KU-348-024

Bird Atlas 2007–11

The breeding and wintering birds of Britain and Ireland

Dedicated to the Regional Organisers, Regional Validators and volunteers who made *Bird Atlas 2007–11* possible.

Also dedicated to the memory of Oscar Merne (1943–2013) who passed away in January 2013. The breadth and depth of Oscar's ornithological legacy are inestimable. Professionally, but more significantly personally, his contribution influenced every significant area of Irish ornithology.

Published by BTO Books, Thetford
BTO Books is the trading name of BTO Services Limited,
The Nunnery, Thetford, Norfolk, IP24 2PU

First published in 2013

ISBN 978-1-908581-28-0

Designed by: O'Connor Design Consultants
Tom Sayer & Chris O'Connor
www.o-connordesign.co.uk
info@o-connordesign.co.uk

Chapter illustrations: Dave Daly

Printed by: Printer Trento

Environment: The paper used for this book has been certified as coming from well-managed forests and other controlled sources according to the rules of the Forest Stewardship Council.

This book was printed and bound in Italy by Printer Trento, an FSC certified company for printing books on FSC mixed paper in compliance with the chain of custody and on-product labelling standards. Printer Trento has an ISO14001 certified environmental management system.

Recommended citation:
Balmer, D.E., Gillings, S., Caffrey, B.J., Swann, R.L., Downie, I.S. & Fuller, R.J. 2013. *Bird Atlas 2007–11: the breeding and wintering birds of Britain and Ireland*. BTO Books, Thetford.

Citation for chapters, for example Chapter 2:
Gillings, S., Balmer, D.E., Caffrey, B.J. & Swann, R.L. 2013. Survey methods and data sources. In *Bird Atlas 2007–11: the breeding and wintering birds of Britain and Ireland* (Balmer, D.E., Gillings, S., Caffrey, B.J., Swann, R.L., Downie, I.S. & Fuller, R.J.), pp 33–45. BTO Books, Thetford.

CONTENTS

CHAPTER 5
Coverage and effort.
Simon Gillings, Dawn E. Balmer, Robert L. Swann & Brian J. Caffrey

CHAPTER 6
Pattern and change in the British and Irish avifaunas over a 40-year period.
Robert J. Fuller, Simon Gillings, Alan W. Lauder & Olivia Crowe

CHAPTER 7
Interpretation of species accounts.
Simon Gillings & Dawn E. Balmer

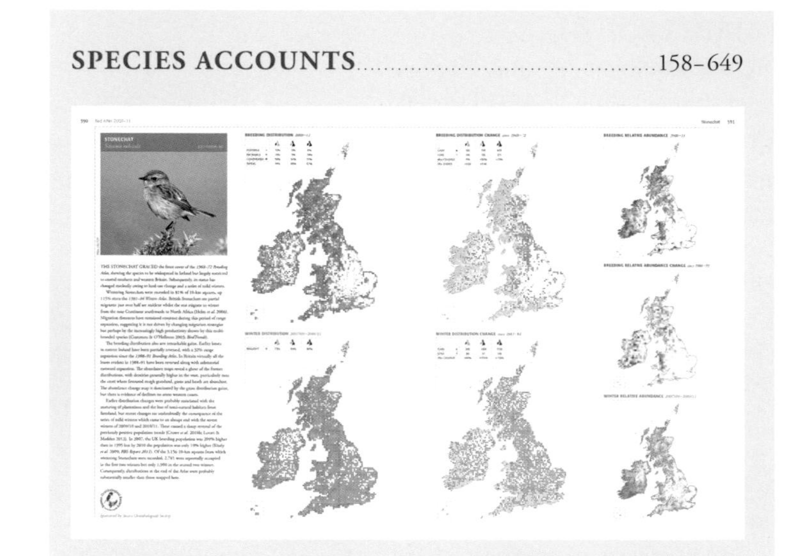

The species accounts form stand-alone material that should be largely self-explanatory, though the reader should first be familiar with Chapter 7, which provides information that will help to interpret the different types of maps provided. It is not necessary to read Chapters 1 to 6 before consulting the species accounts. These chapters provide detailed contextual information on the project development, methodology and patterns that emerge from the Atlas.

However, Chapter 5 includes a detailed examination of how recording effort has varied through the atlases and its conclusions should be borne in mind when reading the species accounts.

Appendices

PREFACE

BY NOW, EVERYONE involved in *Bird Atlas 2007–11* must appreciate that their contributions and this book will set the agendas for bird conservation in Britain and Ireland for the next twenty years. *Bird Atlas 2007–11* presents the results of Britain and Ireland's most ambitious atlas project to date. By combining in one project the data collection and reporting of the third breeding atlas and the second winter atlas there is, for the first time, an overview of wintering and breeding bird distributions and their relative abundances in one volume. Bringing the two seasons together makes sense because winter distributions are just as important in conservation terms as breeding ones and, for our resident birds, winter survival plays a key role in determining subsequent breeding population sizes. The book allows comparisons of range size over two consecutive 20-year periods for our breeding birds and over an approximately 30-year period for our wintering birds. Furthermore, the maps of relative abundance and of change in breeding abundance reveal interesting patterns worthy of further exploration by amateurs and professional scientists alike.

Bird Atlas 2007–11 is the product of a major cooperative effort extending over more than a decade and involving many thousands of people. It is appropriate, in these opening pages, to thank each and every one of the dedicated volunteers without whom the project could not have succeeded. The names of all those who have contributed their time and expertise directly to the Atlas are listed in the Appendices—we also thank all those who have contributed records indirectly through their local bird club or by taking part in monitoring schemes organised by BTO and BirdWatch Ireland (abbreviations and partners of collaborative schemes and projects can be found in the glossary on page 14). The Atlas is a wonderful example of what can be achieved through citizen science and clearly illustrates the critical role that volunteer observers can play in documenting changing patterns of biodiversity on large scales. Apart from carrying out the familiar atlas inventory surveys, they counted every individual bird encountered in their tetrads for a minimum of two hours in both breeding and winter seasons. It has been estimated that the kilometres walked during these Timed Tetrad Visits alone would have been sufficient to walk around the equator twenty times, and would have kept a single observer occupied full-time for 100 years! In addition, some eight million casual records were collected, on a scale which was only possible because of the fantastic uptake of the online data entry facilities. The fruits of all this labour can be seen in the quality of the maps of species distribution, abundance and change presented in the following pages.

The project began as a proposal from BTO scientists in 2001. Like previous atlases, it soon became a cooperative venture with BirdWatch Ireland and the Scottish Ornithologists' Club coming on board as partners, leading to the formation of two overview groups fully committed to making *Bird Atlas 2007–11* a success. The groups comprised members (see Acknowledgements) from a range of interest groups including, in addition to BTO, BirdWatch Ireland and SOC, local bird clubs, BTO Regional Representatives, previous local and national atlas organisers, academic experts and conservation professionals. As Chairmen of the Atlas Working Group and Irish Atlas Steering Group we saw the project evolve and mature. Throughout, we were hugely impressed both by its scale, and the skills and dedication of all those involved in bringing it to fruition, especially the Regional Organisers, who rallied to find observers to cover every last corner of Britain and Ireland. We would like to take this opportunity to thank each of them for their contributions and support, which made our roles both uncomplicated and rewarding.

The input from many BTO, BirdWatch Ireland and SOC staff members was crucial to the successful operation and outcome of the Atlas. Although space does not allow us to mention everyone by name, all the authors of this book were key players throughout the lifetime of the project. We are sure you will agree with us that they have done a fantastic job.

The Atlas Working Group held its first meeting in November 2004, three years prior to fieldwork commencing. Although a great deal of preparation and planning had been carried out prior to that opening session, numerous questions about methods, costs, timing and publication of the results remained to be answered. Thanks to the generous support provided through the BTO Atlas Appeal and to the work of BirdWatch Ireland to secure funding in Ireland, the project was never compromised by lack of funds. The overall aim was clear: to obtain results in both winter and the breeding season that would be directly comparable with the earlier atlases. This required that the core of the survey methodology remained, as far as possible, unchanged, although lessons learned from the previous projects would be incorporated where appropriate.

THE FIELDWORK BEGAN with a pilot survey in 2005 which allowed the methods to be tested in winter. One new aspect of this atlas, which was particularly helpful and impressive, was the use of the Internet to gather data and feed results back to observers. Although it was still possible to enter records on paper, the vast majority were submitted via the Bird Atlas website or BirdTrack. For the fieldworkers, this was quick and easy to do but, even better, enabled rapid feedback on progress. It allowed gaps in coverage to be identified and filled within a season, rather than waiting for end-of-season results. It also enabled 19 million records to be confirmed as valid, or queried; an enormous job undertaken by Regional Validators. The number of records received is impressive and the fact that only 0.05% of the records remained queried at the end of the Atlas is a testament to the hard work of the team of local validators and the cooperation of observers. Fortunately, it was possible to use records from long-running monitoring schemes, short-term surveys focused on particular species and research projects organised by BTO, BirdWatch Ireland, and many conservation organisations and specialist groups to ensure that distribution maps were as complete as possible.

The Atlas also made an important step forward in further strengthening the relationship between BTO and local bird clubs. For the first time, formal agreements were reached between the Bird Atlas partners and local bird clubs covering ownership and use of Atlas data. Some 45 local atlas projects were carried out alongside the national Atlas, utilising the central online recording and validation systems. These will clearly provide a detailed pattern of local bird distributions and numbers, complementary to the national Atlas.

It was Roger Tory Peterson who described birds as being akin to ecological litmus paper. He explained that, because of their rapid metabolism and wide geographical range, they can reflect changes in the environment quickly. This is undoubtedly true for many species, but accurately assessing such changes across the whole of our islands' avifaunas is far from easy. That is why atlases, repeated over regular intervals, are so powerful. They show bird distributions and densities on visually striking maps which make it easy to see where changes have occurred. By covering all species across a large geographic area, they show whether the changes involve a few or many species and whether they are confined to small, local patches or spread more generally across the range. They can help to tell us if measures are needed to protect important habitats and species, whether earlier actions have been successful or where further work is required to understand the causes of the change. They are, therefore, an invaluable tool for conservation.

SOME OF THE trends in the distributions of our resident birds, which are clear from the results in this book, were already apparent in the *1988–91 Breeding Atlas*, or had been anticipated in the intervening years—the rapid rise of the Little Egret being the most striking example. However, for example, the colonisation of Ireland by Great Spotted Woodpeckers, and the northward spread of Goldfinches and Jays in Britain, have been more of a surprise. These success stories sit alongside stories of worrying range contractions. The ranges of several of our resident farmland birds, including the Grey Partridge and Corn Bunting have continued to shrink. Indeed, the Corn Bunting is now considered extinct as a breeding species in Ireland. Although steps have been taken to reverse these trends and have had some local success, *Bird Atlas 2007–11* shows that they have not been sufficient to halt the declines. These have been joined by woodland species such as Lesser Spotted Woodpeckers and Willow Tits, by migrants from equatorial Africa, including Turtle Doves and Whinchats, and by breeding waders such as Curlews and Redshanks. There are likely to be many factors driving the changing ranges of species that are documented in this book. These valuable data will be further analysed to help elucidate the mechanisms and causes of distributional and population change. No doubt climate change has been involved in some cases and will continue to play a part in future distributional changes. Clearly, work remains to be done and this book helps to point the way.

John Eyre
Chairman of the Atlas Working Group

John O'Halloran
Chairman of the Irish Atlas Steering Group

May 2013

ABOUT THE AUTHORS

THE *BIRD ATLAS 2007–11* project was a large and complex undertaking, and the core team had specific roles which are explained below.

Dawn Balmer – Atlas Coordinator

For birdwatchers, Dawn was the face of the Atlas. Working for BTO, she was responsible for the overall coordination and promotion of the project and worked closely with the network of Regional Organisers, Regional Validators and volunteer observers throughout the United Kingdom, Isle of Man and the Channel Islands to ensure that comprehensive coverage was achieved and the validation of records completed. She coordinated fieldwork for the contract fieldworkers in England, Wales and Northern Ireland.

Simon Gillings – Senior Research Ecologist

As a Senior Research Ecologist at BTO, Simon was involved in the early stages of planning of the Atlas and organised the Pilot Winter Atlas fieldwork in 2005/06. He played a key role in developing the methods for the Atlas, and was responsible for all development of analytical methods, data analysis and the design and production of maps and figures. He worked closely with Iain Downie and made significant contributions to the Atlas website. He served on the Rare Breeding Birds Panel as the BTO representative and led on the mapping of rare breeding birds.

Brian Caffrey – Atlas Organiser in Ireland

Brian, employed by BirdWatch Ireland, was responsible for the organisation of the Atlas in Ireland, and specifically fieldwork in the Republic of Ireland. He worked closely with the Irish Atlas Steering Group to promote the project and secure funding for Atlas work in Ireland. He also worked closely with BirdWatch Ireland branches and observers to ensure that fieldwork was completed and managed a team of contract fieldworkers. He led on organising the validation of records in the Republic of Ireland.

Bob Swann – Atlas Organiser in Scotland

Bob, employed part-time by BTO on behalf of SOC, worked closely with Regional Organisers and observers across Scotland to ensure fieldwork was completed and records checked through validation. He made a significant contribution to fieldwork, particularly in Sutherland and Ross-shire, and helped local teams fill gaps in coverage in the Northern Isles and Outer Hebrides. He coordinated fieldwork for contract fieldworkers, visiting expedition teams and RSPB staff on sabbatical in Scotland. He worked with SOC and BTO Scotland to promote the Atlas to observers through writing articles, giving talks and leading training courses.

Iain Downie – Joint Head of Information Systems

Since joining BTO in 2001, Iain has led on the Trust's development of novel web applications to capture online data for a variety of surveys and monitoring schemes. He was responsible for the building and continued development of the Bird Atlas Online system, incorporating the needs of the Regional Organisers, observers and local atlas projects. He worked closely with Dawn and Simon throughout the project and with them (and others in the BTO Information Systems Team) developed innovative tools to help target fieldwork effort and built the *Review & Validation* module used by Regional Validators.

Rob Fuller – Director of Science

Throughout his career at BTO, Rob has been closely involved in the planning and strategic aspects of large-scale bird surveys. He produced the initial concept of *Bird Atlas 2007–11* and undertook much of the early planning for the project. From inception to completion he was responsible for overseeing the whole Atlas.

Others

Karen Wright (Joint Head of Information Systems, BTO), Alan Lauder (Chief Executive Officer, BirdWatch Ireland) and Olivia Crowe (Senior Conservation Officer, BirdWatch Ireland) assisted with writing chapters. Jeff Baker project managed the production of this book, which included liaison with designers, photographers, printers and distributors.

ACKNOWLEDGEMENTS

Volunteer observers

Over 40,000 volunteer observers have contributed records to *Bird Atlas 2007–11*, either directly to the project, or indirectly through surveys and schemes organised by BTO, BirdWatch Ireland, local bird clubs and specialist groups. Records were submitted directly to the Atlas or BirdTrack by just over 17,000 people, sometimes on behalf of couples, families and small groups. Their names are listed in Appendix 1 in recognition of the time and effort so many have put into the project. Needless to say, we have not been able to capture the names of everyone who took part and we extend our thanks to every volunteer who has submitted a record, whether directly or indirectly.

Organisers, staff, groups and individuals

The network of Regional Organisers managed the allocation of Timed Tetrad Visits, organised coverage of Roving Records and motivated and encouraged fieldworkers in each area. Regional Validators checked all of the data submitted in each area. Regional Organisers and Validators were the backbone of the Atlas and they are named and acknowledged in Chapter 1.

We would like thank members of the Atlas Working Group (AWG: 2004–12) who provided guidance and support throughout the planning and fieldwork stages of the Atlas, particularly John Eyre who was Chairman throughout. The following were members of the AWG for all or part of the period: Graham Appleton (BTO), Jeff Baker (BTO), Dawn Balmer (BTO), Tom Cadwallender (Regional Representative, local atlas organiser), Brian Caffrey (BirdWatch Ireland), Andy Clements (BTO), Martin Cook (SOC, local atlas organiser), John Eyre (Chairman), Rob Fuller (BTO), Joe Furphy, Kevin Gaston (University of Exeter), David Gibbons (RSPB), Simon Gillings (BTO), Jeremy Greenwood (BTO), Bob Harris (Regional Network Committee), Mark Hill (Centre for Ecology and Hydrology), Mark Holling (RBBP, SOC), Alan Lauder (BirdWatch Ireland), John Lloyd (Regional Network Committee 2004–09, Regional Representative), Liam Lysaght (National Biodiversity Data Centre), Stephen Newton (BirdWatch Ireland), David Noble (BTO), Nicki Read (Minuting Secretary, BTO), David Stroud (JNCC) and Bob Swann (BTO, SOC).

The Irish Atlas Steering Group (IASG: 2005–12) was formed to provide guidance on issues specific to Ireland and acted as a platform to provide feedback on progress to the key funding groups. The following were members of the IASG for all or part of the period: Graham Appleton (BTO), Dawn Balmer (BTO), Jim Bowman (EPA), Brian Caffrey (BirdWatch Ireland), Kendrew Colhoun (RSPB Northern Ireland), Niamh Connolly (EPA), Olivia Crowe (BirdWatch Ireland), Rob Fuller (BTO), Joe Furphy, Simon Gillings (BTO), George Henderson (NIEA), Alan Lauder (BirdWatch Ireland), Liam Lysaght (National Biodiversity Data Centre), Neil McCulloch (NIEA), Stephen Newton (BirdWatch Ireland), David Norriss (NPWS), Cliona O'Brien (Heritage Council), John O'Halloran (Chairman, University College Cork), Oran O'Sullivan (BirdWatch Ireland), David Tierney (NPWS), Wayne Trodd (EPA), John Wilson (NPWS) and Shane Wolsey (BTO).

The Rare Breeding Birds Panel (RBBP) provided valuable data for the Atlas and also guidance on appropriate mapping resolutions for rare and scarce breeding birds in the UK. Members during the Atlas period were Mark Eaton (Chairman 2012 to present), Ian Francis, Simon Gillings, Mark Holling (Secretary), Andrew King, David Norman, Judith Smith, Ken Smith and David Stroud (Chairman 2008–12). We would like to extend special thanks to Mark Holling who dealt with so many queries regarding rare breeding birds and gave such sound advice.

The Irish Rare Breeding Birds Panel (IRBBP) provided data and advice on rare breeding birds in Ireland. Members were Ian Enlander, George Gordon, Paul Hillis (Secretary), Neil McCulloch, Oscar Merne, Stephen Newton, Oran O'Sullivan, James Robinson (Chairman) and David Tierney. Particular thanks go to Paul Hillis and Stephen Newton who helped with data capture and many validation queries.

BTO staff have been supportive throughout the whole project and we would particularly like to thank the following for help and advice: Kate Aldridge, Robin Anderson, Graham Appleton, Stephen Baillie, Jeff Baker, Carl Barimore, Bryony Braschi, John Calladine, Ellen Chestney, Jacquie Clark, Andy Clements, Greg Conway, Mandy Cook, Anne Cotton, Ieuan Evans, Sophie Foulger, Rachel Gostling, Su Gough, Mark Grantham, Mary Gray, Jeremy Greenwood, Mark Hammond, Paul Harrup, Chas Holt, Alison Johnston, Andrew Joys, Jacqui Kaye, Maria Knight, Peter Lack, John Marchant, Stuart McHugh, Teresa Miyar, Nick Moran, Chris Morley, Andy Musgrove, Stuart Newson, Nicki Read, Sam Rider, Kate Risely, Staffan Roos, Andrew Scott, Carole Showell, Gavin Siriwardena, Sandra Sparkes, Paul Stancliffe, Mike Toms, Chris Wernham, Shane Wolsey, Karen Wright and Ian Yates.

We are especially grateful to Graham Appleton and colleagues in BTO's Communications Department for their terrific fundraising efforts and for their promotion of the project in a wide variety of ways. Kate Aldridge deserves a special mention for her fundraising efforts, including organising a 'species auction'.

BTO Council and Management have supported the Atlas team throughout, always recognising the key role played by *Bird Atlas 2007–11* in fulfilling the scientific and volunteer engagement aims of the BTO as a whole.

In addition to the hundreds of hours BirdWatch Ireland staff contributed to Atlas fieldwork, staff in the Development, Administration and Conservation teams and the BirdWatch Ireland Board also played an important role in *Bird Atlas 2007–11*. In particular we would like to thank: Dick Coombes, Alex Copland, John Cromie, Olivia Crowe, Jamie Durrant, Triona Franks, Niall

Hatch, Katie Jennings, Michelle Kavanagh, Alan Lauder, Helen McCarthy, Declan Murphy, Stephen Newton, Oran O'Sullivan and Annette Quigley. We would also like to thank volunteers at the BirdWatch Ireland offices that contributed to the Atlas, particularly Stephen McAvoy, Blanaid O'Connell, Matthew Thiebaut and Jeroen Veldman. At SOC, Wendy Hicks has been helpful throughout. Chris Waltho, David Jardine and Ken Shaw provided valuable support in their successive roles as SOC Presidents during the Atlas period and subsequently, and previous SOC Presidents Janet Crummy and Mark Holling gave support to the project during planning.

RSPB staff were helpful in sourcing and extracting valuable data from their reserves and monitoring projects and we would like to thank Steve Blain, William George, Darren Oakley-Martin, Andrew Stanbury and Ellen Wilson for their help. Mark Thomas and the Investigations Team reviewed maps of scarce and rare breeding birds. A number of RSPB staff provided advice on specific species: Bruce Fowkes & Leigh Lock (Stone-curlew), Malcolm Smith (Red-necked Phalarope), Andrew Stanbury (Crane), Andrew Taylor (Great Bustard) and Simon Wotton (Bittern). Mark Eaton, Ian Francis, David Gibbons and Jeremy Wilson provided help and support during the Atlas. RSPB also supported the project by allowing some staff to undertake Atlas fieldwork on sabbaticals.

We would also like to thank county bird recorders and other bird club committee members throughout Britain and Ireland for dealing helpfully with so many queries raised through the validation process and for arranging access to bird records in their archives.

All the paper Roving Records forms were input by volunteers and we thank them for many hours of inputting: Liz Blackwell, Judy Follows, Elizabeth Gill, Rosy Jones, Christine Moss, Nicky Parish and Angela Pooley. BirdWatch Ireland fieldworkers and volunteers also input Roving Records forms in Ireland.

The species accounts were drafted largely by Bob Swann and Dawn Balmer. A small number were drafted by staff from BTO and BirdWatch Ireland and we thank them for their help: Graham Appleton, Phil Atkinson, Helen Boland, Brian Caffrey, Neil Calbrade, John Calladine, Greg Conway, Dick Coombes, Alex Copland, Olivia Crowe, Sinead Cummins, Anita Donaghy, Kathryn Finney, Simon Gillings, Ian Henderson, Chas Holt, Peter Lack, Alan Lauder, Dave Leech, Lesley Lewis, John Lusby, John Marchant, Nick Moran, Andy Musgrove, Stephen Newton, Oran O'Sullivan, Kate Risely, Gavin Siriwardena and Rachel Taylor. All accounts were reviewed thoroughly by Dawn Balmer, Simon Gillings, Rob Fuller, Brian Caffrey and Bob Swann. Olivia Crowe assisted with the early stages of reviews.

Species accounts were improved by comments from the following species specialists: John Calladine, Paul Castle, Olivia Crowe, Andy Dobson, John Eyre, Jenny Gill, Mike Harris, Mark Holling, Liz Humphreys, Peter Lack, Tim Mackrill, John Marchant, Mick Marquiss, Barry O'Donoghue, John O'Halloran, David Stroud, Sarah Wanless and Chris Wernham.

We thank the following for their helpful comments on draft chapters: Graham Appleton, Chris Cheffings, Martin Cook, Olivia Crowe, Chris Dee, John Eyre, Barrie Galpin, Jeremy Greenwood, Mark Holling, Alison Johnston, Andy Musgrove, Ian Newton, John O'Halloran, Gavin Siriwardena, Pat Smiddy and Gerard Troost. Statistical and analytical advice was gratefully received from Lluis Brotons, Greg Conway, Jeremy Greenwood, Sergi Herrando, Mark Hill, Alison Johnston and Gavin Siriwardena.

The following photographers kindly provided photographs for use in this book: Alan Adam, Philip Alexander, Jim Almond, John Anderson, Robin Anderson, Harry Appleyard, Michael Ashforth, Dennis Atherton, Geoff Athey, Jason Atkinson, Jeff Baker, Chris Bale, Dawn Balmer, Paul Bateson, Martin Bennett, Tom Bickerton, John Boorman, Mary Braddock, Colin Brown, Steve Bullen, Brian Caffrey, Neil Calbrade, John Camp, Steve Carter, Christopher Cassidy, Graham Catley, Leigh Caudwell, Keith Chapman, Derek Charles, Colum Clarke, Graham Clarke, Ken Clarke, Mark Coates, Kim Connolly, Shay Connolly, Rob Cross, Liz Cutting, Sarah Dalrymple, Steve Davies, Nico de Vasser, Iain Downie, Allan Drewitt, Desmond Dugan, Mick Durham, Mick Eade, Dean Eades, Ruth Eastwood, Gary Faulkner, Edmund Fellowes, Tom Finch, John Flowerday, Les Foster, Vic Froome, Brian & Sophie Fuller, Simon Gillings, Martin Goodey, Dennis Gurner, Gerry Gutteridge, Thomas Hanahoe, Terry Hancock, Brian Harper, Jeff Hazell, Paul Hillion, Ken Hindmarch, Margaret Holland, Mark Hope, Mark Hows, David Hutchinson, Hugh Insley, Jen Jarratt, Gary Jenkins, Niels Jensen, Simon Johnson, Polina Kasapova, David Kay, Andrew Kelly, Jeff & Allison Kew, David Kjaer, Chris Knights, Norman Mackenzie, Graham Maples, Abbie Marland, Breffni Martin, Alison McArthur, Alan McFadyen, Tony McLean, Tim Melling, Jeremy Moore, David Moreton, Jackie Moreton, Greg Morgan, Dennis Morrison, Paul Morrison, Mark Newsome, Paul Newton, Richard Nicoll, Nottsexminer, Joe O'Hanlon, Jill Pakenham, Garth Peacock, Joe Pender, Martin Pulling, Colin Pumfrett, Mike Rae, Brian Rafferty, Morris Rendall, Colin Richards, Mark Richards, Roger Ridley, Rob Robinson, Craig Round, Martin Rutledge, Richard Saxton, Naomi Scott, Glyn Sellors, Zoe Shreeve, Max Silverman, Anna Simpson, Oliver Smart, Peter D. Smith, Steve Smith, John Spaull, Richard Stonier, David Talbot, David Tipling, Brian Tollitt, Joel Walley, Tom Wallis, Alyn Walsh, Clare Ward, Abi Warner, Derek Washington, Damian Waters, Tom Way, Jim Welford, Simon West, J.B. Wilson, Peter Wilson, Richard Winston, Gary Woodburn, Jack Wright and Gordon Youdale.

We thank Caroline Dudley and Barrie Galpin for copy-editing, and John Marchant for copy-editing and proof-reading. Chris O'Connor and Tom Sayer at O'Connor Design Consultants were responsible for the design and layout of the book. The excellent chapter artwork was produced by David Daly. Jeff Baker collated all the photographic material, arranged the artwork, applied final edits to the text and managed the production of the book.

Fieldwork

John Bell, Paul Collin, Phil Davey, Susan Holoran, Alan Salter and Michael Stinson were employed by BTO as fieldworkers. Thanks go also to BTO staff James Bray and Richard Thewlis for filling in gaps. In Ireland, Jamie Durrant, Alison Phillip, Niall Tierney and Robert Wheeldon were employed by BirdWatch Ireland as fieldworkers. In addition to these full-time fieldworkers, contract staff were employed by BirdWatch Ireland to complete smaller packages of survey work (Timed Tetrad Visits and Roving Records) in areas of low coverage: Michael Bell, John Carey, Padraig Cregg, Davy Farrar, Colin Gallagher, Geoff Hunt, Harry Hussey, Niall Keogh, Breffni Martin, John Meade, Martin Moloney, Tony Nagle, Laura Nuttall, Michael O'Clery, Shane O'Neill, Christian Osthoff, Peter Phillips, Ken Shaw, Pat Smiddy, Jenifer Troake, Paul Troake, Stephen Ward, Neal Warnock and David Watson. We thank them all for their commitment and long hours in the field.

Expeditions were organised to areas of poor coverage or areas difficult to access and we thank all those who organised them and participated in them. Expeditions that received some financial support are listed below:

- Winter: Louise Bacon & team (Wester Ross), Yvonne Benting (Barra), Alan Bull & team (Cornwall, Islay), Mike Duckham (Republic of Ireland), Edmund Fellowes & Mike Martin (Coll), Jacquie Heaton & team (Ross-shire), Mark Holling & team (Uist), Alex Joss (Raasay), Bob McMillan (South Rona), Tom Wells & team (Republic of Ireland).
- Breeding season: Mel ab Owain (Co. Clare, Co. Galway), Anne Brenchley & Ian Spence (Co. Kilkenny), Chris Dee (Coll), Steve Gilbert (Co. Clare, Co. Donegal), Steve Hunt (Islay), Alex Joss (Ross-shire), Sussex Ornithological Society (Mull).
- Winter and breeding season: John Armitage (Jura), Dave Conway (Cornwall), Colin Corse & team (Orkney Islands), Hugh Insley & team (North Argyll, Raasay, Cape Wrath).

In addition, the Royal Air Force Ornithological Society (RAFOS) organised their own expeditions to Argyllshire over three summers, Islay in winter and Lincolnshire in summer and the Army Ornithological Society made trips to Jura in both seasons. Stuart Benn (Co. Donegal) and Ken & Linda Smith (Co. Mayo) organised expeditions to Ireland at their own expense. Barrie & Tricia Galpin (Dumfries & Galloway) organised trips with friends in both seasons over a number of years.

In Ireland many volunteers organised trips to visit areas of poor coverage, locations particularly difficult to access and offshore islands. We are very grateful to all these people for their time and dedication.

A number of RSPB staff used their sabbaticals to carry out fieldwork in Scotland and in Ireland.

Finally, we wish to thank the many landowners and land managers who allowed or facilitated access to land in order that Atlas fieldwork could be undertaken.

Local atlas projects

We thank all the local atlas projects for generating excellent local coverage during *Bird Atlas 2007–11*. The individuals and groups behind these local projects worked closely with the Atlas team and ensured that priorities for *Bird Atlas 2007–11* were met alongside their local fieldwork. All local atlas projects that started are listed below.

Assynt, Avon, Bedfordshire, Berkshire, Brecknock, Buckinghamshire, Caithness, Cambridgeshire & Huntingdonshire, Carmarthenshire, Clyde, Cumbria, Derbyshire, Devon, Durham, Fife, Glamorgan, Gloucestershire, Gower, Greater Manchester, Hampshire, Herefordshire, Hertfordshire, Isle of Arran, Isle of Bute, Isle of Wight, Kent, Kinross, Lancashire & north Merseyside, Leicestershire, London, North Wales (Clwyd, Caernarvonshire, Anglesey, Merioneth), Northumberland, Nottinghamshire, Orkney, Oxfordshire, Pembrokeshire, Shetland, Shropshire, Somerset, Southeast Scotland (Lothian, Borders), Suffolk, Surrey, Sussex, Co. Waterford and Wiltshire.

Data sources

In addition to people and organisations mentioned above we would like to thank the following for providing additional distribution records (see Section 2.4).

We are grateful to the members of the Scottish Raptor Study Groups and other contributors to the Scottish Raptor Monitoring Scheme (a partnership between the Scottish Raptor Study Groups, SNH, BTO Scotland, RSPB Scotland, SOC, RBBP and JNCC, grant-aided by SNH with in-kind contributions from the other partners) for providing information on breeding raptors. Brian Etheridge, Staffan Roos, Bob Swann and Chris Wernham enabled the transfer of records. The Northern Ireland Raptor Study Group, facilitated by Jim Wells, the Irish Raptor Study Group and the Golden Eagle Trust also provided valuable records. David Ramsden (Barn Owl Trust) arranged access to breeding records of Barn Owls in southwest England. In Ireland, EPA, NPWS, Bord na Móna and Bat Conservation Ireland kindly provided records. The Game & Wildlife Conservation Trust (GWCT) provided records of Woodcock collected by their members and staff.

We would like to thank the following bird clubs, branches and groups who kindly extracted records from their databases for the Atlas and all those that facilitated this process: Argyll Bird Club, Banbury Ornithological Society, Bedfordshire Bird Club, Berkshire Ornithological Club, Birds in Central Ireland Bird Report, BirdWatch Ireland East Coast Nature Reserve, Borders SOC, Buckinghamshire Bird Club, Cambridgeshire Bird Club, Cape Clear Bird Observatory, Cheshire and Wirral Ornithological Society, Clwyd Bird Recording Group, Cumbria Bird Club, Devon Bird Watching & Preservation Society, Dungeness Bird Observatory, Durham Bird Club, Dursley Birdwatching & Preservation Society, East Lancashire Ornithologists' Club, East London Birders' Forum, Filey Brigg Ornithological Group and Bird Observatory, Fylde Bird Club, Galway Branch of BirdWatch Ireland, Glamorgan Bird Club, Gloucestershire Ornithological Co-ordinating Committee, Greater Manchester Bird Group, Hampshire Ornithological Society, Herefordshire Ornithological Club, Hertfordshire Bird Club, Isles of Scilly Bird Group, Kent Ornithological Society, La Société Guernesiaise, Lancashire & Cheshire Fauna Society, Leeds Birdwatchers' Club, Leicestershire and Rutland Ornithological Society, Lincolnshire Bird Club, Lincolnshire Wildlife Trust Reserves, London Bird Club, Lothian SOC, Malvern Hills Bird Group, Manx BirdLife, Nar Valley Ornithological Society, Norfolk & Norwich Naturalists' Society, North East Scotland Bird Club, Northern Ireland Birdwatchers' Association, Northumberland & Tyneside Bird Club, Nottinghamshire Birdwatchers, Orkney Bird Club, Oxford Ornithological Society, Pembrokeshire Bird Group, Rochdale Field Naturalists' Society, Rossendale Ornithologists'

Club, RSPB Croydon Local Group, RSPB Rainham Local Group, Sheffield Bird Study Group, Shetland Biological Records Centre, Shetland Bird Club, Shropshire Ornithological Society, Sligo Branch of BirdWatch Ireland, Somerset Ornithological Society, Surbiton & District Birdwatching Society, Sussex Ornithological Society, West Midlands Bird Club (Staffs) and Wiltshire Ornithological Society.

We thank Ron Summers for providing distribution data for the Scottish Crossbill based on playback surveys. Stephen Roberts and Malcolm Cowlard provided breeding records for Honey-buzzard.

Fiona Barclay and Dave Dunsford at BirdGuides facilitated the supply of records into BirdTrack and Irishbirding.com allowed use of records from the website.

Roddy Mavor (JNCC) gave access to data in the Seabird Monitoring Programme.

Funding sources

A total of £1.5 million was raised for Atlas work in the United Kingdom, with the biggest segment of the funding coming from tremendously generous BTO members, BTO Garden BirdWatchers, ringers and other supporters. The project was underpinned by a major donation from the estates of Joy and Ted Danter and the initial development work on methods was funded by the Garfield Weston Foundation.

Atlas work in Ireland was supported by the National Parks & Wildlife Service of the Department of the Arts, Heritage and the Gaeltacht, the Heritage Council, the Northern Ireland Environment Agency (Department of Environment, Northern Ireland), the Environmental Protection Agency, BirdWatch Ireland and BTO. This funding paid for the Irish Atlas Coordinator position, the employment of fieldworkers and short-term contract field staff and also the 'hit squads' organised by Shane Wolsey to complete TTVs and gather Roving Records in Northern Ireland.

The Esmée Fairbairn Foundation paid for the Atlas Coordinator's salary and the J. & J.R. Wilson Trust and AEB Charitable Trust facilitated the employment of Bob Swann as the Atlas Organiser in Scotland. Contributions of similar value were made by the H.B. Allen Charitable Trust, the John Ellerman Foundation, Hobart Charitable Trust and the Ernest Kleinwort Charitable Trust, and we are grateful to David Musson, who sold his book collection to raise over £25,000 for the project. SOC part-funded an extension to the Scottish Organiser's contract during the write-up period. Many of the generous donations by individuals, bird clubs, conservation organisations, businesses and charitable trusts are shown as sponsors in the species accounts.

We thank all the charitable trusts concerned: AEB Charitable Trust, Aberbrothock Charitable Trust, H.B. Allen Charitable Trust, Balmain Charitable Trust, Barbour Trust, Charles & Caroline Barratt Charitable Trust, Rowan Bentall Charity Trust, E.G. & M.A. Bousfield Charitable Trust, A.S. Butler Charitable Trust, Chapman Charitable Trust, Edward Joseph Colclough's Trust, Hamish & Doris Crichton's Charitable Trust, Dennis Curry's Charitable Trust, D'Oyly Carte Charitable Trust, John Ellerman Foundation, Esmée Fairbairn Foundation, Gordon Fraser Charitable Trust, Gatliff Trust, Grocers' Charity, Harris Charitable Trust, Lady Hind Trust, Hobart Charitable Trust, Ernest Kleinwort Charitable Trust, the Lever Trust, John Spedan Lewis Foundation, Lyddon Charitable Trust, The Mercers' Company, The Norman Family Trust, Cecil Pilkington Charitable Trust, Mr & Mrs Philip Rackham Charitable Trust, the Christopher H.R. Reeves Charitable Trust, Sandra Charitable Trust, ShareGift (The Orr Mackintosh Foundation), Shears Foundation, The Slater Foundation, E.J.H. Stephenson Charitable Trust, John Swire 1989 Charitable Trust, Tay Charitable Trust, Uplands Charitable Trust, The Emily Weircroft Charitable Trust, Westminster Foundation, Garfield Weston Foundation, Whitaker Charitable Trust, J. & J.R. Wilson Charitable Trust and The Wixamtree Trust.

We thank Scottish Natural Heritage, and particularly Andy Douse, for the grant-aid supplied for the three-year Building Bird Monitoring in Scotland (BBMS) project, which allowed BTO and SOC to support their volunteer networks in maintaining effort on national long-term monitoring at the same time as delivering *Bird Atlas 2007–11*. We also thank the Gillman Trusts for grant-aiding the BBMS initiative.

The writing of Chapter 6 was partly funded by the Joint Nature Conservation Committee, in a partnership with BTO, on behalf of the statutory conservation bodies: Natural England, Natural Resources Wales and Scottish Natural Heritage, and the Department of the Environment of Northern Ireland.

The Nightingale account was sponsored by the Nightingale Supporters Group who gave generously: J.E. Anderson, Malcolm Anderson, Richard Attenborrow, Birmingham University Ornithology Masters Students, Ann Bowtell, A.S. Butler Charitable Trust, J. Julian Clapham, Dr F. Cole, Mr S.L. Dean, Roger Dickey, Nicholas Donnithorne FLS FRES, Mr G.W. Green, Val Gwynn, Mr C Hall, Claire Halpin, Annie & John Harris, Janet Hatton, Andrew Henderson, Mrs D. Housley, Ruth Jones, Helen Lumley in memory of Phil Lumley, Marine Information Limited, John & Jean McCaig, David Milne QC, Mr G.M.B. Moser, J.A. Newnham, Jon Pavey, Rutland Natural History Society, Mrs P.S. Sharp, Sir John Swire and Mr P. Wilson.

Countryside Council for Wales (now Natural Resources Wales) provided funding for an Atlas fieldworker in Wales during one breeding season. Thanks to Siân Whitehead for making funds available. Defra kindly provided additional funding.

Opticron kindly sponsored the production of the paper Roving Records Forms and the Atlas Field Notebook.

We are grateful to the individuals and groups that sponsored a species and their acknowledgements are shown on the species account pages.

BirdWatch Ireland is grateful for external funding for Bird Atlas work in Ireland, which was provided by:

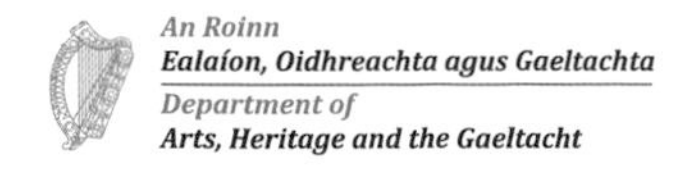

GLOSSARY OF TERMS AND ABBREVIATIONS COMMONLY USED IN THE TEXT

Geographical, mapping and atlas terms

Britain ..... refers to England, Scotland, Wales, Isle of Man and Channel Islands
Ireland ..... refers to the island of Ireland and associated small islands
Britain & Ireland ..... refers to Britain and Ireland as a single geographical entity
Britain and Ireland ..... refers to Britain and Ireland separately
km ..... kilometre
m ..... metre
tetrad ..... a grid square measuring 2 km × 2 km
10-km square ..... a grid square measuring 10 km × 10 km
20-km square ..... a grid square measuring 20 km × 20 km
50-km square ..... a grid square measuring 50 km × 50 km
100-km square ..... a grid square measuring 100 km × 100 km
TTV ..... Timed Tetrad Visit

Organisations and groups

AWG ..... Atlas Working Group
BOU ..... British Ornithologists' Union
BTO ..... British Trust for Ornithology
CCW ..... Countryside Council for Wales (now subsumed into Natural Resources Wales)
EPA ..... Environmental Protection Agency
GWCT ..... Game & Wildlife Conservation Trust
IASG ..... Irish Atlas Steering Group
IRBBP ..... Irish Rare Breeding Birds Panel
JNCC ..... Joint Nature Conservation Committee
NE ..... Natural England
NIEA ..... Northern Ireland Environment Agency
NPWS ..... National Parks & Wildlife Service
NRW ..... Natural Resources Wales (now incorporates CCW)
RBBP ..... Rare Breeding Birds Panel
RSPB ..... Royal Society for the Protection of Birds
SNH ..... Scottish Natural Heritage
SOC ..... Scottish Ornithologists' Club
WOS ..... Welsh Ornithological Society
WWT ..... Wildfowl & Wetlands Trust

Schemes and surveys

BBS ..... Breeding Bird Survey (BTO/JNCC*/RSPB)
BirdTrack ..... BirdTrack (BTO/RSPB/BirdWatch Ireland/SOC/WOS)
CBC ..... Common Birds Census (since 2000 replaced by BBS)
CBS ..... Countryside Bird Survey (BirdWatch Ireland/NPWS)
I-WeBS ..... Irish Wetland Bird Survey (BirdWatch Ireland/NPWS)
NRS ..... Nest Record Scheme (BTO/JNCC)
Ringing Scheme ..... BTO Ringing Scheme (BTO/JNCC*/NPWS and ringers)
WeBS ..... Wetland Bird Survey (BTO/RSPB/JNCC*, in association with WWT)

* on behalf of the statutory conservation bodies: Natural England, Natural Resources Wales and Scottish Natural Heritage, and the Department of the Environment of Northern Ireland.

STANDARD REFERENCES

THE FOLLOWING REFERENCES are commonly used throughout this book. Where used, the abbreviated form is given in italics. In most cases, years detailed in the abbreviated form relate to the period of coverage rather than the year of publication.

1968–72 Breeding Atlas......Sharrock, J.T.R. 1976. *The Atlas of Breeding Birds in Britain and Ireland.* T. & A.D. Poyser, Berkhamsted.

1981–84 Winter Atlas........Lack, P.C. 1986. *The Atlas of Wintering Birds in Britain and Ireland.* T. & A.D. Poyser, Calton.

1988–91 Breeding Atlas......Gibbons, D.W., Reid, J.B. & Chapman, R.A. 1993. *The New Atlas of Breeding Birds in Britain and Ireland: 1988–1991.* T. & A.D. Poyser, London.

BBS Report 2011...............Risely, K., Massimino, D., Johnston, A., Newson, S.E., Eaton, M.A., Musgrove, A.J, Noble, D.G, Procter, D. & Baillie, S.R. 2012. *The Breeding Bird Survey 2011.* BTO Research Report 624. BTO, Thetford.

Birds in England..............Brown, A. & Grice, P. 2005. *Birds in England.* T. & A.D. Poyser, London.

Birds in Ireland................Hutchinson, C.D. 1989. *Birds in Ireland.* T. & A.D. Poyser, Calton.

Birds in Wales...................Lovegrove, R., Williams, G. & Williams, I. 1994. *Birds in Wales.* T. & A.D. Poyser, London.

Birds of Scotland...............Forrester, R.W., Andrews, I.J., McInerny, C.J., Murray, R.D., McGowan, R.Y., Zonfrillo, B., Betts, M.W., Jardine, D.C. & Grundy, D.S. (eds). 2007. *The Birds of Scotland.* SOC, Aberlady.

BirdTrack.........................BirdTrack. 2013. Reports by species. URL: http://blx1.bto.org/birdtrack/results/index.jsp [31 July 2013].

BirdTrends.......................Baillie, S.R., Marchant, J.H., Leech, D.I., Massimino, D., Eglington, S.M., Johnston A., Noble, D.G., Barimore, C., Kew, A.J., Downie, I.S., Risely, K. & Robinson, R.A. 2013. *BirdTrends 2012: trends in numbers, breeding success and survival for UK breeding birds.* BTO Research Report 644. BTO, Thetford. URL: www.bto.org/birdtrends/2012 [31 July 2013].

BWP..............................Cramp, S. *et al.* (eds). 1977–94. *Handbook of the Birds of Europe, the Middle East and North Africa: the birds of the Western Palearctic.* Nine volumes. Oxford University Press, Oxford.

CBS Report 2010..............Crowe, O., Coombes, R.H., Lauder, A., Lysaght, L., O'Brien, C., O'Halloran, J., O'Sullivan, O., Tierney, T.D. & Walsh, A.J. 2011. *Countryside Bird Survey Report 1998–2010.* BirdWatch Ireland, Wicklow.

CBS Trend 2010...............Crowe, O. 2012. CBS Trend 1998–2010. Unpublished.

European Atlas..................Hagemeijer, E.J.M. & Blair, M.J. (eds). 1997. *The EBCC Atlas of European Breeding Birds: their distribution and abundance.* T. & A.D. Poyser, London.

IRBBP 2010.....................Hillis, J.P. & the Irish Rare Breeding Birds Panel. 2011. Rare Irish breeding birds, 2010: the annual report of the Irish Rare Breeding Birds Panel (IRBBP). *Irish Birds* 9: 241–250.

I-WeBS Report 2009..........Boland, H. & Crowe, O. 2012. *Irish Wetland Bird Survey: waterbird status and distribution 2001/02–2008/09.* BirdWatch Ireland, Wicklow.

Migration Atlas.................Wernham, C.V., Toms, M.P., Marchant, J.H., Clark, J.A., Siriwardena, G.M. & Baillie, S.R. (eds). 2002. *The Migration Atlas: movements of the birds of Britain and Ireland.* T. & A.D. Poyser, London.

PECBMS 2010.................PECBMS. 2012. *Population trends of common European breeding birds 2012.* CSO, Prague. URL: www.ebcc.info/pecbm.html [31 July 2013].

RBBP 2010......................Holling, M. & the Rare Breeding Birds Panel. 2012. Rare breeding birds in the United Kingdom in 2010. *British Birds* 105: 352–416.

Seabird 2000....................Mitchell, P.I., Newton, S.F., Ratcliffe, N. & Dunn, T.E. 2004. *Seabird Populations of Britain and Ireland.* T. & A.D. Poyser, London.

SUKB 2012......................Eaton, M.A., Cuthbert, R., Dunn, E., Grice, P.V., Hall, C., Hayhow, D.B., Hearn, R.D., Holt, C.A., Knipe, A., Marchant, J.H., Mavor, R., Moran, N.J., Mukhida, F., Musgrove, A.J., Noble, D.G., Oppel, S., Risely, K., Stroud, D.A., Toms, M. & Wotton, S. 2012. *The state of the UK's birds 2012.* RSPB, BTO, WWT, CCW, NE, NIEA, SNH and JNCC, Sandy.

WeBS Report 2011............Holt, C.A., Austin, G.E., Calbrade, N.A., Mellan, H.J., Hearn, R.D., Stroud, D.A., Wotton, S.R. & Musgrove, A.J. 2012. *Waterbirds in the UK 2010/11: the Wetland Bird Survey.* BTO/RSPB/JNCC, Thetford.

CHAPTER 1

SETTING THE SCENE

Dawn Balmer, Simon Gillings, Brian Caffrey, Bob Swann & Rob Fuller

1.1 HISTORICAL BACKGROUND

FOR MANY NATURALISTS and conservation biologists an atlas is not just a collection of maps, but a book or digital publication that reveals the distribution of the members of a floristic or faunistic group across a defined region. Since the first atlas of British flora (Perring & Walters 1962) atlasing has spread through taxonomic groups, from plankton (Beaugrand 2004) to mammals (Mitchell-Jones *et al.* 1999), and across geographical scales ranging from the localised and regional (e.g. Lysaght 2002; Norman 2008) to the continental (Barrett *et al.* 2003). The aim of atlases has gradually shifted: initially they simply documented *distribution*, but repeat atlases have allowed the documentation of *changes in range*. They have also increasingly reported measures of *abundance* and, more recently, *change in abundance*.

There have now been in excess of 400 bird atlases worldwide (Gibbons *et al.* 2007) of which there are three previous atlases covering Britain and Ireland. The first was a breeding bird atlas, referred to here as the *1968–72 Breeding Atlas* (Sharrock 1976). The second was an atlas of birds in winter, with data collated for three winters, 1981/82 to 1983/84: it is referred to here as the *1981–84 Winter Atlas* (Lack 1986). The third was a repeat breeding atlas—the *1988–91 Breeding Atlas* (Gibbons *et al.* 1993). The first two of these atlases were collaborative projects run between the British Trust for Ornithology (BTO) and the Irish Wildbird Conservancy (IWC). The third atlas was undertaken by the BTO, the IWC and the Scottish Ornithologists' Club (SOC). A brief description of the methods used in these earlier atlases is provided in Section 4.1.2.

In December 2001, BTO staff proposed to the BTO Council that a new combined winter and breeding atlas be undertaken covering the years 2007–11. Between 2002 and 2004 negotiations took place between the BTO, BirdWatch Ireland (formerly IWC) and the SOC, and in 2004 the project was announced to the national network of local bird clubs and natural history societies. Key stages in the planning and organisation of what became known as *Bird Atlas 2007–11* are shown in Figure 1.1.

1.2 GEOGRAPHICAL SCOPE

BIRD ATLAS 2007–11 covers all 10-km squares containing land throughout Great Britain, Ireland, the Isle of Man, and the Channel Islands. This geographical area was first used by the Botanical Society of the British Isles in their ground-breaking and influential *Atlas of the British Flora* (Perring & Walters 1962).

In the interest of brevity this entire geographical region is referred to as 'Britain and Ireland' in the title of this atlas. In the text 'Britain & Ireland' is used when referring to the whole area as a single entity, and 'Britain and Ireland' when referring to the main island groups individually. For example, we would say that the Nuthatch occurs in 'Britain & Ireland' (i.e. somewhere within Britain and/or Ireland) but it would be incorrect to say it occurs in 'Britain and Ireland' (i.e. implying that it occurs in both Britain and Ireland). Sometimes in Chapters 5 and 6 the term 'island' is used when referring to analyses that summarise data separately for the island of Britain (and associated offshore islands) and for the island of Ireland (and associated offshore islands).

In many analyses and summaries the Isle of Man and Channel Isles are combined with Britain under a single title of 'Britain'. No separate statistics are presented for the Isle of Man or the Channel Isles, nor for the separate countries forming Britain & Ireland.

1.3 WHY ANOTHER ATLAS?

ATLASES FORM ONE of several complementary approaches for assessing the numbers, trends and distributions of birds (Donald & Fuller 1998; Bibby 2003). They are unique in providing an overview of distribution and spatial abundance of the entire avifauna and represent an opportunity to take stock of the status and distribution of all bird species. When repeated at intervals they have the potential to provide information on changes both in distribution and in abundance for all species in a region. In these respects atlases complement annual monitoring schemes whose purpose is to provide information on population trends of bird species, but they do so by gathering data from a sample of locations and invariably for only the more widespread species. Taken together, the complete distribution and range-change maps of the atlases and the population trends from the monitoring schemes, provide powerful tools for scientists and conservationists. Atlas data therefore provide one key element in reporting on the broader health of our wildlife, in assessing how well we are protecting it and in informing us of conservation priorities.

Atlases provide various other advantages over routine monitoring. As well as generating large-scale abundance patterns, they also allow regional changes in range to be identified. They can, in some situations, give an opportunity to assess national population size (see Section 1.6). They are a prime example of how 'citizen science' can make major contributions to both conservation and academic research (Greenwood 2007).

The *1968–72 Breeding Atlas* and the *1988–91 Breeding Atlas* were carried out 20 years apart—a sufficient period for changes in populations to potentially become manifested as changes in distribution. To continue the 20-year cycle, the BTO proposal was that a breeding atlas should be planned for 2008–11 and that this should be combined with winter fieldwork. Winter distributions are just as important in conservation terms as breeding ones, so it was sensible to give them equal prominence. Whilst the breeding season is a crucial time of year for all species, many species are limited by factors affecting overwinter survival (Newton 1998). In addition to these scientific benefits of measuring winter and breeding ranges at the same time, a combined winter and breeding-season atlas would also be more cost-effective than running separate atlases.

FIGURE 1.1 TIMELINE OF KEY DATES IN *BIRD ATLAS 2007–11*

Timeline showing key events, meetings and employed staff during the planning, fieldwork and post-fieldwork phases of the Atlas. Blue and orange shaded bars show the main winter and breeding-season fieldwork periods.

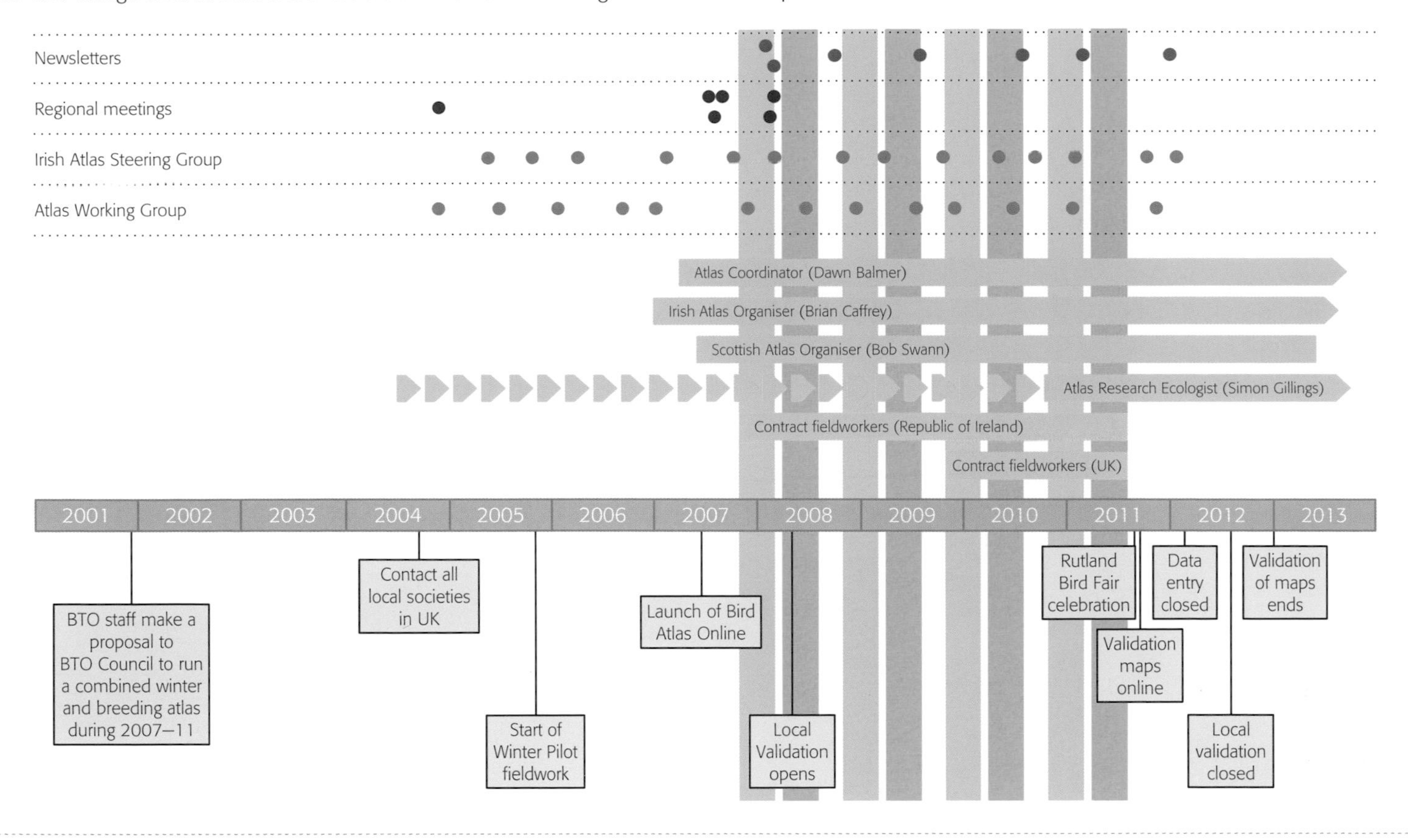

1.3.1 Research uses of atlas data

The research significance of *Bird Atlas 2007–11* is yet to be realised but is likely to be high when one considers that data from the *1988–91 Breeding Atlas* are still being used in high quality publications addressing contemporary conservation issues (e.g. Anderson *et al.* 2009; Franco *et al.* 2009). More than 50 peer-reviewed journal articles published since the early 1990s have made major use of Britush & Irish atlas data, including landmark papers such as those on farmland bird declines (Fuller *et al.* 1995) and on range shifts in response to climate change (Thomas & Lennon 1999). Much of this work is also of considerable policy relevance, especially in the areas of identifying priority conservation sites and strategies (Brown *et al.* 1995; Araújo *et al.* 2002), predicting the effects of environmental changes such as climate and land-use (Bellamy *et al.* 1998; Chamberlain & Fuller 2001) and measuring human use and management of natural resources ('ecosystem services') (Holland *et al.* 2011).

To illustrate the breadth of use of atlases, an Internet search reveals the following numbers of citations in articles and books: 708 for *1968–72 Breeding Atlas*, 311 for *1981–84 Winter Atlas* and 856 for *1988–91 Breeding Atlas* (Google Scholar 2013).

1.4 ORGANISING BIRD ATLAS 2007–11

THIS SECTION DETAILS the key considerations when planning and undertaking *Bird Atlas 2007–11*, ranging from structural organisation, through publicity, to validation of the incoming data.

1.4.1 Overview groups

Serious planning for the Atlas began in 2004 with the formation of the Atlas Working Group (AWG). This was made up of BTO staff, BirdWatch Ireland staff, SOC representatives, previous national and local atlas organisers, experienced BTO Regional Representatives, academics and data users (see Acknowledgements for group members). From its first meeting in November 2004 the AWG met approximately twice per year until its final meeting in November 2011 and oversaw all major decisions concerning the Atlas. The Irish Atlas Steering Group (IASG) was formed to oversee the implementation of the Atlas methods and to handle any issues specific to the island of Ireland. The IASG comprised BirdWatch Ireland staff, BTO staff, academics, representatives from conservation NGOs and funders (see Acknowledgements for group members).

1.4.2 National and regional organisation

An Atlas Coordinator (Dawn Balmer) was appointed from 1st April 2007 and was responsible for the overall coordination and promotion of the project. A part-time Scottish Atlas Organiser (Bob Swann) was appointed from 1st June 2007 to promote the project in Scotland and to liaise with the network of Regional Organisers and SOC branches to organise fieldwork. An Irish Atlas Organiser (Brian Caffrey) was appointed from 1st January 2007 to promote the project throughout Ireland, and in particular to organise fieldwork in the Republic of Ireland. The Senior Research Ecologist (Simon Gillings) began working on aspects of

FIGURE 1.2 ATLAS REGIONS

Regions used for organisation of fieldwork and validation in Britain & Ireland.

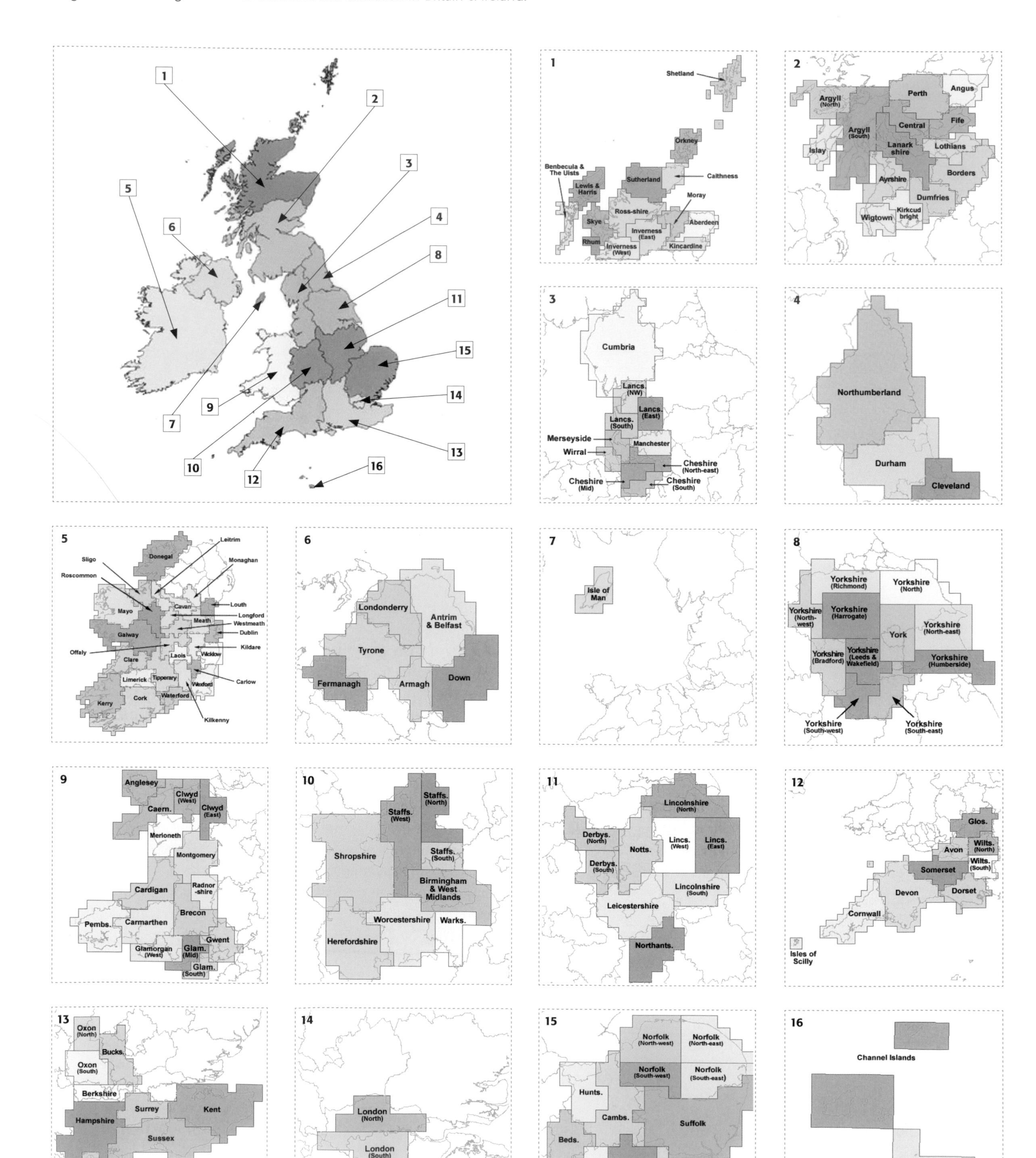

the field methodology (see Chapter 2) and analytical techniques (see Chapter 4) in 2004, and the principal web developer (Iain Downie) starting designing and programming Bird Atlas Online in August 2006 (see Chapter 3). Rob Fuller was responsible for the planning and supervising the entire Atlas project, from conception to publication. The support of the BTO's and BirdWatch Ireland's fundraisers was of paramount importance to ensure that the project was fully funded.

Britain & Ireland was divided into 152 blocks of 10-km squares to break down into more manageable units the tasks of organisation of fieldwork and validation of data. In the UK these 'Atlas Regions' followed the existing BTO regions with the associated network of Regional Representatives. Some of the larger Scottish BTO regions were subdivided, for example the North Argyll BTO region was split into two Atlas Regions. In the Republic of Ireland, new regions were created by amalgamating groups of 10-km squares to approximately follow Irish county boundaries. The Atlas Regions are illustrated in Figure 1.2.

Within each region there was one nominated Regional Organiser whose tasks were to recruit observers for Timed Tetrad Visit (TTV) coverage and to direct effort for Roving Records (see Chapter 2 for details of approach and methods). Largely, Regional Organisers decided for themselves how best to achieve the target coverage. In some large regions they took on one or more Assistant Regional Organisers to help divide the task further, and in others 'square stewards' were identified to act as organisers of fieldwork in individual 10-km squares. Regional Organisers and their assistants were invited to join an online email forum to discuss problems and share good practice. Each region also had between one and nine Regional Validators who had the task of checking the tens of thousands of records submitted to the Atlas in their region. These regional tasks were fundamental to the success of the Atlas and we are extremely grateful to the dedicated organisers and validators who undertook them (Table 1.1).

1.4.3 Training for Regional Organisers, Validators and observers

Three training days for Regional Organisers in England were organised prior to the start of fieldwork in mid to late 2007, with further days in Wales and Scotland organised during early 2008. The main focus of these days was training in the Atlas methods and familiarisation with the interactive Bird Atlas Online system (see Chapter 3), particularly the *Tetrad Management* module (see Section 3.7.1).

ALYN WALSH

For some of the remotest and least populated parts of Britain and Ireland, such as Eirk Bog, Co. Kerry, *Bird Atlas 2007–11* relied upon records provided by visiting walkers and special visits by Atlas volunteers.

In Scotland the Building Bird Monitoring Scheme (BBMS) funded training sessions for Regional Organisers and Validators. These were instrumental in explaining how to use the online system and in the later stages were used to suggest strategies for targeting uncovered areas. The scheme also funded courses for fieldworkers in Scotland, providing training in fieldwork methods and how to use the online system. The courses also generally built up observer confidence levels.

Other BTO training courses in bird monitoring covered the Atlas field methods too. Additionally, a number of Regional Organisers arranged local training at the start of the project and also held regular social events to encourage observer participation and to provide local feedback of coverage and results. These events included a morning gathering Roving Records in a specific area followed by a pub lunch or afternoon tea in a village hall.

In the Republic of Ireland, the Irish Atlas Organiser held numerous Atlas workshops and training events, focusing on recruiting volunteers and explaining the Atlas methods and online components. Specific Atlas events were organised, mainly by BirdWatch Ireland Branches to gather Roving Records from 10-km squares that were under-represented.

1.4.4 Bird Clubs and local tetrad atlases

Early in the planning for *Bird Atlas 2007–11*, BTO wrote to all regional bird clubs to explain the proposed fieldwork timetable and enquire whether they were planning to undertake a local atlas project alongside the national one. Many had previously completed a tetrad atlas concurrent with the *1988–91 Breeding Atlas* and were keen to run a repeat atlas to update information on distribution and assess local changes. Others were keen to embark on their first tetrad atlas. We explained our plans to collect data through the Internet and hoped that this logistical support and technological capacity would make some smaller atlases feasible.

Over the course of *Bird Atlas 2007–11*, 44 counties or regions in Britain and one in Ireland planned to run a local atlas, either from the outset in 2007/08, or later as it became clear that complete coverage was a realistic goal. It was mutually advantageous to the local atlases and *Bird Atlas 2007–11* to run concurrently. The local atlas team (usually a local society) and BTO had a shared understanding, in the form of an agreement, about the ownership and exchange of data. BTO supplied data for the local atlas at the end of each season, four months after the end of a fieldwork period. Two one-day meetings were organised at BTO HQ to discuss how the local and national atlases could work together to best advantage.

Though a small number of local atlas projects were abandoned part way through *Bird Atlas 2007–11*, many others completed on the same timescales as the national atlas and a small number continued fieldwork beyond 2011. Excellent local organisation, skilled data handling and mapping, and efficient writing teams have enabled some projects to publish soon after completion of fieldwork (Bland & Dadds 2012; Ferguson 2012; Forrester *et al.* 2012; Newnham & Crabtree 2012; Brenchley *et al.* 2013; Kirk & Phillips 2013).

1.4.5 Use of the Internet for data capture and volunteer engagement

The planning for *Bird Atlas 2007–11* took place at a time when the Internet was already a proven platform for the collection of data and feedback of results, although the use of applications on mobile telephones for data collection had yet to become firmly established. From an early stage in planning the Atlas it was decided that the Internet should be promoted as the main method for the submission of bird records, and that it should be run alongside traditional paper forms. This was an important decision to make early so that adequate funds could be raised to finance the advance planning and development of online modules for allocation of tetrads for TTVs, data collection, editing of records and the review and validation of records (see Chapter 3). It was necessary for much of this work to be completed ahead of the start of fieldwork in November 2007.

It turned out that over 97% of Atlas records were submitted online or uploaded through the website, illustrating the value of using the Internet for data collection. Its role for providing feedback and motivation to observers is difficult to gauge, although some figures on use of feedback elements of the Bird Atlas Online application can be found in Section 3.6.4.

The Internet will continue to play an important role in disseminating information, including providing an online resource of maps and tables from this and previous atlases. These online resources can be accessed at www.birdatlas.net.

1.4.6 Contracted fieldworkers and expeditions

Although the aim was for the majority of fieldwork to be undertaken by volunteer observers, it was recognised from the outset that some parts of Britain & Ireland have a low density of resident observers and that it would be necessary to arrange and promote coverage of these areas, either using paid contract fieldworkers or by supporting expeditions. This approach had also been adopted in previous atlases.

Contracted fieldworkers

BTO employed two full-time fieldworkers in year three of Atlas fieldwork to cover parts of western Scotland (Islay, Jura, Mull, Argyllshire, particularly the Kintyre peninsula) and Northern Ireland in both winter and the breeding season. In addition, five part-time fieldworkers were employed to cover parts of South Uist, Perthshire, Dunbartonshire, Lanarkshire, Ayrshire, Kirkcudbrightshire, Wigtownshire, Caithness, Wester Ross and Lewis. In year four, one fieldworker was employed full-time (winter and breeding season) in Wales, based in Carmarthenshire, and part-time fieldworkers undertook additional top-up fieldwork in Cornwall, Lincolnshire, North Yorkshire, Shetland and Northern Ireland, filling in a small number of gaps.

Bob Swann (Scottish Organiser) also carried out a large number of TTVs in Sutherland, Wester Ross, Lewis, Orkney, Shetland and Caithness.

TABLE 1.1 LIST OF ATLAS REGIONS AND THEIR REGIONAL ORGANISERS (INDICATED WITH AN ASTERISK) AND VALIDATORS

In the Republic of Ireland Brian Caffrey led on organising fieldwork and coverage in most counties, however many of the Validators (listed below) also greatly assisted with this process.

Code	Region	Organisers and Validators
CHANNEL ISLANDS		
CHAN	Guernsey, Alderney, Sark	Phil Alexander*, Phil Atkinson, Melanie Broadhurst, J P Down, Mark Lawlor
CHJE	Jersey	Tony Paintin*
ISLE OF MAN		
IMAN	Isle of Man	Pat Cullen*, Dawn Balmer*
ENGLAND		
AVON	Avon	Richard Bland*
BEDS	Bedfordshire	Nigel Willits*
BERK	Berkshire	Chris Robinson*, Ken & Sarah White
BIRM	Birmingham & West Midlands	Steve Davies*, Kevin Clements
BUCK	Buckinghamshire	David Lee*, John Gearing*, David Ferguson, Andy Harding
CAMB	Cambridgeshire	Louise Bacon*, Tony Fulford*, Dawn Balmer, Simon Gillings
CHMI	Cheshire (Mid)	Paul Miller*
CHNE	Cheshire (North & East)	John Headon*, Tony Coatsworth, Mark Eddowes
CHSS	Cheshire (South)	Charles Hull*
CLEV	Cleveland	Vic Fairbrother*
CORN	Cornwall	Stephen Jackson*, Dawn Balmer, Darrell Clegg, Greg Conway
CUMB	Cumbria	Clive Hartley*, Colin Gay*, David Piercy*, Stephen Westerberg*
DERN	Derbyshire (North)	David Budworth*, Dawn Balmer, Rod Key
DERS	Derbyshire (South)	David Budworth*, Dawn Balmer, Rod Key
DEVN	Devon	John Woodland*, Mark Blacksell*, John Twyford*, Stella Beavan*, Mark Darlaston, Mike Langman, Peter Robinson
DORS	Dorset	Mike Pleasants*, Dawn Balmer*, Jim Winsper*, Mark Darke, Su Gough, Kevin Lane, Jol Mitchell
DURH	Durham	David Sowerbutts*, Dawn Balmer, Tim Dean
EXNE	Essex (Northeast)	Simon Cox*
EXNW	Essex (Northwest)	Simon Cox*
EXSO	Essex (South)	Simon Cox*
GLOS	Gloucestershire	Gordon Kirk*, Richard Baatsen, Andy Lewis, John Phillips
HAMP	Hampshire	Glynne Evans*, Keith Betton, John Clark, John Eyre, John Shillitoe
HERE	Herefordshire	Steven Coney*, Chris Robinson*, Peter Eldridge, Nick Smith
HERT	Hertfordshire	Chris Dee*
HUNT	Huntingdon & Peterborough	Louise Bacon*, Martin Coates*, Dawn Balmer*, Simon Gillings, Derek Langslow
IWGT	Isle of Wight	James Gloyn*
KENT	Kent	Sally Hunter*, Tim Hodge
LANE	Lancashire (East)	Tony Cooper*, Bill Aspin, Jim Ormerod
LANN	Lancashire (Northwest)	Jean Roberts*, Chris Batty, Peter Crooks, Graham Hulme, Pete Marsh, John Wilson
LANS	Lancashire (South)	Stuart Piner*, Chris Batty, Robert Pyefinch, Neil Southworth, Frank Walsh, Steve White
LEIC	Leicestershire & Rutland	Tim Grove*
LINE	Lincolnshire (East)	Dawn Balmer*, Phil Espin*, Mark Grantham*, Anne Goodall
LINN	Lincolnshire (North)	Dawn Balmer*, Chris Gunn*
LINS	Lincolnshire (South)	Richard & Kay Heath*
LINW	Lincolnshire (West)	Peter Overton*
LOND	London South	Richard Arnold*
LONN	London (North)	Ian Woodward*
MANC	Manchester	Steve Suttill*, Steve Atkins, Judith Smith
MERS	Merseyside	Bob Harris*, Steve White
NBLD	Northumberland	Muriel & Tom Cadwallender*, John Day, Tim Dean, Mike Hodgson
NHAM	Northamptonshire	Barrie Galpin*, Bob Bullock
NKNE	Norfolk (Northeast)	Chris Hudson*
NKNW	Norfolk (Northwest)	Fred Cooke*
NKSE	Norfolk (Southeast)	Rachel Warren*, John Marchant, Steve Piotrowski
NKSW	Norfolk (Southwest)	Vince Matthews*
NOTT	Nottinghamshire	Lynda Milner*, Andy Hall, David Parkin
OXFN	Oxfordshire (North)	Frances Buckel*
OXFS	Oxfordshire (South)	John Melling*, Ben Carpenter
SCLY	Isles of Scilly	Will Wagstaff*, Dawn Balmer
SHRP	Shropshire	Allan Dawes*, John Arnfield, John Tucker
SOMT	Somerset	Eve Tigwell*, David Balance, Rob Grimmond, Stephen Moss, Julian Thomas
STFN	Staffordshire (North)	Martin Godfrey*, Nick Pomiankowski
STFS	Staffordshire (South)	Martin Godfrey*, Nick Pomiankowski
STFW	Staffordshire (West)	Martin Godfrey*, Nick Pomiankowski
SUFF	Suffolk	Mick Wright*, Andrew Easton, Andrew Green, Gi Grieco, Colin Jakes, Peter Lack, Nick Mason, Steve Piotrowski, David Walsh
SURR	Surrey	Hugh Evans*, Steve Chastell, Richard Denyer, Jeffrey Wheatley
SUSS	Sussex	Helen Crabtree*, Richard Cowser, Paul James, John Newnham
WARW	Warwickshire	Mark Smith*
WILN	Wiltshire (North)	David Wall*, Bill Quantrill
WILS	Wiltshire (South)	Bill Quantrill*, David Wall
WIRR	The Wirral	Paul Miller*
WORC	Worcestershire	Harry Green*
YKBR	Yorkshire (Bradford)	Mike Denton*, Dawn Balmer*, Dudley Hind, Lucy Yates
YKEA	Yorkshire (East)	Dawn Balmer*, Geoff Dobbs*, Richard Baines, Nick Moran, Lucy Yates
YKHA	Yorkshire (Central)	Mike Brown*
YKHU	Yorkshire Hull	Dawn Balmer*, Geoff Dobbs*, Ian Henderson
YKLW	Yorkshire (Leeds & Wakefield)	Dawn Balmer*, Andy Musgrove, Mike Denton, Chris Thaxter
YKNE	Yorkshire (Northeast)	Mick Carroll*, Tom Denney*
YKNW	Yorkshire (Northwest)	Gerald Light*, John Flood
YKRH	Yorkshire (Richmond)	John Edwards*, Nick Morgan
YKSE	Yorkshire (Southeast)	Dawn Balmer*, David Gains*, Mike Archer
YKSW	Yorkshire (Southwest)	Dawn Balmer*, David Gains*, Mike Archer
YORK	Yorkshire (York)	Rob Chapman*, Phil Bone, Peter Watson

continued ▶

◀ continued

SCOTLAND		
ABER	Aberdeen	Paul Doyle*, Ian Francis, Dave Gill, Hywel Maggs, Bob Swann
ANGU	Angus	Ken Slater*, Bruce Lynch*, Alan Barclay, Jon Cook, Ron Lawie
ARGN	Argyll (Mull, Coll, Tiree & Morvern)	John Bowler (Coll & Tiree)*, Shaun McCullagh (Mull & Morven)*, Paul Daw
ARGS	Argyll (main & Gigha) & Bute	Richard Allan*, Ron Forrester (Bute)*, Jim Cassels (Arran)*, Paul Daw, Bob Swann
AYRS	Ayrshire	Brian Broadley*, Bob Swann
BENB	Benbecula & the Uists	Brian Rabbitts*
BORD	Borders	Ray Murray*, David Parkinson, Graham Pyatt
CAIT	Caithness	Donald Omand*
CENT	Central	Neil Bielby*, Chris Pendlebury
DUMF	Dumfries	Edmund Fellowes*, Paul Collin
FIFE	Fife & Kinross	Norman Elkins*
IMAY	Isle of May	Norman Elkins*
INVE	Inverness (East & Speyside)	Hugh Insley*
INVW	Inverness (West)	Hugh Insley*
ISLA	Islay, Jura & Colonsay	John Armitage*, James How, David Jardine, Bob Swann
KINC	Kincardine & Deeside	Graham Cooper*
KIRK	Kirkcudbright	Andrew Bielinski*
LANA	Lanark, Renfrew & Dumbarton	John Knowler*, Iain Gibson
LEWH	Lewis & Harris	Chris Reynolds*, Brian Rabbitts
LOTH	Lothian	Mark Holling*, Ian Andrews, Neville Crowther, Alan Heavisides
MORA	Moray & Nairn	Bob Proctor*, Martin Cook
ORKN	Orkney	Colin Corse*, Bob Swann
PERT	Perthshire	Richard Paul*, Michael Bell, Bob Swann
RHUM	Rum, Eigg, Canna & Muck	Bob Swann*
ROSS	Ross-shire	Bob Swann*, Simon Cohen*, Paul Copestake (NW Ross)*
SHET	Shetland	Dave Okill*, Martin Heubeck, Mike Pennington, Bob Swann
SKYE	Skye	Bob McMillan*
SUTH	Sutherland	Bob Swann*
WIGT	Wigtown	Geoff & Jean Sheppard*, Paul Collin
WALES		
ANGL	Anglesey	Tony & Karin White*, Peter Stuttard
BREC	Brecknock	John Lloyd*, Andrew King, Graham Rees
CAER	Caernarfon	Geoff Gibbs*, Kelvin Jones, David Lee, Rhion Pritchard
CARD	Cardigan	Moira Convery*, Russell Jones
CARM	Carmarthen	John Lloyd*, Sally Hall*
CLWE	Clwyd (East)	Anne Brenchley*, Jane Hemming, Ian Spence
CLWW	Clwyd (West)	Mel ab Owain*
GLAM	Glamorgan (Mid)	Wayne Morris*, Rob Nottage*, David Jenkins-Jones
GLAS	Glamorgan (South)	Wayne Morris*, Rob Nottage*, David Jenkins-Jones
GLAW	Glamorgan (West)	Bob Tallack*, Derek Thomas
GWEN	Gwent	Jerry Lewis*, Richard Clarke, Chris Jones
MERI	Merioneth	Dave Anning*, Tom McCanna, Rhion Pritchard
MONT	Montgomery	Jane Kelsall*, Dawn Balmer
PEMB	Pembroke	Bob & Annie Haycock*
RADN	Radnor	Carlton Parry*, Dawn Balmer*

NORTHERN IRELAND		
ANTR	Antrim & Belfast	Shane Wolsey*, Dawn Balmer, David Galbraith, Adam McClure
ARMA	Armagh	Shane Wolsey*, Dawn Balmer, Stephen Hewitt, David Knight
DOWN	Down	Shane Wolsey*, Dawn Balmer, Stephen Hewitt, Kerry Leonard
FERM	Fermanagh	Shane Wolsey*, Dawn Balmer, Michael Stinson
LDRY	Londonderry	Shane Wolsey*, Dawn Balmer, Charles Stewart
TYRO	Tyrone	Shane Wolsey*, Dawn Balmer, Michael Stinson
REPUBLIC OF IRELAND		
CARL	Carlow	Brian Caffrey*, Oscar Merne
CAVA	Cavan	Brian Caffrey*, Heather Bothwell, John Hand, Elizabeth McKenna
CLAR	Clare	Brian Caffrey*, Oscar Merne, Paul Troake, Stephen Ward
CORK	Cork	Brian Caffrey*, Dawn Balmer, Harry Hussey, Oscar Merne, Tony Nagle, Patrick Smiddy, Paul Walsh
DONE	Donegal	Brian Caffrey*, Dermot Breen, John Cromie, Dave Duggan, Martin Moloney, Neil Sharkey, Ralph Sheppard, Rob Wheeldon
DUBL	Dublin	Brian Caffrey*, Tom Kealy, Oscar Merne, Niall Tierney
GALW	Galway	Brian Caffrey*, Dermot Breen, Neil Sharkey, Paul Troake, Neal Warnock
KERR	Kerry	Brian Caffrey*, Jill Crosher, Geoff Hunt, Michael O'Clery
KILD	Kildare	Brian Caffrey*, Bernard King, Oscar Merne, Niall Tierney, Neal Warnock
KILK	Kilkenny	Brian Caffrey*, Edmund Foyle, Liam Lysaght, Paul Walsh
LAOI	Laois	Brian Caffrey*, Alex Copland, Charles Tweney
LEIT	Leitrim	Brian Caffrey*, Seamus Feeney, Fiona & Rob Wheeldon,
LIME	Limerick	Brian Caffrey*, Geoff Hunt, Tom Tarpey
LONG	Longford	Brian Caffrey*, John Hand, Rob Wheeldon
LOUT	Louth	Brian Caffrey*, Brefni Martin, Oscar Merne, Peter Phillips
MAYO	Mayo	Brian Caffrey*, Eoin McGreal, Neil Sharkey, Dave Suddaby, Rob Wheeldon
MEAT	Meath	Brian Caffrey*, Oscar Merne
MONA	Monaghan	Brian Caffrey*, Joe Shannon
OFFA	Offaly	Brian Caffrey*, Dermot Breen, Alex Copland
ROSC	Roscommon	Brian Caffrey*, Michael Bell, Neil Sharkey, Rob Wheeldon
SLIG	Sligo	Brian Caffrey*, Michael Bell, Seamus Feeney, Connor Mcelhinney, Neil Sharkey
TIPP	Tipperary	Brian Caffrey*, Kevin Collins, Edmund Foyle, Paul Walsh
WATE	Waterford	Paul Walsh*, Declan McGrath
WEST	Westmeath	Brian Caffrey*, Triona Finnen
WEXF	Wexford	Brian Caffrey*, Dawn Balmer, Simon Collins, Tom Kealy, Oscar Merne, Christian Osthoff, Paul Walsh
WICK	Wicklow	Brian Caffrey*, Oscar Merne, Stephen Newton, Christian Osthoff

BirdWatch Ireland employed three full-time fieldworkers in the first year of Atlas fieldwork to cover parts of the northwest (Counties Donegal, Sligo, Leitrim, Longford, Cavan and Roscommon). In the second year of fieldwork, Counties Tipperary, Offaly, Westmeath, Laois, Limerick, Kerry and north Cork were targeted. During the third and fourth years of fieldwork, the system of using the fieldworkers as a team of three travelling as one unit was altered, using instead contract staff to complete survey work independently in selected areas of low coverage. These contract fieldworkers completed TTVs in the priority chequerboard 10-km squares (see Section 2.2.4) and Roving Records throughout Ireland to ensure that minimum coverage was achieved.

Expeditions

As with previous atlases, BTO and BirdWatch Ireland provided financial support for volunteer observers willing to visit areas with poor coverage or areas with difficult access that required an expeditionary approach. Funds were made available, primarily to teams, to cover aspects of transport including ferry fares and basic accommodation. We thank everyone who organised and participated in expeditions (see Acknowledgements).

1.4.7 Publicity, promotion and resources

It was imperative that details of the Atlas field methods and means of data submission were circulated to as many British and Irish birdwatchers as possible. In the lead up to the Atlas and at key stages during it, the Atlas staff wrote articles for BTO, BirdWatch Ireland and SOC publications. Pieces were also written for the mainstream national and regional press, as well as articles for the main birding magazines and specialised magazines targeting groups such as mountaineers, anglers and environmental survey workers. Staff and Regional Organisers attended and spoke at many regional and county conferences and bird club meetings. They also provided much local publicity.

FIGURE 1.3 EXAMPLES OF ATLAS NEWSLETTERS

Six Atlas Newsletters in Britain and one in Ireland were produced between 2008 and 2012 to provide feedback to volunteer observers on progress with coverage and provisional results, and to encourage further participation. The front covers of two Newsletters are shown.

Bird Atlas 2007-11

February 2008

BTO SOC BirdWatch Ireland

Tremendous start

Twenty-four million bird records, twelve thousand tetrads visited, £189,000 raised from supporters, over 24 species with sponsors and great news for species such as Buzzard and Little Egret. It has been a dream start to Bird Atlas 2007-11.

In the late 1960s, during the planning of the first breeding atlas, there was great debate whether an atlas of distribution could be achieved by volunteer birdwatchers. Fortunately, optimistic BTO Council members supported the idea and the rest is history. Now, thanks to modern communication and a bigger pool of birdwatchers upon whom we can call, we have made tremendous progress with Bird Atlas 2007-11 in just a few months. In this newsletter we are delighted to be able to reveal some of the new distributions for species, based upon Atlas visits and BTO/RSPB/BWI BirdTrack results. Records from other schemes, such as BTO/CJ Garden BirdWatch and the BTO/JNCC/RSPB Breeding Bird Survey, will be used to 'top-up' the database in the near future. There is still plenty of birdwatching to do, especially in counties where bird clubs have challenged people to visit every tetrad and in the more remote areas of Ireland, Scotland, Wales and England, but we are confident of success.

Thank you!

We are grateful to everyone who has supported Bird Atlas 2007-11 so far, the most ambitious and important project ever taken on by the BTO. Most members of the BTO family are part of the Atlas, contributing by sending in their records from gardens, birdwatching excursions and Timed Tetrad Visits (TTV). In addition, 2,400 Members, Garden BirdWatchers and other supporters have already raised £189,000 for the appeal, through their donations and promises of long-term support. This number constitutes about 1 in 10 of the people to whom the appeal mailing was sent last April. To make sure that we have the funding we need for this project, we would like to increase that fraction to a quarter ... or even a third ... We appreciate every donation, however small.

Andy Clements (BTO Director)

Extra benefits from the tax man

We are delighted that charities can still claim an extra 28% Gift Aid to add to donations from tax payers, as revealed in the Chancellor's Budget Statement on Wednesday, 12 March. Anyone who is paying sufficient personal income tax or capital gains tax can add over a quarter to the value of each donation to the Atlas Appeal. Simply tick the box which asks us to check that we already have a Gift Aid declaration set up in your name and we will make sure that full advantage is being taken of this facility.

giftaid it

Graham Appleton (Head of Fundraising & Publicity)

Percent of expected species richness
Winter
Over 90%
75 - 89%
50 - 74%
1 - 49%
No coverage

Figure 1. There are still gaps in many areas but, with your help, we have found plenty of birds. Red circles show where we have received records for over 90% of the species we expect to find in a 10-km square, with orange showing over 75% and yellow over 50%.

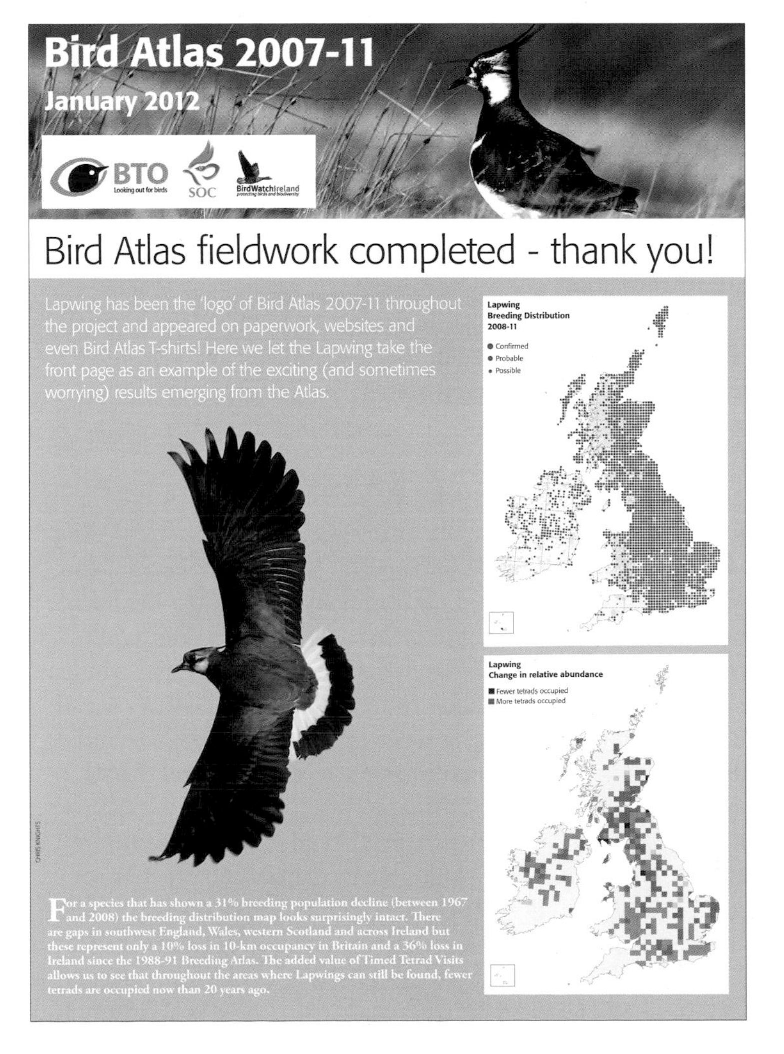

Bird Atlas 2007-11

January 2012

BTO Looking out for birds SOC BirdWatch Ireland

Bird Atlas fieldwork completed - thank you!

Lapwing has been the 'logo' of Bird Atlas 2007-11 throughout the project and appeared on paperwork, websites and even Bird Atlas T-shirts! Here we let the Lapwing take the front page as an example of the exciting (and sometimes worrying) results emerging from the Atlas.

CHRIS KNIGHTS

For a species that has shown a 31% breeding population decline (between 1967 and 2008) the breeding distribution map looks surprisingly intact. There are gaps in southwest England, Wales, western Scotland and across Ireland but these represent only a 10% loss in 10-km occupancy in Britain and a 36% loss in Ireland since the 1988-91 Breeding Atlas. The added value of Timed Tetrad Visits allows us to see that throughout the areas where Lapwings can still be found, fewer tetrads are occupied now than 20 years ago.

Lapwing
Breeding Distribution
2008-11
Confirmed
Probable
Possible

Lapwing
Change in relative abundance
Fewer tetrads occupied
More tetrads occupied

FIGURE 1.4 A PAGE FROM A BIRD ATLAS FIELD NOTEBOOK

Example of part of a completed winter-season TTV using one of the Bird Atlas field notebooks.

DAWN BULMER

During the fieldwork years, six Atlas Newsletters were sent to all participants and one was distributed in Ireland only (Figure 1.3). Throughout the project there was a huge and relentless email correspondence between national organisers, Regional Organisers and volunteer observers that required constant attention; this continued up to a year after the completion of fieldwork, when validation issues dominated.

In addition to regular articles and newsletters, a number of online and paper resources were made available to observers. A large number of paper Roving Record and TTV forms and instructions were printed and distributed. At least one paper Roving Records form was sent to as many birdwatchers as possible via the joint memberships of BTO, BirdWatch Ireland and SOC. Paper TTV forms were only available upon request to ensure that TTVs were only completed after being allocated by the appropriate Regional Organiser. A printed 'Letter of Introduction' was made available on request to help observers introduce themselves, and the aims of the Atlas, to landowners. Laminated A4 and A3 posters were also available for display in bird hides, visitor centres and other locations to publicise *Bird Atlas 2007–11* and the Bird Atlas Online system. Copies of the instructions and forms can be seen in Appendix 3.

Specially designed Bird Atlas field notebooks were produced (sponsored by Opticron) for recording TTVs in the field (Figure 1.4) and in a format that could be readily transferred to web submissions. These were well received and available for a small fee. Also produced were small information cards, tri-folded to the size of a credit card, containing the breeding evidence codes (see Table 2.2) and national grid systems with tetrad grid (see Box 2.1). Transparent overlays of this grid, to help in the identification of tetrads on 1:50,000 scale maps, were also made available on request and proved popular.

The Bird Atlas Online system provided feedback to observers but in addition, the Bird Atlas website was used as the central place to access information and resources, including:

- a list of Frequently Asked Questions;
- downloadable electronic copies of forms and instructions;
- general fieldwork advice and specific tips on methods;
- breeding evidence codes;
- help on reading maps and giving grid references;
- a 'Tip of the Week' during the breeding season drawing attention to the key species active at that time;
- links for printing maps of tetrads in Britain;
- a 'tetrad calculator' to help convert a grid reference to a tetrad code;
- protocols for mapping rare breeding birds;
- links to contact Regional Organisers throughout Britain and Ireland;
- Health and Safety advice for fieldworkers;
- a list of local atlases.

1.4.8 Validation of records

The validation of records collected by volunteer observers is considered an important step in ensuring records are fit for purpose and are widely accepted by the scientific community (Dickinson *et al.* 2010). Validation has been carried out for bird projects in the USA such as eBird (Sullivan *et al.* 2009) and Project FeederWatch (Bonter & Cooper 2012), and also for other taxa. For example, Gardiner *et al.* (2012) showed that species richness of ladybirds in the UK and USA was over-estimated when using data that had not been validated. Previous Britain & Ireland atlases had informal validation through the course of the projects in addition to formal validation of the draft maps prior to publication. Also, existing BTO online systems already used automated validation at the point of data entry (e.g. BirdTrack) and it was envisaged that this technology would be incorporated into the planned Bird Atlas Online system.

Nevertheless, simple typographic errors or incorrect species codes could still occur and these would need to be captured and corrected manually. We anticipated that some grid reference errors might arise, even with the use of point-and-click maps on the website. We envisaged that breeding evidence codes, which were unfamiliar to many observers, might be used inappropriately on occasion. Furthermore, we had to deal with the sensitive issue of occasional species misidentification.

A complex *Review and Validation* module (see Section 3.7.2) was built into the Bird Atlas Online system to allow a carefully selected pool of knowledgeable Regional Validators to check every record submitted to the Atlas. Validators either confirmed records or marked them with one of four query types: location, count, breeding evidence or identification. Observers with queries were encouraged to check their records to correct simple errors, or to provide additional information (including descriptions where necessary) to corroborate unusual records. In the case of the latter, the AWG felt that the Atlas should only use records of local rarities that had been accepted by county rarities committees and national rarities that had been accepted by the British Birds Rarities Committee or the Irish Rare Birds Committee.

The validation process was an essential one and rapidly resolved a number of simple errors that nonetheless would have impacted upon the accuracy of the final maps. Common issues included:

- incorrect letter codes for 100-km squares, especially near 100-km borders;
- central grid references for large BirdTrack sites spanning multiple 10-km squares;
- general grid reference errors;
- breeding evidence added to winter visitors still on wintering ground or on migration (e.g. Goldeneye, Redwing, Brambling);
- breeding evidence added in absence of any suitable breeding habitat in the square (e.g. terns, gulls, Grey Heron);
- lack of Flying (F) code for species flying over unsuitable habitat or commuting between feeding and roosting areas (e.g. Bewick's or Whooper Swans over cities, Pink-footed Goose flocks).

Whilst these issues were relatively easy to resolve, some identification issues were not. An atlas inevitably leads to a greater degree of coverage than most counties receive outside atlas periods, often by surveyors who are not part of the local birding scene. Consequently, atlases can generate records of species in hitherto unknown locations for which descriptions are rarely forthcoming. However, it is also a reality of any form of birdwatching that misidentification errors do arise and these cannot always be resolved. Three species topped the list for the most unresolved identification issue: the Willow Tit (2% queried), Tree Sparrow (1% queried) and Common Redpoll (9% queried). For the Willow Tit and Tree Sparrow, acute recent range contractions mean that in several counties these species are reputedly extinct or confined to a handful of well-known locations. In such counties contemporary records away from known sites are treated with a high degree of scepticism. Problems with the Common Redpoll stem from two sources. Firstly, it is a scarce winter visitor in some counties and requires a description owing to the difficulty of distinguishing some individuals from the Lesser Redpoll. Secondly, its unfortunate common name may have led to the misreporting of some Lesser Redpolls.

Some Goldeneyes displaying in early spring whilst still on their wintering grounds were erroneously coded as *probable* breeding and were corrected through the validation process.

For a tetrad to be eligible for TTV coverage its centre had to fall on land, ruling out coverage of some seabird islands such as the Farnes.

A minority of queried records (0.05%) remained in a state of query because they could not be resolved for one of several reasons. In total there were 9,391 outstanding queries broken down as follows: breeding status = 652 (7%), identification = 6,943 (74%), location = 1,275 (16%) and number = 521 (6%). Such records formed a tiny proportion of the total records submitted and were not used in the production of the Atlas.

1.5 AVAILABILITY OF THE ATLAS DATA

FINALISED DATA FROM *Bird Atlas 2007–11*, along with data from previous atlases are available for use by amateurs, professionals and consultants. Data can be requested from BTO (for Britain) and from BirdWatch Ireland (for Ireland) and applications are welcomed through the respective data-request systems of the two organisations. It should be noted that where data requests include sensitive species mapped at 20-km or 50-km for all or part of their range in this book, only the data as published will be provided, i.e. tetrad or 10-km information will not be disclosed. This is in accordance with the conditions of use of the providers of confidential data (e.g. the UK and Irish Rare Breeding Bird Panels).

As with previous atlases, data will also be made available to appropriate non-governmental and statutory organisations so that the fullest use of these data can be made for conservation purposes.

1.6 POPULATION ESTIMATES

PREVIOUS BRITAIN & IRELAND atlases have taken the opportunity to assess population size and to publish revised population estimates within the books. These have complemented other attempts to estimate national population size (Parslow 1973; Hudson & Marchant 1984; Marchant *et al.* 1990).

The population estimate figures published in the *1968–72 Breeding Atlas* were based largely on informed guesswork, typically using estimates of the number of pairs breeding in a 10-km square and multiplying this by the number of occupied squares and then applying a correction factor to take into account assumptions about how densities of different species varied across their ranges. Those published in the *1981–84 Winter Atlas* were also based on informed guesswork. The *1988–91 Breeding Atlas* represented an improvement in that it adopted a range of clearly defined methods, notably a modified version of the methods used by Hudson & Marchant (1984) using densities from the Common Birds Census (Marchant *et al.* 1990).

Whether to derive new population estimates as part of *Bird Atlas 2007–11* was discussed at length and several times by the AWG. A significant amount of work would be involved in producing new population estimates and Atlas data would be just one component of an exercise involving large volumes of data from other monitoring schemes and sources. Furthermore, for Britain and the United Kingdom there are now formal processes for producing and revising population estimates for all common wintering and breeding species (Stone *et al.* 1997; Baker *et al.* 2006; Musgrove *et al.* 2013) and separately for waterbirds in Britain (Musgrove *et al.* 2011) and Ireland (Crowe *et al.* 2008). None of these existed at the time of the previous atlases. Population estimates for non-waterbird species are not available for the Republic of Ireland, except for a small number of well-studied species (Madden *et al.* 2009). The UK Rare Breeding Birds Panel (RBBP) and the Irish Rare Breeding Birds Panel (IRBBP) are also important sources of population estimates for rare and scarce breeding birds (*RBBP 2010*; *IRBBP 2010*). Data from long-term monitoring schemes such as the Breeding Bird Survey (*BBS Report 2011*) and Countryside Bird Survey (*CBS Report 2010*), that were not running at the time of previous atlases, can be analysed using distance sampling methods to derive estimates of national population sizes (Newson *et al.* 2008a; Crowe 2011). These methods will continue to make an important contribution to producing population estimates, potentially in the future using the Atlas distribution data to produce more refined estimates.

On balance, weighing up the availability of contemporary population estimates and the tasks that would be required to produce robust new figures, the AWG opted not to include population estimates in the printed book. Instead, the plan is to provide links to existing published information and new information as it becomes available at www.birdatlas.net.

1.7 LESSONS LEARNT AND RECOMMENDATIONS FOR THE FUTURE

THIS ATLAS USED new methods and new technologies. It combined data from the winter and breeding seasons and from multiple sources, generating a vastly greater quantity of data than the previous three atlases. Based on our experience with *Bird Atlas 2007–11*, several important points can be made that should help in planning and designing future atlases.

1.7.1 Planning well ahead

- Start planning an atlas early, particularly if online data entry will be used, as time is required to set this up in advance of the start of fieldwork.
- Give careful thought to the funding of the project and try to secure funds upfront. Ensure that funding covers all aspects of the project including data collection, analysis, mapping and writing.
- Consider whether professional fieldworkers will be required to achieve coverage and how these will be funded. Unless it is clear that additional help will be required from the outset, do not deploy contract fieldworkers until the latter half of the project, by which time the difficult gaps will be evident.
- Consider whether stratified or systematic regular sampling can help to alleviate some coverage issues in areas with low observer

IAIN DOWNIE

density. In the case of this atlas we adopted a chequerboard grid of High- and Low-priority squares for TTV coverage in Ireland (Section 2.2.4).

- Consider timing in relation to other surveys. For example, if seabirds are an important component of the avifauna, can the atlas be timed to coincide with national seabird surveys so as to capitalise on efforts to visit remote seabird colonies?

1.7.2 Field methods

- Tetrad Population Estimates were found to be controversial and not as useful as hoped, despite the success of a similar method used in the *Catalan Breeding Bird Atlas* (Estrada *et al.* 2004). For further information about TPEs and analysis of data see Sections 2.2.4 and 5.4.4.
- Ensure that atlas instructions and promotion of atlas methods stress the importance of collecting the highest possible level of breeding evidence for each species in a 10-km square. Although *possible* breeding was recorded by many volunteer observers, it proved more difficult than expected to gather records of *probable* and *confirmed* breeding. Breeding evidence is not routinely recorded by birdwatchers and it took time to get the message across that recording signs of breeding was important. Many observers did not appreciate that breeding evidence should only be recorded when birds are in suitable nesting habitat.
- From the outset, encourage the collection of breeding-season records of birds not showing signs of breeding. Some of the most thought-provoking breeding-season maps were for species that do not breed in Britain & Ireland (e.g. Great Northern Diver).
- Colonial nesting species—given there was no national seabird census during *Bird Atlas 2007–11*, we developed methods to capture the minimum information required for the Atlas. It was difficult to find the balance between making the methods not too onerous for the volunteer observers yet gathering accurate information; using TTV data alone we found the encounter rate with seabird colonies was low (see Section 5.4.3).
- When a visit was made to a 10-km square in the breeding season but no nocturnal species recorded, a specific code ('Nilnoct') could be submitted. This was designed to help interpret the distribution maps of nocturnal and crepuscular species. The take-up was relatively small (see Section 5.3.6) and not as successful as expected, perhaps because it was not sufficiently emphasised and publicised.

Data from bird clubs and online bird information services were invaluable for improving the recording of scarce or nomadic wintering species such as Short-eared Owls.

1.7.3 Data sources and analysis

- Overall, records from online rarity services were very useful in adding species to 10-km lists, especially in winter. They did, however, cause some potential validation issues at a local level, particularly with the many records that referred to local and national rarities. Descriptions were not received by the County Recorder for many records reported to BirdGuides and IrishBirding.com and therefore could not be used in the Atlas.
- The value of records from some ringers and nest recorders could have been increased if accurate grid references had been provided. Some recording sites had a single central grid reference with a 'grid reference accuracy' value (e.g. ±2 km) to denote that records spanned a large area. Consequently, up to c.10% of these records could not be used in the Atlas as the record could not be pin-pointed to a single 10-km square. Some records did not have a grid reference at all and the location was submitted as the county name only. This problem also applied to some records received by the RBBP from some Schedule 1 licence holders and consequently none of these records could be used in the Atlas either. If records are to be used for conservation purposes then it is essential that accurate, high resolution grid references are provided and that, to maintain security, measures are put in place to present data at the most appropriate resolution for the species concerned (see Sections 2.3 and 4.1.3).
- It is important to weigh up carefully the potential benefits of different data sources: we had envisaged that data from the Ringing Scheme and Nest Record Scheme would be of high value for improving breeding evidence but the number of records that *confirmed* breeding proved to be surprisingly low (see Section 5.3.4). This may have been because participants were already entering their records directly to the Atlas, or because the species included in those schemes were often those that were relatively easy to confirm by conventional means.
- Validation was an essential task but took far longer than anticipated because of the immense quantities of data involved. The use of online data screening helped reduce data inputting errors but it was still vital that every record was checked by local experts. Nevertheless, short-cuts could have been taken. For example, breeding-season records lacking breeding evidence could from the outset have been excluded from validation, or the number of species included in the auto-validation routine could have been increased. The magnitude of this online module also caused problems from a hardware and software perspective (see Section 3.9).
- Records can be gleaned from a wide range of sources including groups not usually involved in bird recording. Examples are foresters, gamekeepers and local bat groups who will be out at night and may be able to record nocturnal species. Records of upland species may come from mountaineers and hill walkers who reach high-tops, especially in winter.
- It is important that all records can be traced back to the original observer if they need to be checked during validation.
- The use of out-of-season records with breeding evidence codes has been carefully documented in this atlas (see Appendix 4) and it will be important to do so in future atlases. It is unclear to what extent out-of-season records have been used for individual species in the compilation of distribution maps in previous atlases.
- If using a live online system to manage the atlas dataset, consider storing the date each record was submitted. This will help in the closing stages of the atlas when there is a need to finalise a dataset for analysis and writing yet some observers may still be adding or editing late data, especially in parts of the country with ongoing local atlases.

1.8 USING THIS BOOK

THE BULK OF the book is taken up with the species accounts and these have been written to be stand-alone—there is no requirement to read any introductory chapters beforehand, although Chapter 7 provides simple guidance on the interpretation of maps and figures along with key cross-references to the other chapters should readers wish to explore further. Each species account is supported by many statistics and these are given in Appendix 2.

However, there are five more chapters that give full details of the process of creating this atlas, from planning the fieldwork through to a discussion of what the Atlas has revealed. These chapters may be of particular interest to ornithologists and others engaged in conservation science but they have been written for a much wider audience and we hope that they will be particularly informative for the tens of thousands of citizen scientists who helped to gather the data for the Atlas.

The following chapter concerns the field methods and the wide range of data sources used to make the maps as comprehensive as possible (Chapter 2). Chapter 3 details how the Internet was used to collect data and engage with observers by providing personalised feedback. The methods used to analyse these data and to produce the maps are explained in Chapter 4. In Chapter 5 we review the patterns of coverage achieved in *Bird Atlas 2007–11* and tackle important questions concerning how recording effort differs among atlases and what impact this may have when assessing changes in range. Chapter 6 provides an overview of patterns and change in the British and Irish avifaunas for the c.40-year period spanned by atlases, detailing the emerging conservation implications and suggested areas for further work.

The appendices towards the end of the book provide a range of supplementary information including the names of over 17,000 volunteer observers, copies of the survey paperwork and statistics of range size and change for the 514 species and subspecies recorded during *Bird Atlas 2007–11*.

BRIAN RAFFERTY

CHAPTER 2

SURVEY METHODS AND DATA SOURCES

Simon Gillings, Dawn Balmer, Brian Caffrey & Bob Swann

2.1 TWO COMPLEMENTARY FIELDWORK COMPONENTS

THE METHODOLOGICAL DESIGN of any repeat atlas has to balance two opposing requirements: the first, to maintain consistency with previous techniques so as to allow direct comparability and the calculation of change; the second, to increase the effort so that, as far as possible, the places in which a species was not recorded are indeed places from which it was absent rather than simply places where the effort was insufficient to detect it. These requirements are incompatible unless the atlas survey has two complementary field methods, one of them systematic and, so far as possible, identical with what was used in previous atlases, the other designed to achieve the maximum possible effort.

This division of field effort into two components began with the *1988–91 Breeding Atlas*. For its systematic component, observers were asked to submit lists of species encountered in two 1-hour visits to a sample of eight tetrads in each 10-km square (see Box 2.1 for an explanation of tetrad and square definitions). The second component, to provide the additional effort needed to increase species lists, involved the submission of supplementary records for 10-km squares. These methods provided a strong foundation for the design of the breeding-season fieldwork for *Bird Atlas 2007–11*, with just some aspects of the field methods needing to be reviewed (e.g. whether to provide counts of individuals rather than merely list species). However, more major changes were necessary for the winter surveys.

The *1981–84 Winter Atlas* did not include a standardised field component. Instead, observers submitted species lists with maximum counts for each 10-km square, and indicated how long they had spent surveying that square; standardisation was attempted only at the mapping stage. Whilst this method is repeatable for mapping abundance, it cannot be considered repeatable for producing statistical measures of change because effort per 10-km square was highly variable. Furthermore, the *1981–84 Winter Atlas* methods did not provide fine-scale tetrad data, which are useful for conservation and research purposes. We were therefore keen to apply the two-component approach of the *1988–91 Breeding Atlas* to the winter methods. A pilot study was carried out over the winter of 2006/07 to determine the number and duration of timed visits needed to provide basic 10-km species lists and precise relative abundance data (Gillings 2008).

The two-component approach proved very popular and the phrases 'Timed Tetrad Visits', or 'TTVs', and 'Roving Records' soon became embedded in the vocabulary of Atlas observers. This chapter describes these field methods in detail, along with how they were implemented in practice.

2.2 FIELD METHODS

2.2.1 Grid references and defining tetrads

The respective national grids of Great Britain, Ireland and the Channel Islands were used as the basis for all recording (see Box 2.1). All records were submitted with at least 10-km precision and with tetrad (2-km) precision where possible. Gathering data at a finer spatial scale (e.g. 1-km square) was considered but rejected on the grounds that it would be too onerous and would exacerbate local concerns about recording scarce species. Consistent with the *1988–91 Breeding Atlas*, but contrary to the *1968–72 Breeding Atlas* and the *1981–84 Winter Atlas*, records from the Channel Islands and Fair Isle were submitted with actual grid references rather than being amalgamated under a single nominal grid reference at the level of the island.

Having decided on the grid system and resolution of squares to be used, an extensive Geographical Information System (GIS, using ESRI ArcMap, version 10) was constructed to hold high-resolution digital information. It included outlines of countries and counties, national grids at various resolutions and a number of environmental attributes used in subsequent analyses (e.g. altitude, land cover; see Section 4.4.3). The GIS was used to calculate geographical metrics such as the area of land and the coordinates of the centre point of each square at each grid resolution. As in the *1988–91 Breeding Atlas*, tetrads were deemed eligible for TTV coverage when the centre fell on land or in fresh water, with the low-water line used to define the boundary between land and sea. The GIS significantly speeded up the process of identifying eligible tetrads, though it proved too precise because it sometimes excluded terrestrial tetrads with centres in tidal creeks and included inaccessible offshore tetrads falling on sandbars. These anomalies were rectified following feedback from Regional Organisers. Another adjustment was that some ineligible coastal tetrads were reclassified as eligible if they contained habitat that was not otherwise represented by eligible tetrads in the 10-km square. This ensured that TTV coverage did not miss significant populations of coastal habitat specialists.

In total, 84,993 tetrads (in 3,896 10-km squares) containing land were identified. After making the adjustments detailed above, 79,675 tetrads (in 3,718 10-km squares) were eligible for TTV coverage (Table 2.1). These tetrads formed the basis of database tables used by the Bird Atlas Online application (Chapter 3), for example to allow observers to select tetrads for coverage. Though the 5,318 tetrads whose centre did not fall on land were not technically eligible for TTV coverage for the national atlas, they were included as additional tetrads in these tables so that observers participating in local atlases could achieve complete coverage.

2.2.2 Recording periods

Fieldwork was scheduled to span the four winters 2007/08 to 2010/11, and the four breeding seasons 2008 to 2011.

There was much debate within the AWG about what period constituted 'winter'. One option was to remain consistent with the *1981–84 Winter Atlas* and define winter as the middle of November to the end of February. Not only might this facilitate comparisons with previous data, but it would also avoid the late-autumn passage migration that continues into early November, when the distributions of some species are still not established—though, of course, the distributions of many wintering species are in constant flux throughout the winter months as they respond to

BOX 2.1 GRID SQUARES AND DEFINITIONS

The traditional recording unit in British and Irish national atlases of flora and fauna is a square measuring 10 km × 10 km, commonly referred to as a '10-km square'. Similarly, a '1-km square' measures 1 km × 1 km, a '20-km square' measures 20 km × 20 km, etc. Those terms are used in this atlas and also 2 km × 2 km squares are described as 'tetrads'. The term 'tetrad' refers to the fact that the square is made up of four 1 km × 1 km squares ('monads'). Similarly, a 10-km square is a 'hectad' (100 monads) and a 100-km square a 'myriad' (10,000 monads), with a 50-km square being a 'quarter myriad'; but these other technical terms are rarely used and are avoided here.

The identification of 100-km squares, 10-km squares and tetrads follows common usage. These, and new notation for 20-km and 50-km squares, are detailed below and illustrated in the figures:

- **100-km square:** in Britain a two-letter prefix applies to all grid references; in Ireland a single-letter prefix, to which a leading 'I' was added to facilitate data processing;
- **50-km square:** the prefix letters of the 100-km square followed by the ordinal compass direction (NE, NW, SE, SW) to identify a quadrant of 25 10-km squares, e.g. SPNE;
- **20-km square:** the prefix letters of the 100-km square followed by a single letter from the DINTY grid (see below) to identify the array of four 10-km squares, e.g. SP_G;
- **10-km square:** the prefix letters of the 100-km square followed by the easting and northing of the southwest corner of the 10-km square, e.g. SP11;
- **Tetrad:** the reference of the 10-km square plus a suffix letter (A–Z, omitting O) describing the location of the tetrad of four 1-km squares, e.g. SP11S. The arrangement of tetrad letters is often colloquially known as the 'DINTY grid' on account of the letters of the second horizontal row.

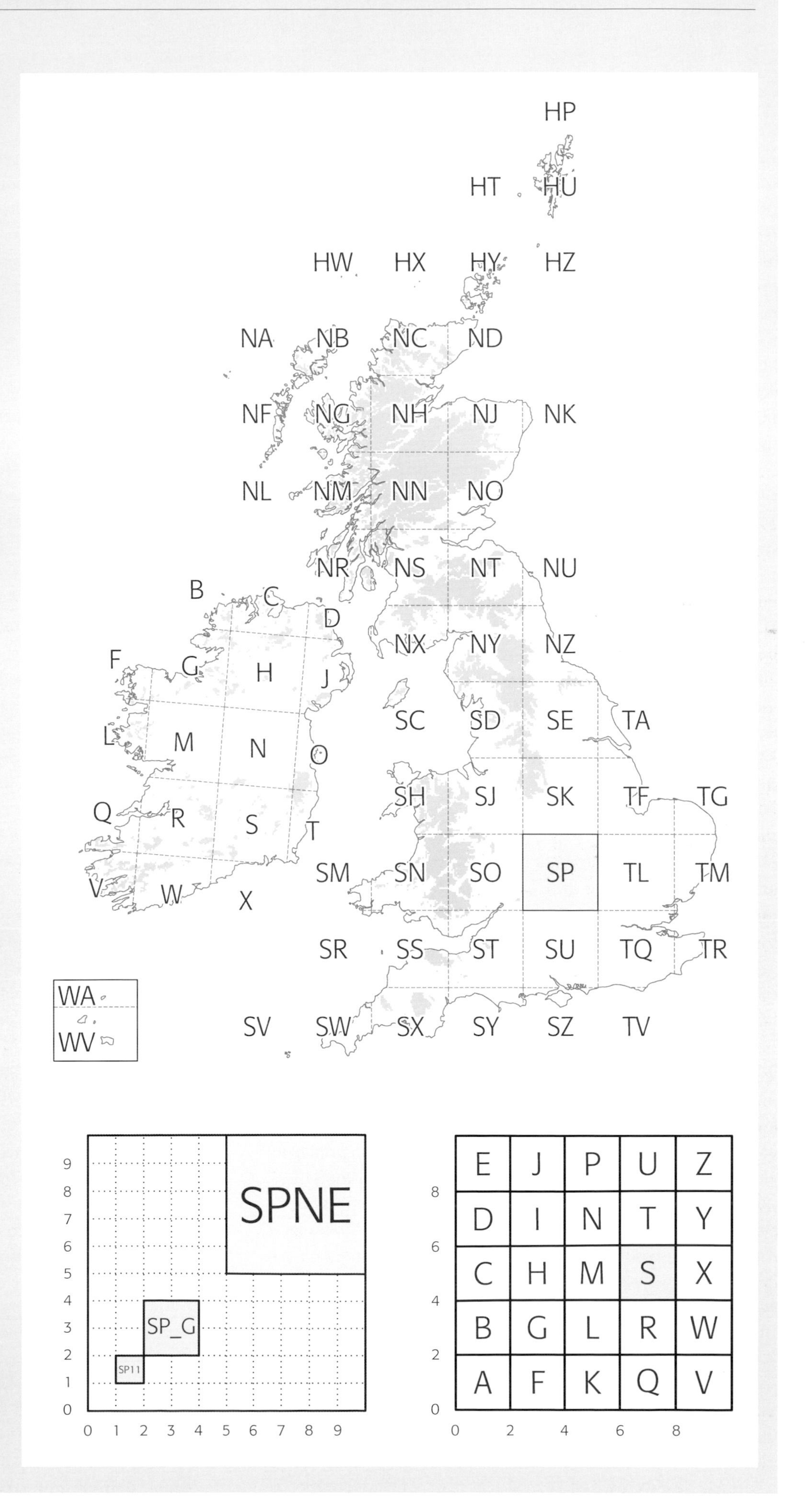

In remote areas, a single 2-hour Timed Tetrad Visit could be undertaken provided the timing of breeding-season visits was optimised to coincide with peak activity patterns.

changing food supplies and weather conditions. There was also concern that, during the recent mild winters leading up to the Atlas, some breeding species had returned to vacated inland or upland breeding areas as early as February, blurring the distinction between winter and breeding ranges. Given that relatively few species were thought to behave in this way, we decided that it was not a major impediment to continue the recording period up to the end of February. The final decision was partly swayed by the need to provide sufficient time for observers to complete pairs of TTVs, so we eventually opted for 1st November to 28th/29th February inclusive, giving a four-month window consistent with the breeding season.

In keeping with previous breeding atlases in Britain & Ireland, the core recording period for the breeding season was 1st April to 31st July, principally to span the main breeding periods of most species, and to accommodate the north–south trend in the onset of breeding. However, the breeding seasons of many species extend beyond these months. Of the 146 species treated in *A Field Guide to Monitoring Nests and Eggs* (Ferguson-Lees *et al.* 2011), 47 can begin nesting earlier: four species in January, 13 in February and 30 in March. For those species that begin nesting early, such as the Goshawk, Long-eared Owl and crossbills, the likelihood of detecting singing or displaying birds is much reduced by April. Moreover, the nesting season of 116 species can extend beyond July, into August (46 species), September (50), October (11), November (4) and even December (5). For certain seabirds, notably the burrow-nesting petrels and the Manx Shearwater, confirmation of breeding comes largely from ringing chicks, which can take place as late as September. For these reasons no strict seasonal limit was imposed on when distribution records and breeding evidence could be submitted. This open policy for data submission necessitated careful validation (see Section 1.4.8) and required a selective and species-specific approach to which out-of-season records were used in map production (see Section 4.2.2).

2.2.3 Building species lists ~ 'Roving Records'

The essence of atlasing is the production of species lists for every grid square containing land, which can be transposed into comprehensive species distribution maps. This was the principal aim of 'Roving Records', a name given to the casual records submitted at 10-km or tetrad resolution, and equivalent to the 'supplementary records' collected in the *1988–91 Breeding Atlas*. A single Roving Record comprised a grid reference (at least 10-km resolution), location name (to aid validation), date and species name. During the breeding season, observers were strongly encouraged to submit evidence of breeding (see Section 2.2.5) and, at the request of local atlases, space was provided so that observers could enter a count of the number of individuals seen or heard if they wished. Records of dead birds and signs (e.g. droppings) were not permitted.

This was the non-systematic component of the Atlas fieldwork: observers were free to visit as many grid squares as they wished, as many times as they wished, ideally surveying all the habitat types present in each square to ensure that species lists were as complete as possible. We also asked that each 10-km square should receive at least one visit at dusk or nocturnally during the breeding season to check for owls and other nocturnal species. Observers undertaking nocturnal visits on which no nocturnal species were found were asked to submit a Roving Record with 'Nilnoct' in place of a species name. The purpose of this was to improve the assessment of nocturnal coverage and to ascertain whether gaps in the distribution maps of nocturnal species were real or due to incomplete coverage. For mostly logistical reasons it was felt that the same target for nocturnal coverage could not be insisted upon for winter, nor would it be as beneficial as in the breeding season.

At the start of the Atlas, large numbers of paper Roving Records forms (see Appendix 3) were sent directly to the members and volunteers of the partner organisations, and to Regional Organisers who distributed them locally to potential surveyors. After receiving these initial paper forms, observers were strongly encouraged to submit the Roving Record data online through the Bird Atlas Online application (see Section 3.6.2). They were also encouraged when planning future roving effort to use a number of online tools on the Atlas results pages to identify squares with short species lists or with little breeding evidence (see Section 3.6.4). Some Regional Organisers arranged 'Hit Squads' or went 'Square Bashing' with groups of observers, to complete species lists and TTVs. In Ireland, many BirdWatch Ireland Branches successfully organised such events to help fill gaps in remote parts of their counties.

2.2.4 Standardised measures of detection and abundance ~ 'Timed Tetrad Visits'

Species distribution maps are highly informative in telling us where species occur but they obscure variations in abundance: a dot on a map could represent one pair or many. Representing geographic patterns in density adds a highly informative dimension to an atlas, and has become a common feature of many second-generation atlases (Gibbons *et al.* 2007). In practice, it is impossible to count all individual birds of all species in a 10-km square—not only would the effort required be prohibitive, but species differ greatly in detectability. Instead data can be collected using a standardised method, and then be used to generate a 'relative abundance' statistic that indicates not the absolute number of birds present but is an index to compare abundance of the species between different areas. The key to success is in the standardisation: we must be certain that an area of the map showing greater abundance truly does have more birds, rather than arising, for example, because an observer spent more time searching that square. Visits can be standardised either by using a fixed search area, as with the BBS and CBS, which use 1-km length transects, or by using a fixed time. For consistency with the *1988–91 Breeding Atlas*, and because it is easier in the field to measure time than area searched, fixed-time surveys of tetrads, known as 'Timed Tetrad

MARTIN ROUTLEDGE

Visits', were used. In addition to providing the data for relative abundance mapping, TTVs fulfilled the need for a systematic fieldwork component to allow statistical calculation of changes in distribution and abundance between atlases. Full instructions and copies of survey forms provided to observers can be found in Appendix 3, and the following sections discuss the detail and reasoning behind these methods.

How many tetrads to survey?

As in the *1988–91 Breeding Atlas*, to ensure an even and geographically representative spread of TTV coverage and a minimum sample size of coverage in each area, a coverage target was set of a minimum of eight eligible tetrads in each 10-km square. On the coast and on islands where the 10-km square was largely in the sea and where there were fewer than eight eligible tetrads (see Section 2.2.1), the target was to survey all eligible tetrads. For ease of reference, hereafter this is called the 'Rule-8' target. The Rule-8 target was applied to all 10-km squares with any eligible tetrads in Britain, the Isle of Man and the Channel Islands. An adjustment was made in Ireland, where the human population density is roughly one quarter of that in Britain and there are considerably fewer birdwatchers to achieve the same level of systematic TTV coverage. There, a representative distribution of TTVs was achieved by adopting a chequerboard of High- and Low-priority 10-km squares. High-priority 10-km squares were defined as those 10-km squares where the sum of the easting and northing numerals was an odd number (e.g. IH01, IH03, etc). The Rule-8 target was then applied to all High-priority 10-km squares with any eligible tetrads. TTV coverage in Low-priority 10-km squares was still encouraged but without an expectation of meeting the Rule-8 target.

Applying the Rule-8 target in this way gave an overall target of 24,444 tetrads to be surveyed in Britain & Ireland (Table 2.1). In many areas the Rule-8 target was actually exceeded, especially in the 45 counties and regions that embarked on local tetrad atlases. In those areas, observers aimed to survey all tetrads with any land. In Ireland many TTVs were also completed in Low-priority 10-km squares, particularly in the more densely populated areas. For these reasons the intensity of tetrad sampling was geographically highly variable (see Section 5.4) and analytical procedures were developed to produce unbiased maps and statistics (see Section 4.4–4.6).

For the *1988–91 Breeding Atlas*, observers were permitted to choose which eight tetrads to survey, rather than having a predetermined regular or random selection imposed upon them. The AWG discussed whether a random selection should be implemented, and also whether repeat surveys of tetrads covered in the *1988–91 Breeding Atlas* should be encouraged to facilitate future analyses. The AWG concluded that, as there was no evidence of bias in coverage during 1988–91, and that, since habitats and land access that might determine observer choice of tetrads were likely to be similar, it was best to use the same advice as given in the *1988–91 Breeding Atlas*. It was expected that this

TABLE 2.1 NUMBERS OF 10-KM SQUARES AND TETRADS, BY COUNTRY, ACCORDING TO ELIGIBILITY FOR TTV COVERAGE

The last column gives the number of tetrads needed to satisfy the Rule-8 target. The Rule-8 target figures for Irish Low-priority squares are given in parentheses as these were not part of the overall target.

Country	Number of 10-km squares		Number of tetrads		
	With 1+ eligible tetrads	With 0 eligible tetrads	Eligible for TTV (centre on land)	Ineligible for TTV (centre in sea)	Rule-8 minimum coverage
England	1,430	37	32,898	1,069	11,108
Scotland	1,031	81	20,021	2,569	7,582
Wales	253	13	5,304	327	1,904
Channel Islands	11	6	57	50	50
Isle of Man	13	1	149	40	77
Northern Ireland chequerboard High-priority	79	2	1,737	107	611
Northern Ireland *chequerboard Low-priority*	*79*	*2*	*1,734*	*95*	*(612)*
Republic of Ireland chequerboard High-priority	408	17	8,895	540	3,112
Republic of Ireland *chequerboard Low-priority*	*414*	*19*	*8,880*	*521*	*(3,123)*

would most likely result in a high overlap with *1988–91 Breeding Atlas* coverage without the need for additional rules. So observers were simply asked to select eight eligible tetrads approximately in accordance with landscape and habitat availability within the 10-km square.

Visit timing and duration

For each selected tetrad, the aim was to make one pair of timed visits during the breeding season and one pair during the winter. Each pair of visits consisted of an early-season visit, which had to be completed during the first two months of the season (November–December or April–May), and a late-season visit to be completed in the remaining two months (January–February or June–July). Individual breeding-season and winter visits did not necessarily have to be completed during a single season and could be carried out in any order.

The choice of date of visits within a period was left to the observer to decide, though general advice was given to synchronise visits with regional peaks in bird activity. In southern areas this meant aiming to complete late breeding-season visits before July; in northern areas, early breeding-season visits were often delayed to allow for the arrival of spring migrants. It was strongly recommended that periods of inclement weather likely to affect bird detectability were avoided. Observers were encouraged to undertake breeding-season TTVs during mornings, while winter TTVs could be conducted at any time of day except the first and last hours of daylight when roost movements could increase the probability of double counting. The recommendation was that coastal tetrads be surveyed between mid and high tide, when birds on mudflats and on the sea would be more concentrated closer to the shoreline, and thus easier to identify and count.

As a minimum, each TTV should have lasted exactly one hour, during which all birds encountered within the tetrad were to be counted. The route through the square should have been planned to represent the main habitats present and conducted at a walking pace. Optionally, observers could add a second hour to cover additional ground within the tetrad. The purpose of the second hour was to give greater precision to relative abundance estimates and to improve tetrad species lists for local atlases. In remote areas an exception was made in that a single 2-hour TTV could be made rather than two separate 1-hour visits. In such cases, observers were encouraged to make the visit during mid-season, when most species ought to be present and active, in order to maximise the opportunity to detect the species present and to record breeding evidence.

First- and second-hour counts were recorded separately to allow the greatest flexibility for subsequent analyses, especially to allow direct comparison with the 1-hour visits from the *1988–91 Breeding Atlas*. An allowance was made for observers to 'stop the clock' if they encountered large or mixed flocks that were time-consuming to identify and count. In those tetrads where it was not possible to survey continuously for a whole hour, for example because a river limited access to part of the square, stopping the clock was permitted to enable the observer to move to the other part of the tetrad. However, this was an exception and clock stopping and relocating were strongly discouraged to avoid any TTV becoming a series of disconnected point counts, which might have led to 'cherry-picking' of habitats. It might also have led to under-recording of species that are detected mainly by being flushed by the observer.

What to count?

During winter visits, observers were asked to count all live individuals of all species encountered, irrespective of age, provided they were 'using the square'. This meant that birds flying over at height (e.g. migrating Pink-footed Geese and gulls flying to a roost outside the tetrad) should have been excluded but those hunting on the wing included (e.g. Sparrowhawks). During breeding-season visits,

Although the main focus of TTVs was to record abundance, behaviour such as food-carrying or the presence of fledged young could be noted to confirm breeding.

observers were instructed to count all live full-grown birds. Immature individuals of species such as Cormorants, eagles and gulls from previous seasons should have been counted but not birds that hatched in the same year. Distinguishing adults from juveniles almost certainly became more difficult later in the season for species such as pipits and larks on moorland, and passerines (e.g. tits, warblers, finches) in woodland and farmland. As required for the winter TTVs, birds flying over and not associating with the tetrad were to be omitted. Observers were asked to pay special attention to ensure that Swifts and hirundines, which are most often recorded in flight, were included where they were obviously using the area. On the coast, birds could be counted as far offshore as the observer was confident in identifying them, provided they were still within the tetrad.

During the breeding season there was no requirement that the counted birds must be breeding in the square. Therefore, TTV counts could include, for example, Fieldfare flocks prior to departure, Wheatears stopping off on northward passage and Black-headed Gulls feeding in fields away from colonies (which were counted separately, see below). See Section 4.4 for details on how these data were treated in the production of abundance maps and other statistics based on TTV data.

Space was provided on paper and online TTV forms for observers to record additional species encountered after the allotted time but within the tetrad (to be treated as additional Roving Records). A column was also provided for recording incidental breeding evidence (see Section 2.2.5), but TTV instructions stressed that the main purpose of TTVs was to produce counts and that observers should not use up the time to explicitly seek evidence of breeding. However, revisiting parts of the tetrad on the way out, or on a subsequent visit, to amass breeding evidence was encouraged.

Counts of colonial species at colonies

The fifth page of the breeding-season TTV form (see Appendix 3) permitted observers to indicate the presence of colonies of the following species: Fulmar, Manx Shearwater, Storm Petrel, Leach's Petrel, Gannet, Cormorant, Shag, Little Egret, Grey Heron, Rook, Sand Martin and all skua, gull, tern and auk species. Except for three species that require specialised techniques (Manx Shearwater, Storm Petrel and Leach's Petrel), optional counts were requested of the size of the colony. The count units were species-specific to make data consistent with existing monitoring schemes. For the Fulmar, Gannet, Cormorant, Shag, Little Egret, Grey Heron, Rook and Sand Martin observers entered the number of 'Apparently Occupied Nests' (AONs); for Guillemots and Razorbills they entered the number of individuals in the colony. For the remaining species the preferred method was the AON, but observers could enter the number of individuals if they wished. Observers were asked to submit the maximum colony count from their pair of breeding-season visits to the tetrad. If any of these species were encountered away from colonies (e.g. off-duty gulls feeding in fields), they were to be counted as normal on the main part of the form.

How many birds were really present? ~ 'Tetrad Population Estimates'

With very few exceptions, one or two hours are insufficient to cover all the land area of a tetrad to count all the birds present. Moreover, even in those parts of a tetrad surveyed adequately, for the majority of species, some individuals will remain hidden and go undetected. There are a number of field methods, backed up with complex analytical techniques, that allow bird detections in the field to be converted to formal estimates of density by attempting to correct for imperfect detection (Thompson 2002). The AWG felt that though it would be a great advance to have absolute density estimates for all species, the increased complexity of these field methods might impact upon the level of observer uptake required to achieve the Rule-8 target coverage in Britain & Ireland. Furthermore, the data derived by these methods would not be comparable to those from the *1988–91 Breeding Atlas*.

Instead, observers were asked, if willing, to estimate the total number of adult birds likely to be present during the breeding season or in winter in each tetrad they surveyed. The *Catalan Breeding Bird Atlas* (Estrada *et al.* 2004) adopted a similar approach by asking its observers to estimate the number of pairs within order-of-magnitude bands (1–9, 10–99, etc) for each surveyed 10-km square. In subsequent analyses, these estimates were reasonably well correlated with independent, more formal estimates (though there was greater disparity for the more abundant species). This approach was extended to the tetrads surveyed for TTVs rather

Skeins of Pink-footed Geese passing over a tetrad or 10-km square were recorded as 'flying' as they were not using the resources of the square.

than 10-km squares. The general premise of this method is that the observer covering a tetrad has the best knowledge of which patches of habitat they have visited and, with some rough calculations, is best placed to convert the number of birds seen to an estimate of the number of individuals that may have been present. To reiterate the approximate nature of these figures, we stated that estimates should be given to the nearest 1, 10, 100, and so on. These calculations require a reasonable knowledge of the habitat preferences of species encountered during a TTV and some appreciation of how the detection of individuals differs among species (see Box 2.2). We fully accepted that not all TTV observers would be confident in providing such estimates, and accordingly made the submission of Tetrad Population Estimates (TPEs) optional.

2.2.5 Breeding evidence

Recording evidence of breeding has always been a standard feature of atlasing, with the result that the terms *possible* breeding, *probable* breeding and *confirmed* breeding have become commonplace, though their meaning and interpretation is sometimes confusing (see Section 7.2.2). For this atlas observers were requested to submit the codes shown in Table 2.2, reflecting the evidence they saw in the field, rather than just the level of breeding confirmation. This was done to aid the validation process and so that, if needed, we could upgrade or downgrade the breeding category for the behaviour of individual species. For example, observations of nest building, which would normally constitute only *probable* breeding, were upgraded to *confirmed* breeding for many species in the *1968–72 Breeding Atlas*.

BOX 2.2 TETRAD POPULATION ESTIMATES

The following worked example was given to observers, to indicate a suitable approach that they might consider when producing Tetrad Population Estimates in their tetrad.

This hypothetical tetrad was covered for two 1-hour visits in the breeding season. The route walked on each visit is shown as a red dashed line and follows a road with substantial hedges either side and also follows a footpath alongside the block of woodland in the west of the tetrad. The rest of the tetrad is mostly open farmland with good visibility across the flat fields. The mature woodland in the west is dense and most of the trees are covered in Ivy. Along the hedgerow it is likely that most birds were detected but small birds in the middle of the field could have been missed, as may others in the small patches of woodland. In the forest block, only birds audible from the path were detected. The likely detection distance for birds (especially songbirds) is within 100 m either side of the route walked (shown approximately by the blue-shaded area) although singing Skylarks could be detected across a wider area (shown approximately by the yellow-shaded area).

On the early visit, 30 Chaffinches were recorded (of which approximately 25 were singing males), and 14 were recorded on the late visit (of which approximately 10 were singing males). Given the route taken in the tetrad and habitat covered (good quality habitat for Chaffinches along the hedgerow), the habitat preferences of the species (hedges, woods) and the area of the tetrad not visited (mostly open farmland with no wooded boundary and the dense woodland), the TPE was estimated to be 100. This value was formed by taking the highest count of singing males (25) and doubling to take into account females not detected (total = 50) and then doubling again to take into account the suitable habitat (for Chaffinch) not visited, which gives an estimate of 100. This may be an underestimate or overestimate of the real number

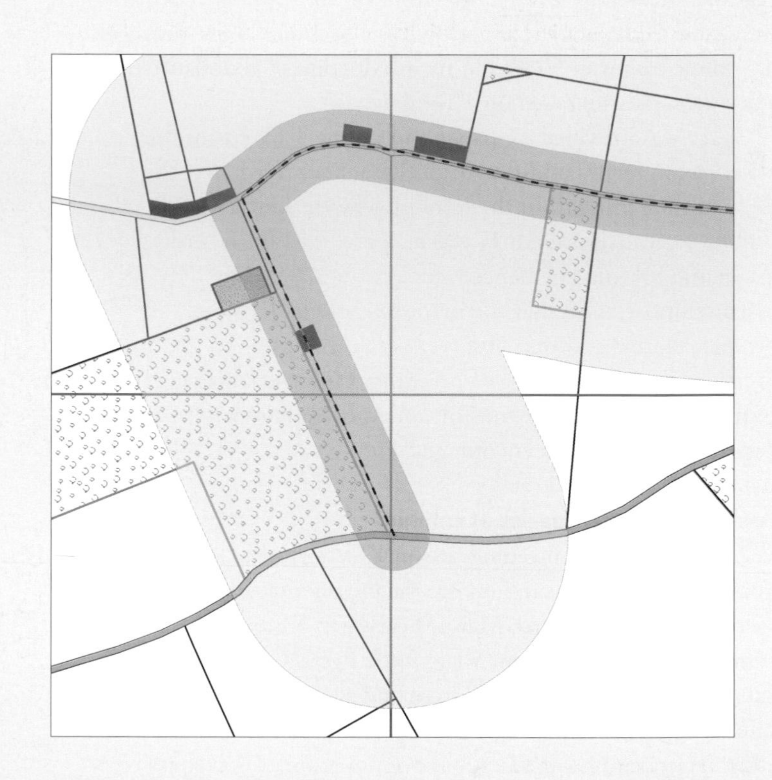

but it is likely to be broadly correct.

For Skylark (where singing birds are readily detectable over a greater distance than Chaffinch) 16 were recorded singing on the early visit and 15 on the late visit and a TPE of 40 seemed reasonable. This number was estimated by taking the maximum count of singing birds (16) and then doubling to take into account females (total = 32) and then adding a few more that may have been in the suitable habitat not covered in the southeast corner of the tetrad. It is likely that this is still an underestimate of the population of Skylarks in the tetrad as not all singing birds would have been detected on the visits.

DAWN BALMER

Whilst we had no intentions at the outset of the Atlas to follow this precedent, we decided that it was advisable to record unsummarised data to afford the greatest flexibility in subsequent analyses.

To aid the recording of passage or summering birds, and to distinguish them from breeding individuals, we added two new codes which could be applied to birds thought to be migrants, or summering, respectively (Table 2.2). Making a distinction between singing *possible* breeders and birds singing during migration is not always easy, although there is a prerequisite for *possible*, *probable* and *confirmed* breeding codes that individuals must be present in suitable breeding habitat. This is important because 10-km squares registering only migrant or summering status are not counted in estimates of range size or change statistics (see Section 4.2.2). Interpretation of breeding evidence is discussed further in Section 7.2.2.

Birds passing over the tetrad or 10-km square and not using its resources were to be coded as 'flying over' (Table 2.2). This code was not to be applied to species such as raptors and hirundines if they were hunting or feeding within the square. Whether birds were flying over or using the tetrad also dictated whether they were included on TTVs (see Section 2.2.4). Discussion of the treatment of flyover records for distribution mapping can be found in Section 4.2.

2.3 TREATMENT OF SENSITIVE RECORDS

THERE ARE A number of reasons that may make the locations of certain breeding and wintering species sensitive (see Section 4.1.3, Box 4.2). We were keen to maximise the submission of records of these species with high accuracy and precision, and to publish their distribution patterns as informatively as possible, but without increasing the risk of disturbance or persecution. The advantage of being able to distribute real-time feedback across the Internet, based on the data from thousands of observers, could have become a serious disadvantage if sensitive data were perceived to be freely available at an inappropriate scale. To alleviate this concern the Atlas team worked closely with the Rare Breeding Birds Panel (RBBP, covering the UK) and the Irish Rare Breeding Bird Panels (IRBBP, covering Ireland) to set a minimum resolution of 10-km or 100-km for each rare species, thus limiting the precision with which spatial information was publicly presented during the data collection phase of the project. In addition, a small number of species falling outside the remit of RBBP/IRBBP but locally scarce in some parts of their range (e.g. Pied Flycatcher) were set a minimum resolution of 10-km. These decisions dictated whether distribution information was used in publicity and feedback material on paper and online, and if so, at what resolution.

Most restrictions applied to the breeding season (March to August inclusive) but a handful of highly sedentary breeding species were also assigned restrictions during non-breeding periods (e.g. White-tailed Eagle and Crane). Sometimes the resolution differed markedly between Britain and Ireland. For example, there were no restrictions to mapping the Corn Bunting in Britain but at the start of the Atlas the species was thought to be extinct in Ireland, so a resolution of 100-km was assigned to protect any remnant populations discovered during the project.

2.4 OTHER DATA SOURCES

THOUGH DATA WERE collected from all parts of Britain & Ireland by Atlas observers, there was still the possibility that some species or breeding confirmations might have been missed. This was likely to be especially true for localised species requiring specialist knowledge. As in previous atlases, the Atlas distribution data were supplemented using casual records compiled from bird clubs and using surveys and research organised by BirdWatch Ireland, BTO, JNCC and RSPB. Combining data sources in this way would result in some distribution maps with variable recorder effort in different parts of the map, but the aim was to produce comprehensive maps with as near as possible saturation of recorder effort, at least for the regularly occurring species.

The datasets listed below were analysed and summarised to produce data conforming to the Roving Record format (Section 2.2.3) for incorporation into the main Atlas dataset. Those likely to have the greatest impact on evidence of breeding (e.g. the Nest Record Scheme and the Ringing Scheme) were loaded annually as soon as they became available. This ensured that feedback included the most complete assessment of breeding distributions currently available, so that observers could target areas and species most effectively. Other datasets were loaded only at the end of the project as they became available, and as time permitted. During the first two years of the Atlas we tended to load all records from each source. Towards the end of the Atlas project, and especially for datasets requiring validation, only 'gap-fillers' were loaded, i.e. those records that either added a new distribution dot, or improved the breeding evidence of an existing dot. Datasets that were added to *Bird Atlas 2007–11* are listed in the next two sections, drawing attention to any issues that were particular to each.

TABLE 2.2 BREEDING EVIDENCE CODES AND THEIR DEFINITIONS

Note that codes for *possible* breeding, *probable* breeding and *confirmed* breeding should be applied only when the species is observed in suitable nesting habitat.

Level	Code	Description
Non-breeding	**F**	**F**lying over and not using the resources of the square
	M	Species observed but suspected to be still on **M**igration
	U	Species observed but suspected to be a s**U**mmering non-breeder
***Possible* breeding**	**H**	Species observed in breeding season in suitable nesting **H**abitat
	S	**S**inging male present (or breeding calls heard) in breeding season in suitable breeding habitat
***Probable* breeding**	**P**	**P**air observed in suitable nesting habitat in breeding season
	T	Permanent **T**erritory presumed through registration of territorial behaviour (song, etc) on at least two different days a week or more apart at the same place, or many individuals on one day
	D	Courtship and **D**isplay (judged to be in or near potential breeding habitat; be cautious with wildfowl)
	N	Visiting probable **N**est site
	A	**A**gitated behaviour or anxiety calls from adults, suggesting probable presence of nest or young nearby
	I	Brood patch on adult examined in the hand, suggesting **I**ncubation
	B	Nest **B**uilding or excavating nest hole
***Confirmed* breeding**	**DD**	**D**istraction **D**isplay or injury feigning
	UN	**U**sed **N**est or eggshells found (occupied or laid within period of survey)
	FL	Recently **FL**edged young (nidicolous species) or downy young (nidifugous species). Careful consideration should be given to the likely provenance of any fledged juvenile capable of significant geographical movement. Evidence of dependency on adults (e.g. feeding) is helpful. Be cautious, even if the record comes from suitable habitat
	ON	Adults entering or leaving nest site in circumstances indicating **O**ccupied **N**est (including high nests or nest holes, the contents of which cannot be seen) or adults seen incubating
	FF	Adult carrying **F**aecal sac or **F**ood for young
	NE	**N**est containing **E**ggs
	NY	**N**est with **Y**oung seen or heard

2.4.1 Survey and scheme datasets from BTO and BirdWatch Ireland

BTO and BirdWatch Ireland undertake a number of surveys and schemes as part of their core monitoring activities, plus many projects targeted on specific species, habitats or issues. These have the potential to provide valuable distribution records, often with breeding evidence attributes. Records were extracted from the following datasets. The survey abbreviation, geographical scope and partner organisations are indicated as appropriate (see also page 14 for abbreviations).

BirdTrack (Britain & Ireland; BTO, RSPB, BirdWatch Ireland, SOC, WOS)

By far the single largest source of additional data was BirdTrack. During the four years of the Atlas, over 10,000 observers submitted 8.1 million records from 1.1 million visits to 126,000 sites throughout Britain & Ireland. We used records from complete species lists, incomplete species lists and casual records, provided they came from sites defined as falling entirely within a single 1-km square, tetrad or 10-km square. In months outside the formal Atlas seasons, only records with breeding evidence were used.

BTO Nest Record Scheme (NRS, Britain & Ireland; BTO, JNCC)

Every year c.40,000 Nest Record cards are submitted, detailing nesting attempts for c.170 species (Dadam *et al.* 2012). They provide a wealth of evidence of breeding, including for some species that are difficult to confirm without finding nests (e.g. owls). We converted nest activity and outcome codes to breeding evidence to derive new Atlas records. The majority of NRS cards could be used but for some scarce breeding species, observers conceal exact nesting locations by providing a displaced grid reference with a grid accuracy value to indicate the proximity of the real location. Such cards were used only if the grid reference and grid accuracy unambiguously identified a single 10-km square or tetrad.

Ringing Scheme (Britain & Ireland; BTO, JNCC, NPWS)

Just over 1 million birds are ringed annually in Britain & Ireland (Dadam *et al.* 2012) and from these records we can extract three forms of breeding evidence: adults with a well-formed brood patch (brood patch codes 2 & 3; Redfern & Clark 2001) translated to *probable* breeding; chicks in the nest (pulli) and very recently fledged young (age code 1J; Redfern & Clark 2001) translated to *confirmed* breeding. These data are especially valuable for species whose skulking habits make detecting food provisioning visits or fledged young very difficult (e.g. petrels, owls, certain warblers) but they require careful validation to eliminate, for example, spurious evidence arising from post-breeding dispersal of juveniles. Ringing data can also provide information on birds present during winter, especially for skulking species and for habitats that are difficult to search (e.g. reedbeds). As with the Nest Record Scheme, records are provided with a grid reference and an accuracy value; in addition to masking sensitive nesting sites, these can also indicate that a precise grid reference describes the centre of a larger ringing site. A combination of grid reference and grid accuracy was used to identify the records that could be assigned unambiguously to a single 10-km square or tetrad.

BTO Heronries Census (UK; BTO)

At the end of the Atlas the Heronries Census (Marchant *et al.* 2004) was checked for additional gap-filling records of Grey Heron and Little Egret. All Grey Heron records were loaded as *confirmed* breeding (Occupied Nest) records. However, because the Little Egret is spreading rapidly, not all of these records related to *confirmed* breeders, so records were assigned breeding evidence in accordance with observers' comments.

Breeding Bird Survey (BBS, UK; BTO, JNCC, RSPB)

Countryside Bird Survey (CBS, Republic of Ireland; BirdWatch Ireland, NPWS)

More than 3,200 1-km BBS squares and over 300 1-km CBS squares are surveyed each spring to record relative abundance along transects (*CBS Report 2010*; *BBS Report* 2011). Since there is no associated evidence of breeding, nor presumption that birds are on territory, these records could be used only to provide presence data. BTO contacted BBS participants to highlight where they had seen species as yet not reported to the Atlas and requested they submit Roving Records. BirdWatch Ireland provided all CBS participants with additional forms for submitting breeding evidence.

Wetland Bird Survey (WeBS, UK; BTO, RSPB, JNCC, in association with WWT)

Irish Wetland Bird Survey (I-WeBS, Republic of Ireland; BirdWatch Ireland, NPWS)

All estuaries and major inland wetlands are surveyed monthly between September and March (and for many sites, all year round) for waterbirds through WeBS and I-WeBS Core Counts in the UK (*WeBS Report 2011*) and in the Republic of Ireland (*I-WeBS Report 2009*) respectively. On a rolling basis, a sample of estuaries is also surveyed at low tide in the UK (*WeBS Report 2011*). These datasets are not limited to waterbirds—they can also include raptors and coastal passerines—and provide many winter distribution records, and breeding-season presence-only records. The main constraint on their use was whether the count sectors used for reporting purposes fell within grid squares. Distribution records were extracted only where at least 95% of a count sector fell within a single tetrad or 10-km square. For example, of the 8,348 sectors registered for WeBS Core Counts, 7,025 (84%) could be assigned to a 10-km, and 4,248 (51%) to a tetrad.

BTO Garden BirdWatch (GBW, UK; BTO)

Garden Bird Survey (GBS, Republic of Ireland; BirdWatch Ireland)

Winter gap-filling records at 10-km resolution were sourced from the weekly submissions from GBW and GBS participants. No breeding-season data were used because evidence of breeding is not collected by either survey. Most breeding in gardens was likely to have been confirmed by conventional Atlas methods. However, in the Republic of Ireland, BirdWatch Ireland emailed forms to all GBS volunteers appealing for sightings in and around their gardens of birds displaying evidence of breeding.

Special surveys and datasets organised by BirdWatch Ireland

During the period of the Atlas, BirdWatch Ireland conducted the following species-focused or area-focused research and monitoring projects, and data from these were converted into Roving Records where possible:

- Whooper Swan Survey 2010 (funded by NPWS)
- Greylag Goose Survey 2007–08 (funded by NPWS and NIEA)
- Barnacle Goose Aerial Surveys 2008, 2011 (funded by NPWS)
- Red Grouse Survey 2008 (funded by NPWS)
- Merlin 2010 Survey (funded by NPWS and BirdWatch Ireland)
- Raptor Conservation Project (funded by NPWS, Department of Agriculture, Food & the Marine, the Heritage Council, Irish Council for Science, Engineering and Technology, Cork County Council and Kerry County Council)
- Corncrake Project 2008–2009 (funded by NPWS)
- Breeding Wader Machair Surveys (funded by NPWS)
- Shannon Callows Breeding Wader Management Project (funded by NPWS)
- Waterways Birds Project (funded by NPWS and the Office of Public Works)
- Phoenix Park Survey (funded by the Office of Public Works)
- Baseline surveys within coastal SPA sites (funded by NPWS)
- BirdWatch Ireland reserve and observatory data

All data covered the core Atlas survey months with the exception of the Barnacle Goose Aerial Surveys. These are conducted in March but since the marine islands surveyed are not visited at any other time, and because the geese are thought to be present throughout the winter, these data were given a nominal date at the end of February.

2.4.2 External data sources

We are grateful to the following groups and organisations that extracted data from their archives and recent monitoring and research activities.

County and regional clubs and reports (Britain & Ireland)

59 county and regional bird clubs and report compilers (see Acknowledgements) either manually entered data from their archives or provided data to the Atlas team to be processed and loaded centrally. As far as possible, breeding evidence was coded using observers' comments. Where data had already been through local validation processes, records were loaded pre-validated.

JNCC Seabird Monitoring Programme (Britain & Ireland)

Each year a sample of colonies is monitored for population size and breeding success (JNCC 2012). Depending on the species, counts take the form of number of Apparently Occupied Nests, Apparently Occupied Burrows, etc. Data were available for the first three breeding seasons of the Atlas. Most data were assumed to relate to *confirmed* breeding, but where mixed pairs were apparent, e.g. for Roseate × Common Tern pairs, only *probable* breeding was assumed. Data from colonies that extended across 10-km square boundaries could not be used owing to uncertainties regarding the nesting locations of individual species within the broader colony.

RSPB (UK)

Every year RSPB reserve managers compile an Annual Reserve Monitoring Report detailing the species breeding on each reserve; they also collate winter bird sightings. Approximately 50% of RSPB reserves fall entirely within a single 10-km square and these data could be used as provided. For reserves where it was not possible to assign reserve-level records to a single square, we worked closely with RSPB staff to identify records that could fill gaps for the Atlas.

For several scarce and localised species the RSPB Data Unit also provided access to invaluable records from RSPB research and monitoring activities. These included annual conservation science projects such as Black Grouse and Capercaillie lek counts, wintering and booming Bittern monitoring, Stone-curlew nest monitoring, Slavonian Grebe, White-tailed Eagle, Corncrake, Red-necked Phalarope and Cirl Bunting annual surveys. RSPB also provided the results of the following special surveys: 2008 Scottish Crossbill, 2010 Hen Harrier, 2011 Montane Survey (mainly Ptarmigan, Dotterel and Snow Bunting).

National Parks & Wildlife Service (NPWS, Republic of Ireland)

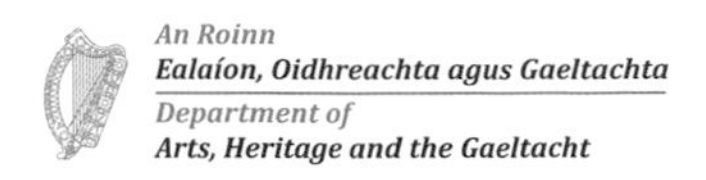

In addition to the many thousands of records submitted by NPWS staff across Ireland, specific datasets were also provided for: Greenland White-fronted Goose, Grey Partridge (supported by the Irish Grey Partridge Conservation Trust and members of the National Association of Regional Game Councils in Ireland), The Hen Harrier Winter Survey, Corncrake (2010–2011) and Waterbirds on Lough Ree.

Environmental Protection Agency (EPA, Republic of Ireland)

EPA provided records of riparian species recorded during surveys and monitoring of Irish rivers.

Rare Breeding Birds Panel (RBBP, UK)

Irish Rare Breeding Birds Panel (IRBBP, Ireland)

Both Panels collate and archive breeding evidence and population information for a suite of scarce and rare breeding birds. Records with 10-km resolution grid references were provided for the Atlas on the understanding that their visibility and any onward data flows would be in accordance with agreements regarding the sensitivity of species (see Section 4.1.3).

Scottish Raptor Monitoring Scheme (SRMS, Scotland)

The Scottish Raptor Monitoring Scheme is a partnership between the Scottish Raptor Study Groups, SNH, BTO Scotland, RSPB Scotland, SOC, RBBP and JNCC, grant-aided by SNH with in-kind contributions from the other partners. Most of the data are collected by members of the Scottish Raptor Study Groups and have helped to produce the most comprehensive assessment of the distribution of breeding raptor species in any Britain & Ireland atlas. The SRMS officer converted SRMS nest outcome codes to standard breeding evidence codes and summarised data at 10-km resolution before provision to the Atlas. The dataset covered raptors, owls and Raven.

Northern Ireland Raptor Study Group (NIRSG, Northern Ireland)

NIRSG is a volunteer organisation that exists to promote the knowledge of raptor biology and populations through volunteer field-based research, working in partnership with landowners, industry, statutory agencies and non-governmental organisations. NIRSG provided breeding-season 10-km distribution records for raptors, owls and Raven derived from their research projects.

Irish Raptor Study Group (IRSG, Republic of Ireland)

IRSG is a voluntary organisation specialising in research, monitoring, conservation and protection of raptors in Ireland. IRSG provided access to 10-km distribution records from their breeding-season and winter surveys and research activities. These records related to raptors, owls and Raven.

Golden Eagle Trust (Republic of Ireland)

The Golden Eagle Trust is a charity dedicated to the conservation and restoration of Ireland's native birds and their habitats. It manages reintroduction programmes for Red Kite, White-tailed Eagle and Golden Eagle. It provided winter and breeding-season distribution records for these species, and for Ring Ouzels.

Barn Owl Trust (Britain)

Members of the Barn Owl Trust monitor Barn Owl nests in southwest England, mostly in Cornwall and Devon. The Barn Owl Trust provided breeding-season distribution records with breeding evidence.

BirdGuides (Britain & Ireland)

IrishBirding.com (Ireland)

BirdGuides and Irish Birding are Internet-based sightings services for nationally and locally rare birds. BirdGuides provided full access to their 'Bird News Extra' database and records were imported into BirdTrack on a nightly basis. Subject to site definition, these records were used in the Atlas. In Ireland, records were manually extracted from the IrishBirding.com website and added as Roving Records.

Bat Conservation Ireland (Republic of Ireland)

Bat Conservation Ireland provided records of owls, plus a number of records of waterbirds encountered during Daubenton's Bat monitoring.

Bord na Móna (Republic of Ireland)

BORD NA MÓNA

Staff from Bord na Móna (Irish Turf Board) provided bird records from the company's land holdings (c.80,000 ha), which comprise mostly industrial and cutaway peat bog in the Irish midlands.

CHAPTER 3

ONLINE DATA CAPTURE AND ENGAGEMENT

Iain Downie, Simon Gillings & Karen Wright

3.1 WHY PROVIDE AN ONLINE SURVEY?

WHEN THE ATLAS Working Group (AWG) was convened in 2004 to discuss the design and implementation of *Bird Atlas 2007–11*, more than 50% of people in the UK and Republic of Ireland had Internet access, a figure that was growing at c.4.5% per annum. The proportion using broadband-speed connections capable of handling content-rich applications by 2006 was c.70%, also growing annually (Office for National Statistics 2012; Central Statistics Office 2013). Prior to *Bird Atlas 2007–11*, BTO had already built online data-submission facilities for some of the existing, long-term bird monitoring schemes (e.g. BBS, Garden BirdWatch, WeBS), plus a bespoke year-round bird recording package (BirdTrack). The year-on-year growth in uptake of these Internet applications was a good indication that web technology would have a major part to play in the forthcoming Bird Atlas. In particular the following key advantages that a web application offered were identified:

- **Improved data provenance** Allowing observers to enter and manage their own data maintains the link between observer and data. This is a major advantage over paper forms, which were typically sent to third-party companies for data entry, and makes it considerably easier to track and resolve issues with data.
- **Faster feedback and engagement** Feedback is critical to the success of any survey, especially long and iterative projects such as this atlas. A key advantage of using the Internet over paper for data submission is that feedback can be instantaneous and responsive to the rate of data entry. Evidence from BBS submissions for 2007 shows that 18% of forms were submitted online on the day of the final survey, and 51% of forms had been submitted within 21 days; in contrast, some paper forms were delayed by over 200 days. Furthermore, feedback online can be personalised to each observer to a much greater extent than is the case with hard copy, for example it can be personalised to the observer's local area.
- **Better management of effort** For surveys that involve one-to-one matching of observers to sample squares, an Internet system is an effective way of managing sample-square allocations and granting access accordingly so as to avoid effort duplication. Provided data are submitted online in a timely fashion, it also allows rapid assessments of coverage leading to more engaging feedback and more effective targeting of future effort.
- **Improved data quality** Interactive tools such as clickable maps and validation of values entered into forms at point of entry rigorously control the type of data submitted and reduce many common data-entry errors. Internet systems also allow context-specific validation in real time, for example checking that certain species or counts are appropriate to the time of year of the record.
- **Cost saving** Although developing and maintaining a web application has inherent financial costs, we estimated that, overall, these would be considerably lower than the costs associated with printing and posting paper forms, professional data entry and manual data checking.

Although there were strong arguments in favour of using web applications, there were also a number of potential challenges that had to be overcome during the lifetime of the project:

- **Unique requirements** Although BTO had already built online data-collection systems, the specific requirements of the Atlas field methods required the building of a new bespoke system from scratch.
- **Initial costs and risks** Many of the modules were built prior to or early in the main data-recording period, which resulted in greater effort to raise adequate funds early in the project. Additional investment in specialist web server hardware was also required early in the project. Staggering the development and release of the remaining modules where possible helped to spread the cost to some extent, but there was an inherent risk in this front-loading of resources that is reduced in paper-based surveys, should uptake be less than expected.
- **Managing success** With limited resources, yet an unknown but potentially high uptake by the birdwatching community, the online systems had to be able to cope with the volume of data, the peaks in use and the users' expectations. Systems needed to be scalable in all these areas, which added complexity and increased costs.
- **Project longevity** Giving surveyors access to their own records during the lifetime of a project increases the expectation of continued access post-project, as these records become a part of the online user's 'complete birding experience'. For previous paper-based surveys, this had never been an expectation. This requires extended maintenance.
- **Connectivity issues** A significant number of observers live in areas with slow Internet connections (narrow bandwidth), such as rural parts of Ireland or the Scottish Highlands and Islands. This necessitated more careful development and presentation of information in smaller pages, a trade-off resulting in more navigation and sub-setting of information online than would be optimal for areas with faster access.
- **Integrating web technology with paper forms** Experience from previous surveys had shown that some observers would begin the project using paper forms and subsequently migrate to using the online system, whilst others would never use the online system. This meant that two parallel data entry systems (web and paper) had to be maintained and fully integrated at all times to facilitate users' transitions from paper to online, and so that feedback represented the real state of coverage and completeness from all participants.

On balance, the AWG agreed in 2006 that a major bespoke web application was needed and that it should be the primary (but not sole) means by which observers should submit and manage their data. The rest of this chapter details the timings of developments, the various modules that were built by the BTO Information Systems (IS) Team working with the Atlas Team, how the data flowed between observers and local experts, and how the data were summarised for public presentation. It details the technological problems that were encountered and how they were overcome. Many of these elements and solutions may be specific to this particular atlas project or to the technologies available at the

FIGURE 3.1 TIMELINE OF MAJOR MODULE DEVELOPMENT AND DEPLOYMENT ALONGSIDE RECORDING SEASONS

The red line shows the rate of data accumulation.

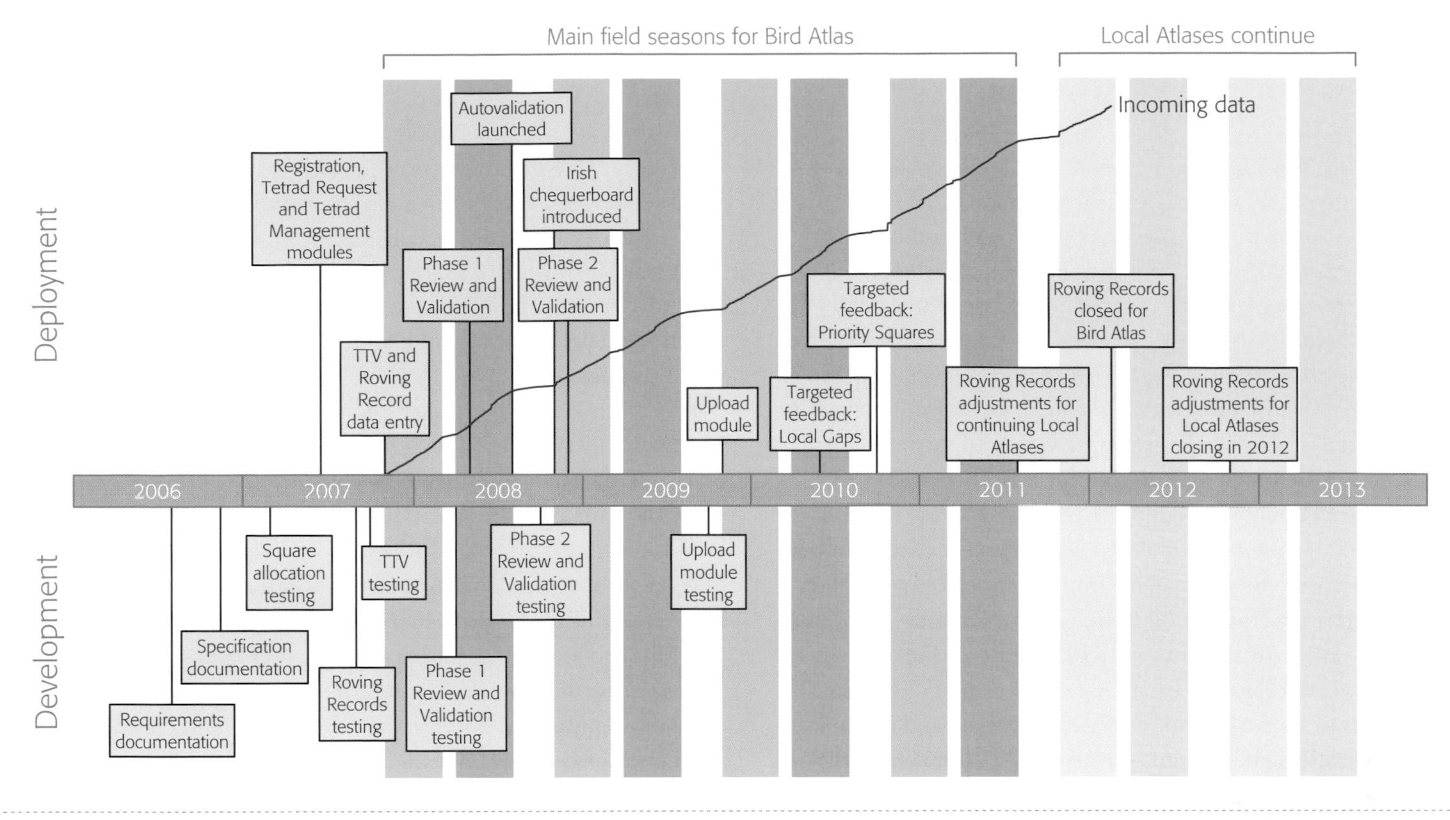

time. However, we hope that this chapter will be useful to others contemplating building an online atlas application. It may also be of interest in years to come in explaining the rationale for why the Bird Atlas Online system was developed in the way that it was.

3.2 ONLINE DEVELOPMENT TIMINGS

THE BIRD ATLAS Online system was built as a series of 'modules' so that various milestones could be met, and to facilitate the addition of new features throughout the project. With fieldwork due to start in earnest on 1st November 2007, the AWG was keen to ensure that modules were in place to allow observers to add, review and edit Roving Records and TTV data from this day onwards. However, prior to this, observers had to be able to register their interest to participate and be able to request tetrads to survey (a process that continued throughout the project too). At the same time, Regional Organisers had to be able to process these requests so that fieldwork could commence on 1st November. Therefore, the first modules to be built by the BTO IS Team were the '*Registration*', '*Tetrad Request*' and '*Tetrad Management*' modules. Planning for this portion of online development started in 2006 alongside the drafting of specifications of the other longer-term requirements (see Chapter 1 for timings of the overall Atlas project). The software development began in January 2007 (see Figure 3.1 for a timeline of notable development dates) and *Registration*, *Tetrad Request* and *Tetrad Management* modules were available from June 2007, giving plenty of time for the Regional Organisers to engage with, motivate and allocate tetrads to their observers. These modules also included feedback systems (see Section 3.6 and 3.7) to allow observers and Regional Organisers to gauge the uptake of tetrads. By 31st October 2007, over 3,400 observers had already registered on the Bird Atlas website, ready for the first day of recording (see Figure 3.2).

Developing in a modular fashion and launching these first three modules early worked well, as it provided confidence in the website and database operations prior to the onset of fieldwork and data entry. This long lead-in time is typical of online developments, and future atlases with an online component will need to consider

FIGURE 3.2 GRAPH SHOWING THE CUMULATIVE NUMBER OF USERS REGISTERED FOR THE BIRD ATLAS ONLINE SYSTEM

Shading indicates the main field seasons.

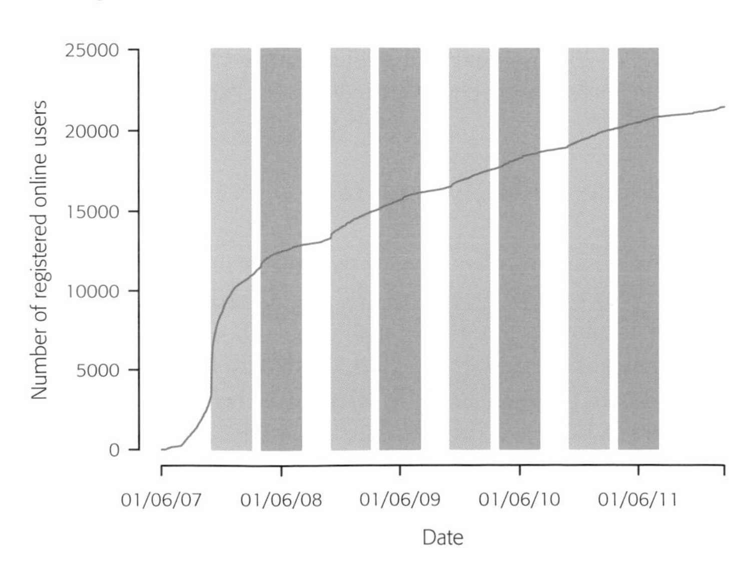

this during their planning stages. It represents a shift in personnel resources from typically science- and support-based staff to a dependency on specialist Information Systems staff in this early period (in this case, database and web application developers).

Following the completion of pilot fieldwork over winter 2006/07, the AWG finalised the Roving Record and TTV field methodologies and paper forms during early 2007. These forms provided the basis for the layout of the online versions, although extra features could be added to the online version. For example, a helpful interactive map was added to the online Roving Record form to allow users to more accurately find or double-check grid references. The first feedback module, called '*My Square Summaries*', was built at the same time so that live feedback would be available from day one. These modules were available for use from 00:00:01am on 1st November 2007 and the first observation contributed was a Roving Record of a Tawny Owl, entered at 02:16am.

Once the core *Tetrad Request*, *Tetrad Management* and data entry modules had been established and data started accumulating, the next major development was a '*Review and Validation*' module to give Regional Validators the ability to view summaries of the data in their region and identify unusual records potentially needing correction or supporting descriptions. This was a significant module for the developers, which changed the data presentation relationship from the normal one of a single observer viewing his or her own data in real time to an expert interrogating the data from multiple observers across an entire region, also in real time. To illustrate the scale of this shift, note that more than 1,300 observers submitted data in the Atlas region of Suffolk.

During the lifetime of the project, numerous other interactive modules were developed or updated, delivering novel ways to summarise and feed data back to observers to engage them into filling gaps, and to Regional Organisers to allow them to target local observers more effectively to fill gaps in coverage. Similarly, several manual processes were developed and used during the project on an ad-hoc basis, such as scripts to load paper-based data and other useful datasets (e.g. Ringing records, Rare Breeding Birds Panel data) into the database to fill gaps. It should also be mentioned that various observers and Regional Organisers tested most modules prior to release, and some of the major testing phases are indicated in Figure 3.1.

3.3 SYSTEM REQUIREMENTS

BTO HAS ORGANISED several surveys where online data capture and processing is the only method used (e.g. BirdTrack—www.birdtrack.net) and several with a combination of paper and online data capture and processing (e.g. BBS, Garden BirdWatch, WeBS). These had been running online for several years before the start of *Bird Atlas 2007–11*, and they provided a good basis for gauging the hardware and software required for running the Atlas, which was perceived to be much larger in scope than previous Britain & Ireland atlases and the first delivered online. Those earlier projects also meant that many of the data resources such as lists of species, counties and grid references already existed as online database tables.

The key system issues to consider were availability, performance and scalability, which together provide users with a reliable system able to cope with both expected and unexpected growth in the number of users and the volume of their data. The figures on BBS submissions (see Section 3.1) suggested that a large number of observers could be expected to be active on a daily basis, concurrently submitting their day's sightings at the end of the day, and the system needed to be able to cope with such peaks in demand. The ethos of rapid data entry was heavily promoted, and so successfully that 35% of all TTVs were submitted online on the day of the visit, and 64% had been submitted within a week of the visit. This meant that observers and Regional Organisers could rapidly, and accurately, gauge coverage and identify gaps to target, but this was only possible because the scalability and capacity to allow hundreds of observers to submit data simultaneously had been successfully built in. The busiest day for online Atlas submissions was 27th December 2007, when more than 7,000 users submitted records.

Other important Information Systems areas to consider were data security and backup, testing, and hardware maintenance and software 'patching'. It should be noted that although data security was important and was delivered through restricted or individual-surveyor access, we did not use Secure Hypertext Transfer Protocol (HTTPS, the typical secure/encrypted web protocol), just HTTP. It was agreed that although we were dealing sometimes with data on sensitive breeding species, for example, this did not merit the same security that financial websites require. This also helped to reduce costs. Backups of the data were taken nightly as described in Box 3.1.

All development and testing was carried out on parallel production and development platforms, with the latter being provided by less powerful, cheaper servers but with the same software and facilities, and with comparable data content and volume. Box 3.1 shows the main servers that were used throughout the development and maintenance of *Bird Atlas 2007–11*, and how the main web applications relevant to the Atlas were built, tested and deployed to the observers. As is common in projects such as this, we used an Integrated Development Environment (IDE), in our case NetBeans (netbeans.org) for the development of the interactive web pages. This IDE provided a locally installed web server with linkages to the test database for rapid and accurate development, and its use was essential to building an online project of this scale.

Finally, having an accessible and responsive online support mechanism (a web support email address with a team behind it) was fundamental in resolving unexpected issues and user feedback, where necessary leading to rapid bug fixing and improvements through further development.

More detail on the specific hardware and software systems, and application development life cycle used by BTO during *Bird Atlas 2007–11* is included in Box 3.1.

BOX 3.1 HARDWARE, PRIMARY SOFTWARE AND THE DEVELOPMENT CYCLE LEADING TO THE RELEASE OF *BIRD ATLAS 2007–11* ONLINE APPLICATION

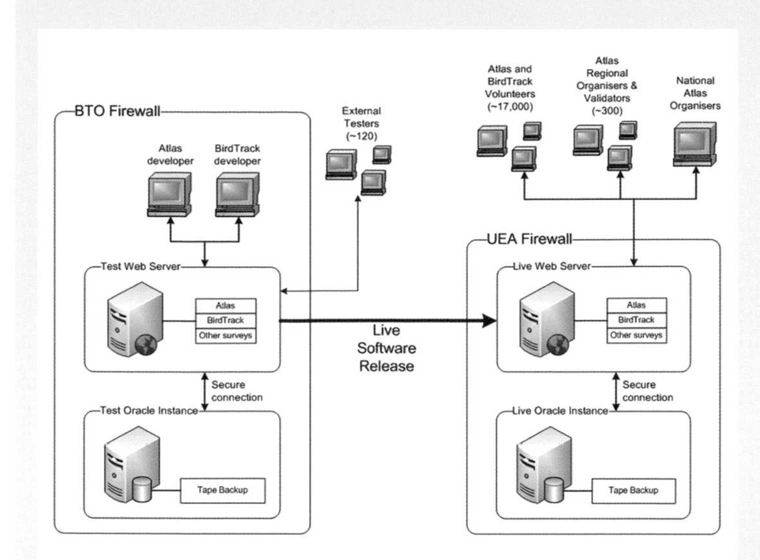

It is currently standard practice to separate the software architecture into Database and Web Servers within a Client-Server model, in order to improve data security and performance. Typically this is a three-tier architecture (Eckerson 1995) where the Database Server holds all data and associated lookup tables (e.g. definitive species lists and spatial data such as grid squares with land eligible for survey coverage) and the Web Server processes incoming and outgoing data to generate the various web pages that are displayed to the Client (the user's web browser).

Oracle Relational Database (www.oracle.com) is one of the industry standard databases and used by many sectors. It comes with strong support, regular patching, and is highly scalable to cater for growth. BTO's Oracle Database hosts the survey data from all BTO online projects, and is its main scientific data repository. Since the Web Server was mainly a service machine and not storing data, we used free open-source software solutions that were equivalent to supported products used in the commercial sector, thus reducing running costs. For example, Glassfish (now an Oracle development) acts as an open-source Java-based web server and CentOS is an open-source version or clone of the popular RedHat operating system. These free versions typically have only community support.

SAS software was used to interrogate data collected and extracted from Oracle on a nightly basis. This software has powerful analysis capabilities, and the ability to create reports and summaries direct to web-browser-compatible outputs (HTML web pages).

It should be noted that any atlas project looking to closely replicate any of these systems in the future may encounter difficulties due to out-dated hardware or software, but anyone attempting to do so would presumably use hardware and software current to the time.

DATABASE SERVER	
Hardware	Sun Microsystems® Sun Fire X4440, 32 Gb Ram, 146 Gb HDD, two AMD Dual core @ 3.2 Ghz
Operating System	Solaris® 10
Software	Oracle® 10g Standard Edition
Backup	DAT 72 tape

WEB SERVER	
Hardware	Dell® PowerEdge 2950, 8 Gb Ram, 250 Gb HDD, two Intel® Xeon® CPU 5148 @ 2.33 GHz
Operating System	CentOS® 5
Software	Open-source Glassfish®, Java® 1.6 on Apache®, SAS®
Backup	LTO2 tape

The hardware and software combinations used for the live installation were hosted at the University of East Anglia (UEA). BTO has a long-standing agreement with UEA and JANET (Joint Academic Network) to host our online systems within the UEA server infrastructure. This has advantages of greater Internet bandwidth and improved environment for the physical storage of the hardware and data. At the time of launch, BTO did not have the bandwidth capacity to cope with a project of this magnitude. In addition, in 2001/02 when BTO invested in online data capture with the development and launch of Migration Watch (now BirdTrack), the 'Cloud' as we know it now was in its infancy and too unreliable or expensive. This resulted in BTO investing in its own hardware and systems, and UEA was used simply as 'somewhere to store them'.

3.4 CORE DATABASE TABLES

ANY BIOLOGICAL RECORD typically has four common elements, which can be summarised as "who saw it?", "where did they see it?", "when did they see it?" and "what did they see?". These elements provide the logical basis for storage of bird records in a relational database (see Figure 3.3). Since these attributes are common to many bird surveys, database tables to hold similar information from existing BTO online surveys had already been built prior to Atlas development. In addition to these four core tables, a number of bespoke tables, indexes and other database structures had to be built to accommodate the particular requirements of the Bird Atlas study design. This provided a framework for individual observers to add, view, edit and track their own records throughout the entire project and beyond, which was vital for engaging the observers and the ultimate success of the project. At the close of data collection, the four core tables contained approximately 18,000 users, 600,000 locations, 7.4 million submissions and 19 million observations relating to Atlas and BirdTrack. From these, data were extracted according to various filters such as season, provision of breeding evidence and validation status for use in the production of this book.

FIGURE 3.3 CORE ORACLE DATABASE TABLES (LIMITED TO ATLAS PROJECT ESSENTIALS ONLY)

Primary keys are marked with an *, relevant foreign keys are italicised.

USERS	User details
USER_ID*	Login name
PASSWORD	Login password
NAME	6 × name fields
ADDRESS	7 × address fields
POSTCODE	UK only
REGION	Atlas-specific regions
EMAIL	Email address
SOURCE_ADDED	Online or other route
COUNTRY	UK, Ireland etc
DOB	Date of birth if under 16 years old

SUB	Submission details
PROJ_ID	Project identifier (ATLAS)
SUB_ID*	Unique identifier
PROTOCOL_ID	Roving Record or TTV component (1st Hour Count, 2nd Hour Count, Extra Species, Colony Count or Tetrad Population Estimate)
USER_ID	Login name
LOC_ID	Unique location identifier
CREATION_DT	Date record created
LAST EDITED_DT	Date record last edited
OBS_DT	Observation date
YEAR	Observation year
DATA_SOURCE	Web, Upload or Paper
VISIT	Early or Late (TTV only)
SEASON	Winter, Breeding or Other

LOCATION	Location details
LOC_ID*	Unique identifier
NAME	Name of location
UNCERTAINTY_RADIUS	Defines size of location
CREATION_DT	Date location created
USER_ID	Login name
GRIDREF	Britain or Ireland grid
COUNTRY_CODE	One of 6 countries
PROJ_ID	Project identifier (ATLAS)
DATA_SOURCE	Web, Upload or Paper
TETRAD	Tetrad identifier
TENKM	10-km square identifier

OBS	Observation details
PROJ_ID	Project identifier (ATLAS)
SUB_ID	Unique submission identifier
OBS_ID*	Unique identifier
SPECIES_CODE	Species code
HOW_MANY	Count
VALID	Valid or type of query
REVIEWED	Status and reviewer code
BREED_STATUS	Breeding evidence code

Other relevant tables include:
List of tetrads eligible for coverage with USER_ID of allocated observer: Lookup of species with confidentiality ratings; Audit tables of records edited, deleted and queried

3.5 THE INTERACTIVE BIRD ATLAS ONLINE SYSTEM

THIS AND THE subsequent two sections describe in more detail the overall aims, structure and modular development of the online system built for *Bird Atlas 2007–11*. The two main areas of development are split into 'Options for Observers' (Section 3.6) and 'Options for Regional Organisers and Validators' (Section 3.7). However, fundamental to these two areas of detail was a framework and core set of modules common to all online surveys and projects organised by BTO. These include registration of the observer (e.g. name, address, contact details and relevant agreements) to use Bird Atlas Online and individual login protocols and branding involving a suite of dedicated banners, logos and colours to identify the project as a distinct entity online.

Although not a functional requirement, individual design and branding for Bird Atlas Online was very important for publicity, promotion and the online 'experience'. As a partnership project, it needed to be distinctly unique so a single public-facing web address (www.birdatlas.net) was used, with neutral, Atlas-specific banners and colours (all agreed by the partners at an early stage in the project).

Before any tangible user interfaces or modules were built, a framework for the project had to be created, primarily using the Java, JavaScript, HTML and CSS programming languages. A fuller description of the development technologies and methods used is outlined in Box 3.1. The main role of the framework was to provide a browser 'session' that an online observer could utilise to allow them to move from page to page, retaining relevant credentials and data, with only one log in. The concept of a session is not unique to Java web applications (e.g. it is used in PHP and Microsoft® Active Server Pages extensively), and more detail on its implementation within a Java context can be found in Hunter and Crawford (2001). Once this framework was established, the main elements described below were added and tested sequentially (see Figure 3.1 for approximate deployment dates). Testing with a realistic look and feel provided the testers with a better perspective of the final product, rather than a non-contextual blank canvas.

Although designed to be brand neutral, the functionality of the online system was designed to be familiar to existing online surveyors of projects such as BBS, Garden BirdWatch, WeBS and BirdTrack. Indeed, Bird Atlas Online was built to run alongside BirdTrack, and was designed not only to absorb as many BirdTrack records as possible to fill in gaps in coverage, but also to be compatible with BirdTrack for users who wanted to take part in the Atlas Timed Tetrad Visits. Bird Atlas Online was therefore built to be familiar to 'BirdTrackers' when they switched between projects. Both projects relied on a '*Data Home*' page that acted as a hub to the main elements of the project (for example, TTV or List data entry, Roving Record or Casual Record data entry, results viewing, help, etc).

3.5.1 Registration

The importance of being able to attribute every record in the Atlas to an identifiable individual was paramount. Therefore, each and every online observer involved in *Bird Atlas 2007–11* had to create an online identity for adding, organising and reviewing their own records. The *Registration* module enabled surveyors to create their own unique USER_ID and PASSWORD for repeated access, against which their contact details were stored. Contact details were essential for day-to-day administration of the project by the National and Regional Organisers, and were also used in some of the dynamic feedback modules to tailor the results to each observer's home 10-km square. Registration is typical of most websites, collecting user details and permissions, and although the page text related to the Atlas, the process is common across all BTO online surveys and projects (and indeed individual observers can use the same access details across different surveys). Having successfully registered, logged-in users landed on the primary page for accessing all Atlas functionality, known as *Data Home*.

3.5.2 Data Home

Data Home represented the centre of Bird Atlas Online (Figure 3.4). It provided surveyors with access to modules for contacting their Regional Organiser, requesting tetrads, adding, reviewing and editing their records and accessing feedback in a number of novel ways. For users with Regional Organiser or Regional Validator privileges (as set within the main database by the National Organisers), *Data Home* also provided access to additional management and validation modules. Although used primarily for navigation, *Data Home* included a message system to alert users to possible issues with individual records (see Section 3.7.2), and any other ad-hoc messages that the National Organisers wished to promote throughout the recording period. For example, each October a message was added reminding observers that the next season's winter fieldwork was due to begin on 1st November.

FIGURE 3.4 SCREENSHOT OF *DATA HOME* SHOWING OPTIONS AVAILABLE TO THE DIFFERENT TYPES OF USERS

The screenshot also shows the common branding that was used throughout the online systems (see Section 3.5) and an example of a queried record alert (see Section 3.7.2).

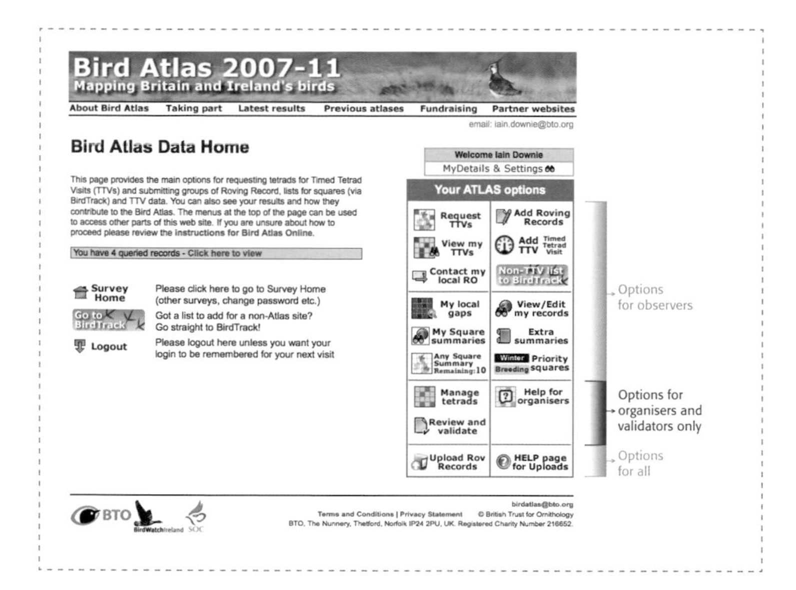

3.6 OPTIONS FOR OBSERVERS

THIS SECTION DESCRIBES some of the core features that helped observers to target squares they wished to survey, then subsequently submit, view and edit records for their squares and other areas (through a combination of 'allocated' tetrads, and more widespread Roving Records). Another helpful feature provided to observers was the ability to view results in areas where they themselves were active, which proved useful in targeting effort to fill in gaps (see Section 3.6.4).

3.6.1 Administration ~ requesting tetrads, contacting Regional Organisers

In order to achieve the minimum target coverage of tetrads in 10-km squares in Britain and Ireland (see Section 2.2.4) it was critical that observers could see which 10-km squares had not yet reached the minimum allocation, and then within such squares, which tetrads were available to be requested (Ireland employed a chequerboard approach, see Section 2.2.4). The *Request Tetrads* module comprised a navigation system where observers could sequentially drill down through maps of countries and regions to interactive maps showing the 10-km squares in a chosen Atlas Region. From there the observer could select an individual 10-km square to see another interactive map showing its constituent tetrads. These last two maps were built with embedded Google Maps. At the time this was the best solution for presenting dynamic maps that allowed the user to zoom in and out and to toggle between simple road/feature maps or satellite images to aid navigation.

Colour coding was used at the Atlas Region level to show the 10-km squares that had achieved different degrees of allocation (below target, at target, or 100%), and also at the individual 10-km square level (Figure 3.5) to show which tetrads were available for surveying (separated into 'available', 'allocated' or 'visited'). Symbols were used to indicate whether tetrads were allocated in winter, the breeding season, or both. Symbols and colour coding were also used to denote 'Low-priority' 10-km squares and tetrads on the Irish chequerboard.

The aims of this module were twofold: firstly, to allow observers to navigate quickly to tetrads they wished to survey, and secondly, to display current uptake to avoid duplication and maximise coverage. For example, if an observer wished to cover their local tetrad or perhaps a more remote one whilst on holiday, they could quickly see if it was available, and if not, select an alternative. This maintained engagement and made it possible to measure allocation and uptake as the project progressed.

Once observers had found an unallocated tetrad, they were directed to a web form to send an electronic request to the Regional Organiser. They in turn could allocate the tetrad to the observer in their *Tetrad Management* module (see Section 3.7.1).

FIGURE 3.5 SCREENSHOT OF STEP 4 OF THE *TETRAD REQUEST* MODULE

The tetrads of the 10-km square TF20 are colour-coded to indicate their allocation and coverage status in real time, and options are presented for the observer to contact the relevant Regional Organiser to request tetrads.

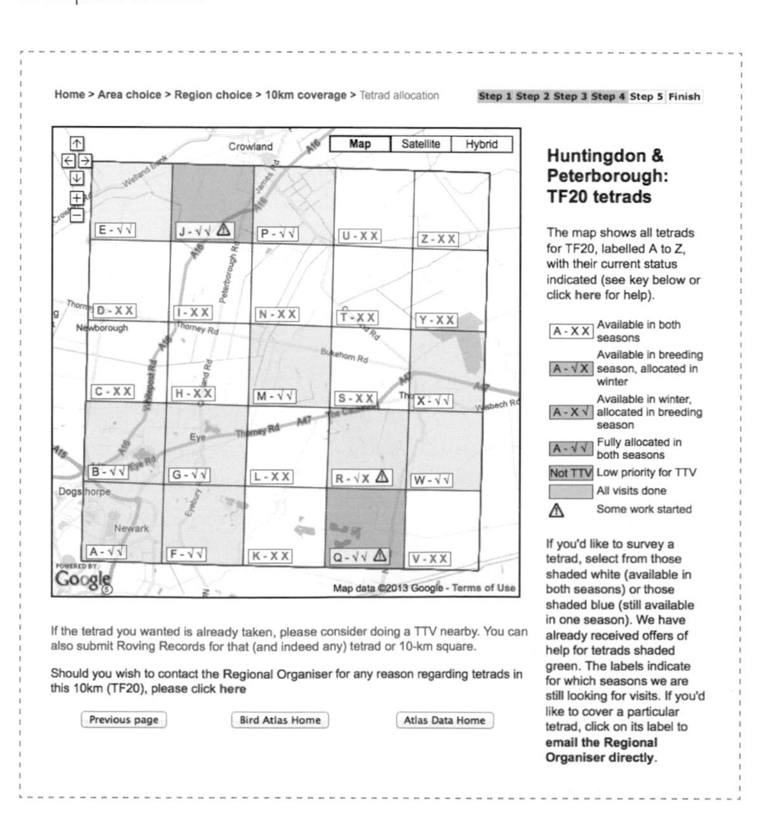

It should be remembered that Bird Atlas Online was just one way to request a tetrad, and Regional Organisers had to deal with multiple means of communication. This was the main reason why this module was built on a request rather than booking basis, in case the requested tetrad had already been promised to another surveyor via offline means. It also gave Regional Organisers the opportunity to discuss the survey with prospective surveyors to check they had the necessary skills to complete a TTV.

This entire module was closely tied into the rest of the database. It could reflect both real-time allocation and actual completion of TTVs, and had numerous links to connect observers to their local Regional Organiser or the Regional Organiser in any area that they might be interested in surveying. A simple '*View my TTVs*' module enabled observers to see a list of the tetrads allocated to them in each season, a record of coverage to date, and links to view maps. Buttons on the *Data Home* screens allowing access to *View my TTVs* and the *'Add TTV'* module (see below) were presented only if observers had at least one tetrad allocated to them.

3.6.2 Submitting data

For the individual observer, this area of the website was the most important—allowing them to add, review and, if necessary, edit their records. To simplify the transfer of records from paper forms carried in the field to the online forms, the appearance of online record submission pages closely mirrored the paper forms, but with the addition of features that online technologies allow to improve data entry.

Roving Records ~ direct data entry

The '*Add Roving Records*' module allowed all individual Roving Records to be collected by a single web form, which could capture up to twenty records at a time. Since a single Roving Record form could collate records from a variety of locations and dates, each Roving Record had to be treated as an individual entity. Each row in the form was set to be completely open, in that an observer had to enter a location (grid square, optional tetrad and place name), date, species, an optional count and a breeding evidence code. All fields could be entered manually, but several features were included both to make it easier for the observer and to reduce errors in the format of data being entered into the different fields:

- An inset interactive map (using Google Maps technology) was provided to assist with location entry. It allowed the observer to zoom in to where they saw particular birds to pinpoint and populate the exact 10-km square and tetrad. This was a major advance because provision of accurate grid references was crucial and a difficult skill for some observers.
- A list of place names previously used on the observer's Roving Records was provided to help to correctly populate the place-name field and reduce unnecessary duplication of identical (but differently named) locations in the database.
- A 'pop-up' calendar function was provided to help with adding the date field in the expected 'dd/mm/yyyy' format (records could be from the entire year).
- Autocomplete technology was widely used to limit the values that could be entered into key fields, and to speed up the data entry process. This worked by presenting an ever-diminishing list of possible values as the user typed each letter or number into a form field. For example, this technology was used to limit permissible 10-km square letter and number permutations; the location name field was pre-populated with any location names used by the observer on previous Roving Record forms; the species name could be selected once the user had typed the first three letters of any word in a species name; and breeding evidence codes could be selected after typing the first letter. The species name autocomplete was especially helpful in allowing the user to select the correct species without typing full names (see Figure 3.6). In other electronic data sources miskeyed species names are one of the most common errors, so this was an enormous help in reducing errors and standardising names entered (although this was by no means perfect; see Section 3.9).
- Helpful 'copy previous information' functions were provided to allow the observer to repeat the location and date from the row above (if the record was at the same place and date during a typical bout of Roving Record collection), or to copy just the date from the row above (if the observer was entering records from a variety of different locations visited on the same day).
- Mouse 'hover-over' features were strategically placed within the form to assist with the different codes that could be used or to remind users what a particular button or function did.

FIGURE 3.6 SCREENSHOT OF THE MAIN DATA ENTRY FORM OF THE *ADD ROVING RECORDS* MODULE

The page includes an inset dynamic map to help populate the form, and other help features improve record entry speed and validation. The user has started to type Wigeon and the drop-down menu is presenting valid species names.

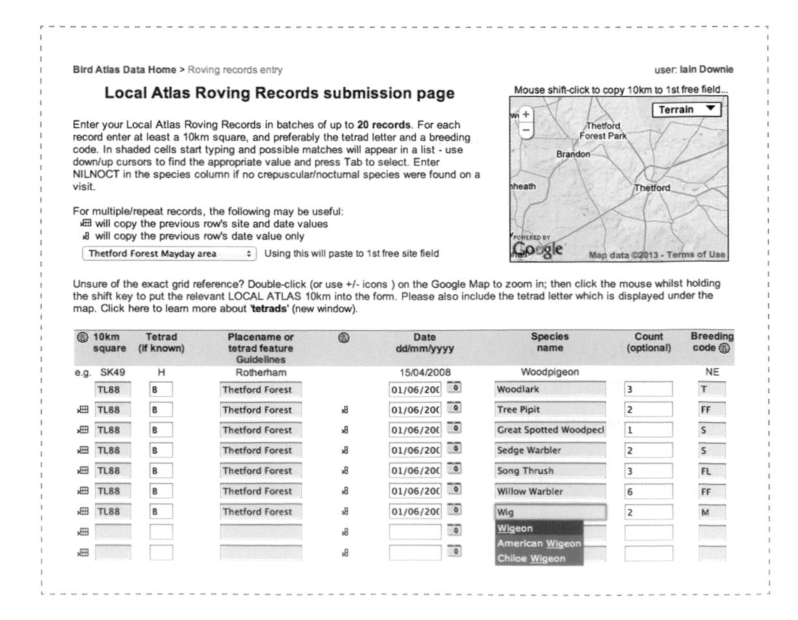

Bird Atlas Data Home > Roving records entry

user: Iain Downie

Local Atlas Roving Records submission page

Enter your Local Atlas Roving Records in batches of up to **20 records**. For each record enter at least a 10km square, and preferably the tetrad letter and a breeding code. In shaded cells start typing and possible matches will appear in a list - use down/up cursors to find the appropriate value and press Tab to select. Enter NILNOCT in the species column if no crepuscular/nocturnal species were found on a visit.

For multiple/repeat records, the following may be useful:
will copy the previous row's site and date values
will copy the previous row's date value only

Thetford Forest Mayday area — Using this will paste to 1st free site field

Unsure of the exact grid reference? Double-click (or use +/- icons) on the Google Map to zoom in; then click the mouse whilst holding the shift key to put the relevant LOCAL ATLAS 10km into the form. Please also include the tetrad letter which is displayed under the map. Click here to learn more about **'tetrads'** (new window).

10km square	Tetrad (if known)	Placename or tetrad feature Guidelines	Date dd/mm/yyyy	Species name	Count (optional)	Breeding code
e.g. SK49	H	Rotherham	15/04/2008	Woodpigeon		NE
TL88	B	Thetford Forest	01/06/200	Woodlark	3	T
TL88	B	Thetford Forest	01/06/200	Tree Pipit	2	FF
TL88	B	Thetford Forest	01/06/200	Great Spotted Woodpecl	1	S
TL88	B	Thetford Forest	01/06/200	Sedge Warbler	2	S
TL88	B	Thetford Forest	01/06/200	Song Thrush	3	FL
TL88	B	Thetford Forest	01/06/200	Willow Warbler	6	FF
TL88	B	Thetford Forest	01/06/200	Wig	2	M
				Wigeon		
				American Wigeon		
				Chiloe Wigeon		

JOHN WELLS

Whilst out surveying, records were noted down on forms or in specially designed Atlas notebooks, and over 90% were then submitted via Bird Atlas Online using the *Add Roving Records* and *Add TTV* modules.

Web form validation was also used during record entry and submission, ensuring that only certain characters or data types could be entered into the fields. For example, the tetrad field could accept only alphabetic characters, but not the 'O' character since it is not used in naming tetrads; dates could not be before 1st November 2007 nor in the future; and only numbers could be entered into the count field. Further automated validation of species, dates and counts for ornithological correctness took place when the page of records was submitted (see section on Data Entry Threshold Validation below).

Roving Records ~ bulk upload

Part way through the Atlas project, it became clear there was a demand for an upload system to accept large volumes of Roving Records that may have come from various electronic sources or from observers who found the online Roving Record system too labour-intensive for large numbers of records collected. The '*Upload Roving Records*' module was devised and implemented during the 2009/10 winter season. At the time of development, with the limited resources and time available, the module was designed only to accept Microsoft Excel files containing data adhering to a strict format. Microsoft Excel was chosen since it was the most popular spreadsheet system available at the time, most electronic systems could export in this format, and it was popular amongst observers (therefore no need to provide additional tutorials etc). The strict format was a simple table of records that mirrored the Roving Record paper forms. Uploaded records were placed in a holding table after initial screening and validation for common errors in formatting or field order. A second-stage process skimmed the records for potential biological and geographical errors, and only if there was a small and manageable number of records (e.g. maximum 20 records) that broke this type of validation (see section on Data Entry Threshold Validation below) were the data entered into the database, with any record raising a warning presented for more thorough checking.

Timed Tetrad Visit ~ direct data entry

Compared to the entry of Roving Records, the *Add TTV* module required a more complex navigation system to cater for the fact that individual observers could have any number of tetrads allocated to them (range = 1–652 tetrads; mean = 7; number of observers = 8,231), with each tetrad potentially having up to four visits, two in winter and two in the breeding season. This could result in a large matrix of individual TTVs that would be difficult to keep track of, so a primary navigation page was created to show the observer's allocated tetrads, split into the four visits. In addition to the different visits, the data submission for each TTV was split into four or five screens, or 'pages', that mirrored the different pages of the paper TTV forms. This was done so that the observer was not presented with a single long page of all c.200 species of a TTV form. Experience shows that such large pages can lead to significant errors and failed submissions due to pages 'timing out'. The primary navigation page also indicated the status of data entry of each TTV as 'No pages done', 'Some pages' or 'All pages', and traffic-light colour-coding was used to reinforce these states (Figure 3.7A). Once the observer clicked on a particular tetrad, the system checked which pages of a TTV had been completed, and which were still outstanding (Figure 3.7B).

When an observer started to enter records for a TTV, they were guided through the pages in order. Even if none of the species on a page were observed, that page still had to be submitted to confirm this. The user only had to enter the visit date (validated to be within the appropriate TTV season) and whether the visit was for one or two hours on Page 1. The rest of the form then mirrored the paper form (see Appendix 3), presenting all the species in the same order as on paper, with cells to enter 1st- and 2nd-hour counts (the latter only presented if a 2-hour visit had been selected on Page 1), a checkbox for the 'extra' sightings, a breeding evidence field (with similar autocomplete prompting and validation as described for Roving Records) and an optional tetrad population estimate field (Figure 3.7C).

FIGURE 3.7 SCREENSHOT EXTRACTS OF STAGES OF THE *ADD TTV* MODULE

A the state of TTV submissions for this observer, indicating that they are part way through submitting the late breeding-season visit to TL88T;

Tetrad	Region	Status of data entry			
		Winter season		Breeding season	
		Early visit	Late visit	Early visit	Late visit
TL88 B	Suffolk	All pages	All pages	All pages	No pages done
TL88 T	Suffolk	All pages	All pages	All pages	Some pages

B the state of individual page submissions for TL88T, indicating that pages 1–3 are complete but 4–5 have yet to be completed;

Tetrad	Region	Visit details	Breeding season: Access and Status				
			Page 1 Mute Swan to Buzzard	Page 2 Golden Eagle to Ring-necked Parakeet	Page 3 Cuckoo to Spotted flycatcher	Page 4 Pied Flycatcher to Corn Bunting...	Page 5 Colonies
TL88 T	Suffolk	Fri, Jul 24, 2009, 2 hours	✓	✓	✓	✕	✕

C an extract from the fourth page of TTV data entry for the breeding season, late visit, 2-hour count, including the early visit counts for reference.

Species (partial list)		Early visit 1st hour count	Early visit 2nd hour count	Early visit Extra (check)	Early visit Breed Code	Late visit 1st hour count	Late visit 2nd hour count	Late visit Extra (check)	Late visit Breed Code	Tetrad Popn Estimate
Pied Flycatcher	PF									
Bearded Tit	BR									
Long-tailed Tit	LT	4			N	10			FF	
Marsh Tit	MT									
Willow Tit	WT									
Crested Tit	CI									
Coal Tit	CT	22	16		N	17	21		FL	

In the example shown, the corresponding early visit to the tetrad had been completed and entered, and the counts and other relevant records were displayed to the observer as a reminder of the records they had previously entered.

Unusual species that were not on the paper forms (vagrants, escapes, etc) were handled by Page 4 of the web form, with a free-form species name field complete with autocomplete validation. A total of 15 extra species records could be entered, which was considered ample. The primary difference between winter and breeding-season TTV forms was the inclusion of a fifth page on the latter to collect colony records. Again, this web form reflected the paper form (see Appendix 3) in showing the 28 colonial species and boxes for entering counts.

One late addition to the TTV data entry module was a system to present an on-screen dialogue box reminding users to try to add breeding evidence for records where possible. This sort of technology was responsive, easy to code and could be used in preference to larger-scale changes to the data entry pages when the priorities required focusing users more.

Data Entry Threshold Validation

All records entered into the Atlas via the *Add Roving Records*, *Upload Roving Records* or *Add TTV* modules, and all BirdTrack submissions (via defined sites, see Section 2.4.1), were subject to further automated ornithological validation, called 'threshold breaking validation'. This utilised a matrix of nominal counts, first and last dates of occurrence and rarity status, independently compiled for every species in each UK and Republic of Ireland county by local experts or the BirdTrack Organiser. Once records had passed the earlier logical validation (e.g. count fields only containing numbers), records were assessed against the county thresholds corresponding to each record's grid reference. Records that broke thresholds were displayed back to the observer during the confirmation stage with a 'Points to check' warning in red text, displaying the threshold that had been broken. Figure 3.8 shows a hypothetical example with explanation of some likely error checking.

Flagging records in red text with a warning was only a rough indication that there may be an error when entering records. The observer could easily override these warnings and continue to add their records, as the threshold values were merely approximate values to catch the majority of errors.

3.6.3 Viewing and editing records

Providing observers with a data entry system is of limited value if they cannot view and, if necessary, edit their records. For example,

FIGURE 3.8 EXAMPLE OF THE DATA ENTRY THRESHOLD VALIDATION APPLIED TO RECORDS AT POINT OF ENTRY

10km/Tetrad	Town, village or feature	Date	Species: Points to check	Count	Breeding code
TF63 L	Snettisham RSPB (L)	Sun, Jun 1, 2008	Scottish Crossbill: Unusually high bird count	2	H
TF63 L	Snettisham RSPB (L)	Sun, Jun 1, 2008	Fieldfare: Out of Season	20	M
TF63 L	Snettisham RSPB (L)	Sun, Jun 1, 2008	Spotted Redshank: Rare or scarce species	3	H
TF63 L	Snettisham RSPB (L)	Sun, Jun 1, 2008	Coal Tit: Unusually high bird count	500	FF
TF63 L	Snettisham RSPB (L)	Sun, Jun 1, 2008	Avocet	140	ON

Here five hypothetical Roving Records for a summer visit to a Norfolk birdwatching site are shown. The Scottish Crossbill record failed because any number of individuals for this range-restricted species in Norfolk will flag a warning based on its regional count—this might indicate that the observer meant to enter Common Crossbill. The Fieldfare, a typical winter visitor, is flagged as out-of-season. Spotted Redshank is not unknown in Norfolk, but is deemed scarce in summer, so flagged as a 'Rare or scarce species'. The Coal Tit failed on its unusually high count of 500 individuals, where presumably the observer meant 5, which would be acceptable. Finally, a record of 140 Avocet was not flagged in red, as for this county and time of year it is quite normal.

JIM ALMOND www.shropshirebirder.co.uk

Dynamic predictive text fields made entry of species names quick and easy but validation was needed to catch incorrect species selections. For example, upon typing 'snipe' a user could then accidentally select Jack Snipe.

knowing what records they have entered and which species they have seen when and where reduces data duplication and effort. Additionally, if local experts queried any records, then there needed to be a mechanism for observers to navigate to assess and edit those records. The '*View/Edit my records*' module was available from *Data Home* and provided access to all the observer's Atlas records. It was divided into the following three areas for ease of navigation.

First, for general information, personalised species lists were presented for winter, breeding-season and out-of-season records, each showing the number of Roving Record and TTV locations where the observer had reported each species, along with the maximum count entered for that season. Drilling down to view a particular species presented a summary of dates and locations for that species (for both Roving and TTV records entered). It was hoped that observers might use these simple lists to spot unfamiliar species, which might be pointers to earlier data entry errors.

The second area dealt with Roving Records. For general information, a graph presented the number of records submitted per month. Navigating to individual Roving Records was complex owing to the potential breadth of locations and seasons, and the possibility of repeat records. To aid navigation, a summary table presented the total numbers of records in each of the eight recording seasons from 2007 to 2011. Separate totals were provided for records entered directly and those uploaded (see Section 3.6.2) so that users could quickly find uploaded records that had broken validation thresholds. In either summary table, clicking on the total for a season further summarised that subset of records into a matrix of 10-km squares and the four months in the season. Finally, clicking on the total for a particular 10-km square in a particular month presented a date-ordered list of those Roving Records. Each record could then be edited or deleted as desired. Any records that had a pending validation query were marked with a red dot.

The third area dealt with TTVs. The structured nature of TTVs made navigation simpler and a table displayed all allocated TTVs and the number of pages entered for each visit. Clicking on a visit displayed all TTV counts for the species encountered, ordered by page number. Options to delete and edit individual species were available. Only once all pages were entered for a TTV did the option to add a species (if missed during normal record entry) become available—this was to avoid adding species outside of the normal page order. The option to delete the entire TTV was also available, but only a page at a time, working backwards from the last page. This was important because the visit duration (and date) was set on Page 1, and it was critical that subsequent pages did not have different attributes. As above, records with validation queries were marked with a red dot.

All edits and deletions were logged in an audit table within the database as a precaution against accidental loss of data, but in practice these were rarely used.

3.6.4 Dynamic personalised feedback

A key aim of the Atlas was to achieve complete 10-km resolution coverage of Britain and Ireland. To facilitate this, a suite of feedback modules was created to help to direct observers into covering poorly recorded areas, finding apparently missing species and confirming breeding, whilst at the same time minimising duplication and wasted effort. Great efforts were made to develop modules that delivered feedback that was personalised to the observer, to make the feedback more appropriate to them, and therefore more effective and engaging.

Personal Results as part of the Local Effort

The first module to provide this was *My Square Summaries*, which gave access to a list of all 10-km squares and tetrads where the observer had entered winter or breeding records. Observers could

GRAHAM CATLEY

Feedback modules were developed throughout the fieldwork period to help observers to identify species, such as the Grasshopper Warbler, not yet recorded or *confirmed* to breed in their square.

drill into a 10-km square or tetrad to view a list of species submitted for that square by the entire Atlas and BirdTrack recording community, with ticks to indicate the species they had contributed themselves. Breeding-season lists showed the highest breeding evidence they had submitted plus the highest submitted from all contributors and incorporating other data sources (see Section 2.4). Species that only they had seen were indicated with a smiley face icon. Although essentially a fun thing to show, it was well received by observers, and added a modicum of competition and increased engagement. To safeguard the locations of rare or sensitive breeding and wintering species (see Section 2.3), such species were omitted from the species list of a square if reported only by other observers; if the observer had reported a sensitive species, only their breeding evidence code was shown, and smiley face icons were replaced with 'anonymous' yellow dots. A 'source' field was also shown to indicate which survey or data source the records came from (often multiple data sources added data for the same species). Records collected out-of-season were not shown within *My Square Summaries*.

The benefit of this system was that it helped observers to prioritise future recording effort by identifying species that were perhaps missing from the Atlas, or lacking *confirmed* breeding evidence, in squares they had already visited.

Promoting Recording in Other Areas

My Square Summaries helped to steer observers into doing more targeted recording in the areas they were already visiting. However, it was vital for the success of the Atlas that surveyors visited unfamiliar squares rather than focusing only on their home area or popular birding spots. The aim of the '*Any Square Summary*' module was to allow observers to see a species list and maximum breeding evidence information for squares they had not yet contributed to. Observers chose a 10-km square, either from a list or via an interactive map, optionally selected a tetrad, then selected the season, and were provided with an impersonalised list of species and maximum breeding evidence (even if the records were their own). This was clearly a popular module and was used by almost 5,000 surveyors to view species lists on over 225,000 occasions. Observers were restricted to 10 views of squares or tetrads per day, to minimise data mining using Internet robots and to prevent misuse of the data before they had been completed and validated.

Identifying Gaps and Priority Areas

The rapidity of online data submission allowed us to monitor the type and quantity of data being submitted and to identify possible problems. One area of concern was that many surveyors were not providing evidence of breeding, so during 2010, we launched a module to promote the recording of breeding evidence, and to identify the 10-km squares most needing extra effort. The '*My local gaps*' module was designed to present information on species that had been recorded in the nine 'local' 10-km squares but not yet *confirmed* to breed. 'Local' was determined by the observer's postcode or, if they had not provided one, by where they were most actively recording using an algorithm based on weighting the locations of (in decreasing order of most useful for Bird Atlas) TTVs, BirdTrack lists, Roving Records and BirdTrack casual records. Whilst there was no prior expectation regarding what was an acceptable percentage of species to have *confirmed* breeding, observers could still use the fact that few species in a square had been *confirmed* breeding as indicative of poor coverage in a particular local 10-km square. With this in mind, the calculation of the percentage figure excluded any species classed as a migrant or summering; otherwise, coastal migration hotspots featured as having apparently poor breeding confirmation. For each of the nine 10-km squares shown, the module presented a list of the species that had yet to have any breeding evidence added, and those that had lower levels of evidence. Observers could also view similar information for any area in Britain and Ireland by re-centring the mapping window.

Finally, later in 2010, the seasonal '*Priority Squares*' module was introduced, which allowed observers to view lists of species that had been seen in the eight surrounding 10-km squares, but not yet recorded from their home 10-km square (again initially defined by their postcode or most active recording area). This operated on the assumption that if a species had been seen in the eight surrounding squares but not in the middle square, the chances were that it was widespread in the area, but under-recorded in the central 10-km square and a prime target for field effort. In contrast, if a species had been reported in few of the neighbouring 10-km squares, the chances were that it was a habitat-specialist or highly localised and might not be worth the effort of searching for in the central 10-km square.

All these modules were designed to both immediately inform and motivate observers to better target areas, thus improving coverage during *Bird Atlas 2007–11* more effectively than is likely to have been the case in previous atlases that could only rely on periodic and impersonalised coverage summaries in newsletters.

3.7 OPTIONS FOR REGIONAL ORGANISERS AND VALIDATORS

TWO IMPORTANT MODULES were provided for the Regional Organisers and Regional Validators. The first was designed in parallel with the *Tetrad Request* module (see Section 3.6.1), allowing Regional Organisers to assign individual tetrads to observers and to monitor progress towards TTV targets in their region. The second was a complex module for Regional Validators to view and validate all records in their region.

3.7.1 Management of TTV coverage

The *Tetrad Management* module allowed the Regional Organiser to allocate individual TTVs to observers in response to emails that had been generated through the *Tetrad Request* module (see Section 3.6.1) or following more traditional forms of communication. Access to this system was restricted to Regional Organisers. Only one Regional Organiser was allowed to administer tetrad allocation per region, but individual Regional Organisers could be responsible for more than one region.

Each Regional Organiser potentially had many hundreds of tetrads to administer, so a simple navigation system was developed to allow them to see at a glance the 10-km squares that were under their control, and how many tetrads had been allocated out of the maximum number available in each 10-km square. Figure 3.9 shows an extract of the navigation system for Anglesey, with an inset map on the left showing the general location and shape of the Atlas region. The map on the right shows the detail of each 10-km square, and the numbers allocated out of those available. Anglesey is an interesting example as the figure demonstrates that while the inland 10-km squares comprise 25 potential tetrads, the coastal 10-km squares contained fewer tetrads, some eligible for coverage, and some varying in management ownership (Figure 3.9).

Drilling down into a 10-km square presented a table showing the constituent tetrads, plus editable cells where the Regional Organiser could assign an observer for the winter and/or breeding season. Assignment was done by entering either a recognised online username or, if the username was unknown or the observer was not an online participant, the observer's surname and postcode (if available). This was felt to be enough information to manage the allocation of tetrads. During the course of the Atlas, this system also showed the Regional Organiser which TTVs had been completed and which were outstanding. Once records had been added to a TTV, this portion of this system became uneditable (see Section 3.9).

Clearly, the advantages of this system were that it was immediate and easy to use. It was one of the first modules to go live, getting tetrads allocated early so that there were no delays when fieldwork started. It also helped to prevent blockages in the progress of the Atlas as Regional Organisers could see who might be slow to complete their fieldwork, or might be allocated what was beyond a sensible number of tetrads, as well as helping to spread the load across an optimal number of observers.

FIGURE 3.9 EXTRACT FROM THE *TETRAD MANAGEMENT* MODULE FOR THE ATLAS REGION OF ANGLESEY

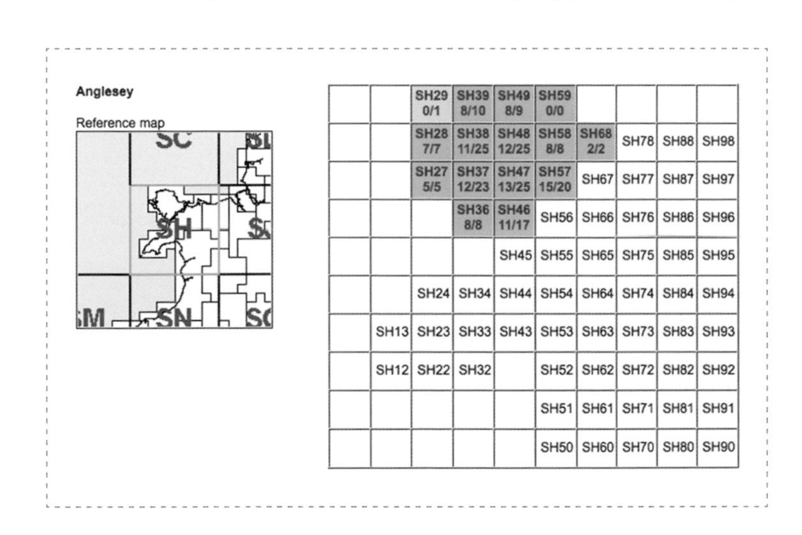

Green shading of 10-km squares indicates that the minimum Rule-8 allocation has been attained whereas brown shading indicates squares currently below minimum allocation; yellow shading indicates 10-km squares outside the region. Numbers beneath the 10-km square labels indicate the number of tetrads fully allocated in both seasons out of the number available. In SH48, for example, 12 of the possible 25 have been fully allocated, exceeding the target of 8 and the square is shaded green. SH57 has 24 tetrads where the centre point is on land (see Section 2.2.1), but only 20 were under the control of the Anglesey organiser, four being under the control of the adjacent region (Caernarfon). Dividing the tetrads between organisers was done for pragmatic reasons where geographic features (here the Menai Strait) bisected 10-km squares. In SH28 all seven on-land tetrads are fully allocated, and all within Anglesey region. Lastly, SH29 had only one tetrad where the centre was on land, and this had not been allocated at this time so the square is shaded brown.

FIGURE 3.10 FLOW CHART SHOWING THE *REVIEW AND VALIDATION* MODULE

These are the routes by which Regional Validators could browse and filter records to produce a manageable subset to check by eye.

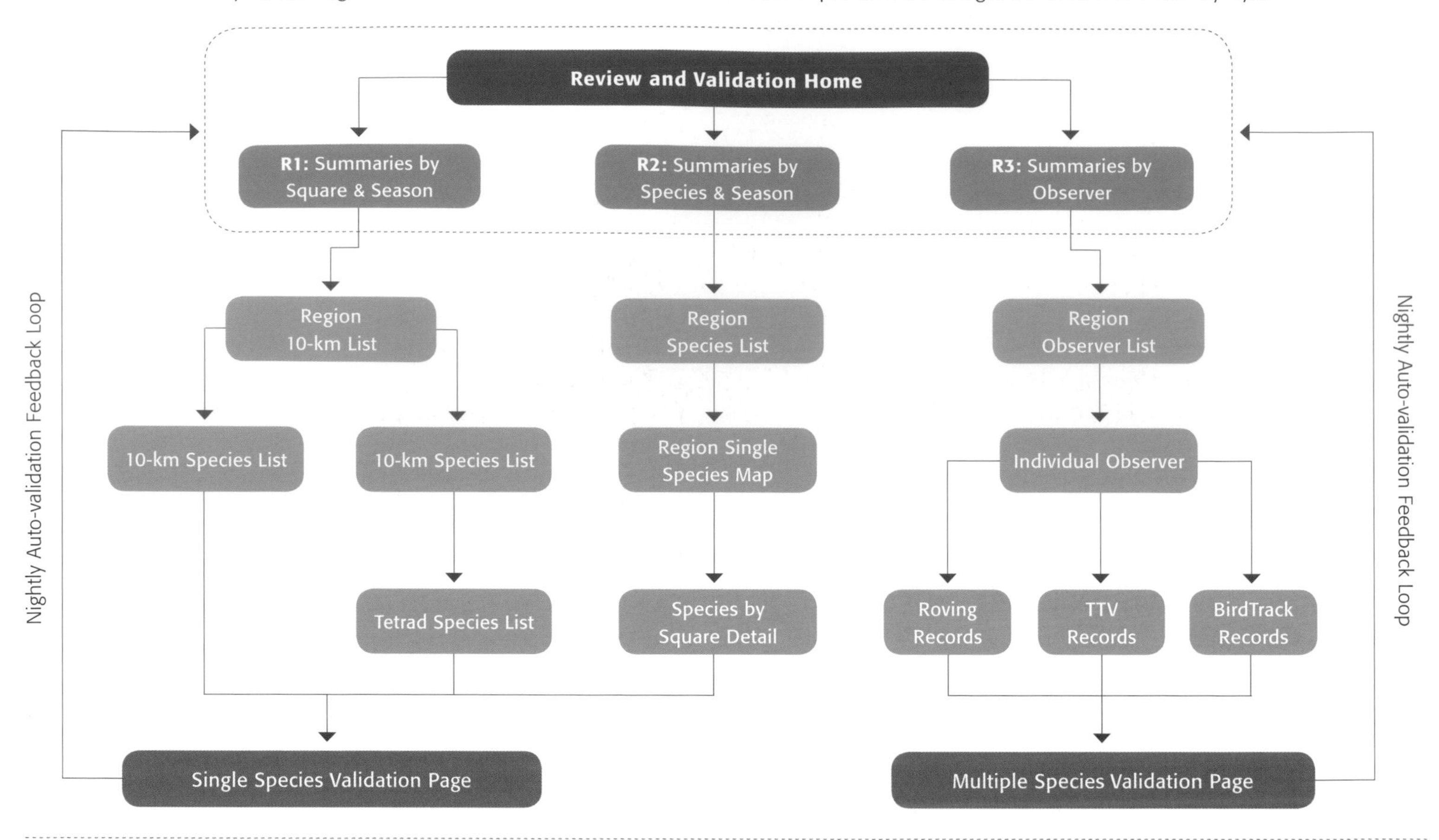

3.7.2 Reviewing and Validation of records

The *Review and Validation* module was the largest, most complex and problematic module built for Bird Atlas Online. The basic requirements were to allow local experts the ability to view all Atlas and BirdTrack records entered within an Atlas region, to confirm which records were valid to use as data for the Atlas, and to query those that needed additional supporting information or correction. The largest problem faced was the sheer volume of records that had to be summarised and navigated, requiring a fast and logical system to arrive at manageable subsets that could be checked by eye. To enable this, a variety of navigation routes were created, based on summaries by location, species or observer. Each of these is briefly described below, and their place within the system is shown in Figure 3.10. The first and second routes were season dependent (winter, breeding season or out-of-season); the third operated across seasons. The routes were:

- **Summaries via 10-km squares and tetrads** This route allowed Regional Validators to view the number of records per 10-km square, or tetrad, and drill down through a list of species entered per square, finally arriving at the 'species view' (Figure 3.11).
- **Summaries via Species** This route showed a list of species entered for the entire region. Selecting a species generated an interactive map showing 10-km squares with data, shaded to indicate maximal breeding evidence. Selecting a 10-km square revealed a table that allowed access to a final 'species view' page, either for the whole 10-km square or for individual tetrads.
- **Summaries via Observers** This presented a list of all observers who contributed data to the region, whether they resided in the region or not. It indicated the number of records and whether BirdTrack records were involved. Drilling down to an observer presented a summary of their Roving Records, TTVs and BirdTrack records, very much like that presented to observers in their own *View/Edit my records* module (see Section 3.6.3). For Roving Records and BirdTrack records, validation was via final pages like the 'species view' described above but potentially encompassing any number of species seen in the square or month selected. For each TTV, validation could be accomplished by accepting the whole visit, or by stepping into each page to accept or query individual records as required.

Regardless of which route the Regional Validator used to access the final 'species view', at the record level, every record had to be set either as 'Valid' or to one of four types of query (Figure 3.11):

- **Identification** Used if there was some uncertainty about the species identification, possibly indicating a data entry error or misidentification, or because the species required a description in the region;
- **Number** An unusually high count that could indicate a data entry error;
- **Breeding status** Either an unusual or incorrect breeding status (most often due to the suitable breeding habitat criterion being ignored);
- **Grid reference** Used if the site name (or sometimes species) hinted strongly that the grid reference had been incorrectly given.

There were several common elements within the *Review and Validation* module that helped summarise and navigate to records, and the eventual validation of records. Throughout navigation, 'progress bars' were used to indicate the proportion of records in the current subset that were unchecked, in a queried state or already set as valid. This allowed the Regional Validator to assess instantly if new records required checking, or if they were on top of their validation role.

The *Review and Validation* module was linked directly to each observer's *Data Home* pages so if there were any queried records, an alert would appear on the respective Atlas or BirdTrack *Data Home* pages (see Figure 3.4 for an example alert). These alerts linked directly to a summary showing the type of query and gave the observer quick links either to edit or delete queried records, or to enter into an email discussion with the Regional Validator to resolve valid records.

All queried records were also 'audited' in the database, so that the Regional Validator could easily see summaries of the records that were outstanding (number of records, type of query, identity of observer), and those that had been resolved. An additional 'email slow observers' function was provided to send a one-off reminder email to all observers who had not resolved queried records within a 14-day period.

Even though the *Review and Validation* module was designed to be as broad and interactive as possible to enable Regional Validators to validate records quickly, regularly and easily, the volume and repetition of records for some species made it a much bigger task than envisaged. To reduce the volume of records that needed targeting by a trained eye, a system of 'auto-validation' was devised.

FIGURE 3.11 A TYPICAL 'SPECIES VIEW' WITHIN THE *REVIEW AND VALIDATION* MODULE

Date	Observer	Count (+nests)	Breed code	Tetrad	Source	Type	Place name	Current	Options				
									OK	ID	No.	BS	GR
							Check this box to set/unset all records below as Okay >			Set/unset all GR >			
5 Apr, 2010	Moran, Nick	2	F	A	web	BT List	2012NM L...	GR					
25 Apr, 2010	Balmer, Dawn Elizabeth	2	FF	D	web	Roving	Barnham ...	BS					
26 Apr, 2009	Dataset (bbs), Bto	18		D	web	BBS	BBS squa...	✓A					
1 Apr, 2010	Carey, Steven	12		A	web	BT List	Lackford...	✓A					
9 Apr, 2009	Regnault, Margaret	10		A	web	Roving	Lackford...	✓A					
22 Apr, 2009	Collins, David	1	S	C	web	1st hour	TTV TL87C	✓A					
25 Apr, 2009	Fairley, Malcolm	1		A	web	BT List	Lackford...	✓A					
1 Apr, 2010	Fairley, Malcolm	-		A	web	BT List	Lackford...	✓A					
4 Apr, 2010	Austin, T S	-		H	web	Roving	Wordwell	✓A					
15 Apr, 2009	Fairley, Malcolm	4		A	web	BT List	Lackford...	✓					
15 Apr, 2009	Regnault, Margaret	4		A	web	Roving	Lackford...	✓					
Number of records: 11													

Here breeding-season records for Lesser Redpoll in TL87 are shown. All records have been reviewed but they have different states. Seven have been set to valid (green tick) by the Regional Validator. Two records have been auto-validated (purple tick) based on the validity of the 1st April 2010 record (see Auto-validation in Section 3.7.2). One record has been queried with the 'GR' flag because the Validator considers the location to be incorrect. Lastly, one record has been queried with the 'BS' flag because the Validator suspects an error because April is potentially too early for the 'Adults carrying Faecal sac or Food for young' code.

Auto-validation

Auto-validation was a database script run nightly, that took advantage of validation effort to date, to batch process new records. The rules were relatively simple, and outlined below, but essentially worked on the principle of changing unchecked records to valid if the Regional Validator had already set a comparable record as valid. For example, within a tetrad, there may be many records of common species such as the Blackbird added by numerous different observers (especially when BirdTrack records were included). If a Regional Validator set a record of 10 Blackbirds as valid for tetrad TL88T, in June, and accepted breeding evidence of Nest with Young, then it could be assumed that all other June records with Blackbird counts no higher than 10 and with breeding evidence no higher than Nest with Young were also valid. Note also:

- Not all species were included. Initially auto-validation involved a subset of 60 common species, but as the volume of valid data from which to 'learn from' became more complete, this was extended to cover some 258 species. Most of these species were auto-validated year-round, but some, for example species that were common winter visitors but rare breeders, were auto-validated just in the one season.
- In most areas auto-validation operated at the 10-km-square resolution, effectively ignoring tetrad letters when comparing records. However, in regions undertaking tetrad-resolution local atlases, auto-validation took place only at tetrad resolution.
- Auto-validation operated within each month because distributions can change within seasons and it was not safe to assume, for example, that a record in February was acceptable on the basis of a prior valid record in November.

This process happened at approximately 01:00am each morning and had a huge impact on reducing the volume of records to validate, allowing the local experts to target more vital and individual records. Clearly, auto-validation was based on assumptions and was not infallible, and for that reason, Regional Validators were also able to query records that had been auto-validated. Generally though, the assumptions did not significantly affect the maps and results being fed back to the Regional Organisers and Validators.

3.8 NIGHTLY FEEDBACK OF RESULTS

THE FEEDBACK DESCRIBED in Section 3.6.4 available to registered Bird Atlas Online users was dynamically generated, and was personalised to what they had submitted and where. In addition to those feedback mechanisms, nightly extracts of all submitted Atlas and BirdTrack records were used to produce generic feedback that was made available publicly. After extracting the current data at 01:00am each morning, bespoke SAS® programs were run to analyse the data, generate graph and map image files, XML data files and write HTML web pages which were then posted to the website www.birdatlas.net ready for the next day's viewers. The web pages produced included:

- **Keystats** A page summarising all the main headline statistics, including the total number of observers, records and species to date; the number of tetrads allocated to observers and the number completed, tallying progress towards the Rule-8 targets (see Section 2.2.4).
- **TTV Progress Maps** Maps showing the 10-km squares that had achieved Rule-8 tetrad allocation or coverage in red, versus those still in need of tetrad allocation and coverage. These maps showed the chequerboard in Ireland. National Organisers used these maps and talked about 'turning the map red'.
- **Expected Species Richness Maps** For every 10-km square previous atlas data were used to estimate how many species might be expected. Since there was no wish to publicise a figure that might be taken as a definitive target for each 10-km square, species richness to date in a square was instead accumulated as a percentage of the expected figure and mapped up to an upper category of 'over 90%' thereby making it impossible to know when a square had met its target. Species were counted towards the breeding-season species richness only if they had at least *possible* breeding evidence. This made the breeding-season map appear to lag behind the winter map, for which species were much easier to add to the list.
- **'Bird of the Day'** Every day a species was chosen at random and a distribution map corresponding to the current season was created. This simple page was very popular and generated many followers and email engagement. The selection of species was automated and unsupervised, so rare breeding species were excluded (see Section 2.3).
- **Percentage of Species *Confirmed* to Breed** A map was produced showing the percentage of species reported in the breeding season for which *confirmed* breeding evidence had been supplied. Whilst we had no ideal or expected figure in mind, a low percentage of species *confirmed* to breed was taken as indicative of likely under-recording.
- **Example Species Maps** Rather than show current species maps for all species, we presented daily updating maps for a suite of 12 species that were strongly associated with particular habitats or landscape types (e.g. the Red Grouse for uplands, the House Sparrow for urban areas). The hope was that these would be good indicators of coverage of key habitats and landscapes. The suite of species was modified through the course of the Atlas, for example to add owl species when there was a focus on nocturnal recording, and to add seabirds to encourage adequate coverage of colonies.
- **'Priority' and 'Gaps' Maps** Before the closing seasons, new interactive maps were produced to draw attention to the relatively small number of 10-km squares that were deficient in TTV coverage, species richness or breeding confirmation and therefore high priorities for coverage.
- **Regional Results** All of the maps described above were also reproduced for single Atlas regions and presented as static images and interactive maps. Since these maps showed smaller areas, it was possible to show information at tetrad resolution as well as for 10-km squares. This allowed us to show precisely which tetrads had been allocated and surveyed, and for regions participating in local atlases, tetrad-resolution species richness and example species maps were produced (static only). Regional pages also included tables summarising progress towards Rule-8 coverage.

In addition to these public pages, similar nightly analyses were used to generate limited-access pages for Atlas administration. In the latter stages of validation, final sets of national distribution maps for all but the rarest breeding species were produced and made available via password-protected web pages to Regional Validators to make final checks. Password-protected web pages were made for the National Organisers and the team of species-account writers. There were daily updates on validation progress, various data diagnostics and a complete set of species pages including all distribution, distribution change, abundance and abundance change maps and associated statistics. Aside from the value of the Internet for the survey itself, its value for organising, distributing and accessing results of this kind from an administration perspective cannot be understated.

3.9 PROBLEMS ENCOUNTERED AND LESSONS LEARNT

NO PROJECT OF this magnitude runs without problems, and a number of unanticipated challenges arose during the course of *Bird Atlas 2007–11*. Some of these are described below.

Hardware and Software

- The *Review and Validation* module could have been designed to cope better with the high volume of data. Drawing on records for an entire region resulted in poor database and server performance, and several remedial patches to the web application were necessary. As a result, new hardware needed to be purchased in early 2008 to improve performance. Auto-validation was a partial solution, though it generated problems itself owing to the huge volume of data it had to process every night.

Observer Management

- The *Tetrad Management* module was designed on the assumption that a single observer would complete both visits in a season. We had not anticipated and catered for the inevitable situations where observers were unable to do the second visit (e.g. illness, moving from local area). In these situations, allocations had to be changed manually directly in the database.

MICK DURHAM www.wildlife-photographer.net

The Fulmar is widely distributed around the coasts of Britain and Ireland so daily updates to its breeding-season distribution map could be used by observers to gauge coverage of rocky coasts and help target future fieldwork.

Data Entry

- For a small number of species pairs, the autocomplete technology actually increased the incidence of data entry errors rather than reducing them. These errors arose when a rarer species occurred earlier in the taxonomic order than a common species. For example, an observer typing 'buzzard' would be presented with Honey-buzzard before Buzzard.
- Colony data collection proved awkward to fit in an online model. It was not compulsory to complete, but if the observer did complete this section, there was a mandatory connection within the database to the same species in earlier pages (e.g. Rook was on Page 4 of the TTV online form, and also on Page 5, the 'Colonies' page, and records on these two pages were stored as separate but connected records in the database). This meant that restrictions in editing had to be included, since an observer could not delete the normal TTV Rook entry without considering the colony entry.
- Uploaded records could not be bulk handled easily to include Threshold Validation. In addition, the lack of resources that imposed a single-upload format (Microsoft Excel) meant that formats had to be more rigid and mistakes at the processing stage were more commonly encountered. This had a knock-on effect of having to adapt feedback systems to allow observers easily to locate uploaded records for their own checking.

Local Atlases

- Thirty county and regional bird clubs continued fieldwork for their tetrad atlases by one to two years, taking advantage of the momentum in observer effort in the relevant areas. This meant that there were differences in the allowed end date of Roving Record data collection depending on region. An additional Roving Record submission module had to be built to allow for continuing regions, while the national *Add Roving Records* module set a hard cut-off date of 31st October 2011. This complexity was often compounded by differences in the boundaries between BTO regions and areas being recorded after the end of the national recording period.
- Ideally, observers would enter their records soon after observation but this did not always occur as some sat on their records for a while. It was therefore necessary to keep record entry open for several months after the end of the final field season, necessitating tighter validation on dates. This was exacerbated by variation in local atlas continuation dates and boundary changes. This also impacted on the *Review and Validation* process, as 'new' records trickled in late.
- Holding the definitive Atlas dataset in the live database caused unforeseen difficulties in the closing stages of book production when there was a need for a stable dataset from which to derive text and maps. However, the dataset was in constant change owing to the addition of new and late data for local atlases, ongoing validation in those regions, and BirdTrack observers entering their old records. Such additions and edits had the potential to cause small changes to maps and statistics for some of the scarcer species.

CHAPTER 4

DATA ANALYSIS AND MAP PRODUCTION

Simon Gillings

4.1 INTRODUCTION

ANALYTICAL TECHNIQUES AND map production methods have changed markedly since the pioneering atlases of the 1960s–70s. Whilst the availability of high-performance computers and data storage technology has made it considerably easier to store, access and manipulate huge volumes of data, analytical techniques have become increasingly complex to allow greater insights to be drawn from the data. Many atlases now use Geographical Information Systems (GIS) and forms of statistical modelling to analyse the spatial data and create attractive maps. This chapter explains the methods used to convert volunteers' bird records into maps and statistics. Firstly, however, there is an important discussion of terminology and the interpretation of atlas data (see also Chapter 7).

4.1.1 False absences and the interpretation of gaps and changes

It is tempting to look at an atlas distribution map and infer that gaps in the map indicate an absence of the species in question. Whilst this interpretation is probably correct in many instances, it may not always be true. 'False absences' arise where a species that is actually present is missed by fieldwork. For example, a map for a nocturnal species might contain a mix of true absences where the species was not present, and false absences in areas that did not receive night-time visits. In clear-cut cases like this, allowances can be made in the interpretation, but false absences can also arise in more subtle ways. For example, observers rely on territorial behaviour to detect many species but this behaviour may be much reduced at low population density (e.g. McShea & Rappole 1997), giving fewer opportunities to detect a species and accrue breeding evidence. Atlas field methods are tailored to minimise such issues by targeting field effort to the main song periods, by making visits early in the day, and by avoiding bad weather. Whilst this reduces the potential for false absences, they can never be truly eliminated, so it is more appropriate to refer to gaps in maps as indicating where the species was 'not detected' rather than that it was 'absent'.

These considerations also extend to the interpretation of change maps and statistics because the likelihood of registering a false absence is inversely related to the amount of recording effort an area receives, and this may vary through time. We might want to describe 'gains' and 'losses' on a species' change map, but such interpretation assumes that the species really was absent from some squares in one of the time periods and ignores the possibility that the species has remained present throughout but exhibited false absences in those squares during atlas fieldwork. To convey this uncertainty it is more appropriate to refer to changes as 'apparent gains' and 'apparent losses'.

Wherever possible, distribution and change patterns have been interpreted in accordance with these principles.

4.1.2 Previous atlas data

The range of potential change analyses and maps is determined partly by the taxonomic status of species at the time of different atlases (see Box 4.1) and also by the type and availability of data. It is worth recalling the type of data each previous atlas collected as well as detailing what was archived at the time and is thus still available for analysis. Full field methods are given in the introductions and appendices of the respective atlas publications.

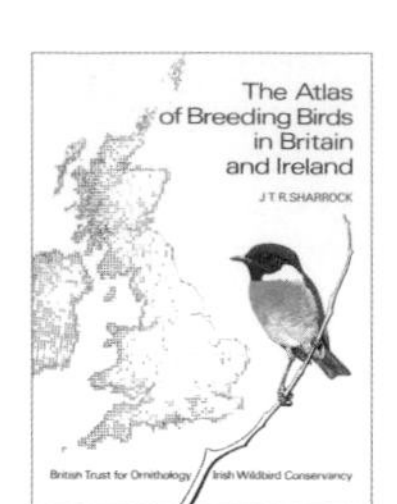

1968–72 Breeding Atlas

Lists of species with *possible*, *probable* or *confirmed* breeding evidence were submitted for 10-km squares but data were not requested at tetrad resolution. No estimates of abundance were made. There was no standardisation of recording effort.

Only the final summarised 'map' data are computerised; the numbers of visits to individual squares and of participants are not available.

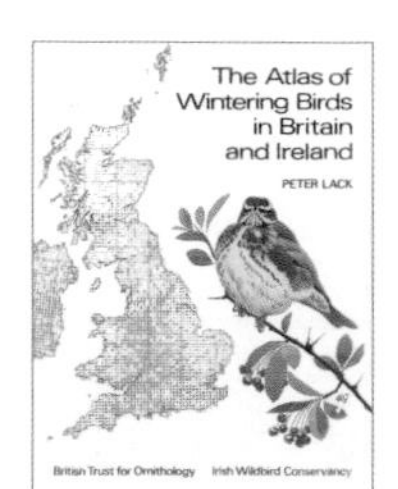

1981–84 Winter Atlas

Lists of species with counts were submitted for 10-km squares along with the length of time spent in the square. Length of time varied from 20 minutes up to 16 hours and counts were standardised to 6 hours during map production. Supplementary records at 10-km resolution were submitted without effort data. No data were submitted at tetrad resolution.

All the original and summarised map data are available, though some errors found during the final stages of map production were never traced back and corrected in the original data.

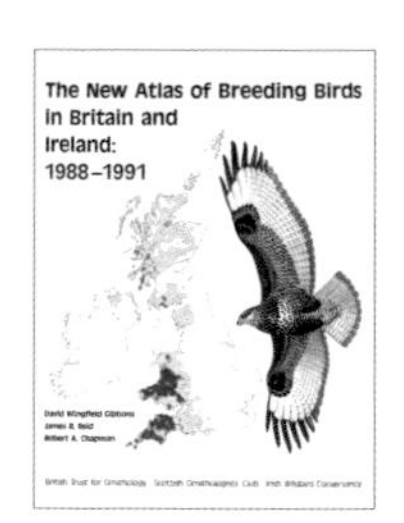

1988–91 Breeding Atlas

Tetrads were used for the first time and, although they were referred to differently, the equivalent of TTVs and Roving Records were included. They were called 'timed counts' and 'supplementary records', but for simplicity in later sections we will refer to them as TTVs and Roving Records (but see below). Each tetrad received two 1-hour TTVs and the majority of common and widespread species were not counted, only listed as present on either visit; 117 scarce, colonial and wetland species were counted. Visit dates were not recorded, nor was the visit(s) on which each species was present or counted. Tetrad coverage followed the same Rule-8 approach (see Section 2.2.4) throughout Britain and Ireland (where, unlike in 2007–11, there was no chequerboard; see Section 2.2.4). Supplementary records were collated at 10-km resolution but they did not receive such a high profile as Roving Records, leading to probable under-reporting of some species. Aside from the TTVs, no other tetrad-resolution data were collected. Breeding evidence was simplified to two levels: *seen* and *breeding*. These broadly translate to *possible* breeding and *probable* & *confirmed* breeding,

JIM ALMOND www.shropshirebirder.co.uk

Following the decision to recognise the Caspian Gull as a distinct species, *Bird Atlas 2007–11* presents new winter and breeding-season distribution maps.

BOX 4.1 TAXONOMIC CHANGES

The years spanning the breeding and winter atlases have seen changes to the taxonomic status of several species and subspecies. Those accepted by the relevant authorities are detailed below, indicating which forms were recognised at the time of each atlas. Only the taxa included in the main species accounts are detailed here. These changes are important to note because they have the potential to limit the permutations of change that can be calculated for some species in some or all seasons (see Section 4.3). In addition, Hooded Crow (*Corvus cornix*) was officially recognised as a species separate from Carrion Crow (*C. corone*) in 2002 (Parkin *et al.* 2003), but the two forms were recorded separately in all previous atlases.

<table>
<tr><th>Taxon</th><th>1968–72 Breeding Atlas</th><th>1981–84 Winter Atlas</th><th>1988–91 Breeding Atlas</th><th>Bird Atlas 2007–11</th></tr>
<tr><td rowspan="2">Teal
Anas spp
(Sangster et al. 2002)</td><td rowspan="2">Teal
A. crecca</td><td rowspan="2">Teal
A. crecca</td><td rowspan="2">Teal
A. crecca</td><td>Teal
A. crecca</td></tr>
<tr><td>Green-winged Teal
A. carolinensis</td></tr>
<tr><td rowspan="2">Manx Shearwater
Puffinus spp
(Sangster et al. 2002)</td><td rowspan="2">Manx Shearwater
P. puffinus</td><td rowspan="2">Manx Shearwater
P. puffinus</td><td rowspan="2">Manx Shearwater
P. puffinus</td><td>Manx Shearwater
P. puffinus</td></tr>
<tr><td>Balearic Shearwater
P. mauretanicus</td></tr>
<tr><td rowspan="3">Herring Gull
Larus spp
(Sangster et al. 2005, 2007)</td><td rowspan="3">Herring Gull
L. argentatus</td><td rowspan="3">Herring Gull
L. argentatus</td><td rowspan="3">Herring Gull
L. argentatus</td><td>Herring Gull L. argentatus</td></tr>
<tr><td>Yellow-legged Gull L. michahellis</td></tr>
<tr><td>Caspian Gull L. cachinnans</td></tr>
<tr><td rowspan="2">Rock Pipit
Anthus spp
(BOU 1986)</td><td rowspan="2">Rock Pipit
A. spinoletta</td><td rowspan="2">Rock Pipit
A. spinoletta
(An unknown proportion of Water Pipits was recorded to subspecies level)</td><td>Water Pipit
A. spinoletta</td><td>Water Pipit
A. spinoletta</td></tr>
<tr><td>Rock Pipit
A. petrosus</td><td>Rock Pipit
A. petrosus</td></tr>
<tr><td rowspan="2">Redpoll
Carduelis spp
(BOU 2001)</td><td rowspan="2">Redpoll
C. flammea</td><td rowspan="2">Redpoll
C. flammea</td><td rowspan="2">Redpoll
C. flammea</td><td>Lesser Redpoll
C. cabaret</td></tr>
<tr><td>Common Redpoll
C. flammea</td></tr>
<tr><td rowspan="2">Crossbill
Loxia spp
(BOU 1980)</td><td rowspan="2">Common Crossbill
L. curvirostra</td><td>Common Crossbill
L. curvirostra</td><td>Common Crossbill
L. curvirostra</td><td>Common Crossbill
L. curvirostra</td></tr>
<tr><td>Scottish Crossbill
L. scotica</td><td>Scottish Crossbill
L. scotica</td><td>Scottish Crossbill
L. scotica</td></tr>
</table>

CHRIS KNIGHTS www.chrisknightswildlife.com

respectively, though a weaker requirement to have suitable breeding habitat for the *seen* category complicates change analyses (see Chapter 7).

TTV data are held as originally submitted but the original supplementary records are not held, having been summarised into final map data in the form of a list of the highest-level breeding status for each species in each 10-km square.

4.1.3 Safeguarding sensitive species

Section 2.3 describes how, in feedback material, the resolution of distribution information for rare breeding species was limited during the data-collection phase to protect sensitive populations from persecution and disturbance. In 2011, discussions between the Atlas team, RBBP and IRBBP turned to the question of how to publish informative maps without impacting upon the conservation status of these species. We reviewed how previous Britain and Ireland atlases solved this question and formulated a policy detailing the circumstances under which accurate 10-km resolution mapping would not be used, and listing three alternative presentation methods (Box 4.2). We were keen to move away from the practice of shifting dots because this obscures the shape of ranges and involves changing accuracy rather than precision. Instead new maps were produced with some or all of the range shown at a resolution of 20-km, 50-km or 100-km; downgrading evidence or removing dots were a last resort. Provisional maps were assessed by BTO, RBBP and IRBBP to identify preferred mapping resolutions and candidate dots were checked for downgrading or removal. During subsequent months these preferences were circulated for comment among key stakeholders (e.g. RSPB Investigations Team, Raptor Study Groups and species specialist groups). In April–May 2012, with all data entered and validation at an advanced stage, a final check was made to ensure that the preferred mapping options were still acceptable in light of the latest data.

4.1.4 Revised TTV targets for analyses

Section 2.2.4 describes how the target for TTV coverage was set as eight tetrads per 10-km square, or all tetrads if fewer than eight were present; for ease of reference this was called the 'Rule-

Threats to rare birds such as the Red-necked Phalarope were carefully assessed and distributions are mapped at the most appropriate resolution to protect individuals and sites.

8' target. From a fieldwork perspective, this was a simple and easily communicated method to ensure a minimum level of TTV coverage across the countries. From an analytical perspective, it is a potential benchmark for determining sample sizes. However, for many analyses, Rule-8 actually delivers a biased sample of tetrads. This is because it states that coverage of an inland 10-km square should reach eight of the 25 tetrads (a 32% sample), but coverage for a coastal square with, for example, approximately half the area of land should also aim for eight out of 12 tetrads, achieving 67% coverage. In this way, the total sample of tetrads will over-represent coastal areas (see Box 5.2). Therefore, many analyses described in Sections 4.4–4.6 use an adjusted target that is proportional to land area to ensure a 32% sample regardless of the land area of the 10-km square. This target, called 'Rule-8_{adj}', generates an unbiased sample of tetrads.

4.2 DISTRIBUTION MAPPING

THE AIM OF distribution maps is to show all the 10-km squares in which each species was recorded as present in any of the four winters and breeding seasons during 2007–11. Only data from 10-km squares with at least some land or freshwater were included, and records of birds reported from entirely marine squares excluded. A summary of which squares were surveyed and any gaps in coverage can be found in Chapter 5. All data sources were used in the production of distribution maps. Below is an explanation of how the different types of distribution maps were created.

4.2.1 Winter distributions

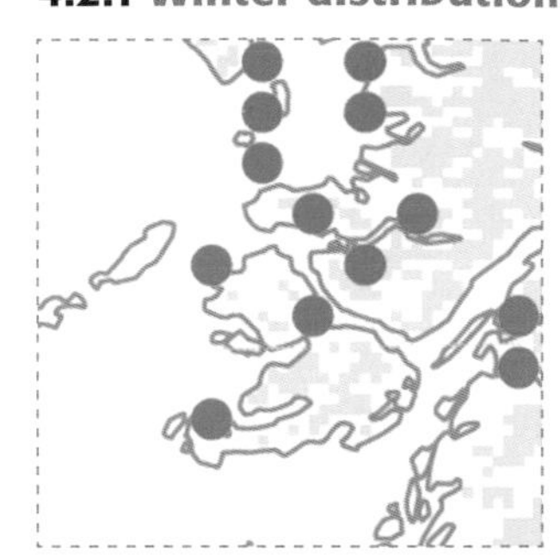

Winter distribution maps were compiled by identifying all 10-km squares where a species was recorded during November to February inclusive, having first excluded all records coded as 'Flying'. Observers should only have applied the Flying code to birds flying over at height in order to exclude species not using the resources of the square. For example, at the country scale, tens of thousands of Pink-footed Geese migrate between their favoured wintering areas in East Anglia and northwest England, overflying many inland counties that they rarely use. At a local scale, corvids, gulls, Starlings and many other species commute between feeding and roosting habitats on a daily basis. Usually it was relatively straightforward for observers to separate these cases of flyovers (and assign them a Flying code) from instances of birds hunting on the wing (e.g. raptors) that were using the resources of the square and did not require a Flying code.

BOX 4.2 POLICY FOR PRESENTATION OF SENSITIVE BREEDING AND WINTERING SPECIES

The presumption will be for accurate presentation (mapping or listing) of breeding range at 10-km square resolution except for:

1. sensitive sites whose accurate identification may result in increased uncontrolled visitation by birdwatchers/photographers with consequent risk of damaging disturbance;
2. breeding sites for species of high conservation concern whose eggs are highly attractive to egg collectors, where identification of particular locations would materially increase the risk of egg theft;
3. sites, the identification of which would jeopardise relations with important stakeholders to the detriment of the conservation of species of high conservation status at specific locations;
4. species where breeding has occurred recently or is thought likely to occur soon and where the breeding habitat is restricted or where presence in the breeding season is indicative of a potential breeding occurrence.

However, where breeding sites of species of high conservation concern are already widely available within the public domain, the presumption is that data will be accurately presented at 10-km square resolution. In winter the presumption will be for accurate presentation at 10-km square resolution except for certain highly sedentary species where wintering locations pin-point breeding locations that fall under one of more of the points above.

A risk-based approach will be used with assessments of sensitivity being particularly informed by:

1. conservation status (population size, range size, trend);
2. known sensitivity to disturbance;
3. traditional use of specific breeding locations (e.g. certain wader species) versus one-off use of breeding locations (e.g. certain warblers, shrikes).

The following three options are considered for each species:

- change dot resolution—i.e. present at 20-km, 50-km or 100-km resolution;
- downgrade breeding evidence;
- remove dot.

A combination of approaches may be considered for an individual map. Note that moving dots, as undertaken in previous atlases, has been discounted.

BTO, RBBP & IRBBP,
March 2011

4.2.2 Breeding distributions

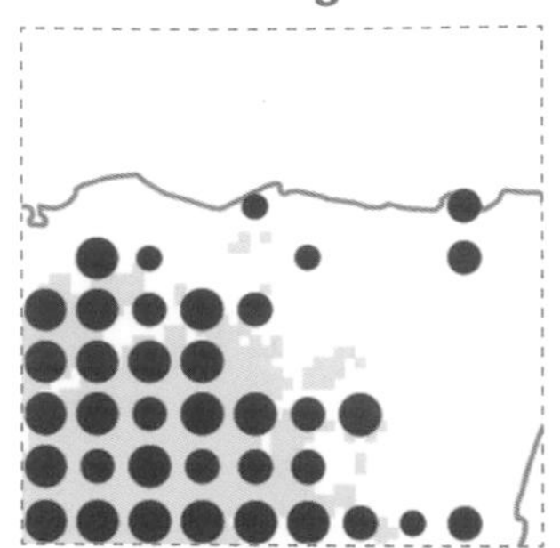

Breeding distribution maps were compiled by determining the highest breeding code (the strongest breeding evidence) for each 10-km square, from among *possible*, *probable*, or *confirmed* breeding statuses (see Table 2.2). As well as records from the main April–July breeding-season fieldwork period, certain out-of-season breeding evidence records were used for 159 species that have a breeding season either starting in March or extending into August–October (see Section 2.2.2). The inclusion of these records was governed by a set of rules that were species-, month- and breeding-evidence-specific (see Appendix 4). The rules ensured the exclusion of false evidence, such as from birds displaying during migration or birds that had already departed from their true breeding area. Though no upgrades to breeding evidence were implemented for individual species, two minor adjustments were made to records. Firstly, records of old, Used Nests submitted in winter (e.g. House Martin nests) were back-dated to the previous breeding season to ensure that these did not register as false winter records. Secondly, presence of colonies on TTVs (Section 2.2.4) was assumed to confer *confirmed* breeding and such records were assigned an Occupied Nest code by default.

Records lacking breeding evidence

Contrary to previous breeding atlases for Britain and Ireland, we chose to include records lacking breeding evidence on distribution maps, provided they could not be confused with the true breeding range and were not used in the calculation of change. Records lacking any code or assigned Flying, Migrant and Summering codes were combined under a single title of '*non-breeding*' records and shown as small grey dots.

There was no original intention to map these records and their collection was not systematic (see Section 7.2.2), but as the Atlas took shape it became clear that these records showed interesting aspects of some species' distributions. For example, adult gulls will forage widely over terrestrial habitats some distance from their discrete colonies (Thaxter *et al.* 2012) and, for the larger species of gull that take several years to reach maturity, large populations of non-breeders are an important component of the avifauna of lowland farmland (Schwemmer *et al.* 2008). Other species for which breeding areas tell only part of the story are the passage migrants that either breed in only part of our region, or only occur on passage to northern breeding areas. For example, *Bird Atlas 2007–11* generated spring records of Great Northern Divers in a large number of 10-km squares off the west and north coasts of Ireland and Scotland, and the scattering of spring records of Great White Egrets has proved to be the precursor to the first successful breeding in Britain (Holt 2013).

In practice, when breeding-season records with the Flying code were examined, in many instances there was little distinction among records attributed that code and those lacking any code or given the Summering or Migrant codes. For example, some records of Gannets outside breeding areas were given a Flying code, some a Summering code, and many no code at all; they probably all involved breeding birds on foraging trips or summering non-breeders. Similarly, southern England was peppered with records of northbound Whimbrels of which the majority were probably flyovers but many were coded as Migrant.

4.2.3 Distribution maps for sensitive species

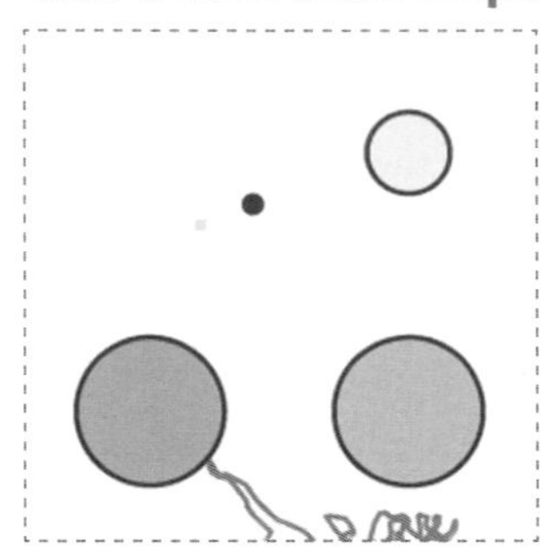

For species where all or part of the range was thought to be sensitive (Section 4.1.3), the 10-km distribution data were summarised within the parent 20-km, 50-km or 100-km squares. So, if a species was to be mapped at a 50-km resolution, the 10-km data were summarised to identify which 50-km squares were occupied, how many of the constituent 10-km squares were occupied and, for the breeding season, the highest breeding evidence across the constituent 10-km squares. Maps were then produced to show a dot for each occupied 50-km square—a large dot for winter maps, and dots of three different sizes to indicate the level of breeding evidence for breeding-season maps. In both seasons, the intensity of shading of each dot was used to indicate the number of occupied 10-km squares (regardless of their breeding evidence). Maps were produced in a similar way for 20-km and 100-km species, and for species where only part of the range was sensitive. As an example, Figure 4.1 shows the breeding distribution of Red Kites at its final 10-km resolution, and as it would have appeared had we opted for one of the other three resolutions. The 100-km map is not particularly informative because almost all mainland British 100-km squares are occupied, though the shading identifies central southern England as the core area in terms of overall occupancy. The 50-km and 20-km resolution versions are more informative without revealing the more precise pattern shown on the 10-km resolution map. For some sensitive species these maps, especially the 20-km and 50-km versions, are a good compromise, remaining informative without drawing potentially damaging attention to particular sites of conservation sensitivity.

4.2.4 Species issues associated with distribution mapping

In the majority of cases this book only presents maps for taxa with full species status as of January 2012 (BOU 2012; see also Box 4.1). Exceptions were made for the Blue-headed *flava* race of Yellow Wagtail and the identifiable forms of Bean Goose, White-fronted Goose and Brent Goose because populations of subspecies are of general interest to birdwatchers and some have different conservation status (Wetlands International 2013).

Whereas virtually all species are identified on sight, sound or both, there is one species, the Scottish Crossbill, which is often

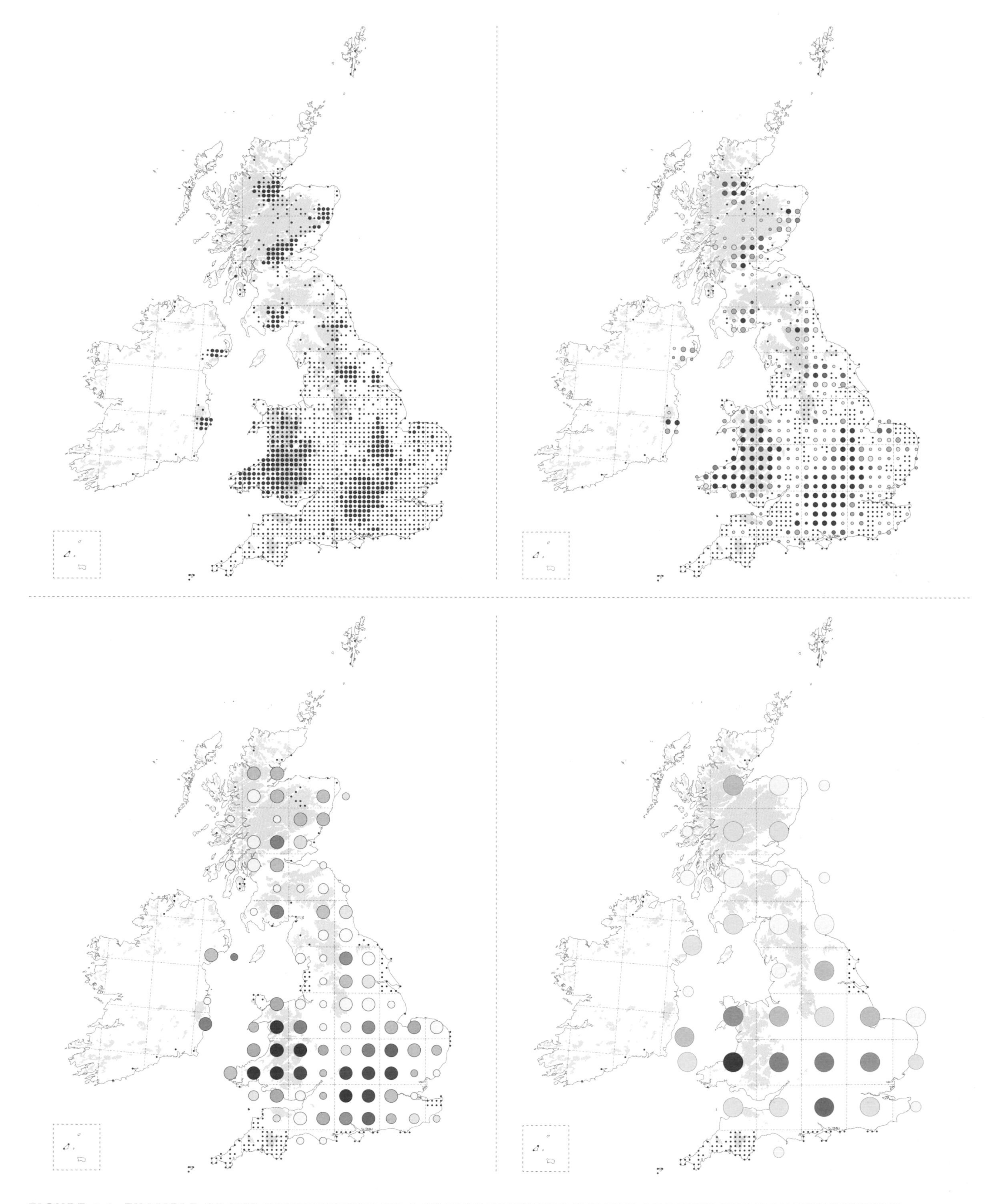

FIGURE 4.1 EXAMPLE OF THE DISTRIBUTION OF A SCARCE BREEDING SPECIES AT DIFFERENT MAP RESOLUTIONS

Maps show the breeding distribution of the Red Kite at 10-km (top left) as used in the species account and as it would have appeared if the mapping resolution had been 20-km, 50-km or 100-km. In all cases the size of a red dot indicates the highest evidence of breeding in the grid square and darker shading indicates a greater number of underlying 10-km squares reported with breeding evidence. Grey dots show 10-km squares with only *non-breeding* records. Note how dots for larger squares with little land can appear in the sea.

In areas of overlap, Common and Scottish Crossbills are difficult to separate unless diagnostic calls are heard. For this reason, a composite species account is presented.

identified partly by where it is seen. Recent work has suggested that recording distinct call-types is the only definitive way to identify the species in the field (Summers *et al.* 2002). The implication of this is that an unknown proportion of past and present records cannot be reliably assigned to this species. By extension, an unknown number of Common Crossbill records may be missing from that species' maps. Because of this uncertainty, crossbill data are presented as follows. As in the *1988–91 Breeding Atlas*, there is a combined species account for Common and Scottish Crossbill in which all data are merged to produce composite maps of crossbill records. This is followed by two Scottish Crossbill maps, one showing the records from 2008–11 as submitted to the Atlas and the other showing only data from the RSPB Scottish Crossbill survey (Summers & Buckland 2011), which used the recommended playback technique for Scottish Crossbill identification (Summers *et al.* 2002). The sample nature of that survey means that the full range for this species cannot be inferred and no distribution statistics are shown, but the benefits in terms of certainty of identification merit including a separate map.

A similar problem arises with the Feral Pigeon and the Rock Dove. Pure Rock Doves are thought to exist only in the extreme north and west of Britain and Ireland. When birds resembling pure individuals are seen in other areas, especially in coastal cliff environments, there can be considerable debate regarding what to call them. Therefore, the practice of previous atlases was followed and maps are shown of both forms in combination.

4.3 DISTRIBUTION CHANGE MAPPING

THE PURPOSE OF distribution change maps is to show how the pattern of occupancy of 10-km squares by a species has changed over time. Usually these maps only show apparent gains and apparent losses since previous atlases; it is then necessary to cross-reference these with distribution maps if the reader wishes to identify areas of apparent stability. Recently, the *Norfolk Bird Atlas* (Taylor & Marchant 2011) showed areas of 'no change' on its change maps, and we have adopted this approach to show 'apparent stability' in addition to the apparent gains and losses.

The starting point for change mapping was the data used for the distribution maps, plus any corresponding data from previous atlases. An initial analysis of these data examined which 10-km squares had evidence of coverage in each atlas (see Section 5.5) as changes could only be calculated for 10-km squares visited in all relevant periods. Adjustments had to be made for the Channel Isles and Fair Isle, where records in the *1968–72 Breeding Atlas* and the *1981–84 Winter Atlas* were only collected at the whole-island level, rather than for constituent 10-km squares. In these places, change statistics were calculated only after amalgamating data at the island level for the *1988–91 Breeding Atlas* and *Bird Atlas 2007–11*.

4.3.1 Winter distribution change

Since the three winter fieldwork periods of the *1981–84 Winter Atlas* ran from the 12th, 13th and 14th of November respectively until the end of February, the winter recording periods for *Bird Atlas 2007–11* were appromimately two weeks longer (as well as there being a fourth winter). Initially, change maps and statistics were produced using all the 2007–11 winter data but it became apparent that a number of autumn migrants showed many apparent gains and, on closer inspection, a large proportion of these could be attributed to 10-km squares that had only been occupied in the early November periods of 2007–10. The Swallow, for example, showed an apparent range expansion of over 200%, but when early November records from 2007–10 were excluded to make the data consistent with the recording period of the *1981–84 Winter Atlas*, the range expansion was a more modest 16%. For this reason all winter distribution change maps and associated statistics only use 2007–11 data collected on or after 14th November. Using these reduced data, each 10-km square was scored for whether it had apparently retained, gained, or lost each species in winter between 1981–84 and 2007–11. Table 4.1 gives an explanation of the symbols used on winter distribution change maps.

4.3.2 Breeding distribution change

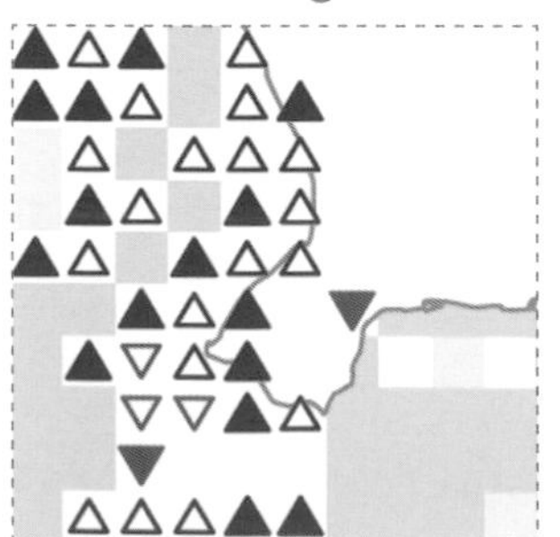

These maps summarise how breeding distributions have changed over four decades since the first Britain & Ireland breeding atlas in 1968–72, through the second atlas in 1988–91 until *Bird Atlas 2007–11*. The maps concentrate on showing the patterns of apparent stability, gains and losses, rather than on presenting changes in the level of breeding evidence. For the majority of regularly occurring species, if a square registers *possible* breeding in suitable nesting habitat, the species is almost certainly attempting to breed.

Whether observations accumulate to satisfy the strict criteria of *probable* or *confirmed* breeding will depend on how easy that species is to observe, with increased recorder effort and increased numbers of the species in question also improving the chances of detecting breeding. Therefore, if there are subtle changes in observer field effort or in species' population density between atlases, change in recorded level of breeding evidence may not convey much of ornithological relevance.

Even considering only changes in gross breeding occupancy, with three atlas periods to consider, there are eight potential permutations of change (Table 4.1). As it would be too confusing to show all of these in detail we have concentrated on the long-term pattern, identifying long-term stability, long-term apparent gains and long-term apparent losses. The apparent gains and losses can be divided into those where the change is recent, i.e. it occurred between 1988–91 and 2008–11, or is old, i.e. it occurred between 1968–72 and 1988–91. Which of these is more important is a matter of context and personal preference. These maps give greater emphasis to recent changes on the basis that most of the old changes have already been reported in the *1988–91 Breeding Atlas*.

As with winter changes, it was necessary to consider whether any changes in methods could skew the breeding-season change maps and statistics. Previous breeding atlases relied largely on April–July fieldwork. Breeding records were accepted from other months but it is unclear how many were submitted, or for which species. *Bird Atlas 2007–11* used strict criteria for 159 species to dictate which out-of-season breeding evidence could add or upgrade breeding evidence on distribution maps (Section 4.2.2). Owing to the uncertainty concerning the use of out-of-season records in previous atlases, distribution-type information for 2008–11 was recalculated without any out-of-season data and then used for all breeding-season change maps and statistics.

As in the *1988–91 Breeding Atlas*, distribution changes for breeding seabirds were calculated after excluding any squares with only *possible* breeding evidence (*seen* in 1988–91 breeding evidence terminology). Table 4.1 gives an explanation of the symbols used on breeding distribution change maps.

4.3.3 Distribution change for sensitive species

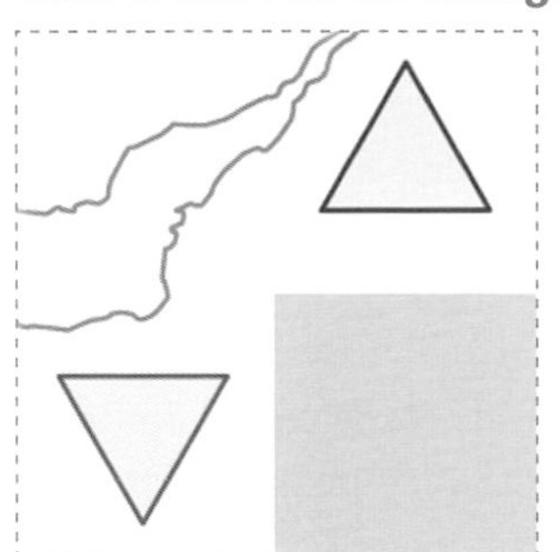

The same concerns that governed the production of distribution maps for rare breeding species had to be reflected in distribution change maps so that changes did not reveal sensitive locations. The mapping resolutions and exclusions identified for distribution maps were also applied to the change maps. Thus, if a species had a 50-km resolution distribution map, the change map was created by calculating the difference in the number of occupied 10-km squares in each 50-km square. Then the 50-km square was shaded where there was no change and marked with an upward or downward triangle if the number of occupied 10-km squares increased or decreased respectively. Shading intensity of the

TABLE 4.1 PERMUTATIONS OF PRESENCE FOR DISTRIBUTION CHANGE MAPPING

For winter and the breeding season the tables show permutations of apparent occupancy across atlases, indicating how these can be interpreted and the map symbols used.

A Winter

1981–84 Winter Atlas	*Bird Atlas 2007–11*	Interpretation	Map symbol
Present	Present	Always present	■
Present	Not detected	Apparent loss	▼
Not detected	Present	Apparent gain	▲
Not detected	Not detected	Apparently never present	Not mapped

B Breeding season

1968–72 Breeding Atlas	*1988–91 Breeding Atlas*	*Bird Atlas 2007–11*	Interpretation	Map symbol
Present	Present	Present	Always present	■
Present	Not detected	Present	Old apparent loss then recovery	■
Present	Present	Not detected	Recent apparent loss	▼
Present	Not detected	Not detected	Old apparent loss	▽
Not detected	Present	Present	Old apparent gain	△
Not detected	Not detected	Present	Recent apparent gain	▲
Not detected	Present	Not detected	Old apparent gain then apparent loss	Not mapped
Not detected	Not detected	Not detected	Apparently never present	Not mapped

JILL PAKENHAM

For well-established non-native species such as the Little Owl, atlases record distribution and changes accurately. However, for more exotic species, apparent changes may reflect shifts in attitudes to recording as well as real status changes.

triangles was varied to indicate the magnitude of the change in the number of occupied 10-km squares and for some species different resolutions were applied to different parts of the range.

4.3.4 Species issues associated with change mapping

Over the four decades encompassed by the series of Britain and Ireland atlases there have been a number of taxonomic changes (Collinson 2006) that affect how change maps are produced (Box 4.1). For 'splits' that have resulted in one widespread species and one or more rare or localised sister species (e.g. the 'Herring Gull' complex), it is safe to assume that the majority of records from earlier atlases relate to the widespread form. For example, Yellow-legged Gulls and Caspian Gulls are scarce and typically occur where Herring Gulls are found. It is a fair assumption that all 10-km squares with previous records of 'Herring Gull' will have had individuals of the current Herring Gull species, so a change map can be generated. For the scarcer forms such as Yellow-legged Gull, no such assumptions can be made and no change map can be produced. No data were combined across species to produce change maps with the exception of the following two cases:

- Redpolls have caused considerable difficulty following the decision in Britain to split Redpoll (*Carduelis flammea*), naming the resulting species the Lesser Redpoll (*C. cabaret*) and the Common Redpoll (*C. flammea*) (BOU 2001). The recording of these species during 2007–11 has been hampered by these confusing new vernacular names: the Lesser Redpoll is by far the commoner species in Britain yet the Common Redpoll was widely reported. Furthermore, the identification of the two species is not always straightforward and records of 'unidentified Lesser/Common Redpoll' accounted for 7% of all redpoll records. To permit change calculations we made the assumption that all 2008–11 breeding-season records of unidentified redpolls were Lesser Redpolls; this seemed reasonable given that the Common Redpoll has only been recorded breeding in very small numbers in a few locations in the extreme north and west of Scotland (*RBBP 2010*), all of which also have breeding Lesser Redpolls. Similarly, we assumed that all redpoll records in previous breeding atlases related to Lesser Redpolls. The same assumption could not be made in winter because Common Redpolls are more widely distributed than in the breeding season. For simplicity, no combining of records was undertaken in winter, so the winter change map uses all redpolls from 1981–84 versus only Lesser Redpolls from 2007–11.
- Records of the Scottish Crossbill and the Common Crossbill were combined to produce a 'Common/Scottish Crossbill' change map in recognition of the identification issues (see Section 4.2.4) and changed taxonomic status since the *1968–72 Breeding Atlas*.

The recording of non-native species has varied markedly through time and among species. Species with recognised self-sustaining populations, such as the Canada Goose, Pheasant and Little Owl, have been recorded consistently in all atlases. Others, such as the Black Swan and Indian Peafowl, have been less routinely documented, especially in previous atlases, and were rarely given the breeding evidence codes they may have merited. Generally, no change maps are presented for such species but change statistics (Section 4.6), where they can be calculated, are presented in Appendix 2; these should be interpreted with caution so as not to conflate trends in surveyor behaviour and trends in status.

4.4 RELATIVE ABUNDANCE MAPPING

IN ONE SENSE, distribution maps only tell part of the story because they make no distinction between 10-km squares occupied by few breeding pairs and those with many more. They do not allow the identification of hotspots where it might be easiest to see a species, or areas where conservation action may be best directed. Abundance maps fill this gap by showing how numbers of a species vary spatially. Ideally maps would show spatial variation in density (absolute abundance) but fieldwork to make this possible was not practical (see Section 2.2.4). However, the counts of birds per hour in a tetrad are relative measures of abundance that can be compared among different parts of Britain and Ireland.

There are several ways that relative abundance maps can be produced:

1. The simplest method is to map the actual counts that observers made in each square. This is more or less how relative abundance was represented in the *1981–84 Winter Atlas*. One disadvantage is that it is difficult to distinguish on the maps between actual zero counts and gaps in coverage (false absences; Section 4.1.1). More importantly, the counts depend not just on the absolute abundance of the species but also on the simple chance of whether the observer happens to see most of the individuals present or just a few of them. This means that some of the variation within a map will be 'noise', making it difficult to pick out the true patterns in abundance.
2. The data can be 'smoothed' to reduce the effects of noise or chance variations in the counts. This is a common technique and was used in the production of relative abundance maps in the *1988–91 Breeding Atlas*. Essentially, it is a form of spatial modelling by which the mapped value for each square is adjusted according to counts in adjacent squares, on the assumption that abundance varies across the landscape in some spatially gradual way. The disadvantage is that this assumption is quite likely to be incorrect: sudden changes in habitat, for instance, can cause abundance to change abruptly rather than gradually.
3. More refined predictive models can be produced that incorporate habitat and other relevant variables. Statistical analysis of the data is used to determine the form of the relationship between these variables and observed bird counts so that predictions of abundance can be made.

All three approaches have been used in *Bird Atlas 2007–11* to produce relative abundance maps, using the most refined approach that the data for a species will permit. For widespread species the third approach was used where possible to model counts to produce detailed tetrad-resolution maps of relative abundance (see Section 4.4.3). For coastal species the second approach was used to smooth counts around the coastline (see Section 4.4.2). For localised species and those where the statistical models were a poor fit to the data, simple count averages were mapped (see Section 4.4.1).

All relative abundance maps are based only on systematic count data from TTVs; they do not use any count data from Roving Records because the variation in search effort that went into obtaining those counts would be likely to bias the maps. TTV data were summarised to obtain, for every tetrad visit, the number of birds detected per hour; for 2-hour TTVs the first- and second-hour counts were averaged. Generally, breeding-season TTVs were adjusted to set to zero any counts in 10-km squares where there was no evidence of a species breeding. This was done because some spring passage migrants (e.g. Wheatear) were counted on TTVs well outside their breeding range and we did not want abundance maps to show a confusing mix of breeding and passage hotspots. The only exception to this was for the smoothed coastal relative abundance maps of colonial seabirds where all records were retained, regardless of breeding evidence.

4.4.1 Simple mean abundance maps at 20-km resolution

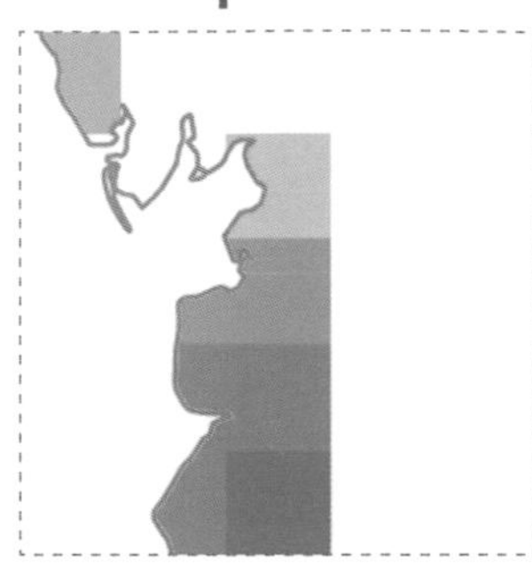

Simple mean abundance maps were produced by calculating the average count per hour across visits for each surveyed tetrad, then averaging these values across all tetrads in each 20-km square to give a local estimate of relative abundance for that square. Rather than 10-km squares, 20-km squares were used as the mapping resolution because up to half of 10-km squares in Ireland would have been deficient in data owing to the chequerboard pattern of TTV coverage (Section 2.2.4). Using a 20-km resolution also decreased the likelihood of spurious patterns emerging because the abundance estimates were based on larger samples of tetrads than would have been the case with a 10-km resolution. This risk was further reduced by excluding those 20-km squares (47 squares in winter and 51 in the breeding season) in which fewer than five tetrads had been surveyed. Rather than areas of poor coverage, these were coastal and marine 20-km squares containing on average just 2% land (up to a maximum of 9% land).

Which species and seasons?

These maps were typically used for species that were scarce or localised in a particular season (e.g. wintering White-fronted Geese and breeding Hawfinches) for which insufficient data were available for statistical modelling. They were also used for species for which the predictive performance (see Section 4.4.3) of abundance models was poor (e.g. breeding Pintails and Quails).

4.4.2 Smoothed maps for coastal species at tetrad resolution

Since a tetrad was only eligible for the TTV method if its centre fell on land, no entirely marine TTVs were completed. However, many coastal tetrads yielded counts of divers, grebes, seaducks, intertidal waders and seabirds that provided information on patterns of abundance around our shorelines. It was not possible to use a complex predictive model for these species because of a lack of environmental variables describing the marine environment in comparable detail and resolution to the terrestrial environment. Instead, a simple spatial model was used to smooth counts and represent abundance as a 'ribbon' of colour around the coastline. The width of the ribbon (8 km) was optimised for the size at which maps would be reproduced in the book, and so that ribbons around Britain and Ireland did not merge.

VIC FROOME

Previous atlas abundance maps for seabirds such as Gannets have used different dots to indicate colony size. Here a 'ribbon' of shading around the coastline is used to indicate variation in abundance in winter and the breeding season.

The ribbon was produced by first creating a grid of 2-km squares across Britain and Ireland. This was slightly offset from the national grids so that no ribbon grid square had precisely the same coordinates as a surveyed tetrad; this ensured that it was always possible to calculate a non-zero distance between any tetrad and any coastal grid square. Centroid coordinates were determined for every grid square. In the GIS, a 'buffer' polygon was created for the Britain & Ireland outline, stretching from the low tide line to 8 km out to sea. The buffer polygon was used to crop the grid of 2-km squares to create a set of 28,353 whole and partial 2-km squares in the shape of an 8-km ribbon around the coast. For each of these whole and partial squares the smoothed abundance value for a given species was calculated by taking an inverse-distance weighted mean (power parameter = 1) of the abundance values for all surveyed coastal and inland tetrads within a radius of 15 km. We had previously experimented with a range of different values for the power parameter and radius and selected 1 and 15 km as giving the best balance of a satisfactory representation of general patterns of abundance without excessive 'bleeding' of abundance information across peninsulas or the mouths of major estuaries.

Which species and seasons?

This method was used for a suite of wintering or breeding species where at least 80% of the 10-km squares occupied by a species in a season could be classed as coastal (defined as 10-km squares containing any sea). These maps were not used for species such as the Oystercatcher which, although most numerous on the coast in winter, also occurred in a large number of inland 10-km squares.

4.4.3 Modelled abundance maps at tetrad resolution

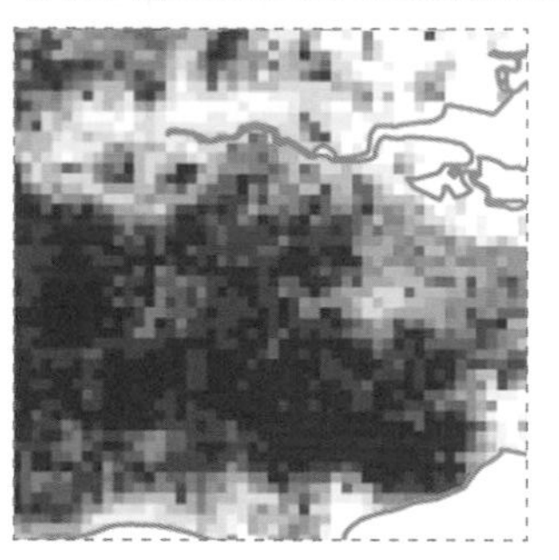

Originally, TTV coverage was designed to provide a minimum of eight counts per 10-km square which could be averaged within each 10-km square, then smoothed across 10-km squares, to produce maps similar to those in the *1988–91 Breeding Atlas*. In addition to the disadvantage described earlier (that smoothing assumes gradual gradients in abundance), the first stage of averaging within 10-km squares discards valuable information about fine-scale variation in bird numbers. For example, habitat can vary strongly among tetrads, and some TTVs took place before certain migrants had arrived or after the main song period of some residents. A better analysis of these data would use all the individual TTVs and account for the habitat in the square, and the timing of the visit, to produce an overall picture of relative abundance that reflected the environmental conditions across Britain and Ireland. This is the essence of the statistical modelling used to produce many of the abundance maps in this atlas.

There are many analytical frameworks that are capable of modelling the relationship between abundance of species and their environment. Based on published reviews (e.g. Elith *et al.* 2006; Prasad *et al.* 2006; Cutler *et al.* 2007; Franklin 2009) and our own testing of different model frameworks, we opted for the family of machine-learning techniques known as 'Regression Trees' (Breiman *et al.* 1984). These rely on finding thresholds in explanatory variables

BOX 4.3 REGRESSION TREES: A HYPOTHETICAL EXAMPLE FOR THE RED GROUSE

EDMUND FELLOWES

Abundance of Red Grouse is determined by altitude, climate and the availability of preferred habitat.

Classification trees are familiar to many people as decision flow charts, for example as dichotomous keys for identifying species which ask a series of 'yes/no' type questions: 'Are the flowers white?', 'Are there five petals?'. By answering the questions we can classify a plant. Regression Trees are an extension of this idea but rather than giving the result in the form of a class such as a species name, they give a numerical quantity, such as a number of birds. Just as a taxonomist works out which sequence of attributes most accurately identifies the plants, so we can use regression tree analysis to find thresholds in environmental attributes that most accurately describe how many birds are reported on surveyed tetrads.

A hypothetical regression tree analysis might find that two questions—'Is altitude greater than 214 m?' and 'Is heather cover greater than 72%?'—best describe variation in observed counts of Red Grouse in a sample of 123 tetrads. In this simple example, for 58 tetrads the answer was 'no' to the altitude question, and in those the average grouse count was 0.0; for 65 tetrads the answer was 'yes' to the altitude question and these could be further subdivided according to their heather cover. On average, tetrads at high altitude and with high heather cover had 9.8 grouse.

Having 'trained' our Regression Tree using the observed counts to identify these questions and critical thresholds, we can now use the tree to make predictions for new tetrads. So, for each unsurveyed tetrad, we could ask these questions about altitude and heather cover and assign each a predicted count of 0.0, 2.1 or 9.8. This is a simplified example—a real tree would be considerably more complex, having more than two questions—but the principle is the same.

Regression Trees were chosen for a number of reasons:

1. Classification Trees have proved highly effective in the field of species distribution modelling (Franklin 2009). Regression Trees are used in medical research (e.g. Özçift 2011) but, as far as we are aware, application to abundance-type data is limited to two studies of the abundance of trees (Iverson & Prasad 1998; Moisen *et al.* 2006) and one of species richness of oceanic fish (Leathwick *et al.* 2006).
2. Regression Trees are non-parametric, meaning they are well suited to skewed abundance data.
3. Extreme predictions, such as those that can arise from poorly fitted curvilinear responses in some other modelling frameworks, are avoided because predictions are bounded by the range of the observed counts.
4. Regression Trees handle complex interactions between factors more effectively than generalised linear or additive modelling frameworks. Indeed, they are conceptually attractive in allowing a hierarchy of factors, potentially at different scales, to describe patterns of abundance.

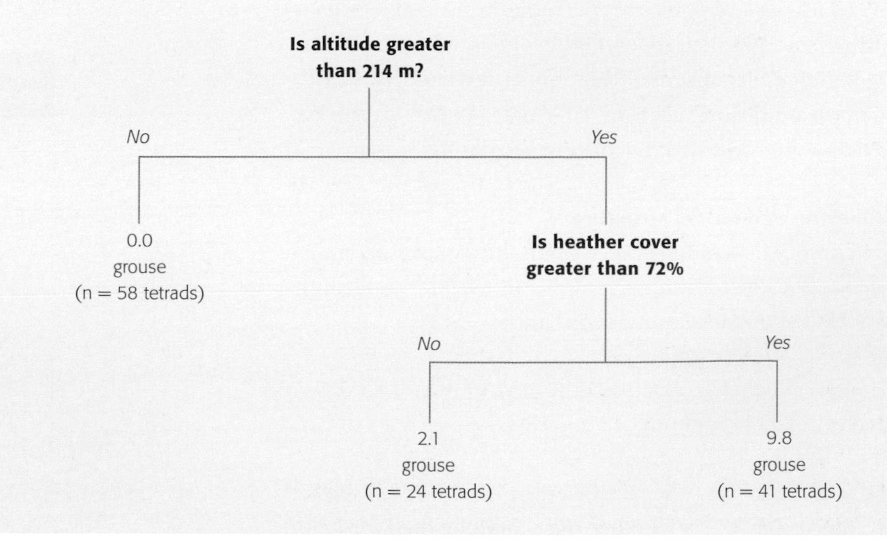

(e.g. more than 50% of tetrad under woodland, mean winter temperature less than 2°C, surveyed before 26th April) to account for variation in the quantity of interest, in our case bird abundance. Box 4.3 explains the principles of a Regression Tree analysis, showing how a tree can be 'trained' to identify thresholds that describe observed variation in bird counts.

In reality, in addition to trees being much larger (i.e. having very many more questions) than the example given in Box 4.3, it is usual to produce many trees and average the results to give an 'ensemble model' which is better than any one of its parts (Araújo & New 2007). This is done because a single tree can become so finely tuned (or 'over-fitted') to its training data that it is unlikely to yield robust predictions when making predictions for the whole of a country. So, for the Red Grouse example, a second tree using slightly different data might find that thresholds in summer temperature and the percentage cover of forest are also good predictors of grouse abundance. The averaged predictions from these two trees should provide more robust predictions than either of the trees in isolation. We used a common ensemble-averaging procedure for Regression Trees called 'Random Forest' (Breiman 2001; Breiman & Cutler 2004). In a Random Forest model, hundreds or thousands of trees are produced; each tree is trained on the same data but different trees are permitted to use only a proportion of the explanatory variables, chosen at random. This means that every tree has the potential to be slightly different and although a single tree in the Random Forest may not perfectly describe the data, if predictions are made from every tree and then averaged, these ensemble predictions are a good fit to the observed data and predictions made for new data are then robust.

The ensemble principle of averaging across models also helped with an atlas-specific issue: TTVs were so popular that in some areas, especially those running a local atlas, coverage far exceeded the Rule-8_{adj} target leading to a geographically biased sample of TTV data. Whilst it would be possible to remove the bias by randomly sampling from the available tetrads to have just the Rule-8_{adj} target number in each 10-km square, many valuable TTVs would then be wasted and the model outcome might be reliant on the chance choice of tetrads. To overcome this we took the ensemble principle one step further to produce an 'ensemble of ensembles' in which we averaged the results of several Random Forest models, each one trained on a different subset of TTV data. In this way all the available data were used, without risking geographic biases.

Cross-validation and model structure

The proven ability of Random Forests in published studies gave us confidence that the technique would work well for modelling patterns of bird abundance, provided that:

a. we had sufficient data for a species in a season;

b. the variables adequately described the environment and resources that determined bird abundance in tetrads.

A method was required to check the performance of each species' models to objectively assess whether they could be used for map

BOX 4.4 FLOW DIAGRAM SHOWING THE STEPS

Here the aim is to produce a map of Ptarmigan abundance in winter.

A Split the data ready to make many Training Datasets and one

1. Get all TTVs for the chosen season.

2. Save a copy of all the TTV data for possible use in Panel E.

3. Take c.25% all 10-km squares by choosing those with both an even easting digit and an even northing digit. TTVs from these 10-km squares will be used in making one Evaluation

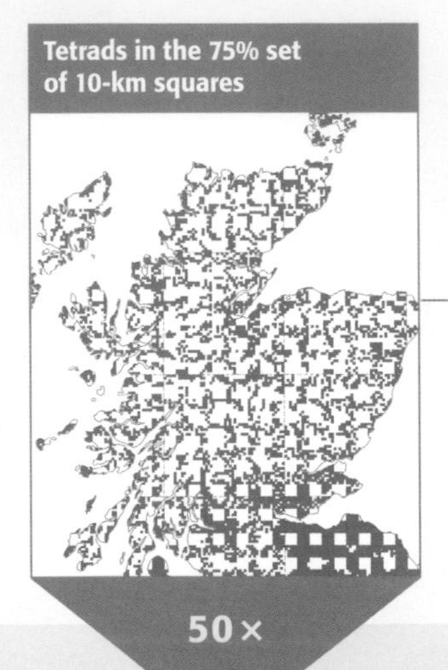

From this 75% set of 10-km squares, tetrads will be repeatedly drawn by random sampling to produce 50 Training Datasets, each of which will contain a slightly different set of TTV data. Data from a high proportion of all TTVs will have been used at least once in the 50 Training Datasets (Figure 4.2).

C Making and analysing *one* of the 50 Training Datasets

1. Remove the geographic bias by randomly selecting n = Rule-8_{adj} tetrads from each 10-km square (this is what generates the variation across the 50 Training Datasets).

2. Restrict data to the species' range using its contagion variable. The TTV data from these tetrads are Training Dataset *i*.

3. Get the environmental variables (see Figure 4.3) for every surveyed tetrad in Training Dataset *i*.

4. Analyse with a Random Forest consisting of 50 trees. Each tree uses the same TTV dataset (below) but is permitted to use only one third of the environmental variables drawn at random (Breiman & Cutler 2004).

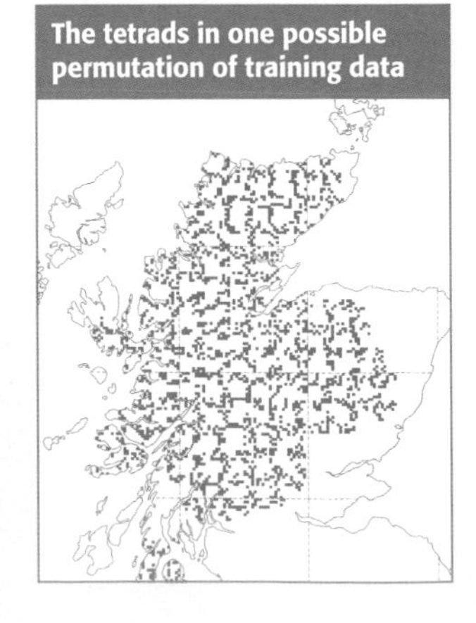

The tetrads in one possible permutation of training data

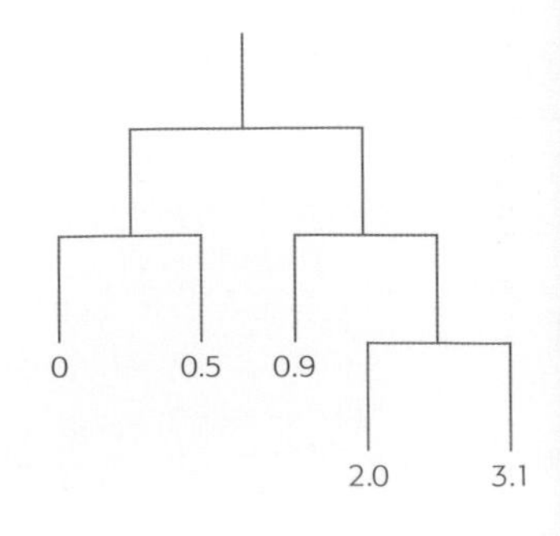

IN ENSEMBLE MODELLING OF TTV DATA FOR ABUNDANCE MAPPING

The inset maps have been cropped to cover just part of Scotland for illustrative purposes only.

Evaluation Datase

Dataset (see Panel B).

4. Use the remaining 75% of 10-km squares as the basis for making multiple Training Datasets (see Panel C).

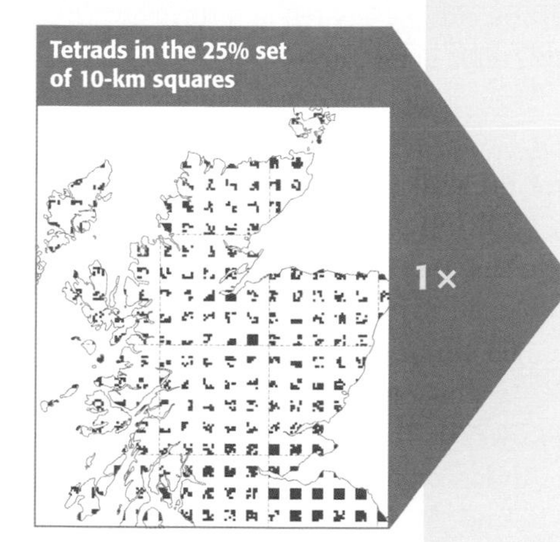

B Make one Evaluation Dataset for this species

1. Remove the geographic bias by randomly selecting n = Rule-8_{adj} tetrads in each 10-km square.
2. Restrict the data to the vicinity of the species' range by excluding all tetrads more than 20 km from the centre of occupied 10-km squares. The retained tetrads are those for which the 'contagion' variable (see 'Variables to explain bird relative abundance') is greater than zero.
3. The TTV data from the retained tetrads form the Evaluation Dataset.

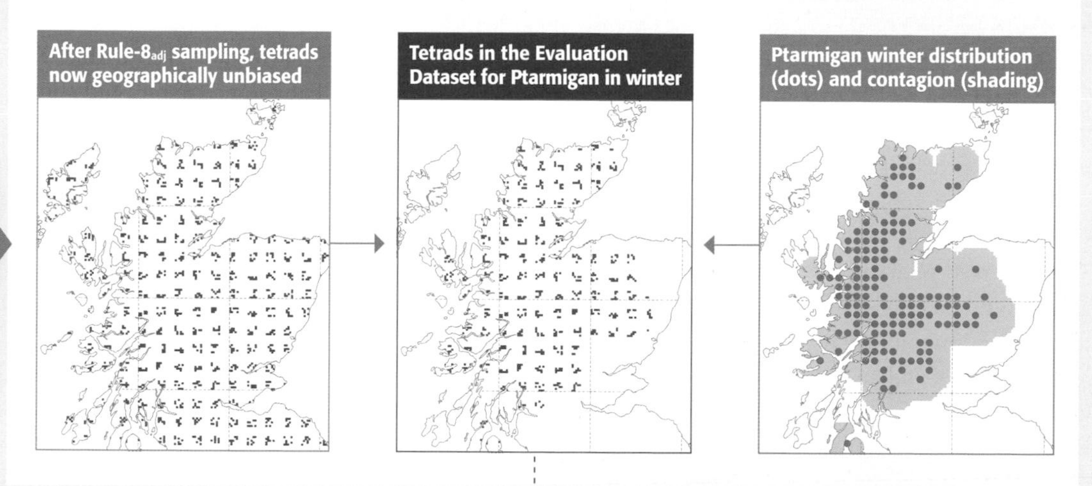

5. Use the Random Forest model to make predictions, p_i, of the expected number of birds for each TTV in the Evaluation Dataset.
6. Calculate the correlation coefficient between the observed bird counts on each TTV and these single-model predictions. This is the Single-model Performance Statistic (below).

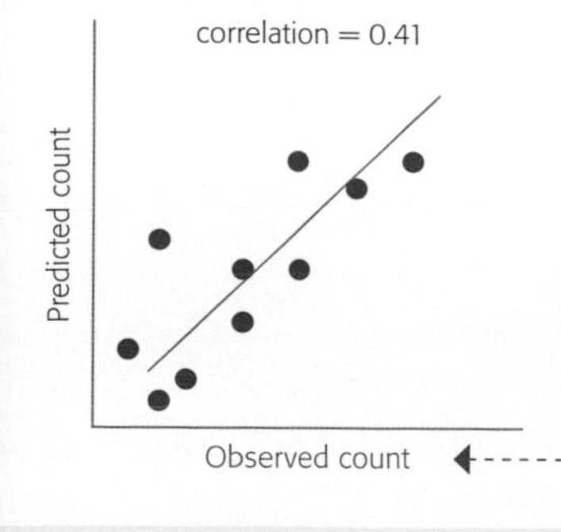

D Making and testing the Ensemble

1. Collate the 50 sets of predictions, p_1 to p_{50}, and for each TTV in the Evaluation Dataset calculate the median prediction. These are the Ensemble Predictions from the training data.

TTV	p_1	p_2	p_3	p_4	...	p_{50}	$p_{ensemble}$	Observed
NH50A-early	0.5	1.0	0.7	1.0	...	1.2	1.0	1
NH50A-late	0.9	1.1	2.6	0.0	...	1.0	1.0	1
NH50C-early	3.1	3.4	2.1	5.2	...	3.3	3.3	4
NH50C-late	2.0	2.5	1.9	7.0	...	4.1	2.5	3
...								
NS89J-late	0	0	0	0	...	0	0.0	0

2. Calculate the correlation coefficient between the observed bird counts and the ensemble predictions to give the tetrad-resolution Ensemble Performance Statistic which gives a measure of ensemble performance at a fine scale.
3. For each 10-km square, calculate the mean ensemble prediction and the mean observed bird count. Then calculate a new correlation coefficient which is the 10-km resolution Ensemble Performance Statistic and gives a measure of ensemble performance at a coarser scale.

E Making the final Full Models

1. Provided the 10-km Ensemble Performance Statistic is ≥0.5, produce an ensemble of 50 Random Forests trained on the full data: i.e. repeat steps C1 to C4 but base each of the 50 Training Datasets on 100% of 10-km squares, rather than the 75% sample set.
2. From each model make predictions of the expected number of birds for every tetrad in the species' range. Collate the 50 sets of predictions and calculate the median to give the final Ensemble Predictions.
3. Create the map.

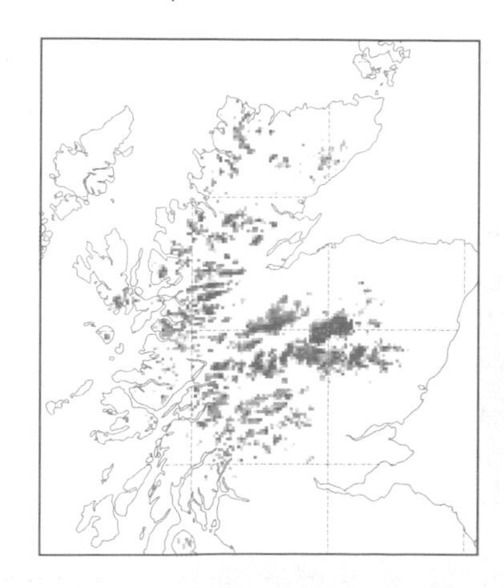

production. Cross-validation is a widely used technique for assessing model fit prior to making final predictions. It involves dividing a dataset into two parts: models are then 'trained' on the first part and 'evaluated' against the second independent part. Provided the evaluation is acceptable, a final model is then created using all the data (Fielding & Bell 1997).

The cross-validation procedure that was used here had to be complex to accommodate the ensemble of ensembles modelling framework. In outline it involved splitting the whole sample of TTV data into a 75% part from which data were drawn repeatedly for model training, and a 25% part that was set aside for evaluating each of the models built from the 75% part. Provided the ensemble predictions (calculated by averaging the predictions from all the training models) were sufficiently similar to the observed values in the withheld 25% of the data, final models could be built using 100% of the data and then used for map production. Full details of how this was accomplished are given in Box 4.4 and an example of the sample sizes involved at different stages of the process is given in Table 4.2. Ultimately, the final ensemble predictions were based on 2500 trees (50 Random Forests each containing 50 trees) and this led to a genuine increase in model performance for all species. On average, across species and seasons, ensemble performance statistics averaged c.30% higher (range 5% to 192%) than their constituent single-model performance statistics. For example, all of the 50 single-model performance statistics for Pheasant breeding abundance in Britain were in the range 0.41–0.42 but the ensemble performance statistic was 0.63 (Figure 4.2). That figure also illustrates another success of the ensemble framework: as the number of single models contributing to the ensemble increased, progressively more of the total sample of surveyed tetrads was used in the training of the models, with the result that the final ensemble was based on more than 99% of the available data.

Bespoke programmes were written in the R language (R Core Team 2013) for the tetrad selection, modelling framework and cross-validation routines, incorporating the R package *randomForest* (Liaw & Wiener 2002).

Variables to explain bird relative abundance

The modelling detailed in the previous sections relied on a set of variables that described the environment in every tetrad in Britain and Ireland. The variables fell into three main categories—habitat, geography and climate. They are described below and summarised in Figure 4.3.

The habitat variables described the availability of 18 habitat types and were derived from the CORINE Land Cover Map (European Commission 1994) because this source covered every tetrad of the UK, Isle of Man and Republic of Ireland. Unfortunately, land cover data could not be sourced for the Channel Islands and abundance mapping on the islands had to be handled separately (see Section 4.7.3). The CORINE dataset contains several cover types that, though common elsewhere in Europe, are scarce in Britain & Ireland. Also, certain distinctions were irrelevant from an ornithological perspective so the 36 CORINE land cover types that were present in Britain & Ireland were reassigned to 18 relevant categories (Figure 4.3). To account for the fact that bird abundance in a tetrad may be determined both by local and by landscape-scale habitat availability (see Fuller 2012a), for each habitat type two variables were produced, one giving the percentage cover of the habitat in each tetrad, and one giving the percentage cover calculated for each tetrad together with its eight neighbours.

Two geographical variables were used. The first described the mean altitude of every tetrad in Britain & Ireland and was derived from a 90 m resolution digital elevation model (CGIAR-CSI 2008). The second divided the individual islands or groups

TABLE 4.2 SAMPLE SIZES AT DIFFERENT STAGES OF THE ENSEMBLE MODELLING OF TTV DATA FOR ABUNDANCE MAPPING

These values are for a combined Britain & Ireland model and are taken prior to the stage of restricting the data to the vicinity of the species' range. Values marked with an asterisk are only estimates because, although the procedure to randomly select n = Rule-8_{adj} tetrads generates a fixed number of tetrads, the number of these with one or two visits varies, giving a variable total number of TTVs.

Data selection stage	Winter	Breeding season
Total TTV sample	**91,883 visits (47,292 tetrads)**	**89,876 visits (46,268 tetrads)**
Number of 10-km squares removed for the 25% Evaluation Dataset	866	826
Number of TTVs available in the Evaluation Dataset's 10-km squares	21,928 visits	21,242 visits
Number of TTVs in the Evaluation Dataset after Rule-8_{adj} sampling	10,645* visits (5,531 tetrads)	10,349* visits (5,359 tetrads)
Number of 10-km squares remaining in the 75% set used for the production of Training Datasets	2,722	2,689
Number of TTVs available for selection for the Training Datasets	69,955 visits	68,634 visits
Example number of TTVs in one Training Dataset after Rule-8_{adj} sampling	33,650* visits (17,391 tetrads)	33,732* visits (17,384 tetrads)
Number of 10-km squares available for Full Models	3,588	3,515
Number of TTVs available for Full Models	91,883 visits	89,876 visits
Example number of TTVs in one Full Model Dataset after Rule-8_{adj} sampling	42,734* visits (22,098 tetrads)	42,843* visits (22,094 tetrads)

FIGURE 4.2 THE USAGE OF TTV DATA AND PREDICTIVE PERFORMANCE ENABLED BY THE ENSEMBLE MODELLING FRAMEWORK

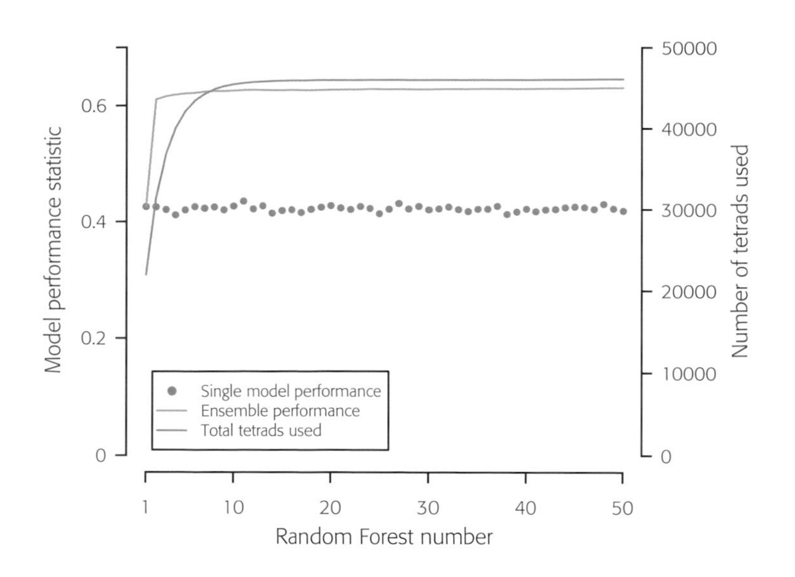

The graph shows diagnostics of the 50 training Random Forest models and the overall ensemble for the abundance of Pheasants in the breeding season. A major aim of the ensemble approach was to make good use of the thousands of TTVs. The brown line indicates that the number of tetrads used at least once increased rapidly as more Random Forests were produced, reaching a total of 99.7% of all tetrads with breeding-season TTVs. The green points show that the predictive performance of each of the single training Random Forest models was slightly greater than 0.4. The blue line shows that producing an ensemble of even just two Random Forest models yielded a major increase in performance which continued to rise to a maximum of 0.63.

of islands into nine different categories (Figure 4.3). This was included because it was clear from distribution maps that many species showed abrupt changes in patterns of occupancy corresponding to islands and island groups. Previous work has shown that this is due to a combination of habitat diversity, island area and isolation from mainland populations (Reed 1981). The inclusion of habitat variables should have helped to describe these observations, but the island variable was included to encompass any remaining variation in bird occupancy of islands.

To characterise the climate of tetrads, temperature and precipitation data were obtained for the five years prior to the start of the fieldwork from Haylock *et al.* (2008). Daily data at a resolution of one quarter degree of latitude and longitude were summarised as follows:

- For mean temperatures, we took the grand mean of all daily mean temperatures of all days during either the winter or the breeding season;
- For minimum temperatures, we first calculated the average of the daily minima within each month, then calculated the minimum value across all months, so the variable describes the average daily minimum temperature of the coldest month;
- For rainfall, we summed the precipitation within each breeding season and winter, and then averaged across years.

Finally, the data for all the quarter-degree cells were interpolated by inverse-distance weighting to generate a continuous surface of climate data from which values for individual tetrads could be extracted (Figure 4.3).

Two other variables were included in models. First, the week number in which a TTV was conducted (week number 1 indicated 1–7 November in winter and 1–7 April in the breeding season) was included as a discrete variable to describe variation in apparent abundance through the season (Figure 4.3). Second, a variable was included to describe the shortest distance from each tetrad to an occupied part of the distribution map. We did this because it became clear from trial models that, for some species, our ability to identify suitable tetrads based only on satellite-derived habitat data, climate and geography was limited. This implied that there might be other factors (e.g. dispersal) that prevented the species from reaching apparently suitable habitats that were distant from areas it occupied. This variable was called 'contagion' after a similar technique used by Estrada *et al.* (2004) in the *Catalan Breeding Bird Atlas*. It had to be calculated for each species and season separately as follows. For the breeding season, all 10-km squares with evidence of breeding (*possible* or higher) by a species were identified, and all constituent tetrads assigned an initial starting value of 1; all remaining tetrads outside the breeding range were given an initial starting value of 0. For every tetrad in Britain & Ireland, a contagion value was calculated as the inverse-distance weighted mean of the 1s and 0s in the surrounding tetrads. By trial and error and visually comparing contagion surfaces with distribution maps, the power parameter was set to 0.75 and the maximum distance over which to average was set to 20 km. Aside from the breeding evidence criterion, the same approach was applied to winter distribution data to derive winter contagion values for each tetrad for each species.

In summary, 46 variables were used for modelling: 18 tetrad habitat variables, 18 landscape habitat variables, six climate variables, one altitude variable, one 'island' variable, one contagion variable and one variable describing the timing of the TTV (Figure 4.3).

Which species and seasons?

Whether an ensemble regression tree model was produced for a species in a season depended partly on the volume of data available in Britain and in Ireland. If a species was reported on at least 100 TTVs in Britain but none in Ireland (e.g. Nuthatch), a Britain-only model was produced. The opposite scenario, of sufficient data in Ireland but not Britain did not arise. Where a species was reported on at least 100 TTVs in both Britain and in Ireland, separate models were run for each island to better account for the differences in sampling arising from the chequerboard. Some species were recorded on more than 100 TTVs in Britain and though they occurred in Ireland they were recorded on fewer than 100 TTVs (e.g. Mandarin Duck); in these cases a single combined Britain & Ireland model was produced. In instances where separate Britain and Ireland models were produced but only one was robust (see below), a single combined Britain & Ireland model was attempted.

Once a model had been produced for a species in a season, whether the results were considered sufficiently robust for map production depended on a careful interpretation of the performance statistics. Models were rejected outright if the 10-km resolution ensemble performance statistic was less than 0.5. Next, the tetrad-resolution ensemble performance statistic was considered. In this case there was no fixed criterion for acceptance because a low statistic could have arisen through false negatives in the TTV data rather than false positives in the model predictions (see Section 4.1.1). Instead these statistics were interpreted alongside a visual check of the congruence of predictions with occupied areas on the distribution map and with areas of high and low relative abundance on simple mean abundance maps at 20-km resolution (Section 4.4.1).

4.5 BREEDING-SEASON RELATIVE ABUNDANCE CHANGE

JUST AS THE distribution maps are a simplification of the status of a species, so the distribution change maps are a simplification of how status is changing because they only indicate extreme local colonisation and extinction events (or at least apparent gains and losses that approximate these changes). Consider, for example, that only one singing bird needs to colonise a 10-km square (and be detected) to register a gain; if numbers then increase further there is no indication of this on a distribution change map. Similarly, a population of thousands can dwindle to just one detected bird without a loss registering on a change map. Between the *1968–72 Breeding Atlas* and *1988–91 Breeding Atlas* the breeding range of the Skylark contracted by only 1.6% in Britain. However, Skylarks showed one of the greatest population declines among farmland species monitored by the Common Birds Census, declining by 54% between 1968 and 1991 (Fuller *et al.* 1995). If we had been able to map changes in local abundance we would have been able to see whether the population decline had taken place throughout the Skylark's range, or only in certain areas, which might have given us a head start on identifying the causes of the decline.

Unlike *Bird Atlas 2007–11*, during the *1988–91 Breeding Atlas* Skylarks and other common species were not counted, only listed as whether detected in each visited tetrad. In the *1988–91 Breeding Atlas* the abundance maps for many common and widespread species were based on a 'Frequency Index', defined as the proportion of surveyed tetrads in a 10-km square found to be occupied by a species. Whilst it is straightforward to calculate in each square a similar Frequency Index for 2008–11 and then to calculate by how much the Frequency Index has changed since 1988–91 as an estimate of abundance change, this relies on one key assumption: that the Frequency Index is a valid measure of bird abundance. Pilot work for the *1988–91 Breeding Atlas* confirmed that a positive relationship does exist between density and the Frequency Index, but the curved shape of the relationship varies among species. For species that are widespread but do not attain locally high densities, the Frequency Index increases gradually throughout the observed range of abundances. In the example in Figure 4.4 a doubling of the density of Nuthatches corresponds to a near doubling of its Frequency Index, meaning that the Index is a sensitive measure of spatial and temporal change in abundance. However, for species that can reach locally high densities, the species can be recorded in a high proportion of tetrads before density reaches its maximum. Using the Wren as an example, the Frequency Index can reach 0.9–1.0 at densities of less than half the observed maximum, meaning that the Frequency Index is less sensitive to changes at the upper end of the abundance spectrum. For both species the inset box-whisker plots show that the Frequency Index values have tended to increase between the *1988–91 Breeding Atlas* and *Bird Atlas 2007–11*, but it should be noted that for locally abundant species such as the Wren, areas with the maximum Frequency Index in both atlas periods could still have undergone abundance change. Box 4.5 describes the process by which Abundance Change Indices were calculated as the difference between Frequency Indices from 1988–91 and 2008–11.

It should be noted that for the 117 less abundant species that were counted in the *1988–91 Breeding Atlas* it was possible to calculate a direct measure of abundance change using the observed counts, rather than relying on the Frequency Index. However, computing an abundance change metric based on the difference in average count between atlases was not straightforward and, since these showed the same geographical pattern of abundance changes as the change in Frequency Index, only the latter are presented in the Atlas.

No attempt was made to produce maps of change in abundance for winter because the *1981–84 Winter Atlas* did not provide tetrad-resolution information comparable to *Bird Atlas 2007–11*. The new TTV count data will form a strong baseline for future assessments of winter abundance change.

4.6 STATISTICS OF RANGE SIZE AND CHANGE

VARIOUS STATISTICS WERE developed to aid the interpretation of the maps. The number of 10-km squares in which a species was detected in a season was totalled separately for Ireland, for Britain (including the Channel Isles and Isle of Man) and Britain & Ireland overall. For the breeding season, separate figures were calculated for each level of breeding evidence. As with the distribution maps, breeding-season totals

FIGURE 4.3 VARIABLES USED IN THE REGRESSION TREE MODELLING OF RELATIVE ABUNDANCE DATA

Each tetrad in Britain and Ireland was assigned values for each of the 44 habitat, geographical and climate variables. The species contagion variable was specified per model according to the species and season. The visit timing variable was specific to each TTV.

36 × Habitat variables

CORINE land cover types were summarised into 18 cover types to describe the habitat availability in each tetrad. Percentage cover was also calculated across each tetrad and its immediate neighbourhood of eight tetrads to give a measure of habitat availability in the immediate landscape.

- Bare & burnt ground
- Coastal - beach & dunes
- Coastal - saltmarsh
- Coastal - sea & ocean
- Coastal - tidal flats & lagoons
- ◀ **Farmland - crops**
- Farmland - pasture
- Human - urban & industrial
- Human - mineral extraction & construction
- Inland water bodies & water courses
- Inland marsh
- Moorland & heathland
- Natural grassland
- Peat bogs
- Scrub
- Woodland - broad-leaved
- Woodland - coniferous
- Woodland - mixed

2 × Geographical variables

▶ The mean altitude of each tetrad.

▶ Island group: nine categories for the individual islands or group of islands.

1 × Species variable

▶ The variable 'contagion' described each tetrad's position relative to the species' range in the relevant season.

6 × Climate variables

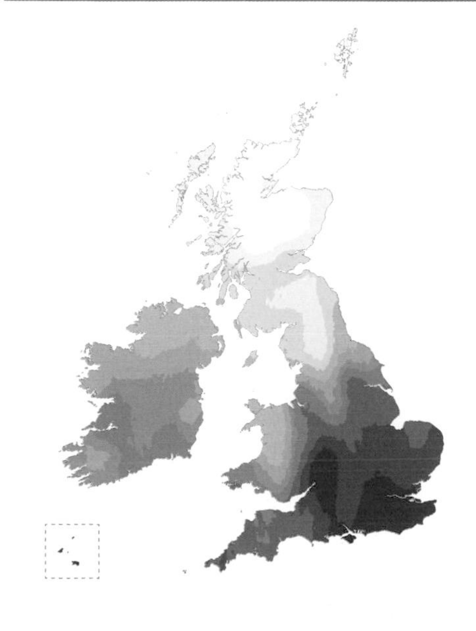

Quarter-degree latitude/longitude gridded climate data were summarised for the five years prior to the Atlas to give the following variables for each tetrad:

- Minimum winter temperature
- Minimum summer temp.
- Mean winter temp.
- ◀ **Mean summer temp.**
- Total winter rainfall
- Total summer rainfall

1 × Visit timing variable

The number of the week in which the TTV was undertaken to help describe seasonal trends in detection and abundance (e.g. week 1 = 1st–7th April).

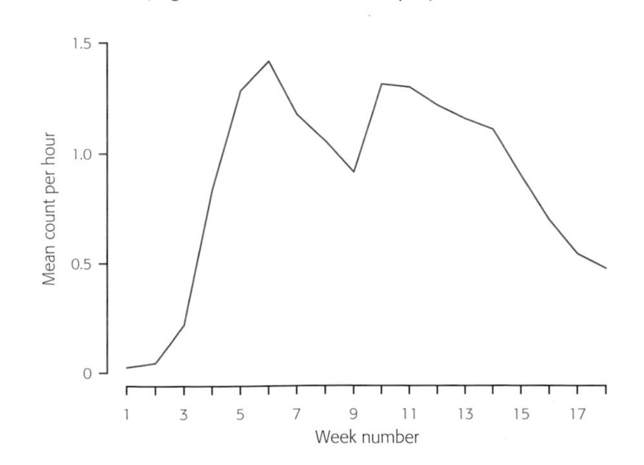

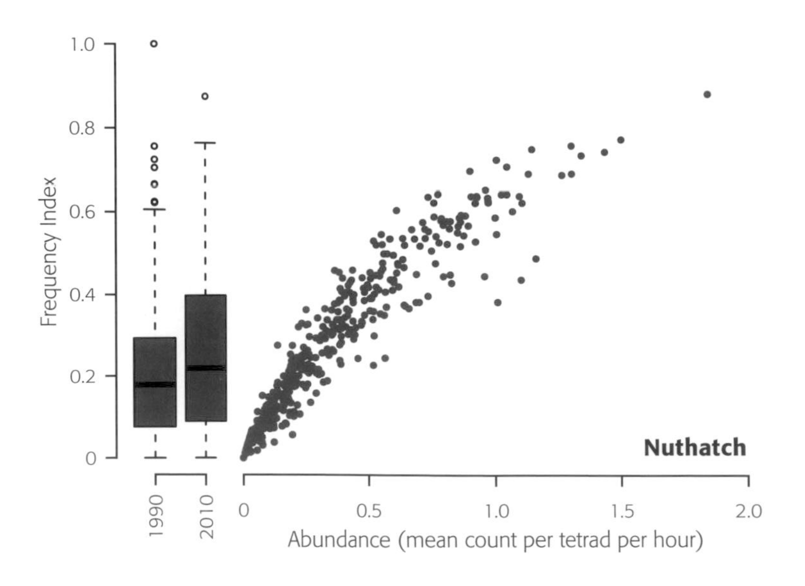

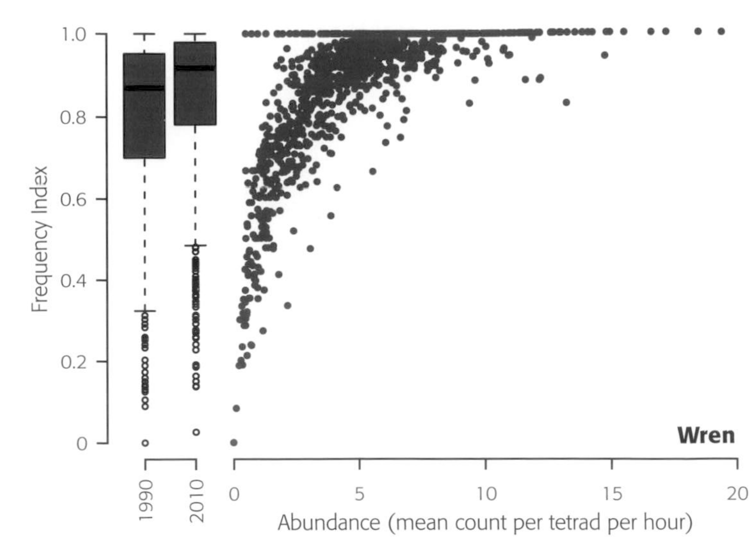

FIGURE 4.4 EXAMPLES OF THE RELATIONSHIP BETWEEN RELATIVE ABUNDANCE AND THE FREQUENCY INDEX

The box-whisker plots show the distribution of breeding-season Frequency Index values for the *1988–91 Breeding Atlas* (labelled 1990) and for *Bird Atlas 2007–11* (labelled 2010) calculated for 20-km squares. Frequency Index values have tended to increase for both species. The scatter plots show for 2008–11 the relationship between abundance, as measured by the mean hourly count across tetrads per 20-km square, and the Frequency Index, as measured by the proportion of tetrads occupied within the 20-km square. Both species have a curvilinear relationship between local abundance and the Frequency Index but they differ in shape.

included out-of-season records with breeding evidence, and winter totals included early November records. In the species accounts, these figures are expressed as a percentage of the potential maximum range size in each country (2,876 10-km squares with land in Britain, the Channel Isles and Isle of Man; 1,018 in Ireland). The actual figures can be found in Appendix 2.

Figures were produced to quantify by how much species' ranges changed between pairs of atlases. In addition to the traditional 'percentage range change' figure common to many second-generation atlases, the numbers of 10-km squares showing apparent gains and losses were also tallied separately. These are useful because some species can show little overall change in range size, but still show appreciable range shift or range turnover.

For winter, we totalled for Ireland, for Britain (including the Channel Isles and Isle of Man) and Britain & Ireland overall, the number of 10-km squares showing apparent gains or losses between the *1981–84 Winter Atlas* and 2007–11; as with the change maps, all records of fly-overs were excluded, as were records from early November 2007–10 to match the recording period with the *1981–84 Winter Atlas*. For the breeding season, similar gain/loss statistics were calculated spanning the entire 40-year period from 1968–72 to 2008–11 and for the 20-year period from 1988–91 to 2008–11; these figures omitted out-of-season breeding evidence records and, for the majority of species, were based only on squares with at least *possible* breeding evidence. For colonial seabirds, squares with only *possible* breeding evidence (or the *seen* category in the *1988–91 Breeding Atlas*) were excluded to be consistent with the changes described in the *1988–91 Breeding Atlas*.

The same basic data were used to produce percentage change figures for the change in range size from the earlier atlases to the current one. For most species the breeding-season change map and associated statistics are for the 40-year period, but there were a small number of cases where a species was most widespread during 1988–91 (e.g. Ruddy Duck) and the 20-year results were more informative. Regardless of which were used for species accounts, all figures are presented in Appendix 2.

These range-change figures are informative but they could suffer from bias if changes in methodology or recorder effort affected the probability that a species that was present in a square was detected. For example, an increase in the number of observers visiting remote upland areas in winter could have led to larger estimates of winter range size. Differences in observer effort are difficult to estimate, especially since for the earlier atlases few primary data were retained (see Section 5.6). However, the TTVs from the *1988–91 Breeding Atlas* and 2008–11 offer a simple way to produce estimates of breeding-season range change that are free from effort biases. In outline, this method (full details in Box 4.6) first determines whether a 10-km square was found to be occupied in each atlas period by taking the data only from a standard number of TTVs within the 10-km square. By fixing the number of TTVs contributing data to each 10-km square in both the early and later atlases, the searching effort is held constant. Differences between these fixed-effort distribution data can then be used to estimate range changes that are not biased by changes in the number and activity of observers. Unfortunately, because the TTV methodology was not used in the *1968–72 Breeding Atlas* or in the *1981–84 Winter Atlas*, it is not possible to calculate similar unbiased estimates of breeding range change for longer periods, or for winter range change.

4.7 MAP PRODUCTION

ALL MAPS WERE produced using the program R (version 2.14.0) using custom-written scripts employing the packages *maptools* (Lewin-Koh & Bivand 2011) and *sp* (Bivand *et al.* 2008) to process spatial data. A distinction is made between categorical and continuous data in their presentation, using symbols where possible to show categories (e.g. levels of breeding evidence) and colour shading to show continuous data (e.g. abundance). This necessitates two different approaches to map production, with symbols being plotted at the centre coordinates of a grid square, and shading relying upon GIS layers of grid squares cropped to the shape of the coastline (Section 2.2.1) and coloured accordingly.

Most maps are seasonally colour-coded, with those that relate to winter using a palette of cool blue colours and those for the breeding season a palette of warm orange-reds. All symbols and colours used for mapping were extensively tested for readability by a range of users. In particular, red and green are never used on the same map to limit problems for red-green colour-blind readers.

4.7.1 Distribution maps

Central coordinates of 10-km, 20-km or 50-km squares were used to locate the dot symbols on maps. Since many species' distributions were related to altitude, a shaded background was included for upland areas. This was derived from the covariate data used in abundance modelling (Section 4.4.3) and comprised all tetrads with mean altitude equal to or greater than 200 m.

4.7.2 Distribution change maps

Central coordinates of 10-km, 20-km or 50-km squares were used to locate the triangle symbols on maps. Stable areas on change maps were plotted by shading the entire relevant squares or polygons cropped to the coastline (see Section 2.2.1).

4.7.3 Relative abundance maps

These used the GIS grid square layers (Section 2.2.1), either at tetrad resolution for widespread species with an ensemble regression tree model, at 20-km resolution for the simple mean abundance maps for scarce species, or the offshore ribbon of polygons for coastal species. Note that because it was not possible to produce regression tree models in the Channel Isles owing to a lack of environmental data, figures for the Channel Isles were substituted from the simple mean abundance at 20-km resolution.

For relative abundance maps it was necessary to find an appropriate way to divide the wide spectrum of modelled or

BOX 4.5 CALCULATING THE BREEDING ABUNDANCE CHANGE INDEX

The maps of change in abundance were produced at the 20-km resolution because it was impossible to show change in every 10-km square owing to the use of the chequerboard in Ireland. The change index was calculated as follows:

1. Rule-8_{adj} targets of individual 10-km squares were summed within 20-km squares to determine the target number of tetrads for each 20-km square, giving a target for each land-locked 20-km square of 32 tetrads in Britain and, because of the chequerboard, 16 tetrads in Ireland. To reduce the occurrence of spurious patterns, 20-km squares with a target of fewer than five tetrads were excluded from the analysis, leaving 918 20-km squares in which 39,936 tetrads were surveyed in 1988–91 and 45,374 tetrads surveyed in 2008–11. We chose not to restrict the analysis to the tetrads surveyed in both periods, but there was a high degree of overlap: 70% of tetrads surveyed in 1988–91 were also surveyed in 2008–11.
2. Presence/non-detection data were obtained for every species in each surveyed tetrad in 1988–91. Comparable data were generated for 2008–11 by taking only first-hour counts from the two visits in a season and determining presence/non-detection information for each species in each surveyed tetrad.
3. In either period, if a species had been recorded on a TTV but there was no evidence of at least *possible* breeding in the 10-km square, the record was re-coded as not detected. This was done to exclude species that were not part of a 10-km square's breeding avifauna, e.g. late-departing Fieldfares.
4. Having prepared the data, they were analysed as follows:
   a. In each 20-km square, and in each period, select tetrads randomly without replacement according to the Rule-8_{adj} targets (see step 1);
   b. In each 20-km square, using only the data from those selected tetrads, calculate the proportion of tetrads occupied by each species in each atlas period.
5. To reduce the influence of precisely which of the surveyed tetrads were included in the random selection in step 4a, steps 4a–b were repeated 1,000 times, resulting in 1,000 estimates of the Frequency Index for each species in each 20-km square in 1988–91 and in 2008–11.
6. After taking the mean Frequency Index values for each period across the 1,000 runs, the Abundance Change Index for a species in a 20-km square was taken as the arithmetic difference between the 2008–11 Frequency Index and the 1988–91 Frequency Index.

For Golden Plover and other species of remote areas, differences in recording effort could affect standard change measures. Fixed-effort change statistics provide robust indications of change.

observed mean abundance values into discrete categories, each of which had a different tint of orange-red or blue on the map. As in the *1988–91 Breeding Atlas*, we chose to represent abundance in 'deciles', dividing the values into ten equal-sized groups, with the simple interpretation that the deepest colour tint indicates the 10% of squares with the highest abundance, ranging through to the palest colour tint indicating the 10% of squares with the lowest abundance.

4.7.4 Relative abundance change maps

These maps used GIS layers of 20-km squares, with polygons shaded according to the Abundance Change Index value. Before these indices could be mapped, a simple rescaling of the indices was needed so that species varying in baseline abundance could be mapped on the same scale and colour scheme. The need for this can be illustrated by comparing two species towards the extreme ends of baseline abundance. Wrens are common, widespread and occupy a range of habitats. They reach a Frequency Index of 1 (i.e. present in 100% of surveyed tetrads in a square) in many 20-km squares. Abundance Change Index values for the Wren could potentially vary between -1 if Wrens disappeared from all tetrads surveyed in a 20-km square, to 1 if they colonised all surveyed tetrads. Contrast this with the figures for the Black Grouse which, like many scarce habitat-specialists, never achieves a Frequency Index of 1. Over the period of the last two breeding atlases, within 20-km squares in the core of its range, the Black Grouse was detected in only c.20% of surveyed tetrads, thereby limiting the observable range of Abundance Change Index values to -0.2 to 0.2. Nevertheless, for the Black Grouse a decrease from 20% of tetrads to 10% of tetrads may be just as significant as a decrease from 100% of tetrads to 50% of tetrads for the Wren.

To put scarce and common species on an even footing for mapping, all Abundance Change Index values for a species were divided by the 90th percentile of the species' non-zero Frequency Index values, pooled across atlas periods. This meant that, regardless of the abundance of the species, its modified Abundance Change Index values mostly ranged between -1 and 1 so that there was a common interpretation of the maps. The 90th percentile was selected so that the transformed Frequency Index values were better distributed from -1 to 1; had the maximum value been used instead to transform the indices then they might have been aggregated in just part of the -1 to 1 range and the map would have shown few discernible patterns.

BOX 4.6 CALCULATION OF CONTROLLED ESTIMATES OF RANGE CHANGE

The index of breeding-season range change between 1988–91 and 2008–11 was calculated using a bootstrapping method that used randomly selected tetrads to maintain geographically representative coverage. The data and procedures were as follows:

1. For each 10-km square, we identified the tetrads that were surveyed in 1988–91 and those surveyed during the breeding seasons 2008–11. In total, 3,231 10-km squares achieved at least Rule-8_{adj} coverage in both periods. After excluding Irish 10-km squares that were not part of the chequerboard, 3105 10-km squares remained: 463 in Ireland (containing 4,028 tetrads in 1988–91 and 3,837 tetrads in 2008–11) and 2,642 in Britain (34,425 tetrads in 1988–91 and 40,353 tetrads in 2008–11).
2. We then identified the species recorded in each tetrad in 1988–91 and in 2008–10. The latter were based only on the first hour of TTVs to maintain consistent effort with the 1988–91 data. Also, if a species was recorded on a TTV but there was no evidence of breeding in the 10-km square, the record was re-coded as an apparent absence. This was done to exclude species that were not part of a 10-km square's breeding avifauna e.g. late departing Fieldfares.
3. Having prepared the data, they were analysed as follows:
   a. In each 10-km square, select randomly without replacement Rule-8_{adj} tetrads from the list of surveyed tetrads in each period.
   b. Using only the data from those selected tetrads, determine whether each 10-km square was occupied in 1988–91 or 2008–11.
   c. Assign each 10-km square a change value of 0, -1 or +1 according to whether the species was detected in both periods, only in the first (i.e. apparent loss), or only in the last (i.e. apparent gain). 10-km squares where the species was apparently absent in both periods were excluded.
   d. Average these change values across British 10-km squares and Irish 10-km squares separately. Each average figure (multiplied by 100) is a single estimate of the percentage change in range size between periods, based on effort-controlled data.
4. To reduce the influence of precisely which of the surveyed tetrads were included in the initial random selection in step 3a, steps 3a–d were repeated 1,000 times, resulting in 1,000 estimates of range change for each species, in Britain and in Ireland.
5. The index quoted in the species accounts is the median of these values. The 25th and 975th ranked values were taken as the bounds of the level of certainty of the apparent range change.

MICK DURHAM www.wildlife-photographer.net

CHAPTER 5

COVERAGE & EFFORT

Simon Gillings, Dawn Balmer, Bob Swann & Brian Caffrey

Records from ringers and nest recorders were invaluable in helping to confirm breeding for scarce and localised species such as the Hobby.

5.1 INTRODUCTION

BIRD ATLAS 2007–11 was highly successful thanks to tens of thousands of volunteer observers who worked together to produce a dataset containing 19 million records. They visited virtually every 10-km square containing land in Britain & Ireland and provided systematic data on relative abundance of birds in more than 50,000 tetrads. In addition, data were incorporated from a wide range of external data sources to make distribution maps as comprehensive as possible. But as impressive as these figures are, serious questions can be raised about comparability between this atlas and previous ones, which relied upon different field methods and involved fewer observers (Section 5.2) and fewer data sources and records (Section 5.3).

Differences in coverage and effort arise because there is an inevitable tension in repeat atlases. On the one hand, maintaining consistent methods and coverage is crucial to allow valid comparisons to be made. Yet, on the other hand, there is a constant desire to maximise effort and allow the evolution of field and analytical methods to deliver ever more refined results. Indeed, even when field methods remain unchanged, one cannot guarantee constant effort without a systematic (e.g. timed) component to the methods (Greenwood *et al.* 1997; Section 2.1). This atlas followed the *1988–91 Breeding Atlas* in using timed counts to enable robust effort-controlled comparisons to be made in the breeding season (see Sections 2.1 and 4.6), and to facilitate the same for future winter atlases. However, the majority of comparisons drawn in this atlas concern changes in gross distribution patterns and these are subject to potential differences in recording effort among atlases.

Recording effort is not simply a matter of whether or not a 10-km square was visited. Prendergast *et al.* (1993) defined the recorder effort problem as "variation in the distribution and behaviour of recorders [that] creates data sets which may be heavily biased in favour of recorder, rather than species, distributions". If either or both the distribution and behaviour of observers change through time, some of the emergent patterns in atlas change maps could partly reflect these changes rather than true changes in species' distributions (Prendergast *et al.* 1993; Dennis *et al.* 1999; Hill 2012). There was much discussion of this upon completion of the *1988–91 Breeding Atlas* (see discussion therein and Bircham & Jordan 1996; Greenwood *et al.* 1997) and these issues can affect subsequent research that uses atlas data (Rocchini *et al.* 2011; Kujala *et al.* 2013).

Recording effort could have changed in various ways. The Britain & Ireland atlases have involved different field methods with a varying emphasis on the collection of distribution and breeding evidence data versus quantitative relative abundance data (Section 4.1.2). Also, since the *1988–91 Breeding Atlas*, the membership of BirdWatch Ireland has more than doubled and that of BTO has increased by more than 80%, leading to a significantly larger pool of potential observers. More subtle aspects of participation could also have changed. Has an observer's willingness to travel to distant squares changed? Has it become more difficult to gain access to private property? Have observers' equipment, skills or acuity for detecting songs and calls changed (e.g. Porter 2012)? One of the most significant changes since previous atlases has been the use of the Internet to collate data. This has made it considerably easier to gather larger numbers of records but it does not necessarily follow that effort has changed. It is quite likely that observers were more selective in the past about which records they submitted by, for example, only submitting records with breeding evidence.

These possible changes make it essential to test whether there have been localised or more broad-scale changes in recording effort between previous atlases and *Bird Atlas 2007–11*, so that readers and future users of atlas data can correctly interpret regional patterns of gains and losses and the overall change statistics. The latter part of this chapter uses a new analytical procedure (Hill 2012) for estimating recording effort, which can be applied to quantify how effort may have varied across Britain & Ireland and among the different atlases. We strongly recommend that the conclusions of these analyses (Section 5.7) be remembered when reading the species accounts and overview (Chapter 6).

Various statistical analyses are performed in this chapter, including correlations, least-squares linear regression and generalised linear models. Where appropriate the latter used log links and Poisson errors for count data. For the sake of readability, generally only the significance of the tests is shown (P values).

5.2 NUMBERS, DISTRIBUTION AND ACTIVITY PATTERNS OF VOLUNTEER OBSERVERS

VOLUNTEER OBSERVERS WERE the backbone of *Bird Atlas 2007–11* and it pays to understand patterns of distribution and activity of observers so that the potential for biases can be evaluated. The total number of observers participating in the Atlas is difficult to quantify precisely. Overall, c.18,000 people registered to use Bird Atlas Online (Chapter 3) but not all submitted data, or they submitted only out-of-season records lacking breeding evidence that were not used for the Atlas. Approximately 9,000 observers submitted data to BirdTrack during the Atlas period, although a proportion of records were out-of-season and lacked breeding evidence and some BirdTrack users also submitted via Bird Atlas Online.

The records from all sources used in the production of the Atlas—that is validated records in winter, the breeding season, or out-of-season with appropriate breeding evidence—were stored in the main project database against 17,427 usernames. This does not equate directly to 17,427 observers because at least 20 of these were schools, c.210 were for birdwatching families and many individuals submitted data on behalf of couples and groups (see Appendix 1). Moreover, for the many observers who took part in other surveys and schemes (e.g. WeBS, CBS), their non-Atlas records were incorporated into the Atlas dataset under single

DAWN BALMER

usernames (Sections 2.4 and 5.3.4). For example, the Breeding Bird Survey, Garden BirdWatch, Heronries Census, Nest Record Scheme, Ringing Scheme and Wetland Bird Survey accounted for only six of the 17,427 usernames, but BTO records show that c.18,200 birdwatchers and surveyors participated in those six schemes and did not contribute directly to the Atlas or BirdTrack. To these c.35,000 observers could be added the number of participants in Irish monitoring schemes (CBS, GBS, I-WeBS), an unknown number of county bird club members, RSPB reserve staff and other observers who contributed to the external data sources that were incorporated into the Atlas. On this basis we estimate that records from c.40,000 observers went into the production of this atlas.

The number of observers is undoubtedly higher than for previous atlases. The number of participants for the *1988–91 Breeding Atlas* is not known, but the *1981–84 Winter Atlas* stated that fieldwork was carried out by "over 10,000 dedicated observers" and for the *1968–72 Breeding Atlas* it was thought that 10–15,000 observers took part, but that "the bulk of the work" was carried out by about 1,500 observers. As with all atlases, the amount of data submitted by observers involved in *Bird Atlas 2007–11* varied considerably. Of the c.17,000 named observers, 50% of records were submitted by 5% of observers, and 90% were submitted by 27%. However, quantity does not necessarily equate to quality because not all records included breeding evidence (see Section 5.3.3).

5.2.1 Where did observers live?

When registering online or when returning paper survey forms, all observers had to provide a postal address. This made it possible to assign a 'home' 10-km square to 17,232 observers, having first excluded a small number of 'group users' (e.g. bird clubs) for which the home address was potentially misleading. The numbers of resident observers per country/region were: England = 12,508, Scotland = 2,300, Wales = 1,175, Republic of Ireland = 952, Northern Ireland = 240, Channel Islands = 43 and Isle of Man = 14. Figure 5.1A shows the density of observers per 10-km square and highlights the marked difference between Britain and Ireland: only 27% of British 10-km squares lacked a resident observer compared to 56% of Irish 10-km squares. The number of resident observers per 10-km square was significantly greater in Britain than Ireland (mean ± SE: Britain = 5.6 ± 0.1; Ireland = 1.2 ± 0.1; $P < 0.0001$). Not surprisingly, within Britain the density of observers was greatest in the lowlands and in the major urban centres. Little geographic variation could be discerned in the density of observers in Ireland except that densities were higher around Belfast, Dublin and Cork. Overall, 15% of Britain & Ireland's 10-km squares had just one resident observer and 53% of squares had 10 or fewer resident observers. In Britain the 10-km square with most resident observers was NT27 (Edinburgh) with 119 observers; in Ireland, IO13 (Dublin) had the most with 59 resident observers.

FIGURE 5.1 MAPS SHOWING THE DISTRIBUTION OF RESIDENT AND SURVEYING OBSERVERS

Inishmore, Aran Islands (Co. Galway), had two resident Atlas observers.

Map A shows the numbers of observers resident in each 10-km square in Britain & Ireland; white areas indicate 10-km squares with no resident observers. **Maps B** and **C** show the numbers of observers submitting data per 10-km square in winter and in the breeding season. On all maps, darker colours indicate a greater number of observers per 10-km square.
In **A** the scale is 1, 2, 3–5, 6–10, 11–20, 21–50, >50.
In **B** and **C** the scale is 1–10, 11–20, 21–30, 31–40, 41–50, 51–100, >100.

A Resident Observers

5.2.2 Where did observers undertake fieldwork?

The numbers of observers submitting any Atlas records (Roving Records or TTV data) per country were: England = 13,087, Scotland = 3,350, Wales = 2,146, Republic of Ireland = 1,179, Northern Ireland = 409, Channel Islands = 68 and Isle of Man = 44. It is interesting to compare these figures with those from Section 5.2.1 on numbers of resident observers. They suggest that twice as many observers submitted data in Wales as lived there and that 70% more submitted data in Northern Ireland than lived there.

Figures 5.1B and 5.1C show spatial variation in the numbers of observers who submitted data for each 10-km square. The major urban centres are still visible but reassuringly, the gaps in Figure 5.1A have been filled, indicating that observers travelled to the squares where there were no resident observers. Of the 3,894 10-km squares containing any land, all but 47 in winter and 23 in the breeding season were visited by at least one observer (Table 5.1; see Section 5.3 for further discussion of spatial coverage). The number of observers contributing data per 10-km square differed significantly between Britain and Ireland in both seasons (both $P < 0.0001$). Across Britain & Ireland, the number of observers visiting a square was similar in winter and the breeding season ($r_s = 0.86$, $P < 0.0001$, $n = 3{,}894$), reaching a maximum of 227 observers in winter and 232 in the breeding season. In both these cases the square was TF74, which includes the Norfolk birding hotspots of Holme and Titchwell. The Irish 10-km square receiving the most observers was the same square as had the most resident observers, IO13 (Dublin). In general, the number of resident observers was positively correlated with the number of observers submitting data (winter: $r_s = 0.67$, $P < 0.0001$; breeding: $r_s = 0.78$, $P < 0.0001$. Both $n = 3{,}894$ 10-km squares).

5.2.3 Patterns in observer behaviour

With information on the home 10-km square of each observer, and of the 10-km squares they visited, it was possible to gain some insights into how far afield observers travelled. The maximum straight-line distance between an observer's home 10-km square and one they surveyed was 1,154 km (Devon to Shetland). This was, of course, an extreme and 75% of observers visited only 10-km squares within 100 km of their home square. Approximately one third of observers submitted data only from their home 10-km square. A further 19% submitted data from their home square plus its immediate neighbours.

We can also examine the different ways in which observers contributed to the Atlas. After excluding group users, 57% of observers submitted at least one Roving Record and 47% completed at least one TTV; 36% of observers undertook both survey types. Data were submitted via BirdTrack by 46% of observers. That figure could be divided into the 14% who submitted both Atlas and BirdTrack records, and 32% that used BirdTrack exclusively. The volumes of records and the contribution of these different sources is covered in Section 5.3.4.

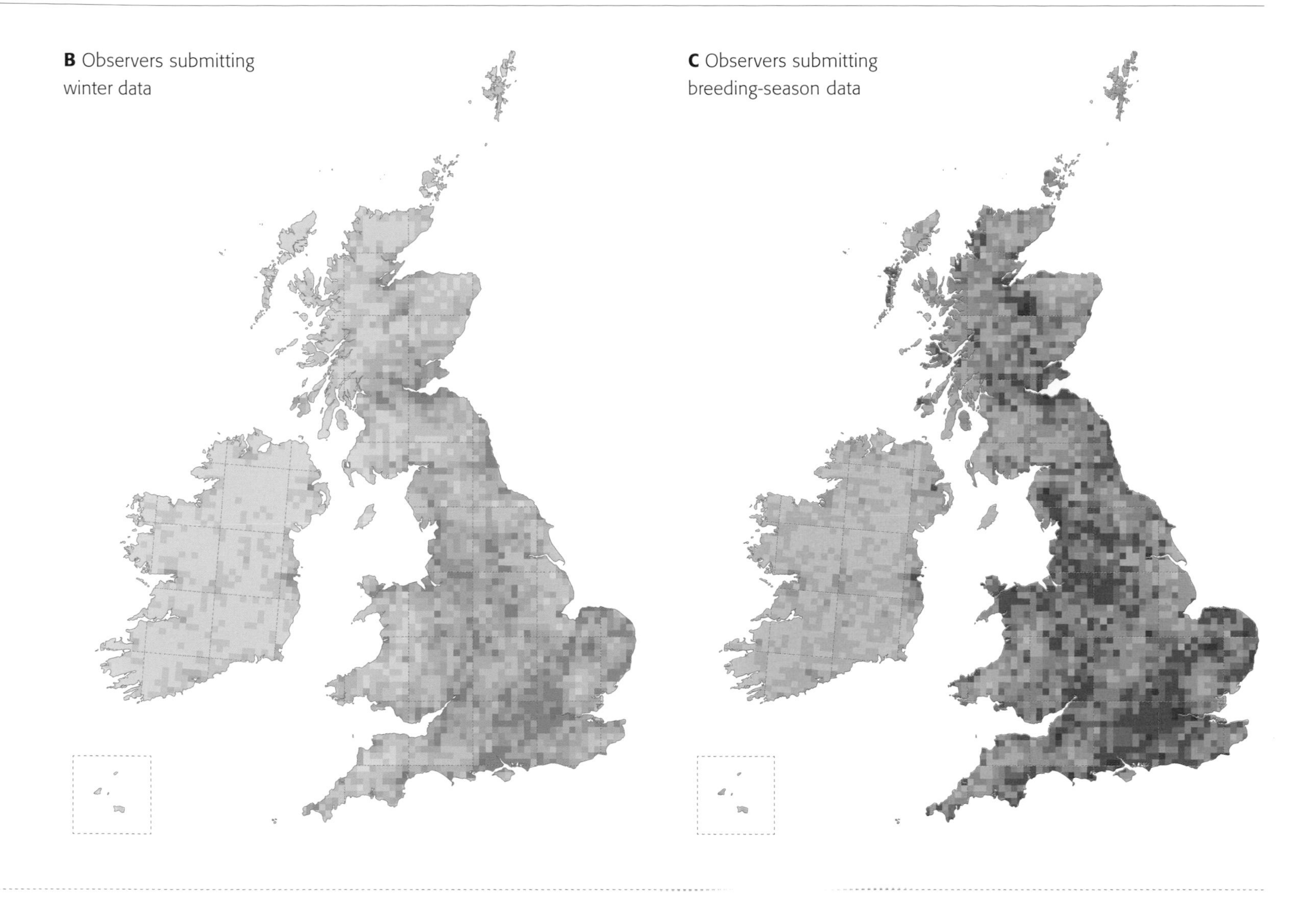

TABLE 5.1 COVERAGE OF 10-KM SQUARES BY COUNTRY OR REGION

Only a small number of 10-km squares were not visited in each country or region and these tended to contain very little land. In winter, unsurveyed 10-km squares amounted to just 0.02% of the land area of Britain & Ireland. In the breeding season just 0.004% of the land area was not visited.

Country/region	Season	Surveyed		Not surveyed	
		Number of squares	Combined area (km²)	Number of squares	Combined area (km²)
England	Winter	1,459	131,928	6	33.5
	Breeding	1,459	131,952	6	9.6
Scotland	Winter	1,080	80,083	34	35.0
	Breeding	1,101	80,115	13	3.4
Wales	Winter	263	21,263	3	0.1
	Breeding	264	21,263	2	0.02
Channel Islands	Winter	16	208	1	0.3
	Breeding	16	208	1	0.3
Isle of Man	Winter	14	585	0	-
	Breeding	14	585	0	-
Britain	**Winter**	**2,832**	**234,068**	**44**	**68.9**
	Breeding	**2,854**	**234,123**	**22**	**13.3**
Republic of Ireland	Winter	853	70,595	3	0.1
	Breeding	855	70,595	1	0.01
Northern Ireland	Winter	162	14,067	0	-
	Breeding	162	14,067	0	-
Ireland	**Winter**	**1,015**	**84,662**	**3**	**0.09**
	Breeding	**1,017**	**84,662**	**1**	**0.01**

5.3 COVERAGE FOR DISTRIBUTION MAPPING

THE MAIN AIM of any atlas is to survey every grid square in sufficient detail to be able to produce a comprehensive species list. At the simplest level, *Bird Atlas 2007-11* was a resounding success in that 3,844 (98.7%) of the 3,894 10-km squares in Britain & Ireland containing any land were visited at least once in both seasons. Indeed the 50 10-km squares with coverage in neither, or only one, season contained only small amounts of coast, offshore sandbars and rocks and they amounted to only 0.02% of the land area of Britain & Ireland. Of these squares, 47 were not visited in winter compared to 23 not visited in the breeding season (Table 5.1). The difference can be attributed to the difficulty of accessing many of these squares in winter and because several received dedicated seabird survey visits in the summer months.

5.3.1 Spread of visits to 10-km squares across seasons

The aim was to accumulate coverage gradually over the four winter field seasons and the four breeding field seasons. However, the very cold winters of 2009/10 and 2010/11 (Box 5.1) caused high mortality among small-bird populations (Balmer & Vickery 2010, 2011; Lovatt & Madden 2012). Therefore there was a concern that the distributions of some small passerines susceptible to cold weather (e.g. Stonechat) might have been under-recorded if a significant number of 10-km squares had been visited only in the two field seasons that followed.

During the first winter of fieldwork, 2,756 10-km squares in Britain and 868 in Ireland were visited as inferred by at least one completed TTV or Roving Record for each 10-km square in that field season (Table 5.2). The following winter, similar numbers of squares were visited (2,754 and 877). This pattern continued through the course of the Atlas. With such a high proportion of squares visited per winter, inevitably a high proportion of 10-km squares were visited at least once in multiple winters. By the close of the final winter, 2,832 10-km squares in Britain and 1,015 in Ireland had been visited at least once, and of these, 2,645 and 733 respectively had been visited in all four winters. Similar patterns of annual and cumulative coverage were apparent in the breeding season.

Table 5.2 also shows that few 10-km squares received visits in only a single field season. In Britain, 60 10-km squares in winter and 22 in the breeding season were only visited in either or both of the final two field seasons; corresponding figures for Ireland were 55 and 16 squares respectively. These figures are encouraging because they suggest that a high proportion of 10-km squares had received some survey attention prior to the cold winters. Nevertheless, it is possible that for species such as the Stonechat, small numbers of 10-km squares may have been occupied during 2007–09 but by the time they were visited in the last two winters, the birds had been lost.

TABLE 5.2 ANNUAL AND CUMULATIVE VISITS TO 10-KM SQUARES

The table shows the numbers of squares that were visited in each separate field season. It also gives the numbers that were visited in only that season and the numbers that were visited in all seasons. These figures do not indicate that a 10-km square had adequate coverage in a field season, only that it was visited at least once as judged by the submission of at least one Roving Record or TTV.

Island	Season	Field season	Number visited in this field season	Number visited in this field season *only*	Number visited during *all* field seasons so far	Number visited in *at least one* field season so far
BRITAIN 2,876 available 10-km squares	**Winter**	2007/08	2,756	7	2,756	2,756
		2008/09	2,754	6	2,713	2,797
		2009/10	2,763	11	2,678	2,820
		2010/11	2,755	12	2,645	2,832
	Breeding	2008	2,800	3	2,800	2,800
		2009	2,822	4	2,787	2,835
		2010	2,814	5	2,773	2,844
		2011	2,819	10	2,759	2,854
IRELAND 1,018 available 10-km squares	**Winter**	2007/08	868	3	868	868
		2008/09	877	2	783	962
		2009/10	929	4	751	998
		2010/11	972	17	733	1,015
	Breeding	2008	949	0	949	949
		2009	970	2	918	1,001
		2010	974	1	890	1,012
		2011	1005	5	882	1,017

BOX 5.1 WEATHER DURING *BIRD ATLAS 2007–11*

The following points summarise the key features of the weather during the field seasons from winter 2007/08 to breeding season 2011. These are based on a combination of the UK bimonthly 'Weather Notes' from the relevant issues of the journal *British Wildlife* (e.g. Branson 2011), the UK monthly 'Weather Log' and annual report in the journal *Weather* (e.g. National Climate Information Centre 2012) and Irish seasonal and annual reports from the website of the Irish Meteorological Service (2013).

HUGH INSLEY

Heavy snowfall in the second two winters of the Atlas presented access difficulties for fieldworkers.

Winter 2007/08: mild, wet

- After a dry autumn, it stayed dry and warm generally until mid December.
- Late December and much of January wet with heavy snow in southern Britain in early January.
- February mostly dry and sunny in England but very wet in northwest Scotland.
- January–February mostly mild, but with a few stormy periods.

Breeding season 2008: cool, dull and wet but variable

- More snow in March and April than January and February.
- Very cold in Britain until third week of April.
- Very variable especially though June and July.
- May up to 2°C warmer than average, but June coolest for 10 years or so.
- Driest year in Scotland since 1994, but very wet in west (twice the average in parts of Ireland).

Winter 2008/09: cold, snow, dry in west

- Colder than average, indeed coldest for 10–15 years in Ireland.
- Significant snow in January and early February including southeast England in the latter month.
- Ireland relatively dry all through winter (though with January about average) and Wales only had about a third of normal rain in February.

Breeding season 2009: wet

- Unsettled through most of period.
- Mild (April averaged 2°C above average in Britain and 1°C above in Ireland).
- Dry in east until July (e.g. Kent only a third of normal rainfall in June) although some stormy weather in June; very wet in parts of west (up to twice average rainfall in Ireland).
- July very wet everywhere especially in west (parts of southwest and northwest Britain up to three times average rainfall).
- About average temperatures.

Winter 2009/10: very cold with snow but mostly dry and often sunny

- Mild until mid December though with heavy rain at times.
- Mid December to mid January very cold (down to -18.5°C) and disruptive snow in places but overall dry and sunny.
- An interlude of average temperatures but turned cold again for February with more snow near end of the month.
- Overall the coldest winter since 1962/63 in Ireland and since 1978/79 in Britain (with north Scotland the coldest on record).

Breeding season 2010: dry, warm

- Very dry and sunny through first half of the season, with western Scotland and southern England especially dry.
- From late May quite warm with July being warmest in eastern England since 1976.
- However, twice average rain in July in Scotland, western Wales, Cumbria and parts of Ireland.

Winter 2010/11: cold first half, mild second half

- End of November heralded one of coldest spells for 100 years with December up to 5°C below long-term average, with most precipitation as snow, especially in the first and third weeks. However much sunnier than usual.
- Became fairly mild from mid January.
- Southwest Britain fairly dry but northern England and southern Scotland very wet, and overall rather dull.

Breeding season 2011: very warm and dry start, more typical from May onwards

- Exceptionally warm at start—up to 6°C above average, with April the warmest in England for 350 years.
- Very dry in March and April—parts of England had <1 mm rain in April.
- May onwards more typical although a southeast (dry) to northwest (wet) split with Scotland having wettest May since 1910.
- Quite cool in many places.

FIGURE 5.2 MAPS OF PERCENTAGE OF EXPECTED SPECIES

Using the species richness in each 10-km square from previous atlases as a target, the coloured shading indicates the percentage of the expected species total reported at the completion of *Bird Atlas 2007–11*. There are four shades on each map representing, from lightest to darkest, 1–49%, 50–74%, 75–89% and 90% or more of expected species richness.

5.3.2 Effectiveness of coverage

Table 5.1 shows the numbers of 10-km squares with any evidence of field recording effort but this does not necessarily mean all those squares were surveyed sufficiently to produce comprehensive species lists. A better measure of the effectiveness of coverage would be the completeness of species lists in the 10-km square. During the course of *Bird Atlas 2007–11*, coverage in each 10-km square was monitored by expressing the observed species richness as a proportion of the richness reported in the *1981–84 Winter Atlas* or the *1988–91 Breeding Atlas*. For the breeding season, a species only counted towards a square's richness if it had at least *possible* breeding evidence. The resulting 'percentage of expected species richness' maps were a key part of the online feedback (see Section 3.8).

Overall, most of Britain & Ireland achieved high levels of coverage (Figure 5.2) with expected species richness of 90% or more achieved in 98% of British 10-km squares and 95% of Irish squares in winter. In the breeding season the corresponding figures were 85% and 87% of squares, respectively. Species richness accrued more slowly in the breeding season than in winter owing to the prerequisite for breeding evidence before species could be counted. Squares with reduced percentages of expected species richness in the breeding season were scattered throughout Britain & Ireland. They were not particularly aggregated in areas where poor coverage might be expected (e.g. in the uplands). In many of these areas, observers spent considerable amounts of time looking for additional species but to no avail. In some squares it was clear that major changes in habitat since the *1988–91 Breeding Atlas* meant that species richness really had declined. It is also worth noting that the strict adherence to suitable nesting habitat being a prerequisite for breeding evidence in *Bird Atlas 2007–11* may have meant that some species lists from the *1988–91 Breeding Atlas* (which did not impose this rule) were unattainable.

5.3.3 Provision of breeding evidence

Possible, *probable* and *confirmed* breeding evidence was central to the production of breeding-season distribution maps and was provided on 55% of the c.7.9 million records submitted for the standard April–July breeding season. For distribution mapping purposes a further 0.1 million out-of-season (March, August–

CHRIS BALE www.guernseybirdnerd.com

Breeding by Long-eared Owls can be *confirmed* from the begging calls of chicks which may be heard in late summer.

October) records were added. These were the out-of-season records that conformed to the species-specific and month-specific rules on acceptability of out-of-season breeding evidence (see Appendix 4 and Sections 2.2.2 and 4.2.2). The proportion of records submitted with evidence codes decreased at the higher levels of confirmation: 26% of records had a *possible* breeding code, 16% a *probable* code and 14% a *confirmed* code. The *non-breeding* codes for flying, migrant and summering birds (Table 2.2) were provided on 3.6% of records and are not considered further in this summary.

The provision of breeding evidence was similar among the main types of Atlas survey methods with 79% of records from breeding-season TTVs and 77% of Roving Records having a breeding evidence code. The high provision of breeding evidence across the whole sample of TTVs was also replicated at the level of the individual TTV. Only 4% of TTVs were submitted with no breeding evidence for any species and 29% were submitted with breeding evidence for all species encountered; the average TTV had breeding evidence for 79% of species. Provision also varied among the other data sources: 19% of records from internal data sources (e.g. other monitoring schemes; see Section 2.4), 21% of BirdTrack records, 36% of bird club records and 97% of records from external data sources (e.g. UK and Irish RBBPs; see Section 2.4.1).

The inclusion of breeding evidence does not, on its own, indicate the value of a survey type or dataset because there could be a high degree of duplication (or redundancy) within and among data sources. Section 5.3.4 considers how different data sources measure up in contributing to the production of distribution maps.

5.3.4 Contribution of different data sources to distribution maps

As for previous Britain & Ireland atlases, additional data sources (see Section 2.4) were collated to ensure that distribution maps in *Bird Atlas 2007–11* were as comprehensive as possible. Some datasets contributed comparatively few records but were carefully selected to add or upgrade existing information (e.g. rare breeding bird panel data), whilst others were large and added in their entirety (e.g. BirdTrack). This section considers the relative value of the major datasets, partly to assess the impact of those that were not available to previous atlases and partly to aid future atlases when prioritising the data sources to include.

FIGURE 5.3 CONTRIBUTION OF DIFFERENT TYPES OF DATA SOURCES TO THE DISTRIBUTION MAPS

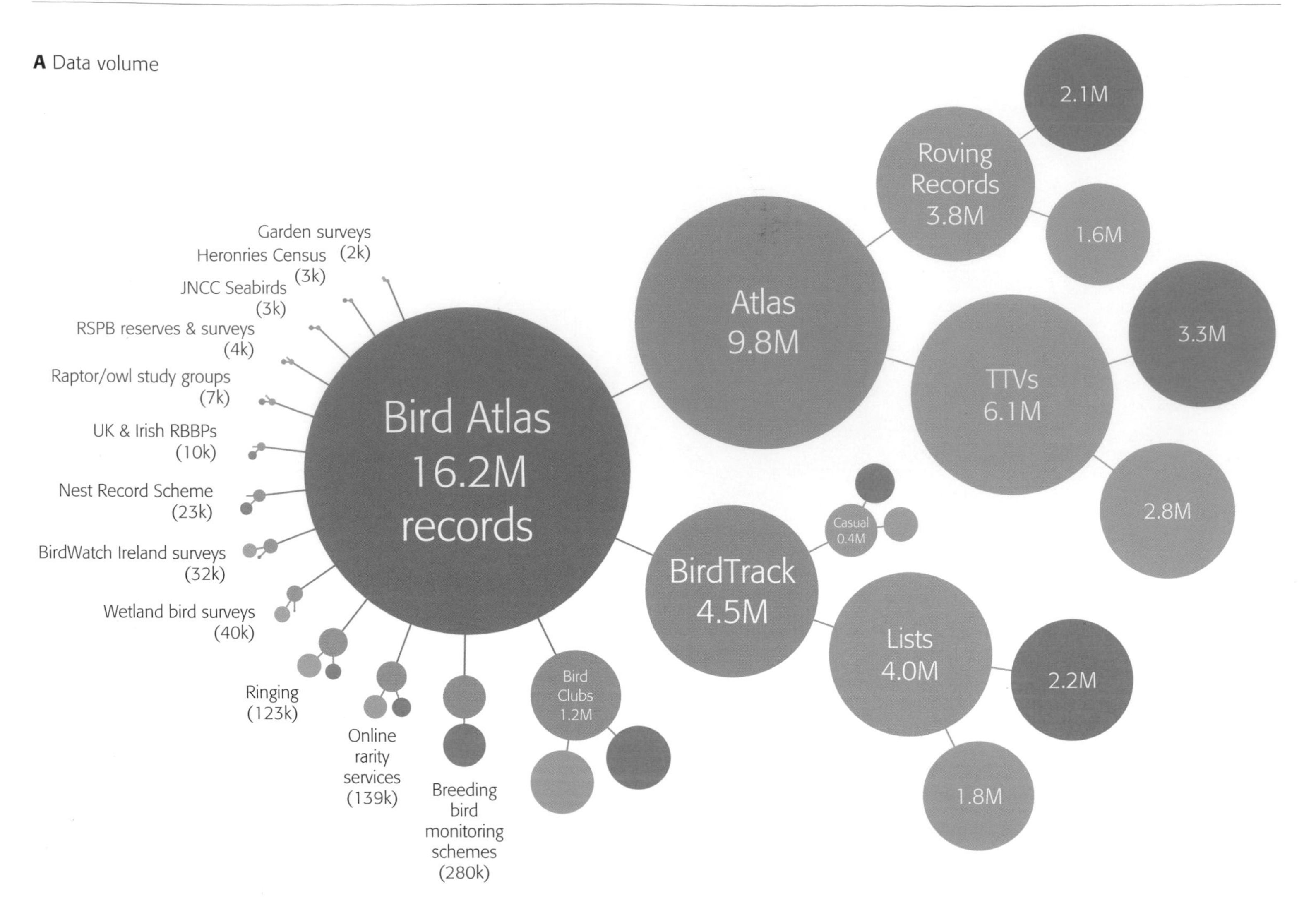

In **A** the area of each circle is proportional to the number of records in the particular subset of data whereas in **B** the area is proportional to the amount of each species' range that was uniquely contributed by the data source. For this analysis, data were grouped according to the type of source rather than geographic scope. For example, 'Wetland bird surveys' includes both the UK Wetland Bird Survey and the Irish Wetland Bird Survey. 'BirdWatch Ireland surveys' encompasses a number of special surveys of individual species, habitats and sites. In **A**, figures indicate numbers of records and are rounded so may not sum perfectly. In **B**, figures indicate the number of maps that the dataset (or data subset)

Of the total dataset of c.19 million submitted and valid records, c.16.2 million records were used for distribution mapping after removing Tetrad Population Estimates (Section 5.4.4) and unused out-of-season records (Section 4.2.2). In Figure 5.3 the 16.2 million records are broken down to assess the contribution of different data source types in the two seasons. Data sources are grouped according to the type of scheme or survey they originated from, rather than geographically or by source organisation. For example, the 'Breeding bird monitoring scheme' data type summarises the contribution from the UK Breeding Bird Survey and the Republic of Ireland Countryside Bird Survey. In Figure 5.3A the area of each circle is proportional to the number of records in that subset of data. For example, 9.8 million records of the 16.2 million grand total could be described as 'Atlas records', having been submitted directly on paper forms or through the Bird Atlas Online system; of these, 3.8 million were Roving Records and 6.1 million were TTV records (numbers appear not to sum owing to rounding errors). These can be subdivided further by season, showing that slightly more records were received by each survey type in the breeding season than in winter.

From Figure 5.3A it is apparent that Atlas and BirdTrack records accounted for the majority of bird records whereas some of the other data sources accounted for as few as c.2,000 records. However, data volume is only one measure of contribution because there can be a high degree of redundancy within and among data sources. Furthermore, a data source could contain a large number of records but unless they have associated breeding evidence, their value for the production of breeding-season distribution maps could be limited.

In Figure 5.3B the order and layout of data sources is identical to that in Figure 5.3A but here the area of each circle indicates the unique contribution of each data source and subset. Contribution was estimated by first finding which species maps (breeding or

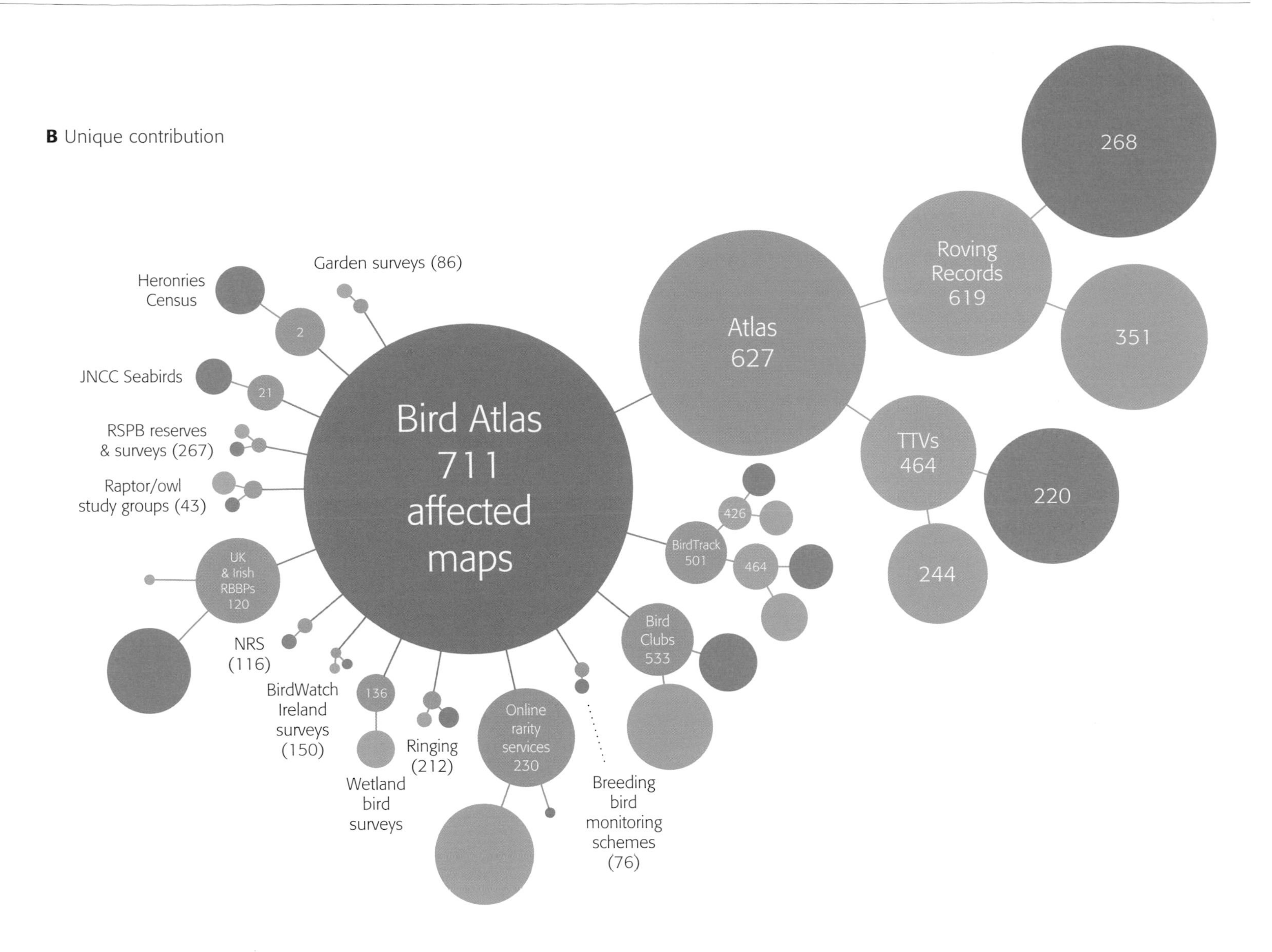

affected. That is, if a dataset affected both the breeding and winter distribution maps for one species it would be counted as affecting two maps. The area of the circle is set according to the median percentage of dots affected by the data source, calculated across all affected maps. For example, the Heronries Census affected breeding-season maps for two species: it affected 27% of the Grey Heron dots and 1% of the Little Egret dots, giving a median unique contribution of 14%. For reference, the equivalent figure for Atlas records was 88%. Blue and orange circles denote values for winter and the breeding season respectively.

winter) were 'affected' by a data source. 'Affected maps' were defined as those where at least one dot on the map (an 'affected dot') would have been missing entirely, or would have been present but with a lower breeding evidence, if the data source had not been included. For each affected map, the percentage of dots that were affected by the data source was determined, and the median calculated across all affected maps; this figure was used to scale the circles. So large circles indicate that a high proportion of the range was affected by a data source (numbers indicate the number of affected maps). This shows, for example, the large difference in the unique contribution of BirdTrack relative to Atlas records: of the 627 maps affected by Atlas records, on average 48% of the range was contributed solely by Atlas records, whereas for the 501 maps affected by BirdTrack data, on average only 3.5% of the range was contributed. These figures are averages and a data source could still make a large impact for some individual species. BirdTrack, for example, made the greatest contribution to winter ranges of rarities (e.g. 64% of the American Herring Gull range), scarcities (e.g. 29% of the Rough-legged Buzzard range) and late-departing summer visitors (e.g. 33% of the House Martin range).

Contrast the contribution of BirdTrack with that from the UK and Irish Rare Breeding Birds Panels. Despite contributing only c.1% of the volume of data, panel data contributed 6.7% of the breeding ranges of 119 species including the Spotted Crake (37% of range), Pintail (33%), Whooper Swan (28%), Bittern (23%) and Goshawk (21%). Data from RSPB reserves and surveys covered many species and on average yielded only 0.3% of ranges. However, there were some species where RSPB data were crucial, including the Capercaillie (43% of range), Stone-curlew (35%), Corncrake (33%), Bittern (28%), Black Grouse (22%) and White-tailed Eagle (22%). The importance of RSPB data and panel data may be underestimated to a degree because, for example, many of the

FIGURE 5.4 DISTRIBUTION OF 10-KM SQUARES WITH 'NILNOCT' VISITS

The map shows the 10-km squares with breeding-season nocturnal visits where the observer found no nocturnal species and submitted a 'Nilnoct' return.

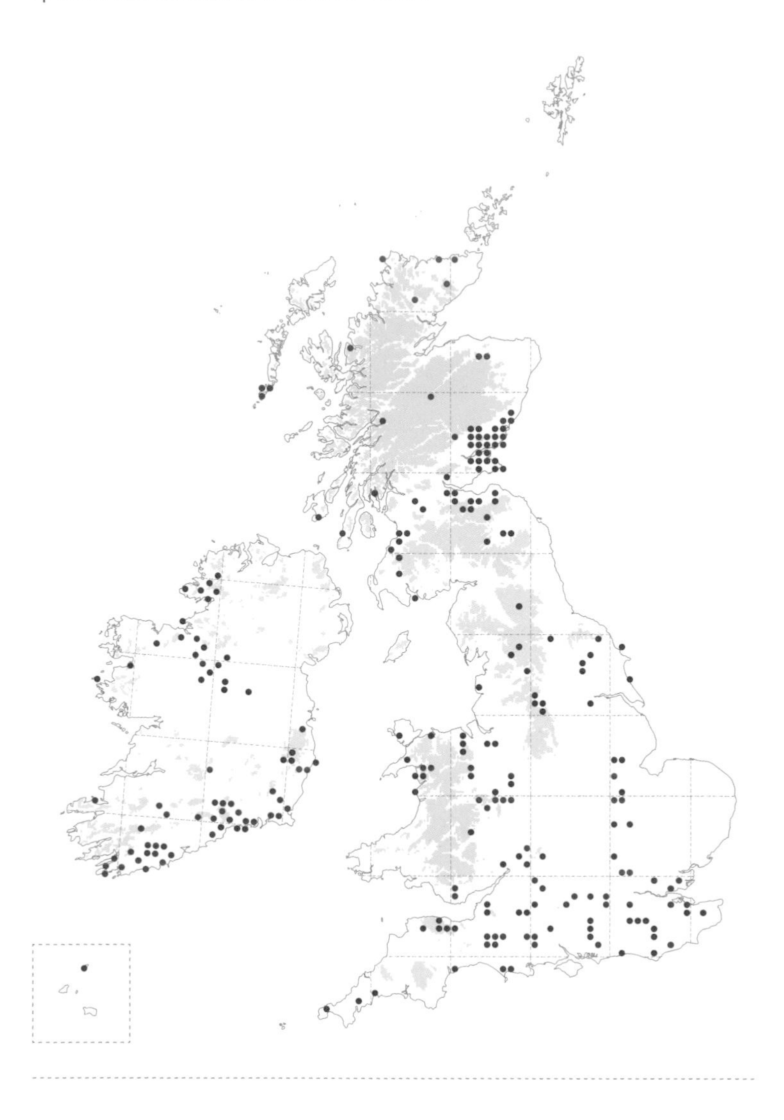

Bittern records in the RBBP archives will have originated from RSPB, and since both sources have the same data, neither will appear to provide these as a unique contribution in this analysis.

It was expected that Nest Record Scheme data and Ringing Scheme data would be especially useful in upgrading breeding evidence, but on average, they accounted for only 0.2% and 0.4% of breeding ranges respectively. Top species were the Pied Flycatcher (1.7% of range from NRS, 3.4% from Ringing) and the Barn Owl (4.6% of range from Ringing). As with RSPB and RBBP data, there is likely redundancy in Ringing and NRS data because a brood monitored for a Nest Record card may subsequently have been ringed. However, when the data were re-analysed after combining Ringing and NRS as a single 'Demographic monitoring' source, the conclusions were the same.

Finally, although this atlas incorporated a large volume of records from BirdTrack, bird clubs and online rarity services that were not available to previous atlases, their unique contribution to range sizes, especially in the breeding season, was actually quite small for the majority of species. Where they do have an impact is in providing records for scarce wintering species and for vagrant species that have not featured in previous atlases. A consequence of this is that the species pool for this atlas is somewhat larger than in previous atlases (see Section 6.3). Breeding-season data from other sources would be more valuable if breeding evidence were to be routinely collected.

5.3.5 Impact of the chequerboard on species numbers in Ireland

Efforts were made to survey every 10-km square in Ireland thoroughly for distribution-mapping purposes. However, with the use of the chequerboard to stratify the collection of TTVs for the systematic collection of abundance information (see Section 2.2.4), there was the potential for coverage, and therefore species detection, to differ between adjacent 10-km squares as some would have only Roving Records and others would have both Roving Records and many hours of TTV effort. Also, given the high level of reporting of breeding evidence on TTVs, there was the added possibility that breeding status could be underestimated in Low-priority 10-km squares.

In fact in the breeding season there was a statistically significant difference in species richness between Irish High-priority (i.e. targeted) and Low-priority 10-km squares ($P = 0.0002$) but the difference was very small: the median species richness in High-priority 10-km squares was 60 species (quartiles = 55.5, 66) compared to 59 species (quartiles = 53, 65) in Low-priority squares. There was no difference in the proportion of breeding species with *probable* or higher breeding evidence ($P = 0.9$). In winter there was also a statistically significant difference in species richness ($P < 0.0001$) but again the actual difference was small: the median species richness in High-priority 10-km squares was 62 species (quartiles = 53, 75) compared to 59 species (quartiles = 50, 73) in Low-priority squares.

On balance there is little evidence that the use of the chequerboard made any appreciable impact on the completeness of distribution information across Ireland.

5.3.6 Recording of nocturnal species

During the breeding season there was an aim to make at least one twilight or nocturnal visit to every 10-km square. To help identify squares that had been visited but where no nocturnal species were found, observers were asked to submit a Roving Record with the word 'Nilnoct' in place of a species name. In total, 636 Nilnoct records were submitted for the breeding season, covering 245 10-km squares in Britain & Ireland (Figure 5.4). These amount to just 6% of 10-km squares. Either the majority of twilight/nocturnal visits were successful in finding at least one nocturnal species or, as seems more likely, very few specific twilight/nocturnal visits were actually undertaken. As a result, Figure 5.4 has limited value for interpreting apparent absences on breeding-season distribution maps of nocturnal species. Note that a further 556 Nilnoct records were received for winter and had a similarly limited value for winter distribution maps.

FIGURE 5.5 **DISTRIBUTION OF TETRADS SURVEYED WITH TIMED VISITS**

Every tetrad that received at least one TTV in winter or the breeding season is shown. Many of the areas of solid colour are indicative of local tetrad atlases. The chequerboard can be seen from the regular pattern of coverage in Ireland.

5.4 SYSTEMATIC COVERAGE OF TETRADS BY TTVS

5.4.1 Numbers of surveyed tetrads and completed TTVs

Coverage of tetrads by the TTV method for the systematic component of the Atlas exceeded all expectations. Between one and four TTVs were made to 50,089 tetrads and many had the 'full set' of a pair of winter TTVs and a pair of breeding-season TTVs. In total, 182,228 TTVs were completed, amounting to a staggering 287,149 hours of dedicated timed-count field effort (Table 5.3A). In Britain, more TTVs were completed in the first field seasons (i.e. winter 2007/08 and breeding season 2008) than in any subsequent seasons, whereas in Ireland the peak was reached in the third field seasons (Table 5.3A).

Figure 5.5 shows the distribution of tetrads with at least one TTV in winter or the breeding season. Several areas of southern, central and northern England, plus southeast Scotland, North Wales and Co. Waterford had complete or near complete tetrad coverage in the breeding season (and many in winter also) often because of local atlases that had reached an advanced stage by the end of *Bird Atlas 2007–11*.

Each surveyed tetrad should ideally have had two winter TTVs (one early season, one late) and two breeding-season TTVs (one early, one late). Any or all of these TTVs could have been of 1-hour or 2-hour duration. Of the 182,228 completed TTVs, 57% were of 2-hour duration. The proportion of TTVs with 2-hour visits was far higher in Britain than in Ireland (Table 5.3b), partly because many areas of Britain undertook local atlases that were more likely to use 2-hour TTVs, and also because many TTVs in Ireland were completed by contract fieldworkers who had to make 1-hour visits in order to get the required number of tetrads surveyed. However, in both islands and in both seasons, more than 90% of tetrads received both early and late visits (Table 5.3B). In remote areas where it was impractical to make two TTVs the rules permitted a single visit of 2-hours. Up to 4% of tetrads were surveyed in this way, but a further 1–5% of tetrads received a single visit of just 1-hour duration (Table 5.3B). Regardless of the number and duration of visits, all TTVs were used in the production of abundance maps.

TABLE 5.3 SUMMARIES OF NUMBERS OF TIMED TETRAD VISITS AND TETRADS SURVEYED

The tables summarise the number of completed Timed Tetrad Visits and the number of tetrads the TTVs covered. Any particular tetrad could have received up to four TTVs: in early winter, late winter, early breeding season and late breeding season.

A The number of Timed Tetrad Visits

	Britain		Ireland	
	Winter	Breeding	Winter	Breeding
Total number of completed TTVs	81,334	79,591	10,778	10,525
% of total achieved per field season (1st/2nd/3rd/4th field season)	36/24/22/18	33/25/25/17	21/26/30/23	24/29/31/16
% of TTVs that were of 2-hour duration	62%	65%	14%	13%

B The number of tetrads with different types of TTV coverage

	Britain		Ireland	
	Winter	Breeding	Winter	Breeding
Total number of tetrads with at least one TTV	41,776	40,994	5,622	5,396
Number covered by TTVs in both seasons	39,180		4,519	
Number with an early and a late visit	39,558 (95%)	38,599 (94%)	5,156 (92%)	5,129 (95%)
Number where both visits were of 2-hour duration	22,154 (53%)	23,337 (57%)	549 (10%)	550 (10%)
Number with only one visit of 2-hour duration	1780 (4%)	1778 (4%)	186 (3%)	88 (2%)
Number with only one visit of 1-hour duration	438 (1%)	617 (2%)	280 (5%)	179 (3%)

5.4.2 Were sufficient tetrads surveyed in all regions?

Though the number of TTVs completed is impressive, the true test of success is whether the number of tetrads surveyed by the TTV method in each 10-km square met the Rule-8 coverage targets that were designed to ensure representative coverage throughout Britain & Ireland. To recap, the Rule-8 target stated that a minimum of eight tetrads should be surveyed in every 10-km square, or all of those available if there were fewer than eight. The target applied to all 10-km squares in Britain and alternate 10-km squares in Ireland, giving spatially stratified targets totalling 20,721 tetrads in Britain and 3,723 in Ireland (see Section 2.2.4 and Table 2.1). The figure of 50,089 surveyed tetrads does not guarantee that at least eight tetrads were surveyed in 10-km squares and it was necessary to check TTV coverage for each 10-km square.

For example, IO16 was an Irish High-priority 10-km square and had 23 tetrads on land, giving it a Rule-8 target of eight tetrads. In winter, 12 tetrads were surveyed, so IO16 can be said to have met its Rule-8 target. To test how closely coverage met the target of 3,723 tetrads in Ireland, only eight tetrads from IO16 can be counted towards the total (the other four are 'spare' for this accounting purpose). After repeating this calculation for every High-priority 10-km square in Ireland, it was found that 474 (97.3%) of the 487 squares met their Rule-8 targets in winter (Table 5.4). After discounting tetrads over and above the target of each 10-km square, the overall total of 5,622 surveyed tetrads was adjusted to 3,709 tetrads. This represents 99.6% coverage of the 3,723 target.

Figures for each island's constituent countries and regions are given in Table 5.4. Overall, 99.9% of the combined Britain & Ireland target was achieved in winter and the breeding season, and in each country/region at least 97% of 10-km squares met their targets and at least 98% of the country/region's target number of tetrads were surveyed (Table 5.4). The shortfall totalled just 34 tetrads in winter and 32 tetrads in the breeding season. Apart from two 10-km squares in Co. Cork (IR61 and IW58), no 10-km square missed its Rule-8 target by more than two tetrads and several of the 10-km squares involved contained only small remote islands. As is clear from Table 5.4 and Figure 5.5, the actual number of tetrads surveyed far exceeded the Rule-8 targets in many areas. As discussed in Section 4.1.4, the Rule-8 target is itself slightly biased in favour of coastal areas and for analytical purposes we used an adjusted target, Rule-8_{adj}, in which the number of tetrads required per 10-km square is proportional to land area, up to a maximum of eight tetrads per 10-km square. In Britain and in Ireland over 99% of 10-km squares met their Rule-8_{adj} target in winter and the breeding season. In fact in 90% of British 10-km squares and 40% of Irish squares, winter TTV coverage actually exceeded the Rule-8_{adj} target; corresponding figures for the breeding season were 90% and 38% respectively. This meant that the overall sample of TTVs was geographically biased, especially in Britain. One of the key aims of the complex procedure designed for modelling abundance was to remove this bias whilst making use of the whole dataset (see Section 4.4.3). The procedure involved randomly sampling from among surveyed tetrads within each 10-km square to remove any gross geographical bias before models were made. However, there was still the potential for bias if the actual tetrads that observers chose to survey within each 10-km square failed to accurately reflect the habitats present in the 10-km square. This question is considered in Box 5.2. It shows some degree of bias in the coverage of coastal and upland habitats, although in both cases the habitats are rare and in absolute terms the biases are small. Nevertheless, this fact should be borne in mind when interpreting figures such as the percentage of tetrads occupied by a species. The bias should not have a major impact on abundance maps.

ALYN WALSH

In coastal areas, tetrads were only eligible for TTV coverage if their centre was above the low-water mark. For this reason, the sandy spit of Raven Point, Co. Wexford, was not eligible for TTV coverage.

TABLE 5.4 HOW FAR WERE RULE-8 TARGETS FOR TTV COVERAGE MET?

The table shows the numbers of tetrads required per country/region to meet the Rule-8 target of eight tetrads per 10-km square (copied from Table 2.1, see Section 2.2.1 for details). Winter and the breeding season then have three columns each. The first gives the actual number of tetrads surveyed. The second gives the number of 10-km squares in which the Rule-8 target was met. The third gives the number of tetrads after discounting any above each 10-km square's Rule-8 target. Note that for the Irish Low-priority rows (in italics), some columns are blank because Low-priority 10-km squares were not part of the overall target.

		Winter			Breeding		
Country/region	**Rule-8 target number of tetrads (in number of 10-km squares)**	**Number of tetrads surveyed**	**Number of 10-km squares with tetrad coverage meeting the Rule-8 target (% of 10-km squares)**	**Number of tetrads surveyed after limiting to n = Rule-8 per 10-km square (% of target)**	**Number of tetrads surveyed**	**Number of 10-km squares with tetrad coverage meeting the Rule-8 target (% of 10-km squares)**	**Number of tetrads surveyed after limiting to n = Rule-8 per 10-km square (% of target)**
England	11,108 (1,430 10-kms)	27,261	1,427 (99.7%)	11,105 (100%)	26,768	1,424 (99.6%)	11,101 (99.9%)
Scotland	7,582 (1,034 10-kms)	11,200	1,022 (98.8%)	7567 (99.8%)	10,918	1,030 (99.6%)	7,577 (99.9%)
Wales	1,904 (253 10-kms)	3,142	252 (99.6%)	1,903 (99.9%)	3,133	252 (99.6%)	1,903 (99.9%)
Channel Islands	50 (11 10-kms)	91	10 (90.9%)	49 (98.0%)	94	11 (100%)	50 (100%)
Isle of Man	77 (13 10-kms)	82	13 (100%)	77 (100%)	80	13 (100%)	77 (100%)
Britain total	20,721 (2,741 10-kms)	41,776	2,724 (99.4%)	20,701 (99.9%)	40,993	2,730 (99.6%)	20,708 (99.9%)
Republic of Ireland - chequerboard High-priority	3,112 (408 10-kms)	3,227	395 (96.8%)	3,098 (99.6%)	3,286	400 (98.0%)	3,094 (99.4%)
Republic of Ireland - *chequerboard Low-priority*	-	1,281	-	-	1,046	-	-
Northern Ireland - chequerboard High-priority	611 (79 10-kms)	710	79 (100%)	611 (100%)	685	78 (98.7%)	610 (99.8%)
Northern Ireland - *chequerboard Low-priority*	-	404	-	-	379	-	-
Ireland total – chequerboard High-priority	3,723 (487 10-kms)	3,937	474 (97.3%)	3,709 (99.6%)	3,971	478 (98.2%)	3,704 (99.5%)

BOX 5.2 HOW REPRESENTATIVE OF HABITAT AVAILABILITY IS THE SAMPLE OF TTV-SURVEYED TETRADS?

There are two ways by which bias could arise in the sample of tetrads with TTVs:

1. The sampling protocol itself could cause bias. It has already been mentioned (see Section 4.1.4) that the Rule-8 target could lead to an over-representation of coastal habitats and for this reason an adjusted target (Rule-8_{adj}) based on land area is used in all analyses. Furthermore, it is possible, though unlikely, that the chequerboard could lead to bias if the alternating 10-km squares were systematically deficient in certain habitats.

2. Observers could choose which tetrad(s) to survey in each 10-km square, but with the proviso that selected tetrads should represent the local habitats. This may not always have been easy to achieve, for example owing to access constraints, and if an appreciable number of observers preferentially chose tetrads for their habitat composition, the sample of surveyed tetrads could over- and under-represent certain habitats.

We tested by how much the habitat composition of surveyed tetrads deviated from what would be expected by chance. First, for every tetrad in Britain and every High-priority tetrad on the Irish chequerboard, the availability of 18 habitat types was quantified using the CORINE land cover map (see Section 4.4.3, subsection 'Variables to explain bird relative abundance'). The availability of each habitat type was summarised across all these tetrads in Britain and Ireland, showing that, for example, 32% of Britain and 14% of Ireland was agriculturally cropped land. Ideally, the sample of surveyed tetrads would match these values within some reasonable margin of error owing to sampling only a fraction of the available squares. To determine what would be a reasonable margin of error, we simulated random sampling of available tetrads, taking n = Rule-8_{adj} tetrads in every 10-km square and calculating habitat availability figures. This was repeated 1,000 times and the 25th and 975th ranked values of the availability of each habitat across the random samples were found and expressed as a percentage of the actual habitat availability values. This showed, for example, that if observers picked tetrads at random, the amount of crops within the set of surveyed tetrads in Britain would be in the range of 99–101% of the real value. This process was repeated, but using the tetrads actually chosen by observers and this showed that crop availability in the sample of surveyed tetrads was also in the range 99–100% indicating that the choice of tetrads did not give a biased sample as far as crops are concerned.

The graphs below show the expected and observed ranges of habitat coverage for Britain and Ireland. The real habitat availability, in terms of % cover of the land, is shown in brackets next to the habitat name. Pale green bars show the range of values expected if tetrads were selected randomly. For most habitats the bars span the 100% line, indicating that random sampling would not have yielded a biased sample, although for rare habitats (e.g. saltmarsh in Ireland) the bars are wide indicating that we could expect quite wide deviations in percentage cover for rare habitats even if tetrads were selected at random.

The dark green bars show the corresponding range of habitat cover values from the set of tetrads actually surveyed. For most habitats, coverage is within the range of what could be expected from random sampling but there are some clear cases where the surveyed tetrads are biased. In Britain, coastal habitats are over-represented whereas some upland habitats are under-represented. In Ireland, deciduous and mixed woodland were over-sampled and natural grasslands and peat bogs were under-sampled. It is worth noting that the decision to use the Rule-8_{adj} target rather than Rule-8 significantly reduced the bias in coverage of coastal habitats.

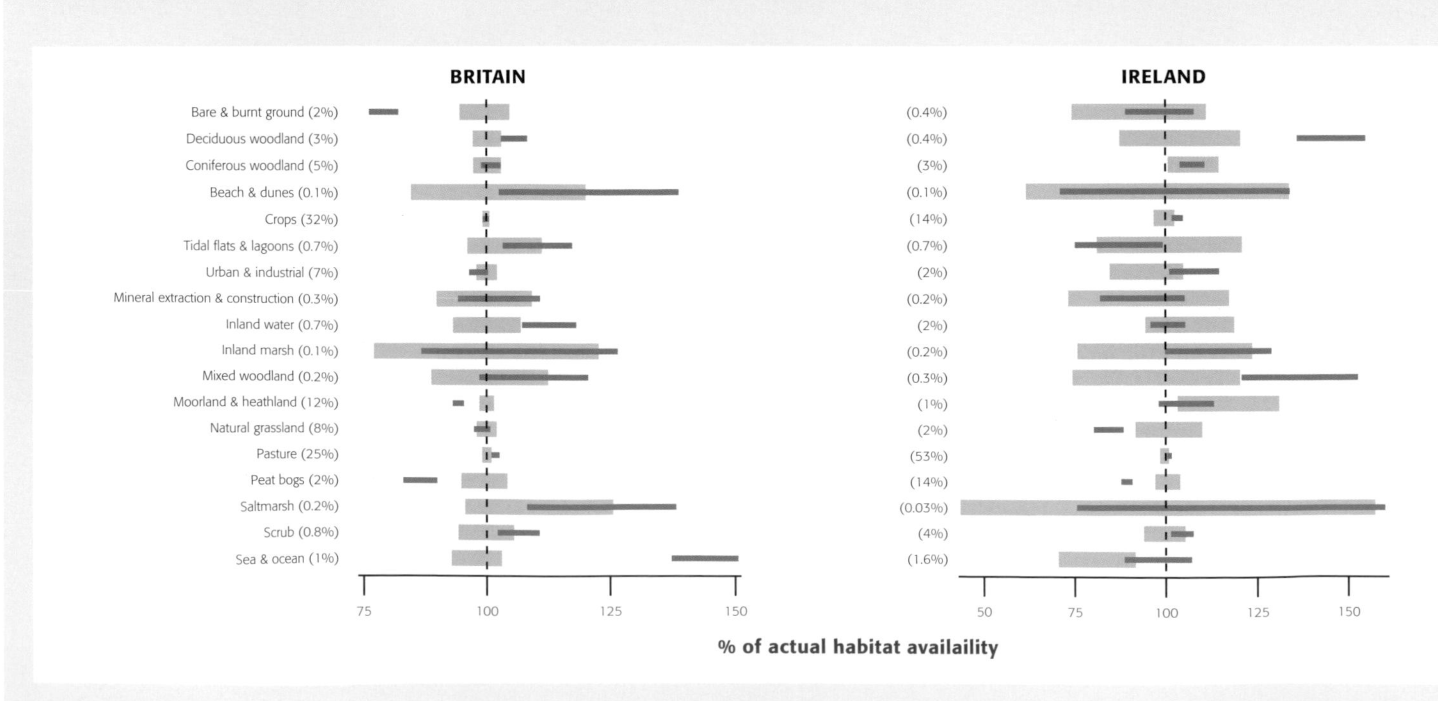

TABLE 5.5 SUMMARY OF THE COLONY INFORMATION SUBMITTED ON TIMED TETRAD VISITS

To illustrate the degree to which the colony data represent the species' ranges, columns two and three give the percentage of 10-km squares with breeding evidence (i.e. the breeding range) in which colony information was also submitted in each island. The Ireland figure was calculated only for chequerboard High-priority 10-km squares. Note that both individual counts and apparently occupied nest (AON) counts were provided for some colonies. n/a entries indicate count methods that were inapplicable for certain species.

Species	% of range with colony returns		Numbers of tetrads with different levels of information				Count totals summed across tetrads	
	Britain	Ireland	Any colony information	Only presence noted	Individuals counted	AONs counted	Sum of Individuals	Sum of AONs
Fulmar	56	44	1,042	382	n/a	660	n/a	48,575
Manx Shearwater	21	40	15	15	n/a	n/a	n/a	n/a
Storm Petrel	19	13	15	15	n/a	n/a	n/a	n/a
Gannet	22	0	10	8	n/a	2	n/a	9,900
Cormorant	14	7	135	53	n/a	82	n/a	2,764
Shag	32	23	328	182	n/a	146	n/a	2,668
Little Egret	10	15	69	18	n/a	51	n/a	406
Grey Heron	18	9	597	125	n/a	472	n/a	3,904
Arctic Skua	25	-	67	24	33	16	165	61
Great Skua	30	0	135	55	66	24	1,285	1,763
Kittiwake	41	20	206	81	55	98	11,945	22,895
Black-headed Gull	28	16	517	110	288	239	63,507	35,144
Mediterranean Gull	17	17	21	6	9	14	74	110
Common Gull	33	15	599	121	349	266	13,304	5,520
Lesser Black-backed Gull	22	13	462	152	183	211	26,565	25,765
Herring Gull	35	17	1,044	371	333	492	29,157	20,737
Great Black-backed Gull	23	15	344	121	114	150	1,806	1,599
Little Tern	26	7	40	10	17	18	711	248
Sandwich Tern	13	13	15	6	6	6	11,621	6,279
Common Tern	20	15	185	41	93	100	4,479	3,358
Roseate Tern	8	33	2	1	1	1	180	90
Arctic Tern	29	20	166	39	86	66	6,383	6,042
Guillemot	41	40	202	107	95	n/a	118,802	n/a
Razorbill	41	34	198	101	97	n/a	7,768	n/a
Puffin	32	30	91	47	38	15	40,039	18,720
Rook	86	75	11,985	3,559	n/a	8,426	n/a	245,190
Sand Martin	51	26	2,191	617	n/a	1,574	n/a	40,615

5.4.3 Counts of colonial species

Observers encountering colonies of herons, seabirds, Rooks or Sand Martins during TTVs were asked to record at least the presence of the colony, and ideally submit counts of individuals present or the number of apparently occupied nests (AONs). In total, 20,681 records of colonial species were submitted from 11,292 tetrads in Britain and 1,727 in Ireland. For most species at least 60% of colony returns included count information (Table 5.5), but it is unclear whether the counts are representative of colonies more broadly. For example, observers may have tended to provide counts for small and easily visible/countable colonies but not large ones that were difficult to view (e.g. seabird colonies on cliffs).

For a widespread colonial species such as the Rook, the colony information appeared to cover the range well because colony information was provided on TTVs in 86% of British and 75% of Irish 10-km squares with breeding Rooks (Table 5.5). Figures were considerably lower for many of the seabirds, suggesting that TTVs were an ineffective means of getting representative sample data for colonies. For seabirds this is unsurprising given the remote nature of many colonies and the requirement for a tetrad's centre to be on land for it to be eligible for TTV coverage. As a further indication of levels of coverage, the total counts for Fulmars, Gannets, Cormorants and Shags corresponded to c.9%, 4%, 20% and 8% of the most recent Britain & Ireland population estimates (*Seabird 2000*). For these reasons, use of the colony data in the Atlas was confined to providing breeding evidence for distribution mapping purposes.

5.4.4 Tetrad Population Estimates

Tetrad Population Estimates (TPEs) were an optional aspect of the TTV method, aimed at generating figures that could be used to help develop population estimates in the future. They relied upon observers being able to scale up the number of a species encountered on a TTV to give an estimate of the number that could have been present in the whole tetrad. A similar method had been used with moderate success in Catalonia (Estrada *et al.* 2004).

This was an optional and experimental aspect of the TTV method which required good knowledge of species, habitats and the tetrad being surveyed (see Section 2.2.4) and it was unclear how many observers would have the necessary knowledge and experience to complete them. Superficially, Britain and Ireland had identical levels of TPE uptake, with 27% of breeding-season tetrads and 21% of winter tetrads having at least one species with a TPE. However, the number of species given a TPE varied widely. For breeding-season TTVs with at least one TPE, a far higher percentage of species on each TTV was given a TPE in Britain (median = 91%, quartiles = 65, 100, n = 11,101 tetrads) compared with Ireland (median = 13%, quartiles = 9, 89, n = 1,502). The same pattern was apparent in winter (Britain: median = 98%, quartiles = 78, 100, n = 9,013; Ireland: median = 18%, quartiles = 13, 88, n = 1,235). The large difference can be attributed to the smaller pool of observers in Ireland (both volunteers and contracted professionals) who had to survey many unfamiliar tetrads in which they may not have felt confident in providing TPEs for all but the most straightforward of species.

The multiplication factor between the maximum number of birds seen by the observer during the TTV and the TPE they provided is a useful way to gauge how figures were scaled up. For the ten species most commonly given a TPE, counts during the TTV were scaled up by a factor of between ×1 and ×6 (Table 5.6). These factors seem on the low side considering that in most cases a significant proportion of the tetrad will not have been visited, and many birds will have gone undetected in the visited parts.

To properly test the TPEs requires independent robust density estimates but these are very difficult to produce. The annual breeding-bird monitoring schemes in the UK and Republic of Ireland (BBS and CBS) use a transect method in which birds are recorded within distance bands and then distance-sampling theory is used to estimate how many individuals may have gone undetected at different distances from the transect. Estimates derived in this way have their own weaknesses, but they are generally thought to be a useful way of deriving population estimates (Newson *et al.* 2008a). In Table 5.6, for comparative purposes, density estimates from the BBS have been converted to numbers of birds per tetrad and shown with the TPEs for the top ten species. If the BBS figures are close to reality, they indicate that TTV counts were not scaled up sufficiently to produce realistic TPE figures.

We acknowledge that TPEs proved highly controversial, with some observers objecting to their use on the grounds that they were non-scientific and subjective. The preliminary analyses above support those objections. Nevertheless, the TPE data merit further analysis, for example to determine whether the figures for scarce species, habitat-specialists or highly detectable species produce sensible estimates of population size. However, for the purposes of the Atlas, no further use is made of TPE data.

TABLE 5.6 TETRAD POPULATION ESTIMATES FOR 10 COMMON AND WIDESPREAD SPECIES

For the ten species most often given a breeding-season Tetrad Population Estimate, the table gives, for each season, the average Tetrad Population Estimate (number of individuals) and the lower and upper quartiles of multiplication factors between the observer's maximum TTV count and their TPE (based on data from Britain & Ireland). With the exception of the Swallow in winter, all these species were given TPEs in at least 7,000 tetrads. The Distance-corrected density column gives an independent estimate of breeding-season densities based on BBS data from 2006 that were analysed to account for some aspects of under-detection (S. Newson pers comm).

Species	Mean TPE (birds per tetrad)	Count to TPE multiplier	Mean TPE (birds per tetrad)	Count to TPE multiplier	Distance-corrected breeding-season density (birds per tetrad)
Wren	29	2.3, 6.0	42	2.0, 5.0	152
Blackbird	58	1.7, 3.8	50	1.8, 4.0	225
Robin	43	1.9, 4.4	40	2.0, 5.0	169
Chaffinch	56	1.5, 3.3	47	1.7, 3.8	255
Woodpigeon	182	1.2, 2.6	73	1.4, 3.0	152
Blue Tit	50	1.8, 4.0	38	2.0, 5.0	263
Great Tit	35	1.9, 4.3	29	2.0, 5.0	132
Swallow	-	-	23	1.3, 2.5	36
Dunnock	25	2.1, 6.0	24	2.2, 5.5	91
Carrion Crow	31	1.3, 2.3	23	1.3, 2.5	47

5.5 COVERAGE FOR CHANGE MAPPING

WHEN PRODUCING MAPS and statistics of distribution change, it is important to know that areas showing apparent change have had similar or at least adequate levels of coverage in the different time periods being considered. At the very least, change should be calculated only for 10-km squares that were surveyed in each time period. Figure 5.6 shows that 3,834 10-km squares were surveyed in all three breeding atlases and it is these squares that are used for the majority of breeding-distribution change maps in this book. For a small number of species, changes are calculated for the period since the *1988–91 Breeding Atlas* and for them the sample size is 3,837 10-km squares. In winter, 3,790 10-km squares were surveyed in both periods (Figure 5.6).

The 13 squares surveyed in previous breeding atlases but not during 2008–11 were either very small stretches of coastline or remote islands. Together they totalled less than c.3 km² of land. Similarly, the 11 10-km squares surveyed in the *1981–84 Winter Atlas* but not in 2007–11 were mostly remote islands and totalled c.12 km² of land. The large number of 'new' squares in winter in 2007–11 were also mostly marine or coastal: 35 of them totalled just 31 km² of land and only one square (NC93) was a complete 10-km square.

5.6 DID RECORDING EFFORT VARY ACROSS BRITAIN & IRELAND AND AMONG ATLASES?

DURING THE FIELDWORK phase of the Atlas, the percentage of the 'expected' species total reported for a 10-km square was widely used as a measure of coverage (Section 5.3.2). The number of expected species was based on the previous atlas data for that particular square which has one key drawback: it assumes that coverage in the earlier atlas was perfect when, in reality, some squares could have been poorly covered, leading to artificially low targets.

Comparing observed species lists with expected ones is the basis of a novel method (Hill 2012) for assessing whether both distribution and change patterns are biased because of underlying variations in effort. Hill proposed that for any 10-km square it is possible to use the species lists of 'nearby similar squares' to identify a set of 'benchmark species' that ought to be always present in the 10-km square. If in one atlas period, all of a square's benchmark species are detected but in a later period some of them are not, this is taken as an indication of low recording effort of the square in the second period.

It follows from this that the proportions of benchmark species that are actually reported in the 10-km square in each atlas can then be taken as estimates of recording effort in that square in each atlas. Here 'recording effort' is a somewhat intangible quantity. It does not have units such as the number of observers or hours spent in the field per square. Rather it is an overall assessment of how well the local species pool was detected, and so should incorporate the major factors like number of observers and time spent in the field, as well as the more subtle factors such as observer skill and land access.

FIGURE 5.6 NUMBERS OF 10-KM SQUARES SURVEYED DURING DIFFERENT ATLASES

For either season, only squares covered in all periods under consideration can be used for change mapping.

A Breeding season

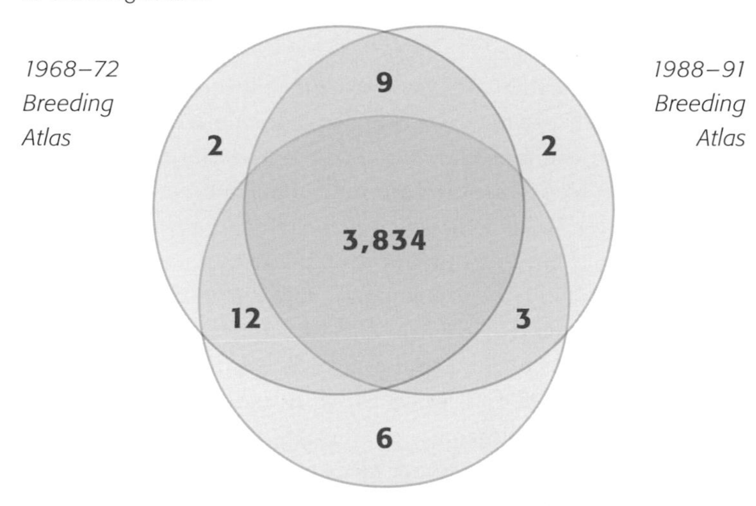

B Winter

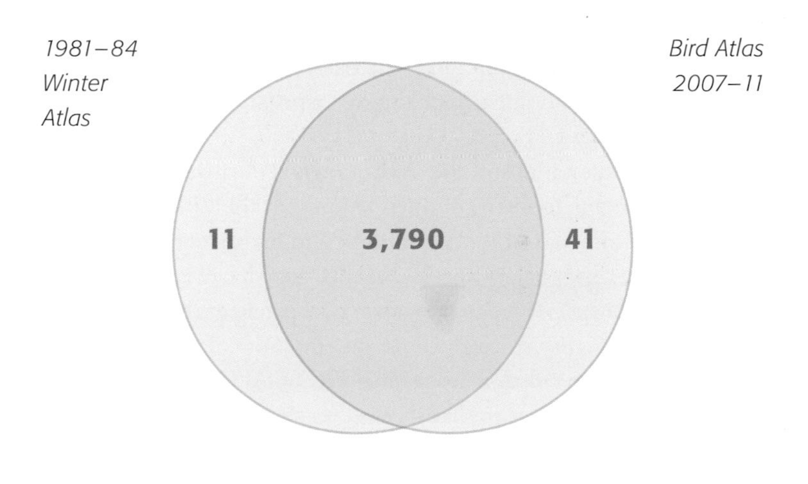

Hill's method is implemented in FRESCALO ('FREquency SCAling LOcal'; Hill 2012), a free programme available from the Biological Records Centre (www.brc.ac.uk/resources.htm). We used FRESCALO in two ways:

- When applied to a single atlas it can be used to produce a map of relative recording effort to show parts of Britain & Ireland where effort may be poor (Section 5.6.1). This goes some way towards meeting the call of Rocchini *et al.* (2011) for a better appreciation of uncertainty by the production of 'maps of ignorance' to show where knowledge may be poor.
- Secondly, FRESCALO produces 'time factors' that describe each species' range size in each atlas period after correcting for variation in effort in each period. For individual species, changes in time factors were compared with changes in range size as measured by the atlases to test the degree to which atlas range-change statistics may be biased (Section 5.6.2).

5.6.1 Estimating patterns of recording effort in different atlases

The first stage of FRESCALO was to define the 'nearby similar squares' for each 10-km square using environmental attributes. This analysis used the altitude and land-cover variables already described and used in the production of abundance models (see Section 4.4.3). For land cover the presence or absence of each of the 18 land-cover types in each 10-km square was recorded. For altitude we first assigned an altitude class (up to 50 m; 51–100 m; 101–200 m; 201–500 m; 501–1,000 m; or >1,000 m) to each tetrad according to its mean altitude and then summarised these to list which altitude classes were present in each 10-km square.

Next, four species distribution files were created: two breeding-season files, one for Britain and one for Ireland, containing the distribution of each species in the three breeding atlases, and likewise two files for winter. Separate files were made for Britain and Ireland so that neighbourhoods (see below) could not contain squares from both islands. Exotic non-native species (see Section 6.5.2) were excluded because they suffer from inconsistent reporting through atlases. Likewise, species with changing taxonomic status were excluded (see Box 4.1).

The distribution files and the environmental data were analysed according to the recommendations of Hill (2012), using neighbourhoods of the 100 nearest and most similar 10-km squares; benchmark species for a 10-km square are the top 27% of species ordered by their occupancy of the neighbourhood squares, weighted by the similarity of each square. As recommended, a provisional FRESCALO analysis was used to identify species that showed very pronounced range changes that could be indicative of significant and real range changes, as opposed to changes resulting from effort variation. Examples were the Buzzard, Avocet, Turtle Dove and Willow Tit. These and others had to be prevented from being selected as benchmark species because true absences from a square in one period could be mistaken by the programme for under-recording (Hill 2012). In the final FRESCALO analysis we followed Hill's recommendation to exclude species from becoming benchmark species if their time factors from the provisional analysis changed by a factor of two or more (i.e. range doubled or halved between periods). In practice, most of the species excluded were too scarce to be selected as benchmark species so this made little impact to the final results.

At this stage of the analysis we had derived a winter benchmark species list and a breeding-season benchmark species list for every 10-km square in Britain and Ireland. The next stage was to compare the list of species observed in each 10-km square in the *1981–84 Winter Atlas* with each square's winter benchmark species list and calculate the percentage of benchmark species that had been detected. This was repeated for the species observed in each square in winter during *Bird*

FIGURE 5.7 ESTIMATED PATTERNS OF UNDER-RECORDING IN DIFFERENT ATLASES

Some Purple Sandpiper gains in northwest Scotland could be artefacts.

Darker shading indicates 10-km squares where a higher proportion of 'benchmark species' were not recorded (as estimated by the FRESCALO analysis) which is a measure of possible under-recording. Areas where no shading is apparent are classed as having high coverage, i.e. at least 90% of benchmark species were recorded. Figures annotating the map give the percentage of 10-km squares in Britain and Ireland with high coverage.

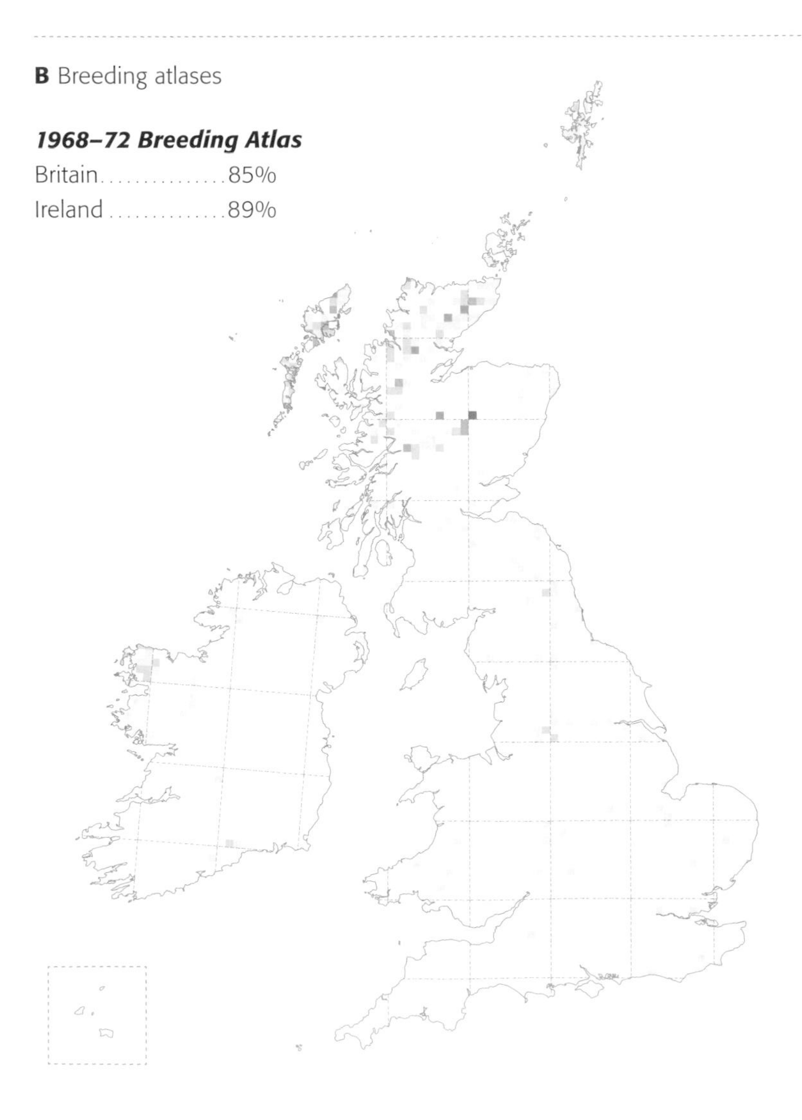

A Winter atlases

1981–84 Winter Atlas

Britain 63%

Ireland 63%

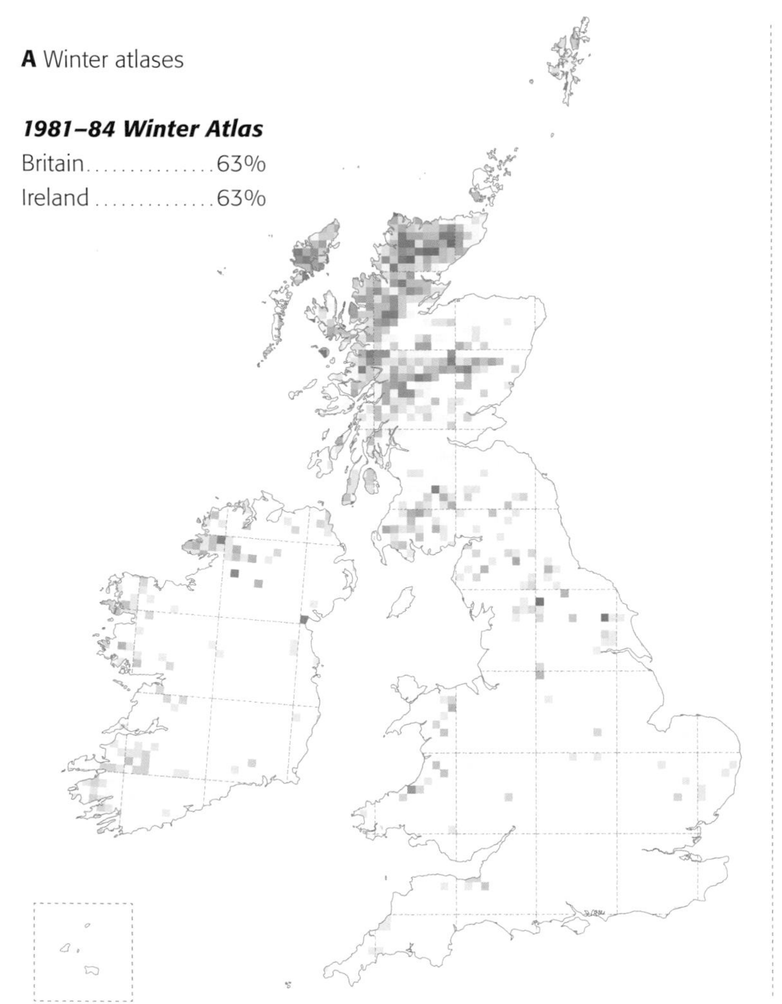

Bird Atlas 2007–11

Britain 86%

Ireland 89%

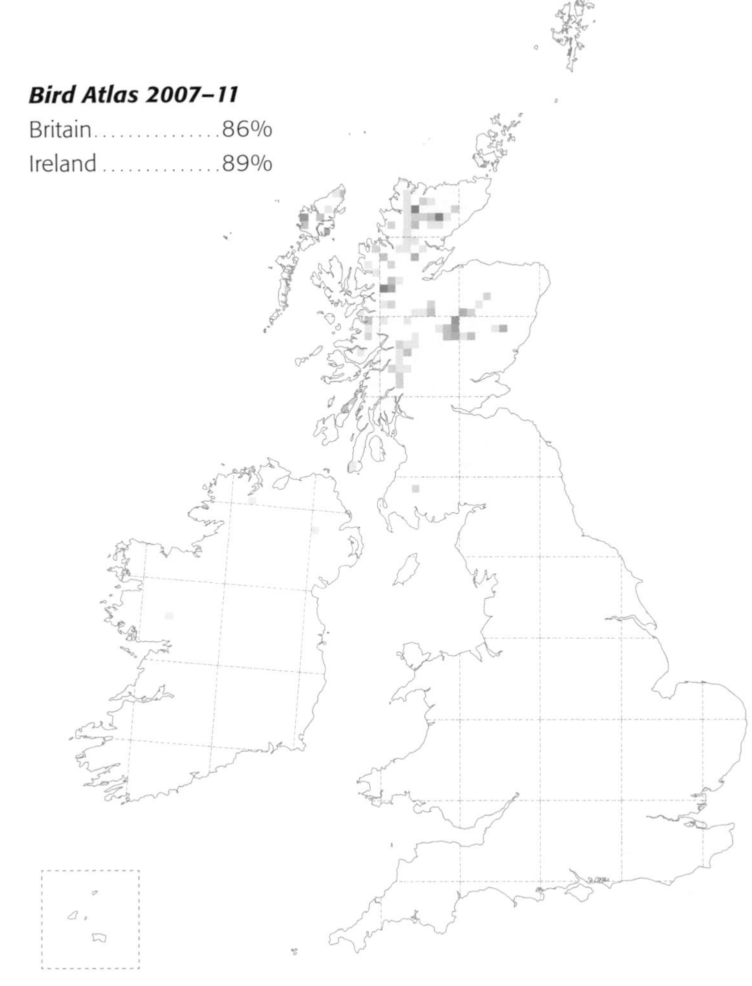

1988–91 Breeding Atlas

Britain 81%

Ireland 70%

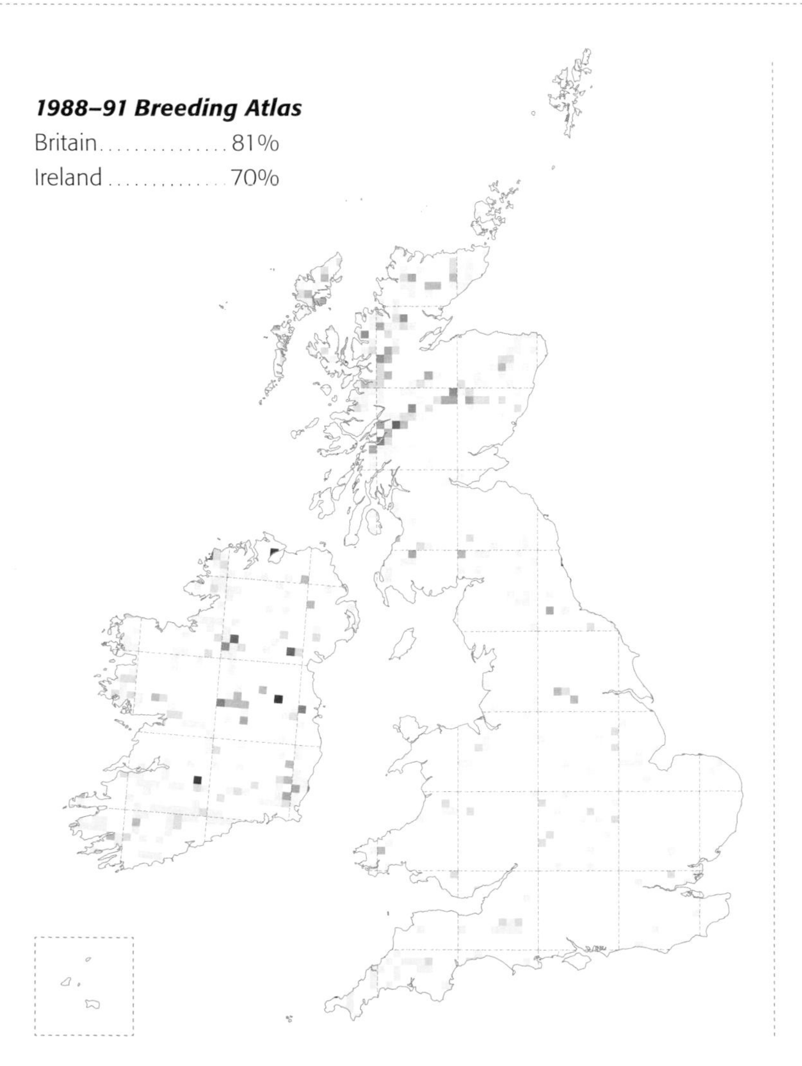

Bird Atlas 2007–11

Britain 85%

Ireland 86%

PAUL HILLION

Apparent absences in the distribution of common species such as the Dunnock can be used to infer the levels of recording effort.

Atlas 2007–11. Similarly, data from the three breeding atlases were compared with the breeding-season benchmark lists.

The figures for the percentage of benchmark species recorded in each 10-km square in each atlas were used as a measure of recorder effort. To make it easier to identify areas of low coverage, the maps in Figure 5.7 show the inverse of those figures, that is the percentage of benchmark species *not* recorded in a 10-km square in a particular atlas. To varying degrees, all five maps show apparently low recording effort in the highest peaks of the Scottish Highlands, but this could be an artefact of FRESCALO. For example, the Cairngorms form an island of unusual habitat with a characteristic suite of montane bird species. It is possible that the 'nearby similar squares' from which FRESCALO identifies benchmark species were actually not sufficiently similar. As a result the benchmark lists included some species that did not occur in montane habitats and were truly absent from the Cairngorms, rather than undetected and indicative of low effort. The same phenomenon may affect the Northern Isles and Outer Hebrides where the neighbourhood of similar squares included mainland Scotland, where species richness and composition are different. Nevertheless, for any given area, this problem applies to all time periods, so any differences in recording effort should still be apparent.

In winter, the most significant result is that recording effort of the western and central Scottish Highlands was poor relative to the rest of Britain in the *1981–84 Winter Atlas* (Figure 5.7A). This pattern is less apparent in the winter 2007–11 map, though some under-recording is still apparent in the same regions. To a lesser extent there were areas of relatively low effort in northwest and southwest Ireland, and in southwest Scotland and the north Pennines in 1981–84 but effort across these areas was more uniform in 2007–11. When averaged across all squares, the median percentage of benchmark species found per 10-km square increased from 94% in the *1981–84 Winter Atlas* to 100% in *Bird Atlas 2007–11*, and the percentage of squares with at least 90% of benchmark species detected (here defined as squares with high effort) also increased (Figure 5.7A).

In the breeding season, recording effort was slightly poorer and patchier in the *1988–91 Breeding Atlas* than in either the *1968–72 Breeding Atlas* or *Bird Atlas 2007–11*, especially in Ireland (Figure 5.7B). The decrease in recording effort from 1968–72 to 1988–91 may be partly attributable to the strong emphasis given in the *1988–91 Breeding Atlas* to the systematic timed-visit component at the expense of casual recording of breeding evidence. Indeed calling the latter 'supplementary' may have given the wrong impression of their importance.

The maps in Figure 5.7 are useful aids when interpreting apparent range changes in the species accounts. For example, the large change in recording effort in winter in northwest Scotland may explain some of the apparent gains seen in that area, and this is discussed where appropriate in the species accounts.

FIGURE 5.8 TESTING FOR BIAS IN ATLAS RANGE-CHANGE ESTIMATES

Examples of plots comparing FRESCALO and atlas range-size figures, one for winter changes in Britain (A) and one for 40-year breeding-season changes in Ireland (B). On each plot, dots represent individual species and are plotted on a $\log_{10}$ scale. The solid line shows the regression line of best fit (see Table 5.7). Where it crosses the black dotted 'no-change' line a red dotted line is marked. If the red line is above or below $y = 1$ this indicates bias in atlas range figures owing to recording effort variation. See text for full explanation.

A Britain winter changes from 1981–84 to 2007–11

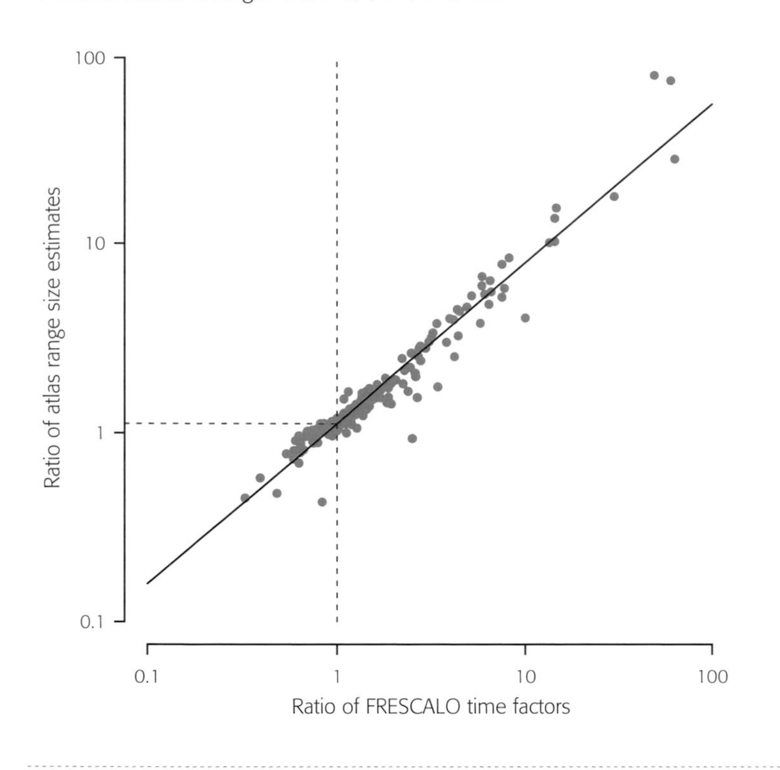

B Ireland breeding-season changes from 1968–72 to 2008–11

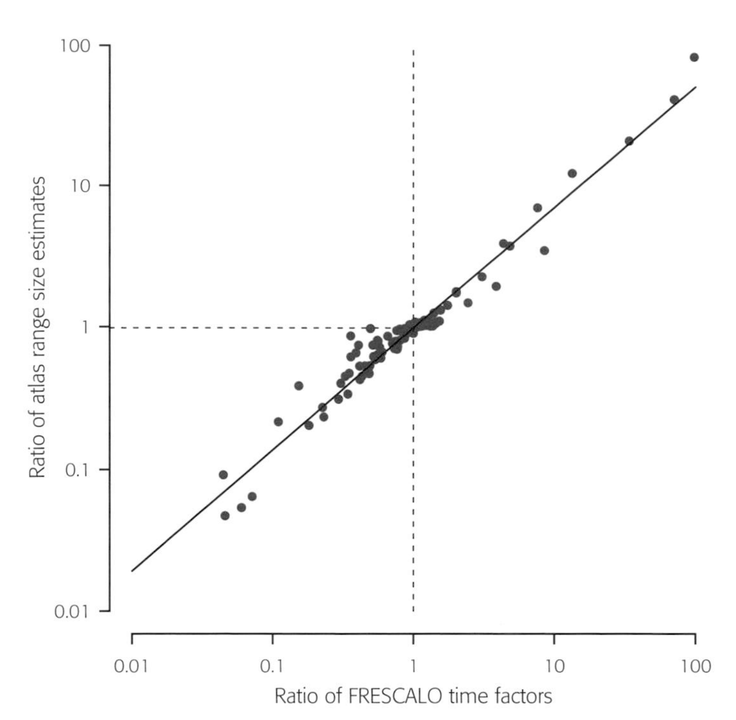

5.6.2 Did effort vary sufficiently to affect distribution change statistics?

The key question that arises from the maps in Figure 5.7 is whether the changes in effort shown are sufficiently large to bias our standard range-change estimates. To test this, we compared measures of species' range change derived directly from Atlas data with those produced by FRESCALO. The logic of this is as follows. If, for example, recording effort is low in the first of two atlases, the range sizes of species will be systematically under-estimated in the first period relative to the second. The implications of this for Atlas range change estimates are:

- species that have actually remained stable will appear to have expanded in range;
- real range expansions will be exaggerated;
- real range contractions will appear less serious than they really are.

Assuming that FRESCALO accurately compensates for the degree of under-recording in the first period, its estimates of change would be unbiased. If there is no appreciable under-recording, a graph of the Atlas estimates of change for species plotted against the corresponding FRESCALO estimates of change would reveal a positive relationship with both methods agreeing on which species have expanding, contracting or stable ranges. If under-recording in one period or the other has significantly affected the atlas range sizes and thus the range changes, the relationship would deviate from this expected pattern.

We tested for bias in Atlas changes by producing separate graphs for Britain and Ireland for each pair of atlases and testing the form of the relationship between Atlas and FRESCALO changes using linear regression. The y-axis of each plot is the ratio of the range sizes for each species in the two atlases. For example, breeding-season range sizes for the Nuthatch in Britain during the *1968–72 Breeding Atlas* and *Bird Atlas 2007–11* were 1,147 and 1,589 10-km squares, giving a ratio of 1.39 (= 1,589/1,147). The x-axis is the ratio between the FRESCALO 'time factors' for each species. Full details of how the time factors are derived can be found in Hill (2012) but essentially they give an unbiased estimate of the range of a species in each atlas period having corrected for any variation in effort. The FRESCALO time factors for the Nuthatch in Britain were 0.248 and 0.430, giving a change ratio of 1.73 (= 0.430/0.248). These pairs of ratios were plotted on $\log_{10}$-scale axes. Example graphs for changes in Britain between the *1981–84 Winter Atlas* and *Bird Atlas 2007–11*, and in Ireland between the *1968–72 Breeding Atlas* and *Bird Atlas 2007–11* are shown in Figure 5.8.

Both graphs show the general expected pattern of a strong positive relationship, with Atlas and FRESCALO broadly agreeing on which species have expanded or contracted in range. The dotted 'no-change' lines show the points of no change (i.e. ratio = 1.0) on the FRESCALO axis. If the Atlas figures are unbiased the regression line should intersect the no-change line at the point corresponding to no-change on the y-axis, i.e. at $y = 1.0$. In the case of winter changes in Britain (Figure 5.8A) the intersection of the regression and no-change lines is above $y = 1.0$, indicating that atlas changes are generally positively biased. For the 40-year breeding-season changes in Ireland (Figure 5.8B) the regression line passes very close to $y = 1.0$. This can be tested formally with linear regression to determine the y-value where $x = 1.0$ (on a log-scale this is where $x = 0$ and the value of interest is the graph's intercept).

Table 5.7 shows that there was evidence in both islands of positive bias for winter range changes, with regression lines intersecting no-change lines (where FRESCALO suggests no change) at 1.12 (Britain) and 1.14 (Ireland). In Britain in the breeding season there were no biases for any pair of atlases. In Ireland in the breeding season there was evidence of negative bias for changes between 1968–72 and 1988–91 followed by positive bias for changes between 1988–91 and 2008–11. Interestingly, the magnitude of bias was the same but in opposite directions (-0.07 then +0.07) and there was no evidence of bias in Ireland for overall changes between 1968–72 and 2008–11.

The conclusion that can be drawn from this is that in winter, as might be expected from the estimated recording effort maps (Figure 5.7A), there was significant bias in change statistics in Britain and in Ireland. These results suggest that lower recording effort in the *1981–84 Winter Atlas* is capable of generating spurious range expansions of up to +12% in Britain and up to +14% in Ireland for species that have actually seen no range change. We also conclude that breeding-season range sizes were estimated relatively well in 1968–72 and 2008–11 but were slightly under-estimated in 1988–91 in Ireland. This means that in Ireland, apparent range changes of approximately -7% between 1968–72 and 1988–91, or approximately +7% between 1988–91 and 2007–11 could be attributable entirely to differences in effort. It is encouraging, however, that no significant biases are apparent in Britain, nor for either island between 1968–72 and 2007–11.

It is worth noting that these analyses rely on a new technique that has seen few applications to date. The resulting maps of effort variation and graphs of its impact on distribution and change metrics appear sensible and match anecdotal evidence but some uncertainties remain. FRESCALO relies on differences in species distributions between atlases in order to estimate variation in effort, but if the differences are actually real, effort variations could be overestimated. We have followed the recommendation of Hill (2012) to reduce this problem but some errors may still exist. For example, is the method less robust in winter when distributions are less static or for species undergoing shifts in range without associated changes in range size? Despite these uncertainties we think these results are an important statement about the need to understand effort variations and offer useful guidance for the interpretation of species accounts and information presented in Chapter 6.

5.7 CONCLUSIONS

CHANGES IN METHODS, recording effort and data volume are significant issues for any repeat atlas but the impact of these changes upon range-size estimates and measures of distribution change can be exceedingly difficult to quantify. In view of the significant increase in the number of participating observers, the breadth of data sources used and the order of magnitude growth in the number of records contributing to this Britain & Ireland atlas compared to previous ones, it was essential to assess the implications for estimating range changes.

The analyses in this chapter show that long-term comparisons between the 1968–72 and 2008–11 breeding atlases are remarkably robust in both Britain and Ireland, raising two interesting points. Firstly *Bird Atlas 2007–11* did not benefit in the breeding season as

TABLE 5.7 TESTING FOR BIAS IN RANGE CHANGE ESTIMATES BETWEEN PAIRS OF ATLASES

Results of regression tests to determine the intercept of log-log plots of FRESCALO time-factor ratios and atlas range-size ratios for individual species calculated between different pairs of atlases on each island. The plots for the 1981–84 to 2007–11 changes in Britain, and for the 1968–72 to 2008–11 changes in Ireland, are shown in Figure 5.8. Values in the Intercept column correspond to the value on the atlas ratio axis where the FRESCALO ratio equals 1.0 (the dotted no-change lines in Figure 5.8). These values are on a $\log_{10}$ scale. To aid interpretation, for those comparisons where the intercept was significantly different from zero on the $\log_{10}$ scale (indicated in column *P*), the Bias column gives the intercept on a linear scale. For example, a bias value of 1.12 means that Atlas figures suggest a 12% range expansion when FRESCALO figures suggest no change. See text for further discussion. *N* is the number of species in each test.

Season	From	To	Britain				Ireland			
			Intercept	*P*	Bias	*N*	Intercept	*P*	Bias	*N*
Winter	1981–84	2007–11	0.0490	<0.0001	1.12	211	0.0574	<0.0001	1.14	143
Breeding	1968–72	1988–91	-0.0011	0.73	n/a	179	-0.0317	<0.0001	0.93	104
	1988–91	2008–11	0.0049	0.19	n/a	182	0.0278	<0.0001	1.07	108
	1968–72	2008–11	0.0056	0.24	n/a	177	-0.0078	0.39	n/a	106

SIMON GILLINGS

Using records from mountaineers and hill walkers helped to ensure that upland areas received adequate coverage in the winter months.

significantly as one might expect from the large volume of external data, mostly owing to the lack of breeding evidence. Secondly, these findings reiterate the invaluable contribution made by observers in 1968–72 to get the atlas in Britain and Ireland underway. These analyses also confirm the perception that range sizes may have been underestimated in the *1988–91 Breeding Atlas*, although only significantly so in Ireland. This helps to explain the large number of species in Ireland that show an apparent range contraction from 1968–72 to 1988–91 followed by an apparent range expansion from 1988–91 to 2008–11 (see Section 6.5.1). This was in no way a fault of the efforts of observers but an unanticipated consequence of introducing a new field method to the atlases which, in other respects, has provided immense benefits in the shape of abundance maps and systematic measures of distribution and change. The Atlas Working Group, Irish Atlas Steering Committee and organisers of *Bird Atlas 2007–11* paid particular attention to the equal promotion of TTVs and Roving Records and it is encouraging to see that this has resulted in improved coverage. Continuing the use of TTVs has paid dividends in terms of the number of species for which we can present maps of relative abundance, but also in providing effort-controlled statistics of breeding-season range change between 1988–91 and 2008–11. The extent of TTV coverage will also be invaluable for future research examining the environmental drivers of patterns of distribution, abundance and change.

It is reassuring that the breeding-season changes between 1968–72 and 2008–11 are unbiased. More problematic, however, are the biases that arise from the improvements in winter coverage in parts of Scotland, northern England and Ireland. It is perhaps unsurprising that changes in effort are so pronounced in winter because the vagaries of weather can impact especially severely upon fieldwork opportunities and access to remote areas. Furthermore, in winter there is no 'filter' comparable to breeding evidence that can limit what can be added to lists, possibly introducing more volatility into the results of winter atlas work. The analysis of bias does not mean that all winter range change statistics should be adjusted by subtracting 12–14 percentage points, but they should prompt readers to consider whether apparent gains for a species in northwest Scotland, or western Ireland, could be the result of changing effort. We strongly encourage the organisers of the next winter atlas to repeat the use of systematic TTVs so that effort-controlled statistics of winter range change can be produced to reduce some of the inherent uncertainties associated with winter fieldwork.

CHAPTER 6

PATTERN AND CHANGE IN THE BRITISH AND IRISH AVIFAUNAS OVER A 40-YEAR PERIOD

Rob Fuller, Simon Gillings, Alan Lauder & Olivia Crowe

6.1 INTRODUCTION

ATLASES HAVE BECOME a powerful tool for documenting and understanding the shifting nature of the avifaunas of Britain and Ireland. By far the largest part of this book is devoted to accounts for individual species, mapping their distribution during 2007–11 and charting changes in distribution and abundance since the *1968–72 Breeding Atlas*, the *1981–84 Winter Atlas* and the *1988–91 Breeding Atlas*. This chapter, however, draws out some of the patterns and changes that are evident over the last 40 years in the broader composition and character of the avifaunas in both the breeding and winter seasons (broadly defined as April to July and November to February respectively). In doing so, several comparisons are made between Britain and Ireland (see Section 1.2 for geographical definitions), asking whether similar, or somewhat different, processes are likely to have affected birds on the two islands over recent decades. Compared with Britain, Ireland is poorer in bird species and the ecology and behaviour of some bird species appear to differ (O'Halloran *et al.* 2012). These biological differences are probably underpinned by the position of Ireland at the western edge of Europe, its smaller size, narrower band of latitude and more limited range of habitat types relative to Britain.

This is not the place to attempt detailed explanations of the emerging patterns, though some general observations about potential processes are made. The analyses presented here should be regarded as preliminary; the Atlas data represent a huge opportunity for further research, including understanding the causes of change in bird distributions.

In this chapter we summarise various broad patterns relating to changes in both range size (or extent of distribution) and abundance of birds. Some initial comments on the relationships between these two attributes of species are, therefore, appropriate. Atlases have traditionally measured and represented distribution as simple detection/non-detection in grid squares, but increasingly they are designed to generate more sophisticated maps of spatial variation in relative abundance (Gibbons *et al.* 2007). To some extent, this blurs the distinction between distribution and abundance. Spatial changes in abundance are, arguably, biologically more interesting than absolute changes in range size, because local extinction is essentially the extreme outcome of abundance change. Indeed, range size across species is typically strongly correlated with both total population size and local density; these relationships have been demonstrated for many taxonomic groups and at different scales.

The mechanisms driving relationships between abundance and range size are somewhat unclear but several hypotheses have been proposed (Gaston 2003). Not surprisingly, changes in the abundance and distribution of birds are nearly always positively correlated. When the population of a species declines, one may reasonably expect it to become less widespread, with the change in range typically lagging behind the change in overall numbers. Although a species could in theory become more localised in distribution while increasing its local density, or vice versa, this seldom happens. Perhaps the most likely circumstance where this might occur concerns colonial breeding or gregarious wintering birds, where the loss of a few small outlying concentrations or colonies, results in a contraction of range and coincides with growth of the remaining concentrations or colonies.

6.2 APPROACH

MANY OF THE analyses in this chapter are primarily concerned with distribution, and are based largely on the numbers and proportions of occupied 10-km squares in different atlases. Proportional change in the number of apparently occupied squares gives a simple measure of range change that has been widely used (e.g. Fuller *et al.* 1995; Chamberlain & Fuller 2000; Donald & Greenwood 2001). This simple measure has the drawback that it can be affected by differences in observer effort in space and time (see Chapter 5). It should, therefore, be borne in mind that 'square occupancy' or 'occupied squares', refer to the number of squares where a species was detected, which may be less than the number of squares in which it actually occurs (see Section 4.1.1). Furthermore, depending on variation in effort, estimates of range change could be overestimated or underestimated.

A thorough discussion of the evidence for changes in effort through the series of Britain & Ireland atlases can be found in Chapter 5, and its conclusions should be remembered when reading this chapter. The use of TTVs in the *1988–91 Breeding Atlas* and *Bird Atlas 2007–11* has made it possible to produce a more robust measure of change based solely on the timed observations for these two periods, which controls for spatial and temporal changes in observer effort (see Chapters 2 and 4). The resulting 'fixed effort' or 'effort-controlled' distributions therefore remove the potentially confounding trends in observer effort. TTVs have the disadvantage, however, that sparsely distributed and nocturnal species may be difficult to detect, with the result that their distributions are likely to be underestimated. Nevertheless, the degree to which species were missed from TTVs ought to be constant through time, with the exception of those in which a change in abundance is associated with a change in detectability due to reduced or increased territorial defence. In several analyses of the most recent 20-year breeding-season changes, effort-controlled data are used for drawing conclusions in preference to simple proportional occupancy data.

6.3 THE SPECIES POOL

THE ATLAS GENERATED records of 502 species (see Appendix 2). There were 289 and 171 species with breeding evidence in Britain and Ireland respectively, and 410 and 268 in winter, though not all of these can be considered regular

FIGURE 6.1 THE SPECIES POOL IN DIFFERENT BREEDING ATLASES IN BRITAIN AND IRELAND

The Venn diagrams show the number of species shared or unique to the three breeding atlases in Britain (A) and Ireland (B). Only species recorded with *probable* or *confirmed* breeding evidence in at least one of the three breeding atlases are considered. Species that were unique to any particular atlas are listed. Those whose presence was the result of reintroduction programmes are marked with an asterisk; those that have only bred in a hybrid pair are marked with a superscript H; exotic non-natives are marked in italics.

A Britain

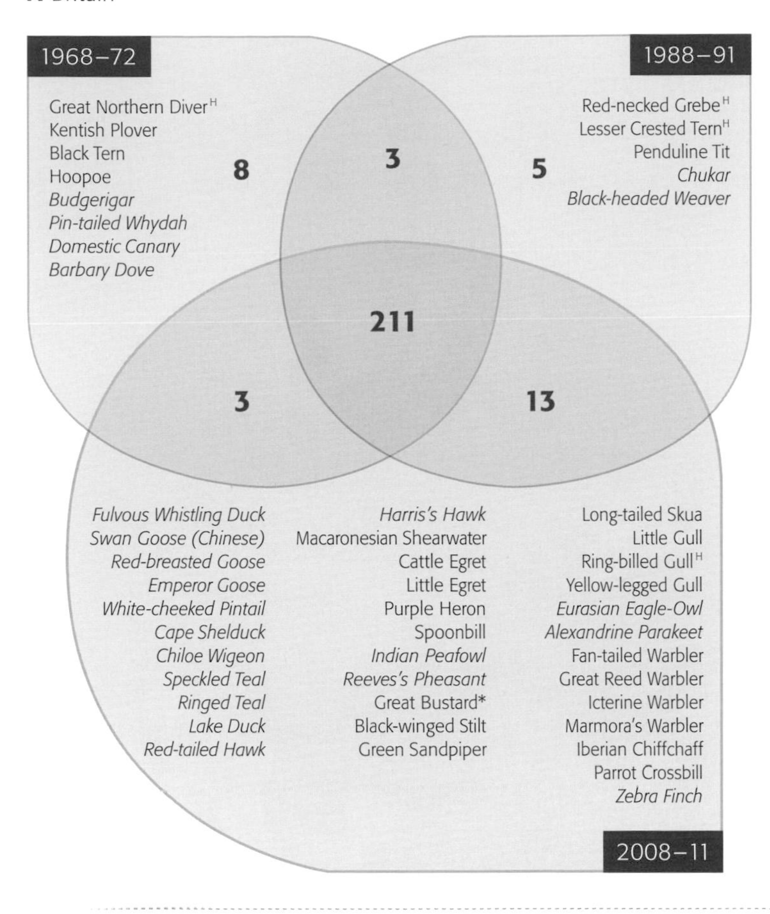

B Ireland

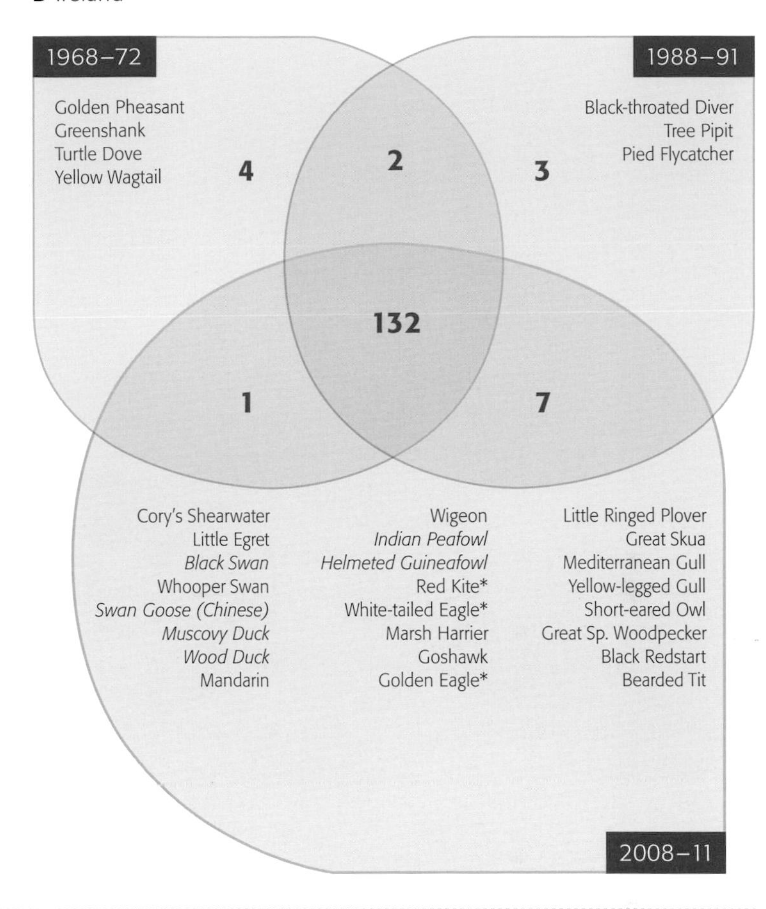

breeding or wintering species. Roughly similar numbers of species were recorded with at least *probable* breeding evidence in each of the previous breeding atlases, although the number of species involved was consistently more than 50% higher in Britain than in Ireland (Figure 6.1). In both Britain and Ireland the number of breeding species unique to any particular atlas period was highest in 2007–11. To some extent this was a consequence of relatively large numbers of 'exotic' non-native species (see Box 6.1 for definitions), which have undoubtedly increased in diversity and abundance but have also been recorded more thoroughly in the latest atlas (see Section 6.5.2).

Most of the 'extinctions' have involved very 'marginal' or 'erratic' breeders such as the Kentish Plover and Red-necked Grebe. However, over the last 40 years there have been some major declines to near extinction of previously widespread species (e.g. Red-backed Shrike) or actual extinction in the case of the Corn Bunting in Ireland where breeding evidence was obtained in 69 10-km squares in the *1968–72 Breeding Atlas*. For information on species that bred in the years between atlas periods but not during these periods, see reports of the UK and Irish Rare Breeding Bird Panels, including Holling *et al.* (2010b) which documents the year of colonisation for species new to the UK during 1973–2008.

Similar patterns were apparent in winter—c.50% more species were reported in Britain than Ireland, and the number of species unique to an atlas period was considerably higher in 2007–11 (Britain 117, Ireland 72) than in 1981–84 (Britain 14, Ireland 10). The apparently greatly increased number of winter species exclusive to *Bird Atlas 2007–11* was partly a consequence of improved data capture methods, for example the automated inclusion of records of rare species (see Sections 2.4 and 5.3.5).

An analysis of range sizes showed a preponderance of species with very small range sizes in winter or the breeding season. During the breeding season, 30% of all species were reported from fewer than 25 10-km squares in Britain & Ireland; the corresponding figure for winter was 42%. In contrast, ubiquitous species, defined as ones occurring in more than 90% of 10-km squares, made up just 6% of species in the breeding season and 3% in winter. Species with small ranges were especially apparent in winter when there was no prerequisite for breeding evidence, enabling more rarities to be considered.

More generally, many of the species with small range sizes were transient visitors or 'exotic' non-native species that do not have established populations. Small fluctuations (in absolute terms) in the size of the ranges of these species can generate anomalously

CHRIS KNIGHTS

One of the most widespread species, Wrens were recorded in at least 96% of 10-km squares in both seasons and on both islands.

TABLE 6.1 THE TEN MOST WIDESPREAD SPECIES BY ISLAND AND SEASON

For each season species are ordered by the number of occupied 10-km squares in 2007–11 in Britain (A) or Ireland (B). The ranked position and range size (occupied squares in parentheses) in previous atlases is shown for comparison. The total numbers of 10-km squares with land are 2,876 in Britain and 1,018 in Ireland.

A Britain

Rank	Winter			Breeding			
	Species	Range size in 2007–11	Rank and range size in 1981–84	Species	Range size in 2007–11	Rank and range size in 1968–72	Rank and range size in 1988–91
1	**Wren**	2,774	2 (2,587)	**Wren**	2,787	2 (2,762)	1 (2,761)
2	**Blackbird**	2,742	1 (2,639)	**Skylark**	2,753	1 (2,783)	2 (2,742)
3	**Robin**	2,741	3 (2,568)	**Pied Wagtail**	2,745	7 (2,662)	4 (2,674)
4	**Chaffinch**	2,694	4 (2,548)	**Blackbird**	2,719	3 (2,727)	3 (2,678)
5	**Mallard**	2,671	5 (2,509)	**Swallow**	2,698	12 (2,605)	6 (2,640)
6	**Buzzard**	2,661	61 (1,508)	**Chaffinch**	2,679	15 (2,593)	9 (2,614)
7	**Meadow Pipit**	2,651	18 (2,319)	**Mallard**	2,675	8 (2,633)	11 (2,606)
8	**Dunnock**	2,649	9 (2,422)	**Robin**	2,675	10 (2,614)	5 (2,643)
9	**Fieldfare**	2,637	8 (2,457)	**Willow Warbler**	2,671	13.5 (2,595)	10 (2,612)
10	**Grey Heron**	2,627	14 (2,390)	**Song Thrush**	2,666	5 (2,685)	7.5 (2,634)

B Ireland

Rank	Winter			Breeding			
	Species	Range size in 2007–11	Rank and range size in 1981–84	Species	Range size in 2007–11	Rank and range size in 1968–72	Rank and range size in 1988–91
1	**Hooded Crow**	1,005	3 (987)	**Wren**	1,003	2 (993)	1 (987)
2	**Wren**	996	4 (986)	**Swallow**	1,000	4 (987)	2 (982)
3	**Robin**	994	1 (989)	**Hooded Crow**	997	6.5 (983)	4 (970)
4	**Blackbird**	993	2 (988)	**Blackbird**	987	5 (985)	3 (976)
5	**Song Thrush**	993	8 (965)	**Robin**	985	9 (977)	5 (967)
6	**Starling**	987	7 (966)	**Starling**	985	8 (981)	8 (957)
7	**Pied Wagtail**	986	9 (959)	**Pied Wagtail**	982	6.5 (983)	6 (966)
8	**Magpie**	986	5 (973)	**Magpie**	982	17 (954)	7 (962)
9	**Dunnock**	984	12 (936)	**Dunnock**	974	10.5 (970)	10.5 (947)
10	**Chaffinch**	980	6 (968)	**Meadow Pipit**	973	1 (1,001)	12.5 (945)

high percentage range-change figures. Consequently, analyses including these species could have been problematic because their range changes could appear to be more significant than those shown by what we might think of as 'regular' species. Also, many of the exotic non-native species were not consistently reported in previous atlases so range expansions may be over-estimated. For many analyses, therefore, we excluded exotic non-natives and highly range-restricted species, defined as species present in fewer than 25 10-km squares in the periods under consideration. Where necessary, 'breeding species' were defined as those that have been recorded with either *probable* or *confirmed* breeding evidence in Britain or Ireland in at least one of the three breeding atlases (excluding those that have only been present in hybrid pairs). This helped to exclude some scarce and rare passage migrant species that sporadically sang in apparently suitable habitat (e.g. River Warbler).

Note, however, that having selected the species pool on the above basis, the default approach when calculating breeding-season range size has been to use all *possible*, *probable* and *confirmed* breeding evidence records.

6.4 RANGE SIZE AND ABUNDANCE PATTERNS

THE MOST WIDESPREAD breeding species have largely remained the same in the last 40 years. In Britain, the Wren, Skylark and Blackbird have been amongst the five most widespread breeding species in the three atlases (Table 6.1). In Ireland, the Wren, Swallow, Hooded Crow and Blackbird have consistently had very large ranges (Table 6.1). Five species featured in the largest ten breeding ranges in both islands in 2008–11: the Wren, Pied Wagtail, Blackbird, Swallow and Robin. In winter there has also been considerable constancy in the widespread species with the Wren, Blackbird, Robin, Chaffinch and Mallard having the largest ranges in both atlases in Britain. In Ireland, the Hooded Crow, Wren, Robin and Blackbird had the largest winter ranges in both atlases. Five species appeared amongst the ten largest winter range sizes for both islands in 2007–11 namely the Wren, Blackbird, Robin, Chaffinch and Dunnock.

When species were ranked according to the percentage of tetrads per 10-km square where each was detected, slightly different patterns emerged to those evident from simple 10-km square

TABLE 6.2 THE TEN MOST FREQUENTLY OCCURRING SPECIES RANKED BY TETRAD OCCUPANCY

For each season species are ordered by the mean percentage of tetrads occupied per 10-km square, within the occupied range, in 2007–11 in Britain (A) and Ireland (B). *N* is the number of 10-km squares used for the analysis, and is smaller than the range sizes given in Table 6.1 owing to the exclusion of a small number of 10-km squares where few TTVs were available (Section 2.2.1); furthermore in Ireland, tetrad occupancy could only be calculated for chequerboard squares (see Chapter 2).

A Britain

Rank	Winter			Breeding		
	Species	% tetrads occupied	*N*	Species	% tetrads occupied	*N*
1	**Carrion Crow**	84.0	2,444	**Chaffinch**	86.8	2,661
2	**Blackbird**	82.0	2,707	**Carrion Crow**	84.2	2,405
3	**Robin**	80.7	2,711	**Wren**	83.4	2,751
4	**Blue Tit**	79.9	2,557	**Woodpigeon**	82.2	2,574
5	**Chaffinch**	79.6	2,673	**Blackbird**	82.0	2,696
6	**Wren**	76.8	2,737	**Robin**	80.3	2,657
7	**Great Tit**	75.6	2,552	**Blue Tit**	77.0	2,554
8	**Woodpigeon**	75.2	2,473	**Swallow**	74.3	2,679
9	**Magpie**	71.2	2,114	**Great Tit**	74.1	2,548
10	**Dunnock**	65.6	2,623	**Dunnock**	66.9	2,613

B Ireland

Rank	Winter			Breeding		
	Species	% tetrads occupied	*N*	Species	% tetrads occupied	*N*
1	**Robin**	94.8	488	**Wren**	95.6	490
2	**Blackbird**	93.9	487	**Robin**	94.8	488
3	**Wren**	91.6	488	**Blackbird**	94.4	487
4	**Chaffinch**	90.7	485	**Swallow**	94.2	490
5	**Magpie**	89.5	485	**Chaffinch**	93.0	478
6	**Hooded Crow**	89.0	489	**Woodpigeon**	90.7	478
7	**Blue Tit**	86.3	483	**Magpie**	87.3	484
8	**Jackdaw**	82.9	477	**Hooded Crow**	86.3	489
9	**Rook**	81.3	476	**Blue Tit**	85.4	476
10	**Dunnock**	81.0	484	**Willow Warbler**	84.4	479

occupancy (compare Table 6.1 and Table 6.2). For example, the Skylark and Pied Wagtail were no longer in the top ten species for Britain in the breeding season but the Carrion Crow and Woodpigeon were near the head of the list. Similarly, in Ireland, Woodpigeons featured for tetrad occupancy but not 10-km square occupancy, while the opposite was the case for Pied Wagtails. Differences in 10-km square and tetrad occupancy may partially reflect detectability but these are measures of rather different aspects of range. Occupancy at 10-km square resolution measures the extent to which a species is distributed across a wide geographical area, whereas tetrad occupancy is a finer-grain measure of distribution *within* the occupied range. Hence, these findings suggest, for example, that whilst Skylarks and Pied Wagtails were more widely distributed within Britain in the breeding season than were Carrion Crows and Woodpigeons, the latter two species were more evenly distributed than the former two species within their ranges. Importantly, they illustrate how the conclusions that might be drawn from such distributional data are affected by the resolution of the analysis, in this case the size of the grid squares.

The raw counts of species undertaken in TTVs were, of course, biased towards highly detectable and gregarious species. Despite the inherent difficulties in comparing these counts across species, they are of interest from a 'cultural service' perspective. Cultural services are one of the four principal services that ecosystems are deemed to provide to human well-being (Millennium Ecosystem Assessment 2005) and, in this context, can be broadly defined as the benefits that biodiversity provides to wider society through, for example, amenity or health benefits. In this context, TTVs indicated the species that birdwatchers and the general public may expect to encounter in the greatest numbers on a typical one-hour walk. This is reflected in Table 6.3, which shows the median counts of the more widely distributed species. Interestingly, the lists of the ten species with the highest counts were very similar for Britain and Ireland, having nine species in common in the breeding season and seven in winter. Counts of Rooks, Jackdaws, Swallows, Wrens and Starlings tended to be higher in Ireland, whereas those of Woodpigeons appeared higher in Britain. In both seasons the Robin featured in the Irish but not the British list.

Finally, there is the question of how range size, or extent of distribution, relates to the abundance of bird species. As explained

TABLE 6.3 SPECIES RECORDED IN THE LARGEST NUMBERS ON TIMED TETRAD VISITS

For each season species are ordered by the median count per hour (and quartiles) in surveyed tetrads in occupied 10-km squares in 2007–11 for Britain (A) and Ireland (B). The focus is on the more widespread species and the use of the median count avoids over-emphasising highly aggregated and localised species; nonetheless, colonial seabirds were excluded from the breeding counts. *N* is as defined in Table 6.2.

A Britain

Rank	Winter				Breeding			
	Species	Median count	Quartiles	*N*	Species	Median count	Quartiles	*N*
1	Woodpigeon	17.1	3.4, 48.2	2,473	Woodpigeon	8.2	2.5, 16.8	2,574
2	Starling	16.6	7.1, 34.9	2,558	Rook	6.4	2.7, 11.3	2,156
3	Rook	11.5	3.8, 21.4	2,286	Chaffinch	5.9	3.8, 7.9	2,661
4	Jackdaw	9.8	4.7, 16.4	2,381	Jackdaw	5.5	2.8, 8.6	2,326
5	Chaffinch	7.4	4.1, 11.3	2,673	Blackbird	5.3	1.9, 7.9	2,696
6	Blackbird	6.5	2.5, 9.9	2,707	House Sparrow	5.2	2.5, 8.4	2,582
7	Black-headed Gull	6.4	0.9, 17.8	2,212	Carrion Crow	4.4	2.6, 6.1	2,405
8	Carrion Crow	6.2	3.4, 9.1	2,444	Swallow	3.7	2.1, 5.5	2,679
9	Fieldfare	5.5	1.2, 13.7	2,621	Starling	3.5	1.2, 7.3	2,559
10	Blue Tit	5.0	2.5, 7.2	2,557	Wren	3.5	1.9, 5.2	2,751

B Ireland

Rank	Winter				Breeding			
	Species	Median count	Quartiles	*N*	Species	Median count	Quartiles	*N*
1	Starling	25.4	12.3, 42.3	485	Rook	11.9	4.9, 19.6	450
2	Rook	16.9	5.6, 29.5	476	Swallow	8.1	5.3, 11.6	490
3	Jackdaw	13.8	6.8, 20.9	477	Jackdaw	7.9	4.7, 13.0	472
4	Redwing	9.8	3.8, 23.0	479	Wren	7.1	4.2, 10.0	490
5	Blackbird	8.5	5.4, 11.5	487	Starling	6.6	3.4, 10.9	486
6	Chaffinch	7.6	4.9, 11.9	485	Blackbird	6.1	3.9, 8.8	487
7	Robin	6.3	4.2, 9.0	488	Chaffinch	5.8	3.7, 8.0	478
8	Woodpigeon	5.6	2.2, 12.9	472	Robin	5.5	3.3, 7.6	488
9	Fieldfare	4.6	1.6, 9.8	472	Woodpigeon	5.5	3.0, 8.4	478
10	House Sparrow	3.8	1.8, 7.0	472	House Sparrow	3.9	2.0, 7.8	474

above, one generally expects a broadly positive relationship—the more abundant a species, the more widely distributed it is likely to be. This has previously been demonstrated to be the case for birds in Britain (e.g. Fuller 1982; Newton 1997; Gaston *et al.* 1997). However, there are many exceptions as illustrated in Figure 6.2 which uses breeding-season range sizes in Britain from *Bird Atlas 2007–11* in combination with national population estimates for birds in Britain (Musgrove *et al.* 2013). Formal population estimates for the majority of species in Ireland were unavailable at the time of writing.

As one might expect, two groups of species effectively formed outliers within what was otherwise a linear positive relationship. Group 1 consisted of 10 highly gregarious and localised colonial-nesting seabirds: the Arctic Tern, Fulmar, Gannet, Guillemot, Kittiwake, Leach's Petrel, Manx Shearwater, Puffin, Razorbill and Storm Petrel. Whilst the breeding ranges of these species, defined by terrestrial 10-km squares, are small, obviously their feeding ranges are extremely large. Group 2 includes more than 20 rare breeding species that were confined to very few squares with small numbers of breeding pairs. Examples include the Slavonian Grebe, Common Scoter, Crane, Honey-buzzard, Marsh Warbler, Red-backed Shrike and Red-necked Phalarope.

Despite the fact that a positive relationship existed once these two groups were excluded, there was still considerable scatter, with individual species being more or less abundant than the average trend might predict for a given range size. For example, the variation in population size for a species with a range size of about 2,000 10-km squares was from approximately 7,000 pairs for the Raven to 700,000 pairs for the Yellowhammer. It is likely that much of this variation related to the ranging and territorial behaviour, and possibly the body mass, of individual species.

6.5 CHANGES IN RANGE SIZE AT THE SPECIES LEVEL

6.5.1 Range expansions and contractions: the winners and losers

Reaching any conclusion about the overall numbers of species showing range expansions or contractions is complicated by the fact that the direction of range change shown by a species has not necessarily been consistent over the last 40 years. Table 6.4 tallies the numbers of breeding species showing different trajectories of range change from 1968–72 to 1988–91, and from 1988–91 to 2008–11, and what the net effect of those changes has been. The first row of each table shows that the ranges of 42 species in Britain (10 in Ireland) expanded sequentially in both the 20-year periods, leading, inevitably, to a net gain in range size over the whole period. The last row indicates that 31 species in Britain (24 in Ireland) showed the opposite pattern, their ranges consistently shrinking over this period. The fifth row shows that for 28 species, British range sizes stayed more or less constant (within ±5%) from one atlas to the next. Interestingly, however, (row 3 of the tables) 28 species in Britain (39 in Ireland) showed an apparent range contraction in the first period, followed by expansion in the second. By contrast, (row 7) 20 species in Britain (2 in Ireland) showed an apparent range expansion followed by contraction. It would be informative to look at the identity of these species and test whether these fluctuations correspond with known population trends because, as discussed in Chapter 5, there were some variations in effort between the different atlases that could account for some apparent range changes.

The totals at the bottom of the tables show that, overall, a higher percentage of species in Britain showed a 40-year net expansion of breeding range (38%) than in Ireland (18%). Proportionally, slightly fewer species in Britain appeared to show net stability (24% vs. 35%) or net contractions (37% vs. 47%). When we consider just the recent changes (right-hand totals columns) since the *1988–91 Breeding Atlas*, the range expansions were more similar, involving 45% of British species and 47% of Irish species (Table 6.4).

Assessment of range changes in winter is simpler because there have been just two atlases (1981–84 and 2007–11). In Britain, 169 species expanded in range between these periods (76%), 35 were stable (16%) and 18 contracted in range (8%). The equivalent figures for Ireland were 91 (60%), 37 (24%) and 24 (16%) respectively. Hence, as with the 40-year breeding-season patterns, a higher proportion of species showed apparent range expansion and a lower proportion showed apparent range contraction in Britain compared with Ireland. In both countries there is evidence that observer effort in winter has increased over this period and that this may explain some of the apparent gains (see Section 5.6).

The simplified bar charts shown in Figure 6.3 show an interesting seasonal contrast in the pattern of absolute range expansions and contractions. Considering the overall picture for both Britain and Ireland, the numbers of breeding species expanding in range by at least one square versus contracting by at least one square were relatively evenly balanced and the magnitudes of range change in each season were similar. In winter, however, range expansions outnumbered contractions and the magnitude of the former was greater than the latter. This seasonal contrast was stronger in Britain than in Ireland. In Britain 89%, and in Ireland 74%, of species increased their winter range size. The respective figures for breeding ranges between 1988–91 and 2008–11 were 61% and 70%. The median changes across species in the breeding season were +33 10-km squares in Britain and +23 squares in Ireland; the median changes in winter were +120 squares in Britain and +27.5 in Ireland. It should, however, be borne in mind that improvements in winter coverage may in part account for the high proportion of species showing apparent range expansion (see Chapter 5).

Species showing the largest breeding range expansions and contractions are listed in Tables 6.5 and 6.6. Comparable data for winter are in Table 6.7. The species listed embraced a diversity of taxonomy, ecology and life history. Nonetheless, some broad patterns could be identified which are summarised below. Note that these comments relate only to species with the most strongly expanding and contracting ranges. Changes in selected species groups are examined further in Section 6.6, which gives a wider perspective on patterns of change.

There was a strong pattern of expansion in large wading waterbirds (herons, etc.), especially in Britain. The three egret species showed the greatest British range expansions of any species in winter. Since first breeding in Britain in 1996 and Ireland in 1997, the Little Egret has hugely expanded its breeding range in

FIGURE 6.2 THE RELATIONSHIP BETWEEN RANGE SIZE AND POPULATION SIZE IN THE BREEDING SEASON IN BRITAIN

Range size (apparently occupied 10-km squares in 2008–11) and its relationship with the latest estimates of population size (log_{10} transformed) for breeding bird species in Britain. Estimated population size was derived from Musgrove *et al.* (2013). Exotic non-native species were excluded from the analysis (see Section 6.3). Groups 1 and 2 are discussed in the text.

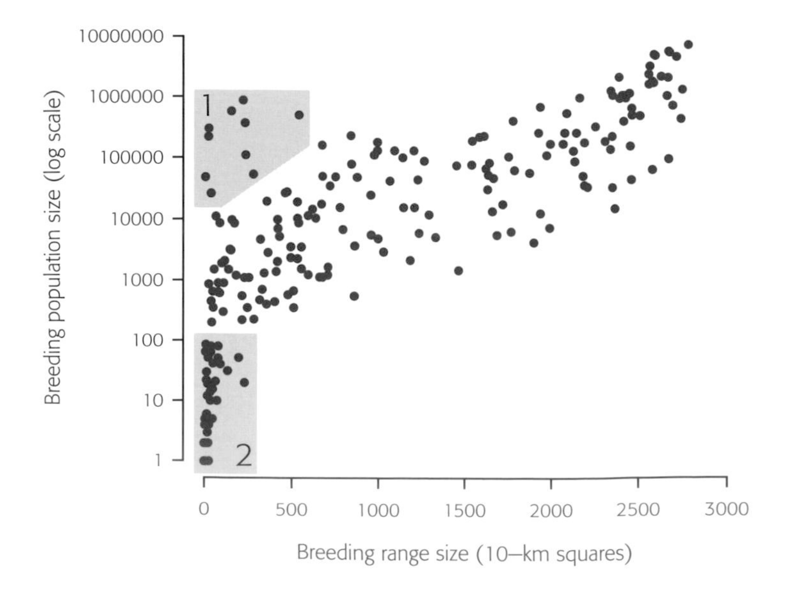

SHAY CONNOLLY www.flickr.com/photos/shayc

During the course of the Britain & Ireland atlases, Little Egrets have undergone a remarkable change in status, going from being an annual vagrant requiring a written description to a widespread wintering species and localised breeder.

TABLE 6.4 NUMBERS OF SPECIES SHOWING DIFFERENT PATTERNS OF SHORT- AND LONG-TERM CHANGE IN BREEDING RANGE

The tables show, for Britain (A) and Ireland (B) separately, numbers of species with different patterns of short- (c.20-year) and long-term (c.40-year) change in breeding-range size. Symbols indicate range expansion (↗), range stability (→) and range contraction (↘) based on % change in the numbers of 10-km squares with breeding evidence. For this purpose a range was considered stable if the change between two atlases was <±5%. The analysis excludes exotic non-natives and species with ranges smaller than 25 10-km squares (see Section 6.3 for further explanation).

A Britain

Change from 1968–72 to 1988–91	Change from 1988–91 to 2008–11	Net change (40-year) 1968–72 to 2008–11			Totals for recent change (20-year)
		↗	→	↘	
↗	↗	42	0	0	↗ 87 (45%)
→	↗	11	6	0	
↘	↗	7	9	12	
↗	→	4	2	0	→ 45 (23%)
→	→	2	23	3	
↘	→	0	3	8	
↗	↘	8	4	8	↘ 61 (32%)
→	↘	0	0	10	
↘	↘	0	0	31	
Totals for net change (40-year)		**↗ 74 (38%)**	**→ 47 (24%)**	**↘ 72 (37%)**	

B Ireland

Change from 1968–72 to 1988–91	Change from 1988–91 to 2008–11	Net change (40-year) 1968–72 to 2008–11			Totals for recent change (20-year)
		↗	→	↘	
↗	↗	10	0	0	↗ 55 (47%)
→	↗	6	0	0	
↘	↗	5	20	14	
↗	→	0	0	0	→ 33 (28%)
→	→	0	19	0	
↘	→	0	2	12	
↗	↘	0	0	2	↘ 28 (24%)
→	↘	0	0	2	
↘	↘	0	0	24	
Totals for net change (40-year)		**↗ 21 (18%)**	**→ 41 (35%)**	**↘ 54 (47%)**	

FIGURE 6.3 DIAGRAMMATIC SUMMARY OF SPECIES' RANGE CHANGES BY ISLAND AND SEASON

The shaded parts of the figure consist of many narrow vertical bars, each one representing the change in range of one species between two atlas periods in Britain (A) or Ireland (B). The bars are ordered from the species that showed the greatest gains, in terms of the absolute numbers of 10-km squares occupied, on the left to those with the greatest losses on the right. For each graph, species that did not occupy at least 25 10-km squares in either period were excluded. The dashed line shows the transition from species that showed gains to ones that showed losses.

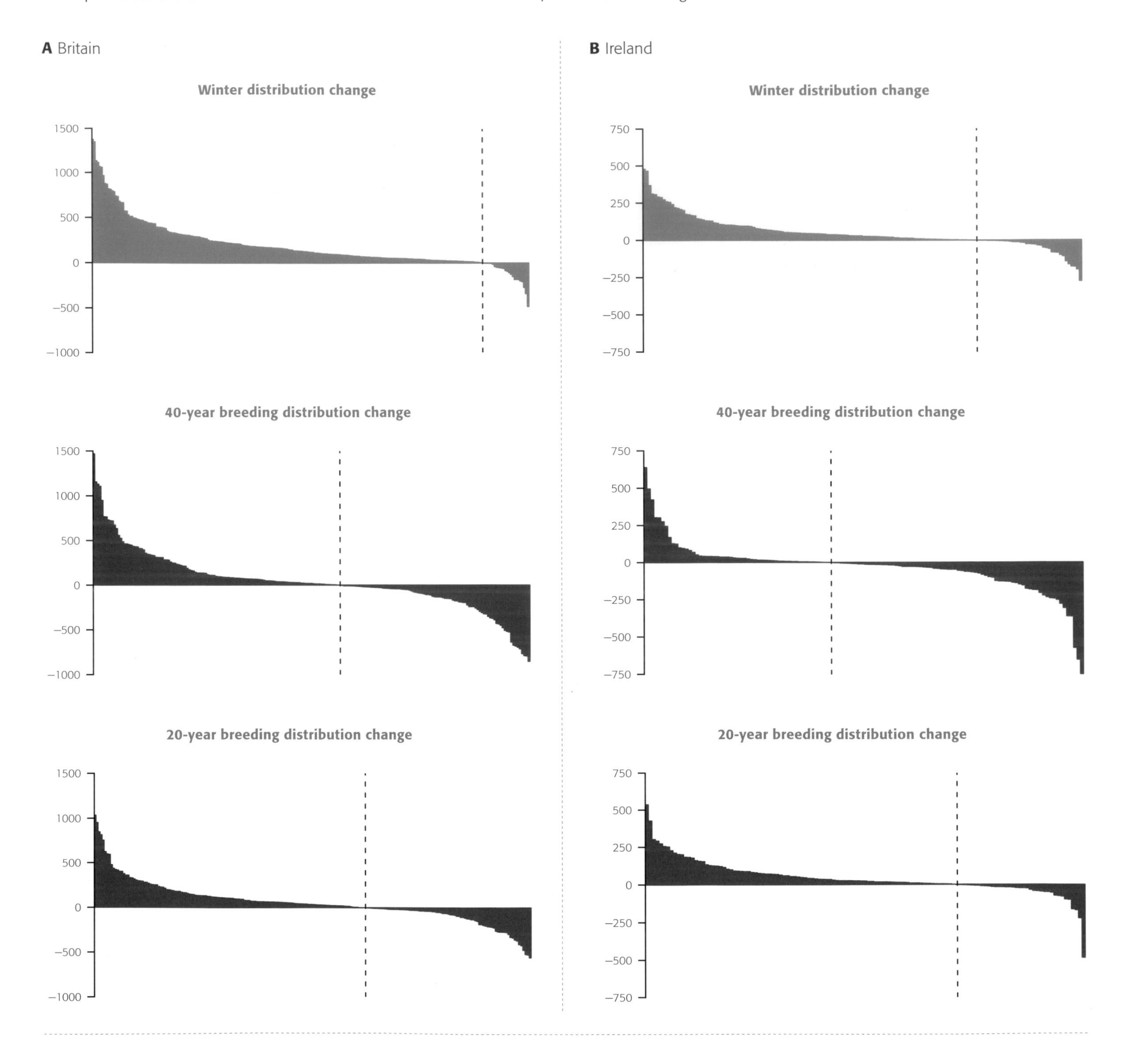

both islands. Subsequently, the 2000s have seen a remarkable series of breeding records in Britain: the first modern-day breeding of the Spoonbill, the first ever of the Purple Heron and Cattle Egret, the second of the Little Bittern and, in 2012, the first breeding of the Great White Egret (see Holling *et al.* 2010b, 2011b; Holt 2013). The breeding ranges and populations of several of these species have also increased in recent years on mainland Europe (e.g. van Turnhout *et al.* 2010; Holt 2013). Additionally there have been winter influxes into Britain and Ireland of Glossy Ibises, with many individuals lingering into the breeding season though as yet showing no signs of breeding.

Among other waterbirds, ducks and geese featured strongly in the lists of large changes of breeding range (Tables 6.5 and 6.6). Many of the range expansions concerned non-native species (see also Section 6.5.2) or ones where native populations have been augmented by introductions (notably the Greylag Goose and Gadwall). However, breeding range contractions were evident in several native wildfowl, including the Pochard and Common Scoter in both Britain and Ireland.

Birds of prey also tended to show range expansions. Some (notably Red Kites and White-tailed Eagles) have been significantly assisted by deliberate reintroductions but most have not. In Britain, since 1968–72, six species appeared on the top 20 lists of species with strongly increasing breeding-range sizes, and this did not include Buzzards and Peregrines, which expanded by 81% and 184% respectively. In Ireland, Buzzards (1976%), Peregrines (276%) and Long-eared Owls (12%) appeared on the equivalent breeding lists and at least five species of birds of prey have re-colonised or been reintroduced since 1988–91 (the Golden Eagle, Marsh Harrier, Red Kite, Short-eared Owl, White-tailed Eagle; Figure 6.1). The one bird of prey with a strongly contracting range was the Short-eared Owl in Britain.

Another striking pattern concerned the status of breeding waders, but unlike other waterbirds and birds of prey, this was predominantly a picture of range contraction both in Britain and Ireland. Only one species, the Avocet in Britain, showed large expansion, whereas seven species showed especially strong range contractions in Britain or Ireland, or in both.

The species with largest winter-range changes (Table 6.7) constituted a mixed bag of late-staying summer visitors, winter visitors and residents. Nevertheless, a high proportion of the winter-range contractions were of resident species also undergoing substantial breeding-range contractions. Also, several seabird species appeared to have contracted their winter ranges—three auks and the Kittiwake were amongst the 20 species that have shown the greatest range contractions in Britain, while five species of gulls featured in Ireland. On the other hand, apparent range expansions for the Sooty Shearwater, Pomarine Skua and Leach's Petrel may have partly reflected increased seawatching effort and increased collation of incidental records for the latest atlas. Obviously most of the seabirds mentioned here are pelagic species whose real winter ranges are far more extensive.

6.5.2 Changes in range size of non-native species

Non-native species cut across several taxonomic and ecological groupings and, since many show strong apparent or real patterns of change, they potentially disrupt our ability to look for more general patterns among native species. Furthermore, introduced species are frequently considered to impact negatively on native species (see Wright *et al.* 2010 for a discussion of non-native waterbirds). For these reasons, range-size changes for non-natives are discussed separately here, and omitted from the analyses presented in Section 6.6.

Changes in the ranges of non-native species are difficult to interpret because some of these species were definitely under-recorded in earlier atlases. Birdwatchers have sometimes ignored non-native species (Wright *et al.* 2010), but a major effort has recently been made to ensure that they are recorded in the national monitoring schemes of the UK and Republic of Ireland. A further problem is that it is not always possible to distinguish native from non-native individuals. A case in point is the Red-crested Pochard. While birds have been released by wildfowlers in some areas of Britain and some may have escaped from collections, there also appears to be a trend of natural expansion in Europe. For example, the large increase of breeding birds in the Netherlands is thought to be mainly attributable to natural expansion (van Turnhout *et al.* 2010). The merging of native and re-established populations of Greylag Geese within Britain is another example (Mitchell *et al.* 2012). Nonetheless, there appears to be a genuine strong pattern of increasing numbers of breeding non-native species and of range expansion in several since the 1970s. So far most have occupied only a small part of the potential habitat available to them so, with suitable conditions, further expansions are to be expected.

Waterbirds feature especially strongly in the list of established non-native species in Britain. Based on a review covering 89 Eurasian and African countries, Wright *et al.* (2010) concluded that the UK has an exceptionally large number of introduced waterbird species, even compared with other western European countries, and that this was probably a legacy of the strong tradition and long history in the UK (especially in Britain) of maintaining waterbird collections.

We can be reasonably confident that non-native species with established self-sustaining populations were well recorded in previous atlases and refer to them as 'established non-natives'. However, most of the species introduced relatively recently do not have self-sustaining populations and are more likely to have been overlooked or ignored in earlier atlases. These species are referred to as 'exotic non-natives'. The distinction between species in these two groups is explained further in the Box 6.1.

In Britain, over the last 40 years, eight of the ten established non-native species expanded their breeding ranges; only the Little Owl (-11%) and Lady Amherst's Pheasant (-46%) contracted. Over the last 20 years, six of the ten species increased range size; the largest contractions were for the Golden Pheasant (-45%) and Ruddy Duck (-21%), the latter due to the active policy of population eradication (Henderson 2009).

The number of exotic non-native species recorded breeding by *Bird Atlas 2007–11* but not by previous atlases was more than twice as high in Britain as in Ireland (Figure 6.1). Particularly in Britain, there were also very large apparent range expansions in several exotic non-natives. Increases of more than 1,000% were recorded between 1968–72 and 2008–11 for breeding Reeves's Pheasants, Muscovy Ducks and Red-crested Pochards. Between 1988–91 and 2008–11, the Black Swan and Helmeted Guineafowl were amongst the three species showing the largest breeding range expansions in Britain, exceeded only by the Little Egret.

TABLE 6.5 SPECIES SHOWING THE LARGEST CHANGES IN SIZE OF BREEDING RANGE BETWEEN 1968–72 AND 2008–11

Percentage changes in numbers of 10-km squares with breeding evidence are shown for the 20 species that underwent the largest range expansions (A) and contractions (B) in Britain and Ireland. The following species were excluded: those that occupied fewer than 25 10-km squares in both periods; those that have never attained *probable* or *confirmed* breeding; those that have only bred in hybrid pairs; exotic non-natives. Established non-native species are shown in italics. Species that benefitted significantly from a reintroduction programme are marked with an asterisk.

Note that the Little Egret, White-tailed Eagle, Icterine Warbler and Common Rosefinch were not recorded in 1968–72 but occurred in >25 British 10-km squares in 2008–11; relevant species in Ireland are the Little Egret and Reed Warbler. Because these species had zero values in 1968–72 it was not possible to calculate a percentage change. Note that the Great Spotted Woodpecker is included for Ireland in Table A on the basis of a migrant male in one 10-km square during 1968–72; breeding did not take place until 2006.

A Species with greatest breeding range expansions between 1968–72 and 2008–11

Britain species	% expansion	Ireland species	% expansion
Cetti's Warbler	6,783%	Common Crossbill	8,133%
Mediterranean Gull	6,500%	Great Spotted Woodpecker	4,000%
Ring-necked Parakeet	4,400%	Buzzard	1,976%
Red Kite*	1,971%	Greylag Goose	1,125%
Avocet	1,663%	*Canada Goose*	600%
Goshawk	1,291%	Gadwall	293%
Egyptian Goose	1,172%	Peregrine	276%
Mandarin Duck	1,144%	Blackcap	249%
Ruddy Duck	1,121%	Siskin	127%
Osprey	1,032%	Raven	95%
Firecrest	935%	Tree Sparrow	78%
Marsh Harrier	876%	Eider	71%
Greylag Goose	733%	Collared Dove	48%
Honey-buzzard	555%	Rock Dove/Feral Pigeon	42%
Black-necked Grebe	382%	Jay	31%
Whooper Swan	375%	Carrion Crow	25%
Dartford Warbler	352%	Long-eared Owl	12%
Gadwall	351%	Garden Warbler	11%
Hobby	295%	*Pheasant*	10%
Common/Scottish Crossbill	253%	Great Crested Grebe	9%

B Species with greatest breeding range contractions between 1968–72 and 2008–11

Britain species	% contraction	Ireland species	% contraction
Red-backed Shrike	-88%	Corn Bunting	-100%
Cirl Bunting	-84%	Grey Partridge	-95%
Hawfinch	-76%	Nightjar	-95%
Wryneck	-73%	Turtle Dove	-94%
Capercaillie	-73%	Corncrake	-91%
Corncrake	-72%	Twite	-80%
Corn Bunting	-56%	Curlew	-78%
Willow Tit	-55%	Whinchat	-76%
Turtle Dove	-51%	Woodcock	-73%
Woodcock	-50%	Dunlin	-69%
Short-eared Owl	-48%	Red Grouse	-66%
Whinchat	-47%	Little Tern	-62%
Ring Ouzel	-43%	Yellowhammer	-61%
Common Scoter	-43%	Ring Ouzel	-57%
Nightjar	-43%	Redshank	-55%
Redshank	-43%	Black-headed Gull	-55%
Nightingale	-43%	Pochard	-53%
Stone-curlew	-42%	Lapwing	-53%
Lesser Spotted Woodpecker	-41%	Golden Plover	-50%
Tree Sparrow	-41%	Red-breasted Merganser	-50%

TABLE 6.6 SPECIES SHOWING THE LARGEST CHANGES IN SIZE OF BREEDING RANGE BETWEEN 1988–91 AND 2008–11

Percentage changes in numbers of 10-km squares with breeding evidence are shown for the 20 species that underwent the largest range expansions (A) and contractions (B) in Britain and Ireland. The following species were excluded: those that occupied fewer than 25 10-km squares in both periods; those that have never attained *probable* or *confirmed* breeding; those that have only bred in hybrid pairs; exotic non-natives. Established non-native species are shown in italics. Species that benefitted from a reintroduction programme are marked with an asterisk.

Note that the Little Egret is included in Table A on the basis of two 10-km squares in Britain and one in Ireland occupied in 1988–91, although breeding did not take place until 1996 and 1997 respectively. Note also that the Great Spotted Woodpecker was not recorded in Ireland in 1988–91 but occurred in >25 Irish 10-km squares in 2010, hence it was not possible to calculate a percentage change.

A Species with greatest breeding range expansions between 1988–91 and 2008–11

Britain species	% expansion	Ireland species	% expansion
Little Egret	16,350%	**Little Egret**	12,600%
Icterine Warbler	2,600%	**Buzzard**	458%
Mediterranean Gull	843%	**Greylag Goose**	326%
Red Kite*	728%	***Red-legged Partridge***	200%
Bittern	531%	**Reed Warbler**	177%
Common Rosefinch	440%	**Blackcap**	150%
Avocet	404%	***Canada Goose***	121%
Cetti's Warbler	375%	**Gadwall**	120%
White-tailed Eagle*	318%	**Jay**	94%
Marsh Warbler	220%	**Siskin**	90%
Dartford Warbler	198%	**Grasshopper Warbler**	74%
Honey-buzzard	167%	**Long-eared Owl**	70%
Egyptian Goose	163%	**Tree Sparrow**	67%
Greylag Goose	132%	**Rock Dove/Feral Pigeon**	66%
Mandarin Duck	121%	**Common Gull**	59%
Marsh Harrier	114%	**Common Crossbill**	58%
Firecrest	109%	**Lesser Black-backed Gull**	53%
Woodlark	107%	**Collared Dove**	52%
Goshawk	106%	**Lesser Redpoll**	51%
Gadwall	100%	**Hen Harrier**	45%

B Species with greatest breeding range contractions between 1988–91 and 2008–11

Britain species	% contraction	Ireland species	% contraction
Fieldfare	-78%	**Curlew**	-73%
Goldeneye	-74%	**Dunlin**	-71%
Hawfinch	-65%	**Corncrake**	-69%
Ruff	-62%	**Grey Partridge**	-66%
Common Scoter	-53%	**Whinchat**	-64%
Willow Tit	-50%	**Twite**	-57%
Whimbrel	-49%	**Ring Ouzel**	-48%
Redwing	-45%	**Redshank**	-47%
Golden Pheasant	-45%	**Little Tern**	-44%
Short-eared Owl	-39%	**Sandwich Tern**	-43%
Golden Oriole	-38%	**Golden Plover**	-42%
Pochard	-38%	**Pochard**	-40%
Whinchat	-37%	**Yellowhammer**	-38%
Wood Warbler	-37%	**Lapwing**	-33%
Corn Bunting	-36%	**Quail**	-32%
Redshank	-35%	**Wood Warbler**	-32%
Turtle Dove	-34%	**Red-breasted Merganser**	-26%
Lesser Spotted Woodpecker	-34%	**Arctic Tern**	-24%
Black-tailed Godwit	-32%	**Woodcock**	-23%
Woodcock	-30%	**Stock Dove**	-21%

TABLE 6.7 SPECIES SHOWING THE LARGEST CHANGES IN SIZE OF WINTER RANGE BETWEEN 1981–84 AND 2007–11

Percentage changes in numbers of occupied 10-km squares are shown for the 20 species that underwent the largest range expansions (A) and contractions (B) in winter in Britain and Ireland. The following species were excluded: those that occupied fewer than 25 10-km squares in both periods, and exotic non-natives. Established non-native species are shown in italics. Species that benefitted from a reintroduction programme are marked with an asterisk.

Recording effort in winter has increased in parts of northwest Scotland, western Ireland and parts of upland Britain (see Chapter 5) to the extent that some apparent range changes may be positively biased. However, for these tables of the largest range expansions and range contractions in winter, the magnitudes of the changes are so large that there can be little doubt that they are real.

A Species with greatest winter range expansions between 1981–84 and 2007–11

Britain species	% expansion	Ireland species	% expansion
Little Egret	112,900%	**Ring-necked Duck**	3,700%
Great White Egret	7,800%	**Waxwing**	1,857%
Cattle Egret	7,300%	**Goosander**	1,800%
Sooty Shearwater	2,700%	**Common Crossbill**	1,150%
Yellow-browed Warbler	1,667%	**Red Kite***	967%
White-tailed Eagle*	1,429%	**Buzzard**	885%
Red Kite*	1,250%	**Mediterranean Gull**	638%
Pomarine Skua	917%	**Pink-footed Goose**	369%
Waxwing	903%	**Green-winged Teal**	289%
Glossy Ibis	733%	**Black-throated Diver**	247%
Spoonbill	669%	**Ring-billed Gull**	230%
Red-breasted Goose	563%	**Blackcap**	225%
Green-winged Teal	531%	**Siskin**	189%
Marsh Harrier	530%	**Velvet Scoter**	147%
Crane	492%	**Gadwall**	143%
Leach's Petrel	473%	**Slavonian Grebe**	118%
Dartford Warbler	449%	**Whimbrel**	111%
Cetti's Warbler	433%	**Greylag Goose**	111%
Egyptian Goose	423%	**Hen Harrier**	100%
Richard's Pipit	414%	**Jay**	86%

B Species with greatest winter range contractions between 1981–84 and 2007–11

Britain species	% contraction	Ireland species	% contraction
House Martin	-57%	**Bewick's Swan**	-90%
Capercaillie	-55%	**Grey Partridge**	-61%
Puffin	-52%	**Pochard**	-49%
Willow Tit	-42%	**Yellowhammer**	-38%
Kittiwake	-31%	**Herring Gull**	-38%
Corn Bunting	-27%	**Carrion Crow**	-32%
Grey Partridge	-23%	**Twite**	-30%
Guillemot	-21%	**Kittiwake**	-26%
Lesser Spotted Woodpecker	-20%	**Curlew**	-23%
Tree Sparrow	-20%	**Glaucous Gull**	-23%
Short-eared Owl	-19%	**Red-breasted Merganser**	-20%
Pochard	-13%	**Great Black-backed Gull**	-20%
Hooded Crow	-13%	**Black-headed Gull**	-19%
Bewick's Swan	-12%	**Coot**	-18%
Cirl Bunting	-12%	**Spotted Redshank**	-16%
Razorbill	-10%	**Stock Dove**	-14%
Yellowhammer	-9%	**Skylark**	-14%
Wheatear	-7%	**Tufted Duck**	-14%
Black Grouse	-5%	**Lapwing**	-12%
Great Black-backed Gull	-5%	**Goldeneye**	-11%

BOX 6.1 SPECIES GROUPED BY TAXONOMY, MIGRATION STRATEGY AND ORIGIN

The following groupings of species are used for analyses in Table 6.9 and Figure 6.4. Note that the 'Resident passerines' and 'Migrant passerines' are small insectivorous species chosen so that direct comparisons can be made between them in terms of range changes. 'Migrants' refers to species wintering south of the Sahara. 'Arid-zone migrants' and 'Humid-zone migrants' are species wintering in West Africa in the Sahel and in more southerly tropical regions, respectively (Ockendon *et al.* 2012).

KEN HINDMARCH/GRAHAM CATLEY http://pewit.blogspot.co.uk

Though closely related, the members of this species pair winter in different parts of West Africa.

Wildfowl
Mute Swan, Whooper Swan, Greylag Goose, Shelduck, Wigeon, Gadwall, Teal, Mallard, Pintail, Garganey, Shoveler, Pochard, Tufted Duck, Eider, Common Scoter, Goldeneye, Red-breasted Merganser, Goosander.

Raptors
Honey-buzzard, Red Kite, White-tailed Eagle, Marsh Harrier, Hen Harrier, Montagu's Harrier, Goshawk, Sparrowhawk, Buzzard, Golden Eagle, Osprey, Kestrel, Merlin, Hobby, Peregrine.

Waders
Oystercatcher, Avocet, Stone-curlew, Little Ringed Plover, Ringed Plover, Dotterel, Golden Plover, Lapwing, Dunlin, Ruff, Snipe, Woodcock, Black-tailed Godwit, Whimbrel, Curlew, Common Sandpiper, Greenshank, Wood Sandpiper, Redshank, Red-necked Phalarope.

Gulls and terns
Kittiwake, Black-headed Gull, Mediterranean Gull, Common Gull, Lesser Black-backed Gull, Herring Gull, Great Black-backed Gull, Little Tern, Sandwich Tern, Common Tern, Roseate Tern, Arctic Tern.

Resident passerines
Goldcrest, Firecrest, Blue Tit, Great Tit, Crested Tit, Coal Tit, Willow Tit, Marsh Tit, Bearded Tit, Cetti's Warbler, Long-tailed Tit, Dartford Warbler, Nuthatch, Treecreeper, Wren, Robin, Stonechat, Dunnock, Grey Wagtail, Pied Wagtail, Meadow Pipit, Rock Pipit, Chaffinch.

Migrant passerines
Sand Martin, Swallow, House Martin, Wood Warbler, Willow Warbler, Garden Warbler, Lesser Whitethroat, Whitethroat, Grasshopper Warbler, Savi's Warbler, Icterine Warbler, Sedge Warbler, Marsh Warbler, Reed Warbler, Spotted Flycatcher, Nightingale, Pied Flycatcher, Redstart, Whinchat, Wheatear, Yellow Wagtail, Tree Pipit.

Arid-zone migrants
Sand Martin, Lesser Whitethroat, Whitethroat, Grasshopper Warbler, Sedge Warbler, Redstart, Wheatear, Yellow Wagtail.

Humid-zone migrants
Common Sandpiper, Turtle Dove, Wood Warbler, Willow Warbler, Garden Warbler, Reed Warbler, Spotted Flycatcher, Nightingale, Pied Flycatcher, Whinchat, Tree Pipit.

All migrants
Includes all the 'migrant passerines' plus the following: Garganey, Honey-buzzard, Hobby, Spotted Crake, Corncrake, Little Ringed Plover, Ruff, Common Sandpiper, Turtle Dove, Cuckoo, Nightjar, Swift, Wryneck, Golden Oriole, Red-backed Shrike.

Established non-native species
These are based on the species classified by BOU (BOU 2013) as Category C and defined as: 'Species that, although introduced, now derive from the resulting self-sustaining populations'. The group comprises: Canada Goose, Egyptian Goose, Mandarin Duck, Ruddy Duck, Red-legged Partridge, Pheasant, Golden Pheasant, Lady Amherst's Pheasant, Ring-necked Parakeet, Little Owl. These species have been relatively consistently recorded through the series of atlases and range-change estimates can be used with confidence.

Exotic non-native species
This group basically comprises the remaining non-native species having excluded the established species mentioned in the previous group. To a large extent these are introductions, human-assisted transportees and escapes from captivity and include the Black Swan and other exotic waterbirds, Indian Peafowl, rare pheasants and various parrots. Their breeding populations (if any) are thought not to be self-sustaining. The list is slightly modified in winter because a handful of these species, notably Barnacle Goose, occur mainly as wild populations in winter. There are inevitably borderline cases in winter. For example, we have treated Red-breasted Goose as native in winter because far more records occur in places where one might expect naturally occurring birds than is the case with Snow Goose which we have treated as an exotic non-native. These exotic non-native species have been inconsistently reported in atlases and range-change estimates should be used with caution.

BOX 6.2 SPECIES GROUPED BY BREEDING-SEASON HABITAT

Species associated with distinct habitat types may have shown common changes in range or abundance which might point to common causes.

The following groupings of species are used for analyses in Table 6.8 and Figure 6.4. Groupings are based on Gibbons *et al.* (1993) and Fuller *et al.* (1995) but updated with major additions such as the Little Egret as appropriate. Note that the Red Kite was treated as an upland species by Gibbons *et al.* (1993) but not in this classification. Species are not necessarily confined to the habitat types under which they are listed. There may be differences in habitat selection of some species between Britain and Ireland (see text) and species marked with an asterisk are excluded from that habitat group for Ireland. Note that very scarce species not included in the analyses are not listed.

Woodland birds

Capercaillie, Honey-buzzard, Goshawk, Sparrowhawk, Woodcock, Tawny Owl, Long-eared Owl, Wryneck, Green Woodpecker, Great Spotted Woodpecker, Lesser Spotted Woodpecker, Golden Oriole, Jay, Goldcrest, Firecrest, Blue Tit, Great Tit, Crested Tit, Coal Tit, Willow Tit, Marsh Tit, Long-tailed Tit, Wood Warbler, Chiffchaff, Willow Warbler, Blackcap, Garden Warbler, Lesser Whitethroat, Nuthatch, Treecreeper, Wren, Blackbird, Fieldfare, Song Thrush, Redwing, Spotted Flycatcher, Robin, Nightingale, Pied Flycatcher, Redstart, Dunnock, Tree Pipit, Chaffinch, Siskin, Lesser Redpoll, Common & Scottish Crossbills, Bullfinch, Hawfinch.

Farmland birds

Grey Partridge, Quail, Montagu's Harrier, Kestrel, Hobby, Corncrake, Stone-curlew, Lapwing, Stock Dove, Woodpigeon, Turtle Dove, Barn Owl, Jackdaw, Rook, Skylark, Whitethroat, Starling, Tree Sparrow, Yellow Wagtail, Greenfinch, Goldfinch, Linnet, Yellowhammer, Cirl Bunting, Reed Bunting, Corn Bunting.

Wetland birds

Mute Swan, Whooper Swan, Greylag Goose, Gadwall, Mallard, Pintail, Garganey, Shoveler, Pochard, Tufted Duck, Goldeneye, Bittern, Little Egret, Grey Heron, Spoonbill, Little Grebe, Great Crested Grebe, Slavonian Grebe, Black-necked Grebe, Marsh Harrier, Osprey, Water Rail, Spotted Crake, Moorhen, Coot, Crane, Little Ringed Plover, Kingfisher, Bearded Tit, Cetti's Warbler, Savi's Warbler, Sedge Warbler, Marsh Warbler, Reed Warbler.

Upland birds

Wigeon, Teal, Common Scoter*, Red-breasted Merganser*, Goosander, Red Grouse, Ptarmigan, Black Grouse, Red-throated Diver, Black-throated Diver, Hen Harrier, Buzzard, Golden Eagle, Merlin, Peregrine, Dotterel, Golden Plover, Dunlin, Snipe, Whimbrel, Curlew, Common Sandpiper, Greenshank, Wood Sandpiper, Red-necked Phalarope, Arctic Skua, Great Skua, Short-eared Owl, Raven, Dipper, Ring Ouzel, Whinchat, Wheatear, Grey Wagtail, Meadow Pipit, Twite, Snow Bunting.

Coastal Birds

Shelduck, Eider, Fulmar, Manx Shearwater, Storm Petrel, Gannet, Cormorant, Shag, Avocet, Ringed Plover, Kittiwake, Lesser Black-backed Gull, Herring Gull, Great Black-backed Gull, Little Tern, Sandwich Tern, Roseate Tern, Arctic tern, Guillemot, Razorbill, Black Guillemot, Puffin, Chough, Rock Pipit.

6.6 RANGE AND ABUNDANCE CHANGES ANALYSED BY SPECIES GROUPS

FURTHER INFORMATIVE PATTERNS in the change of the British and Irish avifauna emerged when range changes were examined for native breeding species grouped in various ways. Table 6.8 shows range changes for species grouped according to broad breeding-season habitat type. Table 6.9 presents range changes for four groups of non-passerines within which the species have broadly similar lifestyles (wildfowl, waders, raptors, gulls & terns) and for groups of species with similar migration characteristics. Boxes 6.1 and 6.2 list the species assigned to each of these groups.

To clarify *where* the changes were most pronounced for different species groups, we examined 20-km resolution abundance change information (see Section 4.5 for derivation) to classify each member of a species group as increasing (scored as +1) or decreasing (scored as -1) in each 20-km square during 1988–91 to 2008–11. These scores were then averaged across the occurring members of the species group to produce a composite change for the group in each 20-km square. The resulting maps (Figure 6.4) show visually arresting geographical differences where the species groups were increasing or decreasing in abundance and occupancy.

In particular over the last 20 years, abundance of both woodland and farmland species in Ireland have increased relative to Britain, while the abundance of migrant passerines also showed contrasts between the two islands. Interestingly, compared with Britain, Ireland presented a relatively uniform picture of increases in abundance change of all species combined. These differences are discussed in more detail below and spatial variations in range change are considered further in Section 6.8.

6.6.1 Main messages for species grouped by habitat type

Both in Britain and in Ireland, the average changes of breeding species associated with five broad habitat types differed considerably (Table 6.8). We have assumed that most species showed the same *broad* patterns of habitat selection in Britain as in Ireland, though we acknowledge that differences are to be expected (Wesołowski & Fuller 2012). We are aware of several species where habitat use differs between the two islands but these differences are mostly subtle and insufficient to invalidate the approach taken here. However, in a few cases the differences are more striking (Nairn & O'Halloran 2012). For example, in Ireland Skylarks use farmland in winter but now avoid cultivated land in the breeding season, being mainly associated with moorland and peatland (Copland *et al.* 2011). Nesting Hen Harriers appear be more dependent on conifer plantations in Ireland than Britain, though they still hunt over adjacent moorland. Unlike in Britain, Red-breasted Mergansers and Common Scoters are not associated with upland water bodies in Ireland.

The main messages that emerge are as follows:

- Woodland birds in Ireland have shown a general pattern of increase in range size over the last 20 years, whereas in Britain, there was a more even balance between expansion and contraction.
- Strong range contractions were evident in many farmland birds in both islands over the last 40 years, but less so when considering only the latter half of that period. Nonetheless, several species continued to show large contractions. Although, farmland species appeared to have generally increased in abundance during this more recent period in Ireland and parts of Scotland (Figure 6.4), this does not indicate a recovery. In Ireland, the suite of farmland species is less diverse than in Britain and perhaps tends to give more weight to relatively successful generalists than to less successful granivores such as Yellowhammers and Corn Buntings. Furthermore, in both islands the population levels of most farmland birds remain at a far lower level than 40 years ago and some of the species continue to decrease.
- In Britain, overall range changes of wetland birds have been strongly positive whereas in Ireland, over the last 40 years, wetland species showed more marked range contractions. However, since 1988–91 wetland species have tended to expand their breeding ranges throughout much of Ireland, though there were some abundance declines, most notably in the east and parts of the southern midlands (Figure 6.4).
- Range contractions in upland birds have been considerably stronger in Ireland than Britain over the past 40 years. There was a widespread pattern of decline in the abundances of upland birds over the last 20 years with especially striking declines in western Ireland, southern Scotland and northern England (Figure 6.4).
- Range change in many coastal bird species has been negative in both Britain and Ireland over the 40 year period and in Britain over the 20-year period.

6.6.2 Possible drivers of patterns for species grouped by habitat type

With a few exceptions, the drivers of the patterns outlined above are generally unclear but some possibilities are indicated below and there is further relevant discussion in Section 6.9.2.

- Several species of woodland birds are severely declining in both population and range size, but the mechanisms are potentially numerous and, unlike farmland birds, this has not translated into a large general pattern of range contraction and there is no obvious overarching cause (Fuller *et al.* 2005).
- The range contractions that have accompanied the large population declines of many farmland bird species over the last 40 years have been attributed mainly to intensification of agriculture (Fuller *et al.*, 1995; Newton 2004; Wilson *et al.* 2009).
- The increase in range sizes of many wetland species, at least in Britain, may in part be due to two factors. First, the increase

FIGURE 6.4 20-YEAR SPATIAL ABUNDANCE CHANGES FOR GROUPS OF BREEDING SPECIES

Patterns of change in breeding abundance are shown across Britain and Ireland for the period 1988–91 to 2008–11. Red or brown shading indicates that the members of the species group, on average, increased in abundance or decreased in abundance, respectively, in a 20-km square. Darker shading indicates that a greater proportion of the species present in the 20-km square showed an increase (or decrease), with the darkest shading indicating that all group members showed the same direction of change. The shading is independent of the number of species in a group and is only shown for 20-km squares with at least four members of a group present. Underlying data are effort-controlled (i.e. TTV data are used). 'All breeding species' includes species with at least *probable* breeding evidence and excludes exotic non-natives. See Boxes 6.1 and 6.2 for the species composition of each group.

in numbers of man-made waterbodies in Britain must have had a profound effect on waterbirds in the latter half of the 20th century, though this has not been fully assessed (Fuller & Ausden 2008). Second, many of the most important wetlands now receive some form of protection through a combination of national and international legislation; conservationists have also become adept at creating and managing high quality habitats for a range of wetland species (Ausden & Fuller 2009). In Ireland, losses of wetlands to drainage may have contributed to the range contractions evident since 1968–72. Whilst similar habitat changes occurred in Britain, these have possibly been offset by larger numbers of man-made wetlands.

- The overall range contractions in upland species over the last 40 years and the abundance declines over the last 20 years were higher than those for the other habitats. The reasons are probably various but, as outlined in Section 6.6.3 for waders, changes in habitat quality and predation pressure are relevant. In addition, altitudinal shifts, which would result in range size change, are to be expected in mountain birds as climate warms (Chamberlain *et al.* 2012) though we are unaware of firm evidence that this happening yet in Britain or Ireland. There is, however, evidence that climate change may be reducing the food supplies of some upland species, notably the Golden Plover (Pearce-Higgins *et al.* 2010).
- The relatively high level of range shrinkage in coastal birds was driven mainly by apparent losses in several auks, gulls and terns. This may be associated with complex changes in quantity and quality of food supplies linked with fisheries and sea temperatures, which affect species in different ways (e.g. Frederiksen *et al.* 2004, 2006; Furness 2007). Other apparent changes may partly reflect coverage where colonies have not benefitted from dedicated seabird expeditionary work (see Chapter 5). Some colonies may have been reduced, or even exterminated, by predators including the American Mink (Craik 1997).

6.6.3 Patterns for selected non-passerine species groups

Breeding raptors have shown general range expansion in Britain over both time periods (Table 6.9). Some of the raptors have benefitted from increased levels of protection from persecution and recovery from organochlorine impacts (Newton 2013), though reintroductions have played a part. Abundance gains for raptors were especially striking in the southern half of Britain. The situation in Ireland was similar since 1988–91.

In contrast, two groups stood out as having undergone large general contractions of breeding range. Waders and gulls & terns both showed large decreases since 1968–72, with the losses being considerably stronger in Ireland than in Britain. It should be noted, however, that the *1988–91 Breeding Atlas* coincided with the second national seabird census, so coverage of seabird colonies then was better compared with 2008–11 for at least some regions within Britain & Ireland. In the period 1988–91 to 2008–11, breeding waders experienced markedly larger contractions than gulls & terns. In terms of abundance, the overall impression was one of widespread declines in breeding waders (Figure 6.4). In Britain this contrasted

TABLE 6.8 BREEDING-SEASON RANGE CHANGES FOR SPECIES BROADLY ASSOCIATED WITH DIFFERENT HABITAT TYPES

Medians (and quartiles) are given for percentage range changes of species in each group by island. Species composition of the groups is given in Box 6.2. Species occurring in fewer than 25 10-km squares were excluded, as were all non-native species. N = number of species in each group.

Kruskal-Wallis tests for differences in median change among habitats were statistically significant for both islands over the 40-year periods ($P < 0.01$). Over the 20-year period, differences in median change were significant for Britain ($P < 0.01$) but not for Ireland. Within each habitat group, differences in range changes between Britain and Ireland were compared and the following were statistically significant: woodland, 20-year period, $P < 0.05$; wetland, 40-year period, $P < 0.001$; upland 40-year period, $P < 0.05$.

Habitat group	Change since	Britain			Ireland		
		Median % range change	Quartiles	*N*	Median % range change	Quartiles	*N*
Woodland species	1968–72	0.7	-22.5, 13.7	43	0.9	-0.4, 6.8	24
	1988–91	3.5	-13.8, 13.9	46	8.0	2.5, 38.6	25
Farmland species	1968–72	-5.3	-39.4, 0.2	26	-2.8	-46.7, 0.4	17
	1988–91	0.2	-10.0, 7.6	26	2.8	-7.9, 19.8	17
Wetland species	1968–72	18.5	1.1, 94.8	24	-6.5	-19.6, -0.6	12
	1988–91	15.6	1.5, 57.7	27	8.6	6.0, 26.5	13
Upland species	1968–72	-10.5	-19.5, 18.7	33	-34.5	-50.0, -7.8	17
	1988–91	-5.3	-17.2, 3.5	34	-4.8	-14.7, 13.8	17
Coastal species	1968–72	-5.9	-17.0, 20.8	20	-4.6	-25.3, 3.8	16
	1988–91	-4.6	-13.8, 18.6	21	4.6	-1.6, 19.9	16

with the general picture of increasing wildfowl abundances. Various kinds of habitat loss and deterioration, both in Britain and in Ireland, may be implicated in the range contractions and abundance declines of waders in lowland and upland contexts, as a result of agricultural changes, peat extraction and afforestation (Fuller & Ausden 2008; Nairn & O'Halloran 2012). However, generalist predators have also increased in recent decades with potential consequences for ground-nesting birds, including waders (Fletcher *et al.* 2010). Changing food availability is likely to be a major driver of change for gulls & terns, although increased pressure from the American Mink, Red Fox and other predators cannot be discounted.

6.6.4 Migrants and residents

Over the period 1968–72 to 2008–11, there was a tendency for range sizes of migrant species (all species combined) to contract in Britain and Ireland (Table 6.9). Over the last 20 years average range sizes of migrants have not changed greatly but there has been a strong geographical divergence in patterns of abundance change. In the southern half of Britain, abundances of migrant passerines have tended to decrease, whereas in the north and in Ireland increases have been more evident (Figure 6.4). A comparison of range changes of migrant and resident passerines in Britain showed that migrants had tended to contract whereas residents had expanded during 1968–72 to 2008–11. The expansion of resident passerines was probably due, in part, to a sequence of relatively mild winters up to the late 2000s enhancing overwinter survival.

Several studies have demonstrated that breeding populations of migrant birds in Britain and elsewhere in Europe are declining (e.g. Böhning-Gaese & Bauer 1996; Sanderson *et al.* 2006; Hewson *et al.* 2007) and it now appears that in Britain these abundance declines are sufficiently strong to register as reductions of range size. The range contractions were especially strong amongst species wintering in the humid tropics of West Africa and, for Britain during 1988–91 to 2008–11, were significantly larger than those for species wintering in the arid zone of West Africa. Migrants that winter in the arid zone showed higher levels of losses over the full

TABLE 6.9 BREEDING-SEASON RANGE CHANGES FOR SELECTED SPECIES GROUPS

Medians (and quartiles) are given for percentage range changes of species in each group by island. Species composition of the groups is given in Box 6.1. Species occurring in fewer than 25 10-km squares were excluded, as were all non-native species. N = number of species in each group.

Kruskal-Wallis tests for differences in median change were conducted for (i) resident and migrant passerines and (ii) arid-zone and humid-zone migrants. Out of eight tests (2 islands, 2 time periods) the following three contrasts were significantly different at $P < 0.05$. Migrant passerines in Britain tended to show larger range contractions than resident passerines over the 40-year period. Humid-zone migrants, in both Britain and Ireland, tended to show larger range contractions than arid-zone migrants over the 20-year period. Within each group, differences in range changes between Britain and Ireland were compared and the only significant difference was for waders over the 40-year period ($P < 0.01$).

Species group	Change since	Britain			Ireland		
		Median % range change	Quartiles	*N*	Median % range change	Quartiles	*N*
Wildfowl	1968–72	8.7	-13.7, 46.3	15	-4.1	-33.1, 2.8	8
	1988–91	3.5	-4.6, 18.6	17	8.9	8.5, 25.1	9
Raptors	1968–72	80.8	10.9, 880.0	13	-3.7	-7.8, 276.1	6
	1988–91	51.5	-1.4, 106.4	14	30.0	10.9, 43.9	6
Waders	1968–72	-15.2	-30.0, 3.2	16	-50.0	-55.0, -34.3	9
	1988–91	-14.1	-25.8, 0.0	17	-23.5	-42.1, -8.3	9
Gulls & terns	1968–72	-8.7	-13.8, 12.5	10	-28.5	-42.5, -5.7	8
	1988–91	9.1	-13.8, 21.2	10	-0.7	-9.3, 33.6	8
Resident passerines	1968–72	2.8	0.5, 18.7	21	0.4	-2.8, 1.6	15
	1988–91	5.5	2.2, 24.7	23	3.7	1.4, 10.1	15
Migrant passerines	1968–72	-9.8	-31.1, 3.2	19	0.1	-20.9, 2.6	11
	1988–91	2.0	-19.7, 12.2	19	5.4	-3.2, 26.0	11
Arid-zone migrants	1968–72	-12.7	-22.7, 1.5	8	-5.6	-7.3, 0.1	5
	1988–91	9.4	-10.6, 13.5	8	26.0	24.4, 33.6	5
Humid-zone migrants	1968–72	-13.8	-42.6, 2.8	11	-20.9	-40.9, 2.4	5
	1988–91	-16.7	-34.4, 2.1	11	-3.2	-14.7, 3.8	5
All migrants	1968–72	-7.3	-31.1, 3.3	29	-7.3	-25.7, 2.4	15
	1988–91	2.1	-16.7, 17.9	31	1.7	-13.2, 24.4	15

FIGURE 6.5 THE RELATIONSHIPS BETWEEN POPULATION TRENDS AND RANGE CHANGES FOR BREEDING SPECIES

Separate figures are shown for Britain (A) and Ireland (B). The British population trends were derived from BBS data and span the period 1994–2011; the Irish population trends were derived by combining CBS data from the Republic of Ireland with BBS data from Northern Ireland and they span the period 1998–2011. In each case, only species occurring in >30 monitoring squares were used. The range change index is an effort-controlled measure of range change lying between -1 and +1 and is based on TTVs undertaken in the *1988–91 Breeding Atlas* and *Bird Atlas 2007–11*.

For presentation purposes, the Ring-necked Parakeet (trend = 942%, range = 0.80) and Red Kite (trend = 727%, range = 0.91) were omitted from A; the Buzzard (trend = 10,948%, range = 0.85), Blackcap (trend = 731%, range = 0.67) and Turtle Dove (trend = 48%, range = -1.00) were omitted from B; neither affected the overall relationships. Correlation coefficients, including the five species not shown in the figures: Britain $r = 0.69$, $P < 0.0001$, $n = 103$; Ireland $r = 0.42$, $P = 0.0011$, $n = 58$.

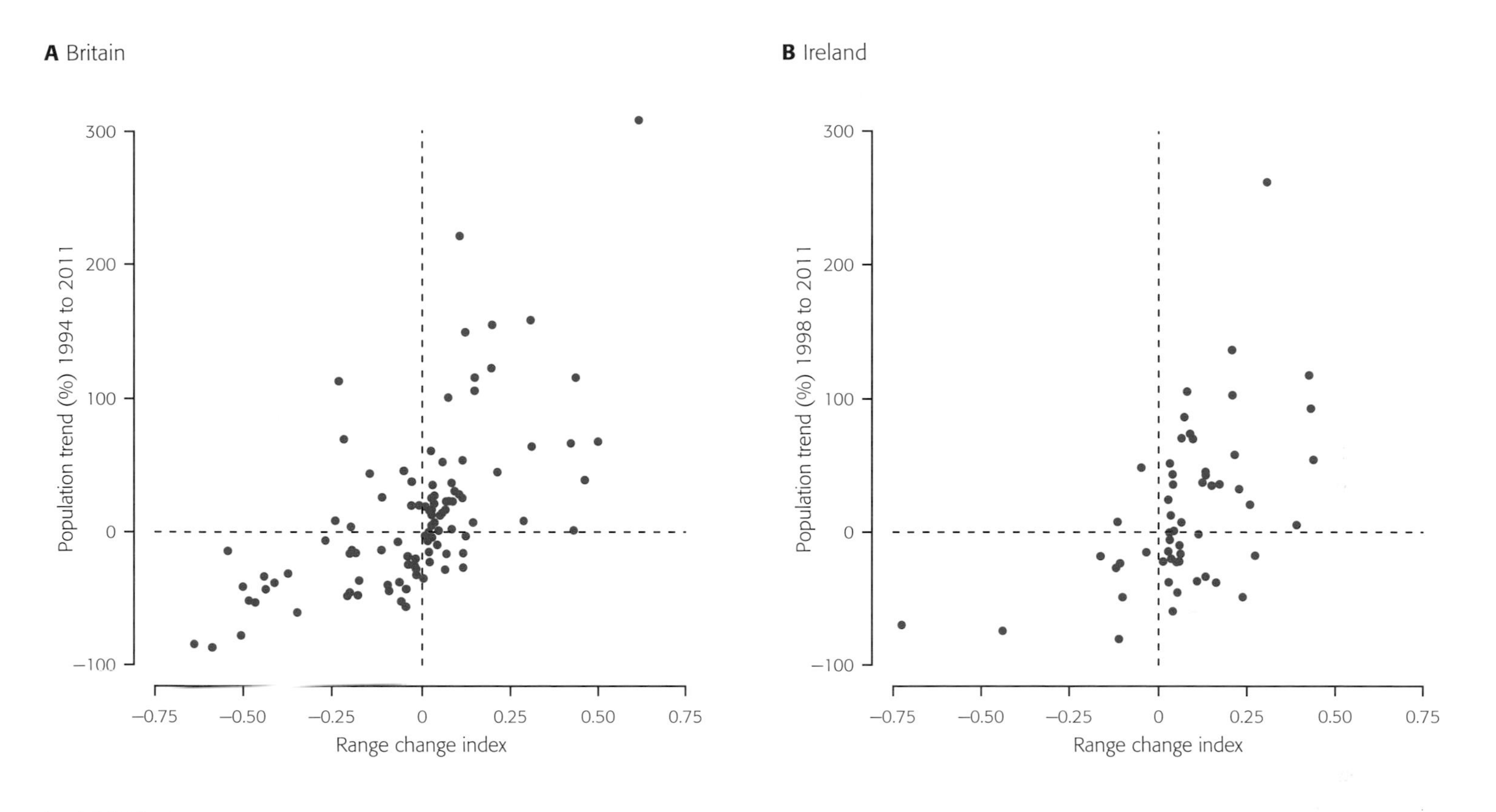

40-year period than between 1988–91 to 2008–11. This relationship between the changing status of migrants and their wintering area has previously been demonstrated using abundance trend data (Ockendon *et al.* 2012).

6.7 RANGE SIZE RELATIONSHIPS

HERE WE CONSIDER whether general patterns can be detected when range changes are examined across large numbers of species. We do so by asking two straightforward questions.

First, how closely did changes in range size correspond to changes in population levels as measured by independent data (largely of relative abundance) from annual bird monitoring schemes? The expectation is that although there should be a close fit between the two measures, range size would not be sensitive to small changes in population level. One would expect that as a species becomes more abundant it should also become more widespread and occur in more 10-km squares (Newton 1997). The relationships for those species for which there was both independent range change and population trend data are presented in Figure 6.5. To limit the potential for observer effort biases, the range-change figures were effort-controlled (i.e. they use TTV data only) and relate to the period 1988–91 to 2008–11. This was the period that most closely matched that for which population trend data were available. Both in Britain and Ireland there were positive relationships between change in breeding-range size and change in breeding-population size but there was considerable scatter. As expected, the majority of species showed the same direction of change in range size and population (79% in Britain and 64% in Ireland). A higher proportion of species showed negative range size changes and negative population size changes in Britain (33%) than in Ireland (14%).

Second, how consistent were range changes between Britain and Ireland? The hypothesis is that species should show similar patterns in the two islands, if they are affected by the same processes. Effort-controlled range change indices were also used in this analysis. For those species occurring in both Britain and Ireland, as one might expect, there was a positive relationship between range changes in the two islands (Figure 6.6). Although there was a large core of species showing very similar changes, there was much scatter around this core, with a small number of species, such as the Barn Owl and Short-eared Owl, showing opposite trends in the two islands. This

FIGURE 6.6 THE RELATIONSHIP BETWEEN BREEDING-SEASON RANGE CHANGE IN BRITAIN AND IRELAND

Changes were calculated between the *1988–91 Breeding Atlas* and *Bird Atlas 2007–11* and are based on effort-controlled data in the form of indices lying between -1 and +1. $r_s = 0.58$, $P < 0.0001$, $n = 160$ species occurring in both islands.

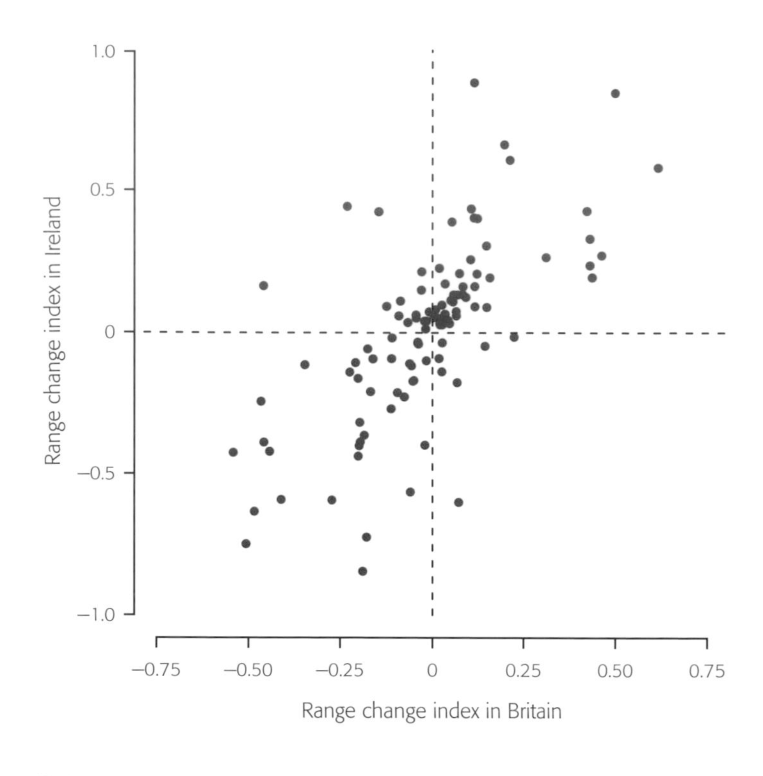

could suggest that different processes have affected some species, or that different land-use trends have occurred on the two islands. However, some of these contrasts were probably due to species that are marginal breeders in Ireland showing relatively large changes (involving colonisation or near extinction of small populations) compared with their larger, more stable populations in Britain.

6.8 SPATIAL ASPECTS OF RANGE CHANGE

THE MAPS OF abundance changes (Figure 6.4) indicated that geographical differences existed in the recent range dynamics of many species. In this section we extend this theme, asking questions about where, across Britain and Ireland, the distributions were changing and whether general patterns could be detected.

6.8.1 Broad patterns of range change

There was enormous variation among species in the types of range changes that have occurred in the breeding season over the past 40 years. Virtually all species that were ubiquitous 40 years ago remained so today but for many less widespread species there was no clear evidence of a systematic change in their range boundaries—losses and gains were roughly evenly distributed within the same broad geographical area occupied in earlier periods. However, substantial numbers of species showed striking geographical patterns of contraction or expansion with a strong directional component.

Based on a purely visual assessment, we draw attention in Box 6.3 to some patterns that appear to be shared by several species. For most resident species the geographical range changes were similar in summer and winter. Whilst populations of several waterbirds have tended to winter further east in recent decades (Austin & Rehfisch 2005; Maclean *et al.* 2008; Lehikoinen *et al.* 2013), this is not strikingly evident from the distribution data probably because, for these highly aggregated species, substantial change in numbers can take place before major range changes are apparent. A clear exception is the Bewick's Swan, which has contracted in range greatly in Ireland.

A relatively large suite of breeding species has expanded northwards or westwards, or both, while, in contrast, another large group of species has lost ground in the west (Box 6.3). The gains and losses in Ireland could be regarded as western extensions of these respective dynamics. The north/west expansions includes species that have expanded in range within lowland England (e.g. Cetti's Warbler), ones that have moved north into Scotland (e.g. Nuthatch) and others that have expanded in range within Scotland (e.g. Goldfinch). Several species that occurred almost everywhere else in Britain have now colonised the extreme fringes of northern Scotland. Species showing these expanding and contracting range dynamics were heterogeneous, sharing no obvious habitat affinities or other traits. Not surprisingly, however, the expanders included several non-natives that are continuing to spread following their

Several waterbirds species such as the White-fronted Goose appear to be changing their migratory patterns in response to climate change.

BOX 6.3 PRONOUNCED PATTERNS IN CHANGE OF BREEDING RANGE

MORRIS RENDALL/DAVID TIPLING/ALISON MCARTHUR

Range dynamics vary among these three widespread breeding wader species (Snipe, Ringed Plover and Oystercatcher).

Several types of spatial pattern of breeding-range expansion or contraction, in terms of apparent occupancy of 10-km squares, were identified based on visual inspection of maps. In each case, the broad pattern was exhibited by several species, though there was considerable variation amongst them. The species listed are apparent examples of ones that showed either (i) clear expansion or contraction, rather than infilling or disappearing from an existing patchy range or (ii) a disproportionate number of gains or losses in a particular region. In most cases, the patterns were evident over the 40-year period. Where expansion was a consequence of recent colonisation this is denoted by a superscript +. Established non-native species are marked with an asterisk (exotic non-native species were excluded). Localised and purely coastal species (e.g. breeding auks) have not been considered.

Expansion in Ireland: Canada Goose*, Little Egret+, Buzzard, Collared Dove, Great Spotted Woodpecker+, Jay, Raven, Great Tit (extreme W), Coal Tit (extreme W), Blackcap, Reed Warbler+.

Inland expansion in Britain: Shelduck, Cormorant (lowland England), Marsh Harrier (E England), Oystercatcher, Avocet (E England), Ringed Plover (N England), Lesser Black-backed Gull, Herring Gull, Common Tern (lowland England), Stonechat (N & W).

Northward/westward expansion within Britain: Canada Goose*, Egyptian Goose*, Mandarin Duck*, Gadwall, Red-legged Partridge*, Quail (Scotland), Little Egret+ (England), Hobby (England), Little Ringed Plover, Kingfisher, Jay (Scotland), Carrion Crow (Scotland), Cetti's Warbler+ (England), Chiffchaff (Scotland), Blackcap (Scotland), Garden Warbler (Scotland), Lesser Whitethroat, Reed Warbler, Nuthatch, Greenfinch (Scotland), Goldfinch (Scotland).

Expansion in extreme north of Scotland: Great Tit, Sand Martin, Swallow, Willow Warbler, Mistle Thrush, Robin, Pied Wagtail, Chaffinch, Linnet.

Contraction in Ireland: Red Grouse, Grey Partridge, Corncrake, Moorhen, Coot, Black-headed Gull, Common Tern, Golden Plover, Lapwing, Dunlin, Snipe, Curlew, Redshank, Stock Dove, Cuckoo (E), Barn Owl, Swift, Skylark, Whinchat, Grey Wagtail, Yellowhammer.

Contraction in western Britain: Grey Partridge, Kestrel (Wales/S Scotland), Corncrake, Moorhen, Coot, Lapwing, Curlew, Black-headed Gull, Stock Dove (Wales), Green Woodpecker (Wales), Yellowhammer.

Contraction in English lowlands (towards the north/west): Snipe, Willow Tit, Wood Warbler, Redstart, Whinchat, Tree Pipit, Lesser Redpoll.

The distribution of the Turtle Dove has shrunk towards southeast England but even in the remnant stronghold areas, densities have tumbled.

establishment in the wild. For many of the expanders, further suitable habitat exists to the north and west of their current ranges so ongoing expansion seems likely.

By no means all the striking spatial patterns of range change fit into those outlined in Box 6.3. The remarkable eastward spread of Buzzards and Ravens in Britain was unique to these species, though Sparrowhawks and Grey Wagtails also gained ground in parts of the east. Only the Turtle Dove and Nightingale have contracted strongly to southeastern England. Few species have shown marked southerly expansions; the Osprey and Goosander are the primary examples, the former partly aided by reintroduction into central England. The Marsh Tit has thinned out greatly at the northern extremity of its range. Some upland species, notably the Red Grouse, Black Grouse and Golden Plover, may have undergone greater losses at the margins than the centres of their upland ranges.

A further point is that many species have undergone quite complex patterns of change, expanding in some regions but contracting in others. For example, Grasshopper Warblers and Sand Martins have lost ground in the south but gained in the north of Britain. Green Woodpecker distribution has contracted in the west but expanded in the east. The range of the Common Tern has contracted in Ireland and Scotland but expanded inland in England whereas the Tree Sparrow showed the opposite pattern. In some cases these changes appeared to involve a systematic spatial shift in distribution. The evidence for range shifts is discussed next and possible drivers of change are discussed in Section 6.9.2.

FIGURE 6.7 SHIFTS IN THE LOCATION OF BREEDING RANGE MARGINS IN BRITAIN AND IRELAND

The shaded boxes show the interquartile range for the margin shift between 1968–72 and 2008–11 shown by species in each group; the solid lines show the median range shifts. For species with a southerly range (A), the shift (km) in the northern range margin is shown; for species with a northerly range (B), the shift in the southern range margin is shown. In both cases positive values indicate the margin has moved further north. Numbers of species in each group are shown on the x-axis labels.

Following Thomas and Lennon (1999), species present in more than 90% of 10-km squares in a country were excluded, as were those present in fewer than 20 squares in either period; a species was classed as southerly (or northerly) if the mean location of occupied squares in an island in 1968–72 was to the south (or the north) of the midpoint of all squares in an island. Margin shift was measured as the difference in the mean location of the 10 most southerly (or northerly) occupied 10-km squares in each time period.

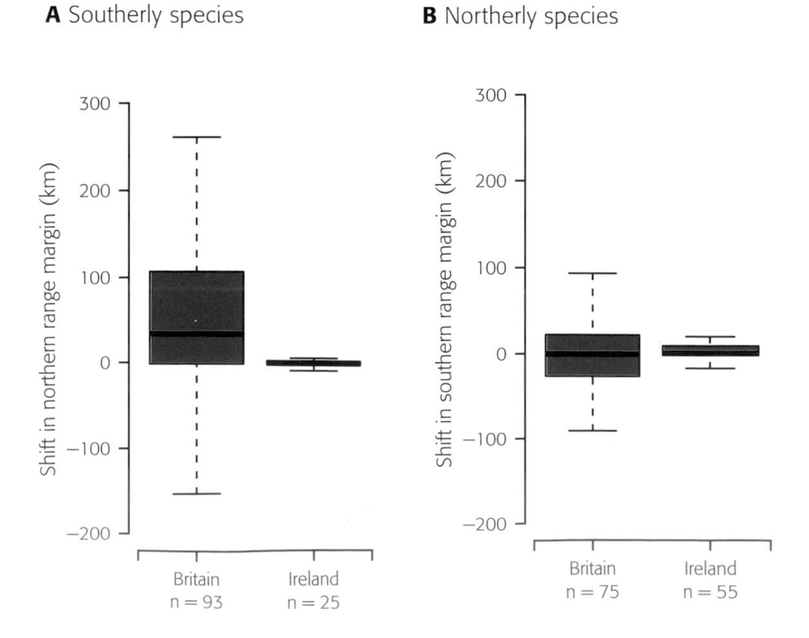

6.8.2 Changes at range margins

The previous section points to some general patterns of distribution change. Here we examine those changes further to begin to consider the question of how climate change may have impacted bird distributions over the last 40 years. Spatial shifts in the ranges of many species are predicted to occur with increasing frequency in response to current climate change (Huntley *et al.* 2007) and there is growing evidence that this indeed is happening across various taxa (Hickling *et al.* 2006; Parmesan 2006).

Several methods, varying in complexity and assumptions, have been proposed to assess changes in the shape and location of species' ranges. Thomas and Lennon (1999) proposed a simple method in which the locations of the 10 most southerly or northerly 10-km squares occupied by a species in Britain were taken as measures of its southern and northern range margins, respectively. Change in the average northing of these margin squares between atlases was used by them as a measure of range shift. Their method, which attempted to control for other confounding trends, found that the range of southerly distributed species had extended northwards, on average, by 18.9 km between 1968–72 and 1988–91. This method could have exaggerated the extent of range shift if changes were in fact due to increased survey effort in the north (Kujala *et al.* 2013), although the analyses in Chapter 5 suggest that survey effort may actually have decreased over that period.

Atlas data can offer insights into the responses of distributions to climate change, though distinguishing between the effects of climate change and other factors may be difficult (see Section 6.9.2). Analyses using the effort-controlled distributions (see Section 4.6) can assess distribution shifts after removing observer-effort biases. Such detailed analyses are beyond the scope of this chapter but here we give a simple summary of the magnitude of apparent range shift for southerly and northerly distributed species in Britain and Ireland for the period 1968–72 to 2008–11 (Figure 6.7). In doing so, no account has been made of other factors such as coverage, species status or population trend. Thomas and Lennon (1999) defined southerly species as those whose mean location of occupied 10-km squares was south of the mean location of all squares in Britain, and vice versa for northerly species. We have classified

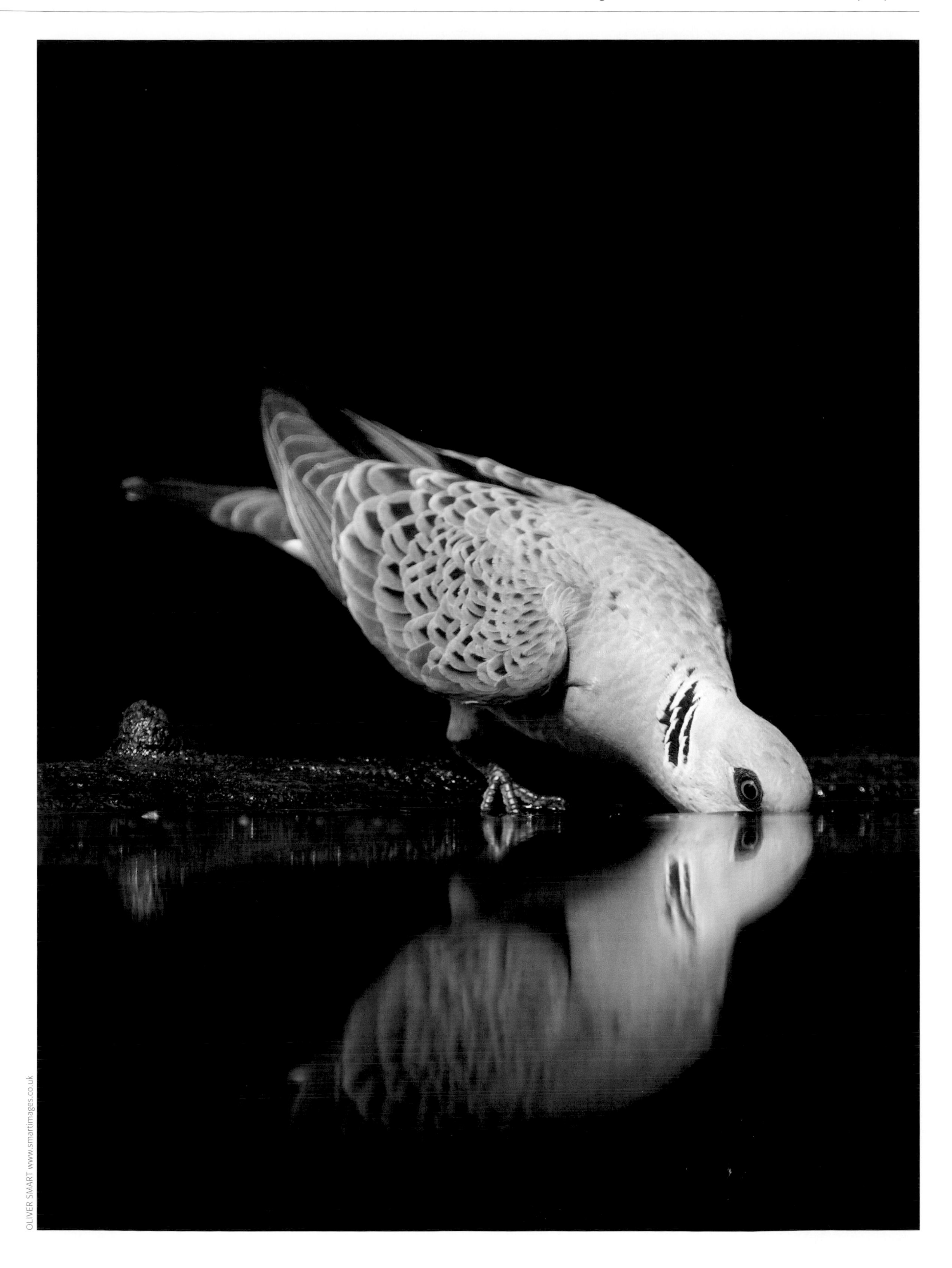

OLIVER SMART www.smartimages.co.uk

FIGURE 6.8 A CLASSIFICATION OF BREEDING SPECIES ACCORDING TO RANGE AND ABUNDANCE CHANGES

The matrix shows where breeding species in Britain & Ireland lie on gradients of range changes versus abundance changes during the period 1988–91 to 2008–11. The range-size change axis was derived from the percentage change in the number of 10-km squares with breeding evidence. The abundance change axis was derived from the percentage of 20-km squares showing increases or decreases in abundance. Only species with abundance change data for at least 50 20-km squares are shown and colonial seabirds are omitted. Species in the highlighted cells show little range change but moderate abundance changes (whether positive, negative, or both).

% CHANGE IN BREEDING RANGE SIZE, 1988–91 TO 2008–11	% OF RANGE SHOWING BREEDING-SEASON ABUNDANCE CHANGE, 1988–91 TO 2008–11				
	Abundance declines dominate (81–100% of range)	Abundance declines common (61–80% of range)	Mixed pattern (declines account for 40–60% of range)	Abundance increases common (61–80% of range)	Abundance increases dominate (81–100% of range)
Large gain (exceeds + 50%)			Water Rail	Canada Goose, Mandarin Duck, Gadwall, Marsh Harrier, Goshawk, Hobby, Woodlark, Stonechat, Firecrest	Greylag Goose, Egyptian Goose, Red Kite, Buzzard, Osprey, Cetti's Warbler, Raven, Siskin
Moderate Gain (+26% to +50%)			Goosander, Little Grebe, Peregrine, Little Ringed Plover, Grasshopper Warbler	Red-legged Partridge, Reed Warbler, Jay	Blackcap
Small Gain (+6% to +25%)		Hen Harrier, Grey Wagtail	Shelduck, Shoveler, Tufted Duck, Quail, Black-throated Diver, Great Crested Grebe, Sparrowhawk, Oystercatcher, Kingfisher, House Martin, Rock Pipit, Sedge Warbler, Garden Warbler, Lesser Whitethroat, Goldcrest, Treecreeper, Chough, Linnet, Bullfinch, Reed Bunting	Mute Swan, Pheasant, Greenshank, Collared Dove, Green Woodpecker, Sand Martin, Whitethroat, Long-tailed Tit, Coal Tit, Nuthatch, Greenfinch	Great Spotted Woodpecker, Chiffchaff, Great Tit, Goldfinch
Little Change (-5% to +5%)		Ptarmigan, Black Grouse, Golden Eagle, Kestrel, Moorhen, Cuckoo, Little Owl, Skylark, Dipper, Starling	Wigeon, Teal, Eider, Merlin, Coot, Stock Dove, Meadow Pipit, Pied Wagtail, Mistle Thrush, Willow Warbler, Magpie, Rook, House Sparrow, Lesser Redpoll	Mallard, Woodpigeon, Swallow, Wren, Dunnock, Robin, Blackbird, Song Thrush, Blue Tit, Jackdaw, Carrion Crow, Hooded Crow, Chaffinch	
Small Loss (-6% to -25%)	Lapwing, Nightingale, Ring Ouzel, Spotted Flycatcher, Yellowhammer	Red-breasted Merganser, Ruddy Duck, Red Grouse, Red-throated Diver, Ringed Plover, Golden Plover, Snipe, Common Sandpiper, Swift, Tree Pipit, Redstart, Wheatear, Marsh Tit, Tree Sparrow, Twite	Grey Heron		
Moderate Loss (-26% to -50%)	Grey Partridge, Curlew, Redshank, Turtle Dove, Short-eared Owl, Lesser Spotted Woodpecker, Yellow Wagtail, Whinchat, Wood Warbler, Pied Flycatcher, Corn Bunting	Pochard, Corncrake, Dunlin, Woodcock, Redwing			
Large Loss (exceeds -50%)	Willow Tit, Hawfinch	Goldeneye			

species in this way both in Britain and in Ireland.

For both groups of species, there is evidence of margin shifts in both directions, though this is far more apparent in Britain than Ireland. However, it is striking that the northern margin of British southerly species has shown the most marked pattern of northwards shift (Figure 6.7A). Some of the largest shifts in Britain are for scarce species; the Marsh Warbler, Marsh Harrier and Black Redstart all showed northward margin shifts in excess of 250 km. For such species the range margin is ill-defined. However, common and widespread species such as the Lesser Whitethroat (shift of 247 km north), Chiffchaff (175 km), Reed Warbler (171 km) and Nuthatch (106 km) are among the many showing northward shifts, some in conjunction with overall range expansion and positive population trends that may, or may not be, climate driven. Some species showed range margin shifts in the opposite direction to that expected if climate change were to be the principal driver. For example, among the southerly species showing apparent southward movement of the northern range margin are the Corn Bunting (-245 km) and Turtle Dove (-301 km)—two species predicted to fare well under a warming climate (Huntley *et al.* 2007) but actually declining rapidly. These emphasise the difficulty of finding clear climate-change signals when other drivers such as changes in land management can have opposing effects (Eglington & Pearce-Higgins 2012). Furthermore, a focus on range margins is interesting but it ignores the wider spatial dynamics of a species, a topic that is explored in the next section.

6.8.3 Broad patterns of abundance change

The previous sections have concentrated on gross patterns of change in occupancy but it is clear that some species can show pronounced changes in abundance within otherwise stable ranges. Analyses of population trend data for example have shown that many generalist and widespread species in Britain are increasing in abundance (Davey *et al.* 2011). The 1988–91 and 2008–11 breeding-season data offered an opportunity to assess spatial patterns of abundance change in breeding birds by using the fixed-effort tetrad occupancy data (see Section 4.5).

Breeding species were classified according to the magnitude of range change and the proportion of the range showing increasing or decreasing relative abundance using combined data from Britain & Ireland (Figure 6.8). As expected, the resulting matrix of species shows that overall range change and abundance change are generally positively correlated, ranging from species such as the Willow Tit with strongly contracting range and declining abundance to the Siskin with strongly expanding range and increasing abundance.

However, the 37 species in the highlighted cells of the matrix have shown little or no overall change in range size but showed widespread abundance decreases, widespread abundance increases, or a complex pattern of increases and decreases. Some, such as the Black Grouse and Little Owl, showed little overall range change but a moderate degree of 'range turnover'. For example, 15% of the current Little Owl range has only recently been occupied and replaces 15% of the former range that has been lost. Whether this is a real pattern of change or an artefact of poor detection in different squares in different atlases is unclear. Nevertheless, even in the part of the range where there have been few apparent gains and losses in 10-km square occupancy, changes in abundance were still prevalent. Of the species showing little range change but increasing abundance across much of their range, all except the Jackdaw occupied more than 90% of 10-km squares (Carrion and Hooded Crows were combined for this purpose) and consequently have little capacity for further range expansion, yet all showed abundance increases across 60–80% of the range.

As with range changes, species differed in their geographical patterns of abundance change (Box 6.4). A substantial number (21) have increased in Ireland and/or northern Britain, but have either decreased or not changed in southern Britain. A well-known example is the Willow Warbler (Morrison *et al.* 2010) but the Cuckoo, House Martin, House Sparrow, Lesser Redpoll and Swallow, all showed similar, but not identical, patterns. By contrast, only two species, the Stock Dove and Green Woodpecker, appeared to show the opposite pattern with a strong decrease in the west and an increase towards the southeast.

While the human eye is remarkably effective at perceiving spatial patterns, a more objective method of assessing change was desirable. The approach used to achieve this is illustrated by the example of the Green Woodpecker as follows. The abundance change map for the Green Woodpecker shows decreases in Wales and southwest England, contrasting with increases throughout central and eastern England. But there are also abundance decreases in southern Scotland, so which direction best explains the gradient in status? To answer this question, a linear regression was computed between the abundance change estimates in each 20-km square and the square's easting and northing coordinates. By combining with trigonometry the easting and northing parameter estimates from the regression it was possible to calculate the compass bearing describing the direction in which there was the most pronounced gradient in abundance change values. For the Green Woodpecker this produced a compass bearing of 98°, confirming the visual interpretation that the species was increasing the most in the east and decreasing the most in the west. The R^2 statistic of the regression provided a relative measure of how much of the observed spatial variation in abundance change values can be attributed to the gradient. For the Green Woodpecker R^2 was 54%, indicating that the west–east gradient explained over half of the variation in abundance change.

This analysis was repeated for any species recorded in at least 25 20-km squares. Species were grouped into those where most 20-km squares were showing declines in abundance, increases in abundance, or a mix of the two (based on whether the lower and/or upper quartiles of the abundance change figures had positive and/or negative values). For the positive-trend and mixed-trend species, the compass bearing pointed in the direction that abundance was increasing the most. To aid interpretation, for the negative-trend species the compass bearing was reversed by 180° so that it pointed in the direction that abundance was decreasing the most.

The numbers of species showing significant compass-directional changes in abundance were: negative-trend species 20 out of 30, positive-trend species 22 out of 31 and mixed-trend species 70 out of 110 species. Figure 6.9 shows the results of the regression tests as a series of radial plots for negative-trend, mixed-trend and positive-trend species. In each plot, each line shows the compass direction of the gradient for one species. The length of each line is based on the R^2 value for the species and gives a relative indication of the strength of the gradient.

In general, the gradients for declining species were not clustered in any particular direction. They were mostly weak with the exception of the Turtle Dove, Spotted Flycatcher and Arctic Skua. Gradients for positive-trend species were also relatively weak and could be divided into those species increasing to the east (e.g. Marsh Harrier and Woodlark) and those increasing to the west (e.g. Blackcap, Coal Tit and Great Tit). The Greylag Goose was a notable outlier, increasing to the northeast. Gradients for mixed-trend species were the strongest and were strongly clustered in the west–northwest direction. This was true for several resident species (Starling, House Sparrow and Lesser Redpoll) and several migrants (Cuckoo, House Martin and Willow Warbler). Only the Green Woodpecker showed a pronounced gradient in the opposite direction.

6.9 GENERAL DISCUSSION

ATLASES PROVIDE INVALUABLE data for tracking avifaunal composition, species status and distributional shifts. Using an atlas to take stock approximately every 20 years would seem an appropriate interval for detecting and measuring these gross dynamics across all species. The current atlas shows that many changes in the geographical breeding ranges and abundance of species have occurred in the last 20 years. In some cases these are continuations of patterns that were evident by the time of the *1988–91 Breeding Atlas* but new patterns have emerged since, sometimes utterly unexpected in their magnitude. An implication of the large numbers of range expansions and contractions is that many species, including ones for which annual monitoring data

BOX 6.4 SPECIES SHOWING CONTRASTING GEOGRAPHICAL PATTERNS OF CHANGE IN BREEDING ABUNDANCE

Some species showed visual evidence of major changes in abundance within their ranges over the period 1988–91 to 2008–11. The species listed below are ones that have been established in Britain and/or Ireland for many decades and excludes those increasing in abundance as a result of colonisation.

Increase in Ireland and in some cases N Britain; no clear change in Britain: Goldcrest, Blue Tit, Coal Tit, Blackbird, Mistle Thrush.

Increase in Ireland and in some cases N Britain; decrease in lowland Britain: Cuckoo, Sand Martin, Swallow, House Martin, Willow Warbler, Sedge Warbler, Starling (decreasing most of Britain), Song Thrush, House Sparrow, Tree Sparrow, Linnet, Lesser Redpoll, Bullfinch.

Increase in N Britain; decrease in S Britain: Garden Warbler, Grasshopper Warbler, Tree Pipit.

Increase in E England; decrease in Ireland and W Britain: Stock Dove.

Increase in E England; decrease in W Britain: Green Woodpecker.

ALLAN DREWITT

The Bullfinch is one of 13 species to have increased in abundance in Ireland and parts of northern Britain yet declined in southern and eastern Britain.

FIGURE 6.9 THE COMPASS DIRECTIONS IN WHICH ABUNDANCE CHANGES WERE GREATEST

Species were divided into those showing mostly negative abundance changes (A), mostly positive abundance changes (C), or a mix of the two (B). On each plot, the coloured lines show different species, with the line pointing in the direction in which abundance changes were largest (from 1988–91 to 2008–11). That is, for species with negative trends the line points in the direction that abundance was declining the most; for mixed trend and positive-trend species it points in the direction that abundance was increasing the most. The length of the line indicates how strongly direction explained spatial variation in abundance change based on the R^2 of each species' linear regression. The circular scale is shown in increments of 10% of R^2, up to a maximum of 60% and key species on each plot are labelled with their two-letter code.

See Section 6.8.3 for details of the calculation. The species codes are as follows: AC Arctic Skua, BC Blackcap, CK Cuckoo, CT Coal Tit, GJ Greylag Goose, GT Great Tit, G. Green Woodpecker, HM House Martin, HS House Sparrow, LR Lesser Redpoll, MR Marsh Harrier, SF Spotted Flycatcher, SG Starling, TD Turtle Dove, WL Woodlark, WW Willow Warbler.

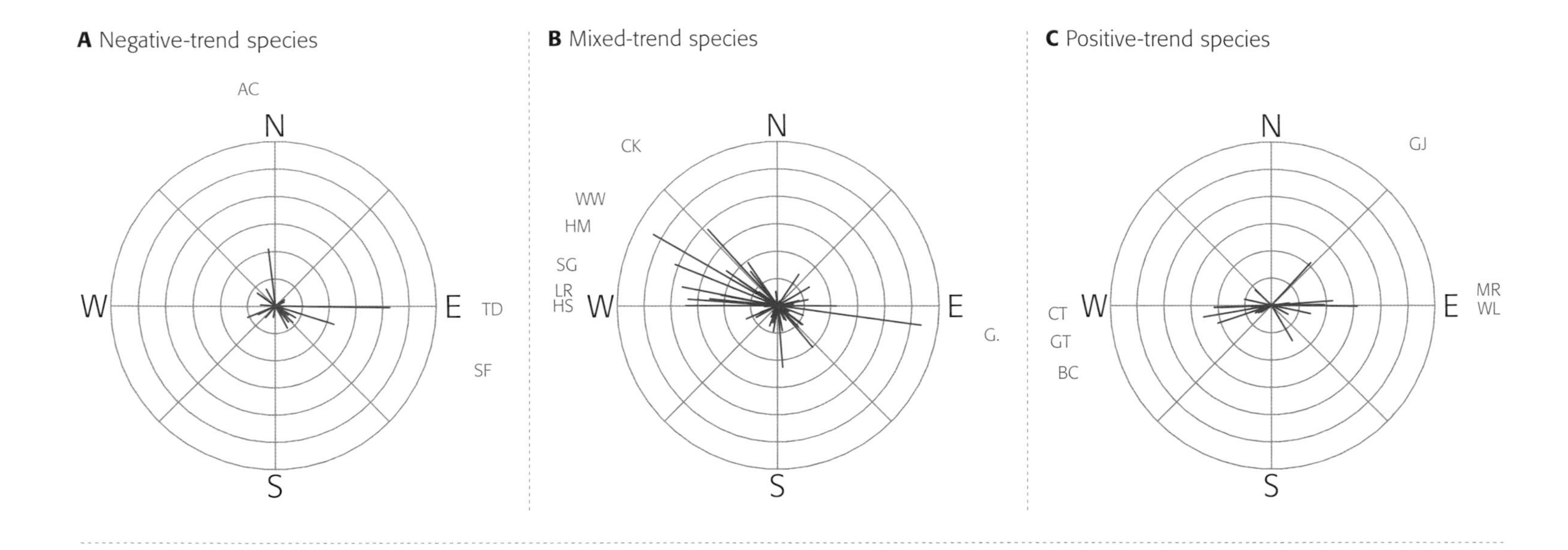

are unavailable, have undergone major population changes over this period. This chapter concludes with some general observations on pattern and change in the British and Irish avifaunas over the last 40 years, identifying several aspects that merit further analysis.

6.9.1 Range dynamics ~ the patterns

The scale on which the range of a species is viewed profoundly affects the conclusions drawn about its dynamics (Gaston 2003). Rarely is it possible to have complete knowledge of distribution or abundance of a species across its full range. However, the greater the proportion of the range studied, the greater the insights are likely to be. By treating Britain & Ireland as a single biogeographical unit, it is possible to discern patterns that would not be evident at smaller scales. Some range changes within Britain may effectively be extensions of ones that are especially prominent within Ireland, and vice versa. For example, this may apply to the range contractions of Grey Partridges, Stock Doves and Yellowhammers in western Britain and to the westward expansion of Buzzards, Great Spotted Woodpeckers and Reed Warblers into Ireland. Indeed for many of the species considered here, Britain and Ireland lie at the extreme edge of a pan-European distribution, and processes such as climate change may apply to different degrees throughout the area. It will be interesting to see whether changes presented here are manifested elsewhere in Europe when the *European Atlas* is repeated.

Naturally, there is great variability in range changes amongst species, but a major geographical gradient of change appears to be emerging in our avifauna. Crudely speaking, this involves expansions and contractions of ranges around two broad axes, one running west to east, the other north to south. Several species are changing on both axes, for instance the Blackcap is increasing in occupancy and abundance towards the north and the west. For some species, these axes embrace both Britain and Ireland, in other species they are evident just within Britain or, more rarely, only within Ireland.

With the development of abundance estimation within atlases, it has become possible to examine how distribution and abundance interact to give a more a complete picture of range dynamics. The expansion or contraction of a species' range may be the outcome of progressive and spatially uneven change in abundance within the former range. Species that appear to fit this model, showing some degree of spatial matching of their range change and abundance change, include the Sand Martin, Garden Warbler, Grasshopper Warbler and Stock Dove. On the other hand, the range may remain unaltered in terms of occupancy, despite considerable changes in abundance within it, resulting in some parts becoming populated at lower or higher density. The latter type of change may be a precursor to future change in range size or a range shift (see below) or it could be a new pattern of geographical variation in density that persists indefinitely. Examples include those species such as the Blackbird, Blue Tit and Goldcrest that have increased in abundance in Ireland but show no clear change elsewhere.

It is interesting to explore the possibility of range shift, a concept relevant to understanding the responses of biodiversity to environmental change (Section 6.8). Range shift is defined as a directional movement of range location, characterised by a

Farmland birds have undergone marked population declines and range contractions as management of lowland landscapes has intensified.

contraction in one area being compensated to some extent by an expansion elsewhere. Change in range size on the other hand implies an overall contraction or expansion. Range shift and change of range size are not necessarily mutually exclusive because there may be shift in location along with a net change in range size. The simple description of range margin shifts in Section 6.8.2 does not fully address this issue because it only considers shifts in a north-south context, whereas range shifts could also be occurring in other directions. Furthermore, there is scope to take account of patterns of abundance change, not just distribution change.

Based on a systematic visual review of the range change and abundance change maps, there is considerable evidence of range shifts occurring over the last 40 years within Britain and Ireland. In identifying candidates for range shift, species have been included where the apparent occupancy of 10-km squares has declined in one region but apparently increased elsewhere. Importantly, we have also included species that showed a striking geographical divergence in abundance change, representing a systematic directional shift in abundance pattern. These species fall into several of the patterns summarised in Boxes 6.3 and 6.4. Species that may be shifting north and/or west include the Dunlin, Garden Warbler, Grasshopper Warbler, House Sparrow, Lesser Redpoll, Nuthatch, Sand Martin, Sedge Warbler, Swallow, Tree Sparrow and Willow Warbler. Species that may be showing an eastward shift in range include the Green Woodpecker, Grey Wagtail and Stock Dove.

For the great majority of species living within Britain & Ireland, these islands form a small part of their global range. This means that range changes within Britain & Ireland may form part of a larger-scale dynamic of which we are unaware. For example, could the contractions of the ranges of the Turtle Dove and Nightingale towards southeast England actually form part of a wider range shift that has not yet been detected? Alternatively, and perhaps more likely, these contractions may reflect a withdrawal to the core of the range as overall population size declines due to some unknown factor (Fuller *et al.* 2007). This seems plausible in the case of species appearing to reach the biological limit of their ranges somewhere within Britain & Ireland (rather than simply 'running out of land'). This may have been a factor in the demise of Wrynecks and Red-backed Shrikes. At least 10 scrub/forest species appear to have their global range boundaries within Britain but none within Ireland (Fuller *et al.* 2007). The species are the Honey-buzzard, Lesser Spotted Woodpecker, Nightingale, Lesser Whitethroat, Firecrest, Marsh Tit, Willow Tit, Nuthatch, Golden Oriole and Hawfinch.

The notion of core and edge of range can also be examined at finer scales. Where populations are concentrated into particular areas within the range, these can be regarded as regional cores. They may be areas with an abundance of high-quality habitat but this is not the only potential explanation of such patterns. When a species is prospering, one might expect to see expansion around such regional cores, but when it is declining there may be shrinkage towards the core. It is difficult to identify such potential patterns unambiguously from the atlas maps without the use of spatial statistics, additional population-size or demographic data and appropriate habitat data. However, it is noticeable that several characteristic upland species appear to have lost ground mainly at the fringes of the main blocks of moorland and mountain—see the breeding maps for the Golden Plover, Black Grouse, Red Grouse and Common Sandpiper.

6.9.2 Range dynamics ~ possible drivers

Many observed changes in range size or range shifts are direct responses to factors causing a reduction or an improvement in the quality of that part of the range. So-called 'niche models' make the assumption that distributions have a direct spatial linkage with the availability of relevant resources or environmental conditions and, therefore, that distribution can be predicted by these factors. This is the principle underlying the modelled abundance maps presented for many species in this book (see Section 4.4.3). Two broad suites of factors are relevant: (i) ones concerning habitat availability and quality, and (ii) ones relating to climate. These factors are not independent because climate affects land use and vegetation. Nonetheless, the Atlas data potentially offer a huge resource for researching the relative importance of these two broad sets of factors in shaping the distribution patterns and changes we see.

Human activities have substantially altered the quantity and quality of habitats available to many bird species in Britain & Ireland over recent decades. The implications have been diverse with some species benefitting, others being adversely affected (Fuller & Ausden 2008; Nairn & O'Halloran 2012). The major measurable impacts of habitat change on birds over the last 40 years probably continue to be within farmland. The massive increase in agricultural production within grassland and arable systems has been accompanied by declines in the ranges and population sizes of several species closely dependent on farmland with a few, such as the Woodpigeon, benefitting from the changes in cropping practices (Fuller *et al.* 1995; Chamberlain *et al.* 2000; McMahon 2007). There are strong causal links associated with loss of food resources and preferred habitat structures both in Britain and Ireland (Newton 2004; Wilson *et al.* 2009; Copland & Lusby 2012). In the case of farmland birds, habitat change is probably the predominant driver of recent range changes rather than climate (Eglington & Pearce-Higgins 2012). For most other groups it is less clear that habitat change is the predominant driver and it is to be hoped that future research will enable conclusions to be drawn. Atlas data may, for example, be extremely valuable in gaining a better understanding of how urbanisation relates to the composition of bird communities.

Whilst habitat deterioration and its consequences are very likely to be implicated in the range contractions of many breeding waders (see Section 6.6.3), it is possible that climate change is also involved either directly or through its effects on habitat suitability, for example by changing soil moisture which might affect species such as the Snipe and Woodcock. Under modelled climate-change scenarios, Huntley *et al.* (2007) have predicted that the ranges of several of these species will shift northward, for example the Common Sandpiper, Curlew, Golden Plover, Lapwing, Redshank and Snipe. Similarly, some upland birds, including waders, Red Grouse, Ring Ouzels and Short-eared Owls are also predicted to contract their British and Irish ranges (Huntley *et al.* 2007); this may come about to some extent through birds being forced to occupy higher altitude areas. The range changes observed over the last 40 years for these species could be initial signs of responses to climate change, but habitat change in various guises, ranging from afforestation to changes in grazing and land abandonment, may be equally significant. Future research may give the insights needed to assess whether there is an overriding driver of these changes.

It is tempting to ascribe many of the range shifts and range expansions to climate change, especially in the case of species moving towards the north and west. In theory, the patterns shown by species such as the Willow Warbler are consistent with climate change, making areas in the north more suitable at the expense of areas in the south (Morrison *et al.* 2010). The northward movement of the range margins of southerly distributed species (Section 6.8.2) is also consistent with the expected responses to a warming climate. There is no question that, at the largest scales, climate sets limits to distribution (Newton 2003).

However, when considering distributions at the scale of Britain & Ireland, it is important to recognise that many other factors can potentially drive range dynamics. Exactly where a species may or may not be found is frequently the outcome of multiple factors, often resulting in apparently suitable habitat not always being occupied (Fuller 2012b). Subtle changes may occur in habitat quality in particular parts of the range. For example, little is understood about the effects of many kinds of pollutants on birds, with notable exceptions including the effects of organochlorines on raptors (Newton 2013) and agrochemicals on some farmland birds (Potts 2012). Predation pressure or interspecific competition may be important drivers in some cases. Population pressure may also drive range expansion and the occupancy of new habitats or landscapes, without any obvious improvement in habitat quality or change in climate in the newly occupied area (Dolman 2012; Wesołowski & Fuller 2012). Equally, contraction at the edge of the range may be

Corn Buntings are now extinct as a breeding species in Ireland, and their breeding range in Britain has shrunk by 58% since 1968–72.

a symptom of reduced population pressure and a retreat to the core of the range (Fuller *et al.* 2007). It is evident that the ranges of some species have moved substantially in the opposite direction to that predicted under climate-change scenarios, suggesting that factors other than warming climate are involved. Goosanders have expanded south, while Marsh Tits have lost ground in the north; in both instances these differed from the pattern predicted by Huntley *et al.* (2007). The Turtle Dove has shown one of the most striking range contractions and, like the Nightingale, is rapidly becoming confined to southeast Britain, contradicting the expected responses to current patterns of climate (Wilson *et al.* 2002; Huntley *et al.* 2007).

Investigation into the causes of range change, and especially of the role of climatic changes, will continue to be a high-priority area of research, with Atlas data playing a significant role. From the perspective of Britain & Ireland, situated in an Atlantic climate, changes in precipitation should be considered. Climate-change research has tended to emphasise the consequences of rising temperatures. However, increasingly volatile and unpredictable precipitation could affect survival and breeding success of some species, even to the extent of countering the effects of temperature. Tingley *et al.* (2012) have suggested that varied long-term responses to climate change are to be expected, deriving from species-specific and region-specific factors that determine the localised character of future climate and its effect on species. Nor should it be forgotten that it is important to disentangle the effects on birds of regional fluctuation in weather from those of longer-term trends in climate. For example, following a run of relatively mild winters, many parts of Britain experienced several cold winters in the late 2000s which greatly reduced populations of Dartford Warblers and possibly other species at the edges of their climatic range.

A final thought is that past human persecution may have excluded some species from substantial parts of their potential ranges. It is possible that recent range expansions of certain expanding species, most notably some raptors and fish-eating birds, represent recolonisation of areas from which they were once eliminated.

6.9.3 Conservation implications

The latest estimates of range change, together with other information on status, are periodically used to determine which species should be identified as 'Birds of Conservation Concern'. Both in Britain and Ireland, 'breeding range decline' is one of several criteria used to identify Red-listed and Amber-listed species. The respective thresholds used in Britain by Eaton *et al.* (2009) were >50% and >25% range contraction over a 25 year period. Similar distribution-change criteria have been adopted to identify 'Birds of Conservation Concern in Ireland' (Lynas *et al.* 2007). Atlas data will play a major role in updating these assessments of conservation status when these lists are next updated.

Atlases are also valuable for drawing attention to particular groups of species that appear to be of conservation concern because a high proportion of their component species are showing range contractions (see Section 6.6 for further discussion). Where these species share similar resource needs, this may point to a potential factor that needs to be understood through research and tackled through conservation action. This was the case with the massive range contractions of farmland birds that were apparent between the *1968–72 Breeding Atlas* and *1988–91 Breeding Atlas*. While these losses have not continued at the same pace over the last 20 years, there is no evidence that the ranges of any of these species are recovering. Indeed the ranges of several farmland species have continued to reduce by at least 25% since the *1988–91 Breeding Atlas*. In Britain this applied to the Corn Bunting, Grey Partridge, Tree Sparrow, Turtle Dove and Yellow Wagtail. In Ireland, there were large recent range losses of the Corncrake, Grey Partridge, Lapwing, Quail and Yellowhammer. Farmland birds, therefore, clearly remain as conservation priorities.

In contrast to farmland birds, woodland birds have shown considerable gains in Ireland over the last 20 years, though the picture is not so positive in Britain, confirming the results of recent surveys (Hewson *et al.* 2007). The Lesser Spotted Woodpecker, Willow Tit, Wood Warbler and Hawfinch are amongst the 20 species showing the greatest range contractions in Britain. The latest range change statistics indicate that migrant birds and coastal birds (auks, gulls and terns) are faring poorly in both Britain and Ireland. Again, this is broadly consistent with other current knowledge.

Perhaps the two main 'new' groups of concern revealed by this analysis are breeding waders and upland birds. There has been recognition for some time that many of these species are under pressure (e.g. Tierney *et al.* 2002; Sim *et al.* 2005; Ausden & Bolton 2012; Grant & Pearce-Higgins 2012), but the Atlas data indicate that the scale of changes may be greater than previously recognised. Both in Britain and Ireland, range contraction in upland birds appears to have been a progressive process over the last 40 years. In Britain, species that have shown especially large contractions (>25%) since 1968–72 are the Common Scoter, Dunlin, Ring Ouzel, Short-eared Owl, Snipe, Whimbrel and Whinchat. In Ireland, losses over the same period are relatively large for the Common Sandpiper, Curlew, Golden Plover, Merlin, Red Grouse, Ring Ouzel, Snipe, Teal, Twite and Whinchat. These Atlas findings suggest that uplands merit considerably more attention to elucidate causes of range change in many of their characteristic bird species.

Finally, we offer a comment concerning range shifts. As we see it, one of the most important findings concerns the emerging evidence that a considerable number of species are not just moving their range margins, but that spatial patterns of abundance are changing too (Section 6.8.3). This emphasises the value of taking a broad geographical perspective wherever possible. What appears to be a decline when viewed from the perspective of one part of the range, may be part of a wider picture of readjustment. In a world of limited conservation resources, it seems increasingly important to take account of these large-scale dynamics when setting priorities for conservation action.

JILL PAKENHAM

CHAPTER 7

INTERPRETATION OF SPECIES ACCOUNTS

Simon Gillings & Dawn Balmer

THIS CHAPTER PROVIDES guidance on the interpretation of the maps that form the core of the book. It also explains the rationale behind the choice of species and maps that are presented. In addition to the graphical key and annotated maps shown here, easy-reference guides to the maps can be found on the front and back inside covers of this book.

7.1 SPECIES INCLUDED IN THE BOOK

RECORDS FOR OVER 500 species and subspecies were submitted during the fieldwork for *Bird Atlas 2007–11*. Many of these were non-breeders or rarities and were recorded in only a few squares. Space would not permit the treatment of all these species with full accounts. We attempted to select species for the book objectively, on the basis of the quantity of information available and status of the species, but inevitably an element of subjectivity and practicality was involved. Initially, species were identified that were present in winter in at least ten 10-km squares, or that had *confirmed* breeding in at least one 10-km square in either this or one of the previous breeding atlases. The final list of 296 species was arrived at after making two further adjustments. Firstly, species were excluded if the majority of squares were only occupied in November, as these tended to be scarce autumn passage migrants (e.g. Sabine's Gull) rather than true wintering species. Secondly, we excluded some non-native species, only including those with relatively large numbers of winter records or which were also breeding in notable numbers.

Range size, abundance and change statistics for all species without a species account can be found in Appendix 2. For such species, and for seasons or change permutations not illustrated in the book, maps will be accessible at www.birdatlas.net via the 'Map Explorer' after publication.

The ordering of species in the book is based upon the recommendations of Sangster *et al.* (2011) with some minor changes to fit the species accounts into the book efficiently. Common names for species follow the recommendations of BOU (2013b) and Gill and Donsker (2013). See Section 4.2.4 and Box 4.1 for discussion about the taxonomy and identification of Feral Pigeon/Rock Doves, redpolls and crossbills and the implications for species accounts.

7.2 SPECIES ACCOUNTS

ALL SPECIES ACCOUNTS start with the common and scientific names of the species and the two-letter BTO code is also shown for those that have one. The majority of species were generously sponsored by individuals, groups, companies and organisations, and their dedications and logos are at the bottom of each species text. We are grateful to all these sponsors.

Each species account consists of a number of elements. In general, maps relating to the winter season use a palette of blue colours and those relating to the breeding season use a palette of orange-red colours. A key to all symbols and shading is given in Box 7.1. Most maps are accompanied by a small table giving key statistics pertinent to the type of map, season or period of change. Full details of the analyses and production methods behind the maps and statistics can be found in Chapter 4. General information and pointers to the interpretation of maps, tables and statistics are given below and particularly in associated boxes containing annotated maps.

7.2.1 Text

In designing this book we have chosen to emphasise the mapped outputs of the Atlas. This has necessitated considerably shorter species texts than in previous atlases. Rather than discuss aspects of the species' natural history and basic biology, which are readily available from other sources (e.g. *BWP*; *Birds in Ireland*; *Birds in Wales*; Snow & Perrins 1998; *Birds in England*; *Birds of Scotland*; Parkin & Knox 2010), the text instead focuses on patterns and trends in distribution and abundance that emerge from the new maps. When discussing breeding evidence collected during atlases, the terms '*non-breeding*', '*possible*', '*probable*' and '*confirmed*' have been italicised, as have the terms '*seen*' and '*breeding*' which were used in the *1988–91 Breeding Atlas*.

As far as possible, the patterns of distribution, abundance and change have been interpreted with reference to recently published works (by July 2013) on status and on the drivers of patterns of distribution and change. These include a number of standard trend statistics from UK and Irish wintering and breeding bird monitoring schemes. It should be noted that these schemes report on different time periods to atlases, so it is not always possible to equate distribution or abundance changes from atlases directly with population trend information. Whilst every care has been taken to use these sources where available, interpretations are inevitably subjective and based on the experience of the authors. The text of each species account has been reviewed by a small number of ornithologists and species experts in Britain and in Ireland to check that a balanced summary is presented.

In the great majority of cases, the results of *Bird Atlas 2007–11* concerning changes in status are consistent with information derived from other monitoring schemes. Occasionally, however, there are apparent contradictions which would merit closer investigation. Such differences are most likely to arise for methodological reasons. Atlases gather information every 20 years from all parts of Britain and Ireland, whereas most monitoring schemes are based on annual sample counts, or rely on periodic counts (e.g. every 10 years) at intervals that are usually different to those of the atlases. It is also the case that certain field methodologies fit the behaviour of some species better than others. For example, mobile or gregarious species, and ones that do not show strong territorial breeding behaviour, can be especially difficult to monitor effectively, often requiring specifically tailored methods.

The full references for most citations given in the accounts can be found in the main reference section at the back of the book. However, a number of commonly cited 'standard references' are presented in an abbreviated form in italics (their full references are given on page 15). Note that, for the majority of these, years indicate the period to which the source relates rather than the year of publication. For example, *RBBP 2010* refers to Holling *et al.* (2011b) which contains information on rare breeding birds in the UK for the year 2010.

Scientific names of all bird species are given in Appendix 2 in a large summary table that also contains statistics of range size and change for all species. Scientific names of other animals and plants mentioned in the species accounts are in Appendix 5.

7.2.2 Distribution maps and tables

Distribution maps show the areas of Britain & Ireland where each species winters or breeds. On most winter distribution maps, blue dots (Box 7.1) show 10-km squares in which each species was recorded at least once during November–February during the winters 2007/08 to 2010/11 (see Section 4.2.1). The table above the map gives the percentage of 10-km squares in Britain, in Ireland or in Britain & Ireland combined that were occupied in winter (see Sections 1.2 and 4.6).

On most breeding distribution maps, small, medium and large red dots show 10-km squares with *possible*, *probable* and *confirmed* breeding evidence respectively for the breeding seasons 2008 to 2011 (see Boxes 7.1 and 7.2). They are based on records from April–July for all species, plus 'out-of-season' breeding evidence (March and August–October) for selected species (see Section 4.2.2). Grey dots show 10-km squares where a species was recorded during April–July but where no breeding evidence was supplied. Such dots could include birds flying over and not using the square, migrants, summering individuals and records with no breeding evidence. These are collectively referred to as *non-breeding* dots (see Section 4.2.2). The table above the map gives the percentage of 10-km squares that were occupied with each level of breeding evidence in Britain, in Ireland, or in Britain & Ireland combined.

For some rare breeding and wintering species, all or part of the range is summarised at the resolution of 20-km or 50-km squares. The size of the dot indicates the maximum breeding evidence recorded in any of the constituent 10-km squares, and the depth of shading of the dot indicates the number of 10-km squares that were occupied (see Box 7.2, Section 4.2.3 and Figure 4.1). In practice, most dots have pale shading owing to the rarity of the species involved. For a very small number of species, certain breeding records have not been mapped at all owing to sensitivities over the location. These are specifically mentioned in the text of the species accounts.

The choice of whether to show distribution maps of the winter, breeding season, or both, was based on the number of occupied squares in the relevant season, on whether the species was recorded as breeding and on space constraints. Where possible,

BOX 7.1 KEY TO MAP SYMBOLS

The shape, size, colour and shade of each symbol convey information about the status of a species in a grid square and Boxes 7.2–7.4 have annotated examples of their use. Distribution maps for rare breeding and wintering species include all or part of the range shown at 20-km or 50-km resolution and in those cases the depth of shading of the dots indicates how many of the constituent 10-km squares were occupied.

WINTER DISTRIBUTION 2007/08–2010/11

10-km	20-km	50-km	
•	○ ● ●	○ ● ●	= Present in winter

BREEDING DISTRIBUTION 2008–11

10-km	20-km	50-km	
•			= *Non-breeding*
•	○ ● ●	○ ● ●	= *Possible* breeding
•	○ ● ●	○ ● ●	= *Probable* breeding
•	○ ● ●	○ ● ●	= *Confirmed* breeding

WINTER DISTRIBUTION CHANGE SINCE 1981–84

▲ = Apparent gain: not detected in 1981–84, present in 2007–11
■ = Stable: present in 1981–84 and 2007–11
▼ = Apparent loss: present in 1981–84 but not detected in 2007–11

BREEDING DISTRIBUTION CHANGE SINCE 1988–91

▲ = Apparent gain: not detected in 1988–91, present in 2008–11
■ = Stable: present in 1988–91 and 2008–11
▼ = Apparent loss: present in 1988–91 but not detected in 2008–11

BREEDING DISTRIBUTION CHANGE SINCE 1968–72

△ = Old apparent gain: not detected in 1968–72, present in 1988–91 and 2008–11
▲ = Recent apparent gain: not detected in 1968–72 or 1988–91 but present in 2008–11
■ = Stable: present in all three breeding atlases
□ = Apparently variable: present in 1968–72 and 2008–11 but not detected in 1988–91
▼ = Recent apparent loss: present in 1968–72 and 1988–91 but not detected in 2008–11
▽ = Old apparent loss: present in 1968–72 but not detected in 1988–91 or 2008–11

WINTER RELATIVE ABUNDANCE 2007/08–2010/11

Absent or very low abundance — High relative abundance

BREEDING RELATIVE ABUNDANCE 2008–11

Absent or very low abundance — High relative abundance

BREEDING RELATIVE ABUNDANCE CHANGE SINCE 1988–91

Relative abundance declines greater — Relative abundance increases greater

white = 'no change' or 'not recorded'

BOX 7.2 ANNOTATED BREEDING-SEASON DISTRIBUTION MAP AND TABLE

The example map and table are for the Golden Eagle, a species with 50-km resolution mapping at the edge of its range in southern Scotland and Ireland. Equivalent distribution maps for winter use a single blue dot to indicate presence in winter (see Box 7.1).

BREEDING DISTRIBUTION 2008–11
Distribution statistics
The percentage of all 10-km squares in Ireland, Britain, and Britain & Ireland overall, that were occupied at each level of breeding evidence, and the total occupied with any breeding evidence.
Coarse-resolution dots for rare species
The Irish and southern Scottish breeding distributions are shown at 50-km resolution to protect breeding sites.
In two of the 50-km squares, the highest breeding evidence recorded in the constituent 10-km squares was possible breeding (small dots). In two 50-km squares, probable breeding was recorded (medium dots) and in a further five 50-km squares, confirmed breeding was recorded (large dots).
All the 50-km dots have pale shading indicating that few of the 10-km squares were occupied in each 50-km square.
Small red dot = Possible breeding
Present or singing in suitable breeding habitat.
Medium red dot = Probable breeding
Prolonged territoriality, pair, courtship, possible nest, nest building, brood patch.
Large red dot = Confirmed breeding
Distraction display, egg shells, old nest, occupied nest, adult with food or faeces, fledged young.
Small grey dot = Non-breeding
Present, migrant, summering, flying over.
Brown shading = Uplands
Altitude at least 200 m.

	Ireland	Britain	Britain & Ireland
POSSIBLE •	<1%	3%	2%
PROBABLE •	<1%	3%	2%
CONFIRMED •	<1%	8%	6%
TOTAL	<1%	14%	10%

breeding-season maps for passage migrants that did not breed (e.g. Turnstone) have been included. Range-size statistics for species, whether mapped or not, are presented in Appendix 2.

Interpretation points for distribution maps

- The Channel Islands and Isle of Man are included within 'Britain' (see Section 1.2 for further discussion of geographical terminology).
- Brown shaded areas show tetrads with a mean altitude of at least 200 m (see Box 7.2).
- Rock Doves/Feral Pigeons, Common/Scottish Crossbills and Lesser/Common Redpolls presented identification challenges and combined maps are produced for these forms and species (Section 4.2.4). For the Scottish Crossbill, independent sample data from playback surveys are shown separately to indicate definitive Scottish Crossbill records.
- For several migratory waterbirds it was not possible to separate all late-departing migrants from breeding individuals. Thus for species such as the Wigeon and Goldeneye, apparent breeding ranges may be slightly exaggerated; species accounts highlight this where appropriate.
- Interpretation of what forms appropriate breeding evidence for some scarce passage migrants (e.g. Icterine Warbler) can vary among observers. Whilst one person may record a bird as *possible* breeding on the basis of a bird in suitable habitat, another may record the same bird as a migrant.

- Non-native species have not always been given a breeding evidence code owing to uncertainties over whether they should be recorded. Though less of a problem in *Bird Atlas 2007–11* compared with previous atlases, breeding range sizes for some non-native species may be under-estimated.
- The Atlas did not set out systematically to collect *non-breeding* records so the distribution of grey dots on breeding-season maps may be incomplete. For the same reason, the percentage of squares occupied by *non-breeding* birds is not presented in the statistics table but the number of squares is given in Appendix 2.
- All dots are placed at the centre of the 10-km, 20-km or 50-km square to which they relate. Consequently, some dots in coastal areas can appear in the sea.

7.2.3 Distribution change maps and tables

Change maps summarise where the distribution of a species has changed or remained stable across atlas periods. The winter distribution change maps show in pale blue shading the 10-km squares that were occupied in both the *1981–84 Winter Atlas* and *Bird Atlas 2007–11* (Box 7.1). Blue upward-pointing triangles show squares recorded as apparently not occupied in 1981–84 but occupied in 2007–11 and are referred to as 'gains'. Grey downward-pointing triangles signify squares recorded as occupied in 1981–84 but apparently not occupied in 2007–11 and are referred to as 'losses'. Stability and changes are only calculated for 10-km squares visited at least once in both atlas periods, and records from early November in 2007–10 were omitted to make the recording period comparable with that of the *1981–84 Winter Atlas* (see Section 4.3.1).

Breeding-season distribution change maps can be of two types. For most species the map summarises changes across all three breeding atlases (labelled as 'since 1968–72'). An example is shown in Box 7.3. These maps focus on the major long-term trends in occupancy, rather than all possible permutations of presence or non-detection (see Table 4.1). They use red upward- and grey downward-pointing triangles to indicate gains and losses respectively over the period 1968–72 to 2008–11. The shading of the triangles indicates whether the changes have been recent (i.e. occurring since 1988–91 and shown as filled) or old (i.e. occurring prior to 1988–91 and shown as unfilled). Squares that were occupied in all three breeding atlases are shaded in pink. Those occupied in 1968–72 and 2008–11 but apparently not in 1988–91 are shown in paler pink.

For a minority of species the breeding-season distribution change maps show only recent changes since the *1988–91 Breeding Atlas*, either because they were declining non-natives that peaked in occurrence around 1988–91 (e.g. Ruddy Duck) or because all or part of the distribution was shown at 20-km or 50-km resolution and therefore it was impossible to summarise 40 years of changes across multiple 10-km squares. These 20-year change maps and tables, like those for winter, simply show and tally stability and changes between the last two atlas periods.

As with the distribution maps (see Box 7.2), change maps for rare breeding and wintering species may have changes in all or part of the range summarised at the resolution of 20-km or 50-km squares. In those cases, triangles are used where there is an overall change in the number of occupied squares, but the whole 20-km or 50-km square is shaded if the number of occupied squares has remained constant, even if the actual squares have changed (see Section 4.3.3 for more details).

Tables give the number of gains and losses together with percentage change in range size (all in terms of 10-km squares) for the relevant season and period of comparison for Britain, Ireland, and for Britain & Ireland combined. Note that in the breeding distribution change tables, for the period 1968–72 to 2008–11, the numbers of gains and losses relate to the full 40-year period and, hence, are the summed totals of recent and old gains or losses respectively. Note also that percentage change figures are asymmetric: a value of +100% indicates that the range size has doubled whereas a value of -100% indicates that the range size has decreased to zero. For colonising species, where range size increased from zero, the percentage change is represented by an infinity symbol (∞). Breeding-season change tables include the 20-year fixed-effort range change index (labelled as '20yr index'). This figure, which varies between -1 (complete loss) to +1 (complete gain) essentially provides an additional indicator of change in breeding-season range size between 1988–91 and 2008–11 based only on TTV data, in order to limit recorder effort biases (see Section 4.6).

Change maps are not shown for species new to Britain & Ireland because they would be identical to the distribution map. All range change statistics can be found in Appendix 2.

One factor is of overriding importance in the interpretation of change maps: differences in recording effort within and among atlases can have major impacts on the completeness of species lists and has the potential to generate spurious range changes. This topic is considered in detail in Chapter 5. We recommend that readers check the main conclusions in Section 5.7 and examine the maps of recording effort in Figure 5.7. The key messages are repeated here:

1. Recording effort improved significantly in winter, especially in northwest Scotland, the Pennines and western Ireland. As a result, range change estimates may be positively biased and apparent gains in these areas may be indicative of improved coverage rather than a real change in status. These biases are mentioned in the species accounts where appropriate.
2. Recording effort in the *1968–72 Breeding Atlas* and in *Bird Atlas 2007–11* was comparable. This is encouraging because the majority of breeding-season change maps summarise the 40-year period. However, effort was lower in the *1988–91 Breeding Atlas*, especially in Ireland, so any changes since 1988–91 may be positively biased. In these cases the 20-year fixed effort change index is a more robust indicator of change, being based on TTV data and therefore free from effort biases.

BOX 7.3 ANNOTATED BREEDING-SEASON DISTRIBUTION CHANGE MAP AND TABLE

The example map and table are for the Grey Wagtail, a species whose breeding distribution has remained broadly stable in most of Ireland and western and central Britain. It is showing minor losses in Ireland, mostly recent, whereas in eastern Britain it is showing pronounced gains mainly over the last 20 years. Equivalent maps for changes in winter distribution since the *1981–84 Winter Atlas* use blue shading for stability, blue upward-pointing triangles for gains and grey downward-pointing triangles for losses.

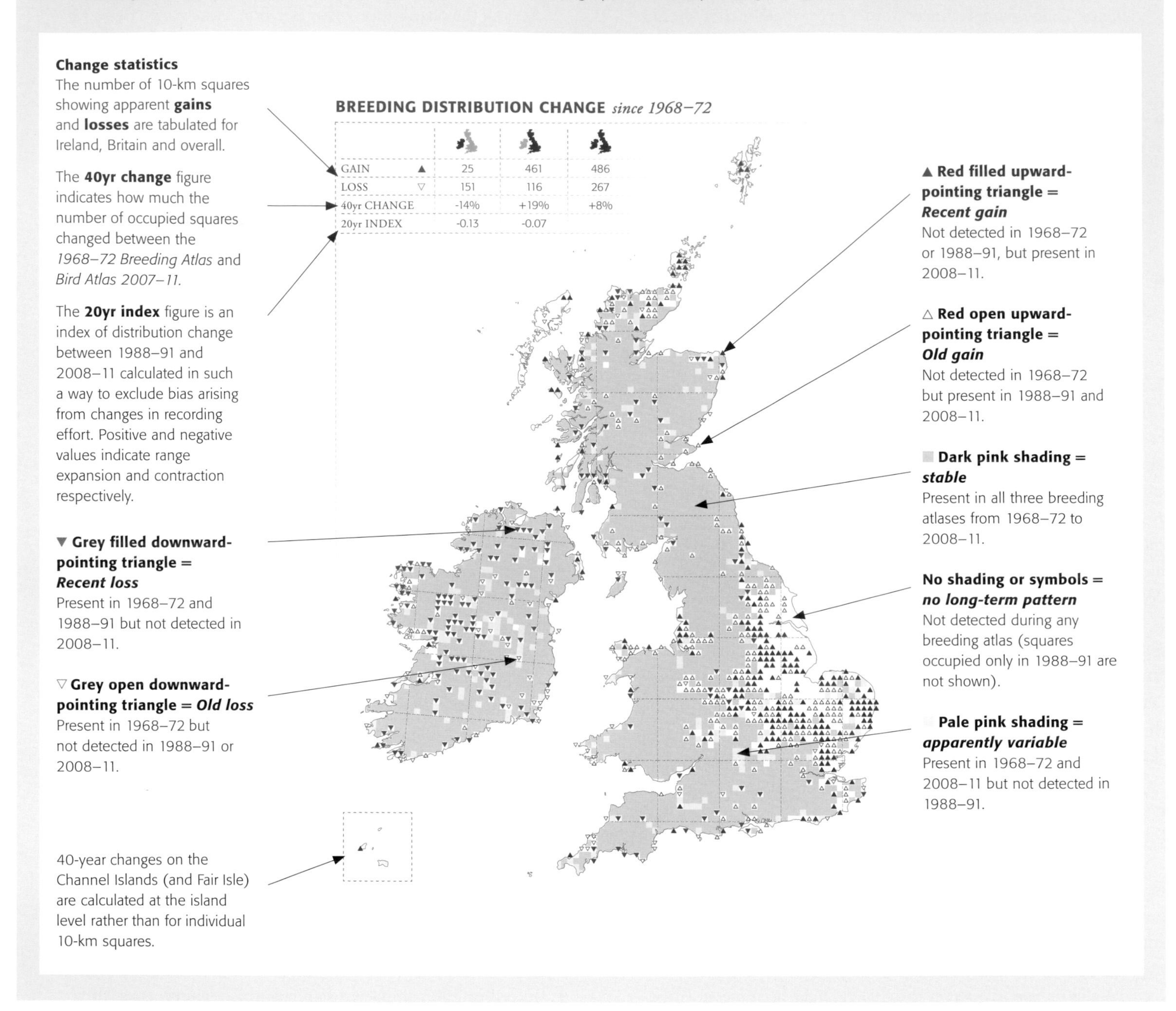

Change statistics
The number of 10-km squares showing apparent gains and losses are tabulated for Ireland, Britain and overall.
The 40yr change figure indicates how much the number of occupied squares changed between the 1968–72 Breeding Atlas and Bird Atlas 2007–11.
The 20yr index figure is an index of distribution change between 1988–91 and 2008–11 calculated in such a way to exclude bias arising from changes in recording effort. Positive and negative values indicate range expansion and contraction respectively.
▼ Grey filled downward-pointing triangle = Recent loss
Present in 1968–72 and 1988–91 but not detected in 2008–11.
▽ Grey open downward-pointing triangle = Old loss
Present in 1968–72 but not detected in 1988–91 or 2008–11.
40-year changes on the Channel Islands (and Fair Isle) are calculated at the island level rather than for individual 10-km squares.
BREEDING DISTRIBUTION CHANGE since 1968–72
GAIN ▲ 25 461 486
LOSS ▽ 151 116 267
40yr CHANGE -14% +19% +8%
20yr INDEX -0.13 -0.07
▲ Red filled upward-pointing triangle = Recent gain
Not detected in 1968–72 or 1988–91, but present in 2008–11.
△ Red open upward-pointing triangle = Old gain
Not detected in 1968–72 but present in 1988–91 and 2008–11.
Dark pink shading = stable
Present in all three breeding atlases from 1968–72 to 2008–11.
No shading or symbols = no long-term pattern
Not detected during any breeding atlas (squares occupied only in 1988–91 are not shown).
Pale pink shading = apparently variable
Present in 1968–72 and 2008–11 but not detected in 1988–91.

Interpretation points for change maps

- Although terms such as 'gain' and 'loss' are routinely used, the potential for false changes owing to under-detection in one atlas period or another should be borne in mind and these terms are often prefixed with 'apparent' (see Section 4.1.1).
- Unlike both previous breeding atlases, *Bird Atlas 2007–11* did not coincide with a major seabird census project (Heubeck 2013), so some remote colonies did not receive comparable breeding-season coverage. Consequently, some apparent losses may be artefacts.
- Changes in taxonomic status affect which maps can be produced (see Section 4.3.4 and Box 4.1).
- Changes in attitudes towards the recording of non-native species mean that change maps and statistics for some of these species may be misleading. Generally, recording of non-native species has been better in *Bird Atlas 2007–11* than in previous atlases (see Section 6.5.2).
- Some dots on distribution maps do not have corresponding symbols on change maps because some distribution records were omitted to make the seasonal recording periods consistent

with those of previous atlases. Specifically, winter changes are calculated from data for the period mid November to February (see Section 4.3.1). For breeding-season changes, as it is unclear to what degree and for which species out-of-season breeding records were used in previous breeding atlases, all out-of-season records were excluded (see Section 4.3.2).

- For breeding-season changes, only records with breeding evidence are used (i.e. no grey dots are used in the calculation of change).
- Breeding-season change maps for seabirds use only 10-km squares with *probable* or *confirmed* breeding evidence (see Section 4.3.2).
- The *1968–72 Breeding Atlas* and *Bird Atlas 2007–11* required that birds be in suitable breeding habitat for the provision of breeding evidence. This was not the case in the *1988–91 Breeding Atlas* with the result that some apparent losses between 1988–91 and 2008–11 may be spurious. This is a particular problem for scarce migrants (e.g. Black Tern) and some waterfowl (e.g. Goldeneye).
- Owing to data for Fair Isle and the Channel Islands being collected at the island scale (rather than for individual 10-km squares) in the *1968–72 Breeding Atlas* and the *1981–84 Winter Atlas*, certain changes are summarised at the island scale.

7.2.4 Relative abundance maps

Relative abundance maps show the parts of Britain & Ireland where species are relatively common or relatively scarce. Winter maps use a scale from white to blue and breeding-season maps a scale from white to red to show the parts of Britain & Ireland that support progressively greater numbers of each species (Box 7.1). On each map there are equal numbers of squares with each of the 10 colour tones. Therefore, each tone represents one tenth of the species range with the darkest tone indicating the top 10% of squares for each species (see Section 4.7.3). The numerical scale (e.g. the number of birds that dark blue represents) is not shown so that readers can concentrate on the emergent geographical pattern in relative abundance and avoid the temptation to use the numbers as densities. For reference purposes, the average percentage of tetrads occupied per 10-km square, and the average number of birds counted per hour per tetrad are given in Appendix 2.

One of three types of map is presented, depending on the availability of data and the distribution of the species:

1. For common and widespread species there are tetrad-resolution maps showing fine-scale variation in abundance derived from statistical models of TTV data (see Section 4.4.3);
2. For scarcer species (e.g. Willow Tit), shading is based on the average number of birds counted per hour per surveyed tetrad in each 20-km square (see Section 4.4.1);
3. For certain highly coastal species (e.g. Eider), abundance is represented in a 'ribbon' around the coastline and was generated by spatial smoothing of counts in surveyed tetrads (see Section 4.4.2).

Interpretation points for relative abundance maps

- The maps do not include corrections for variable detectability among species. Consequently, it is invalid to infer that Species A is more abundant than Species B in a particular area by directly comparing relative abundance on their maps, because the abundance scale is species-specific. However, it is valid to say that a particular region of Britain or Ireland is relatively more important for Species A than Species B. We recommend that any such comparisons focus on blocks of squares, as counts in single squares could be prone to errors.
- No corrections have been made for variation in detectability within species in relation to habitat. We consider it very unlikely that such detection issues will result in biased patterns of relative abundance at the large scales presented here.
- Abundance maps rely solely on TTV data and do not employ counts from Roving Records owing to potential effort biases. For some scarce or localised species, the sample of surveyed tetrads could have missed significant concentrations. This is a particular problem for seabirds, especially on islands lacking tetrads valid for TTV coverage.
- For highly aggregated species (e.g. seabirds), the highest abundance class can encompass a wide range of observed counts.
- Abundance maps are not shown for strictly nocturnal species (these were inadequately surveyed by diurnal TTVs) or for scarce and rare species encountered on few TTVs.

7.2.5 Relative abundance change maps

These maps provide an indication of the parts of Britain & Ireland where species are increasing or decreasing in abundance. They are for the breeding season only and show 20-km squares where the proportion of tetrads that were occupied by a species changed between the *1988–91 Breeding Atlas* and *Bird Atlas 2007–11*. The proportion of tetrads occupied by a species in a 20-km square is positively related to the local density of the species. Therefore, we can infer that regions on the maps with marked changes between 1988–91 and 2008–11 in the proportion of tetrads occupied are likely to have seen changes in local abundance over that period (see Sections 4.5 and 4.7.4).

The 20-km squares where the proportion of occupied tetrads has increased are shown in orange-red tones and those where it has decreased are shown in brown tones (Box 7.4). Abundance change was only calculated for squares with breeding evidence in at least one atlas period.

Interpretation points for relative abundance change maps

- Areas of overall distribution gain will inevitably show as increases in abundance (from zero); likewise losses will show as decreases in abundance. The most interesting situations are where distribution has changed little but abundance is in a state of flux (e.g. Willow Warbler).

BOX 7.4 ANNOTATED BREEDING RELATIVE ABUNDANCE CHANGE MAP

The example map is for the House Martin, a species showing a marked geographical dichotomy in local breeding-season trends, with decreases in abundance apparent since 1988–91 in southern Britain contrasting with increases in abundance in Ireland and northern Britain. The abundance change map should be viewed alongside the breeding-season abundance map to identify how abundance varied spatially in 2008–11.

White areas = *absent* or *no abundance change* Some white areas will correspond with parts of Britain & Ireland where the species has never been present, or present in sufficient numbers to measure changes (see the distribution and change maps).

White area in occupied regions indicate where the proportion of occupied tetrads has remained approximately constant through time. For most species this can be interpreted as broadly stable abundance patterns but for certain very common species (e.g. Wren), changes in abundance can occur without the proportion of occupied tetrads changing.

Red shading = *increasing abundance* The proportion of occupied tetrads in a 20-km square increased between 1988–91 and 2008–11. Darker red indicates the increase has been large.

Brown shading = *declining abundance* The proportion of occupied tetrads in a 20-km square decreased between 1988–91 and 2008–11. Darker brown indicates the decrease has been large.

- For most species, white areas indicate 20-km squares where the species did not breed in either period, or where abundance has remained broadly stable. For a small number of easily-detected and locally abundant species (e.g. Wren), abundance changes in white areas cannot be ruled out. This is because the local abundance of such numerous species could increase or decrease substantially without the proportion of occupied tetrads actually changing (see Figure 4.4 and Section 4.5). For these very common species it helps to check the abundance maps in this book and in the *1988–91 Breeding Atlas* to see if relative abundance was high in both periods. Additional maps of change in abundance for 49 common species in the UK based on annual monitoring data are presented by Massimino *et al.* (2013).
- Areas showing increases in abundance do not necessarily have higher abundance than other areas of the map.

7.3 ADDITIONAL RESOURCES

RANGE SIZE AND change statistics for all species recorded during 2007–11 are presented in a single table in Appendix 2. An online 'Map Explorer' will be available after publication to provide access to digital copies of all the maps presented in the book, as well as the remaining maps that could not be fitted in the book. Links to access published population estimates will also be made available on the web site. Readers should visit www.birdatlas.net to access these additional resources.

Bird Atlas 2007–11
Species Accounts

MUTE SWAN

Cygnus olor

BTO CODE: MS

LES FOSTER www.springerwildlifephotography.co.uk

THE MOST STRIKING change in the distribution of the Mute Swan is its expansion into Shetland, where it now winters and is a confirmed breeder. A pair first bred there in 1992 and by 2010 there were 12 pairs breeding (Shetland Bird Club 2011). The Mute Swan is largely sedentary, and the winter and breeding distributions are very similar. It is widespread across Britain & Ireland, though generally absent from higher ground, and occupies a wide range of lowland habitats. During the breeding season the nests are often conspicuous and it is relatively easy to confirm breeding.

The maps of relative abundance, for both seasons, show the highest concentrations to be throughout landscapes with abundant wetlands and waterways, such as the Somerset Levels, the Fens, the Broads, Orkney and the Uists in Britain, and Lough Neagh, Lough Erne, Lough Oughter, Lough Corrib and along the River Shannon in Ireland. A survey in 2002 produced a population estimate of 31,700 birds in Britain, consisting of 6,150 pairs and 19,400 non-breeders (Ward *et al.* 2007).

There was little change in breeding distribution between the *1968–72 Breeding Atlas* and *1988–91 Breeding Atlas*, but there was variation in population size, partly due to localised habitat degradation but also due to lead poisoning through the ingestion of spent lead shot and fishing weights (Rowell & Spray 2004). A recovery of the population followed the adoption of suitable alternatives and the subsequent banning of lead for fishing weights in the UK in 1987. The distribution change maps show apparent range expansion in both seasons, perhaps as a consequence, although poisoning from lead shot continues to be a significant cause of death (Pitches 2012). Likewise the abundance change map shows density increases in many areas since the *1988–91 Breeding Atlas*. The populations may have benefited also from generally milder winters, which are associated with higher survival and subsequent greater productivity (Esselink & Beekman 1991). The UK wintering population increased by 17% between 1983/84 and 2008/09 (*SUKB 2012*), whilst in the Republic of Ireland numbers increased by 27% between 1994/95 and 2008/09 (*I-WeBS Report 2009*). The creation of new waterbodies will have provided additional wintering and breeding habitats and may account for some of the gains in 10-km-square occupancy.

Sponsored by D. Graham Bell

BREEDING DISTRIBUTION *2008–11*

POSSIBLE •	6%	3%	4%
PROBABLE ●	13%	5%	7%
CONFIRMED ●	42%	54%	50%
TOTAL	61%	62%	61%

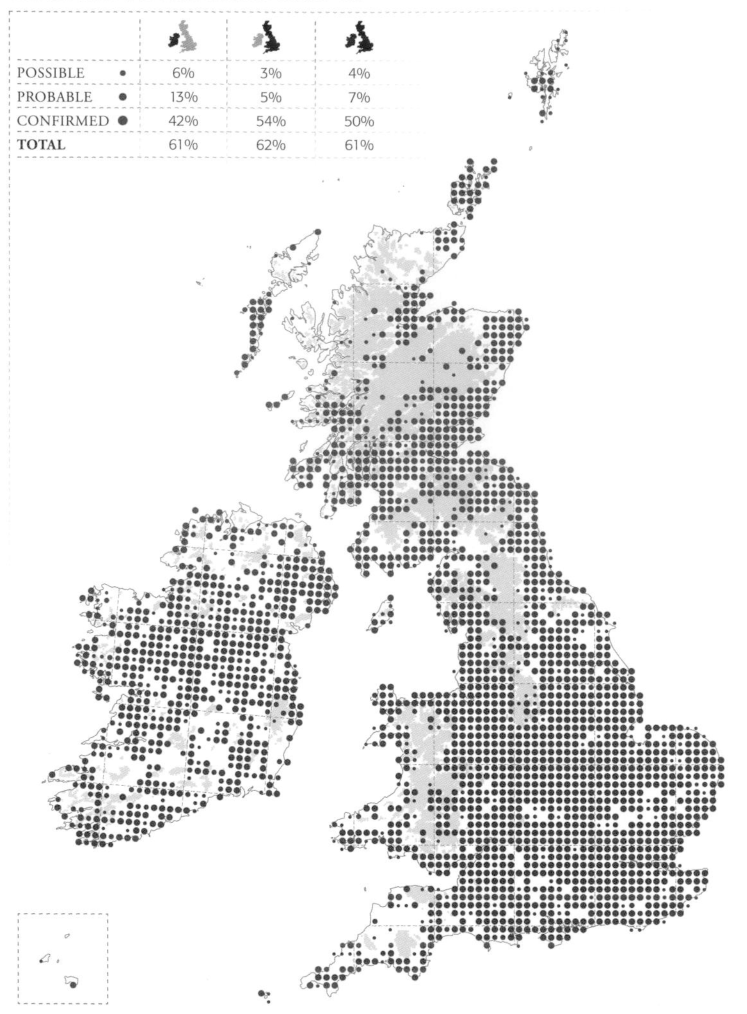

WINTER DISTRIBUTION *2007/08–2010/11*

PRESENT ●	67%	67%	67%

BREEDING DISTRIBUTION CHANGE *since 1968–72*

	Ireland	Britain	Britain and Ireland
GAIN ▲	95	330	425
LOSS ▽	121	186	307
40yr CHANGE	-4%	+9%	+5%
20yr INDEX	-0.07	+0.14	

WINTER DISTRIBUTION CHANGE *since 1981–84*

	Ireland	Britain	Britain and Ireland
GAIN ▲	112	421	533
LOSS ▼	110	96	206
30yr CHANGE	+<1%	+21%	+15%

BREEDING RELATIVE ABUNDANCE *2008–11*

BREEDING RELATIVE ABUNDANCE CHANGE *since 1988–91*

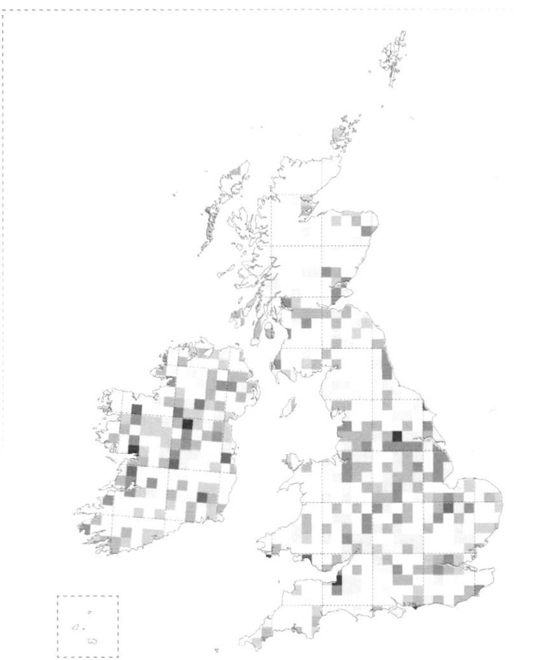

WINTER RELATIVE ABUNDANCE *2007/08–2010/11*

BEWICK'S SWAN

Cygnus columbianus

BTO CODE: BS

STEVE DAVIES www.swindon-birds.co.uk

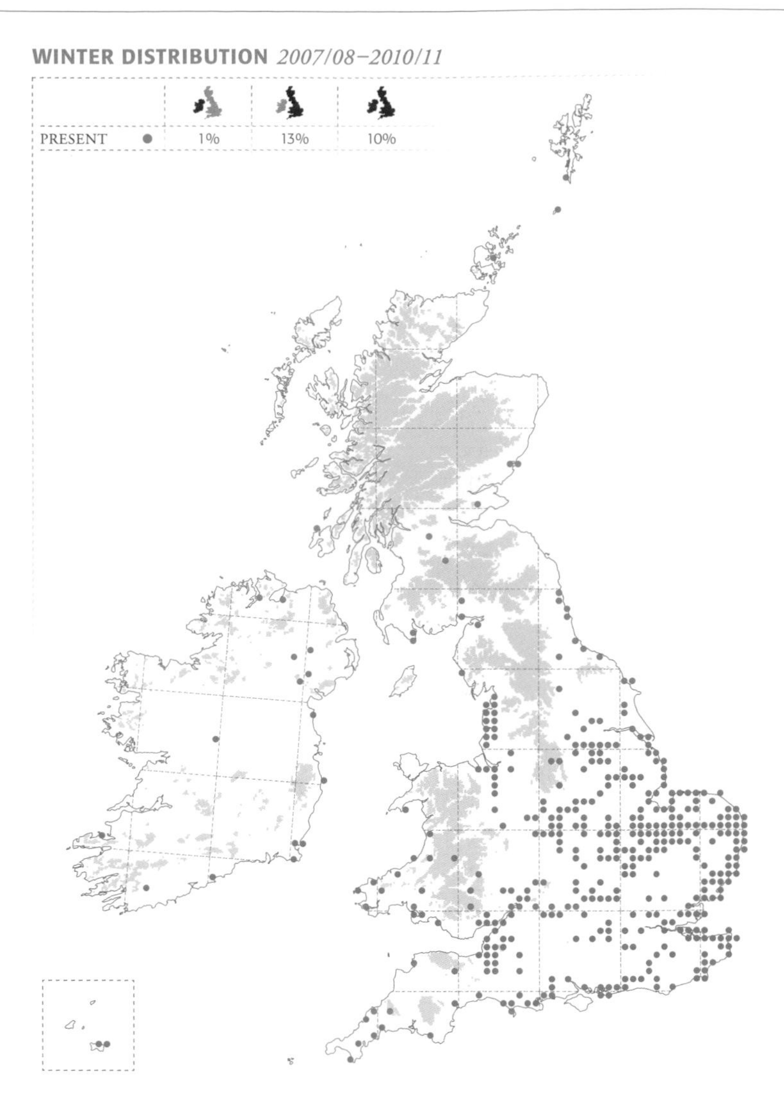

THE BEWICK'S SWAN, the Palearctic form of Tundra Swan, breeds across the tundra of European Russia, and 30–40% of the population winters in Britain & Ireland. The species' winter distribution is patchy, with most flocks occurring in southeastern England, and very few in Ireland, Wales or Scotland. Despite a relatively large number of occupied squares, significant flocks are concentrated in the fenlands of eastern Britain, chiefly around the Ouse Washes and Nene Washes (*WeBS Report 2011*), with smaller concentrations in parts of the west, especially the Severn Estuary. In Ireland, small numbers are concentrated at two sites in the southeast, Wexford Harbour & Slobs and the Cull & Killag (Boland *et al.* 2010).

This flyway population has declined by 27% between 1995 and 2005 (Rees & Beekman 2010). Numbers in Britain & Ireland have declined since reaching a peak in the early 1990s (*I-WeBS Report 2009*; *WeBS Report 2011*). The decline in Ireland has been pronounced, and is reflected in the 90% range contraction shown in the change map. Just 80 individuals were recorded throughout Ireland during the 2010 International Swan Census (Boland *et al.* 2010), representing an 86% decline since 1995. In Britain, the contraction of range has been largely from the north and west, and much of this seems to have been balanced by gains in eastern England, with only a modest 12% range loss apparent overall. The downward trend in numbers has been much shallower in eastern England compared with Ireland, largely because numbers at the key sites of the Ouse Washes and Nene Washes, which together support more than 80% of the total wintering population in Britain & Ireland, have remained broadly stable. Counts during winter 2010/11 were higher than during recent winters, perhaps indicative of a response to the cold winter, forcing birds from the Continent into Britain (*WeBS Report 2011*).

Generally, there seems to have been an eastward shift in the wintering range of this population, as evidenced by the declines in Ireland and western Britain, which are on the western edge of the species' Palearctic range. This is possibly caused by birds 'short-stopping' in continental Europe.

Sponsored by Northumberland & Tyneside Bird Club

WINTER DISTRIBUTION CHANGE *since 1981–84*

GAIN ▲	5	173	178
LOSS ▼	147	224	371
30yr CHANGE	-90%	-12%	-33%

WINTER RELATIVE ABUNDANCE *2007/08–2010/11*

▲ An increasing proportion of Bewick's Swans that winter in Britain & Ireland are now concentrated in the fenlands of eastern England.

WHOOPER SWAN

Cygnus cygnus BTO CODE: WS

EDMUND FELLOWES

THE GENERAL PATTERN of winter distribution of the Whooper Swan remains consistent with the *1981–84 Winter Atlas*, with the largest numbers of occupied squares in more northerly areas of both Britain and Ireland. The winter change map shows that many of the areas in central and north-central Ireland and in south and central Scotland were occupied during both winter atlases. Some range expansion in southern areas is evident, particularly in southeast England. This is also reflected in the results of the 2010 International Swan Census, which reported a significant increase at the Ouse Washes (Hall *et al.* 2012). Overall, range increases of 35% and 16% were identified in Britain and Ireland respectively. These results are consistent with an increase in the Icelandic breeding population of the Whooper Swan, which winters exclusively in Britain & Ireland, to around 30,000 birds during the 2010 census, representing an 11% increase since 2005 (Boland *et al.* 2010; Hall *et al.* 2012).

The winter abundance map reflects the concentrations that have been identified through other monitoring schemes (*I-WeBS Report 2009*; *WeBS Report 2011*). In Ireland these are predominantly in central, northern, western and northeastern counties. In Britain, the highest densities are mainly in parts of northern and eastern England and Scotland.

There are relatively few records of breeding Whooper Swans in Britain & Ireland. The map of breeding distribution shows *confirmed* breeding records in southeast England, the northern part of Ireland and Scotland. Some are known to stem from injured birds remaining to summer and breed, and in at least one case, in Highland, injured birds were joined by wild birds to form a resident breeding population (*RBBP 2010*). The *confirmed* breeding records on Shetland and the Outer Hebrides, however, could reflect an expansion in the species' breeding range. This expansion is possibly due to the increase in the Icelandic breeding population, especially as most of the gains in these areas have occurred since the *1988–91 Breeding Atlas*. The *possible* breeding record in Merioneth is of two birds (one injured) that summered in 2010 and may have bred (*RBBP 2010*).

Sponsored by Offaly County Council

WINTER DISTRIBUTION *2007/08–2010/11*

	Britain & Ireland	Britain	Ireland
PRESENT ●	54%	45%	47%

BREEDING DISTRIBUTION *2008–11*

	Britain & Ireland	Britain	Ireland
POSSIBLE •	<1%	<1%	<1%
PROBABLE ●	<1%	<1%	<1%
CONFIRMED ●	<1%	1%	1%
TOTAL	<1%	1%	1%

WINTER DISTRIBUTION CHANGE *since 1981–84*

GAIN ▲	157	528	685
LOSS ▼	86	233	319
30yr CHANGE	+16%	+35%	+28%

WINTER RELATIVE ABUNDANCE *2007/08–2010/11*

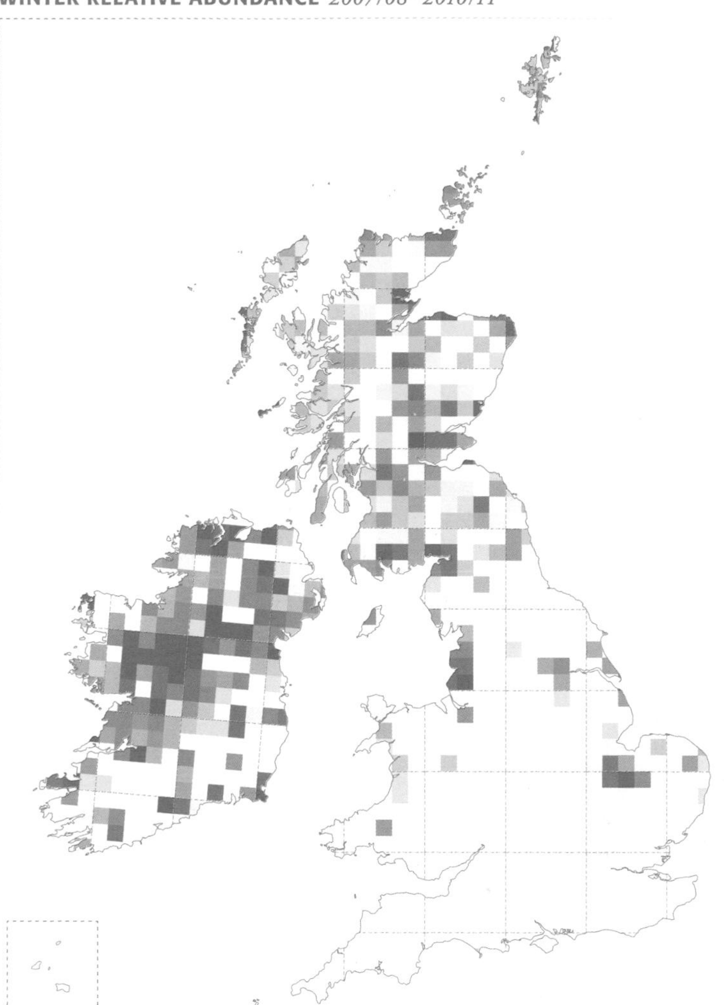

BREEDING DISTRIBUTION CHANGE *since 1968–72*

GAIN ▲	8	37	45
LOSS ▽	0	7	7
40yr CHANGE	∞	+375%	+475%
20yr INDEX	-	+1.00	

BLACK SWAN

Cygnus atratus BTO CODE: AS

MORRIS RENDALL

NATIVE TO AUSTRALIA, Black Swans were imported to Britain around 1791 and first bred in the wild in 1851, at Carshalton in Surrey (Lever 2009). They now occur widely, as introductions or escapes, in a variety of habitats including ornamental ponds, lakes and marshes. When last assessed, the British population was not considered to be self-sustaining (Dudley 2005).

This species was shown to be more widespread than previously thought, mainly due to better recording of non-native species during the Atlas period. The winter distribution map shows Black Swans to be widely distributed in the southern half of Britain. Elsewhere, reports were more scattered and some records may have related to single wandering individuals.

Breeding evidence was recorded in 111 10-km squares and these were mainly in southern and central England, the Isle of Man and Channel Islands; two records came from Ireland. In addition, Black Swans were recorded as *non-breeding* in a further 153 10-km squares. It is likely that many of these *non-breeding* records came from birds in suitable habitat but there are differences amongst birdwatchers in the way that breeding codes are applied to non-native species.

Since 2003, the number of breeding pairs has been consistently in double figures (Holling *et al.* 2011a). There is an established breeding population in the Broads of east Norfolk, where numbers have built up since 1980 and breeding is regular (Taylor *et al.* 1999).

Assessing change in range is difficult due to the likely under-reporting of Black Swans in previous atlases. There were records from 30 10-km squares in the *1981–84 Winter Atlas*, of which 11 were again occupied in 2007–11. Apparently, only one 10-km square was occupied in the *1988–91 Breeding Atlas* and that square was again occupied in 2008–11. Black Swans are known to form hybrid pairs with Mute Swans and produce offspring (known as 'Blute Swans' in captivity) and this was reported from one 10-km square in central England.

Sponsored by John & Jean McCaig (BTO Life Fellows)

BREEDING DISTRIBUTION *2008–11*

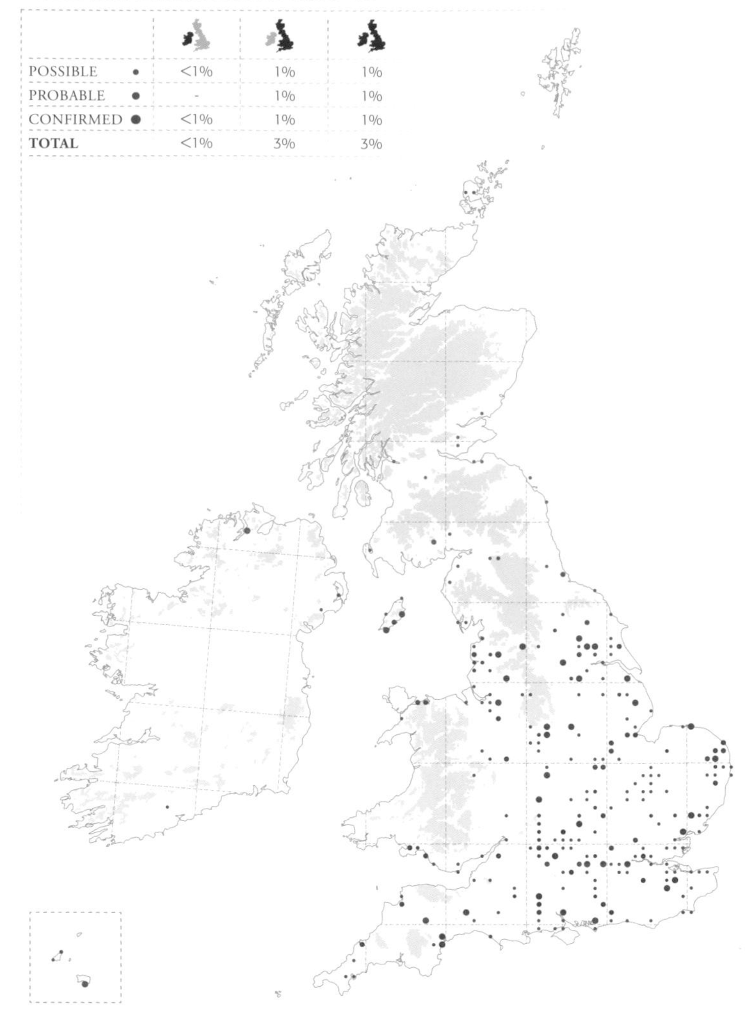

POSSIBLE •	<1%	1%	1%
PROBABLE •	-	1%	1%
CONFIRMED •	<1%	1%	1%
TOTAL	<1%	3%	3%

WINTER DISTRIBUTION *2007/08–2010/11*

PRESENT •	1%	10%	8%

SWAN GOOSE

Anser cygnoides

BTO CODE: HN

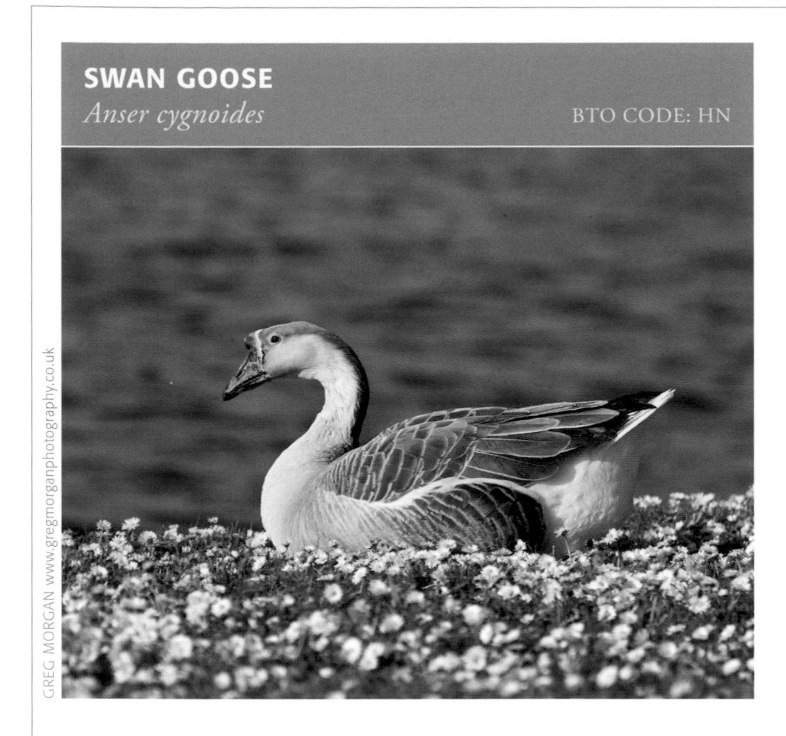

GREG MORGAN www.gregmorganphotography.co.uk

THE NATURAL BREEDING range of the Swan Goose spans Central Asia to southeast Siberia and Mongolia, and it winters in east China and Korea, where it is declining (Zhang *et al.* 2011; Wetlands International 2013). The species is listed as Vulnerable; the rapid population decline is thought to be due to poor breeding success as a result of drought and habitat loss. Unsustainable levels of hunting are also contributing to the species' decline (BirdLife International 2013).

In Britain & Ireland the domesticated form, commonly known as 'Chinese Goose' occurs in a feral state and was reported from a small scattering of 10-km squares. The maps for winter and the breeding season are similar, with the majority of records in England. Breeding was *confirmed* in two 10-km squares: at Scaling Dam in Cleveland in 2008 and 2009, where a pair has been present since 2006, and at Hollybed Common in Herefordshire in 2011. It is unclear how well reported this species has been during 2007–11, but it was certainly under-reported in the past, thereby limiting as assessment of its trends. It was not reported during the *1968–72 Breeding Atlas* and there was only one record during each of the other previous atlases. Eight Swan Geese were recorded in the 1991 survey of introduced and escaped geese, five of them in Norfolk (Delany 1993).

Elsewhere in Europe, non-native records of this species are widespread but no introduced populations are known to be fully established. In the Netherlands, Swan Geese largely associate with feral Greylag Geese and some 10–20 pairs were found during 1998–2000 but only a few were thought to be breeding (SOVON Vogelonderzoek Nederland 2002, 2011). By 2005 there were an estimated 150 breeding pairs, dropping to 13 pairs by 2009 (van der Jeugd *et al.* 2006). Breeding is also known from Germany, where the species first bred in 1967 and around 15–20 pairs now breed (Südbeck *et al.* 2007; Bauer & Woog 2008); ecological damage through hybridization has been noted (Gebhardt 1996). There were five records of probable breeding in Belgium in 2000–02 (Vermeersch *et al.* 2004).

BREEDING DISTRIBUTION *2008–11*

	Ireland	Britain	Britain & Ireland
POSSIBLE ●	-	<1%	<1%
PROBABLE ●	<1%	<1%	<1%
CONFIRMED ●	-	<1%	<1%
TOTAL	<1%	<1%	<1%

WINTER DISTRIBUTION *2007/08–2010/11*

	Ireland	Britain	Britain & Ireland
PRESENT ●	<1%	1%	1%

BEAN GOOSE

Anser fabalis

BTO CODE: BE

AT THE TIME of the *1981–84 Winter Atlas*, the two commonly occurring forms of Bean Goose, 'Taiga' (*fabalis*) and 'Tundra' (*rossicus*), were not routinely separated. Therefore, to make comparisons between atlases, all records, irrespective of form, were combined. Most records come from East Anglia, southeast England and northwest England, with a scatter elsewhere along the east coast and a few inland. There are a few occupied squares in mainland Scotland, but many more on Orkney and Shetland. Bean Geese are rare in Ireland, with records from just four 10-km squares. The winter relative abundance map indicates that the majority of the wintering Bean Goose population is highly concentrated, and the many occupied squares elsewhere usually support few birds, or birds for short periods only.

The change map shows a mix of gains and losses. The areas of most striking gain are Orkney and Shetland, where the Bean Goose is usually a rare winter visitor. The Yare Valley in the Norfolk Broads is one of two traditional wintering areas for the Taiga Bean Goose. The population has declined substantially, both locally (Holt *et al.* 2011a) and at flyway-level (Wetlands International 2013), and may have become more site-faithful to well-managed grasslands. Expected losses here have probably been masked by influxes of Tundra Bean Geese.

Separate maps of Taiga and Tundra Bean show the winter distribution where birds have been identified to subspecies level. In addition to the Yare Valley, the other traditional wintering area on Slamannan Plateau, Stirlingshire (*WeBS Report 2011*), is shown on the Taiga Bean Goose distribution map. Elsewhere, small groups occur at coastal and inland sites and may move around within a winter. Small, but regular, numbers of Tundra Bean Geese are recorded each year and these tend to be related to cold-weather movements. The map shows concentrations in Britain along the east coast, in the Northern Isles, on the Lancashire coast and inland around the Fens. In cases where 'gain' squares could be assigned to subspecies, 66% were due to the presence of the Tundra form, 19% to the Taiga form and 15% to both forms.

In memory of Robert Corrie

WINTER DISTRIBUTION *2007/08–2010/11*

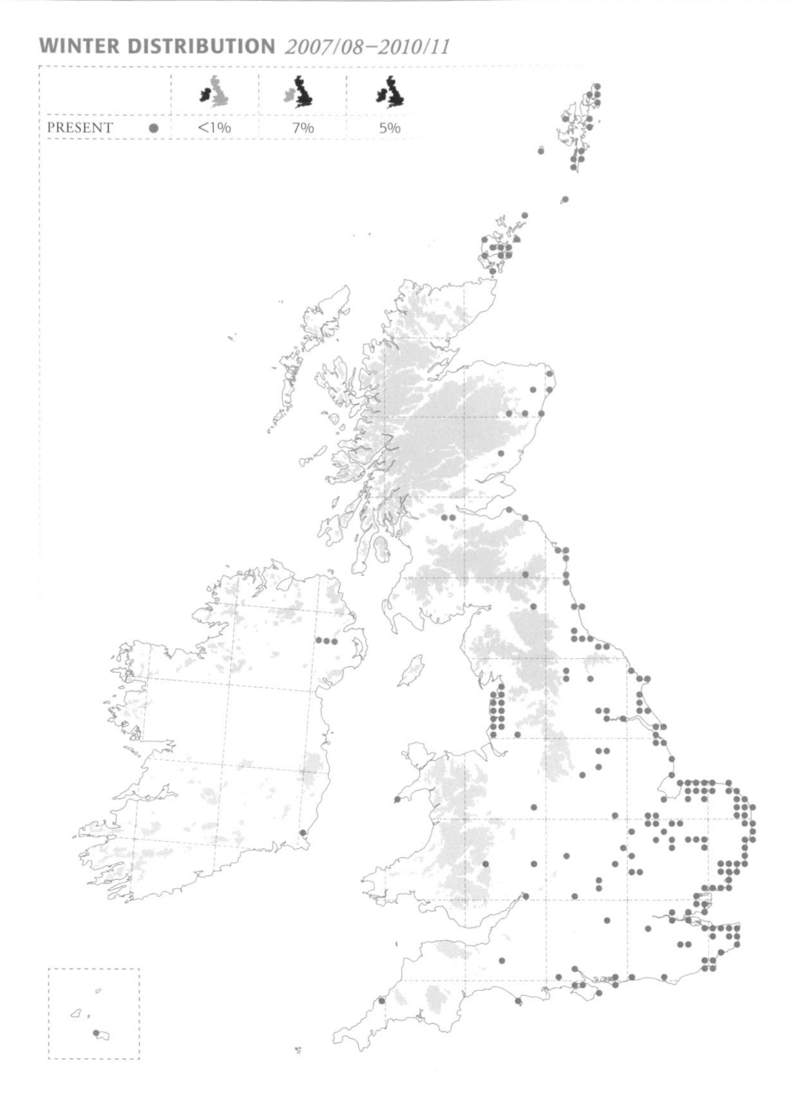

TAIGA BEAN GOOSE WINTER DISTRIBUTION *2007/08–2010/11*

WINTER DISTRIBUTION CHANGE *since 1981–84*

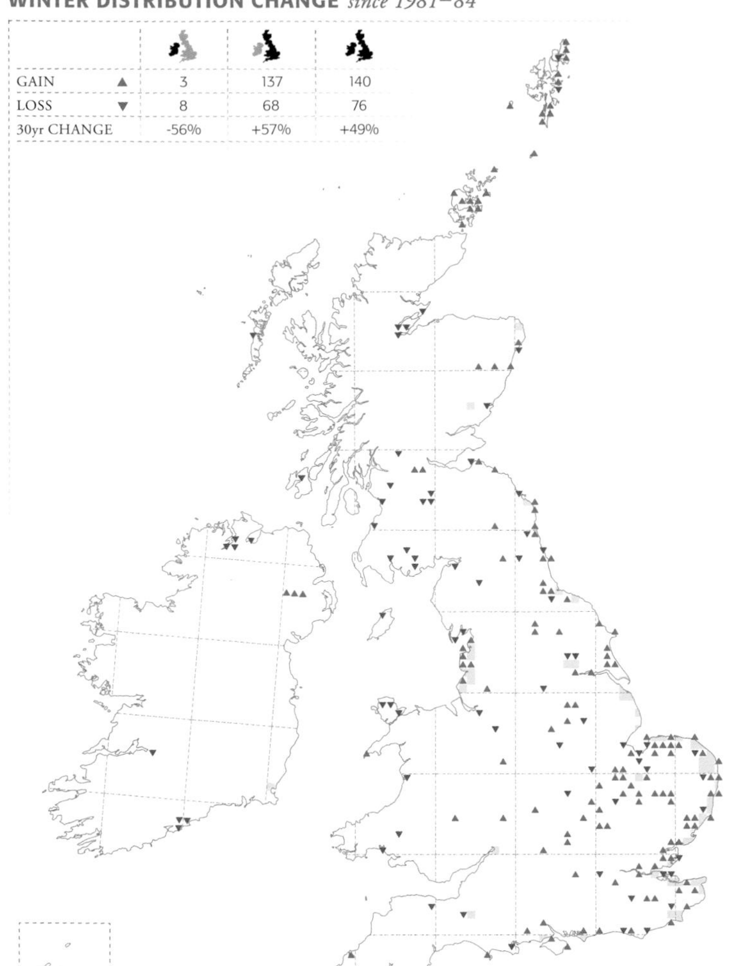

GAIN ▲	3	137	140
LOSS ▼	8	68	76
30yr CHANGE	-56%	+57%	+49%

WINTER RELATIVE ABUNDANCE *2007/08–2010/11*

TUNDRA BEAN GOOSE WINTER DISTRIBUTION *2007/08–2010/11*

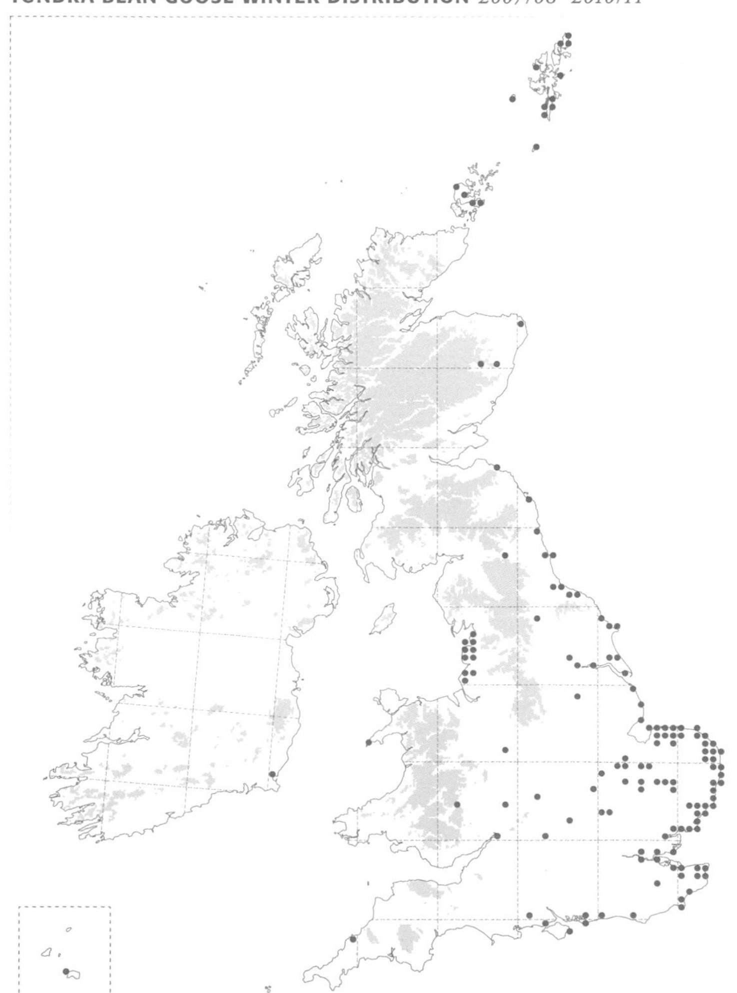

PINK-FOOTED GOOSE

Anser brachyrhynchus

BTO CODE: PG

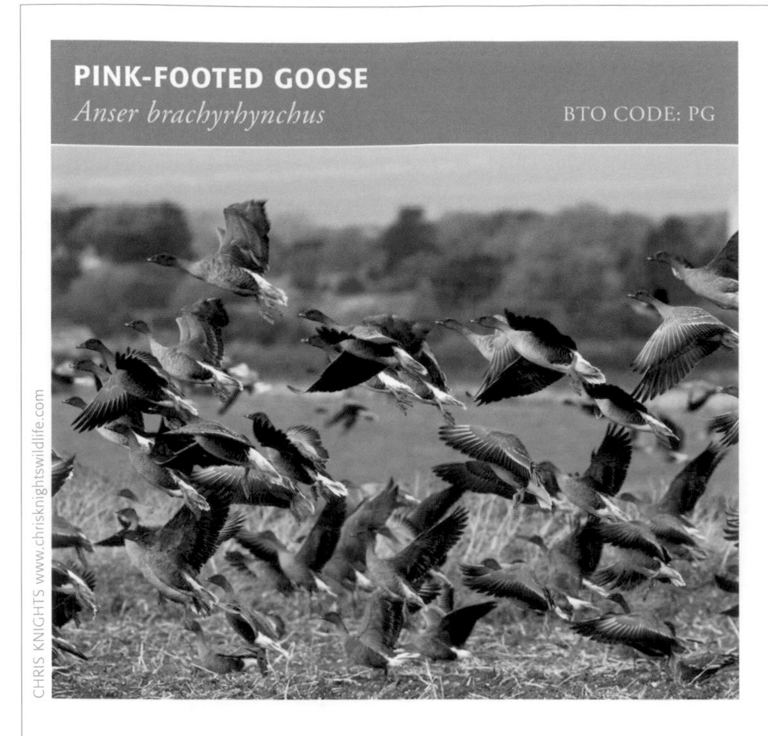

CHRIS KNIGHTS www.chrisknightswildlife.com

EACH AUTUMN THE entire population of Pink-footed Geese breeding in Iceland and eastern Greenland migrates southeast to winter almost exclusively in Britain, where flocks favour intensively farmed lowlands, and generally avoid upland areas. A few individuals, especially in eastern Britain, are from the Svalbard population, which winters mainly in Denmark. This species is uncommon in Shetland, northwest Scotland, Ireland, Wales and southwest and southern England. In Scotland it is found along the eastern coastal plain, through the central lowlands, mainly in the east, and around the Solway Firth. In England it is found mainly in a broad band from Lancashire across to Lincolnshire and Norfolk, though the winter relative abundance map indicates that the highest densities are found close to the coast.

The winter population has increased markedly from around 90,000 at the time of the *1981–84 Winter Atlas* to over 360,000 currently (Mitchell 2010). This increase is considered to be due primarily to improved foraging opportunities on agricultural land in the main wintering areas, which may have increased overwinter survival and body condition prior to return migration and breeding (Fox *et al.* 2005).

This population increase has resulted in a doubling of the British range size since the *1981–84 Winter Atlas*. Gains have taken place throughout the range, but are most noticeable in England, northeast Scotland and the Northern Isles. It should be noted that a significant number of these apparent gains, especially those along the central England flight-line between Lancashire and East Anglia, may be due to the inclusion of some flying birds that were not associating with the square, though some relate to individuals and small flocks that linger with naturalised geese on pits, lakes and reservoirs.

In spring, flocks of Pink-footed Geese move north and stage in central and northern Scotland before returning to Iceland. This results in large numbers of *non-breeding* records in Britain during April and early May. A small number of birds remain each year to summer. Two *confirmed* breeding records were recorded during this atlas in Cumbria and Inverness-shire, with a *probable* breeding record on the Outer Hebrides (*RBBP 2010*).

Sponsored by Sporting Gun

WINTER DISTRIBUTION *2007/08–2010/11*

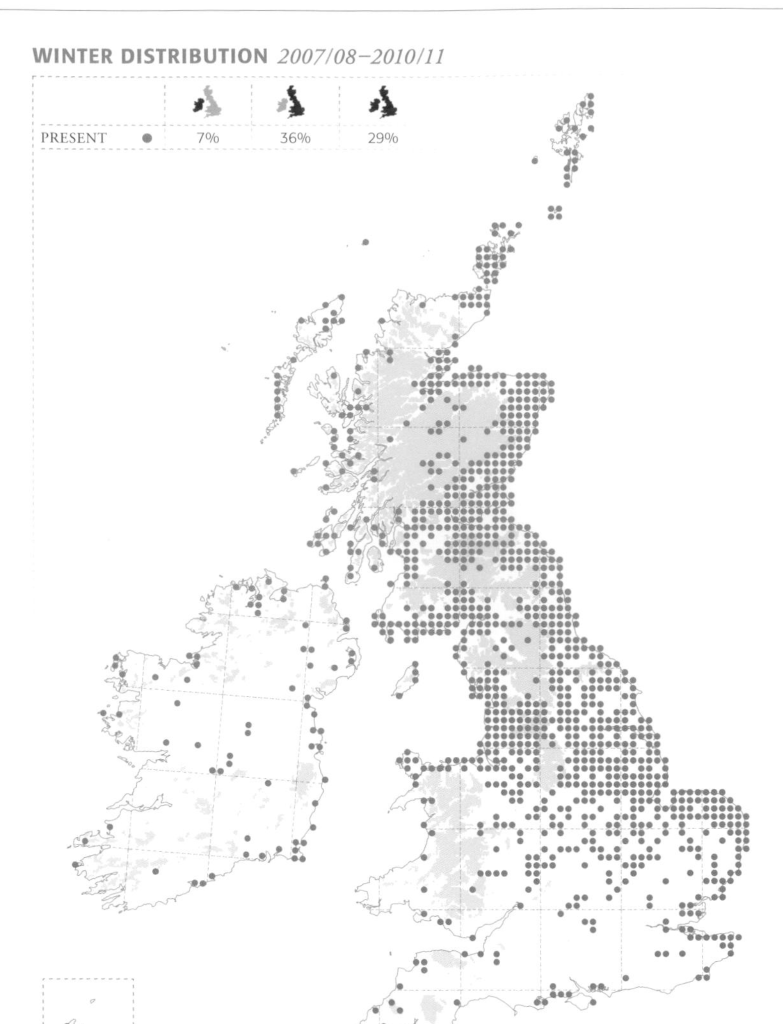

	Ireland	Britain	Britain & Ireland
PRESENT ●	7%	36%	29%

BREEDING DISTRIBUTION *2008–11*

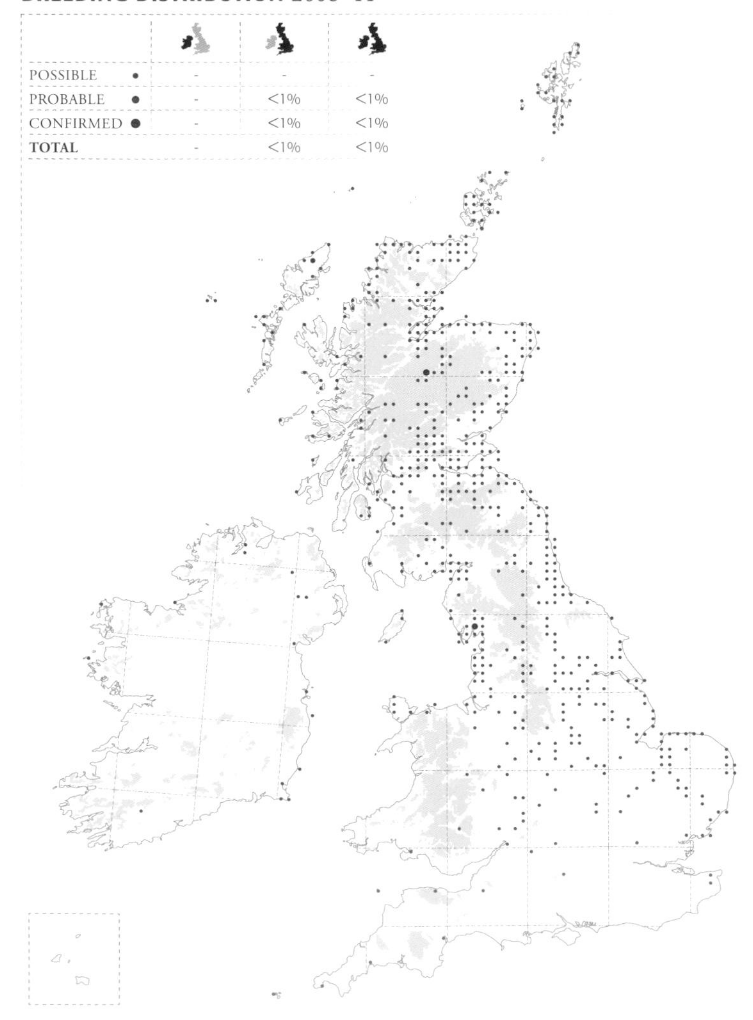

	Ireland	Britain	Britain & Ireland
POSSIBLE •	-	-	-
PROBABLE ●	-	<1%	<1%
CONFIRMED ●	-	<1%	<1%
TOTAL	-	<1%	<1%

WINTER DISTRIBUTION CHANGE *since 1981–84*

GAIN ▲	53	597	650
LOSS ▼	5	117	122
30yr CHANGE	+369%	+93%	+100%

WINTER RELATIVE ABUNDANCE *2007/08–2010/11*

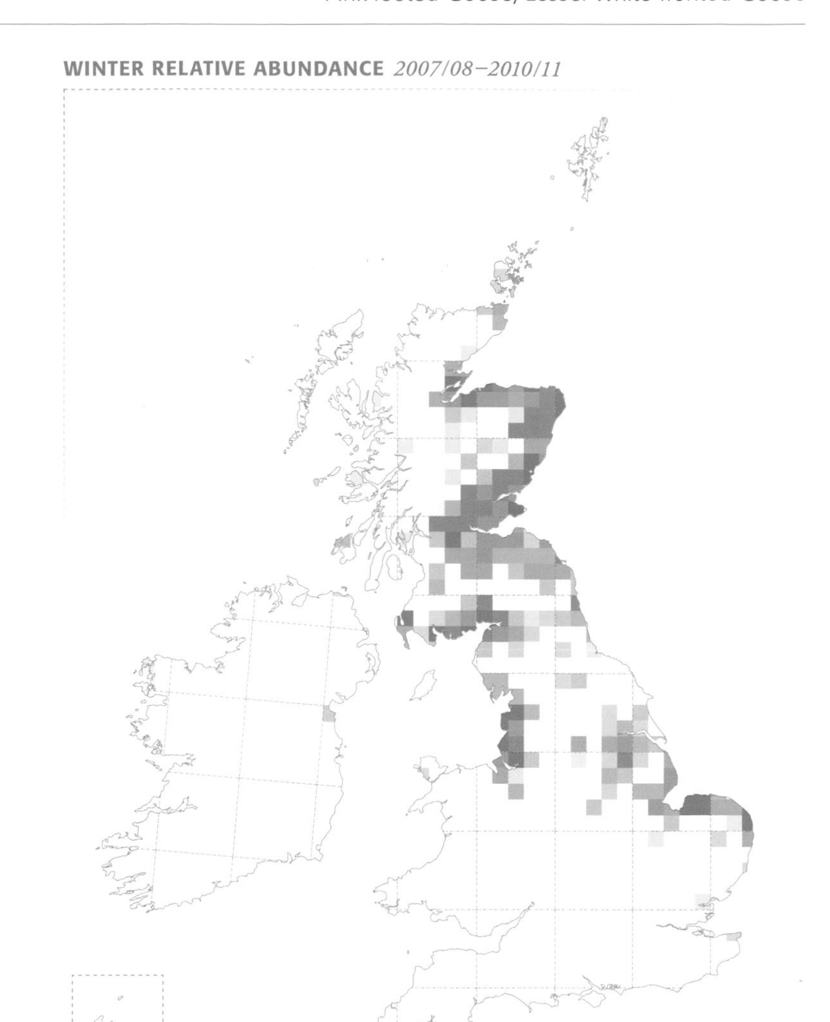

LESSER WHITE-FRONTED GOOSE

Anser erythropus

BTO CODE: LC

DAVID KJAER www.davidkjaer.com

LESSER WHITE-FRONTED GEESE are very rare winter vagrants to Britain, a situation confused by the presence of escapes, some showing evidence of hybridisation. They are extremely rare winter visitors in Ireland. Since the *1981–84 Winter Atlas* the species has more than doubled its winter range size but it is impossible to be certain about the provenance of all individuals. However, the fact that ten 10-km squares were occupied during the breeding season casts doubt on these records being of vagrants. Most records were from the southern half of England, a pattern shared with many non-native species escaping from captivity. A cluster of records in Norfolk includes a possibly wild bird migrating with 'Taiga' Bean Geese, but known escapes are also present there.

WINTER DISTRIBUTION *2007/08–2010/11*

WHITE-FRONTED GOOSE

Anser albifrons

BTO CODE: WG

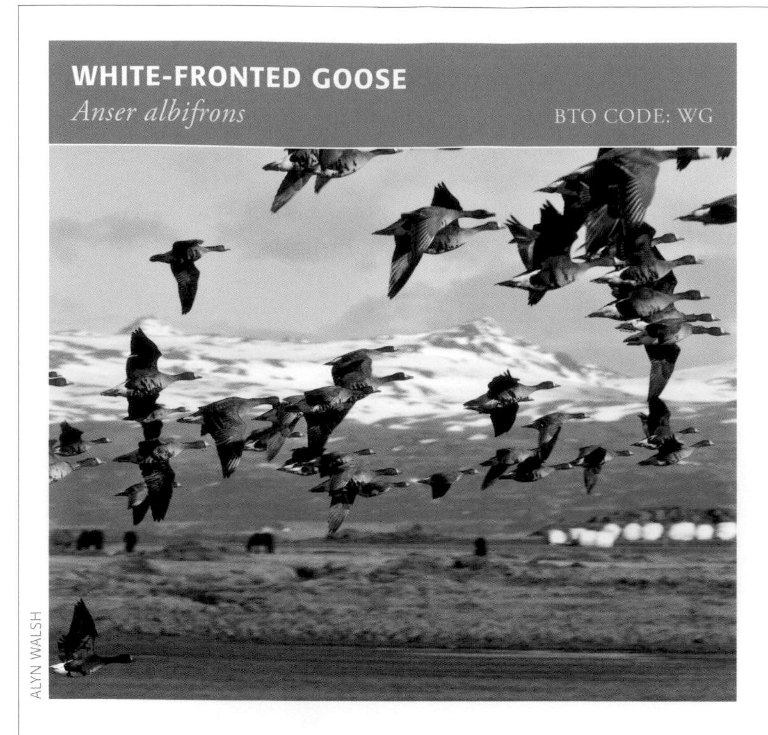

TWO DISTINCTIVE SUBSPECIES of White-fronted Goose winter in Britain & Ireland. There is little range overlap, with the 'European' (*albifrons*) subspecies occurring mainly in southern England, and the 'Greenland' (*flavirostris*) subspecies occurring predominantly in Ireland and western and northern Scotland. These subspecies were not reported separately during the *1981–84 Winter Atlas*, thereby limiting range-change calculations to the species level. Nevertheless, knowledge of the general distribution of the two forms helps when interpreting the change map, which shows a 10% range contraction in Ireland compared with a 36% expansion in Britain, most notably in southeast England, along the east coast of England and in south and east Scotland.

Many of the gains in Britain reflect a distributional shift eastwards in the core wintering range of the European subspecies. While the population size has increased at flyway level (Wetlands International 2013), there have been declines at most sites in Britain (*WeBS Report 2011*), especially at the Severn Estuary which, though still one of the most important sites in Britain, saw an 89% decline in numbers between 1996/97 and 2009/10 (*WeBS Report 2011*). Meanwhile, numbers are increasing farther east, especially in the Netherlands (Hornman *et al.* 2012a). Some gains may also relate to small numbers of European birds arriving during cold-weather influxes.

The story is very different in Ireland and parts of north and west Scotland, where losses reflect a flyway-level decline in the Greenland population, due to chronic low productivity (Stroud *et al.* 2012a). Patterns involve loss of smaller flocks feeding mainly on traditional peatland sites, coincident with a shift towards managed grasslands (Fox & Stroud 2002). More than 60% of the population remains concentrated at two sites, Wexford Slobs in Ireland and Islay in southwest Scotland. Elsewhere, in south and east Scotland, and along the English east coast, birds of the Greenland subspecies occur with other 'grey' geese, especially Pink-footed Geese. These gains may, in part, be due to the greater effort of birdwatchers, during this atlas, in searching flocks of commoner geese for more unusual species.

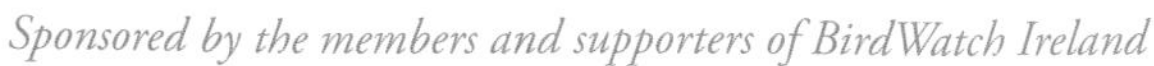

WINTER DISTRIBUTION *2007/08–2010/11*

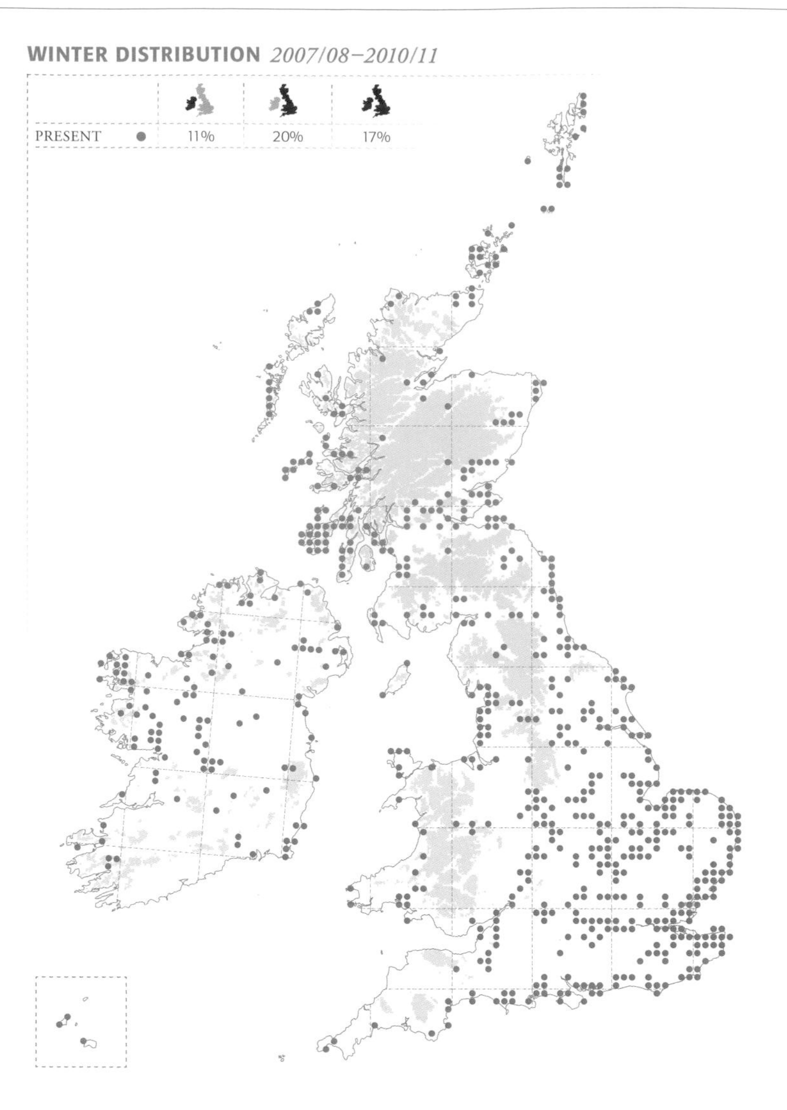

EUROPEAN WHITE-FRONTED GOOSE WINTER DISTRIBUTION *2007/08–2010/11*

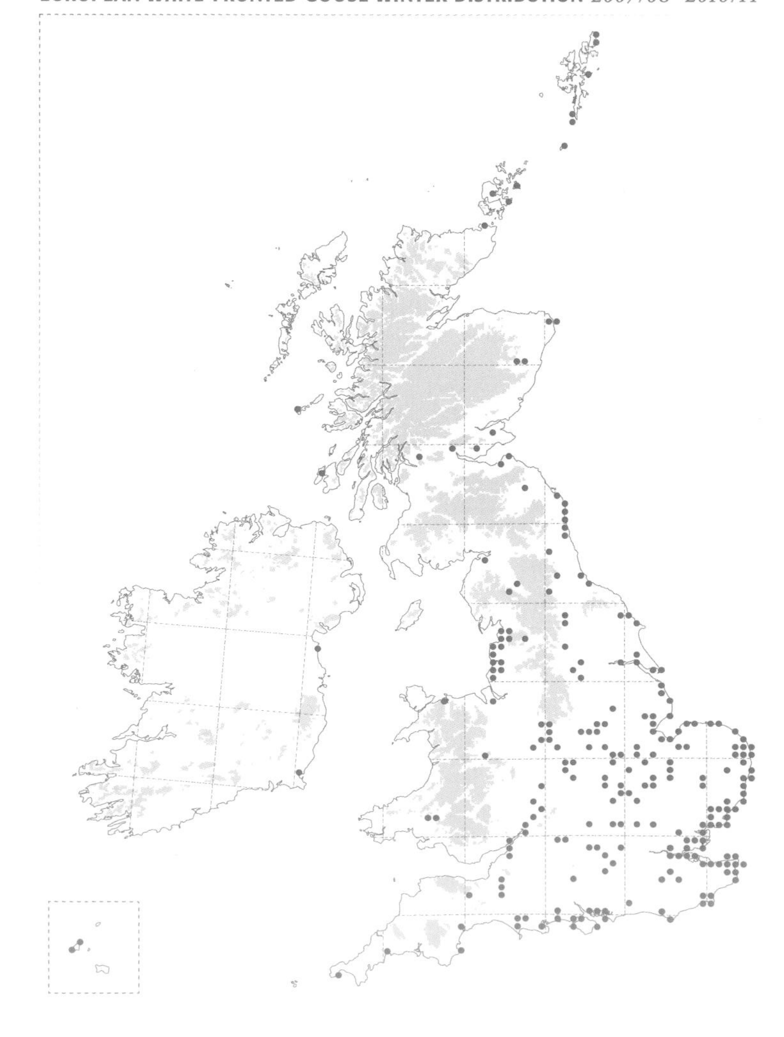

WINTER DISTRIBUTION CHANGE *since 1981–84*

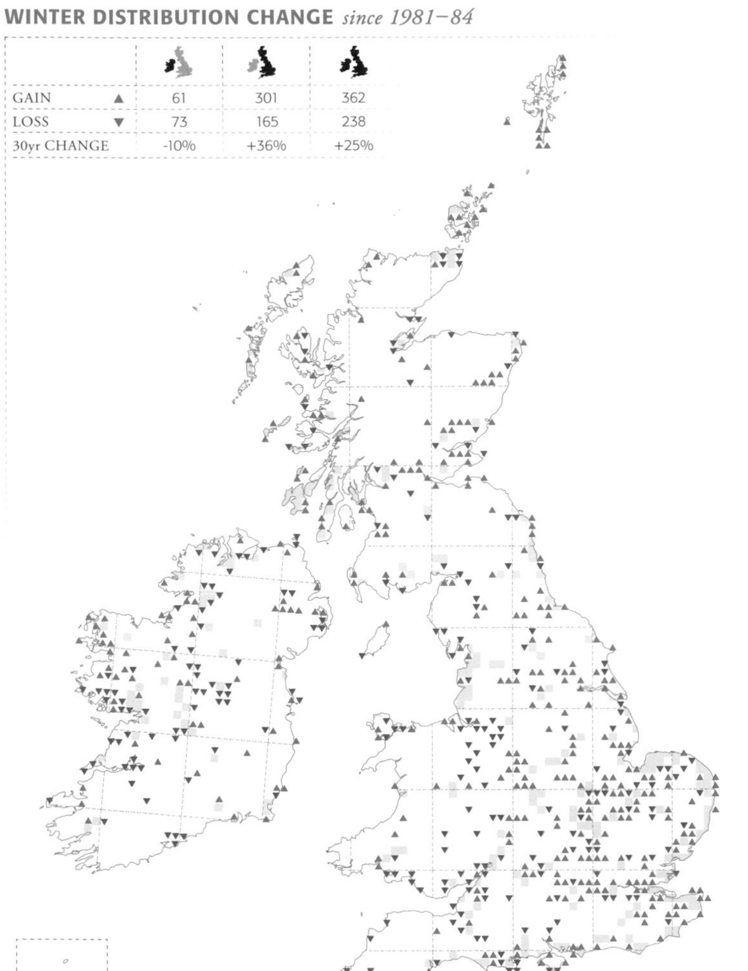

WINTER RELATIVE ABUNDANCE *2007/08–2010/11*

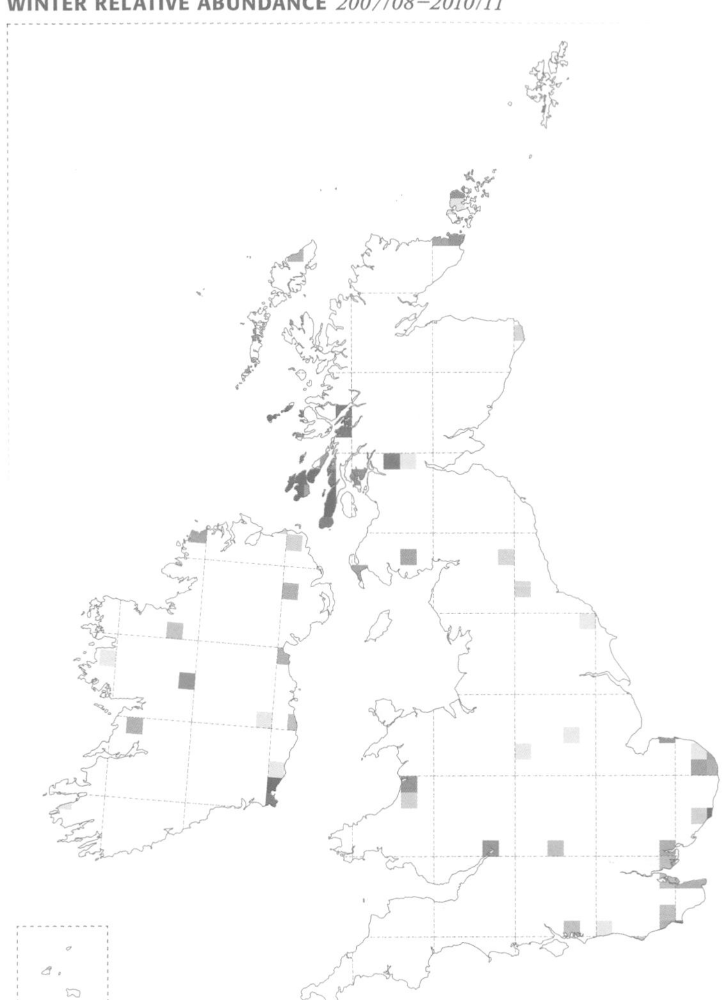

GREENLAND WHITE-FRONTED GOOSE WINTER DISTRIBUTION *2007/08–2010/11*

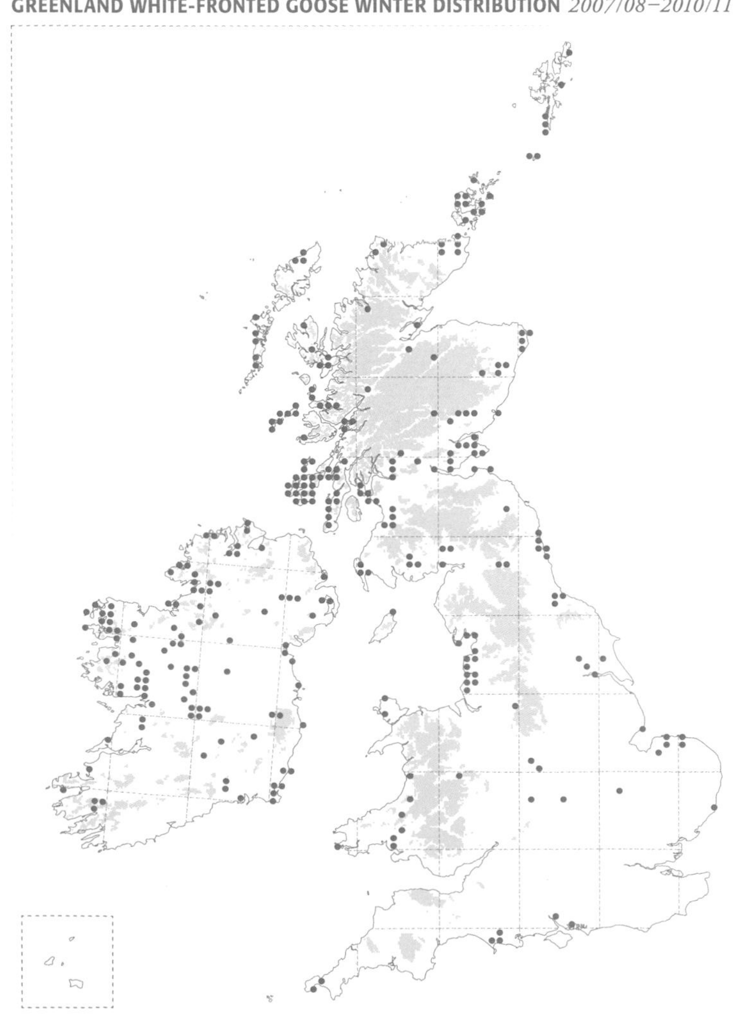

GREYLAG GOOSE

Anser anser

BTO CODE: GJ

DAVID KJAER www.davidkjaer.com

IT IS EVIDENT from the winter distribution that the three populations of the Greylag Goose that occur in Britain & Ireland in winter now overlap to such an extent that in many places it is impossible to separate resident native, re-established and Icelandic wintering birds (Mitchell *et al.* 2010, 2012). All favour low-lying agricultural land, though the early return of some birds to marginal upland breeding sites in February disrupts the pattern. In Ireland the records are well scattered, although it is thought that most birds of Icelandic breeding origin occur in eight flocks in north-, east- and south-coast counties between Lough Swilly, Co. Donegal, and the River Suir, Co. Waterford. The other Irish flocks are typically much smaller, and their origin remains unknown (*I-WeBS Report 2009*).

Although the Icelandic wintering population has shown no significant change in numbers since the *1981–84 Winter Atlas*, there have been significant increases in the resident populations (*I-WeBS Report 2009*; *WeBS Report 2011*), consistent with the observed 91% increase in range size. These increases have taken place throughout the winter range but are particularly noticeable in Orkney and Shetland, western Scotland and much of England. A scattering of losses in Ireland and Scotland may be linked with a northerly redistribution of the Icelandic wintering population (Mitchell 2012).

The increase in range of the resident populations is clearly seen in the breeding distribution change map, which shows a 138% increase in range size since the *1988–91 Breeding Atlas* and a 748% expansion since the *1968–72 Breeding Atlas*. This species is now widespread throughout England, bar the southwest, and in north and southwest Wales. It is widespread in Scotland, apart from the uplands and the northeast, and the ranges of breeding native and re-established populations have now merged. With the exception of eastern Scotland, the breeding relative abundance map shows many similarities with the winter map, reflecting the sedentary nature of the resident populations. The breeding abundance change map largely reflects the gross apparent range changes, but also indicates that Greylag Geese have rapidly increased in density to occupy a high proportion of tetrads in these new areas.

Sponsored by Michael Maggs

BREEDING DISTRIBUTION *2008–11*

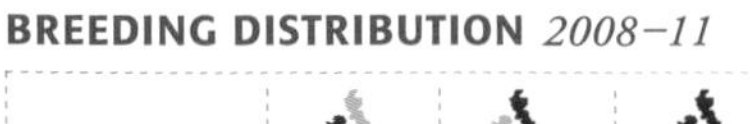

POSSIBLE •	3%	6%	5%
PROBABLE ●	2%	11%	9%
CONFIRMED ●	5%	42%	32%
TOTAL	10%	59%	46%

WINTER DISTRIBUTION *2007/08–2010/11*

PRESENT ●	18%	66%	53%

BREEDING DISTRIBUTION CHANGE *since 1968–72*

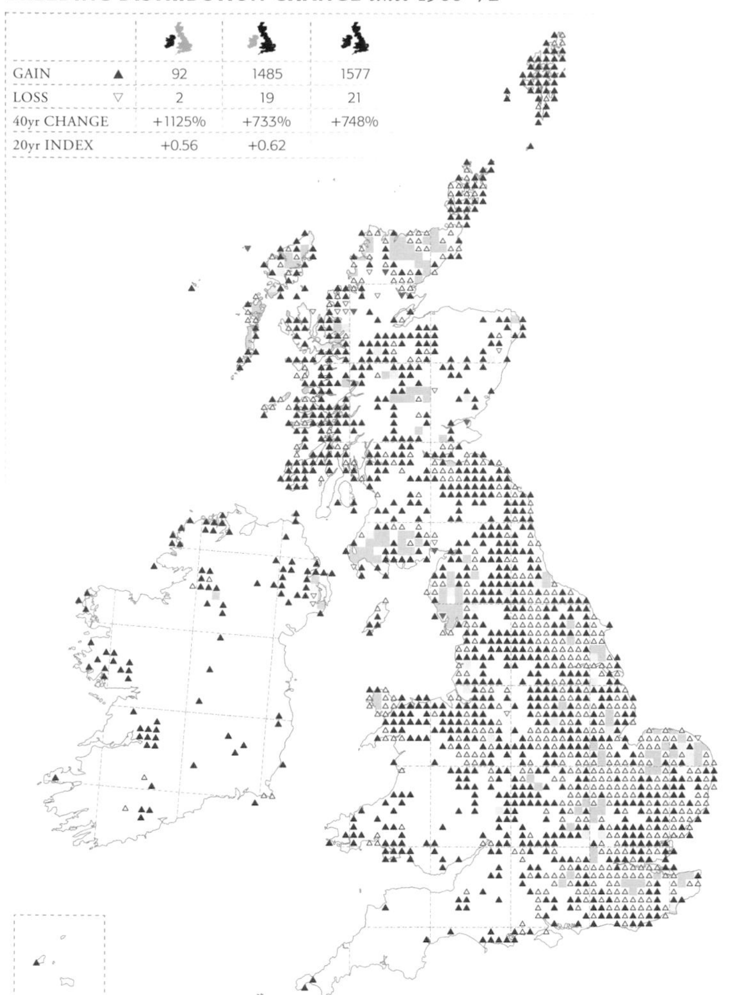

GAIN	▲	92	1485	1577
LOSS	▽	2	19	21
40yr CHANGE		+1125%	+733%	+748%
20yr INDEX		+0.56	+0.62	

WINTER DISTRIBUTION CHANGE *since 1981–84*

GAIN	▲	125	958	1083
LOSS	▼	33	89	122
30yr CHANGE		+111%	+89%	+91%

BREEDING RELATIVE ABUNDANCE *2008–11*

BREEDING RELATIVE ABUNDANCE CHANGE *since 1988–91*

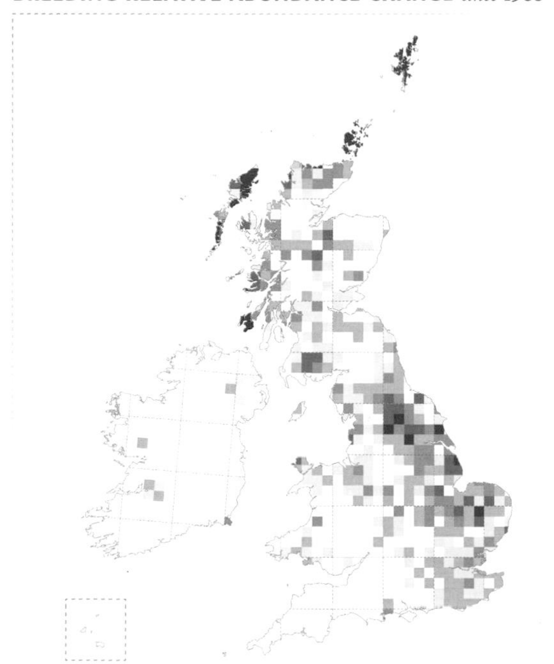

WINTER RELATIVE ABUNDANCE *2007/08–2010/11*

CANADA GOOSE
Branta canadensis

BTO CODE: CG

STEVE CARTER www.colnevalleybirding.co.uk

THIS INTRODUCED SPECIES, now to be found on a wide range of waterbodies throughout much of Britain, and locally in Ireland, is rather sedentary in nature (*Migration Atlas*) and as a result there are strong similarities in distribution and abundance patterns between winter and the breeding season. The species is widespread through England and Wales with relative abundance being highest in a broad band from southeast England to the Irish Sea and parts of northeast England. In Scotland the distribution is patchier with the main concentrations being around the inner Solway Firth and from Perthshire through to the Clyde and into Argyllshire. In Ireland, the distribution is very localised, with more than half of the range centred on the catchment of the River Erne, where birds occur predominantly on inter-drumlin lakes. They have yet to establish themselves in urban areas in Ireland, as they have done in Britain.

The widespread availability of secure predator-free nesting sites on a variety of waterbodies surrounded by good feeding areas has resulted in high chick productivity. This, combined with relatively low hunting pressure, has fuelled spectacular population increases since the previous atlases. In the UK, numbers in winter have more than doubled since 1984 (*WeBS Report 2011*), whilst the numbers in the breeding season have increased by 73% since 1995 (*BBS Report 2011*).

The winter and breeding season change maps show this increase as expansion westwards into Wales and southwest England, and northwards through eastern and northern England and central Scotland through to Argyllshire. This is represented by a 67% increase in winter range size in Britain since the *1981–84 Winter Atlas*, and a 162% increase in breeding range size since the *1968–72 Breeding Atlas*. In addition to these large-scale range expansions, the abundance change map points to large increases in density in the original source areas in central and southeast England. Over the same period this species has increased its Irish range by 80% in winter and 600% in the breeding season, but it still has a very localised distribution.

There is anecdotal evidence of aggression and hybridisation with native species, water eutrophication and agricultural damage but little research has quantified the ecological impact of introduced Canada Geese (Rehfisch *et al.* 2010).

Sponsored by Mrs H. Forrest

BREEDING DISTRIBUTION *2008–11*

POSSIBLE •	1%	4%	3%
PROBABLE •	1%	8%	6%
CONFIRMED •	2%	50%	38%
TOTAL	4%	62%	47%

WINTER DISTRIBUTION *2007/08–2010/11*

BREEDING DISTRIBUTION CHANGE *since 1968–72*

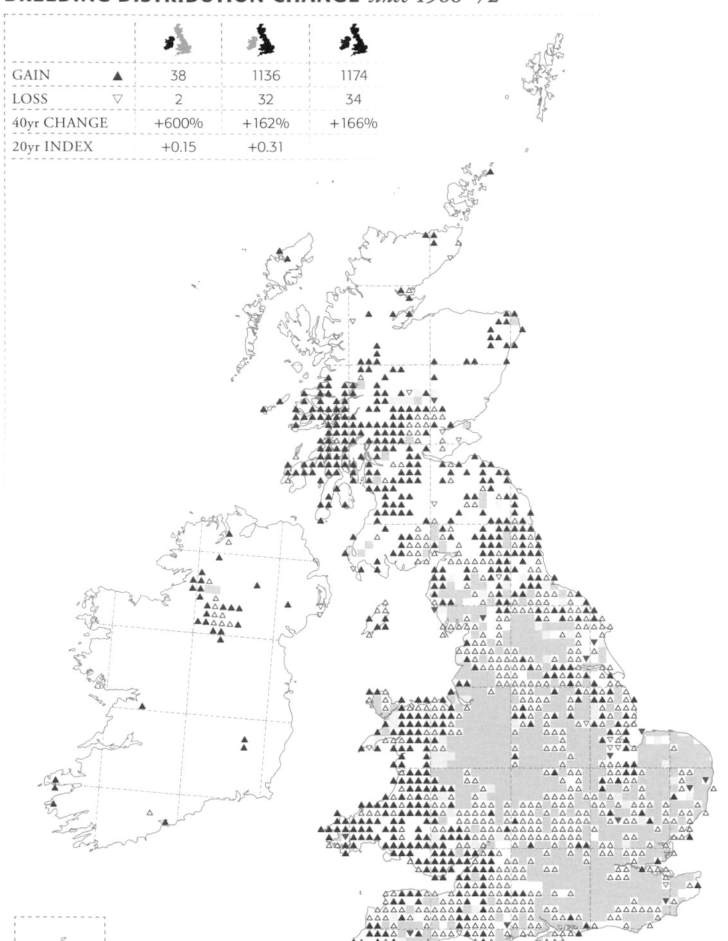

GAIN ▲	38	1136	1174
LOSS ▽	2	32	34
40yr CHANGE	+600%	+162%	+166%
20yr INDEX	+0.15	+0.31	

WINTER DISTRIBUTION CHANGE *since 1981–84*

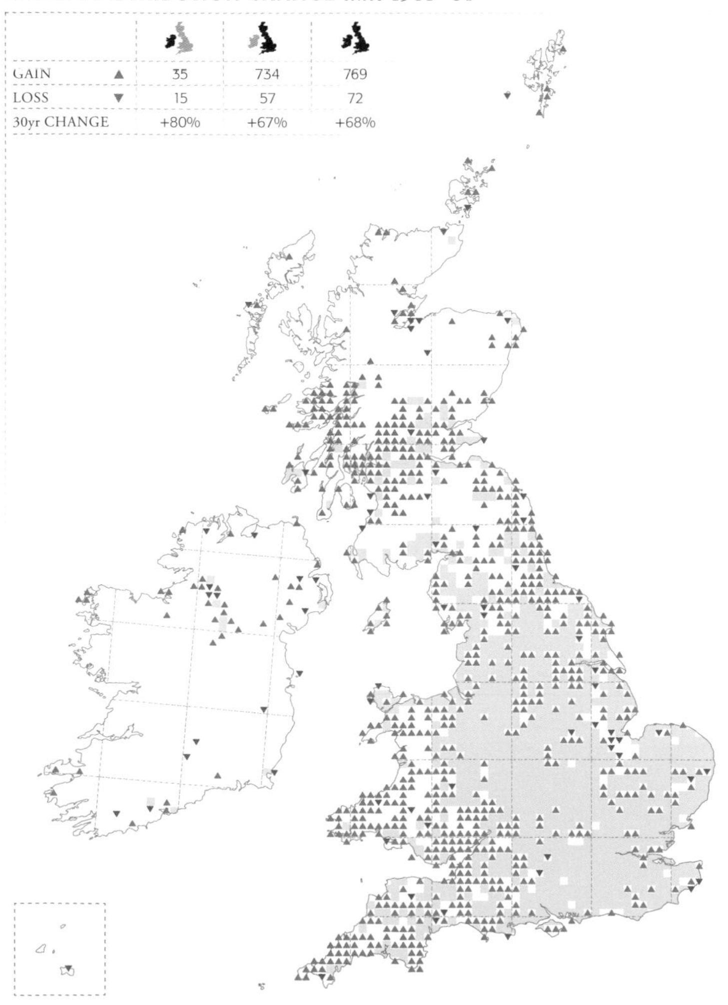

GAIN ▲	35	734	769
LOSS ▼	15	57	72
30yr CHANGE	+80%	+67%	+68%

BREEDING RELATIVE ABUNDANCE *2008–11*

BREEDING RELATIVE ABUNDANCE CHANGE *since 1988–91*

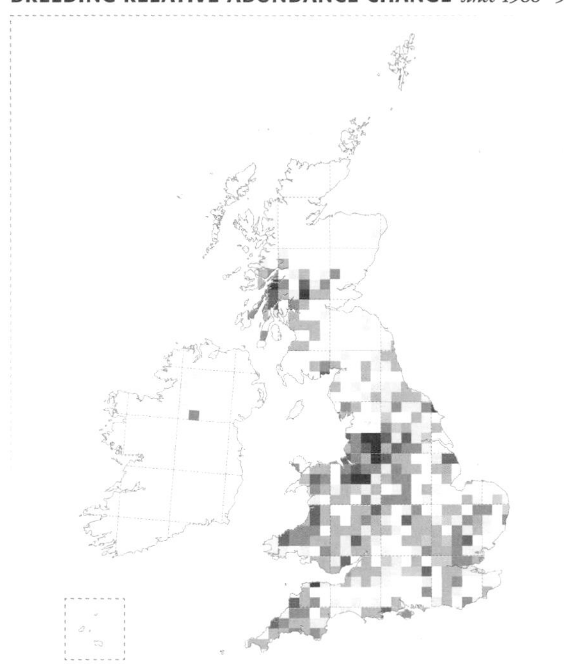

WINTER RELATIVE ABUNDANCE *2007/08–2010/11*

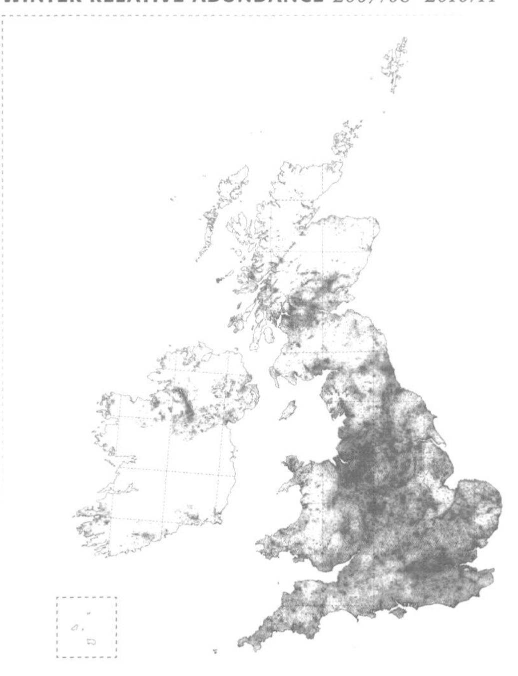

BARNACLE GOOSE

Branta leucopsis

BTO CODE: BY

COLIN PUMFRETT

WHEREAS FOR THE Brent Goose different breeding populations are largely separable in the field, the same cannot be said of the Greenland-breeding and Svalbard-breeding populations of Barnacle Goose that winter in Britain & Ireland, merging with a growing naturalised population. For this reason the maps shown here are a composite of all three populations. However, there is some geographical delineation of these populations in winter. Birds along the coast and islands of northwestern Scotland and western Ireland are largely from the Greenland-breeding population, whilst the birds on the Solway Firth and on the east coast of Britain are largely from the Svalbard population. The naturalised population, which is found mainly in Britain, has a much wider distribution, and accounts for many of the occupied 10-km squares in inland areas, particularly throughout England. This is reflected on both the wintering and the breeding distribution maps. Migratory Barnacle Geese also utilise inland feeding areas but these are generally not more than a few kilometres from their coastal wintering locations.

In Ireland a modest 7% increase in range size is apparent since the *1981–84 Winter Atlas*, with gains occurring particularly in the areas that support the highest densities along the west coast, and little evidence of a growing naturalised population inland. In Britain, a range increase of 79% is evident since the *1981–84 Winter Atlas*, with many gains throughout England being associated with the increasing naturalised population. Along the western coast of Scotland, particularly in the strongholds of the Solway Firth, Islay and the Outer Hebrides, the number of occupied 10-km squares has largely remained stable, despite a population increase (*WeBS Report 2011*). The growth in the Greenland population is reflected in the increasing numbers at a small number of key British and Irish wintering sites that together support approximately 80% of the total population (Mitchell *et al.* 2008).

During the breeding season, the naturalised population of Barnacle Geese has expanded in range by 88% throughout Britain since the *1988–91 Breeding Atlas*, although it was noted then that the distribution map under-represented the species' range. In the mid 2000s this population was thought to number c.900 pairs (Musgrove *et al.* 2013).

Sponsored by Jean Stewart, in memory of the late Donald Stewart

WINTER DISTRIBUTION *2007/08–2010/11*

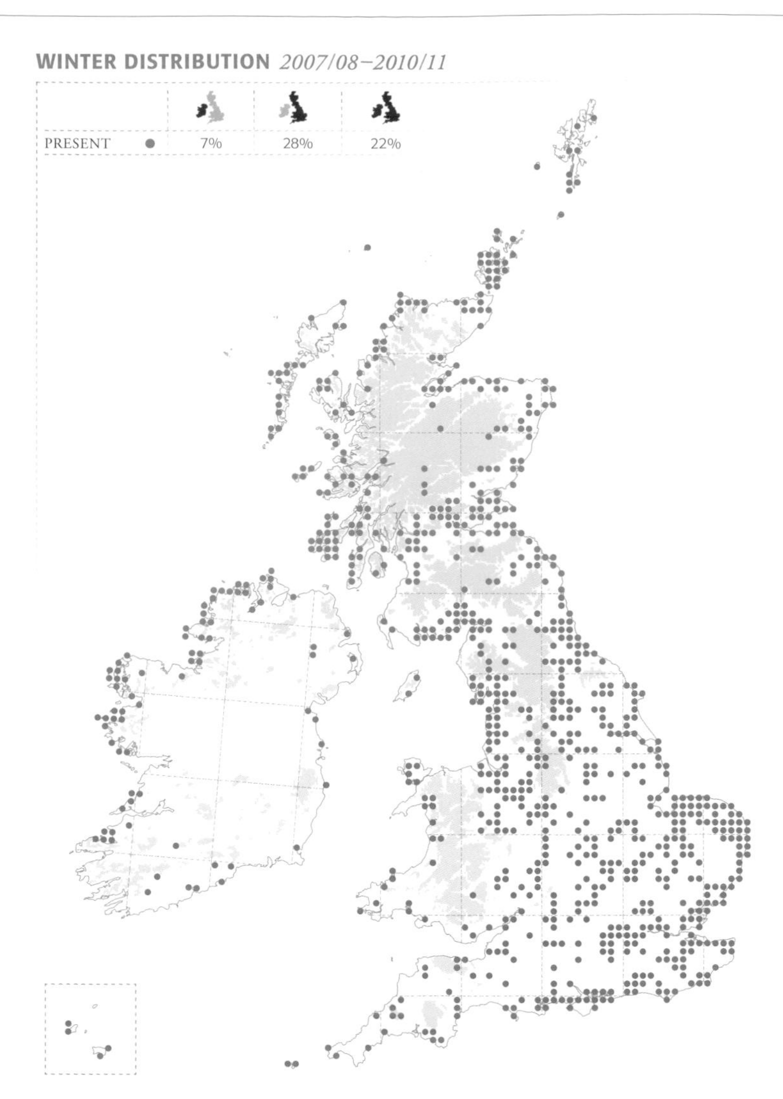

PRESENT ●	7%	28%	22%

BREEDING DISTRIBUTION *2008–11*

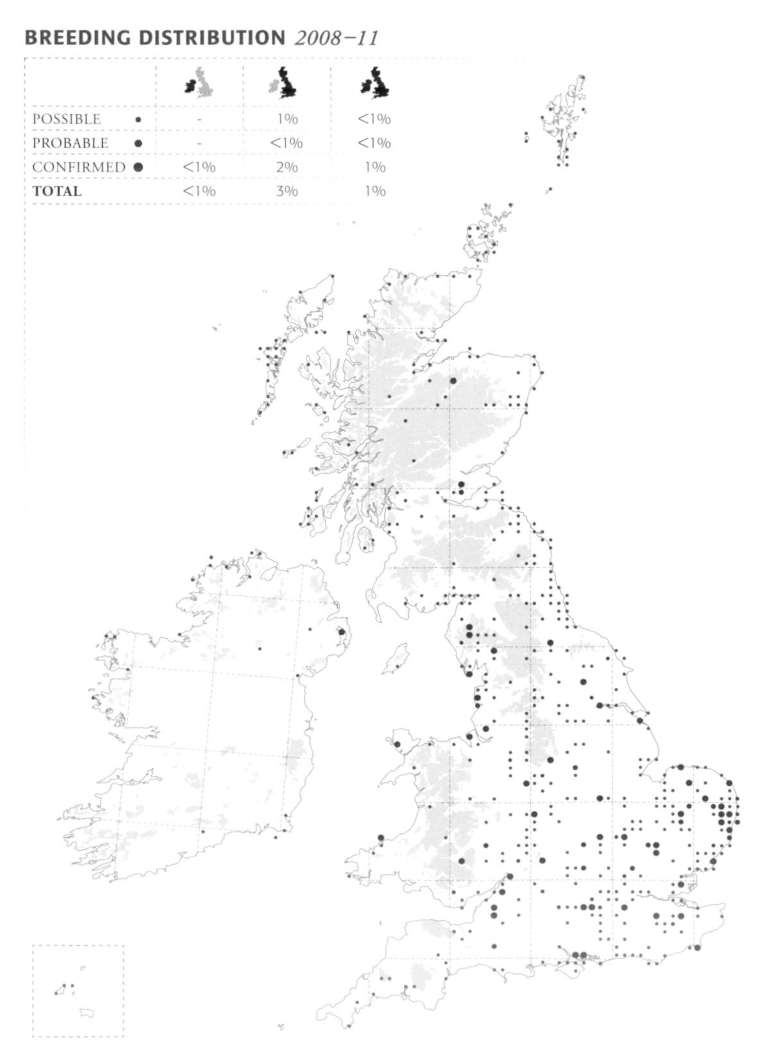

POSSIBLE ●	-	1%	<1%
PROBABLE ●	-	<1%	<1%
CONFIRMED ●	<1%	2%	1%
TOTAL	<1%	3%	1%

WINTER DISTRIBUTION CHANGE *since 1981–84*

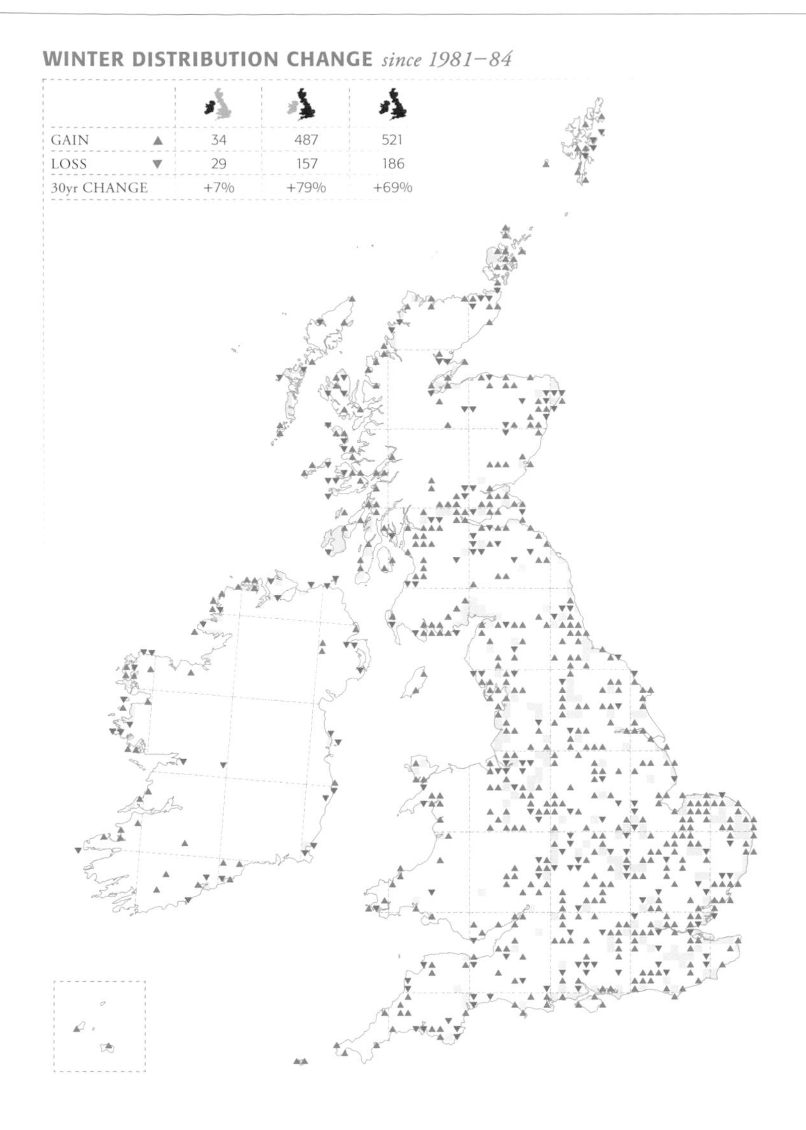

	(Ireland)	(Britain)	(Britain & Ireland)
GAIN ▲	34	487	521
LOSS ▼	29	157	186
30yr CHANGE	+7%	+79%	+69%

WINTER RELATIVE ABUNDANCE *2007/08–2010/11*

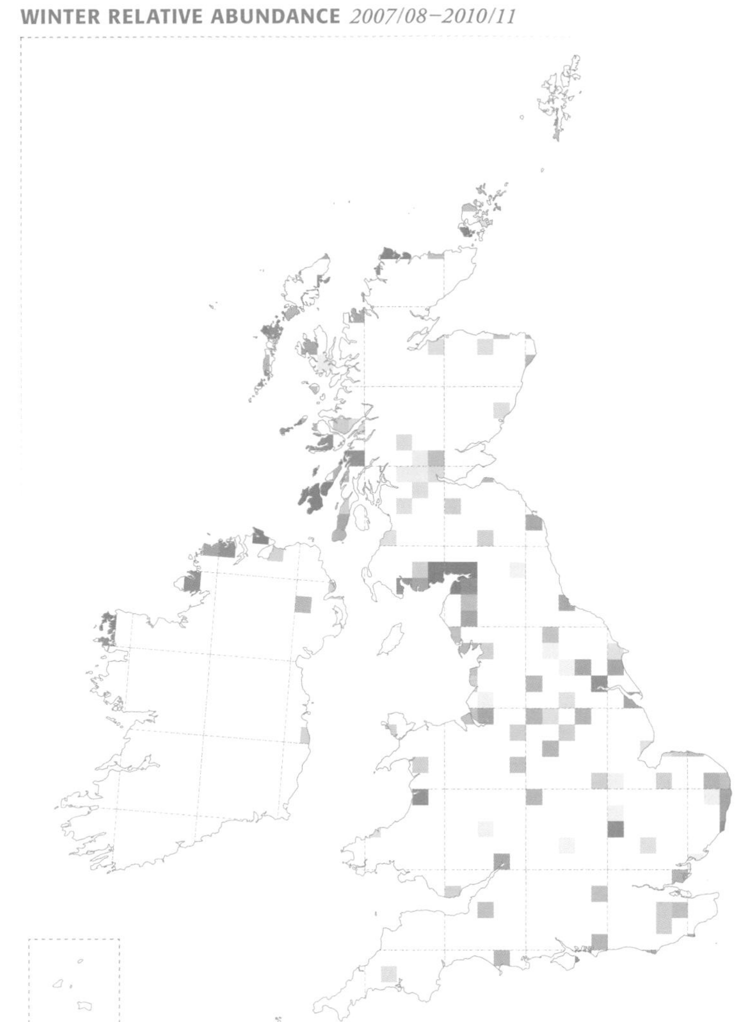

SARAH DALRYMPLE

◀ These Barnacle Geese wintering on the Solway Firth originate from Svalbard. This population has been increasing; the high proportion of young recorded in flocks suggests that high productivity is the cause.

BRENT GOOSE
Branta bernicla

BTO CODE: BG

POLINA KASAPOVA www.thingsarelikethis.com

BRENT GEESE WERE recorded in winter from almost the entire coastline of Britain & Ireland, except for north and west Scotland. This overall distribution conceals separate patterns for birds from four different breeding populations. The Canadian Light-bellied population winters predominantly in Ireland. Although subspecies were not separately recorded in the *1981–84 Winter Atlas*, the scarcity in Ireland of individuals from other populations means that the 67% range expansion of Brent Geese can be safely ascribed to this population and is consistent with a population increase from 15,000 in the early 1980s to 38,000 in 2010 (Colhoun *et al.* 2011). The relative abundance map suggests highest densities were located around estuarine sites in Ireland.

Birds from the Svalbard Light-bellied Brent Goose population winter in Denmark and in Britain, where they winter predominantly around Lindisfarne in Northumberland. Typically, small numbers occur at other British east-coast sites but the many gains shown on the change map, particularly in northern Scotland, probably correspond to heavy snowfalls and frosts in winters 2009/10 and 2010/11 in Denmark, which resulted in birds moving west to irregularly used sites in milder areas (Clausen *et al.* 2012).

The bulk of the Brent Geese that winter elsewhere in Britain are from the arctic Russian-breeding Dark-bellied population. This population has also expanded its range around much of the British coastline, although the highest numbers remain, as in the *1981–84 Winter Atlas*, along the south and east coasts of England between the Humber estuary and Dorset. Within these strongholds the range size has largely remained stable, as has the wintering population since rising to a peak in the early 1990s (*WeBS Report 2011*). Individuals and small groups moving between coastal sites probably account for the scatter of inland records. This population is scarce in Ireland, where counts of only one or two individuals were reported from 26 occupied 10-km squares.

Finally, Black Brants are relatively scarce winter visitors to Britain and Ireland from eastern Siberia, Alaska or western Canada. Most records are confined to the south and east of England with only a few individuals recorded in Ireland.

Sponsored by Tolka Branch of BirdWatch Ireland

DARK-BELLIED BRENT GOOSE WINTER DISTRIBUTION *2007/08–2010/11*

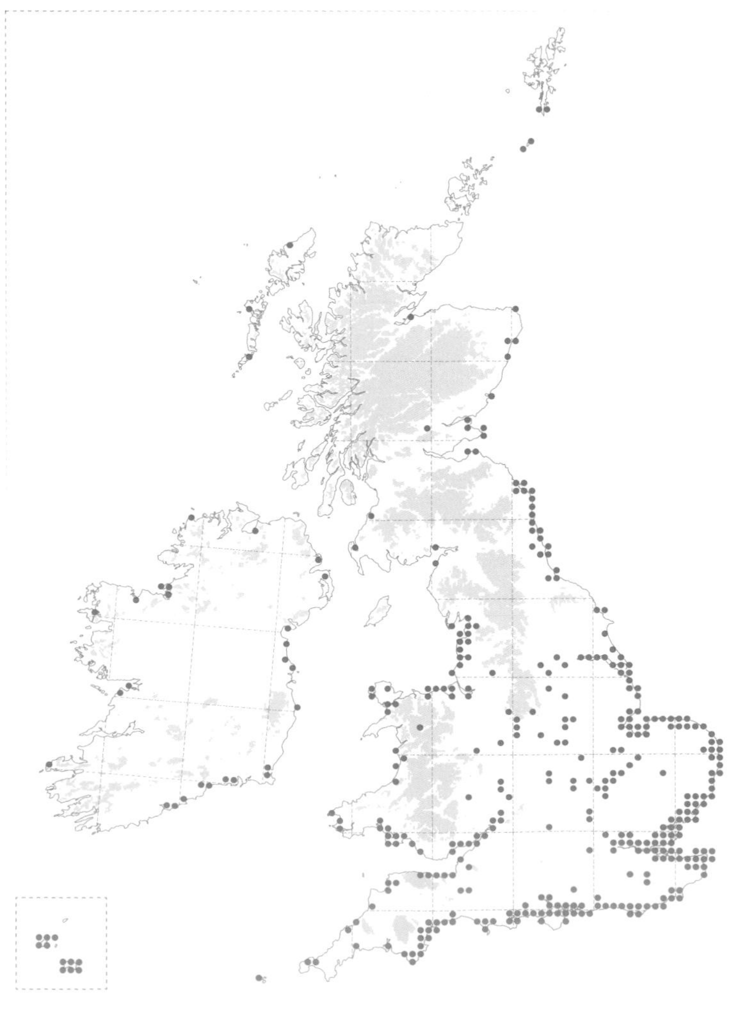

WINTER DISTRIBUTION CHANGE *since 1981–84*

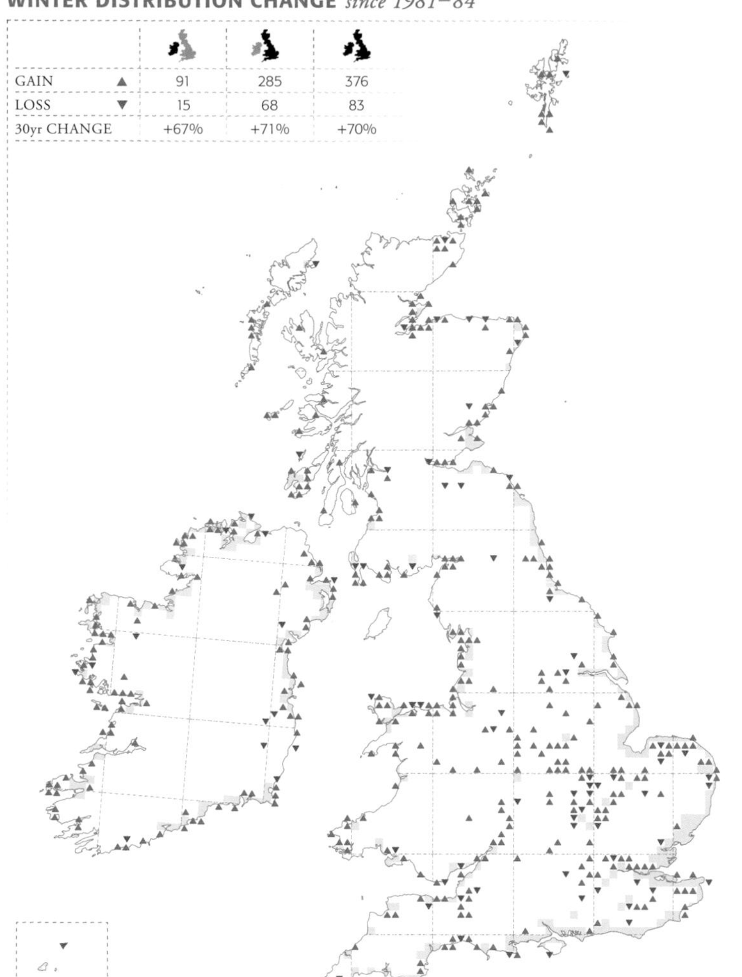

GAIN ▲	91	285	376
LOSS ▼	15	68	83
30yr CHANGE	+67%	+71%	+70%

WINTER RELATIVE ABUNDANCE *2007/08–2010/11*

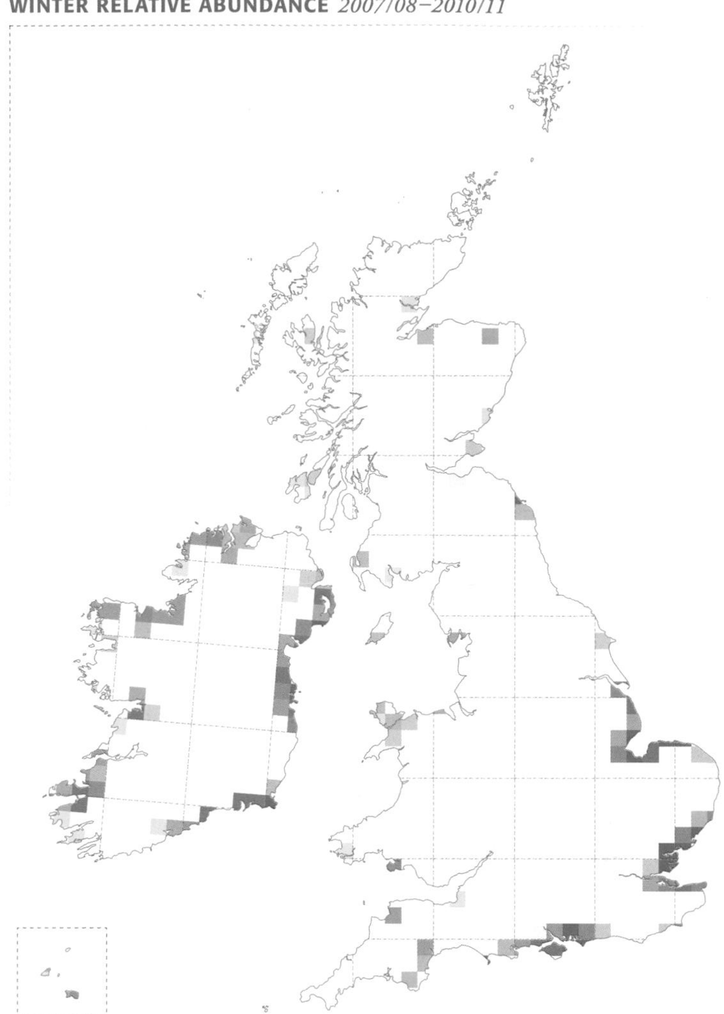

LIGHT-BELLIED BRENT GOOSE WINTER DISTRIBUTION *2007/08–2010/11*

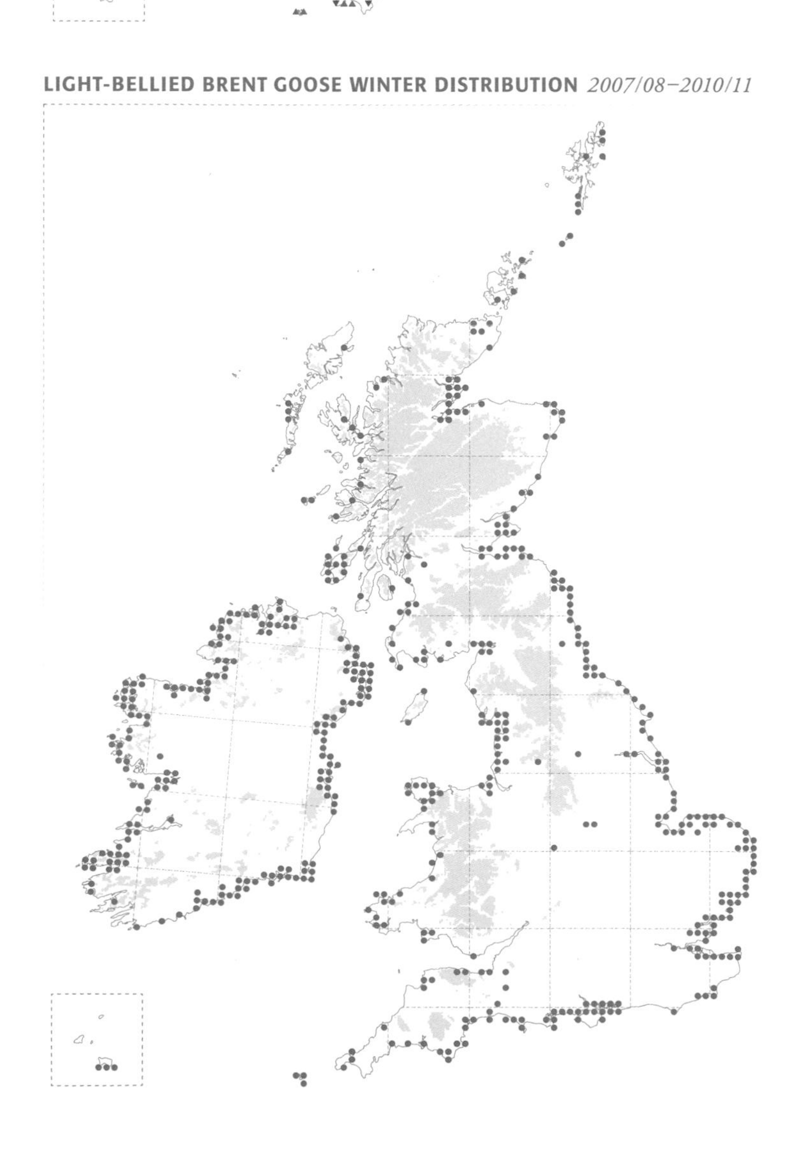

BLACK BRANT WINTER DISTRIBUTION *2007/08–2010/11*

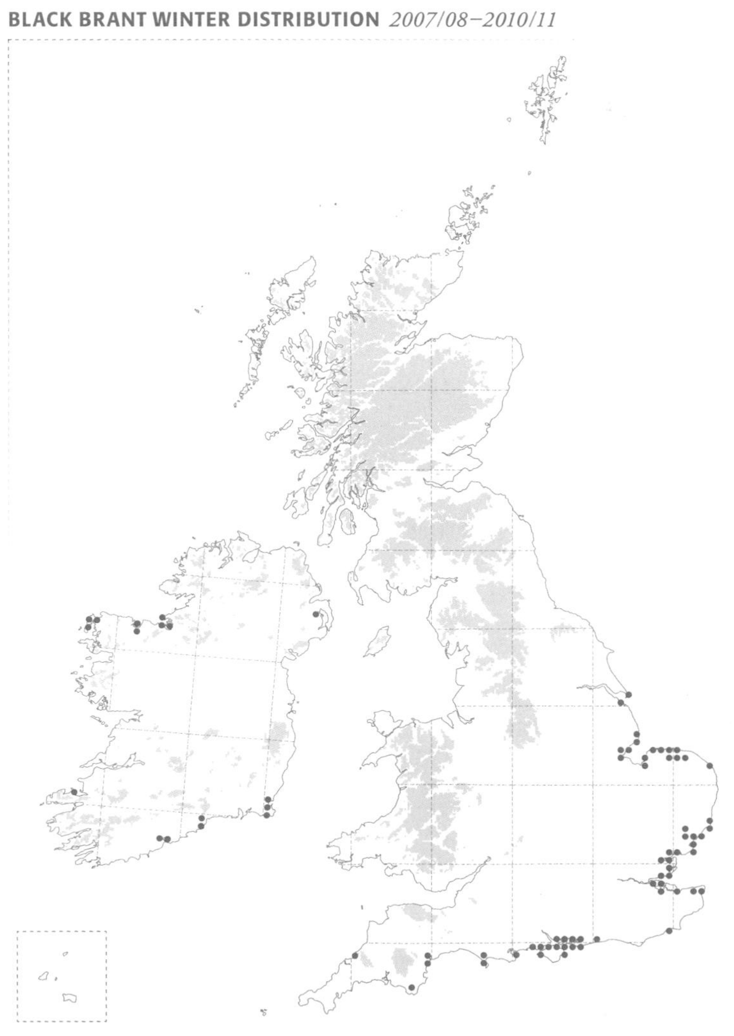

BAR-HEADED GOOSE

Anser indicus

BTO CODE: HD

JIM ALMOND www.shropshirebirder.co.uk

BAR-HEADED GEESE in Britain & Ireland tend to associate with other geese, usually Greylag Geese or Canada Geese. In their native range they breed in the montane regions of Central Asia, Mongolia and China and winter in northern India and northern Myanmar (Lever 2009). They are commonly kept in wildfowl collections in Europe and the birds occurring in Britain & Ireland have all originated from captivity.

The winter distribution map shows records from widespread locations across England, with fewer in Wales and Scotland. Just one 10-km square was occupied in Ireland in the winter, and an adjacent square occupied in the breeding season, suggesting that the same individual was involved. In the breeding season, records with breeding evidence were submitted for only 17 10-km squares, yet there were a further 136 10-km squares where birds were recorded as *non-breeding*. This is partly reflected in the counts submitted, as 78% were just of single birds. Peak counts included up to 10 individuals reported from some of the London parks. These figures indicate a marked increase since the *1981–84 Winter Atlas*, when Bar-headed Geese were recorded in only 22 10-km squares, and also since the *1988–91 Breeding Atlas*, when four 10-km squares were occupied. A survey of introduced and escaped geese in Britain in summer 1991 recorded 85 Bar-headed Geese, mostly in southern England. Successful breeding was recorded in Hampshire where a flock of 19 individuals included nine goslings from three broods (Delany 1993).

The combination of a genuine increase in numbers and range alongside the spread of Greylag and Canada Geese with better recording during the present atlas is the most likely explanation for the wider distribution. The species has the capacity to become locally common, as illustrated by the 100–125 pairs in the Netherlands (Banks *et al.* 2008), and can breed semi-colonially (SOVON Vogelonderzoek Nederland 2011). Monitoring by the Rare Breeding Birds Panel suggests that there have been fewer breeding attempts in recent years compared with the late 1990s when, for example, eight pairs bred in 1999 (Holling *et al.* 2011a).

Sponsored by Dr & Mrs Alder, North Yorkshire

BREEDING DISTRIBUTION *2008–11*

POSSIBLE •	-	<1%	<1%
PROBABLE •	-	<1%	<1%
CONFIRMED ●	-	<1%	<1%
TOTAL	-	<1%	<1%

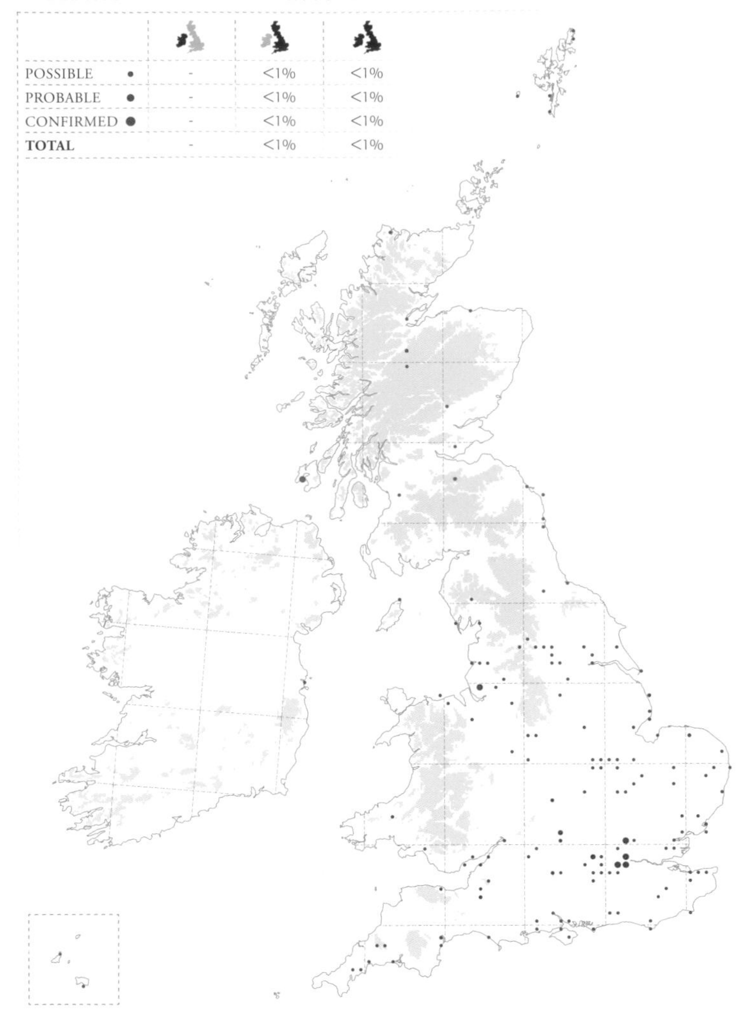

WINTER DISTRIBUTION *2007/08–2010/11*

PRESENT ●	<1%	5%	4%

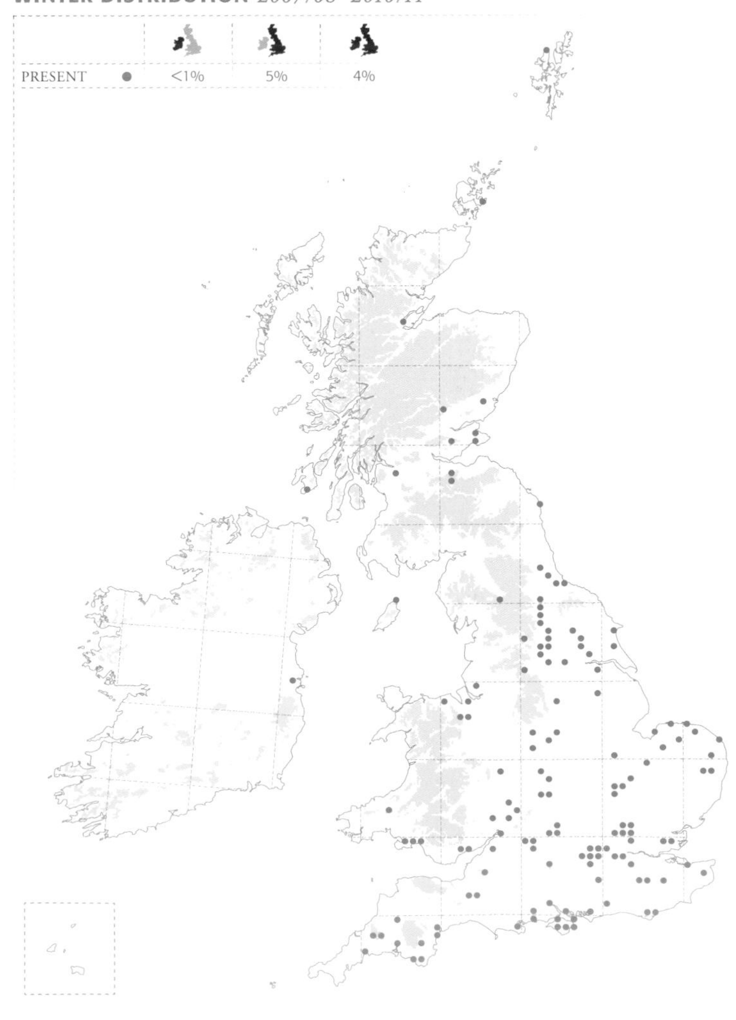

SNOW GOOSE

Anser caerulescens

BTO CODE: SJ

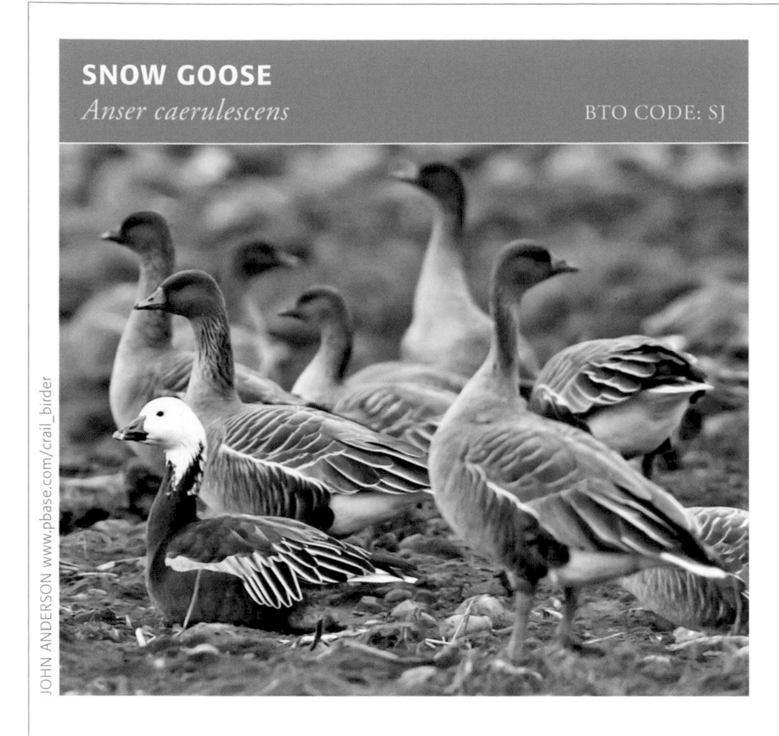

JOHN ANDERSON www.pbase.com/crail_birder

ALTHOUGH SMALL NUMBERS of presumably vagrant Snow Geese are recorded annually, the majority of individuals recorded during 2007–11 were escapes from captivity. In Britain, wild Snow Geese are most likely to occur at traditional staging or wintering areas of Pink-footed Geese, such as around Loch Leven in Scotland, the marshes in Lancashire and the north Norfolk coast. In Ireland, wild Snow Geese tend to be found in flocks of Greenland White-fronted Geese. These records of wild geese are birds that have become displaced during their migration from breeding grounds that extend from northeast Siberia eastwards across the whole of arctic North America to northwest Greenland, and perhaps cross to Greenland or Iceland where they join flocks of geese destined for Britain & Ireland. Once in this flyway, some individuals return year after year with the migrating Pink-footed and Greenland White-fronted Geese.

The winter distribution map also shows many inland records, which most likely relate to escaped birds. Clusters of occupied 10-km squares often reflect the same individuals moving from site to site with introduced geese during the winter.

Since the *1981–84 Winter Atlas* there has been a 33% increase in the number of occupied 10-km squares, probably due to a combination of improved coverage and a genuine increase in the number of birds present. Furthermore, in Britain, the increased population size and range of the Pink-footed Goose means that individuals associating with this species are liable to be reported in many more locations.

The breeding records are all of birds of captive origin and tend to be at sites with ornamental lakes, such as Stratfield Saye, Hampshire, and the University of York campus. The only apparently self-sustaining populations were those in Argyll, on northwestern Mull and Coll in the Inner Hebrides (Lever 2009). Although there have been no recent breeding records from Mull (Holling *et al.* 2011a), the small, sedentary population on Coll continues to breed in small numbers. Breeding was recorded for the first time on Orkney in 2009 and also occurred in the following two summers.

Sponsored by Judith Manners for her mother

BREEDING DISTRIBUTION *2008–11*

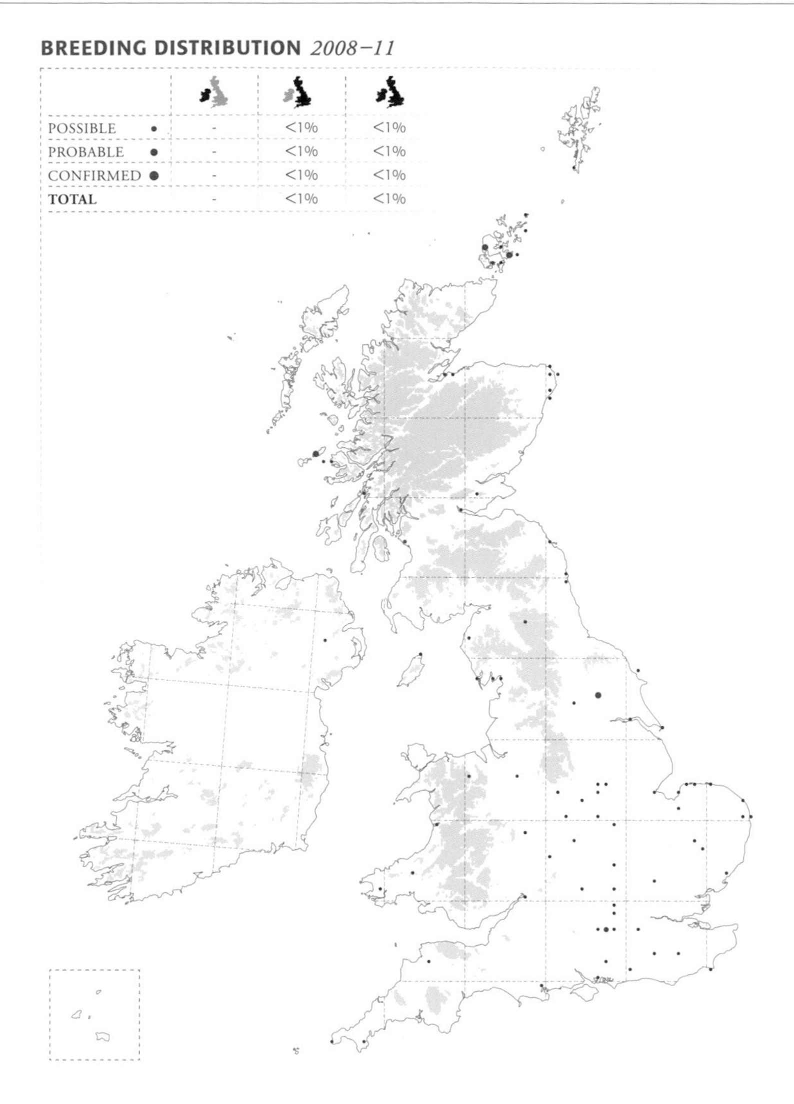

POSSIBLE •	-	<1%	<1%
PROBABLE •	-	<1%	<1%
CONFIRMED ●	-	<1%	<1%
TOTAL	-	<1%	<1%

WINTER DISTRIBUTION *2007/08–2010/11*

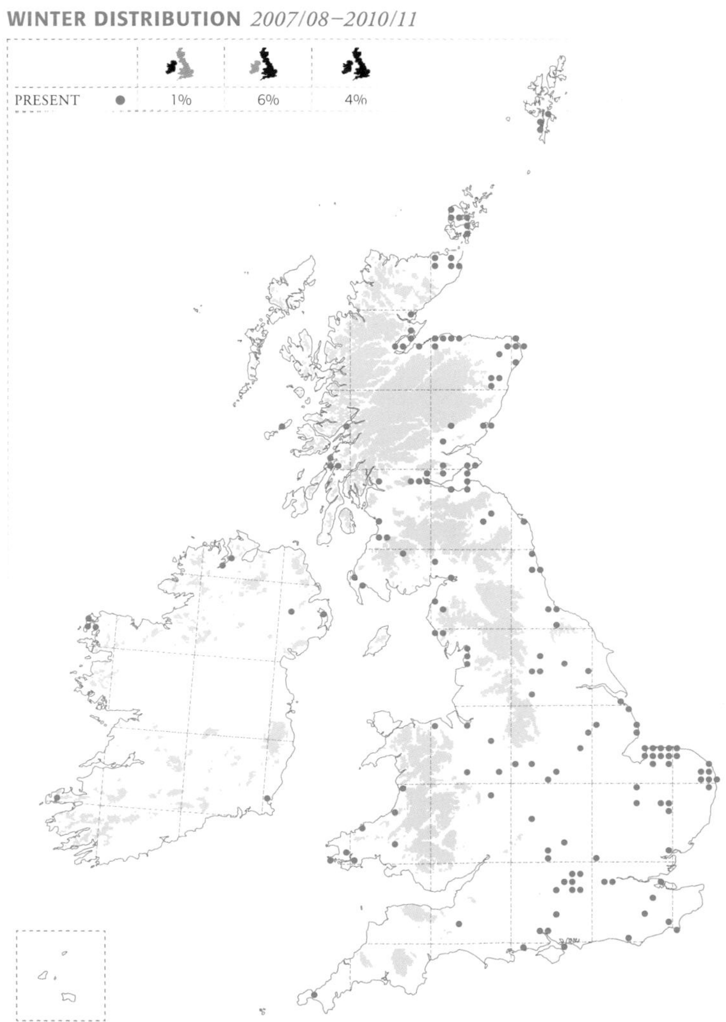

PRESENT ●	1%	6%	4%

ROSS'S GOOSE

Anser rossii

BTO CODE: RJ

JOHN ANDERSON www.pbase.com/crail_birder

THE ROSS'S GOOSE is not considered to have occurred in Britain or Ireland in a wild state and is currently on Category D of the British List. It was not recorded in Ireland during this atlas and is not included on the Irish List. Although some of the birds recorded in Britain are escapes from captivity, it is possible that some of those recorded in large flocks of geese could be genuine vagrants from North America; of around 30 submissions to BBRC, in 2011, around a dozen individuals were placed in Category D and two were considered likely escapes (Hudson *et al.* 2012). The winter distribution shows records from coastal and inland 10-km squares, some of which include staging and wintering areas of other geese, such as Pink-footed Goose.

WINTER DISTRIBUTION *2007/08–2010/11*

PRESENT ●	-	2%	1%

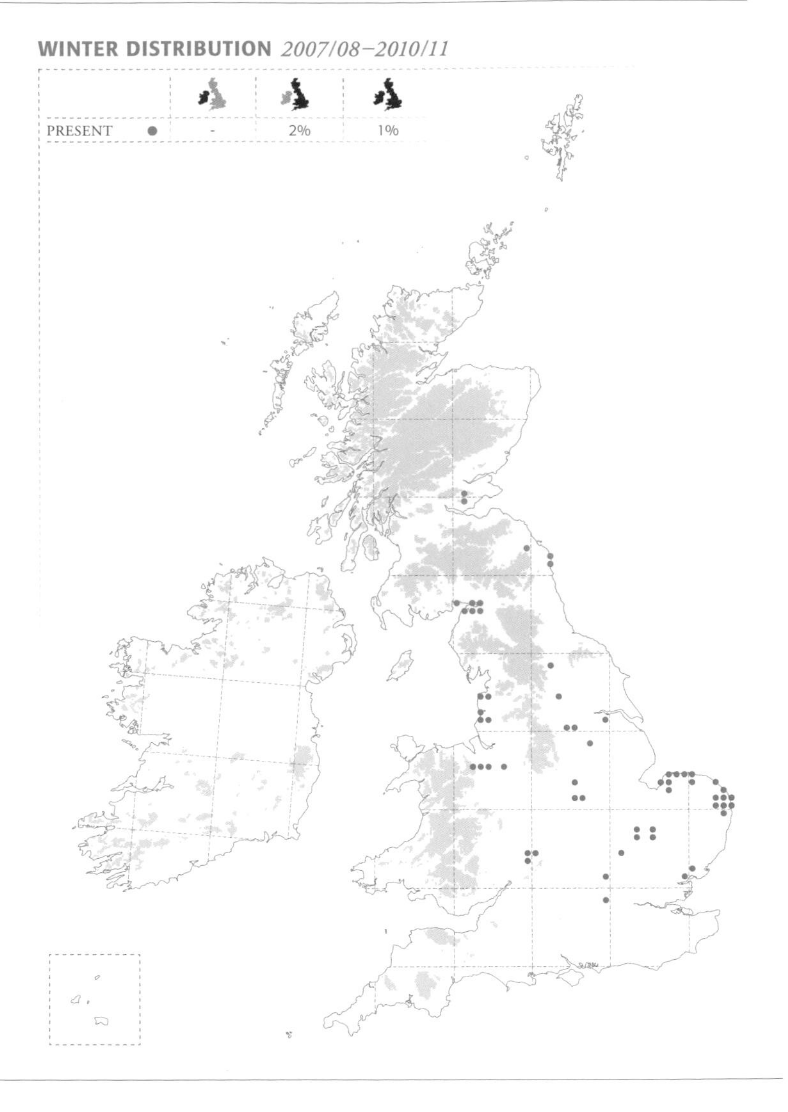

RED-BREASTED GOOSE

Branta ruficollis

BTO CODE: EB

GRAHAM CATLEY http://pewit.blogspot.co.uk

ALTHOUGH A SMALL number of vagrant Red-breasted Geese from the declining arctic Siberian breeding population winter in Britain each year, the species is popular in wildfowl collections and many birds are of captive origin. Vagrants are typically found in flocks of Brent Geese or Barnacle Geese so the winter distribution map mimics the stopover and wintering areas of those species. Brent and Barnacle Geese wintering in Ireland originate from the Nearctic, which may explain the lack of Red-breasted Goose sightings in Ireland. Inland records, and some coastal records, relate to escapes, for example five such birds were reported at several locations along the Suffolk coastline. There were reports from 24 10-km squares in the breeding season, one of these recording *confirmed* breeding.

WINTER DISTRIBUTION *2007/08–2010/11*

PRESENT ●	-	2%	1%

RUDDY SHELDUCK

Tadorna ferruginea

BTO CODE: UD

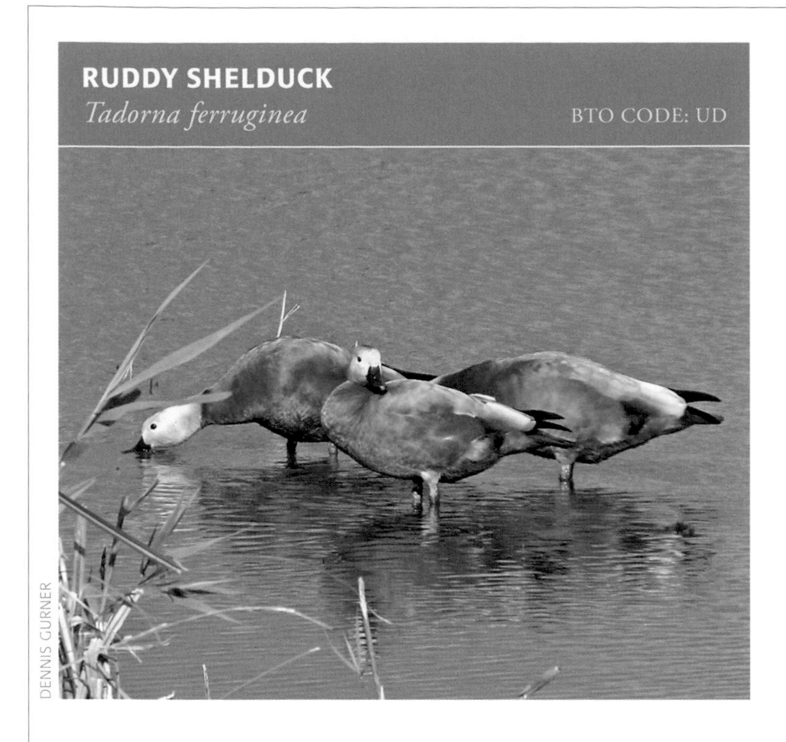

DENNIS GURNER

THERE ARE HISTORICAL records that may relate to genuine vagrants from breeding grounds in southeast Europe, north Africa and Central Asia, but it is thought that most Ruddy Shelducks recorded nowadays are escapes from wildfowl collections, or their descendants. Opinions on origin differ, though: Vinicombe (2002) makes a case for natural origin whilst Harrop (2002) favours captive origin or perhaps vagrancy from the introduced population in Askaniya Nova, Ukraine.

During *Bird Atlas 2007–11*, winter records came from 80 10-km squares in Britain and three 10-km squares in Ireland. Within Britain, the records are mainly in southern and central England, with just three occupied squares in Scotland. In Ireland, the records are from coastal localities in Counties Sligo and Wexford. The observed increase since the *1981–84 Winter Atlas* is in part due to improved recording of non-native species, but also perhaps to an increase in the number of escapes, both here and in continental Europe.

There is a small introduced breeding population in Britain and others elsewhere in Europe (Lever 2005). Records with breeding evidence came from 16 10-km squares in Britain; all of these were in the southern half of England. There were records of *confirmed* breeding from four 10-km squares and of *probable* breeding from a further seven 10-km squares; this is slightly more than recorded by RBBP between 2006 and 2008, which reported no more than three pairs in each year (Holling *et al.* 2011a). Taking into account all records, regardless of breeding status, more 10-km squares were occupied during the breeding season than in the winter. There is a clear pattern of arrival in July, peaking in August (Parkin & Knox 2010), and although these birds are believed to be largely non-natives, this is also the time of year when we might expect vagrants to arrive from their original breeding grounds.

BREEDING DISTRIBUTION *2008–11*

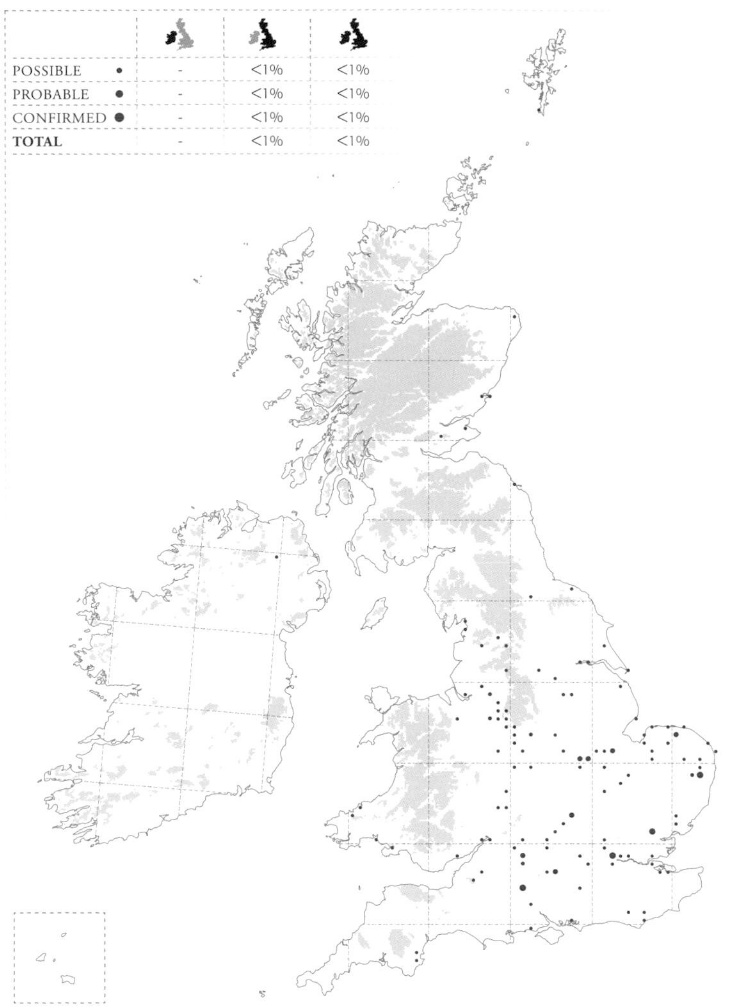

WINTER DISTRIBUTION *2007/08–2010/11*

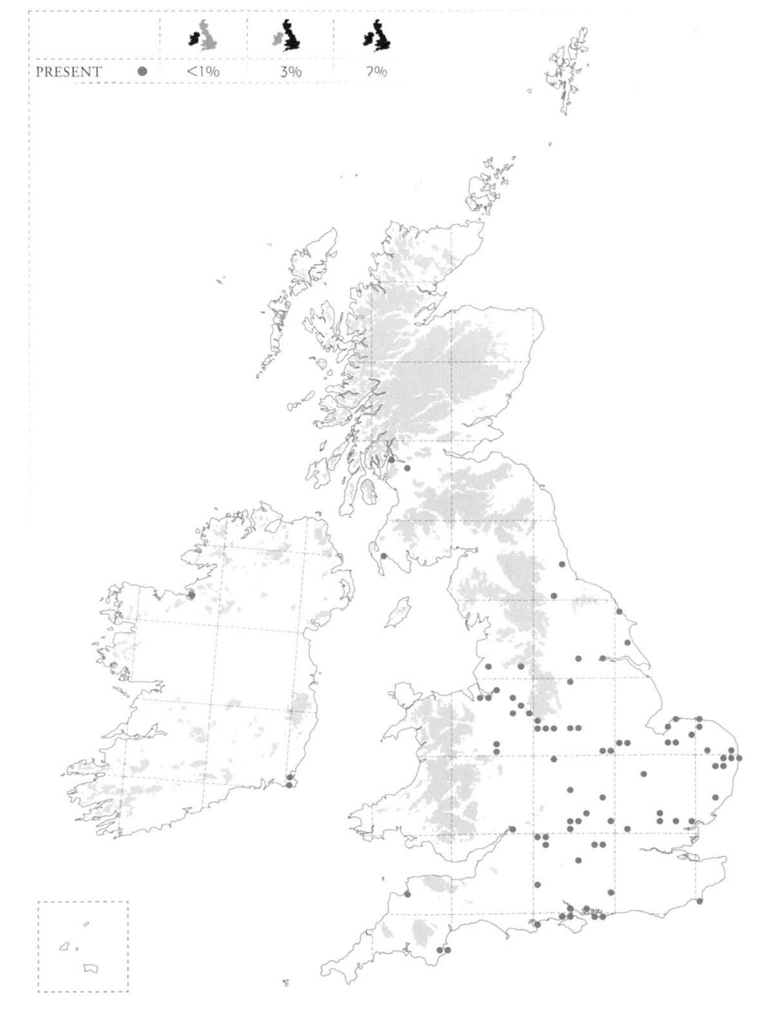

EGYPTIAN GOOSE

Alopochen aegyptiaca

BTO CODE: EG

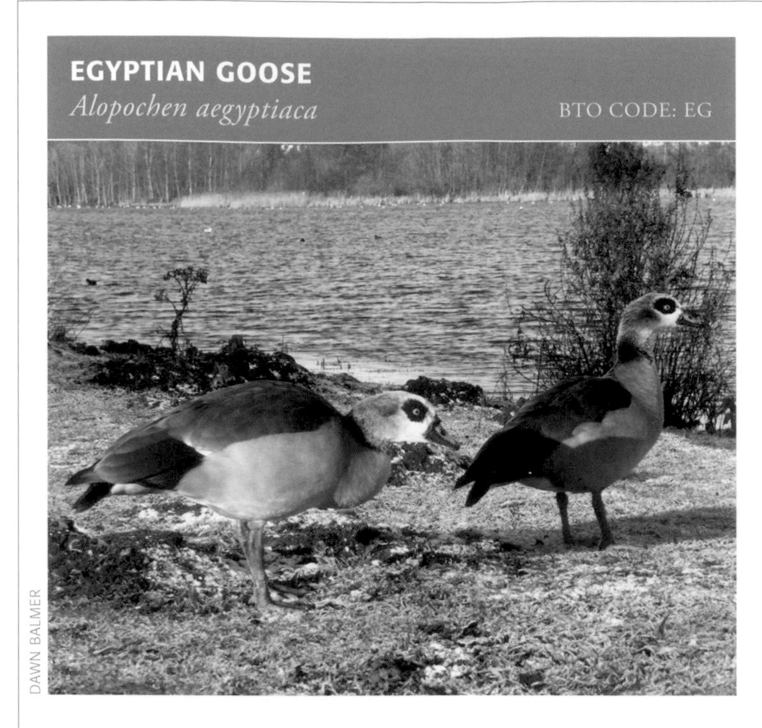

DAWN BALMER

THE EGYPTIAN GOOSE was introduced to Britain, mainly from South Africa, as early as the 17th century and it has bred in a wild state for over 200 years in locations from Devon to East Lothian (Holling *et al.* 2011a). The early distribution was tied to the history of releases, though it was only in Norfolk that a self-sustaining population had become established, as shown by both the *1968–72 Breeding Atlas* and the *1981–84 Winter Atlas*. By the time of the *1988–91 Breeding Atlas* the species had spread into Breckland and east Suffolk as numbers increased. By 1991 at least 900 individuals were present, 91% of which were in Norfolk (Delany 1993). Fieldwork for *Bird Atlas 2007–11* has revealed increases in relative abundance in these core areas as well as further significant range expansion.

The distribution maps show that Egyptian Geese are now found in two main areas: East Anglia and an area centred on Greater London extending into parts of neighbouring counties. There is a suggestion that the East Midlands may develop into a third stronghold, with a wide scattering of records there in both seasons. Away from these key areas there are isolated records as far north as Shetland, west Wales, southwest England and Ireland, with most of these records in winter. The winter map shows dispersal away from the breeding area as birds utilise a wider range of habitats such as winter cereals, root crops or open pig units.

The reason for the expansion over the last 20 years is unknown, though is probably fuelled by an increasing population in East Anglia leading to wider dispersal. There were an estimated 750–900 breeding pairs in Norfolk between 2000 and 2007 (Taylor & Marchant 2011). The key requirements for breeding are open water, adjacent short grass for grazing and large old trees with holes for nesting (Sutherland & Allport 1991). The wide scatter of *non-breeding* records on the breeding map perhaps hints at further expansion to come. There has been considerable expansion in the Netherlands, though the severity of the winters may limit growth in the east of the country (Lensink 1998). By 2009 the Dutch population numbered c.10,000 pairs, or 50,000 individuals post-breeding (Gyimesi & Lensink 2010).

BREEDING DISTRIBUTION *2008–11*

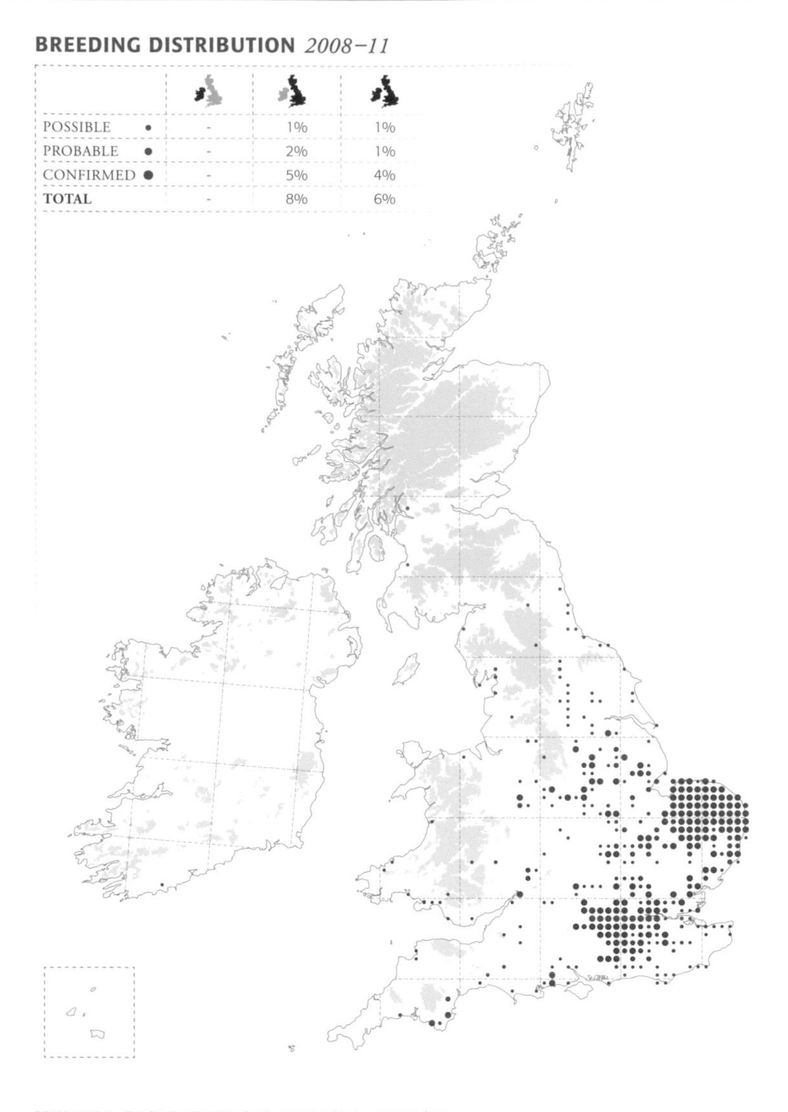

POSSIBLE ●	-	1%	1%
PROBABLE ●	-	2%	1%
CONFIRMED ●	-	5%	4%
TOTAL	-	8%	6%

WINTER DISTRIBUTION *2007/08–2010/11*

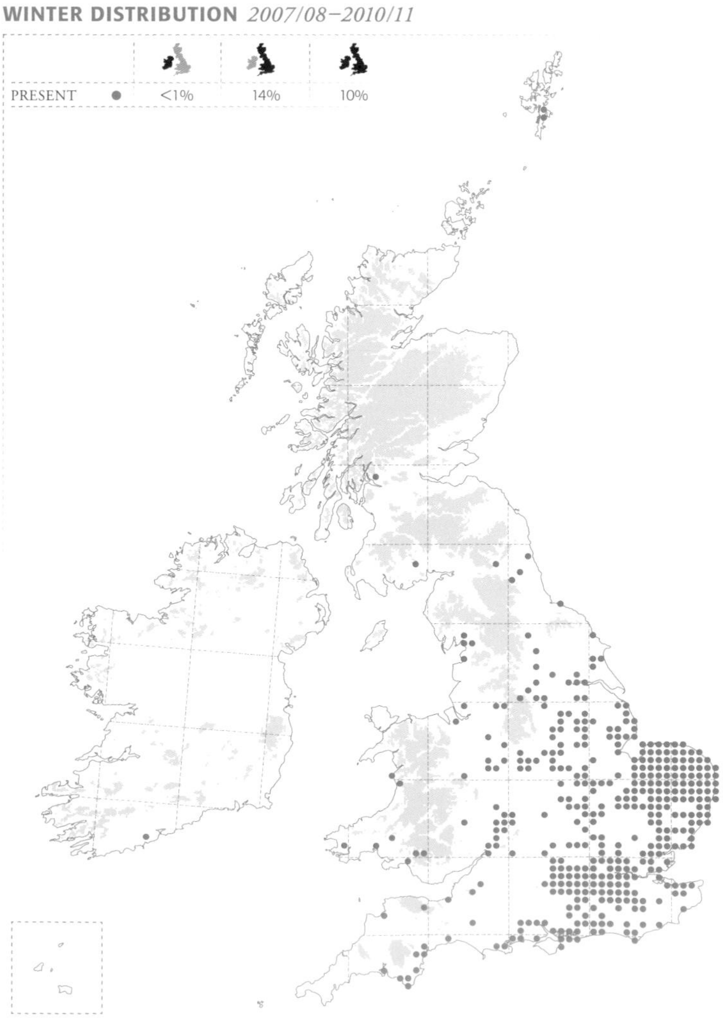

PRESENT ●	<1%	14%	10%

Sponsored by Nicholas Donnithorne FLS FRES

BREEDING DISTRIBUTION CHANGE *since 1968–72*

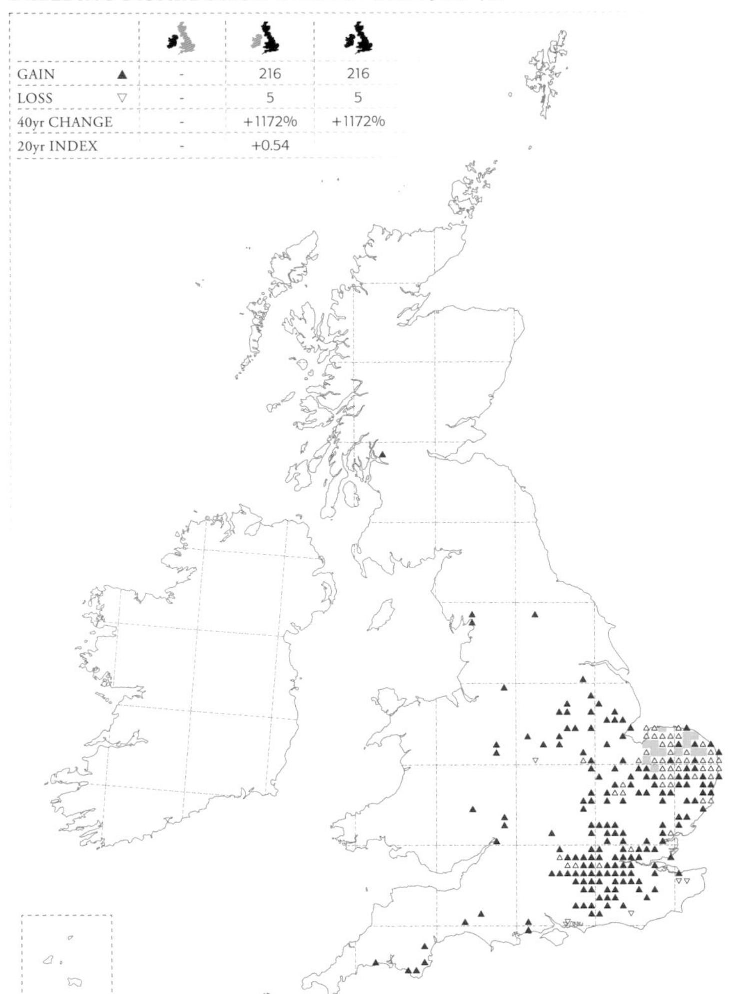

GAIN ▲	-	216	216
LOSS ▽	-	5	5
40yr CHANGE	-	+1172%	+1172%
20yr INDEX	-	+0.54	

BREEDING RELATIVE ABUNDANCE *2008–11*

WINTER DISTRIBUTION CHANGE *since 1981–84*

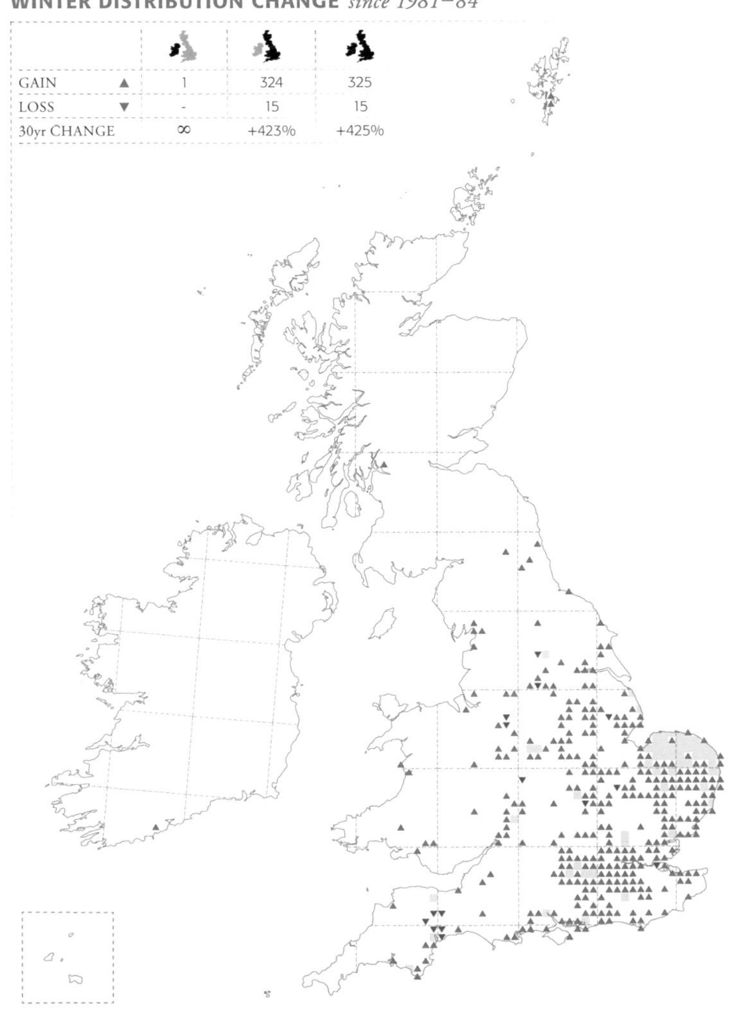

GAIN ▲	1	324	325
LOSS ▼	-	15	15
30yr CHANGE	∞	+423%	+425%

WINTER RELATIVE ABUNDANCE *2007/08–2010/11*

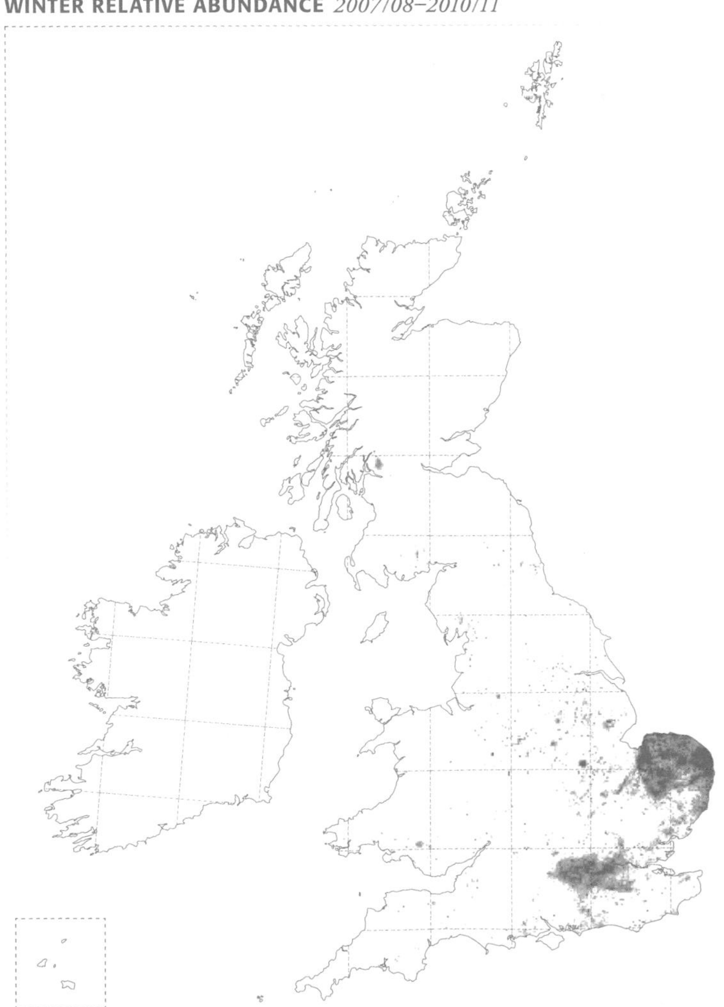

SHELDUCK

Tadorna tadorna

BTO CODE: SU

DAVID KJAER www.davidkjaer.com

THE CONTINUED COLONISATION of inland breeding sites since the *1968–72 Breeding Atlas* is the most striking change for the Shelduck and is also consistent with the change in range that has taken place in winter; increases of 34% and 17% in range size were recorded when compared with the *1968–72 Breeding Atlas* and the *1981–84 Winter Atlas* respectively. Shelducks wintering around the coastline favour muddy estuaries and coastal marshes. Inland, they utilise a range of sites including valley farmland, lakes, reservoirs and pig fields. A closer look at the winter records shows that some 10-km squares inland and on the west coasts of Scotland and Ireland are occupied only in February and these are likely to involve returning breeding birds that have wintered elsewhere.

The breeding distribution map shows birds to be well distributed around the lowland coastal fringe and at scattered inland locations in northern, central and southern England. Inland breeding is rare in Ireland, with most of the records coming from Lough Neagh. In the Channel Islands, Shelducks have recently established a small breeding population. The breeding season relative abundance map shows the importance of locations along the East Anglian coastline, the Lancashire and Cumbrian marshes, the Uists and Orkney, as well as the area inland of the Wash extending into the Fens and Breckland. The change in relative abundance map indicates an interesting pattern of changes in tetrad occupancy, such as the losses in Orkney and gains in Breckland.

A comparison of atlas results with population monitoring reveals intriguing differences during both seasons. Despite the increase in winter range, winter population trends in the Republic of Ireland and the UK show shallow, steady declines since the mid 1990s (*I-WeBS Report 2010*; *WeBS Report 2011*). Causes of some local declines have been identified—for example, the over-exploitation of the cockle and mussel fishery in the Wash (Atkinson *et al.* 2010). The UK breeding population increased by 2% between 1995 and 2010 (*BBS Report 2011*), much less than the range expansion shown by this atlas.

Sponsored by Tay Ringing Group

BREEDING DISTRIBUTION *2008–11*

	Ireland	Britain	Britain & Ireland
POSSIBLE •	3%	3%	3%
PROBABLE ●	9%	13%	12%
CONFIRMED ●	11%	24%	20%
TOTAL	23%	40%	35%

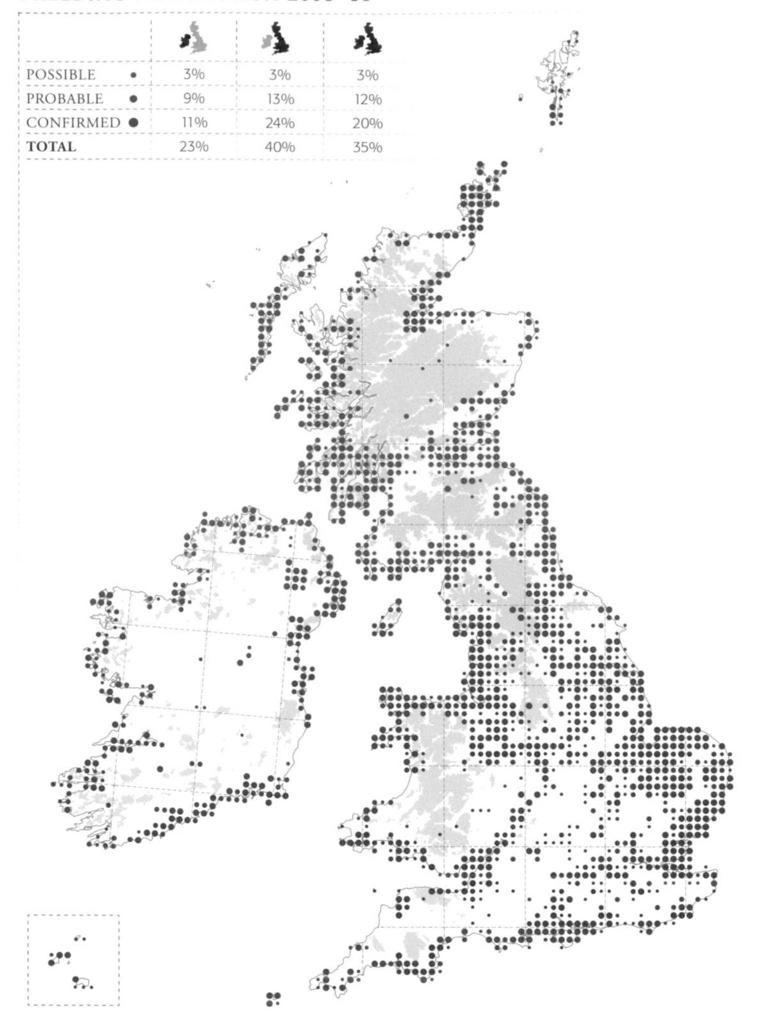

WINTER DISTRIBUTION *2007/08–2010/11*

	Ireland	Britain	Britain & Ireland
PRESENT ●	20%	40%	35%

BREEDING DISTRIBUTION CHANGE *since 1968–72*

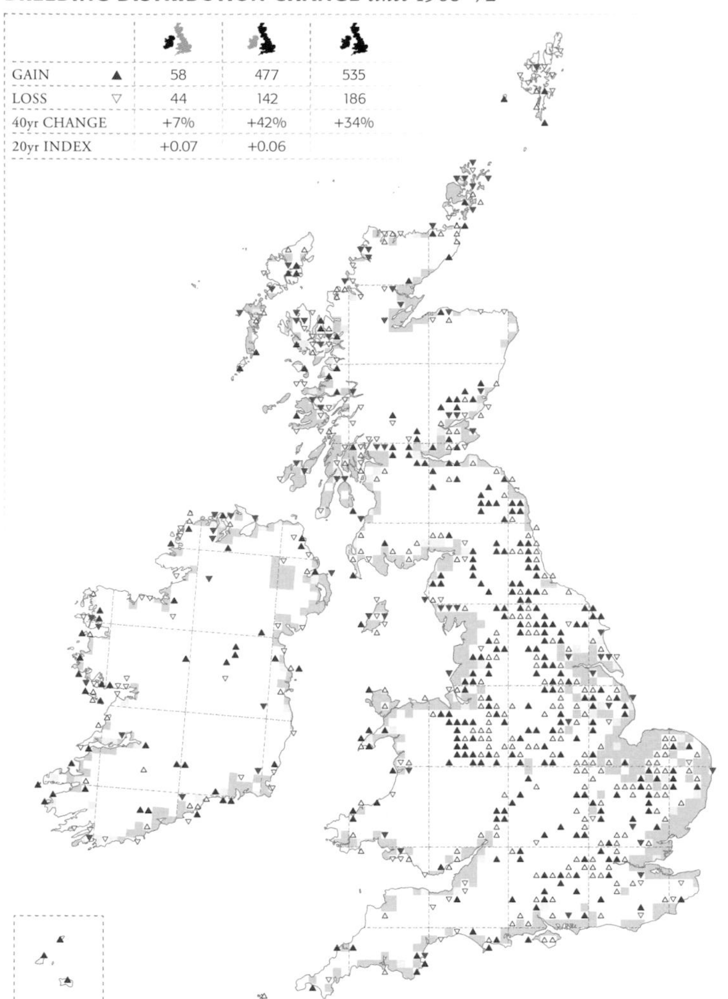

GAIN ▲	58	477	535
LOSS ▽	44	142	186
40yr CHANGE	+7%	+42%	+34%
20yr INDEX	+0.07	+0.06	

WINTER DISTRIBUTION CHANGE *since 1981–84*

GAIN ▲	56	351	407
LOSS ▼	42	174	216
30yr CHANGE	+7%	+19%	+17%

BREEDING RELATIVE ABUNDANCE *2008–11*

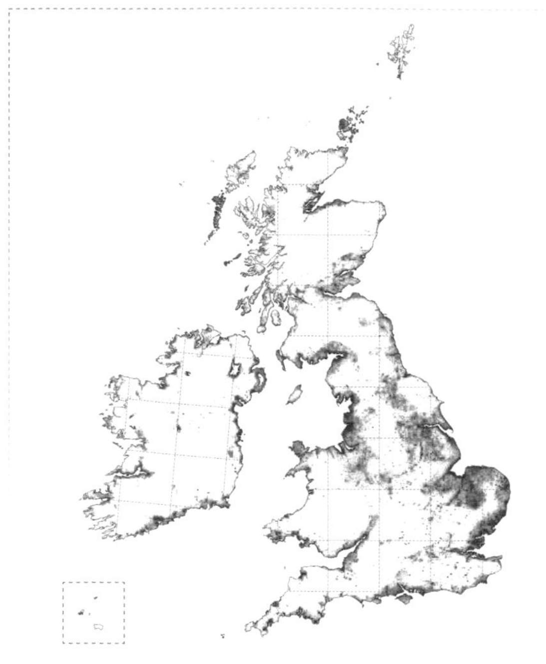

BREEDING RELATIVE ABUNDANCE CHANGE *since 1988–91*

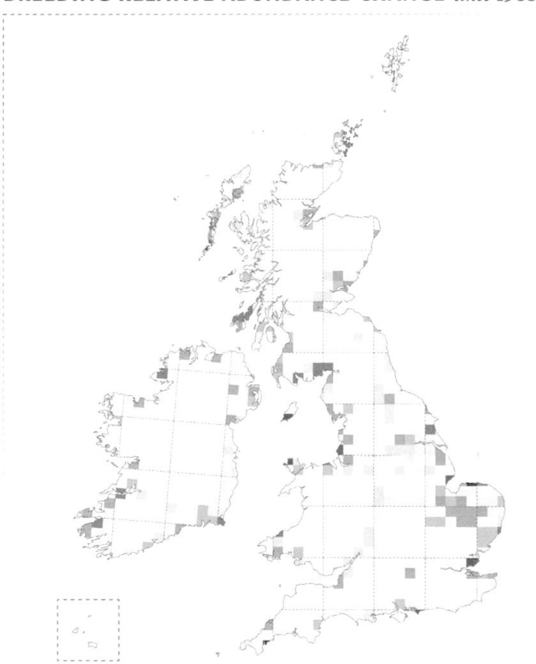

WINTER RELATIVE ABUNDANCE *2007/08–2010/11*

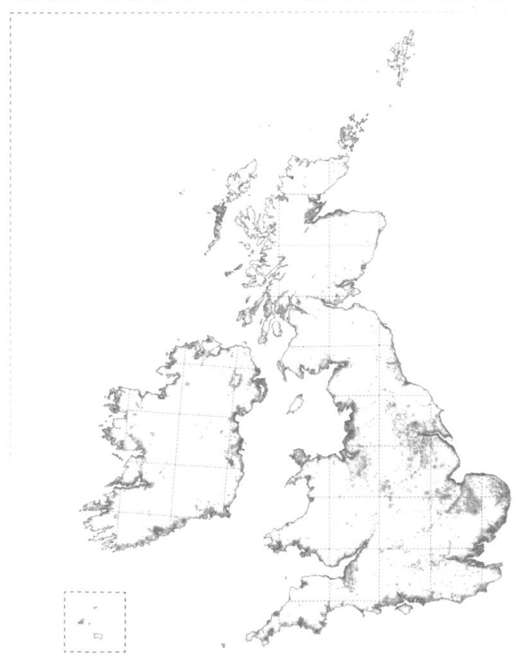

MUSCOVY DUCK
Cairina moschata

BTO CODE: MY

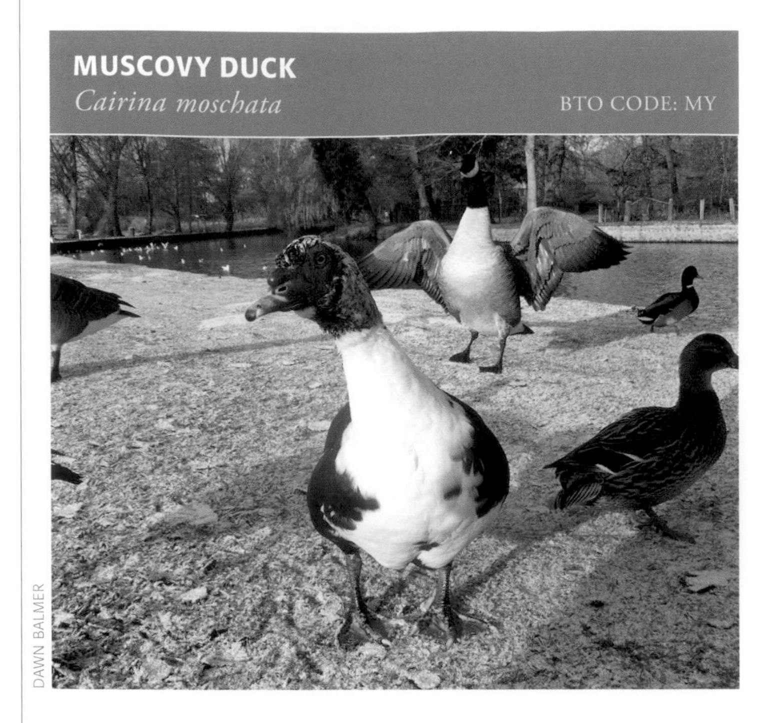

DAWN BALMER

MUSCOVY DUCKS ARE often found near human habitation, prompting uncertainties as to whether they are introduced or semi-domesticated and leading to regular under-reporting (Banks *et al.* 2008). Consequently, reviews such as Holling *et al.* (2011a) will inevitably underestimate numbers and range. The results from *Bird Atlas 2007–11* probably give us the best overview of distribution for many years.

The breeding and winter distribution maps show Muscovy Ducks to be found at scattered locations throughout England, with fewer in Scotland and Wales. They were recorded in a very small number of 10-km squares in Ireland and on the Isle of Man, the latter with a count of nine birds in one square. During the breeding season, evidence of breeding came from 109 10-km squares, and birds were present in a further 91 10-km squares. This implies that there has been a substantial increase in range since the *1968–72 Breeding Atlas* and *1988–91 Breeding Atlas*, when they were recorded in just five 10-km squares in each atlas, although it is likely that increased reporting during *Bird Atlas 2007–11* will account for much of the increase. Counts of over 20 birds submitted to the Atlas came from just three sites: Buxton Pavilion Gardens, Derbyshire; Beccles Quay, Suffolk, and Ely, Cambridgeshire, where maxima of 27, 25 and 22 were recorded respectively in the winter months.

Since the early 1980s there has been an established group present at Ely on the River Ouse. Broods were first noticed in 1987 and the population reached a maximum of 82 birds in January 1996 (Cambridge Bird Club 1997; Lever 2009). Though there have been some attempts at control in the past, for example around 20 were culled in early 2000 (Cambridge Bird Club 2001), they are now thought to be spreading along the River Ouse and appear to be self-sustaining. During the Atlas period a maximum of 46 (including young) were recorded here in May 2009 (Cambridge Bird Club 2011).

In memory of Brunon Stanislaw Zakrzewski, 1926–2010

BREEDING DISTRIBUTION *2008–11*

POSSIBLE •	<1%	2%	2%
PROBABLE •	<1%	1%	1%
CONFIRMED •	-	1%	1%
TOTAL	<1%	4%	4%

WINTER DISTRIBUTION *2007/08–2010/11*

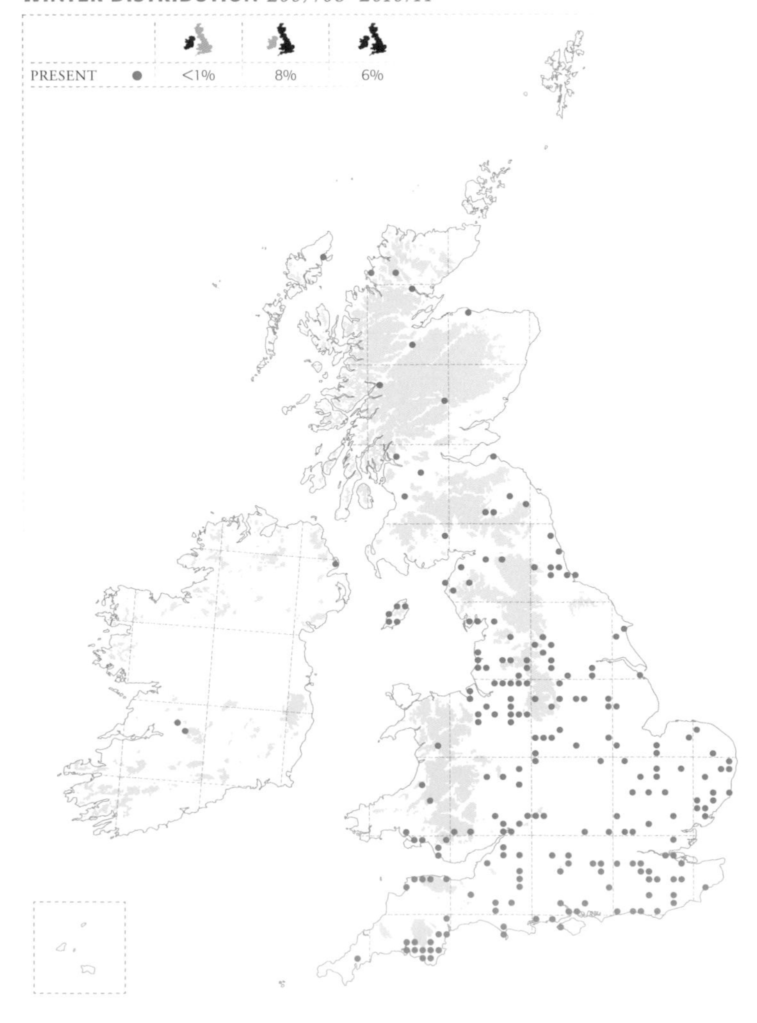

PRESENT •	<1%	8%	6%

WOOD DUCK

Aix sponsa

BTO CODE: DC

DEREK WASHINGTON www.washington.clara.net

THIS NORTH AND Central American species is commonly kept in wildfowl collections and was first introduced into England in the 1870s at Bicton House, near Exeter, Devon (Lever 2009). Free-flying birds were kept at sites such as Woburn, Bedfordshire, and Foxwarren Park, Surrey, during the early 20th century. There is no evidence to suggest that it has ever occurred in a wild state, though it is suspected that genuine vagrants might have reached western Europe (Taylor *et al.* 1999).

Wood Ducks were recorded in 77 10-km squares during the winter months, with most records in central and southern England. This is a more widespread distribution than shown by waterbird monitoring schemes, where they were recorded at just 12 sites in 2010/11 (*WeBS Report 2011*). During the breeding season there were records with breeding evidence from 30 10-km squares, including 10 squares with *confirmed* breeding. Records of birds present during the breeding season also came from a further 44 10-km squares. Atlas fieldwork has uncovered more *confirmed* breeding records than previously thought (Holling *et al.* 2011a). While there has been no overall change in range size since the *1988–91 Breeding Atlas*, there has been a high degree of turnover: 28 of the 30 10-km squares were gains, balancing losses from a different set of squares. Breeding attempts in Britain since the late 1960s have been well documented (*1968–72 Breeding Atlas*; Lever 2009; Holling *et al.* 2011a) and are generally related to isolated nesting records, often near wildfowl collections. Small groups have become established at some sites, such as Stanton Park, Wiltshire, and breeding has been recorded there in several successive years.

Unlike the closely related Mandarin Duck, the Wood Duck has failed to become firmly established as a widespread breeding bird. Lever (2009) suggests a number of reasons including the slightly longer fledging period for Wood Ducks, which makes ducklings vulnerable to predators for a greater period of time, competition for nest sites with Mandarin Ducks and reduced genetic variability.

BREEDING DISTRIBUTION *2008–11*

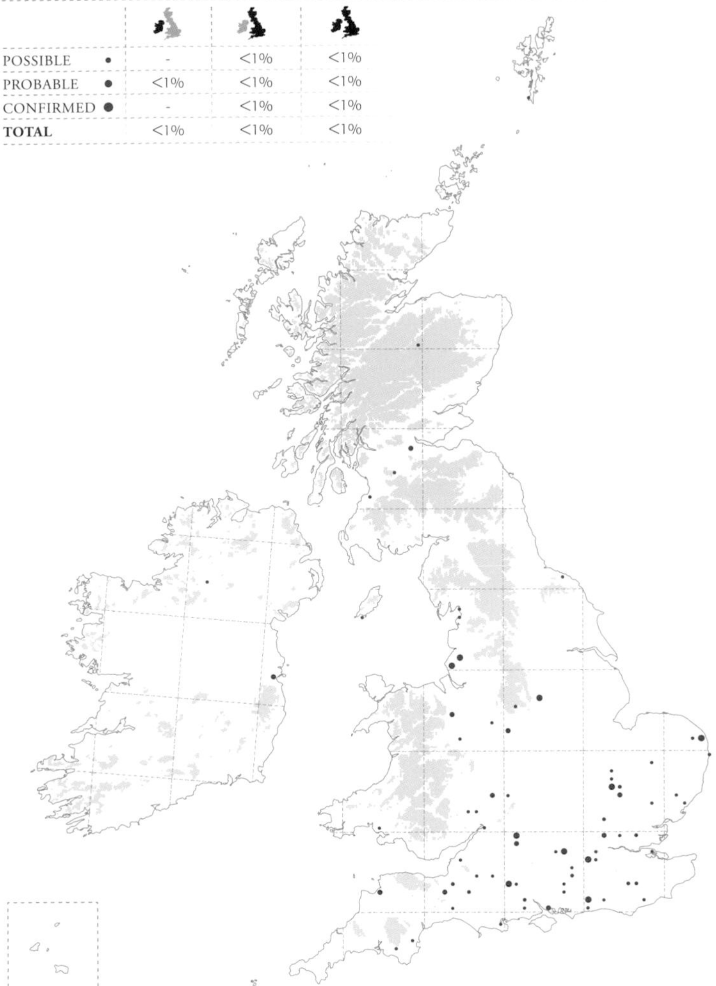

	Ireland	Britain	Britain & Ireland
POSSIBLE •	-	<1%	<1%
PROBABLE ●	<1%	<1%	<1%
CONFIRMED ●	-	<1%	<1%
TOTAL	<1%	<1%	<1%

WINTER DISTRIBUTION *2007/08–2010/11*

	Ireland	Britain	Britain & Ireland
PRESENT ●	<1%	3%	2%

MANDARIN DUCK
Aix galericulata
BTO CODE: MN

MARK COATES

THERE HAS BEEN considerable range expansion since the *1988–91 Breeding Atlas*, particularly through southern and central England, with an overall 123% increase in range size. Away from the swathe of records in southern and central England, the breeding distribution map shows tight clustering in areas with suitable breeding habitat such as Jersey, south Devon, Kielder Forest in Northumberland, Argyll Forest Park and around Inverness and Berwick. The species is scarce in the western half of Wales. In Ireland a small breeding population has established (Mathers 1993) and breeding was *confirmed* in two 10-km squares.

The expansion in breeding range is clear to see on the breeding distribution change map, which illustrates the small number of squares, centred on the Berkshire/Surrey border, where Mandarin Ducks have been present in all three breeding atlases. The winter and breeding distribution maps are similar, although in winter there is a little more spread in the records, with more 10-km squares occupied.

Although generally a shy and secretive species inhabiting areas with a dense growth of marginal trees and shrubs, the Mandarin Duck has been well recorded in *Bird Atlas 2007–11* and enough birds were registered on Timed Tetrad Visits for maps of relative abundance to be produced in both seasons. These show the main strongholds to be in southeast England, in the Severn Vale and around the Peak District, with smaller hotspots elsewhere.

Although the Mandarin Duck may have been under-recorded in the *1968–72 Breeding Atlas*, there is no doubt that there has been a genuine increase in range and numbers over the last 40 years. The number of sites recording this species during wintering waterbird surveys continues to increase and the monthly maxima of 703 birds recorded in the 2010/11 winter represents the highest-ever total (*WeBS Report 2011*). These counts include 245 in the Forest of Dean Ponds and 135 at Linacre Reservoirs, Derbyshire. Similarly, BBS has recorded a small increase in the number of squares supporting Mandarin Ducks (*BBS Report 2011*). They remained under-recorded in Ireland during the Atlas period; the estimated population in Northern Ireland in 2005–06 was 20–30 pairs (Northern Ireland Birdwatchers' Association 2008).

Sponsored by Lady Baird

BREEDING DISTRIBUTION *2008–11*

POSSIBLE •	<1%	3%	2%
PROBABLE ●	<1%	5%	4%
CONFIRMED ●	<1%	10%	7%
TOTAL	<1%	18%	13%

WINTER DISTRIBUTION *2007/08–2010/11*

PRESENT ●	1%	21%	15%

BREEDING DISTRIBUTION CHANGE *since 1968–72*

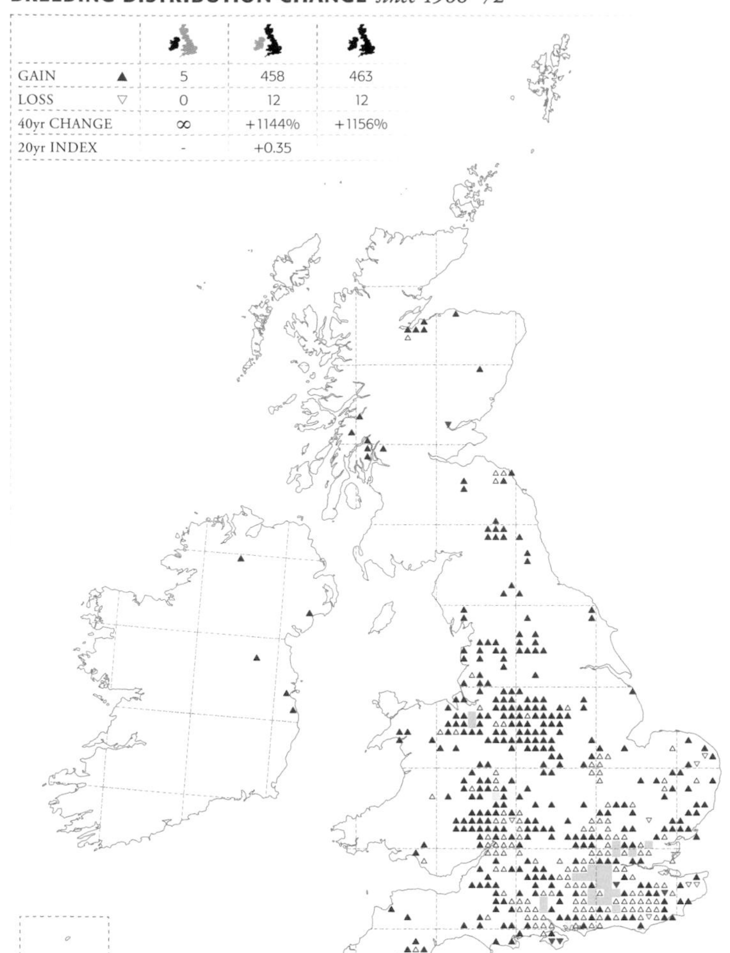

GAIN ▲	5	458	463
LOSS ▽	0	12	12
40yr CHANGE	∞	+1144%	+1156%
20yr INDEX	-	+0.35	

BREEDING RELATIVE ABUNDANCE *2008–11*

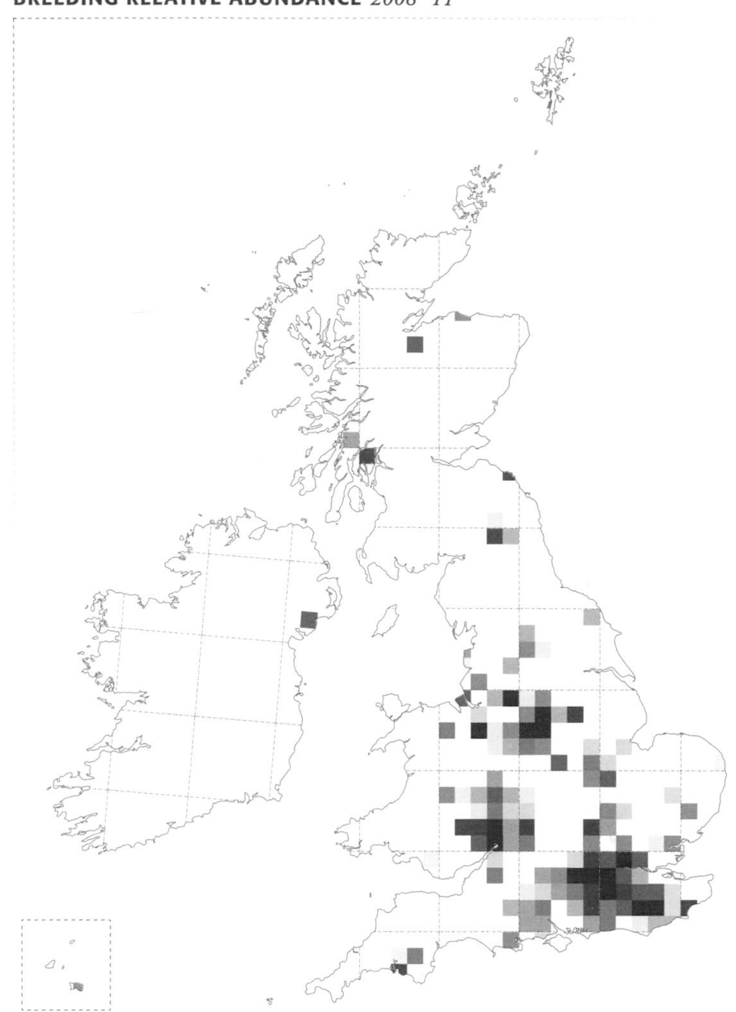

WINTER DISTRIBUTION CHANGE *since 1981–84*

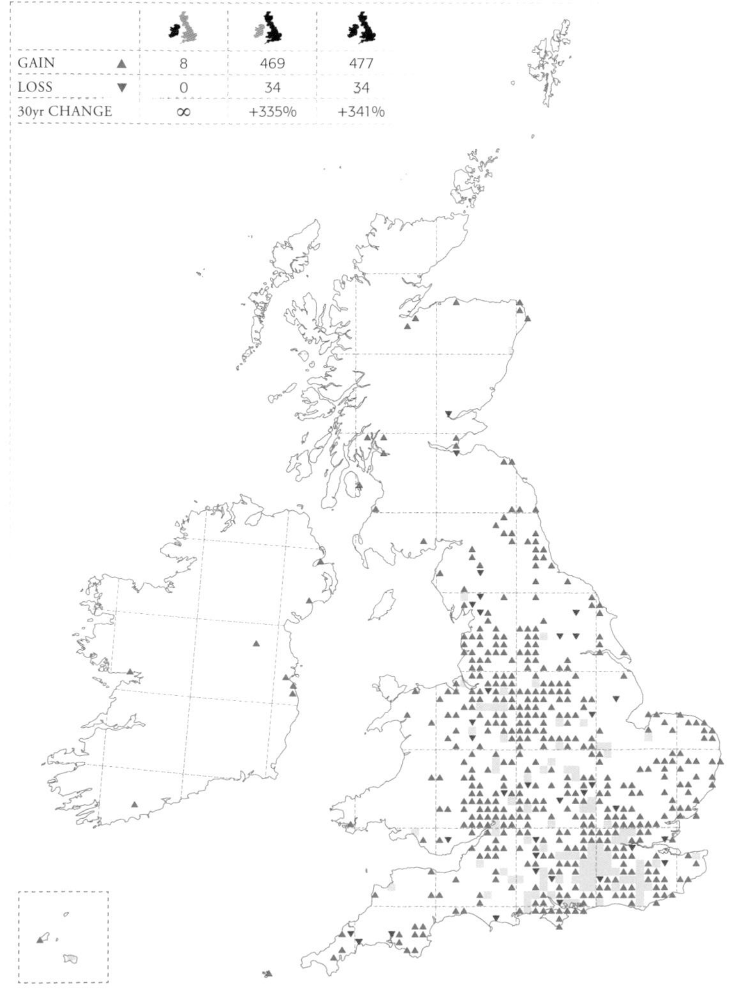

GAIN ▲	8	469	477
LOSS ▼	0	34	34
30yr CHANGE	∞	+335%	+341%

WINTER RELATIVE ABUNDANCE *2007/08–2010/11*

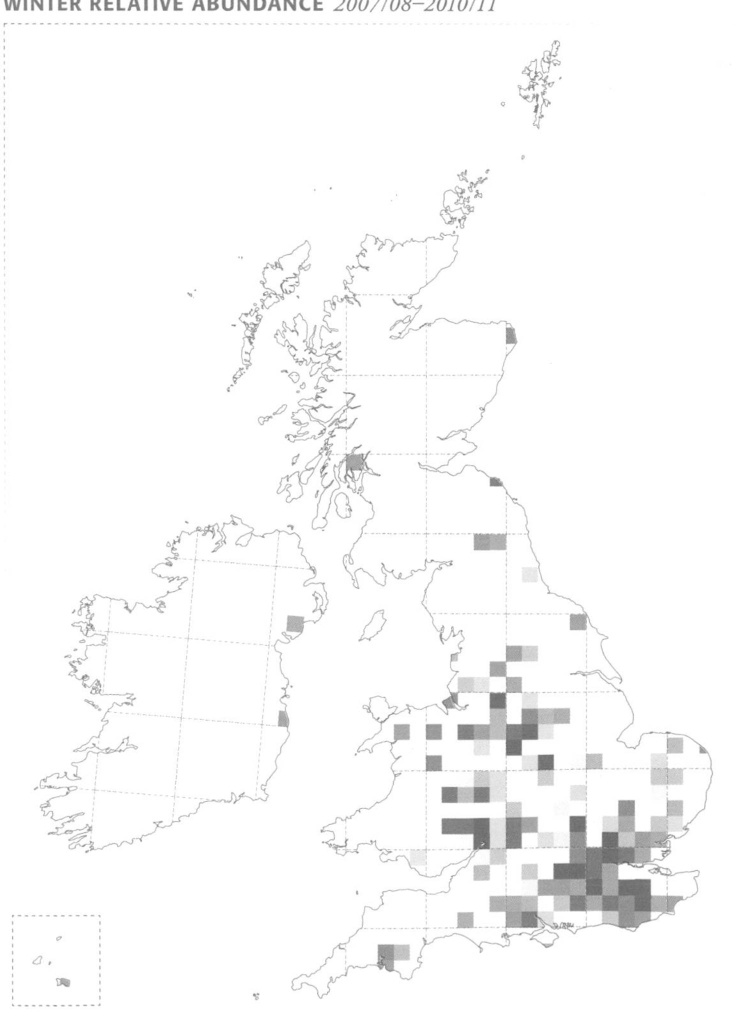

WIGEON

Anas penelope

BTO CODE: WN

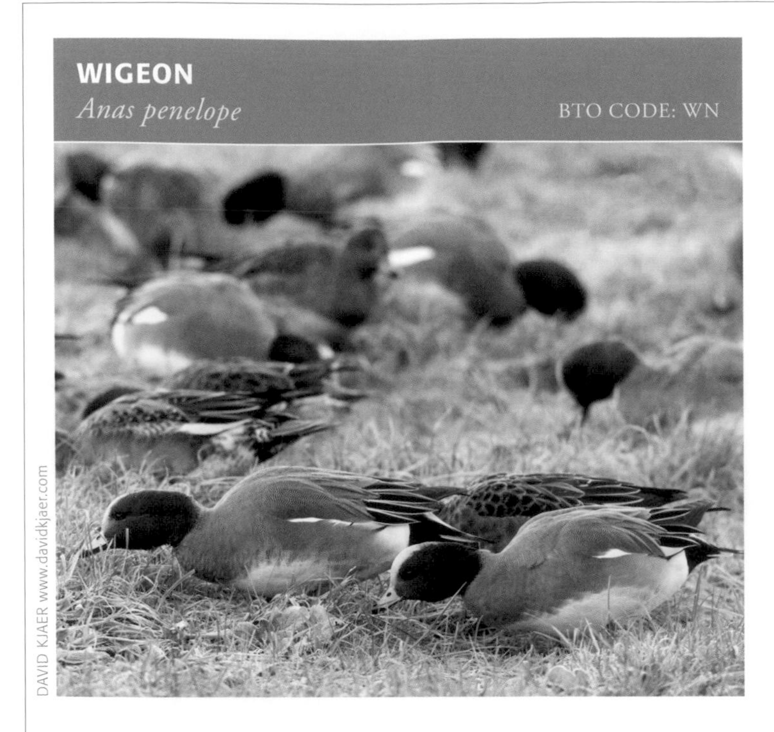

THE SMALL BREEDING population of the Wigeon in Britain & Ireland is augmented by the autumn arrival of vast numbers of birds from Iceland, Fennoscandia and Russia. Through the winter this species is widespread in coastal and lowland areas and largely absent from the uplands. The winter abundance map shows particular concentrations in the Northern Isles, the inner Moray Firth, parts of central Scotland, the Irish Sea coastal fringe, major river valleys in eastern England, estuaries of south and southeast England and the rivers, lakes and turloughs of the west midlands of Ireland.

The UK wintering population increased by 63% between 1983/84 and 2008/09 (*SUKB 2012*) and this is reflected by a 27% increase in the number of occupied 10-km squares in Britain since the *1981–84 Winter Atlas*, with gains most evident in lowland areas of England. In Ireland, there has been a modest 6% increase in range, despite reported declines in numbers since the mid 1990s (*I-WeBS Report 2009*; *WeBS Report 2011*). It has been suggested that the decline in Northern Ireland may be due to a shift in distribution in response to climate change (Holt *et al.* 2011a). The 2010/11 winter saw high December and January counts at a number of sites in Britain, presumably associated with the frozen conditions prevalent across northwest Europe at the time (*WeBS Report 2011*).

Some wintering birds may remain into late April making it difficult to differentiate them from potential breeders. As a result, some records of *possible* breeding should be treated with caution. In Britain, the main breeding areas are in the Pennines and in northern Scotland from Fife through the eastern Highlands north to Sutherland and Caithness, the Northern Isles and the Uists. There are also a few *confirmed* breeding records in central and eastern England and southern Scotland. The breeding distribution change map suggests a decline in the Scottish uplands and gains in the islands, but inaccurate coding of records hinders proper analysis. In Ireland, nine 10-km squares had pairs in suitable breeding habitat but none were *confirmed* as breeding.

Sponsored by Jon Pavey

BREEDING DISTRIBUTION *2008–11*

	Ireland	Britain	Britain & Ireland
POSSIBLE •	1%	4%	3%
PROBABLE •	1%	5%	4%
CONFIRMED •	-	4%	3%
TOTAL	2%	13%	10%

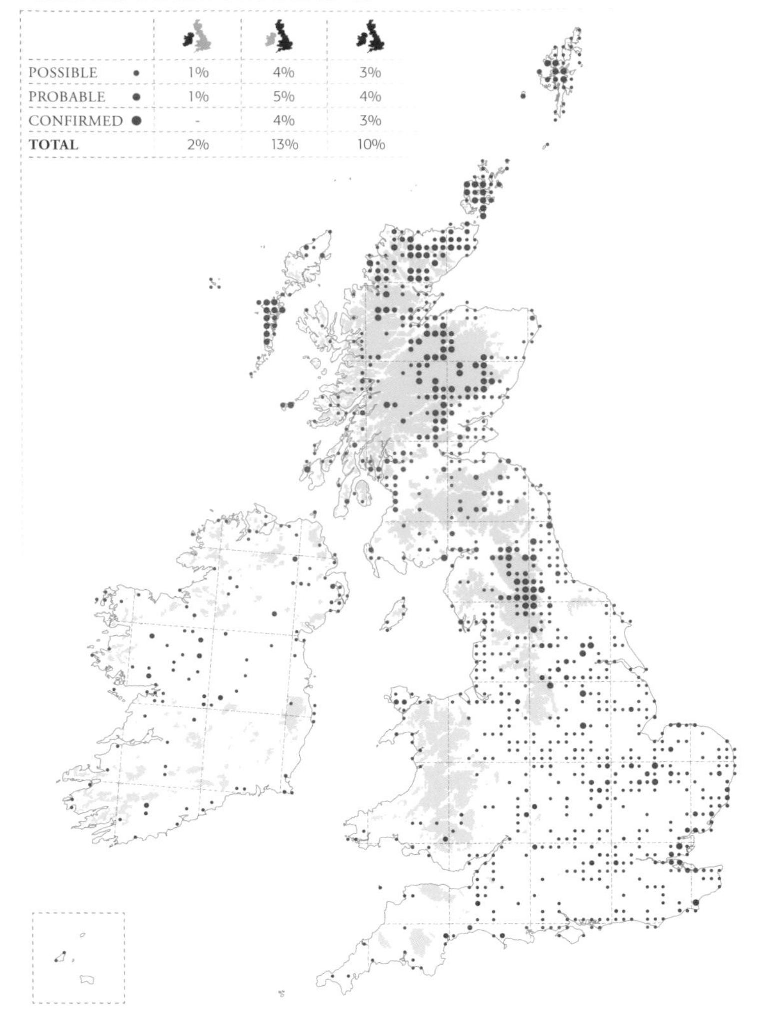

WINTER DISTRIBUTION *2007/08–2010/11*

	Ireland	Britain	Britain & Ireland
PRESENT •	49%	65%	61%

BREEDING DISTRIBUTION CHANGE *since 1968–72*

GAIN ▲	24	216	240
LOSS ▽	0	141	141
40yr CHANGE	∞	+27%	+35%
20yr INDEX	-	-0.07	

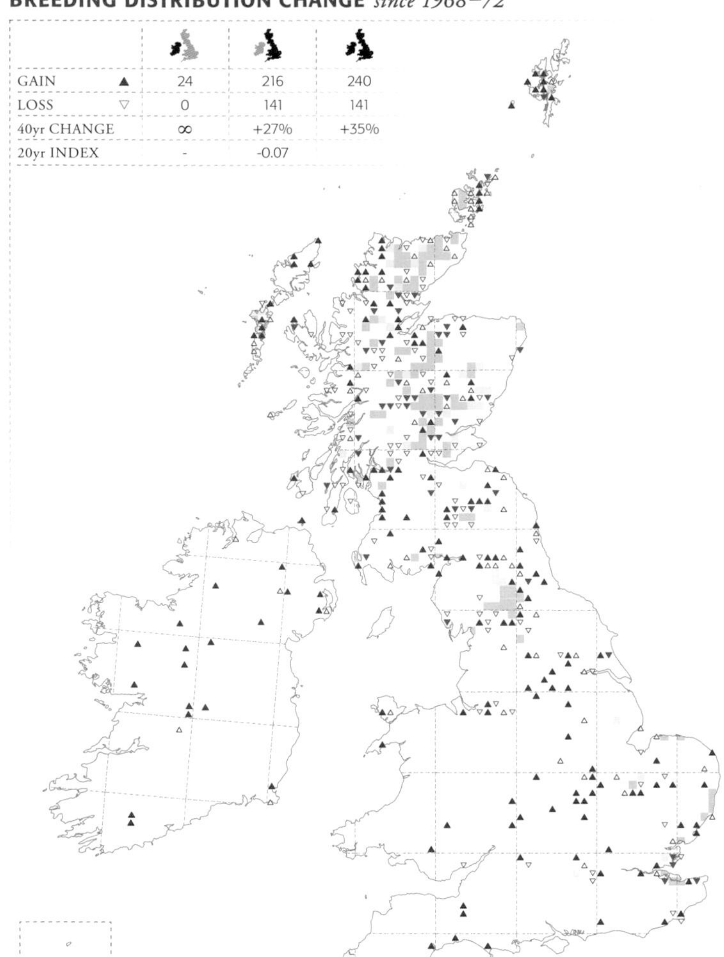

BREEDING RELATIVE ABUNDANCE *2008–11*

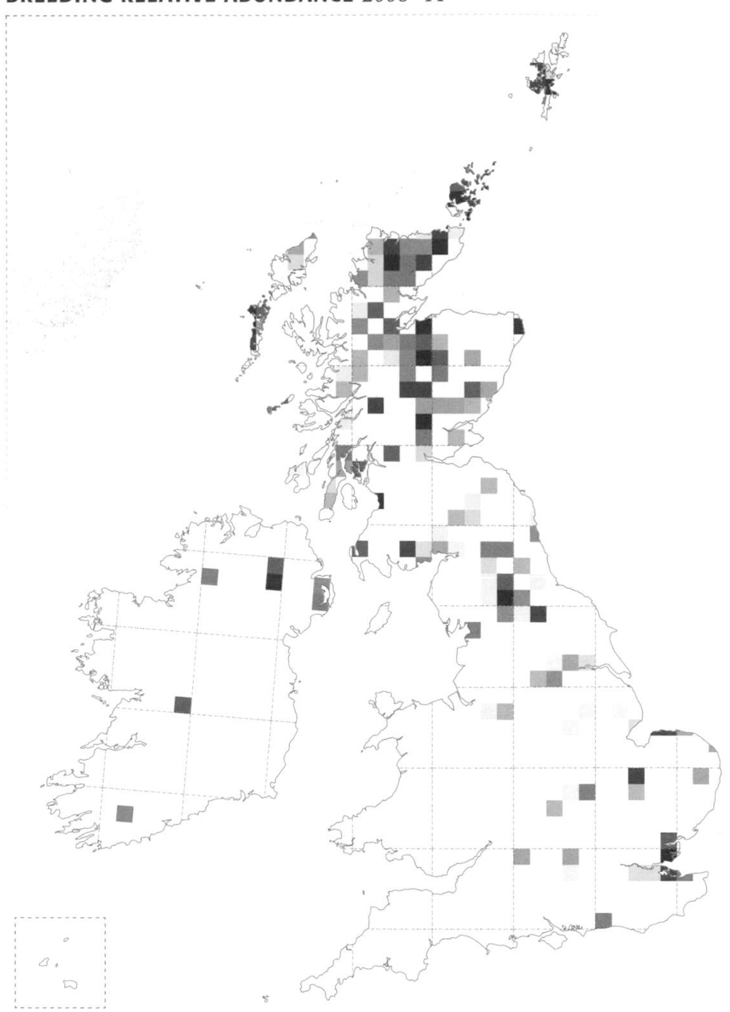

WINTER DISTRIBUTION CHANGE *since 1981–84*

GAIN ▲	131	529	660
LOSS ▼	106	146	252
30yr CHANGE	+5%	+27%	+22%

WINTER RELATIVE ABUNDANCE *2007/08–2010/11*

GADWALL

Anas strepera

BTO CODE: GA

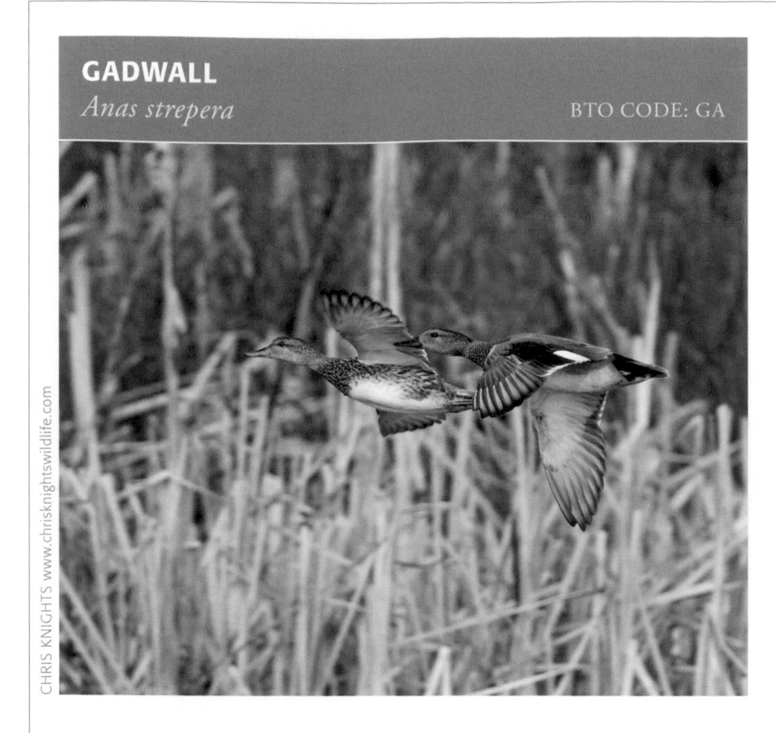

CHRIS KNIGHTS www.chrisknightswildlife.com

FROM THE FORMER strongholds in East Anglia, Fife and Lough Neagh recorded during the *1968–72 Breeding Atlas*, the breeding range of the Gadwall has expanded throughout much of the lowlands of central, eastern and northwest England, eastern Scotland, Orkney and the Uists. In Ireland, breeding is still concentrated around Lough Neagh, with scattered breeding records elsewhere through the island, particularly in the south and west. The distribution of birds is wider in winter than in the breeding season owing to dispersal from natal areas and the arrival of migrants from Iceland and the near Continent. This is particularly evident in southwest England, coastal Wales and Ireland, and Shetland. The relative abundance maps are broadly similar between seasons, although the wider inland distribution in winter is apparent.

The original breeding stock originates largely from pinioned wild birds caught at Dersingham Decoy, Norfolk, and introduced into Breckland around 1850. Much of the present population is descended from this source, although genuinely wild birds wintering here may have stayed to breed and thus contributed to the breeding population. Birds colonised Loch Leven, Fife, in 1909 and that area remains a stronghold. The increase in the number of lowland reservoirs and gravel pits in Britain has aided the spread of this species and provided suitable wintering and breeding areas. The UK breeding population increased by 83% between 1995 and 2010 (*BBS Report 2011*) but the species remains too scarce as a breeding species in Ireland for annual monitoring to produce trends.

The increase in wintering numbers has been tracked by winter waterbird monitoring schemes, which show a 312% increase in the UK between 1983/84 and 2008/09 (*SUKB 2012*), and a 200% increase in the Republic of Ireland between 1994/95 and 2008/09 (*I-WeBS Report 2009*). This is in contrast to other species of duck, which are showing apparent short-stopping. The overall increase may mask some local changes. For example, numbers in Northern Ireland declined after the early 1990s but showed substantial recovery in 2009/10 and 2010/11 (*WeBS Report 2011*). Monitoring schemes across Europe, for example in the Netherlands (Hornman *et al.* 2012b) and Switzerland (Keller & Burkhardt 2012), have also documented a rapid increase in winter numbers.

Sponsored by Paul Stanbrook

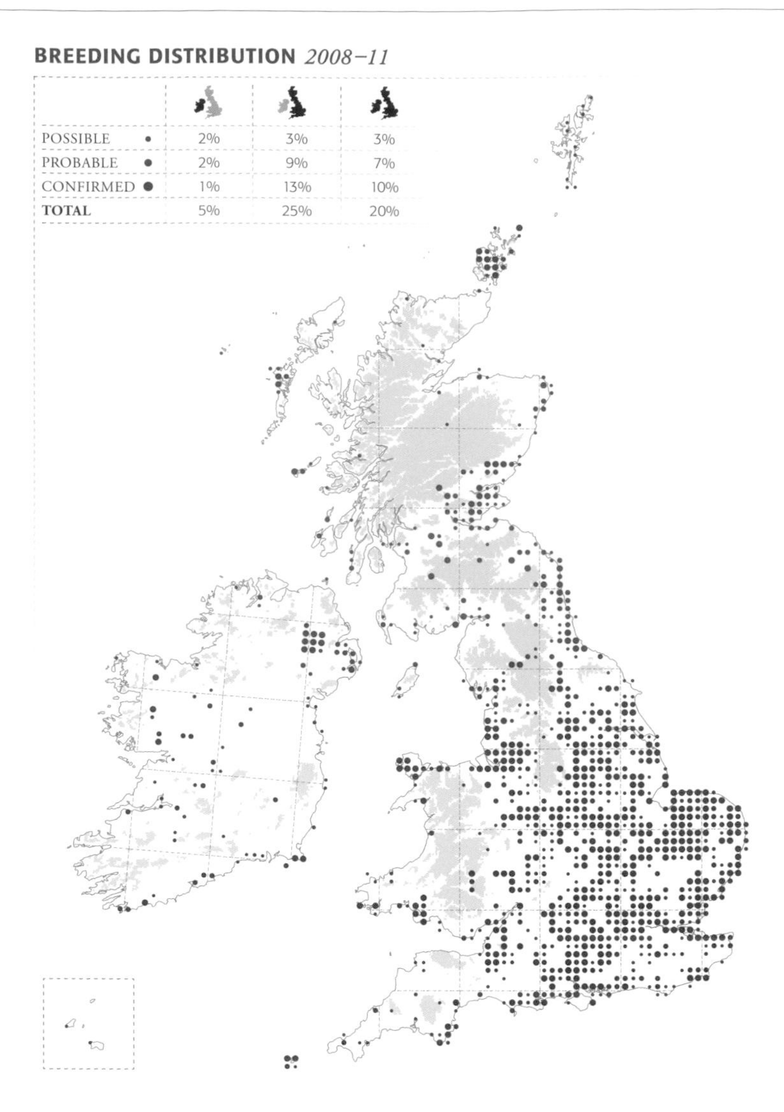

BREEDING DISTRIBUTION *2008–11*

POSSIBLE •	2%	3%	3%
PROBABLE •	2%	9%	7%
CONFIRMED •	1%	13%	10%
TOTAL	5%	25%	20%

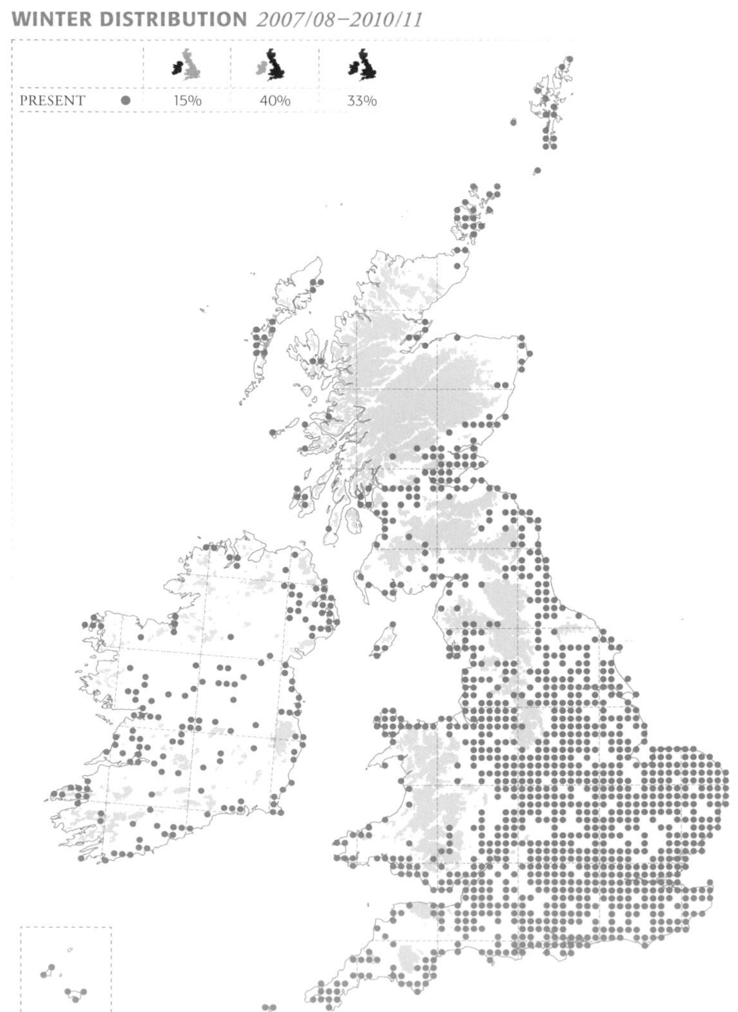

WINTER DISTRIBUTION *2007/08–2010/11*

PRESENT •	15%	40%	33%

BREEDING DISTRIBUTION CHANGE *since 1968–72*

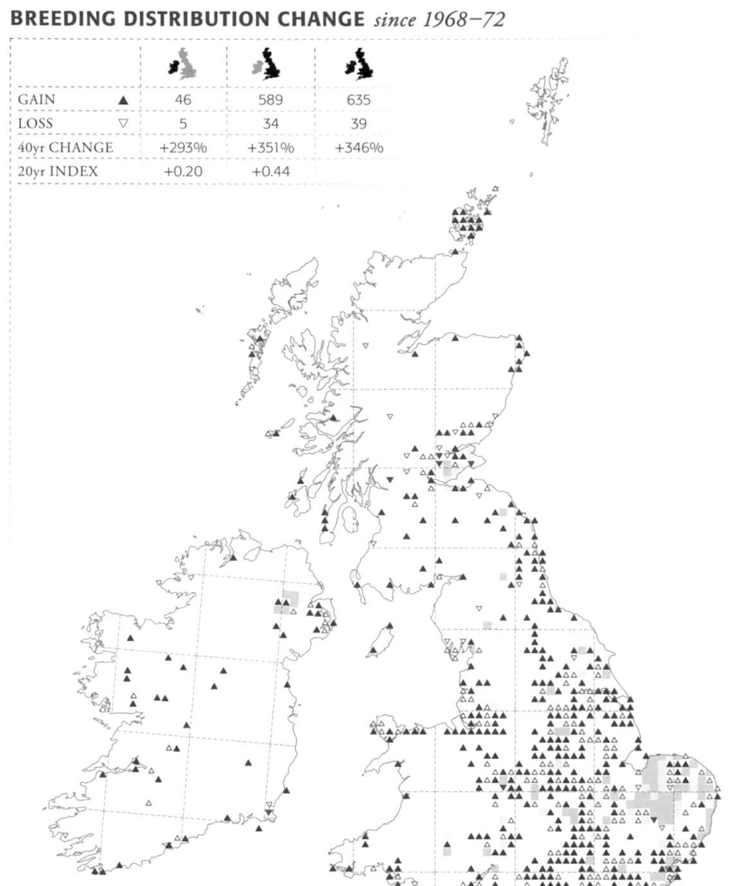

GAIN ▲	46	589	635
LOSS ▽	5	34	39
40yr CHANGE	+293%	+351%	+346%
20yr INDEX	+0.20	+0.44	

BREEDING RELATIVE ABUNDANCE *2008–11*

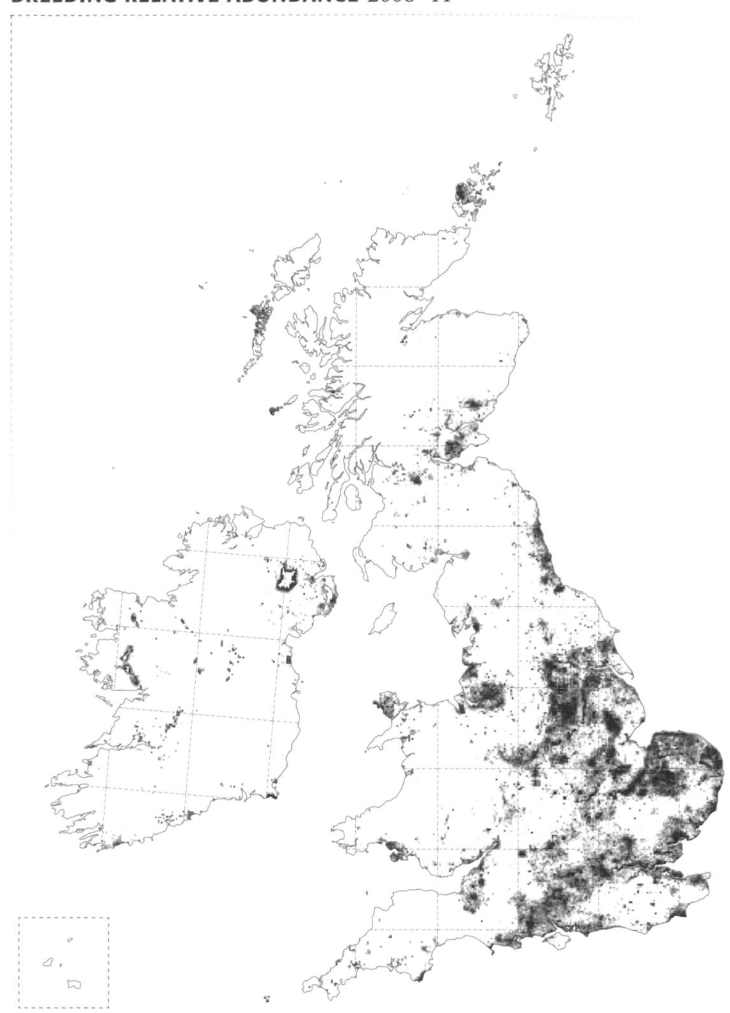

WINTER DISTRIBUTION CHANGE *since 1981–84*

GAIN ▲	109	599	708
LOSS ▼	23	74	97
30yr CHANGE	+143%	+90%	+95%

WINTER RELATIVE ABUNDANCE *2007/08–2010/11*

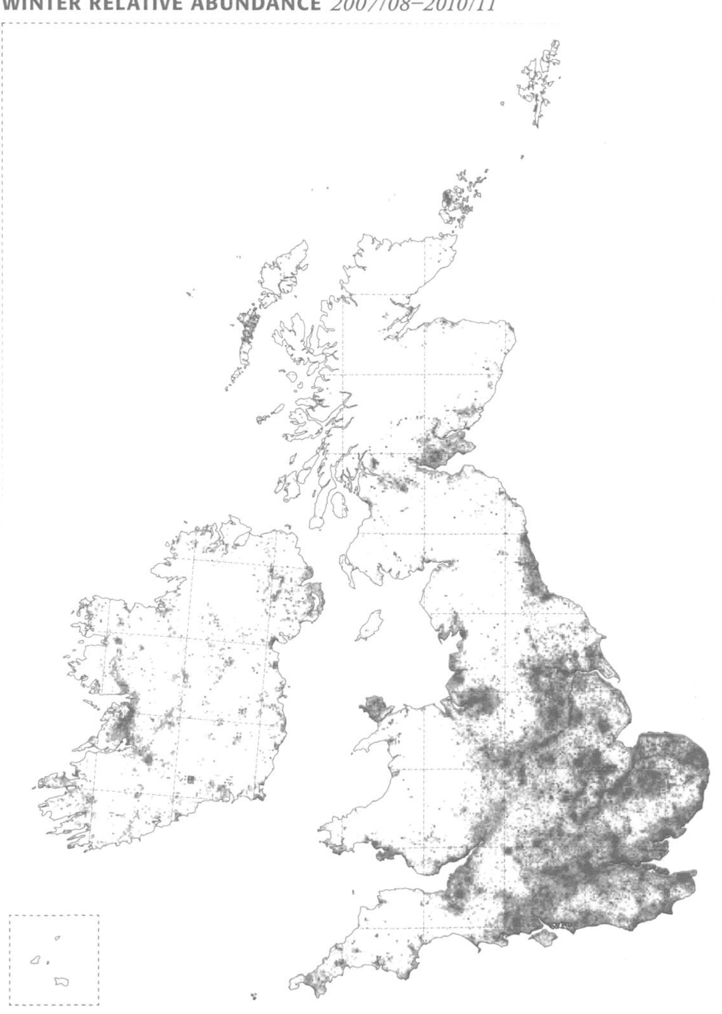

TEAL

Anas crecca

BTO CODE: T.

DESMOND DUGAN www.AbernethyImages.co.uk

IN WINTER, THIS species is widespread throughout Britain & Ireland, being found in a variety of coastal and freshwater sites from small pools to large estuaries. Large numbers of winter immigrants arrive from Iceland, the near Continent, Scandinavia and northwest Russia each autumn (*Migration Atlas*). Gaps in the winter distribution map mostly coincide with upland areas, whilst the winter relative abundance map shows that high densities are associated with major river valleys and coastal areas, especially around the major estuaries.

A 41% increase in numbers wintering in the UK between 1983/84 and 2008/09 (*SUKB 2012*) is reflected in a 13% increase in the number of occupied 10-km squares in Britain since the *1981–84 Winter Atlas*. There is no obvious pattern in the squares where this species has been lost or gained, probably due to its use of a wide range of small wetlands, including ephemeral sites, which may vary from one winter to the next. Movements may have been especially pronounced in the 2009/10 and 2010/11 winters because Teals are particularly vulnerable to severe weather and move in search of ice-free conditions (Ridgill & Fox 1990). The 2010/11 winter saw a record count of 45,884 Teals on the Somerset Levels (*WeBS Report 2011*). In Ireland, despite an apparent 28% increase in numbers between 1994/95 and 2008/09 (*I-WeBS Report 2009*), there was only a negligible increase in the number of occupied 10-km squares, with losses in inland areas being balanced by gains in coastal counties.

The breeding season distribution map is distorted by the continued presence of wintering birds through to the end of April, many of which may have been incorrectly coded as *possible* or *probable* breeders. Although Teals breed throughout Britain & Ireland, their distribution is patchy south of a line from Anglesey to Norfolk, and in Ireland. The breeding season relative abundance map shows Teals to be most abundant in parts of northern Scotland.

Since the *1968–72 Breeding Atlas* there has been a 14% decrease in range size in Britain. In Scotland, gains in the north and west, and in Shetland, contrast with declines in the south and east. In Ireland, range size has decreased by 46% since the *1968–72 Breeding Atlas*, resulting in large gaps in breeding distribution, especially in the north and west of the island. The predominant loss in range in Ireland is likely to have resulted from land-use change, especially land drainage.

Sponsored by Anne Reid

BREEDING DISTRIBUTION *2008–11*

POSSIBLE •	7%	8%	8%
PROBABLE •	10%	23%	20%
CONFIRMED •	3%	10%	8%
TOTAL	20%	41%	36%

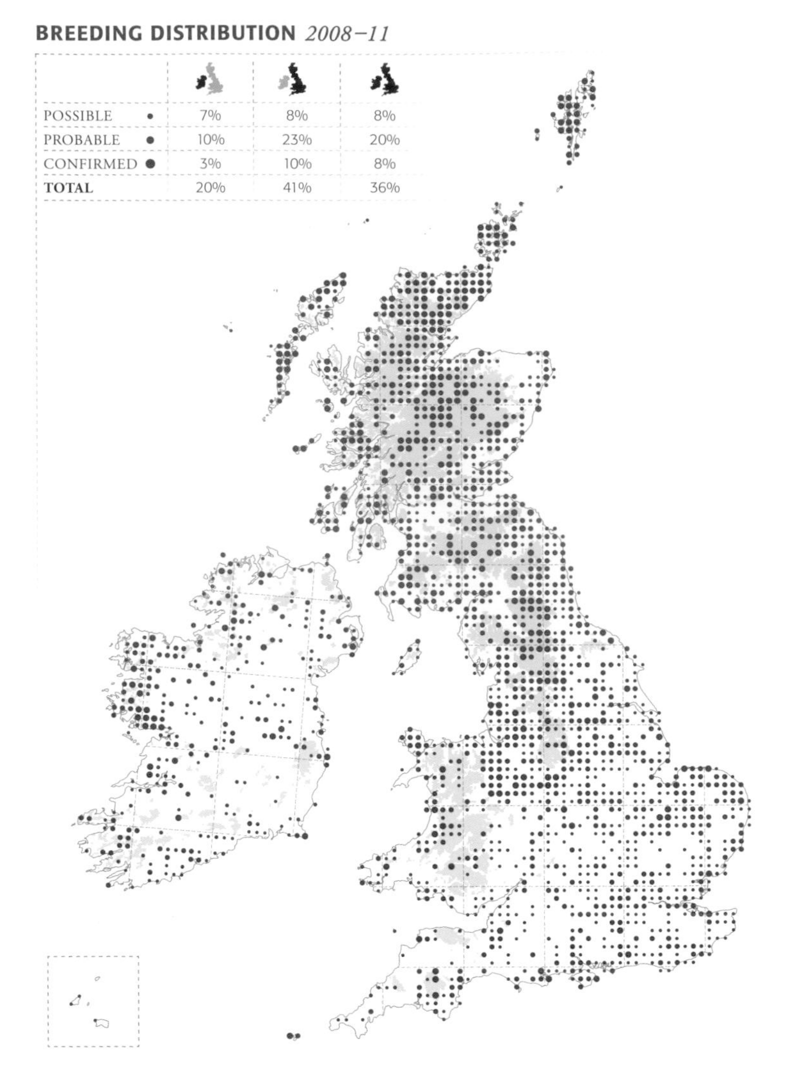

WINTER DISTRIBUTION *2007/08–2010/11*

PRESENT •	67%	75%	73%

BREEDING DISTRIBUTION CHANGE *since 1968–72*

GAIN ▲	89	374	463
LOSS ▽	269	562	831
40yr CHANGE	-46%	-14%	-21%
20yr INDEX	-0.25	-0.08	

BREEDING RELATIVE ABUNDANCE *2008–11*

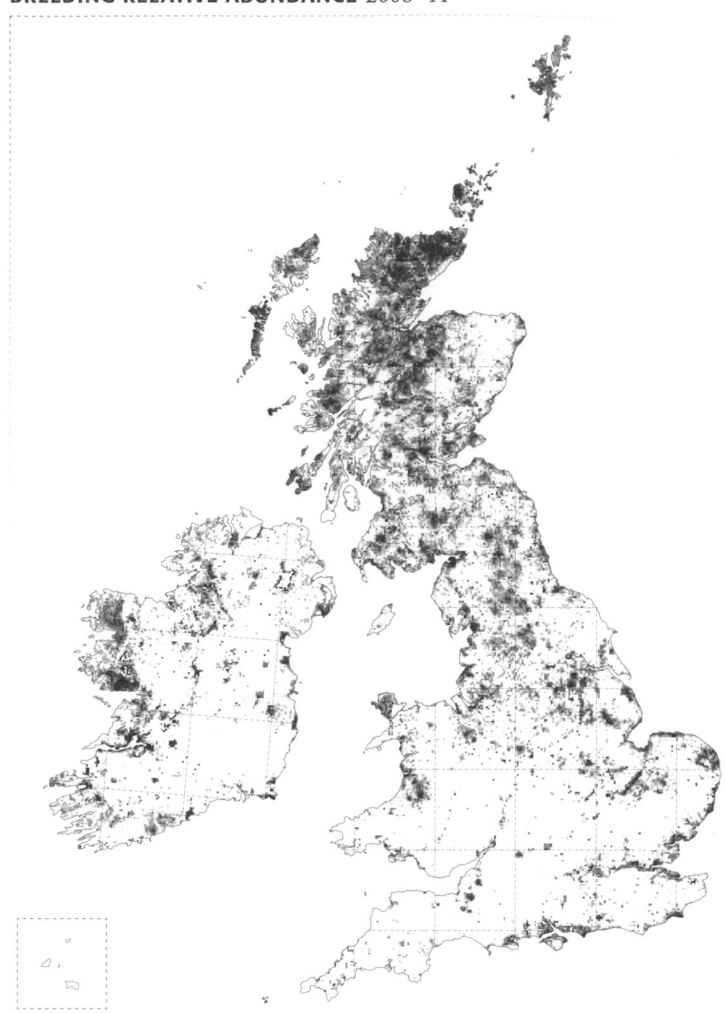

WINTER DISTRIBUTION CHANGE *since 1981–84*

GAIN ▲	143	421	564
LOSS ▼	142	177	319
30yr CHANGE	+<1%	+13%	+10%

WINTER RELATIVE ABUNDANCE *2007/08–2010/11*

MALLARD

Anas platyrhynchos

BTO CODE: MA

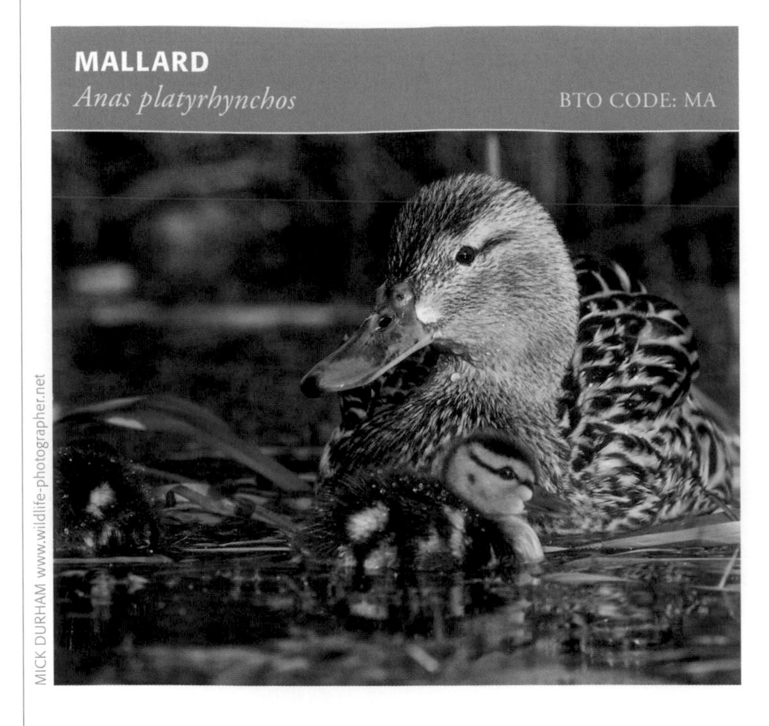

MICK DURHAM www.wildlife-photographer.net

THE MALLARD IS among the most widespread species recorded, both in the winter and in the breeding season, with records in 91% and 92% of 10-km squares respectively. This duck is really only absent from the most remote mountainous areas and from squares with no suitable aquatic habitats. This high level of occupancy has changed only modestly throughout the series of atlases. A recent 2% increase in occupied 10-km squares in Britain and 8% in Ireland has largely reversed the small contraction in occupancy noted in the *1988–91 Breeding Atlas*.

Since the *1988–91 Breeding Atlas*, there also appears to have been a significant and widespread increase in breeding abundance, as indicated by the breeding abundance change map. This is mirrored by a 20% increase in the UK breeding population during 1995–2010 (*BBS Report 2011*), although monitoring in the Republic of Ireland shows no significant change in breeding numbers (*CBS Report 2010*). In some areas, numbers are boosted by annual releases of large numbers of captive-bred birds for shooting purposes (*Birds in England*; *Birds of Scotland*). Birds submitted as 'domestic Mallard' were reported from c.20% of 10-km squares.

These breeding-season trends are not matched in winter despite evidence suggesting that British and Irish breeders are largely sedentary (*Migration Atlas*). The widespread nature of the species belies the fact that there has been a general reduction in the size of the wintering population across Britain & Ireland since around 1990. Winter waterbird monitoring schemes document a 39% decline in the UK between 1983/84 and 2008/09 (*SUKB 2012*) and a 15% decline between 1994/95 and 2008/09 in the Republic of Ireland (*I-WeBS Report 2009*). The cause of these declines is thought to be a decrease in the proportion of the northwest European breeding population that migrates to Britain & Ireland, perhaps due to milder winters (Sauter 2010). The small 5% increase in range size in winter mostly reflects some expansion within upland areas.

Though numbers are higher in winter, the actual pattern of relative abundance is very similar to that in summer, as might be expected of such a generalist wetland species. The highest abundance coincides with more fertile lowland areas where aquatic productivity is highest. Northwest England, parts of eastern England and the Thames basin hold what appear to be the largest hotspots for the species.

Sponsored by Joan New

BREEDING DISTRIBUTION *2008–11*

POSSIBLE •	10%	3%	5%
PROBABLE •	20%	8%	11%
CONFIRMED ●	58%	83%	76%
TOTAL	88%	94%	92%

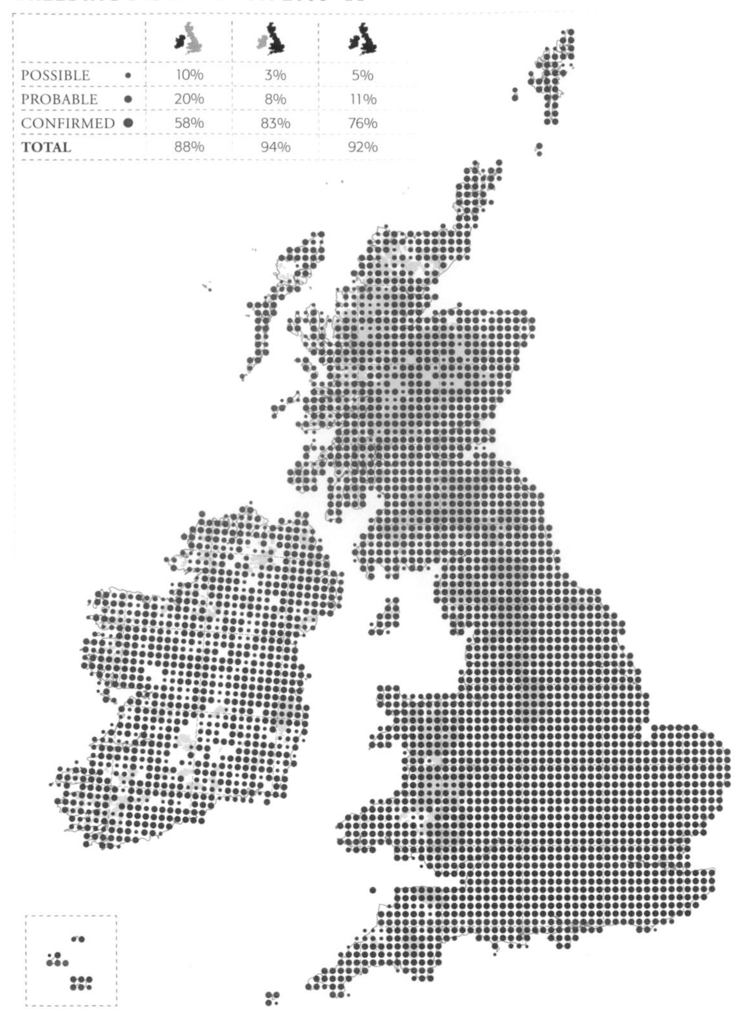

WINTER DISTRIBUTION *2007/08–2010/11*

PRESENT ●	86%	93%	91%

BREEDING DISTRIBUTION CHANGE *since 1968–72*

GAIN ▲	51	93	144
LOSS ▽	63	62	125
40yr CHANGE	-1%	+1%	+1%
20yr INDEX	+0.04	+0.02	

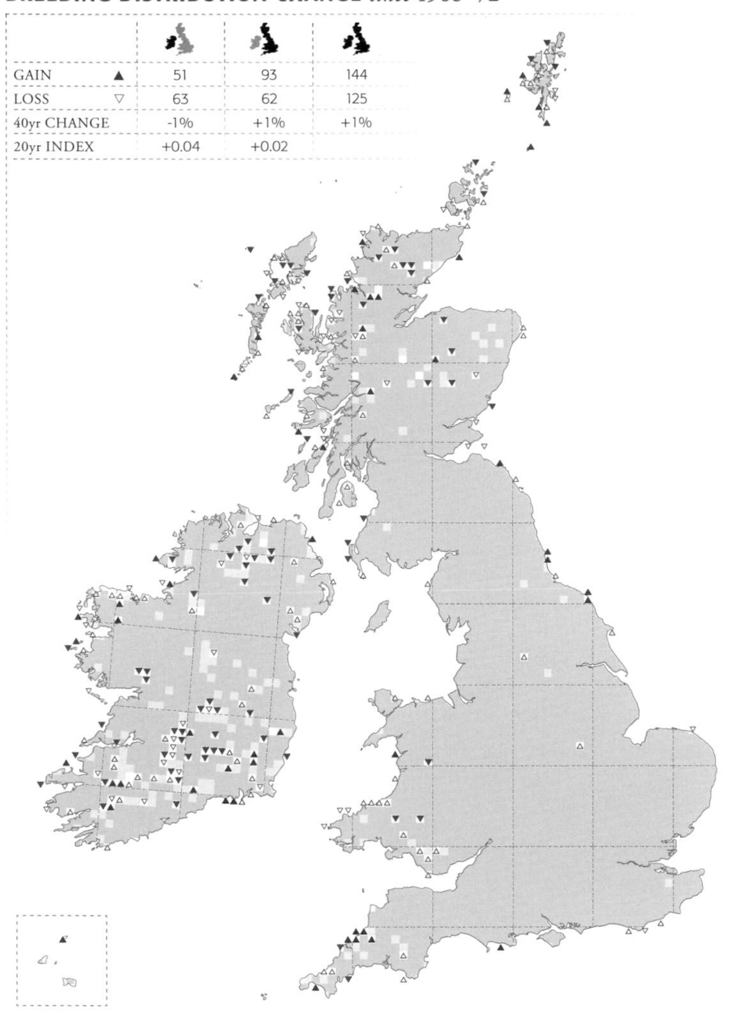

BREEDING RELATIVE ABUNDANCE *2008–11*

BREEDING RELATIVE ABUNDANCE CHANGE *since 1988–91*

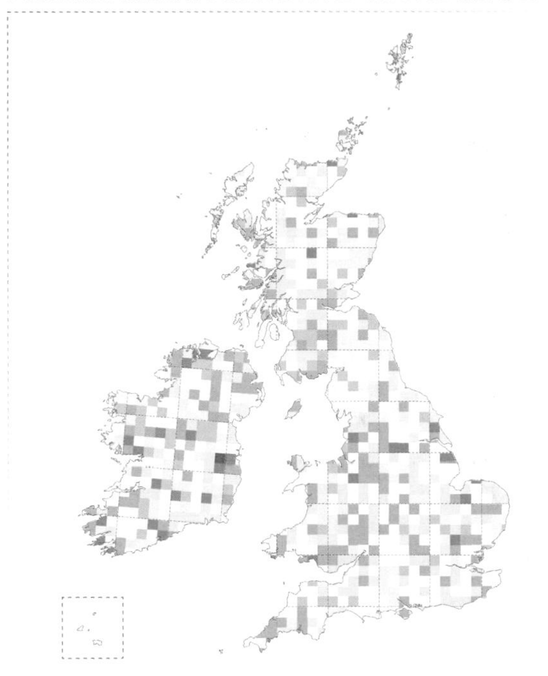

WINTER DISTRIBUTION CHANGE *since 1981–84*

GAIN ▲	93	170	263
LOSS ▼	79	52	131
30yr CHANGE	+2%	+5%	+4%

WINTER RELATIVE ABUNDANCE *2007/08–2010/11*

PINTAIL

Anas acuta

BTO CODE: PT

MICK DURHAM www.wildlife-photographer.net

THE BULK OF the European wintering population of this species is found in Britain, France and the Netherlands (Gilissen *et al.* 2002). Pintails are unusual among dabbling ducks in aggregating at relatively few sites in winter in Britain & Ireland. This behaviour, coupled with their ability to exploit temporary wetland habitats and respond quickly to adverse weather, as illustrated during recent cold winters (2009/10 and 2010/11), can have a profound effect on their occurrence and abundance.

The winter distribution map reveals preferences for grazing marshes, river floodplains, sheltered coasts and estuaries while the relative abundance map confirms the importance of estuaries throughout most of Britain & Ireland for this species. The Burry Inlet, south Wales, and Dee Estuary are the key sites in the UK, each holding on average over 4,000 birds (*WeBS Report 2011*). In Ireland, Lower Brosna Callows averages c.450 birds (*I-WeBS Report 2009*).

There has been a 34% increase in range size in Britain in winter since the *1981–84 Winter Atlas*, and notable gains include the Northern Isles and eastern fringes of the Scottish coast, with more extensive gains farther south into England and Wales. Meanwhile, there has been a long-term increase in numbers wintering in Britain since the early 1970s, yet a notable decline since the mid 2000s (*WeBS Report 2011*). Reasons for the decline are unclear, although an increase in the Netherlands in recent years may imply a shift in core wintering range (Hornman *et al.* 2011). In Ireland losses and gains are more evenly balanced, with a 4% decline in the number of occupied 10-km squares possibly reflecting here, as elsewhere, the changing availability of suitable wetland habitats.

The Pintail is a rare and localised breeder, with clusters of records only on Orkney, North Uist, Tiree, the East Anglian coast and the Ouse Washes and isolated records elsewhere. There has been a 32% reduction since the *1968–72 Breeding Atlas* in the range size occupied by birds showing breeding evidence, while no *confirmed* breeding records were reported in Ireland. The number of *confirmed* records overall dropped from 35 to 15. As with many duck species, an unknown proportion of records of *possible* and *probable* breeding are likely to relate to late wintering or passage birds remaining in April and into early May.

WINTER DISTRIBUTION *2007/08–2010/11*

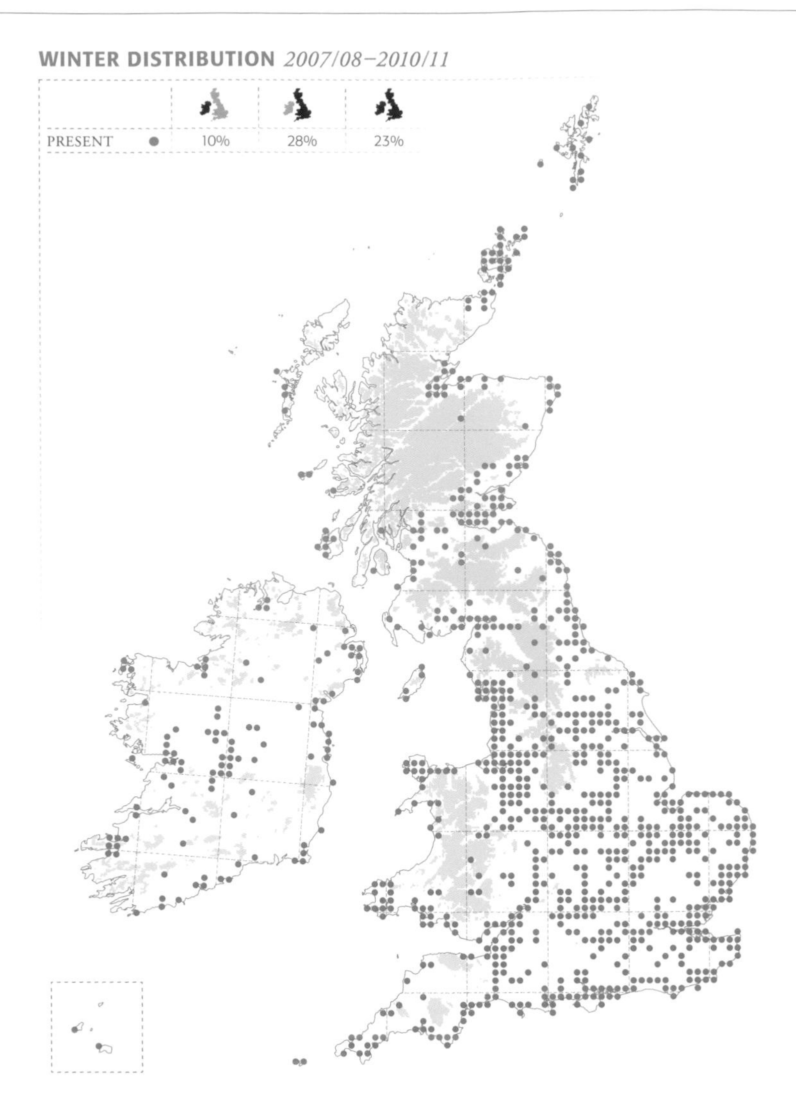

BREEDING DISTRIBUTION *2008–11*

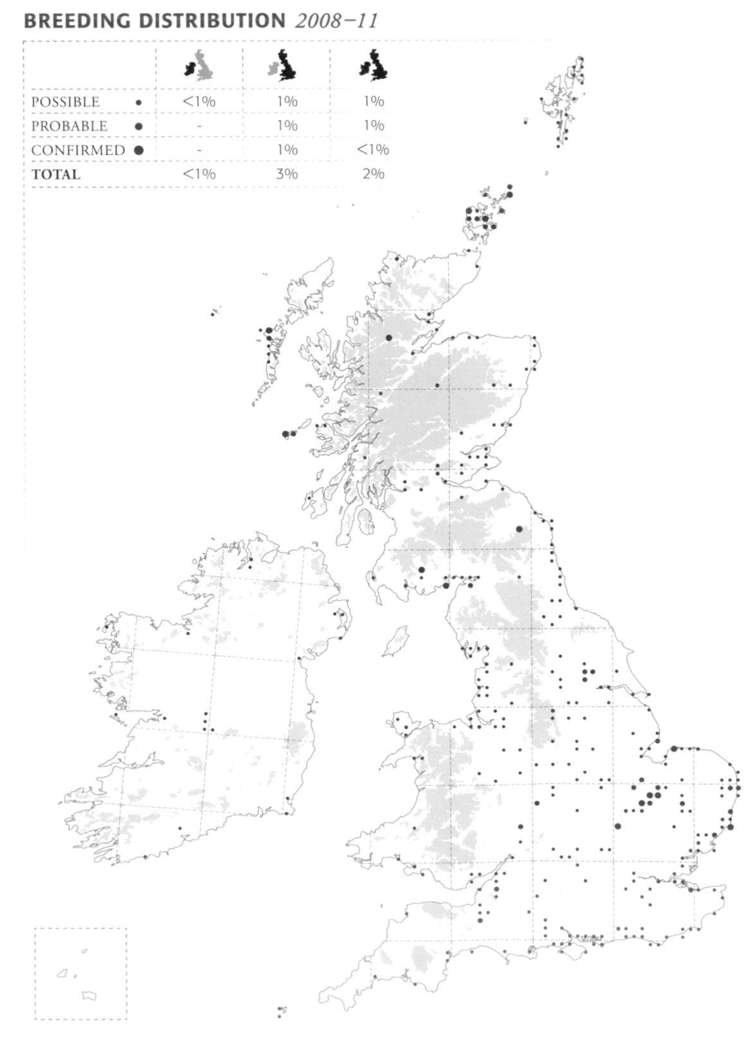

WINTER DISTRIBUTION CHANGE *since 1981–84*

GAIN ▲	47	352	399
LOSS ▼	51	161	212
30yr CHANGE	-4%	+34%	+28%

WINTER RELATIVE ABUNDANCE *2007/08–2010/11*

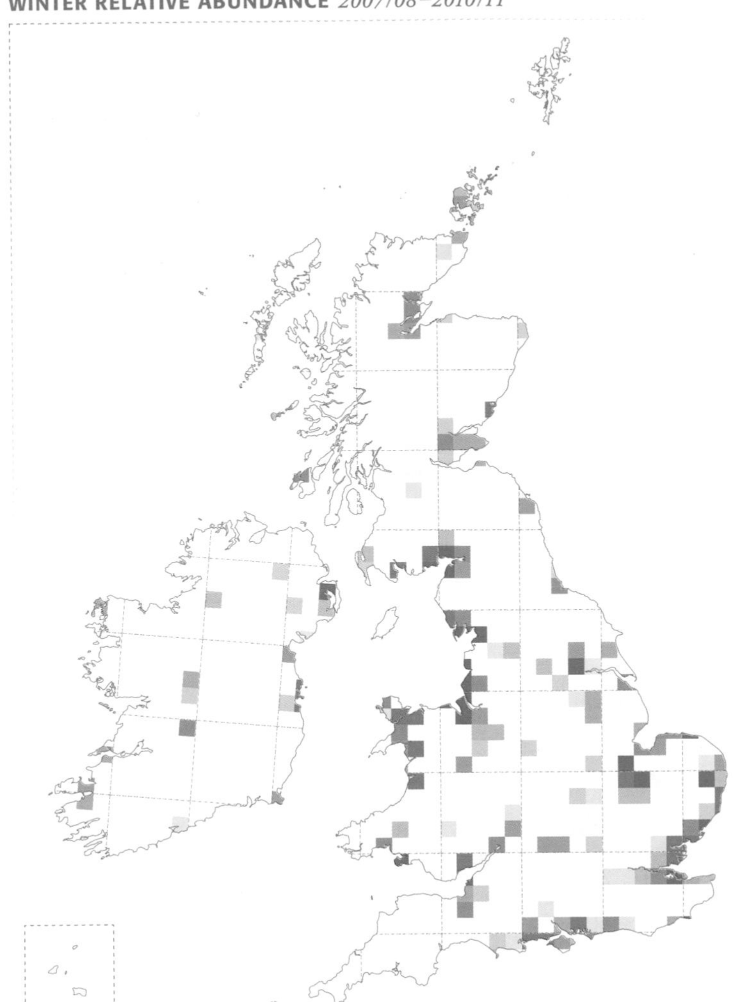

BREEDING DISTRIBUTION CHANGE *since 1968–72*

GAIN ▲	1	43	44
LOSS ▽	12	63	75
40yr CHANGE	-92%	-23%	-32%
20yr INDEX	-	-0.41	

GARGANEY

Anas querquedula

BTO CODE: GY

SIMON JOHNSON simon@purpleheron.demon.co.uk

BREEDING DISTRIBUTION *2008–11*

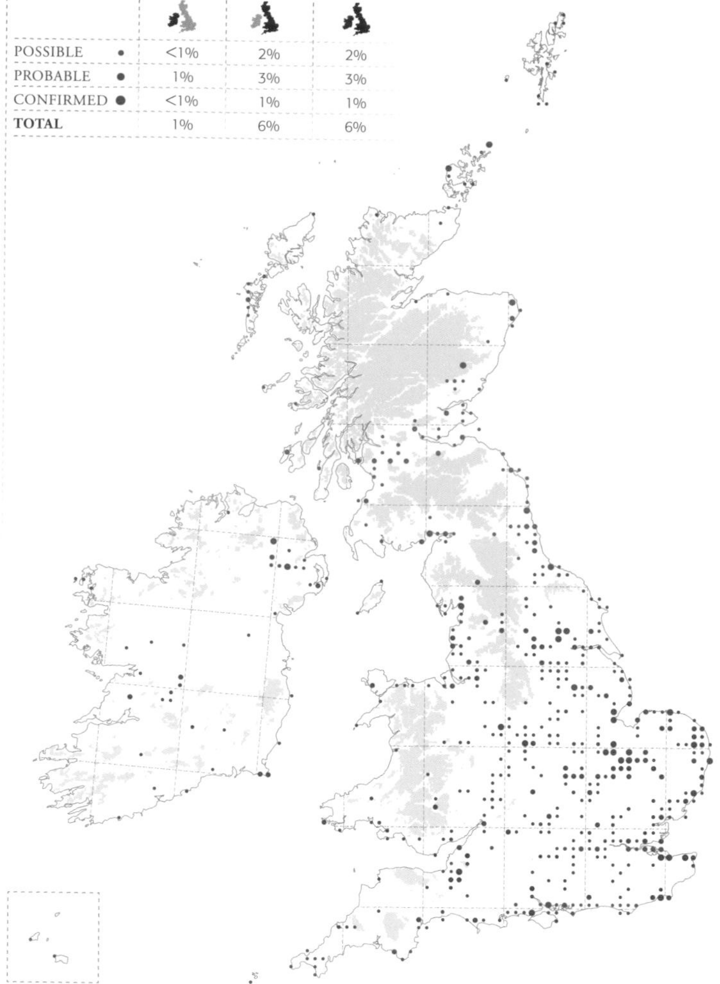

WINTER DISTRIBUTION *2007/08–2010/11*

WIDELY RECORDED IN spring on lakes, marshlands and wet meadows, the Garganey is surely among the most difficult ducks to census as a breeding bird. The breeding distribution map shows that records with breeding evidence were received from 199 10-km squares in Britain and 13 in Ireland. Favoured areas are around the Ouse Washes and nearby fenland sites, Norfolk Broads, Somerset Levels, and English northeast and northwest coastlines. Furthermore, migrants and other non-breeding records were reported from almost 300 10-km squares. The Garganey is a rare breeding bird in Ireland, so *confirmed* breeding records in 2008 and 2009 in Co. Antrim are notable (Hillis 2010). Migrants were reported from 26 10-km squares in Ireland.

One of the problems when recording breeding evidence is that many records are of pairs in early spring. Pair formation occurs mainly in the wintering areas and birds migrate together to their breeding grounds (*BWP*), so the presence of a pair at a site in early spring does not necessarily indicate breeding. It is likely that a number of the *probable* breeding records relate to pairs on passage. Confirming breeding is equally tricky, and although the number of occupied 10-km squares increased by 50% since the *1968–72 Breeding Atlas*, the number of 10-km squares with *confirmed* breeding fell from 46 to 35. Holling *et al.* (2010b) surmised that during the 1990s the maximum number of pairs (on average 121 per annum) was higher than in the 2000s, but the current trend shows numbers rising again. This contrasts with a widespread decline in the large breeding population in eastern and northern Europe (Snow & Perrins 1998; Wetlands International 2013).

There are winter records from 55 10-km squares in Britain & Ireland, though this greatly exaggerates the number in any one winter. Most are from scattered locations in southern and central England, where they are found amongst other dabbling ducks. About 30% of these 10-km squares were occupied in early November only and so most likely by late-departing birds, while around 25% of 10-km squares were occupied in February only, suggesting that the birds had been early returning migrants.

In memory of Gerald Percy and Lord William Percy

AMERICAN WIGEON

Anas americana

BTO CODE: AW

GEOFF ATHEY www.picasaweb.google.com/geoff.athey

THE NUMBER OF American Wigeons wintering in Britain & Ireland has increased steadily since the late 1950s, with an annual mean of 20 birds reported in the period 2000–03 in Britain (Fraser & Rogers 2006a).

During the winter fieldwork period, records came from 33 10-km squares in Britain and 10 in Ireland; a large increase from the 13 occupied during the *1981–84 Winter Atlas*. Except for two together in Shetland in November–December 2007, all were singles. They are found amongst wintering Wigeons, and some return to the same location in consecutive winters. There were records from 24 10-km squares during the breeding season, including birds lingering through to July.

GREEN-WINGED TEAL

Anas carolinensis

BTO CODE: TA

GRAHAM CATLEY http://pewit.blogspot.co.uk

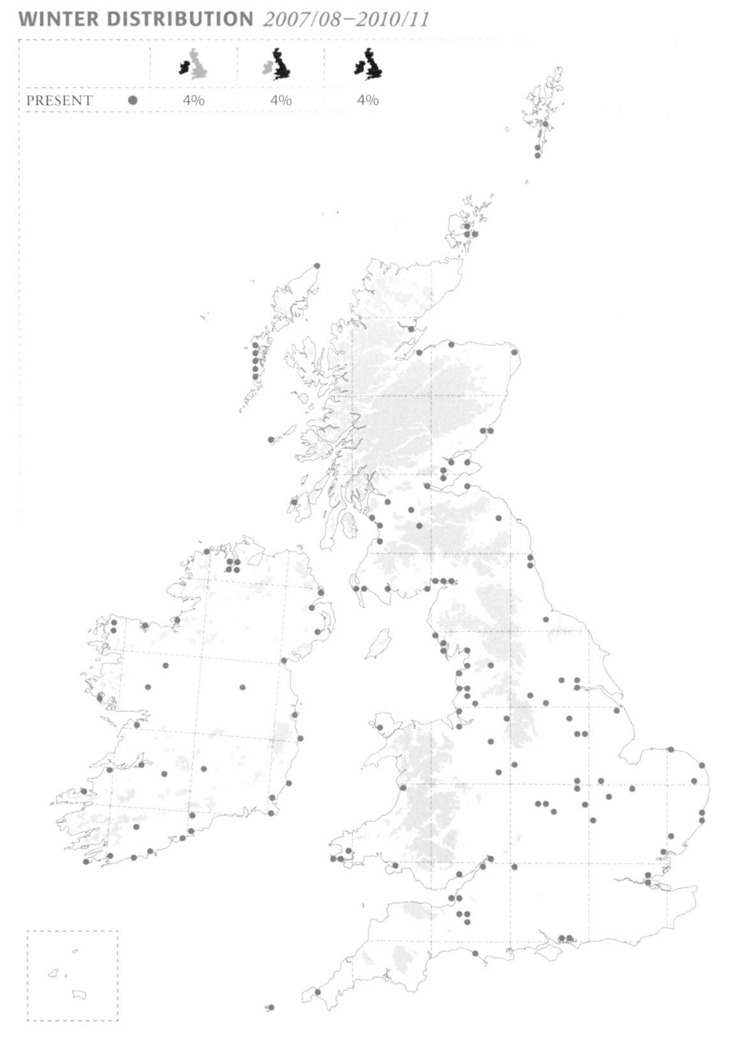

GREEN-WINGED TEALS are the commonest of the Nearctic ducks to be recorded in Britain & Ireland, with an annual mean of 41 birds in Britain in the period 2000–03 (Fraser & Rogers 2006a). The species was recorded in 108 10-km squares in Britain and 36 10-km squares in Ireland, from freshwater lakes, brackish marshes and estuaries. A proportion of records include returning birds from previous winters, either to the same site or nearby. The species is likely to have been significantly under-reported because females and juveniles are extremely difficult to identify in the field. In the breeding season, records of late-wintering or summering birds came from 55 10-km squares in Britain and eight 10-km squares in Ireland.

SHOVELER

Anas clypeata

BTO CODE: SV

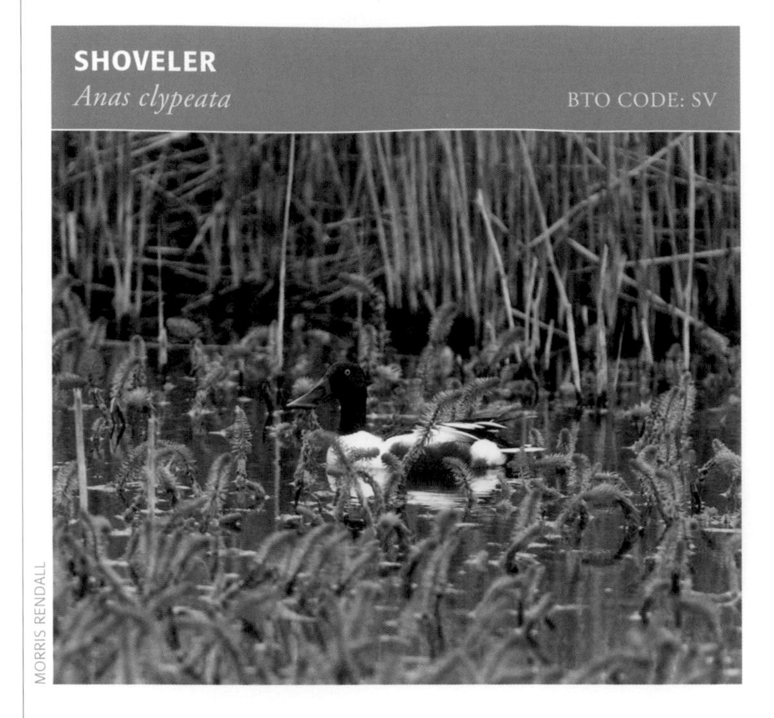

MORRIS RENDALL

THE WINTER DISTRIBUTION map shows that Shovelers are present mainly in lowland areas, favouring a range of inland wetland sites and some coastal estuaries. They are most widely distributed and abundant in parts of southern Britain and in mid-western Ireland, as well as in parts of the Outer Hebrides and Northern Isles. They are largely absent across much of Scotland, except in the central lowlands and on base-rich lochs in the Uists and Orkney. In England, they are widespread, and hotspots are found along major waterways such as the Severn, Trent, Thames and Great Ouse. In Ireland and Wales, discrete distributions are found in low-lying and coastal areas.

The overall range size has increased by 36% since the *1981–84 Winter Atlas*, and the majority of these gains have been in Britain, most notably in Orkney. This corresponds with a 70% increase in numbers in the UK reported by winter waterbird monitoring schemes between 1983/84 and 2008/09 (*SUKB 2012*). While there has been a smaller increase in range in Ireland, Shoveler numbers have fluctuated widely since the 1990s, showing a substantial decline in Northern Ireland (*WeBS Report 2011*) and an increase in the Republic of Ireland (*I-WeBS Report 2009*).

Shovelers are relatively scarce and local breeders in Britain & Ireland. The breeding distribution and abundance maps show a dispersed population with records concentrated along low-lying floodplains and associated wetlands. Overall there has been little change in the number of occupied 10-km squares since the *1968–72 Breeding Atlas*. Gains in many areas are balanced by losses in adjacent 10-km squares, probably reflecting the changing availability of suitable breeding wetlands. Past losses have been reversed somewhat in Scotland and Ireland. In Scotland, gains are evident in the Outer Hebrides and in Orkney. RBBP suggests that this species is well covered and reports that Cambridgeshire, Norfolk and Yorkshire are the strongholds, with over 100 pairs in each (*RBBP 2010*). The Lower Derwent Valley, Yorkshire, and Ouse and Nene Washes are the key breeding sites (*Birds in England*). Numbers and breeding success fluctuate markedly with the extent and timing of spring flooding. In Ireland, IRBBP reports that a rather small proportion of breeding events are detected (*IRBBP 2010*).

Sponsored by Philip Douglas, Douglas Employment Law

BREEDING DISTRIBUTION *2008–11*

POSSIBLE •	1%	3%	3%
PROBABLE •	2%	8%	6%
CONFIRMED •	1%	7%	5%
TOTAL	4%	18%	14%

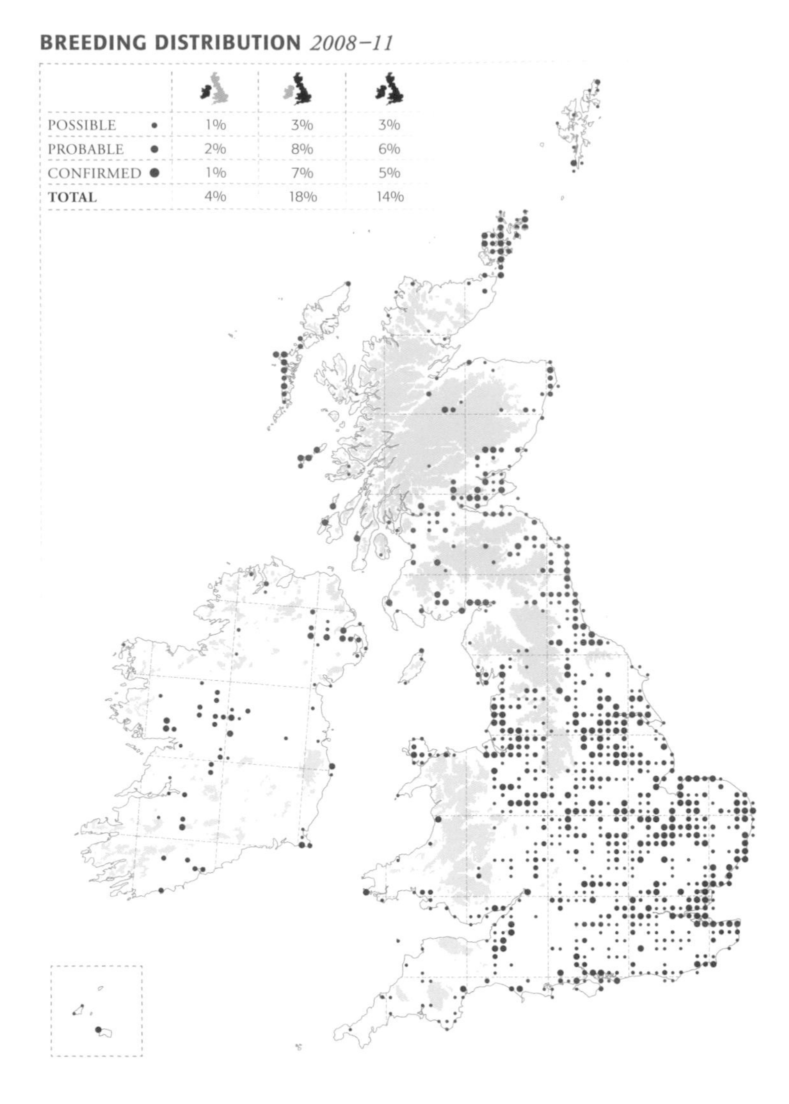

WINTER DISTRIBUTION *2007/08–2010/11*

PRESENT •	20%	37%	33%

BREEDING DISTRIBUTION CHANGE *since 1968–72*

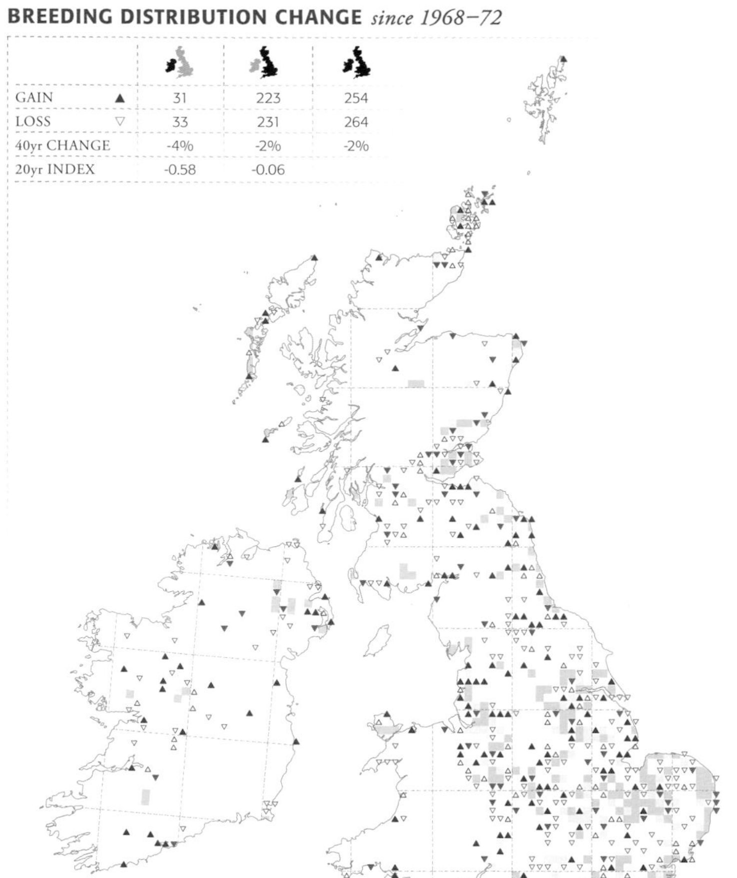

GAIN ▲	31	223	254
LOSS ▽	33	231	264
40yr CHANGE	-4%	-2%	-2%
20yr INDEX	-0.58	-0.06	

BREEDING RELATIVE ABUNDANCE *2008–11*

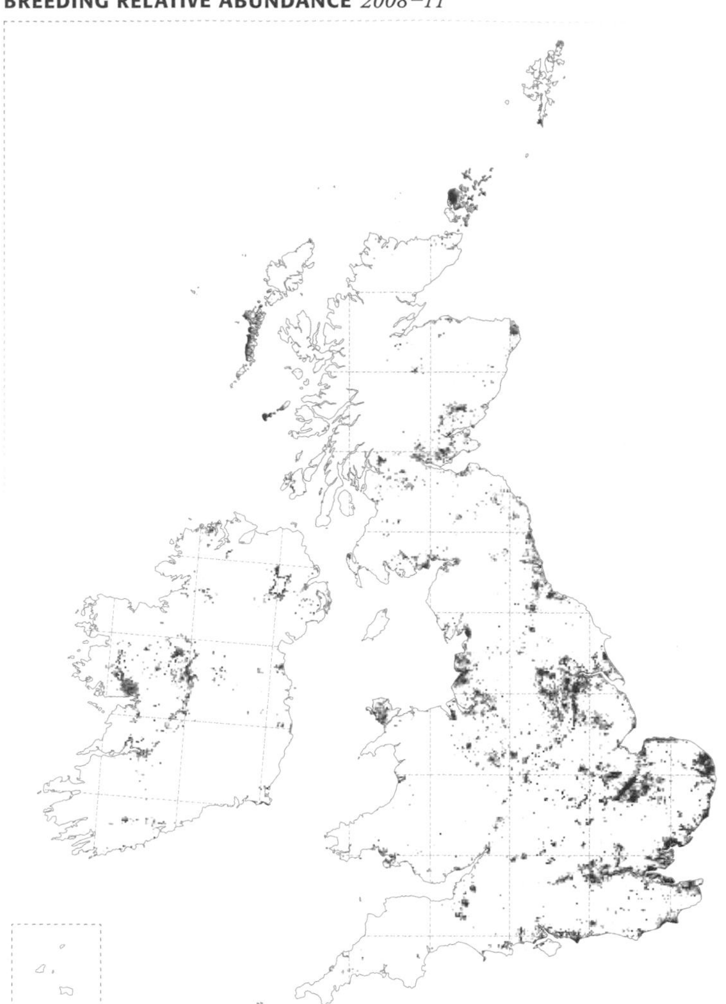

WINTER DISTRIBUTION CHANGE *since 1981–84*

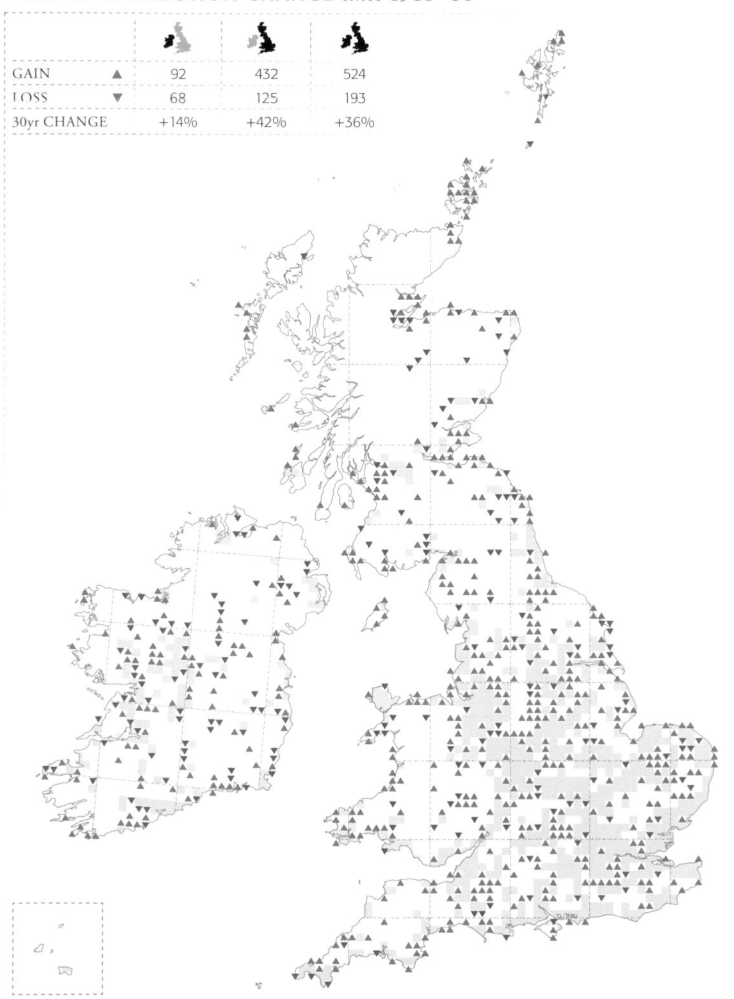

GAIN ▲	92	432	524
LOSS ▼	68	125	193
30yr CHANGE	+14%	+42%	+36%

WINTER RELATIVE ABUNDANCE *2007/08–2010/11*

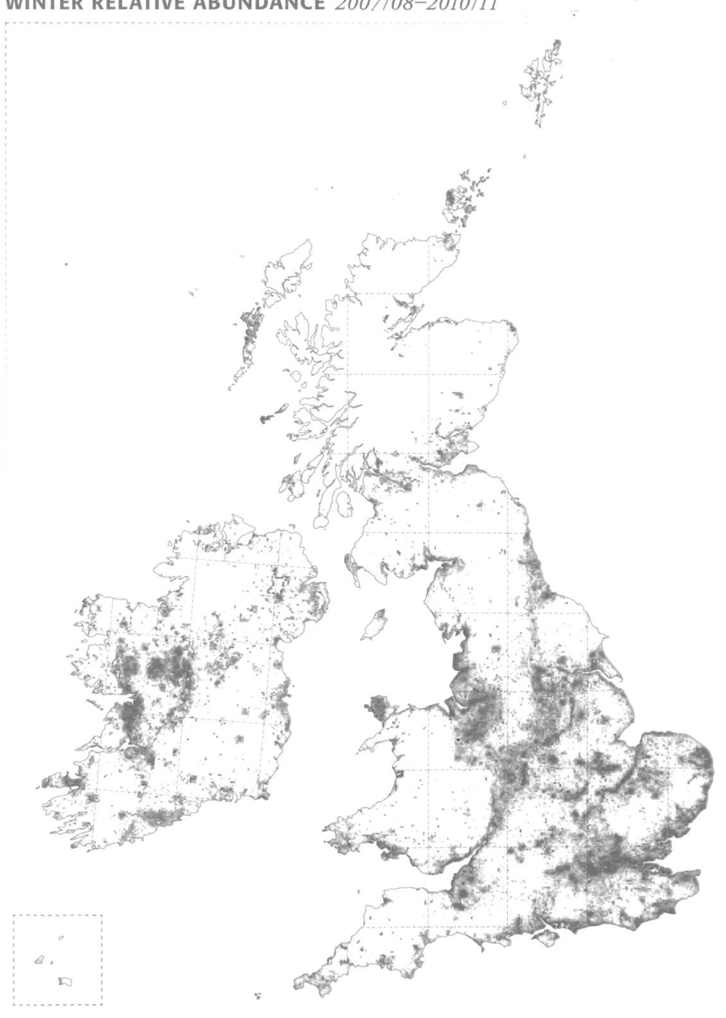

RED-CRESTED POCHARD

Netta rufina BTO CODE: RQ

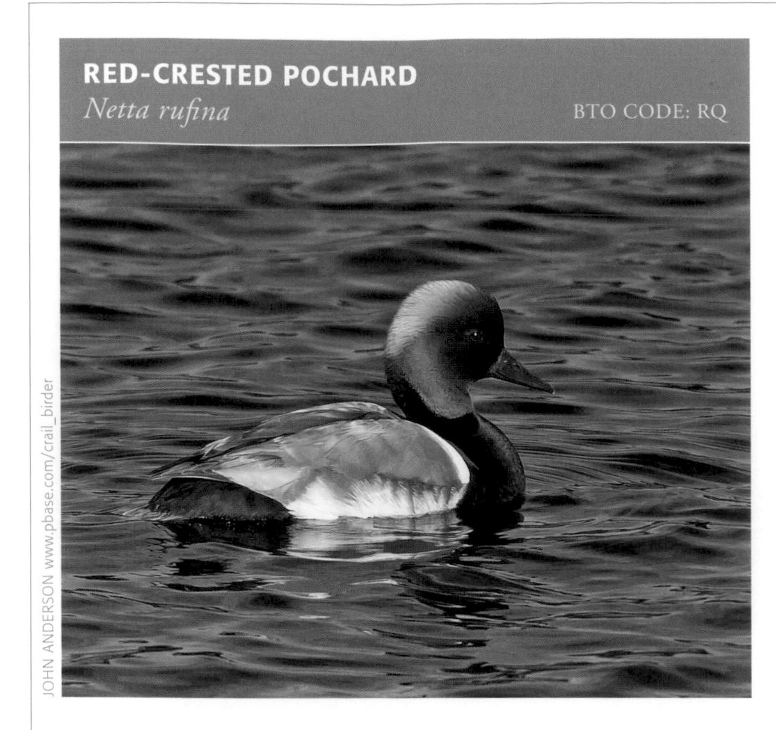

JOHN ANDERSON www.pbase.com/crail_birder

SINCE THE FIRST British breeding record, in Lincolnshire in 1937, numbers of Red-crested Pochards have increased and there is now a naturalised, self-sustaining population centred on the Cotswold Water Park on the Gloucestershire/Wiltshire border, where 20 pairs bred in 2008 (Lever 2009; Holling *et al.* 2011a). The population in Britain & Ireland largely originates from escapes or releases of individuals from waterfowl collections; it is likely that the birds at Cotswold Water Park originated from nearby Slimbridge (Baatsen 1990). Natural vagrancy from continental Europe no doubt occurs but the true status is masked by introduced birds (Parkin & Knox 2010). The species breeds patchily in Europe and east into central Asia and there have been years when influxes have been recorded in Britain (Pyman 1959; Harrop 1991).

In the breeding season, records with breeding evidence came from a much more restricted area, mainly in central and southern England and East Anglia, than in the winter. The breeding populations centred on the Cotswold Water Park, the central and lower Thames Valley and the East Midlands are evident. Red-crested Pochards were present in a further 100 10-km squares in Britain during the breeding season but without showing any breeding evidence, suggesting that there is further scope for an expansion in their breeding range. There were no records from Ireland in the breeding season.

In winter, Red-crested Pochards are widely distributed in England south of a line from Morecambe Bay to Teesside, where they are found mainly in freshwater habitats. There were a small number of occupied 10-km squares in south Wales and Scotland and just three in Ireland. In winter 2010/11 the Cotswold Water Park held c.400 birds and all other sites supported fewer than 100 individuals (*WeBS Report 2011*).

The large range expansion in Britain over the last 40 years is shown on the change maps for winter and the breeding season. Interestingly, the species has shown a parallel increase in numbers in central Europe following a range shift from the western Mediterranean to Switzerland and Germany and the use of a greater number of sites (Keller 2000, 2006), so it is conceivable that some increases in Britain may include birds of wild origin.

BREEDING DISTRIBUTION *2008–11*

POSSIBLE •	-	<1%	<1%
PROBABLE •	-	1%	<1%
CONFIRMED •	-	1%	1%
TOTAL	-	2%	1%

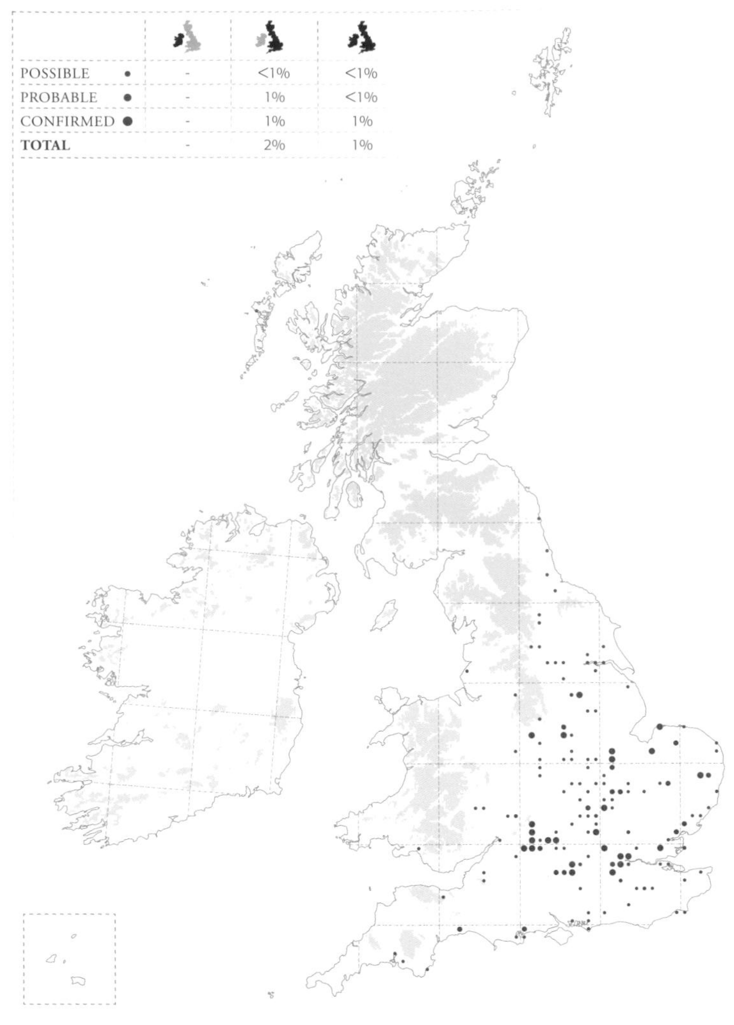

WINTER DISTRIBUTION *2007/08–2010/11*

PRESENT •	<1%	10%	7%

BREEDING DISTRIBUTION CHANGE *since 1968–72*

GAIN ▲	-	48	48
LOSS ▽	-	2	2
40yr CHANGE	-	+1150%	+1150%
20yr INDEX	-	+0.43	

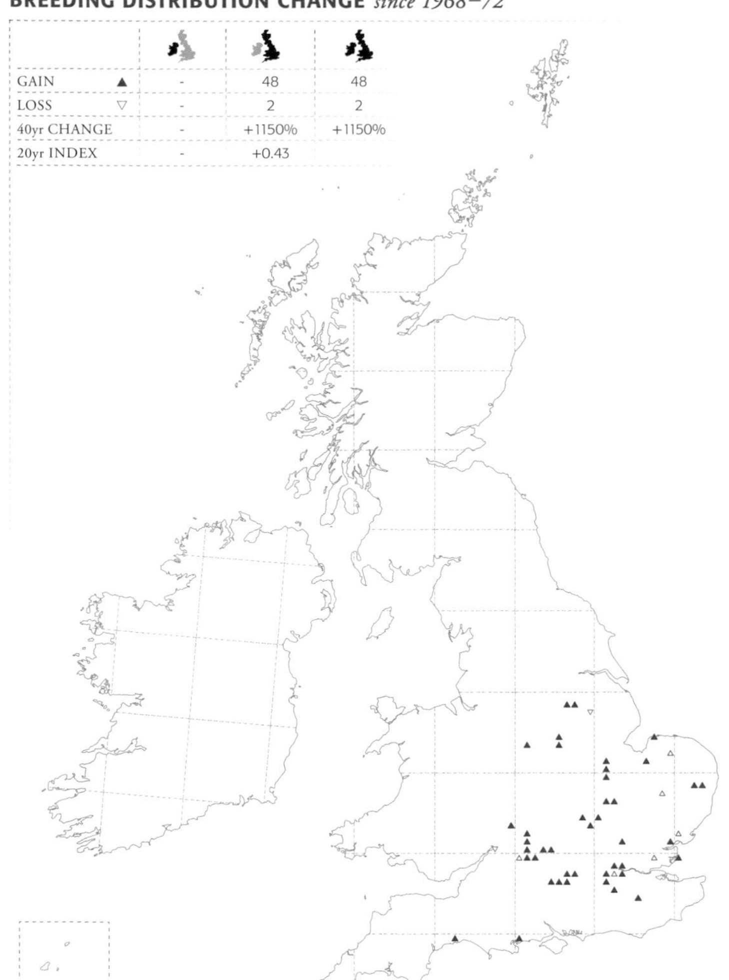

BREEDING RELATIVE ABUNDANCE *2008–11*

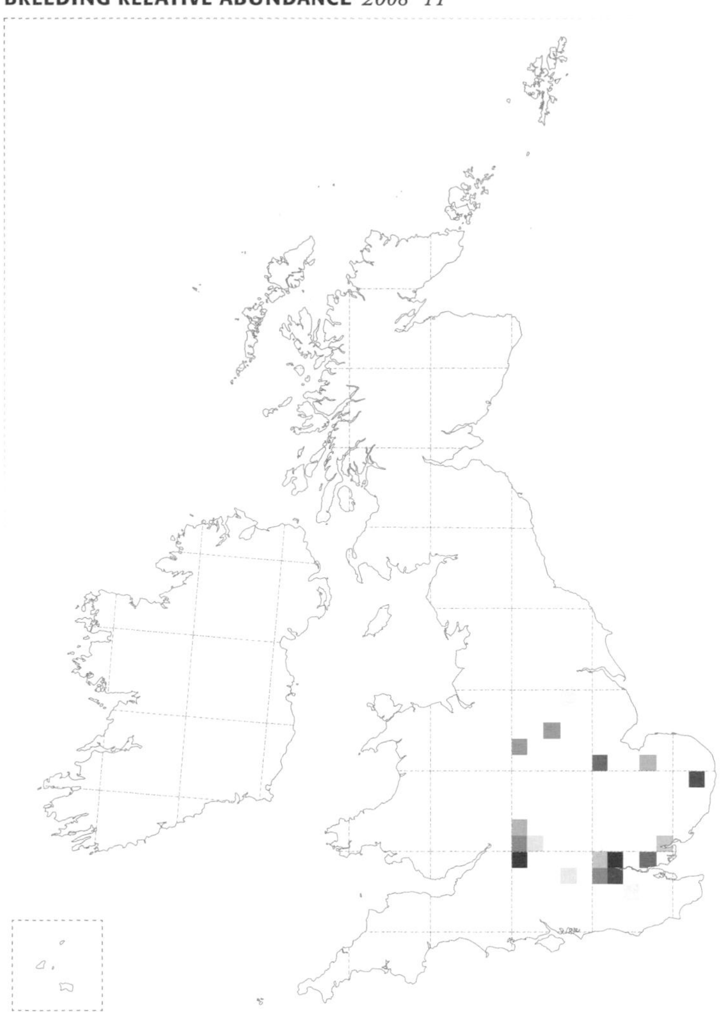

WINTER DISTRIBUTION CHANGE *since 1981–84*

GAIN ▲	1	223	224
LOSS ▼	6	39	45
30yr CHANGE	-71%	+245%	+218%

WINTER RELATIVE ABUNDANCE *2007/08–2010/11*

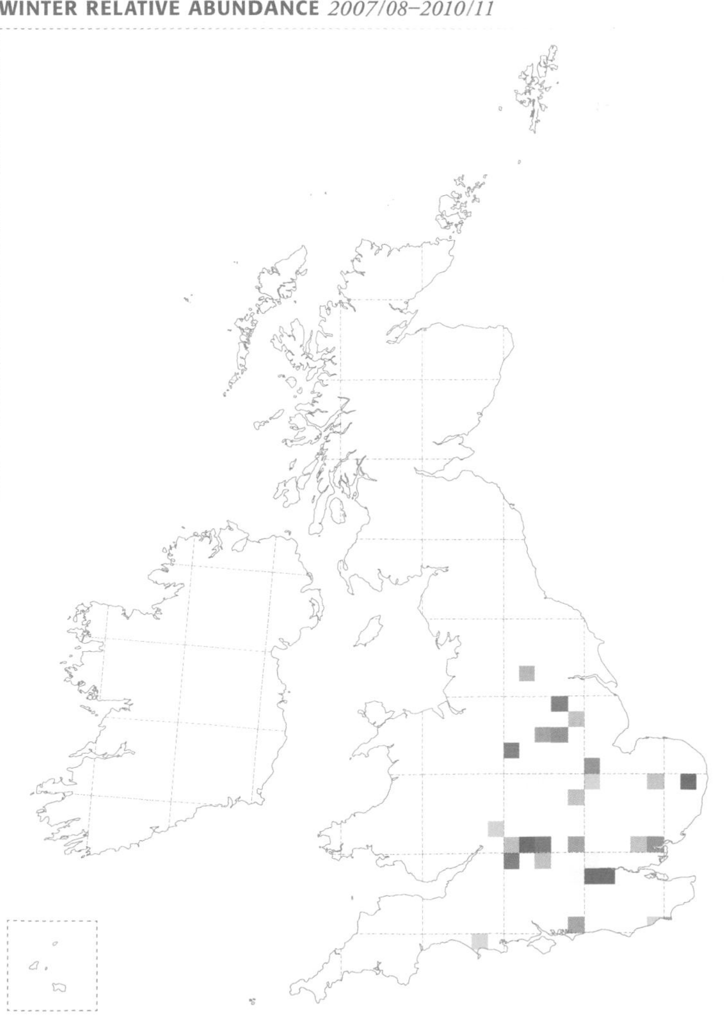

FERRUGINOUS DUCK

Aythya nyroca

BTO CODE: FD

JIM ALMOND www.shropshirebirder.co.uk

THIS SPECIES IS primarily a winter visitor, with most records in the period October–January, reflecting dispersal from the breeding grounds in central Europe east to central Asia (Vinicombe 2000). It is also widely kept in captivity and some of the records in Britain & Ireland are likely to be escapes from waterfowl collections. Winter records in Britain were all south of a line from Morecambe Bay to the Humber with clusters of occupied 10-km squares around Chew Valley Lake in Avon, the Thames Valley, Norfolk Broads and the East Midlands. Ferruginous Ducks generally associate with Pochards, with which they are likely to make local movements, so these clusters of sightings may relate to single birds visiting nearby sites within and across winters. In Ireland, winter records came from three 10-km squares to the southeast of Lough Neagh and one 10-km square in southeast Wexford. The total of 39 occupied 10-km squares in Britain suggests a significant range expansion since the *1981–84 Winter Atlas*, when there were records from 21 10-km squares, all in Britain.

Most breeding-season records within the Atlas fieldwork period were of birds classified as *non-breeding* and these were scattered across 12 10-km squares in England, mostly in the south, and three in Scotland. At one 10-km square in Norfolk, a drake was seen displaying to female Pochards in June 2011 and this was recorded as evidence of *possible* breeding. This observation follows a series of records at Chew Valley Lake in Avon, where a pair was present during 2003–07 (Davis & Vinicombe 2011); breeding may have occurred in 2003 and 2004, when the male was seen displaying with a female Ferruginous Duck, and copulation was noted, although no nest or young were seen. Then in 2006 the pair was again seen displaying, and in the autumn of that year a juvenile male, of unknown origin, was also recorded. The original male apparently returned in 2008 but no breeding evidence was noted during the Atlas period. There have been rapid declines in the European breeding population largely due to habitat degradation (BirdLife International 2013).

BREEDING DISTRIBUTION *2008–11*

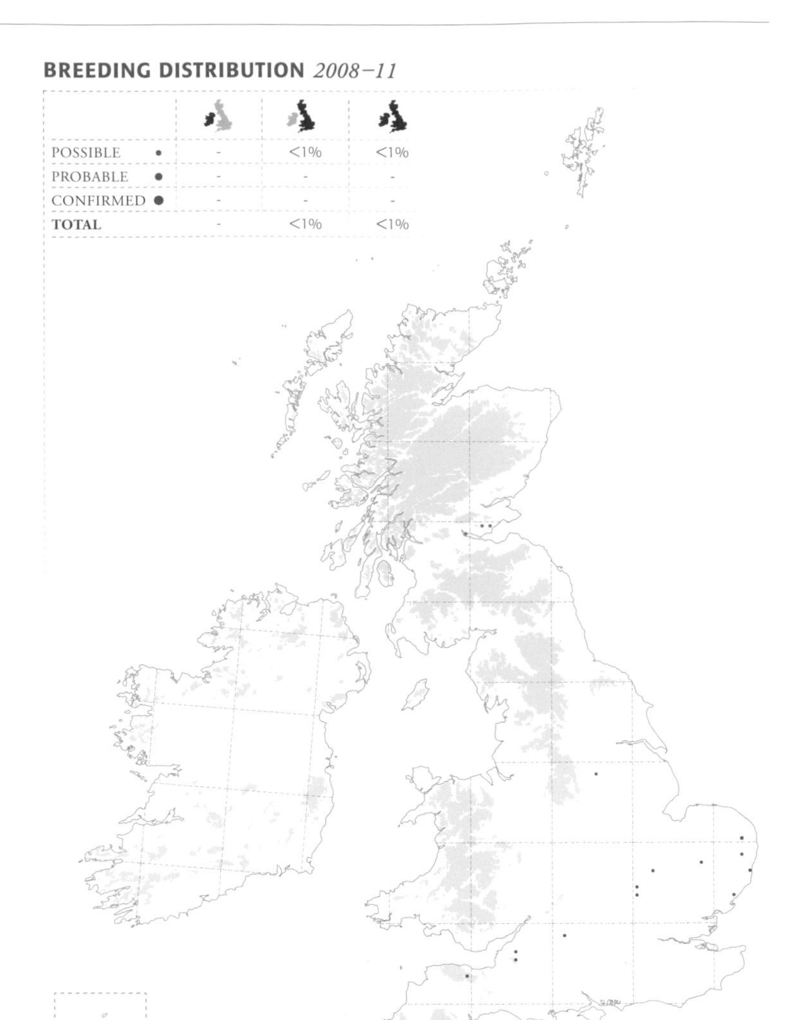

WINTER DISTRIBUTION *2007/08–2010/11*

RING-NECKED DUCK
Aythya collaris

BTO CODE: NG

DEREK CHARLES

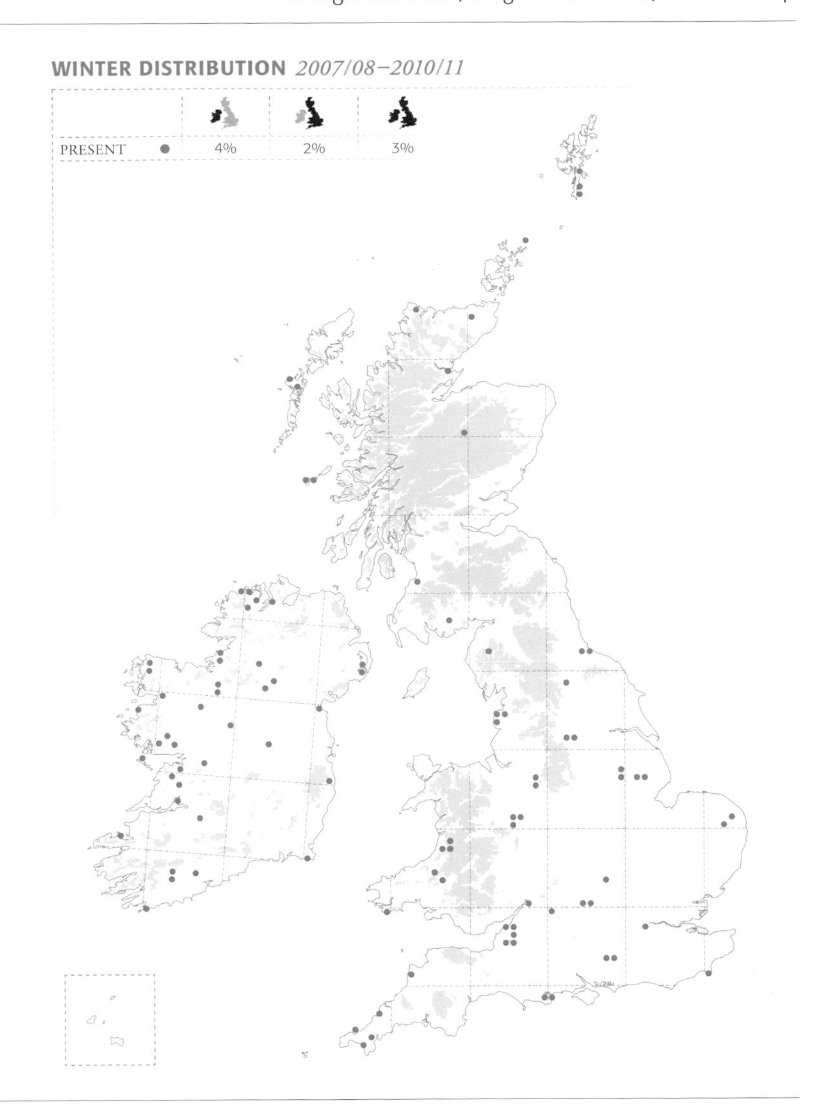

SMALL NUMBERS OF this Nearctic duck occur in Britain & Ireland annually. Many are long-staying individuals, returning to favoured freshwater wetlands in consecutive winters. In winter, 100 10-km squares were occupied in Britain & Ireland, which is a marked increase on the 16 occupied in the *1981–84 Winter Atlas*. The rise in numbers may be partly due to improved detection, although it probably also reflects a genuine increase in vagrancy. There was one record of *possible* breeding, in May 2010, relating to a male paired with and defending a female Tufted Duck in Fife (Fife Bird Club 2012). Elsewhere there were 45 10-km squares with *non-breeding* birds, most of which had departed by May.

LESSER SCAUP
Aythya affinis

BTO CODE: AY

JOHN ANDERSON www.pbase.com/crail_birder

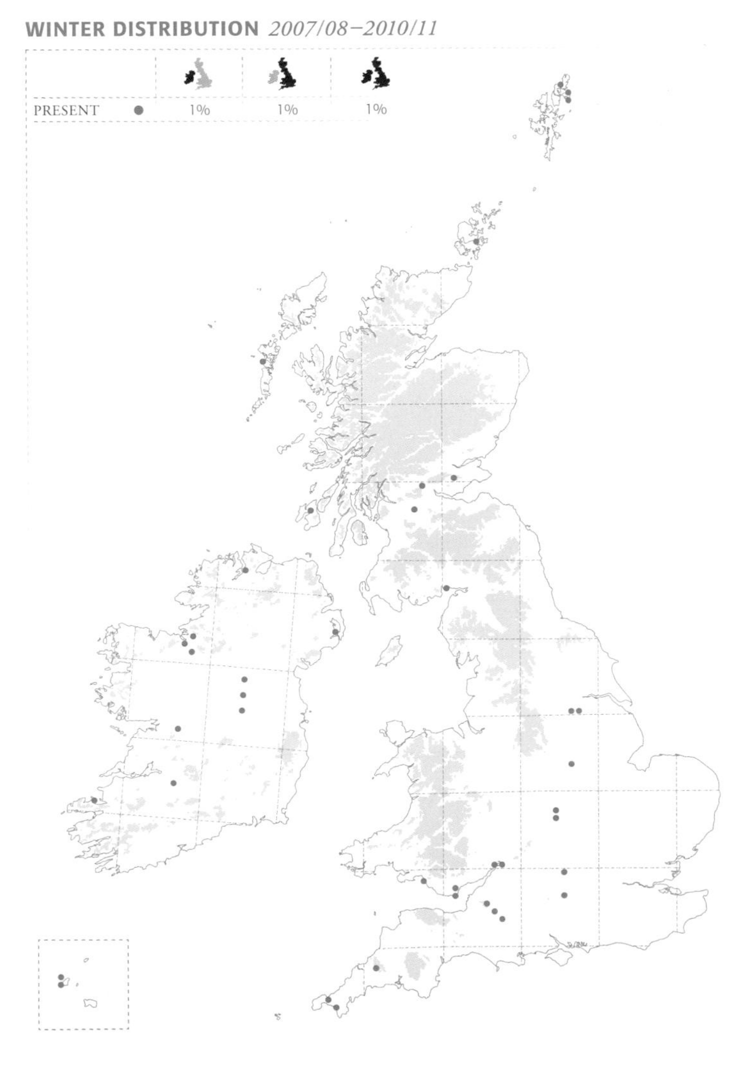

THE LESSER SCAUP is a scarce winter visitor and numbers recorded in Britain & Ireland have increased since the *1981–84 Winter Atlas*, when none were recorded. This is at least in part due to better awareness of the key identification features. There is little pattern to the scatter of 41 10-km squares occupied in winter; single birds of this species usually associate with other *Aythya* ducks, and may move around local waterbodies during the winter. Some records no doubt relate to returning individuals, though the arrival of first-winter birds indicates that new individuals continue to cross the Atlantic. There were records from 17 10-km squares during the breeding season, all in Britain: birds were seen mostly in April but a small number remained into May and June.

POCHARD

Aythya ferina

BTO CODE: PO

STEVE CARTER www.colnevalleybirding.co.uk

THE WINTER DISTRIBUTION of the Pochard is concentrated in lowland areas, where waterbodies have high nutrient status, and the species is generally absent from the acidic, nutrient-poor waterbodies of the uplands of Scotland, Ireland and Wales. There has been a 21% contraction in winter range in Britain & Ireland since the *1981–84 Winter Atlas*. The loss is most prominent in Ireland, where there were 49% fewer 10-km squares occupied, and is also reflected in a long-term decline in the wintering population, driven by a large-scale decline at the key wintering sites, Loughs Neagh and Beg (Allen *et al.* 2004; *WeBS Report 2011*). The reasons for this decline in Britain & Ireland are still unclear. Europe-wide changes such as 'short-stopping' relating to milder winters farther east may be partly responsible and disturbance or changes in food availability through eutrophication may be implicated (Maclean *et al.* 2006). In addition to the recent loss of around 20,000 birds from Lough Neagh, there has been a low-level decline in many other areas and this is reflected in the lower occupancy recorded by the Atlas.

Rich lakes are also preferred for breeding and the map shows the concentration of breeding birds in eastern lowland areas of England and a very sparse distribution farther north and west in Scotland, Wales and Ireland. The Pochard has also declined as a breeding species, with a 39% range contraction since the *1968–72 Breeding Atlas*, with losses occurring in both Britain and Ireland. Despite these losses there are local areas of stability and some recent scattered gains. This may be due to site-level changes, better recording or, perhaps most significantly, the creation of many new wetlands in England in particular. Some apparent gains could also involve late-departing winter visitors falsely recorded in the *possible* breeding category. Although the Atlas has recorded a range contraction, over the last 10 years the mean number of pairs in the UK reported to RBBP has increased, with the change in status largely driven by increasing numbers in England (Holling *et al.* 2011b).

In memory of Christine Simmons 1940–2009

BREEDING DISTRIBUTION *2008–11*

POSSIBLE •	1%	3%	3%
PROBABLE •	<1%	3%	2%
CONFIRMED •	1%	5%	4%
TOTAL	2%	11%	9%

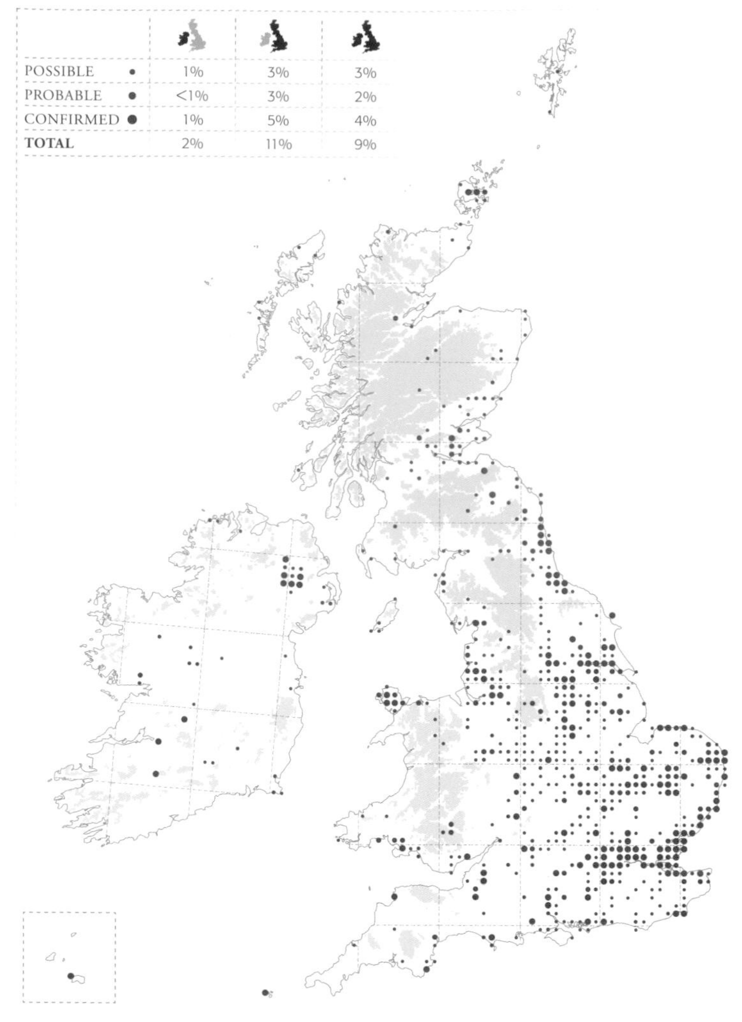

WINTER DISTRIBUTION *2007/08–2010/11*

PRESENT •	20%	44%	37%

BREEDING DISTRIBUTION CHANGE *since 1968–72*

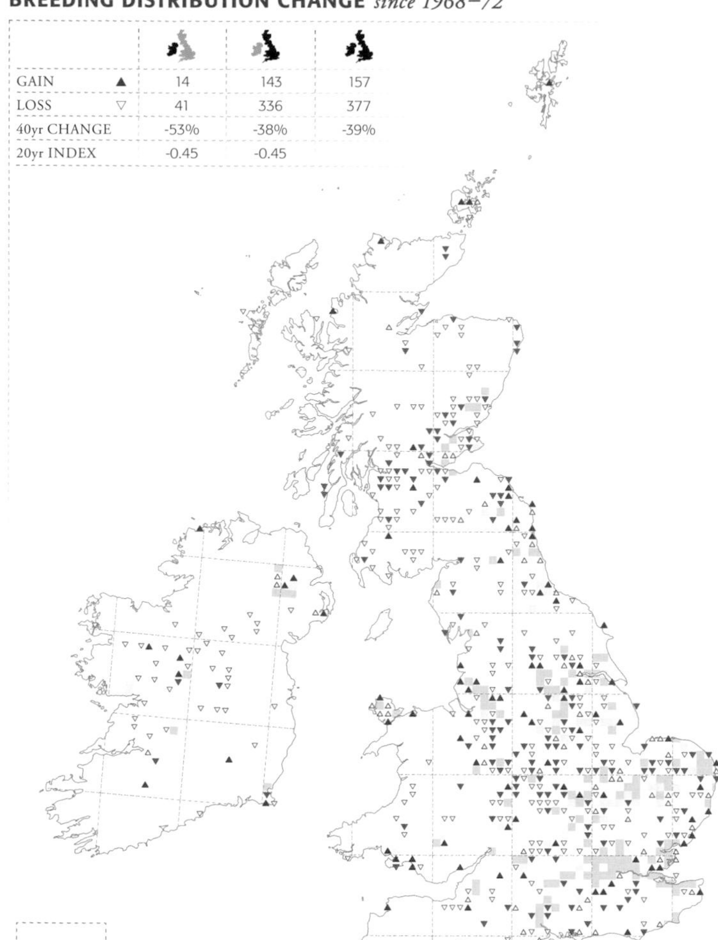

GAIN ▲	14	143	157
LOSS ▽	41	336	377
40yr CHANGE	-53%	-38%	-39%
20yr INDEX	-0.45	-0.45	

BREEDING RELATIVE ABUNDANCE *2008–11*

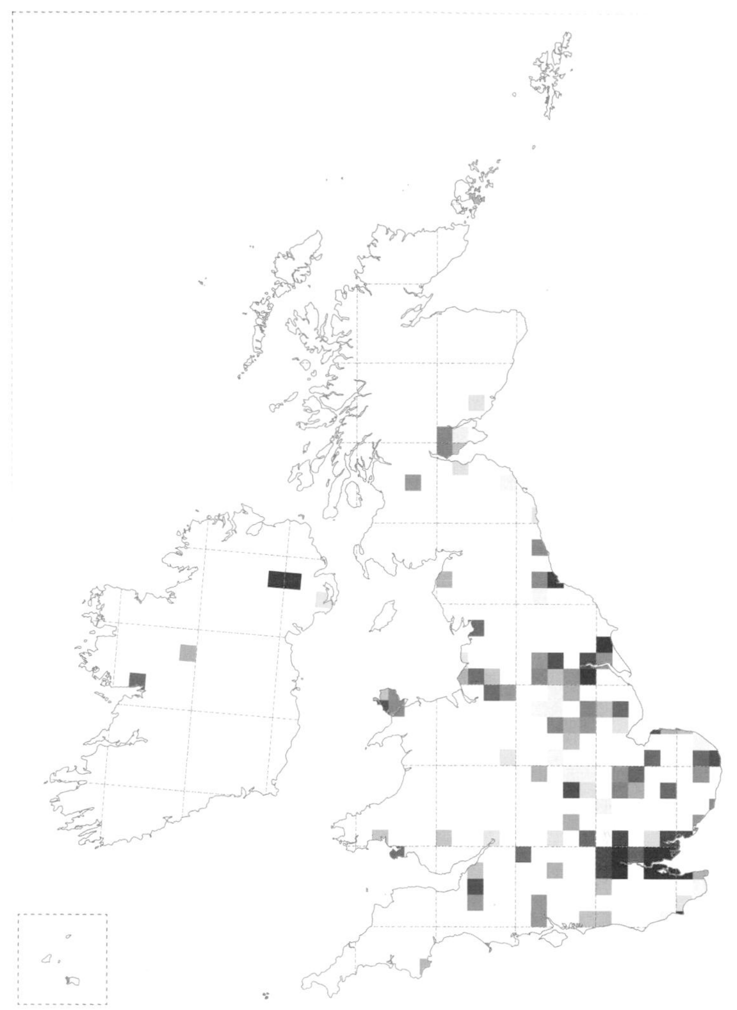

WINTER DISTRIBUTION CHANGE *since 1981–84*

GAIN ▲	41	207	248
LOSS ▼	238	391	629
30yr CHANGE	-49%	-13%	-21%

WINTER RELATIVE ABUNDANCE *2007/08–2010/11*

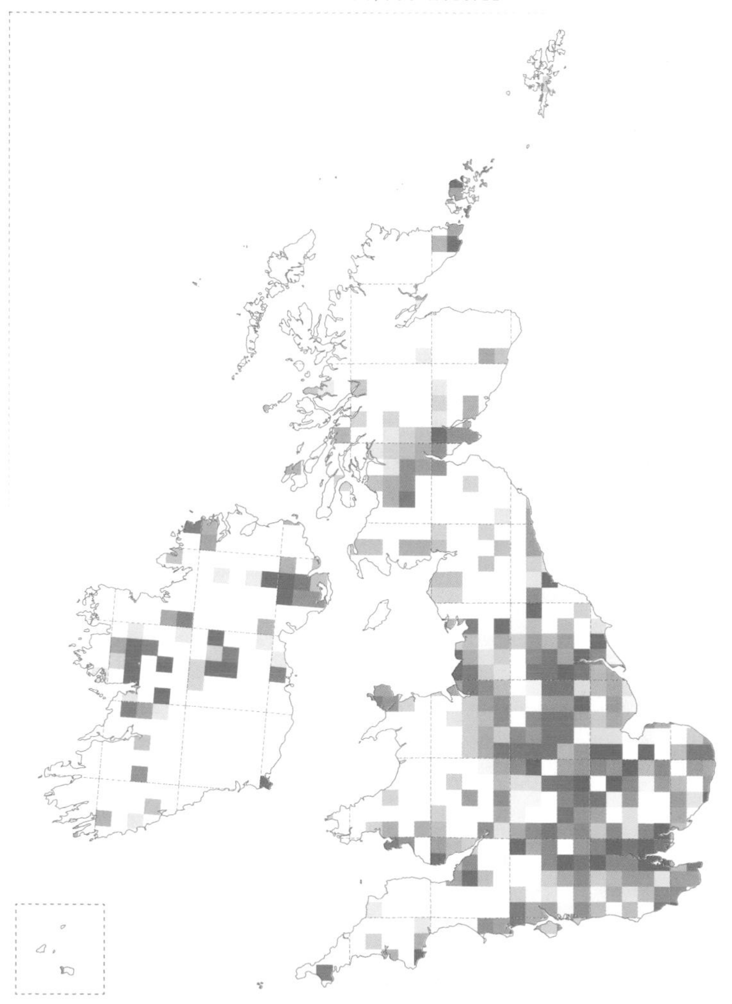

TUFTED DUCK
Aythya fuligula

BTO CODE: TU

MORRIS RENDALL

ONE OF OUR most familiar ducks, the Tufted Duck is widespread in much of Britain during winter. While the distribution is more localised in Ireland, and also in northern Scotland and Wales, at many of these sites large numbers are present. The areas of greatest abundance lie across most of lowland England, in east-central Scotland and in the northern half of Ireland, particularly on the largest loughs. Five sites in Britain regularly hold over 2,000 birds, with Rutland Water the key site with an average of over 6,000 birds. Loughs Neagh, Beg, Corrib and Erne headed the list in Ireland (*I-WeBS Report 2009*; *WeBS Report 2011*).

Bird Atlas 2007–11 shows a 10% increase in range in Britain in winter since the *1981–84 Winter Atlas*. The pattern is different in Ireland, where a winter range contraction of 14% is evident. There has been a shallow increase in wintering numbers in Britain and the Republic of Ireland (*I-WeBS Report 2009*; *WeBS Report 2011*). In Northern Ireland, a very large decline is evident at Loughs Neagh and Beg, paralleling the decline shown by the Pochard (Allen *et al.* 2004; *WeBS Report 2011*). While the reasons for these losses are still unclear, 'short-stopping' relating to milder winters farther east may be partly responsible, and changes in food availability may be implicated (Allen *et al.* 2004; Maclean *et al.* 2006). Competition for food with introduced Roach has also been cited as a possible reason (Winfield & Winfield 1994).

The breeding range of the Tufted Duck has expanded in Britain since the *1968–72 Breeding Atlas*, but in Ireland, although there has been an overall contraction during that period, there has been a small gain of 6% since the *1988–91 Breeding Atlas*. Increases in range are most noticeable along the northern and western margins of the 1968–72 range in Britain, with much infilling elsewhere. Proving breeding is usually relatively straightforward as broods are easily detected and therefore there is a high incidence of *confirmed* breeding. The proportion of occupied 10-km squares with *confirmed* breeding is lower in Ireland and in northern Scotland, however, where observers are fewer, the density is lower and repeat visits to check for broods may have occurred less often. The change in relative abundance shows a mix of gains and losses which could reflect changes such as habitat availability and water quality (Lauder 2007).

Sponsored by Sue Rowe

BREEDING DISTRIBUTION *2008–11*

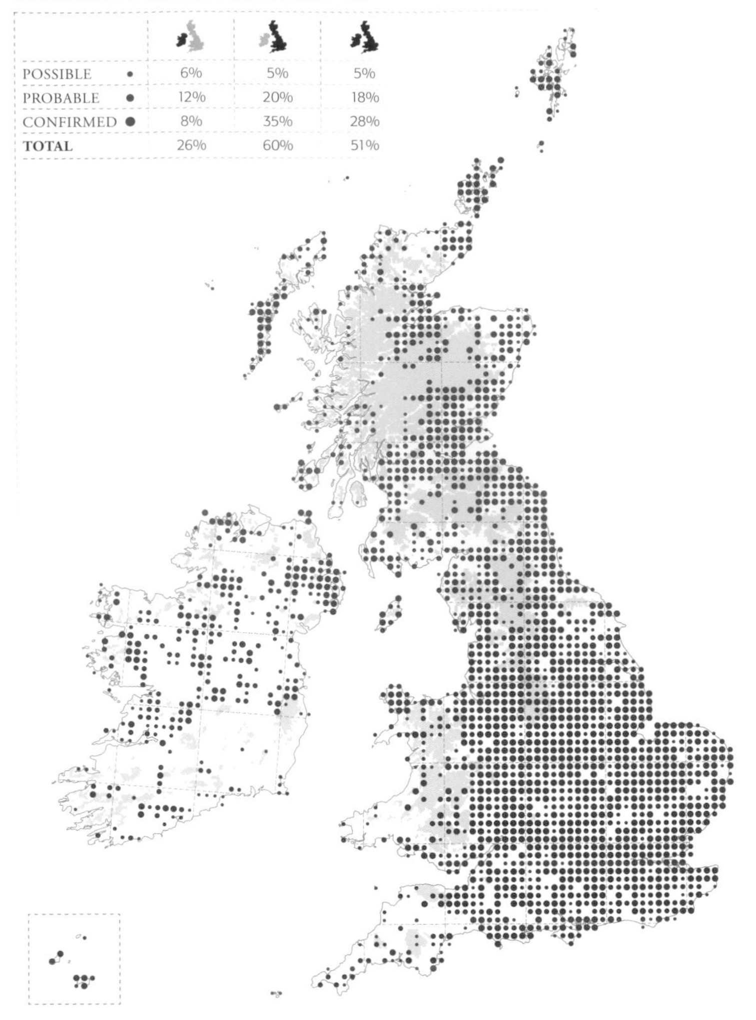

POSSIBLE •	6%	5%	5%
PROBABLE ●	12%	20%	18%
CONFIRMED ●	8%	35%	28%
TOTAL	26%	60%	51%

WINTER DISTRIBUTION *2007/08–2010/11*

PRESENT ●	37%	63%	56%

BREEDING DISTRIBUTION CHANGE *since 1968–72*

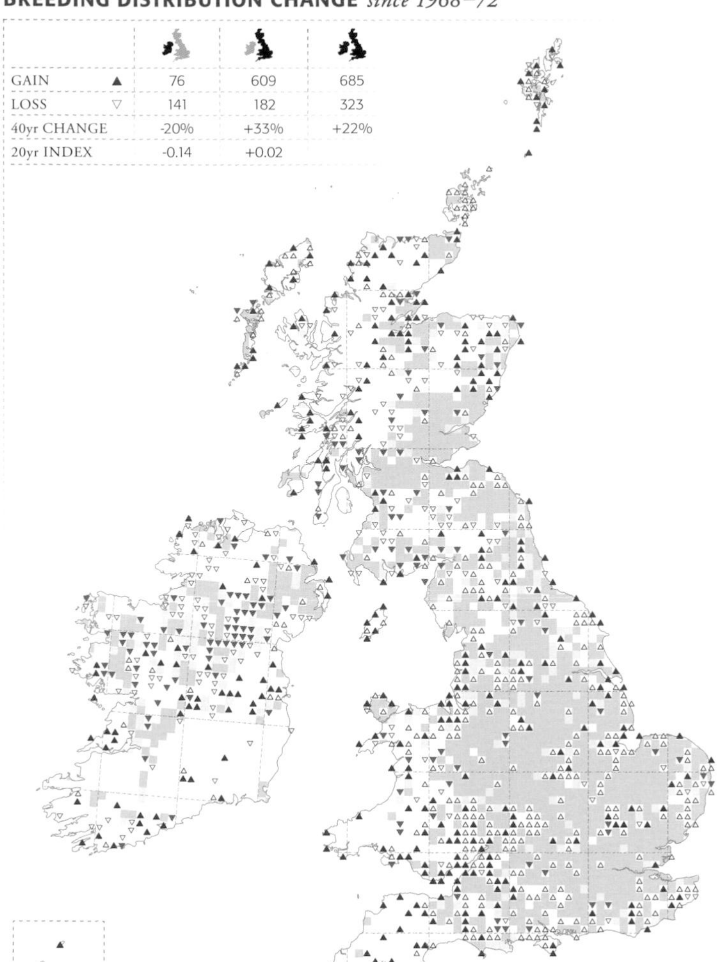

GAIN ▲	76	609	685
LOSS ▽	141	182	323
40yr CHANGE	-20%	+33%	+22%
20yr INDEX	-0.14	+0.02	

BREEDING RELATIVE ABUNDANCE *2008–11*

WINTER DISTRIBUTION CHANGE *since 1981–84*

GAIN ▲	77	344	421
LOSS ▼	135	180	315
30yr CHANGE	-14%	+10%	+5%

WINTER RELATIVE ABUNDANCE *2007/08–2010/11*

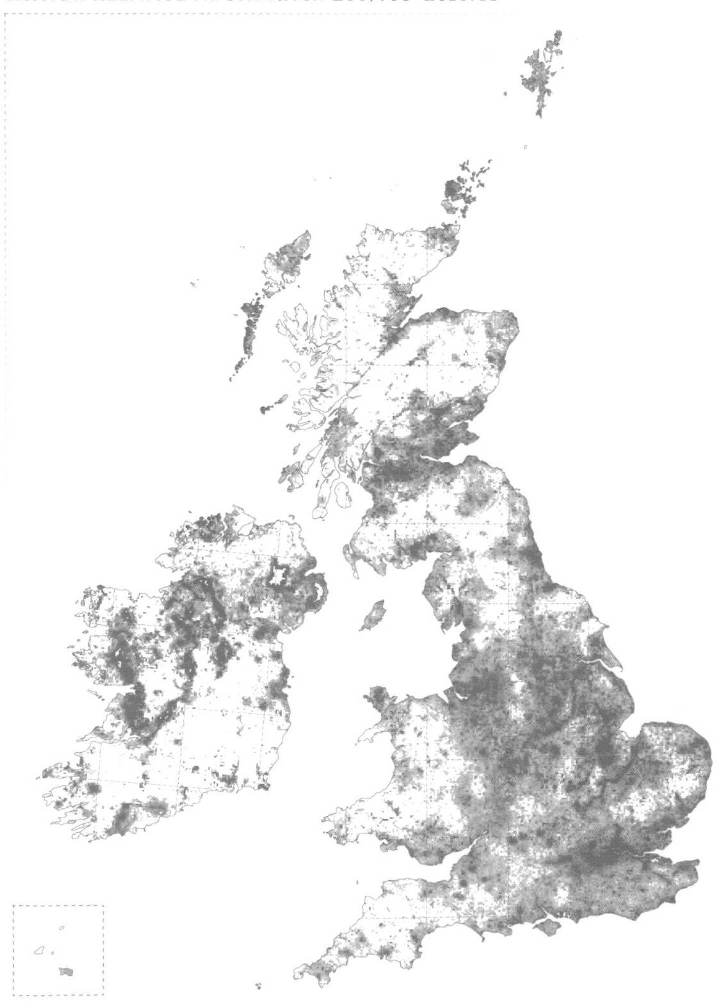

SCAUP
Aythya marila

BTO CODE: SP

PAUL HILLION

THE SCAUP OCCURS in Britain & Ireland predominantly during the winter, originating from breeding populations that stretch from western Siberia through European Russia to northern Fennoscandia and Iceland (*Migration Atlas*). Birds were reported from a wide range of coastal and inland 10-km squares, though the greatest numbers were found on the coast in northern and western Britain and northeastern and southwestern Ireland. A clear exception to this is the large concentrations at Lough Neagh and Lough Beg in Northern Ireland, where a peak of 6,335 birds were counted in the 2008/09 winter (*WeBS Report 2011*).

Overall, there has been a 57% range expansion in Britain & Ireland since the *1981–84 Winter Atlas*. Gains have occurred in some coastal areas, including the Northern Isles, Moray Firth and southwest England, and also at inland sites. Despite these gains in range, numbers in the Republic of Ireland declined significantly from the mid 1990s to 2002/03, since when they have been stable (*I-WeBS Report 2009*). Numbers in Britain have seen a shallow decline since the mid 1970s (*WeBS Report 2011*), though there was a massive decline in the Firth of Forth in the early 1970s (*Birds of Scotland*). Local losses have been linked to changes in food supply and water quality (Campbell 1984; Quinn *et al.* 1997). In contrast, numbers in Northern Ireland have shown a large increase due to an increasing trend at Loughs Neagh and Beg, thought to be associated with the recovery of the Icelandic breeding population (*WeBS Report 2011*). However, there was a significant drop in numbers during the winter of 2009/10, which may have been related to a corresponding increase at the Solway Firth.

Small numbers can be found in Britain & Ireland throughout the year, with most *non-breeding* records believed to be late migrants or summering birds. Unlike during previous breeding atlases, no evidence of *confirmed* breeding was received, though *possible* breeding was recorded at five 10-km squares, three centred on Lough Neagh. The last documented breeding in Britain & Ireland was in 1999, in Co. Armagh (Holling *et al.* 2011b).

Sponsored by Bill Barkworth

WINTER DISTRIBUTION *2007/08–2010/11*

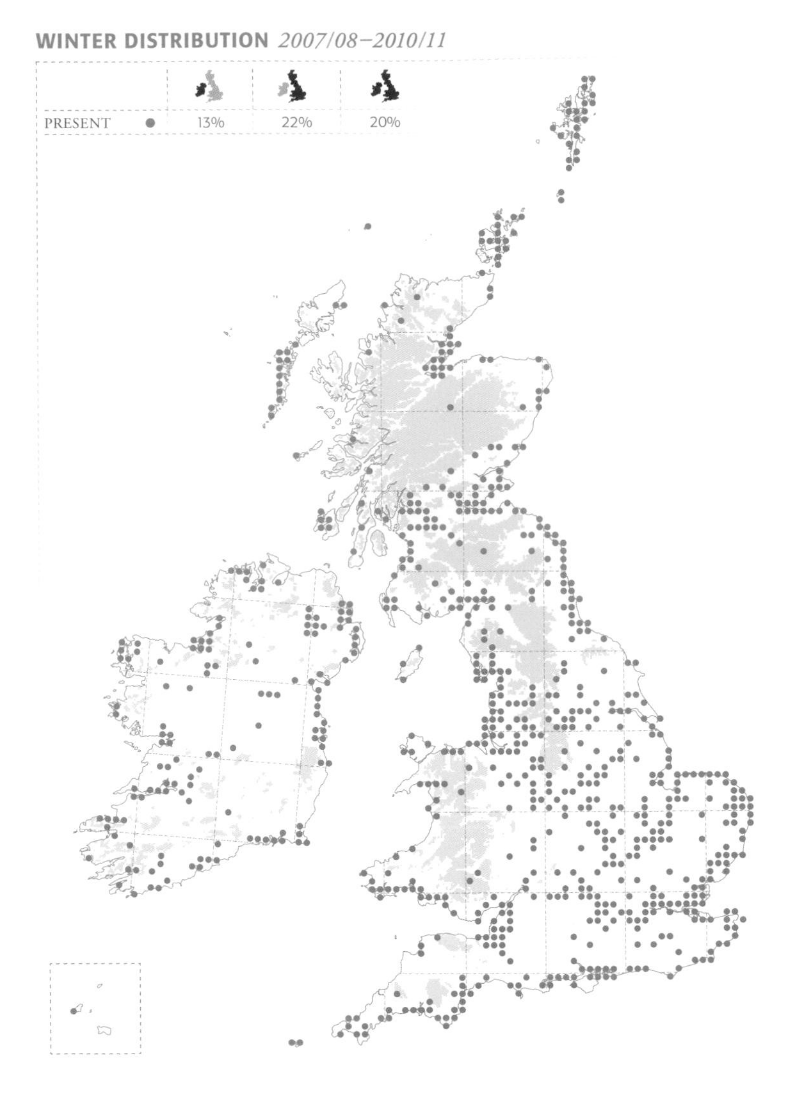

PRESENT ●	13%	22%	20%

BREEDING DISTRIBUTION *2008–11*

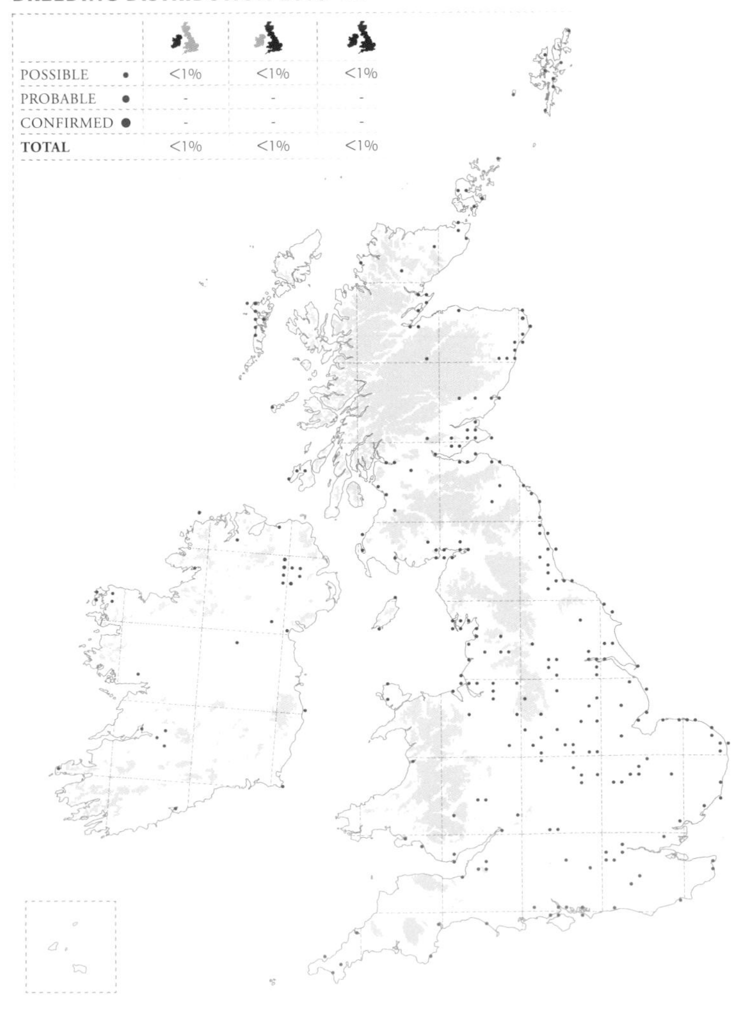

POSSIBLE •	<1%	<1%	<1%
PROBABLE ●	-	-	-
CONFIRMED ●	-	-	-
TOTAL	<1%	<1%	<1%

WINTER DISTRIBUTION CHANGE *since 1981–84*

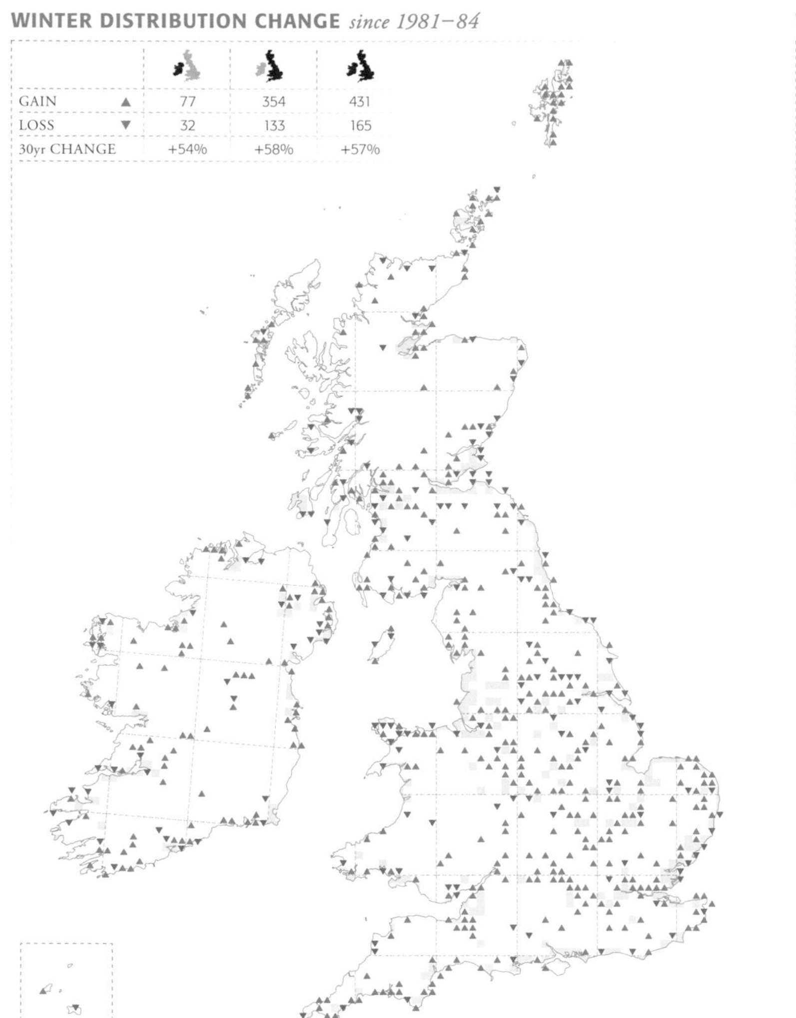

WINTER RELATIVE ABUNDANCE *2007/08–2010/11*

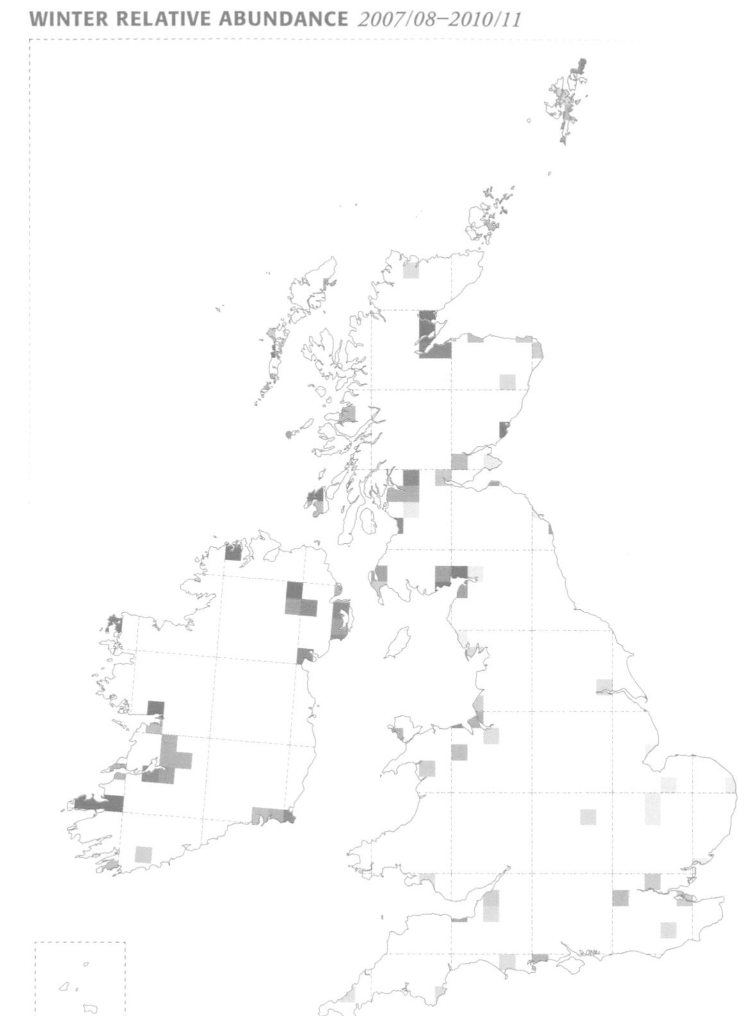

MORRIS RENDALL

◀ The majority of the Icelandic breeding population are thought to winter in Ireland and Scotland. Loughs Neagh and Beg (Northern Ireland) and the Solway Firth (Scotland/England) are sites of international importance.

EIDER

Somateria mollissima

BTO CODE: E.

JOHN ANDERSON www.pbase.com/crail_birder

THE EIDER IS very much a seaduck and this is evident from the distribution maps. In winter it is found all around the coastline of Britain with the exception of much of the Solway Firth, Cardigan Bay and the Bristol Channel. In Ireland it is restricted largely to the northern coasts. The greatest concentrations are to be found along the North Sea coast from Shetland to North Yorkshire, and in the northwest in Argyllshire and the Firth of Clyde, northern parts of Ireland and Morecambe Bay. The overall winter range size is largely unchanged since the *1981–84 Winter Atlas*, but this conceals an increase in numbers in Ireland and the Isle of Man (Sharpe 2007; *I-WeBS Report 2009*) and a decline at several key sites in Britain (*WeBS Report 2011*). The losses in central England relate to there having been fewer sporadic storm-driven individuals.

Since the *1968–72 Breeding Atlas* there have been interesting changes in the distribution of breeding Eiders. There have been notable gains in northwest Wales, Morecambe Bay, the Isle of Man (since the *1988–91 Breeding Atlas*), and throughout the north coast of Ireland, especially Co. Down in the northeast and also in Co. Mayo (Murray & Cabot 2002). A survey of breeding Eiders in Northern Ireland in 2009 revealed the population to be ten times greater than in the previous survey in 1977, mostly likely due to an increase in food abundance, although there had been little change in range since the *1988–91 Breeding Atlas* (Leonard 2010). More surprisingly, since 2003, Eiders have attempted to breed in the west Solent, with three young fledged in 2010 (Hampshire Ornithological Society 2011). Range losses in western Scotland and Shetland, combined with declining numbers and significant reductions in breeding abundance, are possibly the result of predation by introduced American Mink (Craik 1997). Conflict with mussel farms and oil-pollution incidents, particularly in Shetland, have also been implicated in these declines (*Birds of Scotland*).

Declines in breeding numbers have been noted elsewhere in northern Europe (Ekroos *et al.* 2012) with possible causes including food shortages (Coulson 2010) and thiamine deficiency (Balk *et al.* 2009).

Sponsored by The Crown Estate (Marine)

BREEDING DISTRIBUTION *2008–11*

POSSIBLE •	1%	2%	2%
PROBABLE •	2%	3%	2%
CONFIRMED ●	4%	12%	10%
TOTAL	7%	17%	14%

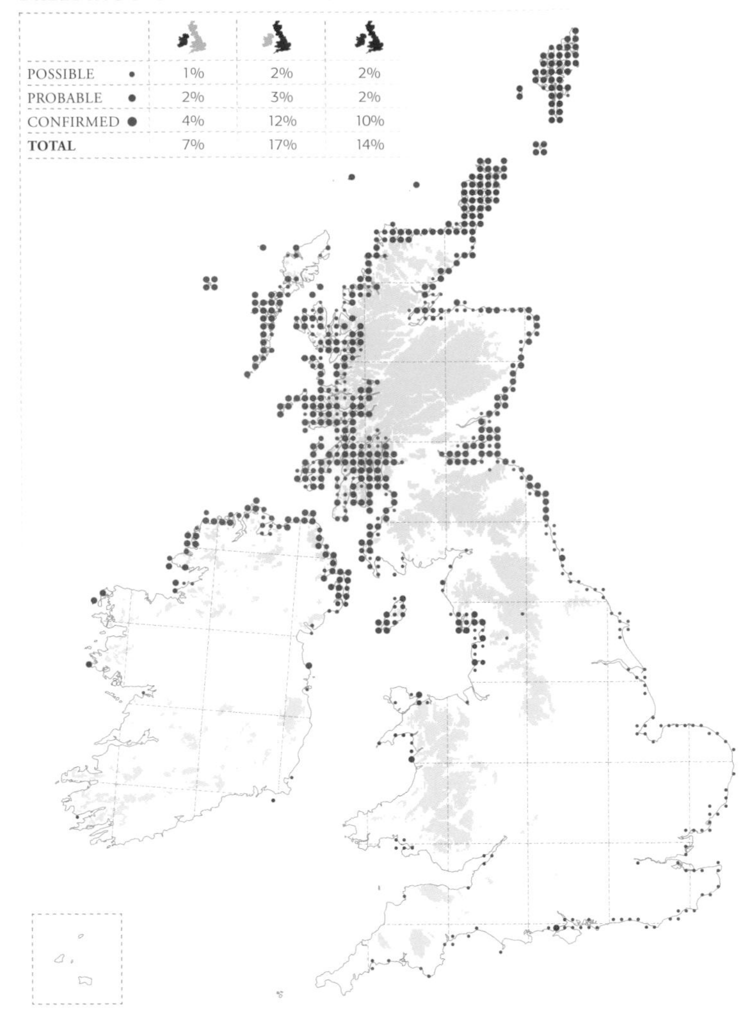

WINTER DISTRIBUTION *2007/08–2010/11*

PRESENT •	8%	23%	19%

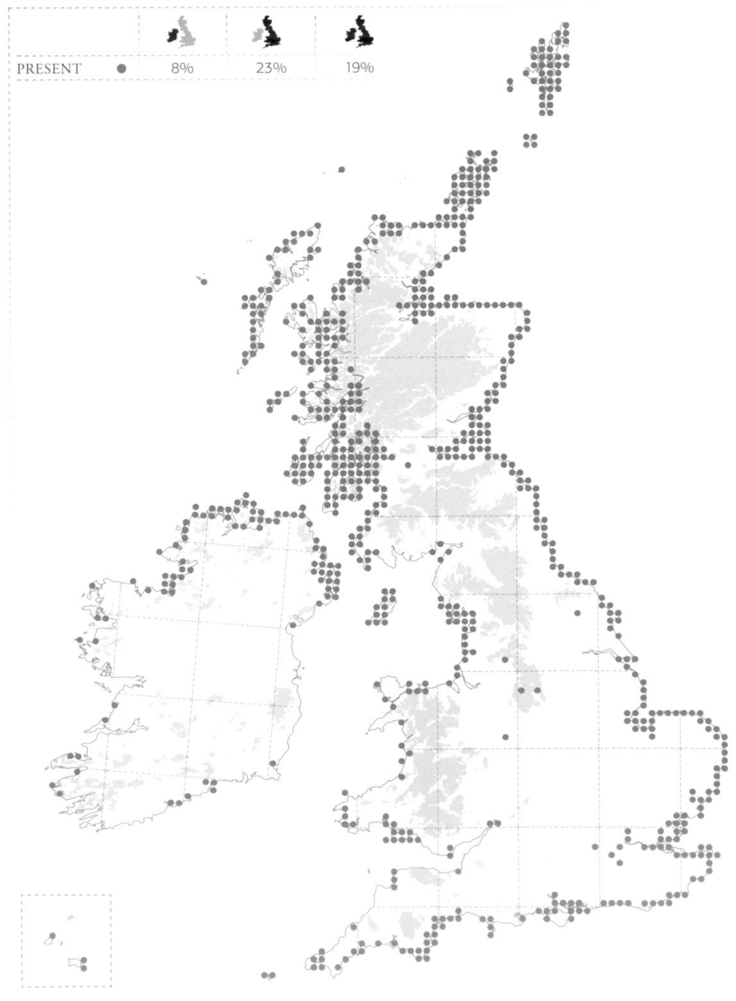

BREEDING DISTRIBUTION CHANGE *since 1968–72*

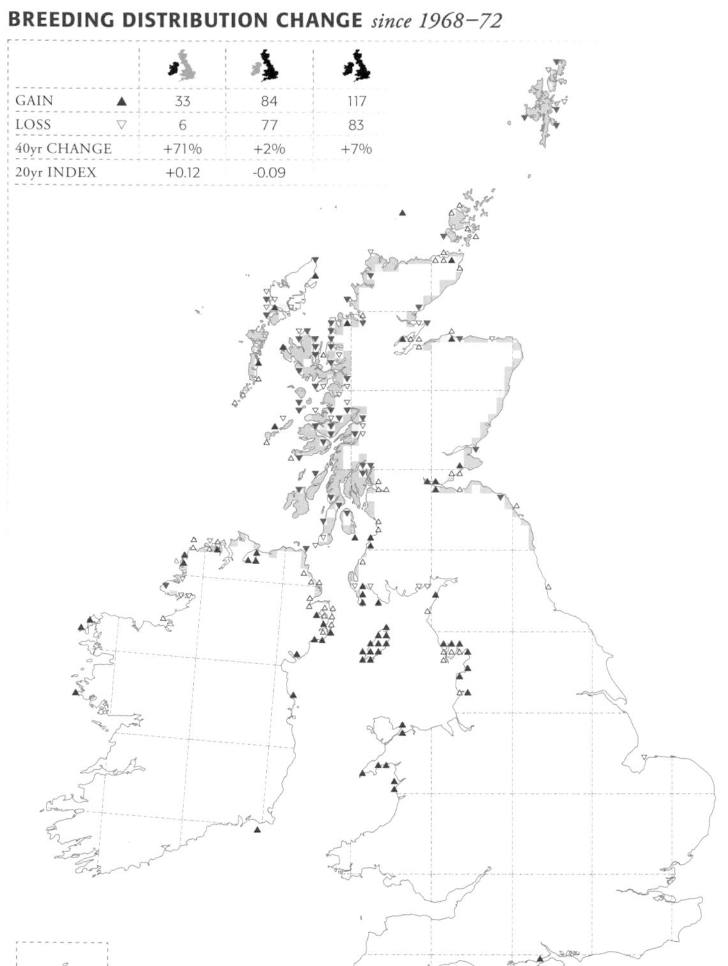

GAIN ▲	33	84	117
LOSS ▽	6	77	83
40yr CHANGE	+71%	+2%	+7%
20yr INDEX	+0.12	-0.09	

BREEDING RELATIVE ABUNDANCE *2008–11*

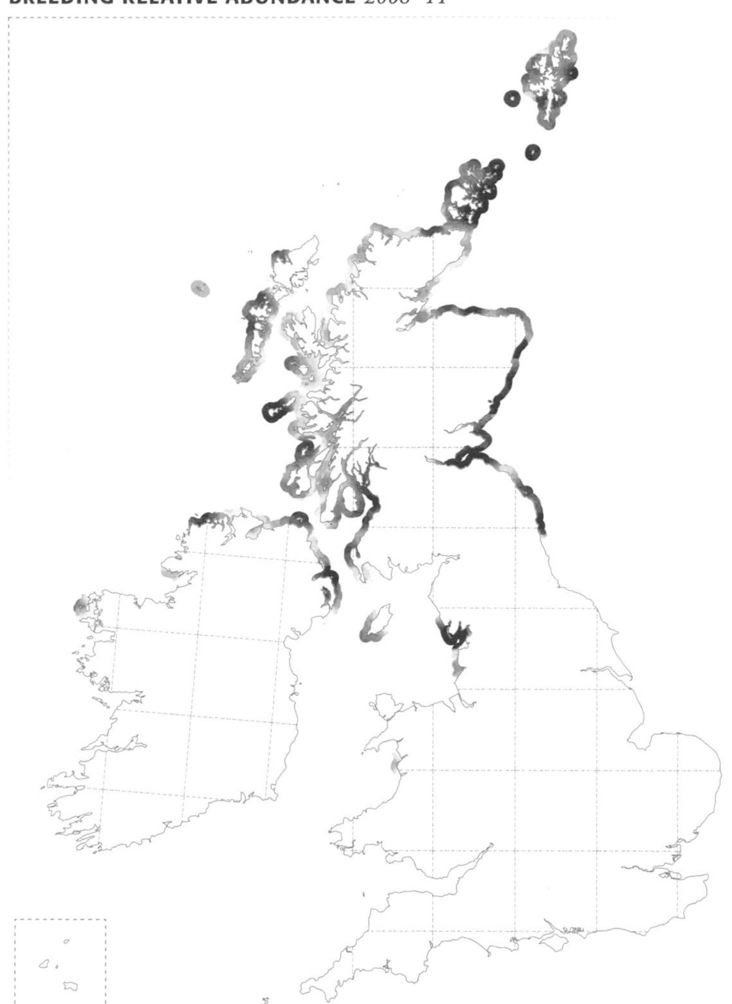

WINTER DISTRIBUTION CHANGE *since 1981–84*

GAIN ▲	28	136	164
LOSS ▼	29	141	170
30yr CHANGE	-1%	-1%	-1%

WINTER RELATIVE ABUNDANCE *2007/08–2010/11*

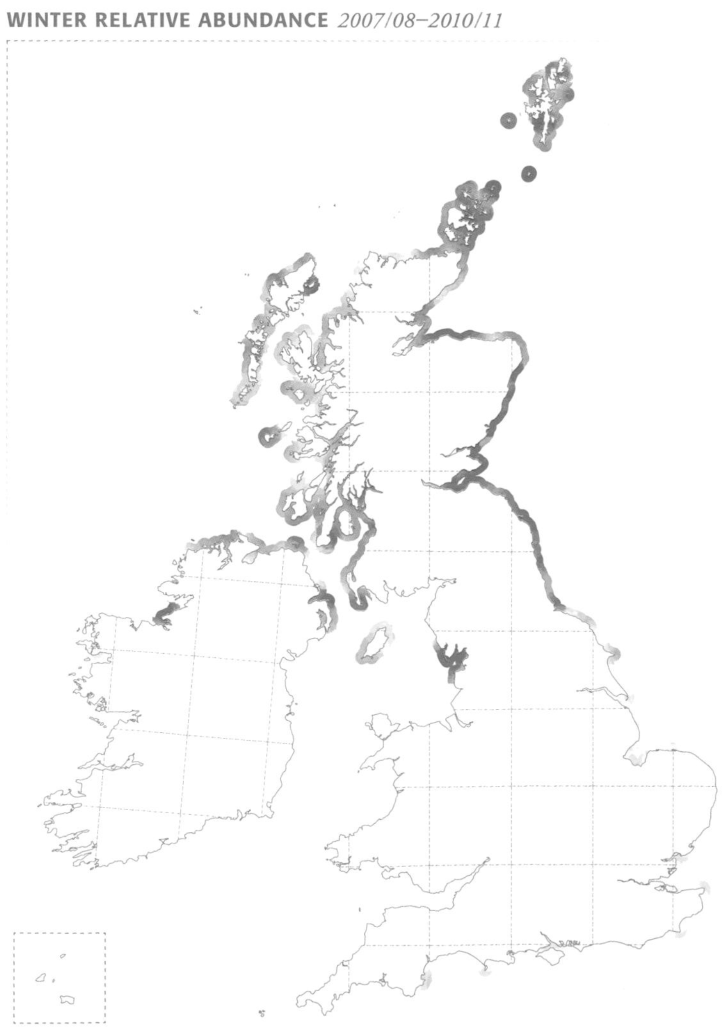

LONG-TAILED DUCK

Clangula hyemalis

BTO CODE: LN

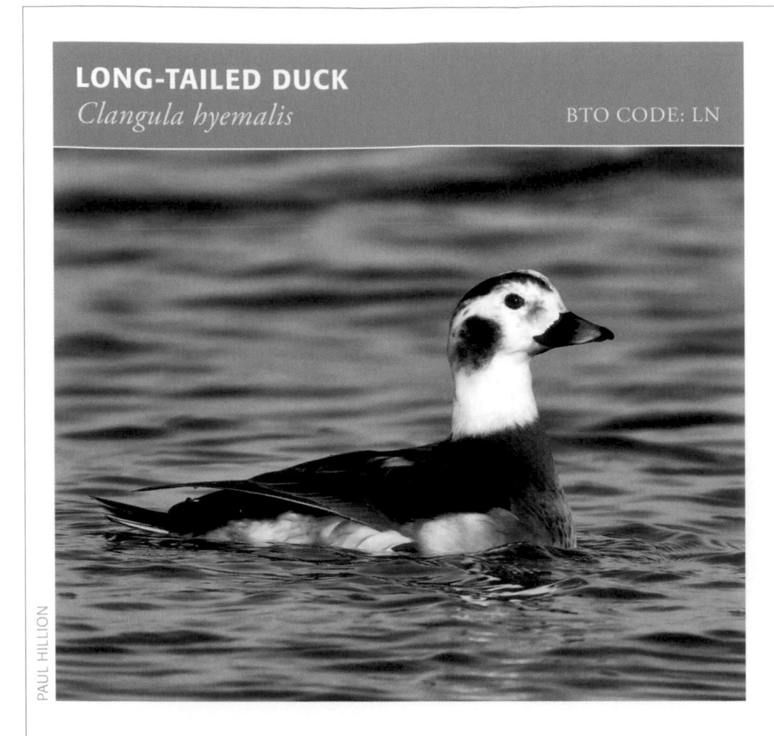

PAUL HILLION

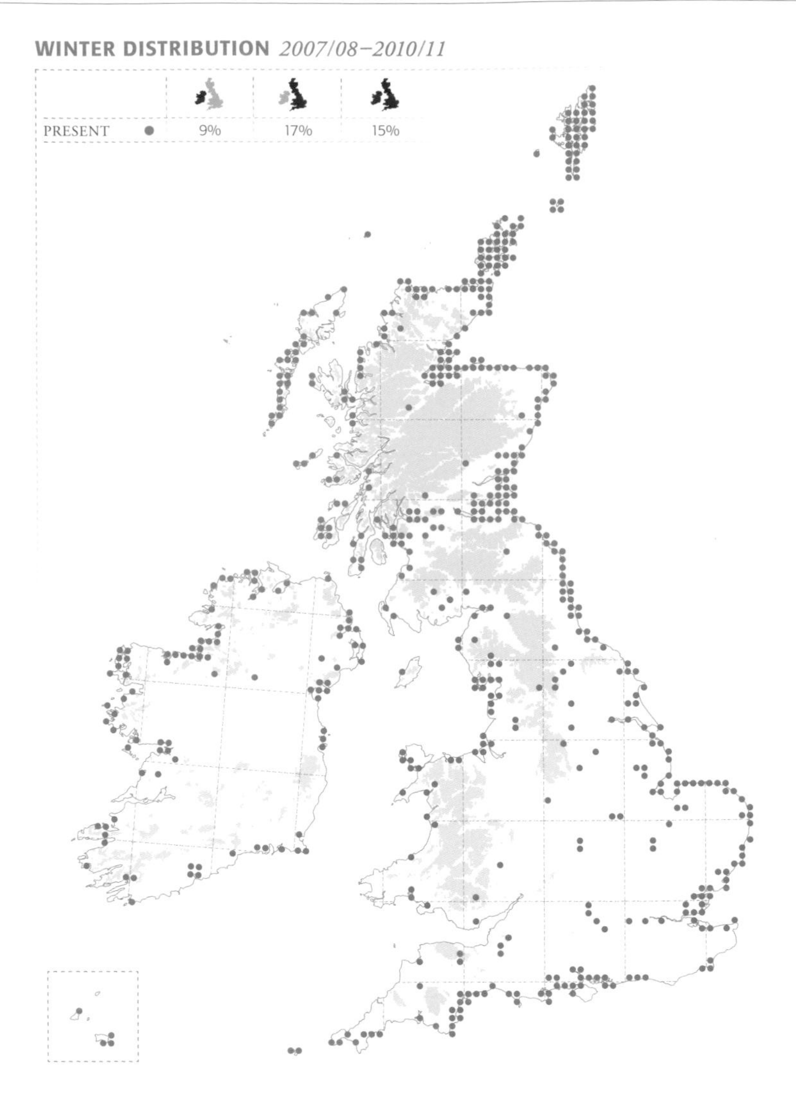

LONG-TAILED DUCKS are winter visitors to Britain & Ireland and tend to be found well offshore in deep waters, so the numbers detected from shore may give a misleading impression of distribution and abundance. During the winter they were found all along the North Sea coast from Shetland south, though the winter abundance map indicates that the largest numbers were in the Firth of Forth, Firth of Tay, the Moray Firth, and around the Outer Hebrides and the Northern Isles. Smaller numbers are patchily distributed along the coasts of southern England, Wales, western Scotland and parts of Ireland. Individual birds also turn up occasionally at inland wetlands.

The Long-tailed Duck is listed as Vulnerable by IUCN following the detection of a significant decline in the Baltic wintering population (BirdLife International 2013). The marine habits of this species make reliable monitoring difficult but it is clear that numbers wintering in the UK have declined over the last 20 years (*WeBS Report 2011*), with losses noted in the key wintering areas of Shetland and Orkney (*Birds of Scotland*). Numbers in Ireland are too small to draw meaningful trends (*I-WeBS Report 2009*). There has been an apparent range expansion of 7% in Britain & Ireland since the *1981–84 Winter Atlas*. This, however, is highly influenced by apparent changes in occupancy in the more marginal parts of the range, partly as a result of better winter coverage during *Bird Atlas 2007–11*. In addition, many of the losses and an appreciable number of gains are at inland sites and reflect the unpredictable distribution of storm-driven individuals. Changes affecting its food supply, disturbance from dredging and recreation, and marine pollution are listed as possible threats (*Birds of Scotland*).

A number of historic records of breeding have been claimed in Scotland, although there is considerable doubt over their accuracy (*Birds of Scotland*). During this atlas there was a scattering of *non-breeding* records of late-departing and passage birds along the North Sea coast and around the Outer Hebrides. The 13 10-km squares reportedly occupied during the *1988–91 Breeding Atlas* were all assigned the *seen* category and probably also relate to *non-breeding* records so should not be considered as breeding losses.

Sponsored by Mark Penty

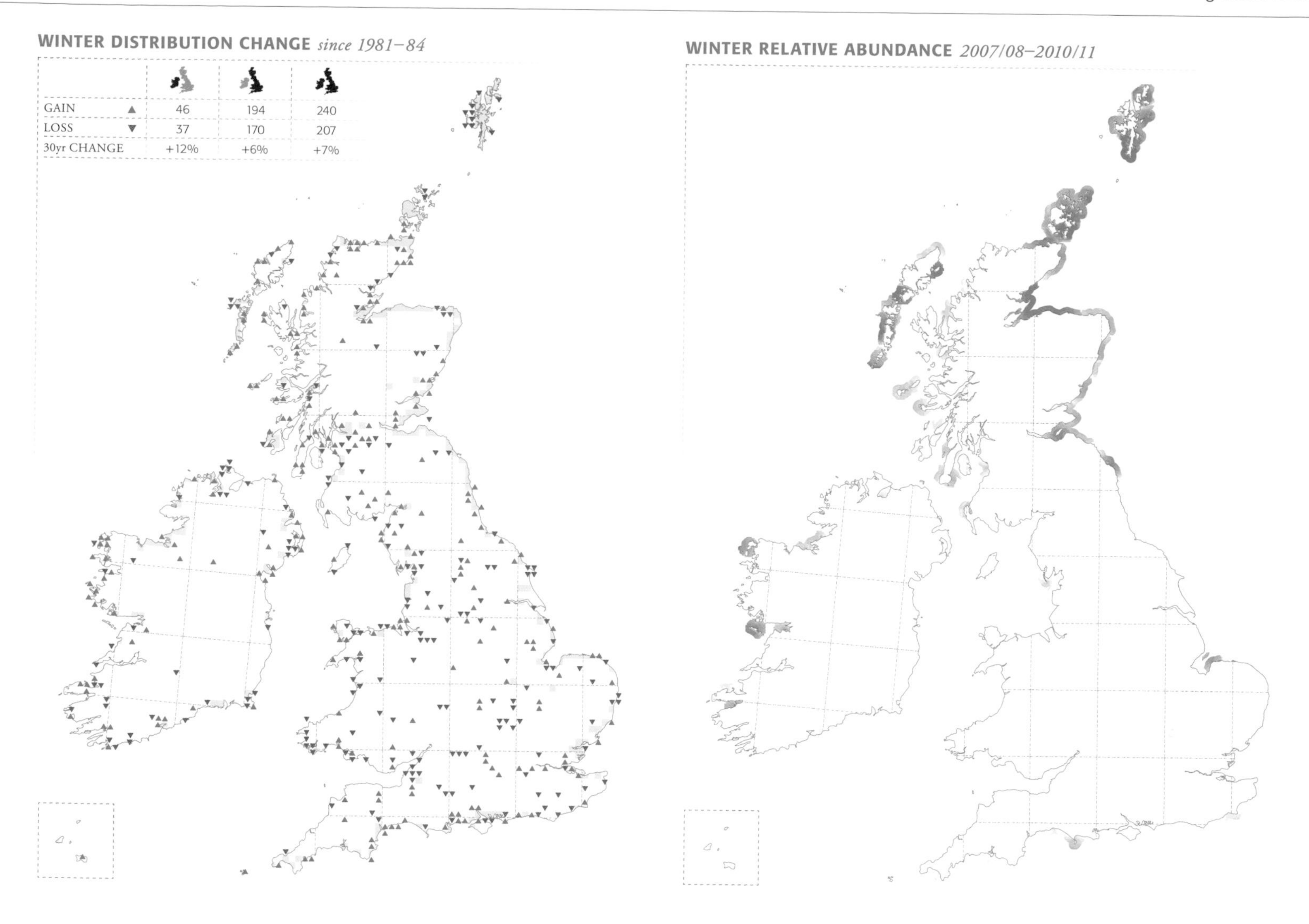

GAIN ▲	46	194	240
LOSS ▼	37	170	207
30yr CHANGE	+12%	+6%	+7%

▲ Long-tailed Ducks are unusual in that they commute between daytime feeding areas and night roosts farther out at sea.

COMMON SCOTER

Melanitta nigra

BTO CODE: CX

GRAHAM CATLEY http://pewit.blogspot.co.uk

ALTHOUGH SMALL NUMBERS breed in both Britain and Ireland, the Common Scoter is to most birdwatchers a familiar winter visitor. The Atlas shows it is present around much of the British & Irish coastline, though patchily distributed in northwest Ireland and western Scotland. There is also a scatter of inland records throughout central and southeast England that relates to singles or small groups that may turn up following poor weather. The highest concentrations recorded by the Atlas were in the Moray Firth, the coast from Angus south to County Durham, off Norfolk, from Carmarthen Bay and the Irish Sea, and off the southwest coast of Ireland.

There has been a 39% range expansion since the *1981–84 Winter Atlas*, with gains occurring throughout the range, particularly evident in areas of lower wintering abundance, such as in Shetland. The mixed pattern of gains and losses at inland sites reflects the chance events of migrant or storm-driven individuals dropping in at particular sites.

The small breeding population is restricted to the Flow Country of Caithness and Sutherland and larger lochs in west Inverness-shire and Perthshire in Scotland, and to a few scattered loughs in western Ireland. Dedicated surveys in Scotland indicate that there has been a decline from 95 pairs in 1995 to 52 in 2007, accompanied by a 25% reduction in the number of occupied lochs and a 17% reduction in occupied 10-km squares (Holling *et al.* 2010a). By 2010, the numbers had fallen further, to 42 pairs in Scotland (*RBBP 2010*).

In Ireland, a small breeding population remains in Counties Galway, Mayo and Sligo but a significant decline is evident, especially around Lough Erne (Tierney *et al.* 2000). A 2012 pre-breeding and brood survey carried out at the four known breeding sites reported a 50% decline since 1999, producing an estimated national breeding population of 39 pairs (Hunt *et al.* 2012). The breeding population in Northern Ireland became extinct in 1993 and eutrophication of waters, predation by American Mink and the introduction of non-native coarse fish were given as possible reasons for its demise (Whilde 1993). In Scotland, disturbance through increased human access may also be a problem at some sites (*Birds of Scotland*).

Sponsored by Scottish Environment Protection Agency (SEPA)

BREEDING DISTRIBUTION *2008–11*

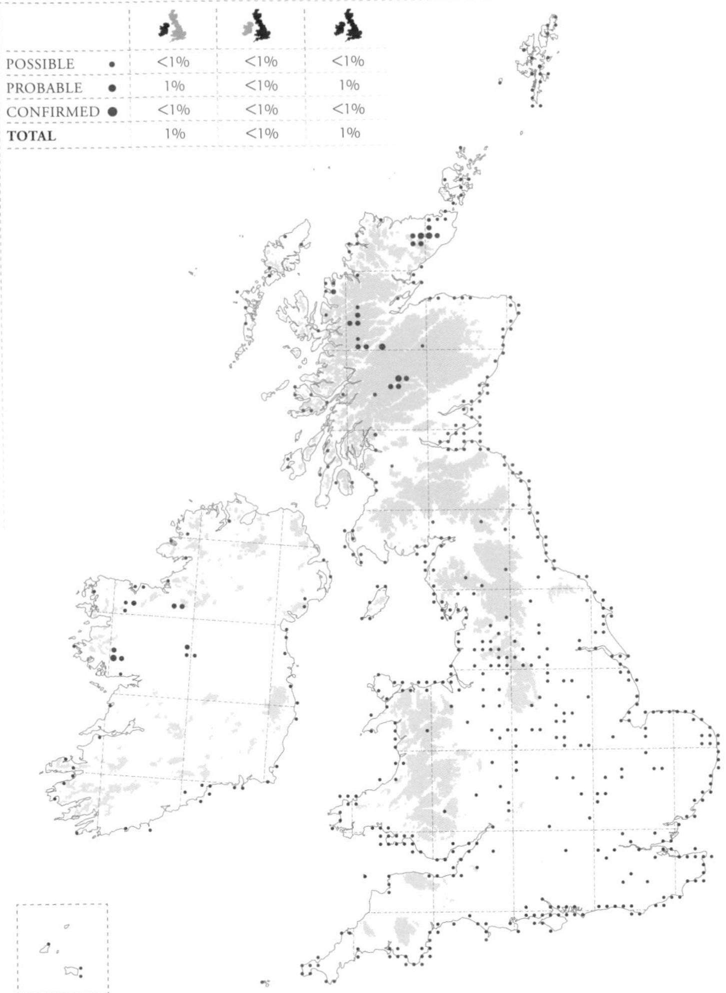

POSSIBLE •	<1%	<1%	<1%
PROBABLE •	1%	<1%	1%
CONFIRMED •	<1%	<1%	<1%
TOTAL	1%	<1%	1%

WINTER DISTRIBUTION *2007/08–2010/11*

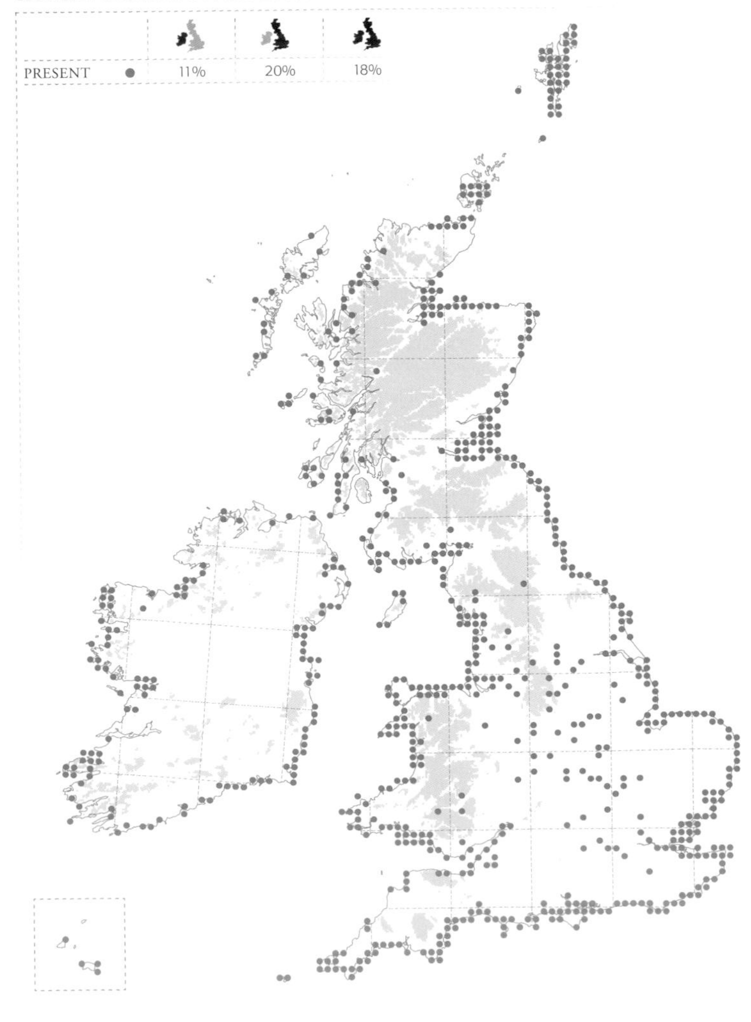

PRESENT •	11%	20%	18%

BREEDING DISTRIBUTION CHANGE *since 1968–72*

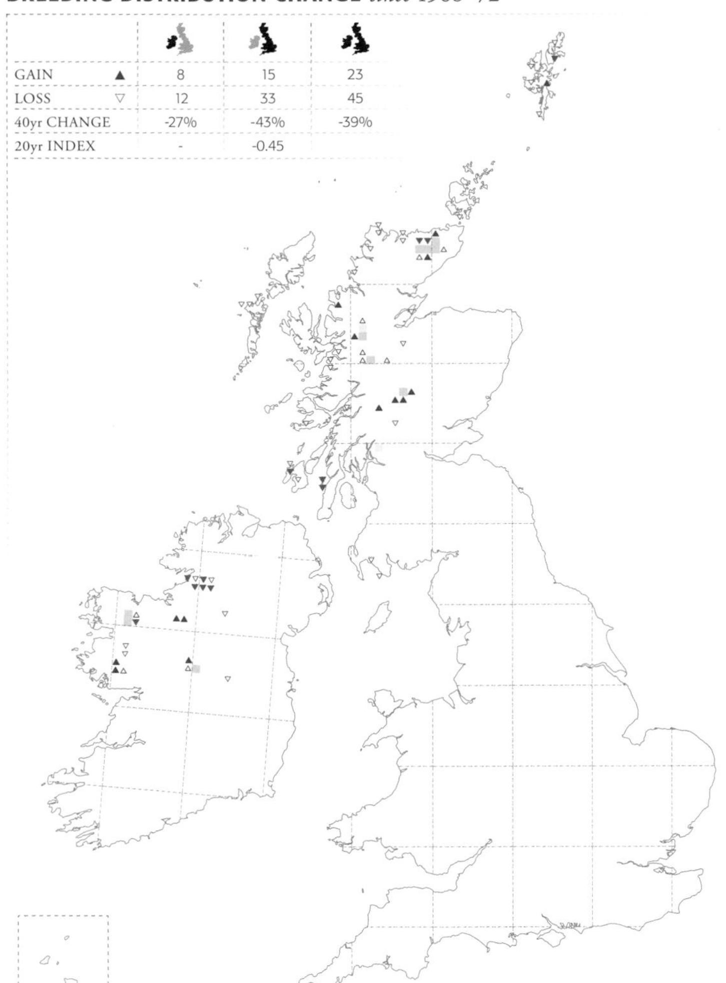

GAIN ▲	8	15	23
LOSS ▽	12	33	45
40yr CHANGE	-27%	-43%	-39%
20yr INDEX	-	-0.45	

BREEDING RELATIVE ABUNDANCE *2008–11*

WINTER DISTRIBUTION CHANGE *since 1981–84*

GAIN ▲	52	247	299
LOSS ▼	35	87	122
30yr CHANGE	+20%	+43%	+39%

WINTER RELATIVE ABUNDANCE *2007/08–2010/11*

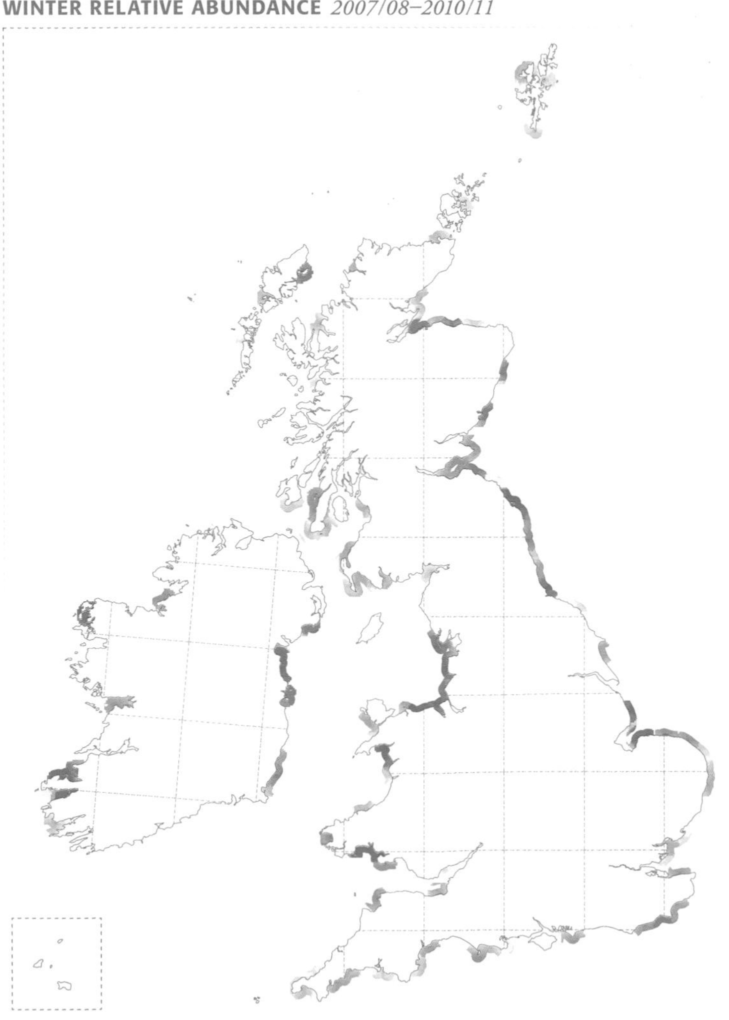

VELVET SCOTER

Melanitta fusca

BTO CODE: VS

JOHN ANDERSON www.pbase.com/crail_birder

IN WINTER THE Velvet Scoter has a coastal distribution that is almost continuous from Shetland southwards along the North Sea coast and along the south coast of England, but this conceals much geographic variation in abundance. It has a much patchier distribution along other coasts and is absent from large stretches in Ireland, Wales and western Scotland. It is more rarely recorded inland than the Common Scoter and there were only 10 occupied inland 10-km squares. Though some individuals arrive in late summer to moult in Scottish waters (*Birds of Scotland*), most are present from November onwards and the largest concentrations tend to occur around Orkney, and in the Moray Firth and Firth of Forth. Reported numbers vary considerably annually (*WeBS Report 2011*), possibly as a consequence of local movements in relation to food supply, but also the ease with which offshore flocks are seen from the coast (*Birds of Scotland*).

There has been a 65% increase in the number of occupied 10-km squares since the *1981–84 Winter Atlas*. It is probable that some gains reflect improved levels of coverage during this atlas, whereby individual birds and small flocks have been recorded outside the core wintering areas. However, the increases in Shetland and the Firth of Forth and the losses in Orkney are probably genuine. Counts of Velvet Scoters from annual waterbird monitoring schemes are difficult to interpret owing to wide fluctuations and coverage problems in the core areas. Nevertheless, a long-term population decline is clear (*WeBS Report 2011*). Putting this into context, wintering numbers in the Baltic have declined by over 60% during 1993–2009 (Skov *et al.* 2011) and autumn migration through Finland is progressively later (Lehikoinen & Jaatinen 2012), suggesting that significant changes are taking place at a flyway level.

Velvet Scoters may linger into the spring and there were records of migrant or summering birds from 117 10-km squares in Britain and three in Ireland. There is one record of *possible* breeding on the Isle of May, Fife, where a male Velvet Scoter was seen accompanying a female Eider to her nest in May 2010 (Brockie 2011).

Sponsored by Fife SOC—celebrating 60 years (1950–2010)

WINTER DISTRIBUTION *2007/08–2010/11*

PRESENT ●	4%	10%	8%

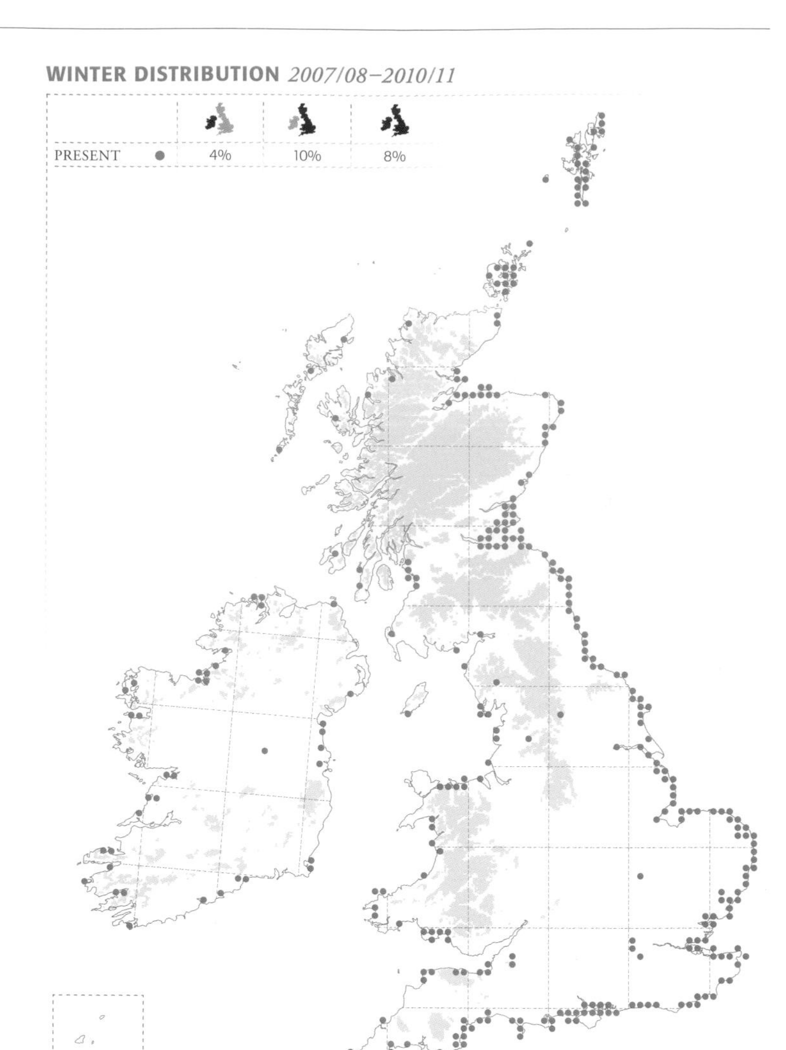

WINTER DISTRIBUTION CHANGE *since 1981–84*

GAIN ▲	33	154	187
LOSS ▼	11	59	70
30yr CHANGE	+147%	+58%	+65%

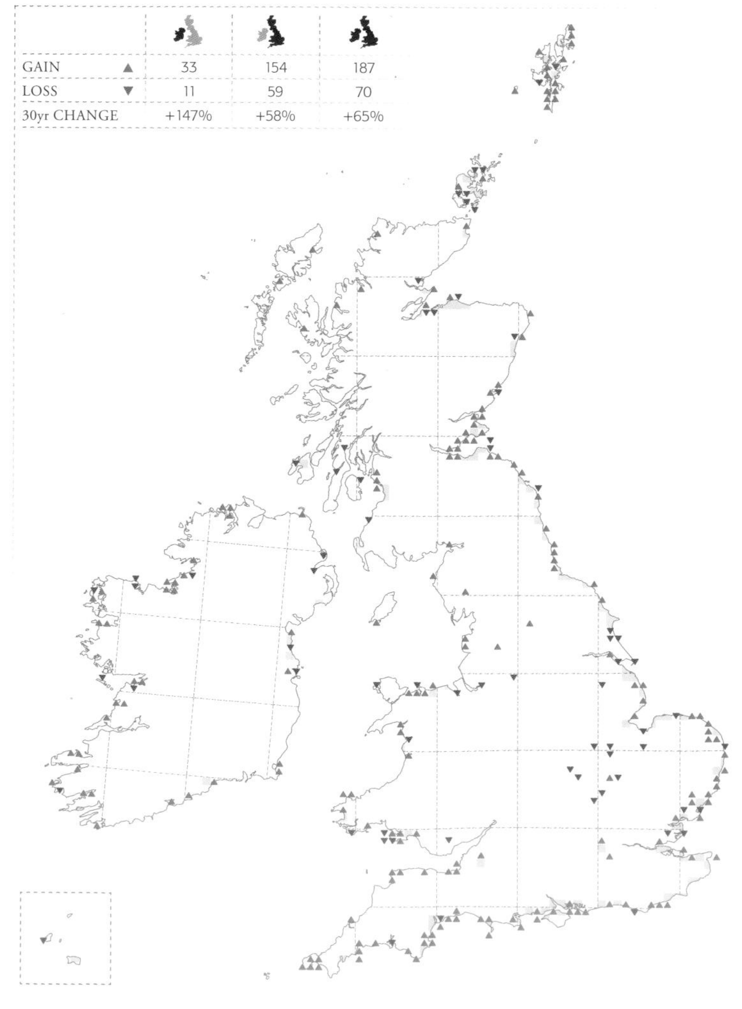

SURF SCOTER
Melanitta perspicillata

BTO CODE: FS

MARK NEWSOME

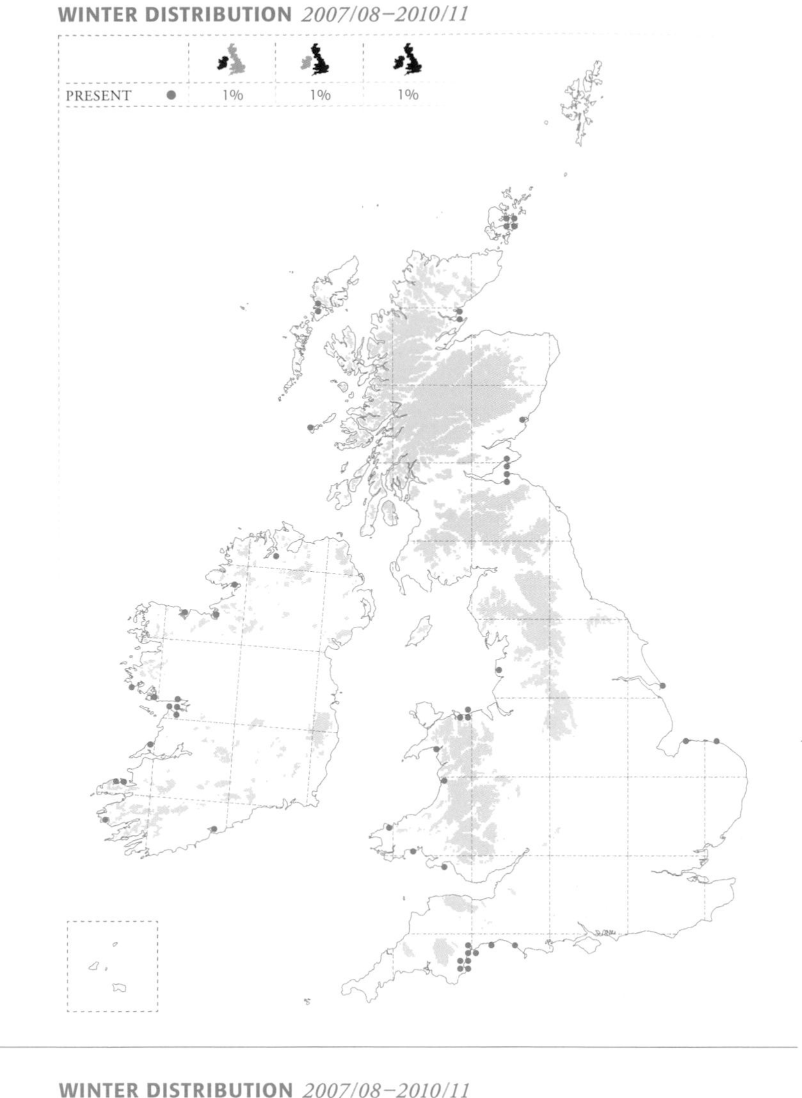

SURF SCOTERS ARE rare winter visitors from North America, often found associated with large flocks of Common Scoters. Some individuals are known to return to the same general area in subsequent winters, which may have resulted in the clusters of occupied squares in areas such as the Firth of Forth, North Wales, Devon and at Galway Bay in western Ireland.

In winter, they were recorded in 35 10-km squares in Britain and 15 in Ireland, which represents an increase from 10 and seven squares respectively in the *1981–84 Winter Atlas*. The increase is probably associated with improved coverage. There were records of *non-breeding* birds in 27 10-km squares in Britain & Ireland during the breeding season.

KING EIDER
Somateria spectabilis

BTO CODE: KE

JOHN ANDERSON www.pbase.com/cruz_birder

KING EIDERS ARE scarce annual winter visitors from breeding grounds on the high-arctic coastline. In Britain & Ireland, winter records are widely scattered, with 23 occupied 10-km squares in Britain, mostly on the east coast, and two squares in Ireland. These constitute more than a doubling in range size since the *1981–84 Winter Atlas*. A review of records during 1958–90 suggested that a minimum of 61 individuals had occurred in Britain & Ireland, once returning individuals had been accounted for, fewer than the published figure of 153 for the same period (Suddaby *et al.* 1994). This species was also recorded in 17 10-km squares during the breeding season, including individuals summering in consecutive years in Scotland.

GOLDENEYE

Bucephala clangula

BTO CODE: GN

DESMOND DUGAN www.AbernethyImages.co.uk

LARGE NUMBERS OF Goldeneyes from Fennoscandian and Russian breeding grounds winter in Britain & Ireland (*Migration Atlas*), where they are found in coastal areas and a wide variety of freshwater habitats. They are widely distributed, particularly throughout Scotland and northern England, with the exception of some upland areas. Farther south, in England, Wales and Ireland, the distribution is patchier and centred on suitable coastal areas, river valleys and wetland networks.

In winter, the number of apparently newly occupied 10-km squares in Britain since the *1981–84 Winter Atlas* has been almost entirely balanced by losses, while in Ireland there has been a predominant contraction of range (11%). Monitoring of numbers has shown a 28% decline between 1983/84 and 2008/09 in the UK (*SUKB 2012*) and a 37% decline between 1994/95 and 2008/09 in the Republic of Ireland (*I-WeBS Report 2009*), which are thought to be due to a northeasterly shift in the core wintering range due to milder winters (Lehikoinen *et al.* 2013).

Records of *confirmed* breeding are concentrated in Strathspey and around the Great Glen in Inverness-shire, along the River Dee in Aberdeenshire, and around Loch Tay in Perthshire. Elsewhere, the only *confirmed* breeding records are from Northumberland (20-km dot; *RBBP 2010*) and Avon (Holling *et al.* 2010b). The pair in Avon constitutes the first *confirmed* breeding by apparently wild birds in England. Distribution changes since the *1988–91 Breeding Atlas* include 'losses' south of the breeding range which almost certainly relate to migrants assigned the *seen* category in 1988–91. A better indication of change is that the number of 10-km squares with *probable* or *confirmed* breeding increased from 13 to 38, and included colonisation of Perthshire and Aberdeenshire. There were at least 195 breeding females in northern Scotland in 2010 (Goldeneye Study Group 2010), where one possible limiting factor may be competition for nest cavities with Mandarin Ducks (Cosgrove 2003). However, Langridge (1996) cites predation by Pine Martens as the single most important factor limiting breeding success. In Ireland, records of *possible* breeding follow records of confirmed breeding in 2000 on Lough Neagh (Perry & Kennedy 2002) and on Lough Sheelin, Co. Cavan, in 2003 (Hillis 2007).

Dedicated to Alison Stuck

BREEDING DISTRIBUTION *2008–11*

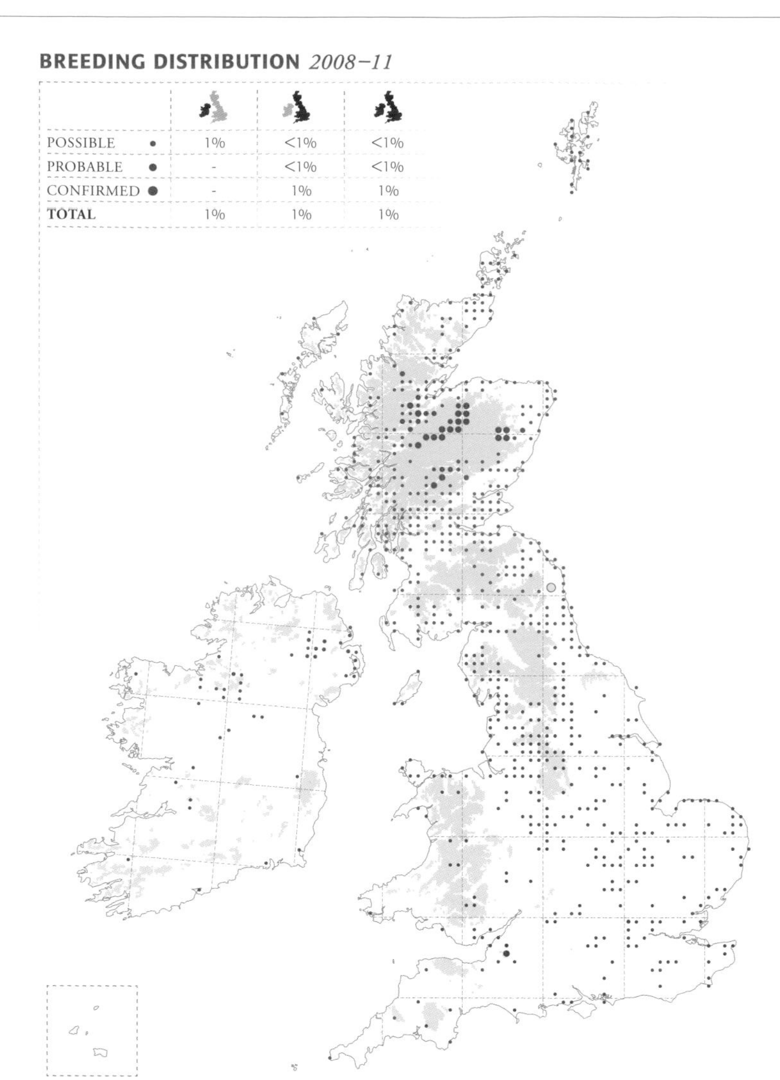

POSSIBLE •	1%	<1%	<1%
PROBABLE ●	-	<1%	<1%
CONFIRMED ●	-	1%	1%
TOTAL	1%	1%	1%

WINTER DISTRIBUTION *2007/08–2010/11*

PRESENT ●	28%	58%	50%

BREEDING DISTRIBUTION CHANGE *since 1988–91*

GAIN ▲	6	26	32
LOSS ▽	11	154	165
20yr CHANGE	-38%	-74%	-72%
20yr INDEX	-	-0.51	

BREEDING RELATIVE ABUNDANCE *2008–11*

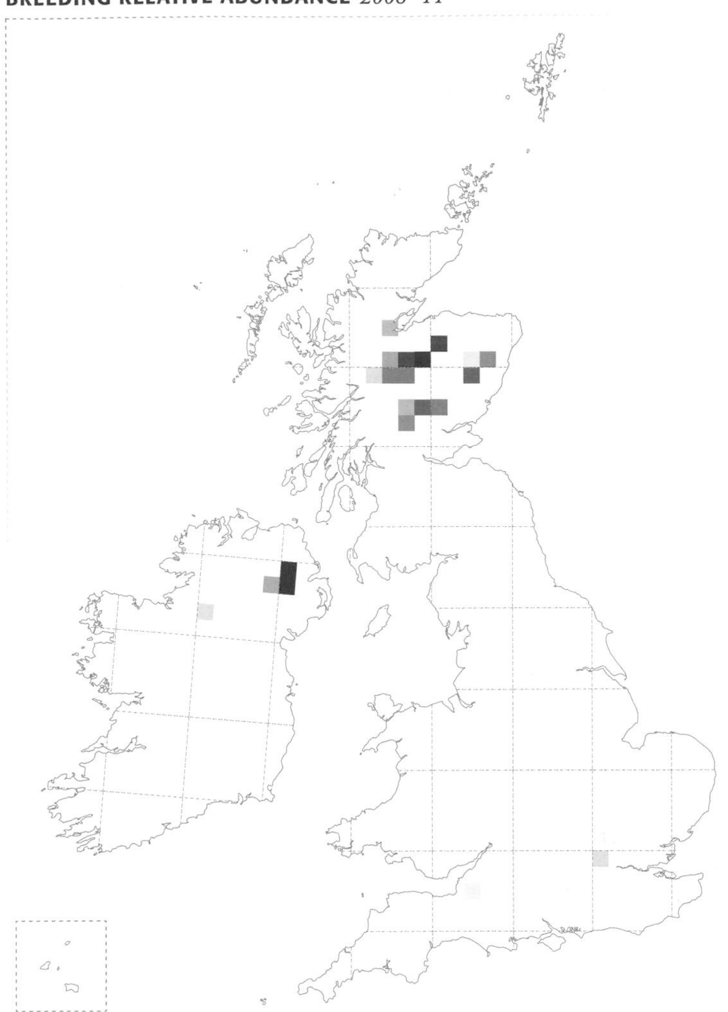

WINTER DISTRIBUTION CHANGE *since 1981–84*

GAIN ▲	74	304	378
LOSS ▼	109	297	406
30yr CHANGE	-11%	+<1%	-1%

WINTER RELATIVE ABUNDANCE *2007/08–2010/11*

SMEW

Mergellus albellus

BTO CODE: SY

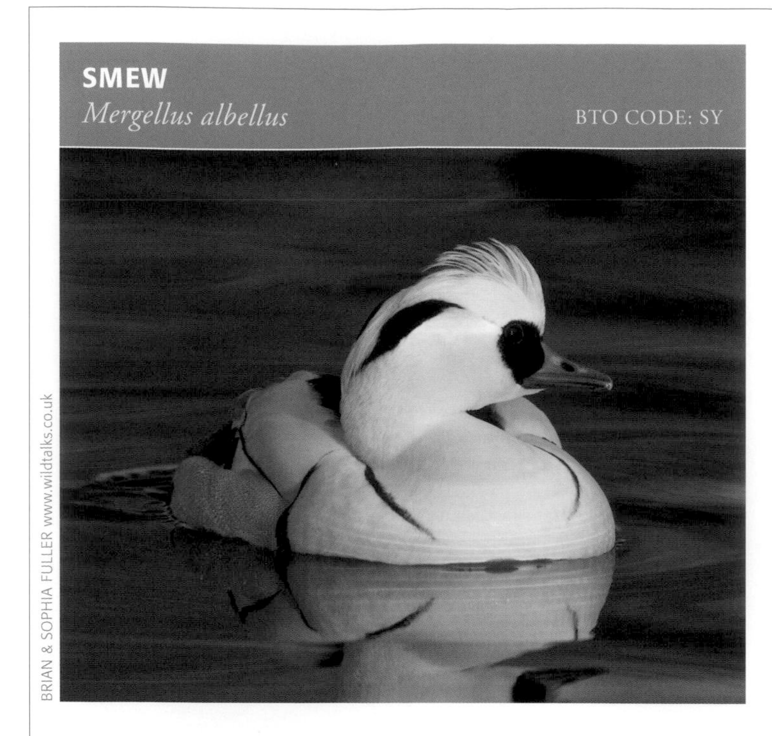

BRIAN & SOPHIA FULLER www.wildtalks.co.uk

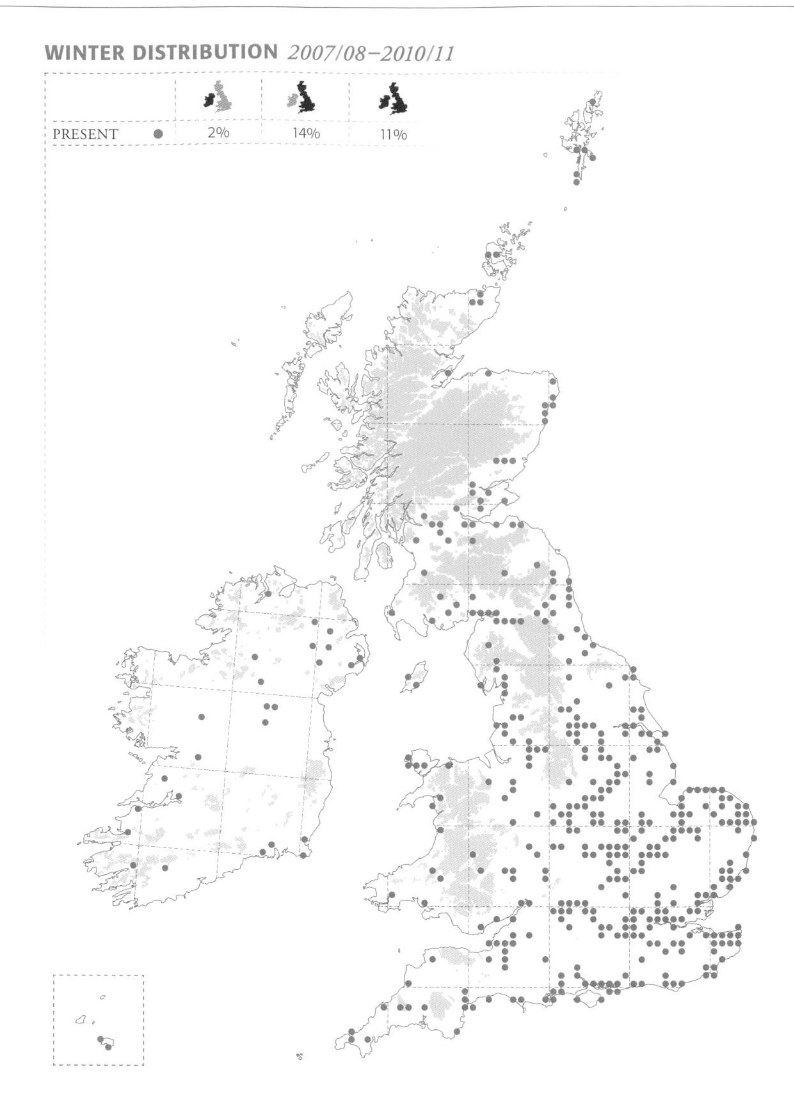

OF THE 40,000 Smews wintering in northwest and central Europe (Wetlands International 2013), a variable but very small proportion winters in Britain & Ireland. The winter distribution map shows that they were recorded widely across lowland England, Scotland and Wales. There were also records from the Isle of Man and the Channel Islands, where they are scarce winter visitors. In Ireland, records were relatively few and well scattered.

In general, the number of Smews wintering in Britain is thought to have been in decline since the 1950s. However, wintering numbers are heavily influenced by weather conditions on the near Continent, as evident in the cold winter of 1978/79 when nearly 400 birds were recorded (Chandler 1981). During this atlas, cold spells during January–February 2010, and again during December 2010–January 2011, resulted in small influxes into the UK. In the former winter, the number of occupied sites counted for waterbird monitoring increased from a recent average of 58 to 91 and the number of birds doubled to 199 (Holt *et al.* 2011a); and in the 2010/11 winter, Smews were recorded at 103 sites (*WeBS Report 2011*). The maximum count in the Atlas was 32, in TL37 (near Huntingdon) in February 2011. These influxes probably explain why the number of occupied 10-km squares was 72% higher than during the *1981–84 Winter Atlas.* Changes in the extent and suitability of gravel pits also may be important.

Winter weather patterns also may be affecting the distribution of Smews at a larger scale. For example, in Sweden wintering numbers increased from 400 in 1971 to 3,800 in 2004 (Nilsson 2008). This pattern of apparently wintering nearer to the breeding grounds ('short-stopping') is common to several waterbird species. Declines in Smew numbers wintering in the Netherlands (Hornman *et al.* 2012a) are consistent with short-stopping, which also could help to explain the long-term decline in numbers wintering in Britain & Ireland.

Smews do not breed in Britain or Ireland but in the breeding season there were records of *non-breeding* birds from 48 widely scattered 10-km squares in Britain and eight 10-km squares in Ireland. These were mainly records of birds remaining beyond the winter period, some of which were recorded summering throughout June and July.

Sponsored by Michael and Helen Cox

WINTER DISTRIBUTION CHANGE *since 1981–84*

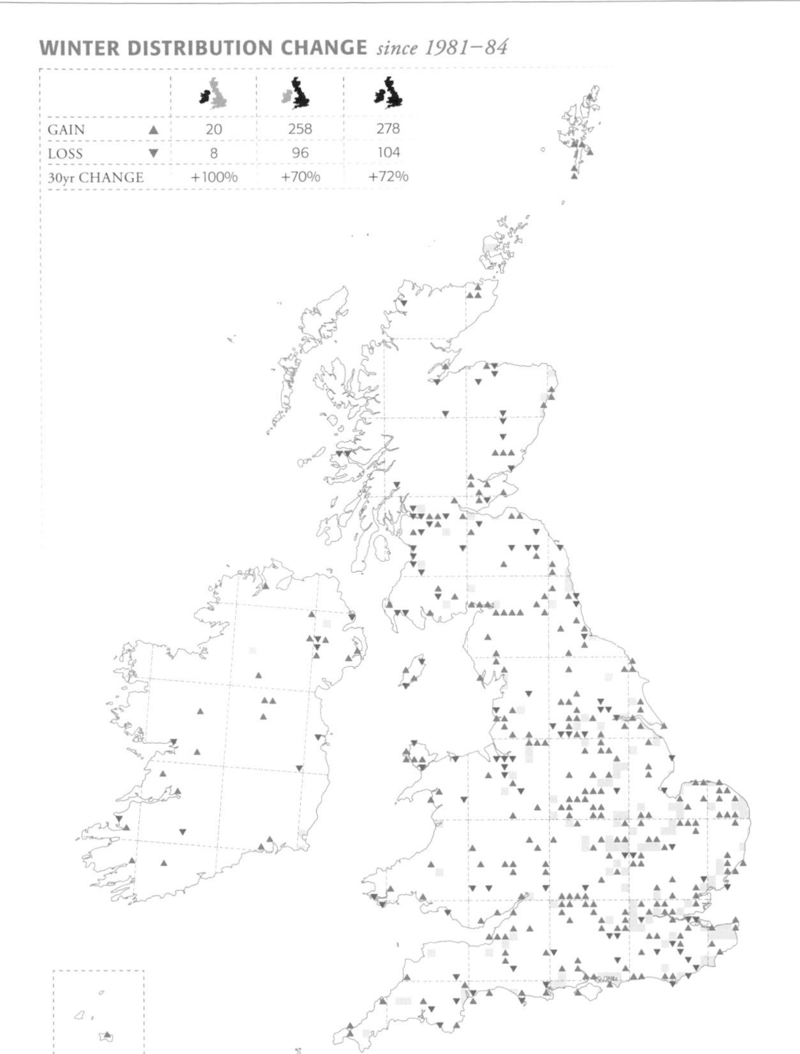

GAIN ▲	20	258	278
LOSS ▼	8	96	104
30yr CHANGE	+100%	+70%	+72%

WINTER RELATIVE ABUNDANCE *2007/08–2010/11*

▲ **Smews are often site-faithful, returning from mid October to favoured fresh or slightly brackish waters year after year.**

RED-BREASTED MERGANSER

Mergus serrator

BTO CODE: RM

JOHN ANDERSON www.pbase.com/crail_birder

RED-BREASTED MERGANSERS are widely distributed around the coast of Britain & Ireland in winter, though there are distinct concentrations in western Scotland, the Northern Isles, northwest and southwest Ireland, north Wales and the Solent. Small numbers are recorded on inland waters, particularly in Britain.

Since the *1981–84 Winter Atlas*, the range has expanded by 15% in Britain but contracted by 20% in Ireland. The expansion in Britain seems at odds with a documented long-term population decline (*WeBS Report 2011*), though some wintering sites are not covered by waterbird monitoring schemes, and, since most of the British gains were at inland sites, in reality they probably relate to very few individuals. Furthermore, in the more remote parts of Scotland, apparent gains could reflect improved winter coverage since 1981–84. In Ireland, the losses were concentrated along the west and south coasts, as well as inland, and parallel a reported shallow decline (*I-WeBS Report 2009*). It is currently unclear how changes in Britain & Ireland fit into the broader European picture, where trends vary by country, for example, increasing in Sweden and the Netherlands but declining in the Baltic Sea (Nilsson 2008; Skov *et al.* 2011; Hornman *et al.* 2012a).

The breeding distribution is more concentrated than in winter, and includes the Scottish Isles, northwest Scotland, parts of north and west Ireland, northern England and northwest Wales. In addition to these breeding areas, there are late-summer moulting assemblages on the British east coast, such as round the Firth of Forth.

There has been a 28% range contraction since the *1968–72 Breeding Atlas*, with losses concentrated in western Ireland, southwest Scotland and many inland parts of Scotland, particularly in the east and far north. Few population monitoring data exist: the Breeding Sawbill Surveys in 1987 and 1997 only targeted riverine systems and did not detect significant temporal trends in density (Armitage *et al.* 1997). The fact that numbers can vary markedly from year to year makes monitoring difficult (Marquiss & Duncan 1993). Two possible explanations for the observed range contraction are predation by American Mink and shooting, particularly on salmonid rivers where there is perceived predation upon fish stocks (Russell *et al.* 1996; Harris *et al.* 2008).

BREEDING DISTRIBUTION *2008–11*

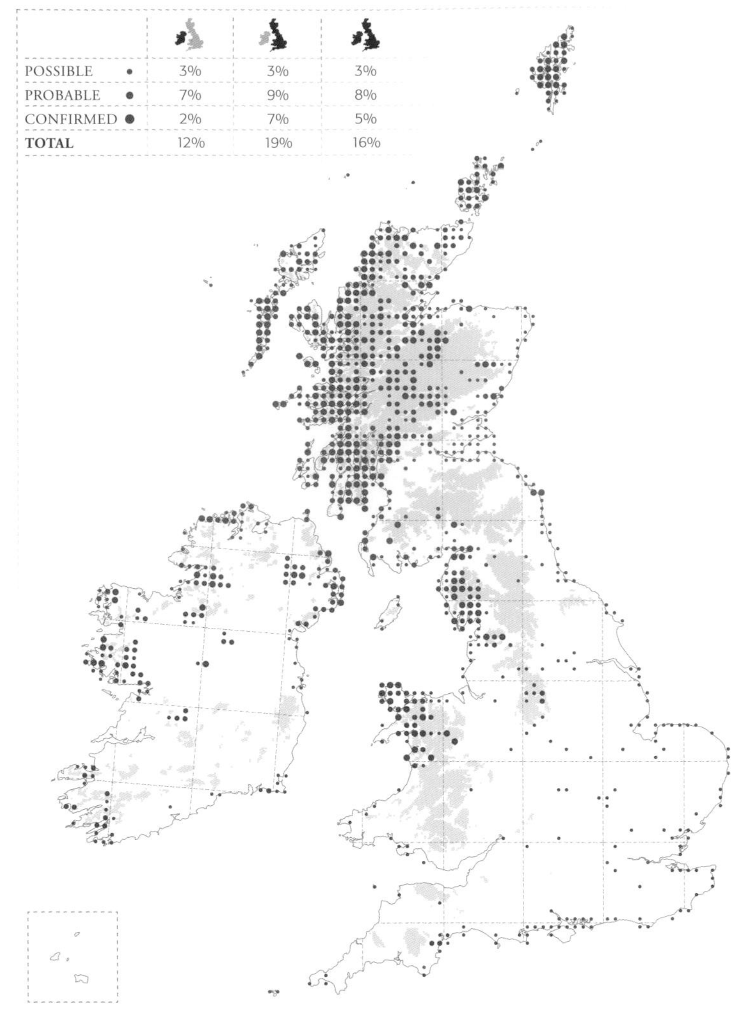

POSSIBLE •	3%	3%	3%
PROBABLE •	7%	9%	8%
CONFIRMED •	2%	7%	5%
TOTAL	12%	19%	16%

WINTER DISTRIBUTION *2007/08–2010/11*

PRESENT •	21%	35%	31%

Sponsored by Anne Reid

BREEDING DISTRIBUTION CHANGE *since 1968–72*

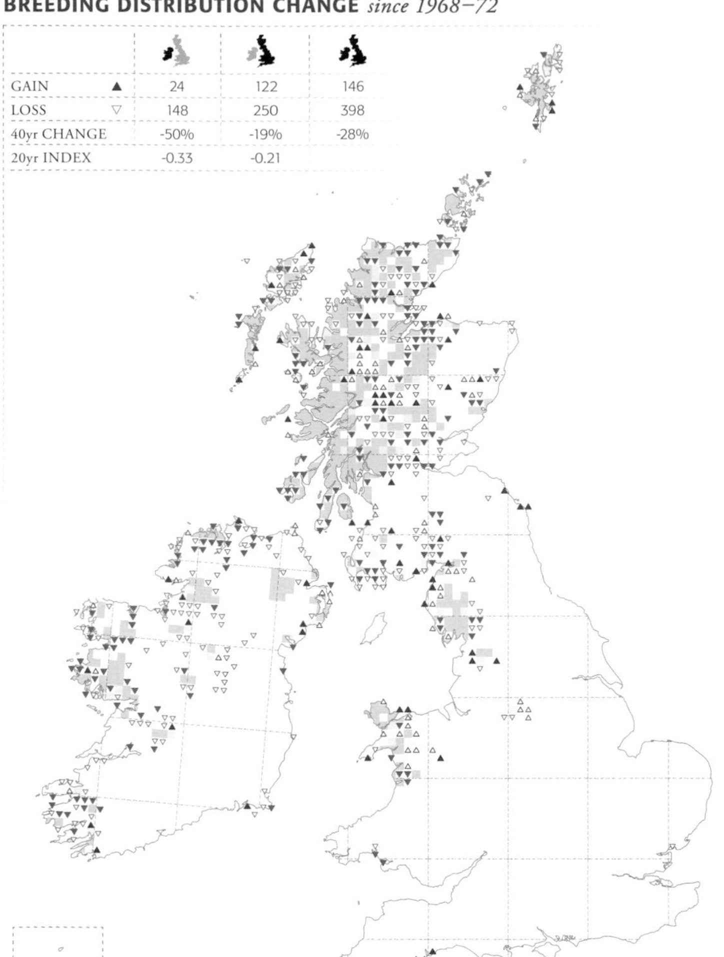

GAIN ▲	24	122	146
LOSS ▽	148	250	398
40yr CHANGE	-50%	-19%	-28%
20yr INDEX	-0.33	-0.21	

BREEDING RELATIVE ABUNDANCE *2008–11*

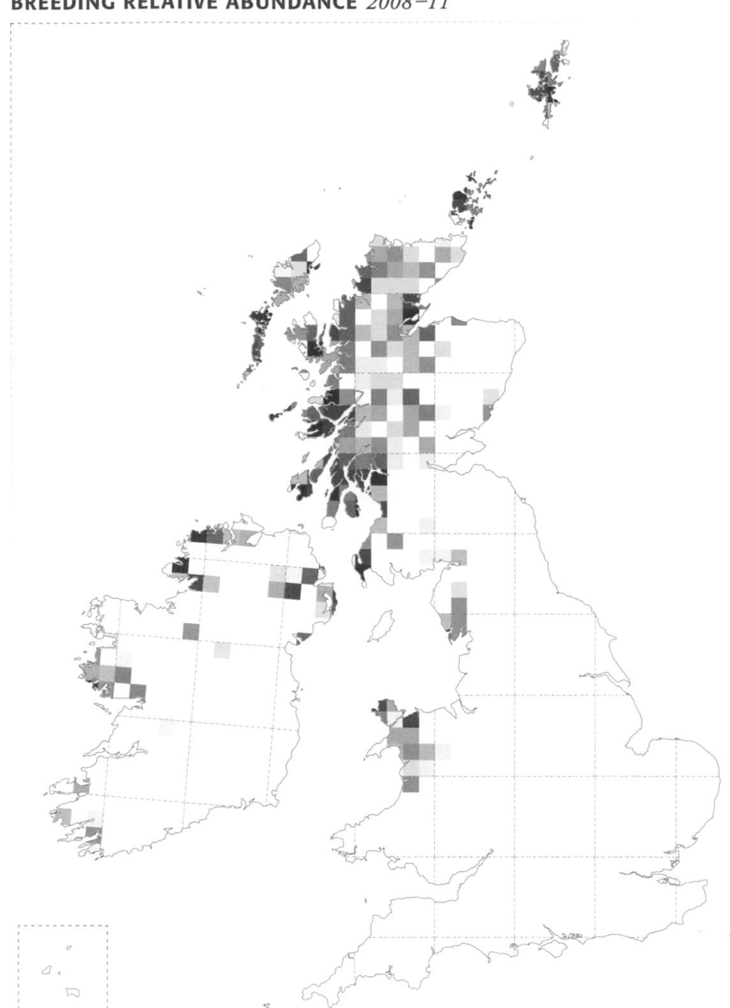

WINTER DISTRIBUTION CHANGE *since 1981–84*

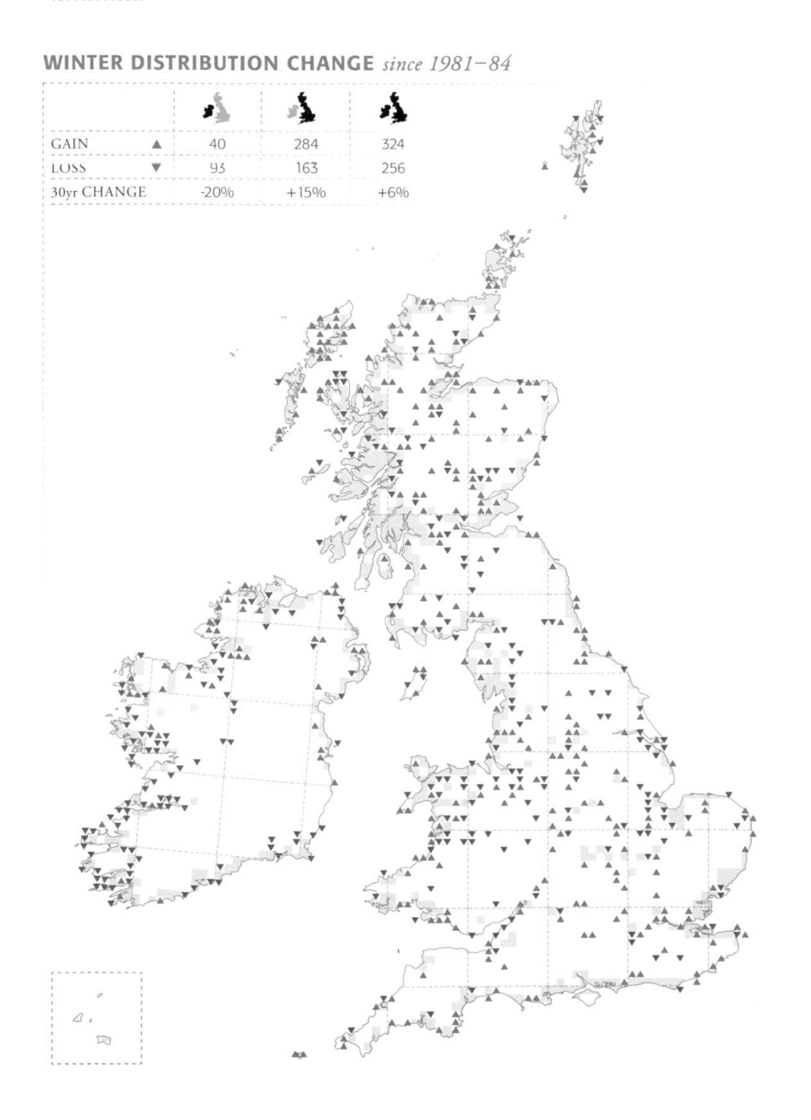

GAIN ▲	40	284	324
LOSS ▼	93	163	256
30yr CHANGE	-20%	+15%	+6%

WINTER RELATIVE ABUNDANCE *2007/08–2010/11*

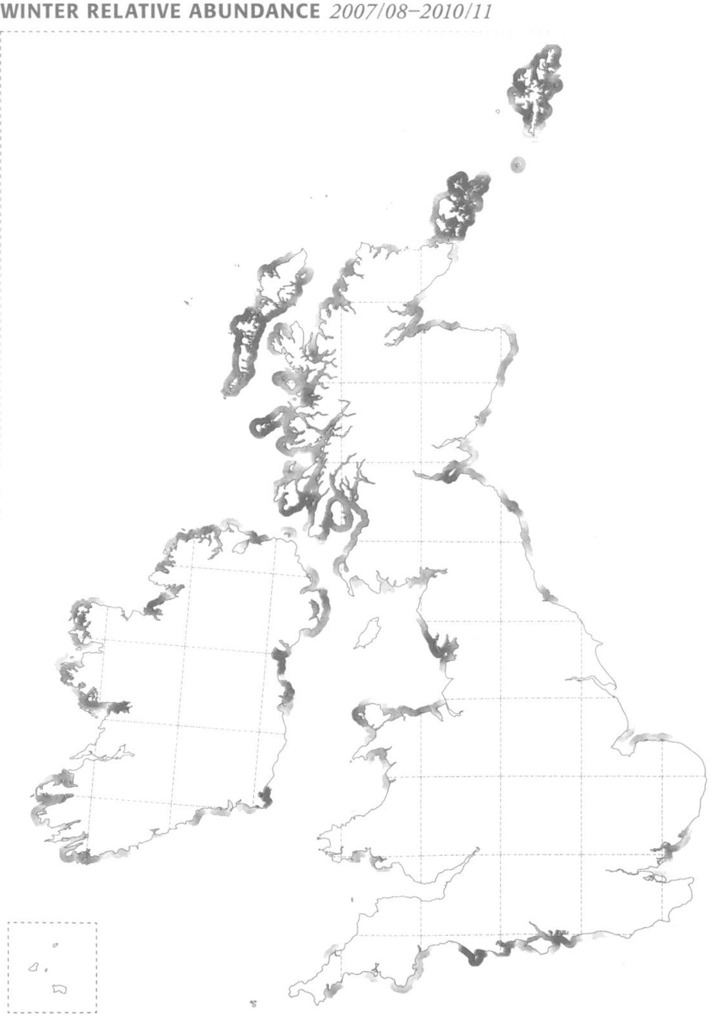

GOOSANDER

Mergus merganser

BTO CODE: GD

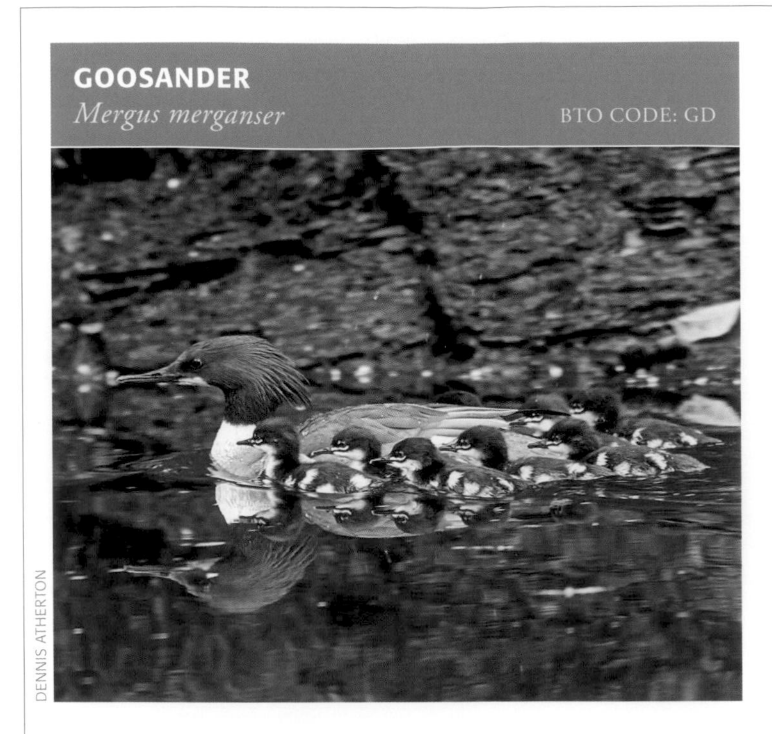

DENNIS ATHERTON

GOOSANDERS BREED THROUGHOUT Scotland, and from northern England south to Derbyshire, Wales, the Marches and Devon. There were just two *confirmed* breeding records in Ireland. In Britain, breeding Goosanders have shown a significant range expansion over the last 40 years, a result supported by dedicated merganser surveys (Armitage *et al.* 1997; Rehfisch *et al.* 1999). There was an 87% increase in numbers during 1981–2010 (*BirdTrends*), whilst Atlas data reveal a 112% range expansion since the *1968–72 Breeding Atlas*.

The almost complete colonisation of Wales over the last 40 years is striking, as is ongoing southward expansion in northern England and the colonisation of southwest England. There has also been some extralimital expansion with *confirmed* breeding in Northamptonshire, East Anglia and Hampshire. Within the core areas there is little variation in abundance, although high-altitude areas tend to be avoided. Reasons for the increases are unclear, though the provision of nestboxes, warmer springs as a result of climate change, and the decline of the Pine Marten during the colonisation and expansion period in the late 1800s have been suggested as important factors (Kear 2003).

In winter, Goosanders occupy rivers and a wide range of freshwater bodies throughout Britain, as well as some western and northern estuaries. They are largely absent from upland areas in northern Scotland and from parts of southern and eastern England. They are relatively scarce in Ireland, being found mostly along the coastal fringe. Relative abundance was highest in central and eastern Scotland, from the Solway and northeast England down to the West Midlands and throughout much of Wales and the Marches. Over 200 individuals are regularly reported from each of the Firths of Solway, Tay and Forth (*WeBS Report 2011*).

The winter range has expanded by 87% since the *1981–84 Winter Atlas*. Gains are apparent throughout and are particularly noticeable in Ireland, Wales and central and southern England. Some losses have also occurred, mostly in northern Scotland and to a lesser extent in north and southeast England. Licensed control of Goosanders is permitted where there are perceived impacts on fisheries (Harris *et al.* 2008); the effects of this and illegal killing on Goosander populations are unknown.

In memory of Roy Jessop
Staffordshire Branch of the West Midland Bird Club

BREEDING DISTRIBUTION *2008–11*

POSSIBLE ●	-	7%	5%
PROBABLE ●	<1%	9%	7%
CONFIRMED ●	<1%	15%	11%
TOTAL	<1%	31%	23%

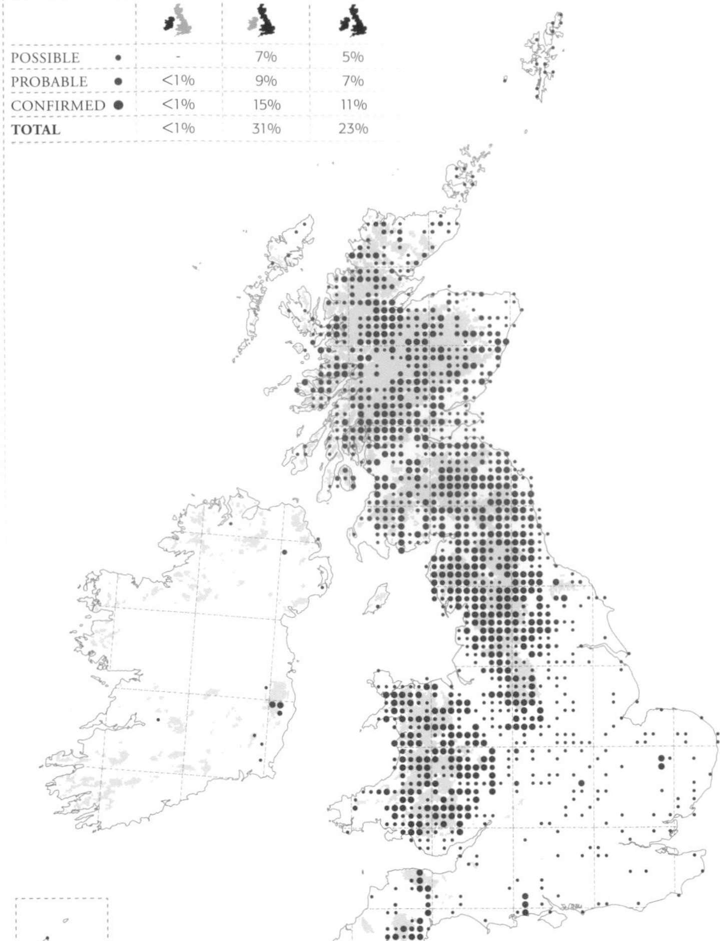

WINTER DISTRIBUTION *2007/08–2010/11*

PRESENT ●	8%	65%	50%

BREEDING DISTRIBUTION CHANGE *since 1968–72*

GAIN ▲	4	571	575
LOSS ▽	1	112	113
40yr CHANGE	+300%	+112%	+112%
20yr INDEX	-	+0.06	

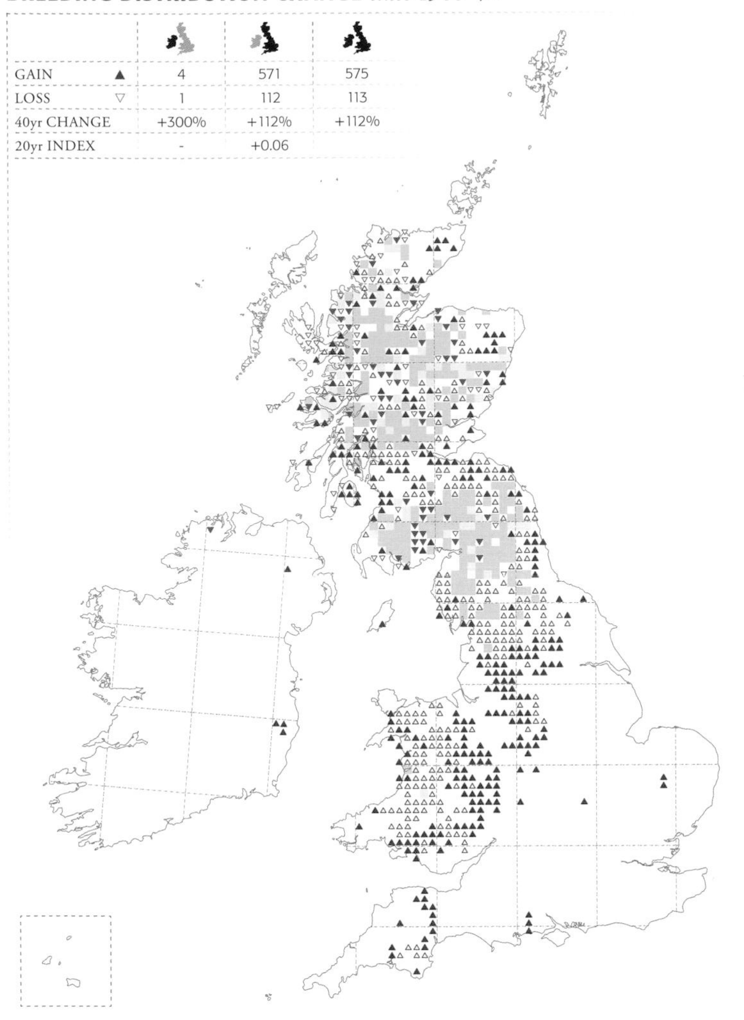

WINTER DISTRIBUTION CHANGE *since 1981–84*

GAIN ▲	76	940	1016
LOSS ▼	4	130	134
30yr CHANGE	+1800%	+81%	+87%

BREEDING RELATIVE ABUNDANCE *2008–11*

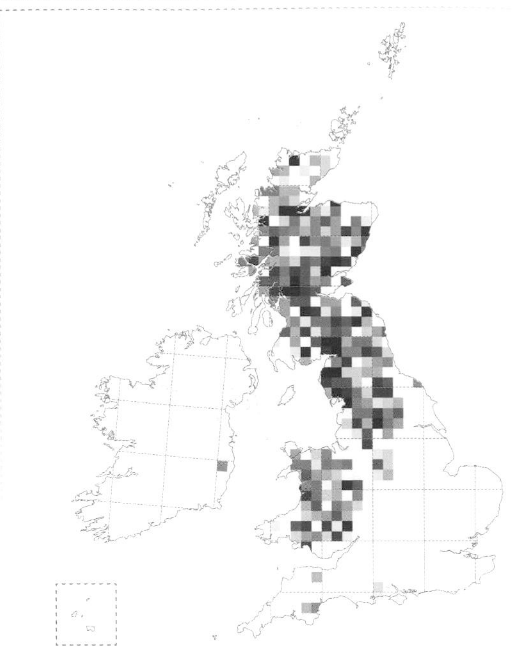

BREEDING RELATIVE ABUNDANCE CHANGE *since 1988–91*

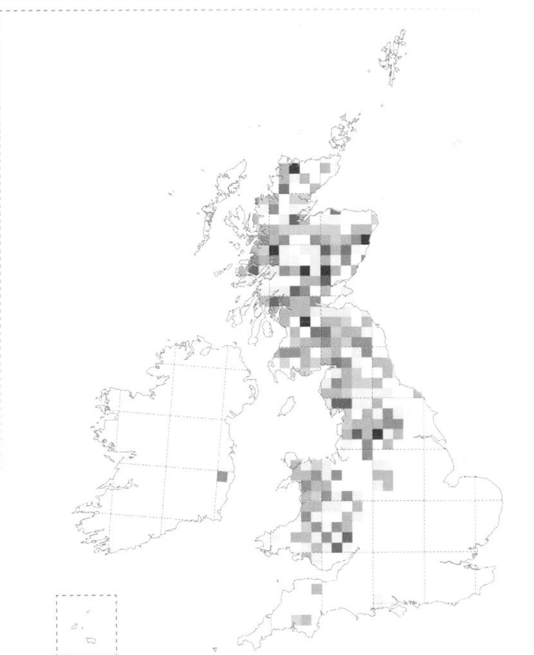

WINTER RELATIVE ABUNDANCE *2007/08–2010/11*

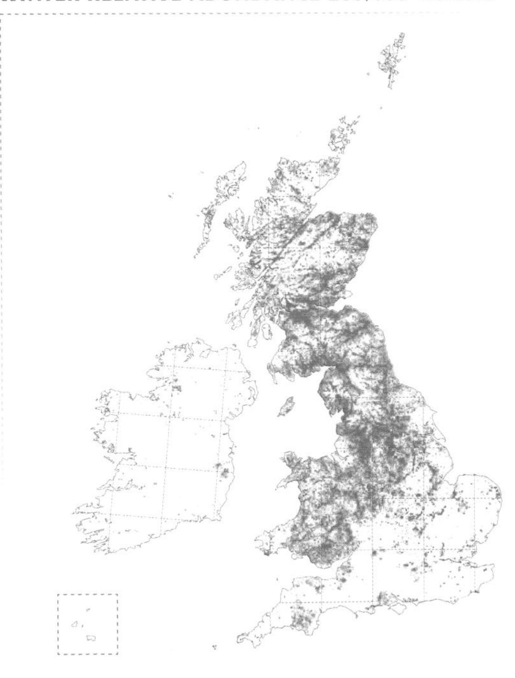

RUDDY DUCK
Oxyura jamaicensis

BTO CODE: RY

OLIVER SMART www.smartimages.co.uk

RUDDY DUCKS FORMED a small introduced population in southwest England following a series of escapes from WWT Slimbridge, Gloucestershire, in the late 1950s (Henderson 2006). Following breeding at Chew Valley Lake, Avon, in the 1960s, the breeding range spread to the West Midlands, where it was concentrated during the *1968–72 Breeding Atlas*, and then expanded rapidly during the 1980s and 1990s. The species was breeding in Northern Ireland, Scotland and Wales by the time of the *1988–91 Breeding Atlas*. The population peaked at c.6,000 birds in 2000 (Kershaw & Hughes 2002).

The appearance of populations in several European countries, and concerns over hybridisation with native White-headed Ducks in Spain, prompted plans for eradication of the introduced Ruddy Duck population in Britain & Ireland. Following research to identify effective culling methods, an eradication programme started in 2005 and by the second winter of the Atlas, the UK population had been reduced by almost 90% (Henderson 2009), and by 2012 to around 60 individuals (Henderson 2012). The distribution maps shown here largely reflect the status of the Ruddy Duck in the first seasons of Atlas fieldwork, revealing a patchy distribution throughout central and eastern Britain, together with scattered records elsewhere. It is likely that many of these areas were unoccupied by the final years of fieldwork.

The breeding change map is calculated from the *1988–91 Breeding Atlas* to illustrate changes since around the time of peak abundance. It shows many apparent gains, some of which will be breeding areas that were colonised after the *1988–91 Breeding Atlas* and still occupied in 2008. Others may reflect new locations occupied by birds disturbed from traditional sites during the eradication. Overall, the change map shows a 21% range contraction, but by the conclusion of Atlas fieldwork the contraction was far greater. Indicative of the ongoing eradication, the BirdTrack reporting rate for the Ruddy Duck dropped from c.4% of lists in 2005 to less than 0.5% in 2012 (*BirdTrack*). The winter change map shows a range expansion because at the time of the *1981–84 Winter Atlas* the population was still at a relatively low level.

In memory of Sandra Cooper

BREEDING DISTRIBUTION *2008–11*

POSSIBLE •	<1%	2%	2%
PROBABLE •	<1%	4%	3%
CONFIRMED •	<1%	2%	2%
TOTAL	<1%	8%	7%

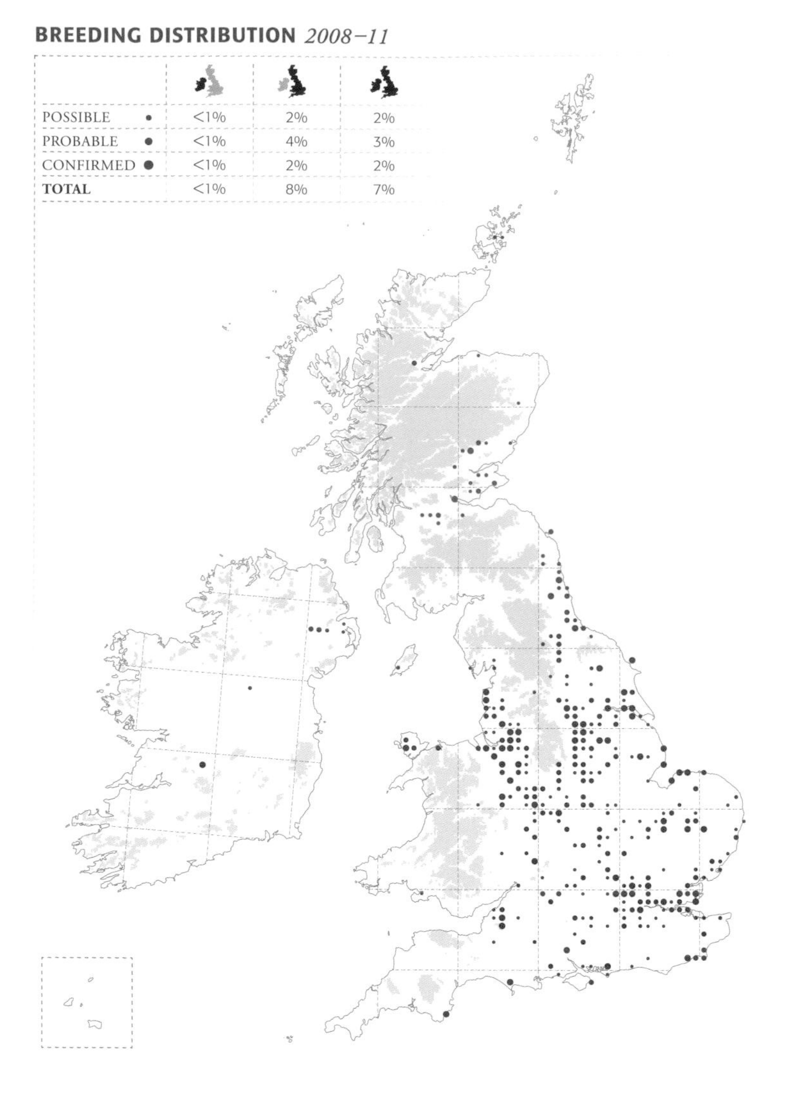

WINTER DISTRIBUTION *2007/08–2010/11*

PRESENT •	2%	11%	9%

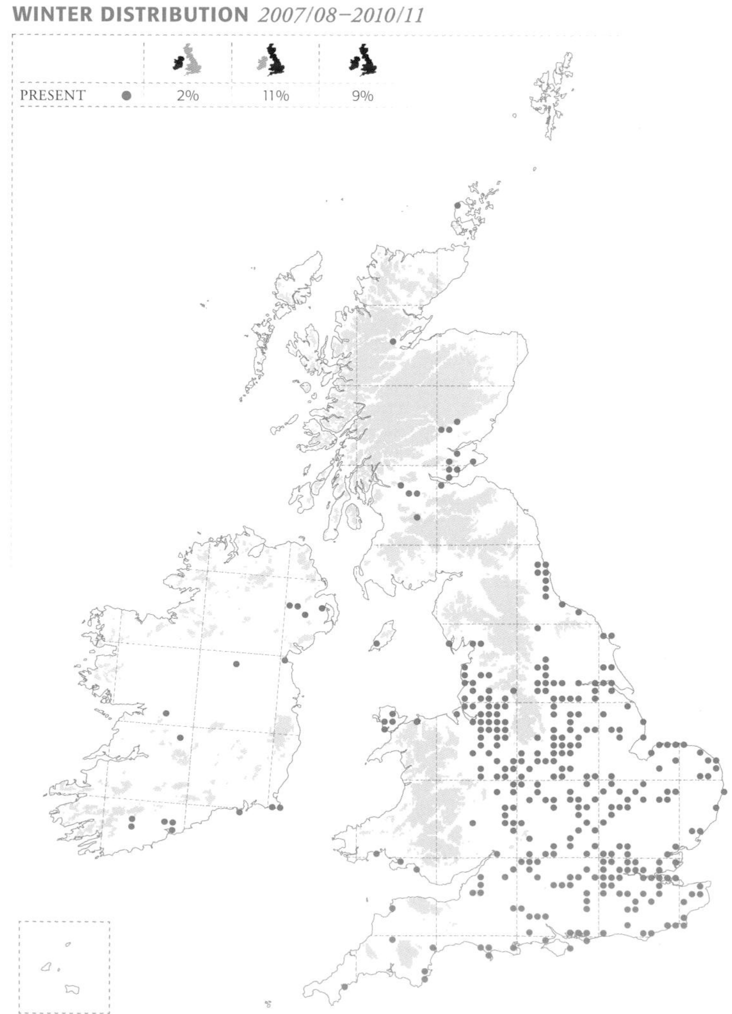

BREEDING DISTRIBUTION CHANGE *since 1988–91*

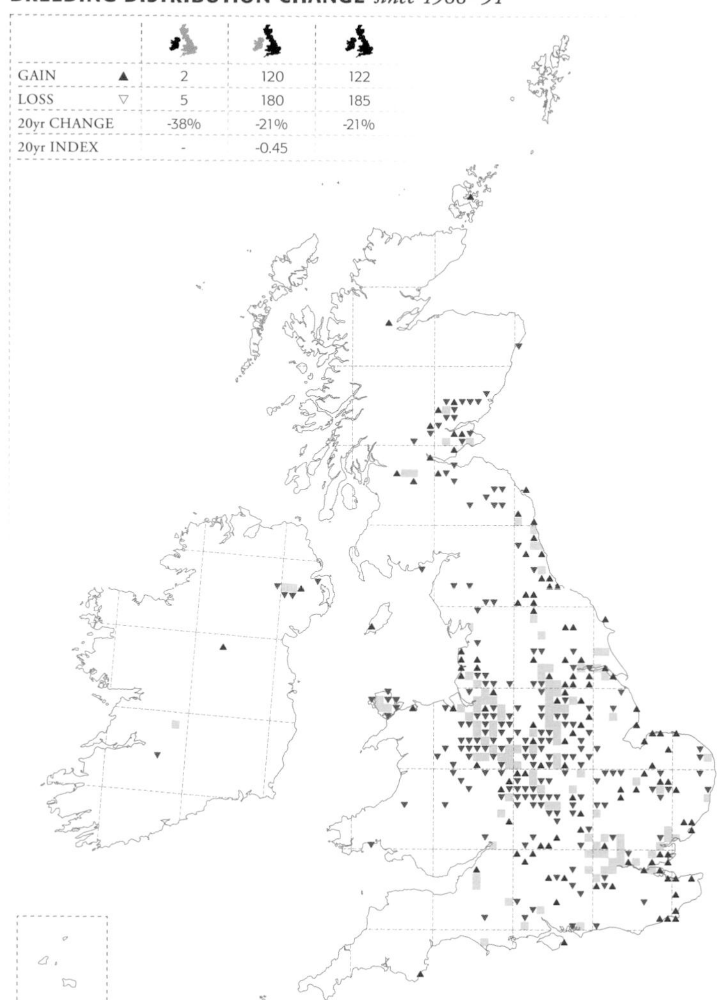

GAIN ▲	2	120	122
LOSS ▽	5	180	185
20yr CHANGE	-38%	-21%	-21%
20yr INDEX	-	-0.45	

BREEDING RELATIVE ABUNDANCE *2008–11*

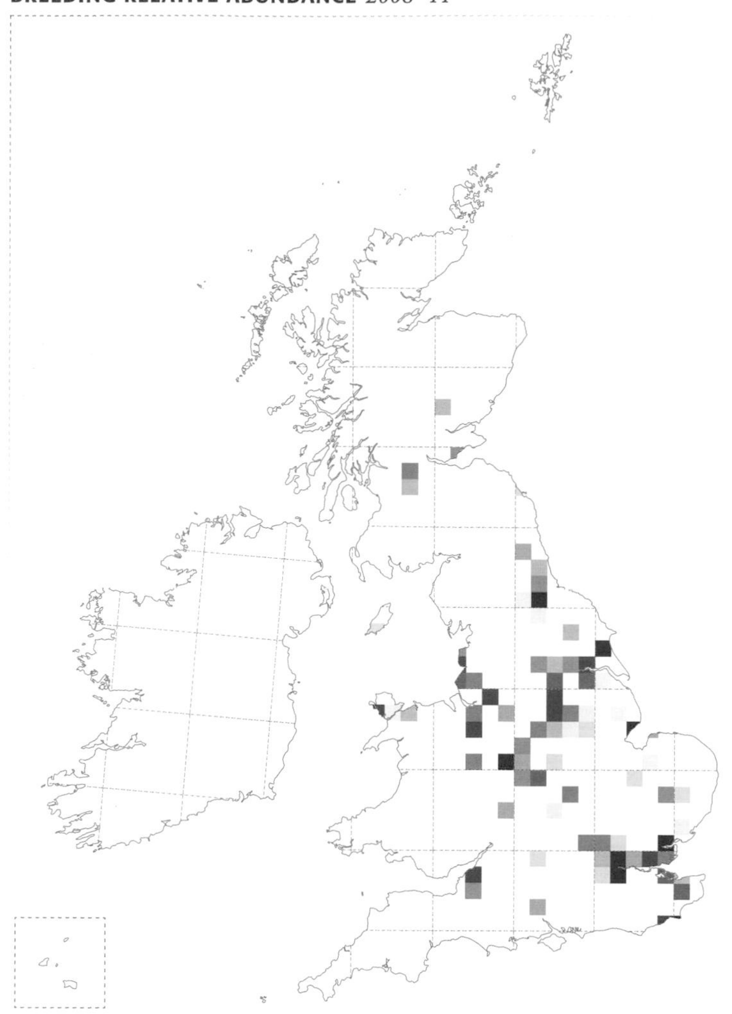

WINTER DISTRIBUTION CHANGE *since 1981–84*

GAIN ▲	12	159	171
LOSS ▼	4	96	100
30yr CHANGE	+133%	+26%	+29%

WINTER RELATIVE ABUNDANCE *2007/08–2010/11*

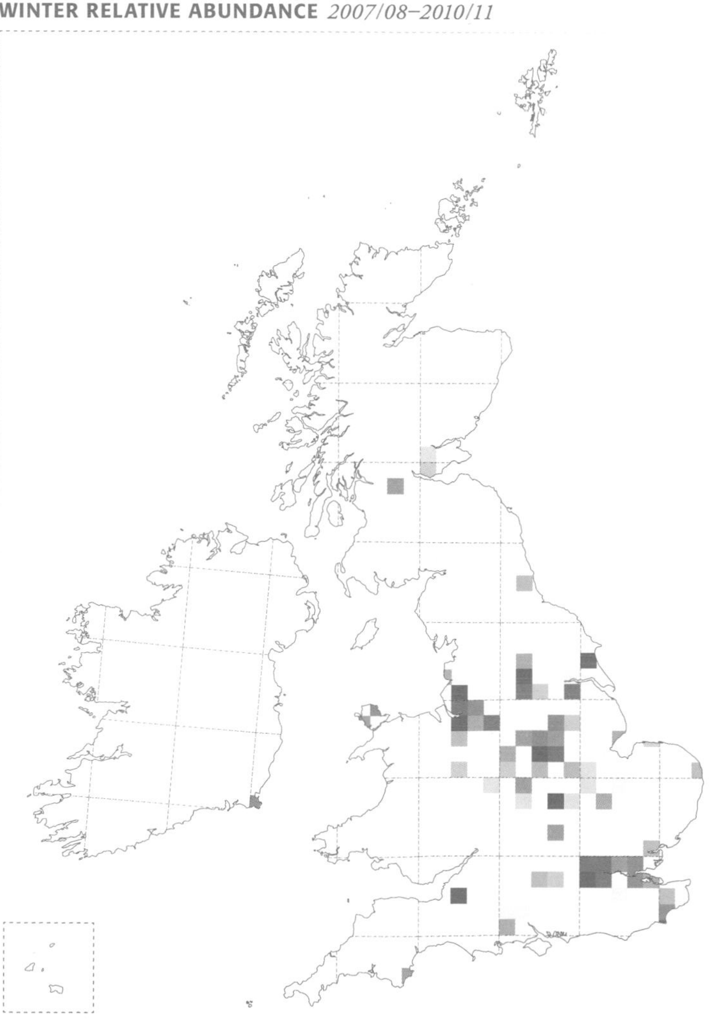

RED GROUSE
Lagopus lagopus

BTO CODE: RG

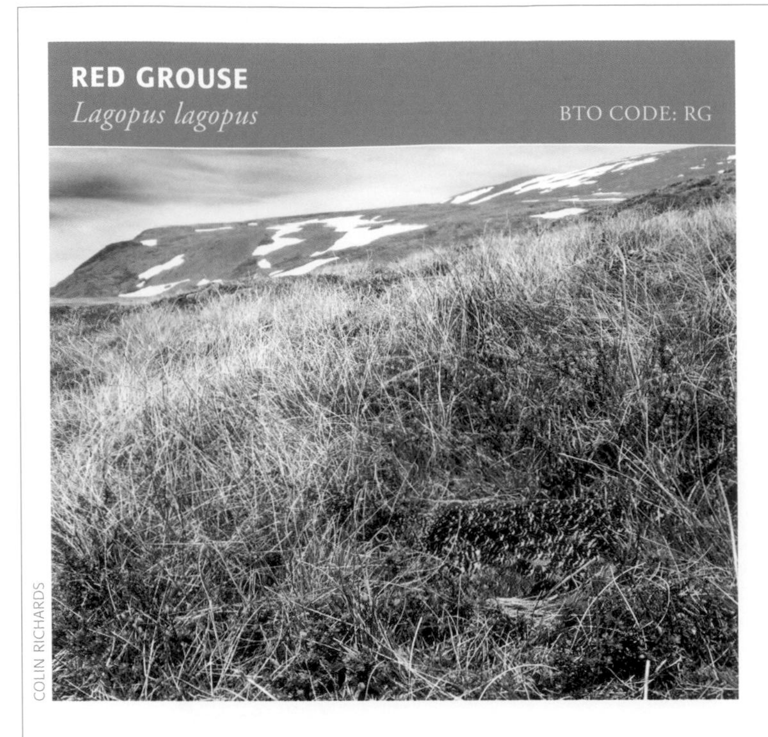

COLIN RICHARDS

THE RED GROUSE is characteristic of heather-dominated moorland in Britain and of raised and blanket bogs in Ireland. As a result of its highly sedentary nature, the winter and breeding distribution maps are similar and show presence where suitable upland habitat occurs. Likewise, the winter and breeding-season relative abundance maps are similar and show the highest densities occurring in the Pennines, North York Moors, Cheviot Hills, eastern Borders and the eastern Highlands of Scotland. Areas of high density are due largely to the management of heather moorlands on large estates to create a harvestable surplus for shooting purposes. On unmanaged moorlands, highest densities occur above the natural treeline and on areas with moorland soils derived from basic rocks rather than from acidic rocks (*Birds in Scotland*).

Within Britain there are regional differences in Red Grouse trends. Although in winter a 12% range expansion is shown, many of the gains were in areas where grouse occur at low density, and probably reflect improved coverage. They have increased in Shetland as reductions in stock levels have allowed heather to become better established (Shetland Bird Club 2007). However, in both seasons, significant losses are apparent in many areas. Eastern, southern and western Scotland and Wales show losses, particularly along the moorland edge. In southwest England the Exmoor population may be extinct: no records were received during Atlas fieldwork and the last report was in September 2009 (Devon Birdwatching & Preservation Society 2010). Many of the core areas also show reductions in abundance. Several factors may be implicated in the losses including overgrazing leading to a shift from heather to grass-dominated moors, afforestation reducing moorland habitats, and increased predator abundance (Thirgood *et al.* 2000; Allen *et al.* 2005).

In Ireland, Atlas data suggest a 66% breeding range contraction since 1968–72, which accords well with a 50% contraction over 40 years documented by dedicated grouse surveys (Cummins *et al.* 2010) and surveys in Northern Ireland suggesting a population of only 202–221 territories (Allen *et al.* 2005). Losses have taken place across the raised bogs of the midlands and the blanket bogs of the west and southwest, associated with habitat losses from afforestation, large-scale peat extraction and inappropriate burning (Cummins *et al.* 2010).

Sponsored by Gillian Stewart in memory of her husband John

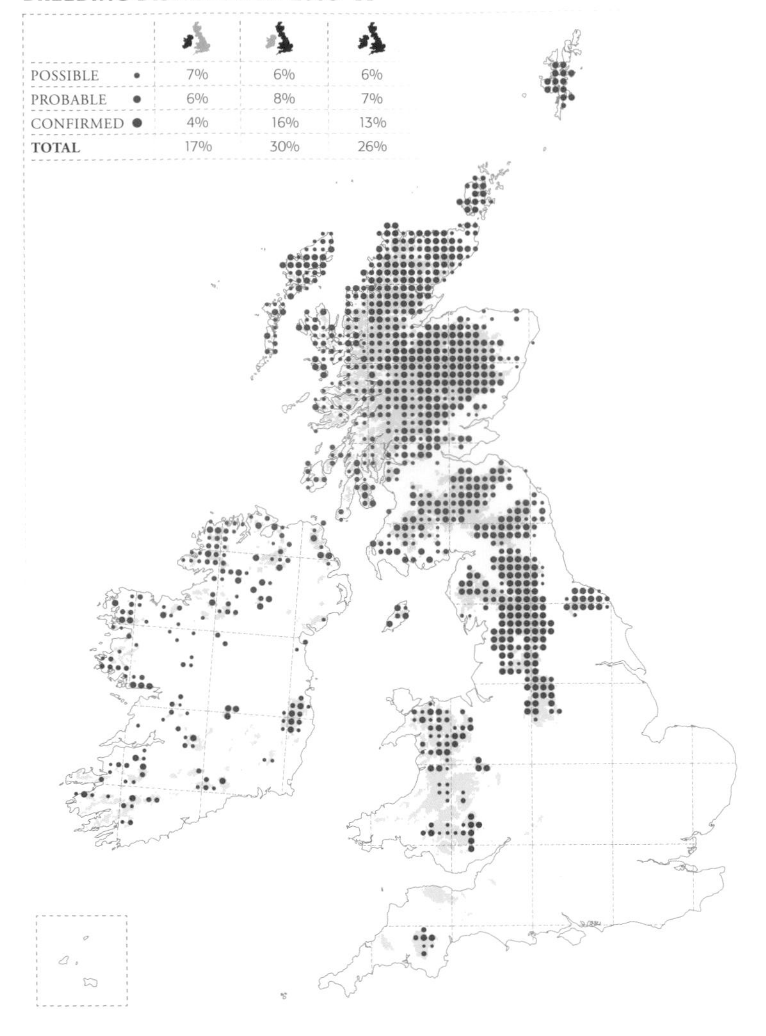

BREEDING DISTRIBUTION *2008–11*

	Ireland	Britain	Britain & Ireland
POSSIBLE •	7%	6%	6%
PROBABLE •	6%	8%	7%
CONFIRMED ●	4%	16%	13%
TOTAL	17%	30%	26%

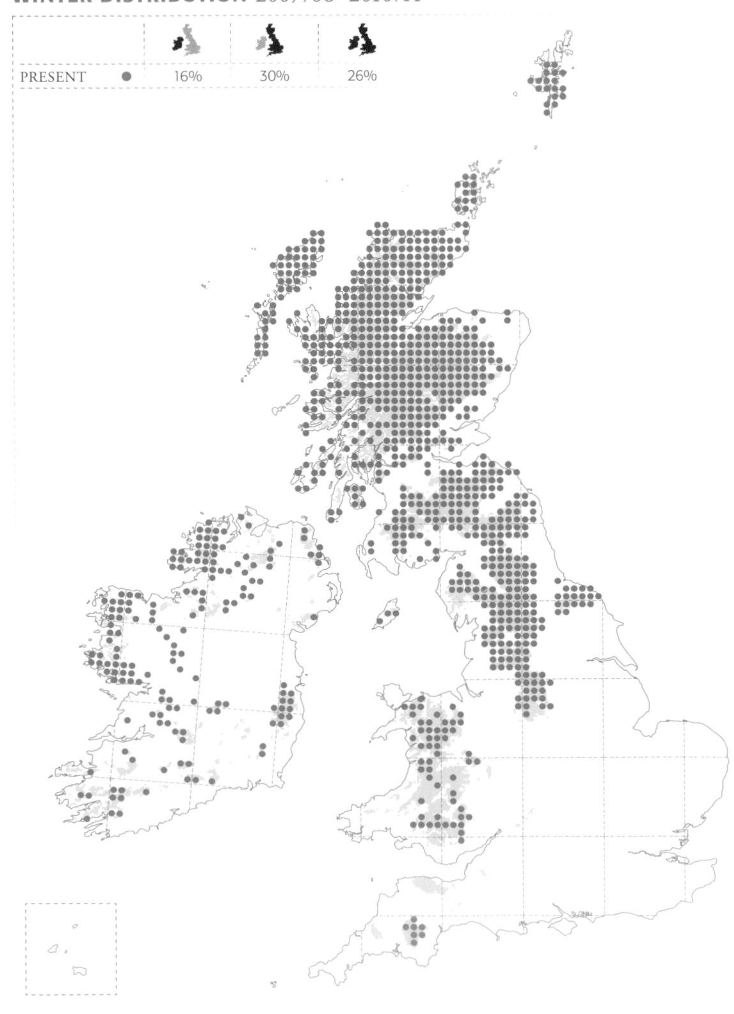

WINTER DISTRIBUTION *2007/08–2010/11*

	Ireland	Britain	Britain & Ireland
PRESENT ●	16%	30%	26%

BREEDING DISTRIBUTION CHANGE *since 1968–72*

GAIN ▲	19	62	81
LOSS ▼	296	302	598
40yr CHANGE	-66%	-22%	-34%
20yr INDEX	-0.40	-0.22	

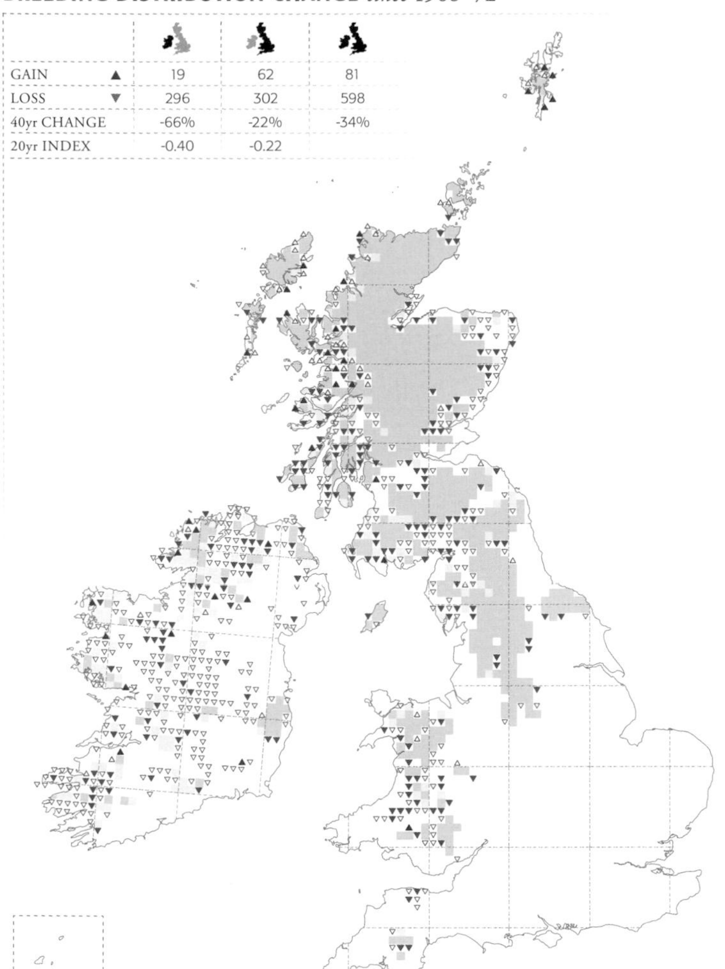

WINTER DISTRIBUTION CHANGE *since 1981–84*

GAIN ▲	83	240	323
LOSS ▼	86	153	239
30yr CHANGE	-2%	+12%	+9%

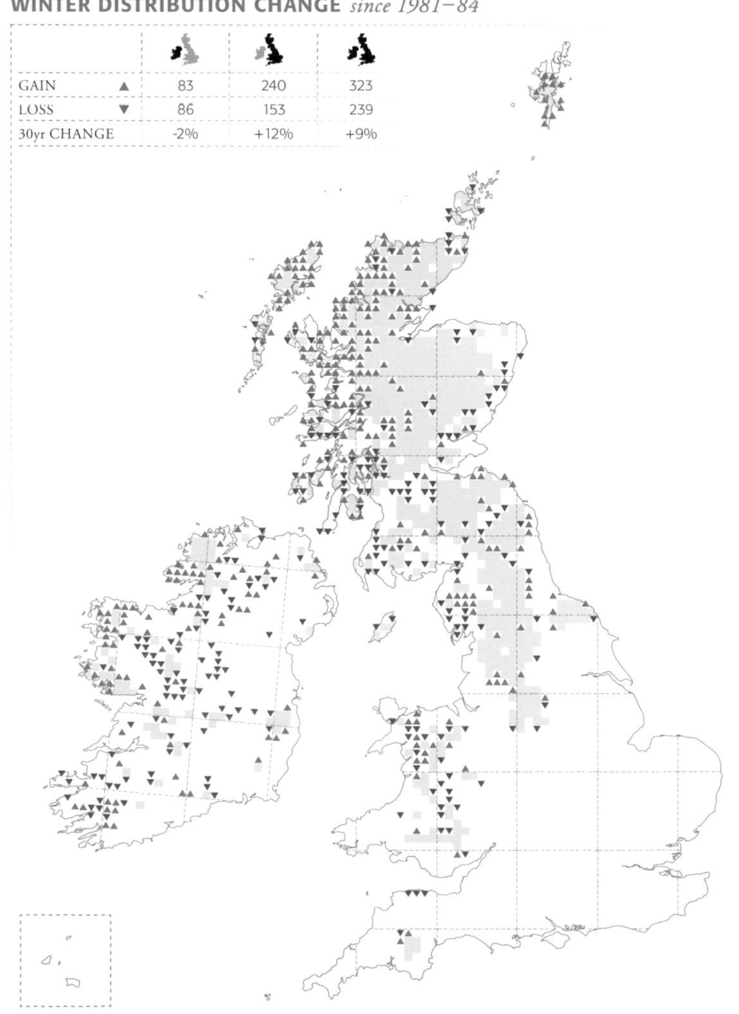

BREEDING RELATIVE ABUNDANCE *2008–11*

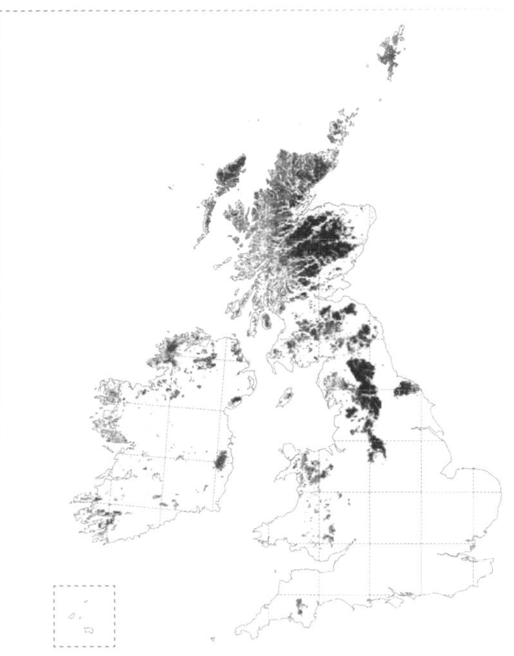

BREEDING RELATIVE ABUNDANCE CHANGE *since 1988–91*

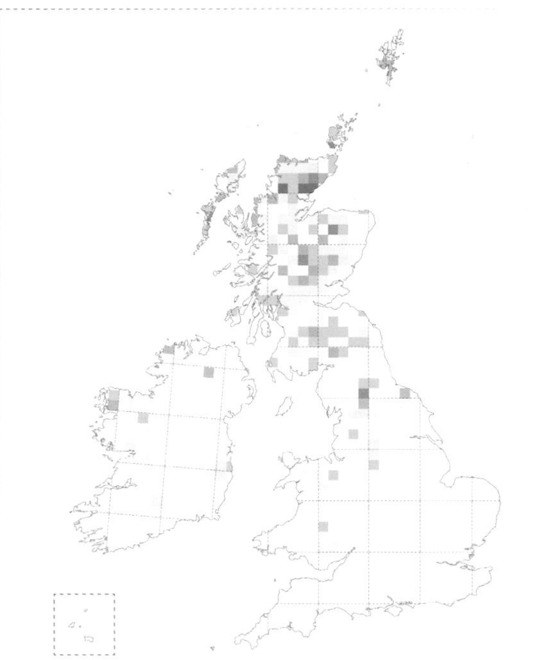

WINTER RELATIVE ABUNDANCE *2007/08–2010/11*

PTARMIGAN

Lagopus muta

BTO CODE: PM

JOHN CAMP

THE PTARMIGAN OCCURS throughout the high tops of the Scottish Highlands, favouring montane habitats. Birds are resident in this harsh environment, though during prolonged periods of snowfall they are sometimes forced down to lower altitudes. Some records from lower altitudes were reported during the Atlas, for example, in NH83 (East Inverness-shire) in December 2009.

There is little difference between the winter and breeding-season distribution maps, which show presence in the Grampian Mountains and northwest Highlands. There are also outliers in Caithness and Moray and on Skye, Mull and Arran; it is possible that where they occur at very low densities they have not been detected by atlas methods.

The winter change map shows a 64% increase in the number of occupied 10-km squares, but this is likely to be due to improved coverage during *Bird Atlas 2007–11*, particularly through help from mountaineers, who were encouraged to submit sightings. In contrast, the breeding change map shows an 11% range contraction since the *1968–72 Breeding Atlas*. This was the net result of apparent gains in 16 10-km squares, which are probably a result of improved coverage, and loss from 38 squares. Losses have mostly occurred in more marginal areas on lower hills towards the edge of the range, particularly in the southwest. Loss of food plants may have occurred through overgrazing by sheep (Francis & Cook 2011). The effects of changes in climate on high-altitude specialists like the Ptarmigan are currently unknown.

Habitat changes have the potential to lead to local losses. For example, Watson & Moss (2004) reported adverse impacts on numbers and breeding success during 1967–96 at a ski resort in the Cairngorms, where Ptarmigans normally show 10-year population cycles (Watson *et al.* 2000). On four areas they studied breeding success, population cycles and predation rates by Carrion Crows, which colonised following the development. On the most developed area near the main car park, where Ptarmigans occurred at high density, predation rates by crows were high, they reared few broods and birds died flying into the ski-lift wires. Breeding success improved at study sites farther away from the development.

Sponsored by Scottish Countryside Alliance

BREEDING DISTRIBUTION *2008–11*

POSSIBLE •	-	1%	1%
PROBABLE •	-	1%	1%
CONFIRMED •	-	3%	2%
TOTAL	-	5%	4%

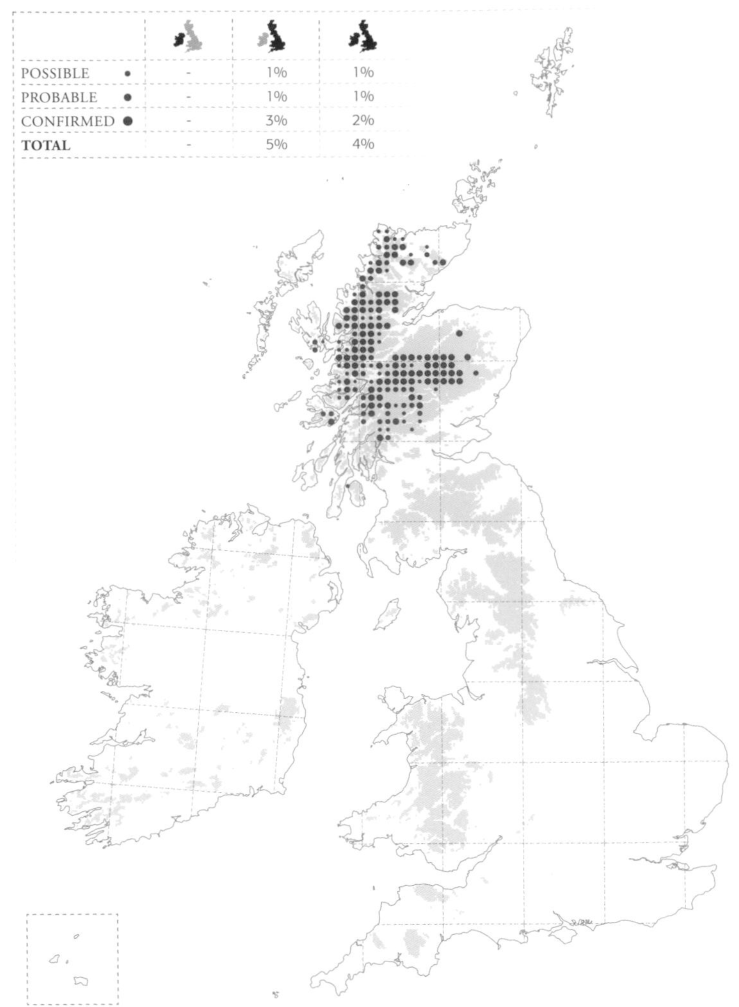

WINTER DISTRIBUTION *2007/08–2010/11*

PRESENT •	-	5%	4%

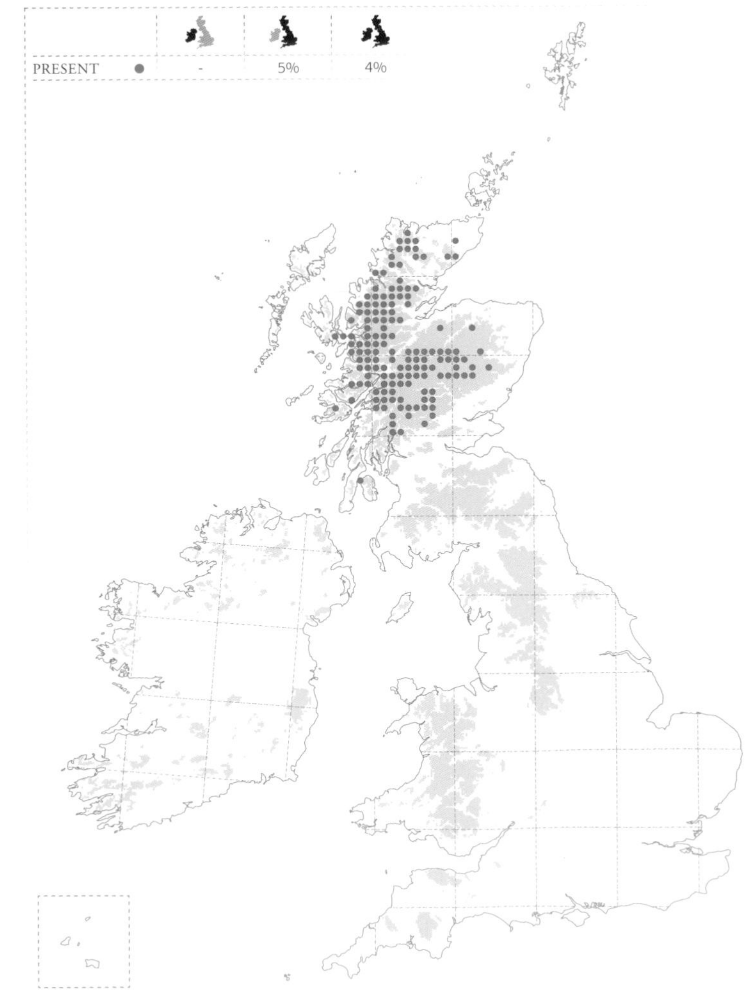

BREEDING DISTRIBUTION CHANGE *since 1968–72*

GAIN ▲	-	16	16
LOSS ▽	-	38	38
40yr CHANGE	-	-11%	-11%
20yr INDEX	-	-0.20	

WINTER DISTRIBUTION CHANGE *since 1981–84*

GAIN ▲	-	73	73
LOSS ▼	-	20	20
30yr CHANGE	-	+64%	+64%

BREEDING RELATIVE ABUNDANCE *2008–11*

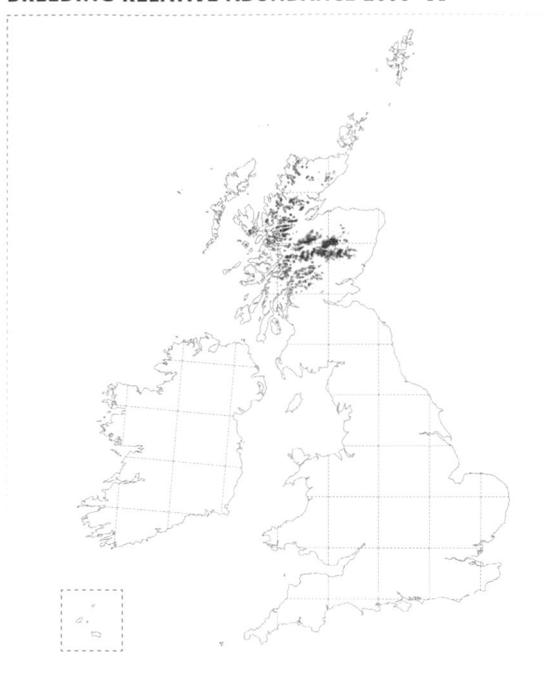

BREEDING RELATIVE ABUNDANCE CHANGE *since 1988–91*

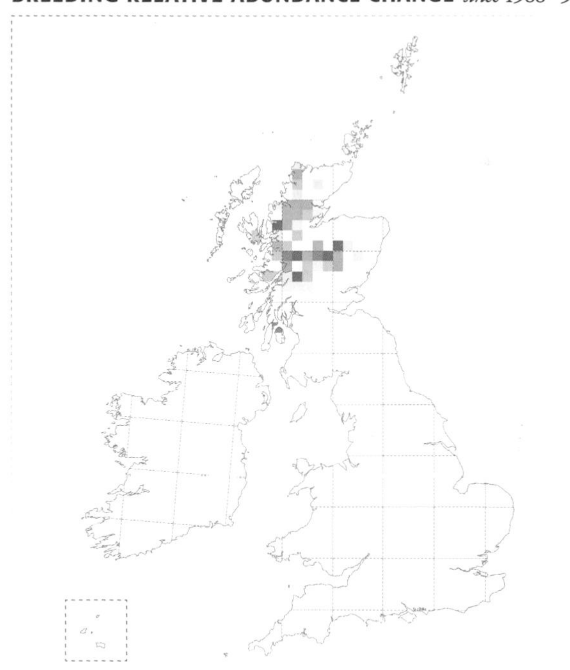

WINTER RELATIVE ABUNDANCE *2007/08–2010/11*

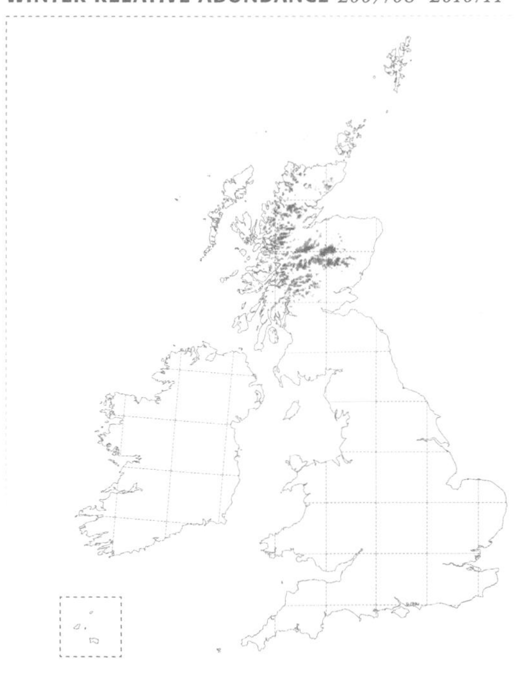

BLACK GROUSE

Tetrao tetrix

BTO CODE: BK

DAVID KJAER www.davidkjaer.com

THE BLACK GROUSE is a resident species found in upland areas of Britain, where it favours a mix of moorland fringes, marginal farmland and woodland edge. It declined steeply in Britain throughout the 20th century and as a result has been the subject of much research and conservation action to reverse these declines (Grant *et al.* 2009).

Black Grouse are largely sedentary, and the distribution and relative abundance maps show much similarity between winter and the breeding season. Winter and breeding change maps again show similar patterns. The 29% contraction in breeding range since the *1968–72 Breeding Atlas* is especially notable, while losses are approximately balanced by gains when compared with the *1988–91 Breeding Atlas* and there is a small 5% loss since the *1981–84 Winter Atlas*. Losses in both seasons are evident from much of Wales, parts of northern England, and southern and central Scotland. A local extinction on Islay is particularly notable. The population in the Peak District became extinct in 2000, though a large reintroduction project started in 2003. Losses are also evident from tetrads within the core breeding range.

Wide-scale habitat and land-use changes have been the main cause of long-term declines of the Black Grouse in Britain (*Birds of Scotland*; Sim *et al.* 2008). Young plantations on moorland can be particularly attractive to Black Grouse, though second and subsequent rotation plantings are not nearly so readily used. Recent schemes to encourage the planting of native woodlands or to encourage the regeneration of upland woodlands have undoubtedly benefited this species and have resulted in range gains, particularly in northern Scotland. As plantations mature they become less suitable (Pearce-Higgins *et al.* 2007) and this may then result in losses, as is evident on the winter change map in parts of southwest and eastern Scotland. Black Grouse respond well to conservation prescriptions such as grazing modifications and predation management (Grant *et al.* 2009) and this has led to recent gains in marginal areas such as North Wales and increases in the North Pennines stronghold (Calladine *et al.* 2002; Warren & Baines 2008).

Sponsored by RSPB

BREEDING DISTRIBUTION *2008–11*

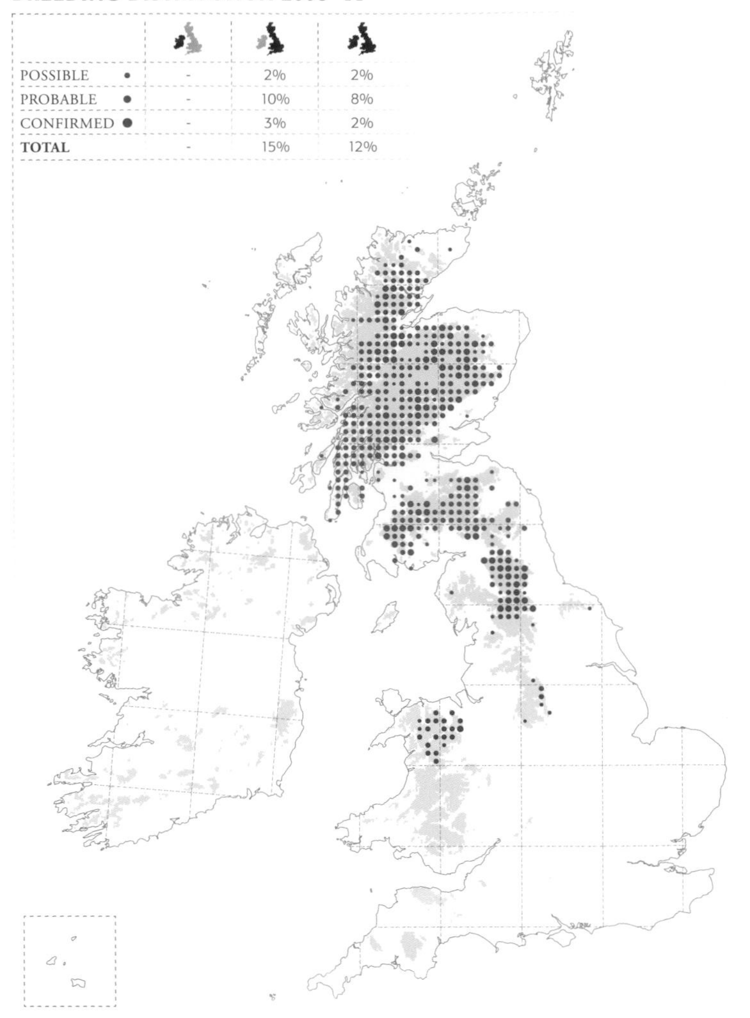

POSSIBLE ●	-	2%	2%
PROBABLE ●	-	10%	8%
CONFIRMED ●	-	3%	2%
TOTAL	-	15%	12%

WINTER DISTRIBUTION *2007/08–2010/11*

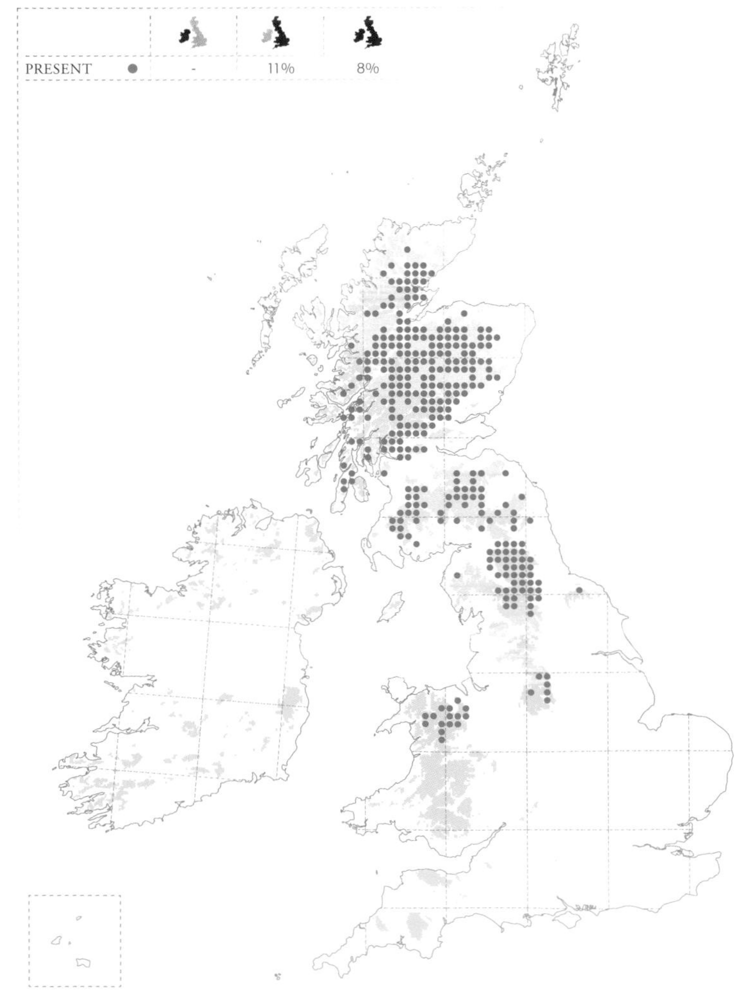

PRESENT ●	-	11%	8%

BREEDING DISTRIBUTION CHANGE *since 1968–72*

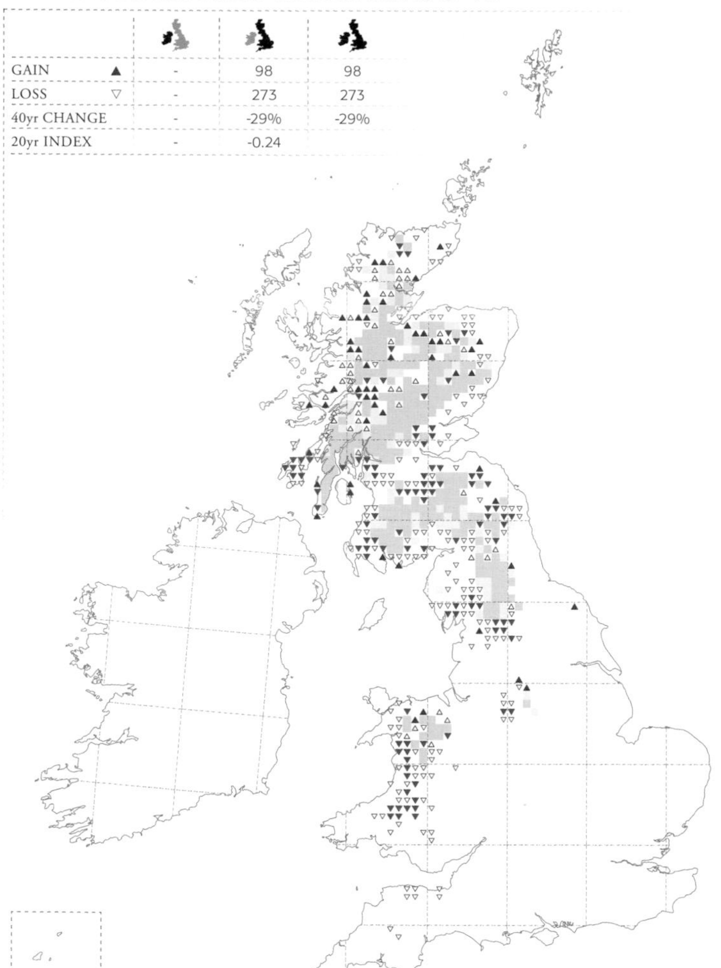

GAIN ▲	-	98	98
LOSS ▽	-	273	273
40yr CHANGE	-	-29%	-29%
20yr INDEX	-	-0.24	

WINTER DISTRIBUTION CHANGE *since 1981–84*

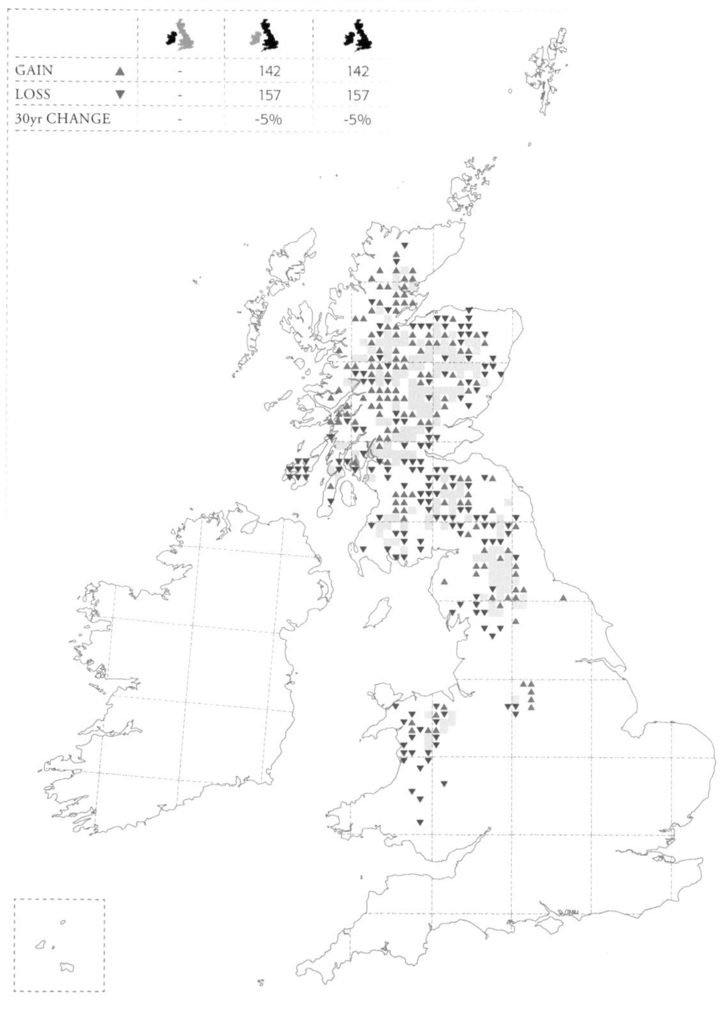

GAIN ▲	-	142	142
LOSS ▼	-	157	157
30yr CHANGE	-	-5%	-5%

BREEDING RELATIVE ABUNDANCE *2008–11*

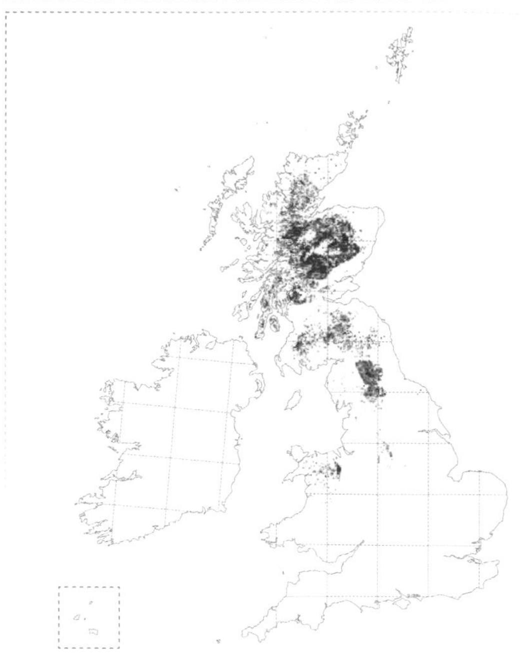

BREEDING RELATIVE ABUNDANCE CHANGE *since 1988–91*

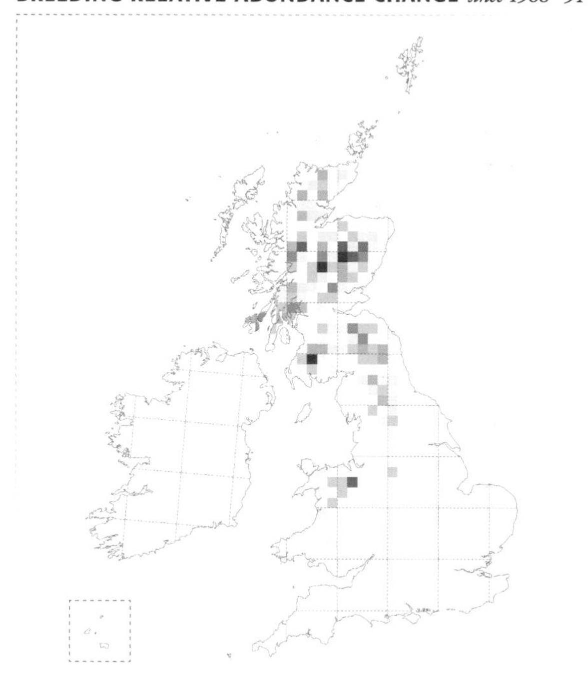

WINTER RELATIVE ABUNDANCE *2007/08–2010/11*

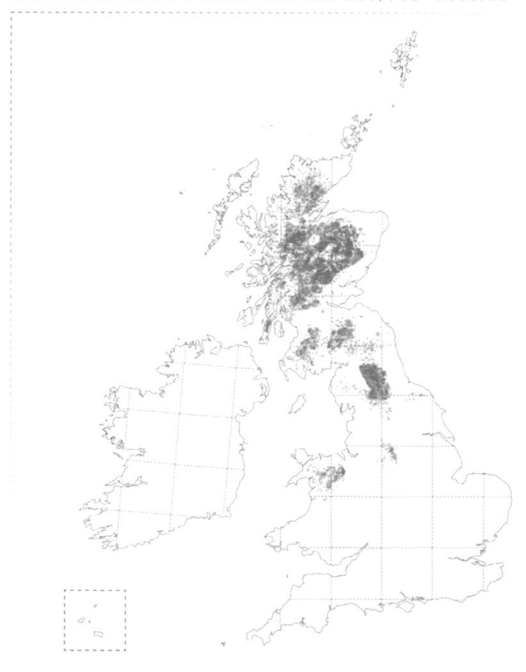

CAPERCAILLIE

Tetrao urogallus

BTO CODE: CP

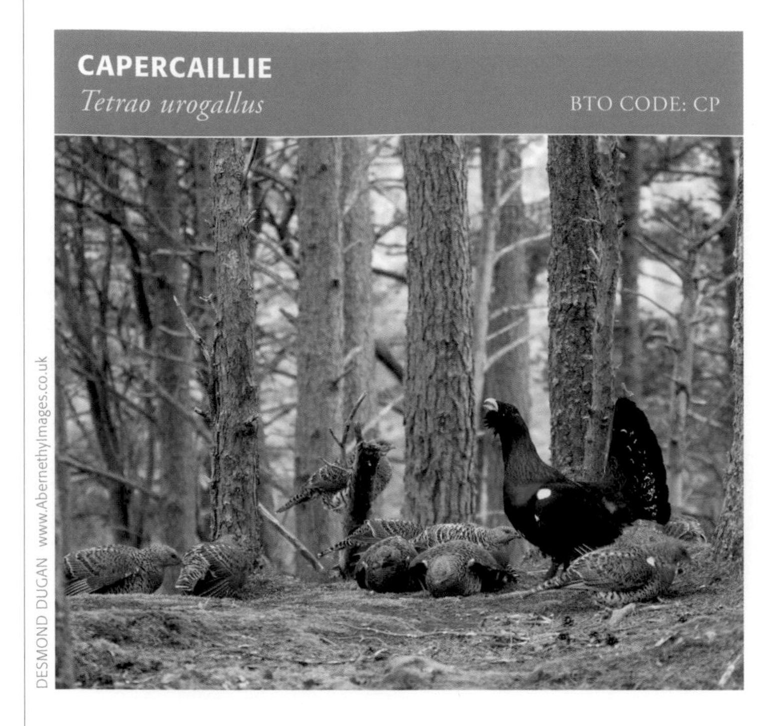

DESMOND DUGAN www.Abernethyimages.co.uk

THIS WOODLAND GROUSE is confined to pine forests in the north of Scotland. It is a resident species and is fairly sedentary, so it is no surprise that the distribution maps for winter and the breeding season are broadly similar. They highlight the core areas in Easter Ross, Strathspey and Aberdeenshire. Only a few other occupied sites remain outwith this area, in Inverness-shire, Perthshire and Dunbartonshire. Despite being large, Capercaillies can be difficult to locate in extensive forested areas, particularly when present at very low densities. This probably accounts for the discrepancy in apparent occupancy between the two distribution maps.

Capercaillie numbers have undergone a huge decline since the 1970s. This is reflected in a 55% decrease in the number of occupied 10-km squares since the *1981–84 Winter Atlas*. Through the three breeding atlases the range size has decreased from 182 10-km squares in 1968–72, through 66 in 1988–91 to 51 currently, an overall decline of 73% in squares surveyed in all atlases. Apparent losses are most noticeable in the southern, western and eastern parts of the former range. The first of four national surveys, undertaken during 1992–94, gave an estimated 2,200 birds (Catt *et al.* 1998). The *1988–91 Breeding Atlas* had just found a 64% contraction in range since 1968–72 which, along with evidence from bag returns, indicated that the population was already much reduced. The fourth survey, conducted in winter 2008/09, estimated a population size of only 1,285 individuals, of which 75% were found in the Badenoch and Strathspey regions (Ewing *et al.* 2012). Recent climate change has resulted in a preponderance of cooler and wetter summers, and this, together with increased predation, changes in forest habitat and collisions with forest fences, has been blamed for the decline (Moss *et al.* 2000, 2001; Baines *et al.* 2004).

Given the difficulty of detecting Capercaillies, it is little surprise that they were encountered on few TTVs and the relative abundance maps show the core areas only.

Sponsored by RSPB

BREEDING DISTRIBUTION *2008–11*

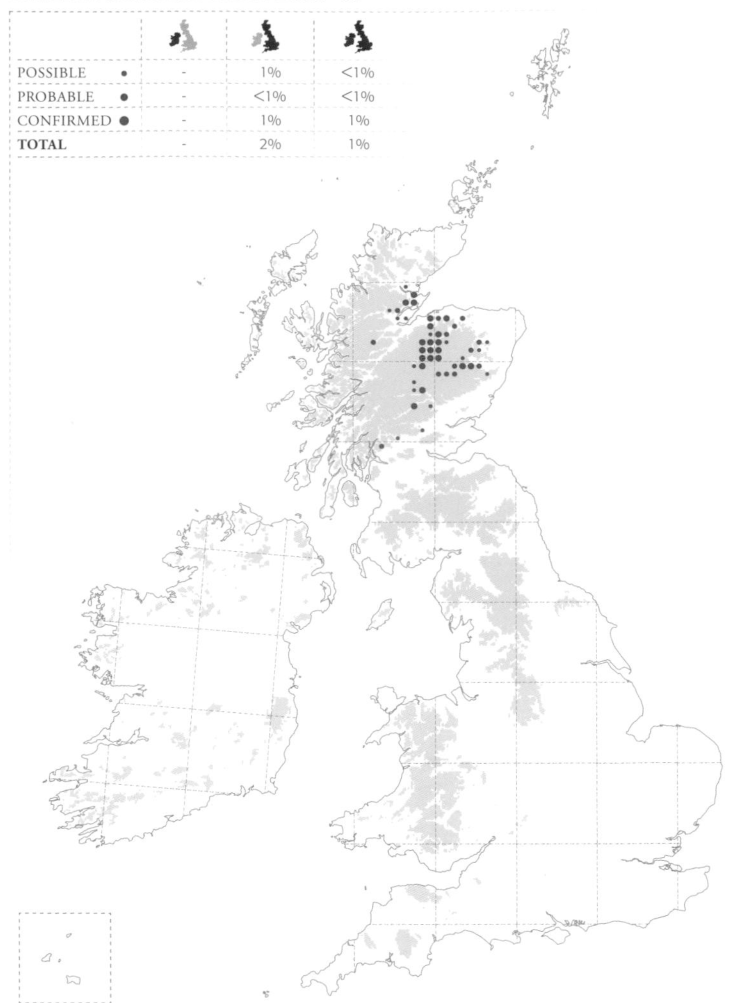

POSSIBLE •	-	1%	<1%
PROBABLE •	-	<1%	<1%
CONFIRMED •	-	1%	1%
TOTAL	-	2%	1%

WINTER DISTRIBUTION *2007/08–2010/11*

PRESENT •	-	1%	1%

BREEDING DISTRIBUTION CHANGE *since 1968–72*

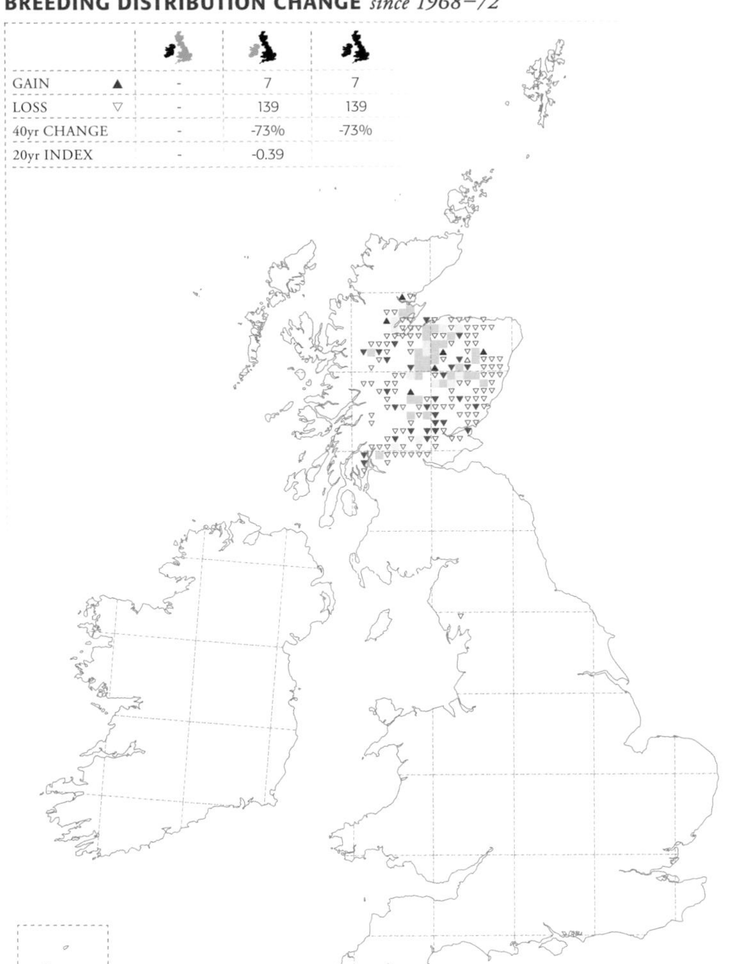

GAIN	▲	-	7	7
LOSS	▽	-	139	139
40yr CHANGE		-	-73%	-73%
20yr INDEX		-	-0.39	

BREEDING RELATIVE ABUNDANCE *2008–11*

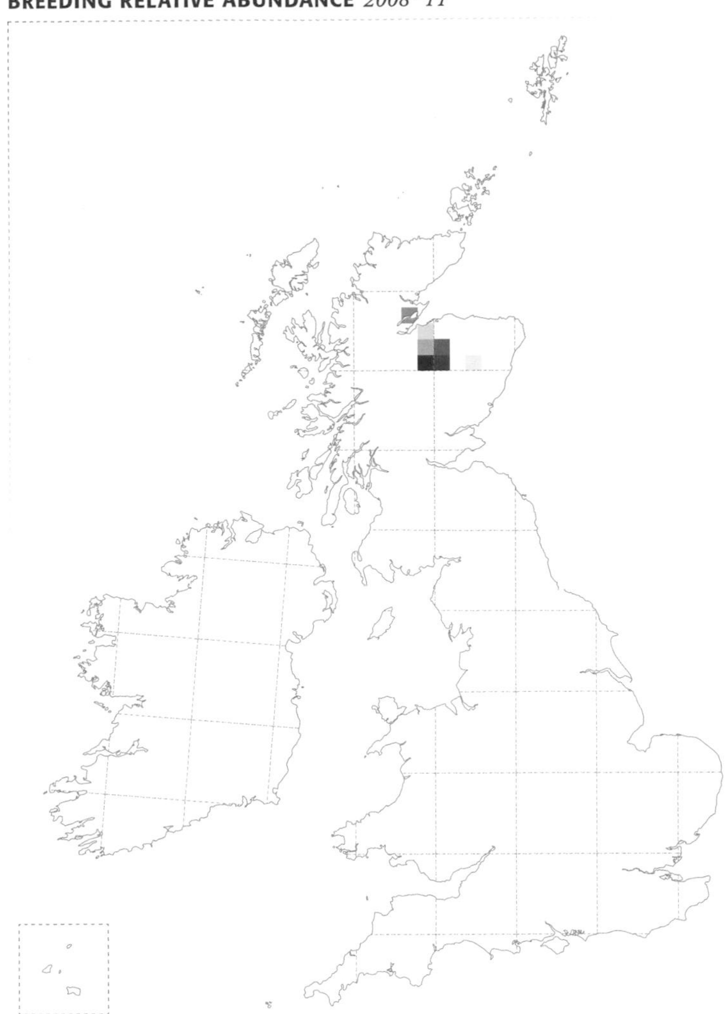

WINTER DISTRIBUTION CHANGE *since 1981–84*

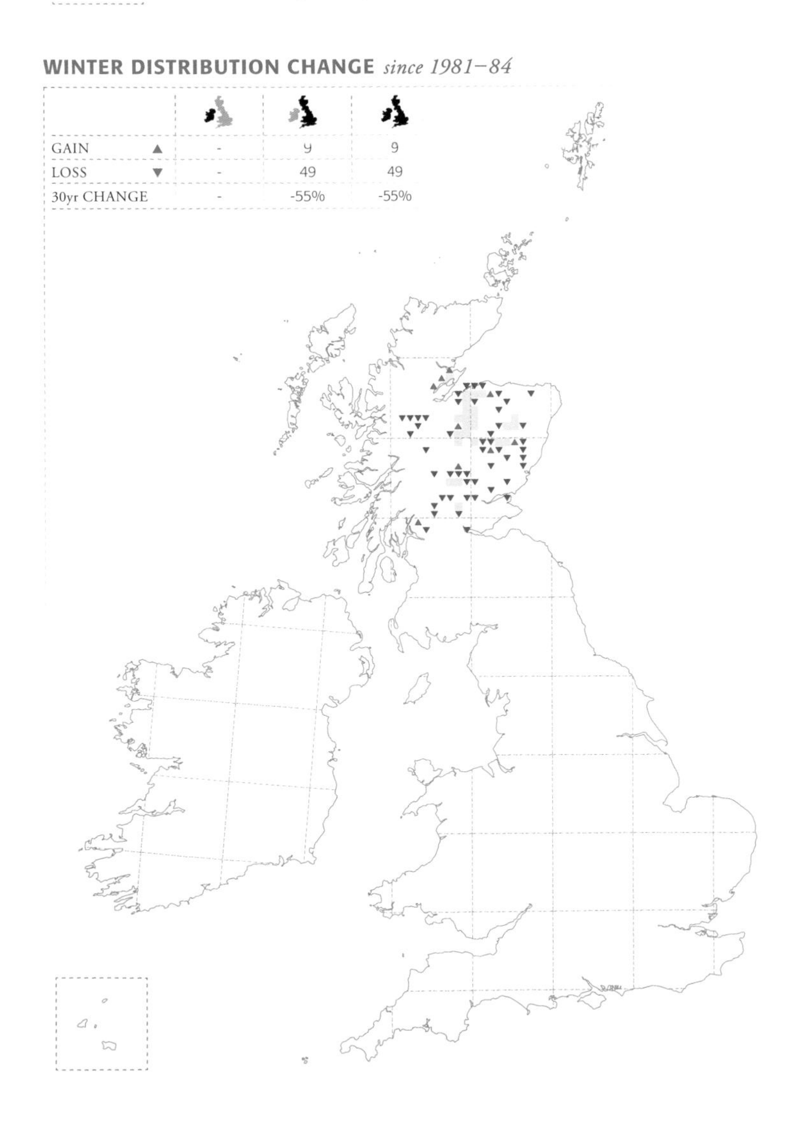

GAIN	▲	-	9	9
LOSS	▼	-	49	49
30yr CHANGE		-	-55%	-55%

WINTER RELATIVE ABUNDANCE *2007/08–2010/11*

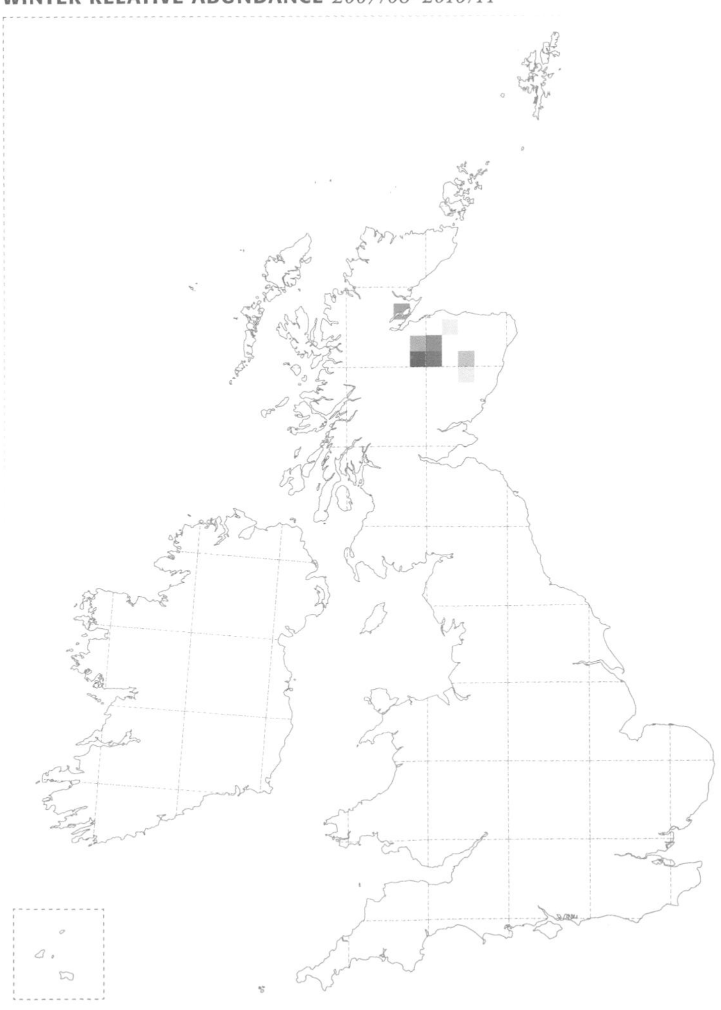

RED-LEGGED PARTRIDGE

Alectoris rufa

BTO CODE: RL

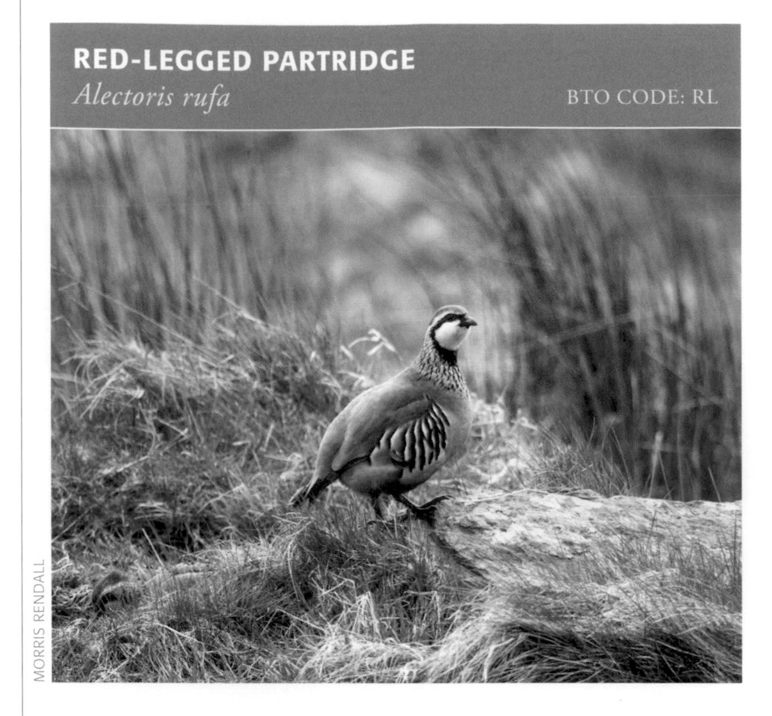

MORRIS RENDALL

RED-LEGGED PARTRIDGES are distributed widely across England, the eastern half and southwest of Scotland and the Isle of Man. In Wales they are found mainly in Anglesey and Pembrokeshire with scattered records elsewhere in the northeast and southeast. Records in Ireland are relatively scarce and come from a small scattering of 10-km squares in agricultural areas. They were recorded in the Channel Islands for the first time in any of the atlases. The distribution and relative abundance maps show that they are largely absent from upland and built-up areas and that abundance is greatest in East Anglia and the east Midlands.

The change maps for winter and the breeding season show an increase in the number of occupied 10-km squares in north, northwest and southwest England, Scotland, Wales and Ireland. These gains are likely to be due mainly to local releases, with some natural range expansion. As well as the increase in the number of occupied 10-km squares, there has been an increase in relative abundance in central southern England, northern England and parts of Scotland. However, within the stronghold areas of central and eastern England there is an indication of a decline in abundance. Between 1967 and 2010, the UK population has decreased by 19% (*BirdTrends*), although this conceals a recent small upturn between 1995 and 2010 (*BBS Report 2011*).

As more farms diversify into shooting, the number of Red-legged Partridges released has increased and this is illustrated by the National Gamebag Census, where numbers shot quadrupled between 1990 and 2005 (Game & Wildlife Conservation Trust 2013). It is estimated that 6.5 million partridges (Grey and Red-legged) were released across the UK in 2004, and 2.6 million were shot (Public and Corporate Economic Consultants 2006). There has been little research on the impacts of released birds on native species, but there is some evidence that shooting operations based on large-scale releases of Red-legged Partridges could be implicated in local extinctions of Grey Partridges (Watson *et al.* 2007).

BREEDING DISTRIBUTION *2008–11*

POSSIBLE •	2%	8%	6%
PROBABLE ●	1%	25%	19%
CONFIRMED ●	<1%	24%	18%
TOTAL	3%	57%	43%

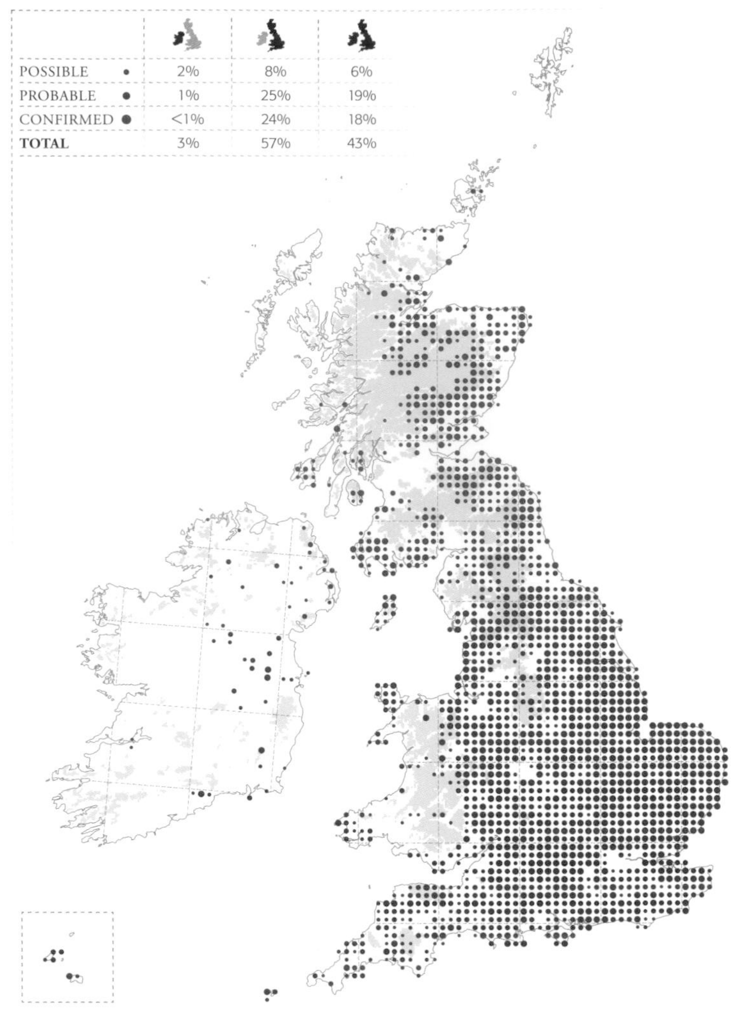

WINTER DISTRIBUTION *2007/08–2010/11*

PRESENT ●	2%	55%	41%

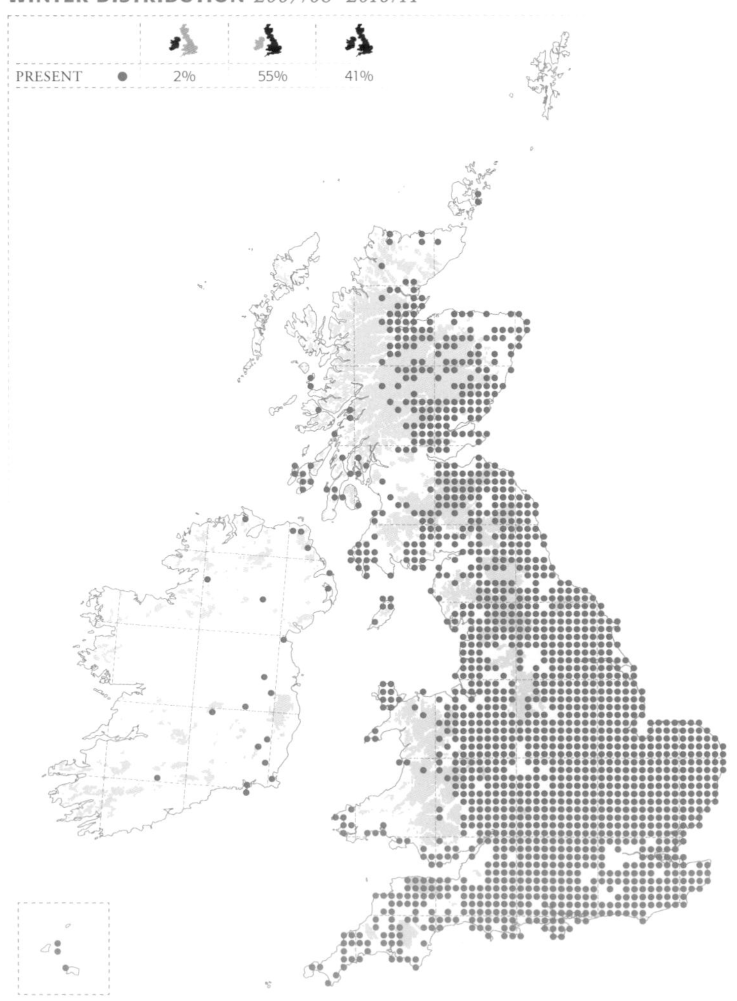

BREEDING DISTRIBUTION CHANGE *since 1968–72*

GAIN ▲	36	763	799
LOSS ▽	0	49	49
40yr CHANGE	∞	+78%	+82%
20yr INDEX	-	+0.21	

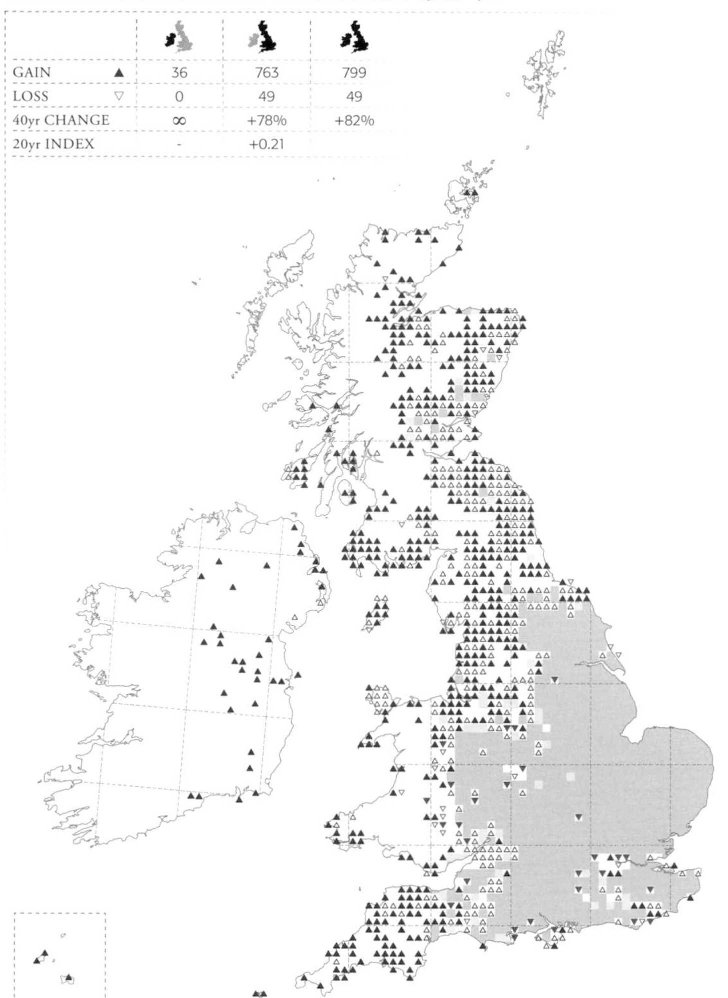

WINTER DISTRIBUTION CHANGE *since 1981–84*

GAIN ▲	18	710	728
LOSS ▼	3	49	52
30yr CHANGE	+500%	+75%	+76%

BREEDING RELATIVE ABUNDANCE *2008–11*

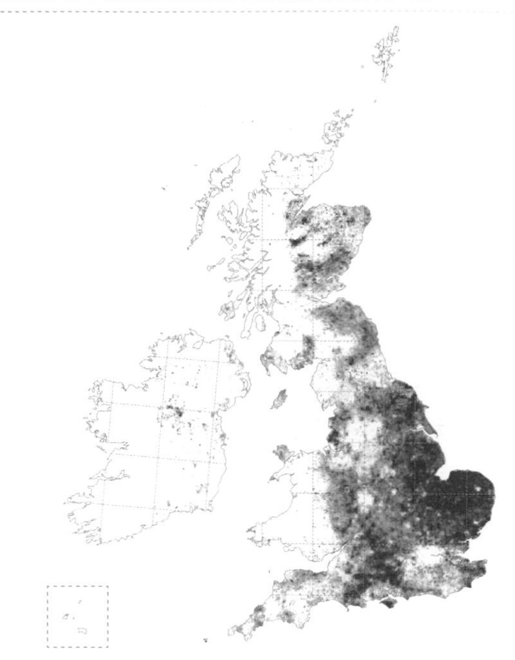

BREEDING RELATIVE ABUNDANCE CHANGE *since 1988–91*

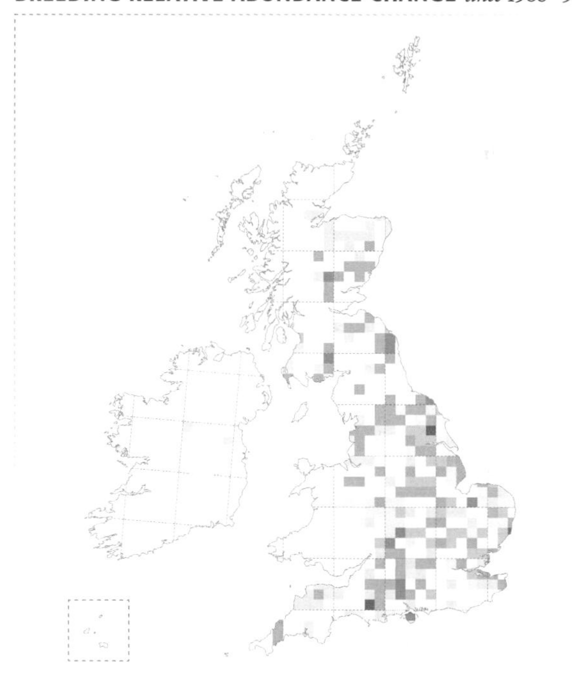

WINTER RELATIVE ABUNDANCE *2007/08–2010/11*

GREY PARTRIDGE

Perdix perdix

BTO CODE: P.

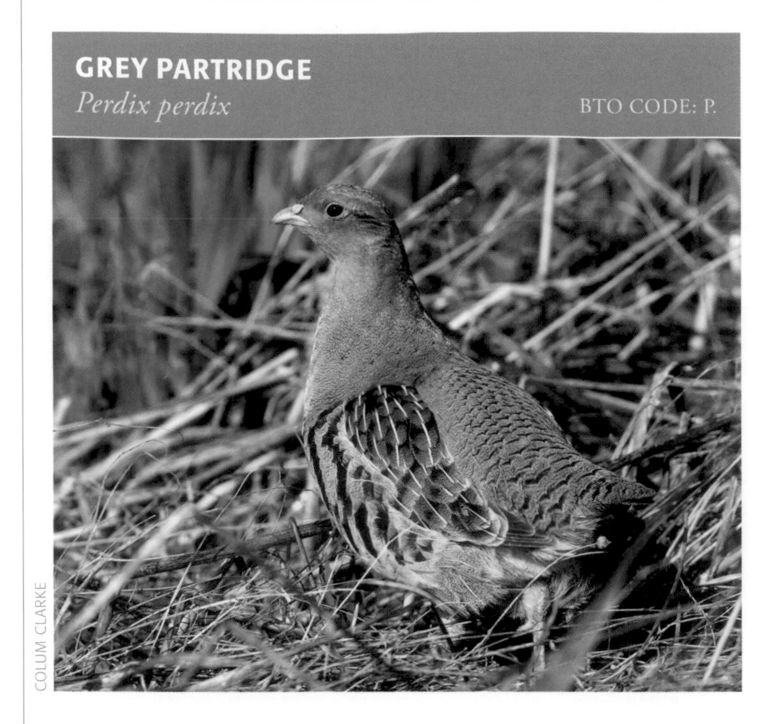

COLUM CLARKE

THE WINTER AND breeding distribution maps show that Grey Partridges are widespread across lowland England, except for the southeast and southwest. In Scotland they occupy a wide strip through the lowlands in the east but are now very scarce in the north and southwest. They are absent from most of Wales, except Anglesey and along the Welsh Marches. In Ireland, the cluster of dots in Co. Offaly marks the location of an Irish Grey Partridge Conservation Trust project where birds have been released since 2002 (Potts 2012).

There has been a 46% decline in the number of occupied 10-km squares since the *1968–72 Breeding Atlas* and this is coupled with a 91% population decline in the UK during 1967–2010 (*BirdTrends*). The breeding distribution change map shows that the loss of range in Ireland, Wales, southwest England and southwest Scotland took place largely between the *1968–72 Breeding Atlas* and the *1988–91 Breeding Atlas*, and that the main areas of loss in the last 20 years are in southeast England, the west Midlands and on the western edge of the range in Scotland. Local extinctions may be masked in some areas by the release of captive-bred birds onto shooting estates; about 100,000 captive-reared Grey Partridges are released in Britain each year (Public and Corporate Economic Consultants 2006).

The Grey Partridge is one of the most strongly decreasing bird species in Europe (*PECBMS 2010*), due to the effects of agricultural intensification. One of the key reasons for the decline appears to be the application of herbicides on the food plants of the young chicks' invertebrate prey (Potts 2012). Recent research indicates that infection with caecal nematodes from farm-reared Pheasants may also be contributing to the decline (Tompkins *et al.* 2002).

The abundance change map suggests that densities are decreasing in virtually all areas. As a result, Grey Partridges are thinly spread throughout their current range, although they remain relatively abundant in northwest Norfolk, Lancashire, Lincolnshire and eastern Scotland, perhaps due to the presence of weedy stubble fields, a key winter foraging habitat (Moorcroft *et al.* 2002).

Sponsored by the Holkham Estate, Norfolk

BREEDING DISTRIBUTION *2008–11*

POSSIBLE ●	<1%	6%	5%
PROBABLE ●	-	19%	14%
CONFIRMED ●	1%	18%	13%
TOTAL	1%	43%	32%

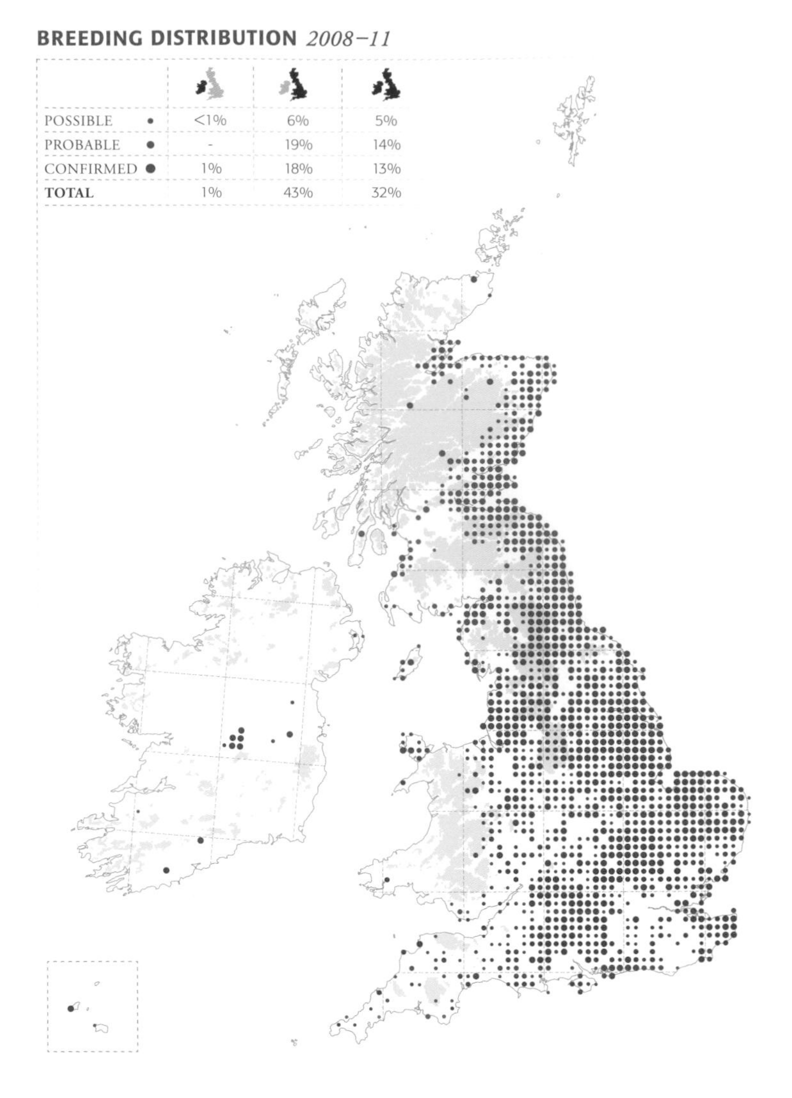

WINTER DISTRIBUTION *2007/08–2010/11*

PRESENT ●	1%	42%	31%

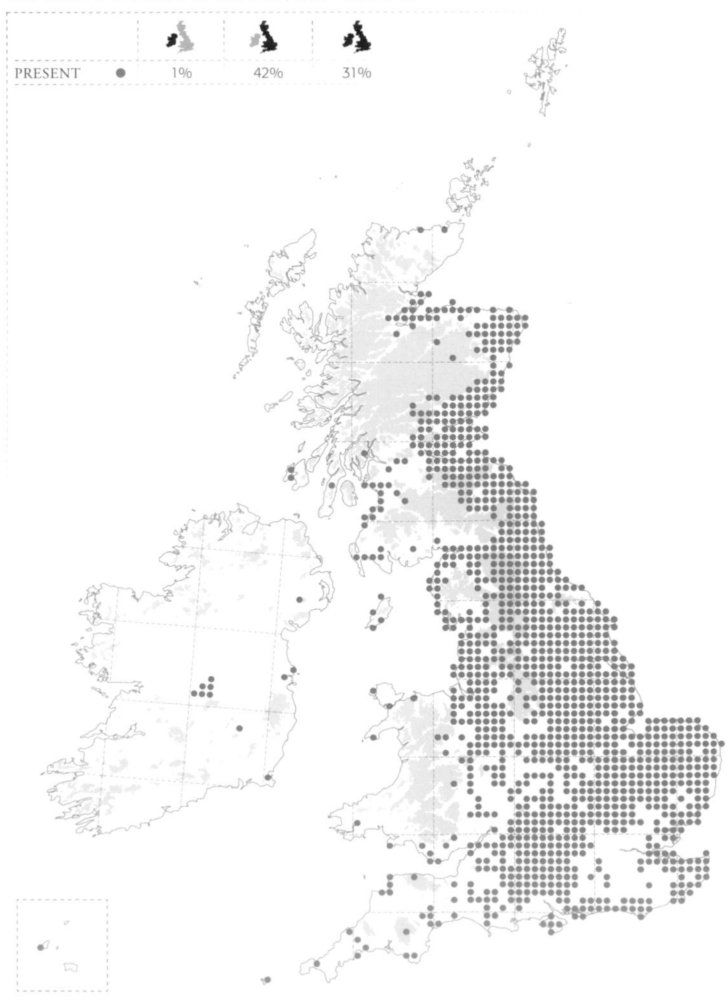

BREEDING DISTRIBUTION CHANGE *since 1968–72*

GAIN ▲	5	24	29
LOSS ▽	248	813	1061
40yr CHANGE	-95%	-39%	-46%
20yr INDEX	-	-0.47	

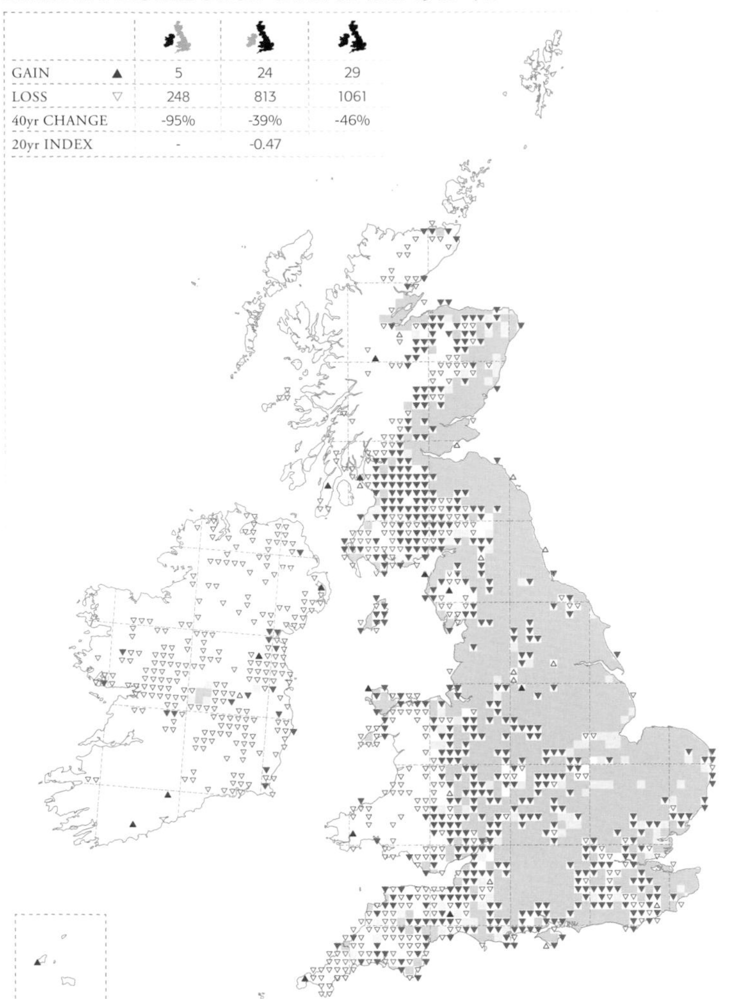

WINTER DISTRIBUTION CHANGE *since 1981–84*

GAIN ▲	9	154	163
LOSS ▼	26	497	523
30yr CHANGE	-61%	-23%	-23%

BREEDING RELATIVE ABUNDANCE *2008–11*

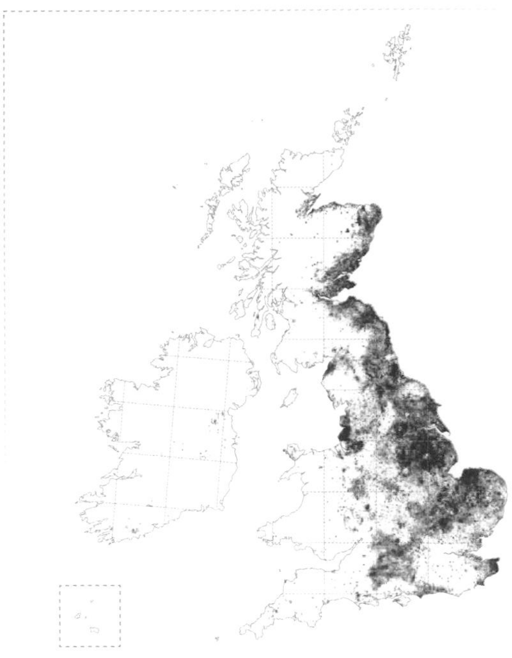

BREEDING RELATIVE ABUNDANCE CHANGE *since 1988–91*

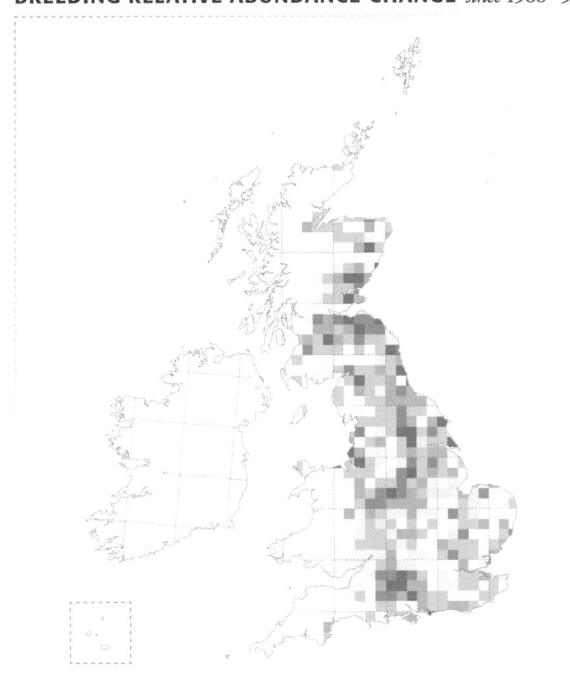

WINTER RELATIVE ABUNDANCE *2007/08–2010/11*

QUAIL

Coturnix coturnix

BTO CODE: Q.

GRAHAM CATLEY http://pewit.blogspot.co.uk

THE BREEDING DISTRIBUTION map shows records throughout lowland England, where arable farmland and open grassland predominate, and in a narrow band along the eastern side of Scotland as far north as Shetland. Records elsewhere, including in Ireland, are more scattered. Information from BirdTrack suggests that 2011 was a good 'Quail year', with records widespread across Britain & Ireland (*BirdTrack*), and there was a localised influx in 2009 to East Lothian (Welch 2010) and Northumberland (Northumberland & Tyneside Bird Club 2010). This is reflected in the number of 10-km squares occupied per year: 286, 371, 265 and 462 in 2008 to 2011 respectively. Record number of Quails were reported to RBBP in 2011, with a total of 1,992 males including over 100 in each of Angus & Dundee, Borders, Fife and Yorkshire (Holling *et al.* 2013). A Quail year was also recorded in each of the previous breeding atlases: in 1970 and 1989. Reasons for invasions are poorly understood but may be associated with warm, dry springs and southeasterly winds. In some years there is a late influx of birds, as in East Lothian in 2009, and these may have attempted to breed elsewhere in Europe before continuing their migration and undertaking further breeding activity here.

Traditional strongholds on calcareous soils in southern England are evident on the breeding distribution change map as areas where Quails have been recorded in all three breeding atlases. Gains and losses elsewhere are generally scattered, reflecting the sporadic nature of occurrence. The total of 890 10-km squares with reported breeding evidence is only marginally greater than the number of squares occupied during the *1988–91 Breeding Atlas*, but double that in the *1968–72 Breeding Atlas*. In addition, Quails were reported from 132 10-km squares without breeding evidence. Given that most birds are detected by song, we may speculate how many of these *non-breeding* records may have warranted breeding evidence.

Proving breeding is exceptionally difficult and there were only 32 10-km squares where breeding was *confirmed*. Most records refer to singing males but this is not necessarily an indication of breeding, as some males move farther north after calling for a few days and failing to attract a mate. This behaviour, and the tendency for Quails to change markedly in number and distribution between years, means that distribution maps and range-size estimates over emphasise presence in any single year; only 11% of squares were recorded as occupied in three or more years.

Sponsored for S. Harvey

BREEDING DISTRIBUTION *2008–11*

POSSIBLE •	1%	19%	14%
PROBABLE ●	1%	10%	8%
CONFIRMED ●	<1%	1%	1%
TOTAL	2%	30%	23%

BREEDING DISTRIBUTION *in 2011*

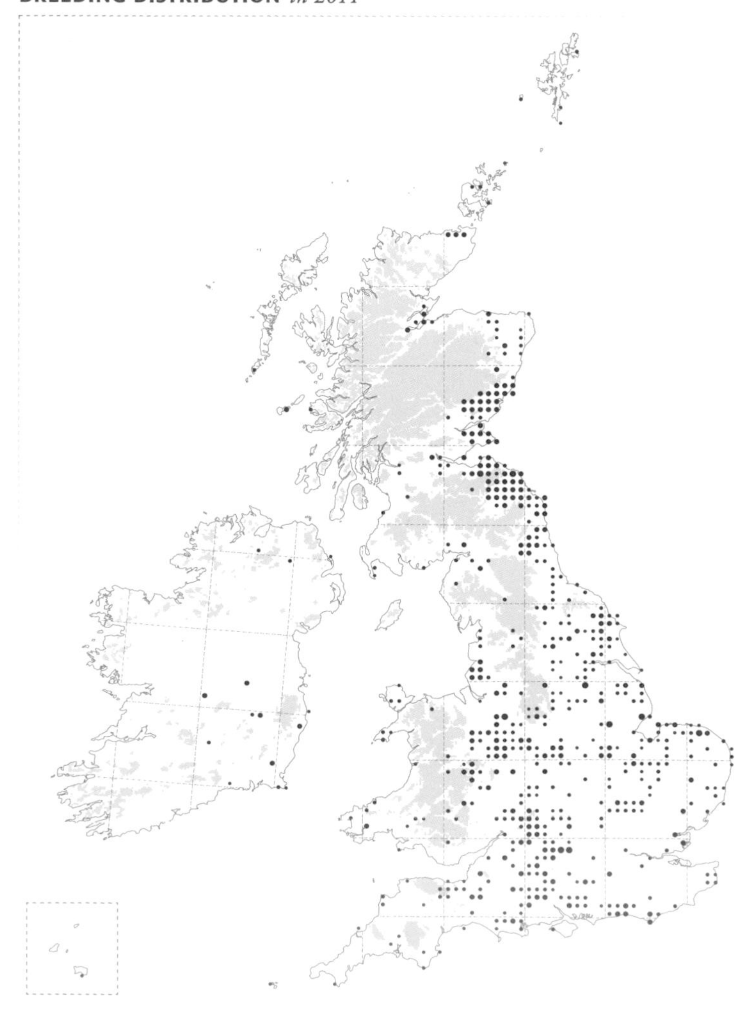

BREEDING DISTRIBUTION CHANGE *since 1968–72*

GAIN ▲	19	638	657
LOSS ▽	29	178	207
40yr CHANGE	-30%	+114%	+103%
20yr INDEX	-	-0.12	

BREEDING RELATIVE ABUNDANCE *2008–11*

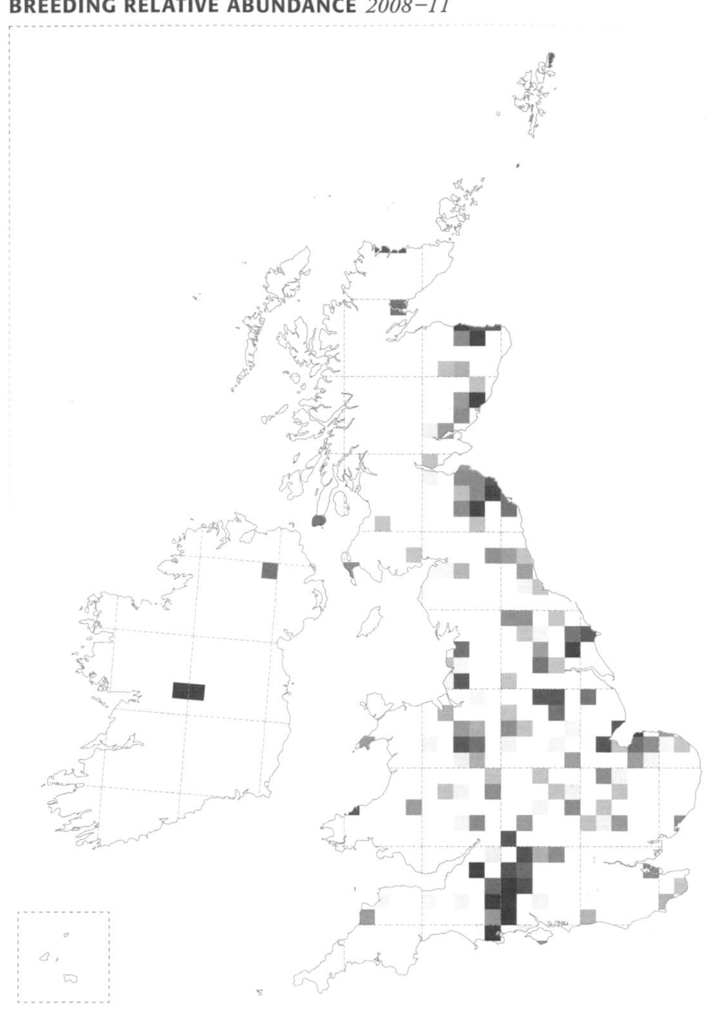

HELMETED GUINEAFOWL

Numida meleagris

BTO CODE: FW

NATIVE TO SUB-SAHARAN Africa, Helmeted Guineafowls are widely kept in captivity and many roam freely around farms and estates. Following the first documented record of feral breeding at Houghton Park, Norfolk, in 2001 (Ogilvie *et al.* 2003a), they were confirmed to have bred in four further counties by 2008 (Holling *et al.* 2011a). Most records in the breeding season are south of the Severn–Wash line, with a few records scattered elsewhere in Britain & Ireland. With breeding-season presence in 196 10-km squares, 89 of which recorded breeding evidence, and 106 10-km squares occupied in winter, a significant expansion is apparent since previous atlases. It is unlikely that this is due entirely to better recording of non-natives.

BREEDING DISTRIBUTION *2008–11*

POSSIBLE •	<1%	1%	1%
PROBABLE •	<1%	1%	1%
CONFIRMED ●	-	<1%	<1%
TOTAL	<1%	2%	2%

PHEASANT

Phasianus colchicus

BTO CODE: PH

DESMOND DUGAN www.abernethyimages.co.uk

PHEASANTS WERE INTRODUCED to England in the late 11th century and into Scotland, Wales and Ireland in the late 16th century (Lever 2005). The distribution maps are similar for the winter and the breeding season and show that, although widely distributed, Pheasants are absent from much of northwest Scotland and are scarce on the Outer Hebrides, Skye and Shetland. Since the previous winter and breeding atlases, there have been small gains in range in the north and west of Scotland, Wales and parts of Ireland, particularly in the west and southwest.

Although Pheasants are widespread throughout Ireland, their relative abundance is lower than in Britain. The relative abundance maps for both seasons are similar and show lower abundance in Wales, in upland areas and around large conurbations in Britain. In autumn, numbers are greatly increased by the release of captive-reared birds for shooting. Such releases have increased approximately fivefold since the early 1960s, to around 35 million birds annually, with some 15 million shot annually (Public and Corporate Economic Consultants 2006). Some of the gains in range may be due to local releases.

High densities of Pheasants potentially have negative effects on native species, but these have been poorly studied. One study found no direct impacts, but mixed indirect effects on native vertebrates due to management practices (Davey 2008). Indirect effects possibly include modification of the structure of the field layer, the spread of disease and parasites, and competition for food (Fuller *et al.* 2005). Work by the Game & Wildlife Conservation Trust has focused on changes in ground flora caused by Pheasant poults in woodland release pens and the influence of release density on biodiversity in the wider countryside. The results suggested that the interiors of many lowland woods managed for Pheasants have a more open structure, creating favourable conditions for the growth of herbs and brambles and supporting higher densities of songbird species requiring dense low cover for nesting (Game Conservancy Trust 2004).

BREEDING DISTRIBUTION *2008–11*

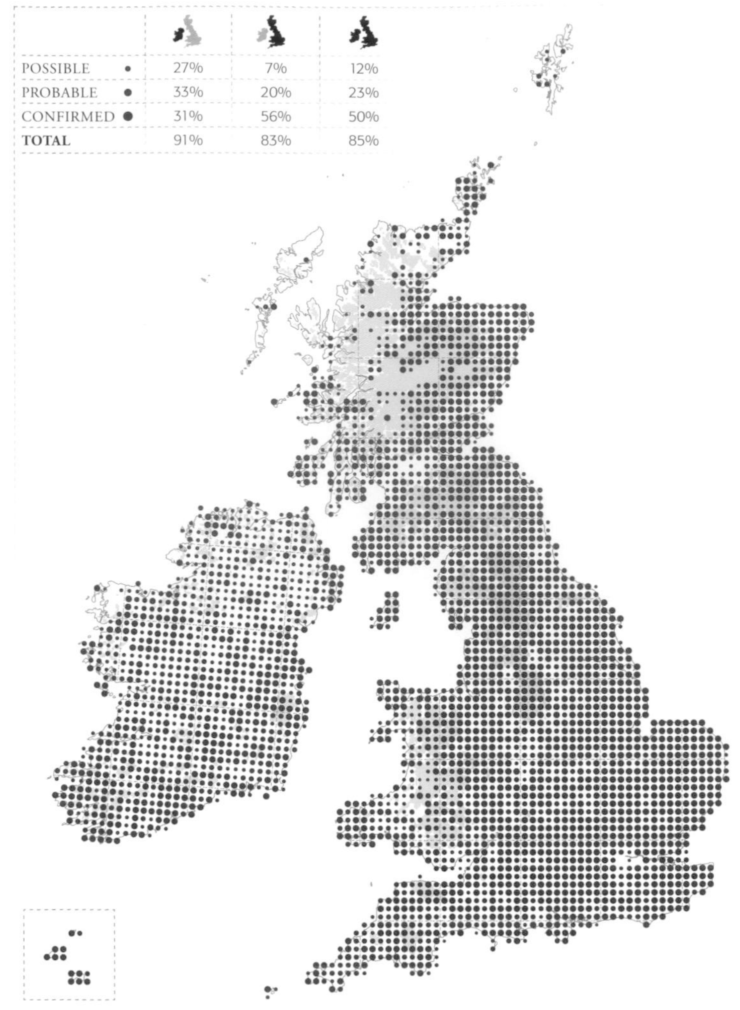

POSSIBLE •	27%	7%	12%
PROBABLE ●	33%	20%	23%
CONFIRMED ●	31%	56%	50%
TOTAL	91%	83%	85%

WINTER DISTRIBUTION *2007/08–2010/11*

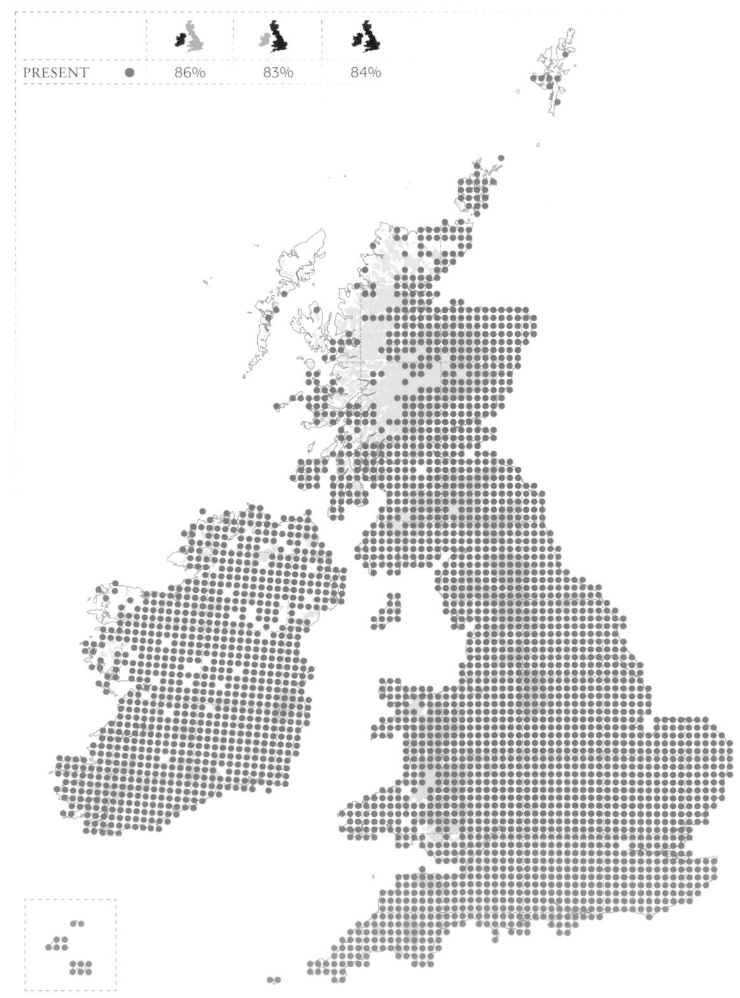

PRESENT ●	86%	83%	84%

Sponsored by Trevor McCann

BREEDING DISTRIBUTION CHANGE *since 1968–72*

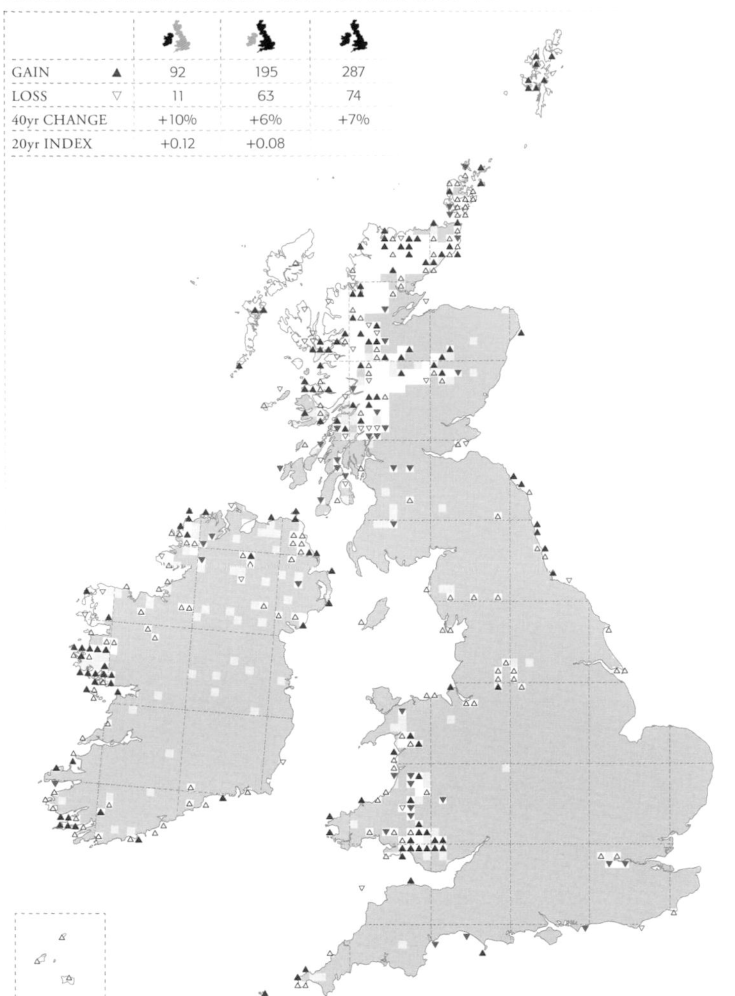

GAIN ▲	92	195	287
LOSS ▽	11	63	74
40yr CHANGE	+10%	+6%	+7%
20yr INDEX	+0.12	+0.08	

WINTER DISTRIBUTION CHANGE *since 1981–84*

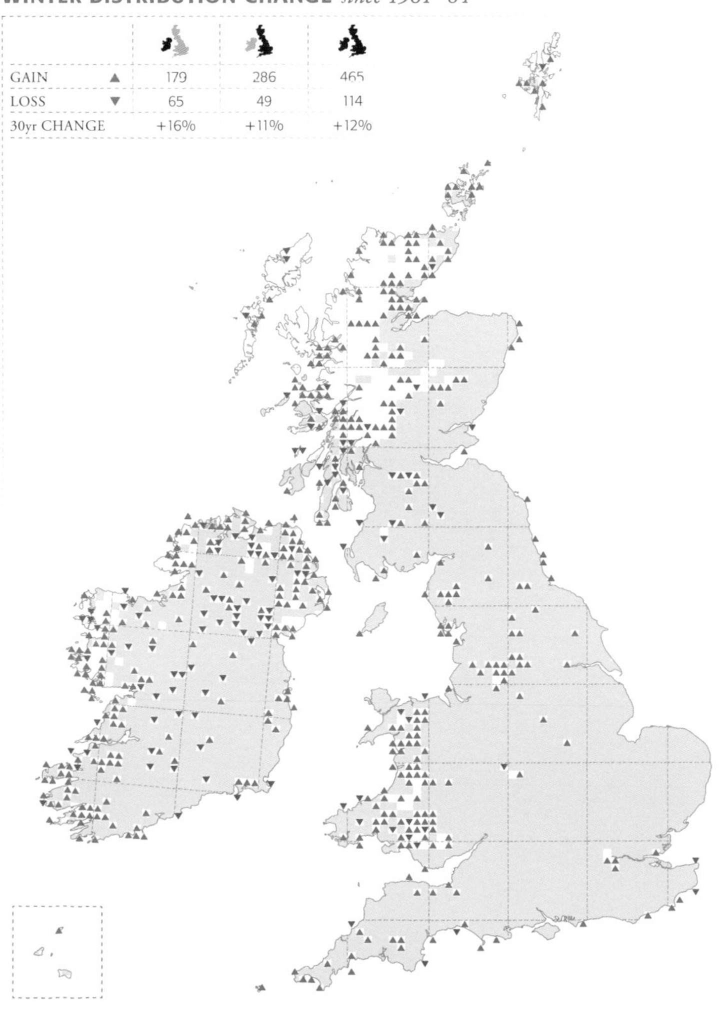

GAIN ▲	179	286	465
LOSS ▼	65	49	114
30yr CHANGE	+16%	+11%	+12%

BREEDING RELATIVE ABUNDANCE *2008–11*

BREEDING RELATIVE ABUNDANCE CHANGE *since 1988–91*

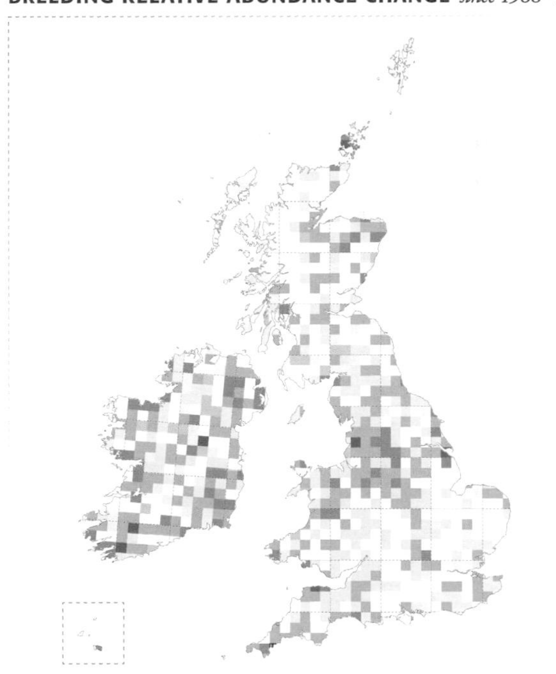

WINTER RELATIVE ABUNDANCE *2007/08–2010/11*

GOLDEN PHEASANT

Chrysolophus pictus

BTO CODE: GF

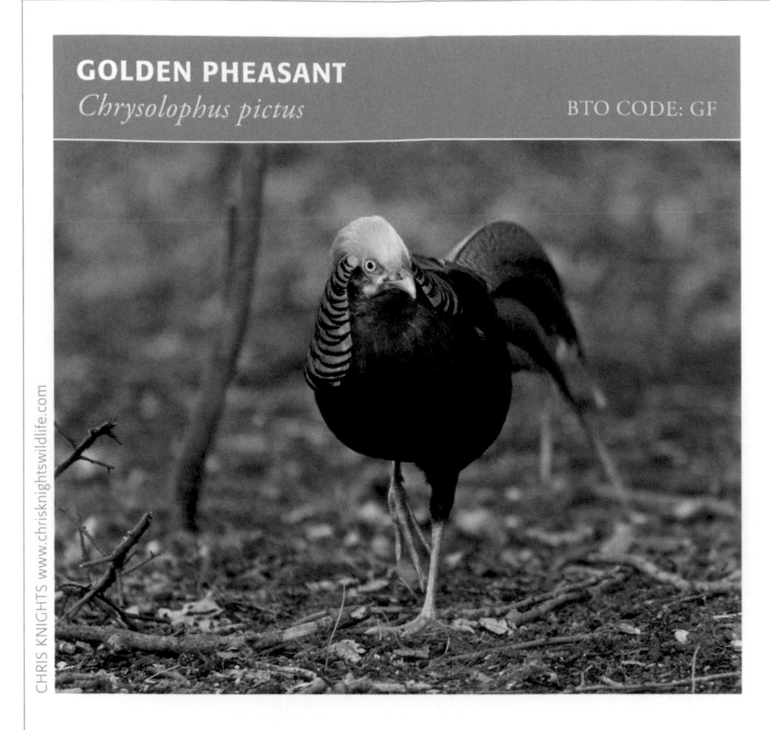

CHRIS KNIGHTS www.chrisknightswildlife.com

AT THE TIME of the *1988–91 Breeding Atlas*, the introduced population of Golden Pheasants in Britain was thought to be the only self-sustaining population outside their native uplands of central China. From 1725 onwards, Golden Pheasants were released in scattered locations from the Isles of Scilly to Kent and northwards to the Isle of Gigha off the west coast of Argyll, although many of these populations died out after just a few years (Lever 2009). During the *1968–72 Breeding Atlas* the strongholds of the population in Britain were in Breckland, northwest Norfolk, Dumfries & Galloway and on Anglesey. By the *1988–91 Breeding Atlas*, 20 years later, two more populations had become established: on the South Downs in Hampshire and West Sussex.

The current winter and breeding distribution maps show that the stronghold remains in Breckland, with another very small population in northwest Norfolk. Elsewhere, there are many isolated records of local escapes or recent introductions. Both the breeding change map since the *1988–91 Breeding Atlas* and the winter change map show losses in the former strongholds in Hampshire, West Sussex, Dumfries & Galloway and Anglesey.

Within Breckland, Golden Pheasants apparently have disappeared from some of their traditional sites and numbers have declined throughout the region. At some of these sites there has been no apparent change in habitat, whilst at others coniferous plantations have matured or have been felled, or habitat management in non-forest sites has been undertaken to the detriment of this species. In Norfolk, some losses in range were noted west of Thetford between 1980–85 and 1999–2007, but some gains in the eastern part of Thetford Forest (Taylor & Marchant 2011). The reasons for the decline in the stronghold populations are not fully understood but may include predation by the Red Fox and raptors, inbreeding, disturbance and a reduction in the availability of suitable habitat (Balmer *et al.* 1996; *Birds of Scotland*; Lever 2009; Wearing 2011).

Mark Rehfisch, sponsor of the Golden Pheasant, with thanks and in recognition of 21 years' service to BTO

BREEDING DISTRIBUTION *2008–11*

POSSIBLE •	-	1%	<1%
PROBABLE •	-	<1%	<1%
CONFIRMED ●	-	<1%	<1%
TOTAL	-	1%	<1%

BREEDING DISTRIBUTION CHANGE *since 1988–91*

GAIN ▲	-	19	19
LOSS ▽	-	40	40
40yr CHANGE	-	-45%	-45%
20yr INDEX	-	-0.33	

CHRIS KNIGHTS www.chrisknightswildlife.com

◀ **Breeding was *confirmed*, on the basis of fledged young, in just three 10-km squares. Some of the 24 squares with records that lack breeding evidence could also be part of the breeding range.**

LADY AMHERST'S PHEASANT

Chrysolophus amherstiae

BTO CODE: LM

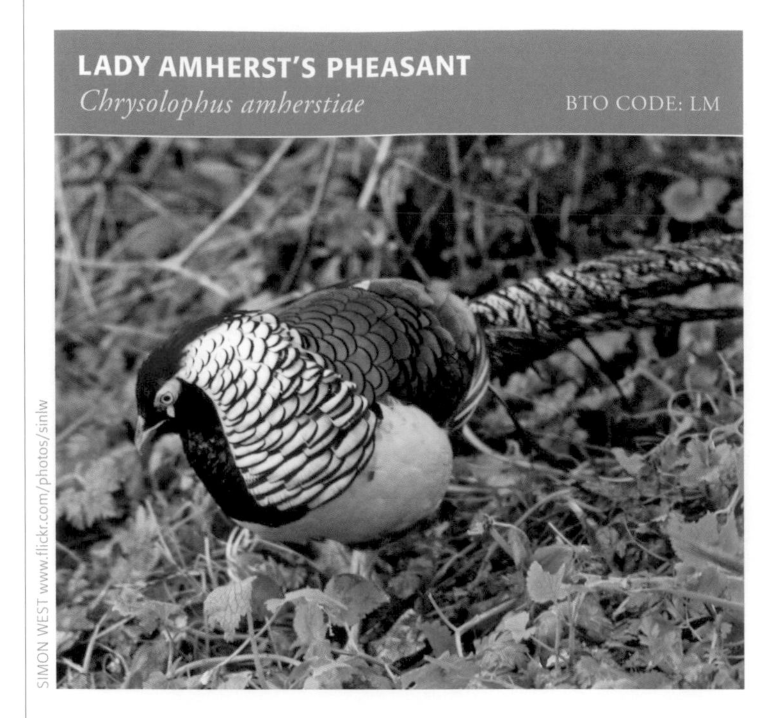

SIMON WEST www.flickr.com/photos/sinlw

THERE HAS BEEN an established population of Lady Amherst's Pheasants in Bedfordshire since the 1890s, following an introduction at Woburn, with later introductions at Whipsnade and at Mentmore, Buckinghamshire, in the 1930s. Following further releases in the 1950s, a separate colony became established at Luton Hoo in south Bedfordshire. Further releases of birds at Woburn in 1984 resulted in birds spreading some 5 km west to woods around Bow Brickhill, Buckinghamshire. Elsewhere there were widespread other releases at localities such as Richmond Park in Surrey, Halkyn churchyard in Clwyd, Elveden in Suffolk and Beaulieu Manor Woods in Hampshire between 1925 and 1955, though none of these populations survived for more than a few years (Lever 2009).

The winter distribution map shows records from just seven 10-km squares, all in Britain, with no winter records from the previous stronghold populations in Bedfordshire or Buckinghamshire. Despite the males being highly distinctive, they are difficult to census in their preferred woodland habitat and may have been overlooked. The breeding-season map shows records of *probable* breeding in just two adjacent 10-km squares in the former Bedfordshire stronghold, while isolated records elsewhere relate to escaped birds. The winter and breeding distribution change maps illustrate the loss of range from this core area.

Local atlas fieldwork in Bedfordshire between 1968 and 1977 found this species to be present in 32 tetrads (Harding 1979), which dropped to 24 tetrads during the second local atlas undertaken between 1988 and 1992 (Dazley & Trodd 1994). Fieldwork during *Bird Atlas 2007–11* revealed that fewer than five tetrads were occupied in the same recording area. Numbers have dropped from around 250 birds during 1968–77 to just five males in 2008 (Holling *et al.* 2011a). The decline has been rapid, yet the reasons are poorly understood. Suggestions include habitat loss, change in woodland structure, increased predation (by Red Foxes, Magpies and Goshawks), an increase in Chinese Muntjac deer affecting the understorey, disturbance by birdwatchers and perhaps inbreeding due to a lack of interchange between fragmented populations (Nightingale 2005).

BREEDING DISTRIBUTION *2008–11*

POSSIBLE •	-	<1%	<1%
PROBABLE •	-	<1%	<1%
TOTAL	-	<1%	<1%

BREEDING DISTRIBUTION CHANGE *since 1968–72*

GAIN ▲	-	5	5
LOSS ▽	-	11	11
40yr CHANGE	-	-46%	-46%
20yr INDEX	-	-	

MARK HOWS www.hows.org.uk

◀ Numbers in the former stronghold of Bedfordshire and neighbouring Buckinghamshire have declined significantly and extinction within Britain is thought to be almost inevitable.

REEVES'S PHEASANT

Syrmaticus reevesii

BTO CODE: RV

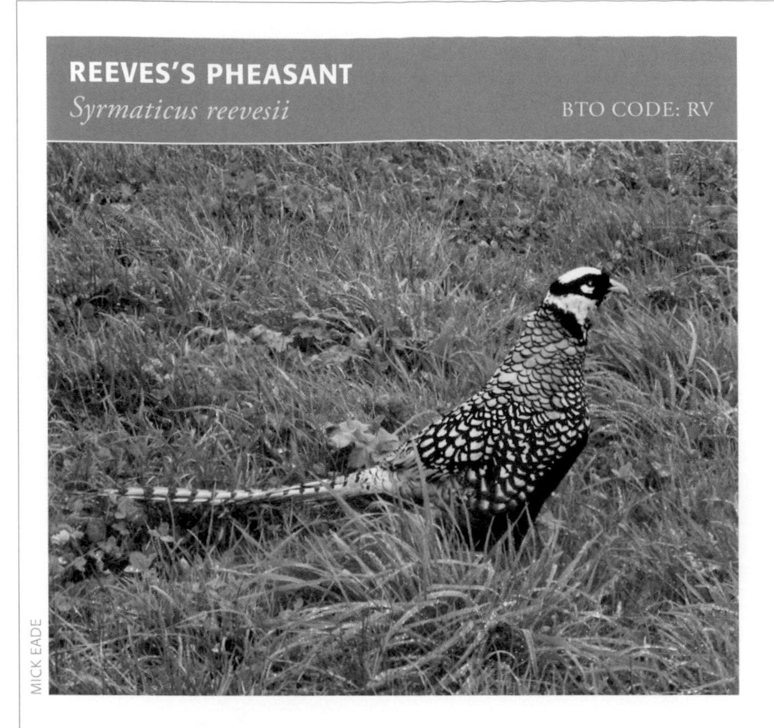

MICK EADE

A NATIVE OF north central China, the Reeves's Pheasant is widely kept in captivity and most records relate to occasional escapes and the remnants of planned introductions. The species was introduced, with varying success, to a number of estates between 1870 and 1890, including Lilford Park, Northamptonshire; Woburn Abbey, Bedfordshire; Tortworth, Gloucestershire; Bedgebury, Kent; Guisachan, Inverness-shire; Duff House, Aberdeenshire; and Markree Castle, Co. Sligo. More recent releases include Elveden Hall, Suffolk, in 1950; Cumbria in 1969; and Kinveachy Forest, Inverness-shire, in 1970 (Lever 2009).

The winter and breeding distribution maps are similar, though this species is more widely recorded in the winter. The distribution maps largely agree with the known patterns of release, with most records coming from Wales and southern England. There are also two distinct clusters in Scotland, on the Cowal Peninsula in Argyll and east of Grantown-on-Spey on the Inverness-shire/Moray border. Most records with breeding evidence relate to birds in suitable habitat and there are six 10-km squares where *probable* breeding has been recorded, based on pairs or birds holding territory. There were no records of *confirmed* breeding and none were reported to the Rare Breeding Birds Panel within the Atlas period (Holling *et al.* 2011a; M. Holling pers comm).

Lever (2009) suggested that the failure of the Reeves's Pheasant to become established in Britain & Ireland may partly be attributable to the fact that the males disperse too widely for successful pairing. It is a strong flier and lacks a far-carrying call, unlike the Golden Pheasant and Lady Amherst's Pheasant, and males may become isolated from females at the release location.

Elsewhere in Europe, Reeves's Pheasants have been successfully introduced in the former Czechoslovakia for sporting purposes (Kokes 1977 in Lever 2005) and in France (*European Atlas*). They were also introduced to parts of Austria, Germany and Hungary, though stable populations failed to establish (Lever 2005).

In its native range, the species is listed as Vulnerable because its small and mostly unprotected population is severely fragmented and is declining rapidly in the face of continuing habitat loss and over-hunting (BirdLife International 2013).

BREEDING DISTRIBUTION *2008–11*

	Ireland	Britain	Britain & Ireland
POSSIBLE •	-	1%	<1%
PROBABLE ●	-	<1%	<1%
CONFIRMED ●	-	-	-
TOTAL	-	1%	<1%

WINTER DISTRIBUTION *2007/08–2010/11*

	Ireland	Britain	Britain & Ireland
PRESENT ●	-	2%	1%

INDIAN PEAFOWL

Pavo cristatus

BTO CODE: PX

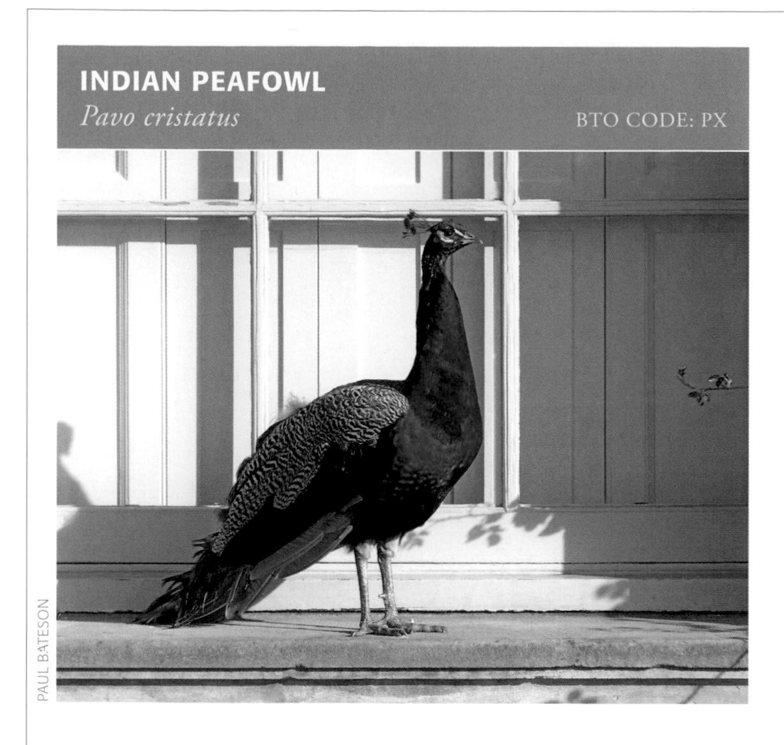

PAUL BATESON

THE INDIAN PEAFOWL is a native of the Indian subcontinent and Sri Lanka but is familiar to many as an ornamental species associated with stately homes, ornamental parks and country estates. These birds are usually free-flying and many individuals wander some distance and attempt to breed in the wild (Holling *et al.* 2011a). Historically, such birds have not been routinely documented because birdwatchers tend not to record exotic species, and judging when birds are naturalised is problematic. The results presented here are a step forward in describing the status of this species in the wild, but are still likely to suffer from reporting issues, both of the birds themselves and of the inclusion of appropriate breeding evidence.

Records in the winter came from 108 10-km squares in Britain and from three squares in Ireland. Birds were widely scattered south of a line from the Wash to the Severn, with a distinct group in north Devon. Elsewhere, there were isolated winter records in northern England, Wales, Scotland and the Isle of Man. The three records in Ireland were all from the east coast.

Breeding evidence was recorded in 136 10-km squares in Britain and eight in Ireland, while there were a further 153 10-km squares in Britain and three in Ireland where Indian Peafowl were recorded but for which no breeding evidence was provided. An unknown proportion of these potentially also relates at least to birds in suitable breeding habitat. There are clusters of records with breeding evidence in Bedfordshire, Norfolk, Suffolk, Herefordshire, Yorkshire, Lothian, Borders and east Inverness-shire. There were 22 10-km squares with records of *confirmed* breeding, all in Britain. Many of these areas correspond to breeding locations published by RBBP (Holling *et al.* 2011a), though they acknowledge that breeding in the wild is under-recorded.

No previous atlas records exist for this species so it is impossible to assess changes in range.

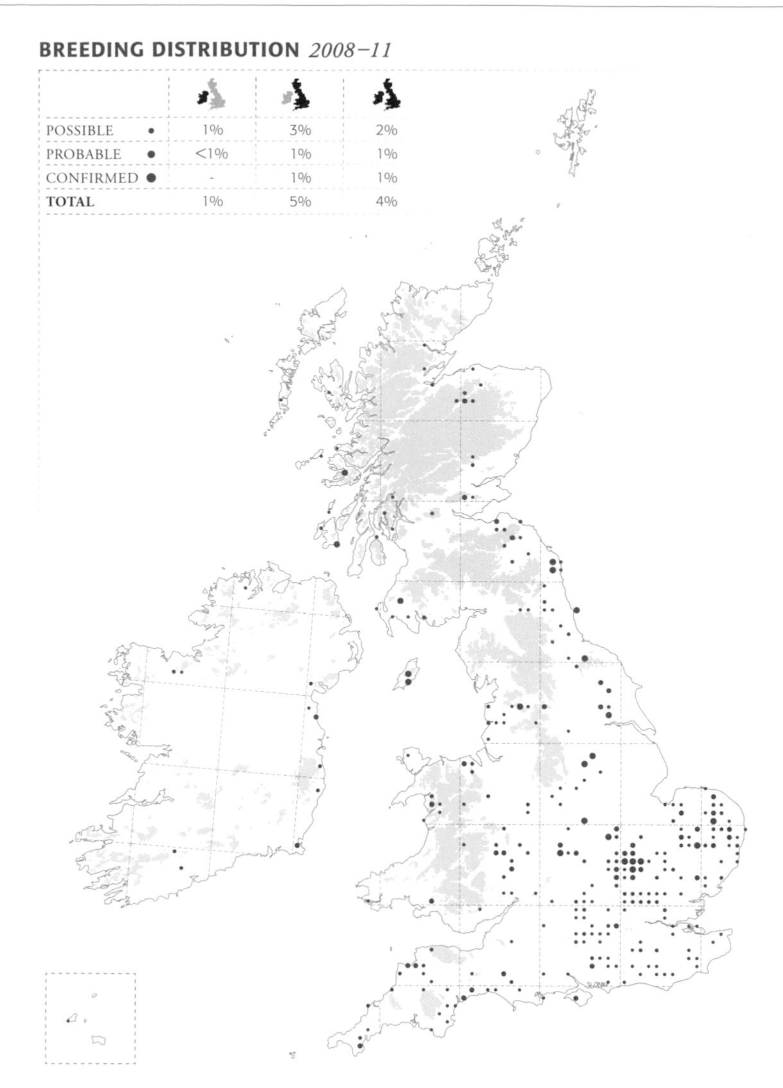

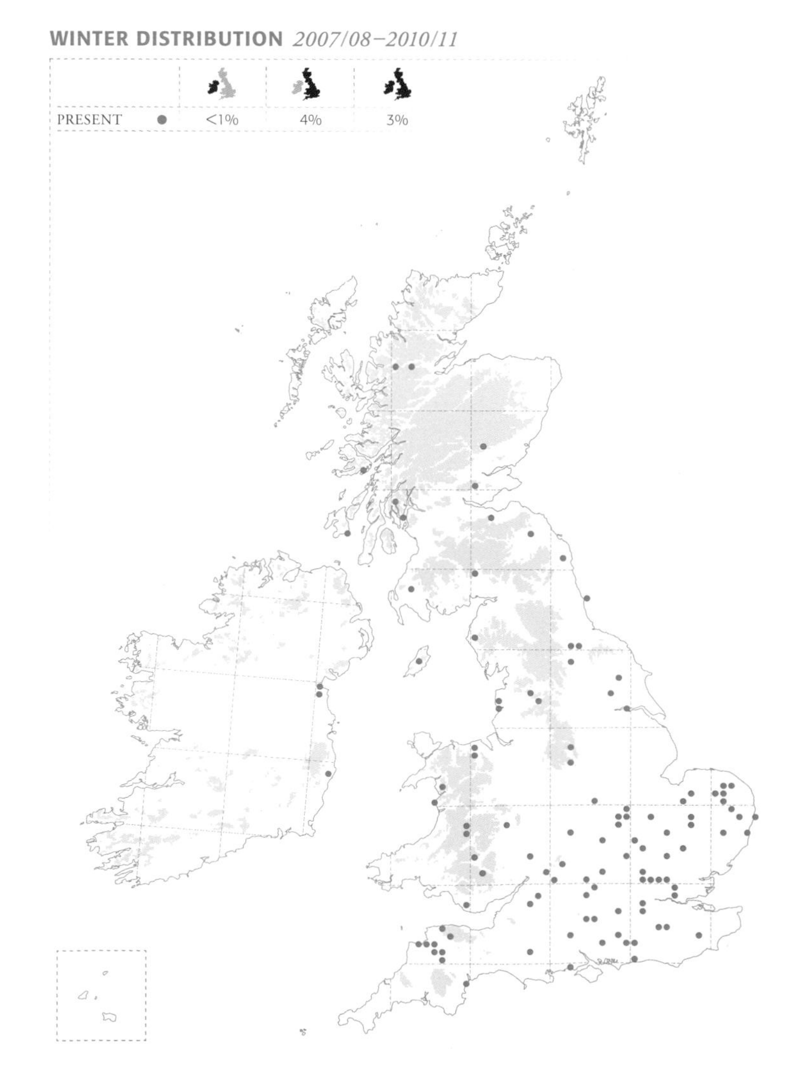

Sponsored by Tom and Muriel Cadwallender

RED-THROATED DIVER

Gavia stellata

BTO CODE: RH

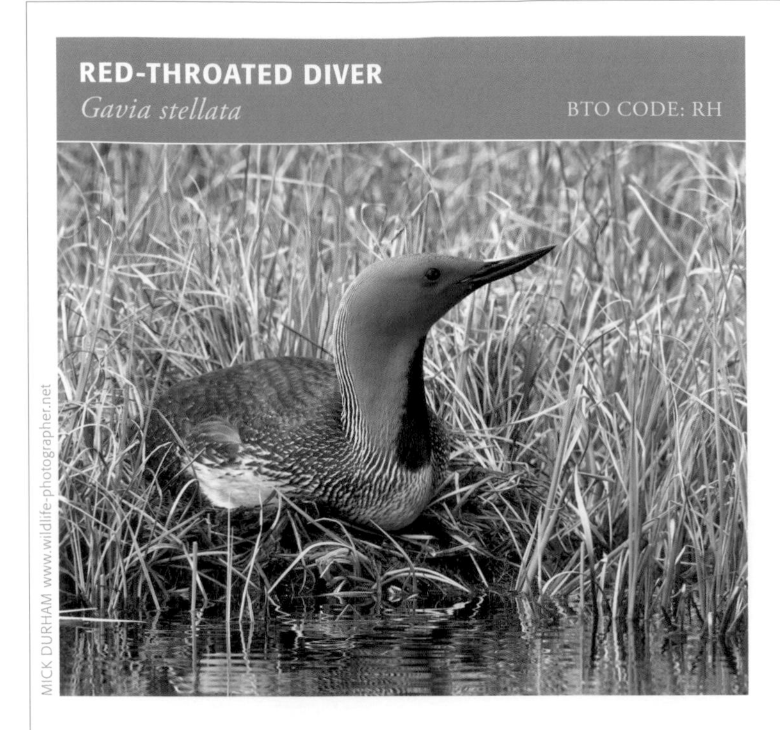

MICK DURHAM www.wildlife-photographer.net

BREEDING DISTRIBUTION *2008–11*

	Ireland	Britain	Britain & Ireland
POSSIBLE •	-	2%	1%
PROBABLE •	<1%	4%	3%
CONFIRMED •	1%	6%	5%
TOTAL	1%	12%	9%

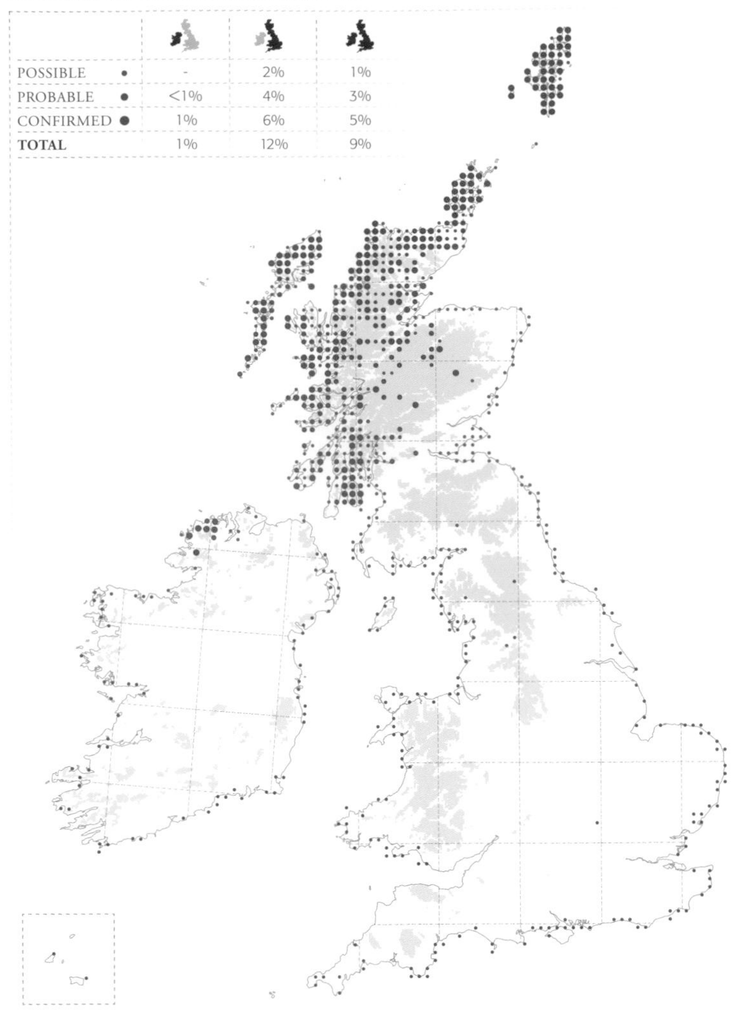

WINTER DISTRIBUTION *2007/08–2010/11*

	Ireland	Britain	Britain & Ireland
PRESENT •	20%	26%	25%

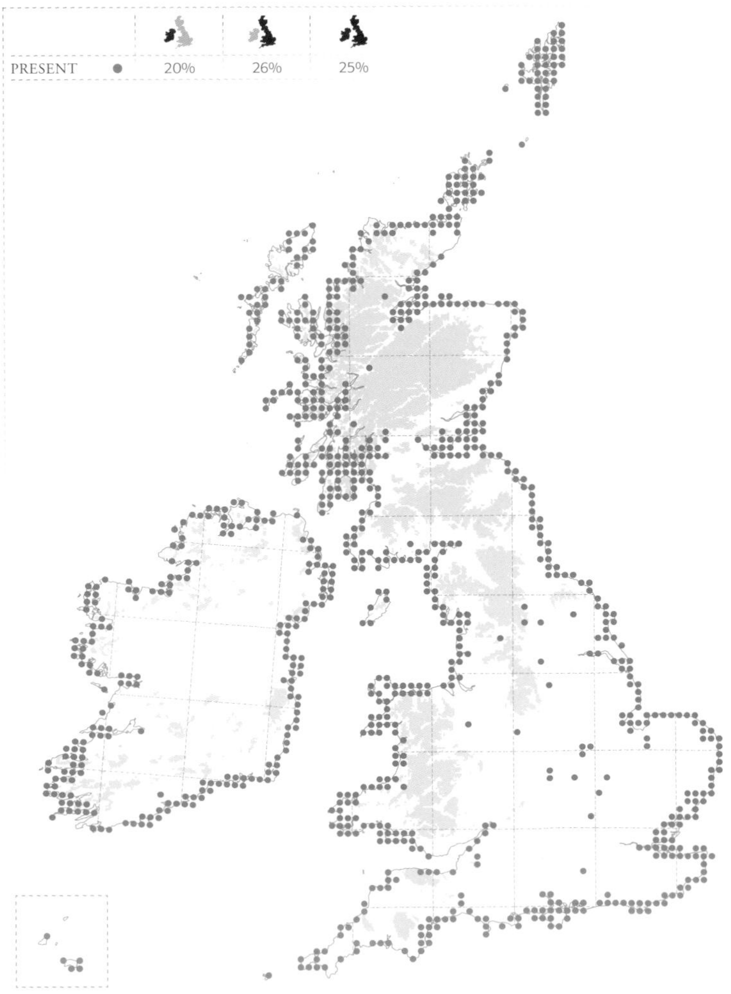

RED-THROATED DIVERS were recorded in winter all round the coast of Britain & Ireland, and at a scatter of inland sites, mainly in northern and central England. The highest concentrations were recorded along North Sea coasts, in southwest Scotland and in southwest Ireland. In Britain, this pattern partly agrees with the results of detailed offshore aerial surveys that show large concentrations off southeast England, especially in the Greater Thames where over 6,000 individuals have been recorded (Webb *et al.* 2009). Taking these offshore birds into account, the British wintering population may number c.17,000 individuals (O'Brien *et al.* 2008).

There has been a 32% winter range expansion since the *1981–84 Winter Atlas*, with a large part of this being off northwest Scotland and parts of western Ireland. Some of this increase may be due to better coverage but some gains are likely to be real. Greatest losses are evident from low-density areas in the west and also from inland sites. Losses from inland sites are balanced by gains and are likely be due to chance events relating to storm-driven individuals.

Breeding birds are restricted to freshwater lochs and bog pools in Co. Donegal in Ireland and in north and west Scotland. The Irish breeding population is small: six pairs bred in 2010 (*IRBBP 2010*), which is a small increase over recent years (Cromie 2002). Within Scotland, the highest densities are found in Shetland, parts of Orkney, Caithness, the western fringe of the Scottish Highlands and the Outer Hebrides. A Scotland-wide survey in 2006 showed that 33% of the breeding population was on Shetland and 26% on the Outer Hebrides, and that numbers overall had increased by 38% since a similar survey in 1994 (Dillon *et al.* 2009). Range change figures suggest an 11% expansion since the *1968–72 Breeding Atlas*, yet a 9% contraction since the *1988–91 Breeding Atlas*. These numbers may misrepresent changes because divers commute large distances to feed and may have been mis-coded as *possible* breeders. However, there has been a 21% increase in the number of squares with *probable* or *confirmed* breeding since 1988–91.

Sponsored by the Gay Birders Club

BREEDING DISTRIBUTION CHANGE *since 1968–72*

GAIN ▲	5	118	123
LOSS ▽	4	83	87
40yr CHANGE	+14%	+11%	+11%
20yr INDEX	-	-0.20	

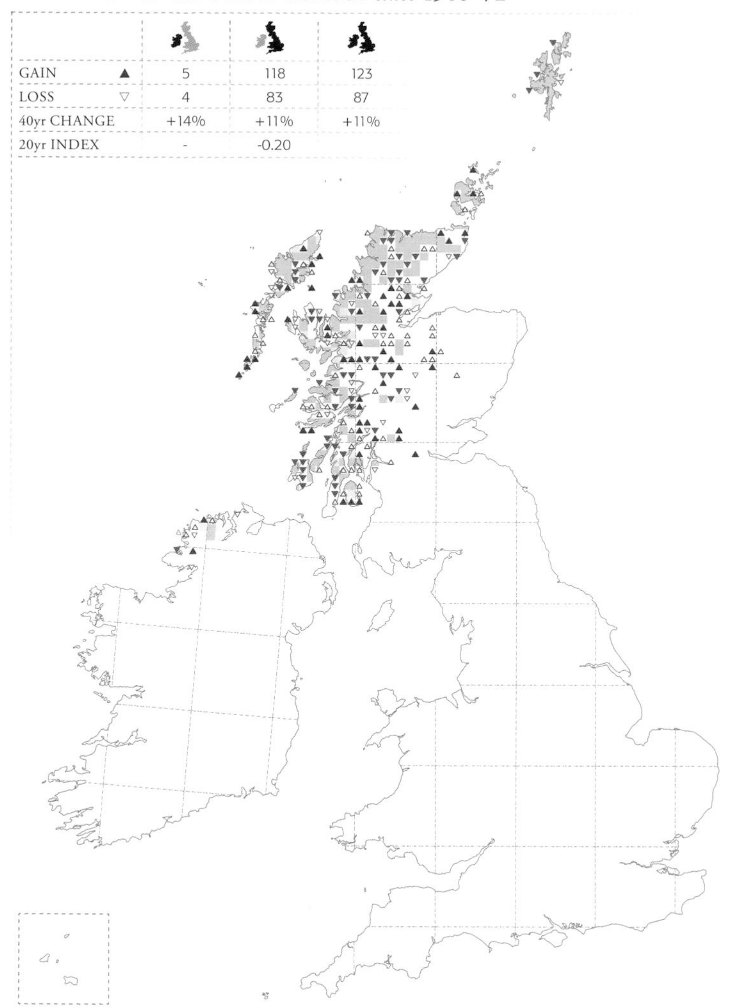

WINTER DISTRIBUTION CHANGE *since 1981–84*

GAIN ▲	77	278	355
LOSS ▼	36	98	134
30yr CHANGE	+26%	+33%	+32%

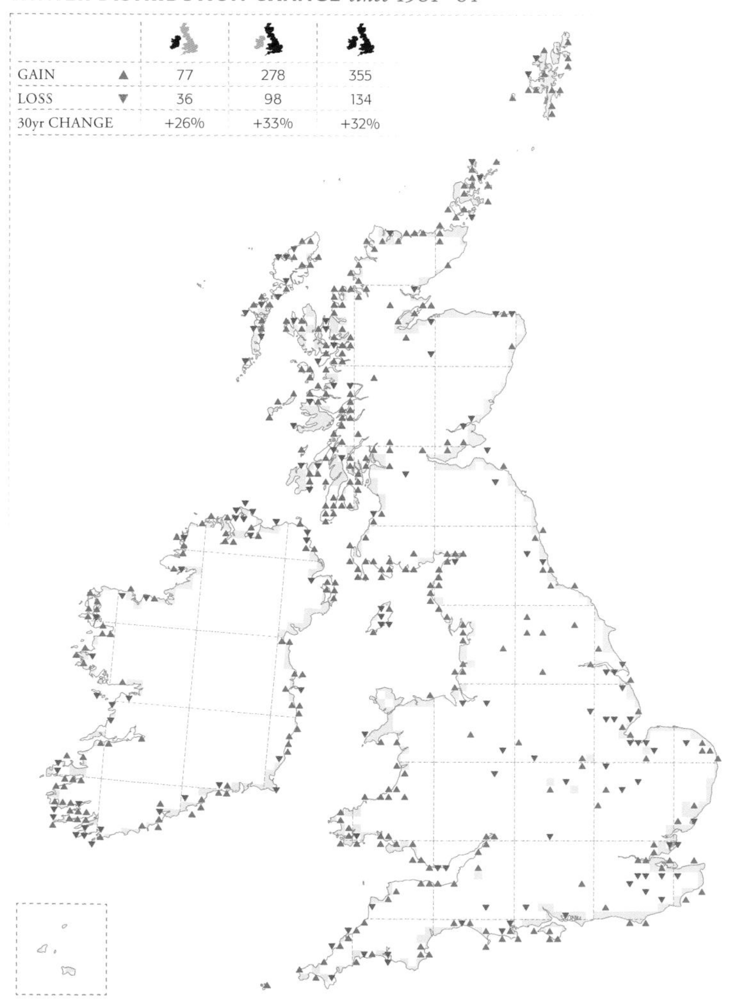

BREEDING RELATIVE ABUNDANCE *2008–11*

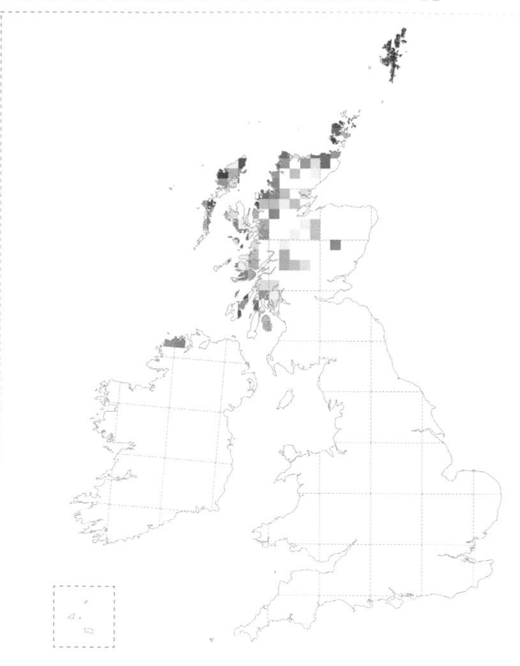

BREEDING RELATIVE ABUNDANCE CHANGE *since 1988–91*

WINTER RELATIVE ABUNDANCE *2007/08–2010/11*

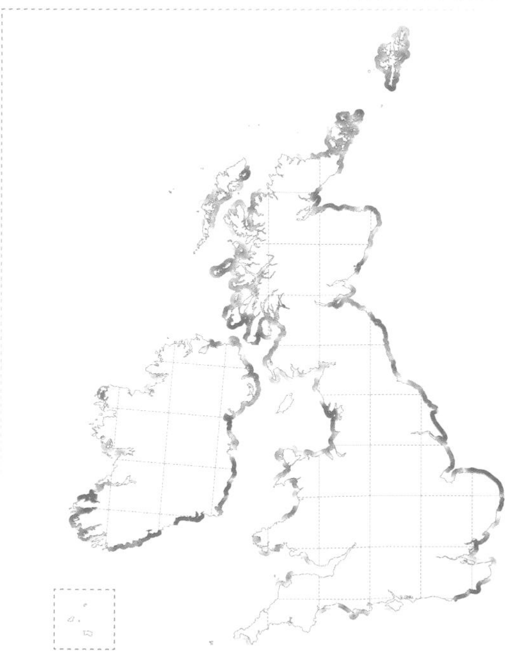

BLACK-THROATED DIVER

Gavia arctica

BTO CODE: BV

OLIVER SMART www.smartimages.co.uk

BLACK-THROATED DIVERS are present round much of the coast of Britain in winter, though are virtually absent from Shetland and are scarce around northeast Scotland and Wales. Given the strong westerly emphasis to the winter abundance map, with hotspots in Cornwall and northwest Scotland, it is surprising that the species is so scarce around the Irish coastline and Irish Sea. There has been a 51% range expansion since the *1981–84 Winter Atlas* but in some areas gains are undoubtedly due to improved coverage, for example in northwest Scotland.

The breeding distribution of Black-throated Divers is confined mainly to north and west Scotland and the Outer Hebrides. One *confirmed* record in Galloway is outside the core range. Breeding in Ireland has never been proven; although an individual with young was recorded on the sea during the *1988–91 Breeding Atlas*, the birds' origin was never confirmed, and there is no evidence to suggest that they have previously bred there (*Birds in Ireland*). Loch occupancy is associated with the abundance of small salmonids and complex shorelines (Jackson 2005). Highest breeding densities are in Sutherland, in Wester Ross and on the Outer Hebrides, where the species favours large lochs with suitable islets or, increasingly, artificial rafts to nest upon. The breeding distribution map is supplemented by *non-breeding* birds in 245 10-km squares, many of which were late wintering birds remaining in coastal areas in the early part of the breeding season.

A 10% range expansion in Britain is evident since the *1988–91 Breeding Atlas*, which mirrors national surveys showing an increase from 187 territories in 1994 to 217 territories in 2006 (Holling *et al.* 2009). Gains are concentrated in the eastern and southern parts of the range, though many of these newly occupied 10-km squares refer to single birds or pairs on suitable lochs, rather than *confirmed* breeding events. This contrasts with losses in the more northerly part of the range. Provision of rafts to moderate the effects of unnaturally fluctuating water levels and disturbance from human activities is thought to have increased productivity of the Scottish Black-throated Diver population by around 44% (Hancock 2000), and this may have fuelled the observed range expansion.

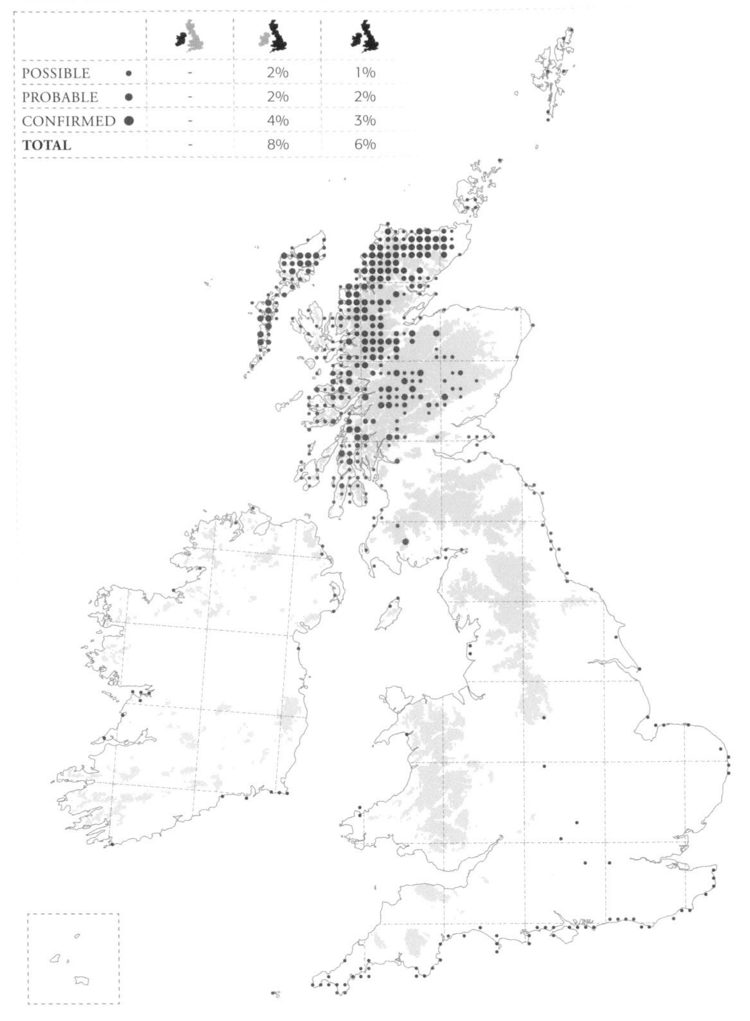

BREEDING DISTRIBUTION *2008–11*

POSSIBLE •	-	2%	1%
PROBABLE •	-	2%	2%
CONFIRMED •	-	4%	3%
TOTAL	-	8%	6%

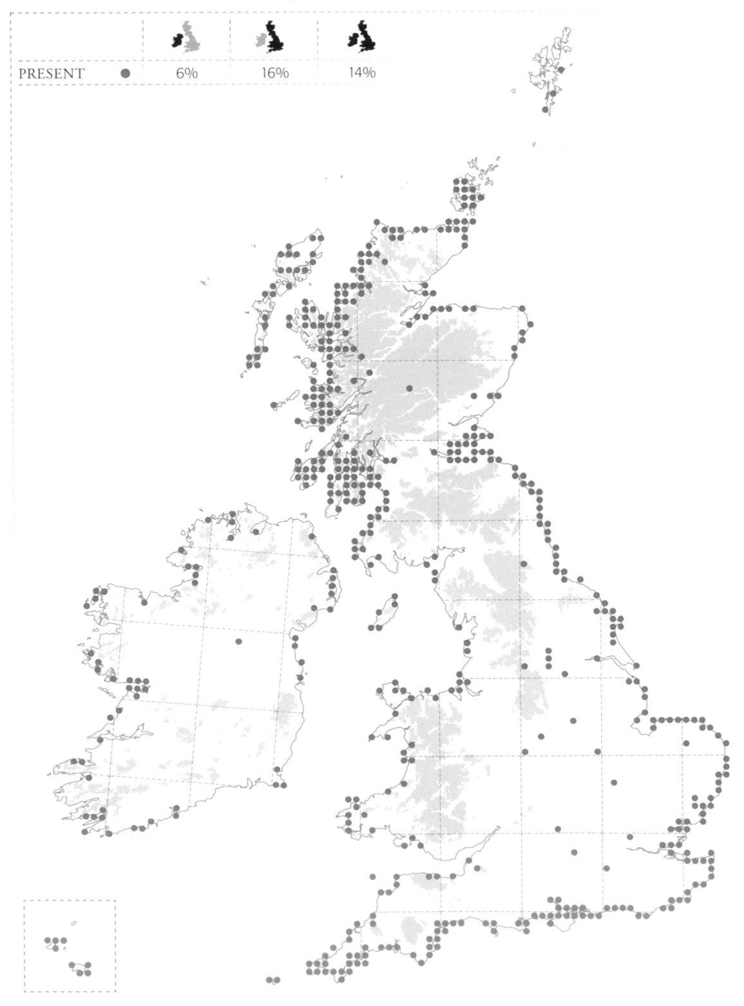

WINTER DISTRIBUTION *2007/08–2010/11*

PRESENT •	6%	16%	14%

BREEDING DISTRIBUTION CHANGE *since 1968–72*

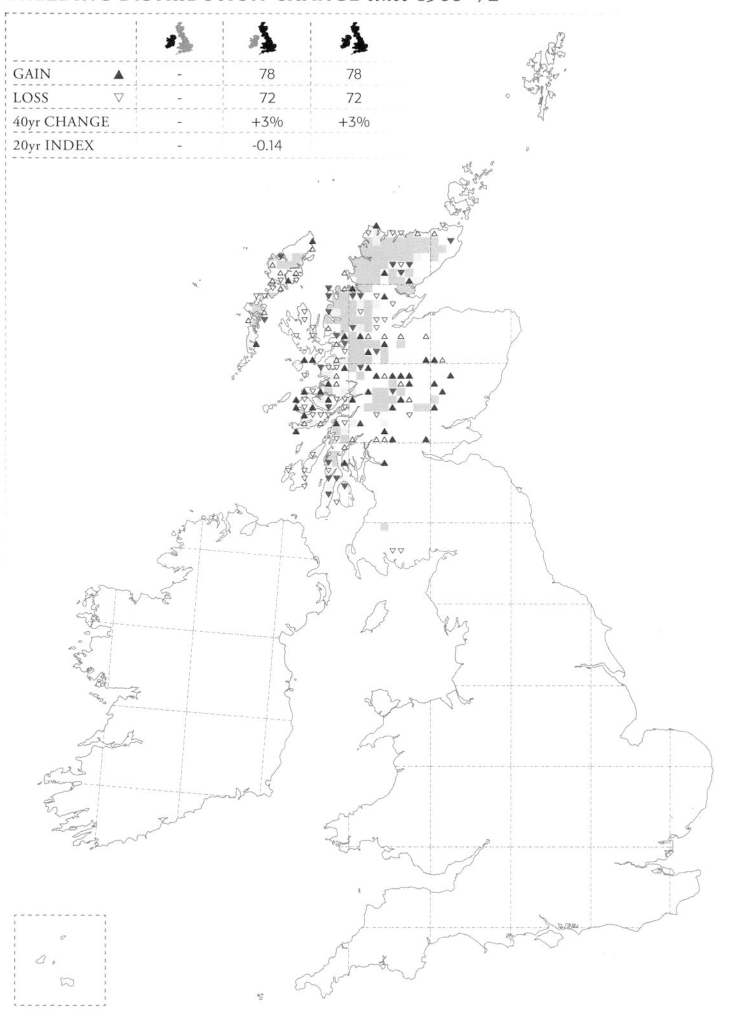

GAIN ▲	-	78	78
LOSS ▽	-	72	72
40yr CHANGE	-	+3%	+3%
20yr INDEX	-	-0.14	

WINTER DISTRIBUTION CHANGE *since 1981–84*

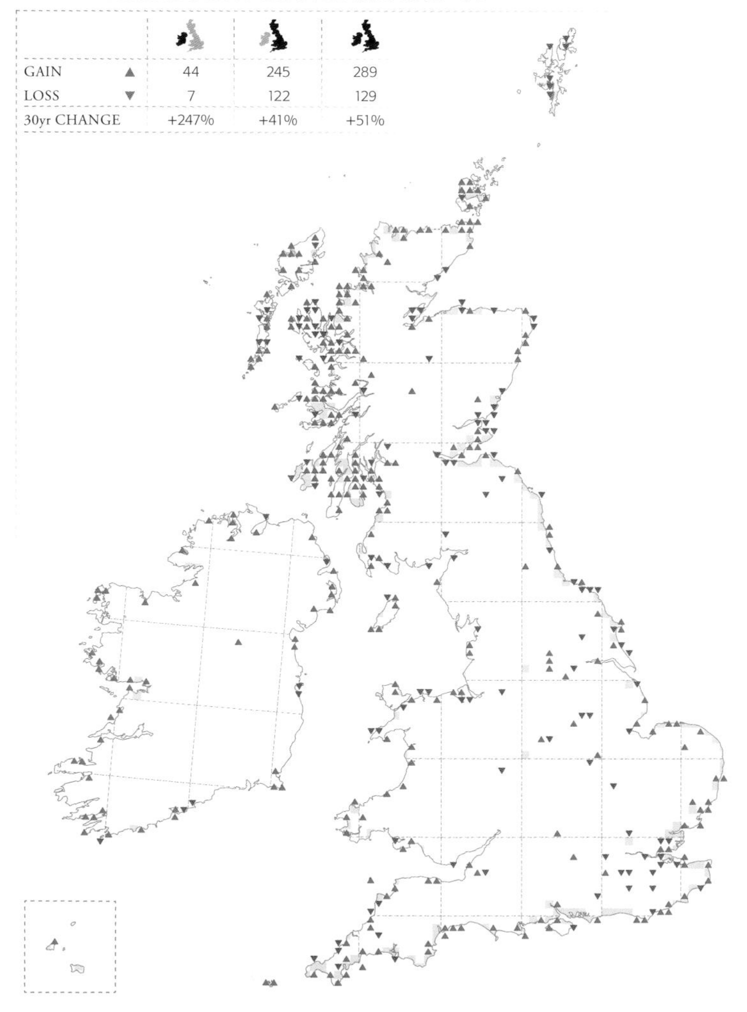

GAIN ▲	44	245	289
LOSS ▼	7	122	129
30yr CHANGE	+247%	+41%	+51%

BREEDING RELATIVE ABUNDANCE *2008–11*

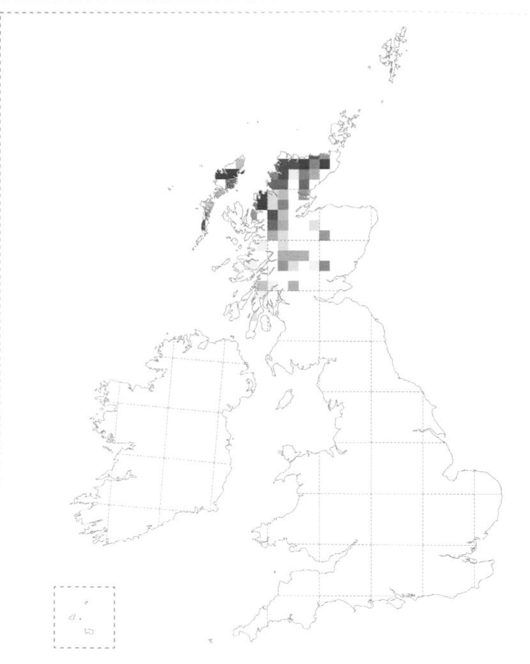

BREEDING RELATIVE ABUNDANCE CHANGE *since 1988–91*

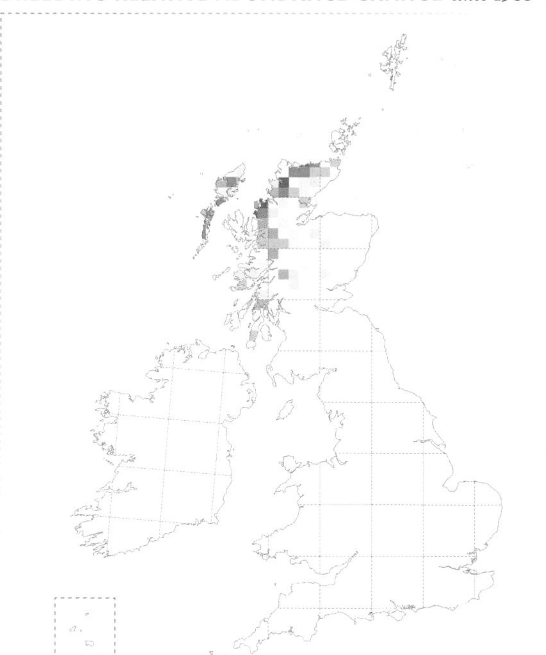

WINTER RELATIVE ABUNDANCE *2007/08–2010/11*

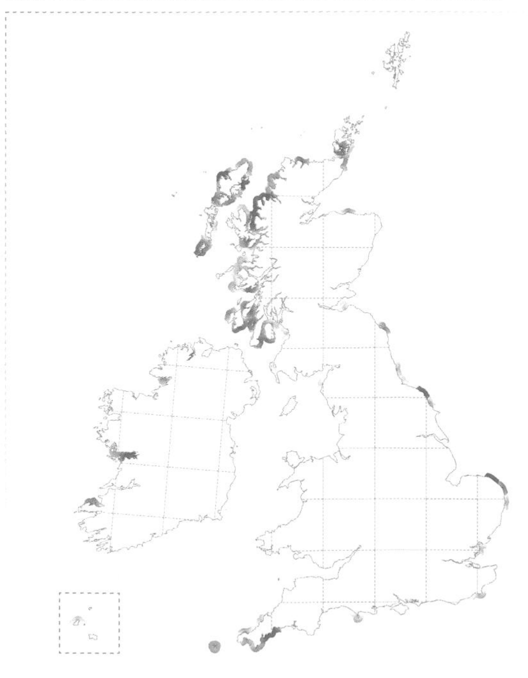

GREAT NORTHERN DIVER

Gavia immer

BTO CODE: ND

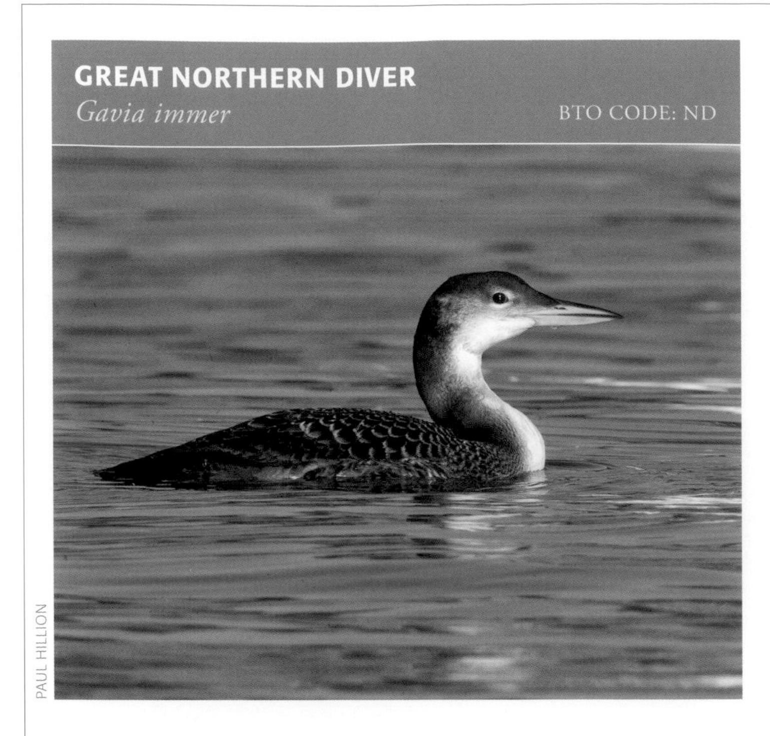

PAUL HILLION

GREAT NORTHERN DIVERS are mainly arctic breeders that move south to winter in our region (*Migration Atlas*) or still farther south. They are recorded in coastal waters around Britain & Ireland as well as occasionally appearing on inland wetlands. The largest concentrations during this atlas were found in the Northern Isles, the Outer Hebrides, northwest Scotland south to Argyll, and western and southern Ireland, which corresponds well to key areas found in a dedicated survey in 2006/07 (Lewis *et al.* 2008). In Ireland, internationally important concentrations of 385, 148 and 136 birds were recorded in winter 2008/09 at Galway Bay, Donegal Bay and Blacksod & Tullaghan Bays respectively (*I-WeBS Report 2009*). Comparable sites in Britain were Sound of Gigha (217 birds in 2008/09) and Rova Head to Wadbister Ness (113 birds in 2009/10) but, in general, systematic winter monitoring does not cover this species well (*WeBS Report 2011*). Elsewhere, records tend to be more of occasional individuals and this is particularly true of the scattering of inland records, although some larger reservoirs regularly attract birds that can stay for several weeks.

Since the *1981–84 Winter Atlas* there has been a 39% increase in the number of occupied 10-km squares in Britain & Ireland. In parts of northwest Scotland, apparent gains may be a consequence of improved coverage since the *1981–84 Winter Atlas*. Elsewhere the explanation is less clear, though Wood (2007) also ascribed increased numbers of Great Northern Divers in Essex to improved coverage.

The single mapped breeding record relates to a bird that paired with a Black-throated Diver for several consecutive seasons up to 2008 (Holling *et al.* 2010b). This mirrors the only previous putative breeding record, from the *1968–72 Breeding Atlas*, which is thought to have related to a hybrid pairing (*Birds of Scotland*). The breeding distribution maps also show that significant numbers of wintering or passage birds are present off northwest Ireland and Scotland into the spring, with some reported until mid or even late May. Birds migrating north may be responsible for Atlas records along the south coast of England in April. Small numbers remain to summer in coastal waters in the north and west.

Sponsored by Andrew Forbes

WINTER DISTRIBUTION *2007/08–2010/11*

		Britain & Ireland	Britain	Ireland
PRESENT	●	28%	27%	27%

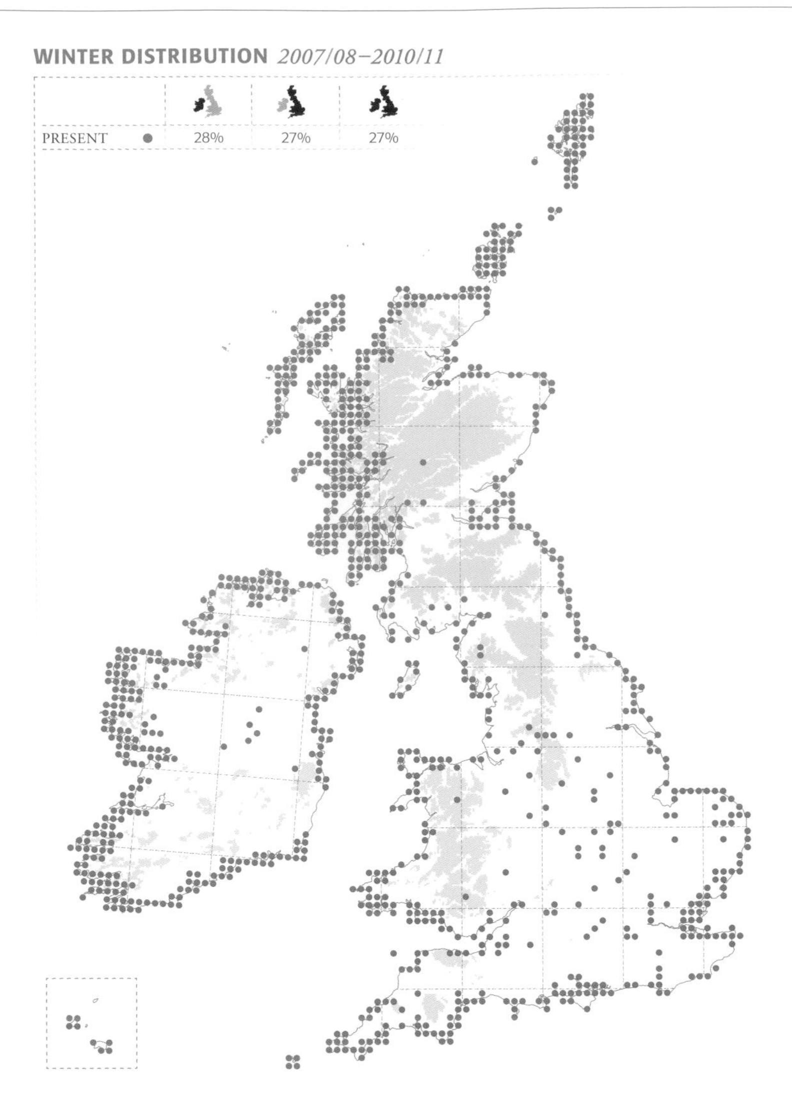

BREEDING DISTRIBUTION *2008–11*

		Britain & Ireland	Britain	Ireland
POSSIBLE	•	-	<1%	<1%
PROBABLE	●	-	-	-
CONFIRMED	●	-	-	-
TOTAL		-	<1%	<1%

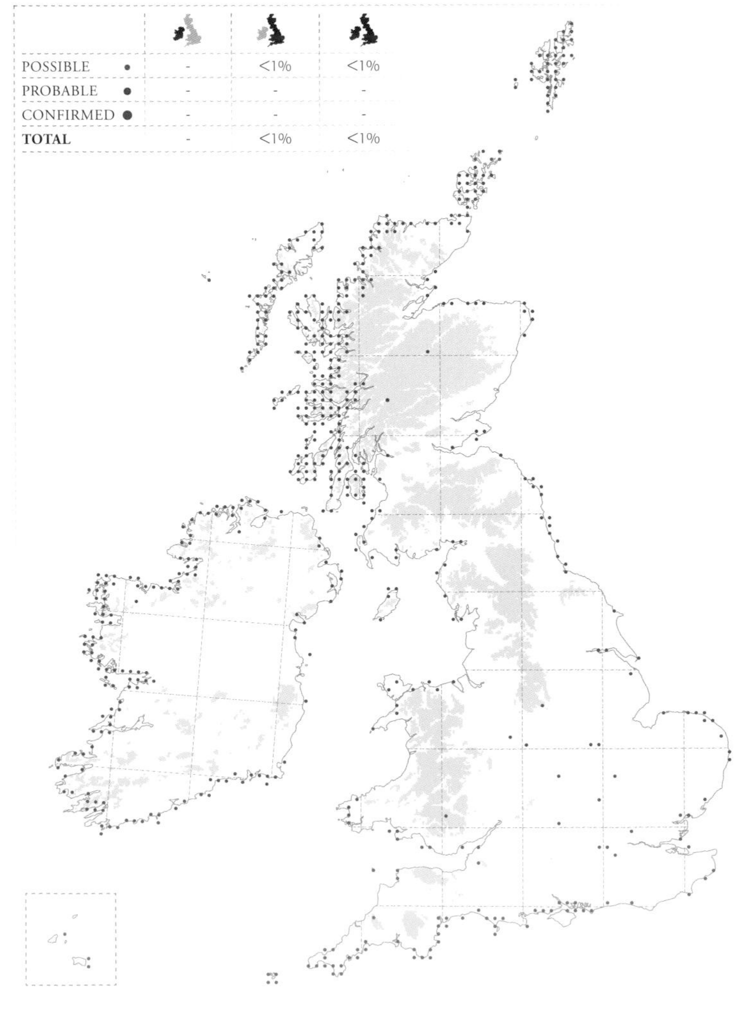

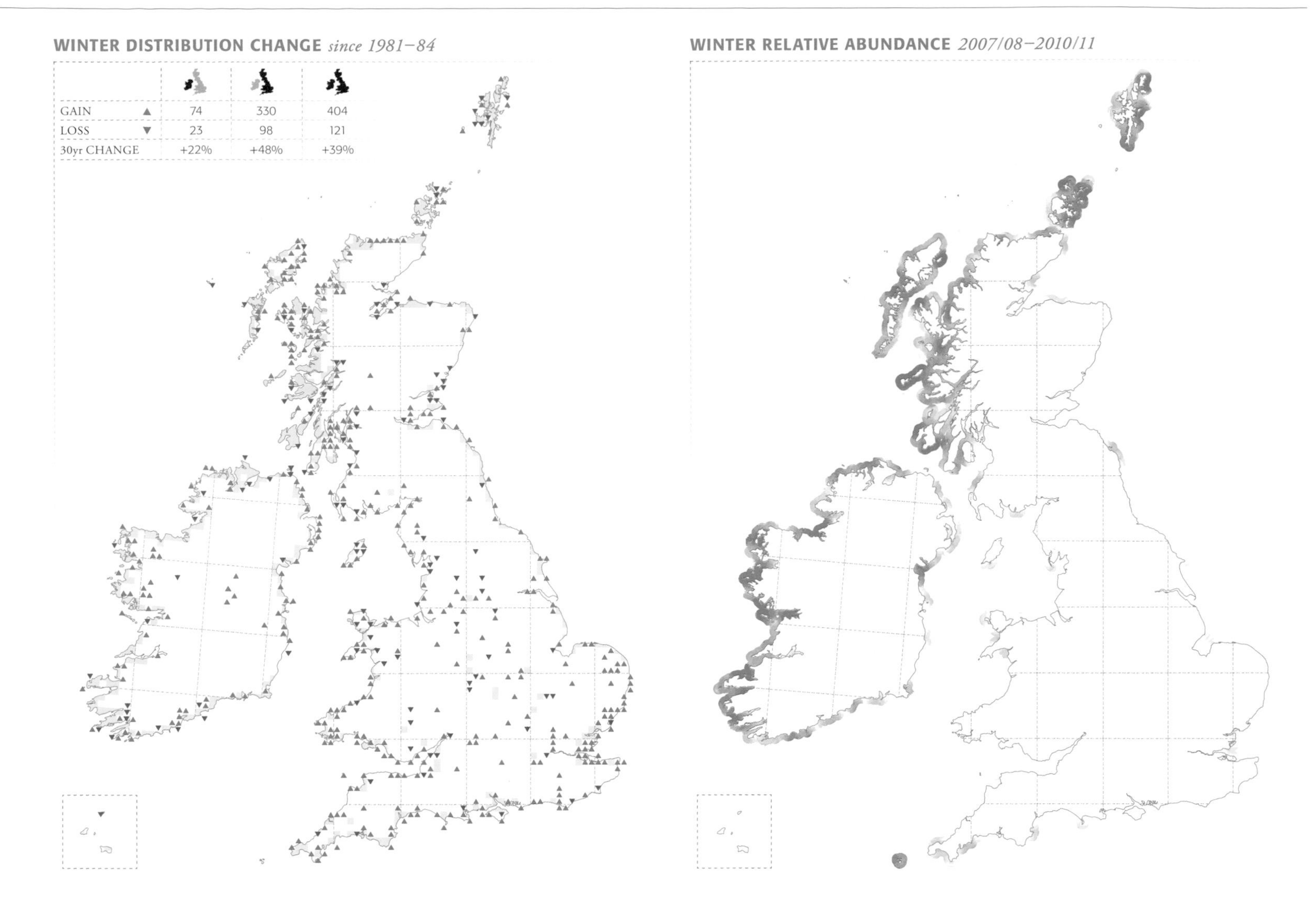

WINTER DISTRIBUTION CHANGE since 1981–84
GAIN ▲ 74 330 404
LOSS ▼ 23 98 121
30yr CHANGE +22% +48% +39%
WINTER RELATIVE ABUNDANCE 2007/08–2010/11

GRAHAM CATLEY http://pewit.blogspot.co.uk

◀ **The spread of records in the breeding season is perhaps surprising; *non-breeding* records came from 168 10-km squares in June and 72 squares in July.**

FULMAR

Fulmarus glacialis

BTO CODE: F.

NAOMI SCOTT

UP TO THE mid 18th century, the Fulmar's European breeding range was restricted to a few colonies in Iceland and St Kilda. The range then expanded, leading to the first British colony outside St Kilda, on Foula, in 1878, and the first Irish colony in 1911 (Fisher 1952). The breeding distribution map shows that Fulmars now breed around almost the entire coastlines of Scotland, Wales and Ireland, and in the southwest and northeast of England. They are absent from long stretches of the east and southeast coasts of England, reflecting the limited availability of suitable seacliffs for nesting. In winter, some of these gaps are filled, since Fulmars disperse widely outside the breeding season.

The breeding abundance map shows concentrations on the southwest and west coasts of Ireland and the north coast of Scotland and its offshore islands, especially Shetland. Over 90% of Britain & Ireland's breeding Fulmars are in Scotland (*Seabird 2000*).

The change maps show range expansion since the *1968–72 Breeding Atlas* and the *1981–84 Winter Atlas*, but a contraction since the *1988–91 Breeding Atlas*, though the magnitudes of these changes were small. Dedicated seabird monitoring shows that numbers increased until the mid 1990s then declined sharply from the turn of the century and that productivity has declined since the mid 1990s (JNCC 2012).

The availability of discards from fishing vessels (Hudson & Furness 1989) is believed to have been a major factor in facilitating the Fulmar's range expansion over many decades. However, the recent decline in breeding numbers has been linked to a reduction in whitefish fishing in the North Sea and a shortage of sandeels (Votier *et al.* 2004; Furness 2007). Climate change may also be affecting the abundance of zooplankton (and higher trophic levels, i.e. prey fish), another important food source. Seabird bycatch in the Grand Sol longline fishery to the west of Ireland may kill considerable numbers of Fulmars but as yet this has not been quantified nor linked to changes in breeding numbers at nearby colonies (Anderson *et al.* 2011).

Sponsored by David Milne QC

BREEDING DISTRIBUTION *2008–11*

POSSIBLE •	<1%	1%	1%
PROBABLE •	1%	1%	1%
CONFIRMED •	15%	17%	17%
TOTAL	16%	19%	19%

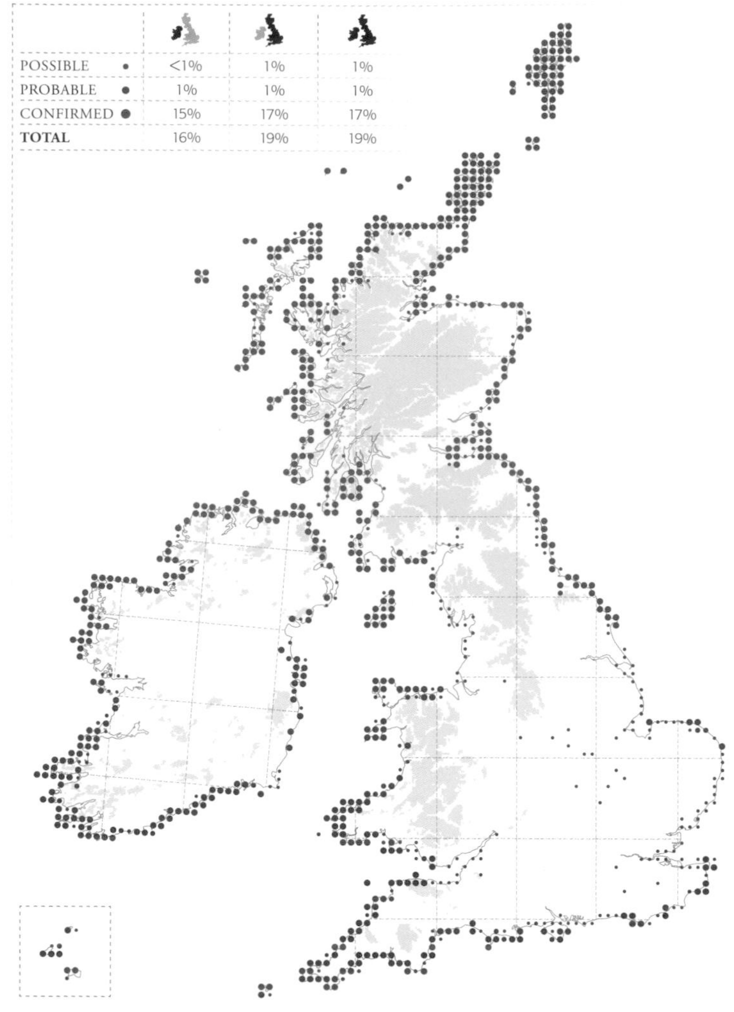

WINTER DISTRIBUTION *2007/08–2010/11*

PRESENT •	16%	19%	18%

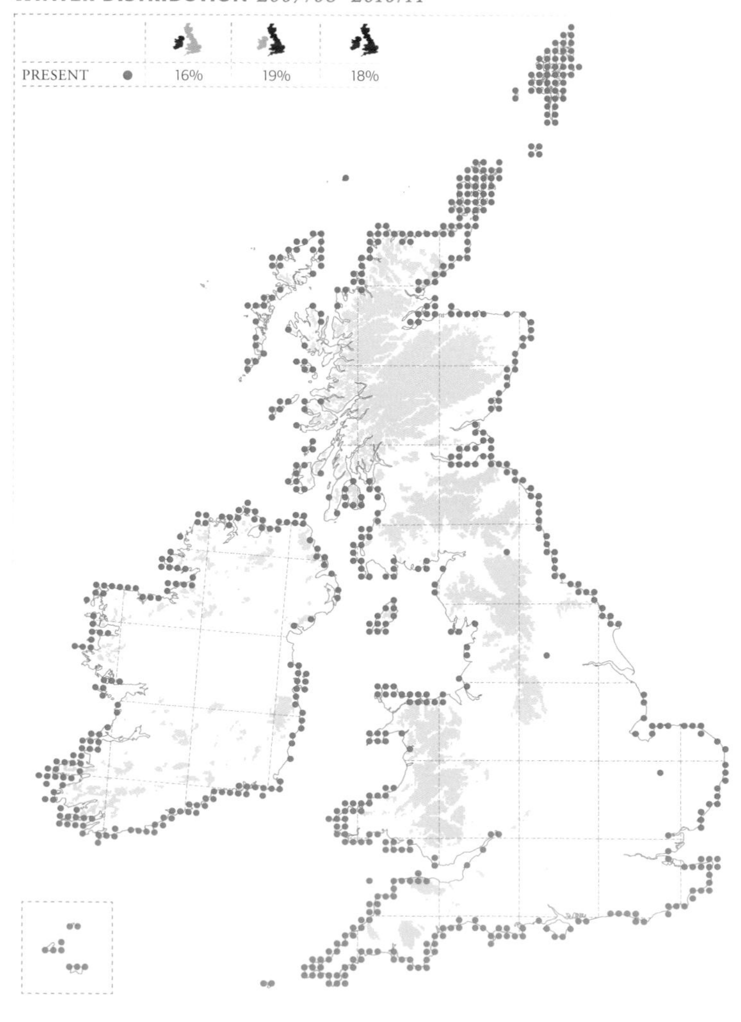

BREEDING DISTRIBUTION CHANGE *since 1968–72*

GAIN ▲	31	66	97
LOSS ▽	19	55	74
40yr CHANGE	+8%	+2%	+4%
20yr INDEX	-0.11	-0.09	

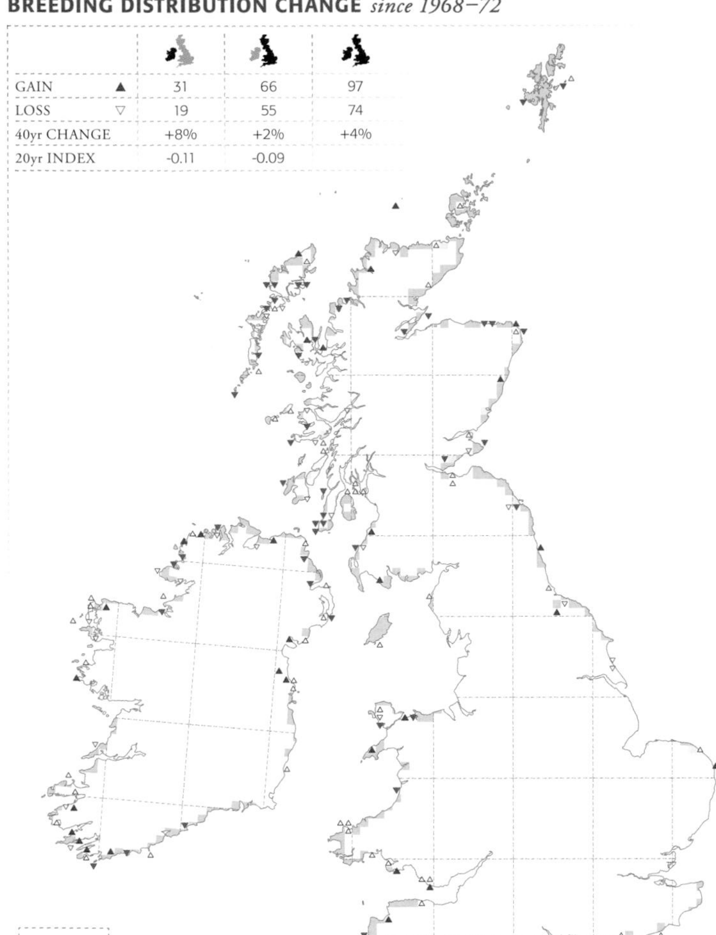

BREEDING RELATIVE ABUNDANCE *2008–11*

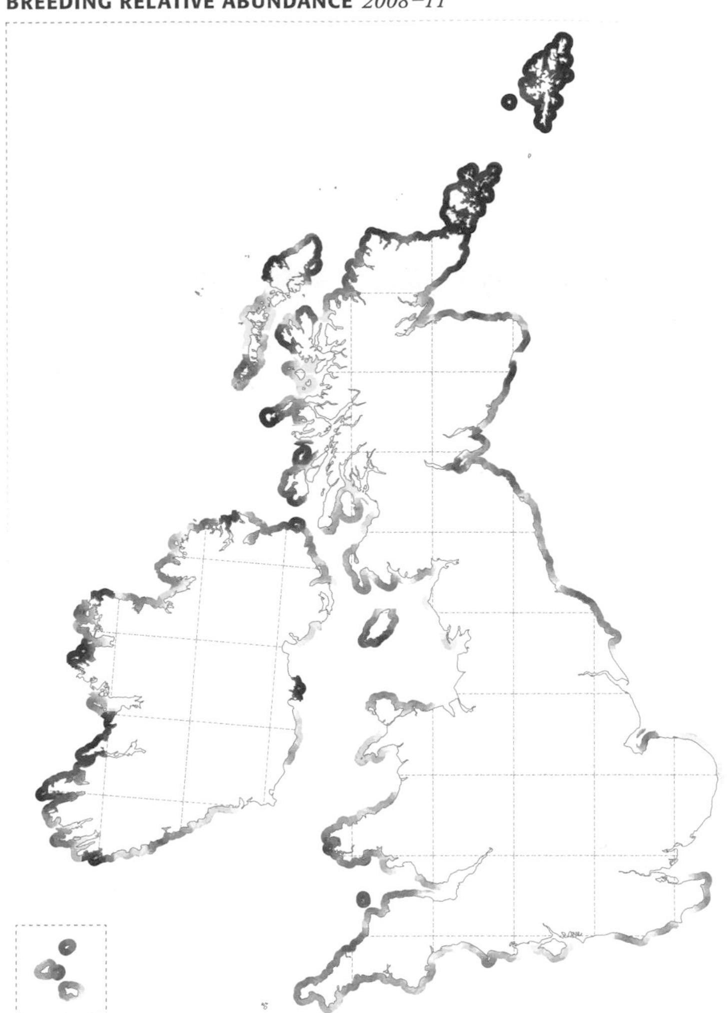

WINTER DISTRIBUTION CHANGE *since 1981–84*

GAIN ▲	47	133	180
LOSS ▼	39	91	130
30yr CHANGE	+5%	+9%	+8%

WINTER RELATIVE ABUNDANCE *2007/08–2010/11*

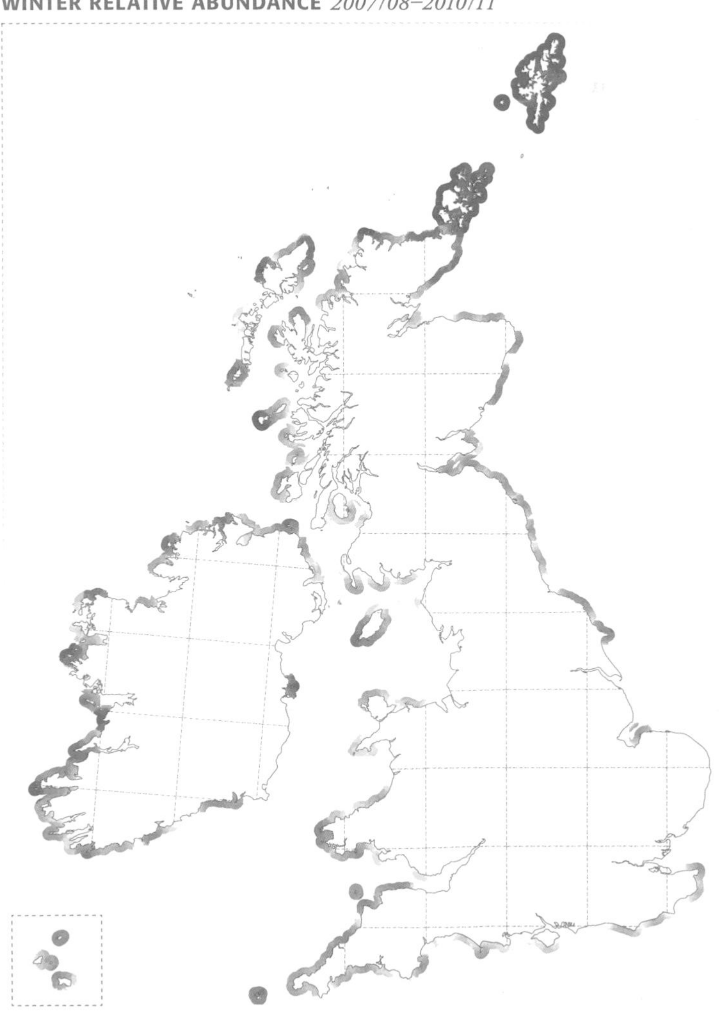

MANX SHEARWATER

Puffinus puffinus BTO CODE: MX

GARY FAULKNER

MANX SHEARWATERS ARE summer visitors to Britain & Ireland, breeding in roughly 50 colonies of which the largest, on Rum in Scotland and Skomer in Wales, account for over 75% of the total British & Irish population, itself accounting for 80% of the global population (*Seabird 2000*; JNCC 2012). The colonies tend to be located on steep grassy slopes on offshore islands, mainly along the western coastlines of both Britain and Ireland, and this is reflected in the breeding distribution map. However, they are wide-ranging during the summer with foraging individuals seen widely around the coastlines of Britain and Ireland; birds from Skomer move northwards and westwards into the Irish Sea to feed (Guilford *et al.* 2008).

Surveying Manx Shearwaters is difficult owing to their preference for remote islands, which they visit only at night. Unlike previous atlases, *Bird Atlas 2007–11* did not benefit from a major seabird census project so it is likely that some apparent losses, particularly in western Ireland, shown on the breeding change map reflect variation in coverage. There were no records of *confirmed* breeding during this atlas on Foula, Shetland, Rathin Island, Co. Antrim, nor Sark, Little Sark or Jethou, Channel Islands, despite visits to these colonies. Likewise, some apparent gains may be due to colony expansion into new 10-km squares or to better coverage, though gains on the Isle of May and Isles of Scilly are real. Currently there are no national population trends available for this species (JNCC 2012). On Canna, Calf of Man and Lundy, numbers declined markedly due to predation by rats, though extinction was avoided through timely eradication projects (Bell *et al.* 2011; Brown *et al.* 2011). In Ireland, the American Mink has recently spread to Manx Shearwater colonies in Counties Kerry and Wexford, with consequences as yet unknown (Newton 2011).

Following fledging in September, most depart for their South American wintering grounds (Guilford *et al.* 2009) but some remain into November, leading to a scattering of records around Britain, south of the Firth of Forth, and Ireland. There has been a marked increase in winter records since the *1981–84 Winter Atlas*.

BREEDING DISTRIBUTION *2008–11*

POSSIBLE •	<1%	<1%	<1%
PROBABLE ●	-	<1%	<1%
CONFIRMED ●	1%	1%	1%
TOTAL	1%	1%	1%

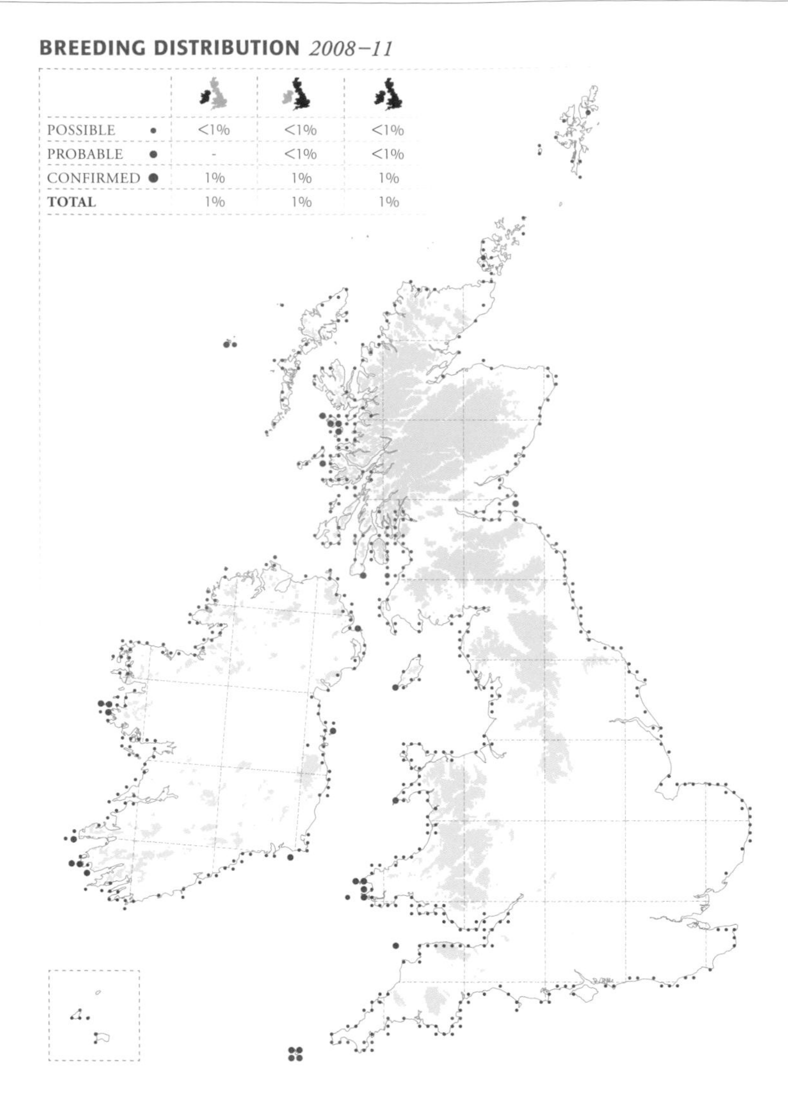

Sponsored by Mike Archer

BREEDING DISTRIBUTION CHANGE *since 1968–72*

GAIN ▲	2	6	8
LOSS ▽	8	15	23
40yr CHANGE	-38%	-28%	-31%
20yr INDEX	-	-	

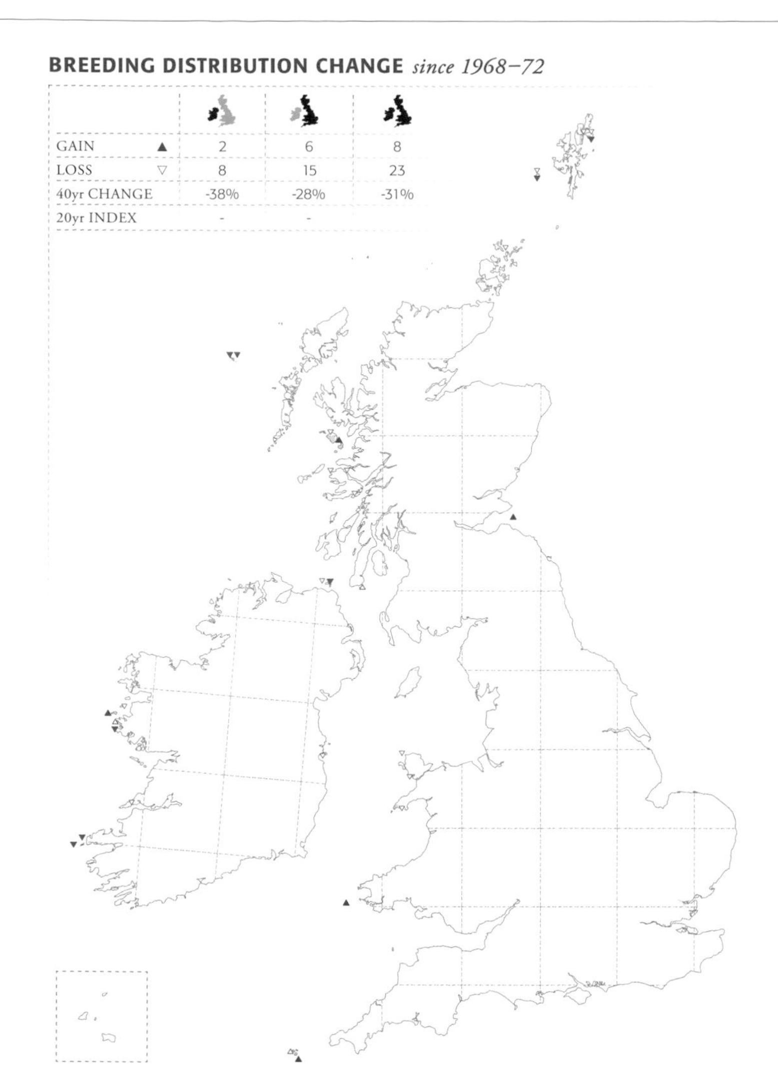

WINTER DISTRIBUTION *2007/08–2010/11*

PRESENT ●	1%	2%	1%

BALEARIC SHEARWATER

Puffinus mauretanicus BTO CODE: YQ

RICHARD STONIER www.birdsonline.co.uk

OVER THE LAST decade the Balearic Shearwater has become more regularly recorded in British and Irish coastal waters, possibly due to a northward shift in post-breeding movement into the Atlantic (Wynn 2009). This change is thought to relate to increased sea-surface temperatures (Yésou 2003). Occurrences in British and Irish waters are largely between July and October and occasionally into early winter. There also appears to be a mid winter influx into these waters, with winter records most concentrated off the coast of southwest Britain and the Channel Islands, and further records patchily distributed elsewhere along the southern and eastern British coast and off southwest Ireland. They were also reported as *non-breeding* from 94 10-km squares in the breeding season.

WINTER DISTRIBUTION *2007/08–2010/11*

PRESENT ●	<1%	2%	1%

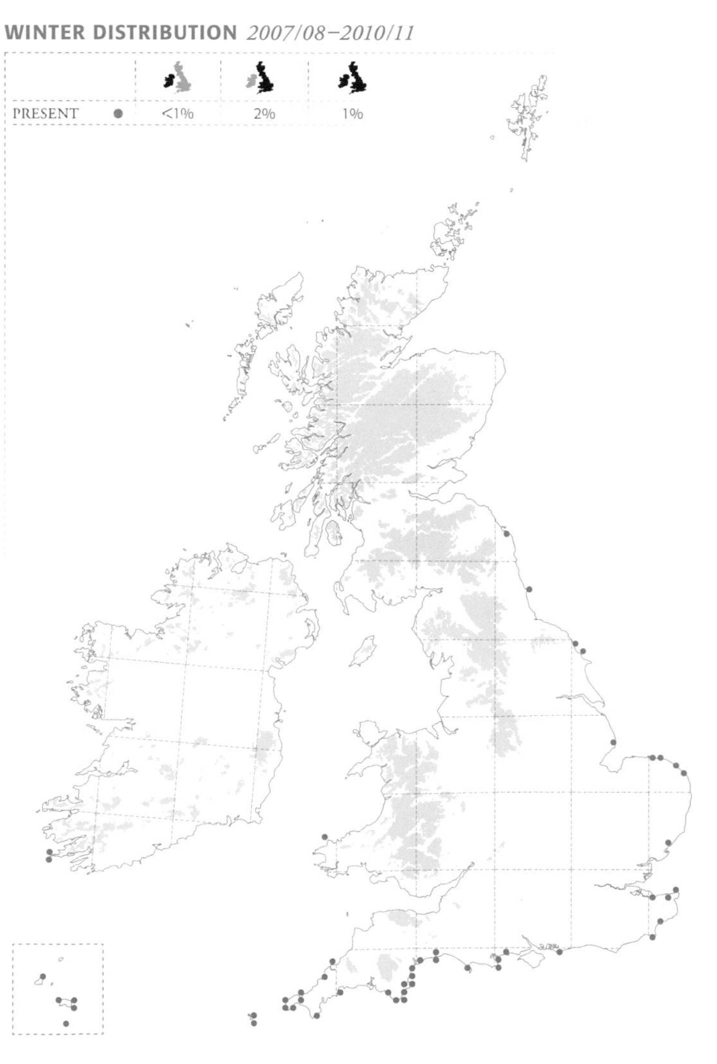

STORM PETREL

Hydrobates pelagicus

BTO CODE: TM

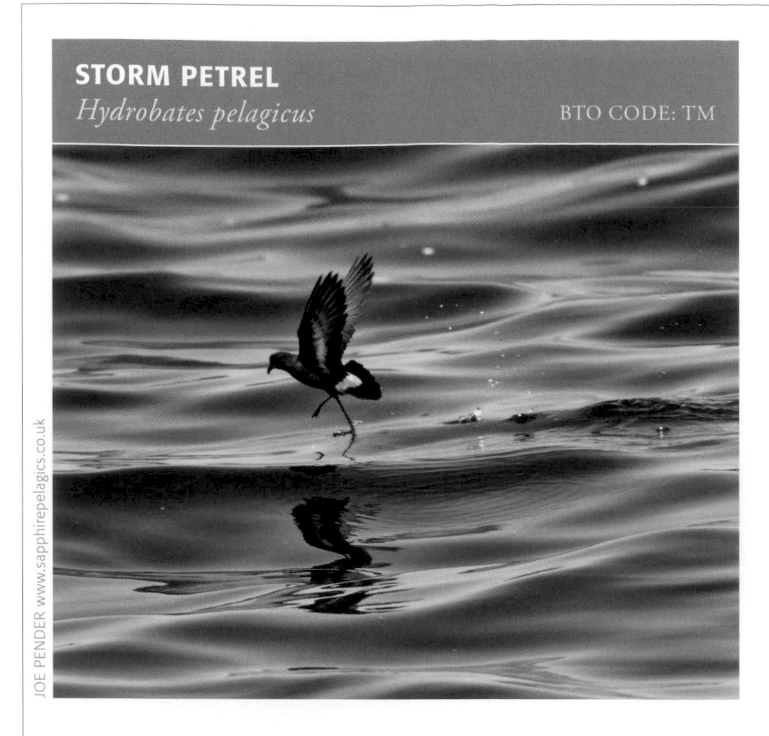

JOE PENDER www.sapphirepelagics.co.uk

STORM PETRELS NEST on remote, small, rat-free, offshore islands on the Atlantic fringe of north and west Britain & Ireland from Shetland to Burhou in the Channel Islands (*Seabird 2000*) and the distribution map presented here accords well with this pattern.

Unlike previous atlases, *Bird Atlas 2007–11* did not benefit from a major seabird census project so it is likely that some apparent losses shown on the breeding change map reflect variation in coverage; thus drawing conclusions on trends is difficult. The Irish distribution documented in the *1968–72 Breeding Atlas* was thought to be under-recorded. Of the atlases to date, perhaps the most complete was the *1988–91 Breeding Atlas*, which used data from the recently completed Seabird Colony Register (1985–87; Lloyd *et al.* 1991) during which visits were made to most known colonies. Compared to those atlases, *Bird Atlas 2007–11* figures suggest range contractions of 22% since 1968–72 and 28% since 1988–91 but these are unlikely to be accurate, given the variable nature of coverage.

What can be gleaned from these data is that colonies on the Calf of Man and Foula seem to have become extinct. In Ireland there are no known colony extinctions and the losses are likely to be coverage artefacts. As with the other burrow-nesting species, predation by invasive mammalian predators must be a serious issue. Monitored colonies in Ireland on Great Skellig, Co. Kerry, and Inishglora, Co. Mayo, show population fluctuations but no observable trend, whilst in Scotland monitoring on Priest Island reveals a continuing decline in numbers (Highland SOC 2011) Apparently new colonies were located on Unst and Yell in Shetland, Copinsay in Orkney, Islay in the Inner Hebrides and Ramsay Island in Pembrokeshire.

Away from known colonies there is a scatter of sightings of *non-breeding* birds around the coasts of Britain and Ireland.

In winter, Storm Petrels were recorded in 50 10-km squares, mostly along the North Sea coast and off southwest England. This represents a significant increase compared to the 11 squares occupied in the *1981–84 Winter Atlas*, possibly as a consequence of the popularity of seawatching.

Sponsored by Harry Green, in memory of his inspirational biology teacher, Arnold Darlington

BREEDING DISTRIBUTION *2008–11*

POSSIBLE •	1%	<1%	<1%
PROBABLE •	-	<1%	<1%
CONFIRMED ●	1%	1%	1%
TOTAL	2%	1%	1%

BREEDING DISTRIBUTION CHANGE *since 1968–72*

GAIN ▲	5	15	20
LOSS ▽	10	23	33
40yr CHANGE	-31%	-18%	-22%
20yr INDEX	-	-	

LEACH'S PETREL

Oceanodroma leucorhoa

BTO CODE: TL

JASON ATKINSON www.at2h.blogspot.com

LEACH'S PETREL BREEDING colonies are limited to remote offshore islands in the far north and west of Scotland and southwest Ireland (*Seabird 2000*), close to the deep-water areas beyond the continental shelf where birds are thought to feed (*Seabird 2000*). St Kilda alone supported more than 90% of the British & Irish breeding population, while the Flannan Islands and North Rona, both off northwest Scotland, each supported over 1,000 apparently occupied sites at the time of the last survey (*Seabird 2000*). These colonies constitute a small proportion of the estimated 10 million pairs globally.

Surveying Leach's Petrels is difficult owing to their preference for remote islands, which they visit only at night. Unlike previous atlases, *Bird Atlas 2007–11* did not benefit from a major seabird census project so it is likely that some apparent losses shown on the breeding change map reflect variation in coverage. Nevertheless, during this atlas, two new colonies were found on Shetland (Miles *et al.* 2010). Singing or displaying birds noted on Fair Isle and on Great Skellig in Co. Kerry may indicate breeding attempts at these sites.

There have been declines in numbers at existing colonies (Newson *et al.* 2008b; Murray *et al.* 2010), and increased predation by cats (Foula) and Great Skuas (St Kilda) has been implicated. Since the *1988–91 Breeding Atlas*, three colonies, Sula Sgeir, Foula and Sule Skerry, appear to have been lost, although there are doubts that breeding actually occurred on the last island. The Stags of Broadhaven in Co. Mayo currently register as a loss but this is because they were not visited during the current atlas period, though Leach's Petrels have been recorded in the area by seawatchers. Away from known colonies, the small number of records scattered round the coast are possibly influenced by ringers attracting birds by sound playback.

Leach's Petrels can be seen from coastal headlands and seawatching hotspots during autumn and early winter gales. Many of the 100 10-km squares on the winter distribution map were occupied only in early November. Nevertheless, Atlas data show a 500% range expansion since the *1981–84 Winter Atlas* when just 11 squares were occupied. The increase is probably a consequence of the growing popularity of seawatching.

Sponsored by David Milne QC

BREEDING DISTRIBUTION *2008–11*

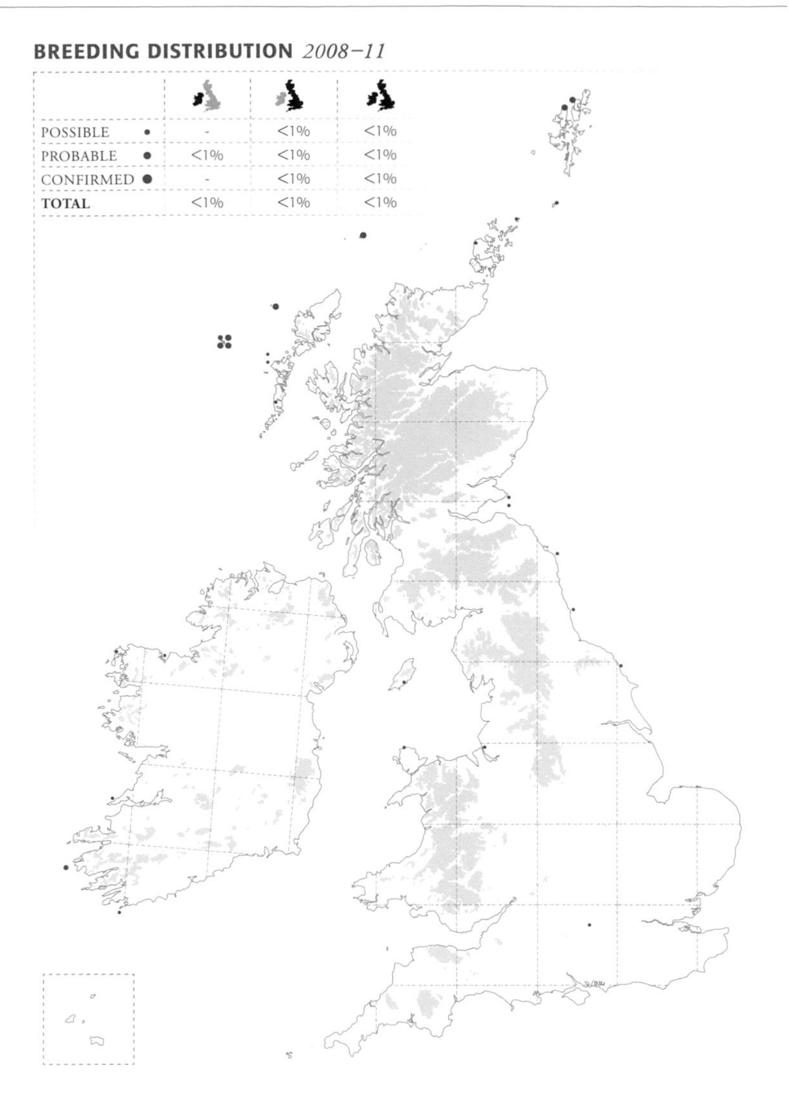

POSSIBLE •	-	<1%	<1%
PROBABLE •	<1%	<1%	<1%
CONFIRMED •	-	<1%	<1%
TOTAL	<1%	<1%	<1%

WINTER DISTRIBUTION *2007/08–2010/11*

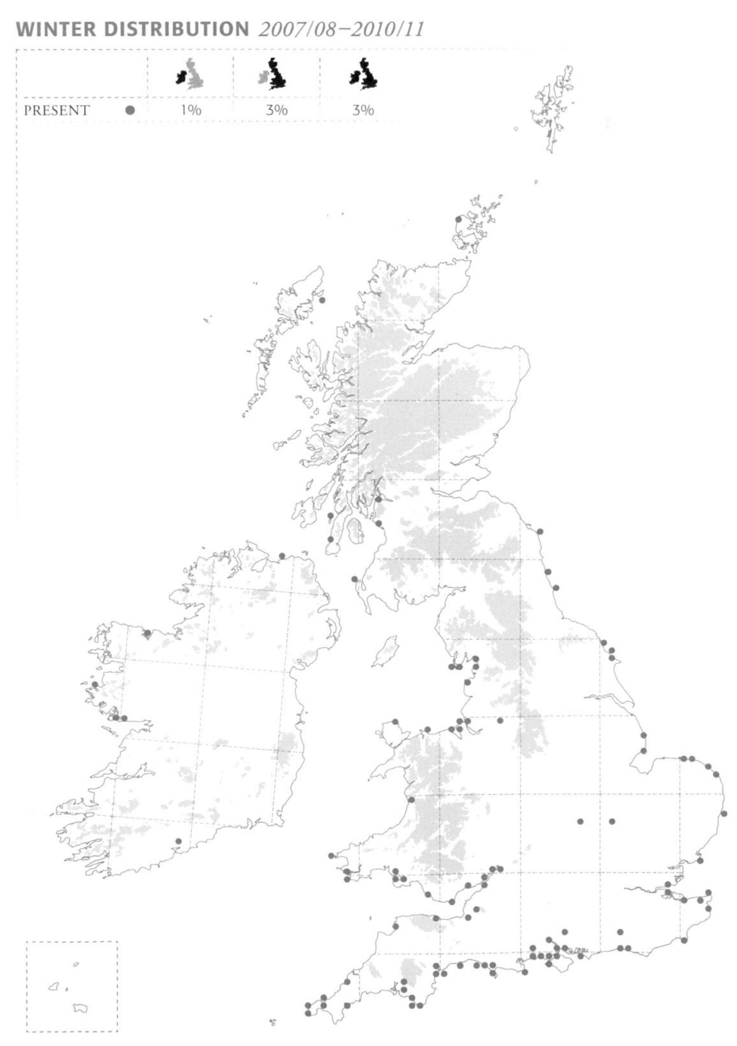

PRESENT •	1%	3%	3%

GANNET

Morus bassanus

BTO CODE: GX

TOM FINCH www.tomfinchphotography.co.uk

PERHAPS THE MOST striking feature of the Gannet breeding distribution map is that birds were recorded from virtually every coastal 10-km square in Britain and Ireland. These represent a combination of adult birds ranging widely on foraging trips from colonies, non-breeding adults (Hamer *et al.* 2001; Camphuysen 2011) and immature birds.

In contrast, *confirmed* breeding was confined to 26 colonies in 33 10-km squares. During the *1988–91 Breeding Atlas* Gannets occupied 22 colonies, an increase of six compared with the *1968–72 Breeding Atlas*. By 2011 another four colonies had become established, on Westray and Sule Skerry, both in Orkney (2003), Lambay Island, Co. Dublin (2006), and most recently Berneray, Outer Hebrides (2009). Records of *possible* breeding came from Copinsay, Orkney, in 2008 and 2009, where birds have been seen on the island but have never bred.

The last complete survey in Britain & Ireland, in 2003/04, found c.261,000 apparently occupied nests (Wanless *et al.* 2005a). Gannets can travel great distances from their nest site to forage and also eat a wide range of live fish and discards from fishing vessels. They thus seem buffered against food shortages. This, and high adult survival rates, may have enabled the population to increase (JNCC 2012). However, the rate of population increase may be slowing, possibly due to a slight reduction in survival rates (Wanless *et al.* 2006). The breeding abundance map shows high concentrations located close to colonies, but also significant numbers along many northern and western coasts.

In winter, Gannets were widely recorded around the entire coastlines of Britain and Ireland, with highest concentrations around the Northern Isles and off southeast Scotland and northeast England, southwest England, southwest Wales, and southwest Ireland. The winter change map shows a 52% range expansion in Britain and a 1% contraction in Ireland since the *1981–84 Winter Atlas*. Numbers are small, however, compared to those in the summer months, since most Gannets winter in the Bay of Biscay, in the Mediterranean or off the coast of West Africa (Kubetzki *et al.* 2009; Veron & Lawlor 2010).

Sponsored by BTO

BREEDING DISTRIBUTION *2008–11*

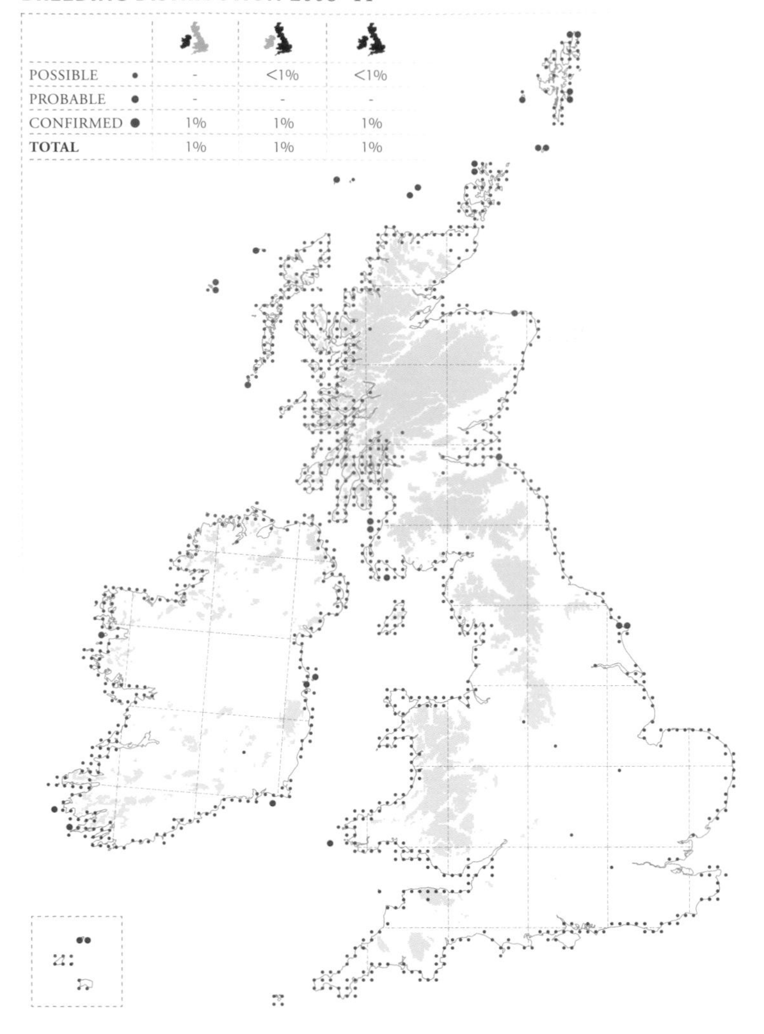

WINTER DISTRIBUTION *2007/08–2010/11*

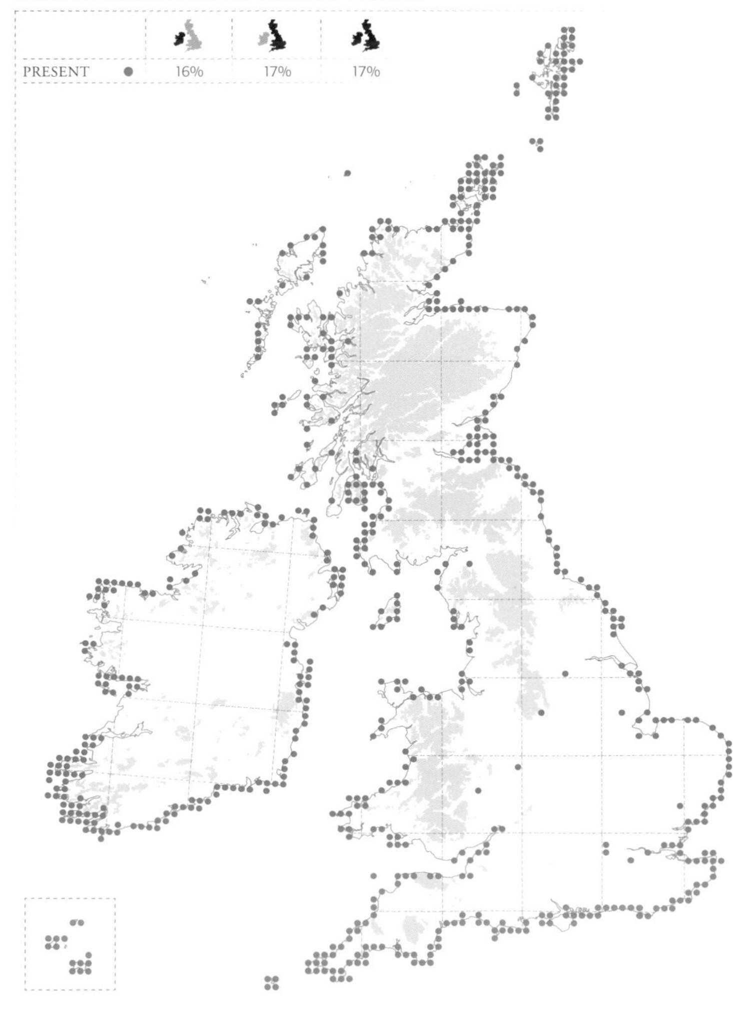

BREEDING DISTRIBUTION CHANGE *since 1968–72*

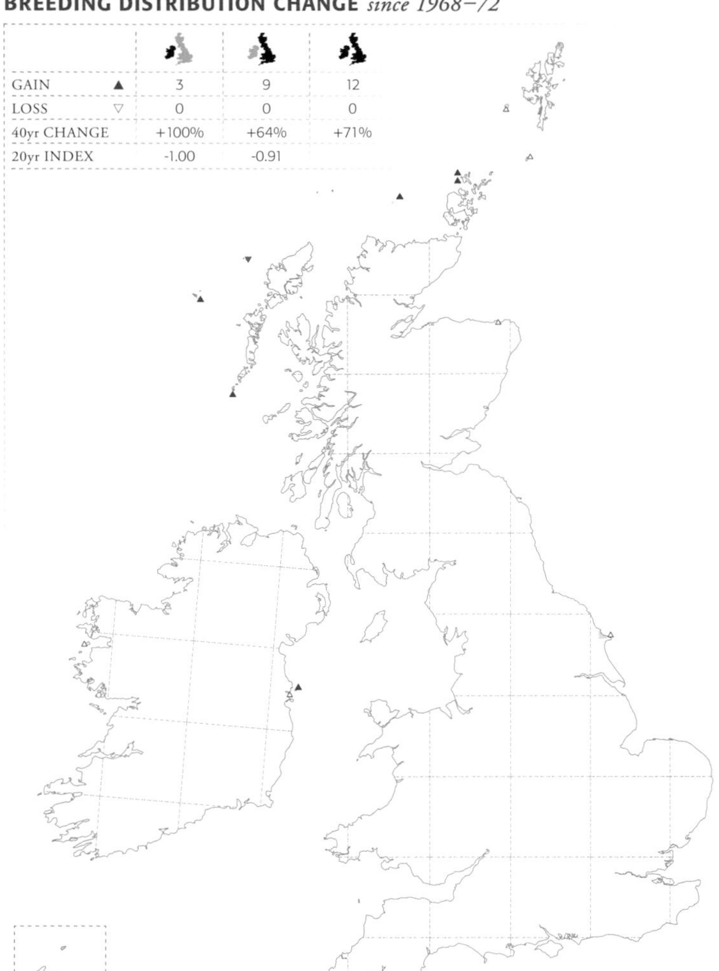

GAIN ▲	3	9	12
LOSS ▽	0	0	0
40yr CHANGE	+100%	+64%	+71%
20yr INDEX	-1.00	-0.91	

BREEDING RELATIVE ABUNDANCE *2008–11*

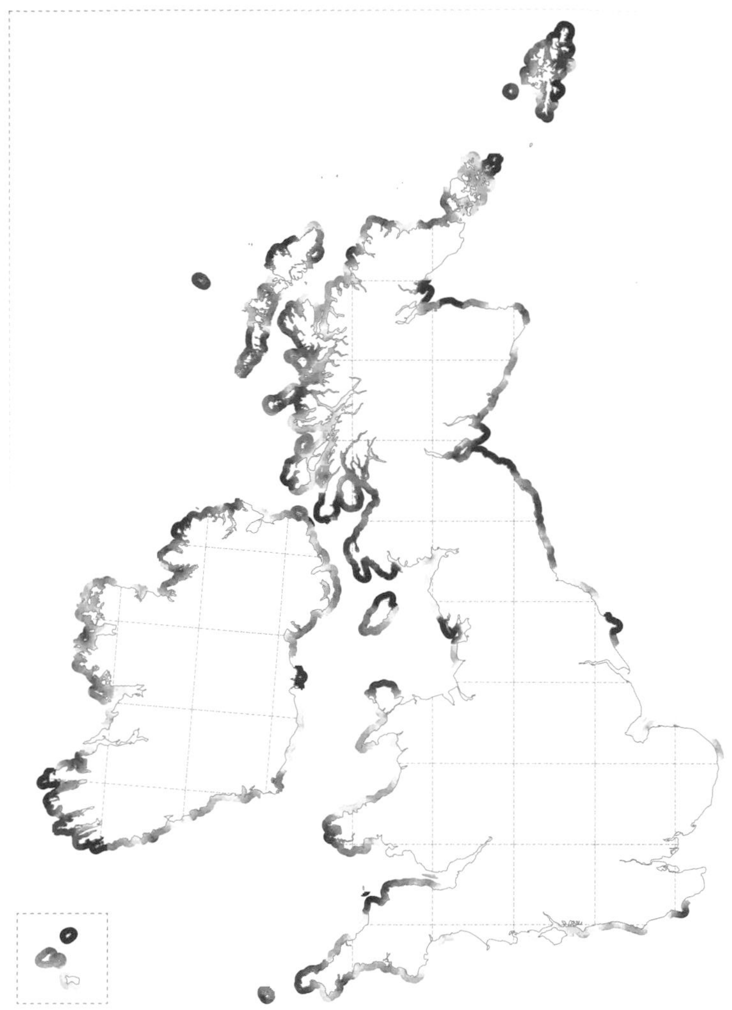

WINTER DISTRIBUTION CHANGE *since 1981–84*

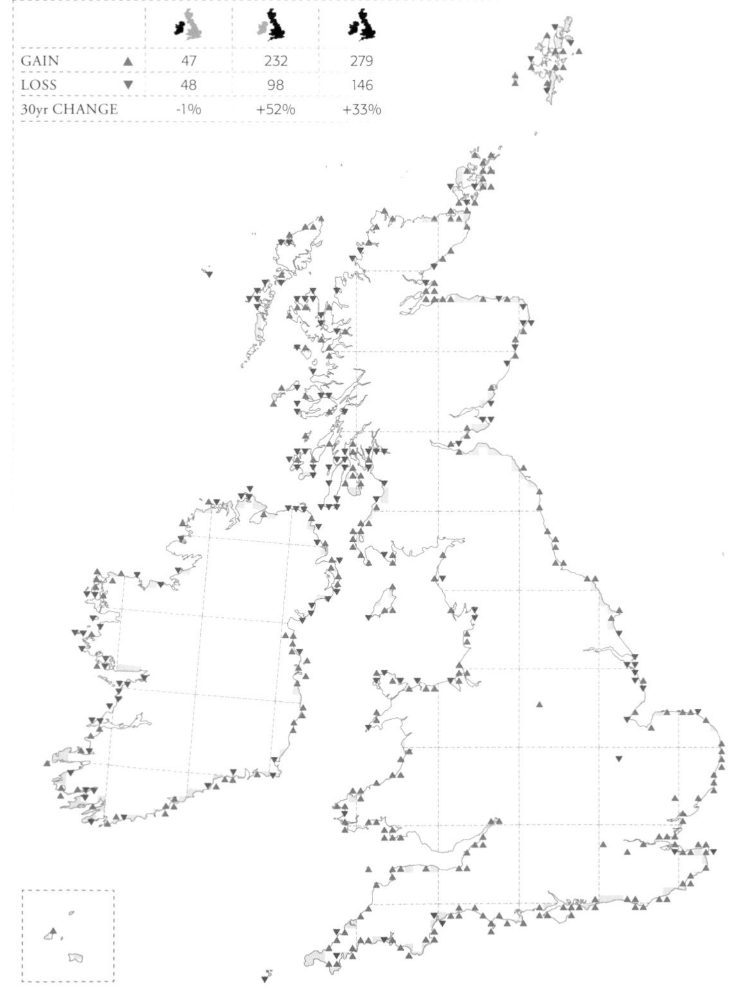

GAIN ▲	47	232	279
LOSS ▼	48	98	146
30yr CHANGE	-1%	+52%	+33%

WINTER RELATIVE ABUNDANCE *2007/08–2010/11*

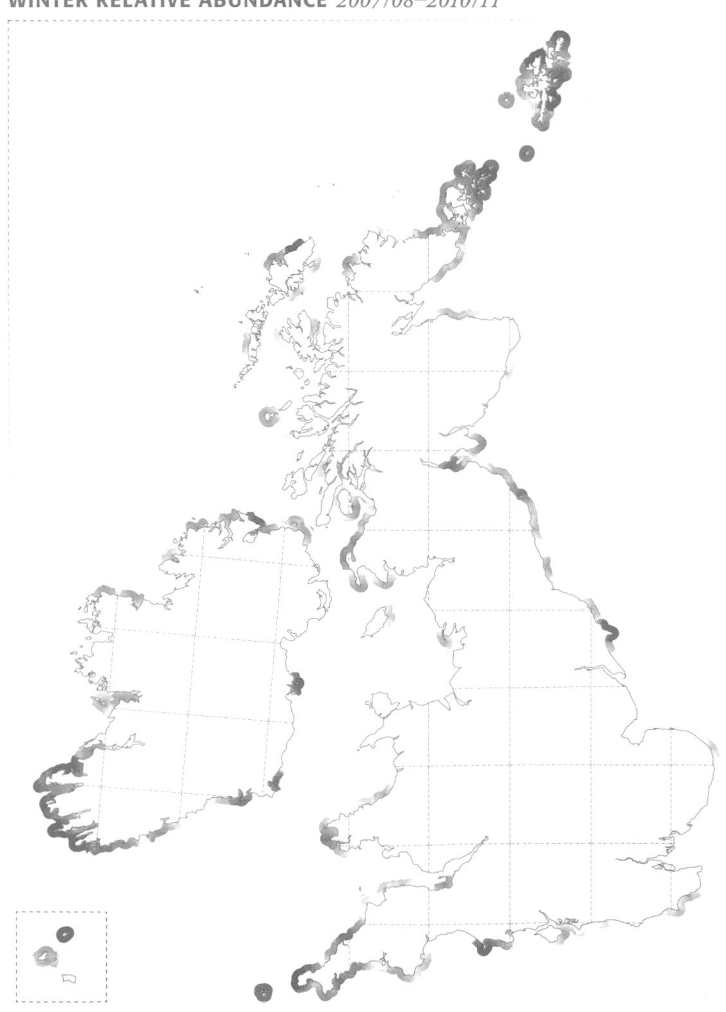

CORMORANT
Phalacrocorax carbo

BTO CODE: CA

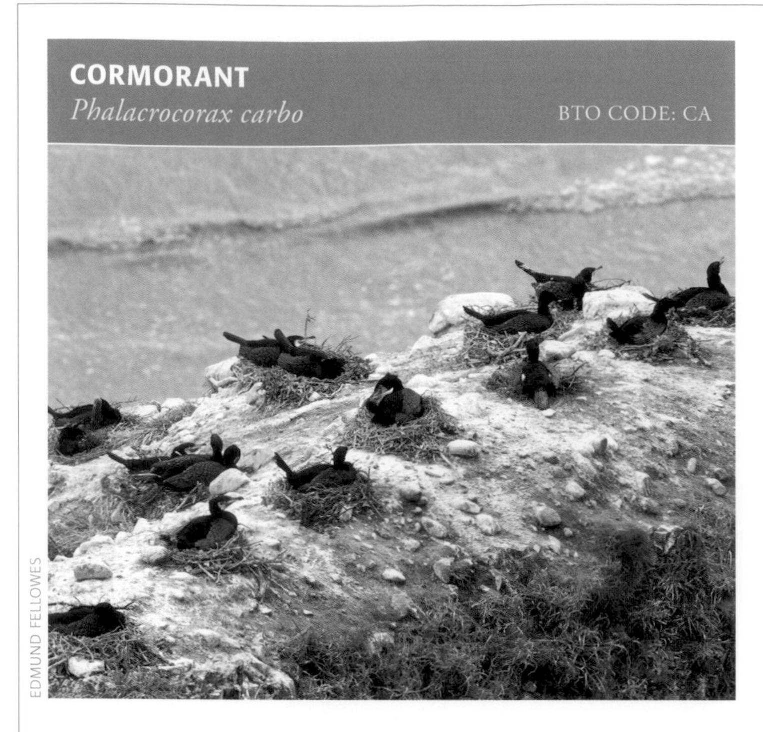
EDMUND FELLOWES

IN WINTER, CORMORANTS are widely distributed throughout Britain & Ireland, being found in 80% of 10-km squares. Low densities are associated with upland areas, while the highest densities are along the coasts, particularly around major estuaries, and along major lowland river systems. Since the *1981–84 Winter Atlas* there has been a 53% range expansion in Britain and an 18% range expansion in Ireland, both mainly in inland, lowland areas. Over the same period, wintering populations have increased steadily in the UK and the Republic of Ireland (*I-WeBS Report 2009*; *WeBS Report 2011*).

Breeding colonies are now widely distributed across Britain & Ireland. The breeding distribution map shows that *non-breeding* birds also remain widely distributed along the coasts and throughout the lowlands with records from 1,547 10-km squares in Britain and 430 10-km squares in Ireland. Inland colonies exist in western Ireland and central and eastern England, and the breeding distribution change map indicates that many of these have established since the *1988–91 Breeding Atlas* (Sellers *et al.* 1997). During the same period, colonies have been lost from some coastal areas, particularly in western Scotland.

Comparisons with previous atlas data are complicated because the expansion of the species inland to tree-fringed wetlands means that potentially any Cormorant perched in trees may be coded as being in suitable nesting habitat. The species' breeding change map is therefore derived from *probable* and *confirmed* breeding records only, to help overcome this problem. Some gains shown on the change map in England are associated with the expansion of the continental race *P. c. sinensis*, which is more typical of fresh water (Newson *et al.* 2007) and exploits fish populations in stocked lakes and reservoirs. Small numbers of individuals of the *sinensis* population have also been recognised in Ireland. The expansion of this population inland has led to conflict with inland fisheries, prompting demands for licensed control. The gains inland in winter are fuelled by the increasing proportion of British & Irish coastal breeders that have been using these freshwater habitats (Rehfisch *et al.* 1999), as well as by birds from the recently established inland breeding population (Hughes *et al.* 2000).

Sponsored by Essex & Suffolk Water

BREEDING DISTRIBUTION *2008–11*

POSSIBLE •	7%	8%	8%
PROBABLE •	1%	2%	2%
CONFIRMED •	9%	9%	9%
TOTAL	17%	19%	19%

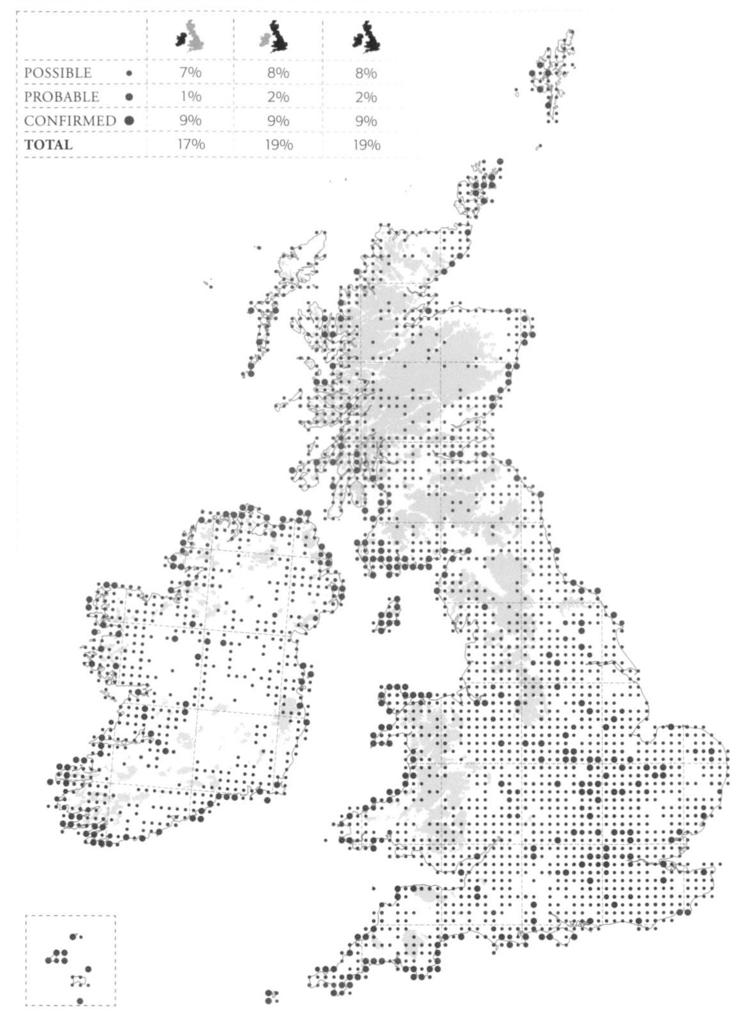

WINTER DISTRIBUTION *2007/08–2010/11*

PRESENT •	77%	82%	80%

BREEDING DISTRIBUTION CHANGE *since 1968–72*

GAIN ▲	60	195	255
LOSS ▽	60	110	170
40yr CHANGE	0%	+39%	+27%
20yr INDEX	-0.45	-0.21	

BREEDING RELATIVE ABUNDANCE *2008–11*

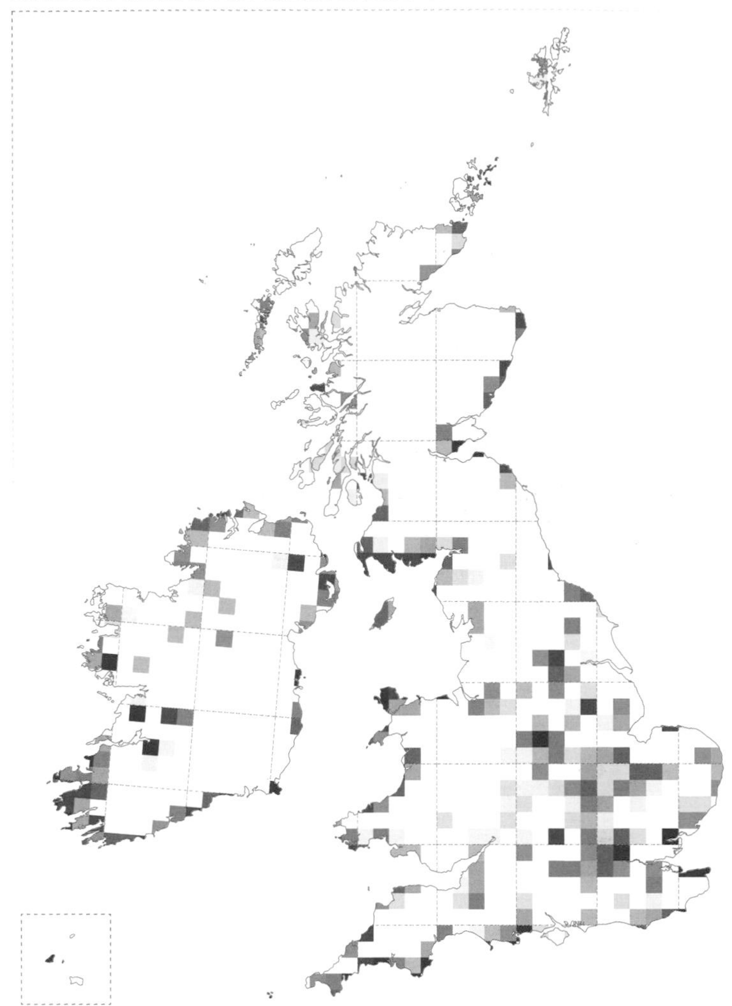

WINTER DISTRIBUTION CHANGE *since 1981–84*

GAIN ▲	180	861	1041
LOSS ▼	64	67	131
30yr CHANGE	+18%	+53%	+43%

WINTER RELATIVE ABUNDANCE *2007/08–2010/11*

SHAG
Phalacrocorax aristotelis

BTO CODE: SA

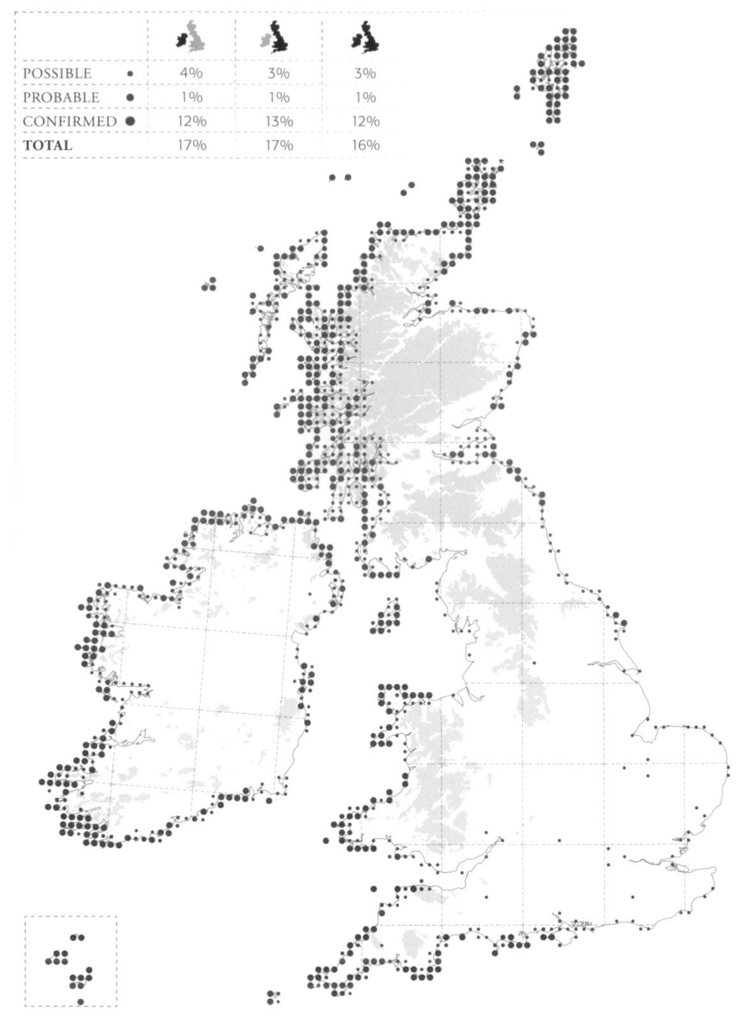

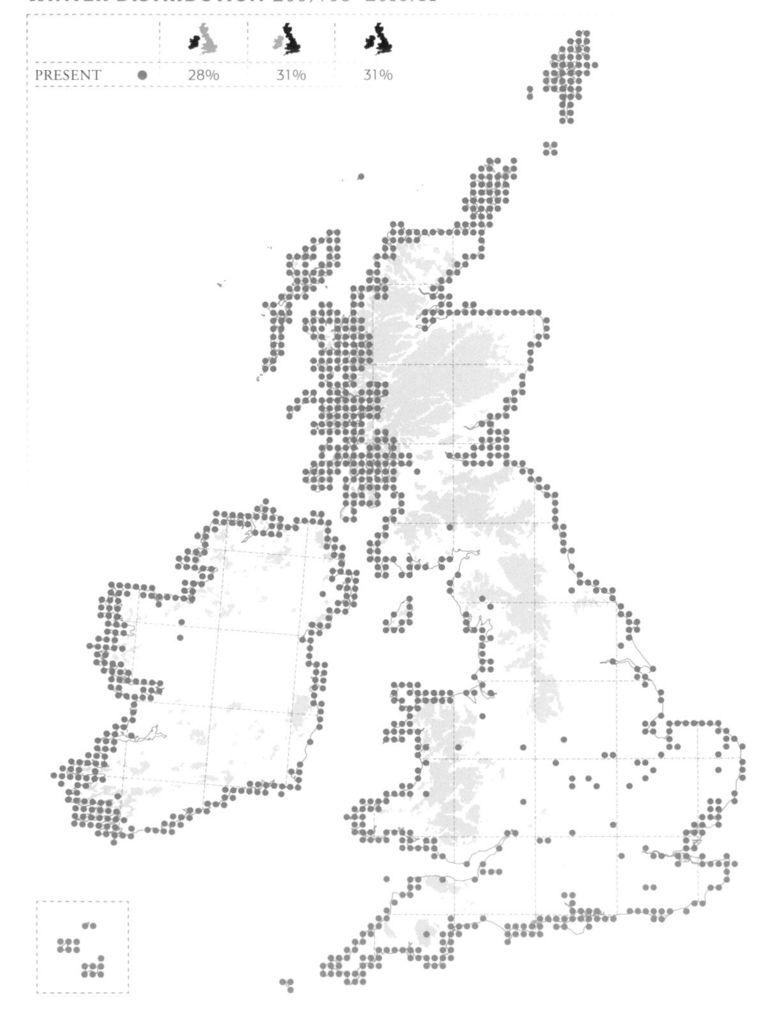

IN WINTER, SHAGS were recorded around much of the coast of Britain & Ireland with the highest densities in northern and western Scotland and western Ireland. An overall 9% range expansion since the *1981–84 Winter Atlas* is noticeable throughout their winter range. Inland areas are characterised by a mixture of gains and losses, reflecting the unpredictable distribution of these displaced individuals. Breeding and winter relative abundance maps are similar because adult birds are fairly sedentary (*Migration Atlas*): immature birds occur at low density along the North Sea coast, accounting for the wide spread of records, but are not systematically detected by TTVs.

Britain & Ireland supports c.40–45% of the world's breeding Shags (*Seabird 2000*). Difficulties in coverage, and the fact that in some years many experienced adults do not breed (Harris *et al.* 1998), make assessing trends difficult but it is clear that populations have fluctuated historically. In the UK, a 21% population increase occurred between 1969–70 and 1985–88. This was followed by a 27% decline to 1998–2002, and a further 26% decline to 2011 (JNCC 2012). A similar pattern has been observed in Ireland. The magnitude of these population declines is not reflected in the 9% range contraction in Britain & Ireland since the *1968–72 Breeding Atlas*, although this figure conceals a mix of localised gains and losses. However, differing interpretations of breeding evidence codes applied to birds away from colonies make it difficult to assess changes in breeding range from the Atlas data set.

Reasons for any declines are speculative. Large wrecks associated with prolonged periods of onshore gales result in increased mortality, for example in the North Sea in 1994 and 2005 (Harris & Wanless 1996; Frederiksen *et al.* 2008). Predictions of increased incidences of extreme weather suggest that such mortality events may become more frequent (Frederiksen *et al.* 2008). Predation of eggs and chicks by invasive mammalian predators has had localised effects (Craik 1997; Martin 2002). Shags are heavily reliant on sandeels (Harris & Wanless 1991); food shortages could have contributed to periodic low productivity and population decline (JNCC 2012) but critical data are lacking.

Sponsored by The Crown Estate (Marine)

BREEDING DISTRIBUTION CHANGE *since 1968–72*

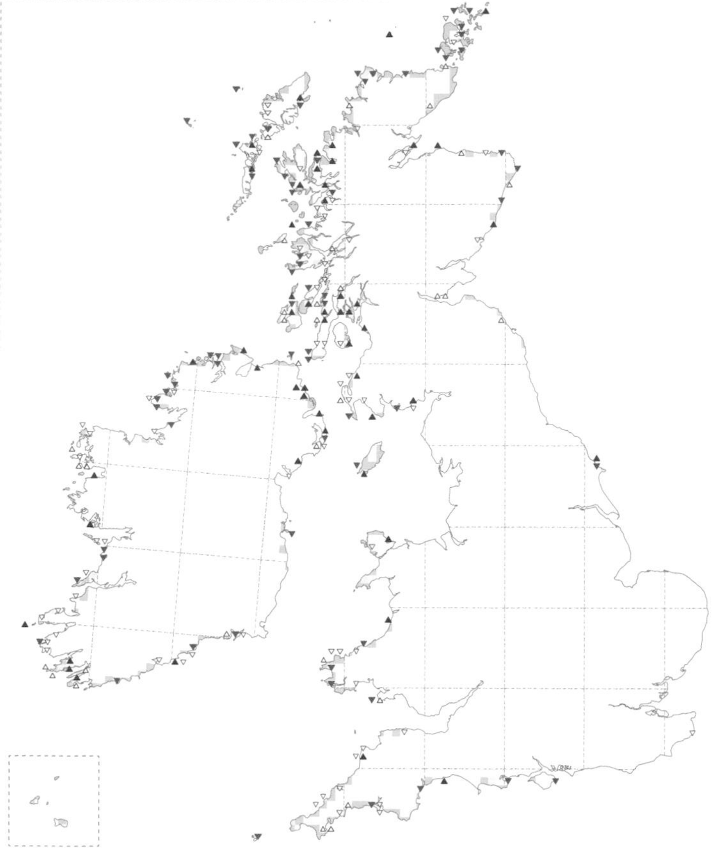

GAIN ▲	26	61	87
LOSS ▽	40	97	137
40yr CHANGE	-10%	-9%	-9%
20yr INDEX	+0.13	+0.04	

BREEDING RELATIVE ABUNDANCE *2008–11*

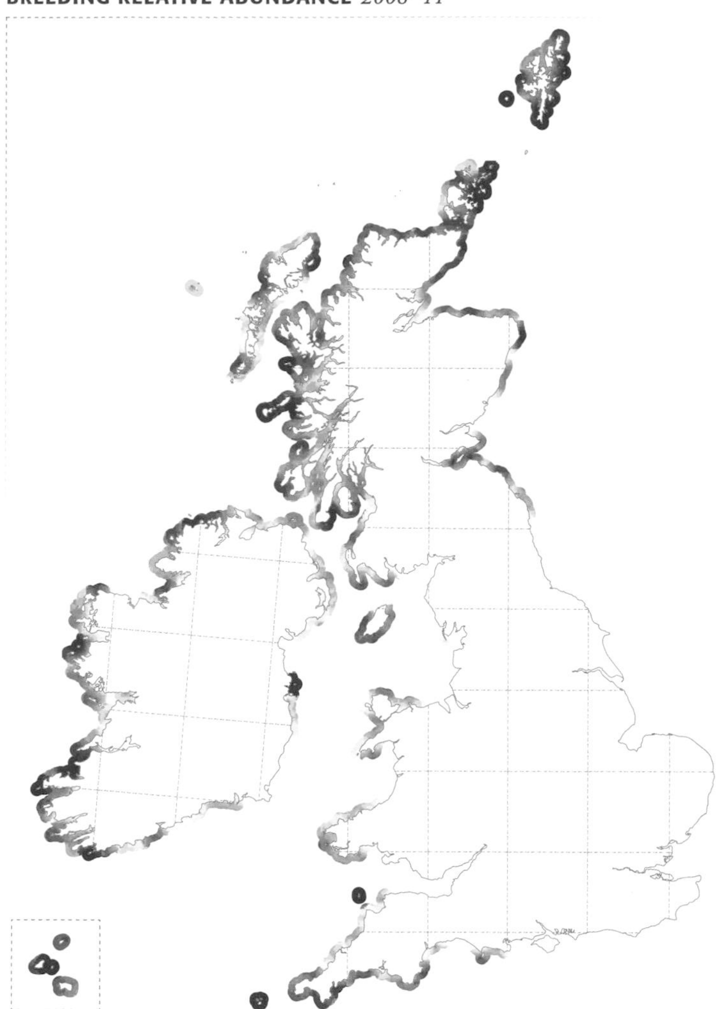

WINTER DISTRIBUTION CHANGE *since 1981–84*

GAIN ▲	36	195	231
LOSS ▼	30	105	135
30yr CHANGE	+2%	+12%	+9%

WINTER RELATIVE ABUNDANCE *2007/08–2010/11*

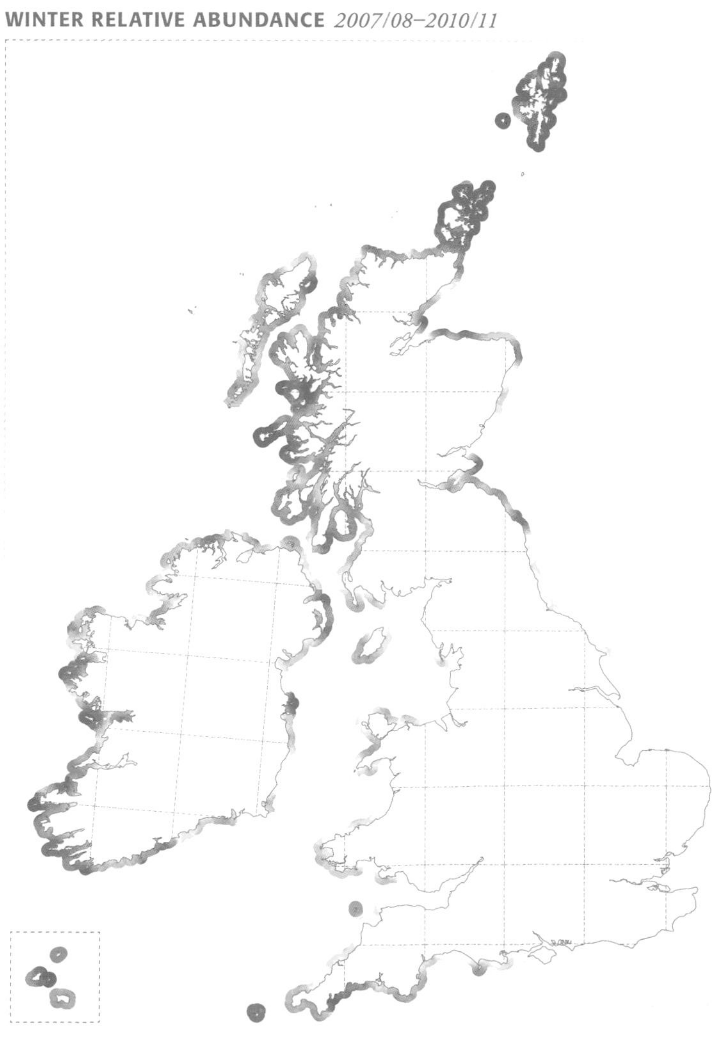

BITTERN
Botaurus stellaris

BTO CODE: BI

GORDON YOUDALE

TOTAL NUMBERS OF breeding Bitterns in Britain increased from a low point in 1997, when just 11 booming males were recorded, to 104 booming males in 2011 (Brown *et al.* 2012) thanks to the restoration of existing reedbeds and creation of new ones. Bitterns require extensive wet reedbeds with areas of open water and a sustainable fish population for breeding. Clusters of sites providing these conditions explain the hotspots in wetland landscapes in Kent, East Anglia, Somerset and Yorkshire. Many of these areas show as gains on the breeding-season change map and highlight the success of much-needed conservation action over the last 40 years.

Just a handful of 10-km squares along the Norfolk and Suffolk coastlines, in the Norfolk Broads and in north Kent have recorded breeding evidence for Bitterns in all three breeding atlas periods. Many of the gains at inland sites since the *1988–91 Breeding Atlas* correspond with extensive habitat creation schemes. There were two reports of *non-breeding* birds in western Ireland in April 2008; Bitterns bred in Ireland up to about 1840, and only occasional booming has been recorded since (*Birds in Ireland*).

In winter, Bitterns are much more widespread and can be found at many smaller sites with a mix of reedbeds and pools and also along riverbanks. During harsh winters, including those of 2009/10 and 2010/11, freezing conditions force many Bitterns to forage more widely, and during the winter of 2009/10 there were at least 600 Bitterns at nearly 400 sites (Wotton *et al.* 2011). Although the majority of wintering individuals reported during the Atlas came from England and Wales, there was also a small number of records from Scotland and Ireland. The gains shown on the winter change map reflect an increasing breeding population since the early 1980s and also influxes from the Continent during cold spells in the winter recording period.

Although the key requirements of breeding Bitterns are now better understood (Brown *et al.* 2012), threats such as sea-level rise and drying out of reedbeds, possibly enhanced by climate change, mean that continued research and management are required.

Sponsored by RSPB

BREEDING DISTRIBUTION *2008–11*

POSSIBLE ●	-	1%	1%
PROBABLE ●	-	1%	1%
CONFIRMED ●	-	1%	1%
TOTAL	-	3%	3%

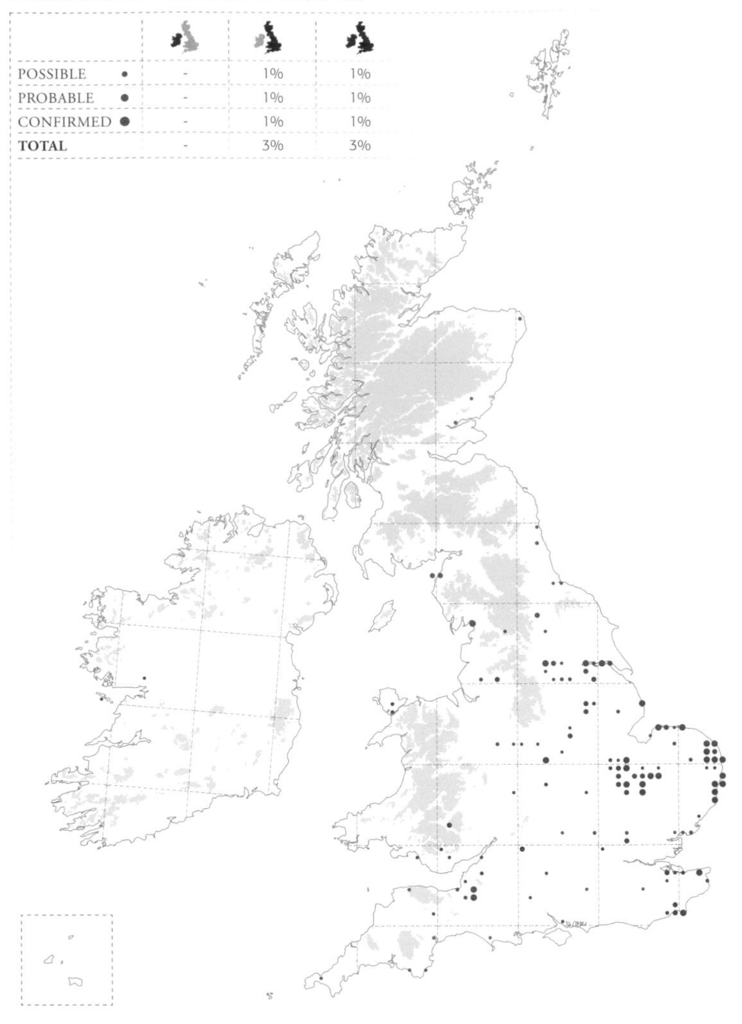

BREEDING DISTRIBUTION CHANGE *since 1968–72*

GAIN ▲	-	56	56
LOSS ▽	-	9	9
40yr CHANGE	-	+134%	+134%
20yr INDEX	-	+0.71	

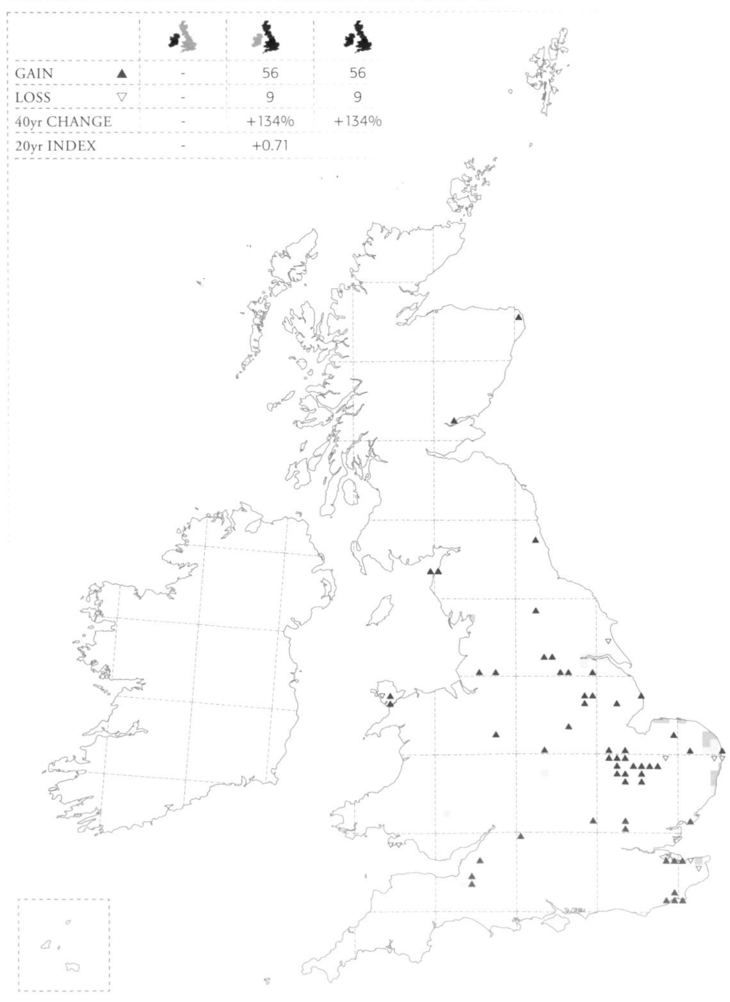

WINTER DISTRIBUTION *2007/08–2010/11*

PRESENT ●	<1%	17%	13%

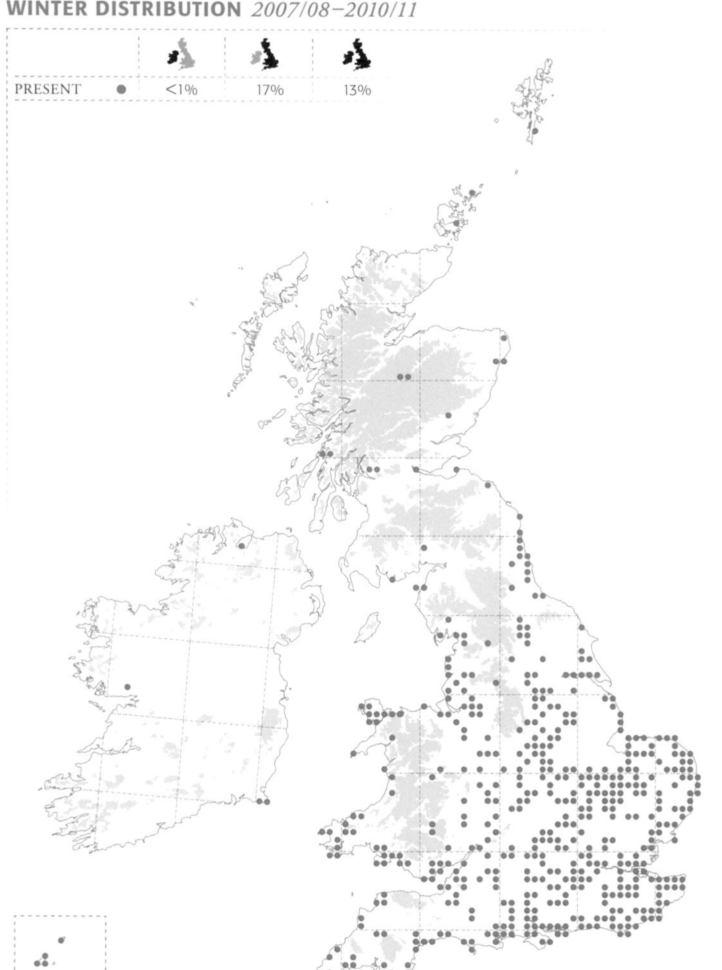

WINTER DISTRIBUTION CHANGE *since 1981–84*

GAIN ▲	4	367	371
LOSS ▼	5	69	74
30yr CHANGE	-20%	+161%	+156%

LITTLE BITTERN

Ixobrychus minutus

BTO CODE: LL

JEFF HAZELL www.freewebs.com/levels-birder

THE LITTLE BITTERN is a scarce migrant to Britain & Ireland but following a decline in its European breeding range (*European Atlas*) it has become an increasingly rare visitor. However, following the presence of a calling male in 2009, a pair nested successfully at Ham Wall, Somerset, in 2010, fledging at least one young (*RBBP 2010*). A pair also nested in 2011, possibly producing chicks on the second nesting attempt in late summer (Holling *et al.* 2013). The only previous reported instance of confirmed breeding was in 1984, when a pair raised at least three young at Potteric Carr, Yorkshire (*Birds in England*). Previously, pairs have lingered several times and booming males have been recorded, such as one summering for six weeks in June and July 1970 during the *1968–72 Breeding Atlas*.

BREEDING DISTRIBUTION *2008–11*

POSSIBLE ●	-		-
PROBABLE ●	-	-	-
CONFIRMED ●	-	<1%	<1%
TOTAL	-	<1%	<1%

LITTLE EGRET
Egretta garzetta BTO CODE: ET

LES FOSTER www.springerwildlifephotography.co.uk

THE COLONISATION AND range expansion of the Little Egret represent one of the most phenomenal shifts in abundance and distribution of any bird in Britain & Ireland over the past 20 years. The species was recorded in only a single 10-km square during the *1981–84 Winter Atlas*, and as a vagrant in just three 10-km squares during the *1988–91 Breeding Atlas*. Now, a winter distribution encompassing over 1,500 10-km squares, and *confirmed* breeding in more than 160 10-km squares, emphasises the extent to which the Little Egret population has flourished over the relatively short period since its initial establishment.

Increases in wintering and breeding populations throughout Spain and France during the 1980s and 1990s (Hafner & Fasola 1997) coincided with the expansion of Little Egrets into Britain & Ireland, when their status changed from a rare and sporadic visitor to an annual visitor and subsequently a permanent resident. Such movements northwards and westwards are likely to have been facilitated by subtle climatic changes. Successful breeding was first documented in Britain on Brownsea Island in Dorset in 1996 (Lock & Cook 1998), and was confirmed in Ireland, in Co. Cork, the following year (Smiddy & Duffy 1997). Since then the range has expanded considerably (Smiddy 2002; *RBBP 2010*), although the evidence for some *possible* breeding records may be interpreted as *non-breeding* by other surveyors, therefore perhaps overstating the current breeding distribution.

Despite wide juvenile dispersal, the Little Egret winter distribution is still strongly biased towards southern and eastern areas of both Britain and Ireland, where the species is now regularly observed at coastal and inland wetlands. In comparison, the population is still relatively sparse in north Ireland and particularly in Scotland, where birds are recorded predominantly on offshore islands. The breeding distribution shows similar patterns. The majority of breeding colonies have been established in existing Grey Heron colonies, mainly in close proximity to the coastlines of southern England, Wales and Ireland. However, inland colonies at freshwater sites have also been confirmed. These recent trends indicate that the impressive expansion throughout Britain & Ireland and the establishment of new colonies are likely to continue apace.

Sponsored by Thames Water Utilities Ltd

BREEDING DISTRIBUTION *2008–11*

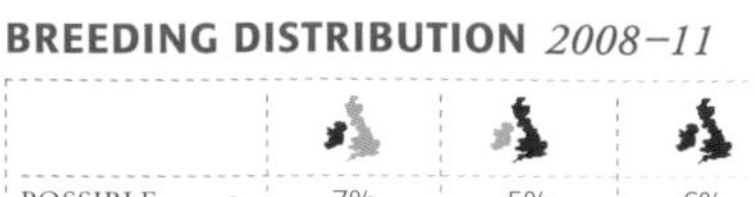

POSSIBLE •	7%	5%	6%
PROBABLE ●	2%	2%	2%
CONFIRMED ●	4%	4%	4%
TOTAL	13%	11%	12%

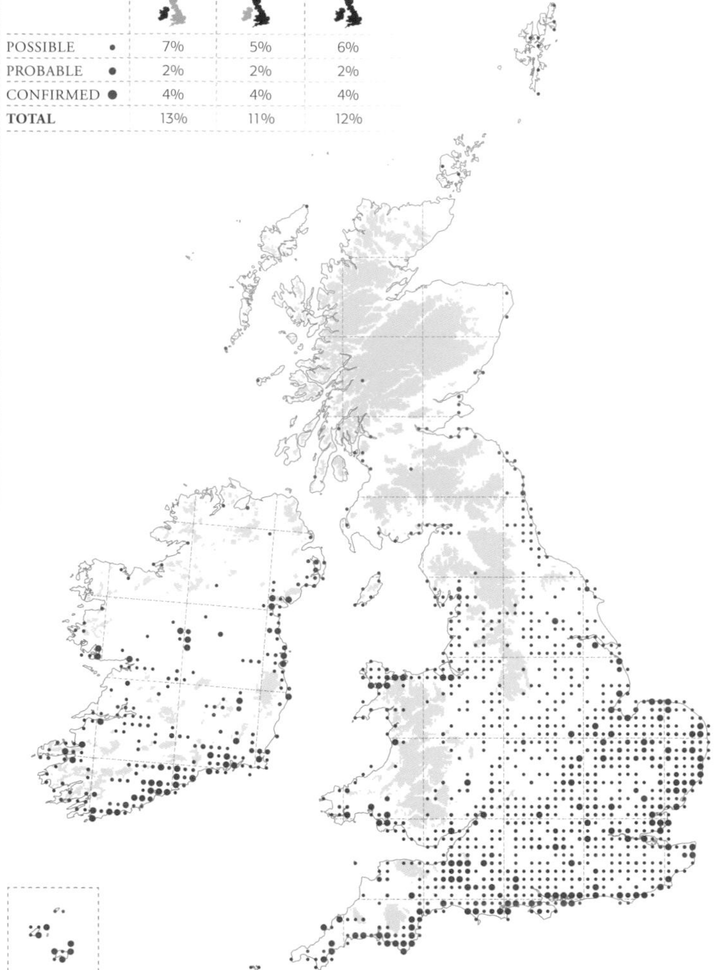

BREEDING RELATIVE ABUNDANCE *2008–11*

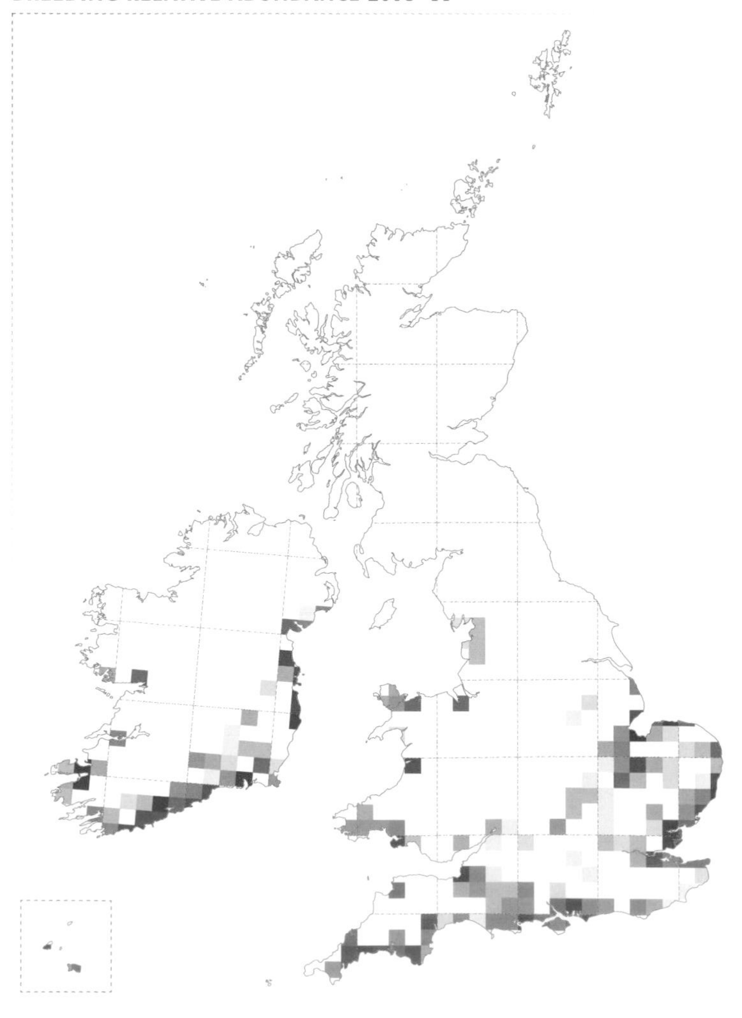

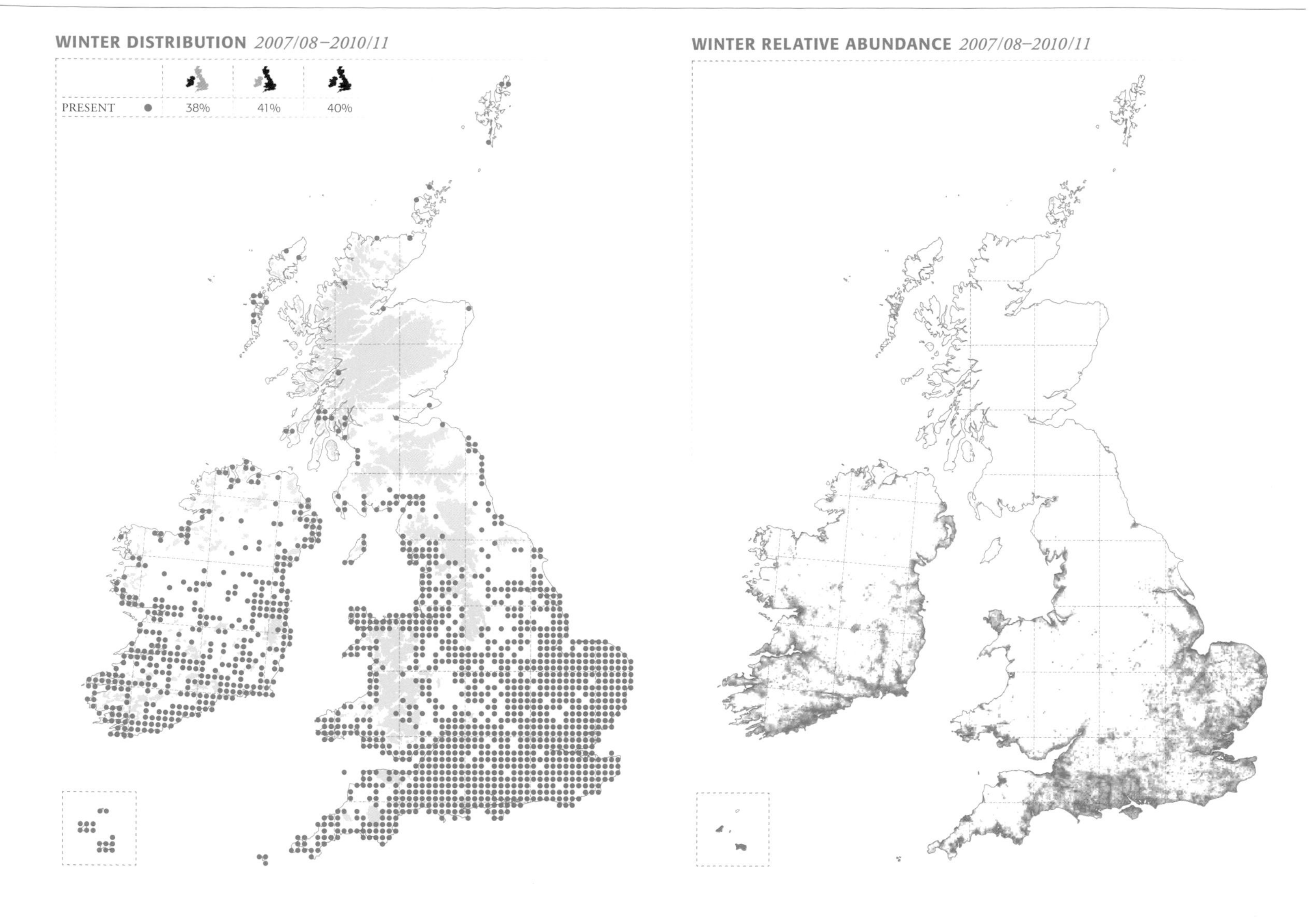

◀ The Little Egret has shown the greatest range expansion of any wintering species in the Atlas, and in the breeding season was recorded as present or with breeding evidence in 1,214 10-km squares.

CATTLE EGRET

Bubulcus ibis

BTO CODE: EC

JIM ALMOND www.shropshirebirder.co.uk

THE START OF Atlas fieldwork coincided with an unprecedented influx of Cattle Egrets into Britain & Ireland in late 2007. In Britain, initial sightings were in October, followed by the main influx into southwest England in early November, with arrivals continuing throughout December and into January 2008. At least 90 and 168 individuals were recorded in Britain in 2007 and 2008 respectively (Hudson *et al.* 2008, 2009). In Ireland, where there had been just 10 previous records, the timing of arrival was much the same as in Britain, and by the end of December 2007 there was a minimum of 29 birds (Milne & McAdams 2009). Although the initial arrival centred on southwest England and southwest Ireland, birds quickly moved northwards and eastwards. There was evidence of a smaller influx at the end of 2008, with birds lingering into 2009.

Cattle Egrets were recorded breeding in Britain for the first time in 2008, when breeding was *confirmed* in two 10-km squares in Somerset, where two or three pairs fledged at least two young (Holling *et al.* 2010b). Although birds returned to Somerset in 2009 and 2010, breeding was not confirmed. The Cattle Egrets nested within a heronry, where it was difficult to observe them. Records of *possible* breeding came from Sussex in 2008 and from Dorset and Avon in 2009. The breeding-season map also shows that *non-breeding* birds were recorded in 99 10-km squares in Britain and 11 in Ireland during the Atlas period. The majority of these records came from coastal areas, though there was a scatter of inland records in the East Midlands, Lancashire and Merseyside.

Within Europe, the Cattle Egret was originally native only to Spain and Portugal, but being an opportunistic species readily able to adapt to natural and man-made landscapes, it expanded its range rapidly during 1970–90 and continued to move northwards (BirdLife International 2004). Numbers recorded in Britain increased during the 1990s and 2000s, with a record total of 29 in 2006 (Hudson *et al.* 2009): this was a precursor to the influx in late 2007, which may have been sparked by meteorological conditions in the western Mediterranean.

Sponsored by Middletown House Nursing Home & Retirement Village

BREEDING DISTRIBUTION *2008–11*

	Ireland	Britain	Britain & Ireland
POSSIBLE •	-	<1%	<1%
PROBABLE ●	-	-	-
CONFIRMED ●	-	<1%	<1%
TOTAL	-	<1%	<1%

WINTER DISTRIBUTION *2007/08–2010/11*

	Ireland	Britain	Britain & Ireland
PRESENT ●	4%	6%	5%

GREAT WHITE EGRET

Ardea alba

BTO CODE: HW

JIM ALMOND www.shropshirebirder.co.uk

FOLLOWING HOT ON the heels of colonisation by the Little Egret, the Great White Egret was widely considered to be on the verge of doing the same, with a marked increase in numbers in Britain & Ireland since the beginning of the 21st century. Numbers had risen to more than 25 birds per year in Britain since 2000 and from 2006 it was removed from the list of species assessed by the British Birds Rarities Committee. In Ireland, 2009 was the third year in succession that the species occurred in record numbers (Fahy 2011).

Great White Egrets are recorded in Britain throughout the year, though there is a spring peak reflecting the arrival of birds on northward migration from wintering grounds in western and central Europe (*Birds in England*; *Birds of Scotland*). During the Atlas breeding period, 134 10-km squares in Britain and 12 in Ireland were recorded as occupied by *non-breeding* birds. In Britain most records were south and east of a line running from the Wash to the Severn. Records of birds in suitable breeding habitat submitted with a breeding evidence code from two 10-km squares were downgraded to *non-breeding*, because at the time the species had not been proven to breed in Britain or Ireland. However, with up to six birds during the late-winter period and several summering on the Somerset Levels in 2011, it was little surprise that breeding began there in spring 2012, with two pairs fledging a total of four young (Holt 2013).

More 10-km squares were occupied in the winter than in the breeding season, though the overall pattern of distribution was similar. There were more records from northwest England and Scotland during the winter period, some of which may relate to a handful of wandering individuals.

The recent expansion of the Great White Egret's breeding range into northwest Europe, and the increasing tendency for birds to overwinter in western and central Europe, probably accounts for the rise in records in Britain & Ireland. In the Netherlands the breeding population started increasing from 2000 and by 2010 there were around 150 breeding pairs (SOVON 2013).

Sponsored by Graham Appleton

BREEDING DISTRIBUTION *2008–11*

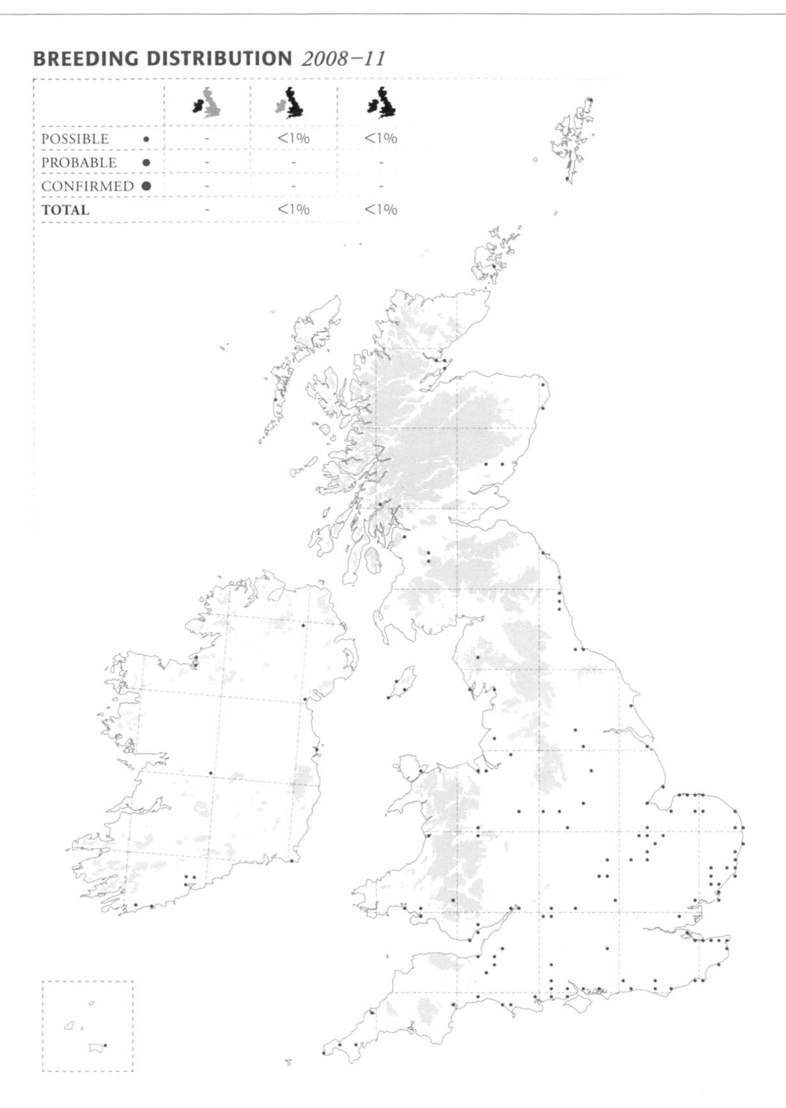

	Britain	Ireland	Britain & Ireland
POSSIBLE	-	<1%	<1%
PROBABLE	-	-	-
CONFIRMED	-	-	-
TOTAL	-	<1%	<1%

WINTER DISTRIBUTION *2007/08–2010/11*

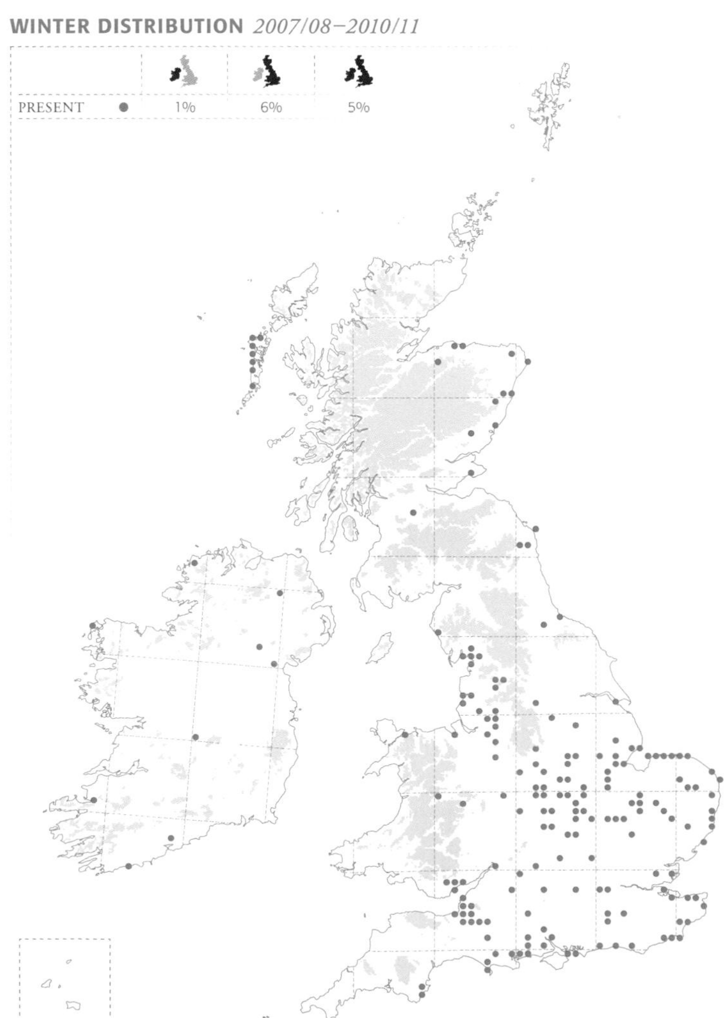

	Britain	Ireland	Britain & Ireland
PRESENT	1%	6%	5%

GREY HERON

Ardea cinerea

BTO CODE: H.

LIZ CUTTING www.lizcuttingphotos.com

WINTERING GREY HERONS are widely distributed across Britain & Ireland except for the most mountainous areas. There may be some local movements away from breeding areas in the winter but many birds return to breeding colonies by late February and are shown within the winter distribution map. Many colonies in Britain are remarkably well monitored, though small colonies and isolated pairs may go undetected, especially in Scotland, where heronries tend to be smaller than in England or Wales. There is relatively little monitoring of heronries in Ireland. Recording breeding evidence for Grey Herons can be problematic. Many of the *possible* breeding codes will be of birds in suitable habitat where it is difficult to confirm breeding, but a proportion are likely to show the presence of *non-breeding* or feeding birds at apparently suitable sites but where colonies have never been present.

The small overall gain shown by the winter change map, particularly in northwest Scotland and in Ireland, is consistent with good coverage during the winter months, and increasing numbers (*I-WeBS Report 2009*; *SUKB 2012*). The breeding change map shows a mix of gains and losses and is more difficult to interpret. Some gains may reflect the discovery of new colonies whilst others may reflect better coverage. However, some gains and losses could reflect subtle differences among atlases in the way that birds outside known colonies, yet still in suitable habitat, have been assigned breeding evidence. Grey Herons last nested on Orkney in 2005 (Orkney Bird Report Committee 2009), and similarly there has been no recent breeding on Shetland.

The BTO Heronries Census has shown Grey Herons to be more abundant in the UK in the early 2000s than at any time in the last 80 years (Marchant *et al.* 2004). Harsh winters have caused severe mortality and these periods are clearly visible in the long-term trend (*BirdTrends*). The general increase may stem from reduced persecution, improvements in water quality, the provision of new habitat as new lakes and gravel pits mature, and increased feeding opportunities at freshwater fisheries (*BirdTrends*). These lowland wetlands and coastal habitats account for the areas of highest density on both the breeding and the winter relative abundance maps.

Sponsored by attendees of BTO Swanwick Conference 2008

BREEDING DISTRIBUTION *2008–11*

POSSIBLE •	40%	26%	30%
PROBABLE •	12%	6%	8%
CONFIRMED ●	26%	36%	33%
TOTAL	78%	68%	71%

WINTER DISTRIBUTION *2007/08–2010/11*

PRESENT ●	89%	91%	91%

BREEDING DISTRIBUTION CHANGE *since 1968–72*

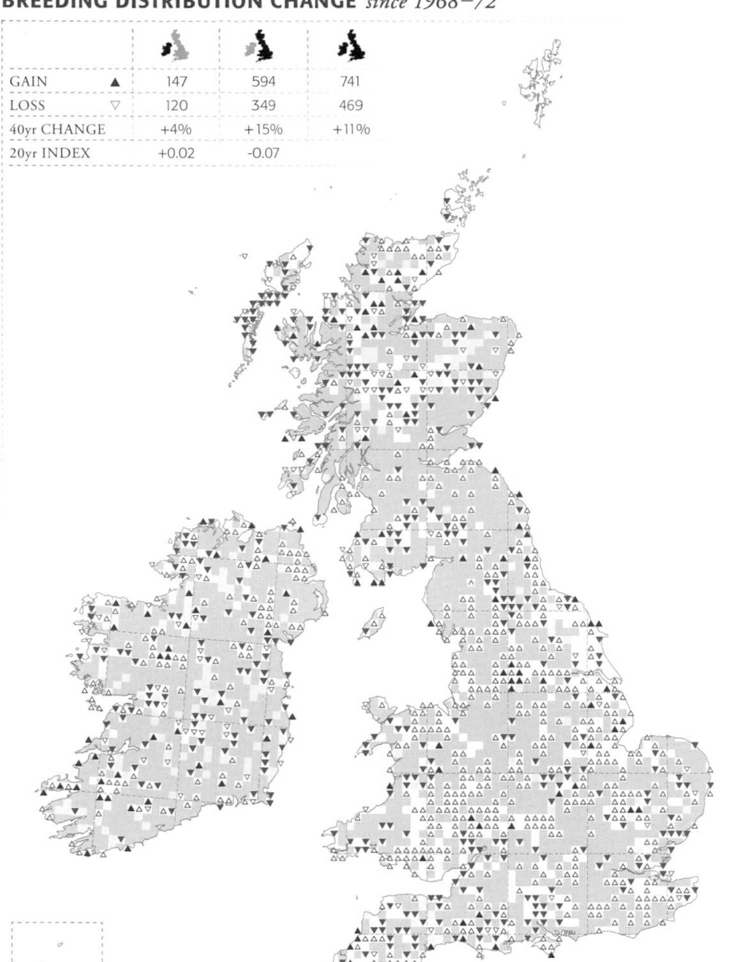

GAIN ▲	147	594	741
LOSS ▽	120	349	469
40yr CHANGE	+4%	+15%	+11%
20yr INDEX	+0.02	-0.07	

BREEDING RELATIVE ABUNDANCE *2008–11*

WINTER DISTRIBUTION CHANGE *since 1981–84*

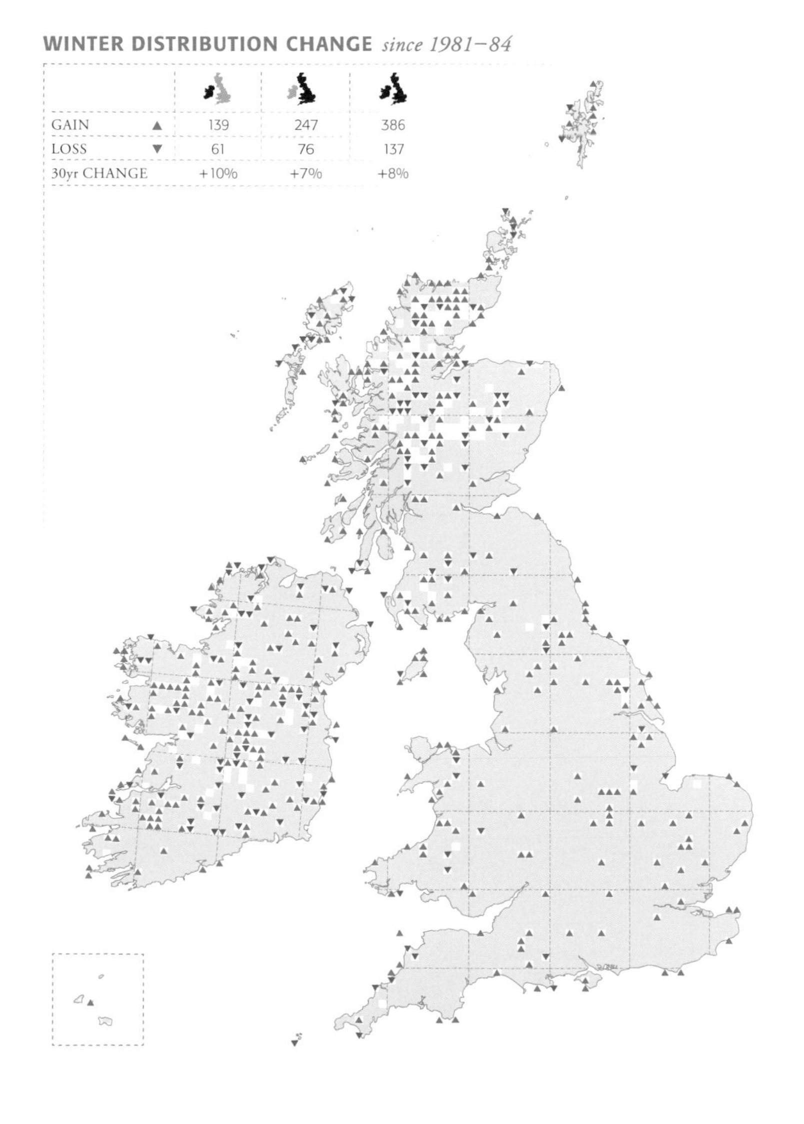

GAIN ▲	139	247	386
LOSS ▼	61	76	137
30yr CHANGE	+10%	+7%	+8%

WINTER RELATIVE ABUNDANCE *2007/08–2010/11*

PURPLE HERON
Ardea purpurea BTO CODE: UR

BRIAN HARPER

THOUGH EXTREMELY RARE in Ireland, Purple Herons are regularly recorded in Britain in the spring, typically arriving during periods of warm southerly winds between April and May and occasionally staying into the summer. During 2008–11, *non-breeding* birds were reported from 56 10-km squares in Britain, and two in Ireland, including the first Northern Irish record. In 2010 the first *confirmed* breeding occurred when a pair successfully raised two young at Dungeness RSPB reserve, Kent (*RBBP 2010*); this follows an attempt in Suffolk in 2007 (Holling *et al.* 2010a). During the *1968–72 Breeding Atlas* there was one record of *possible* breeding at a site in southeast England in 1972. There were winter records from two 10-km squares in England, both in early November.

BREEDING DISTRIBUTION *2008–11*

	Ireland	Britain	Britain & Ireland
POSSIBLE •	-	-	-
PROBABLE ●	-	-	-
CONFIRMED ●	-	<1%	<1%
TOTAL	-	<1%	<1%

GLOSSY IBIS
Plegadis falcinellus BTO CODE: IB

SHAY CONNOLLY www.flickr.com/photos/shayc

EXCEPTIONAL NUMBERS WERE recorded throughout 2007–11, following a major arrival into Britain & Ireland in April 2007 (c.29 Britain, six Ireland), some of which remained into the winter, and further influxes in the autumns of 2009 (c.38 Britain, 31 Ireland) and 2010 (c.21 Britain, 17 Ireland) (Fahy 2011; Hudson *et al.* 2008, 2010, 2011). In the winter months they occupied 28 10-km squares in Britain and four in Ireland. Some birds remained into spring, when they were recorded in 20 10-km squares in Britain and three in Ireland. Sightings of colour-ringed birds were traced back to the expanding colony in the Coto Doñana in Spain and to the Camargue in France.

WINTER DISTRIBUTION *2007/08–2010/11*

	Ireland	Britain	Britain & Ireland
PRESENT ●	<1%	1%	1%

SPOONBILL

Platalea leucorodia

BTO CODE: NB

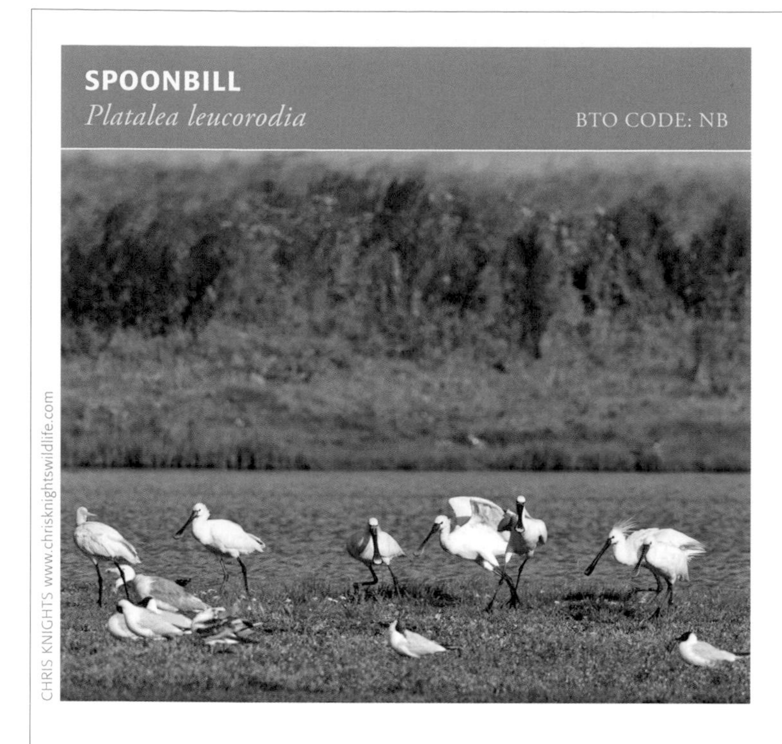

CHRIS KNIGHTS www.chrisknightswildlife.com

THE SPOONBILL IS a rare breeding bird in Britain. Breeding was *confirmed* in two 10-km squares during the Atlas recording period, continuing the pattern of recent sporadic breeding activity (Holling *et al.* 2010b). A pair was found with three juveniles in Dumfries & Galloway in August 2008 and is the first proven breeding record for Scotland. In 2010, six pairs fledged 10 young at Holkham National Nature Reserve in Norfolk, and the following year eight pairs fledged 14 young. Colour-ringed individuals amongst the flock in Norfolk showed that these birds had come from various countries in Europe, including the Netherlands, Germany and Spain.

Elsewhere in Britain, birds in suitable breeding habitat were recorded at sites along the Suffolk coastline and in Kent. They bred in 1998 and 2007 in Suffolk (eggs laid), and in 1999 in Lancashire & North Merseyside (two young fledged) (Holling *et al.* 2010b). These are the only breeding records since about 1668, when birds in East Anglian fenland, the last of a former population, were wiped out by hunting and land drainage. Spoonbills have been observed displaying and carrying nest material at other sites in Britain in recent years, suggesting that future range expansion is possible. Indeed, *non-breeding* birds, usually immatures, were recorded in 233 10-km squares in Britain and in 11 in Ireland.

This species has shown a major increase in its winter range since the *1981–84 Winter Atlas*. All winter records in Britain, except for one in Scotland, came from the southern half of England, with the majority from coastal sites. In Ireland, birds were recorded at 10-km squares along the south and southwest coasts.

Populations are increasing within western Europe, with new colonies being established, for example in France, the Netherlands, Belgium and Germany (Triplet *et al.* 2008). Some populations are increasing in central and southeast Europe (Hungary and Italy), although others are decreasing sharply (Turkey). The loss and degradation of wetland habitats is a major threat, although overfishing, water pollution, illegal hunting and collisions with power lines all contribute.

Sponsored by Amelia and Beatrice Bradshaw

BREEDING DISTRIBUTION *2008–11*

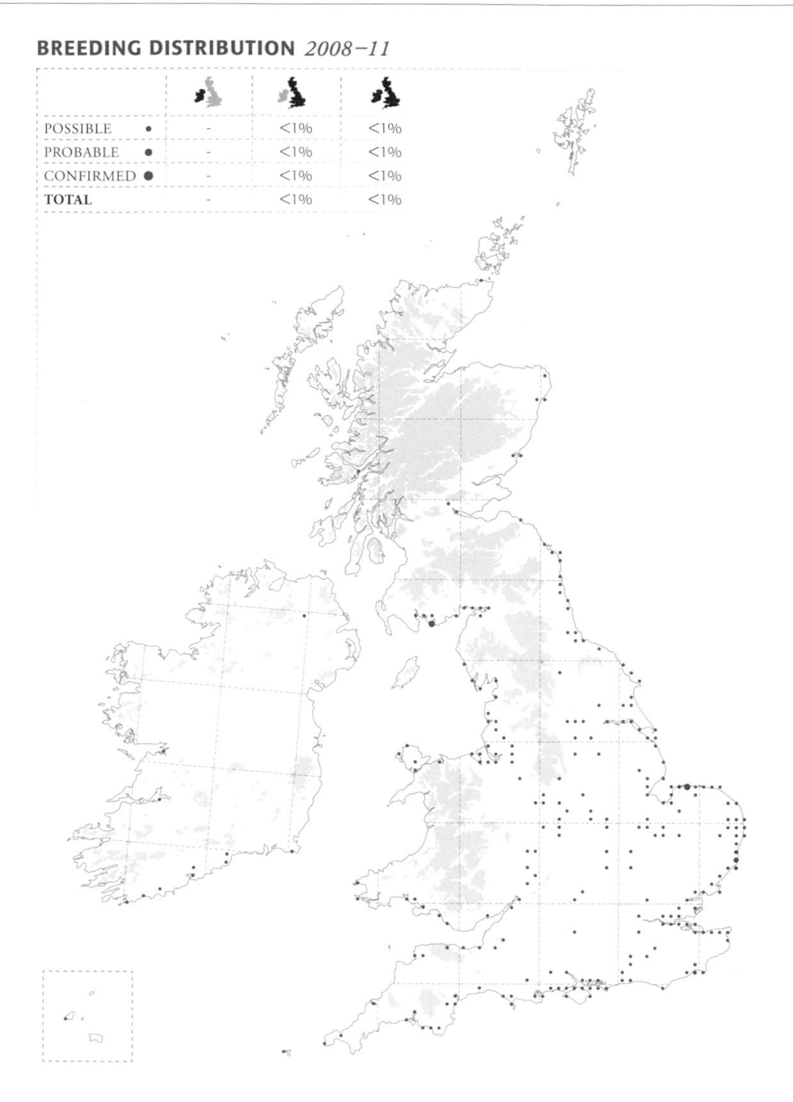

POSSIBLE •	-	<1%	<1%
PROBABLE •	-	<1%	<1%
CONFIRMED •	-	<1%	<1%
TOTAL	-	<1%	<1%

WINTER DISTRIBUTION *2007/08–2010/11*

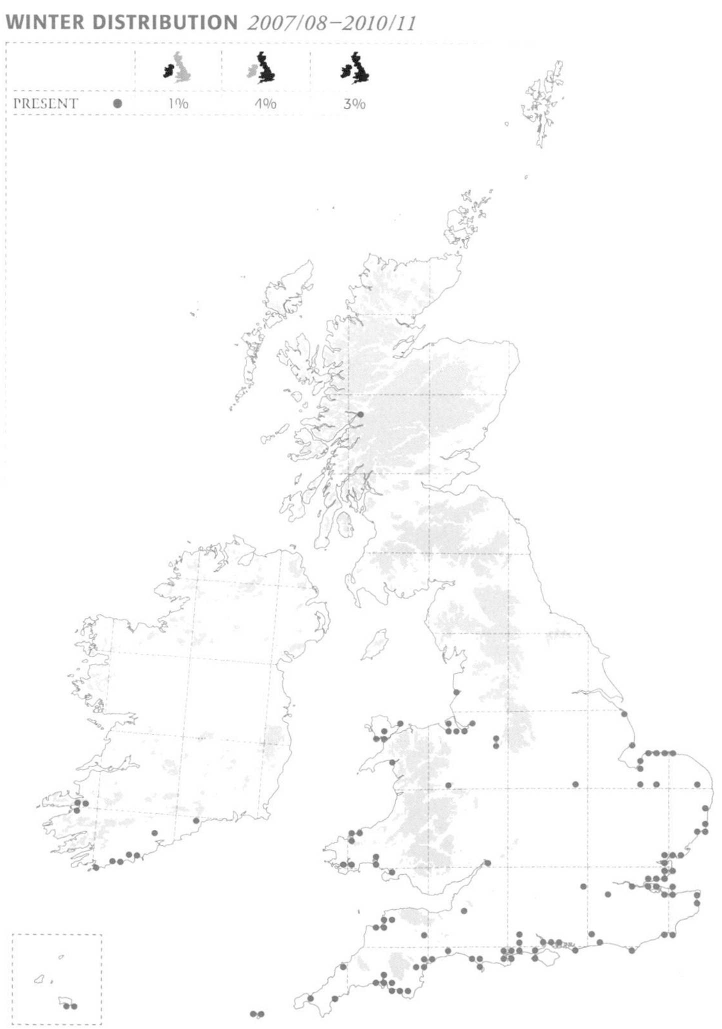

PRESENT •	1%	4%	3%

LITTLE GREBE
Tachybaptus ruficollis

BTO CODE: LG

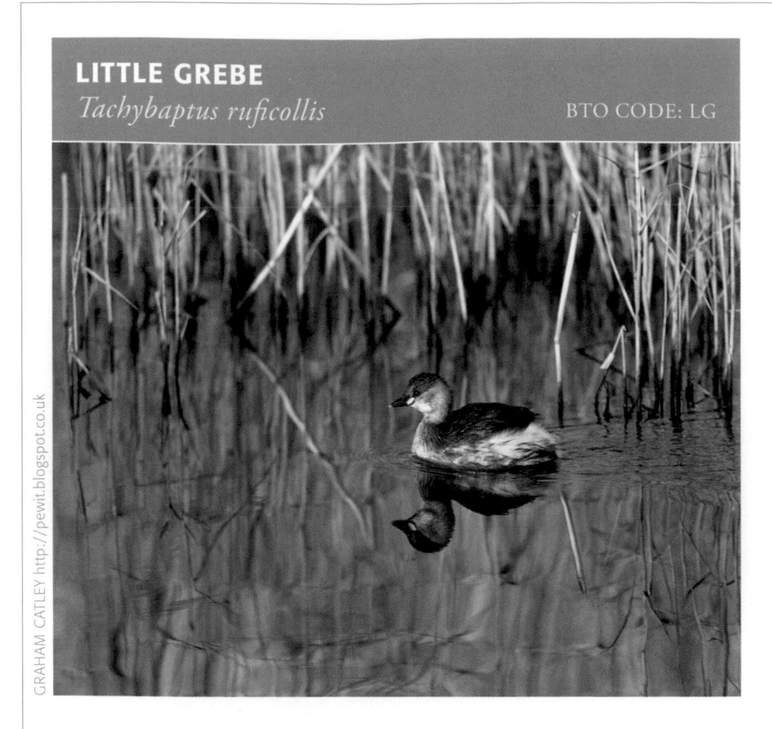

GRAHAM CATLEY http://pewit.blogspot.co.uk

THE LITTLE GREBE is a common and widespread inhabitant of small and shallow freshwater wetlands throughout lowland Britain & Ireland. Most breeders probably remain in Britain & Ireland in winter, with some post-breeding dispersal (Vinicombe 1982), and some winter immigration from the Continent is likely (*Migration Atlas*). This explains very similar breeding-season and winter distribution and abundance patterns, with birds widespread across most of the low-lying regions of Britain & Ireland, absent from parts of northern Scotland, the Pennines, Wales and southern Ireland, and most abundant along the major river valleys.

In Ireland, a large apparent range contraction between the *1968–72 Breeding Atlas* and the *1988–91 Breeding Atlas* has been largely reversed since. In Britain, over the same periods, a small range contraction was followed by a large range expansion. In Ireland the gains and losses are scattered; in Britain, recent gains are most notable in southwest England, much of Wales and west and north Scotland. These range changes are difficult to explain. Though highly vocal during the breeding season, Little Grebes are quite shy, and can remain hidden in emergent vegetation. Hence it is possible that subtle changes in coverage could explain these patterns, although recent increases are still apparent when calculated using standardised TTV data only. Loss of wetlands, mortality due to cold winters (Marchant *et al.* 1990) and, locally, the introduction of American Mink (Cunningham 1987; Smith 1988; Ferreras & Macdonald 1999) have all been suggested as potential causes of change.

Since the *1981–84 Winter Atlas,* there has been a notable extension of range throughout Britain & Ireland with a 27% increase in the number of occupied 10-km squares. Winter numbers are poorly monitored by existing schemes because Little Grebes occupy many small wetlands and waterways that are not routinely surveyed. Trend analyses suggest that there has been a steady increase in numbers in Britain and stability in Northern Ireland since the early 1990s (*WeBS Report 2011*). In the Republic of Ireland, numbers have been largely stable over a similar period but underwent a recent increase in 2007/08 and 2008/09 (*I-WeBS Report 2009*).

Sponsored by the Gay Birders Club

BREEDING DISTRIBUTION *2008–11*

POSSIBLE ●	11%	8%	9%
PROBABLE ●	11%	9%	10%
CONFIRMED ●	25%	41%	37%
TOTAL	47%	58%	56%

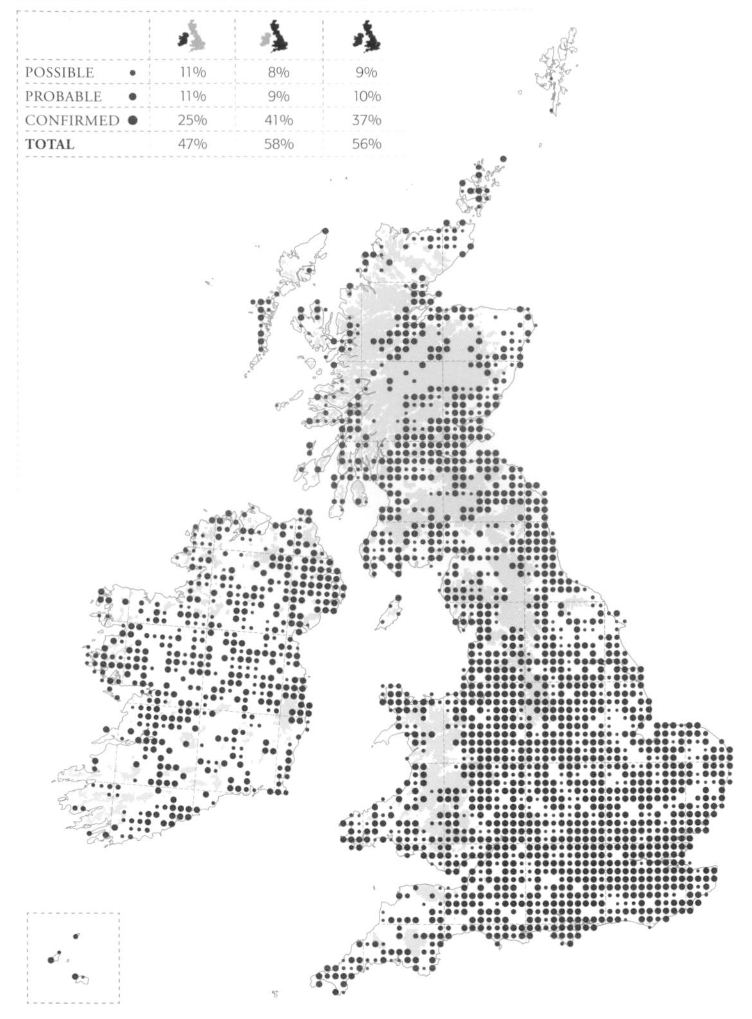

WINTER DISTRIBUTION *2007/08–2010/11*

PRESENT ●	52%	64%	61%

BREEDING DISTRIBUTION CHANGE *since 1968–72*

GAIN ▲	128	547	675
LOSS ▽	172	238	410
40yr CHANGE	-9%	+23%	+14%
20yr INDEX	+0.15	+0.08	

BREEDING RELATIVE ABUNDANCE *2008–11*

BREEDING RELATIVE ABUNDANCE CHANGE *since 1988–91*

WINTER DISTRIBUTION CHANGE *since 1981–84*

GAIN ▲	192	538	730
LOSS ▼	95	151	246
30yr CHANGE	+24%	+27%	+27%

WINTER RELATIVE ABUNDANCE *2007/08–2010/11*

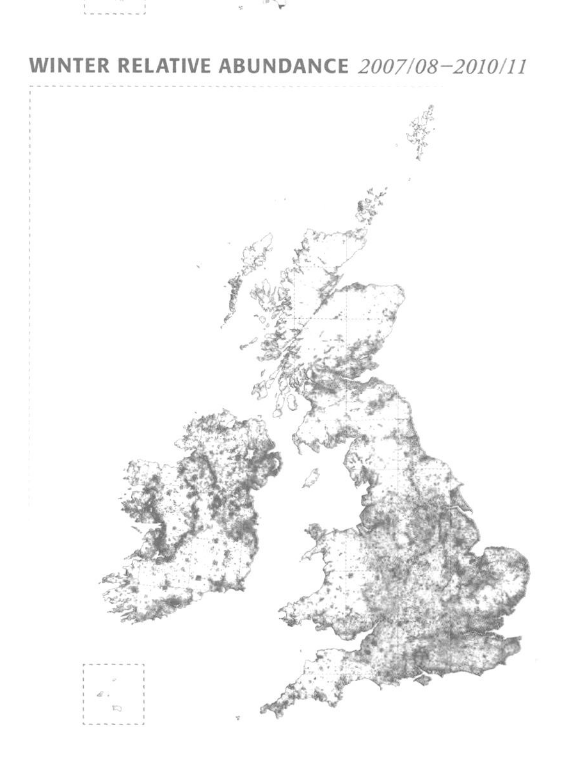

GREAT CRESTED GREBE

Podiceps cristatus

BTO CODE: GG

DAVID KJAER www.davidkjaer.com

THROUGHOUT THE WINTER, Great Crested Grebes are found on lowland lakes, on estuaries and on shallow sea coasts. Inland, birds in winter are most concentrated in central and southern England, central Scotland and in Ireland on the lakes that are north of a diagonal band from Co. Kerry in the west to Co. Antrim and Co. Down in the east. On the coast, winter abundance is highest in the major estuaries and at other shallow coastal sites.

There has been a 22% increase in winter range size since the *1981–84 Winter Atlas*. This gain is broad, with no obvious focal points, and is likely to relate to a continued long-term recovery that has occurred since cessation of persecution in the early 20th century (Harrison & Hollom 1932; Holloway 1996). Some coastal gains could reflect hard-weather movements associated with the two colder-than-normal winters during the Atlas period. The observed range expansion corroborates an increase in winter numbers in Britain since the mid 1980s, though there is evidence of a shallow decline since the mid 2000s (*WeBS Report 2011*). Numbers have declined in the Republic of Ireland by 31% during 1994/95–2008/09 (*I-WeBS Report 2009*), with a recent decline also noted in Northern Ireland (*WeBS Report 2011*). Some of this may relate to fewer continental immigrants from declining northwest European breeding populations (Wetlands International 2013).

The breeding distribution mirrors the winter distribution though it is restricted to fresh water. The areas of highest breeding abundance appear to be Ireland's midland and western lakes, Lough Neagh, the meres of northwest England, the fens of East Anglia and in the Thames Basin. A survey of Lough Neagh in 1998 revealed 1,827 breeding pairs on 138 km of shoreline, establishing it as the primary breeding stronghold in Britain & Ireland (Perry *et al.* 1999).

There was a parallel breeding range expansion of 26% since the *1968–72 Breeding Atlas*. This most likely reflects the continued historical increase in the population as well as localised effects such as the increased availability of gravel pits, though breeding grebes may be susceptible to local disturbance from watersports and angling activities at such sites.

Sponsored by South Dublin Branch of BirdWatch Ireland

BREEDING DISTRIBUTION *2008–11*

POSSIBLE ●	3%	2%	3%
PROBABLE ●	9%	5%	6%
CONFIRMED ●	12%	28%	24%
TOTAL	24%	35%	33%

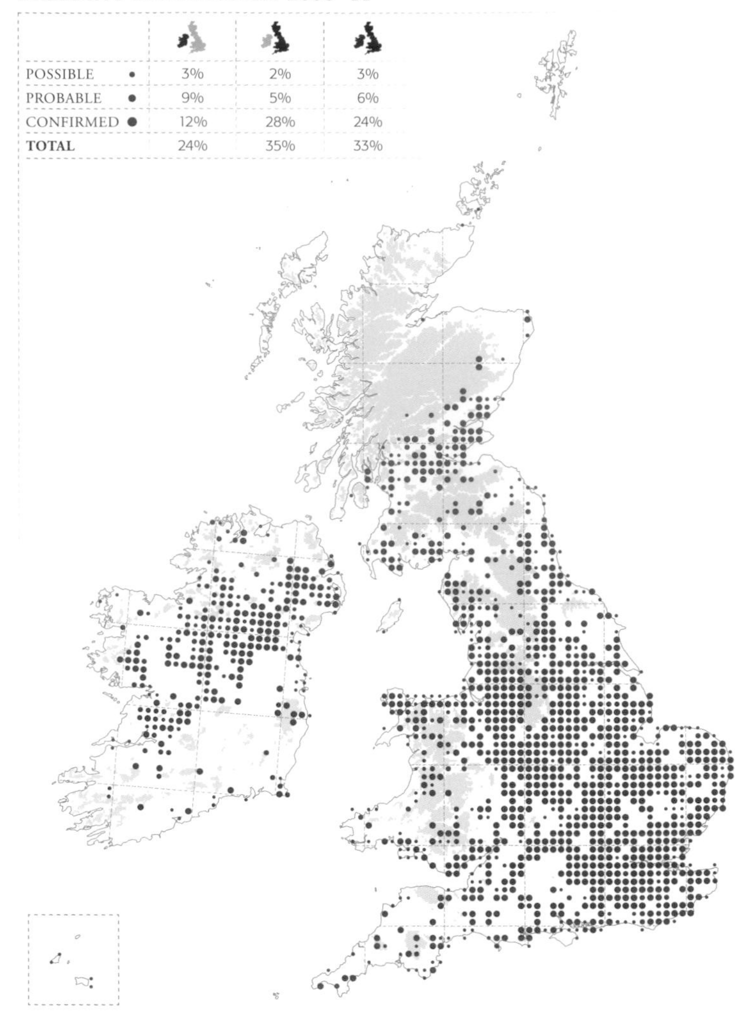

WINTER DISTRIBUTION *2007/08–2010/11*

PRESENT ●	32%	41%	38%

BREEDING DISTRIBUTION CHANGE *since 1968–72*

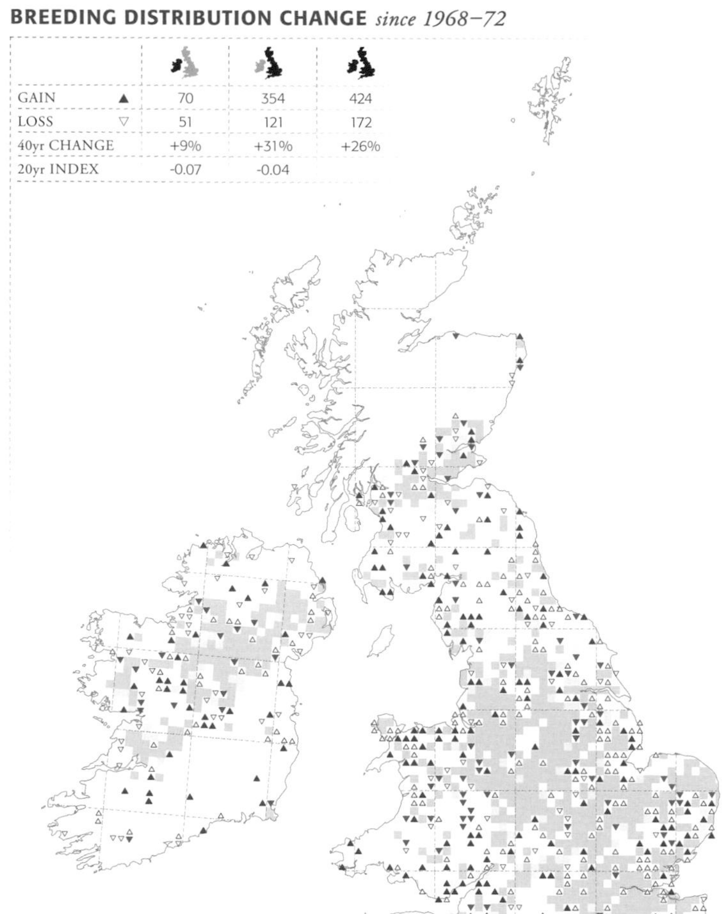

GAIN	▲	70	354	424
LOSS	▽	51	121	172
40yr CHANGE		+9%	+31%	+26%
20yr INDEX		-0.07	-0.04	

BREEDING RELATIVE ABUNDANCE *2008–11*

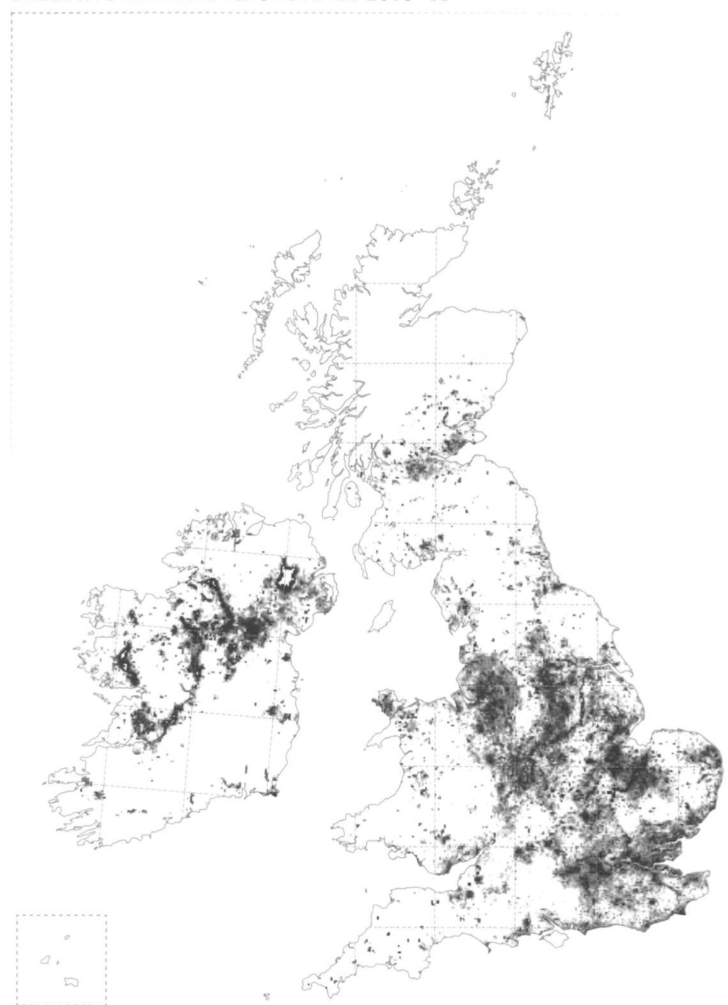

WINTER DISTRIBUTION CHANGE *since 1981–84*

GAIN	▲	110	347	457
LOSS	▼	60	136	196
30yr CHANGE		+19%	+23%	+22%

WINTER RELATIVE ABUNDANCE *2007/08–2010/11*

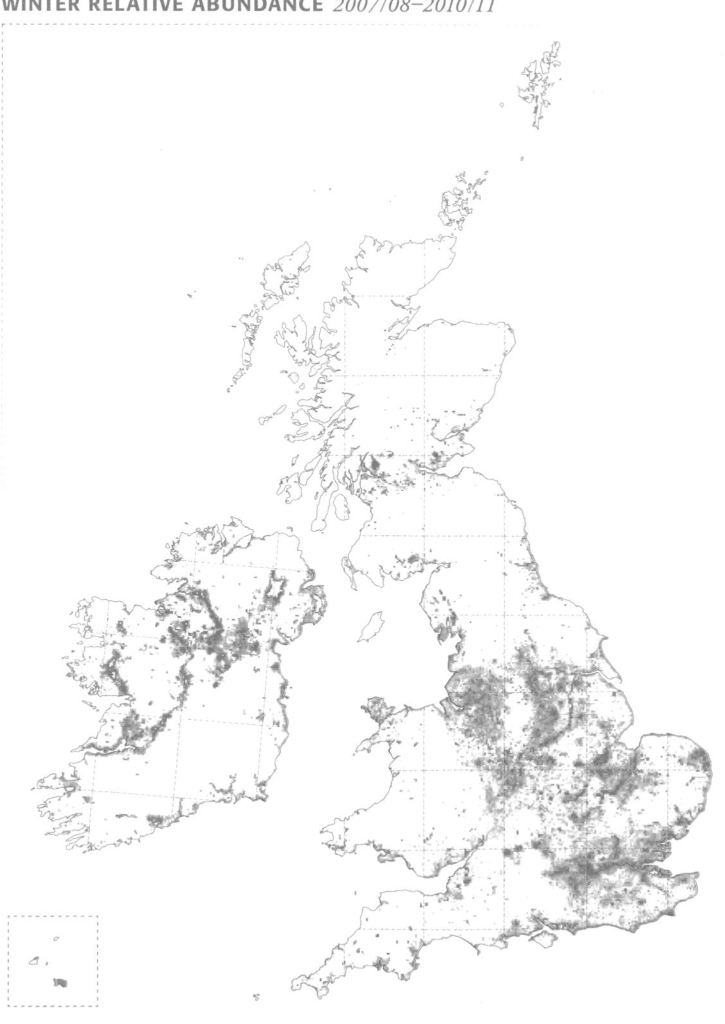

RED-NECKED GREBE
Podiceps grisegena

BTO CODE: RX

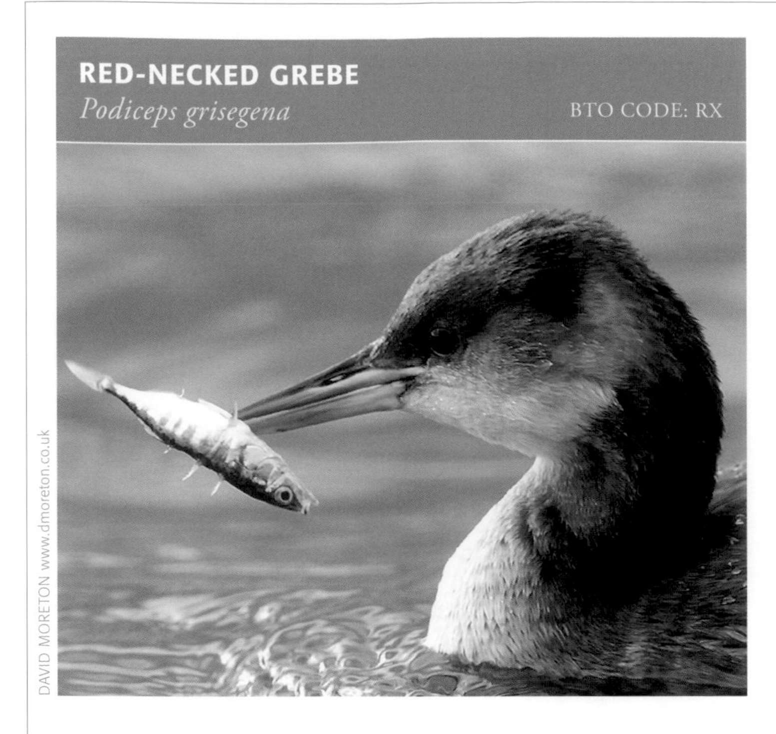

DAVID MORETON www.dmoreton.co.uk

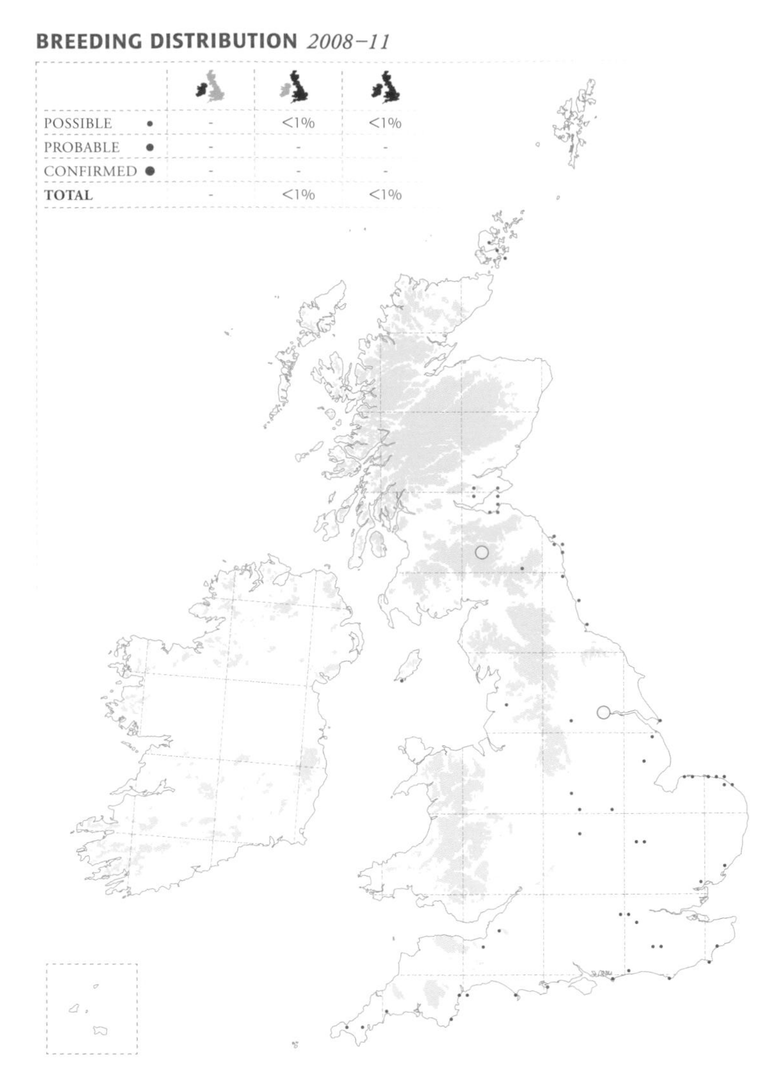

BREEDING DISTRIBUTION *2008–11*

	Ireland	Britain	Britain & Ireland
POSSIBLE •	-	<1%	<1%
PROBABLE •	-	-	-
CONFIRMED •	-	-	-
TOTAL	-	<1%	<1%

RED-NECKED GREBES are regular but rather scarce winter visitors to Britain & Ireland from continental Europe. Musgrove *et al.* (2011) estimated a British winter population of approximately 55 birds for the period 2004/05 to 2008/09. The winter distribution map shows a strong bias to the North Sea coast from Angus to Kent and along the south coast of England. Records are also scattered at inland sites, mainly in eastern England. In Ireland, it is found mainly at a small number of coastal locations.

At most sites, records often refer to just one or two individual birds. The Firth of Forth is the only area to support larger numbers in winter on a regular basis, though numbers there have declined in recent years. For example, in 1994/95 up to 100 birds were recorded compared with just four in 2010/11 (*WeBS Report 2011*). Since the *1981–84 Winter Atlas* there has been a 7% increase in the number of occupied 10-km squares in Britain and an increase from four to 18 squares in Ireland (excluding early November records). In many areas, this increase probably reflects improved coverage. However, there were genuine increases in well-covered areas such as south and southwest England. The mix of gains and losses, especially inland, reflects the high turnover in site use through time.

Breeding attempts by Red-necked Grebes in Britain are very rare and sporadic; the only successful breeding was in Borders in 2001 (Ogilvie *et al.* 2003b). During 2008–11 there were no *confirmed* breeding records, though at one site a bird returned to suitable breeding habitat for several consecutive years and exhibited territorial behaviour. However, most records of birds present during the breeding season represented *non-breeding* individuals, including wintering birds remaining at coastal and inland sites in April and early May and summering birds in June. In Ireland a single bird was present in one 10-km square in 2008 (not mapped); breeding was last confirmed in 2003 (Hillis & O'Sullivan 2012). Records from mid July at the Firth of Forth are wintering birds returning to begin their moult (*Birds of Scotland*).

JIM WELFORD

▲ **A male has summered annually at Hatfield Moors, South Yorkshire, since 2004 and was observed making territorial calls during the Atlas period.**

SLAVONIAN GREBE

Podiceps auritus

BTO CODE: SZ

JOHN ANDERSON www.pbase.com/crail_birder

WINTERING SLAVONIAN GREBES occur around much of the coast of Scotland, with particular concentrations in sheltered waters in the Northern Isles, northwest Scotland, the Moray Firth, the Firth of Forth and Kintyre. In England they are found mainly in Northumberland and farther south along the coast from East Anglia to Cornwall. In Ireland they are coastally distributed with the highest concentrations at Strangford Lough, Lough Foyle, Lough Swilly and Blacksod Bay (*I-WeBS Report 2009*; *WeBS Report 2011*). Elsewhere, particularly at inland sites in Britain, records are more sporadic and mostly of one or two birds.

There has been a 69% winter range expansion in Britain & Ireland since the *1981–84 Winter Atlas*. This may partly stem from improved coverage, but some increases are real: numbers wintering around Shetland have doubled since 2000/01 and there have been gains in Orkney and the Outer Hebrides, perhaps in response to an increase in the Icelandic breeding population (Harvey & Heubeck 2012).

Slavonian Grebes do not breed in Ireland, and in Britain are restricted to a shrinking part of the eastern Scottish Highlands. Historically, the population rose from 49 pairs in 1971 to 73 pairs in 1993, but has since declined substantially. In 2010 there were just 22 breeding pairs (*RBBP 2010*). Reasons for this decline and their restricted breeding distribution are unclear. A diverse suite of predators can cause nest failure but clutch predation rates are not unusually high and not a cause for concern; losses due to flooding and wave damage may be greater (Perkins *et al.* 2005). Brood survival may be key and is lower where introduced Pike are present (Summers *et al.* 2009). Alternatively, productivity may be linked to the availability of chironomid larvae, which is in turn linked to loch productivity (Brooks *et al.* 2012). Being at the extreme south of the species' range, the Scottish population may also be susceptible to impacts related to climate change.

Outside the Highlands, Slavonian Grebes have been recorded summering on Orkney and Shetland and in the East Midlands a female paired with a Great Crested Grebe and nested in three summers from 2006 to 2008, failing each time at the egg or chick stage (Toon 2007; Holling *et al.* 2010b).

Sponsored by RSPB

WINTER DISTRIBUTION *2007/08–2010/11*

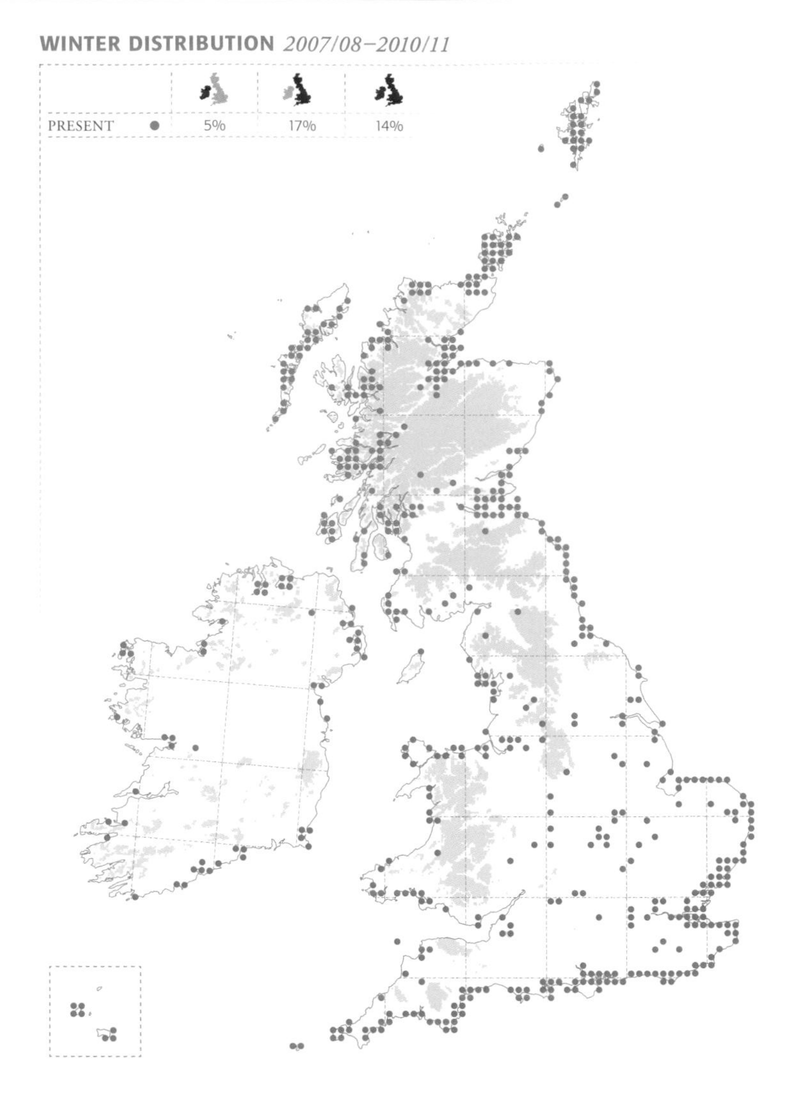

PRESENT ●	5%	17%	14%

BREEDING DISTRIBUTION *2008–11*

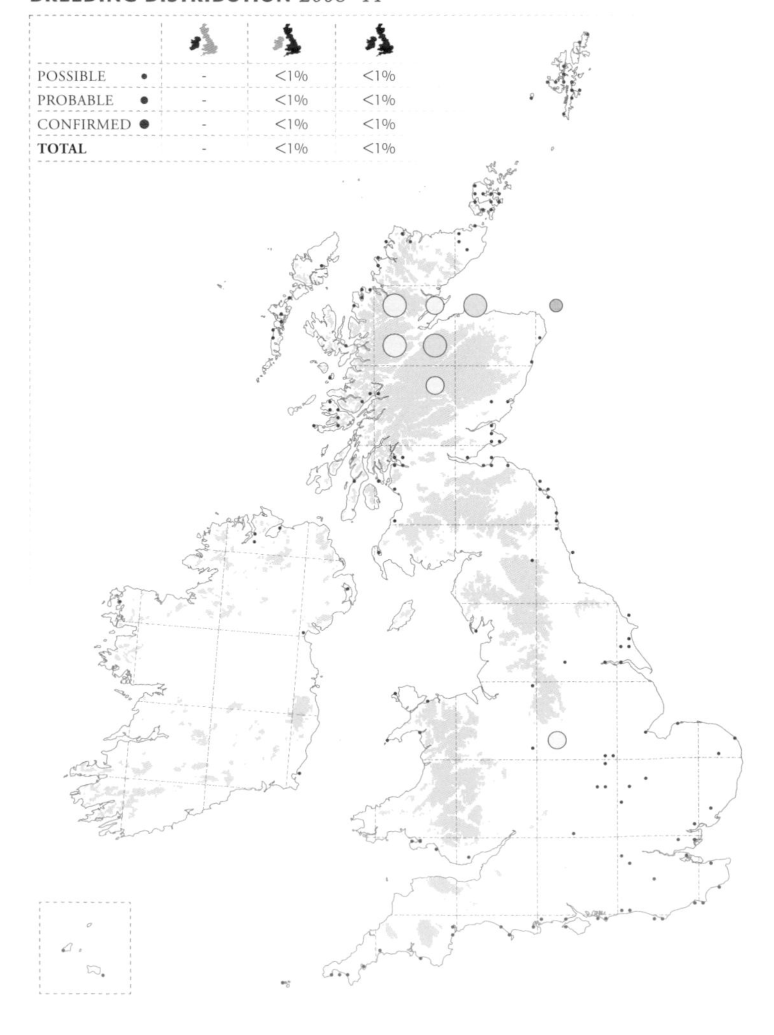

POSSIBLE ●	-	<1%	<1%
PROBABLE ●	-	<1%	<1%
CONFIRMED ●	-	<1%	<1%
TOTAL	-	<1%	<1%

WINTER DISTRIBUTION CHANGE *since 1981–84*

	Ireland	Britain	Britain & Ireland
GAIN ▲	39	277	316
LOSS ▼	13	91	104
30yr CHANGE	+118%	+65%	+69%

WINTER RELATIVE ABUNDANCE *2007/08–2010/11*

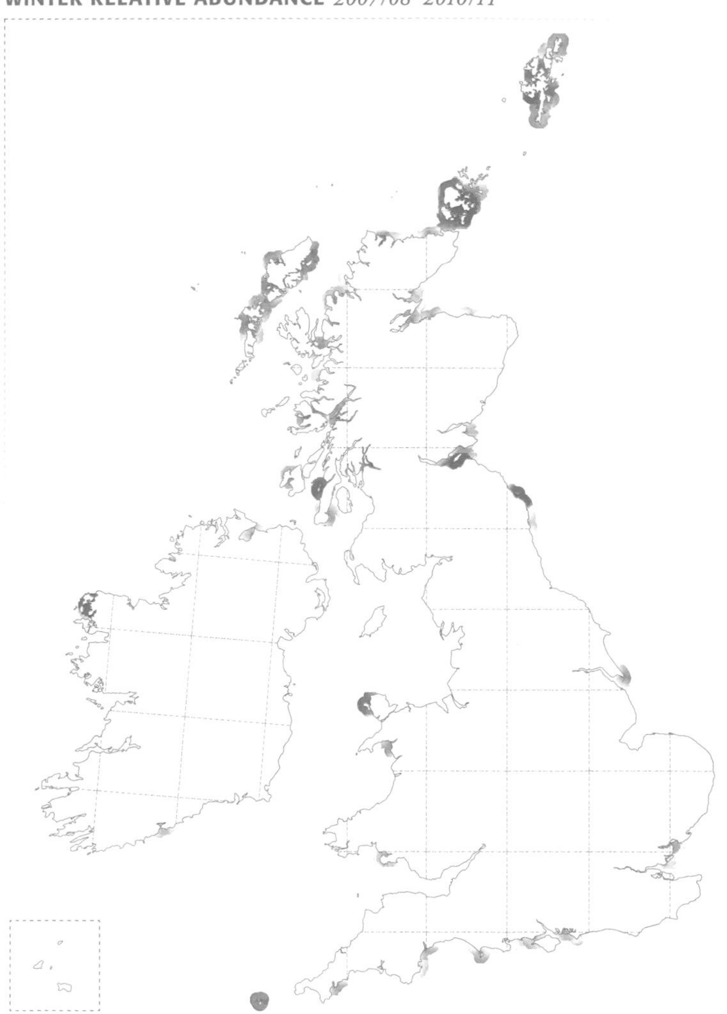

BREEDING DISTRIBUTION CHANGE *since 1988–91*

	Ireland	Britain	Britain & Ireland
GAIN ▲		7	7
LOSS ▽	-	15	15
20yr CHANGE	-	-33%	-33%
20yr INDEX	-	-	

BLACK-NECKED GREBE

Podiceps nigricollis

BTO CODE: BN

OLIVER SMART www.smartimages.co.uk

DURING THE WINTER this small grebe prefers sheltered coastal waters and large open reservoirs. It has a southerly distribution, being most abundant around the Thames Estuary and along the south coast of England. Here, small groups occur at favoured locations such as Studland Bay, Dorset, and Langstone Harbour, Hampshire. It is also found on inland waters from Lancashire through the East Midlands to the large reservoirs around London, including sites such as William Girling and Staines Reservoirs. Elsewhere, records often refer to just one or two birds. In Ireland, there were reports from ten 10-km squares, most being on the coast.

Since the *1981–84 Winter Atlas* there has been a marked 69% increase in the number of occupied 10-km squares in Britain & Ireland. The winter change map shows apparent gains across inland sites as well as along the south and southeast coasts of England. Most of the records in Ireland represent gains. The cause of the gains is unclear although, in England at least, improved coverage seems an unlikely reason given the areas and sites in question.

Black-necked Grebes are scarce breeders in England, occurring in a wide scatter of 10-km squares, mainly in central and eastern England. The species is a rare breeder elsewhere in Britain & Ireland. Following recent losses of sites with *confirmed* breeding in Scotland, it clings on in only two 10-km squares: a *probable* breeding record at one site in the Borders and a *possible* record in a nearby square. *Non-breeding* birds were found in 111 10-km squares, mostly in England. Breeding was *confirmed* in one 10-km square in the Republic of Ireland in 2011 and *non-breeding* records came from one further square (not mapped).

In their review of records up to 2004, Martin and Smith (2007) showed that the breeding population had risen during the late 1990s to a peak of 50 confirmed pairs in 2002. However, the number of sites occupied and the number of confirmed pairs has since declined, and by 2010 there were 38 confirmed pairs spread out between sites in north, central and southeast England (*RBBP 2010*). Despite the recent decline, the number of 10-km squares where evidence of breeding was recorded is still higher than during the *1988–91 Breeding Atlas*.

BREEDING DISTRIBUTION *2008–11*

	Ireland	Britain	Britain & Ireland
POSSIBLE ●	-	<1%	<1%
PROBABLE ●	-	1%	<1%
CONFIRMED ●	<1%	1%	1%
TOTAL	<1%	2%	1%

BREEDING DISTRIBUTION CHANGE *since 1968–72*

	Ireland	Britain	Britain & Ireland
GAIN ▲	1	52	53
LOSS ▽	1	10	11
40yr CHANGE	0%	+382%	+350%
20yr INDEX	-	-	

Sponsored by Friends of Tring Reservoirs

◀ For breeding, Black-necked Grebes favour sheltered, shallow, eutrophic waters, usually with extensive fringing vegetation and floating aquatic plants. Woolston Eyes in Cheshire & Wirral is the main site in Britain.

RED KITE

Milvus milvus

BTO CODE: KT

DAWN BALMER

THE RE-ESTABLISHMENT OF the Red Kite as a breeding species in Britain & Ireland, outside Wales, has been a major conservation success. Reintroduction began during the *1988–91 Breeding Atlas* but breeding did not occur until 1992 (Wotton *et al.* 2002a). Since then, the range has increased sevenfold, and 20-fold since the *1968–72 Breeding Atlas*, resulting in what must be the most extensively human-facilitated distribution of any native species covered in this atlas.

The remnant Welsh population increased from around 80 pairs in 1992 to over 1,000 pairs by 2011 (Welsh Kite Trust 2011), accompanied by expansion from its former heartlands into most of the rest of Wales and into Shropshire and Herefordshire.

From 1989 onwards, Red Kites were reintroduced to the Chilterns in England and the Black Isle in Scotland. By 2011 the Chiltern population had increased to over 800 pairs. As breeding pairs became established, they occupied 10-km squares close to the release areas, and then gradually spread into adjacent squares. From this release area, kites have colonised much of central southern England, and established satellite populations in Wiltshire, Hampshire and most recently Sussex (Welsh Kite Trust 2011). Despite comparable nesting success, the Black Isle population has increased at a much slower pace, with only 55 pairs recorded in 2011. Illegal killing through poisoning is thought to be limiting this population (Smart *et al.* 2010).

Since 1995, six further reintroduction projects have taken place in England and Scotland. These are clearly evident as clusters of occupied 10-km squares on the breeding distribution map. Outside these core areas, breeding is localised but the wide spread of *non-breeding* records suggests that other founder populations could appear in time.

In Ireland, the first reintroductions were in Co. Wicklow in 2007 (now 16 pairs) and Co. Down in 2008 (now nine pairs). Breeding in 2010 represented the first in Ireland since c.1790 (Straughan 2010; Welsh Kite Trust 2011).

The reintroduction programmes have led to a 14-fold expansion of the winter range. The release areas are clearly visible in the winter maps owing to short average dispersal, in part a result of supplementary winter feeding (Evans *et al.* 1999).

Sponsored by Northumbrian Water; a leading environmental business and proud partner of the Northern Kites project

BREEDING DISTRIBUTION *2008–11*

POSSIBLE •	1%	10%	8%
PROBABLE •	<1%	5%	3%
CONFIRMED ●	1%	10%	7%
TOTAL	2%	25%	18%

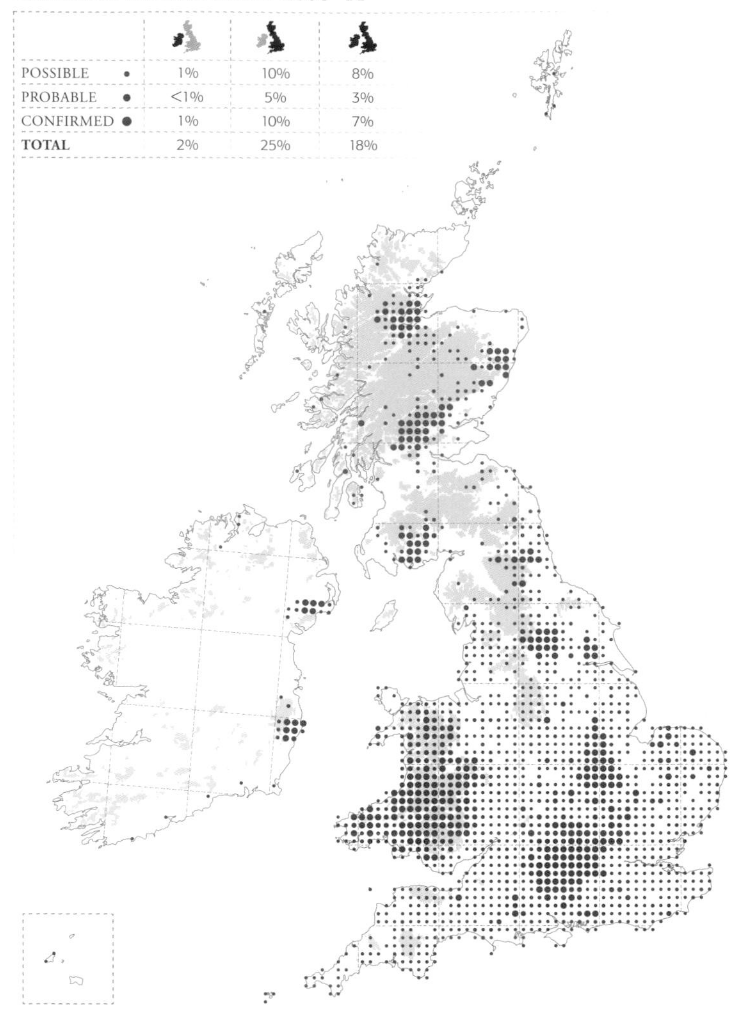

WINTER DISTRIBUTION *2007/08–2010/11*

PRESENT ●	3%	34%	26%

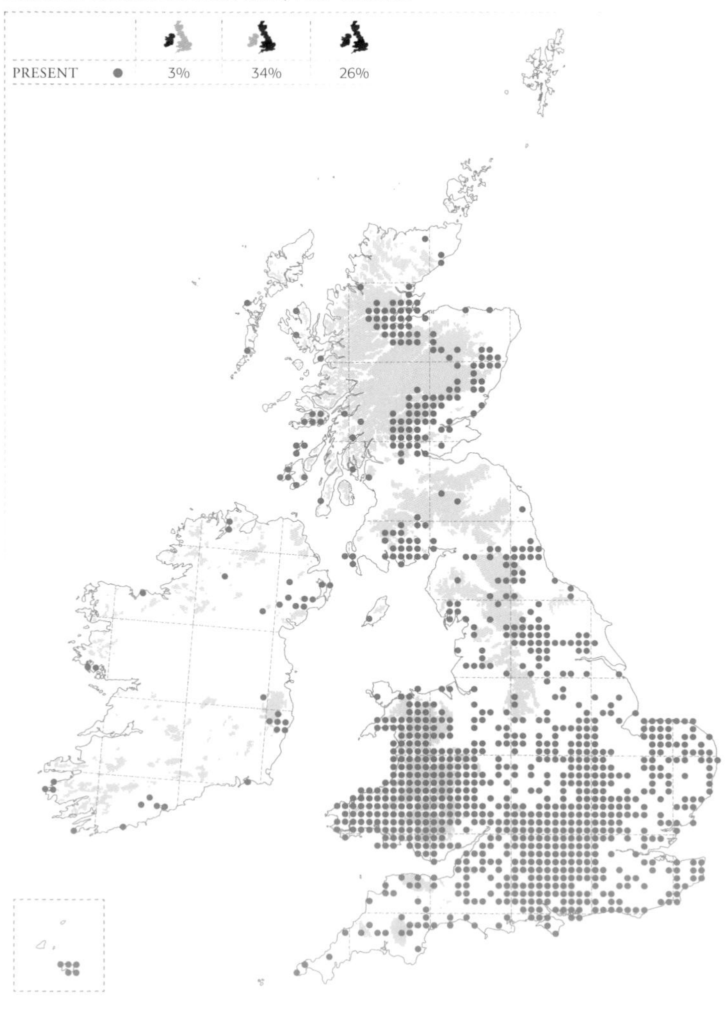

BREEDING DISTRIBUTION CHANGE *since 1968–72*

GAIN ▲	22	673	695
LOSS ▽	0	3	3
40yr CHANGE	∞	+1971%	+2035%
20yr INDEX	-	+0.91	

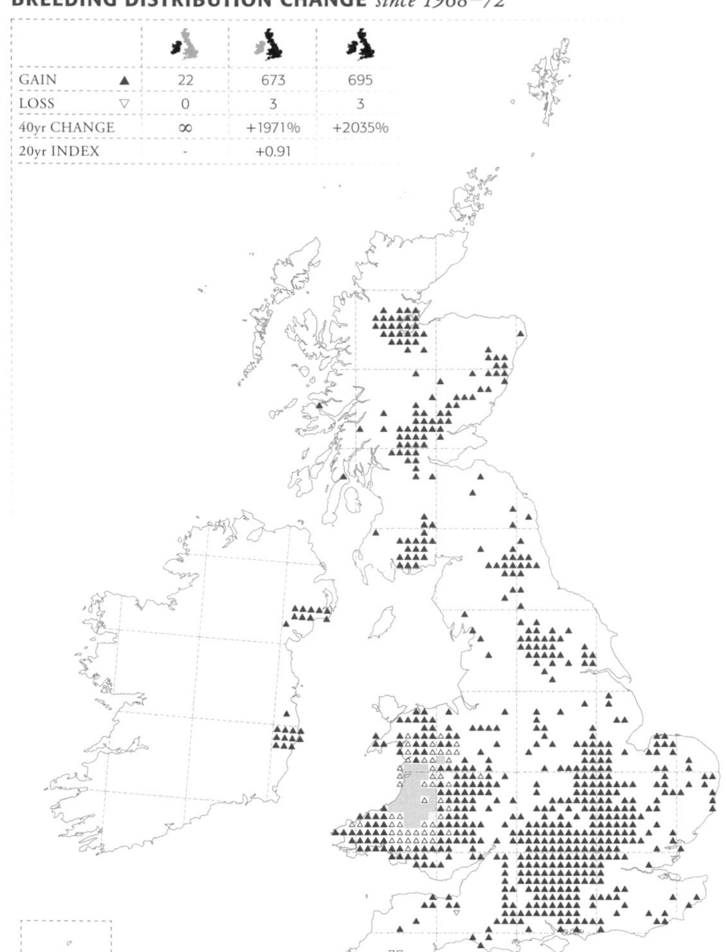

BREEDING RELATIVE ABUNDANCE *2008–11*

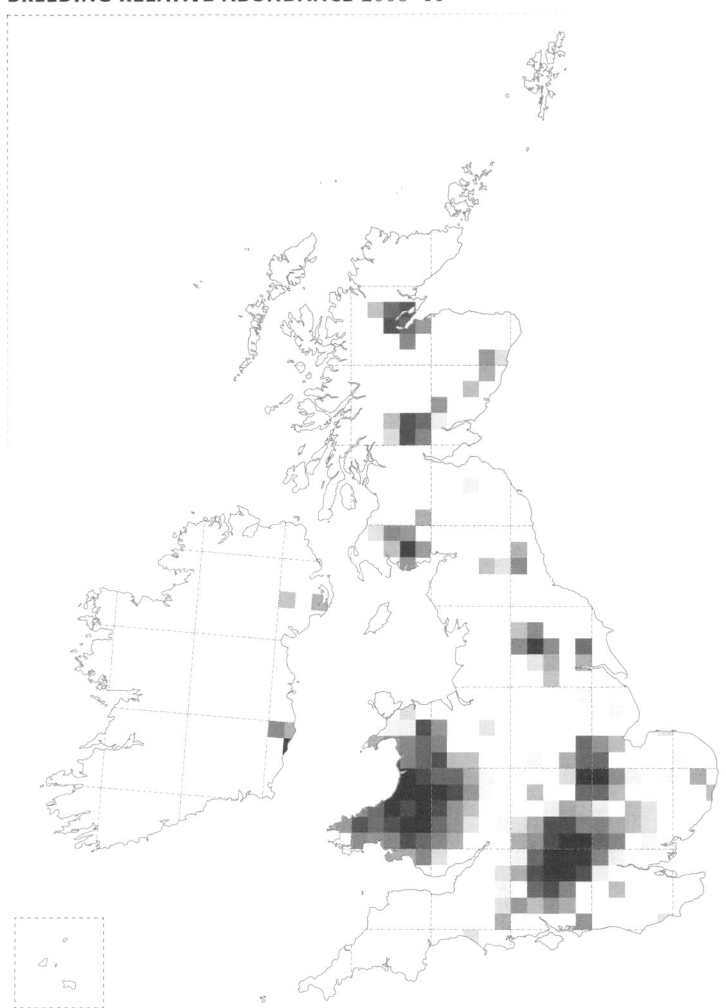

WINTER DISTRIBUTION CHANGE *since 1981–84*

GAIN ▲	32	887	919
LOSS ▼	3	12	15
30yr CHANGE	+967%	+1250%	+1238%

WINTER RELATIVE ABUNDANCE *2007/08–2010/11*

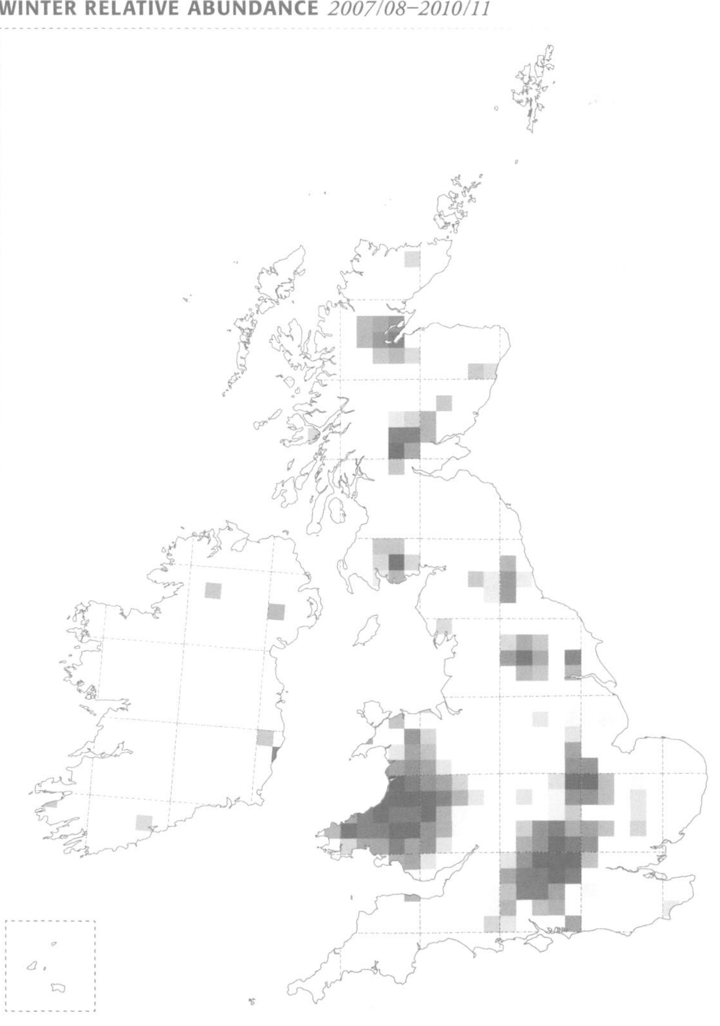

WHITE-TAILED EAGLE
Haliaeetus albicilla

BTO CODE: WE

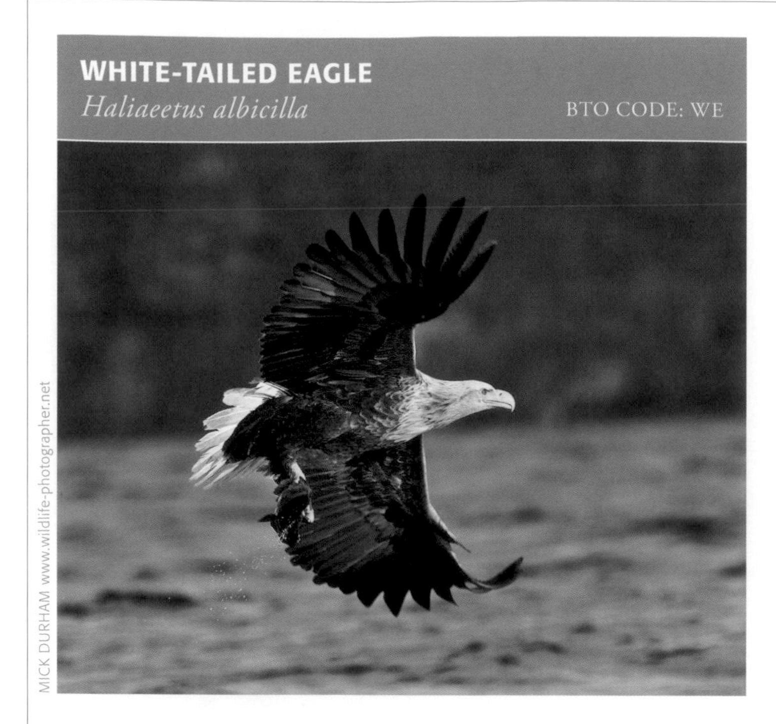

MICK DURHAM www.wildlife-photographer.net

EVIDENCE FROM PLACE names suggests that White-tailed Eagles were formerly widespread in Britain & Ireland (Evans *et al.* 2012), before they were driven to extinction by humans early in the 20th century (Green *et al.* 1996). They have since been successfully reintroduced to the western Highlands of Scotland in two phases: a total of 140 birds released on Rum during 1975–85 and in Wester Ross during 1993–98 (Whitfield *et al.* 2009). The first breeding attempt took place in 1983, although it was unsuccessful, and by the *1988–91 Breeding Atlas* there were 11 territorial pairs. The population has continued to increase and spread so that by 2010 at least 52 pairs fledged 46 young (Etheridge *et al.* 2012). The 94 occupied 10-km squares fall into four key breeding areas: Outer Hebrides, Wester Ross, Skye and the Small Isles, and north Argyll centred on Mull. During 2007–12, a further 85 birds were released in east and central Scotland (RSPB 2012). These did not breed before the end of fieldwork for this atlas but they, together with wandering immature birds, explain the wide scattering of *non-breeding* records.

In Ireland, 100 birds were released in Killarney National Park, Co. Kerry, during 2007–11 (Golden Eagle Trust 2013). Though these did not successfully breed during the Atlas period, a pair laid eggs at Lough Derg in Co. Clare in 2012; the nest was later abandoned.

In winter, most breeding adults remain on their territories, while other birds, particularly immatures, move considerable distances. Released birds in Ireland have reached central and western Ireland and some in Scotland travel inland following highland glens until they reach the east coast. Wandering continental immigrants account for some of the occupied squares, particularly in England.

Breeding success in Scotland is low, though comparable with some Scandinavian populations, and is slowly increasing as individuals gain experience and more wild-reared birds enter the population (Evans *et al.* 2009). Although some birds are known to have been illegally killed, such deliberate human interference does not appear to be a limiting factor at present.

Sponsored by Natural Research Projects Ltd in memory of Mike Madders

BREEDING DISTRIBUTION *2008–11*

POSSIBLE •	<1%	1%	1%
PROBABLE ●	<1%	<1%	<1%
CONFIRMED ●	-	2%	2%
TOTAL	<1%	3%	3%

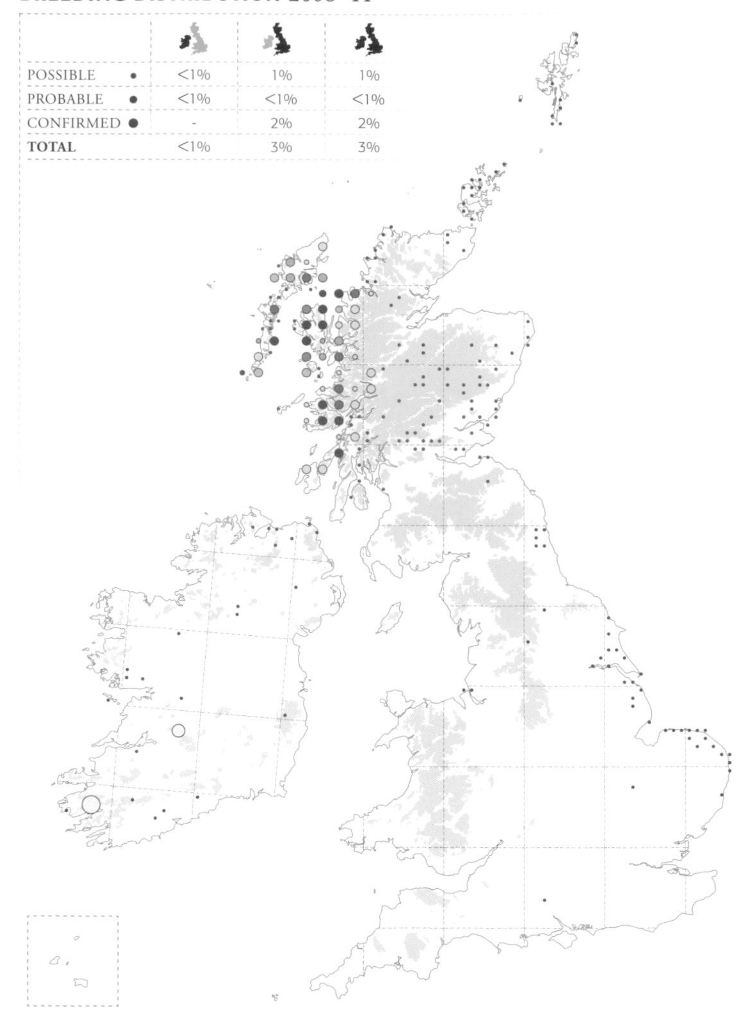

WINTER DISTRIBUTION *2007/08–2010/11*

PRESENT ●	3%	10%	8%

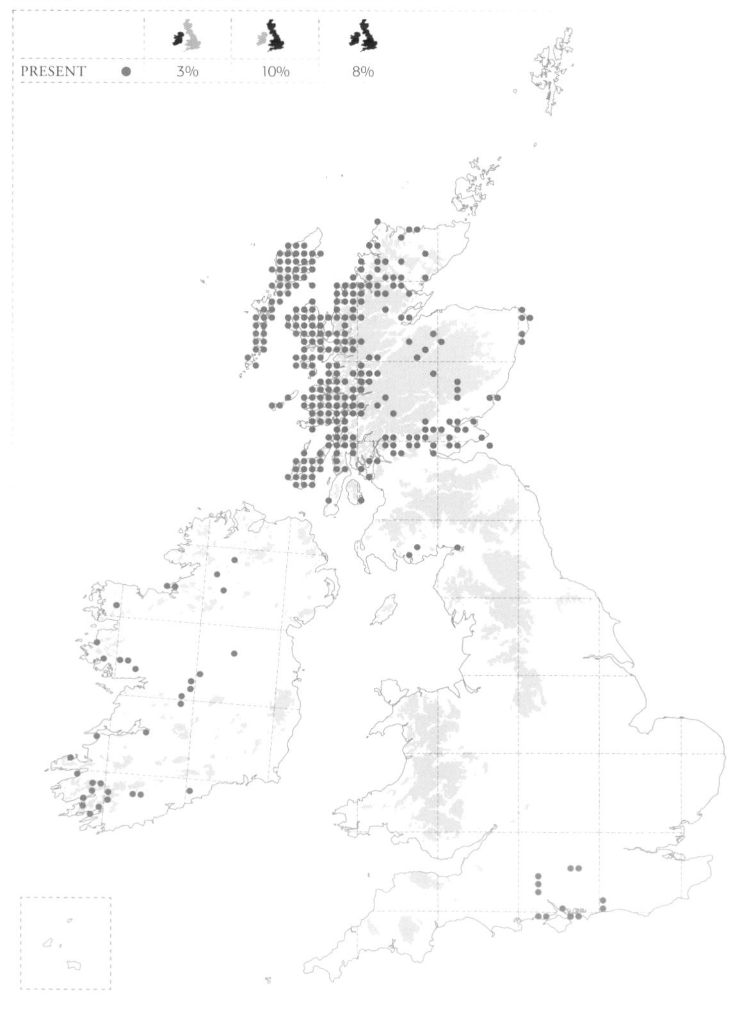

BREEDING DISTRIBUTION CHANGE *since 1988–91*

GAIN	▲	5	79	84
LOSS	▽	0	9	9
20yr CHANGE		∞	+318%	+341%
20yr INDEX		-	+0.81	

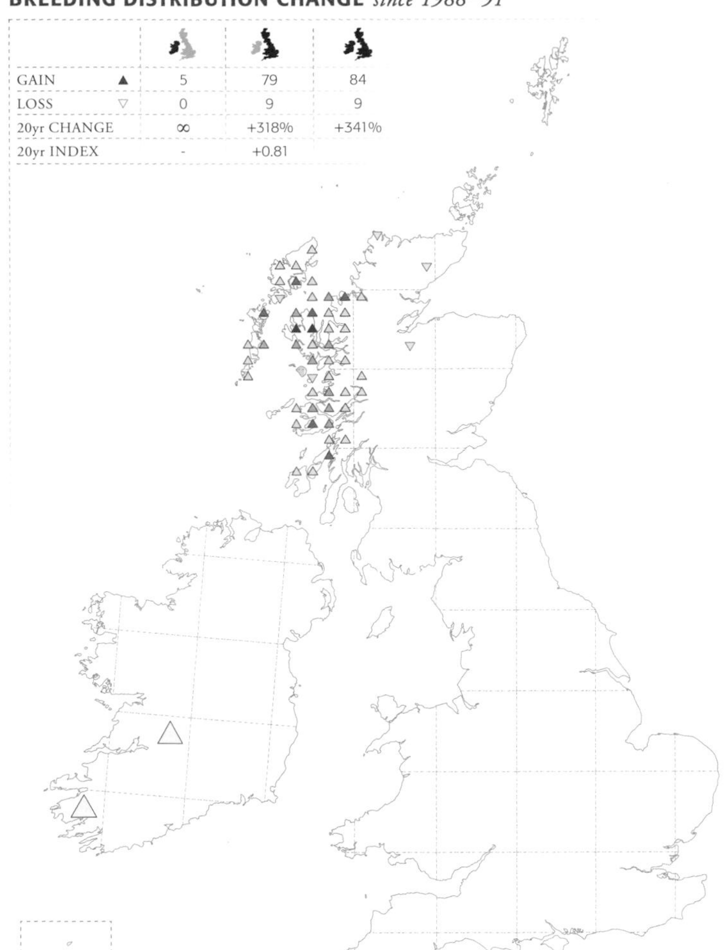

BREEDING RELATIVE ABUNDANCE *2008–11*

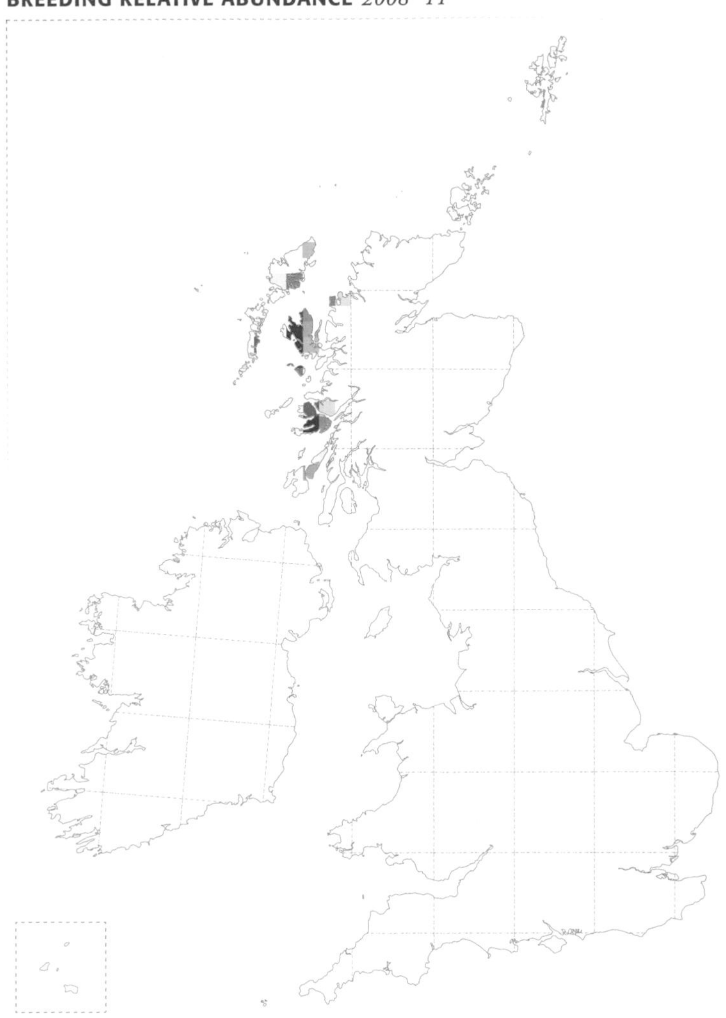

WINTER DISTRIBUTION CHANGE *since 1981–84*

GAIN	▲	30	251	201
LOSS	▼	0	8	8
30yr CHANGE		∞	+1429%	+1606%

WINTER RELATIVE ABUNDANCE *2007/08–2010/11*

MARSH HARRIER

Circus aeruginosus

BTO CODE: MR

CHRIS BALE www.guernseybirdnerd.com

MARSH HARRIERS HAVE a chequered history as breeding birds, with temporary extinction in Britain at the end of the 19th century (Clarke 1995) and in Ireland in 1917 (*Birds in Ireland*). Numbers subsequently rose before a crash to just one breeding pair in Suffolk in 1971. Abundance and range have since increased as a result of the ban on organochlorine pesticide use, reduced interference from humans (Underhill-Day 1998) and the creation and restoration of wetland habitat.

Since the *1988–91 Breeding Atlas*, when there were c.90 breeding female Marsh Harriers (Underhill-Day 1998), the breeding range has doubled. This has involved consolidation of the core range in the east and southeast, and range extension to northwest England, the Channel Islands, Isles of Scilly and several sites in Scotland. They bred in Northern Ireland in 2009 and 2011, the first breeding records there since the 1840s (Scott *et al.* 2009). A survey in 2005 produced a UK breeding population estimate of 363–429 pairs (Holling *et al.* 2008), and showed that the number of breeding females increased by 131% between 1995 and 2005, while there was an increase from 350 young fledged in 1995 to 800 in 2005 (Pitches 2006). In 2005, Lincolnshire, Norfolk and Suffolk held 68% of the total population, with Kent and Cambridgeshire holding a further 21% (Holling *et al.* 2008).

In winter, Marsh Harriers are found in a broad coastal band along the east and south coast of England and at more scattered locations along the Welsh coast. There are also clusters of records around the Dee Estuary in northwest England, in the fens of East Anglia, on the Somerset Levels and on the Channel Islands, and scattered inland records. They remain scarce winter visitors in Scotland and Ireland. This picture contrasts markedly with that from the *1981–84 Winter Atlas*, when as few as 10 individuals were thought to winter, mostly in East Anglia and South Wales. The number of birds wintering in Norfolk is ten times higher than in the 1980s (Taylor & Marchant 2011) and several communal roost sites now hold double figures and, exceptionally, as many as 100 individuals.

Sponsored by White Maple Consulting Ltd with Bob & Sally Empson

BREEDING DISTRIBUTION *2008–11*

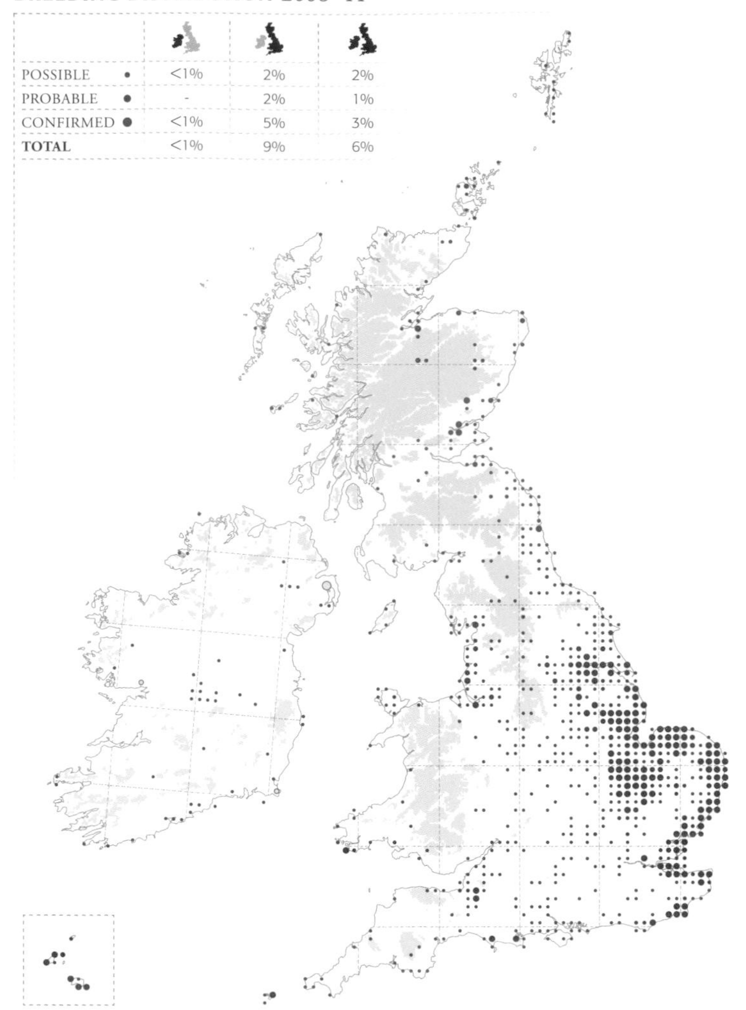

WINTER DISTRIBUTION *2007/08–2010/11*

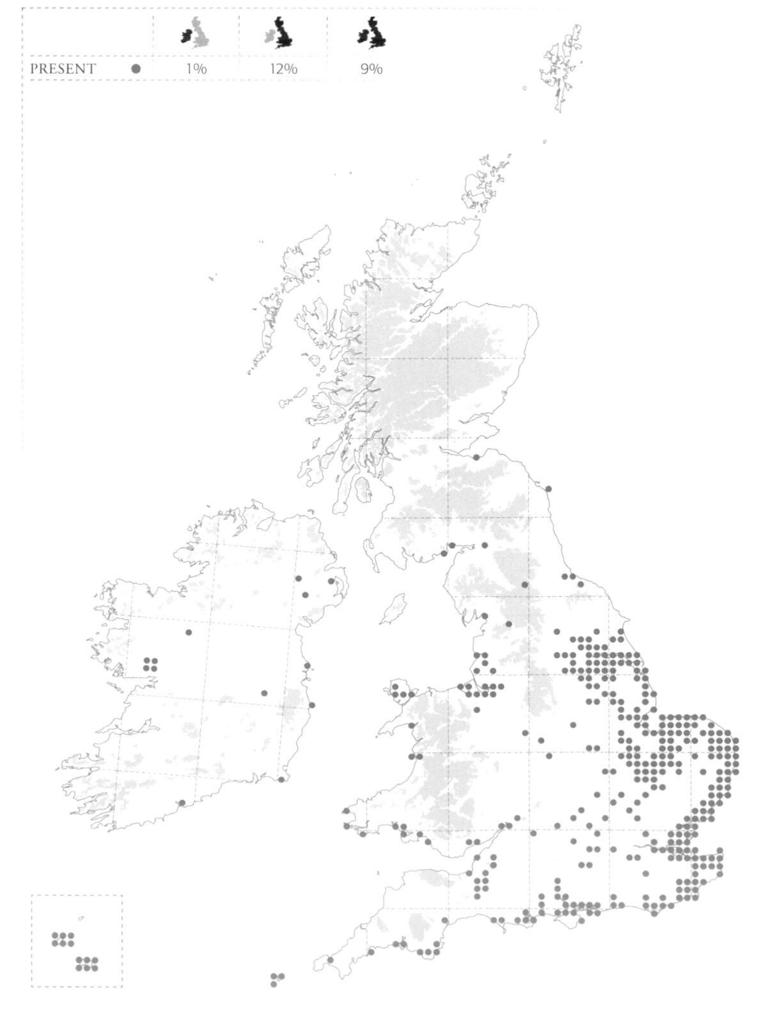

BREEDING DISTRIBUTION CHANGE *since 1988–91*

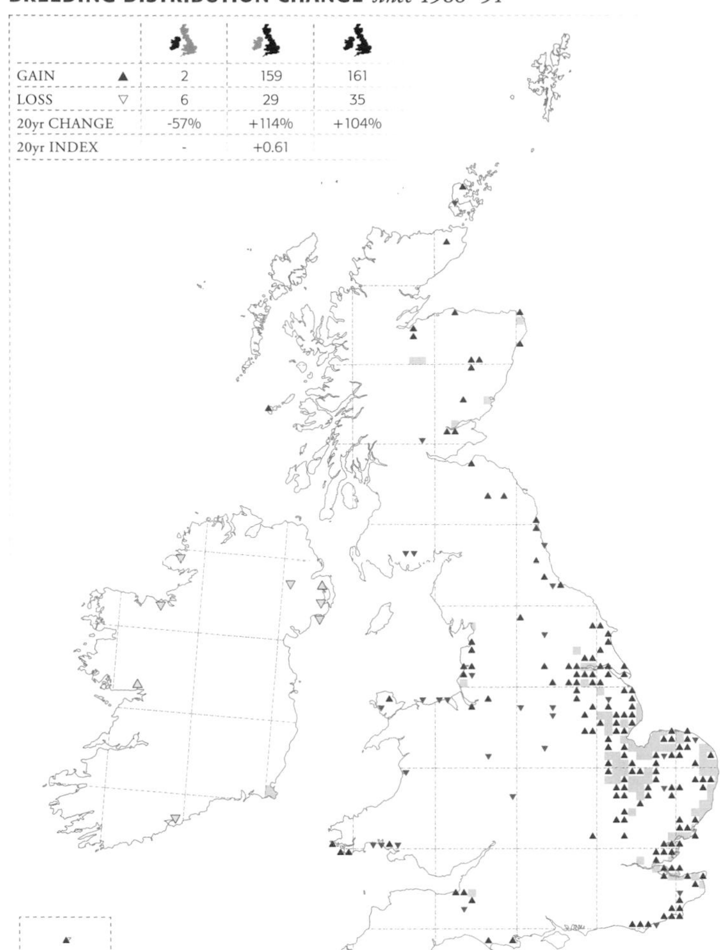

GAIN ▲	2	159	161
LOSS ▽	6	29	35
20yr CHANGE	-57%	+114%	+104%
20yr INDEX	-	+0.61	

BREEDING RELATIVE ABUNDANCE *2008–11*

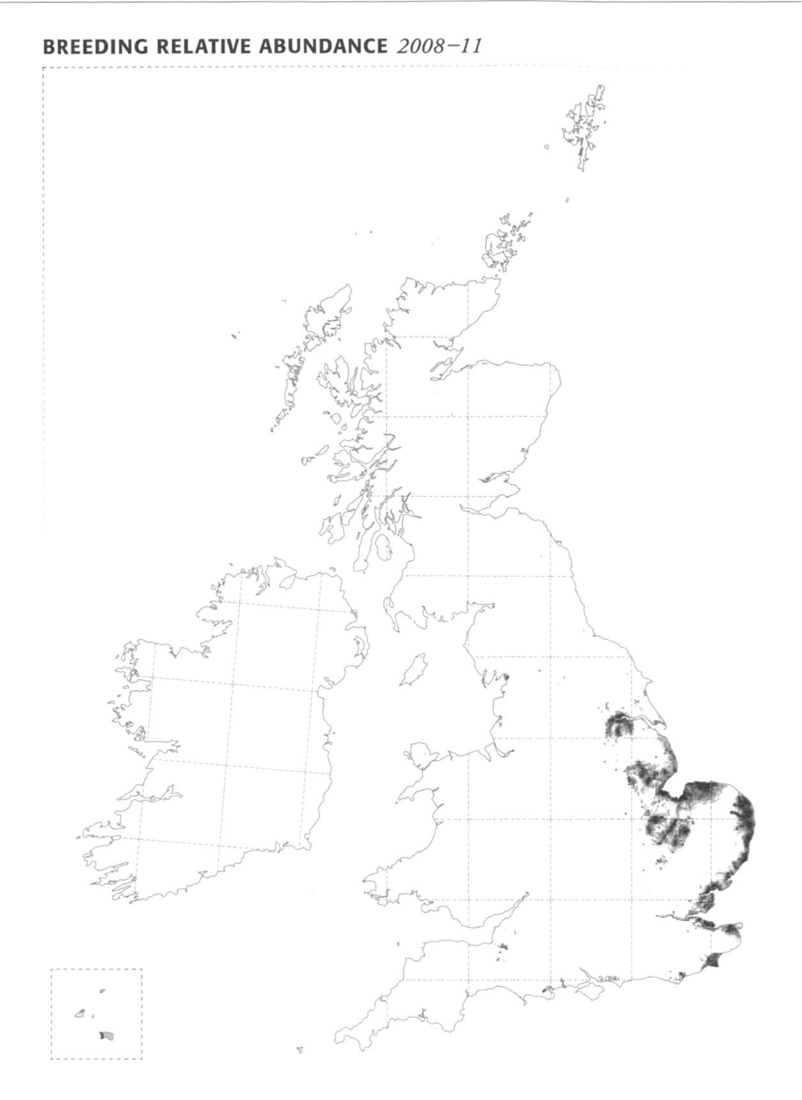

WINTER DISTRIBUTION CHANGE *since 1981–84*

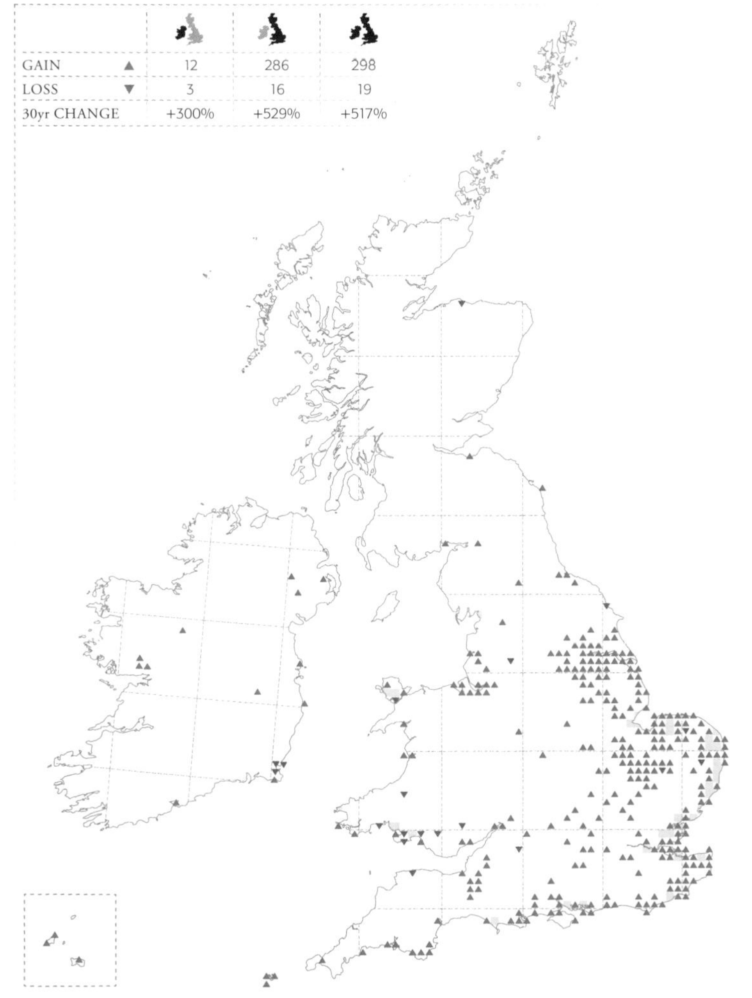

GAIN ▲	12	286	298
LOSS ▼	3	16	19
30yr CHANGE	+300%	+529%	+517%

WINTER RELATIVE ABUNDANCE *2007/08–2010/11*

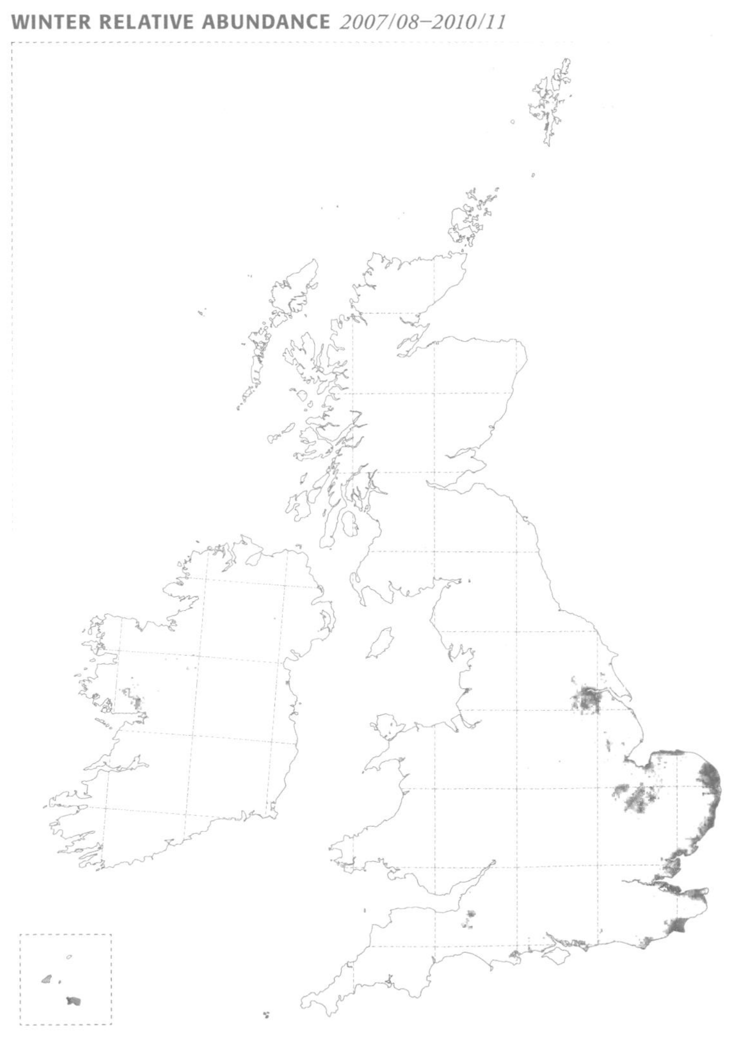

HEN HARRIER

Circus cyaneus

BTO CODE: HH

MORRIS RENDALL

DURING THE BREEDING season, Hen Harriers frequent heather moorland, extensive farmland and young forest plantations. In Scotland they are found in Orkney, the Uists and Inner Hebrides, parts of the Highlands and locally in the Southern Uplands. Smaller numbers are present in northern England, Wales and the Isle of Man. Notably, one pair bred in southwest England in 2009 (not mapped). In Ireland they breed on uplands mainly in the southwest, midlands and northwest.

In 2010 there were c.662 breeding pairs in Britain (*RBBP 2010*) and 158–205 pairs in Ireland (Ruddock *et al.* 2012). In Britain, numbers increased from 630 pairs in 1988–89 to 806 pairs in 2004, but an 18% decline followed (*RBBP 2010*). Atlas data show a 29% range expansion in Britain since the *1968–72 Breeding Atlas*, including major gains in Wales which parallel observed population increases (*RBBP 2010*). The decline on the Isle of Man (*RBBP 2010*) has not yet led to local range losses but major losses are evident in southwest Scotland. In Ireland, the range loss and population decline evident between 1968–72 and 1988–91 (O'Flynn 1983) now appears to have been partly offset by gains in the west and north, giving an 8% contraction since 1968–72. Gains may partly reflect improved coverage and reporting, because in many areas steep population declines are evident (O'Donoghue 2012; Ruddock *et al.* 2012).

Forestry is influencing population trends: while useful in pre-thicket stage (<15 years), it is deleterious thereafter (15–45 years) (O'Flynn 1983; Madders 2000; O'Donoghue *et al.* 2011). Other factors include deliberate human interference and illegal killing, which in Britain is still prevalent on driven grouse moors (Fielding *et al.* 2011), and limited food availability caused by intensified grazing and reduction of rough grass, and by afforestation (Amar *et al.* 2003, 2005; O'Donoghue 2012).

The wintering population is probably largely composed of British & Irish breeders (Dobson *et al.* 2012). Birds were reported in 38% more 10-km squares than during the *1981–84 Winter Atlas*. Over that time, the breeding population has increased, but improved winter coverage may also have increased the odds of detecting wide-ranging foraging birds.

BREEDING DISTRIBUTION *2008–11*

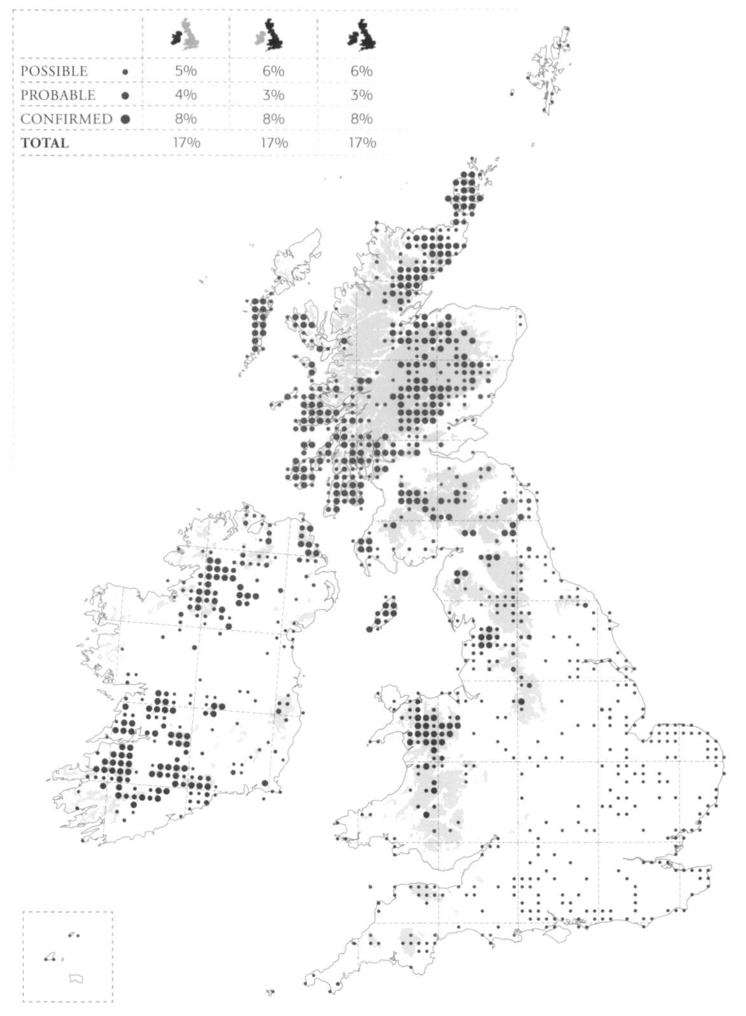

WINTER DISTRIBUTION *2007/08–2010/11*

BREEDING DISTRIBUTION CHANGE *since 1968–72*

GAIN ▲	90	262	352
LOSS ▽	104	154	258
40yr CHANGE	-7%	+29%	+17%
20yr INDEX	-0.25	-0.18	

WINTER DISTRIBUTION CHANGE *since 1981–84*

GAIN ▲	297	550	847
LOSS ▼	83	311	394
30yr CHANGE	+100%	+24%	+38%

BREEDING RELATIVE ABUNDANCE *2008–11*

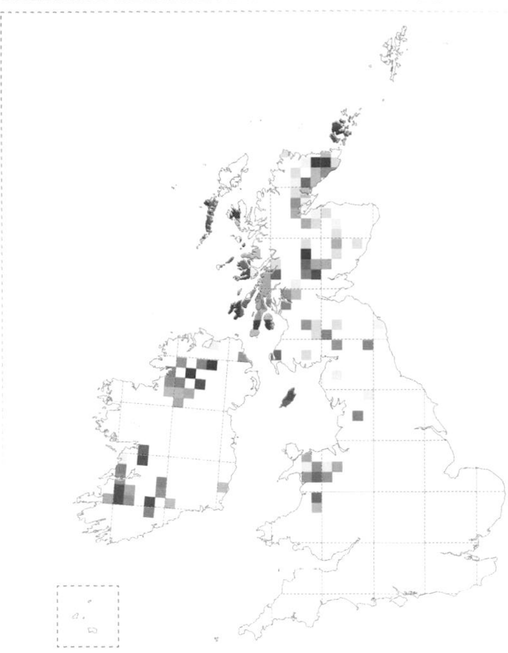

BREEDING RELATIVE ABUNDANCE CHANGE *since 1988–91*

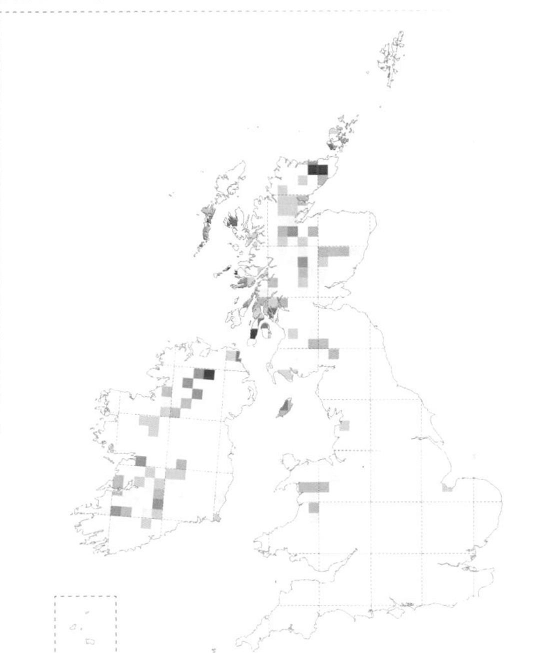

WINTER RELATIVE ABUNDANCE *2007/08–2010/11*

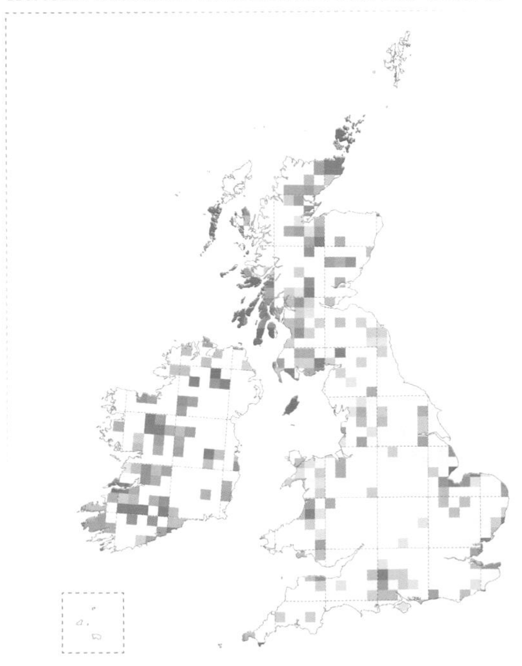

MONTAGU'S HARRIER

Circus pygargus

BTO CODE: MO

GRAHAM CATLEY http://pewit.blogspot.co.uk

MONTAGU'S HARRIERS OCCUR as scarce passage migrants between late April and early May and *non-breeding* birds were reported in 121 10-km squares in southern and eastern Britain; there were just eight squares occupied by non-breeders in Ireland. Breeding evidence was provided from 36 10-km squares, the majority falling into one of two core areas. In the east, along the Fenland fringes and arable areas of the Wash in Norfolk and Lincolnshire, around eight pairs breed annually (*RBBP 2010*). In the south, the chalk downlands and open arable farmland of Wiltshire and neighbouring counties are an important stronghold with around seven pairs recorded annually (*RBBP 2010*).

The change map shows shifts in range since the *1988–91 Breeding Atlas*; by that time Montagu's Harriers had already been lost from southwest Wales and the young conifer plantations and heathlands of Devon and Cornwall that were occupied during the *1968–72 Breeding Atlas*. The map shows losses from the Norfolk Broads and coastal Suffolk since the *1988–91 Breeding Atlas* and what appears to be a high degree of turnover in the southern population, with gains only slightly outweighing losses. The current breeding range is 13% larger than that reported in 1988–91 but still 28% smaller than in 1968–72. Breeding was at a peak in the mid 1950s, when up to 30 pairs nested in a year; numbers then dropped rapidly and there were no breeding records in 1974. Since then numbers have slowly increased to a maximum of 13 confirmed breeding pairs in 2006, 2009 and 2010.

Pesticide ingestion leading to the production of thin-shelled eggs, maturing conifer plantations and the drought in the Sahel, which began in 1968, are all likely to have affected breeding numbers (Clarke 1996). Persecution, particularly from egg collectors and gamekeepers, has been an influential factor historically (*Birds in England*). Montagu's Harriers have an increasing tendency to nest in agricultural habitats, particularly arable crops, making intervention by conservationists difficult. Better site monitoring by local experts, including minimising disturbance from birdwatchers and deterring egg collectors, and active protection of nests and young from farming operations should improve the fortunes of this species.

Sponsored by Devon Birdwatching & Preservation Society

BREEDING DISTRIBUTION *2008–11*

	Ireland	Britain	Britain & Ireland
POSSIBLE •	-	1%	<1%
PROBABLE •	-	<1%	<1%
CONFIRMED •	-	<1%	<1%
TOTAL	-	1%	<1%

BREEDING DISTRIBUTION CHANGE *since 1988–91*

	Ireland	Britain	Britain & Ireland
GAIN ▲	-	30	30
LOSS ▽	-	26	26
20yr CHANGE	-	+13%	+13%
20yr INDEX	-	-	

HONEY-BUZZARD

Pernis apivorus

BTO CODE: HZ

DAMIAN WATERS www.drumimages.co.uk

HONEY-BUZZARDS ARE secretive and are less detectable than other raptors because they spend a lot of time on the forest floor at the nests of bees and wasps, and relatively little time in the air. The information presented here is probably more complete than that presented in previous atlases, thanks to the targeted efforts and improved data sharing of local Honey-buzzard researchers.

Breeding evidence was reported from 81 10-km squares, all in Britain, and *non-breeding* birds were reported from 186 squares. To protect sensitive breeding locations, the distribution is summarised at 50-km resolution. The largest concentration of occupied squares was in the English south-coast counties of Dorset through to Kent; these counties accounted for more than half of the pairs reported in 2010 (*RBBP 2010*). Elsewhere, the main breeding areas were in Wales, Norfolk, North Yorkshire and Scotland. In 2010, 47 known territories were occupied and at least 40 young fledged (*RBBP 2010*). This was the highest number of occupied territories recorded since surveys undertaken in 2000 and 2001 found up to 51 pairs (Batten 2001; Holling *et al.* 2011b).

The change map summarises differences in the number of 10-km squares occupied per 50-km square since the *1988–91 Breeding Atlas*. Overall, there were 16 losses, including the formerly well-known site at Haldon Forest, Devon, and 61 gains. These must be interpreted with caution, however, due to variable reporting of Honey-buzzards through time. Gains are evident in most areas. Honey-buzzards were first confirmed breeding in Wales in 1992 (*Birds in Wales*), in Cumbria in 1997 (Cumbria Raptor Study Group 2002) and in southwest Scotland in the early 1990s (*Birds of Scotland*). They have since been found in many other locations throughout these regions, reflecting ongoing colonisation and improved efforts to find the birds.

The apparent increase has been much debated (Combridge *et al.* 2003, 2005; Wiseman 2004; Clements 2005). Roberts *et al.* (1999) identified a significant expansion into upland forests in western and northern Britain. Furthermore, they are successful breeders: one long-term study in the New Forest showed that 91% of pairs fledge young successfully (Wiseman 2012).

Sponsored by Pamela Rhodes

BREEDING DISTRIBUTION *2008–11*

POSSIBLE ●	-	1%	1%
PROBABLE ●	-	1%	<1%
CONFIRMED ●	-	1%	1%
TOTAL	-	3%	2%

BREEDING DISTRIBUTION CHANGE *since 1988–91*

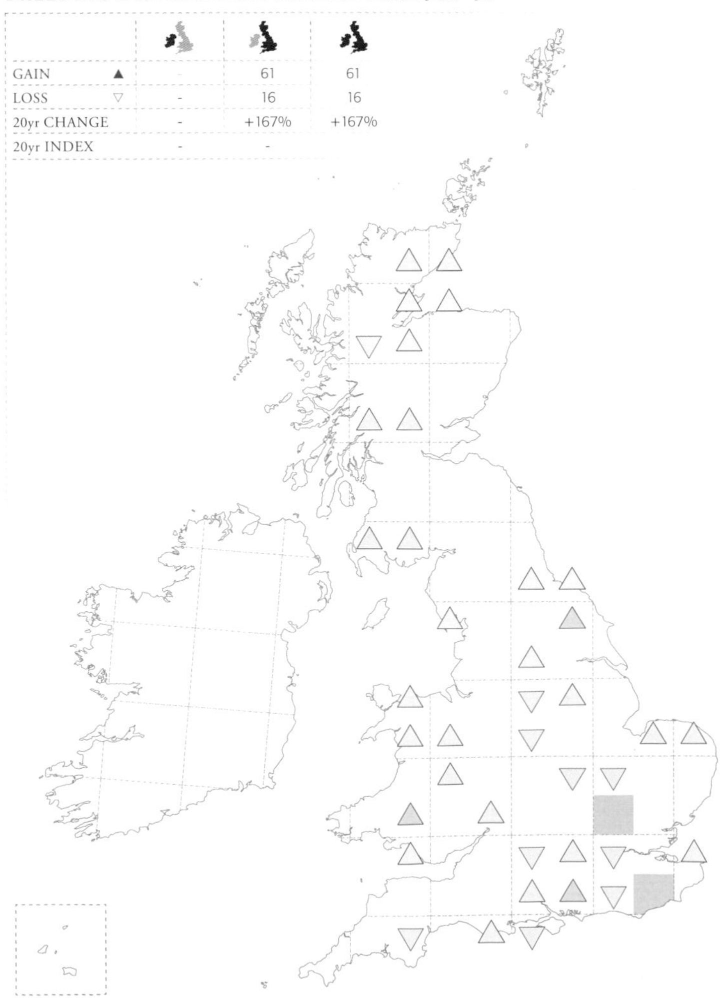

GAIN ▲	-	61	61
LOSS ▽	-	16	16
20yr CHANGE	-	+167%	+167%
20yr INDEX	-	-	

GOSHAWK

Accipiter gentilis

BTO CODE: GI

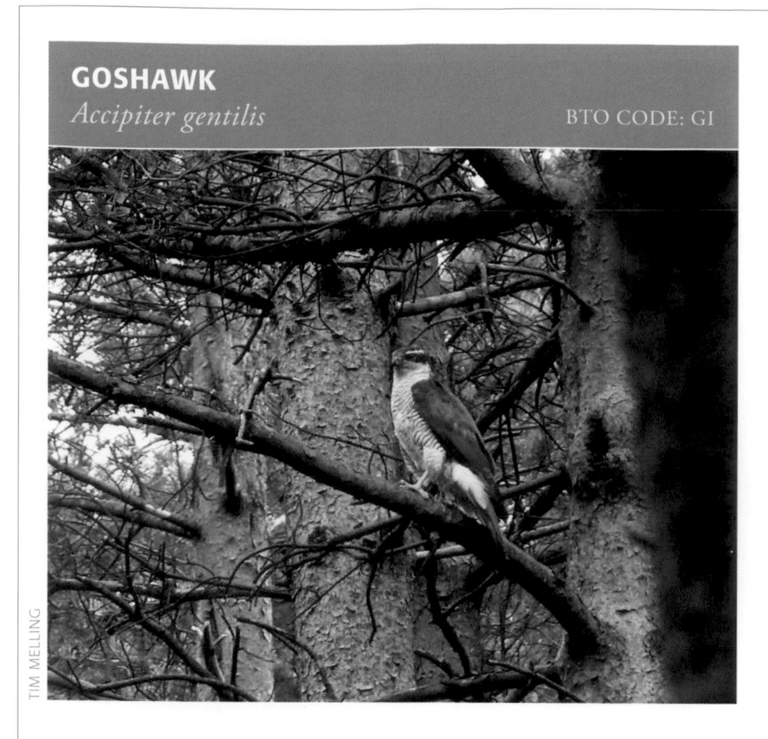

TIM MELLING

THE GOSHAWK WAS extinct in Britain & Ireland, or almost so, by the late 19th century as a result of deforestation and persecution (Hollom 1957). Continued sporadic records of breeding were bolstered by Goshawks reintroduced to Britain in the 1960s and 70s through accidental and deliberate releases by falconers (Marquiss & Newton 1982). The combination of the wide separation of release sites and locally high levels of illegal killing is considered to have had a major impact on the subsequent development of Goshawk distribution (Marquiss *et al.* 2003); Goshawks disperse only slowly (*Migration Atlas*).

Goshawks are resident, leading to similar patterns of distribution and abundance between seasons; wandering immature birds probably explain the scatter of winter records beyond the key breeding areas. Outside Wales and the Borders, the distribution comprises a number of clusters of occupied 10-km squares. Goshawks are rare in Ireland, though probably under-recorded.

In many areas Goshawks are associated with extensive state-owned forests. Marquiss *et al.* (2003) contrasted the performance of Goshawks inhabiting state-owned forests in the Borders with those in private forests in northeast Scotland. Despite comparable productivity, population growth in northeast Scotland was markedly lower than in the Borders, leading them to conclude that this difference was attributable to illegal killing.

Both change maps show large-scale increases in range. Some of the gains may result from improved coverage and data availability, but a marked population increase is likely to be under way in many areas. In northeast Scotland, in 2,500 km^2 of terrain, the population increased from initial colonisation in 1973 at roughly 5% per annum, to 30 breeding pairs in 2006 (Francis & Cook 2011) with subsequent further increases. In the Borders, the first successful breeding took place in 1977 and by 1996 there were 87 pairs (Marquiss *et al.* 2003). Goshawks can occur at quite high density: Squires *et al.* (2009) found 14 pairs per 100 km^2 of forest in Wales, whilst in northeast Scotland the maximum potential density is 9 pairs per 100 km^2 (M. Marquiss pers comm). During 2000–04, there were 100 known occupied home ranges in Scotland (*Birds of Scotland*) and in 2010 up to 435 pairs in the UK (*RBBP 2010*). The relative abundance maps highlight Wales as a major stronghold of the species.

Sponsored by Forestry Commission (England)

BREEDING DISTRIBUTION *2008–11*

	Ireland	Britain	Britain & Ireland
POSSIBLE •	1%	4%	3%
PROBABLE •	<1%	5%	4%
CONFIRMED •	<1%	9%	7%
TOTAL	1%	18%	14%

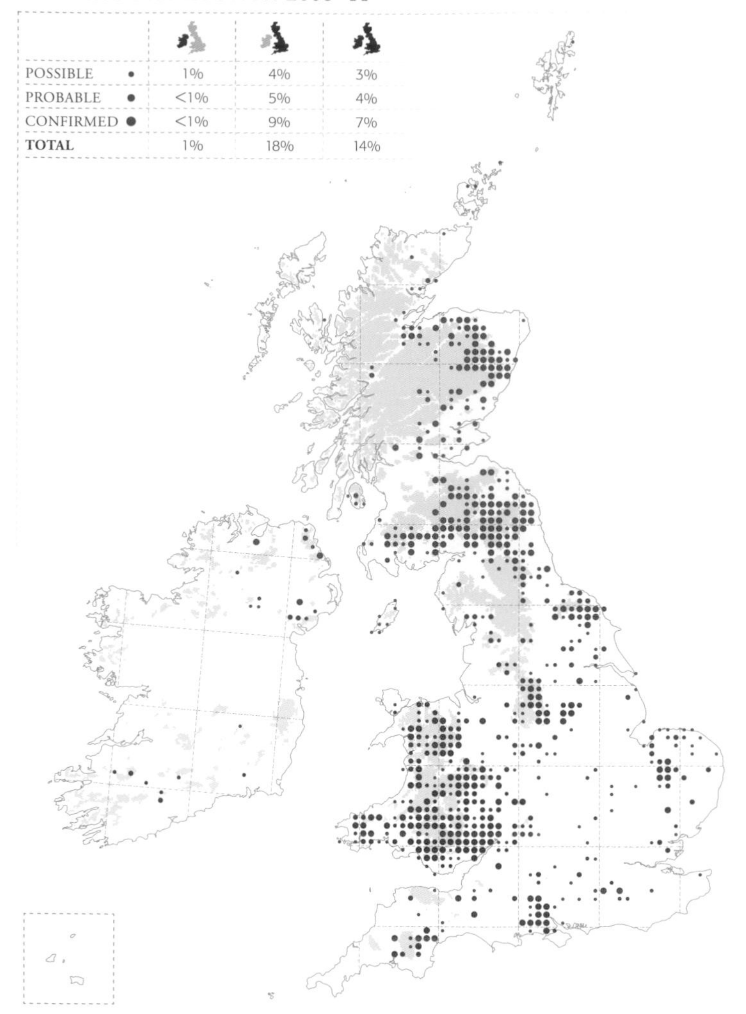

WINTER DISTRIBUTION *2007/08–2010/11*

	Ireland	Britain	Britain & Ireland
PRESENT •	<1%	18%	13%

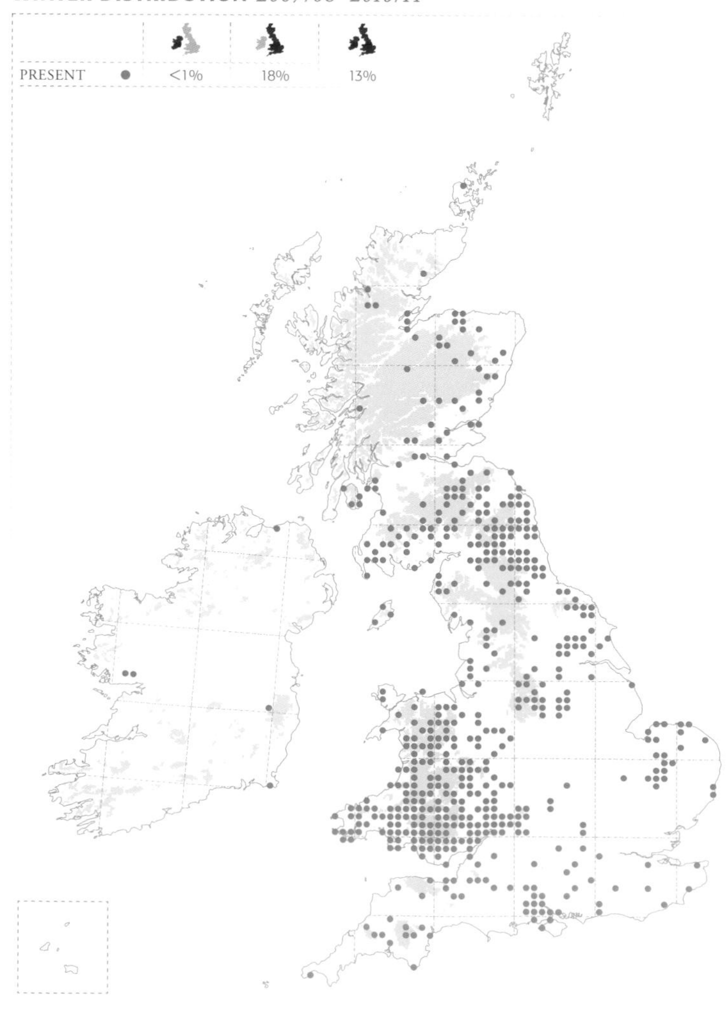

BREEDING DISTRIBUTION CHANGE *since 1968–72*

GAIN ▲	19	467	486
LOSS ▽	0	15	15
40yr CHANGE	∞	+1291%	+1346%
20yr INDEX	-	+0.41	

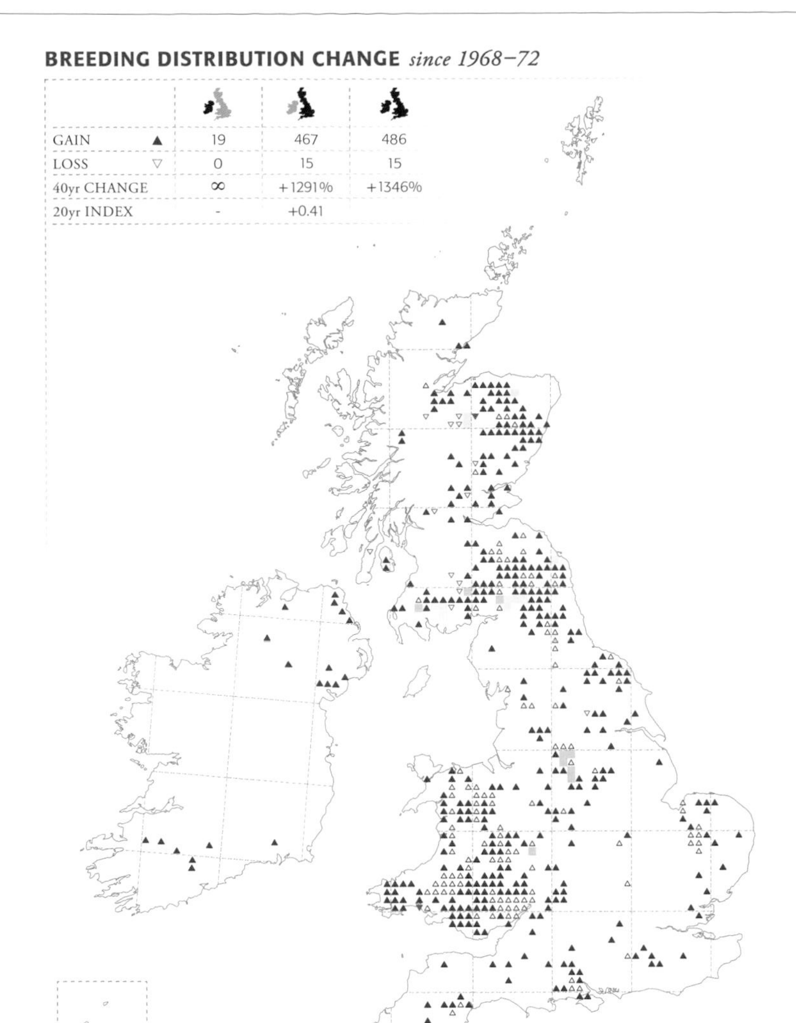

BREEDING RELATIVE ABUNDANCE *2008–11*

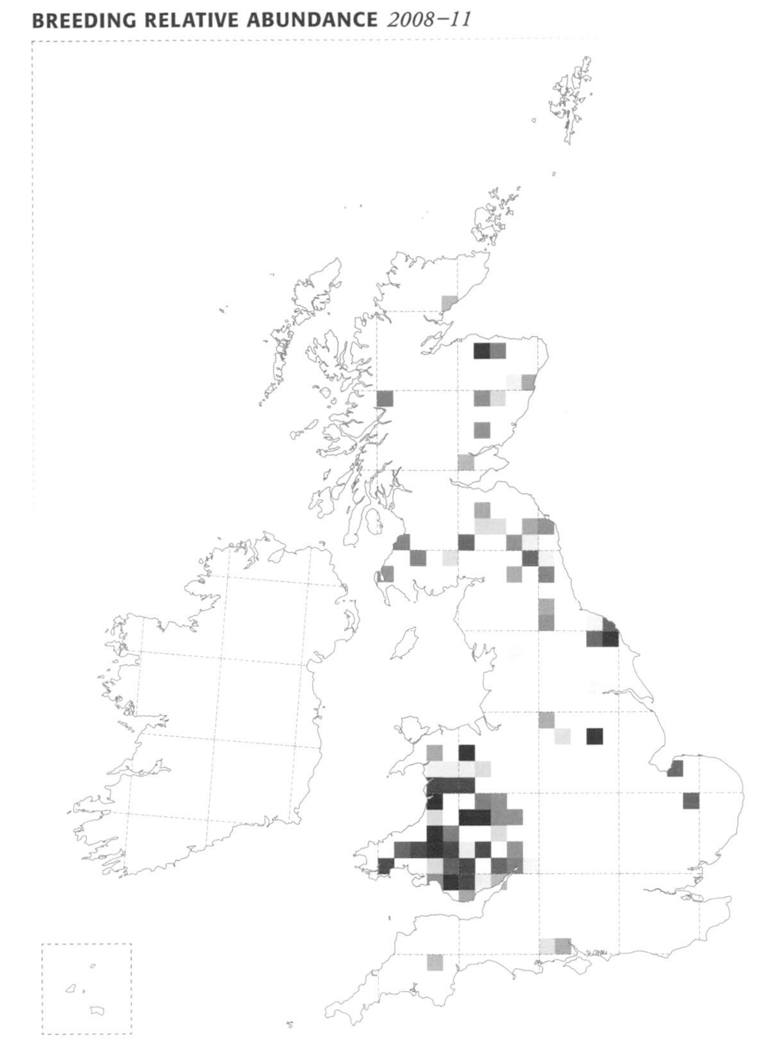

WINTER DISTRIBUTION CHANGE *since 1981–84*

GAIN ▲	5	447	452
LOSS ▼	0	69	69
30yr CHANGE	∞	+344%	+348%

WINTER RELATIVE ABUNDANCE *2007/08–2010/11*

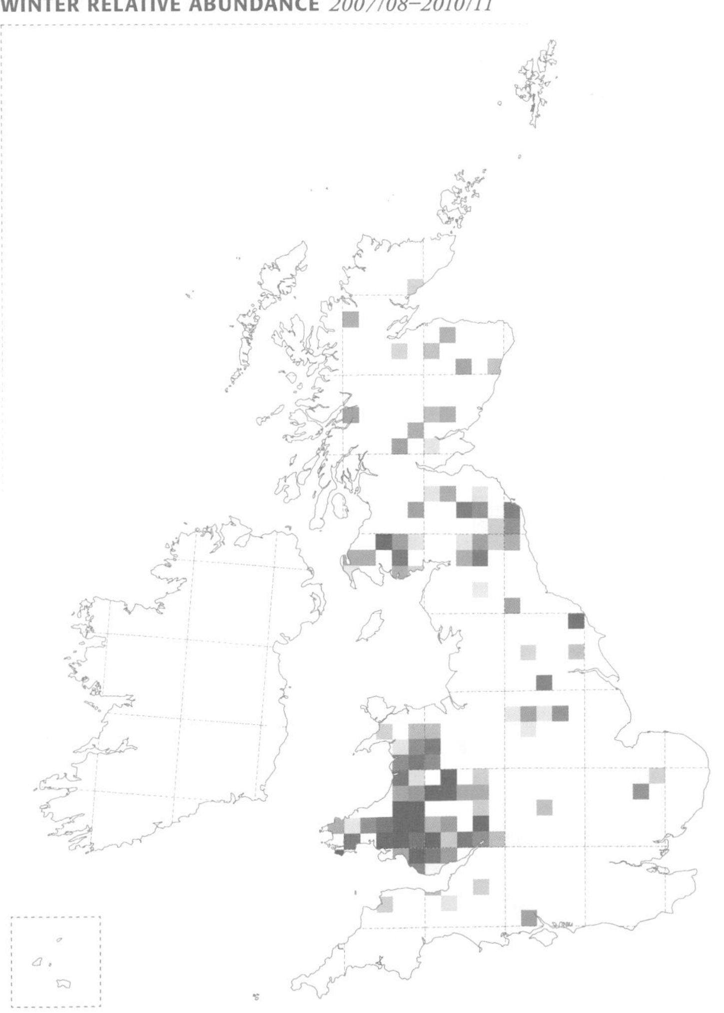

SPARROWHAWK

Accipiter nisus

BTO CODE: SH

ALAN MCFADYEN www.naturephotographyhides.co.uk

IN THE BREEDING season, Sparrowhawks are widespread across most of Britain & Ireland, with the exception of the northern Scottish uplands and some island groups. Abundance is highest in lowland areas such as eastern England, and particularly low in northwest Scotland and other upland areas. The change maps tell quite different stories in Britain and Ireland.

The series of breeding atlases charts the recovery of British Sparrowhawks from the low point in the early 1960s caused by the widespread agricultural use of organochlorine pesticides (Newton 1986). A 29% range expansion has occurred since the *1968–72 Breeding Atlas*, most of which had already taken place by the *1988–91 Breeding Atlas*: only a 7% expansion is apparent since then. Gains in eastern England probably reflect the completion of population recovery, whereas those in northern and western Scotland may be associated with expansion into maturing conifer plantations. Breeding population trends also document this recovery, with a steep increase from the 1970s to the early 1990s, followed by stability until c.2005, then a small decline (*BirdTrends*). The abundance change map shows the recovery of Sparrowhawks in eastern England since 1988–91. It also indicates that abundances have decreased in many areas in western and northern Britain. In support of this, annual monitoring data indicate that numbers in northwest England decreased by 40% during 1995–2010 (*BBS Report 2011*).

In Ireland, the effect of organochlorine poisoning was less marked (*Birds in Ireland*); overall a 2% range contraction is evident since the *1968–72 Breeding Atlas*. This conceals a 20% contraction between 1968–72 and 1988–91, followed by recovery, perhaps suggesting that observer coverage in 1988–91 had been incomplete. However, fixed-effort comparisons and the abundance change map confirm the recent increase. Monitoring data suggest a stable population, though a decline is apparent in recent years (*CBS Trend 2010*).

The winter distribution is very similar to the breeding one. Continental immigrants probably account for winter records on the Isles of Scilly and Shetland, where Sparrowhawks do not breed. The 19% winter range expansion mirrors the gains seen during the breeding season.

BREEDING DISTRIBUTION *2008–11*

POSSIBLE ●	36%	18%	23%
PROBABLE ●	13%	13%	13%
CONFIRMED ●	28%	51%	45%
TOTAL	77%	82%	81%

WINTER DISTRIBUTION *2007/08–2010/11*

PRESENT ●	84%	86%	86%

Sponsored by Nature Photographers' Portfolio

BREEDING DISTRIBUTION CHANGE *since 1968–72*

GAIN ▲	112	648	760
LOSS ▽	125	126	251
40yr CHANGE	-2%	+29%	+19%
20yr INDEX	+0.12	+0.06	

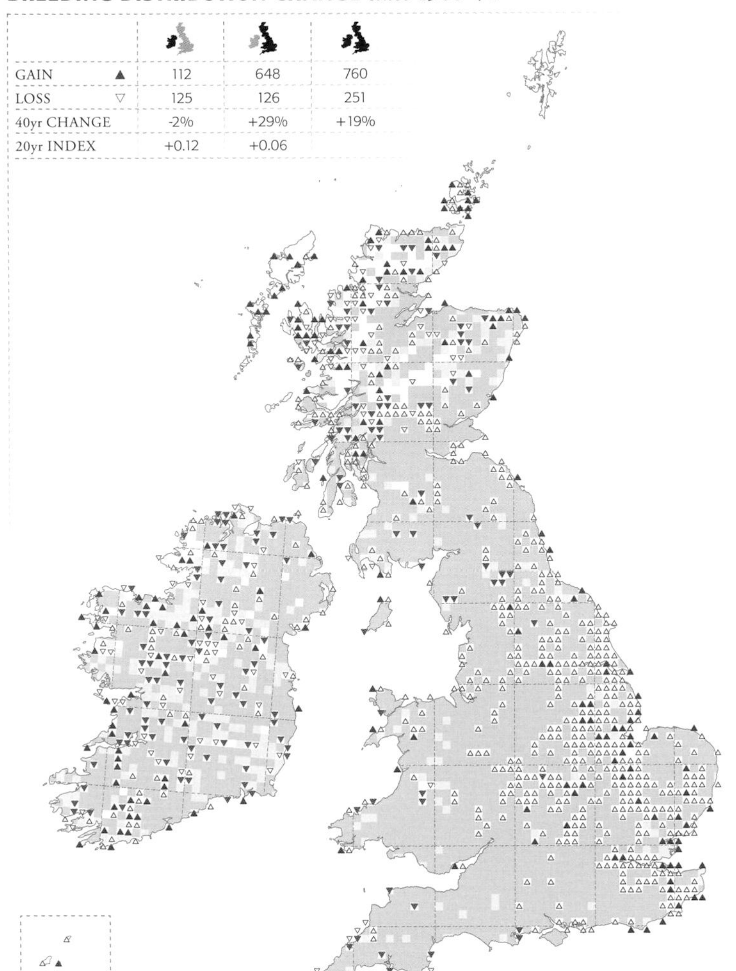

WINTER DISTRIBUTION CHANGE *SINCE 1981–84*

GAIN ▲	260	440	700
LOSS ▼	96	91	187
30yr CHANGE	+25%	+17%	+19%

BREEDING RELATIVE ABUNDANCE *2008–11*

BREEDING RELATIVE ABUNDANCE CHANGE *since 1988–91*

WINTER RELATIVE ABUNDANCE *2007/08–2010/11*

BUZZARD

Buteo buteo

BTO CODE: BZ

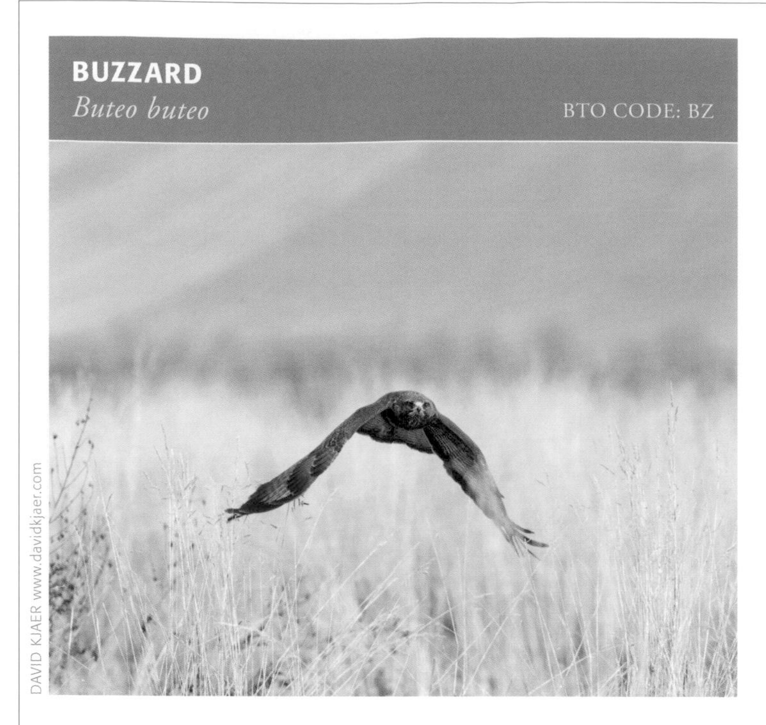

DAVID KJAER www.davidkjaer.com

ONE OF THE most extraordinary changes in our avifauna over recent decades has been the spread of the Buzzard. Following more than a doubling of its range, it is now widespread year-round all across Britain and in the eastern half of Ireland. In most areas this represents a recolonisation of former range, mainly following contraction caused by persecution in the 18th, 19th and early 20th centuries, prey reduction following myxomatosis in the 1950s, and the impacts of organochlorine pesticides in the 1950s and 1960s (Parkin & Knox 2010).

There are slightly more 10-km squares occupied in winter than in the breeding season, owing mainly to dispersal of juveniles. The change maps show that the main areas of gain have been in eastern Ireland and eastern Britain, together with the Isle of Man and Channel Islands. Buzzards have yet to colonise Shetland and the western half of Ireland.

The relative abundance maps reveal that, although the species is now widespread, it is still most abundant in Wales, southwest and northern England, southern Scotland and the low ground of eastern Scotland. High densities are associated with the availability of unimproved pasture and mature woodland (Sim *et al.* 2001). Although the breeding abundance change map is dominated by the wholesale eastward expansion of Buzzards, increases in abundance are also apparent in western areas, where territory density has risen markedly (Clements 2002; Driver & Dare 2009; Roberts & Jones 2009; Prytherch 2013).

The reasons for the recovery of the Buzzard are not fully understood, though reduction in persecution is considered to have played a key role (Prytherch 2013). The species has a broad diet, but in many areas European Rabbits are a key prey and their availability can directly affect Buzzard densities and breeding success (Graham *et al.* 1995; Sim *et al.* 2001). In some areas, such as in Ireland, Buzzards have a relatively unspecialised diet (Rooney & Montgomery 2013), and this may have facilitated their recolonisation. New nesting habitat, in the form of woodlands, shelterbelts and plantations, has also provided good opportunities for breeding. Elsewhere, the removal of hedgerows, improvement of pastureland and increased use of pesticides may have limited the extent of local recovery (Sim 2003).

Sponsored by Shropshire Ornithological Society

BREEDING DISTRIBUTION *2008–11*

	Ireland	Britain	Britain & Ireland
POSSIBLE •	13%	5%	7%
PROBABLE •	18%	14%	15%
CONFIRMED •	20%	71%	57%
TOTAL	51%	90%	79%

WINTER DISTRIBUTION *2007/08–2010/11*

	Ireland	Britain	Britain & Ireland
PRESENT •	54%	93%	82%

BREEDING DISTRIBUTION CHANGE *since 1968–72*

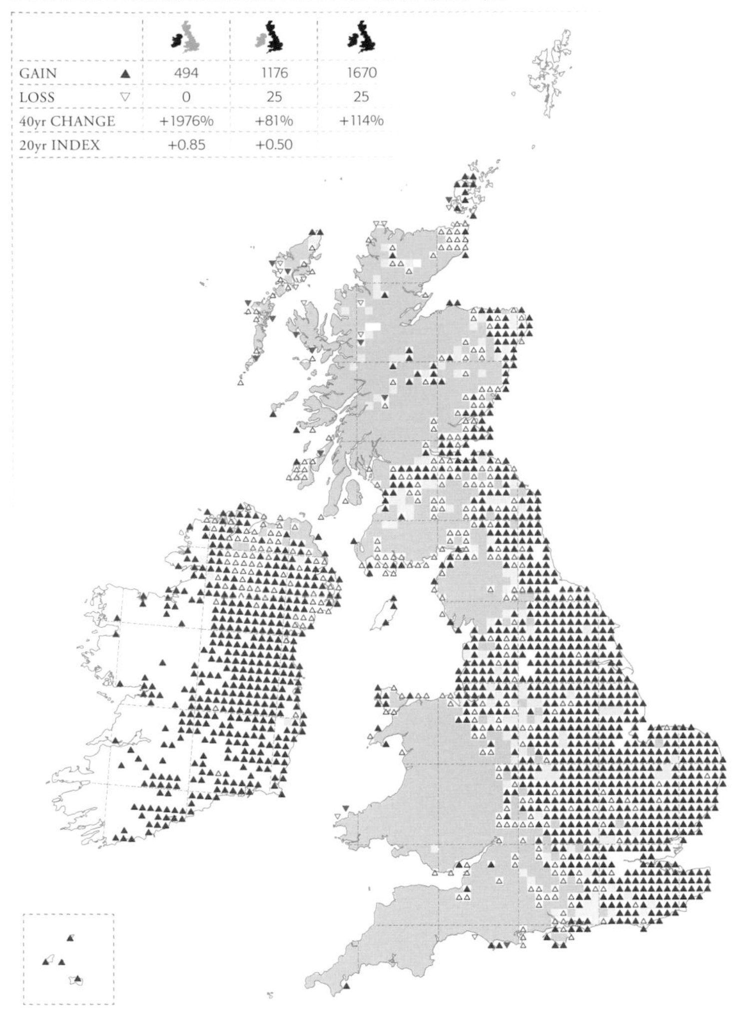

GAIN ▲	494	1176	1670
LOSS ▽	0	25	25
40yr CHANGE	+1976%	+81%	+114%
20yr INDEX	+0.85	+0.50	

WINTER DISTRIBUTION CHANGE *since 1981–84*

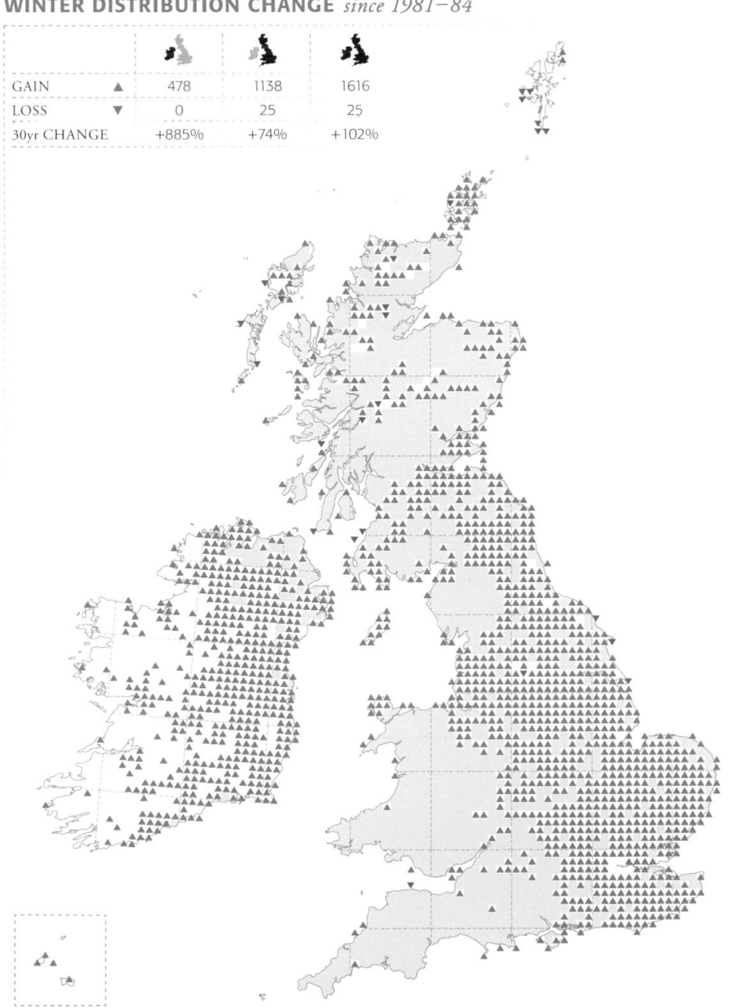

GAIN ▲	478	1138	1616
LOSS ▼	0	25	25
30yr CHANGE	+885%	+74%	+102%

BREEDING RELATIVE ABUNDANCE *2008–11*

BREEDING RELATIVE ABUNDANCE CHANGE *since 1988–91*

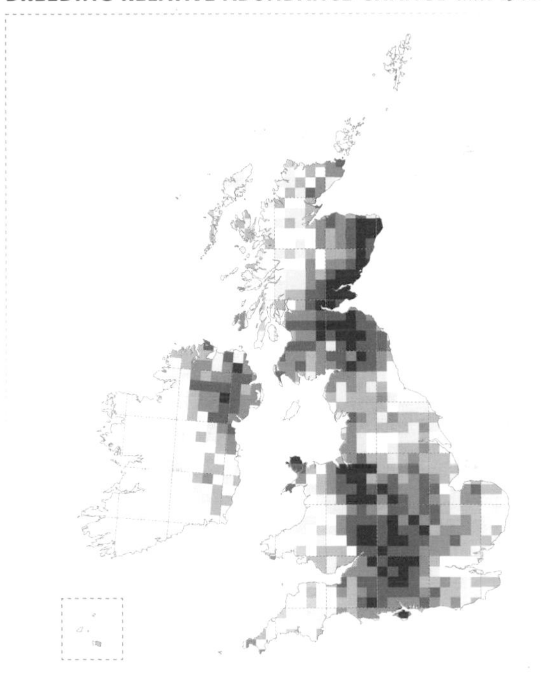

WINTER RELATIVE ABUNDANCE *2007/08–2010/11*

GOLDEN EAGLE

Aquila chrysaetos

BTO CODE: EA

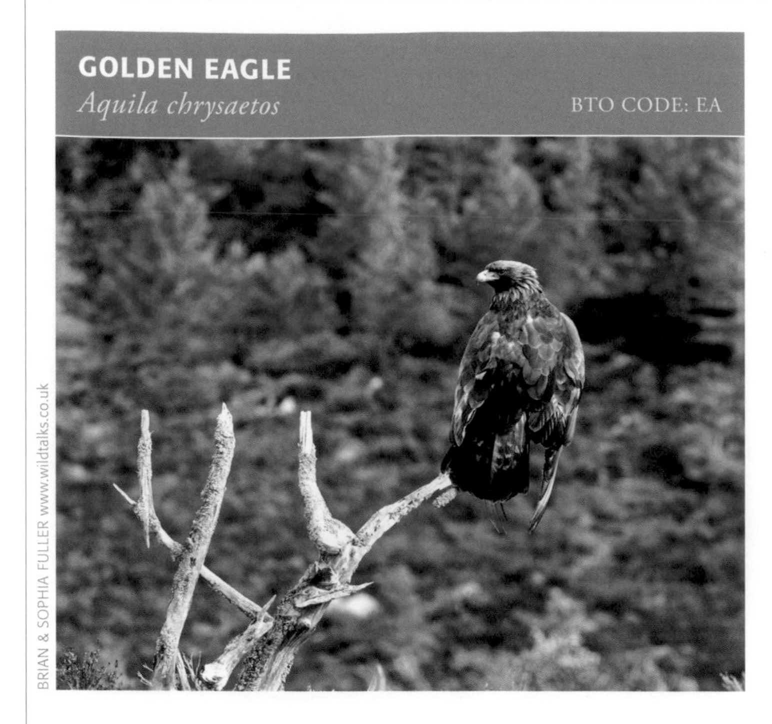

BRIAN & SOPHIA FULLER www.wildtalks.co.uk

ADULT GOLDEN EAGLES remain on their upland territories throughout the year, whilst immature birds roam widely within the uplands, though there is little difference between the breeding season and winter maps shown here. They are found throughout the Scottish Highlands and on most Hebridean islands. In England, a lone bird remains on territory in the Lake District. In Ireland, Golden Eagles became extinct in the Republic around 1912 and in Northern Ireland in the 1960s (O'Toole *et al.* 2002). A reintroduction project began in Co. Donegal in 2002, resulting in three breeding pairs by 2010 (*IRBBP 2010*). Due to site sensitivity, edge-of-range territories in Galloway and in the Borders and the reintroduced pairs in Ireland are presented at 50-km resolution.

The Golden Eagle population in Britain has been broadly stable since at least 1982 (Eaton *et al.* 2007). The latest survey in 2003 produced an estimate of 442 pairs and the pattern of abundance agrees with that presented here, showing the highest densities on the Scottish islands, declining at the eastern fringes of the range and in parts of Argyll (Eaton *et al.* 2007).

Breeding-season gains in Scotland marginally outweighed losses and are scattered throughout the Highlands with no obvious pattern. Old losses in Galloway and Argyllshire may be due to large-scale afforestation having reduced the availability of open moorland, but Whitfield *et al.* (2007) found little recent evidence of this. Elsewhere, the two main constraints on the Scottish Golden Eagle population are low food availability in western Scotland, associated with high levels of grazing and burning, and persecution in the central and eastern Highlands, mainly associated with grouse-moor management (Whitfield *et al.* 2004, 2006; Eaton *et al.* 2007). There is no evidence of competition for food with White-tailed Eagles (Whitfield *et al.* 2013).

The 11% expansion in winter range in Britain since the *1981–84 Winter Atlas* is probably a consequence of improved observer coverage. Some records reflect young birds wandering widely beyond the regular breeding areas, and those in southern England and Wales were believed to relate to escaped birds. Gains in Ireland, in both seasons, follow the recent reintroduction there.

Scottish Natural Heritage
Dualchas Nàdair na h-Alba
All of nature for all of Scotland
Nàdar air fad airson Alba air fad

Sponsored by Scottish Natural Heritage

BREEDING DISTRIBUTION *2008–11*

POSSIBLE •	<1%	3%	2%
PROBABLE ●	<1%	3%	2%
CONFIRMED ●	<1%	8%	6%
TOTAL	<1%	14%	10%

WINTER DISTRIBUTION *2007/08–2010/11*

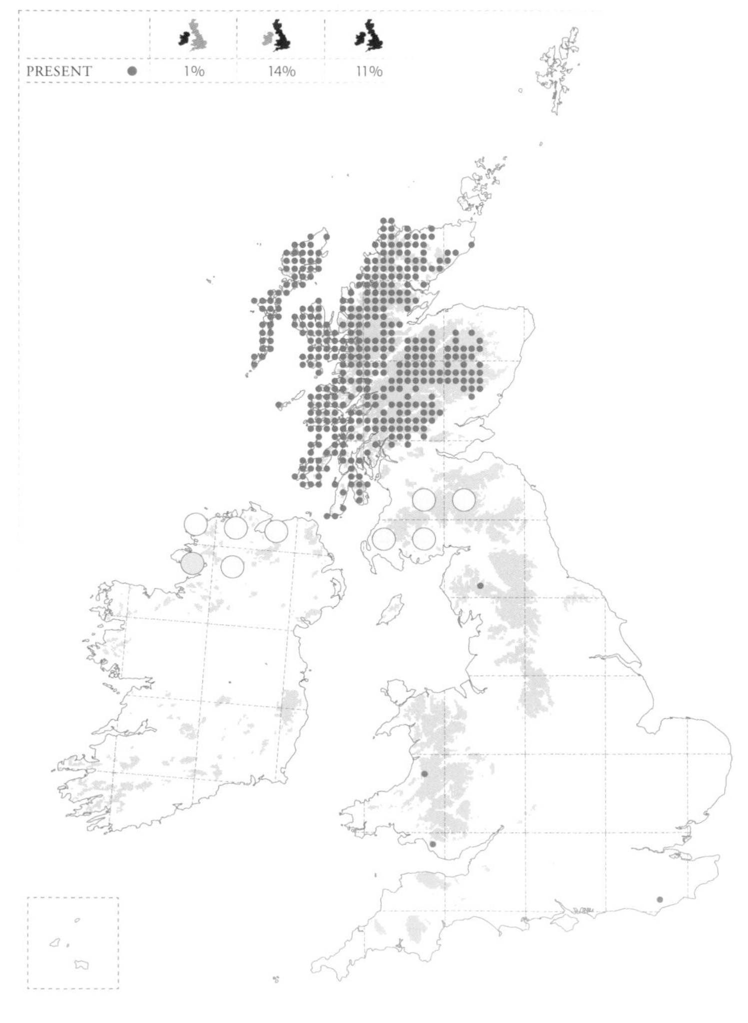

PRESENT ●	1%	14%	11%

BREEDING DISTRIBUTION CHANGE *since 1988–91*

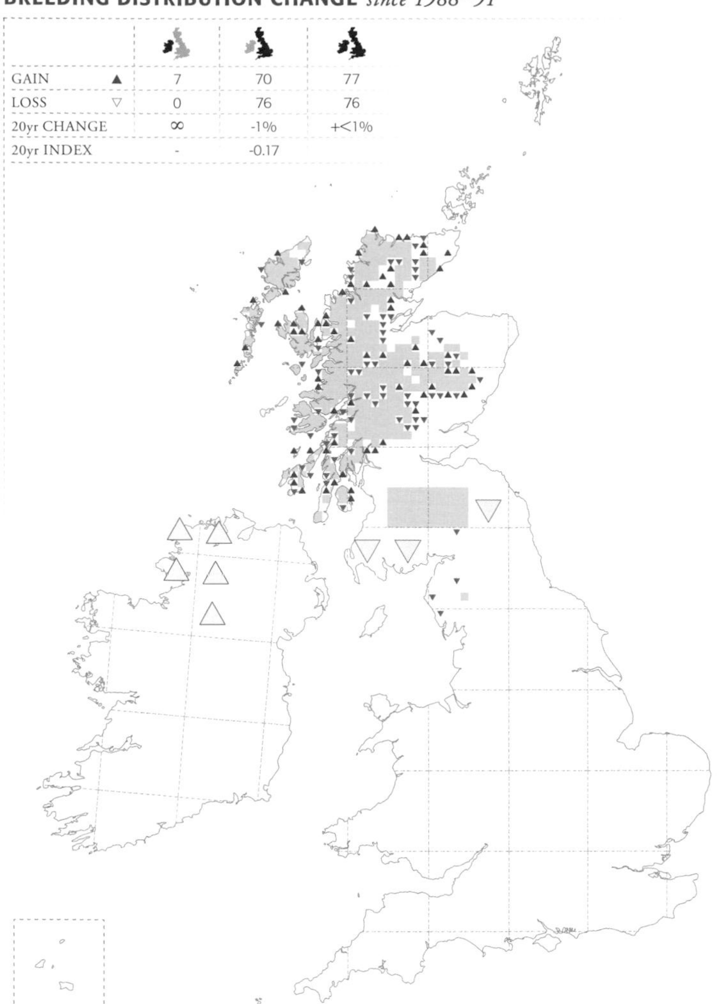

GAIN ▲	7	70	77
LOSS ▽	0	76	76
20yr CHANGE	∞	-1%	+<1%
20yr INDEX	-	-0.17	

WINTER DISTRIBUTION CHANGE *since 1981–84*

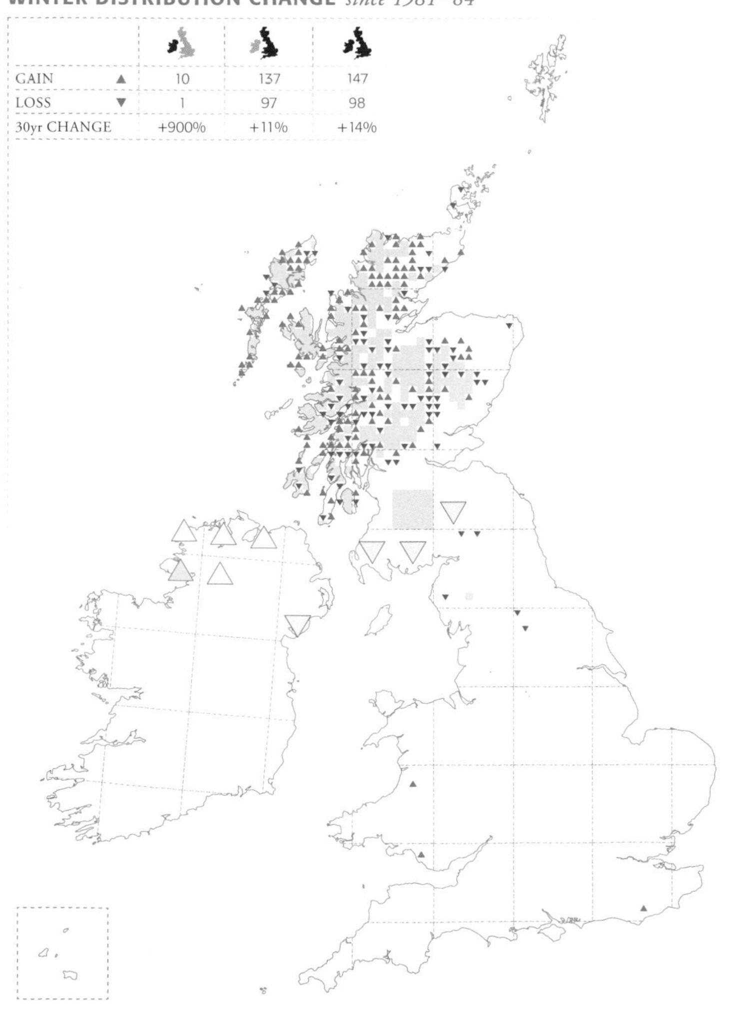

GAIN ▲	10	137	147
LOSS ▼	1	97	98
30yr CHANGE	+900%	+11%	+14%

BREEDING RELATIVE ABUNDANCE *2008–11*

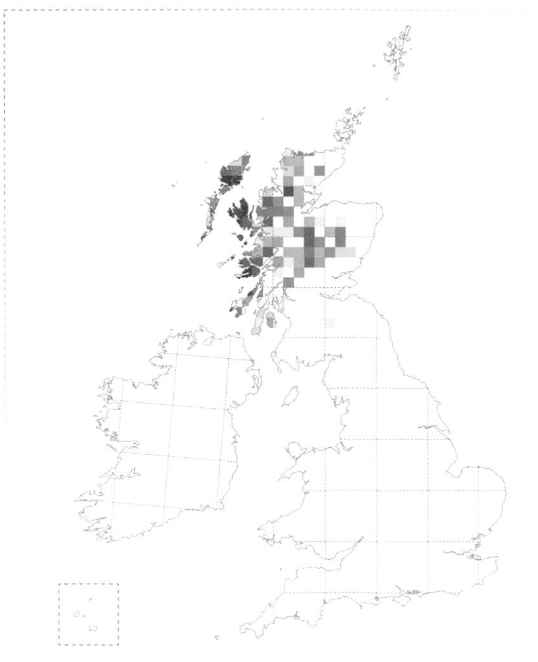

BREEDING RELATIVE ABUNDANCE CHANGE *since 1988–91*

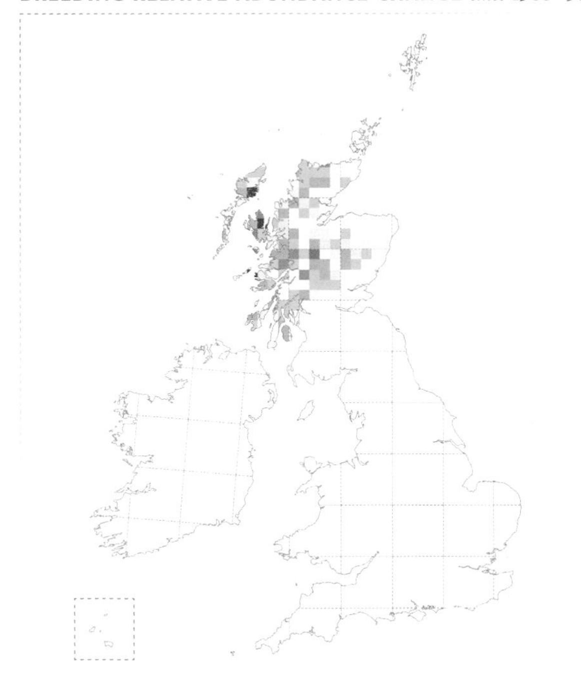

WINTER RELATIVE ABUNDANCE *2007/08–2010/11*

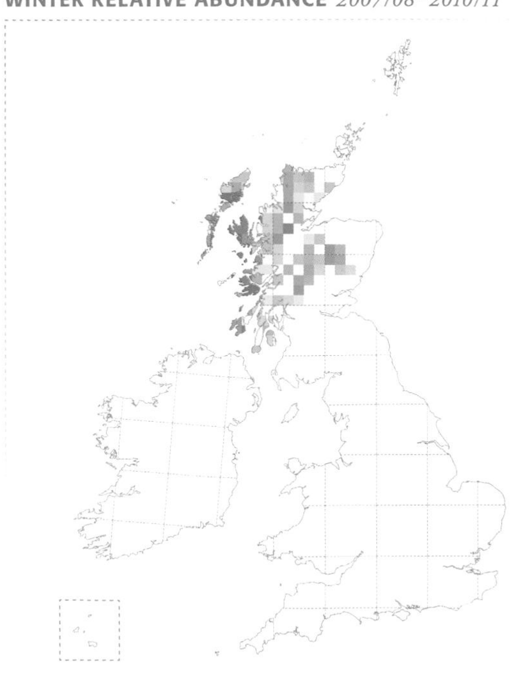

ROUGH-LEGGED BUZZARD

Buteo lagopus

BTO CODE: RF

GRAHAM CATLEY http://pewit.blogspot.co.uk

ROUGH-LEGGED BUZZARDS breed from Fennoscandia eastwards and in the autumn they migrate south or southeast. The British east coast is at the extreme northwestern fringe of their normal wintering range (Fraser & Rogers 2006a) and they are scarce but regular winter visitors there. During 2007–11 most records were from the eastern half of England, from Yorkshire to Kent, and on the Northern Isles. Like many raptors, Rough-legged Buzzards can roam widely and some clusters of occupied 10-km squares may refer to a single individual; equally, certain areas can hold several individuals, including north Norfolk, which hosted up to three individuals in the 2010/11 winter. The few records in western and southern England probably reflect onward movement by some birds.

Rough-legged Buzzards are rare in Ireland and, although a handful of birds was reported during the Atlas winter periods, none had been accepted by the relevant rarities committees at the time of publication.

The change map illustrates a predominant gain in range, with 15% more 10-km squares occupied during 2007–11 than in the *1981–84 Winter Atlas*. The only area with an overall reduction in range was eastern Scotland. The number of records of Rough-legged Buzzards reaching Britain has varied historically, possibly owing to temporary shifts in breeding range in response to changing prey abundance in the Arctic or reflecting changes of wintering grounds, associated with a more westerly autumn migration, particularly by western breeders (Scott 1978). Mean annual totals were 10 for 1974–79, 28 and 73 in the 1980s and 1990s and 35 for 2000–03 (Fraser & Rogers 2006a). Although counts of birds migrating through Falsterbo, Sweden, declined from the early 1980s to the late 1990s (Kjellén & Roos 2000), they have since recovered somewhat (Kjellén 2013), possibly indicating the trends of the population from which British winterers originate.

Some birds remained in their wintering areas into spring and were reported as non-breeders in 49 10-km squares. Rough-legged Buzzards have not bred in Britain & Ireland; the one 'loss' in the breeding season concerns a remarkable record of a female that spent 10 months, from early summer 1988 to spring 1989, on Lundy (Davis & Jones 2007) during the *1988–91 Breeding Atlas*.

Sponsored by David J. Montier

WINTER DISTRIBUTION *2007/08–2010/11*

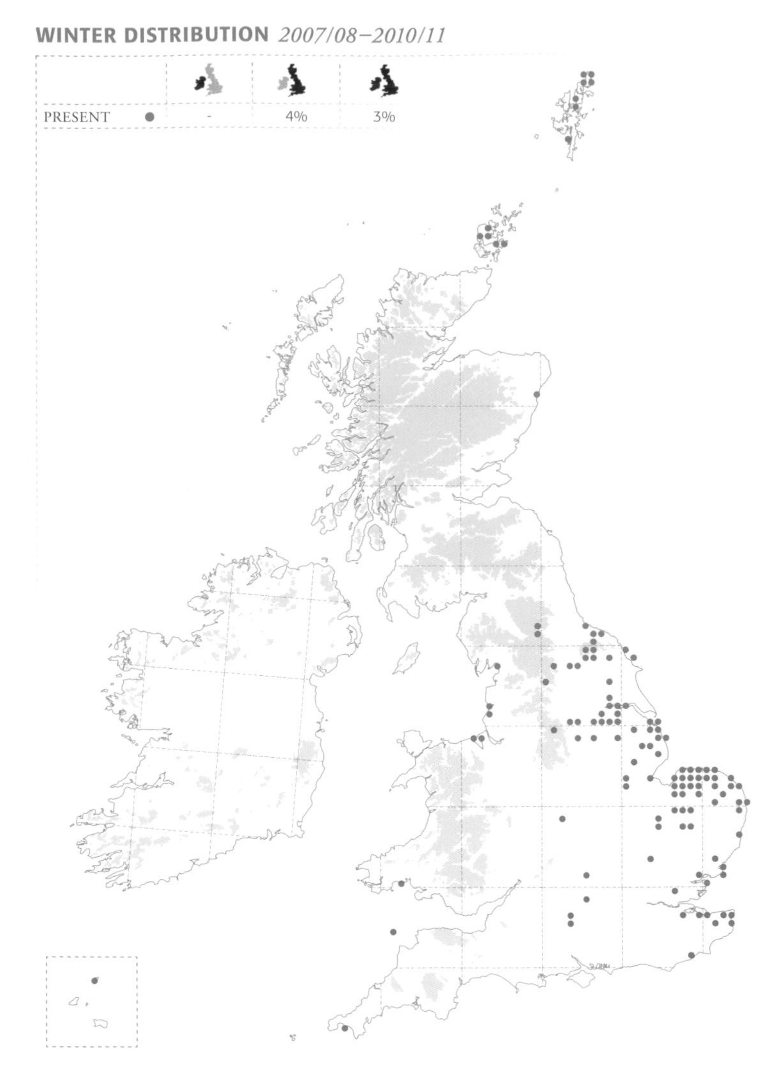

WINTER DISTRIBUTION CHANGE *since 1981–84*

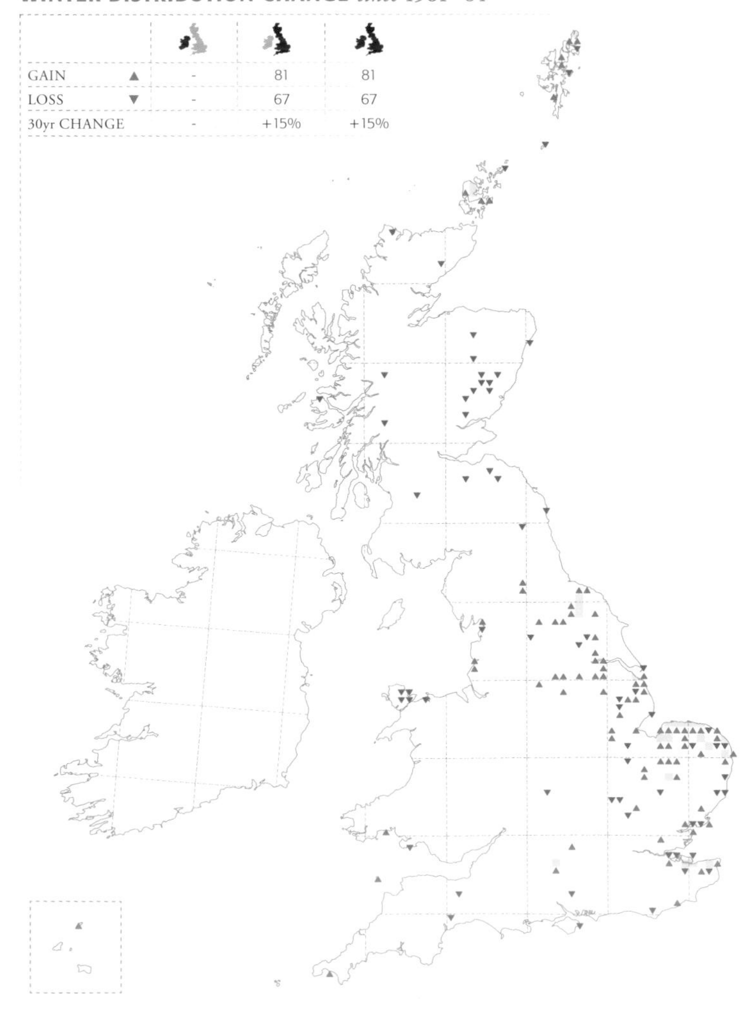

HARRIS'S HAWK

Parabuteo unicinctus

BTO CODE: HA

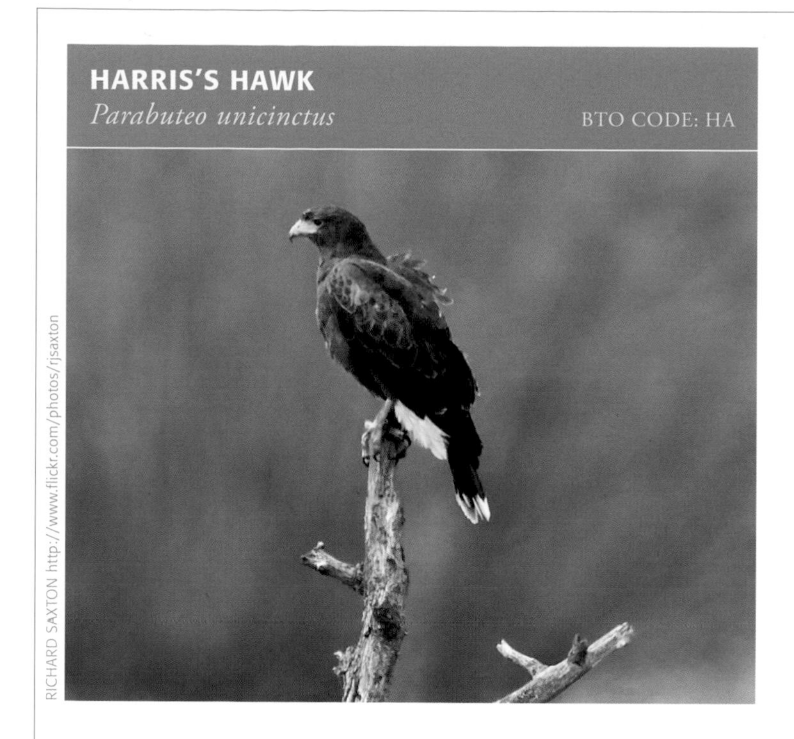

RICHARD SAXTON http://www.flickr.com/photos/rjsaxton

THE HARRIS'S HAWK is native to the Americas where it breeds between southwestern USA and central Chile and Argentina. It is a popular subject for falconry, as it is easy to train and highly social. All records of Harris's Hawks in Europe relate to escaped birds.

During this atlas there were winter records from 59 10-km squares in Britain. Occupied 10-km squares were widely scattered, extending from central Scotland to the south coast of England. *Confirmed* breeding involving pairs of Harris's Hawks was recorded in two 10-km squares in northwest England. In one square, a male that had been present for three years paired with a female in 2008 and eggs were laid in an old Magpie nest. The eggs were subsequently taken, under licence, by Natural England and hatched in captivity and the female was captured (Holling *et al.* 2011a). The other record of *confirmed* breeding involved a pair that nested in 2010 and 2011 (Lancashire & Cheshire Fauna Society 2012). Of the four 10-km squares where *probable* breeding was recorded, those in Devon and Yorkshire involved Harris's Hawks paired with Buzzards which went on to raise hybrid young. Those in Derbyshire and Kent involved one or more birds nest building. There were records from a further 36 10-km squares, all in Britain, where no evidence of breeding was submitted. It is likely that some of these could have merited at least *possible* breeding status owing to presence in potential breeding habitat. The distribution of *non-breeding* records was similar to the scattering of winter records.

Individual free-living Harris's Hawks have long been present in Britain but they have been poorly documented and no records were submitted to earlier atlases. Previous records include mixed pairings between a male Harris's Hawk and a Buzzard in Cheshire & Wirral from at least 1997 to 2001. They produced two hybrid young in 2000 and possibly also in 1998 (Holling *et al.* 2011a). The increasing interest in non-native species, and the recent breeding attempts, should prompt more attention being given to this species.

BREEDING DISTRIBUTION *2008–11*

POSSIBLE ●	-	-	-
PROBABLE ●	-	<1%	<1%
CONFIRMED ●	-	<1%	<1%
TOTAL	-	<1%	<1%

WINTER DISTRIBUTION *2007/08–2010/11*

PRESENT ●	-	2%	2%

OSPREY

Pandion haliaetus

BTO CODE: OP

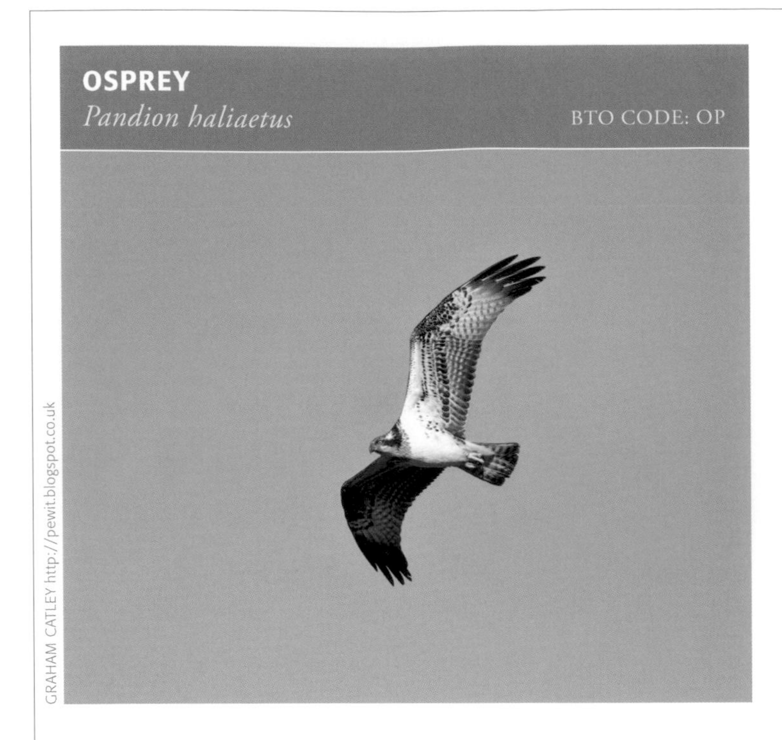

GRAHAM CATLEY http://pewit.blogspot.co.uk

OSPREYS BECAME VIRTUALLY extinct as a breeding species in Britain during the early 1900s, due to killing and egg collecting by humans (Dennis 2008). With increased protection, there has been a steady increase in numbers and range in Scotland since successful breeding began at Loch Garten, Inverness-shire, in the 1950s (*Birds of Scotland*). From 25 occupied 10-km squares and c.7 breeding pairs in the *1968–72 Breeding Atlas*, numbers increased to 73 pairs in 1991 (Ogilvie *et al.* 1994); 168 squares were occupied in the *1988–91 Breeding Atlas*, although that figure includes some where only migrants were recorded. The range has now increased by a further 68% to 286 10-km squares and in 2010 there were at least 211 breeding pairs (*RBBP 2010*).

The straths and lowlands of the eastern and central Highlands remain the stronghold for Britain's Ospreys, but a further significant proportion breeds in Tayside and central Scotland. Since the *1988–91 Breeding Atlas*, Ospreys have spread north into Caithness and west into Lochaber, and have become established in Aberdeenshire, Kincardineshire, Argyllshire, the Borders and, more recently, Dumfries & Galloway. They expanded over the border into Cumbria in 2001 and Northumberland in 2010. Birds summer annually in some other historical breeding areas in northwest Scotland where recolonisation has not yet occurred.

A translocation programme to re-establish a breeding population in England began at Rutland Water in 1996. After the release of 64 chicks, the first successful breeding took place in 2001 (Fray *et al.* 2009) and in 2010 there were five active nests from which 12 chicks fledged (Mackrill 2013). Birds from this programme have also settled in Wales, where since 2004 a pair has bred at Glaslyn, Meirionnydd, in North Wales (Holling *et al.* 2007a) and in 2011 a second pair bred at Cors Dyfi in Mid Wales.

Records of *non-breeding* Ospreys were reported in 832 10-km squares widely distributed throughout Britain and from 28 squares in Ireland, where the species remains a scarce visitor.

Most of the 24 winter-season records of Osprey were of late passage birds in November. Records in four 10-km squares in February were probably of birds returning early.

BREEDING DISTRIBUTION *2008–11*

POSSIBLE ●	-	3%	2%
PROBABLE ●	-	1%	1%
CONFIRMED ●	-	6%	4%
TOTAL	-	10%	7%

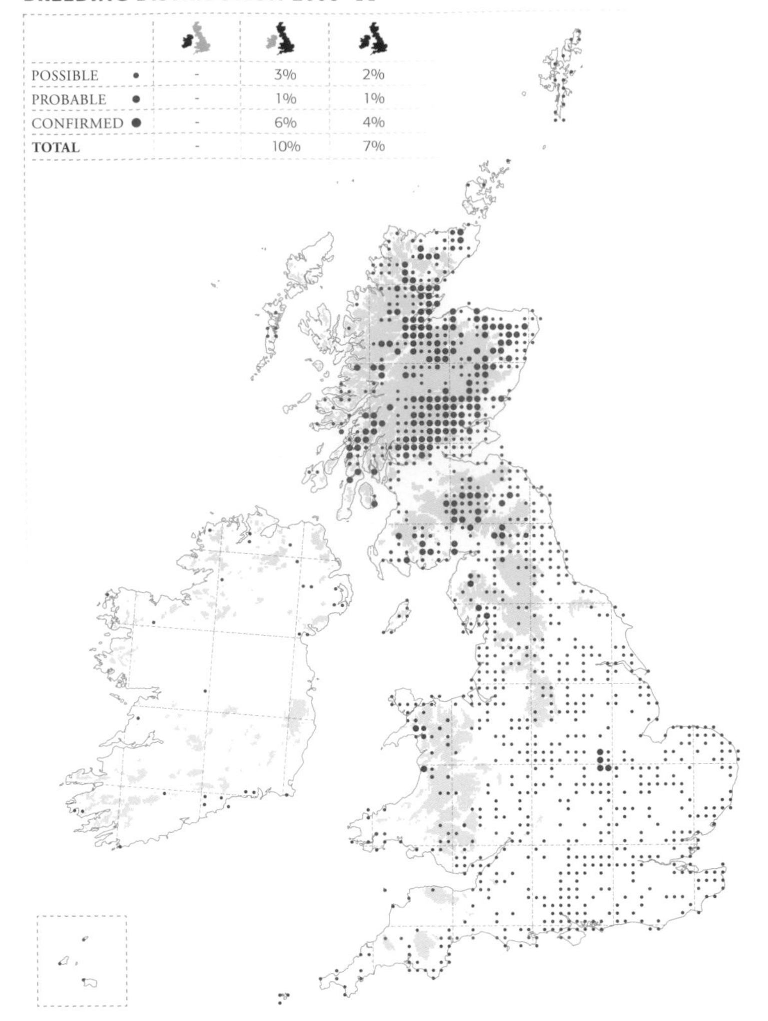

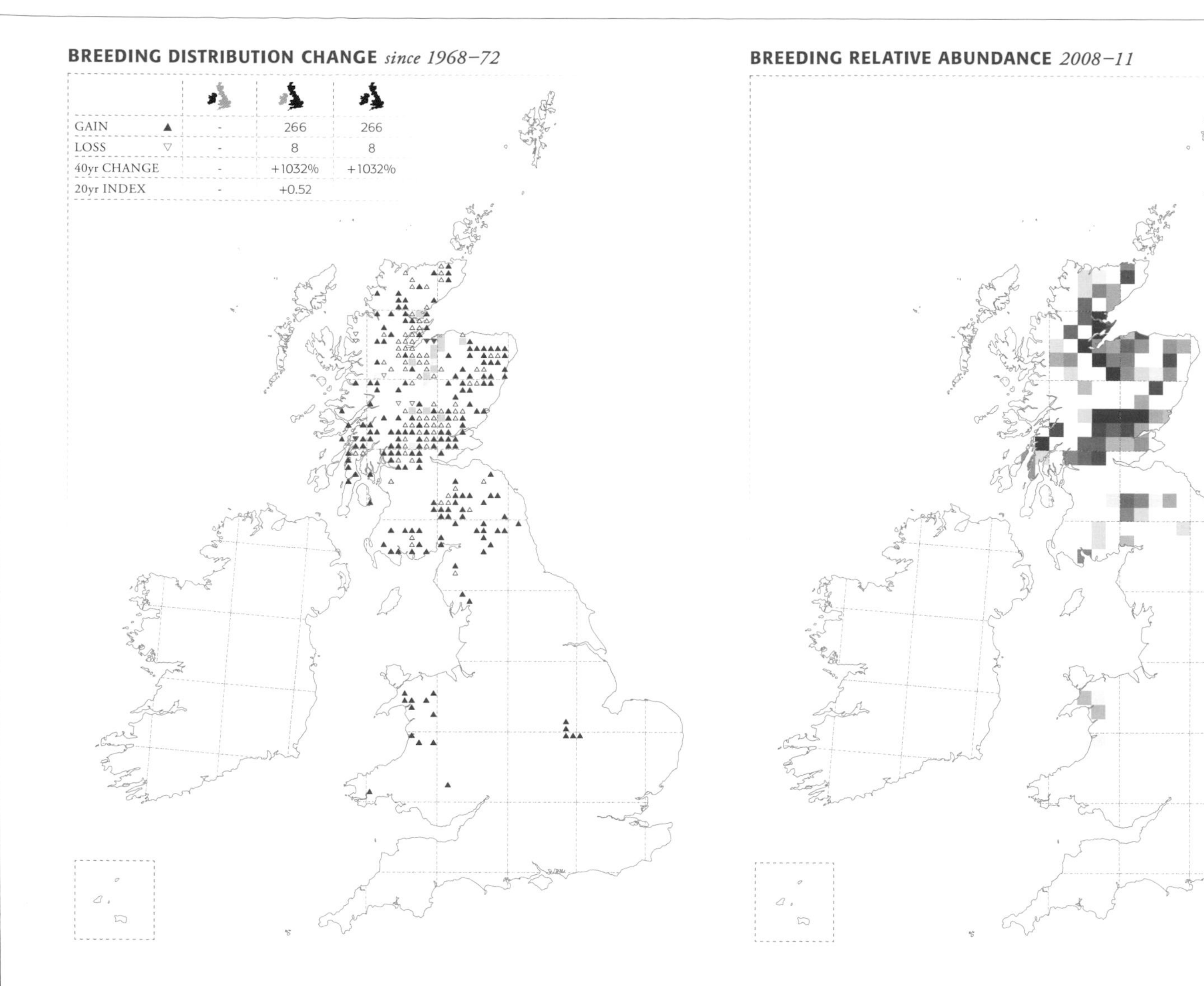

BREEDING DISTRIBUTION CHANGE *since 1968–72*

GAIN ▲	-	266	266
LOSS ▽	-	8	8
40yr CHANGE	-	+1032%	+1032%
20yr INDEX	-	+0.52	

BREEDING RELATIVE ABUNDANCE *2008–11*

GARTH PEACOCK www.garthpeacock.co.uk

◀ There are many large reservoirs, gravel pits and river systems in central England with nearby mature trees that would provide suitable habitat for expansion from the reintroduced population.

KESTREL

Falco tinnunculus

BTO CODE: K.

PAUL HILLION

THE KESTREL IS one of the most widespread and abundant raptors in Britain & Ireland, being present in almost 90% of 10-km squares in both winter and the breeding season. It is absent only from small areas of Northern Ireland and southwest and central Wales, from certain upland areas of western Scotland and mostly from Shetland. Based on this atlas, in Britain the Kestrel has lost its position as the most widespread raptor to the Buzzard, but in Ireland it currently retains that accolade. Densities are highest in central and eastern England and southwest Ireland; elsewhere, especially in southwest England, Wales and western Scotland, Kestrels are noticeably scarcer.

The breeding range has contracted by 6% since the *1968–72 Breeding Atlas*, with losses occurring in southwest and central Wales, in western Scotland and sparingly through the midlands and north of Ireland. These relatively small losses conceal a larger population decline: numbers in the UK declined by 32% during 1995–2010 (*BBS Report 2011*), part of an overall decline of 44% since 1970 (*SUKB 2012*). In Ireland, numbers fell by 34% during 1998–2010 (*CBS Trend 2010*). Kestrels can breed at densities of 10–50 pairs per 10-km square (Village 1990; Clements 2008) so a substantial decline can take place before 10-km squares appear as losses. The abundance change map shows that Kestrels have declined in abundance in many parts of Britain and Ireland, including areas where no observable 10-km square losses have taken place. Abundance increases are confined to southwest Ireland and parts of central England.

The winter distribution, which is very similar to that during the breeding season, has expanded by 7% since the *1981–84 Winter Atlas*. A large part of the expansion took place in Ireland, and may be an artefact of improved winter coverage.

Causes of decline are unclear and possibly complex, although agricultural intensification, the impact of second-generation rodenticides, damping of the vole cycles (Cornulier *et al.* 2013) and competition with other raptors have all been proposed. Petty *et al.* (2003) gave good evidence that direct predation of Kestrels by Goshawks had contributed to the decline in Kielder Forest, Northumberland, and Shrubb (2004) considered this likely to be occurring also in Wales.

BREEDING DISTRIBUTION *2008–11*

POSSIBLE ●	35%	12%	18%
PROBABLE ●	18%	12%	13%
CONFIRMED ●	35%	61%	55%
TOTAL	88%	85%	86%

WINTER DISTRIBUTION *2007/08–2010/11*

BREEDING DISTRIBUTION CHANGE *since 1968–72*

GAIN ▲	39	74	113
LOSS ▽	93	218	311
40yr CHANGE	-6%	-6%	-6%
20yr INDEX	+0.05	-0.10	

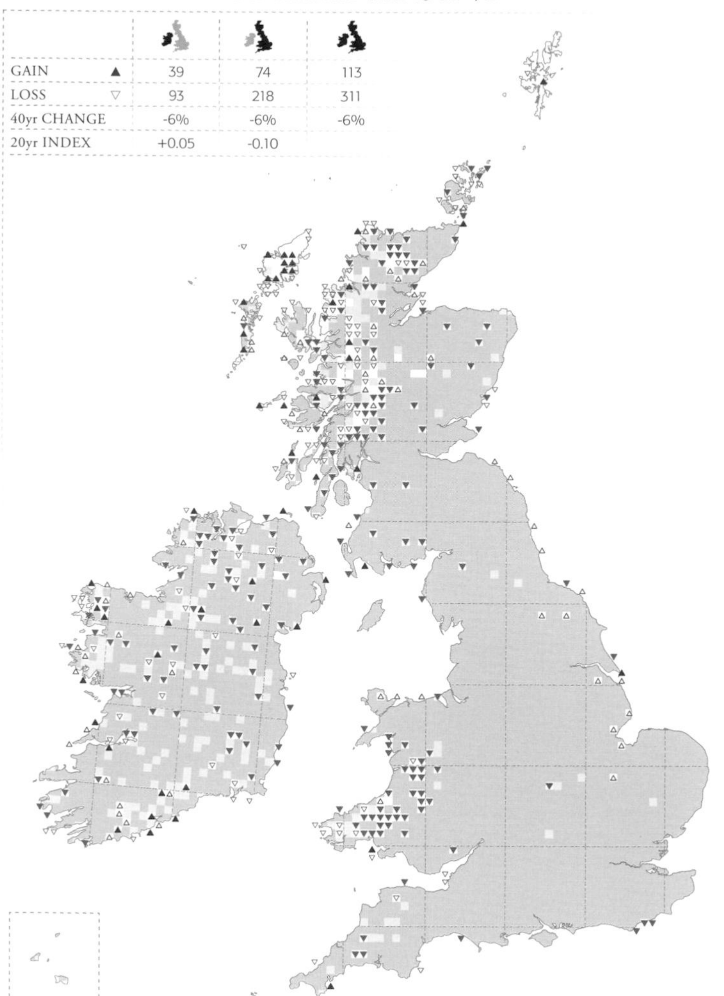

WINTER DISTRIBUTION CHANGE *since 1981–84*

GAIN ▲	216	186	402
LOSS ▼	79	91	170
30yr CHANGE	+19%	+4%	+7%

BREEDING RELATIVE ABUNDANCE *2008–11*

BREEDING RELATIVE ABUNDANCE CHANGE *since 1988–91*

WINTER RELATIVE ABUNDANCE *2007/08–2010/11*

MERLIN

Falco columbarius

BTO CODE: ML

MICK DURHAM www.wildlife-photographer.net

THE MERLIN'S BREEDING preference for uplands, and especially heather-dominated moorland (Rebecca 2011), is clearly evident on the distribution map. The abundance map shows that the Scottish islands, northern and eastern Highlands, north Pennines and northwest Ireland held the highest densities. Scotland held 65% of the British breeding population in 2008 (Ewing *et al.* 2011).

Apparent patterns of range change since the *1968–72 Breeding Atlas* differ between Britain (11% expansion) and Ireland (40% contraction), whereas since the *1988–91 Breeding Atlas* changes are similar (4% and 6% contractions respectively). All areas show a mix of recent gains and losses but the species is easily overlooked (Lusby *et al.* 2011), especially where nesting in plantations (Norriss *et al.* 2010).

The Merlin breeding population in Britain has been the subject of three national surveys. Numbers increased from 1983–84 to 1993–94 but figures from 2008 suggest a subsequent 13% decline, although this is not statistically significant (Ewing *et al.* 2011). Repeated counts from areas with good coverage, however, confirm large declines in several upland regions of England, Wales and northeast Scotland (Ewing *et al.* 2011). Despite these recent declines, breeding numbers in 2008 were significantly larger than those in 1983–84, and numbers may have been lower still in 1968–72 when depressed by organochlorine pesticides. Possible reasons for the recent declines include changes in prey availability and in nesting habitat through more frequent heather burning and the maturing of plantations (Ewing *et al.* 2011). Northern Ireland was covered for the first time in the 2008 survey and numbers and trends in Ireland are poorly known (Lusby *et al.* 2011).

Winter numbers are supplemented by immigrants from Iceland (*Migration Atlas*). In Britain, wintering Merlins largely avoid the higher uplands of Scotland and Wales, with most occupied squares being in lower-lying habitats. In contrast, in Ireland many of the breeding areas are occupied in winter, in addition to parts of the midlands. The reasons underlying the 52% winter range expansion are unclear, although improved coverage in Ireland may partly explain the greater apparent expansion there, as may a reported breeding population increase in Iceland (O. Nielsen pers comm).

Sponsored by The Moorland Association

BREEDING DISTRIBUTION *2008–11*

POSSIBLE •	8%	7%	7%
PROBABLE ●	2%	3%	3%
CONFIRMED ●	4%	13%	11%
TOTAL	14%	23%	21%

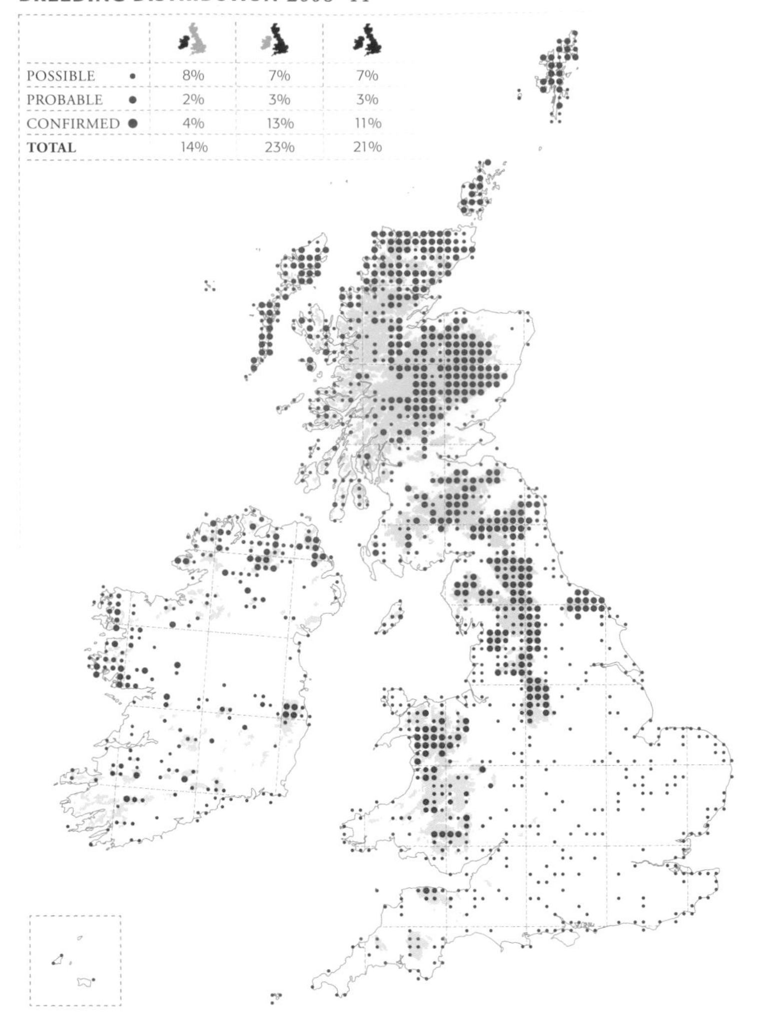

WINTER DISTRIBUTION *2007/08–2010/11*

PRESENT ●	39%	55%	51%

BREEDING DISTRIBUTION CHANGE *since 1968–72*

	Ireland	Britain	Britain & Ireland
GAIN ▲	89	310	399
LOSS ▽	186	245	431
40yr CHANGE	-40%	+11%	-4%
20yr INDEX	-0.09	-0.18	

WINTER DISTRIBUTION CHANGE *since 1981–84*

	Ireland	Britain	Britain & Ireland
GAIN ▲	269	773	1042
LOSS ▼	103	305	408
30yr CHANGE	+80%	+46%	+52%

BREEDING RELATIVE ABUNDANCE *2008–11*

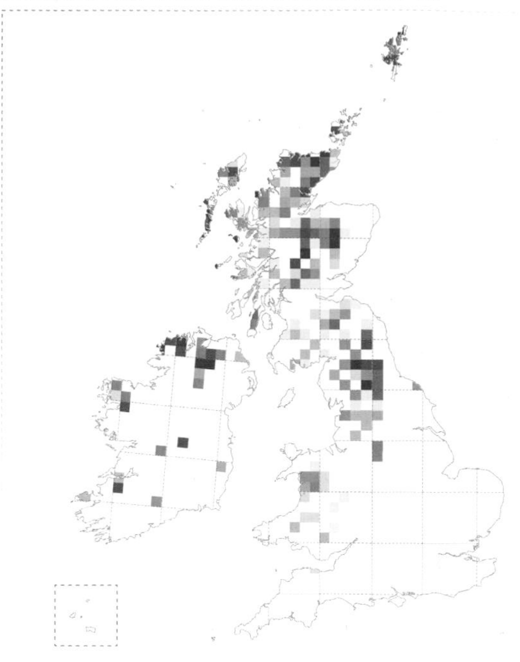

BREEDING RELATIVE ABUNDANCE CHANGE *since 1988–91*

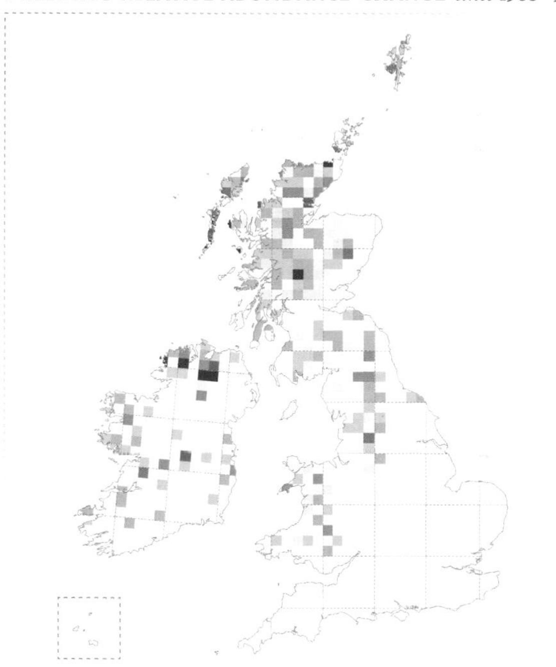

WINTER RELATIVE ABUNDANCE *2007/08–2010/11*

HOBBY
Falco subbuteo

BTO CODE: HY

MIKE RAE www.mikerae.com

THE HOBBY IS a breeding migrant occupying a wide range of landscapes throughout southern Britain. Numbers can be difficult to assess, particularly in farmland areas seldom visited by birdwatchers, and its wide-ranging behaviour makes assigning breeding evidence difficult. It is a scarce migrant to Ireland and does not breed (*Birds in Ireland*), although the number of records has increased and the presence of adults in the breeding season may be a precursor to future colonisation (*IRBBP 2010*).

Hobbies have undergone a large-scale expansion in range, consolidating their range in southern England and spreading north, west and east, bringing about a 16% population increase during 1995–2010 (*BBS Report 2011*). They now occupy four times as many 10-km squares as in the *1968–72 Breeding Atlas* and are widespread south of a line from the Humber to the Mersey, with the exception of west Wales and Cornwall where they remain scarce. North of this line, they are thinly distributed as far north as southern Northumberland. There is a small cluster of squares with breeding evidence in the Scottish Highlands, where birds have been recorded at various times since the 1970s but breeding was not proven until 2001 (*Birds of Scotland*). It is possible that sporadic nesting elsewhere in Scotland may have been overlooked. The highest densities of Hobbies occurred in eastern England, with notable hotspots apparent in the Fens of East Anglia, the Norfolk Broads, North Kent Marshes and the New Forest.

Population estimates range from a minimum of 993 pairs reported to RBBP (*RBBP 2010*) to perhaps a more realistic estimate of approximately 2,200 pairs based on breeding densities and the availability of suitable breeding habitat (Clements 2001).

Reasons for the increases are only speculative. They may be linked to increases in the Hobby's dragonfly prey supplies associated with a greater number of gravel pits and reservoirs (Prince & Clarke 1993). Increased availability of prey was thought to be the main reason for good breeding success in a study in lowland farmland in Derbyshire (Messenger & Roome 2007). Climate change could also have benefited this species and its prey (Burton 1995).

BREEDING DISTRIBUTION *2008–11*

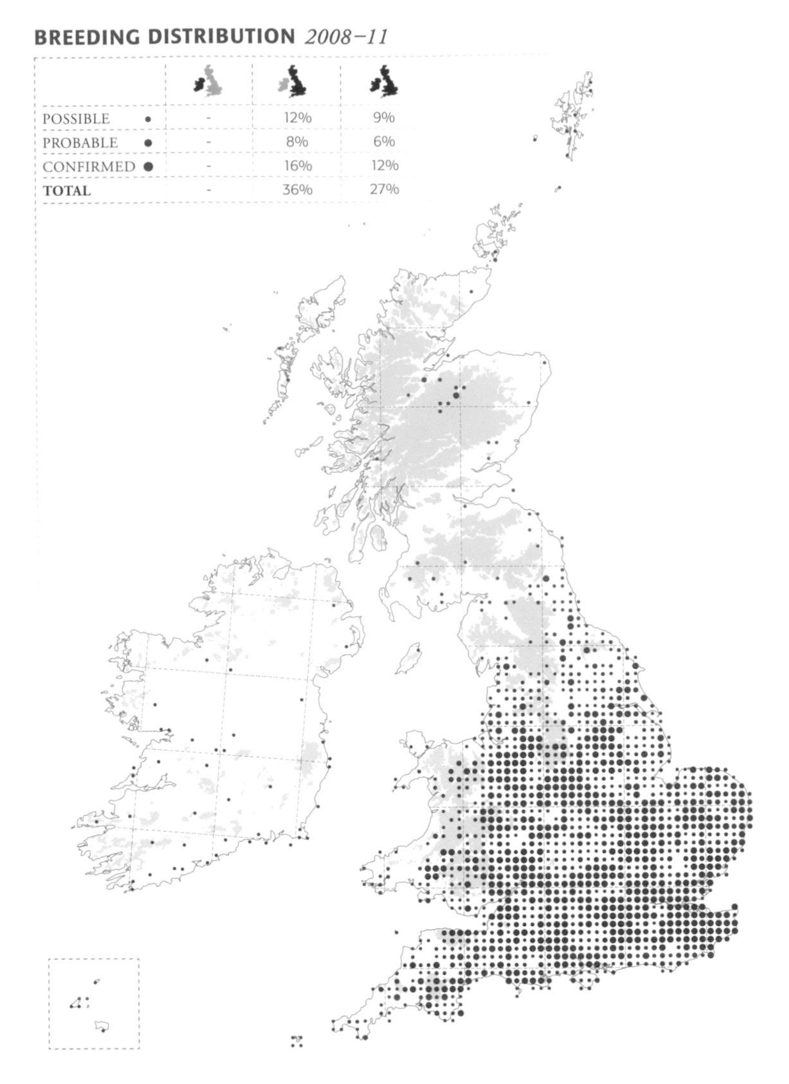

	Ireland	Britain	Britain & Ireland
POSSIBLE •	-	12%	9%
PROBABLE ●	-	8%	6%
CONFIRMED ●	-	16%	12%
TOTAL	-	36%	27%

BREEDING RELATIVE ABUNDANCE CHANGE *since 1988–91*

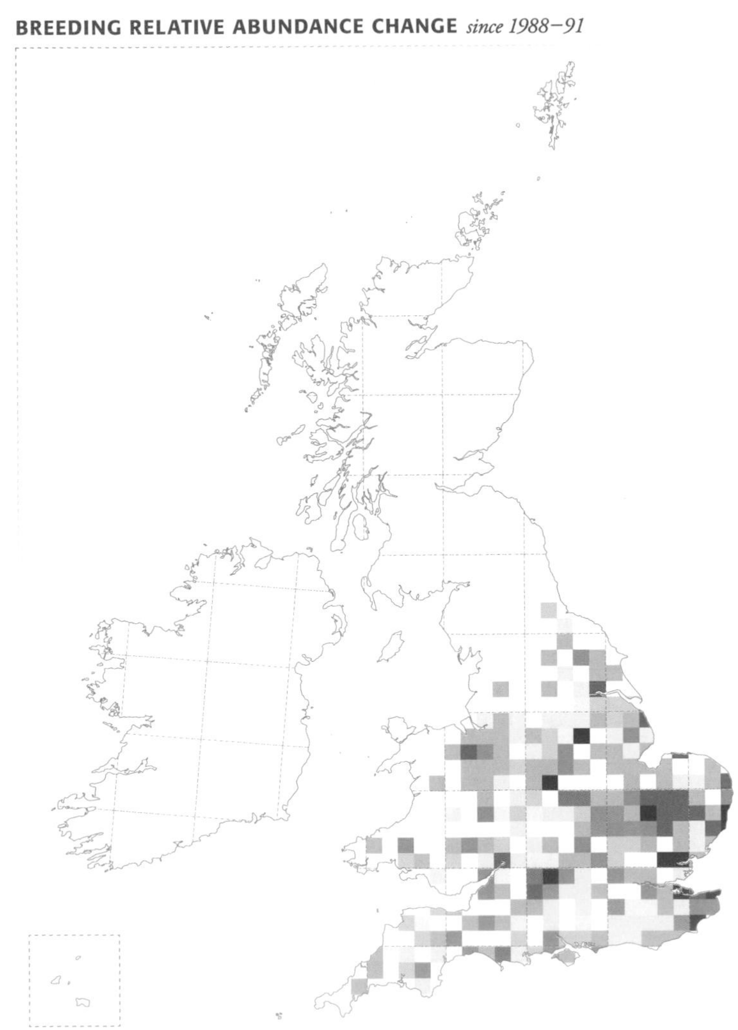

Sponsored by The Royston Scroggs Memorial Fund

BREEDING DISTRIBUTION CHANGE *since 1968–72*

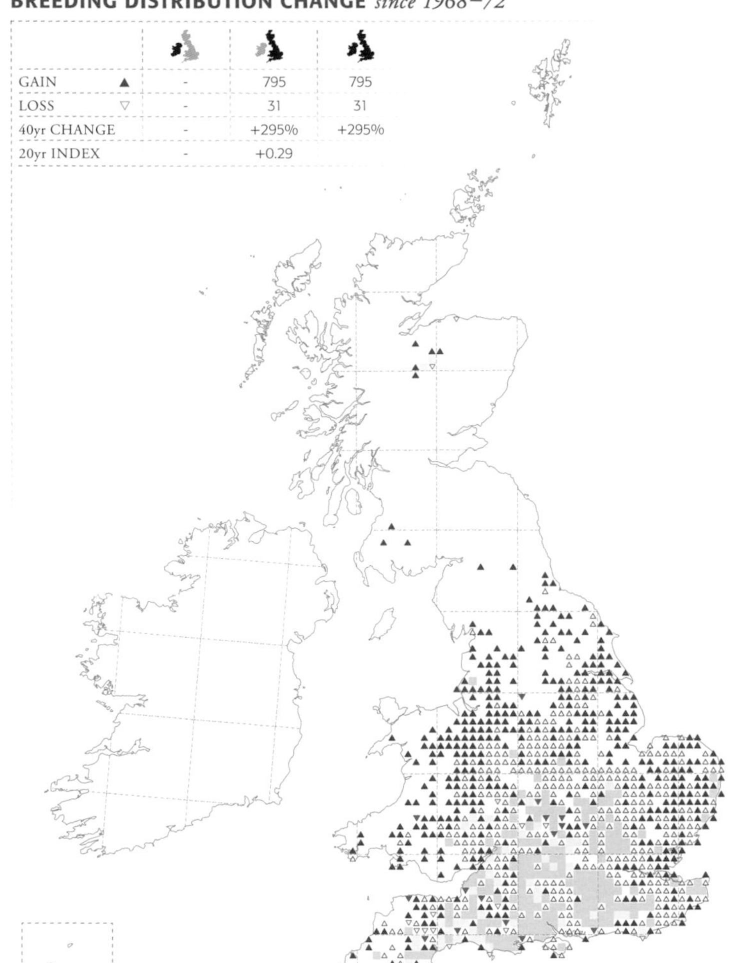

GAIN ▲	-	795	795
LOSS ▽	-	31	31
40yr CHANGE	-	+295%	+295%
20yr INDEX	-	+0.29	

BREEDING RELATIVE ABUNDANCE *2008–11*

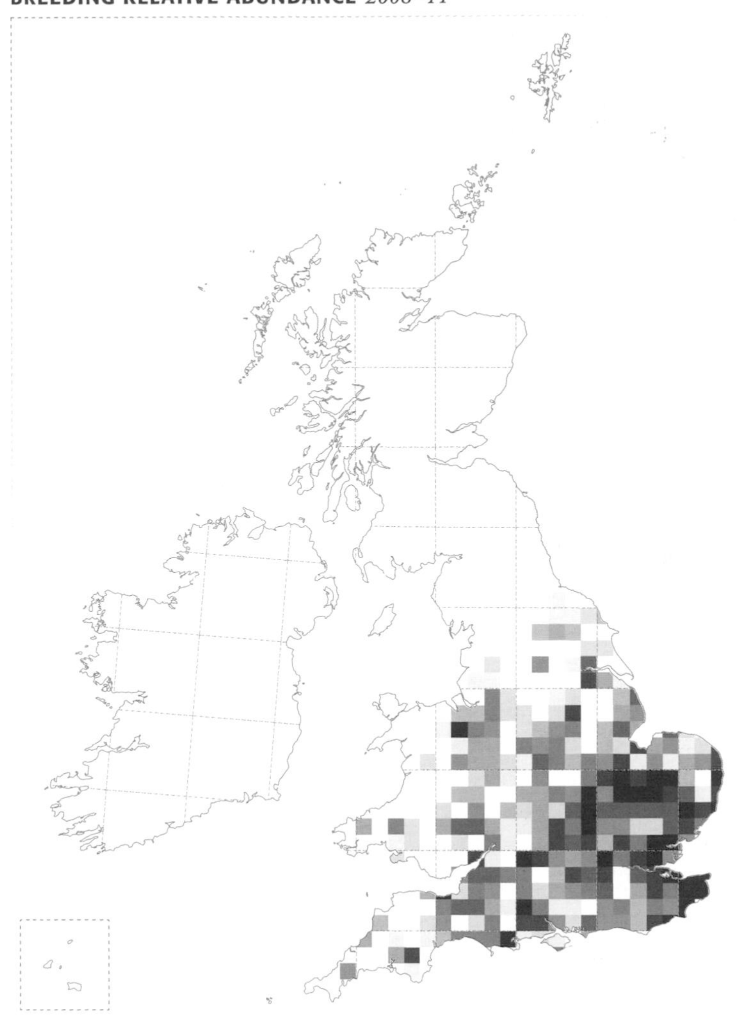

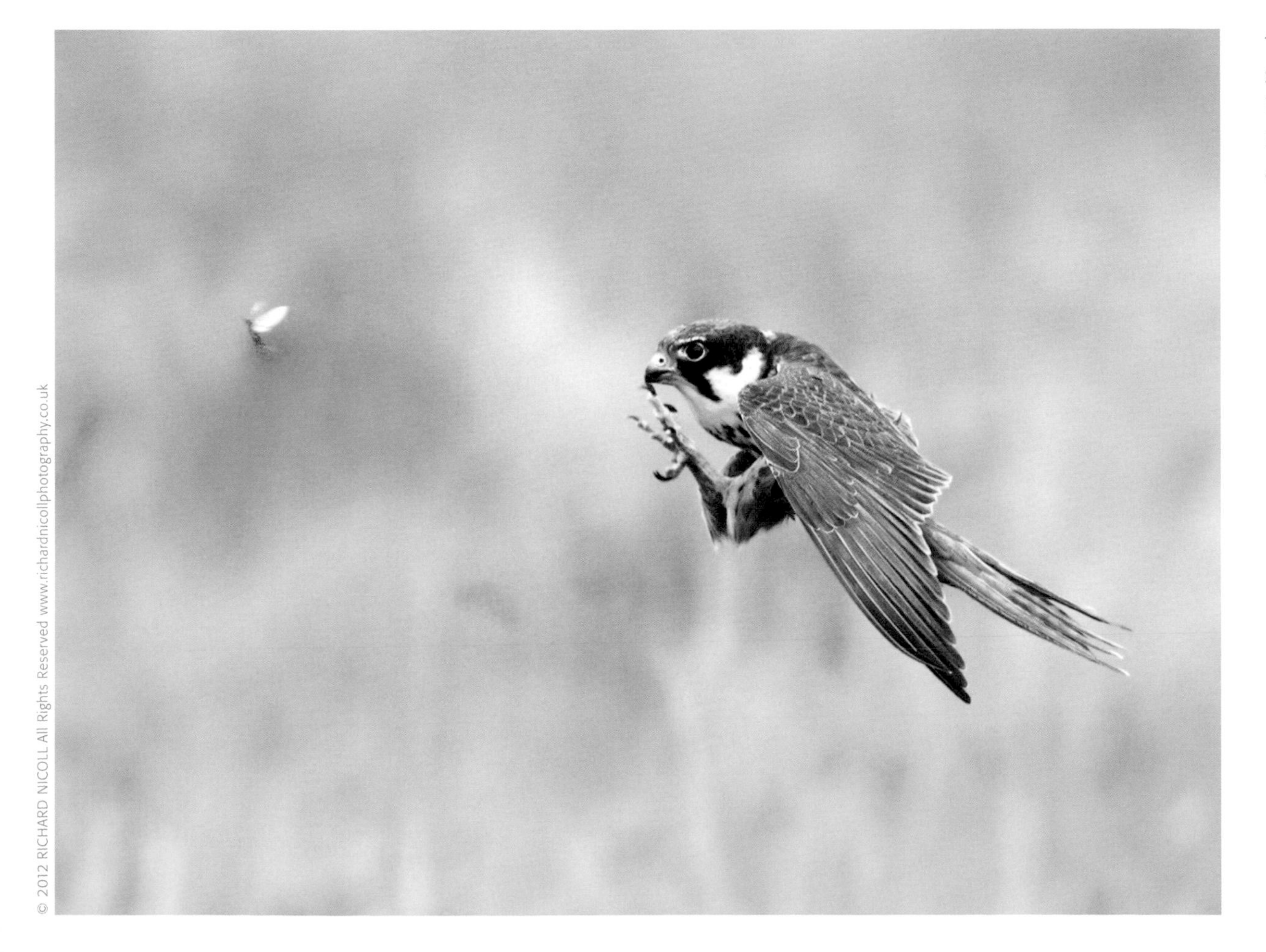

◀ **Hobbies have greatly expanded their breeding range in Britain but have yet to colonise Ireland.**

PEREGRINE
Falco peregrinus

BTO CODE: PE

MICK DURHAM www.wildlife-photographer.net

FOLLOWING THE BAN on organochlorine pesticide usage in agriculture, there has been a remarkable recovery in the Peregrine population and the species is now widespread in both Britain and Ireland. It breeds in upland and coastal areas with suitable cliffs, and across much of the lowlands, where it utilises quarries and man-made structures (Ratcliffe 2003). The species is still a localised breeder in eastern England and the interior of Ireland. The winter range in Britain has doubled since the *1981–84 Winter Atlas*, with gains primarily across lowland England. In Ireland a 51% winter range expansion has seen birds spread from the coast to the midlands.

Overall, the breeding distribution change map reveals a 200% range expansion since 1968–72 and a 40% expansion since 1988–91, with gains taking place in both islands. In Ireland the gains are scattered throughout, though especially concentrated in the southwest. The Channel Islands have also been colonised since the *1988–91 Breeding Atlas*. Breeding populations are now at an all-time high in both the UK (Dixon *et al.* 2008; Banks *et al.* 2010) and the Republic of Ireland (Madden *et al.* 2009).

Marked variation in population trends has been demonstrated among British regions, corroborating the divergent patterns of gain versus loss now seen in the breeding change map. Whilst the pattern is generally one of population recovery and expansion around the coast and into the lowlands and urban areas of England, declines had occurred in parts of Scotland by 2002 (Banks *et al.* 2010) and local extinctions are now apparent in western and northern Scotland. Indeed, compared to 1,104 10-km squares showing range gain in Britain since the *1968–72 Breeding Atlas*, there were 158 showing range loss and all but six were in Scotland. Gains are generally considered to be associated with reduced illegal killing, abundant prey and increased tolerance of humans, whereas losses are thought to be associated with local reductions in food availability, habitat degradation and in some instances continuing illegal killing (Ratcliffe 2003; Banks *et al.* 2010; Amar *et al.* 2012). In Ireland, coastal populations have remained stable since the early 1990s, with the major changes being at inland quarries and man-made sites (Madden *et al.* 2009).

An RWE company

Sponsored by Fawley Power Station & the Environment Management Dept, RWE npower

BREEDING DISTRIBUTION *2008–11*

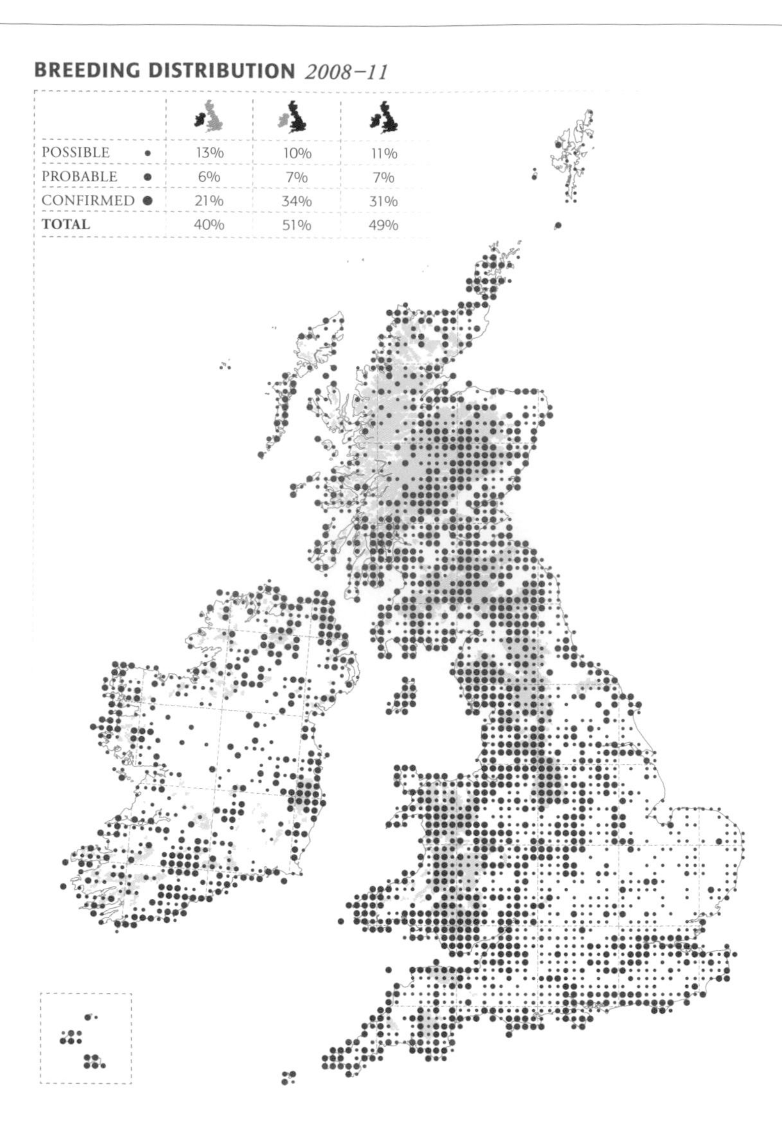

	Ireland	Britain	Britain & Ireland
POSSIBLE •	13%	10%	11%
PROBABLE •	6%	7%	7%
CONFIRMED •	21%	34%	31%
TOTAL	40%	51%	49%

WINTER DISTRIBUTION *2007/08–2010/11*

	Ireland	Britain	Britain & Ireland
PRESENT •	44%	75%	67%

BREEDING DISTRIBUTION CHANGE *since 1968–72*

GAIN ▲	327	1104	1431
LOSS ▽	26	158	184
40yr CHANGE	+276%	+184%	+200%
20yr INDEX	+0.03	-0.05	

WINTER DISTRIBUTION CHANGE *since 1981–84*

GAIN ▲	257	1265	1522
LOSS ▼	116	207	323
30yr CHANGE	+51%	+105%	+93%

BREEDING RELATIVE ABUNDANCE *2008–11*

BREEDING RELATIVE ABUNDANCE CHANGE *since 1988–91*

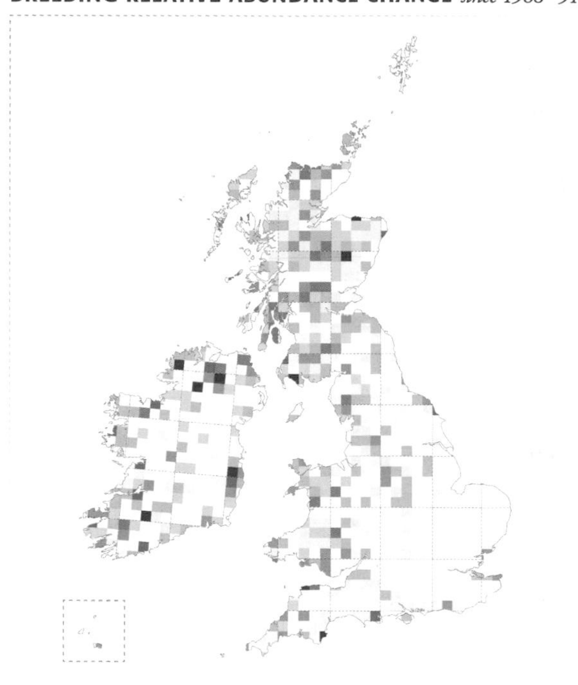

WINTER RELATIVE ABUNDANCE *2007/08–2010/11*

WATER RAIL

Rallus aquaticus BTO CODE: WA

VIC FROOME

THE WATER RAIL is among the most elusive of Britain & Ireland's waterbirds, being skulking and secretive when not calling. Consequently, these maps may under-represent the true range of the species and apparent changes could reflect chance patterns of detections in different periods rather than true changes in square occupancy.

Breeding Water Rails are patchily distributed throughout the lowlands of Britain and Ireland. The winter distribution includes the same general areas but the species is far more widespread, being detected in almost twice as many 10-km squares in Britain than during the breeding season. This may be because resident breeders are supplemented by migrants from continental Europe (*Migration Atlas*). This certainly accounts for the large increase in occupancy in Shetland (*Birds of Scotland*). Also cold winter weather may force birds out of frozen marshes to seek food in more-open sites where they are more readily detected.

The abundance map for the breeding season suggests strongholds in parts of Ireland, East Anglia and southern Scotland. In winter, hotspots are more widely spread and often coastal. However, these results should be interpreted with caution, being based on sparse data owing to the low likelihood of detecting Water Rails during TTVs.

The breeding-season change map reports a 6% range expansion in Britain contrasting with a 17% contraction in Ireland since the *1968–72 Breeding Atlas*. Throughout Ireland and England, gains are interspersed with losses, perhaps indicative of coverage and detection issues. Noticeably, from Northumberland northwards the pattern is almost exclusively one of gain. It is unclear to what extent the species has been overlooked in the past in Scotland or whether changes are real, perhaps reflecting a response to mild winters prior to this atlas (*Birds of Scotland*). In winter, gains predominate, giving an overall 54% range expansion since the *1981–84 Winter Atlas*. Gains may partly reflect improved coverage and perhaps increased population size following a run of relatively mild winters in the 1990s and 2000s. Unfortunately, no trend information is available for either season and the species has recently been added to those covered by RBBP (Holling *et al.* 2009) in a bid to increase knowledge of this under-recorded species (Murray 2013).

Sponsored by Robert and Linda Maddock

BREEDING DISTRIBUTION *2008–11*

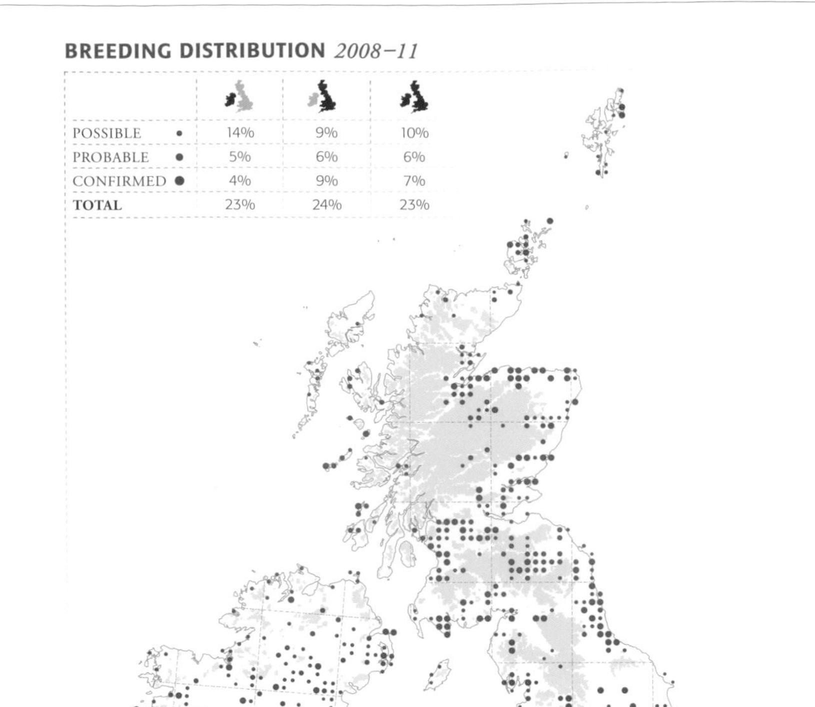

POSSIBLE •	14%	9%	10%
PROBABLE •	5%	6%	6%
CONFIRMED •	4%	9%	7%
TOTAL	23%	24%	23%

WINTER DISTRIBUTION *2007/08–2010/11*

PRESENT •	28%	45%	40%

BREEDING DISTRIBUTION CHANGE *since 1968–72*

GAIN ▲	127	378	505
LOSS ▽	174	342	516
40yr CHANGE	-17%	+6%	-1%
20yr INDEX	-0.07	-0.04	

BREEDING RELATIVE ABUNDANCE *2008–11*

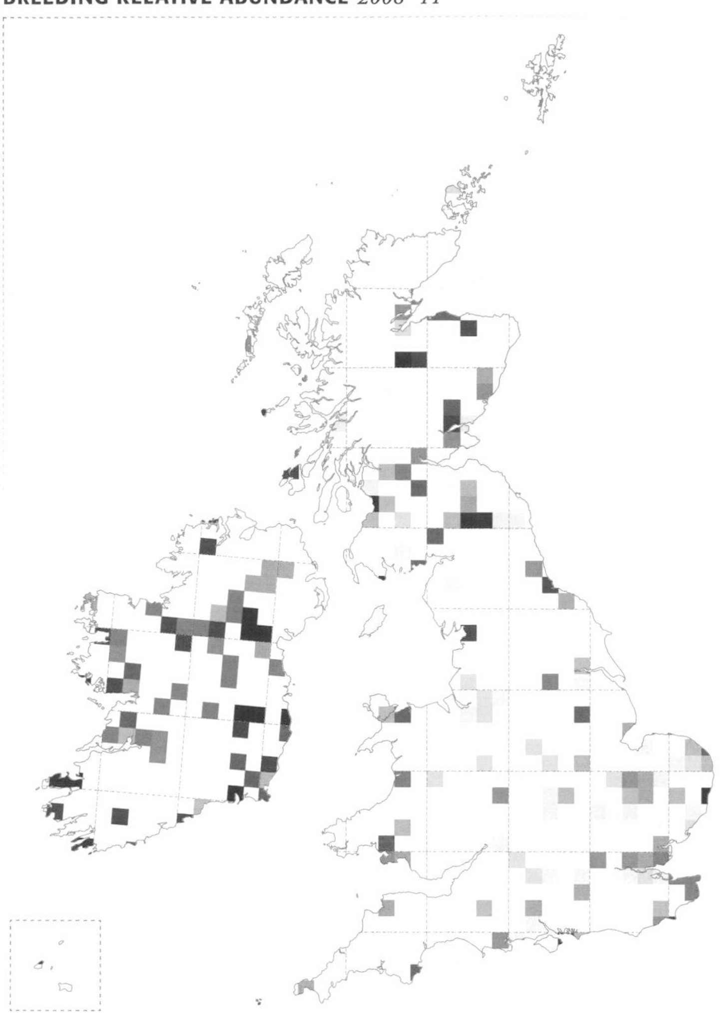

WINTER DISTRIBUTION CHANGE *since 1981–84*

GAIN ▲	182	592	774
LOSS ▼	82	162	244
30yr CHANGE	+59%	+53%	+54%

WINTER RELATIVE ABUNDANCE *2007/08–2010/11*

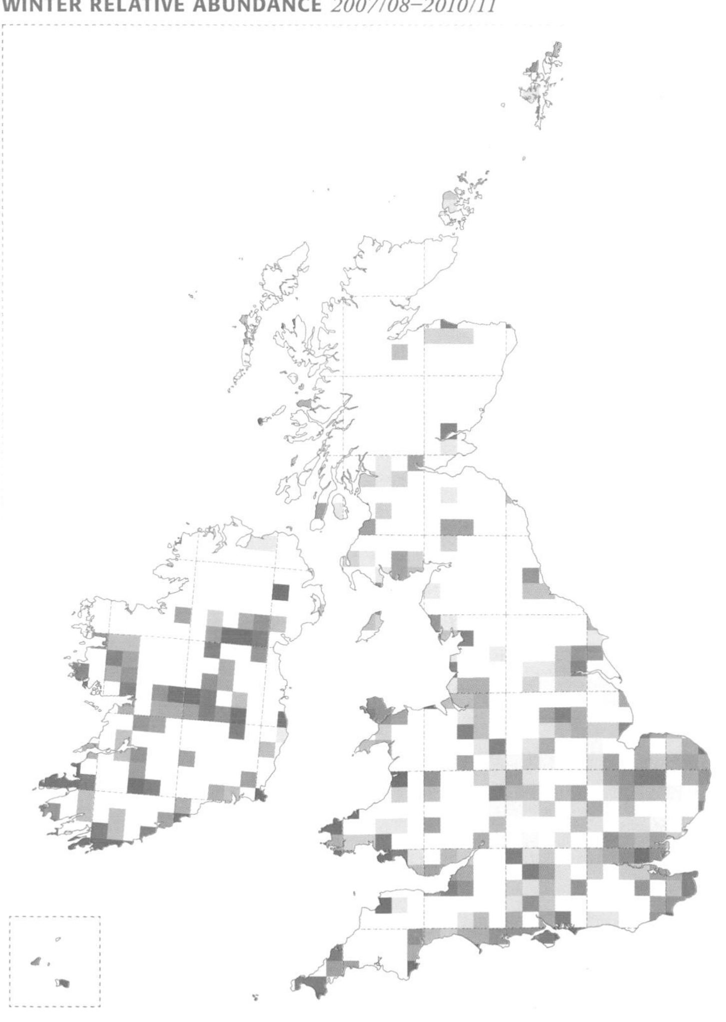

SPOTTED CRAKE
Porzana porzana

BTO CODE: AK

JIM ALMOND www.shropshirebirder.co.uk

THE SPOTTED CRAKE is a very rare, but regular breeding migrant to Britain and Ireland, although its status is very difficult to confirm for several reasons. Individuals are highly secretive and difficult to observe, they favour wet, herb-rich grassland and swamps, which can present access difficulties, and they are usually detected only by the male's nocturnal song (Gilbert 2002; Stroud *et al.* 2012b).

Nevertheless, breeding was *confirmed* in two 10-km squares in southwest England and Yorkshire and *probable* breeding was reported from Somerset, East Anglia, Yorkshire, the Inner and Outer Hebrides and the Northern Isles. *Possible* breeding birds were recorded in a further ten 10-km squares in England, 12 in Scotland and two in Ireland. Records lacking breeding evidence were received from ten 10-km squares, some of which could have involved birds in suitable breeding habitat.

Estimating abundance and trends for Spotted Crakes is beset by problems, partly because many records are not reported to RBBP, and the annual total reported varies with search effort (*RBBP 2010*). Perhaps only half the annual total is documented by RBBP reports (Stroud *et al.* 2012b). The best estimate of the British breeding population was derived from the 1999 national survey, which estimated 73 singing males at 29 sites (Gilbert 2002). However, Stroud *et al.* (2012b) found a further 11 breeding records for that year and after a reappraisal of records they produced a revised total of 80 males. They also found records additional to those reported to previous atlases and it remains to be seen whether the range shown here is lacking other records from the 2008–11 period.

Despite these difficulties, it is clear that the Spotted Crake population has declined since the early 2000s, in terms of both breeding birds and passage migrants (Stroud *et al.* 2012b). Provisional results from a national survey in 2012 suggest that the UK population probably numbers fewer than 30 calling males, although the unusually wet spring may have influenced numbers, making some sites unsuitable but possibly enhancing others (S. Schmitt pers comm).

BREEDING DISTRIBUTION *2008–11*

POSSIBLE •	<1%	1%	1%
PROBABLE •	-	1%	<1%
CONFIRMED •	-	<1%	<1%
TOTAL	<1%	2%	1%

BREEDING DISTRIBUTION CHANGE *since 1968–72*

GAIN ▲	2	37	39
LOSS ▽	0	35	35
40yr CHANGE	∞	+5%	+10%
20yr INDEX	-	-	

CORNCRAKE

Crex crex

BTO CODE: CE

GARY FAULKNER

EVEN UP TO the *1968–72 Breeding Atlas* the Corncrake was an abundant and widespread breeding bird in Britain and Ireland but it is now confined to a small number of coastal and island strongholds in Scotland and Ireland. Intensive management of meadows, particularly the change from late-cut hay to early cut silage, combined with the mechanisation of meadow mowing are the main factors causing these overall declines (Green & Williams 1994; Green 1996; Green *et al.* 1997).

Since the *1988–91 Breeding Atlas* there have been significant gains in the Scottish islands, and in the Nene Washes in eastern England where a successful reintroduction project started in 2002 (Carter & Newbery 2004). The scatter of records in eastern Scotland and eastern England probably relates to migrating birds, some of which occasionally call for a few days before moving on. The gains in Britain have largely been the result of intensive conservation measures, pioneered by RSPB and adopted in agri-environment schemes (O'Brien *et al.* 2006; Wilson *et al.* 2009). These measures include payments to farmers to delay mowing hay until August and to mow fields from the centre outwards, allowing chicks to escape to margins (Tyler *et al.* 1998). Maintenance of suitable tall herbaceous vegetation (such as Common Nettle and irises) to provide early cover for returning birds in April and May has also been important. The application of these measures has been sufficient to lead to a 14% increase in range size in Britain since the *1988–91 Breeding Atlas* and a 141% population increase in the UK during 1993–2009 (Holling *et al.* 2011b).

In contrast, there have been continued losses in Ireland, with only a few gains in the strongholds of Co. Donegal and West Connaught, where some conservation measures have been deployed which have allowed the populations to persist (Copland 2002; Donaghy *et al.* 2011). Regular summer flooding in the 2000s led to the virtual loss of the Shannon Callows population. Overall the loss from the wider countryside far eclipses any gains, with a 69% reduction in the number of occupied 10-km squares in Ireland since the *1988–91 Breeding Atlas*.

BREEDING DISTRIBUTION *2008–11*

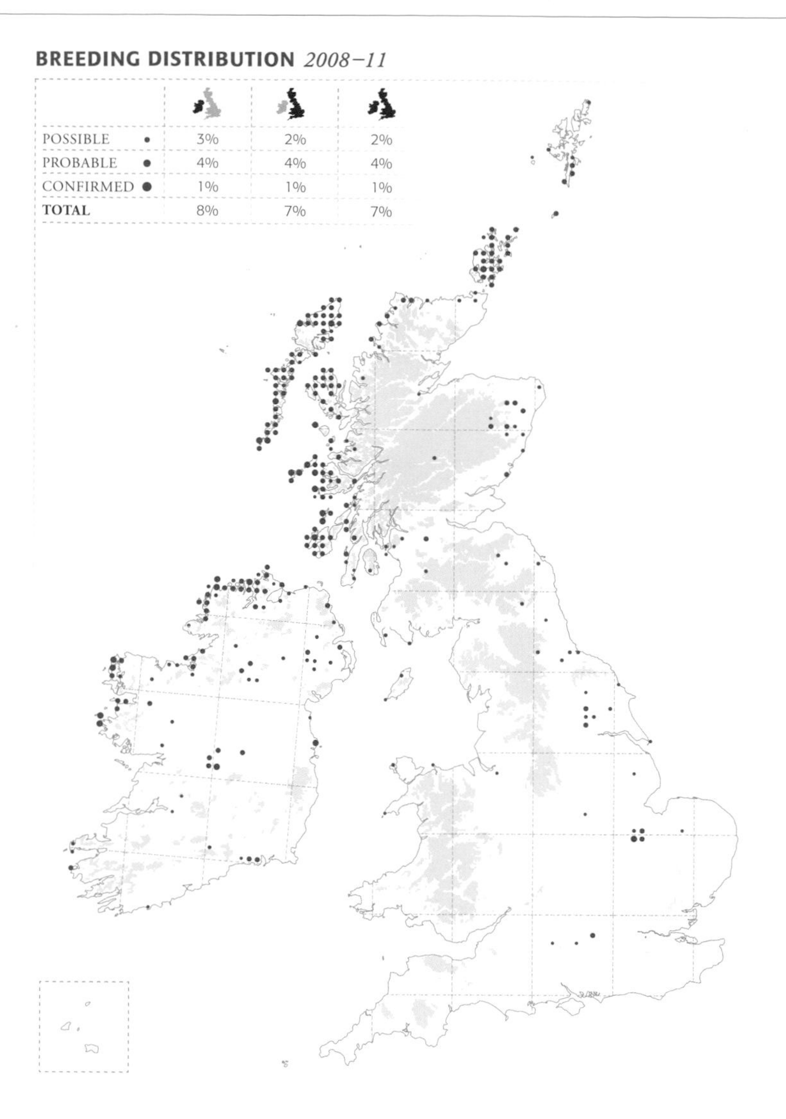

POSSIBLE •	3%	2%	2%
PROBABLE •	4%	4%	4%
CONFIRMED •	1%	1%	1%
TOTAL	8%	7%	7%

BREEDING DISTRIBUTION CHANGE *since 1968–72*

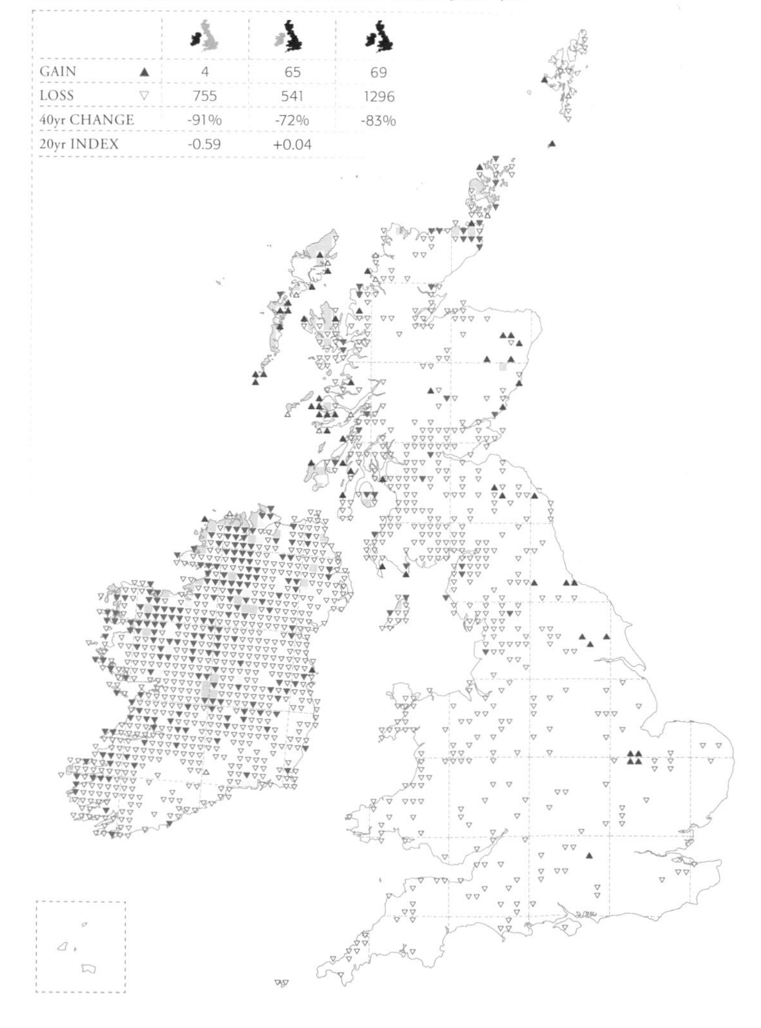

GAIN ▲	4	65	69
LOSS ▽	755	541	1296
40yr CHANGE	-91%	-72%	-83%
20yr INDEX	-0.59	+0.04	

Sponsored by a supporter of BirdWatch Ireland

MOORHEN
Gallinula chloropus

BTO CODE: MH

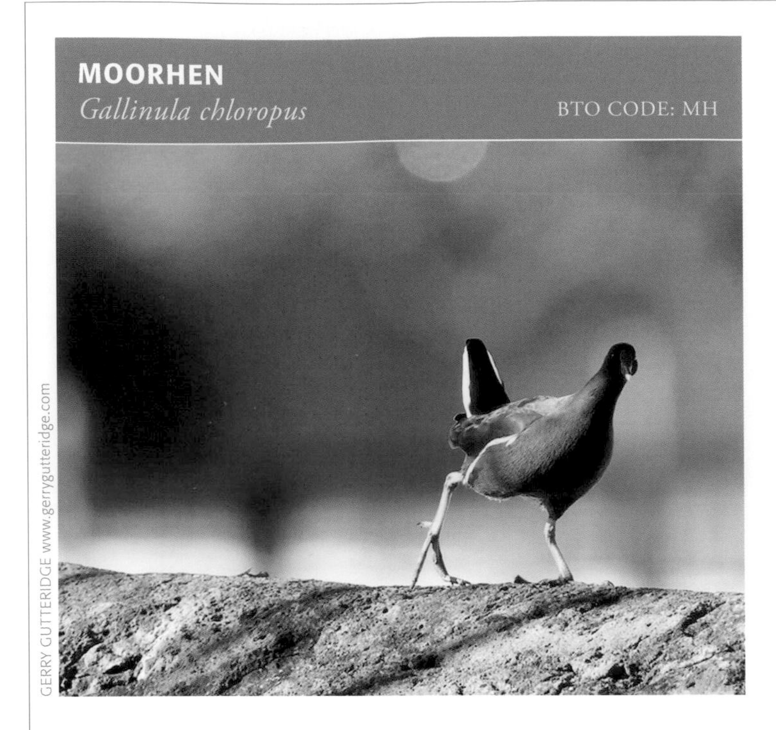

GERRY GUTTERIDGE www.gerrygutteridge.com

BRITISH AND IRISH Moorhens are highly sedentary, and their numbers are enhanced in winter by birds migrating from elsewhere in northwest Europe (*Migration Atlas*). Both the distribution and the pattern of relative abundance are broadly similar during the breeding season and winter. Moorhens are widely distributed throughout most of Britain and Ireland on a wide variety of lowland freshwater wetlands and agricultural areas with abundant drainage ditches. They are absent from many upland areas of Wales, northern England and northern and western Scotland.

The change maps indicate that Moorhens have contracted in range in Ireland, in winter by 11% since the *1981–84 Winter Atlas* and in the breeding season by 19% since the *1968–72 Breeding Atlas*. A decline in breeding-season relative abundance is also evident throughout Ireland. These losses may have been due to large-scale drainage of wetlands owing to increasing agricultural demands. Paradoxically, breeding-season population monitoring suggests a 76% increase but these data cover the period 1998–2010 only (*CBS Trend 2010*).

In Britain the pattern of change is mixed. Gains and losses are mostly on the edge of the range, such as in the margins of the uplands, and could reflect subtle changes in abundance in these low-density areas, or simply chance patterns of detection. There have been notable gains in both seasons in Shetland. The abundance change map suggests that there has been a predominant decline in relative abundance since the *1988–91 Breeding Atlas* throughout central England and southern Scotland. Long-term annual monitoring schemes show that the Moorhen population decreased markedly from a peak in the mid 1970s to a low in the mid 1980s, then partly recovered up to the mid 2000s before declining again. Overall, numbers in 2010 were 26% lower than in 1975 (*BirdTrends*). Spatial analysis has detected declines in the fringes of the range in Scotland over the period 1994–96 to 2007–09 and small increases on the upland margins in England and Wales (Massimino *et al.* 2013). Some of the long-term fluctuations may be related to cold winters, though local losses may be due to the filling in of farm ponds and the spread of the American Mink.

Sponsored by Joy & Ian Castle

BREEDING DISTRIBUTION *2008–11*

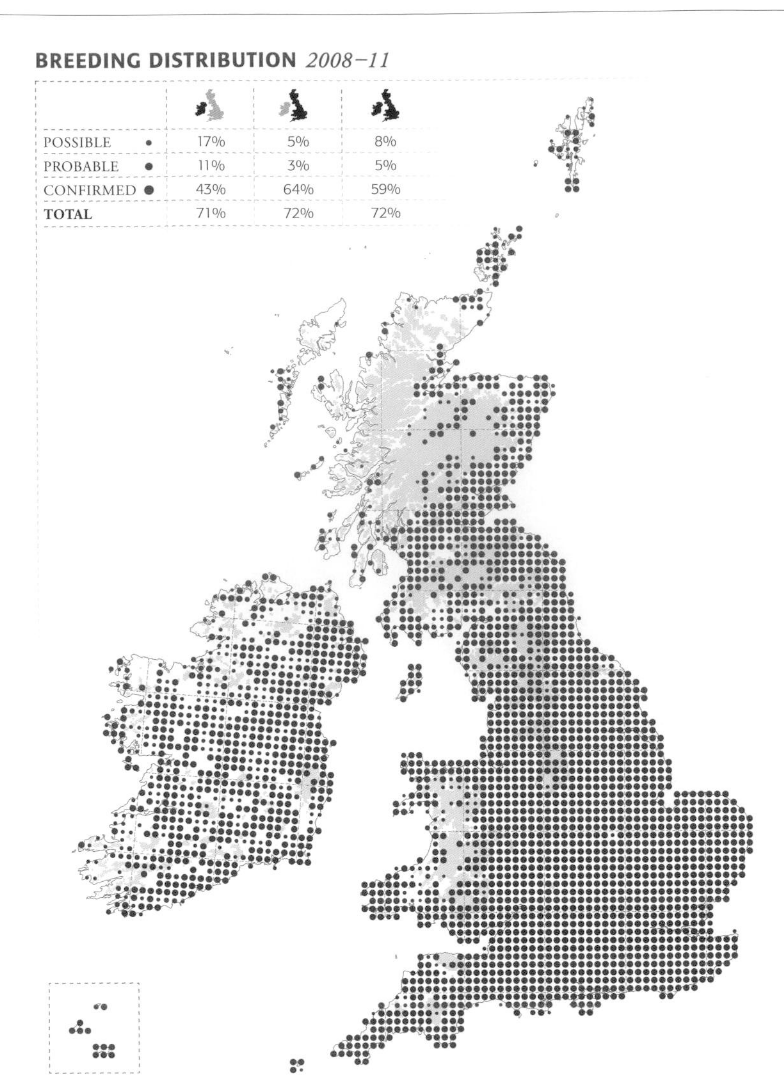

POSSIBLE •	17%	5%	8%
PROBABLE •	11%	3%	5%
CONFIRMED ●	43%	64%	59%
TOTAL	71%	72%	72%

WINTER DISTRIBUTION *2007/08–2010/11*

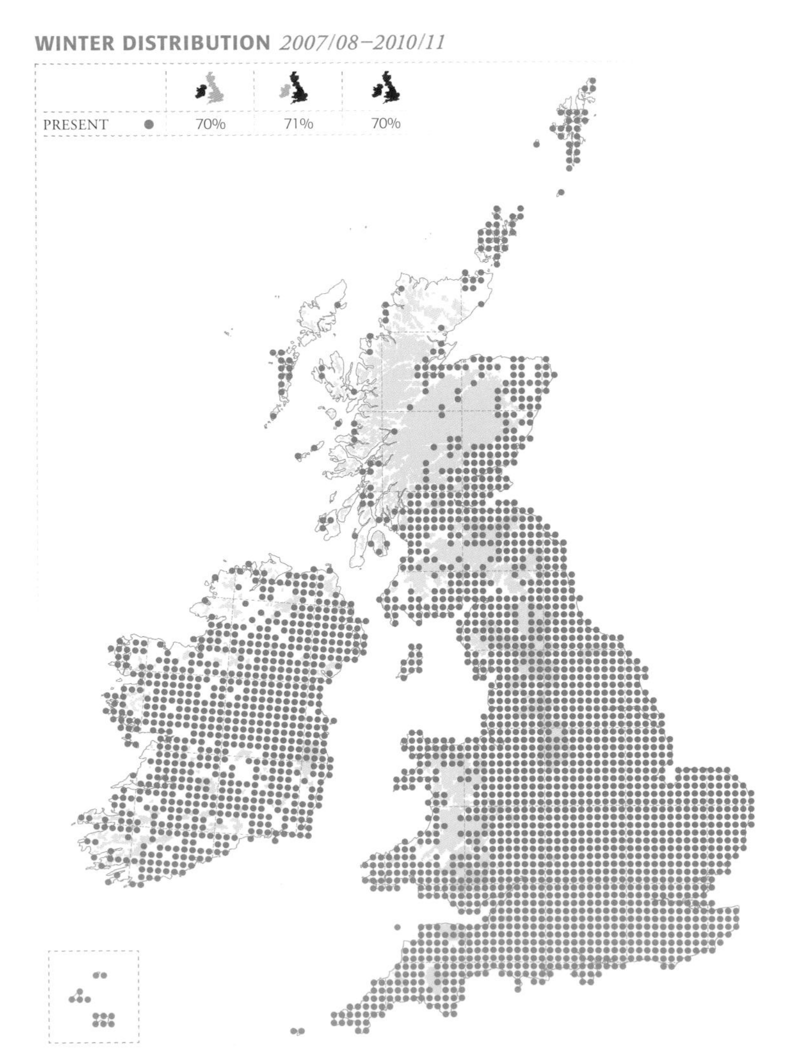

PRESENT ●	70%	71%	70%

BREEDING DISTRIBUTION CHANGE *since 1968–72*

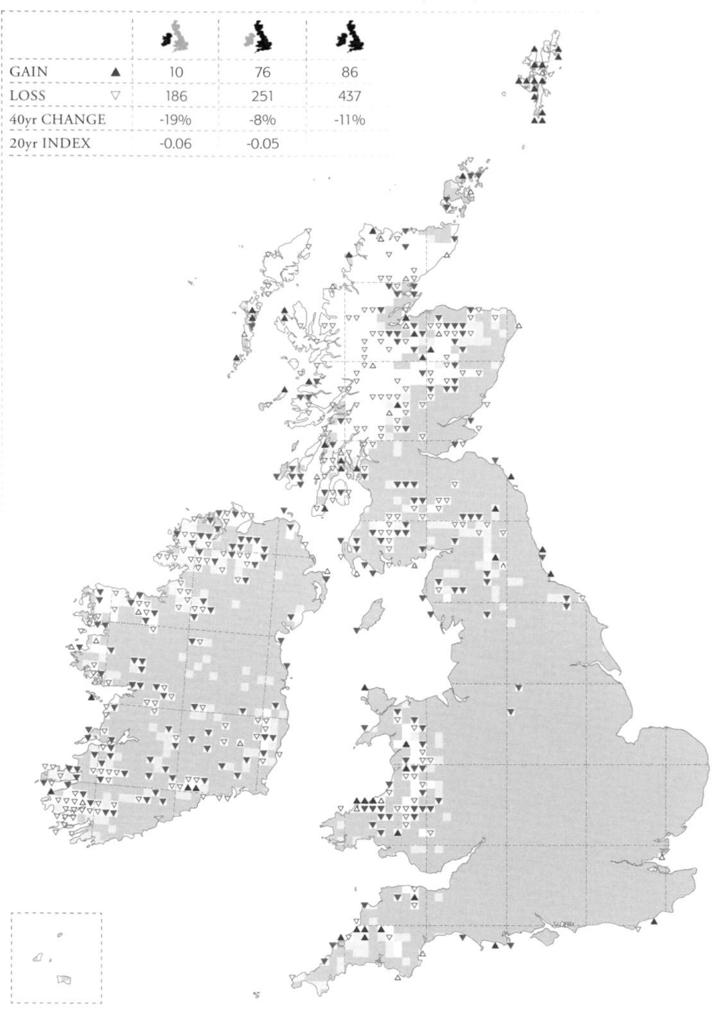

GAIN ▲	10	76	86
LOSS ▽	186	251	437
40yr CHANGE	-19%	-8%	-11%
20yr INDEX	-0.06	-0.05	

WINTER DISTRIBUTION CHANGE *since 1981–84*

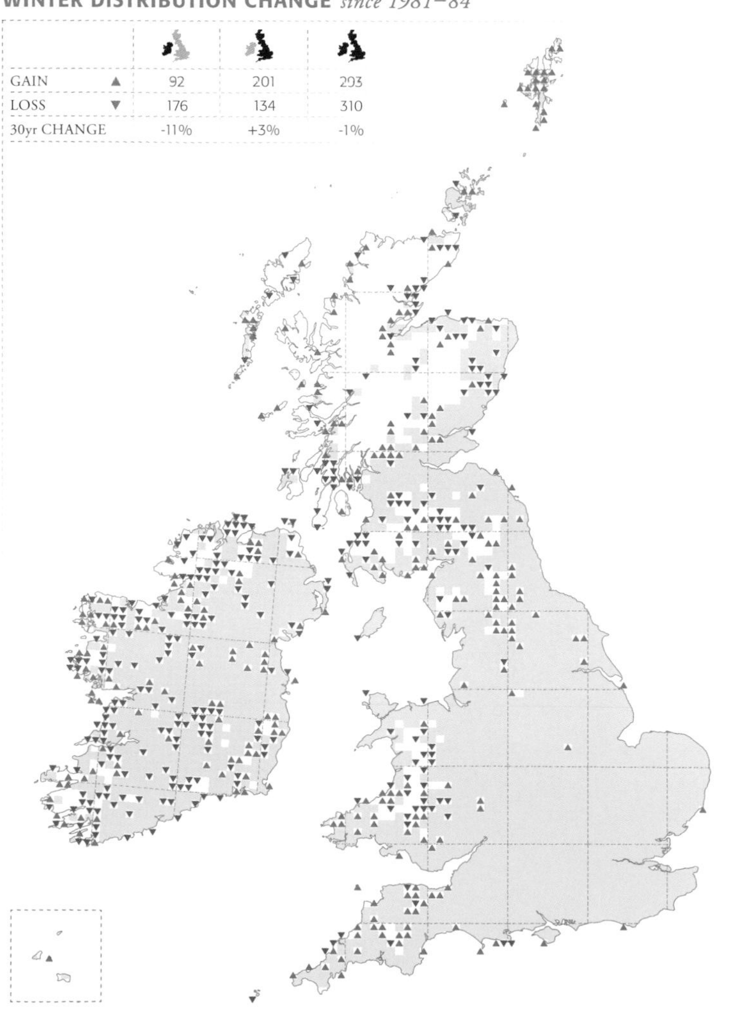

GAIN ▲	92	201	293
LOSS ▼	176	134	310
30yr CHANGE	-11%	+3%	-1%

BREEDING RELATIVE ABUNDANCE *2008–11*

BREEDING RELATIVE ABUNDANCE CHANGE *since 1988–91*

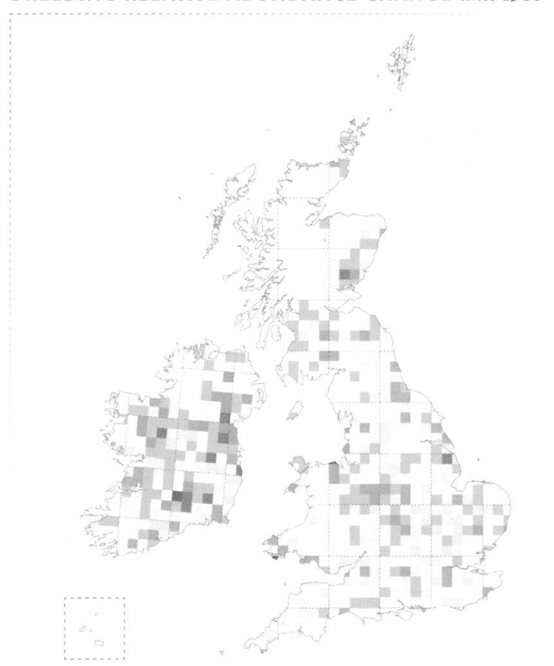

WINTER RELATIVE ABUNDANCE *2007/08–2010/11*

COOT
Fulica atra

BTO CODE: CO

STEVE CARTER www.colnevalleybirding.co.uk

THE COOT IS a resident waterbird that is widely distributed on low-lying wetlands across Britain and Ireland. Numbers are boosted every winter by birds migrating from elsewhere in northern and eastern Europe (*Migration Atlas*) and they become highly gregarious, forming large flocks that are often associated with diving ducks.

Patterns of distribution and abundance remain broadly similar during winter and the breeding season. Coots occur throughout most of England, only becoming patchily distributed in upland regions. The same is true in Ireland and Wales, but in Scotland they are highly localised outside the Central Belt. They require larger, more open wetlands for nesting than Moorhens do (Taylor & van Perlo 1998), which may explain their absence from parts of the west and southwest of Ireland (*Birds in Ireland*). Concentrations are highest throughout much of southern Britain, and are much lower in Ireland away from the wetland complexes in the midlands and west.

The breeding change map shows extensive losses in Ireland contributing to a 36% range contraction since the *1968–72 Breeding Atlas*, possibly as a consequence of drainage of wetlands since the 1970s. The range contraction in Britain is less severe and losses in the upland margins of Scotland and Wales are almost balanced by gains elsewhere in England. Despite the slight range contraction, long-term monitoring of numbers reveals a 74% population increase in Britain during 1970–2010 (*SUKB 2012*). The abundance change map supports this, showing increases in relative abundance throughout the large populations in south–central England but declines in the smaller populations in the north.

The winter change map also illustrates a substantial range contraction in Ireland, parts of Scotland and Wales since the *1981–84 Winter Atlas*. Although winter numbers in Britain have remained largely stable, there has been a significant decline in Ireland since the early 2000s (*I-WeBS Report 2009*; *WeBS Report 2011*), including a large decline at Loughs Neagh & Beg. This may be part of an eastward shift in the winter range of the flyway population, or may have been caused by a deterioration of the condition of this site.

Sponsored by Helen Baker, Hertfordshire

BREEDING DISTRIBUTION *2008–11*

	Ireland	Britain	Britain & Ireland
POSSIBLE •	7%	3%	4%
PROBABLE •	6%	3%	3%
CONFIRMED •	23%	51%	44%
TOTAL	36%	57%	51%

WINTER DISTRIBUTION *2007/08–2010/11*

	Ireland	Britain	Britain & Ireland
PRESENT •	38%	58%	53%

BREEDING DISTRIBUTION CHANGE *since 1968–72*

GAIN ▲	54	203	257
LOSS ▽	259	268	527
40yr CHANGE	-36%	-4%	-12%
20yr INDEX	-0.18	-0.06	

WINTER DISTRIBUTION CHANGE *since 1981–84*

GAIN ▲	89	247	336
LOSS ▼	171	197	368
30yr CHANGE	-18%	+3%	-2%

BREEDING RELATIVE ABUNDANCE *2008–11*

BREEDING RELATIVE ABUNDANCE CHANGE *since 1988–91*

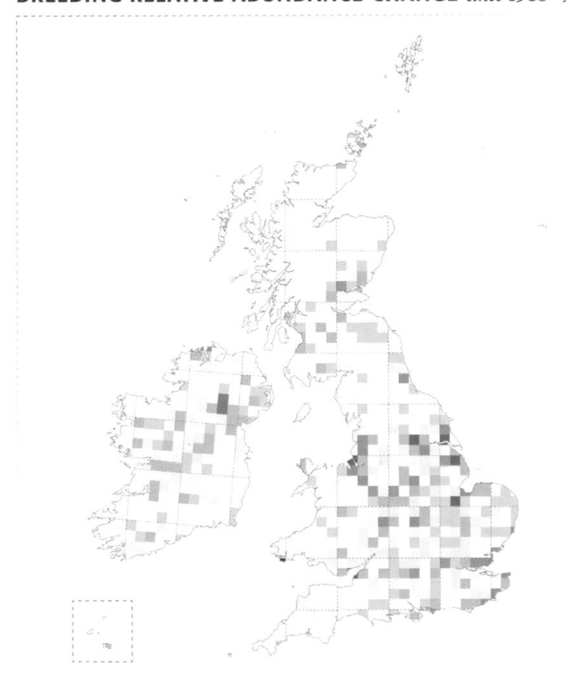

WINTER RELATIVE ABUNDANCE *2007/08–2010/11*

CRANE

Grus grus

BTO CODE: AN

DAVID KJAER www.davidkjaer.com

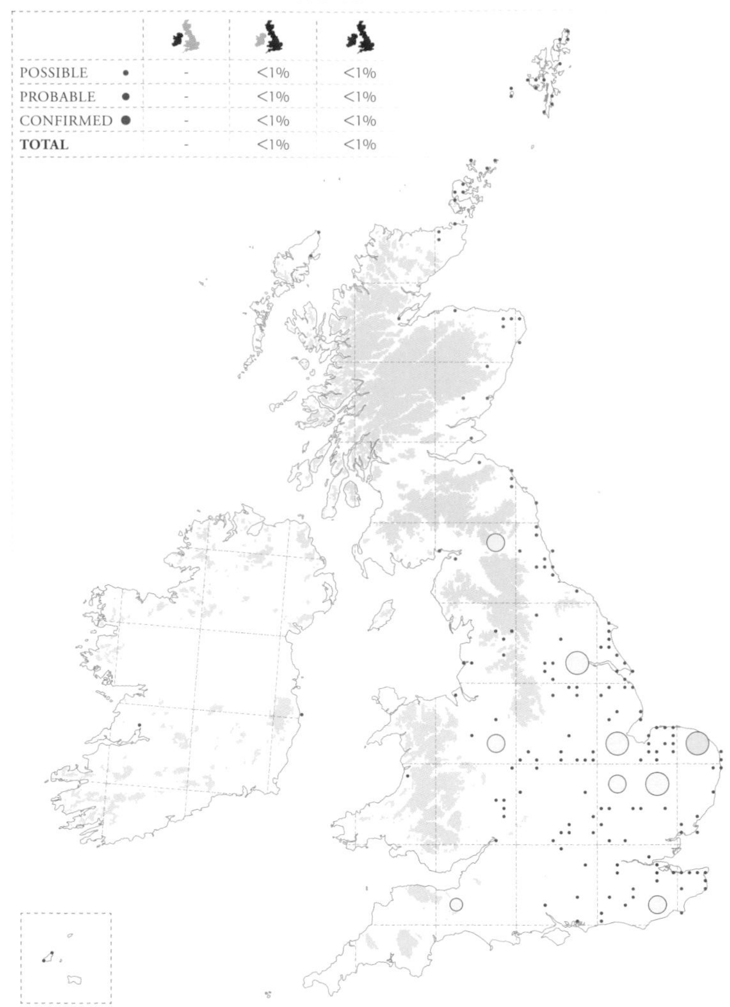

BREEDING DISTRIBUTION *2008–11*

POSSIBLE •	-	<1%	<1%
PROBABLE ●	-	<1%	<1%
CONFIRMED ●	-	<1%	<1%
TOTAL	-	<1%	<1%

CRANES ARE AGAIN regular winter visitors to Britain, although not as widespread as place names and bone remains would suggest they were historically (Boisseau & Yalden 1998). Their former breeding status is less clear: a note referring to a young Crane for sale in Norfolk in 1542 was the last documented breeding record until the 20th century (Gurney 1921).

Cranes are highly prone to disturbance and so the breeding distribution is displayed at 50-km resolution to protect the less well-known locations. Breeding evidence was received for 21 10-km squares, most falling in four extensive areas of relatively undisturbed wetlands in Britain, while *non-breeding* records came from a further 219 10-km squares in Britain and two in Ireland.

Stanbury *et al.* (2011) reviewed the recolonisation of Britain by Cranes, which began with two sub-adults that arrived in the Norfolk Broads in autumn 1979, where breeding has occurred every year since 1981 (Buxton & Durdin 2011). Nine pairs produced six young in the Broads in 2010 (*RBBP 2010*). A pair summered in Yorkshire in 2001, first bred in 2002, and had fledged a total of five young by 2009. Colonisation of the East Anglian Fens began in 2007, and up to three pairs now breed annually at two sites (*RBBP 2010*). It is unclear whether the new colonisations arose by dispersal from the Broads or from continental Europe (Stanbury *et al.* 2011). Elsewhere there are five 10-km squares with breeding evidence, including one 10-km square on the Somerset Levels where 20 juvenile Cranes were released in August 2010 in a bid to establish a population of at least 20 pairs by 2030 (Great Crane Project 2013).

The wintering population is thought to number c.50 birds (Stanbury *et al.* 2011) and although records came from both Britain (77 10-km squares) and Ireland (five squares) the majority were in southern England, particularly the Broads and Fens. The Yorkshire birds are thought to winter in the Norfolk Broads. Post-breeding dispersal and autumn migrants from the Continent may account for some records. Restoration of extensive wetlands, coupled with protection of breeding and wintering areas, particularly minimising disturbance, are likely to benefit Cranes.

BREEDING DISTRIBUTION CHANGE *since 1988–91*

GAIN ▲	-	20	20
LOSS ▽	-	1	1
20yr CHANGE	-	+950%	+950%
20yr INDEX	-	-	

Sponsored by Baroness Young of Old Scone

LEIGH CAUDWELL

▲ In winter, the Cranes in the Norfolk Broads are sometimes joined by birds from elsewhere, and feed together in the surrounding fields and grazing marshes.

GREAT BUSTARD
Otis tarda

BTO CODE: US

DAVID KJAER www.davidkjaer.com

HISTORICALLY, GREAT BUSTARDS probably bred in 11–13 British counties, all but one in England, and last nested in 1832. Their extinction was probably a result of agricultural change and persecution (Waters & Waters 2005; Shrubb 2011). It is thought that the factors that caused the loss are no longer operating, raising the possibility of a reintroduction (Osborne 2005). Assessments of habitat availability have suggested that pockets of suitable habitat still exist across England, notably on Salisbury Plain in Wiltshire where a large tract has been set aside for military training and conservation purposes. The Plain supports a mosaic of habitats that potentially meet the species' requirements: short grass areas for lekking, arable land for nesting and long grassland supporting arthropods at densities sufficient to rear chicks (Osborne 2005).

A programme to reintroduce bustards to Salisbury Plain began in 2004 with the translocation of chicks from a stable breeding population in southern Russia (Great Bustard Project 2013). Approximately 20 chicks have been imported each year since, and 86 were released during 2004–09 (Burnside *et al.* 2012). Mortality is high in their first year, probably through predation and collisions, but low thereafter (Burnside *et al.* 2012). Releases have formed a core population of around 15 birds close to the release site. The first breeding attempts were recorded in 2007 and 2008, though these proved unsuccessful. In 2009 two nests were found and these produced the first chicks to be reared in Britain for 177 years. Breeding was also *confirmed* in 2010 and 2011. To protect the breeding location, the *confirmed* breeding record is plotted at 20-km resolution. The breeding-season distribution map also shows *non-breeding* records in a further ten 10-km squares in southern England, which relate to sightings of wing-tagged individuals that had dispersed away from the release area.

During the winter, birds disperse farther afield from the core area and there were records from 20 10-km squares, mainly in Wiltshire, Dorset, Somerset and Gloucestershire. Vagrant continental Great Bustards were recorded in the *1981–84 Winter Atlas*; a lone bird was present in two adjacent 10-km squares in north Kent and three birds in two adjacent squares in south Kent, all in December 1981.

BREEDING DISTRIBUTION *2008–11*

	Ireland	Britain	Britain & Ireland
POSSIBLE •	-	-	-
PROBABLE •	-	-	-
CONFIRMED •	-	<1%	<1%
TOTAL	-	<1%	<1%

WINTER DISTRIBUTION *2007/08–2010/11*

	Ireland	Britain	Britain & Ireland
PRESENT •	-	1%	1%

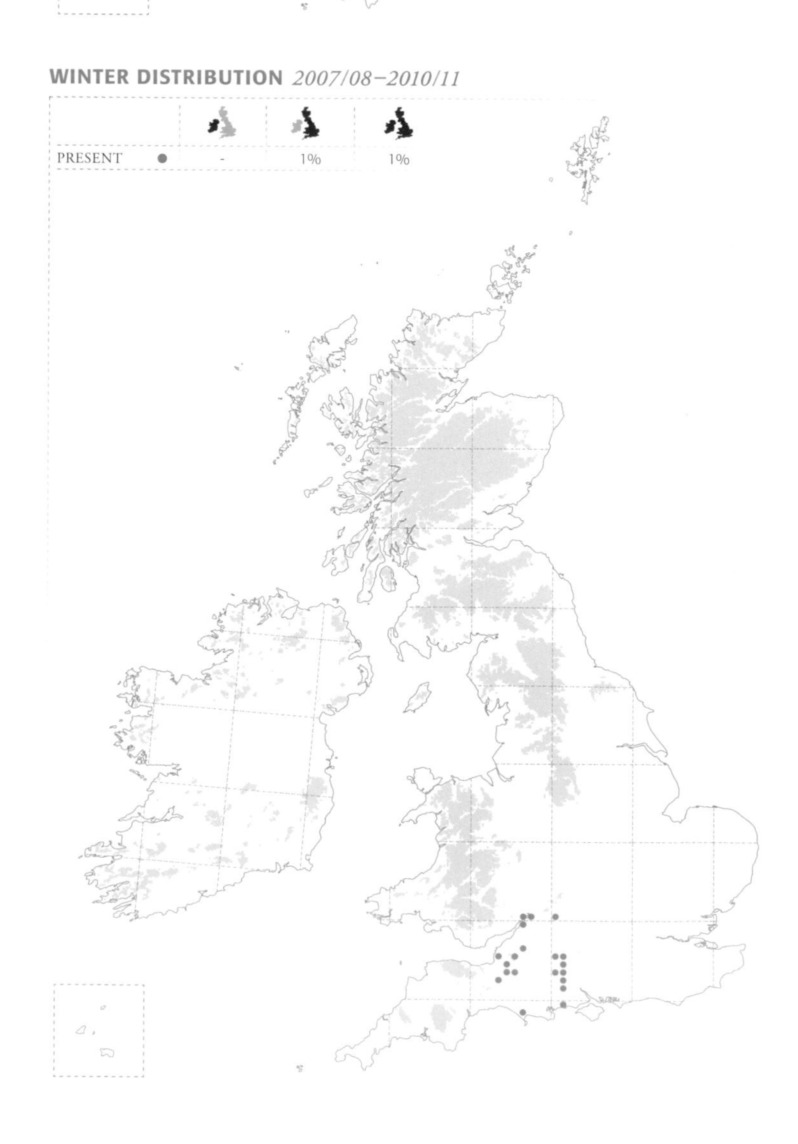

Sponsored by Dr Derek Rose, Newcastle upon Tyne, via Devizes

BLACK-WINGED STILT
Himantopus himantopus BTO CODE: IT

JOHN ANDERSON www.pbase.com/crail_birder

BREEDING DISTRIBUTION *2008–11*

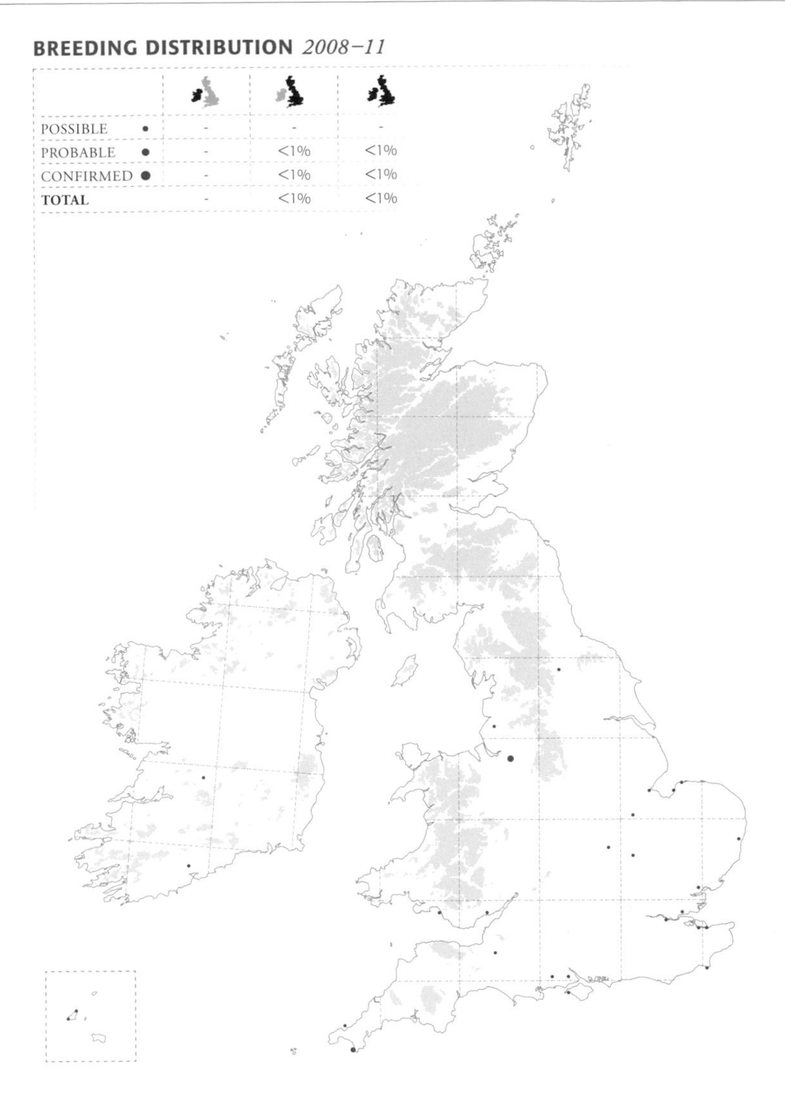

POSSIBLE ●	-	-	-
PROBABLE ●	-	<1%	<1%
CONFIRMED ●	-	<1%	<1%
TOTAL	-	<1%	<1%

THE EUROPEAN BREEDING range of the Black-winged Stilt is largely centred on the Mediterranean, with small populations as far north as the Netherlands (*European Atlas*). The majority winter south of the Sahara and it is during return spring migration that most records of Black-winged Stilts occur in Britain & Ireland (Parkin & Knox 2010). The majority of records involve wandering individuals or small groups. Such *non-breeding* birds were recorded in 24 10-km squares in southern Britain and two squares in Ireland. Some records related to pairs or trios but, where these did not remain for long, or the habitat was deemed unsuitable, no breeding evidence was assigned. Many records relate to the same nomadic individuals. This nomadism may be related to the species' dependence on ephemeral habitats which in its core breeding range can lead to substantial population fluctuations as changing water levels affect the suitability of sites (*Migration Atlas*).

Prior to the Atlas there had been six previous breeding attempts in Britain (Holling *et al.* 2010b). In 2008 a seventh attempt took place, at Neumann's Flash in Cheshire, but this was ultimately unsuccessful as all three chicks disappeared before fledging (Gregory 2008). In addition to the Cheshire record, there was also one record of *probable* breeding involving three birds displaying at a site in Cornwall in April 2008. Only two of the previous breeding attempts have proved successful. In 1945, two pairs bred at Nottingham sewage works, fledging three young, and probably a third pair laid eggs which were subsequently taken (Staton 1945; *Birds in England*). In 1987 a pair bred at Holme-next-the-Sea, Norfolk, fledging two chicks (Boyd 1987). The failed attempts were in Cambridgeshire in 1983, Cheshire in 1993, Suffolk in 2005 and Lancashire in 2006. Just after the Atlas, an eighth breeding attempt took place in Somerset but the birds deserted after laying eggs (Thomas 2012).

There were no breeding records of this species during previous atlases, although migrants were recorded in 1968 and, following the successful breeding in 1987, in each year during 1988–91. There were no winter records following the disappearance of 'Sammy', the Black-winged Stilt that took up residency at Titchwell in north Norfolk during 1993–2005.

Sponsored by John S. M. Bower MBE

OYSTERCATCHER

Haematopus ostralegus

BTO CODE: OC

ANDREW KELLY www.akellyphoto.com

IN WINTER, OYSTERCATCHERS are widely distributed on rocky and estuarine shores, with the largest concentrations forming on the major estuaries. Atlas data suggest a 40% winter range expansion in Britain since the *1981–84 Winter Atlas* but the majority of gains are inland 10-km squares that were occupied only in February, fitting a pattern of earlier return to breeding sites (Jenkins & Sparks 2010). A 5% range contraction is evident in Ireland. Long-term monitoring shows a gradual increase in winter numbers overall in the UK until c.1990, followed by decline (*WeBS Report 2011*), though trends vary among regions (Austin *et al.* 2000). Locally, declines may be linked to the activities of commercial shellfisheries (Atkinson *et al.* 2010). Numbers in the Republic of Ireland continue to increase (*I-WeBS Report 2009*).

Breeding Oystercatchers are found around most of the coast of Britain & Ireland, with the exception of southeast Ireland. In all three atlases, breeding Oystercatchers have been nearly ubiquitous in Scotland except for some northwestern upland areas. In recent decades the species has bred increasingly inland and it is now widespread across northern and eastern England. High densities are associated with the upland margins in eastern Scotland and northern England, as well as with the Northern Isles; concentrations around some of the larger estuaries, such as the Wash, Thames Estuary and Morecambe Bay, reflect flocks of summering non-breeders. In Ireland, the breeding distribution remains predominantly coastal.

Since the *1968–72 Breeding Atlas* there has been a 28% range expansion in Britain, but little change in Ireland. Gains in Britain are almost exclusively at inland sites, though there are some gains along the south coast of England. Some losses have occurred inland in northwest Scotland. Indeed, the abundance change map shows declines since the *1988–91 Breeding Atlas* in the proportion of occupied tetrads across eastern and southern Scotland, contrasting with increases in northern and eastern England. These mirror annual population monitoring results which show a 29% decline in Scotland during 1995–2010, contrasting with a 48% increase in England (*BBS Report 2011*). The causes of these changes are unclear but, perhaps significantly, the declines are shared by several wader species in Scotland.

Sponsored by Daphne Watson

BREEDING DISTRIBUTION *2008–11*

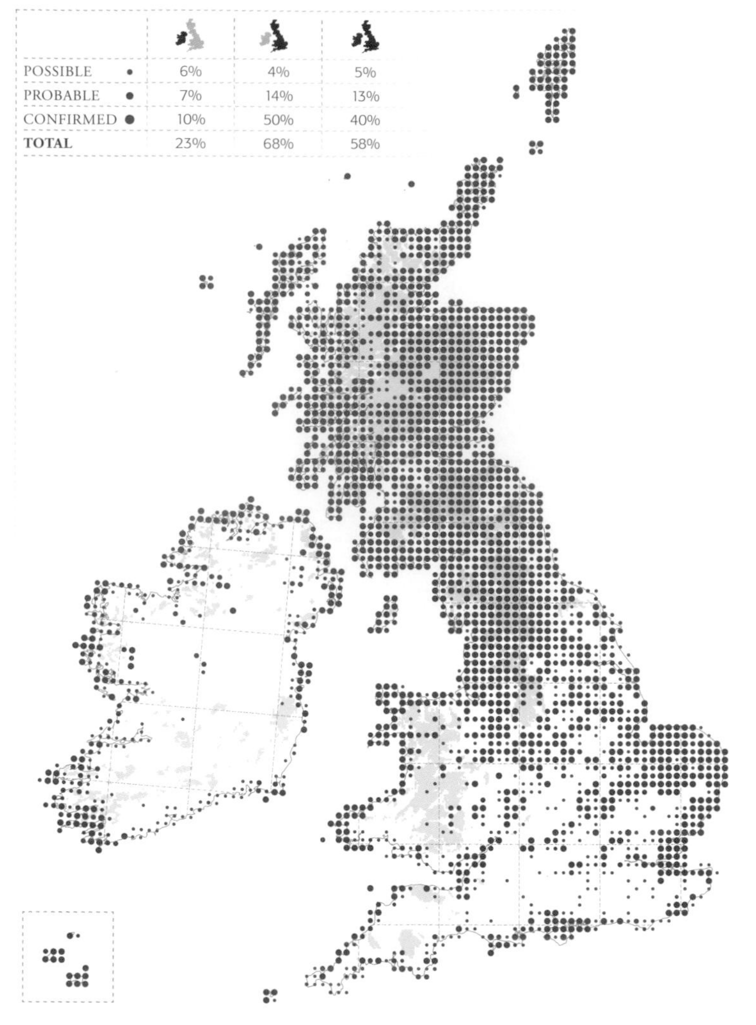

POSSIBLE 6% 4% 5%
PROBABLE 7% 14% 13%
CONFIRMED 10% 50% 40%
TOTAL 23% 68% 58%

WINTER DISTRIBUTION *2007/08–2010/11*

BREEDING DISTRIBUTION CHANGE *since 1968–72*

GAIN ▲	45	525	570
LOSS ▽	59	94	153
40yr CHANGE	-5%	+28%	+23%
20yr INDEX	+0.07	+0.11	

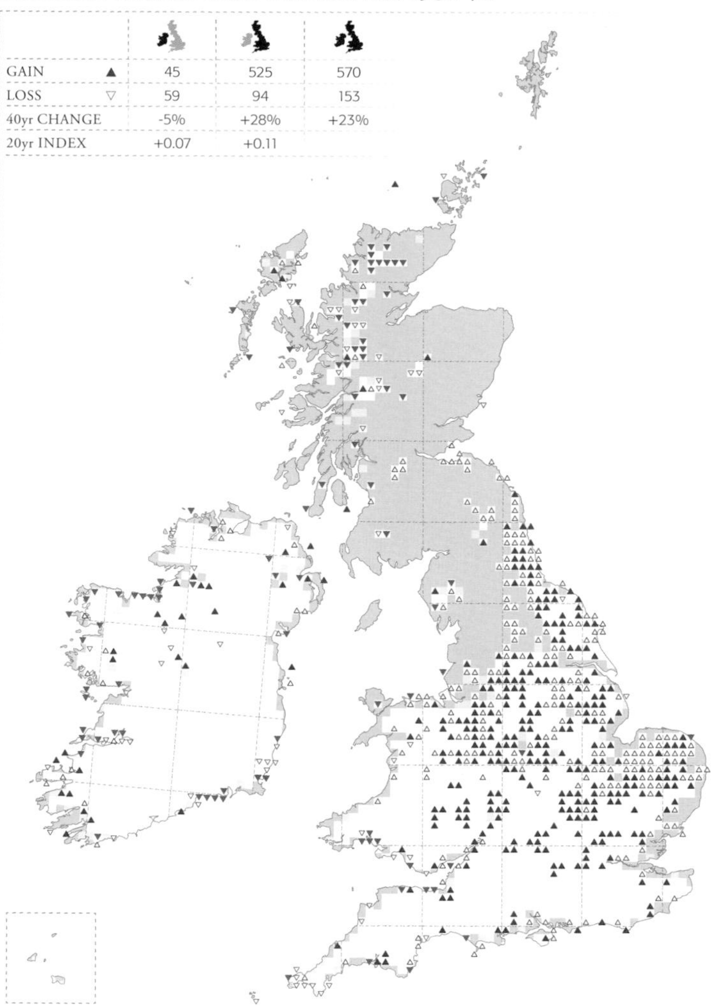

WINTER DISTRIBUTION CHANGE *since 1981–84*

GAIN ▲	21	561	582
LOSS ▼	37	95	132
30yr CHANGE	-5%	+40%	+29%

BREEDING RELATIVE ABUNDANCE *2008–11*

BREEDING RELATIVE ABUNDANCE CHANGE *since 1988–91*

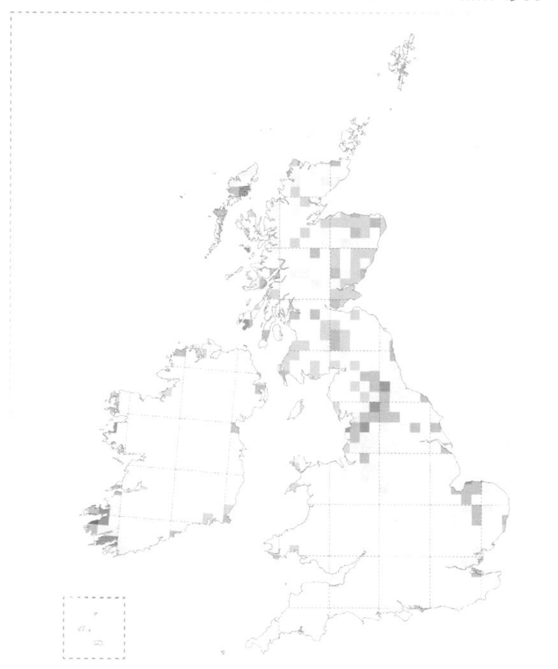

WINTER RELATIVE ABUNDANCE *2007/08–2010/11*

AVOCET

Recurvirostra avosetta

BTO CODE: AV

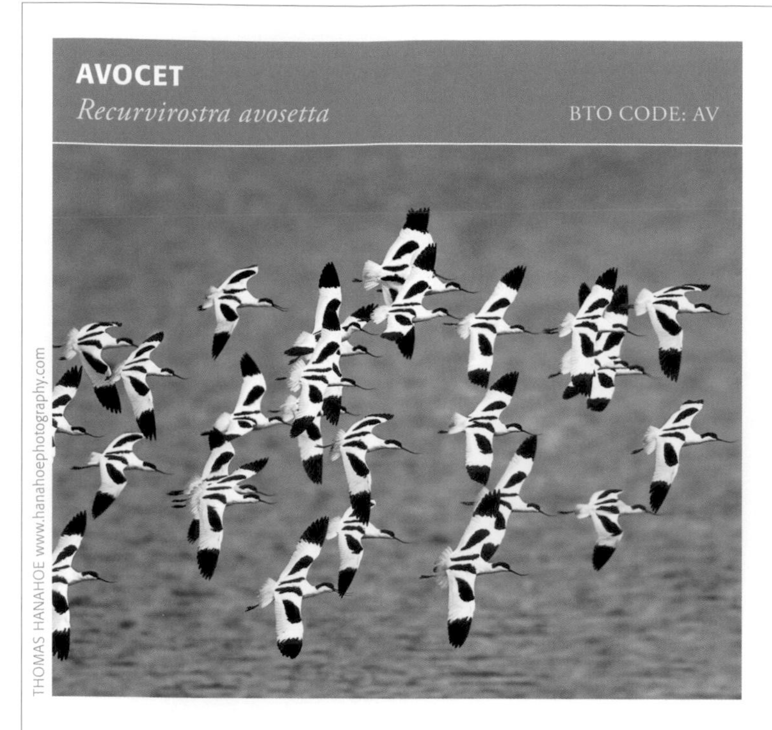

THOMAS HANAHOE www.hanahoephotography.com

THE SPREAD OF the Avocet has been a major feature on many southern British wetlands. They bred regularly on the east coast of England until the mid 1840s. Following sporadic breeding in Co. Wexford, Norfolk and Essex, they re-established themselves in Suffolk in 1947 (Cadbury & Olney 1978). The Avocet breeding population in Britain has since increased steadily to at least 1,600 pairs (*RBBP 2010*) and the winter population has swelled to 7,500 individuals (Musgrove *et al.* 2011).These increases have been accompanied by significant expansions of range in both winter, by 165% since the *1981–84 Winter Atlas*, and the breeding season, by 1,663% since the *1968–72 Breeding Atlas*. The continuing rate of expansion can also be seen from the number of occupied breeding sites, which by 2010 had increased to at least 95 from 66 just five years earlier (*RBBP 2010*). Despite the expansion within Britain, Avocets remain scarce in Ireland.

The winter distribution and abundance maps show the largest concentrations to lie around the Thames and Medway Estuaries, on the Suffolk coast and along the south coast between the Solent and Poole Harbour. Long-term monitoring shows a steady increase in the wintering population since the mid 1980s (*WeBS Report 2011*). There is no evidence of the trend slowing, with a 75% increase during 1998/99–2008/09 (*SUKB 2012*), and the range continues to extend northwards (*WeBS Report 2011*).

The breeding-season distribution is concentrated into eastern and southeast England, although there was one pair in Wales (*RBBP 2010*), and several pairs in northwest England. The most northerly breeding site was in Northumberland. All of the occupied squares outside East Anglia and the southeast have been colonised since the *1988–91 Breeding Atlas*. Breeding has yet to be proven in Scotland, where most records related to wandering migrants, although some birds displayed in suitable breeding habitat before moving on. The relative abundance map shows the importance of coastal sites such as those along the East Anglian coastline. Although still predominantly a coastal species, Avocets have started to breed inland in the Fens, the east Midlands and South Yorkshire. *Non-breeding* birds were reported from 159 10-km squares and there is much scope for further expansion, especially to inland sites (*RBBP 2010*).

Sponsored by RSPB

BREEDING DISTRIBUTION *2008–11*

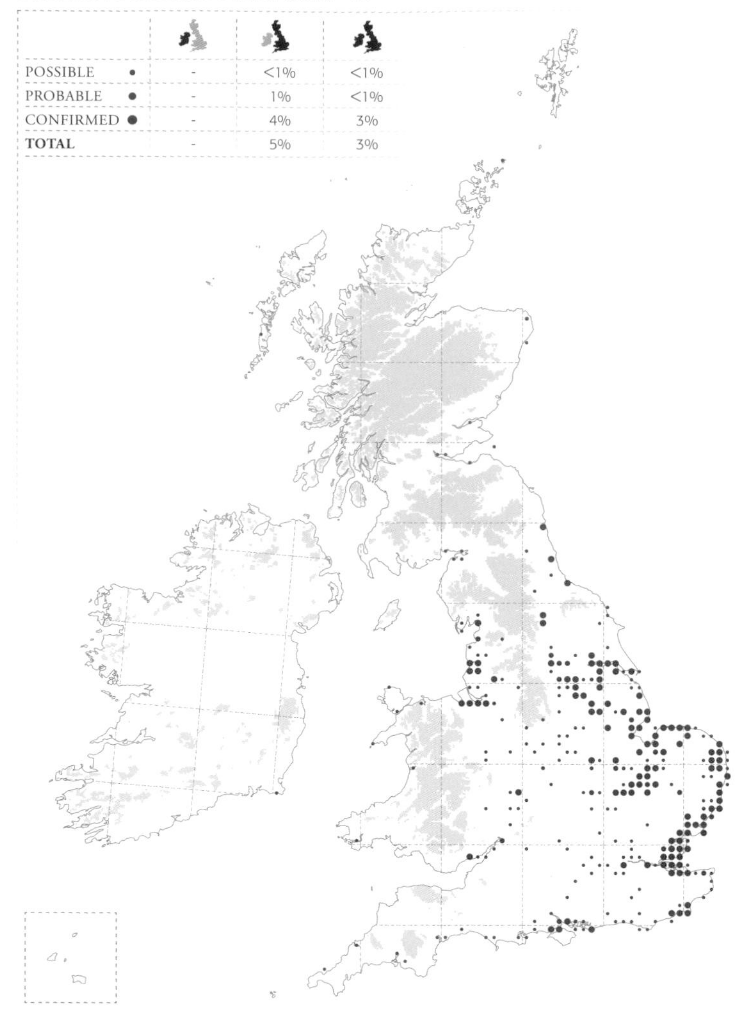

POSSIBLE •	-	<1%	<1%
PROBABLE •	-	1%	<1%
CONFIRMED •	-	4%	3%
TOTAL	-	5%	3%

WINTER DISTRIBUTION *2007/08–2010/11*

PRESENT •	<1%	5%	4%

BREEDING DISTRIBUTION CHANGE *since 1968–72*

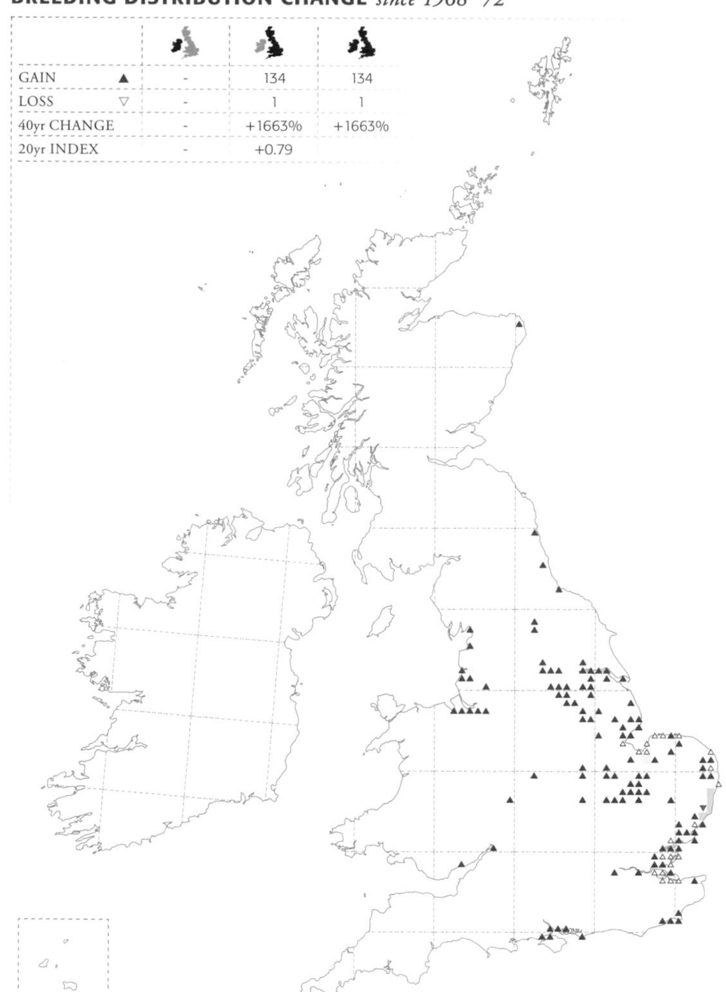

GAIN ▲	-	134	134
LOSS ▽	-	1	1
40yr CHANGE	-	+1663%	+1663%
20yr INDEX	-	+0.79	

BREEDING RELATIVE ABUNDANCE *2008–11*

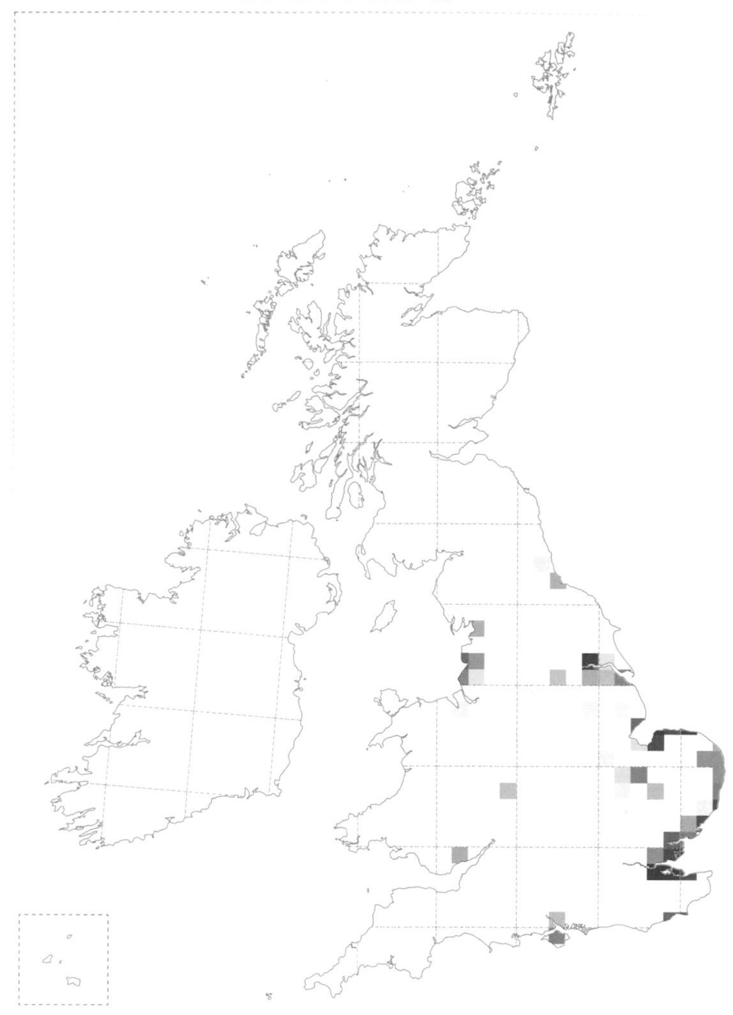

WINTER DISTRIBUTION CHANGE *since 1981–84*

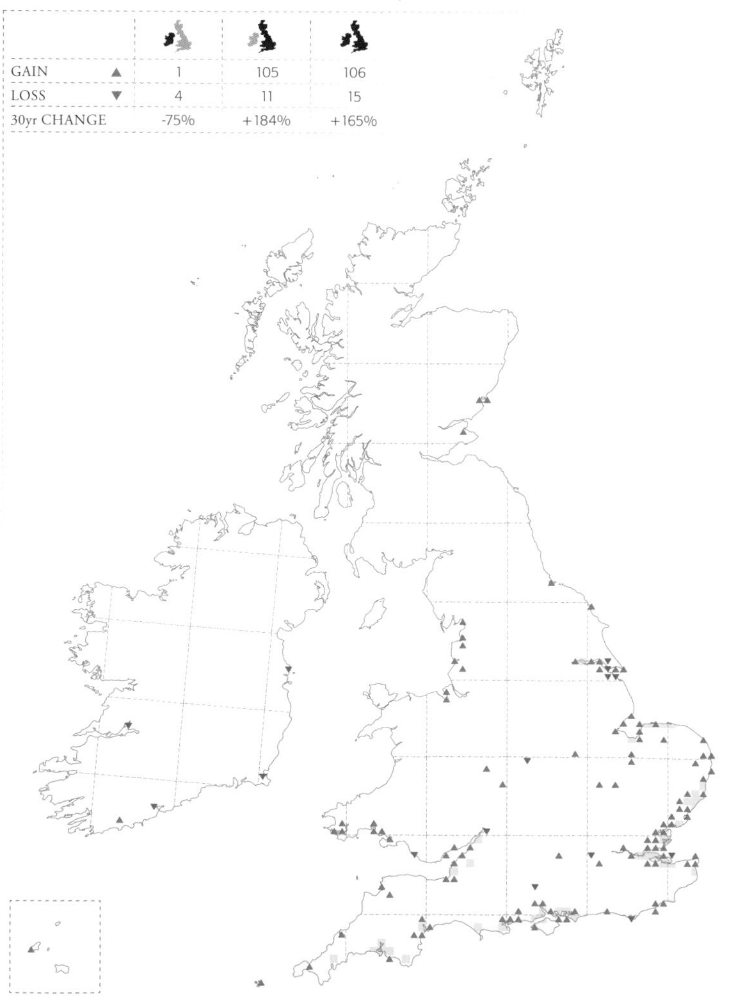

GAIN ▲	1	105	106
LOSS ▼	4	11	15
30yr CHANGE	-75%	+184%	+165%

WINTER RELATIVE ABUNDANCE *2007/08–2010/11*

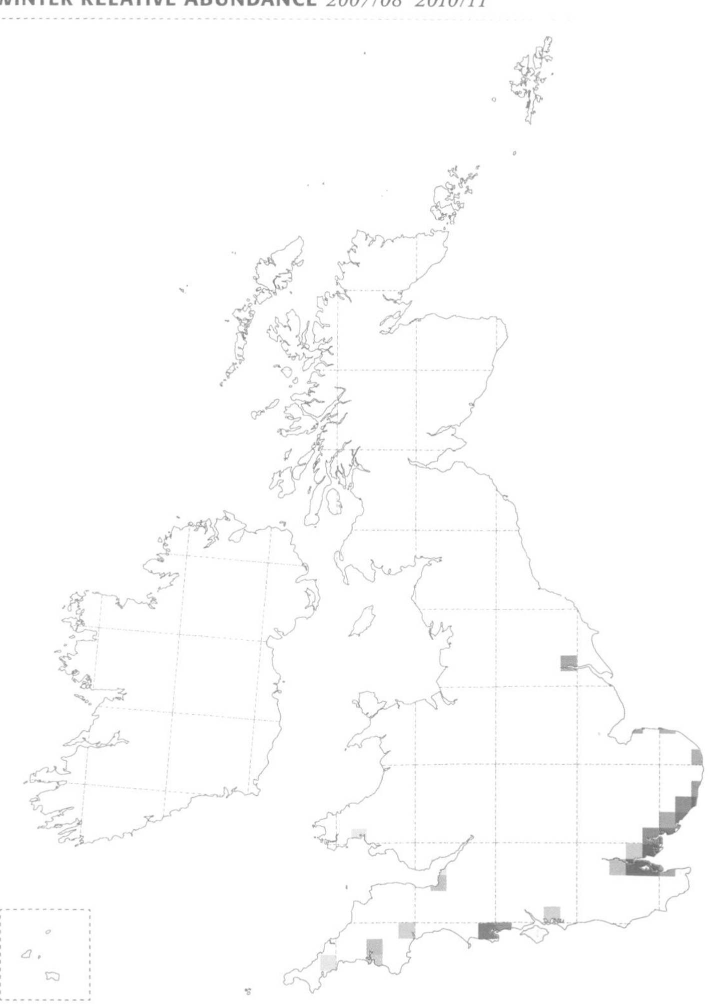

STONE-CURLEW

Burhinus oedicnemus

BTO CODE: TN

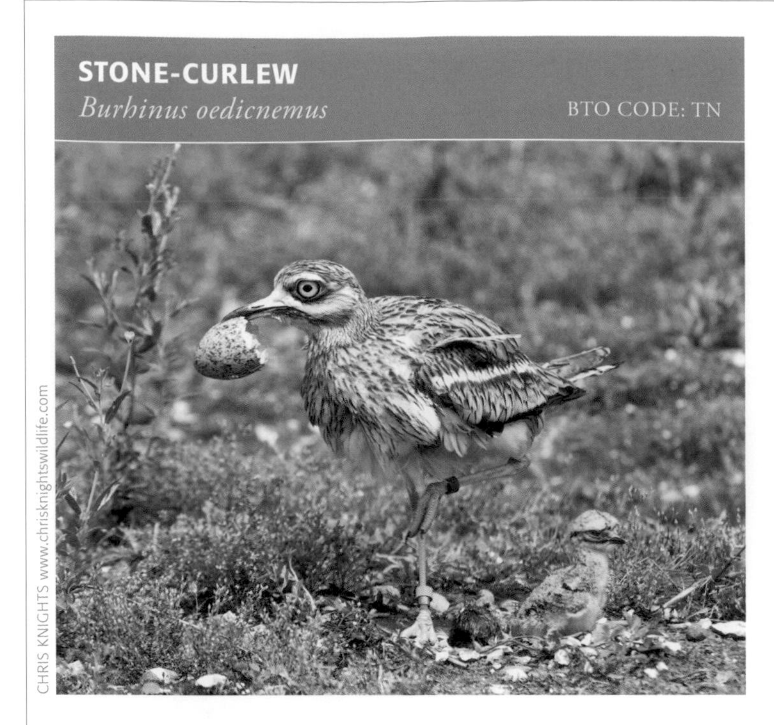

CHRIS KNIGHTS www.chrisknightswildlife.com

OWING TO RANGE contraction since the 1850s, the Stone-curlew is considerably more localised in Britain than the free-draining stony soils that it prefers. Losses since 1850 have been associated with land enclosure, afforestation, lack of grazing, the cessation of European Rabbit warrening and the conversion of permanent pasture to arable farmland. Virtually all of the c.375 breeding pairs are monitored annually by RSPB staff and landowners (*RBBP 2010*) and these maps benefit from their records. The two main breeding areas apparent in the *1988–91 Breeding Atlas*, the East Anglian Breckland and Salisbury Plain in Wiltshire, remain the strongholds but the population in the Suffolk Sandlings has expanded and in recent years two pairs have bred in Sussex (*RBBP 2010*; plotted at 20-km resolution).

After a range contraction of 42% between the *1968–72 Breeding Atlas* and the *1988–91 Breeding Atlas*, no net change was apparent up to 2008–11. The range was not static, however, with 22 10-km squares gained and another 22 lost. There is a slight tendency towards losses on the fringes of the Salisbury Plain population and towards gains around the Brecks population. Indeed, the latter has expanded to the north and west beyond Breckland, but there is still some way to go to recover the extent of occupancy there in 1968–72.

Stone-curlews prefer sparse vegetation and bare ground. In Britain this combination is provided either by short semi-natural grassland or by spring-sown crops, with densities generally higher on grassland (Green & Griffiths 1994; Green *et al.* 2000). In both habitats, Stone-curlews depend heavily on management and intervention (Newton 2004). In grasslands, densities and breeding success are highest where European Rabbits are abundant (Green & Griffiths 1994; Green & Taylor 1995; Bealey *et al.* 1999). For pairs using arable farmland, nest finding is essential to safeguard eggs and chicks from machinery, whilst agri-environment schemes promote spring sowing of crops to deliver the preferred nesting habitat (Aebischer *et al.* 2000).

Non-breeding Stone-curlews were recorded from 54 widely scattered 10-km squares in Britain, together with one in Ireland. In winter, late-departing and early returning breeders accounted for most of the 14 occupied 10-km squares, though birds do occasionally overwinter.

BREEDING DISTRIBUTION *2008–11*

POSSIBLE •	-	<1%	<1%
PROBABLE ●	-	<1%	<1%
CONFIRMED ●	-	2%	1%
TOTAL	-	2%	1%

WINTER DISTRIBUTION *2007/08–2010/11*

PRESENT ●	-	<1%	<1%

Sponsored by Hugh van Cutsem, Hilborough Estate

BREEDING DISTRIBUTION CHANGE *since 1988–91*

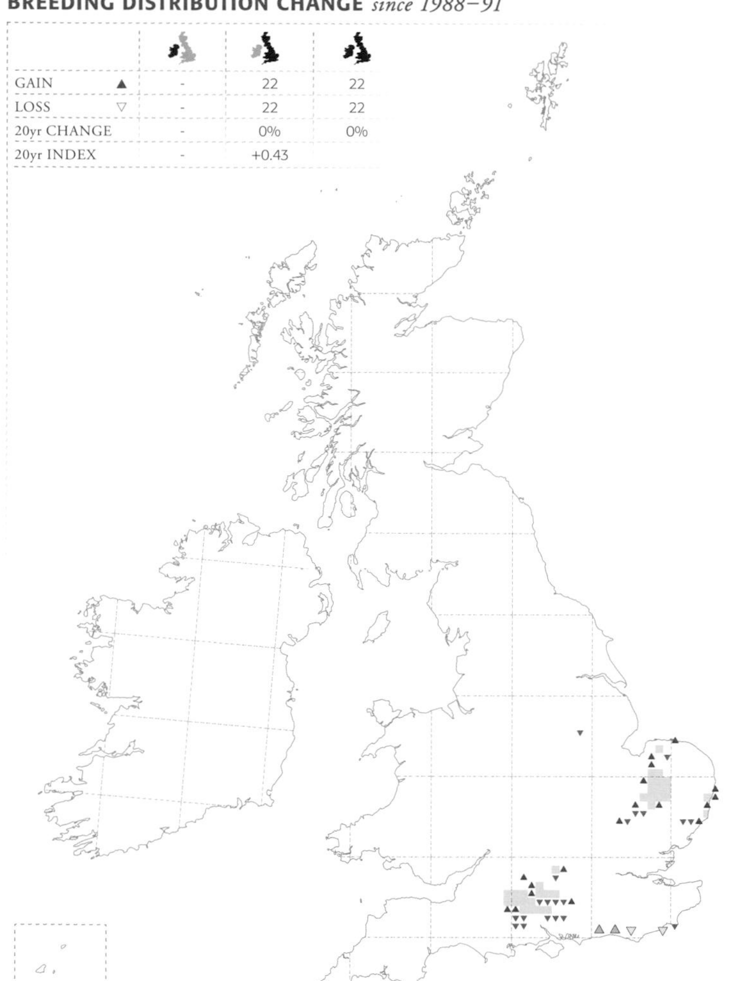

GAIN ▲	-	22	22
LOSS ▽	-	22	22
20yr CHANGE	-	0%	0%
20yr INDEX	-	+0.43	

BREEDING RELATIVE ABUNDANCE *2008–11*

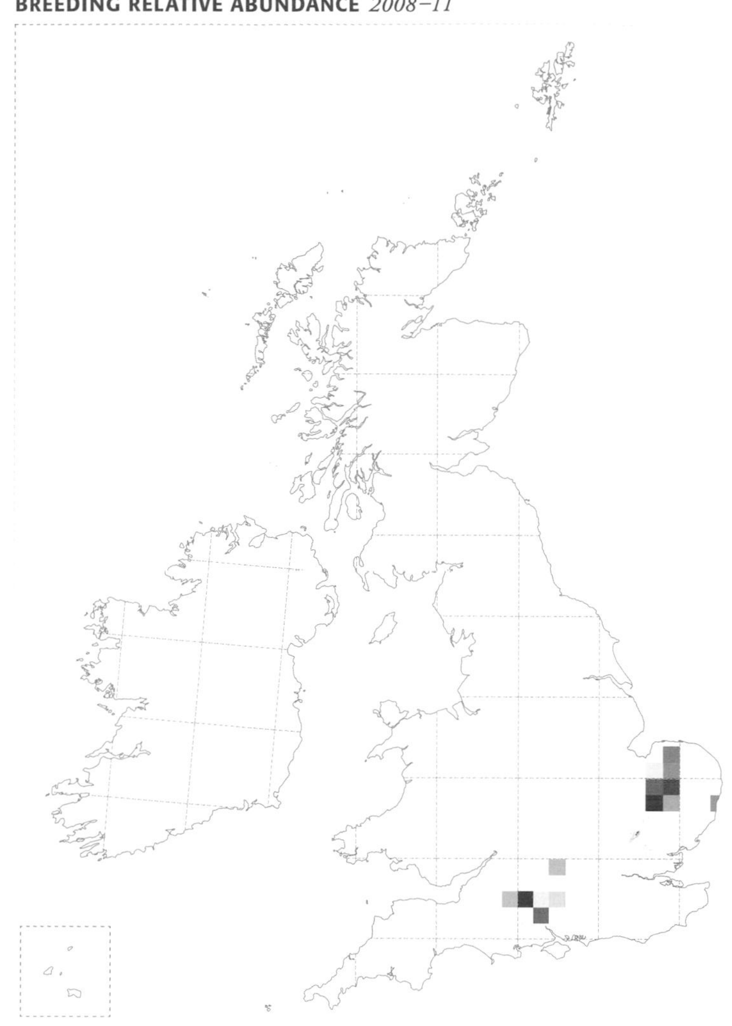

◀ Stone-curlews feed on tightly grazed grasslands and take a range of prey including earthworms, woodlice, millipedes and beetles. Such animals are more likely to be available on sparsely vegetated ground at night.

BRIAN TOLLITT Briantollitt@googlemail.com

LITTLE RINGED PLOVER

Charadrius dubius

BTO CODE: LP

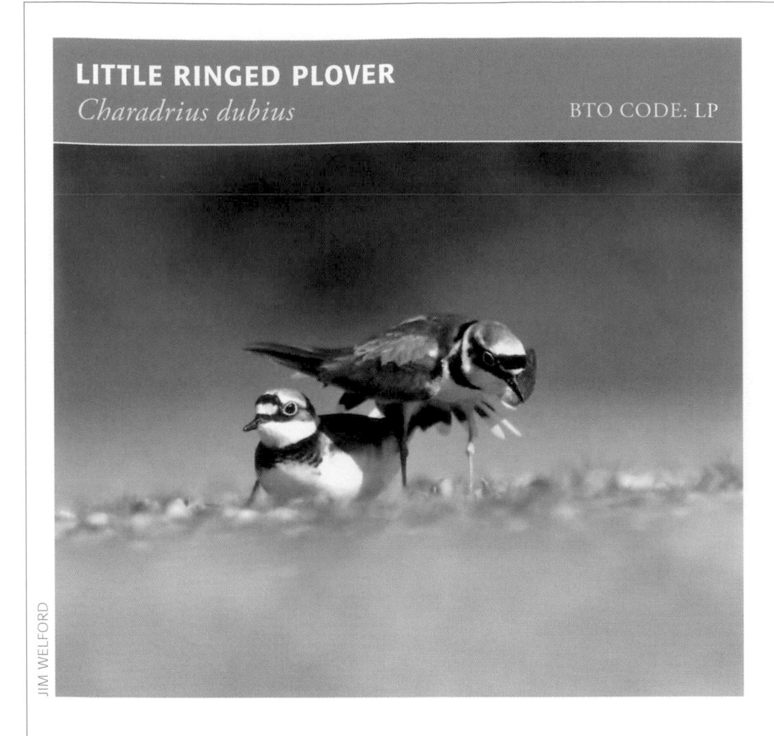

JIM WELFORD

THE BRITISH BREEDING range of the Little Ringed Plover is centred on northern-central and southeast England but the species now also breeds sparsely in Wales and the lowlands of central and eastern Scotland. It is a rare new colonist in Ireland (Collins 2008) and the three 10-km squares with breeding evidence are mapped at 20-km resolution. The species is almost exclusively a summer visitor; records from five 10-km squares in winter probably related to late migrants or early arriving birds.

A national survey in 2007 estimated a British population of 1,239 pairs (G. Conway pers comm), almost double the figure estimated by the previous survey, in 1984 (Parrinder 1989). Abundance is highest in northwest England, though there are local concentrations in South Wales and southeast England. Wales accounts for c.16% of the numbers reported to RBBP, and Scotland just 2% (*RBBP 2010*).

Little Ringed Plovers bred for the first time in Britain in 1938 (Ledlie & Pedler 1938) and their range has expanded progressively through the series of atlases, doubling overall since the *1968–72 Breeding Atlas*. Most gains between 1968–72 and the *1988–91 Breeding Atlas* amounted to infilling in southeast and central England, although there was some expansion towards the Channel coast, into southern and central Wales and southern Scotland. Since 1988–91 this pattern of expansion has continued, with many more gains in northern England and Scotland and breeding *confirmed* in Ireland for the first time in 2008 (Collins 2008).

The main habitats used in Britain in 2007 were sand and gravel pits (30%) and the margins of other static water bodies (28%). Colonisation of Wales and Scotland has seen increasing use of river shingle, and accordingly the relative use of this habitat rose from an overall 2% of pairs in 1984 to 18% in the 2007 survey (Parrinder 1989; G. Conway pers comm). Further range expansion and population increase is expected, given the momentum of the recent changes and the considerable area of potential breeding habitat yet to be occupied. Threats include flooding, to which nest sites are susceptible (Lloyd & Friese 2013), and human disturbance, which can cause birds to redistribute to other sites, making the assessment of breeding numbers problematic.

Sponsored by Biocensus Ltd

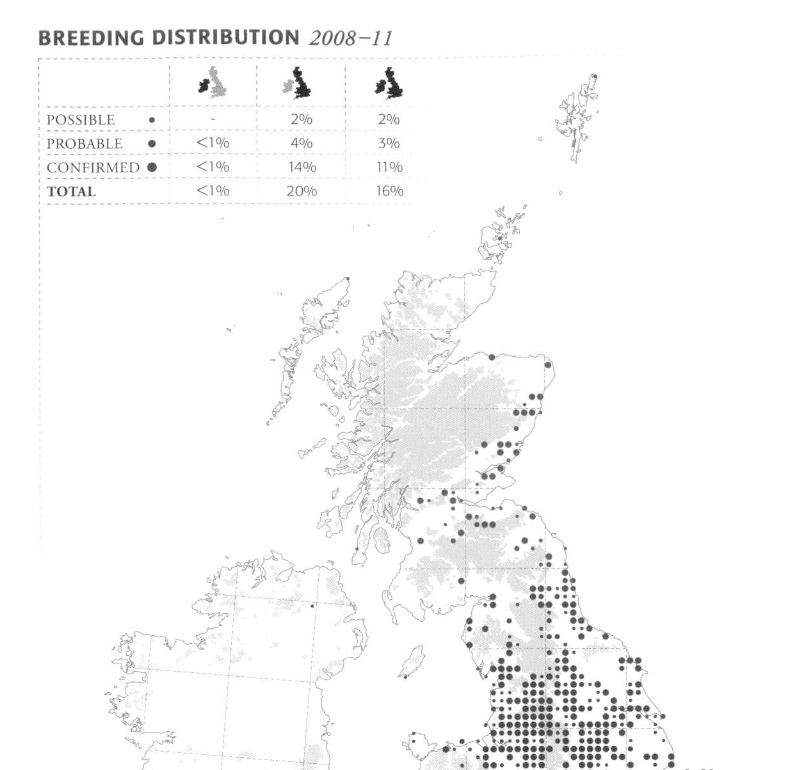

BREEDING DISTRIBUTION *2008–11*

POSSIBLE •	-	2%	2%
PROBABLE •	<1%	4%	3%
CONFIRMED •	<1%	14%	11%
TOTAL	<1%	20%	16%

WINTER DISTRIBUTION *2007/08–2010/11*

PRESENT •	-	<1%	<1%

BREEDING DISTRIBUTION CHANGE *since 1988–91*

	Ireland	Britain	Britain & Ireland
GAIN ▲	3	285	288
LOSS ▽	1	108	109
20yr CHANGE	+200%	+42%	+43%
20yr INDEX	-	-0.10	

BREEDING RELATIVE ABUNDANCE *2008–11*

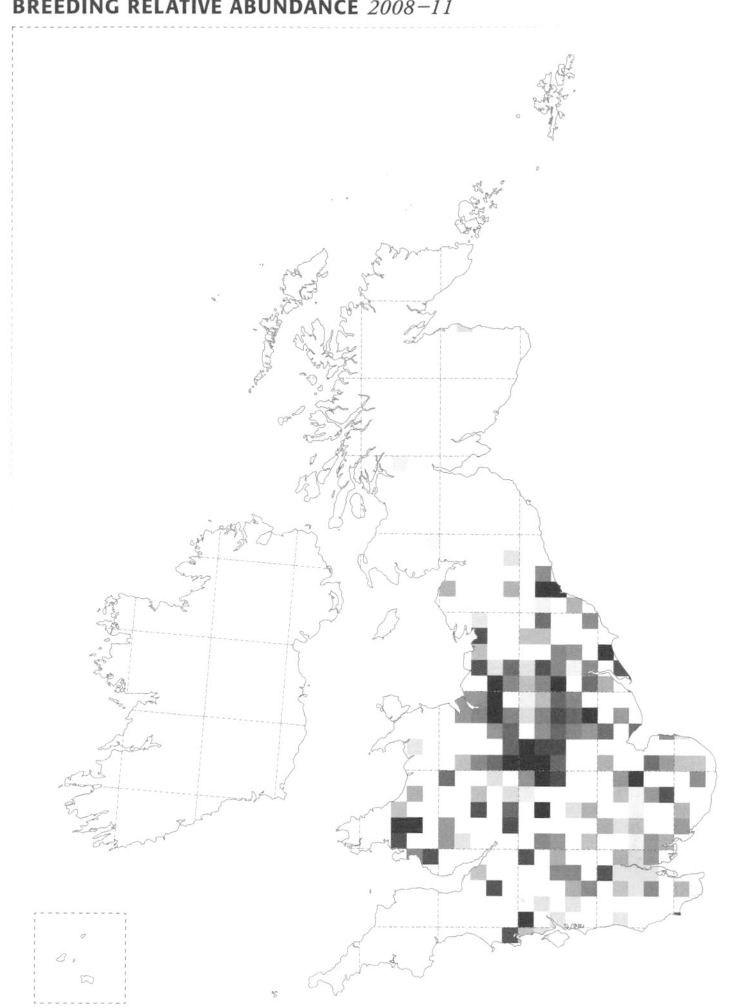

KENTISH PLOVER

Charadrius alexandrinus

BTO CODE: KP

GARTH PEACOCK www.garthpeacock.co.uk

PERSECUTION AND DISTURBANCE are thought to have led to the demise of the Kentish Plover as a British breeding species (*Birds in England*). They last bred in England in 1979, at Gibraltar Point in Lincolnshire (Archer *et al.* 2010), and in the Channel Islands on Guernsey and Alderney in 1974 (Young *et al.* 2013). During the *1968–72 Breeding Atlas* there were *confirmed* breeding records on three of the Channel Islands: Guernsey, Alderney and Herm. The distribution map presented here shows records of *non-breeding* birds in 22 10-km squares in Britain and one in Ireland. There were exceptional winter records in two 10-km squares; one in Co. Cork and one on the Outer Hebrides.

BREEDING DISTRIBUTION *2008–11*

	Ireland	Britain	Britain & Ireland
POSSIBLE •	-		-
PROBABLE •	-	-	-
CONFIRMED •	-	-	-
TOTAL	-	-	-

RINGED PLOVER
Charadrius hiaticula

BTO CODE: RP

LIZ CUTTING www.lizcuttingphotos.com

IN WINTER THIS shorebird is widely distributed round the coasts of Britain and Ireland. Small numbers were recorded inland, mainly across northern and eastern England, though probably most were early returning breeders. There has been negligible range change since the *1981–84 Winter Atlas*. Apparent gains in northwest Scotland may be an artefact of improved coverage. It is surprising there are not more losses at this scale, given that the UK wintering population declined by 24% during 1983/84–2008/09 (*SUKB 2012*): this is thought to be part of a redistribution within Europe in response to changing climate (Maclean *et al.* 2008). In contrast, there has been a notable increase in numbers in the Republic of Ireland since the mid 1990s (*I-WeBS Report 2009*).

The breeding distribution also is mainly coastal, with significant gaps only in southwest England, Yorkshire, and southwest Wales. Inland breeding occurs in a range of mainly wetland habitats including along rivers, by lochsides and gravel pits and, in the midlands of Ireland, in harvested peat bogs (Cooney 1998). Gains since the *1988–91 Breeding Atlas* have mostly been inland at such wetlands, whereas the Breckland population has continued to decline, despite little apparent loss of habitat. Overall, there has been a 23% range contraction in Ireland and a 5% expansion in Britain since the *1968–72 Breeding Atlas*. Despite this small range expansion, the British breeding population declined by c.37% during 1984–2007 (Conway *et al.* 2008). Disturbance can have a major impact on population size at coastal sites (Liley & Sutherland 2007). Other factors, not fully understood, are involved in the recent declines of the breeding populations on the Uist machair (Jackson *et al.* 2004).

Highest breeding-season abundance was recorded on the Northern Isles and Outer Hebrides and in parts of western Ireland, though not all records in these places were of breeders. Large concentrations at some major estuaries and at some breeding sites in northwest Scotland are a reflection of the heavy spring passage of high-arctic breeders. Migrants in spring and dispersing juveniles in July may also explain the scatter of inland records of *non-breeding* birds in southern and central England.

BREEDING DISTRIBUTION *2008–11*

POSSIBLE •	3%	4%	4%
PROBABLE ●	6%	8%	7%
CONFIRMED ●	15%	22%	20%
TOTAL	24%	34%	31%

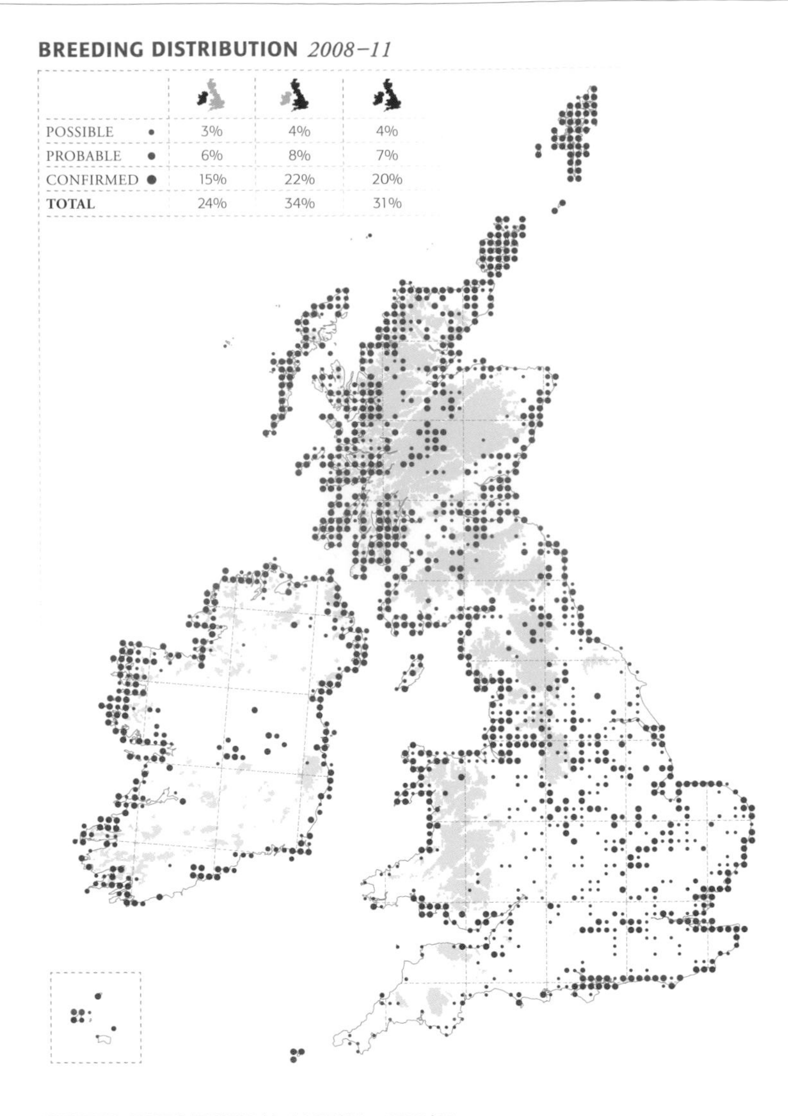

WINTER DISTRIBUTION *2007/08–2010/11*

PRESENT ●	25%	29%	28%

BREEDING DISTRIBUTION CHANGE *since 1968–72*

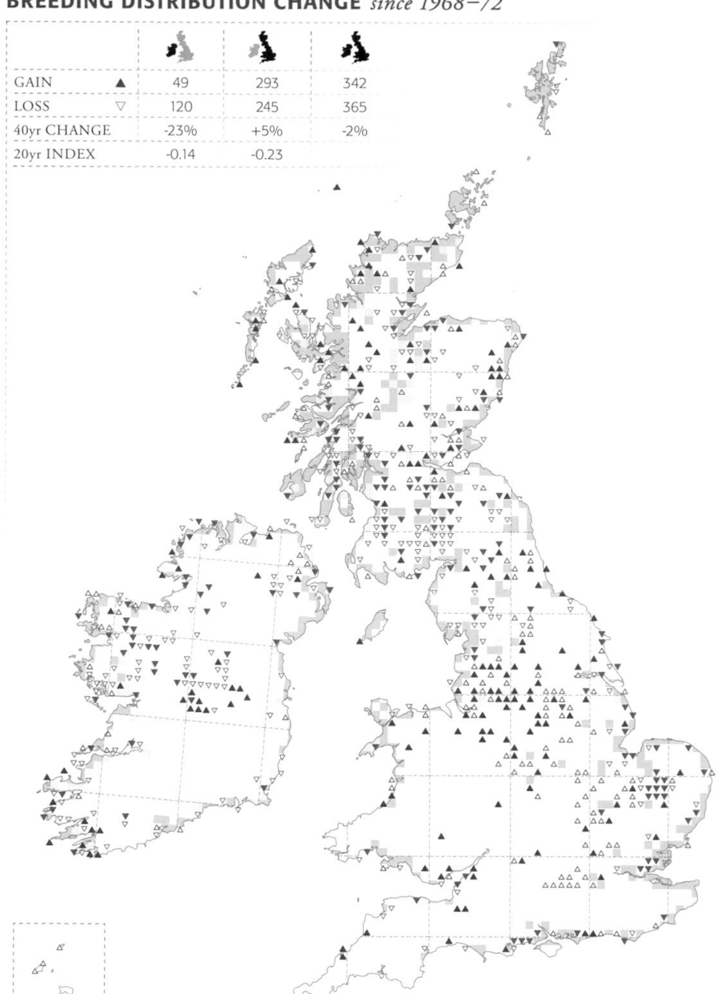

	Ireland	Britain	Britain & Ireland
GAIN ▲	49	293	342
LOSS ▽	120	245	365
40yr CHANGE	-23%	+5%	-2%
20yr INDEX	-0.14	-0.23	

BREEDING RELATIVE ABUNDANCE *2008–11*

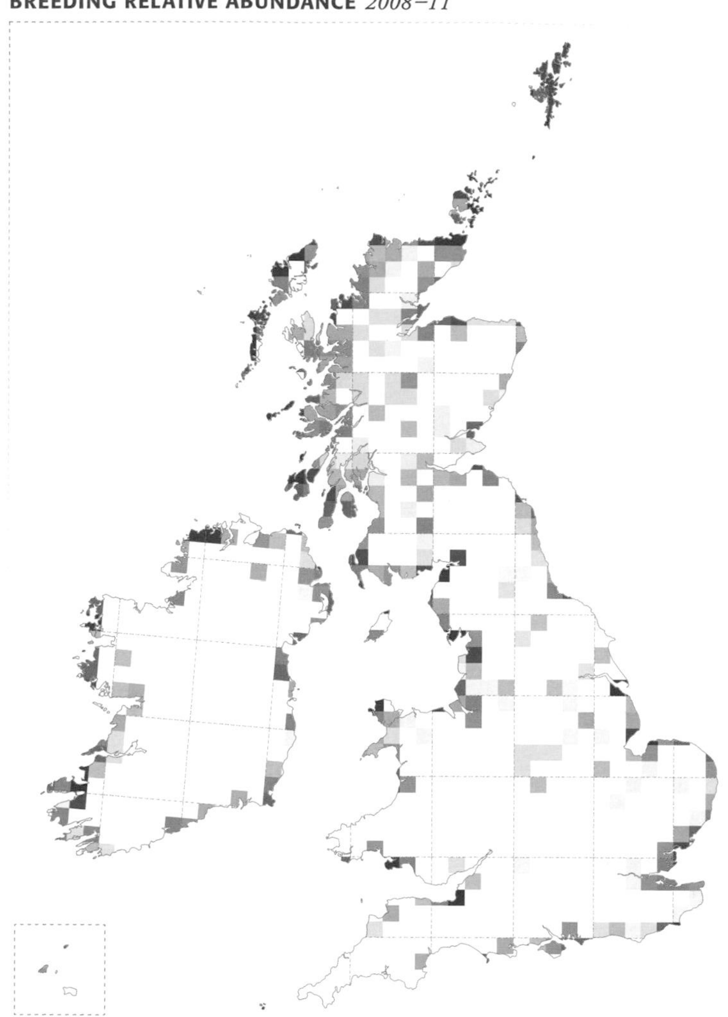

WINTER DISTRIBUTION CHANGE *since 1981–84*

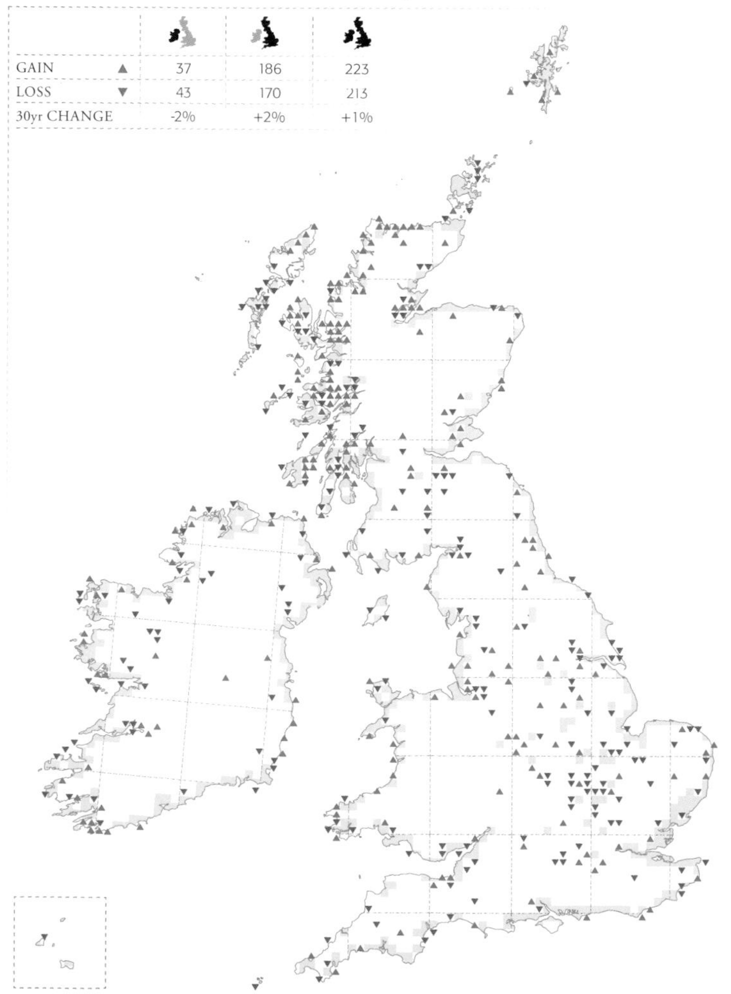

	Ireland	Britain	Britain & Ireland
GAIN ▲	37	186	223
LOSS ▼	43	170	213
30yr CHANGE	-2%	+2%	+1%

WINTER RELATIVE ABUNDANCE *2007/08–2010/11*

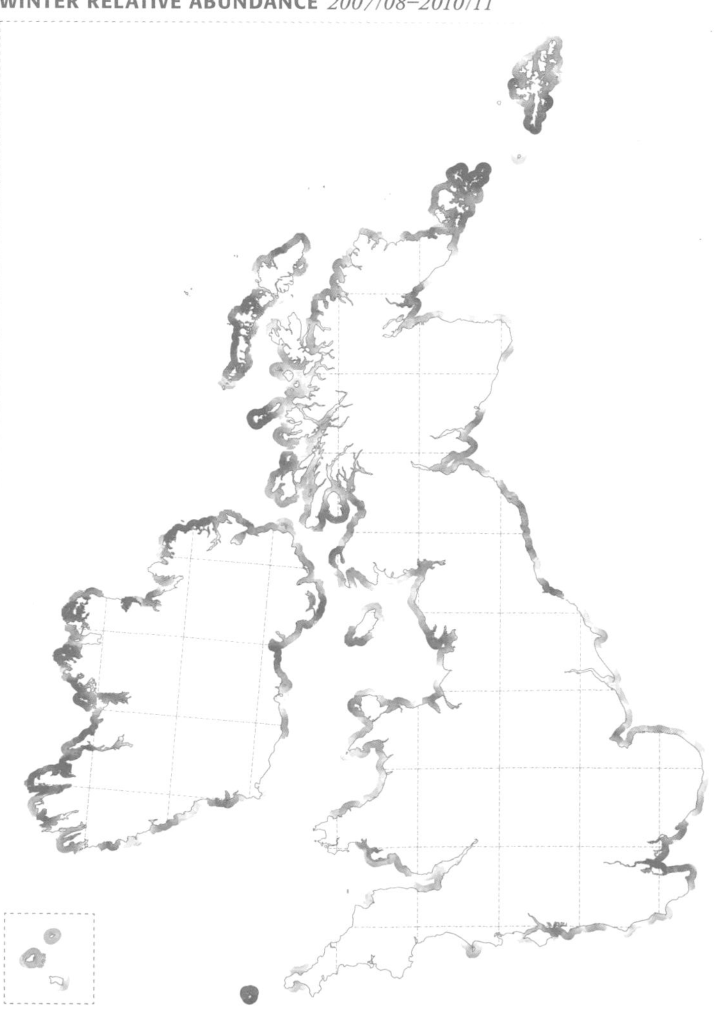

DOTTEREL

Charadrius morinellus

BTO CODE: DO

VIC FROOME

THE DOTTEREL IS a summer visitor to montane alpine breeding areas. In Scotland it is mainly restricted to the Grampian Mountains, with smaller numbers in the northwest Highlands (Galbraith *et al.* 1993). In the Grampians they are present in areas of suitable habitat at altitudes over 700 m. Further north and west they will breed at lower altitudes, for example down to 500 m in northwest Sutherland. During *Bird Atlas 2007–11*, breeding was also recorded in Galloway (one pair) and Cumbria: these and other southern records are mapped at 50-km resolution. The breeding abundance map shows an incomplete picture, owing to low encounter rates during TTVs. There were no breeding records for Ireland, where the species is very rarely found nesting (*Birds in Ireland*).

The apparent 78% range expansion since the *1968–72 Breeding Atlas*, most evident in northwest Scotland, is probably largely due to improved coverage. These data also suggest a 17% range contraction since the *1988–91 Breeding Atlas*, but part of that is an artefact of the codes used for breeding evidence, with many of the birds classed as *seen* in 1988–91 being migrants rather than *possible* breeders.

A better assessment of status can be drawn from sample surveys that returned British population estimates of 840, 630 (recently revised to 747) and 423 males in 1987/88, 1999 and 2011 respectively (Galbraith *et al.* 1993; Whitfield 2002; D. Hayhow pers comm; Holling *et al.* 2013). These figures suggest a halving of the breeding population since the *1988–91 Breeding Atlas*. The causes of the decline are unclear. Losses in England have been attributed to overgrazing, disturbance and deposition of atmospheric nitrogen (*Birds in England*). In Scotland, redistribution to Fennoscandia and problems in the wintering grounds have been suggested as alternative causes (Whitfield 2002).

Migrants, passing through to breeding sites in Scotland and Fennoscandia, were reported from 162 10-km squares throughout Britain, and from four squares in Ireland.

Dotterel were also recorded in nine 10-km squares in winter. Six of the English records were during early November and are best regarded as late migrants.

BREEDING DISTRIBUTION *2008–11*

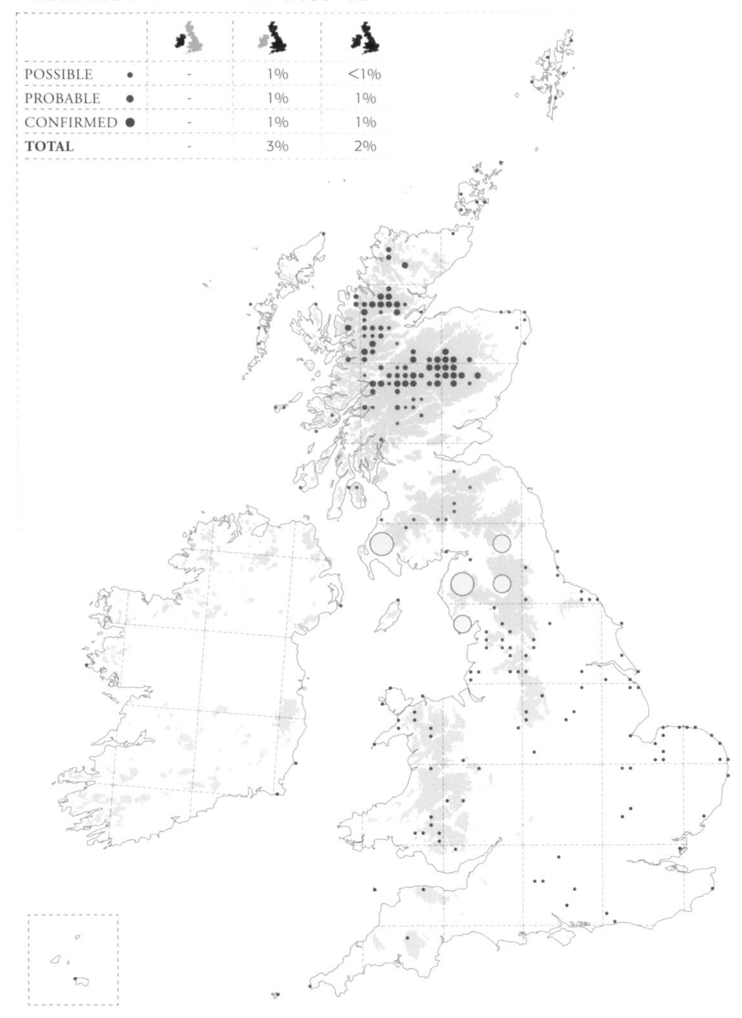

POSSIBLE •	-	1%	<1%
PROBABLE •	-	1%	1%
CONFIRMED •	-	1%	1%
TOTAL	-	3%	2%

WINTER DISTRIBUTION *2007/08–2010/11*

PRESENT •	<1%	<1%	<1%

BREEDING DISTRIBUTION CHANGE *since 1988–91*

GAIN ▲	-	35	35
LOSS ▽	-	52	52
20yr CHANGE	-	-17%	-17%
20yr INDEX	-	-0.16	

BREEDING RELATIVE ABUNDANCE *2008–11*

COLIN RICHARDS

◀ **Remarkably, after laying a clutch for a male in Scotland, some 80–90% of females then relocate to Scandinavia or elsewhere in the same year in search of further nesting opportunities.**

GOLDEN PLOVER
Pluvialis apricaria

BTO CODE: GP

PAUL HILLION

WINTERING GOLDEN PLOVERS are recorded through most of the lowlands of Britain and the midlands and coastal regions of Ireland. The scattered records in upland areas result from birds briefly returning to the margins of their breeding areas in February. There are few clear patterns in the gains and losses, though perhaps a suggestion of apparent gains around the western and northern fringes of Scotland, which could result from increased coverage. The number of Golden Plovers recorded on winter waterbird surveys, especially on eastern British estuaries, has increased significantly since the mid 1980s (Gillings *et al.* 2006), and a comparison of the winter abundance map with that from the *1981–84 Winter Atlas* suggests a subtle increase in density in Lincolnshire and East Anglia. Decreases in western Britain (e.g. Lock 1994) may have been temporarily reversed as a consequence of the two cold winters during the Atlas period. The pattern of abundance in Ireland is relatively unchanged.

During the breeding season, birds are to be found in the uplands of the north and west of Britain and of Ireland. Within the core of the breeding range in Scotland, highest densities occur on the Outer Hebrides, Shetland and the flows of Caithness and Sutherland, although comparable densities still exist in parts of the English Pennines. In Ireland and Wales breeding densities are low. The breeding-season range loss that was detected in the *1988–91 Breeding Atlas* has accelerated, such that half of the Irish range and one fifth of the British range have been lost over the last 40 years, mirroring the 13% UK population decline (*SUKB 2012*). Losses have been greatest in the southern uplands of Scotland, and around the margins of the smaller outlying British and Irish populations, fragmenting them further. The small population in southwest England appears all but extinct, with no *confirmed* breeding during 2008–11. Possible causes of decline in the uplands include afforestation, increases in generalist predators and changes to hill farming (*Birds of Scotland*; Pearce-Higgins *et al.* 2009a). Drying of peatland soils, as a result of artificial drainage and summer warming, is also detrimental (Pearce-Higgins *et al.* 2010).

Sponsored by Scottish Countryside Alliance

BREEDING DISTRIBUTION *2008–11*

POSSIBLE •	<1%	4%	3%
PROBABLE ●	1%	8%	6%
CONFIRMED ●	2%	12%	9%
TOTAL	3%	24%	18%

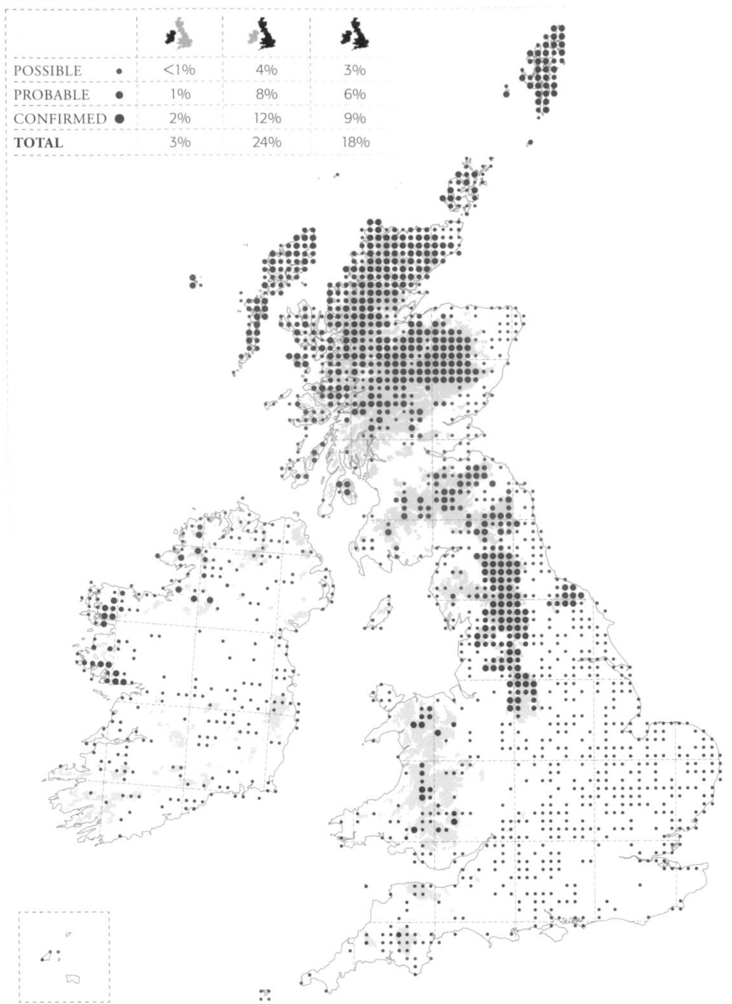

WINTER DISTRIBUTION *2007/08–2010/11*

PRESENT ●	51%	66%	62%

BREEDING DISTRIBUTION CHANGE *since 1968–72*

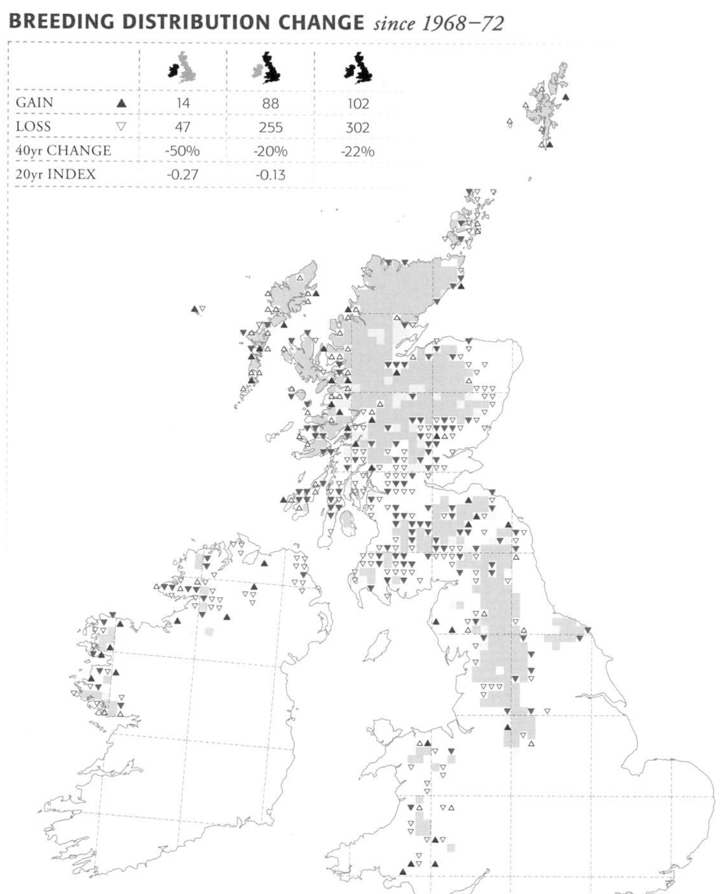

GAIN ▲	14	88	102
LOSS ▽	47	255	302
40yr CHANGE	-50%	-20%	-22%
20yr INDEX	-0.27	-0.13	

WINTER DISTRIBUTION CHANGE *since 1981–84*

GAIN ▲	205	531	736
LOSS ▼	167	242	409
30yr CHANGE	+8%	+19%	+17%

BREEDING RELATIVE ABUNDANCE *2008–11*

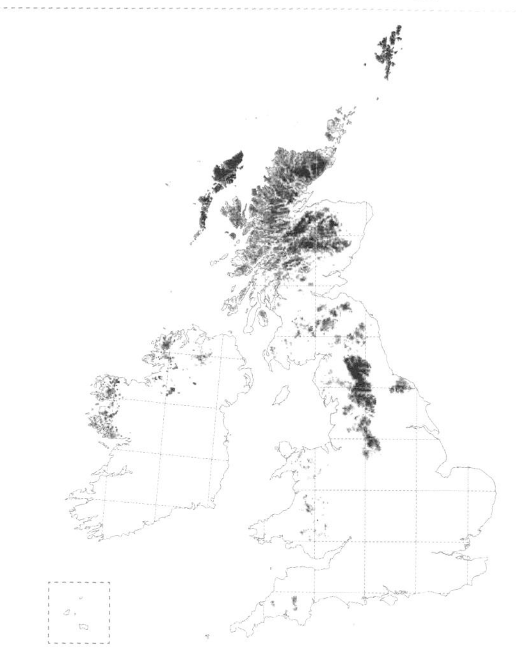

BREEDING RELATIVE ABUNDANCE CHANGE *since 1988–91*

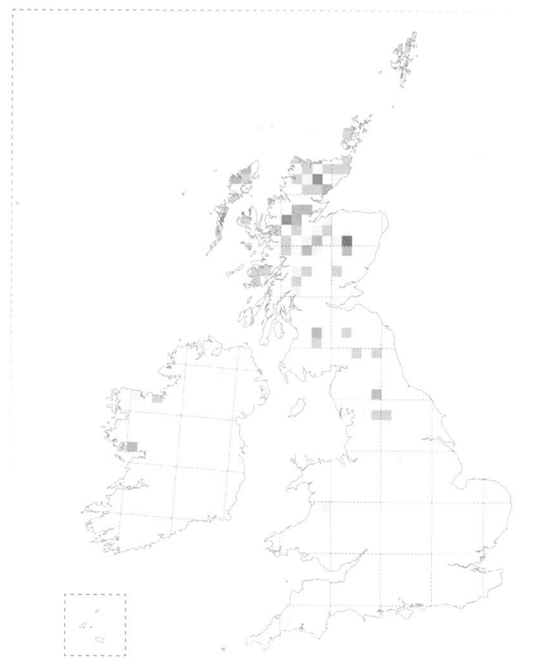

WINTER RELATIVE ABUNDANCE *2007/08–2010/11*

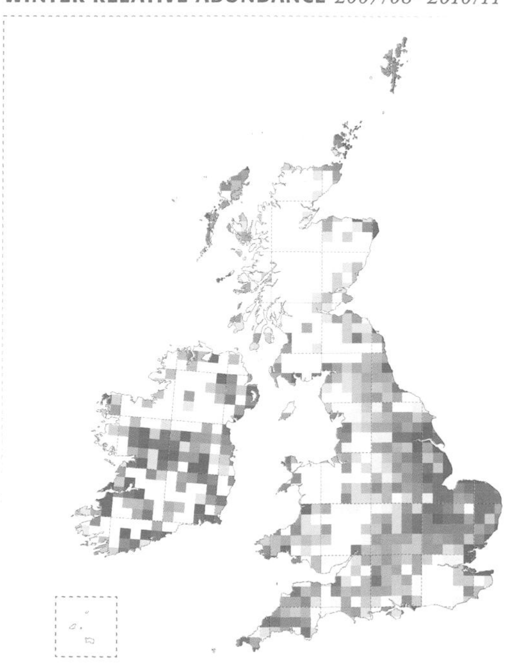

GREY PLOVER

Pluvialis squatarola

BTO CODE: GV

JIM ALMOND www.shropshirebirder.co.uk

GREY PLOVERS ARE winter visitors and passage migrants to Britain & Ireland from their breeding grounds on the Siberian tundra. They are widely distributed around the coasts, with a particular preference for areas with intertidal mud and sandflats. Grey Plovers are consequently scarcer in northern and western Scotland and the north of Ireland, where these habitats are less available. Birds were occasionally recorded inland, mostly in eastern England at wetlands or on farmland associating with flocks of Golden Plovers.

The largest concentrations in winter occurred on the major estuaries, in Britain and to a lesser extent in Ireland. During 2010/11, seven sites passed the 2,500-bird threshold to qualify as sites of international importance, headed by the Wash in eastern England with a five-winter average peak of over 10,000 birds (*WeBS Report 2011*).

There has been little change in range since the *1981–84 Winter Atlas*, with range expansions of 3% in Britain and 1% in Ireland. Most of the gains have been at more marginal sites, particularly inland ones, where birds are present in only very low and variable numbers. Exceptions are in Orkney and the Moray Firth, where there have been significant gains.

Wintering numbers in the UK increased from the 1970s to a peak in the mid 1990s and have since declined. So although numbers in the UK in 2008/09 were 72% higher than in 1983/84, they were 15% lower than in 1988/89 (*WeBS Report 2011*; *SUKB 2012*). In the Republic of Ireland there was a 36% decline between 1994/95 and 2008/09 (*I-WeBS Report 2009*). It is unclear whether these changes reflect large-scale redistribution in response to climate change (Maclean *et al.* 2008) or population declines (Delany *et al.* 2009). At the flyway level the population is thought to have declined during 1990–2010 (Wetlands International 2013).

During the breeding season, *non-breeding* birds were reported in 361 10-km squares in Britain and 28 in Ireland. These include spring and early autumn passage migrants and summering first-year birds. Seven 10-km squares coded as *seen* in the *1988–91 Breeding Atlas* will also have related to passage or summering birds.

Sponsored by Wash Wader Ringing Group

WINTER DISTRIBUTION *2007/08–2010/11*

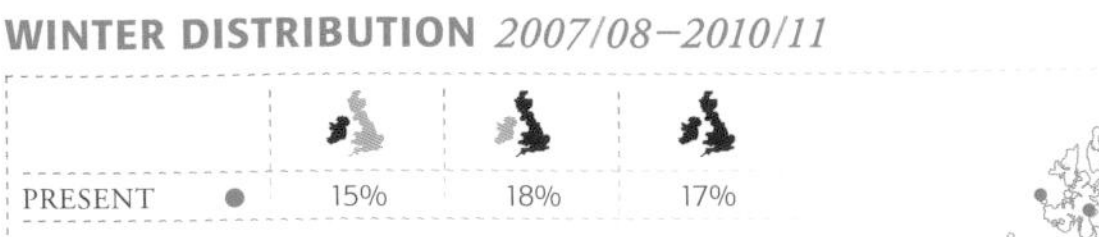

PRESENT ●	15%	18%	17%

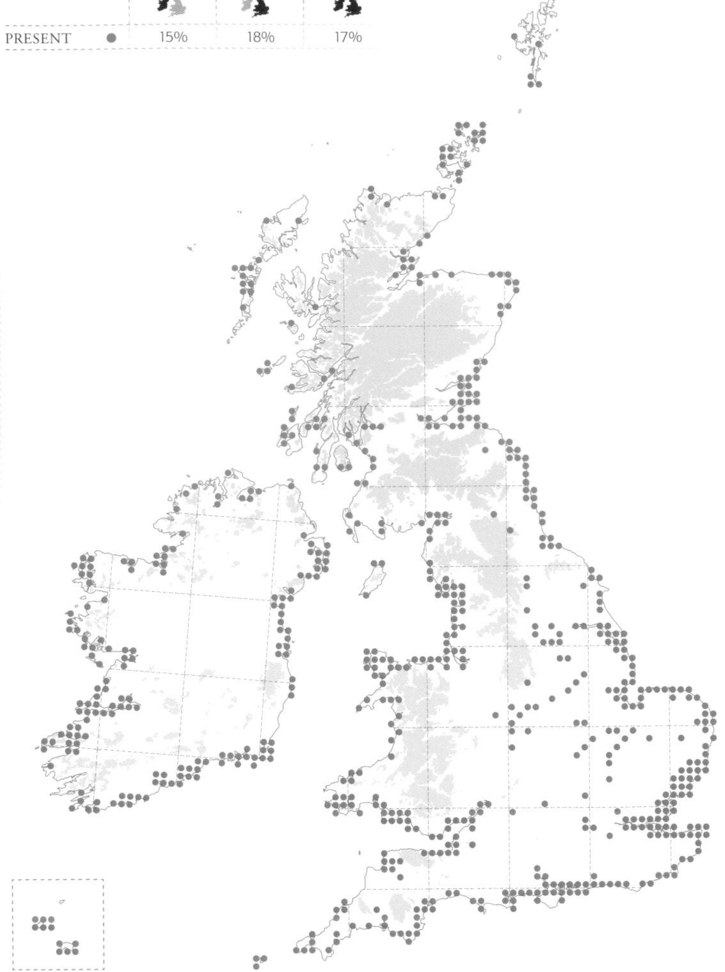

BREEDING DISTRIBUTION *2008–11*

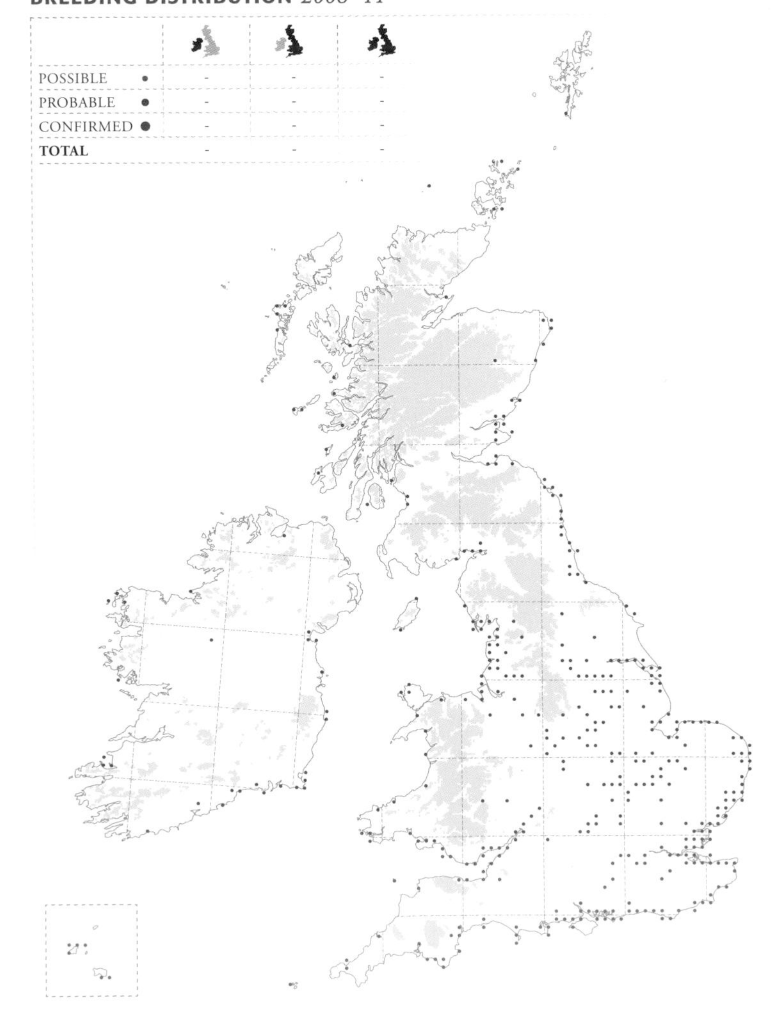

POSSIBLE ●	-	-	-
PROBABLE ●	-	-	-
CONFIRMED ●	-	-	-
TOTAL	-	-	-

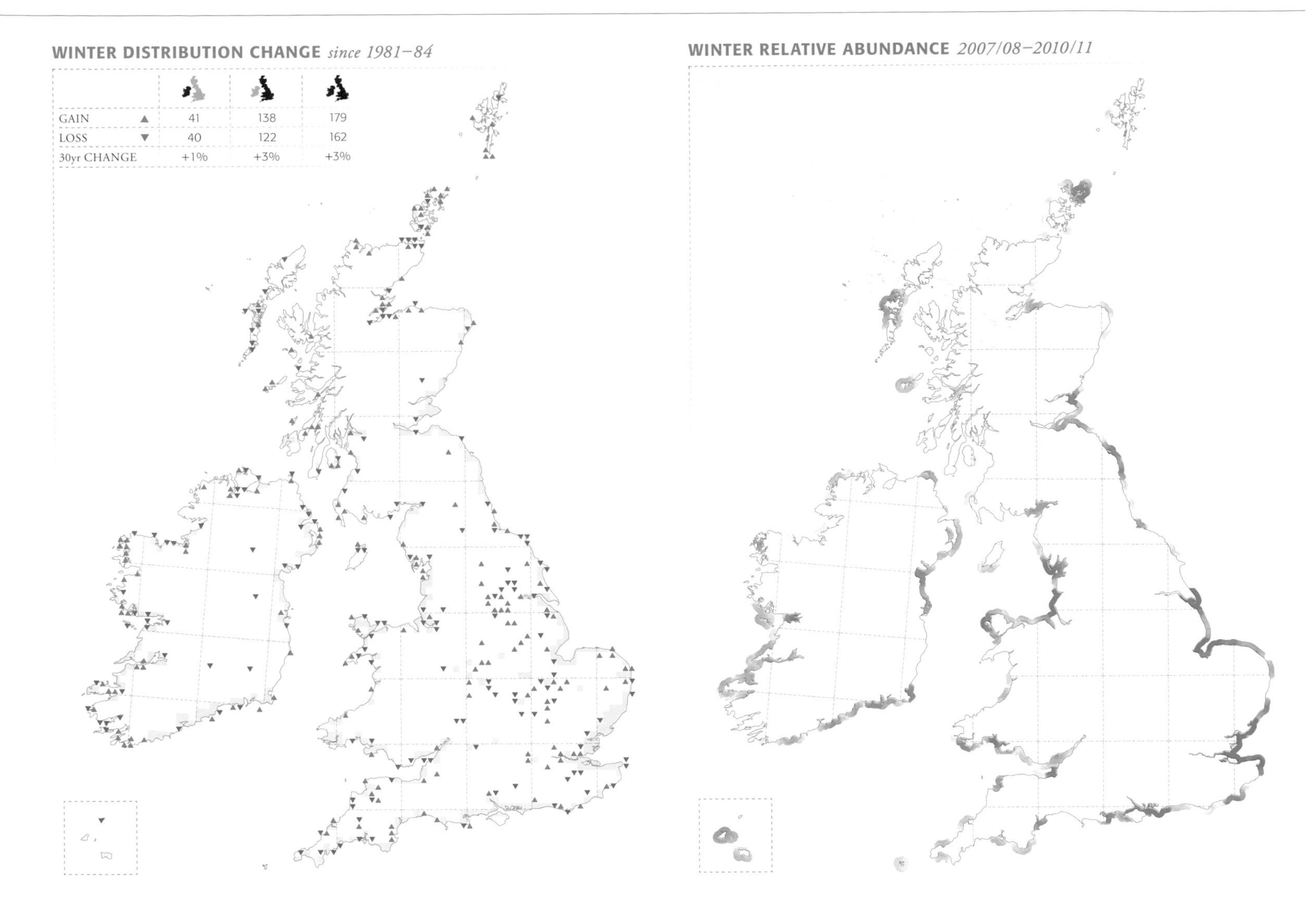

WINTER DISTRIBUTION CHANGE since 1981–84
GAIN ▲ 41 138 179
LOSS ▼ 40 122 162
30yr CHANGE +1% +3% +3%
WINTER RELATIVE ABUNDANCE 2007/08–2010/11

◀ The Wash and Dengie Flats, Essex, are the two most important sites in Britain and hold approximately half of all British Grey Plovers in winter between them.

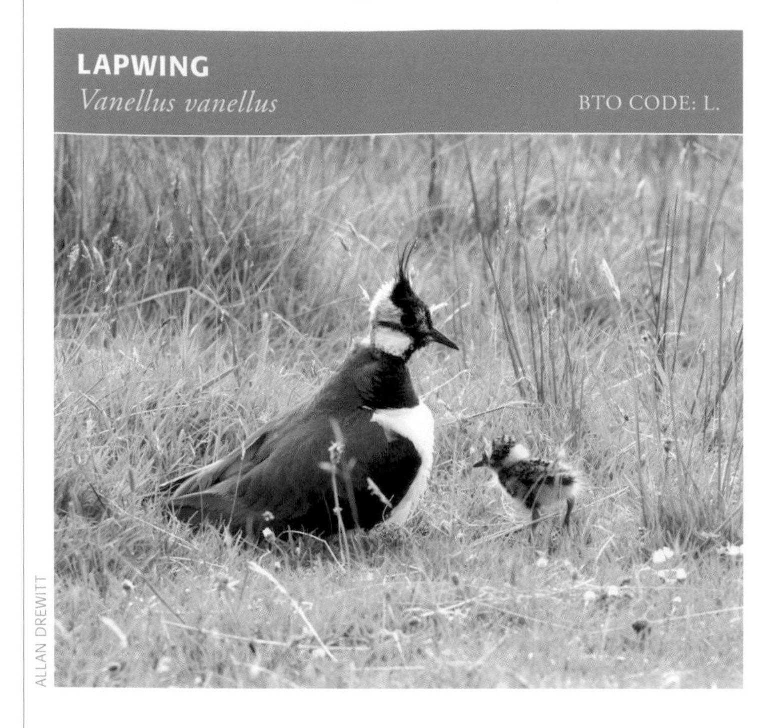

LAPWING

Vanellus vanellus

BTO CODE: L.

ALLAN DREWITT

DESPITE DECADES OF population decline (Shrubb 2007), the Lapwing is still the most widespread breeding wader species in Britain & Ireland. Over the last 40 years, it has been lost as a breeder from most of southwest England, western Wales and western mainland Scotland, at a rate of c.110 10-km squares per decade. Losses throughout Ireland mean that the current range is less than half that in the *1968–72 Breeding Atlas*. Numbers at the Shannon Callows, formerly an important stronghold for breeding waders, declined by more than 80% during 1987–2002 (Tierney *et al.* 2002). The abundance change map indicates that even in those parts of Britain & Ireland where Lapwings still breed, densities have declined. Changes in agricultural practice are the main drivers (Baines 1988; Sheldon *et al.* 2004), though predation pressure may be important in some areas (Bolton *et al.* 2007). Currently, densities remain highest in northwest England, Orkney, Shetland and the Outer Hebrides; those in Ireland are low throughout.

After the Woodcock and Snipe, the Lapwing is the third most widespread wintering wader in Britain & Ireland, occupying 78% of squares. This figure is very similar to that from the *1981–84 Winter Atlas*, although the Lapwing's propensity to move in response to cold weather means that distribution patterns may differ between winters. Range change is negligible in Britain, but in Ireland there has been a modest 12% range contraction, though with no clear spatial pattern. Contractions might have been greater had the run of mild winters continued through the full Atlas period. The two cold winters during *Bird Atlas 2007-11* probably resulted in a greater westward movement from the Continent and from Britain into Ireland. Wintering concentrations are highest in the Shannon and western midlands regions of Ireland, and in Britain in the Somerset Levels, Fens, northwest England, the Uists, Caithness and Orkney. The former core area in central England is less dominant and Fenland and the east coast now more prominent than in the *1981–84 Winter Atlas*. Like many wader species, there is evidence of increased wintering on and near eastern British estuaries (Gillings *et al.* 2006), possibly in response to changing climate (Maclean *et al.* 2008).

Sponsored by Philip Merricks—Elmley National Nature Reserve

BREEDING DISTRIBUTION *2008–11*

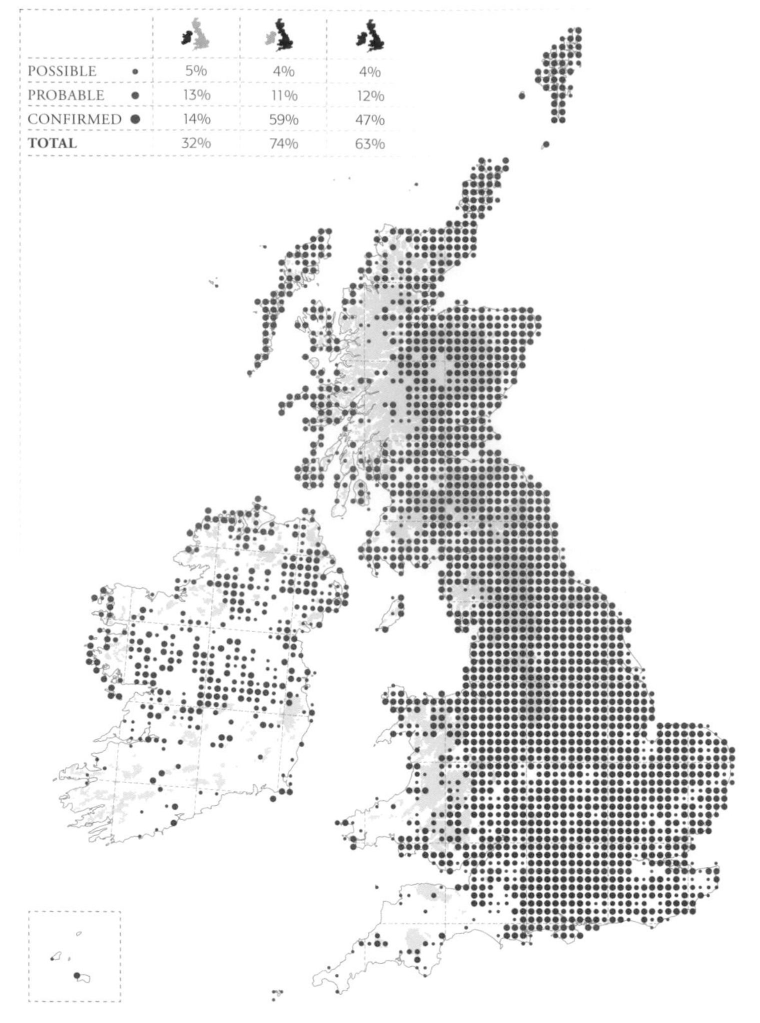

	Ireland	Britain	Britain & Ireland
POSSIBLE •	5%	4%	4%
PROBABLE •	13%	11%	12%
CONFIRMED •	14%	59%	47%
TOTAL	32%	74%	63%

WINTER DISTRIBUTION *2007/08–2010/11*

	Ireland	Britain	Britain & Ireland
PRESENT •	72%	80%	78%

BREEDING DISTRIBUTION CHANGE *since 1968–72*

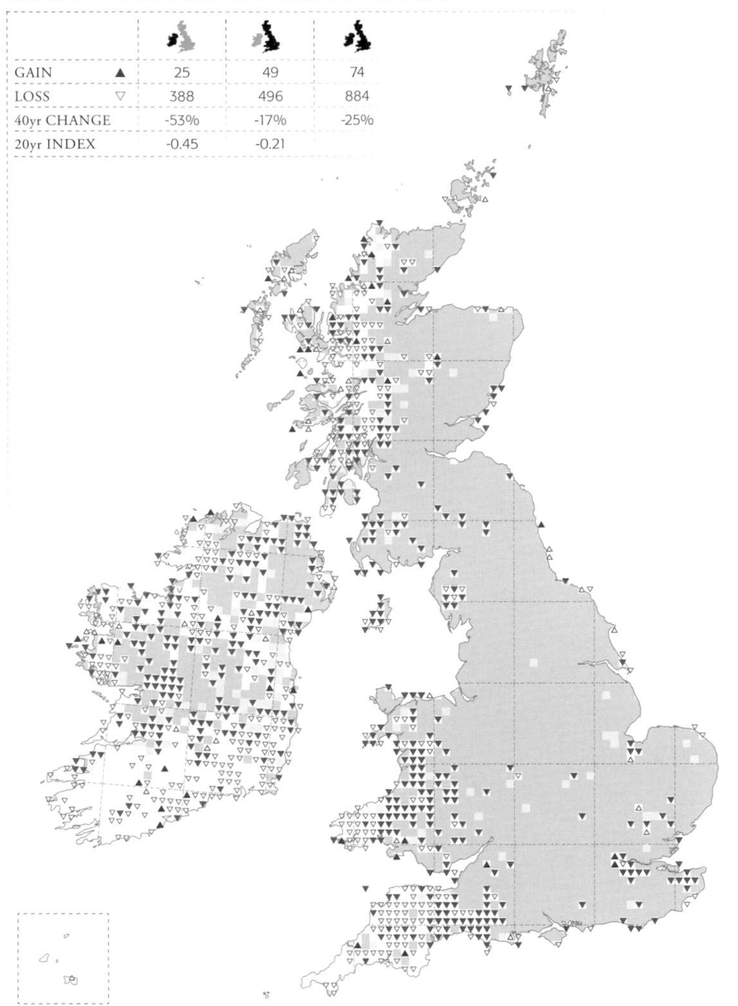

GAIN ▲	25	49	74
LOSS ▽	388	496	884
40yr CHANGE	-53%	-17%	-25%
20yr INDEX	-0.45	-0.21	

WINTER DISTRIBUTION CHANGE *since 1981–84*

GAIN ▲	60	193	253
LOSS ▼	159	171	330
30yr CHANGE	-12%	+1%	-3%

BREEDING RELATIVE ABUNDANCE *2008–11*

BREEDING RELATIVE ABUNDANCE CHANGE *since 1988–91*

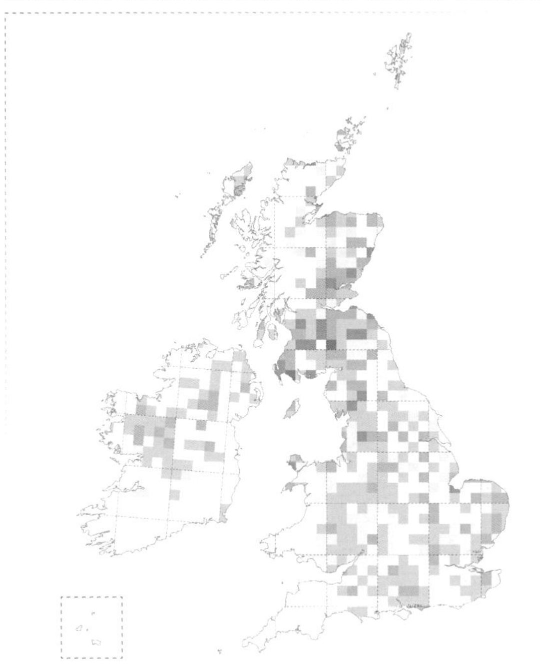

WINTER RELATIVE ABUNDANCE *2007/08–2010/11*

KNOT
Calidris canutus

BTO CODE: KN

ALLAN DREWITT

KNOT OF THE race *islandica* breeding in Greenland and the Canadian Arctic commute to northwest Europe and over 65% winter around the coasts of Britain & Ireland (Delany *et al.* 2009), where they feed on muddy and sandy shores, especially in estuaries. They are generally absent from northern and western Scotland and from other stretches of rocky coastline. The Knot's preference for estuaries is readily apparent from the winter abundance map: most of the sites of international importance are estuarine, with the Wash, where the recent five-winter average peak has been 138,000 birds, topping the list (*WeBS Report 2011*). Dundalk Bay, with an average of 6,750 birds, is the most important Irish site (*I-WeBS Report 2009*).

The number of occupied 10-km squares has increased in Britain by 27% and in Ireland by 58% since the *1981–84 Winter Atlas*. A concurrent 15% increase in numbers was measured in the UK between 1983/84 and 2008/09 (*SUKB 2012*). Wintering numbers in the Republic of Ireland were stable from 1994/95 until 2005/06 but there has since been a marked increase, with an overall increase of 42% between 1994/95 and 2008/09 (*I-WeBS Report 2009*). Recent increases have also been reported in the Netherlands (Hornman *et al.* 2011) but the overall flyway population has not shown any clear trend (Wetlands International 2013).

Coastal range gains are scattered throughout, with little discernible pattern. Changes to habitat and prey supply, including human overexploitation of shellfisheries, can cause local fluctuations at individual wintering and passage sites (Atkinson *et al.* 2010). Interchange between nearby sites, such as the Dee and Ribble Estuaries in northwest England, can also lead to large fluctuations (*WeBS Report 2011*). The many inland gains are of marginal population significance, given that most were occupied fleetingly and by fewer than five individuals.

Whereas some individuals depart for continental pre-breeding moulting areas in mid or late March, others fatten and moult into breeding plumage prior to departure in early May for staging areas in Iceland and Norway (Delany *et al.* 2009). These moulting assemblages, passage birds and individuals that summer on some estuaries account for the 423 10-km squares occupied by *non-breeding* birds.

Sponsored by William Dick

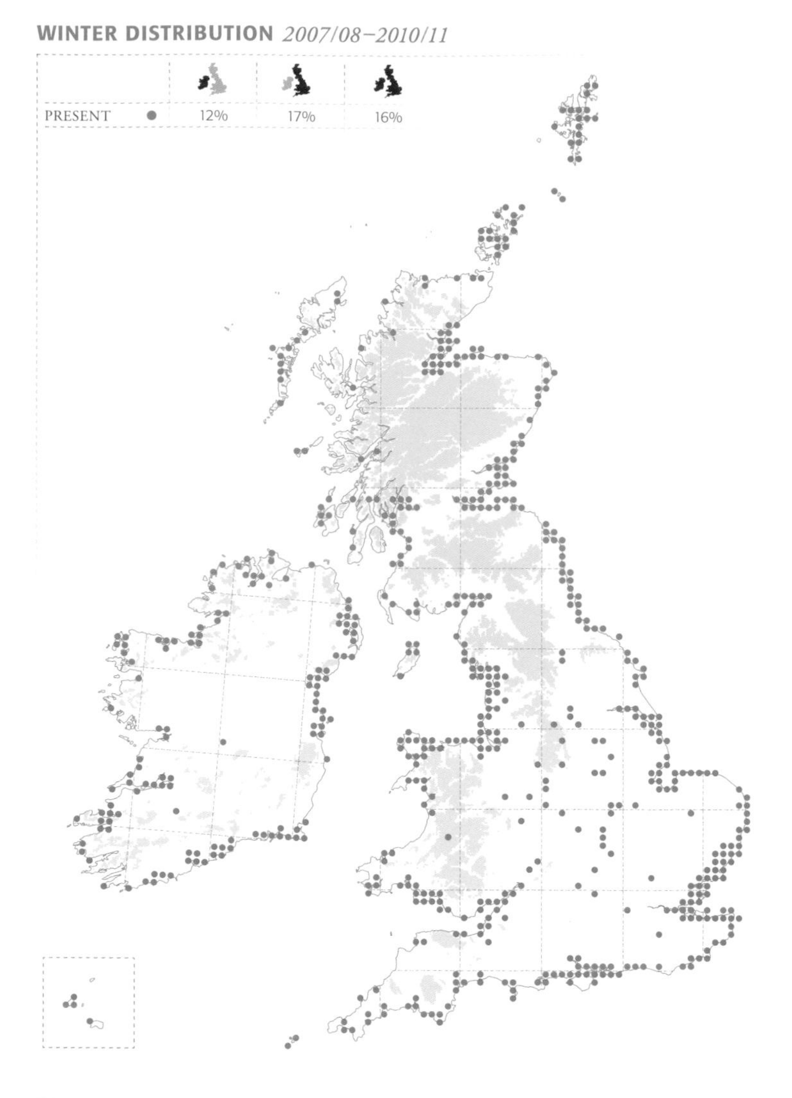

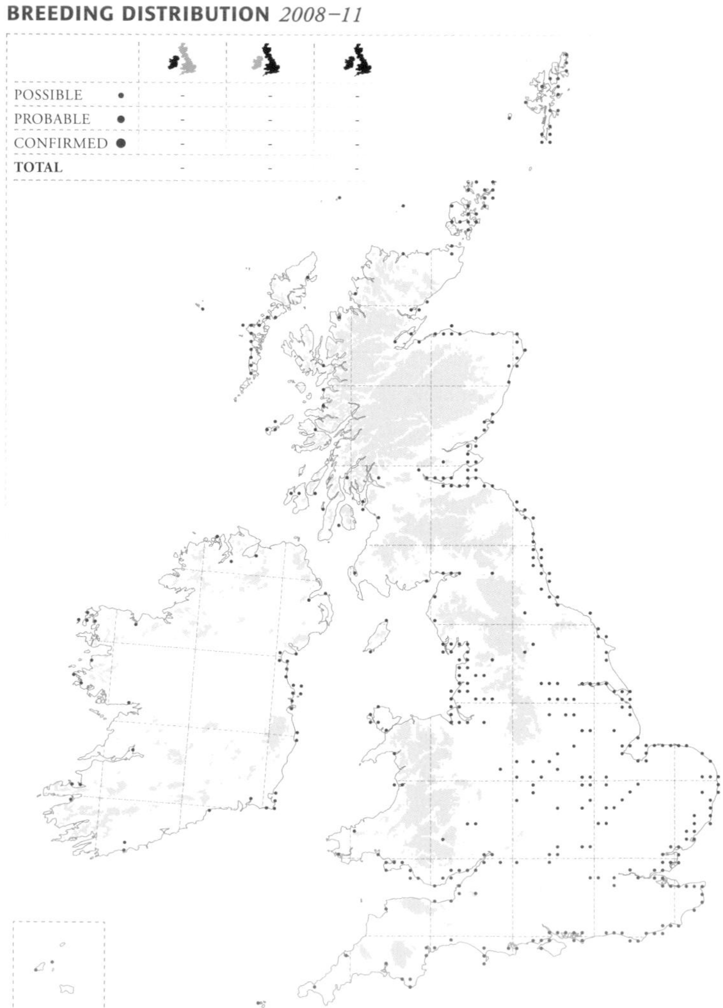

WINTER DISTRIBUTION CHANGE *since 1981–84*

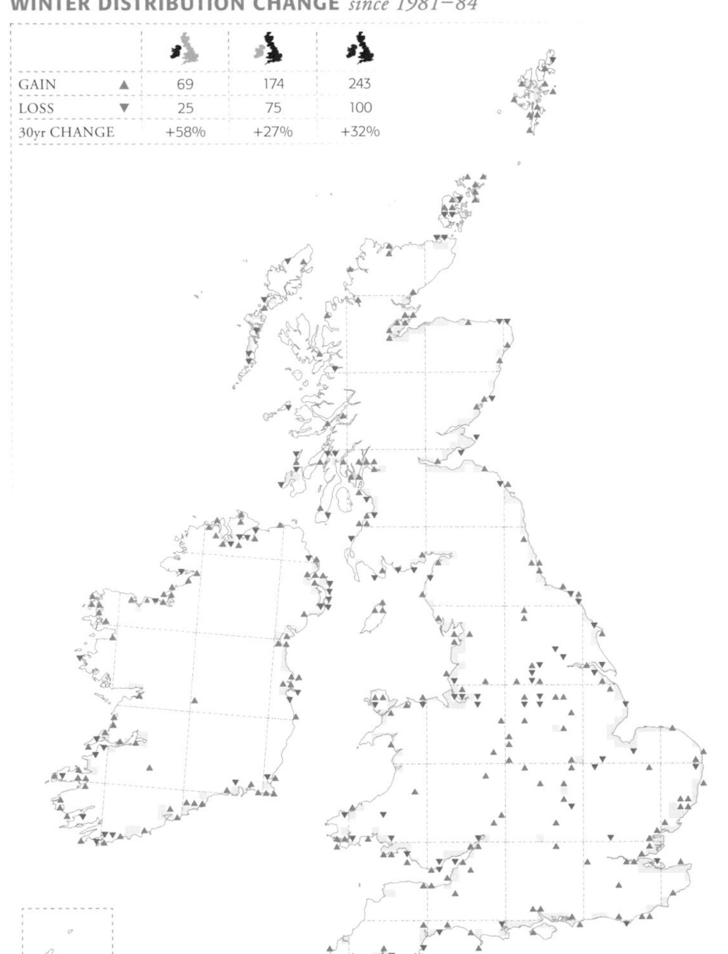

GAIN	▲	69	174	243
LOSS	▼	25	75	100
30yr CHANGE		+58%	+27%	+32%

WINTER RELATIVE ABUNDANCE *2007/08–2010/11*

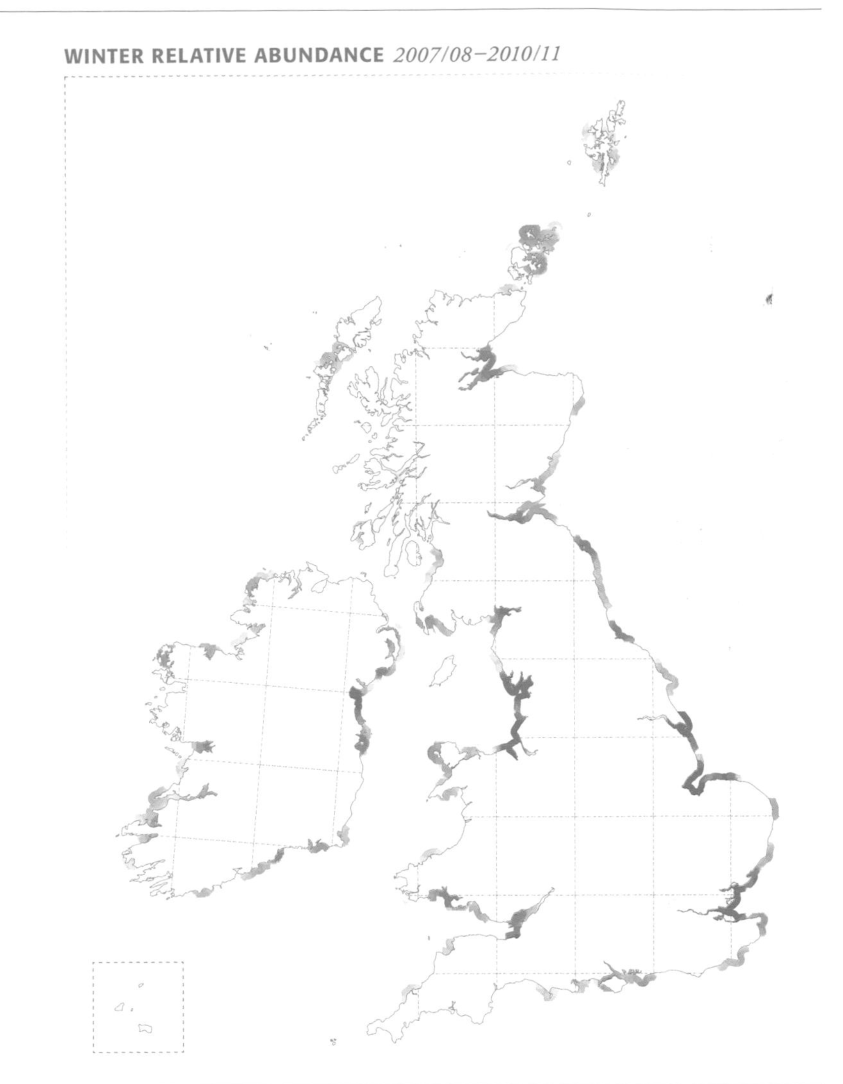

LITTLE STINT
Calidris minuta

BTO CODE: LX

LITTLE STINTS PASS through Britain & Ireland en route from their breeding areas on arctic tundra in Scandinavia and Siberia to wintering areas in Africa and around the Mediterranean, mainly during late July–October, and there is a smaller return movement in late spring. During the winter period birds were recorded from 108 10-km squares, mostly in the southern half of England and at coastal sites, though inland reservoirs and gravel pits were also utilised. Records from 45 of these squares were in November only and may have involved birds on late passage, rather than wintering. There has been a 78% increase in the number of occupied squares since the *1981–84 Winter Atlas*.

Sponsored by a Highland SOC Member

WINTER DISTRIBUTION *2007/08–2010/11*

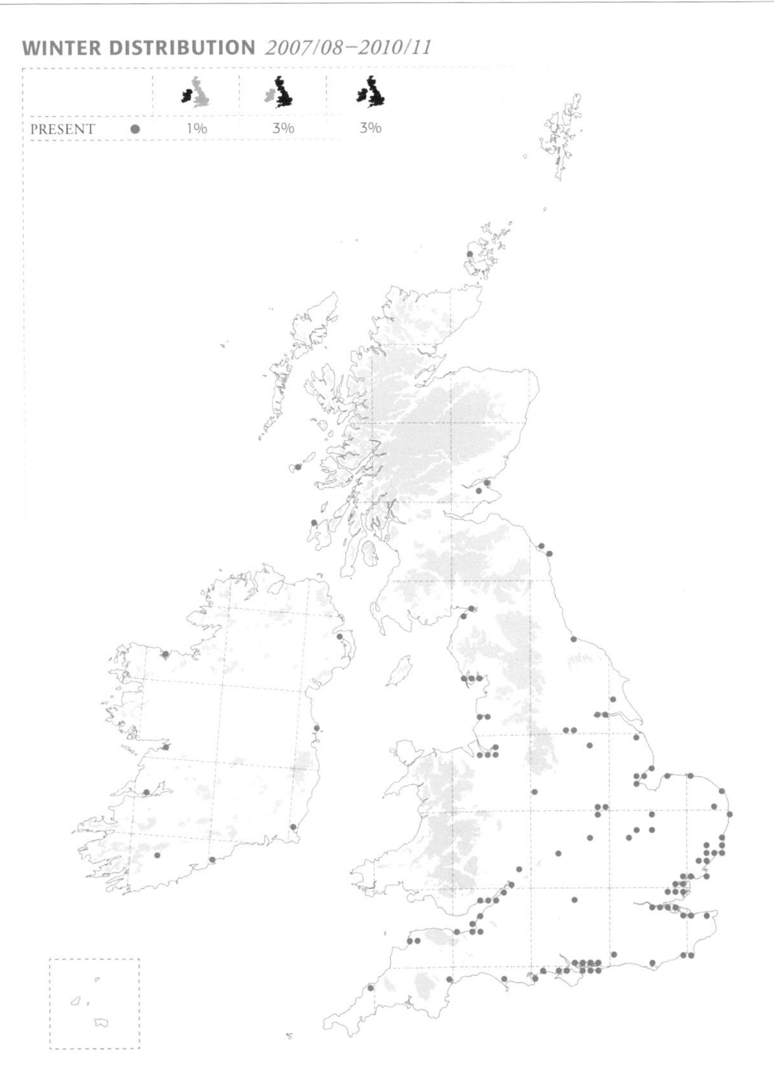

PRESENT	●	1%	3%	3%

SANDERLING
Calidris alba

BTO CODE: SS

ALLAN DREWITT

THE SANDERLING IS a high-arctic breeder and occurs in Britain & Ireland on passage and during winter. The winter distribution is predominantly coastal and Sanderlings are scarcer inland than most other waders. The distribution map shows winter records along most stretches of coast but the abundance map confirms that Sanderlings are most abundant away from estuaries, instead preferring areas where sandy beaches and sandbars prevail. The North Sea coast, Outer Hebrides, Ribble Estuary, Carmarthen Bay and Tralee Bay (Co. Kerry) held the highest densities.

The species has expanded in range, with 31% and 21% range expansions in Britain and Ireland respectively since the *1981–84 Winter Atlas*. The gains in Britain have occurred throughout most of the coastline, and at inland sites, although distribution along the North Sea coast has remained stable. In Ireland, gains have been predominantly in the west and northwest. Gains are consistent with a 76% increase in winter numbers in the UK during 1983/84–2008/09 and a 141% increase in the Republic of Ireland during 1994/95–2008/09 (*I-WeBS Report 2009*; *SUKB 2012*). These figures might not represent the population as a whole, however, because many birds on open coasts are not routinely monitored (Rehfisch *et al.* 2003; Crowe *et al.* 2008). Approximately 64% of the Irish population occurs on non-estuarine coasts, where they have also increased, but by only 5% during 1994/95–2003/04 (Crowe *et al.* 2008). Increases have been reported in several other northwest European countries and not all can be attributed to improved count coverage. Patterns of increase differ between winter and spring passage periods and among habitat types (Reneerkens *et al.* 2009).

It was previously thought that Sanderlings wintering in Britain & Ireland were all of Siberian origin, but colour-marking studies have since shown that some Greenland-breeding birds not only pass through, but also overwinter here (Reneerkens *et al.* 2009). Birds on passage, shown as *non-breeding*, were reported from 528 10-km squares in Britain and 67 10-km squares in Ireland.

Sponsored by Joan Harvey

WINTER DISTRIBUTION *2007/08–2010/11*

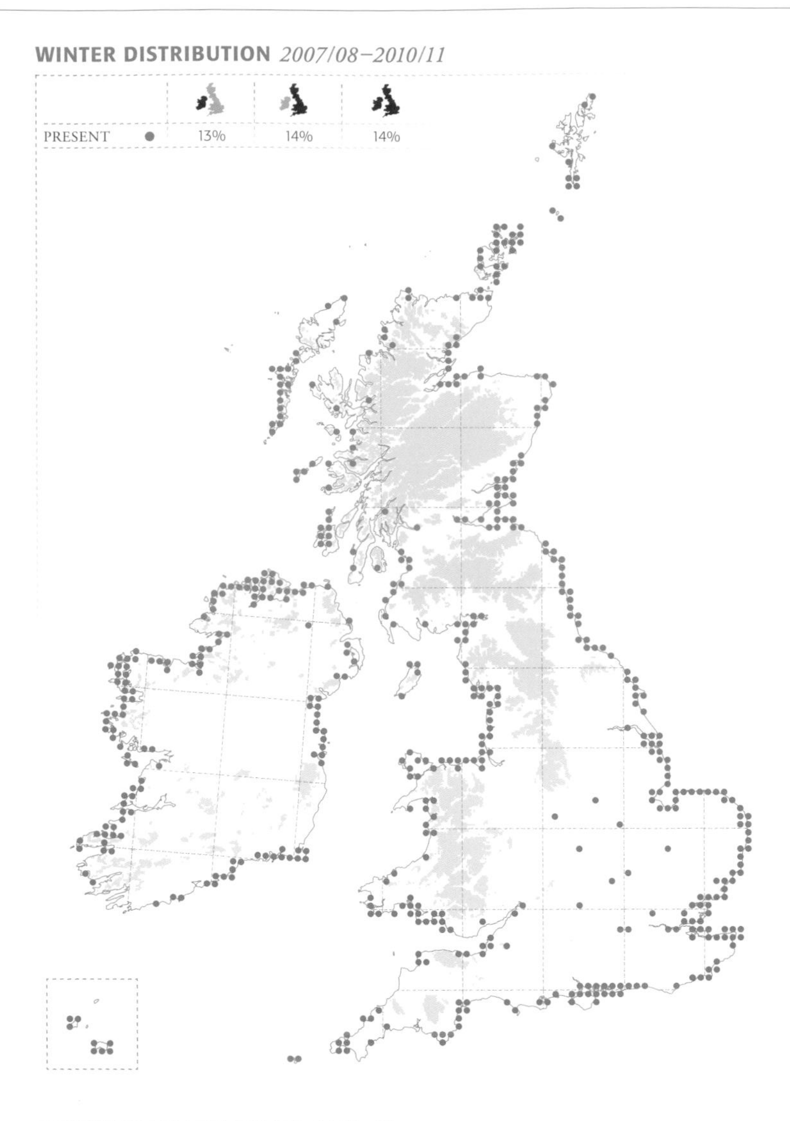

BREEDING DISTRIBUTION *2008–11*

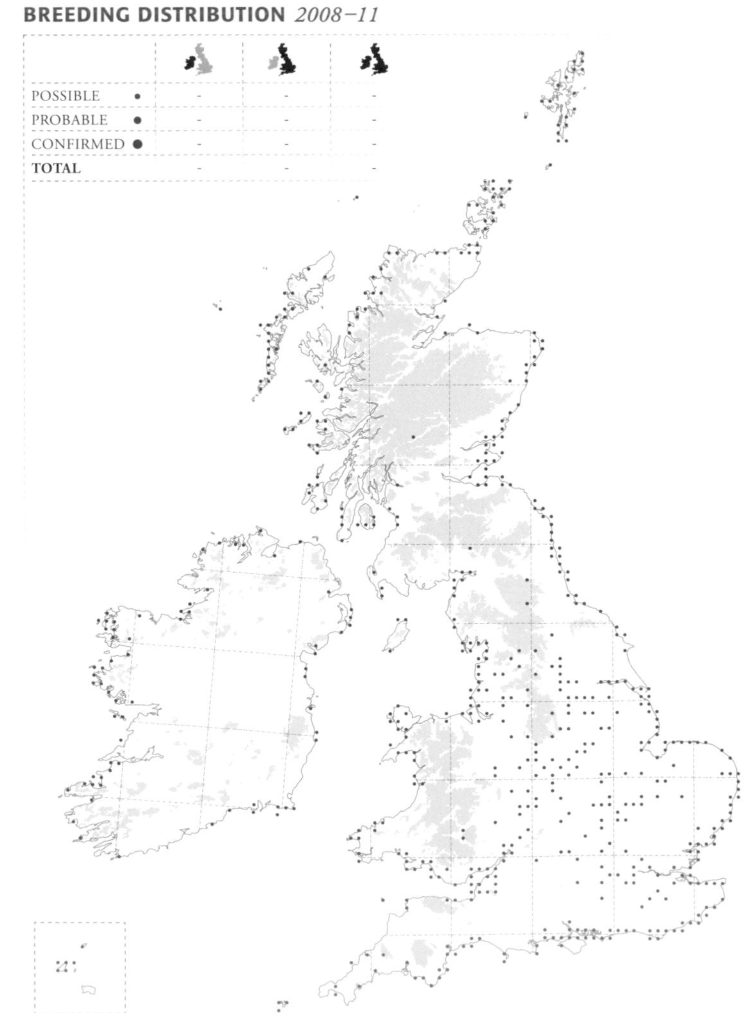

WINTER DISTRIBUTION CHANGE *since 1981–84*

GAIN ▲	48	143	191
LOSS ▼	25	54	79
30yr CHANGE	+21%	+31%	+28%

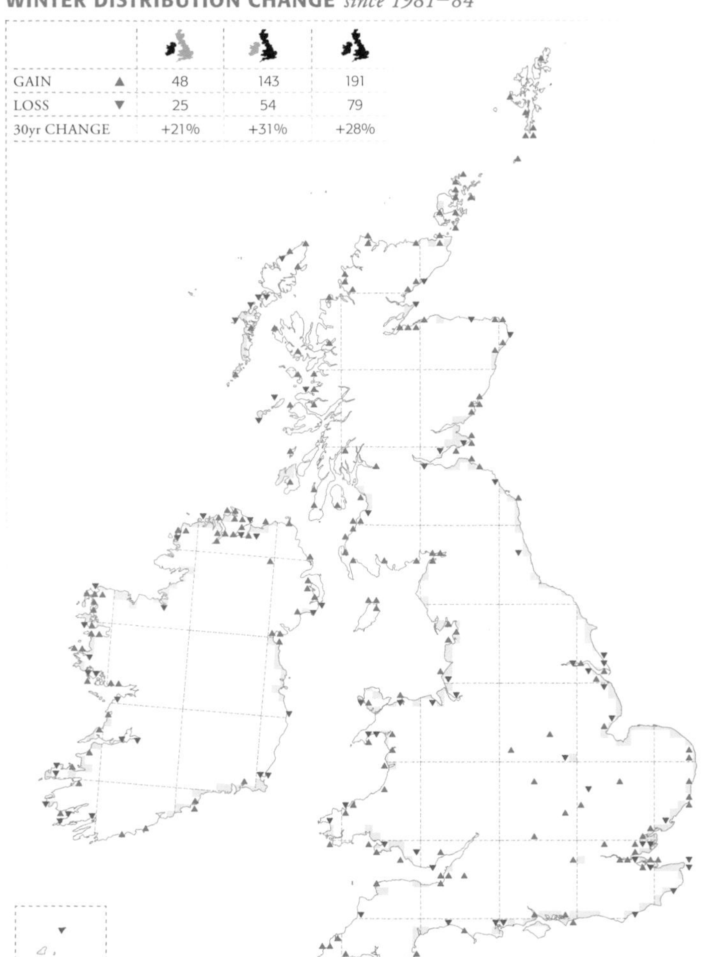

WINTER RELATIVE ABUNDANCE *2007/08–2010/11*

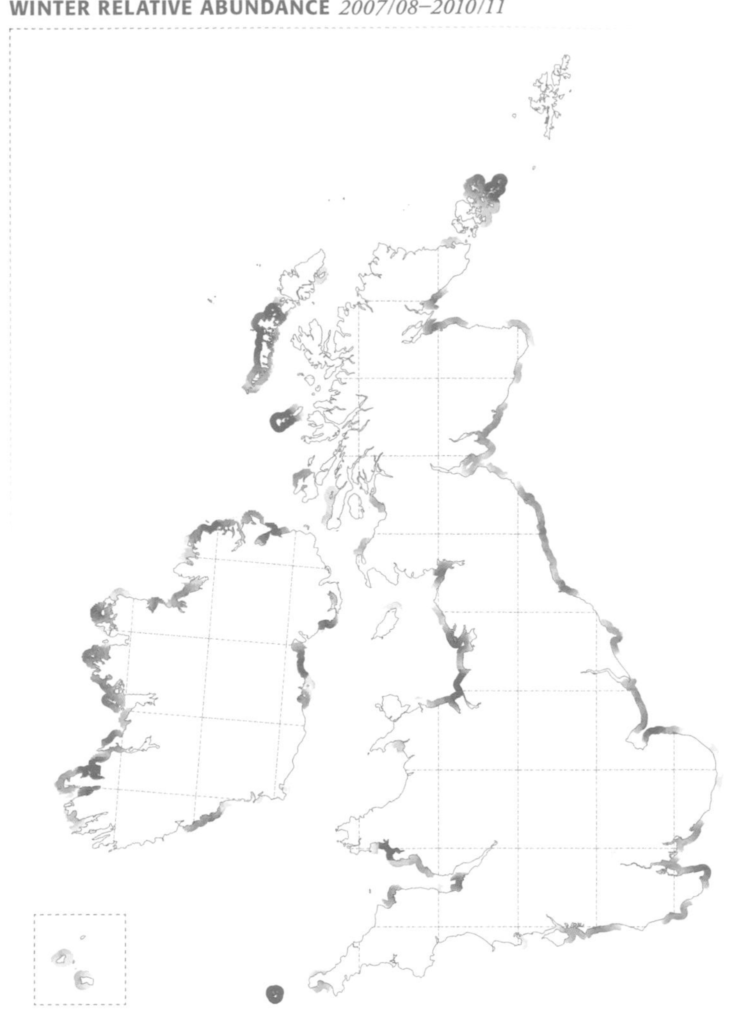

TEMMINCK'S STINT

Calidris temminckii

BTO CODE: TK

DEAN EADES www.birdmad.com

THERE WAS JUST one record of the Temminck's Stint exibiting breeding evidence during 2008–11. It involved a single bird singing and displaying in June 2010. Owing to the ephemeral nature of the habitat and absence of birds in subsequent years the record of this exceptionally rare breeder has been mapped at 10-km resolution. Breeding-season records in the remaining 133 occupied 10-km squares involved passage birds. The last successful breeding in Britain was in 1993, though birds were reported in suitable habitat until at least 2007 when there is some evidence that another breeding attempt took place (Holling *et al.* 2010a; Francis & Cook 2011). There was just one winter record, in Gloucestershire in January–February 2008.

BREEDING DISTRIBUTION *2008–11*

POSSIBLE •	-	<1%	<1%
PROBABLE •	-	-	-
CONFIRMED •	-	-	-
TOTAL	-	<1%	<1%

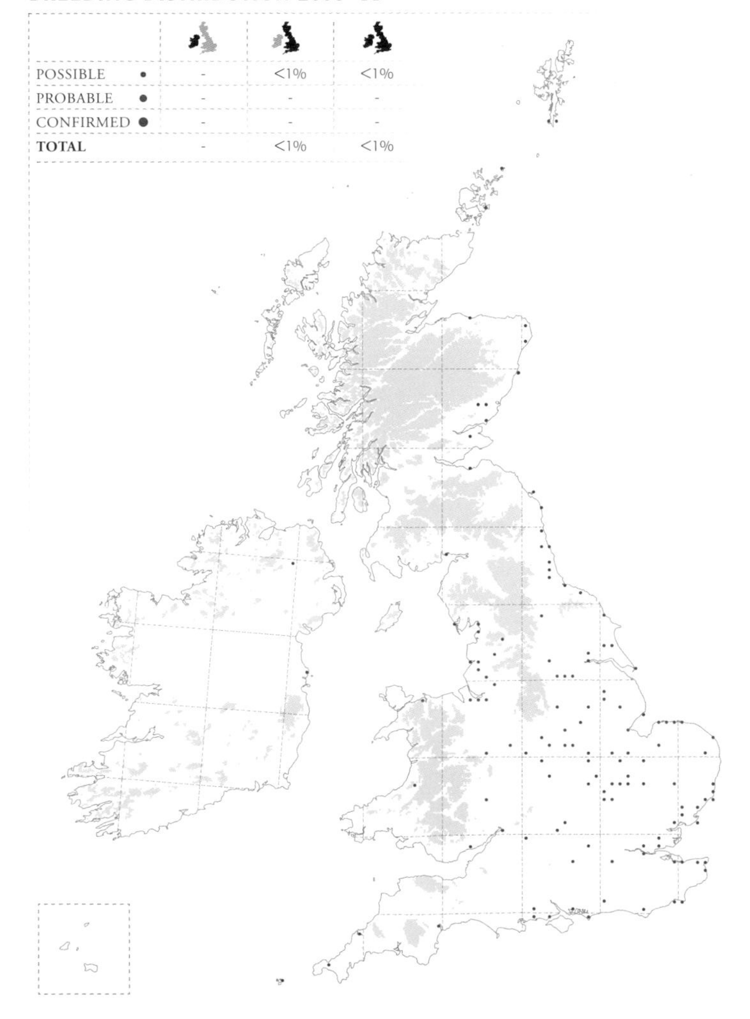

PURPLE SANDPIPER

Calidris maritima

BTO CODE: PS

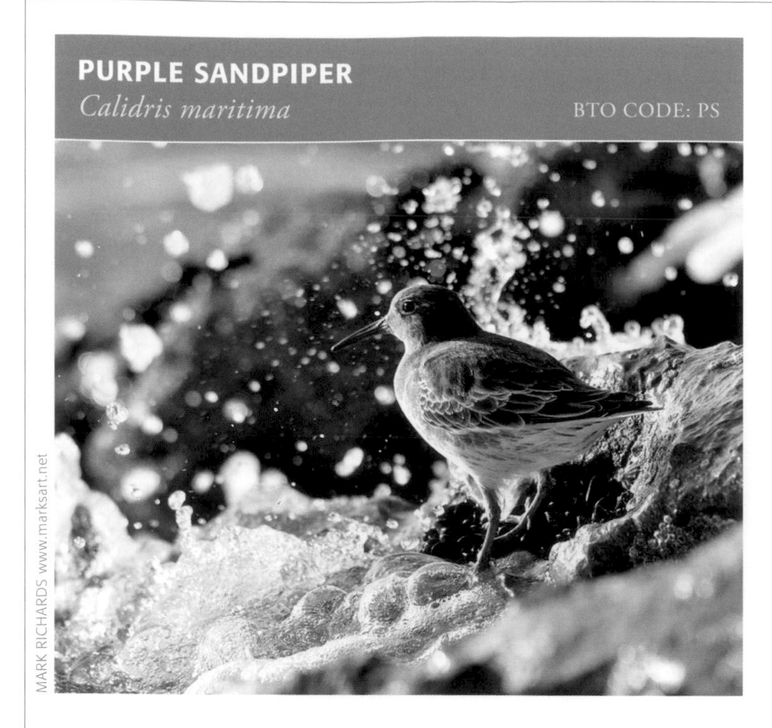

MARK RICHARDS www.marksart.net

WINTERING PURPLE SANDPIPERS prefer northern, exposed, shallow rocky shores. This is evident from the relative abundance map, which shows high concentrations along the coasts of the northern North Sea, Northern Isles and Outer Hebrides and around exposed headlands in Ireland. Elsewhere, many of the occupied squares around Wales and southeastern and southern England hold very few birds (Rehfisch *et al.* 2003).

Determining trends for this species is difficult. Comparisons with the *1981–84 Winter Atlas* suggest a 13% range expansion in Britain and a 10% contraction in Ireland, although some of these changes, such as the gains in northern and western Scotland, may reflect variations in coverage. Annual estuarine counts have declined by 52% in the UK during 1983/84–2008/09 (*SUKB 2012*) yet increased by 40% in the Republic of Ireland during 1994/95–2008/09 (*I-WeBS Report 2009*). However, relatively few birds use estuaries. Decennial non-estuarine surveys provide the best trends, showing that the UK open-coast population declined by 27% during 1984/85–2006/07 (Austin *et al.* 2003). Irish densities declined by 33% during 1987/88–1997/98 (Colhoun *et al.* 2008) and subsequently there was a further small decline (Crowe *et al.* 2008).

Purple Sandpipers are the most northerly of wintering waders (*BWP*). Three separate breeding populations winter in Britain & Ireland. Most in the north and west originate from the declining northeast Canadian population and those in eastern Britain from populations in Scandinavia and Svalbard (Delany *et al.* 2009; Hallgrimsson *et al.* 2012; Wetlands International 2013). In the UK, regional analyses of the non-estuarine coast surveys show that Purple Sandpipers have increased only in the northwest, which is consistent with the hypothesis that recent climate change has permitted waders to winter nearer to their breeding grounds (Rehfisch *et al.* 2004; Maclean *et al.* 2008).

Most Purple Sandpipers depart in April or May, and records during that period account for the 271 10-km squares with *non-breeding* records. The first record of breeding was in the Scottish Highlands in 1978, and between one and five pairs have nested annually since (Smith & Summers 2005; *RBBP 2010*). During 2008–11, *confirmed* breeding was recorded in two 10-km squares in the Cairngorms National Park.

WINTER DISTRIBUTION *2007/08–2010/11*

	Ireland	Britain	Britain & Ireland
PRESENT ●	11%	17%	16%

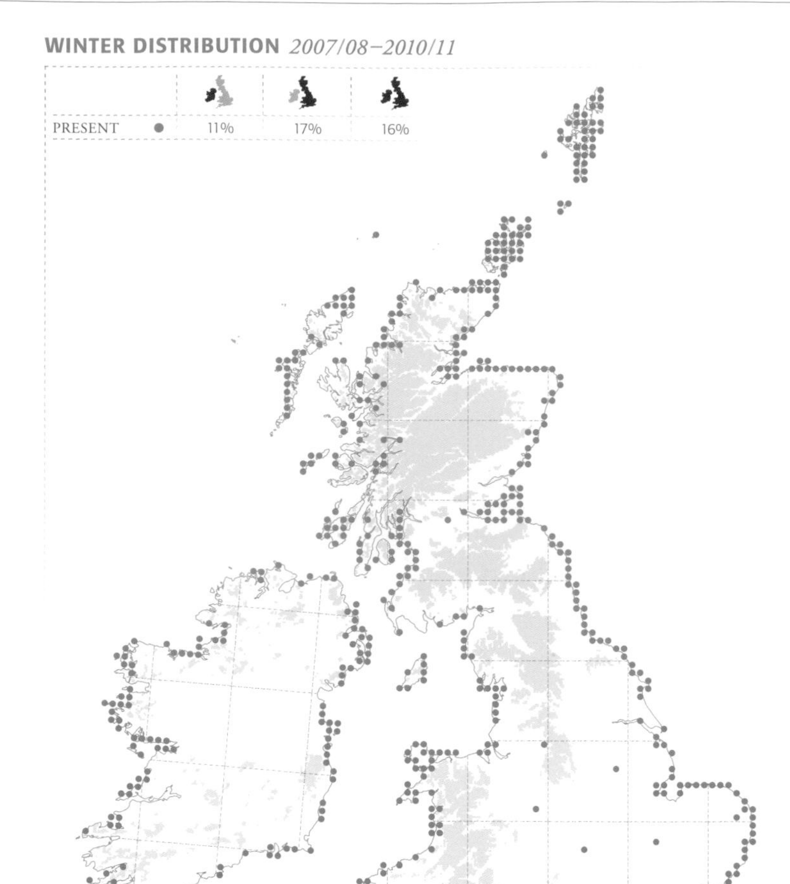

BREEDING DISTRIBUTION *2008–11*

	Ireland	Britain	Britain & Ireland
POSSIBLE •	-	-	-
PROBABLE ●	-	-	-
CONFIRMED ●	-	<1%	<1%
TOTAL	-	<1%	<1%

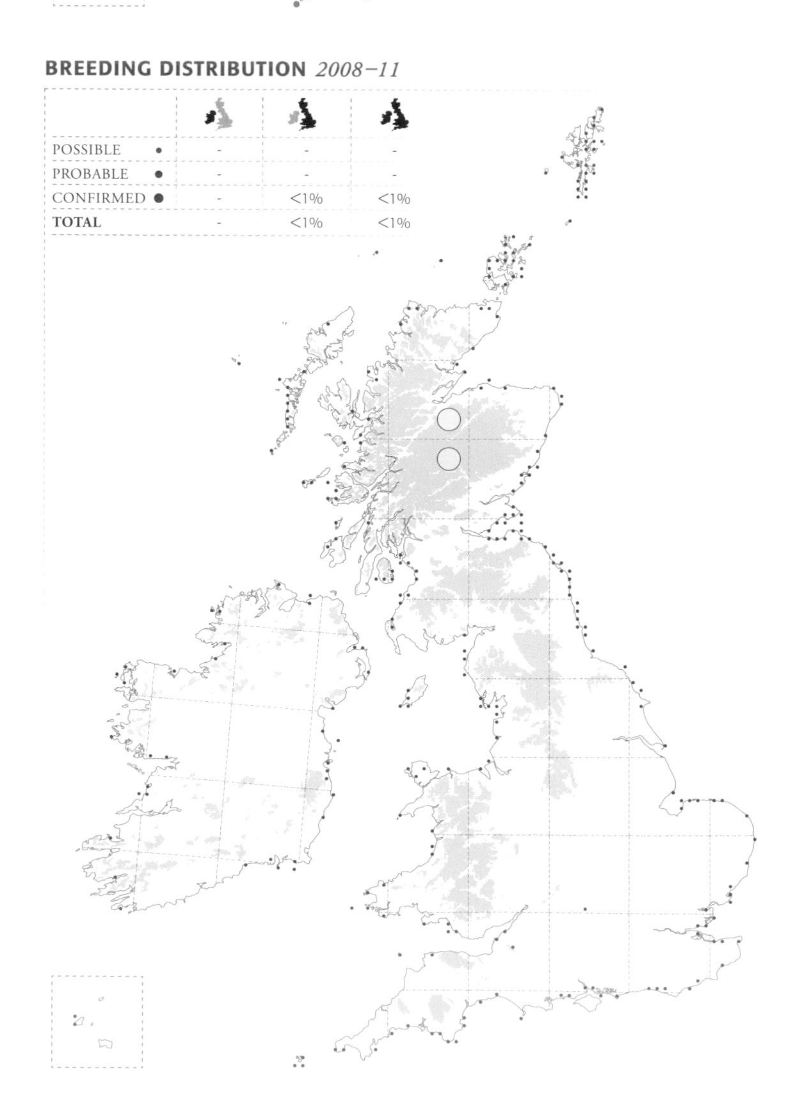

Sponsored by David Marshall Evans

WINTER DISTRIBUTION CHANGE *since 1981–84*

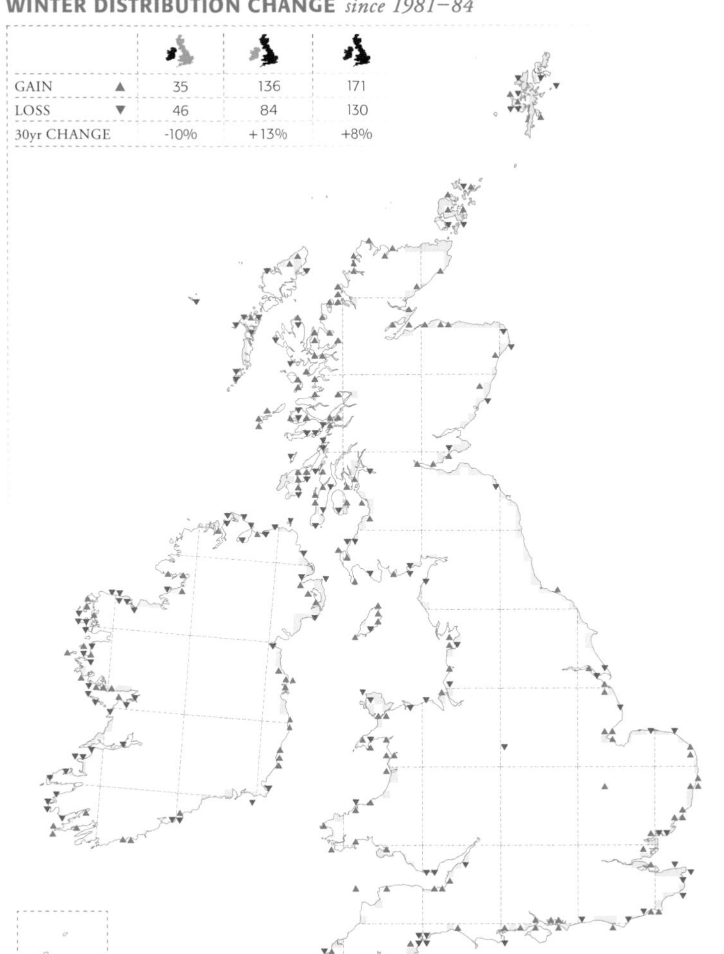

GAIN ▲	35	136	171
LOSS ▼	46	84	130
30yr CHANGE	-10%	+13%	+8%

WINTER RELATIVE ABUNDANCE *2007/08–2010/11*

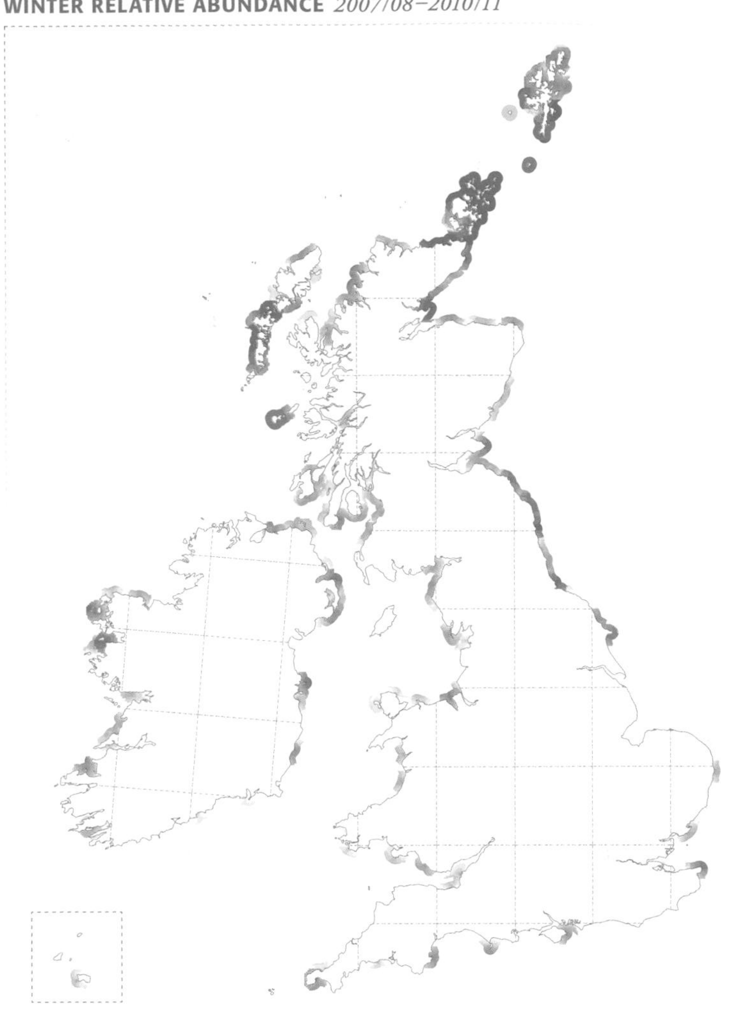

CURLEW SANDPIPER

Calidris ferruginea

BTO CODE: CV

JIM ALMOND www.shropshirebirder.co.uk

MOST RECORDS OF Curlew Sandpipers in Britain & Ireland relate to birds on passage between their central Siberian breeding grounds and wintering areas in central and southern Africa. They are most frequently reported while on autumn passage in July–October and are scarce in spring. Over half the occupied 10-km squares on the winter distribution map are late-autumn migrants recorded in November. The few records of true wintering birds are all from coastal marshes or estuaries. Though the species remains rare in winter, three times as many 10-km squares were occupied as in the *1981–84 Winter Atlas*. During the breeding season, *non-breeding* birds were recorded in 161 10-km squares in Britain and 18 10-km squares in Ireland.

WINTER DISTRIBUTION *2007/08–2010/11*

PRESENT ●	1%	2%	1%

DUNLIN
Calidris alpina

BTO CODE: DN

RICHARD WINSTON www.flickr.com/photos/honeycreeper

BREEDING DISTRIBUTION *2008–11*

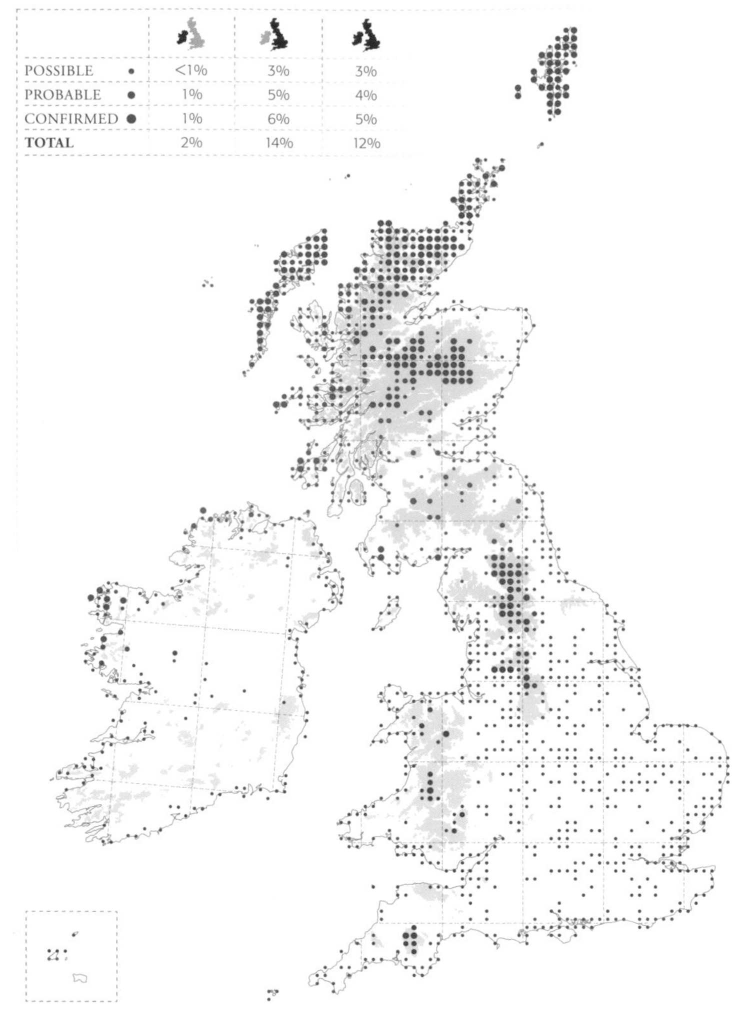

	Ireland	Britain	Britain & Ireland
POSSIBLE •	<1%	3%	3%
PROBABLE •	1%	5%	4%
CONFIRMED •	1%	6%	5%
TOTAL	2%	14%	12%

WINTER DISTRIBUTION *2007/08–2010/11*

	Ireland	Britain	Britain & Ireland
PRESENT •	26%	31%	29%

WINTERING DUNLINS ARE widely distributed throughout the coastlines of Britain and Ireland, with the exception of steep rocky shores, as predominate in western Scotland. They also occur regularly in small numbers at inland sites, especially in England. The largest concentrations occur on the major estuaries of Britain and Ireland, and in Orkney and the Uists.

The winter range has changed little in Britain & Ireland, with a 2% expansion recorded since the *1981–84 Winter Atlas*. This is despite a 28% decline in numbers in the UK between 1983/84 and 2008/09 and a 52% decline in the Republic of Ireland between 1994/95 and 2008/09 (*I-WeBS Report 2009*; *SUKB 2012*). Some gains may reflect improved coverage or the creation of new inland wetlands.

During April and May there is a large movement of Dunlins through Britain & Ireland en route to their arctic breeding grounds, with a return movement starting in mid July. This has resulted in many breeding-season records of passage birds and an exaggerated impression of breeding abundance in parts of northern and western Scotland.

British and Irish breeding Dunlins have returned to their territories by April. The strongholds are all in Scotland, in the Northern Isles and Outer Hebrides, the flow country of Caithness and Sutherland, the hills of the northwest Highlands and the Grampian Mountains. Further south, good numbers remain in the Pennines, but only a few pairs are to be found in the Southern Uplands, in central Wales and on Dartmoor. In northwest Ireland, a small population remains thinly scattered on uplands, blanket bogs and coastal machair; a survey in 2009 found birds at just seven sites (Suddaby *et al.* 2009).

The breeding change map shows losses in marginal upland areas, particularly in western Ireland, northern England and southern Scotland. Formerly the machair of the Outer Hebrides was estimated to support a quarter of the British breeding population but numbers there declined by 50% during 1983–2000 and by a further 15% during 2000–07 (Fuller *et al.* 2010). Predation by introduced Hedgehogs has contributed to these declines (Jackson & Green 2000). Elsewhere, causes of decline include habitat loss, particularly the afforestation of moorland sites (Lavers & Haines-Young 1997).

Sponsored by Ian Bradshaw

BREEDING DISTRIBUTION CHANGE *since 1968–72*

GAIN ▲	9	157	166
LOSS ▽	53	207	260
40yr CHANGE	-69%	-11%	-18%
20yr INDEX	-0.81	-0.20	

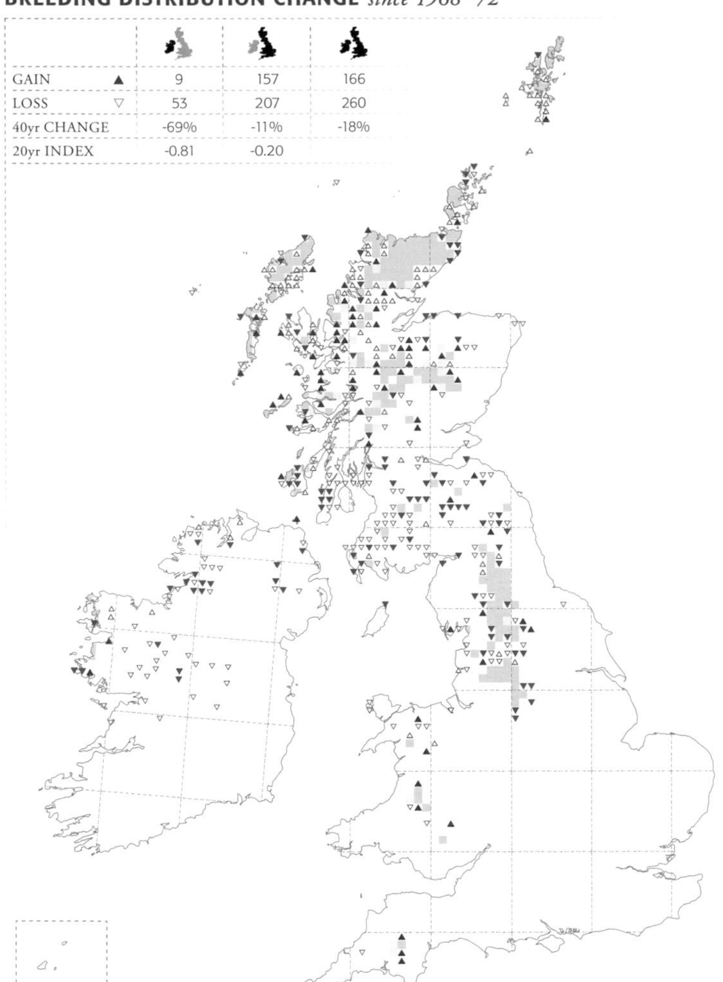

BREEDING RELATIVE ABUNDANCE *2008–11*

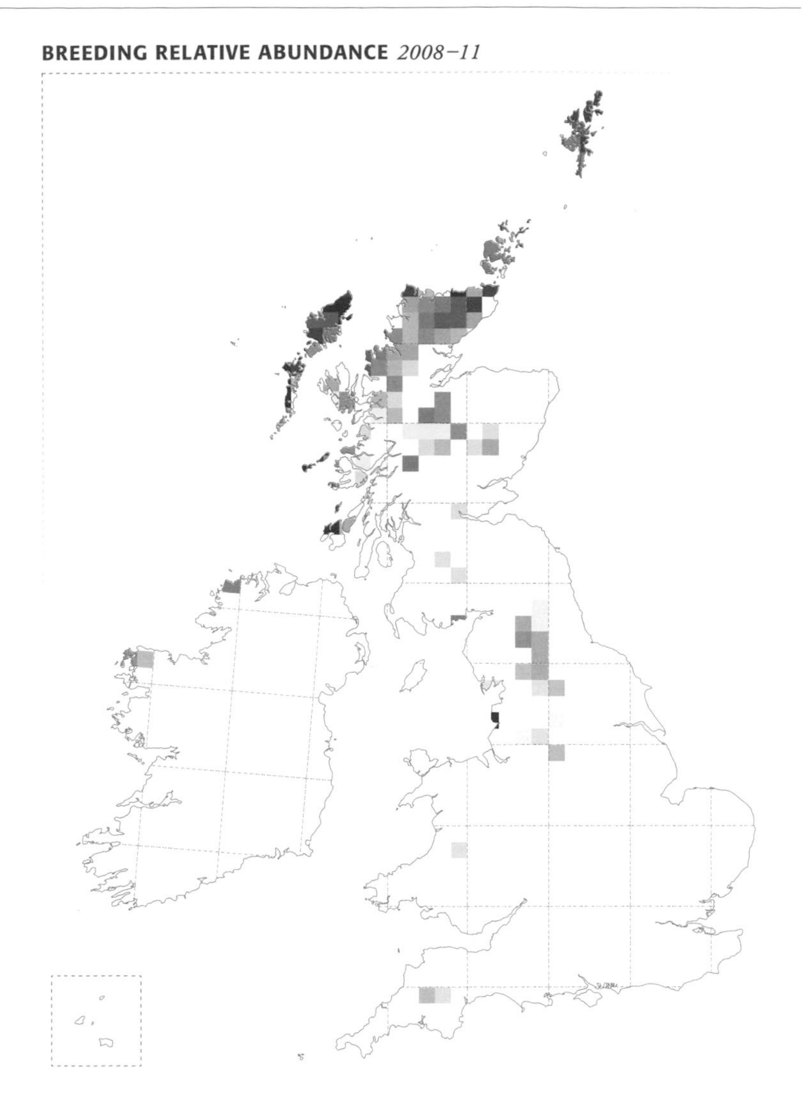

WINTER DISTRIBUTION CHANGE *since 1981–84*

GAIN ▲	60	209	269
LOSS ▼	53	197	250
30yr CHANGE	+3%	+1%	+2%

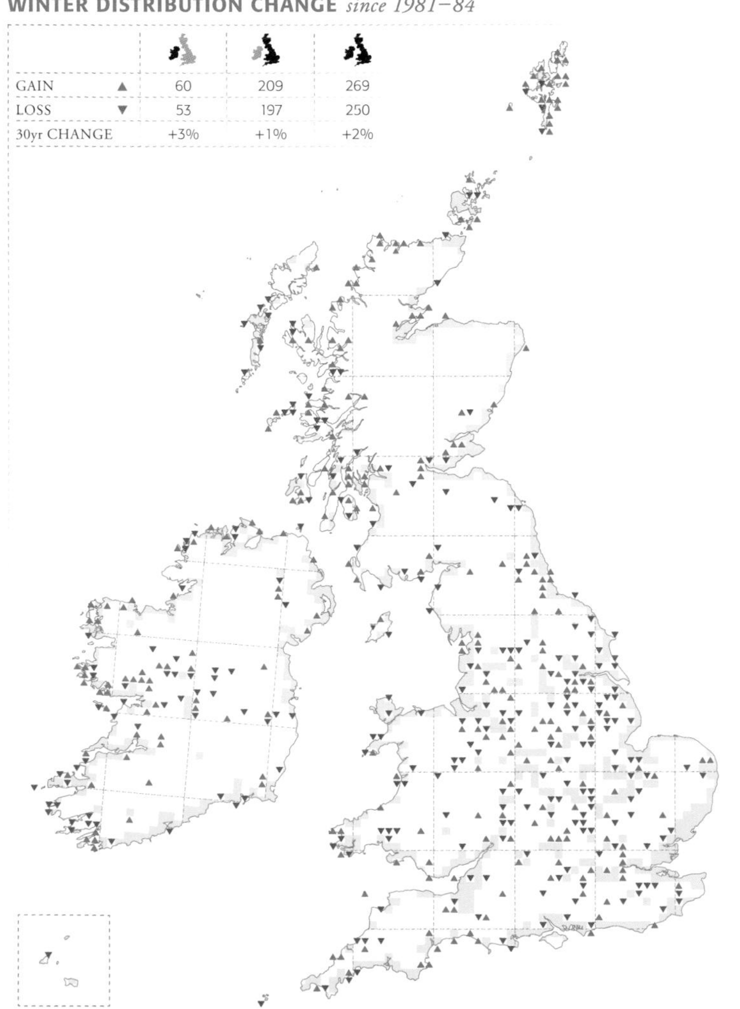

WINTER RELATIVE ABUNDANCE *2007/08–2010/11*

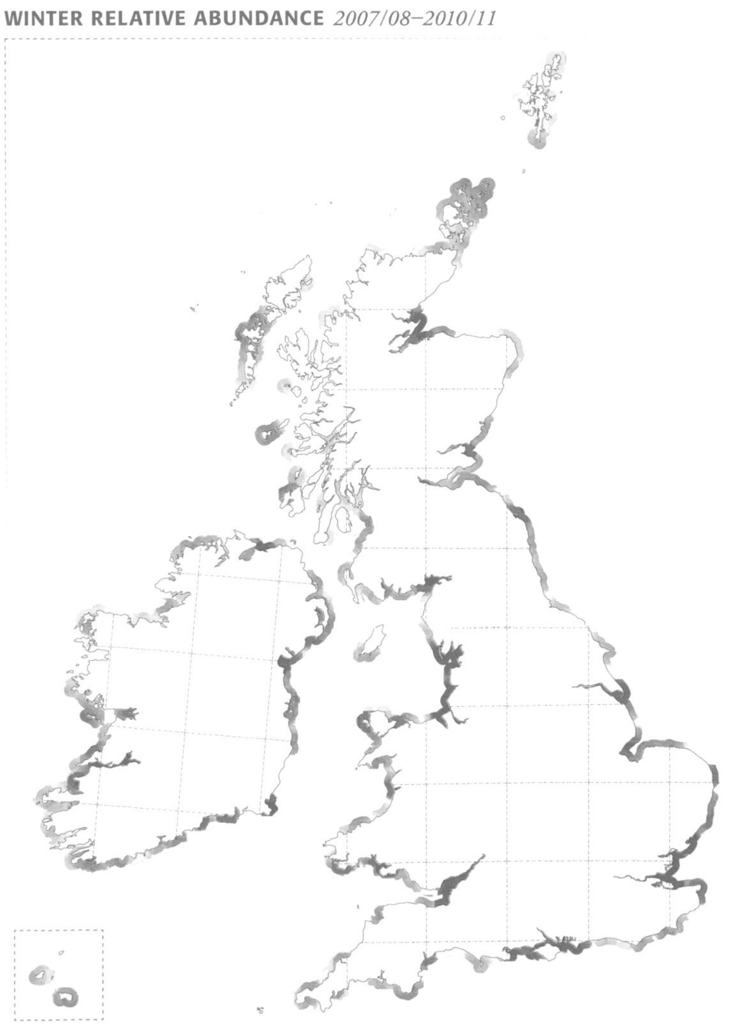

RUFF

Calidris pugnax

BTO CODE: RU

TOM WALLIS

IN WINTER, RUFFS are found at inland and coastal locations, predominantly in England, and are scarce elsewhere. A currently estimated wintering population of 800 birds occurs mostly on freshwater marshes, wet grassland or other suitable habitats adjacent to estuaries (Musgrove *et al.* 2011). Since the *1981–84 Winter Atlas*, Ruffs have become slightly more widespread at coastal sites throughout the existing range. This corresponds with a marked increase in winter numbers from the 1970s to the early 2000s, although numbers have subsequently dropped (*WeBS Report 2011*). The two most important areas in terms of abundance are the Ouse Washes and north Norfolk. Cold winters during the second half of the Atlas period may have resulted in a redistribution of birds from inland to coastal sites in Britain and Ireland, and larger arrivals of birds from continental Europe.

Ruffs have been very scarce breeders in Britain since major declines occurred in the 18th and 19th centuries, brought about by the intensification of grassland management and by hunting (*Birds in England*). Breeding behaviour is restricted to areas of extensive damp grassland habitats, where the species forms small leks. Some gains and losses on the change map reflect birds lekking on passage, but the trends also match fluctuations in numbers reported by RBBP (*RBBP 2010*). Fenland has remained a favoured area, but elsewhere breeding is now confined to northwest England and the Hebrides. The last confirmed record was in 2006, when a pair nested in northwest England; prior to that, breeding was confirmed in Shetland in 2003 and the Outer Hebrides in 2004 (Holling *et al.* 2007a, 2009).

Deterioration in the quality of staging areas in the Netherlands has caused a global redistribution of migrating and breeding Ruffs over the last two decades: passage numbers in the Netherlands have declined substantially whilst the eastern migration route through Belarus has become more important and the breeding population has shifted from the European Arctic to western Siberia (Rakhimberdiev *et al.* 2011; Verkuil *et al.* 2012). Consequently, even on established wetland nature reserves with beneficial management, the future does not look promising for Ruffs breeding in Britain.

Sponsored by John & Viv Phillips in memory of Linford Rose

WINTER DISTRIBUTION *2007/08–2010/11*

PRESENT ●	3%	9%	8%

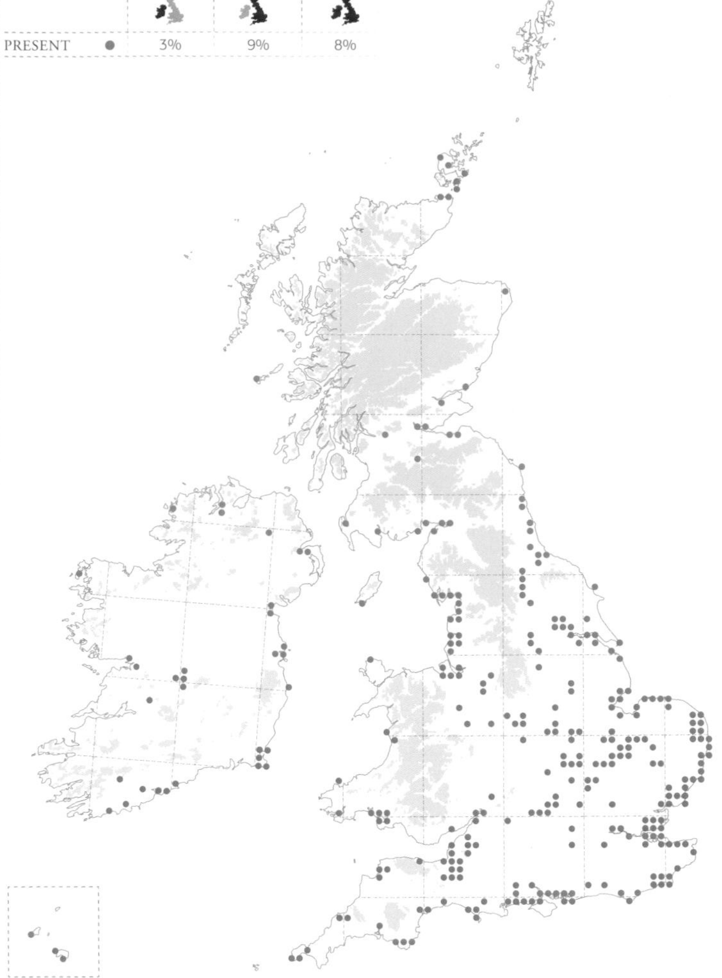

BREEDING DISTRIBUTION *2008–11*

POSSIBLE •	-	<1%	<1%
PROBABLE ●	-	<1%	<1%
CONFIRMED ●	-	-	-
TOTAL	-	<1%	<1%

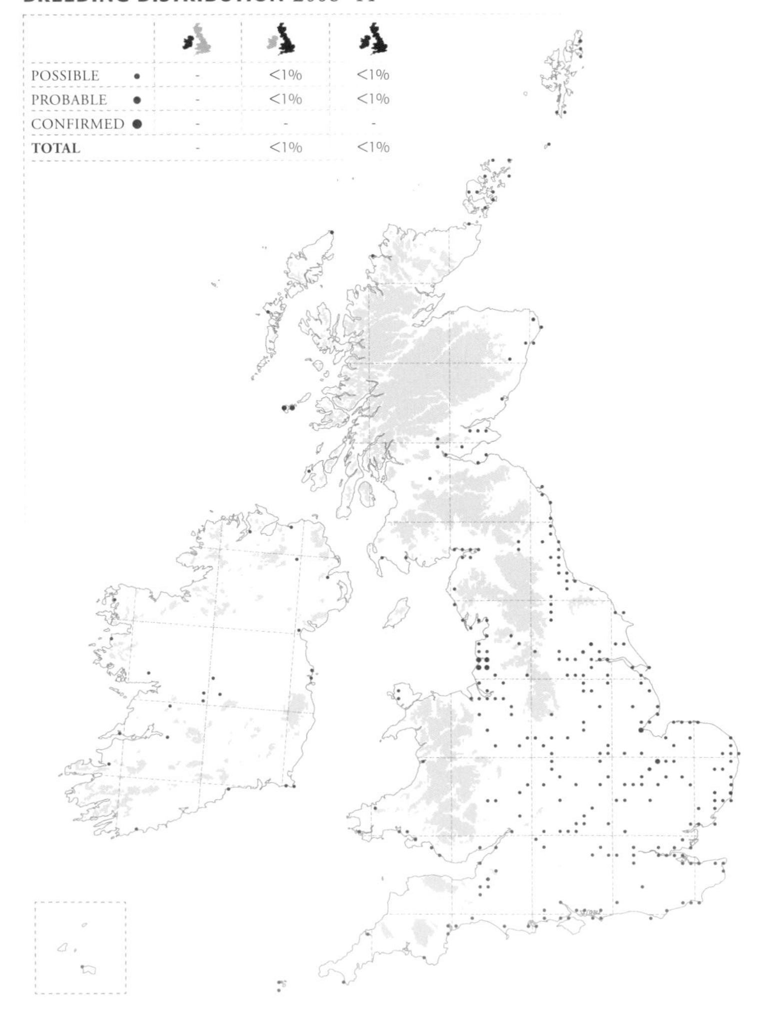

WINTER DISTRIBUTION CHANGE *since 1981–84*

GAIN ▲	20	127	147
LOSS ▼	17	97	114
30yr CHANGE	+12%	+14%	+13%

WINTER RELATIVE ABUNDANCE *2007/08–2010/11*

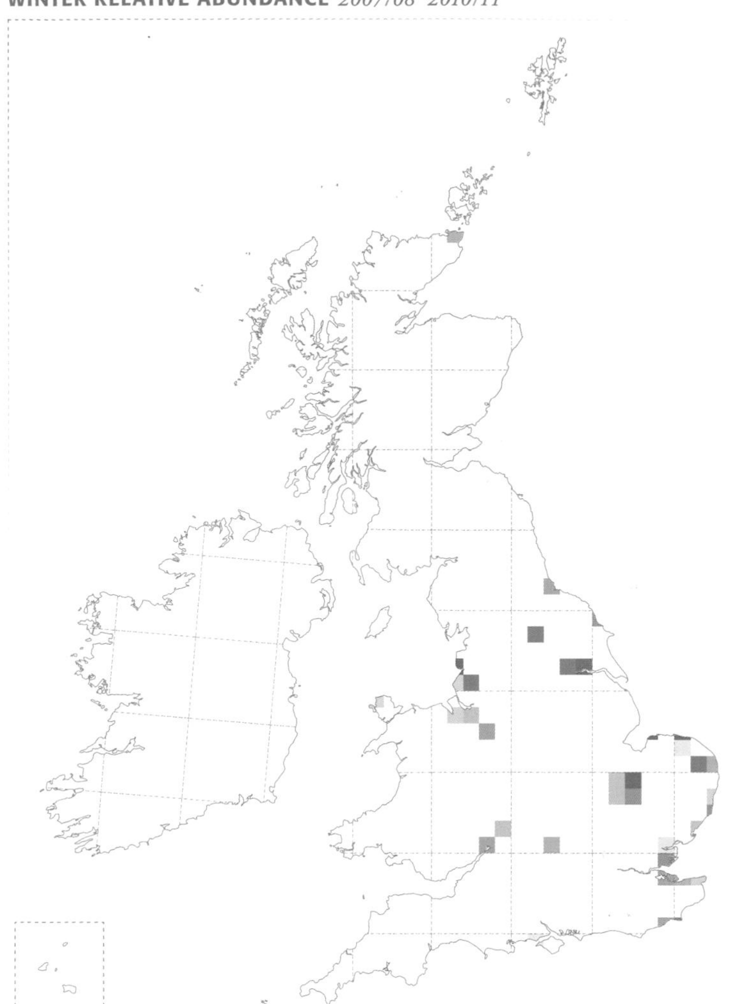

BREEDING DISTRIBUTION CHANGE *since 1968–72*

GAIN ▲	-	12	12
LOSS ▽	-	10	10
40yr CHANGE	-	+14%	+14%
20yr INDEX	-	-1.00	

JACK SNIPE

Lymnocryptes minimus

BTO CODE: JS

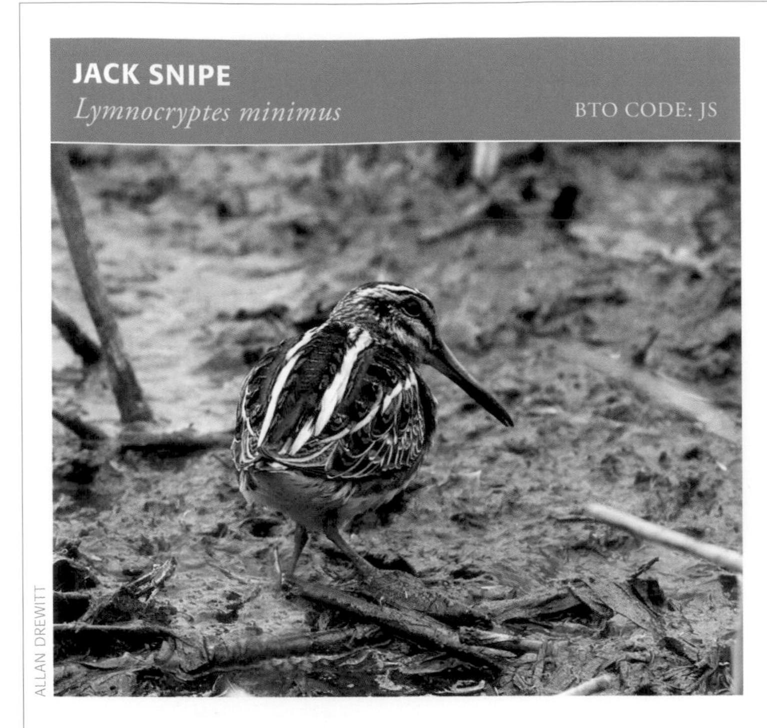
ALLAN DREWITT

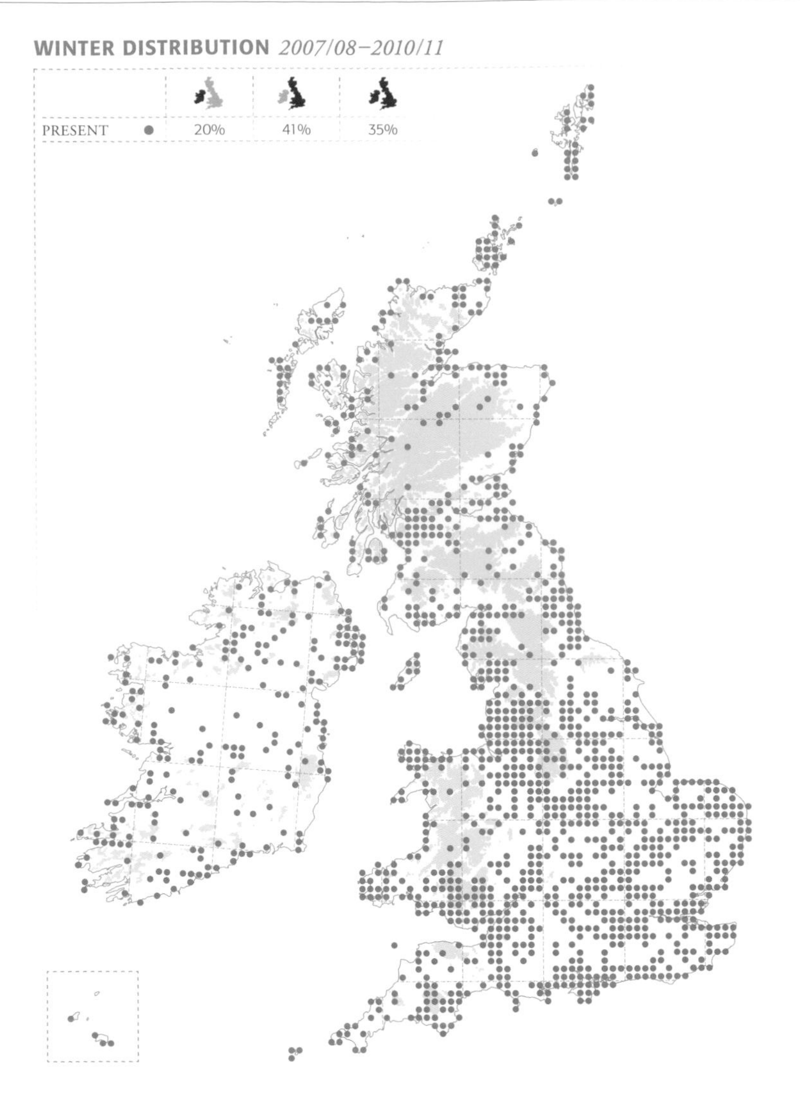

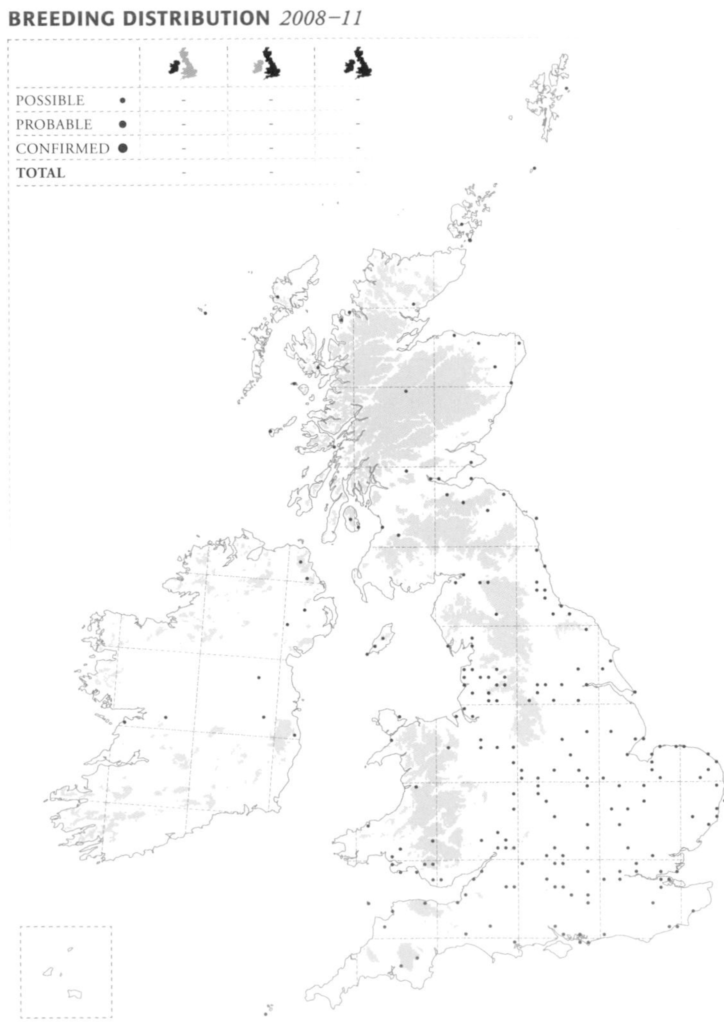

THE JACK SNIPE is a regular winter visitor and passage migrant to Britain & Ireland from northeast European and Siberian breeding grounds. Because it is a very secretive bird, however, there is considerable uncertainty concerning population sizes, trends and distribution patterns throughout its Afro-Palearctic flyway (Delany *et al.* 2009). The winter distribution map is likely to underestimate the true level of 10-km square occupancy, especially in areas with relatively low levels of coverage.

The Jack Snipe is associated with the coastal fringe and damp lowland wetlands, where it favours areas with shallow fresh or brackish water and short vegetation. In Scotland it is most widespread in the Central Lowlands, whilst in England it is widely distributed in Northumberland and southward from Cumbria and Lincolnshire. In Wales it favours lower ground in the south and along the north coast. It has a wide but very patchy distribution throughout Ireland, with highest concentrations close to the western coastline.

Since the *1981–84 Winter Atlas*, the number of occupied 10-km squares has increased by 71% in Britain, but by only 29% in Ireland. Counts of three birds or fewer are usual but exceptional counts included 35 at Craigmarloch, Lanarkshire, in November 2007, 21 at Bickershaw Rucks, Manchester, in November 2008 and 18 at Carrigower Bog, Co. Wicklow. It is unclear to what extent range gains are real or due to increased effort during *Bird Atlas 2007–11*. That gains are more prevalent in Britain than in Ireland might reflect the substantially higher observer effort in Britain.

Scarcely more than 100 individuals are counted during annual waterbird monitoring in any single winter and any trends are unknown (*I-WeBS Report 2009*; *WeBS Report 2011*). The British wintering population has recently been estimated at around 100,000 individuals, but with very low reliability (Musgrove *et al.* 2013), and no Irish figure is available.

There were *non-breeding* records of late-departing wintering individuals and passage migrants from 206 10-km squares in Britain and nine in Ireland. There have been no breeding records for Britain & Ireland, despite many claims in Scotland in the past (*Birds of Scotland*).

Sponsored in memory of Margaret Alison Smith (1957–2006)

WINTER DISTRIBUTION CHANGE *since 1981–84*

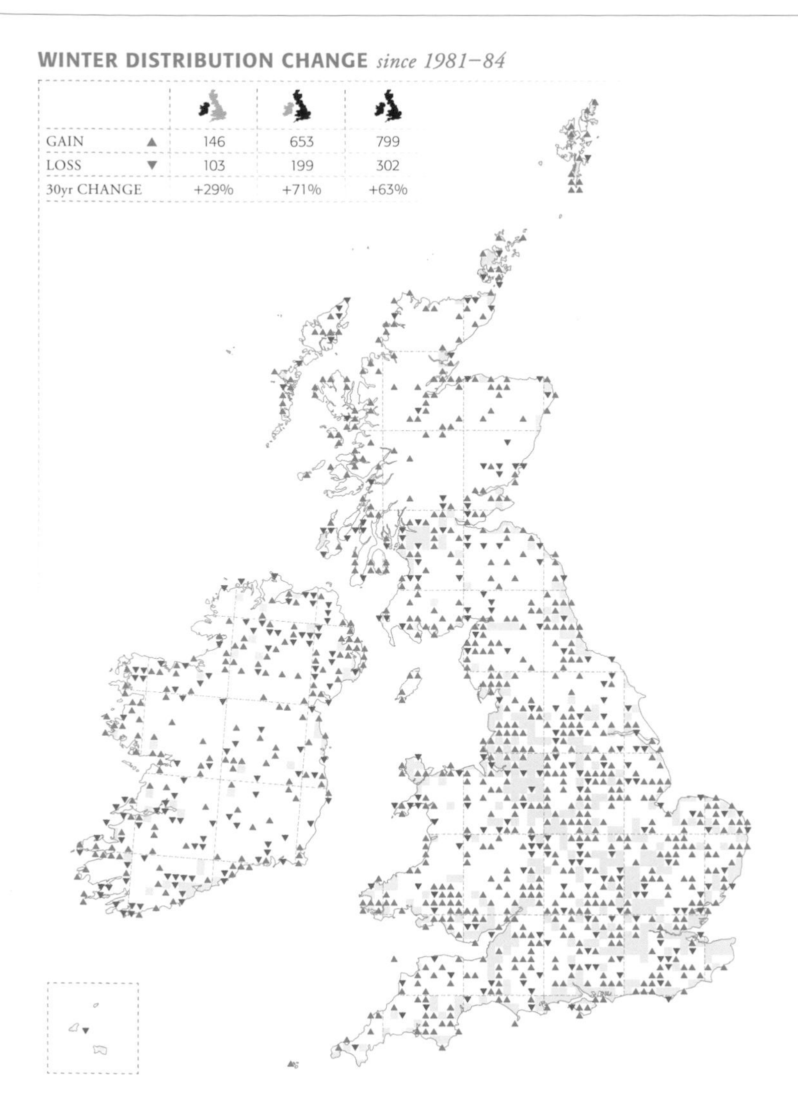

WINTER RELATIVE ABUNDANCE *2007/08–2010/11*

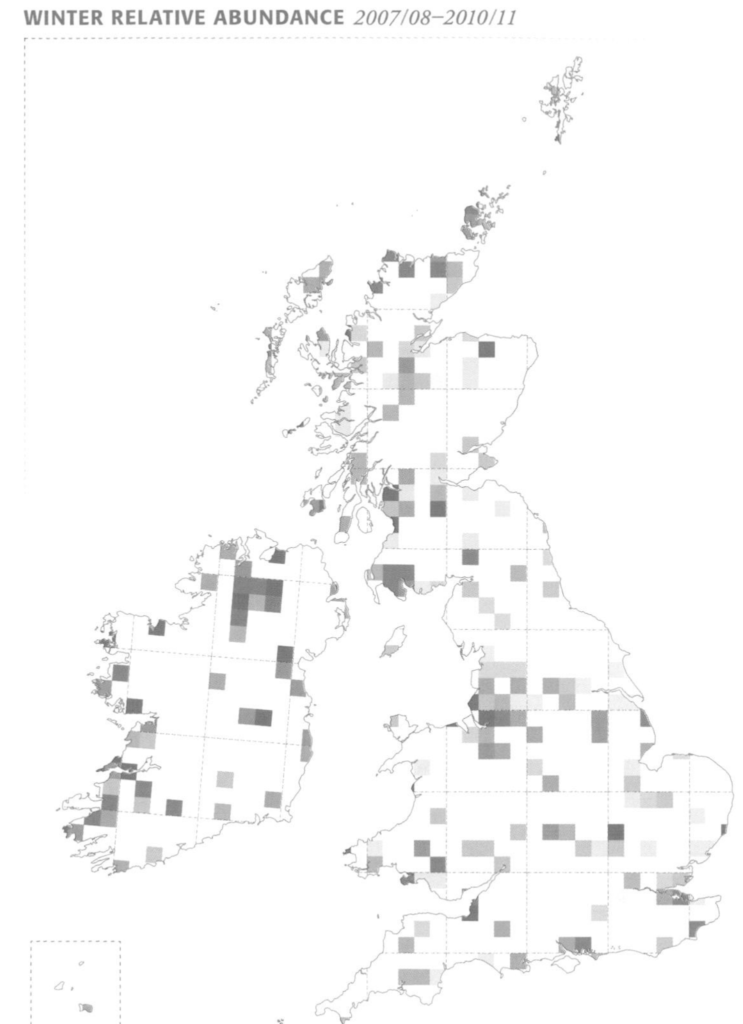

▲ Migrants arrive in Britain from mid September, but in Ireland most birds arrive in October or November.

SNIPE

Gallinago gallinago

BTO CODE: SN

ROGER RIDLEY

THE SNIPE IS the most widespread wintering wader in Britain & Ireland, with records from 88% of 10-km squares, though the population size is hard to determine (Musgrove *et al.* 2011) and trends are poorly monitored. Resident birds are joined by others from northern continental Europe, Iceland and the Faeroes (*Migration Atlas*). They winter in a wide range of lowland and upland habitats, except for the highest parts of Scotland. Densities are highest in southwest Ireland, the Northern Isles and in the coastal margins of Britain & Ireland. Since the *1981–84 Winter Atlas* there has been a 13% range expansion, although gains in northwest Scotland may be attributable to improved coverage.

Since the *1968–72 Breeding Atlas* a significant population decline has taken place, mostly in lowland areas (Henderson *et al.* 2002; Wilson *et al.* 2005), leading to a range contraction towards the uplands and generating a polarised pattern of occupancy and abundance. The current breeding range and highest abundances are focused in Scotland, northwest England, Wales and in the midlands and west of Ireland. These are the areas with the greatest remaining extent of its favoured marshy wet pasture and moorland habitats. Although numbers have fallen by 13% in England during 1995–2010, there has been a 30% increase in Scotland (*BBS Report 2011*). The abundance change map reveals a contrasting pattern of declines in the proportion of occupied tetrads since the *1988–91 Breeding Atlas*, particularly in lowland areas, but gains in some northern and western upland areas.

Lowland breeding population declines are linked to agricultural intensification and the drainage of wet tussocky grassland, resulting in the loss of damp pasture (Siriwardena *et al.* 2000a; Henderson *et al.* 2002; Newton 2004). Drainage dries the ground, leading to shortened breeding seasons and reduced productivity (Green 1988), though management of water levels and habitat can increase breeding densities (Baines 1988; Ausden & Hirons 2002). However, despite habitat improvements at many lowland wetland reserves, numbers have not increased, suggesting that other key aspects of habitat quality, such as prey abundance, may also be important (Smart *et al.* 2008).

BREEDING DISTRIBUTION *2008–11*

POSSIBLE •	23%	14%	16%
PROBABLE •	30%	28%	28%
CONFIRMED •	5%	12%	10%
TOTAL	58%	54%	54%

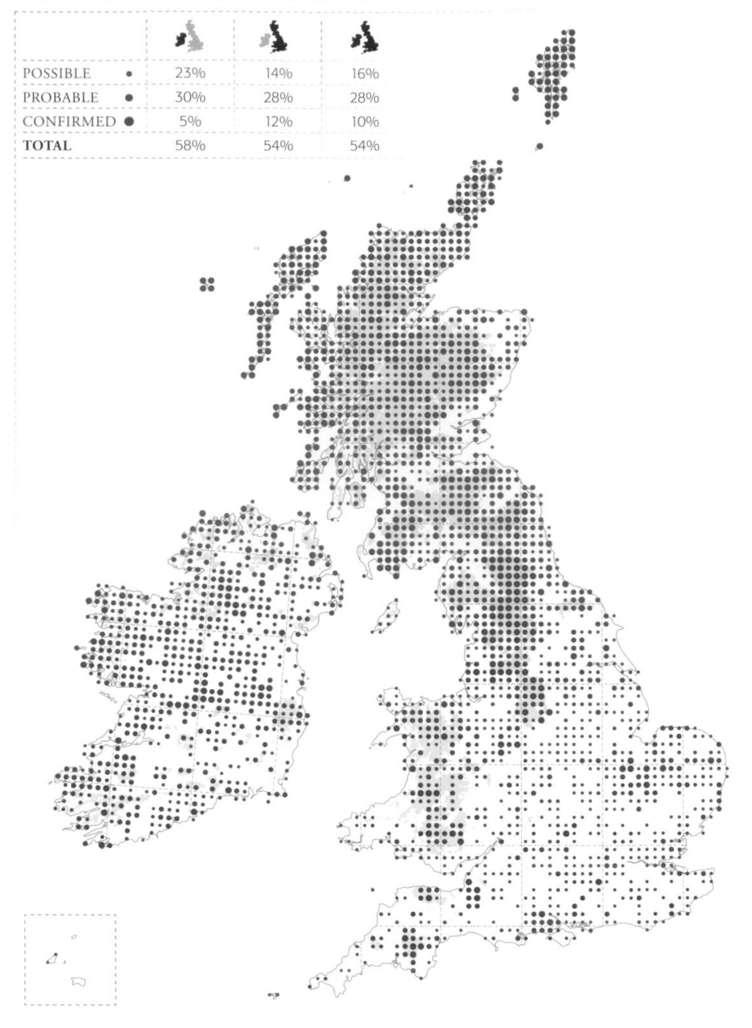

WINTER DISTRIBUTION *2007/08–2010/11*

PRESENT •	90%	87%	88%

BREEDING DISTRIBUTION CHANGE *since 1968–72*

GAIN ▲	33	131	164
LOSS ▽	340	826	1166
40yr CHANGE	-34%	-31%	-32%
20yr INDEX	-0.19	-0.21	

WINTER DISTRIBUTION CHANGE *since 1981–84*

GAIN ▲	132	473	605
LOSS ▼	83	142	225
30yr CHANGE	+6%	+16%	+13%

BREEDING RELATIVE ABUNDANCE *2008–11*

BREEDING RELATIVE ABUNDANCE CHANGE *since 1988–91*

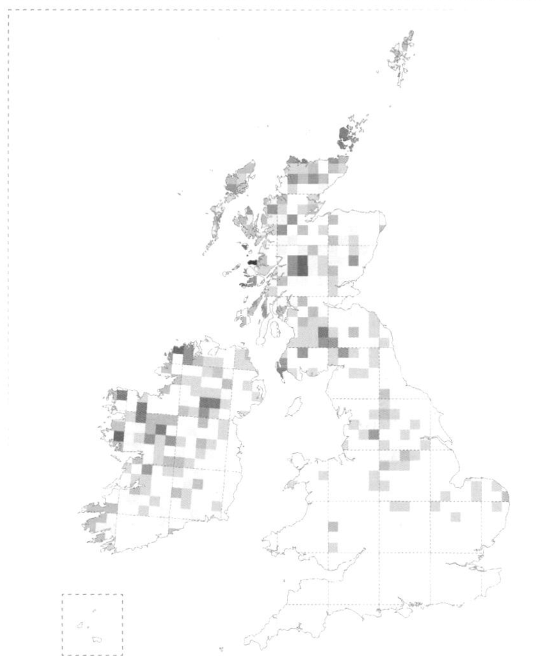

WINTER RELATIVE ABUNDANCE *2007/08–2010/11*

WOODCOCK

Scolopax rusticola

BTO CODE: WK

PETER D SMITH

DURING THE AUTUMN there is a large influx of Woodcocks from the Continent, outnumbering the British & Irish breeding population five to one (Hoodless & Powell 2010) and leading to a wide winter distribution throughout Britain and much of Ireland. Relatively low winter densities occur in much of central and southern England and the highest uplands in Scotland. In Ireland the highest densities are associated with the uplands in the west and with the Wicklow Mountains. A range expansion of 37% since the *1981–84 Winter Atlas* is suggested but the secretive Woodcock is a species for which improved coverage can generate spurious range gains. Nonetheless, the Atlas coincided with some exceptional years for Woodcocks: BirdTrack reporting rates were 2–3 times higher in 2009/10 and 2010/11 than in previous winters (*BirdTrack*) so many of the apparent gains could be real. Losses are most notable in central Ireland where raised bogs have been cut away and surrounding woodland lost.

Breeding Woodcocks have a much more restricted distribution, being absent from many of the Scottish islands, much of central and southwest England and large swathes of Wales and western Ireland. They are also absent from the Channel Islands. Breeding-season TTVs were undertaken during mornings, so are inappropriate for measuring Woodcock abundance, and therefore no map is shown here. A dedicated survey in 2003 produced a population estimate of 78,346 males, based on counts of birds roding at dusk, with the highest densities recorded in eastern England and the lowest in Wales (Hoodless *et al.* 2009).

Since the *1988–91 Breeding Atlas* the Woodcock's breeding range has contracted by 29%. There was uncertainty whether the 37% range contraction between the *1968–72 Breeding Atlas* and *1988–91 Breeding Atlas* was real or an artefact of changing recording methods and coverage. This latest atlas strongly suggests that the current breeding range is less than half its 1960s extent, supporting other evidence of a long-term population decline (Marchant *et al.* 1990; Hoodless & Hirons 2007). The reasons for the decline are unclear but may include recreational disturbance, the drying out of woodlands, increased browsing by deer, declining woodland management, and the maturing of new plantations (Fuller *et al.* 2005; Hoodless & Hirons 2007).

Sponsored by Game & Wildlife Conservation Trust

WINTER DISTRIBUTION *2007/08–2010/11*

	Ireland	Britain	Britain & Ireland
PRESENT ●	57%	89%	80%

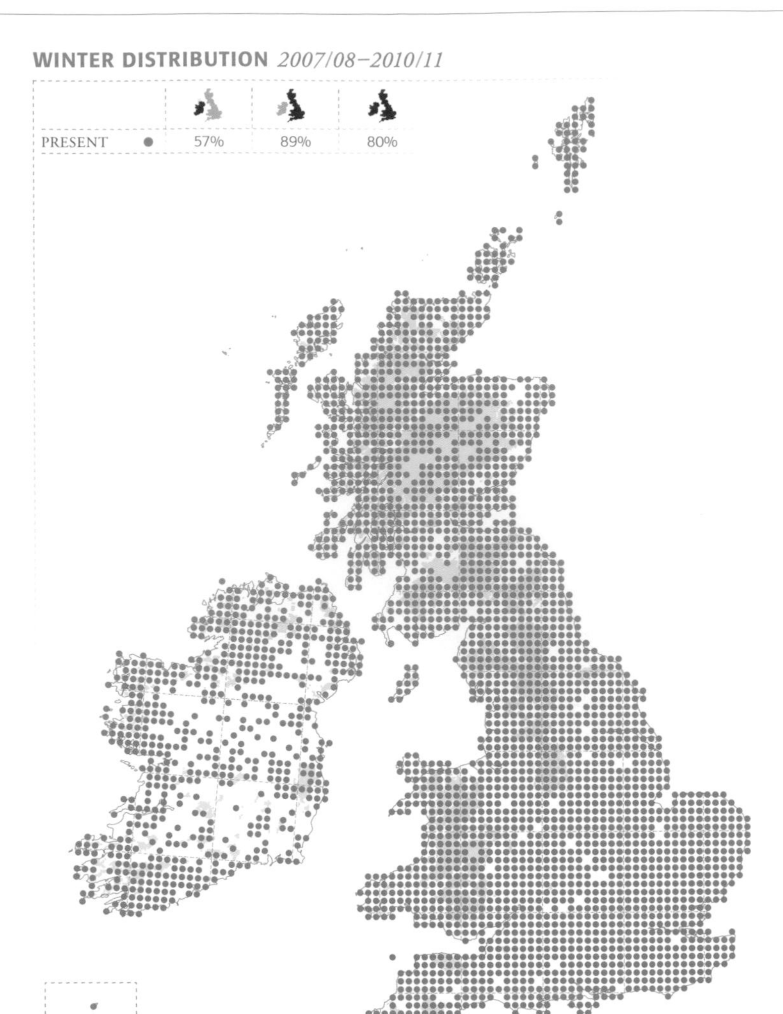

BREEDING DISTRIBUTION *2008–11*

	Ireland	Britain	Britain & Ireland
POSSIBLE •	5%	8%	7%
PROBABLE ●	7%	18%	15%
CONFIRMED ●	1%	4%	3%
TOTAL	13%	30%	25%

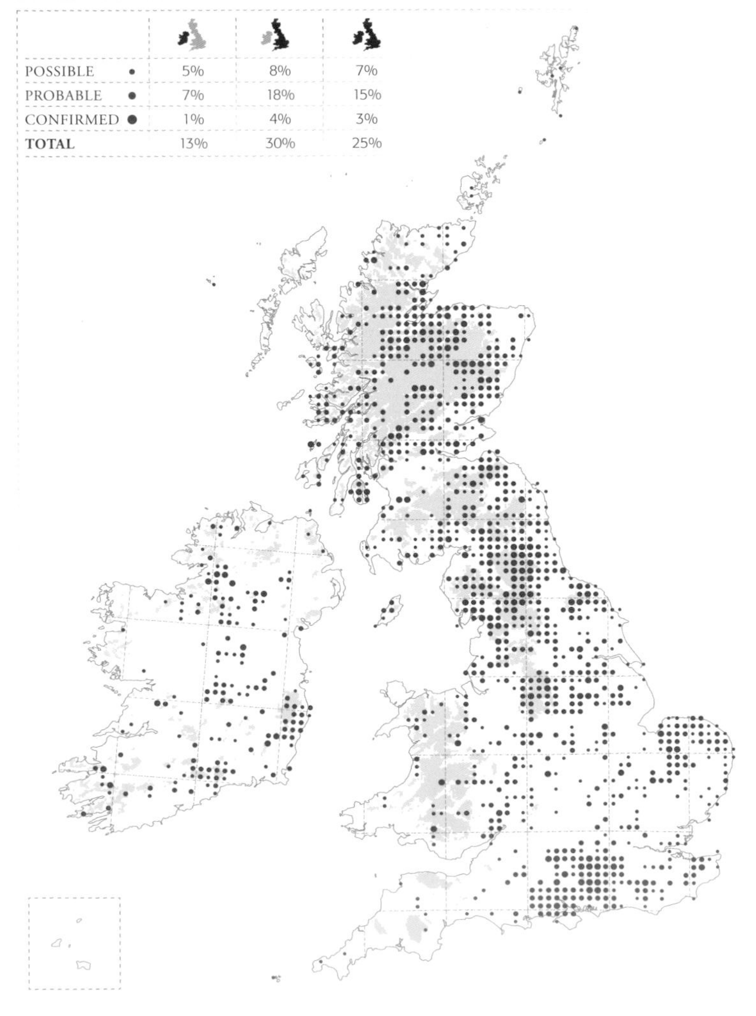

WINTER DISTRIBUTION CHANGE *since 1981–84*

GAIN ▲	269	833	1102
LOSS ▼	175	98	273
30yr CHANGE	+20%	+42%	+37%

WINTER RELATIVE ABUNDANCE *2007/08–2010/11*

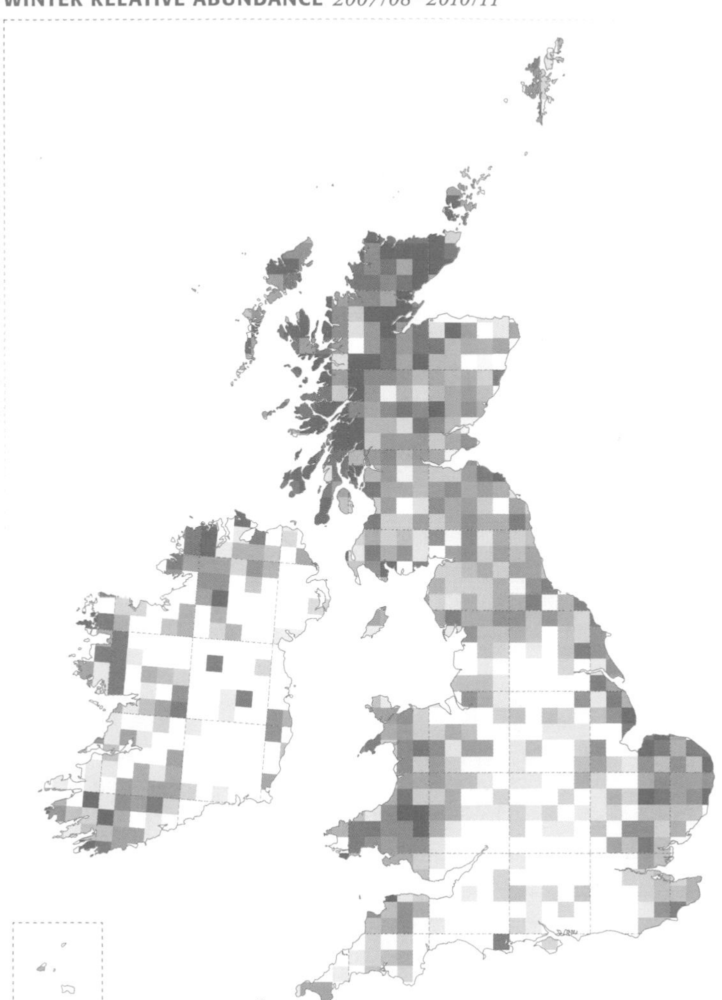

BREEDING DISTRIBUTION CHANGE *since 1968–72*

GAIN ▲	42	103	145
LOSS ▽	404	955	1359
40yr CHANGE	-73%	-50%	-55%
20yr INDEX	-0.43	-0.55	

BLACK-TAILED GODWIT

Limosa limosa

BTO CODE: BW

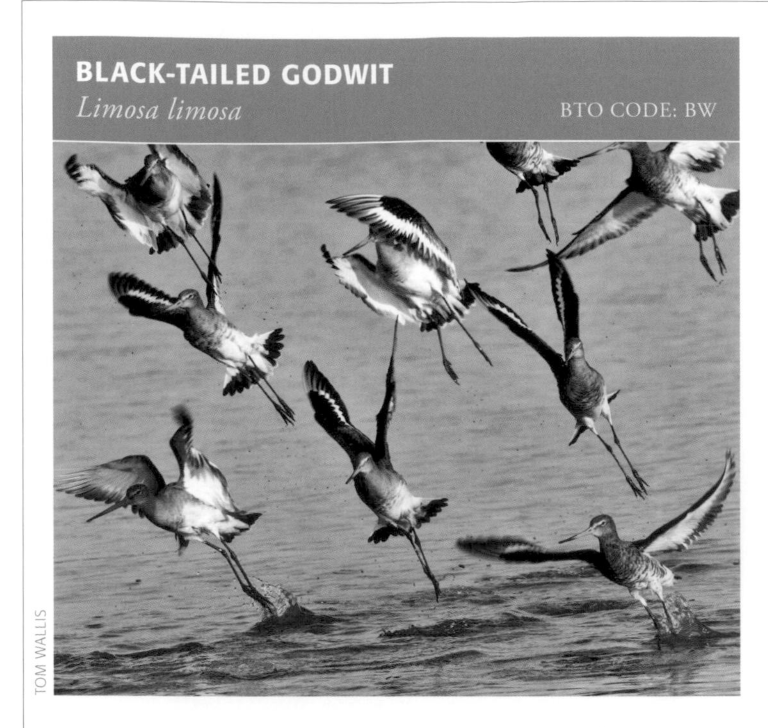

TOM WALLIS

VIRTUALLY ALL OF the Black-tailed Godwits that winter in Britain & Ireland breed in Iceland (Gunnarsson *et al.* 2005). The subspecies *islandica* has undergone a sustained breeding population increase over the past century, which has translated into a long-term increase in the numbers wintering in Britain and Ireland (Gill *et al.* 2001; Crowe *et al.* 2008; *WeBS Report 2011*). Coupled to this increase has been a 177% winter range expansion in Britain and a 55% expansion in Ireland since the *1981–84 Winter Atlas*. Gains are geographically widespread but there are distinct clusters in East Anglia, the Thames Basin, North Wales, northwest England, the east and south Irish coasts and the Shannon Estuary. The greater use of inland sites during winter in Britain is notable. The increase in the *islandica* population is not fully understood but may be linked to climatic and habitat changes operating on both the breeding and the wintering grounds (Gill *et al.* 2007).

Britain supports small breeding populations of the nominate subspecies *limosa* and the Icelandic subspecies *islandica*. In contrast to *islandica*, the west European population of *limosa*, which breeds mostly in lowland grassland areas, has declined at an alarming rate (Jensen *et al.* 2008; Lourenço & Piersma 2008; Wetlands International 2013). Changes in productivity as a consequence of agricultural intensification are most likely to be driving the decline in *limosa* (Gill *et al.* 2007).

In England, breeding was *confirmed* in eight 10-km squares and these records more than likely refer to *limosa*. The Ouse Washes and nearby Nene Washes in East Anglia remain the centre of the British breeding population; other *confirmed* squares likely to relate to nominate birds are localised and limited to Lancashire, Yorkshire and Kent. Breeding was also *confirmed* in southwest England but is not mapped here owing to site sensitivities. *Confirmed* breeding records in Orkney and Shetland, and at a site in Ireland are thought to involve individuals of the *islandica* subspecies. The numbers of *limosa* breeding in Britain and their range size have decreased and fluctuated since the 1970s, and breeding populations currently number 50–60 pairs of *limosa* and 5–10 pairs of *islandica* (Holling *et al.* 2010b; *RBBP 2010*).

BREEDING DISTRIBUTION *2008–11*

POSSIBLE •	-	<1%	<1%
PROBABLE ●	-	1%	1%
CONFIRMED ●	<1%	<1%	<1%
TOTAL	<1%	1%	1%

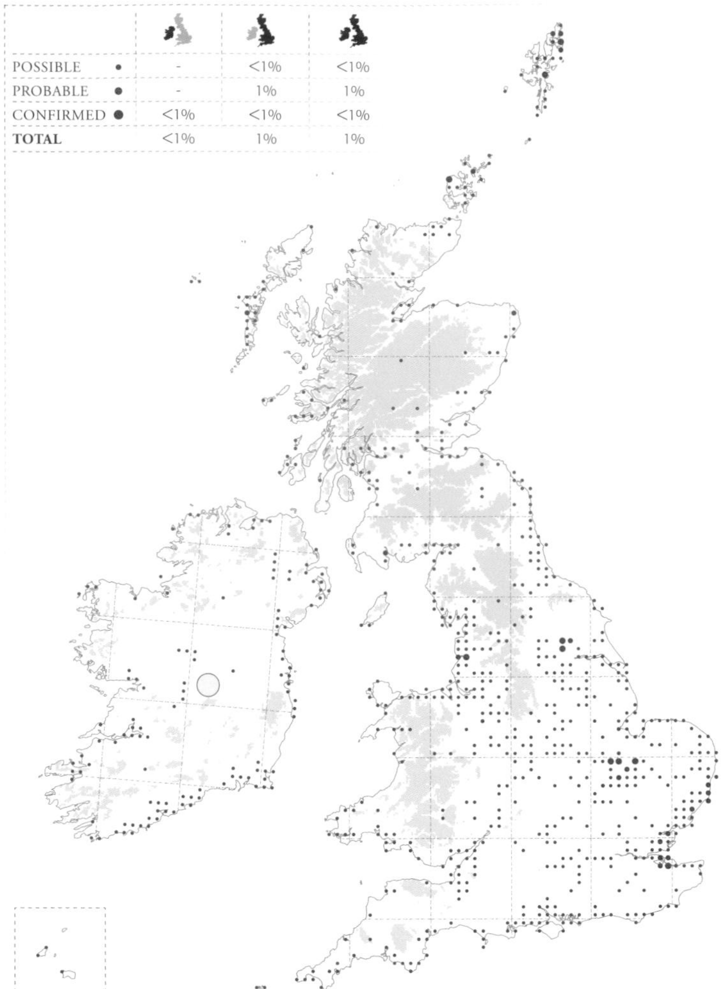

WINTER DISTRIBUTION *2007/08–2010/11*

PRESENT ●	16%	16%	16%

Sponsored by Associated British Ports

BREEDING DISTRIBUTION CHANGE *since 1988–91*

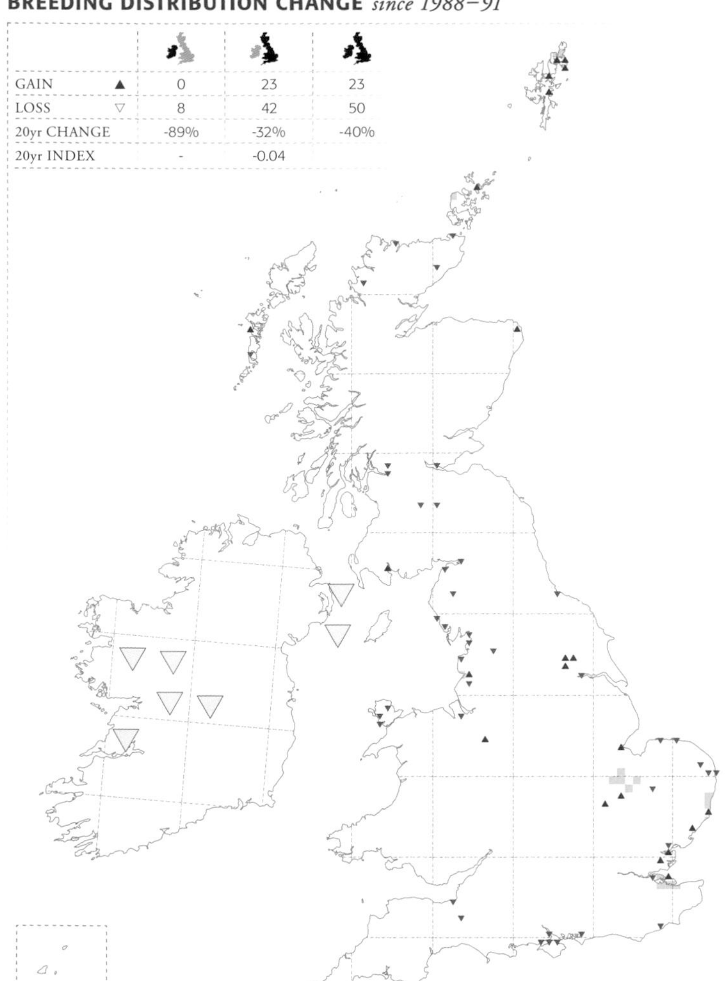

GAIN ▲	0	23	23
LOSS ▽	8	42	50
20yr CHANGE	-89%	-32%	-40%
20yr INDEX	-	-0.04	

BREEDING RELATIVE ABUNDANCE *2008–11*

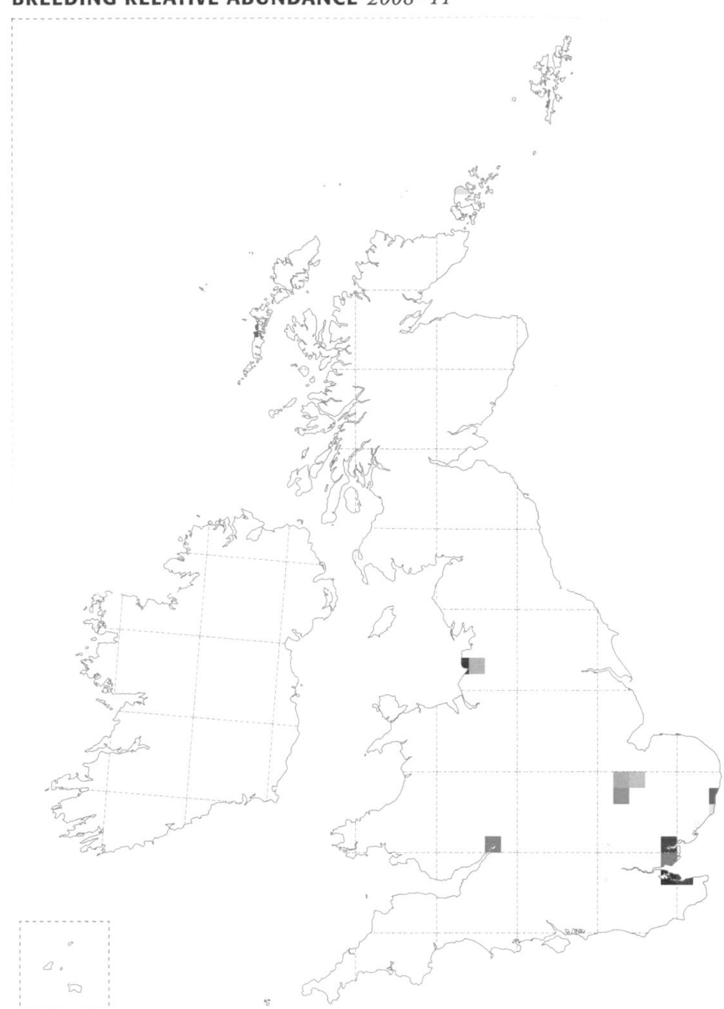

WINTER DISTRIBUTION CHANGE *since 1981–84*

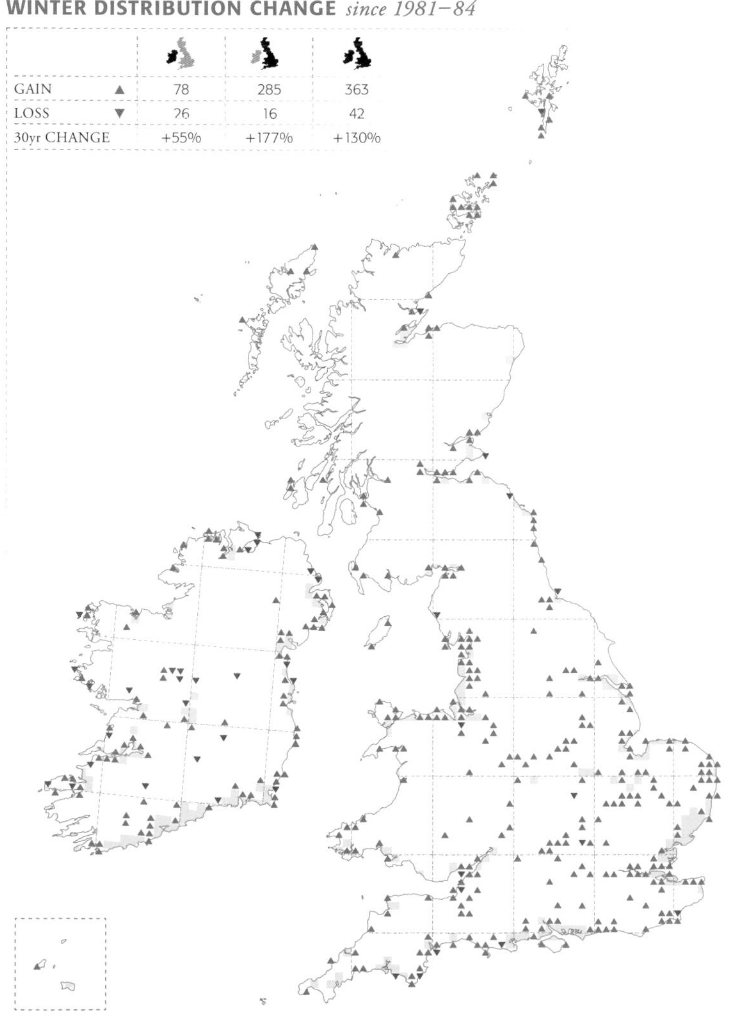

GAIN ▲	78	285	363
LOSS ▼	26	16	42
30yr CHANGE	+55%	+177%	+130%

WINTER RELATIVE ABUNDANCE *2007/08–2010/11*

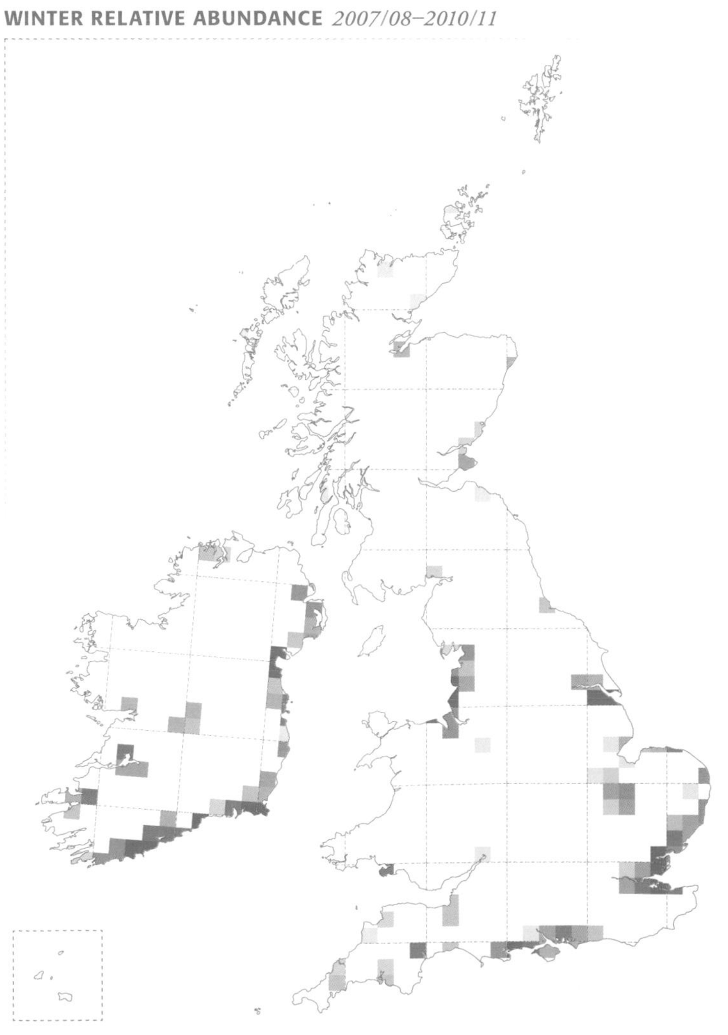

BAR-TAILED GODWIT
Limosa lapponica

BTO CODE: BA

OLIVER SMART www.smartimages.co.uk

BAR-TAILED GODWITS from arctic breeding areas in Fennoscandia and Russia arrive in late summer and winter around the coasts of Britain & Ireland, on suitable low-lying shores. The exceptions are much of northern and western Scotland and elsewhere where there are sections of steep cliff coastline. There is a scatter of inland records in southern Britain. The winter abundance map shows that the main concentrations are associated with the major British and Irish estuaries, together with the low-lying shorelines of Northumberland, the Outer Hebrides and Orkney.

Overall, range size has increased by 19% since the *1981–84 Winter Atlas*, although this masks a notable difference between Britain and Ireland where there were increases of 26% and 2% respectively. Many of the gains were in marginal, low-density areas such as Shetland, western Scotland, western Ireland, Wales and southwest England and may correspond to only small changes in the number of birds. Furthermore, some of these areas have seen improved coverage since the *1981–84 Winter Atlas*. The number of inland records has also increased, although these are usually of single birds only. The UK wintering population has remained largely stable over this period, with the exception of a steep drop in numbers during the mid 2000s, which has now been reversed (*WeBS Report 2011*). In the Republic of Ireland there was a 1% increase in wintering numbers during 1994/95–2008/09 (*I-WeBS Report 2009*), which is in line with the change in range size. The flyway population of this species is estimated to be stable (Delany *et al.* 2009), though numbers have been rising steadily in the Netherlands (Hornman *et al.* 2012a).

The majority of Bar-tailed Godwits that have wintered in Britain & Ireland depart in February or March, staging in the Wadden Sea before migrating to breeding grounds in Fennoscandia and Russia. During April–May there is a second movement, involving birds from West Africa en route to Siberian breeding grounds (*Migration Atlas*; Scheiffarth *et al.* 2002). These southern migrants account for many of the squares occupied by *non-breeding* birds, of which there were 540 in Britain and 69 in Ireland. Additionally, a proportion of *non-breeding*, mainly first-year birds remain to summer.

Sponsored by Guy Phillipson and Swale Wader Group

WINTER DISTRIBUTION *2007/08–2010/11*

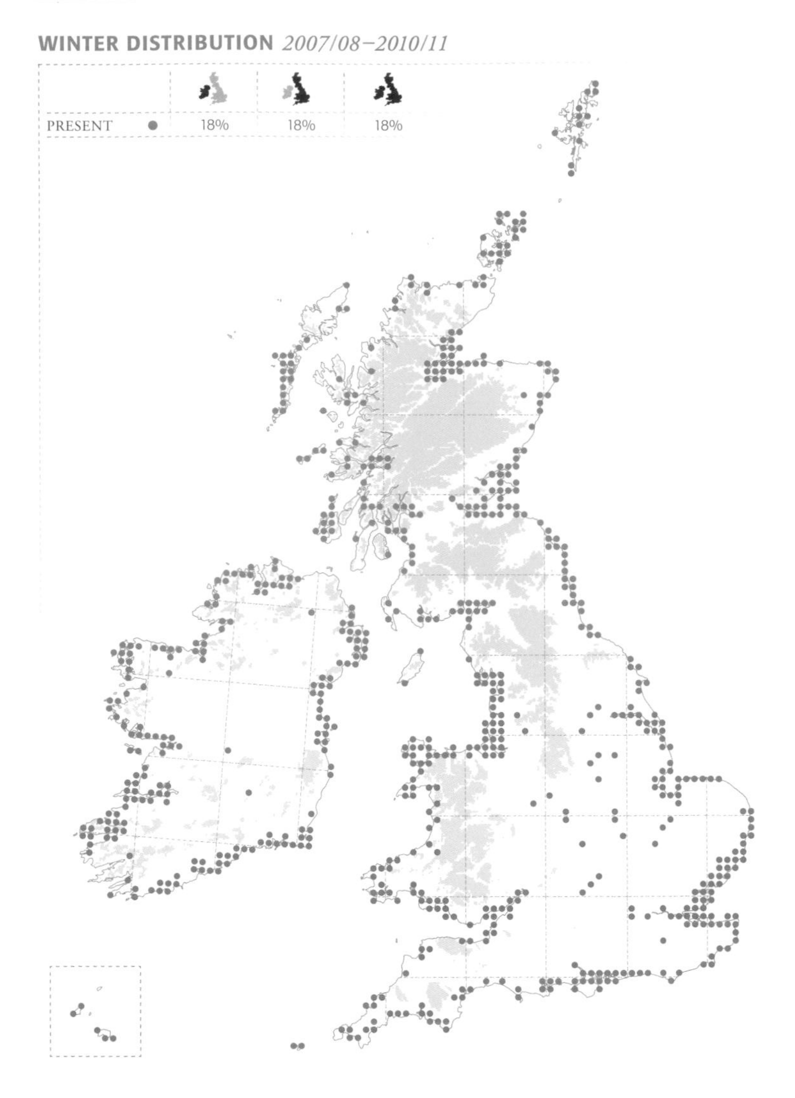

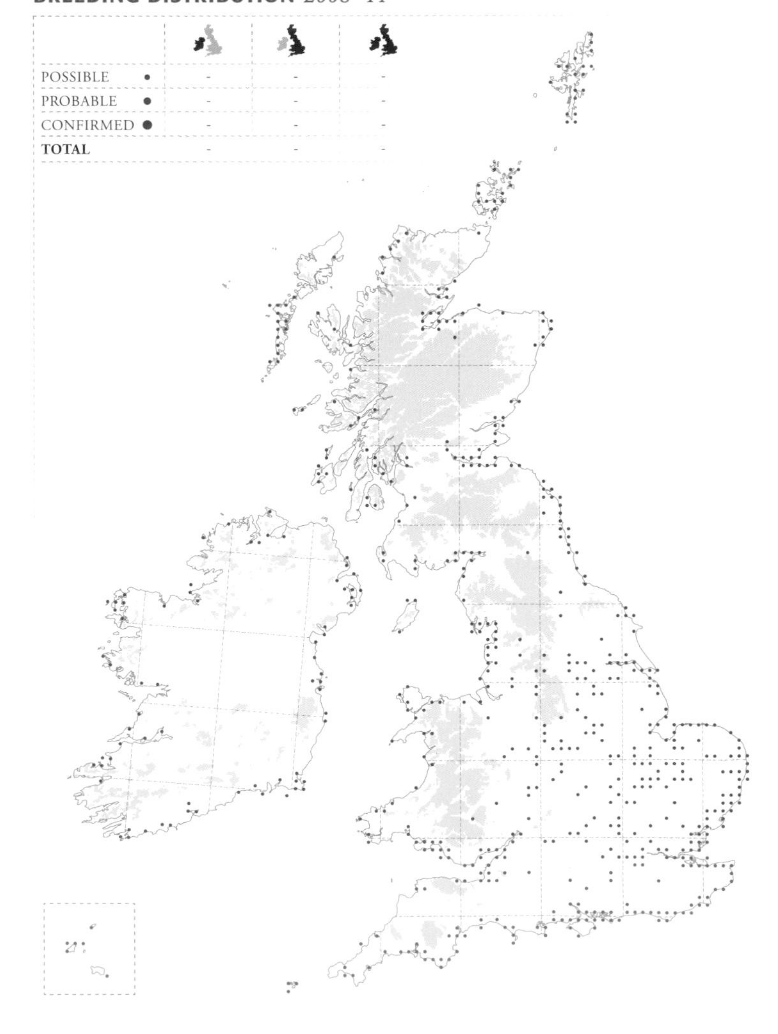

GAIN ▲	35	176	211
LOSS ▼	31	74	105
30yr CHANGE	+2%	+26%	+19%

TOM WALLIS

◀ **In Britain, the Wash, Alt Estuary and Thames Estuary each hold over 5,000 birds. In Ireland, Dundalk Bay, Dublin Bay, Lough Foyle and the Wexford Slobs all hold internationally important numbers.**

WHIMBREL
Numenius phaeopus

BTO CODE: WM

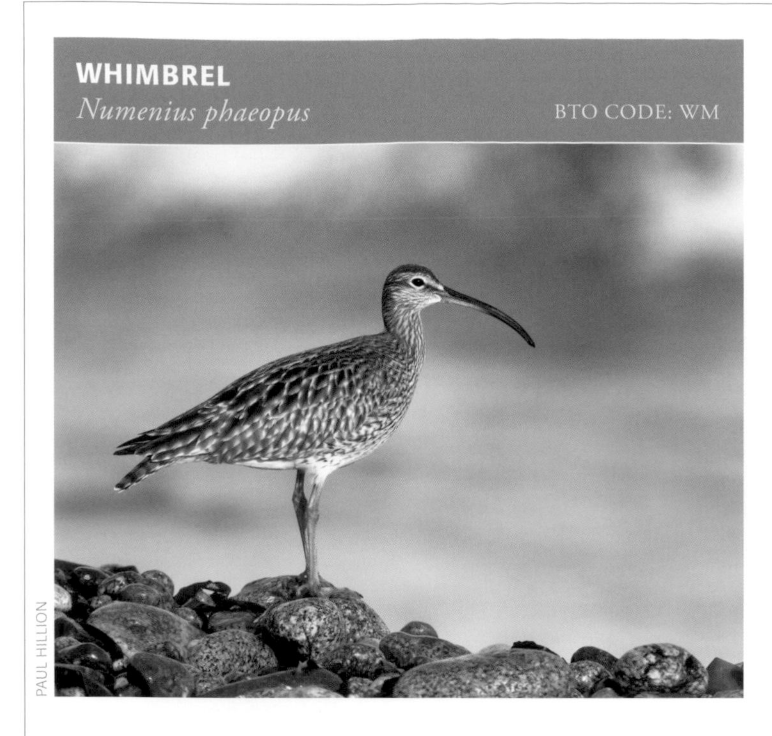

PAUL HILLION

THE WHIMBREL HAS a very restricted breeding distribution in Britain, with most of the population and 76% of the range confined to the Shetland Islands. The breeding population on Shetland was estimated at c.150 pairs during the *1968–72 Breeding Atlas* and had increased to 410–470 pairs by the *1988–91 Breeding Atlas,* with an extension of their distribution within the archipelago (Dore *et al.* 1996). *Bird Atlas 2007–11* shows further increases in range within Shetland, though surveys of Fetlar, Unst and Yell have revealed significant declines in numbers since the 1980s (Holling *et al.* 2010a) and a large-scale survey in 2009 produced a population estimate of only c.290 pairs (Jackson 2009).

Outside Shetland, breeding was *confirmed* on Orkney and *probable* breeding was recorded on Lewis, North Uist and in Caithness. No breeding evidence was reported on St Kilda, where Whimbrels bred during 1968–72 and were also recorded in 1988–91. Overall, there has been a 29% range contraction since the *1968–72 Breeding Atlas.* The change map indicates a mix of gains and losses, though it is possible that some of the losses relate to unhelpful coding of migrants in the previous atlases. Pressures on the breeding grounds from afforestation, peat extraction, agricultural improvements and climate change may threaten some sites (*Birds in England*; *Birds of Scotland*).

Non-breeding Whimbrels were reported from 1,430 10-km squares. Many will have been spring migrants bound for Greenland, Iceland, Fennoscandia and Russia (*Migration Atlas*). These pass through Britain & Ireland from early April through to late May, when they are widely recorded from coastal sites and, particularly in England, from inland localities. Return passage starts during early July, peaking in August and then declining through September (*Migration Atlas*).

Since the *1981–84 Winter Atlas* there has been a large increase in wintering records, with a 212% increase in the number of occupied 10-km squares in Britain & Ireland, probably as a result of milder winters. These are predominantly at coastal sites, with a concentration along the south coasts of England and Ireland, and tend to involve only a small number of individual birds.

Sponsored by Tom & Julia Lawson

BREEDING DISTRIBUTION *2008–11*

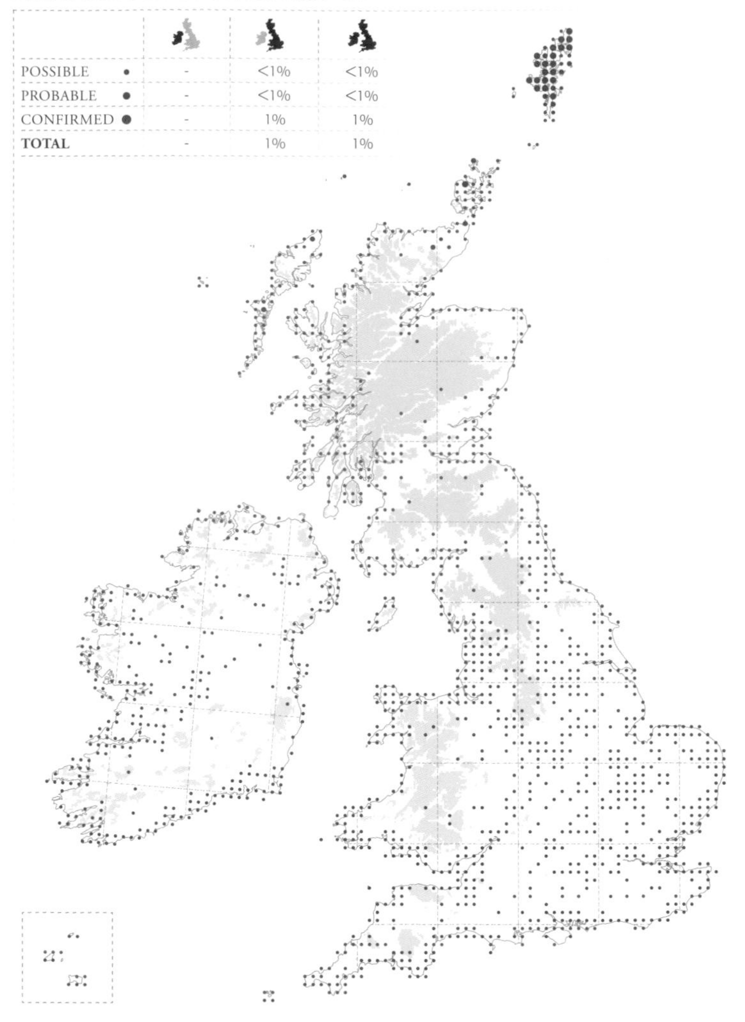

POSSIBLE •	-	<1%	<1%
PROBABLE •	-	<1%	<1%
CONFIRMED •	-	1%	1%
TOTAL	-	1%	1%

BREEDING DISTRIBUTION CHANGE *since 1968–72*

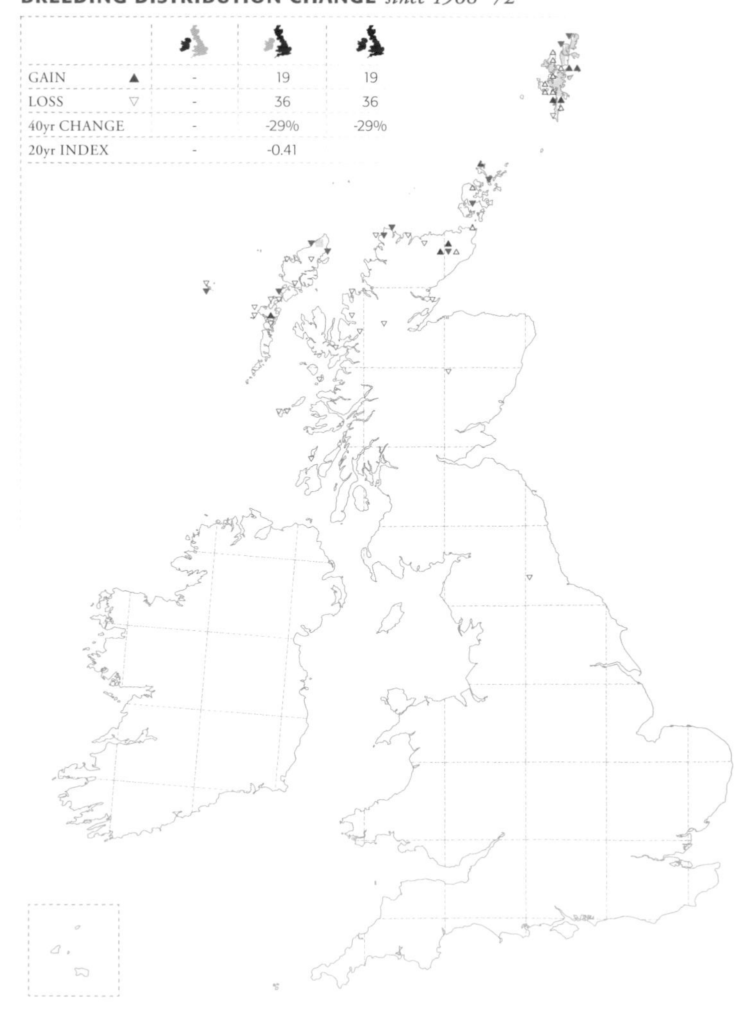

GAIN ▲	-	19	19
LOSS ▽	-	36	36
40yr CHANGE	-	-29%	-29%
20yr INDEX	-	-0.41	

▲ Whimbrels arrive back on their breeding territories from the last week of April, with arrivals peaking in May.

CURLEW

Numenius arquata

BTO CODE: CU

PAUL HILLION

THE LOSS OF breeding Curlews from most of Ireland and parts of western Britain over the last 40 years is a key finding from *Bird Atlas 2007–11*: since the *1968–72 Breeding Atlas* the range has contracted by 78% in Ireland and 17% in Britain. In Ireland losses have occurred mostly within the last 20 years. Gains in eastern and southeast England do not compensate for losses in western Scotland, Wales and southwest England. The abundance change map emphasises these losses, showing that, even in areas such as southern and eastern Scotland, fewer tetrads are occupied now than during 1988–91. These changes match the 44% population decline in the UK during 1995–2010 (*BBS Report 2011*) and a 78% decline in Ireland during 1998–2010 (*CBS Trend 2010*). The highest breeding concentrations remain in northern England, especially the Pennine Hills, eastern Scotland and the Northern Isles.

In the British uplands, declines have been steepest in heather-dominated areas (Amar *et al.* 2011). More generally across Britain and Ireland, breeding habitat has been affected on higher ground by the destruction of peat bogs, afforestation, farmland intensification and land abandonment and in the lowlands by drainage of wetlands and intensive management of lowland grasslands (Wilson *et al.* 2004; *Birds of Scotland*). Increased nest predation has also been implicated in this decline (Grant *et al.* 1999; Johnstone *et al.* 2007a). Climate change may be a problem in the future (Renwick *et al.* 2012).

In winter the breeding population moves to the coast and adjacent farmland, where it is joined by large numbers of migrants from Fennoscandia (*Migration Atlas*). The highest densities are on the major estuaries, in the Northern Isles and in western Ireland. Early returning breeders account for occupancy and low densities at some inland areas in late winter.

Contrasting fortunes between Britain and Ireland are also apparent in winter. The 15% range increase in Britain since the *1981–84 Winter Atlas* is corroborated by a 19% increase in UK wintering numbers between 1983/84 and 2008/09 (*SUKB 2012*). In Ireland a 23% range contraction has occurred, mostly inland, whilst counts from mainly coastal locations suggest a 40% population decline between 1994/95 and 2008/09 (*I-WeBS Report 2009*).

Sponsored by Anne Reid

BREEDING DISTRIBUTION *2008–11*

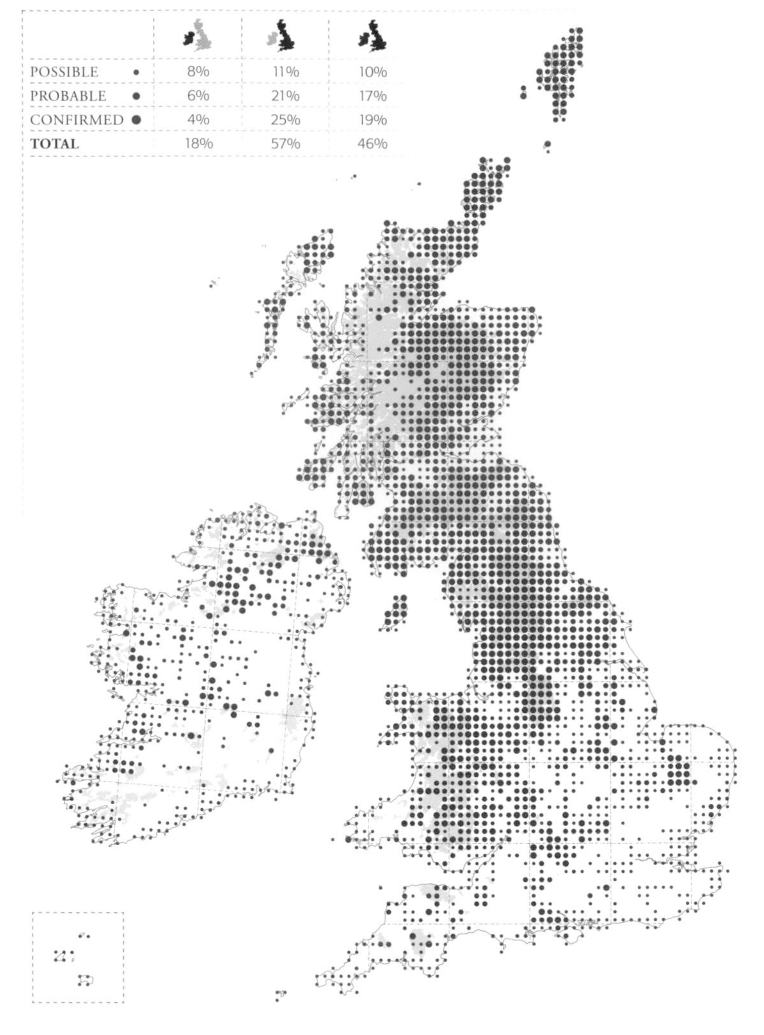

POSSIBLE	8%	11%	10%
PROBABLE	6%	21%	17%
CONFIRMED	4%	25%	19%
TOTAL	18%	57%	46%

WINTER DISTRIBUTION *2007/08–2010/11*

PRESENT	60%	59%	59%

BREEDING DISTRIBUTION CHANGE *since 1968–72*

GAIN ▲	6	167	173
LOSS ▽	660	490	1150
40yr CHANGE	-78%	-17%	-35%
20yr INDEX	-0.73	-0.19	

WINTER DISTRIBUTION CHANGE *since 1981–84*

GAIN ▲	56	470	526
LOSS ▼	234	257	491
30yr CHANGE	-23%	+15%	+2%

BREEDING RELATIVE ABUNDANCE *2008–11*

BREEDING RELATIVE ABUNDANCE CHANGE *since 1988–91*

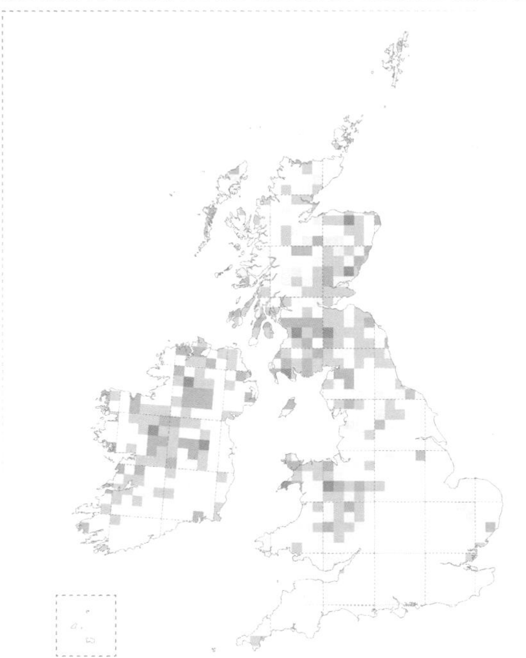

WINTER RELATIVE ABUNDANCE *2007/08–2010/11*

COMMON SANDPIPER

Actitis hypoleucos

BTO CODE: CS

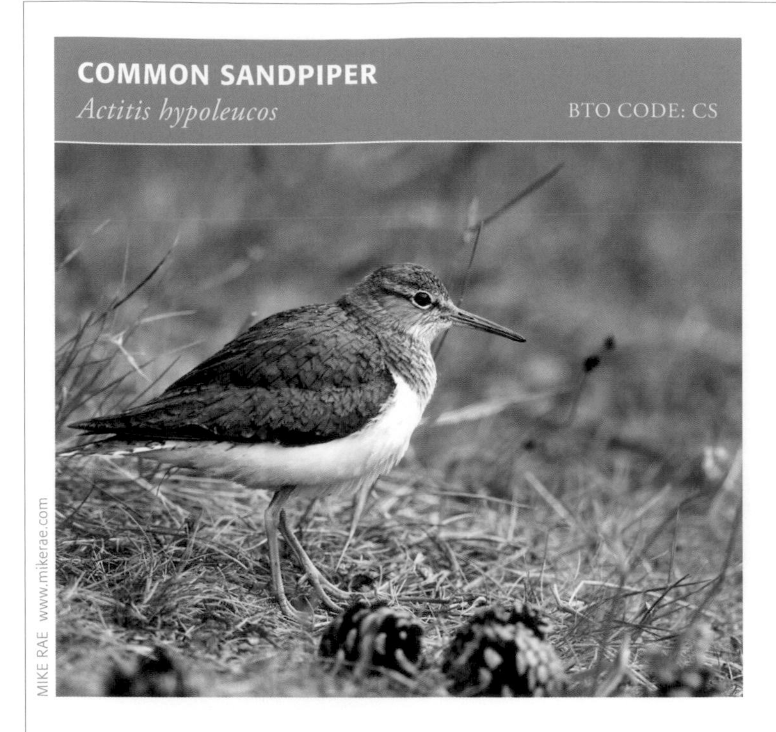

MIKE RAE www.mikerae.com

COMMON SANDPIPERS ARE summer visitors to breeding sites in the uplands of Wales, northern England, much of Scotland and western Ireland, with only small numbers outside these areas. On migration, birds are more widespread, particularly throughout southern and eastern England. The breeding abundance map shows that, within the breeding areas, densities are greatest in Scotland, particularly in the far northwest.

There has been a 20% contraction of the breeding range across Britain & Ireland since the *1968–72 Breeding Atlas*. Losses have been greatest in Ireland, in central Wales and on lower ground adjacent to the uplands in England and Scotland. The relative abundance change map indicates that declines are evident even in the areas of northern England and southern Scotland where range loss has so far been minimal. There was a 7% decline in the UK breeding population during 1995–2010 (*SUKB 2012*). Reasons for these losses are unclear but, in the Peak District, poor breeding success and reduced overwinter survival of first-year birds may have limited the recovery of the population after a hard-weather event in 1989 (Holland & Yalden 2002). Further analysis of data from the same study area suggested that lower adult survival rates were the likely cause of the 59% population decline during 1997–2004, and that variation in climatic conditions on the breeding grounds alone does not therefore appear to be responsible for Common Sandpiper declines (Pearce-Higgins *et al.* 2009b). Little is known of migratory stopover or wintering locations (Dougall *et al.* 2010), although recent work using geolocators has revealed that one bird from Scotland stopped over in Morocco and wintered in southern Senegal or the Gambia (Bates *et al.* 2013).

The UK wintering population is estimated at 73 individuals (Musgrove *et al.* 2011). This may be an underestimate given that, over the four winters of the Atlas, Common Sandpipers were reported from 344 10-km squares in southern parts of Ireland, England and Wales. Most records were of single birds, though some sites supported small groups, usually only in November. Since the *1981–84 Winter Atlas* there has been a 33% increase in winter range size, presumably in response to milder winters.

BREEDING DISTRIBUTION *2008–11*

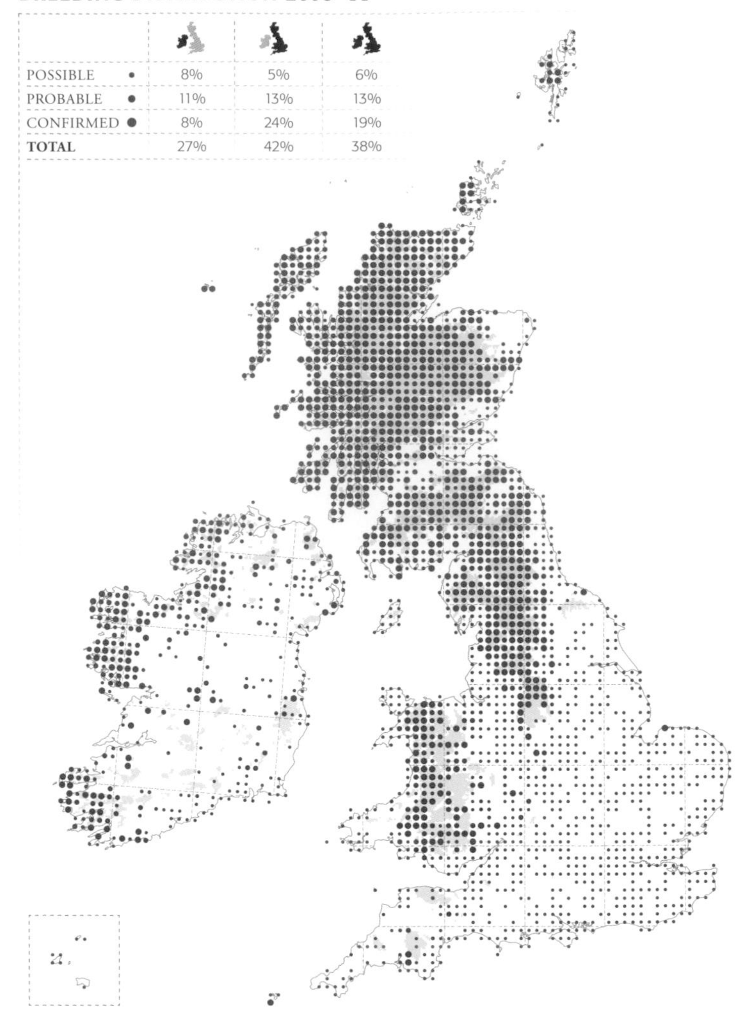

POSSIBLE •	8%	5%	6%
PROBABLE •	11%	13%	13%
CONFIRMED •	8%	24%	19%
TOTAL	27%	42%	38%

BREEDING RELATIVE ABUNDANCE CHANGE *since 1988–91*

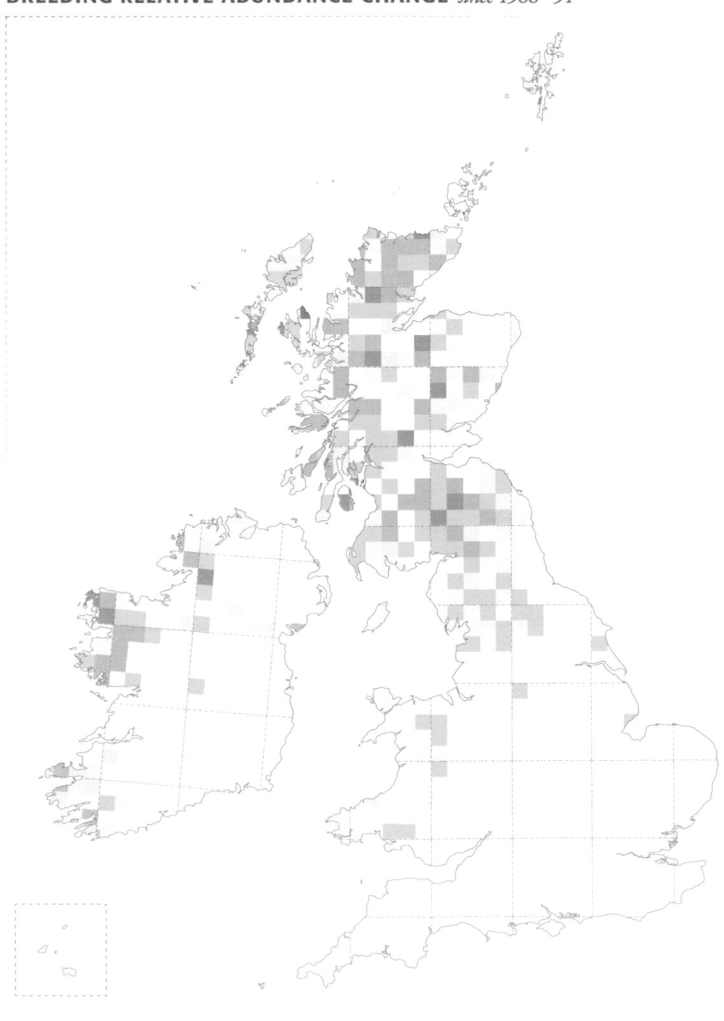

BREEDING DISTRIBUTION CHANGE *since 1968–72*

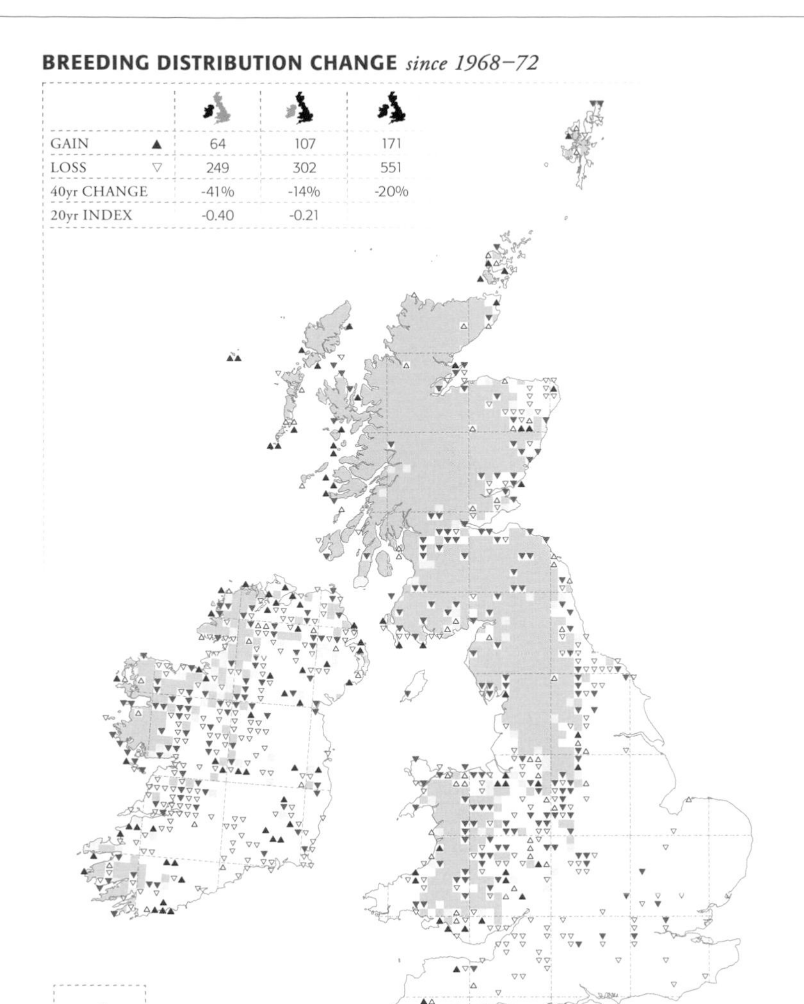

GAIN ▲	64	107	171
LOSS ▽	249	302	551
40yr CHANGE	-41%	-14%	-20%
20yr INDEX	-0.40	-0.21	

BREEDING RELATIVE ABUNDANCE *2008–11*

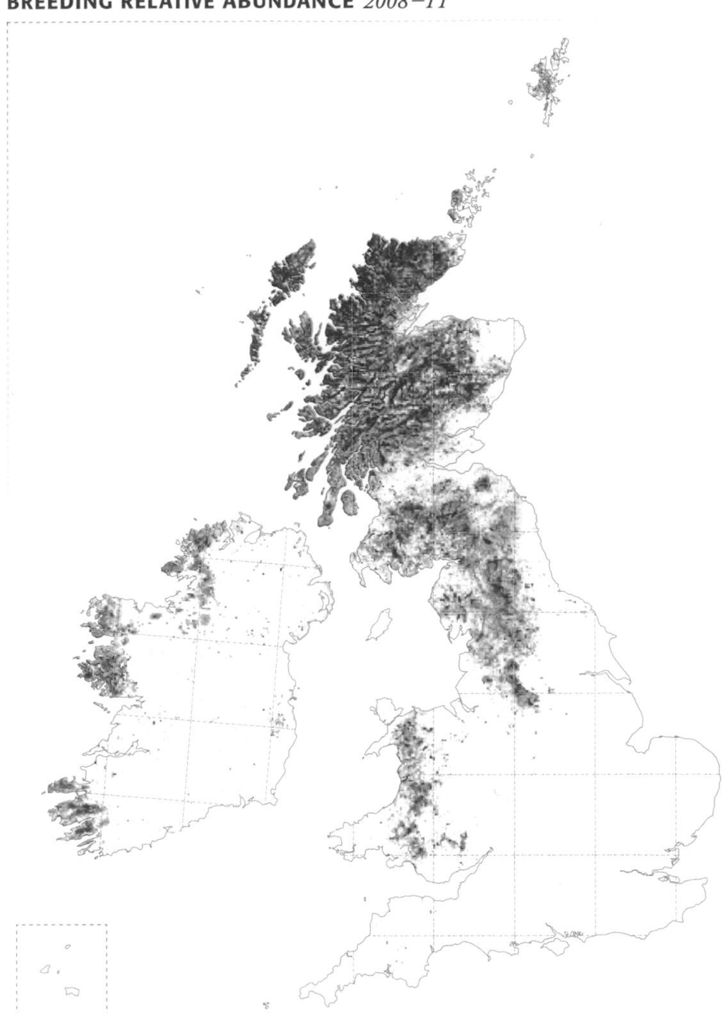

WINTER DISTRIBUTION *2007/08–2010/11*

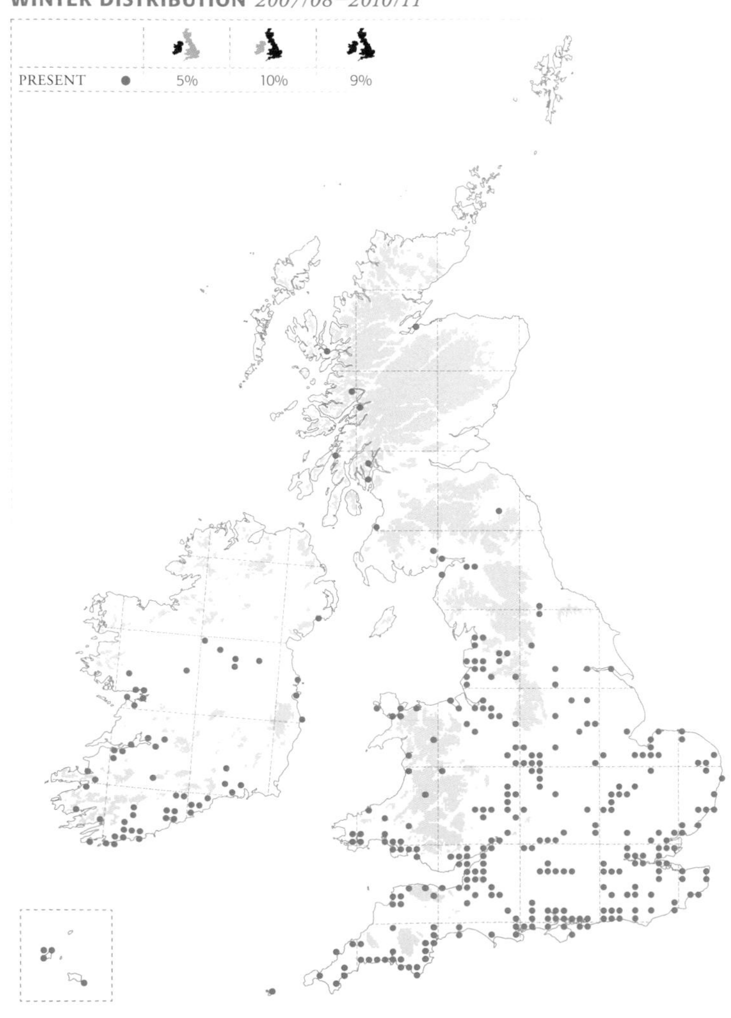

PRESENT ●	5%	10%	9%

WINTER DISTRIBUTION CHANGE *since 1981–84*

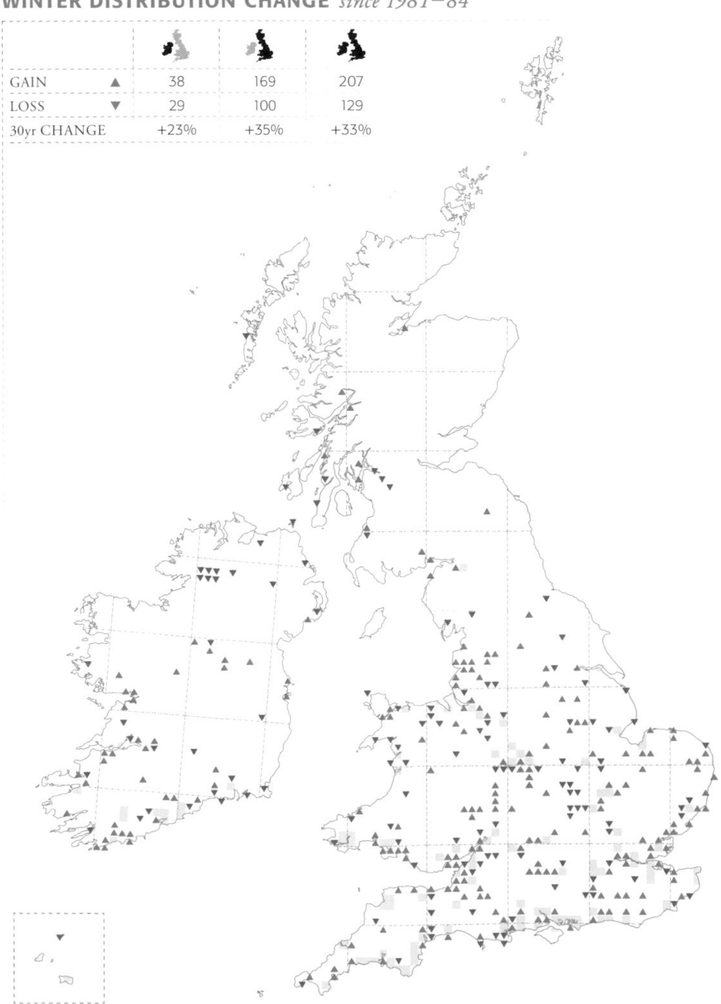

GAIN ▲	38	169	207
LOSS ▼	29	100	129
30yr CHANGE	+23%	+35%	+33%

GREEN SANDPIPER
Tringa ochropus

BTO CODE: GE

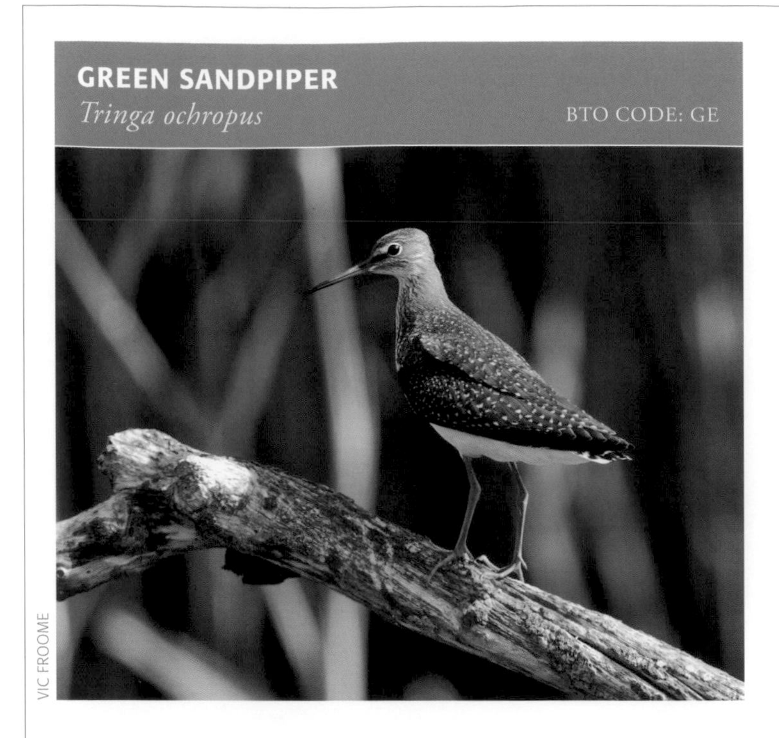

VIC FROOME

THE GREEN SANDPIPER is primarily an autumn migrant in Britain and Ireland from breeding grounds in Fennoscandia, passing through in good numbers during July–September, with smaller numbers of individuals remaining to winter. It favours freshwater sites such as ditches, streams and watercress beds as well as the edges of coastal marshes and estuaries (*BWP*). During the winter, Green Sandpipers are widely spread throughout much of England with the exception of upland areas. The importance of coastal lowlands, floodplains and river valleys is highlighted on the winter relative abundance map. Perhaps as a result of recent milder winters, the number of birds overwintering appears to have increased, with 56% more 10-km squares now occupied than in the *1981–84 Winter Atlas*. The gains have taken place mostly through the lowlands of England, Wales and southern Ireland, though there has also been a notable extension of the winter range into northern England and the Central Belt of Scotland. Losses within the core range probably reflect changes in local habitat suitability due to factors such as variable water levels.

Individual birds may return repeatedly to the same 10-km squares: a colour-ringing study in Hertfordshire showed an overall return rate from one winter to the next of 83.5% (Smith *et al.* 1992). At a finer scale, Green Sandpipers can be mobile, moving up to 3.2 km between diurnal feeding areas and roost sites (Smith *et al.* 1999). The full range of sites and squares used may be underestimated because Green Sandpipers are also nocturnal foragers, especially in cold weather (Smith *et al.* 1999).

The Green Sandpiper is a very rare breeding bird in Britain; it was first confirmed breeding in Inverness-shire in 1959 (*Birds of Scotland*), though there is an unsubstantiated claim from Cumbria in 1917 (Robinson 1917; *Birds in England*). During *Bird Atlas 2007–11* breeding was *confirmed* in three 10-km squares and there was evidence of *possible* breeding in one square. Three pairs bred in 2009 and 2010 (*RBBP 2010*). The breeding-season map shows that *non-breeding* birds are widespread on spring passage through much of lowland England and Wales, and are more thinly scattered in eastern Scotland.

Sponsored by Sir John Baird

WINTER DISTRIBUTION *2007/08–2010/11*

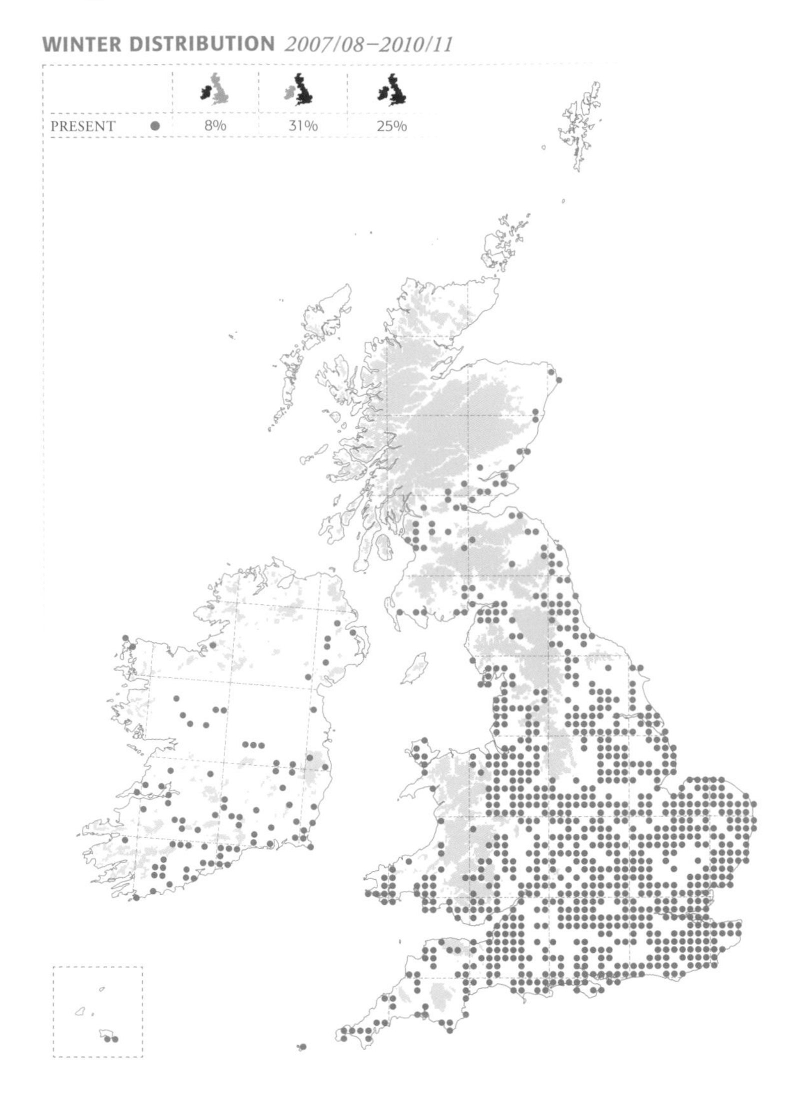

PRESENT ●	8%	31%	25%

BREEDING DISTRIBUTION *2008–11*

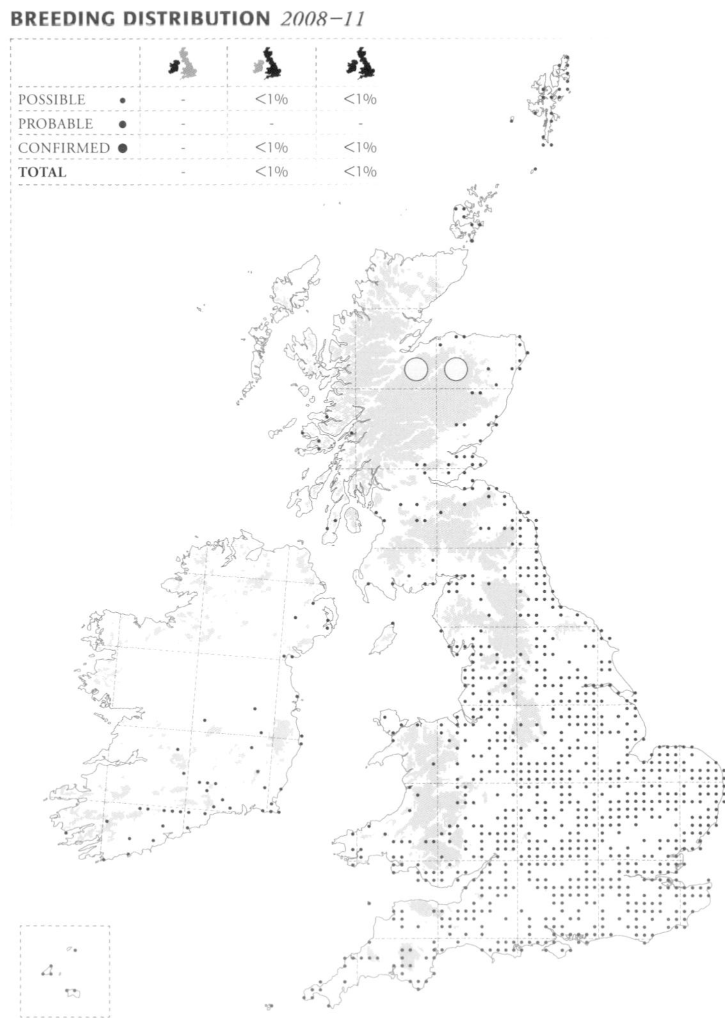

POSSIBLE ●	-	<1%	<1%
PROBABLE ●	-	-	-
CONFIRMED ●	-	<1%	<1%
TOTAL	-	<1%	<1%

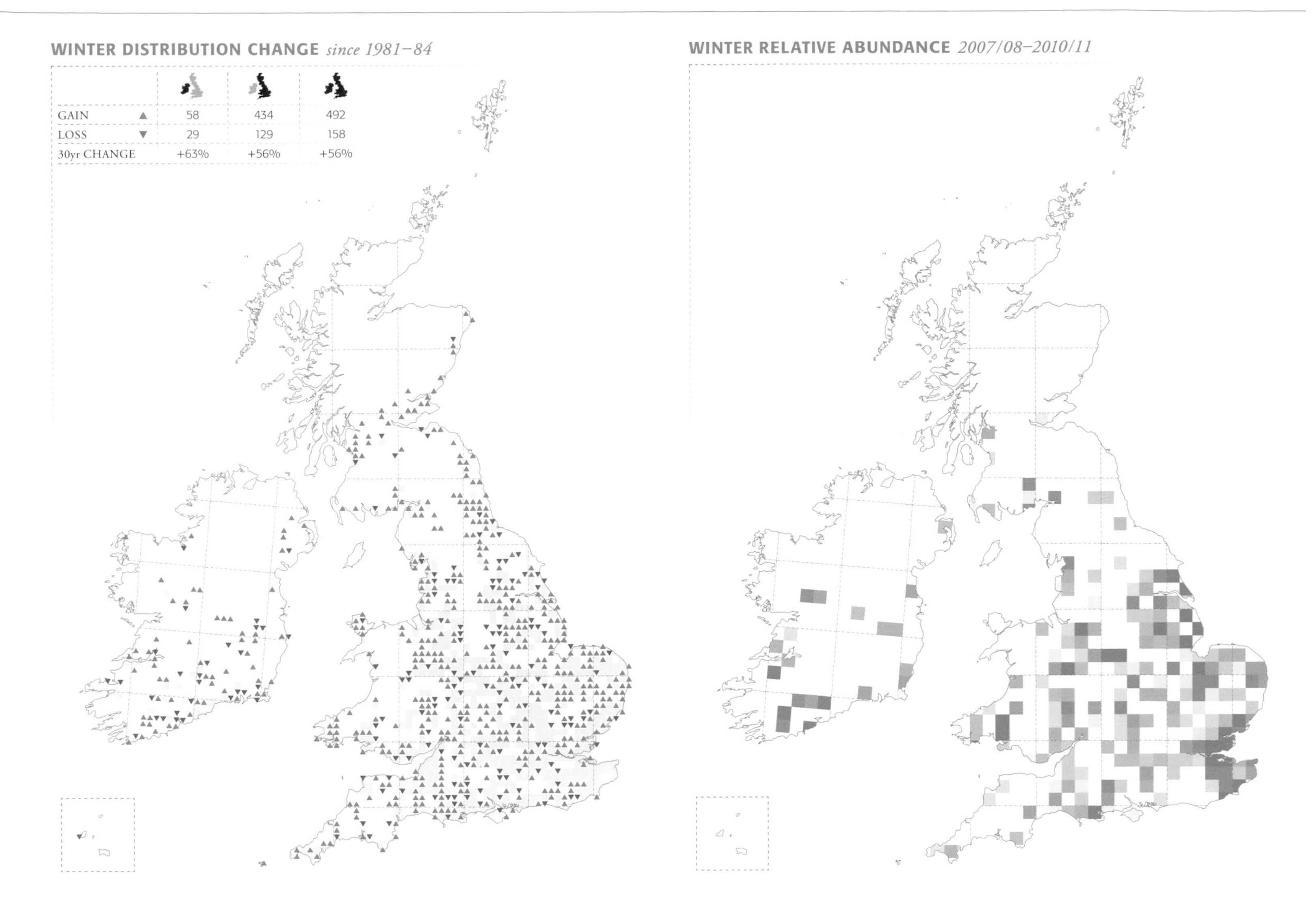

GAIN ▲	58	434	492
LOSS ▼	29	129	158
30yr CHANGE	+63%	+56%	+56%

◀ **Although Green Sandpipers are often seen singly, the peak counts for the breeding season and winter were 65 at Cantley Beet Factory, Norfolk, in July 2009 and 14 at Mirelake, Shropshire, in November 2007.**

SPOTTED REDSHANK

Tringa erythropus

BTO CODE: DR

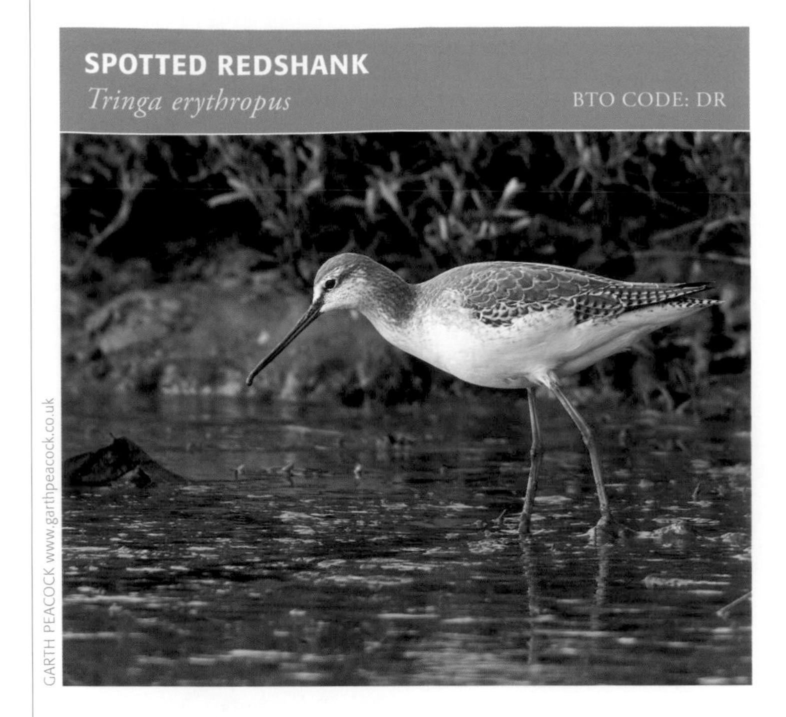

GARTH PEACOCK www.garthpeacock.co.uk

SPOTTED REDSHANKS BREED from northern Scandinavia through subarctic Russia and eastward to Siberia, but have never been proven to breed in Britain & Ireland. The species is most familiar as a passage migrant and, although most individuals winter in equatorial Africa (Wetlands International 2013), a small proportion winters in western Europe, including up to c.100 birds in Britain (Musgrove *et al.* 2011) and tens of individuals in Ireland (*I-WeBS Report 2009*).

The winter distribution map shows that Spotted Redshanks are largely, but not exclusively, found on the coast, where there can be subtle temporal changes in site use during autumn and winter. For example, in East Anglia numbers peak at the Wash, Lincolnshire/Norfolk, during the main autumn passage period, while counts on the north Norfolk coast increase later in the autumn, and those at the Blackwater Estuary, Essex, during early winter (*WeBS Report 2011*). They are scarce in the north and west of Scotland and of Ireland. During the Atlas period the highest winter counts were recorded around the Solent, Hampshire, where a peak of 22 was counted in November 2011, and on the north Norfolk coast a peak of 15 was recorded in January 2008. In Ireland, most counts were of single birds, but three were reported in Wexford Harbour and in Galway Bay. These counts corresponding well with the peak counts reported by waterbird monitoring schemes (*I-WeBS Report 2009*; *WeBS Report 2011*).

Since the *1981–84 Winter Atlas* there has been a contraction of the winter distribution of Spotted Redshanks in Ireland, contrasting with an expansion of range in Britain. The latter has taken place in conjunction with a slow increase in the local wintering population, which may be linked to the general trend for milder winters in recent decades. Climate-driven shifts in core winter distributions have been demonstrated for other wader species in northwest Europe (e.g. Maclean *et al.* 2008) and Spotted Redshanks are likely to have responded similarly.

During the breeding season, migrant Spotted Redshanks were reported from 236 10-km squares in Britain and 25 in Ireland. These reflect a mix of northbound spring migrants and early returning adults in July. The peak count in Britain was 47 birds at Minsmere, Suffolk, in July 2008, and in Ireland only single birds were recorded at any one location.

Sponsored by Roger Jones

WINTER DISTRIBUTION *2007/08–2010/11*

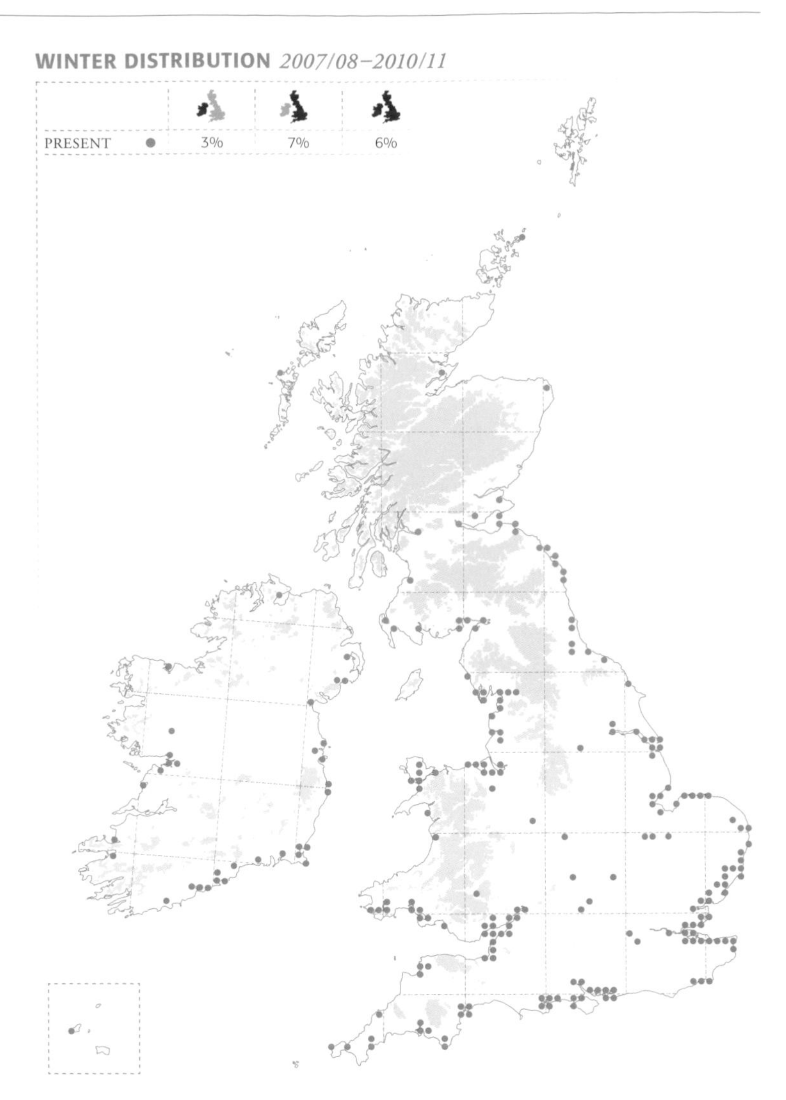

	Ireland	Britain	Britain & Ireland
PRESENT ●	3%	7%	6%

WINTER DISTRIBUTION CHANGE *since 1981–84*

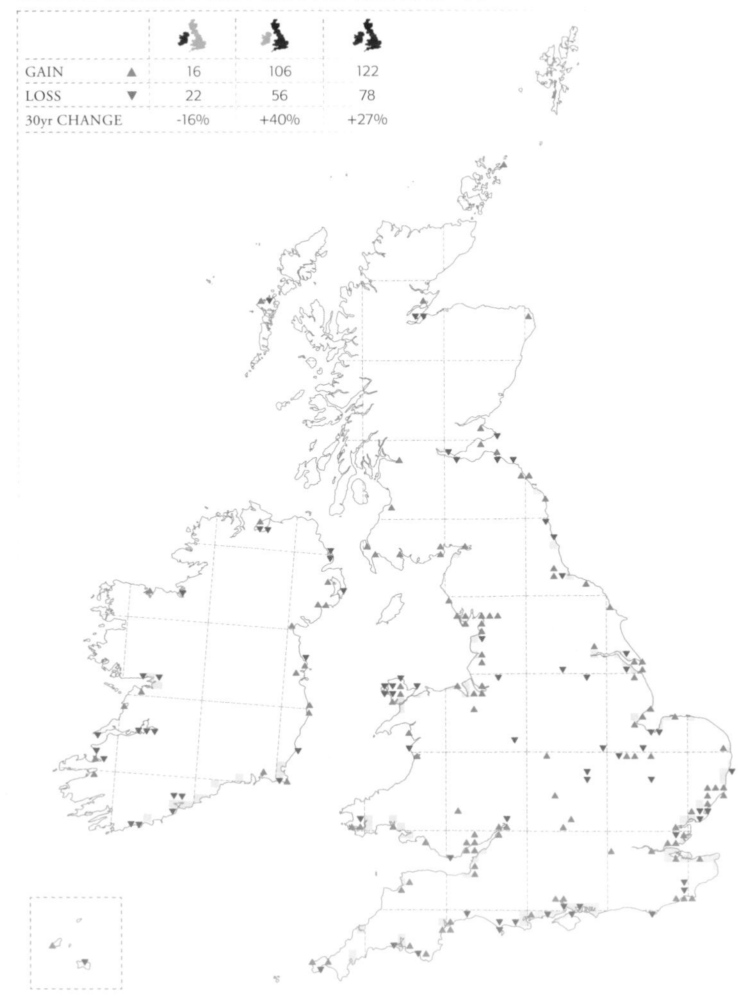

	Ireland	Britain	Britain & Ireland
GAIN ▲	16	106	122
LOSS ▼	22	56	78
30yr CHANGE	-16%	+40%	+27%

WOOD SANDPIPER

Tringa glareola

BTO CODE: OD

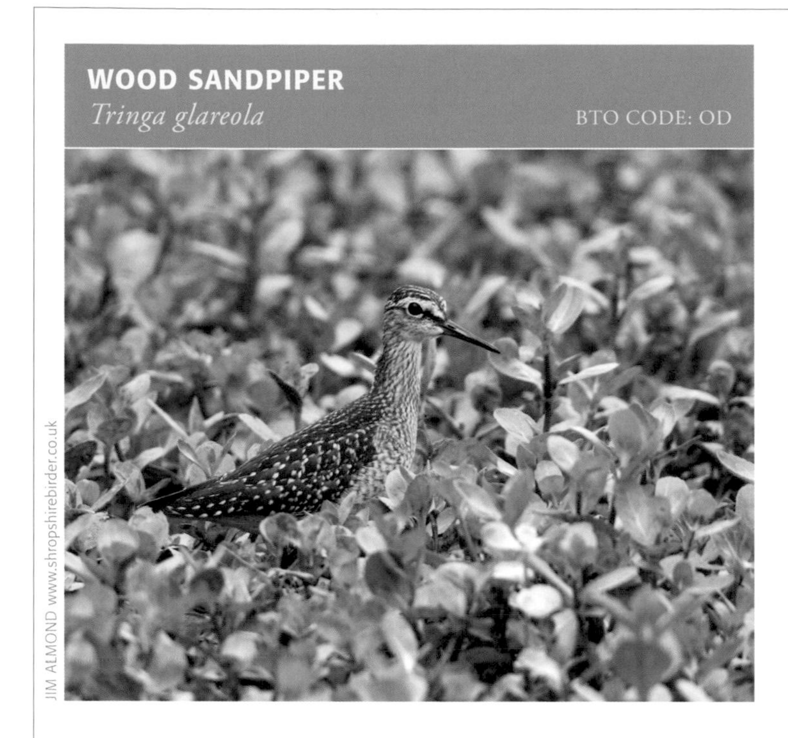

JIM ALMOND www.shropshirebirder.co.uk

THE WOOD SANDPIPER is a very rare breeding wader confined to boggy habitats in Scotland, although it is widely recorded on spring passage throughout the southern half of Britain. Owing to the sensitivity of some of its breeding haunts, its breeding distribution and changes since the *1988–91 Breeding Atlas* are shown here at 50-km resolution.

Breeding evidence was recorded in 22 10-km squares, with most in Sutherland and Caithness, but others in Inverness-shire, Wester Ross and the Outer Hebrides. The breeding change map shows that most 50-km squares have seen an increase in the number of underlying occupied 10-km squares since the *1988–91 Breeding Atlas*: three 10-km squares have lost breeding Wood Sandpipers and 17 squares have gained them. Since the *1968–72 Breeding Atlas*, when 18 10-km squares were occupied, there has been a gain of only five 10-km squares with breeding evidence, however, suggesting that there has been a population recovery since 1988–91, when the maximum annual total was just six pairs (Ogilvie *et al.* 1994).

Breeding Wood Sandpipers can be difficult to locate and the remoteness of their breeding areas means that some pairs may have been missed, especially in those 10-km squares that received single Timed Tetrad Visits and little Roving effort. Nevertheless, the main sites in Highland are regularly surveyed so this species is reasonably well monitored (Chisholm 2007). All the main breeding sites were surveyed during the *1988–91 Breeding Atlas* so the range expansion that has since occurred is not an artefact of coverage. During 2010, 27 pairs were reported from at least 11 sites, equalling the highest previously documented total from the national survey, in 2007, when 27 pairs were reported from 20 sites (Holling *et al.* 2010a; *RBBP 2010*).

Wood Sandpipers are scarce passage migrants in Ireland, with nine 10-km squares occupied during 2008–11, whilst in Britain there were records from 404 10-km squares. Most Wood Sandpipers have departed by winter; there were just two 10-km squares occupied in England, both in early November only, compared with four in the *1981–84 Winter Atlas*. There were none in Ireland in either winter atlas period.

BREEDING DISTRIBUTION *2008–11*

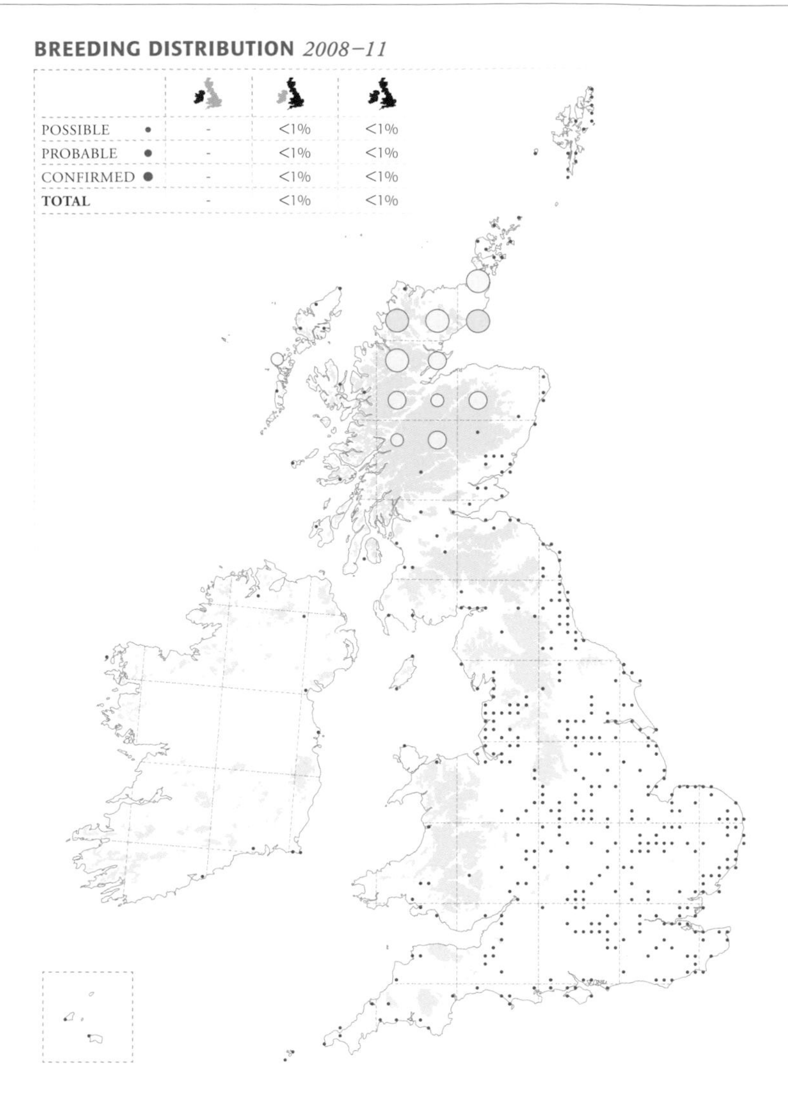

POSSIBLE •	-	<1%	<1%
PROBABLE •	-	<1%	<1%
CONFIRMED ●	-	<1%	<1%
TOTAL	-	<1%	<1%

BREEDING DISTRIBUTION CHANGE *since 1988–91*

GAIN ▲	-	17	17
LOSS ▽	-	3	3
20yr CHANGE	-	+175%	+175%
20yr INDEX	-	-	

Sponsored by White Rose Property Management (Peter Watson & Jenny Dixon)

GREENSHANK
Tringa nebularia

BTO CODE: GK

MARK HOPE

BREEDING GREENSHANKS ARE largely restricted to the bogs and moors of the northwest Highlands and Hebridean islands, with the highest densities in Sutherland, Wester Ross, Lewis, Harris and North Uist.

There has been a 6% range expansion in Britain since the *1988–91 Breeding Atlas* and a 2% range expansion since the *1968–72 Breeding Atlas*. Gains have been on Shetland, Lewis, North Uist and in West Inverness-shire. Losses are most evident in the eastern fringes of the range, particularly in Badenoch and Strathspey. In explaining losses on Deeside, Watson & Francis (2012) cited acidic deposition, overgrazing and trampling, disease from sheep ticks, disturbance, predation and reforestation as possible causes. Population trends are lacking but a survey of the Caithness and Sutherland Peatlands Special Protection Area in 2009 suggested an increase since the 1990s (Bellamy & Eaton 2010). Such an increase may be more widespread as the relative abundance change map indicates many areas where the proportion of occupied tetrads has increased since the *1988–91 Breeding Atlas*. Greenshanks are very rare breeders in Ireland and, unlike the previous two breeding atlases, where one 10-km square was occupied, no breeding Greenshanks were recorded during 2008–11.

Records of *non-breeding* birds in 870 10-km squares reflect a mixture of birds on spring passage and females and failed breeders that leave the moors and turn up at inland and coastal sites in late June and July.

The wintering range coincides with key estuarine complexes throughout the coasts of Britain & Ireland. Relative abundance is highest in the west, especially in Ireland and parts of western Scotland, where birds are more widely distributed. There have been range expansions of 48% in Britain and 13% in Ireland since the *1981–84 Winter Atlas*, with gains most evident in Scotland and eastern England and at several locations around the Irish coastline, most notably in the west, as well as at a small selection of inland sites in Britain. This expansion reflects a continuing increase in abundance (*I-WeBS Report 2009*; *WeBS Report 2011*), probably as a result of milder climatic conditions (Maclean *et al.* 2008).

Sponsored by Barbara Boize

BREEDING DISTRIBUTION *2008–11*

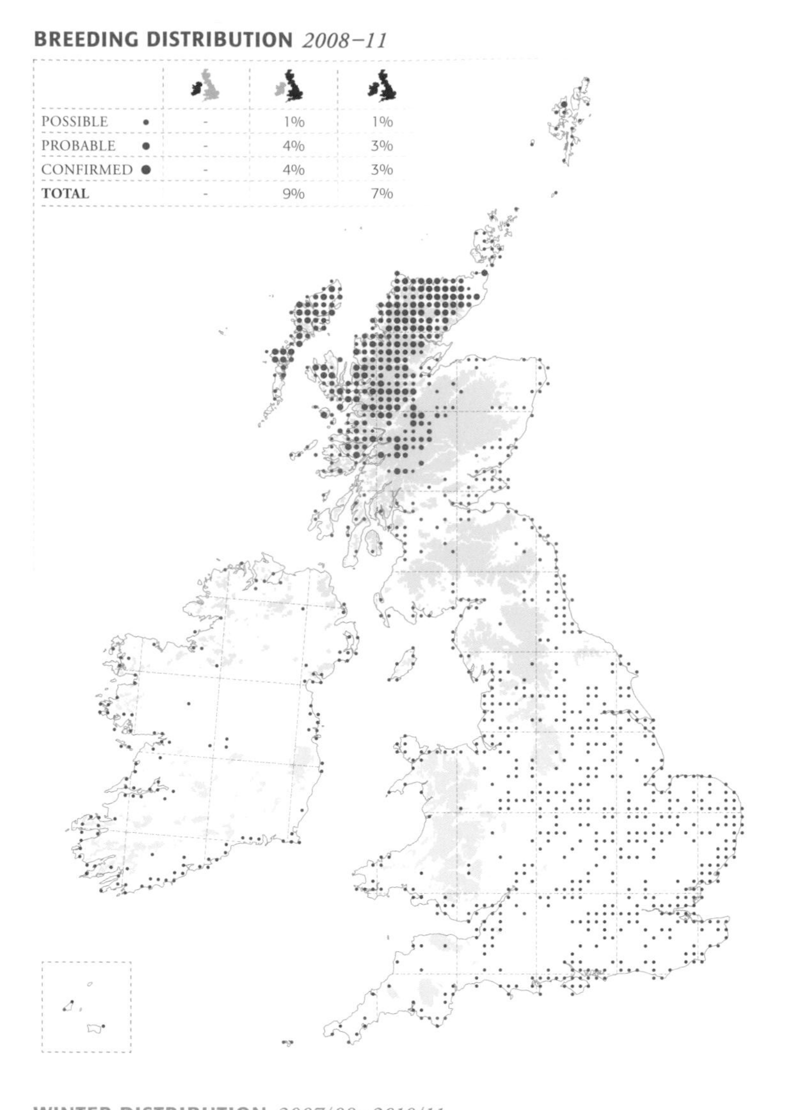

	Ireland	Britain	Britain & Ireland
POSSIBLE •	-	1%	1%
PROBABLE •	-	4%	3%
CONFIRMED •	-	4%	3%
TOTAL	-	9%	7%

WINTER DISTRIBUTION *2007/08–2010/11*

	Ireland	Britain	Britain & Ireland
PRESENT •	26%	14%	17%

BREEDING DISTRIBUTION CHANGE *since 1968–72*

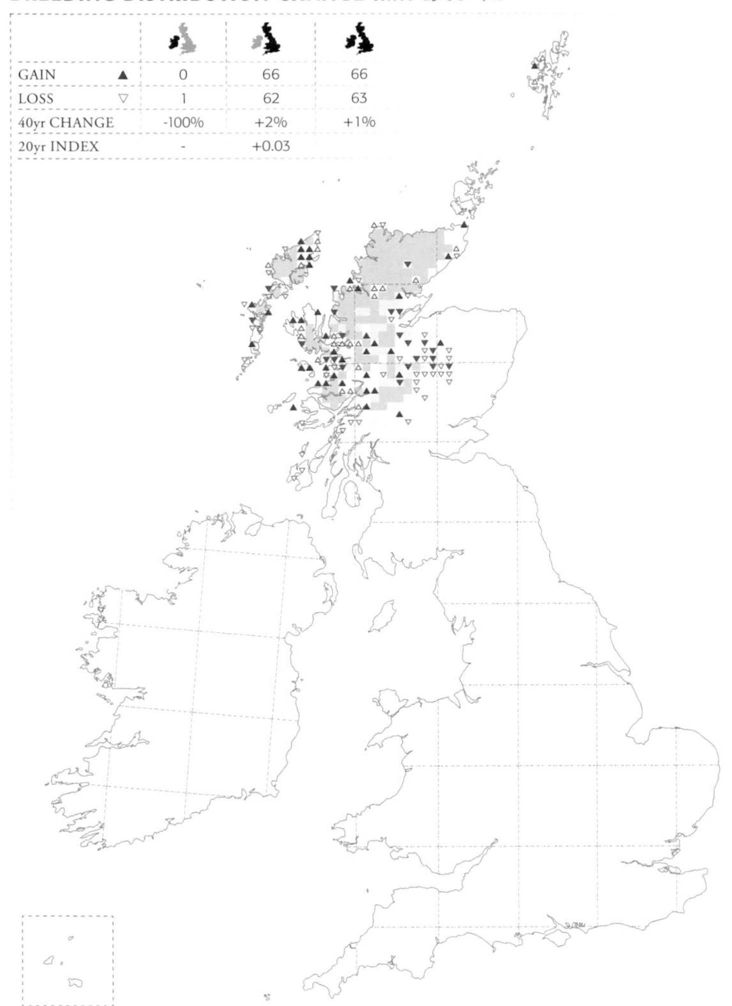

GAIN ▲	0	66	66
LOSS ▽	1	62	63
40yr CHANGE	-100%	+2%	+1%
20yr INDEX	-	+0.03	

WINTER DISTRIBUTION CHANGE *since 1981–84*

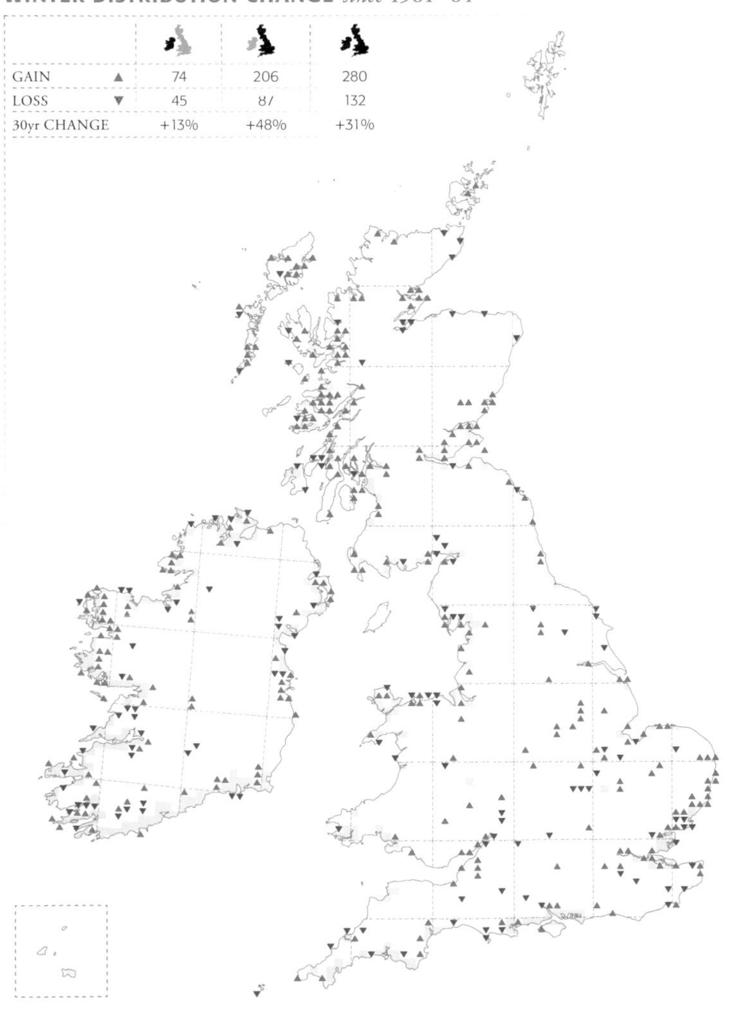

GAIN ▲	74	206	280
LOSS ▼	45	87	132
30yr CHANGE	+13%	+48%	+31%

BREEDING RELATIVE ABUNDANCE *2008–11*

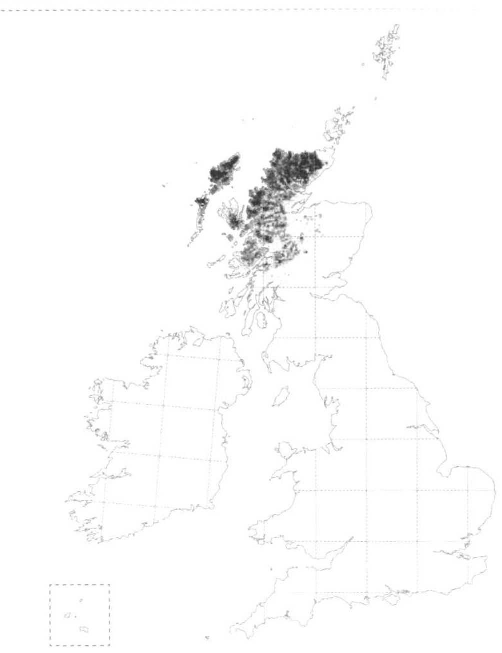

BREEDING RELATIVE ABUNDANCE CHANGE *since 1988–91*

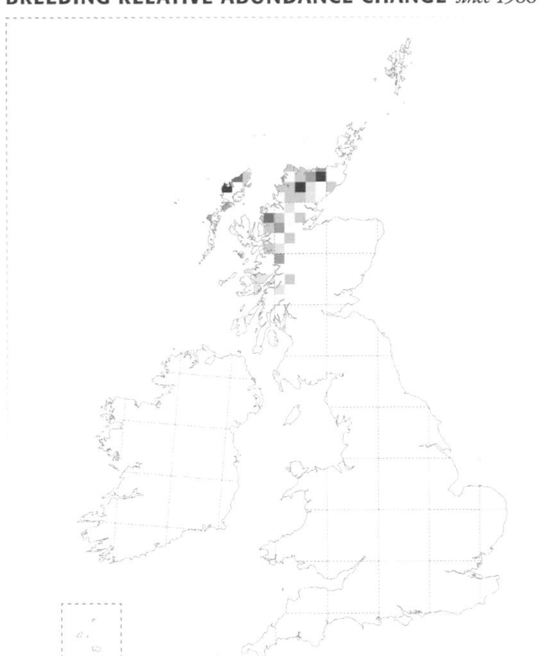

WINTER RELATIVE ABUNDANCE *2007/08–2010/11*

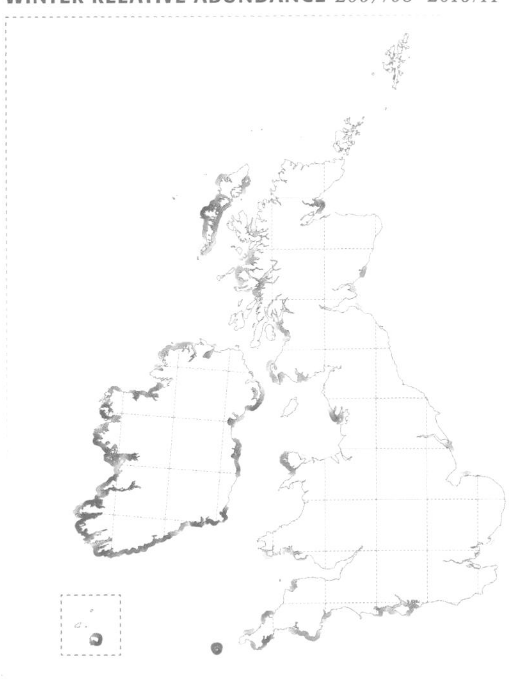

REDSHANK
Tringa totanus

BTO CODE: RK

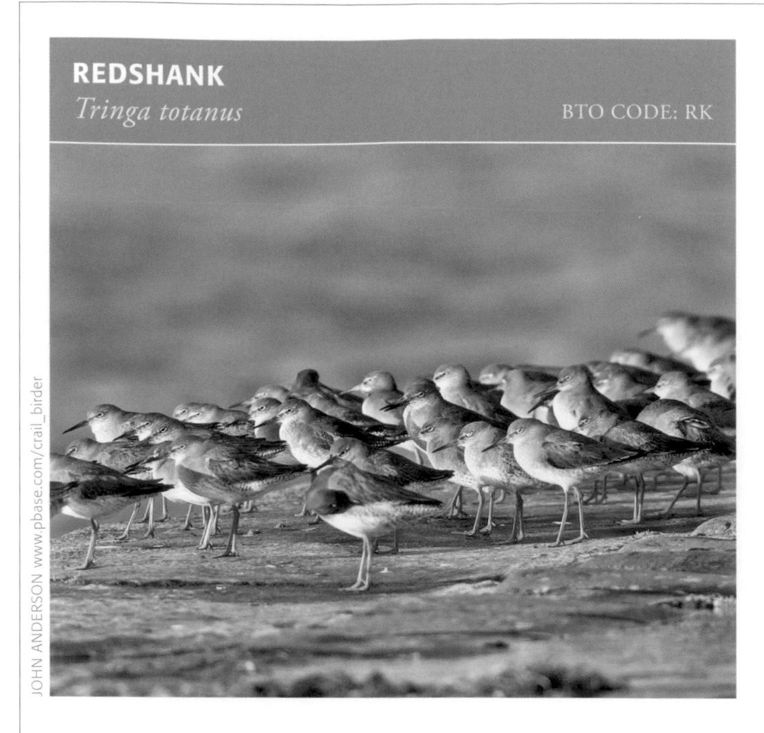
JOHN ANDERSON www.pbase.com/crail_birder

IN WINTER, REDSHANKS are widely distributed round the coasts of Britain & Ireland and in low-lying inland river valleys and callows. The winter relative abundance map suggests that the highest densities are associated with major estuaries and the Northern Isles. Since the *1981–84 Winter Atlas* there has been a 3% increase in range size across Britain & Ireland. Gains in northwest Scotland probably reflect increased recording effort in remote areas. Elsewhere in Britain the greatest changes affect inland 10-km squares. The loss of suitable wetlands through drainage affects some areas whilst in others gains result from the creation of new sites such as gravel pits. In addition, improved coverage is more likely to pick up occasional individuals at these inland sites. In Ireland, winter gains and losses are largely confined to inland areas and are more or less balanced.

During the breeding season the Redshank favours a variety of damp habitats including coastal marshes, lowland wet grasslands and rough pasture on moorland fringes, resulting in a rather fragmented distribution. The highest densities were in Caithness, the Northern Isles, Outer Hebrides, the Pennines and Lancashire, the coastal marshes of southeast England and the river callows and turloughs of the midlands and west of Ireland.

There has been a 44% contraction of the breeding range across Britain & Ireland since the *1968–72 Breeding Atlas*, much of which occurred since 1988–91. Losses are evident throughout the range, with reductions in abundance also apparent within core areas. These losses in range and abundance corroborate a 39% population decline in the UK during 1995–2010 (*BBS Report 2011*). Agricultural intensification and the drainage of wetland sites are considered to have driven declines (Brindley *et al.* 1998; Wilson *et al.* 2005) and factors such as grazing pressure and predation may hinder recovery (Norris *et al.* 1998; Jackson *et al.* 2004; Bodey *et al.* 2010). Afforestation of moorland edges has had an influence in some areas (Thompson *et al.* 1995) and in Ireland changes in land-use practice, increased predation and adverse weather may be contributory factors to the decline on the machair and wet grasslands (Suddaby *et al.* 2009).

Sponsored by Dún Laoghaire–Rathdown County Council

BREEDING DISTRIBUTION *2008–11*

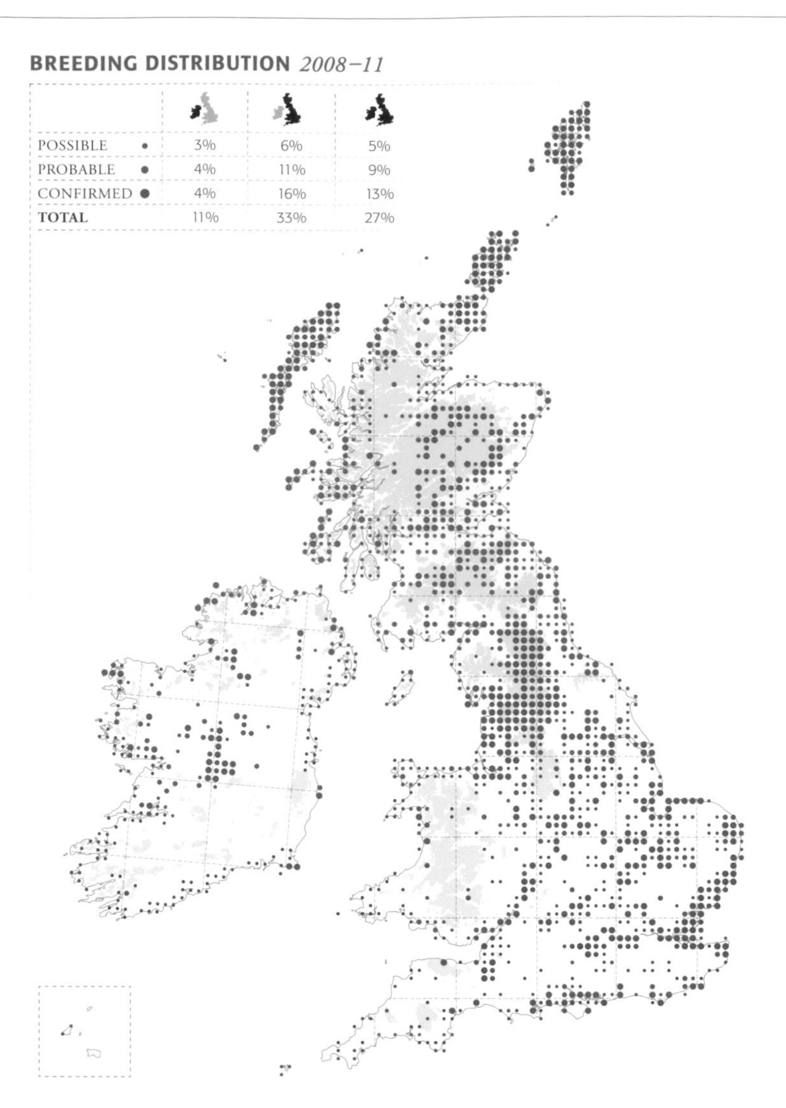

	Ireland	Britain	Britain & Ireland
POSSIBLE	3%	6%	5%
PROBABLE	4%	11%	9%
CONFIRMED	4%	16%	13%
TOTAL	11%	33%	27%

WINTER DISTRIBUTION *2007/08–2010/11*

	Ireland	Britain	Britain & Ireland
PRESENT	40%	45%	44%

BREEDING DISTRIBUTION CHANGE *since 1968–72*

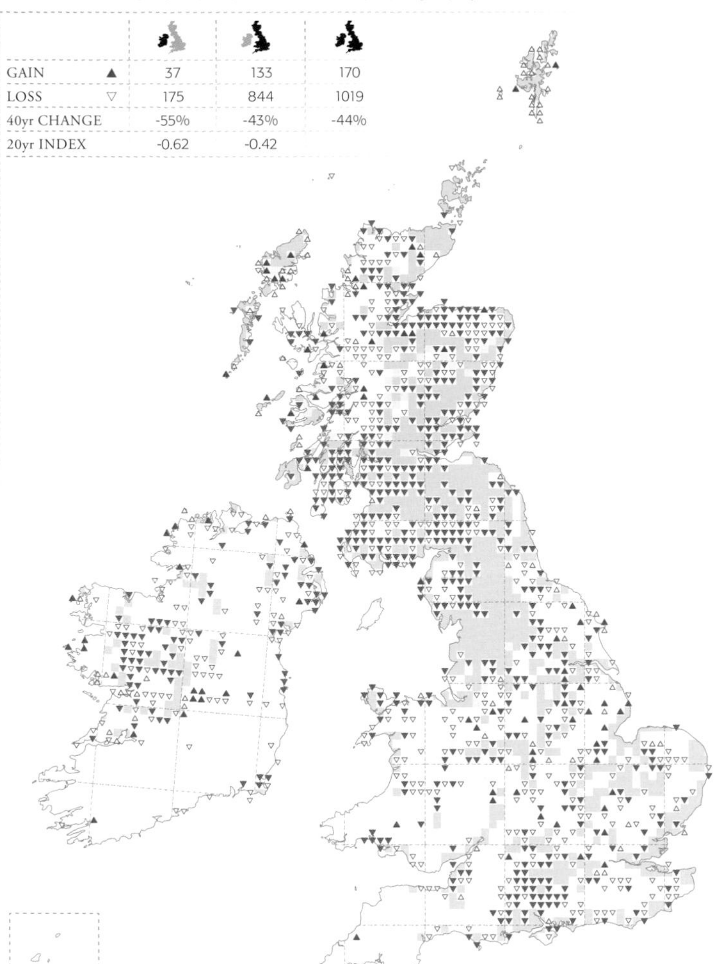

GAIN ▲	37	133	170
LOSS ▽	175	844	1019
40yr CHANGE	-55%	-43%	-44%
20yr INDEX	-0.62	-0.42	

WINTER DISTRIBUTION CHANGE *since 1981–84*

GAIN ▲	75	273	348
LOSS ▼	77	230	307
30yr CHANGE	-1%	+4%	+3%

BREEDING RELATIVE ABUNDANCE *2008–11*

BREEDING RELATIVE ABUNDANCE CHANGE *since 1988–91*

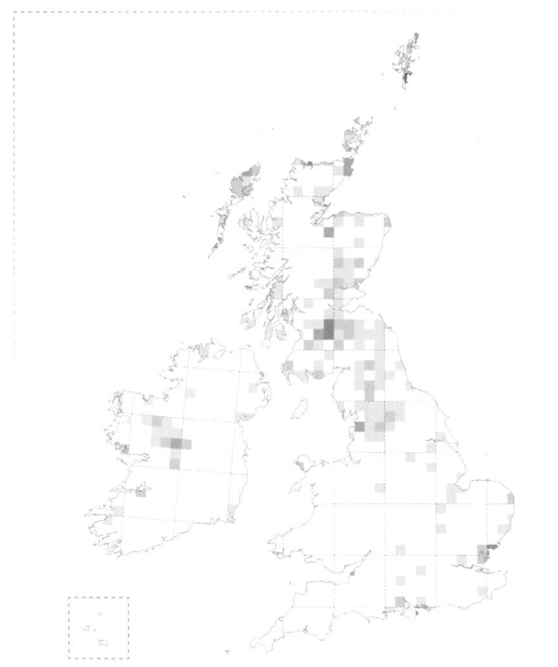

WINTER RELATIVE ABUNDANCE *2007/08–2010/11*

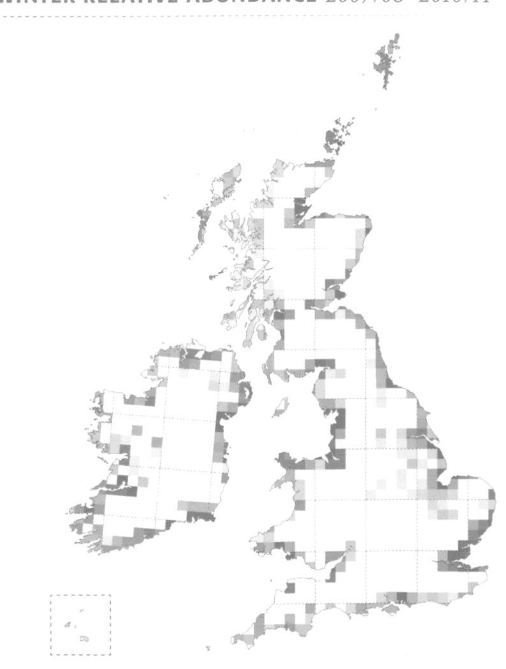

TURNSTONE

Arenaria interpres

BTO CODE: TT

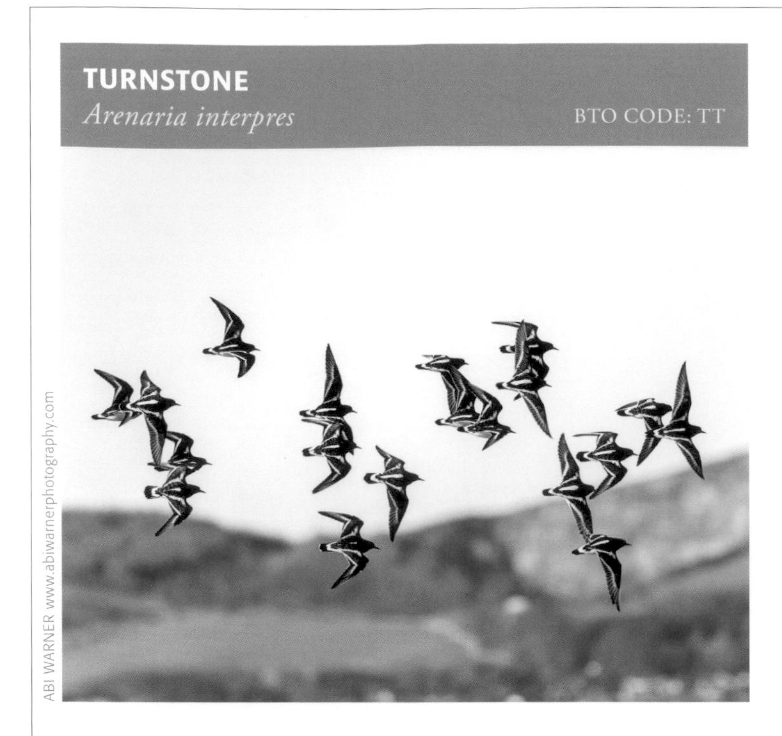

ABI WARNER www.abiwarnerphotography.com

THE TURNSTONE IS a winter visitor to Britain & Ireland, favouring sandy beaches, estuaries and rocky shores. It is widespread and occupies a high proportion of coastal 10-km squares. Since it is so widely distributed, standard shorebird monitoring schemes do not adequately reflect total numbers and additional surveys of open coasts are required. Results of the 2006/07 Non-estuarine Coastal Waterbird Survey showed that numbers had declined by 19% since the 1984/85 Winter Shorebird Count (Austin *et al.* 2007). This is thought to be due to changing climate facilitating a northerly range shift (Rehfisch *et al.* 2004), a pattern now found for several wader distributions in Europe (Maclean *et al.* 2008).

The winter change map, however, shows that there has been an expansion in range across Britain & Ireland since the *1981–84 Winter Atlas*, with a 9% increase in range size overall. Increases were most notable in southwest Ireland, northwest Scotland, and at a few inland sites in Britain. It is likely that some of the gains have resulted from improved coverage of remote coasts during *Bird Atlas 2007–11*. Compared with the *1981–84 Winter Atlas* abundance pattern, the main concentrations appear largely unchanged, still being in the Northern Isles, the Uists, Tiree, western and northeast Ireland and the larger British estuaries.

Most Turnstones remain on their coastal wintering grounds throughout the winter, not departing for their breeding areas until April or May. There were records of *non-breeding* birds from 660 10-km squares in Britain and 124 squares in Ireland during the breeding season. Birds that have wintered farther south also pass through Britain & Ireland at this time while on northbound passage to breeding grounds in the high Arctic. These passage birds possibly account for the scattering of records at inland sites, particularly in England, in early summer.

A small number of birds, mainly immatures, remain to summer, particularly at sites located farther north. These and other *non-breeding* birds may have been coded as *seen* during the *1988–91 Breeding Atlas* and account for the artificial 'breeding' range contraction since that time. Turnstones have never been proven to breed in Britain or Ireland, though some have shown signs of attempting to do so (*Birds of Scotland*).

Sponsored by Neil Metcalfe

WINTER DISTRIBUTION *2007/08–2010/11*

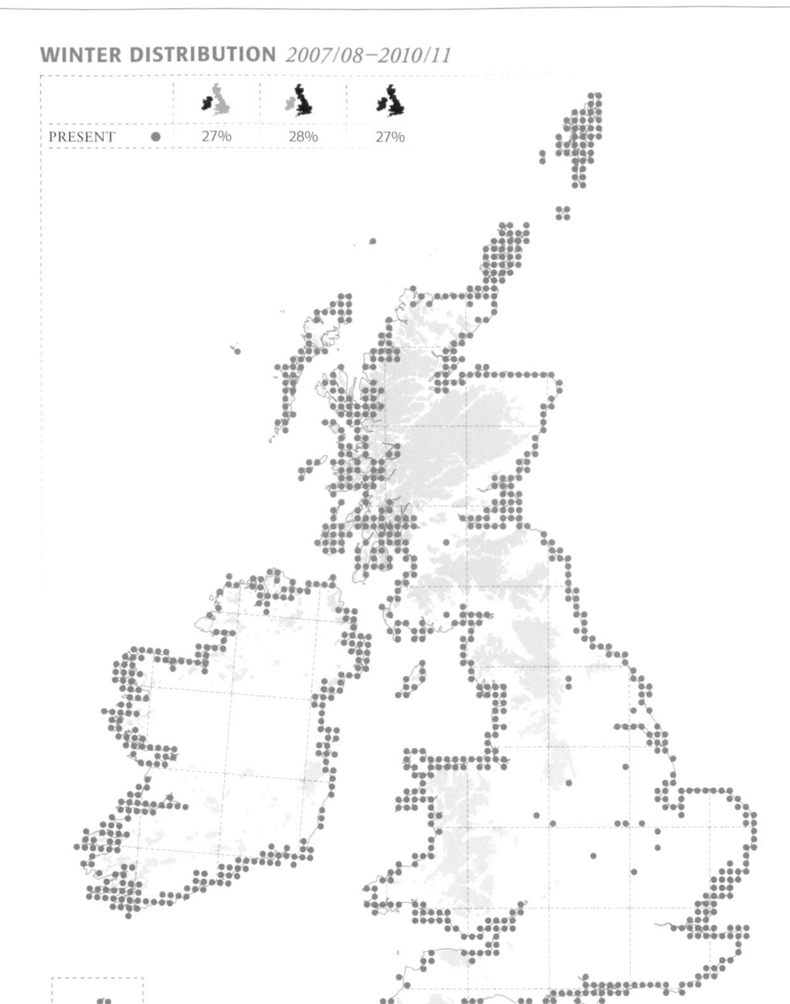

BREEDING DISTRIBUTION *2008–11*

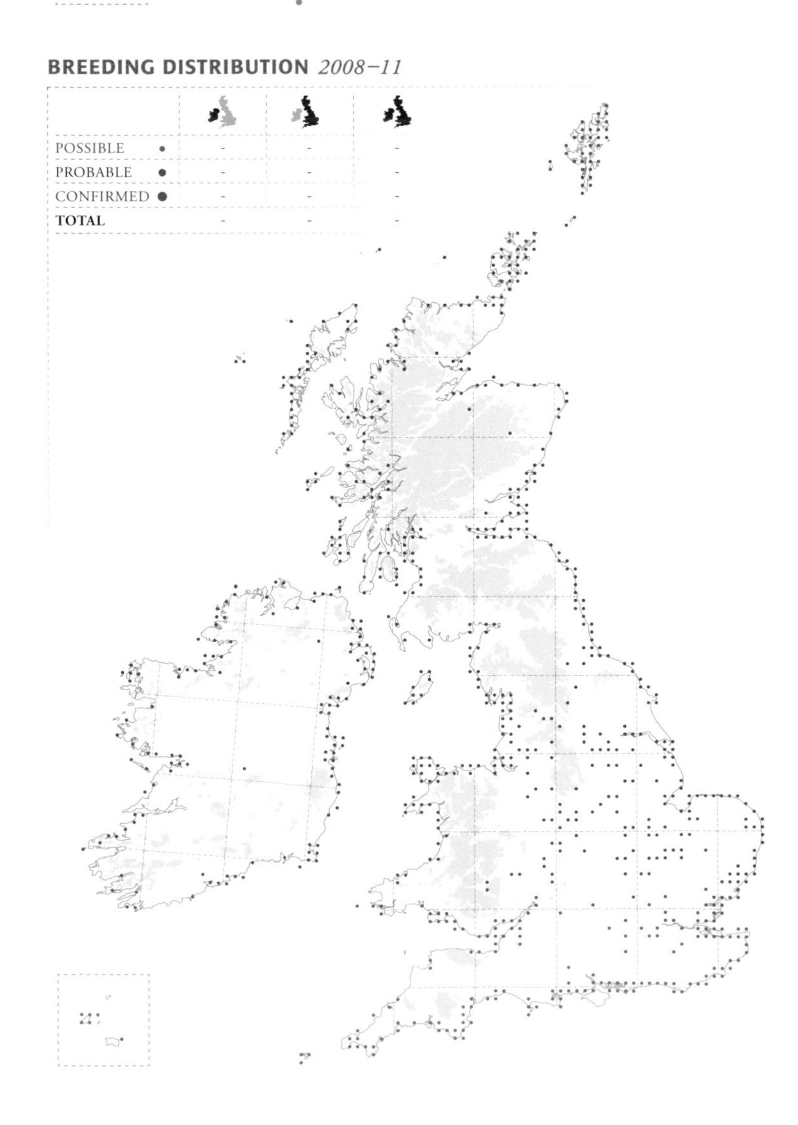

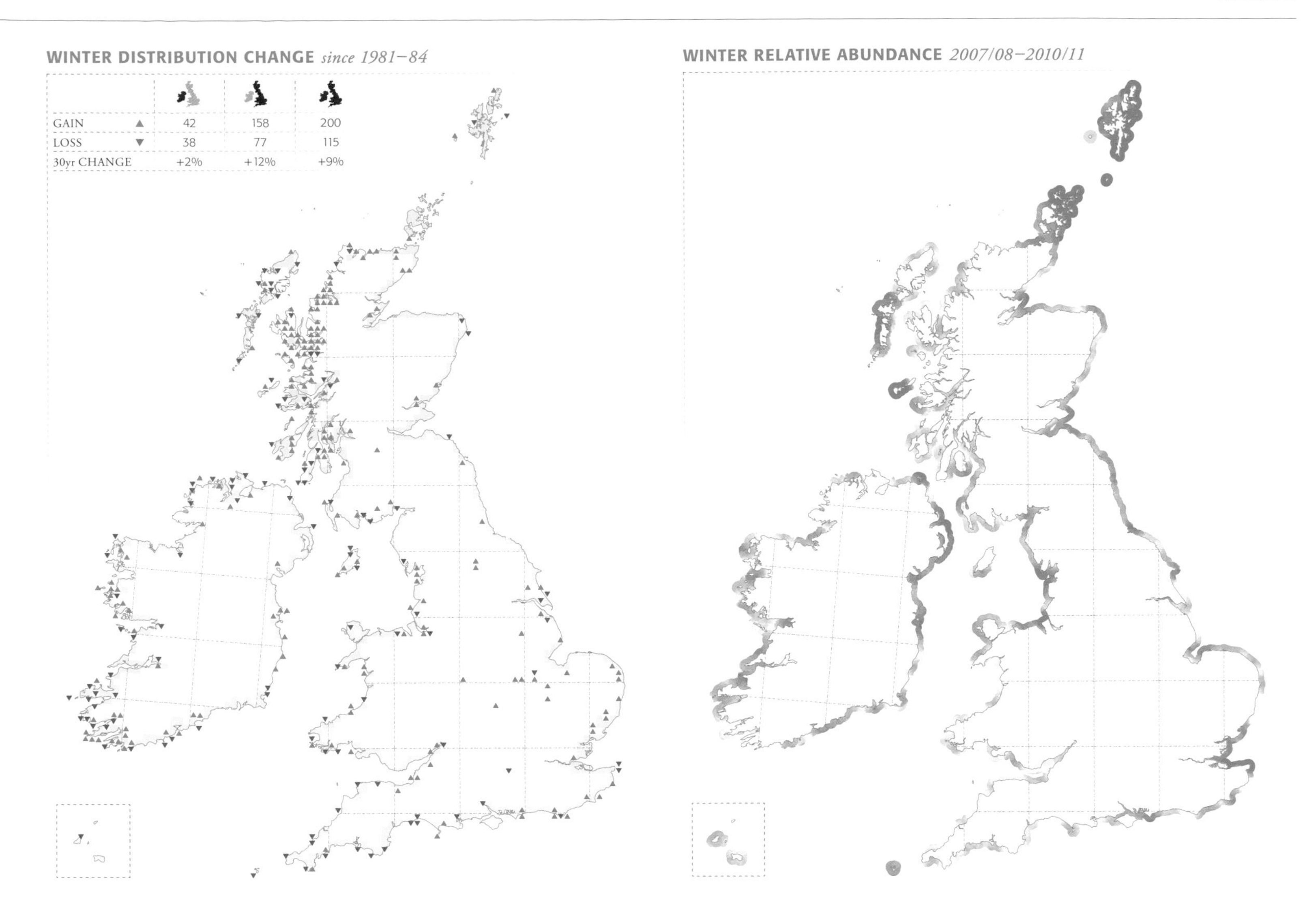

GAIN ▲	42	158	200
LOSS ▼	38	77	115
30yr CHANGE	+2%	+12%	+9%

GERRY GUTTERIDGE www.gerrygutteridge.com

◄ **Birds wintering in Britain & Ireland are drawn from a wide breeding range encompassing arctic northeast Canada, Greenland, Fennoscandia and central Siberia.**

RED-NECKED PHALAROPE

Phalaropus lobatus

BTO CODE: NK

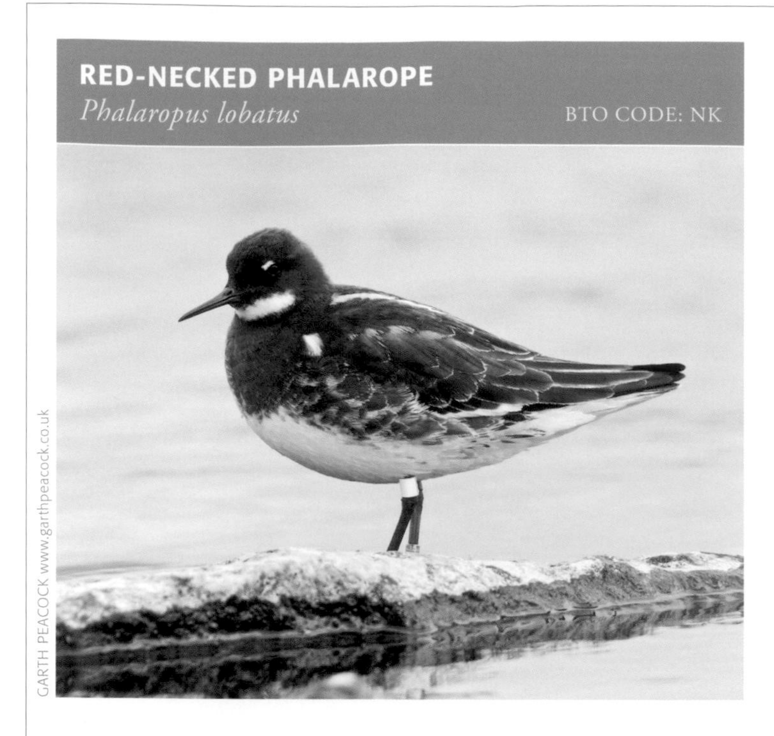

GARTH PEACOCK www.garthpeacock.co.uk

RED-NECKED PHALAROPES are very rare breeders in Scotland and Ireland, where they occur on the southern edge of their circumpolar breeding range. The Britain & Ireland breeding distribution is summarised and mapped at 50-km resolution to protect key sites and breeding birds. In Scotland the main stronghold is Shetland, where birds were located in suitable breeding habitat in seven 10-km squares. Birds were also present at suitable breeding sites in five squares in the Outer Hebrides, one in the Inner Hebrides and one in northeast Scotland. In Ireland there is a record of *possible* breeding in Co. Offaly in 2010 and one of *probable* breeding in Co. Mayo in 2011 (not mapped).

Since the *1968–72 Breeding Atlas* the Scottish breeding population of Red-necked Phalaropes has shown considerable fluctuation in numbers (Ogilvie *et al.* 2004; *Birds of Scotland*) with the current range being larger than during the *1988–91 Breeding Atlas* but smaller than during the *1968–72 Breeding Atlas*. Peak counts in 1978, 1996 and 2005 reached over 30 breeding males, whilst during the intervening periods numbers dropped to around 15 breeding males. In 2010 the UK population numbered between 19 and 27 breeding males and these were recorded at 15 sites, with Shetland accounting for 19 males at 10 sites (*RBBP 2010*). There was an increase on Shetland in 2011 with 24 males recorded as *probable* or *confirmed* breeders reflecting successful habitat creation and management work by RSPB (M. Smith pers comm). Whilst a few sites have been lost since the *1988–91 Breeding Atlas*, other sites have been gained. Red-necked Phalarope sites are known to be vulnerable to a number of factors that can affect their suitability for this species, including succession of vegetation, changes in water levels, eutrophication and perhaps predation (Ellis 2004; JNCC 2010).

Migrant Red-necked Phalaropes are recorded at a small number of coastal and inland wetland sites each spring, mostly in central and eastern England, whilst on their way to breeding grounds in the north. There were records from 50 10-km squares in Britain and five squares in Ireland. In the winter, all but one of the five 10-km squares were occupied only in November, the exception being an individual present with two Grey Phalaropes in the Solent in January 2008.

BREEDING DISTRIBUTION *2008–11*

POSSIBLE •	<1%	<1%	<1%
PROBABLE •	<1%	<1%	<1%
CONFIRMED •	-	<1%	<1%
TOTAL	<1%	<1%	<1%

BREEDING DISTRIBUTION CHANGE *since 1988–91*

GAIN ▲	2	10	12
LOSS ▽	1	5	6
20yr CHANGE	+100%	+56%	+60%
20yr INDEX	-	-	

Sponsored by Baroness Young of Old Scone

GREY PHALAROPE

Phalaropus fulicarius

BTO CODE: PL

JOE PENDER www.sapphirepelagics.co.uk

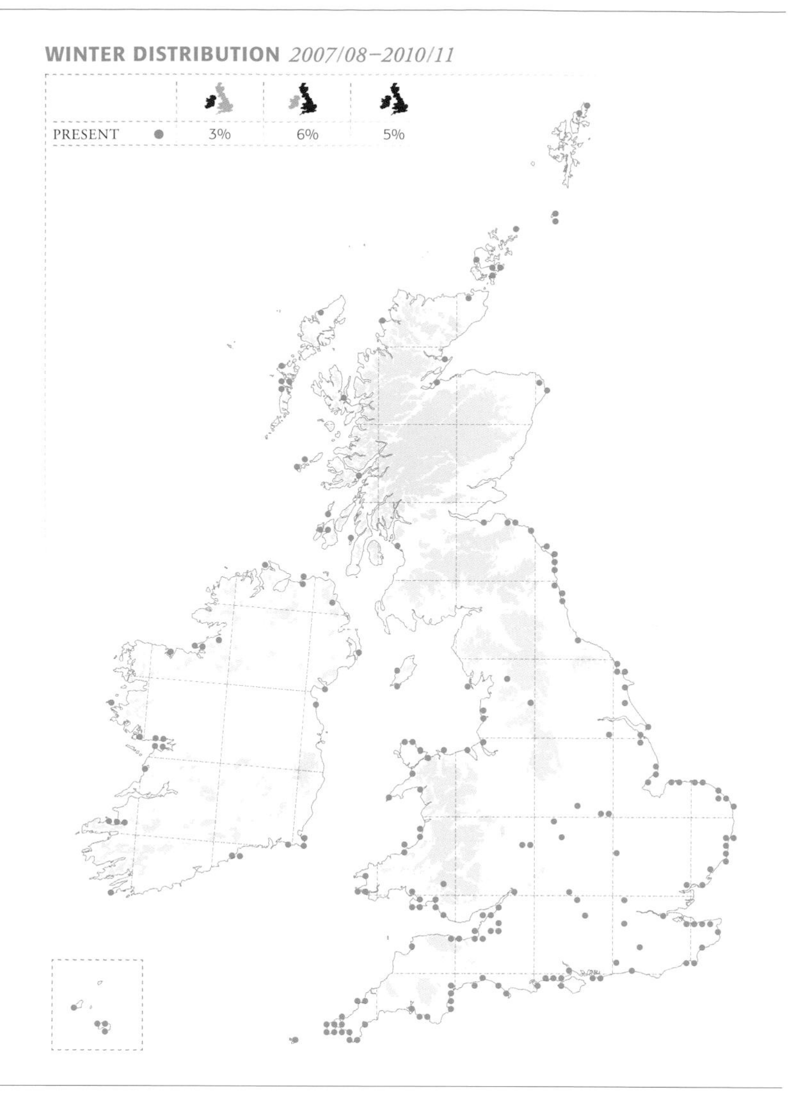

GREY PHALAROPES WERE recorded from 203 10-km squares in winter in Britain & Ireland during 2007–11, representing a large increase from the 41 10-km squares recorded during the *1981–84 Winter Atlas*. Even discounting many early November records, this still amounts to a 229% increase. Birds were recorded mainly in North Sea coastal squares south from the Firth of Forth, along the south coast of England, and north to the southern Irish Sea. Elsewhere the distribution was patchier and included a scattering of inland records in southern and central England. The majority of 10-km squares were occupied in November only, with just 63 10-km squares occupied between December and February.

POMARINE SKUA

Stercorarius pomarinus

BTO CODE: PK

JOE PENDER www.sapphirepelagics.co.uk

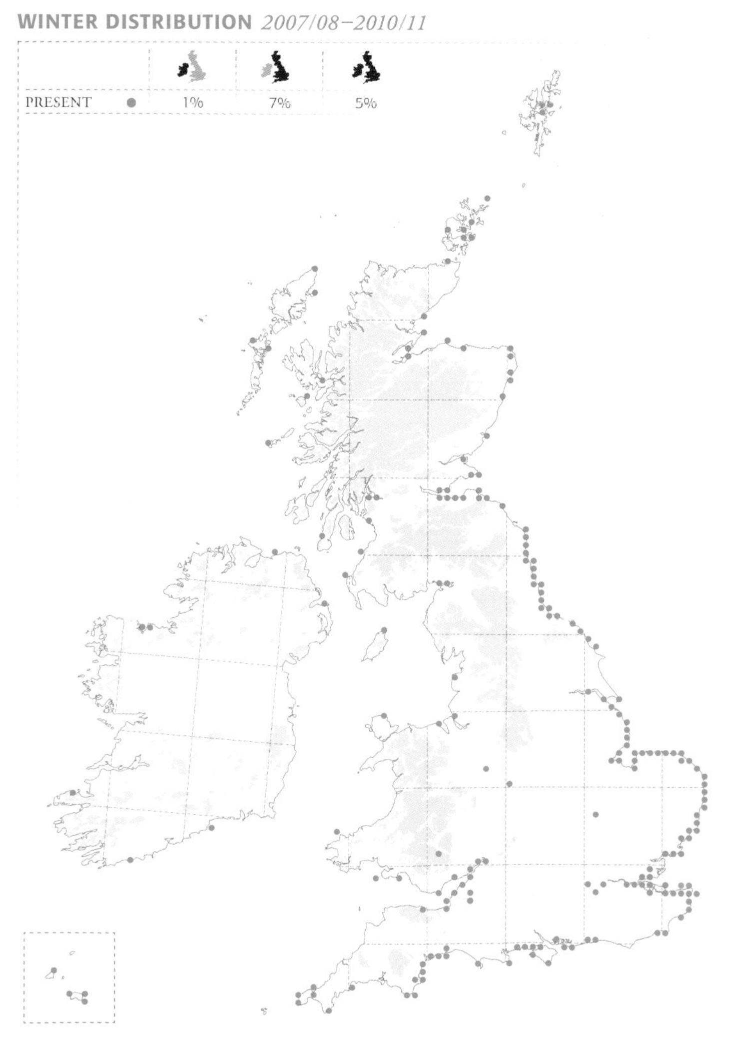

POMARINE SKUAS BREED in the high Arctic tundra where their populations are thought to be stable (BirdLife International 2013). They winter in the tropics so are encountered as passage visitors to our shores. There is a distinct easterly bias to winter records, with birds recorded in 190 10-km squares in Britain and only seven in Ireland. Overall, 57% of winter 10-km squares were occupied in November only, including some inland squares and most northern and western squares. The winter range of the Pomarine Skua has shown a remarkable increase from the 14 10-km squares occupied during the *1981–84 Winter Atlas*. Birds on spring and early autumn passage were recorded in 169 widely scattered coastal 10-km squares.

ARCTIC SKUA

Stercorarius parasiticus

BTO CODE: AC

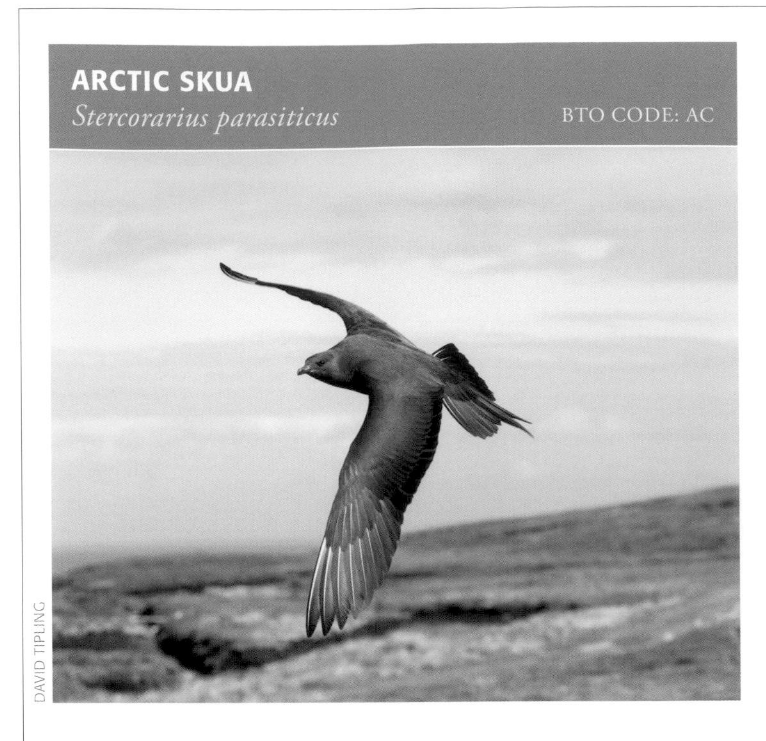

DAVID TIPLING

JUST 1% OF the Arctic Skua's circumpolar population breeds in Scotland (JNCC 2012). Here, at the southwestern extremity of its Eurasian breeding range, birds are restricted to the Northern Isles, Caithness and Sutherland, the Outer Hebrides, St Kilda and a few southern Inner Hebridean islands. Though the species has summered in Ireland, it has never bred there (*Birds in Ireland*). The highest numbers are found on Shetland, Orkney and the Outer Hebrides. Elsewhere, between 13 and 24 pairs breed on Handa, Sutherland (Smith *et al.* 2010), a small number breed on the flows of Caithness and Sutherland, whilst farther south there were records of *probable* breeding from Coll, the Treshnish Islands and Jura.

Atlas data show that there has been a 7% range expansion in Britain since the *1968–72 Breeding Atlas*, though a 6% contraction since the *1988–91 Breeding Atlas* with losses particularly notable from Islay and Jura. Data from the JNCC Seabird Monitoring Programme show that Arctic Skua populations have declined at a greater rate than those of any other UK seabird species: a 73% decline between 1986 and 2011 (JNCC 2012) has led to the species being Red-listed as a species of conservation concern (*SUKB 2012*). These declines have been driven mainly by low food availability. Predation by Great Skuas, due to the scarcity of fish prey such as sandeels, is likely to have played an additional role (Jones *et al.* 2008; Meek *et al.* 2011).

In addition to the breeding areas shown on the distribution map, *non-breeding* Arctic Skuas were reported from many coastal 10-km squares in Britain, and from a relatively small scattering of squares in Ireland. Migrants are recorded in spring, particularly on the west coast during May, whilst by July returning birds are evident at British east-coast sites. The autumn southbound migration continues into early winter and is illustrated by the early November records, which constitute most of the Irish and c.40% of the British 'winter' range. The 55% range expansion, mostly in Britain, probably reflects later migration and a growing interest in seawatching during the winter months.

BREEDING DISTRIBUTION *2008–11*

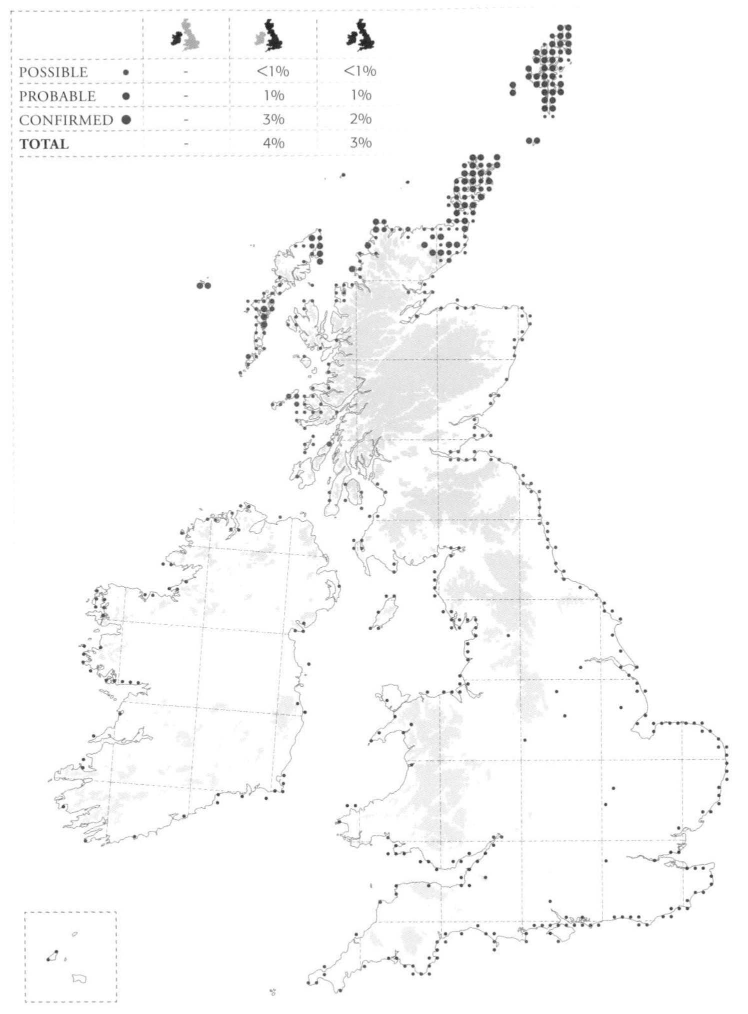

	Ireland	Britain	Britain & Ireland
POSSIBLE •	-	<1%	<1%
PROBABLE •	-	1%	1%
CONFIRMED •	-	3%	2%
TOTAL	-	4%	3%

WINTER DISTRIBUTION *2007/08–2010/11*

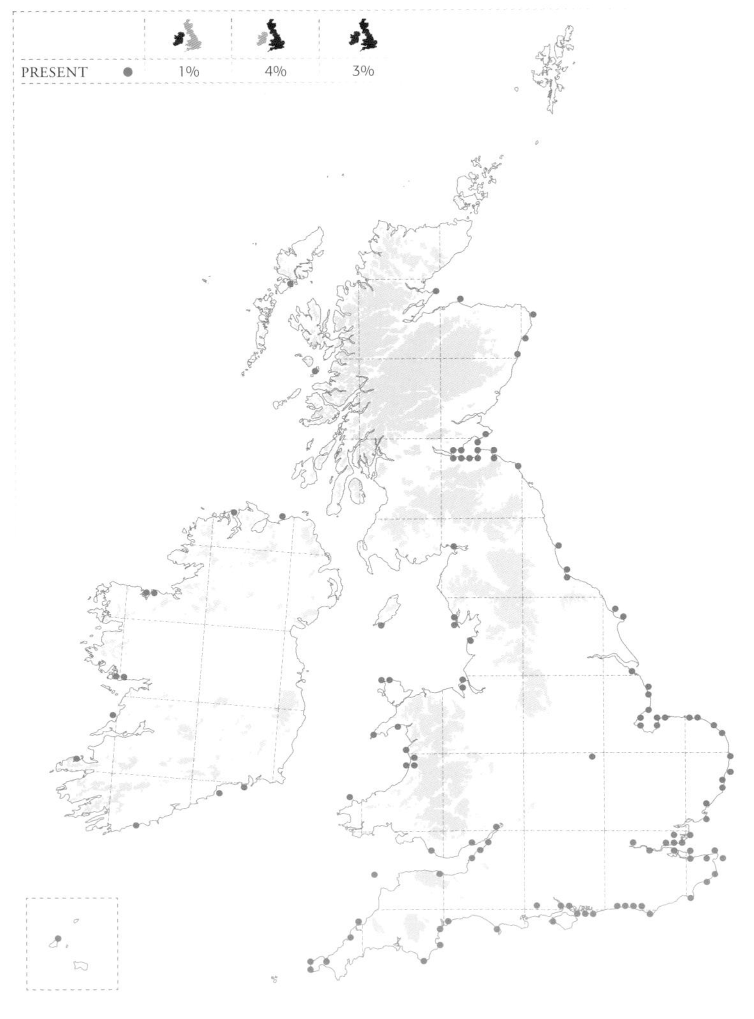

	Ireland	Britain	Britain & Ireland
PRESENT •	1%	4%	3%

BREEDING DISTRIBUTION CHANGE *since 1968–72*

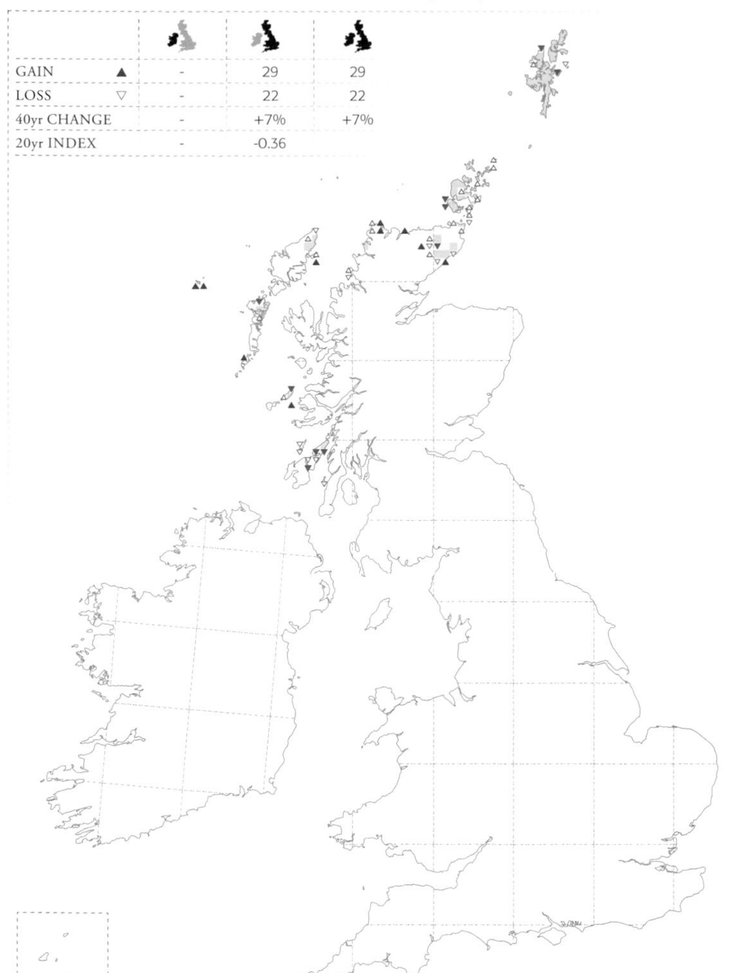

GAIN ▲	-	29	29
LOSS ▽	-	22	22
40yr CHANGE	-	+7%	+7%
20yr INDEX	-	-0.36	

BREEDING RELATIVE ABUNDANCE *2008–11*

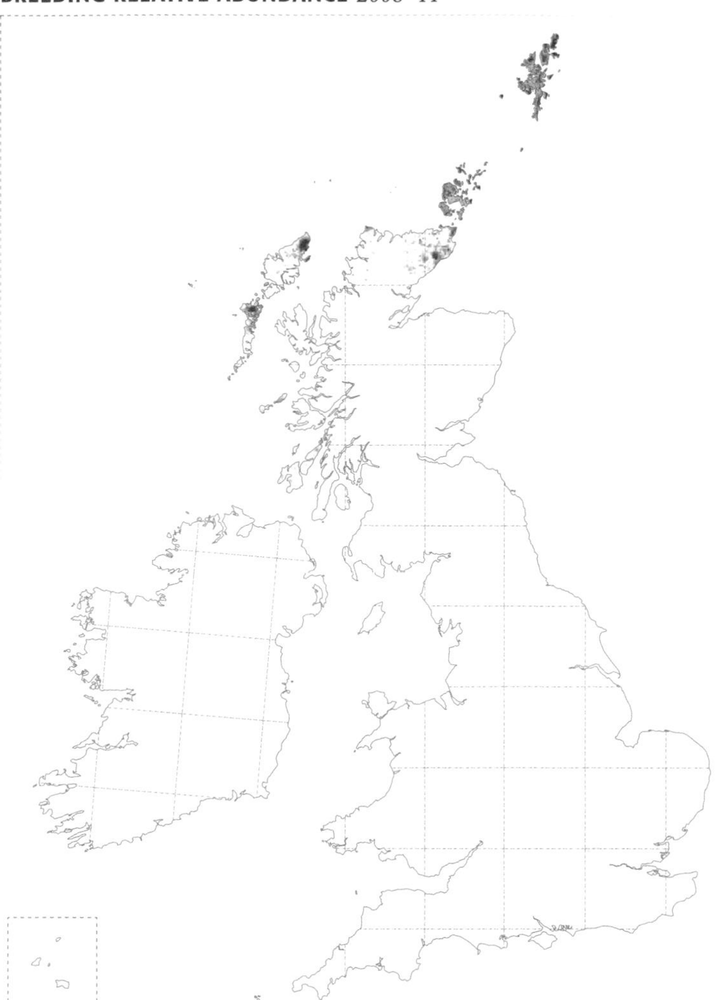

WINTER DISTRIBUTION CHANGE *since 1981–84*

GAIN ▲	2	60	62
LOSS ▼	4	34	38
30yr CHANGE	-50%	+65%	+55%

GREAT SKUA

Stercorarius skua

BTO CODE: NX

DESMOND DUGAN www.AbernethyImages.co.uk

MORE THAN HALF the world population of Great Skuas is found along the northwest seaboard of Scotland and Ireland. The highest concentrations were found on Shetland and Orkney; colonies on Handa, Sutherland, and St Kilda are not shown on the abundance map owing to a lack of eligible tetrads and small sample size respectively.

Breeding populations have increased historically, although there were periods of low productivity during the late 1980s, early 2000s and 2011 in the large Shetland colonies as a result of a decreased availability of sandeels and discards from fishing vessels (*Seabird 2000*; JNCC 2012). This is likely to have affected the size of the breeding population, but few data are available. A complete survey of colonies in Orkney in 2010 showed that the number of occupied territories had dropped by 23% since the *Seabird 2000* surveys. The large colony on Hoy, formerly holding almost 2,000 territories, declined by 32% between *Seabird 2000* and 2010. This is perhaps due to food shortages, maybe even including intraspecific predation of chicks in high-density nesting areas. There was also a marked redistribution with many smaller colonies increasing and new ones being established (Meek *et al.* 2011). Despite these problems in the Northern Isles, there has been a 111% range expansion since the *1968–72 Breeding Atlas*; much of this due to the colonisation of new sites in the west of Scotland and in northwest Ireland since 2000. This has been probably partly fuelled by recruits from the population of c.200 territories on Handa. Although breeding productivity has declined there, it is still relatively high compared to other colonies in Scotland (Jones *et al.* 2008).

By November there were a few scattered records in northern Scotland and along the western coasts of Britain & Ireland, but most were from the Firth of Forth down to the south coast of England and around to the Severn Estuary. From December to February there were records from 94 10-km squares. Overall there has been a 139% increase in the number of occupied 10-km squares since the *1981–84 Winter Atlas*, suggesting later passage, an increase in the number of overwintering birds and perhaps better recording of seabirds during the winter.

BREEDING DISTRIBUTION *2008–11*

POSSIBLE •	<1%	1%	1%
PROBABLE •	<1%	1%	1%
CONFIRMED •	1%	3%	3%
TOTAL	1%	5%	5%

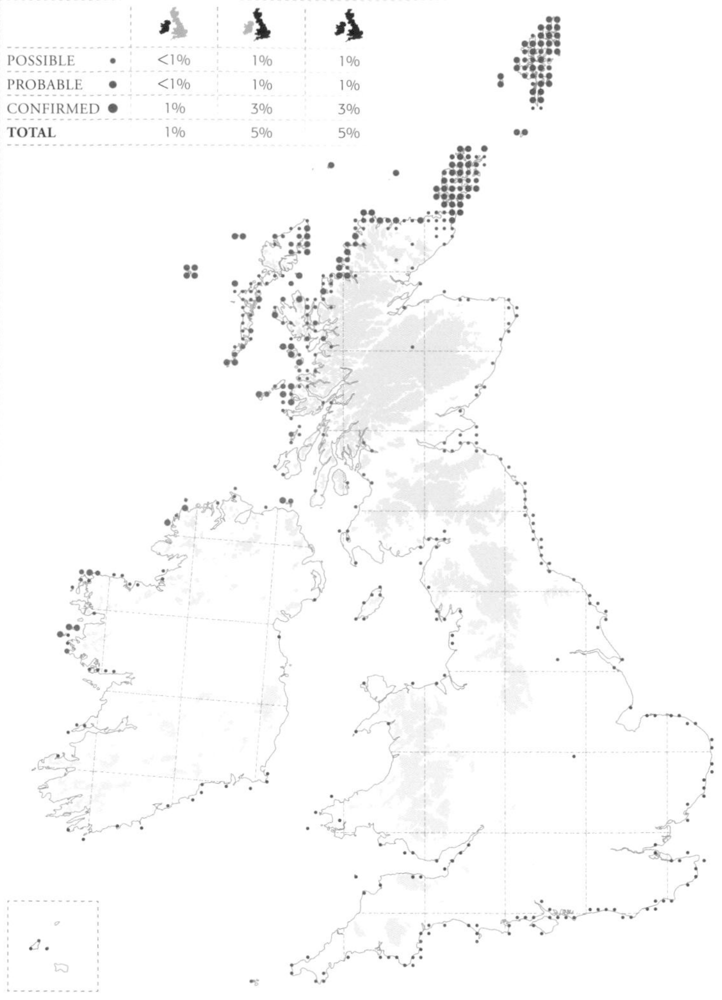

WINTER DISTRIBUTION *2007/08–2010/11*

PRESENT •	2%	6%	5%

Sponsored by Val Gwynn

BREEDING DISTRIBUTION CHANGE *since 1968–72*

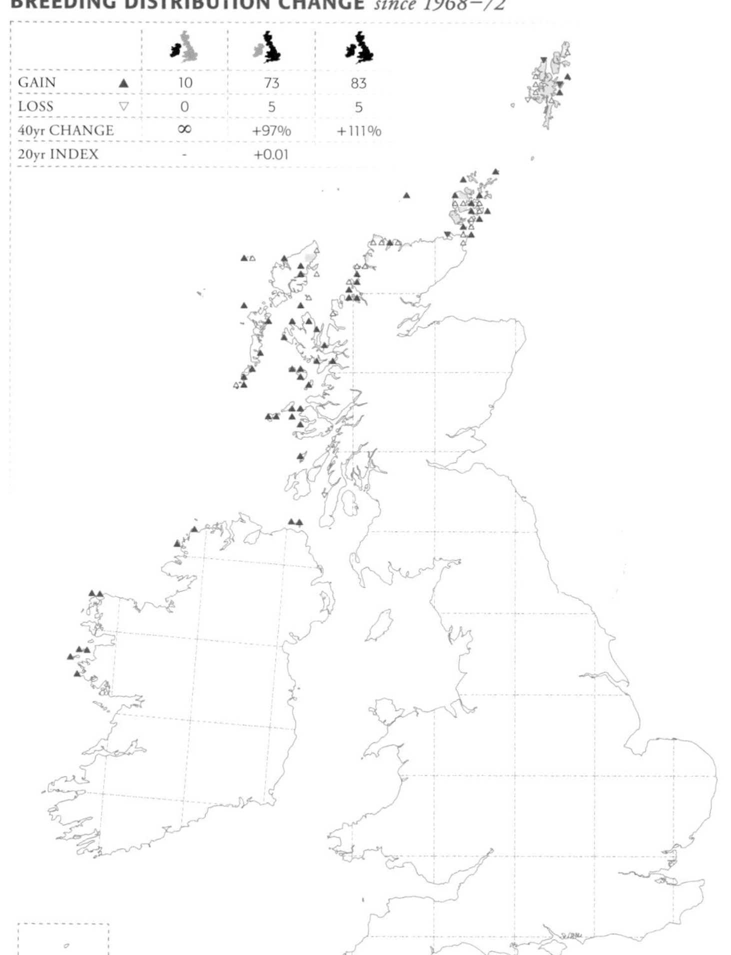

GAIN ▲	10	73	83
LOSS ▽	0	5	5
40yr CHANGE	∞	+97%	+111%
20yr INDEX	-	+0.01	

BREEDING RELATIVE ABUNDANCE *2008–11*

WINTER DISTRIBUTION CHANGE *since 1981–84*

GAIN ▲	10	107	117
LOSS ▼	4	31	35
30yr CHANGE	+150%	+138%	+139%

KITTIWAKE

Rissa tridactyla

BTO CODE: KI

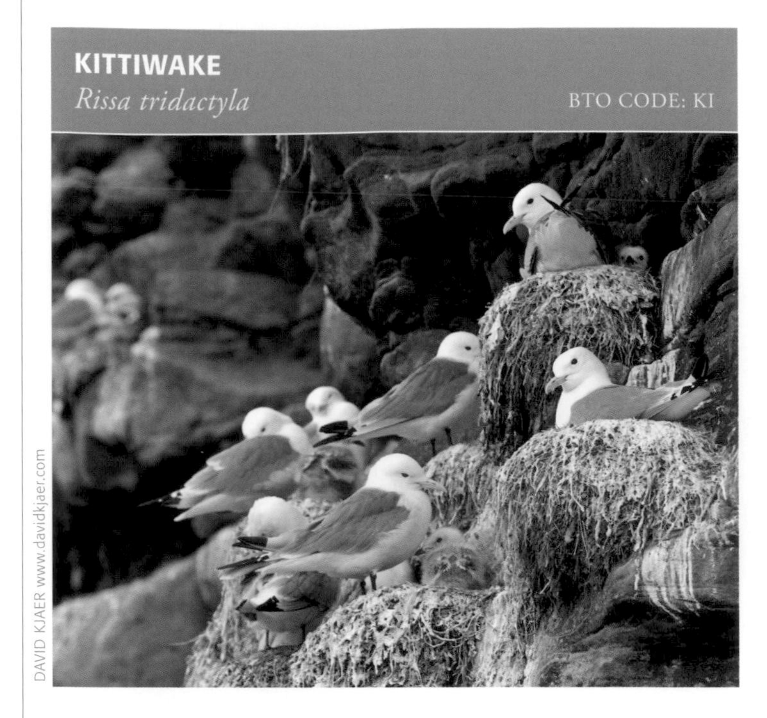

DAVID KJAER www.davidkjaer.com

KITTIWAKES ARE HIGHLY pelagic when not breeding; some British and Irish breeders cross to the west Atlantic but others remain in the eastern Atlantic and North Sea where they are joined by birds from Norway (Frederiksen *et al.* 2012). Winter records are mainly from vantage points along the British and Irish coastlines, but Kittiwakes are generally scarce or present only intermittently within sight of the coast. Concentrations were greatest in southwest Ireland, southwest Wales and the Northern Isles. There are fewer occupied squares now than during the *1981–84 Winter Atlas*, with losses most pronounced in western Scotland, particularly from Islay to Skye, and around the northern and western Irish coastline, despite these areas generally seeing increased recording effort. Kittiwakes are scarce inland and the distribution and change maps reflect the random nature of their occurrence during the winter.

Kittiwake breeding colonies are largely restricted to coasts with rocky cliffs, though man-made structures can be used such as buildings, bridges and offshore rigs. The breeding-season relative abundance map highlights the broad regions where Kittiwake abundance was greatest, particularly between Flamborough Head and Orkney. These areas approximately correspond to the largest colonies, some numbering tens of thousands of pairs (*Seabird 2000*; Coulson 2012).

There has been a mix of some gains and significant losses since the *1968–72 Breeding Atlas*, amounting to a 10% range contraction overall. In the Northern Isles the collapse of Lesser Sandeel populations has resulted in many years of breeding failure (*Seabird 2000*). Recent losses in southwest England may be related to local changes in prey abundance (Heaney *et al.* 2008). In the UK, a 44% population decline was evident during 1986–2011, while numbers in the Republic of Ireland were more or less stable (JNCC 2012). Key factors in the decline appear to be changes in the marine environment that affect abundance of prey, which may in turn have consequences for breeding success, body condition and survival rates of adults (Harris & Wanless 1997; *Seabird 2000*). Predation by Great Skuas of adults and their chicks is also thought to be important in some colonies in Shetland and on St Kilda (Heubeck *et al.* 1997; Phillips *et al.* 1997).

Sponsored by a BTO Member from Yorkshire

BREEDING DISTRIBUTION *2008–11*

	Ireland	Britain	Britain & Ireland
POSSIBLE •	1%	1%	1%
PROBABLE ●	1%	<1%	<1%
CONFIRMED ●	6%	7%	7%
TOTAL	8%	8%	8%

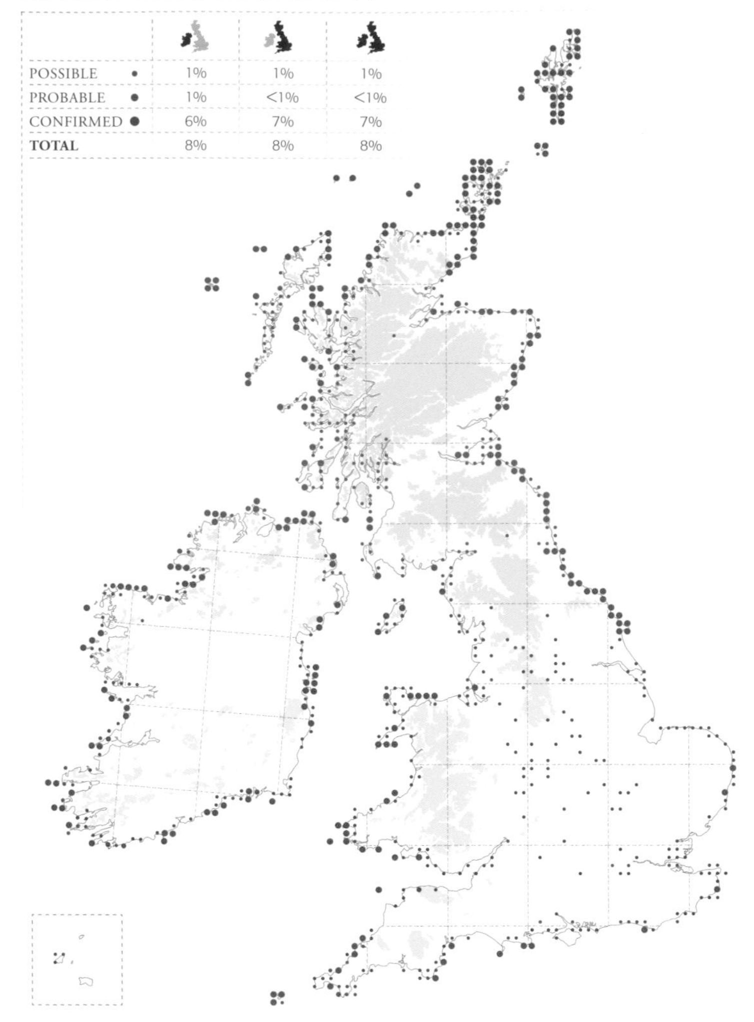

WINTER DISTRIBUTION *2007/08–2010/11*

	Ireland	Britain	Britain & Ireland
PRESENT ●	15%	20%	18%

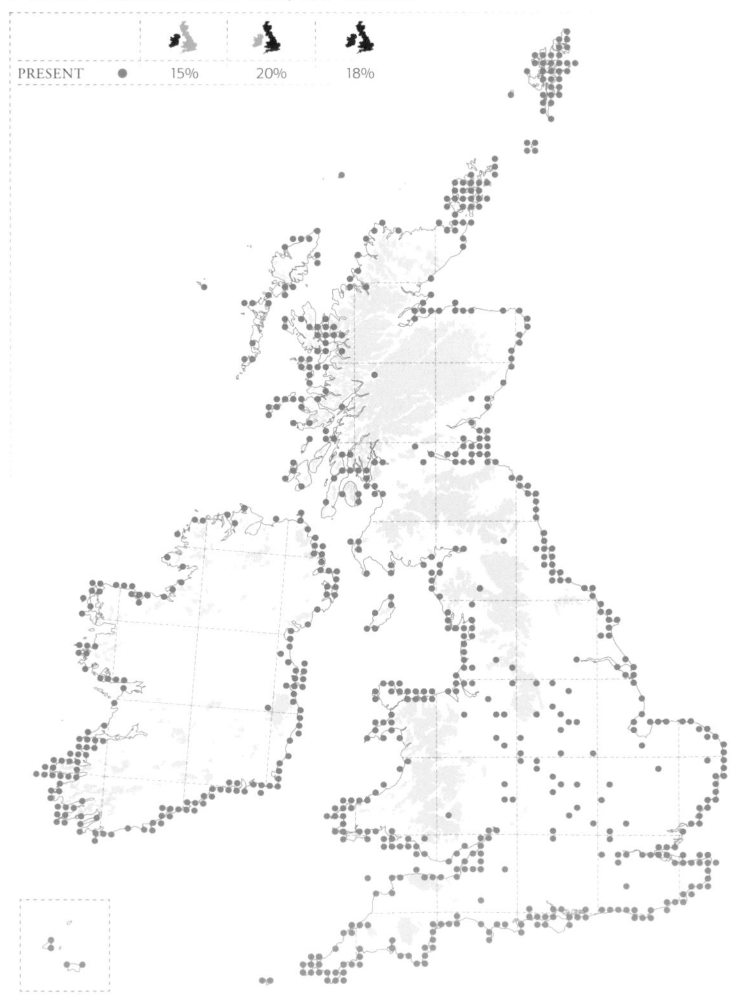

BREEDING DISTRIBUTION CHANGE *since 1968–72*

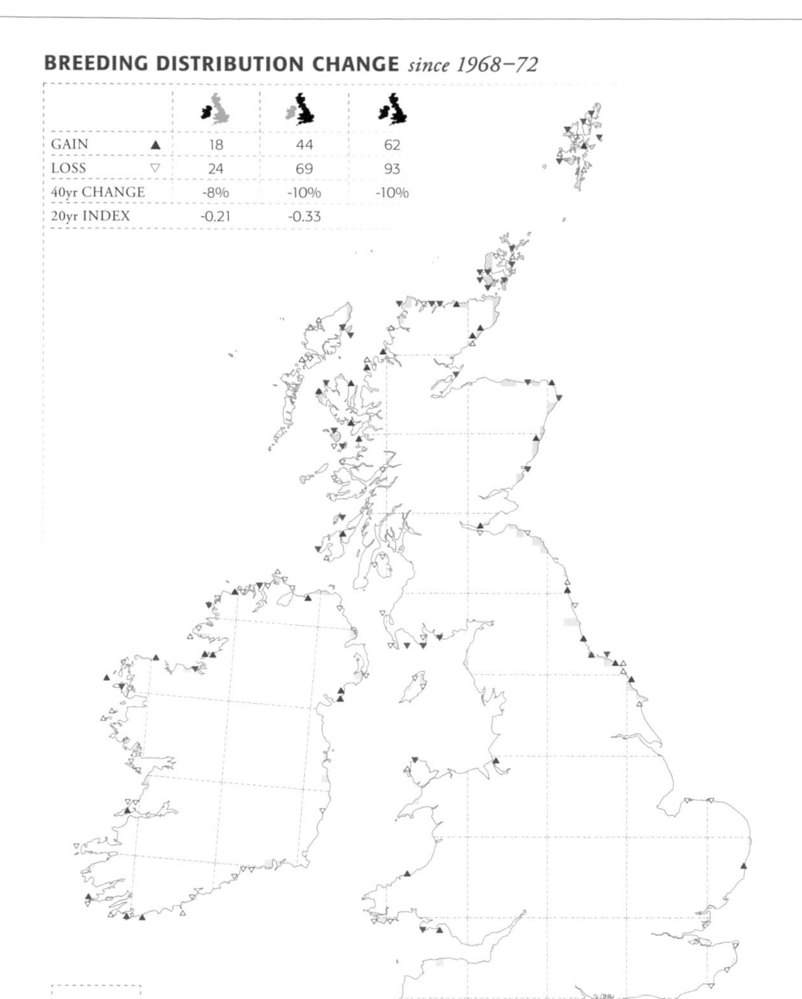

	Ireland	Britain	Britain & Ireland
GAIN ▲	18	44	62
LOSS ▽	24	69	93
40yr CHANGE	-8%	-10%	-10%
20yr INDEX	-0.21	-0.33	

BREEDING RELATIVE ABUNDANCE *2008–11*

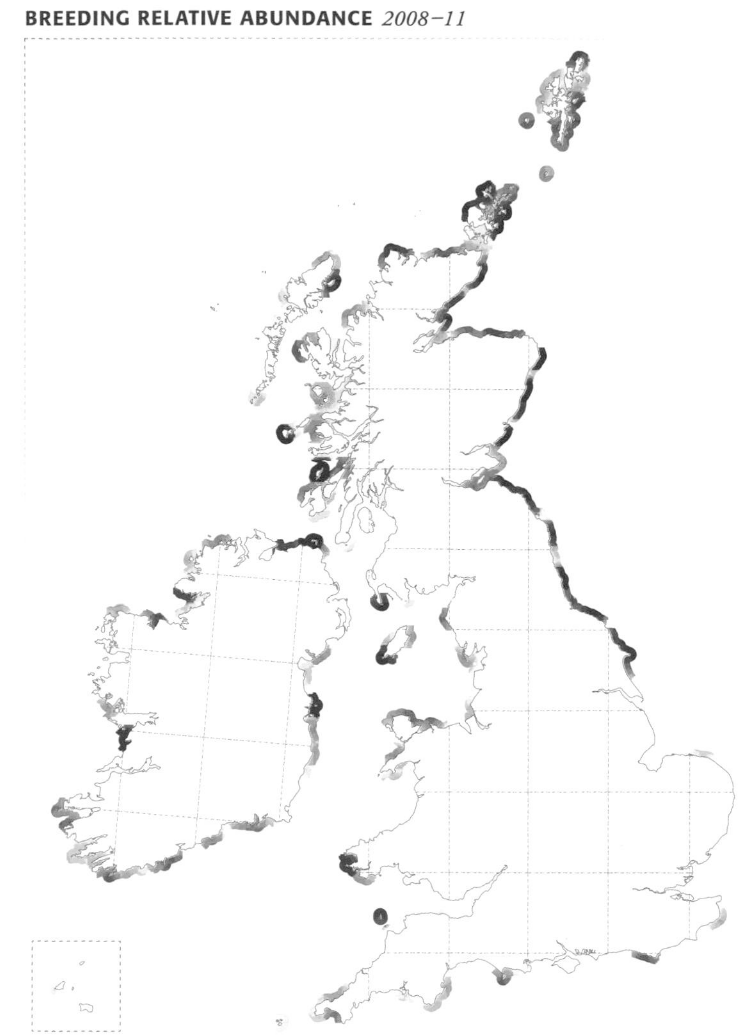

WINTER DISTRIBUTION CHANGE *since 1981–84*

	Ireland	Britain	Britain & Ireland
GAIN ▲	45	154	199
LOSS ▼	95	362	457
30yr CHANGE	-26%	-31%	-30%

WINTER RELATIVE ABUNDANCE *2007/08–2010/11*

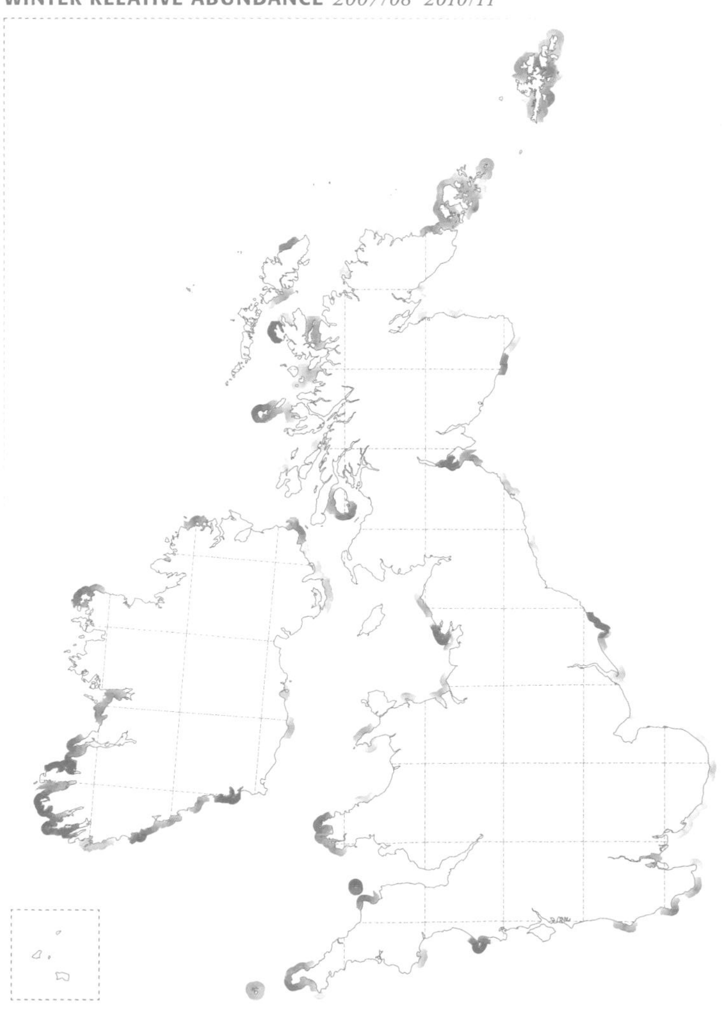

BLACK-HEADED GULL

Chroicocephalus ridibundus

BTO CODE: BH

ZOE SHREEVE

BLACK-HEADED GULLS have a near-ubiquitous wintering distribution in lowland Britain & Ireland. In Britain the only significant gaps are in the Scottish and Welsh uplands, although the range is fairly fragmented in coastal western Scotland. The largest gaps in Ireland are in the northwest, between Antrim and Connemara, while other, smaller inland gaps are in the southwest and eastern midlands. They are most abundant in northeast, northwest, central and eastern England, the Central Belt of Scotland, eastern coastal Ireland and the lower Shannon.

The winter change map shows that there has been a significant contraction since the *1981–84 Winter Atlas*, with the greatest losses apparent in Ireland and Wales. Owing to its wide distribution, only a fraction of the winter population is routinely monitored, but those figures indicate a steady decline over recent decades (*I-WeBS Report 2009*; *WeBS Report 2011*). Analyses of decadal roost counts in Britain indicate that, although numbers have increased significantly since 1953, they reached a peak during 1973–93, and have since declined (Banks *et al.* 2009).

Black-headed Gulls nest in a wide range of coastal and inland, natural and man-made wetlands, allowing a wide breeding distribution of 1,177 10-km squares, with *non-breeding* birds present in a further 1,670 squares. Areas of highest densities in Britain are in Orkney, northern England, East Anglia, the Thames Estuary, and the Solent, and in Ireland are more fragmented around Lough Neagh, Strangford Lough, and at several large wetland complexes scattered throughout the west and northwest.

The breeding distribution change map can be divided roughly diagonally, with the north and west characterised by range losses contrasting with a predominance of range gains in the southeast. Gains in the south follow a similar pattern to some other colonial waterbirds, such as the Common Tern, with more records of nesting at newly created habitats at inland locations. Losses, which contribute to a 22% range contraction overall, are as yet unexplained but suggestions include loss of safe nesting sites, drainage, food depletion (Benton *et al.* 2002) and increased predation, in some areas by introduced American Minks (Craik 1997, 1998).

BREEDING DISTRIBUTION *2008–11*

POSSIBLE •	5%	9%	8%
PROBABLE ●	2%	5%	4%
CONFIRMED ●	10%	21%	18%
TOTAL	17%	35%	30%

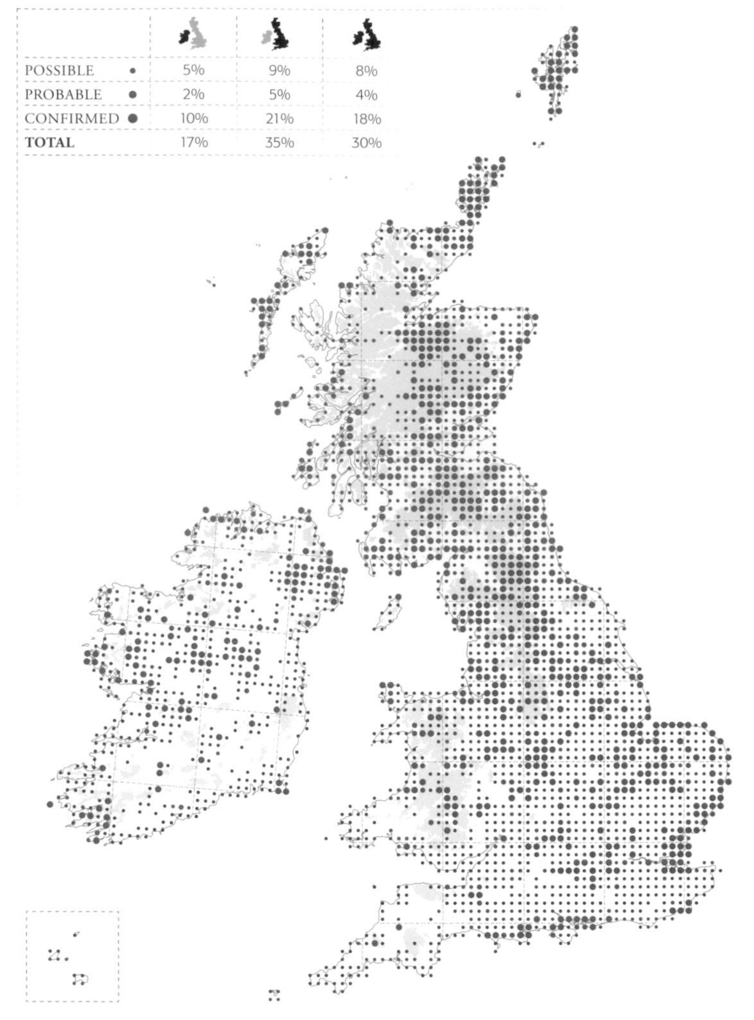

WINTER DISTRIBUTION *2007/08–2010/11*

PRESENT ●	70%	78%	76%

BREEDING DISTRIBUTION CHANGE *since 1968–72*

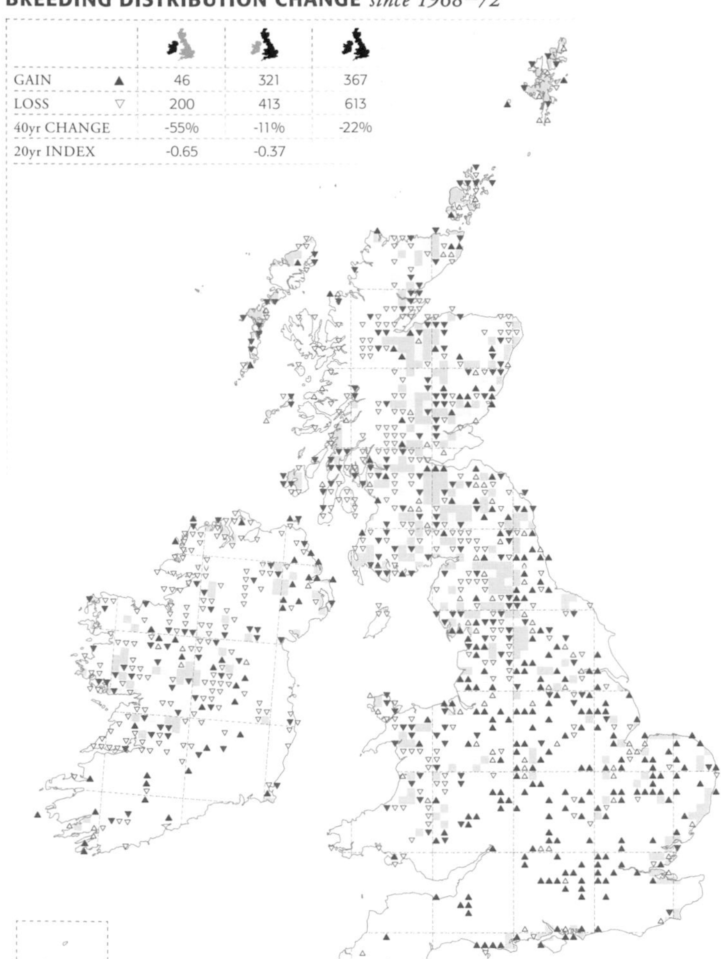

GAIN ▲	46	321	367
LOSS ▽	200	413	613
40yr CHANGE	-55%	-11%	-22%
20yr INDEX	-0.65	-0.37	

BREEDING RELATIVE ABUNDANCE *2008–11*

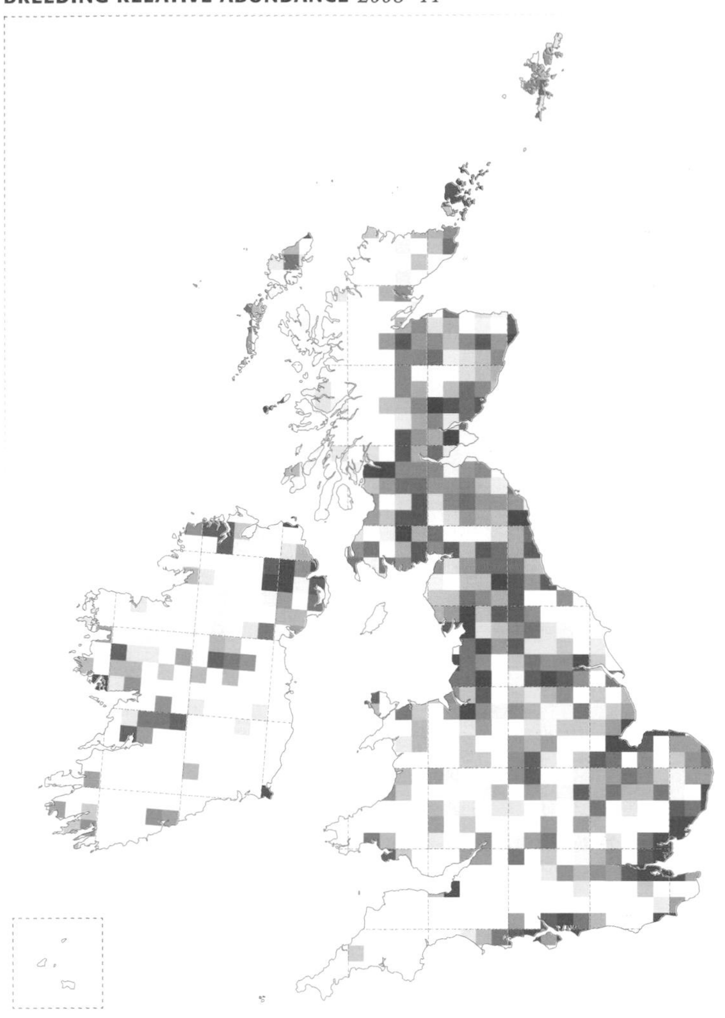

WINTER DISTRIBUTION CHANGE *since 1981–84*

GAIN ▲	28	113	141
LOSS ▼	190	199	389
30yr CHANGE	-19%	-4%	-8%

WINTER RELATIVE ABUNDANCE *2007/08–2010/11*

MEDITERRANEAN GULL

Larus melanocephalus

BTO CODE: MU

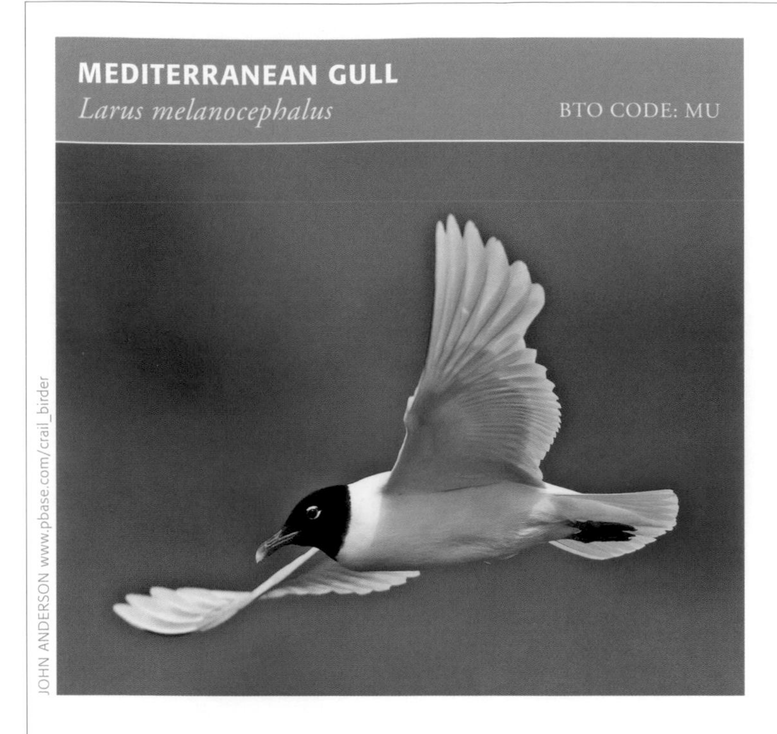

JOHN ANDERSON www.pbase.com/crail_birder

RAPIDLY EXPANDING COLONIES in the Netherlands, France, Germany, Belgium and Hungary have helped to fuel the increase in the numbers and range of the Mediterranean Gull in Britain & Ireland. During the winter months Mediterranean Gulls have a largely coastal distribution, though careful scanning through flocks of Black-headed Gulls at inland gull roosts and refuse tips probably accounts for the band of occupied squares from northwest England, through the midlands to London and the southeast. On the coast they generally occur singly or in small numbers, though counts in excess of 100 are regularly reported from the British south coast, especially around the Solent, with the highest counts generally made in September. In Ireland, Mediterranean Gulls were reported from widely distributed coastal sites, with recent peak counts of up to 70 individuals in Dublin Bay and on Achill Island and up to 48 at Cork Harbour (*I-WeBS Report 2009*). Although still uncommon in much of Scotland, they are now regular in winter in the Firths of Forth and Clyde (*Birds of Scotland*).

Mediterranean Gulls were first recorded breeding in Britain at Needs Ore Point, Hampshire, in 1968. They have bred annually since 1976, and by 1990 there were 11 pairs in Britain (*Birds in England*). The first breeding record for Ireland was in 1995 in Co. Antrim (Allen & Tickner 1996), and was followed by another the next year in Co. Wexford (Milne & O'Sullivan 1997).

Breeding in Britain is concentrated along the south and east coasts between Dorset and Norfolk but the expansion to inland sites in Britain, where they breed amongst Black-headed Gulls, is one of the most striking changes since the *1988–91 Breeding Atlas*. The UK breeding population is still increasing rapidly and jumped from 573 confirmed pairs in 2009 to at least 1,016 in 2010 (*RBBP 2010*). The main breeding site in Ireland is at Lady's Island Lake, Co. Wexford, with sporadic breeding of one or two pairs at a small number of other sites (*IRBBP 2010*). There has not yet been any confirmed breeding in Scotland, although a number of nesting attempts have been observed (*Birds of Scotland*).

Sponsored by Mr Anthony Roberts OBE and Dr Vivian Roberts

BREEDING DISTRIBUTION *2008–11*

POSSIBLE •	<1%	1%	1%
PROBABLE •	<1%	1%	1%
CONFIRMED •	1%	2%	1%
TOTAL	1%	4%	3%

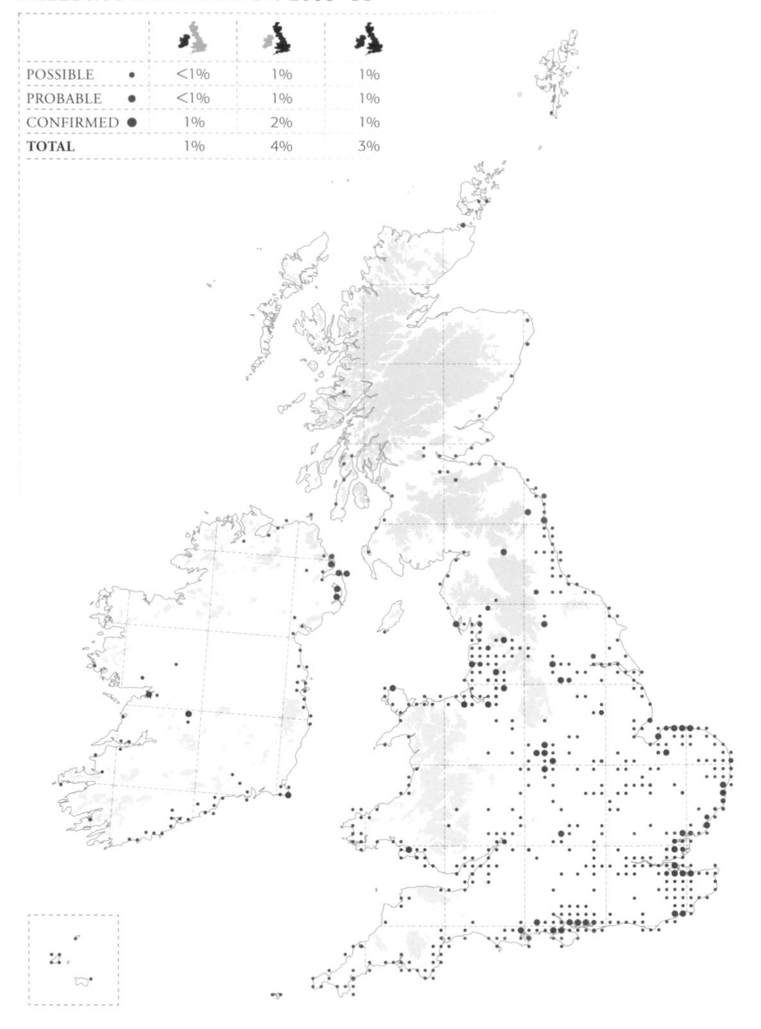

WINTER DISTRIBUTION *2007/08–2010/11*

PRESENT •	13%	24%	21%

BREEDING DISTRIBUTION CHANGE *since 1968–72*

GAIN ▲	9	65	74
LOSS ▽	0	0	0
40yr CHANGE	∞	+6500%	+7400%
20yr INDEX	-	+0.59	

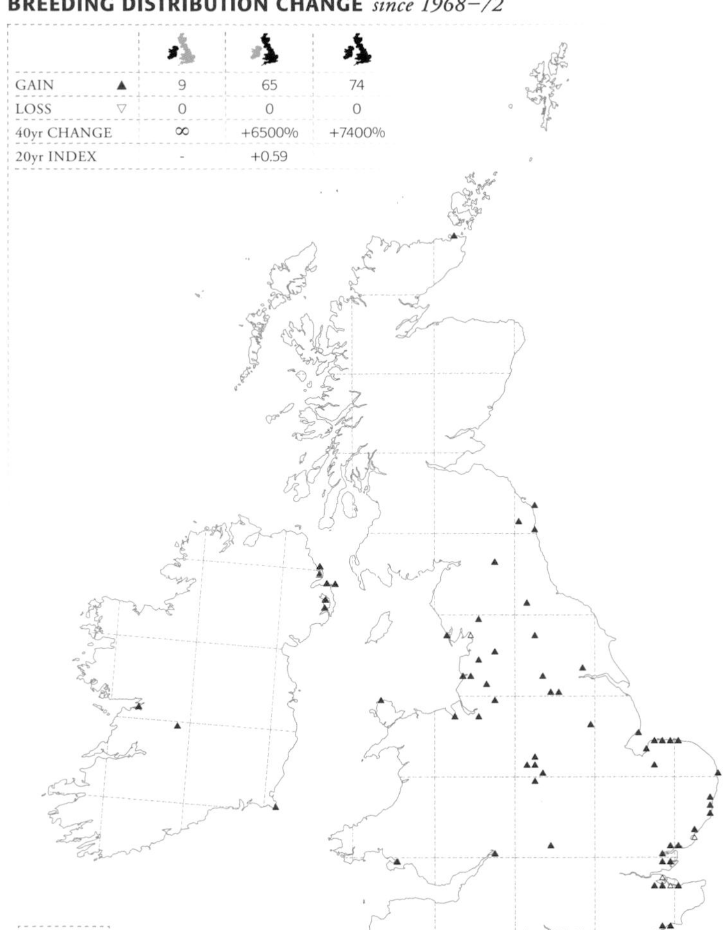

BREEDING RELATIVE ABUNDANCE *2008–11*

WINTER DISTRIBUTION CHANGE *since 1981–84*

GAIN ▲	109	529	638
LOSS ▼	7	20	27
30yr CHANGE	+638%	+356%	+384%

WINTER RELATIVE ABUNDANCE *2007/08–2010/11*

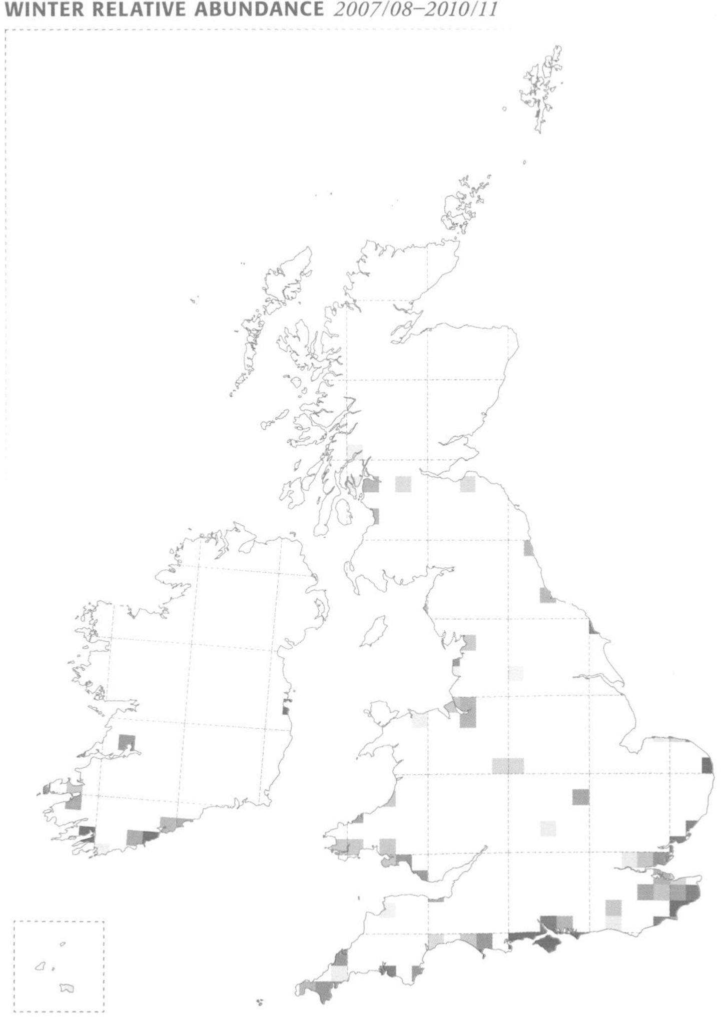

COMMON GULL

Larus canus

BTO CODE:CM

RICHARD WINSTON www.flickr.com/photos/honeycreeper

DURING THE AUTUMN and early winter there is a large influx of continental Common Gulls into Britain & Ireland (*Migration Atlas*), which results in a very wide winter distribution in Britain, with birds being absent only from upland areas. This contrasts with the Irish distribution, where the species generally occurs at few inland locations. The highest winter densities occur on the low ground of eastern Scotland and eastern England, and there are significant concentrations around parts of the Irish coast. Since the *1981–84 Winter Atlas* there has been virtually no change in overall range. Gains in northern Scotland may be a result of improved coverage but those in central England are likely to be genuine.

Continental Common Gulls begin their return movement during March and April, so many were still present in their wintering range at the start of the breeding-season fieldwork period, partly explaining the 1,565 10-km squares occupied by *non-breeding* birds. Breeding Common Gulls are found mostly in the north and west of both Scotland and Ireland. However, in Scotland, the highest densities are found on the eastern half of the country from Angus to the Moray Firth, Caithness and the Northern Isles as well as in many Highland straths and glens. In Ireland, the distribution is strongly coastal away from Counties Mayo and Galway. Outside these areas, Common Gulls are very scarce breeders, with small isolated colonies in northern England and at coastal sites in East Anglia, Kent and Hampshire (*Birds in England*). Remarkably, a pair nested, unsuccessfully, in Derbyshire in 2008 (Carrington & Bradley 2009).

The breeding change map highlights recent losses in many parts of Scotland and western Ireland. In Scotland, many of these losses are linked with increased levels of predation, with surviving colonies tending to remain in or relocate to predator-free areas (*Birds of Scotland*). Losses in parts of western Ireland broadly correspond with the spread of the American Mink, but other causal factors may also be involved (McGreal 2011). Potentially erroneous breeding evidence coding of summering and late wintering birds makes it difficult to assess whether some gains in Ireland and England are real.

Sponsored by Felicity & Howard Woods

BREEDING DISTRIBUTION *2008–11*

POSSIBLE	3%	5%	4%
PROBABLE	4%	4%	4%
CONFIRMED	10%	17%	15%
TOTAL	17%	26%	23%

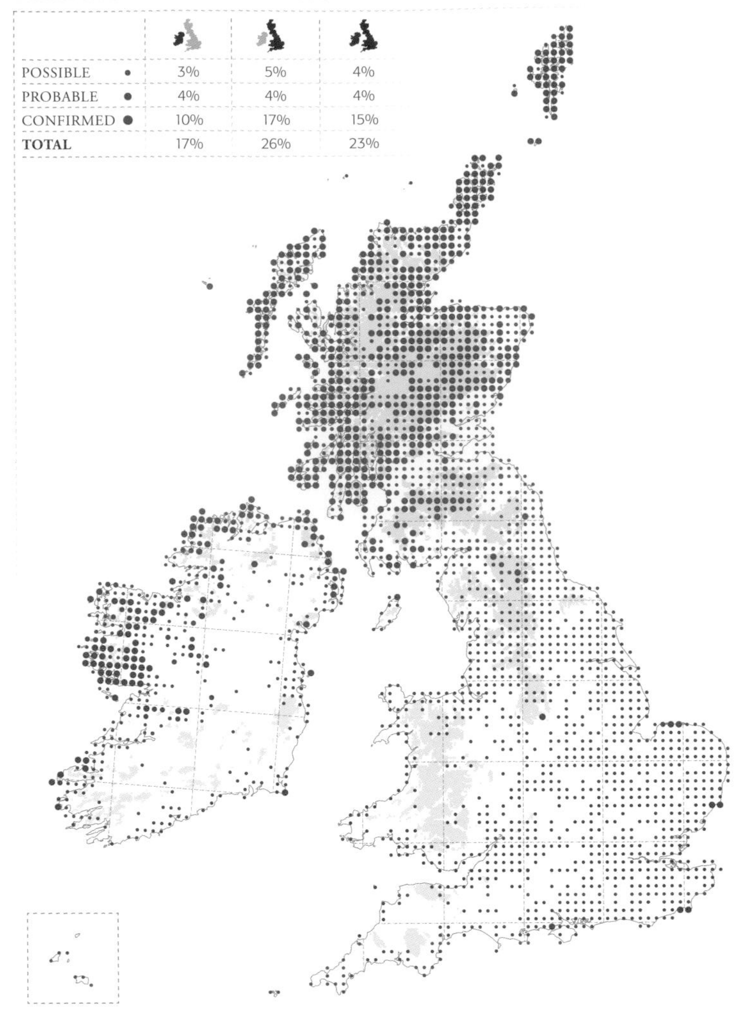

WINTER DISTRIBUTION *2007/08–2010/11*

PRESENT	50%	81%	73%

BREEDING DISTRIBUTION CHANGE *since 1968–72*

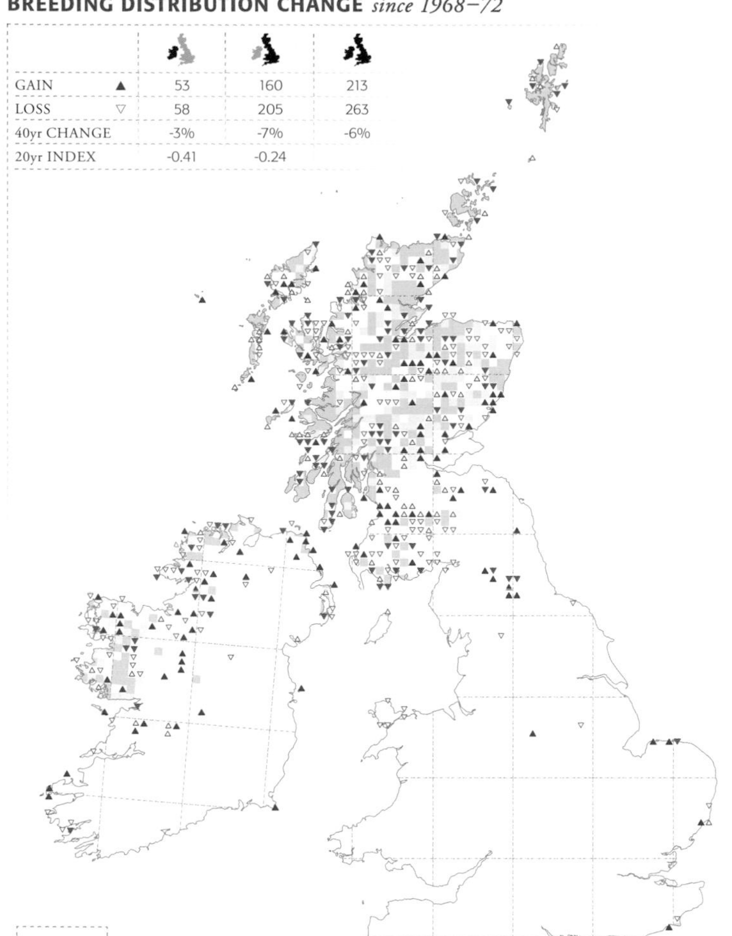

GAIN ▲	53	160	213
LOSS ▽	58	205	263
40yr CHANGE	-3%	-7%	-6%
20yr INDEX	-0.41	-0.24	

BREEDING RELATIVE ABUNDANCE *2008–11*

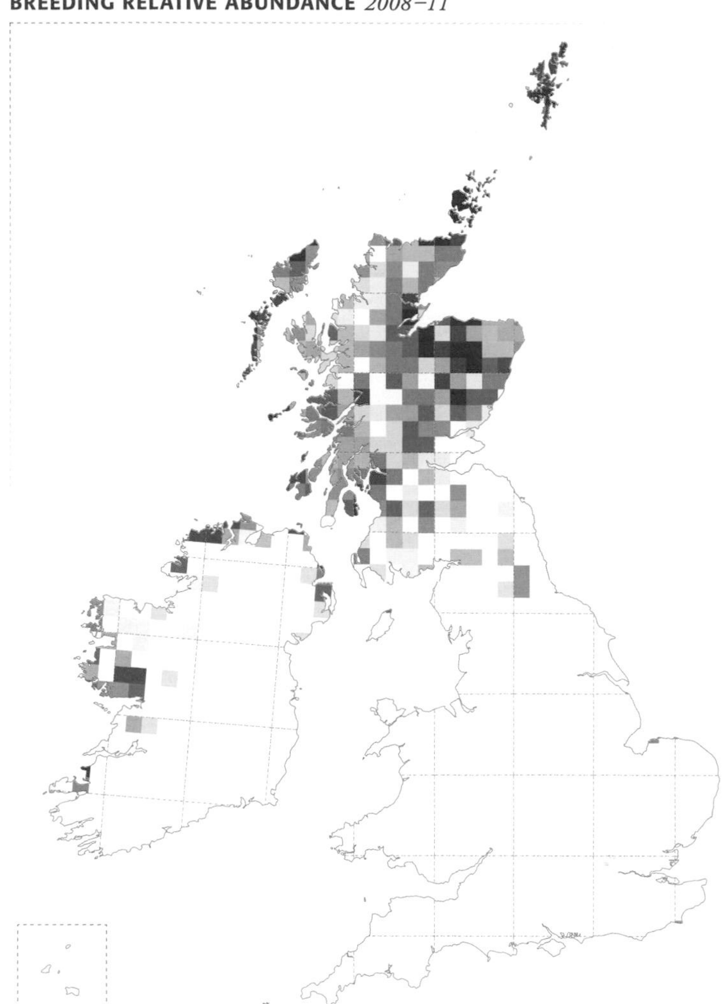

WINTER DISTRIBUTION CHANGE *since 1981–84*

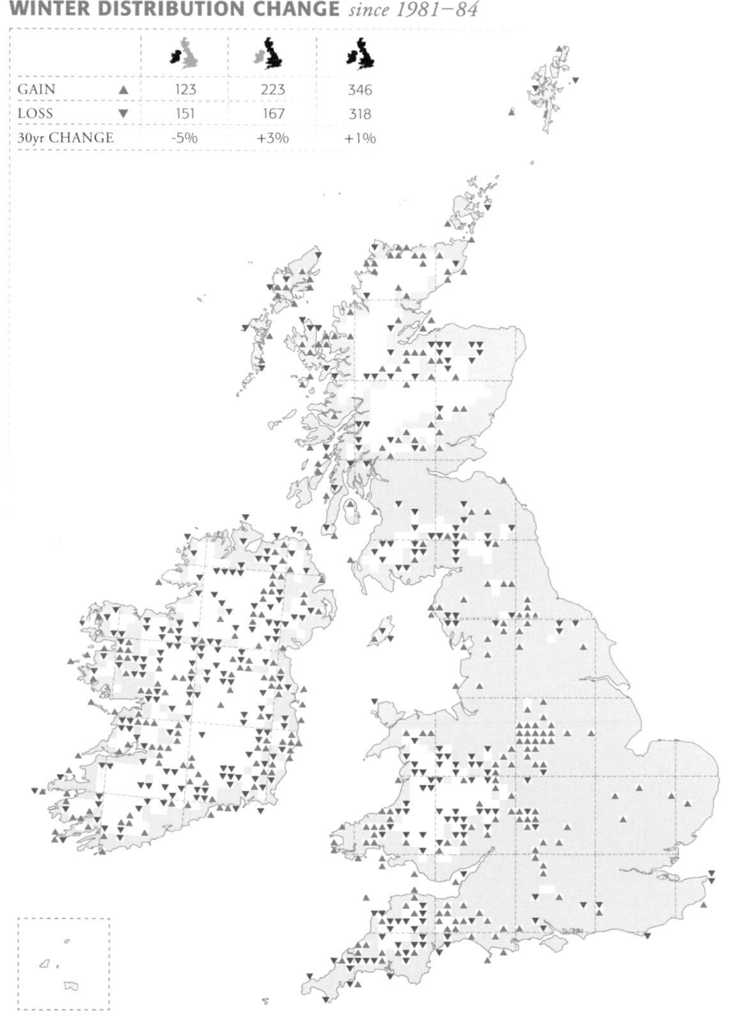

GAIN ▲	123	223	346
LOSS ▼	151	167	318
30yr CHANGE	-5%	+3%	+1%

WINTER RELATIVE ABUNDANCE *2007/08–2010/11*

LITTLE GULL

Hydrocoloeus minutus

BTO CODE: LU

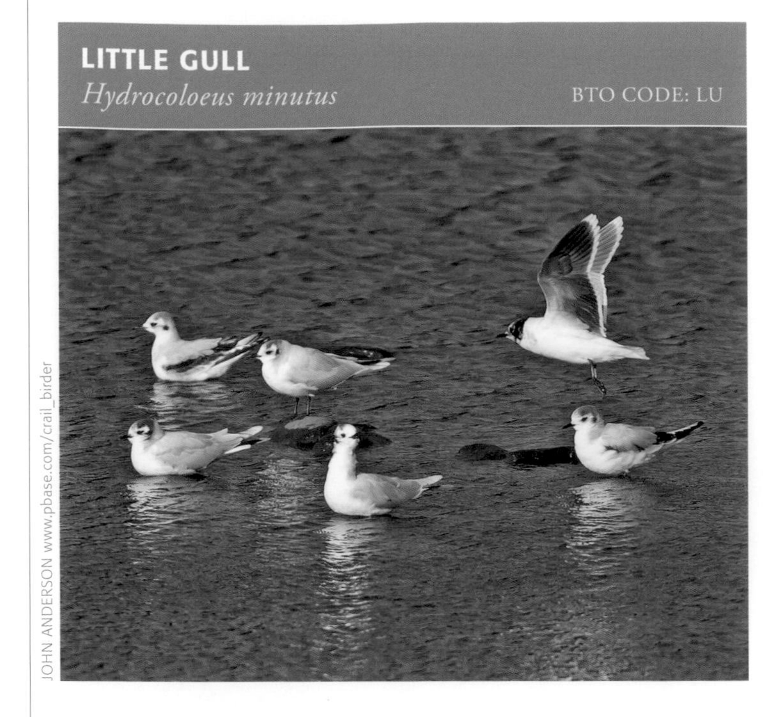

JOHN ANDERSON www.pbase.com/crail_birder

RECORDS OF THE Little Gull have increased considerably in Britain & Ireland since the 1950s and a 44% winter range expansion is apparent since the *1981–84 Winter Atlas*. This is likely to be linked to the westward extension of the species' breeding range to the Netherlands, Denmark, Sweden, Norway and Finland (BirdLife International 2004) and a corresponding increase in breeding numbers (Wetlands International 2013). In parallel, many birds now 'stop over' in the North Sea to undergo their post-breeding moult before continuing to their wintering grounds (Hartley 2004).

Little Gulls winter south to the Mediterranean and the Atlantic Ocean off North Africa, though small numbers remain in the Irish Sea (O'Sullivan & Smiddy 1992), English Channel and North Sea; there is, however, uncertainty over the numbers offshore. Observed numbers vary from year to year and within a winter, depending on the weather conditions; during onshore gales they may be reported from nearby coasts and occasionally inland. This explains the coastal pattern to the current winter distribution map, with the addition of a scattering of inland records, which usually involve just single birds. The winter change map shows an apparent overall gain in Britain & Ireland, which consists of a mixture of apparent gains and losses around the British and Irish coastlines. In Orkney, Shetland and inland England, gains outnumber losses.

During the breeding season there were records from 418 10-km squares in Britain and 26 in Ireland, involving mainly birds on spring passage, with some summering. Most are coastal, but there is a wide scatter of inland records in southern Britain, probably of birds passing over land (Messenger 1993). There have been five breeding attempts in England, most recently in Norfolk in 2007 (Holling *et al.* 2010a), and one or two breeding attempts in Scotland (*Birds of Scotland*), though none has been successful. In 2009, there were *probable* breeding records in north Norfolk (a pair of adults displaying and holding territory) and West Yorkshire (three second-summer individuals displaying and carrying nesting material with one apparently sitting for a week) and in 2011 *possible* breeding was recorded in Caithness (first-summer birds displaying and nest building).

In memoriam Dave Hills

WINTER DISTRIBUTION *2007/08–2010/11*

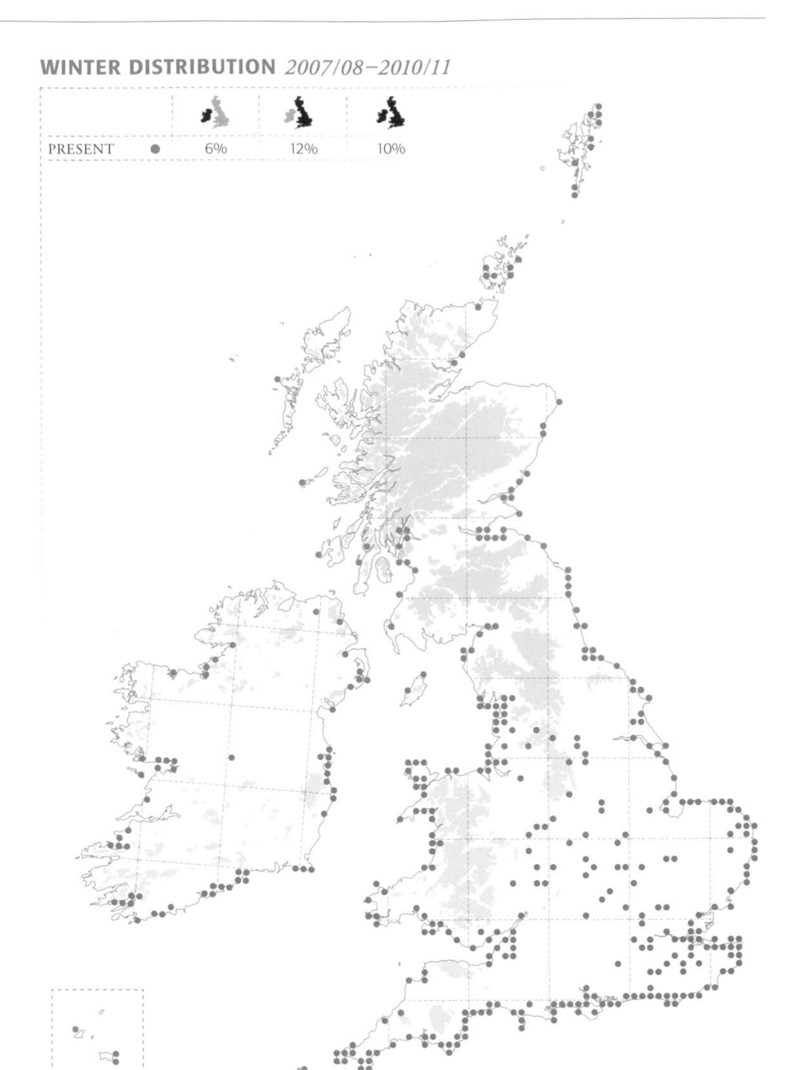

	Ireland	Britain	Britain & Ireland
PRESENT ●	6%	12%	10%

WINTER DISTRIBUTION CHANGE *since 1981–84*

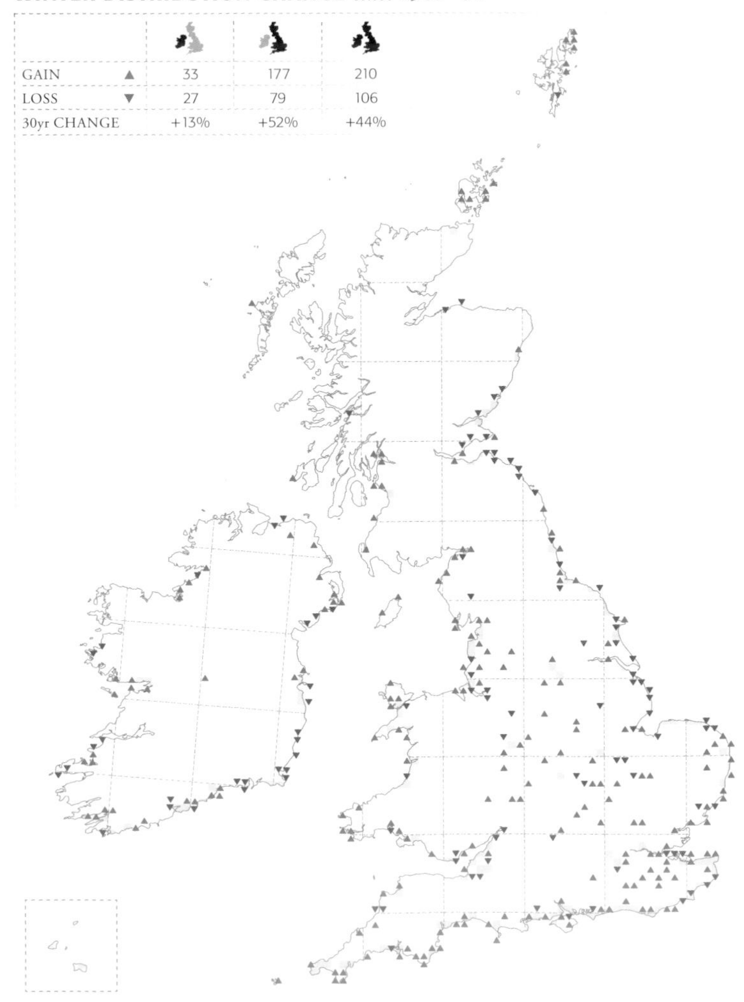

	Ireland	Britain	Britain & Ireland
GAIN ▲	33	177	210
LOSS ▼	27	79	106
30yr CHANGE	+13%	+52%	+44%

RING-BILLED GULL

Larus delawarensis

BTO CODE: IN

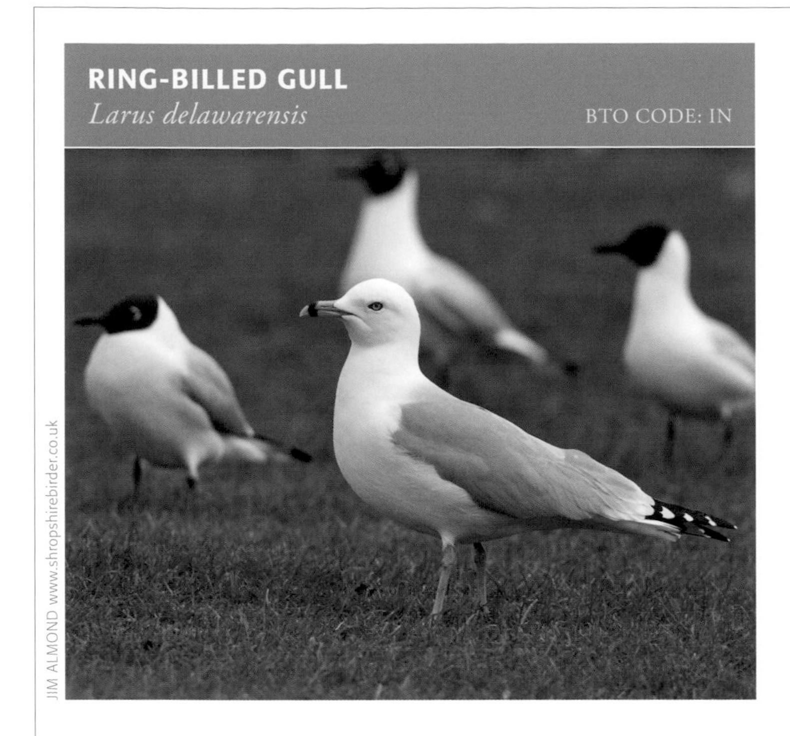

JIM ALMOND www.shropshirebirder.co.uk

THE GENERAL INCREASE in the number of records of Ring-billed Gulls since the first records in Britain in 1973 and in Ireland in 1979 (Parkin & Knox 2010) reflects a genuine increase in population size and breeding range in the species' native North America (Sauer *et al.* 2011). There has also been increasing interest in the identification of gulls and improved knowledge of identification features, especially of immature plumages, which has probably increased detection rates of this species.

The current winter distribution map shows a westerly bias, with most records in Ireland, southwest England and southern Wales, also with a scattering across central England and in ten 10-km squares in Scotland. The range has expanded from 49 10-km squares during the *1981–84 Winter Atlas* to 142 during *Bird Atlas 2007–11*.

The timing of records suggests that a late-winter influx takes place, with some birds remaining into the spring and a few summering. Some remain for many months, or even years, in the same location. Gulls on spring passage across Britain & Ireland may bring with them Ring-billeds that had crossed the Atlantic during the previous autumn and wintered elsewhere in Europe or Africa (Fraser & Rogers 2005). Such individuals, and late-departing winter visitors, were recorded as *non-breeding* in 20 10-km squares in Britain and 26 10-km squares in Ireland during the breeding season. There was one record with breeding evidence, from 2009, when an adult Ring-billed Gull was found paired with a Common Gull in a colony in Scotland and was seen sitting on a nest, though the contents were never seen (Barden 2010; Holling *et al.* 2011b). The only other record of breeding in Britain or Ireland comes from Copeland Island, Co. Down, in 2004, where a Ring-billed Gull was paired with a Common Gull. Breeding was confirmed only when the ring number was read on a hybrid Ring-billed Gull × Common Gull seen at Millisle on the Co. Down coast in February 2008, which indicated it was a chick ringed in that colony in 2004 (Charles 2009).

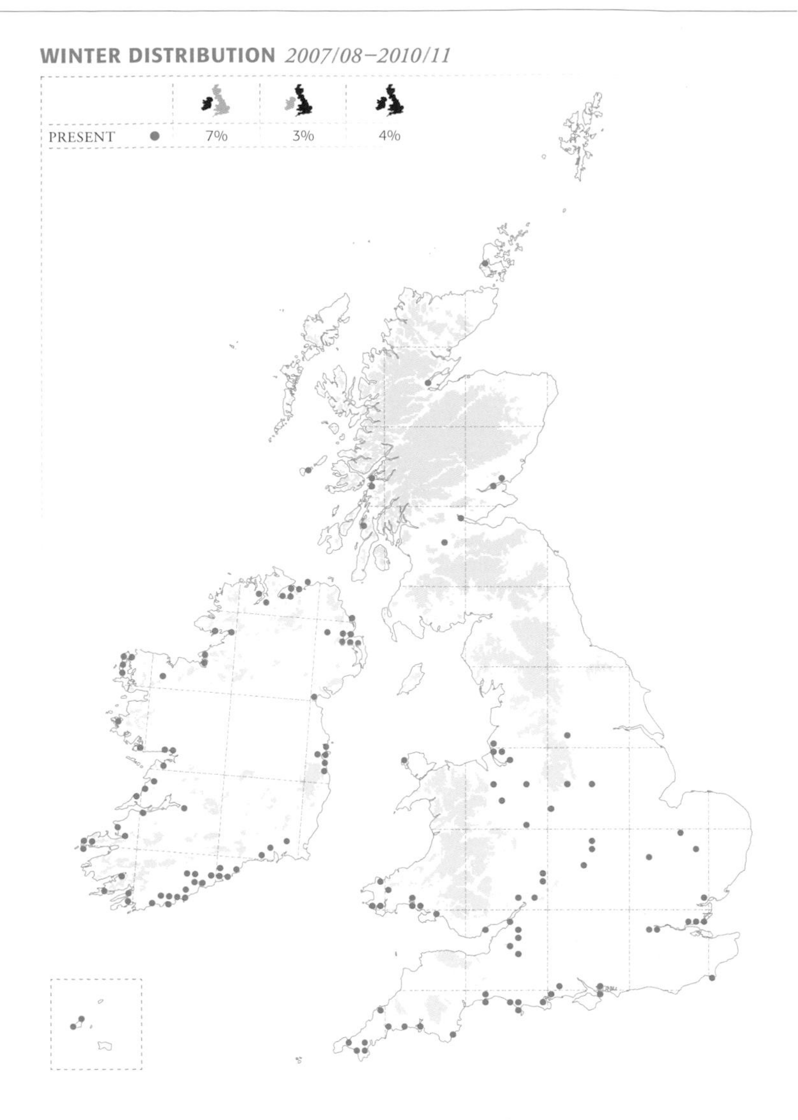

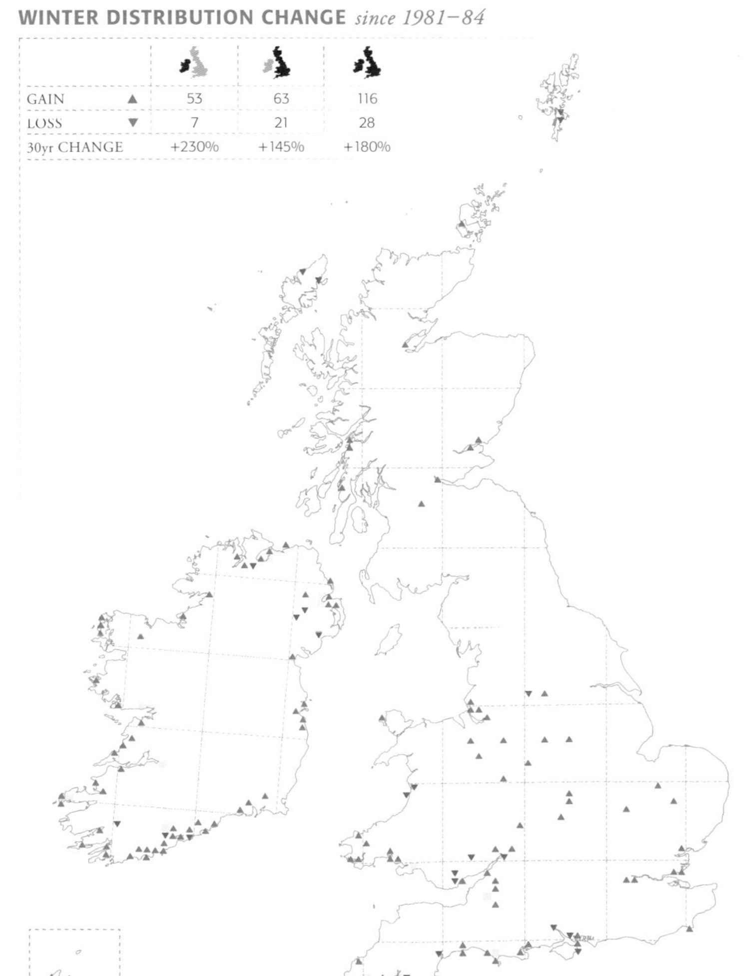

WINTER DISTRIBUTION CHANGE since 1981–84
GAIN ▲ 53 63 116
LOSS ▼ 7 21 28
30yr CHANGE +230% +145% +180%

Sponsored by a BTO Member from Yorkshire

LESSER BLACK-BACKED GULL

Larus fuscus

BTO CODE: LB

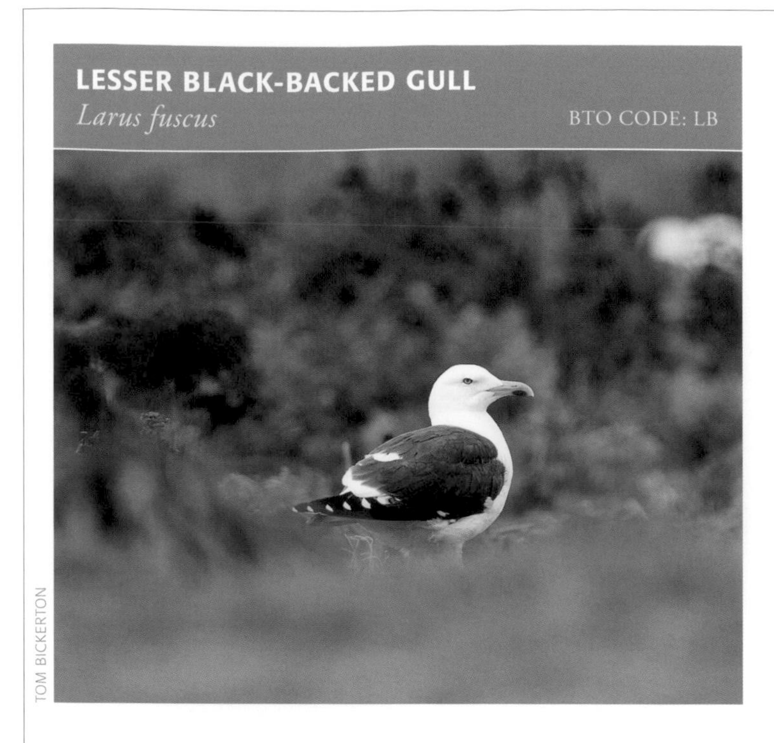

TOM BICKERTON

THERE HAS BEEN a remarkable change in the status of this species, with increases in range in both the winter and the breeding season.

The expansion in winter range since the *1981–84 Winter Atlas* is consistent with a dramatic increase in winter abundance since 1953 (Banks *et al.* 2009; Burton *et al.* 2013). The increase is particularly notable in Ireland, where there has been a 55% expansion in range, and in Wales, southwest England and East Anglia as well as in Scotland, though many of the records from Orkney and Shetland are passage birds recorded in November or February. Much of this expansion is inland and has occurred in different areas to where breeding-season changes have taken place. An ability to exploit a wide range of feeding opportunities and the availability of safe roost sites could be amongst factors contributing to an increased tendency to remain to winter in Britain & Ireland. Some birds disperse post-breeding to other parts of Britain & Ireland prior to moving further south. Those that remain are joined by visitors from growing breeding populations in Iceland, the Faeroes and northwest Europe (Wetlands International 2013).

Counts in UK breeding colonies increased by 29% between 1969–70 and 1985–88 and by a further 40% to 1998–2002, but then declined by 32% between 2000 and 2011; in the Republic of Ireland a 41% increase was evident between 1969–70 and 1998–2002 (JNCC 2012). The increases were attributed to greater legal protection, reduced exploitation and better feeding opportunities, though subsequent declines may have been due to culling, reduced food availability arising from changes in landfill and fisheries practices, predation and habitat change (*Seabird 2000*). The greatest change in breeding distribution has been the increase in inland nesting, mainly on man-made structures such as rooftops (Raven & Coulson 1997; Allard 2007). Losses are most evident in northern and western Scotland, where many colonies have shown a decline in numbers, mainly attributed to reduced food supply (*Birds of Scotland*). Colony counts suggest that declines have become more widespread in recent years, a finding repeated in Counties Galway and Mayo between 1977–78 and 2010 (McGreal 2011). Although the breeding range is smaller than the winter range, *non-breeding* birds are widespread in the breeding season.

Sponsored by Richard & Margaret Bentley

BREEDING DISTRIBUTION *2008–11*

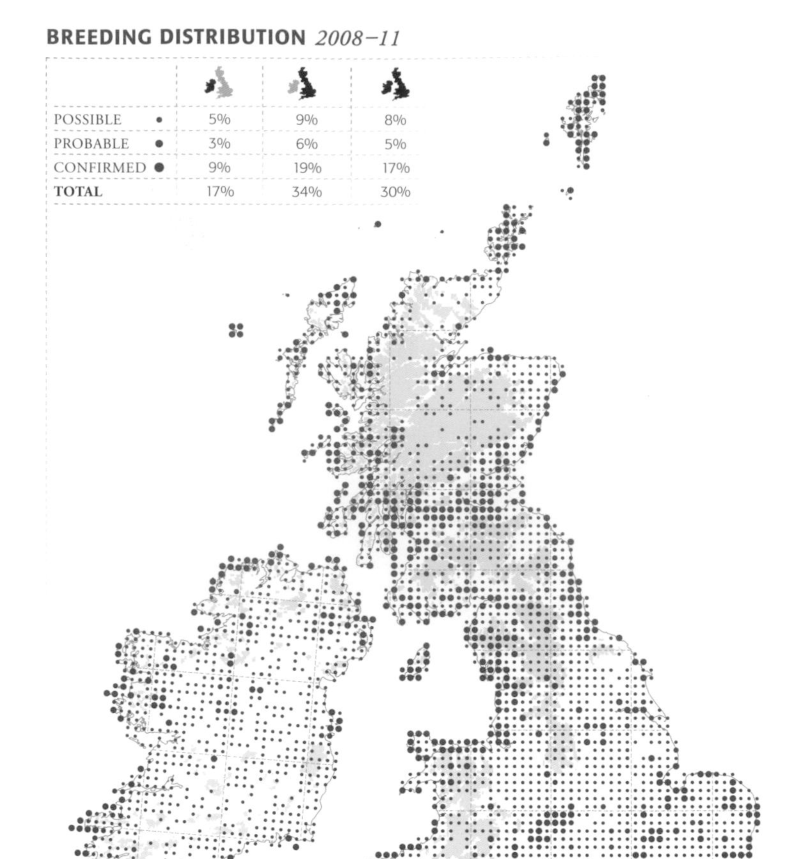

POSSIBLE •	5%	9%	8%
PROBABLE •	3%	6%	5%
CONFIRMED ●	9%	19%	17%
TOTAL	17%	34%	30%

WINTER DISTRIBUTION *2007/08–2010/11*

PRESENT ●	38%	64%	57%

BREEDING DISTRIBUTION CHANGE *since 1968–72*

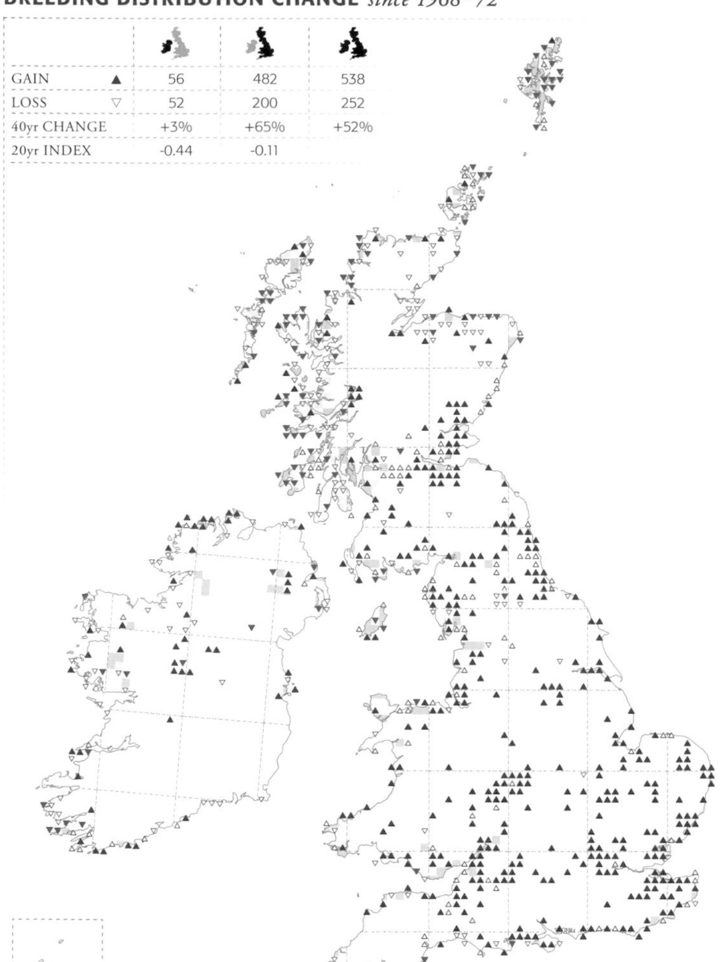

GAIN ▲	56	482	538
LOSS ▽	52	200	252
40yr CHANGE	+3%	+65%	+52%
20yr INDEX	-0.44	-0.11	

BREEDING RELATIVE ABUNDANCE *2008–11*

WINTER DISTRIBUTION CHANGE *since 1981–84*

GAIN ▲	222	517	739
LOSS ▼	91	199	293
30yr CHANGE	+55%	+22%	+27%

WINTER RELATIVE ABUNDANCE *2007/08–2010/11*

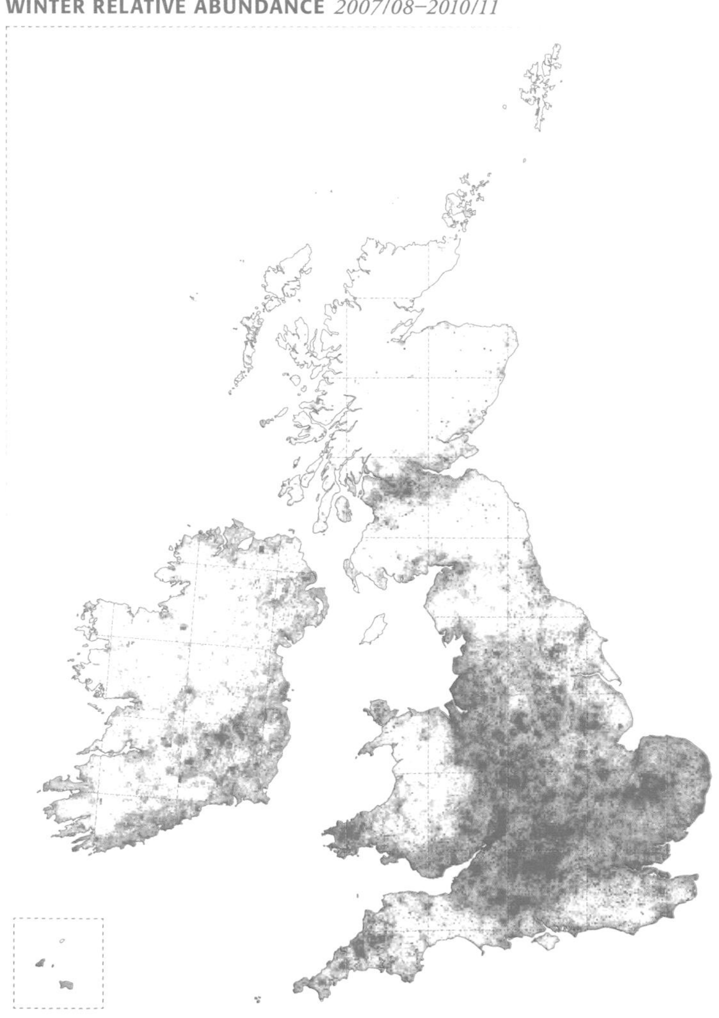

HERRING GULL

Larus argentatus

BTO CODE: HG

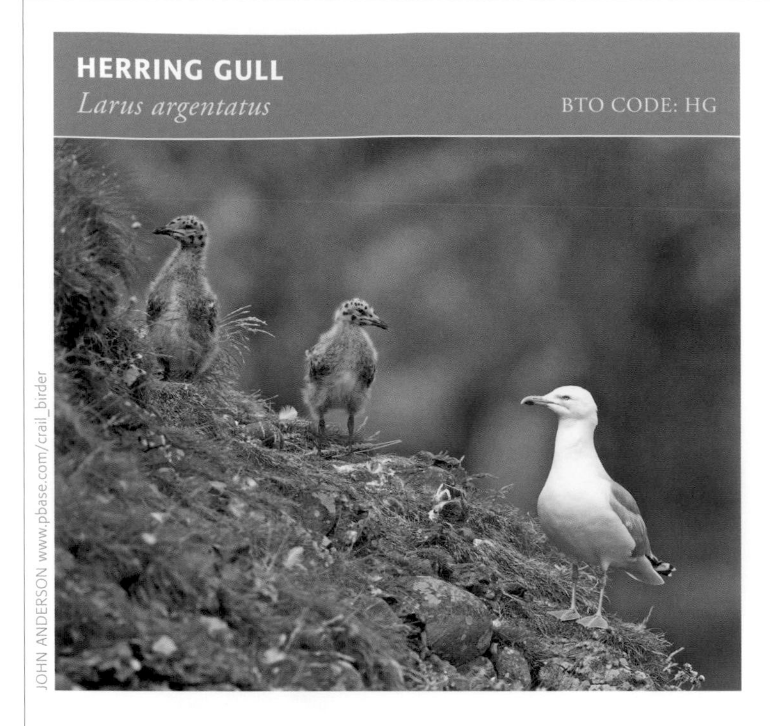

JOHN ANDERSON www.pbase.com/crail_birder

DURING THE WINTER, Herring Gulls are widely distributed throughout lowland areas of Britain, with the highest concentrations near the coast. In Ireland, the distribution is more coastal, although this has not always been the case. There has been a 38% contraction in range in Ireland since the *1981–84 Winter Atlas* and losses have been most marked in inland areas. Losses are also apparent in southern Scotland. This coincides with declines in breeding numbers and poor productivity in Ireland and declines at Scottish colonies. In Britain there was a small, 3% gain in the number of occupied 10-km squares in winter, mainly at inland sites.

During the summer, the breeding distribution is predominantly coastal. However, readiness of this species to nest on buildings has allowed it to colonise urban areas (Raven & Coulson 1997; Rock 2005). Outside breeding areas, large numbers of *non-breeding* birds were recorded in more than 1,200 inland 10-km squares. These were distributed throughout Britain, excluding the higher uplands, though only locally in Ireland.

Seabird colony monitoring in the UK shows that, following a 48% decline in numbers between 1969–70 and 1985–88, and a 13% decline between 1985–88 and 1998–2002, a further rapid decline of 33% took place between 2000 and 2011 (JNCC 2012). Data from six colonies monitored in Ireland suggest an even larger decline of 60% over the same period (JNCC 2012). However, populations in Britain, at least, remain higher than in the early 20th century (Chabrzyk & Coulson 1976). Reductions in feeding opportunities at refuse sites and from fishing-industry discards, changes in intertidal ecology affecting food supplies, continuing effects of botulism plus increased rates of mammalian predation are implicated in the declines (*Seabird 2000*). Though many colonies have decreased greatly in size, few have been completely lost. The breeding change map implies that the greatest losses were in coastal sites in western Ireland and western Scotland. Gains shown in urban areas, particularly in central and eastern Scotland, North Wales and northern and southern England, are indicative of the divergence in trends between urban (generally increasing) and rural (generally decreasing) breeding populations (Calladine *et al.* 2006).

Sponsored by Daniel M. Turner

BREEDING DISTRIBUTION *2008–11*

POSSIBLE •	5%	6%	6%
PROBABLE ●	3%	5%	4%
CONFIRMED ●	13%	31%	26%
TOTAL	21%	42%	36%

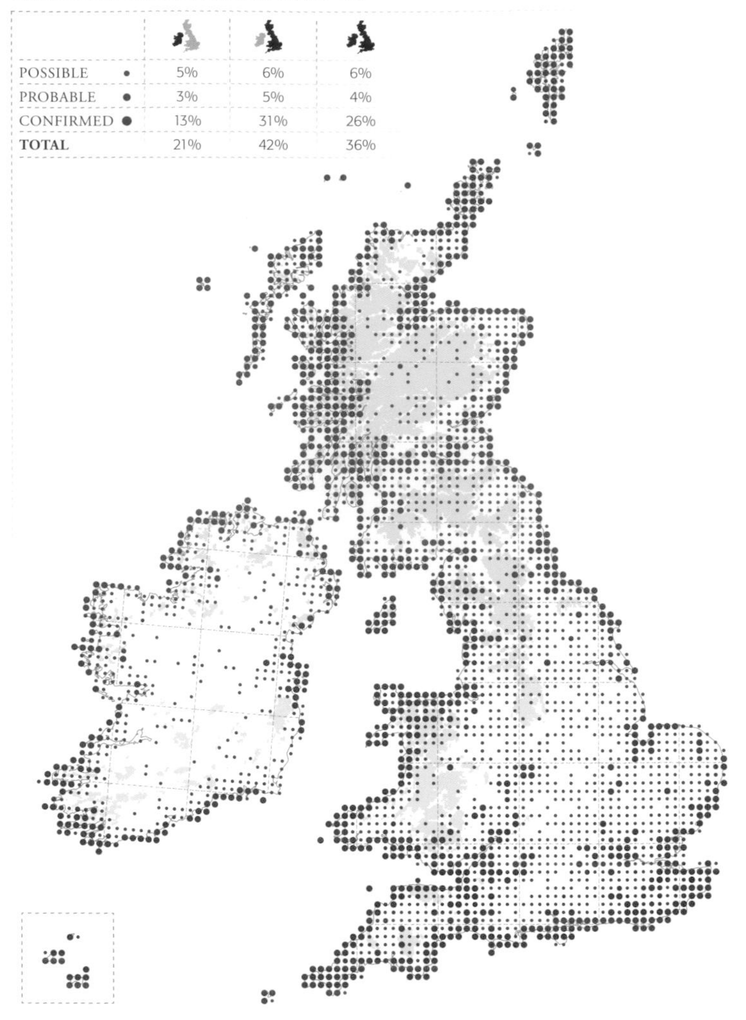

WINTER DISTRIBUTION *2007/08–2010/11*

PRESENT ●	47%	84%	74%

BREEDING DISTRIBUTION CHANGE *since 1968–72*

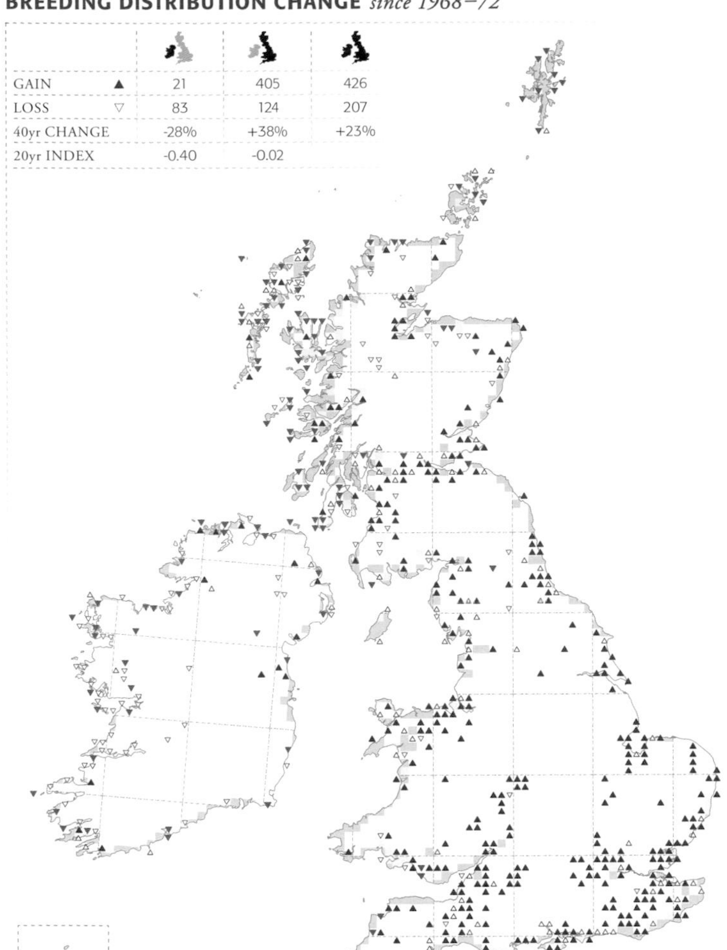

GAIN ▲	21	405	426
LOSS ▽	83	124	207
40yr CHANGE	-28%	+38%	+23%
20yr INDEX	-0.40	-0.02	

BREEDING RELATIVE ABUNDANCE *2008–11*

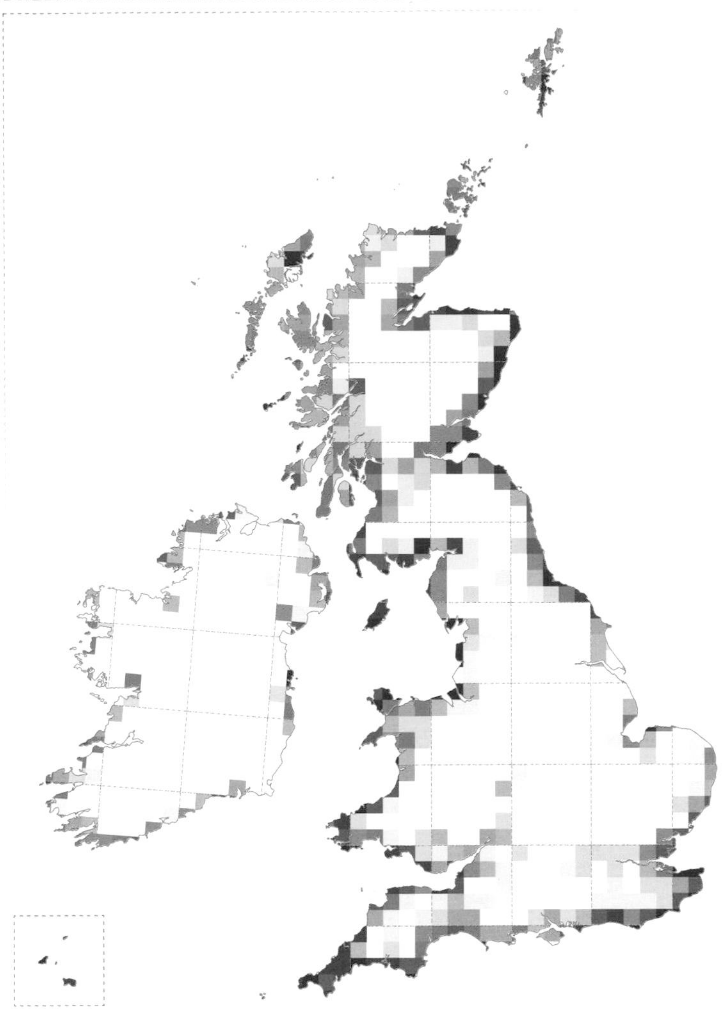

WINTER DISTRIBUTION CHANGE *since 1981–84*

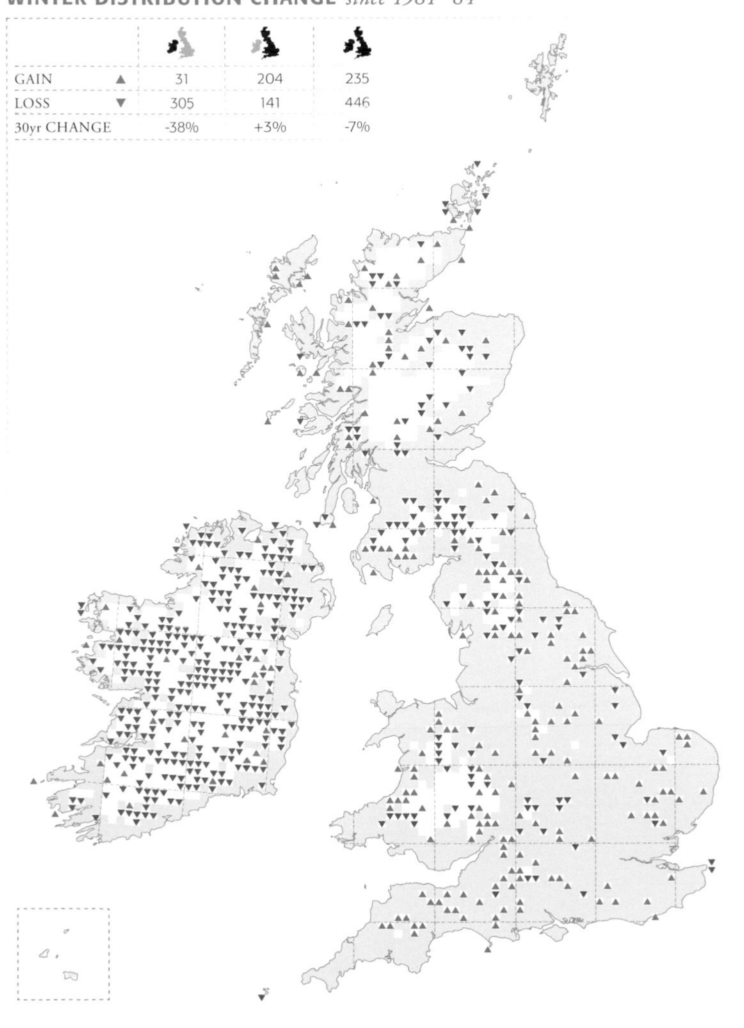

GAIN ▲	31	204	235
LOSS ▼	305	141	446
30yr CHANGE	-38%	+3%	-7%

WINTER RELATIVE ABUNDANCE *2007/08–2010/11*

YELLOW-LEGGED GULL

Larus michahellis

BTO CODE: YG

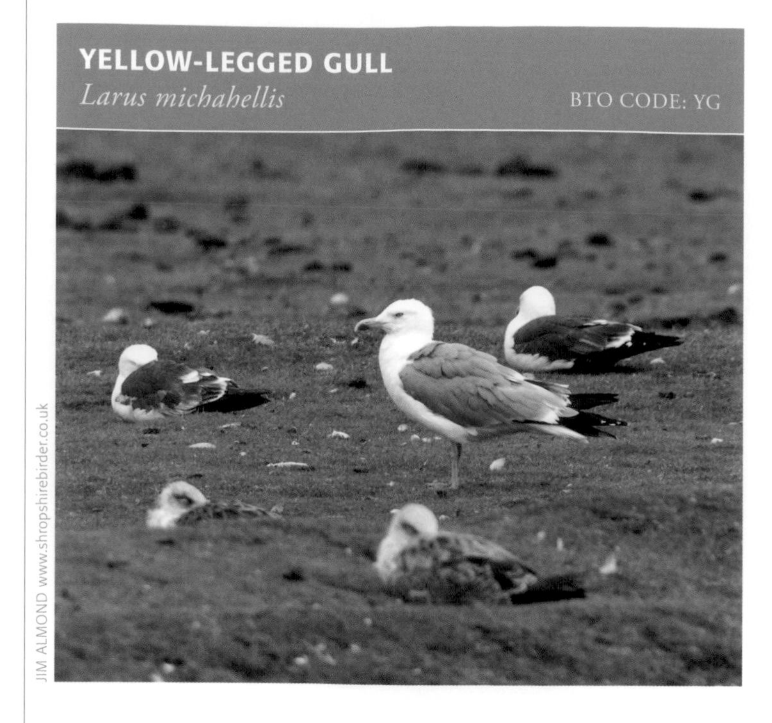

JIM ALMOND www.shropshirebirder.co.uk

SINCE THE EARLY records of Yellow-legged Gulls in the 1960s and 1970s (Hume 1978; Melville 1991), the numbers recorded in Britain & Ireland have increased considerably, especially since the mid 1980s (Dennis 1992; Walker 1995). The Yellow-legged Gull was formally recognised as a full species, separate from Herring Gull, by BOU in 2005 (Sangster *et al.* 2005), though keen gull watchers had been recording its occurrence and distribution for many years. Nevertheless, no previous atlas data exist for this species so no change maps are possible. The increase in numbers is due mainly to the expanding breeding populations in western, central and southern Europe and North Africa, but also to increased awareness by birdwatchers of advances in identification (Garner & Quinn 1997; Garner *et al.* 1997).

Yellow-legged Gulls are found throughout the year but there is a distinct peak in numbers between mid July and the end of August (Wilson & Balmer 2002), when there is an influx of juvenile birds from mainland Europe. The winter distribution map shows that most records came from central, eastern and southern Britain, where they use a range of inland and coastal sites, such as waterbodies, pig fields and landfill sites. In Ireland they were recorded mainly on the east and south coasts.

Breeding-season fieldwork recorded birds showing breeding evidence in six 10-km squares in Britain and one in Ireland, including *confirmed* breeding in one 10-km square in Dorset, and *probable* breeding in two squares in southern and eastern England and in one square in Ireland. *Non-breeding* birds are widespread across the southern half of England and there were records in 307 and 13 squares in Britain and Ireland respectively. Following a hybrid pairing with a Lesser Black-backed Gull in 1992 (Ogilvie *et al.* 1995), the first confirmed breeding of a pure pair was in Dorset in 1995, where one pair bred and another pair was present (Ogilvie *et al.* 1998). This remains a very rare breeding bird, with typical annual totals of just one pure pair breeding, and two or three mixed pairs with Herring or Lesser Black-backed Gulls (*RBBP 2010*).

Sponsored by Dorset Bird Club

WINTER DISTRIBUTION *2007/08–2010/11*

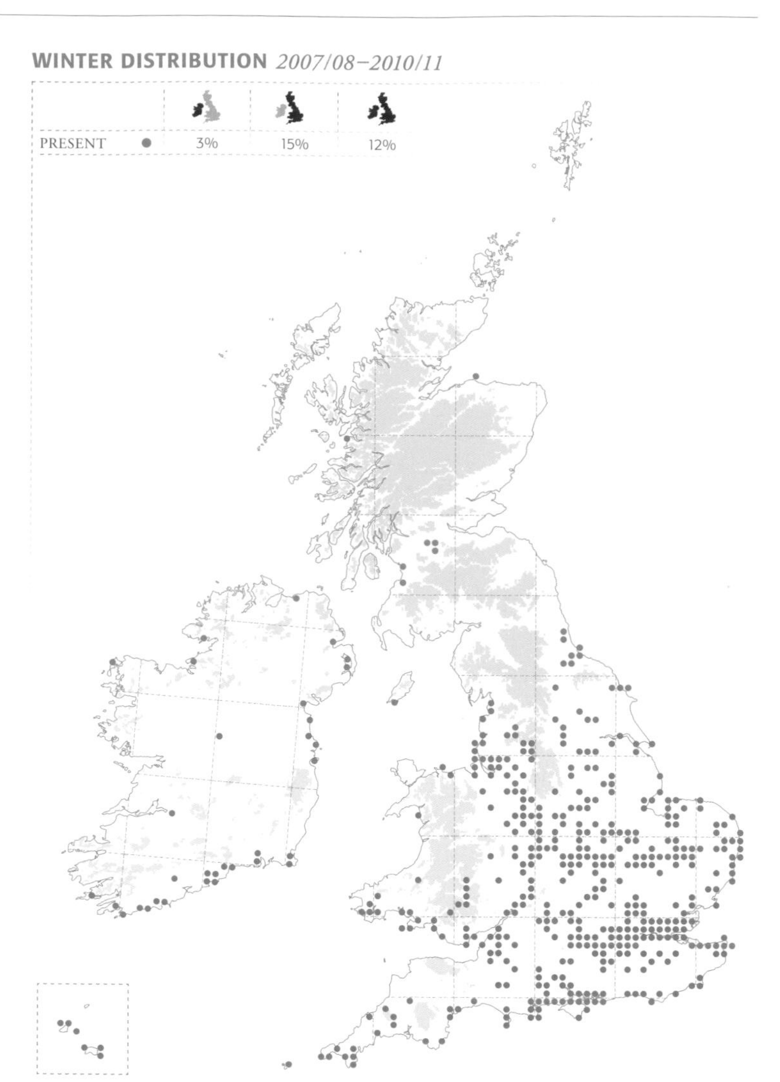

PRESENT	●	3%	15%	12%

BREEDING DISTRIBUTION *2008–11*

POSSIBLE	•	-	<1%	<1%
PROBABLE	●	<1%	<1%	<1%
CONFIRMED	●	-	<1%	<1%
TOTAL		<1%	<1%	<1%

CASPIAN GULL

Larus cachinnans

BTO CODE: YC

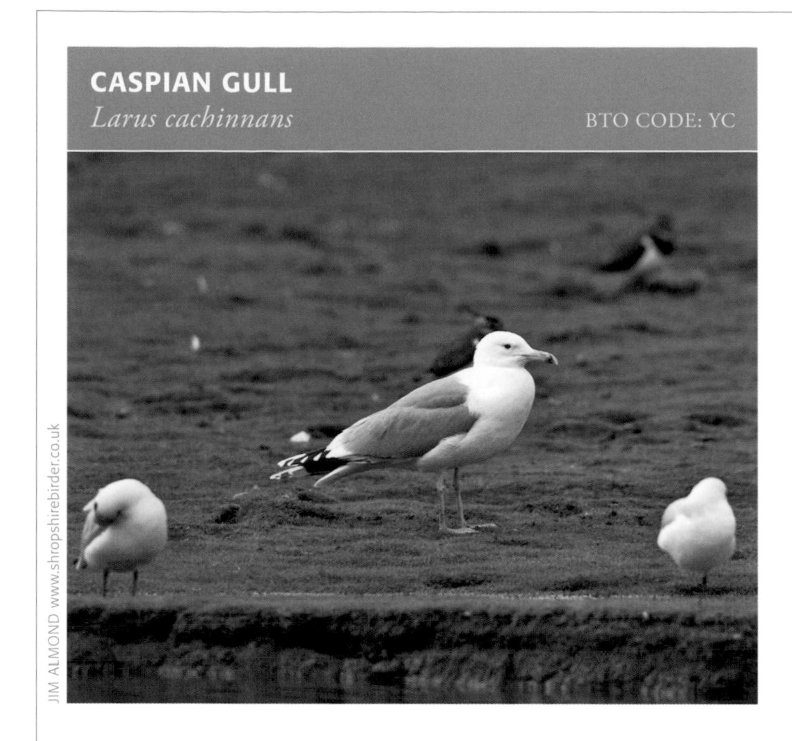

JIM ALMOND www.shropshirebirder.co.uk

THIS IS THE first time that the Caspian Gull has featured in any atlas of British and Irish birds. The species breeds around the Black and Caspian Seas and eastwards to northwest China (Olsen & Larsson 2004). In Europe it has been spreading north and west and now breeds in Poland and eastern Germany (Skórka *et al.* 2005). Mixed-species colonies in central Europe (Neubauer *et al.* 2009) are the source of hybrid birds that also reach Britain.

The first documented record in Britain was from Mucking, Essex, in September 1995 and the first in Ireland was from North Foreshore Tip, Belfast in Co. Antrim, in March 1998 (Parkin & Knox 2010). The Caspian Gull was recognised as a separate species from Herring Gull only in 2007 (Sangster *et al.* 2007; Collinson *et al.* 2008). There has been a significant increase in records over the last 15 years, in line with the expanding breeding colonies, but also because of improved observer awareness thanks to important identification papers (Garner & Quinn 1997; Jonsson 1998; Gibbins *et al.* 2010, 2011).

There is a distinct arrival in late summer and early autumn into Britain, mainly to the southeast and east coast, with birds then moving inland and dispersing north and west as the autumn and winter progress. Juveniles arrive in early August and post-breeding adults later in the autumn, which further boosts numbers. Numbers peak in winter and Caspian Gulls were reported from 169 10-km squares, mostly in southern and central England. There were just two records from Scotland, and none from Wales or Ireland, where this species remains very rare. It should be noted that the range may have been greater still, but Caspian Gulls are a local rarity in many counties and some records had not been submitted and accepted by the completion of the Atlas.

Caspian Gulls were recorded as *non-breeding* from 46 10-km squares in central and eastern England during the breeding season. Displaying birds have not, as yet, been recorded within colonies of Herring or Lesser Black-backed Gulls.

WINTER DISTRIBUTION *2007/08–2010/11*

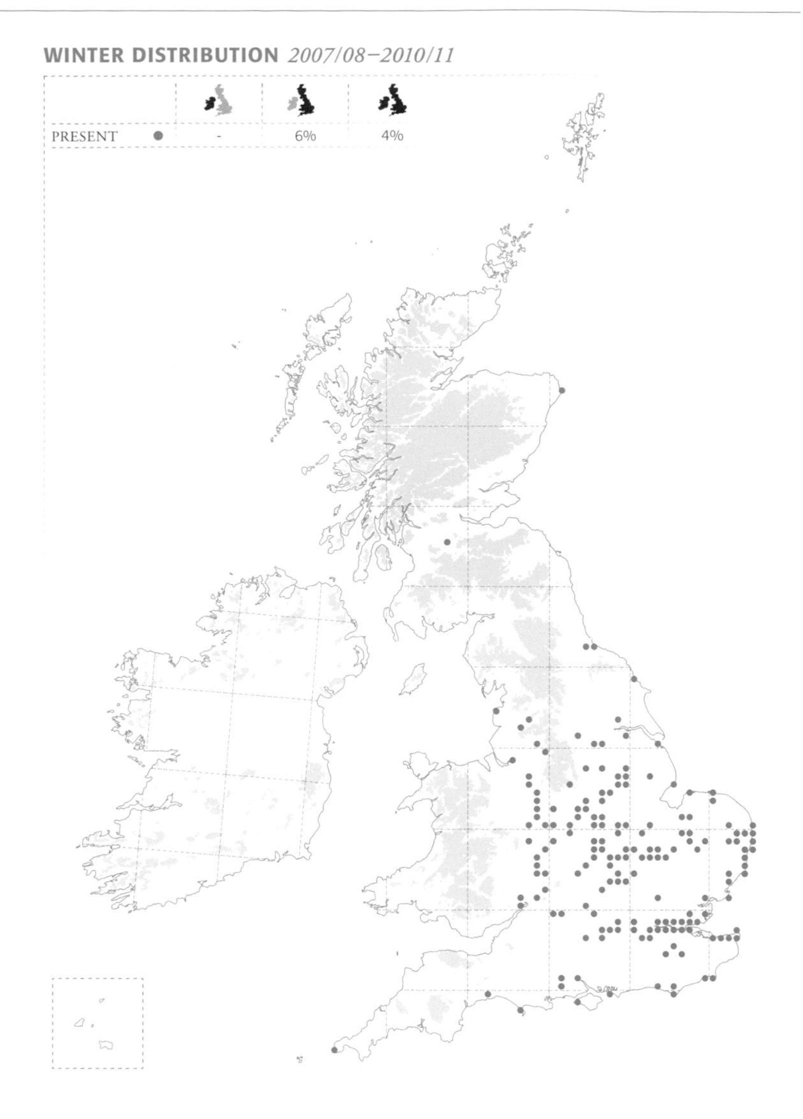

BREEDING DISTRIBUTION *2008–11*

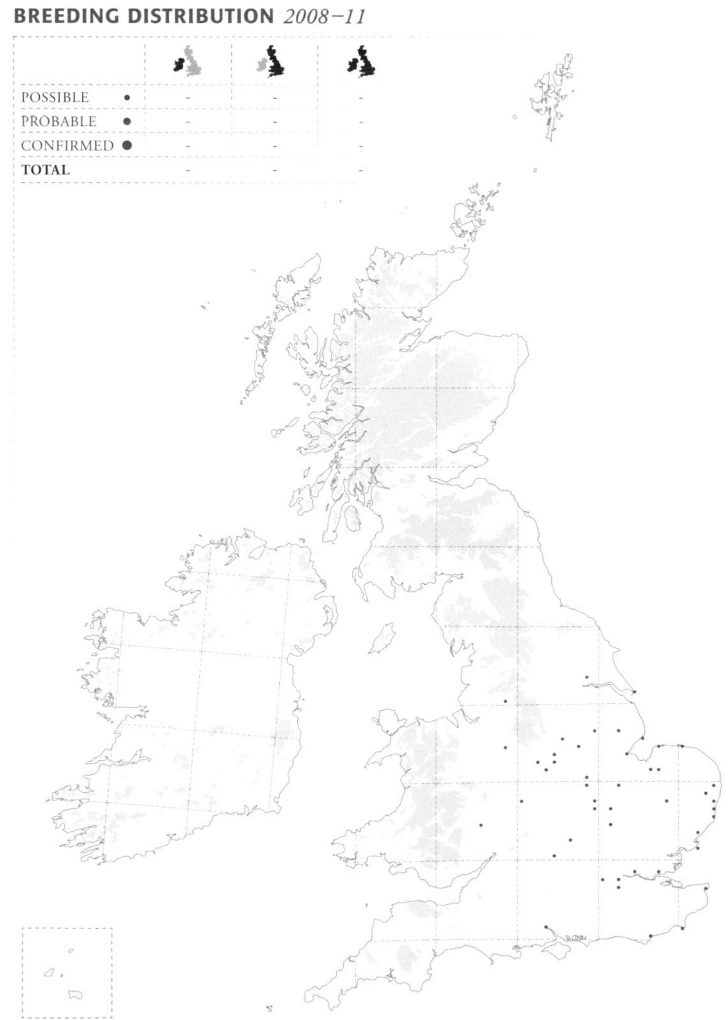

ICELAND GULL

Larus glaucoides

BTO CODE: IG

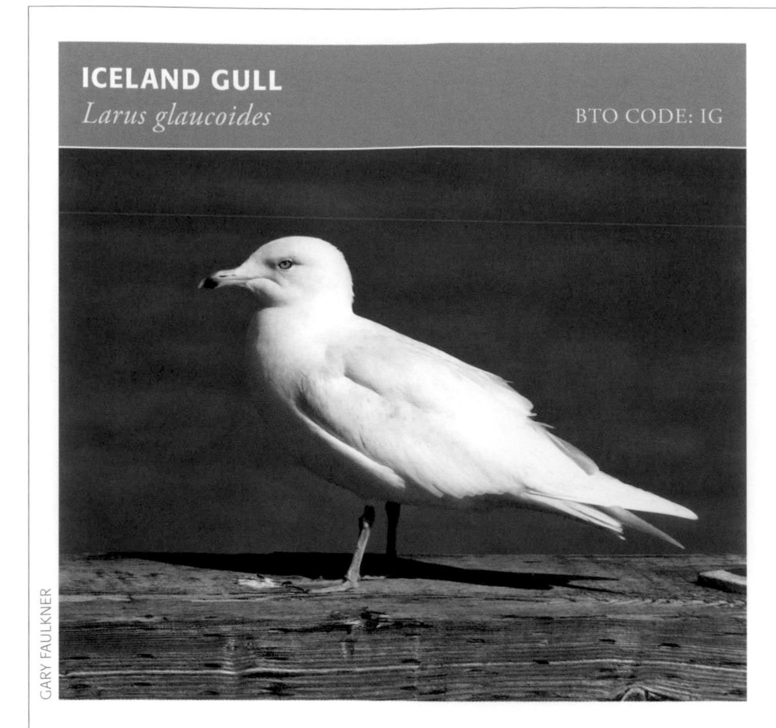

GARY FAULKNER

BREEDING IN ARCTIC regions of Canada and Greenland, the Iceland Gull is a scarce but regular passage migrant and winter visitor to Britain & Ireland (Weir *et al.* 2000). Birds tend to arrive from October or November into northern and western Britain & Ireland and then move south as the winter progresses (*Migration Atlas*), with numbers peaking between January and March. Although there are hotspots around fish quays, harbours and landfill sites, they can be found by carefully searching gull flocks at loafing and roost sites.

During the winter months they are scattered around the coast of Britain with concentrations evident in west Cornwall, northeast and northwest England, Caithness, the Northern Isles and the Outer Hebrides. There are records from many inland locations, particularly in central England where the gull roosts are well watched. In Ireland the distribution is predominantly coastal, with records from throughout the coastline, though relative abundance is highest in the west.

The distribution differs markedly from the westerly biased pattern shown by the *1981–84 Winter Atlas*, and the winter change map shows 28% and 61% range expansions in Ireland and Britain respectively, with many gains in square occupancy in inland parts of England. Interestingly, the 1982/83 winter, and to a lesser extent the 1983/84 one, were considered exceptional for this species, which makes this increase even more notable. There has been growing enthusiasm for gull watching since the 1980s, which may partly explain the increase. The four winters during the Atlas period were good for 'white-winged gulls' overall, particularly the winter of 2008/09 (*BirdTrack*). Numbers are influenced by weather patterns, with numbers appearing after strong northerly or northwesterly gales.

Some Iceland Gulls remain into the spring, and April and May are good months to search for them as they pass through inland and coastal sites. *Non-breeding* birds were recorded from 231 10-km squares in Britain and from 61 10-km squares in Ireland during the breeding season. The distribution was similar to that in winter, with a scatter of records from around the coastline of Britain & Ireland and from inland locations in central England.

Sponsored by Peter Mansell

WINTER DISTRIBUTION *2007/08–2010/11*

	Ireland	Britain	Britain & Ireland
PRESENT ●	12%	16%	15%

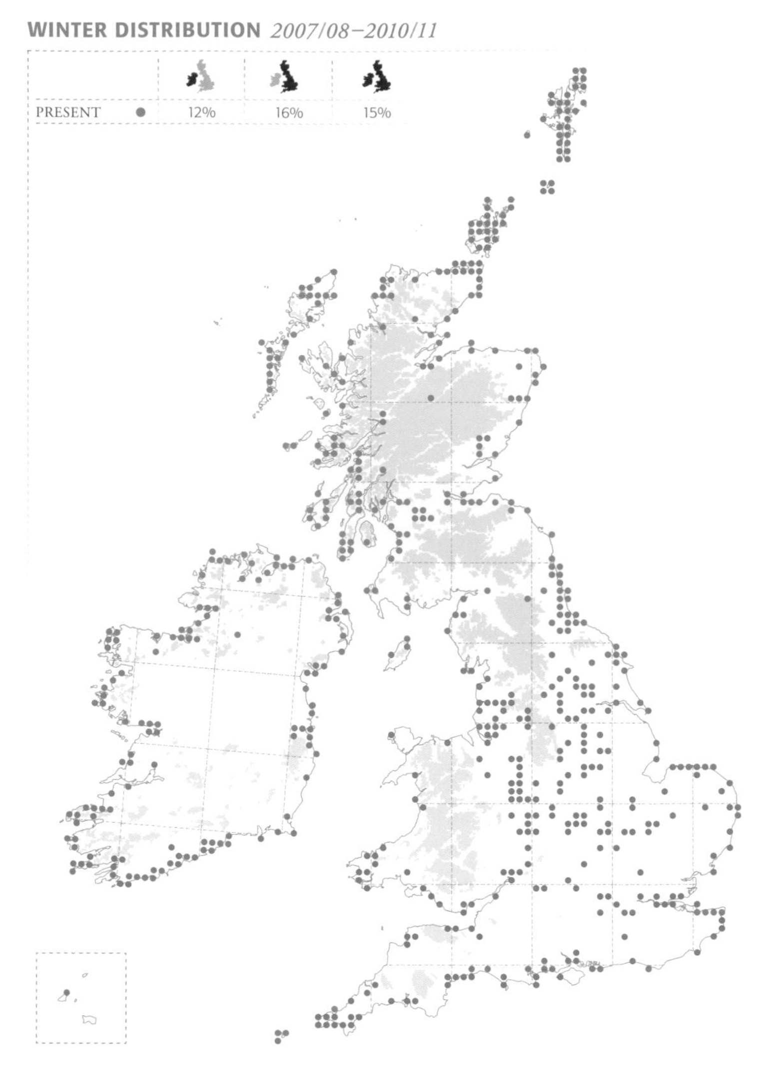

WINTER DISTRIBUTION CHANGE *since 1981–84*

	Ireland	Britain	Britain & Ireland
GAIN ▲	61	296	357
LOSS ▼	35	123	158
30yr CHANGE	+28%	+61%	+53%

GLAUCOUS GULL

Larus hyperboreus

BTO CODE: GZ

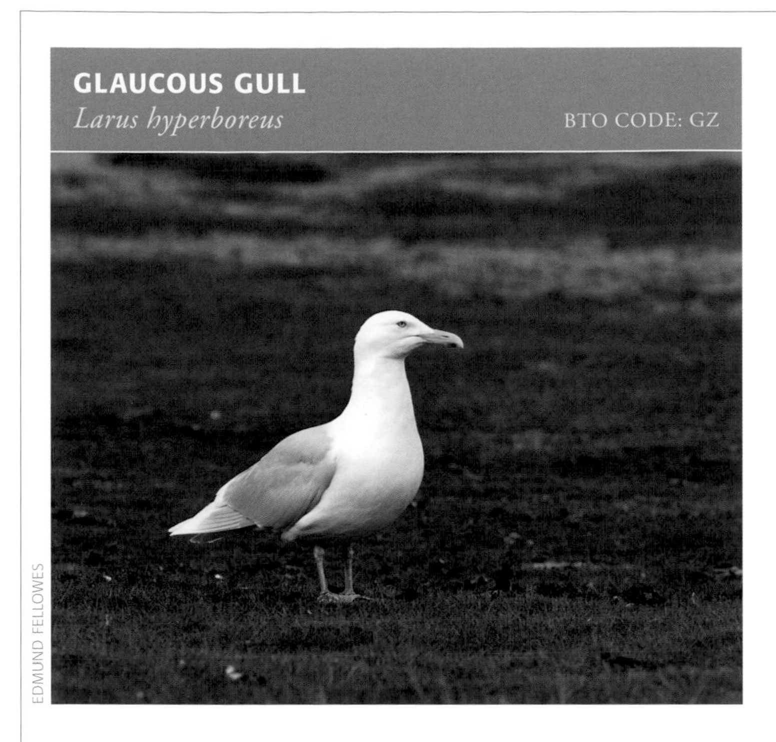

EDMUND FELLOWES

GLAUCOUS GULLS BREED on islands and cliffs around the Arctic south to Iceland and southern Greenland (*BWP*). They are regular winter visitors to Britain & Ireland with most records in the period October to late April. The winter distribution map shows a scatter of records around the coast of Britain, though they are scarcer on the west coast of England and Wales. They are widely recorded across the Northern Isles and the Outer Hebrides. In Ireland they are found mainly around the coast, especially in the west, with few records inland. There has been little overall change in their winter range size, though this masks differences between Britain and Ireland where the changes have been +9% and -23% respectively. The most significant change in distribution is the increased occurrence inland in central England, where they are found on landfill sites and nearby, loafing in fields, and at roost sites. This probably also reflects greater interest and scrutiny of gull flocks by birdwatchers at inland locations.

Numbers reported vary from year to year, and influxes tend to occur following a period of northerly or northwesterly gales, which may displace birds feeding around fishing boats in Icelandic or Greenland waters (*Birds of Scotland*). Exceptional numbers were recorded in winter 1980/81, just prior to the *1981–84 Winter Atlas*, and there is some evidence of another influx in winter 2008/09 (*BirdTrack*).

Historically, Glaucous Gulls were more numerous than Iceland Gulls (*Birds in Ireland*; *Birds in England*), although in recent decades Iceland Gulls have dominated. A review of the numbers of gulls wintering in the West Midlands showed that fewer Glaucous Gulls and more Iceland Gulls have been recorded since 1990 (Dean 2002). In Scotland, similar numbers of each species have been seen since 1990 (*Birds of Scotland*).

During the summer months, *non-breeding* Glaucous Gulls were recorded from 150 10-km squares in Britain and from 41 10-km squares in Ireland. The majority were occupied in April only, presumably by late-departing individuals, and there was a coastal bias to the records, with many coming from Orkney, Shetland and the Outer Hebrides.

WINTER DISTRIBUTION *2007/08–2010/11*

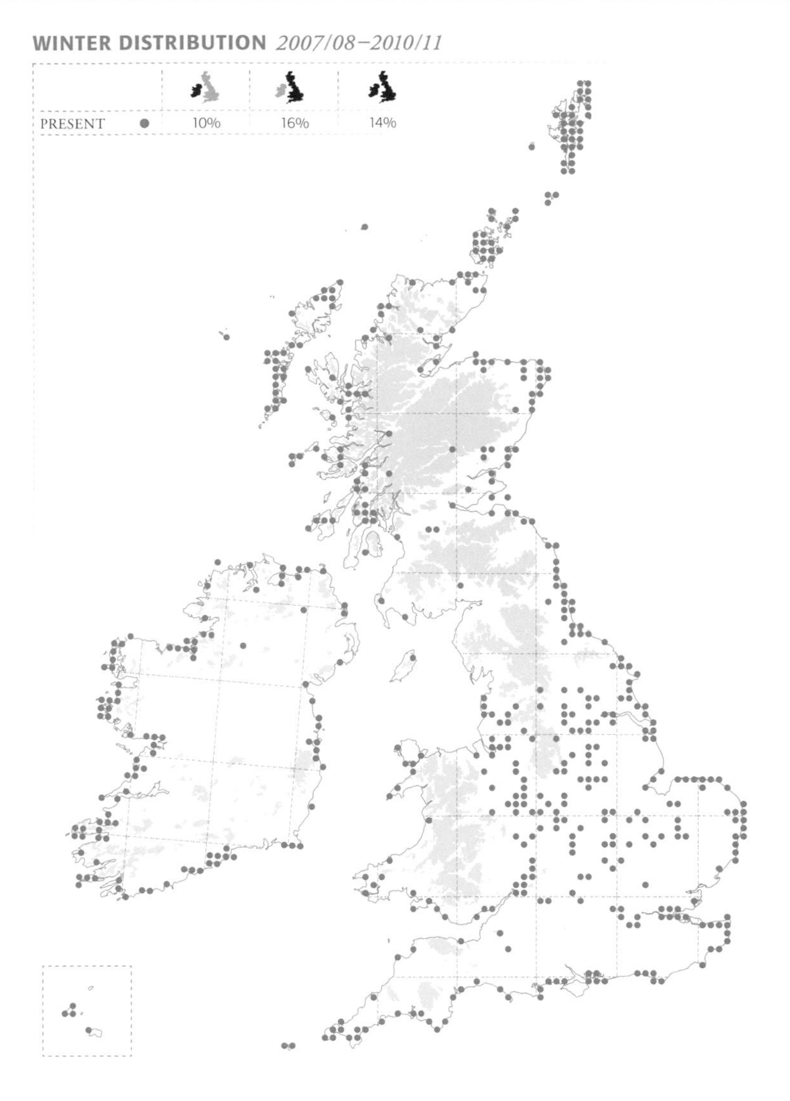

WINTER DISTRIBUTION CHANGE *since 1981–84*

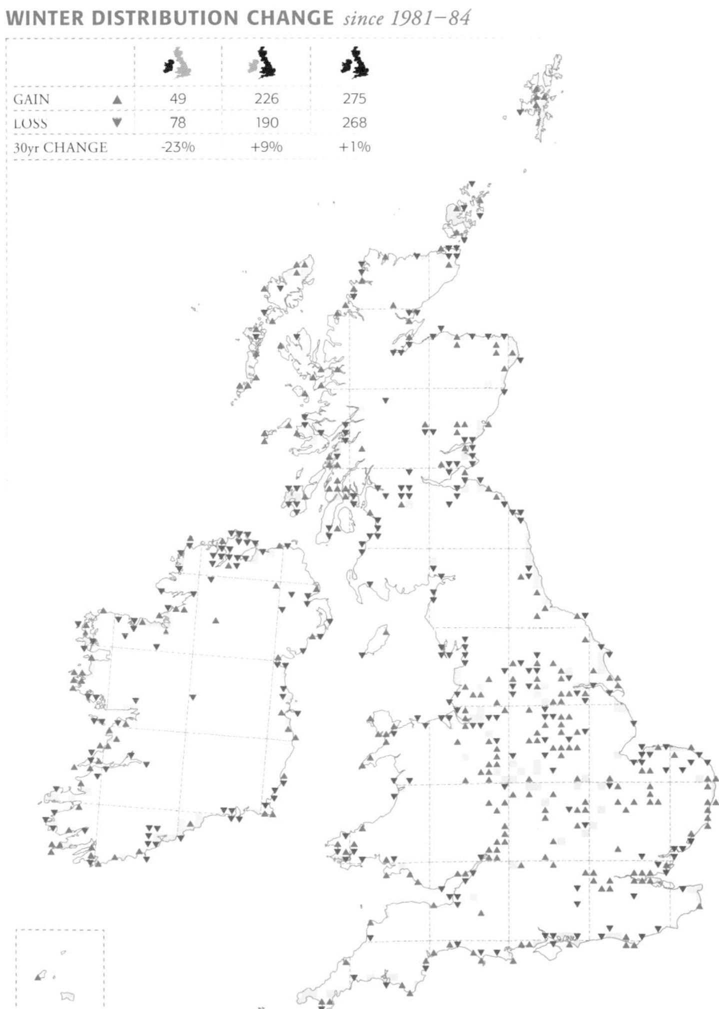

In memory of Cynthia & Jim Turner

GREAT BLACK-BACKED GULL

Larus marinus

BTO CODE: GB

TOM BICKERTON

IN WINTER THE Great Black-backed Gull is to be found all around the coast of Britain & Ireland, as well as at inland, lowland sites in England and Scotland. The winter distribution change map shows a 20% reduction in the number of occupied 10-km squares in Ireland, mainly in inland areas, since the *1981–84 Winter Atlas*, which is in agreement with declining annual peak counts (*I-WeBS Report 2009*). In Britain, losses slightly outweigh gains, giving a 5% range contraction overall, also largely from inland areas. This pattern of range change is inconsistent with decadal gull roost counts, which indicate an increasing winter population in southwest Scotland, northwest England and the English midlands contrasting with a decreasing population in east and southeast England (Banks *et al.* 2009). The inland range losses may reflect the sporadic nature of the occurrence of wandering gulls as they use ephemeral feeding resources.

During the summer, breeding birds have a predominantly coastal distribution around both Britain and Ireland. They are largely absent from most of the North Sea coast between Lothian and Kent. Records of *non-breeding* birds are widespread, particularly in coastal areas, but also inland in central England. The breeding-season relative abundance map highlights the high densities found in the Northern Isles, northwest Scotland and western Ireland.

There has been a substantial 30% decline in range size in Ireland since the *1968–72 Breeding Atlas*, though much of this loss occurred between 1968–72 and 1988–91. A recent range contraction is apparent in western Scotland but is outweighed by gains elsewhere in Britain, especially along the English south coast. Since the *1988–91 Breeding Atlas* there has been a 7% increase in the number of 10-km squares occupied in Britain & Ireland. This is despite monitoring in the UK showing a population decline of 35% during 1986–2011 (*SUKB 2012*). This decline in abundance was associated with a decline in productivity, most evident in Scotland (*Seabird 2000*), though the cause of the decline in the last decade is unknown (JNCC 2012).

BREEDING DISTRIBUTION *2008–11*

POSSIBLE •	4%	5%	4%
PROBABLE •	2%	4%	4%
CONFIRMED ●	11%	15%	14%
TOTAL	17%	24%	22%

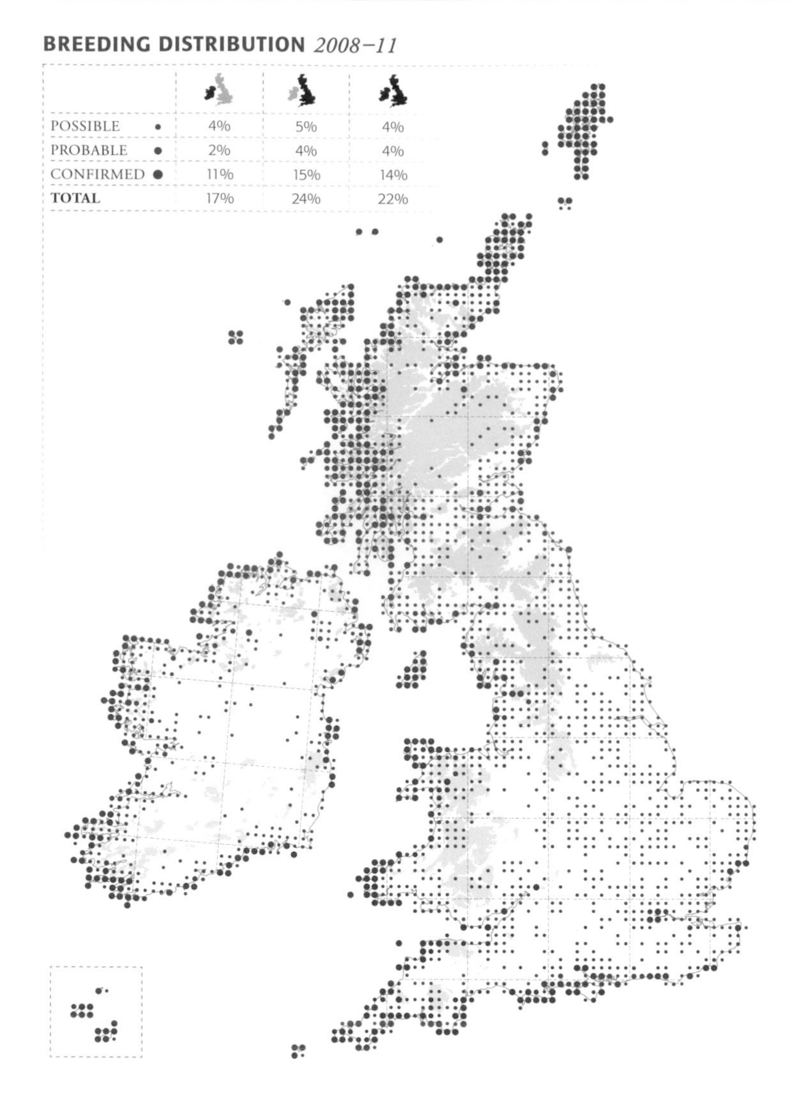

WINTER DISTRIBUTION *2007/08–2010/11*

PRESENT ●	44%	69%	62%

Sponsored by Alastair Whitelaw & Merlin Magnus Baker-Whitelaw

BREEDING DISTRIBUTION CHANGE *since 1968–72*

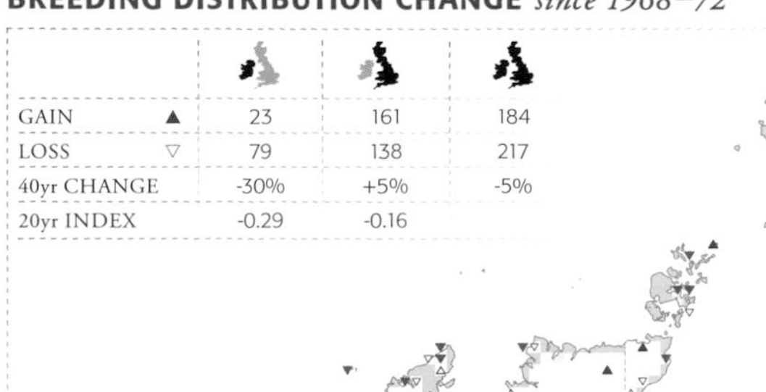

GAIN ▲	23	161	184
LOSS ▽	79	138	217
40yr CHANGE	-30%	+5%	-5%
20yr INDEX	-0.29	-0.16	

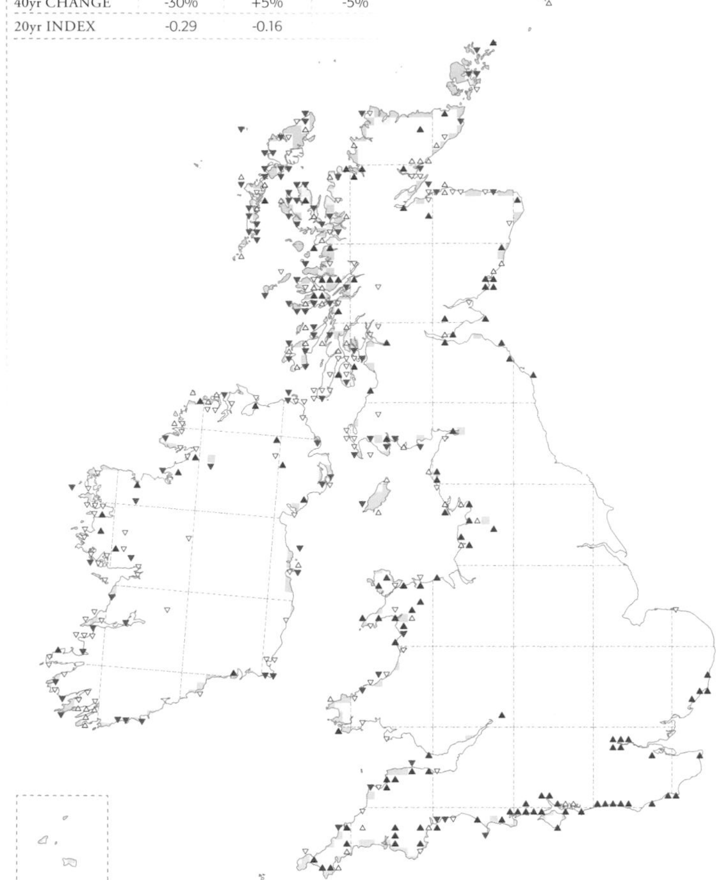

BREEDING RELATIVE ABUNDANCE *2008–11*

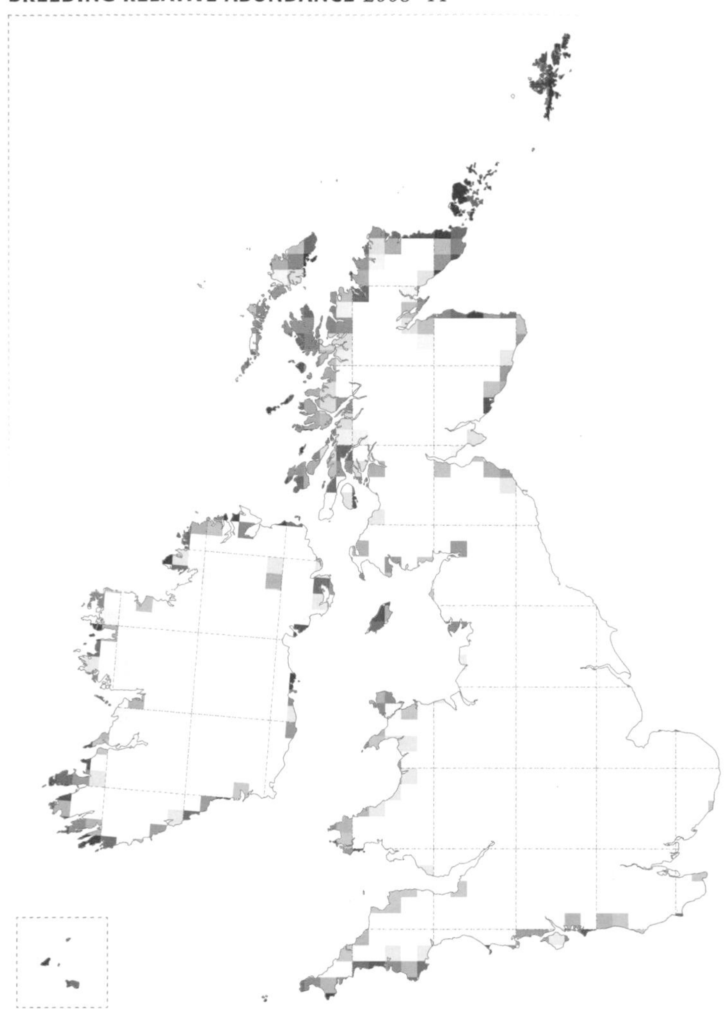

WINTER DISTRIBUTION CHANGE *since 1981–84*

GAIN ▲	46	267	313
LOSS ▼	153	359	512
30yr CHANGE	-20%	-5%	-8%

WINTER RELATIVE ABUNDANCE *2007/08–2010/11*

LITTLE TERN
Sternula albifrons

BTO CODE: AF

MARGARET HOLLAND

LITTLE TERNS NEST mostly on beaches on mainland Britain & Ireland. In Britain, c.75% of the c.2,000 pairs breed in England (JNCC 2012), the majority on three sections of coast: the Humber/ Lincolnshire, East Anglia and the Solent. Gronant is the sole extant Welsh colony and the Isle of Man supports a single colony. They have a patchy distribution around Scotland, including Orkney, but do not breed on Shetland. In western Scotland they are most common on the Atlantic coast of the southern Outer Hebrides and the Inner Hebrides. The Irish population, numbering c.200 pairs, occurs on mainland sandy and shingle beaches in the east and in the west, and especially on islands off Counties Galway, Mayo and Donegal.

While Little Terns are known to shift colonies from one year to the next in response to predation, disturbance and habitat change (JNCC 2012), the breeding range change map principally captures a picture of colony loss, particularly from the Thames Estuary and Kent, Anglesey, the Solway, the outer parts of the Firths of Forth and Tay, and throughout the Irish coast. The most significant gains since the *1988–91 Breeding Atlas* have been the cluster of new colonies in northeast England and northeast Scotland. Offshore sandbars, for example Scroby Sands, Norfolk, are occupied when suitable conditions permit.

The long-term small population decline (JNCC 2012) does not match the magnitude of the range loss, perhaps indicating that there has been a shift into fewer, larger colonies. These are managed areas often fenced and wardened solely for the benefit of Little Terns, and many are nature reserves. Wardening is a significant factor in the conservation of colonies, as mainland beaches increasingly become disturbed by human recreation and frequented by a wide range of potential ground and aerial predators, though declines are also recorded at these guarded sites. Years of poor productivity, perhaps due to predation, disturbance, flooding by high tides and poor weather, and the resultant lower recruitment into the breeding population are likely to be driving the population decline (Ratcliffe *et al.* 2000; *Seabird 2000*).

BREEDING DISTRIBUTION *2008–11*

POSSIBLE •	<1%	<1%	<1%
PROBABLE •	<1%	<1%	<1%
CONFIRMED ●	2%	3%	3%
TOTAL	2%	3%	3%

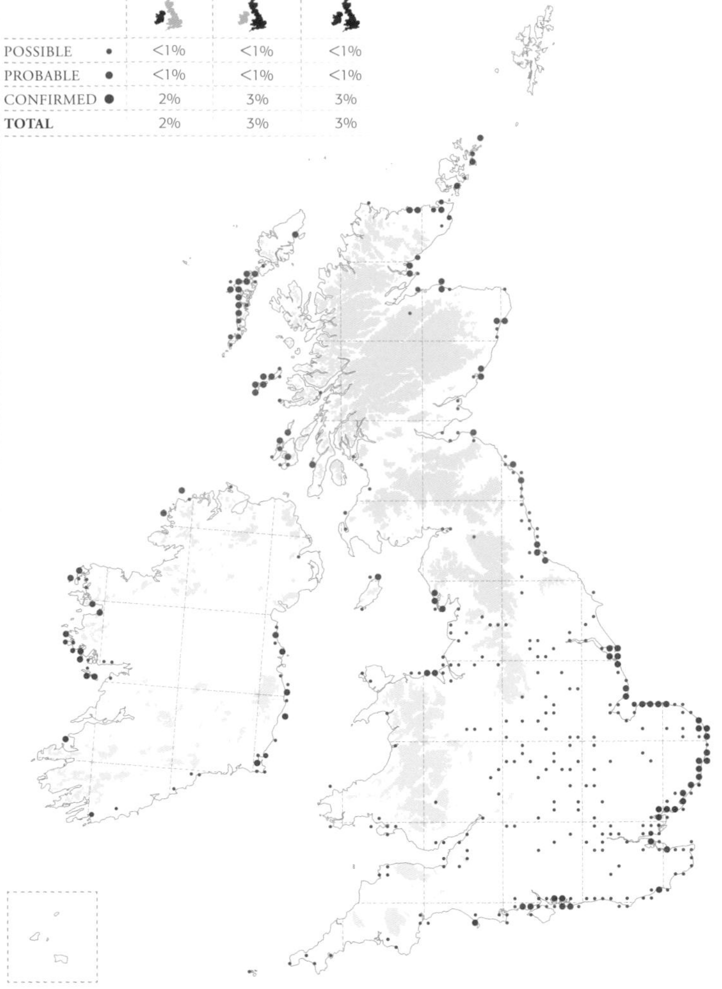

Sponsored by Louth Local Authorities

BREEDING DISTRIBUTION CHANGE *since 1968–72*

	Ireland	Britain	Britain & Ireland
GAIN ▲	7	27	34
LOSS ▽	39	68	107
40yr CHANGE	-62%	-30%	-39%
20yr INDEX	-	-0.50	

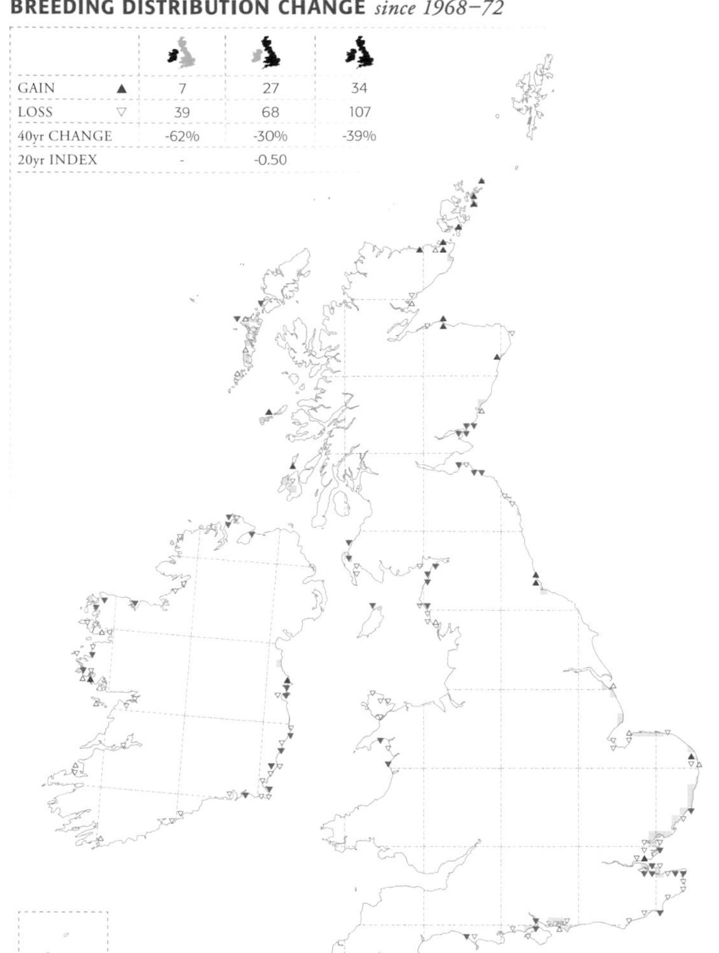

BREEDING RELATIVE ABUNDANCE *2008–11*

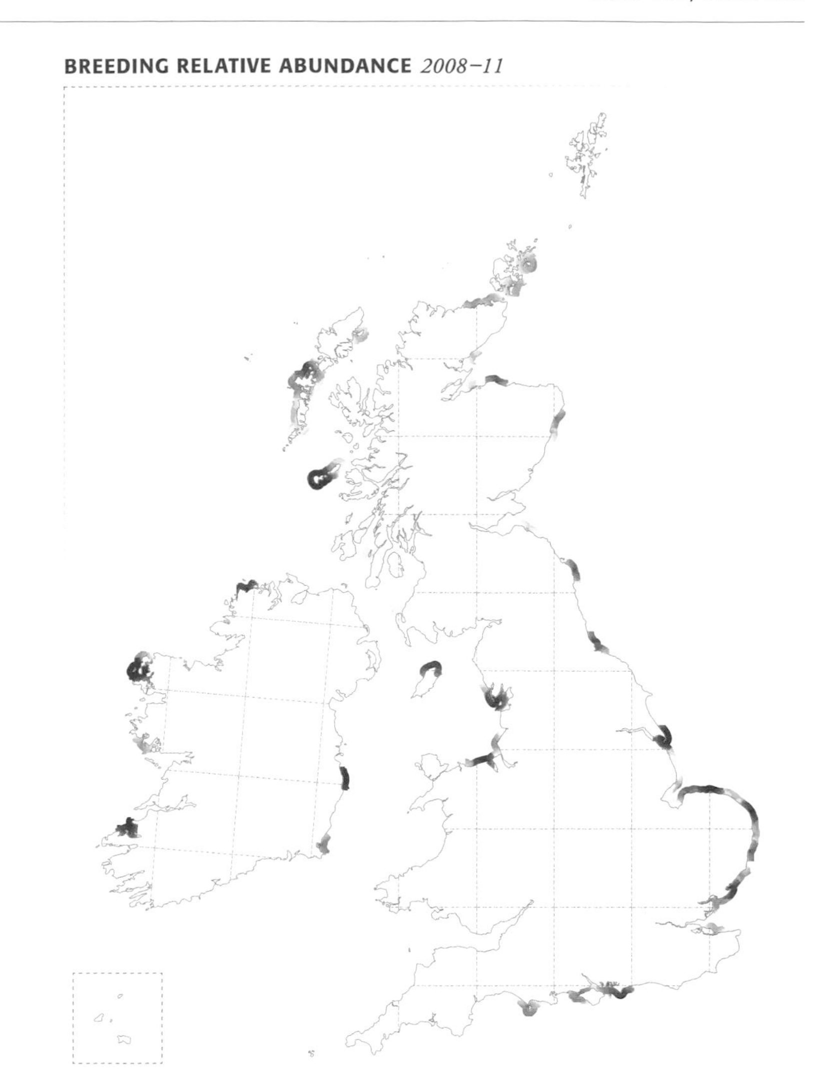

BLACK TERN

Chlidonias niger

BTO CODE: BJ

GRAHAM CATLEY http://pewit.blogspot.co.uk

BLACK TERNS ARE regular migrants in Britain, and scarce migrants in Ireland. Large spring influxes can occur if birds heading to continental breeding grounds are forced west by easterly winds. Passage birds were recorded in 397 10-km squares, mostly in southern, central and eastern Britain, with just six in Ireland. They bred regularly in England until the mid 19th century but have not bred since 1978 (*Birds in England*), though they probably did so in 1983 (Spencer *et al.* 1986). In Ireland, they last bred in 1975 (*Birds in Ireland*). Birds recorded as *seen* in eight 10-km squares in the *1988–91 Breeding Atlas* were probably also migrants but breeding was *confirmed* in two 10-km squares in the *1968–72 Breeding Atlas*.

BREEDING DISTRIBUTION *2008–11*

	Ireland	Britain	Britain & Ireland
POSSIBLE •	-	-	-
PROBABLE •	-	-	-
CONFIRMED ●	-	-	-
TOTAL	-	-	-

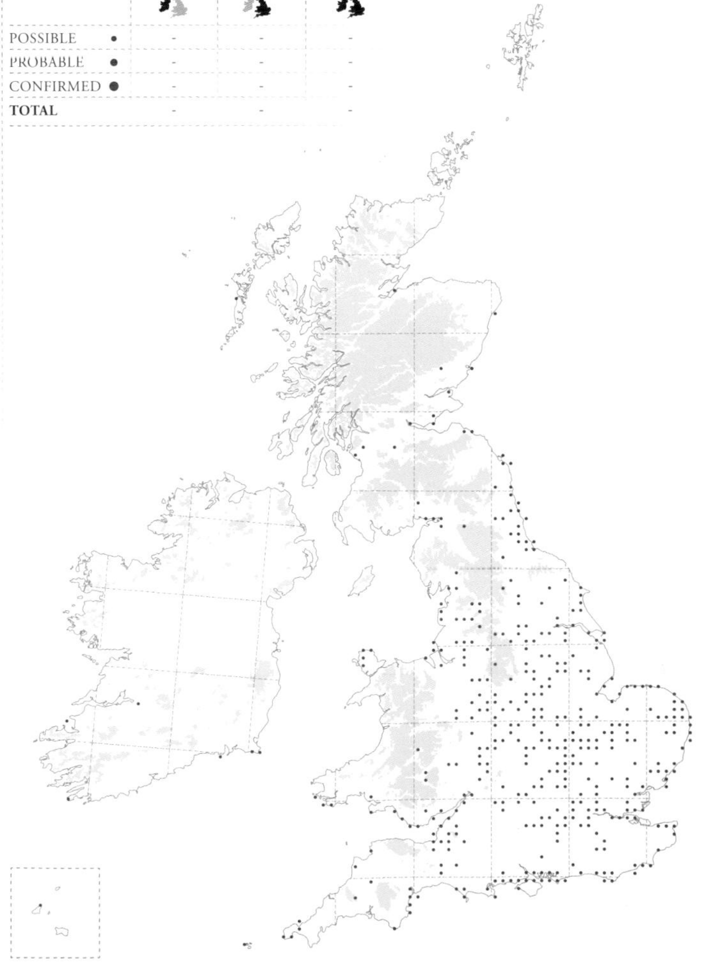

SANDWICH TERN

Sterna sandvicensis BTO CODE: TE

SANDWICH TERNS BREED in a relatively small number of large colonies scattered around the coasts of Britain and Ireland. The breeding abundance map picks out most of the key breeding areas, although the large offshore colonies in Northumberland are under-represented because the tetrads were not eligible for TTVs.

The distribution of colonies can vary from year to year as birds respond to changing conditions including restoration and creation of breeding habitats. In addition, colony sizes are renowned for wide fluctuations as the proportion of adults nesting varies annually (JNCC 2012). Consequently, the distribution change map is marked by a mix of gains and losses, although overall a 23% range contraction is apparent since the *1968–72 Breeding Atlas*. Colonies have been lost throughout the range but particularly in eastern Scotland and along the coast of western Ireland. Losses may be attributable to mammalian predators such as the Red Fox, American Mink and Brown Rat, changes in food availability and competition for nesting areas with large gulls (Ratcliffe *et al.* 2000; *Seabird 2000*; *Birds of Scotland*).

Non-breeding birds, including migrants and commuting foraging adults, were also reported widely around British and Irish coasts during the breeding season, with the main gaps being Shetland and much of the Hebrides. There was a scattering of inland occurrences at gravel pits and reservoirs in England south of a line from the Humber to the Ribble. In Ireland, the only inland records reported were from near the only inland colony, at Lough Erne in the northwest.

Alone among Britain & Ireland's highly migratory terns, the Sandwich Tern is a regular wintering species, albeit it in very small numbers. Sandwich Terns were recorded in November on the North Sea coast southwards from the Firth of Forth and were present for a greater part of the winter along the English south coast and around the Channel Islands. They were virtually absent from Scotland. In Ireland records were clustered around Strangford Lough in the northeast and inner Galway Bay in the west. The winter change map and associated statistics show the expected one-sided story of range expansion.

Sponsored by Kent Ornithological Society

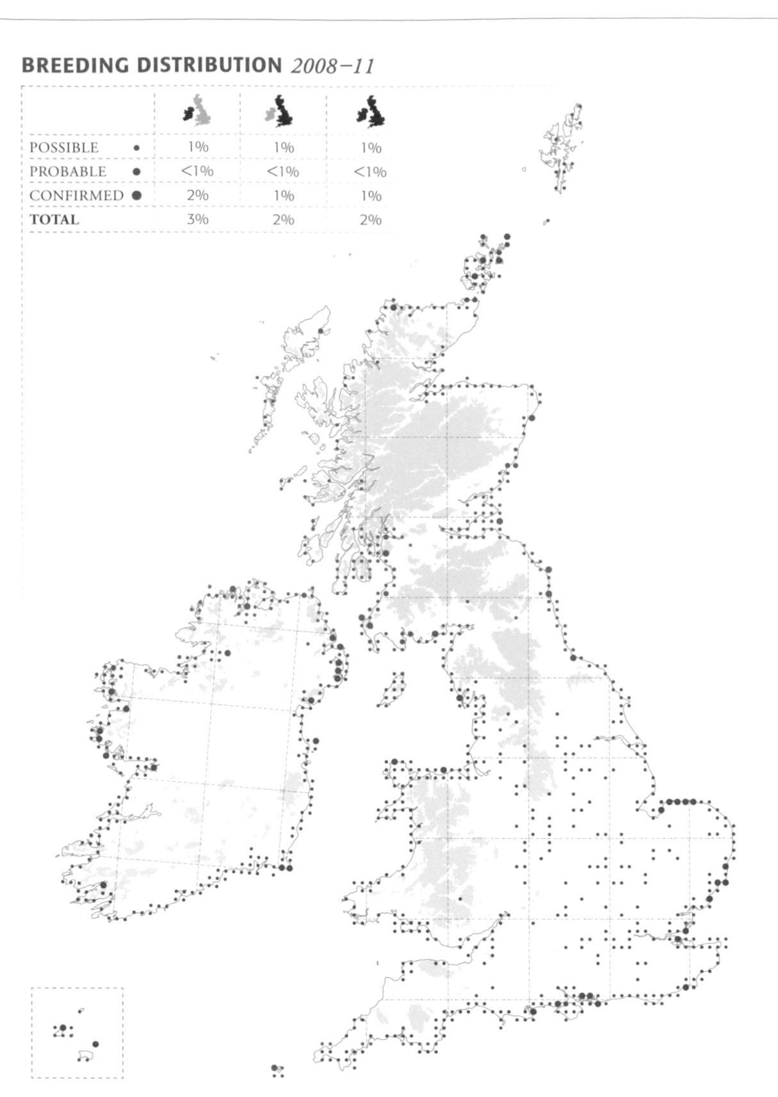

BREEDING DISTRIBUTION *2008–11*

POSSIBLE •	1%	1%	1%
PROBABLE •	<1%	<1%	<1%
CONFIRMED ●	2%	1%	1%
TOTAL	3%	2%	2%

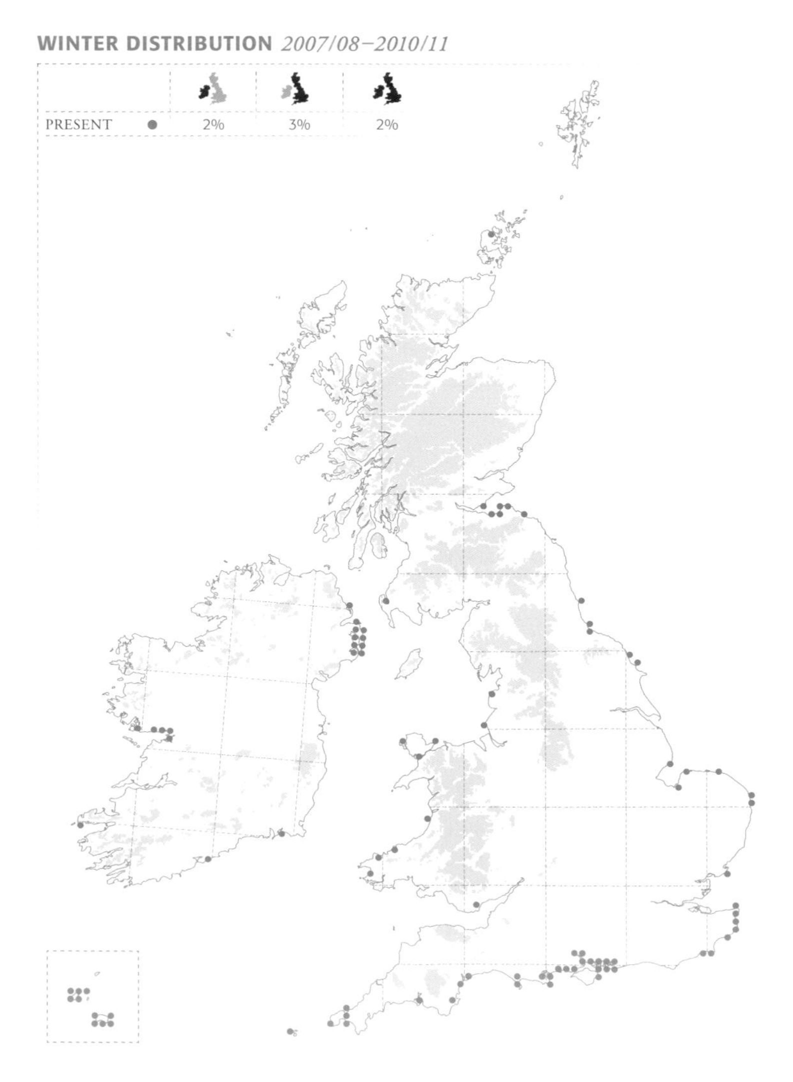

WINTER DISTRIBUTION *2007/08–2010/11*

PRESENT ●	2%	3%	2%

BREEDING DISTRIBUTION CHANGE *since 1968–72*

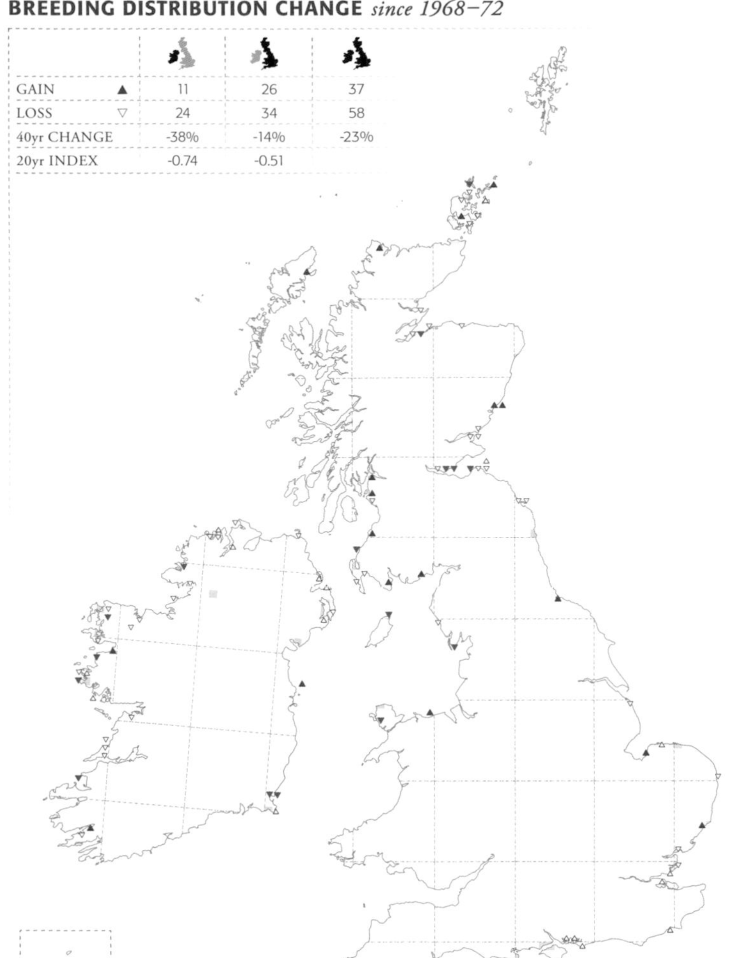

GAIN ▲	11	26	37
LOSS ▽	24	34	58
40yr CHANGE	-38%	-14%	-23%
20yr INDEX	-0.74	-0.51	

BREEDING RELATIVE ABUNDANCE *2008–11*

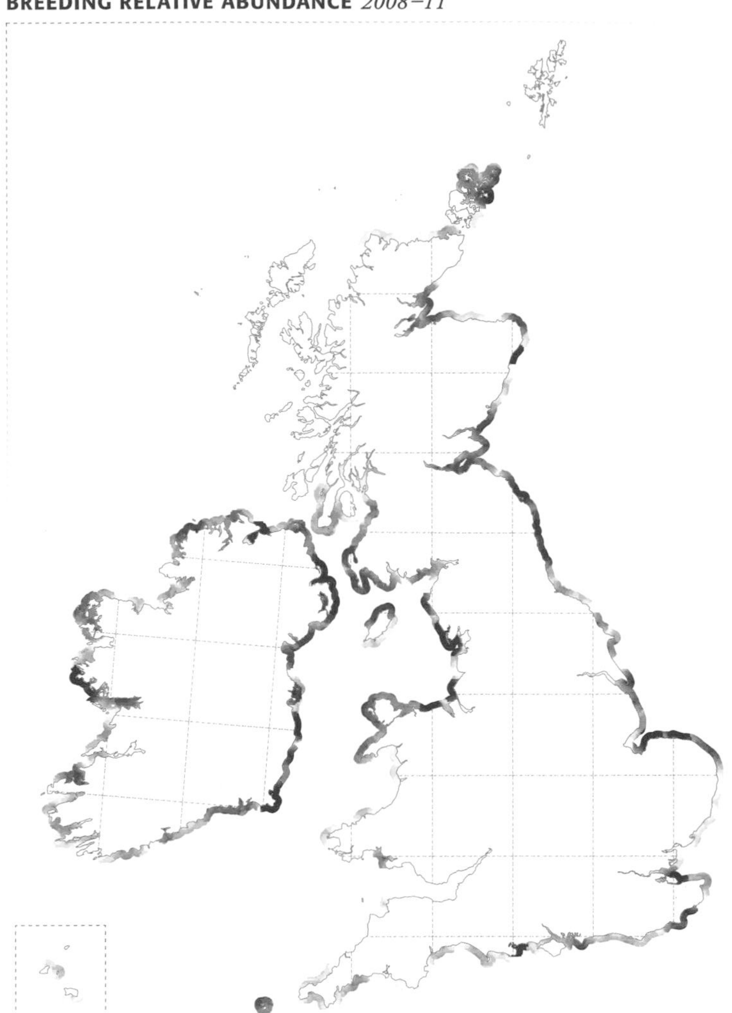

WINTER DISTRIBUTION CHANGE *since 1981–84*

GAIN ▲	15	46	61
LOSS ▼	1	16	17
30yr CHANGE	+1400%	+120%	+169%

COMMON TERN

Sterna hirundo

BTO CODE: CN

GRAHAM CATLEY http://pewit.blogspot.co.uk

BREEDING DISTRIBUTION *2008–11*

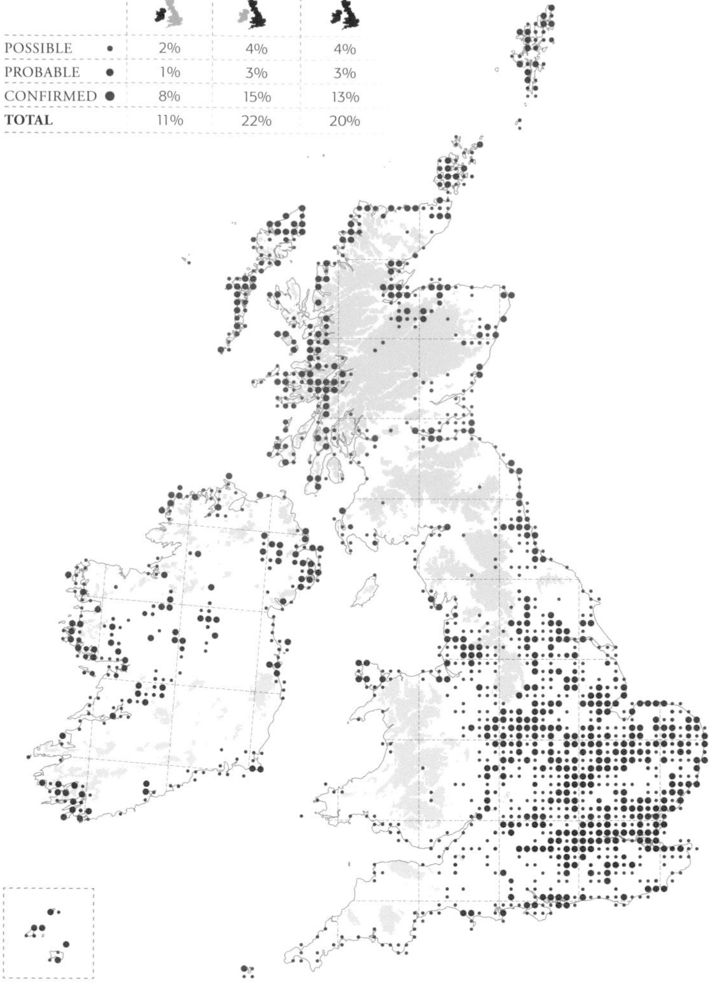

POSSIBLE •	2%	4%	4%
PROBABLE ●	1%	3%	3%
CONFIRMED ●	8%	15%	13%
TOTAL	11%	22%	20%

THE COMMON TERN is a summer visitor to Britain and Ireland. Breeding Common Terns are widespread and primarily coastal in Scotland and are found on the lochs and islands of the west coast, Outer Hebrides, Northern Isles and around the inner Moray Firth. This contrasts with England, where the species' range is dominated by inland colonies. In Ireland, colonies are more clustered, both on the coast and inland where most nest on islands in the largest lakes. These stretch from Lough Neagh in the northeast to Lough Derg in the southwest, and include lakes in the northwest, especially in Counties Mayo and Galway.

The breeding distribution change map shows a clear pattern of losses in Scotland and Ireland contrasting with gains in eastern and central England since both previous breeding atlases. These contrasting trends in distribution are also apparent in annual monitoring of the size of the breeding population (*Seabird 2000*; JNCC 2012). Overall, the breeding range has virtually halved in Ireland since the *1968–72 Breeding Atlas*, whilst in Britain a 13% expansion is apparent. The overriding pattern of gain in inland England is likely to have resulted from the creation of man-made waterbodies, with considerable effort focused on the provision of nesting platforms. This range expansion is reflected in a 59% increase in the number of Common Terns detected on English BBS squares during 1995–2010 (*BBS Report 2011*). Nevertheless, inland colonies tend to be small and to date these gains are outweighed in population terms by losses elsewhere (JNCC 2012). The losses in Scotland have been attributed to increases in predation (Craik 1997) and this factor may be important also in inland and western Ireland. Food shortages reducing breeding productivity have also been implicated in the losses in Scotland (JNCC 2012); there is no evidence of food shortage in Ireland, though the ecology of lakes and their fish populations could be changing in response to other factors including invasive aquatic plants and Zebra Mussels.

True winter records of Common Terns were few and 11 of the 17 10-km squares occupied in winter held birds in the first half of November only.

Sponsored by Alton Water Volunteers

BREEDING DISTRIBUTION CHANGE *since 1968–72*

GAIN ▲	29	310	339
LOSS ▽	110	251	361
40yr CHANGE	-47%	+13%	-3%
20yr INDEX	-0.27	-0.10	

BREEDING RELATIVE ABUNDANCE *2008–11*

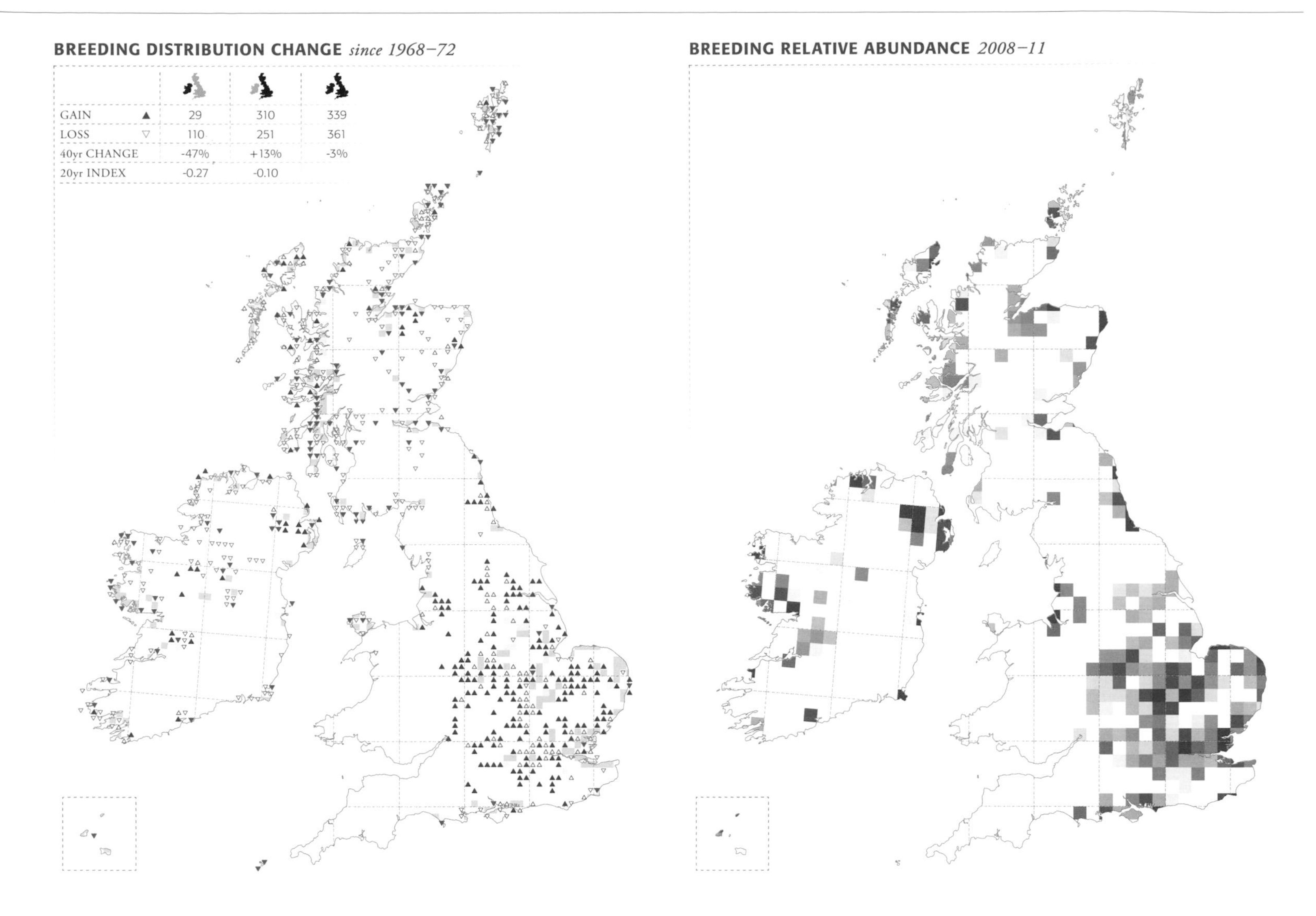

COLIN BROWN

◀ The provision of tern rafts at inland lakes and gravel pits has contributed to the range expansion of the Common Tern.

ARCTIC TERN

Sterna paradisaea

BTO CODE: AE

JEREMY MOORE

BREEDING DISTRIBUTION *2008–11*

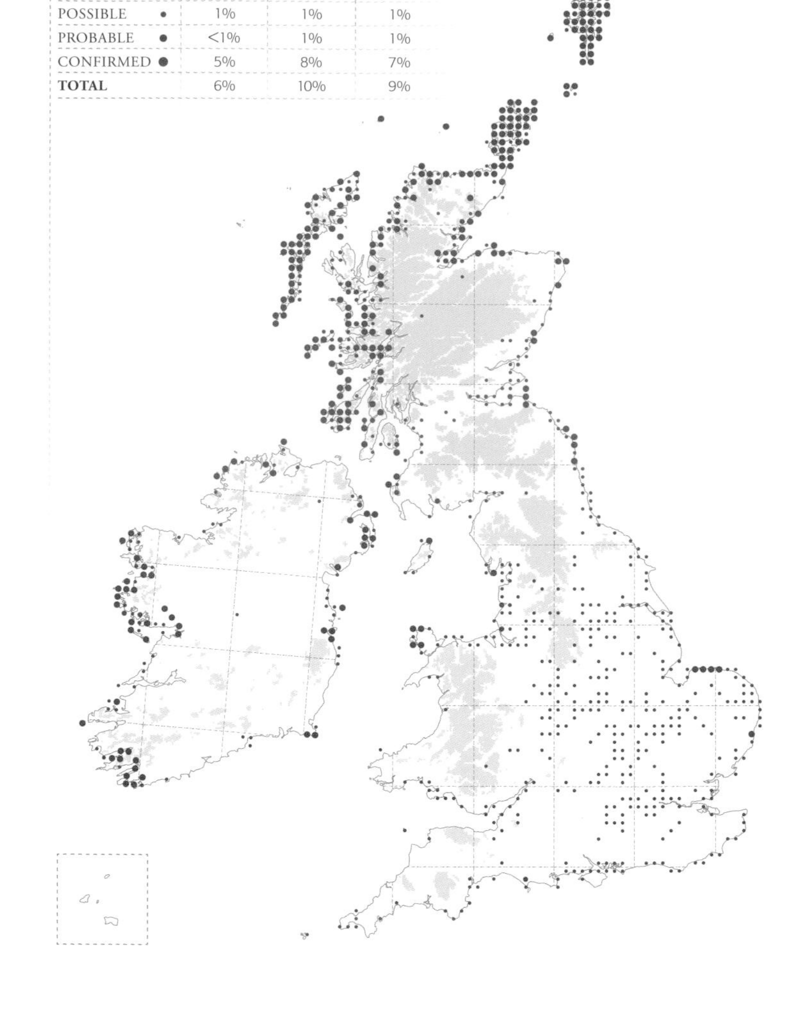

POSSIBLE •	1%	1%	1%
PROBABLE •	<1%	1%	1%
CONFIRMED •	5%	8%	7%
TOTAL	6%	10%	9%

ARCTIC TERNS ARE at the southern limit of their breeding distribution in Britain & Ireland. They breed predominantly in coastal areas of Scotland and Ireland, with major colonies found in Shetland and Orkney. In Wales they are restricted to Anglesey and its offshore islands and there is a small colony on the Isle of Man. Breeding colonies in Ireland are located in five main clusters. Very few were recorded breeding inland, with exceptions in northern Scotland and the Galway area of western Ireland. *Non-breeding* birds were recorded widely in central, southern and eastern England, where there is a pronounced spring overland migration in some years, with individuals and small flocks stopping off at gravel pits, lakes and reservoirs (Kramer 1995).

The breeding range change map shows an overall range contraction of 31% since the *1968–72 Breeding Atlas*. Losses are greatest in western Scotland, where predation by American Minks has been implicated (Craik 1997; Clode & MacDonald 2002), and in eastern Scotland, where food shortages have resulted in very low breeding productivity (JNCC 2012). There was an extensive programme of eradication of the American Mink on the southern islands of the Outer Hebrides during 2001–06 (Roy 2011) and attempts on Lewis and Harris are currently under way (Hebridean Mink Project 2011). The continuity of breeding range in the Northern Isles is perhaps surprising, given the fluctuations in the sandeel population (Monaghan 1992), though there is a high rate of intercolony movement in response to food availability. In Ireland, losses were more noticeable in the northwest. Arctic Tern numbers are difficult to monitor due to movements among colonies, though a 29% decline was recorded for Britain & Ireland between 1985–88 and 1998–2002 (*Seabird 2000*) and a 15% decline during 2000–11 (JNCC 2012), with years of poor productivity and poor recruitment noted as reasons for the decline.

Most winter observations were coastal, particularly along the English North Sea, and most were from early November, indicating late departure of a few birds. Away from the North Sea, records were very scattered, but included inland occurrences in three English 10-km squares.

Sponsored by the Royal Naval Birdwatching Society

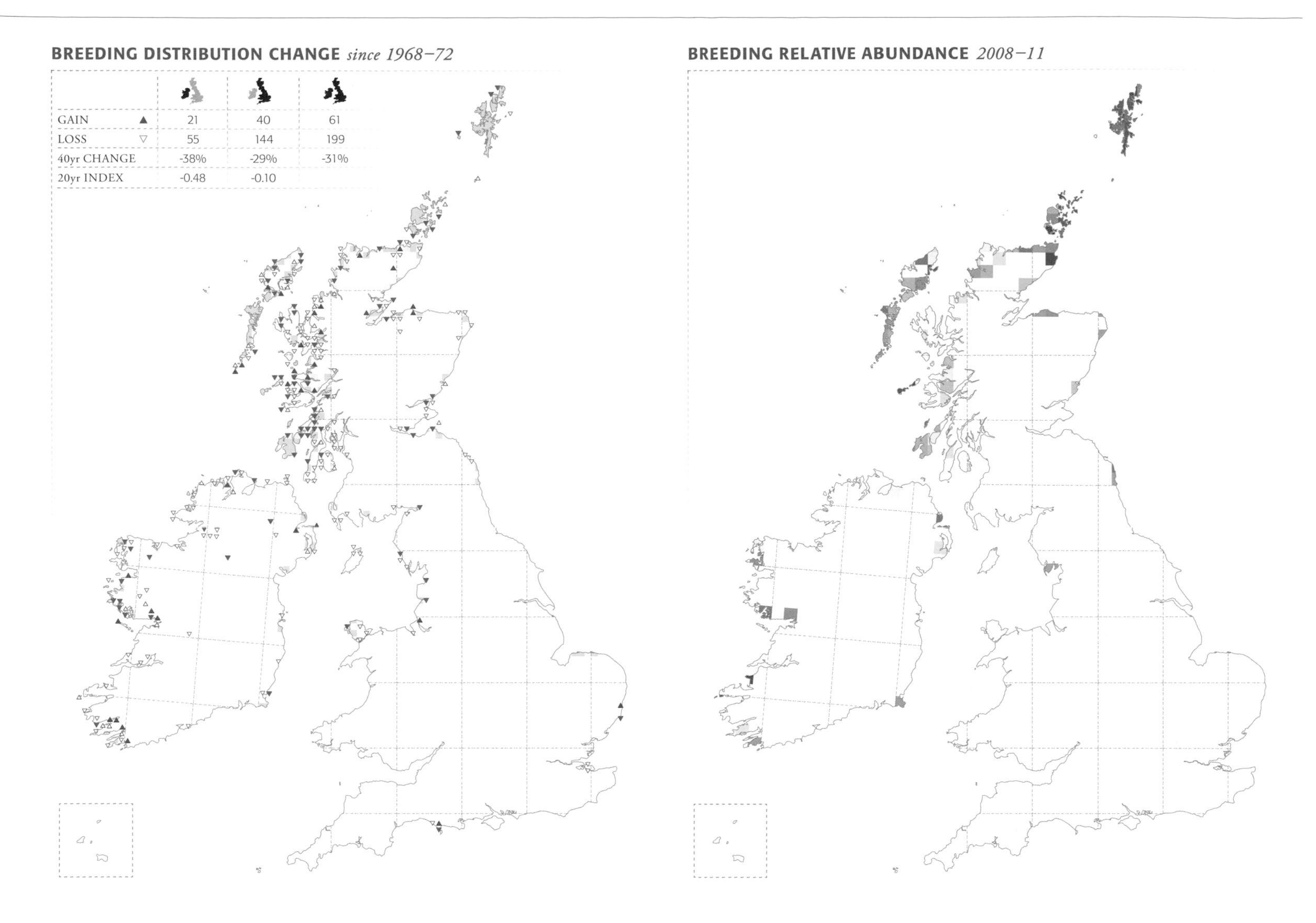

BREEDING DISTRIBUTION CHANGE *since 1968–72*

GAIN ▲	21	40	61
LOSS ▽	55	144	199
40yr CHANGE	-38%	-29%	-31%
20yr INDEX	-0.48	-0.10	

BREEDING RELATIVE ABUNDANCE *2008–11*

COLIN BROWN

▲ Arctic Terns are almost exclusively coastal breeders in Britain & Ireland, nesting on islands and remote beaches.

ROSEATE TERN
Sterna dougallii

BTO CODE: RS

PAUL MORRISON

AFTER WINTERING OFF the West African coast, Roseate Terns return to the shores of Britain & Ireland mainly during May, and at this time some individuals can be seen outside their normal breeding range, around the Irish Sea, along the south and east coasts of Britain, and very rarely inland.

In Britain & Ireland the Roseate Tern has a high conservation profile because its small breeding population is highly localised and has undergone a decline of over 50% over 25 years. Hunting in its wintering grounds is thought to be the key driver of the long-term decline (Cabot 1996), although at a local scale habitat loss, predation, disturbance and competition for nest sites with gulls are cited as causes of colony contraction and loss (Avery *et al.* 1995; JNCC 2012). During 2008–11, individuals were present in 135 10-km squares in Britain & Ireland, although records with breeding evidence were restricted to just 18 squares. Ninety-seven percent of the British & Irish population breeds at just three colonies—Coquet Island in northeast England and Rockabill and Lady's Island Lake in the east of Ireland (*Seabird 2000*). Numbers occurring at the other colonies are very small, despite focused tern-management programmes having been run at most of them for many years. Some outlying 10-km squares with breeding evidence relate to single Roseate Terns paired with Common Terns (e.g. in Lancashire and on Anglesey; *RBBP 2010*).

In Britain, colony loss has been greatest in the Firth of Forth, where there has been competition with Herring Gulls, low productivity, intermittent breeding and emigration to other areas. In Ireland, losses have been greatest at Strangford Lough and along the west coast. The 33% range contraction in Britain & Ireland between 1968–72 and 1988–91, followed by a further 50% contraction between 1988–91 and 2008–11, has led to the concentration of the species into a small number of key colonies where the management has been most intensive. These key colonies have shown an ongoing increase in breeding numbers, and are thereby contributing to the recovery of this species.

Sponsored by Wexford Naturalists' Field Club

BREEDING DISTRIBUTION *2008–11*

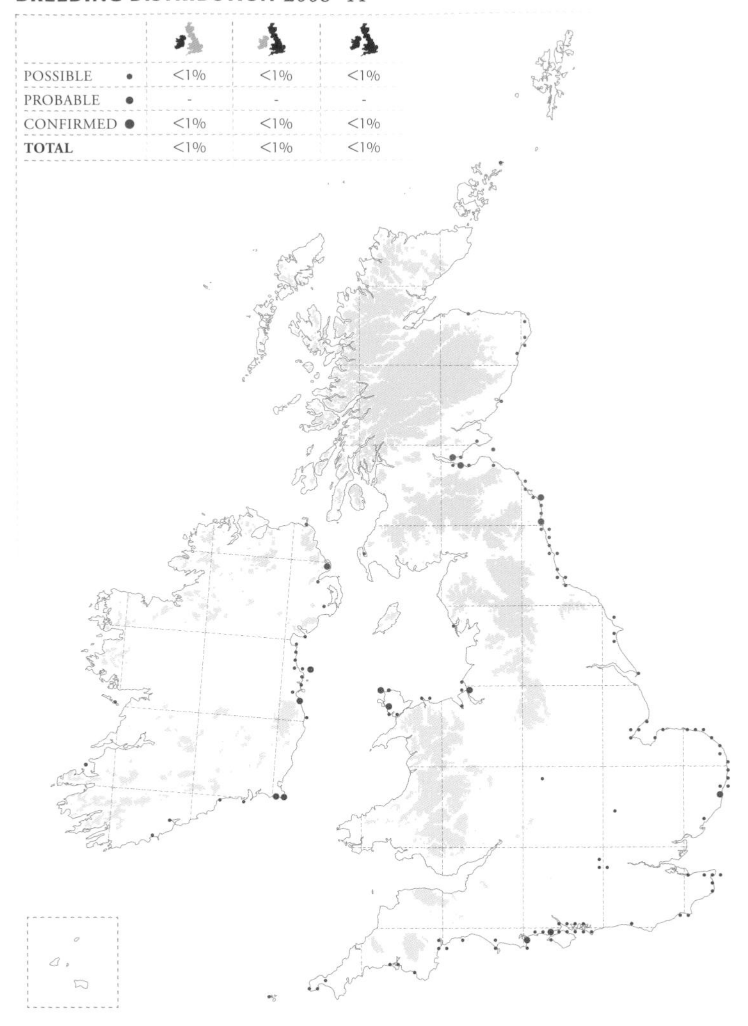

POSSIBLE •	<1%	<1%	<1%
PROBABLE ●	-	-	-
CONFIRMED ●	<1%	<1%	<1%
TOTAL	<1%	<1%	<1%

BREEDING DISTRIBUTION CHANGE *since 1968–72*

GAIN ▲	1	5	6
LOSS ▽	18	18	36
40yr CHANGE	-77%	-57%	-67%
20yr INDEX	-	-	

LITTLE AUK
Alle alle

BTO CODE: LK

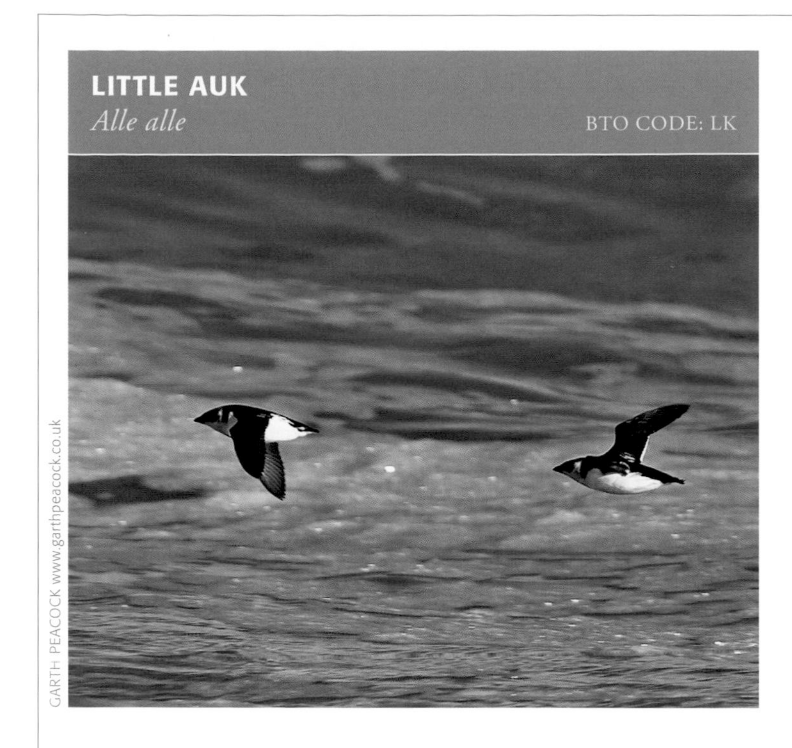

GARTH PEACOCK www.garthpeacock.co.uk

THE LITTLE AUK is a scarce but regular autumn passage migrant and winter visitor from the high Arctic, often driven south after strong northerly gales (Pollock *et al.* 1996). Most records were from North Sea coastal squares from Shetland south to Kent, though they also occurred along the south coast of England. Little Auks are relatively scarce in west-coast and Irish waters and the few inland records were restricted to sites in eastern England. There is a heavy preponderance of early winter records with 66% of the 332 10-km squares reporting birds during November only.

The winter change map excludes squares occupied in early November only, to match the recording period of the *1981–84 Winter Atlas*. This reveals a 3% range contraction in Britain & Ireland, though there was a high degree of turnover of occupied squares. Gains were widespread down the North Sea coast and along the south coast of England, whilst losses were mainly in Shetland, the North Sea coast, west-coast waters and inland sites in eastern England. In Ireland, a 41% range contraction is evident, although in both winter atlas periods the number of occupied 10-km squares was small.

The pattern of gains and losses is likely to reflect whether or not storms occurred in each atlas period during the main Little Auk passage period, rather than any population-level changes in abundance. The distribution of Little Auks documented during the *1981–84 Winter Atlas* was ascribed to a severe auk 'wreck' in February 1983, most likely a result of a fast-moving deep depression that passed from southern Greenland into the North Sea, followed by strong to gale-force northeasterly winds (Underwood & Stowe 1984); post-mortem results revealed that birds were short of food (Hope-Jones *et al.* 1984). It is notable, therefore, at least in Britain, that a comparable distribution was recorded during 2007–11 in the absence of severe wrecks.

There were records of *non-breeding* birds in just three 10-km squares in the breeding season, all April sightings, from the Outer Hebrides, Isle of Man and Borders, illustrating the scarcity of this species in spring.

WINTER DISTRIBUTION *2007/08–2010/11*

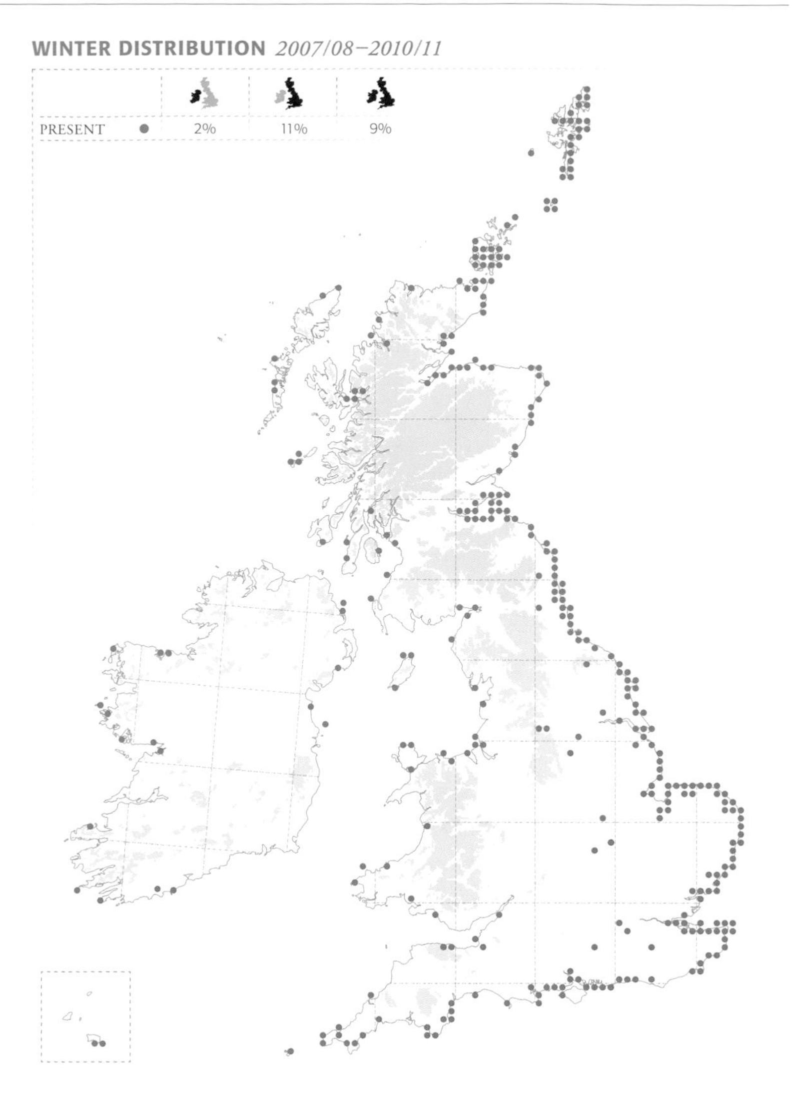

WINTER DISTRIBUTION CHANGE *since 1981–84*

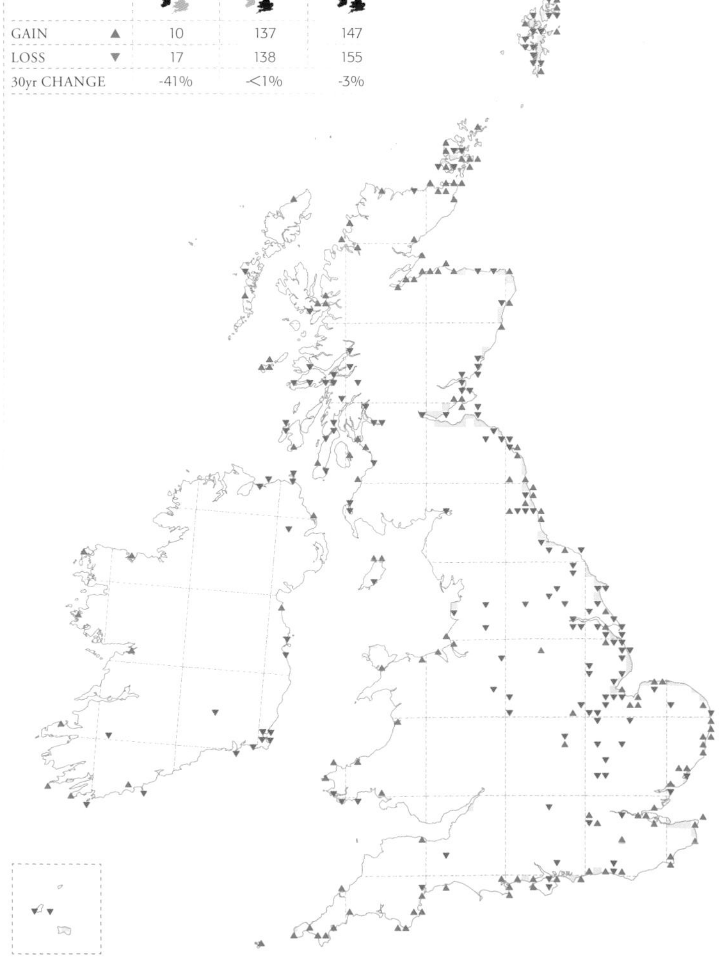

Sponsored by Dave and Moira Brooks

GUILLEMOT

Uria aalge

BTO CODE: GU

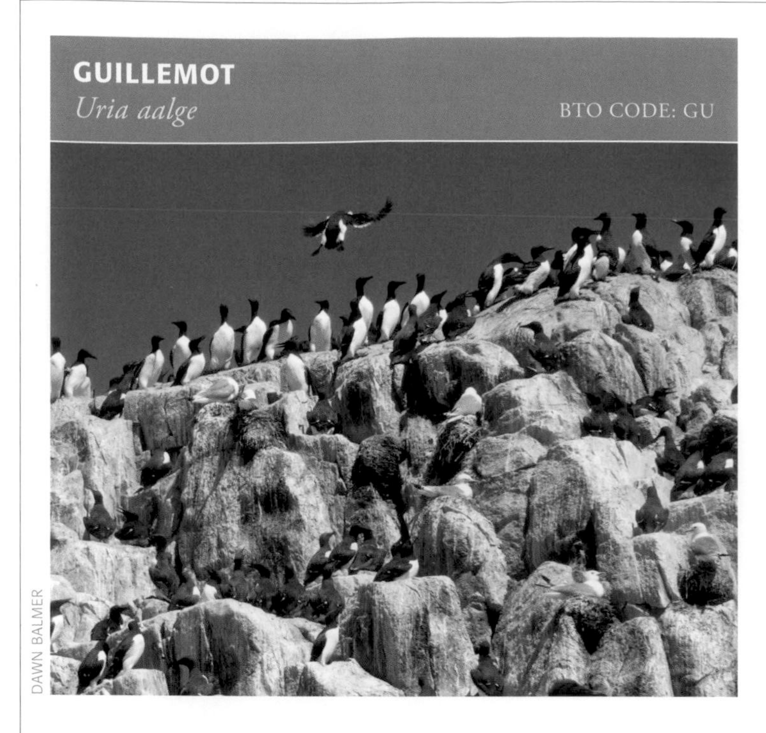

DAWN BALMER

DURING WINTER, GUILLEMOTS are found around much of the coast of Britain & Ireland with the highest densities in west-coast waters and the North Sea coast of southeast Scotland and northeast England. Indeed, by October (North Sea) and November (west coast) breeding birds occasionally appear at their breeding colonies. Since the *1981–84 Winter Atlas* there has been a 21% range contraction in Britain, yet a 7% expansion in Ireland. The losses in Britain are most evident in many parts of northern and western Scotland and southern North Sea coasts around the Wash. Gains are most notable in the west around the Irish Sea, south Wales and southwest England. These changes affect a small part of the largely oceanic winter range of the species.

Guillemot colonies are associated with steep cliffs and rocky offshore stacks and islands, and the paucity of these habitats between Flamborough Head, Yorkshire, and Freshwater, Isle of Wight, is mirrored in the breeding distribution map. Colony size varies hugely in Britain & Ireland and this variation is not shown clearly on the relative abundance map.

Auk range size and population size at national and colony level show complex and sometimes contradictory trends which cannot be explored fully here. Atlas data indicate an overall 11% reduction in the number of 10-km squares with *probable* or *confirmed* breeding since the *1988–91 Breeding Atlas* and an overall 19% reduction since the *1968–72 Breeding Atlas*, with losses in Orkney, western Scotland and western Ireland. Regular monitoring at four of Ireland's largest colonies over the last decade has shown a population increase of 22%.

Seabird colony monitoring in the UK suggests that Guillemots were 35% more abundant in 2011 than in the mid 1980s (*SUKB 2012*) and that numbers increased from 611,000 individuals in 1969/70 to over one million in 1985–88, with apparent stabilisation since 2001 (JNCC 2012), though a decline in numbers in Scotland is evident (Wanless & Harris 2012). In the UK, Guillemot breeding productivity has declined in recent years, possibly linked to reduced food availability and reduced energy content of fish brought to some colonies (Wanless *et al.* 2005b; JNCC 2012).

Sponsored by The Crown Estate (Marine)

BREEDING DISTRIBUTION *2008–11*

POSSIBLE •	1%	1%	1%
PROBABLE •	<1%	<1%	<1%
CONFIRMED ●	5%	6%	6%
TOTAL	6%	7%	7%

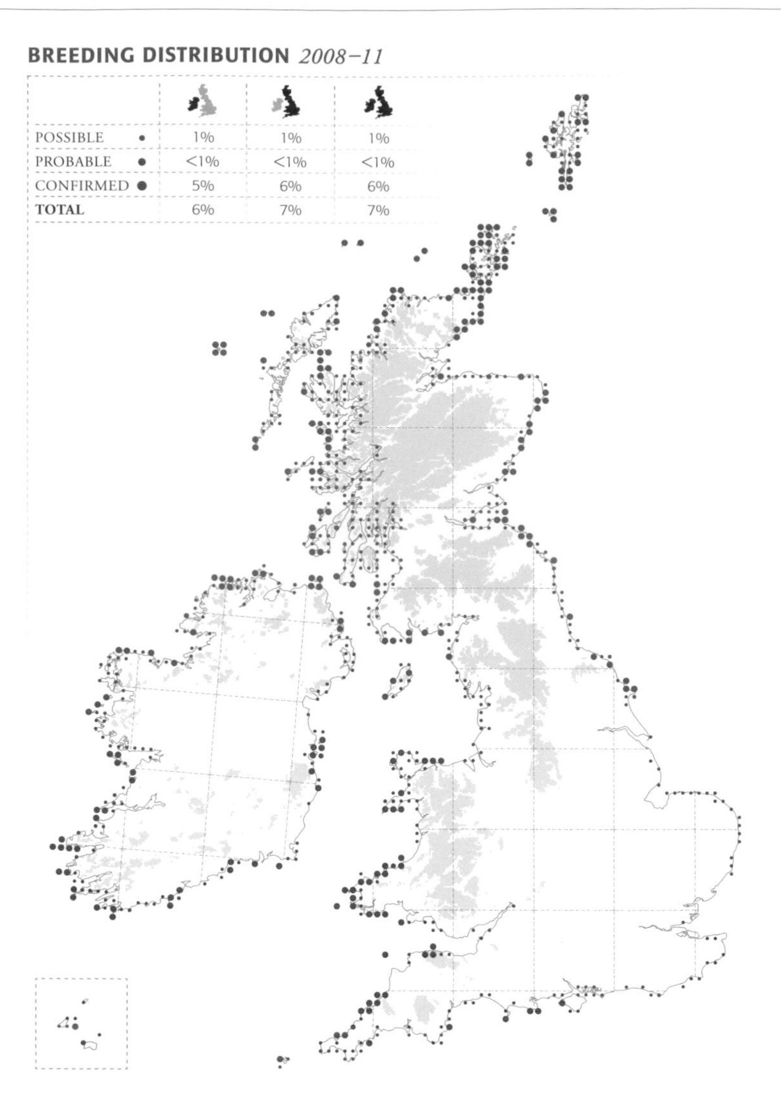

WINTER DISTRIBUTION *2007/08–2010/11*

PRESENT ●	17%	18%	18%

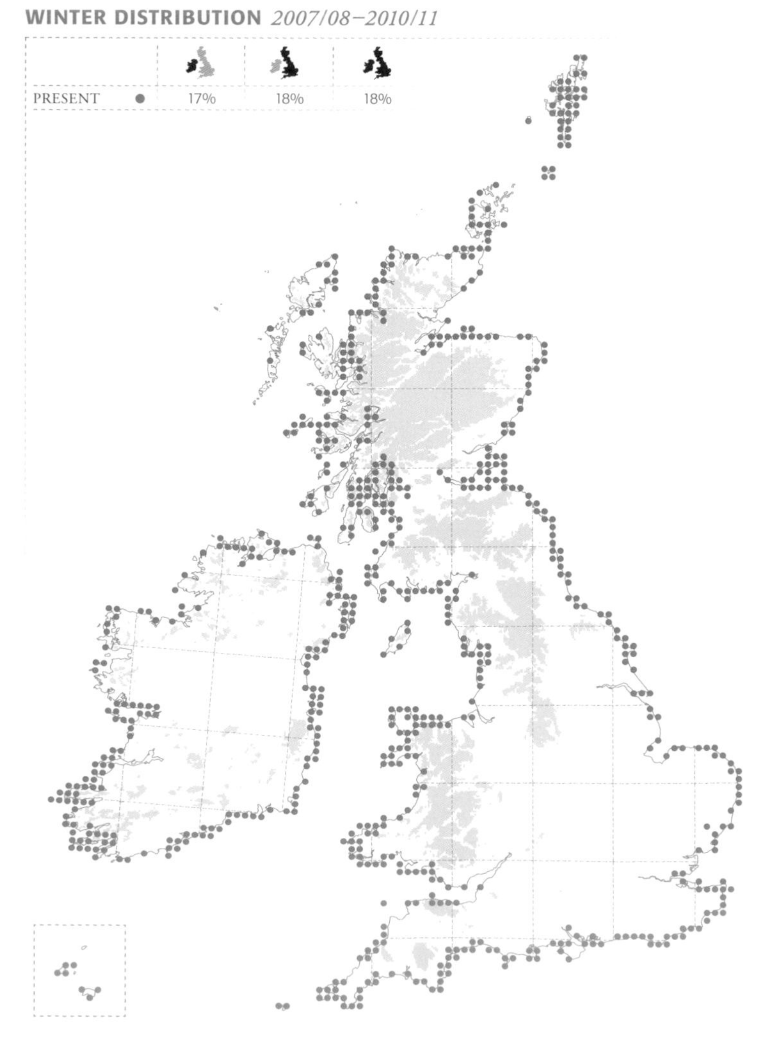

BREEDING DISTRIBUTION CHANGE *since 1968–72*

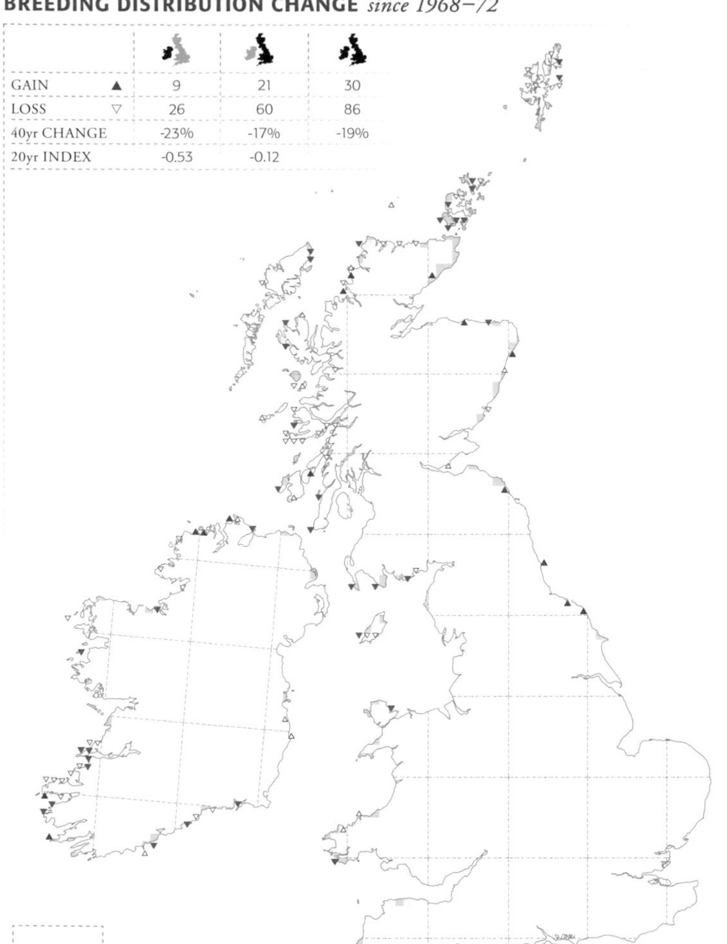

GAIN ▲	9	21	30
LOSS ▽	26	60	86
40yr CHANGE	-23%	-17%	-19%
20yr INDEX	-0.53	-0.12	

BREEDING RELATIVE ABUNDANCE *2008–11*

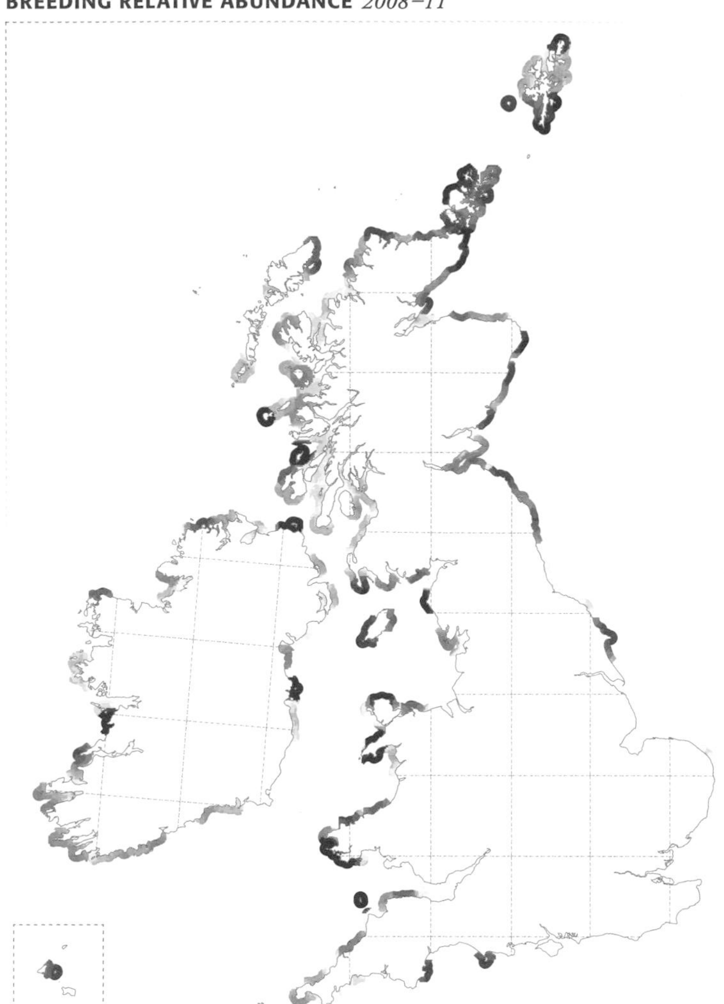

WINTER DISTRIBUTION CHANGE *since 1981–84*

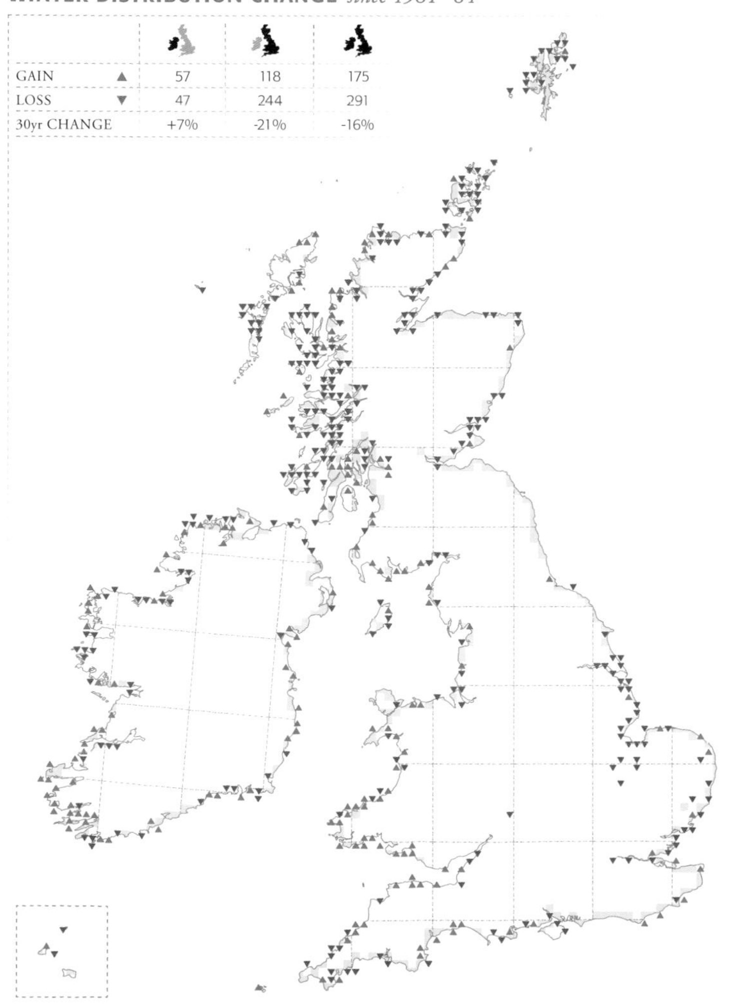

GAIN ▲	57	118	175
LOSS ▼	47	244	291
30yr CHANGE	+7%	-21%	-16%

WINTER RELATIVE ABUNDANCE *2007/08–2010/11*

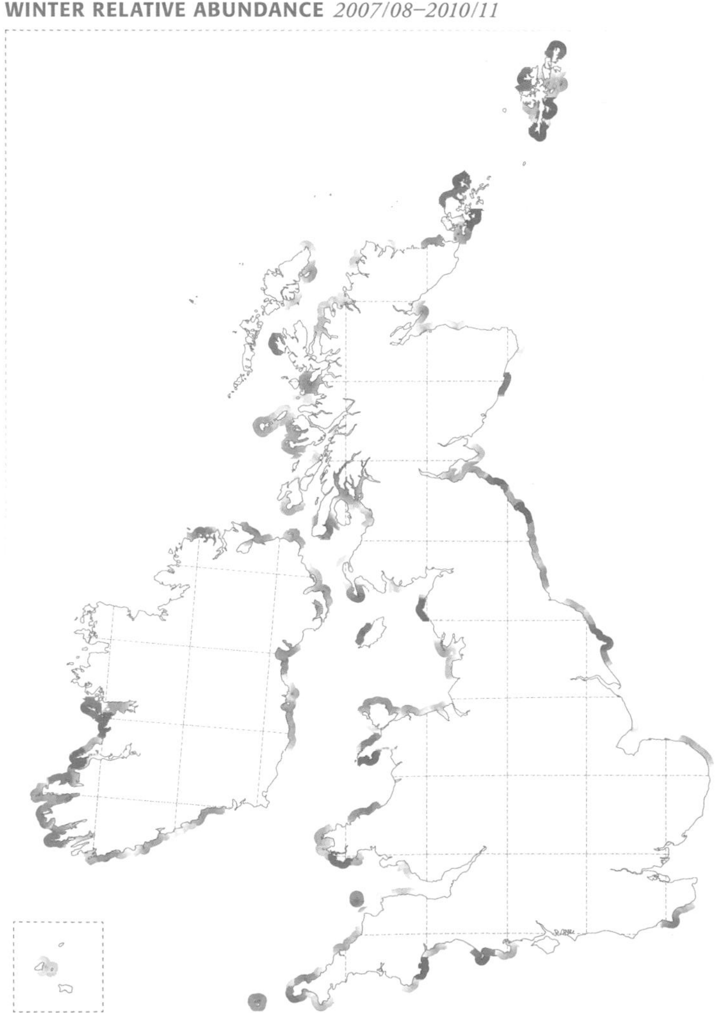

RAZORBILL

Alca torda

BTO CODE: RA

DAVID KJAER www.davidkjaer.com

DURING WINTER, RAZORBILLS were reported around most of the coast of Britain & Ireland; however, this is likely to be just a small part of the pelagic winter distribution of the species (Stone *et al.* 1995). In parts of the north and west of Britain they were absent during much of the winter until February, when they started returning to breeding colonies. Since the *1981–84 Winter Atlas*, gains and losses in Ireland have been balanced, whereas there has been a 10% range contraction in Britain. Losses apparent in northern and western Scotland and in eastern England north of the Wash have been partly compensated for by gains in the Firth of Clyde, Irish Sea, southwest approaches and the south coast of England.

The breeding-season distribution is similar to that of the Guillemot, with breeding birds absent between Yorkshire and Dorset, where there is almost no suitable nesting habitat. The breeding relative abundance map shows that the largest concentrations occur from Flamborough Head to the Northern Isles and at many locations up the western seaboards of both Britain and Ireland.

Razorbill colonies differ widely in their trends in abundance, breeding success and survival (*Seabird 2000*) but space does not permit an in-depth review of these in relation to the national range changes shown here. Overall, seabird monitoring shows that the Razorbill breeding population increased by 59% in the UK between 1986 and 2011 (*SUKB 2012*), whilst monitoring at four of Ireland's largest colonies over the last 10–12 years has shown a stable population. Despite this large increase in numbers in the UK, Atlas data show a 12% reduction in the number of 10-km squares with *probable* or *confirmed* breeding since the *1988–91 Breeding Atlas*, with the greatest losses apparent in northern and western Scotland and in western Ireland. The impact of severe mortality events such as wrecks on breeding populations appear to be buffered by non-breeding birds (Heubeck *et al.* 2011; Wanless & Harris 2012). Although overall numbers have increased, low productivity, coinciding with food shortages in the north and east of Britain, could lead to declines in the future (JNCC 2012).

Sponsored by Janet Atkinson

BREEDING DISTRIBUTION *2008–11*

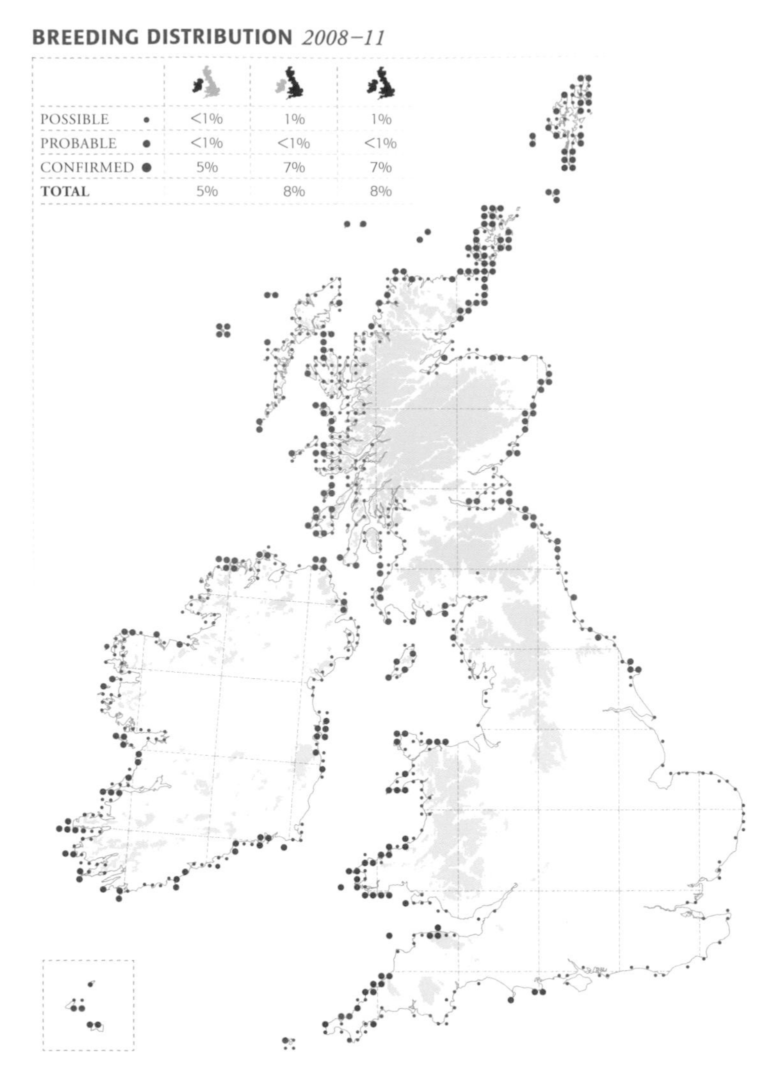

POSSIBLE •	<1%	1%	1%
PROBABLE •	<1%	<1%	<1%
CONFIRMED ●	5%	7%	7%
TOTAL	5%	8%	8%

WINTER DISTRIBUTION *2007/08–2010/11*

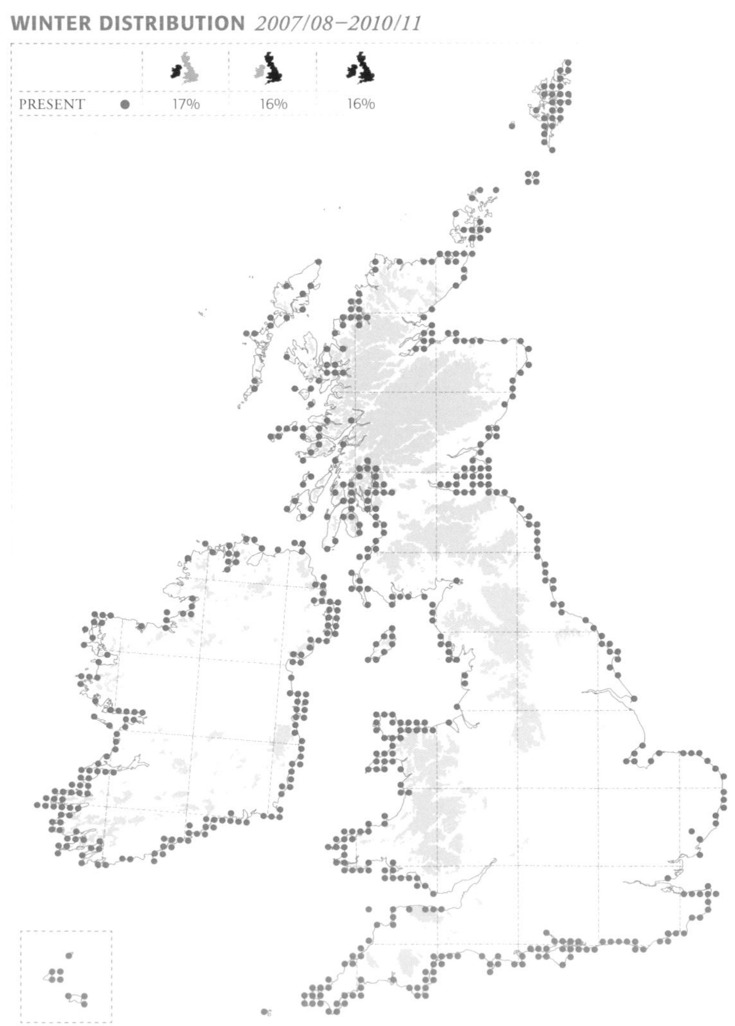

PRESENT ●	17%	16%	16%

BREEDING DISTRIBUTION CHANGE *since 1968–72*

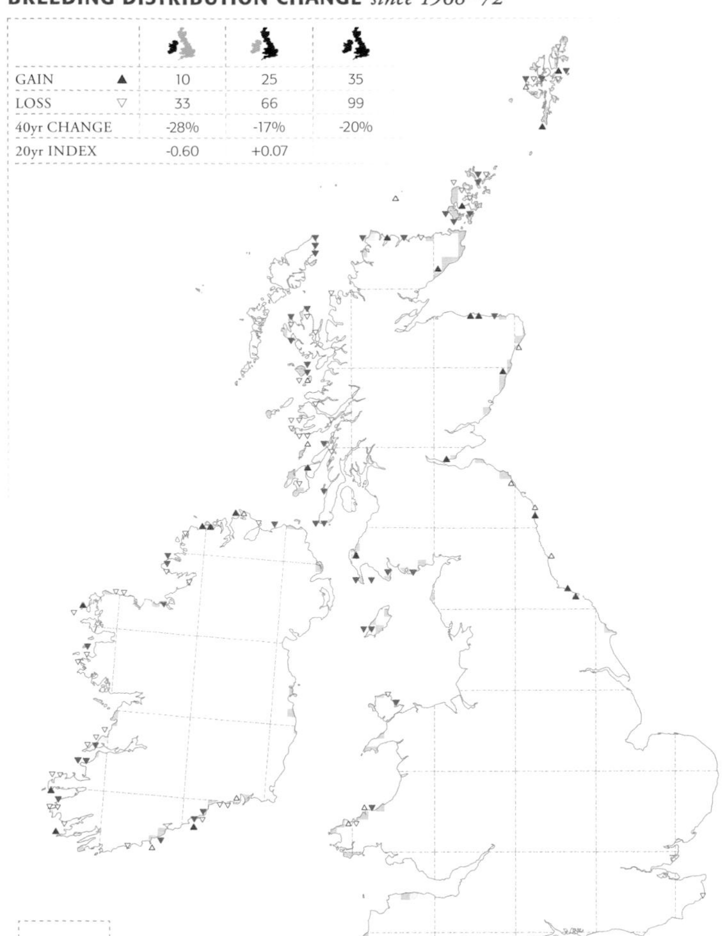

GAIN ▲	10	25	35
LOSS ▽	33	66	99
40yr CHANGE	-28%	-17%	-20%
20yr INDEX	-0.60	+0.07	

BREEDING RELATIVE ABUNDANCE *2008–11*

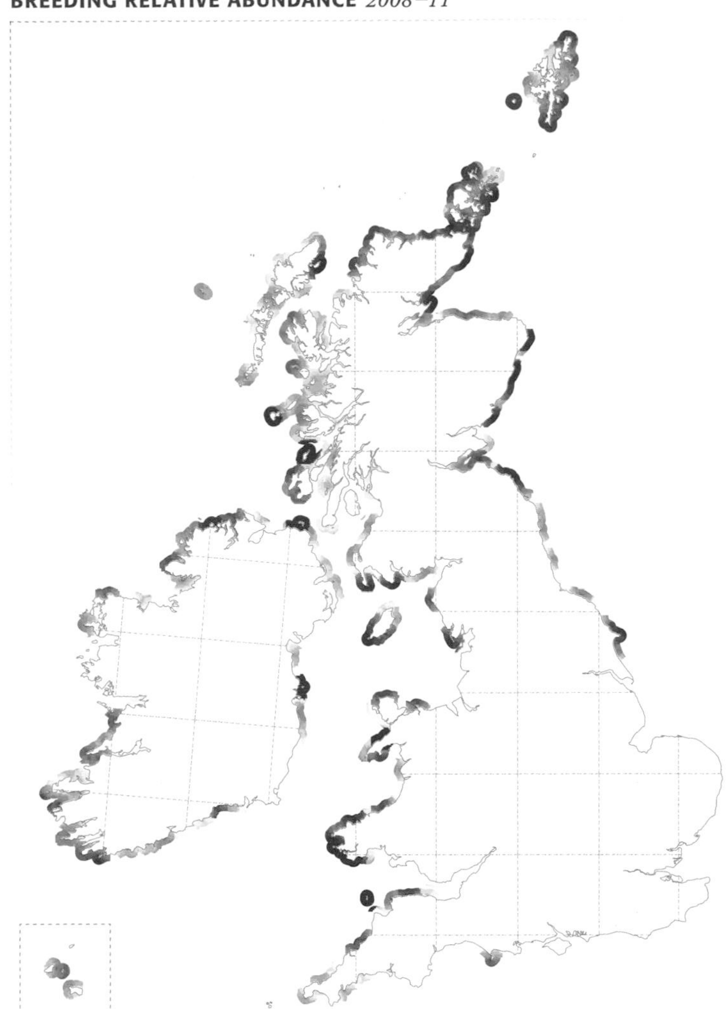

WINTER DISTRIBUTION CHANGE *since 1981–84*

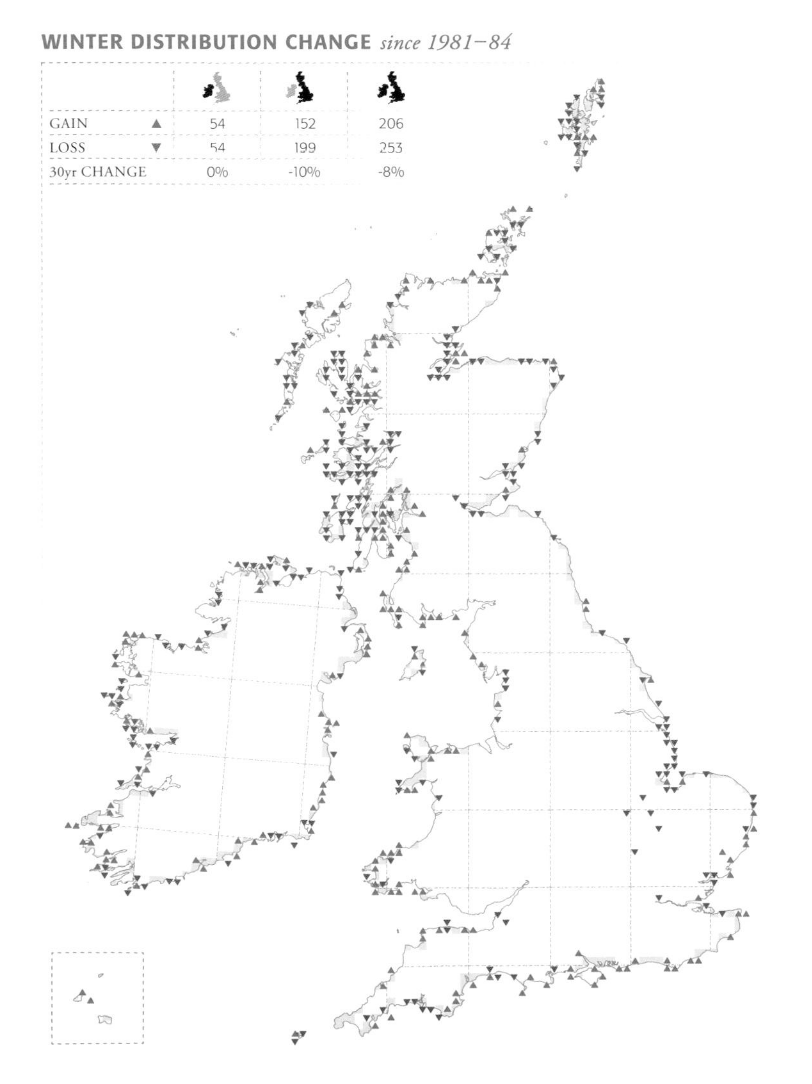

GAIN ▲	54	152	206
LOSS ▼	54	199	253
30yr CHANGE	0%	-10%	-8%

WINTER RELATIVE ABUNDANCE *2007/08–2010/11*

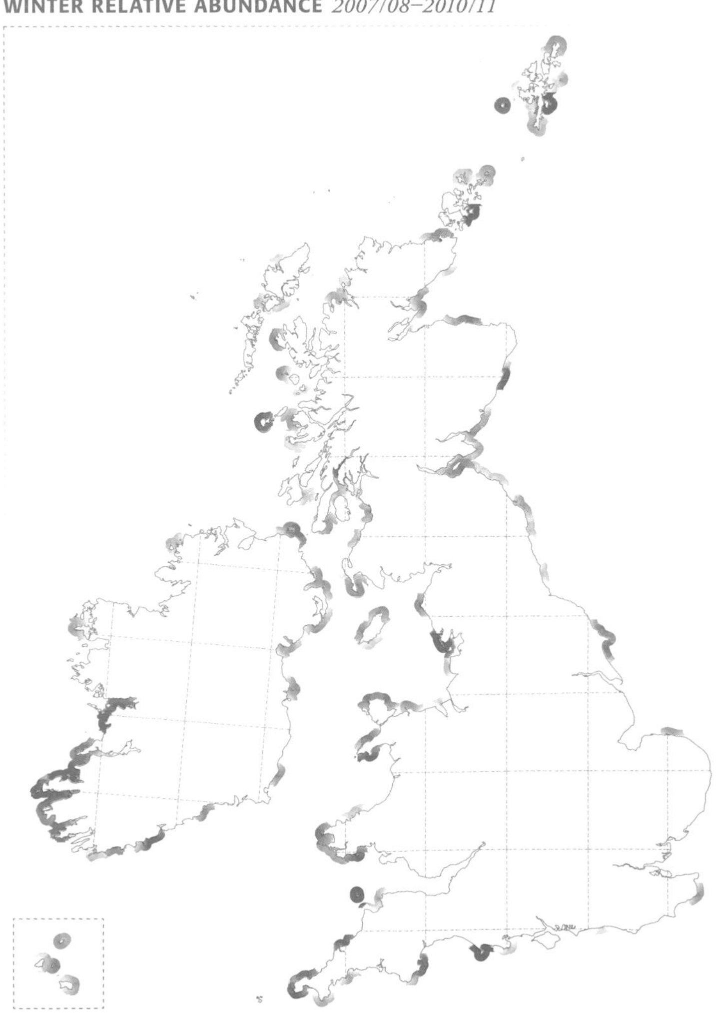

BLACK GUILLEMOT

Cepphus grylle

BTO CODE: TY

GARTH PEACOCK www.garthpeacock.co.uk

THE BLACK GUILLEMOT is one of the most sedentary seabird species breeding in Britain & Ireland and its principal range in both seasons is the coasts of Ireland, Anglesey, the Isle of Man and northern and western Scotland. In winter there are additional records scattered along the coasts of the North Sea and southwest and southern England, perhaps reflecting limited dispersal away from breeding areas.

The breeding and winter abundance maps show the highest densities to lie in the core of the breeding distribution in the Northern Isles. In comparison to previous breeding atlases, the distribution has changed most in western Scotland, with losses predominating in the Outer and Inner Hebrides, but striking gains in the inner north Clyde area and on the east side of the Kintyre Peninsula, especially since the *1988–91 Breeding Atlas.* There has been a 29% range contraction in Britain since the *1988–91 Breeding Atlas*, contrasting with a 42% expansion in Ireland over the same period. Apparent losses in northwest Scotland may be due to small colonies surveyed for *Seabird 2000* not being covered during 2008–11. Seabird Monitoring Programme data indicates that the status of the species has probably changed little since *Seabird 2000* and, unlike other auks, there has been no significant change in breeding productivity at monitored colonies between 1987 and 2011 (JNCC 2012).

Winter range size has increased overall by 4% since the *1981–84 Winter Atlas*. In some areas, such as southwest Scotland, the Irish Sea and the North Sea, these increases are real, but the mixture of gains and losses in western Scotland and Ireland is likely to reflect the difficulty of detecting individuals or small groups in remote coastal areas, although local extinctions may have occurred.

Reasons for the changing fortunes of this species, particularly along the western seaboard, are poorly understood. In some areas increased predation by American Minks (Craik 1997) and Brown Rats (Ewins & Tasker 1985) has led to local extinctions. In Ireland recent increases have been aided by further exploitation of artificial nest sites in harbour walls (Greenwood & Tickner 2002), and this may also explain some of the gains around the Clyde (*Seabird 2000*).

BREEDING DISTRIBUTION *2008–11*

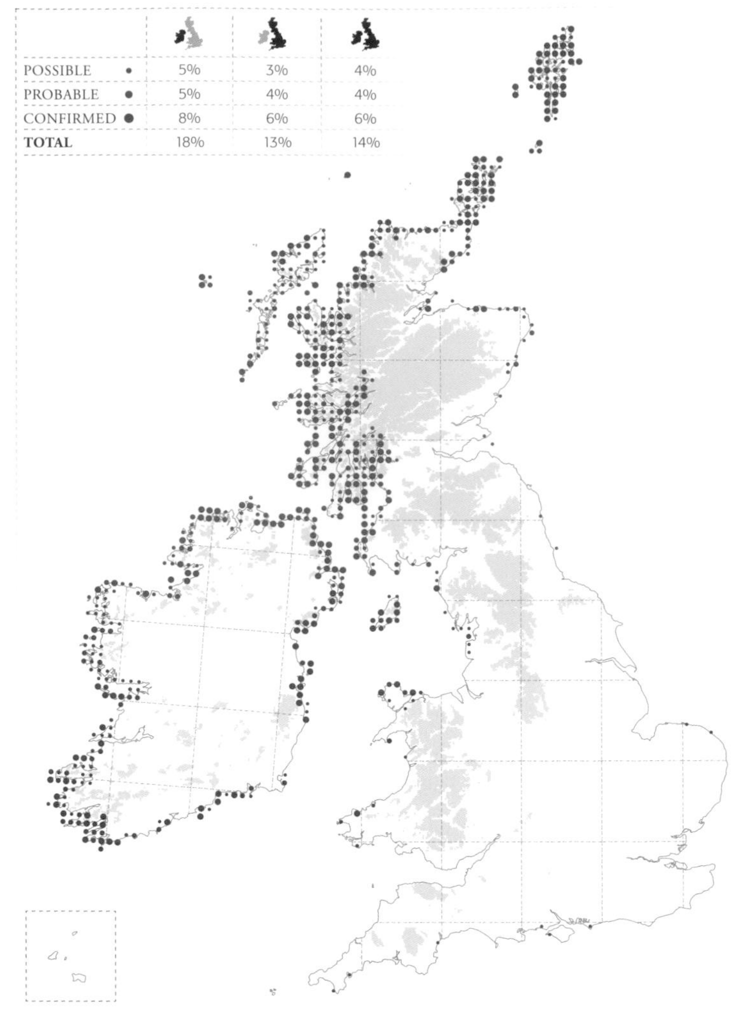

	Britain & Ireland	Britain	Ireland
POSSIBLE •	5%	3%	4%
PROBABLE ●	5%	4%	4%
CONFIRMED ●	8%	6%	6%
TOTAL	18%	13%	14%

WINTER DISTRIBUTION *2007/08–2010/11*

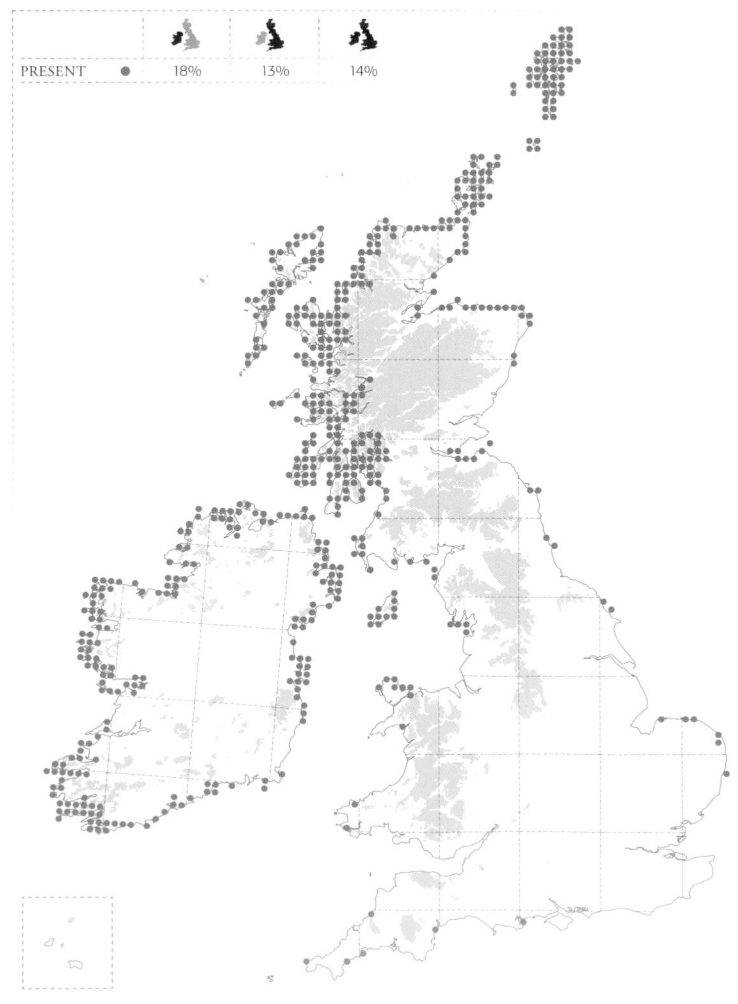

	Britain & Ireland	Britain	Ireland
PRESENT ●	18%	13%	14%

BREEDING DISTRIBUTION CHANGE *since 1968–72*

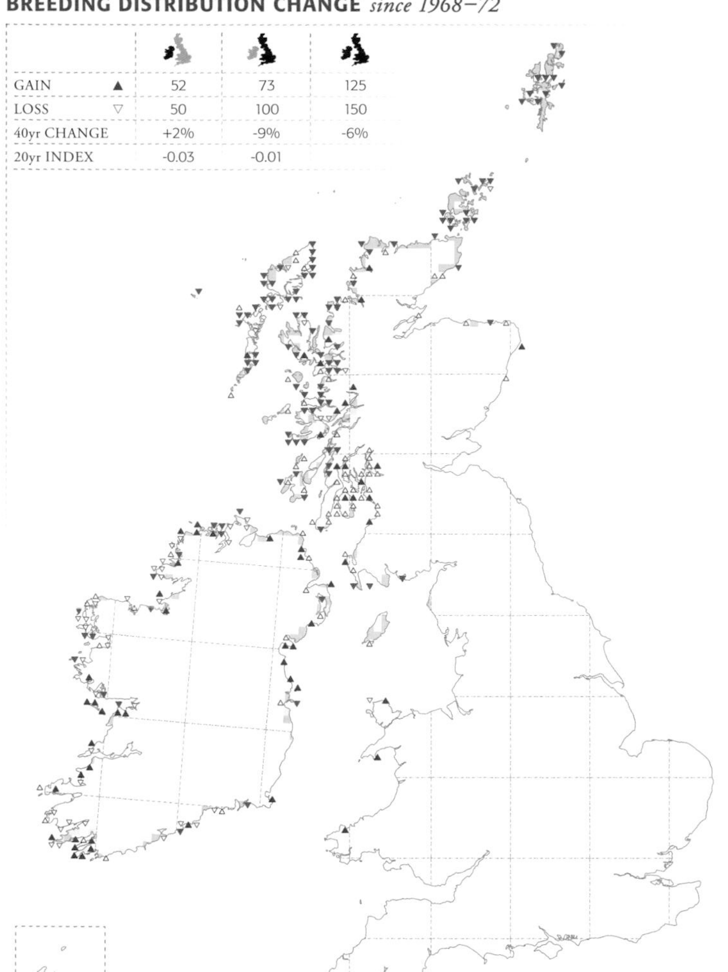

GAIN ▲	52	73	125
LOSS ▽	50	100	150
40yr CHANGE	+2%	-9%	-6%
20yr INDEX	-0.03	-0.01	

BREEDING RELATIVE ABUNDANCE *2008–11*

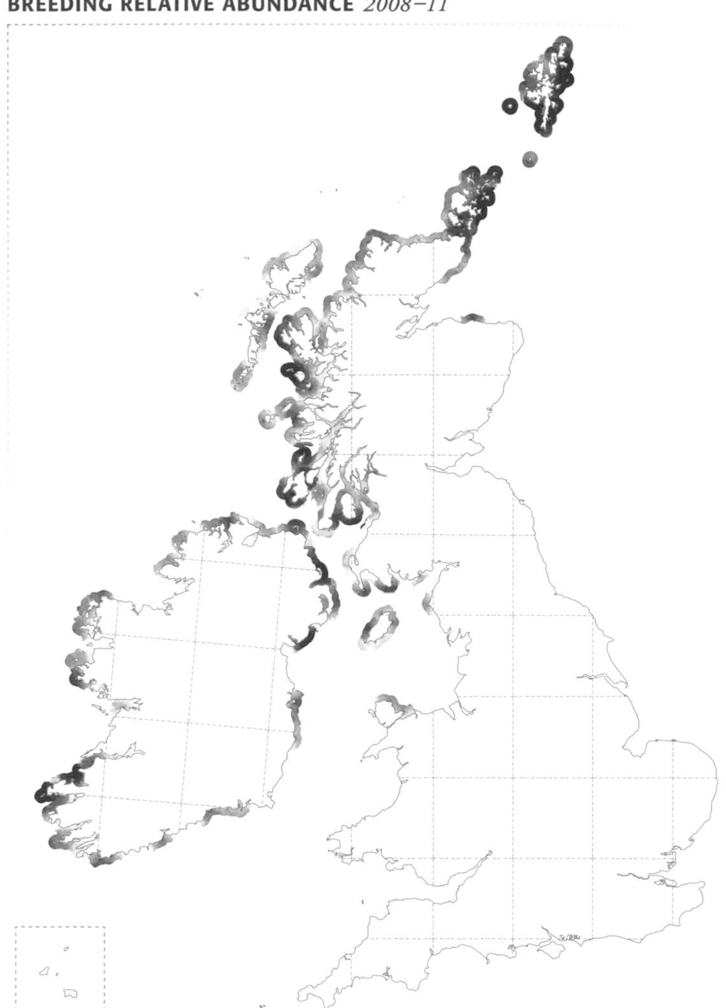

WINTER DISTRIBUTION CHANGE *since 1981–84*

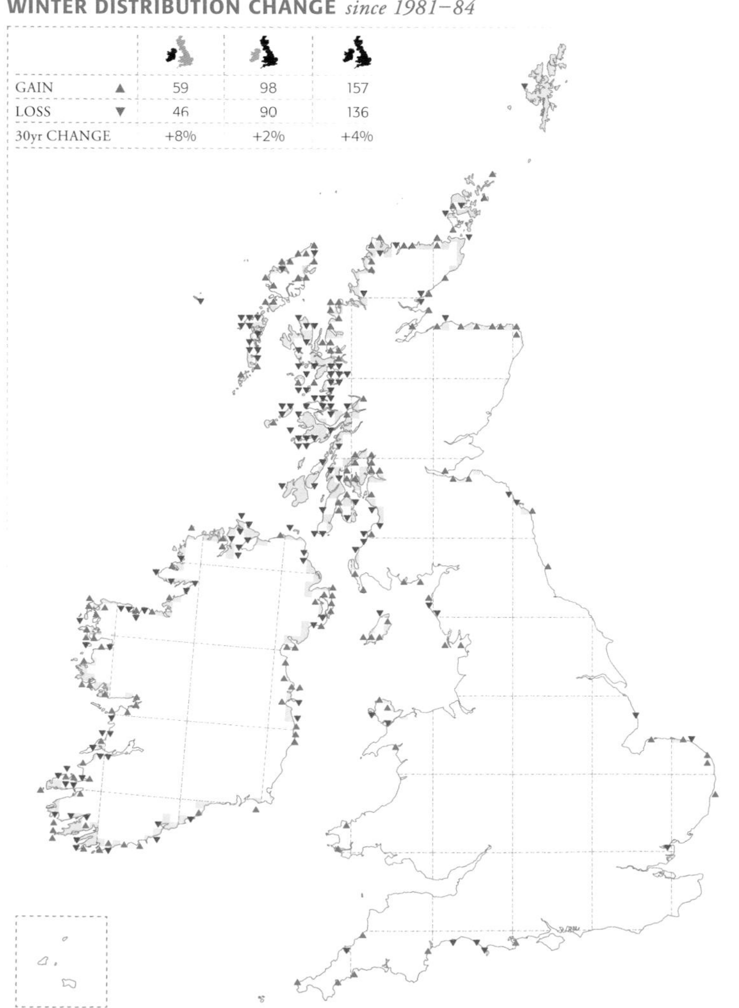

GAIN ▲	59	98	157
LOSS ▼	46	90	136
30yr CHANGE	+8%	+2%	+4%

WINTER RELATIVE ABUNDANCE *2007/08–2010/11*

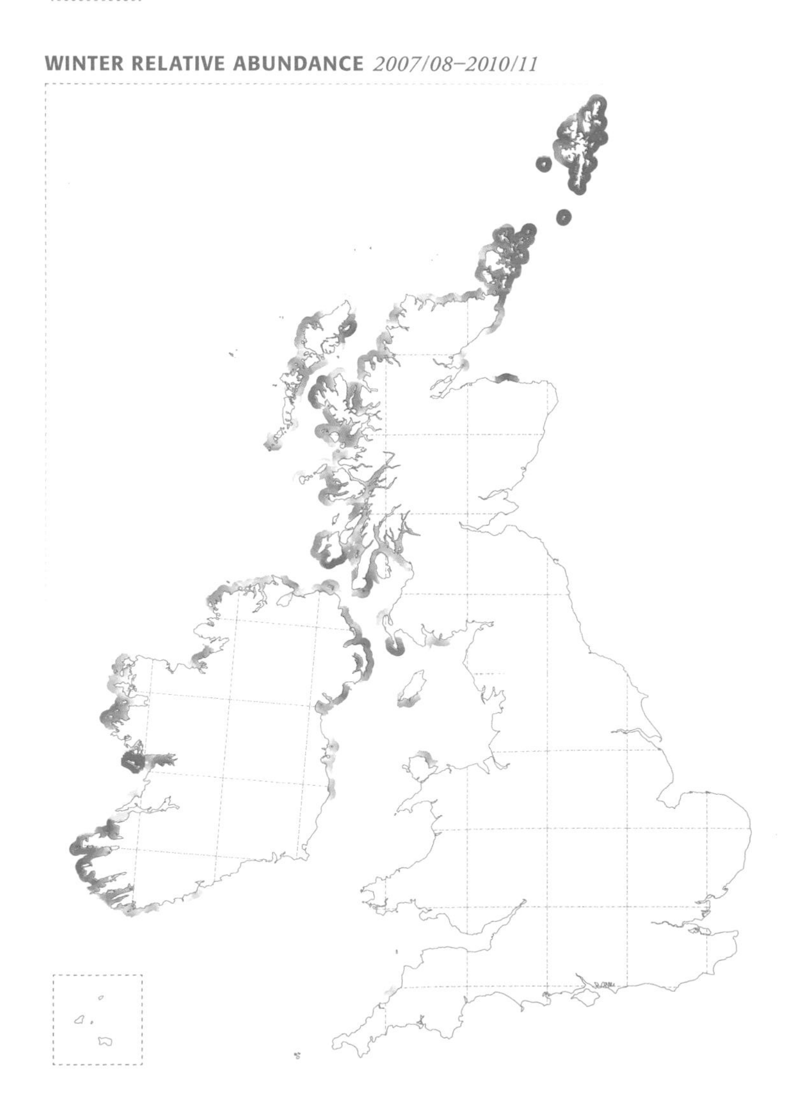

PUFFIN

Fratercula arctica

BTO CODE: PU

ALLAN DREWITT

THE BREEDING DISTRIBUTION map is a good reflection of the distribution of colonies, with the bulk of the range, and of the breeding population, occurring in the Northern Isles, St Kilda, along the North Sea coast south to Yorkshire, in southwest Wales and western Ireland (*Seabird 2000*). The distribution of the main colonies has changed little since the *1968–72 Breeding Atlas*, but the 28% range contraction documented here may reflect the difficulty of detecting birds in small colonies. These losses in range are more than offset by increases in numbers elsewhere, including a 21-fold increase in colony size on the Isle of May between 1969–70 and 1998–2002 (*Seabird 2000*). However, several of the larger colonies, including those on the Farne Islands and Isle of May, have since seen substantial declines and associated low productivity (Harris *et al.* 2009; Harris & Wanless 2011). The future prospects of these colonies are uncertain in the face of climate-induced sea-temperature rises and corresponding changes in marine communities, in addition to pressures from commercial fisheries (Frederiksen *et al.* 2006; Harris & Wanless 2011).

For a pelagic-wintering species such as the Puffin, land-based atlas field techniques can only record those exceptional individuals that pass close to land. Knowledge of the at-sea distribution of Puffins has advanced significantly with the deployment of 'geolocator' tags (Harris *et al.* 2010; Guilford *et al.* 2011), which show that birds from some colonies winter at sea thousands of kilometres from the colonies, although a considerable proportion of birds from at least the Isle of May remain within the North Sea, where small numbers can sometimes be seen from coastal vantage points, particularly in November.

The winter change map shows a 54% range contraction since the *1981–84 Winter Atlas*, with the main area of loss being along the North Sea coast. Rather than loss of a traditional wintering area, the reduction reflects an exceptional pattern of inshore records following a major auk wreck in February 1983 (Underwood & Stowe 1984). As shown in the *1981–84 Winter Atlas*, the coastal distribution in 'normal' winters is considerably more restricted and more in keeping with that documented during 2007–11.

BREEDING DISTRIBUTION *2008–11*

POSSIBLE •	<1%	1%	1%
PROBABLE •	<1%	1%	<1%
CONFIRMED ●	2%	4%	4%
TOTAL	2%	6%	5%

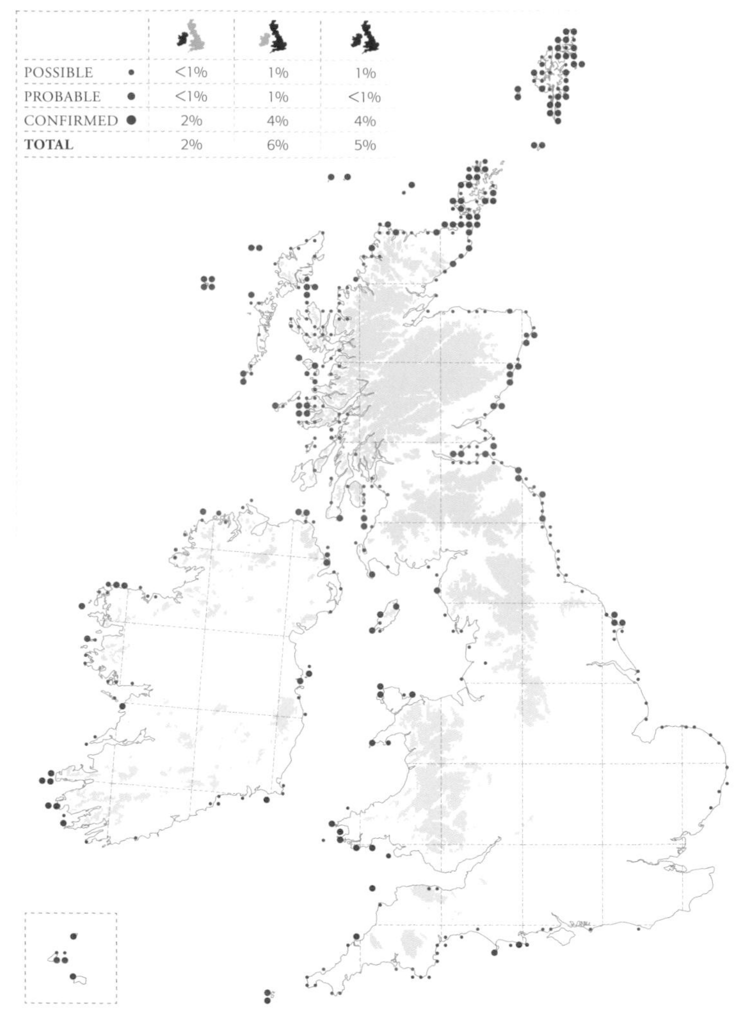

WINTER DISTRIBUTION *2007/08–2010/11*

PRESENT ●	<1%	3%	2%

Sponsored by Opticron

BREEDING DISTRIBUTION CHANGE *since 1968–72*

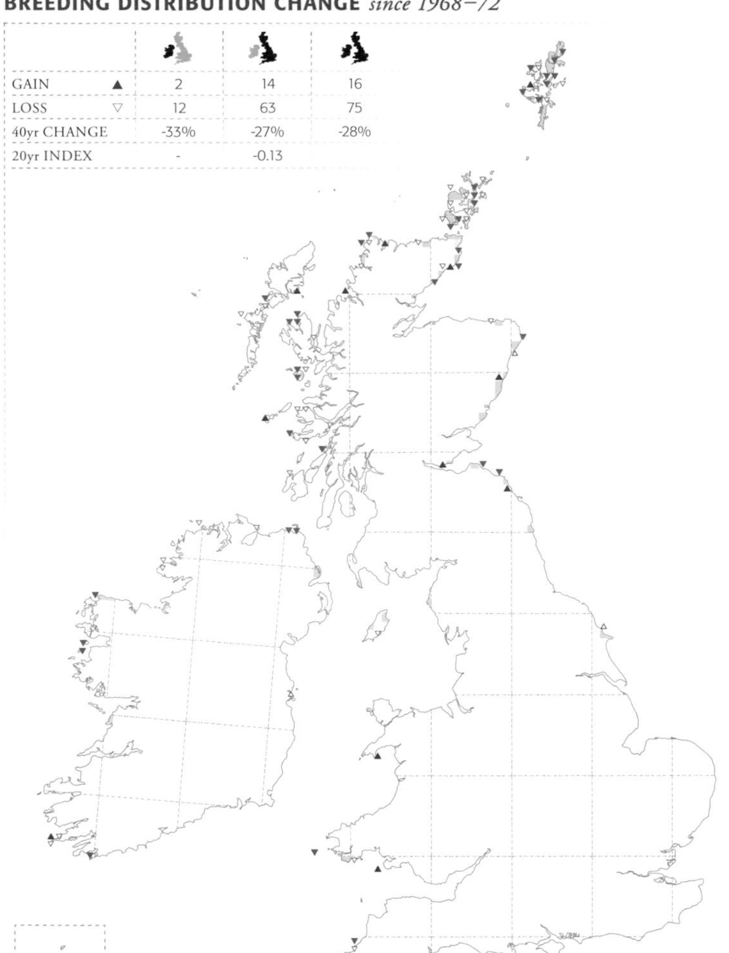

GAIN ▲	2	14	16
LOSS ▽	12	63	75
40yr CHANGE	-33%	-27%	-28%
20yr INDEX	-	-0.13	

BREEDING RELATIVE ABUNDANCE *2008–11*

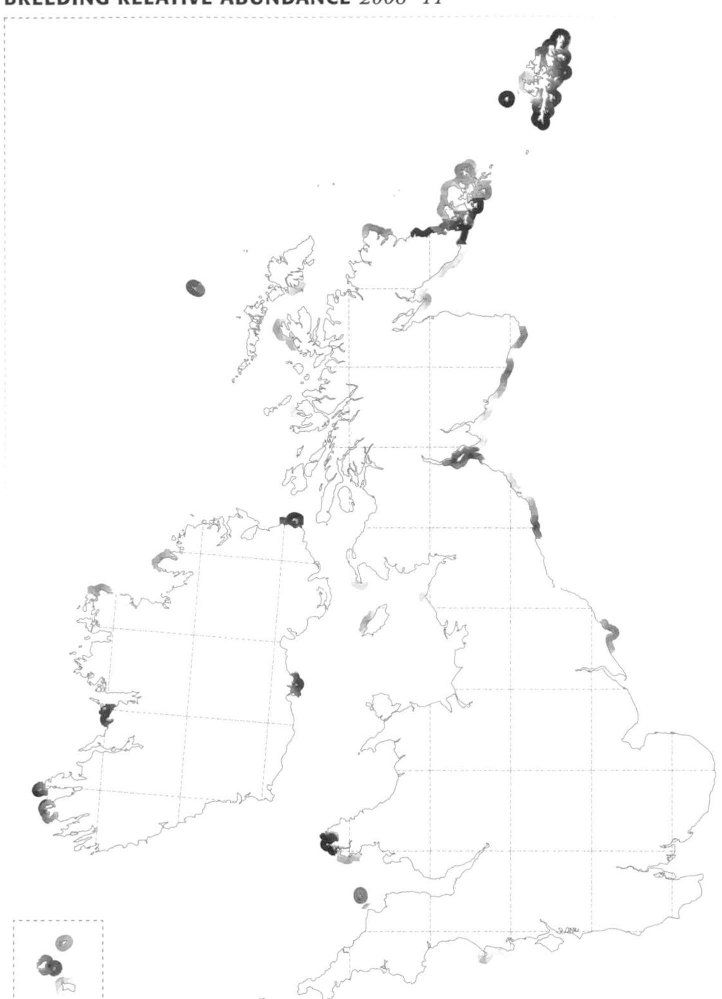

WINTER DISTRIBUTION CHANGE *since 1981–84*

GAIN ▲	3	33	36
LOSS ▼	11	93	104
30yr CHANGE	-73%	-52%	-54%

ROCK DOVE/FERAL PIGEON

Columba livia

BTO CODE: DV/FP

ROB ROBINSON

PURE ROCK DOVES are thought to be restricted to remote coastlines of northern and western Scotland and northern, western and southern Ireland. Elsewhere, birds resembling Rock Doves occur in many Feral Pigeon flocks and those on cliffs, such as in eastern Scotland and northeast England, present identification problems. Owing to the difficulty of assigning such individuals to each form, and to permit comparisons with previous atlas data, all records of Rock Doves and Feral Pigeons were combined. The resulting maps show this resident species to be widespread in Britain, though largely absent from the uplands of central Wales, northwest England, the Southern Uplands and Scottish Highlands. In Ireland it is found mostly in the east and south, and is absent from large parts of the midlands.

The winter and breeding-season relative abundance maps are very similar and show high densities in conurbations, in low-lying arable-dominated areas of southeast Ireland and eastern Britain, and in Orkney, Shetland and the Outer Hebrides.

The change maps show considerable gains in both seasons, mainly in rural areas, such as the margins of the uplands, lowland agricultural areas of eastern England, and much of eastern and southern Ireland. The majority of the change in Britain occurred between 1968–72 and 1988–91, but the expansion in Ireland was later, occurring mostly between 1988–91 and 2008–11. The abundance change map confirms the overall gain in Ireland, but also suggests that densities may have declined in parts of England since 1988–91.

These patterns of range and abundance change correspond well with observed breeding population trends. In the UK, Feral Pigeon breeding population trends are mixed, showing a 20% decline in England during 1995–2010 contrasting with increases in Scotland and Wales (*BBS Report 2011*), which concurs with the observed pattern of abundance change. Monitoring in the Republic of Ireland shows a 188% increase during 1998–2010 (*CBS Trend 2010*), which agrees with the observed range changes there. Causes of these diverging trends are unclear. In Bristol, closure of animal feed mills was implicated in the decline from 11,000 to 9,500 birds (Tully 2006).

Rock Dove sponsored by Shetland Bird Club, in memory of Calum Mackenzie

Feral Pigeon sponsored by Mike Wildish—Dedicated to the good townsfolk of Andover, Hampshire

BREEDING DISTRIBUTION *2008–11*

POSSIBLE •	17%	14%	15%
PROBABLE ●	19%	23%	22%
CONFIRMED ●	21%	40%	35%
TOTAL	57%	77%	72%

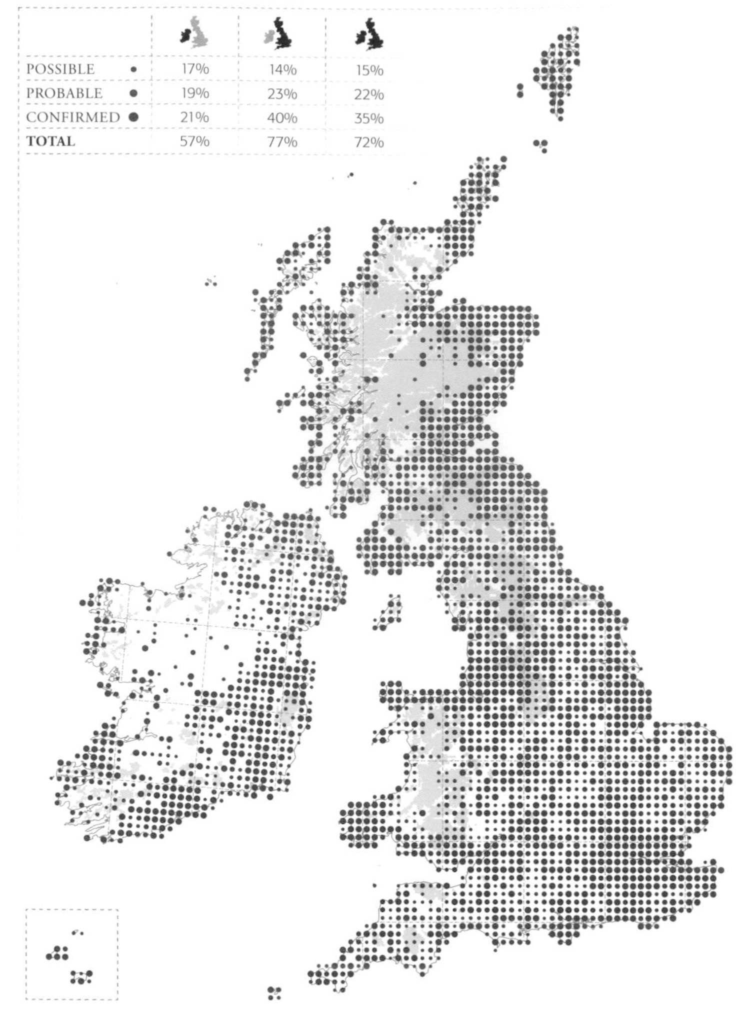

WINTER DISTRIBUTION *2007/08–2010/11*

PRESENT ●	56%	81%	75%

BREEDING DISTRIBUTION CHANGE *since 1968–72*

GAIN ▲	266	891	1157
LOSS ▽	98	170	268
40yr CHANGE	+42%	+48%	+47%
20yr INDEX	+0.40	+0.05	

WINTER DISTRIBUTION CHANGE *since 1981–84*

GAIN ▲	269	489	758
LOSS ▼	69	99	168
30yr CHANGE	+57%	+21%	+26%

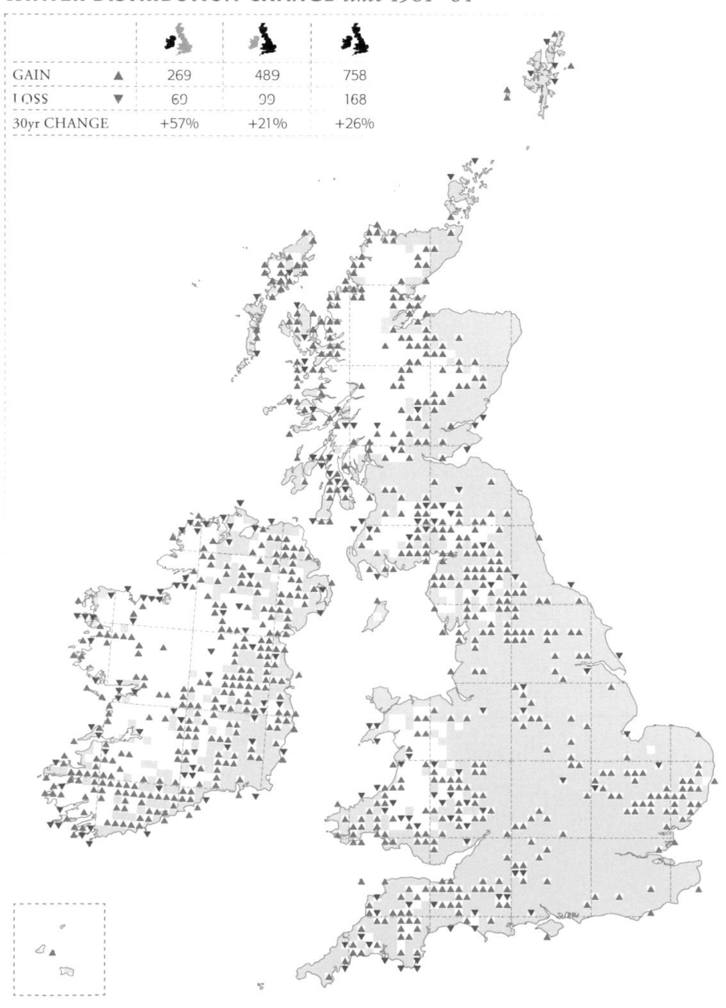

BREEDING RELATIVE ABUNDANCE *2008–11*

BREEDING RELATIVE ABUNDANCE CHANGE *since 1988–91*

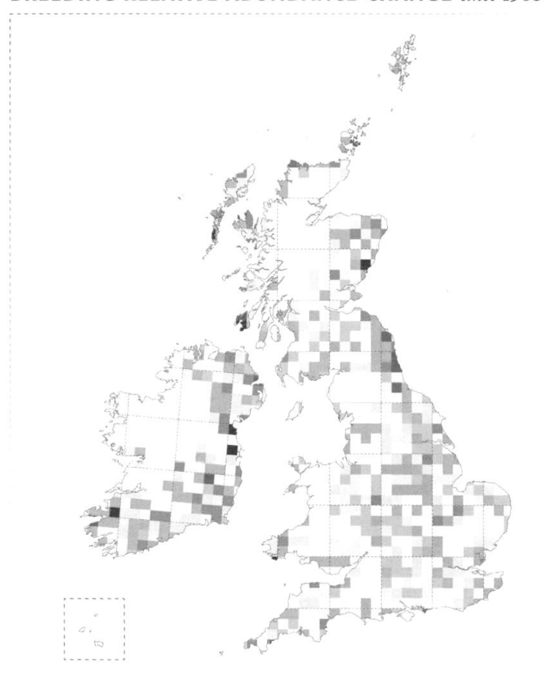

WINTER RELATIVE ABUNDANCE *2007/08–2010/11*

STOCK DOVE
Columba oenas

BTO CODE: SD

TERRY HANCOCK

THE STOCK DOVE is distributed widely across most of England in winter and summer, but is absent in both seasons from most of northwest Scotland and from northern and western Ireland. Other gaps include the uplands of northwest Wales and southern Scotland. Within Britain & Ireland, Stock Doves are largely sedentary (*Migration Atlas*) and summer and winter patterns of relative abundance are similar. The abundance maps suggest that higher densities occur especially around low-altitude alluvial arable farmland and open woodland, as found throughout much of lowland England. The current distribution in Ireland coincides with the availability of arable farmland.

Patterns of change are markedly different in Britain and Ireland. In Britain the small range contraction apparent between the *1968–72 Breeding Atlas* and the *1988–91 Breeding Atlas* has been almost recouped by a small expansion up to 2008–11. In contrast, in Ireland the species has undergone a considerable and continuous range contraction over the last 40 years, shrinking into the southeast. During 1995–2010, there was a 9% breeding population increase in the UK (*BBS Report 2011*). This followed a strong increase during the 1970s, but some levelling off during the 1980s and early 1990s (*BirdTrends*). This overall increase implies that the range contraction recorded in the north and west of the UK has been offset by increases in breeding density in the south and east, an assertion supported by the breeding abundance change map. That map also suggests declining abundance in southwest England, though not yet to the extent that local extinctions are apparent on the distribution change map. In Ireland, where range contraction has been greatest, there was a 50% breeding population decline in the Republic of Ireland between 1998 and 2010 (*CBS Trend 2010*).

In the UK, both breeding productivity and overwinter survival have increased in the last 40 years, following a recovery from the effects of organochlorine seed-dressings in the 1950s (O'Connor & Mead 1984). The recent decline across western Ireland is due to the loss of mixed farming, meaning that the species is now largely confined to the arable belt in the east and southeast.

Sponsored by Ted Forsyth

BREEDING DISTRIBUTION *2008–11*

POSSIBLE •	10%	5%	7%
PROBABLE •	11%	18%	16%
CONFIRMED •	7%	43%	34%
TOTAL	28%	66%	57%

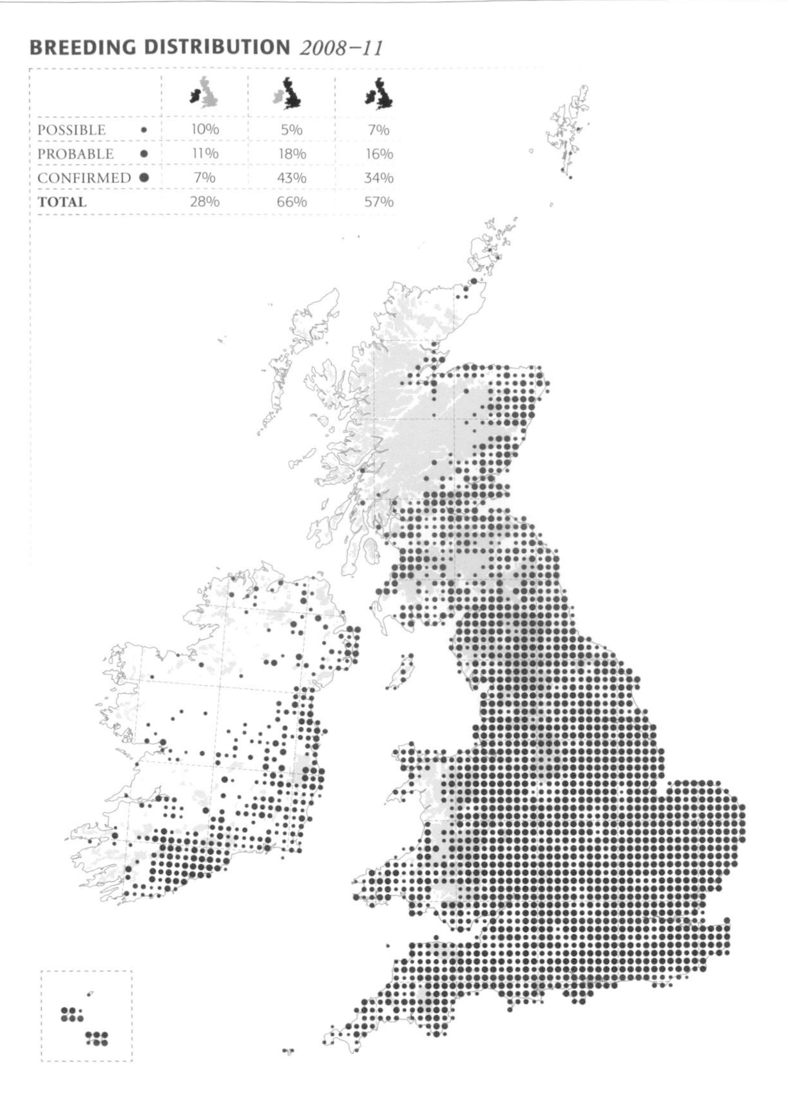

WINTER DISTRIBUTION *2007/08–2010/11*

PRESENT •	23%	65%	54%

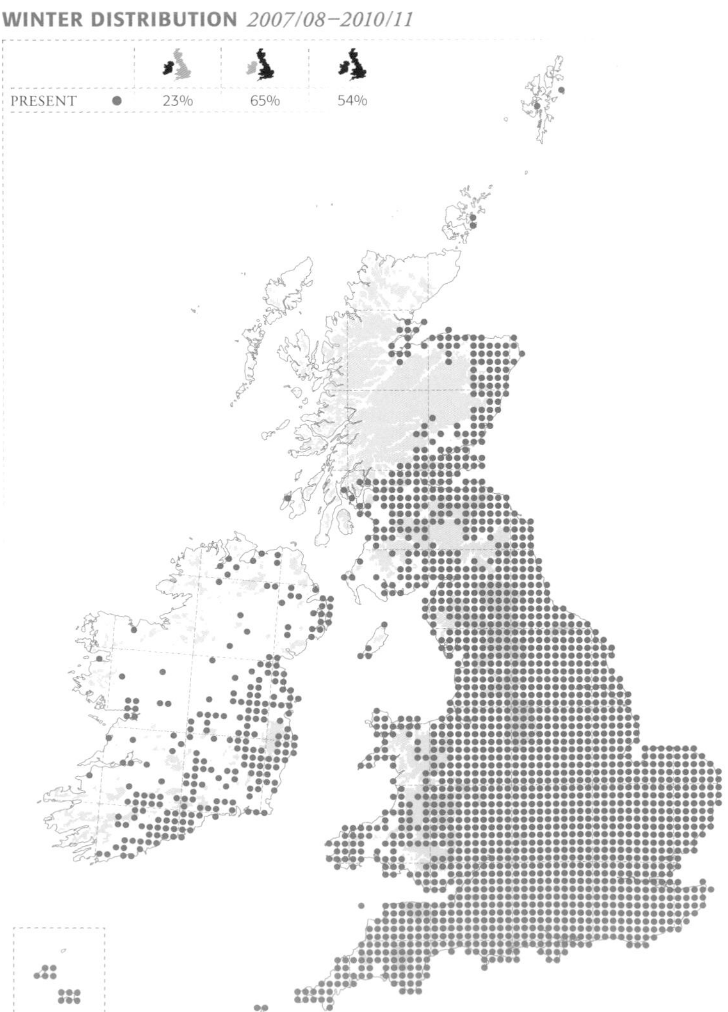

BREEDING DISTRIBUTION CHANGE *since 1968–72*

GAIN ▲	76	161	237
LOSS ▽	327	203	530
40yr CHANGE	-47%	-2%	-12%
20yr INDEX	-0.18	+0.06	

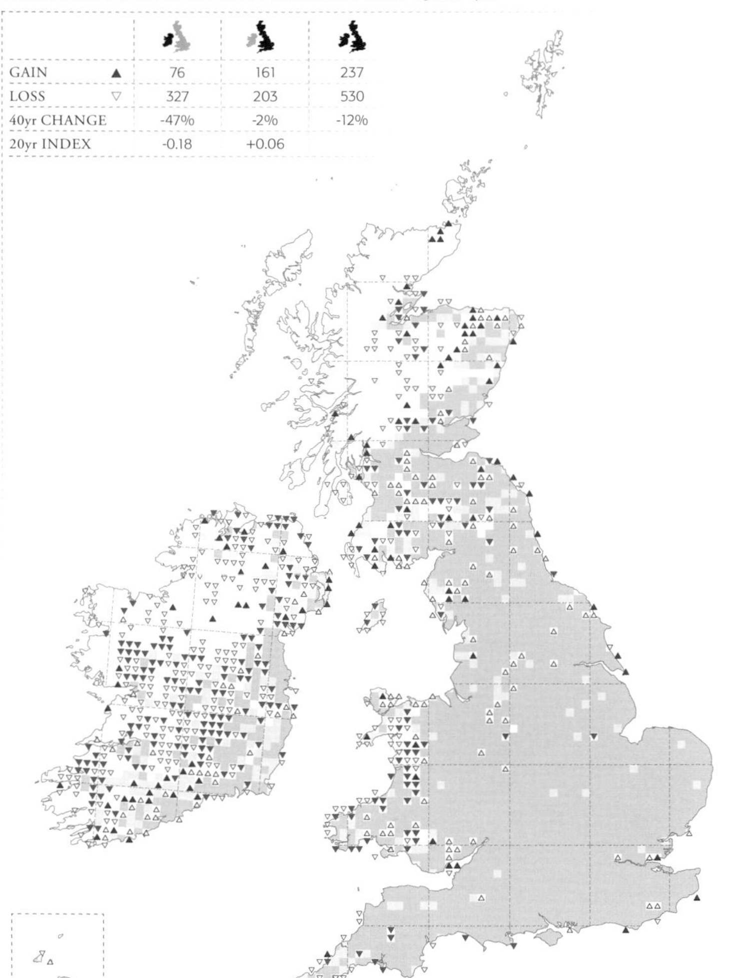

WINTER DISTRIBUTION CHANGE *since 1981–84*

GAIN ▲	108	282	390
LOSS ▼	145	107	252
30yr CHANGE	-14%	+10%	+7%

BREEDING RELATIVE ABUNDANCE *2008–11*

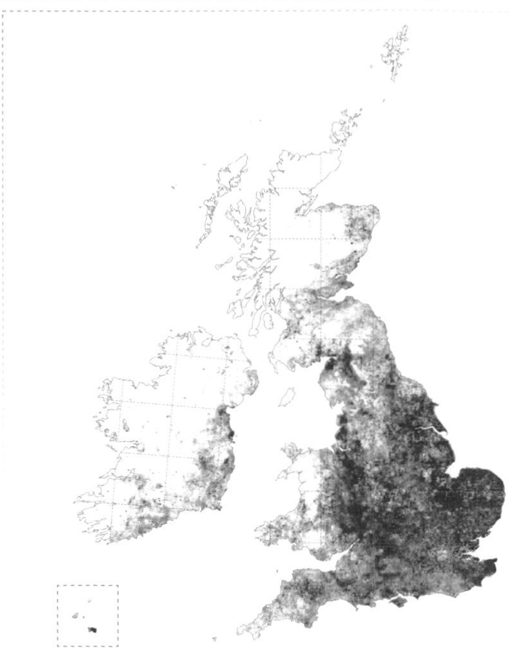

BREEDING RELATIVE ABUNDANCE CHANGE *since 1988–91*

WINTER RELATIVE ABUNDANCE *2007/08–2010/11*

WOODPIGEON
Columba palumbus

BTO CODE: WP

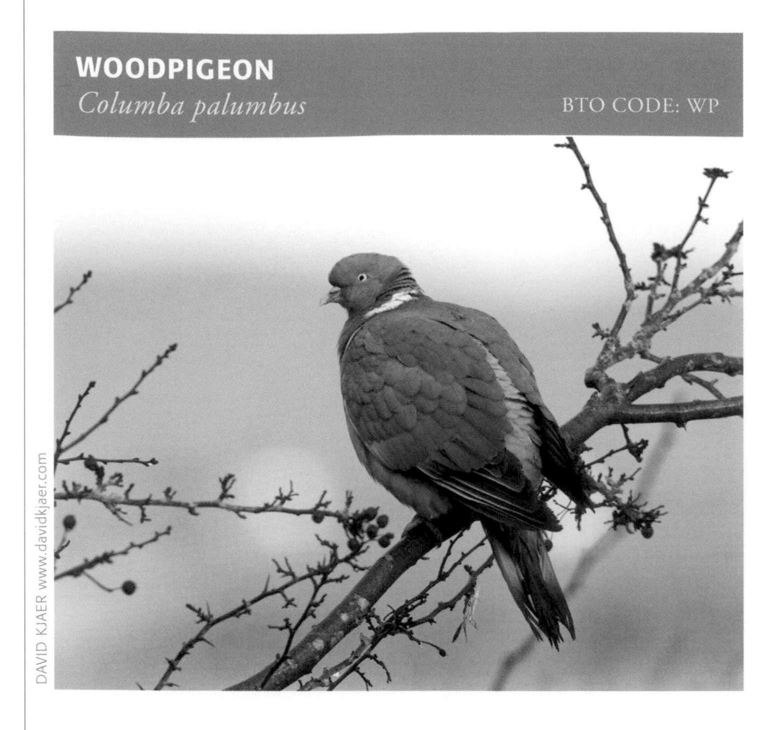

DAVID KJAER www.davidkjaer.com

THE WOODPIGEON OCCUPIES virtually every 10-km square within Britain and Ireland with the exception of the higher montane zones of Scotland and some northern and western islands offering little woodland cover.

Though mobile, Woodpigeons are largely sedentary within Britain & Ireland (*Migration Atlas*), and consequently show very similar patterns of abundance in winter and the breeding season. The highest concentrations occur in the arable eastern lowland regions of both Britain and Ireland, with marginally lower abundance in the more heavily wooded regions of southeast England and still lower abundance in pastoral areas. Previous analyses of atlas data suggested that Woodpigeon winter abundance was highest in areas with a 3:1 mix of arable and pasture (Atkinson *et al.* 2002). Within Ireland, the highest concentrations were in the arable-dominated southeast.

Relatively few gains and losses have occurred in either season and these are mainly along the northwestern periphery of the range. Gains are most notable in the extreme western Highlands, the Outer Hebrides and Northern Isles. Some gains may be due to the recent exploitation of shelter belts and isolated woodlands as Woodpigeons expanded their range. The breeding abundance change map suggests that densities have increased in northwest Britain and western Ireland since the *1988–91 Breeding Atlas*. Densities could have increased in central and eastern England but since virtually all tetrads in those areas were already occupied by 1988–91, this map has little sensitivity to detect further increases. The UK breeding population increased by 169% during 1967–2010 (*BirdTrends*), whilst in the Republic of Ireland there was a 34% increase during 1998–2010 (*CBS Trend 2010*). There is also a strong increasing trend for the use of gardens by Woodpigeons in Britain (Garden BirdWatch 2013). These patterns are probably attributable to higher overwinter survival and the exploitation of abundant food available for grazing in late winter in the form of brassicas, such as Oil-seed Rape (Inglis *et al.* 1997). This remarkable species, however, is very much overlooked and neither its ecology nor its movements are fully understood. For example, the degree of integration between populations within Britain & Ireland or between Britain & Ireland and continental Europe is uncertain.

In memory of Chris Mansell

BREEDING DISTRIBUTION *2008–11*

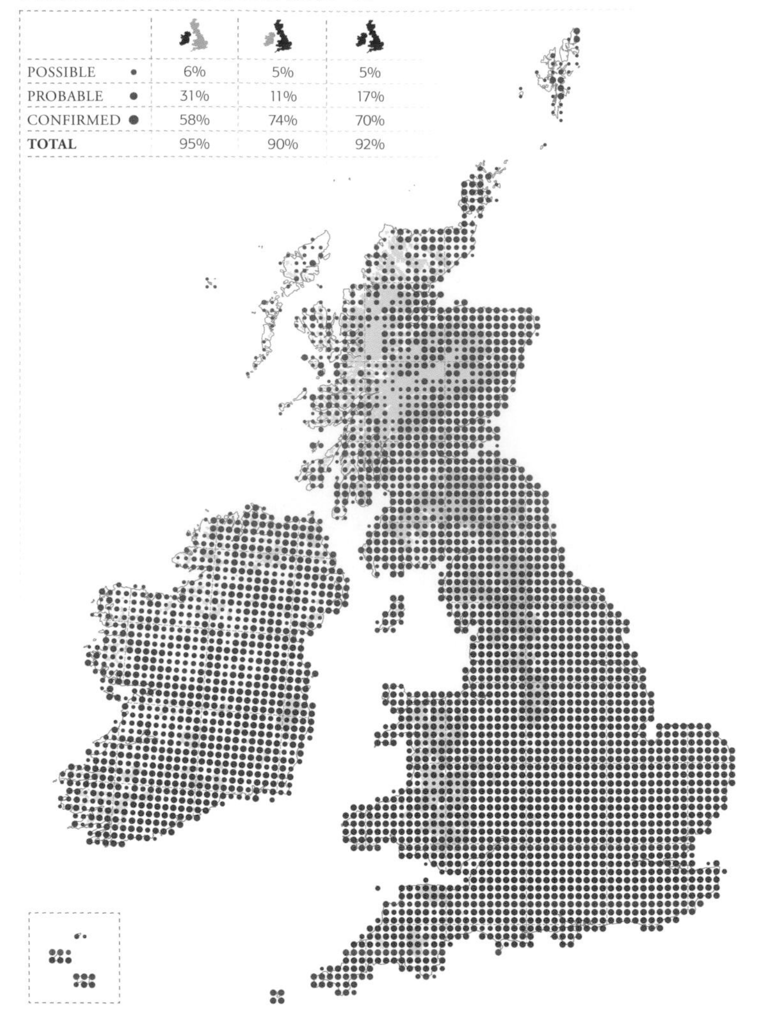

POSSIBLE •	6%	5%	5%
PROBABLE ●	31%	11%	17%
CONFIRMED ●	58%	74%	70%
TOTAL	95%	90%	92%

WINTER DISTRIBUTION *2007/08–2010/11*

PRESENT ●	92%	87%	88%

BREEDING DISTRIBUTION CHANGE *since 1968–72*

GAIN ▲	15	62	77
LOSS ▽	10	60	70
40yr CHANGE	+1%	+<1%	+<1%
20yr INDEX	+0.02	+0.02	

WINTER DISTRIBUTION CHANGE *since 1981–84*

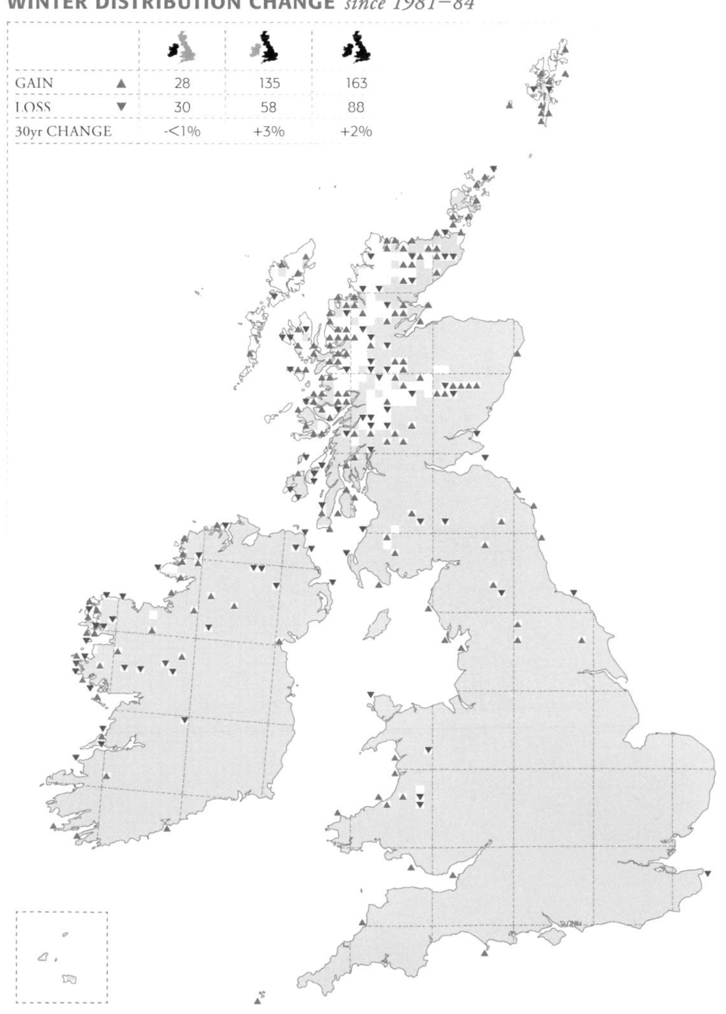

GAIN ▲	28	135	163
LOSS ▼	30	58	88
30yr CHANGE	-<1%	+3%	+2%

BREEDING RELATIVE ABUNDANCE *2008–11*

BREEDING RELATIVE ABUNDANCE CHANGE *since 1988–91*

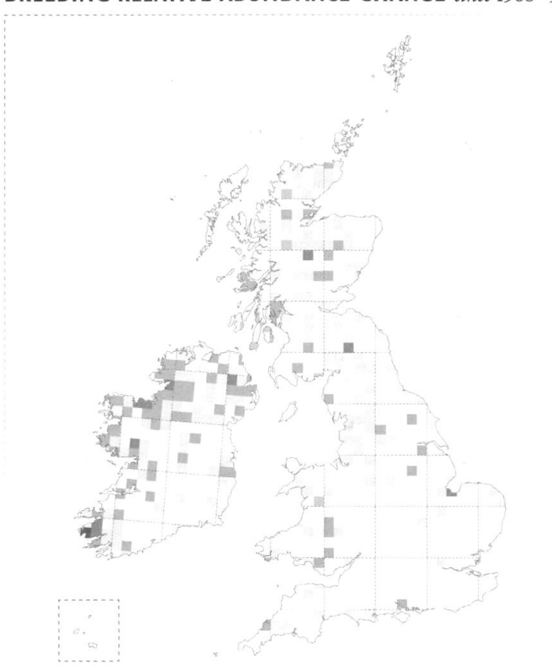

WINTER RELATIVE ABUNDANCE *2007/08–2010/11*

COLLARED DOVE

Streptopelia decaocto

BTO CODE: CD

MORRIS RENDALL

BREEDING DISTRIBUTION *2008–11*

POSSIBLE ●	14%	5%	7%
PROBABLE ●	35%	21%	25%
CONFIRMED ●	35%	57%	51%
TOTAL	84%	83%	83%

THE COLLARED DOVE is a widespread breeding and wintering species all across Britain & Ireland, with the exception of 10-km squares covering higher ground such as the Scottish Highlands, as well as those in remote treeless expanses of bogs or flows. A rapid expansion in range occurred throughout western Europe during the 1930s (Fisher 1953), reaching Britain by the 1950s (Hudson 1965). Breeding was first confirmed in Britain in 1955, in Norfolk (*Birds in England*). Its range in the southeast of Britain has remained relatively stable since the *1968–72 Breeding Atlas*. Elsewhere, gains across northern and western Britain and most of central and western counties of Ireland, as reported between the *1968–72 Breeding Atlas* and *1988–91 Breeding Atlas,* have continued. These increases were particularly notable in Ireland, where there has been a 52% increase in range size since the *1988–91 Breeding Atlas* compared with an 8% increase in Britain.

Strong similarity between the winter and breeding-season patterns of range and relative abundance reflects the sedentary nature of Collared Doves. Relative abundance increased from northwest to southeast, being especially high in the east of England, in the Fens, and near other concentrations of arable or horticultural land use such as in northwest England. Overall, the population trend for this species within the UK is still increasing, though there are signs it may be levelling off (*BirdTrends*). In Ireland, numbers increased by 44% between 1998 and 2010 (*CBS Trend 2010*). This continued rise in numbers is reflected not only in the change maps but also in the breeding abundance change map, which shows that there has been an increase in relative abundance since the *1988–91 Breeding Atlas* throughout the range, particularly in lowland areas. There has been a widespread moderate increase across Europe since 1980 (*PECBMS 2010*).

The species' ecology and demography are poorly studied and consequently the causes of these changes are not fully understood. Collared Doves are known to exploit grain, weed-seeds and discarded or broadcast livestock feed and to be abundant where such food is freely available, around farmyards, grain-stores, allotments, parks and gardens. Productivity appears to have increased and the species has probably benefited from milder winters (*BirdTrends*).

WINTER DISTRIBUTION *2007/08–2010/11*

PRESENT ●	77%	80%	79%

Sponsored by Mike Wildish—Dedicated to all BTO Garden BirdWatch members in Hampshire

BREEDING DISTRIBUTION CHANGE *since 1968–72*

GAIN ▲	318	434	752
LOSS ▽	43	127	170
40yr CHANGE	+48%	+15%	+22%
20yr INDEX	+0.40	+0.11	

WINTER DISTRIBUTION CHANGE *since 1981–84*

GAIN ▲	343	428	771
LOSS ▼	32	38	70
30yr CHANGE	+70%	+21%	+30%

BREEDING RELATIVE ABUNDANCE *2008–11*

BREEDING RELATIVE ABUNDANCE CHANGE *since 1988–91*

WINTER RELATIVE ABUNDANCE *2007/08–2010/11*

TURTLE DOVE

Streptopelia turtur BTO CODE: TD

GARY FAULKNER

THE TURTLE DOVE is one of the latest summer migrants to arrive in Britain and, once here, is a difficult species to confirm breeding. The breeding distribution is restricted to the eastern half of England as far north as Yorkshire and west to the Severn. Overshooting migrants may linger outside the normal range and account for some extralimital records from Scotland, Wales and southwest England. The highest densities are found in Suffolk, Essex and Kent, with further hotspots scattered throughout Lincolnshire, but densities are much lower in the west and north of the range. They are very scarce in Ireland, where there were records of *possible* breeding from two 10-km squares, together with *non-breeding* records from a further 26 10-km squares.

The change map illustrates one of the most striking changes of any species in the Atlas, with progressive loss of range in the north, west and southwest of Britain over the last 40 years. The loss in the Home Counties is also notable. In Ireland, though there was no repeat of the small-scale *confirmed* breeding of the *1968–72 Breeding Atlas*, the change map probably exaggerates changes there because birds coded as *non-breeding* during 2008–11 may have been coded as *seen* or *possible* breeding in earlier atlases and hence appear as losses. The breeding abundance change map shows that, even across the remaining core range, Turtle Dove densities have tumbled, as indicated by a reduction in the proportion of tetrads occupied.

These losses tie in with the large population decline during 1967–2010 (*BirdTrends*). This is not a localised phenomenon; across Europe a 73% decline occurred during 1980–2010 (*PECBMS 2010*). There are several possible explanations for these declines, including conditions on the wintering grounds, hunting pressure during migration, agricultural intensification in the breeding range and a shorter breeding season leading to fewer breeding attempts (Browne & Aebischer 2001, 2004, 2005).

There were winter records from 19 10-km squares: 15 in Britain and four in Ireland. Whilst most of these were records of late migrants in early November, there were a small number of midwinter records, usually of birds found in or around gardens.

Sponsored by Guy Phillipson

BREEDING DISTRIBUTION *2008–11*

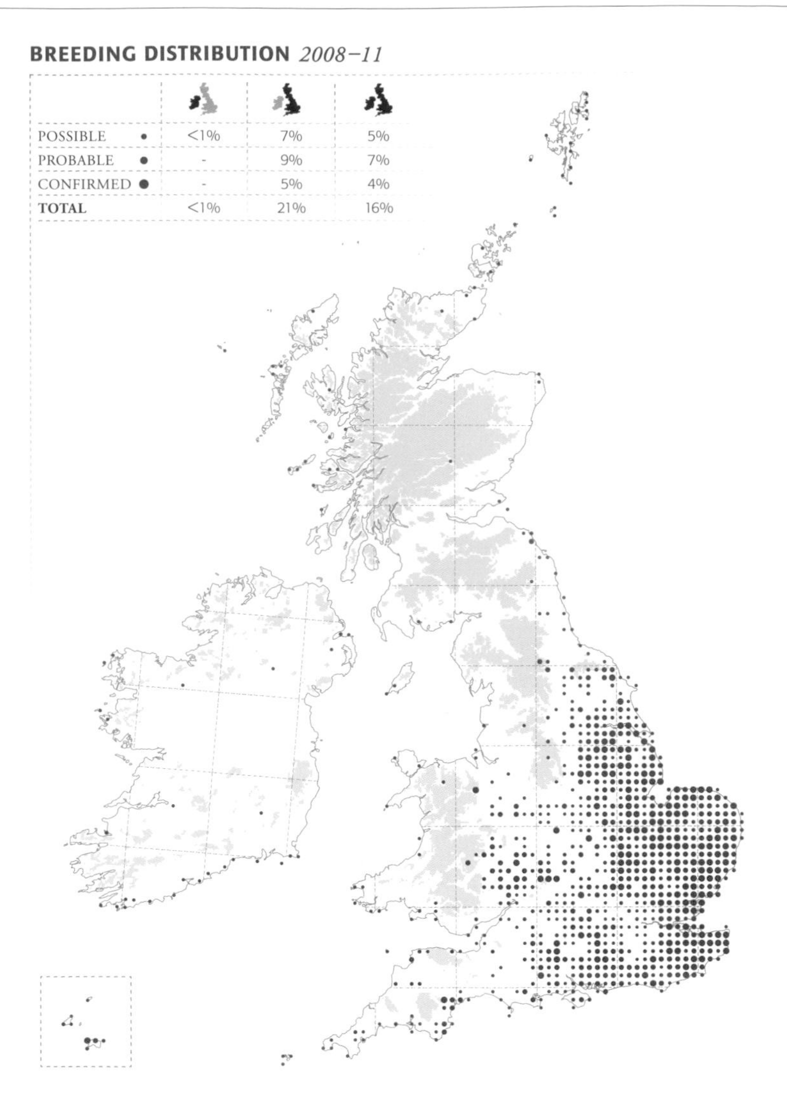

POSSIBLE •	<1%	7%	5%
PROBABLE •	-	9%	7%
CONFIRMED •	-	5%	4%
TOTAL	<1%	21%	16%

BREEDING RELATIVE ABUNDANCE CHANGE *since 1988–91*

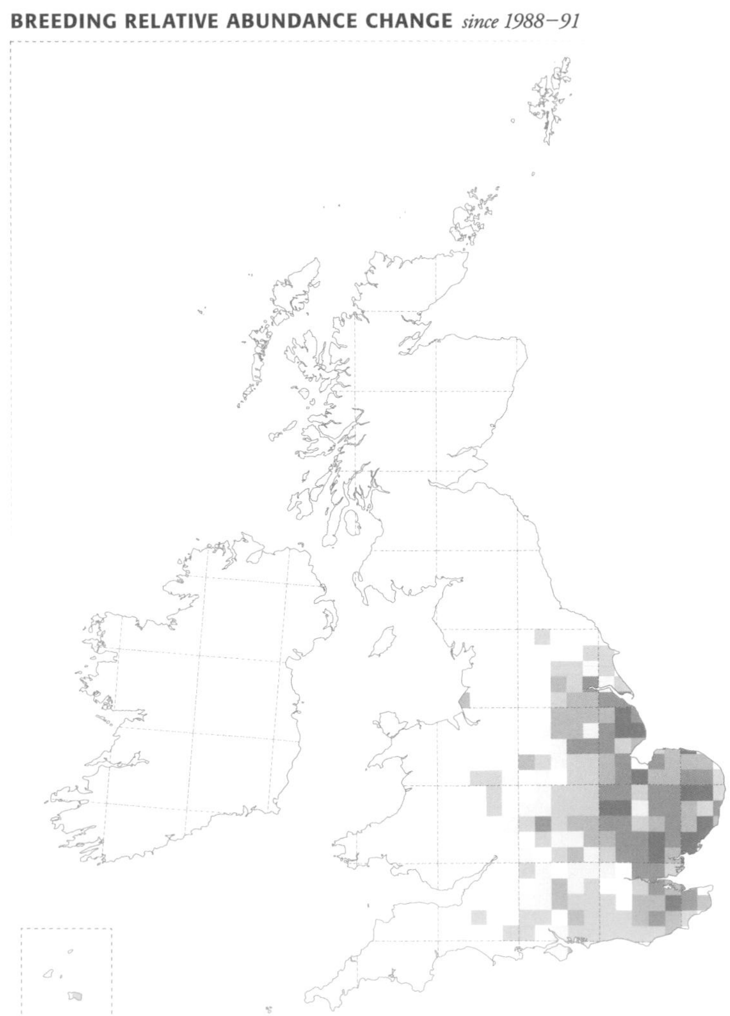

BREEDING DISTRIBUTION CHANGE *since 1968–72*

GAIN ▲	2	11	13
LOSS ▽	31	648	679
40yr CHANGE	-94%	-51%	-52%
20yr INDEX	-	-0.59	

BREEDING RELATIVE ABUNDANCE *2008–11*

WINTER DISTRIBUTION *2007/08–2010/11*

PRESENT ●	<1%	1%	<1%

RING-NECKED PARAKEET

Psittacula krameri

BTO CODE: RI

RICHARD WINSTON www.flickr.com/photos/honeycreeper

AFTER SUSPECTED INSTANCES in Norfolk, Northamptonshire and Kent, breeding by released or escaped Ring-necked Parakeets was first confirmed in Kent in 1971 (*Birds in England*). It was recorded in just four 10-km squares in the *1968–72 Breeding Atlas*. By the *1988–91 Breeding Atlas* a significant range expansion had occurred, particularly in Greater London and parts of eastern Kent, but the species was still largely restricted to southeastern England. Subsequently, the northern range boundary has advanced and parakeets have colonised several large conurbations, including Liverpool, Manchester, Sheffield and Birmingham and *non-breeding* birds were reported from several squares in southern Scotland. Non-native populations elsewhere in Europe are expanding in a similar manner (Strubbe & Matthysen 2009a). Apparent range contraction along the English south coast mostly reflects different treatment of *non-breeding* birds in the *1988–91 Breeding Atlas*, although breeding pairs have disappeared from Sussex (James 1996). Parakeets are generally more mobile during winter and the distribution map reflects this, with sightings scattered across much of England, excluding the southwest. The only record in Ireland was in winter.

Annual monitoring indicates a tenfold increase in numbers during 1995–2010 (*SUKB 2012*). The current population estimate stands at 8,600 breeding pairs or c.30,000 individuals post-breeding (Musgrove *et al.* 2013; Project Parakeet 2013). Given this rapid population growth it is surprising that the species has not spread further. Butler (2003) estimated that, over the previous decade, the Greater London population had been increasing at an annual rate of 30% yet spreading by only 0.4 km per year, leading to a sharp increase in density. The abundance change map shows that increases in tetrad occupancy since the *1988–91 Breeding Atlas* have been restricted to the core range in Greater London and eastern Kent.

Parakeets are still closely associated with areas of human habitation. Were they to colonise other habitats, detrimental impacts on native species might become evident (Strubbe *et al.* 2010). In Belgium, Nuthatch numbers are lower in areas containing more parakeet pairs, suggesting that there may be competition for nest cavities (Strubbe & Matthysen 2007, 2009b). No such relationship is currently apparent in Britain (Newson *et al.* 2011).

Sponsored by Derek Gilby

BREEDING DISTRIBUTION *2008–11*

POSSIBLE •	-	1%	1%
PROBABLE •	-	1%	1%
CONFIRMED •	-	2%	1%
TOTAL	-	4%	3%

WINTER DISTRIBUTION *2007/08–2010/11*

PRESENT •	<1%	8%	6%

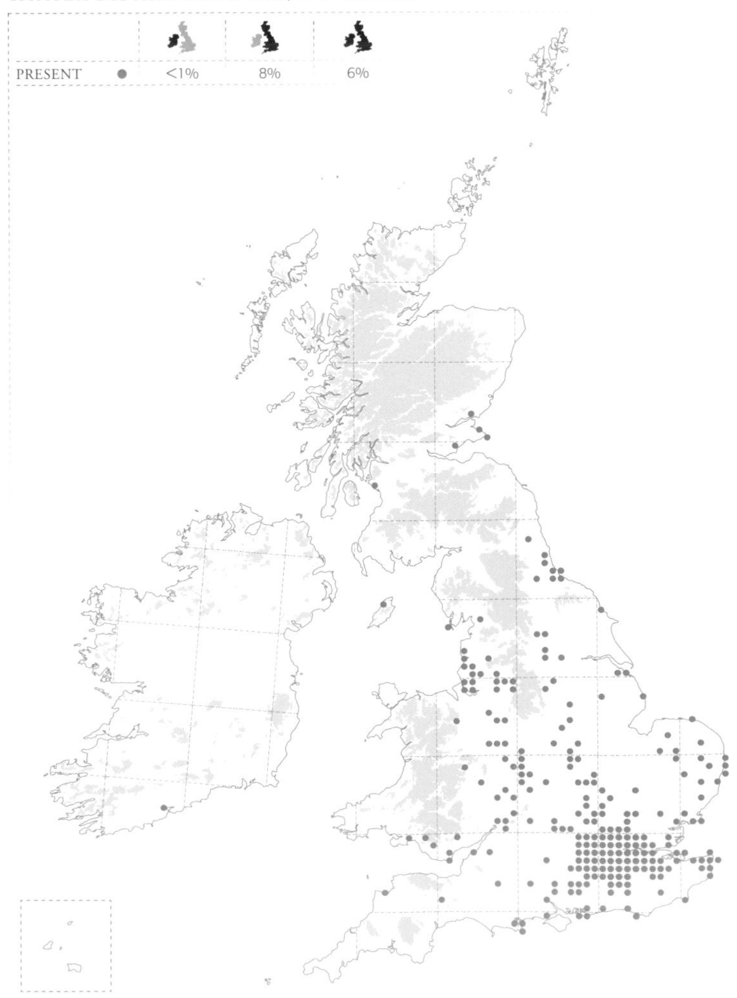

BREEDING DISTRIBUTION CHANGE *since 1968–72*

GAIN ▲	-	88	88
LOSS ▽	-	0	0
40yr CHANGE	-	+4400%	+4400%
20yr INDEX	-	+0.79	

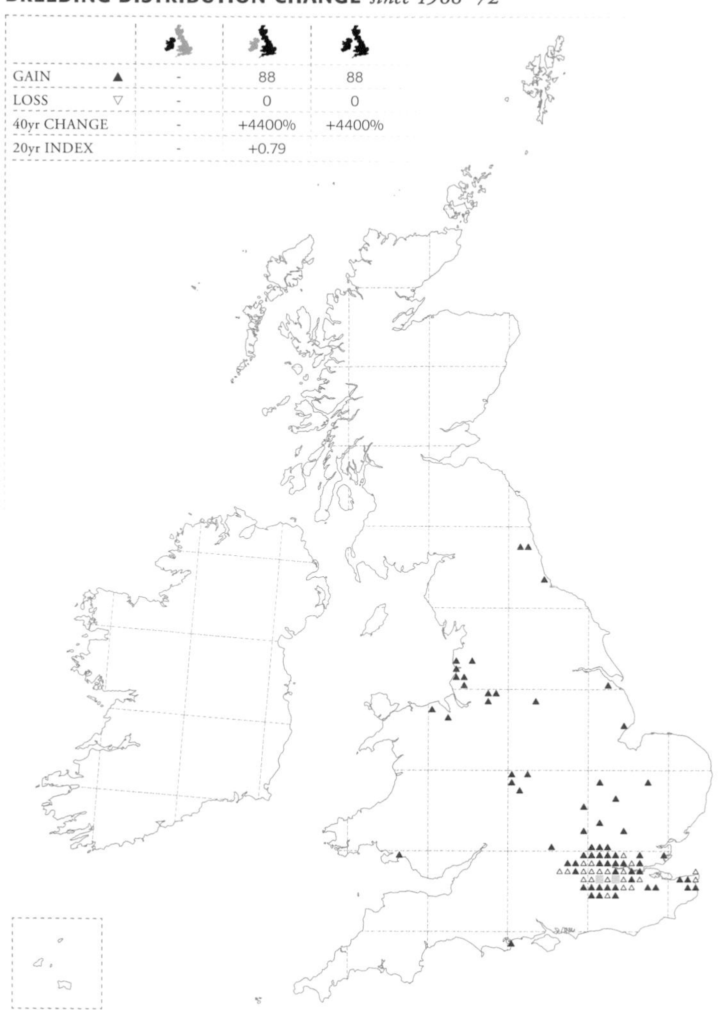

BREEDING RELATIVE ABUNDANCE *2008–11*

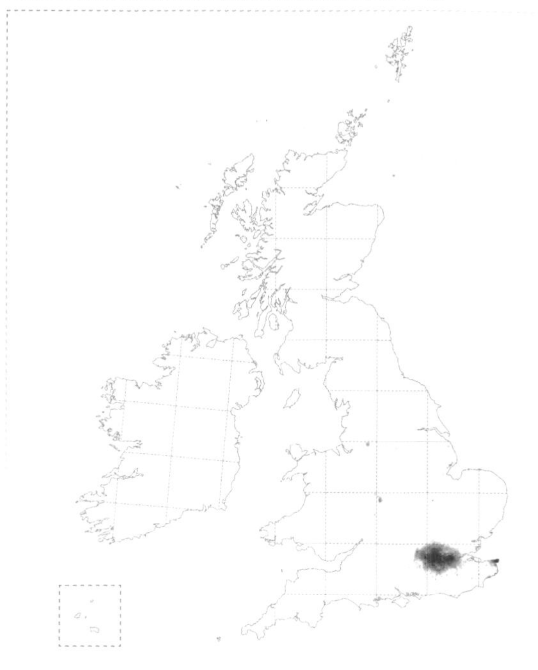

WINTER DISTRIBUTION CHANGE *since 1981–84*

GAIN ▲	1	184	185
LOSS ▼	0	34	34
30yr CHANGE	∞	+214%	+216%

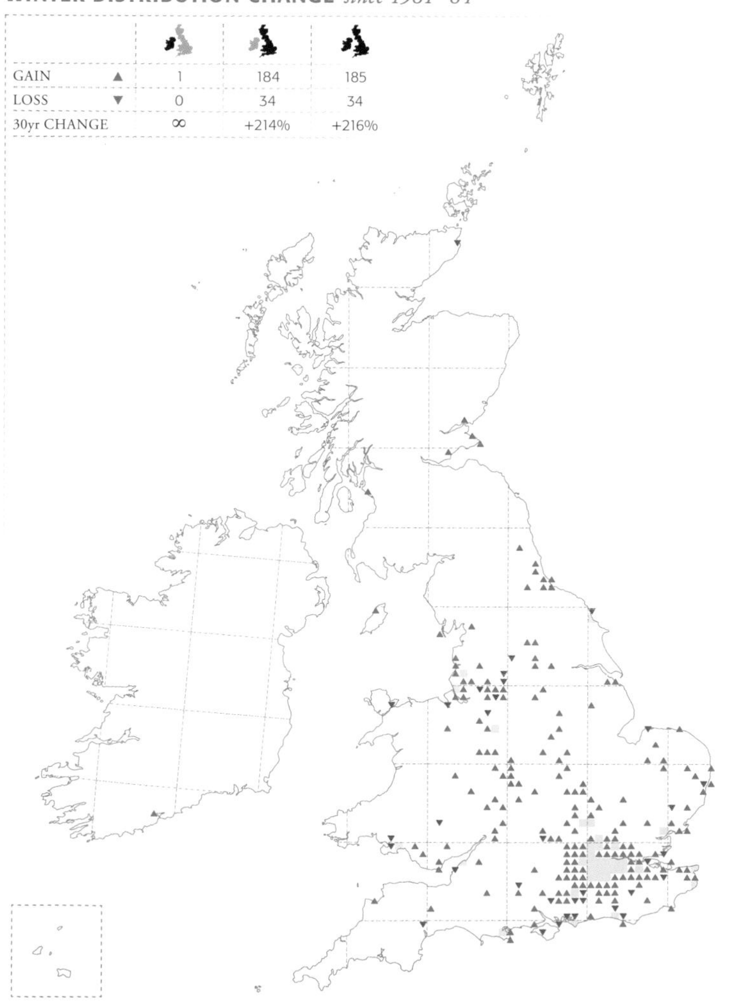

BREEDING RELATIVE ABUNDANCE CHANGE *since 1988–91*

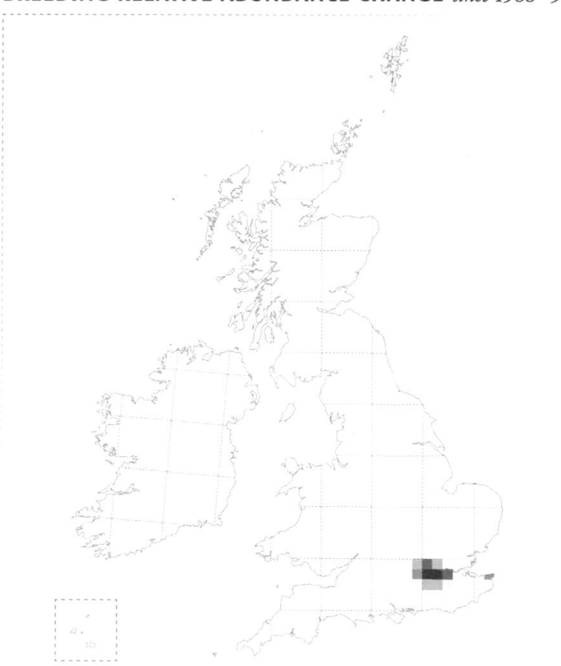

WINTER RELATIVE ABUNDANCE *2007/08–2010/11*

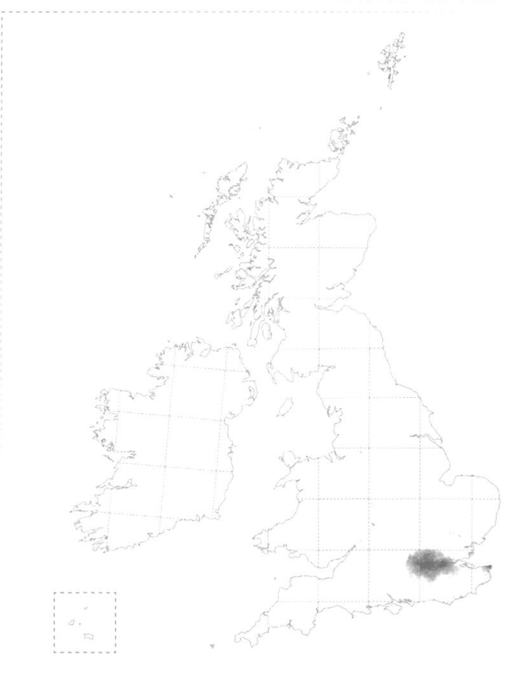

CUCKOO

Cuculus canorus

BTO CODE: CK

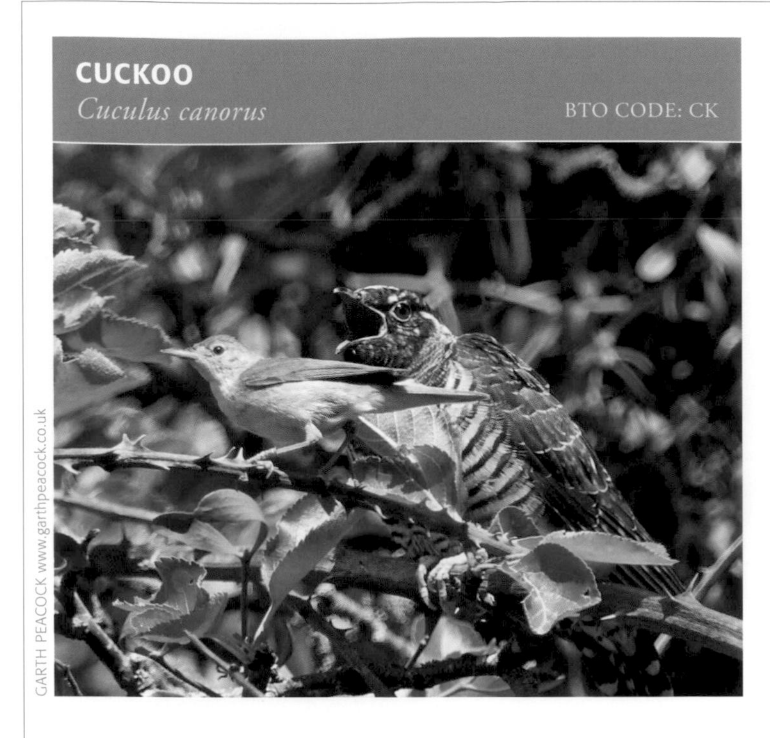

GARTH PEACOCK www.garthpeacock.co.uk

THE FIRST CALL of the Cuckoo is perhaps the most anticipated sound of a British or Irish summer. Breeding Cuckoos are found in 78% of 10-km squares in a range of habitat types, although they reach highest densities in upland and marginal habitats. Dartmoor, Exmoor and the New Forest in England, and the Brecon Beacons in Wales, hold relatively high densities but abundance peaks in western and northern Scotland and in western Ireland.

The Cuckoo's British & Irish range contracted by 10% between 1968–72 and 1988–91, and by a further 2% up to 2008–11. The majority of losses were in eastern Ireland, with a scattering through low-lying and coastal areas of Britain. More significant changes are apparent from the abundance change map. In Britain, areas to the south of the Scottish central lowlands experienced marked declines in tetrad occupancy since the *1988–91 Breeding Atlas*, whereas areas to the north have seen increases. In Ireland the picture is similar, with increases in the west outweighed by decreases in Northern Ireland and in the south and east.

These trends are matched by those in population size. The Republic of Ireland recorded a modest 6% decline during 1998–2010 (*CBS Trend 2010*). A 49% decline in the UK during 1995–2010 comprises declines of 5% in Scotland, 34% in Wales and 63% in England (*BBS Report 2011*). Furthermore, a 21% decline is evident since 1980 across Europe as a whole (*PECBMS 2010*).

Cuckoos are among several declining migrants that winter in the humid zone of Africa (Ockendon *et al.* 2012). Cuckoos have not advanced spring arrival times to the same extent as their hosts, raising the possibility of decreased availability of nests for this brood parasite (Saino *et al.* 2009). Douglas *et al.* (2010) found instead, however, that in Britain earlier nesting by Reed Warblers may have increased nest availability, which may explain the heavier parasitism of that species (Brooke & Davies 1987). Alternative possible contributors to decline include reduced breeding-season prey availability, owing to falling British moth populations (Fox *et al.* 2006), and unknown factors acting on the wintering grounds or during migration.

Sponsored by Harris, Elsie, David & Peter Glue, New Forest, Hampshire

BREEDING DISTRIBUTION *2008–11*

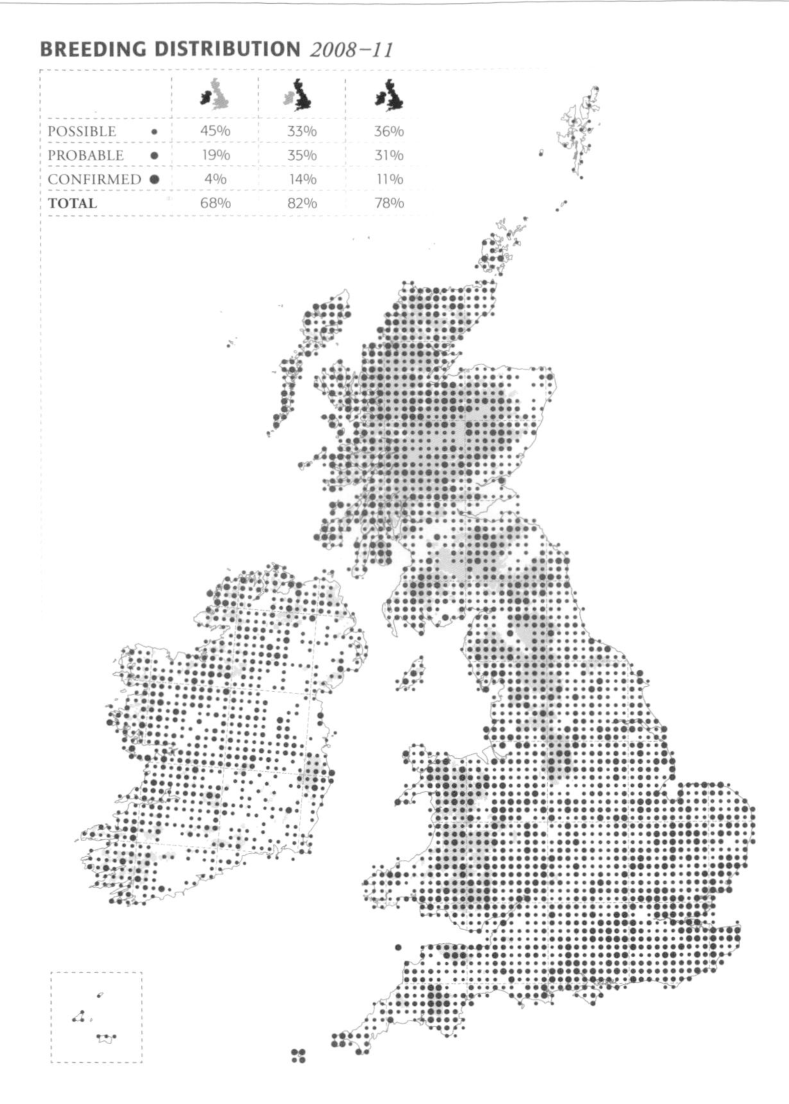

POSSIBLE ●	45%	33%	36%
PROBABLE ●	19%	35%	31%
CONFIRMED ●	4%	14%	11%
TOTAL	68%	82%	78%

BREEDING RELATIVE ABUNDANCE CHANGE *since 1988–91*

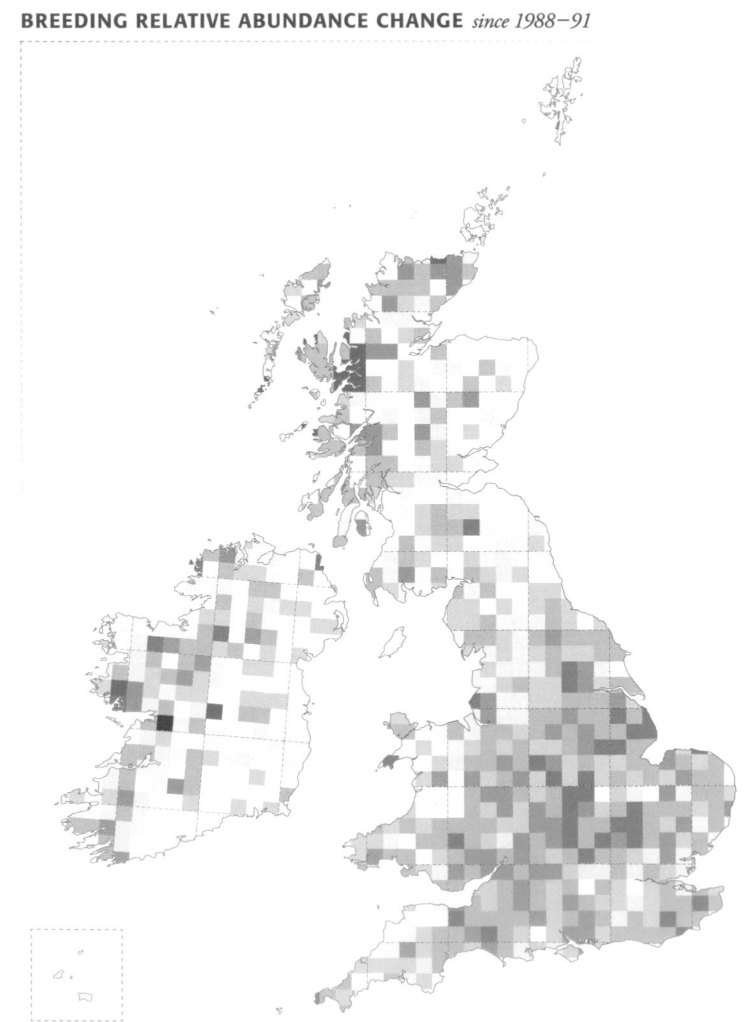

BREEDING DISTRIBUTION CHANGE *since 1968–72*

BREEDING RELATIVE ABUNDANCE *2008–11*

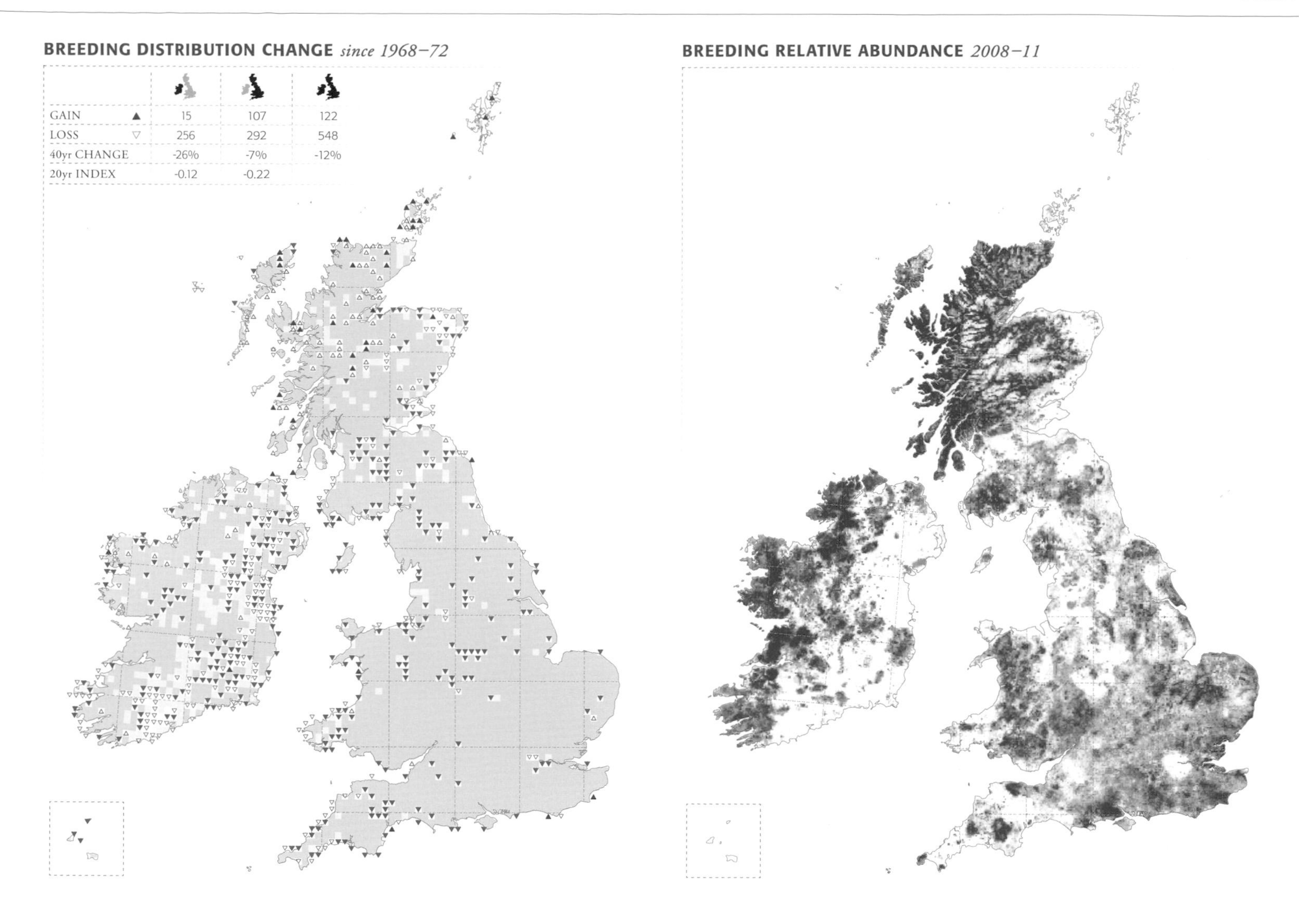

GAIN ▲	15	107	122
LOSS ▽	256	292	548
40yr CHANGE	-26%	-7%	-12%
20yr INDEX	-0.12	-0.22	

◀ Work to track the migration routes and wintering locations of Cuckoos from different parts of the breeding range, in a bid to understand the causes of declines, is already producing valuable results.

DAVID KJAER www.davidkjaer.com

BARN OWL
Tyto alba

BTO CODE: BO

PAUL NEWTON www.paulnewtonphotography.com

BARN OWLS HAVE been in decline since the mid 19th century (Shawyer 1987) owing to changes in agriculture, to mortality associated with road traffic and to loss of nest sites (Ramsden 1998; Toms *et al.* 2001; Nagle 2007).

Year-round, Barn Owls are widely distributed in Britain, avoiding only high-altitude and urban areas and being absent from remoter islands, including the Outer Hebrides and Northern Isles. Diurnal surveys are inefficient for this largely nocturnal species and change estimates may be especially sensitive to biases in effort. Maps of abundance suggest appreciably higher densities in eastern England than elsewhere, but may be confounded by behavioural differences (Palmer 2013). In Ireland the distribution is sparser, with highest densities in the southwest, perhaps attributable to the presence of introduced small mammals.

Large range expansions are shown for both islands in winter. In Scotland, gains could partly reflect improved coverage but also include real gains, probably facilitated by recent mild winters and conservation initiatives (*Birds of Scotland*). By contrast, the apparent 79% winter expansion in Ireland contradicts evidence of a long-term population decline and is likely to be an artefact of low coverage during the *1981–84 Winter Atlas* (*Birds in Ireland*) and recent improvements in Barn Owl monitoring in Ireland.

In Britain, atlas data suggest a 38% breeding range contraction from 1968–72 to 1988–91, followed by a major expansion (67%). Squares that were apparently unoccupied only in 1988–91 are shown in pale pink on the change map: some may be artefacts of the low emphasis on supplementary recording in the *1988–91 Breeding Atlas*. However, mild winters (Dadam *et al.* 2011), nestbox provision, captive release programmes (Meek *et al.* 2003) and the provision and management of field margins through agri-environment schemes (Askew *et al.* 2007) may have facilitated recovery in some areas. The pattern is very different in Ireland, where declines since the 1950s (Bunn *et al.* 1982) have resulted in a 47% range contraction since the *1968–72 Breeding Atlas*, with the greatest losses in northern and eastern Ireland. Declines may be related to agricultural intensification and greater exposure to rodenticides in Ireland than in Britain (Lusby 2012).

Sheepdrove Trust

Sponsored by The Sheepdrove Trust and Wildlife Conservation Partnership

BREEDING DISTRIBUTION *2008–11*

	Ireland	Britain	Britain & Ireland
POSSIBLE •	12%	7%	9%
PROBABLE ●	2%	4%	3%
CONFIRMED ●	12%	55%	44%
TOTAL	26%	66%	56%

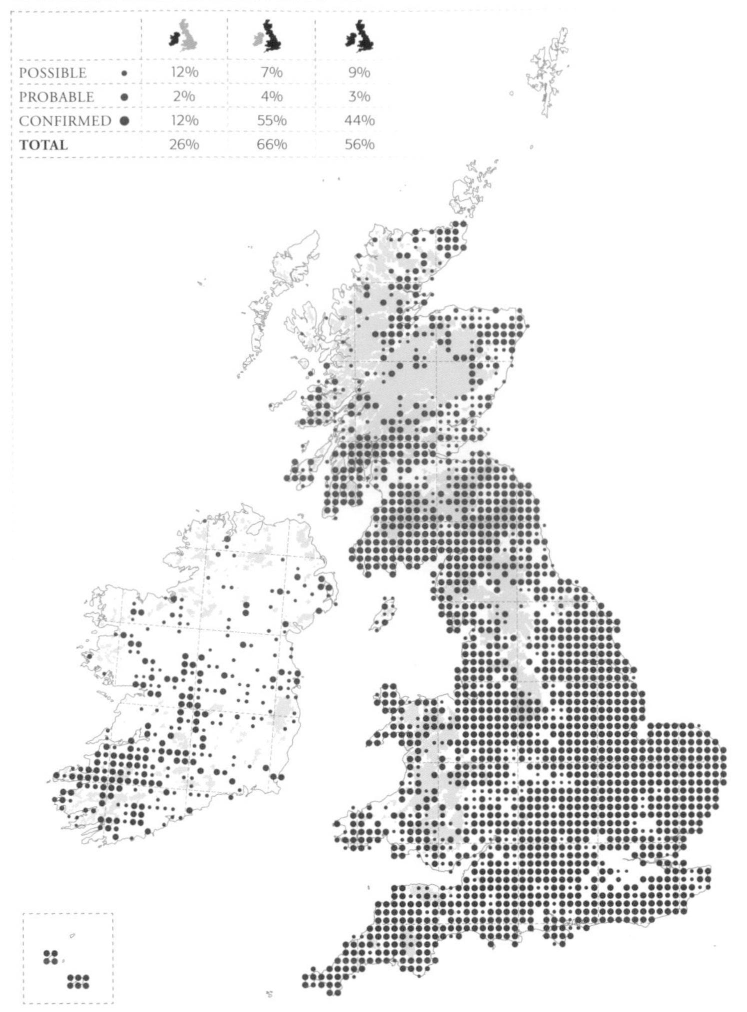

WINTER DISTRIBUTION *2007/08–2010/11*

	Ireland	Britain	Britain & Ireland
PRESENT ●	25%	70%	58%

BREEDING DISTRIBUTION CHANGE *since 1968–72*

GAIN ▲	79	378	457
LOSS ▽	313	294	607
40yr CHANGE	-47%	+5%	-7%
20yr INDEX	-0.44	+0.26	

BREEDING RELATIVE ABUNDANCE *2008–11*

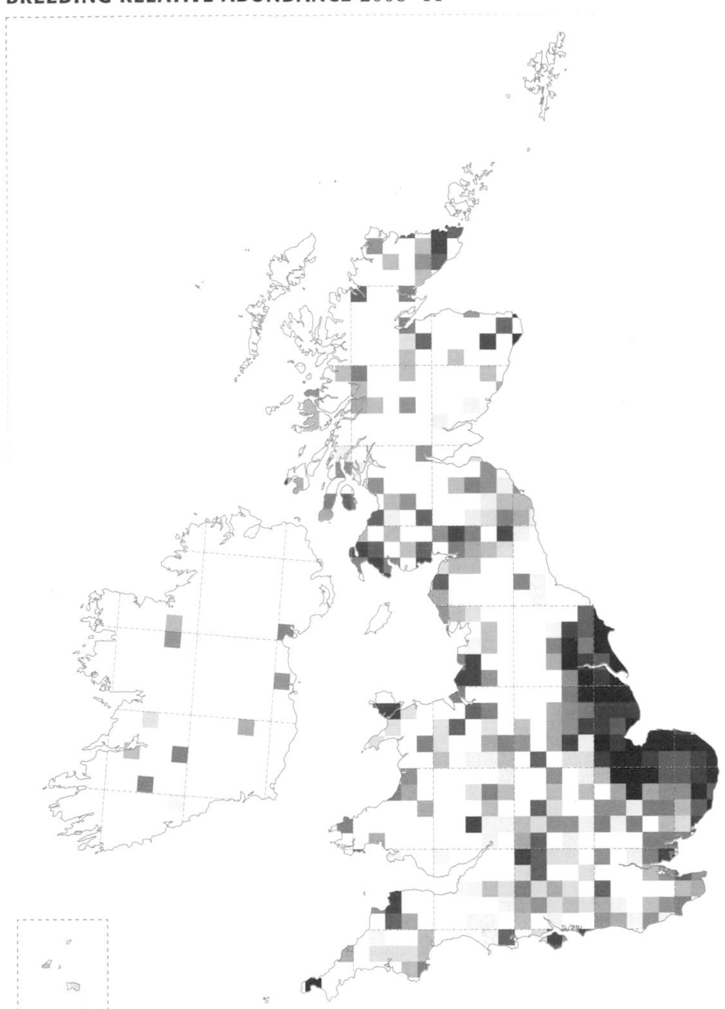

WINTER DISTRIBUTION CHANGE *since 1981–84*

GAIN ▲	184	925	1109
LOSS ▼	78	109	187
30yr CHANGE	+79%	+71%	+72%

WINTER RELATIVE ABUNDANCE *2007/08–2010/11*

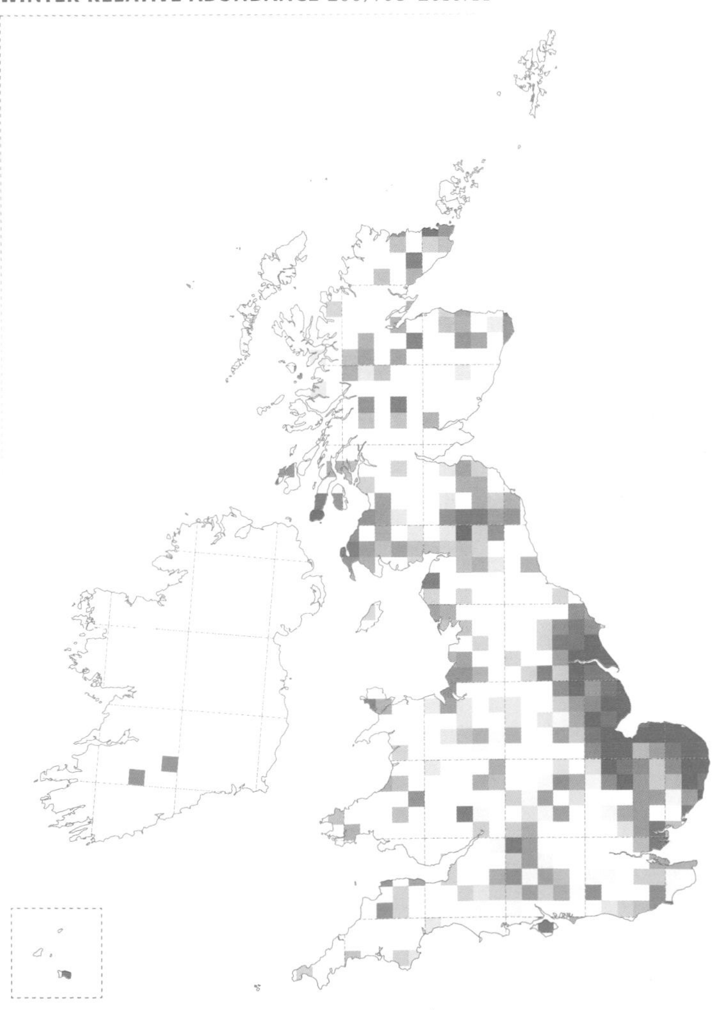

SNOWY OWL
Bubo scandiacus

BTO CODE: SO

CHRIS BALE www.guernseybirdnerd.com

THE SNOWY OWL has a circumpolar breeding distribution on the high-arctic tundra. In winter the birds are largely nomadic, with dispersal a response to food availability. In irruption years, some individuals may wander southward from the breeding range and these birds probably account for the bulk of the 400 or so records in Britain and 70 or so in Ireland. Migrants in Scotland are thought to originate mainly from the European Arctic, an idea supported by the recovery of one Norwegian-ringed bird in the Outer Hebrides (*Birds of Scotland*). Birds might also arrive from the Canadian Arctic, with dispersive birds taking refuge on ships en route to Europe.

Influxes in the 1960s may have facilitated the colonisation by Snowy Owls of Fetlar, in Shetland. A pair bred there each year from 1967 to 1975, with a second nesting female present in 1973–74, and fledged between one and five chicks annually, except in 1972 (Robinson & Becker 1986). Following the death or departure of the breeding male in winter 1975/76, the remaining lone females laid infertile eggs during 1980–90 (*Birds of Scotland*). In Ireland the only known breeding attempt concerned a pair that laid eggs in 2001 but subsequently abandoned them (Hillis 2002).

Since 2001 there have been no known breeding attempts in Britain & Ireland. During the 2008–11 breeding seasons a *non-breeding* adult female was reported from various locations in nine 10-km squares on the west coast of Ireland. In Britain *non-breeding* birds were reported from 14 10-km squares. At least one mobile male summered each year during 2008–11 in the Outer Hebrides, resulting in a scatter of occupied squares. Another female summered on Lihou in the Channel Islands in 2009, where it fed on gull chicks. Most other records were of short-staying migrants.

During the Atlas period, Snowy Owls were recorded during the winter in 13 10-km squares in Britain and five in Ireland. This was a marked increase on the two squares occupied during the *1981–84 Winter Atlas*. The map shows six groups of occupied 10-km squares, which probably related to six or seven mobile individual birds. Two were present in the Channel Islands in winter 2008/09.

BREEDING DISTRIBUTION *2008–11*

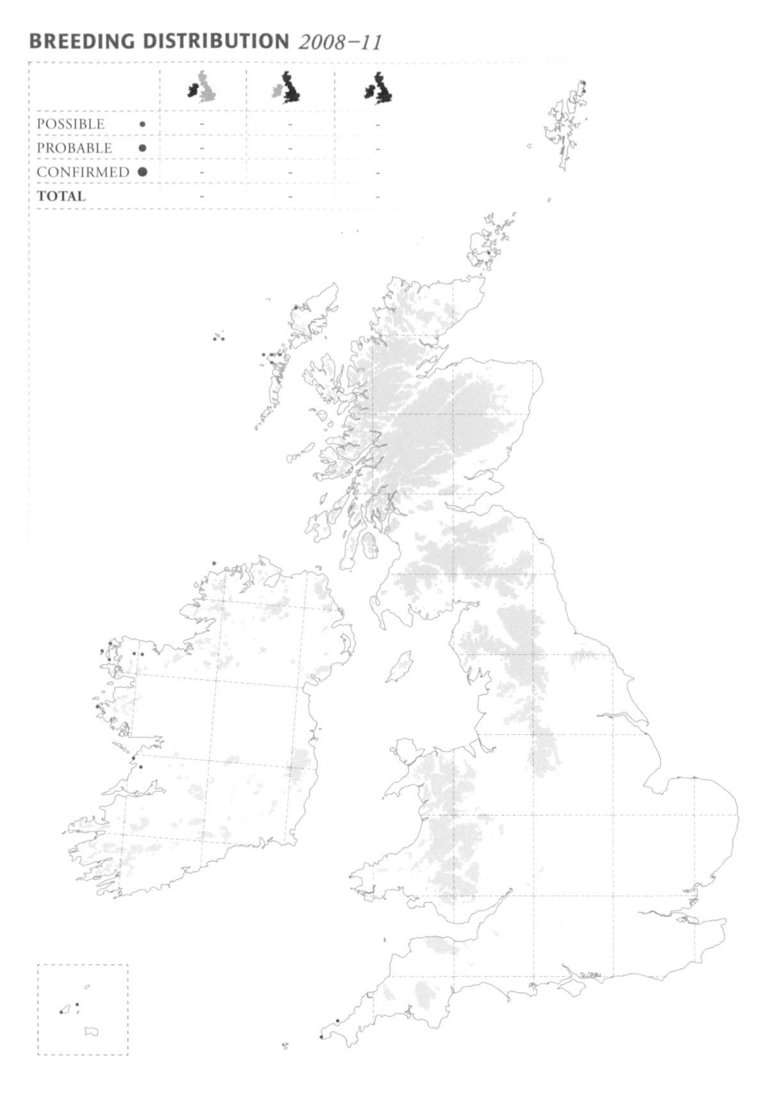

	Ireland	Britain	Britain & Ireland
POSSIBLE •	-	-	-
PROBABLE ●	-	-	-
CONFIRMED ●	-	-	-
TOTAL	-	-	-

BREEDING DISTRIBUTION CHANGE *since 1968–72*

	Ireland	Britain	Britain & Ireland
GAIN ▲	-	0	0
LOSS ▽	-	5	5
40yr CHANGE	-	-100%	-100%
20yr INDEX	-	-	

Sponsored by Mary Waller

WINTER DISTRIBUTION *2007/08–2010/11*

PRESENT ●	<1%	<1%	<1%

WINTER DISTRIBUTION CHANGE *since 1981–84*

GAIN ▲	4	12	16
LOSS ▼	0	2	2
30yr CHANGE	∞	+500%	+700%

◀ This mobile individual accounted for several of the occupied 10-km squares on the Hebrides. Others wandered around Orkney, the Cairngorms, Counties Mayo and Galway, southwest England and the Channel Islands.

GARY JENKINS www.garyjenkinsbirdphotography.blogspot.co.uk

LITTLE OWL
Athene noctua

BTO CODE: LO

LIZ CUTTING www.lizcuttingphotos.com

THE BRITISH POPULATION of the Little Owl derives from introductions made during the late 1800s (Parkin & Knox 2010). By 1900, Little Owls were regular breeders in several English counties and a period of rapid expansion followed, lasting into the 1930s. The first Welsh breeding record came in 1916 and that for Scotland, where the species has subsequently lost its foothold, came in 1958. There have been only four records from Ireland (*Birds in Ireland*), none during 2007–11.

The winter and breeding-season distributions are similar, reflecting the sedentary nature of these owls in Britain (*Migration Atlas*). The species is largely restricted to England and the Welsh borders, with isolated populations in northwest and coastal Wales. In parts of the Little Owl's native range, low temperatures and lying snow reduce survival (Schönn 1986) and this may explain why the British uplands remain unoccupied.

Abundance maps show that the Little Owl is more abundant within East Anglia, the Midlands and central northern England than in the more westerly parts of its breeding range. Little Owls generally are most abundant within areas of mixed habitat, where there are small parcels of farmland, woodland edge and hedgerows (Van Nieuwenhuyse *et al.* 2008); the Atlas data seem to indicate this is true also in Britain.

Whereas there is evidence of expansion within northeast England since 1988–91, this was outweighed by the marked range contraction throughout southwest England and across much of Wales between 1968–72 and 1988–91. Since the 1980s, Little Owls have undergone declines of 47% in Gwent and 81% in Pembrokeshire in the numbers of occupied tetrads (Venables *et al.* 2008; Rees *et al.* 2009). The abundance change map shows similar losses throughout the western part of the range since the *1988–91 Breeding Atlas*, in line with a population decline of 40% during 1995–2010 (*BBS Report 2011*). This pattern is repeated in several other western European countries (Van Nieuwenhuyse *et al.* 2008). It is thought that changing farming practices might have depressed juvenile survival by reducing their food supply (*BirdTrends*). The decreased fledging success reported in Denmark (Thorup *et al.* 2010) is not currently apparent in Britain (*BirdTrends*).

Sponsored by Athene Communications Ltd

BREEDING DISTRIBUTION *2008–11*

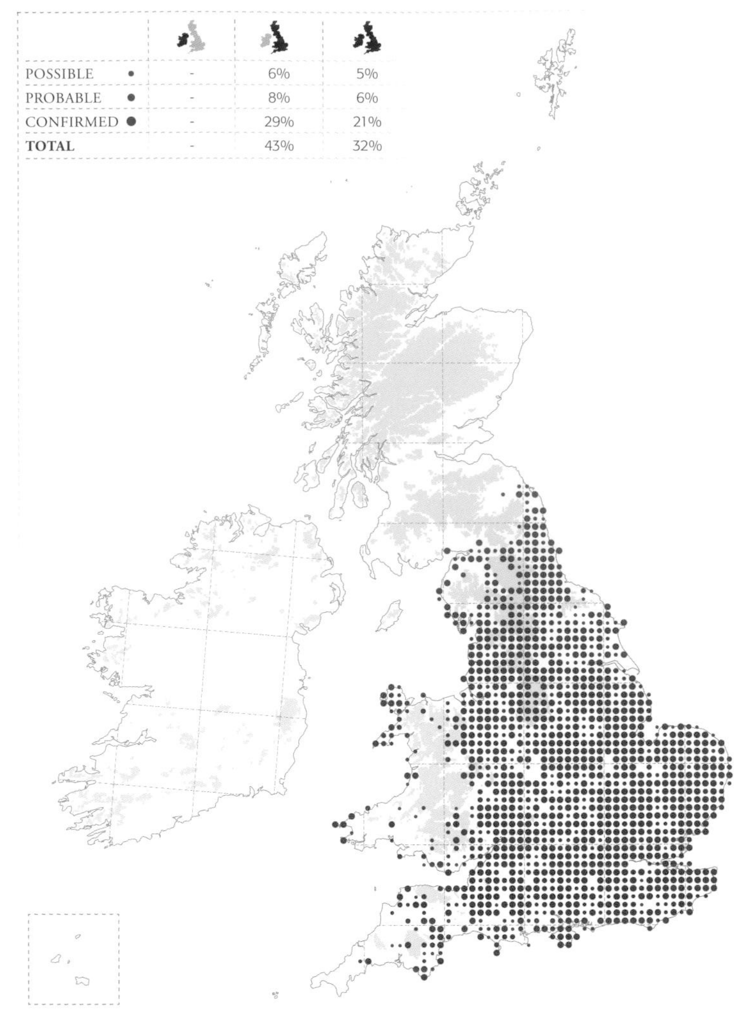

POSSIBLE •	-	6%	5%
PROBABLE •	-	8%	6%
CONFIRMED •	-	29%	21%
TOTAL	-	43%	32%

WINTER DISTRIBUTION *2007/08–2010/11*

PRESENT •	-	41%	30%

BREEDING DISTRIBUTION CHANGE *since 1968–72*

GAIN ▲	-	102	102
LOSS ▽	-	250	250
40yr CHANGE	-	-11%	-11%
20yr INDEX	-	-0.22	

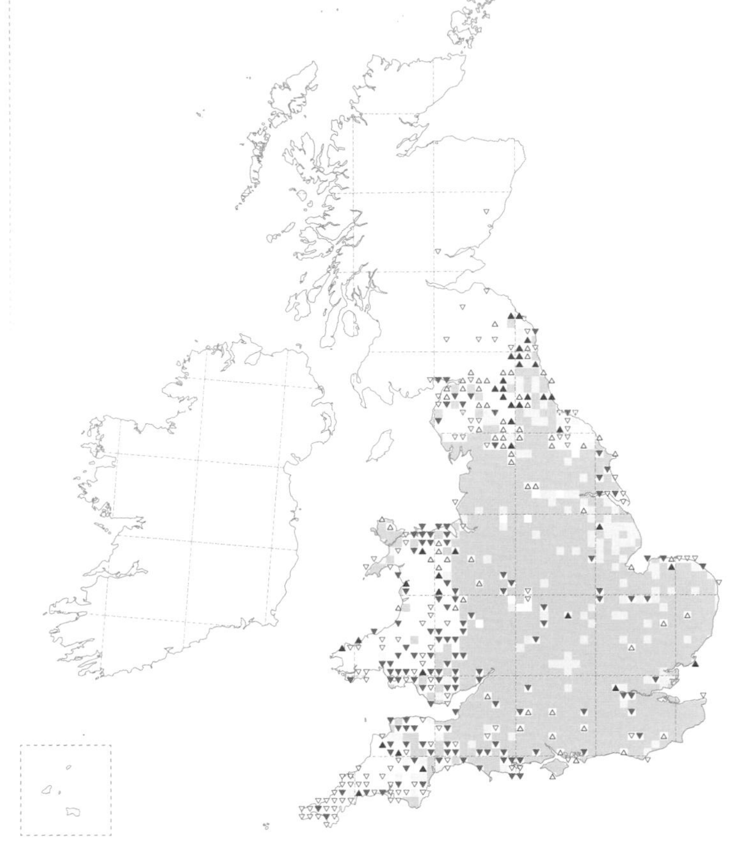

WINTER DISTRIBUTION CHANGE *since 1981–84*

GAIN ▲	0	300	300
LOSS ▼	1	168	169
30yr CHANGE	-100%	+13%	+13%

BREEDING RELATIVE ABUNDANCE *2008–11*

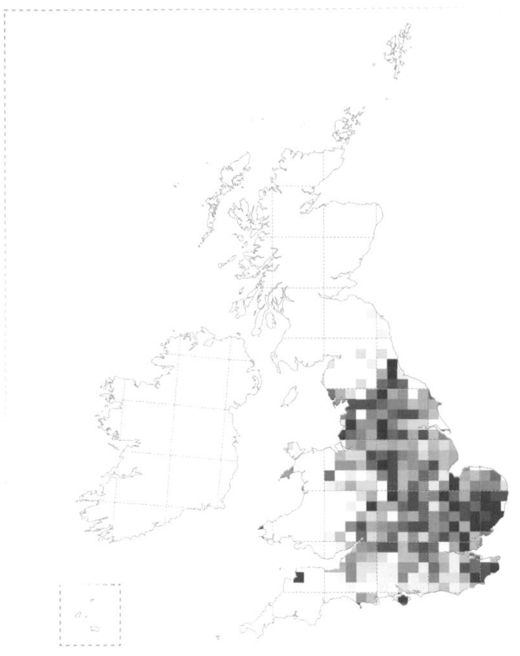

BREEDING RELATIVE ABUNDANCE CHANGE *since 1988–91*

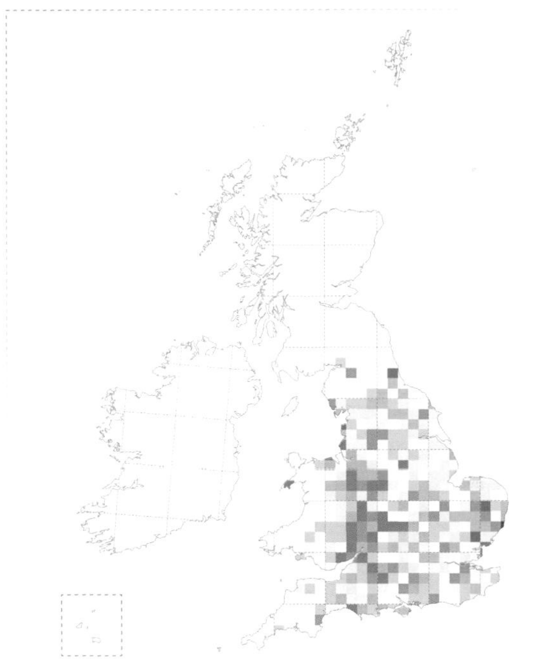

WINTER RELATIVE ABUNDANCE *2007/08–2010/11*

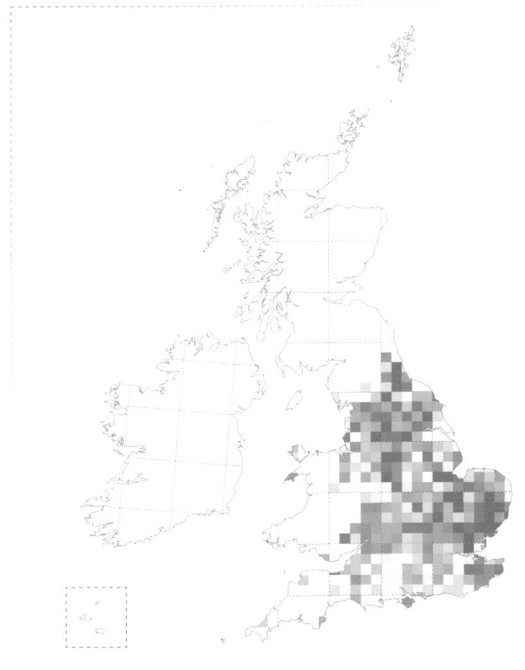

TAWNY OWL

Strix aluco

BTO CODE: TO

NIELS JENSEN www.flickr.com/photos/20952345@N03

TAWNY OWLS ARE resident in woodlands throughout Britain but absent from the Channel Islands and from Ireland, although a bird presumed to be wild was present in Co. Down in 2013. Within Britain they are absent from the Northern Isles, Outer Hebrides, some of the smaller Inner Hebridean islands, the Isles of Scilly and treeless upland areas elsewhere, particularly in northern Scotland. Drawn from diurnal TTVs, abundance maps for this nocturnal species show only the crudest of patterns, suggesting lower densities in major conurbations and open agricultural areas. As previous atlases have shown, distribution patterns vary little by season.

The change maps show a 25% increase in the number of occupied 10-km squares since the *1981–84 Winter Atlas*, contrasting with a 6% decrease since the *1968–72 Breeding Atlas*. There are some areas where gains have definitely occurred. These include the far north and west of Scotland, where maturing plantations have created suitable new habitat. Tawny Owls have recently colonised the Isle of Man, where a bird had been reported regularly since autumn 2000 and breeding was confirmed in 2005 at a site in the north (Sharpe 2007). Since then, birds have become established in three 10-km squares in winter and four in the breeding season. Elsewhere it would appear that the large increases in the numbers of occupied 10-km squares in winter are due primarily to better coverage, aided by nationwide press appeals for Atlas records of calling Tawny Owls.

Standard atlas methods are not ideal for Tawny Owls, owing to their low rates of detection and because they are most vocal in autumn, outside the standard field seasons (Freeman *et al.* 2006). Many of the changes in the breeding season are in low-density areas at range margins and, especially where effort was also low, may reflect failures in detection. However, the 6% range contraction suggested by these data agrees with local atlases (e.g. Norman 2008) and with the shallow 17% decline during 1967–2010 reported from long-term monitoring (*BirdTrends*). The causes of this apparent decline are unknown; Walker et al. (2008) found no evidence of a link with exposure to rodenticides.

In memory of Dilys Breese

BREEDING DISTRIBUTION *2008–11*

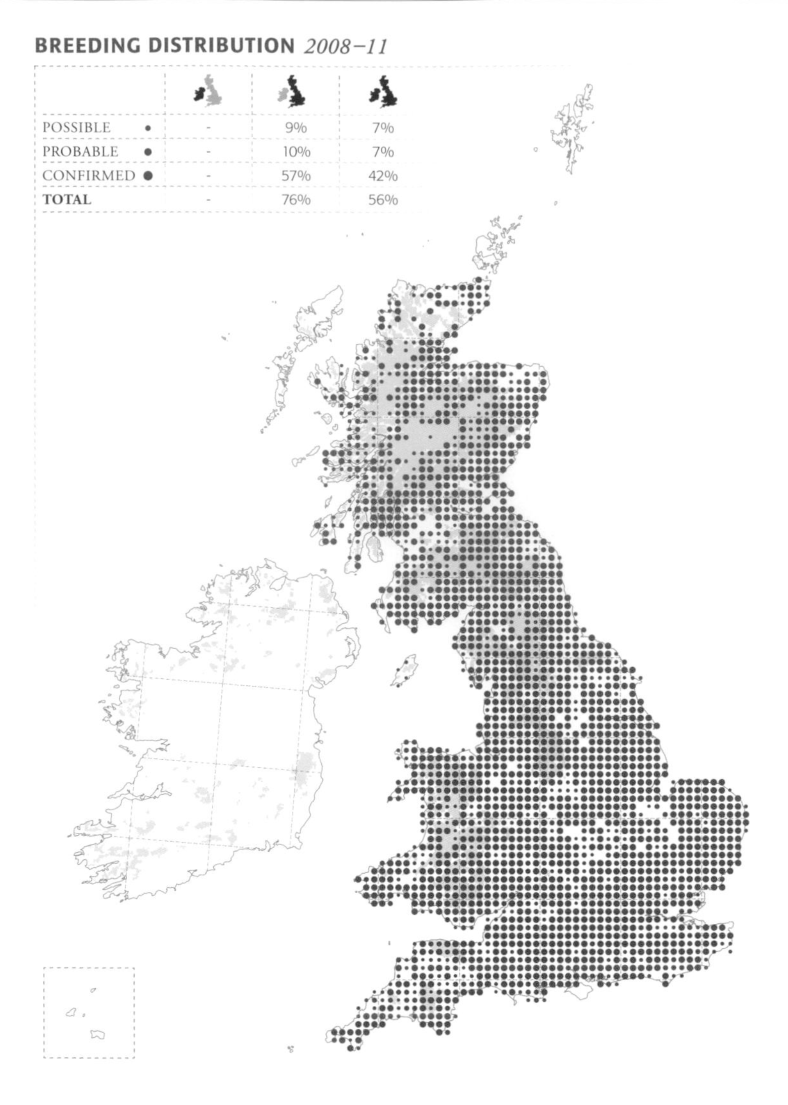

	Ireland	Britain	Britain & Ireland
POSSIBLE •	-	9%	7%
PROBABLE •	-	10%	7%
CONFIRMED •	-	57%	42%
TOTAL	-	76%	56%

WINTER DISTRIBUTION *2007/08–2010/11*

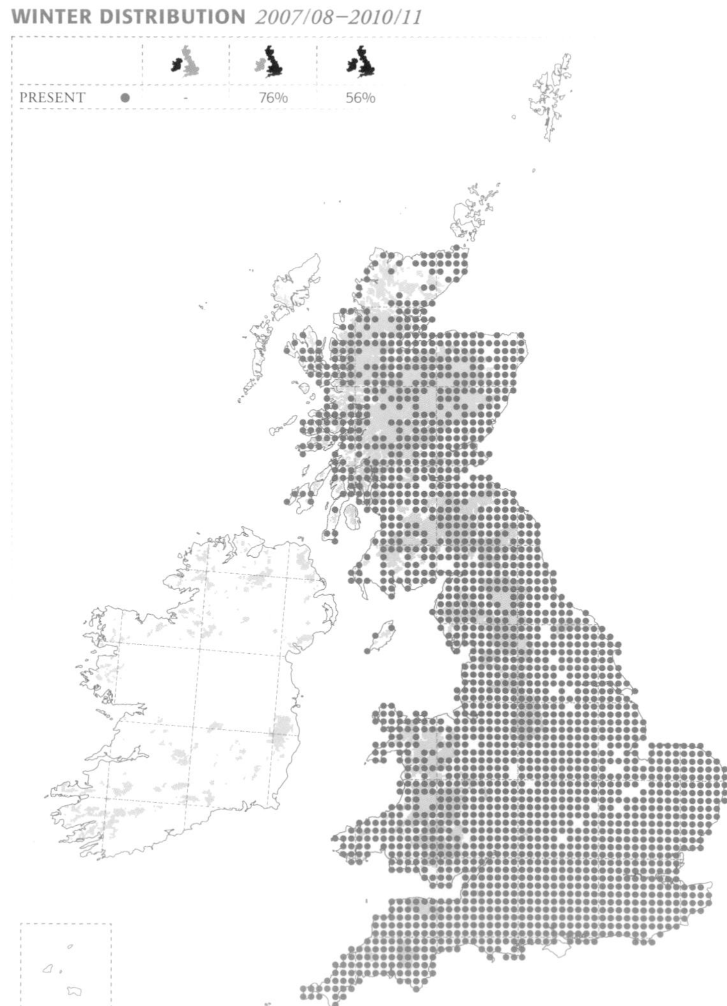

	Ireland	Britain	Britain & Ireland
PRESENT •	-	76%	56%

BREEDING DISTRIBUTION CHANGE *since 1968–72*

GAIN ▲	-	103	103
LOSS ▽	-	236	236
40yr CHANGE	-	-6%	-6%
20yr INDEX	-	-0.33	

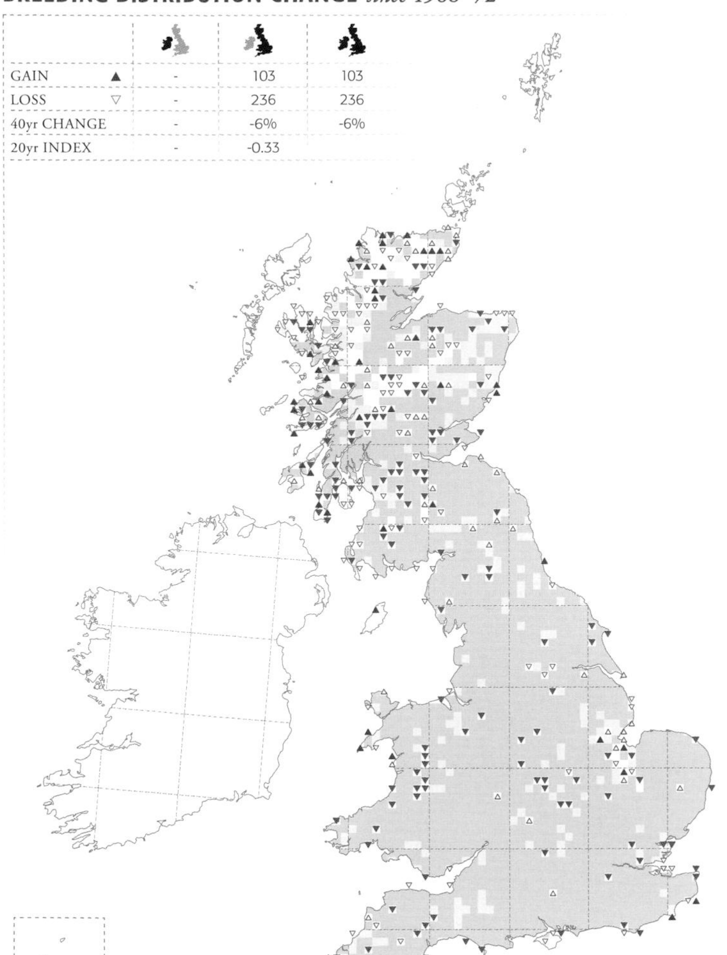

BREEDING RELATIVE ABUNDANCE *2008–11*

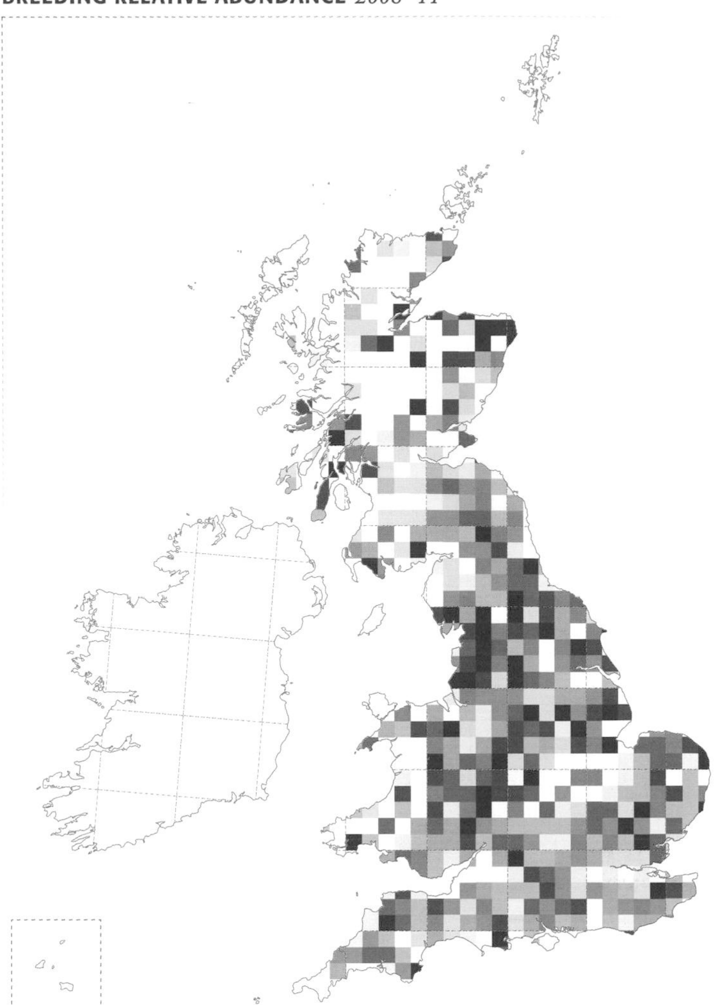

WINTER DISTRIBUTION CHANGE *since 1981–84*

GAIN ▲	-	571	571
LOSS ▼	-	142	142
30yr CHANGE	-	+25%	+25%

WINTER RELATIVE ABUNDANCE *2007/08–2010/11*

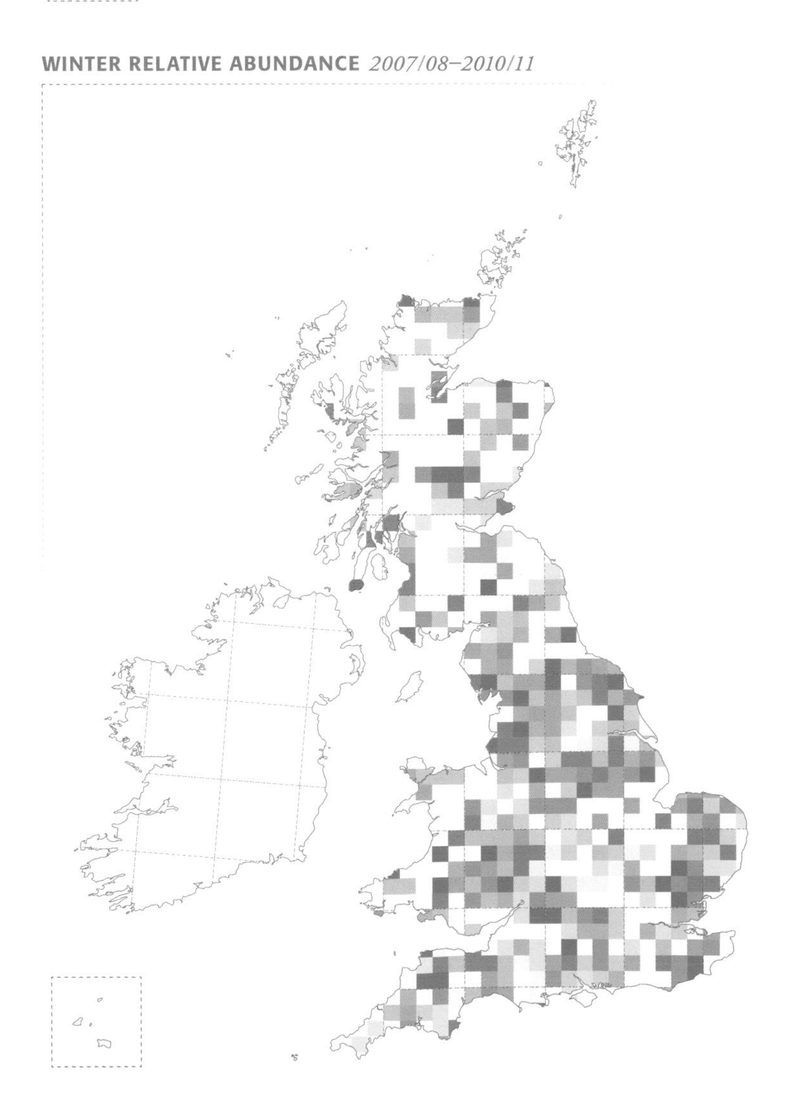

LONG-EARED OWL

Asio otus

BTO CODE: LE

DAVID KAY

THE RANGE OF the Long-eared Owl has changed little since the *1968–72 Breeding Atlas*, with breeding detected at scattered locations in Britain and throughout Ireland. In Ireland it is the most abundant owl and probably benefits from the absence of the competitively dominant Tawny Owl (*Birds in Ireland*). Distributions are subject to under-recording owing to the species' discreet nature and preference for dense woodland. Although territorial behaviour can indicate occupation in early spring, presence is frequently revealed by the penetrating and distinctive food-begging calls of fledglings. Unsuccessful breeders are often overlooked and specific nocturnal fieldwork can introduce biases associated with observer distribution, particularly in remoter areas.

The British range has apparently contracted by 18% since the *1968–72 Breeding Atlas* but there is also marked turnover, with more than half of squares showing gains or losses since 1968–72. This may be partly a consequence of birds being missed in one atlas or another, but the significant gains in northern England and clusters of losses in southeast England, southwest Scotland and the Black Isle are notable. A range expansion of 12% has been recorded in Ireland for the same period, with marked apparent spread in the southwest, particularly in Co. Cork. In some areas, gains may stem from improved knowledge following local surveys (e.g. Piercy 2007) or from the expansion of conifer plantations (*Birds in Ireland*).

Long-eared Owls are sedentary throughout Britain & Ireland but are joined by annually variable numbers of autumn immigrants from Fennoscandia (*Migration Atlas*). The winter distribution is patchy throughout both Britain and Ireland and likely to suffer from under-recording. Overall, Long-eared Owls were recorded in 51% more 10-km squares than in the *1981–84 Winter Atlas*, but again there are many gains and losses that are indicative of variable but generally increasing coverage. The influence of cold winters in making owls more active and detectable, or prompting larger influxes, cannot be discounted. Abundance maps are not shown, because few Long-eared Owls were encountered on TTVs, but double-figure winter counts were received of 18 in Durham, 12 in West Yorkshire and 10 in Lancashire. The highest count in Ireland was of five near Coleraine, Co. Londonderry.

BREEDING DISTRIBUTION *2008–11*

	Ireland	Britain	Britain & Ireland
POSSIBLE •	7%	5%	5%
PROBABLE •	4%	2%	2%
CONFIRMED •	29%	11%	16%
TOTAL	40%	18%	23%

BREEDING DISTRIBUTION CHANGE *since 1968–72*

	Ireland	Britain	Britain & Ireland
GAIN ▲	245	287	532
LOSS ▽	203	396	599
40yr CHANGE	+12%	-18%	-7%
20yr INDEX	-0.44	-0.45	

Sponsored by Ecological Planning and Research Ltd

◀ To help resolve the considerable uncertainty regarding the status of the Long-eared Owl in the UK, it has recently been added to the list of species for which breeding data are collated by the RBBP.

SHORT-EARED OWL
Asio flammeus

BTO CODE: SE

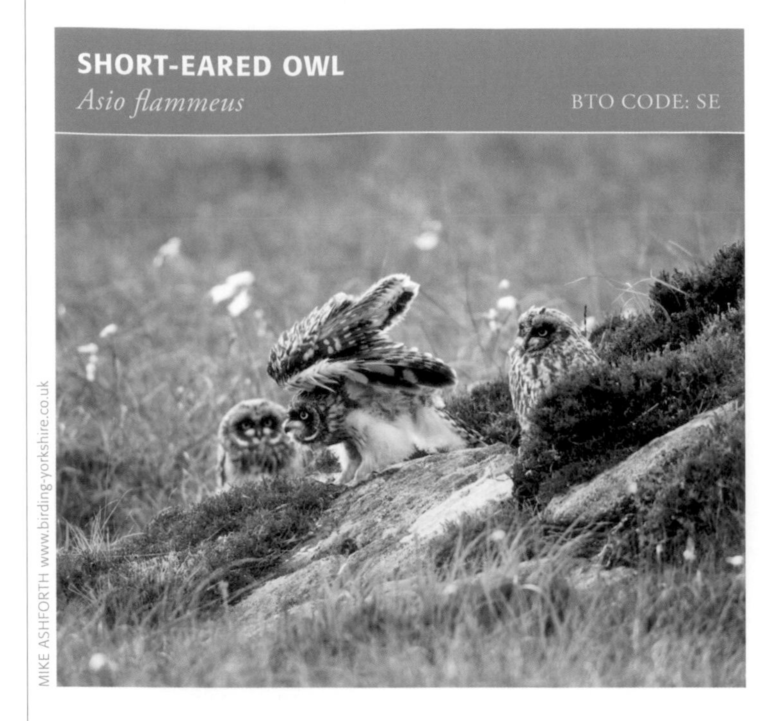
MIKE ASHFORTH www.birding-yorkshire.co.uk

THE CORE BREEDING range of the Short-eared Owl is the open rough country of northern England and Scotland, where they favour heather moorland, rough grassland, bogs and young forestry plantations. Elsewhere breeding records are widely scattered and involve a small number of pairs in lowland coastal marshes and extensive grasslands. Abundance is greatest on Orkney, on the Uists and in the Pennines. Breeding seems to be irregular in the south of Ireland, with the most recent documented case being from 1977 (Jones 1979). However up to five pairs nested in the northeast during 2008–11 (Hillis 2012).

The European population declined overall during 1990–2000 (Burfield 2008) and there have been widespread declines also in Britain alone. Determining trends is problematic, however, owing to these owls' fluctuations and nomadism, associated with small-mammal population cycles, and the inherent difficulties in surveying them (Greenwood *et al.* 2003; Calladine *et al.* 2010). Since the *1968–72 Breeding Atlas*, the breeding range has almost halved and the species is now Amber-listed in the UK and being covered by RBBP (*RBBP 2010*). Losses have affected all areas but have been especially pronounced in central Wales, eastern England and southern and eastern Scotland. The Scottish losses in particular can be linked to forest growth: plantations are used for their first 12 years but owl densities peak at between three and seven years (Shaw 1995).

During the winter, an influx of Fennoscandian breeders greatly boosts numbers in Britain & Ireland, although part of the native population departs for southern Europe (*Migration Atlas*). Wintering occurs widely along the British east coast from Fife to Kent, in large river valleys in southern and eastern England, the lowlands of Lancashire and the downlands of Hampshire and Wiltshire. Birds in Orkney, the Uists and the Pennines appear to winter close to their breeding grounds. Irish records are usually along the east coast (*Birds in Ireland*) but there were several records farther west during 2007–11. Winter range size in Ireland remains unchanged but in Britain there has been a 19% contraction since the *1981–84 Winter Atlas*. There is a high degree of turnover, with the many adjacent gains and losses indicating the species' tendency to wander.

Sponsored by Hamish & Doris Crichton Charitable Trust

BREEDING DISTRIBUTION *2008–11*

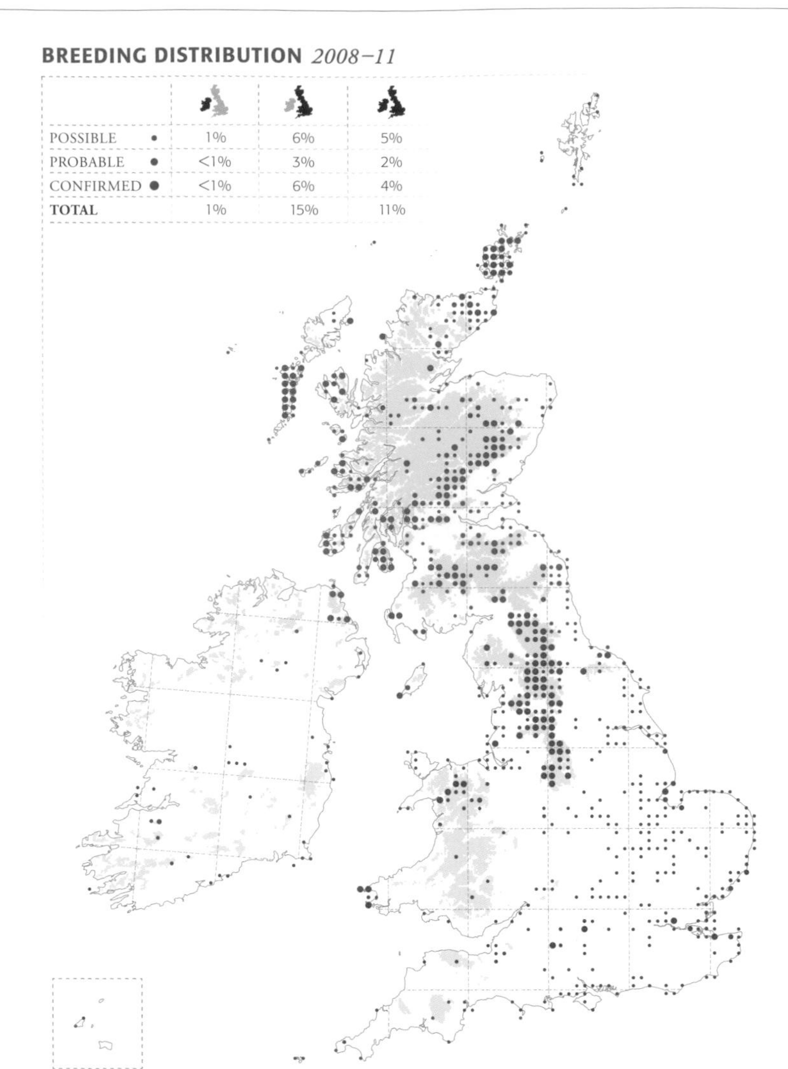

POSSIBLE •	1%	6%	5%
PROBABLE •	<1%	3%	2%
CONFIRMED ●	<1%	6%	4%
TOTAL	1%	15%	11%

WINTER DISTRIBUTION *2007/08–2010/11*

PRESENT ●	5%	30%	24%

BREEDING DISTRIBUTION CHANGE *since 1968–72*

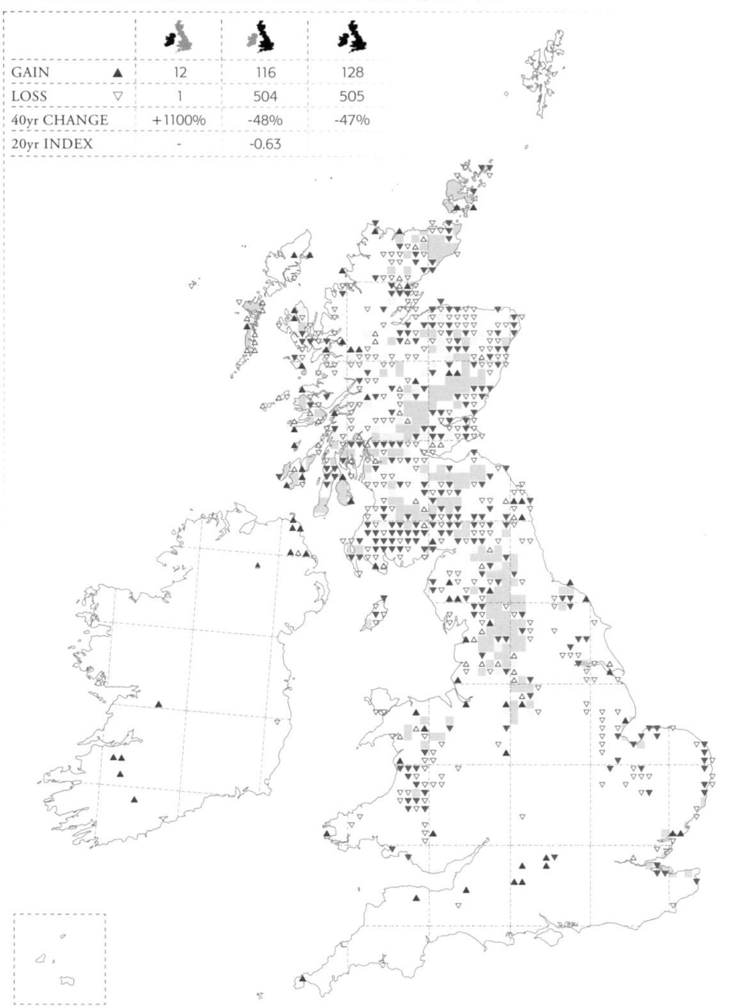

GAIN ▲	12	116	128
LOSS ▽	1	504	505
40yr CHANGE	+1100%	-48%	-47%
20yr INDEX	-	-0.63	

WINTER DISTRIBUTION CHANGE *since 1981–84*

GAIN ▲	38	301	339
LOSS ▼	38	485	523
30yr CHANGE	0%	-19%	-18%

BREEDING RELATIVE ABUNDANCE *2008–11*

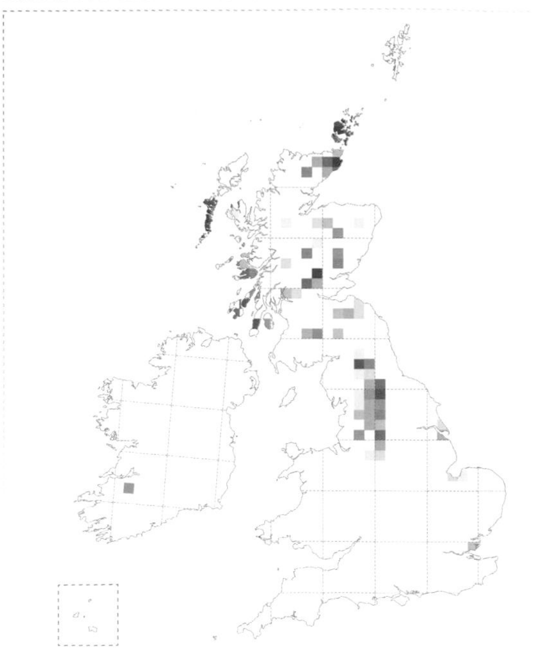

BREEDING RELATIVE ABUNDANCE CHANGE *since 1988–91*

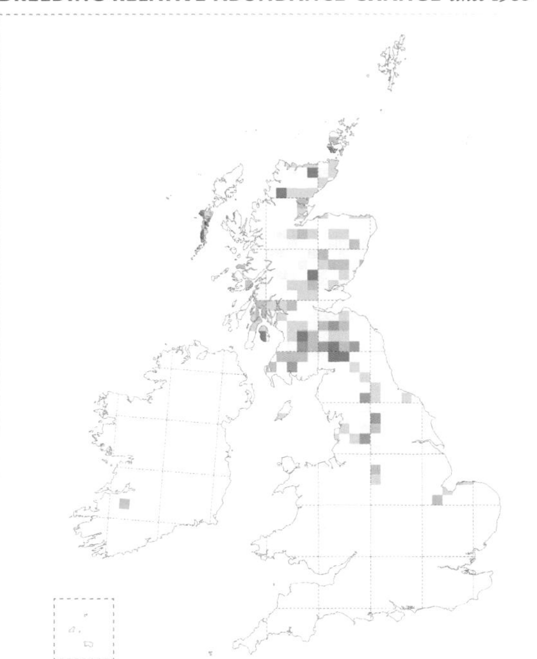

WINTER RELATIVE ABUNDANCE *2007/08–2010/11*

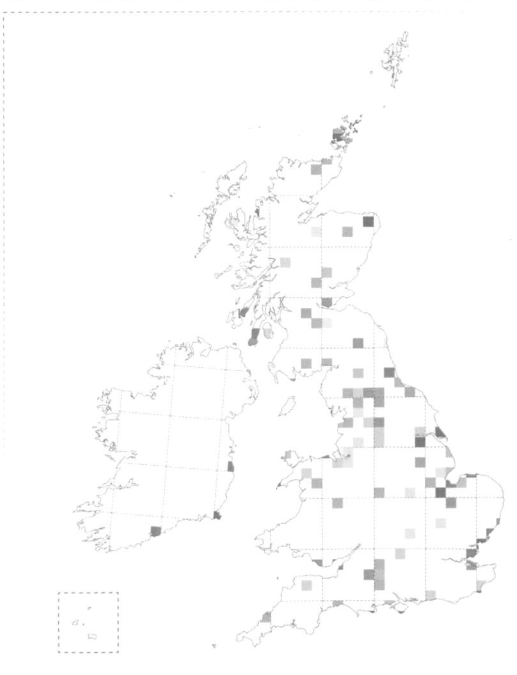

EURASIAN EAGLE-OWL

Bubo bubo

BTO CODE: EO

TOM WAY www.tomwayphotography.co.uk

ARCHAEOLOGICAL EVIDENCE SUPPORTS the 'native' presence of Eurasian Eagle-Owls in prehistoric Britain (Stewart 2007), though the current breeding population is considered to be the result of aviary escapes and deliberate introductions (Melling *et al.* 2008). The arrival of genuine vagrants is thought to be unlikely but not impossible (Aebischer *et al.* 2010; Kelly *et al.* 2010).

Since 1984 more than 200 records of eagle-owls have been published in county and regional bird reports in Britain, sometimes involving birds with jesses or belonging to races or sister species not native to Europe, whose captive origins are clear. The first documented breeding record came from Moray & Nairn in 1984 (Dennis & Cook 1996) and breeding has been reported almost annually since 1997. Breeding records come from a small number of sites, largely centred in northern England and Scotland, but the species has been reported breeding in southern England (Holling *et al.* 2007b; Toms 2009).

The few records collected during Atlas fieldwork also reflect this pattern, with most centred on northwest England. Among these were just two 10-km squares with *confirmed* breeding: in Lancashire & North Merseyside in 2008–11 and in Cumbria in 2010. Other pairs almost certainly bred during the Atlas period but sensitivities surrounding this species mean that not all records are available with sufficient spatial resolution for mapping.

The species is largely sedentary as an adult but young birds may disperse over substantial distances before settling into suitable habitat. Four individuals ringed as chicks at British nests have been recovered, the two longest movements being 160 km and 209 km (Robinson & Clark 2012).

It appears that Britain holds a very small but potentially increasing breeding population. This may have consequences for other bird-of-prey populations. The Eurasian Eagle-Owl is a predator of other birds of prey, including Hen Harriers, Goshawks and Peregrines, whose breeding productivity may be reduced in areas where these owls become established (Brambilla *et al.* 2006; Sergio *et al.* 2007). Efforts should be directed towards monitoring aviary escapes and subsequent breeding attempts in the wild (Penteriani *et al.* 2012).

BREEDING DISTRIBUTION *2008–11*

	Ireland	Britain	Britain & Ireland
POSSIBLE •	-	<1%	<1%
PROBABLE ●	-	<1%	<1%
CONFIRMED ●	-	<1%	<1%
TOTAL	-	<1%	<1%

WINTER DISTRIBUTION *2007/08–2010/11*

	Ireland	Britain	Britain & Ireland
PRESENT ●	-	1%	<1%

Sponsored by Hazel Douglas for My Birds!

NIGHTJAR
Caprimulgus europaeus

BTO CODE: NJ

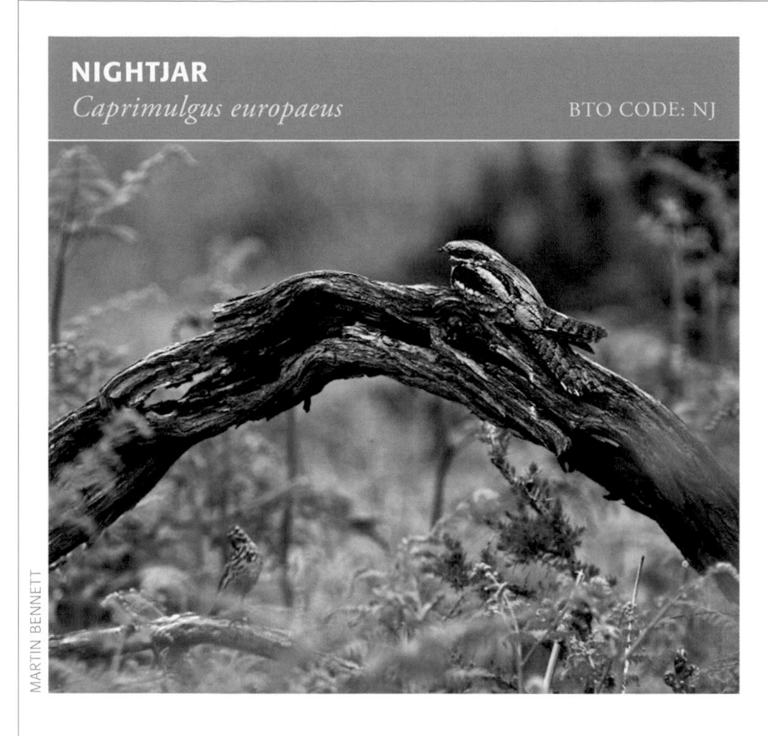
MARTIN BENNETT

CONSOLIDATION AND EXPANSION of the Nightjar's breeding range across England and Wales has been the most notable change since the *1988–91 Breeding Atlas*. Across much of its range, the breeding distribution is closely associated with lowland heathland and felled or recently planted conifer plantations, though coastal moorland (Cornwall), Sweet Chestnut coppice (Kent) and sand dunes (Suffolk) may also be occupied. In Ireland they mainly occupy clear-felled conifer plantations.

Historically, Nightjars were widely distributed throughout Britain & Ireland but between the *1968–72 Breeding Atlas* and the *1988–91 Breeding Atlas* the range contracted by 51% in Britain and by 88% in Ireland. The decline in Ireland seems to have set in during the 1950s and 1960s and by 1981 there was just one report, from Co. Wicklow (*Birds in Ireland*), although the species may be under-reported (*IRBBP 2010*). Dedicated surveys in Britain show that the number of churring males increased from 2,100 in 1981 to 3,400 in 1992 and 4,600 in 2004 (Gribble 1983; Morris *et al.* 1994; Conway *et al.* 2007). Despite this doubling in numbers, there was relatively little expansion: range size increased from 241 to 268 and 275 10-km squares, mostly involving consolidation in southwest and southeast England. The 18% range expansion in Britain from 1988–91 to 2008–11 sees gains beyond the previous distribution and includes a novel broadening of habitat use whereby moorland conifer plantations are now being occupied; this may assist future range expansion.

Despite the recent upward trends for population size and distribution, there are still parts of the 1968–72 range that remain unoccupied or depleted in numbers (Langston *et al.* 2007), notably in Ireland, Scotland, western Wales and northwest and central England. Reasons for this are unclear but may include a shortage of suitable habitat, owing to a reduced area of young plantations in England and Wales, and a move towards continuous-cover forest management. The implications of climate change are uncertain (Langston *et al.* 2007): predictions of increased spring rainfall (Hulme *et al.* 2002) could reduce the success of early nests, whereas warmer temperatures may facilitate northward range expansion (Huntley *et al.* 2007), provided that suitable habitat exists.

Sponsored by Forestry Commission (England)

BREEDING DISTRIBUTION *2008–11*

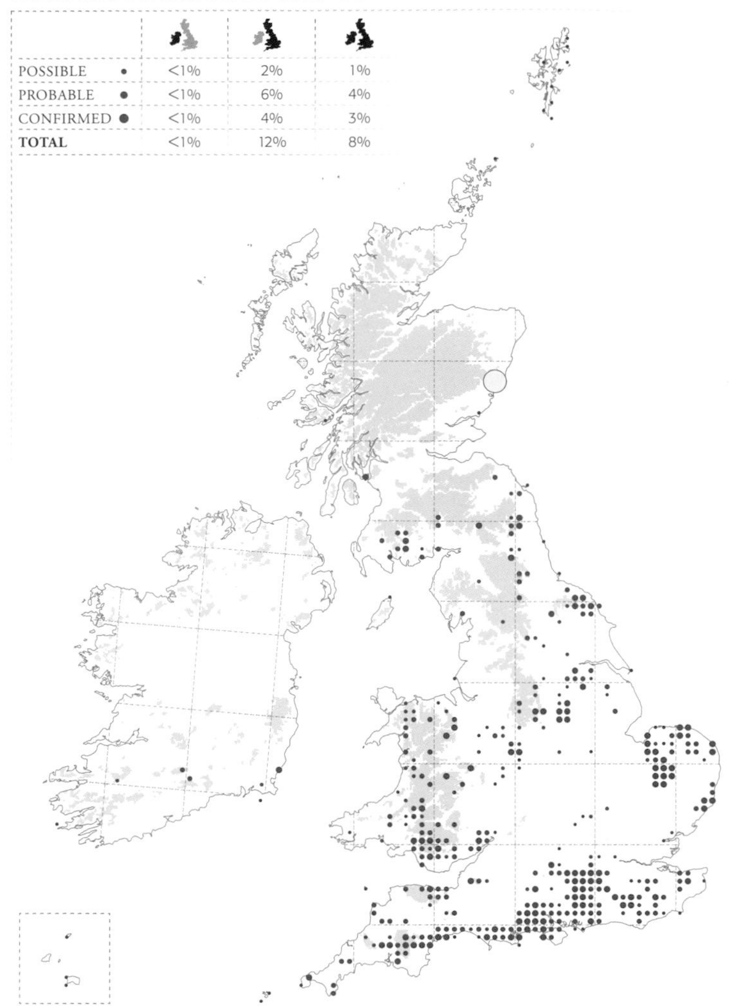

	Ireland	Britain	Britain & Ireland
POSSIBLE •	<1%	2%	1%
PROBABLE ●	<1%	6%	4%
CONFIRMED ●	<1%	4%	3%
TOTAL	<1%	12%	8%

BREEDING DISTRIBUTION CHANGE *since 1988–91*

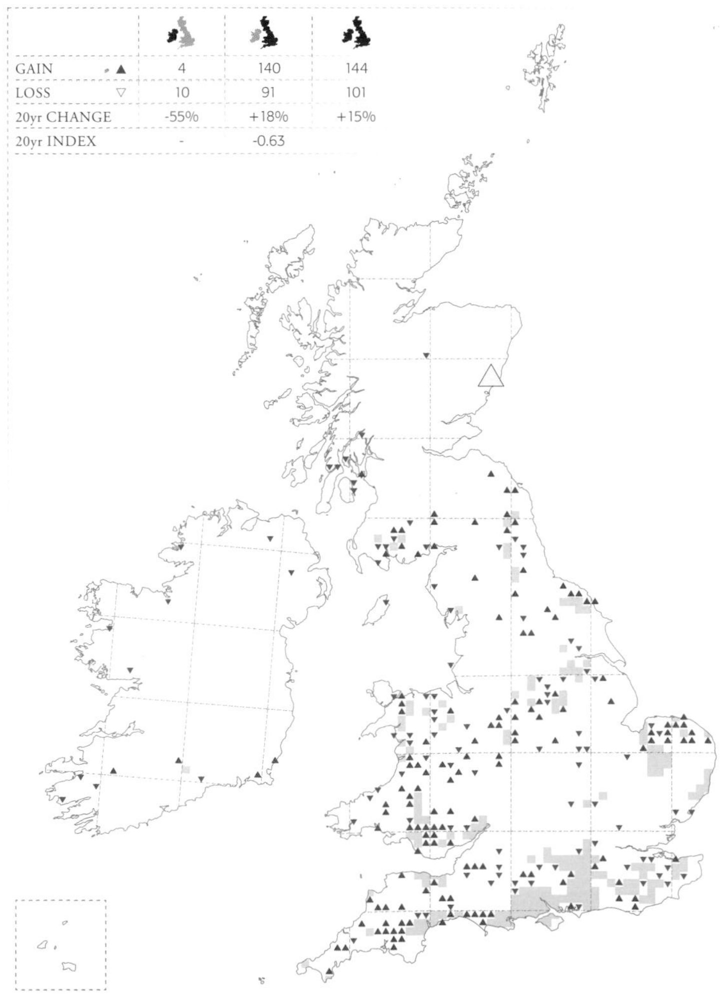

	Ireland	Britain	Britain & Ireland
GAIN ▲	4	140	144
LOSS ▽	10	91	101
20yr CHANGE	-55%	+18%	+15%
20yr INDEX	-	-0.63	

SWIFT
Apus apus

BTO CODE: SI

BREFFNI MARTIN

SWIFTS HAVE A broad breeding distribution, with higher densities in warm, dry areas such as East Anglia, and lower densities in northern and western regions. In Britain & Ireland Swifts nest almost exclusively in buildings, and are able to breed successfully in urban areas, provided that insect-rich skies are available within their foraging range. This is clear in the map of relative abundance, which closely matches that of the built environment, with concentrations in towns and cities.

Distribution is patchy in the western half of Ireland and across northwest Scotland, and patterns of occupancy have shifted over time. In Britain the overall number of occupied 10-km squares has shown little change, but in Ireland there has been a 26% decline in this measure since the *1968–72 Breeding Atlas*. Although Swifts were *confirmed* to have bred in nearly every 10-km square in England and Wales, the abundance change map shows widespread declines in tetrad occupancy rates within 20-km squares since the *1988–91 Breeding Atlas*. In areas of low breeding density or where there was a low intensity of coverage, it can be more difficult to obtain evidence of breeding.

Swifts do not defend territories; their reliance on aerial insects as a food supply means they can nest semi-colonially and range many kilometres in search of food. This behaviour, together with the sometimes-large number of non-breeding birds in the population, makes monitoring their numbers unusually difficult. However the estimated 38% population decline in the UK during 1995–2010 and the estimated 46% decline in the Republic of Ireland during 1998–2010 appear to tally with the change in abundance recorded by the atlases (*CBS Trend 2010*; *BBS Report 2011*).

A possible reason for declines is changes to construction techniques that restrict the availability of nesting cavities in new buildings (Concern for Swifts 2013). Breeding success may be related to summer weather (Thomson *et al.* 1996; Rajchard *et al.* 2006) and, as long-distance migrants wintering south of the equator, Swifts are susceptible to changes they encounter in conditions on migration and in their wintering areas.

Sponsored by Swift Conservation

BREEDING DISTRIBUTION *2008–11*

POSSIBLE •	17%	5%	8%
PROBABLE •	16%	7%	10%
CONFIRMED ●	30%	62%	54%
TOTAL	63%	74%	72%

BREEDING RELATIVE ABUNDANCE CHANGE *since 1988–91*

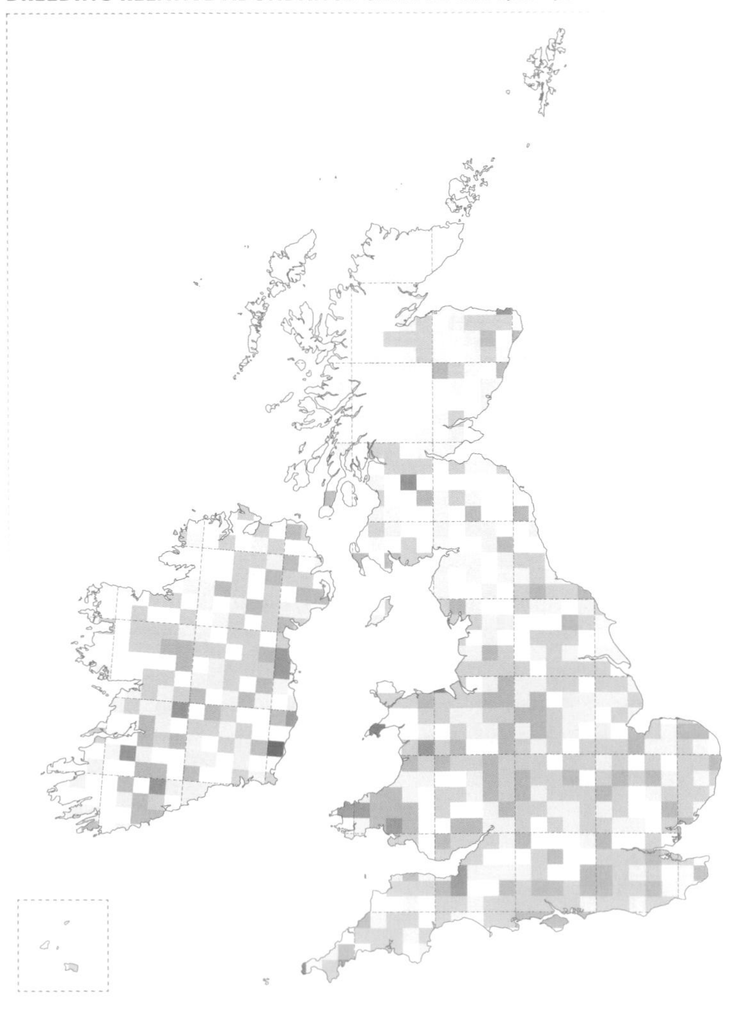

BREEDING DISTRIBUTION CHANGE *since 1968–72*

GAIN ▲	23	90	113
LOSS ▽	246	165	411
40yr CHANGE	-26%	-3%	-10%
20yr INDEX	-0.13	-0.07	

BREEDING RELATIVE ABUNDANCE *2008–11*

HOOPOE

Upupa epops

BTO CODE: HP

MICK DURHAM www.wildlife-photographer.net

DURING THE BREEDING season there were *non-breeding* records in 203 10-km squares in Britain and a further nine squares in Ireland. These were typically spring-migration overshoots and their distribution fits the established pattern (Fraser & Rogers 2006a). Breeding was confirmed in two 10-km squares during the *1968–72 Breeding Atlas*: in Cornwall in 1968 and 1969, and in Sussex in 1971, and *possible* breeding in two further squares. The most recent confirmed breeding in Britain & Ireland occurred in 1996 when a pair fledged three young in Montgomeryshire (Ogilvie *et al.* 1999). Winter records were limited to 12 10-km squares in Britain and three in Ireland, and included four records in early November and two in February.

BREEDING DISTRIBUTION *2008–11*

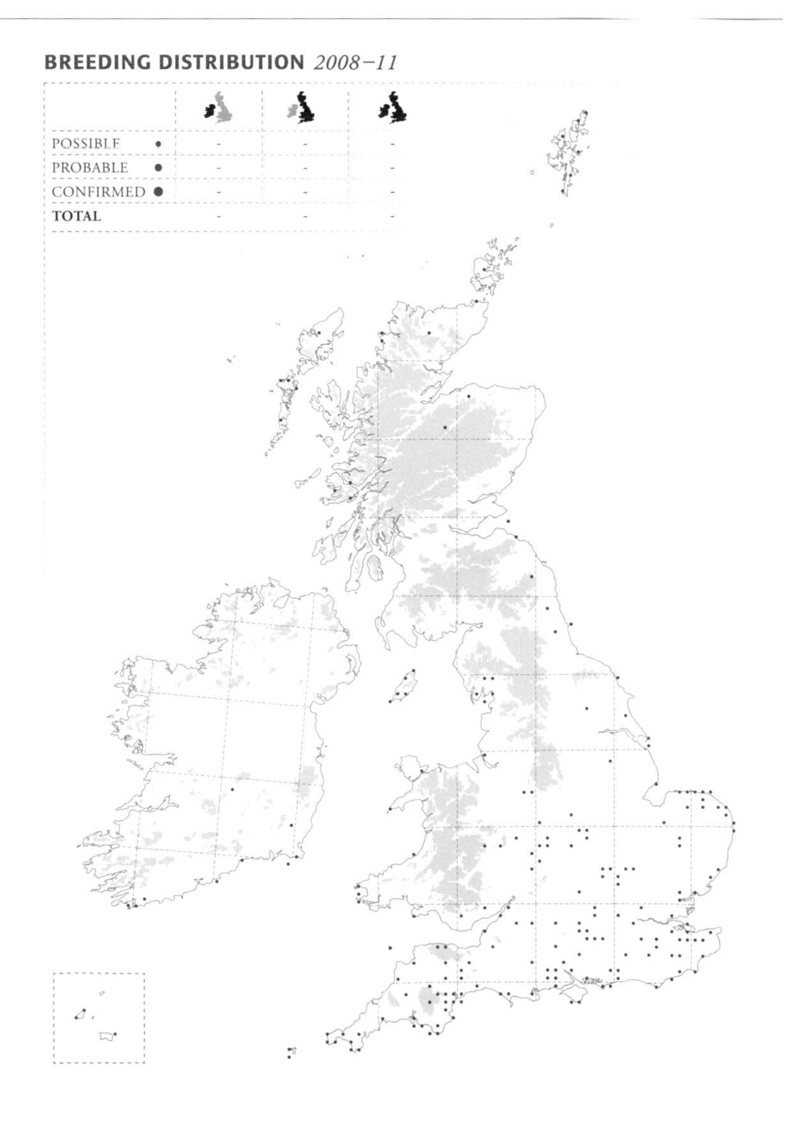

POSSIBLE •	-	-	-
PROBABLE •	-	-	-
CONFIRMED ●	-	-	-
TOTAL	-	-	-

KINGFISHER

Alcedo atthis

BTO CODE: KF

ANNA SIMPSON www.wildenvironmentalist.blogspot.co.uk

THE KINGFISHER IS widely distributed on the lowland rivers of Britain & Ireland. It is resident, with some dispersal away from breeding territories outside the breeding period, especially by juvenile birds. In Britain this may explain the greater number of 10-km squares occupied in winter than in the breeding season. Despite dispersal, the breeding and winter distributions do not differ greatly. Kingfishers are especially prevalent and numerous in England and Wales, whereas in Scotland a small population is concentrated into the lowlands of the south and east. They have a patchy distribution in Ireland, especially in the west and northwest where their low density can make them difficult to detect (Crowe *et al.* 2010a).

Patterns of distribution change indicate large gains in winter range in both Britain and Ireland since the *1981–84 Winter Atlas*, when numbers were at a low point following several cold winters (*BirdTrends*). These gains are notable in southern Scotland, eastern and southwest England, and parts of Wales and Ireland and in some areas may be linked to the run of mild winters prior to the Atlas period.

Over the last 40 years there has been a mix of gains and losses in occupancy in the breeding season, with losses generally outweighing the gains, although gains are evident overall in parts of eastern England and Scotland. In Ireland there has been a 25% contraction of range since the *1968–72 Breeding Atlas*, despite an apparent 26% expansion since the *1988–91 Breeding Atlas*. This pattern is likely to be an artefact of poor coverage in Ireland during 1988–91 (Crowe *et al.* 2010a), because comparisons using fixed-effort surveys indicate decreases in occupancy and abundance since that time.

Breeding numbers in the UK fell sharply through the late 1970s to a low point in the mid 1980s and have since recovered, though a downward trend is evident from around 2005 onwards (*BirdTrends*). Populations may have been impacted by the cold winters during *Bird Atlas 2007–11*, although there was no evidence of this in some Irish river systems (Crowe *et al.* 2010a).

Sponsored by Canal & River Trust

BREEDING DISTRIBUTION *2008–11*

POSSIBLE •	20%	15%	16%
PROBABLE •	7%	9%	9%
CONFIRMED •	11%	22%	19%
TOTAL	38%	46%	44%

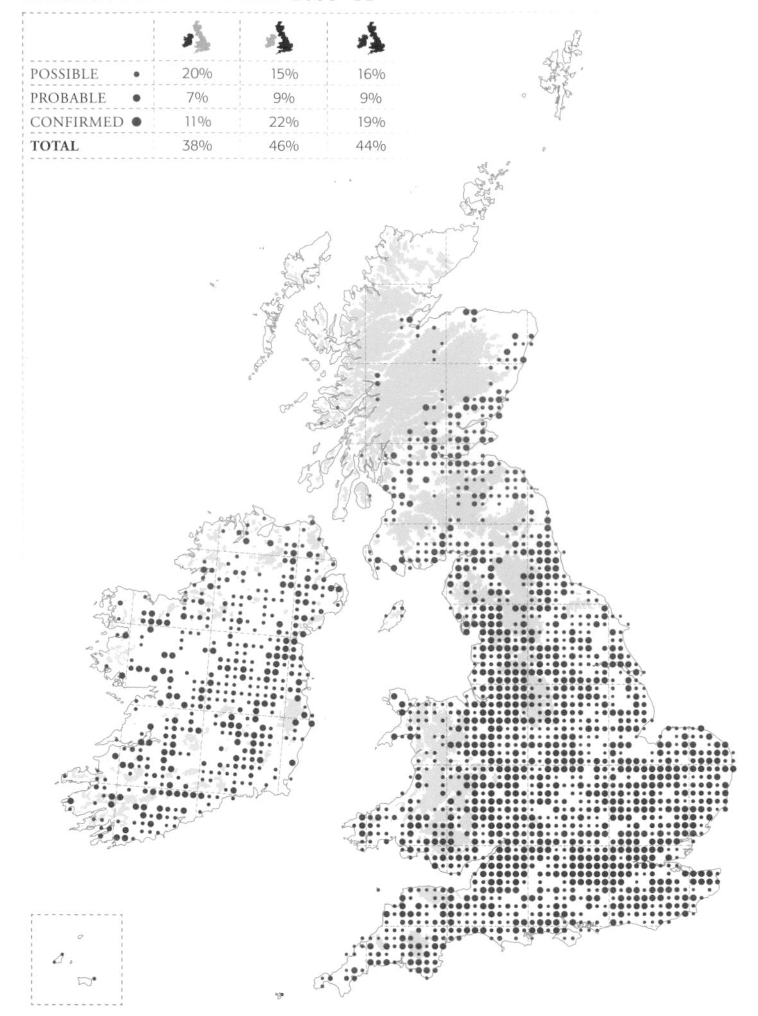

WINTER DISTRIBUTION *2007/08–2010/11*

PRESENT •	34%	56%	50%

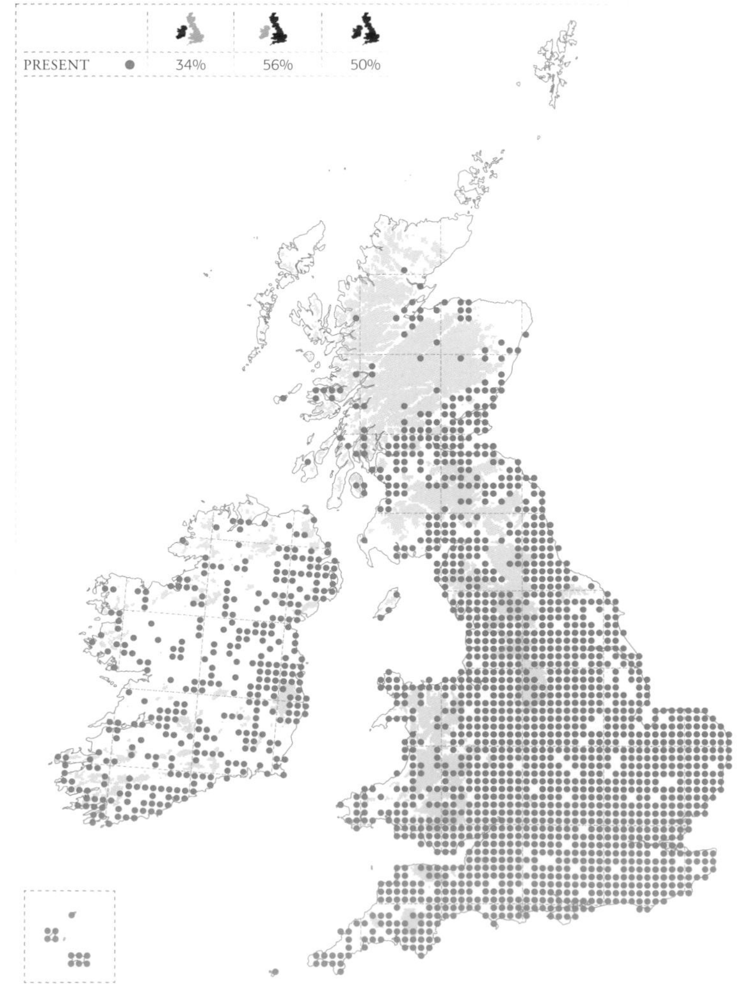

BREEDING DISTRIBUTION CHANGE *since 1968–72*

GAIN ▲	109	301	410
LOSS ▽	235	288	523
40yr CHANGE	-25%	+1%	-6%
20yr INDEX	-0.23	-0.10	

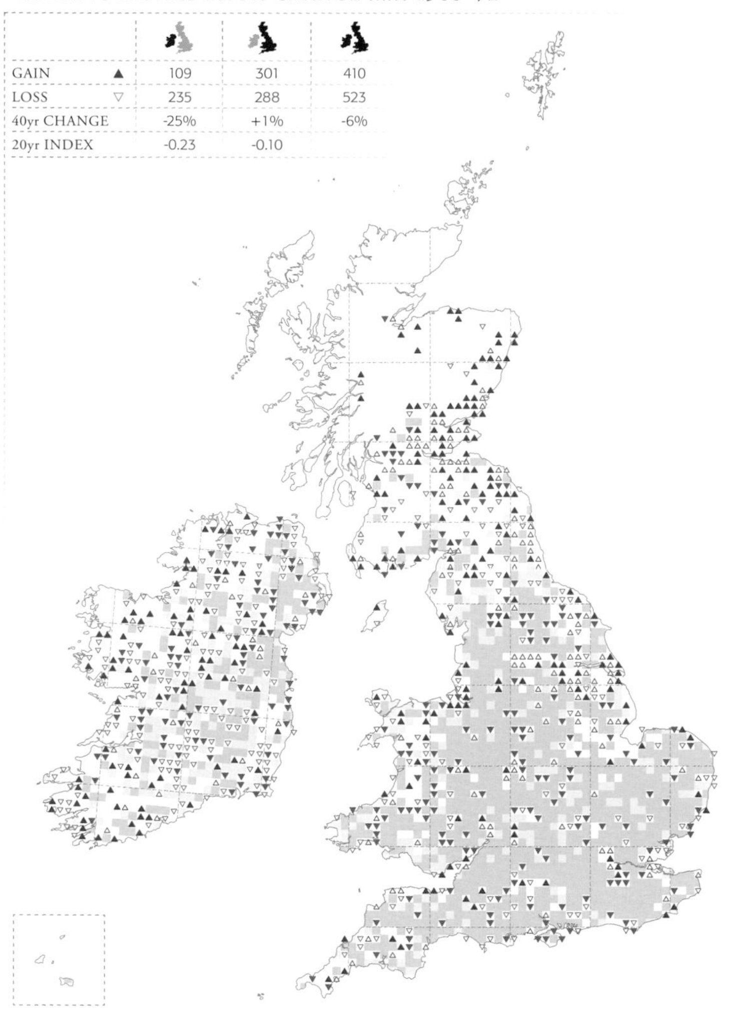

WINTER DISTRIBUTION CHANGE *since 1981–84*

GAIN ▲	163	590	753
LOSS ▼	106	81	187
30yr CHANGE	+22%	+49%	+44%

BREEDING RELATIVE ABUNDANCE *2008–11*

BREEDING RELATIVE ABUNDANCE CHANGE *since 1988–91*

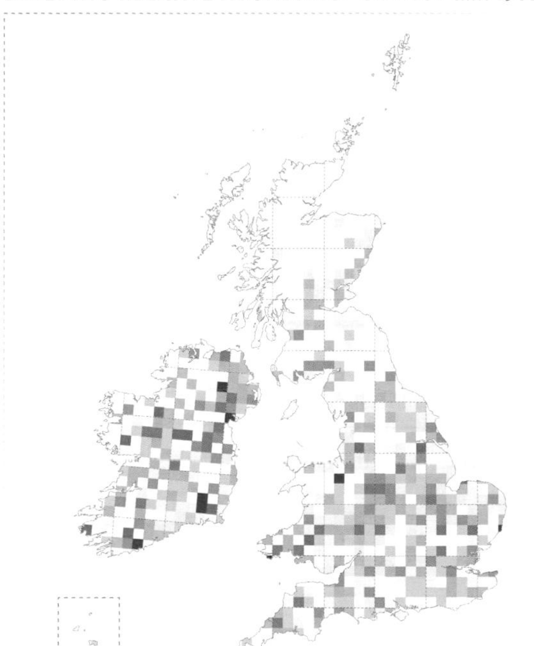

WINTER RELATIVE ABUNDANCE *2007/08–2010/11*

GREEN WOODPECKER

Picus viridis

BTO CODE: G.

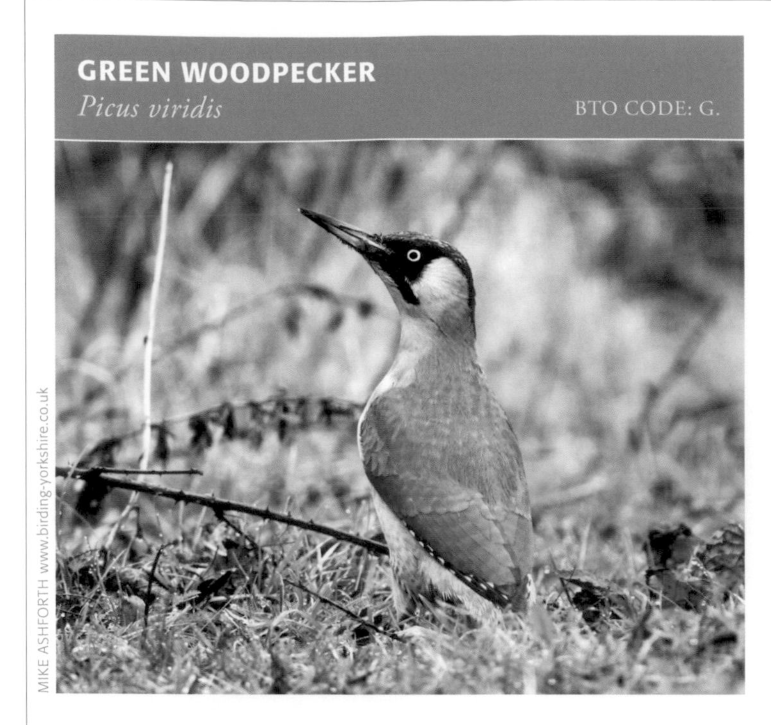

MIKE ASHFORTH www.birding-yorkshire.co.uk

THE GREEN WOODPECKER'S distribution is largely continuous across southern England but becomes patchy in western Wales, northern England and in south-central Scotland; relative abundance shows the same pattern. The species is absent from the Channel Islands, the Isle of Man and from Ireland: three birds shot during the 19th century stand as the only Irish records (*Birds in Ireland*).

There is little evidence of any seasonal movement (*Migration Atlas*) and the breeding and winter maps are essentially the same. Slightly more 10-km squares were occupied in the breeding season than in winter, probably as a result of easier detection owing to the species' far-carrying breeding calls.

Overall, the change in breeding range size is relatively small, but this conceals remarkable regional differences in gains and losses. The *1988–91 Breeding Atlas* had revealed some losses in Cornwall, western Wales and around the Solway Firth and gains in northern Scotland. Since then, the species has disappeared almost entirely from Pembrokeshire and Anglesey and failed to consolidate its range in northern Scotland. Further losses are concentrated across northwest England and southwest Scotland. By complete contrast, the range has spread in parts of eastern Britain, most conspicuously at the 10-km scale in the Fens, but also at finer scales. The abundance change map shows a clear east–west divide, with declining tetrad occupancy in Wales and southwest England contrasting strongly with increasing tetrad occupancy in central and eastern England. Changes in density show the same spatial pattern (Massimino *et al.* 2013). More than twice as many Norfolk tetrads were occupied in the early 2000s as two decades earlier (Taylor & Marchant 2011) and more widely this essentially woodland species has become familiar also as a bird of gardens and farmland (Alder & Marsden 2010). Annual monitoring shows a 115% population increase during 1970–2010 (*SUKB 2012*).

The reasons for the population increase and spread in the east of the British range, and range loss in the far west, are not understood. The availability of ants and their pupae, which are the species' main food, is an important factor governing Green Woodpecker distribution and might also be playing a part in these recent changes.

BREEDING DISTRIBUTION *2008–11*

POSSIBLE ●	-	12%	9%
PROBABLE ●	-	8%	6%
CONFIRMED ●	-	36%	27%
TOTAL	-	56%	42%

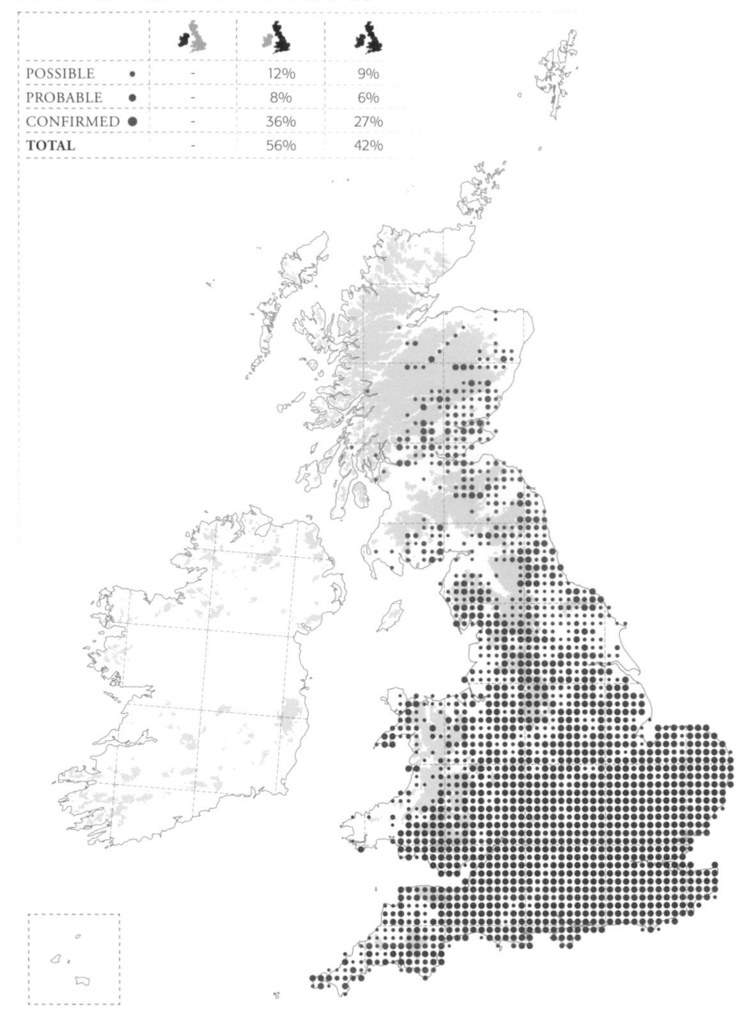

WINTER DISTRIBUTION *2007/08–2010/11*

PRESENT ●	-	54%	40%

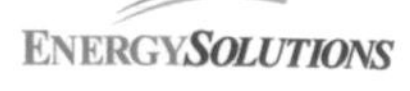

Sponsored by Energy Solutions working with Oldbury Power Station

BREEDING DISTRIBUTION CHANGE *since 1968–72*

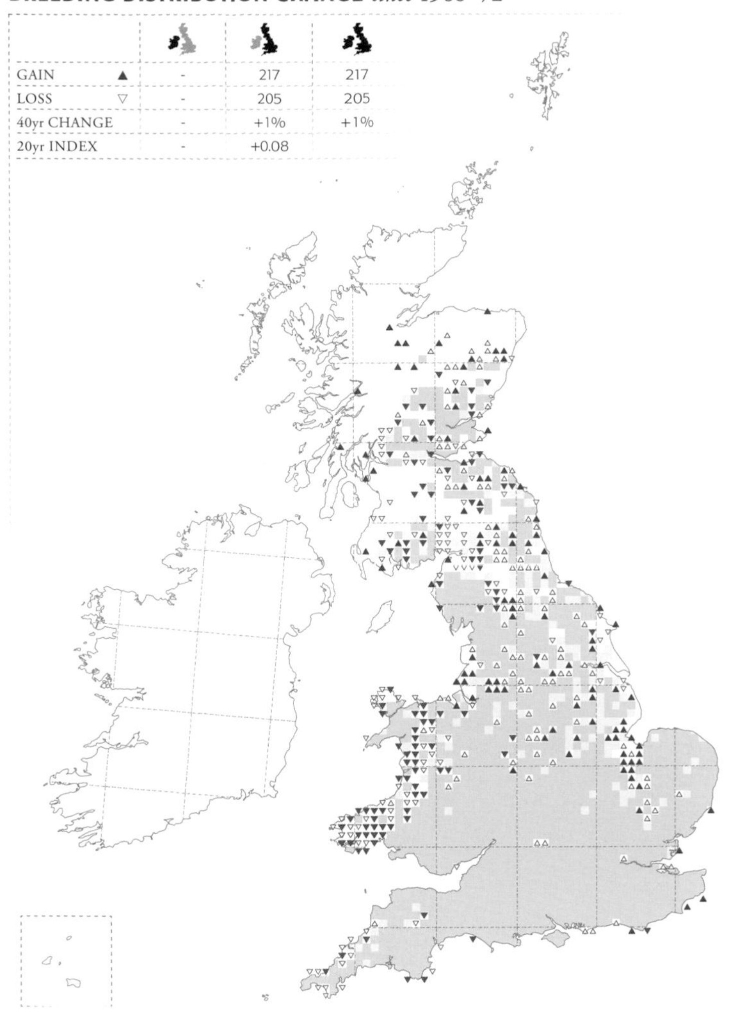

GAIN ▲	-	217	217
LOSS ▽	-	205	205
40yr CHANGE	-	+1%	+1%
20yr INDEX	-	+0.08	

WINTER DISTRIBUTION CHANGE *since 1981–84*

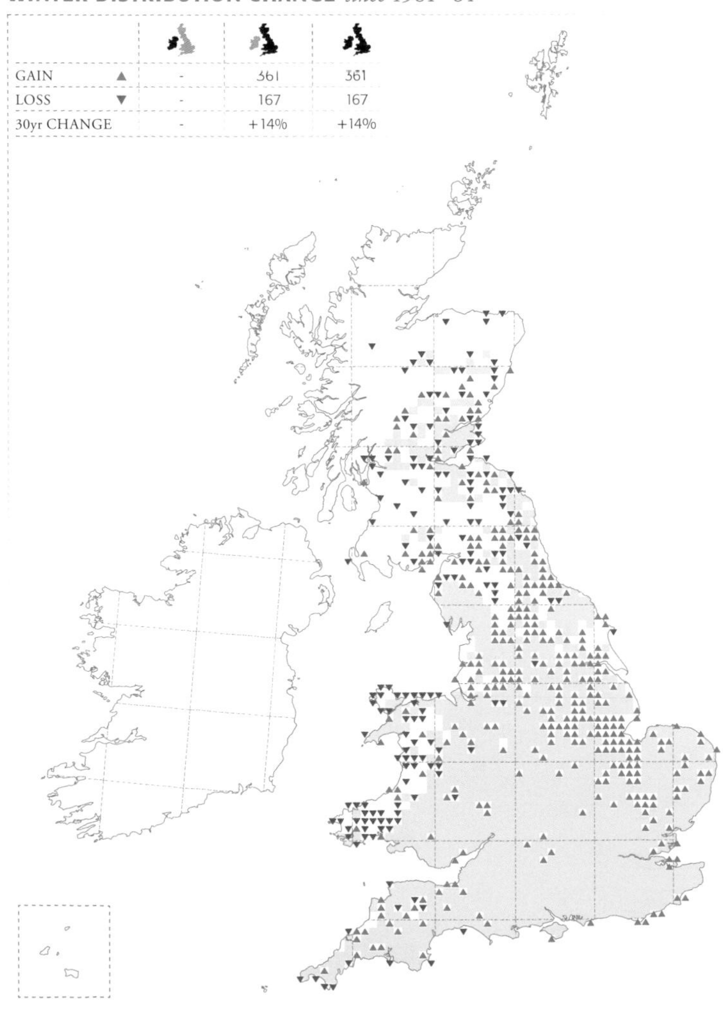

GAIN ▲	-	361	361
LOSS ▼	-	167	167
30yr CHANGE	-	+14%	+14%

BREEDING RELATIVE ABUNDANCE *2008–11*

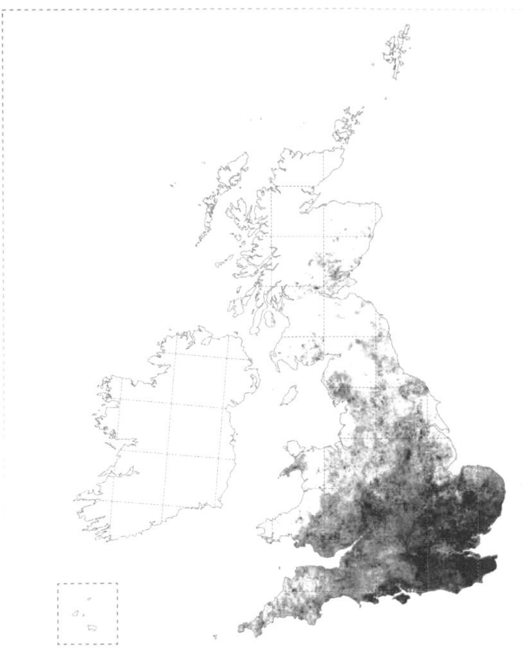

BREEDING RELATIVE ABUNDANCE CHANGE *since 1988–91*

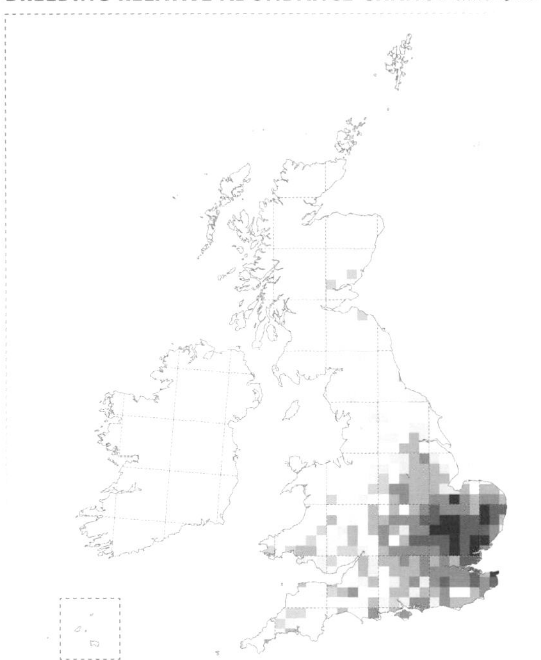

WINTER RELATIVE ABUNDANCE *2007/08–2010/11*

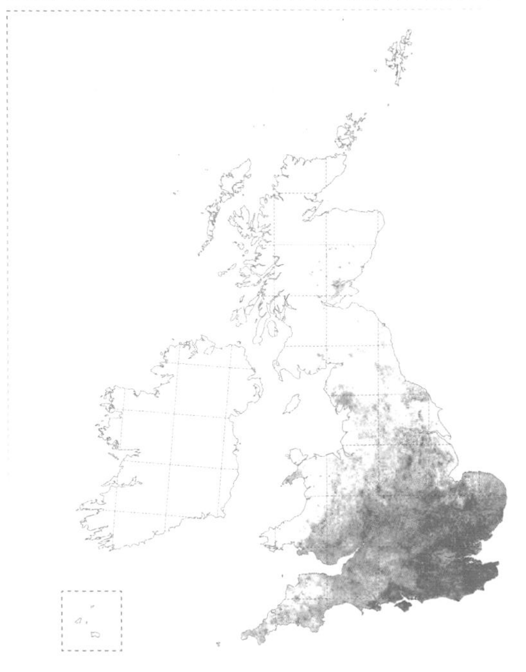

GREAT SPOTTED WOODPECKER

Dendrocopos major BTO CODE: GS

MICK DURHAM www.wildlife-photographer.net

GREAT SPOTTED WOODPECKERS have colonised Ireland since the *1988–91 Breeding Atlas*. Breeding was first proved in Northern Ireland in 2006 (Murphy 2007) and the Republic of Ireland in 2009 (Hillis *et al.* 2010). The Irish distribution is presently concentrated into eastern counties and is probably advancing westwards. Genetic analysis of feathers obtained from nests has shown that these early colonists most probably originated from Britain (McDevitt *et al.* 2011). The majority of previous Irish records occurred during winter irruptions of continental birds (*Birds in Ireland*).

This species is currently undergoing a substantial range expansion in Britain too, most notably towards the north and west, with gains in occupancy evident in Scotland and Wales. The Northern Isles, Outer Hebrides and most of the Inner Hebrides remain unoccupied. The change maps also show infilling within the core range in England, particularly on the eastern fringe from the Fens north to Yorkshire. Records have increased on the Isle of Man since the mid 1990s, with breeding suspected in 2005 (Sharpe 2007) and then *confirmed* in three 10-km squares in 2010 (C. Sharpe pers comm). Densities appear to be highest in southeast England and the border counties of England and Wales, and lowest in upland areas and the north of Scotland.

The range expansion is in line with the rapid increase in population documented in the 1970s, with further increases from the early 1990s: a 408% population increase was recorded in the UK during 1967–2010 (*BirdTrends*). This is also reflected in the abundance change map, which shows an increase across most of the range since the *1988–91 Breeding Atlas* in the proportion of occupied tetrads per 20-km square.

A national decline in Starling numbers and the consequent reduced competition for nest sites (Smith 2005, 2006), increases in the availability of dead and decaying wood (Amar *et al.* 2010a), a resource important for both nesting and feeding (Smith 2007) and the provision of supplementary food at garden feeders (Smith & Smith 2013) have all been suggested as factors potentially contributing to these increases in Britain.

Sponsored for Joseph Smitherman

BREEDING DISTRIBUTION *2008–11*

POSSIBLE •	2%	5%	4%
PROBABLE •	1%	6%	5%
CONFIRMED ●	2%	71%	53%
TOTAL	5%	82%	62%

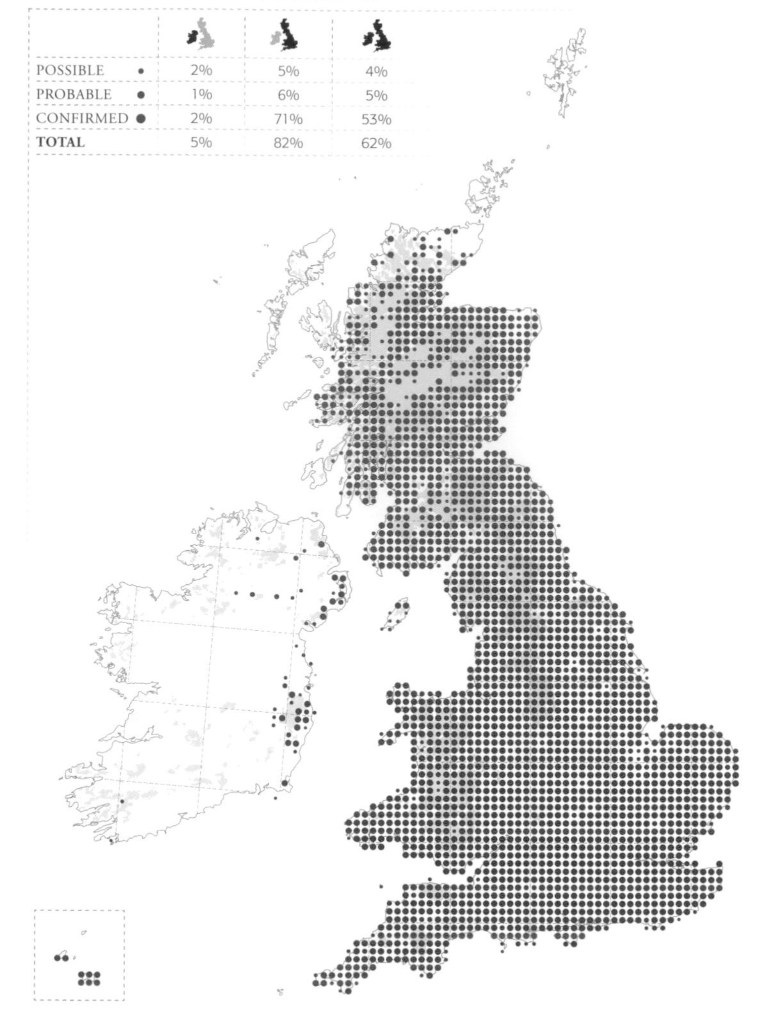

WINTER DISTRIBUTION *2007/08–2010/11*

PRESENT ●	4%	81%	61%

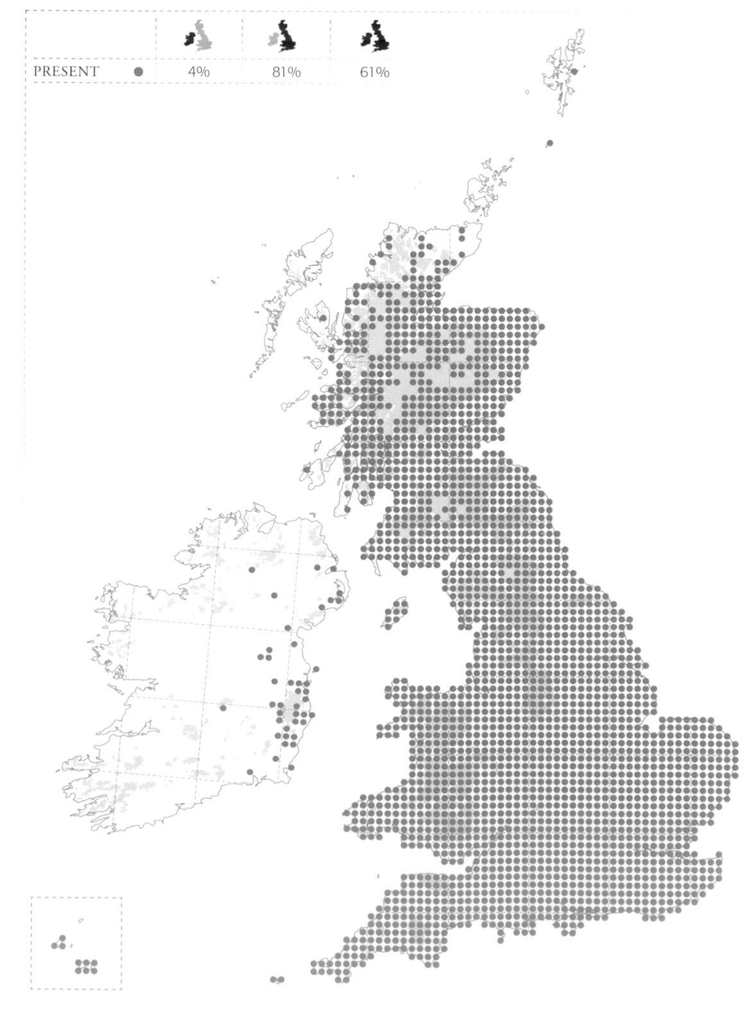

BREEDING DISTRIBUTION CHANGE *since 1968–72*

GAIN ▲	41	315	356
LOSS ▽	1	34	35
40yr CHANGE	+4000%	+14%	+16%
20yr INDEX	-	+0.30	

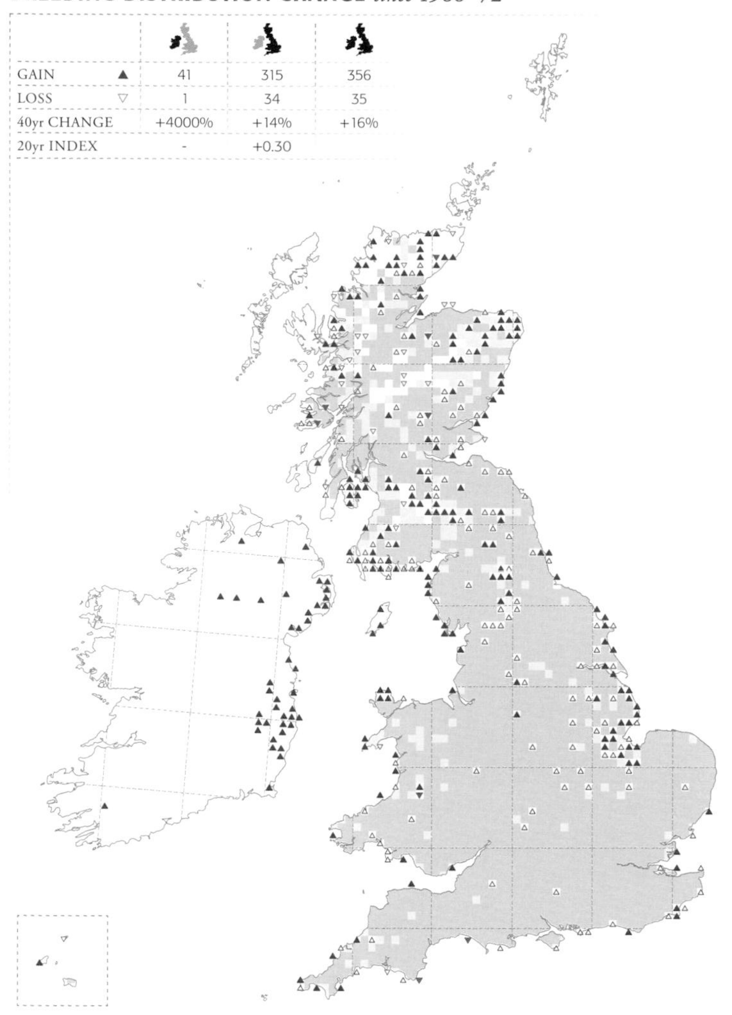

WINTER DISTRIBUTION CHANGE *since 1981–84*

GAIN ▲	36	588	624
LOSS ▼	0	19	19
30yr CHANGE	∞	+33%	+35%

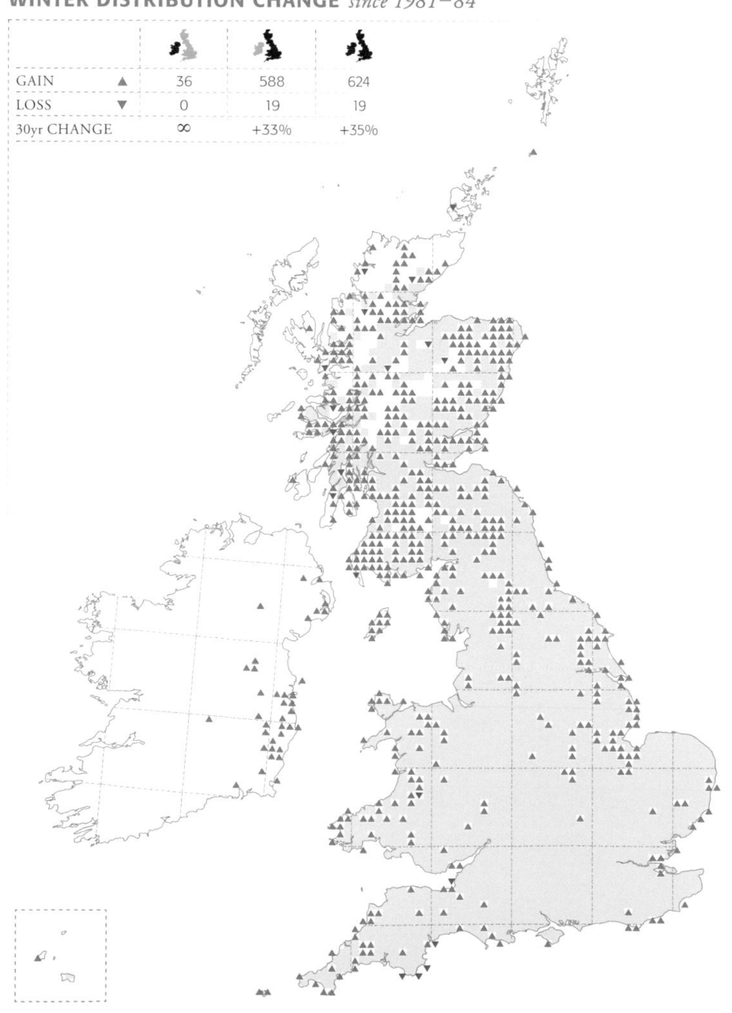

BREEDING RELATIVE ABUNDANCE *2008–11*

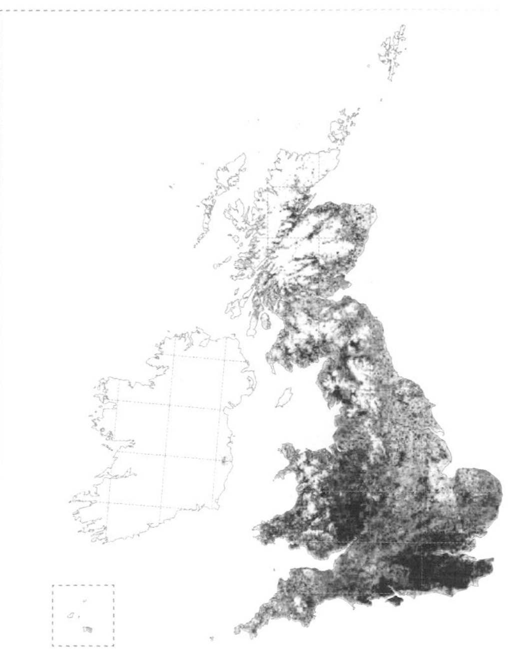

BREEDING RELATIVE ABUNDANCE CHANGE *since 1988–91*

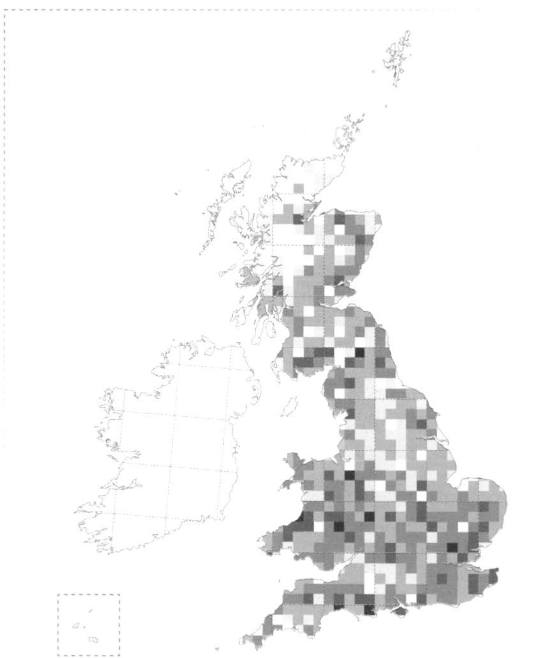

WINTER RELATIVE ABUNDANCE *2007/08–2010/11*

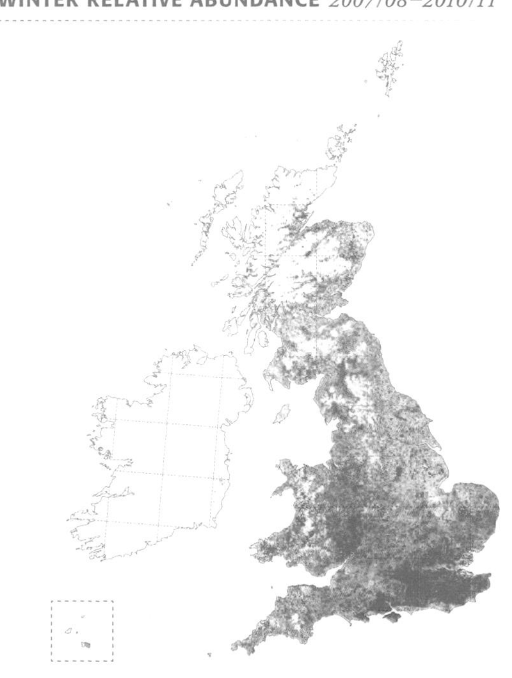

LESSER SPOTTED WOODPECKER

Dendrocopos minor

BTO CODE: LS

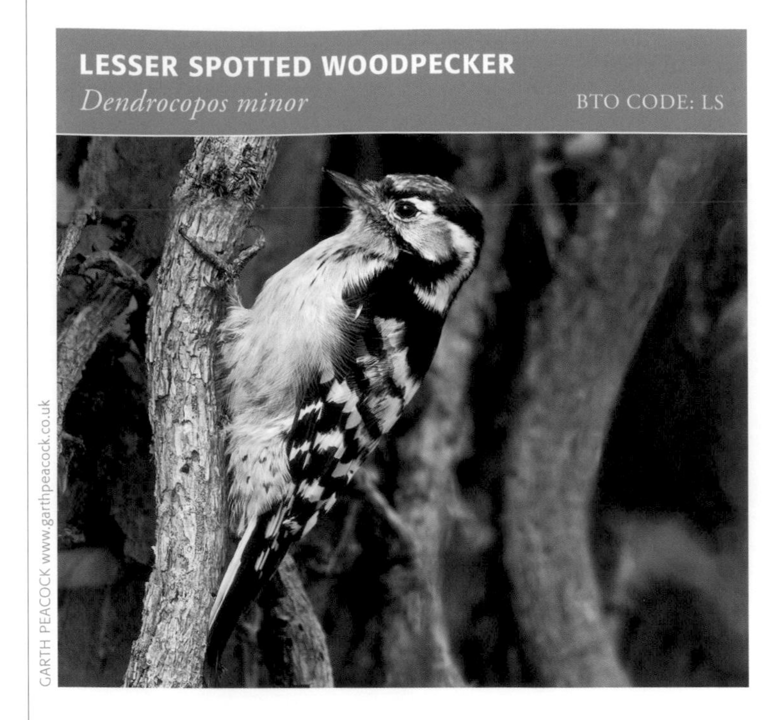

GARTH PEACOCK www.garthpeacock.co.uk

ACCELERATING RANGE CONTRACTION by our smallest woodpecker is one of the most striking findings from *Bird Atlas 2007–11*. The British endemic race *comminutus* has undergone many local losses during recent decades and is beginning to show absences at the whole-county level.

The winter and breeding distributions are very similar: 55 more 10-km squares were found to be occupied in winter than the breeding season, owing mostly to the species' greater mobility at that season and perhaps a greater detectability in February when birds are displaying. The species has never been found in Ireland and, since the Scottish records in the *1968–72 Breeding Atlas* have recently been reviewed and rejected (ap Rheinallt *et al.* 2012), they do not appear on the change maps. The first acceptable record for Scotland occurred in October 2012, on Shetland (Fray 2013).

Losses of range have been extensive in all regions, but especially in East Anglia and in northern and westernmost England. Overall, the breeding range contracted by 11% up to the *1988–91 Breeding Atlas* and subsequently by a further 34%. Only minor local gains have been noted, mostly in northeast England and in Wales. The distribution appears increasingly fragmented, with London and surrounding counties forming the largest remaining area of continuous occupation at this scale. The more isolated occupied 10-km squares might hold no more than a single pair.

The population of Lesser Spotted Woodpeckers in Britain last peaked during the outbreak of Dutch Elm Disease but has been in decline since around 1980. Numbers halved in just the ten years to 1999, since when the species has become too scarce to be monitored by annual schemes (Hewson *et al.* 2007; *BirdTrends*). Causes of the decline are unclear but probably relate to competition, predation or declining woodland quality: squares that retained these woodpeckers between the first two breeding atlases had significantly greater woodland cover than those that lost the species (Charman *et al.* 2010; Broughton *et al.* 2013). Breeding success is currently very low, with chick starvation appearing to be a major problem (Charman *et al.* 2012; Smith & Charman 2012), suggesting that food availability may be a factor limiting the population.

In memory of John Woodland & Mark Blacksell

BREEDING DISTRIBUTION *2008–11*

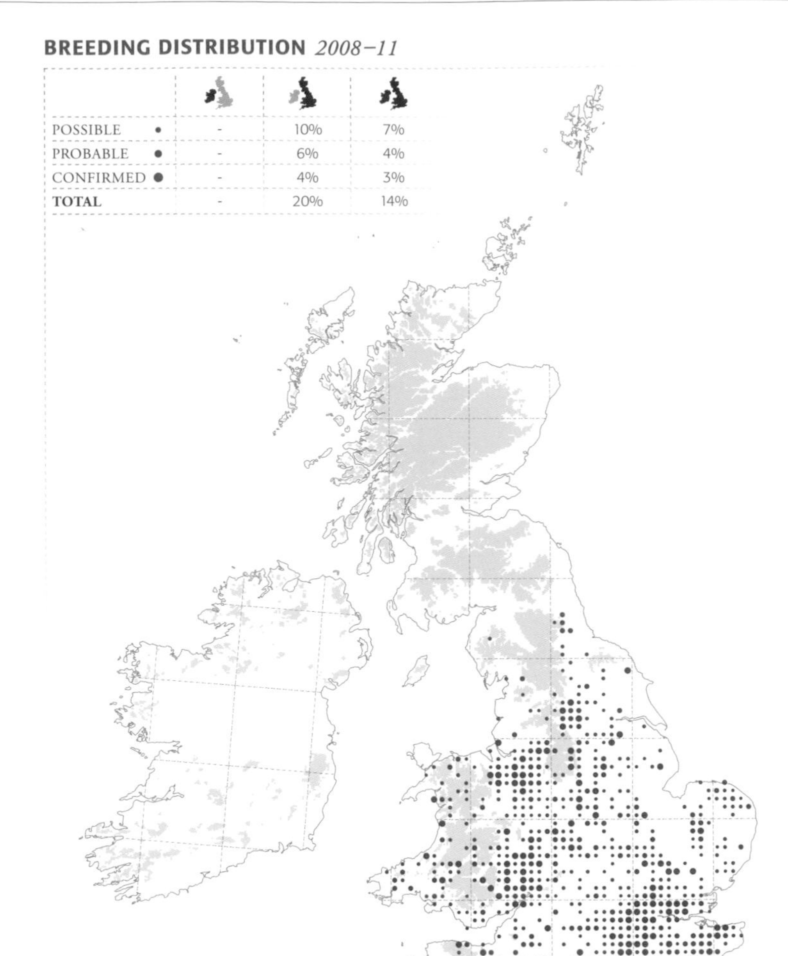

POSSIBLE •	-	10%	7%
PROBABLE •	-	6%	4%
CONFIRMED •	-	4%	3%
TOTAL	-	20%	14%

WINTER DISTRIBUTION *2007/08–2010/11*

PRESENT •	-	21%	16%

BREEDING DISTRIBUTION CHANGE *since 1968–72*

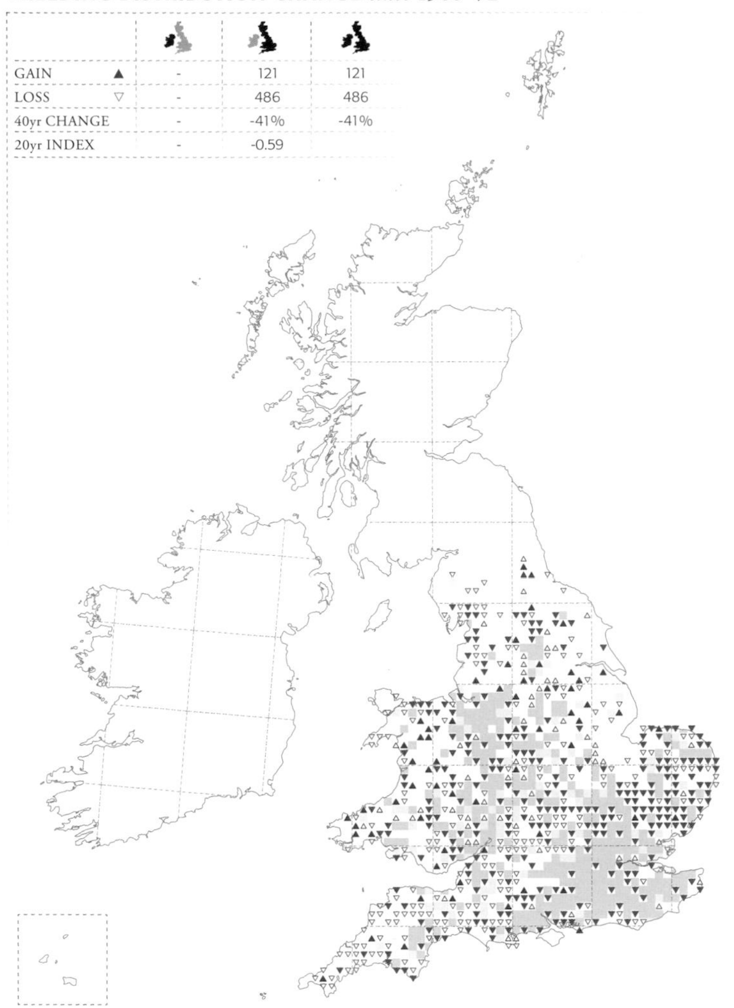

GAIN ▲	-	121	121
LOSS ▽	-	486	486
40yr CHANGE	-	-41%	-41%
20yr INDEX	-	-0.59	

WINTER DISTRIBUTION CHANGE *since 1981–84*

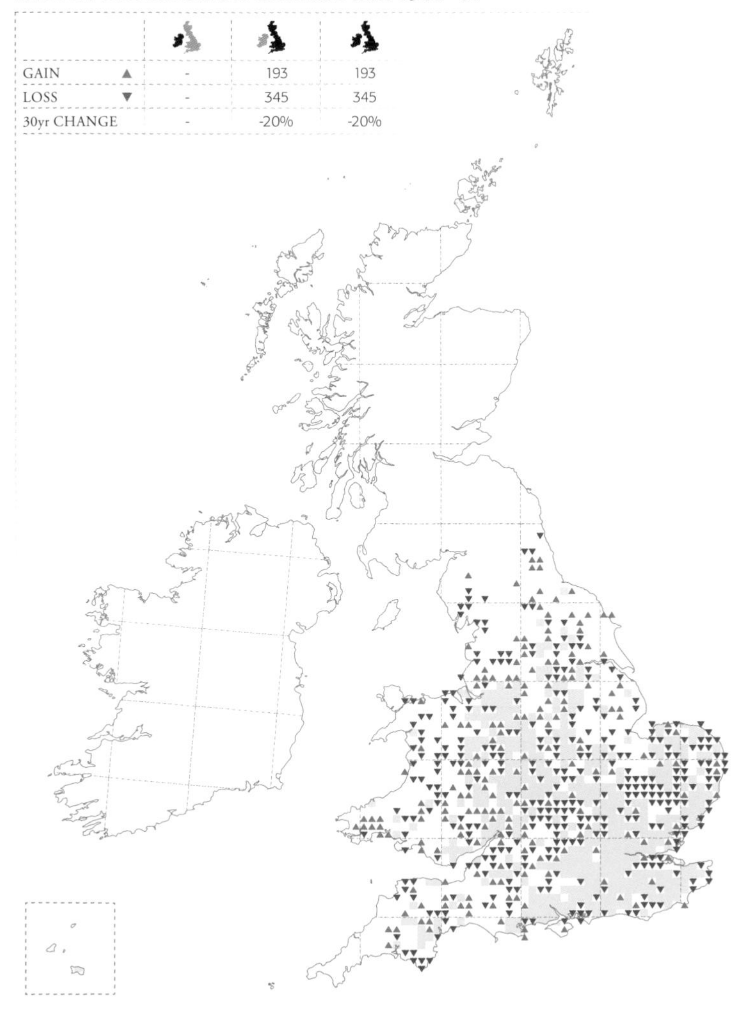

GAIN ▲	-	193	193
LOSS ▼	-	345	345
30yr CHANGE	-	-20%	-20%

BREEDING RELATIVE ABUNDANCE *2008–11*

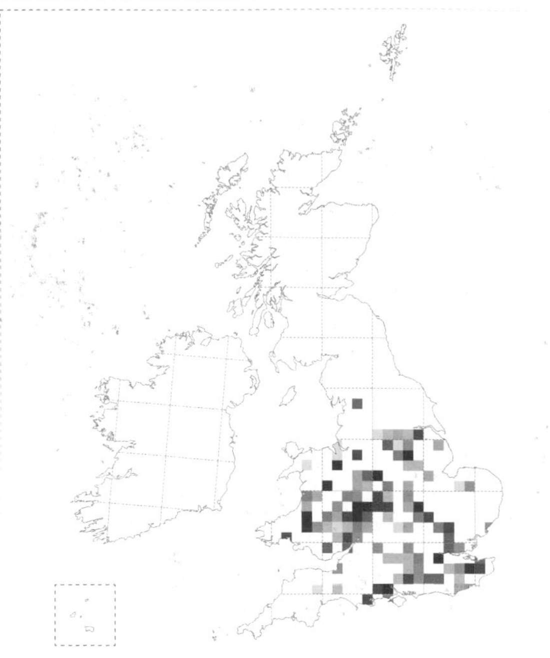

BREEDING RELATIVE ABUNDANCE CHANGE *since 1988–91*

WINTER RELATIVE ABUNDANCE *2007/08–2010/11*

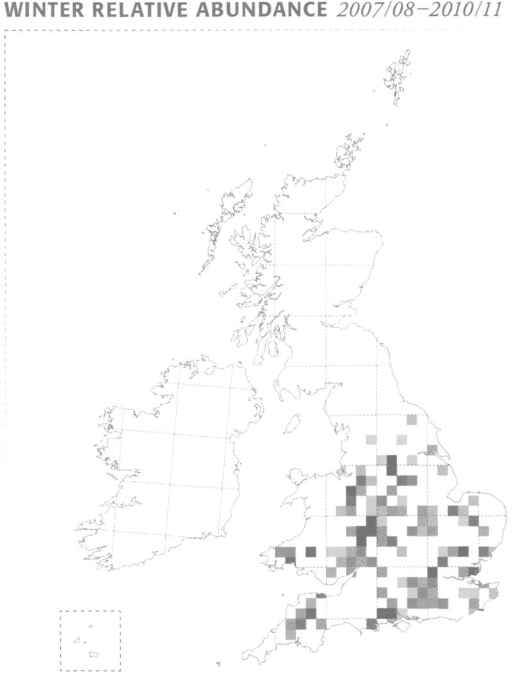

WRYNECK

Jynx torquilla

BTO CODE: WY

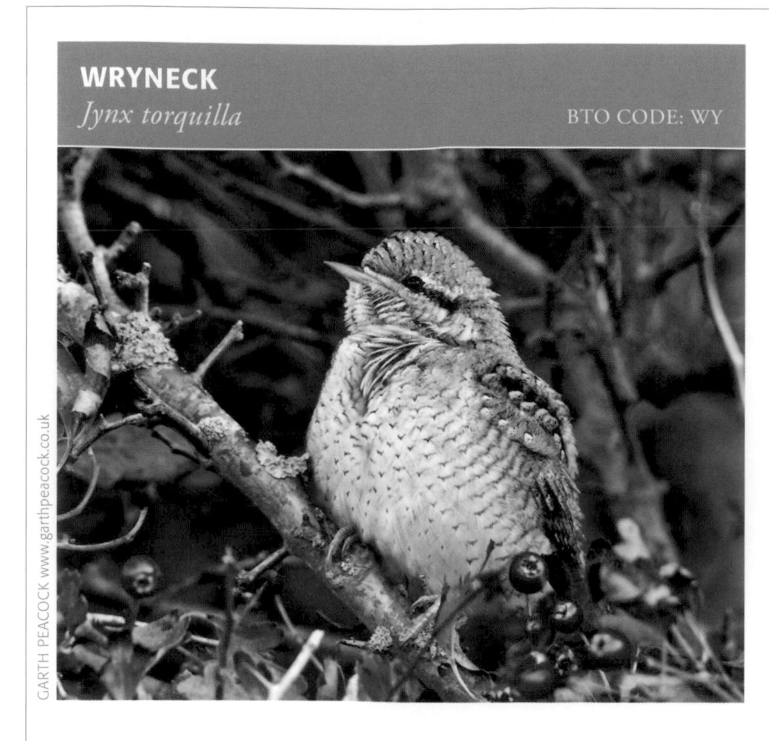

GARTH PEACOCK www.garthpeacock.co.uk

WRYNECKS WERE RECORDED in 48 10-km squares during the *1968–72 Breeding Atlas*, mostly in southeast England, where breeding was *confirmed* in 12 squares. Breeding was also *confirmed* in two squares in a then-expanding population in the Scottish Highlands. By the *1988–91 Breeding Atlas* the species was recorded from only six 10-km squares, four of which were in the Highlands and only one of which supported *confirmed* breeding. The last confirmation of breeding in England was in Buckinghamshire in 1985 and the last in Scotland was in Ross-shire in 2002 (*RBBP 2010*).

During *Bird Atlas 2007–11*, birds were recorded with *possible* breeding evidence in 12 10-km squares and *probable* breeding was noted in just one square. Most records were in the Scottish Highlands, with a marked concentration in Strathspey, and refer to singing males. Due to the scarcity of the species, and the abundance of potentially suitable habitat in Scotland, it is possible that the Atlas underestimates the true status of the species.

Non-breeding birds, not thought to be in suitable breeding habitat, were recorded from a further 87 10-km squares in Britain, mostly in southern and eastern England, and from four squares in Ireland. There were November records from two 10-km squares in Britain and one in Ireland.

Wrynecks have suffered a long-term population decline and range contraction since the mid 19th century in western and central Europe and it has been suggested that the decline is linked with a drop in food availability due to a shortage of bare ground and short vegetation (Tucker & Heath 1994; Mermod *et al.* 2009; Weisshaupt *et al.* 2011). Other factors implicated in this decline include increased rain during the breeding season as part of climatic changes, agricultural improvement, loss of orchards and unimproved meadows, replacement of hardwoods with conifer plantations and the widespread use of pesticides and herbicides (del Hoyo *et al.* 2002; Coudrain *et al.* 2010). Since 1980 there has been a further moderate decline of around 10% in Europe (*PECBMS 2010*). These continuing declines on the near Continent make it even less likely that this species will re-establish itself as a regular breeder in Britain in the near future.

Sponsored by John and Alison Baker

BREEDING DISTRIBUTION *2008–11*

POSSIBLE •	-	<1%	<1%
PROBABLE •	-	<1%	<1%
CONFIRMED •	-	-	-
TOTAL	-	<1%	<1%

BREEDING DISTRIBUTION CHANGE *since 1968–72*

GAIN ▲	-	10	10
LOSS ▽	-	45	45
40yr CHANGE	-	-73%	-73%
20yr INDEX	-	-	

GOLDEN ORIOLE
Oriolus oriolus

BTO CODE: OL

CHRIS KNIGHTS www.chrisknightswildlife.com

GOLDEN ORIOLES ARE scarce spring migrants and are recorded mainly at coastal localities in a range of arboreal habitats. They are now very rare breeders with a recent five-year mean count of just five pairs (2006–10; *RBBP 2010*), whereas at least 16 pairs were *confirmed* breeding during 1988–91 (Ogilvie *et al.* 1994). During 2008–11 breeding was *confirmed* in two 10-km squares in the Suffolk Fens and *possible* breeding was reported in 27 squares. As far as is known, Golden Orioles were last confirmed breeding outside Suffolk in 2003 and since then breeding has occurred at only one site, which is RSPB Lakenheath Fen. Many records away from the Fens probably refer to migrants; *non-breeding* birds were reported in a further 93 10-km squares in Britain and five in Ireland.

Whether to classify extralimital records of singing birds as *possible* breeders or migrants is problematic and affects estimates of range change. Superficially, the 180% range expansion since the *1968–72 Breeding Atlas* appears encouraging but the majority of the 'gains' relate to transient singing males. The number of squares with *probable* or *confirmed* breeding changed from six in 1968–72 to 14 in 1988–91 and to two in 2008–11. The change map highlights losses in the Fens and at other historical breeding sites. The hope, expressed in the *1988–91 Breeding Atlas*, that the population would consolidate and expand in the coming decades has clearly not been realised.

The British population is thought not to be self-sustaining (Milwright 1998); rather it is linked to population trends in the Low Countries (Mason & Allsop 2009). Although the European population increased by 26% during 1990–2010 (*PECBMS 2010*), populations on the northwestern fringe of the range are declining or, at best, are stable (BirdLife International 2004). The preference for poplar plantations in Britain is a restriction on their range, as rates of felling exceed those of planting (Dagley 1994). Orioles are adversely affected by poor weather in June and breeding success is generally low (Mason & Allsop 2009). Encouragingly, however, distribution modelling in relation to future climate predicts a significant expansion of range in southeast England (Huntley *et al.* 2007).

BREEDING DISTRIBUTION *2008–11*

	Ireland	Britain	Britain & Ireland
POSSIBLE ●	-	1%	1%
PROBABLE ●	-	-	-
CONFIRMED ●	-	<1%	<1%
TOTAL	-	1%	1%

BREEDING DISTRIBUTION CHANGE *since 1968–72*

	Ireland	Britain	Britain & Ireland
GAIN ▲	-	26	26
LOSS ▽	-	8	8
40yr CHANGE	-	+180%	+180%
20yr INDEX	-	-	

RED-BACKED SHRIKE

Lanius collurio

BTO CODE: ED

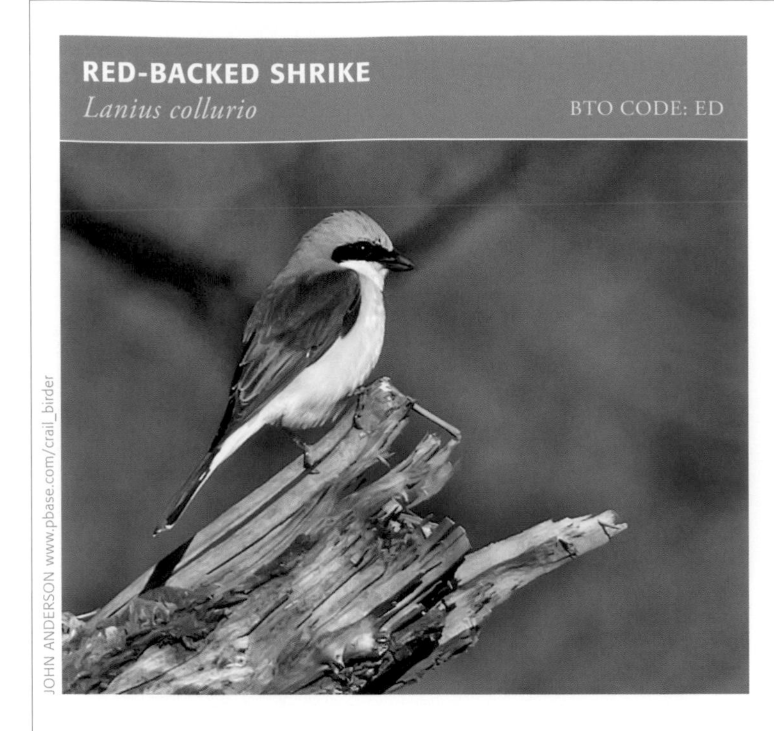

JOHN ANDERSON www.pbase.com/crail_birder

NUMBERS OF BREEDING Red-backed Shrikes in Britain decreased steadily after the middle of the 19th century, with a strong decline noted after 1945 involving range contraction mainly into southeast England. The decline continued rapidly during the 1970s and 1980s, to the point of extinction in 1989 (*Birds in England*). One pair bred in England in 1991 and 1992. Regular breeding commenced again in Scotland in 2004 and one pair nested in Wales during 2005–07 (Holling *et al.* 2007a; *RBBP 2010*). In Scotland, breeding attempts tend to coincide with high numbers on spring passage (*Birds of Scotland*). In Britain & Ireland, the majority of records are now of non-breeding birds observed only on passage, typically scattered across eastern and southern locations in Britain.

During *Bird Atlas 2007–11* potential breeding records came from several widely scattered British locations. Their distribution, here mapped at 50-km resolution and encompassing north-central Scotland, eastern England and southwest England, implies incidental recolonisation on a broader front than might have been predicted from the historical pattern of range contraction. Records of *confirmed* breeding came from southwest England, where breeding took place in Devon in 2010–11 following three males holding territory in 2008–09, and the Scottish Highlands (*RBBP 2010*; K. Rylands pers comm). Whether a viable breeding population can re-establish, given continuing declines on the near Continent (*PECBMS 2010*), remains to be seen. Many of the losses shown on the change map reflect different interpretation of migrants in the *1988–91 Breeding Atlas*.

Breeding Red-backed Shrikes require extensively managed transitional habitats, typically scrub–grassland mosaics with an abundance of large insects such as crickets and beetles (Vanhinsbergh & Evans 2002; Tryjanowski *et al.* 2003). However, their decline also fits a pattern shared with other long-distance migrants wintering in Africa (Sanderson *et al.* 2006), implying there may be effects of land-use practice or of climate, either in Africa or at staging locations en route, that have contributed to the declining European breeding populations. Closer scrutiny of demographic change is required to identify where in the life-cycle the factors limiting recovery are operating, though poor breeding performance is not thought to have driven the decline in Britain (Tryjanowski *et al.* 2006).

BREEDING DISTRIBUTION *2008–11*

POSSIBLE ●	-	<1%	<1%
PROBABLE ●	-	<1%	<1%
CONFIRMED ●	-	<1%	<1%
TOTAL	-	<1%	<1%

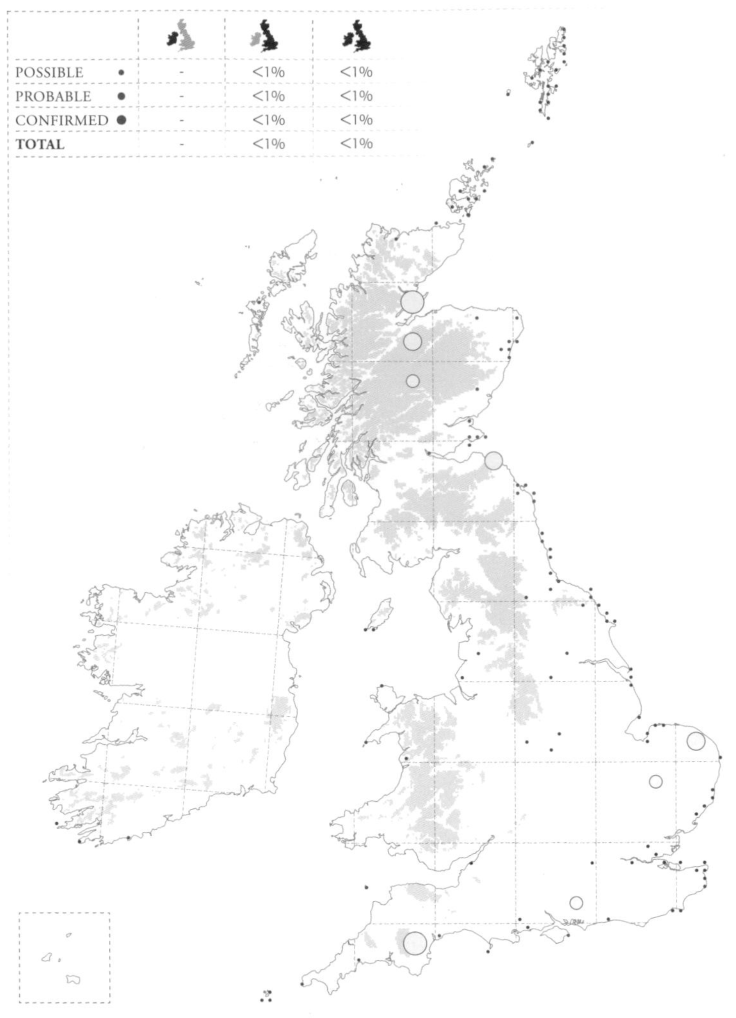

BREEDING DISTRIBUTION CHANGE *since 1988–91*

GAIN ▲	-	12	12
LOSS ▽	-	14	14
20yr CHANGE	-	-13%	-13%
20yr INDEX	-	-	

GREAT GREY SHRIKE

Lanius excubitor

BTO CODE: SR

JIM ALMOND www.shropshirebirder.co.uk

THE WINTER MAP shows a wide but strangely clumped distribution, extending through most of Britain except western Scotland; there were no accepted winter records in Ireland. The birds are winter visitors from Fennoscandia, mainly from Norway: they arrive from late September onwards and by November are mostly settled into their winter territories, where they typically remain until March or April. Some British east-coast squares were occupied only in November, perhaps by birds that then wintered further inland or continued southward into France.

Winter territories, each occupied by a single bird, may span a number of favoured feeding areas several kilometres apart, each holding a 'larder' of stored food to which the bird regularly returns. Thus it is not unusual for an individual to be recorded in two or more 10-km squares during a single winter. Birds may return to the same or partly overlapping winter territories in subsequent years, sometimes adding further adjoining squares. These habits contribute to the clumped nature of the 10-km distribution, as does the patchiness of the species' favoured habitat of large, undisturbed tracts of lowland heath. Extensive areas of prime habitat, such as the New Forest, Dartmoor, Exmoor and the Thames Basin Heaths, regularly hold more than one active winter territory. Other exposed and open habitats may also be occupied, including young forestry plantations.

Analysing bird report data, Fraser & Ryan (1995) estimated that a maximum of 75 birds wintered in Britain during 1986–92. The *1981–84 Winter Atlas* had estimated the population to be 150 individuals and it is likely that numbers in the 1960s and 1970s had been higher still. Henderson (2012) reported a peak in wintering numbers in Dumfries & Galloway during the 1970s and attributed this to the creation of young forestry plantations. The change map, however, indicates that losses in Scotland and the north and east of England were more than outweighed by the gains distributed widely across England and Wales.

During the breeding season there were *non-breeding* records in 87 10-km squares in Britain and one in Ireland; most such records were in April, though five squares were occupied in May and one square in June and July.

Sponsored by Surrey Bird Club

WINTER DISTRIBUTION *2007/08–2010/11*

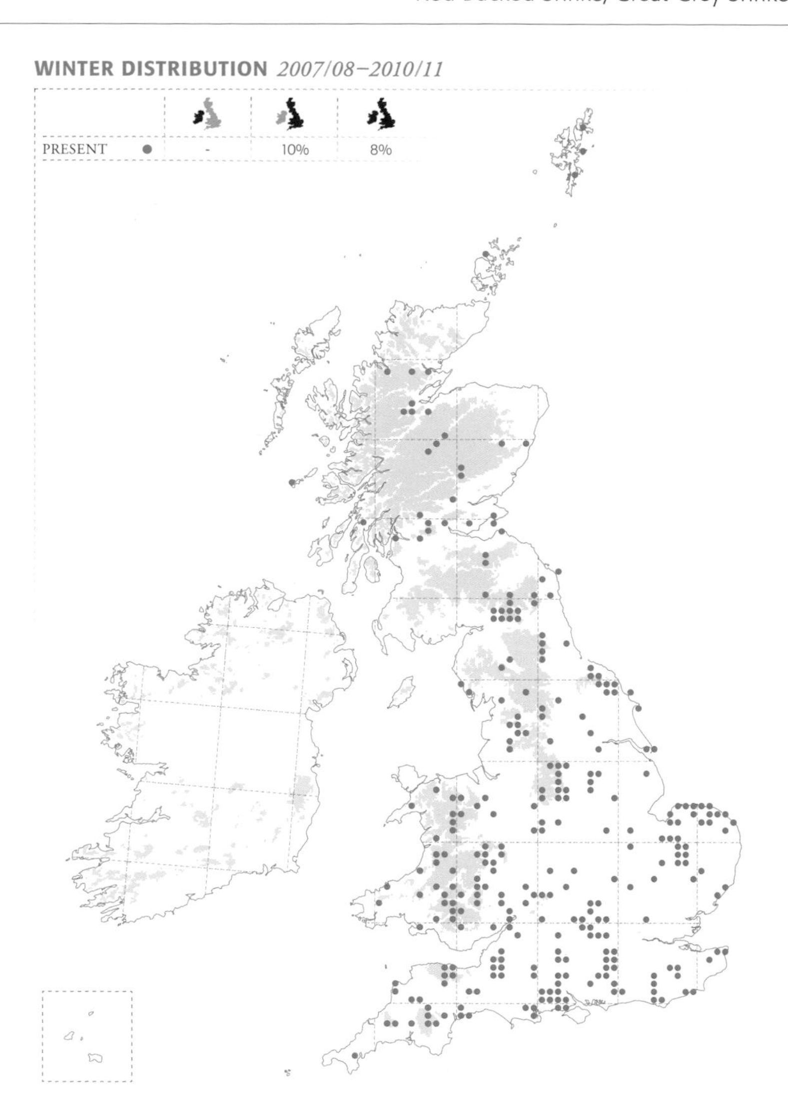

WINTER DISTRIBUTION CHANGE *since 1981–84*

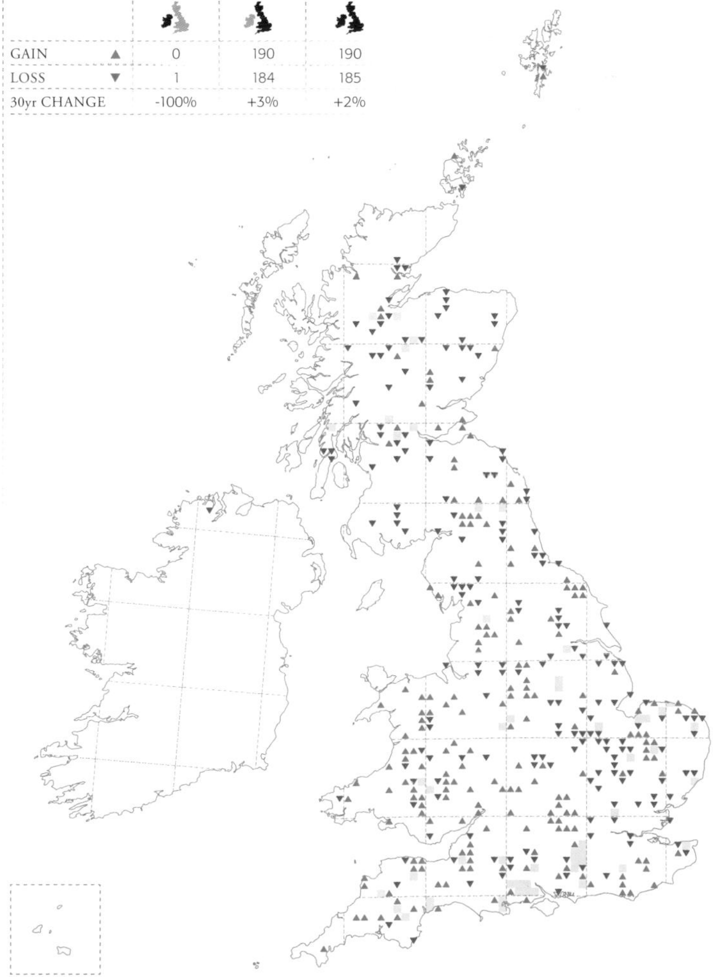

CHOUGH

Pyrrhocorax pyrrhocorax

BTO CODE: CF

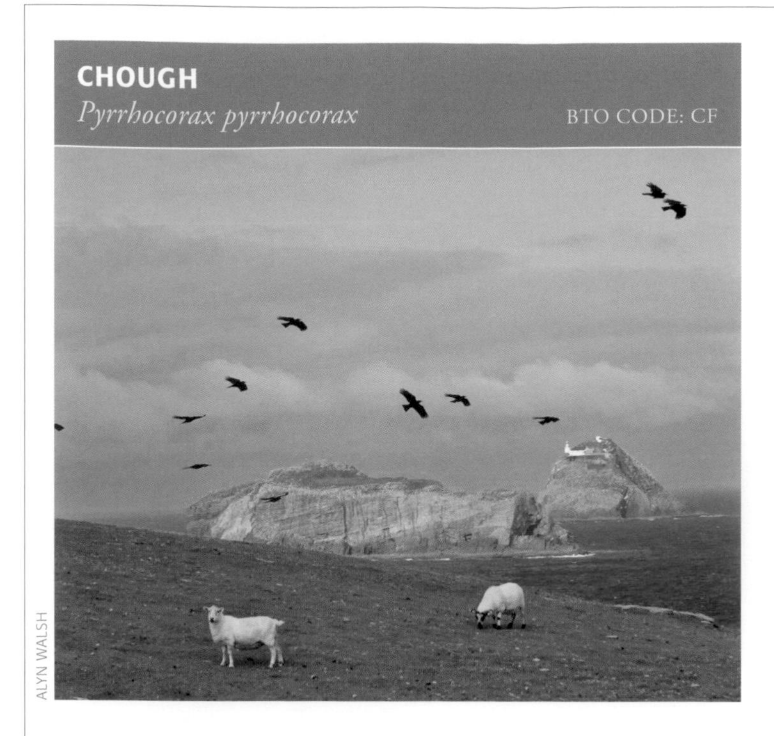

ALYN WALSH

THE MOST STRIKING change over the last 40 years has been the recolonisation of west Cornwall. Birds arrived in 2001, fledged young in 2002 for the first time since 1947, and increased to six pairs by 2011 (Johnstone *et al.* 2011). Genetic analysis shows that Ireland was the most likely source (Wenzel *et al.* 2012).

Chough distribution in Britain & Ireland is limited to regions with very mild winters, suitable nest cavities and grazed pastures. Choughs are essentially sedentary and there are few seasonal differences in distribution. Birds present in the breeding season in southwest Scotland had disappeared by the closing field seasons of the Atlas and were not recorded in the winter. With their loss, Scottish Choughs are now restricted to Colonsay, Islay and Jura. On Islay, variations in population size were accounted for mainly by variation in pre-breeding survival rates (Reid *et al.* 2004). Agricultural change, including loss of outdoor-wintered cattle and permanent coastal pastures, can cause declines in Chough populations (*Birds of Scotland*; Johnstone *et al.* 2007b).

In Ireland, except where deep bays are flanked by low-lying shores, the range is near continuous from Co. Donegal to Co. Wexford. There is a small outlying population on Rathlin Island, Co. Antrim. Chough surveys suggest that Irish numbers have remained broadly stable during 1992–2003 (Berrow *et al.* 1993; Gray *et al.* 2003). The westerly coastal distribution is also noticeable in Wales, although some pairs nest inland at quarries. Choughs are present throughout the Isle of Man, whilst in England they are confined to Cornwall apart from two exceptional winter records in Cumbria and on the Isle of Wight.

In winter, losses in range were recorded on the Mull of Kintyre and Gigha in Scotland and nearby in Northern Ireland. These changes may be connected because wintering birds on the Mull of Kintyre may have originated from Northern Ireland (*Birds of Scotland*). Elsewhere in Ireland there has been a scatter of winter gains and losses, giving an overall expansion of 10%. There are notable gains in South Wales, following breeding which took place in Glamorgan in 2006 for the first time for c.150 years (Holling *et al.* 2009).

Sponsored by Ecology Matters

BREEDING DISTRIBUTION *2008–11*

	Britain & Ireland	Britain	Ireland
POSSIBLE •	1%	<1%	1%
PROBABLE •	4%	1%	2%
CONFIRMED •	11%	3%	5%
TOTAL	16%	4%	8%

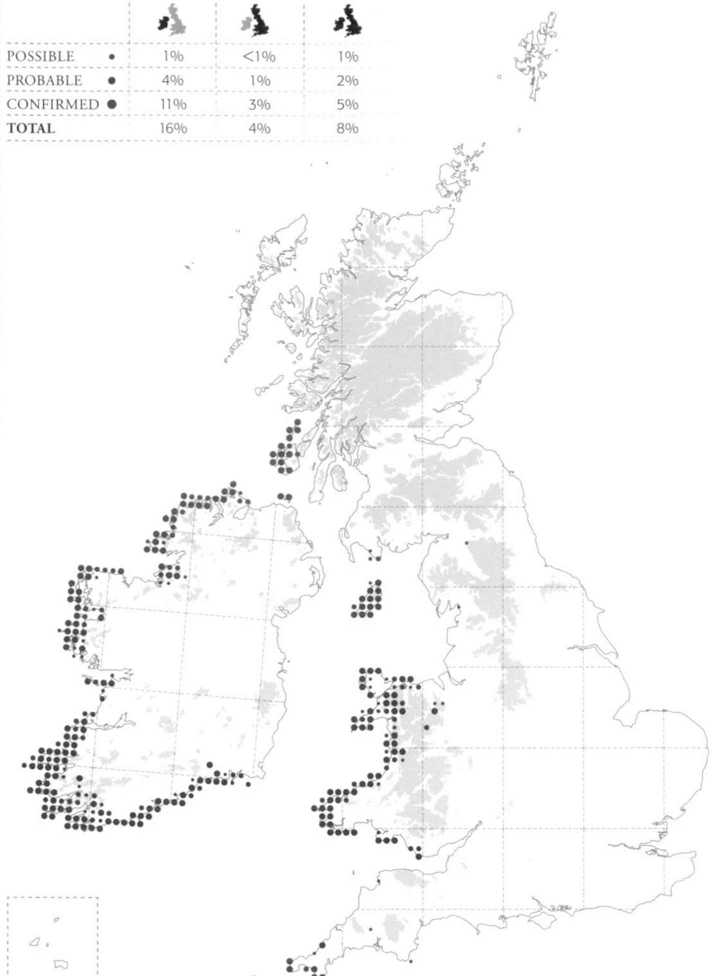

WINTER DISTRIBUTION *2007/08–2010/11*

	Britain & Ireland	Britain	Ireland
PRESENT •	18%	4%	8%

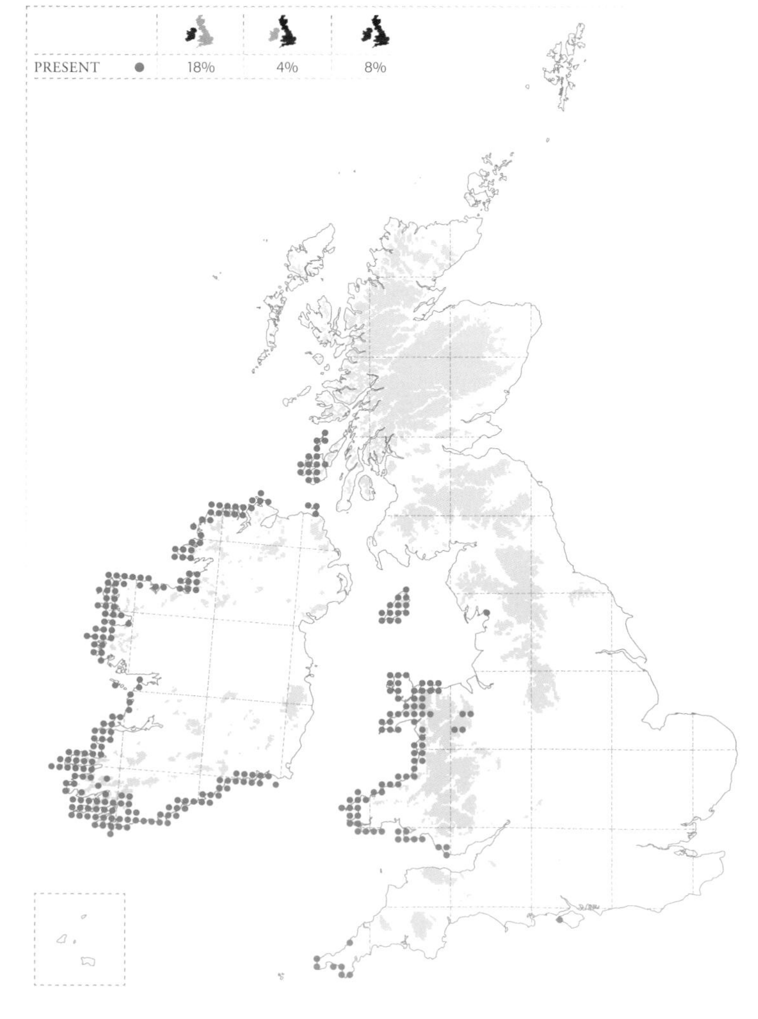

BREEDING DISTRIBUTION CHANGE *since 1968–72*

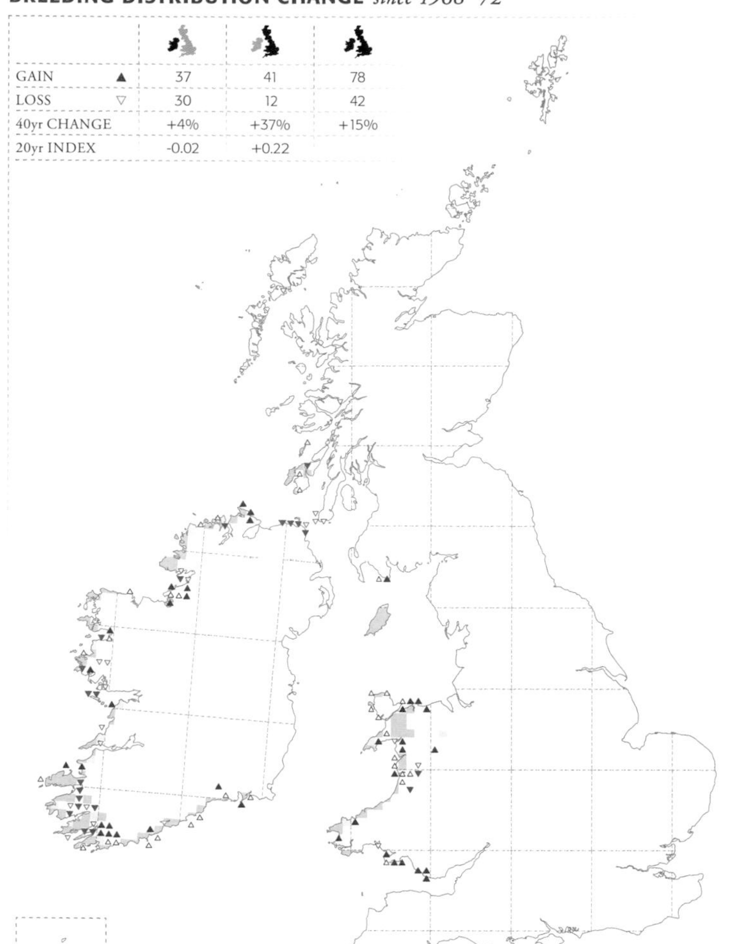

GAIN ▲	37	41	78
LOSS ▽	30	12	42
40yr CHANGE	+4%	+37%	+15%
20yr INDEX	-0.02	+0.22	

BREEDING RELATIVE ABUNDANCE *2008–11*

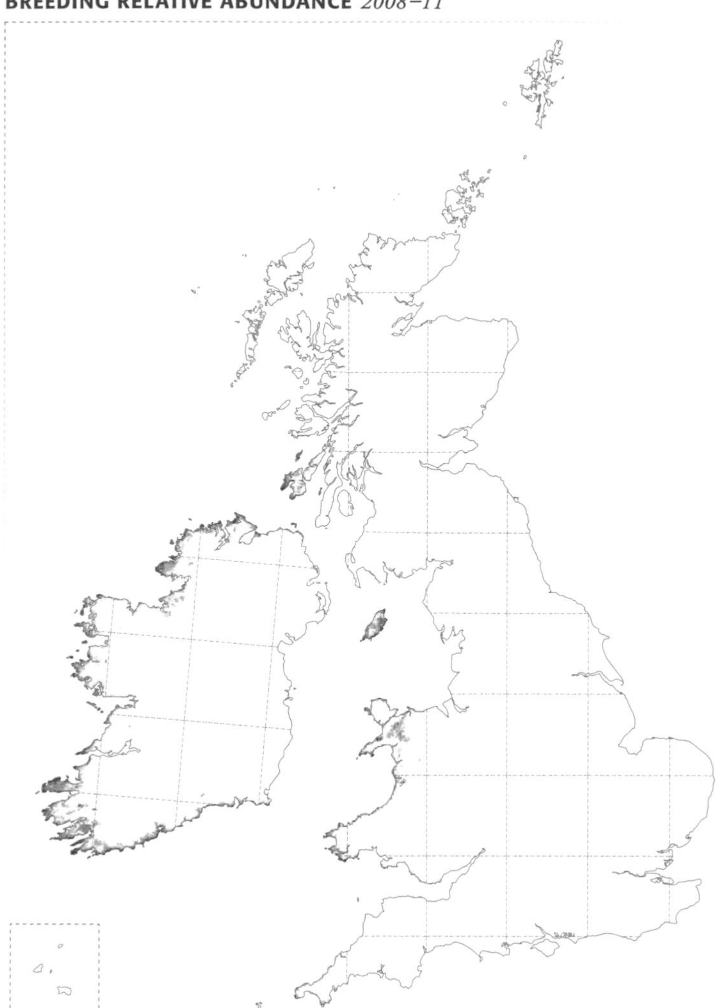

WINTER DISTRIBUTION CHANGE *since 1981–84*

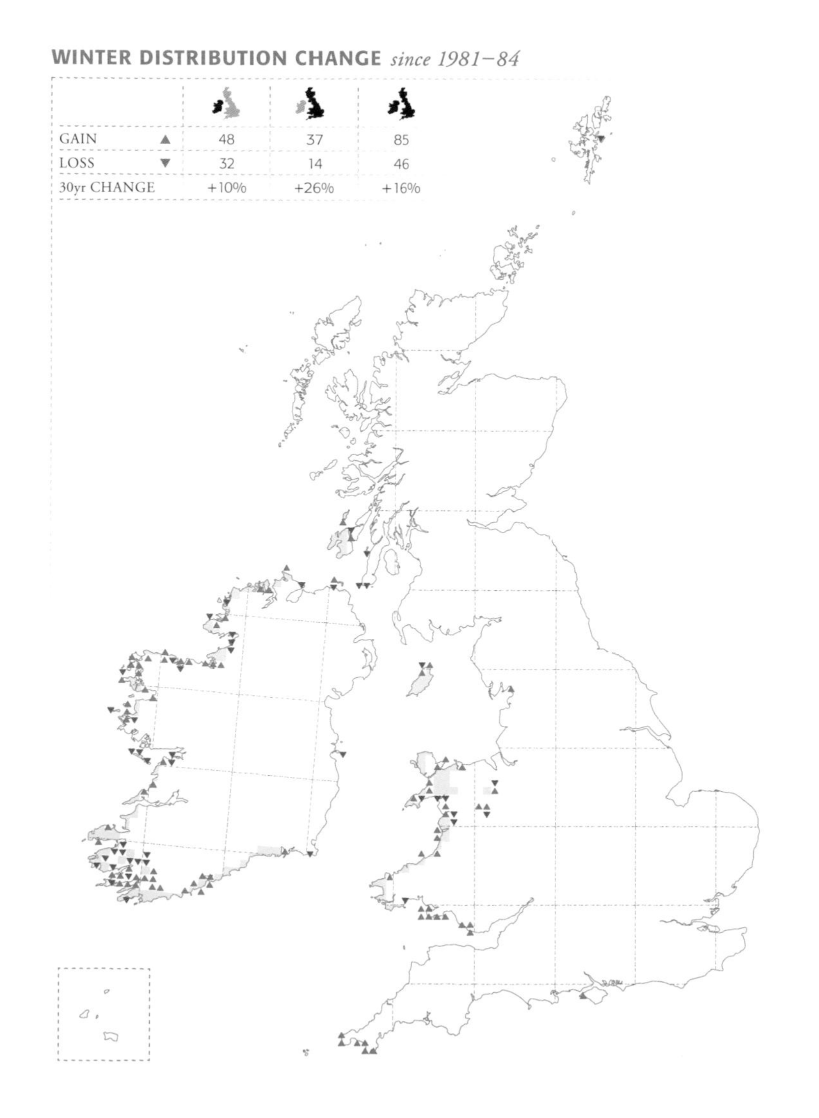

GAIN ▲	48	37	85
LOSS ▼	32	14	46
30yr CHANGE	+10%	+26%	+16%

WINTER RELATIVE ABUNDANCE *2007/08–2010/11*

MAGPIE

Pica pica

BTO CODE: MG

GARTH PEACOCK www.garthpeacock.co.uk

DURING THE WINTER, Magpies were widespread in both Ireland and Britain, being absent only from the northwestern half of Scotland and its islands. They have colonised many of the upland areas of southern Scotland since the *1981–84 Winter Atlas*, as well as spreading into new areas of eastern Scotland and the fringes of the Highlands. The breeding distribution and changes since previous atlases are almost exactly the same as described for winter, which is understandable, given the strong territoriality of pairs of birds at every stage of the year (*BWP*).

The relative abundance maps show that the highest densities are associated with urban and suburban areas from southeast England through the Midlands to Lancashire and West Yorkshire. Other high-density areas include much of northeast England, the central lowlands of Scotland, Northern Ireland and the greater Dublin area; a study in Dublin found 16.6 pairs per km^2 (Kavanagh 1987). In the uplands of Wales, northern England and Scotland, Magpies occur at low density, or are absent entirely.

There have been large and widely distributed abundance increases in Ireland, as indicated by an increased proportion of occupied tetrads since the *1988–91 Breeding Atlas*. This is despite a reported 18% decline in numbers in the Republic of Ireland during 1998–2010 (*CBS Trend 2010*). In the UK, the situation is more mixed, with abundance increases in Scotland and parts of eastern England, but declines elsewhere, particularly in the west. Although overall there was a 97% population increase during 1970–2010 (*SUKB 2012*), there is evidence of a plateau or slight decline since 1995 (*BirdTrends*), which hides regional differences shown both in the Atlas abundance change map and by annual monitoring.

In the areas of gain, reduced control by humans is probably the main driver of change and probably also explains the significant increase in nest survival and productivity (*BirdTrends*). In some areas, increased availability of suitable nesting sites in conifer plantations may have aided colonisation. Magpies may be more difficult to detect at the edge of their range, and especially in areas where keepering may still persist, as they become unobtrusive.

BREEDING DISTRIBUTION *2008–11*

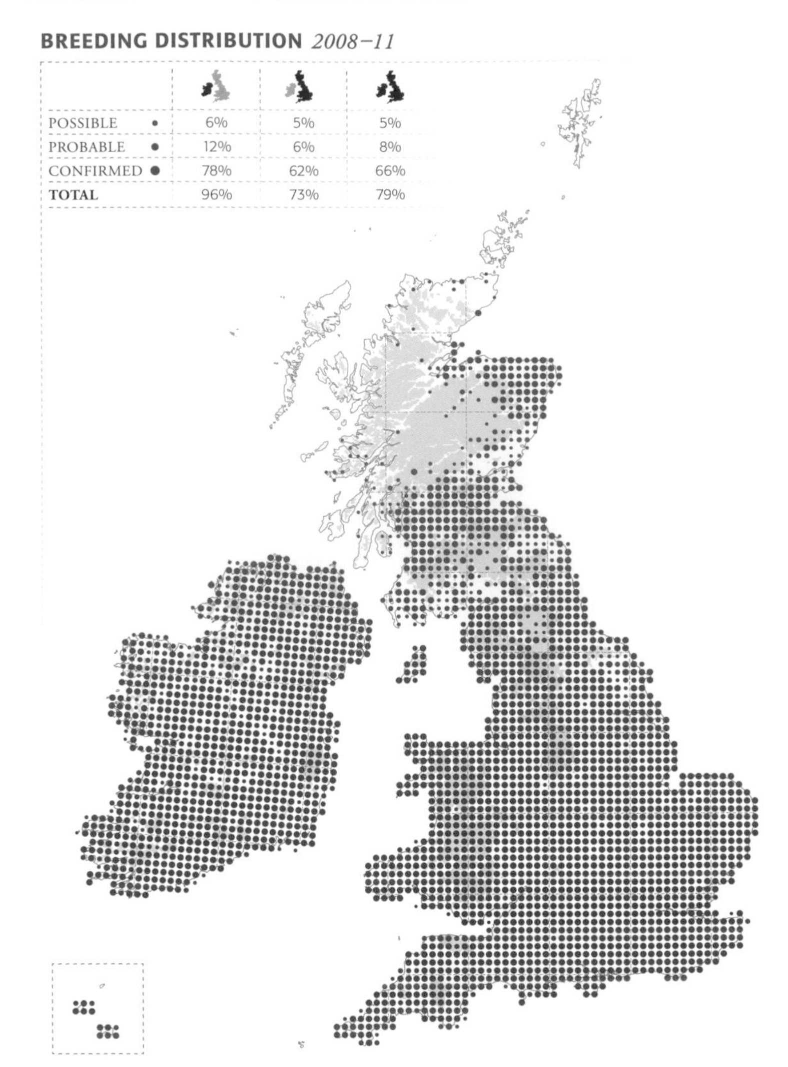

POSSIBLE •	6%	5%	5%
PROBABLE •	12%	6%	8%
CONFIRMED ●	78%	62%	66%
TOTAL	96%	73%	79%

WINTER DISTRIBUTION *2007/08–2010/11*

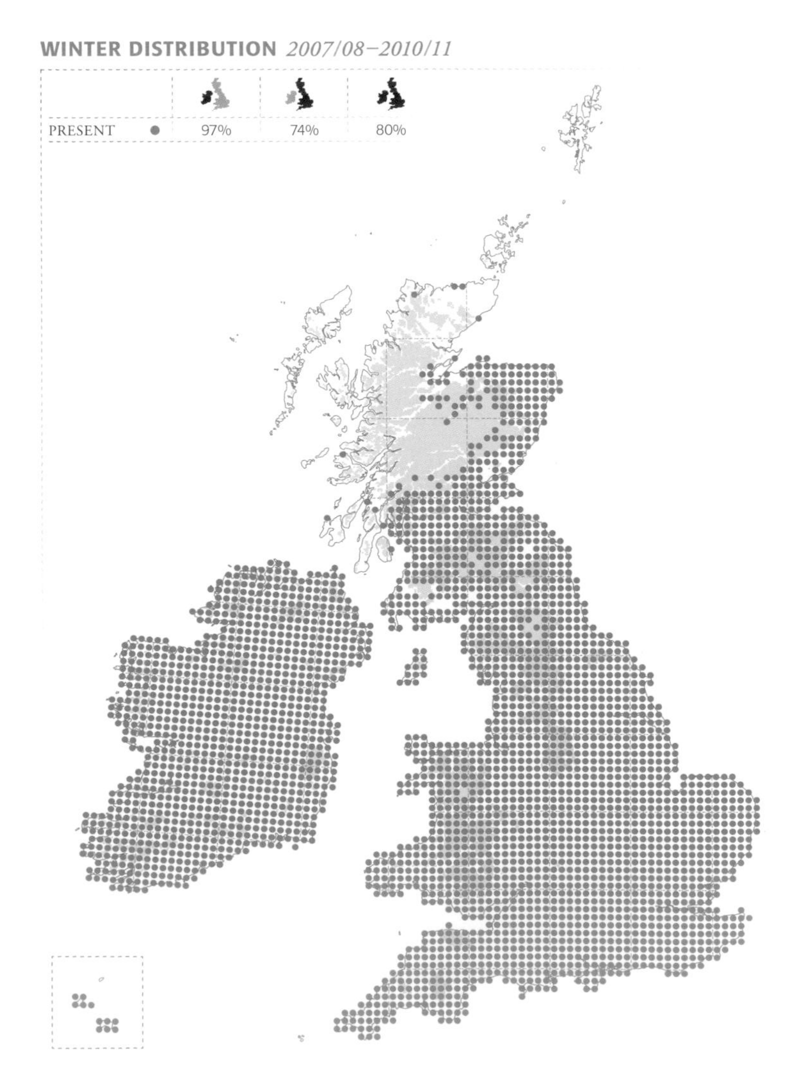

PRESENT •	97%	74%	80%

Sponsored by Peter Wilkinson

BREEDING DISTRIBUTION CHANGE *since 1968–72*

GAIN ▲	29	202	231
LOSS ▽	3	70	73
40yr CHANGE	+3%	+7%	+5%
20yr INDEX	+0.01	+0.04	

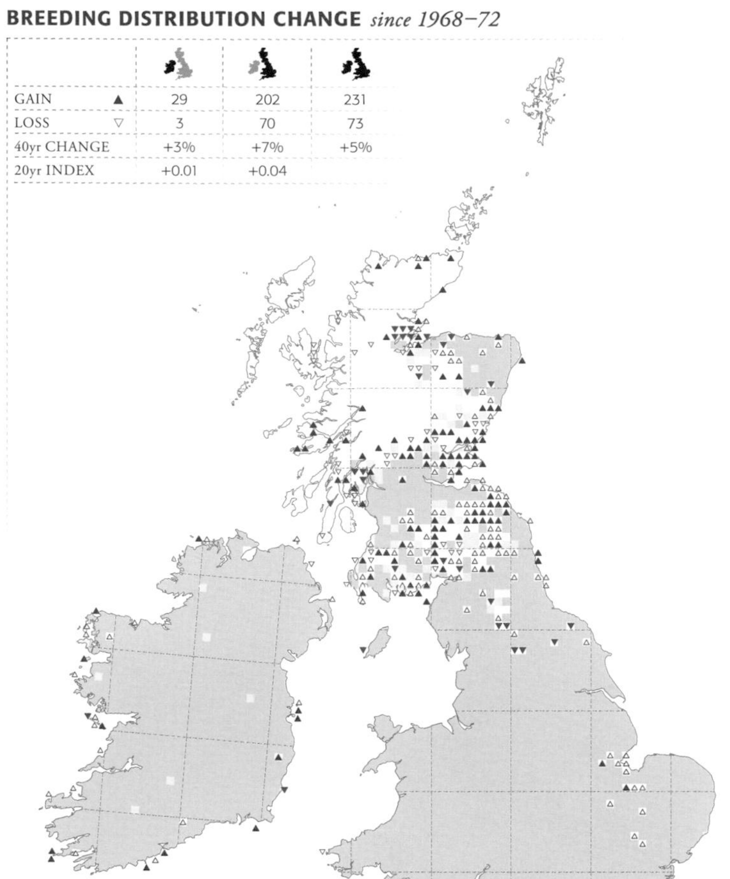

WINTER DISTRIBUTION CHANGE *since 1981–84*

GAIN ▲	18	219	237
LOSS ▼	20	35	55
30yr CHANGE	-<1%	+10%	+6%

BREEDING RELATIVE ABUNDANCE *2008–11*

BREEDING RELATIVE ABUNDANCE CHANGE *since 1988–91*

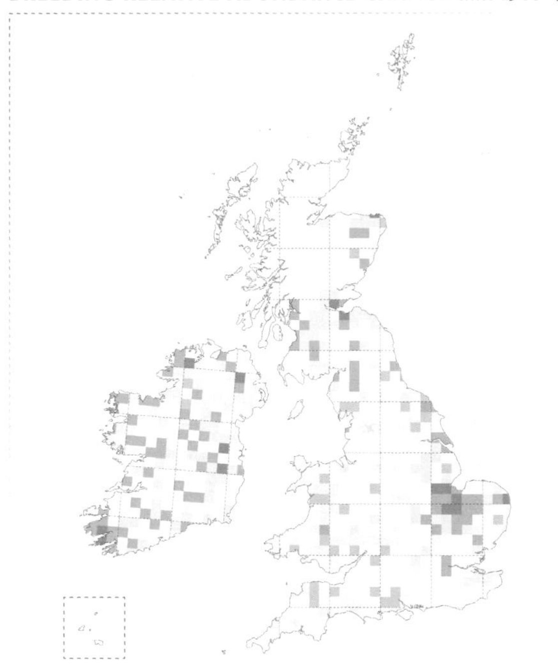

WINTER RELATIVE ABUNDANCE *2007/08–2010/11*

JAY

Garrulus glandarius

BTO CODE: J.

LIZ CUTTING www.lizcuttingphotos.com

JAY DISTRIBUTION WITHIN Britain & Ireland is thought to be constrained by woodland cover and climatic conditions and, at least historically, numbers have been controlled by gamekeepers in many areas. The winter distribution has expanded since the *1981–84 Winter Atlas*, with an 86% range expansion in Ireland and a 16% expansion in Britain, where there have been major gains in Scotland. In the *1981–84 Winter Atlas* there was only one occupied 10-km square northwest of the Grampian Mountains, but Jays can now be found widely within the Great Glen and northeast Scotland. In Ireland there has been infilling of the patchy distribution of the *1981–84 Winter Atlas* but also westward expansion. Southeast England is a major stronghold, with other hotspots of abundance in western England and in Wales.

The breeding distribution map is similar to the winter map, as expected for a resident species. In Britain the 14% range expansion between 1968–72 and 2008–11 has virtually all occurred since the *1988–91 Breeding Atlas*. Increases in relative abundance are apparent in many areas since the *1988–91 Breeding Atlas*. The growth in Scotland is attributed to reduced keepering and an increase in forest cover, particularly of plantation forests, many of which are not keepered (*Birds of Scotland*). Over the period spanning these atlases, the UK Jay population has fluctuated but shown no long-term trend (*BirdTrends*), though there has been a 15% increase between 1995 and 2010 (*BBS Report 2011*). This conceals habitat differences, with Jay populations being stable in woodland but increasing in farmland (Gregory & Marchant 1996).

Ireland has seen substantial breeding-range expansion, particularly in the last 20 years, with a near doubling of the range since the *1988–91 Breeding Atlas*. This follows an apparent 32% range contraction between the *1968–72 Breeding Atlas* and *1988–91 Breeding Atlas*. At the time, those losses were not ascribed to coverage issues and many of the recent gains have been farther north and west, but Jays are still too scarce to be monitored effectively on an annual basis. As in Scotland, the increases may be linked to increased forest cover.

Sponsored by Bristol Ornithological Club

BREEDING DISTRIBUTION *2008–11*

POSSIBLE •	19%	12%	14%
PROBABLE •	16%	21%	19%
CONFIRMED •	16%	36%	31%
TOTAL	51%	69%	64%

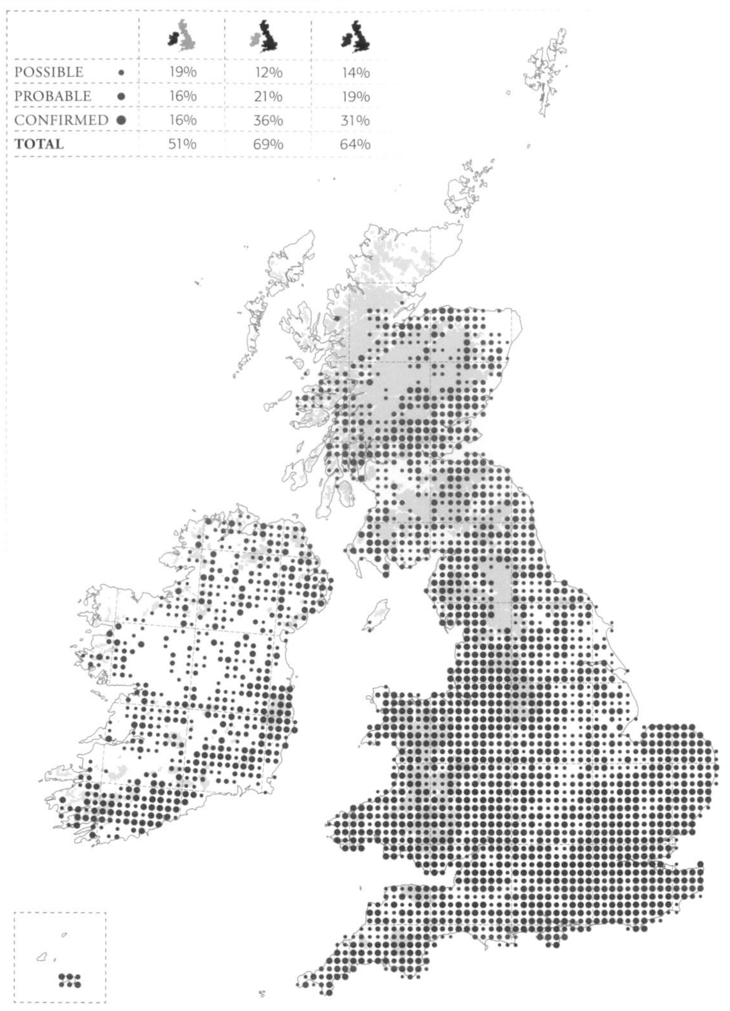

WINTER DISTRIBUTION *2007/08–2010/11*

PRESENT •	55%	73%	68%

BREEDING DISTRIBUTION CHANGE *since 1968–72*

GAIN ▲	244	318	562
LOSS ▽	120	71	191
40yr CHANGE	+31%	+14%	+17%
20yr INDEX	+0.39	+0.12	

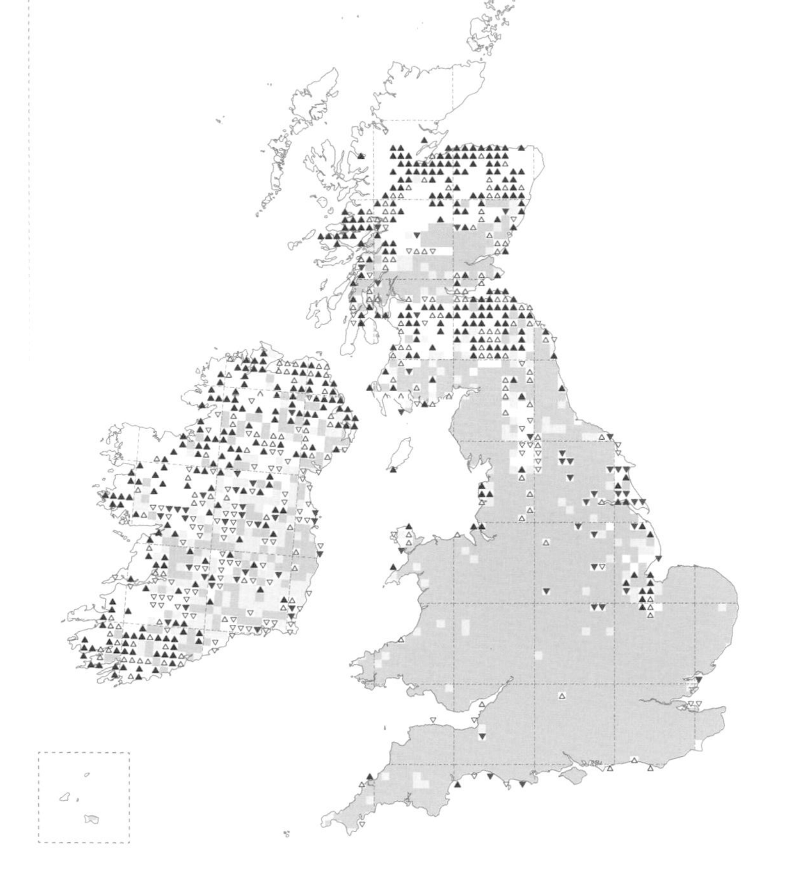

WINTER DISTRIBUTION CHANGE *since 1981–84*

GAIN ▲	291	336	627
LOSS ▼	51	49	100
30yr CHANGE	+86%	+16%	+26%

BREEDING RELATIVE ABUNDANCE *2008–11*

BREEDING RELATIVE ABUNDANCE CHANGE *since 1988–91*

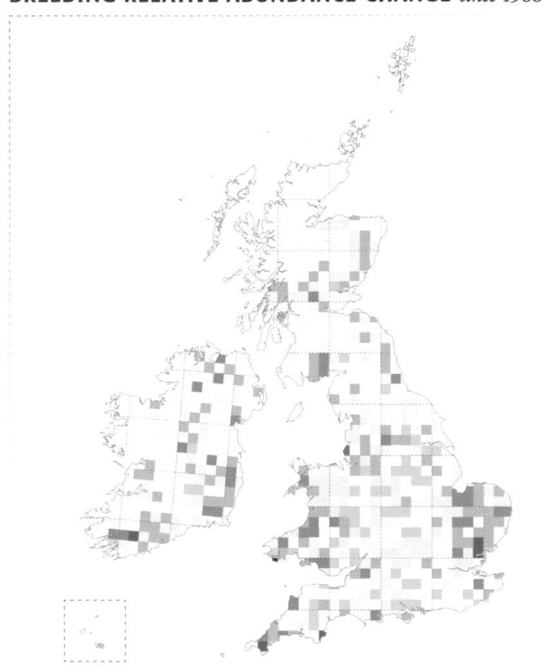

WINTER RELATIVE ABUNDANCE *2007/08–2010/11*

JACKDAW

Corvus monedula

BTO CODE: JD

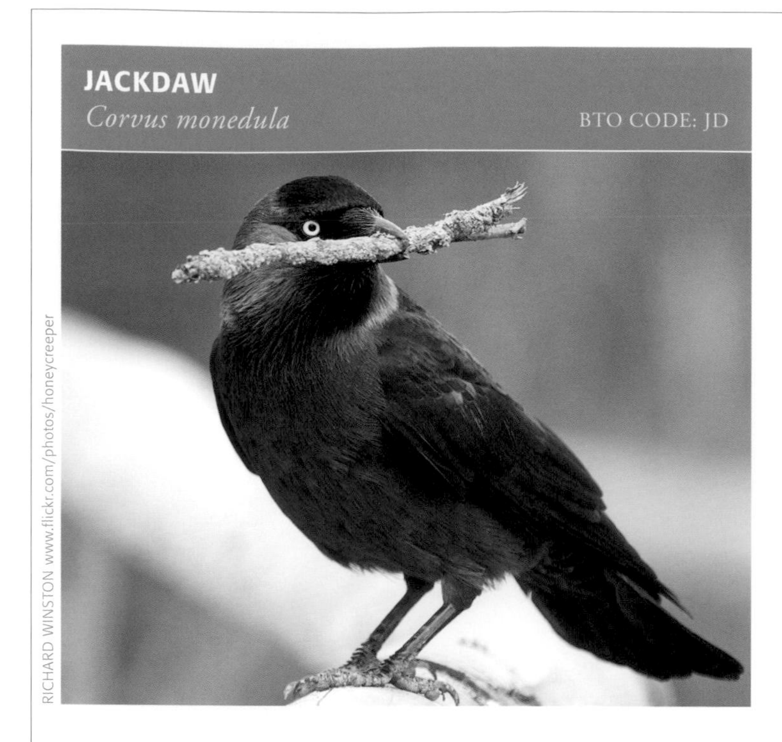

RICHARD WINSTON www.flickr.com/photos/honeycreeper

JACKDAWS ARE WIDESPREAD in both seasons and absent only from large areas of northwest Scotland and some small pockets of northwest Ireland. Since the *1981–84 Winter Atlas*, there has been a mixture of gains and losses at the edge of the range in Scotland, and to a lesser extent in Ireland, though the gains on Shetland are notable, and there has been some infilling of upland areas of northern England and southern Scotland. As this is a gregarious species, any distributional changes that occur are perhaps more obvious than would otherwise be the case, as flocks of birds move into or out of areas.

There has been little overall change in breeding-range size over the sequence of atlases. Throughout the period there have been a few gains and losses along the southeast edge of the Scottish Highlands. In these upland margins Jackdaws occur at low densities, so some changes could reflect detection issues. Alternatively, loss of suitable agricultural land due to afforestation or reductions in livestock stocking densities could have influenced occupancy, because the species is highly dependent on farmland for food. The breeding-season abundance change map shows evidence of increasing tetrad occupancy throughout England, the lowlands of Scotland and parts of Ireland. Intriguingly, only north-central Wales showed consistent declines in abundance.

The Republic of Ireland population increased by 35% during 1998–2010 (*CBS Trend 2010*). The UK breeding population grew by 131% during 1970–2010 (*SUKB 2012*), although most of the increase was from the early 1980s onwards, with the east of England showing the greatest increases since 1995 (*BBS Report 2011*). Overall nest failure rates in the UK have declined since the 1960s and there has been a corresponding increase in breeding success per nesting attempt (*BirdTrends*). Reasons for these changes are unclear but may be linked to changes in patterns of cultivation. In the past, increases were greatest in pasture-dominated farmland (Gregory & Marchant 1996), but the results presented here show that trends are also positive in arable and mixed farming areas. Some gains in the east could be associated with the increased rearing of pigs, which provides year-round food resources.

Sponsored for Peggy Smitherman

BREEDING DISTRIBUTION *2008–11*

POSSIBLE •	4%	3%	3%
PROBABLE •	8%	4%	5%
CONFIRMED •	83%	75%	77%
TOTAL	95%	82%	85%

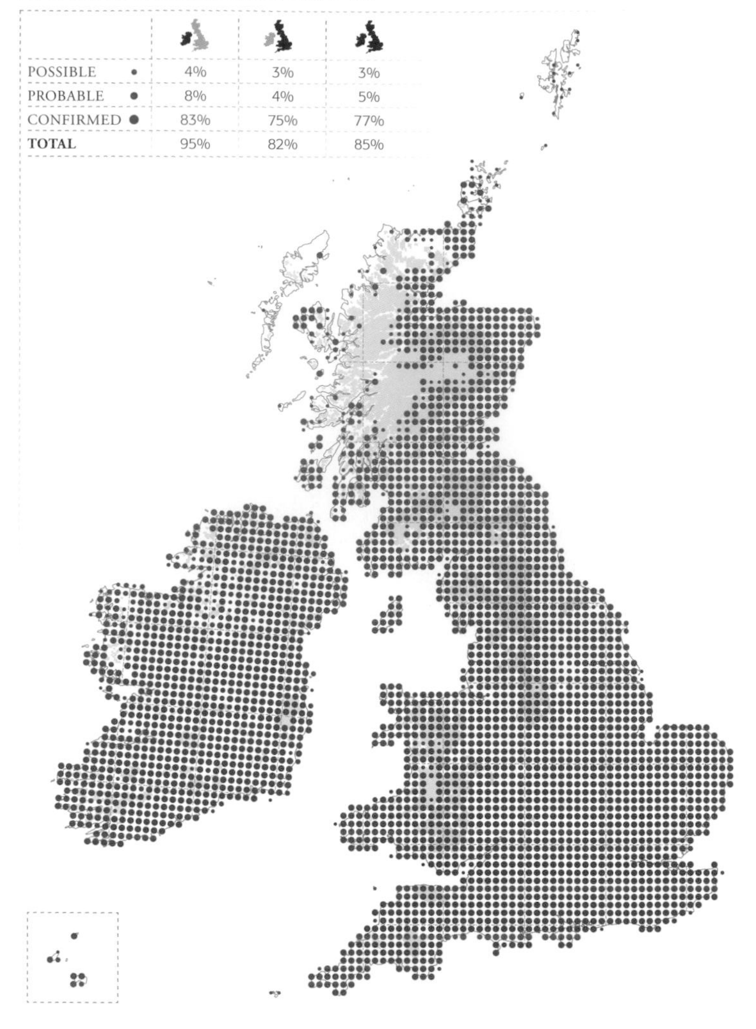

WINTER DISTRIBUTION *2007/08–2010/11*

PRESENT •	95%	83%	86%

BREEDING DISTRIBUTION CHANGE *since 1968–72*

GAIN ▲	14	44	58
LOSS ▽	22	125	147
40yr CHANGE	-1%	-3%	-3%
20yr INDEX	+0.01	+0.02	

WINTER DISTRIBUTION CHANGE *since 1981–84*

GAIN ▲	28	126	154
LOSS ▼	26	77	103
30yr CHANGE	+<1%	+2%	+2%

BREEDING RELATIVE ABUNDANCE *2008–11*

BREEDING RELATIVE ABUNDANCE CHANGE *since 1988–91*

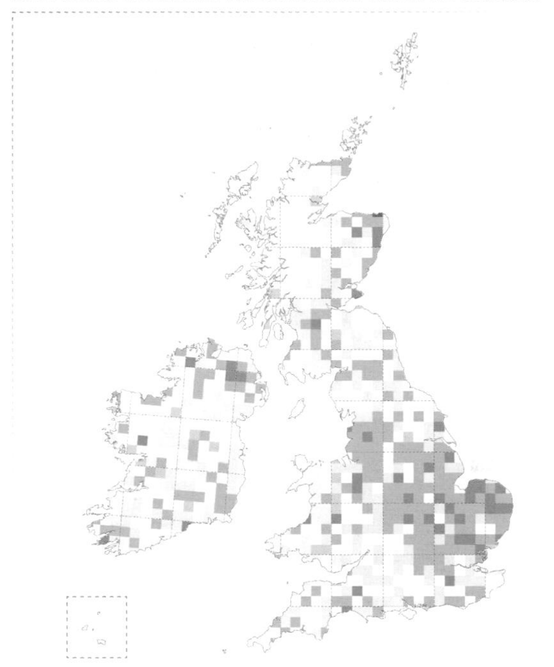

WINTER RELATIVE ABUNDANCE *2007/08–2010/11*

ROOK

Corvus frugilegus

BTO CODE: RO

DAWN BALMER

IN BOTH WINTER and the breeding season, Rooks maintain a nearly ubiquitous distribution across much of Britain & Ireland. Though they are absent from larger conurbations, such as London, the lowland distribution of Rooks includes virtually every 10-km square within the Atlas region. Major gaps occur over higher ground in Scotland, parts of northern England, Wales and the rugged far northwest of Ireland. The winter and breeding-season distribution and abundance patterns are remarkably similar, reflecting the largely sedentary nature of this species.

Rooks favour open, lowland countryside, being less abundant through the wooded Wealden counties of England, for example, than in the arable east of Britain or the eastern and southern counties of Ireland. In general, Rooks are more abundant in Ireland than in Britain.

There have been only relatively small changes in the distribution of Rooks over the last 40 years. Since the *1968–72 Breeding Atlas* some losses have occurred in and around the northerly fringes of the species' range, in the Scottish Highlands, and also in northwest Ireland, and these outnumber the small number of gains. Some of these apparent changes may reflect subtle differences in coding of breeding evidence of birds away from obvious colonies.

Within these broadly stable range limits there is mixed evidence of abundance changes, partly depending on whether estimated trends are based on counts at colonies or of birds in the wider countryside. The abundance change map shown here reveals increases in southwest Ireland, northeast Ireland and across fenland England contrasting with decreases in many other areas. Similarly, surveys of rookeries in the UK found a 40% population increase from 1975–77 to 1996 (Marchant & Gregory 1999). However, annual population monitoring suggests shallow population declines in the last decade (*CBS Trend 2010*; *BBS Report 2011*). Trends could vary geographically due to variations in land-use practice, especially in grassland management and arable crops. A 33% decline in the number of nests and a 20% decrease in the number of rookeries between 1975 and 1998 in Easter Ross was partly due to high levels of persecution, with rookeries in farm woodlands declining and those in gardens and villages increasing (McGhee 2000).

Sponsored for Jack & Mary Hazzard

BREEDING DISTRIBUTION *2008–11*

POSSIBLE •	7%	3%	4%
PROBABLE •	4%	1%	2%
CONFIRMED •	79%	71%	73%
TOTAL	90%	75%	79%

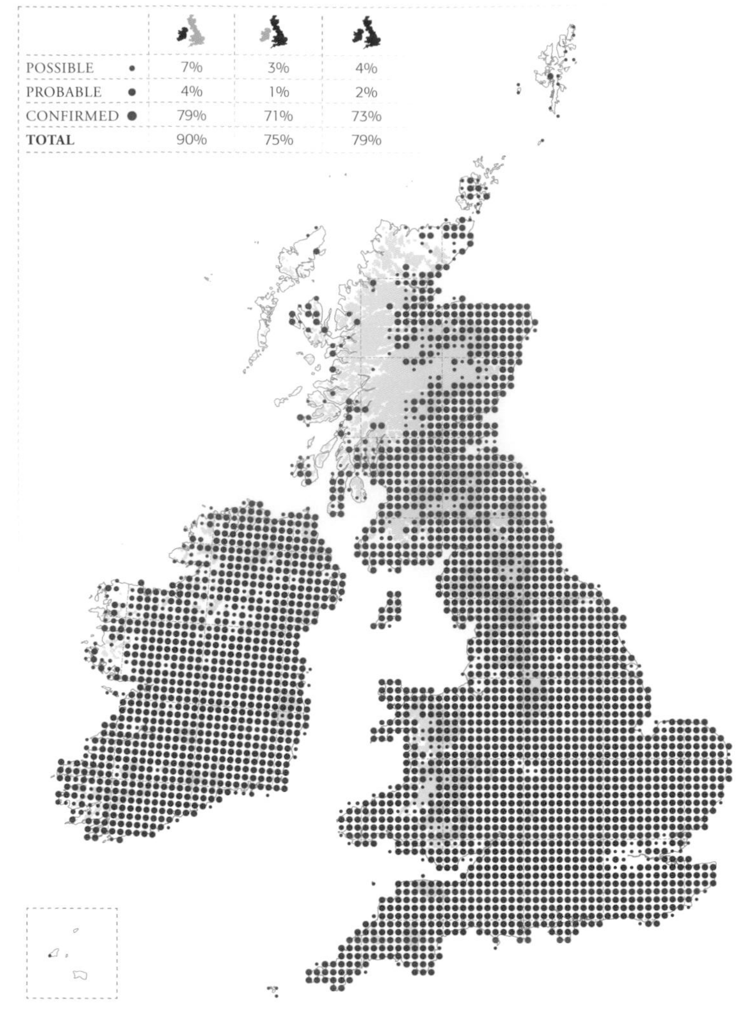

WINTER DISTRIBUTION *2007/08–2010/11*

PRESENT •	94%	80%	83%

BREEDING DISTRIBUTION CHANGE *since 1968–72*

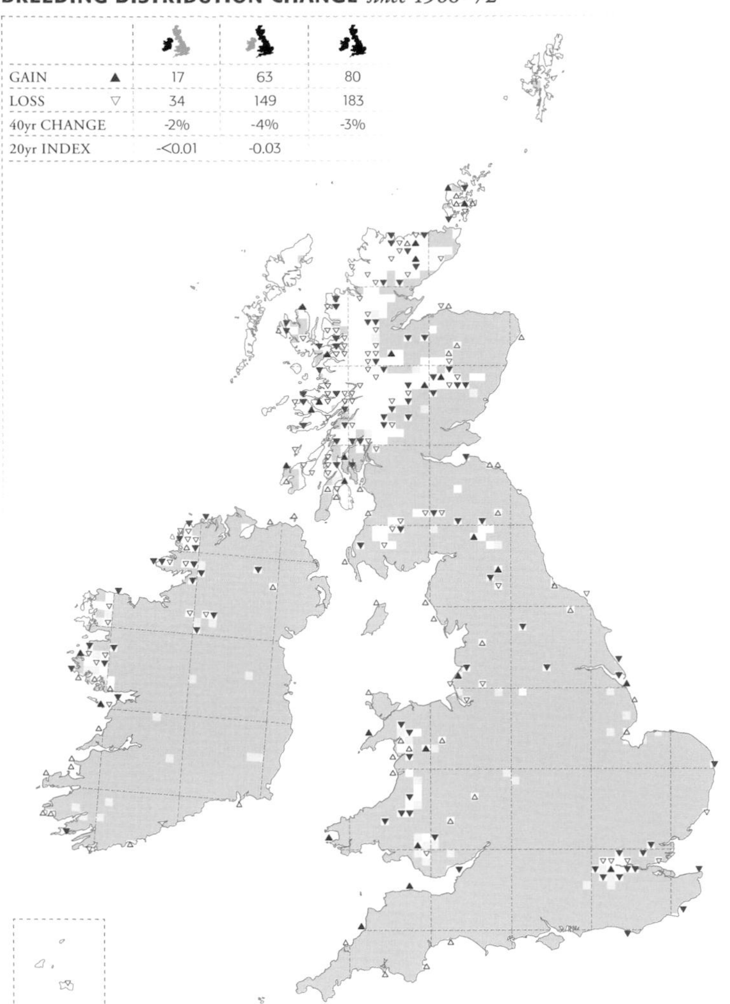

GAIN ▲	17	63	80
LOSS ▽	34	149	183
40yr CHANGE	-2%	-4%	-3%
20yr INDEX	-<0.01	-0.03	

WINTER DISTRIBUTION CHANGE *since 1981–84*

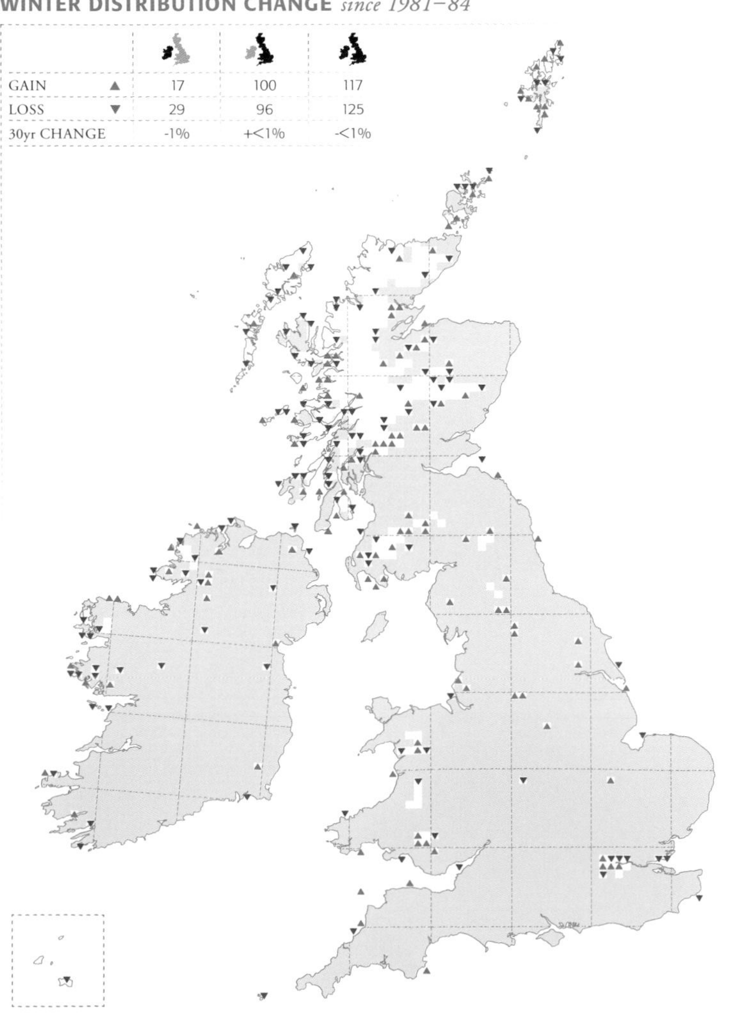

GAIN ▲	17	100	117
LOSS ▼	29	96	125
30yr CHANGE	-1%	+<1%	-<1%

BREEDING RELATIVE ABUNDANCE *2008–11*

BREEDING RELATIVE ABUNDANCE CHANGE *since 1988–91*

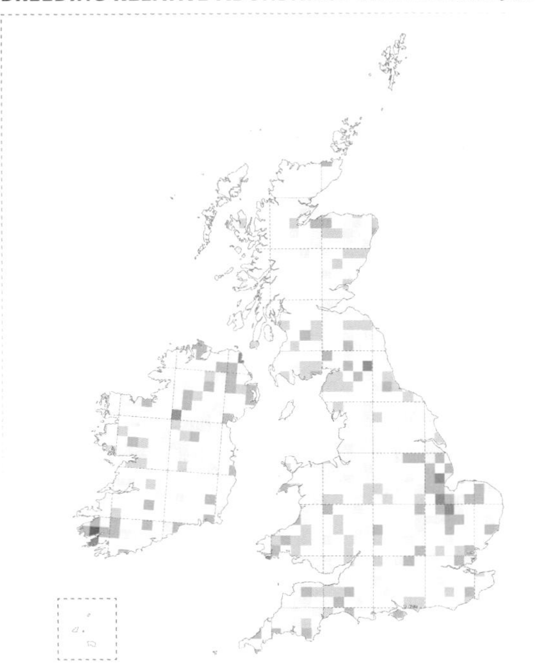

WINTER RELATIVE ABUNDANCE *2007/08–2010/11*

CARRION CROW

Corvus corone

BTO CODE: C.

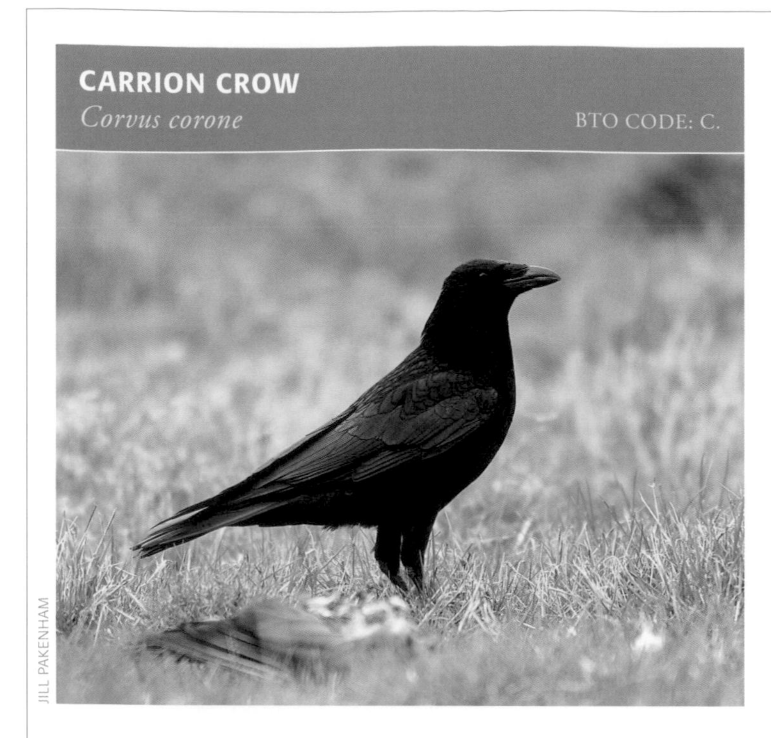
JILL PAKENHAM

THE CARRION CROW is widely distributed throughout England, Wales, the Channel Islands and southern and eastern Scotland. It is replaced by the Hooded Crow in Ireland and in western Scotland from Argyll to Sutherland, including the Hebrides and Northern Isles. Densities are highest on low ground with mixed and pastoral farmland, in southeast and central England into eastern Wales, the central lowlands of Scotland and northeast Scotland.

There has been a 6% range expansion since the *1968–72 Breeding Atlas*. Most gains were in western Scotland and Ireland, though many involve 10-km squares with only single birds and the occasional pair. The rapid movement of the Carrion/Hooded Crow hybrid zone documented by Cook (1975) does not appear to have continued. On the Isle of Man Carrion Crows were present in all atlases but not recorded as pure pairs in the *1968–72 Breeding Atlas*. They have increased there, but their status is complicated because, over many generations of hybridisation, Hooded Crow/Carrion Crow hybrids can appear like pure individuals of either species (Sharpe 2007). In northeast Ireland, Carrion Crows were once rarities but are now regular albeit scarce breeders in the north and east (*Birds in Ireland*; O'Donoghue *et al.* 1996), undergoing a 25% range expansion since 1968–72. Whether these all involved pure Carrion Crows is impossible to ascertain.

During winter, there are more records of wandering birds in the Northern Isles, northwest Scotland and Ireland. Range has changed little in Britain, whilst in Ireland there has been a 32% range contraction since the *1981–84 Winter Atlas*, with losses scattered throughout the northeast and southwest.

The relative abundance change map shows marked increases in breeding abundance in eastern England since 1988–91. Farther west the species has been largely stable, although there is some evidence of declines in southwest England and central Wales (Massimino *et al.* 2013). Overall, the Carrion Crow population increased by 89% during 1970–2010 (*SUKB 2012*), primarily as a consequence of decreased control (Gregory & Marchant 1996) but perhaps also due to increased productivity (*BirdTrends*). However, there are indications that the population has stabilised since the early 2000s (*BirdTrends*).

Sponsored by Mike Wildish—In memory of my parents, Frank and Joan Wildish

BREEDING DISTRIBUTION *2008–11*

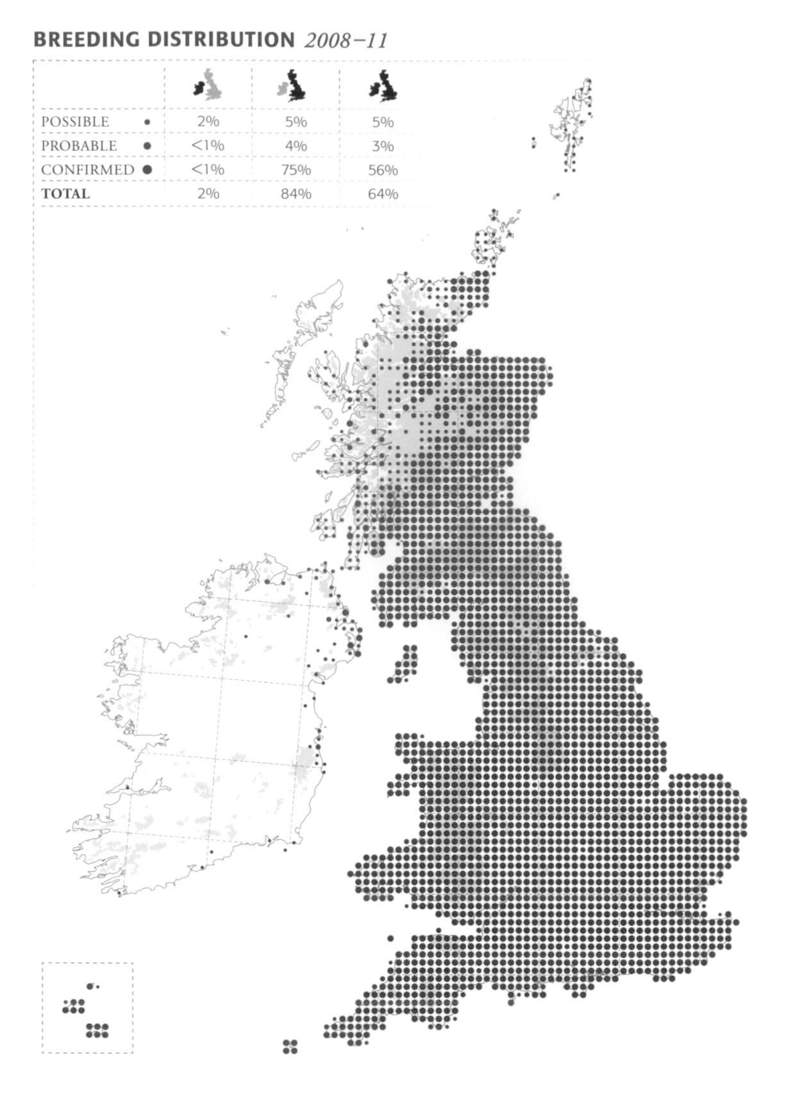

	Ireland	Britain	Britain & Ireland
POSSIBLE •	2%	5%	5%
PROBABLE •	<1%	4%	3%
CONFIRMED •	<1%	75%	56%
TOTAL	2%	84%	64%

WINTER DISTRIBUTION *2007/08–2010/11*

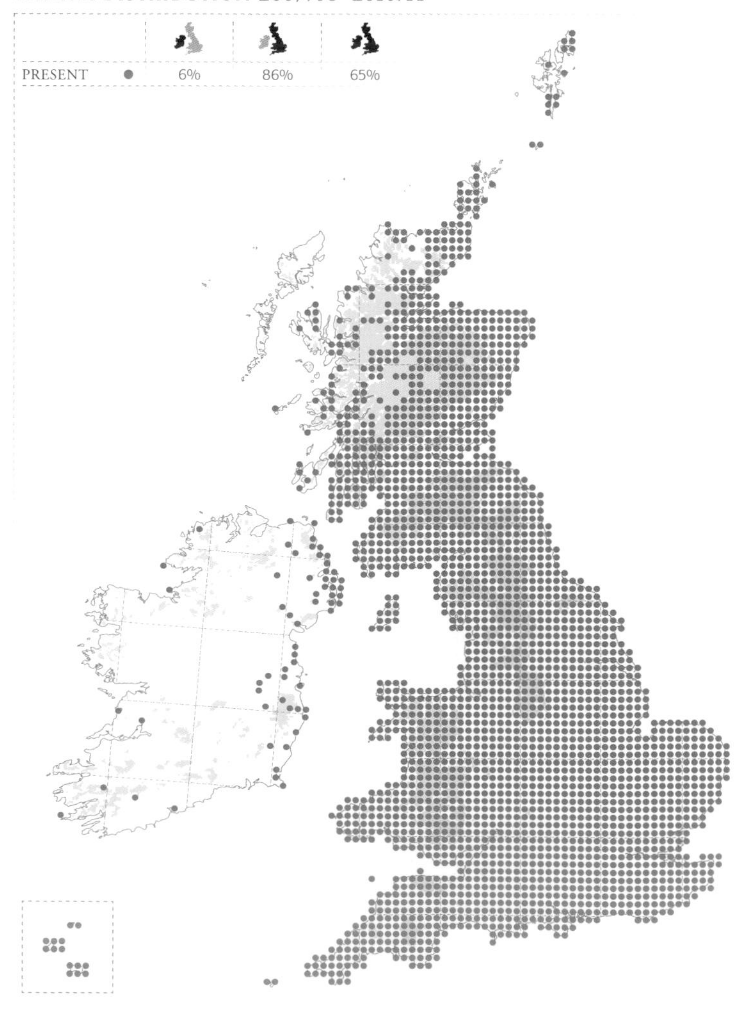

	Ireland	Britain	Britain & Ireland
PRESENT •	6%	86%	65%

BREEDING DISTRIBUTION CHANGE *since 1968–72*

GAIN ▲	21	161	182
LOSS ▽	16	34	50
40yr CHANGE	+25%	+6%	+6%
20yr INDEX	0	+0.02	

WINTER DISTRIBUTION CHANGE *since 1981–84*

GAIN ▲	37	118	155
LOSS ▼	63	56	119
30yr CHANGE	-32%	+3%	+1%

BREEDING RELATIVE ABUNDANCE *2008–11*

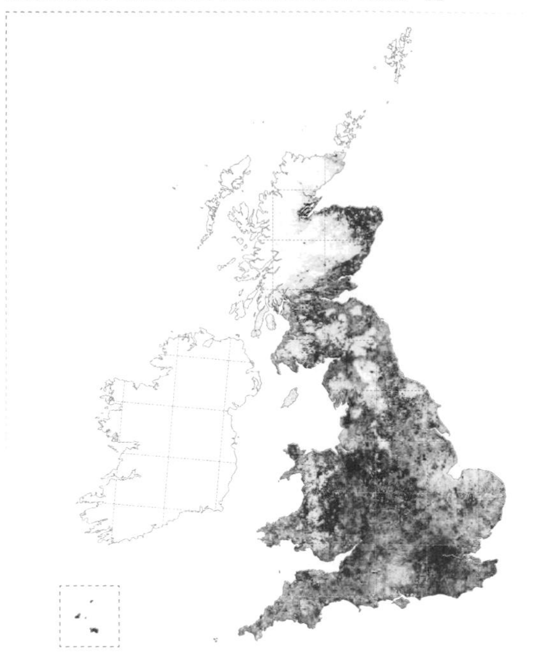

BREEDING RELATIVE ABUNDANCE CHANGE *since 1988–91*

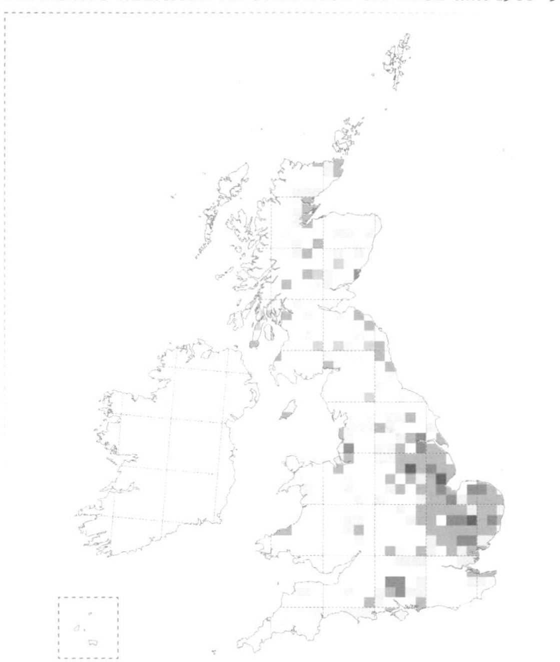

WINTER RELATIVE ABUNDANCE *2007/08–2010/11*

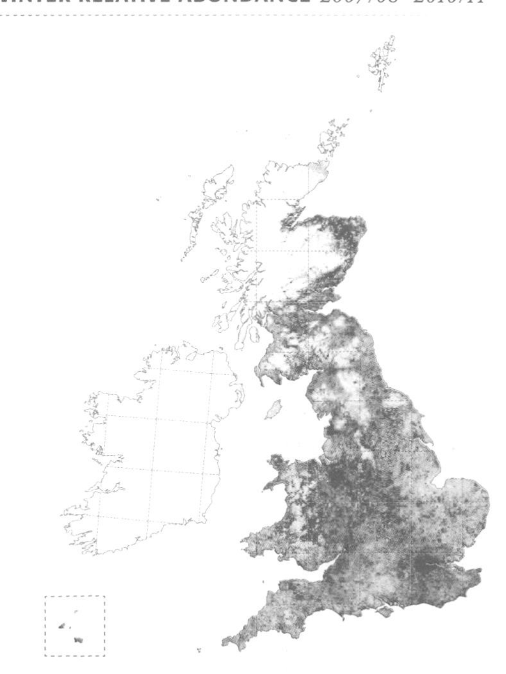

HOODED CROW

Corvus cornix

BTO CODE: HC

JOHN ANDERSON www.pbase.com/crail_birder

THE HOODED CROW replaces the Carrion Crow in Ireland and most of northern and western Scotland. In the eastern part of its Scottish range it overlaps with the Carrion Crow and hybrids are common in a narrow hybrid zone, which had been shifting north and west as Carrion Crows gained ground in the lowlands (Cook 1975; *Birds of Scotland*). Hybridisation also occurs on the Isle of Man, although there is a degree of segregation, with Hooded Crows tending to occur at higher altitudes (Sharpe 2007). Highest densities in Scotland are found in the Northern Isles, on Inner Hebridean islands from Skye to Islay and on the neighbouring Kintyre Peninsula. In Ireland densities are generally greater than in Scotland, particularly in the southwest, northeast, Co. Wicklow and Co. Kildare.

During the breeding season there has been little change in distribution since the *1988–91 Breeding Atlas*. Apparent gains on the eastern edge of the range, particularly in Aberdeenshire, may be due to confusion with hybrid birds; a study of 1,090 pairs in 2007 showed that 96.1% were Carrion Crows, 3.6% were hybrids and 0.3% were Hooded Crows (Francis & Cook 2011). The breeding abundance change map shows that increases in tetrad occupancy have occurred throughout Ireland since the *1988–91 Breeding Atlas* and correspond with a 28% population increase during 1998–2010 (*CBS Trend 2010*). This positive trend contrasts with declines in many parts of Scotland, in particular the Outer Hebrides and towards the eastern fringe of the mainland range, mirroring a 21% population decline in Scotland between 1995 and 2010 (*BBS Report 2011*).

As this is a mainly resident species, the winter distribution is similar to that recorded during the breeding season. The main exception is a scatter of records down the east side of Scotland and England, probably relating to birds arriving from the Continent (*Migration Atlas*). The winter change map shows little change in Ireland, but a 13% decline in the number of occupied 10-km squares in Britain since the *1981–84 Winter Atlas*. This decline is most noticeable down the eastern coast of Britain, as the winter influx of continental birds has continued to diminish.

BREEDING DISTRIBUTION *2008–11*

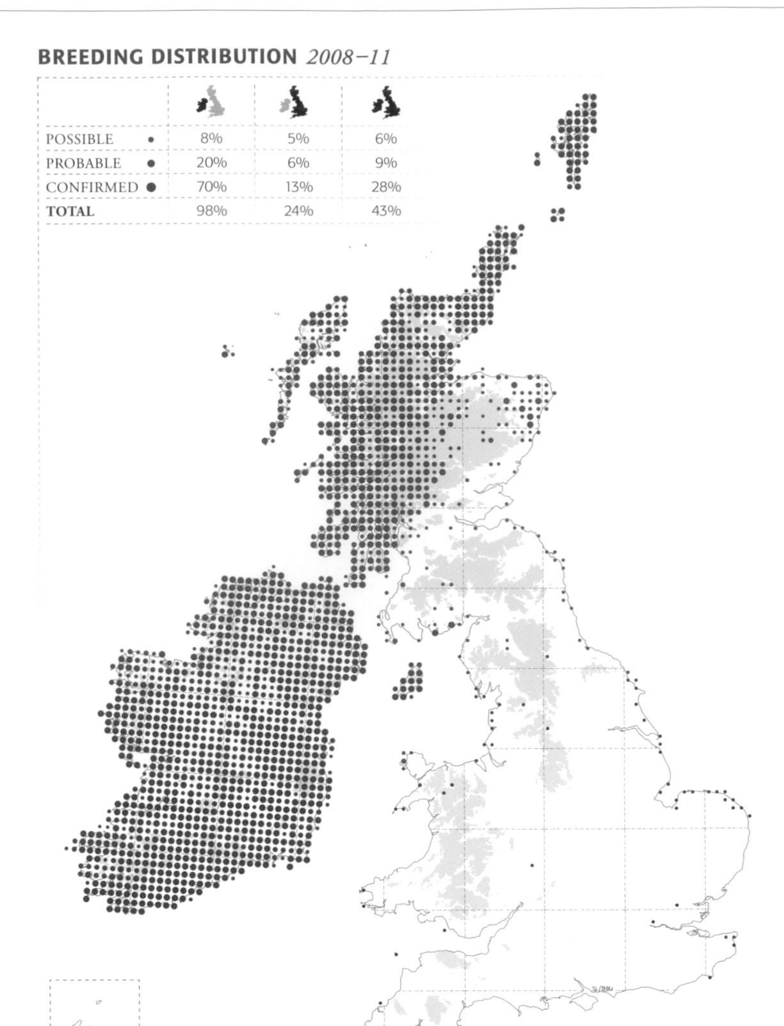

WINTER DISTRIBUTION *2007/08–2010/11*

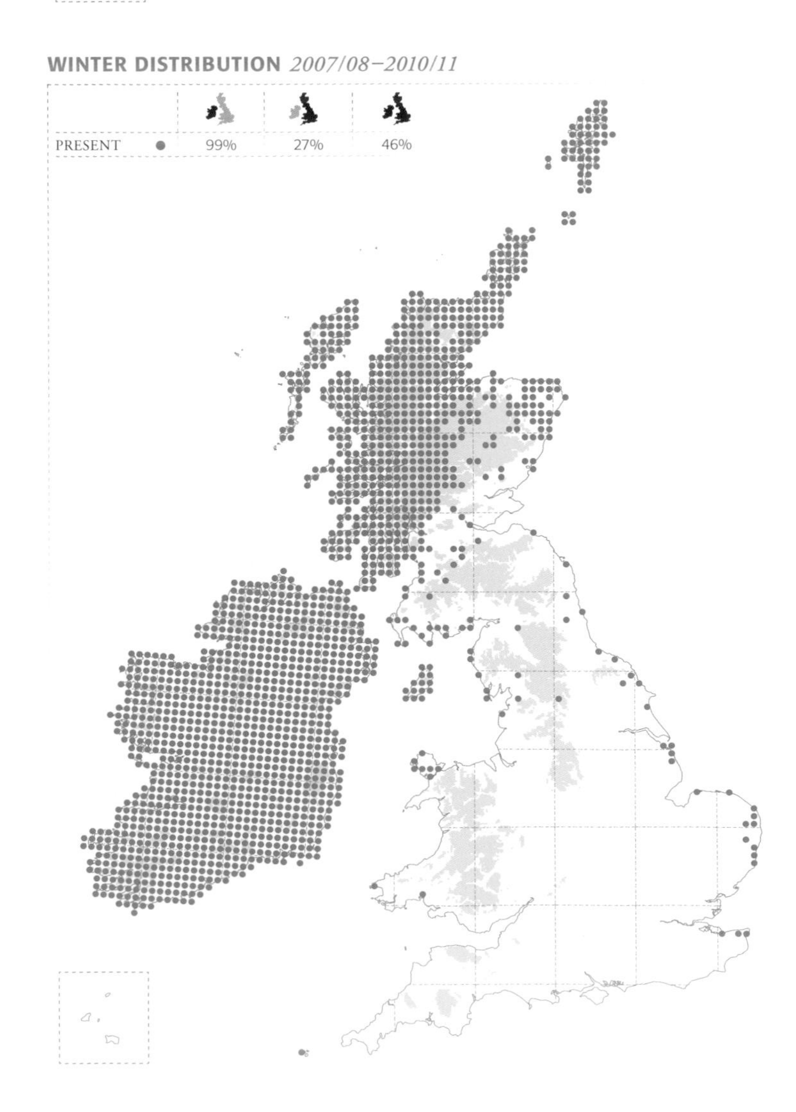

BREEDING DISTRIBUTION CHANGE *since 1968–72*

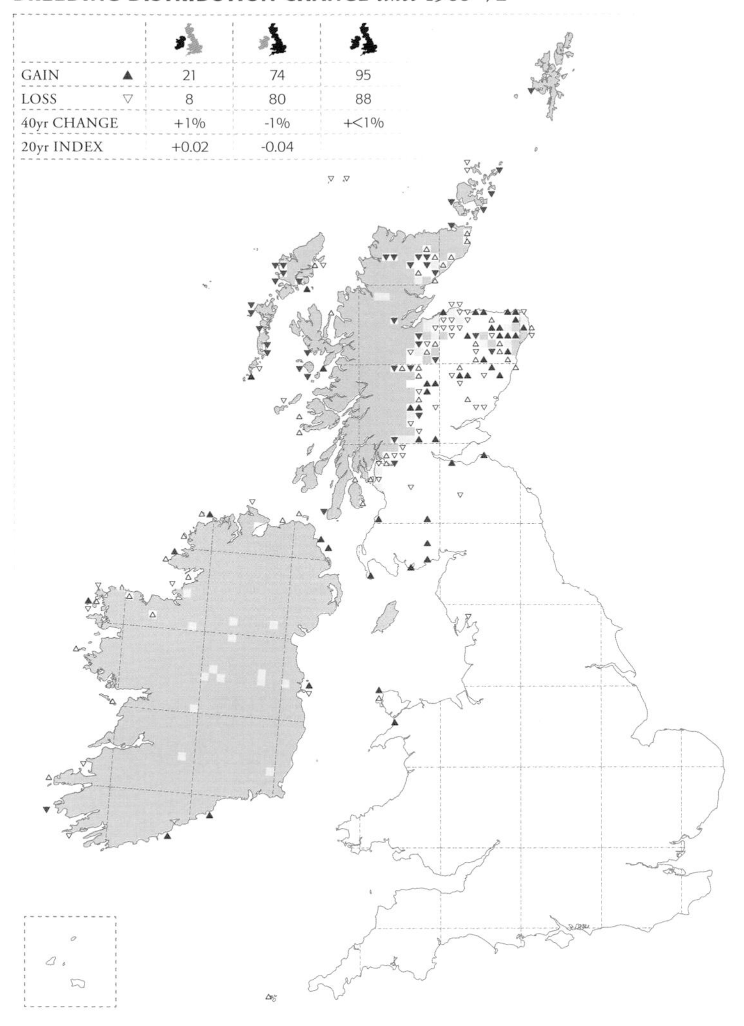

GAIN ▲	21	74	95
LOSS ▽	8	80	88
40yr CHANGE	+1%	-1%	+<1%
20yr INDEX	+0.02	-0.04	

WINTER DISTRIBUTION CHANGE *since 1981–84*

GAIN ▲	16	106	122
LOSS ▼	15	213	228
30yr CHANGE	+<1%	-13%	-6%

BREEDING RELATIVE ABUNDANCE *2008–11*

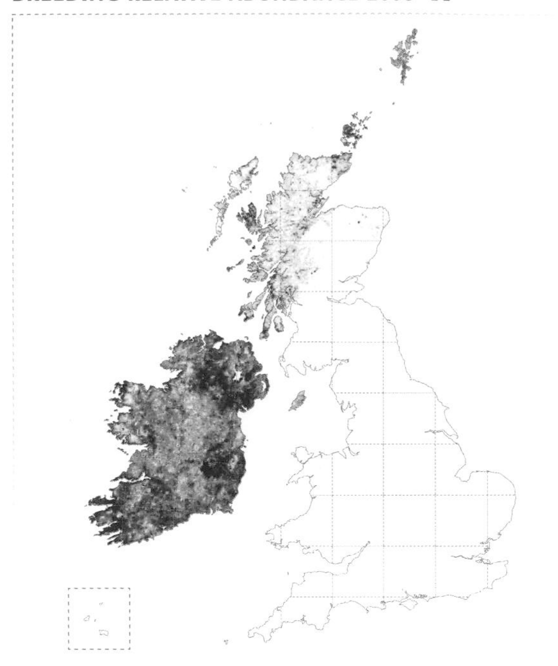

BREEDING RELATIVE ABUNDANCE CHANGE *since 1988–91*

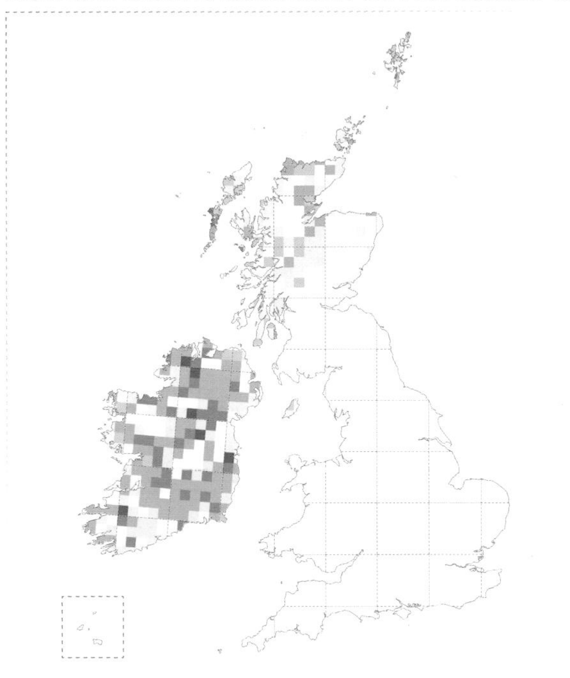

WINTER RELATIVE ABUNDANCE *2007/08–2010/11*

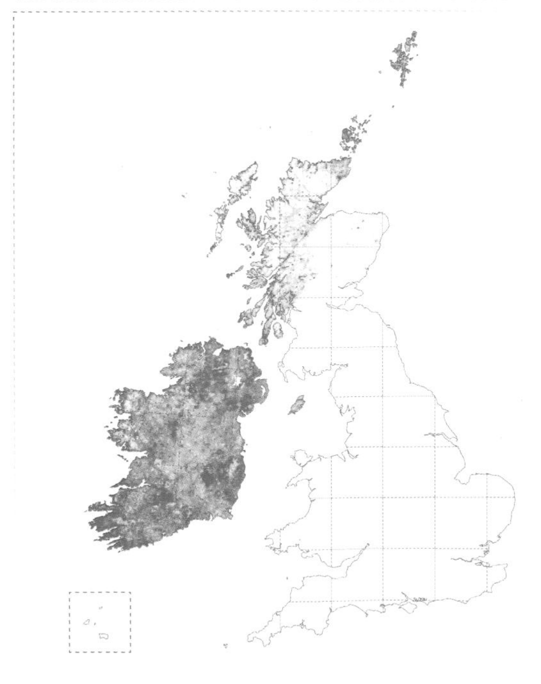

RAVEN

Corvus corax BTO CODE: RN

DESMOND DUGAN www.AbernethyImages.co.uk

THE RAVEN IS now as much a bird of pastoral or mixed lowland farmland and forestry as it is of the uplands, having expanded its British & Irish range by 79% in winter and by 68% in the breeding season since the *1981–84 Winter Atlas* and *1968–72 Breeding Atlas*. The present distributions include virtually all of Ireland and most of Britain, aside from a band running down the east from Inverness to Essex. The only unoccupied upland area is the North York Moors.

Both seasons' change maps show the extent of the gains in Britain, from a narrow strip on the eastern edge of the Scottish uplands, widening to a broad band through central and southern England. Only the Raven and Buzzard share this pattern of eastward range expansion in Britain. In Ireland the change is equally dramatic, with the colonisation of the midlands and northeast. Expansion into the lowlands has been accompanied by the use of quarries for nesting, though the majority of birds in these landscapes nest in trees. The spread into the lowland interior of Ireland has been proceeding steadily since the 1970s, whereas the range gain in Britain has taken place mostly since the 1990s.

Densities of Ravens appear to remain highest in those regions in the north and west of Britain and upland and coastal Ireland that formed the strongholds before the recent range expansion. In all these areas the Atlas data show that abundance of Ravens has increased strongly, which contradicts annual monitoring data suggesting a stable UK breeding population (*BBS Report 2011*) and a declining Irish one (*CBS Trend 2010*).

Raven populations are vulnerable to legal and illegal control and to reductions in food supply, particularly the availability of carrion. Long-term gains in Scotland have been driven by a reduction in both legal and illegal control and a prolonged increase in sheep numbers (*Birds of Scotland*). Gains in Ireland may also be associated with increased carrion availability owing to increased overwintering of sheep (McGreal 2007). Amar *et al.* (2010b) considered whether increased Raven abundance could be having an impact on breeding wader populations but found limited evidence for adverse effects.

BREEDING DISTRIBUTION *2008–11*

POSSIBLE •	21%	7%	11%
PROBABLE •	25%	14%	17%
CONFIRMED •	40%	48%	46%
TOTAL	86%	69%	74%

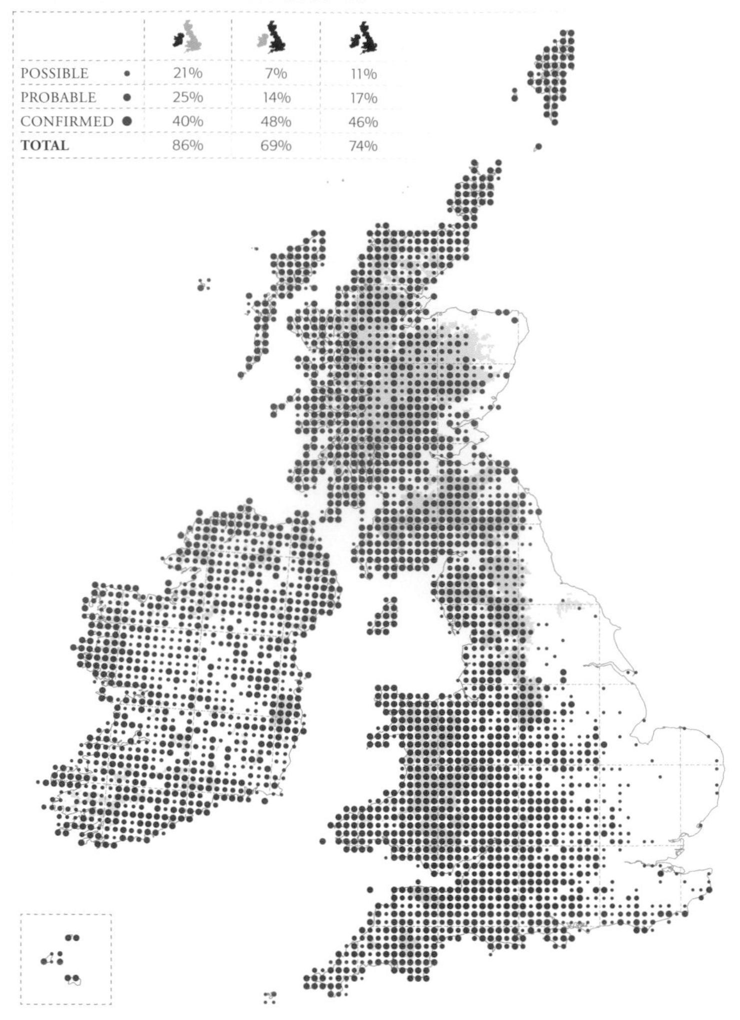

WINTER DISTRIBUTION *2007/08–2010/11*

PRESENT •	91%	77%	81%

BREEDING DISTRIBUTION CHANGE *since 1968–72*

GAIN ▲	450	768	1218
LOSS ▽	28	36	64
40yr CHANGE	+95%	+59%	+68%
20yr INDEX	+0.26	+0.46	

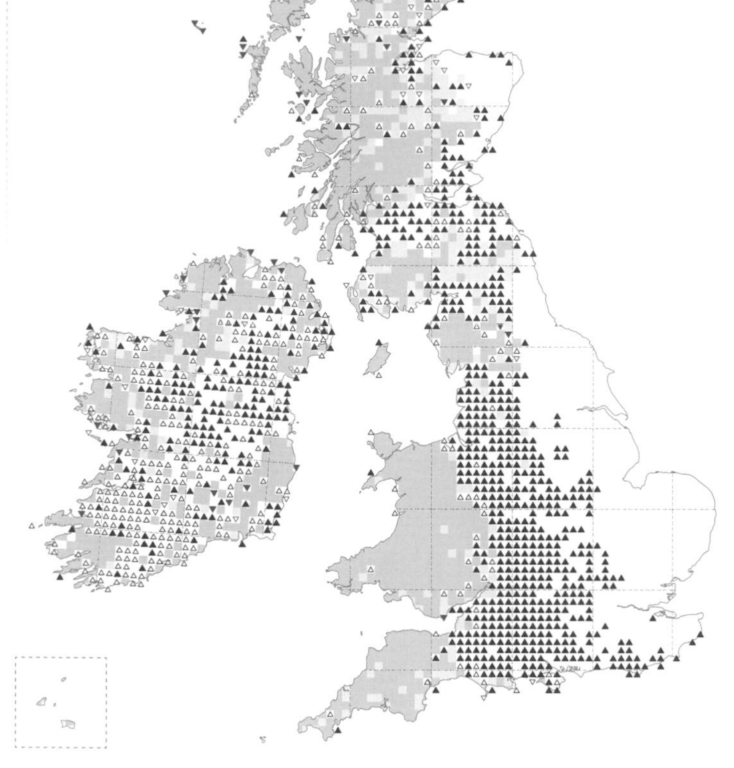

WINTER DISTRIBUTION CHANGE *since 1981–84*

GAIN ▲	335	1078	1413
LOSS ▼	44	13	57
30yr CHANGE	+48%	+97%	+79%

BREEDING RELATIVE ABUNDANCE *2008–11*

BREEDING RELATIVE ABUNDANCE CHANGE *since 1988–91*

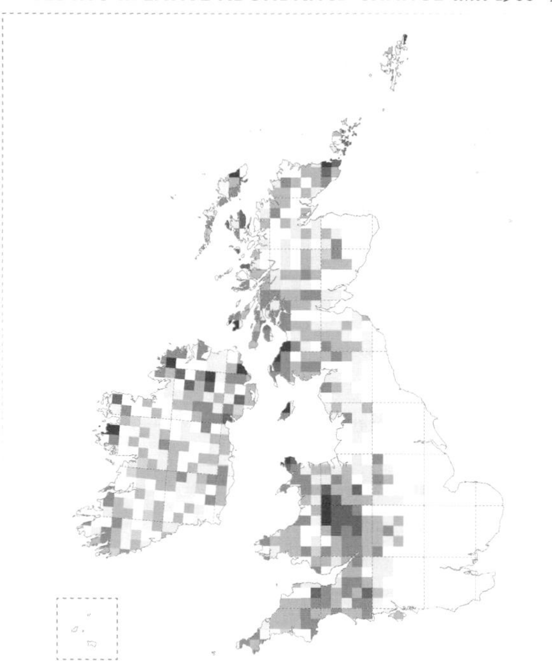

WINTER RELATIVE ABUNDANCE *2007/08–2010/11*

GOLDCREST

Regulus regulus

BTO CODE: GC

STEVE CARTER www.colnevalleybirding.co.uk

THE WINTER AND breeding-season distributions of the Goldcrest are similar, with birds found in 90% of 10-km squares in winter and 88% of squares in the breeding season. The main gaps in distribution occur in the treeless landscapes of the Scottish Highlands, Outer Hebrides, Northern Isles and the Fens in England. The highest densities in both seasons are found in Ireland, particularly in the southeast, southwest and northwest. In Britain the abundance pattern is more fragmented, with pockets of high density concentrated into heavily forested areas; Thetford Forest, the New Forest and Kielder Forest are discernible on the map. In Ireland, Goldcrests occur at relatively high abundance in broad-leaved woodlands as well as in coniferous forests (Nairn & O'Halloran 2012).

Modest gains in the number of occupied 10-km squares have been recorded in both winter and the breeding season. There has been a 12% increase since the *1981–84 Winter Atlas*, most prominent on the margins of the species' range in western Ireland, western Scotland and eastern England. A significant number of passage migrants arrive in the autumn and many stay to overwinter. The increase in breeding range is more modest, 4% since the *1968–72 Breeding Atlas*. Despite these range expansions, the abundance change map shows a rather mixed pattern of change in the proportion of tetrads occupied, with declines more prominent than increases in northern England and Scotland but a more even spread of increases and decreases in the rest of England and Wales, whilst increases predominate in Ireland.

The gains in Ireland are surprising since breeding population monitoring shows a 50% decline in numbers in the Republic of Ireland during 1998–2010 (*CBS Trend 2010*), whilst in the UK there has been a 15% decline during 1995–2010 (*BBS Report 2011*). Goldcrest abundance can be affected by winter weather, and severe spells of cold weather during the Atlas period resulted in high mortality (Kirk 2010) and a decline in numbers (*BBS Report 2011*). This suggests that other factors such as increased availability of suitable habitat, perhaps through the maturing of conifer plantations, is driving the observed range expansion and abundance change (Wilson *et al.* 2006; *Birds of Scotland*).

Sponsored by the readers of Birdwatching magazine

BREEDING DISTRIBUTION *2008–11*

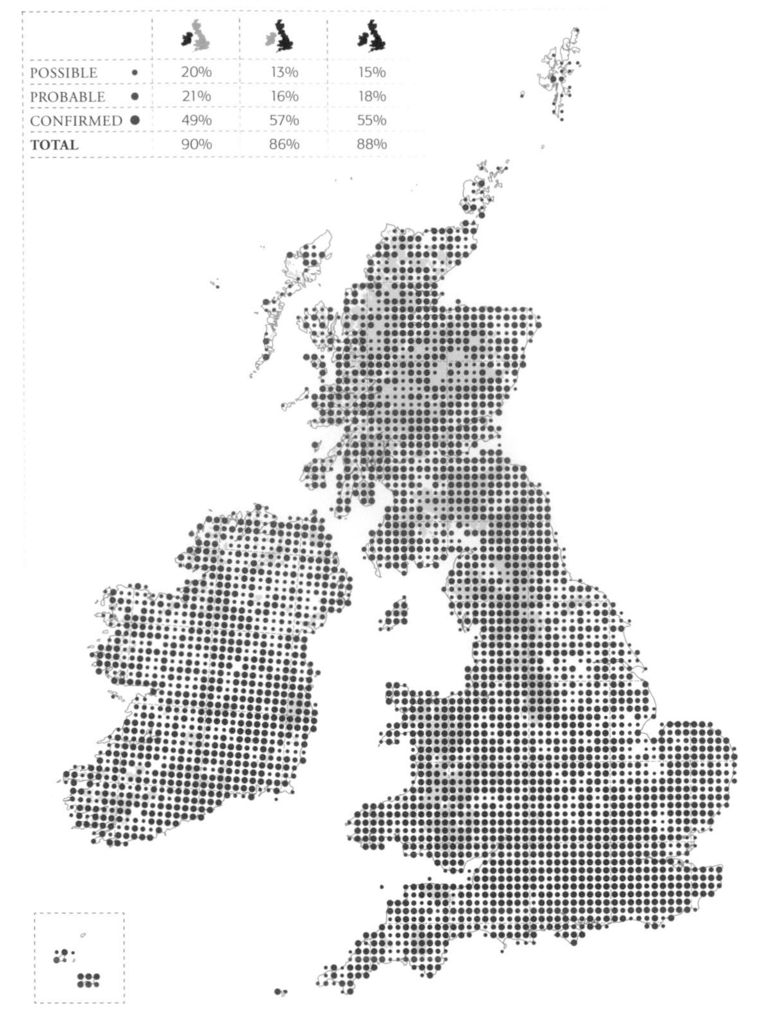

	Ireland	Britain	Britain & Ireland
POSSIBLE •	20%	13%	15%
PROBABLE •	21%	16%	18%
CONFIRMED •	49%	57%	55%
TOTAL	90%	86%	88%

WINTER DISTRIBUTION *2007/08–2010/11*

	Ireland	Britain	Britain & Ireland
PRESENT •	92%	89%	90%

BREEDING DISTRIBUTION CHANGE *since 1968–72*

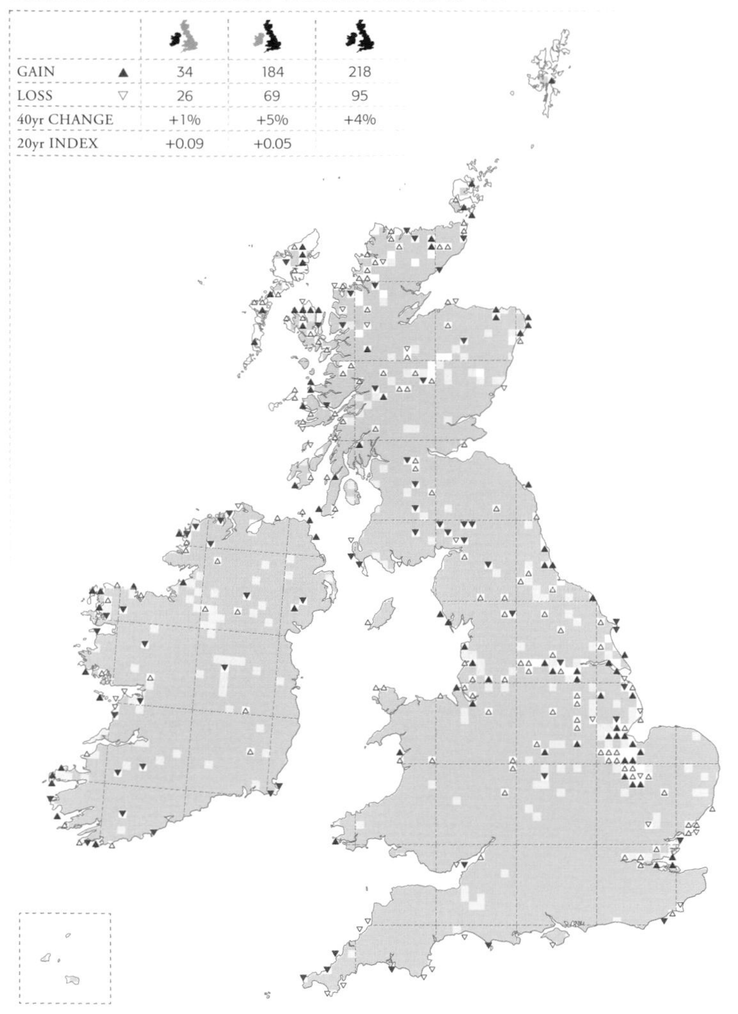

GAIN	▲	34	184	218
LOSS	▽	26	69	95
40yr CHANGE		+1%	+5%	+4%
20yr INDEX		+0.09	+0.05	

WINTER DISTRIBUTION CHANGE *since 1981–84*

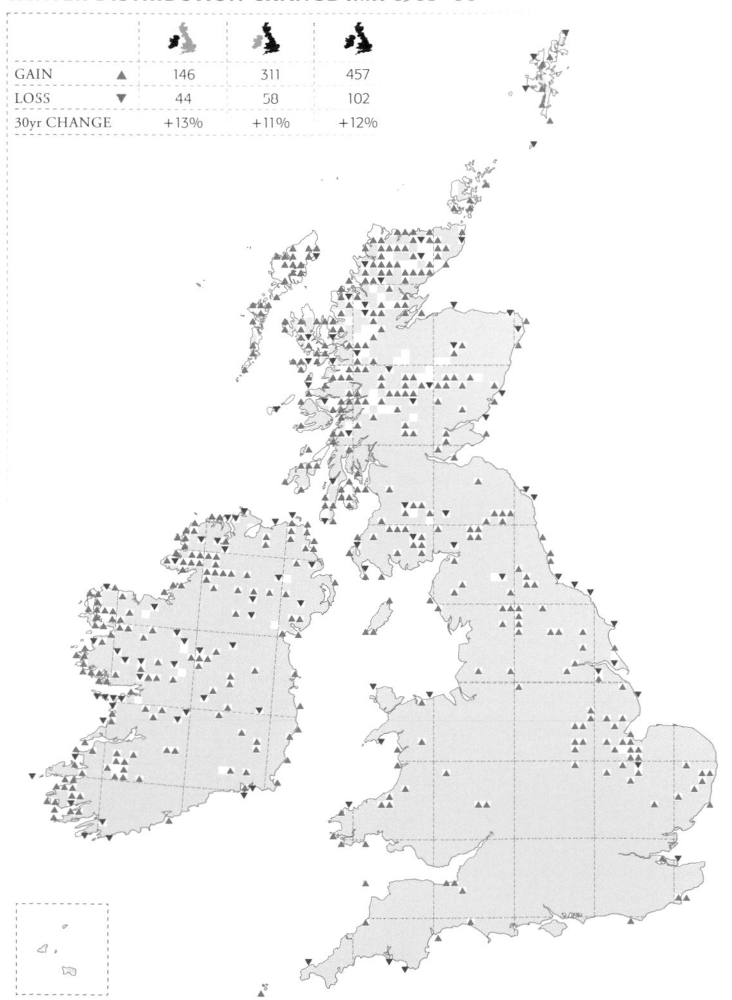

GAIN	▲	146	311	457
LOSS	▼	44	58	102
30yr CHANGE		+13%	+11%	+12%

BREEDING RELATIVE ABUNDANCE *2008–11*

BREEDING RELATIVE ABUNDANCE CHANGE *since 1988–91*

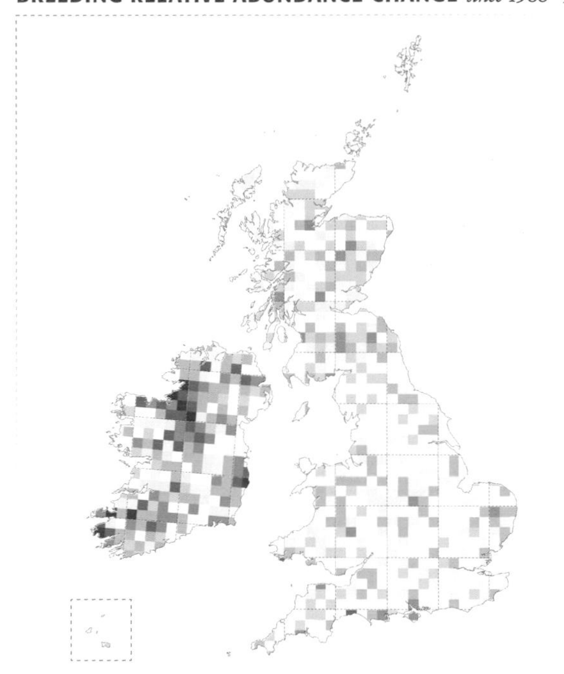

WINTER RELATIVE ABUNDANCE *2007/08–2010/11*

FIRECREST

Regulus ignicapilla

BTO CODE: FC

GRAHAM CLARKE www.grahamsphoto.blogspot.ie

THE WINTER DISTRIBUTION is concentrated into southern Britain and associated with coastal or lowland inland locations. Elsewhere, widely scattered records in Scotland, Ireland and northern England are virtually all in coastal areas. The winter distribution has more than doubled in extent since the *1981–84 Winter Atlas*, involving a substantial shift inland and also northwards to the extremities of Scotland. Squares occupied only in November account for 12% of the 'winter' distribution and these include all records in the Northern Isles and most of those in Ireland. These are likely to be passage birds, involving mainly continental immigrants, as the autumn migration period continues until late November (Riddiford & Findley 1981). The greatest winter abundance was found in southern Britain, associated mainly with coastal areas between southwest Wales and East Anglia.

The breeding distribution is heavily concentrated within southeast and eastern England between Hampshire and Norfolk, with scattered records in Wales, southern and northern England. The breeding abundance map highlights the high density in southern England, with further local concentrations in East Anglia, Wales and Gloucestershire. The breeding distribution is typically inland compared with the much more coastal winter distribution, indicating that the breeding sites are largely vacated in winter. However, some locations are occupied in both summer and winter and there are ringing records confirming that some individuals remain at breeding territories throughout the year (Conway 2010).

The change in breeding distribution is astounding, with a 935% increase in occupation of 10-km squares since the *1968–72 Breeding Atlas*. Following the first proven breeding in Britain, in Hampshire in 1962 (Adams 1966), there has been a near-continual increase in breeding numbers logged by RBBP, with a peak estimate of 1,000 pairs in 2010 (*RBBP 2010*). Around a third of all territories in Britain are in the New Forest (Wynn *et al.* 2012). Despite recent severe winter weather, the breeding population there was little affected, suggesting that most individuals were migratory. Further population growth and range expansion is anticipated in Britain, given the ongoing northward and eastward breeding range expansion through mainland Europe since the 1960s, including colonisation of the Netherlands in the 1970s and subsequent rapid population growth there (Hustings 2002).

Sponsored by Penny Hollow

BREEDING DISTRIBUTION *2008–11*

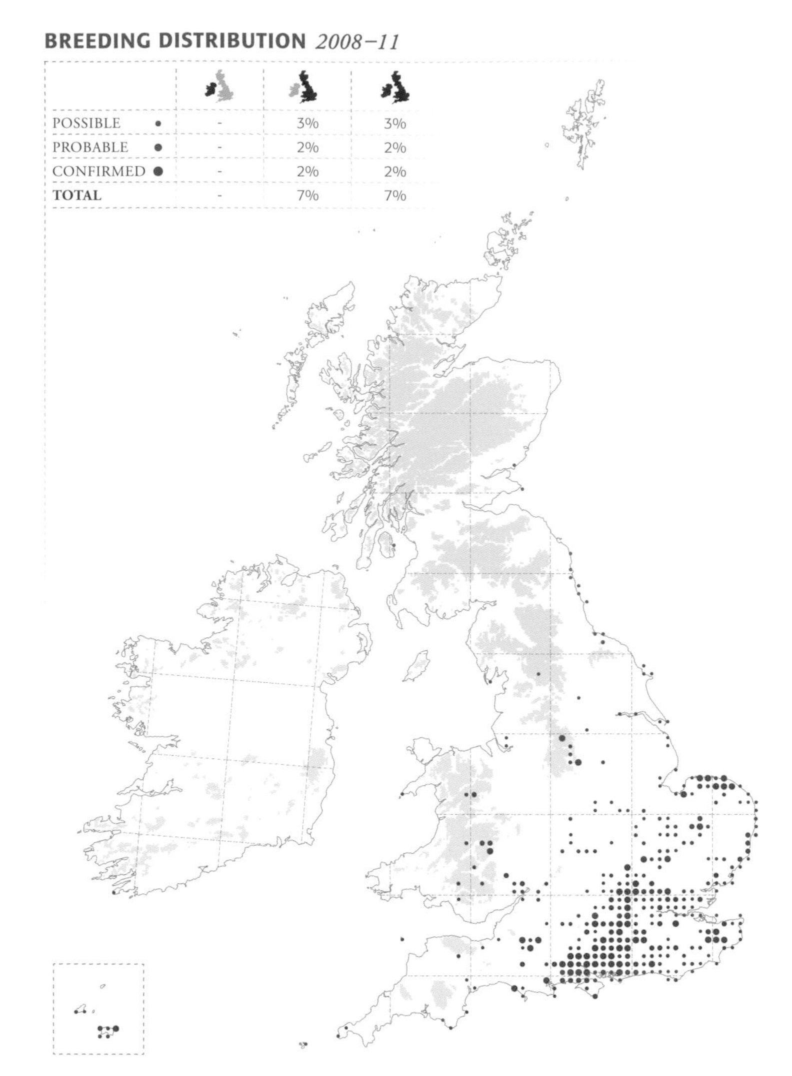

POSSIBLE ●	-	3%	3%
PROBABLE ●	-	2%	2%
CONFIRMED ●	-	2%	2%
TOTAL	-	7%	7%

WINTER DISTRIBUTION *2007/08–2010/11*

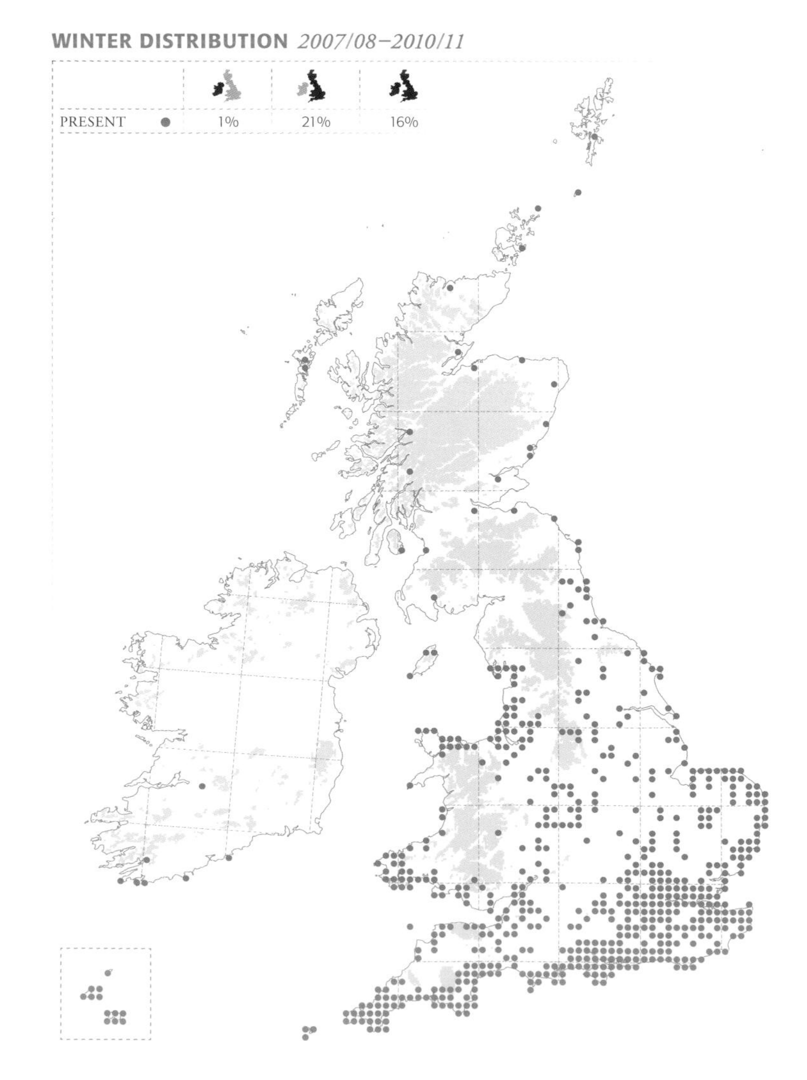

PRESENT ●	1%	21%	16%

BREEDING DISTRIBUTION CHANGE *since 1968–72*

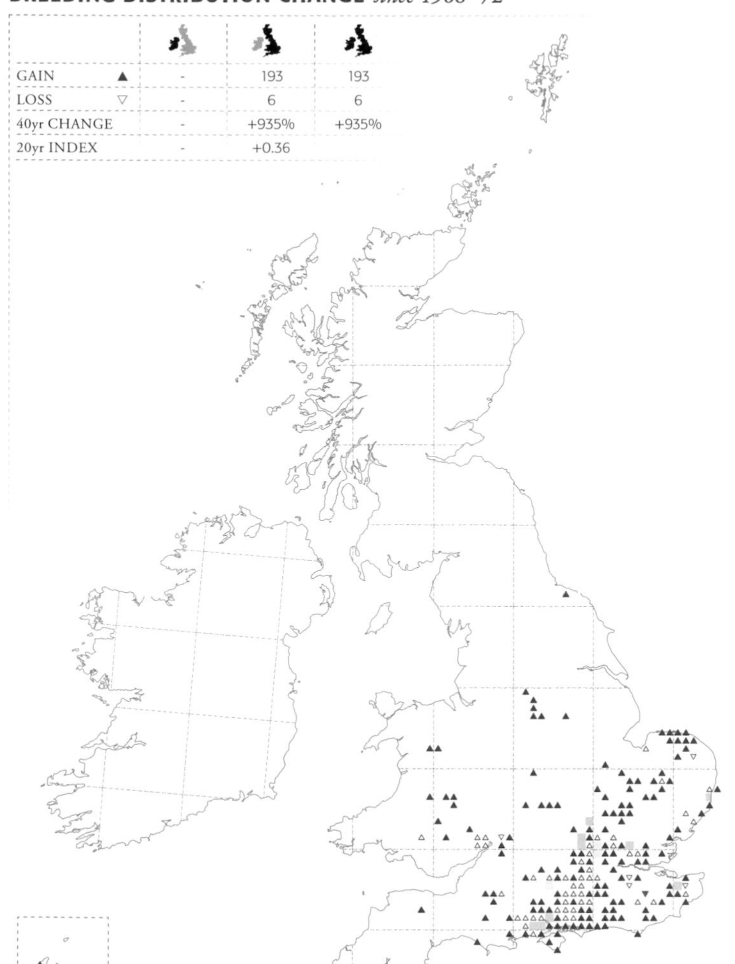

GAIN ▲	-	193	193
LOSS ▽	-	6	6
40yr CHANGE	-	+935%	+935%
20yr INDEX	-	+0.36	

BREEDING RELATIVE ABUNDANCE *2008–11*

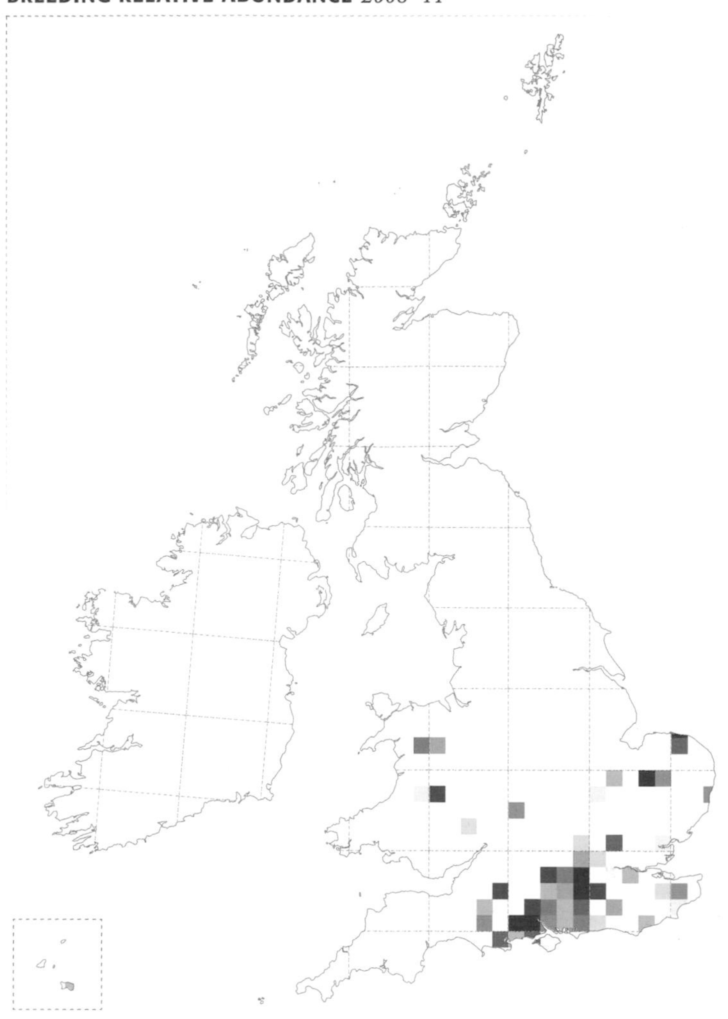

WINTER DISTRIBUTION CHANGE *since 1981–84*

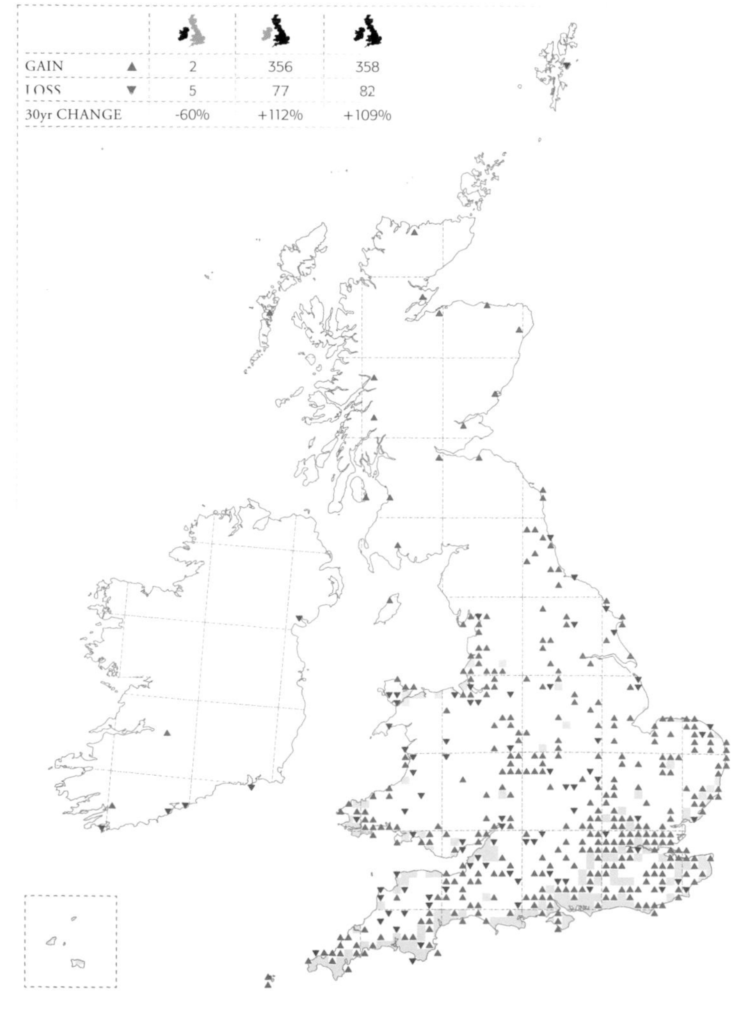

GAIN ▲	2	356	358
LOSS ▼	5	77	82
30yr CHANGE	-60%	+112%	+109%

WINTER RELATIVE ABUNDANCE *2007/08–2010/11*

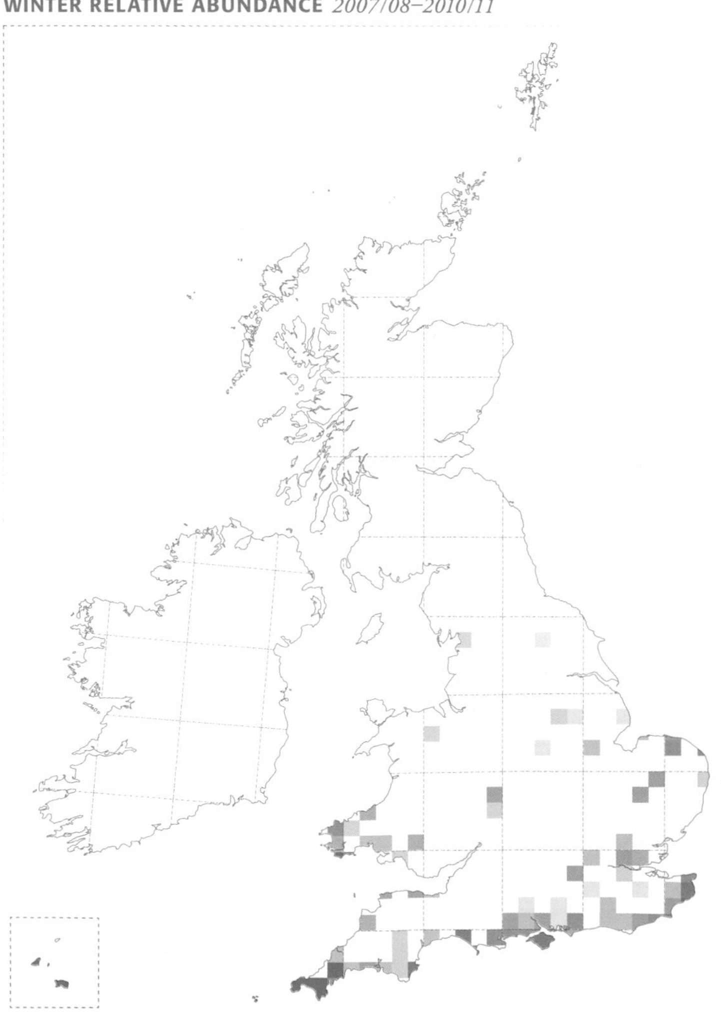

BLUE TIT
Cyanistes caeruleus

BTO CODE: BT

DESMOND DUGAN www.AbernethyImages.co.uk

BREEDING DISTRIBUTION *2008–11*

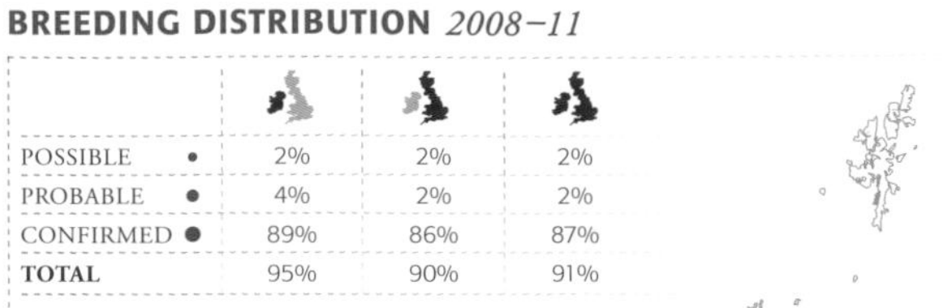

	Ireland	Britain	Britain & Ireland
POSSIBLE •	2%	2%	2%
PROBABLE ●	4%	2%	2%
CONFIRMED ●	89%	86%	87%
TOTAL	95%	90%	91%

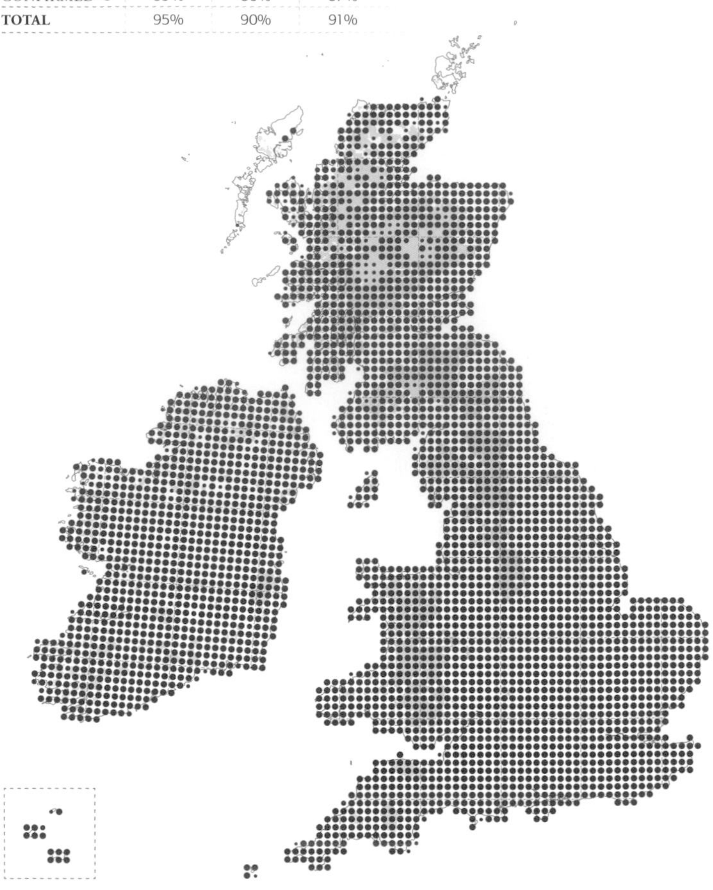

WINTER DISTRIBUTION *2007/08–2010/11*

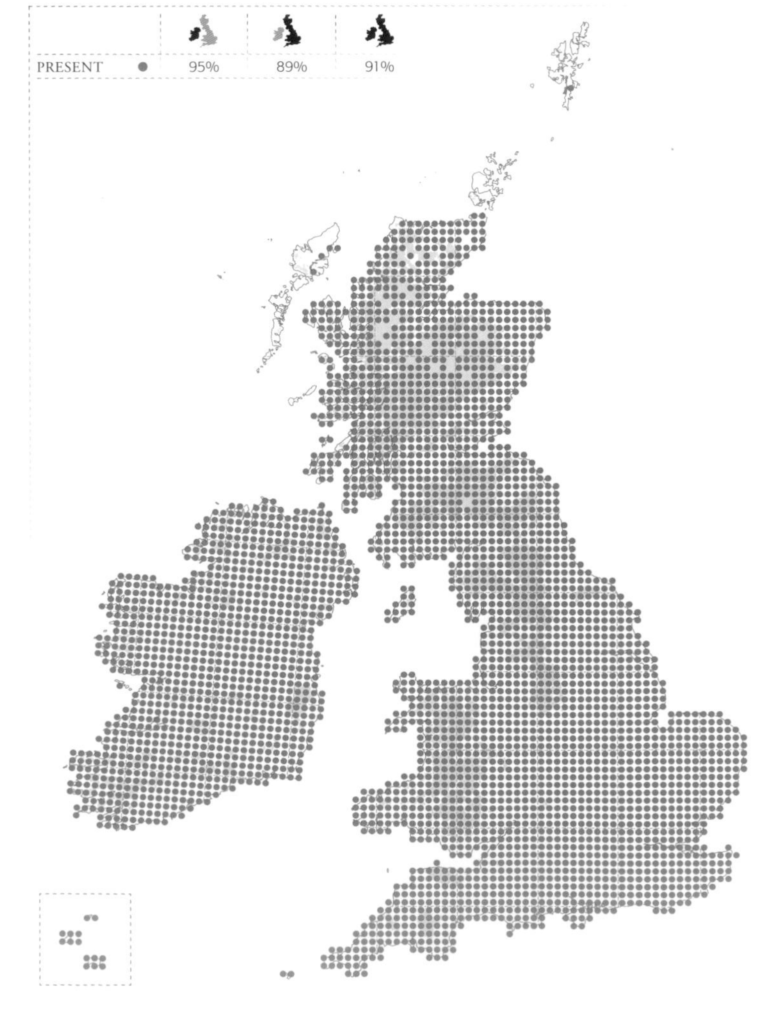

	Ireland	Britain	Britain & Ireland
PRESENT ●	95%	89%	91%

THE BLUE TIT is a widespread breeding and wintering species throughout Britain and Ireland, absent only from the highest ground in Scotland, the Northern Isles, most of the Outer Hebrides and a few Inner Hebridean islands. It is chiefly a bird of broad-leaved woodland, though abundant in a wide range of other woodland, garden and scrub habitats.

Patterns of abundance are virtually identical during winter and the breeding season, owing to high site fidelity and minimal continental immigration (*Migration Atlas*). Densities are low throughout Ireland, in upland areas of Britain and in other landscapes with few woodlands, such as the Fens of East Anglia. The highest densities occur in central and southeast England and in lower-lying areas of Wales, where the Blue Tit is one of the commonest passerines. Pockets of high density are apparent in urban areas, possibly as a result of high nestbox availability and supplementary feeding (Minot & Perrins 1986; Fuller *et al.* 2008). High levels of garden occupancy are found, with 94% of Irish gardens and 98% of British gardens reporting Blue Tits in winter (O'Sullivan 2011; Garden BirdWatch 2013).

Distribution changes in both seasons are minor, involving a few gains in marginal coastal and upland areas in Scotland and Ireland. Some gains may reflect a move towards a greater proportion of broad-leaved tree species in plantation forests (Forestry Commission 2013) but many of the coastal gains are probably related to coverage.

The breeding abundance change map shows marked increases in Scotland and especially Ireland since 1988–91 and no change in much of central England, although the sensitivity of this method for detecting changes in such areas of high tetrad occupancy is limited. Massimino *et al.* (2013) also found density increases in Scotland. Blue Tit numbers in the UK increased by 37% during 1967–2010 (*BirdTrends*), though there have been fluctuations linked to cold winters and poor breeding seasons (Marchant *et al.* 1990; Coiffait *et al.* 2009). Populations in the Republic of Ireland increased by an average of 2% per year during 1998–2008 (Crowe *et al.* 2010b). Food provision in gardens and availability of nestboxes may have contributed to increases in lowland areas but causes of increases in western Scotland are unclear.

Sponsored by Gardman

BREEDING DISTRIBUTION CHANGE *since 1968–72*

GAIN ▲	21	57	78
LOSS ▽	6	23	29
40yr CHANGE	+2%	+1%	+1%
20yr INDEX	+0.05	+0.02	

WINTER DISTRIBUTION CHANGE *since 1981–84*

GAIN ▲	38	105	143
LOSS ▼	17	24	41
30yr CHANGE	+2%	+3%	+3%

BREEDING RELATIVE ABUNDANCE *2008–11*

BREEDING RELATIVE ABUNDANCE CHANGE *since 1988–91*

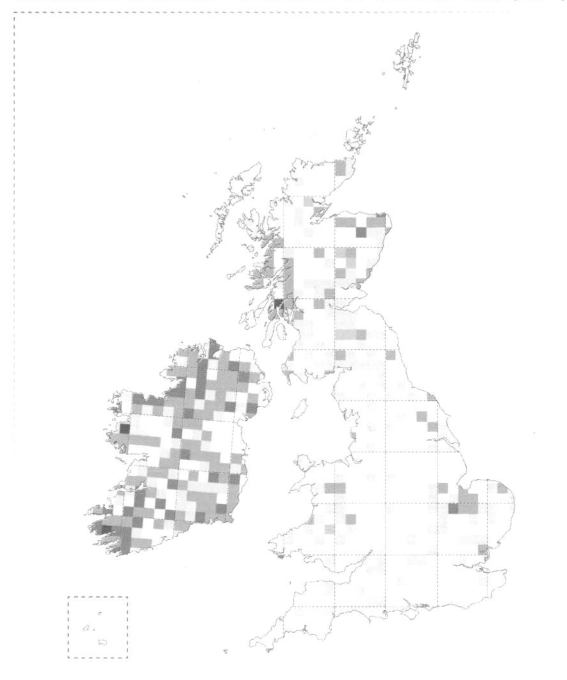

WINTER RELATIVE ABUNDANCE *2007/08–2010/11*

GREAT TIT
Parus major BTO CODE: GT

JOE O'HANLON

THE GREAT TIT was found in 90% of 10-km squares in Britain & Ireland in both winter and the breeding season. As with the Blue Tit, the only significant range gaps are in treeless moorland areas of the Scottish Highlands, the Outer Hebrides and the Northern Isles (*Birds of Scotland*). The relative abundance maps show a very similar pattern for both seasons, with the highest densities found in lowland wooded landscapes of England, Wales and parts of eastern Ireland. Densities are low throughout the uplands, where numbers are presumably limited by the availability of broad-leaved woodland.

Change in range size has been small over the last 40 years. Range expansions of 7% in winter since the *1981–84 Winter Atlas* and 4% in the breeding season since the *1968–72 Breeding Atlas* have arisen mostly through gains on the western and northern margins of the range in Ireland and Scotland. The abundance change map shows widespread and significant increases in tetrad occupancy since the *1988–91 Breeding Atlas*. These are most prominent in Ireland, where tetrad occupancy has increased in almost every 20-km square. These abundance increases may partly explain the observed range gains, although improved coverage in some remote areas may also be involved.

Great Tit breeding numbers in the UK have risen steadily since the 1960s, with a 112% increase recorded during 1967–2010 (*BirdTrends*). In the Republic of Ireland, a 27% increase occurred during 1998–2010 (*CBS Trend 2010*). Great Tits are common visitors to gardens in winter and were reported in 93% and 79% of Irish and British gardens respectively (O'Sullivan 2011; Garden BirdWatch 2013), although fewer birds use gardens in years when beechmast abundance is high in woodland (Chamberlain *et al.* 2007).

On the Continent, population increases (*PECBMS 2010*) have been related to enhanced survival, linked to supplementary feeding and milder winters (Hõrak & Lebreton 1998; Ahola *et al.* 2009). There is some evidence, albeit limited, for similar drivers in the UK (*BirdTrends*). Nairn & O'Halloran (2012) suggested that the colonisation of Ireland by Great Spotted Woodpeckers might be beneficial for species such as the Great Tit by increasing the availability of nest holes.

Sponsored by Gardman

BREEDING DISTRIBUTION *2008–11*

POSSIBLE •	4%	1%	2%
PROBABLE ●	6%	2%	3%
CONFIRMED ●	84%	86%	85%
TOTAL	94%	89%	90%

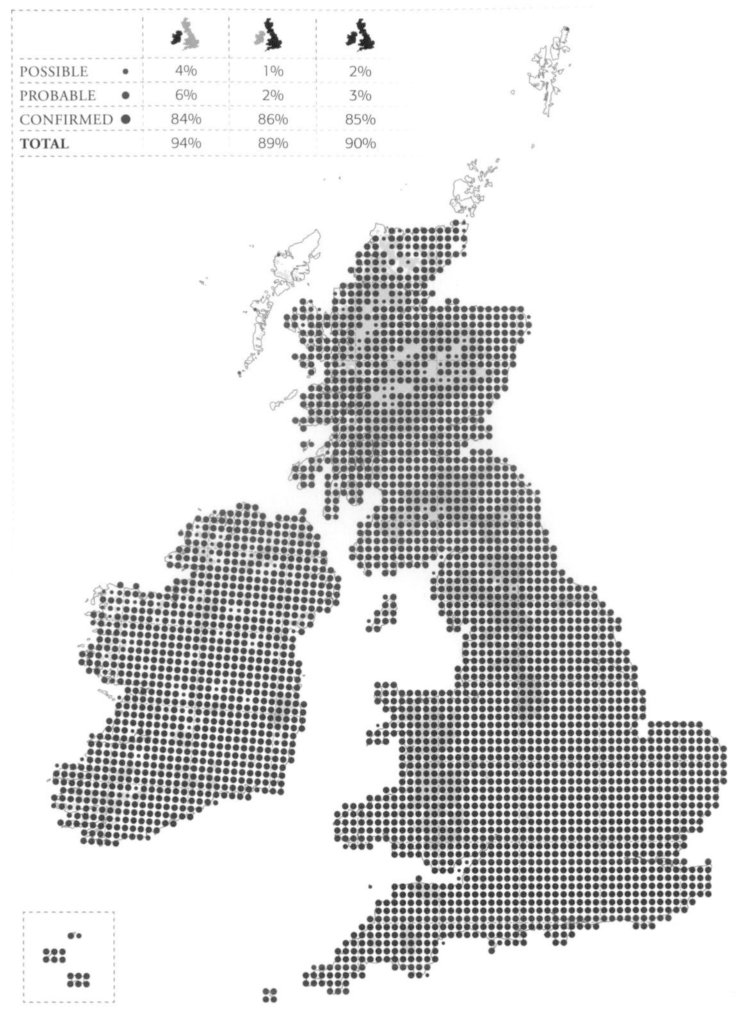

WINTER DISTRIBUTION *2007/08–2010/11*

PRESENT ●	94%	89%	90%

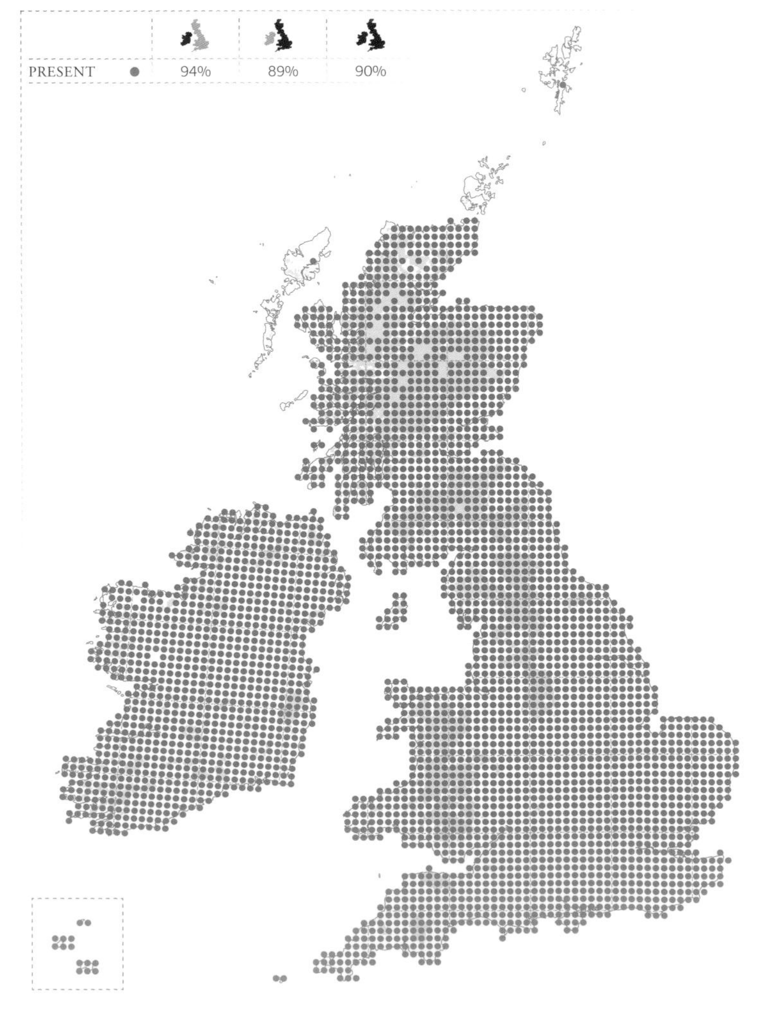

BREEDING DISTRIBUTION CHANGE *since 1968–72*

GAIN ▲	56	112	168
LOSS ▽	5	23	28
40yr CHANGE	+6%	+4%	+4%
20yr INDEX	+0.11	+0.05	

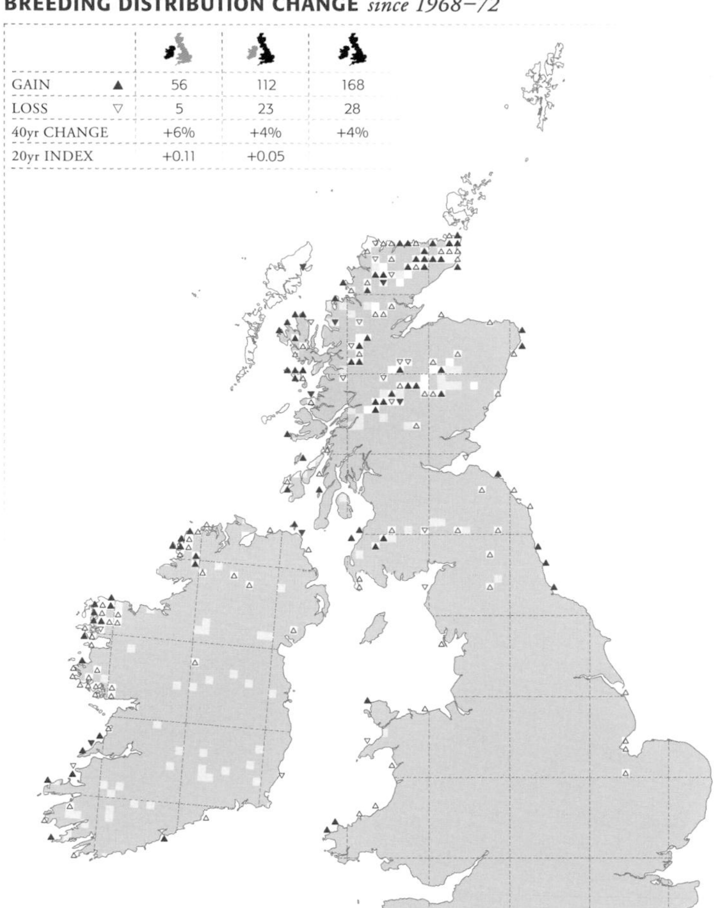

WINTER DISTRIBUTION CHANGE *since 1981–84*

GAIN ▲	115	155	270
LOSS ▼	13	28	41
30yr CHANGE	+12%	+5%	+7%

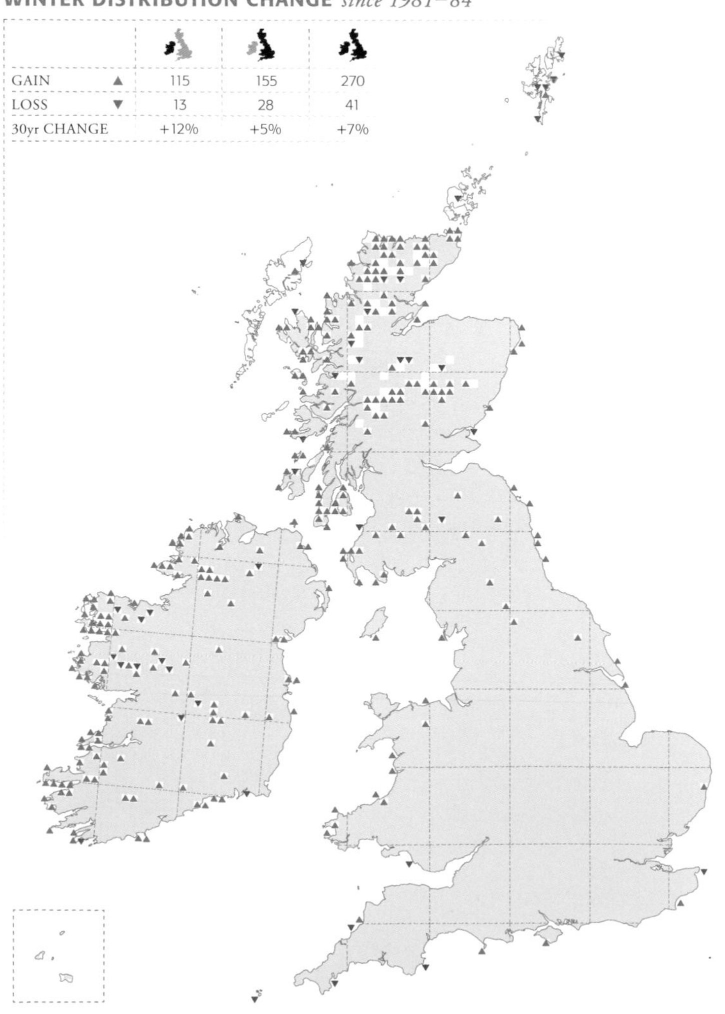

BREEDING RELATIVE ABUNDANCE *2008–11*

BREEDING RELATIVE ABUNDANCE CHANGE *since 1988–91*

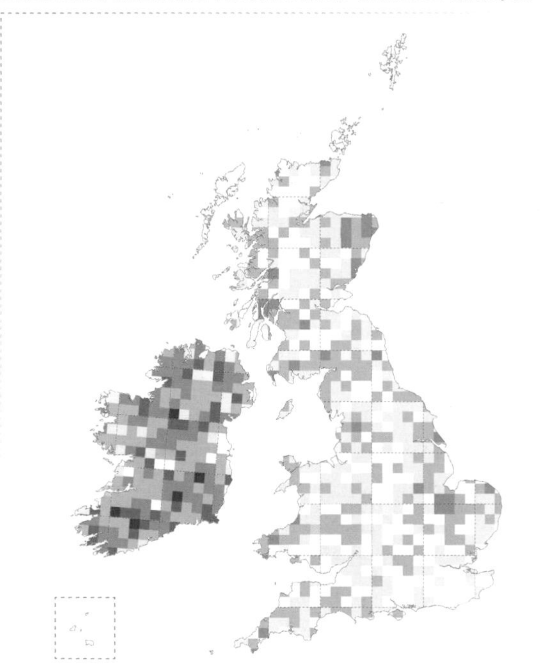

WINTER RELATIVE ABUNDANCE *2007/08–2010/11*

CRESTED TIT

Lophophanes cristatus

BTO CODE: CI

DESMOND DUGAN www.AbernethyImages.co.uk

THE CRESTED TIT is an iconic species of the pinewoods of northern Scotland, where it has a very restricted distribution. Its core range covers the Caledonian pinewoods of upper Strathspey, the pine plantations of lower Strathspey and the coastal plains of Moray and Nairn. In Easter Ross and east Inverness-shire, Crested Tits are found at low densities in pine plantations, especially those containing patches of relict forest, on the low hills above the firths. They also occur in the remnant pine forests of the glens from Strathbran and Strathfarrar south to Glen Garry, west of the Great Glen. Summers *et al.* (1999) found that the density of winter social groups was ten times higher in ancient native pinewoods than in other woods that comprised planted Scots Pine, Lodgepole Pine and Sitka Spruce. A comprehensive survey in 1992–99, covering all months, found Crested Tits in 79 10-km squares (Summers & Canham 2001). In comparison, Atlas data from 2007–11 span 82 10-km squares.

Adults are resident in their territories throughout the year whereas juveniles normally undertake short dispersive movements (Summers 1998; *Migration Atlas*). This means that there is much similarity between the distributions in the breeding season and the winter. Post-breeding dispersal of young birds is probably responsible for the slightly wider scatter of winter records and for records well outside the core area, such as the individuals visiting garden feeding stations on Skye in November 2010 and at Gairloch, Wester Ross, in mid July 2011.

The winter change map shows a 50% increase in the number of occupied 10-km squares since the *1981–84 Winter Atlas*, whilst the breeding change map indicates a 28% increase since the *1968–72 Breeding Atlas*. Most of these gains have been in the lower-density areas west of the Great Glen and are probably influenced by increased observer effort, particularly in winter. New plantings, especially of Scots Pines, and sympathetic forest management may also be leading to local increases in numbers and range expansion (Cook 1982; Summers *et al.* 1999; *Birds of Scotland*). Given the low densities of birds in plantations, there is scope to increase densities through suitable management (Denny & Summers 1996; Summers 2000).

Sponsored by Scottish Ornithologists' Club

BREEDING DISTRIBUTION *2008–11*

POSSIBLE •	-	<1%	<1%
PROBABLE ●	-	<1%	<1%
CONFIRMED ●	-	1%	1%
TOTAL	-	1%	1%

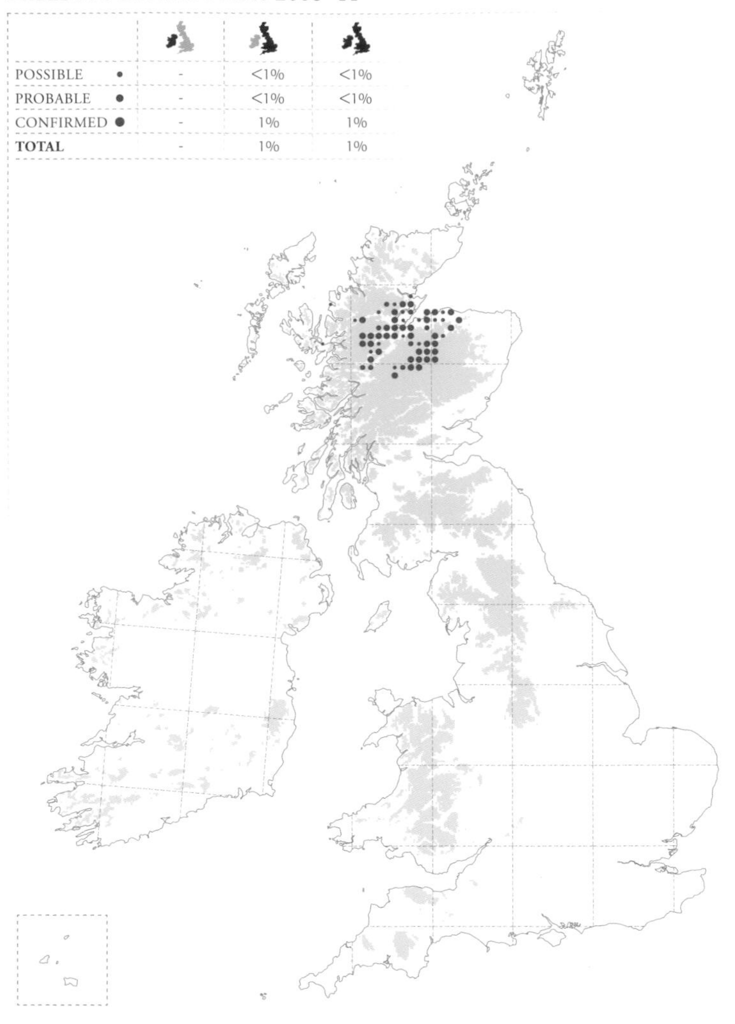

WINTER DISTRIBUTION *2007/08–2010/11*

PRESENT ●	-	3%	2%

BREEDING DISTRIBUTION CHANGE *since 1968–72*

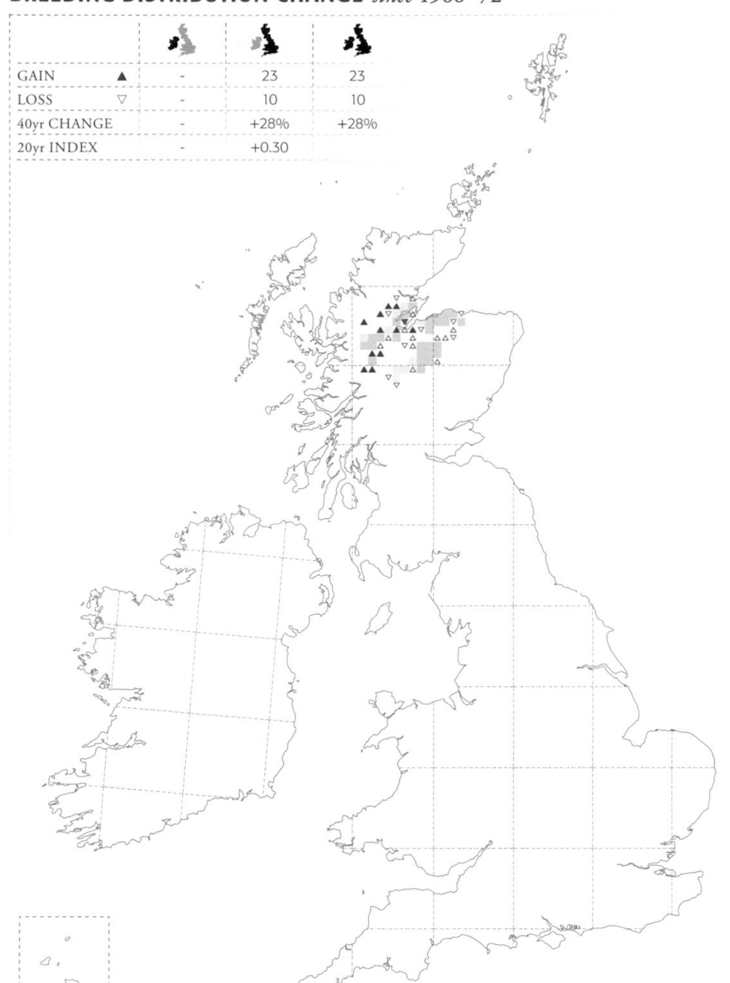

GAIN ▲	-	23	23
LOSS ▽	-	10	10
40yr CHANGE	-	+28%	+28%
20yr INDEX	-	+0.30	

BREEDING RELATIVE ABUNDANCE *2008–11*

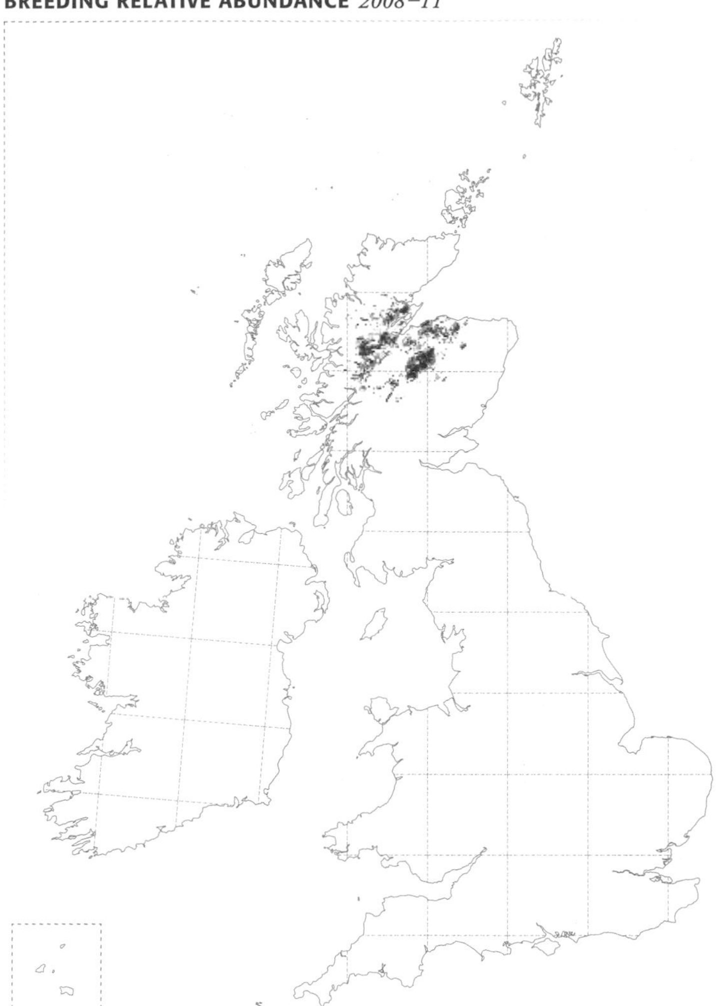

WINTER DISTRIBUTION CHANGE *since 1981–84*

GAIN ▲	-	31	31
LOSS ▼	-	8	8
30yr CHANGE	-	+50%	+50%

WINTER RELATIVE ABUNDANCE *2007/08–2010/11*

COAL TIT

Periparus ater

BTO CODE: CT

A SPECIALIST OF coniferous woodland, the Coal Tit is widely distributed across Britain and Ireland, where subspecies *ater* and *hibernicus* occur respectively (Parkin & Knox 2010). Coal Tits are largely site-faithful, with movements occurring in response to changing conditions, such as food shortages in their preferred woodland habitat (*Migration Atlas*). Due to their relatively sedentary nature, their distribution and patterns of abundance are generally similar in winter and the breeding season.

The Coal Tit is most numerous in conifer plantations, although it occurs abundantly in Irish broad-leaved woods. The highest densities are recorded throughout Ireland, in heavily forested areas in Scotland and Wales, and in localised pockets in England, particularly Thetford Forest and the New Forest. The most notable gaps in distribution occur on the highest ground in Scotland, in the Northern Isles, on some Hebridean islands and in the Fens in eastern England.

Since the *1981–84 Winter Atlas* the species has expanded its range by 12%, with many gains in the western half of Ireland, western and northern Scotland and eastern England.

The breeding abundance change map shows a general pattern of gains since 1988–91 in the proportion of occupied tetrads, most noticeably in Ireland. Since the *1988–91 Breeding Atlas* there has also been a 7% range expansion, most evident in coastal areas of western Ireland, some Hebridean islands, Orkney and parts of eastern England. These increases are mirrored by annual monitoring schemes, which show a 17% population increase in the UK during 1995–2010 and a 6% increase in the Republic of Ireland during 1998–2010 (*CBS Trend 2010*; *BBS Report 2011*).

These increases in range and abundance have been linked to the spread of commercial softwood plantations and also the popularity of exotic evergreens in gardens (Wilson *et al.* 2006; *Birds in England*; *Birds of Scotland*). It is also quite likely that the offer of winter sustenance in gardens has benefited the species. Coal Tits appear to be influenced by changes in cone abundance, switching to supplementary food in gardens more often in years with few spruce cones than in those with many (McKenzie *et al.* 2007).

Sponsored by CJ Wildlife

BREEDING DISTRIBUTION *2008–11*

POSSIBLE •	6%	5%	5%
PROBABLE •	9%	5%	6%
CONFIRMED ●	76%	75%	75%
TOTAL	91%	85%	86%

WINTER DISTRIBUTION *2007/08–2010/11*

PRESENT ●	92%	88%	89%

BREEDING DISTRIBUTION CHANGE *since 1968–72*

GAIN ▲	83	138	221
LOSS ▽	15	71	86
40yr CHANGE	+8%	+3%	+4%
20yr INDEX	+0.11	+0.08	

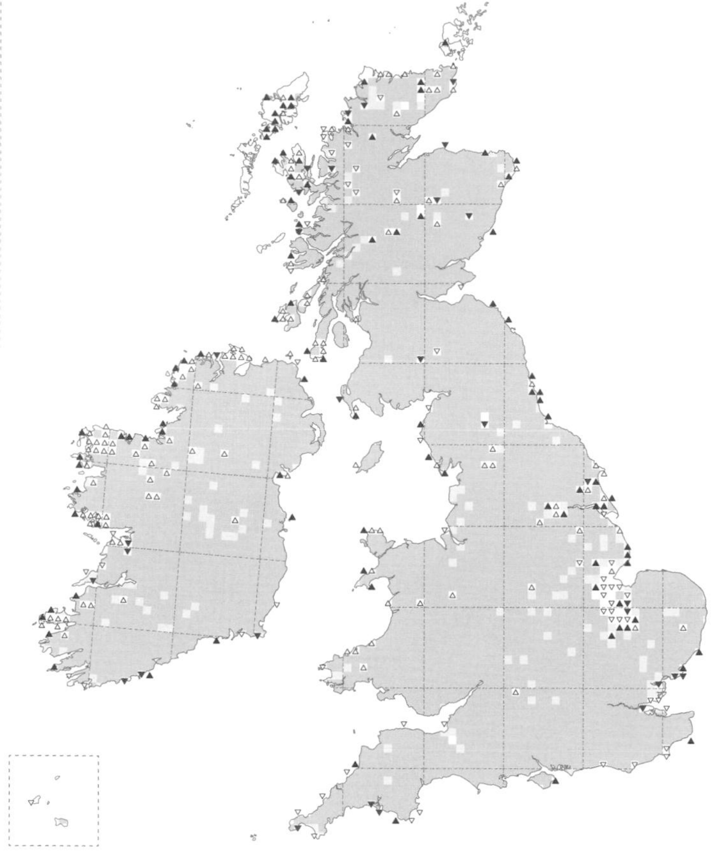

WINTER DISTRIBUTION CHANGE *since 1981–84*

GAIN ▲	162	268	430
LOSS ▼	18	34	52
30yr CHANGE	+19%	+10%	+12%

BREEDING RELATIVE ABUNDANCE *2008–11*

BREEDING RELATIVE ABUNDANCE CHANGE *since 1988–91*

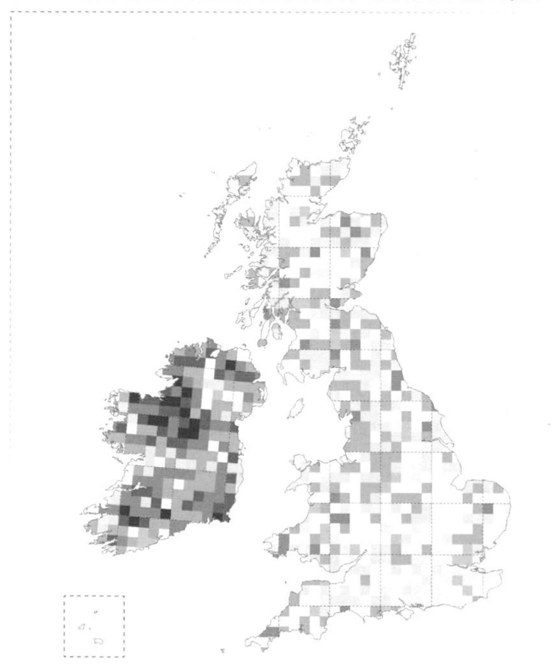

WINTER RELATIVE ABUNDANCE *2007/08–2010/11*

WILLOW TIT

Poecile montana

BTO CODE: WT

GRAHAM CATLEY http://pewit.blogspot.co.uk

THE DISTRIBUTION OF Willow Tits is now concentrated in a crescent from northeast England to South Wales, with several scattered outposts. The species is very sedentary so, not surprisingly, the winter and breeding distributions closely match one another. Fine-scale seasonal differences probably reflect detectability, given that Willow Tits can be inconspicuous and densities are often low.

The distributional change of the Willow Tit is one of the most striking revealed by *Bird Atlas 2007–11*. The species has virtually disappeared from the southeastern part of its English range since 1988–91. Former hotspots in southern England, East Anglia and the southern Midlands have vanished in the space of just two decades. Despite a rather constant population decline, range loss accelerated from 10% between the *1968–72 Breeding Atlas* and the *1988–91 Breeding Atlas* to 50% subsequently. This presumably represents the aggregation of many local extinctions following prolonged decline. There has been an estimated overall 79% decline in numbers during 1995–2010 (*BBS Report 2011*).

Although they occupy a range of wooded and scrubby habitats, Willow Tits in Britain prefer young, damp woodland with suitable dead wood for nesting (Siriwardena 2004; Lewis *et al.* 2007, 2009a; Mollet *et al.* 2009). Wet woodland probably remains one of the core habitats and drying soils may have been an important factor in some losses, though curiously the species is now absent from large areas of wet woodland in East Anglia. In Wales, and possibly elsewhere, Willow Tits occasionally inhabit coniferous woodland but, unlike birds in Scandinavia and central Europe, the British subspecies (*kleinschmidti*) does not use this habitat widely. The suitability of some woods may have been reduced by deer browsing and shading, both leading to thinning of the low vegetation that birds depend on for feeding (Lewis *et al.* 2009b; Newson *et al.* 2012). Competition for nest sites with more dominant tit species may drive local Willow Tit declines (Maxwell 2002, 2003), but this does not explain the large-scale population changes, and only weak evidence links declines with numbers of avian nest predators (Siriwardena 2004; Lewis *et al.* 2007). Many mysteries remain about the causes of the recent range contraction.

Sponsored by N. R. Aker

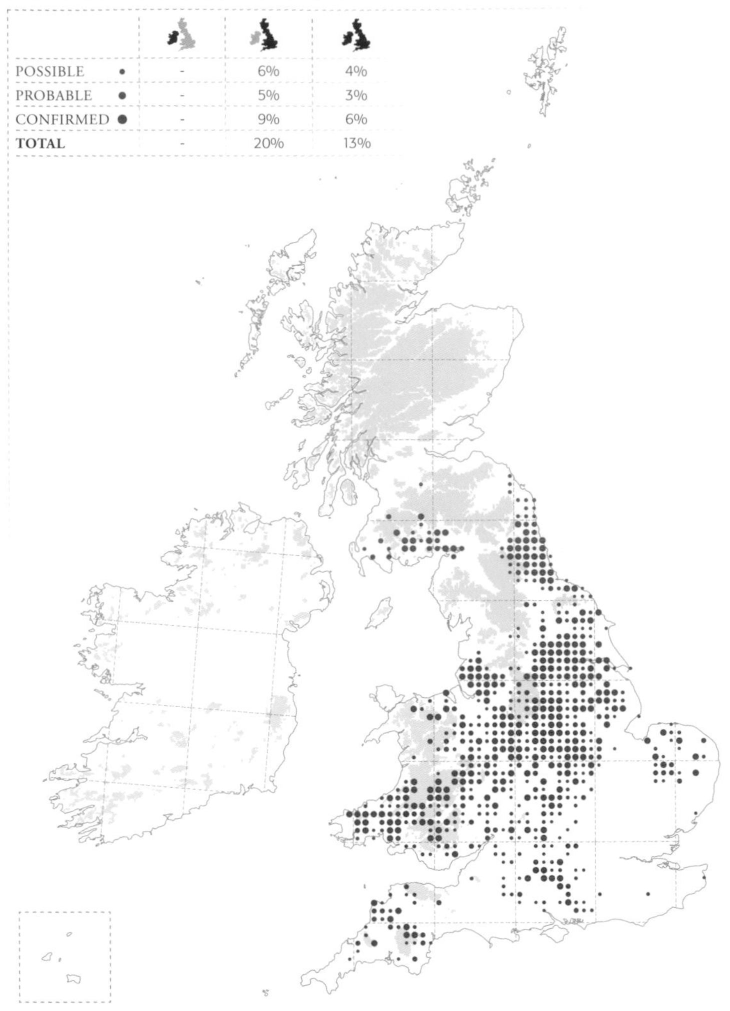

BREEDING DISTRIBUTION *2008–11*

	Ireland	Britain	Britain & Ireland
POSSIBLE •	-	6%	4%
PROBABLE •	-	5%	3%
CONFIRMED ●	-	9%	6%
TOTAL	-	20%	13%

WINTER DISTRIBUTION *2007/08–2010/11*

	Ireland	Britain	Britain & Ireland
PRESENT ●	-	24%	18%

BREEDING DISTRIBUTION CHANGE *since 1968–72*

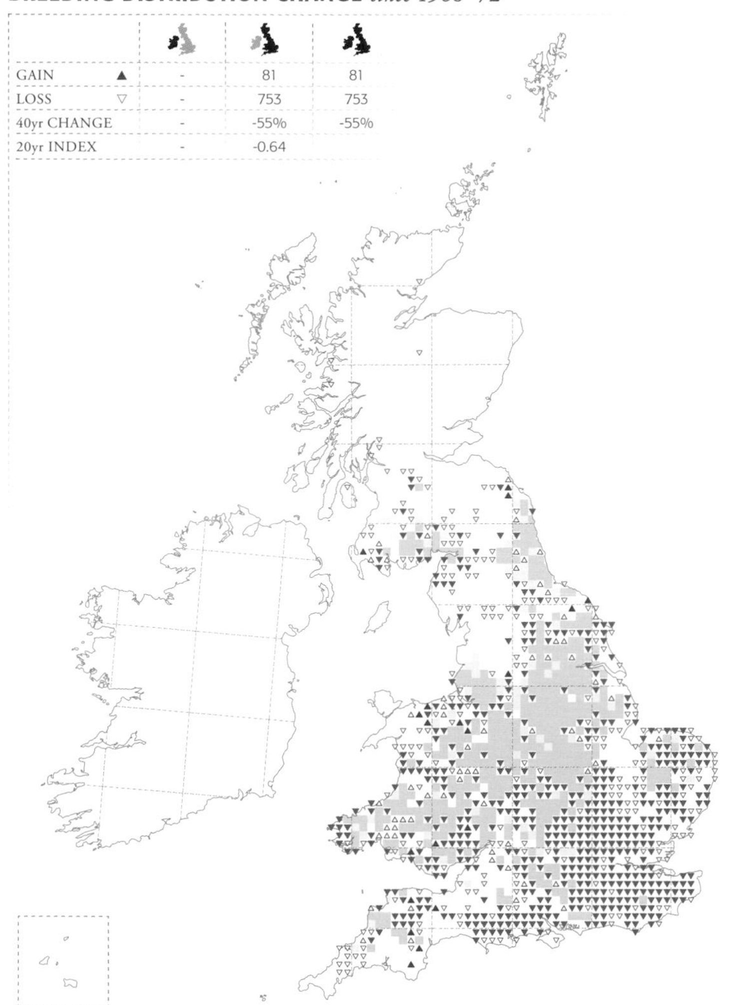

GAIN ▲	-	81	81
LOSS ▽	-	753	753
40yr CHANGE	-	-55%	-55%
20yr INDEX	-	-0.64	

WINTER DISTRIBUTION CHANGE *since 1981–84*

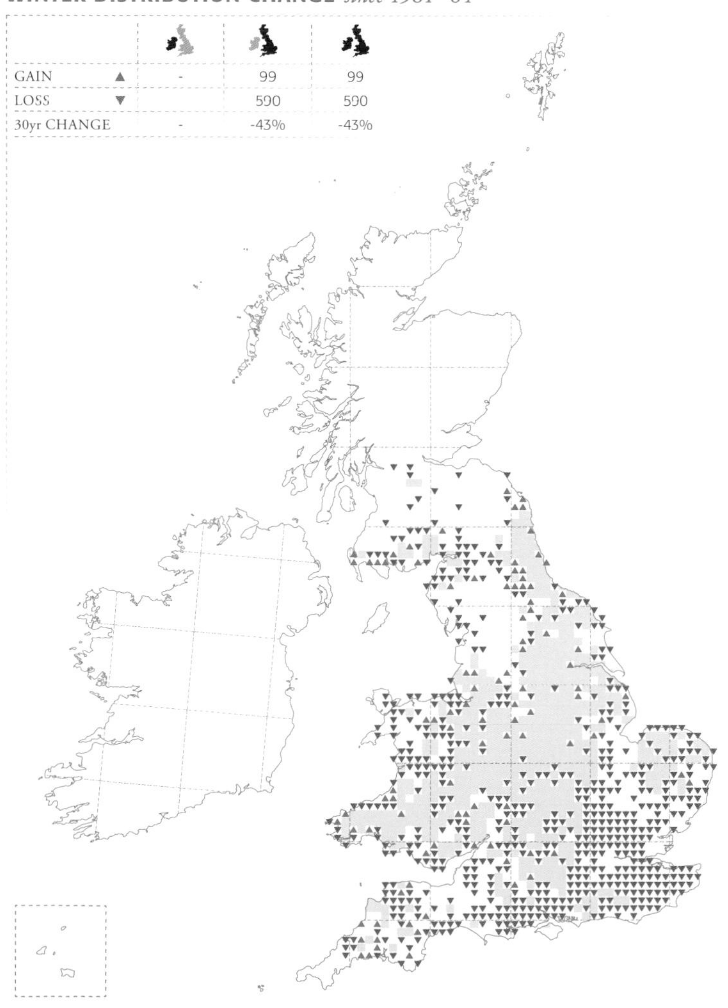

GAIN ▲	-	99	99
LOSS ▼	-	590	590
30yr CHANGE	-	-43%	-43%

BREEDING RELATIVE ABUNDANCE *2008–11*

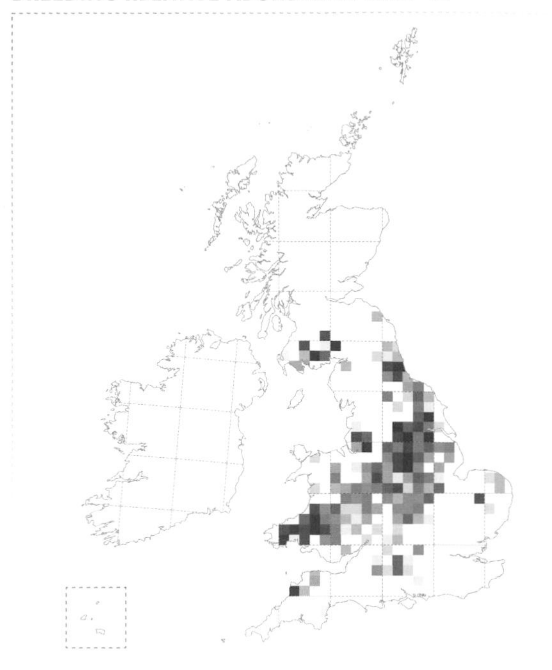

BREEDING RELATIVE ABUNDANCE CHANGE *since 1988–91*

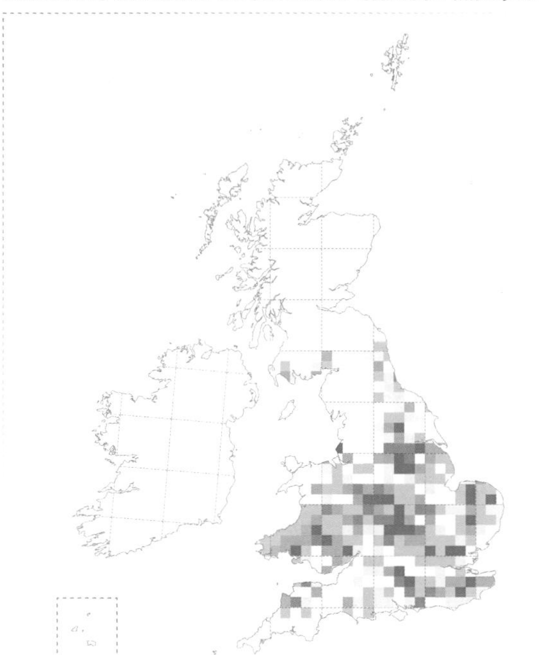

WINTER RELATIVE ABUNDANCE *2007/08–2010/11*

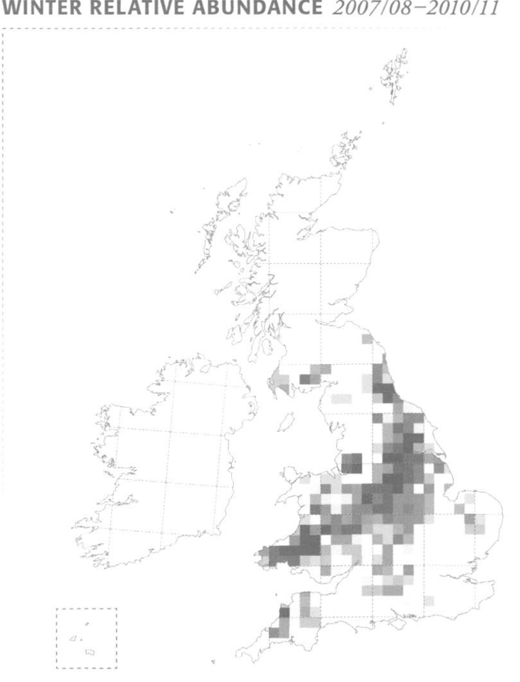

MARSH TIT

Poecile palustris

BTO CODE: MT

STEVE SMITH

MARSH TITS ARE absent from Ireland and most of Scotland, but are widespread in England and Wales, albeit with significant gaps in much of northwest England, northwest Wales, the north Midlands and the Fens. Their winter and breeding distributions are similar, reflecting the species' low propensity to disperse (Broughton *et al.* 2010).

The population is undergoing a continuing decline in Britain, reflected in a tendency for shrinkage at the edges of the gaps in the range, where distributions are less concentrated and changes in apparent occupancy are more likely to occur. Net losses since the *1981–84 Winter Atlas* have actually been quite small, with rather similar numbers of 10-km squares apparently being colonised by Marsh Tits as losing them. However, loss of breeding range has been significant, with a 22% range contraction since the *1968–72 Breeding Atlas*. There have been fewer losses since the *1988–91 Breeding Atlas* than in the earlier 20-year period, reflecting a reduction in the rate of population decline nationally (*BirdTrends*). There have also been local gains. Marsh Tits are often found at low densities at the 10-km-square scale and can be inconspicuous in winter, so some apparent changes may reflect chance encounters in one atlas or another. Changes in the detail of the distribution are probably best interpreted as the consequences of changes in abundance rather than of local changes in habitat suitability. The abundance change map shows that tetrad occupancy in 20-km squares has declined since the *1988–91 Breeding Atlas* throughout virtually the whole range in Britain.

The decline does not appear to be due to competition with Blue and Great Tits nor to increased avian nest predation, and woodland habitat degradation is a more likely explanation (Siriwardena 2006; Broughton *et al.* 2011). Historically, woodland fragmentation has probably been important because Marsh Tits require moderately large woodland patches in which to breed (Hinsley *et al.* 1995). They also prefer a complex understorey below tall, closed tree canopies (Broughton *et al.* 2012). Declining active woodland management causing shading of the preferred shrub-layer structure is a possible mechanism for population decline and range contraction (Carpenter *et al.* 2010; Broughton *et al.* 2012).

Sponsored by Mike Wildish—Dedicated to the Hampshire Wildlife Trust - North West District Committee and Volunteers

BREEDING DISTRIBUTION *2008–11*

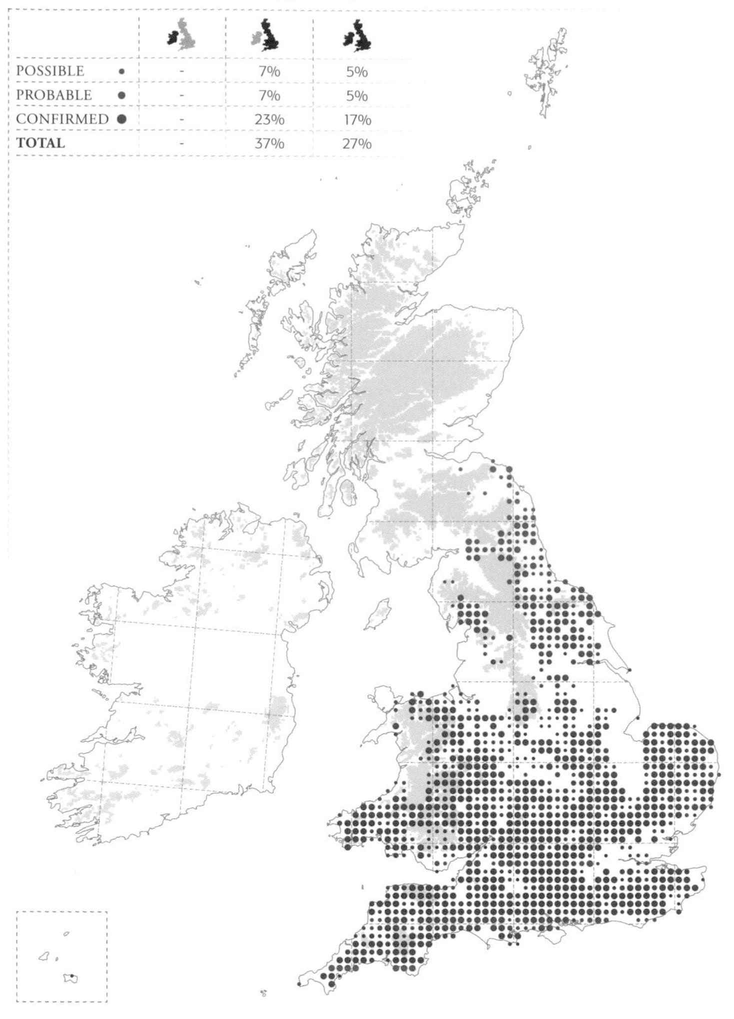

POSSIBLE •	-	7%	5%
PROBABLE •	-	7%	5%
CONFIRMED ●	-	23%	17%
TOTAL	-	37%	27%

WINTER DISTRIBUTION *2007/08–2010/11*

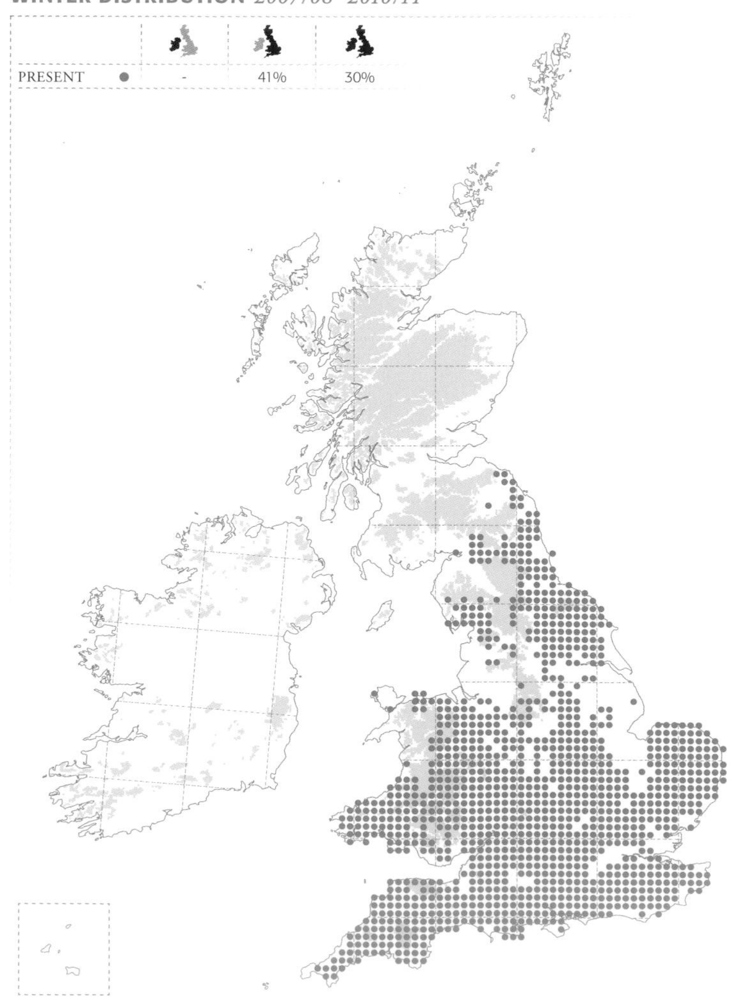

PRESENT ●	-	41%	30%

BREEDING DISTRIBUTION CHANGE *since 1968–72*

GAIN ▲	-	62	62
LOSS ▽	-	369	369
40yr CHANGE	-	-22%	-22%
20yr INDEX	-	-0.27	

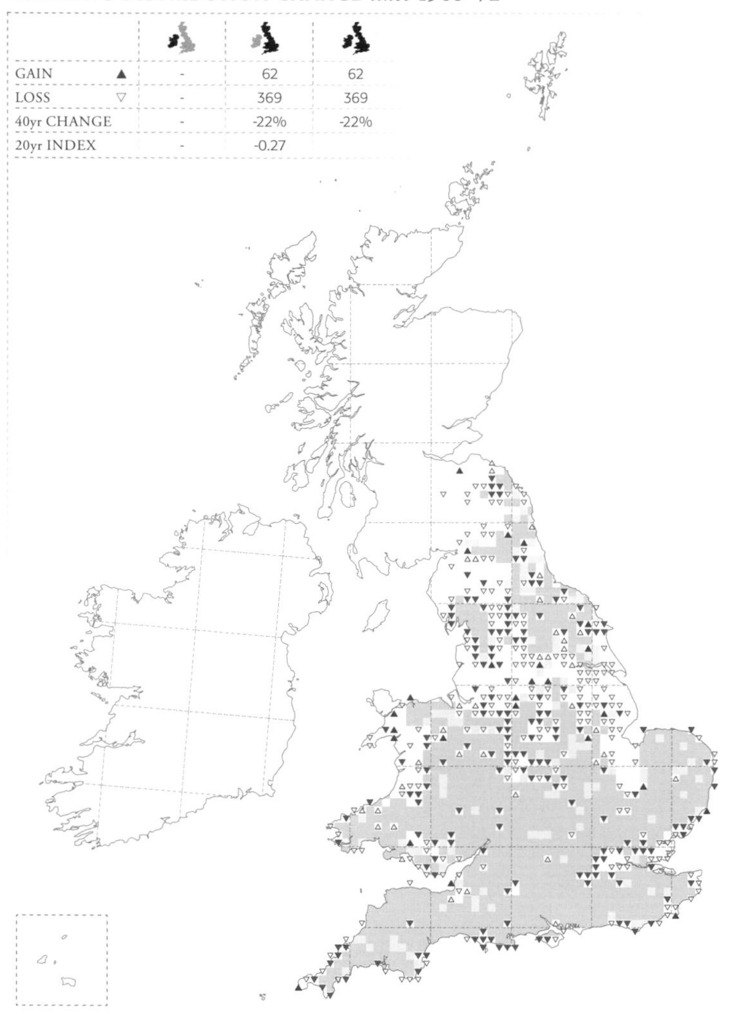

WINTER DISTRIBUTION CHANGE *since 1981–84*

GAIN ▲	-	144	144
LOSS ▼	-	196	196
30yr CHANGE	-	-4%	-4%

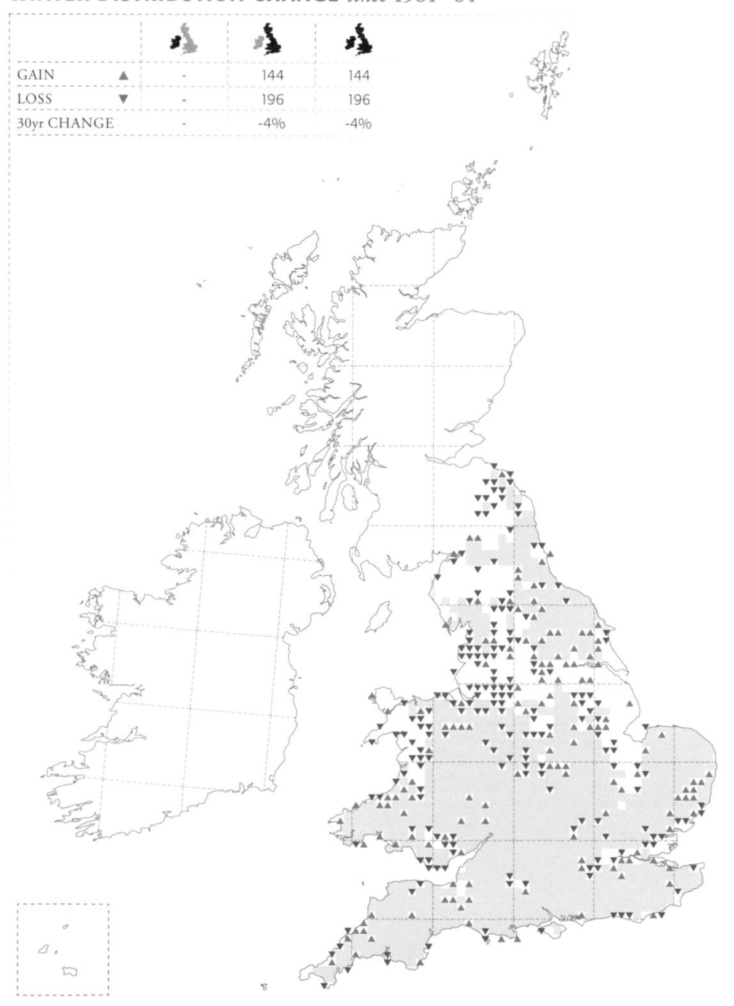

BREEDING RELATIVE ABUNDANCE *2008–11*

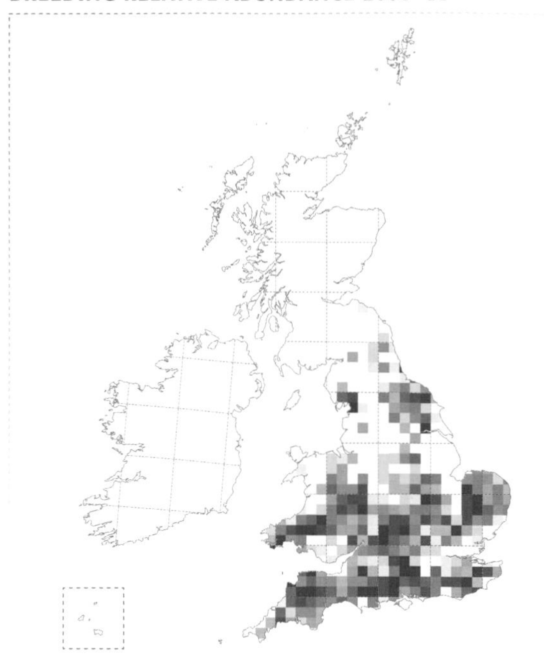

BREEDING RELATIVE ABUNDANCE CHANGE *since 1988–91*

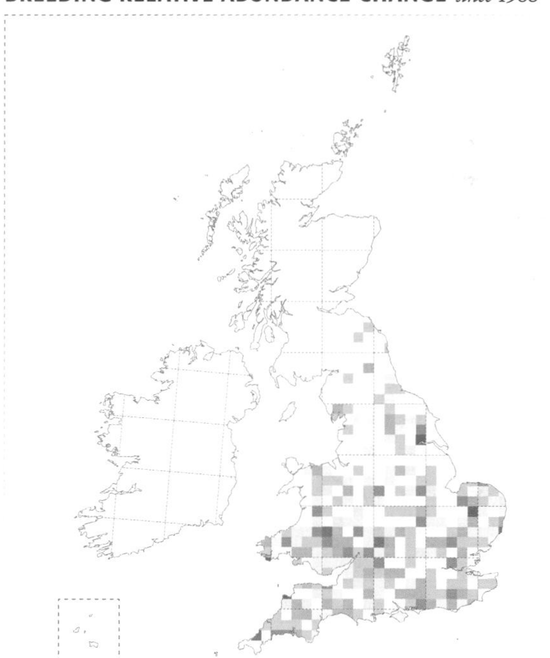

WINTER RELATIVE ABUNDANCE *2007/08–2010/11*

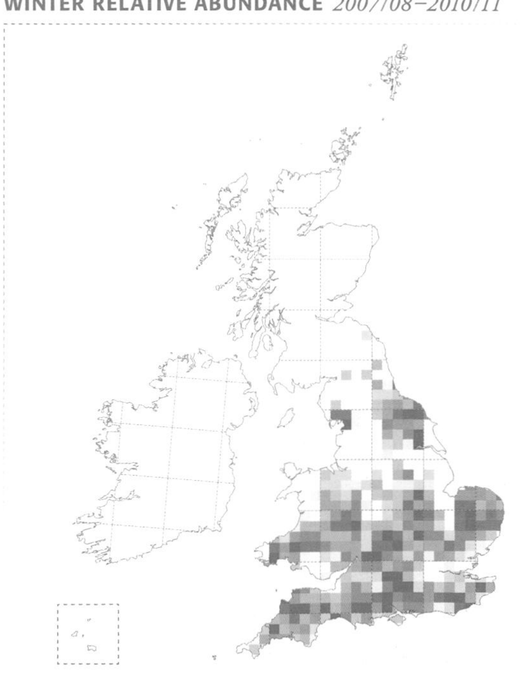

BEARDED TIT

Panurus biarmicus

BTO CODE: BR

ALLAN DREWITT

BECAUSE BEARDED TITS depend on extensive *Phragmites*-dominated reedbeds, habitat availability is likely to be the key factor limiting their distribution. Nonetheless, their range has increased markedly in recent decades. The breeding range expanded by 40% between the *1968–72 Breeding Atlas* and the *1988–91 Breeding Atlas*. This pattern has continued, with a further 32% range expansion by 2008–11, including a return to Wales, expansion in southwest England and apparent colonisation of Scotland. Bearded Tits were not recorded in Scotland during the *1988–91 Breeding Atlas* but bred in the Tay reedbeds probably from 1991 onwards (*Birds of Scotland*). Birds have bred sporadically in Ireland (*Birds in Ireland*) and *confirmed* breeding was reported from one site in the southeast.

Based on reports received by RBBP, the UK population was estimated at 718 pairs in 2010 (*RBBP 2010*). This is likely to be an underestimate owing to the difficulty of detecting Bearded Tits in reedbeds and the omission of certain key sites. Suffolk, Norfolk and Yorkshire accounted for 78% of the *confirmed* or *probable* breeding pairs in 2010 (*RBBP 2010*) but no figures were available for the Tay reedbeds which held 250 pairs in 2004 (*Birds of Scotland*).

Newly created or expanded reedbeds in Somerset, South Yorkshire and the East Anglian Fens have been readily colonised. However, setbacks have occurred. Exceptional flooding followed by severe cold caused high mortality and the near-extinction of the Leighton Moss population (Wilson & Peach 2006) and saltwater inundation and increased water nutrient loading caused local declines in Norfolk (Cadbury 2005). In contrast, brackish water in Scottish reedbeds may help populations survive cold winters (*Birds of Scotland*).

Outside the breeding season, Bearded Tits are highly gregarious and are prone to dispersive movements (*BWP*). This is reflected in the winter distribution, with approximately double the number of 10-km squares occupied as in the breeding season, including a scatter of sites in central-eastern England. Nonetheless, the reporting rate of Bearded Tits in winter is lower than that in the breeding season (*BirdTrack*), indicating under-recording in winter. This, and the sporadic occurrence of dispersing individuals, may explain the mixture of gains and losses in central England.

Sponsored by RSPB

BREEDING DISTRIBUTION *2008–11*

POSSIBLE ●	-	<1%	<1%
PROBABLE ●	-	<1%	<1%
CONFIRMED ●	<1%	2%	2%
TOTAL	<1%	2%	2%

WINTER DISTRIBUTION *2007/08–2010/11*

PRESENT ●	<1%	5%	4%

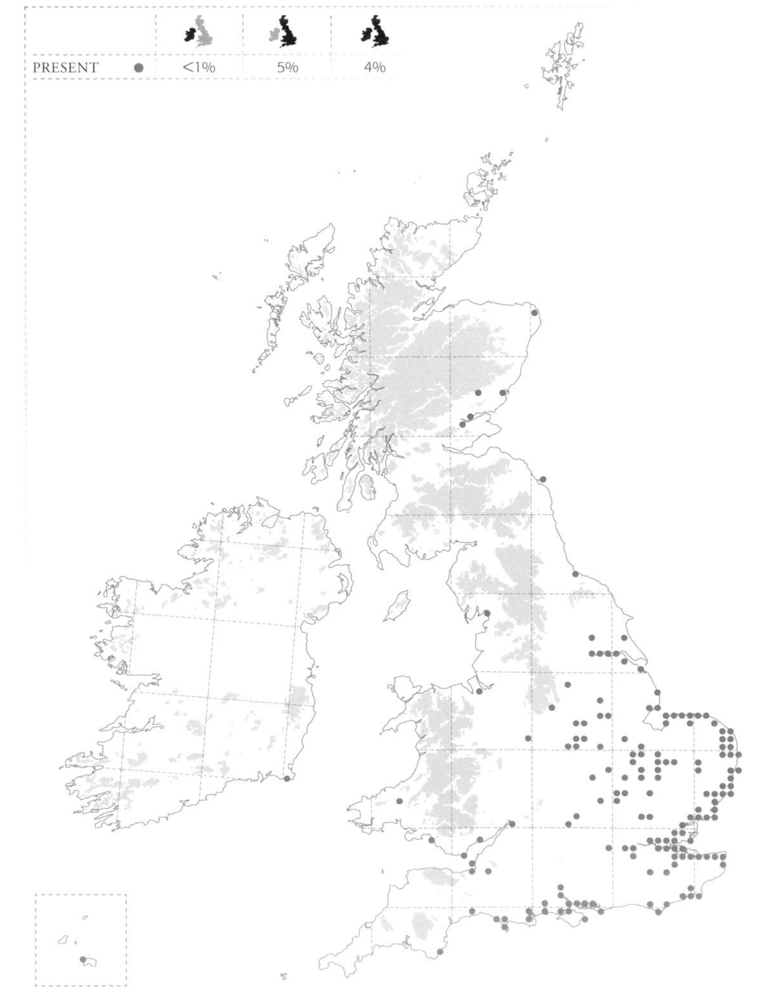

BREEDING DISTRIBUTION CHANGE *since 1968–72*

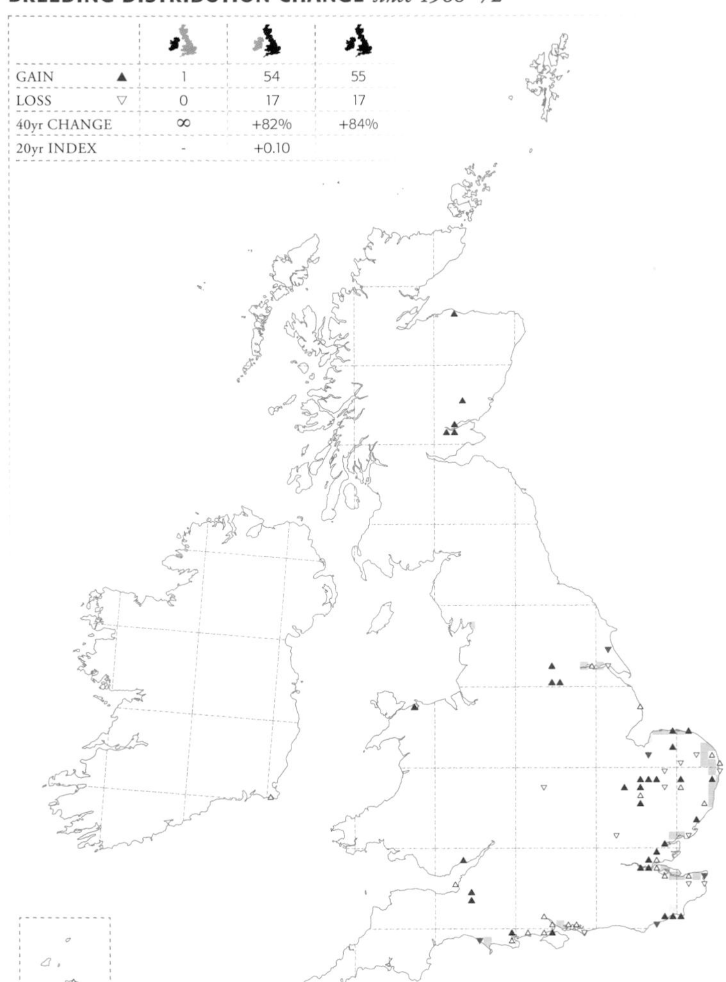

GAIN ▲	1	54	55
LOSS ▽	0	17	17
40yr CHANGE	∞	+82%	+84%
20yr INDEX	-	+0.10	

BREEDING RELATIVE ABUNDANCE *2008–11*

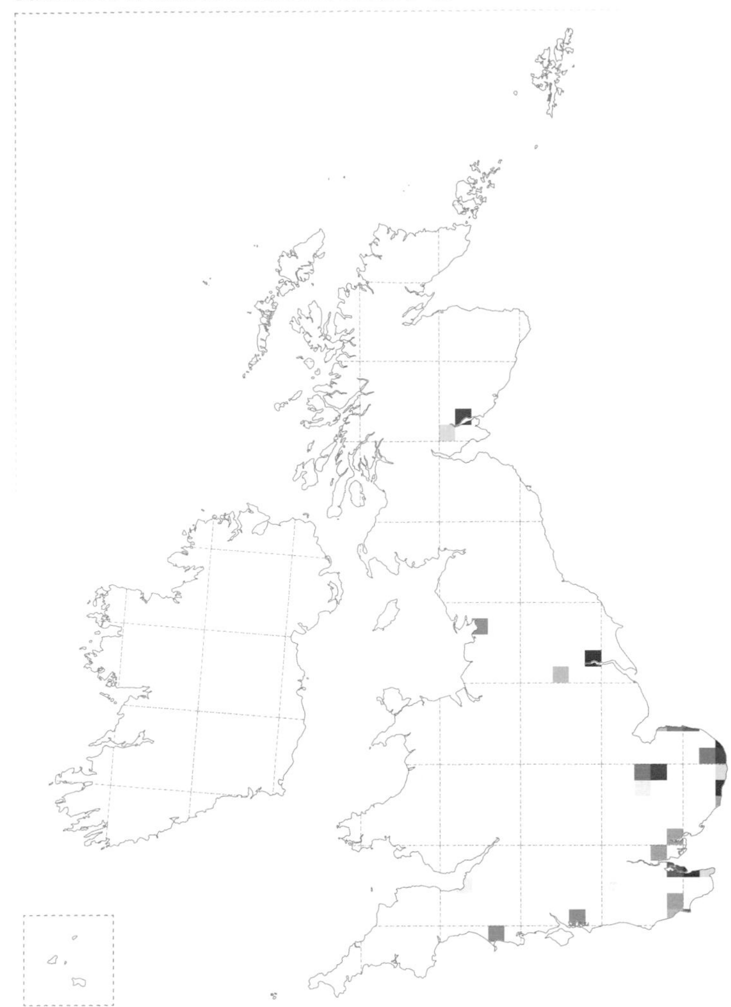

WINTER DISTRIBUTION CHANGE *since 1981–84*

GAIN ▲	1	69	70
LOSS ▼	1	50	51
30yr CHANGE	0%	+16%	+16%

WINTER RELATIVE ABUNDANCE *2007/08–2010/11*

WOODLARK
Lullula arborea

BTO CODE: WL

CHRIS KNIGHTS www.chrisknightswildlife.com

THE STATUS, DISTRIBUTION and behaviour of Woodlarks in Britain have changed considerably during the last 40 years. Only 56 10-km squares were occupied during the *1981–84 Winter Atlas*, a period when birds deserted East Anglia after breeding. Winter range has since increased fourfold and birds now regularly winter near breeding sites in East Anglia and central and northern England. Expansion has also occurred in winter around the core breeding area in southern England. Scattered coastal observations in northeast England and Scotland, mostly in November, relate to dispersing birds or continental migrants (*Migration Atlas*; *Birds of Scotland*). Relative abundance in winter is strongly associated with the breeding population centres, together with a small concentration in Cornwall.

The current breeding distribution broadly matches that of the historic breeding strongholds in Devon, the New Forest, the Thames Basin and East Anglia, but there are also new populations in the West Midlands, northern England and South Wales. The pattern of breeding distribution change is complex. Between 1968–72 and 1988–91 the range contracted by 63%, largely through losses on farmland in Wales and southwest England. Subsequently, a 107% range expansion took place up to 2008–11, involving consolidation in the Thames Basin and East Anglia and expansion into central and northern England.

These changes are consistent with the 46% increase in range size and 88% increase in population size between the 1997 and 2006 national surveys, the latter giving a population estimate of 3,064 territories (Wotton & Gillings 2000; Conway *et al.* 2009). Much of the expansion is associated with restoration and re-creation of heathland (Langston *et al.* 2007), with which 67% of the population was associated in 2006. A third of territories were associated with sympathetically managed plantation forestry. Increases in Thetford Forest were not directly related to climate change and the effects of weather on productivity were of secondary importance to nest predation (Wright *et al.* 2009). Since 1997 numbers have declined in forestry, especially in East Anglia (Conway *et al.* 2009). Possible causes include reduced habitat suitability and increased nest predation (Wright *et al.* 2009). Current levels of recreational disturbance reduce densities, but not breeding success (Mallord *et al.* 2007).

HOS

Sponsored by Hampshire Ornithological Society

BREEDING DISTRIBUTION *2008–11*

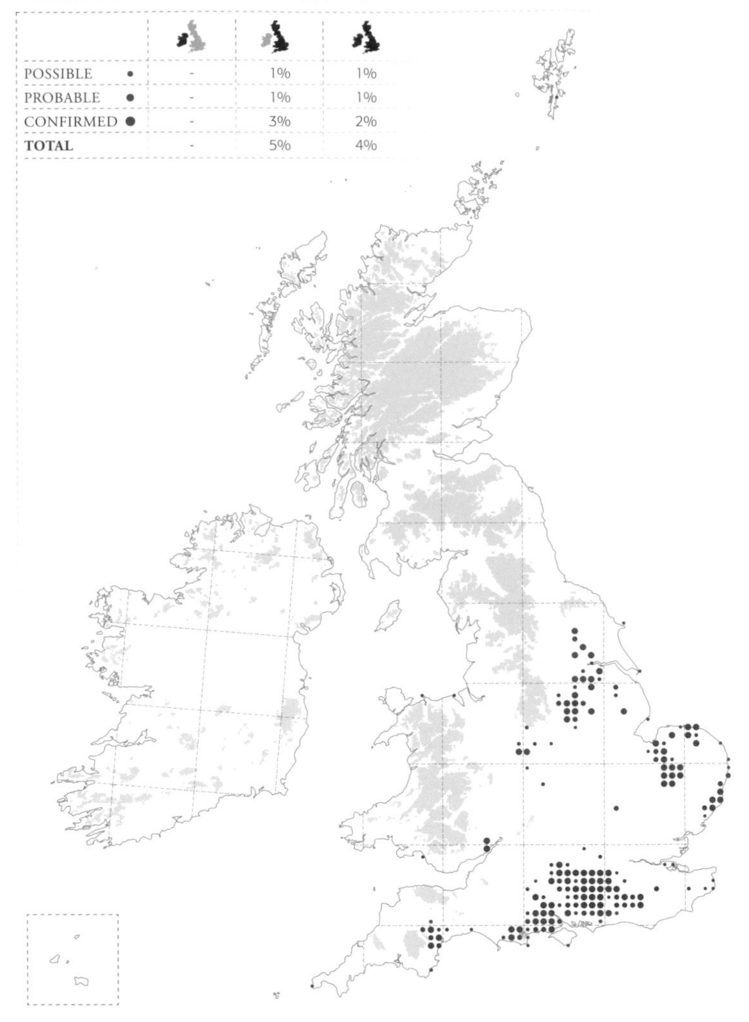

POSSIBLE •	-	1%	1%
PROBABLE •	-	1%	1%
CONFIRMED ●	-	3%	2%
TOTAL	-	5%	4%

WINTER DISTRIBUTION *2007/08–2010/11*

PRESENT ●	-	8%	6%

BREEDING DISTRIBUTION CHANGE *since 1968–72*

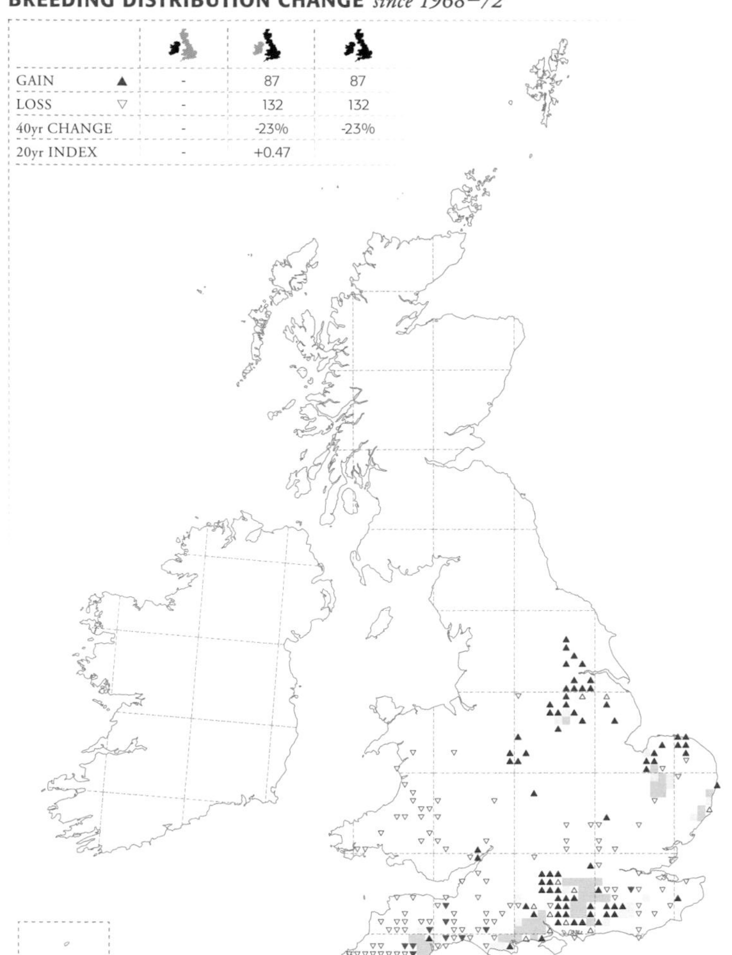

GAIN ▲	-	87	87
LOSS ▽	-	132	132
40yr CHANGE	-	-23%	-23%
20yr INDEX	-	+0.47	

BREEDING RELATIVE ABUNDANCE *2008–11*

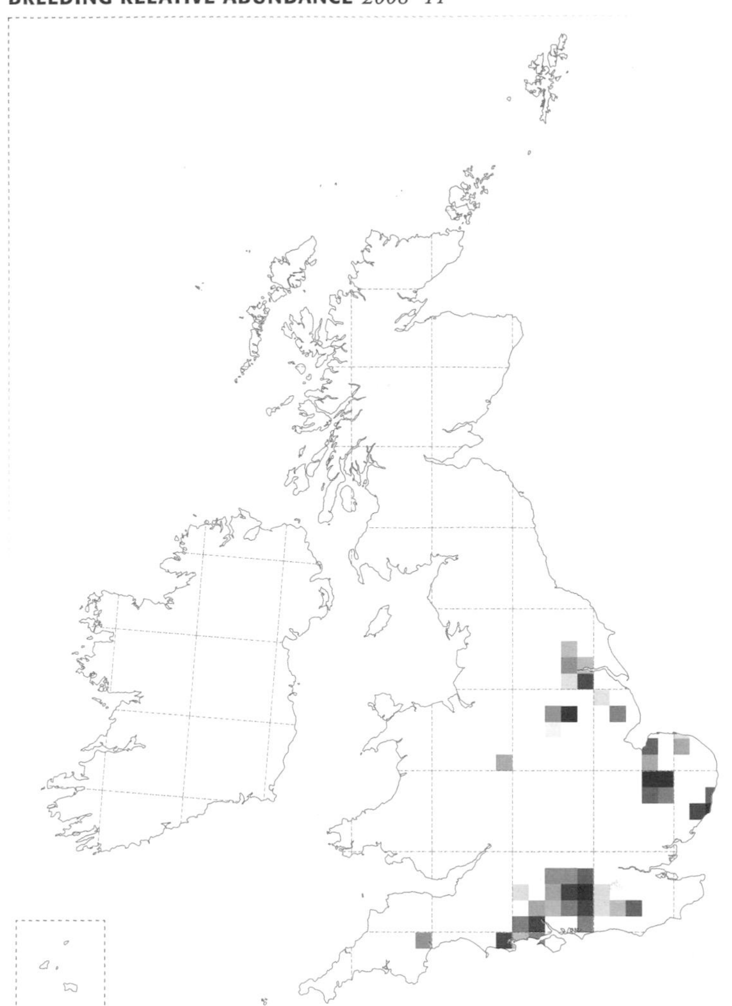

WINTER DISTRIBUTION CHANGE *since 1981–84*

GAIN ▲	-	192	192
LOSS ▼	-	26	26
30yr CHANGE	-	+296%	+296%

WINTER RELATIVE ABUNDANCE *2007/08–2010/11*

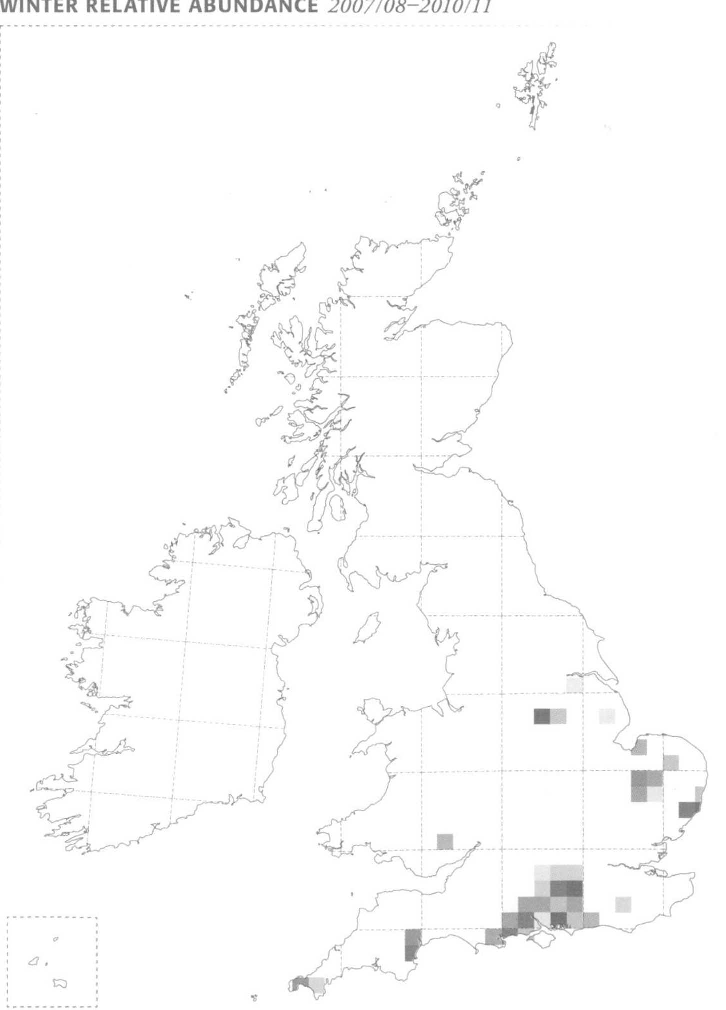

SKYLARK
Alauda arvensis

BTO CODE: S.

NOTTSEXMINER

DURING WINTER, SKYLARKS are widespread throughout Britain, except at higher altitudes. Those upland areas shown as occupied on the winter distribution map were mostly occupied only in February, suggesting that these birds were early returning breeders rather than true winterers. Even in those marginal upland areas occupied throughout the winter, the abundance map indicates that densities are low, as they are around urban areas. Higher densities are found in the south and east of England and near the coast (Gillings & Fuller 2001). In Ireland, birds are sparsely distributed, favouring southern, eastern and coastal regions.

Overall, the number of occupied 10-km squares in Britain in winter has increased by 8% since the *1981–84 Winter Atlas*. Many gains in upland areas might be associated with birds returning earlier to upland breeding sites, however. In Ireland, winter losses outnumber gains, with 14% fewer occupied squares, which probably reflects the decline in the breeding population.

In the breeding season, Skylarks are widespread throughout Britain and most of Ireland. The breeding-season relative abundance map highlights the lower densities in the higher uplands and in built-up areas. Densities are relatively high in the lowlands, particularly in eastern Britain, in the lower hills of western England and Wales, and along the western coastal fringes of Scotland and Ireland.

Despite many decades of breeding population decline there has been very little change in the distribution in Britain. However, in Ireland there has been a substantial range contraction, of 14% since the *1968–72 Breeding Atlas*, with losses in the midlands, south and east. Throughout Ireland there have also been marked declines in the proportions of tetrads occupied since the *1988–91 Breeding Atlas*. Similar declines affect much of Britain at tetrad level but have caused few larger-scale distributional changes. These patterns correspond with known trends in breeding populations, with a 58% decline in the UK during 1970–2010 (*SUKB 2012*) and a 30% decline in the Republic of Ireland during 1998–2010 (*CBS Trend 2010*). Declines in Britain have been linked to changes in farming, particularly intensification of grassland management and the switch from spring to autumn sowing of cereals (Chamberlain & Siriwardena 2000; Donald 2004).

Sponsored by Syngenta

BREEDING DISTRIBUTION *2008–11*

POSSIBLE •	22%	6%	10%
PROBABLE ●	33%	21%	24%
CONFIRMED ●	30%	69%	58%
TOTAL	85%	96%	92%

WINTER DISTRIBUTION *2007/08–2010/11*

PRESENT ●	54%	81%	74%

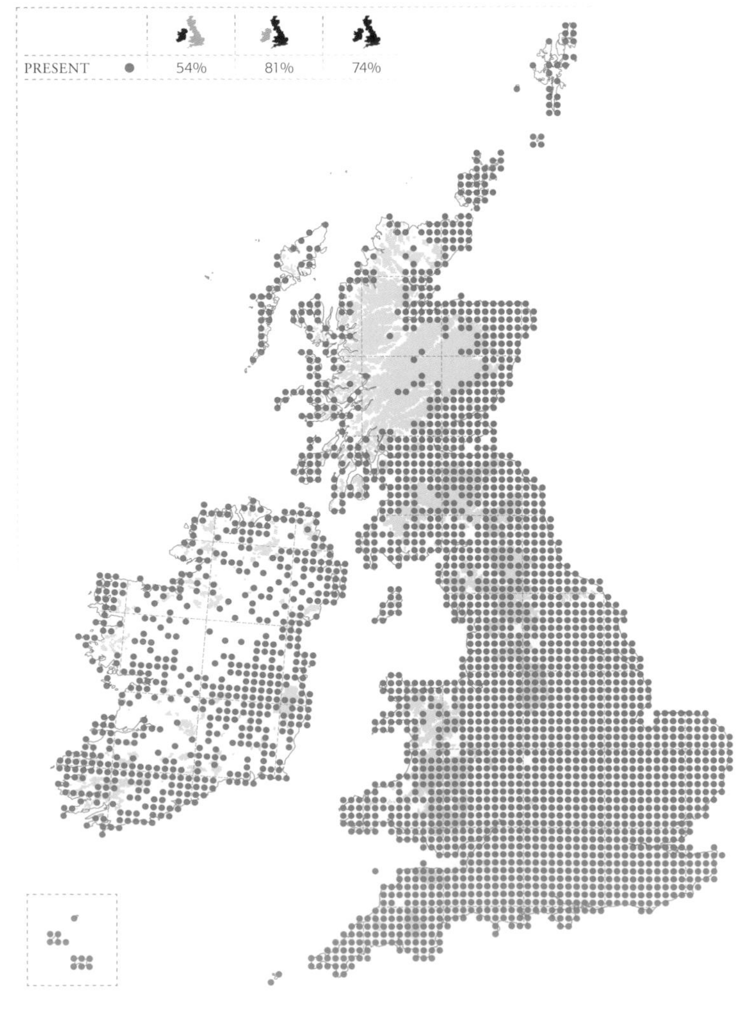

BREEDING DISTRIBUTION CHANGE *since 1968–72*

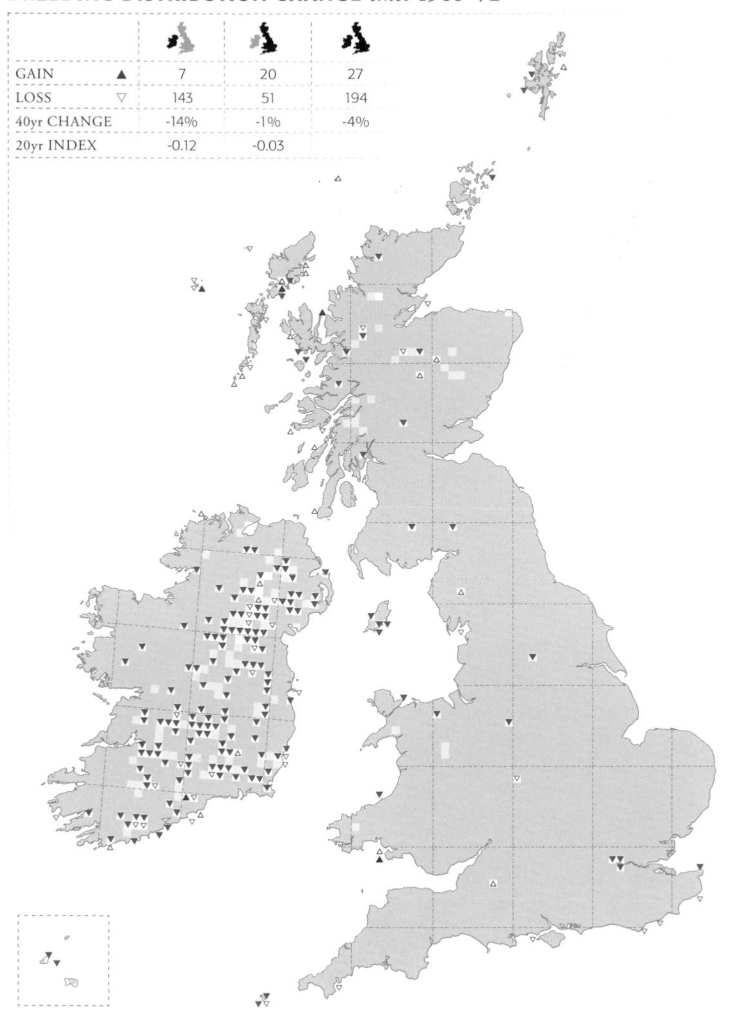

GAIN ▲	7	20	27
LOSS ▽	143	51	194
40yr CHANGE	-14%	-1%	-4%
20yr INDEX	-0.12	-0.03	

WINTER DISTRIBUTION CHANGE *since 1981–84*

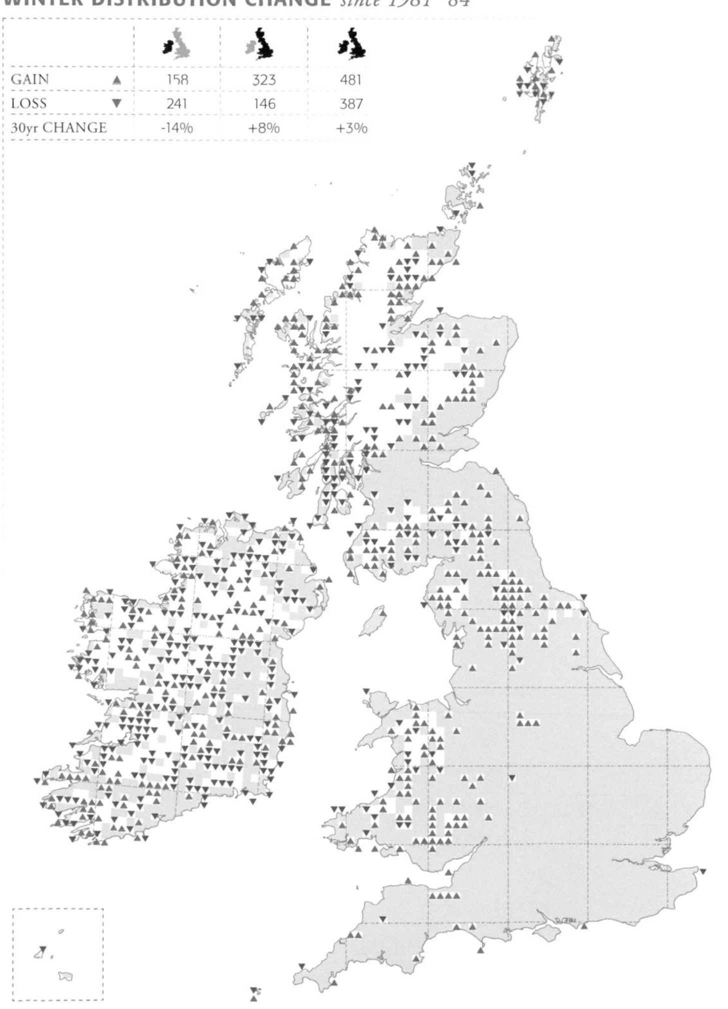

GAIN ▲	158	323	481
LOSS ▼	241	146	387
30yr CHANGE	-14%	+8%	+3%

BREEDING RELATIVE ABUNDANCE *2008–11*

BREEDING RELATIVE ABUNDANCE CHANGE *since 1988–91*

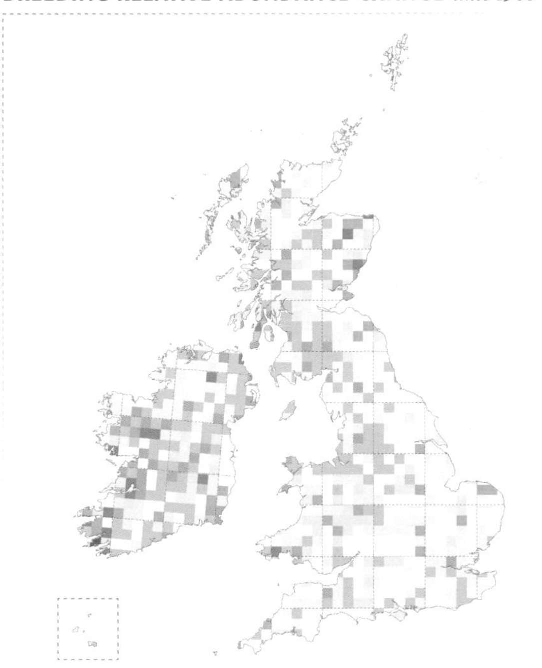

WINTER RELATIVE ABUNDANCE *2007/08–2010/11*

SHORE LARK
Eremophila alpestris

BTO CODE: SX

LIZ CUTTING www.lizcuttingphotos.com

SHORE LARKS ARE scarce winter visitors to Britain, and very rare visitors to Ireland. They were recorded predominantly in coastal habitats on the east coast of England and in southeast Scotland, where they feed on the tideline and amongst saltmarsh and sand-dune vegetation (Brown & Atkinson 1996).

Birds wintering in Britain originate from the breeding population in Fennoscandia, individuals of which winter along North Sea coasts (Dierschke 2001). The majority winter around the Dutch and German Wadden Sea and only a very small, annually variable proportion crosses the North Sea to Britain. Colour ringing has shown that birds wintering in Britain in one year may winter in the Wadden Sea the next (Clark *et al.* 2002a, 2002b). The reasons for the variation in numbers between winters are unknown but may be related to variation in breeding success or conditions in other wintering areas. The single record in Co. Donegal in December 2009 constituted the 23rd record for Ireland; there were none there in winters 1981–84.

The winter distribution in Britain has changed little since the *1981–84 Winter Atlas*; a 4% range contraction was recorded. Gains and losses tend to balance out, particularly along the east coast. Changes in the lower saltmarshes of the Wadden Sea, caused by creation of embankments and by intensified grazing, might have been responsible for the rapid population decline recorded there between the 1960s and 1980s (Dierschke 2002). More recently, numbers there increased approximately threefold during the 1990s owing to the increasing area of pioneer saltmarshes and reduced grazing intensity on the upper saltmarshes (Dierschke 2001). These changes may have affected numbers occurring in Britain.

During several years since the early 1970s, Shore Larks have been found in suitable breeding habitat in parts of the Cairngorms. Breeding has been confirmed in only two years, most recently in 2003 (*Birds of Scotland*), and, although birds were present in seven of the eight years leading up to that event (Holling *et al.* 2007a), none were recorded in suitable habitat during 2008–11. Spring-passage birds were recorded in 26 10-km squares, however, mostly along the English east coast.

WINTER DISTRIBUTION *2007/08–2010/11*

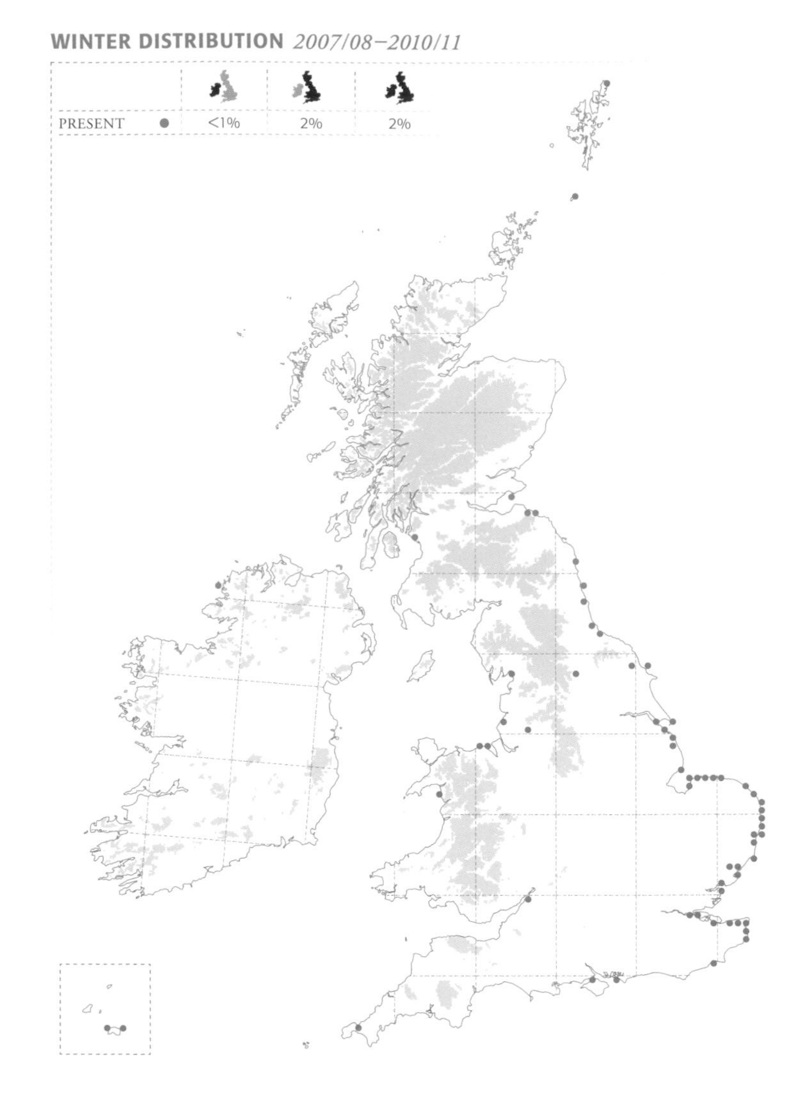

Sponsored by Peggy Heyder

WINTER DISTRIBUTION CHANGE since 1981–84
GAIN ▲ 1 29 30
LOSS ▼ 0 31 31
30yr CHANGE ∞ -4% -2%
BREEDING DISTRIBUTION 2008–11
POSSIBLE ● - - -
PROBABLE ● - - -
CONFIRMED ● - - -
TOTAL - - -

OLIVER SMART www.smartimages.co.uk

▲ The number of Shorelarks recorded varies between winters; the highest number reported to the Atlas was 31 at Holkham in Norfolk in November 2010.

SAND MARTIN

Riparia riparia

BTO CODE: SM

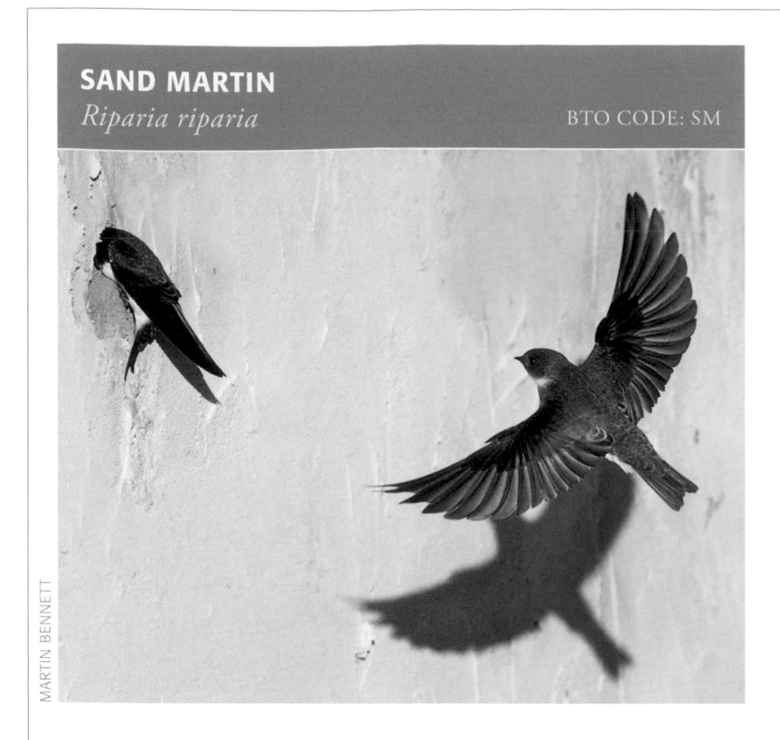

MARTIN BENNETT

BREEDING SAND MARTINS occur throughout Ireland. In Britain, however, they are widespread north of a line from the Wash to the Severn Estuary, but relatively localised further south. This divide was apparent in the *1968–72 Breeding Atlas* but has become more pronounced, owing to relative stability to its north but ongoing losses in southern and southeastern England, through both the 20-year periods between breeding atlases.

In Wales and Ireland there has been little overall change in range between the *1968–72 Breeding Atlas* and the present, but this masks an apparent range contraction between 1968–72 and 1988–91 and something of a recovery since. The range in much of Scotland shows a similar pattern except that, since the *1988–91 Breeding Atlas*, it has pushed further into the northwest and onto some of the Inner Hebrides and into Orkney. Breeding Sand Martins remain scarce in the Outer Hebrides and absent from Shetland.

Atlas data suggest that Sand Martins share the same pattern of regional change in abundance as the Swallow and House Martin. The proportion of tetrads occupied has changed in a systematic manner since the *1988–91 Breeding Atlas*, with marked increases in Ireland, Scotland and northern England but decreases across much of southern England.

At the local level, some gains and losses may have resulted from redistribution within ephemeral nesting habitats, for example owing to erosion of banks and cliffs (*Birds of Scotland*) or changing conditions at aggregate extraction sites. Apparent population levels show considerable fluctuation, partly owing to surveying difficulties (*BirdTrends*). Annual survival rates are related to rainfall, both on the Sand Martin's West African wintering grounds (Robinson *et al.* 2008) and in its breeding areas (Cowley & Siriwardena 2005). Droughts in the Sahel in 1968–69 and 1983–84 caused significant population crashes but it is unclear whether their impact lasted long enough to influence distribution patterns.

Consistent with a documented pattern of earlier arrival (Sparks & Tryjanowski 2007), Sand Martins were recorded in eight times more 10-km squares during the winter period than in the *1981–84 Winter Atlas*, the majority being occupied only in February.

Sponsored by Thelma Sykes in memory of John Fisher

BREEDING DISTRIBUTION *2008–11*

POSSIBLE •	17%	9%	11%
PROBABLE ●	14%	5%	8%
CONFIRMED ●	47%	47%	47%
TOTAL	78%	61%	66%

BREEDING RELATIVE ABUNDANCE CHANGE *since 1988–91*

BREEDING DISTRIBUTION CHANGE *since 1968–72*

GAIN ▲	90	212	302
LOSS ▽	137	506	643
40yr CHANGE	-6%	-14%	-12%
20yr INDEX	+0.19	+0.11	

BREEDING RELATIVE ABUNDANCE *2008–11*

GRAHAM CATLEY http://pewit.blogspot.co.uk

◀ **Sand Martins can form post-breeding flocks and roost communally in late summer so it is important to be certain that fledged young have hatched locally when providing breeding evidence.**

SWALLOW

Hirundo rustica

BTO CODE: SL

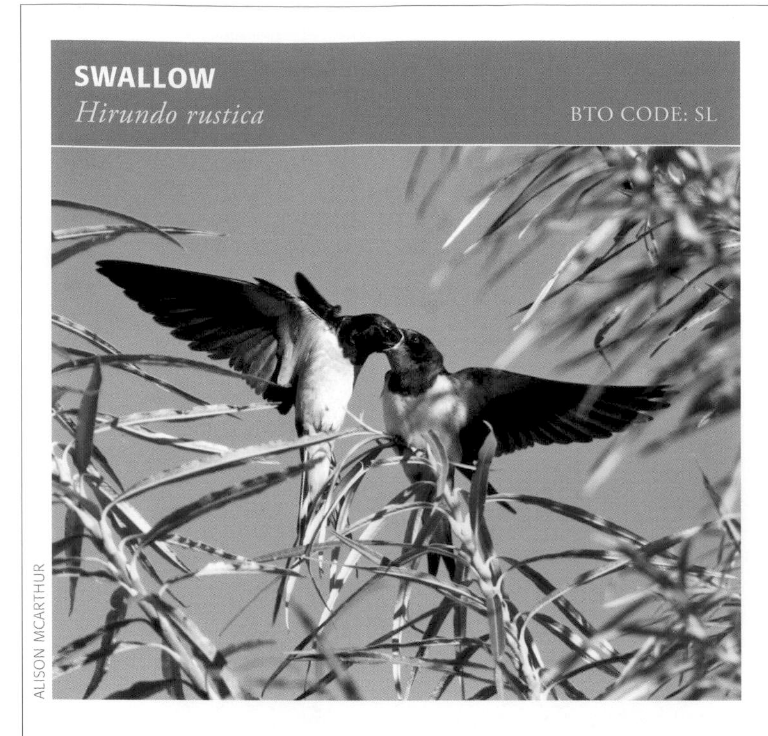

ALISON MCARTHUR

RECORDED BREEDING IN 95% of 10-km squares, the Swallow has the most extensive distribution of any summer migrant in Britain & Ireland. It is absent only from a few areas of northern Scotland and from central London. It occurs at high densities throughout Ireland. Despite breeding across the British uplands it does so at only a low density. Densities are also relatively low throughout arable eastern England. Such patterns may be related to the presence of cattle, which is the best predictor of the distribution of foraging Swallows (Henderson *et al.* 2007). The range changes since the *1968–72 Breeding Atlas* are small in absolute terms but involve steady northward expansion in northwest Scotland, the Northern Isles and the Outer Hebrides, as well as incidental coastal gains in Ireland.

Although Swallow populations are subject to annual fluctuations, the European population has increased overall by 5% during 1980–2010 (*PECBMS 2010*). The annual population trends for the UK and Republic of Ireland both show increases since the mid to late 1990s (*CBS Trend 2010*; *BBS Report 2011*). The abundance change map shows a pattern common to several long-distance migrants whereby tetrad occupancy has increased in many parts of Ireland and northern Britain since the *1988–91 Breeding Atlas* but decreased in southeast England. The causes of these regional differences are unclear. An analysis of CBC data indicated that declines in Swallow numbers in eastern England were not correlated with changing farming practices or decreased nest-site availability (Robinson *et al.* 2003). Given that this regional pattern is shared by several long-distance migrants with differing breeding ecology, it seems likely that factors acting on migration or in the wintering grounds are important (Ockendon *et al.* 2012).

Swallows are comparatively late-departing migrants, featuring on 1.4% of BirdTrack lists in the second week of November during the Atlas years (*BirdTrack*). Despite being a traditional harbinger of spring, Swallows are departing later with growing frequency and in some areas attempting to overwinter. After excluding the many early November records that would not have been covered by the *1981–84 Winter Atlas* methods, a 52% winter range expansion since 1981–84 is still apparent.

In memory of Margaret Hayter

BREEDING DISTRIBUTION *2008–11*

POSSIBLE •	1%	2%	2%
PROBABLE •	3%	2%	2%
CONFIRMED ●	94%	90%	91%
TOTAL	98%	94%	95%

BREEDING RELATIVE ABUNDANCE CHANGE *since 1988–91*

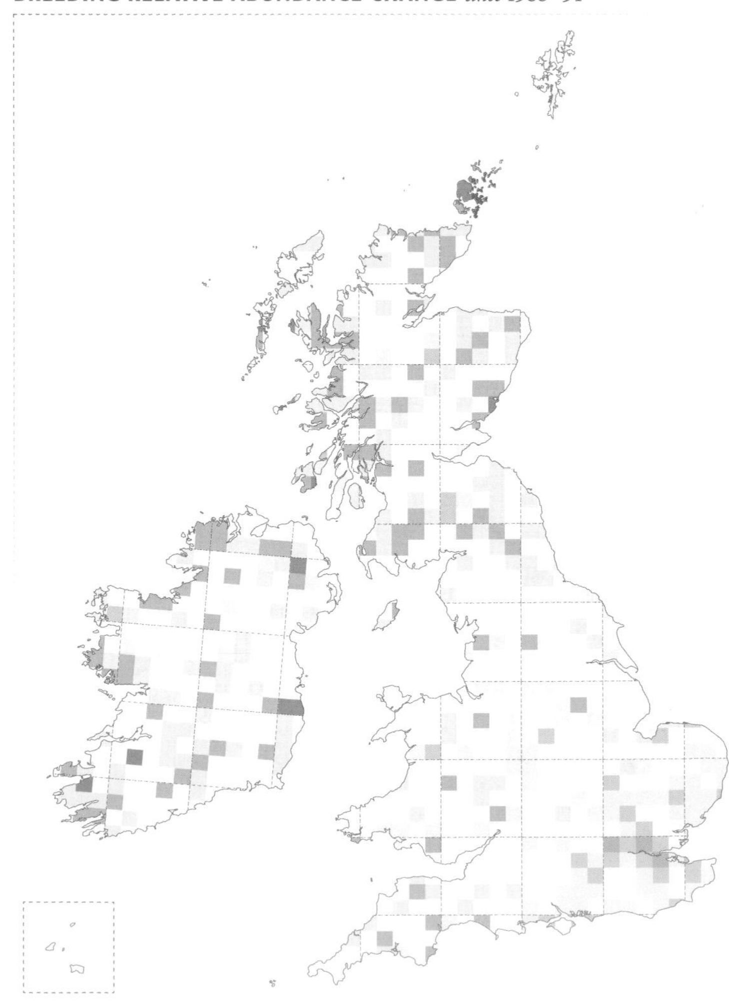

BREEDING DISTRIBUTION CHANGE *since 1968–72*

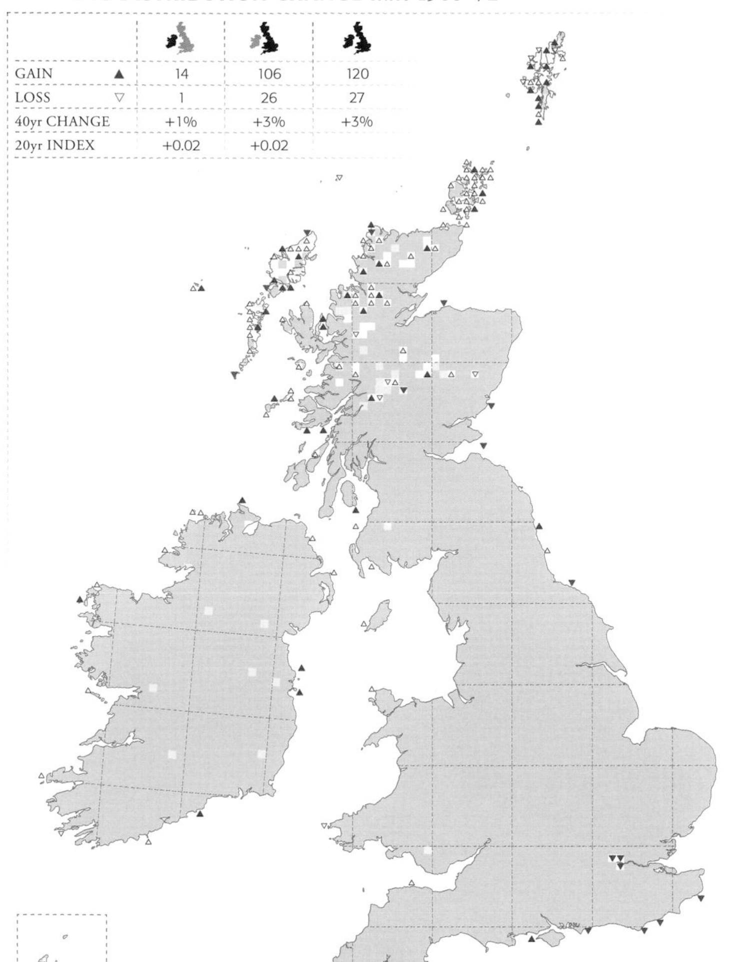

GAIN ▲	14	106	120
LOSS ▽	1	26	27
40yr CHANGE	+1%	+3%	+3%
20yr INDEX	+0.02	+0.02	

BREEDING RELATIVE ABUNDANCE *2008–11*

WINTER DISTRIBUTION *2007/08–2010/11*

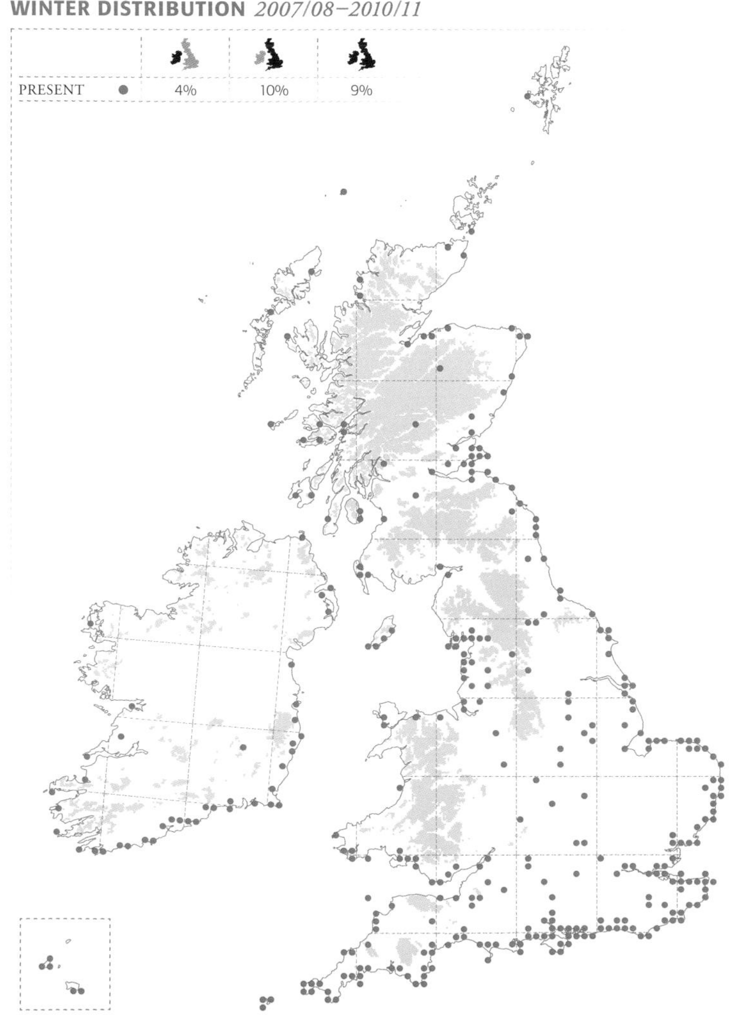

PRESENT ●	4%	10%	9%

WINTER DISTRIBUTION CHANGE *since 1981–84*

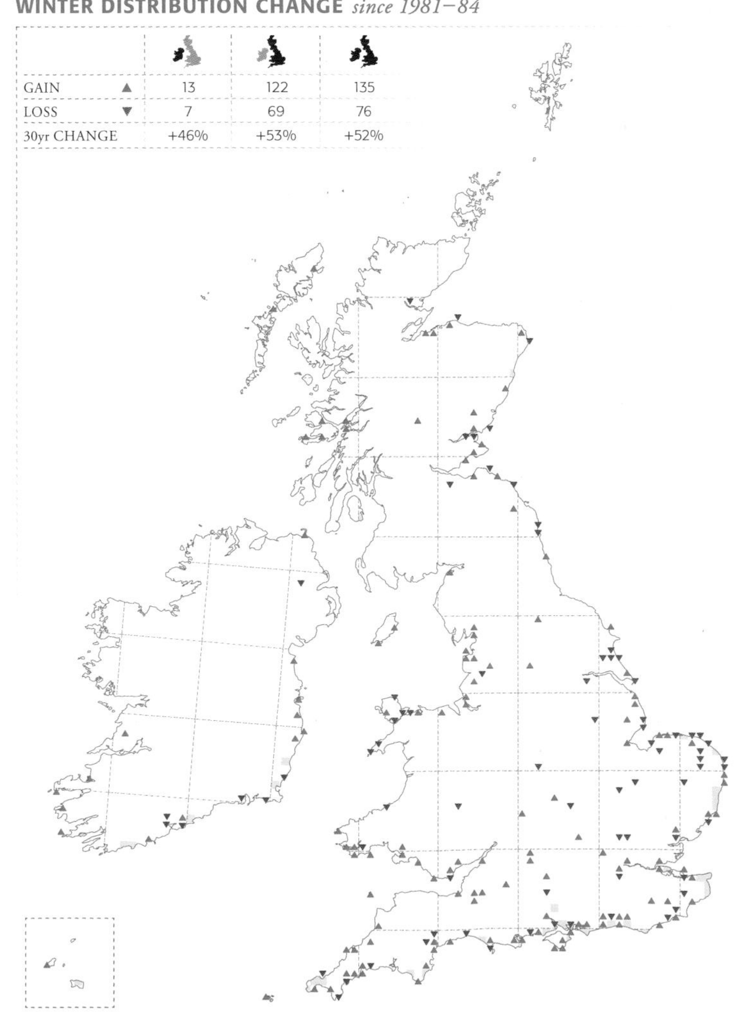

GAIN ▲	13	122	135
LOSS ▼	7	69	76
30yr CHANGE	+46%	+53%	+52%

HOUSE MARTIN
Delichon urbicum

BTO CODE: HM

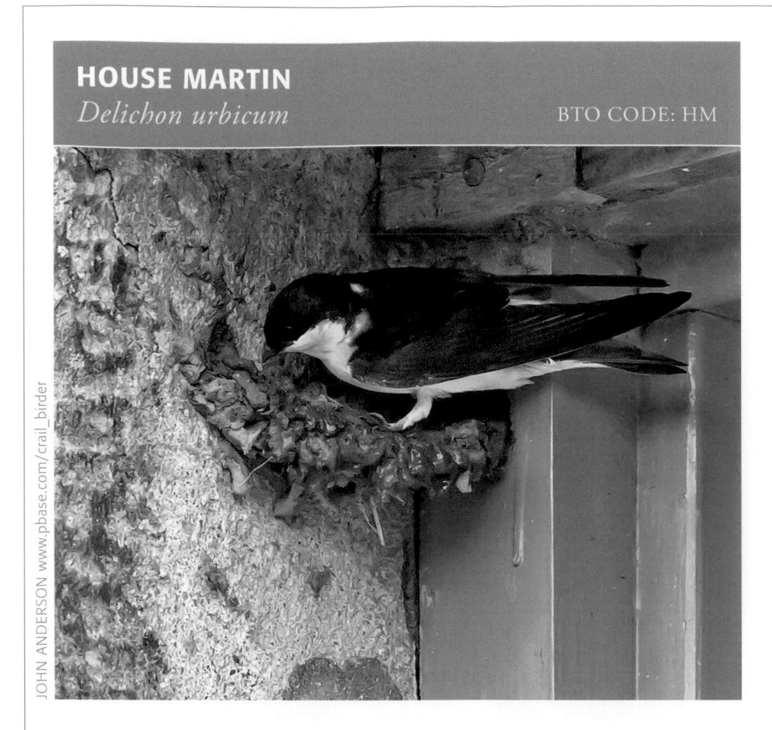

JOHN ANDERSON www.pbase.com/crail_birder

THE ONLY PLACES IN Britain & Ireland now with no, or few, breeding House Martins are the highest parts of the Scottish Highlands, Shetland, the Outer Hebrides and some exposed parts of the west coasts of Scotland and Ireland. This follows some range extension into all these areas during the last 20 years, although some of the recent gains are simply a recovery from apparent losses noted between the *1968–72 Breeding Atlas* and the *1988–91 Breeding Atlas*.

Despite this ubiquity it is clear from the relative abundance map that House Martins are much commoner in the eastern half of Ireland than in the rest of Britain & Ireland. They are least common on all high ground and in a swathe of southeast England from Kent and Sussex northwards to the Wash. Abundance is especially low in some urban areas, including London, Birmingham, Dublin and Belfast, despite the species' reliance on buildings for nesting. The abundance change map shows striking decreases in tetrad occupancy in southern Britain, contrasting with increases in Scotland and throughout Ireland.

These patterns correspond well with population trends. In the Republic of Ireland there was a 23% increase during 1998–2010 (*CBS Trend 2010*). Although the overall UK trend shows only a 2% decline during 1995–2010, it conceals a 14% decline in England and a 104% increase in Scotland (*BBS Report 2011*).

Across Europe, populations are in shallow decline (*PECBMS 2010*), and so these regional differences are particularly interesting. House Martins winter in the humid zone of West Africa and rainfall patterns there are correlated with annual survival rates in Britain (Robinson *et al.* 2008). Of ten species that winter in this humid zone, the House Martin is the only one not showing a strong UK decline (Ockendon *et al.* 2012), but clearly more work is needed to understand the regional differences, including whether regional populations experience different conditions in winter or on passage.

A scattering of House Martins were recorded during winter fieldwork but the majority were in November and, as in the *1981–84 Winter Atlas*, few if any birds genuinely overwintered.

Sponsored by Dr Ruth M. Jones

BREEDING DISTRIBUTION *2008–11*

POSSIBLE	5%	2%	3%
PROBABLE	4%	3%	3%
CONFIRMED	82%	83%	83%
TOTAL	91%	88%	89%

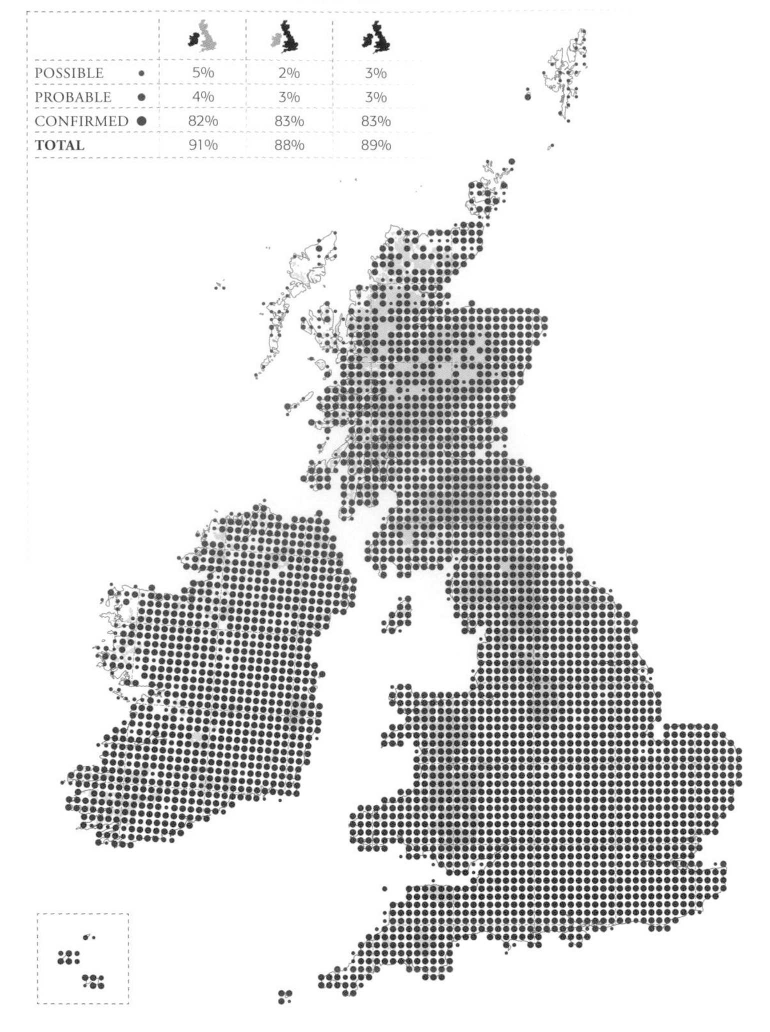

BREEDING RELATIVE ABUNDANCE CHANGE *since 1988–91*

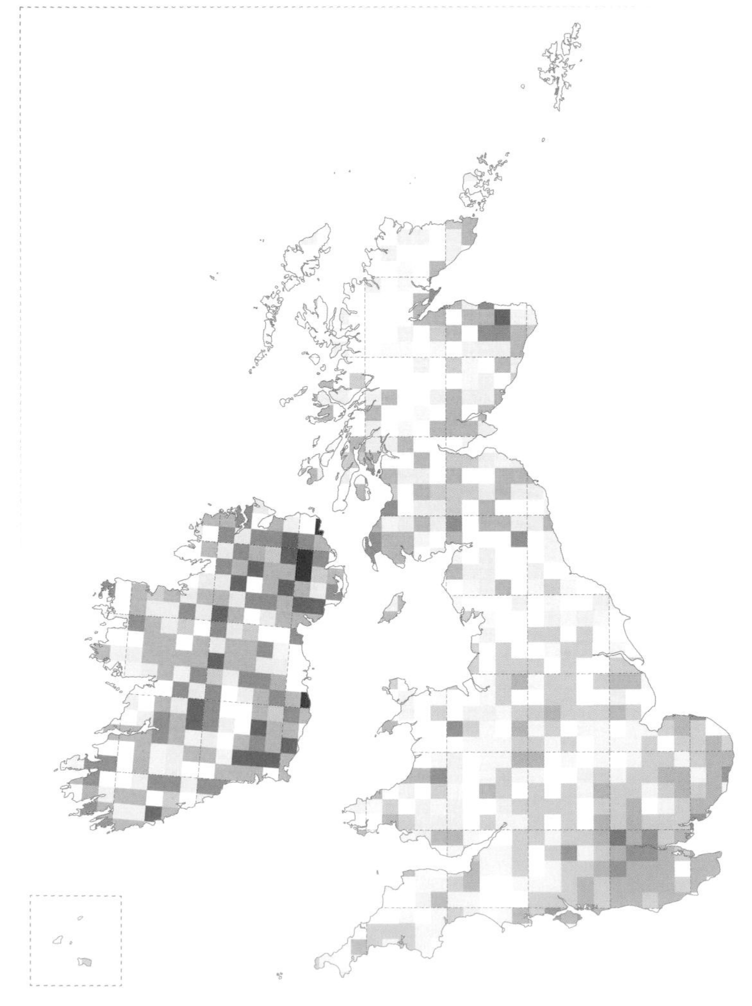

BREEDING DISTRIBUTION CHANGE *since 1968–72*

GAIN ▲	67	142	209
LOSS ▽	27	69	96
40yr CHANGE	+4%	+3%	+3%
20yr INDEX	+0.16	+0.03	

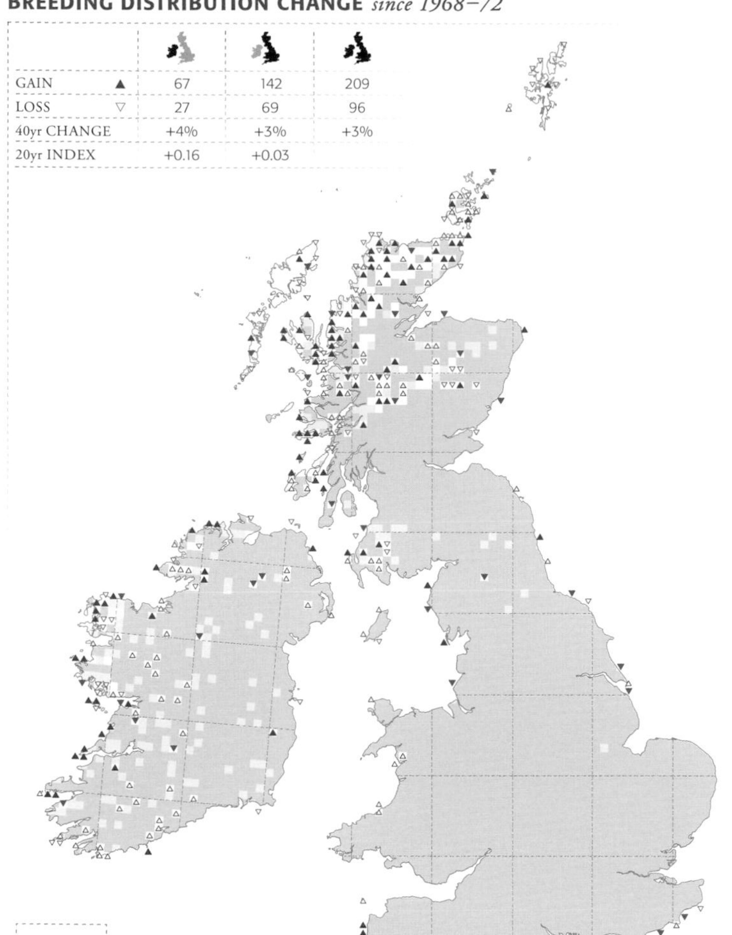

BREEDING RELATIVE ABUNDANCE *2008–11*

WINTER DISTRIBUTION *2007/08–2010/11*

PRESENT ●	<1%	3%	3%

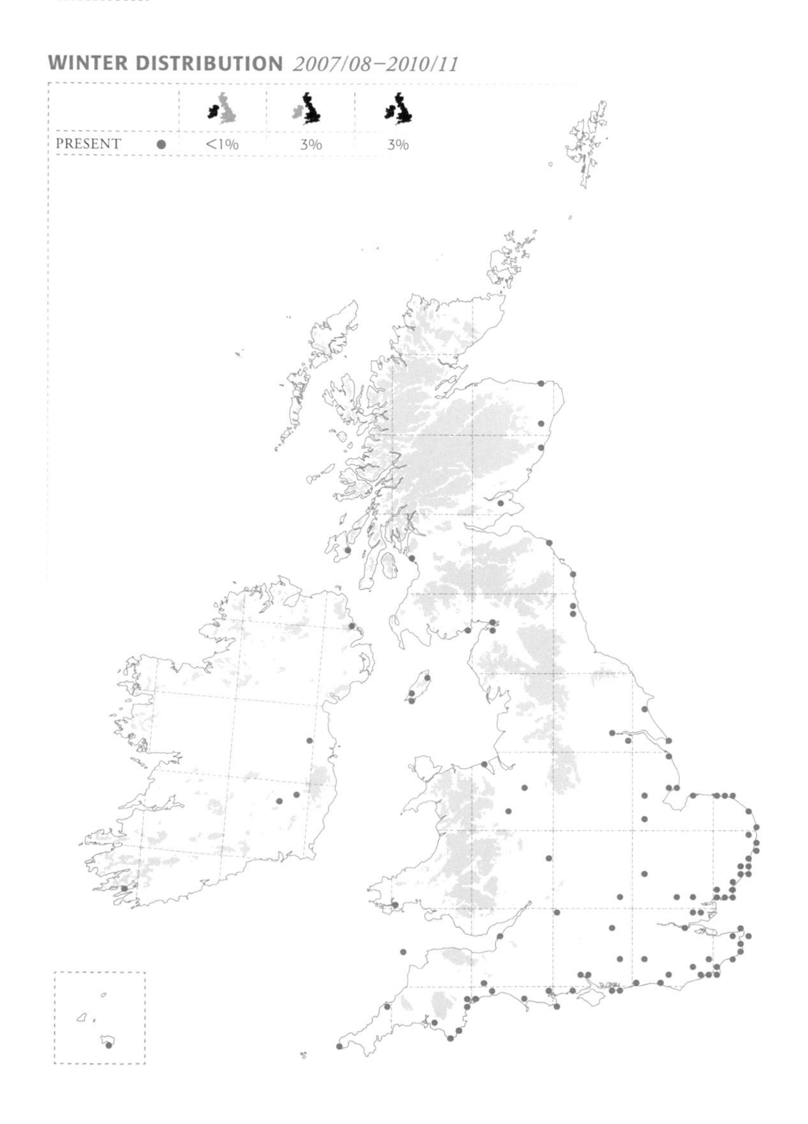

WINTER DISTRIBUTION CHANGE *since 1981–84*

GAIN ▲	2	29	31
LOSS ▼	1	86	87
30yr CHANGE	+100%	-57%	-55%

CETTI'S WARBLER

Cettia cetti

BTO CODE: CW

JIM ALMOND www.shropshirebirder.co.uk

THE REMARKABLE COLONISATION by the Cetti's Warbler has progressed strongly across England and Wales, while in Ireland and in Scotland the species remains a very rare vagrant.

The explosive song means that fieldworkers will have overlooked few breeding sites but, because birds not in song can be unobtrusive, the winter maps may be incomplete. Little difference is apparent, however, between the breeding and wintering distributions. Proving breeding is difficult, because many sites are impenetrable, but probably nearly all summer records refer to breeding birds. Cetti's Warblers now occupy lowland wetlands in a broad but discontinuous band around southern British coasts and are relatively common in certain river basins, such as the Thames and the Great Ouse. Other lowland areas of England and Wales are also occupied, but at lower density.

Range expansion since the first breeding record in Kent in 1973 (Holling *et al.* 2011b) has been documented thoroughly by atlases. In 1968–72, Cetti's Warblers were recorded in six 10-km squares, including a single instance of *confirmed* breeding. Occupancy had risen to 89 squares by the *1988–91 Breeding Atlas*, despite substantial losses in Kent and East Anglia due to severe weather in the mid 1980s. By 1996 there were just under 600 territories in Britain and the Channel Isles (Wotton *et al.* 1998). From a range centred on the English south coast in 1988–91, rapid expansion has since occurred along both west and east coasts, and into Wales and the Midlands. Numbers have risen exponentially, recently doubling in periods of four to seven years, and now total at least 1,900 territories (*RBBP 2010*). Across western Europe, the population has increased almost fourfold since 1990 (*PECBMS 2010*).

The colonisation of Britain was part of a major expansion northwards from Mediterranean France during the 20th century (Bonham & Robertson 1975). There is some evidence that, as the population has increased, density-dependent regulation during cold winters has begun to limit growth rates (Robinson *et al.* 2007a). The hard 2009/10 winter brought about the first drop in reported numbers for 10 years (*RBBP 2010*) but, if the impact of previous cold winters is any indication, Cetti's Warbler numbers will recover rapidly.

Sponsored by Chew Valley Ringing Station

BREEDING DISTRIBUTION *2008–11*

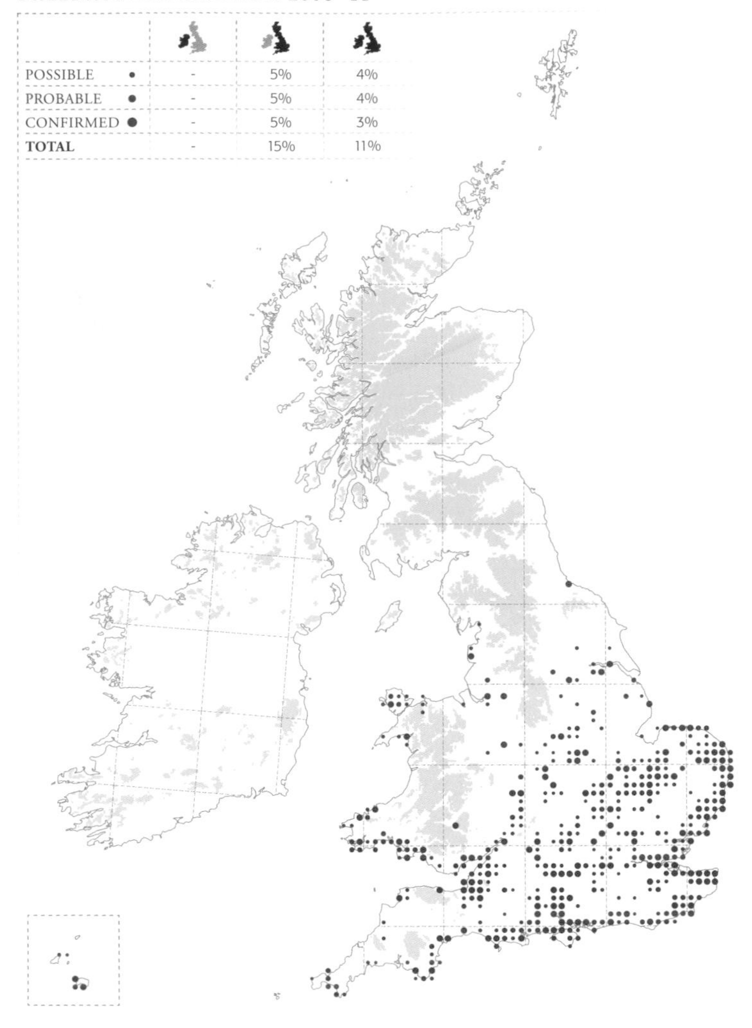

POSSIBLE	●	-	5%	4%
PROBABLE	●	-	5%	4%
CONFIRMED	●	-	5%	3%
TOTAL		-	15%	11%

WINTER DISTRIBUTION *2007/08–2010/11*

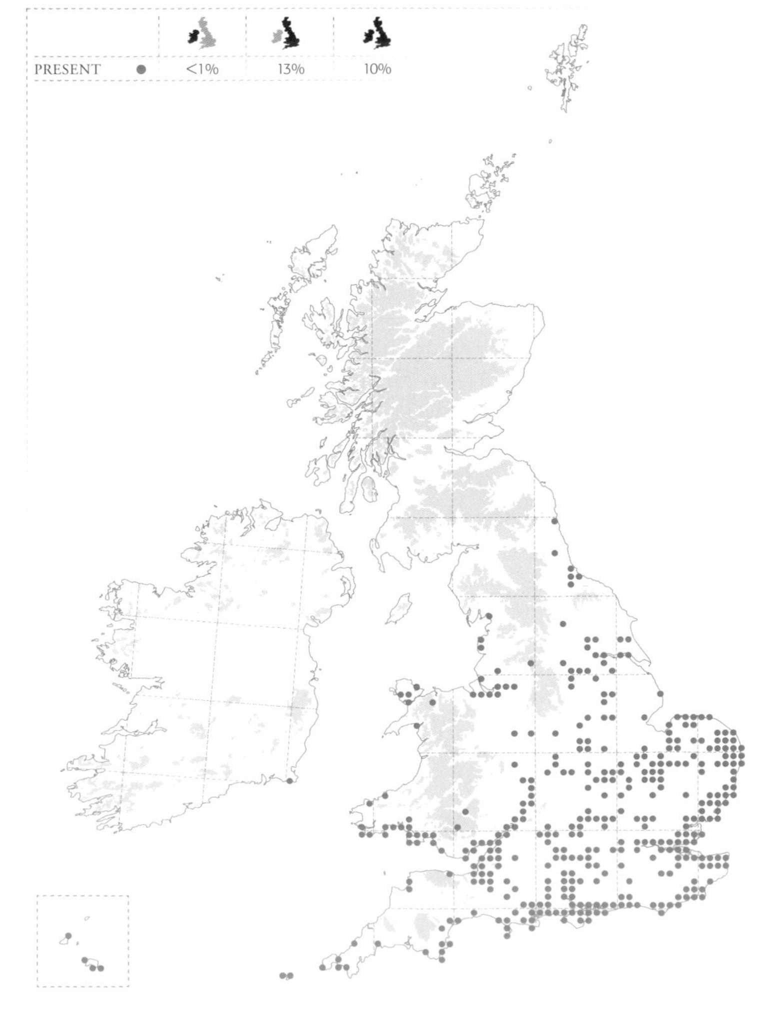

PRESENT	●	<1%	13%	10%

BREEDING DISTRIBUTION CHANGE *since 1968–72*

GAIN ▲	-	407	407
LOSS ▽	-	0	0
40yr CHANGE	-	+6783%	+6783%
20yr INDEX	-	+0.75	

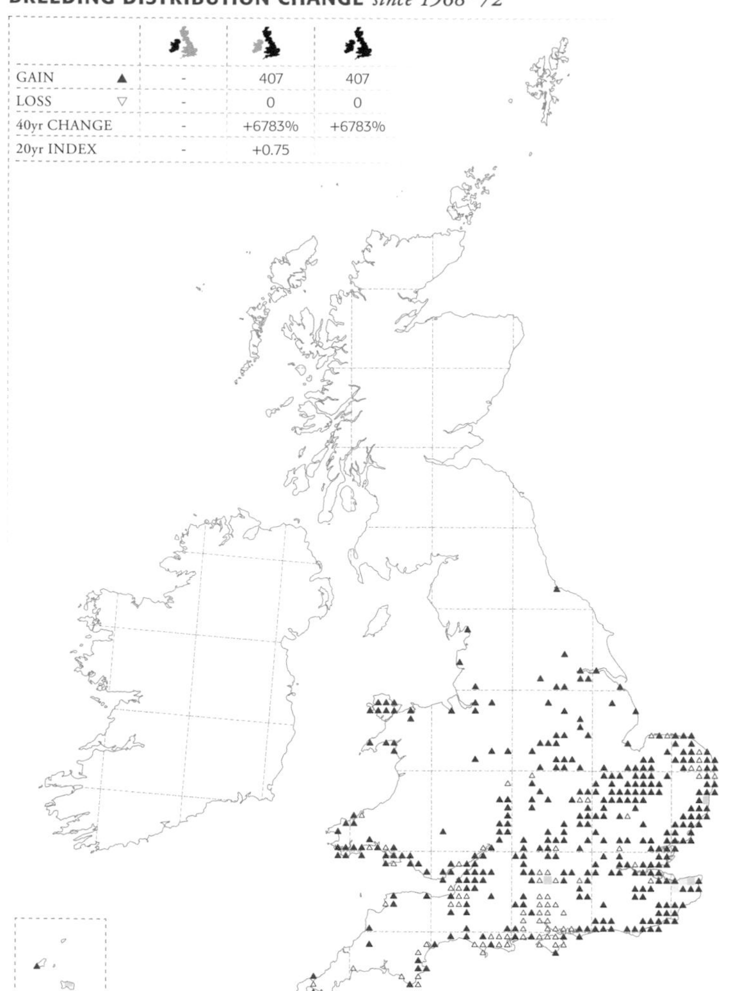

BREEDING RELATIVE ABUNDANCE *2008–11*

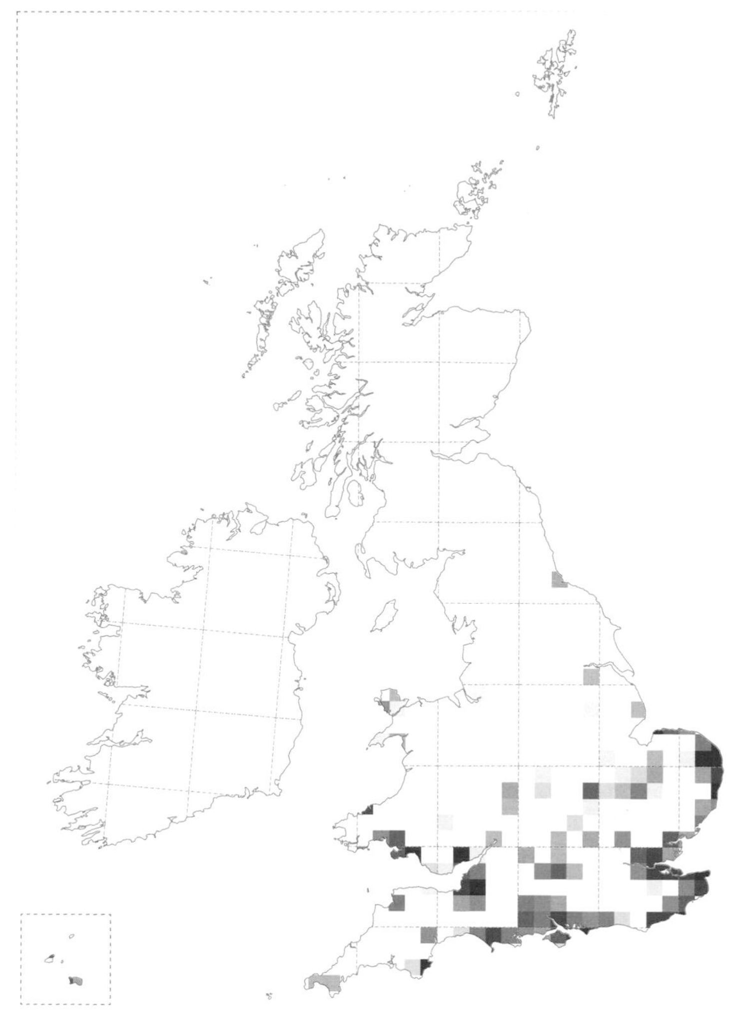

WINTER DISTRIBUTION CHANGE *since 1981–84*

GAIN ▲	1	281	282
LOSS ▼	0	4	4
30yr CHANGE	∞	+433%	+434%

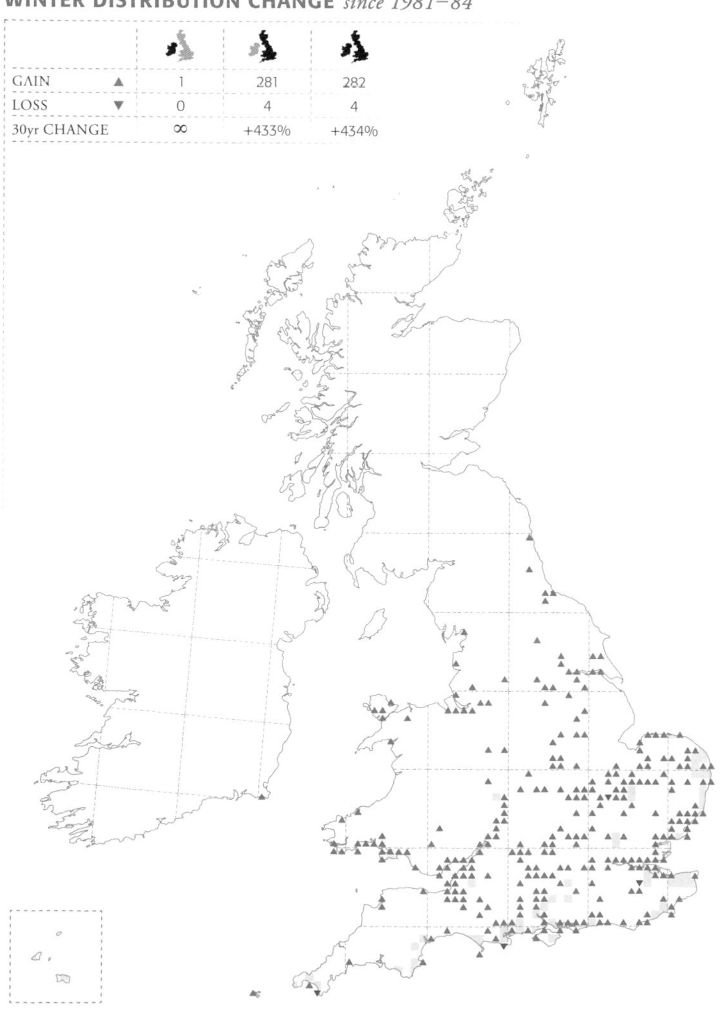

WINTER RELATIVE ABUNDANCE *2007/08–2010/11*

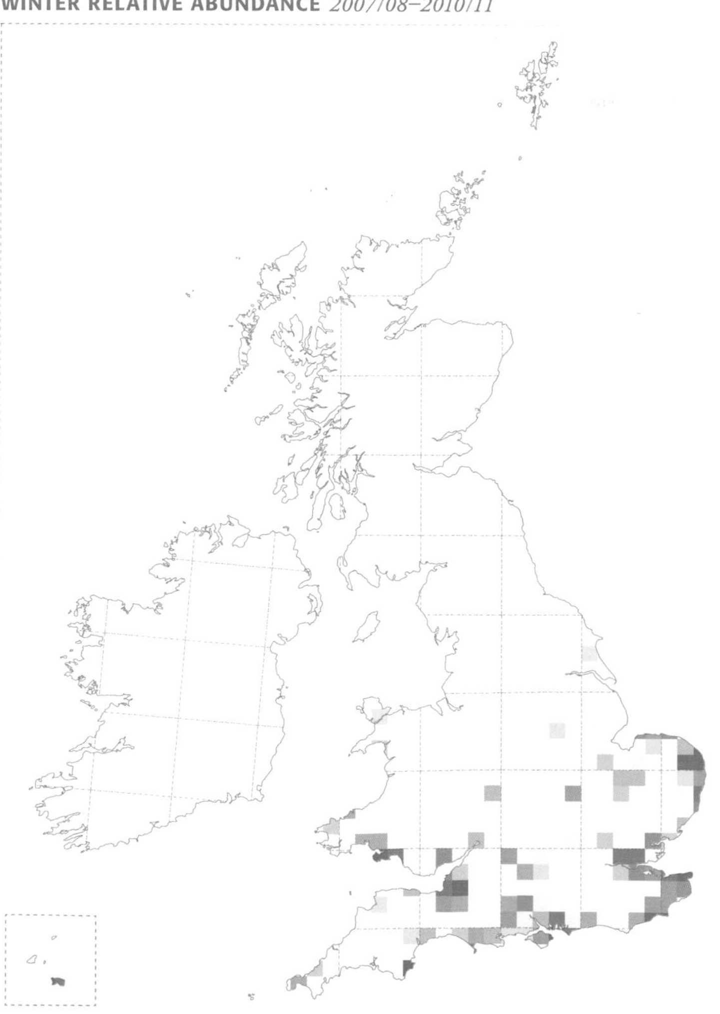

LONG-TAILED TIT
Aegithalos caudatus

BTO CODE: LT

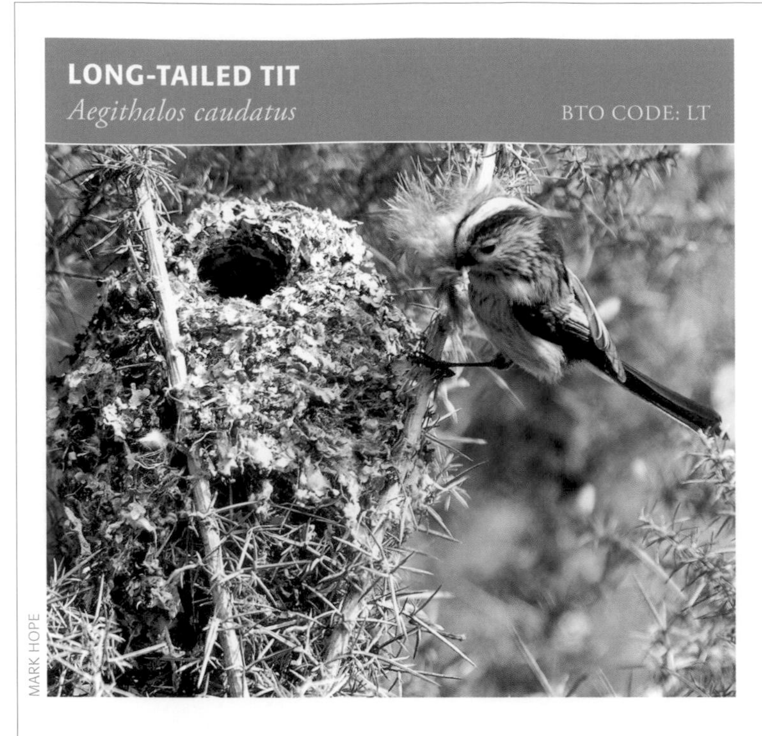

MARK HOPE

LONG-TAILED TITS are widespread in Britain & Ireland, occupying 83% of 10-km squares in winter and 78% in the breeding season. The main gaps in range occur on upland moors in mainland Scotland, on the Outer Hebrides and Northern Isles, and in central and western Ireland.

Their generally sedentary nature within Britain & Ireland is evident from ring recoveries and from detailed population studies (*Migration Atlas*). It is also reflected in the similar patterns shown by the winter and breeding abundance maps. Also evident from these maps is the preference for deciduous woodland, scrub and hedgerow habitats (*BWP*), with Long-tailed Tits occurring at higher densities in the more wooded lowland areas of Britain and Ireland.

The Long-tailed Tit is highly susceptible to cold weather, with numbers fluctuating widely around severe winters. Kennedy *et al.* (1954) considered that the disappearance of Long-tailed Tits from parts of Counties Tipperary, Galway, Mayo, Antrim and Down was a result of the severe weather in January 1945. In Britain, the impacts of cold winters and the subsequent population recoveries have been well documented (Marchant *et al.* 1990; *BirdTrends*). The 22% winter range expansion since the *1981–84 Winter Atlas* could reflect an increase in numbers, with birds spreading farther into upland marginal areas following a series of mild winters, though coverage of such areas also improved.

Long-term monitoring shows progressive increases in Long-tailed Tit numbers since the early 1980s, with a 97% increase in numbers in the UK during the period 1985–2010 (*BirdTrends*). This increase is reflected to a limited degree in the change maps, which show a 3% range expansion since 1968–72; gains are evident mainly in the uplands of Scotland and northern England, in coastal areas of eastern England and in western Ireland. The breeding abundance change map shows increases since the *1988–91 Breeding Atlas* in the proportion of occupied tetrads, particularly in England and Ireland, matching those in population size. Aside from the possible impact of mild winters on population trends, there is evidence that laying dates have become progressively earlier (Crick & Sparks 1999; *BirdTrends*) and for an increase in fledglings produced per nesting attempt.

Sponsored by BTO Garden BirdWatch

BREEDING DISTRIBUTION *2008–11*

POSSIBLE	14%	3%	6%
PROBABLE	11%	4%	6%
CONFIRMED	50%	71%	66%
TOTAL	75%	78%	78%

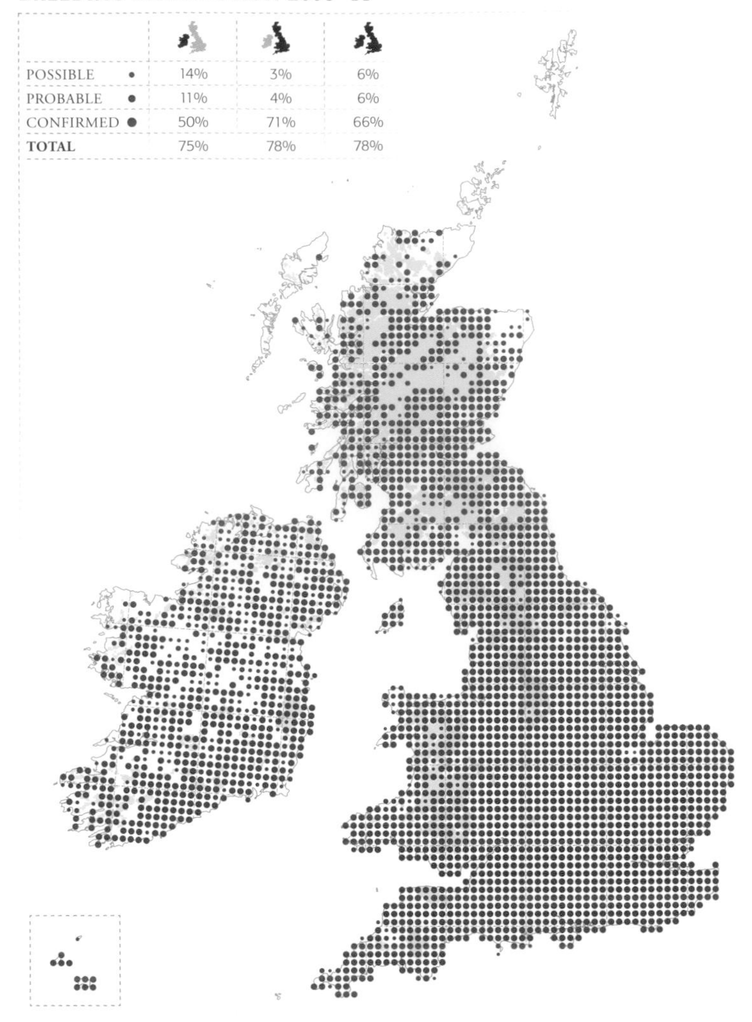

WINTER DISTRIBUTION *2007/08–2010/11*

PRESENT	82%	84%	83%

BREEDING DISTRIBUTION CHANGE *since 1968–72*

GAIN ▲	127	164	291
LOSS ▽	89	120	209
40yr CHANGE	+5%	+2%	+3%
20yr INDEX	+0.25	+0.10	

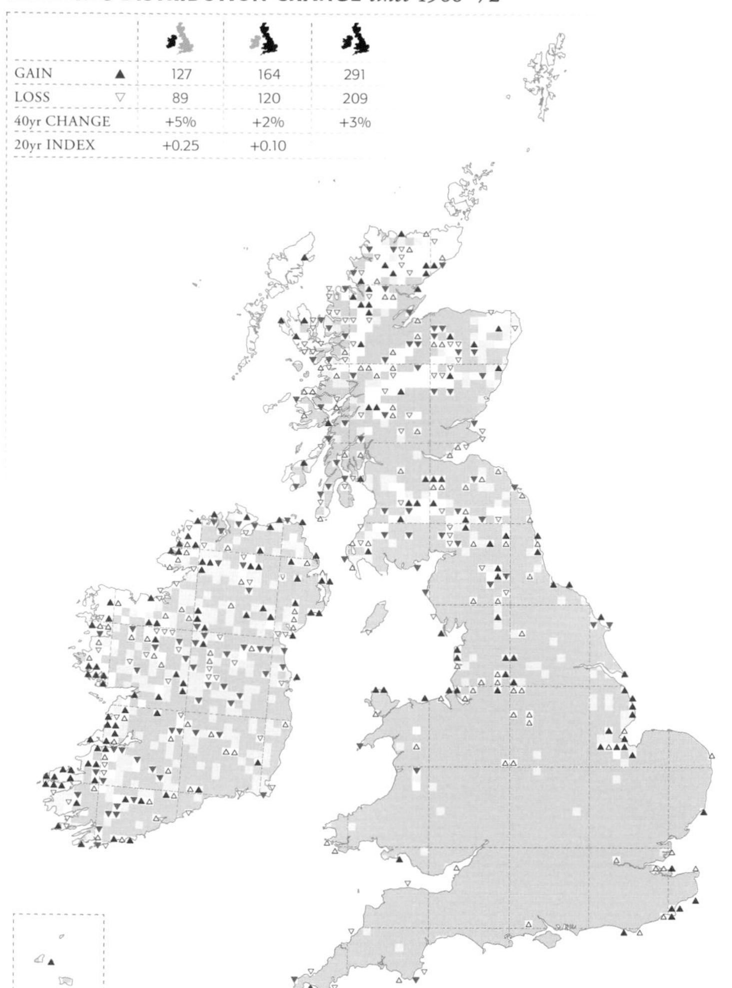

WINTER DISTRIBUTION CHANGE *since 1981–84*

GAIN ▲	301	342	643
LOSS ▼	29	51	80
30yr CHANGE	+52%	+14%	+22%

BREEDING RELATIVE ABUNDANCE *2008–11*

BREEDING RELATIVE ABUNDANCE CHANGE *since 1988–91*

WINTER RELATIVE ABUNDANCE *2007/08–2010/11*

WOOD WARBLER

Phylloscopus sibilatrix

BTO CODE: WO

MARK RICHARDS www.marksart.net

THE WOOD WARBLER favours mature closed-canopy woodland, particularly Sessile Oak and Silver Birch, and occurs mainly in the western half of Britain. The highest densities are associated with wooded uplands, mainly in southwest England, Wales and the Marches, the Peak District, the northern Pennines and Cumbria. In Scotland the main strongholds are in Galloway and in the western Highlands from Argyll to Inverness-shire. Elsewhere, a few small isolated populations remain in suitable habitat in southern and eastern England and eastern and northern Scotland. In Ireland, Wood Warblers are very scarce and irregular breeders (*IRBBP 2010*), now found mainly in the northern half of the island.

Between 1968–72 and 1988–91 there were marked losses in southern and central England, although the overall range size remained almost unchanged. Subsequently there has been a 37% range contraction since the *1988–91 Breeding Atlas*. The losses are particularly evident on lower ground in the south and east, and around the margins of the core areas. The abundance change map shows that tetrad occupancy rates in 20-km squares have decreased throughout the entire range since 1988–91, and this corresponds with a 65% reduction in numbers recorded in the UK during 1995–2010 (*BBS Report 2011*). Longer-term trends are difficult to quantify because the westerly distribution of the species was a poor match to the historical distribution of monitoring plots. Considering data from 1967–99, however, Hewson and Noble (2009) demonstrated that Wood Warblers did not begin to decline significantly until c.1989. There have been widespread declines across Europe since 1980 (*PECBMS 2010*).

Ockendon *et al.* (2012) included Wood Warbler among a range of species wintering in the humid zone of tropical West Africa that are in decline, and suggested that these trends could be driven by changes in climate or land use in the wintering area. Meanwhile, Mallord *et al.* (2012a) found no evidence of increased nest predation having driven the declines in Wales. Mallord *et al.* (2012b) found that territories were associated with a number of structural habitat variables, which could be related to the past management of the study woods, and suggested that a moderate grazing regime would help restore habitat quality.

DJEnvironmental

Sponsored by DJEnvironmental (Tim Davis & Tim Jones)

BREEDING DISTRIBUTION *2008–11*

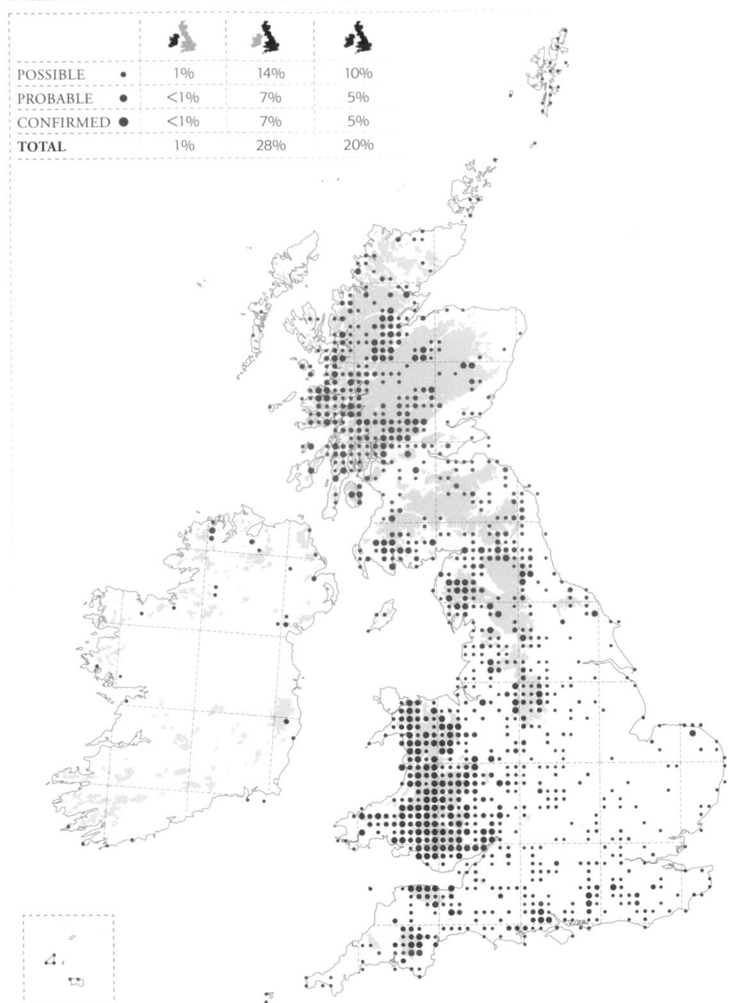

POSSIBLE •	1%	14%	10%
PROBABLE •	<1%	7%	5%
CONFIRMED •	<1%	7%	5%
TOTAL	1%	28%	20%

BREEDING RELATIVE ABUNDANCE CHANGE *since 1988–91*

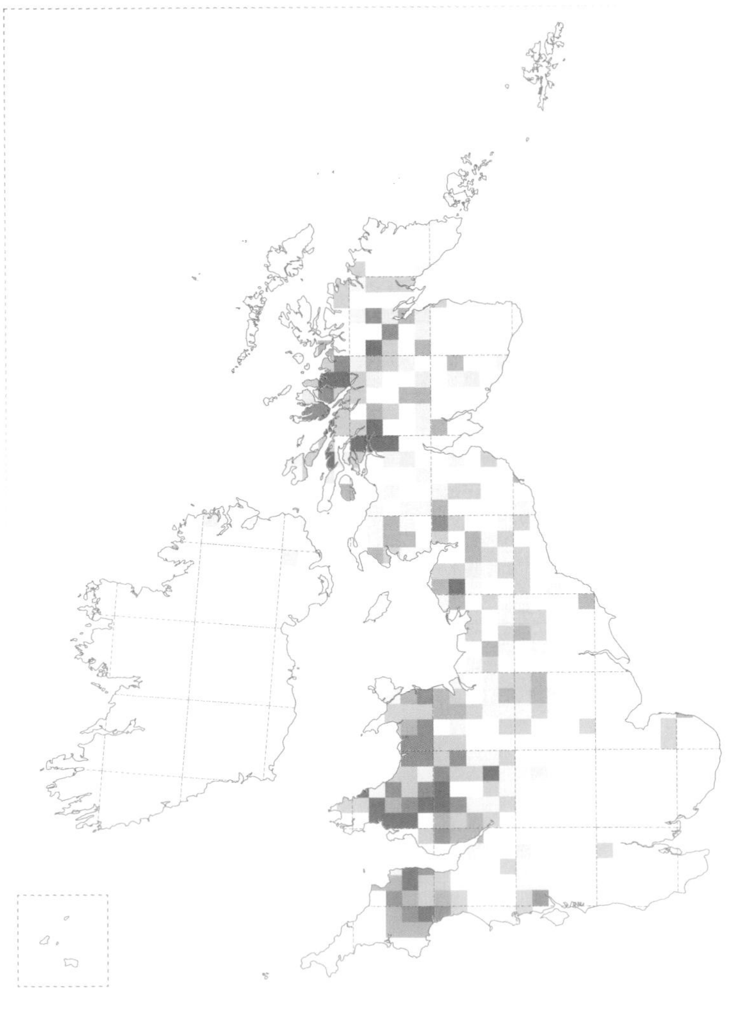

BREEDING DISTRIBUTION CHANGE *since 1968–72*

	Ireland	Britain	Britain & Ireland
GAIN ▲	17	178	195
LOSS ▽	6	607	613
40yr CHANGE	+138%	-35%	-34%
20yr INDEX	-	-0.51	

BREEDING RELATIVE ABUNDANCE *2008–11*

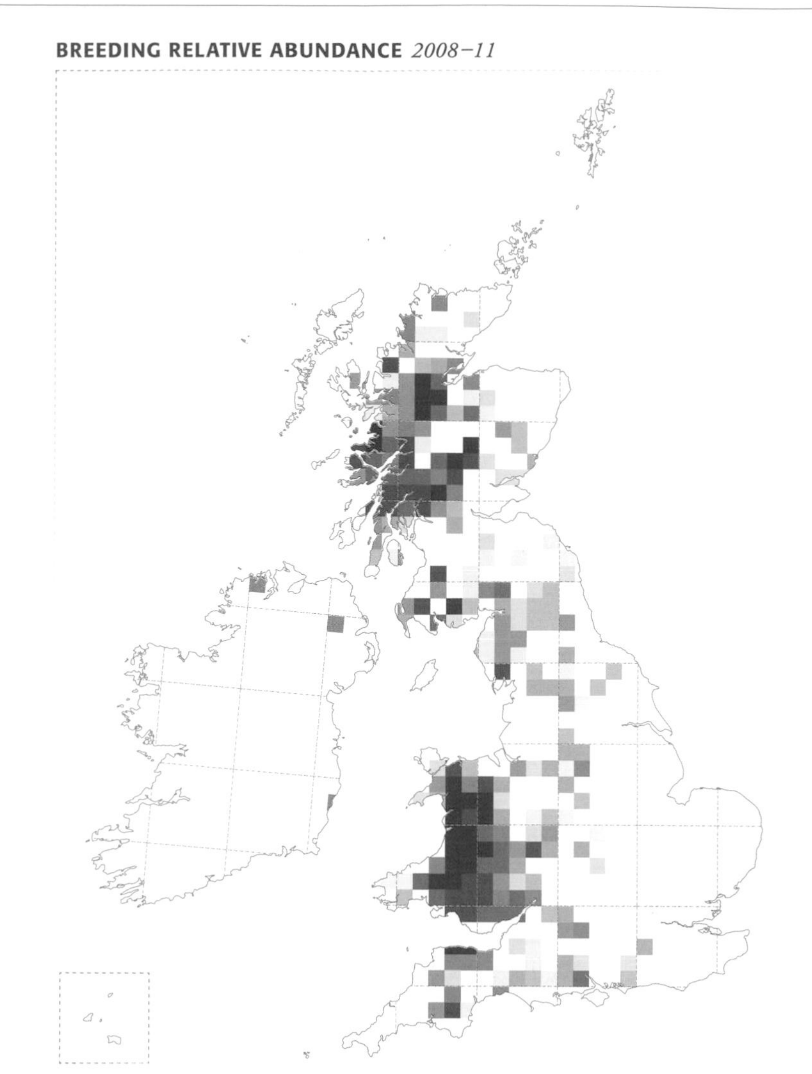

JIM ALMOND www.shropshirebirder.co.uk

◀ **Ireland's position to the west of the African–Eurasian migratory flyway may limit opportunities for the expansion there of migrants such as the Wood Warbler, Redstart and Pied Flycatcher from western Britain.**

CHIFFCHAFF

Phylloscopus collybita

BTO CODE: CC

HARRY APPLEYARD

THE MAJORITY OF the Chiffchaffs that breed in Britain & Ireland winter in the Mediterranean Basin and North Africa. Local birds that remain over winter are joined by nominate-race and *abietinus* individuals from elsewhere in Europe and by small but regular numbers of *tristis* from Asia (*Migration Atlas*; Conway 2011).

In winter, the Chiffchaff is widely distributed throughout the southern half of Britain, and less widely in Ireland, but largely absent from the uplands; winter records farther north in Britain are relatively sparse and occur in coastal or lowland locations. The areas of greatest abundance in Britain and in Ireland are strongly associated with the coast. Birds inland tend to occur by water, at sites where the habitat is rich in invertebrates. Large aggregations may occur at wetlands and waste-water treatment works, some of which regularly hold over 25 wintering individuals (occasionally up to 100: e.g. Penhallurick 1978; Conway 2011). The change in winter distribution since the *1981–84 Winter Atlas* has been substantial, with increases of 85% in Britain and 68% in Ireland in the numbers of occupied 10-km squares. These gains have resulted in a substantial consolidation of range inland and a considerable northward expansion, with wintering now regular in counties north to Shetland and in higher-altitude inland sites.

Breeding Chiffchaffs are widespread throughout Britain & Ireland, being localised on Scottish islands and absent only from upland areas. Breeding abundance is high throughout much of England, Wales and the southern half of Ireland but generally decreases northwards across Ireland, northern England and Scotland. The overall distribution in Ireland has remained relatively unchanged over the last 40 years. In Britain, however, the breeding distribution has expanded by 20% since the *1968–72 Breeding Atlas*, with most of the gains being in northern Britain. This reflects a northward expansion of breeding range and a greater use of high-altitude sites. This pattern is supported by monitoring schemes, which indicate increases of 69% in England and 336% in Scotland between 1995 and 2010 (*BBS Report 2011*) and an increase of 11% in the Republic of Ireland between 1998 and 2010 (*CBS Trend 2010*). Correspondingly, tetrad occupancy rates have increased in many 20-km squares.

Sponsored by Anne Mansell

BREEDING DISTRIBUTION *2008–11*

	Ireland	Britain	Britain & Ireland
POSSIBLE ●	19%	14%	15%
PROBABLE ●	33%	15%	20%
CONFIRMED ●	36%	55%	50%
TOTAL	88%	84%	85%

WINTER DISTRIBUTION *2007/08–2010/11*

	Ireland	Britain	Britain & Ireland
PRESENT ●	17%	42%	35%

BREEDING DISTRIBUTION CHANGE *since 1968–72*

GAIN ▲	32	439	471
LOSS ▽	39	47	86
40yr CHANGE	-1%	+20%	+13%
20yr INDEX	+0.07	+0.14	

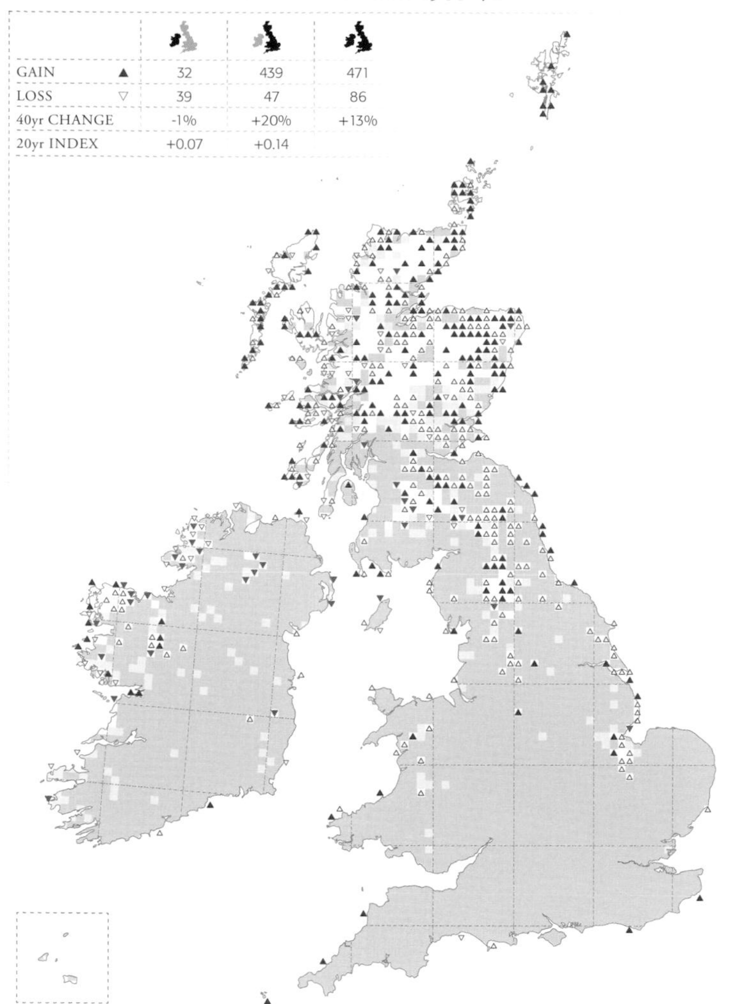

WINTER DISTRIBUTION CHANGE *since 1981–84*

GAIN ▲	104	609	713
LOSS ▼	44	119	163
30yr CHANGE	+68%	+85%	+82%

BREEDING RELATIVE ABUNDANCE *2008–11*

BREEDING RELATIVE ABUNDANCE CHANGE *since 1988–91*

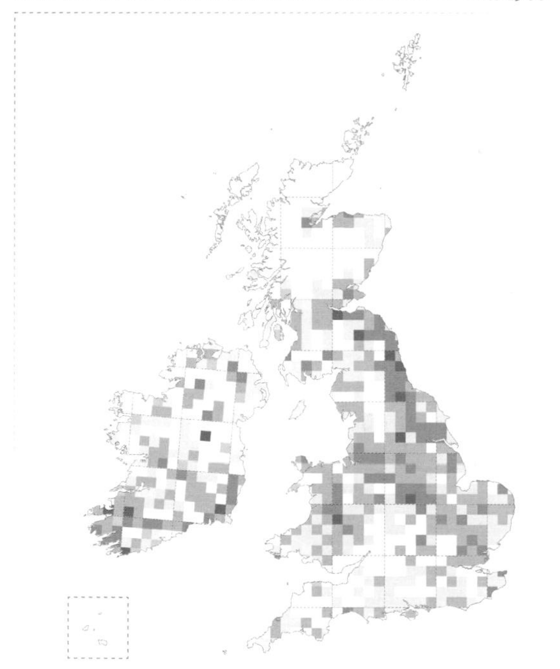

WINTER RELATIVE ABUNDANCE *2007/08–2010/11*

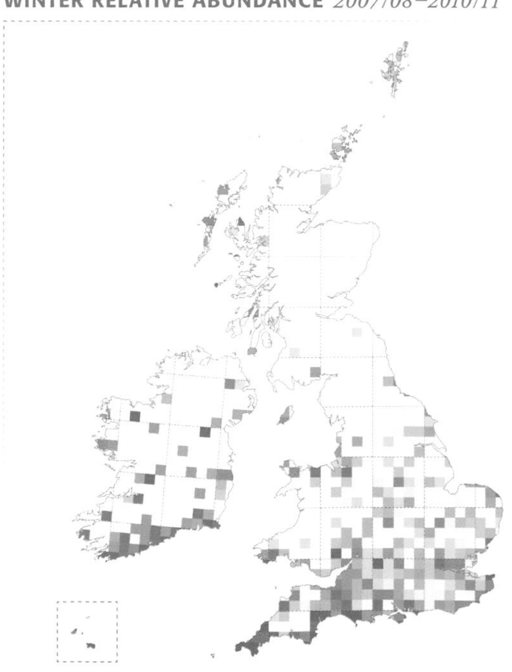

WILLOW WARBLER

Phylloscopus trochilus

BTO CODE: WW

DAVID HUTCHINSON

DECLINES SINCE THE *1988–91 Breeding Atlas* in the proportion of tetrads occupied by Willow Warblers have been widespread across much of England, contrasting with increases in most of Scotland and Ireland. This striking pattern, shown by the abundance change map, is shared with several other long-distance migrants, including the House Martin and Cuckoo.

Despite this pattern of loss at tetrad resolution, the Willow Warbler still has a very wide breeding distribution, with gaps being restricted to a few 10-km squares around the Fens in eastern England and to the Outer Hebrides and Northern Isles. Densities are low in much of southern, central and eastern England, and at higher altitudes in the Scottish Highlands. They are greatest in the northern half of Ireland, Wales, much of northern England and southern Scotland, together with the foothills and glens of the Scottish Highlands.

There has been an increase in range of just 3% in Britain & Ireland since the *1968–72 Breeding Atlas*. These gains are mostly in coastal squares in the west of Ireland or in the Outer Hebrides and Northern Isles, where the creation of new plantations and increasing tree cover around houses in exposed areas has probably facilitated expansion (*Birds of Scotland*).

Monitoring data show a 28% decline in numbers in England during 1995–2010, contrasting with a 33% increase in Scotland (*BBS Report 2011*), whereas in the Republic of Ireland there has been a 64% increase in numbers during 1998–2010 (*CBS Trend 2010*). Between the late 1980s and early 1990s, there was a decline in adult survival in southern Britain, while in northern Britain there was no evidence that survival had changed, suggesting that increased mortality had probably been a major cause of the observed population decline (Peach *et al.* 1995a). Changing climatic conditions in Britain & Ireland, the quality of breeding habitat, and environmental changes in Africa are considered to be driving the changes in abundance at the regional scale (Morrison *et al.* 2010, 2013).

In winter, Willow Warblers were recorded in 31 10-km squares, mostly in England. All except six of these were occupied only in November and presumably therefore by late migrants.

In memory of Dr Elizabeth Ryan

BREEDING DISTRIBUTION *2008–11*

POSSIBLE •	4%	6%	6%
PROBABLE •	12%	13%	13%
CONFIRMED •	79%	73%	75%
TOTAL	95%	92%	94%

BREEDING RELATIVE ABUNDANCE CHANGE *since 1988–91*

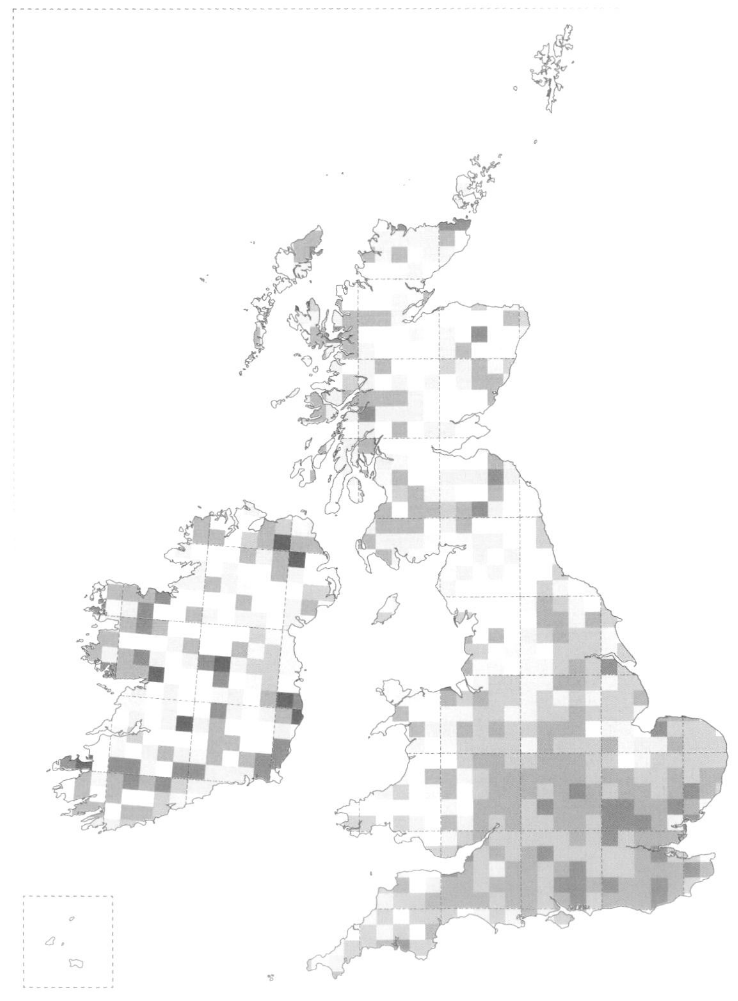

BREEDING DISTRIBUTION CHANGE *since 1968–72*

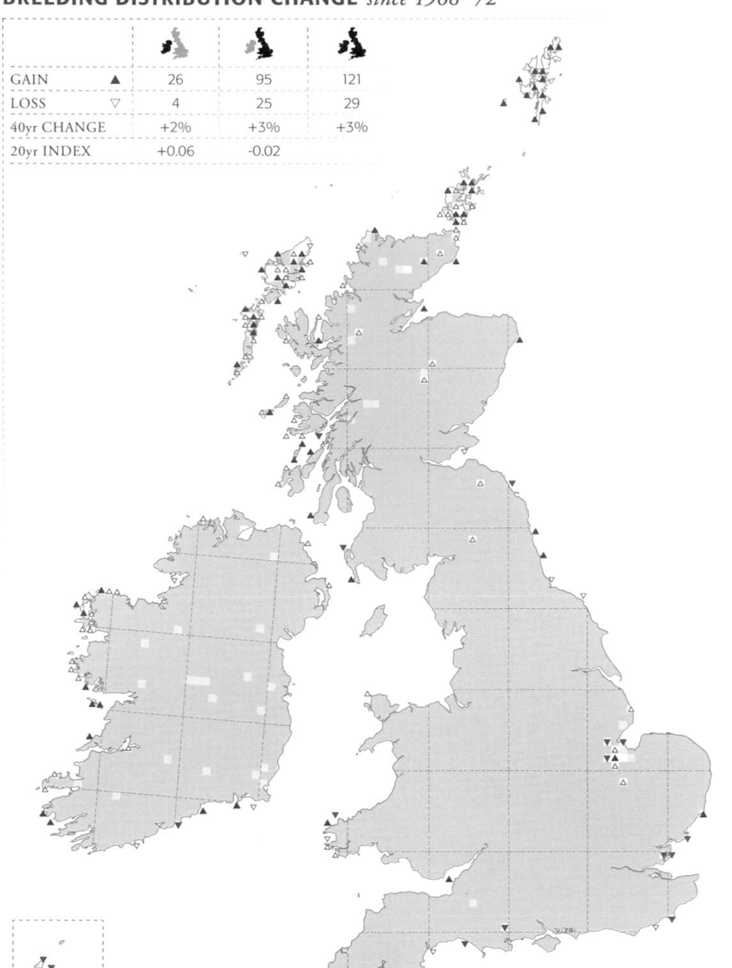

GAIN ▲	26	95	121
LOSS ▽	4	25	29
40yr CHANGE	+2%	+3%	+3%
20yr INDEX	+0.06	-0.02	

BREEDING RELATIVE ABUNDANCE *2008–11*

CLARE WARD

◀ **During the breeding season Willow Warblers were recorded in 92% of 10-km squares in Britain and 95% of Irish squares making them the most widespread of migrant warblers.**

BLACKCAP

Sylvia atricapilla

BTO CODE: BC

JOE O'HANLON

THOUGH THE BLACKCAP is traditionally regarded as a summer visitor, there has been a striking increase since the 1990s in the number of birds wintering, with numbers peaking in gardens in late January or February (Garden Bird Feeding Survey 2013). Accordingly there has been a 77% increase in 10-km square occupancy since the *1981–84 Winter Atlas*. Densities appear to be highest in the milder south and west of Britain & Ireland. This increase has been attributed, at least in part, to milder winters leading to a change in migration patterns. A proportion of Blackcaps from breeding populations in central Europe have adapted their migratory strategy and now winter in gardens in Britain & Ireland (Berthold & Terrill 1988; Bearhop *et al.* 2005). This strategy is thought to increase these birds' winter survival and allow them to return earlier to their central European breeding territories.

During the breeding season, the Blackcap is now widely distributed throughout Britain and Ireland. There have been extensive gains in the breeding range since the *1968–72 Breeding Atlas*; the range has expanded in Ireland by a staggering 249%, and in Britain by 25%, with most of the British gains being in Scotland. The breeding season relative abundance map shows that the highest densities occur in lowland areas in southern Britain and in pockets throughout Ireland.

The breeding abundance change map reveals striking increases in the proportions of occupied tetrads throughout Ireland, and less extensively in Britain, since the *1988–91 Breeding Atlas*. Long-term monitoring shows a similar pattern, with a 439% increase in numbers in the Republic of Ireland during 1998–2010 and a 102% increase in the UK between 1995 and 2010 (*CBS Trend 2010*; *BBS Report 2011*). Separate trends in numbers for Scotland and Northern Ireland reveal even greater percentage increases since 1995 (*BBS Report 2011*).

The Blackcap is a relatively adaptable species that has colonised urban and suburban areas to a far greater degree than most other migrants. Although it prefers areas with dense low vegetation, and is consequently sensitive to factors such as deer browsing (Holt *et al.* 2013), its populations in woodland are also thriving (Hewson *et al.* 2007).

Sponsored for Griselda & B. Harvey

BREEDING DISTRIBUTION *2008–11*

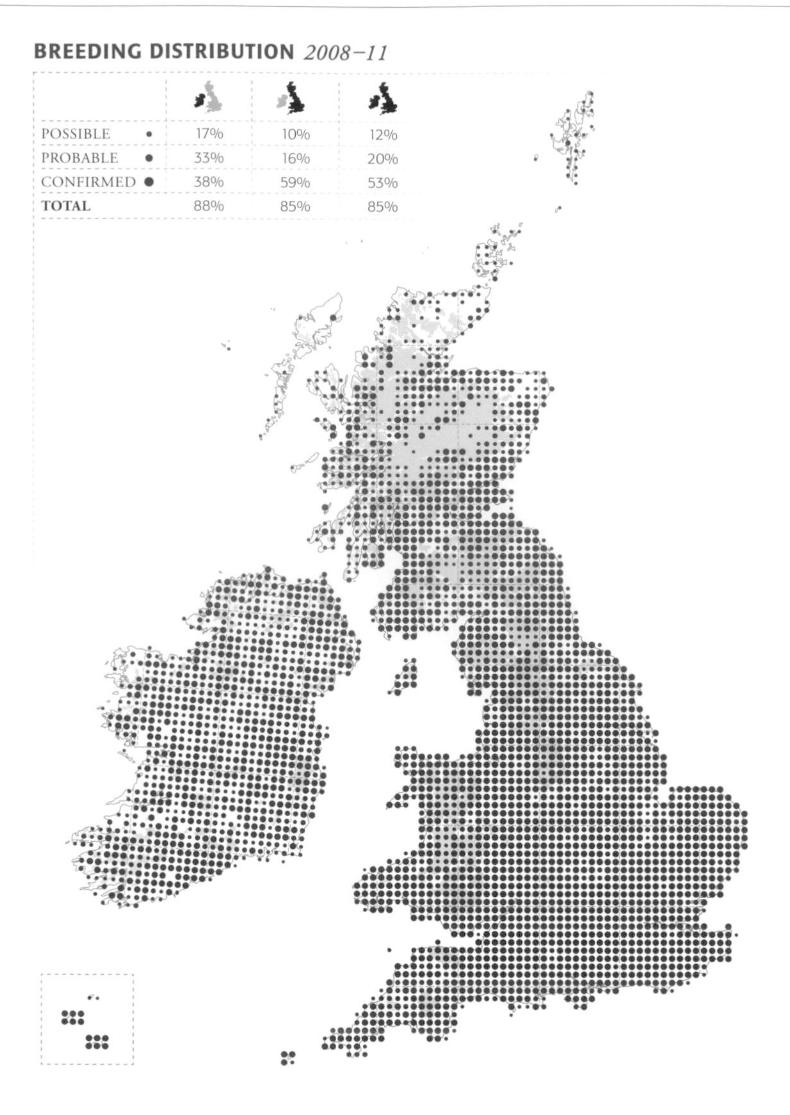

	Ireland	Britain	Britain & Ireland
POSSIBLE •	17%	10%	12%
PROBABLE ●	33%	16%	20%
CONFIRMED ●	38%	59%	53%
TOTAL	88%	85%	85%

WINTER DISTRIBUTION *2007/08–2010/11*

	Ireland	Britain	Britain & Ireland
PRESENT ●	39%	51%	48%

BREEDING DISTRIBUTION CHANGE *since 1968–72*

GAIN ▲	638	493	1131
LOSS ▽	1	12	13
40yr CHANGE	+249%	+25%	+51%
20yr INDEX	+0.66	+0.19	

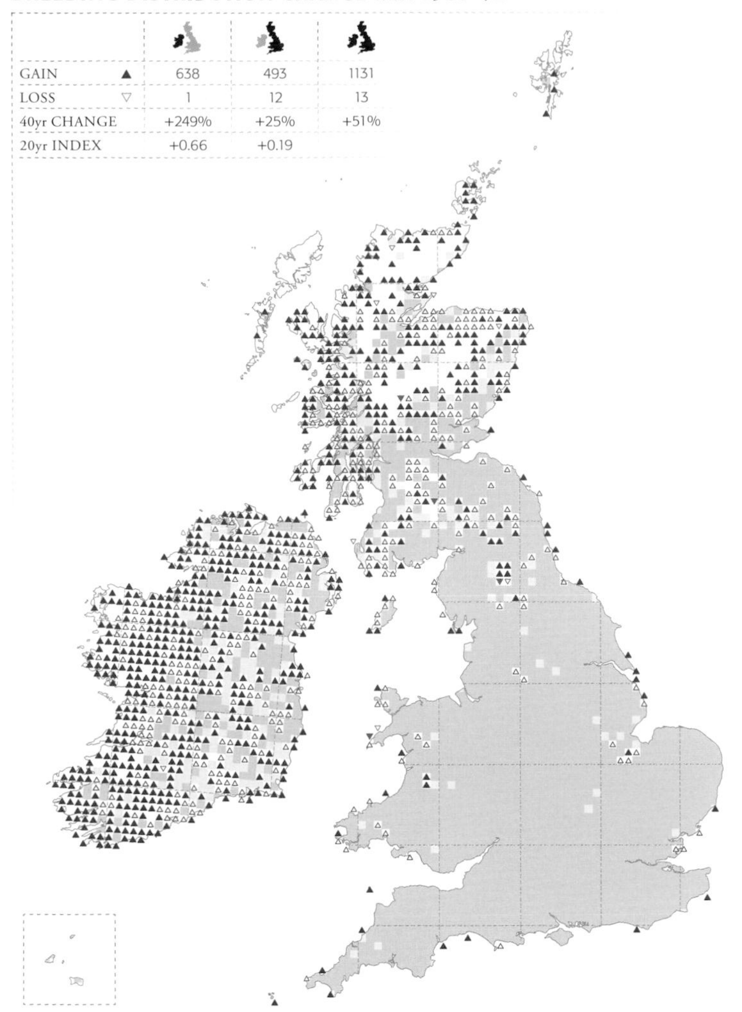

WINTER DISTRIBUTION CHANGE *since 1981–84*

GAIN ▲	290	654	944
LOSS ▼	31	161	192
30yr CHANGE	+225%	+57%	+77%

BREEDING RELATIVE ABUNDANCE *2008–11*

BREEDING RELATIVE ABUNDANCE CHANGE *since 1988–91*

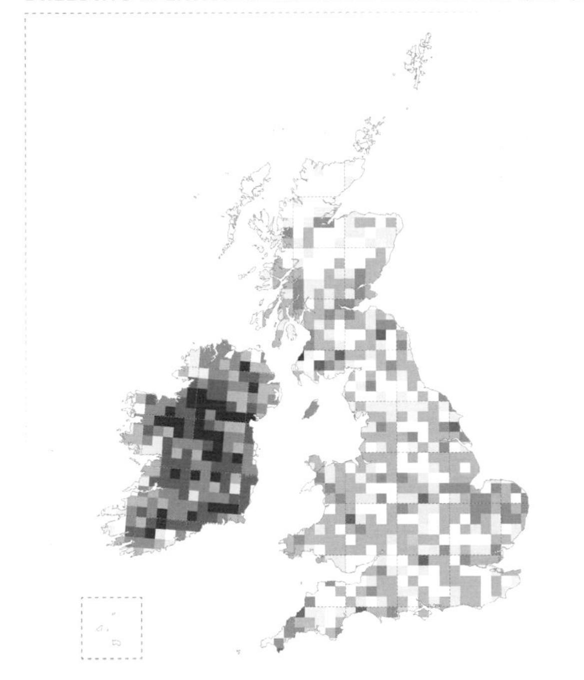

WINTER RELATIVE ABUNDANCE *2007/08–2010/11*

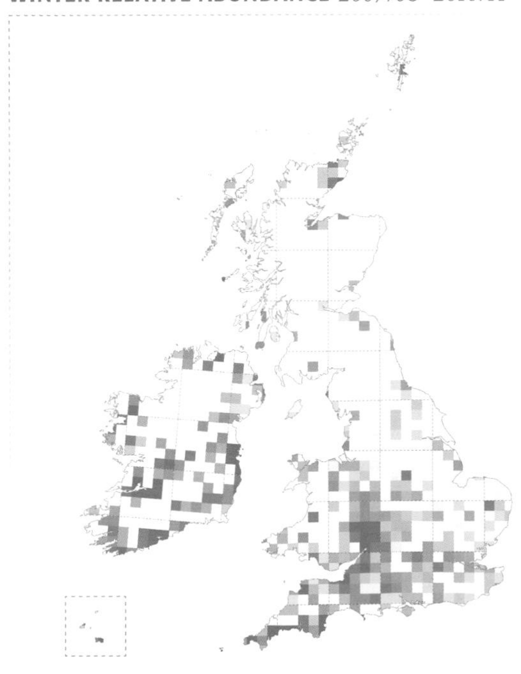

GARDEN WARBLER

Sylvia borin

BTO CODE: GW

JIM ALMOND www.shropshirebirder.co.uk

GARDEN WARBLERS ARE summer visitors that favour scrub, young woodland and woodland edge. The species is widely distributed throughout England, Wales and southern and central Scotland. Farther north they are distributed more patchily, occurring in northeast Scotland, Strathspey and along the Great Glen. In Ireland they are scarce, being concentrated into Counties Fermanagh and Cavan (Herbert 1991). Garden Warblers are absent or scarce also on remote islands, from the Channel Islands to Shetland. Highest densities are found in well-wooded areas, particularly in western Britain from Wales to southwest Scotland. Low densities are associated with uplands in northern England and Scotland and with intensively farmed areas, such as the Fens of eastern England.

Since the *1968–72 Breeding Atlas* there has been a 12% range expansion in Britain & Ireland. Gains are most evident in the west and north of the range, in Ireland, Northumberland and Scotland. Gains are also apparent on low ground around the Fens and in Lancashire.

Garden Warbler numbers have fluctuated since the 1960s, though there is good evidence of a recent decline (Hewson *et al.* 2007; *BirdTrends*). A 16% decline in numbers took place in England between 1995 and 2010 (*BBS Report 2011*) and this corresponds with the pattern shown by the abundance change map for much of England since 1988–91. In contrast, abundance has increased in parts of Wales, and in southwest and central Scotland, suggesting a possible influence of climatic change. Numbers have shown widespread moderate decline across Europe since 1980 (*PECBMS 2010*).

Species such as Garden Warbler that winter in the humid zone of tropical West Africa have generally declined in recent years, whereas those wintering further north in West Africa have not performed so poorly (Ockendon *et al.* 2012); these parallels suggest that changes in climate or land use in these wintering areas may be driving the declines. Processes operating on the breeding grounds may also influence distribution, for example heavy browsing by deer, which can reduce the suitability of woodland as Garden Warbler habitat (Holt *et al.* 2011b).

Garden Warblers were recorded from 19 10-km squares in winter, though all except two records were in November and thus likely to involve late migrants.

Sponsored by Simon Roddis

BREEDING DISTRIBUTION *2008–11*

POSSIBLE •	4%	19%	15%
PROBABLE •	1%	25%	18%
CONFIRMED ●	1%	29%	21%
TOTAL	6%	73%	54%

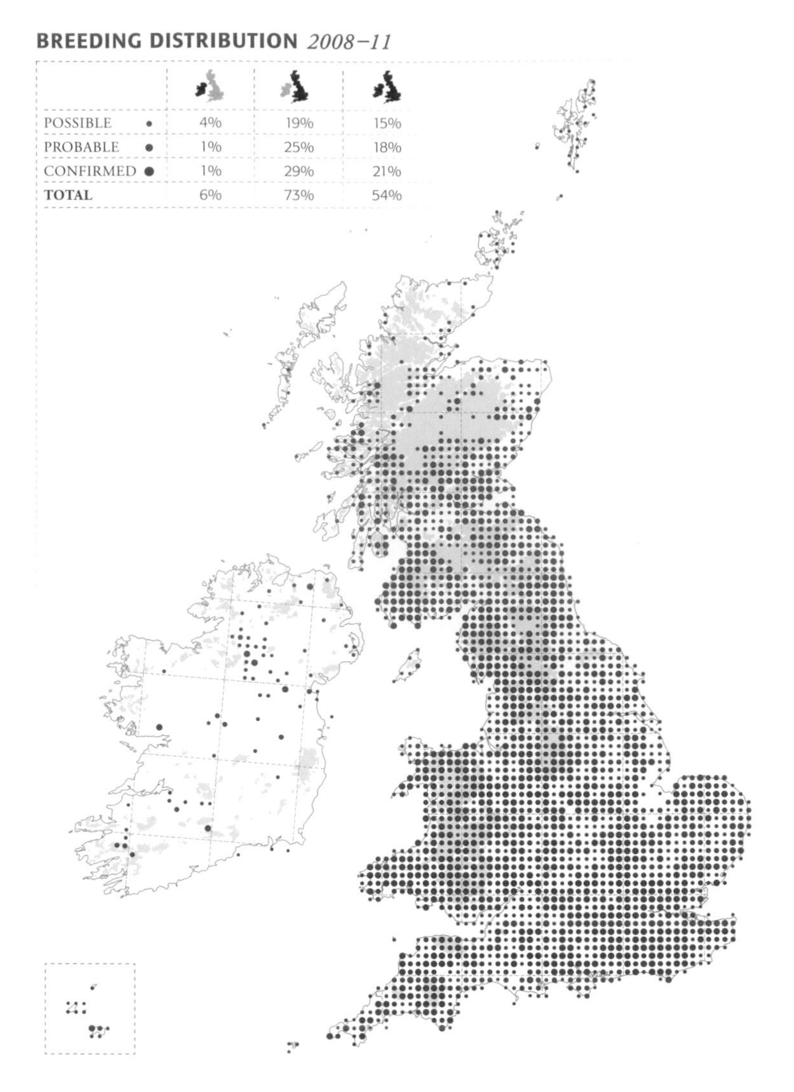

BREEDING RELATIVE ABUNDANCE CHANGE *since 1988–91*

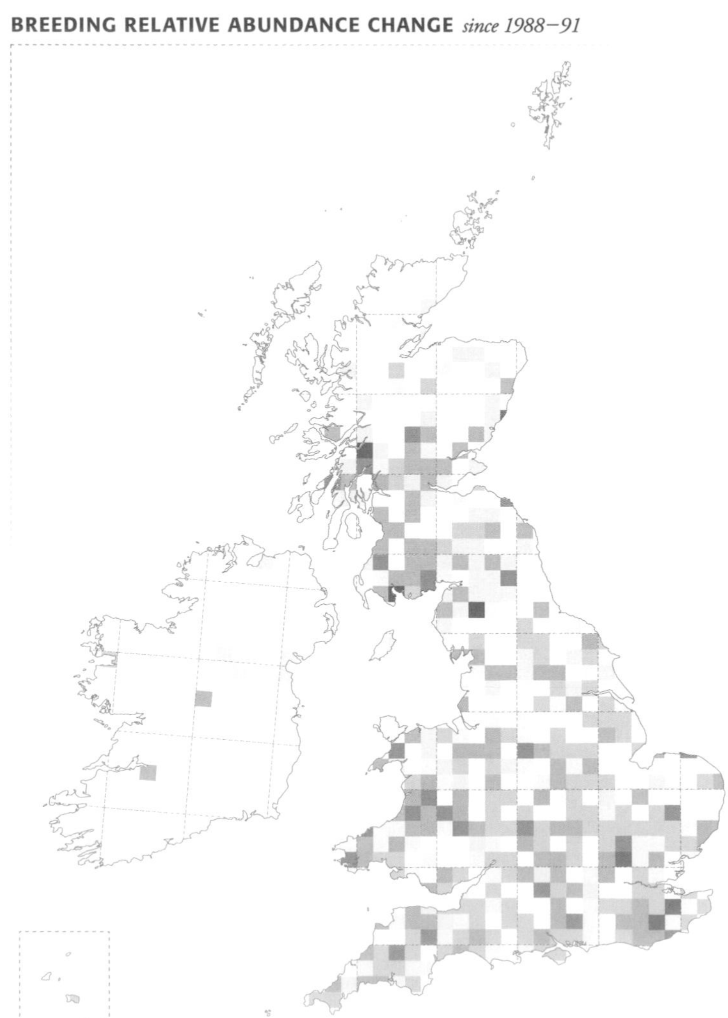

BREEDING DISTRIBUTION CHANGE *since 1968–72*

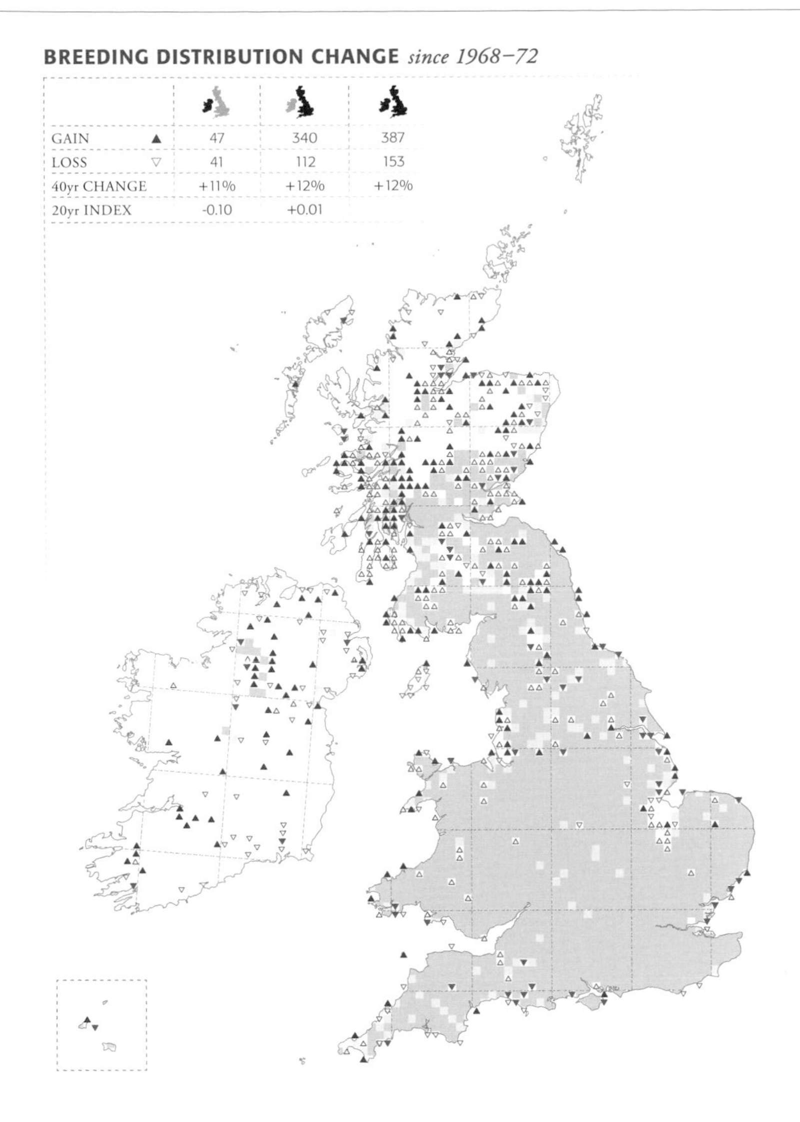

GAIN	▲	47	340	387
LOSS	▽	41	112	153
40yr CHANGE		+11%	+12%	+12%
20yr INDEX		-0.10	+0.01	

BREEDING RELATIVE ABUNDANCE *2008–11*

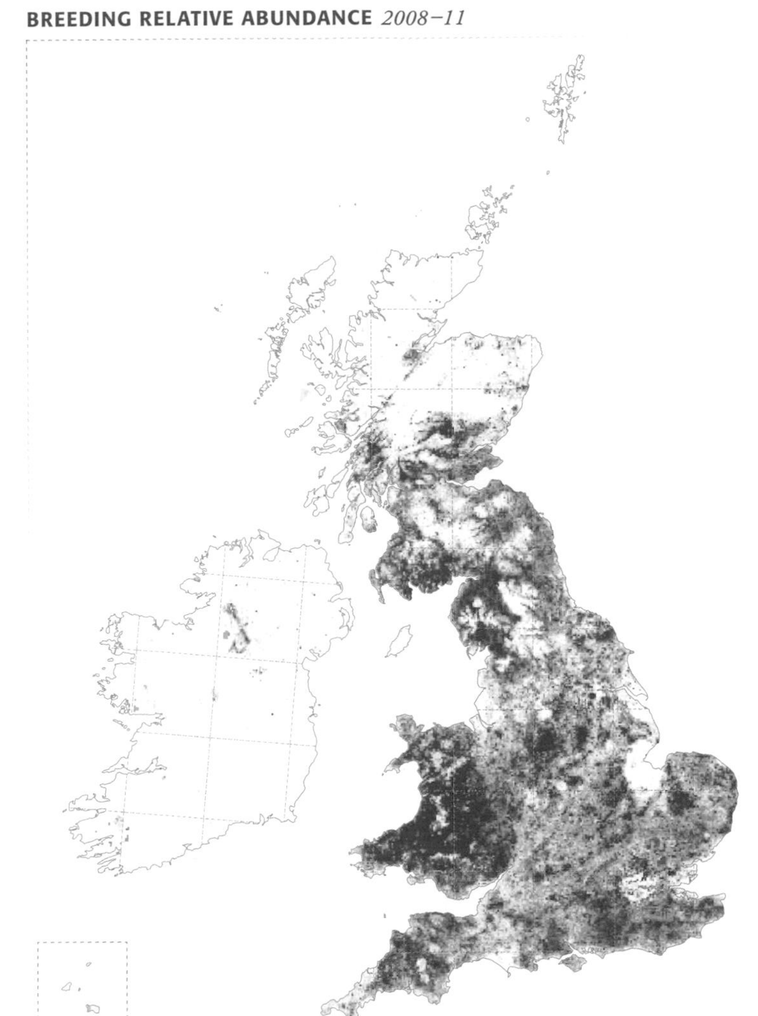

◀ The highest densities in the breeding season are found in well-developed scrub, young coppice and regenerated woodland. A dense understorey is required for nesting and feeding.

NICO DE VISSER www.flickr.com/photos/ndv-photography

LESSER WHITETHROAT

Sylvia curruca

BTO CODE: LW

MORRIS RENDALL

BREEDING LESSER WHITETHROATS in Britain & Ireland are mainly restricted to England, and historically were not known to breed any further north than the Midlands (Holloway 1996). Since the *1968–72 Breeding Atlas*, however, this long-distance migrant has shown a 33% range expansion in the breeding season and has colonised lowland areas of Wales, southern Scotland (Byars 2010; Mearns & Mearns 2012) and northern and southwest England. Gains have been particularly noticeable in western Wales, coastal Dumfries & Galloway, and the British east coast between North Yorkshire and Edinburgh. It is estimated that there are now some 200 pairs in Dumfries & Galloway (Mearns & Mearns 2012).

Breeding abundance is highest in eastern England and the East Midlands. The abundance change map also shows an increase in the proportion of occupied tetrads in many parts of eastern England since the *1988–91 Breeding Atlas*, contrasting with declines further south, perhaps suggesting a northward shift in distribution.

Lesser Whitethroats remain absent from upland areas of England, Wales and Scotland, but have been found nesting in lowlands as far north as Shetland, where breeding was first confirmed in 1993 (*Birds of Scotland*). As a breeding bird, Lesser Whitethroats are very rare in Ireland, with just a handful of singing males recorded annually in recent years (*IRBBP 2010*). No birds were *confirmed* breeding in Ireland during the Atlas period, but two squares recorded *probable* breeding.

Long-term monitoring indicates that the population was stable from the 1960s to the late 1980s, followed by a moderate decline up to the late 1990s but an increase since then (*BBS Report 2011*; *BirdTrends*). Unlike most other warblers in the region, this species migrates across Europe on a southeast–northwest axis, through the eastern Mediterranean, and winters just south of the Sahara in northeast and central Africa. Population trends are likely to be influenced by pressures on migration and in winter (Fuller *et al.* 2005), and so the recent increases may indicate improved conditions at stopover sites or in the winter quarters.

There were 63 10-km squares occupied in winter, mostly by late migrants in November. Only 22 10-km squares held birds during December to February.

Sponsored by the BTO Ringing Scheme

BREEDING DISTRIBUTION *2008–11*

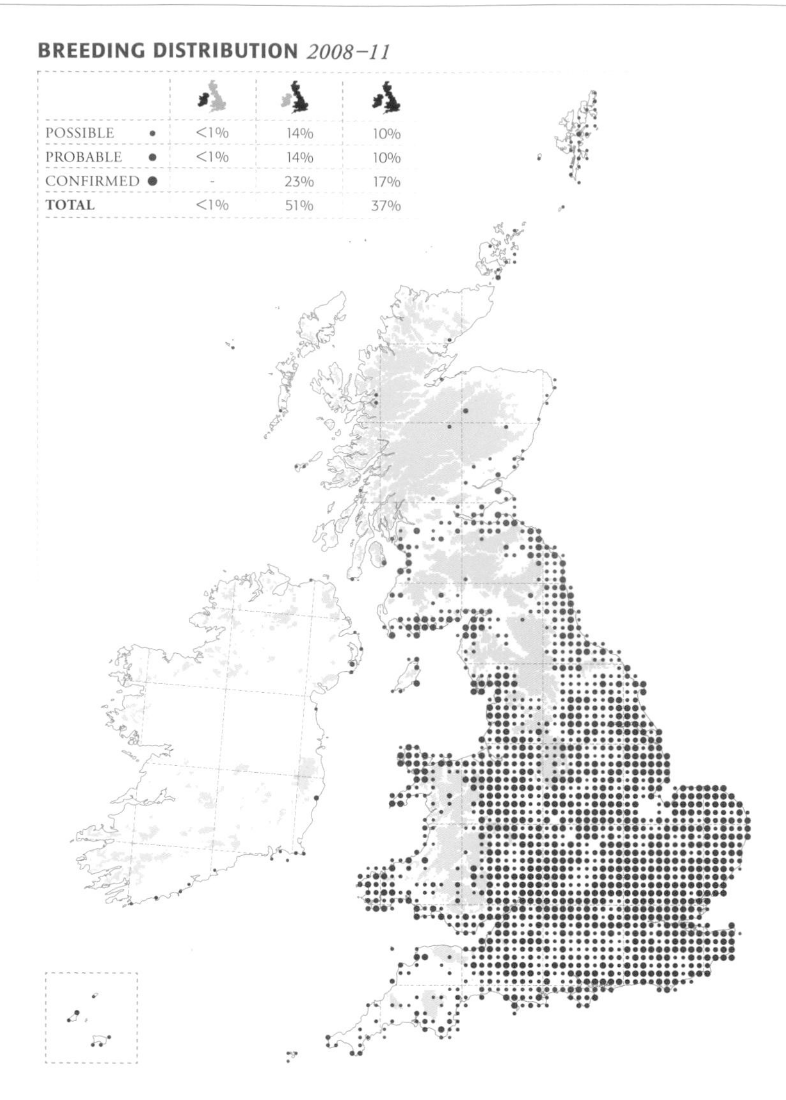

	Ireland	Britain	Britain & Ireland
POSSIBLE •	<1%	14%	10%
PROBABLE ●	<1%	14%	10%
CONFIRMED ●	-	23%	17%
TOTAL	<1%	51%	37%

BREEDING RELATIVE ABUNDANCE CHANGE *since 1988–91*

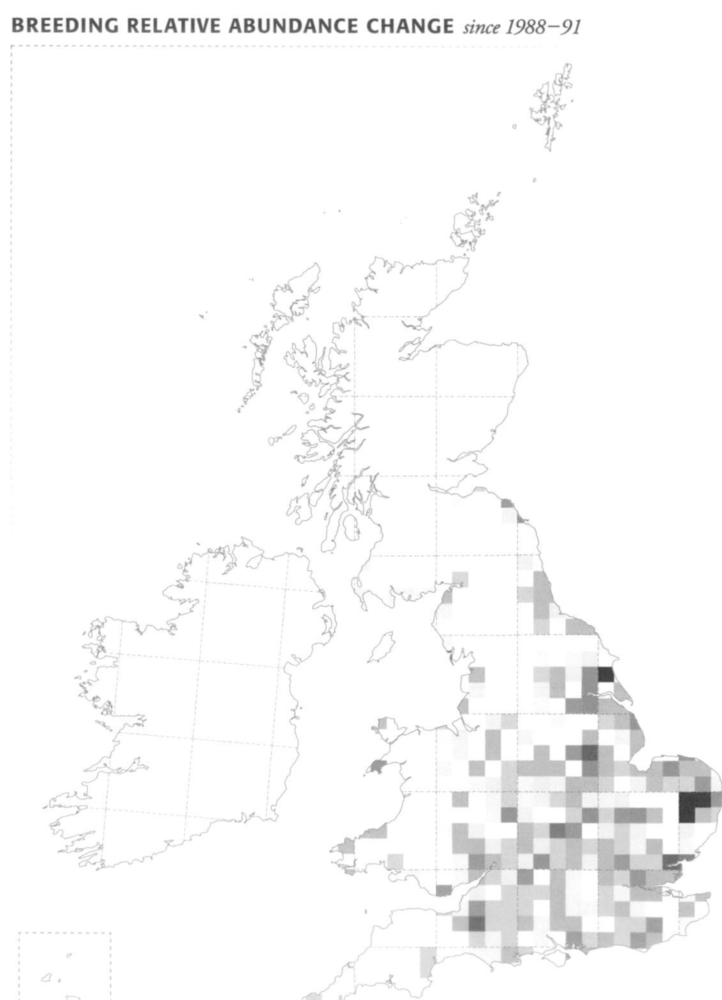

BREEDING DISTRIBUTION CHANGE *since 1968–72*

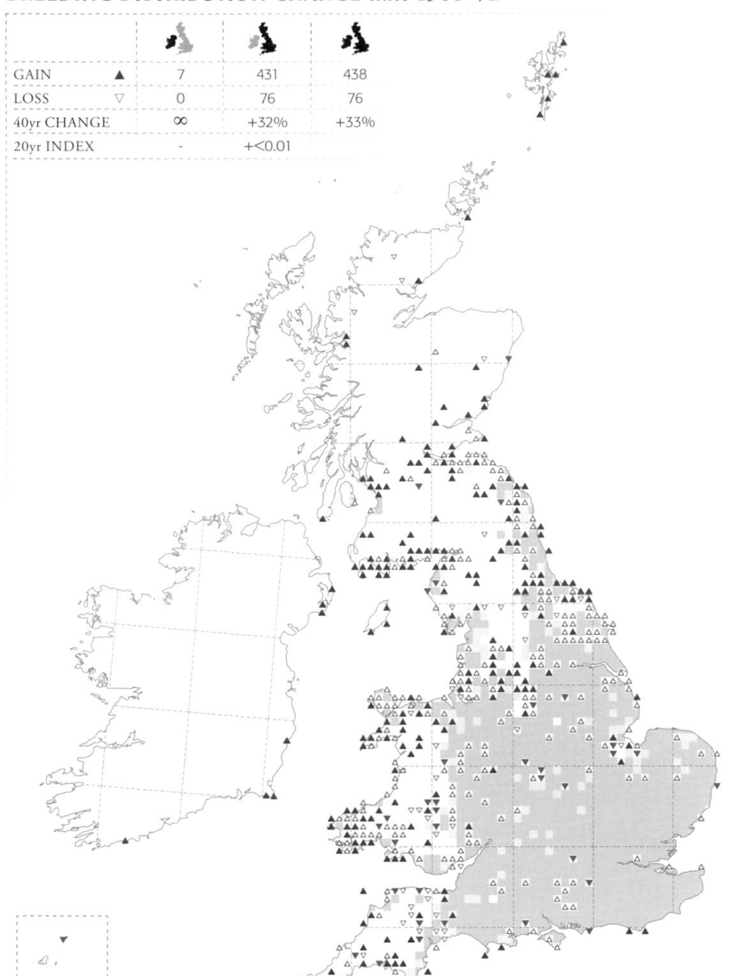

GAIN ▲	7	431	438
LOSS ▽	0	76	76
40yr CHANGE	∞	+32%	+33%
20yr INDEX	-	+<0.01	

BREEDING RELATIVE ABUNDANCE *2008–11*

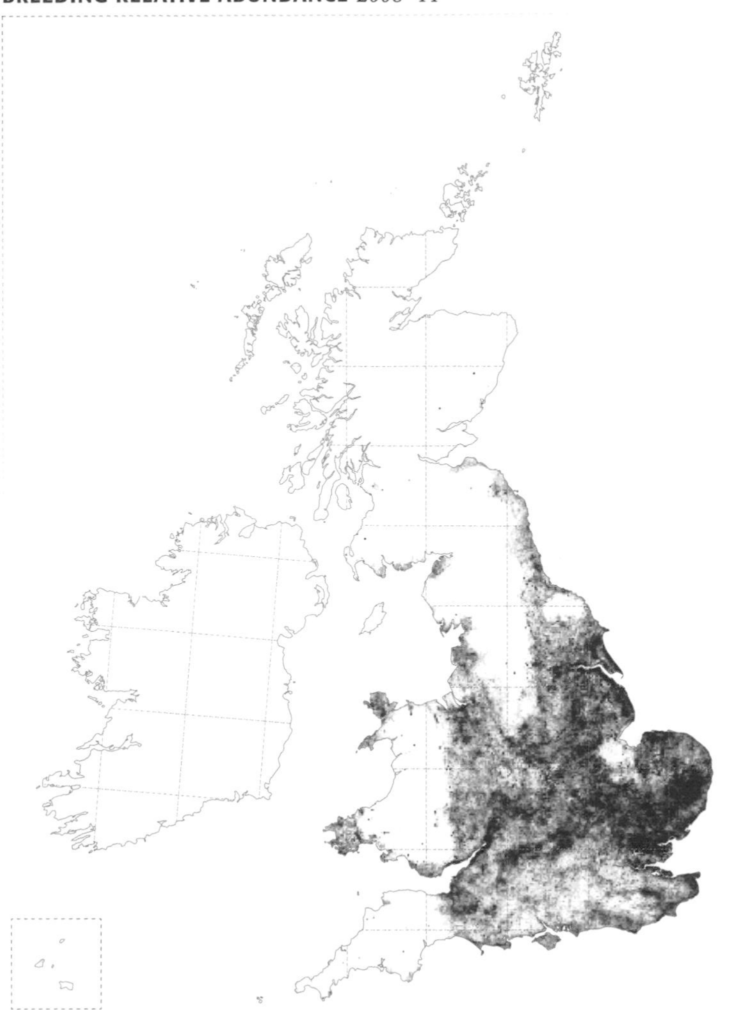

WINTER DISTRIBUTION *2007/08–2010/11*

PRESENT ●	<1%	2%	2%

WINTER DISTRIBUTION CHANGE *since 1981–84*

GAIN ▲	3	23	26
LOSS ▼	0	9	9
30yr CHANGE	∞	+140%	+170%

WHITETHROAT

Sylvia communis

BTO CODE: WH

CRAIG ROUND

WHITETHROATS ARRIVE IN mid to late April and spread throughout Britain and Ireland, excluding the higher uplands particularly of northern England and Scotland. They are thinly distributed in the Outer Hebrides, the Northern Isles and southwest Ireland. The relative abundance map shows that those upland areas and their margins, where they are occupied, support relatively few birds, and densities are also low throughout much of Ireland. The highest densities are found in the lowlands of southern, central and eastern England.

Whitethroat numbers crashed during the 1968/69 winter as a result of severe droughts on their wintering grounds south of the Sahara (Winstanley *et al.* 1974). The British & Irish population has never fully recovered from this crash and the long-term monitoring data for the UK show only a 15% increase in numbers during 1970–2010 (*SUKB 2012*). There was a 43% increase in numbers in the UK, however, during the period 1995–2010 (*BBS Report 2011*).

In Britain, no overall change in distribution has occurred between the *1968–72 Breeding Atlas* and 2008–11, but a 7% range expansion is evident in Britain since the *1988–91 Breeding Atlas*. There has also been a 24% increase in the number of occupied 10-km squares in Ireland since the *1988–91 Breeding Atlas*, corresponding with an 18% increase in numbers in the Republic of Ireland during 1998–2010 (*CBS Trend 2010*). Although long-term losses are still evident in southwest Ireland and in the northern Pennines, recent increases in the number of occupied 10-km squares since the *1988–91 Breeding Atlas* are apparent through much of Ireland and in the marginal upland areas of northern and western Britain.

The abundance change map also highlights recent population increases, with the proportion of occupied tetrads increasing markedly throughout the core range since the *1988–91 Breeding Atlas*. Baillie and Peach (1992) demonstrated that changes in Whitethroat abundance were highly correlated with overwinter survival, suggesting that improved environmental conditions on the wintering grounds may have driven the recent increases.

Only five squares recorded Whitethroats during the winter, all of which were in southern Britain; three of these were occupied only in November and therefore probably by birds still on migration.

Sponsored by Bill Haynes

BREEDING DISTRIBUTION *2008–11*

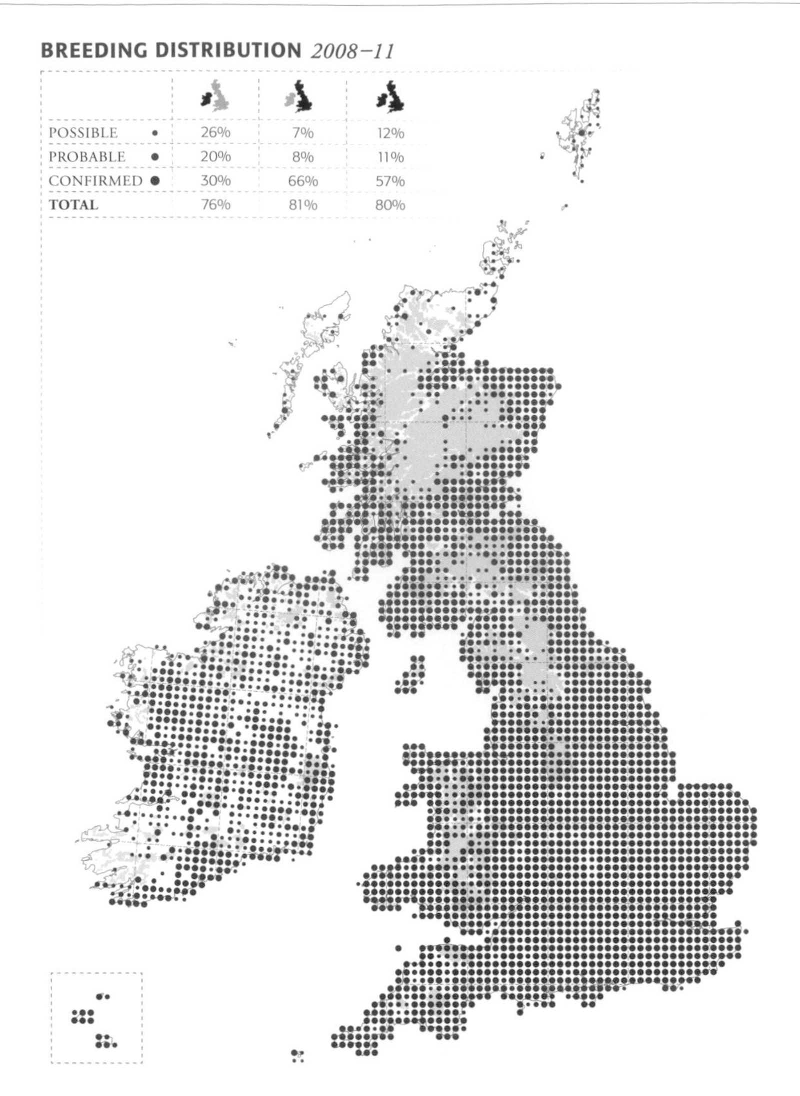

POSSIBLE •	26%	7%	12%
PROBABLE ●	20%	8%	11%
CONFIRMED ●	30%	66%	57%
TOTAL	76%	81%	80%

BREEDING RELATIVE ABUNDANCE CHANGE *since 1988–91*

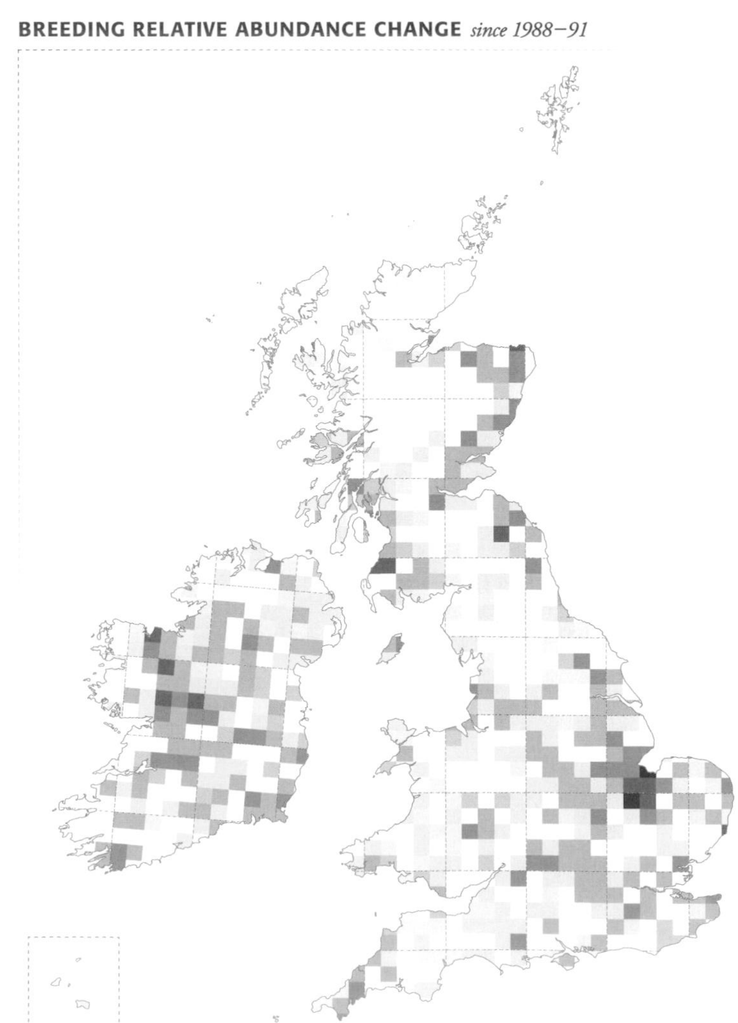

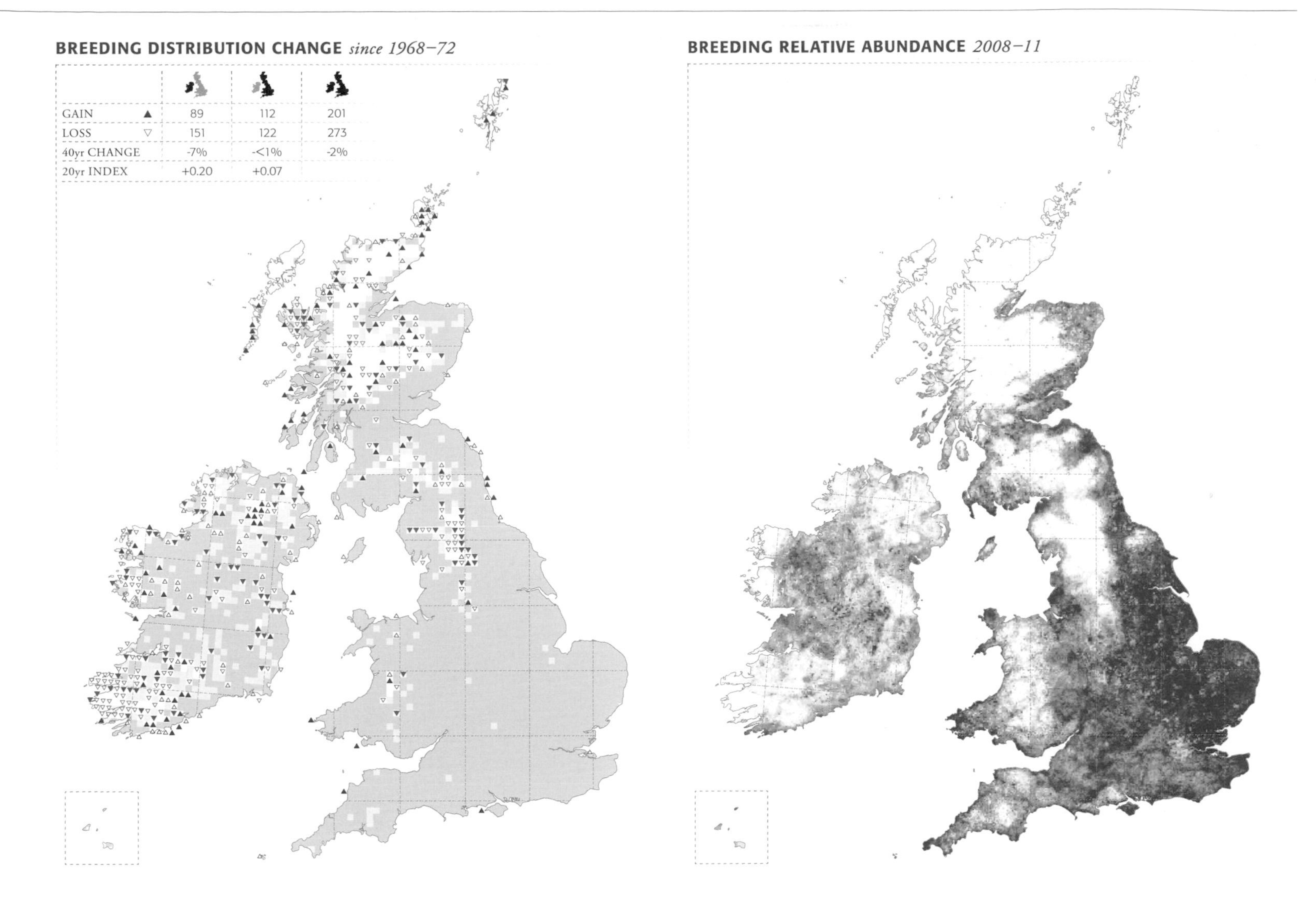

GAIN ▲	89	112	201
LOSS ▽	151	122	273
40yr CHANGE	-7%	-<1%	-2%
20yr INDEX	+0.20	+0.07	

◀ For many passerines, recently fledged young and adults carrying food are the most obvious signs of *confirmed* breeding. For Whitethroats, these accounted for 47% and 43% of *confirmed* breeding records respectively.

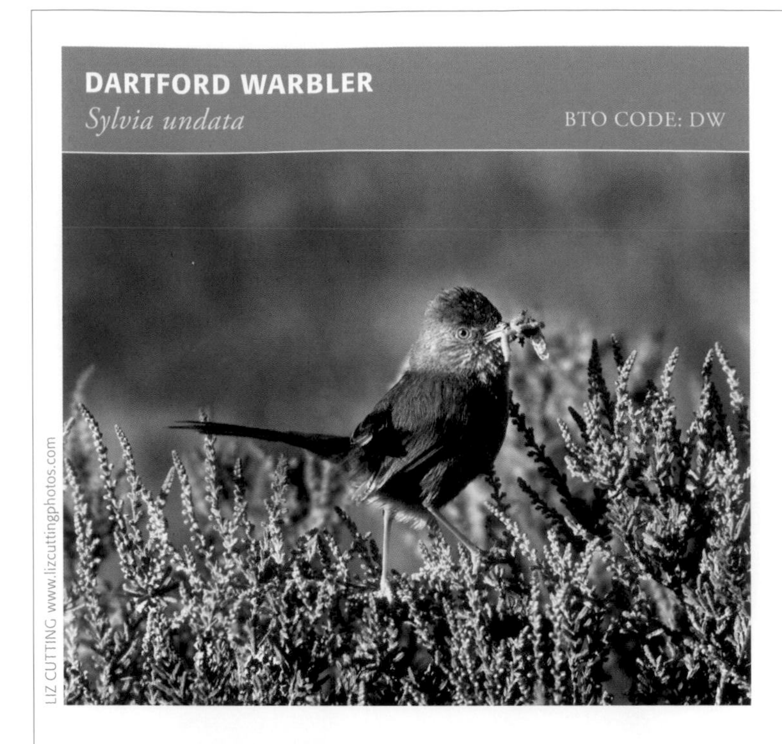

DARTFORD WARBLER

Sylvia undata

BTO CODE: DW

LIZ CUTTING www.lizcuttingphotos.com

THE DARTFORD WARBLER'S winter distribution is rooted firmly in southern Britain, extending from southwest England and Wales to Lincolnshire and East Anglia, but including scattered locations in the Midlands. The number of occupied squares has grown by 449% since the *1981–84 Winter Atlas*, involving extensive expansion from the formerly discrete strongholds in southern England. Many of the newly occupied wintering sites are in mild coastal locations but more remarkable is the use of inland sites, particularly those in the upland moors of southwest England and Staffordshire.

The current breeding distribution is confined to southern Britain, in both coastal and inland locations. The latter comprise lowland heathlands but also upland sites on the moorland fringes. Breeding abundance is greatest on heathlands in Hampshire, Surrey and Dorset, which in 2006 accounted for over 70% of the breeding population (Holling *et al.* 2009).

Breeding range expansion beyond southern England has been recent and rapid, including the colonisation of Wales, central England and East Anglia. The 52% range expansion up to the *1988–91 Breeding Atlas* was modest compared to the subsequent 198% expansion by 2008–11. During the same period there has been an increase in the proportion of occupied tetrads in core areas.

The most recent national survey in 2006 produced a population estimate of 3,214 territories, up 70% on the previous figures from 1994 (Wotton *et al.* 2009). New sites occupied in national surveys after 1984 have been at progressively higher altitudes, in line with climatic warming (Bradbury *et al.* 2011). Historically, there have been devastating population crashes following severe winters (e.g. Tubbs 1967) and most recently, following two successive severe winters, the number of pairs reported in 2010 was reduced to 19% of the 2006 population estimate (*RBBP 2010*). Numbers in the Thames Basin and Wealden Heaths Special Protection Areas, which held 25% of the 2006 population, fell from 978 territories in 2008 to just 50 in 2010 (Clark & Eyre 2012). Consequently, some of the gains shown on these maps may have been in squares that became unoccupied by the final seasons of the Atlas. Despite this setback, this species does have the capacity to recover rapidly and further expand its range.

Sponsored by The Clinton Devon Estates

BREEDING DISTRIBUTION *2008–11*

POSSIBLE •	-	1%	1%
PROBABLE •	-	1%	1%
CONFIRMED ●	-	3%	3%
TOTAL	-	5%	5%

WINTER DISTRIBUTION *2007/08–2010/11*

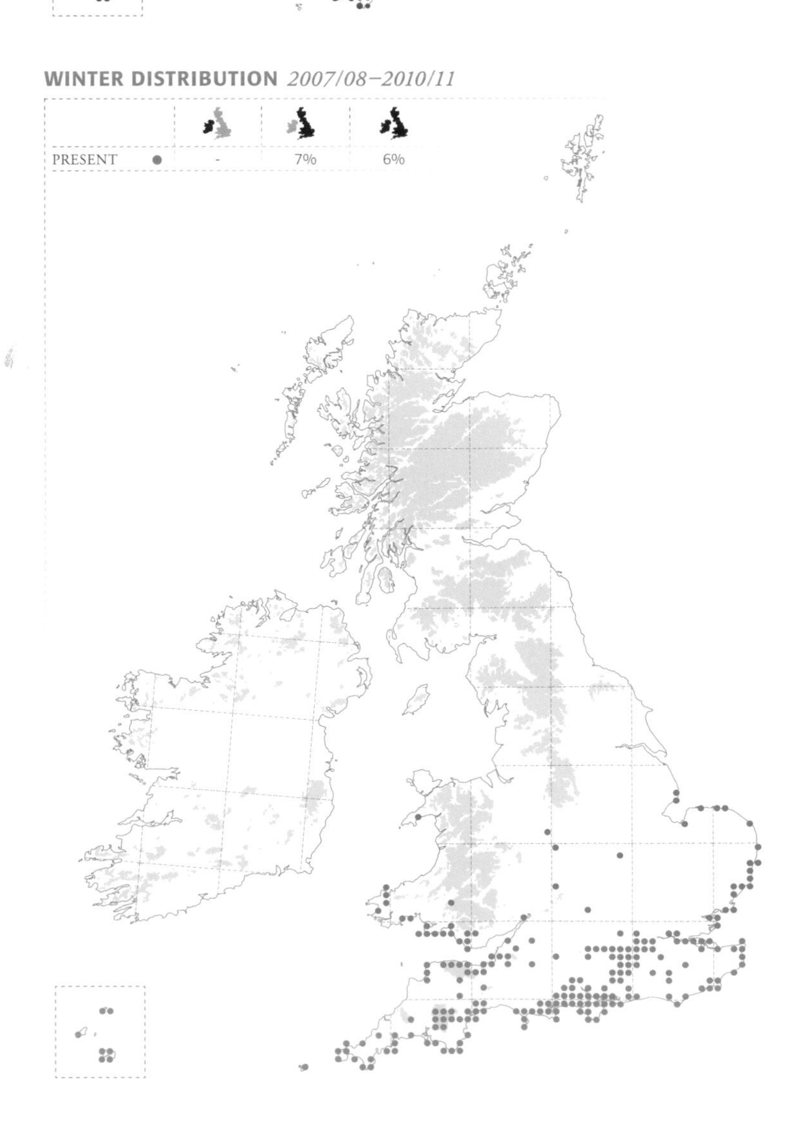

PRESENT ●	-	7%	6%

BREEDING DISTRIBUTION CHANGE *since 1968–72*

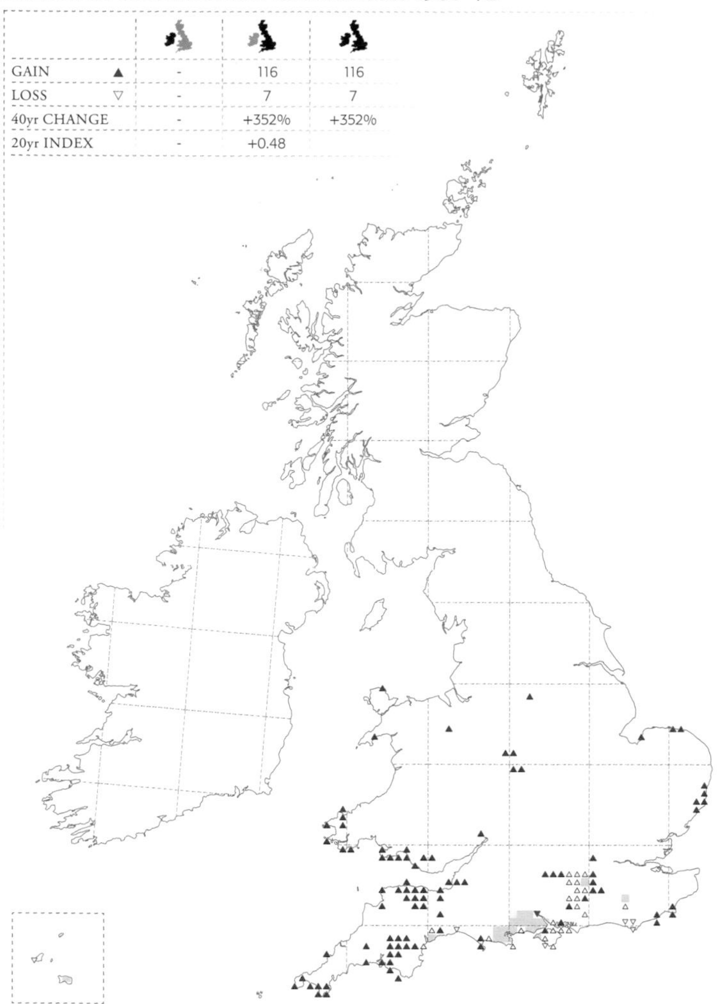

GAIN ▲	-	116	116
LOSS ▽	-	7	7
40yr CHANGE	-	+352%	+352%
20yr INDEX	-	+0.48	

WINTER DISTRIBUTION CHANGE *since 1981–84*

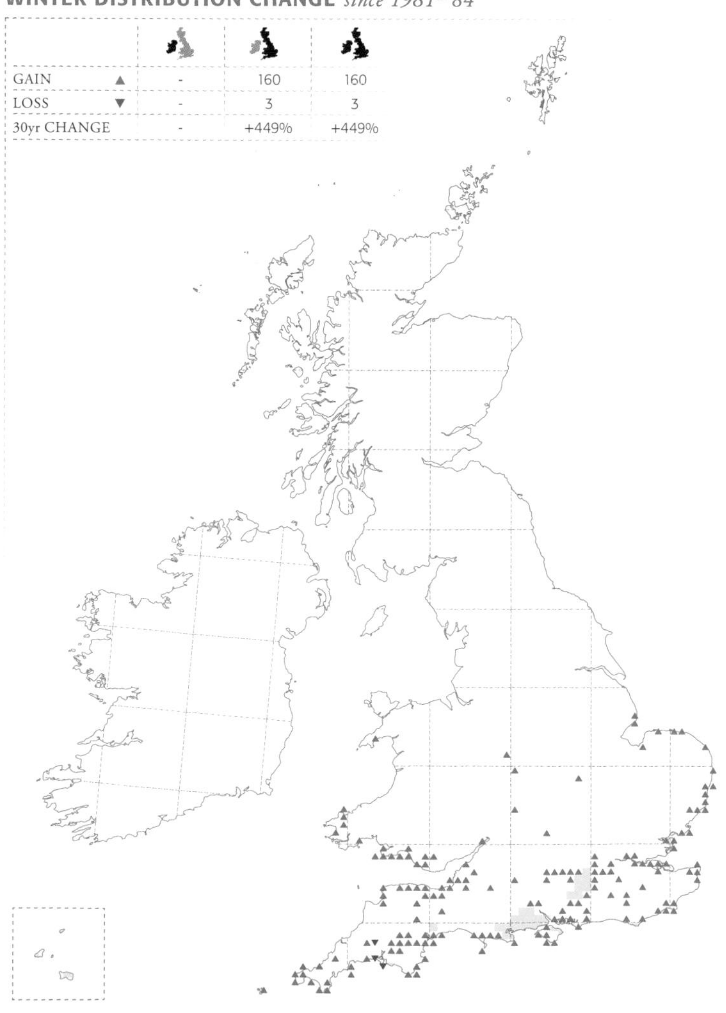

GAIN ▲	-	160	160
LOSS ▼	-	3	3
30yr CHANGE	-	+449%	+449%

BREEDING RELATIVE ABUNDANCE *2008–11*

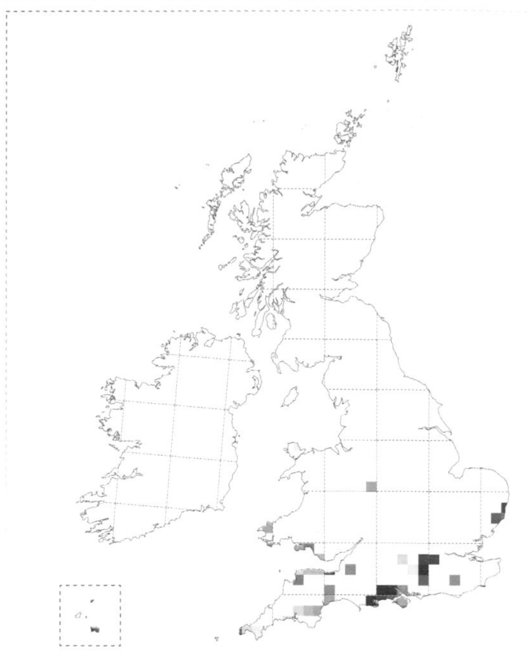

BREEDING RELATIVE ABUNDANCE CHANGE *since 1988–91*

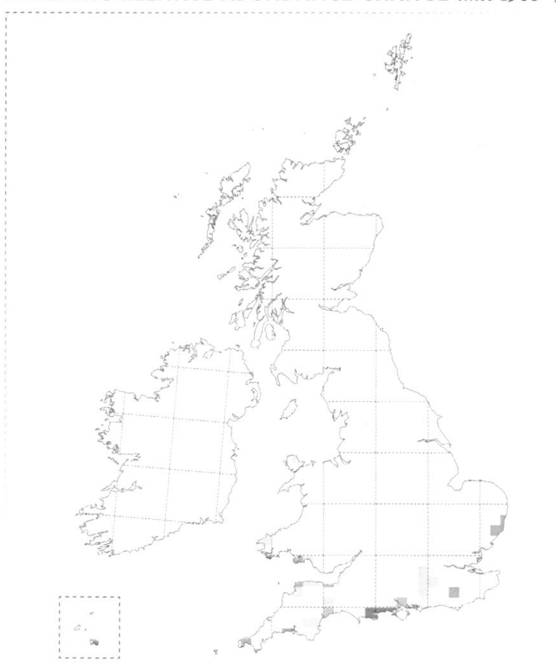

WINTER RELATIVE ABUNDANCE *2007/08–2010/11*

GRASSHOPPER WARBLER

Locustella naevia

BTO CODE: GH

DAVID TIPLING

THE DISTINCTIVE REEL of the Grasshopper Warbler was heard across almost the length and breadth of Britain & Ireland during 2008–11. Although mostly absent from the Outer Hebrides and Shetland, as well as at higher altitudes, this summer visitor is widespread throughout mainland Scotland and the Inner Hebrides. There are gaps in parts of northern England, along the Welsh border, and in southern England. Grasshopper Warblers are widespread in Ireland, except in the southeast. Indeed, the highest densities are found across Ireland, especially the west, as well as in Wales and southwest Scotland.

The change map shows that losses in 10-km square occupancy form a large swathe across English south-coast counties and also pockets in the east and midlands of Ireland. These substantial losses are offset by a marked concentration of gains in Scotland, especially in the northwest, and in the extreme west and southwest of Ireland. The abundance change map also highlights recent density increases in western Ireland, southwest Scotland and Wales.

The species has Red-list status in the UK and Amber-list status in Ireland on the basis of long-term declines and range contractions (Parkin & Knox 2010). There has been a modest recovery since the mid 1990s, however, with a 59% increase in the UK during 1995–2010 (*SUKB 2012*) and a 4% per annum increase in the Republic of Ireland during 1998–2010 (*CBS Trend 2010*). The trend in England amounts to only a 1% increase during 1995–2010 (*BBS Report 2011*) and confirms the regional differences shown by the Atlas.

Reasons for the decline are unclear but may include habitat succession, the maturing of forestry plantations, other forms of habitat loss and degradation, and factors affecting conditions on passage and in the wintering grounds (Marchant *et al.* 1990; Bayly *et al.* 2011). Gilbert (2012) concluded that the availability of suitable habitat was limited in some regions. She found that many formerly occupied sites were no longer suitable owing to canopy closure through woodland succession and to the removal of critical low vegetation through grazing. At present, low rates of ring recovery and nest finding limit the scope for using demographic data to investigate causes further.

BREEDING DISTRIBUTION *2008–11*

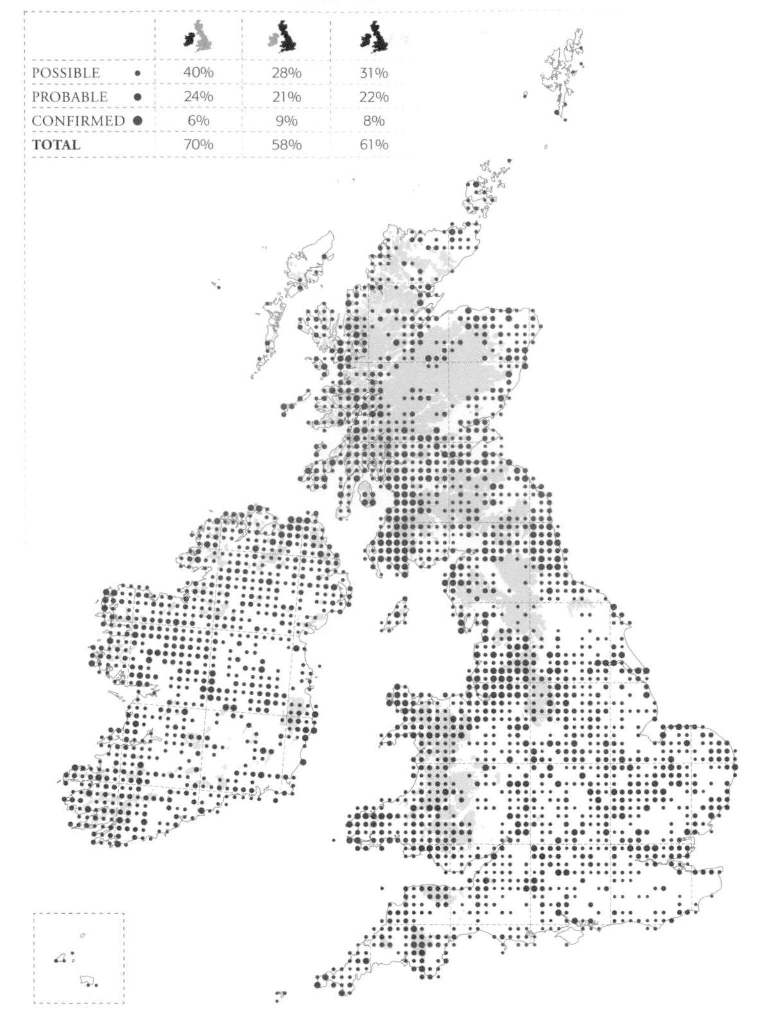

POSSIBLE •	40%	28%	31%
PROBABLE ●	24%	21%	22%
CONFIRMED ●	6%	9%	8%
TOTAL	70%	58%	61%

BREEDING RELATIVE ABUNDANCE CHANGE *since 1988–91*

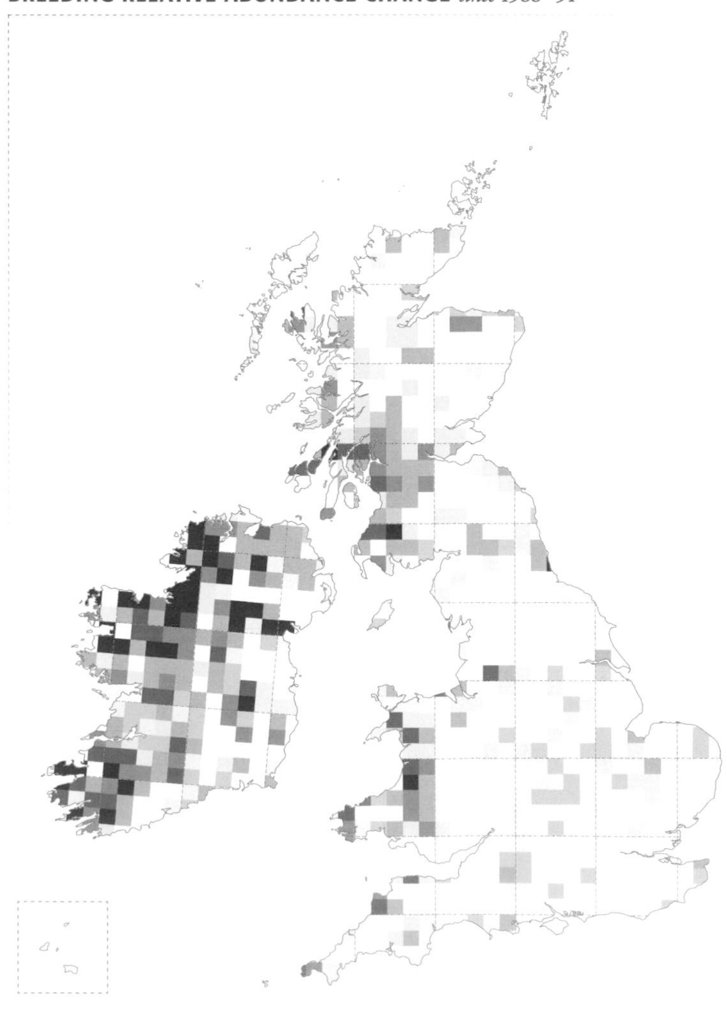

Sponsored by Stephen Harvey for Jill Harvey

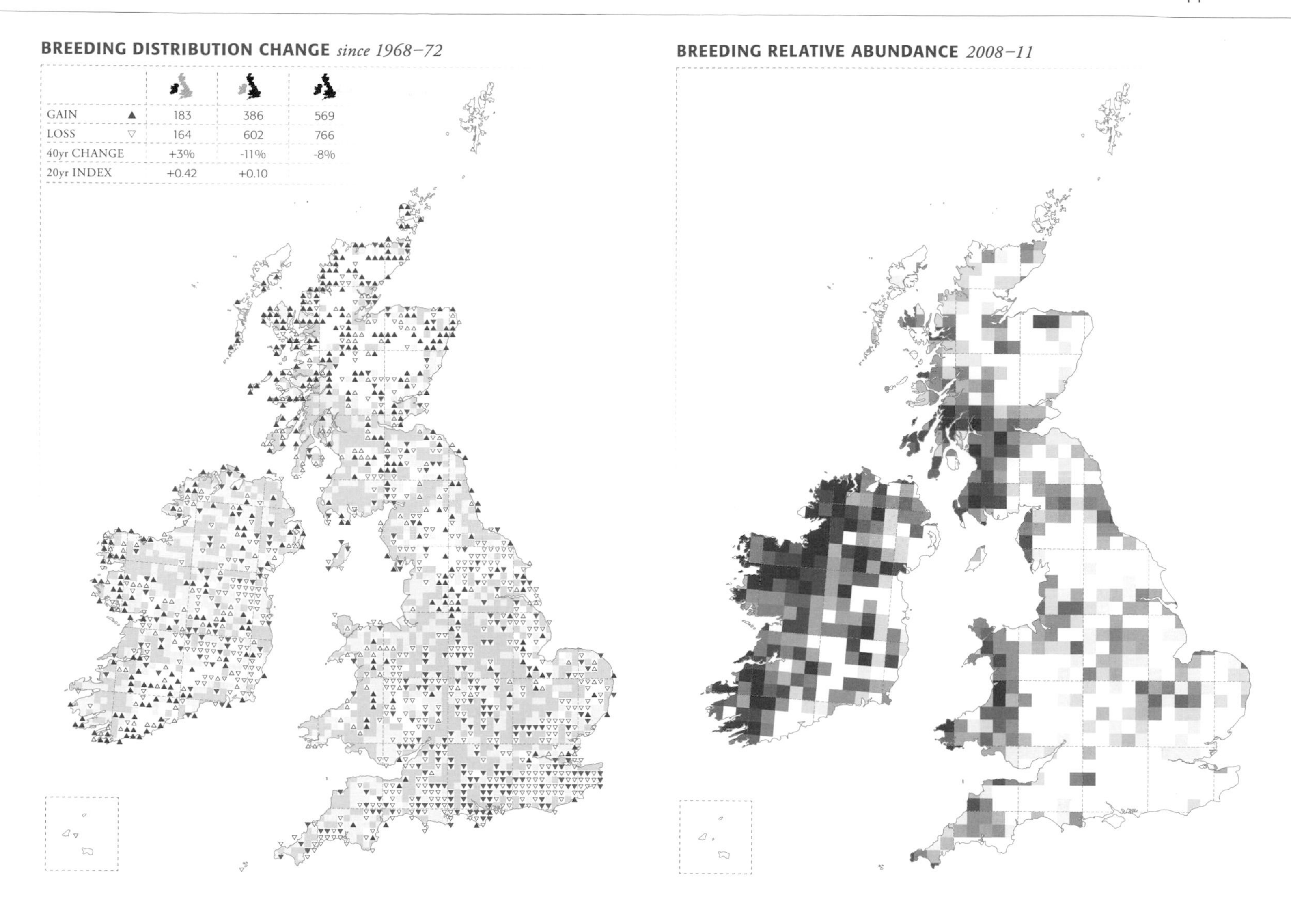

GAIN ▲	183	386	569
LOSS ▽	164	602	766
40yr CHANGE	+3%	-11%	-8%
20yr INDEX	+0.42	+0.10	

GRAHAM CATLEY http://pewit.blogspot.co.uk

◀ The distinctive song of the Grasshopper Warbler means males are easy to detect, yet obtaining evidence of breeding is difficult; just 13% of records with breeding evidence were of *confirmed* breeding.

SEDGE WARBLER

Acrocephalus schoenobaenus BTO CODE: SW

COLIN BROWN

SEDGE WARBLERS BREED throughout Britain & Ireland, though they are scarce or absent in the uplands. In Britain they reach high densities in the lowlands, especially the fens, valleys and coastal marshes of eastern England, the Scottish Central Belt and parts of eastern Scotland. In Ireland, high densities are associated with the Shannon catchment and with inland waterbodies and coasts in the north.

Atlas data suggest that the breeding range contracted by 12% between 1968–72 and 1988–91 and then expanded by 17% up to 2008–11, with a 2% gain overall. Whilst the dip in range size in 1988–91 is consistent with under-recording, it should be borne in mind that Sedge Warbler populations fluctuate widely in response to wet-season rainfall in their West African wintering grounds (Peach *et al.* 1991) and that a major population crash occurred in 1984/85. The long-term pattern in Britain is of range loss in England, especially the south and east, and gains in Wales and northwest Scotland. In Ireland, losses concentrated in the southeast almost exactly balance gains in the northwest. In many areas, however, gains and losses are closely mixed. Some of the complexity of the pattern of gains and losses could be related to habitat changes. Following reed cutting, for example, Sedge Warbler densities are significantly reduced and nest predation rates increased (Graveland 1999). Also, pattern of gains and losses may be related to the size and isolation of habitat patches (Foppen *et al.* 1999).

In addition to the broad changes in distribution, the abundance change map shows declines in the proportions of occupied tetrads in eastern England, contrasting with strong increases in tetrad occupancy in Ireland and in northern and western Britain. Annual monitoring data confirm this pattern, with a 69% population increase in the Republic of Ireland during 1998–2010 and a 36% increase in Scotland during 1995–2010, but only a 5% increase in England over the same period (*CBS Trend 2010*; *BBS Report 2011*). Factors underlying this geographical variation are currently unknown. Sedge Warblers show a trend of earlier spring arrival in Britain (Sparks *et al.* 2007), and thus factors relating to migration and climate change may be implicated.

Sponsored by The Wetland Trust, Porzana & Rye Bay Ringing Group

BREEDING DISTRIBUTION *2008–11*

	Britain & Ireland	Britain	Ireland
POSSIBLE •	22%	17%	18%
PROBABLE ●	26%	16%	18%
CONFIRMED ●	36%	42%	41%
TOTAL	84%	75%	77%

BREEDING RELATIVE ABUNDANCE CHANGE *since 1988–91*

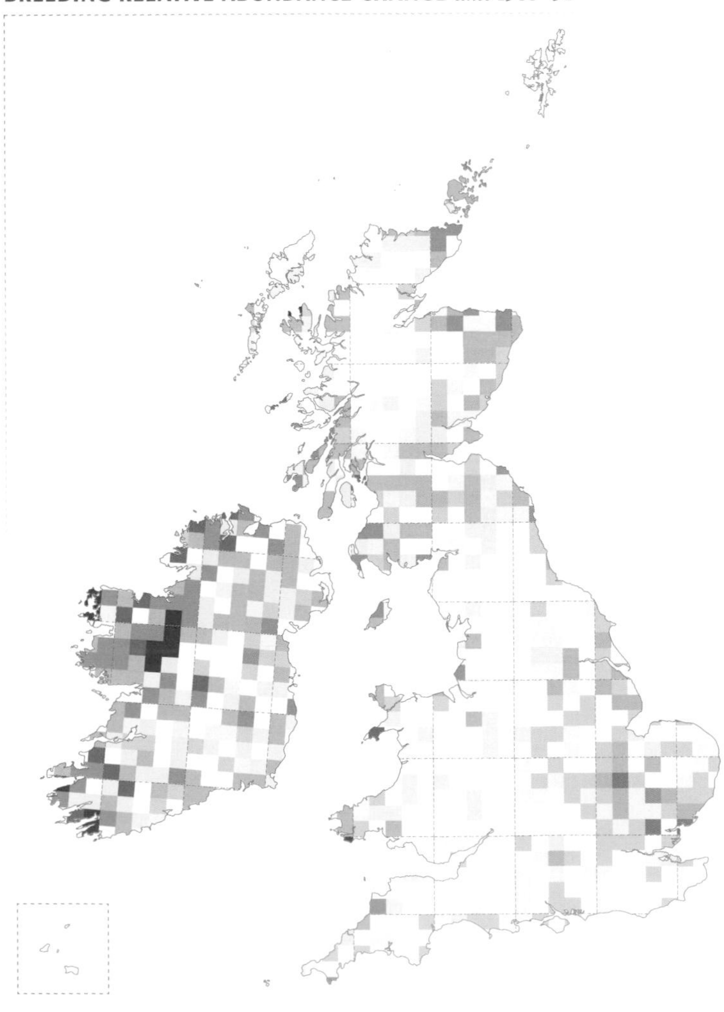

BREEDING DISTRIBUTION CHANGE *since 1968–72*

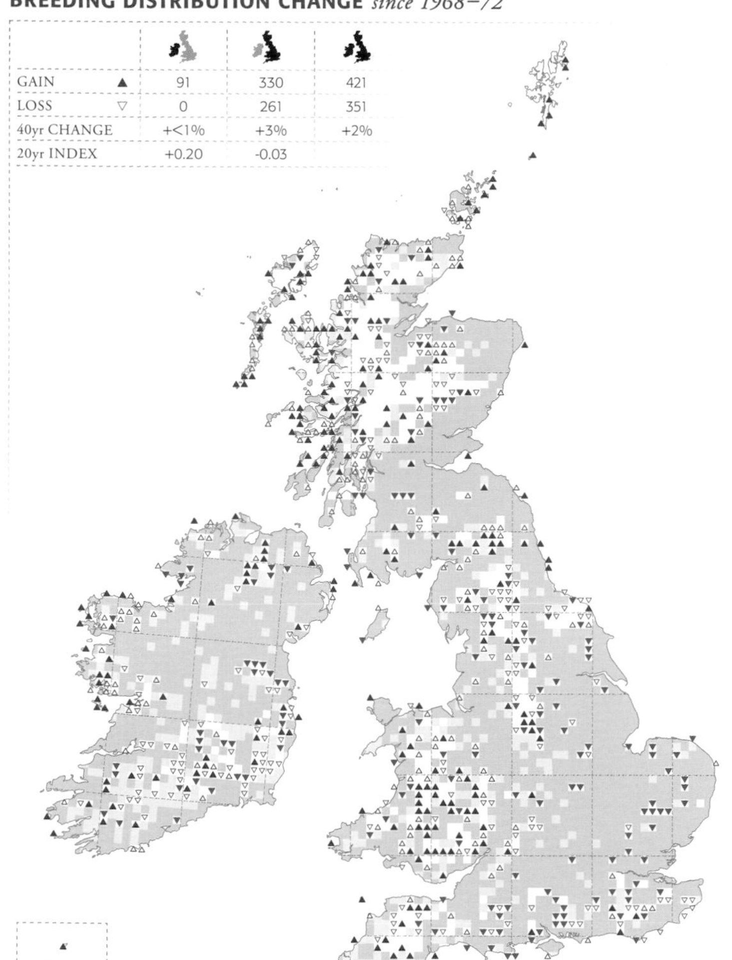

GAIN ▲	91	330	421
LOSS ▽	0	261	351
40yr CHANGE	+<1%	+3%	+2%
20yr INDEX	+0.20	-0.03	

BREEDING RELATIVE ABUNDANCE *2008–11*

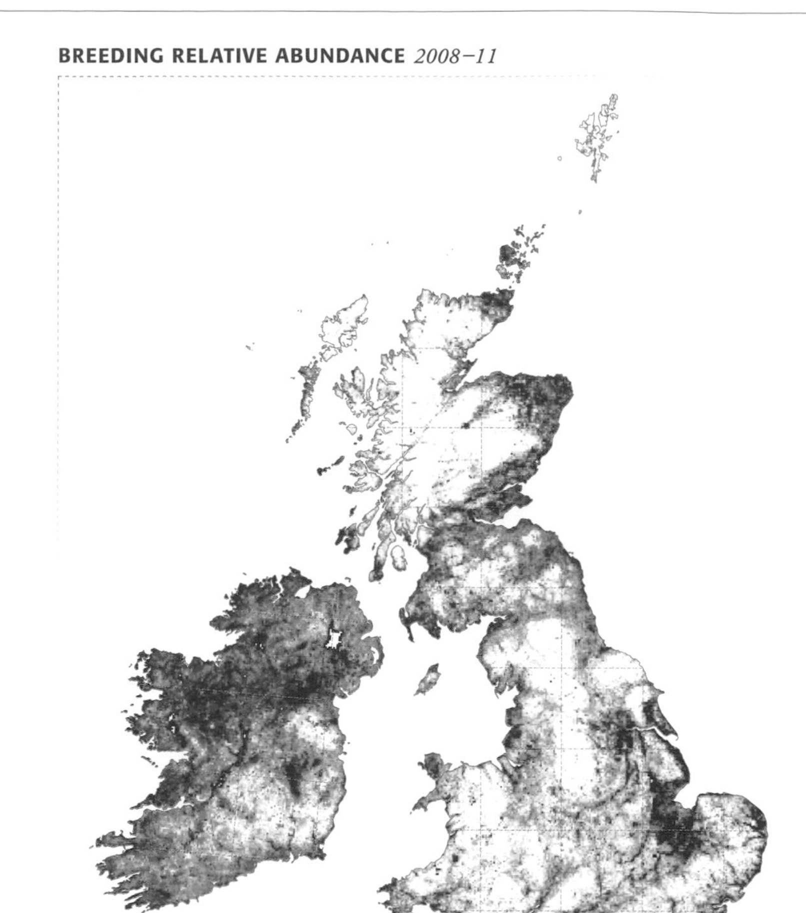

GRAHAM CATLEY http://pewitblogspot.co.uk

◀ Both annual monitoring of breeding populations and the number of birds ringed at Constant Effort Sites show the extent to which populations crash with each drought event in West Africa.

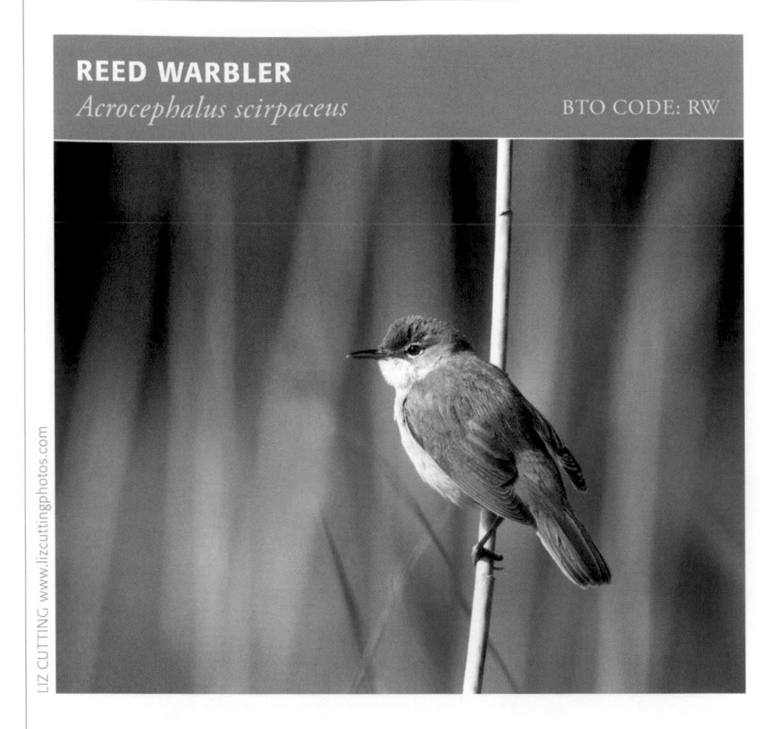

REED WARBLER
Acrocephalus scirpaceus
BTO CODE: RW

REED WARBLERS ARE strongly associated with areas of *Phragmites*, although other tall, robust plants can also be used for nesting. Although the majority of the Reed Warbler population in Britain still breeds to the south of South Yorkshire and Lancashire, the northward and westward range expansion noted in the *1988–91 Breeding Atlas* has continued apace over the last two decades. As a result, the numbers of occupied 10-km squares in the lowlands of both northwestern and northeastern England have increased.

Breeding in mainland Scotland was first proven in 1987 in the large Tay reedbeds (Robertson 2003). Since 1992 populations have become established in southwest Scotland (Bruce 1997), with signs of more recent expansion to east-coast sites. The whole Scottish population is still thought to number only 40–60 pairs (*Birds of Scotland*). The species' range in Wales has also expanded substantially, particularly along the southern and western coast. It has colonised Ireland since the *1968–72 Breeding Atlas*. The first confirmed breeding record in the Republic of Ireland was in 1981 (Smiddy & O'Mahony 1997), since when the population has colonised much of the coastline in the south and southeast.

The highest breeding densities are found in eastern England, particularly in the Fens. This area also shows the greatest increases in tetrad occupancy since the *1988–91 Breeding Atlas*, although increases have also occurred along parts of the southern English and Welsh coasts and in Merseyside, Cheshire and Somerset. Annual monitoring shows that numbers in Britain have more than doubled over the past 40 years and that this growth continues, with a 36% increase during 1995–2010 (*SUKB 2012*), despite results across national schemes suggesting that the European population is stable (*PECBMS 2010*).

Population growth may in part be driven by increased breeding success (*BirdTrends*), possibly relating to an increase in reed growth as the climate warms. In Poland, earlier arrival has been shown to extend the breeding season, allowing more pairs to raise second broods (Halupka *et al.* 2008). With a nine-day advancement of laying in Britain during 1968–2010 (*BirdTrends*), the season may be lengthening here too.

Sponsored by Vine House Farm

BREEDING DISTRIBUTION *2008–11*

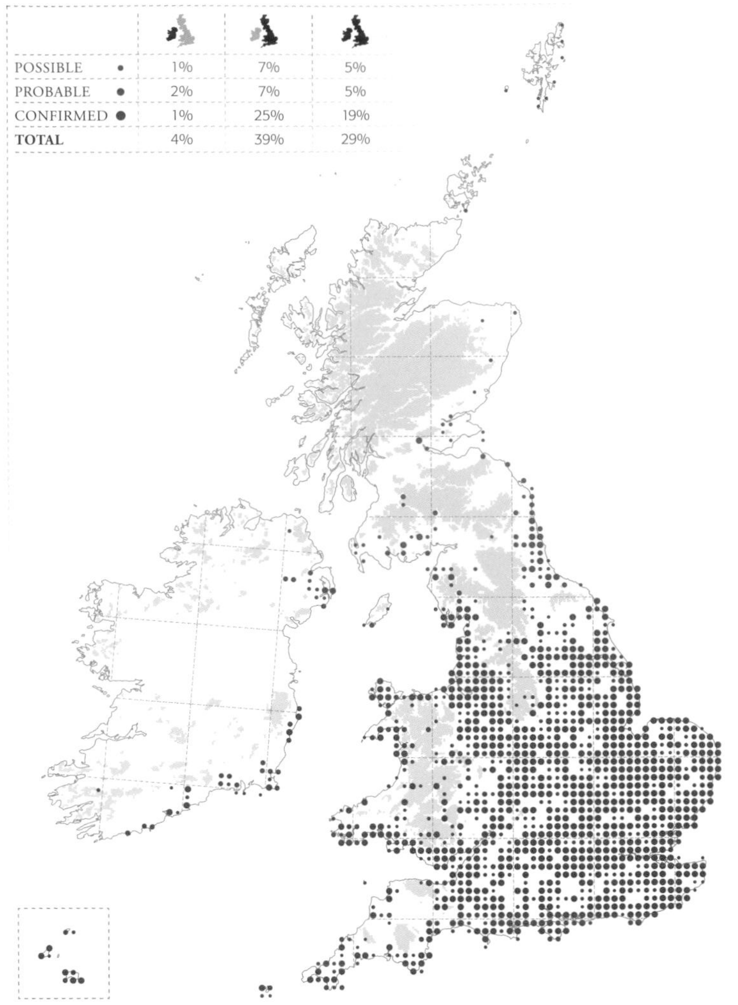

POSSIBLE •	1%	7%	5%
PROBABLE •	2%	7%	5%
CONFIRMED ●	1%	25%	19%
TOTAL	4%	39%	29%

BREEDING RELATIVE ABUNDANCE CHANGE *since 1988–91*

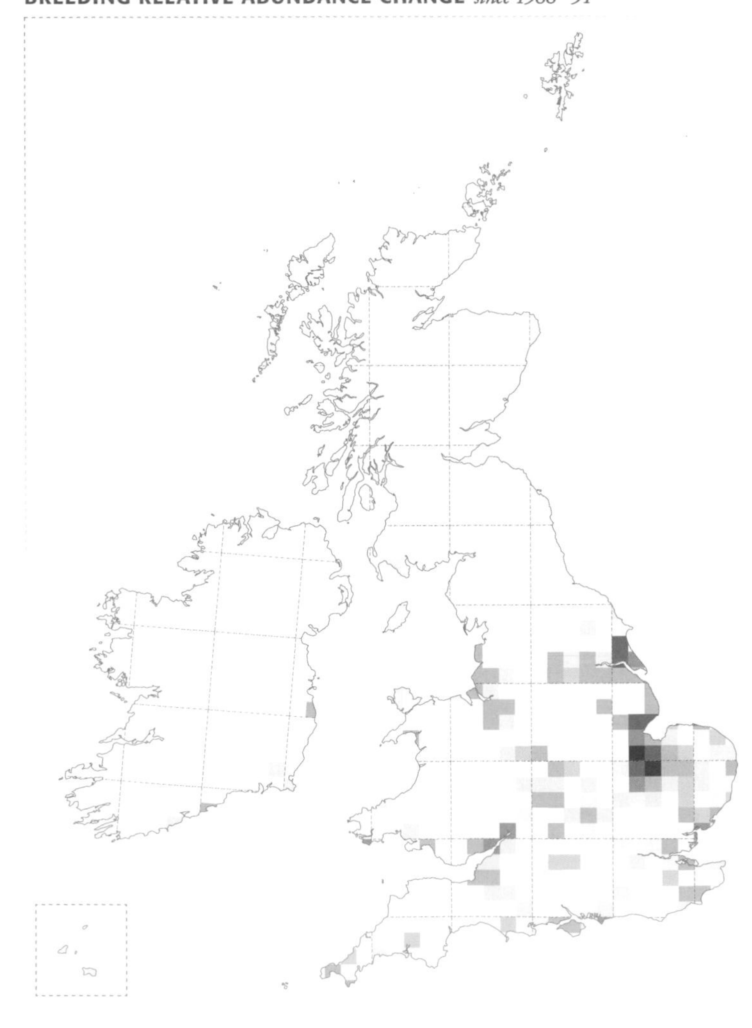

BREEDING DISTRIBUTION CHANGE *since 1968–72*

GAIN ▲	36	376	412
LOSS ▽	0	66	66
40yr CHANGE	∞	+40%	+44%
20yr INDEX	-	+0.11	

BREEDING RELATIVE ABUNDANCE *2008–11*

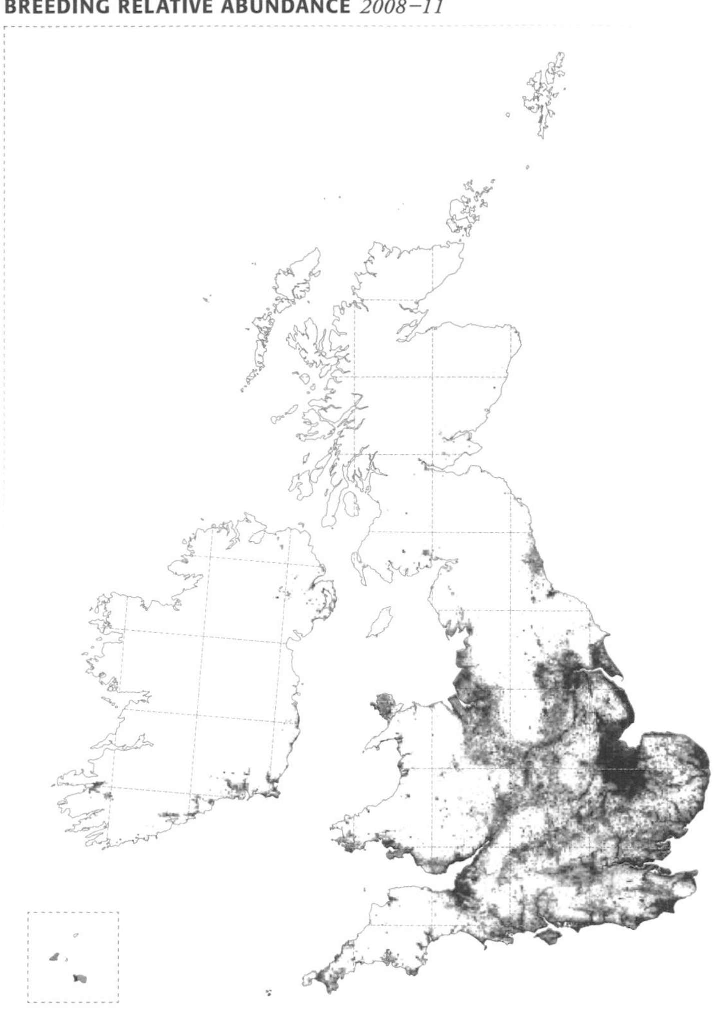

FAN-TAILED WARBLER

Cisticola juncidis

BTO CODE: FZ

PHIL ALEXANDER

THE FIRST KNOWN case of breeding for the region covered by the Atlas was recorded at Port Soif on Guernsey in 2009 (Lawlor *et al.* 2010). There had also been a singing male, and unconfirmed reports of breeding, on Alderney in 2008. There was one winter record within the Atlas recording period, of a migrant at Pegwell Bay in Kent in 2009. Following range expansion in Europe in the 1960s and 1970s, and records of vagrants in Britain and in Ireland during that period, it was predicted that this species would colonise mainland Britain (Ferguson-Lees & Sharrock 1977). As yet, this has not happened, perhaps owing to the species' apparent reluctance to cross large expanses of water.

BREEDING DISTRIBUTION *2008–11*

POSSIBLE •	-	<1%	<1%
PROBABLE •	-	-	-
CONFIRMED ●	-	<1%	<1%
TOTAL	-	<1%	<1%

MARSH WARBLER

Acrocephalus palustris

BTO CODE: MW

JIM ALMOND www.shropshirebirder.co.uk

SINGING MARSH WARBLERS appear from late May onwards in Britain but their locations are unpredictable and only infrequently are sites occupied in successive seasons. Many songsters are present only briefly but some attract a mate and breed. On average around eight territories are reported each year, although in 2009 there were 14 (Holling *et al.* 2011b; *RBBP 2010*).

The breeding-season distribution map shows a strong bias towards the east coast of Britain, reflecting this species' easterly direction of arrival from its East African wintering grounds (*Migration Atlas*). Since most dots represent one bird in just one of the four years of fieldwork, any impression of contiguous occupation of parts of the coast is misleading. Although some of these birds were readily identifiable as migrants, others were given breeding evidence codes and breeding was *confirmed* in Shetland and at several sites on or near the English east coast. Other *possible* or *probable* breeding records were scattered across central England.

As recently as the 1980s, Marsh Warbler distribution in Britain was very different. A robust population was centred on southwest England between Worcestershire and Somerset, perhaps numbering over 180 pairs in the early 1900s, and breeding was then rare or unknown near the east coast (Kelsey *et al.* 1989). The population began to decline after the 1950s. In Worcestershire, 50–60 males remained throughout the 1970s but numbers collapsed to just four by 1987. Uncompensated emigration was probably the main cause of the demise of this population, with Worcestershire lying at the furthest edge of the Marsh Warbler's breeding range at that time (Kelsey *et al.* 1989).

A breeding population in Kent, a major part of the distribution in the *1988–91 Breeding Atlas*, comprised up to c.20 territories in the mid 1990s, but disturbance and egg-collecting are thought to have led to its disappearance (Kent Ornithological Society 1998).

The change map documents the final extinction of the western population and the recent emergence of widespread sporadic breeding in eastern Britain. Coastal arrivals are more evident now than ever before but a regular breeding population has yet to re-establish. Numbers across Europe were stable during 1980–2010 (*PECBMS 2010*).

Sponsored by Harry Green, Ray Bishop and John Hodson in memory of Worcestershire's vanished Marsh Warblers

BREEDING DISTRIBUTION *2008–11*

POSSIBLE ●	-	1%	1%
PROBABLE ●	-	<1%	<1%
CONFIRMED ●	-	<1%	<1%
TOTAL	-	1%	1%

BREEDING DISTRIBUTION CHANGE *since 1968–72*

GAIN ▲	-	47	47
LOSS ▽	-	20	20
40yr CHANGE	-	+129%	+129%
20yr INDEX	-	-	

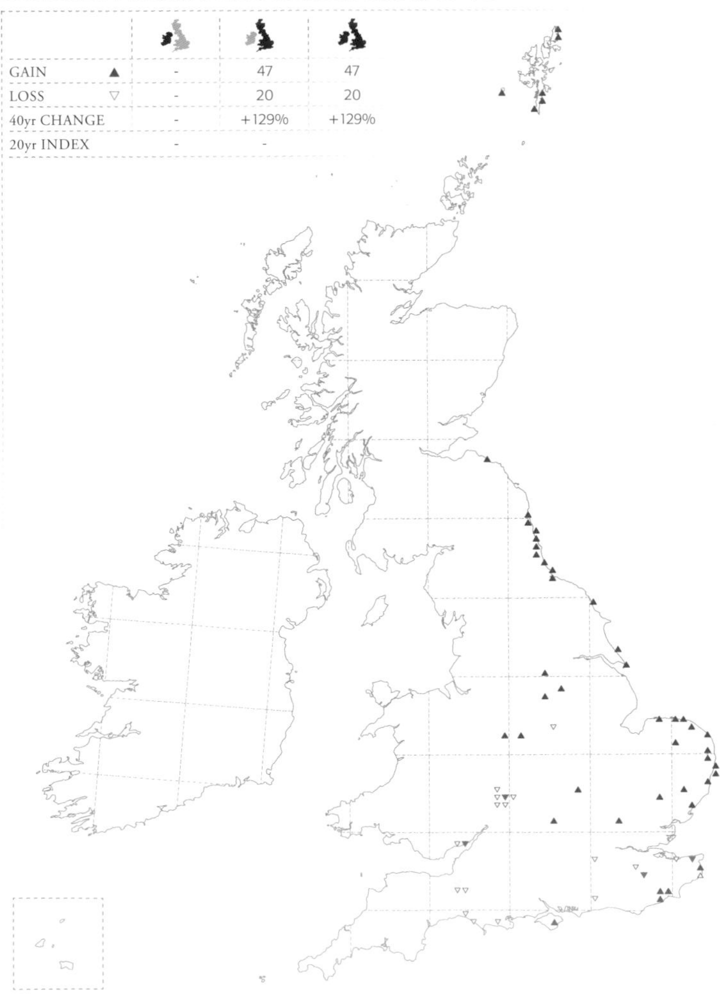

SAVI'S WARBLER

Locustella luscinioides

BTO CODE: VI

MARY BRADDOCK www.flickr.com/photos/nwsurreywildlife

THE MAP SHOWS 19 occupied 10-km squares scattered through southern and eastern England between Lincolnshire and Cornwall. Most involved singing males, some lingering for over a week and so being classed as *probable* breeders. Breeding was *confirmed* in 2010, when two pairs bred in Sussex (*RBBP 2010*). Records in Britain increased during the 1960s and 1970s but since the mid 1980s there has been a steady decline, possibly linked to wintering conditions in the Sahel (Slack 2009; Hudson *et al.* 2010). Recent figures give an average population size of four pairs but 2010 and 2011 were particularly good years, with up to 10 territories established in 2010 (*RBBP 2010*). Savi's Warblers are even rarer in Ireland and there were no records there during 2007–11.

BREEDING DISTRIBUTION *2008–11*

POSSIBLE •	-	<1%	<1%
PROBABLE •	-	<1%	<1%
CONFIRMED •	-	<1%	<1%
TOTAL	-	<1%	<1%

ICTERINE WARBLER

Hippolais icterina

BTO CODE: IC

GARY WOODBURN

ICTERINE WARBLERS, WHICH breed in northern and eastern Europe, declined by 33% during 1980–2010 (*PECMBS 2010*). Although the species is more usually recorded during autumn, spring *non-breeding* individuals were recorded in 48 10-km squares, mostly along the British east coast. In a further 27 10-km squares, birds were assigned breeding evidence, mostly on the basis of song or presence in suitable habitat. Probably many of these records were also of migrants, but four in southeast England and nine in northern Scotland were in suitable, inland breeding habitat. Six of these Scottish records were in 2009, including one of a pair feeding young in the nest in southeast Sutherland (Swann 2010), only the fourth known breeding attempt in Britain. All previous attempts had also been in Scotland (Holling *et al.* 2011b).

BREEDING DISTRIBUTION *2008–11*

POSSIBLE •	-	1%	1%
PROBABLE •	-	<1%	<1%
CONFIRMED •	-	<1%	<1%
TOTAL	-	1%	1%

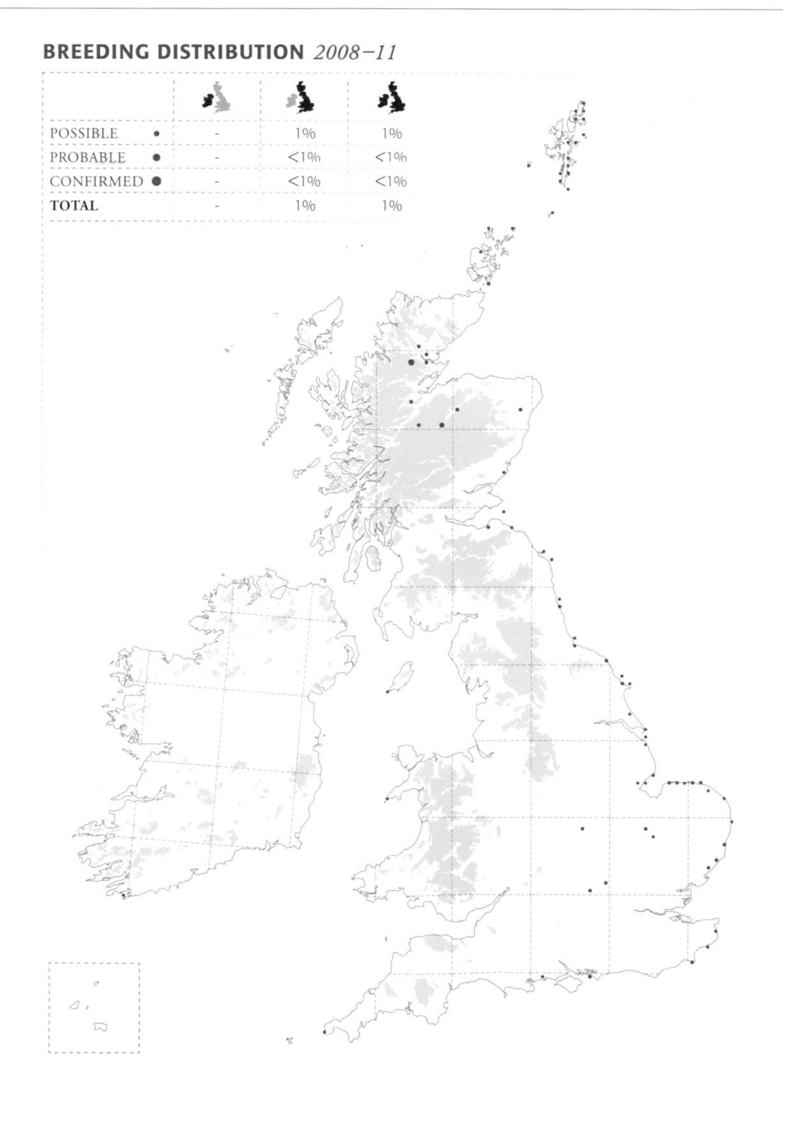

PALLAS'S WARBLER

Phylloscopus proregulus

BTO CODE: PA

GLYN SELLORS www.glynsellorsphotography.com

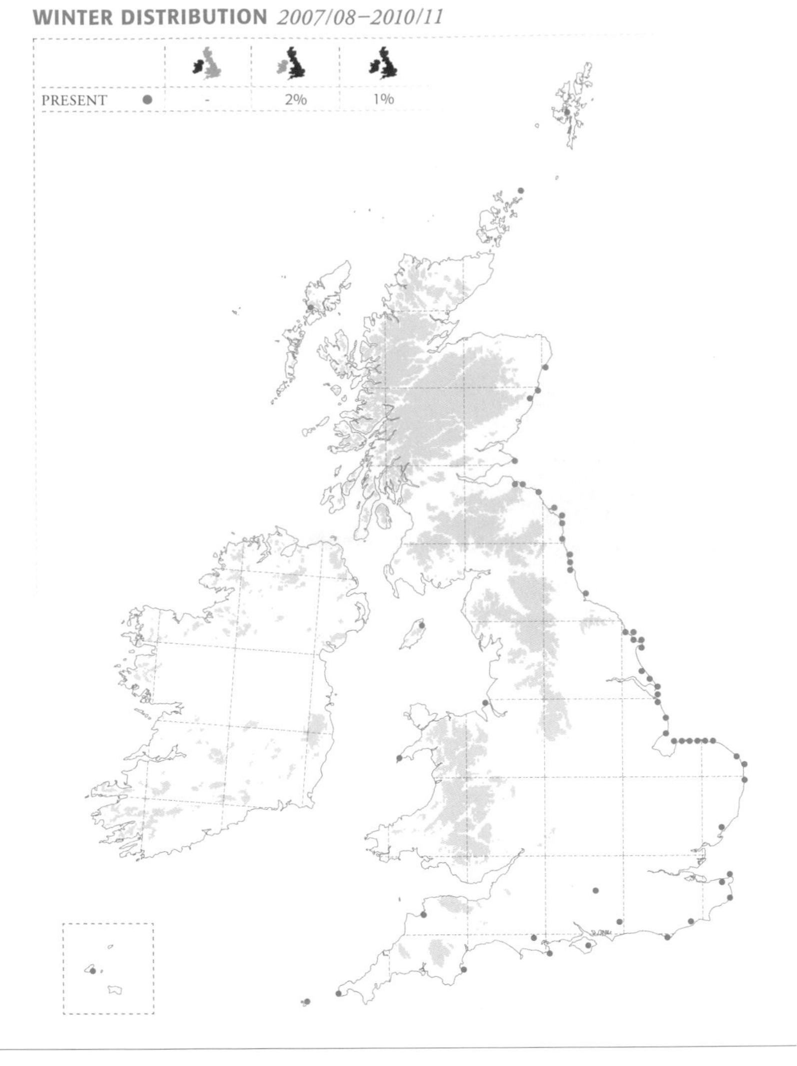

PALLAS'S WARBLERS ARE rare late-autumn visitors from Siberia. During 2007–11 they were recorded from 58 10-km squares in winter, with 52 of these involving sightings in November only. The bulk of the records were from the North Sea coast between Aberdeenshire and north Norfolk. Several of the records in English south-coast counties were made in midwinter. Even discounting squares occupied only in the early November period not covered by the *1981–84 Winter Atlas*, the remaining records represent a large increase on the four squares occupied in 1981–84. This finding is consistent with the significant increase in records from the 1950s to the 2000s (Fraser & Rogers 2006b). Notably, migrants were recorded in five squares in spring.

YELLOW-BROWED WARBLER

Phylloscopus inornatus

BTO CODE: YB

MARTIN GOODEY www.mpgoodeyphotography.com/apps/photos

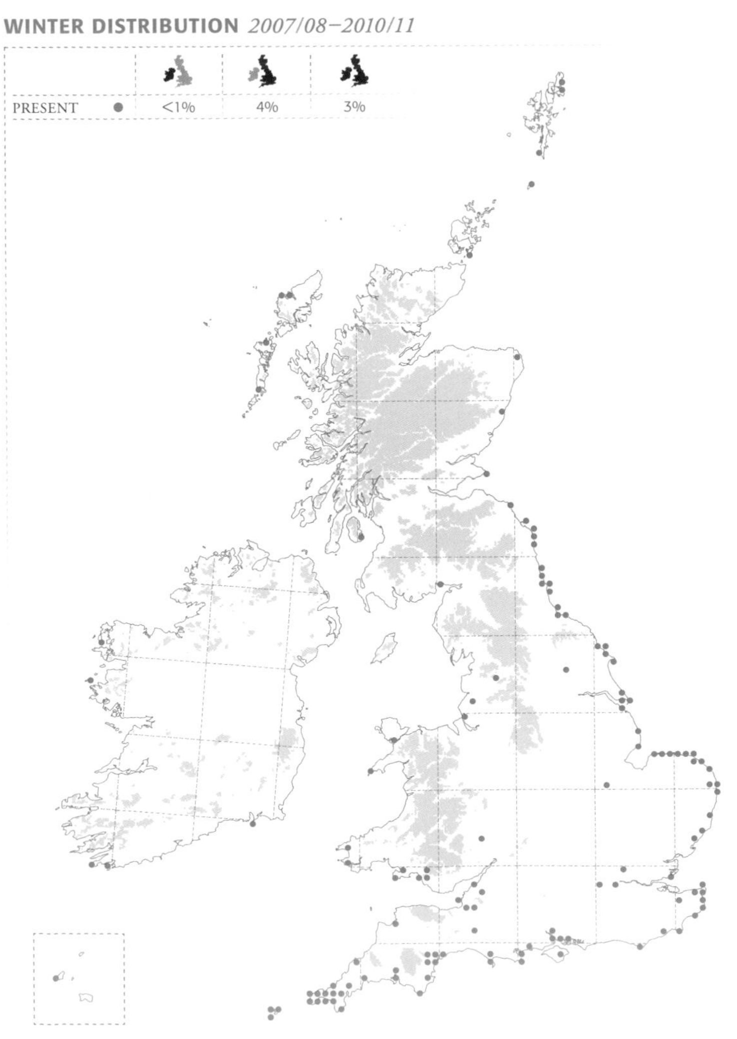

YELLOW-BROWED WARBLERS are scarce but increasing autumn migrants to Britain & Ireland. Most occupied 10-km squares were along the North Sea coast and in southwest England, with just five coastal squares occupied in Ireland. More than three-quarters of squares were occupied only in November, presumably by late migrants, whilst potentially overwintering birds were concentrated into southwest Britain. The number of occupied 10-km squares increased significantly from three in the *1981–84 Winter Atlas* to 130. Mean annual totals of migrant individuals increased from 72 during 1968–79 to 434 during 2000–03 (Fraser & Rogers 2006b). Ongoing debate concerns whether these birds represent migration to undescribed western wintering areas, rather than vagrancy involving reverse migration (Thorup 1998; Gilroy & Lees 2003).

WAXWING

Bombycilla garrulus BTO CODE: WX

GLYN SELLORS www.glynsellorsphotography.com

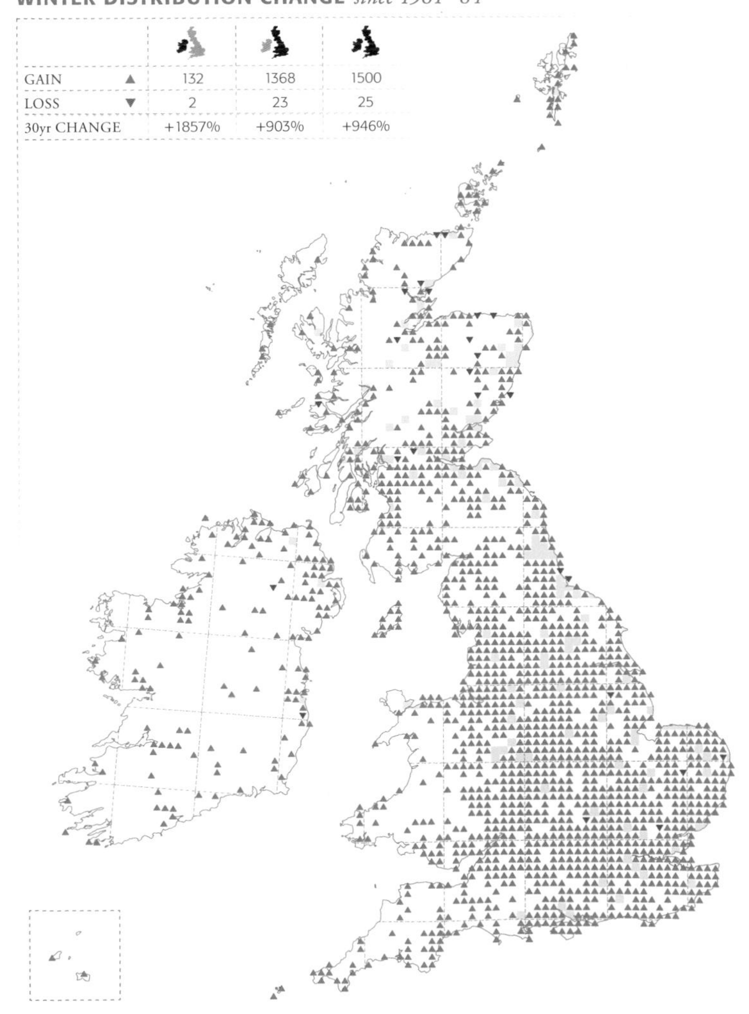

WINTER DISTRIBUTION CHANGE since 1981–84
GAIN 132 1368 1500
LOSS 2 23 25
30yr CHANGE +1857% +903% +946%

FORMERLY IRREGULAR AS immigrants from the Eurasian taiga, it appears that Waxwings are becoming more regular winter visitors to Britain & Ireland. Numbers vary annually, with irruptions occurring when Fennoscandian Rowan berry crops fail. Birds arrive in the Northern Isles and along the North Sea coast, where they can remain in large numbers if berries there are abundant. If the berry crop is poor, or as it becomes depleted, birds move south and west in search of food, as shown by resightings of colour-marked individuals (Murray 2001; *Migration Atlas*; *Birds of Scotland*).

Wintering Waxwings avoid the uplands and particularly favour urban and suburban areas, where they feast on the berries of ornamental trees and shrubs, leading to concentrations in towns and cities in eastern and central Scotland, northern England, the Midlands and southeast England. Numbers thin out to the southwest, resulting in fewer occupied squares in Wales and southwest England, although birds sometimes occur in the Channel Islands. Usually, only small numbers reach Ireland, where records tend to be concentrated in built-up areas in the north and east, with relatively few in central and western regions.

The number of squares holding Waxwings during atlas periods is dependent on the size and frequency of irruptions. During the *1981–84 Winter Atlas* there was only one minor irruption and over the three winters a total of 156 10-km squares were occupied. During 2007–11 there were two major irruptions, in autumns 2008 and 2010. During the latter, severe weather brought about increased competition with resident species for berries and resulted in a mass exodus of Waxwings from parts of Scotland, with many being resighted subsequently as far away as the English south coast (Duncan 2011). These factors contributed to a more-than-tenfold increase in range size at the 10-km scale since 1981–84 and the near-complete distribution through central, southern and eastern England, as well as the scatter of records throughout Ireland. Indeed the number of squares occupied in Ireland was only nine fewer than the entire total for Britain & Ireland in 1981–84.

Two Waxwings entering a Caithness wood in June 2009 constituted the single record of *possible* breeding (Holling *et al.* 2011b).

In memory of Frances (Fran) Blackburn

NUTHATCH
Sitta europaea BTO CODE: NH

THE NUTHATCH SHOWED northward range expansion in the *1988–91 Breeding Atlas* and maps revealing the extent of ongoing expansion have been eagerly awaited. This is very much a resident species (*Migration Atlas*) and the winter and breeding-season maps are virtually identical, with birds widespread in Wales, most of England and southern Scotland but absent from the Channel Islands, Isle of Man, Ireland and much of Scotland. Across England, they are notably absent from the Fens and the arable farmland of East Yorkshire. At least in the Fens, absence is not entirely due to lack of woodland but also because the woods that do exist are too isolated for colonisation to be successful (Bellamy *et al.* 1998).

During the winter, Nuthatches were reported from 38% more 10-km squares than during the *1981–84 Winter Atlas*. Breeding Nuthatches were reported from 25% more 10-km squares than in the *1988–91 Breeding Atlas*, which was in turn 8% more than in 1968–72. These trends partly represent outright colonisation, but the abundance change map highlights that increases since 1988–91 have occurred also in tetrad occupancy within the established range, particularly in western England and Wales. The area experiencing the most active colonisation since 1988–91 is the south of Scotland. Breeding was first proven in Scotland in Borders in 1989 (Murray 1991), since when the population has doubled every two to three years (*Birds of Scotland*).

Breeding numbers have risen by 82% in England and by 52% in Wales during 1995–2010 (*BBS Report 2011*), but small sample sizes hamper the calculation of trends in Scotland. The reporting of Nuthatches in gardens in southern Scotland has increased, however, from virtually zero in 1995 to 10% of gardens by 2013 (Garden BirdWatch 2013). These increases may be related to food supplies becoming more reliable, owing to milder winters and the increased provision of food in gardens (*Birds of Scotland*). Survival data to test this hypothesis are lacking but apparent increases in productivity may be more important (*BirdTrends*). As movements by adults are extremely rare, the spread of the species is probably being facilitated by the dispersal of juveniles in autumn (*Migration Atlas*), this being the period when territories become established (Matthysen 2002).

Sponsored by John & Jean McCaig (BTO Life Fellows)

BREEDING DISTRIBUTION *2008–11*

	Ireland	Britain	Britain & Ireland
POSSIBLE ●	-	6%	4%
PROBABLE ●	-	6%	5%
CONFIRMED ●	-	43%	32%
TOTAL	-	55%	41%

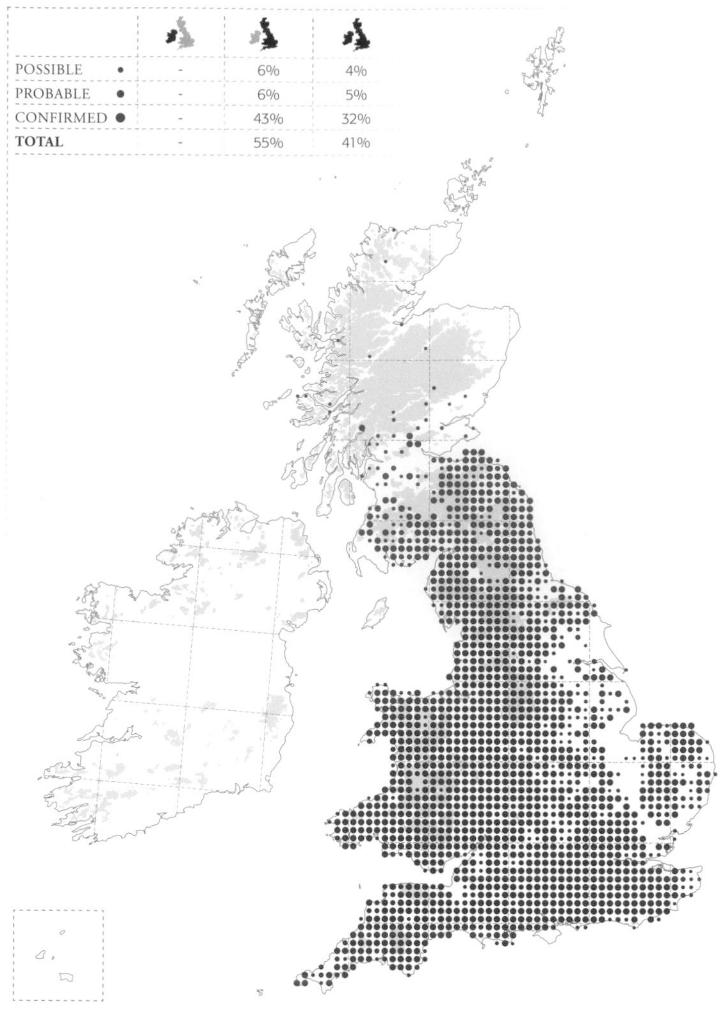

WINTER DISTRIBUTION *2007/08–2010/11*

	Ireland	Britain	Britain & Ireland
PRESENT ●	-	56%	41%

BREEDING DISTRIBUTION CHANGE *since 1968–72*

GAIN ▲	-	469	469
LOSS ▽	-	59	59
40yr CHANGE	-	+35%	+35%
20yr INDEX	-	+0.19	

WINTER DISTRIBUTION CHANGE *since 1981–84*

GAIN ▲	-	477	477
LOSS ▼	-	37	37
30yr CHANGE	-	+38%	+38%

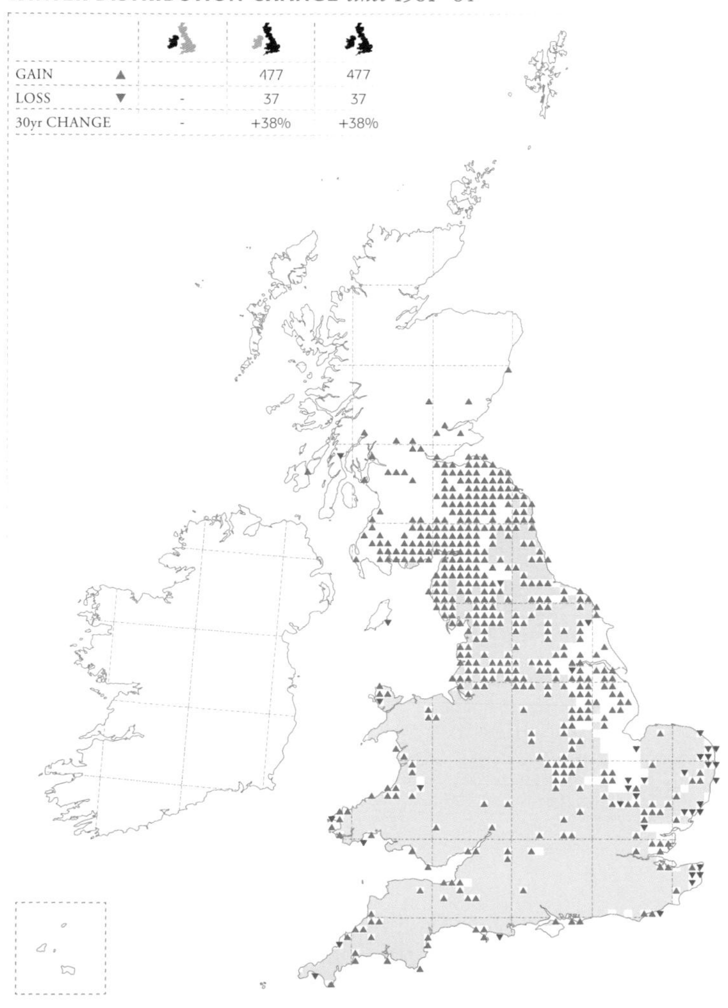

BREEDING RELATIVE ABUNDANCE *2008–11*

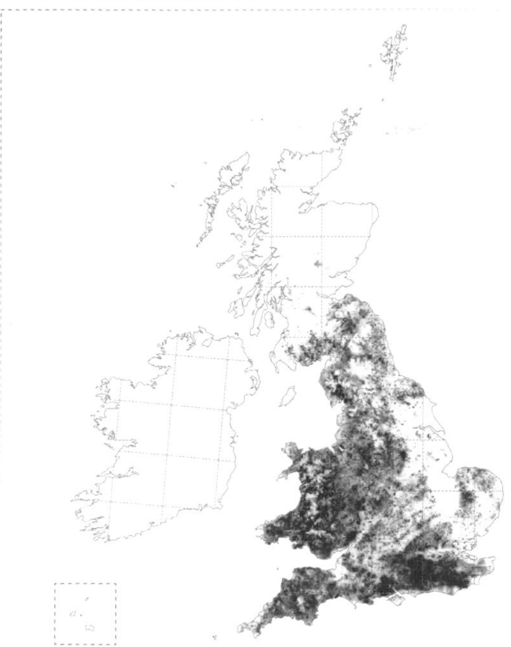

BREEDING RELATIVE ABUNDANCE CHANGE *since 1988–91*

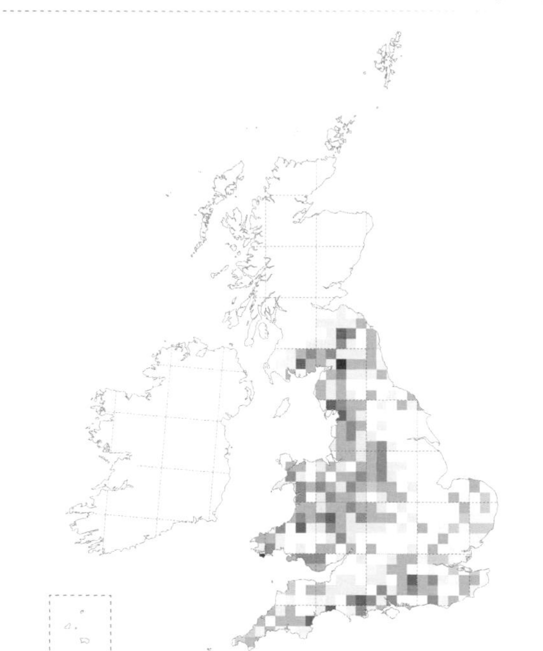

WINTER RELATIVE ABUNDANCE *2007/08–2010/11*

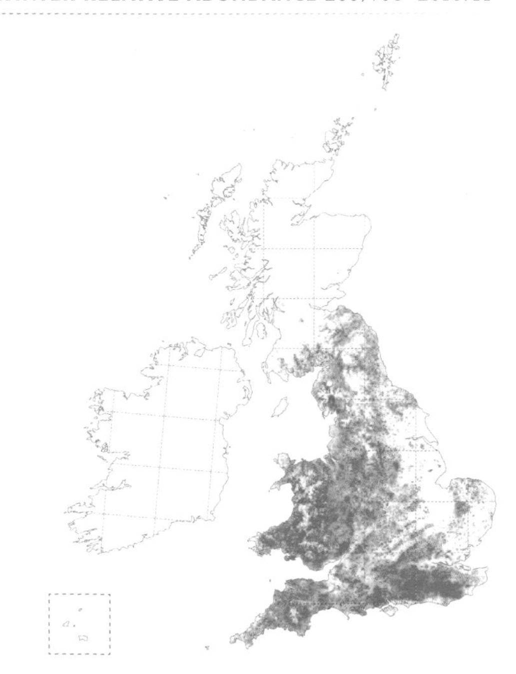

TREECREEPER

Certhia familiaris

BTO CODE: TC

PAUL NEWTON www.paulnewtonphotography.com

TREECREEPERS ARE WIDELY distributed, being recorded in more than 70% of 10-km squares in both seasons, a distribution that broadly accords with the availability of trees. In Britain the only significant gaps were the Northern Isles, Outer Hebrides, the highest peaks of Scotland and the English Fens. Gaps in the midlands and the western half of Ireland correspond with areas of unsuitable habitat (*Birds in Ireland*).

Treecreepers are among our most sedentary bird species (*Migration Atlas*) and, like the distribution maps, the abundance maps were virtually identical in winter and the breeding season. Highest densities tend to occur in the most heavily wooded landscapes, where presence is less constrained by distances between woodlands (Hinsley *et al.* 1995).

The UK breeding population declined by 18% during 1970–2010 (*SUKB 2012*), although numbers have been broadly stable since c.1980 (*BirdTrends*). The 5% range contraction since the *1968–72 Breeding Atlas* is most noticeable in southwest England, eastern England and southwest Ireland and could represent a retreat from marginal areas. Gains in Scotland and northwest Ireland may result from the creation and maturing of new woodland. These may also partly account for the 11% winter range expansion since the *1981–84 Winter Atlas*, although increased observer effort in these areas will also be influential.

Gains since the *1988–91 Breeding Atlas* mostly represent a reversal of the losses apparent between 1968–72 and 1988–91. This effect is particularly pronounced in Ireland where range size decreased by 20% then increased by 18%. The apparent dip in occupancy in 1988–91 may be explained by a combination of the low density of Treecreepers in Ireland at the time and a reduction in recording effort between 1968–72 and 1988–91. The abundance change map suggests that Treecreepers have subsequently increased in abundance in Ireland. In Britain, increases tend to be more frequent in the north than in the south, but this is not currently reflected in annual population monitoring results (*BBS Report 2011*). The causes of these changes are unknown. Compared to the breadth of research on the species in Fennoscandia (e.g. Suorsa *et al.* 2005), relatively little has been undertaken in Britain & Ireland.

Sponsored by Treswell Wood IPM Group

BREEDING DISTRIBUTION *2008–11*

	Ireland	Britain	Britain & Ireland
POSSIBLE	29%	13%	17%
PROBABLE	13%	10%	11%
CONFIRMED	24%	53%	45%
TOTAL	66%	76%	73%

WINTER DISTRIBUTION *2007/08–2010/11*

	Ireland	Britain	Britain & Ireland
PRESENT	69%	79%	77%

BREEDING DISTRIBUTION CHANGE *since 1968–72*

GAIN ▲	120	107	227
LOSS ▽	160	211	371
40yr CHANGE	6%	-5%	-5%
20yr INDEX	+0.13	-0.04	

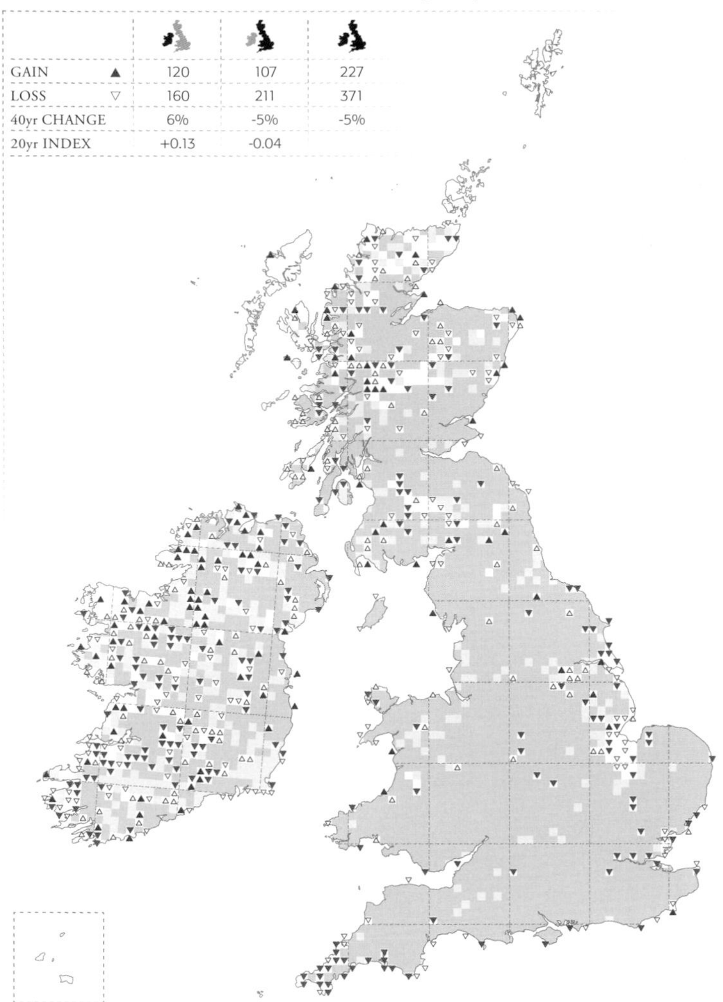

WINTER DISTRIBUTION CHANGE *since 1981–84*

GAIN ▲	212	301	513
LOSS ▼	118	111	229
30yr CHANGE	+16%	+9%	+11%

BREEDING RELATIVE ABUNDANCE *2008–11*

BREEDING RELATIVE ABUNDANCE CHANGE *since 1988–91*

WINTER RELATIVE ABUNDANCE *2007/08–2010/11*

SHORT-TOED TREECREEPER

Certhia brachydactyla BTO CODE: TH

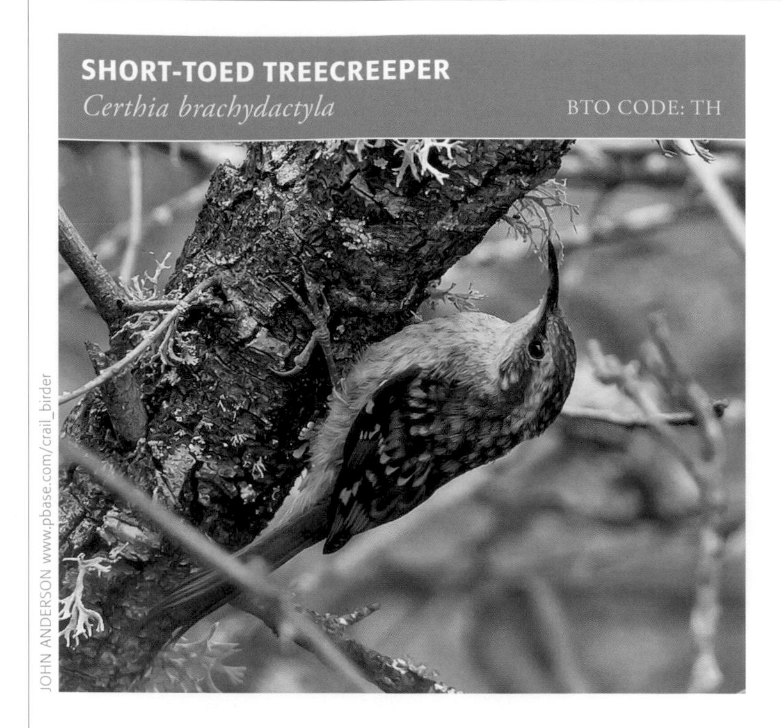

JOHN ANDERSON www.pbase.com/crail_birder

WITHIN THE REGION covered by this atlas the Short-toed Treecreeper is restricted to the Channel Islands. Occasional vagrants are identified on the south and east coasts of Britain, including one record in Suffolk during the Atlas period, but none have yet shown signs of breeding (Parkin & Knox 2010).

The Short-toed Treecreeper is the only treecreeper species to occur regularly in the Channel Islands. It is resident, breeding in woods, parks and large gardens across the main islands of Guernsey and Jersey. It has always been less common on the smaller islands, where limited habitat has probably hampered long-term persistence, and there have been long periods for which no records have been received. In Alderney, since the *1968–72 Breeding Atlas*, the species has gone from being a scarce resident to a vagrant; it was not recorded in the period 1990–2002 and the last record was in November 2002 (Young *et al.* 2013).

On the larger islands, long-term changes in distribution cannot be easily determined from atlas data because earlier data were collected at the island level rather than for individual 10-km squares: hence we report 60% 'range contraction' owing to loss from three of the five islands. This is unfortunate because the distribution of woodland birds like the Short-toed Treecreeper may have been affected by the many changes to the islands' landscapes during the 19th and 20th centuries (Hocart 2010). In the early part of the 19th century, outside the many orchards, tree cover on the islands was sparse and many of the steep-sided valleys which now contain woodland were either grazed by sheep or managed for the collection of gorse for fuel. Most trees were found along earth banks surrounding fields and roads. During the occupation (1940–45) most of the little remaining woodland was cleared and, since its arrival in the late 1970s, Dutch Elm Disease has decimated the islands' elm population. However, since the 1940s, woodland has grown up naturally and has been deliberately planted across the islands resulting in many areas of relatively young woodland. Conditions for woodland birds in the Channel Islands are now probably at their best for many decades.

BREEDING DISTRIBUTION *2008–11*

POSSIBLE •	-	<1%	<1%
PROBABLE •	-	-	-
CONFIRMED ●	-	<1%	<1%
TOTAL	-	<1%	<1%

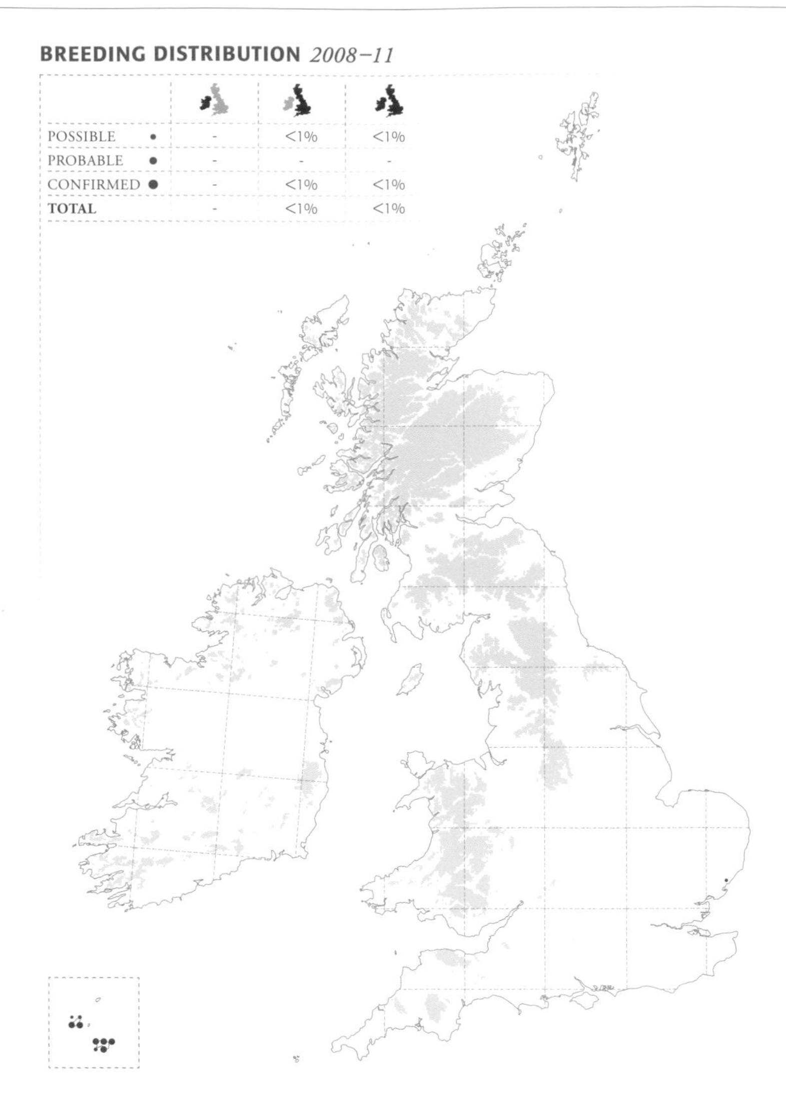

BREEDING DISTRIBUTION CHANGE *since 1968–72*

GAIN ▲	-	0	0
LOSS ▽	-	3	3
40yr CHANGE	-	-60%	-60%
20yr INDEX	-	-0.60	

JEFF & ALLISON KEW

▲ In southern Europe, Short-toed Treecreepers occur in lowland wooded landscapes, and Treecreepers are only found at higher altitude.

WREN
Troglodytes troglodytes

BTO CODE: WR

PAUL HILLION

THE WREN IS among the most widespread species in Britain & Ireland, capable of occupying a wide range of habitats. Wrens were widely recorded during this atlas, being present in at least 97% of squares throughout the year and absent only from some of the highest-altitude areas in Scotland. Of the five races occurring in Britain & Ireland (Parkin & Knox 2011), *fridariensis* on Fair Isle (Aspinall & Aspinall 2011) and *hirtensis* on St Kilda (Miles 2011) are of high conservation concern. The winter and breeding-season relative abundance maps are very similar and highlight the low densities associated with upland areas. Overall densities are higher in Ireland than in Britain.

Given the Wren's widespread nature, there is little scope for further range gains. Data suggest a 4% range expansion overall since the *1981–84 Winter Atlas*, most notably across northern Scotland; this probably reflects increasingly mild winters and improved coverage in these areas of low Wren density. The extent of the breeding range has remained broadly stable throughout the atlases, though this masks fluctuations in population levels. The main driver of Wren population size is winter weather (Peach *et al.* 1995b; Robinson *et al.* 2007b); breeding populations have fluctuated throughout the period spanned by these atlases (*BirdTrends*) but with no discernible impact on range size.

The abundance change map indicates that since the *1988–91 Breeding Atlas* there have been increases in the proportion of occupied tetrads, particularly in many parts of Scotland and north and northwest Ireland; lack of coloration elsewhere on the map probably indicates already high tetrad occupancy for this common species, rather than a lack of abundance trend. In Britain, annual monitoring data confirm this, showing increases in density during 1994–2009 from central England northwards, with increases greatest in Scotland (Massimino *et al.* 2013) where the population rose by 31% during 1995–2009 (Risely *et al.* 2011). Although annual breeding population monitoring data show generally increasing trends throughout Britain & Ireland, a decline is evident in many regions from 2009 onwards following the severe winter of 2009/10 (*BBS Report 2011*; Lovatt *et al.* 2012); such declines have always been short-lived.

Sponsored by WREN

BREEDING DISTRIBUTION *2008–11*

POSSIBLE •	3%	2%	3%
PROBABLE •	8%	6%	7%
CONFIRMED •	87%	88%	88%
TOTAL	98%	96%	98%

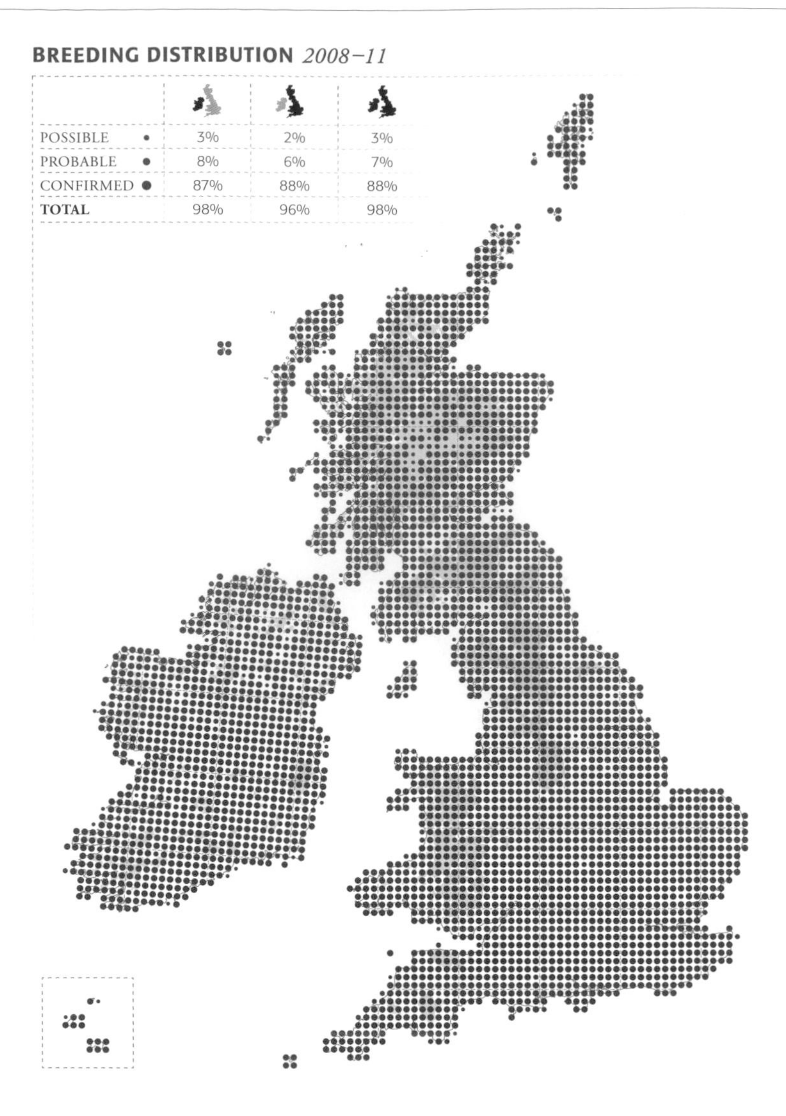

WINTER DISTRIBUTION *2007/08–2010/11*

PRESENT •	98%	96%	97%

BREEDING DISTRIBUTION CHANGE *since 1968–72*

GAIN ▲	12	36	48
LOSS ▽	3	21	24
40yr CHANGE	+1%	+1%	+1%
20yr INDEX	+0.01	+0.01	

WINTER DISTRIBUTION CHANGE *since 1981–84*

GAIN ▲	17	164	181
LOSS ▼	21	19	40
30yr CHANGE	-<1%	+6%	+4%

BREEDING RELATIVE ABUNDANCE *2008–11*

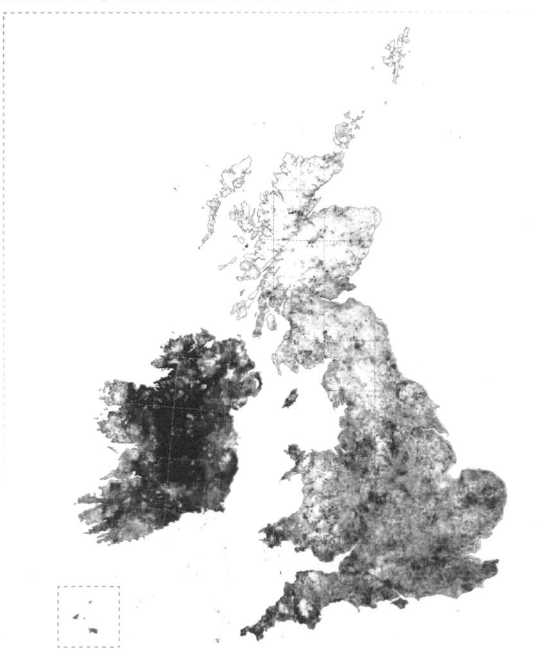

BREEDING RELATIVE ABUNDANCE CHANGE *since 1988–91*

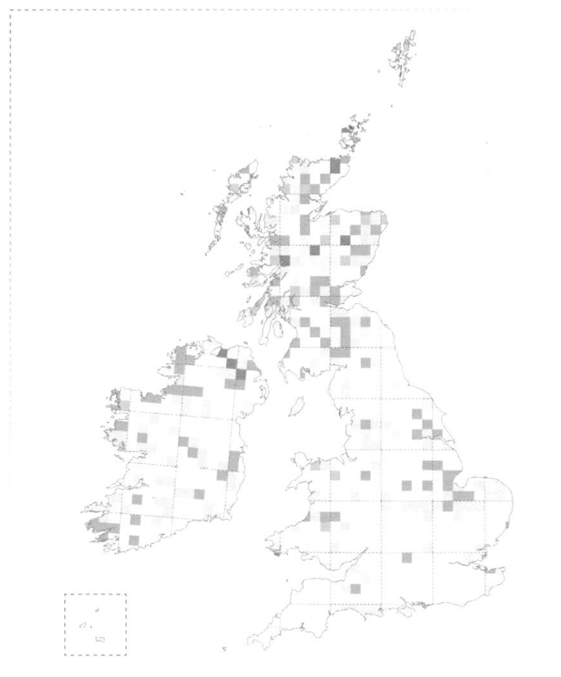

WINTER RELATIVE ABUNDANCE *2007/08–2010/11*

STARLING

Sturnus vulgaris

BTO CODE: SG

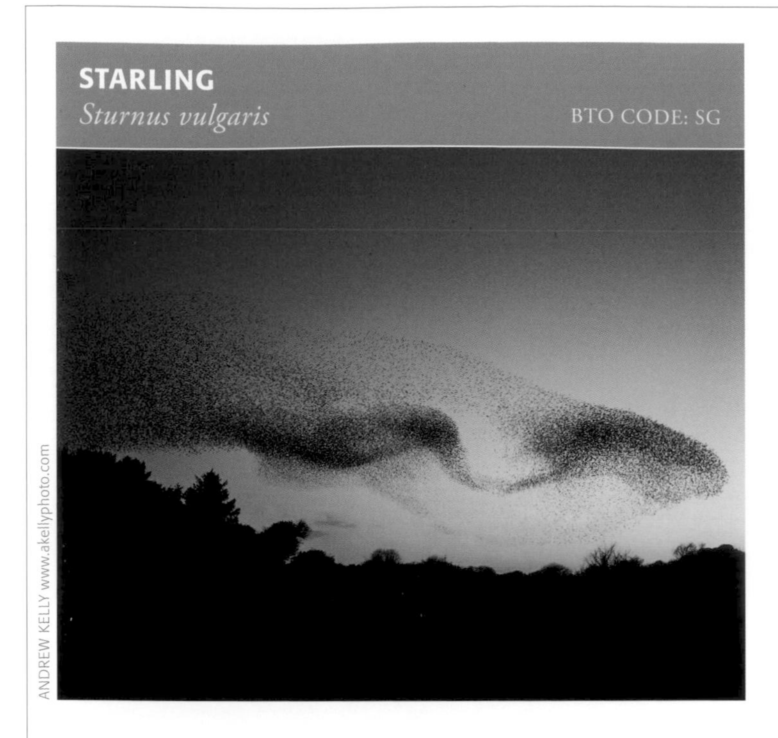

ANDREW KELLY www.akellyphoto.com

WITH THE EXCEPTION of the highest ground in northern and western Scotland, Starlings are widespread throughout Britain & Ireland during winter, with residents joined by migrants from continental Europe (*Migration Atlas*). The breeding-season distribution is largely the same, although some parts of central Wales are unoccupied at this time. Winter abundance is highest in areas of low-lying pastoral farmland in the northeast and midlands of Ireland and in the west and southwest of Britain (Atkinson *et al.* 2002). In the breeding season, abundance is greatest across Ireland and in low-lying areas in eastern Britain and the Northern Isles, with other concentrations associated with urban areas. By comparison, much of Wales, southwest England and the Scottish uplands have very low breeding densities. Range changes for wintering Starlings have been negligible, with a 2% expansion overall since the *1981–84 Winter Atlas*. Expansion is most noticeable on upland margins in northern Scotland, and is possibly a result of improved coverage in the range periphery.

Since the *1968–72 Breeding Atlas* the distribution has remained unchanged in Ireland, but there has been a 5% contraction in Britain with most losses in the Scottish Highlands, western Wales and southwest England. However, the breeding abundance change map highlights major declines in tetrad occupancy through much of Britain since the *1988–91 Breeding Atlas*, consistent with a 50% decline in numbers in the UK during 1995–2010 (Robinson *et al.* 2005a; *SUKB 2012*). A decrease in the survival of first-year birds is likely to be a driver of this decline (Freeman *et al.* 2007). In contrast, Atlas data show a pronounced increase in the proportion of tetrads occupied in Ireland, despite annual monitoring estimating a 7% population decline during 1998–2010 (*CBS Trend 2010*).

Reasons for the contrasting trends of breeding Starlings between Britain and Ireland may be related to feeding preferences and differences in the extent and condition of pastures (Olsson *et al.* 2002; Devereux *et al.* 2004). In grasslands, Starling diet consists largely of cranefly larvae (Whitehead *et al.* 1995) and high-input grassland systems, such as those in much of Ireland, may result in high soil fertility and prey abundance (Vickery *et al.* 2001).

BREEDING DISTRIBUTION *2008–11*

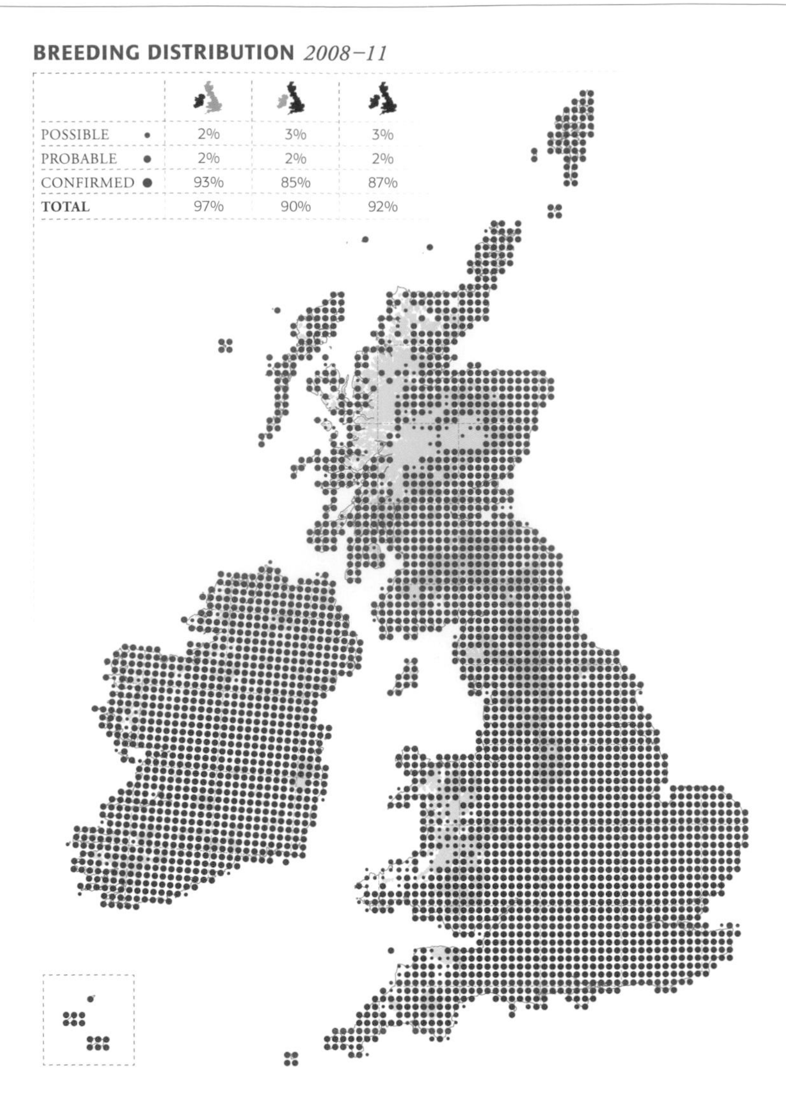

POSSIBLE •	2%	3%	3%
PROBABLE •	2%	2%	2%
CONFIRMED •	93%	85%	87%
TOTAL	97%	90%	92%

WINTER DISTRIBUTION *2007/08–2010/11*

PRESENT •	97%	90%	92%

BREEDING DISTRIBUTION CHANGE *since 1968–72*

GAIN ▲	13	14	27
LOSS ▽	9	161	170
40yr CHANGE	+<1%	-5%	-4%
20yr INDEX	+0.04	-0.05	

WINTER DISTRIBUTION CHANGE *since 1981–84*

GAIN ▲	29	103	132
LOSS ▼	22	56	78
30yr CHANGE	+1%	+2%	+2%

BREEDING RELATIVE ABUNDANCE *2008–11*

BREEDING RELATIVE ABUNDANCE CHANGE *since 1988–91*

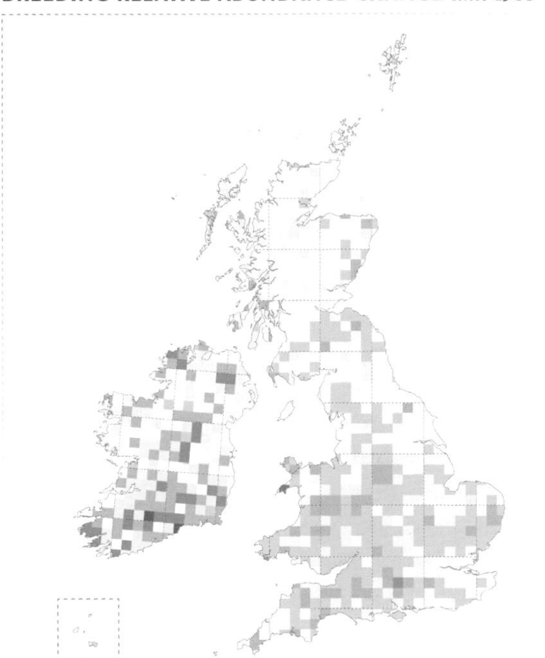

WINTER RELATIVE ABUNDANCE *2007/08–2010/11*

DIPPER

Cinclus cinclus

BTO CODE: DI

GLYN SELLORS www.glynsellorsphotography.com

DIPPERS ARE EXCLUSIVELY associated with fast-flowing, stony-bedded watercourses, whether in winter or in the breeding season. As a result, the species is widespread throughout non-coastal areas in Scotland, Wales, northern and southwestern England, and absent from low-lying areas of central, southern and eastern England. In Ireland they are widespread across the country, with the exception of the vast raised bogs of the midlands and mid-west. The relative abundance maps show that densities tend to be relatively high in Scotland, northern England and the southern half of Wales, and lower in the northern half of Wales, southwest England and Ireland. Outside the core areas, sporadic sightings of birds in winter often relate to continental immigrants of the Black-bellied form (nominate *cinclus*).

There have been small expansions of the winter range in Britain (14%) and Ireland (19%) since the *1981–84 Winter Atlas*, most likely linked to better coverage as these gains are not mirrored on the breeding-season maps. Breeding-range changes since the *1968–72 Breeding Atlas* in Britain have been slight (-11%). Small losses in northern and western Scotland may be related to acidification problems from conifer plantations (*Birds of Scotland*). Farther south, it is notable that losses have taken place on the eastern fringe of the range in Wales and southwest England.

The abundance change map highlights widespread declines in tetrad occupancy across much of the range since the *1988–91 Breeding Atlas,* so it is unsurprising that annual monitoring data reveal a 30% population decline in the UK during 1970–2010 (*SUKB 2012*). The main reason for the decline is thought to be acidification of rivers flowing through conifer plantations (Tyler & Ormerod 1994).

Range contractions reported in Ireland up to the *1988–91 Breeding Atlas* were partly ascribed to decreased coverage but these are still evident and a 21% range contraction from 1968–72 to 2008–11 is apparent. Range changes may conceal local population increases: Perry and Agnew (1993) documented a 116% population increase during 1972–92 in northwest Ireland. However, in Co. Cork the population is largely stable with some losses in squares with apparently poor-quality habitat (P. Smiddy pers comm).

Sponsored by the Environment Agency

BREEDING DISTRIBUTION *2008–11*

POSSIBLE •	15%	8%	10%
PROBABLE •	8%	6%	6%
CONFIRMED ●	26%	32%	30%
TOTAL	49%	46%	46%

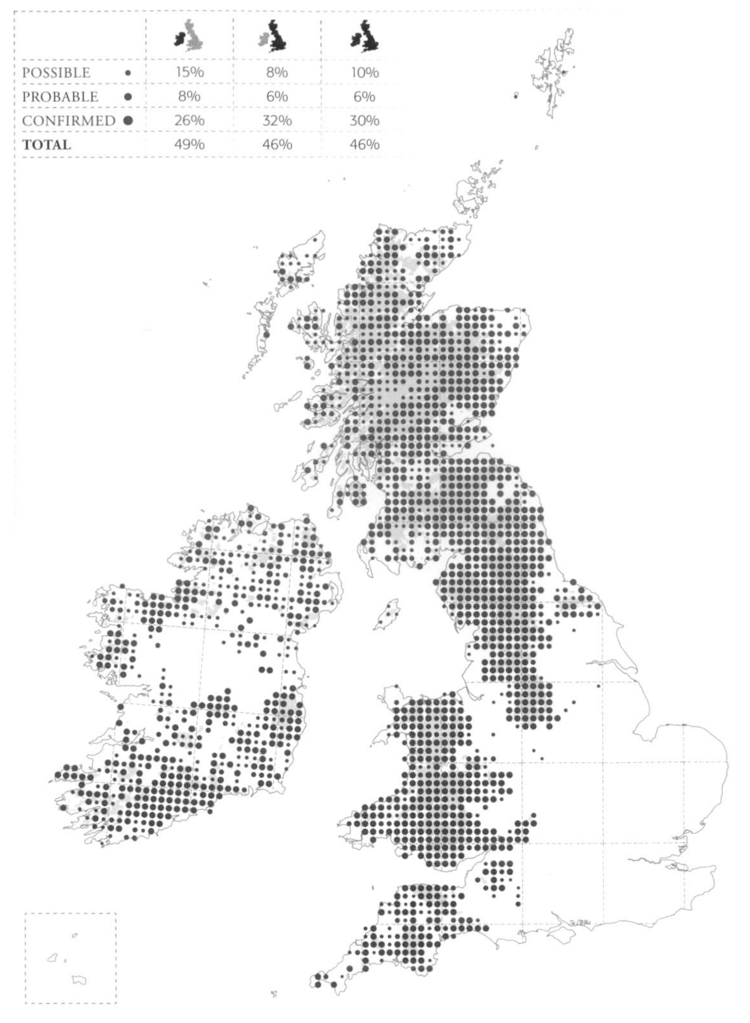

WINTER DISTRIBUTION *2007/08–2010/11*

PRESENT ●	44%	47%	46%

BREEDING DISTRIBUTION CHANGE *since 1968–72*

GAIN ▲	108	119	227
LOSS ▽	236	270	506
40yr CHANGE	-21%	-11%	-14%
20yr INDEX	-0.06	-0.18	

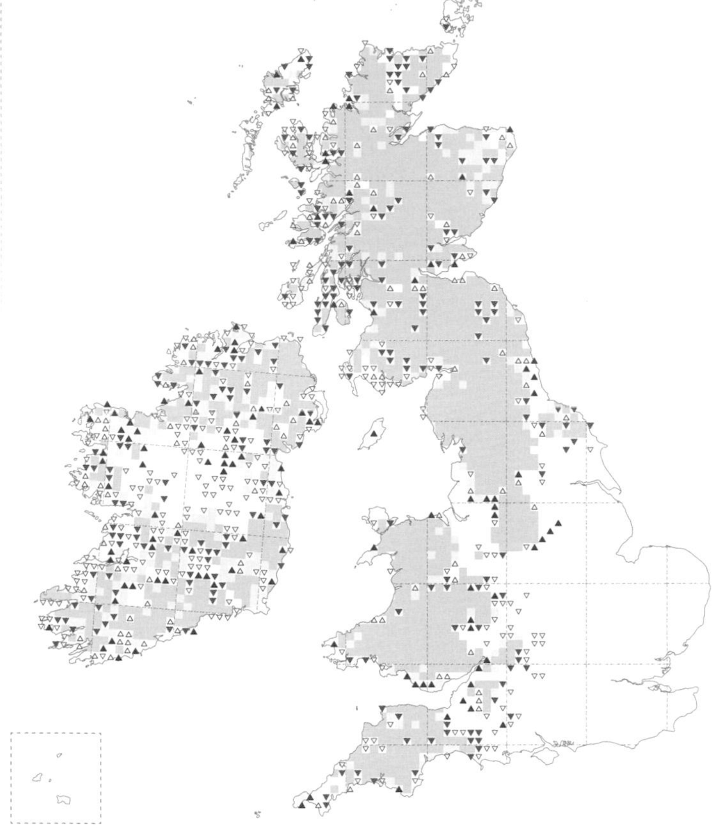

WINTER DISTRIBUTION CHANGE *since 1981–84*

GAIN ▲	172	302	474
LOSS ▼	105	136	241
30yr CHANGE	+19%	+14%	+15%

BREEDING RELATIVE ABUNDANCE *2008–11*

BREEDING RELATIVE ABUNDANCE CHANGE *since 1988–91*

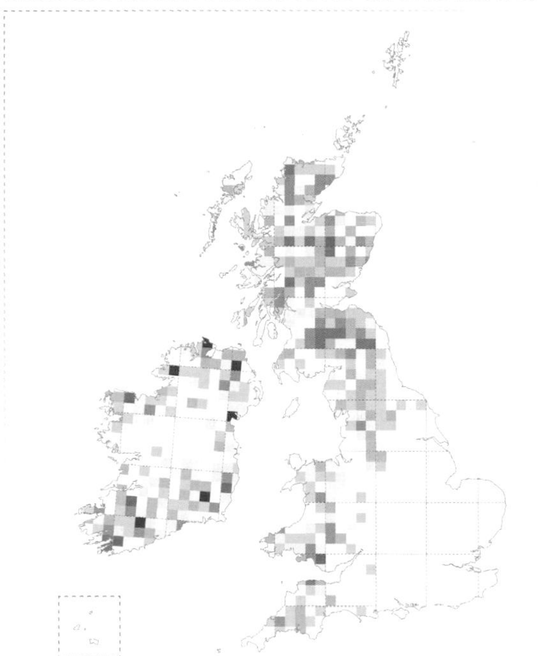

WINTER RELATIVE ABUNDANCE *2007/08–2010/11*

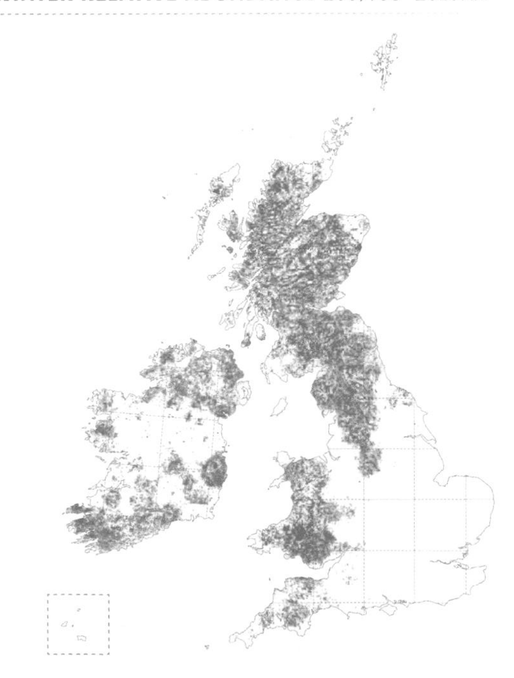

RING OUZEL

Turdus torquatus

BTO CODE: RZ

MARTIN BENNETT

THE RING OUZEL is a summer visitor and its association with high ground in Britain can be clearly seen in the breeding distribution and relative abundance maps. In Ireland it is more restricted, with breeding largely confined to Co. Kerry and the northwest.

The breeding-season distribution change and abundance change maps paint a picture of decline over the last 40 years: range size has decreased by 43% in Britain and by 57% in Ireland since the *1968–72 Breeding Atlas*, with the losses probably accelerating in Ireland. Too few Ring Ouzels are recorded on BBS squares to produce annual trends (*BBS Report 2011*) but periodic repeat surveys and regularly monitored study areas indicate that substantial declines are under way throughout Britain (Wotton *et al.* 2002b; Sim *et al.* 2010).

Recent losses mean that this species no longer breeds in Exmoor and is now largely absent from most of central Wales, the western half of the Scottish Southern Uplands and Co. Wicklow. In Scotland north of the Central Belt, Ring Ouzels have contracted from the range margins, particularly in the southwest and far north. These changes are consistent with research showing contraction to sites at higher altitude, with greater heather cover or more distant from conifer plantations (Sim *et al.* 2010). Low survival rates, particularly of first-year birds, and climate change have been suggested as potential drivers of change (Beale *et al.* 2006; Sim *et al.* 2010, 2011).

The breeding-season distribution map shows 516 10-km squares occupied by *non-breeding* Ring Ouzels, reflecting the many migrants found in April and May on coasts and low ground in southeast and central England. A return passage takes place in autumn and the winter distribution map shows occupied squares along the east coast of England, but more particularly in the far south. The majority of these squares were occupied only in early November, presumably by late migrants. After mid November Ring Ouzels were reported from a wide scattering of 10-km squares, mainly in southern England. The increase in the number of occupied 10-km squares since the *1981–84 Winter Atlas* perhaps reflects a later migration period and an increased tendency for some birds to remain to winter.

Sponsored by North Northumberland Bird Club

BREEDING DISTRIBUTION *2008–11*

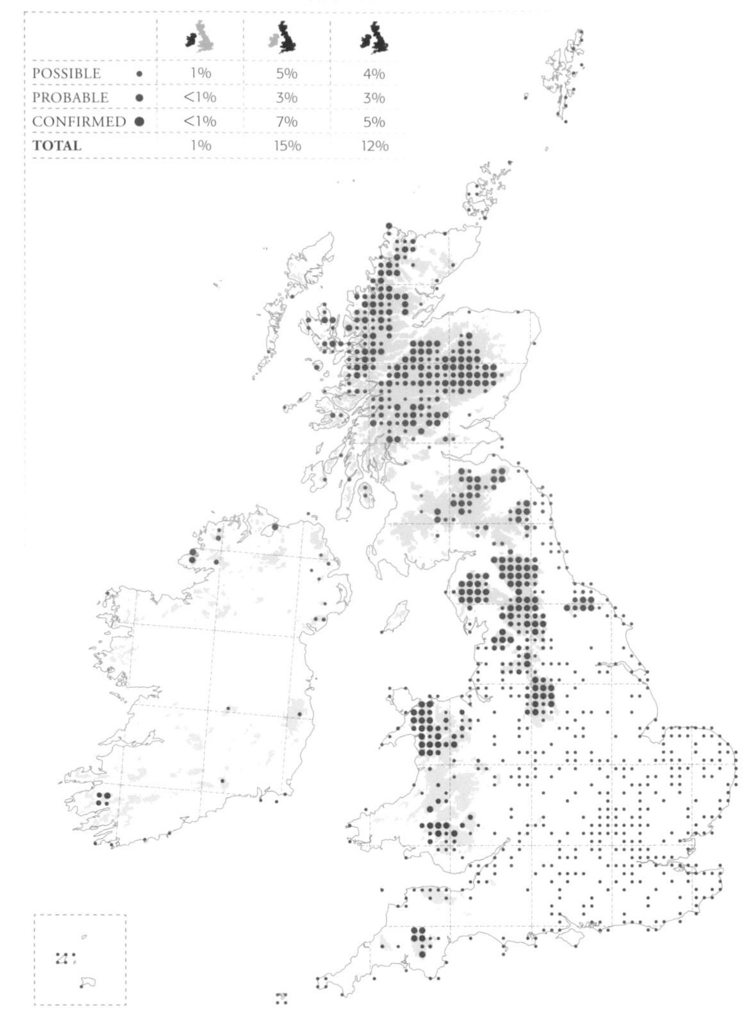

	Ireland	Britain	Britain & Ireland
POSSIBLE •	1%	5%	4%
PROBABLE •	<1%	3%	3%
CONFIRMED •	<1%	7%	5%
TOTAL	1%	15%	12%

BREEDING RELATIVE ABUNDANCE CHANGE *since 1988–91*

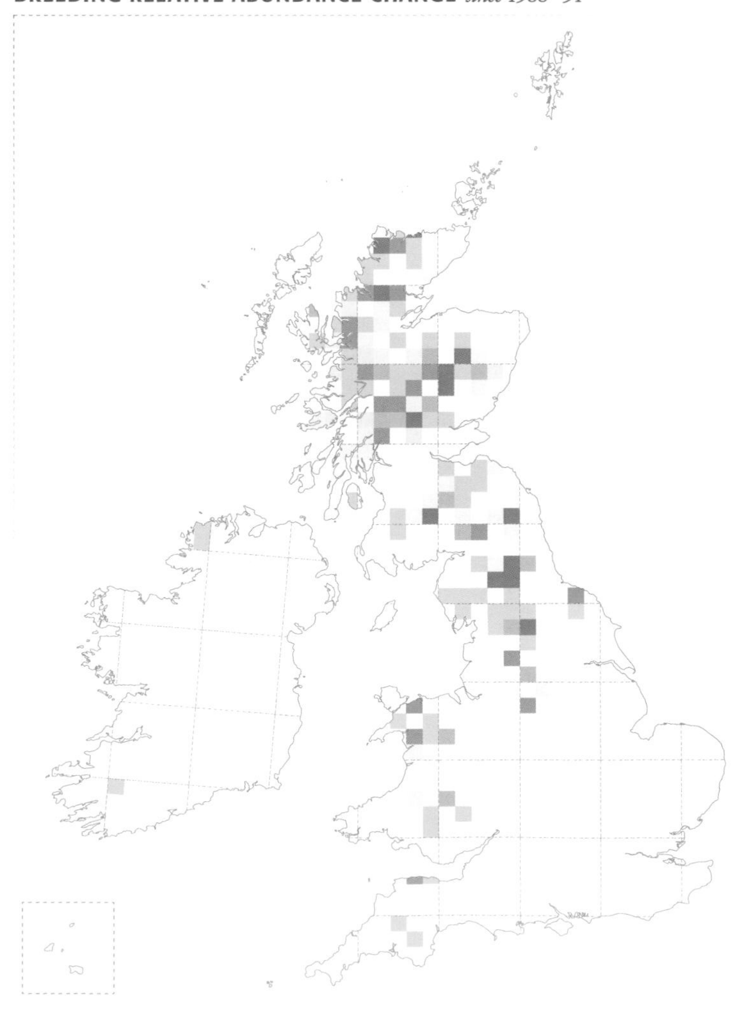

BREEDING DISTRIBUTION CHANGE *since 1968–72*

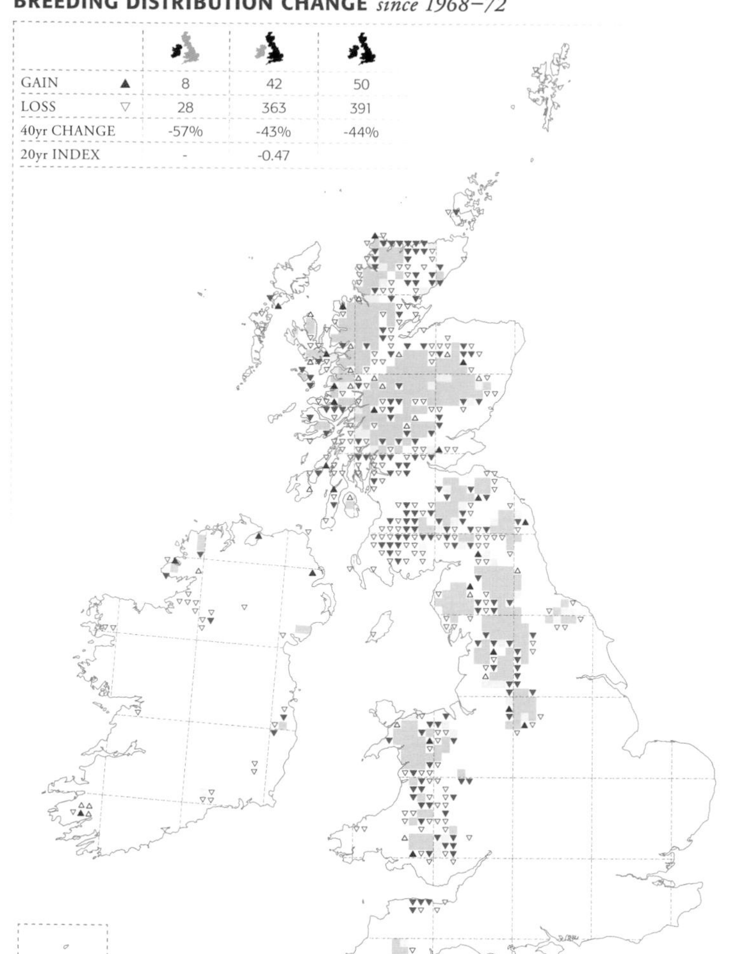

GAIN ▲	8	42	50
LOSS ▽	28	363	391
40yr CHANGE	-57%	-43%	-44%
20yr INDEX	-	-0.47	

BREEDING RELATIVE ABUNDANCE *2008–11*

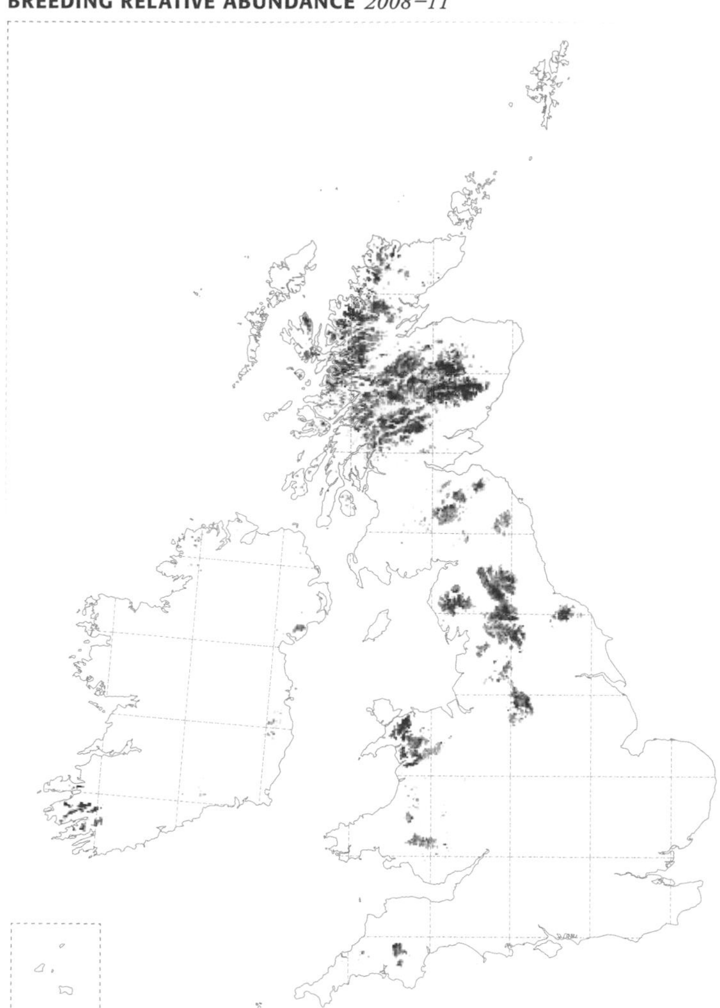

WINTER DISTRIBUTION *2007/08–2010/11*

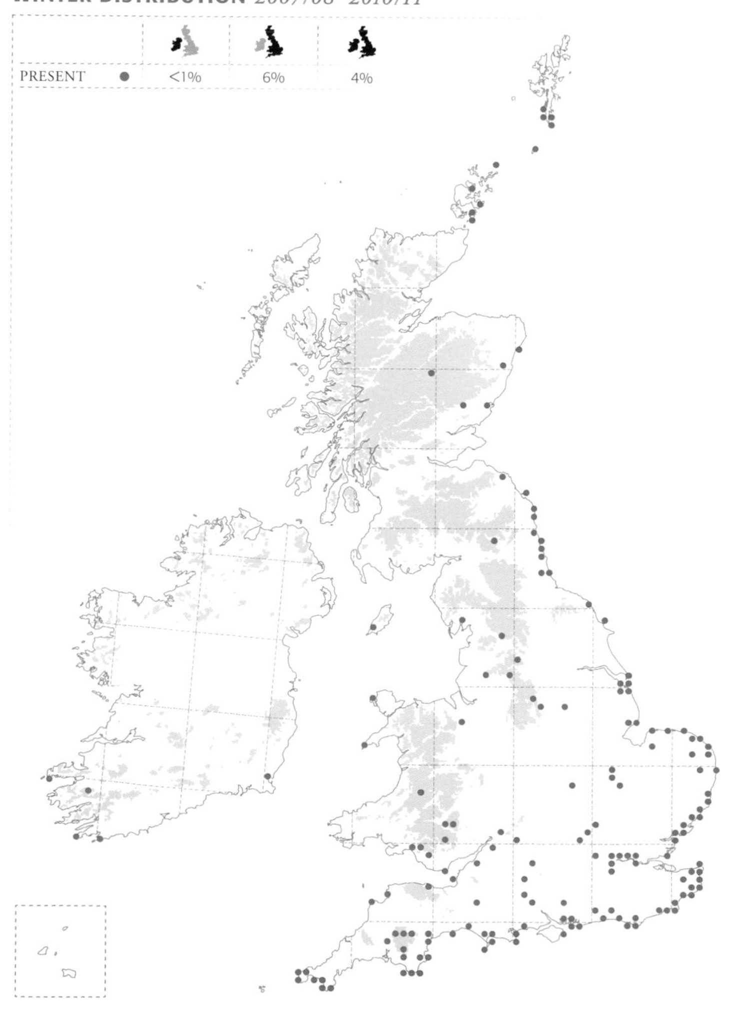

PRESENT ●	<1%	6%	4%

WINTER DISTRIBUTION CHANGE *since 1981–84*

GAIN ▲	4	56	60
LOSS ▼	1	14	15
30yr CHANGE	+300%	+300%	+300%

BLACKBIRD
Turdus merula

BTO CODE: B.

JILL PAKENHAM

THE BLACKBIRD IS amongst the commonest bird species in Britain & Ireland, with records from 96% of 10-km squares in both the winter and the breeding season; it is absent only from a few upland 10-km squares and from some remote islands. Autumn migrants may explain why slightly more squares are occupied in winter than in the breeding season.

Relative abundance maps show similar patterns in the breeding season and winter: high densities in the east and midlands of Ireland and the lowlands of England and Wales contrast with low densities in the uplands of northern England, Scotland, Wales and western Ireland. Interestingly, relative abundance appears lower in winter than in the breeding season in the major urban conurbations of London and northwest England but this could simply reflect differing detectability in the two seasons.

In winter the 2% increase in range size since the *1981–84 Winter Atlas* is largely manifested in northwest Scotland and probably reflects a combination of improved coverage and new habitat creation through woodland plantings. In the breeding season there has been little overall change in range size since the *1968–72 Breeding Atlas*: the few apparent gains and losses are all on the margins of the range in northwest Scotland and at coastal squares with little land. Most are probably the consequence of chance detections in one or more atlases of Blackbirds breeding at low density, although some could also reflect the availability of new habitat.

Although the breeding-season change map emphasises stability, the abundance change map shows increases in the proportion of occupied tetrads per 20-km square in parts of Wales and Scotland since the *1988–91 Breeding Atlas*, a pattern supported by a 23% population increase in the UK during 1995–2010 (*SUKB 2012*). The lack of coloration in England despite a 21% population increase (*BBS Report 2011*) illustrates one of the weaknesses of this mapping method: areas with already high tetrad occupancy can undergo further density increases without corresponding increases in tetrad occupancy. In fact, densities may be rising in all but the extreme southeast of England (Massimino *et al.* 2013). Increasing density in Ireland corroborates a small (5%) population increase during 1998–2010 (*CBS Trend 2010*).

In memory of David Snow

BREEDING DISTRIBUTION *2008–11*

POSSIBLE •	1%	2%	2%
PROBABLE •	4%	2%	3%
CONFIRMED ●	92%	90%	91%
TOTAL	97%	94%	96%

WINTER DISTRIBUTION *2007/08–2010/11*

PRESENT ●	98%	95%	96%

BREEDING DISTRIBUTION CHANGE *since 1968–72*

GAIN ▲	7	30	37
LOSS ▽	6	49	55
40yr CHANGE	+<1%	-1%	-<1%
20yr INDEX	+0.01	+0.01	

WINTER DISTRIBUTION CHANGE *since 1981–84*

GAIN ▲	10	87	97
LOSS ▼	11	27	38
30yr CHANGE	-<1%	+2%	+2%

BREEDING RELATIVE ABUNDANCE *2008–11*

BREEDING RELATIVE ABUNDANCE CHANGE *since 1988–91*

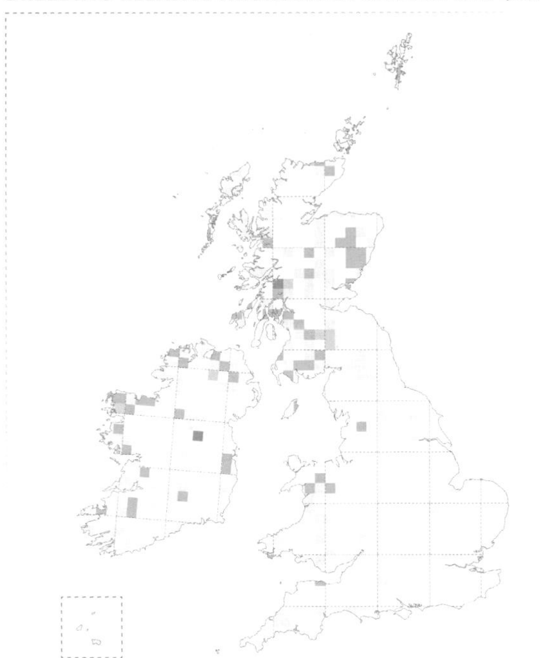

WINTER RELATIVE ABUNDANCE *2007/08–2010/11*

FIELDFARE
Turdus pilaris

BTO CODE: FF

MARTIN BENNETT

FROM THEIR BREEDING grounds in Fennoscandia, up to one million Fieldfares make their way to Britain & Ireland each winter (*Migration Atlas*). They are widely distributed in winter, with the exception of major urban areas and the highest ground in northern Scotland and in northwest Ireland; some marginal upland areas were occupied only in early winter. The winter abundance map clearly shows a preference for low-lying land, where Fieldfares feed in open farmland, hedgerows and woodland, only moving into gardens during harsh winter weather. The highest concentrations are found in southeast and central England.

The winter distribution map shows a 4% range expansion since the *1981–84 Winter Atlas*, with gains confined mainly to the Scottish Highlands, Hebridean islands and the west of Ireland. Although some of this increase is likely to be due to improved coverage in these marginal areas, the two severe winters of 2009/10 and 2010/11 may have forced greater numbers of Fennoscandian immigrants into western parts of Britain and Ireland. Significant increases in Fieldfare numbers in British and Irish gardens were recorded during these two winters (Garden BirdWatch 2010; O'Sullivan 2010).

Mapping the breeding distribution of Fieldfares is difficult owing to migrant birds lingering into May. Once the 1,293 10-km squares with *non-breeding* birds are discounted, Fieldfares are clearly very rare breeders with just a handful of 10-km squares showing evidence of breeding, mostly in Scotland and northern England. Breeding was *confirmed* in just four 10-km squares, in the Cairngorms, Shetland, Scottish Borders and Peak District. Breeding is not always confined to the north: a pair bred in Kent in 1991 (Ogilvie *et al.* 1994).

These data suggest a substantial, 78% contraction of breeding range since the *1988–91 Breeding Atlas* but many of the records in the earlier period probably related to late migrants rather than true breeders. There were 18 10-km squares with *probable* or *confirmed* breeding in both 1968–72 and 1988–91 and 11 10-km squares in 2008–11. The distribution and abundance of breeding Fieldfares is highly unpredictable, as is the number that are actually detected (*RBBP 2010*), making interpretation of trends problematic.

WINTER DISTRIBUTION *2007/08–2010/11*

BREEDING DISTRIBUTION *2008–11*

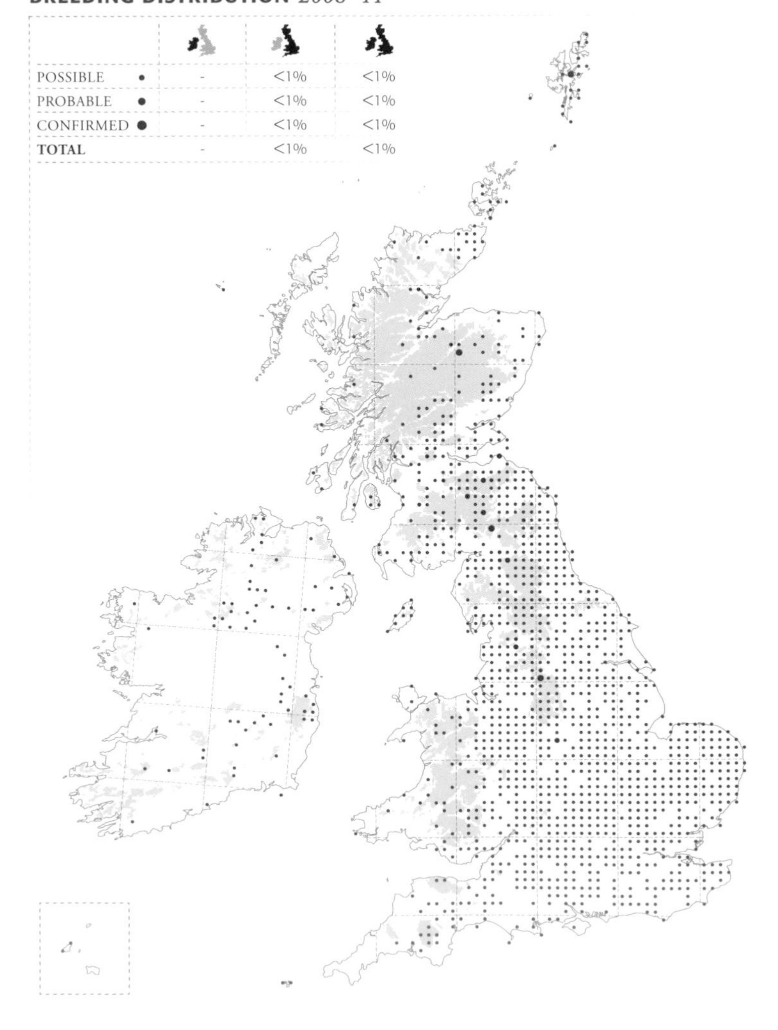

WINTER DISTRIBUTION CHANGE *since 1981–84*

GAIN ▲	75	186	261
LOSS ▼	33	85	118
30yr CHANGE	+5%	+4%	+4%

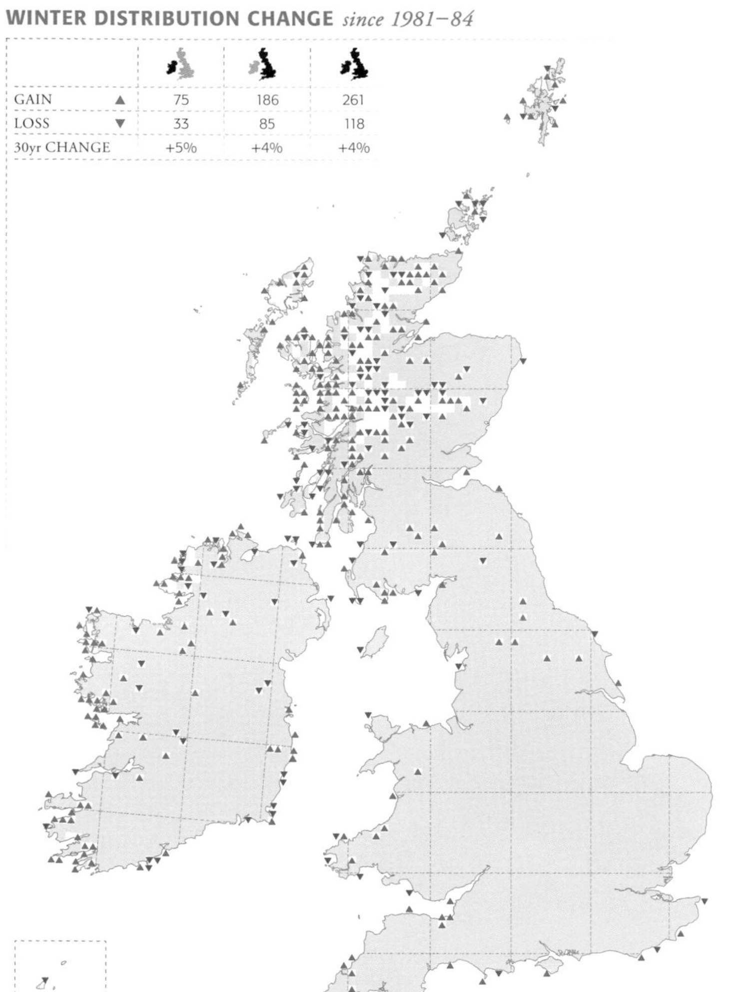

WINTER RELATIVE ABUNDANCE *2007/08–2010/11*

BREEDING DISTRIBUTION CHANGE *since 1968–72*

GAIN ▲	0	23	23
LOSS ▽	1	34	35
40yr CHANGE	-100%	-32%	-34%
20yr INDEX	-	-0.55	

SONG THRUSH

Turdus philomelos

BTO CODE: ST

ALISON MCARTHUR

THE MELODIOUS SONG of the Song Thrush is a familiar sound across British & Irish woodland, farmland and gardens. The species was found in 94% of 10-km squares in the breeding season and 92% of squares in winter, being absent only from parts of the Scottish Highlands and Southern Uplands during the winter months. Relative abundance is distinctly higher in Ireland than in Britain in both seasons and abundance is generally greatest in low-lying areas.

A 6% range expansion in Britain & Ireland since the *1981–84 Winter Atlas* is mostly due to gains in the upland fringes, where many of these records relate to early returning birds in February. Whilst our population is largely resident, small numbers migrate south to winter in France, Spain or Portugal and appreciable numbers from the Low Countries winter in Britain & Ireland (*Migration Atlas*).

There is little change in the breeding distribution since previous atlases; however, the breeding abundance change map shows increases in Ireland, Wales, central and southwestern England since the *1988–91 Breeding Atlas*. In Scotland the situation is less clear, with no obvious pattern of change, while a distinct decline is evident in the east of England.

During 1970–2010 there was an overall 54% population decline in the UK, although the most recent part of that period, from 1995–2010, was characterised by a 13% increase (*SUKB 2012*). However, this partial population recovery is not mirrored in Ireland, despite the increases shown on the abundance change maps, with annual monitoring estimating a 17% decline during 1998–2010 (*CBS Trend 2010*). Historical declines in Britain have been linked to increased mortality during the first year of life; the causes of this are uncertain but likely candidates include changes in farm management, land drainage, pesticide use and predation (Thomson *et al.* 1997; Robinson *et al.* 2004). Thomson *et al.* (1997) found that adult annual survival rates were lower in years with cold winters. The ongoing decline in eastern England may be attributable to the increasing dryness of agricultural soils and the loss of grassland from eastern arable counties (Peach *et al.* 2004).

Sponsored for Charlie & Cath Hazzard

BREEDING DISTRIBUTION *2008–11*

POSSIBLE •	11%	4%	6%
PROBABLE •	14%	4%	7%
CONFIRMED ●	70%	85%	81%
TOTAL	95%	93%	94%

WINTER DISTRIBUTION *2007/08–2010/11*

PRESENT ●	98%	90%	92%

BREEDING DISTRIBUTION CHANGE *since 1968–72*

GAIN ▲	11	33	44
LOSS ▽	9	60	69
40yr CHANGE	+<1%	-1%	-1%
20yr INDEX	+0.03	+0.01	

WINTER DISTRIBUTION CHANGE *since 1981–84*

GAIN ▲	31	228	259
LOSS ▼	18	57	75
30yr CHANGE	+1%	+7%	+6%

BREEDING RELATIVE ABUNDANCE *2008–11*

BREEDING RELATIVE ABUNDANCE CHANGE *since 1988–91*

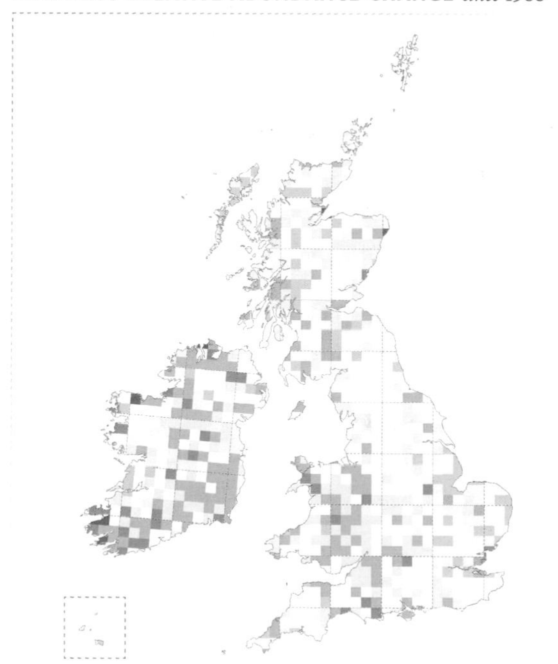

WINTER RELATIVE ABUNDANCE *2007/08–2010/11*

REDWING
Turdus iliacus

BTO CODE: RE

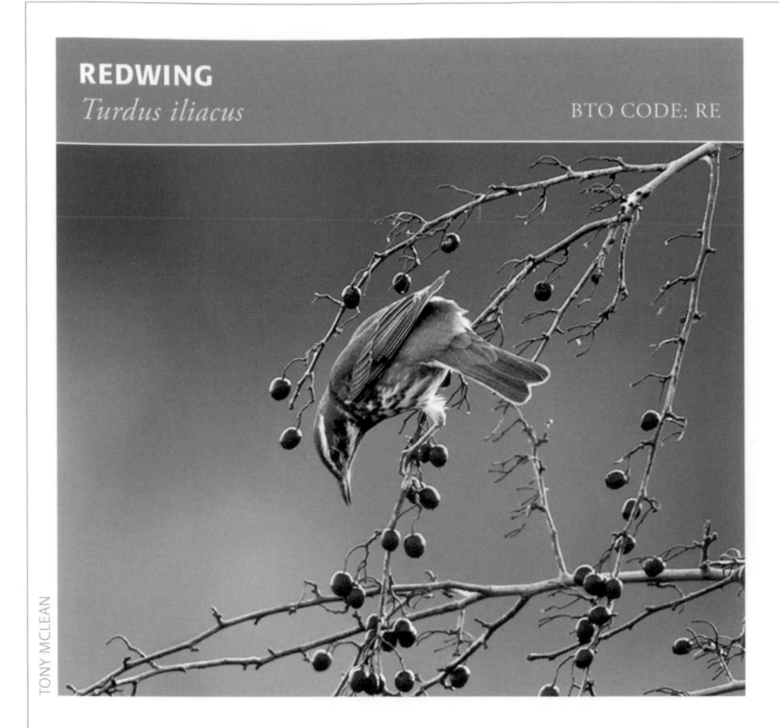

TONY MCLEAN

AUTUMN SEES THE arrival of large numbers of Redwings of the race *coburni* from Iceland and the Faeroes, and of the nominate race from Fennoscandia eastwards. The former dominates in Ireland and perhaps western Scotland, whereas the latter winters throughout Britain, with fewer in Ireland (*Migration Atlas*; Milwright 2002; *Birds of Scotland*). Together they are found in 92% of 10-km squares, with the few significant gaps being in the higher Scottish uplands. Occupied upland areas support very low densities in winter. Highest densities occur in western lowlands, in particular throughout Ireland and in southwest Wales, southwest England and the West Midlands. It is not clear whether this westerly tendency owes more to the milder climate, to the availability of feeding areas, or to the breeding origins of the birds concerned.

Since the *1981–84 Winter Atlas* there has been a 4% increase in the number of occupied 10-km squares. Gains are mostly in the uplands, on Scottish islands or in the far west of Ireland. They are likely to stem from improved recording effort or from milder weather allowing birds to winter further north and at higher altitudes.

Early in the breeding season, Redwings were still present in 780 10-km squares. Most were in eastern Britain and may have included individuals of both races (*Migration Atlas*). Although some migrants remain late into May and even sing, *probable* and *confirmed* breeding records were restricted to northern Scotland, in particular Inverness-shire, Ross-shire, Sutherland and Shetland. Submissions to RBBP give an average breeding population of 16 pairs (*RBBP 2010*) but coverage is known to be incomplete.

The large numbers of breeding-season gains and losses are partly a reflection of the low site fidelity shown by breeding Redwings (*Birds of Scotland*). This also means that the distribution in a single year will be smaller than shown here for the four years combined. Apart from gains in Shetland, a major contraction in range is apparent, particularly in Wester Ross. There are now 32% fewer occupied 10-km squares than during the *1968–72 Breeding Atlas*, and numbers are reduced where birds persist (*Birds of Scotland*). Climate change may be responsible for this recent rapid decline.

Sponsored by Jiggy Lloyd Consulting

WINTER DISTRIBUTION *2007/08–2010/11*

PRESENT ●	96%	91%	92%

BREEDING DISTRIBUTION *2008–11*

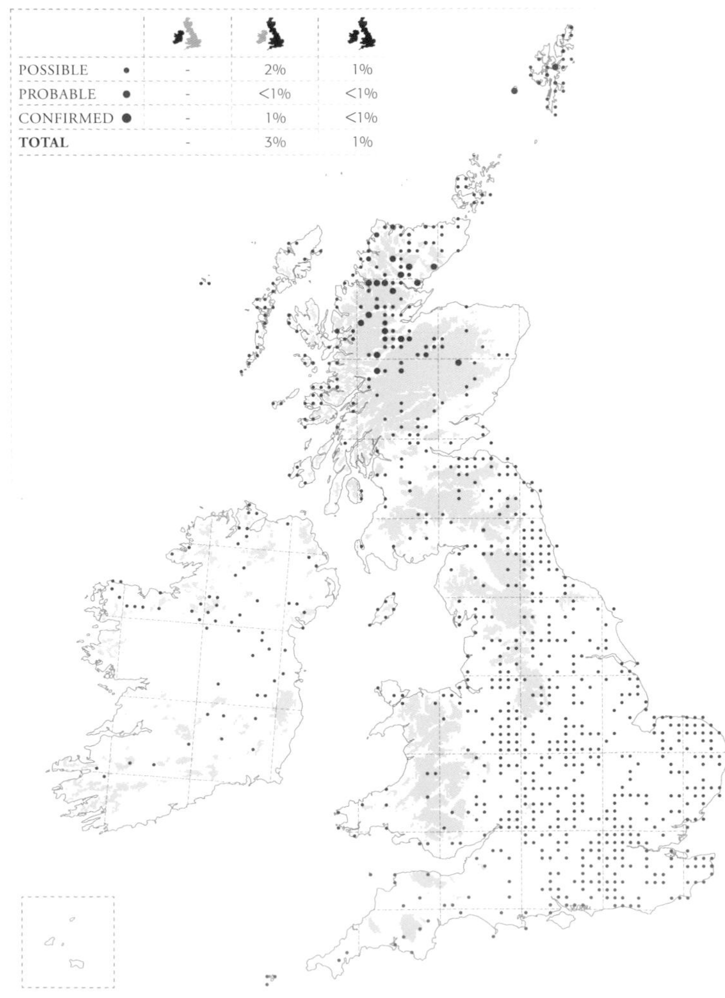

POSSIBLE •	-	2%	1%
PROBABLE ●	-	<1%	<1%
CONFIRMED ●	-	1%	<1%
TOTAL	-	3%	1%

WINTER DISTRIBUTION CHANGE *since 1981–84*

GAIN ▲	63	222	285
LOSS ▼	26	111	137
30yr CHANGE	+4%	+5%	+4%

WINTER RELATIVE ABUNDANCE *2007/08–2010/11*

BREEDING DISTRIBUTION CHANGE *since 1968–72*

GAIN ▲	-	36	36
LOSS ▽	-	72	72
40yr CHANGE	-	-32%	-32%
20yr INDEX	-	-0.46	

MISTLE THRUSH

Turdus viscivorus

BTO CODE: M.

CLARE WARD

THE MISTLE THRUSH is widely distributed throughout Britain & Ireland, being found in 88% of 10-km squares in winter and 86% in the breeding season. Breeding Mistle Thrushes are absent from the Outer Hebrides and Northern Isles and from western parts of Co. Mayo. Some of these areas are occupied in winter by small numbers of immigrants. Otherwise, the only significant gaps are in the higher Scottish uplands.

As expected for a mainly sedentary species (*Migration Atlas*), the winter and breeding-season relative abundance maps are similar. Densities are highest in Ireland. In Britain there are scattered pockets of high density, including some at moderate altitudes on the eastern fringe of the Highlands (*Birds of Scotland*). Higher uplands are sparsely populated in both seasons, but especially in winter.

Since the *1981–84 Winter Atlas*, there have been small increases in the numbers of occupied 10-km squares, mostly in northwest Scotland and the coastal fringes of Ireland. Many are probably associated with improved coverage in areas of low Mistle Thrush density, although year-round gains on Skye may result from colonisation of plantations there (McMillan 2005). Breeding-season changes are also minor, although expansion in northern Scotland, probably associated with maturing plantations, contrasts with unexplained losses in western Ireland and on Guernsey.

These small distribution changes belie marked regional trends in abundance. The abundance change map shows decreases in tetrad occupancy in much of England; these mirror a breeding population decline which has been under way since the 1970s (*BirdTrends*) and encompasses a 35% decline in England during 1995–2010 (*BBS Report 2011*). It is unclear why the increase in tetrad occupancy in Ireland conflicts with a 32% breeding population decline there during 1998–2010 (*CBS Trend 2010*). The population in Scotland had been increasing strongly from 1994 to 2004 but has since declined markedly (*BirdTrends*). The causes of these changes are unclear, although the long-term UK decline is likely to have been driven by reductions in annual survival (Siriwardena *et al.* 1998a). Mason (2000) documented that Mistle Thrushes in eastern England are now largely restricted to villages and towns, but counts in gardens indicate that abundance has declined there also (Chamberlain *et al.* 2005).

Sponsored by Peter Watson & Jenny Dixon (BTO Members)

BREEDING DISTRIBUTION *2008–11*

POSSIBLE •	12%	5%	7%
PROBABLE •	16%	7%	9%
CONFIRMED •	60%	73%	70%
TOTAL	88%	85%	86%

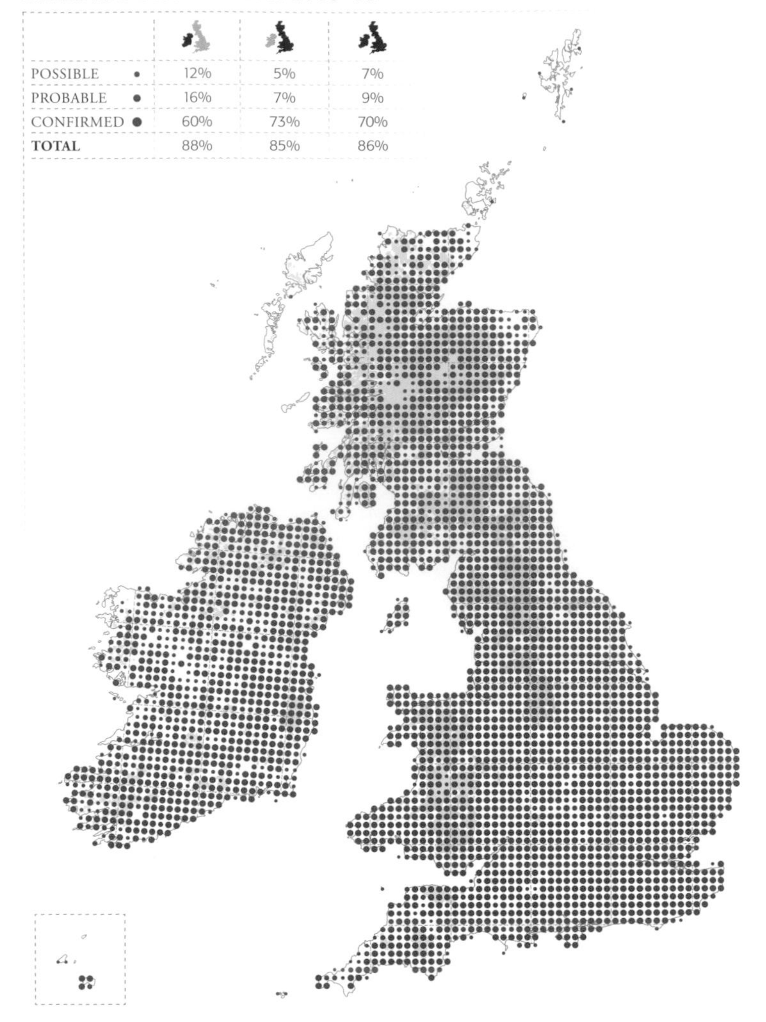

WINTER DISTRIBUTION *2007/08–2010/11*

PRESENT •	93%	87%	88%

BREEDING DISTRIBUTION CHANGE *since 1968–72*

GAIN ▲	12	85	97
LOSS ▽	59	85	144
40yr CHANGE	-5%	0%	-1%
20yr INDEX	+0.03	-0.01	

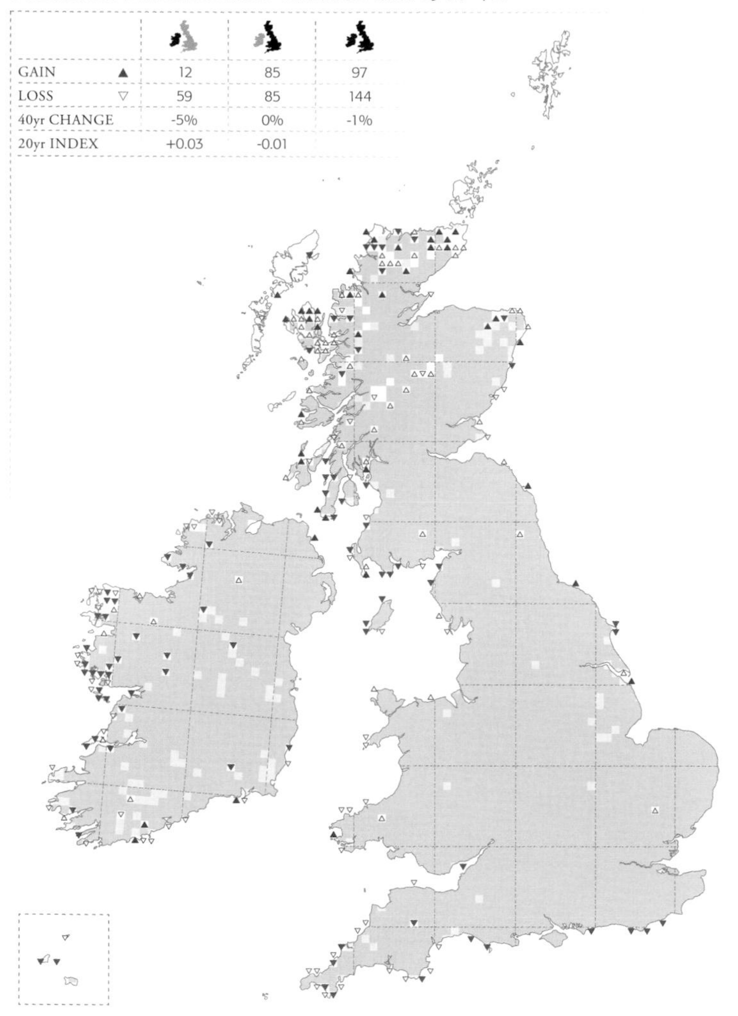

WINTER DISTRIBUTION CHANGE *since 1981–84*

GAIN ▲	70	242	312
LOSS ▼	35	58	93
30yr CHANGE	+4%	+8%	+7%

BREEDING RELATIVE ABUNDANCE *2008–11*

BREEDING RELATIVE ABUNDANCE CHANGE *since 1988–91*

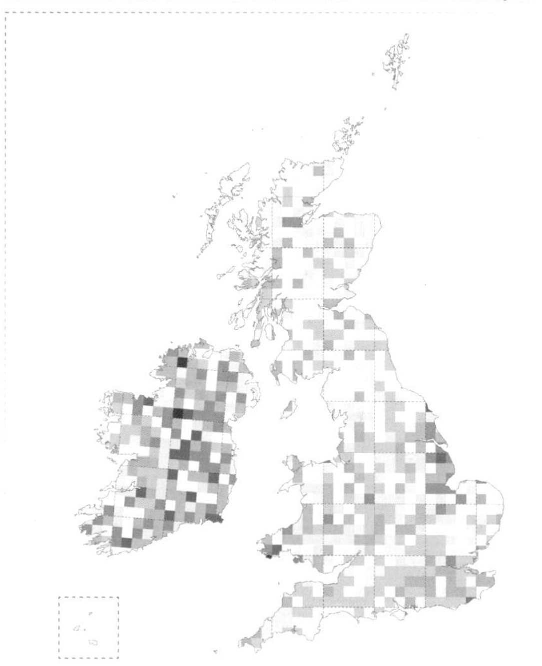

WINTER RELATIVE ABUNDANCE *2007/08–2010/11*

SPOTTED FLYCATCHER

Muscicapa striata

BTO CODE: SF

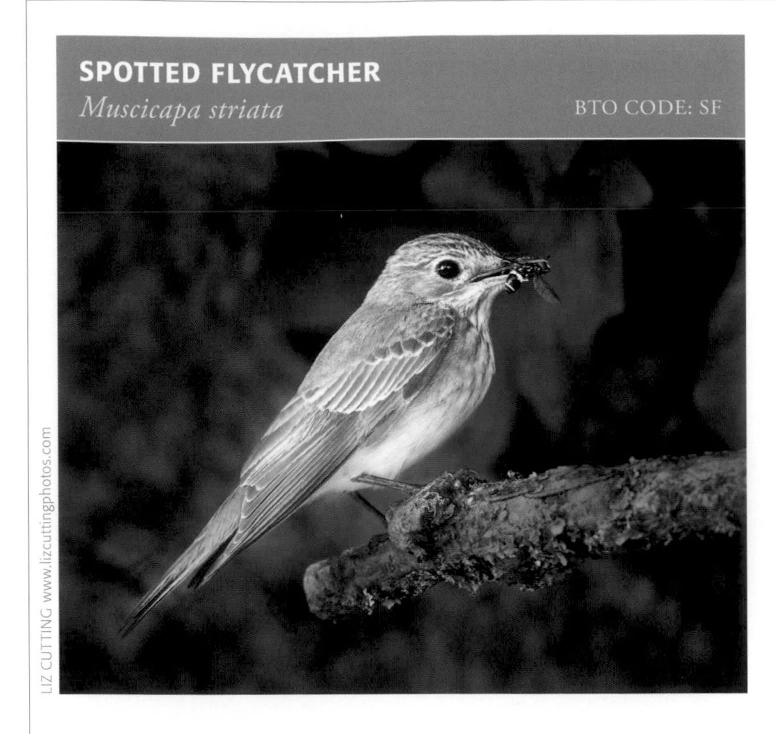

LIZ CUTTING www.lizcuttingphotos.com

THE SPOTTED FLYCATCHER remains a widespread breeding species in Britain & Ireland, despite a long-term population decline, being absent only from the Outer Hebrides, the Northern Isles and major urban centres. Densities are generally greatest in the north and west of the range, excluding the highest uplands, but low in western coastal districts of Ireland, where the distribution is patchy.

Range losses in Britain have been relatively minor, amounting to only 10% at 10-km resolution since the *1968–72 Breeding Atlas*, but largely concentrated in and around major conurbations. In Ireland there has been a more substantial 21% range contraction, with the species now absent from many squares along the west coast. The abundance change map is amongst the most consistently negative map of any species in the Atlas, with declines in tetrad occupancy throughout most of England, Wales and eastern Scotland since the *1988–91 Breeding Atlas*, although there is a suggestion of increases in northern Scotland. In population terms these decreases amount to a 50% decline during 1995–2010, part of an 88% decline recorded since 1970 (*SUKB 2012*). Abundance changes are more mixed in Ireland, where this species is too scarce for annual monitoring to produce a trend. Moderate declines are also evident across Europe (*PECBMS 2010*).

Productivity per nesting attempt has declined, with avian nest predators being the main cause of nest losses in one study (Stevens *et al.* 2008; *BirdTrends*). Higher rates of nest predation in woodland and farmland than in gardens mean that productivity in gardens can be higher than elsewhere (Stevens *et al.* 2007). Reduced survival of first-year birds, however, is thought to be the main demographic factor behind the decline (Freeman & Crick 2003). Lower survival might result from deterioration in woodland quality for young birds, operating in the immediate post-breeding period, or from conditions in winter or on migration (Fuller *et al.* 2005). The Spotted Flycatcher is one of several long-distance migrants wintering in the humid zone of West Africa that are currently in steep population decline (Hewson & Noble 2009), suggesting their declines have a common cause, acting on migration or on the wintering grounds (Ockendon *et al.* 2012).

BREEDING DISTRIBUTION *2008–11*

POSSIBLE •	20%	8%	11%
PROBABLE ●	13%	9%	10%
CONFIRMED ●	37%	60%	54%
TOTAL	70%	77%	75%

BREEDING RELATIVE ABUNDANCE CHANGE *since 1988–91*

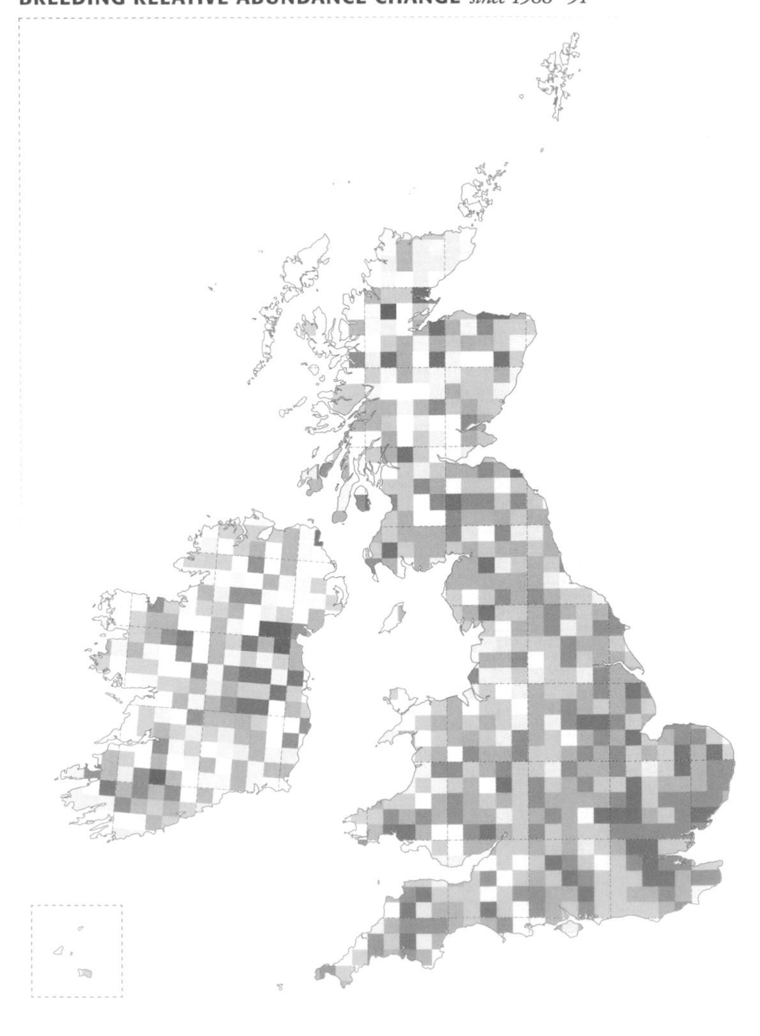

The John Spedan Lewis Foundation

Sponsored by John Spedan Lewis Foundation

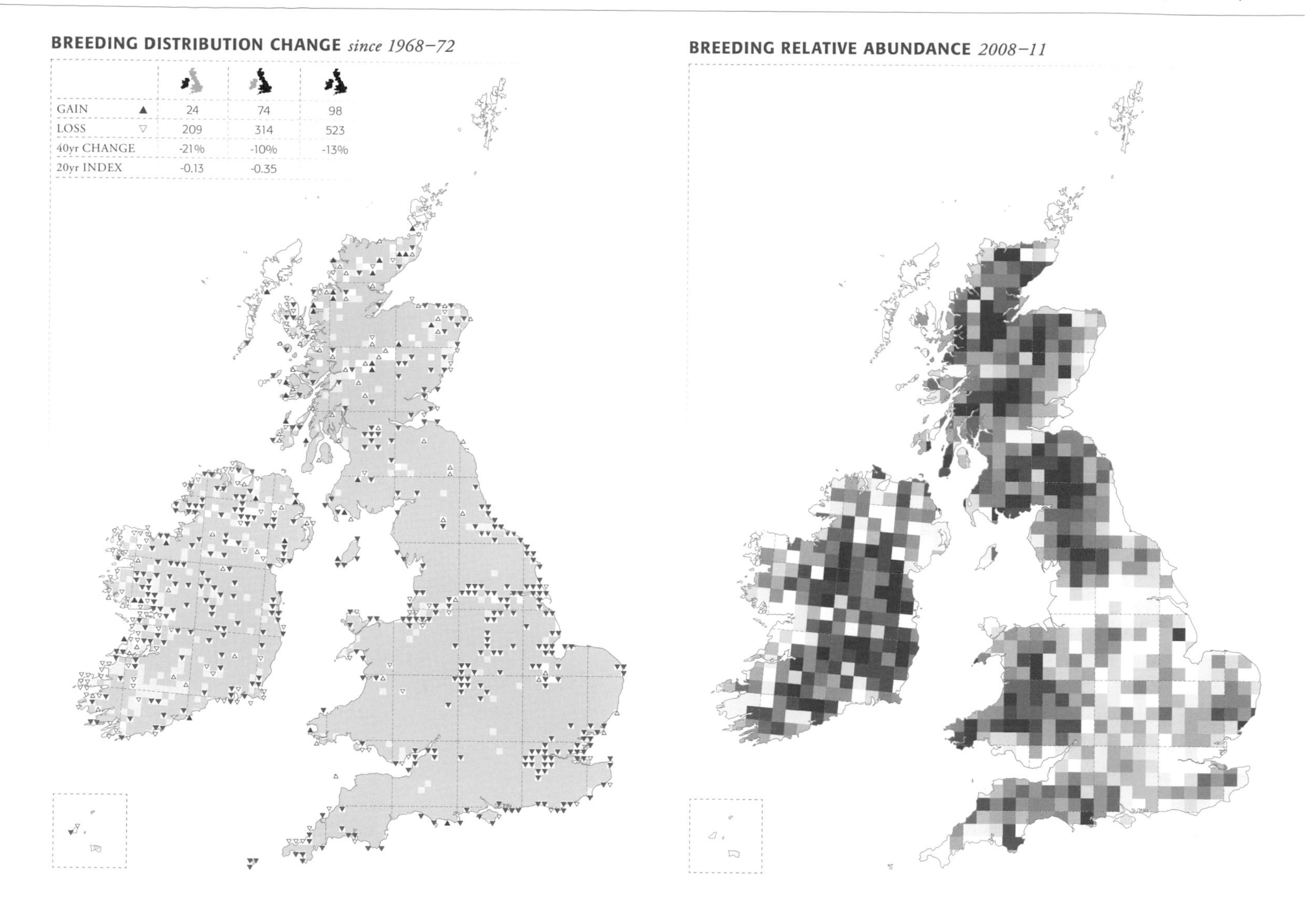

GAIN ▲	24	74	98
LOSS ▽	209	314	523
40yr CHANGE	-21%	-10%	-13%
20yr INDEX	-0.13	-0.35	

◀ The deteriorating status of the Spotted Flycatcher probably has much to do with pressures faced on migration or in its African wintering grounds. However, reduction in invertebrate food during the summer may also be relevant.

ROBIN

Erithacus rubecula

BTO CODE: R.

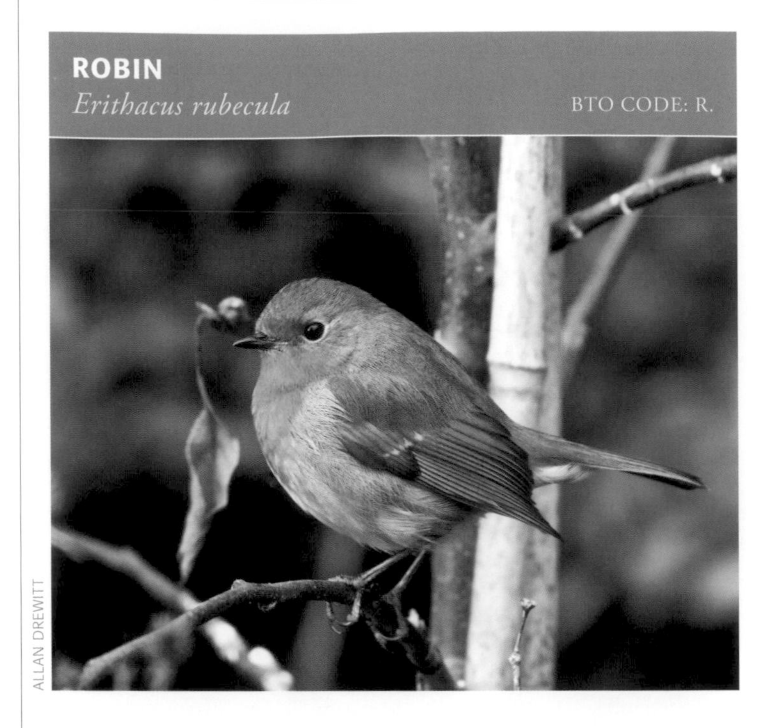

ALLAN DREWITT

ROBINS ARE WIDESPREAD throughout Britain & Ireland, breeding in 94% of all 10-km squares, and are absent only from some of the Scottish islands and from exposed uplands. Their relative abundance is highest in Ireland and southern England and lowest in the uplands, especially above 400 m altitude (*Birds of Scotland*).

There has been a 2% range expansion since the *1968–72 Breeding Atlas*, with gains in the Outer Hebrides, Caithness and the Northern Isles, probably aided by increased tree planting. Between 1995 and the start of the Atlas, the UK breeding population rose by 20% (Risely *et al.* 2009). Robins were recorded in a high proportion of tetrads in Ireland and in southern and central England in the *1988–91 Breeding Atlas* but there can be many pairs per tetrad: consequently, densities can change markedly without trends in tetrad occupancy being apparent on the abundance change map. Maps based on annual monitoring show that Robin abundance increased during 1994–2009 throughout Britain north of the Severn and the Thames (Massimino *et al.* 2013). Causal factors are unclear but may act through concurrent trends towards lower rates of nest failure, higher survival rates and earlier breeding (*BirdTrends*). In contrast, a 25% decline has taken place in Ireland during 1998–2010 (*CBS Trend 2010*). So far this has not been sufficient to impact upon tetrad occupancy, and the only changes apparent are in a number of marginal squares.

Continental immigration is widespread in autumn and explains the greater winter occupancy in the Northern Isles (*Birds of Scotland*). Although birds vacate many upland areas for the winter, most British and Irish Robins are sedentary (*Migration Atlas*). Consequently, the winter abundance map is very similar to the breeding map except that relative abundance is reduced over extensive tracts of the uplands. The small range expansion, of 3%, since the *1981–84 Winter Atlas* has taken place mostly on upland margins and on Scottish islands. It is likely to be due partly to the increased breeding population in these areas and possibly also to milder winters, which might have allowed more birds to overwinter in some of the more upland and exposed squares.

BREEDING DISTRIBUTION *2008–11*

	Britain & Ireland	Britain	Ireland
POSSIBLE •	2%	2%	2%
PROBABLE •	3%	2%	2%
CONFIRMED •	92%	89%	90%
TOTAL	97%	93%	94%

WINTER DISTRIBUTION *2007/08–2010/11*

	Britain & Ireland	Britain	Ireland
PRESENT •	98%	95%	96%

BREEDING DISTRIBUTION CHANGE *since 1968–72*

GAIN ▲	11	74	85
LOSS ▽	3	21	24
40yr CHANGE	+1%	+2%	+2%
20yr INDEX	+0.01	+0.02	

WINTER DISTRIBUTION CHANGE *since 1981–84*

GAIN ▲	7	153	160
LOSS ▼	15	26	41
30yr CHANGE	-1%	+5%	+3%

BREEDING RELATIVE ABUNDANCE *2008–11*

BREEDING RELATIVE ABUNDANCE CHANGE *since 1988–91*

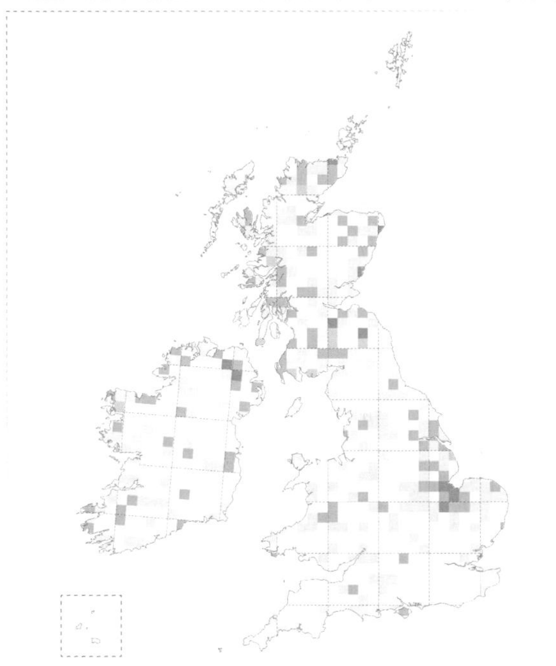

WINTER RELATIVE ABUNDANCE *2007/08–2010/11*

NIGHTINGALE

Luscinia megarhynchos

BTO CODE: N.

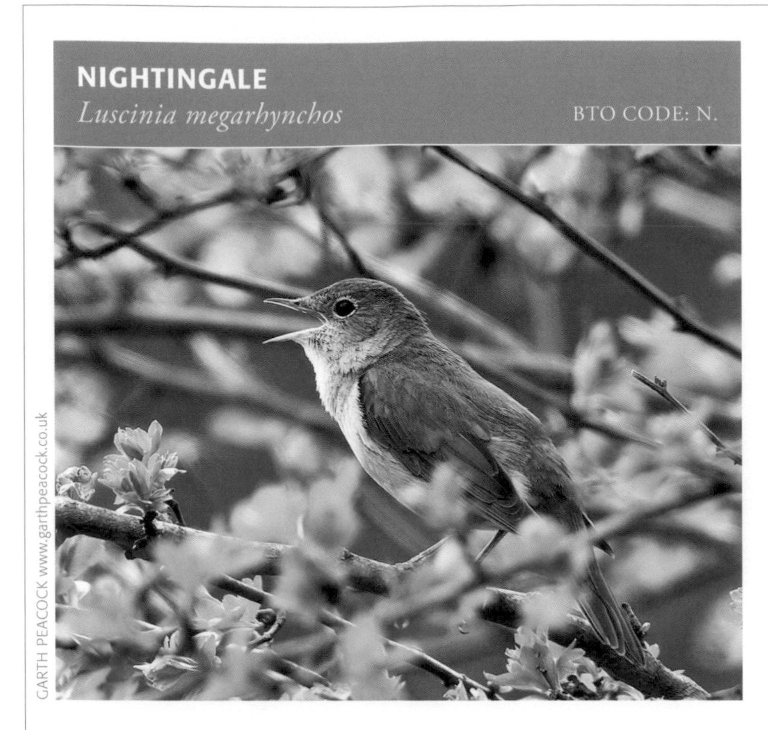

GARTH PEACOCK www.garthpeacock.co.uk

BREEDING DISTRIBUTION *2008–11*

POSSIBLE ●	-	6%	4%
PROBABLE ●	-	4%	3%
CONFIRMED ●	-	3%	2%
TOTAL	-	13%	9%

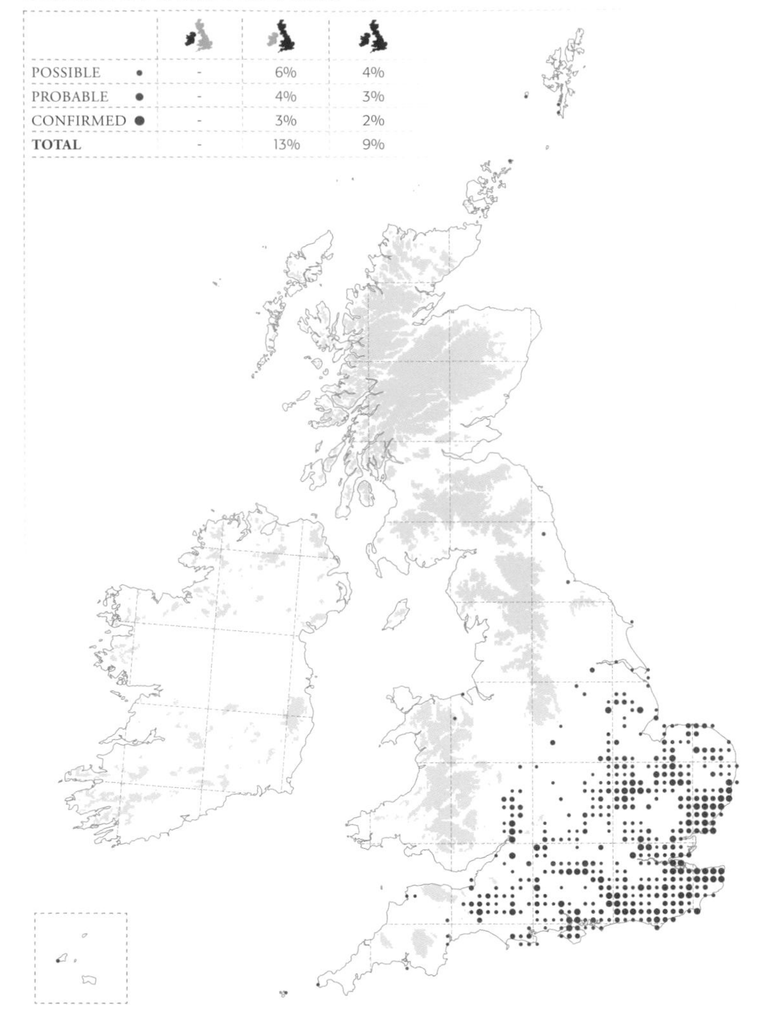

BREEDING RELATIVE ABUNDANCE CHANGE *since 1988–91*

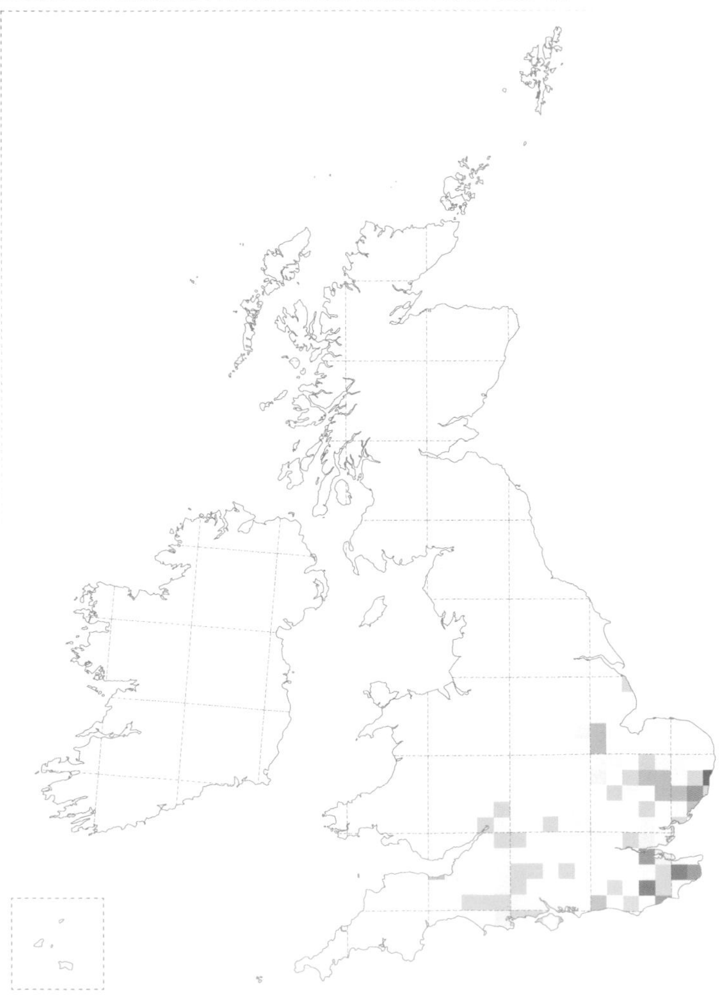

ENGLAND REPRESENTS THE northern edge of the global breeding range of the Nightingale, with the species breeding only to the south of a line from the Severn to the Humber. Despite a projected northward expansion in response to climate change (Wilson *et al.* 2002; Huntley *et al.* 2007) a contraction has occurred towards the southeastern strongholds of Kent, Sussex and Essex in the last four decades, resulting in a range 43% smaller than in 1968–72. These three counties, together with Hampshire, held the highest densities, whereas counties at the range edge are increasingly characterised by a small number of sites of relatively high population density (e.g. Cotswold Water Park, Gloucestershire/Wiltshire, and Paxton Pits, Cambridgeshire). Such sites are becoming increasingly isolated as birds disappear from surrounding countryside.

Numbers of Nightingales in England have declined by approximately 90% in the last 40 years (Holt *et al.* 2012) and in the shorter term by 52% during 1995–2010 (*BBS Report 2011*). The breeding change map and abundance change map together show that the decline is the result of a contraction in range and a reduction in numbers in the core area. Whereas other northwest European countries have also experienced declines, the trend for Europe as a whole is stable, with central and southern countries having recorded increases (*PECBMS 2010*; Holt *et al.* 2012).

The decline in Britain is probably attributable to a combination of factors, acting locally and at large scales. These include habitat loss and reductions in habitat quality at breeding sites, pressures on migration and habitat degradation in the West African winter quarters (Holt *et al.* 2012). The position of Britain at the limit of the species' global range may also be highly relevant (Fuller *et al.* 2007). A national survey in 1999 indicated a reduction in numbers in woodland, including managed coppice, partly offset by increases in scrub habitats (Wilson *et al.* 2002). Deteriorating habitat quality for woodland Nightingales is probably due to management being less active, leading to increased shading, coupled with intensified browsing by deer. Both factors tend to result in loss of the dense understorey vegetation that is required by Nightingales for territory establishment, foraging and nesting (Holt *et al.* 2010).

Sponsored by the Nightingale Supporters' Group (see page 13)

BREEDING DISTRIBUTION CHANGE *since 1968–72*

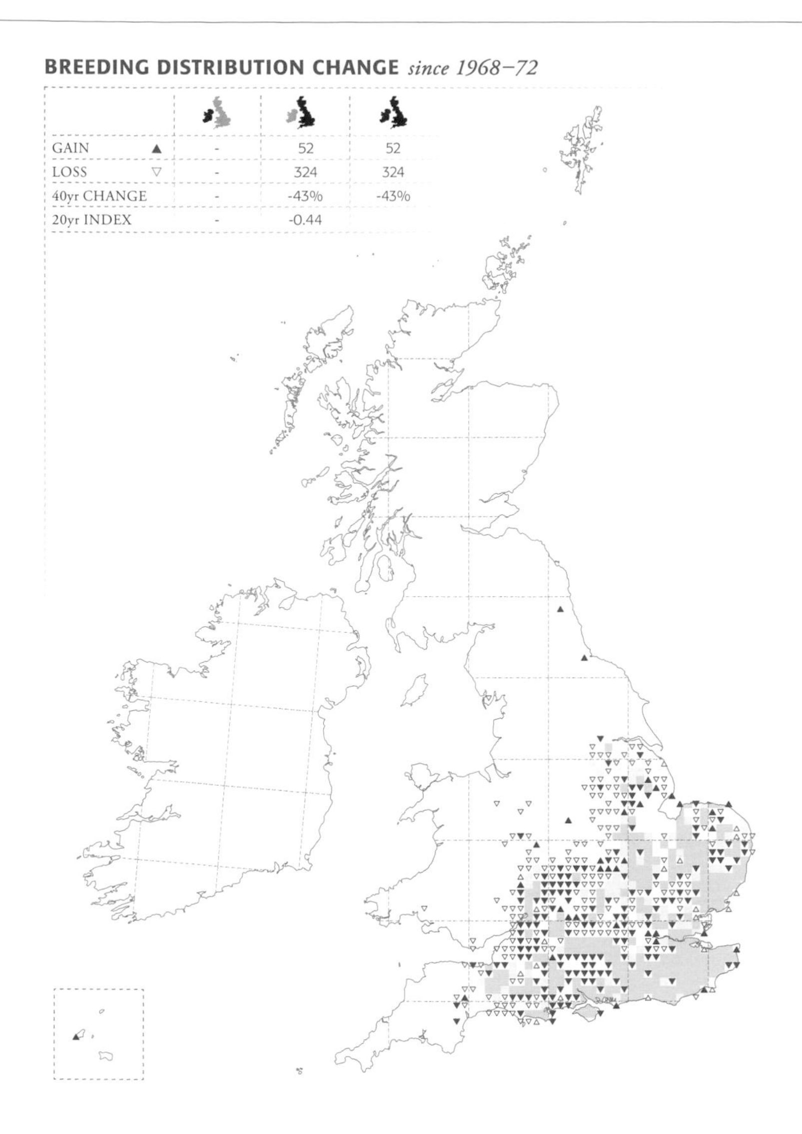

GAIN ▲	-	52	52
LOSS ▽	-	324	324
40yr CHANGE	-	-43%	-43%
20yr INDEX	-	-0.44	

BREEDING RELATIVE ABUNDANCE *2008–11*

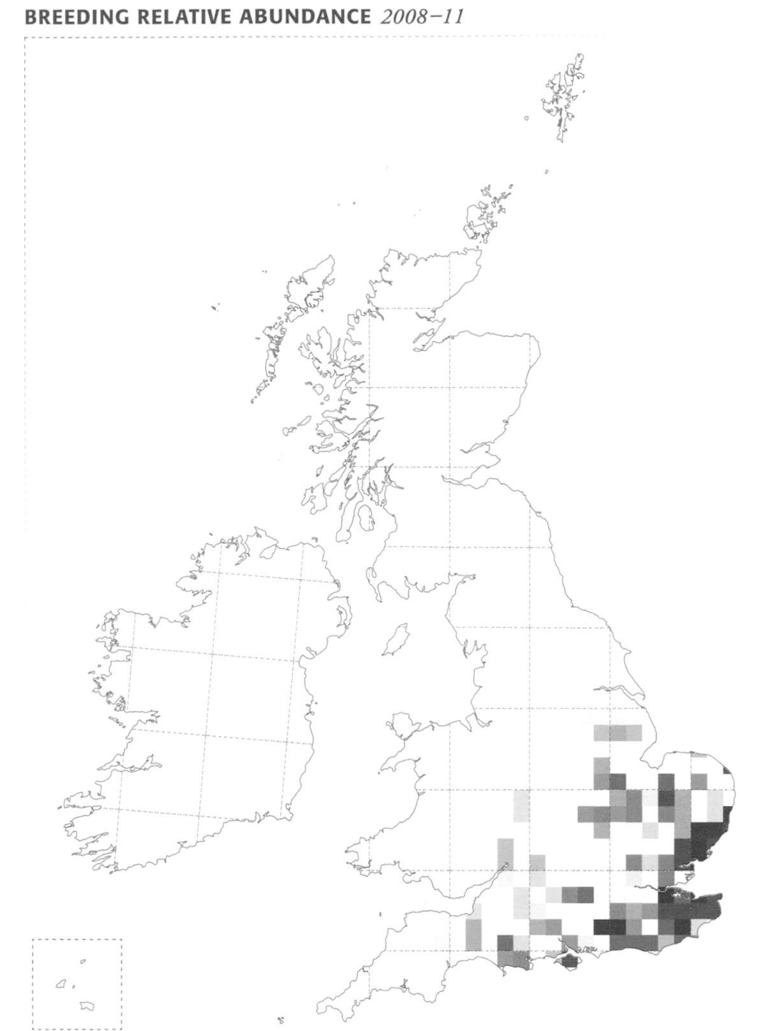

◀ **It can be difficult to confirm breeding for Nightingales owing to the dense understorey habitats they occupy. Singing birds and others holding territory accounted for over 70% of records.**

PIED FLYCATCHER

Ficedula hypoleuca

BTO CODE: PF

EDMUND FELLOWES

AT FIRST GLANCE there appears to be little change in the broad distribution of this summer migrant, which continues to be associated with the mature upland woodlands of western and northern Britain. Pied Flycatchers are widely distributed across most of Wales, parts of Shropshire and Herefordshire, and in northwest England from West Yorkshire through Cumbria to Northumberland, but are more patchily distributed in western Scotland.

In fact, quite marked range changes have occurred over the last 40 years or so. Although a few apparent gains in eastern England in the *1988–91 Breeding Atlas* probably involved migrants, overall there was a 35% range expansion between 1968–72 and 1988–91. Some gains were attributable to the provision of nestboxes during the intervening decades. The subsequent 27% range contraction from 1988–91 to 2008–11 means the net 40-year change amounts to a 2% contraction: there has, however, been a subtle westerly and northerly shift in distribution, with most of the recent losses involving thinning of the range along its eastern fringe and in southwest England. Accompanying the local changes in 10-km occupancy, the abundance change map shows the proportions of occupied tetrads have declined almost everywhere since 1988–91.

During 1995–2010, British breeding numbers decreased steeply by 50% (*BBS Report 2011*), outpacing a wider European decline of 19% during 1980–2010 (*PECBMS 2010*). Nationally there has been no general trend in productivity per breeding attempt (*BirdTrends*) but, at one site on the eastern fringe of the breeding range, productivity has declined (Goodenough *et al.* 2009). There is strong evidence from the Netherlands that local declines can occur where a mismatch has developed between the timing of peak caterpillar abundance and the arrival and breeding times of Pied Flycatchers (Both *et al.* 2006). Changes in climate and in weather on the wintering grounds and at stopover sites also correlate well with European population indices for the species (Chernetsov & Huettmann 2005), suggesting that both local and large-scale factors could be contributing to the observed trends. Indeed, despite finding a decline in breeding performance, Goodenough *et al.* (2009) concluded that effects on the wintering grounds and during migration were more important drivers of local declines.

Sponsored by Lake District National Park Authority

BREEDING DISTRIBUTION *2008–11*

POSSIBLE	-	4%	3%
PROBABLE	-	2%	2%
CONFIRMED	-	12%	9%
TOTAL	-	18%	14%

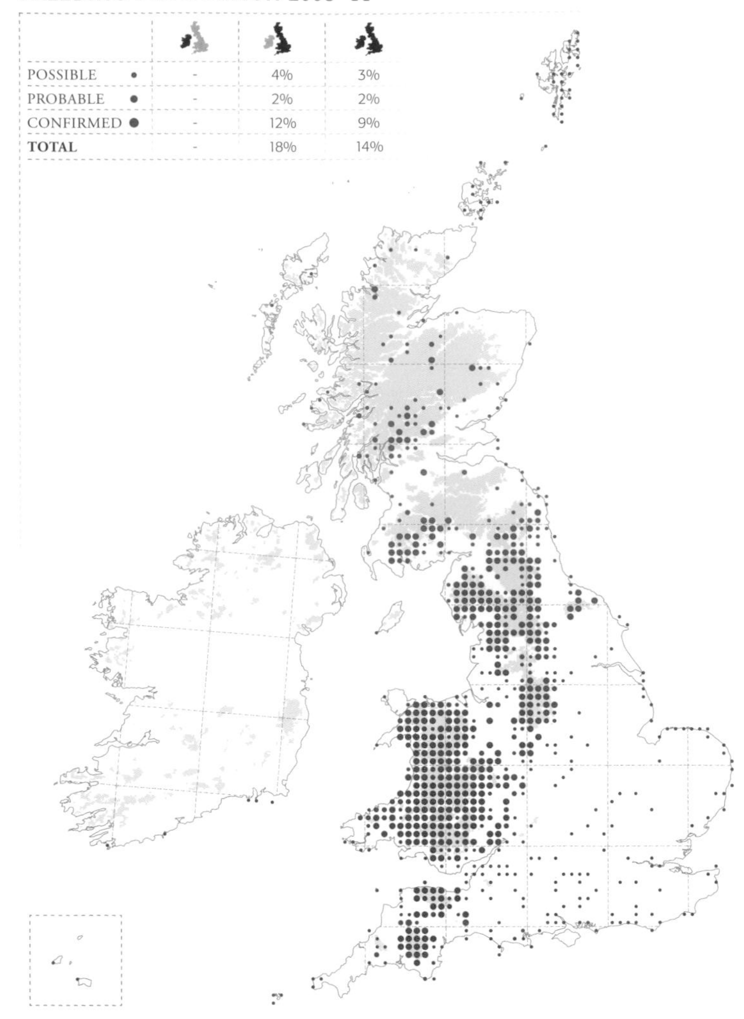

BREEDING RELATIVE ABUNDANCE CHANGE *since 1988–91*

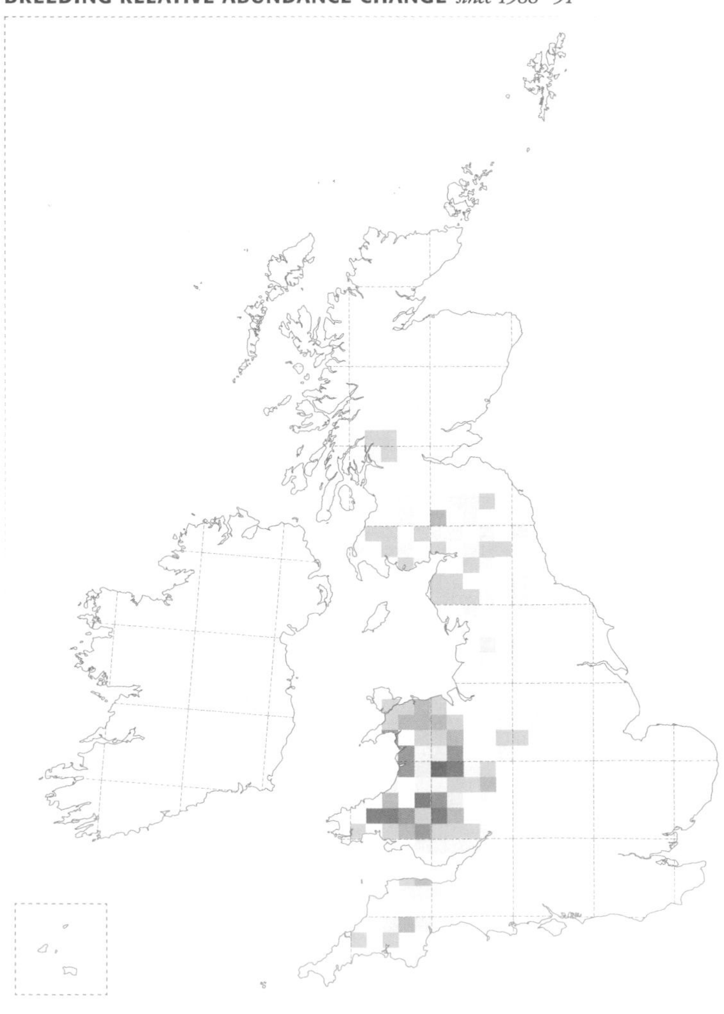

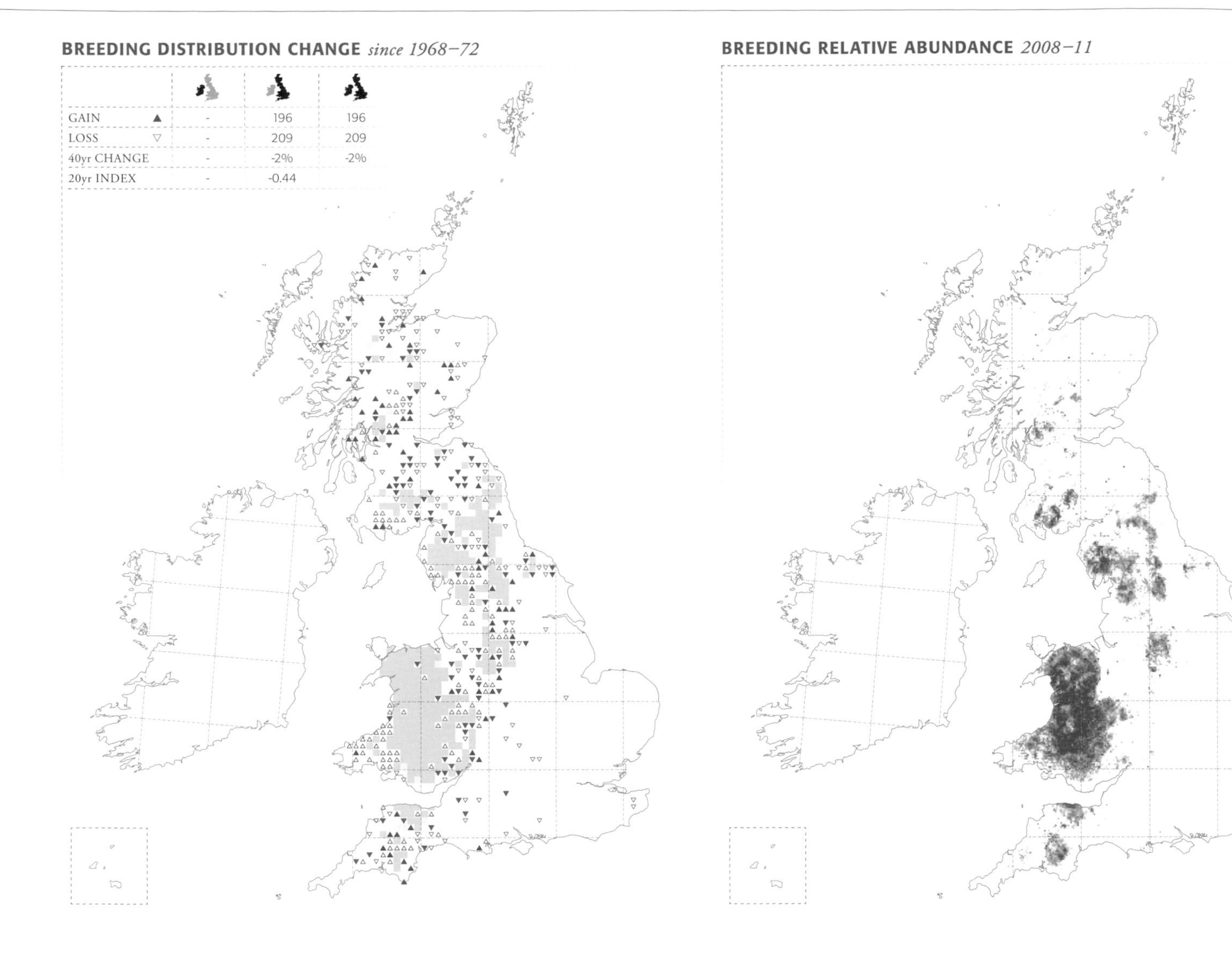

BREEDING DISTRIBUTION CHANGE since 1968–72
GAIN ▲ - 196 196
LOSS ▽ - 209 209
40yr CHANGE - -2% -2%
20yr INDEX - -0.44
BREEDING RELATIVE ABUNDANCE 2008–11

GRAHAM CATLEY http://pewit.blogspot.co.uk

◀ **Pied Flycatchers typically depart Britain between mid August and mid September, with late migrants lingering into mid October. An individual on Flat Holm, Glamorgan, in December 2007 was exceptional.**

BLACK REDSTART

Phoenicurus ochruros

BTO CODE: BX

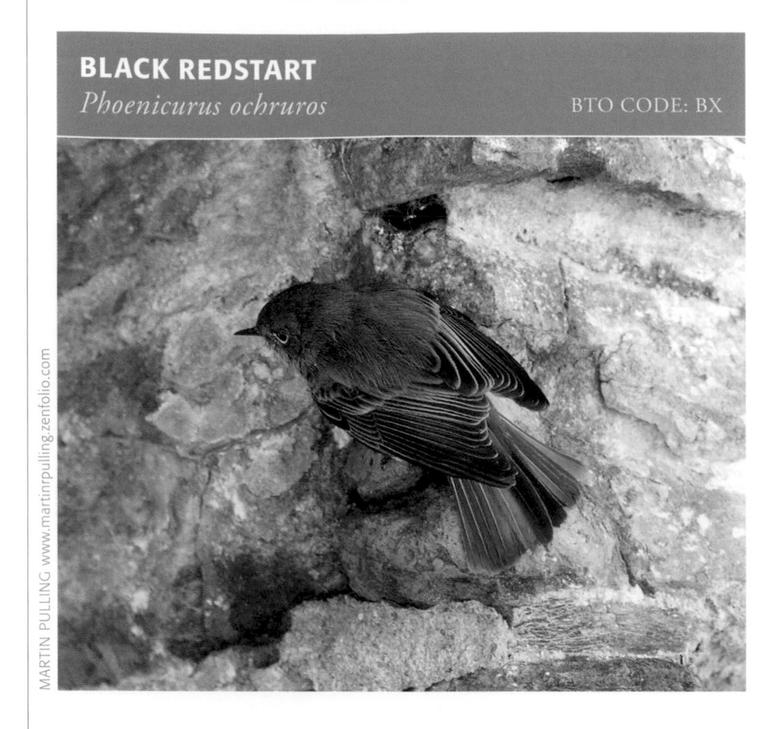

MARTIN PULLING www.martinpulling.zenfolio.com

BLACK REDSTARTS ARE largely coastal birds in winter in Britain & Ireland and were recorded from most squares along the south coasts of Ireland, Wales and England, with records also spread widely further north, particularly on Irish Sea and North Sea coasts. Inland birds were scattered thinly throughout southern and midland England, with a cluster of occupied squares in Somerset and Avon. The winter abundance map suggests that densities are highest in southwest England.

Since the *1981–84 Winter Atlas* the number of occupied 10-km squares has increased by 52% in Britain, though by only 2% in Ireland. Some of the apparent expansion in Britain may be linked to better collation of casual records, but there does seem to have been a clear increase, particularly along the east coast. Numbers of birds wintering in Scotland are thought to have increased since the 1960s (*Birds of Scotland*).

Black Redstarts are more localised in the breeding season. Migrants and other *non-breeding* birds were recorded in 306 10-km squares but breeding evidence was obtained in only a further 135 squares, with a third of those involving *confirmed* breeding. Breeding records were concentrated around London, coastal Kent, East Anglia and northwest England. There were at least 20 territories at 15 sites in Greater London in 2010 (*RBBP 2010*). There were no *confirmed* breeding records in Ireland and the only one in Scotland, on Lewis, involved the sighting of a single recently fledged juvenile, with the exact breeding site not determined (*RBBP 2010*).

Although the core breeding range has changed relatively little, the number of occupied squares with breeding evidence has doubled since 1968–72 and many of the gains were outside southeast England, suggesting a slow range expansion. However, despite the far higher abundance of the species on the Continent, where a modest increase is apparent since the early 1980s (*PECBMS 2010*), there has still been no large-scale colonisation.

The current agreed population estimate for Britain is a mere 19–44 pairs (Musgrove *et al.* 2013). It is quite possible that some breeding pairs are overlooked, however, given their continuing concentration into industrial areas, where song can be difficult to pick out against background noise.

Sponsored by RPS Group PLC

WINTER DISTRIBUTION *2007/08–2010/11*

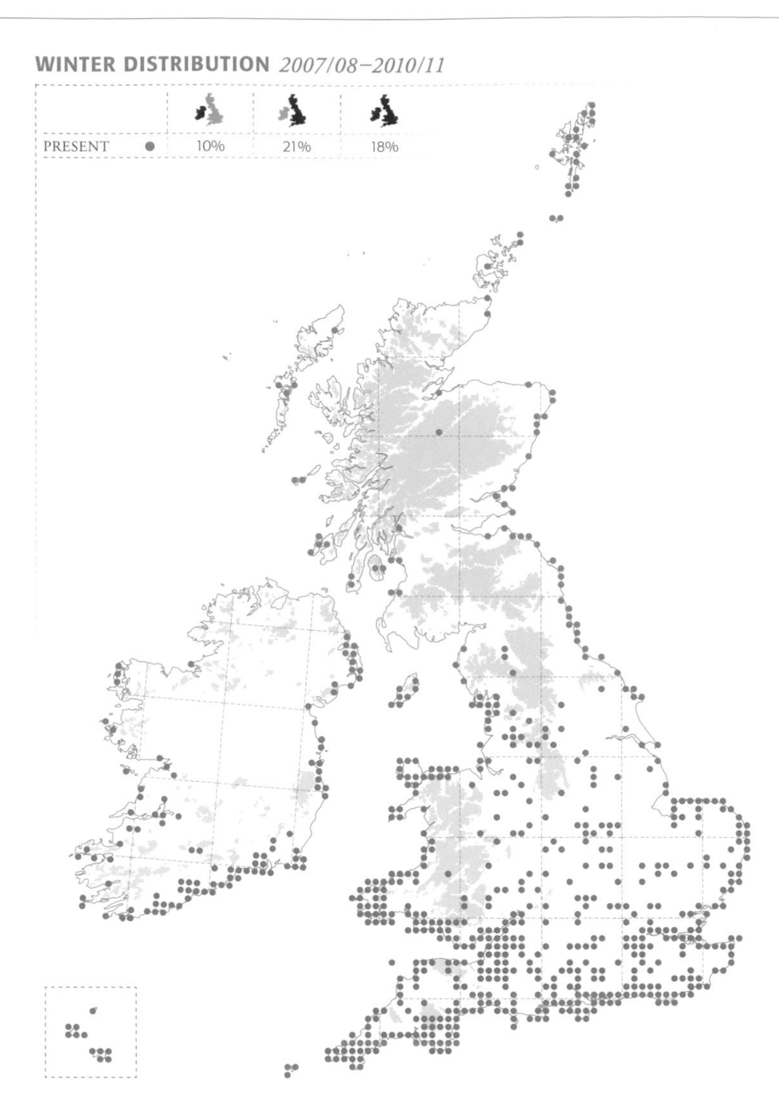

	Ireland	Britain	Britain & Ireland
PRESENT ●	10%	21%	18%

BREEDING DISTRIBUTION *2008–11*

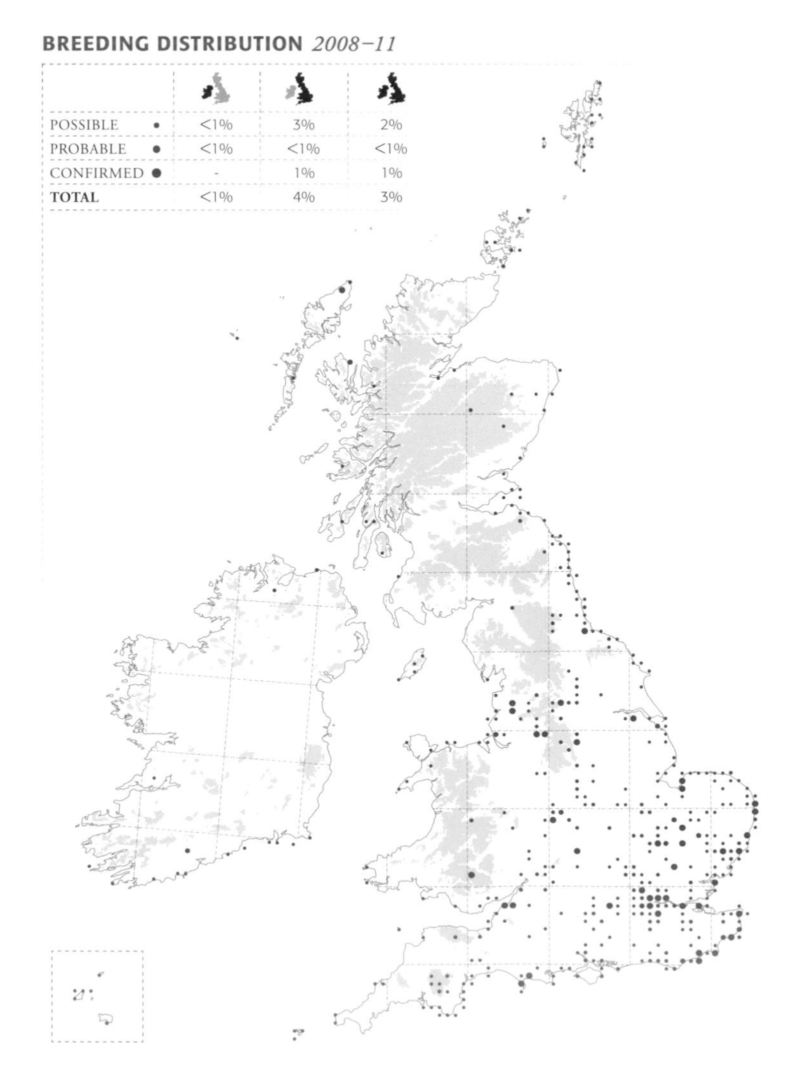

	Ireland	Britain	Britain & Ireland
POSSIBLE •	<1%	3%	2%
PROBABLE ●	<1%	<1%	<1%
CONFIRMED ●	-	1%	1%
TOTAL	<1%	4%	3%

WINTER DISTRIBUTION CHANGE *since 1981–84*

GAIN ▲	45	289	334
LOSS ▼	43	121	164
30yr CHANGE	+2%	+52%	+42%

WINTER RELATIVE ABUNDANCE *2007/08–2010/11*

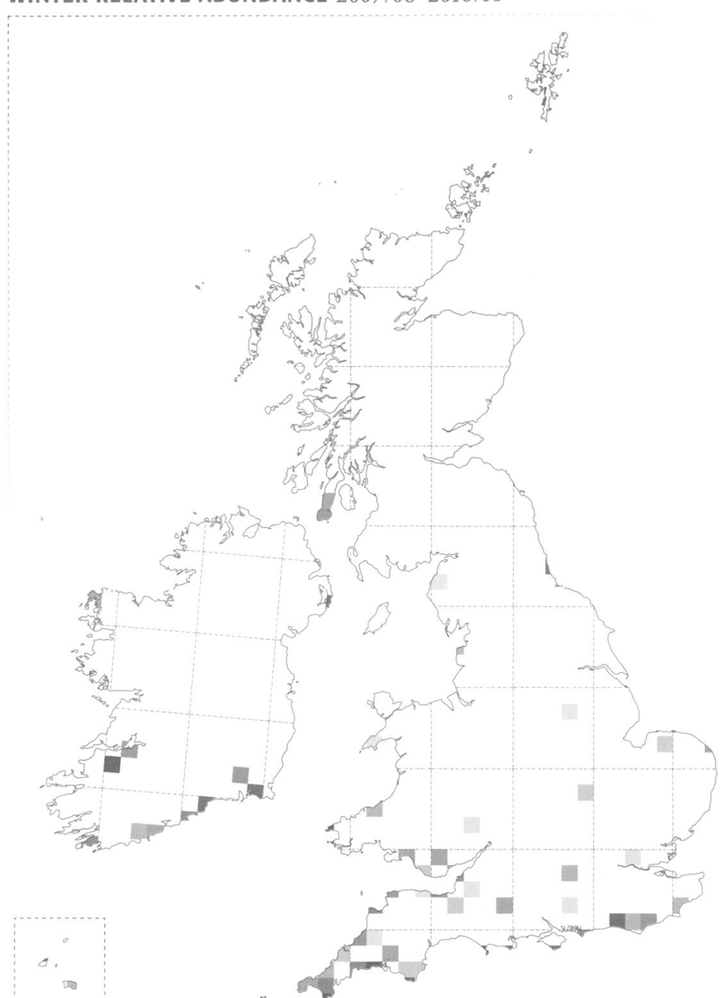

BREEDING DISTRIBUTION CHANGE *since 1968–72*

GAIN ▲	3	93	96
LOSS ▽	0	29	29
40yr CHANGE	∞	+94%	+99%
20yr INDEX	-	-0.63	

REDSTART

Phoenicurus phoenicurus

BTO CODE: RT

TOM WALLIS

THE REDSTART IS a rare and perhaps overlooked annual breeder in Ireland (*Birds in Ireland*; *IRBBP 2010*) and breeding evidence was recorded in two 10-km squares. Breeding Redstarts are absent from the Channel Islands, Isle of Man, Northern Isles and most Hebridean islands, from which suitable wooded habitat is lacking or scarce. Otherwise, they have a wide but patchy distribution in western and northern Britain, with highest abundances in the wooded uplands of Wales, northern England and, to lesser extent, southern Scotland. Small pockets of distribution remain in southern England, mostly associated with heathland.

The British breeding range has contracted by 31% since the *1968–72 Breeding Atlas*. Losses are most apparent in the English lowlands and in Scotland, through the Central Belt and on the northern and western fringes of the range. In the *1988–91 Breeding Atlas*, it was noted that Redstarts were increasing in core areas and decreasing in the range margins; these trends have continued. The abundance change map shows some increases since 1988–91 in Mid Wales but stability or decline more generally.

Drought in the Sahel of West Africa caused a large population decline during c.1968–73 (Marchant *et al.* 1990). Sites surveyed by RSPB in the 1980s and again in 2003–04 showed there had been marked declines throughout Britain over this period, including a 51% decline in Wales (Hewson *et al.* 2007). Loss of mature hedges and changes in woodland management may explain why some lowland populations never fully recovered from the drought (*Birds in England*). Birds have often disappeared where habitat remains apparently suitable, however, and recently intensified deer browsing might have been expected to have increased the area of suitable lowland woodland habitat (Fuller 2001). It remains unclear why Redstarts are faring so poorly, especially in the lowlands. Intriguingly, there has been a small increase since c.2008 and the 19% increase during 1995–2010 reported for the UK (*BBS Report 2011*) is driven largely by a recent steep increase in Wales. Numbers have also increased across Europe, especially since c.2005 (*PECBMS 2010*), suggesting that recent positive trends may be determined by large-scale factors such as rainfall patterns, rather than by factors local to the breeding grounds.

Sponsored by R. R. Langley

BREEDING DISTRIBUTION *2008–11*

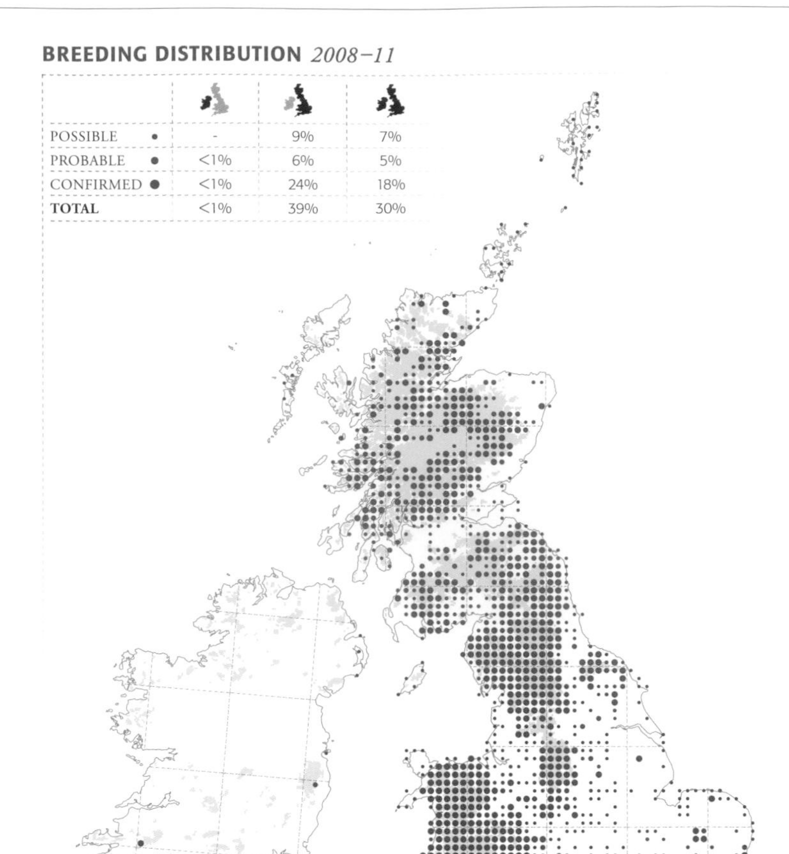

POSSIBLE •	-	9%	7%
PROBABLE •	<1%	6%	5%
CONFIRMED •	<1%	24%	18%
TOTAL	<1%	39%	30%

BREEDING RELATIVE ABUNDANCE CHANGE *since 1988–91*

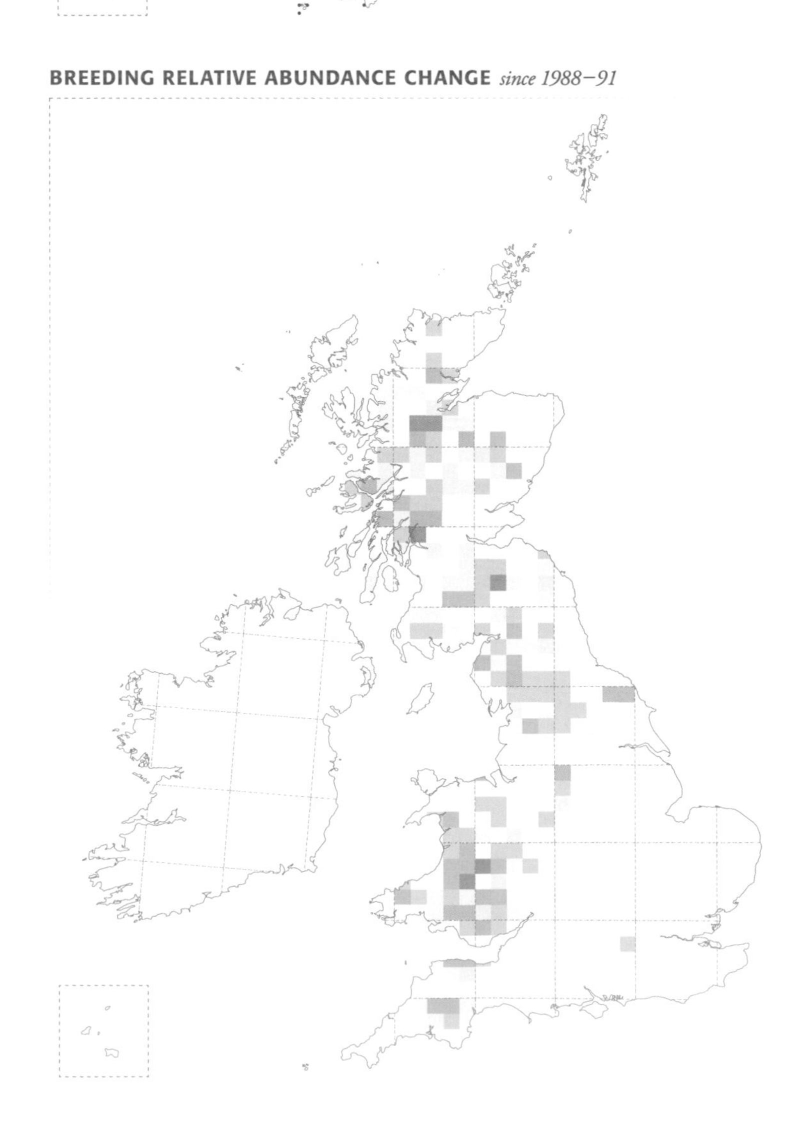

BREEDING DISTRIBUTION CHANGE *since 1968–72*

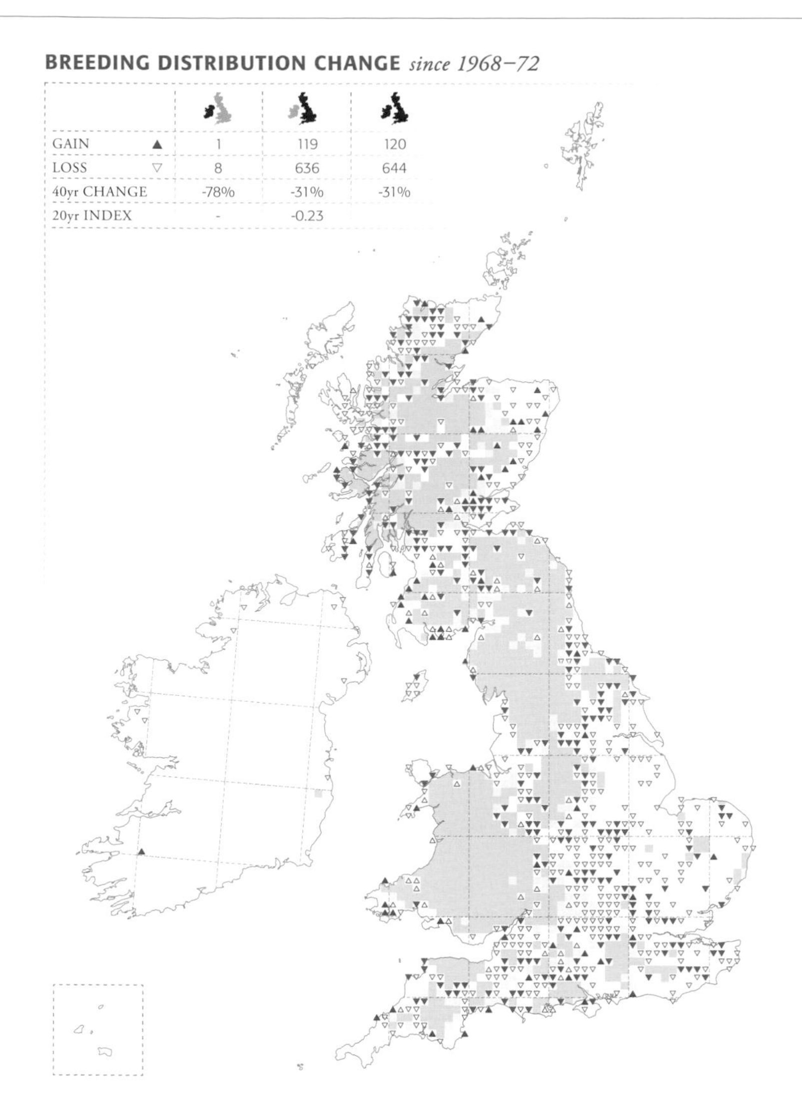

GAIN ▲	1	119	120
LOSS ▽	8	636	644
40yr CHANGE	-78%	-31%	-31%
20yr INDEX	-	-0.23	

BREEDING RELATIVE ABUNDANCE *2008–11*

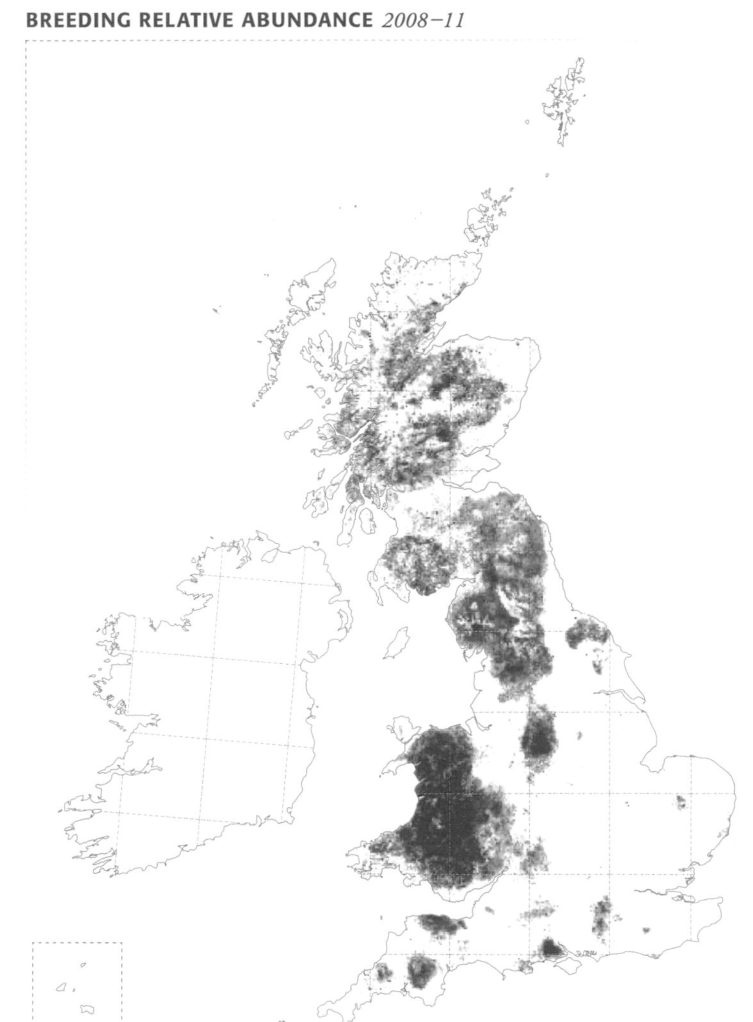

EDMUND FELLOWES

◀ The breeding haunts of Redstarts range from mature broad-leaved woods and copses to hedgerows, wood–pasture landscapes and heathland. In the uplands they nest in stone walls as well as tree holes.

WHINCHAT

Saxicola rubetra

BTO CODE: WC

ROB CROSS www.robcimages.co.uk

IN BRITAIN, THE Whinchat is becoming increasingly confined to the upland margins of Scotland, northern England, central Wales, Exmoor, Dartmoor and the Isle of Man. It occurs in lowland habitats at higher latitudes in Britain but the only significant lowland population in southern Britain is in southwest England and especially on Salisbury Plain. In Ireland, Whinchats are rare breeders, with small populations in the Antrim Hills, Kildare bogs, Wicklow Hills and Shannon Callows. Occupancy of the Shannon area may be facilitated by the management of late-cut hay meadows for Corncrakes. Within occupied regions, there appears to be little geographic variation in abundance, although densities in northern England appear to be lower than elsewhere in Britain, and densities are very low almost everywhere in Ireland.

The breeding range of the Whinchat has been steadily eroding since the early 1900s: already by the *1968–72 Breeding Atlas* it had been lost from much of central and southeast England and successive atlases have charted further losses in the lowlands and upland fringes. Overall, there has been a 47% range contraction in Britain since 1968–72. Over the same period there has been a 76% range contraction in Ireland, with the loss of birds from the northwest and midlands. Throughout the existing range, the abundance change map shows a decrease in the proportion of occupied tetrads since the *1988–91 Breeding Atlas*. Contractions and density declines are reflected in annual population monitoring results, which show a 57% decline during 1995–2010 in the UK (*BBS Report 2011*). A study focusing on upland bird populations found a 95% decline in the abundance of Whinchats between 1968–80 and 1999–2000 (Henderson *et al.* 2004).

Long-term, post-war declines in range were probably a direct consequence of agricultural intensification and the resulting loss of marginal habitats rich in invertebrate prey (e.g. *Birds of Scotland*). Since about 1990, however, declines have accelerated, and reduced overwinter survival has been implicated (Ockenden *et al.* 2012), though this has perhaps been exacerbated by local land-use practices and climate change. In Scotland, there may be altitudinal limits for upland-breeding Whinchats constrained by intensive farming at lower elevations and by suboptimal environmental conditions at higher elevations (Calladine & Bray 2012).

BREEDING DISTRIBUTION *2008–11*

POSSIBLE •	1%	6%	5%
PROBABLE •	1%	5%	4%
CONFIRMED ●	2%	19%	14%
TOTAL	4%	30%	23%

BREEDING RELATIVE ABUNDANCE CHANGE *since 1988–91*

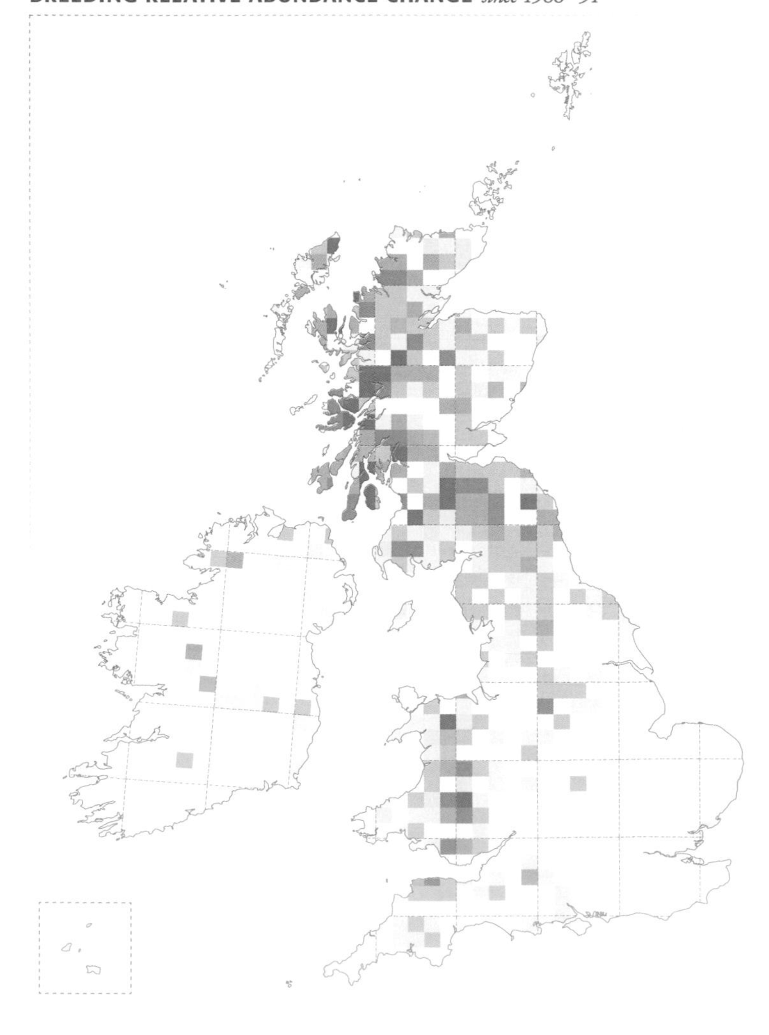

Sponsored by David Musson

BREEDING DISTRIBUTION CHANGE *since 1968–72*

GAIN ▲	21	67	88
LOSS ▽	167	863	1030
40yr CHANGE	-76%	-47%	-50%
20yr INDEX	-0.63	-0.49	

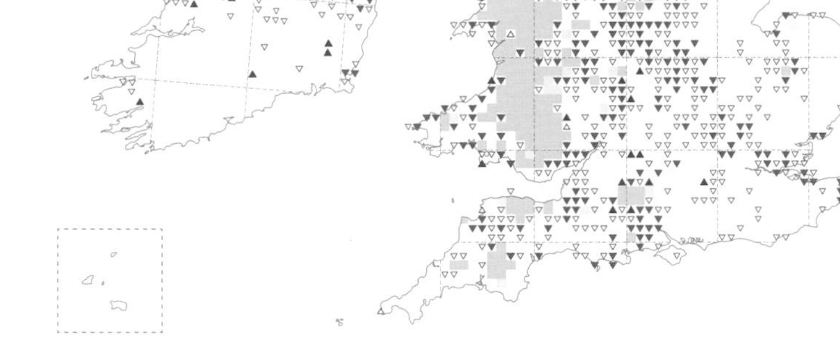

BREEDING RELATIVE ABUNDANCE *2008–11*

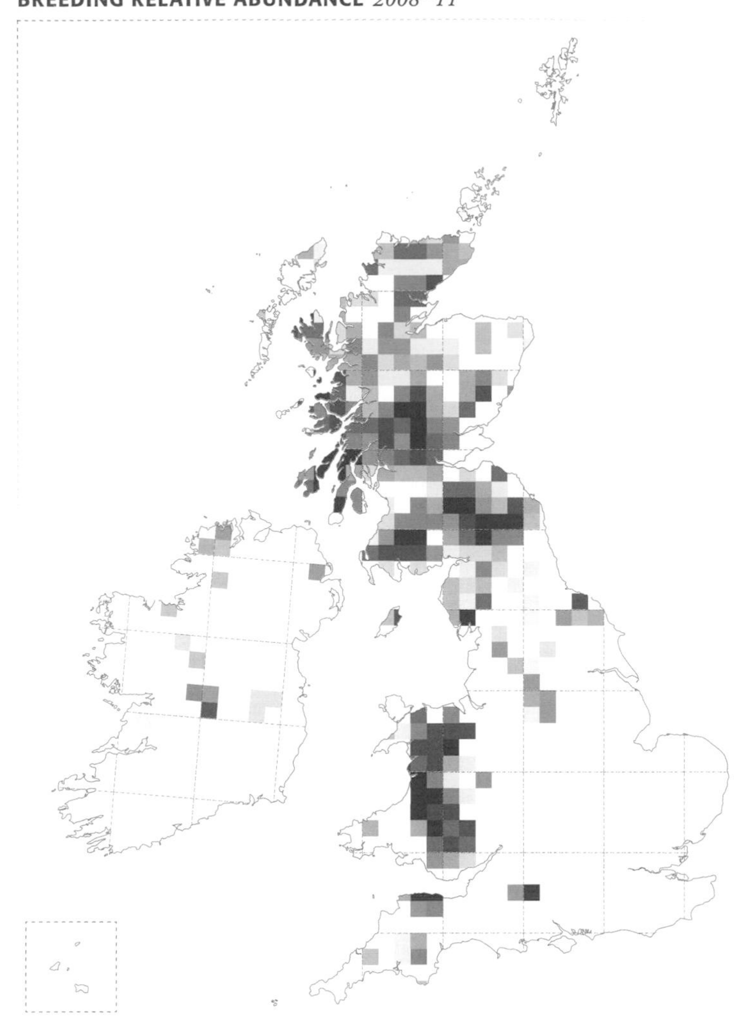

BRIAN RAFFERTY

◀ Marginal upland habitat is important for Whinchats and the declines recorded in the lowland populations in previous atlases have now extended there too.

STONECHAT

Saxicola rubicola

BTO CODE: SC

PAUL HILLION

THE STONECHAT GRACED the front cover of the *1968–72 Breeding Atlas*, showing the species to be widespread in Ireland but largely restricted to coastal southern and western Britain. Subsequently, its status has changed markedly owing to land-use change and a series of mild winters.

Wintering Stonechats were recorded in 81% of 10-km squares, up 115% since the *1981–84 Winter Atlas*. British Stonechats are partial migrants: just over half are resident whilst the rest migrate to winter from the near Continent southwards to North Africa (Helm *et al.* 2006). Migration distances have remained constant during this period of range expansion, suggesting it is not driven by changing migration strategies but perhaps by the increasingly high productivity shown by this multi-brooded species (Cummins & O'Halloran 2003; *BirdTrends*).

The breeding distribution also saw remarkable gains. Earlier losses in eastern Ireland have been partially reversed, with a 32% range expansion since the *1988–91 Breeding Atlas*. In Britain virtually all the losses evident in 1988–91 have been reversed along with substantial eastward expansion. The abundance maps reveal a ghost of the former distributions, with densities generally higher in the west, particularly near the coast where favoured rough grassland, gorse and heath are abundant. The abundance change map is dominated by the gross distribution gains, but there is evidence of declines on some western coasts.

Earlier distribution changes were probably associated with the maturing of plantations and the loss of semi-natural habitats from farmland, but recent changes are undoubtedly the consequence of the series of mild winters which came to an abrupt end with the severe winters of 2009/10 and 2010/11. These caused a sharp reversal of the previously positive population trends (Crowe *et al.* 2010b; Lovatt & Madden 2012). In 2007, the UK breeding population was 209% higher than in 1995 but by 2010 the population was only 19% higher (Risely *et al.* 2009; *BBS Report 2011*). Of the 3,156 10-km squares from which wintering Stonechats were recorded, 2,781 were reportedly occupied in the first two winters but only 1,900 in the second two winters. Consequently, distributions at the end of the Atlas were probably substantially smaller than those mapped here.

Sponsored by Sussex Ornithological Society

BREEDING DISTRIBUTION *2008–11*

POSSIBLE •	10%	5%	6%
PROBABLE •	14%	9%	10%
CONFIRMED •	50%	51%	51%
TOTAL	74%	65%	67%

WINTER DISTRIBUTION *2007/08–2010/11*

PRESENT •	73%	84%	81%

BREEDING DISTRIBUTION CHANGE *since 1968–72*

GAIN ▲	105	758	863
LOSS ▽	146	125	271
40yr CHANGE	-5%	+52%	+29%
20yr INDEX	+0.23	+0.42	

WINTER DISTRIBUTION CHANGE *since 1981–84*

GAIN ▲	305	1430	1735
LOSS ▼	86	57	143
30yr CHANGE	+46%	+151%	+115%

BREEDING RELATIVE ABUNDANCE *2008–11*

BREEDING RELATIVE ABUNDANCE CHANGE *since 1988–91*

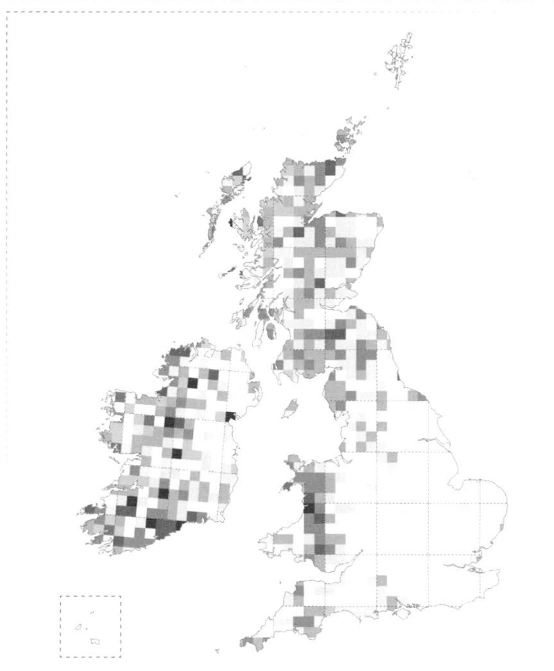

WINTER RELATIVE ABUNDANCE *2007/08–2010/11*

WHEATEAR
Oenanthe oenanthe

BTO CODE: W.

GARY FAULKNER

WHEATEARS ARE SUMMER visitors to the uplands and coasts of northern and western Britain, and to the northern and western counties of Ireland. Elsewhere they are localised, with small outposts in Ireland on the Antrim coast and in the Wicklow Mountains, and in Britain on Salisbury Plain and in a handful of coastal squares in southeast England. In addition, *non-breeding* birds, usually northbound migrants, are recorded in virtually all remaining squares in Britain, outside some major urban centres. Migrants were far less widely recorded in Ireland. Late autumn migrants account for most of the 140 10-km squares on the winter distribution map.

The highest densities occur throughout the uplands and near the coast. Many Scottish islands also support high densities, possibly because on many, most of the land is near the coast (Sellers 2006). In Scotland, occupancy peaks in the 200–500-m elevation range, with birds generally avoiding areas of intensive land management (*Birds of Scotland*).

Since the *1968–72 Breeding Atlas* there has been a 14% range contraction in Britain and a 24% contraction in Ireland. Losses have gradually accumulated, leading to the disappearance of distinct breeding areas, including the southeast coast of Ireland, Breckland and the southern English heaths and downland. Elsewhere, losses of Wheatears have occurred on the edges of the current range, such as in western Ireland and on intensively farmed lowlands in Scotland (*Birds of Scotland*).

The abundance change map shows declines in tetrad occupancy throughout the core breeding range since 1988–91. Long-term UK population trends are not available owing to historically poor coverage of upland habitats. However, on 13 marginal upland CBC plots originally surveyed during 1968–80, numbers had fallen by 96% by 1999–2000 (Henderson *et al.* 2004). In Ireland a 31% decline is apparent during 1998–2010 (*CBS Trend 2010*). Early declines in the lowlands were ascribed to loss of extensive grasslands and declining European Rabbit populations (*BirdTrends*), although some losses occurred in areas with favourable management, suggesting external factors were at play (Taylor & Marchant 2011). The causes of the apparent density reductions across the uplands are unknown, although falling sheep densities may have reduced the suitability of grass swards (Fulton 2010).

Sponsored by Jack Taylor, Liverpool (South West Lancashire Ringing Group)

BREEDING DISTRIBUTION *2008–11*

	Britain & Ireland	Britain	Ireland
POSSIBLE •	11%	8%	9%
PROBABLE •	9%	10%	10%
CONFIRMED ●	21%	38%	34%
TOTAL	41%	56%	53%

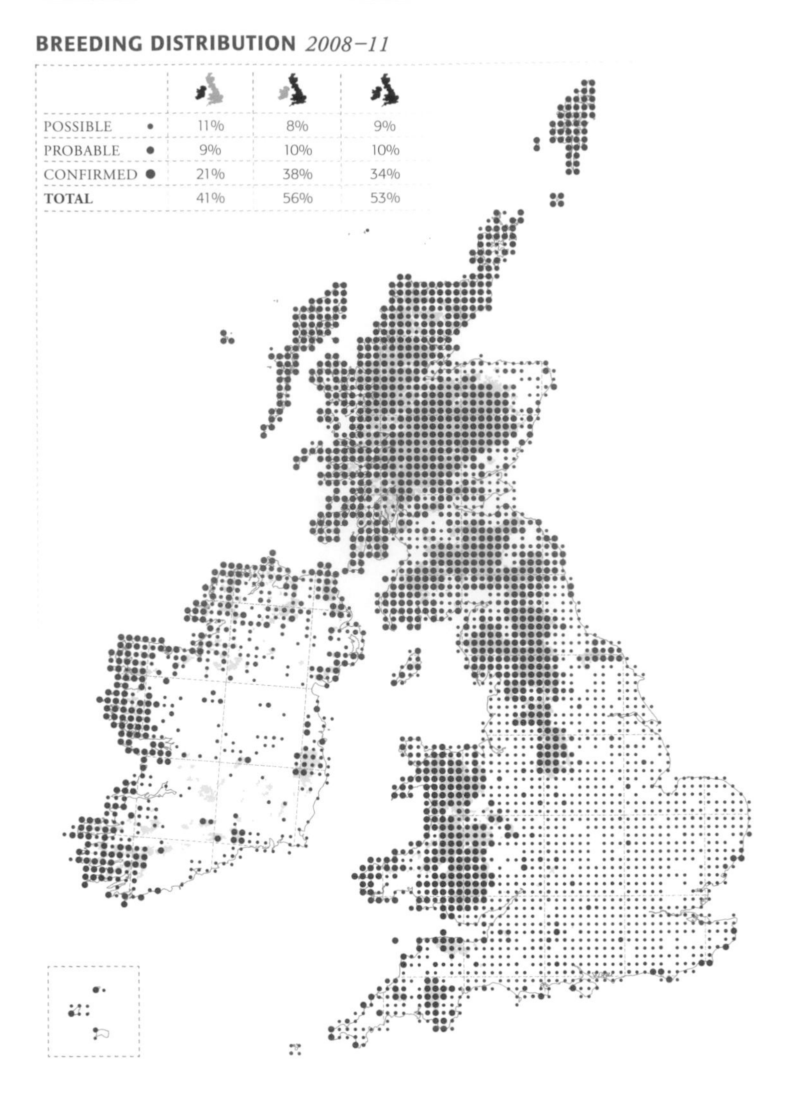

BREEDING RELATIVE ABUNDANCE CHANGE *since 1988–91*

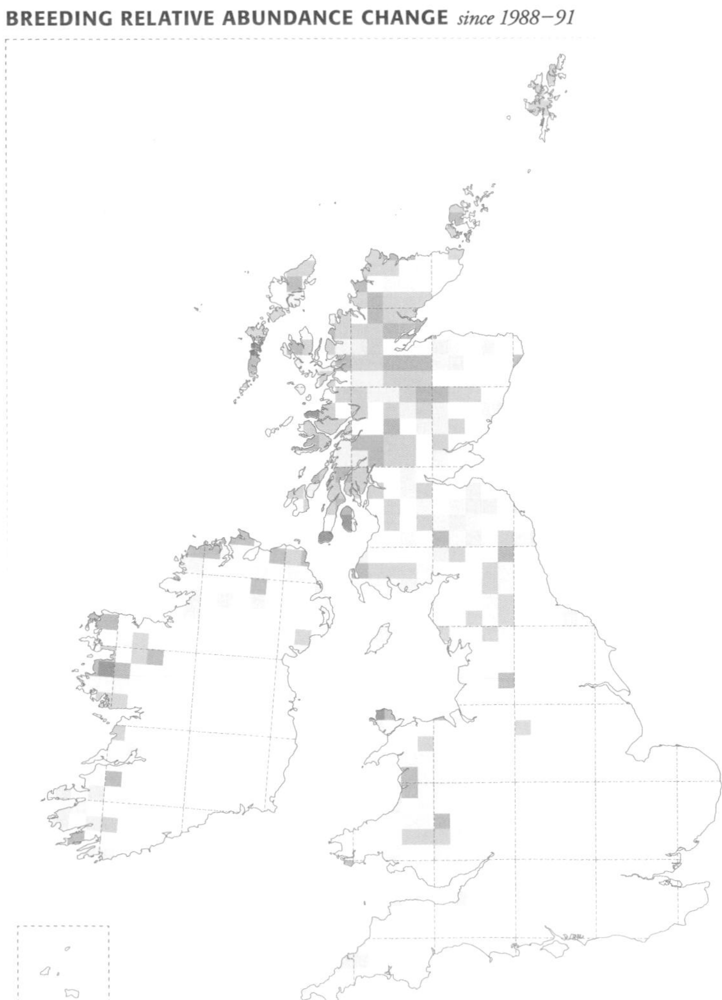

BREEDING DISTRIBUTION CHANGE *since 1968–72*

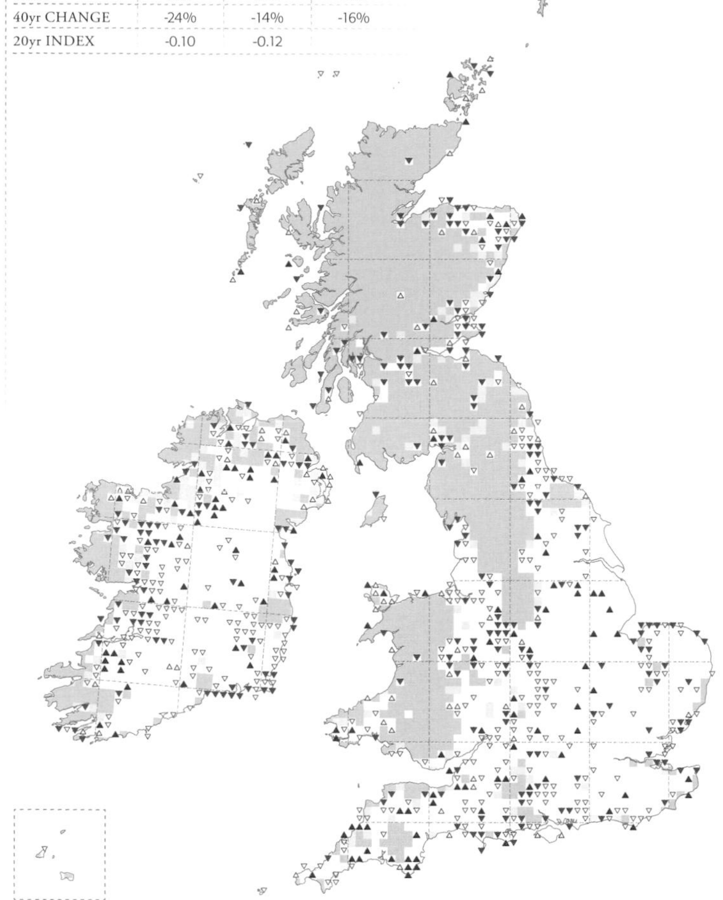

GAIN ▲	73	137	210
LOSS ▽	201	391	592
40yr CHANGE	-24%	-14%	-16%
20yr INDEX	-0.10	-0.12	

BREEDING RELATIVE ABUNDANCE *2008–11*

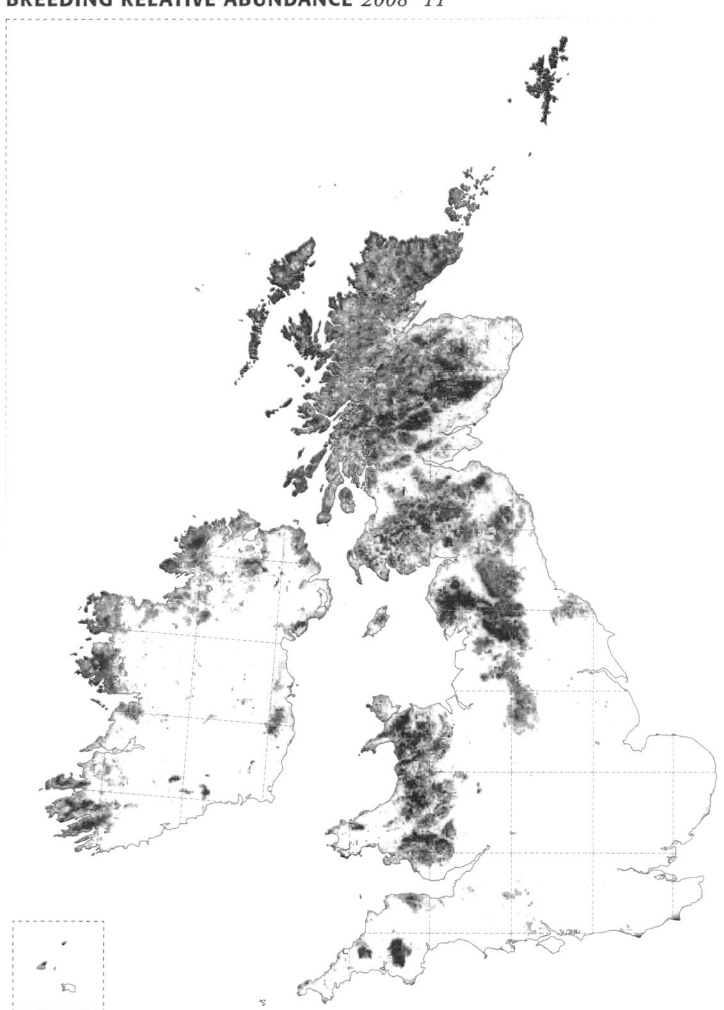

WINTER DISTRIBUTION *2007/08–2010/11*

PRESENT ●	1%	4%	4%

DUNNOCK
Prunella modularis

BTO CODE: D.

CLARE WARD

THE DUNNOCK IS resident in Britain & Ireland and is highly sedentary; it occurs in 93% of 10-km squares both in winter and in the breeding season. The most notable absences are from treeless stretches of the Scottish Highlands, and parts of the Outer Hebrides and Northern Isles. No breeding evidence was reported from Shetland in 2008–11, nor in previous breeding atlases. Confirmation of breeding is also lacking in parts of the northwest of both Ireland and Scotland where Dunnocks occur at low density. Densities in both seasons are highest in southern and eastern Ireland and throughout much of lowland England.

There has been very little change overall in the breeding distribution since the *1968–72 Breeding Atlas*, though some gains are apparent in the uplands of northwest Scotland, on Orkney and in the Outer Hebrides, which may reflect improved coverage. However, the breeding abundance change map reveals widespread increases in the proportion of occupied tetrads since the *1988–91 Breeding Atlas*. In the UK, there was a substantial fall in the Dunnock population between the mid 1970s and mid 1980s, though some recovery has occurred since the late 1990s (*BirdTrends*). In the Republic of Ireland there has been a 13% increase in numbers during 1998–2010 (*CBS Trend 2010*).

The reasons for the initial population decline are unclear, though a reduction in management and increased grazing pressure from deer are likely to have created a more open understorey and reduced the amount of preferred habitat within woodland (Bevington 1991; Fuller *et al.* 2005; Holt *et al.* 2011b). Despite apparent declines in woodland, the Dunnock remains abundant in many urban areas. There is little variation in survival rates through time (Siriwardena *et al.* 1998b) but clutch size, brood size and the number of fledglings per breeding attempt all increased as the population fell (*BirdTrends*).

Compared to the *1981–84 Winter Atlas* there has been a 6% increase in 10-km-square occupancy in winter. Gains are evident in the northern half of Britain, especially in northwest Scotland, as well as in northern and western Ireland. These may be due to a combination of improved coverage and an increase in the area of woodland.

BREEDING DISTRIBUTION *2008–11*

POSSIBLE •	10%	5%	7%
PROBABLE •	21%	6%	10%
CONFIRMED ●	65%	80%	76%
TOTAL	96%	91%	93%

WINTER DISTRIBUTION *2007/08–2010/11*

PRESENT ●	97%	92%	93%

BTO
Looking out for birds

Sponsored by the BTO Ringing Scheme

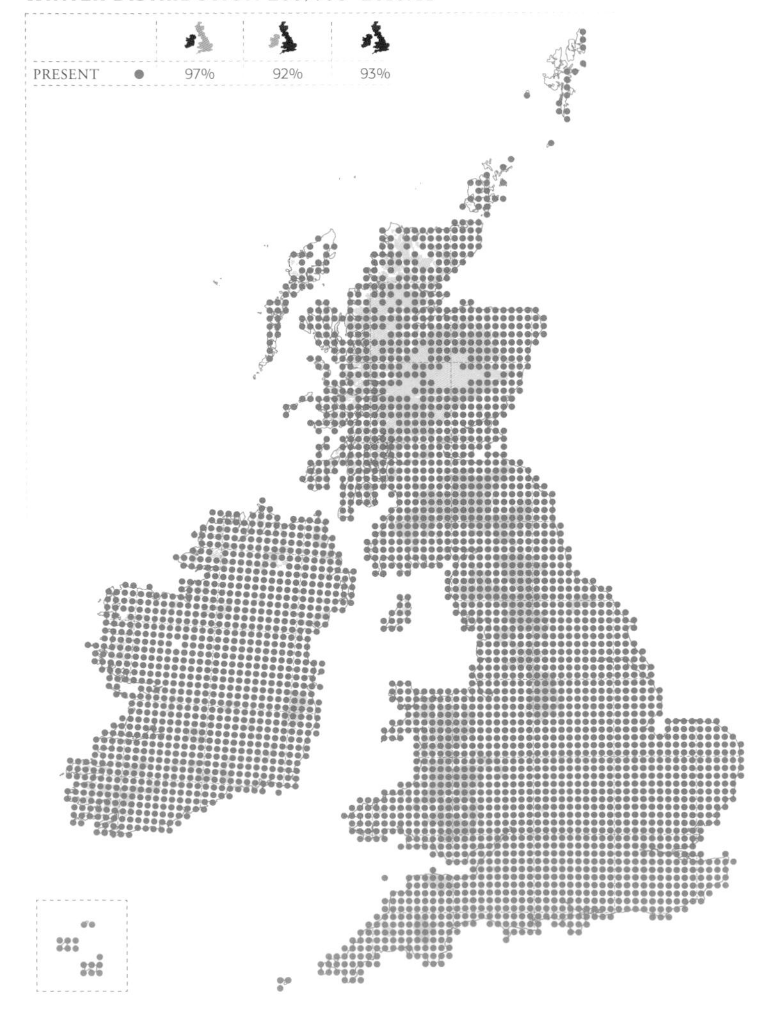

BREEDING DISTRIBUTION CHANGE *since 1968–72*

GAIN ▲	16	66	82
LOSS ▽	13	48	61
40yr CHANGE	+<1%	+1%	+1%
20yr INDEX	+0.04	+0.06	

WINTER DISTRIBUTION CHANGE *since 1981–84*

GAIN ▲	51	207	258
LOSS ▼	26	36	62
30yr CHANGE	+3%	+7%	+6%

BREEDING RELATIVE ABUNDANCE *2008–11*

BREEDING RELATIVE ABUNDANCE CHANGE *since 1988–91*

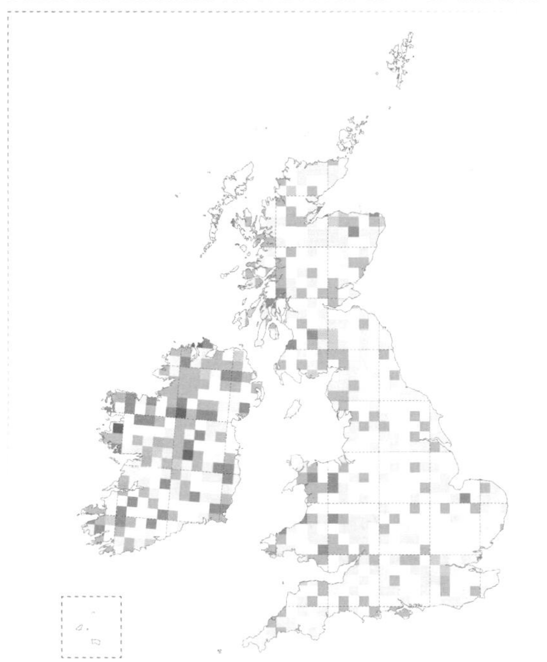

WINTER RELATIVE ABUNDANCE *2007/08–2010/11*

HOUSE SPARROW
Passer domesticus

BTO CODE: HS

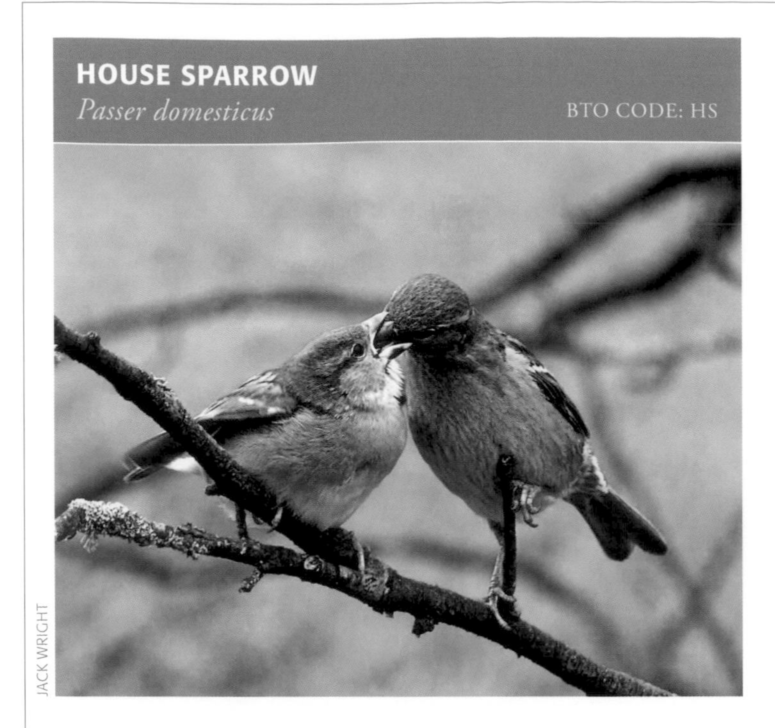

JACK WRIGHT

THE HOUSE SPARROW is among the most widespread species in Britain & Ireland, being found in c.90% of 10-km squares; it is absent only from exposed upland areas of northern Scotland. House Sparrows are highly sedentary so distribution patterns remain virtually unchanged from the breeding season to winter.

A 5% range expansion since the *1981–84 Winter Atlas* has arisen mainly by gains in remote parts of western Ireland, northern and western Scotland and central Wales. In most instances this is probably due to improved coverage allowing the detection of birds in these areas of low sparrow density. The breeding-season change map projects an image of stability; the few gains are mostly restricted to the coastal fringes of Ireland, whilst losses are mainly in upland margins. Overall there has been only a 3% range contraction since the *1968–72 Breeding Atlas*.

These small range changes belie striking changes in population size. Monitoring data indicate a rapid decline of 69% in the UK breeding population during 1977–2010, with numbers perhaps stabilising in the last decade (Siriwardena *et al.* 2002; Robinson *et al.* 2005b; *BirdTrends*). Long-term trends are lacking in the Republic of Ireland but numbers increased by 93% during 1998–2010 (*CBS Trend 2010*). These trends are reflected in the abundance change map, which shows decreases in the proportion of occupied tetrads since the *1988–91 Breeding Atlas* across southeast and central England but increases in Wales, parts of Scotland and most of Ireland. Massimino *et al.* (2013) detected similar density declines in southeast England, particularly in London, a trend also reported in Edinburgh (Dott & Brown 2000).

Extensive research into these changes indicates that different factors are likely to drive trends in rural and urban populations. Low first-year survival rates, connected with reduction in winter food supply caused by agricultural intensification, are probably key factors affecting rural populations (Freeman & Crick 2002; Hole *et al.* 2002). In towns and cities, however, reduced breeding performance has probably been more important, potentially owing to reductions in invertebrate prey availability, air pollution and lack of nest sites (Summers-Smith 2003, 2007; Peach *et al.* 2008; Shaw *et al.* 2008).

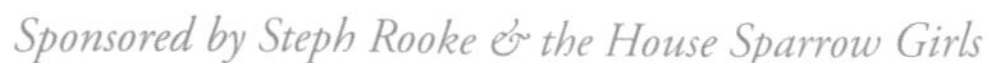
Sponsored by Steph Rooke & the House Sparrow Girls

BREEDING DISTRIBUTION *2008–11*

POSSIBLE	4%	1%	2%
PROBABLE	5%	2%	3%
CONFIRMED	86%	87%	87%
TOTAL	95%	90%	92%

WINTER DISTRIBUTION *2007/08–2010/11*

PRESENT	93%	88%	89%

BREEDING DISTRIBUTION CHANGE *since 1968–72*

GAIN ▲	16	15	31
LOSS ▽	18	104	122
40yr CHANGE	-<1%	-3%	-3%
20yr INDEX	+0.08	+0.02	

WINTER DISTRIBUTION CHANGE *since 1981–84*

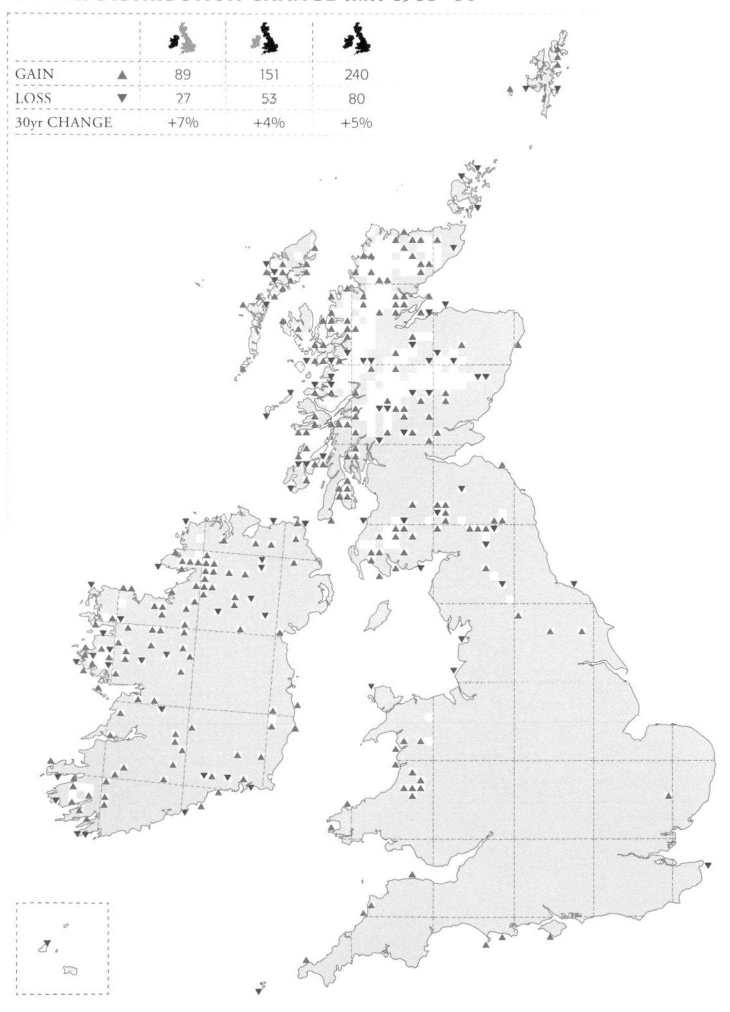

GAIN ▲	89	151	240
LOSS ▼	27	53	80
30yr CHANGE	+7%	+4%	+5%

BREEDING RELATIVE ABUNDANCE *2008–11*

BREEDING RELATIVE ABUNDANCE CHANGE *since 1988–91*

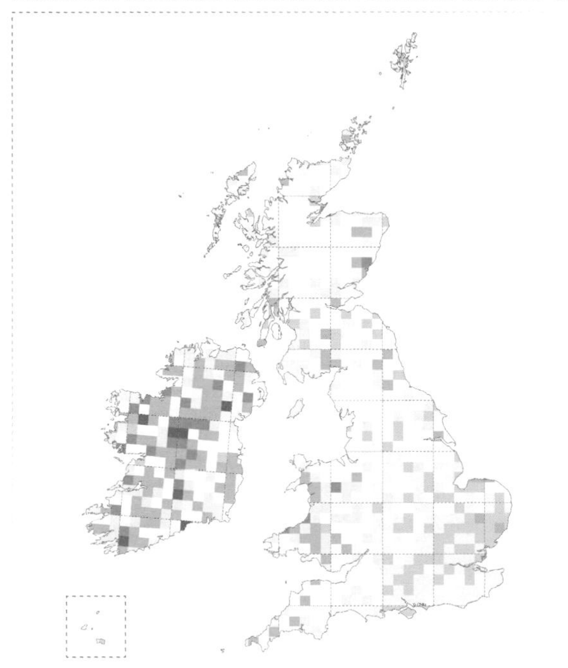

WINTER RELATIVE ABUNDANCE *2007/08–2010/11*

TREE SPARROW

Passer montanus

BTO CODE: TS

KIM CONNOLLY www.kimcomania.co.uk

THE WINTER AND breeding distribution maps show that Tree Sparrows now occur in low-lying parts of England, mostly to the north of a line from the Severn to the Wash, and in eastern Scotland and the Welsh Marches. In Ireland, they are concentrated around Lough Neagh and in eastern counties, with a few scattered around northern and western coasts. In both seasons, densities are highest in northeast England and eastern Scotland.

Trends differ markedly between Ireland and Britain, and to some extent within Britain. In Ireland, the range has expanded, by 71% in winter since the *1981–84 Winter Atlas* and by 78% since the *1968–72 Breeding Atlas*, with most of the gains since 1988–91. Breeding numbers around Loughs Neagh & Beg increased by 65% over just three years in the mid 1990s (Perry & Day 1997). In contrast, the British range has contracted by 20% in winter since 1981–84 and by 41% in the breeding season since 1968–72, involving significant and mostly recent losses from southeast England, South Wales, the Pennines and the western part of the Central Belt in Scotland. These losses are offset slightly by gains in northeast Scotland, where tetrad occupancy in the breeding season increased by 174% between 1981–84 and 2002–06 (Francis & Cook 2011). The abundance change map shows that increases have also occurred in northeast England but that densities have declined in southern England.

The localised increases in abundance shown here may partly explain the 96% population increase apparent in the UK during 1995–2010 (*BBS Report 2011*). However, the decline in the 1970s was so severe that, even with this recent increase, the population still numbers fewer than 10% of those present in the late 1960s (*BirdTrends*). In common with many farmland birds, the substantial decline in Britain has been linked to agricultural changes (Chamberlain *et al.* 2001). As there is evidence of increasing productivity, reduced overwinter survival is the most likely mechanism underlying the decline (*BirdTrends*). There is some evidence of association with winter stubbles and possibly with wetlands for feeding (Hancock & Wilson 2003; Field & Anderson 2004). The causes of the expansion in Ireland and northeast Scotland are unknown.

Sponsored by MKA Ecology Limited, Consultant Ecologists

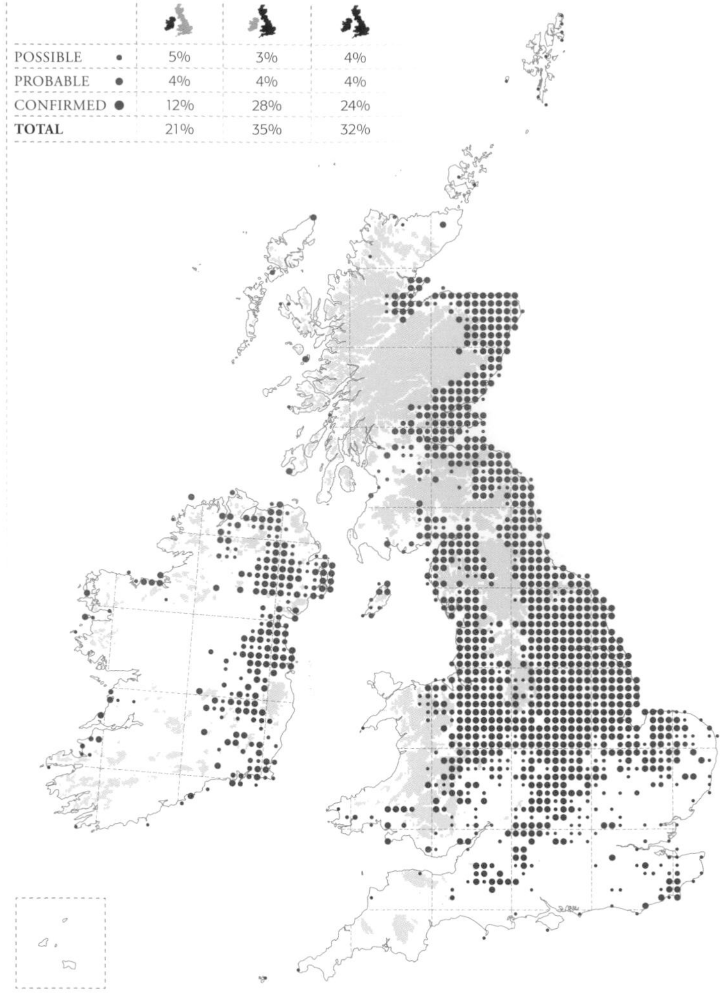

POSSIBLE •	5%	3%	4%
PROBABLE ●	4%	4%	4%
CONFIRMED ●	12%	28%	24%
TOTAL	21%	35%	32%

WINTER DISTRIBUTION *2007/08–2010/11*

PRESENT ●	24%	41%	36%

BREEDING DISTRIBUTION CHANGE *since 1968–72*

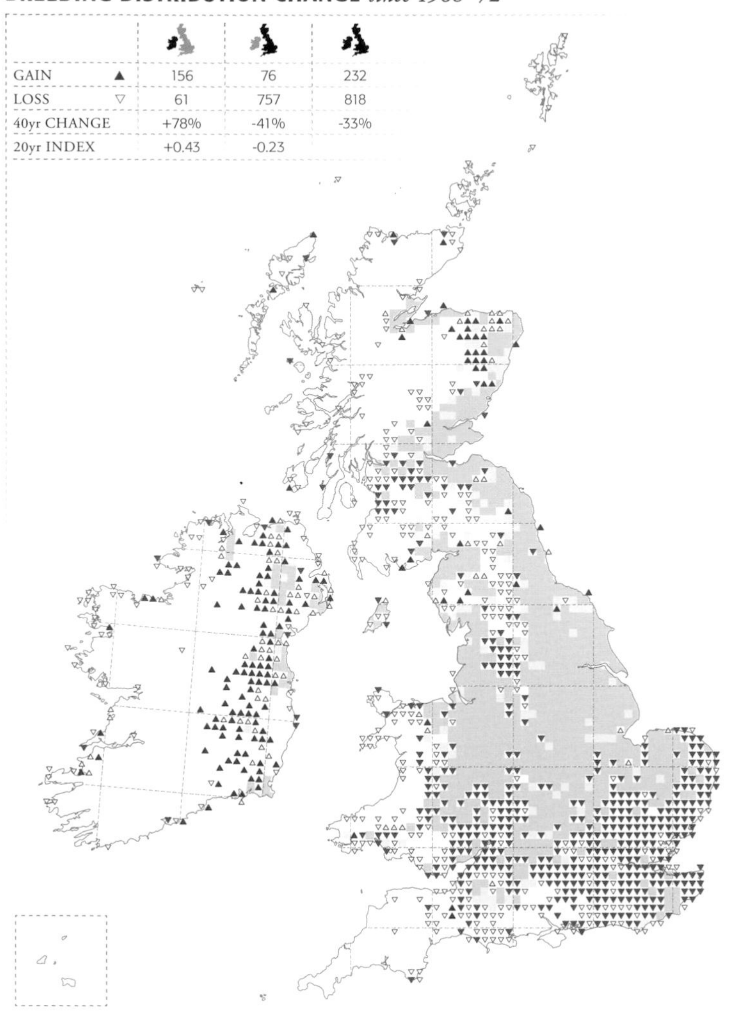

GAIN ▲	156	76	232
LOSS ▽	61	757	818
40yr CHANGE	+78%	-41%	-33%
20yr INDEX	+0.43	-0.23	

WINTER DISTRIBUTION CHANGE *since 1981–84*

GAIN ▲	151	199	350
LOSS ▼	57	473	530
30yr CHANGE	+71%	-20%	-12%

BREEDING RELATIVE ABUNDANCE *2008–11*

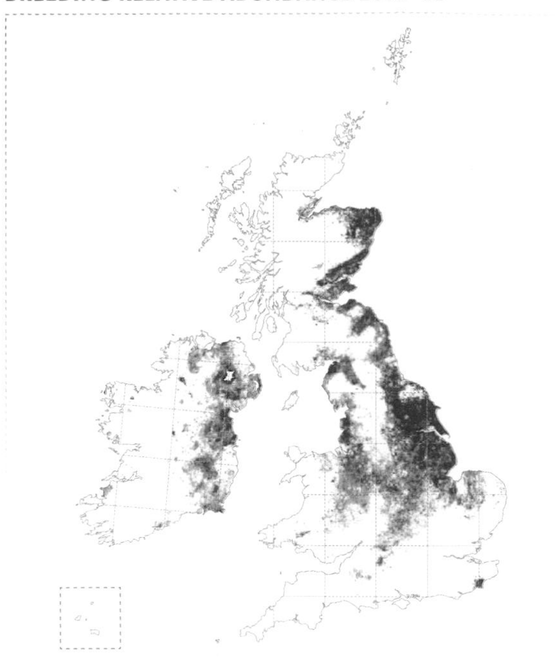

BREEDING RELATIVE ABUNDANCE CHANGE *since 1988–91*

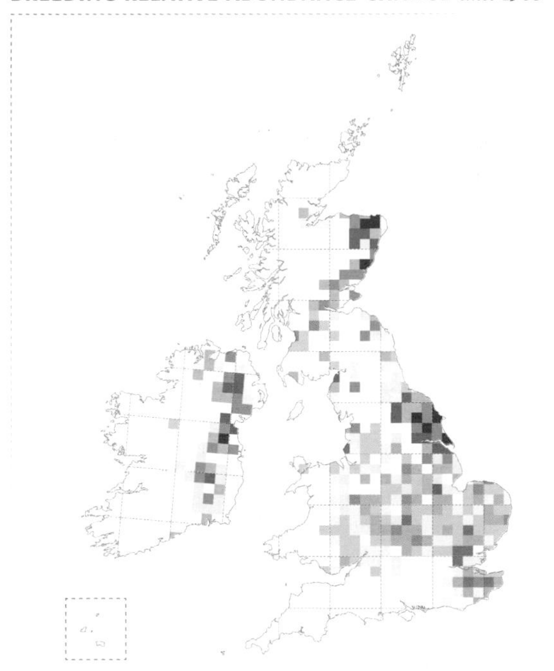

WINTER RELATIVE ABUNDANCE *2007/08–2010/11*

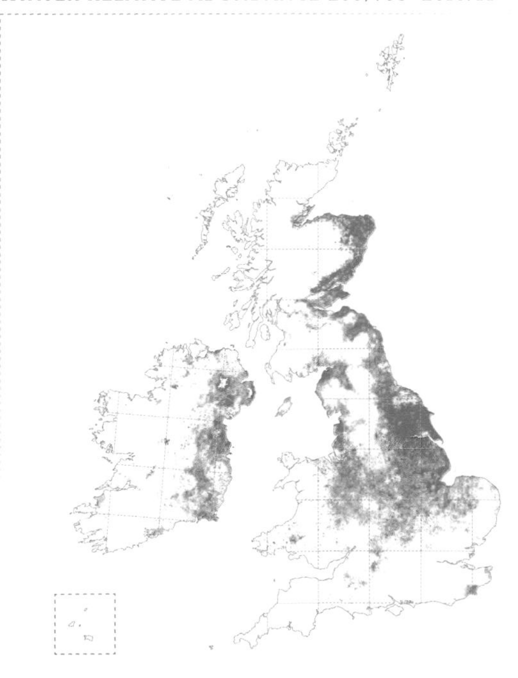

YELLOW WAGTAIL
Motacilla flava

BTO CODE: YW

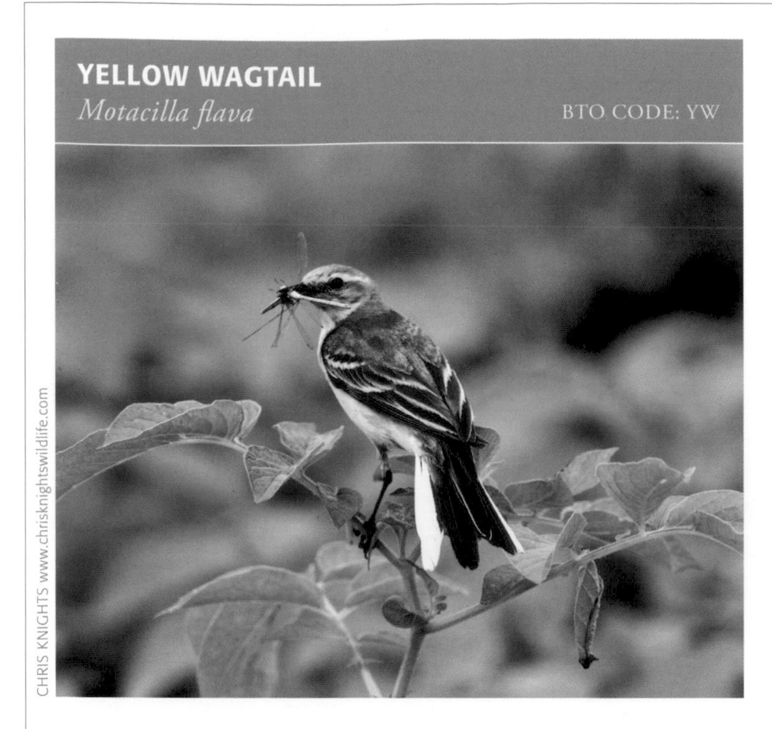

CHRIS KNIGHTS www.chrisknightswildlife.com

BREEDING DISTRIBUTION *2008–11*

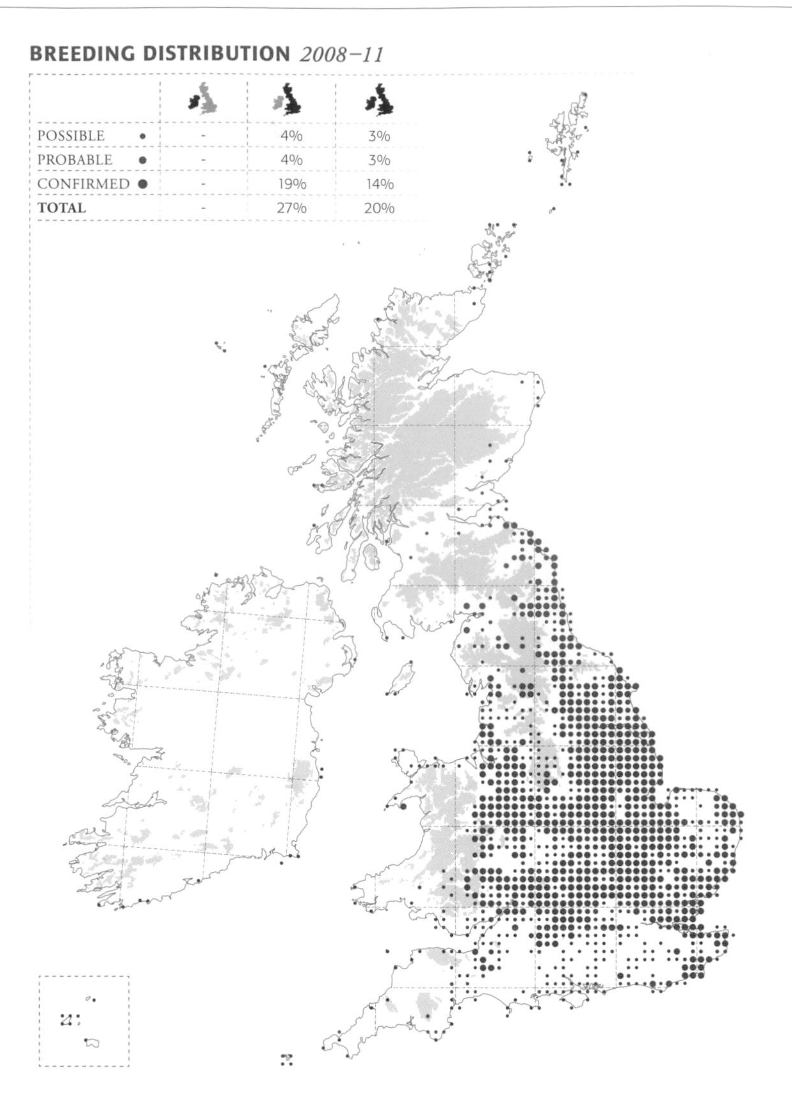

POSSIBLE •	-	4%	3%
PROBABLE •	-	4%	3%
CONFIRMED •	-	19%	14%
TOTAL	-	27%	20%

A MARKED RANGE contraction since the *1968–72 Breeding Atlas* was apparent by the *1988–91 Breeding Atlas*, and this atlas shows a further withdrawal into parts of central, eastern and northeast England. Few squares are now occupied in southern Scotland, and the Somerset Levels is a prominent outlier in southwest England, as is one pair on the Llŷn Peninsula in Wales. None were reported to have bred in Ireland during 2008–11. The highest densities were recorded in East Yorkshire, Lincolnshire, the Fens, Broadland and the Essex and Kent coastal marshes.

The breeding-season change map shows widespread losses in southern England, northwest England, Wales, the Scottish Borders and the Central Belt. Even within the remaining range, the proportion of occupied tetrads has fallen since 1988–91 in many areas. East Lincolnshire and East Yorkshire are among the only areas to show an expansion into new 10-km squares and an associated increase in tetrad occupancy.

As well as considerable range contraction, there has been a 72% population decline in the UK during 1970–2010 (*SUKB 2012*). This decline has been linked to farmland drainage, the conversion of pasture to arable land, the switch from spring- to autumn-sown cereals and a loss of insects associated with cattle (Chamberlain & Fuller 2000, 2001; Vickery *et al.* 2001; Nelson *et al.* 2003). In some arable areas, early nesting attempts are in winter wheat and grass fields but later attempts are in spring-sown broad-leaved crops, especially potatoes (Gilroy *et al.* 2010). Additionally, presence is related to soil penetrability, suggesting a link between soil degradation, invertebrate availability and wagtail abundance (Gilroy *et al.* 2008). Pairs nesting in silage are rarely successful, owing to the frequency of cutting regimes. Those associated with wet grasslands require a combination of recently flooded areas for feeding and tall, dense vegetation for nesting (Bradbury & Bradter 2004).

The continental Blue-headed race of Yellow Wagtail (nominate *flava*) was recorded in 153 10-km squares, of which breeding evidence was obtained in 19 squares, mostly in south and central England. Most breeding records related to mixed pairings with British *flavissima* but a pure pair bred in Cornwall in 2010 (*RBBP 2010*).

BREEDING RELATIVE ABUNDANCE CHANGE *since 1988–91*

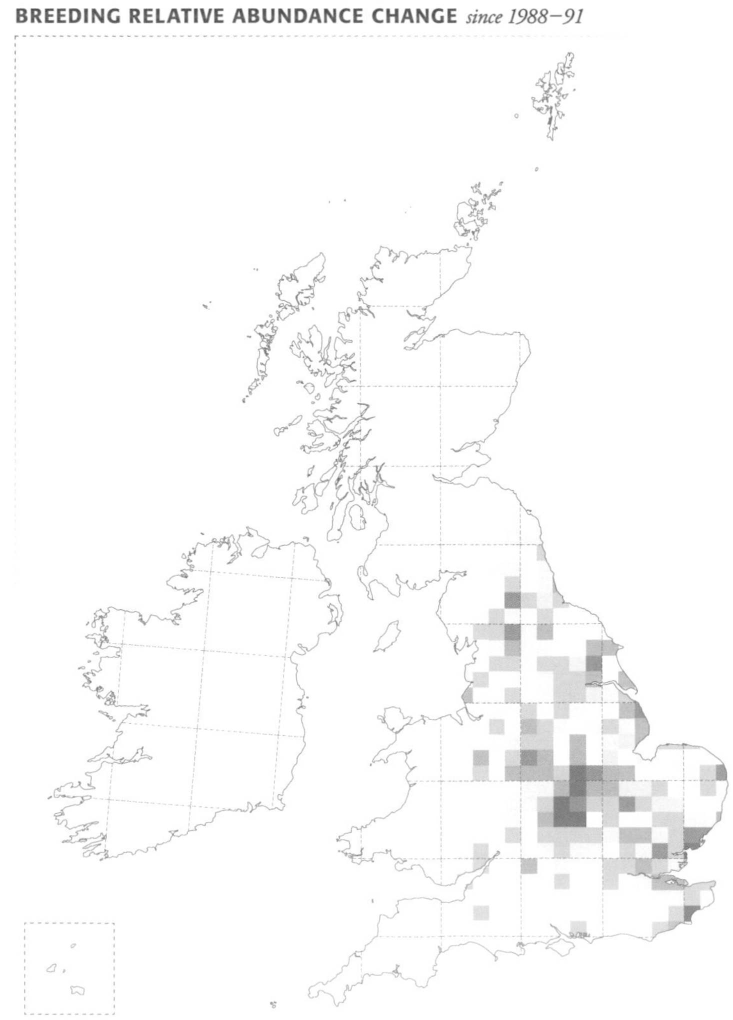

Sponsored by Dr Andrew Harris

BREEDING DISTRIBUTION CHANGE *since 1968–72*

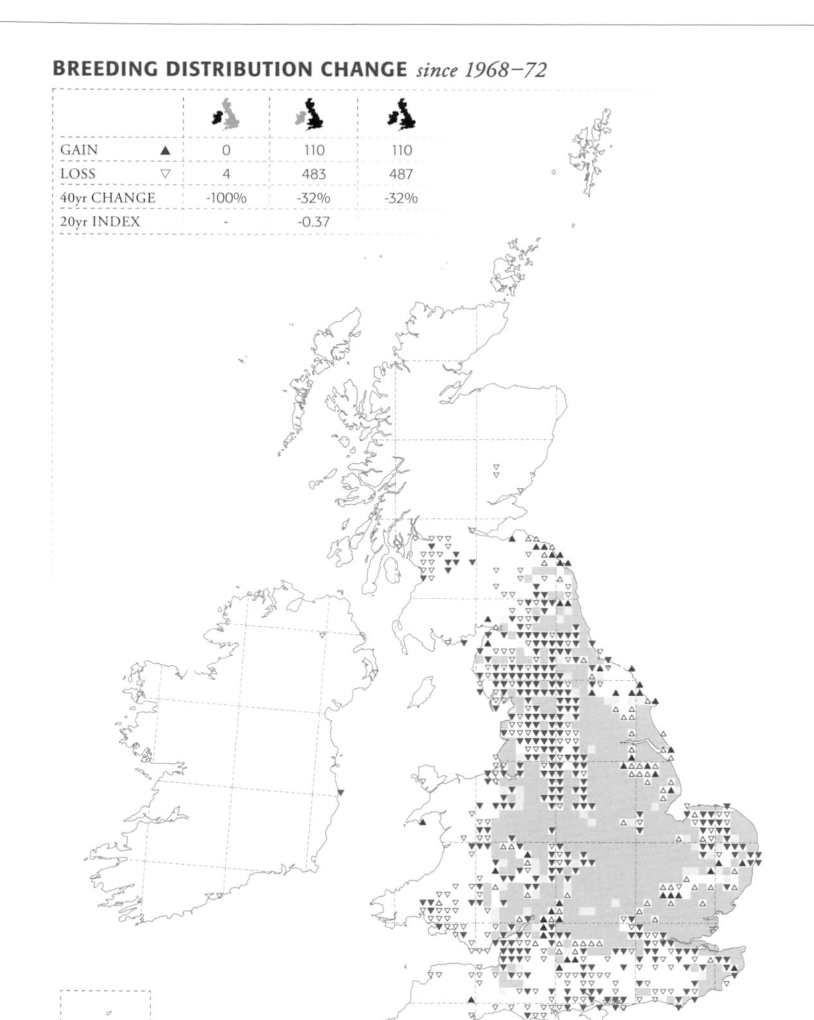

GAIN ▲	0	110	110
LOSS ▽	4	483	487
40yr CHANGE	-100%	-32%	-32%
20yr INDEX	-	-0.37	

BREEDING RELATIVE ABUNDANCE *2008–11*

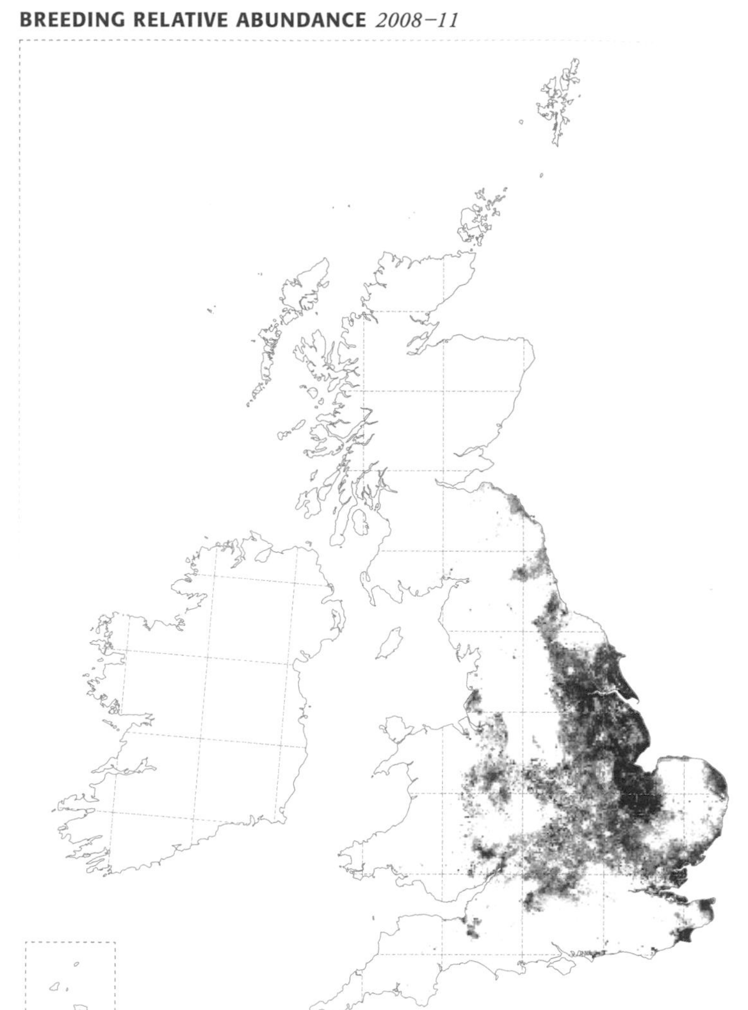

BLUE-HEADED WAGTAIL BREEDING DISTRIBUTION *2008–11*

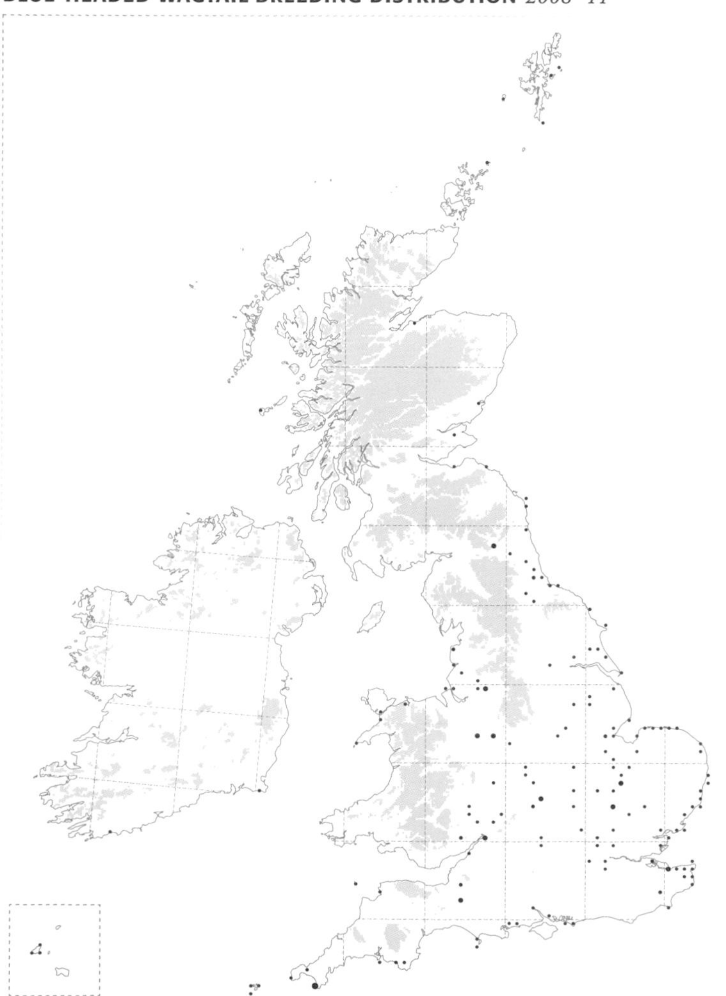

GREY WAGTAIL

Motacilla cinerea

BTO CODE: GL

BRIAN RAFFERTY

WINTERING GREY WAGTAILS are widespread throughout Britain except for parts of eastern England and the uplands. In Ireland, they are particularly widespread in the south and east. There have been 38% and 10% winter range expansions in Britain and Ireland respectively since the *1981–84 Winter Atlas*. Gains have occurred throughout the range, although most continuously in eastern England, on upland fringes and around the Scottish coast. In Ireland gains exceed losses, particularly in central areas. Some gains may be linked to better coverage, although milder winters during the first half of the Atlas may have allowed more birds to remain closer to breeding sites.

After the winter, breeding birds return to the British uplands, giving a distribution that is slightly more extensive than in winter, and 19% larger in this atlas than in the *1968–72 Breeding Atlas*, following expansion into the East Midlands, East Anglia, Caithness and the Northern Isles. Densities are highest in the uplands, particularly in Wales and the Pennines and throughout mainland Scotland. Grey Wagtails prefer wooded streams where the breeding cycle is synchronised with the phenology of insects of broad-leaved woods, a strategy that buffers them against the effects of acidification (Ormerod & Tyler 1991; Buckton & Ormerod 1997).

In Ireland the distribution during the breeding season remains much as in winter. A 14% contraction of the Irish breeding range since 1968–72 may have been exacerbated by the cold winters of 2009/10 and 2010/11. Grey Wagtails are susceptible to cold weather (Dobinson & Richardson 1964) and some breeding birds were probably lost from Irish squares before they could be visited in the latter half of the Atlas. The abundance change map tells a story of breeding-season abundance declines throughout Ireland, as in many parts of western and northern Britain.

The impact of the two severe winters is apparent in breeding population trends. Whereas the UK population showed a 38% increase during 1995–2007, the trend had become a 15% decline by 2010 (Risely *et al.* 2009; *BBS Report 2011*). The Irish population declined by 4% per annum during 1998–2010 (*CBS Trend 2010*). In addition to the effects of cold winters, reductions in clutch and brood sizes are raising concerns (*BirdTrends*).

BREEDING DISTRIBUTION *2008–11*

POSSIBLE •	21%	8%	11%
PROBABLE •	16%	9%	11%
CONFIRMED ●	39%	59%	54%
TOTAL	76%	76%	76%

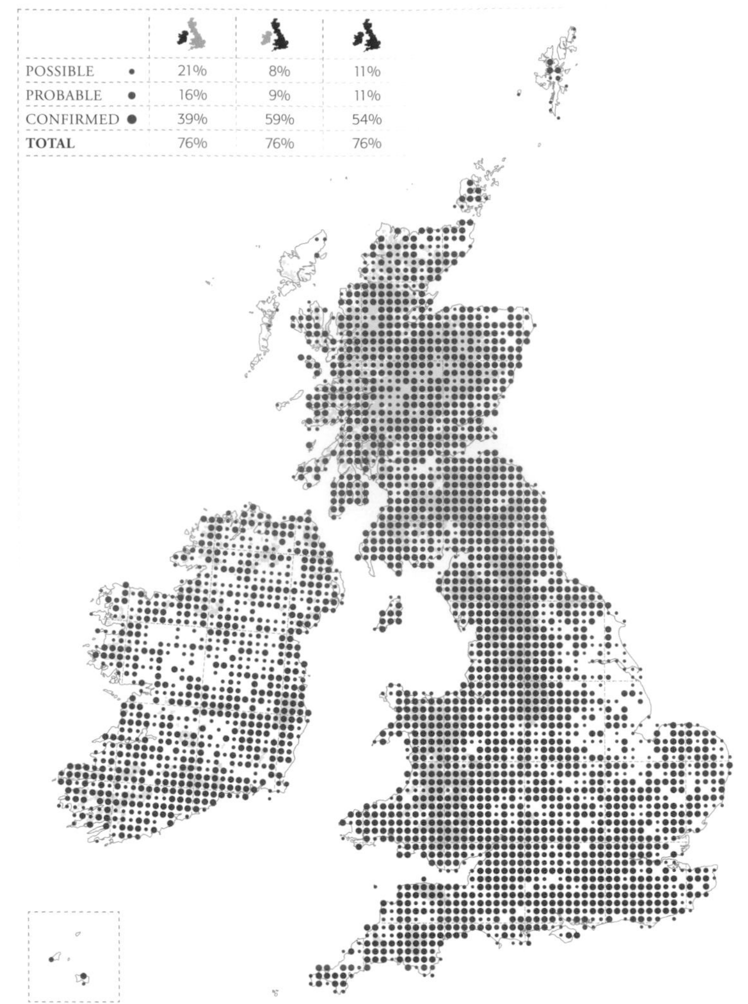

WINTER DISTRIBUTION *2007/08–2010/11*

PRESENT ●	75%	75%	75%

Sponsored by the Environment Agency

BREEDING DISTRIBUTION CHANGE *since 1968–72*

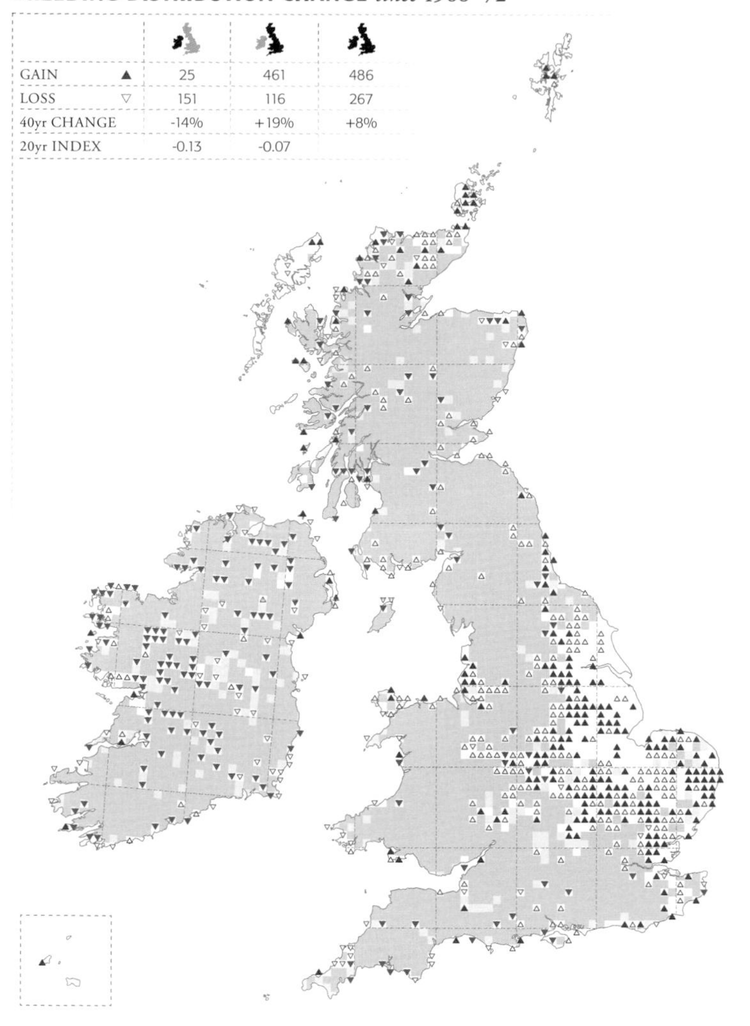

GAIN ▲	25	461	486
LOSS ▽	151	116	267
40yr CHANGE	-14%	+19%	+8%
20yr INDEX	-0.13	-0.07	

WINTER DISTRIBUTION CHANGE *since 1981–84*

GAIN ▲	198	645	843
LOSS ▼	130	78	208
30yr CHANGE	+10%	+38%	+29%

BREEDING RELATIVE ABUNDANCE *2008–11*

BREEDING RELATIVE ABUNDANCE CHANGE *since 1988–91*

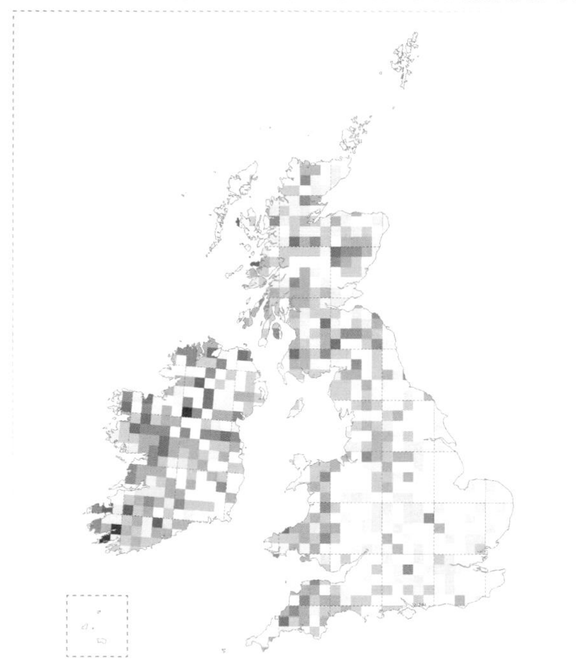

WINTER RELATIVE ABUNDANCE *2007/08–2010/11*

PIED WAGTAIL

Motacilla alba

BTO CODE: PW

CHRISTOPHER CASSIDY

BREEDING PIED WAGTAILS (*M. a. yarrellii*) are almost ubiquitous in Britain & Ireland, being recorded in 95% of 10-km squares, but there is complex variation in abundance. Upland regions have low densities, although associated river valleys and foothills can support higher numbers. Densities are highest in lowland areas of the north and west but relatively low throughout England, even in grass-dominated areas (*contra* Atkinson *et al.* 2002).

Breeding distribution has changed little, with only a 2% range expansion since the *1968–72 Breeding Atlas*, mainly involving gains on Scottish islands. The UK breeding population has undergone marked fluctuations since the 1960s, although abundance has probably increased slightly through the series of breeding atlases (*BirdTrends*). Recently, an 11% decline has been reported in the UK during 1995–2010 (*BBS Report 2011*). This appears to contradict the abundance change map, which suggests increases in tetrad occupancy through a large part of eastern England. However, that area supports relatively few birds and increases there may be outweighed by declines in the larger northern populations (*BBS Report 2011*). Atlas data show widespread abundance increases in Ireland, which correspond with a 14% population increase in the Republic of Ireland during 1998–2010 (*CBS Trend 2010*).

Pied Wagtails vacate upland areas of northern Scotland and northern England in winter. In Britain, winter densities are low to the north of the Humber Estuary, even in occupied lowlands. To the south, densities are highest in valleys, floodplains and along the coast. Densities are high throughout Ireland, where winter distribution has changed little since the *1981–84 Winter Atlas*. In Britain an 11% range expansion has affected mostly upland margins of northern England and Scotland. Often this is a case of individual birds either remaining close to their breeding grounds or returning early in February.

'White Wagtails' (nominate *alba*) were reported in winter from 89 10-km squares, mostly in central and southern England, but rarely breed (*RBBP 2010*). *Confirmed* breeding in three 10-km squares in the Channel Islands and one in Sussex may relate to continental birds, whereas four in the Northern Isles may involve Icelandic birds. Migrant *non-breeding* birds were reported from a further 629 squares.

Sponsored by Janet & Peter Wood

BREEDING DISTRIBUTION *2008–11*

POSSIBLE ●	4%	2%	2%
PROBABLE ●	6%	3%	4%
CONFIRMED ●	87%	90%	89%
TOTAL	97%	95%	95%

WINTER DISTRIBUTION *2007/08–2010/11*

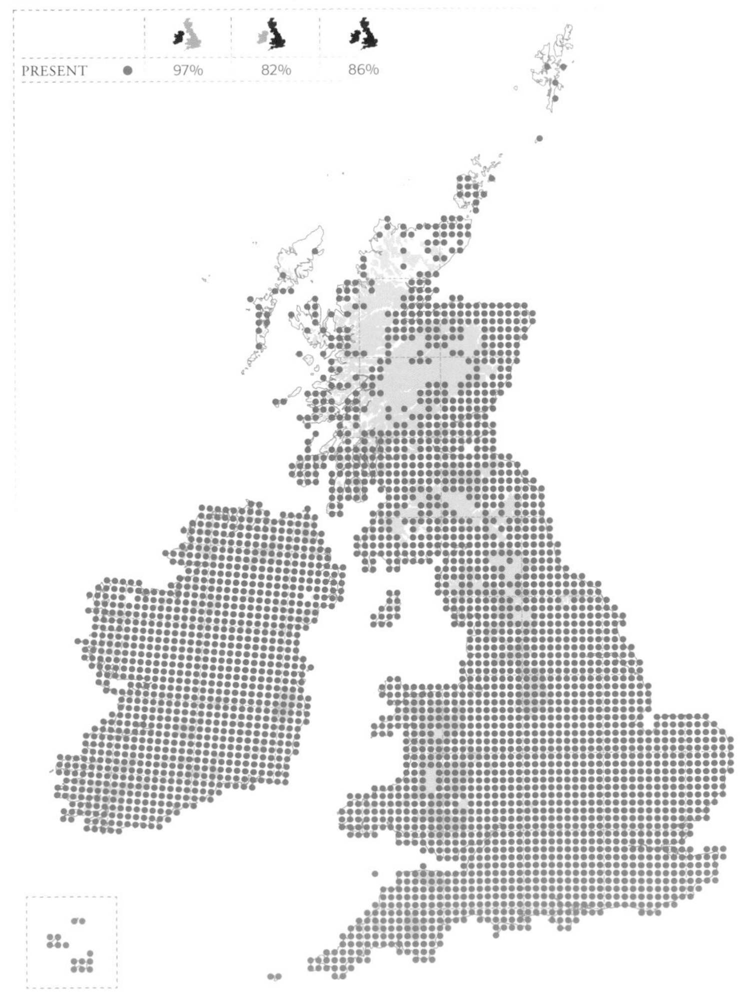

PRESENT ●	97%	82%	86%

BREEDING DISTRIBUTION CHANGE *since 1968–72*

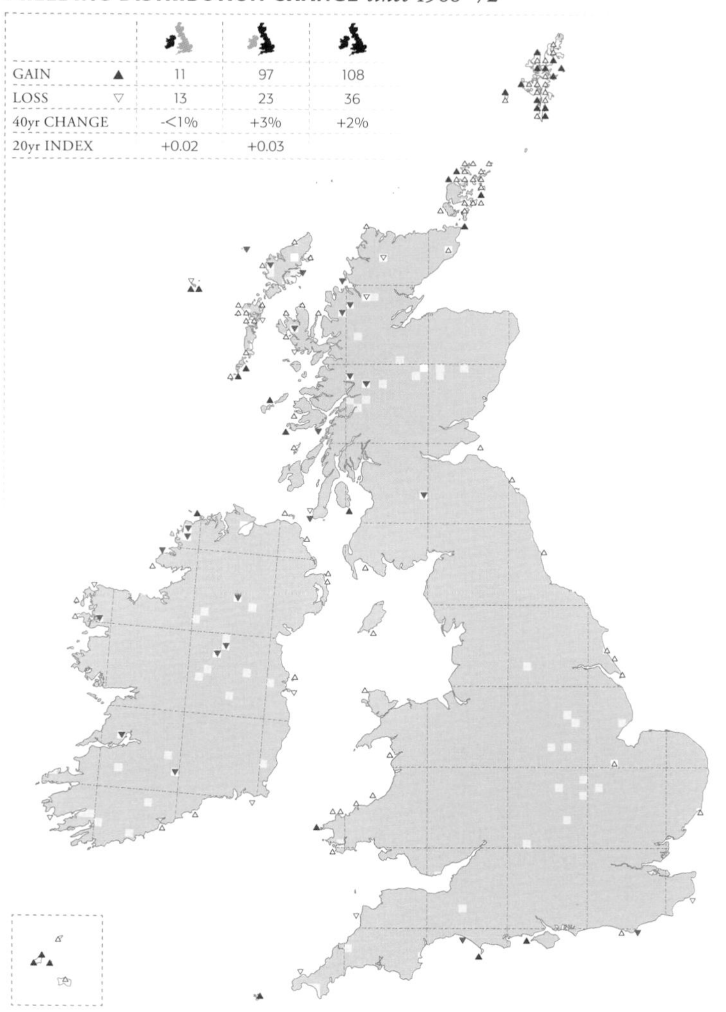

GAIN ▲	11	97	108
LOSS ▽	13	23	36
40yr CHANGE	-<1%	+3%	+2%
20yr INDEX	+0.02	+0.03	

WINTER DISTRIBUTION CHANGE *since 1981–84*

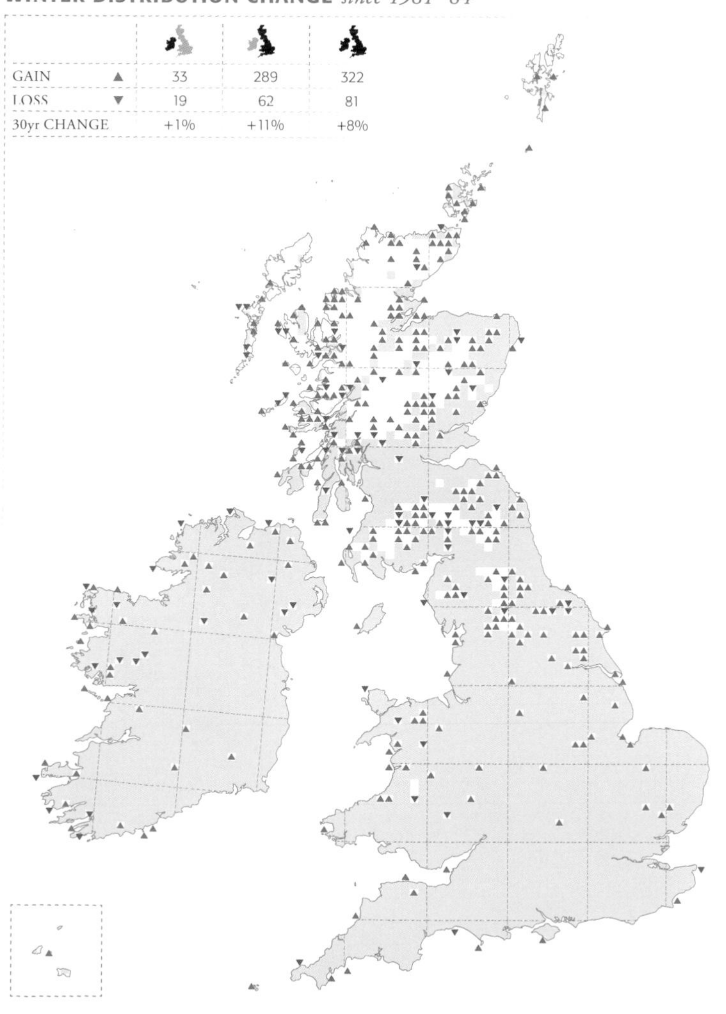

GAIN ▲	33	289	322
LOSS ▼	19	62	81
30yr CHANGE	+1%	+11%	+8%

BREEDING RELATIVE ABUNDANCE *2008–11*

BREEDING RELATIVE ABUNDANCE CHANGE *since 1988–91*

WINTER RELATIVE ABUNDANCE *2007/08–2010/11*

TREE PIPIT

Anthus trivialis

BTO CODE: TP

NEIL CALBRADE www.neilcalbrade.co.uk

TREE PIPITS ARE summer visitors to wooded and scrubby habitats in Britain. Requirements for prominent song-posts coupled with a sparse field and shrub layer are provided by heavily grazed woods in upland areas, early and clear-fell stages of conifer plantations, lowland heaths and scrubby downland. The lack of such habitat explains many range gaps. Tree Pipits are rare breeders in Ireland (*Birds in Ireland*); migrants in four coastal 10-km squares were the only records.

Densities are high in the wooded uplands, more so in birch than oak woods (*Birds of Scotland*). Differences between deciduous woods and coniferous plantations cannot be readily discerned from Atlas data, although abundance in conifer stands in Thetford Forest, Norfolk is comparable to the hotspots of the north and west. In Thetford Forest, densities and productivity are highest in two- to five-year-old stands (Burton 2007, 2009).

Range contractions detected by the *1988–91 Breeding Atlas* have continued and the range is now 29% smaller than that shown by the *1968–72 Breeding Atlas*. Losses that began in central England have deepened and spread to southeast and northern England and the Scottish Central Belt. Outside these areas of outright loss, the abundance change map shows declines in most upland areas since 1988–91. The situation is not wholly negative as there are some increases in eastern Scotland.

Long-term population trends show an overall steep decline in Britain since the late 1980s (*BirdTrends*) and trends during 1995–2010 reveal 46% and 26% declines in England and Wales respectively (*BBS Report 2011*). A reported 72% increase in Scotland should be treated with caution as monitored BBS squares are a poor match to the Tree Pipit's range.

Locally, declines may be related to the maturing of plantations. Losses within woodland have been greatest in areas where vegetation cover has increased and habitat has become less open (Amar *et al.* 2006). Nest failure rates have increased, at least in Thetford Forest, and predation is the main cause (Burton 2009). The Tree Pipit is one of several long-distance migrants undergoing steep declines and therefore factors related to migration or wintering ecology could also be important (Hewson *et al.* 2007; Ockendon *et al.* 2012).

BREEDING DISTRIBUTION *2008–11*

	Ireland	Britain	Britain & Ireland
POSSIBLE •	-	10%	8%
PROBABLE ●	-	14%	10%
CONFIRMED ●	-	20%	15%
TOTAL	-	44%	33%

BREEDING RELATIVE ABUNDANCE CHANGE *since 1988–91*

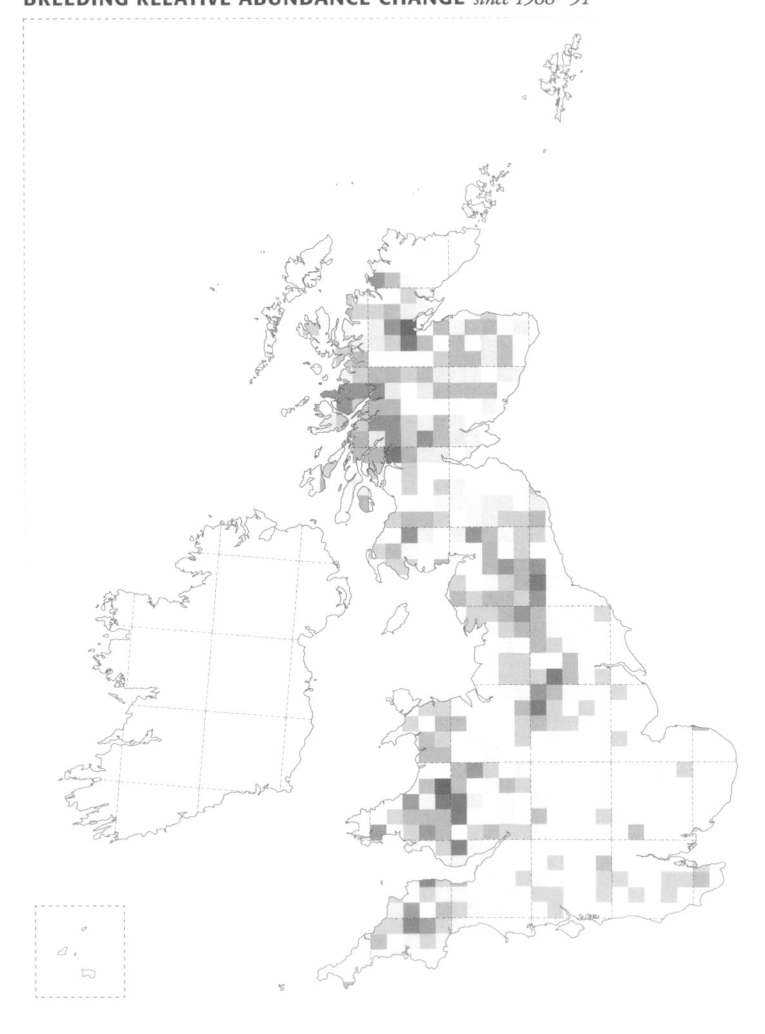

BREEDING DISTRIBUTION CHANGE *since 1968–72*

GAIN ▲	-	158	158
LOSS ▽	-	682	682
40yr CHANGE	-	-29%	-29%
20yr INDEX	-	-0.25	

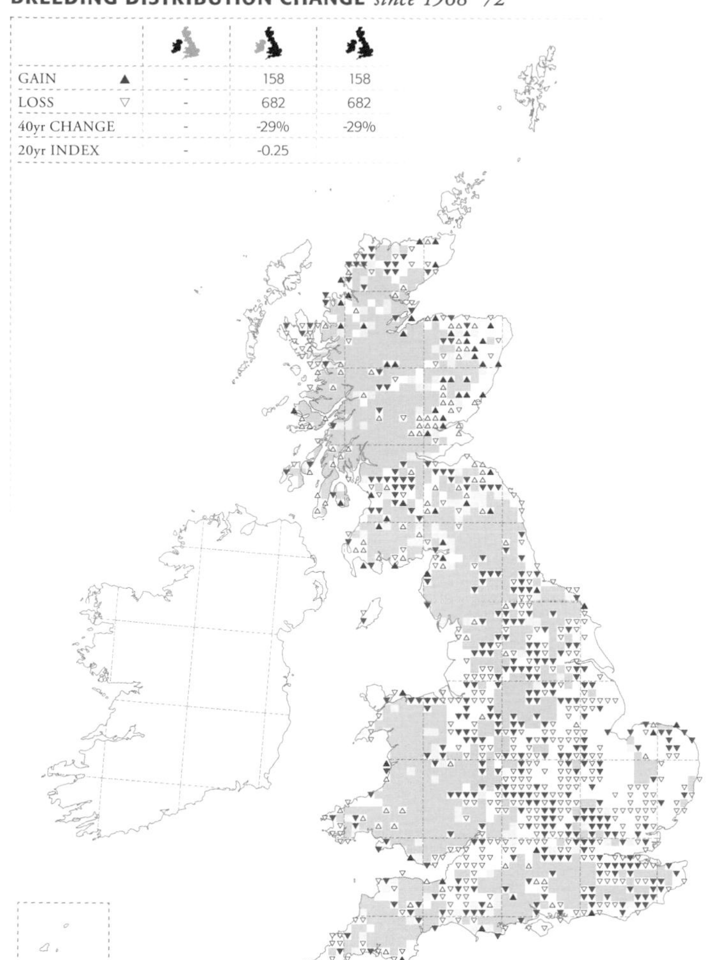

BREEDING RELATIVE ABUNDANCE *2008–11*

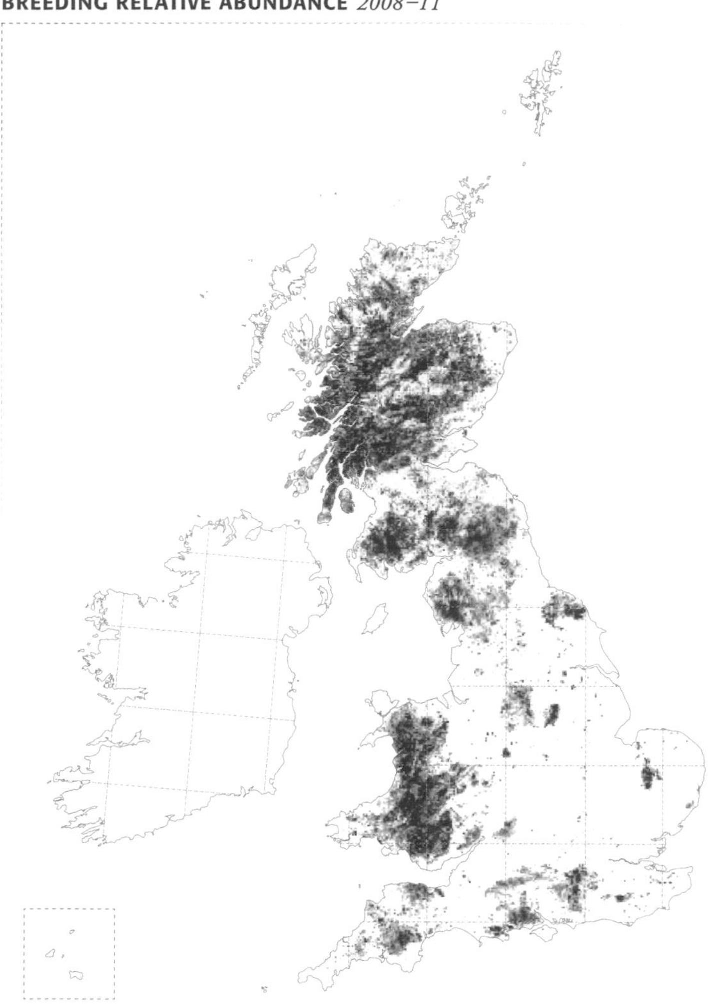

WATER PIPIT
Anthus spinoletta

BTO CODE: WI

GARTH PEACOCK www.garthpeacock.co.uk

WATER PIPITS ARE scarce winter visitors from alpine areas of southern Europe. There is a distinct southern bias to the winter distribution in Britain and Ireland. Most occupied squares are coastal but the many inland records are associated with wetlands and marshes in the larger river valleys. The highest counts were received from the English east coast between the Humber and Suffolk, and from the Fens. Range size appears to have increased markedly, from 94 10-km squares in the *1981–84 Winter Atlas* to 340 squares in 2007–11, but changes cannot be assessed accurately because the Water Pipit did not have full species status in 1981–84 and an unknown proportion of records may have been submitted as Rock Pipits.

Sponsored by The Watercress Company

WINTER DISTRIBUTION *2007/08–2010/11*

PRESENT ●	2%	11%	9%

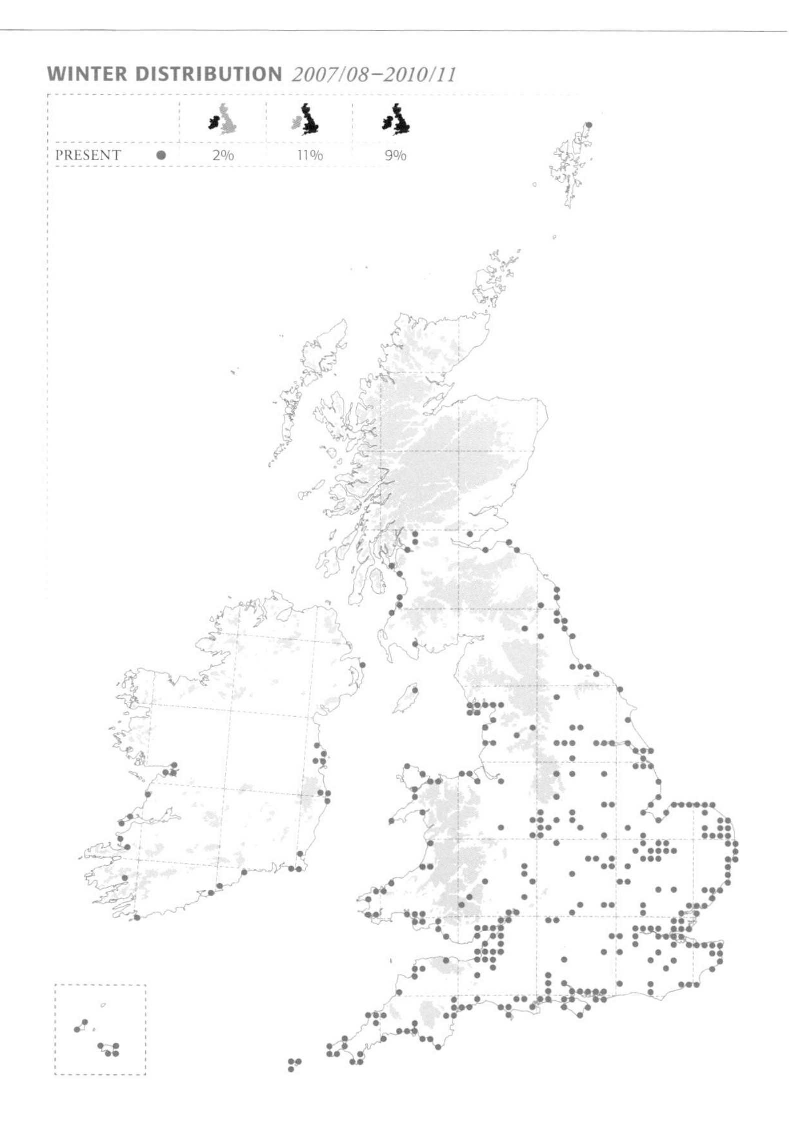

MEADOW PIPIT

Anthus pratensis

BTO CODE: MP

CLARE WARD

WINTERING MEADOW PIPITS are widely distributed across Britain and Ireland, with absences mostly confined to the highest ground. Significant densities were recorded throughout Ireland but in Britain they were largely confined to the coastal fringe and the Fens. Autumn migration of local breeders and Icelandic and Scandinavian migrants follows a southwesterly direction (*Migration Atlas*). This may account for the high densities recorded in Ireland, and on southern and western coasts in Britain, where TTVs carried out in early November may have recorded large numbers of grounded migrants.

Distributional changes in winter have been relatively small, amounting to a 7% expansion overall since the *1981–84 Winter Atlas*. Most changes involve scattered infilling and in Ireland gains exactly balance losses. In Scotland and northern England a more consistent pattern of gains in uplands and islands is apparent, which may stem from a combination of improved coverage and birds lingering in mild winters.

Breeding Meadow Pipits are widespread in the north and west but patchily distributed in the southern English lowlands. The upland–lowland divide is even more striking on the abundance map. Meadow Pipits are the commonest passerine above 500 m altitude in Scotland (*Birds of Scotland*) and a mosaic of heather, bog and grassland provides optimal habitat (Vanhinsbergh & Chamberlain 2001). There has been little overall change in the number of occupied 10-km squares since the *1968–72 Breeding Atlas* and the mix of gains and losses is largely confined to areas of low density in the lowlands.

There has been a steady population decline in the UK since the 1970s, amounting to a 46% decrease during 1970–2010 (*BirdTrends*; *SUKB 2012*) and a 39% decrease in the Republic of Ireland during 1998–2010 (*CBS Trend 2010*). These declines are consistent with a 66% decrease across Europe since 1980 (*PECBMS 2010*). The Atlas abundance change map shows a mix of increases and decreases in the proportions of occupied tetrads but this map may be a relatively insensitive measure of abundance changes for this species because densities in optimal habitat could fall substantially before changes in tetrad occupancy become apparent. Nevertheless, the pattern of increases and decreases in Ireland would benefit closer scrutiny.

Sponsored by Trevor Poyser

BREEDING DISTRIBUTION *2008–11*

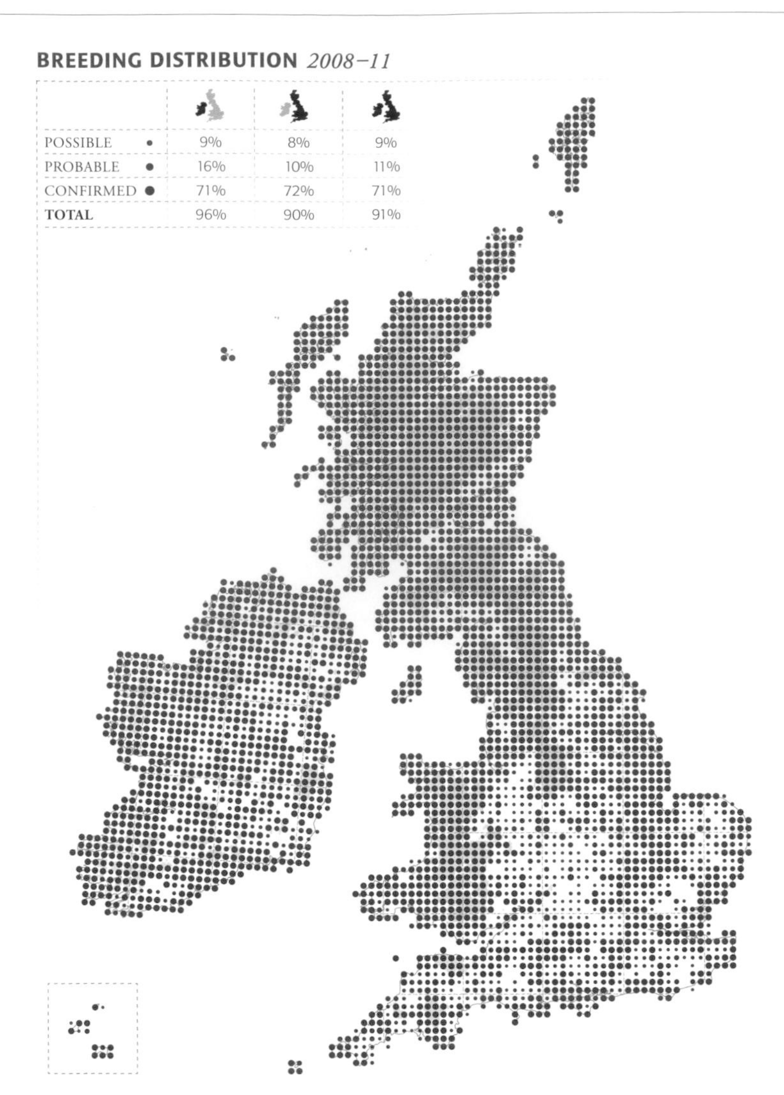

POSSIBLE	•	9%	8%	9%
PROBABLE	•	16%	10%	11%
CONFIRMED	●	71%	72%	71%
TOTAL		96%	90%	91%

WINTER DISTRIBUTION *2007/08–2010/11*

PRESENT	●	94%	92%	93%

BREEDING DISTRIBUTION CHANGE *since 1968–72*

GAIN ▲	6	124	130
LOSS ▽	35	180	215
40yr CHANGE	-3%	-2%	-2%
20yr INDEX	+0.02	-0.04	

WINTER DISTRIBUTION CHANGE *since 1981–84*

GAIN ▲	66	363	429
LOSS ▼	66	124	190
30yr CHANGE	0%	+10%	+7%

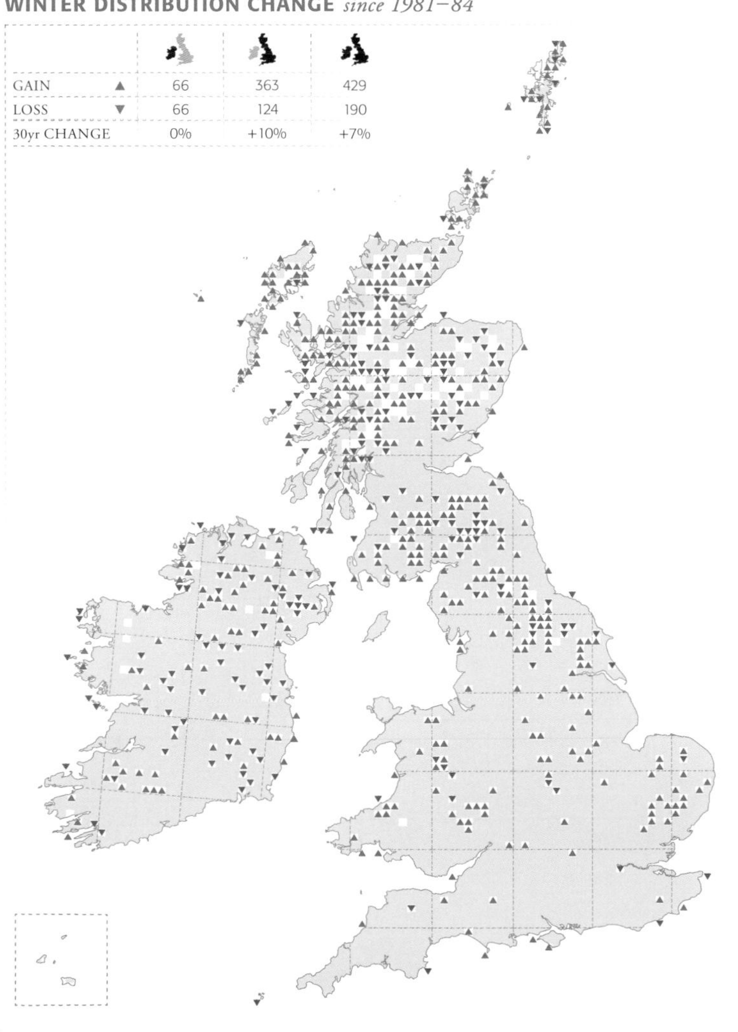

BREEDING RELATIVE ABUNDANCE *2008–11*

BREEDING RELATIVE ABUNDANCE CHANGE *since 1988–91*

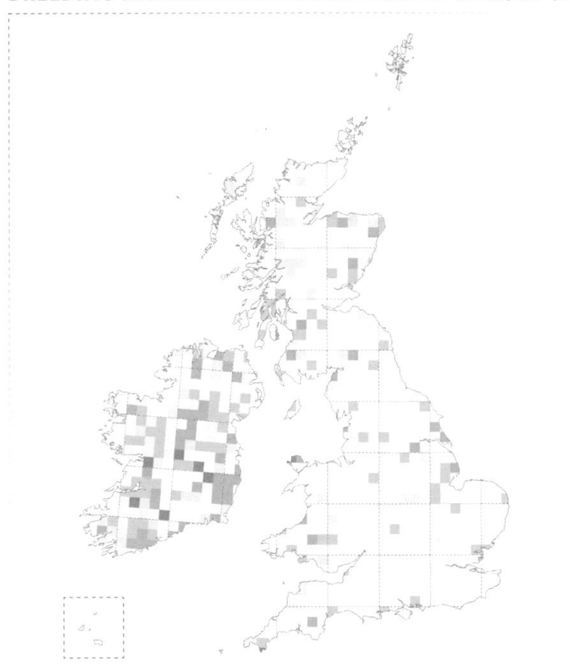

WINTER RELATIVE ABUNDANCE *2007/08–2010/11*

ROCK PIPIT

Anthus petrosus

BTO CODE: RC

GARY FAULKNER

BREEDING DISTRIBUTION *2008–11*

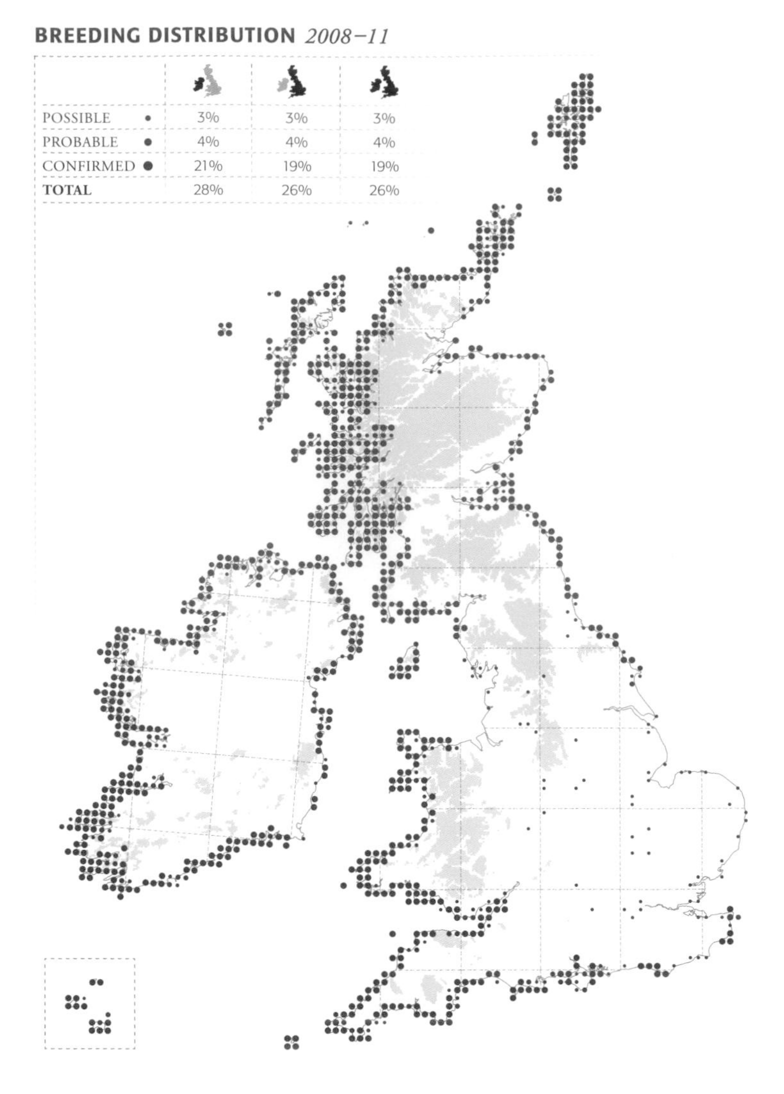

		Britain & Ireland	Britain	Ireland
POSSIBLE	•	3%	3%	3%
PROBABLE	•	4%	4%	4%
CONFIRMED	•	21%	19%	19%
TOTAL		28%	26%	26%

BREEDING ROCK PIPITS are very much restricted to rocky coasts and are absent from long stretches of low-lying coasts characterised by saltmarshes, mudflats or sandy beaches. The highest breeding densities are in western Ireland, southwest Wales, the Hebrides and the Northern Isles.

Since the *1968–72 Breeding Atlas* there has been a small (2%) decrease in the number of occupied 10-km squares, with slightly greater losses in Britain than in Ireland, but little discernible pattern. Some recent gains are a reversal of old losses and may be indicative of under-recording in the *1988–91 Breeding Atlas*, for example on the Cornish coast. Breeding population trend data are lacking but declines have been noted on several Scottish islands; the causes are unknown (*Birds of Scotland*).

British and Irish breeding Rock Pipits are essentially resident (*Migration Atlas*) so the statistic that Rock Pipits occupy 32% of 10-km squares in winter compared to 26% in the breeding season suggests occupancy by continental migrants in winter. The winter distribution remains almost exclusively coastal but unlike the breeding-season distribution includes the saltmarshes of eastern England. The winter abundance map is similar to the breeding-season map with the addition of high densities around estuaries in eastern England. Significant numbers of the *littoralis* race arrive from northern Fennoscandia each autumn to winter in coastal marshes in eastern and southern England and in Wales (*Migration Atlas*). They probably also occur farther north on the North Sea coast but remain undetected when mixed with the resident population of nominate *petrosus* (*Birds of Scotland*).

The winter change map shows gains in core areas in northwest Scotland and on the Irish coast which are likely to be due to improved levels of coverage. Gains elsewhere, such as on the Irish Sea coasts and the coast of south and southeast England, may reflect greater numbers of continental immigrants. Some gains and losses at inland sites will reflect the erratic occurrence of birds at inland wetlands. However, at the time of the *1981–84 Winter Atlas* the Water Pipit had not been split from the Rock Pipit so an unknown proportion of inland 'losses' may actually relate to records of Water Pipits in 1981–84.

WINTER DISTRIBUTION *2007/08–2010/11*

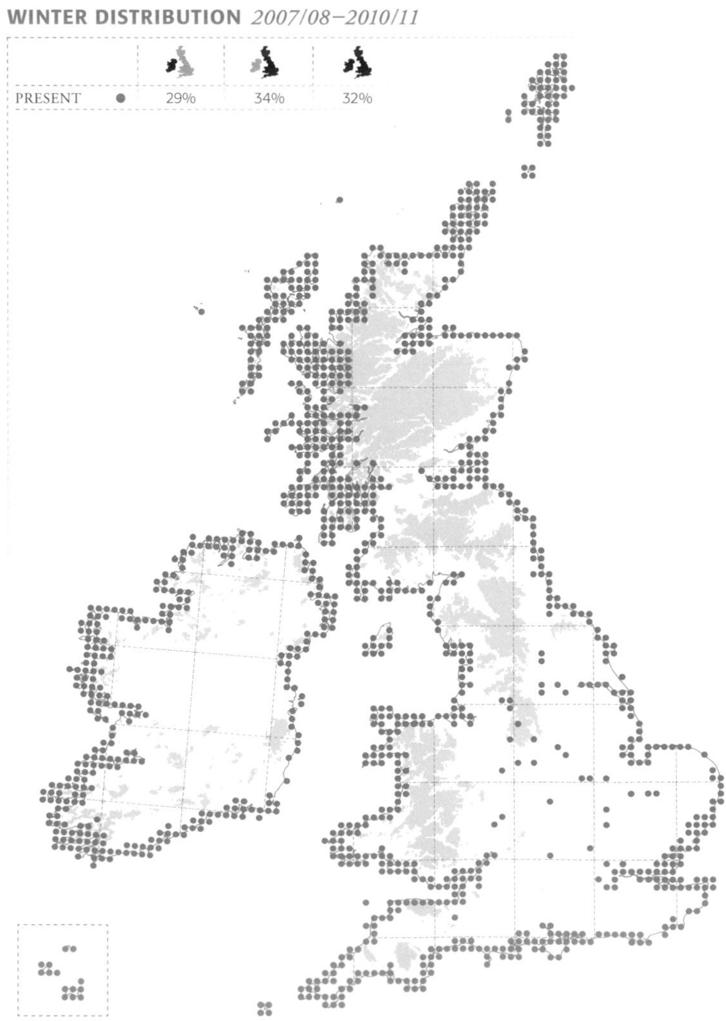

		Britain & Ireland	Britain	Ireland
PRESENT	•	29%	34%	32%

Sponsored by Alastair Whitelaw

BREEDING DISTRIBUTION CHANGE *since 1968–72*

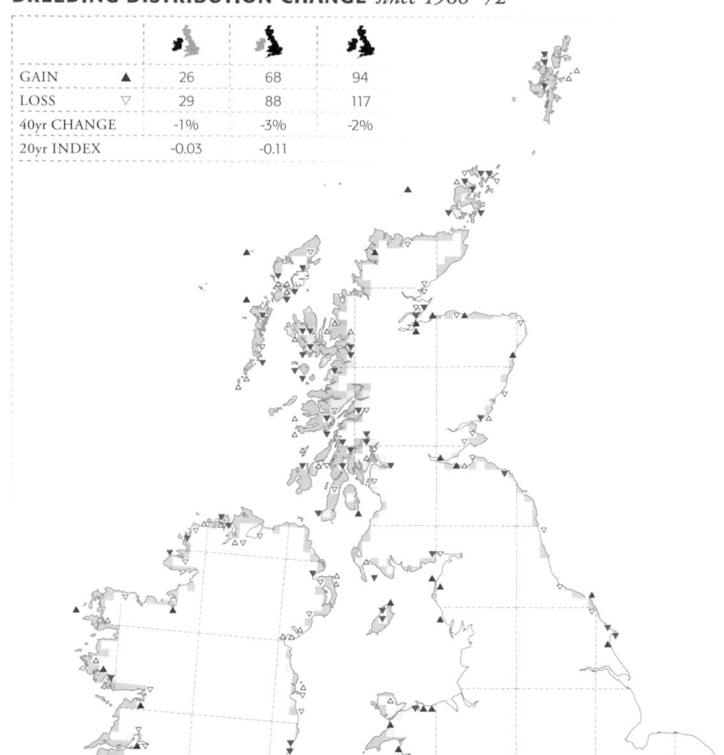

GAIN ▲	26	68	94
LOSS ▽	29	88	117
40yr CHANGE	-1%	-3%	-2%
20yr INDEX	-0.03	-0.11	

BREEDING RELATIVE ABUNDANCE *2008–11*

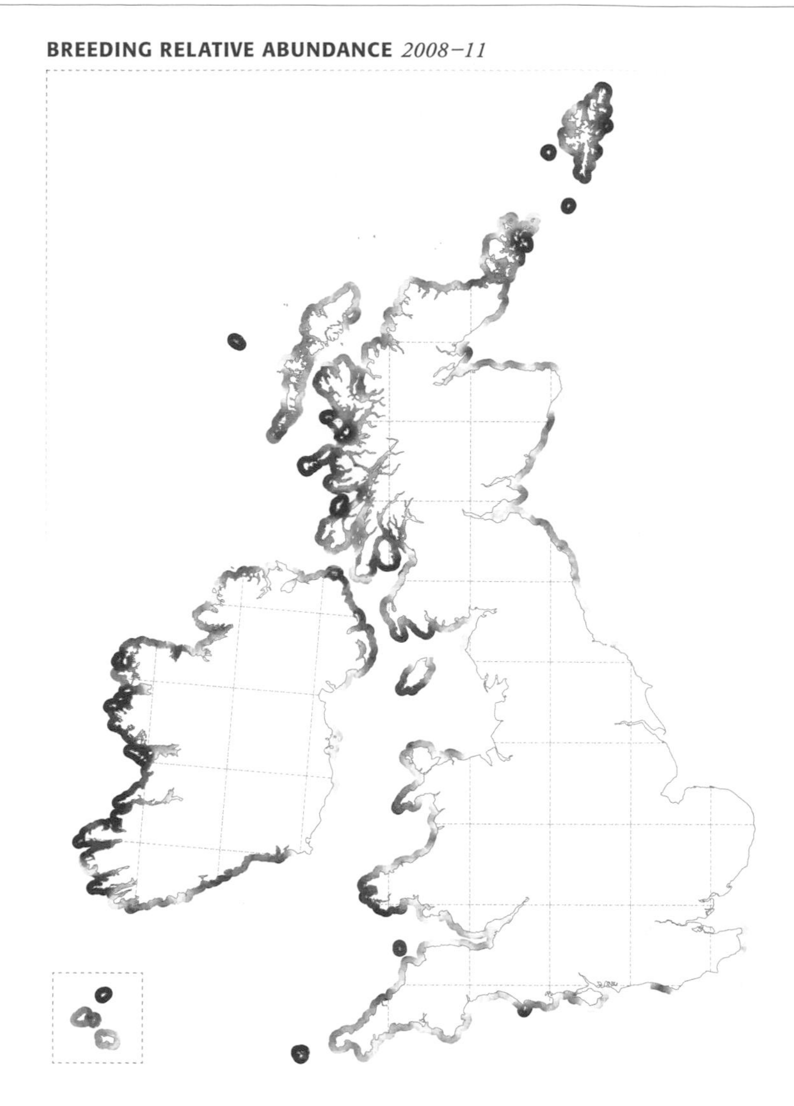

WINTER DISTRIBUTION CHANGE *since 1981–84*

GAIN ▲	43	155	198
LOSS ▼	38	133	171
30yr CHANGE	+2%	+3%	+2%

WINTER RELATIVE ABUNDANCE *2007/08–2010/11*

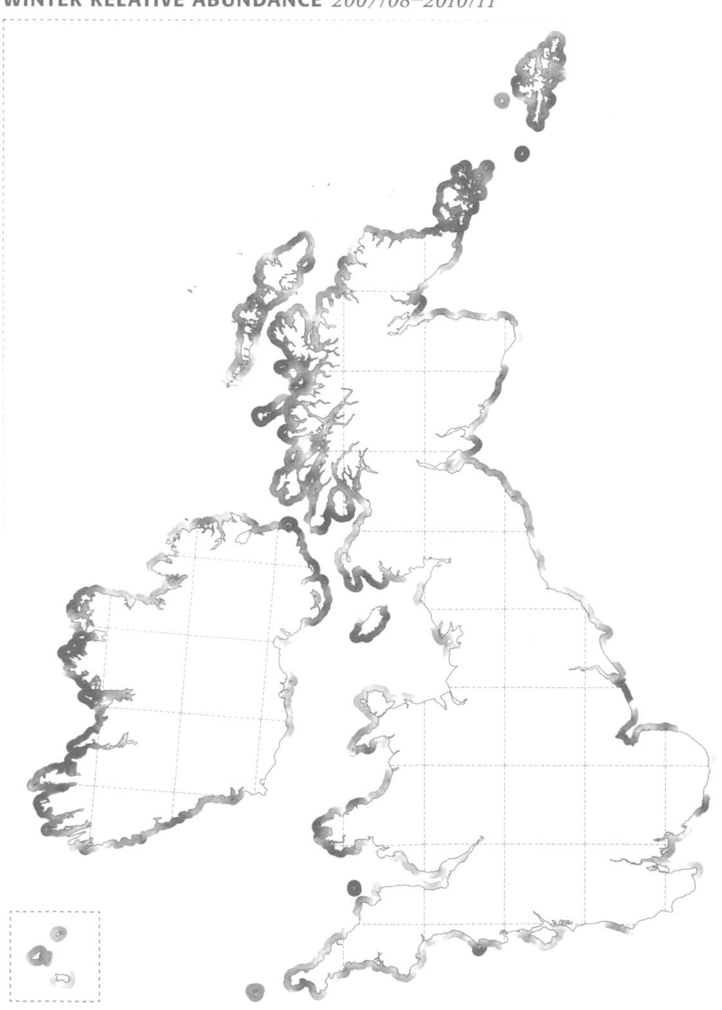

CHAFFINCH

Fringilla coelebs

BTO CODE: CH

ALISON MCARTHUR

THE CHAFFINCH IS one of the commonest birds in Britain & Ireland and was recorded breeding in 94% of all 10-km squares. It is absent only from parts of the Northern Isles, the Outer Hebrides and a few west-coast squares in Ireland. The highest breeding densities in Ireland are found in Ulster, whereas in Britain they are in the lowlands of southern, central and eastern England and on the upland edges in northern England and Scotland. Though occupied, more exposed uplands and islands support low densities, as do urban centres and open agricultural regions such as the Fens and much of the Irish midlands and southwest.

The number of occupied 10-km squares has increased by 2% since the *1968–72 Breeding Atlas*, mainly through gains in coastal 10-km squares and on islands. Gains on the Outer Hebrides and Orkney probably reflect the colonisation of new conifer plantations and shelter belts.

Annual monitoring data show a 17% population increase in the Republic of Ireland during 1998–2010 (*CBS Trend 2010*). A recent 12% increase in the UK during 1995–2010 is part of a longer-term 37% increase during 1970–2010 (*SUKB 2012*). The abundance change map reflects these trends, with gains in western Ireland, western Scotland and the Fens. Massimino *et al.* (2013) also found increases in western Scotland and parts of central and eastern England. These gains would have been greater were it not for a downturn in numbers since 2006 associated with the outbreak of trichomonosis (Robinson *et al.* 2010).

British and Irish Chaffinches are highly sedentary but from autumn onwards they are joined by large numbers of migrants from Scandinavia (*Migration Atlas*). The winter distribution is very similar to the breeding distribution except for the greater occupancy of the Northern Isles by migrants (*Birds of Scotland*). Given the extent of the winter influx the patterns of relative abundance are surprisingly similar in winter and the breeding season although, perhaps indicative of arrivals from the Continent (*Migration Atlas*), southern Ireland gains relative importance in winter. Distribution gains since the *1981–84 Winter Atlas* are largely confined to the Outer Hebrides and are probably linked to colonisation of the area by resident breeders.

Sponsored by Bill Haynes

BREEDING DISTRIBUTION *2008–11*

POSSIBLE •	4%	2%	2%
PROBABLE •	10%	4%	6%
CONFIRMED ●	81%	87%	86%
TOTAL	95%	93%	94%

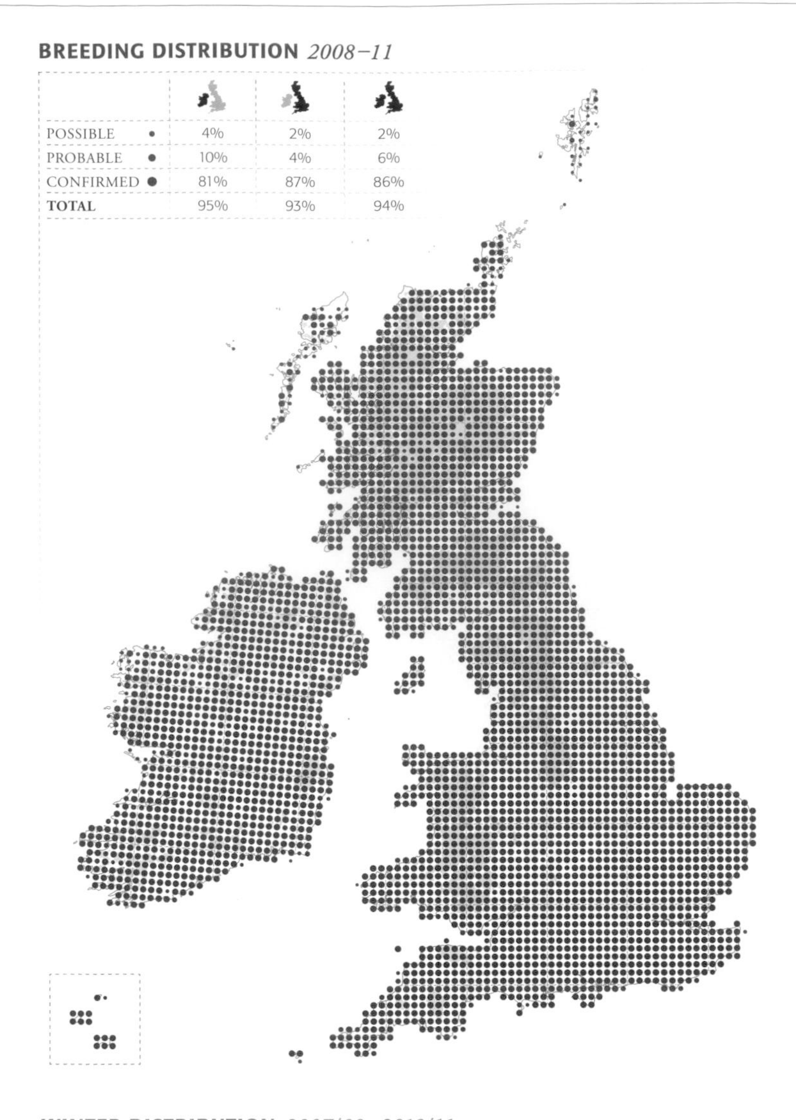

WINTER DISTRIBUTION *2007/08–2010/11*

PRESENT ●	96%	94%	94%

BREEDING DISTRIBUTION CHANGE *since 1968–72*

GAIN ▲	10	82	92
LOSS ▽	11	7	18
40yr CHANGE	-<1%	+3%	+2%
20yr INDEX	+0.01	+0.02	

WINTER DISTRIBUTION CHANGE *since 1981–84*

GAIN ▲	20	125	145
LOSS ▼	21	17	38
30yr CHANGE	-<1%	+4%	+3%

BREEDING RELATIVE ABUNDANCE *2008–11*

BREEDING RELATIVE ABUNDANCE CHANGE *since 1988–91*

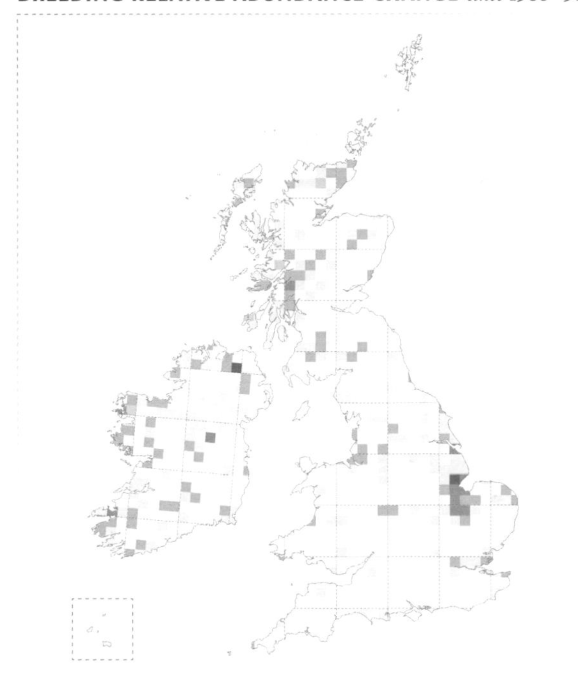

WINTER RELATIVE ABUNDANCE *2007/08–2010/11*

BRAMBLING

Fringilla montifringilla

BTO CODE: BL

DESMOND DUGAN www.AbernethyImages.co.uk

BRAMBLINGS ARRIVE IN large numbers during the autumn from their Fennoscandian breeding grounds. In winter they are widely distributed through most of England, Wales and southern, central and eastern Scotland. In Ireland they are much less widespread, being found in only 32% of 10-km squares compared to 71% in Britain, with greater occupancy in the east and south than in the west. Abundance appears to be relatively high in northeast Scotland, the Southern Uplands, East Anglia and a band running from the Welsh Marches to the Isle of Wight. This pattern largely corresponds with that found in the *1981–84 Winter Atlas*, although abundance now appears to be relatively lower in northeast England and higher in southern England. Patterns of abundance are likely to vary annually according to the size of the autumn influx and the rate at which birds move through the country in search of food.

Since the *1981–84 Winter Atlas* there have been apparent range expansions of 21% in Britain and 53% in Ireland. However, there are few discernible patterns in the gains and losses; this may be a consequence of birds responding to spatially variable food supplies. Some changes may stem from an overall increase in recording effort and the tendency for Bramblings to visit garden feeding stations during years when the beechmast crop is poor (Chamberlain *et al.* 2007).

During April, many Bramblings have yet to depart Britain and such *non-breeding* birds were reported from over 700 10-km squares, mostly in England and eastern Scotland. In most years a few males singing in suitable breeding habitat are reported to RBBP but the Brambling remains a very occasional and possibly declining breeder (*RBBP 2010*; Holling *et al.* 2011b). Most breeding records have been in Scotland but even there no more than one or two pairs are confirmed nesting in a year (*Birds of Scotland*). There were no records of *confirmed* breeding during 2008–11, though there were two 10-km squares with *probable* breeding and five with *possible* breeding, involving birds lingering in suitable habitat. The 40% increase in the number of occupied 10-km squares since 1968–72 is of little significance given the very small number of squares involved.

In memory of my parents and brother by Brian and Pat Martin

WINTER DISTRIBUTION *2007/08–2010/11*

	Ireland	Britain	Britain & Ireland
PRESENT ●	32%	71%	61%

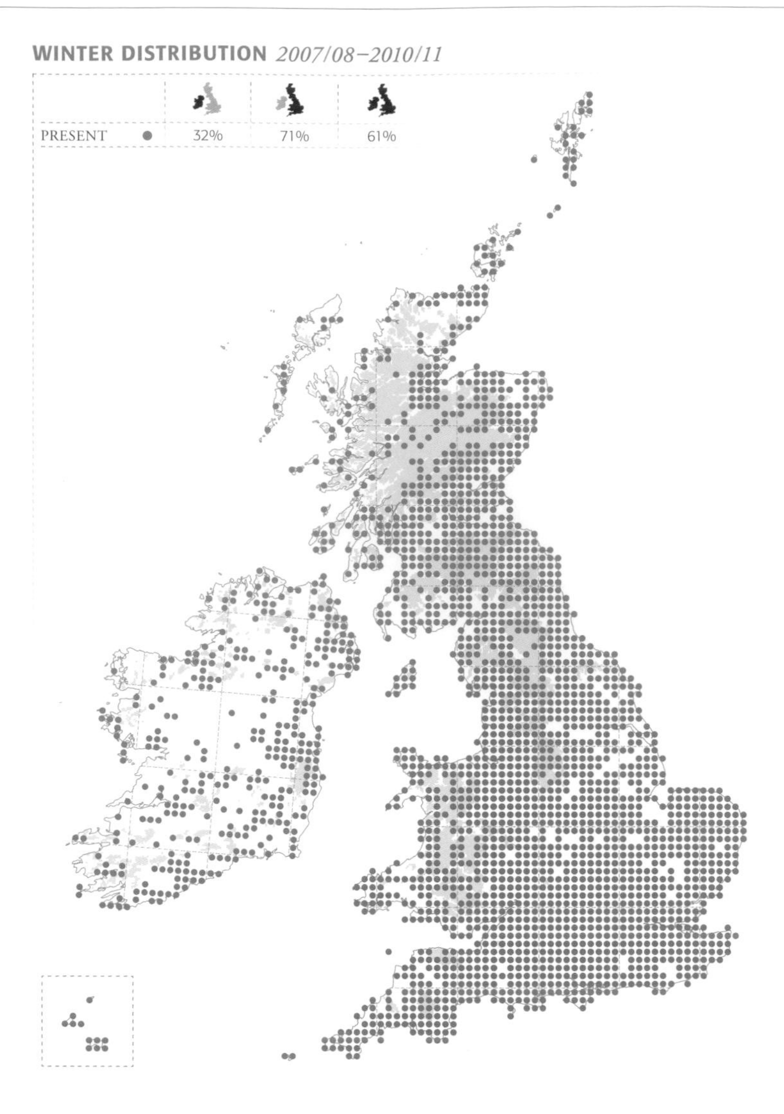

BREEDING DISTRIBUTION *2008–11*

	Ireland	Britain	Britain & Ireland
POSSIBLE •	-	<1%	<1%
PROBABLE ●	-	<1%	<1%
CONFIRMED ●	-	-	-
TOTAL	-	<1%	<1%

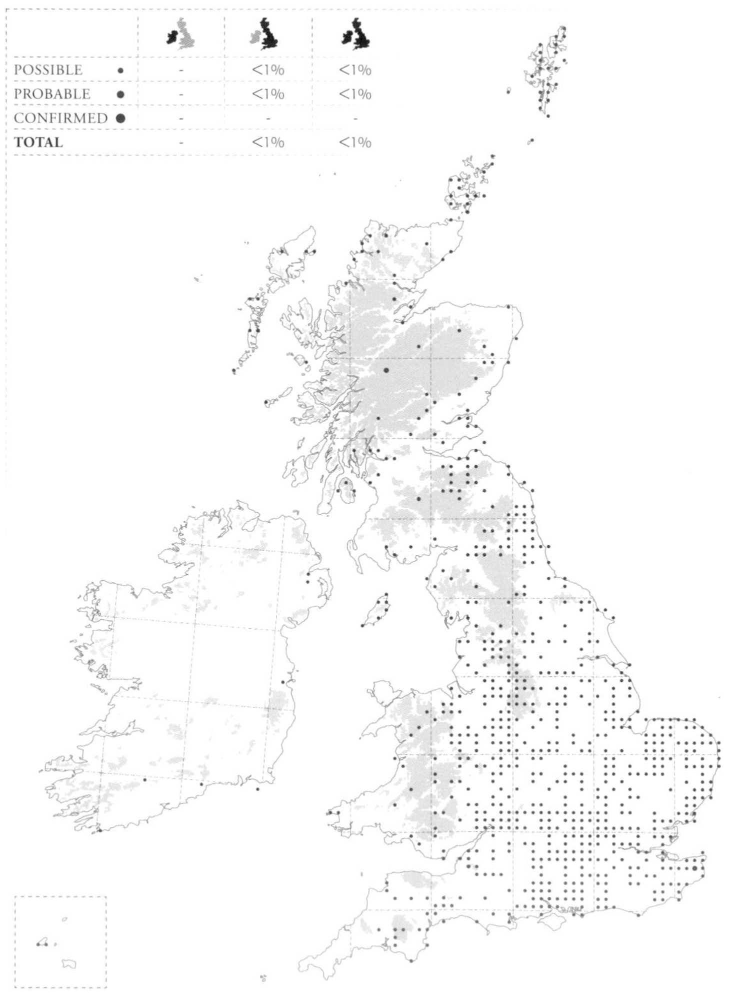

WINTER DISTRIBUTION CHANGE *since 1981–84*

GAIN ▲	206	556	762
LOSS ▼	101	217	318
30yr CHANGE	+53%	+21%	+25%

WINTER RELATIVE ABUNDANCE *2007/08–2010/11*

BREEDING DISTRIBUTION CHANGE *since 1968–72*

GAIN ▲	-	7	7
LOSS ▽	-	5	5
40yr CHANGE	-	+40%	+40%
20yr INDEX	-	-	

GREENFINCH
Chloris chloris

BTO CODE: GR

JOE O'HANLON

BREEDING DISTRIBUTION *2008–11*

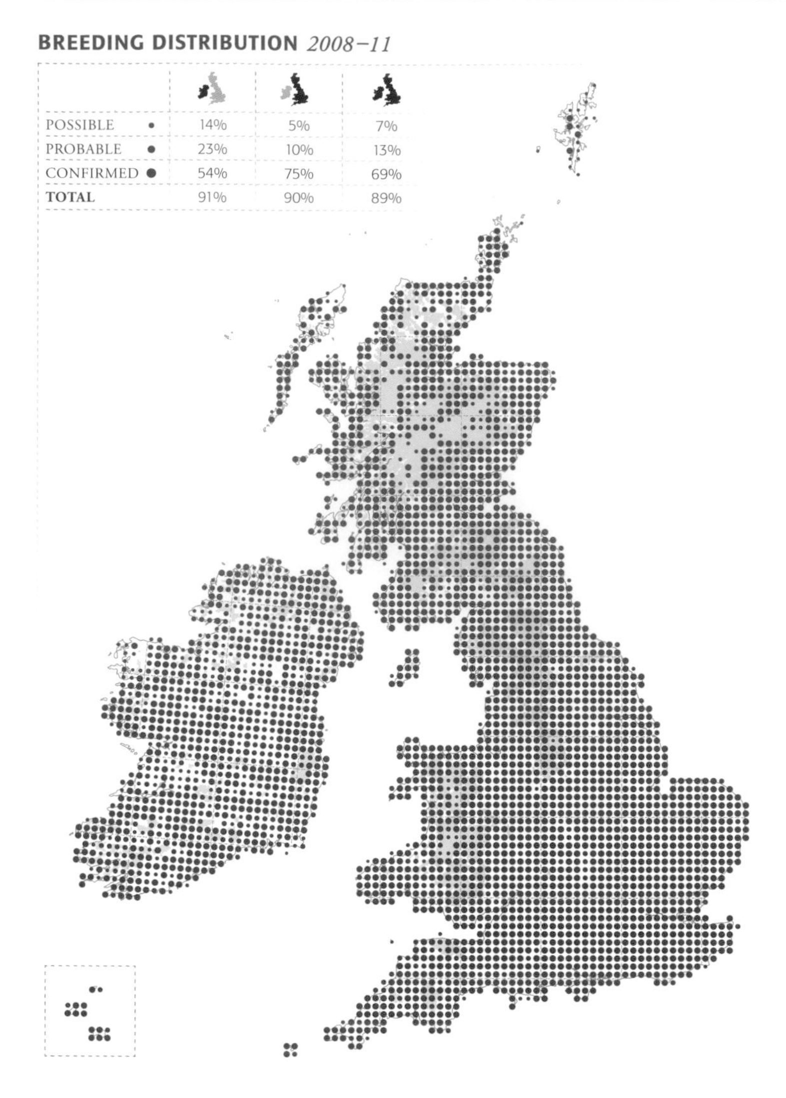

POSSIBLE 14% 5% 7%
PROBABLE 23% 10% 13%
CONFIRMED 54% 75% 69%
TOTAL 91% 90% 89%

WINTER DISTRIBUTION *2007/08–2010/11*

THE GREENFINCH IS widely distributed in winter, perhaps showing especially strong associations with villages and towns across Britain & Ireland. Although this is a mainly resident species, the population is supplemented by small numbers of winter visitors from continental Europe (Main 1999). Like the Goldfinch, its only significant range gaps are in the higher parts of northern Scotland. Since the *1981–84 Winter Atlas* the range has increased by 18%, involving infilling in western Ireland and parts of Wales and northern England, and a more continuous expansion around the coastal–upland fringe in northwest Scotland. The relative abundance map shows that the highest densities are found on lower ground, mainly in southern and southeast England, eastern Scotland and in pockets throughout Ireland.

The breeding-season distribution and patterns of relative abundance are very similar to those in the winter. In Scotland, occupancy and densities decline above 300 m, where agriculture changes from arable to pasture (*Birds of Scotland*). Range expansions similar to those in winter are also apparent, with 5% more 10-km squares occupied since the *1968–72 Breeding Atlas*. Gains in mainland northwest Scotland, parts of the Outer Hebrides and Orkney had begun by the *1988–91 Breeding Atlas* but Shetland has been colonised more recently.

Greenfinch numbers and survival rates have not been affected by agricultural intensification to the same degree as those of other granivorous passerines (Siriwardena *et al.* 2000a) but, since 2005, numbers have been severely affected by the disease trichomonosis (Robinson *et al.* 2010; Lawson *et al.* 2012). In winter, Greenfinches were reported in 74% of British gardens in 2005 but in only 53% in 2011 (Garden BirdWatch 2013); a smaller drop from 90% to 82% occurred between 2003/04 and 2010/11 in Ireland (Crowe 2005; O'Sullivan 2011). In 2006, the UK breeding population was 49% up on 1994 levels (Raven *et al.* 2007) but during 2004–09 it fell by 24% and by 2010 was 10% lower than in 1970 (*BirdTrends*; *SUKB 2012*). In Ireland, breeding numbers were 34% lower in 2010 than in 1998 (*CBS Trend 2010*). Despite these recent declines, the abundance change map shows that Greenfinches are present in more tetrads than during the *1988–91 Breeding Atlas*.

Sponsored by customers of CJ Wildlife

BREEDING DISTRIBUTION CHANGE *since 1968–72*

GAIN ▲	37	194	231
LOSS ▽	30	36	66
40yr CHANGE	+1%	+7%	+5%
20yr INDEX	+0.15	+0.11	

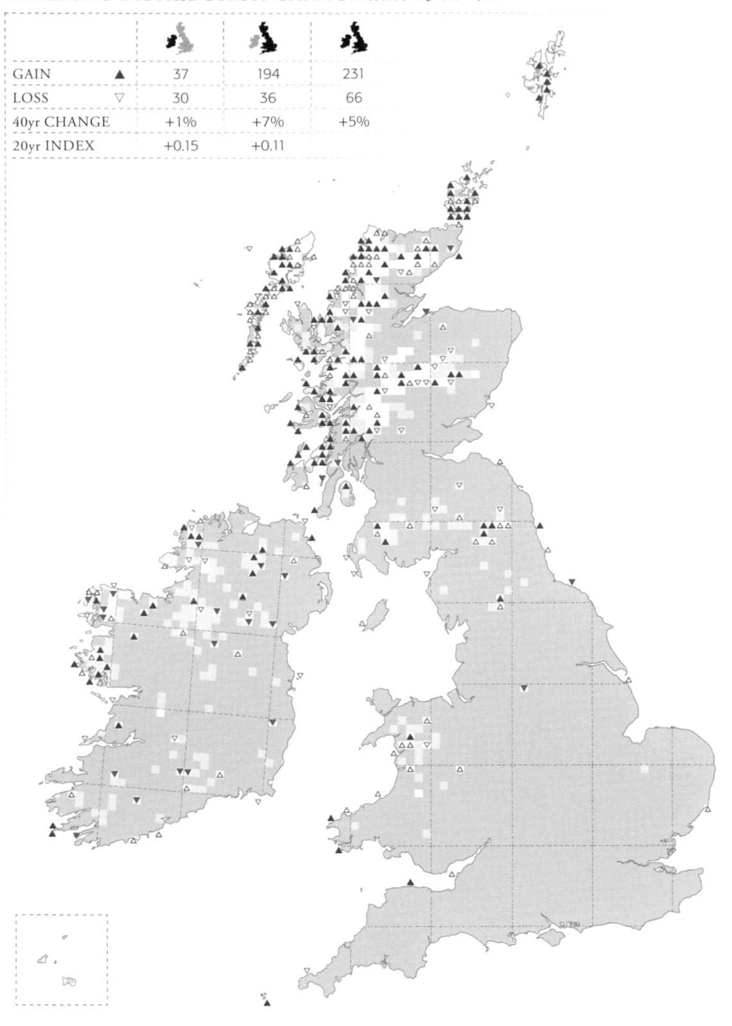

WINTER DISTRIBUTION CHANGE *since 1981–84*

GAIN ▲	211	379	590
LOSS ▼	39	45	84
30yr CHANGE	+24%	+16%	+18%

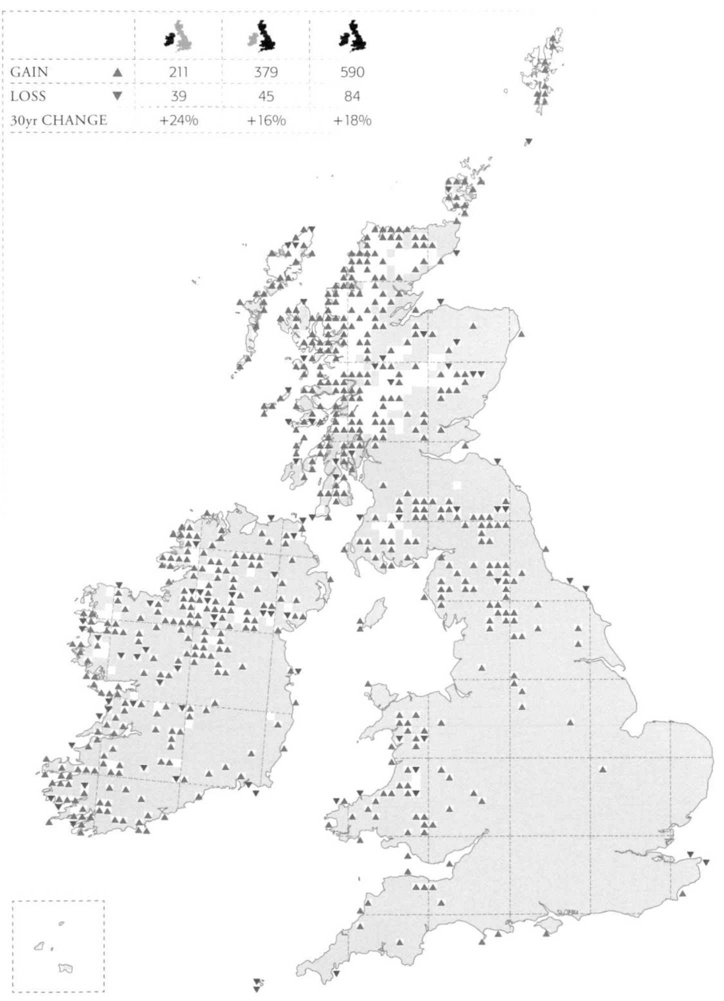

BREEDING RELATIVE ABUNDANCE *2008–11*

BREEDING RELATIVE ABUNDANCE CHANGE *since 1988–91*

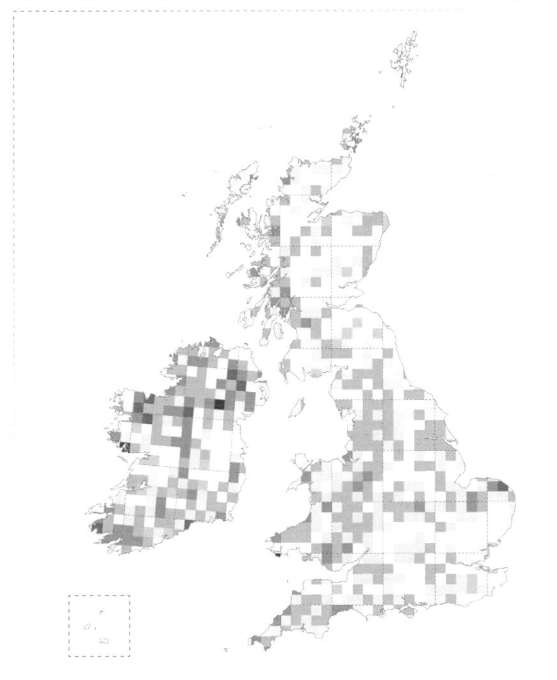

WINTER RELATIVE ABUNDANCE *2007/08–2010/11*

GOLDFINCH
Carduelis carduelis BTO CODE: GO

DESMOND DUGAN www.Abernethyimages.co.uk

THE GOLDFINCH HAS a wide distribution throughout Britain & Ireland in winter but is largely absent from the interior of northwest Scotland. The UK breeding population declined sharply from the mid 1970s to mid 1980s but has since increased significantly (*BirdTrends*). The effect of the recent increase, which is almost entirely explained by improved annual survival (Siriwardena *et al.* 1999), can be seen on the winter distribution change map, which shows a 29% range expansion since the *1981–84 Winter Atlas*, most notably in Ireland and Scotland. The winter relative abundance map emphasises how scarce Goldfinches are outside the lowlands in winter.

Winter and breeding-season patterns of distribution and abundance are similar although, whereas densities are low in all upland areas in winter, progressively more birds breed in upland margins towards the south of Britain. The 91% increase in the UK breeding population during 1995–2010 (*BBS Report 2011*) and the 158% increase during 1998–2010 in the Republic of Ireland (*CBS Trend 2010*) are reflected in 10% and 27% range expansions in Britain and Ireland respectively since the *1988–91 Breeding Atlas*. Gains in Britain are concentrated around the upland fringe in Scotland and continue a pattern that was already under way between the *1968–72 Breeding Atlas* and *1988–91 Breeding Atlas*. The gains in Ireland are a reversal of losses there between the first two breeding atlases; these fluctuations appear to be real and not an artefact of coverage. More significantly, there has been a substantial increase in density throughout Britain & Ireland, as shown by the increased proportions of tetrads occupied on the abundance change map.

Earlier periods of population decline were attributed to agricultural intensification resulting in reduced availability of weed seeds (Siriwardena *et al.* 1999). Subsequent increases partly revolve around the success with which Goldfinches have increasingly exploited specialist seed varieties provided as bird food in gardens. Goldfinches were recorded in 50–60% of gardens in Britain in 2011, up from just under 10% of gardens in 1995 (Garden BirdWatch 2013). A similar pattern is evident in Ireland (Crowe 2005), with Goldfinches recorded in a record 86% of gardens in 2011 (O'Sullivan 2011).

In memoriam Lorna Nickell

BREEDING DISTRIBUTION *2008–11*

POSSIBLE •	7%	3%	4%
PROBABLE ●	22%	7%	11%
CONFIRMED ●	65%	75%	72%
TOTAL	94%	85%	87%

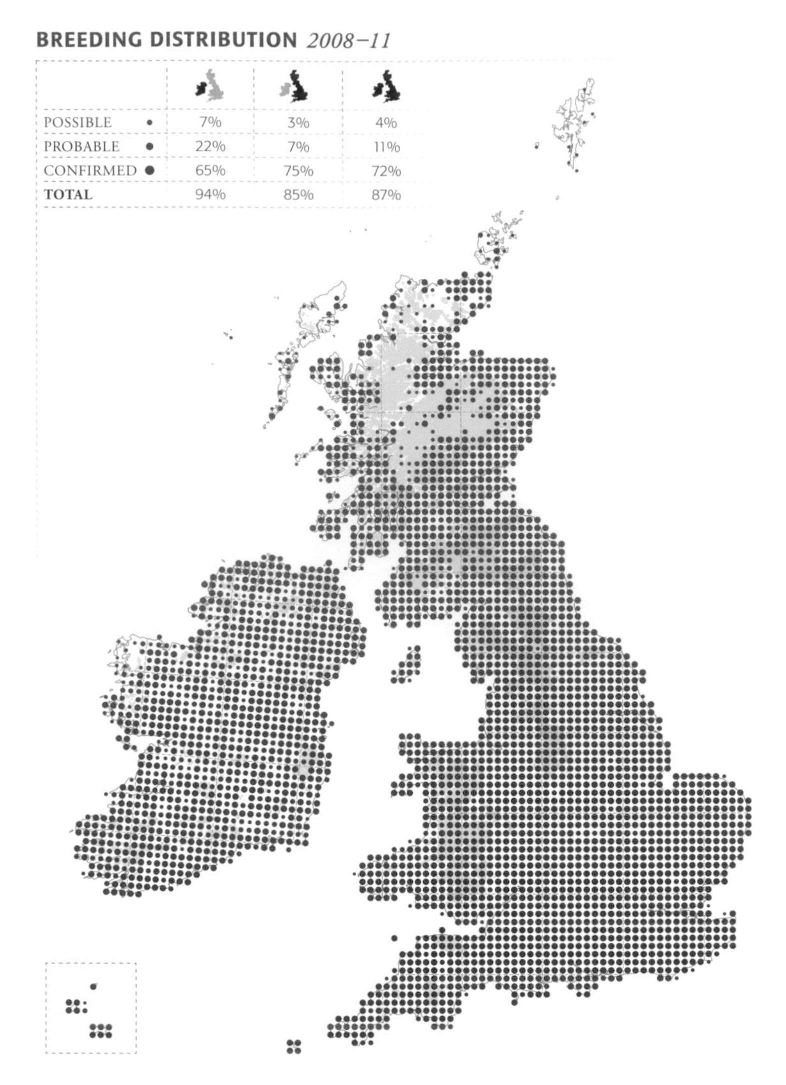

WINTER DISTRIBUTION *2007/08–2010/11*

PRESENT ●	94%	87%	89%

BREEDING DISTRIBUTION CHANGE *since 1968–72*

GAIN ▲	50	353	403
LOSS ▽	11	16	27
40yr CHANGE	+4%	+16%	+12%
20yr INDEX	+0.29	+0.14	

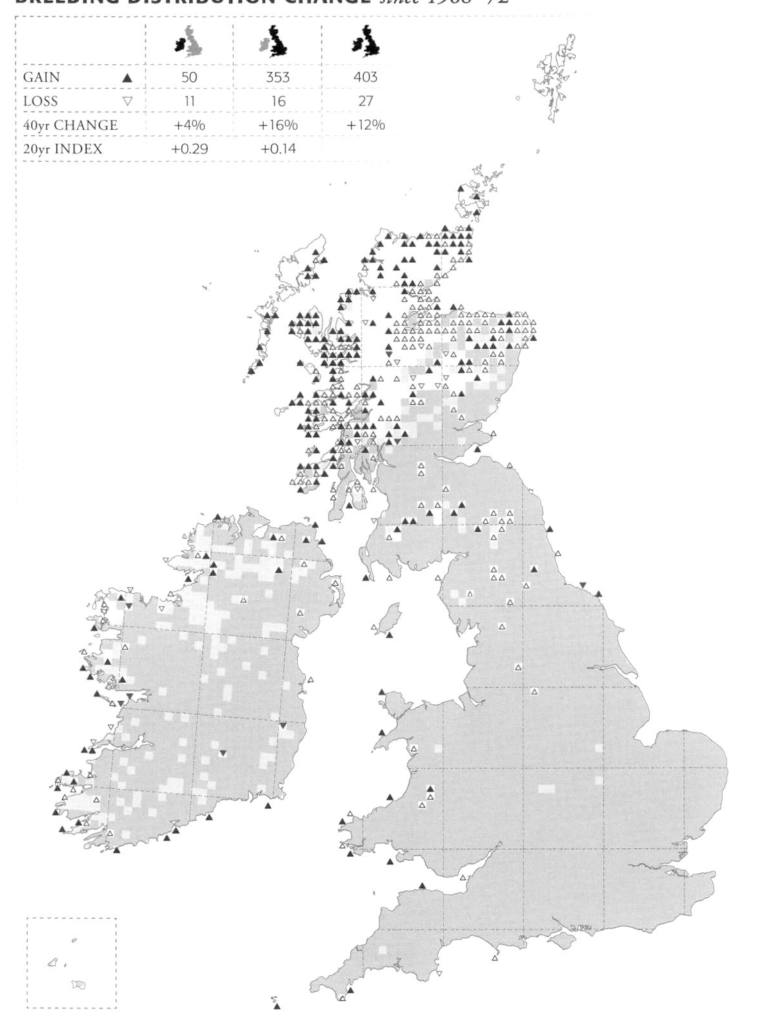

WINTER DISTRIBUTION CHANGE *since 1981–84*

GAIN ▲	321	488	809
LOSS ▼	14	37	51
30yr CHANGE	+49%	+23%	+29%

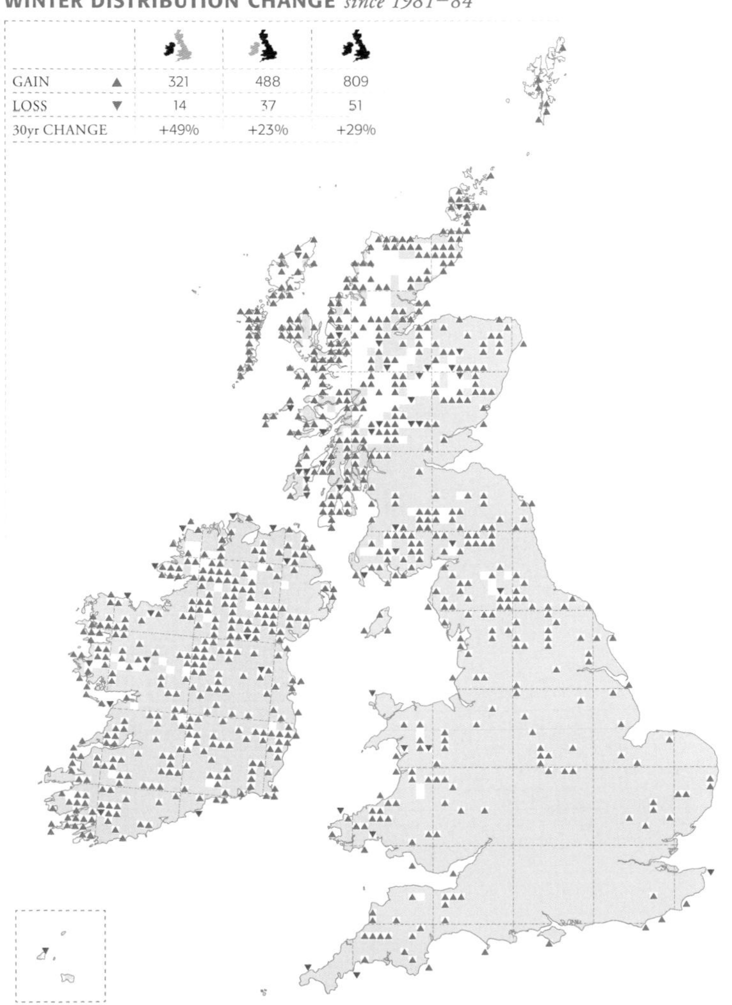

BREEDING RELATIVE ABUNDANCE *2008–11*

BREEDING RELATIVE ABUNDANCE CHANGE *since 1988–91*

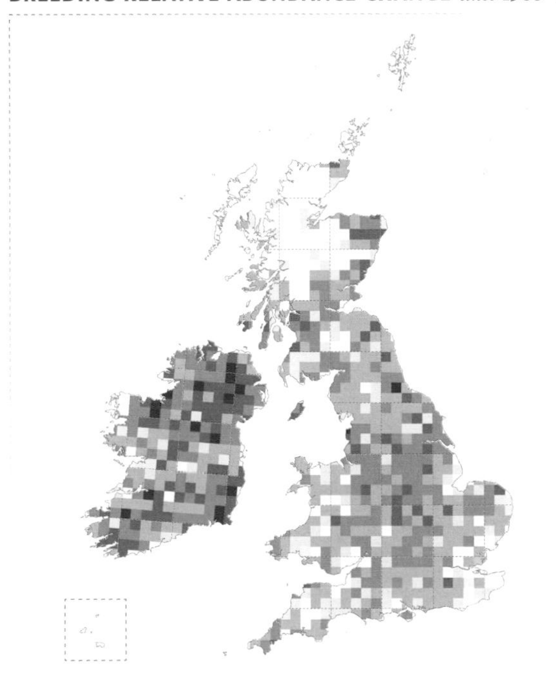

WINTER RELATIVE ABUNDANCE *2007/08–2010/11*

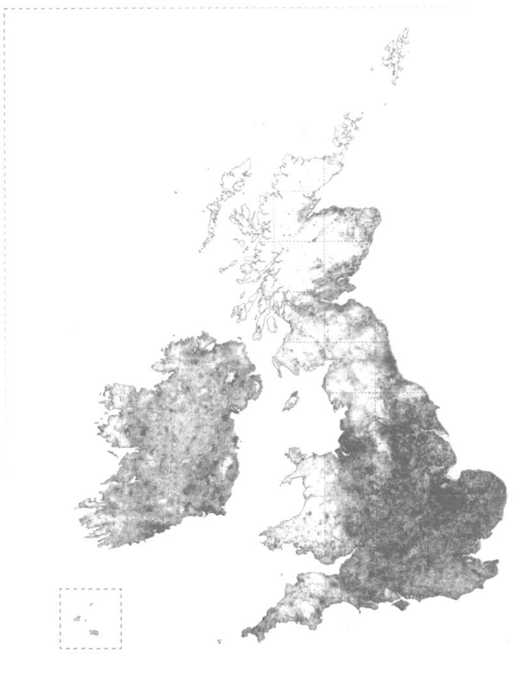

SISKIN

Carduelis spinus

BTO CODE: SK

GRAHAM CATLEY http://pewit.blogspot.co.uk

DURING THE BREEDING season, the preference of the Siskin for extensive coniferous woodland is clearly reflected in the maps. The highest densities are associated with landscapes dominated by conifer plantations, particularly on moors, low hills and the lower slopes of the uplands. Breeding Siskins are found through most of Scotland and Wales and much of northern and southwest England, with a patchier distribution elsewhere. In eastern and southern England, Thetford Forest and the New Forest stand out. In Ireland they are more widespread in the western half of the country, with more-isolated pockets in the east and south. Siskins are absent as breeding birds from the Channel Islands and from much of central and eastern England.

Since the *1968–72 Breeding Atlas* there has been a spectacular increase, of 166%, in the number of occupied 10-km squares. Recent gains have been achieved in Wales, southern and southwestern England and large parts of Ireland. In Scotland, birds have spread into the northeast and onto the Northern Isles and Outer Hebrides. There has also been an increase in abundance since the *1988–91 Breeding Atlas* throughout Wales, Scotland and the western half of Ireland. Annual monitoring data are consistent with this pattern, showing a 55% increase in numbers in the UK during 1995–2010 (*BBS Report 2011*). Exploitation of supplementary food in gardens, particularly when the cone crop has been poor, and the increased availability of maturing conifer plantations as breeding sites may have fuelled the increase (Chamberlain *et al.* 2005; Crowe 2005; McKenzie *et al.* 2007; *Birds of Scotland*).

In winter, Siskins have an even wider distribution, being found in 83% of all 10-km squares, with British & Irish breeders joined by continental immigrants (*Migration Atlas*). The level of migration is likely to depend on the conifer crops in the source and destination areas. In line with the expansion in the British breeding range, there has been a 44% expansion of range in winter, with gains mainly in northern and western Scotland, Wales and southwestern and eastern England. In Ireland there was an even greater and more widespread range expansion, with an increase of 189% in the number of occupied squares.

BREEDING DISTRIBUTION *2008–11*

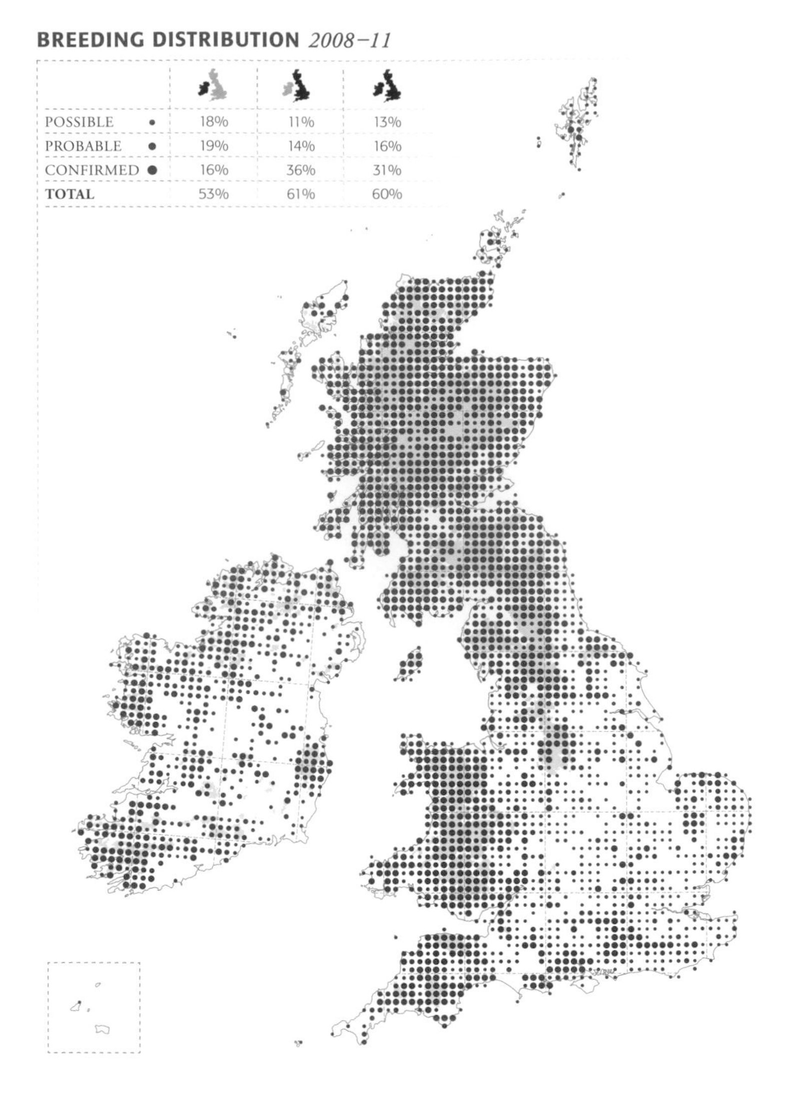

POSSIBLE	18%	11%	13%
PROBABLE	19%	14%	16%
CONFIRMED	16%	36%	31%
TOTAL	53%	61%	60%

WINTER DISTRIBUTION *2007/08–2010/11*

PRESENT	73%	86%	83%

BREEDING DISTRIBUTION CHANGE *since 1968–72*

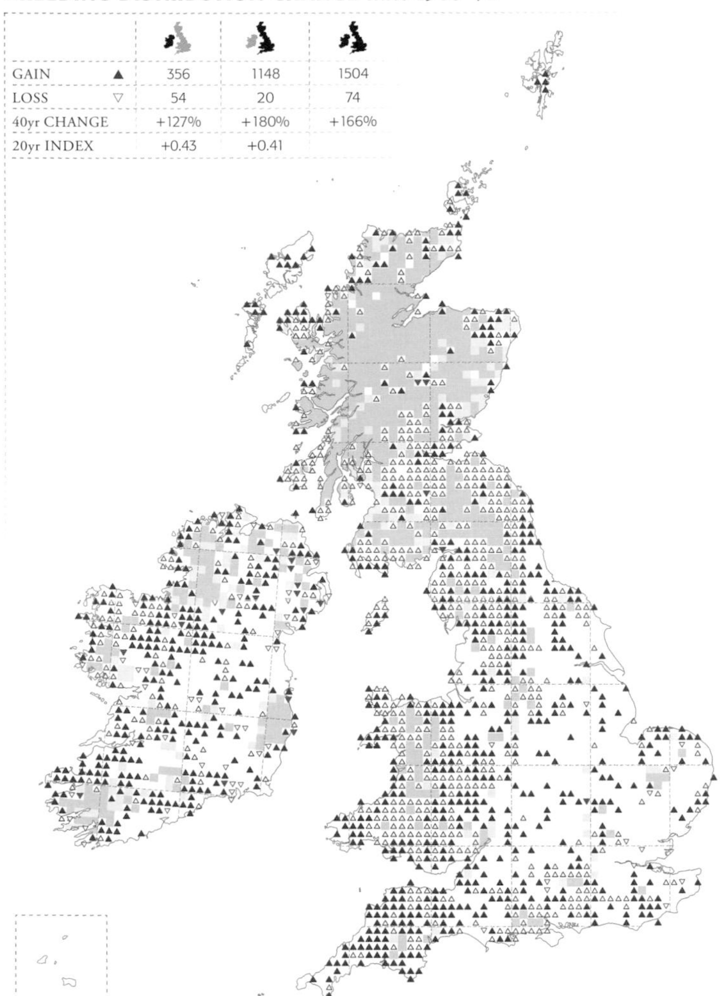

GAIN ▲	356	1148	1504
LOSS ▽	54	20	74
40yr CHANGE	+127%	+180%	+166%
20yr INDEX	+0.43	+0.41	

WINTER DISTRIBUTION CHANGE *since 1981–84*

GAIN ▲	493	783	1276
LOSS ▼	28	53	81
30yr CHANGE	+189%	+44%	+62%

BREEDING RELATIVE ABUNDANCE *2008–11*

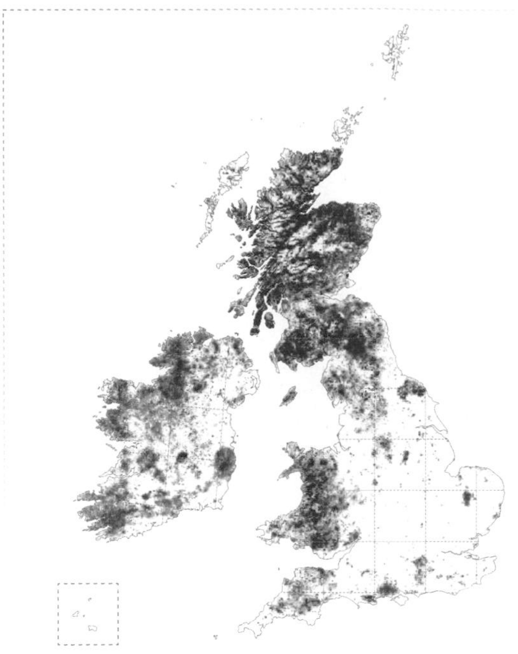

BREEDING RELATIVE ABUNDANCE CHANGE *since 1988–91*

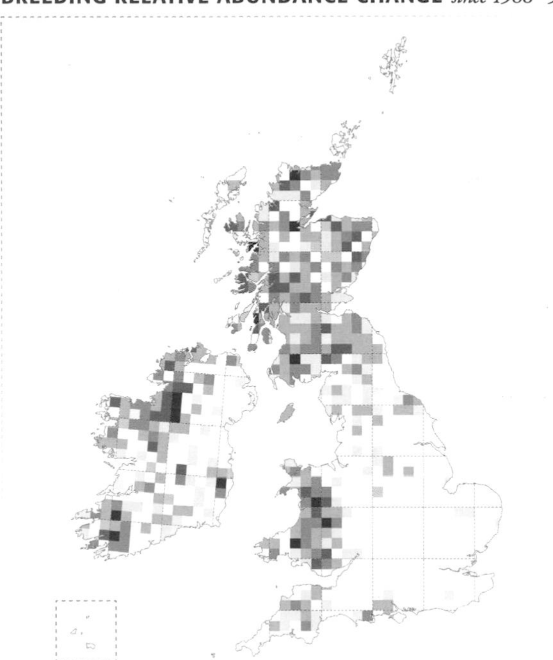

WINTER RELATIVE ABUNDANCE *2007/08–2010/11*

LINNET

Carduelis cannabina

BTO CODE: LI

GRAHAM CATLEY http://pewit.blogspot.co.uk

DURING WINTER, LINNETS withdraw from upland areas and therefore have a slightly more restricted distribution than in the breeding season. Since the *1981–84 Winter Atlas* there has been a 12% increase in the number of occupied 10-km squares. Range gains have been mostly in the more northerly and westerly parts of the range. In some marginal upland areas, gains may reflect birds remaining to spend the winter, possibly as a result of relatively mild weather, at least in the first two winters of the Atlas period.

Linnets require scattered bushes or scrub for nesting and seeds in the surrounding landscape for food. They are widely distributed during the breeding season, being found in virtually every 10-km square in England, Wales and Ireland. In Scotland, however, they are restricted to lower ground. Recent gains in northwest Scotland and throughout Ireland have been balanced by losses on the fringes of the uplands further south in Scotland and in the northern half of Ireland, resulting in no overall change since the *1968–72 Breeding Atlas*. The abundance change map shows large increases in the proportions of occupied tetrads in Ireland and in parts of northern and eastern Scotland, but declines on lower ground elsewhere in Britain. Numbers of Linnets fell rapidly in the UK between the mid 1970s and mid 1980s (Marchant *et al.* 1990), though the subsequent picture has been more mixed, with increases exceeding 50% in Ireland, a small increase in Scotland and declines in England and Wales since the 1990s (*CBS Report 2010*; *BBS Report 2011*).

Loss of weeds in the agricultural landscape, due to the intensification of farming practices, has reduced the availability of invertebrate and seed foods for birds generally (Wilson *et al.* 1999). For Linnet, there has been a marked reduction of several food-plants traditional in the diet of nestlings, which, in some areas, has become dominated by seeds of Dandelion and Oil-seed Rape (Moorcroft *et al.* 2006). Falling breeding success, mediated through increased nest failures during the egg stage, may underpin the decline (Siriwardena *et al.* 2000b). Stubble remaining over winter, supporting high seed abundance and with natural regeneration of weeds, is an important resource for wintering Linnets (Moorcroft *et al.* 2002).

Sponsored by Jen Woodland and her grandsons Robbie, Iain & Alistair, in memory of John

BREEDING DISTRIBUTION *2008–11*

POSSIBLE •	11%	4%	6%
PROBABLE ●	28%	14%	18%
CONFIRMED ●	53%	66%	63%
TOTAL	92%	84%	87%

WINTER DISTRIBUTION *2007/08–2010/11*

PRESENT ●	67%	69%	69%

BREEDING DISTRIBUTION CHANGE *since 1968–72*

	Ireland	Britain	Britain & Ireland
GAIN ▲	31	131	162
LOSS ▽	47	112	159
40yr CHANGE	-2%	+1%	+<1%
20yr INDEX	+0.22	+0.01	

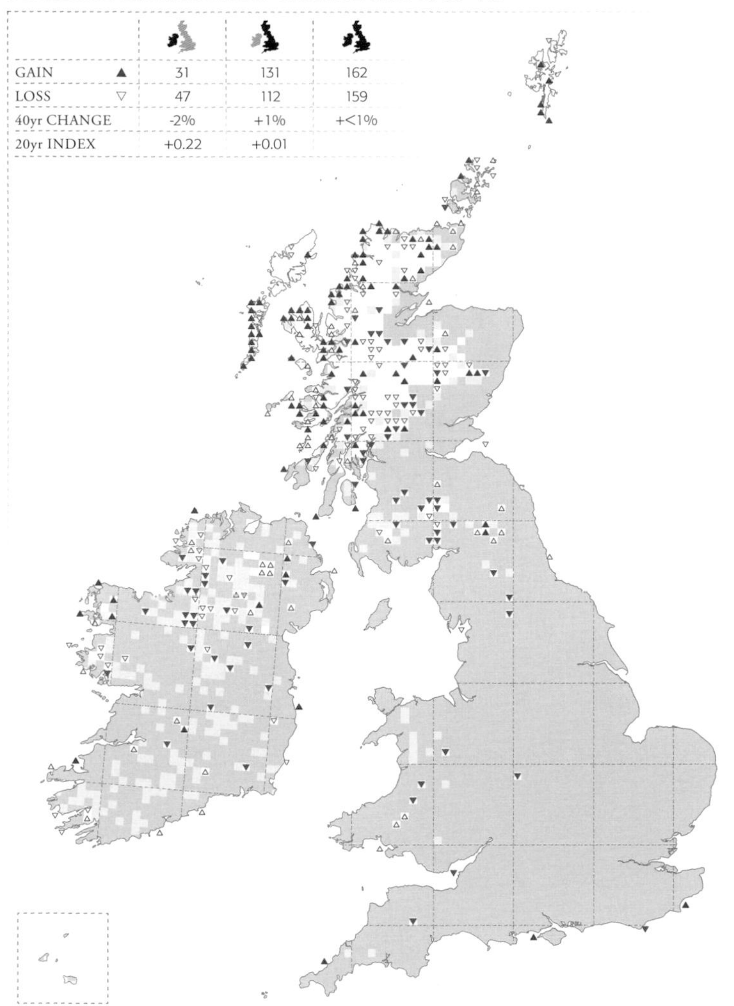

WINTER DISTRIBUTION CHANGE *since 1981–84*

	Ireland	Britain	Britain & Ireland
GAIN ▲	207	362	569
LOSS ▼	155	150	305
30yr CHANGE	+9%	+12%	+12%

BREEDING RELATIVE ABUNDANCE *2008–11*

BREEDING RELATIVE ABUNDANCE CHANGE *since 1988–91*

WINTER RELATIVE ABUNDANCE *2007/08–2010/11*

TWITE

Carduelis flavirostris

BTO CODE: TW

RUTH EASTWOOD

BREEDING TWITES HAVE shown range contractions of 80% in Ireland and 19% in Britain since the *1968–72 Breeding Atlas*. They are now to be found throughout the Northern Isles and close to the western seaboard of Scotland, with inland upland populations primarily in the Scottish Highlands and the Pennines. A few pairs remain in North Wales, the Southern Uplands and the coastal fringes of northwest Ireland. Highest densities are in Shetland, Orkney and Caithness and on the Hebrides and their adjacent mainland coast. In Ireland many of the losses in the west have occurred since the *1988–91 Breeding Atlas* (McLoughlin & Cotton 2008). There have also been recent declines in the southern Pennines and in the Scottish Highlands, mainly in the eastern half of the former range (Langston *et al.* 2006).

Although wintering is largely coastal there are some inland records, particularly in central Scotland. Upland-nesting Twites traditionally move to the coast, though breeding birds in some populations spend the winter locally (Raine *et al.* 2006a; McLoughlin *et al.* 2010). The winter abundance map highlights the importance of the Northern Isles, especially Orkney, and the Hebrides between North Uist and Islay. Further south, large estuaries such as Morecambe Bay, the Humber and the Wash attract substantial flocks to feed in saltmarshes (Brown & Atkinson 2006). Since the *1981–84 Winter Atlas*, 30% fewer 10-km squares have been occupied in Ireland. In Britain, however, there has been a 19% increase, mostly in the north and east, with some losses in southern England. An increase is surprising, given that breeding numbers have declined overall.

Twites require an abundant supply of seeds for feeding throughout the year and are susceptible to changes in agricultural practice that reduce the availability of their food. Breeding population declines have been attributed to the conversion of hay meadows to pasture or silage, early cutting dates for silage, increased stocking levels, a loss of species-rich hay meadows and poor burning practice on heather moorlands (Langston *et al.* 2006). Research confirms that conservation measures are required to safeguard the remaining populations and to provide suitable conditions for recolonisation (Hancock & Wilson 2003; Raine *et al.* 2006b, Wilkinson & Wilson 2010).

Sponsored by Snowdonia National Park Authority

BREEDING DISTRIBUTION *2008–11*

POSSIBLE •	1%	4%	3%
PROBABLE •	<1%	6%	5%
CONFIRMED ●	1%	8%	6%
TOTAL	2%	18%	14%

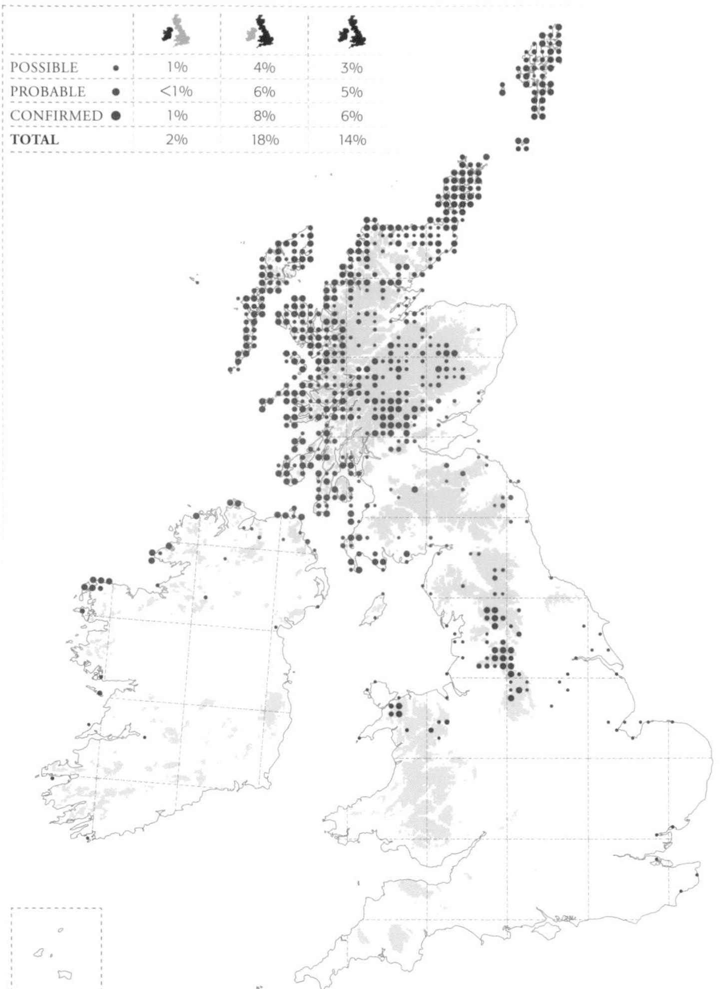

WINTER DISTRIBUTION *2007/08–2010/11*

PRESENT ●	7%	23%	19%

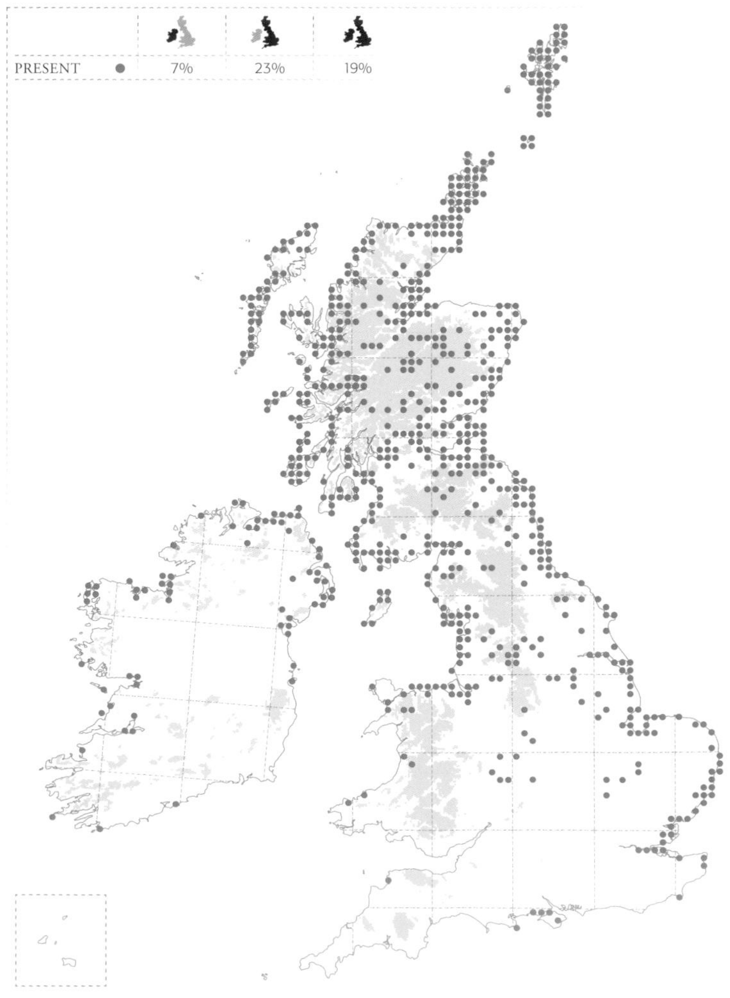

BREEDING DISTRIBUTION CHANGE *since 1968–72*

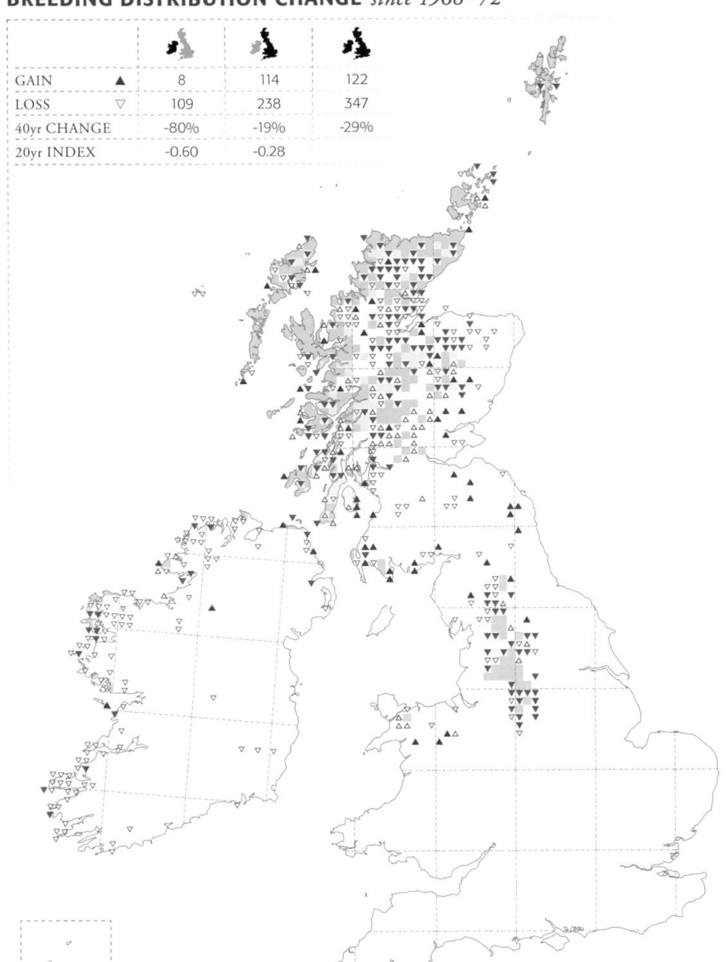

GAIN ▲	8	114	122
LOSS ▽	109	238	347
40yr CHANGE	-80%	-19%	-29%
20yr INDEX	-0.60	-0.28	

BREEDING RELATIVE ABUNDANCE *2008–11*

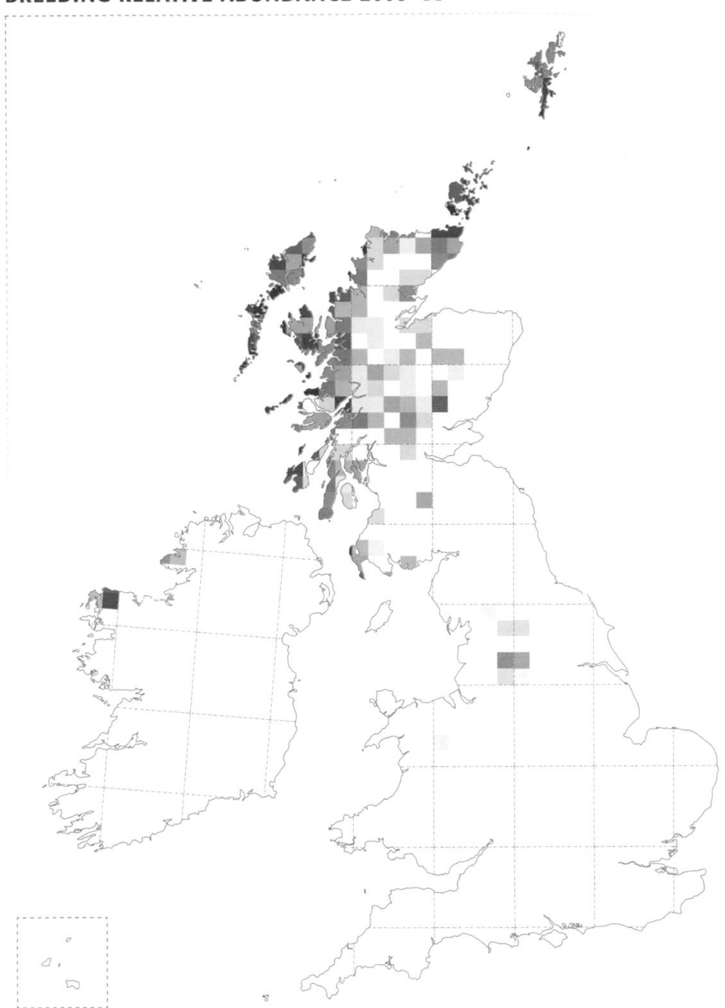

WINTER DISTRIBUTION CHANGE *since 1981–84*

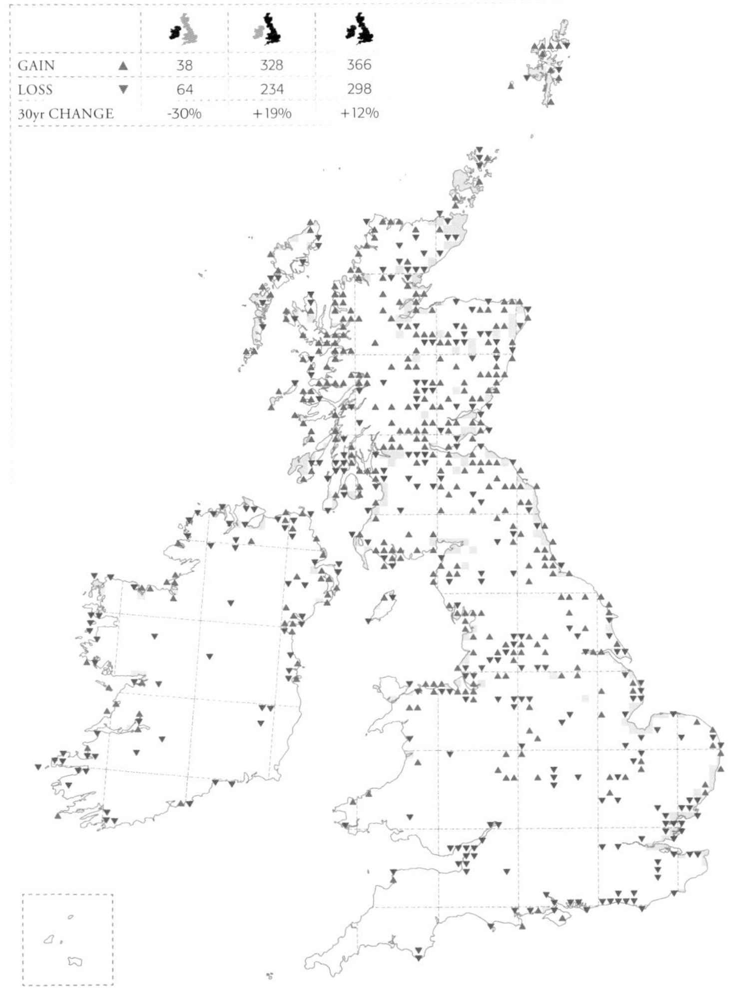

GAIN ▲	38	328	366
LOSS ▼	64	234	298
30yr CHANGE	-30%	+19%	+12%

WINTER RELATIVE ABUNDANCE *2007/08–2010/11*

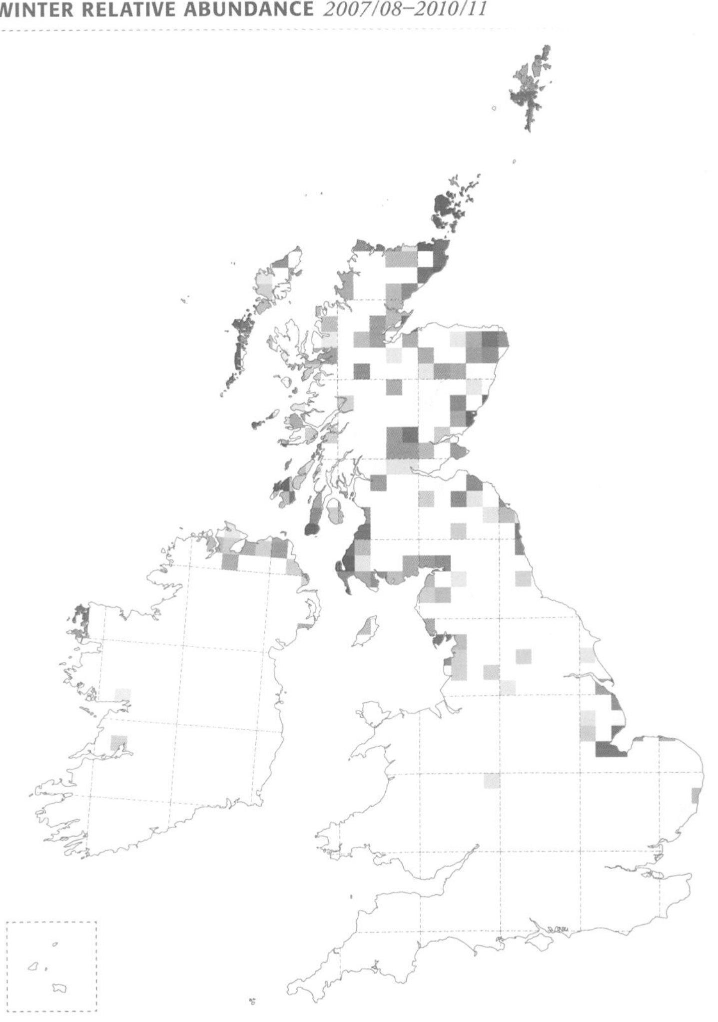

LESSER REDPOLL

Carduelis cabaret

BTO CODE: LR

CLARE WARD

THE LESSER REDPOLL was elevated to full species status by BOU in 2001 (Knox *et al.* 2001), whereas in Ireland and elsewhere it continues to be treated as a race of the Common Redpoll. BOU's taxonomic split renders comparison difficult with the *1981–84 Winter Atlas*, for which data were gathered under the single heading 'Redpoll', but breeding comparisons are little affected. Atlas data were collected also under an 'unidentified Common/Lesser' heading. The breeding-season maps combine records of Lesser Redpolls and birds recorded as 'Common/Lesser', whereas winter maps show Lesser Redpolls only.

Lesser Redpolls breed fairly widely throughout Ireland, Wales, most of Scotland and northern England, and more patchily further south. Since the *1968–72 Breeding Atlas* 16% of the range in Britain & Ireland has been lost, most obviously in eastern England. Recent gains are most evident in Ireland, southern and southwest England and parts of Scotland. The change map may overstate stability, as some records of singing birds in suitable habitat in early spring may refer to birds on passage. The changes in relative abundance are striking, with increases in the west contrasting with declines in the east. Monitoring data support this pattern: during 1995–2010 there was a 10% decline in numbers in England but a 26% increase in Scotland (*BBS Report 2011*). In the Republic of Ireland there has been a 163% increase in numbers during 1998–2010 (*CBS Trend 2010*). New forestry plantations have provided suitable breeding habitat, whilst in other areas the removal of scrub and the maturing of forests may have led to losses (*Birds of Scotland*).

During the winter Lesser Redpolls are widely distributed, but largely absent from sparsely wooded areas, intensively farmed lowlands and some uplands. They occur at relatively high densities throughout Ireland, in wooded valleys and lowlands in northern and eastern Scotland, northern England and Wales, and in parts of southern and southeast England. The 61% range expansion in Ireland since the *1981–84 Winter Atlas* is probably influenced by recent increases in breeding numbers. In Britain there was a smaller, 30% expansion, though some areas of apparent loss, such as on Shetland, relate to birds now classified as Common Redpolls.

In memory of Marjorie Hewes

BREEDING DISTRIBUTION *2008–11*

POSSIBLE ●	21%	13%	15%
PROBABLE ●	33%	18%	22%
CONFIRMED ●	26%	23%	24%
TOTAL	80%	54%	61%

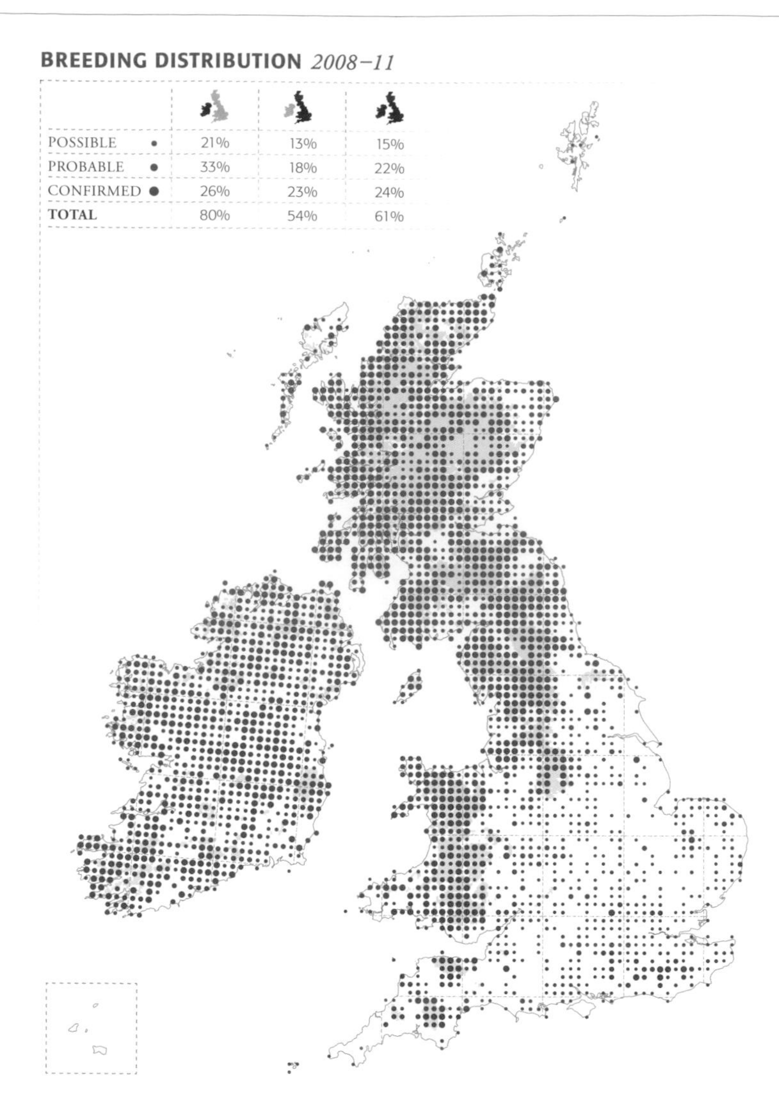

WINTER DISTRIBUTION *2007/08–2010/11*

PRESENT ●	77%	67%	70%

BREEDING DISTRIBUTION CHANGE *since 1968–72*

GAIN ▲	86	221	307
LOSS ▽	114	652	766
40yr CHANGE	-3%	-22%	-16%
20yr INDEX	+0.42	-0.15	

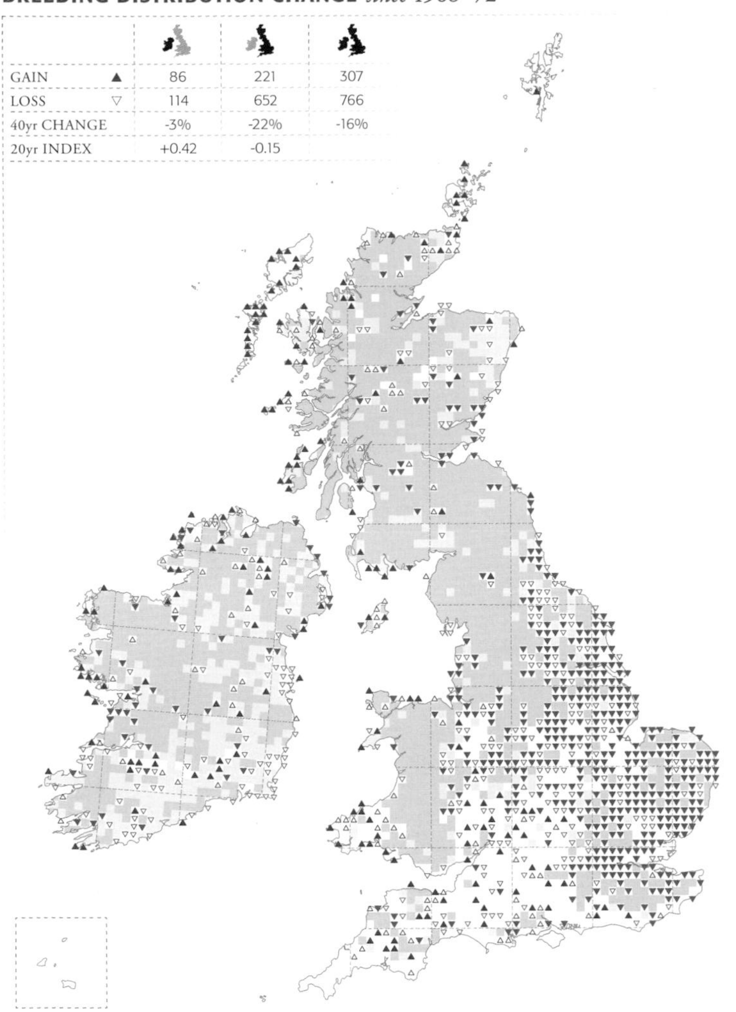

WINTER DISTRIBUTION CHANGE *since 1981–84*

GAIN ▲	338	555	893
LOSS ▼	51	250	301
30yr CHANGE	+61%	+20%	+30%

BREEDING RELATIVE ABUNDANCE *2008–11*

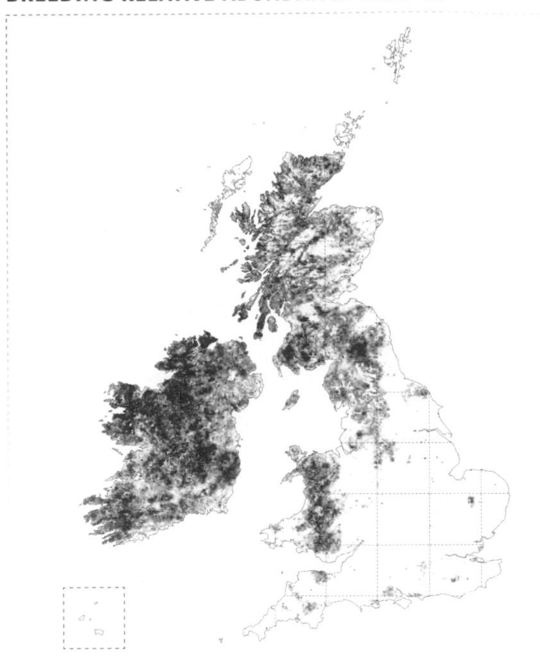

BREEDING RELATIVE ABUNDANCE CHANGE *since 1988–91*

WINTER RELATIVE ABUNDANCE *2007/08–2010/11*

COMMON REDPOLL

Carduelis flammea

BTO CODE: FR

JIM ALMOND www.shropshirebirder.co.uk

THE COMMON REDPOLL is mainly a winter visitor to Britain & Ireland from northern Eurasia, Greenland and possibly Iceland. At least two subspecies have been recognised in Britain & Ireland: the Mealy Redpoll (nominate *flammea*) and the Greenland Redpoll (*rostrata*). There is doubt whether the Iceland Redpoll ('*islandica*') is valid as a separate race and it is currently omitted from the British List (Stoddart 2011; BOU 2013a). Separating the various forms of the Common Redpoll from each other and from the Lesser Redpoll can raise intractable problems (Reid & Riddington 1998; Collinson 2006; Stoddart 2013). The Lesser Redpoll was split from the Common Redpoll by BOU in 2001 (Knox *et al.* 2001), though in Ireland, and elsewhere, they continue to be treated as one species.

Common Redpolls breeding in the boreal zone, across the whole of northern continental Europe, Asia and North America, are highly irruptive. Influxes into the Northern Isles and the east coast of Britain occur occasionally, as in the winter of 2010/11. Many of these birds move south or west, often to inland squares, particularly in central and southern England, where they join flocks of Lesser Redpolls. Small numbers of Greenland Redpolls arrive most autumns in northern Scotland, particularly on the islands from early September, though passage can continue into November (Riddington & Votier 1997; Stoddart 2013).

During 2008–11 Common Redpolls were suspected of breeding in northern Scotland, at sites in Shetland, Orkney, Caithness, Sutherland, the Outer Hebrides and Tiree (*RBBP 2010*). Some of these showed the characteristics of Mealy Redpolls, while others appeared to be Greenland Redpolls.

Extralimital breeding by Mealy Redpolls has been reported previously from the Northern Isles, Sutherland (*Birds of Scotland*) and the Outer Hebrides (Stevenson 2005). There are few documented cases of Greenland Redpolls breeding, though an apparent pair bred in Inverness-shire in 1959 (Murton & Porter 1961) and birds resembling this form, or the Iceland Redpoll, have been recorded in the Outer Hebrides in the breeding season (Stevenson 2005). *Birds of Scotland* suggested that some breeding records of Mealy Redpolls from the Northern Isles may in fact have involved *rostrata* instead and that *rostrata* may go undetected elsewhere in Scotland.

WINTER DISTRIBUTION *2007/08–2010/11*

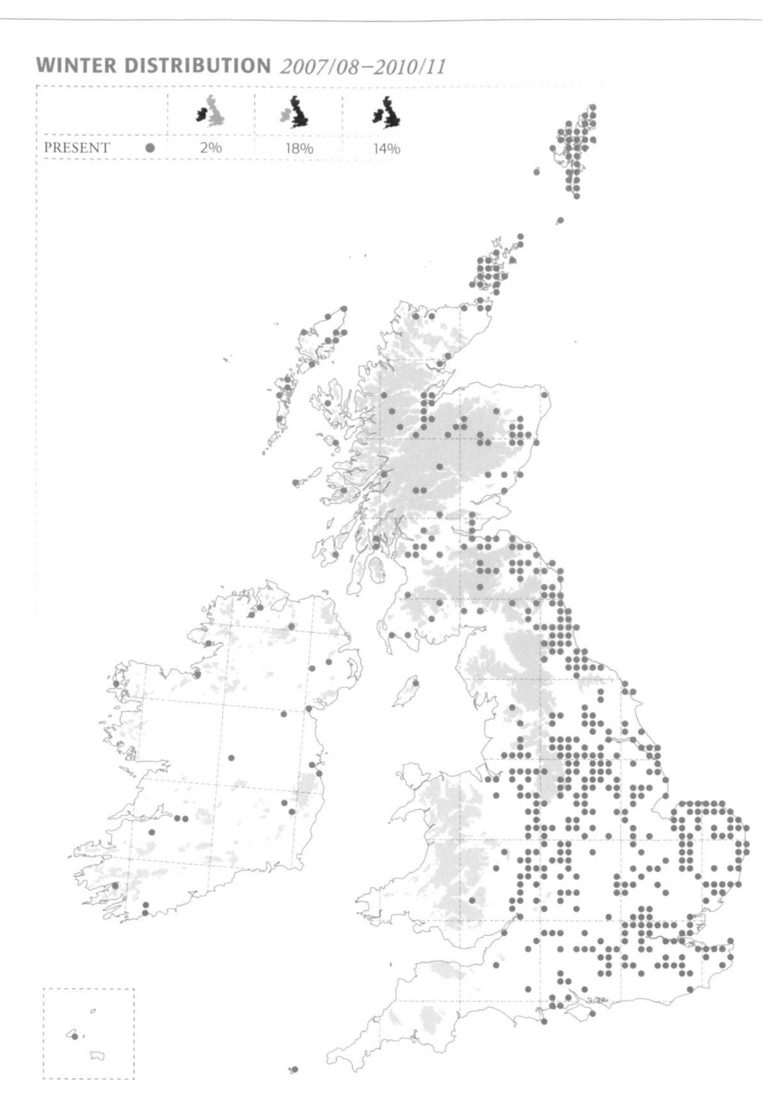

BREEDING DISTRIBUTION *2008–11*

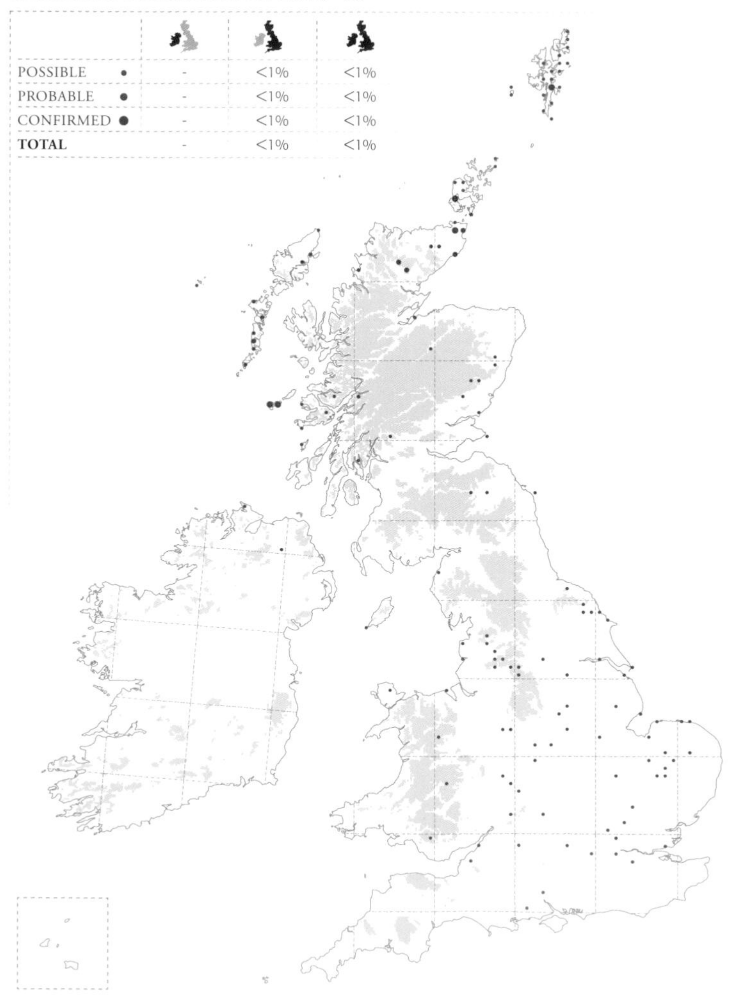

Sponsored by Sarah Mansell

SERIN
Serinus serinus

BTO CODE: NS

J B WILSON

THE SERIN IS a relatively common breeding bird in continental Europe that has been slowly expanding its range northwards. It has bred in southern England on a few occasions and from the 1970s to the 1990s it nested regularly on Jersey. During 2008–11 *possible* breeding was recorded in four 10-km squares, and *probable* breeding was noted in 2009 in a single 10-km square in Dorset (Dorset Bird Club 2011). No birds were *confirmed* breeding and Jersey reported *non-breeding* records only. The most recent confirmed British breeding records were in Norfolk in 2003 and Kent in 1996 (Holling *et al.* 2009).

In winter there were records from 12 10-km squares, mostly in southern England. Half of these squares held birds only in November.

BREEDING DISTRIBUTION *2008–11*

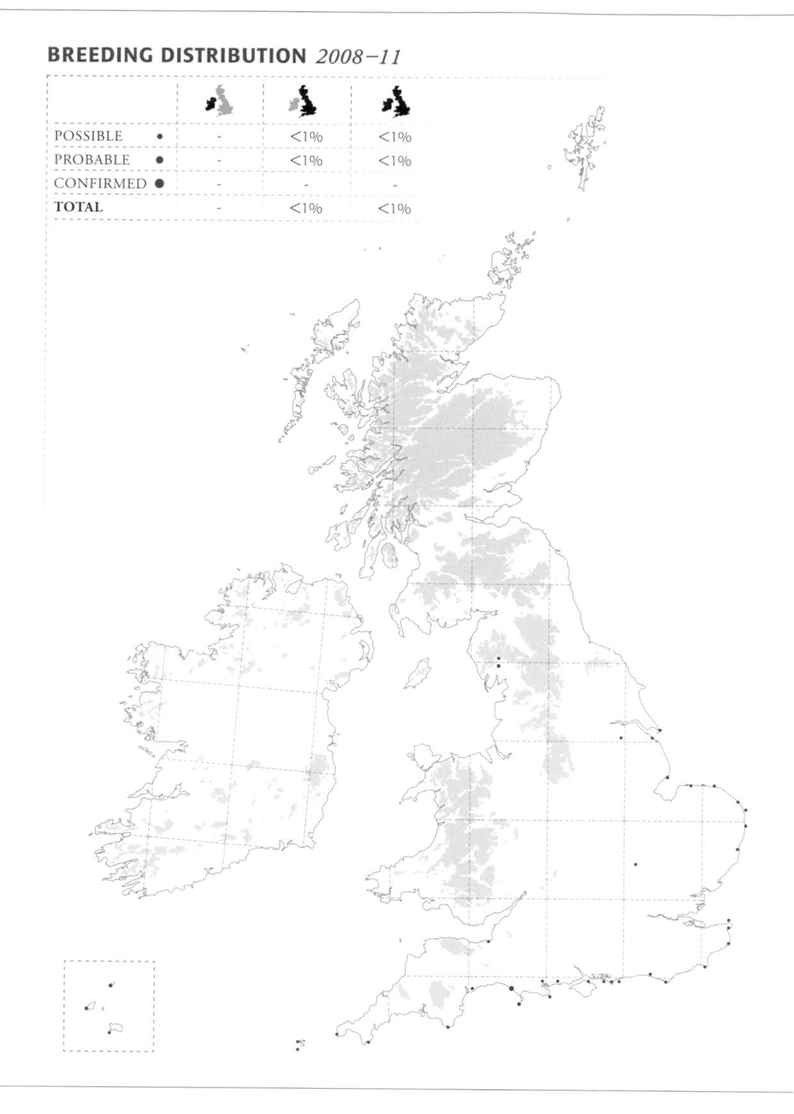

POSSIBLE ●	-	<1%	<1%
PROBABLE ●	-	<1%	<1%
CONFIRMED ●	-	-	-
TOTAL	-	<1%	<1%

ARCTIC REDPOLL
Carduelis hornemanni

BTO CODE: AL

GARY WOODBURN

THE ARCTIC REDPOLL is a rare winter visitor originating from breeding populations in northern Greenland (nominate *hornemanni*) or in northern Scandinavia through to northern Siberia (*exilipes*). During *Bird Atlas 2007–11* it was recorded from 19 10-km squares, mostly in coastal and lowland areas in eastern England. There was a small influx into the Northern Isles in autumn 2010 but the records from the five 10-km squares there all referred to birds present only in November when possibly still on passage. No birds had been recorded wintering during the *1981–84 Winter Atlas*, and arguably the Atlas records represent a genuine increase in occurrence; as further evidence of increase, records of *exilipes* are no longer assessed by the British Birds Rarities Committee (Kehoe *et al.* 2006).

WINTER DISTRIBUTION *2007/08–2010/11*

PRESENT ●	-	1%	<1%

COMMON CROSSBILL/SCOTTISH CROSSBILL

Loxia curvirostra/Loxia scotica BTO CODE: CR/CY

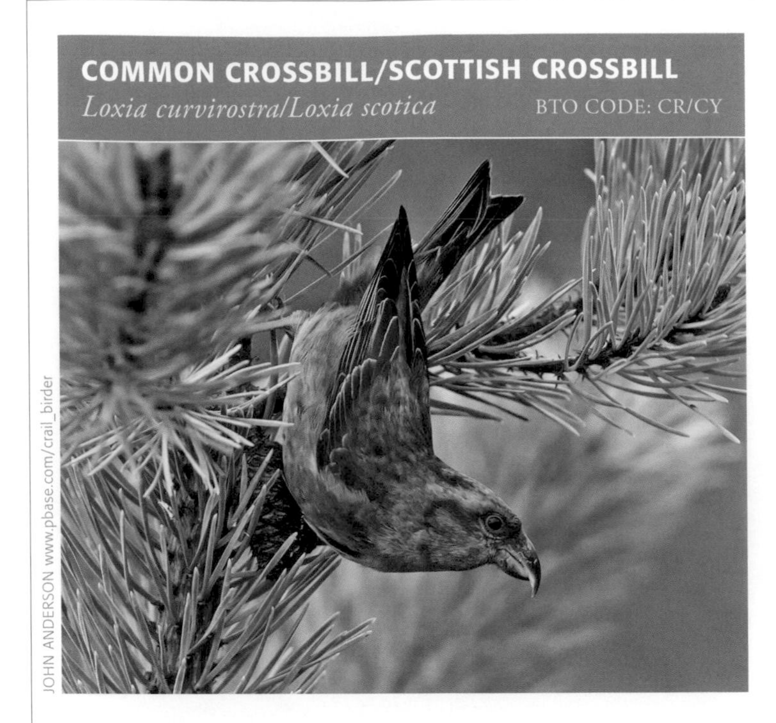

JOHN ANDERSON www.pbase.com/crail_birder

IN AREAS WHERE the Common Crossbill and the Scottish Crossbill co-occur, they cannot be separated reliably without a sonogram of their distinctive calls (Summers *et al.* 2007). As this is unlikely to have been the method of identification for most records submitted to *Bird Atlas 2007–11*, and because many individuals were submitted as 'unidentified crossbill', here we provide a combined account using records of birds reported as the Common Crossbill, Scottish Crossbill and 'unidentified crossbill'. Outside northeast Scotland and the eastern Highlands it is safe to assume that all records refer solely to the Common Crossbill (see Summers & Buckland 2011).

Common Crossbills are highly mobile across Europe and their occurrence in Britain & Ireland varies in response to conifer seed production. During the winters covered by this atlas, crossbills were widely distributed throughout most of Scotland, with the exception of treeless areas such as parts of northwest Scotland, the Northern Isles and some Hebridean islands. In much of Wales and England they were associated with conifer plantations in upland areas and also lowland conifer forests, for example in Norfolk, Hampshire and Dorset. In Ireland the distribution was very patchy and linked with afforestation, mostly on the hills.

Crossbills may start breeding as early as midwinter, depending on the availability of conifer seeds. Consequently the breeding and winter distribution maps are fairly similar. Numbers of crossbills breeding in Britain & Ireland vary widely from year to year; in South Wales they are most abundant in years of high Sitka Spruce cone production (Dixon & Haffield 2013).

There have been significant gains in the numbers of occupied 10-km squares throughout much of Britain & Ireland since the *1981–84 Winter Atlas* and the *1968–72 Breeding Atlas*, associated with the maturing of forestry planted after 1945. The abundance change map also shows increases since the *1988–91 Breeding Atlas* in the proportions of tetrads occupied, particularly in Scotland, Wales and Ireland.

Records of *non-breeding* birds, in areas such as the Northern Isles and central and eastern England, are likely to result from post-breeding irruptions, mainly in June or July, the origins of which may differ between years (Newton 2006; Marquiss *et al.* 2008, 2012).

Sponsored by Forestry Commission (England)

BREEDING DISTRIBUTION *2008–11*

POSSIBLE •	13%	13%	13%
PROBABLE •	7%	8%	8%
CONFIRMED •	5%	17%	14%
TOTAL	25%	38%	35%

WINTER DISTRIBUTION *2007/08–2010/11*

PRESENT •	23%	46%	40%

BREEDING DISTRIBUTION CHANGE *since 1968–72*

GAIN ▲	245	834	1079
LOSS ▽	1	69	70
40yr CHANGE	+8133%	+253%	+331%
20yr INDEX	+0.32	+0.42	

WINTER DISTRIBUTION CHANGE *since 1981–84*

GAIN ▲	211	859	1070
LOSS ▼	4	75	79
30yr CHANGE	+1150%	+177%	+215%

BREEDING RELATIVE ABUNDANCE *2008–11*

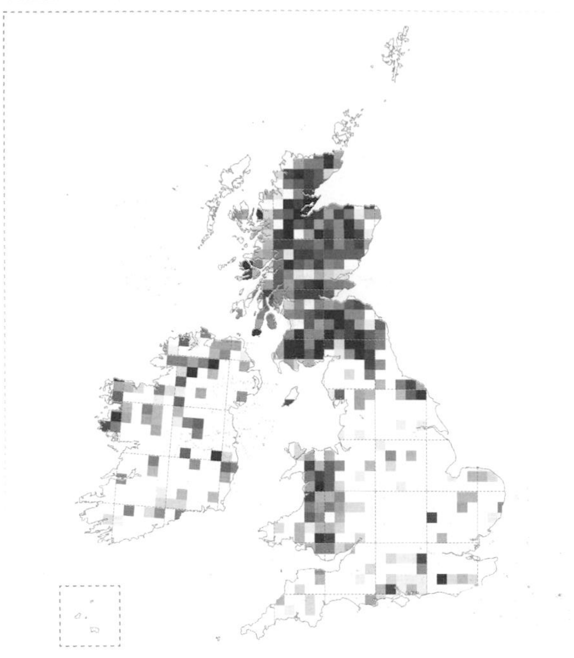

BREEDING RELATIVE ABUNDANCE CHANGE *since 1988–91*

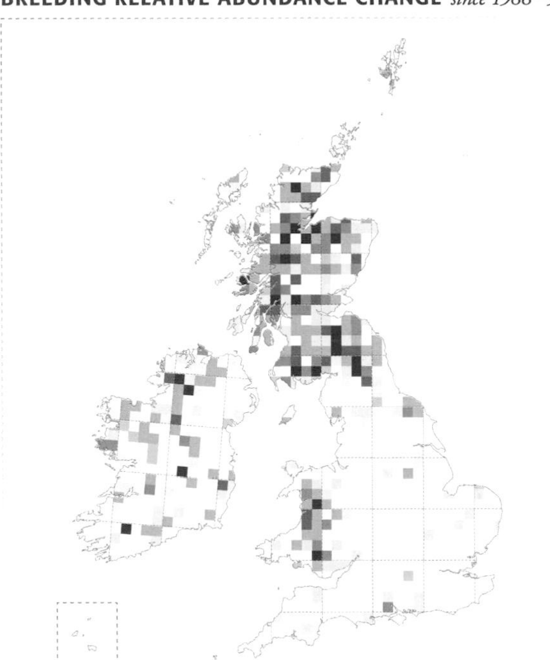

WINTER RELATIVE ABUNDANCE *2007/08–2010/11*

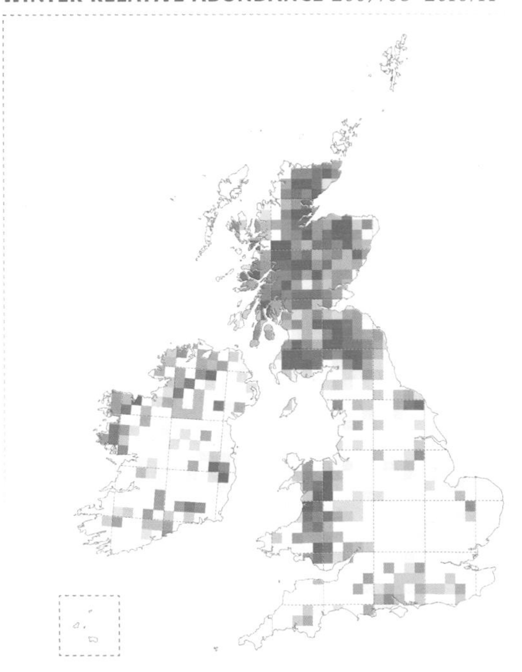

SCOTTISH CROSSBILL

Loxia scotica

BTO CODE: CY

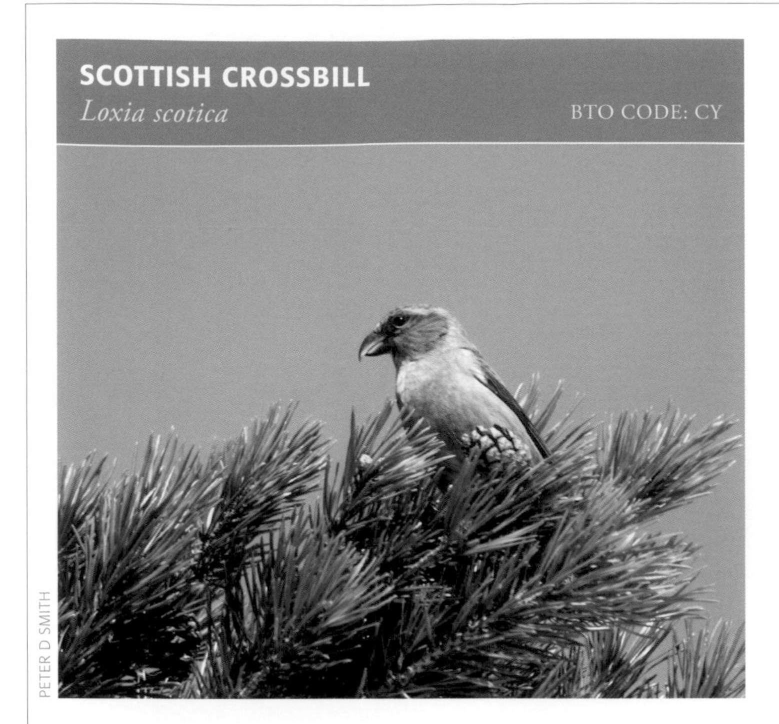

PETER D SMITH

HAVING BEEN SPLIT from the Common Crossbill in 1980 (BOU 1980), the Scottish Crossbill is Britain's only endemic bird species (Knox 1990). It can be reliably separated only on subtle differences in bill measurements or by its distinctive excitement call (Summers *et al.* 2002, 2007) and early information on its distribution and population size might suffer from unreliable field identifications.

Between January and April 2008 a survey was undertaken in northern Scotland to estimate the distribution and population size of this species (Summers & Buckland 2011). At a series of 889 survey points systematically covering the presumed range, sound playback was used to attract crossbills and excitement calls were recorded from the birds attracted, enabling subsequent identification. The resulting map, reproduced here, shows Scottish Crossbills in 72 10-km squares, with a disjunct distribution. Most lay in northeast Scotland, in Nairn, Moray and Banff, extending southward into lower Deeside, and the other main concentration extended from Sutherland to Easter Ross and eastern Inverness-shire. There was a notable absence in Strathspey, where Scottish Crossbills had been recorded in previous surveys (e.g. Summers *et al.* 2004). This gap may have appeared due to changes in the regional and seasonal availability of the various cone crops on which the birds depend (Summers 1999; Marquiss & Rae 2002); Scottish Crossbills utilise a wide range of conifer woods, including plantations of Scots and Lodgepole Pines, Sitka Spruce and larches. The survey estimated a population of 6,800 pairs, considerably higher than previously thought (Summers & Buckland 2011).

Ring recoveries and sightings of colour-ringed individuals show only relatively short-distance movements, within Deeside and between Deeside and Strathspey (Marquiss & Rae 2002); the distribution is thus likely to be similar at all seasons.

A breeding-season map, based on records submitted to *Bird Atlas 2007–11* as 'Scottish Crossbill', is also shown. Records with breeding evidence were received from 60 10-km squares, and *non-breeding* records came from a further 11 squares. Although there is much overlap with the distribution from the playback survey, Common Crossbills occur throughout the same area and identification for some Atlas records might therefore be uncertain, if based more on location than on the diagnostic calls.

Sponsored by Natural Power

CROSSBILL SURVEY *2008*

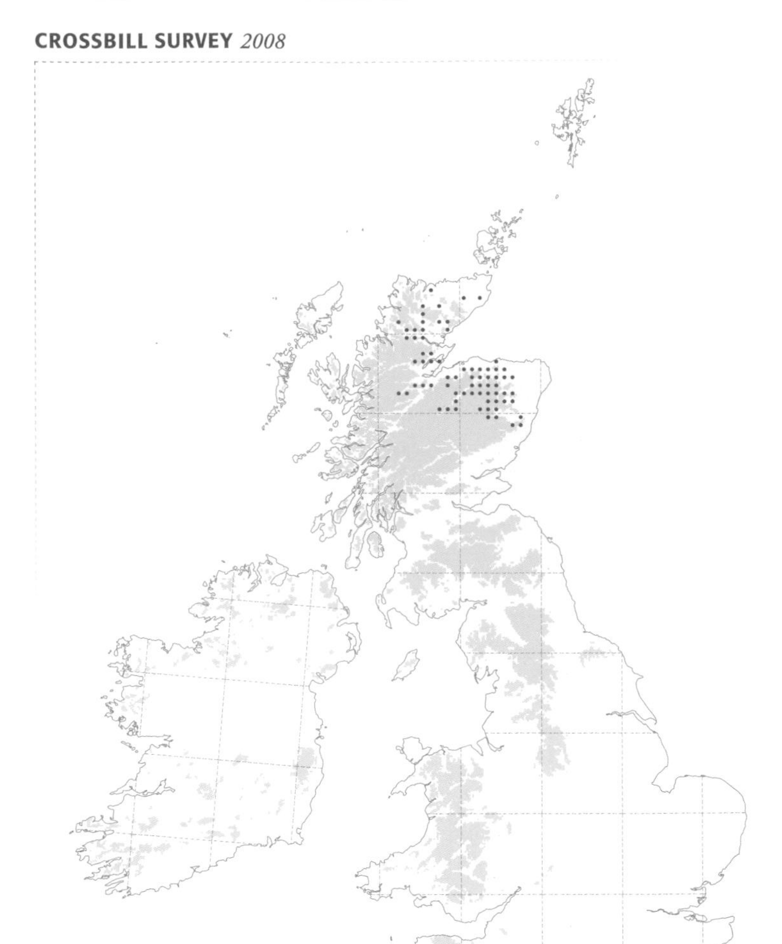

BREEDING DISTRIBUTION *2008–11*

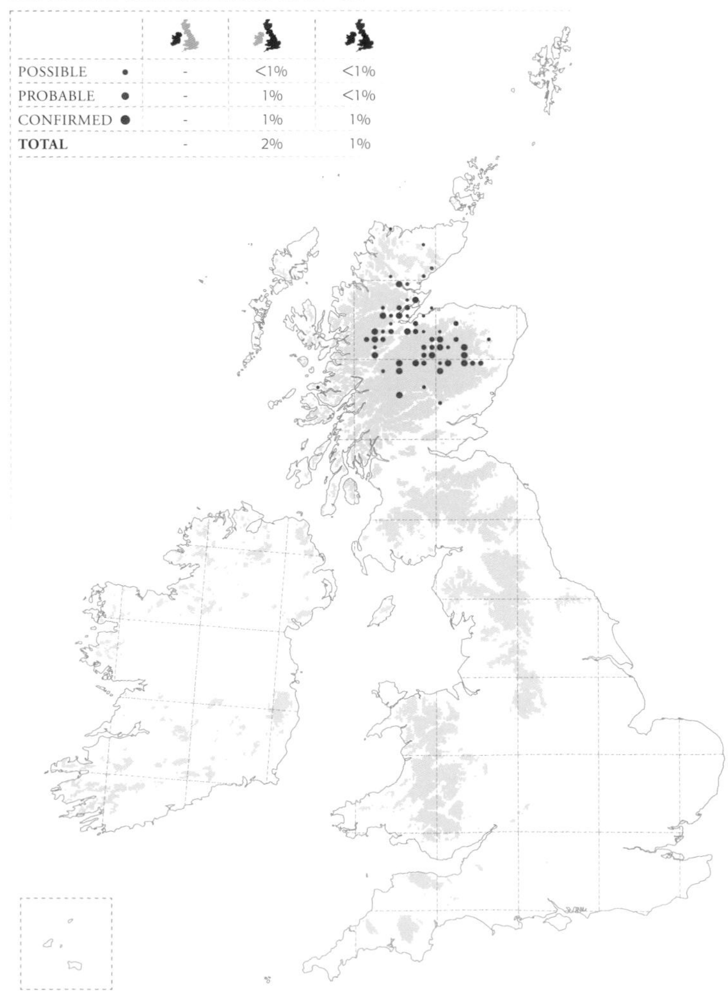

POSSIBLE •	-	<1%	<1%
PROBABLE •	-	1%	<1%
CONFIRMED •	-	1%	1%
TOTAL	-	2%	1%

PARROT CROSSBILL
Loxia pytyopsittacus BTO CODE: PC

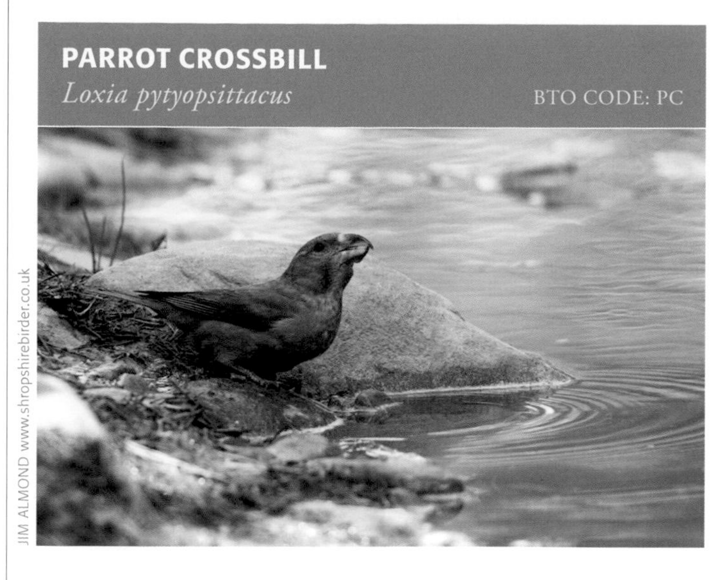

JIM ALMOND www.shropshirebirder.co.uk

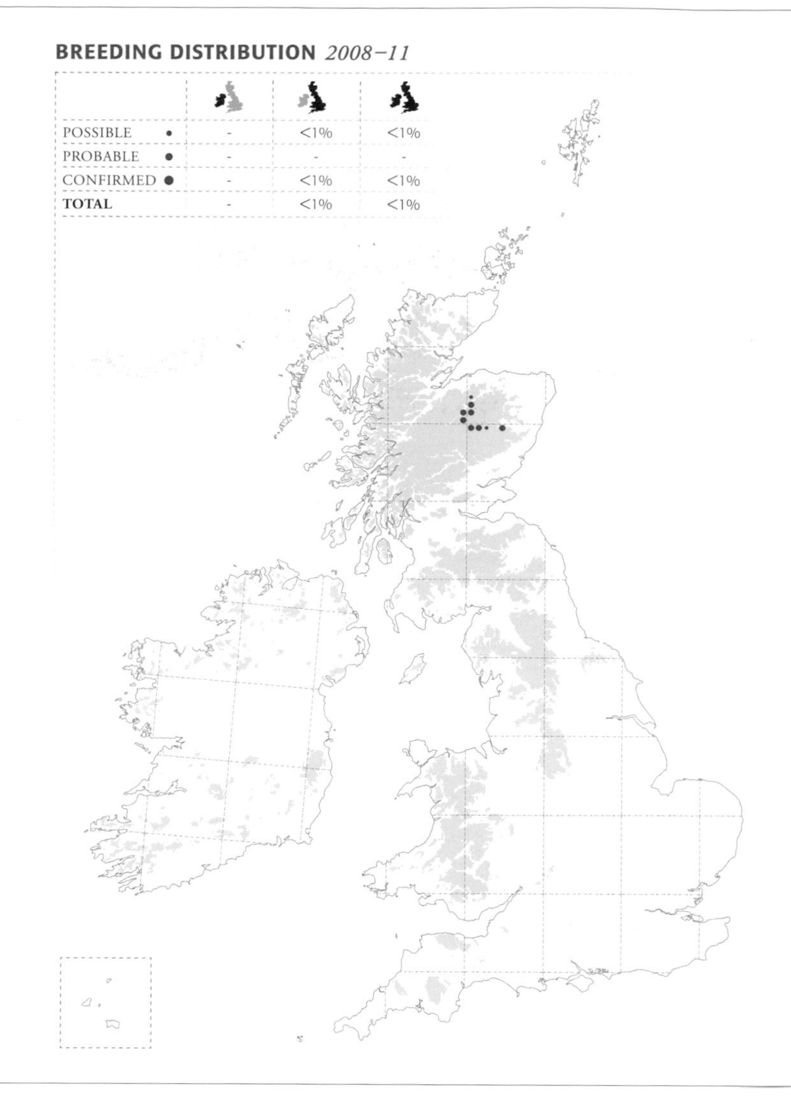

BREEDING DISTRIBUTION *2008–11*

POSSIBLE •	-	<1%	<1%
PROBABLE ●	-	-	-
CONFIRMED ●	-	<1%	<1%
TOTAL	-	<1%	<1%

A RARE RESIDENT breeder, the Parrot Crossbill is found mainly in ancient native pinewoods in northern and northeastern Scotland but also in large plantations of Scots Pine (Marquiss & Rae 2002; Summers 2002). Due to the problems of separating this species from the Scottish Crossbill, it is difficult to get an accurate picture of its breeding distribution. During the breeding season, Parrot Crossbills were recorded from nine 10-km squares, all within Strathspey or Deeside. A survey conducted in spring 2008 estimated a population of 131 birds. The confidence interval around this figure was very wide, however, but no birds were recorded outside the known breeding areas (Summers & Buckland 2011). There were records from just five 10-km squares in winter, all in Scotland.

COMMON ROSEFINCH
Carpodacus erythrinus BTO CODE: SQ

GARTH PEACOCK www.garthpeacock.co.uk

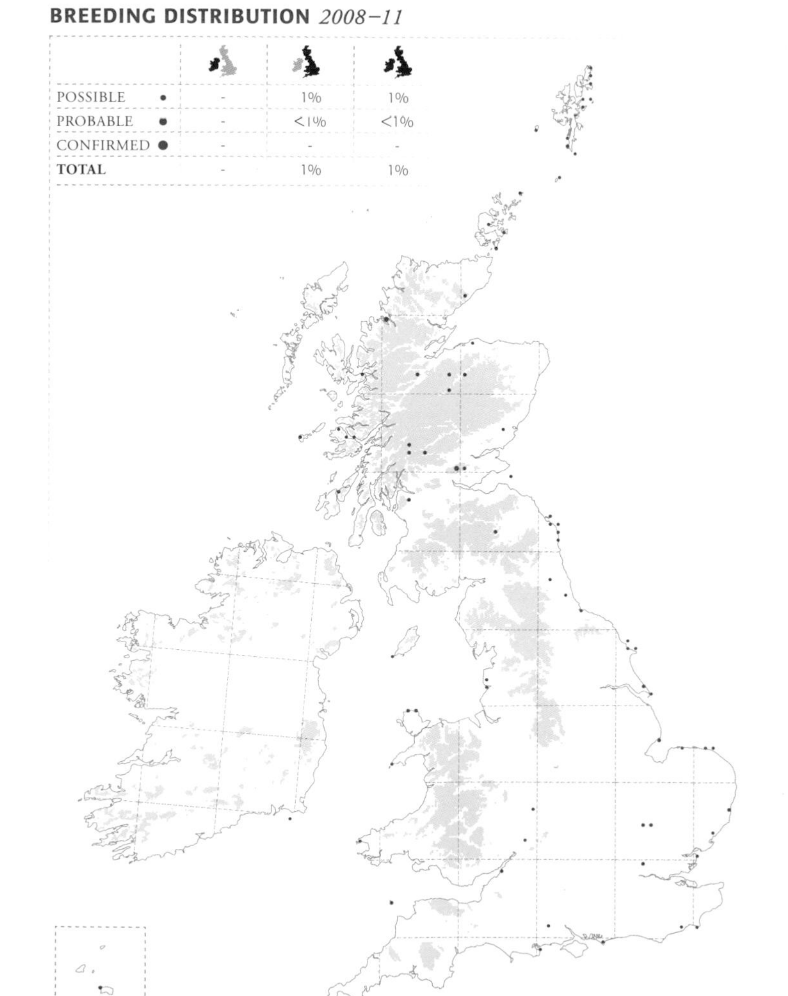

BREEDING DISTRIBUTION *2008–11*

POSSIBLE •	-	1%	1%
PROBABLE ●	-	<1%	<1%
CONFIRMED ●	-	-	-
TOTAL	-	1%	1%

FIRST RECORDED BREEDING in Britain in 1982, in Highland, the Common Rosefinch has bred irregularly since then, with the latest confirmed nesting being in Cumbria in 2001 (Holling *et al.* 2010b). Birds continue to arrive each spring, though these are mainly migrants, remaining for at most a few days. Singing males are occasionally recorded over periods of weeks. During 2008–11, *possible* breeding was recorded from 25 10-km squares and *probable* breeding in two squares. Squares with breeding evidence were widely scattered, with a slight preponderance towards Scotland. They represent a substantial increase from the five squares occupied during the *1988–91 Breeding Atlas.*

There were records from eight 10-km squares during the winter months, five of which reported sightings only in November and presumably therefore of late migrants.

BULLFINCH

Pyrrhula pyrrhula

BTO CODE: BF

ALISON MCARTHUR

THE BULLFINCH IS a relatively sedentary species in Britain & Ireland, with a wide breeding distribution. Densities are highest in lowland wooded landscapes, particularly in Ireland, and lowest in large conurbations and in upland areas. Bullfinches are absent from open country with few trees, such as around the Wash, in exposed coastal regions in western Ireland, and on many Scottish islands.

The overall extent of the breeding range has changed little since the *1968–72 Breeding Atlas*, but there has been some regional variation. Recent gains in western Ireland, on some Inner Hebridean islands and in northern and western Scotland may reflect an increase in tree planting in these exposed locations. Losses are associated mostly with upland edges or with coastal areas.

Bullfinches declined steeply during 1977–82, more so on farmland than in woodland. The decline has eased since the mid 1980s, with evidence of an upturn since 2000, but UK numbers are still 39% lower than in 1967 (*BBS Report 2011*; *BirdTrends*). Since 1988–91, abundance has declined throughout southeast England and northeast Scotland but increased elsewhere, most noticeably in Ireland, where numbers in the Republic of Ireland increased by 39% between 1998 and 2010 (*CBS Trend 2010*).

Patterns of winter distribution and relative abundance are similar to those in the breeding season. The exceptions are in upland areas, where more squares were occupied in winter owing to birds dispersing onto moors to feed on heather seeds, and on the Northern Isles, where migrants arrive, albeit irregularly, from Fennoscandia and beyond (*Migration Atlas*). A 9% range expansion since the *1981–84 Winter Atlas* is apparent, mostly in the north and west. Although some gains are associated with the colonisation of new breeding areas, many are undoubtedly explained by the improvements in coverage, leading to better detection where Bullfinches occur at low density.

Causes of decline potentially include deteriorating habitat quality, constraints on survival outside the breeding season, and the impact of Sparrowhawks on the ability of Bullfinches to exploit resources in some habitats (Proffitt *et al.* 2004; Marquiss 2007). Analyses of survival and breeding success were unable to identify clear demographic mechanisms underlying the decline (Siriwardena *et al.* 1999, 2001).

Sponsored by Nicholas Donnithorne FLS FRES

BREEDING DISTRIBUTION *2008–11*

POSSIBLE •	6%	6%	6%
PROBABLE •	36%	22%	26%
CONFIRMED ●	47%	52%	51%
TOTAL	89%	80%	83%

WINTER DISTRIBUTION *2007/08–2010/11*

PRESENT ●	88%	83%	84%

BREEDING DISTRIBUTION CHANGE *since 1968–72*

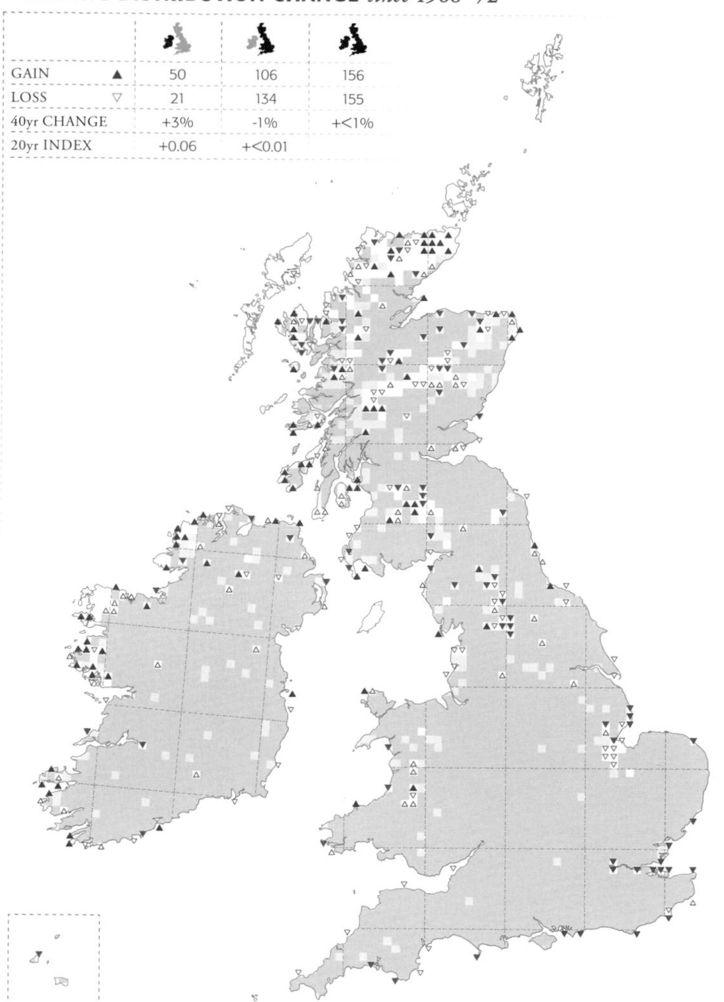

GAIN ▲	50	106	156
LOSS ▽	21	134	155
40yr CHANGE	+3%	-1%	+<1%
20yr INDEX	+0.06	+<0.01	

WINTER DISTRIBUTION CHANGE *since 1981–84*

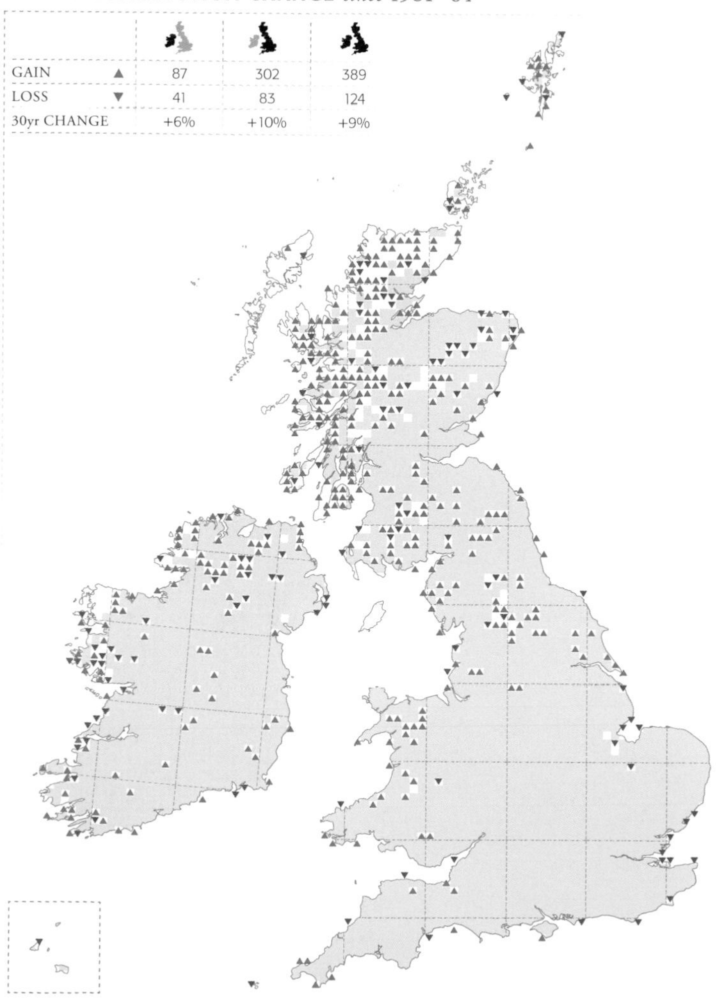

GAIN ▲	87	302	389
LOSS ▼	41	83	124
30yr CHANGE	+6%	+10%	+9%

BREEDING RELATIVE ABUNDANCE *2008–11*

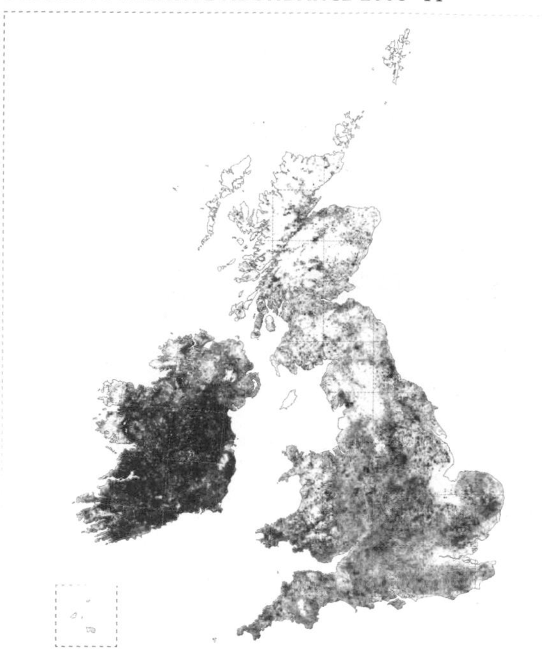

BREEDING RELATIVE ABUNDANCE CHANGE *since 1988–91*

WINTER RELATIVE ABUNDANCE *2007/08–2010/11*

HAWFINCH

Coccothraustes coccothraustes BTO CODE: HF

MARTIN BENNETT

THE HAWFINCH IS an increasingly scarce breeder with a highly fragmented breeding distribution in Britain, associated with larger tracts of mature broad-leaved and mixed woodland; it is more than three times more widespread in winter. In Ireland it is an irregular winter visitor (*Birds in Ireland*).

Over three quarters of the former breeding range in Britain has been lost since 1968–72, with the majority of the deficit arising since the *1988–91 Breeding Atlas*. Losses are most concentrated in southeast England but extend into southeast Scotland. Despite a lack of annual monitoring data, there is compelling evidence of a major breeding population decline. Based on bird-report data, Langston *et al.* (2002) estimated a decline of c.40% between the mid 1980s and late 1990s. Breeding-season numbers in a sample of woods dropped by 74% between the mid 1980s and 2003–04 (Hewson *et al.* 2007). Subsequently, Clements (2013) has produced a new breeding population estimate of 500–1,000 pairs, a major reduction from the 3,000–6,500 pairs estimated from the *1988–91 Breeding Atlas*. The causes of this strong decline, which contrasts sharply with stability in mainland Europe (*PECBMS 2010*), are currently unknown.

The relatively few gains have occurred in the Forest of Dean and in Wales, although Smith (2004) suspected that the population of c.50 pairs in Merioneth had been overlooked. These areas, together with parts of northwest England and the New Forest, now hold the largest concentrations of wintering and breeding birds.

Breeding trends contrast with a 28% winter range expansion in Britain since the *1981–84 Winter Atlas*, with gains most prevalent along the southwestern fringe of the range. Explanations mooted for this disparity include changes in dispersion patterns or habits that have made Hawfinches harder to detect in the breeding season. Although they are certainly inconspicuous, especially when breeding at low density, there is no evidence to suggest any behavioural changes. Alternatively, increased immigration from the Continent may recently have bolstered the winter population. Though there are exchanges of ringed birds between Britain and Scandinavia (Calladine & Morrison 2010; Robinson & Clark 2012), the scale of immigration is unclear, and any temporal trends are not yet understood.

Sponsored by Herts Bird Club

BREEDING DISTRIBUTION *2008–11*

POSSIBLE ●	-	2%	1%
PROBABLE ●	-	1%	1%
CONFIRMED ●	-	1%	1%
TOTAL	-	4%	3%

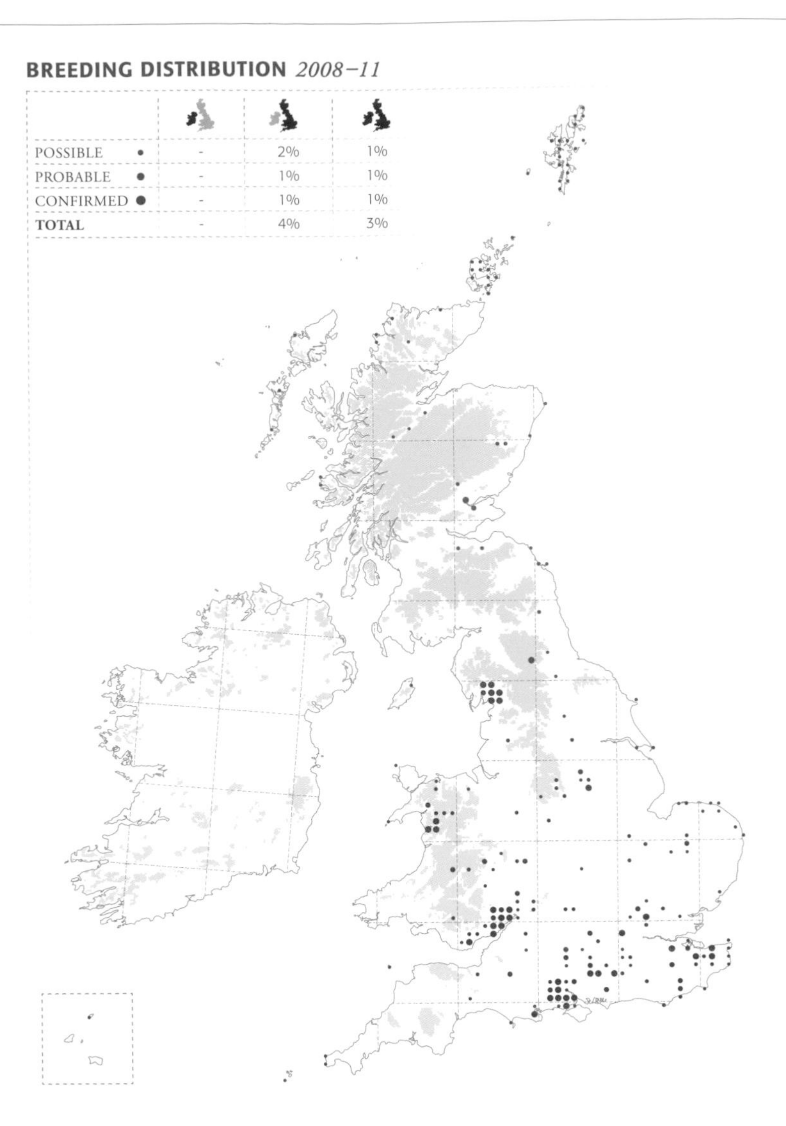

WINTER DISTRIBUTION *2007/08–2010/11*

PRESENT ●	<1%	12%	9%

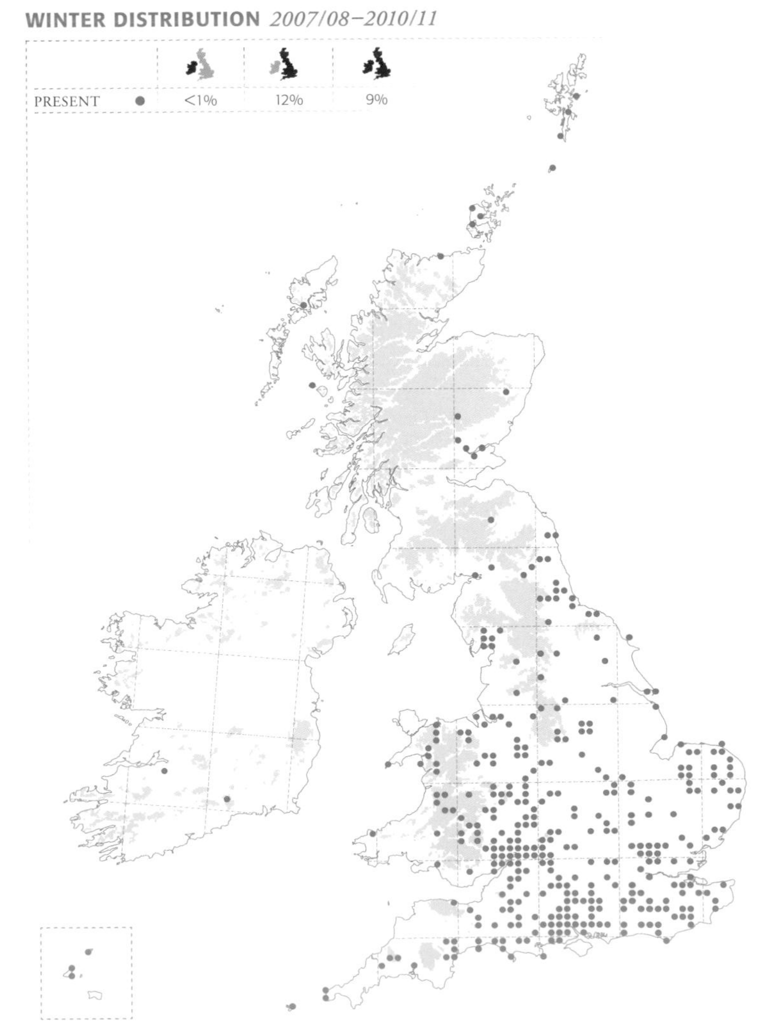

BREEDING DISTRIBUTION CHANGE *since 1968–72*

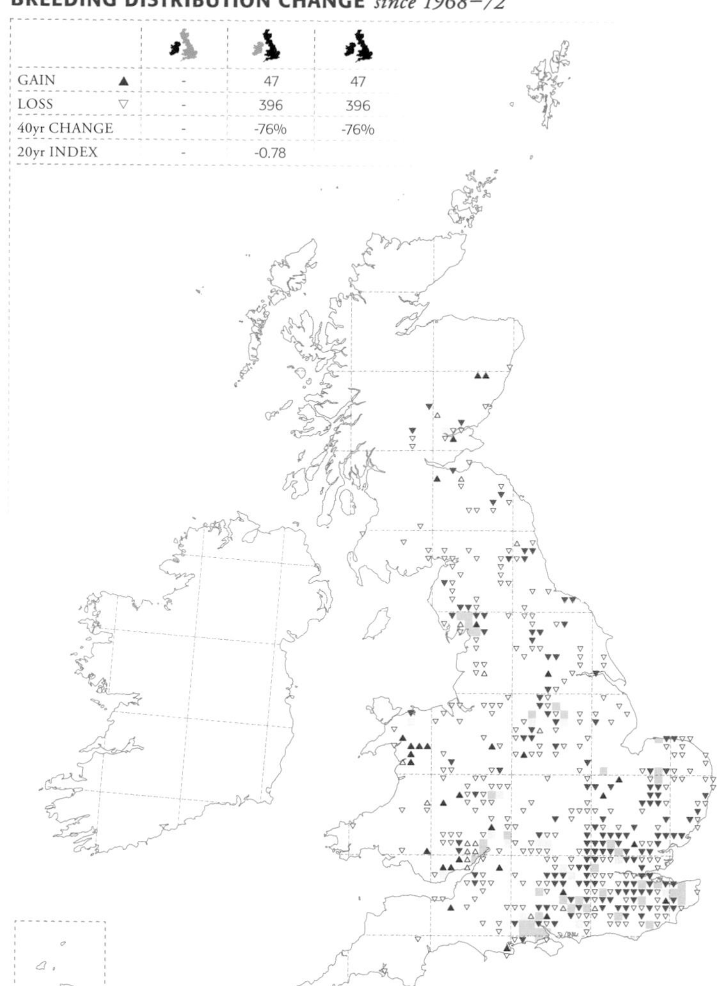

GAIN ▲	-	47	47
LOSS ▽	-	396	396
40yr CHANGE	-	-76%	-76%
20yr INDEX	-	-0.78	

BREEDING RELATIVE ABUNDANCE *2008–11*

WINTER DISTRIBUTION CHANGE *since 1981–84*

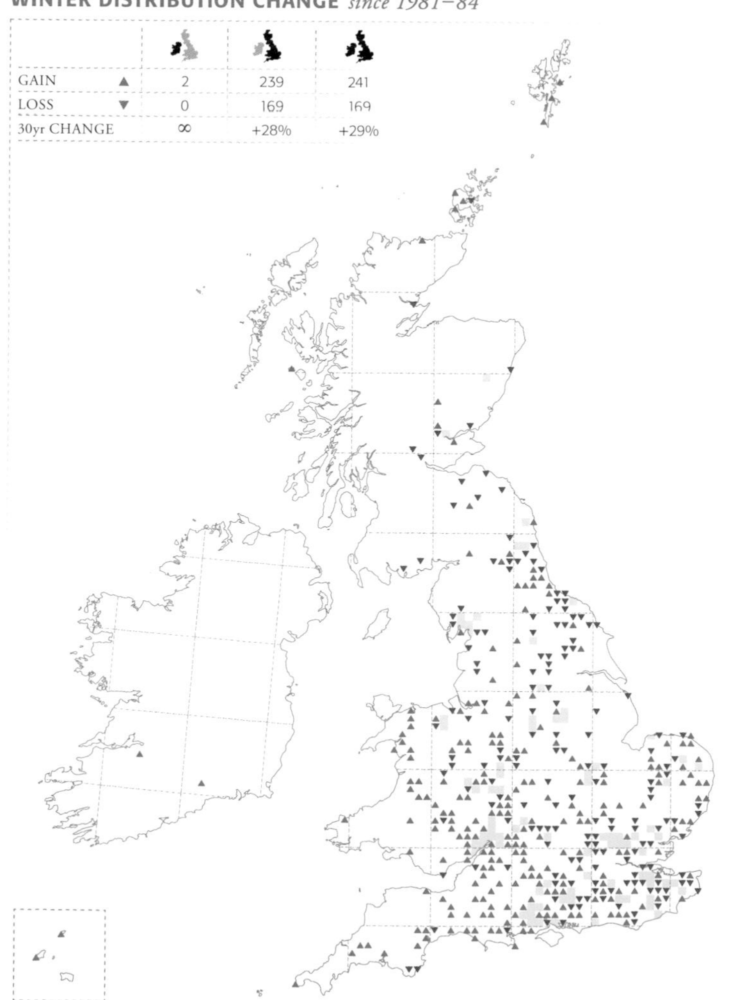

GAIN ▲	2	239	241
LOSS ▼	0	169	169
30yr CHANGE	∞	+28%	+29%

WINTER RELATIVE ABUNDANCE *2007/08–2010/11*

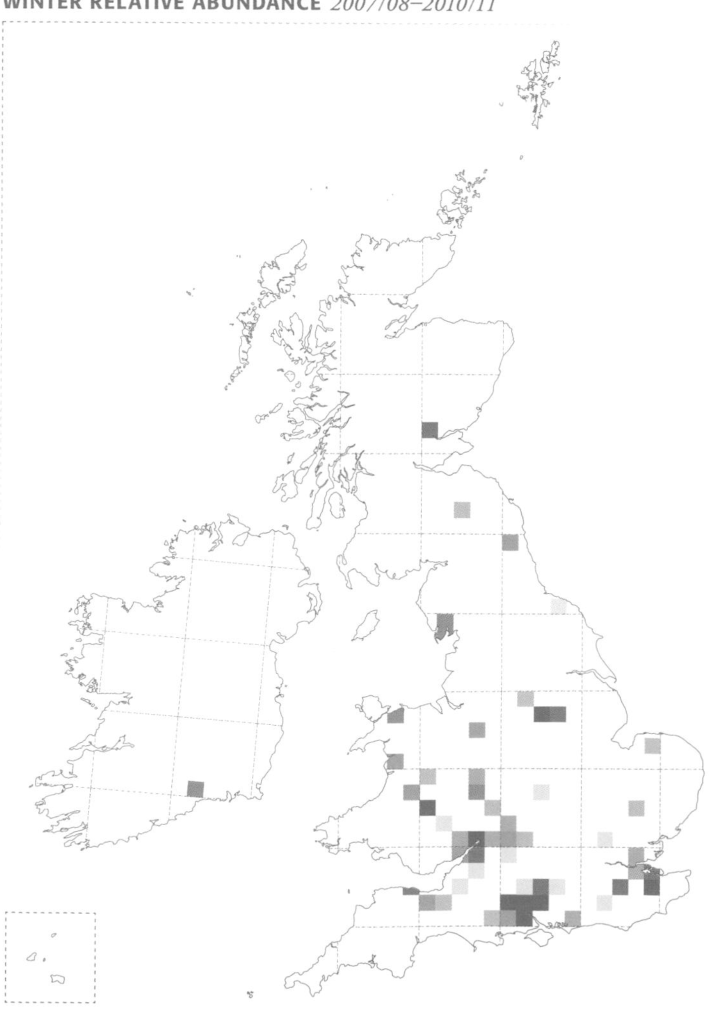

SNOW BUNTING
Plectrophenax nivalis
BTO CODE: SB

PETER D SMITH

A SMALL RELICT population of Snow Buntings still breeds on the high tops of mountains in the Cairngorms and northwest Highlands. It consists mostly of the Icelandic subspecies *insulae* but with small numbers also of nominate *nivalis* (Smith 1996), whose breeding range stretches around the Arctic from Alaska eastward to Scandinavia and northwest Russia (*BWP*).

During 2008–11, breeding was *confirmed* in 11 10-km squares. Atlas data suggest that the current range is substantially larger than in the *1968–72 Breeding Atlas* but smaller than in the *1988–91 Breeding Atlas*. Occupancy figures are likely to be heavily influenced by the intensity of coverage on the high tops inhabited by this species, but other data confirm this trend (Watson & Smith 1991; *Birds of Scotland*). The first national survey of breeding Snow Buntings was conducted in 2011 and found a minimum of 79 possible territories, of which c.80% were in the central Cairngorms area (D. Hayhow pers comm).

The tiny resident population is swamped in winter by visitors from northern breeding grounds. The majority of wintering birds are of Icelandic origin, but *nivalis* individuals from Scandinavia form an increasing proportion of flocks further south, particularly late in the winter when many birds cross the North Sea (North West Norfolk Ringing Group 2013; P. Atkinson pers comm). Large numbers winter throughout the Highlands, with smaller numbers in the uplands of northern England. Otherwise, Snow Buntings are widespread as a coastal wintering species, with the highest coastal densities occurring in the Northern Isles and Outer Hebrides, along the North Sea coast from the Tees to Norfolk, and in northwest Ireland. There is a scattering of inland records, many involving birds on passage in November.

Since the *1981–84 Winter Atlas* there has been a 34% increase in the number of occupied 10-km squares. Winter gains are particularly evident in the uplands, those on the coast being mainly on the margins of the range in southern and southwest England, Wales and Ireland. These trends are in contrast to local declines in wintering numbers in Scotland, as might result were more birds to remain in Iceland, or where agricultural intensification has reduced the suitability of farmland (*Birds of Scotland*).

Sponsored by Scottish Environment Protection Agency (SEPA)

WINTER DISTRIBUTION *2007/08–2010/11*

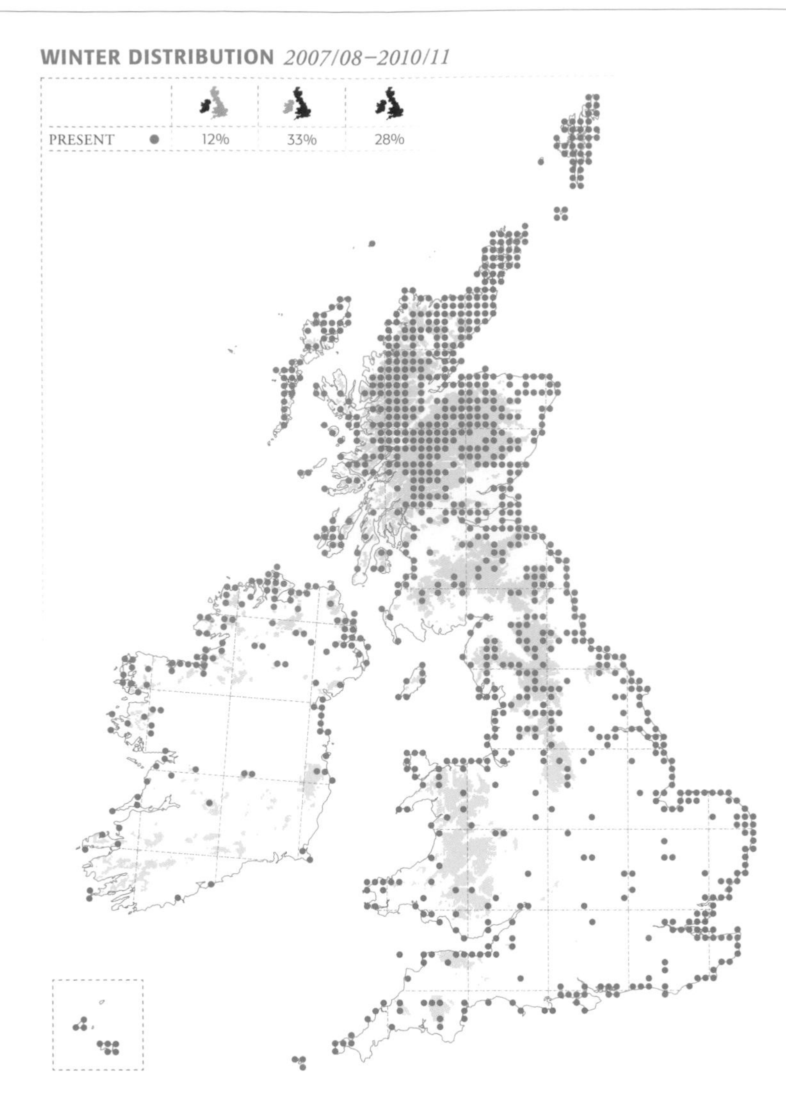

PRESENT •	12%	33%	28%

BREEDING DISTRIBUTION *2008–11*

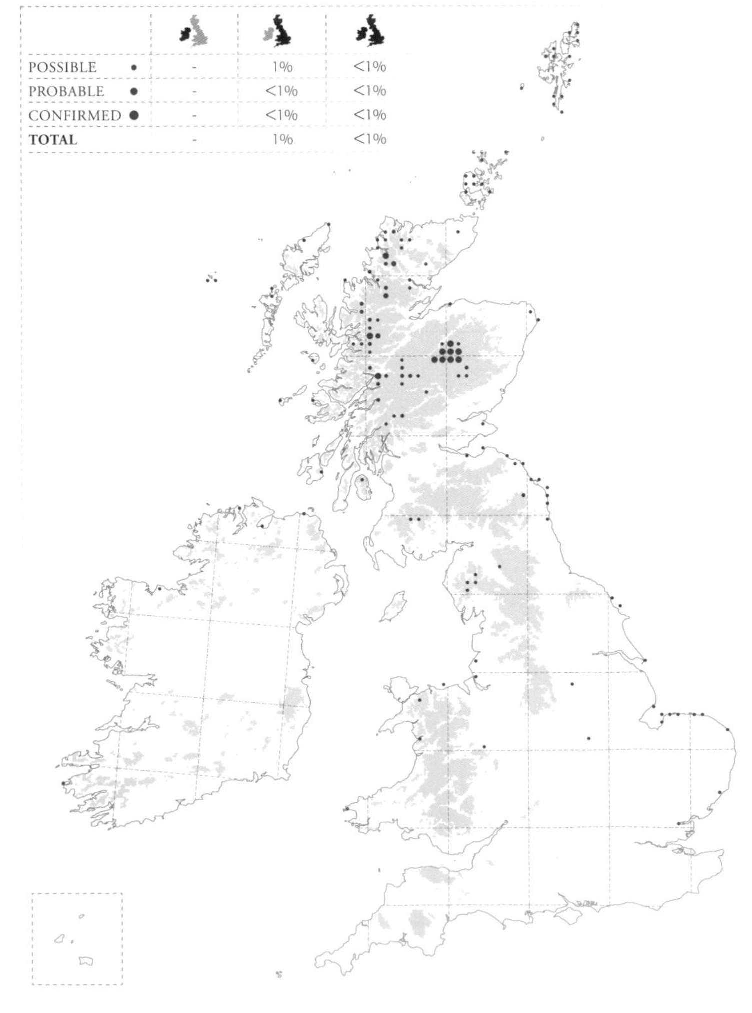

POSSIBLE •	-	1%	<1%
PROBABLE •	-	<1%	<1%
CONFIRMED •	-	<1%	<1%
TOTAL	-	1%	<1%

WINTER DISTRIBUTION CHANGE *since 1981–84*

GAIN ▲	78	446	524
LOSS ▼	45	237	282
30yr CHANGE	+48%	+33%	+34%

WINTER RELATIVE ABUNDANCE *2007/08–2010/11*

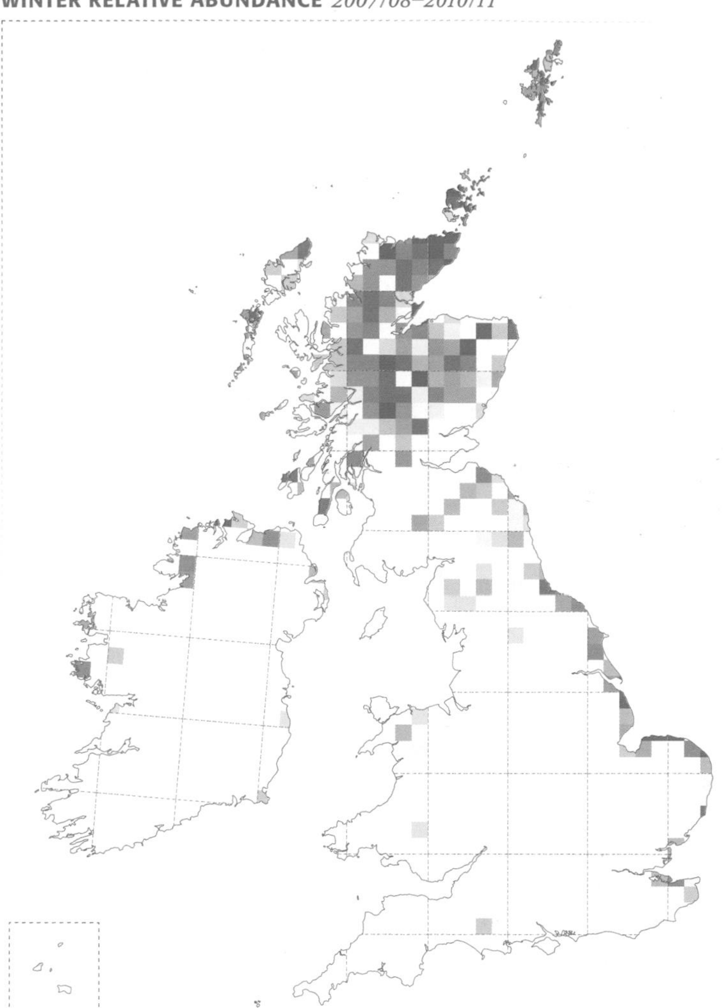

BREEDING DISTRIBUTION CHANGE *since 1968–72*

GAIN ▲	-	25	25
LOSS ▽	-	7	7
40yr CHANGE	-	+129%	+129%
20yr INDEX	-	-	

LAPLAND BUNTING

Calcarius lapponicus

BTO CODE: LA

JIM ALMOND www.shropshirebirder.co.uk

THE LAPLAND BUNTING is a relatively scarce but regular autumn passage migrant and winter visitor. Many arrive from Scandinavia but a variable number may originate from Greenland (*Migration Atlas*).

Most records were from coastal 10-km squares, in particular along the North Sea coast and almost continuously from the Firth of Forth southward to Kent. They were also recorded from many squares along the south coast of England from Kent to Cornwall and on west coasts north to Lancashire, with a scattering further north and around the Irish coast. Lapland Buntings are generally scarce inland. Most inland birds recorded for the Atlas were in central and eastern England, many being present only in early November or in February and probably, therefore, were active migrants.

The distribution recorded during *Bird Atlas 2007–11* represents a 151% range expansion in Britain since the *1981–84 Winter Atlas*, Lapland Buntings being recorded then almost exclusively on the North Sea coast. The gains in Britain are most noticeable on western and southern coasts. In Ireland the number of occupied squares increased from one to 15.

This large increase in the number of occupied squares and the noticeably different distribution can be attributed to a major influx of Lapland Buntings that took place in autumn 2010 (Pennington *et al.* 2012; Rivers & Forsyth 2012). After the first three winters of Atlas fieldwork, Lapland Buntings had been reported in 145 10-km squares. In 2010/11 alone they were reported from 218 squares, 138 of them new squares in Ireland and on western and southern coasts of Britain. Pennington *et al.* (2012) analysed the timing and geographical pattern of arrivals and concluded that the exceptional numbers were most likely to have originated from east Greenland and were a consequence of abundant food enabling high chick production, combined with weather patterns that promoted southeasterly migration.

Lapland Buntings have bred in recent decades in the Scottish Highlands, with breeding confirmed during 1977–80, though single birds have been seen in suitable habitat since (*Birds of Scotland*). During the current atlas period there was only one record in a potential breeding area, of a single bird in suitable habitat in July 2011.

Sponsored by Wincey Willis

WINTER DISTRIBUTION *2007/08–2010/11*

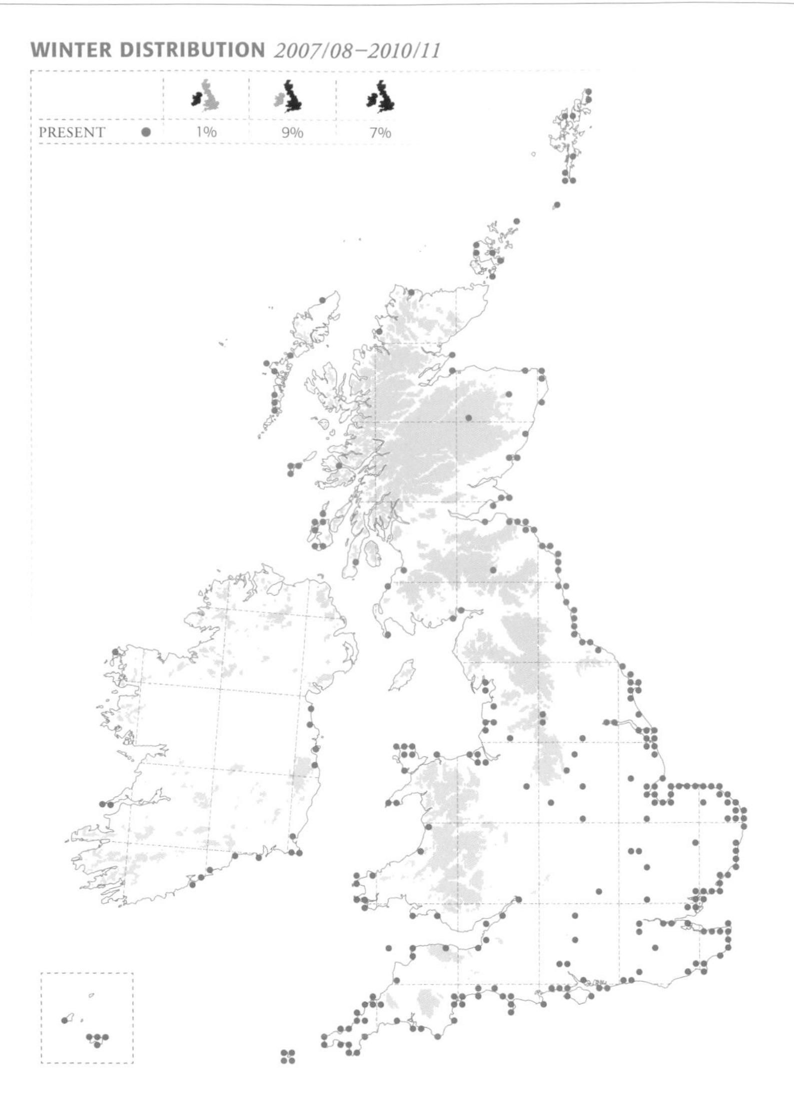

PRESENT ●	1%	9%	7%

BREEDING DISTRIBUTION *2008–11*

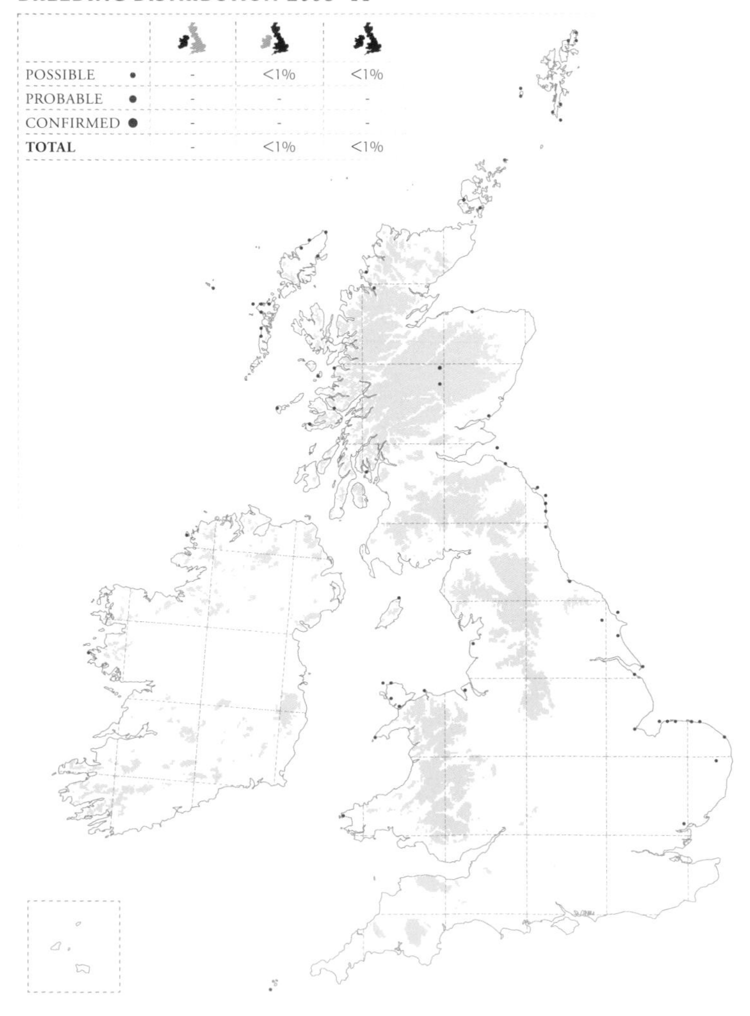

POSSIBLE •	-	<1%	<1%
PROBABLE ●	-	-	-
CONFIRMED ●	-	-	-
TOTAL	-	<1%	<1%

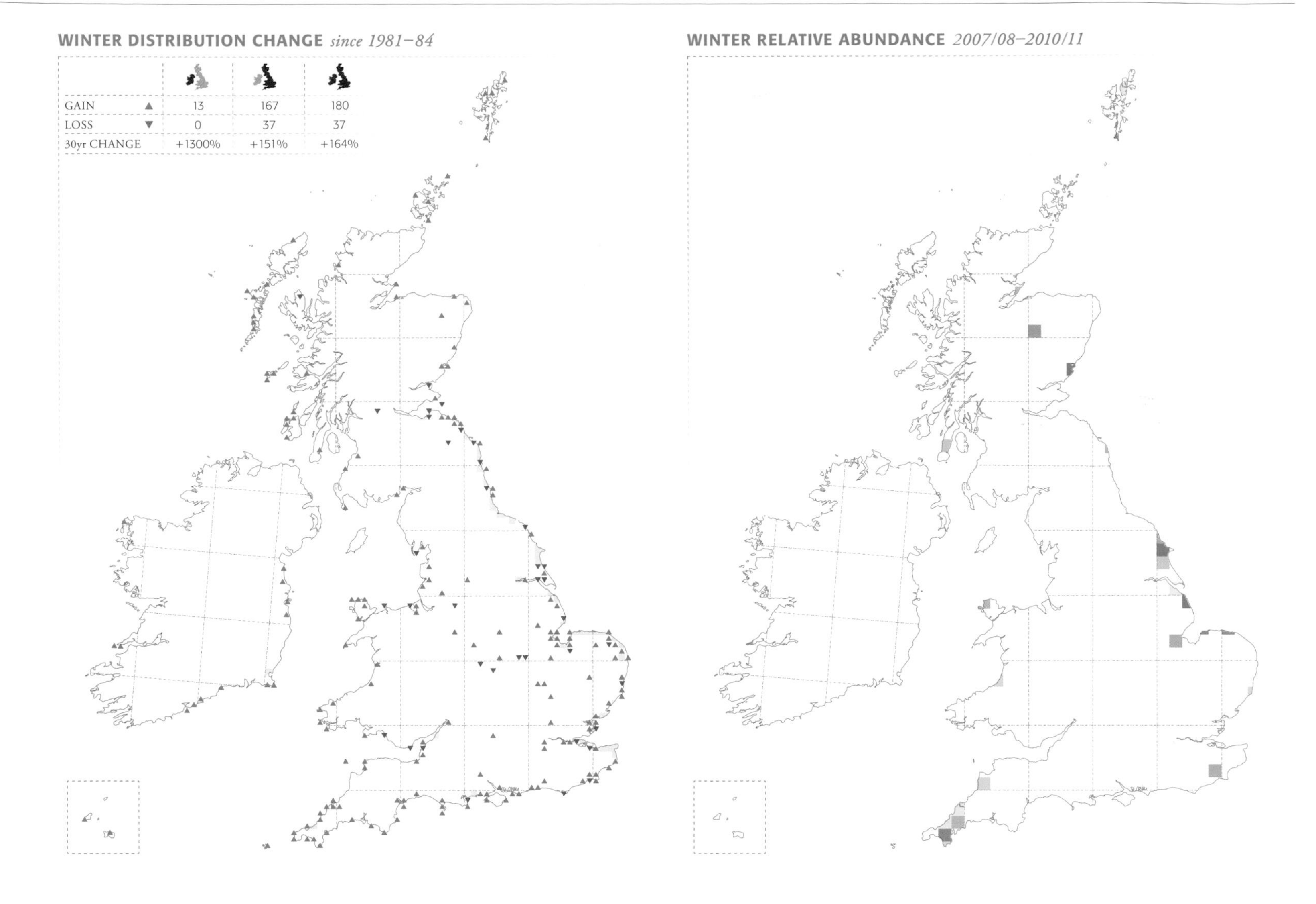

GAIN ▲	13	167	180
LOSS ▼	0	37	37
30yr CHANGE	+1300%	+151%	+164%

◀ **During winter, Lapland Buntings may be found in a range of habitats including extensive saltmarsh, shingle, sand, short grassy turf and stubble fields. They are often in the company of Skylarks or finches.**

YELLOWHAMMER

Emberiza citrinella BTO CODE: Y.

ALISON MCARTHUR

WINTERING YELLOWHAMMERS ARE largely confined to low-lying areas of mainland Britain and southern and eastern Ireland. They rarely breed in the Outer Hebrides or Northern Isles (*Birds of Scotland*). The breeding-season distribution for this largely sedentary species is almost identical to that in winter. Densities are generally highest in eastern Britain and are lower in Ireland, even in the southeast.

The British winter range has contracted by 9% since the *1981–84 Winter Atlas*, with birds withdrawing mostly from upland fringes. In Ireland, however, a far more substantial 38% of the former range has been lost, mainly from the north, the west and parts of the midlands. These losses are part of a longer-term decline, as confirmed by the breeding-season change map which shows a 61% range contraction in Ireland since the *1968–72 Breeding Atlas*. The remnant population in southeast Ireland is still declining (Crowe *et al.* 2010b).

In Britain there have also been declines, with a range contraction of 21% since the *1968–72 Breeding Atlas* and a 55% decline in breeding numbers during 1970–2010 (*SUKB 2012*). Losses have gradually accumulated in the west and in the Pennines but, significantly, the abundance change map indicates that densities have decreased throughout the remaining range since the *1988–91 Breeding Atlas*.

Yellowhammers are closely associated with cereal farming (Kyrkos *et al.* 1998) and require a minimum amount of cereal in the landscape for a population to be maintained, possibly because intensively managed grassland supports few invertebrates and weeds, or because dense swards reduce access to resources (Perkins *et al.* 2000). The almost complete loss of arable farming from northern and western Ireland over the last 40 years probably explains the observed Irish decline.

In Britain, declining overwinter survival, owing to reductions in winter seed availability, has contributed to the declines, and these can be mitigated or reversed by increasing the availability of weedy stubbles or by supplementary feeding (Gillings *et al.* 2005; Siriwardena *et al.* 2007). Losses of Yellowhammers from upland margins may be related to the maturing of plantations, ongoing specialisation in livestock, and the loss of nesting habitat through the management of land to improve agricultural productivity (Fuller *et al.* 2006; *Birds of Scotland*).

Sponsored by SongBird Survival

BREEDING DISTRIBUTION *2008–11*

POSSIBLE •	9%	7%	7%
PROBABLE •	12%	13%	13%
CONFIRMED ●	15%	47%	39%
TOTAL	36%	67%	59%

WINTER DISTRIBUTION *2007/08–2010/11*

PRESENT ●	29%	65%	56%

BREEDING DISTRIBUTION CHANGE *since 1968–72*

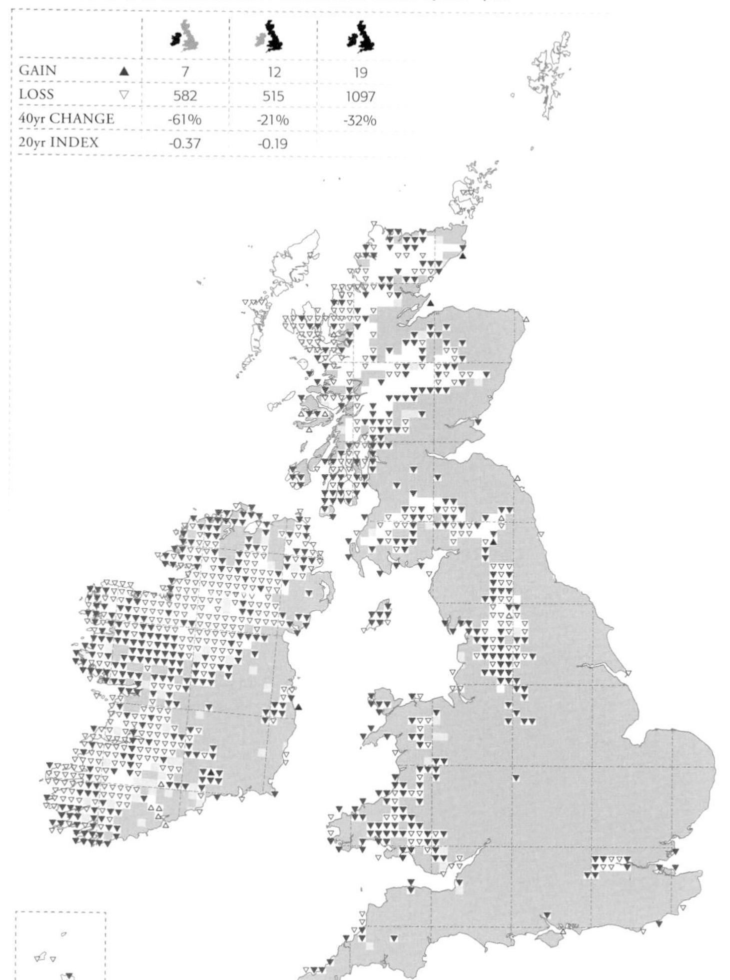

GAIN ▲	7	12	19
LOSS ▽	582	515	1097
40yr CHANGE	-61%	-21%	-32%
20yr INDEX	-0.37	-0.19	

WINTER DISTRIBUTION CHANGE *since 1981–84*

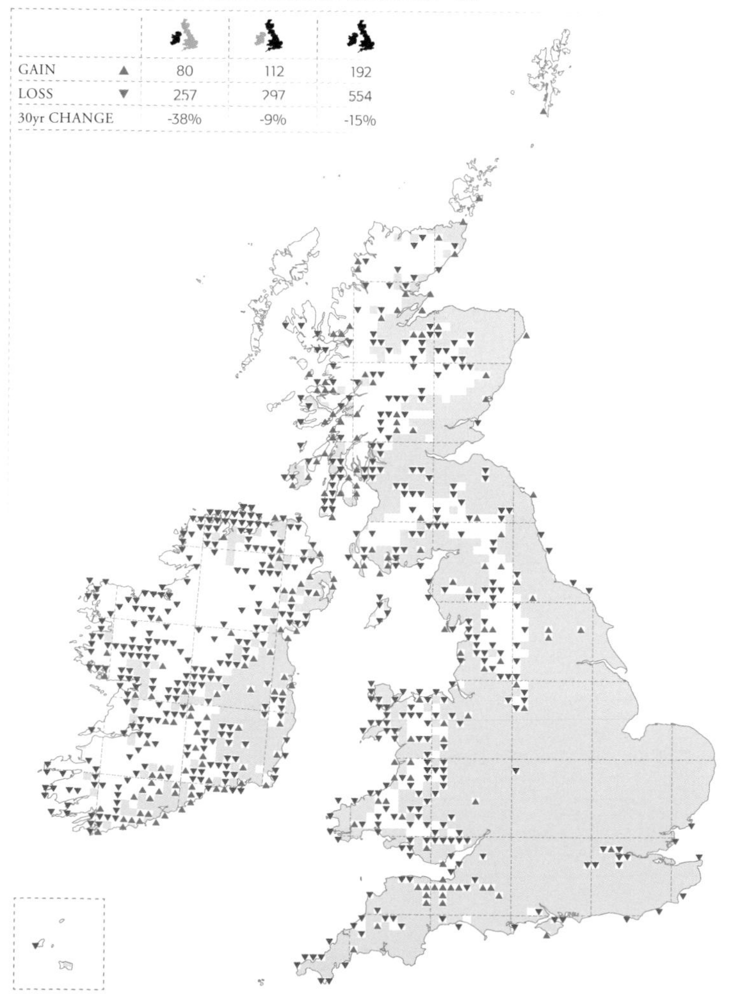

GAIN ▲	80	112	192
LOSS ▼	257	297	554
30yr CHANGE	-38%	-9%	-15%

BREEDING RELATIVE ABUNDANCE *2008–11*

BREEDING RELATIVE ABUNDANCE CHANGE *since 1988–91*

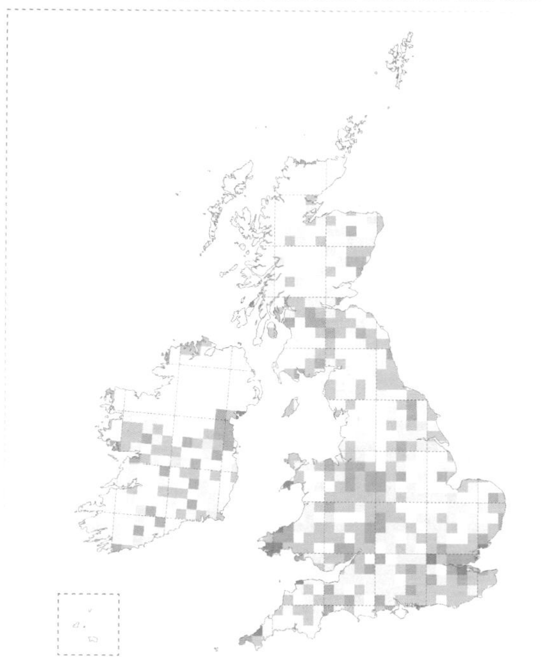

WINTER RELATIVE ABUNDANCE *2007/08–2010/11*

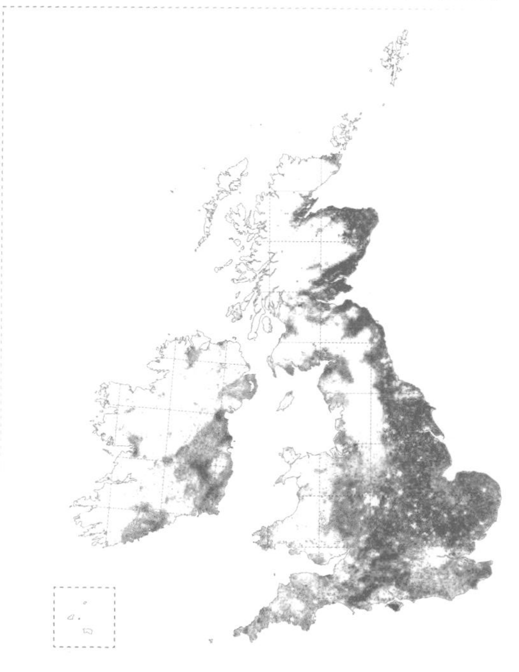

CIRL BUNTING

Emberiza cirlus

BTO CODE: CL

CHRIS BALE www.guernseybirdnerd.com

FORMERLY WIDESPREAD ACROSS southern England and Wales, and a resident also of the Channel Islands, the Cirl Bunting is currently confined to a narrow coastal strip of south Devon and to that part of south Cornwall to which it is being reintroduced (Stanbury *et al.* 2010). In the early 1990s, 19 pairs bred on Jersey (Wotton *et al.* 2000) but breeding in the Channel Islands subsequently ceased and in 2011 *probable* breeding was recorded in just one 10-km square. There is just one record ever for Ireland, in May 2006 (Milne & McAdams 2008). At their zenith, from the late 1800s to around 1930, Cirl Buntings were resident in 39 British counties north to North Wales and Yorkshire but long-term decline and accelerating range contraction took the population to a low point of around 118 territories in 1989, all but four of these in Devon (Evans 1992). Comparison with the *1968–72 Breeding Atlas* documents the final stages of this withdrawal to south Devon.

The species has been the subject of intensive research and targeted conservation measures within Devon, led by RSPB and Natural England. Its marked preference for winter foraging on stubble or fallow fields was quickly discovered, leading to the suggestion that loss of winter stubbles during the 1970s may have been a major factor contributing to the decline (Evans & Smith 1994). Within the remnant range, the provision, through agri-environment schemes, of breeding and wintering habitat designed specifically for Cirl Bunting has benefited both its target species and a range of other taxa (MacDonald *et al.* 2012). Cirl Bunting numbers increased by 83% during 1992–98 in these managed areas but by only 2% elsewhere (Peach *et al.* 2001).

Successive surveys have charted the population's recovery to 862 pairs in 2009 (Stanbury *et al.* 20010). Despite this steep rise, and a 15% increase in occupied tetrads between 2003 and 2009 (Stanbury *et al.* 2010), there is no sign of Cirl Buntings returning naturally to any of their lost ground outside south Devon. Recent cool, wet summers and reductions in the availability of fallow and stubble habitats following the loss of set-aside nationally in 2007 might have limited their potential for recolonisation.

Sponsored by RSPB

BREEDING DISTRIBUTION *2008–11*

POSSIBLE •	-	<1%	<1%
PROBABLE •	-	<1%	<1%
CONFIRMED •	-	1%	<1%
TOTAL	-	1%	<1%

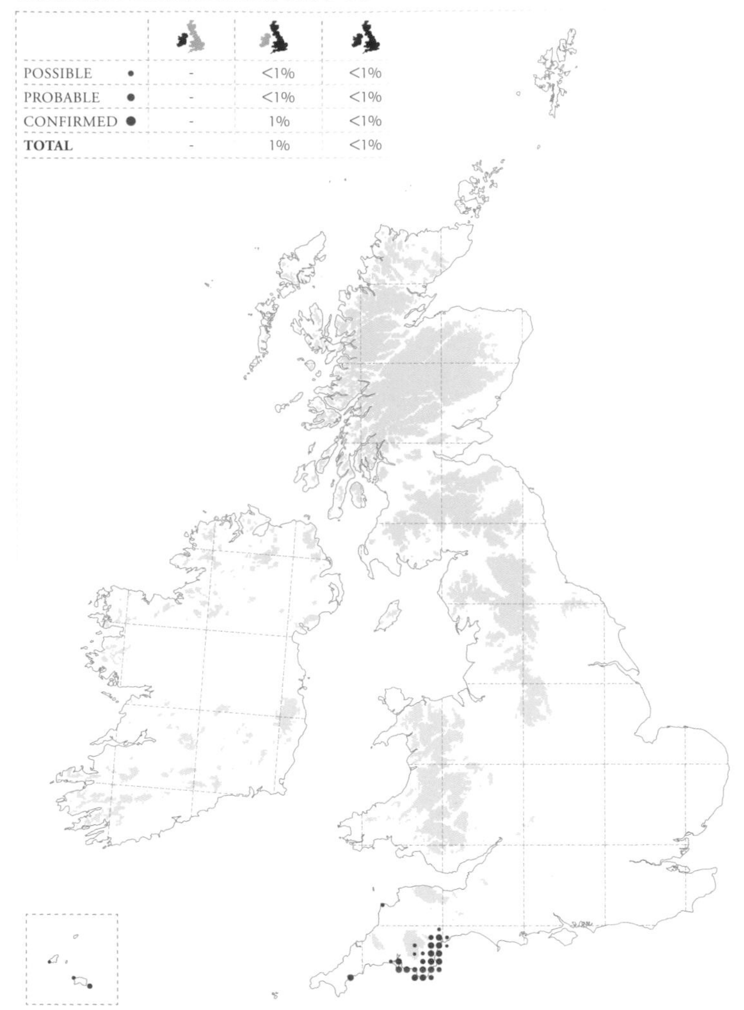

WINTER DISTRIBUTION *2007/08–2010/11*

PRESENT •	-	1%	1%

BREEDING DISTRIBUTION CHANGE *since 1968–72*

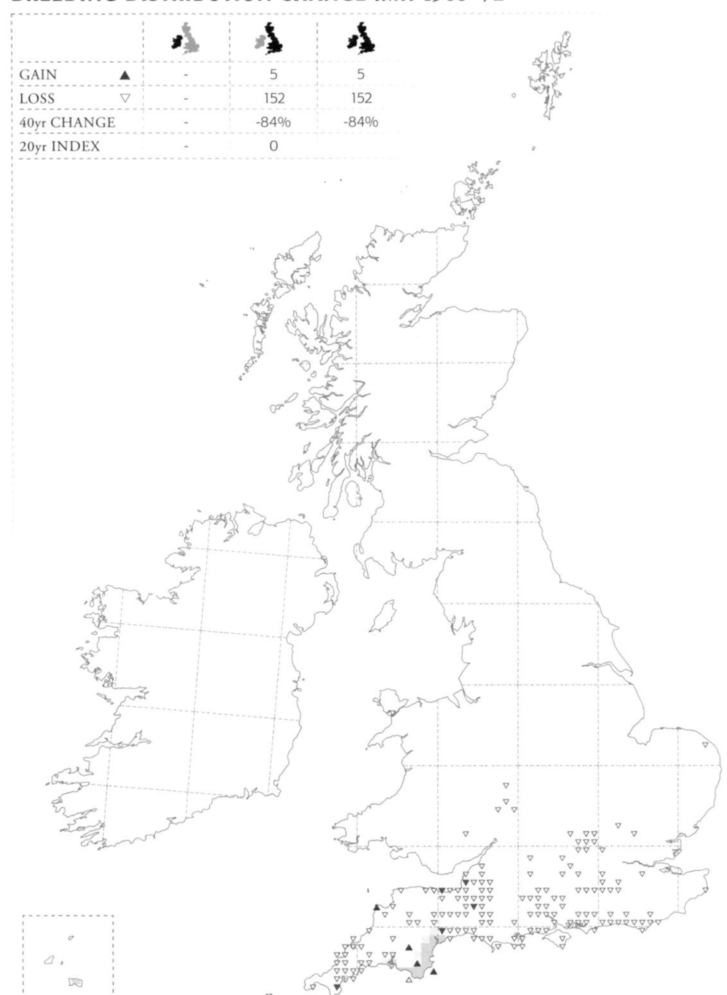

GAIN ▲	-	5	5
LOSS ▽	-	152	152
40yr CHANGE	-	-84%	-84%
20yr INDEX	-	0	

BREEDING RELATIVE ABUNDANCE *2008–11*

WINTER DISTRIBUTION CHANGE *since 1981–84*

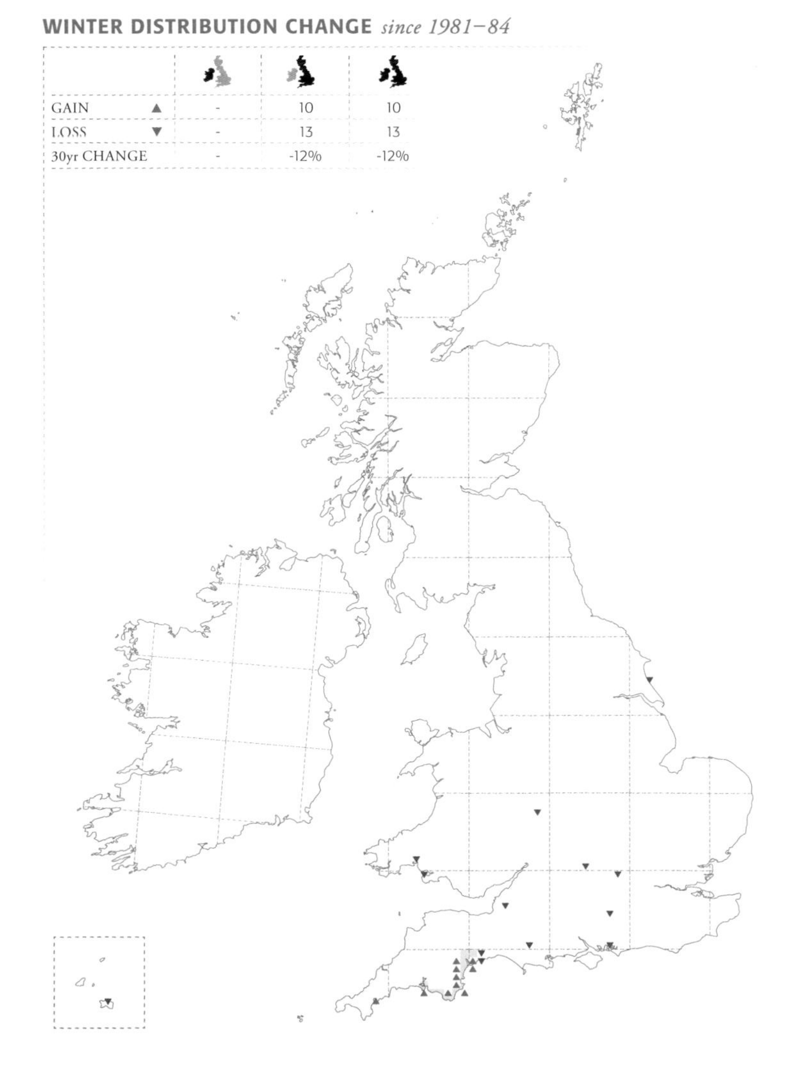

GAIN ▲	-	10	10
LOSS ▼	-	13	13
30yr CHANGE	-	-12%	-12%

WINTER RELATIVE ABUNDANCE *2007/08–2010/11*

REED BUNTING

Emberiza schoeniclus

BTO CODE: RB

JOHN FLOWERDAY

BREEDING REED BUNTINGS are widely distributed across Britain and Ireland. Gaps in the distribution are evident in the more barren uplands of northern Scotland, including Shetland and some Hebridean islands. High-density areas have moved little since the *1988–91 Breeding Atlas*, being associated mostly with large lowland vales and plains.

The breeding distribution has been largely stable since the *1968–72 Breeding Atlas*, despite a major population increase and subsequent decline in the 1970s (*BirdTrends*). In marginal, low-density areas, particularly in northwest Scotland, a mixture of gains and losses probably reflects variation in detection between atlases, though losses in the Outer Hebrides may be real. Reed Buntings have been absent from Shetland since the late 1990s (*Birds of Scotland*). The losses from southeast Ireland apparent by 1988–91 have since been largely reversed, accompanied by modest increases in abundance throughout the island (Crowe *et al.* 2010b).

Increases in abundance are apparent also in Britain, mainly in eastern and northeast England and in southern Scotland, whilst losses are concentrated into southeast England. These changes are consistent with the results of annual monitoring in the UK, where there has been a 24% increase overall during 1995–2010 but a 21% decline in southeast England (*BBS Report 2011*). Earlier declines on farmland were linked to decreased overwinter survival, most likely through reduced food availability (Peach *et al.* 1999). In some arable areas, however, high densities are now associated with Oil-seed Rape (Burton *et al.* 1999; Gruar *et al.* 2006) or with Elephant Grass biomass crops (Bright *et al.* 2013).

Reed Buntings are mostly sedentary (*Migration Atlas*; *Birds of Scotland*) and distribution and abundance patterns in winter and the breeding season differ only where birds abandon upland areas in winter. A small number of continental immigrants arrive in autumn and these account for the Atlas records on Shetland, almost all of which were in early November. Winter gains, mostly in the north and west and around upland margins, amount to a 23% expansion of the winter range. These may reflect wintering at higher elevations during recent mild winters, although improved coverage may be a factor in some, more remote regions.

Sponsored by Lesley J. Nickell

BREEDING DISTRIBUTION *2008–11*

POSSIBLE •	18%	10%	12%
PROBABLE ●	28%	19%	21%
CONFIRMED ●	45%	53%	51%
TOTAL	91%	82%	84%

WINTER DISTRIBUTION *2007/08–2010/11*

PRESENT ●	85%	79%	81%

BREEDING DISTRIBUTION CHANGE *since 1968–72*

GAIN ▲	26	118	144
LOSS ▽	53	252	305
40yr CHANGE	-3%	-5%	-5%
20yr INDEX	+0.10	+0.04	

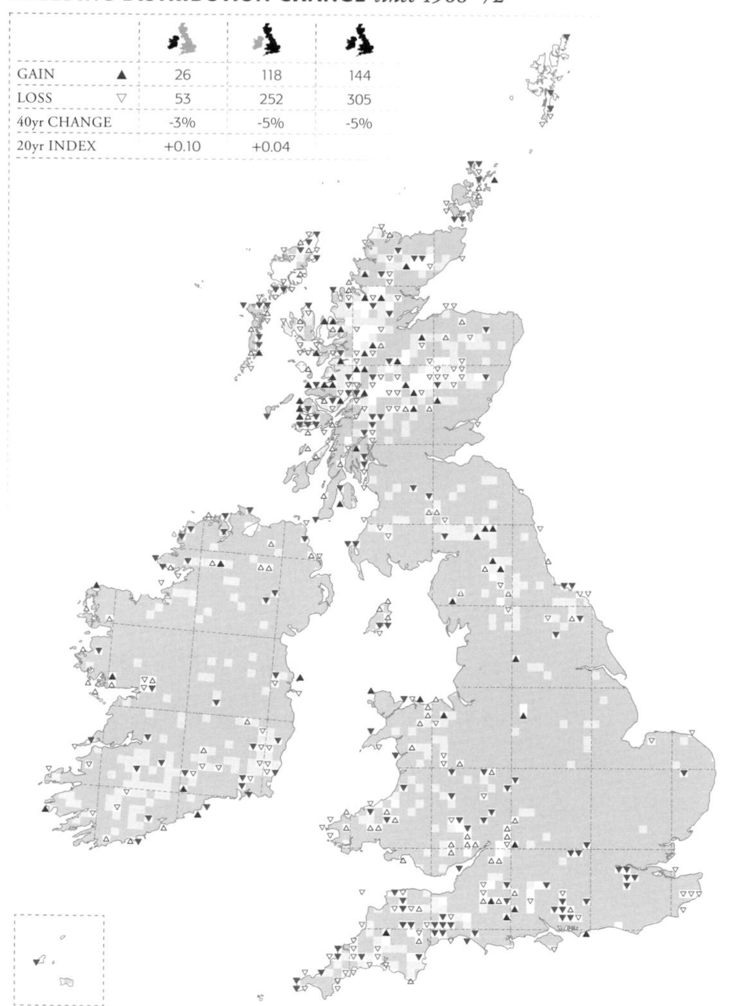

WINTER DISTRIBUTION CHANGE *since 1981–84*

GAIN ▲	213	547	760
LOSS ▼	84	110	194
30yr CHANGE	+18%	+25%	+23%

BREEDING RELATIVE ABUNDANCE *2008–11*

BREEDING RELATIVE ABUNDANCE CHANGE *since 1988–91*

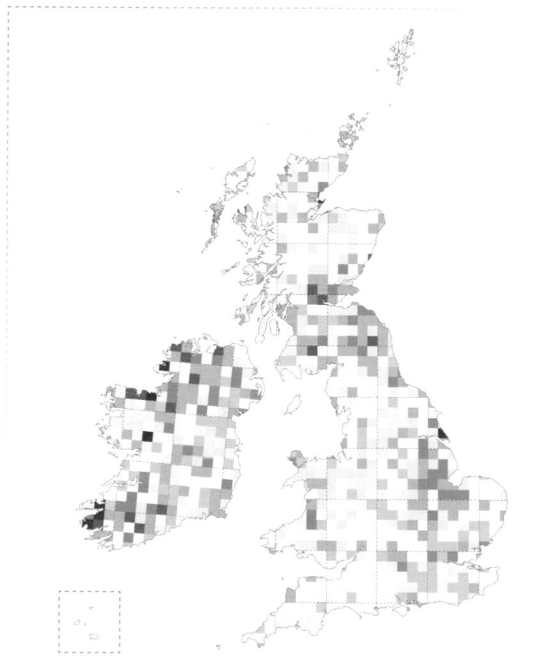

WINTER RELATIVE ABUNDANCE *2007/08–2010/11*

CORN BUNTING

Emberiza calandra

BTO CODE: CB

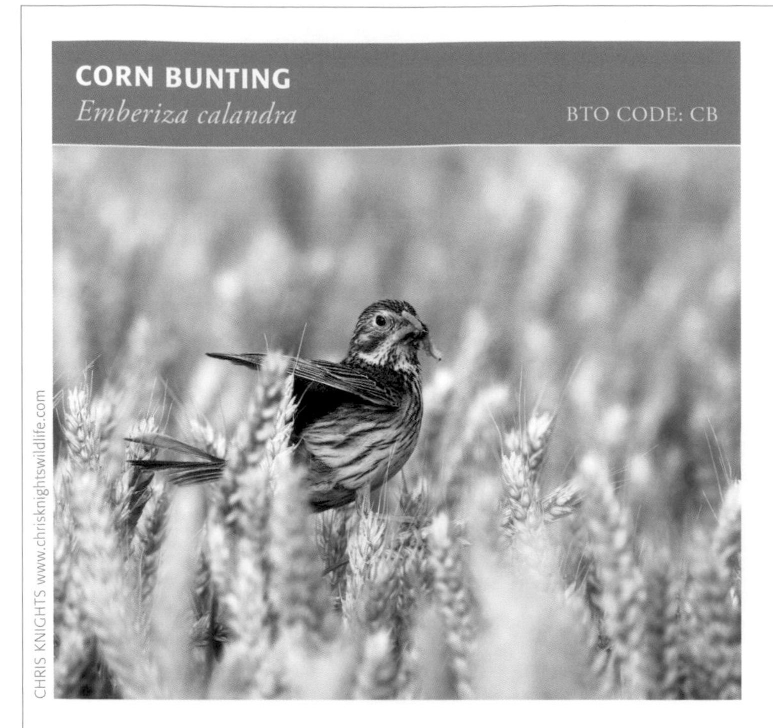

CHRIS KNIGHTS www.chrisknightswildlife.com

BREEDING DISTRIBUTION *2008–11*

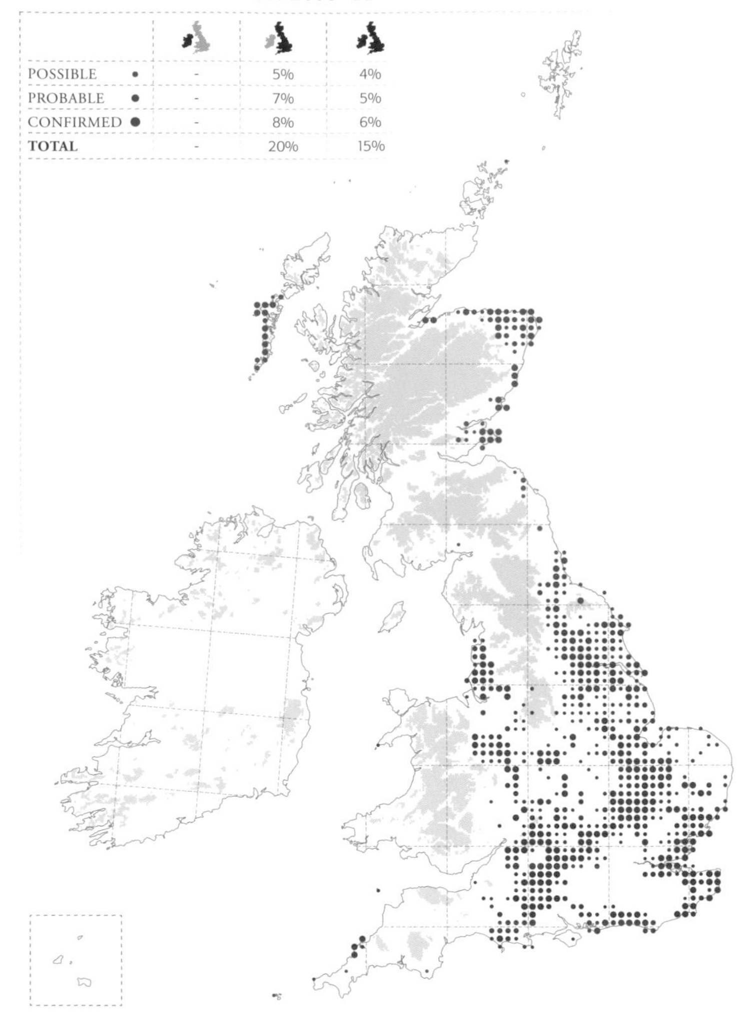

POSSIBLE ●	-	5%	4%
PROBABLE ●	-	7%	5%
CONFIRMED ●	-	8%	6%
TOTAL	-	20%	15%

WINTER DISTRIBUTION *2007/08–2010/11*

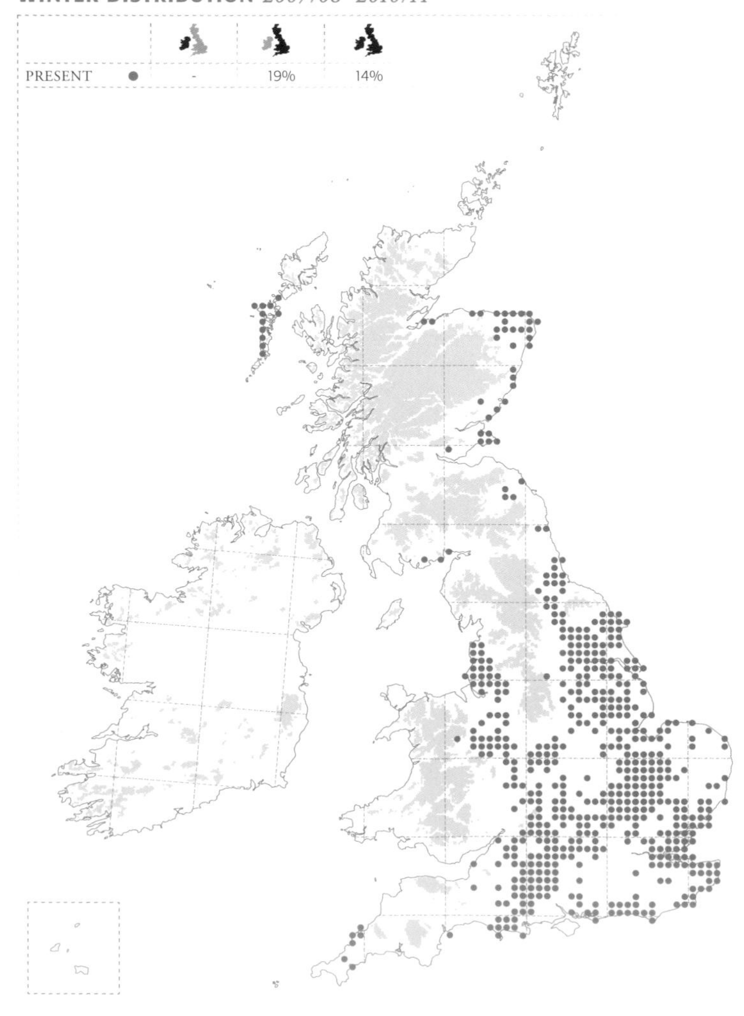

PRESENT ●	-	19%	14%

CORN BUNTINGS HAVE undergone a severe population decline, to the extent that, by the late 1990s, they had ceased to breed in Ireland. Occasional non-breeding records in southeastern Irish counties have been attributed to migrants from Britain or continental Europe (Taylor & O'Halloran 2002) but none were recorded during 2007–11.

In Britain a 90% breeding population decline during 1970–2010 (*SUKB 2012*) has been accompanied by contractions of 27% in winter range, since the *1981–84 Winter Atlas*, and 56% in breeding range, since the *1968–72 Breeding Atlas*. Corn Buntings are so highly sedentary that the heavily fragmented distribution is almost identical in winter and the breeding season. The remnant Scottish populations are isolated from those elsewhere, owing to extinctions in southern Scotland and northern England. With the loss of the species from the Northern Isles, Lewis, Coll and Tiree, the only remaining island population is that on the machair of the Uists.

Major losses in England have fragmented the population into discrete clusters of occupied squares. These include groupings on chalk soils from Dorset to Cambridgeshire, on the coast between Kent and Suffolk and on low-lying arable farmland from the Fens northward to Durham. All these remnant populations have seen declines in abundance since 1988–91 and annual monitoring indicates that total numbers are still declining (*BBS Report 2011*).

Causes of decline vary geographically, owing to the species' use of different habitats. In lowland arable farmland, the switch from spring to autumn sowing of cereals has reduced the availability of weed-rich stubbles for winter feeding (Donald 1997). Changes in cropping may also have caused a decline in the number of birds raising a second brood, with a consequent reduction in overall productivity (Brickle & Harper 2002). Productivity is probably affected also by the intensity of pesticide use (Brickle *et al.* 2000; Ewald *et al.* 2002). In the Hebrides, the practice of harvesting unripe cereal crops for silage has reduced overwinter food availability (Wilson *et al.* 2007). Population recovery can be facilitated by targeted management, for example by leaving crop patches unharvested for winter feeding or by delaying mowing for silage where birds nest in grass (Perkins *et al.* 2011).

Sponsored by Vine House Farm

BREEDING DISTRIBUTION CHANGE *since 1968–72*

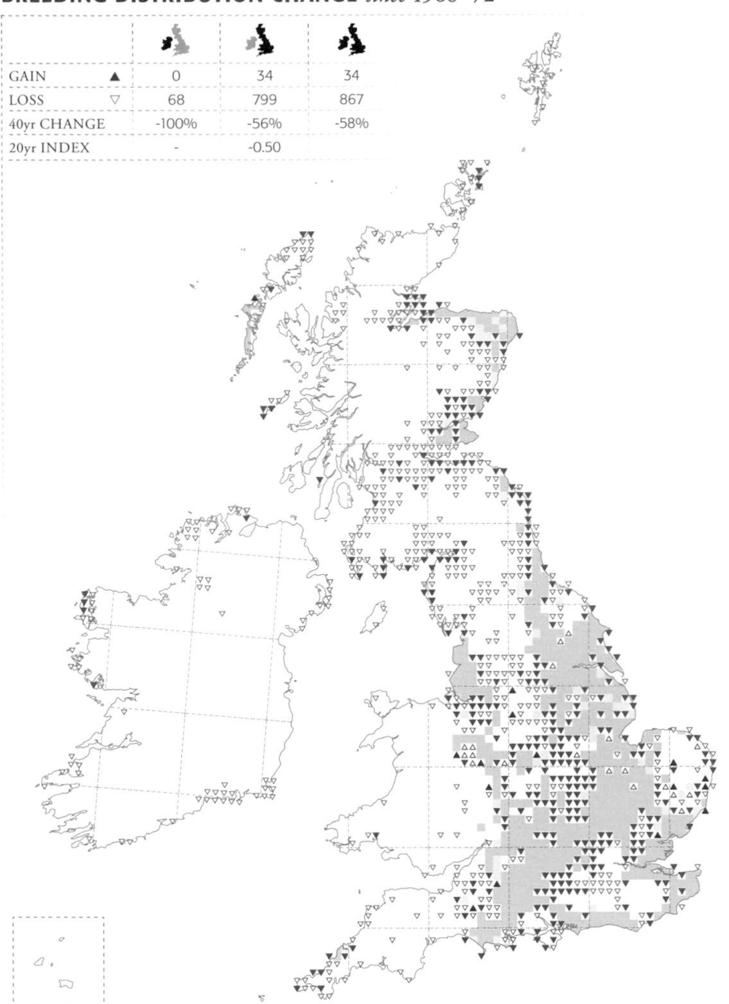

GAIN ▲	0	34	34
LOSS ▽	68	799	867
40yr CHANGE	-100%	-56%	-58%
20yr INDEX	-	-0.50	

WINTER DISTRIBUTION CHANGE *since 1981–84*

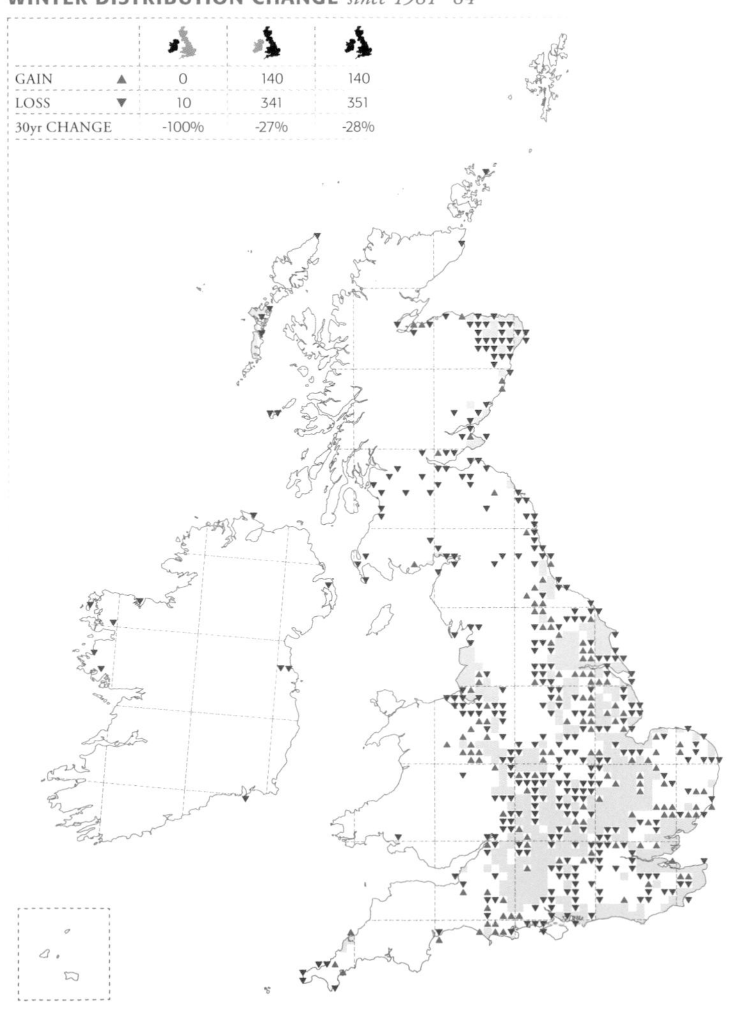

GAIN ▲	0	140	140
LOSS ▼	10	341	351
30yr CHANGE	-100%	-27%	-28%

BREEDING RELATIVE ABUNDANCE *2008–11*

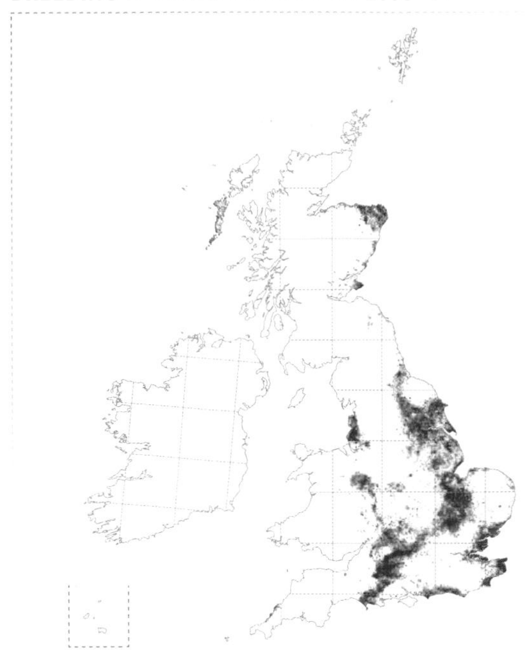

BREEDING RELATIVE ABUNDANCE CHANGE *since 1988–91*

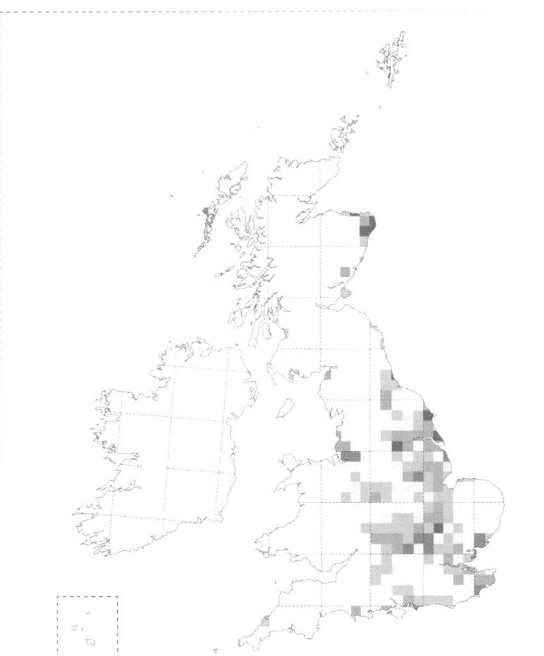

WINTER RELATIVE ABUNDANCE *2007/08–2010/11*

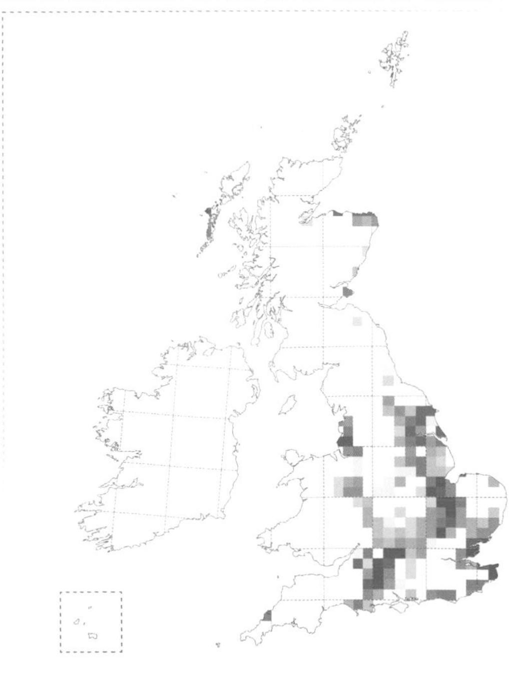

Appendices

APPENDIX 1 – OBSERVERS

THERE WERE SEVERAL ways in which individuals, groups and organisations contributed to *Bird Atlas 2007–11* (see Section 2.4 and Acknowledgements). In the following pages we would like to thank the observers who submitted records directly to the Atlas and BirdTrack. All those named below provided records that were used in the production of this book. To reduce repetition the number of observers with the same name is given in brackets after duplicate names.

A A Rocha UK, Mr L Aaron, Mr E M ab Owain, Mrs S Abbey, Abbey Wood Conservation Group, Abbey Wood Environmental Group, Isobel Abbott, Mr D Abbott (2), Mr F Abbott, Mr M Abbott, Mr P Abbott, Mr P F Abbott, Mr S Abbott (2), Stewart Abbott, Avenue Birders, Mrs N Abbott, Ms A Abbott, Ms J Abbott, Mr M & Mrs A Abbs, Marilyn Abdulla, Mr P Abel, Mr R Abel, Chris Abell, Mrs E Abell, Mrs S Ablitt, Mr D J Abrahams, Mr K Abram, Mr P Abram, Mr C Abrams, Mr T Abrehart, Mr S Abrey, Mr K Abrook, Mr D Acfield, Mrs C Achenbach, Mr N G Acheson, Mr D Ackland, Mr J Ackland, Mr M A'Court, Mr G M Adam, Mr R G Adam, Mrs A Adam, Miss J Adams, Mrs J Adams (2), Mr A Adams, Mr J Adams, Mr M Adams (2), Mr N P Adams, Paul L Adams, Paul Adams, Mr P Adams, Mr R Adams (2), Mr S Adams, Mr T Adams, Mr T D Adams, Mrs A Adams, Mrs Debbie Adams, Mrs H Adams, Ms T Adams, Mrs M Adams RVN, Mr D Adamson, Mrs R Adamson, Mr B Adcock, Mr N Addey, S G Addinall, Mr I M Addis, Mrs L Addison (2), Dr H A Addlesee, Mr J P Adey, Dr K Adkins & Ms J Hanstock, Mr M Adkins, Miss S Adlard, Mr G Adnitt, Mrs J Adrian, T & A Adshead, Miss J Adshead, Dr N J Aebischer, Dr Nigel Agar, Ms S Agnew, Mr D Agombar, Mr S Aguirre, Theresa Ahearne, Mr L Ahlgren, Mrs B Ahmed, Dr J Aidley, Miss D Ainscough, Mrs J Ainsworth, Mrs M P Aird, Mr R Aitchison, Mr F Aitken, Mr G Aitken, Mr S Aitken, Mrs P A Aitken, Mrs R A Aitken, Dr H K Aiton, Mr P Akers, Mr S Alberici, Mr M Albert, Mrs L Albert, Mr T Albone, Mr J Alce, Mrs J Alcolea & J Clover, Joan & Chris Alder, Mr G Alder, Mrs D Alder, Alderney Wildlife Trust, Mr J D Alderson, Mrs D M Alderson, The Aldertons, Mr M Alderton, Mr R J Aldis, Don Aldiss, Mr A Aldous, Dr R Aldridge, Mr John D Aldridge, Ms K J Aldridge, Miss E A Alexander, Mr D Alexander, Mr K Alexander, Mr M Alexander (2), Mr P Alexander, Mr S M D Alexander, Mrs R Alexander, Ms H Alexander, Peter Aley, Mrs A Ali, Mrs R Allaby, Alladale & Deanich, Mr D Allan, Mr J Allan, Mr R G Allan, Mrs J Allan, Mr P R Allard, Mr P Allcock, Mr P S Allday, M G & Eve Allderidge, P Allen & I Forsyth, Miss E Allen, Miss E J I Allen, Miss H Allen, Miss J Allen, Mr A Allen (2), Chris J Allen, Mr C Allen, Mr D Allen (3), Mr G Allen, Mr J Allen, Mr M Allen & volunteers, Mr M Allen (2), Mr N Allen, Mr R Allen, Mr S Allen, Mr T Allen, Alison Allen, Mrs A H Allen, Mrs D Allen, Mrs J Allen, Mrs L A Allen, Mrs M Allen, Ms D Allen, Ms J Allen, Prof David M Allen, Mr E Allenby, Mr H Allenby, Mr J Allenby, Mrs P Allenby, Mrs S Allen-Pearson & Mrs J Wilkinson, Mr M Allerton, Miss S Alliez, Mr C D Allin, Mrs A Allington-Smith, Dr K Allinson, Mr G W Allison, Mr J Allison, Mr M Allison (2), Mr R Allison, Mr Sean Allison, T Allison & V Head, Ms N Allison, D Allison, Mr R M Alliss, Mr D P Allnutt, Mr M G Allott, Mr P Allott, Mr J E Alloway, Dr A L Allsop, Miss J Allsop, Mr B Allsop, Mr P Allsopp, Mr J Allton, The Alltons, Major C H G Allum, Mr T Allwood, Mrs P Allwright, Mr S Al-Mahrouq, Mrs A E Almeida, Mr J Almond, Mr J A Almond, Mr S Almond, Alnwick Birding Group, Ms S Alsbury, Chris Alsop, Mr T Alves, Mr M Ambler, Mrs L E F Ambler, Mrs H Ambrose, Mr R Amer, Mr S Ames, Mr D R Amey, Mrs S A Amies, Mr K W Anckorn, Mr S Andersen, Dr B Anderson, Dr D Anderson, Guy Anderson, Dr J E Anderson, Miss C Anderson, Miss S Anderson, Miss V J Anderson, Mr A Anderson (2), Mr A D Anderson, David B Anderson, Mr D Anderson (5), Mr G Anderson (2), Mr G P Anderson, I D & J Anderson family, Mr I R Anderson, Mr J Anderson & family, Mr J Anderson, Kevin Anderson, Mr M Anderson, Malcolm Anderson, Mr N Anderson (2), Mr P Anderson, Mr P N Anderson, Mr R Anderson (2), Mr S Anderson, Mr S W Anderson, Mr W R Anderson, Mrs D Anderson, Mrs N A B Anderson, Mrs N Anderson, Mrs S Anderson, Mrs W Anderson, Ms H Anderson, Ms J Anderson, Ms K Anderson, Prof M Anderson, Dr E Anderton, Dr S Andrew, Mr P Andrew, Mr P R Andrew, Mrs D L Andrew, Dr C D Andrews, Miss J Andrews, Mr B Andrews, Mr C Andrews (2), Mr D Andrews (2), Mr D J R Andrews, Mr G D M Andrews, Ian J Andrews, Mr I J Andrews, Mr J Andrews, Mr J K Andrews, Mr M Andrews (4), Mr M B Andrews, Mr P Andrews, Mr R Andrews, Mr R D Andrews, Mr S E Andrews, Mr V Andrews, Mrs D M Andrews, Mrs H Andrews, Mr D Angell, Mr W Angell, Mr I Angus, Mrs R Angus, Angus Recorder, Mr J A Ankers, Mr S Ankers, Mr D Annal, Mr C Annall, Mr C R Annan, Mr P & Mrs J Anness, Mr G H Annibal, Mr D Anning, Tessa Anning, Mr T J Ansell, Mr M C Anstee, Mr S Anstey, Mr M Anthony, Mr P Anthony, Mrs J Anthony, Ms M Antram, Mr P Antrobus, Mr D Antropik, Tristan ap Rheinallt, Mr A Aplin, Mr J Applebee, Miss A Appleby, Mr D Appleby, Steve Appleby, Mr T Appleby, Mr A Appleton, Mr D Appleton, Graham Appleton & Jennifer Gill, J R Appleton, Mr T Appleton, Anne V Appleyard, Mrs R Appleyard, Mr S Apps, Mrs H Ap-Rhisiart, Mrs D Aps, Mr F Ar Moenner, Mr P C Arbery, Mr D W Arch, Mr P Archdale, Miss J Archer, Mr B M Archer, Mr D Archer (2), Mr J K Archer, Mike Archer, Mr S Archer, Mrs L Archer, Mrs S Archer, Mr D Arden, Ardstraw Primary School, Argent family, Argyll Bird Club, Argyll Team, Miss A Arkinstall, Mrs M Armandary, Mr J S Armitage, Mr Mike Armitage & Dr Lynsey Robinson, Ms D Armitage, Mr V J Armond, Mr J Armour, Dr A R Armstrong, Miss J Armstrong, Mr A R Armstrong, Mr D Armstrong, Mr Grant Armstrong, Mr I J Armstrong, Mr J Armstrong, Mr P J Armstrong, Mrs H Armstrong, T & L Armstrong, Dr A J Arnfield, Dr A Arnold, Dr M Arnold, Mr B S Arnold, Mr G Arnold, Mr M Arnold (2), Mr N Arnold, Mr P Arnold, Mr R Arnold, Mr S Arnold, Mr T Arnold, Mr V Arnold, Arnold family, Mrs S Arnold, Ms J E Arnold, Arran Ranger Service, Mr J Arrowsmith, Mr J W Arrowsmith, Mr P Arrowsmith, Mr B Arthur, Mr D Arthur, Mr S Arthur, Mr C Arthurton, Mr R Arundale, Dr R Ash, Mr A B Ash, Mr J Ash, Mr T Ash, C & R Ash, Ms L Ash, Kate Ashbrook, Ms M M Ashby, Mr Colin Ashcroft, Mr R Ashcroft, Jean Ashcroft, Mr D Ashdown, Mr P D Asher, Ashfield Rangers, Mr P M & Mrs M Ashford, Mr B Ashley, Mr S Ashley, Mr N Ashman, Mr M Ashpole, The Ashpoles, Mr P Ashton, Mr S Ashton, Mr Sean V M Ashton, Mr R L & Mrs S J Ashton, Ms R Ashton, Mr B Ashton - Wickett, Miss M Ashurst, Dr I A Ashworth, Mr G Ashworth, Mr M Ashworth (2), Mr P Ashworth, Mr R Ashworth, Mrs Patricia M Ashworth, Mr M Askew, Mr N P Askew, Dr P Askham, Mr J R Askins, Mr J Aslett, Dr K Asmussen, Mr W C Aspin, Mr A Asquith, Mr N Astbury, Mr T Astill, Mr D Astin, Mr D J Astins, Mr P Astle, Mr R H Astle, Mr S Astley, Mrs B Aston, Rev K Atchley, Miss M Atherton, Mr G Atherton, Mr S Atherton, Dr G Athey, Mr N D Athey, Mr N Atkey, Mr P T Atkey, Mr G R Atkin, K Atkin, Dr A F J Atkins, Mr C Atkins, Mr D Atkins, Mr J Atkins, Mr R Atkins, Stephen V Atkins, Rosemary Atkins, Dr P W Atkinson, Miss J Atkinson, Miss J E Atkinson, Miss K M Atkinson, Mr A Atkinson, Mr C Atkinson, Mr D R Atkinson, J D Atkinson, Mr M Atkinson, Mr S Atkinson, Mr Tony Atkinson, Ms K Atkinson, Ms V Atkinson, Mr R Atterby, Mr C Attewell, Mr J Attiwell, Ms S Attrell, Dr R P Attrill, Mrs J Attrill, Mr J Atwell, Mr P Attwood, Mrs C Attwood, Paul Aubrey, Dr M Auld, Mr C Auld (2), Russell Auld, Mr J Austen, Dr J Austin, Mr B P Austin, Dave Austin, Mr E Austin, John Austin, Mr John Austin, Mr M Austin, Mr M J Austin, Mr P Austin, Mr S Austin, Mr T S Austin, Mr V J Austin, Mrs A Austin, Mr G C Auty, Mr M B Auvache, Penny Avant, Mr C D C Avanti, Mr M Averill, Mark Avery, Miss M Avery, Mr G Avery, Mr D V Avis, Graham C Avison, Mr J D Avon, Avon Birdwatchers, Avon Valley Adventure & Wildlife Park, Mr D P Ayers, Mr Steven Ayers, Mrs S Ayers, Mr R Aylard, Mr Pete Ayley, Mrs A Ayling, Dr S Aylwin, Mr S Ayre, Miss H Ayshford

B Mr R Baatsen, Mr R J Baber, W J Baber, Mr R Bacciochi, Mr J Bachelor, Mr D Back, Mr D Backhouse, Helen Backhouse, Mrs C Backman, Claire Backshall, Dr C Bacon, Miss L Bacon, John Bacon, Mrs D Bacon, Mrs E Bacon, Bill Badger, Mr J Badger, Dr M Badley, Mr I Baggley, Dr D G Baggott, Mr C Baggott, Mr Paul Bagguley, Mr R Bagguley, Mrs A Bagwell, Mr Tim Bagworth, Ms A Baier, Mr M Baigent, Mr J Bailes, Juliet Bailey, Mr A Bailey (2), Mr C Bailey, Bailey family, Mr G D Bailey, Mr J Bailey, Mick Bailey, Mr M Bailey, Mr M A Bailey, Mr M Bailey (3), Mr P Bailey, Richard G Bailey, Mr R Bailey, Carolyn Bailey, Mrs E Bailey, Dr M R & Mrs S M Bailey, Ms C Bailey (2), Revnd A J Bailey, Mrs M Bailie, Stephen Baillie, Mr H Baillie, Mrs L H Baillie, Ms K Baillie, Revnd C Baillie, Miss C Bain (2), Mr A Bain, Mr D J Bain, Mr D R Bain, Mrs H Bain, Carole & Ian Bainbridge, Dr I P Bainbridge, Miss V Bainbridge, D Bainbridge, Mr S & M Bainbridge, Chris Baines, Mr M & Mr D Baines, Mr M Baines, Mr Stephen Baines, Mr J Baird, Mr P D N Baird, Mr T Baird, Mrs A Baird, Miss S Bairner, Mrs Ann Bairstow, Mr C Baister, Dr J H Baker, Miss C Baker, Miss H Baker, Miss L J Baker, Miss P Baker, Miss S Baker, Mr Tony Baker, Mr A Baker, Mr A D Baker, Alan Baker, Mr A R Baker, Mr C Baker, Mr D Baker (2), Mr E Baker, Mr G Baker, The Bakers, Mr J Baker, Mr J D Baker, Mr J K Baker, Mr M Baker (3), Mr M H Baker, Mike J Baker, Mr N Baker, Mr P Baker (2), Mr Phil J Baker, Mr R Baker, R Baker & L Whitfield, Mr R Baker (3), Mr R T Baker, Mr T Baker, Mr W Baker, Mrs J Baker, Dr M Baker-Schommer, Mr D Bakewell, Mr D Baldock, Mr M Baldock, Mr D Baldwin (2), Joseph Baldwin, Mr J R Baldwin, Mr M Baldwin, Mr T Baldwin, Mrs A E Baldwin, Mr G Balfour, Vanda Balfour, Miss H Ball, Miss L C Ball, Mr A Ball, Mr A R Ball, Mr C Ball, Dean Ball, Mr D Ball, Mr D J Ball, Mr J Ball, Mr J D Ball, Mr K Ball, Mr M J Ball, Mr P Ball, Phil Ball, Mr S Ball, Mr T Ball, Mrs K Ball, Mrs L Ball, Mrs M Ball, Mrs N E Ball, Mrs S Ball, Mrs Ball, Ms E Ball, Ms L Ball, Mr I Ballam, Mr D K Ballance, Miss H Ballantine, Mr P Ballantine, Mr J H Ballantyne, Keith R J Ballantyne, Mr M Ballantyne, Mr S Ballantyne, Mr K Ballard, Judy Ballard, Mrs L Ballard, Mrs R Ballentine, Mrs C B Ballinger, Ballycastle Swifts, Mr A J M Balmer, Dawn Balmer, Mr W Balmont, Balmoral Ranger Service, Mr J Balshaw, Mr B Baltera, Mr T Bamber, M Bamber, Mr T Bambridge, Dr M F M Bamford, Mr R Bamford, J S Bamforth, Banbury Ornithological Society, Mrs S Bancroft, Miss M Band, Mrs C Band, Mr N Banham, Miss S Banks, Mr A Banks, Mr B Banks, Mr C M Banks, Mr D Banks, Mr J Banks (2), Mr M Banks, Peter Banks, Mr S Banks, Mr T Banks, Mrs A Banks, Dr A Bannister, Mr R C A Bannister, Mr A Bannister, Mr K Bannister, Malcolm Bannister, Mrs Y Bannister, Mr E Bannon, Mr John K Bannon, Mr A M & Mrs M G Banthorpe, Dr Helen Bantock, Mr I Banton, Ms G Banton, Mr D Banyard, Mr P Barbagallo, S Barbato & N Gilbert, Mr D Barber, Mr J F Barber, Mr L J Barber, Mr P Barber, S & G Barber, Mrs C J Barber, Mrs E Barber, Mrs G C Barber, Mrs J Barber (2), Mrs R Barber, Mr P Barbour, Alan Barclay, Elizabeth Barclay, Mrs S Barclay, Mr J Bardet, Bardsey Bird Observatory, Mr B Barefield, Chris Barfield, Mr M Barfield, Mr Bob Barfield BEM, Mr AK Barfoot, Mr C Barimore, Dr E Barker, Miss E Barker, Miss J Barker, Mr A Barker, Mr A J Barker, Mr D Barker, Mr D J Barker, Mr G Barker, Mr J R Barker, Kevin Barker, Mr P Barker (2), Simon R J Barker, Terry Barker, Mr T V Barker, Mrs C Barker, Mrs N P Barker, Mrs R Barker, Mrs S Barker, Ms C Barker, Prof A W Barker, Mrs J M Barkla, Mrs C Barkley, Mr J Barley, Dr C J Barlow, Mr D Barlow, Mr G Barlow, Mr G A Barlow, Mr J Barlow (2), Mr N P Barlow, Mr P Barlow, Mr R Barlow, Mr S Barlow, Barn Owl Trust, Mr I Barnard, Mr J D G Barnard, Mr W Barnard, Mr T Barnatt, Mr D J Barnden, Mr R Barnden, Miss P Barnes, Barnes family, Mr A Barnes (3), Mr C R Barnes, Mr D Barnes, David Barnes, Mr D J Barnes, Mr E Barnes (2), Mr G J Barnes, Mr J Barnes, Mr John C Barnes, John E Barnes, K & R A Barnes, Mr P J Barnes, Richard Barnes, Mr R F Barnes, Mr S Barnes, Mrs A I Barnes, Mrs P Barnes, Mrs R Barnes, Ms P Barnes, Dr M Barnett, Miss R Barnett, Mr I C Barnett, Mr J Barnett (2), Mr R G Barnett, Mrs L Barnett, Mrs S M Barnett, Ms C Barnett, Mr M P Barnsley, Mr J Baron, Dr I Barr, Mr A Barr, Mr C J Barr, Mr J Barr, Ms L Barr, Miss C Barrand, Miss N Barrand, Mr R & Mrs A Barratt, Mr R Barratt, Mr T I Barratt, Miss C Barrett, Mr A J Barrett, Mr G Barrett, Mr J Barrett (2), Mr M Barrett, Mr M J Barrett, Mr N Barrett (2), Mr P Barrett, Mr R Barrett, Mr R G Barrett, Mr S Barrett, Barrett family, Mrs F Barrett, Mrs M R Barrett, Ms T F Barrett, Miss K Barrie, Mr H Barringer, Mr J W Barrington, Mr M Barriskill, Mr D Barron, Mr P J Barron, Mr R Barron, Mr D Barrow, Mr M Barrow, Mr P & Mrs C Barrow, Mr P Barrow, The Barrows, Mr D G Barry, Mr H Barry, Mr M J Barsley, Mr M Barstow, Mr J Bart, Mr G Barter, Mr M R Bartle, Mr P Bartle, Mr C Bartlett, Mr J L Bartlett, Mrs H Bartlett, Mr G Bartley, John Bartley, Mrs K Bartman, Mr M Bartolini, Colin Barton, Mr I Barton, Mr M E Barton, Mr N Barton, Mrs L J Barton, Mrs M A Barton, Mrs S Barton, Ms C Barton, Mr C Bartram, Mr C R Bartram, Mrs C Bartram, Jim Barty, Mrs H J Barwick, Mr R I Bashford, Mr R Bason, Mr I H Bass, Mrs V J Bass, Ms L Bass, Mr E Bassett, Mr R Bassett, Mr C Bastian, Mr D W & Mrs S J Bastin, Mr N Bastin, Mr J Bastow, Mrs R S Bastow, Bat Conservation Ireland, Mr J D Batchelder, Mr A Batchelor, Mr G M Batchelor, Bate family, Mr A Bateman, Mr D Bateman, Mr L Bateman, Mr N J Bateman, Mr P Bateman, Mrs L Bateman, Ms D Bateman, Dr A Bates, Mr B Bates, Mr D J Bates, Mr G G Bates, Mr J D Bates, Keith Bates, Mr S M Bates, Mr W T G Bates, Mr P Bateup, Mr G S Batho, Mr J Bathurst, Mr A Batley, Mr B Batson, Mrs H Batson, Mr S Batt, Jennifer K Batten, Peter Batten & Denise Lloyd, Mr S Batten, Ms M Batten, Mr E Battersby, Mr M J Battisson, Mr C G Batty, Mr R E Batty, Mr G Baudoncq, Miss S Baulch (2), Mr B Baverstock, Seb Baverstock, Mr D J Bax, Mrs L Baxendale, Mrs S Baxendell, Mr A Baxter, Mr M Baxter, Mr N Baxter, Sam Baxter, Mr S Baxter, Rowena Baxter, Mrs S Baxter, Mr A R Bayes, Mrs L Bayfield, Mr Sam Bayley, Ms S Baylis, Dr C Bayliss, Mr A Bayliss, Mr J A Bayliss, Mr N Bayliss, Dr N J Bayly, Mr M J Bayne, Mr E Bayton, Mrs B Bazeley, Mr C Beach, Mr M Beacon, Mr C Beadle, Mr S Beal, Mrs Jenny Beal, Mr G Beale, Mrs B M Beales, C E Bealey, Mrs S Beall, Dr D Beamish, Mr D Beamish, Mr & Mrs I Beamish, Dr M Bean, Mr B Bean, Mr S Bean, Mr I C Beaney, Mr V R Beaney, Miss L Beard, Mike Beard, Mr R Beard, Mrs L Beard, Ms L Beard, Mr J Beardsley, Ms S Beardsley, Ms M S Beardsworth, Mr A J & Mrs C A Beasley, Mr G Beasley, Mr R Beasley, Mrs J Beastall, Mr W C Beaton, Mr A A Beattie, Mr A G Beattie, Mr D Beattie, Mr J Beattie, Mrs S Beattie, Ms H Beauchamp, Mrs F Beaumont, Stella Beavan, Mr A Bebbington, Mr B Beck, Ms A Beck, Mr P Beckenham, Dr M Becker, Mr G Beckett, Mr K F Beckett, Mr R Beckett, Mr P Beckingham, Ms S Beckingham, Mr R Beddard, Beddington Collective, Mr R C Beddis, Mr K Beddow, Dr A Bedford, Ian Bedford & Janet Bell, Tom Bedford, Mrs B Bedford, Mrs K Bedford, Bedfordshire Bird Club (2), Mr D M Bednall, Miss H Bedwell, Mr S Beeby, Mr C J Beech, Mr M Beech, Ms L Beech, Mr T Beecher, Mr D V Beeken, Mr A Beeney, Dr R Beer, Dave Beer, Mike Beer, Mrs E Beer, Mrs S A Beer, Mr A Beese, Miss L Beetham, Mr M Beetham, Mrs J Beevers, Miss T Begg, Ms A Begley, Mr W H Beglow, Mr B Behagg, Mrs G Beirne, Mr B R Beischer, Mr C Belcher, Mr G Belcher, Dr C Bell, M V Bell, Miss P Bell, Miss V Bell, Mr A P Bell, Mr C W Bell, Mr D Bell (2), Mr G Bell (2), Mr H Bell, Mr I C Bell, John Bell, Mr J Bell (2), Mr M Bell (6), Mr P Bell (2), Mr R Bell, Mrs J L Bell, Mrs M Bell (2), Mrs R Bell, Mrs S Bell, Ms J Bell, Dr G Bellamy, C & C Bellamy, Mr D Bellamy, Mr P E Bellamy, Mr D Bellis, Mr J Bellorini, Dr L A Belshaw, Mr R Belson, Mr G Belt, Mrs P Belton, Mrs S Belton, Mrs J Benbow, Ms C Bendickson, Mr T Bending, Mr D Benedictus, J & G Benger, Mr P Benham, Mr L Beniston, Mrs J Benito, Mr S Benn & Ms B Brodie, Mr I J Bennallick, Miss D Bennett, Miss E Bennett, Miss K Bennett, Mr A Bennett (2), Mr A J Bennett, Mr B H G & Mrs S E Bennett, Mr D Bennett (3), Mr D I Bennett, Mr G D Bennett, Mr J Bennett (2), Mr J J Bennett, Mr P Bennett, Mrs A Bennett, Mrs J Bennett, Mrs J M Bennett, Mrs K Bennett, Mrs M Bennett, Mrs Y G L Bennett, Ms J Bennett (2), Mr D Bennie, Mr J Bennie, Mr I Bennison, Mr A Benson, Mr C Benson, Mr M Benson, Mr R A Benson, Mr S M Benson, Mr T Benson, Mrs R Benson, Ms L Benson, The Bents, Mrs C M Bent, Mr S Bentall, Ms Y Benting, Miss N Bentley, Mr A Bentley, Mr Grahame A Bentley, Mrs J A Bentley, Mrs M Bentley, Mrs V P Bentley, Mr C J Benton, Rev M Benwell, Mr P Beraet, Mr P M Beraet, Mr A Beresford, Mr S Beresford, Ms B Bergin, Mrs C Bergner, Berkshire Ornithological Society, Mr D Berni, Miss E Berry, Miss J Berry, Miss S Berry, Mr C Berry, Gordon Berry, Mr J Berry, Mr M Berry, Mr R Berry, Mr R J Berry, Mr T Berry, Mr & Mrs Berry, Mrs C Berry, Ms H Berry, Miss C Bertelli, Mr J A Bertenshaw, Mr D Berwick, Mrs C Bessant, Mr A D Best, Mr J Best, Mr J R Best, Mr P Best, Mrs D Best, Mrs K Best, Mrs P Best, Lt Col Nicholas Beswick, Mr I Beswick, Mrs J Beswick, Mr K F Betton, Mr & Mrs Mike Betts, Mr P F Betts, Mrs I Betts, Miss R Bevan, Mr A Bevan, Mr G Bevan, Mr M

Bevan, Mr A V Bevan, Mrs P Bevan, Mr D Beveridge, Mr E A Bew, Mr N Bew, Mr N J R Bewley, Ms J Bezant, Mr A Bezer, Mr D Bezuidenhout, Mr N R Bhatti, Julie Bhatti, Mrs L Bibby, Bicester Garrison Conservation Group, H & E Bickerstaff, Mr D Bickerton (2), Mr M A Bickerton, Mr T Bickerton, Mr P J Bickford, Mr M Bickley, Mr J R Bicknell, Mrs A Bicknell, Martin Biddle, Robin Biddle, Oliver Biddulph, Noranne Biddulph, Mrs K Bidgood, Mr LM Bidwell, Mr N Bielby, Mr A Bielinski, Mr G N Bierton, Mrs S A Bigg, Prof G R Bigg, Ms S Biggane, Dr P Biggar, Ms A Biggins, Dr D T Biggs, Miss K E Biggs, Mr P Biggs, Biggs family, Mr C Bigmore, Mr C E Bignal, Dr S M Bignold, Mr G Bilbao, Mr M E S Bilbe, Mr M Bilbie, Mr D Bilcock, Mrs H Billanie, Mr D Billing, Mr P Billinghurst, Mr R W Billingsley, Dr R W Billington, Mr G Billington, Mr Jonathan Bills, Mr R Billyard, Mrs H E Bilsby, Mr C J Bingham, Paul Bingham, Mr T A Bingham, Mr A Binham, Steve Binney, Miss N Binnis, Mr J Binns, Mr W Binns, Mrs L Binns, Biodiversity Team, Biodiversity Unit - NPTCBC, Dr K Birch, Dr R Birch, Miss J A Birch, Miss S Birch, Mr G D Birch, Mr I Birch, Mr R Birch, Mr R A Birch, Mrs K Birch, Ms T Birch, Mr M Birchall, Mrs K Birchall, Mr R K Bircher, Mr G Bird, Mr J Bird, Mr M Bird (3), Mr R Bird, Mr S F Bird, Mrs B Bird, Bird Recorders 21st Century, Bird Watching & Wildlife Club, BirdGuides, Birds In Belton, BirdWatch Ireland Wicklow Branch, Mr T F Birkbeck, John Birket & Croydon RSPB, Mr G Birkett, Mr J Birkett, Mr P Birkett, Mr M A Birkin, Mr S Birks, Mrs K Birks, Ms W Birks, Mr H Birley, Ms L Birt, Dr R H Bishop, Mr B Bishop, Mr D Bishop, Mr G Bishop, Julian Bishop, Mrs S Bishop, Ms S Bishop (3), Glenn Bishton, Lorne Bissell, Mr A J Bissitt, Mrs A Bissland, Mr A Bisson (2), Mrs G R Bisson, Ms L Bizley, Dr J Black, Dr P J Black, Miss R Black, Mr C Black (2), Mr F Black, Mr G Black, Richard & Ruth Black, Mr R Black, Mrs D Black, Mrs PD Black, Wendy Black, Dr P Blackburn, Miss E Blackburn, Miss J Blackburn, Mr A Blackburn, Mr J R Blackburn, Mr J Blackburn, Mr N Blackburn, Mr R Blackburn, Blackburn Bird Watcher, Fay Blackburn, Mr J Blackburne, Dr J Blacker, Dr T Blackett, Mr S Blackett, Ms B Blackhall, Caroline Blackie, Mr D Blackledge, Mr R Blackler, Mr S Blackley, Mr G M Blackman, Miss L Blackmore, Mr A D Blackmore, Mr D Blackmore, Mr I Blackmore, Mr A Blackshaw, Trevor Blackshaw, Miss E A Blackwell, Mr A Blackwell, Mr H Blackwell, Mr R Blades, Ms J Blades, Mr S Bladwell, Mr S Blain, Miss E Blair, Mr Alisdair Blair, Mr J Blair, Mr A E Blake, Mr K Blake, Mr S Blake, Prof B Blake, Mr A F Blakeley, Mr C Blakeley, Dr D Blakesley, Miss C G Blakey, Mr R Blamey, Mark Blamire & Friedy Luther, Mrs S Blamire, M Blamires & YHA Eskdale, Mr R L Bland (2), David Blandford, Mr M Blane, Mr D Blanning, Mr J Blasco, Mr P Blatcher, Mr C Blaxill, Mr R Bleakley, Mr M F Bleaney, John Bleby, Mrs S Blenkhorn, Mrs S Bletcher, Mr A Bletchly, Mr P Blewitt, Mr M A Blick, Miss D Blight, Mr I Blight, Mr R M Blindell, Mrs A C Blinston, C Blinston, Mr R Block, Mr B Blogg, Mr K P Blomerley, Mr P Bloomfield, Mr S Bloomfield, Mrs C Bloomfield, Mrs D Bloomfield, Dr J W Bloor, Mr D Bloor, Robert & Alison Bloor, Mr G Blow, Mr K D Blowers, Mr M Blowfield, Mrs R J L Bloxam, Mr P Blud, Mr A Bluefield, Mr D Bluemore, Mr D S Bluemore, Mr Andrew Bluett, Mr W G Blumsom, Miss C Blundell, Mr J Blundell, Mr L Blundell, Mr A Blunden, Mr D Blyth, Trevor Blythe, Mr T Blythe, Mr J Boakes, Mr D Boal, Mr R Boan, Mr P Boardman, Mr G Boath, Dr N D Boatman, Mr D Boddington, Mr D Boddy, Mr J Boddy, The Bodilys, Dave Bodley, Anne Bodley, Mr S Boggio, Dr D Boggon, Barbara Boize, Mrs R Bojalil, Mr P L Bolam, Ms H Boland, Mr T Bolderstone, Mr S Bolger, Mr Chris Bollen, Mark Bolton, Mr A Bolton, Mr D Bolton, Mr J Bolton, Mr S Bolton, Mrs L Bolton, Ms A Bolton, L Bond, Mr D A Bond, Mr I Bond, Mr J Bond, Mr R Bond (2), T Bond & D Nevitt, Mrs G Bond, Phil & Sara Bone, Mr R W Bone, Mrs S K Bone, Mr M Bonfield, Dr T J Bonham, Mr M J Bonham, Mr P Bonham, Mr E Bonheim, Mr F Bonney, Mr C Bonnington, Mr R Bono, Mr R Bonser, Mr P Boobyer, Mrs W Booker, Ms J Booker, Ms D Booler, Mr R Boon, Mr H Boorman, Mrs G Boorman, Mr W J G Boot, Mr I Boote, Mr M Boote, Dr C Booth, Dr C M Booth, Mr A Booth (2), Mr A L Booth, Mr C Booth (2), Mr C J Booth, Mr D Booth (2), Mr D J Booth, Mr J Booth, Mr M Booth, Mr S Booth, Mr W H & Miss E Booth, Mrs C Booth, Mr M Boothman, Mr D Boothroyd, Mrs A Boothroyd, Ms A L Borawska, Bord na Móna, Borders Forest Trust, Borders Recorder, Borders SOC, Mr I Boreham, Ms L Boreham, A Borlase, Ms N Borodina, Mrs J Borrill, Mrs R Borrows, Mr R M Borwick, Mr S Bosanquet, Mrs S M Bosley, Dr A Bostock, Mr S Bostock, Ms J Bostock, Revnd M Bostock, Mr C Boston, Mr D Boston, Mrs I Boston, Miss K Boswarva, Mr S A Boswell, Mr K Botham, Mrs H Bothwell, Mr M Botley, Mr M Bott, Mrs C J Botteley, Mr R Botting, Mr J E Bottom, Mr P Bottom, Mr D Bottomer, Mr G Bottomley, Mr R Bottomley, Mr C Bottrell, Mrs G Bottrill, Mr J Boulcott, Mr A Boulton, Mr E Boulton, Mr R Boulton, Mr A Bourn, Dr W R P Bourne, Miss A Bourne, Mrs C Bourne, Ms L Bourne, Mr C Bovis, Dr D Bowden, Mr C G R Bowden, Mr H Bowden, M & L Bowden, Mr P Bowden, The Bowden-Burlinsons, Mr T Bowditch, Dr M Bowen, Mr G S Bowen, Mr J Bowen, Mrs U Bowen, Ms J Bowen, Mr E Bowen-Jones, Mr C Bower, Mr K Bower, Mrs L Bower, Bowerman family, Mr P D Bowerman, Mr R Bowers, Mr R C Bowers, Mr A Bowes, Mr D J Bowes, Mr I R Bowes, Miss H Bowick, Mr N J G Bowie, Mr D J Bowker, Mr K Bowker, Mr W J M Bowker, Mr N Bowland, Dr J Bowler, Mr K Bowler, Mrs H Bowler, Mr R Bowles, Mr T E Bowley, Ms J Bowley, Dr J Bowman, Mr A Bowman, Mr D S Bowman, Mr G Bowman, Mr J Bowman, Mr R N Bowman, Mrs J Bowman, Mr H R Bown, Mr R Bown, Mr R Bowra, Mr D W Bowring, Mrs H Bows, Mr R Box, Mr F Boxell, Mr D Boyce (2), Mr K T Boyce, Mrs P Boyce, Aoife Boyd, Mr D Boyd (2), Mr D A Boyd (2), Mr G F Boyd, Mr M Boyd (2), Mr W Boyd, The Boyds, Miss L Boyer, Mr M Boyer, Mr P R Boyer, Ms D Boyer, Mr G Boyes, Mr S Boyes, Mr P Boylan, Mr B R Boyland, Mr J Boyle, Mrs A Boyle, Ms F Boyle, Mr F Boynton, Mrs K M Boys, Boys School Rathangan, Mr A Boyt, Mr D Brabban, Ms T Bracey, Mr J Bracher, Mr A Brackan, Brackenhurst Birders, Miss C Bracker, Mr D Bradbeer, John B Bradbeer, Dr R B Bradbury, Mr C Bradbury, Mr J Bradbury, Revnd P Bradbury, Mrs M Braddock, Miss E Bradford, Barry Bradford, Mr C Bradford, Mr C A Bradford, Mr D Bradford, Dr S Bradley (2), Mr A Bradley, Mr C Bradley, Mr I Bradley, Mr J Bradley (2), J & B Bradley, Mr W Bradley, Mrs S G Bradley, Ms A Bradley (2), P G Bradley, Mr R F Bradney, Mr B J Bradnum, Mr D Bradnum, Miss S Bradshaw, Mr C G Bradshaw, Mr C J Bradshaw, Mr D W Bradshaw, Mr H Bradshaw, Mr I Bradshaw, Mr L Bradshaw, Philip Bradshaw, Mr R S Bradshaw, Mrs V Bradshaw, Mr A Brady, Mr I M Brady, Mr M Brady, Mr P Brady, Mrs J Brady, Mrs R Brady, Miss G Braes, Mr J A Braggs, Mr J C Brain, Mr T W W Brain, Miss M Braithwaite, Mr A Braithwaite, Mrs J A Braker, Miss E Bramhall, Mr A T Bramhall, Mr P Bramhall, Ms Y Bramley, Mr S J Branch, Mr D Brand, Mr M Brand, Mrs S Brandes, Dr A Brandon, Brandon Country Park, Mrs C Brandon-Lodge, Mr K Brandwood, Dr G Branfield, Mrs K Brannan, Mrs Ida M Brannock, Mr J Branscombe, Miss A Bransden, Mr A Branson, Brant family, Mr P R Brash, Mr P Brasier, Mr D Brassey, Mr R Brassington, Mr R A Brassington, Mr J H Bratton, Mr A Braun, Mr D M Braund, Mrs J A Brause, Mr A Bray (3), Mr B Bray, Mr I Bray, Mr J Bray, Mr M Bray, Ms A Bray, The Braybookes, Mr P Brayshaw, Mr H Brazier, Mr M T Breaks, Mrs M Breakspear, Mr R G Breakwell, Miss S Brealey, Mr B E B Bree, Mr A Breed, Mr C Breen, Mr Dermot Breen, Mr D Breen, Mr M Breeze, Mr S Breeze, Mr D Bremner (2), Dr A Brenchley, Mr C Brene, Mr A Brennan, Mr J Brennan, Mr K Brennan (2), Mr R H Brennan, Mrs M Brennan, Ms M Brennan, Ms S Brennan, Dr A G Brereton, Miss R Bretherick, Mr K J Brett, Mr R Brett, Mr J Brettell, Mr M J Breward, Dr A Brewer, Mr C D Brewer, Mr M Brewer, Mr S Brewer, Mrs J Brewer, Dr M Brewis, Tom Brewis, Mr A Brewster, Mr D Brewster, Ms C Brewster, Mr A Briddon, Mr K Brides, Mr A Bridge, Mr I Bridge, Mr M J Bridge, Mr T Bridge, Mr T E Bridge, Ms G Bridgeman, Miss J Bridger, Mrs J Bridger, D & J Bridges, Mr T F Bridges, Mrs H Bridgland, Mr R Bridson, Mrs E Brien, Mr M Brierley, Mrs S Brierley (2), Dr C Briffett, Dr K B Briggs, Barnaby Briggs, Mr J H Briggs, Mr P Briggs, Mrs G E Briggs, Mrs L Briggs, Ruth Briggs, Miss R Briggs, Mr D Bright, Mr J Bright, Mr P Bright, Ms J Bright, Dr P Bright-Thomas, Mrs M M A Brill, Mr K Brindley, Mr K Brine, Mrs C A Bringloe, Ms L Brinklow, Mr T Briody, Ms D Briscoe, Mr M Britnell, Mr A Britner, Mr S Britstone, Ms A Britt, Brittain/Whitley/ Williams family, Mr K Britten, Mr P Britten, Mr A Britton, Mr C Britton (2), Mr D J Britton, Mr S A Britton, Mr T Britton, Ms J Britton, Rev T Britton, Mr E Brixey, BRNC, Dr G R Broad, Miss E Broad, Miss S Broad, Mr R A Broad, Mr I Broadbent, Mr R C Broadbent, Miss N Broadbridge, Miss M Broadhurst, Mr R G Broadie, Mr B J Broadley, Dr J Brock, Mr P Brock, Mr R W Brock, Mrs H Brock, Mr K Brockie, Miss M Brocklesby, Mr A Brockless, Mr I K Brockway, Ms C Brodribb, Gayle Brogan, Mr D Brogan, Mrs R Brokenshire, Dr R Bromilow, Mr D Bromont, Mr J H Bronner, Mr R Brook, The Brooks, Dr M de L Brooke, Mr S Brooke, Mrs S Brooke, Sheila Brooke, Mr C Brookes, Mr Ian Brookes, Mr M C Brookes, Mrs S Brookes, Mr D Brooke-Taylor, Mr A H Brooking, Mr G Brooking, Mr M Brooking, Mr W Brooking, Mr S Brookland-Beck, Mrs A Brookman, Dr C Brooks, Mr D Brooks, Mr D J Brooks, Mr G R Brooks, Mr K V Brooks, Mr L Brooks (2), Mrs S Brooks, Ms H Brooks, Mr G Broom, Mrs S Broom, Mr C Broome, Mr D P Broome, Mrs S Broome, Miss R Broomfield, Mrs M Broomfield, Mr J Broomhead, Mr J Brophy, Ms J Brophy, Mr J Brosche, Miss A Broszkiewicz, Miss S Brotherton, Mr D Brotherton, Mr M Brotherton, Mr P Brotherton, Ian & Jenny Brotherton, Mr A Brougham, Mr C Brougham, Mr A O Broughton, Mr H E Broughton, Mr M A Broughton, Mr R K Broughton, Mr Broughton, Dr A Brown, Dr J Brown, Dr K Brown, Dr M Brown, Miss E Brown, Miss J E Brown, Miss L Brown, Miss P Brown, Miss S Brown (2), Mr A Brown (2), Dr Allan W Brown, Mr B Brown, Brown family, Mr C Brown (3), Mr D Brown (4), Mr D A Brown, Mr G Brown (5), Mr G B Brown, Mr G E H Brown, Mr I Brown, Mr I A R Brown, Mr J Brown, Mr J A Brown, Mr J H F Brown, Lee Brown, Mr L Brown, Mr L P Brown, Mr Murray JD Brown, S Martin Brown, Mr M Brown (2), Mr M F Brown, Mr M J Brown, Gaddas-Brown, Mr N G Brown, Mr N H Brown, Mr N M Brown, Mr P Brown (3), Mr R Brown (4), Mr R I Brown, Mr R M Brown, Mr Sam Brown & Iona MacGregor, Mr S G Brown, Mr W Brown (2), Anne Brown, Mrs A Brown, Mrs B Brown, Mrs C Brown, Mrs F E Brown, The Browns, Mrs J Brown, Mrs J E Brown, Mrs M Brown, Mrs N Brown, Mrs R Brown, Mrs S A Brown, Mrs T Brown, Mrs V Brown, Mr T R Brown, Wendy Brown, Ms B Brown, Ms J Brown, Ms K Brown, Ms R Brown (2), S J W Brown, Dr S Browne, Miss T Browne, Mr K Browne, Mr M Browne, Mr S Browne, Mr V Browne, Mrs N Browne, Mr M Brownell, Mr J Brownfield, Ms J Brownhill, Mr M Browning, Mr G Brownlow, Mr E Brownsword, Mr P Brownsword, G Bruan, Miss E Bruce, Mr B Bruce, Mr K Bruce, Mr L Bruce, Mr M Bruce, Mr W Bruce, Mrs A M Bruce, Ms L Bruce, Ms V Bruce, Mr J W Brucker, Brue Valley Bird Monitoring, Mrs L Bruges, Mr D Bruin, Mr R Brumby, Mr M K Brummage, Mr A & Mrs M Brunning, Mr A S Brunt, Mr A Brunt, Mr R Bruty, Mr D S F Bryan, Mr P Bryan, Mrs J P Bryan, Mrs K Bryan, Mrs S M Bryan, Ms L Bryan, Dr M J Bryant, Miss E Bryant, Mr J J Bryant, Mr P Bryant, Mr R Bryant, Mrs S Bryant, Prof D M Bryant, Miss R Bryce, Mr J B Bryce, Mr M Bryce, Mr N Bryce, J & D Bryce, Ms L Bryce, James Dunsmuir Bryden, Ms L Brydon, Miss H F Bryett, Mr I Bryn, Mr A Bryson, Mr T Bryson, Mr P Bubb, Rosy Bubb, Buccleuch Ranger Service, Mr D Buchan, Mr M Buchan (2), Dr D J Buchanan, Miss R Buchanan, Duncan Buchanan, Mr J Buchanan, Mr L W Buchanan, Mr M Buchanan, Mr N Buchanan, Mr P Buchanan, Ms C Buchanan, Ms R Buck, Mrs F C & Mr A Buckel, Mrs L Buckell, Miss C Buckerfield, Mr G Buckett, Mr A M Buckingham, Mr D Buckingham, Mr D L Buckingham, Mrs S Buckingham, Buckinghamshire Bird Club, Mr D Buckland, Mr Martin L Buckland, Mr R A Buckland, Prof S T Buckland, Mr I Buckle, Mr S Buckle, Mrs J Buckle, Ms Clare Buckle, Dr P Buckley, Mr D Buckley, Mr E Buckley, Mr I Buckley, Mr M Buckley, Mrs C Buckley, Mrs N Buckley, Mr R P Bucknall, Mrs A E Bucknall, Mr N J Bucknell, Mr A Budd, Mr D J Budd, Mr J Budd, Mr M A Budd, Mr P A Budd, Mr R A Budd, Mr M D Budden, Mr M Budge, Ms D L Buesnel, Mr B D Buffery, Mr R D J Bufton, Revnd Canon P Bugg, Dr R S K Buisson, Mr D K Buisson, Mr R Bulcraig, Mr J Bulfin, Alan Bull, Mr C J Bull, Mr L Bull, Mr P Bull, M & K Bullen, Mr S Bullen, Mr S Bullett, Mr G M Bullivant, Dr R Bullman, Dr D Bullock, Miss N Bullock, Mr C Bullock (2), Mr P Bullock, Mr R W Bullock, Mrs S Bullock, Mr W A Bullough, Mr K Bulpin, Bunchrew Seer's Gang, G F W Bundy, Dr F D Buner, Mrs A Bunker, Mr B Bunn, Mrs S Bunn, Mrs A Bunning, Mr A Bunten, Mr J Bunting, Mrs S Bunting, Mr P Burbage, Mr N J Burchett, Mr M Burdekin (2), Ms H Burden, P Burden, Ms J Burdett, Mr T Burditt, Ian Burfield, Mr N D Burfield, Mrs J Burfield, Mr G D Burfoot, Mr R Burford, Mr M J S Burge, Mrs F S Burge, Mrs V Burger, Dr Malcolm Burgess, Mr C C Burgess, Mr D Burgess, Mr J Burgess, Mr M Burgess (2), Mr R Burgess, Mr R G Burgess, Mr J Burgoine, Mr G Burgon, Mr J J Burgum, Andrew & Nicole Burgum, Mrs H Burkard, Dr N Burke, Miss M Burke, Mr B Burke, Mr J N Burke, Mr M Burke, Mr T Burke, Mr C & Mrs P Burkes, Debbie Burkett, Mr A Burkin, Miss J Burkitt, Mr Richard Burkmar, Mr G & Mrs E Burman, Mr M Burman, Mr R Burman, Mr J Burn, Ms A Burnage, Mr B Burnett, Mr C Burnett, Mr J W Burnett, Mrs S M Burnett, Ms L Burnett, Mr A Burnham (2), Miss M Burnley (2), Miss A Burns, Miss G Burns, Mr A Burns, Mr B Burns, Mr D W Burns, Burns/Roberts, Mr R B Burns, Mr S Burns, Mrs B E Burns, Mrs L Burns, Mr M Burnside, Mr I M Burrell, Mr J Burrell, Mr B Burrows, Mr C Burrows, Mr P C Burrows, Mrs P Burrows, Mr I Burrus, Mr P Burston, Mr P J Burston, Mr R G Burston, Dave Burt, Mr R Burt, Mrs G Burt, Mr J Burtchaell, Dr N H K Burton, Mr A Burton (2), Mr A K Burton, Mr G Burton, Mr G J A Burton, Mr M Burton, Mr Roger Burton, Mr R G Burton, Mr R M Burton, Mr S Burton, Mrs A Burton, Mrs M Burton, Mrs P Burton, Mrs R Burton, Ms C Burton, Ms J M Burt-Smith, Mr R Busby, Mr R Busfield, Mr P J H Bush, Mrs D Bush, Mrs J Bush, Ms C Bushe, Ms H Bushe, Mr C Bushell, Mr R P M Bushell, Mrs C Bushnell, Mr G J Buss, Mrs K Buss, Mr K Bussell, Mr S Busuttil, Dr G I Butcher, J R Butcher, Mr P Butcher, Mr R J Butcher, Dr A J & Mrs E Butler, Miss A Butler, Mr A Butler (2), Mr A K Butler, Mr B Butler (2), Mr David S Butler, Mr D Butler (2), Mr E Butler, Mr G A R & Mrs J M Butler & family, Mr G K Butler, Carlow R G C, Mr J Butler, Mr M J Butler, Mr P Butler, Mr P J Butler, Mr R A Butler, Mr R F Butler, Mr S Butler, Mr T Butler, Jennie Butler, Joyce Butler, Mrs J Butler (2), Mrs L J Butler, Ms C Butler, Ms F C Butler, Ms J Butler, Ms P Butler, Rev J Butler, Mr D Butling, Mr P Butson, Mr C Butt, Mrs S Butt, Miss M Butter, Mr D Butterfield, Mr D P Butterfield, Mr C Butters, Mr T Butters, Mr A M B Butterworth, Mr J M Butterworth, Mr K Butterworth, Mr P G Butterworth, Mr David Butterworth, Mrs K Butterworth & Mr P Kenyon, Mrs S Butterworth, Mr E Buttery, Dr N E Buxton, Mr D Buxton, Mr D N Buxton, Mr I Buxton, Mr N J Buxton, Paul & Lorna Buxton, Mr R R Buxton, Mr T Byars, Mr B Bye, Dr P A F Byle, Mr B Byrne (2), Mr C Byrne (2), Mr J A Byrne, Mr M Byrne, Mrs A Byrne, Mrs C Byrne, Ms R Byrne, Dr P Bysh, Mr R Bysouth, Ms J M Bytheway, Mr D Bywater

C Dr D Cabot, Dr J Caddick, Mr F Cade, Mr M Cade, Dr N Cadee, Mrs S Cadle, Mr D Cadman, Mr J Cadman, Mr R A Cadman, Tom Cadwallender, Muriel Cadwallender, Miss D Cady, Miss L H Cady, Mr B Caffrey (2), Mr N Caffrey, Caffrey family, Mr J Cahalane, Brian Cahill, Mr J Cahill, Mr M J Caiden, Miss S Cain, Miss Thelma S Caine, Mr G D Caine, Mr G Caird, Mr K Cairns, Mr P Cairns, Neil Calbrade, Mr B Calder, Mr J Calder, Mrs J M Calder, Mr A Calderwood, Mr I Calderwood, Mr T Caldicott, Mr C Caldwell, Mrs H E Caldwell, Calf Of Man Bird Observatory Wardens, Miss A Calfe, Mr A S Call, John Calladine, Mike Calladine, Mr D Callaghan (2), Mr J Callaghan, Mr D Callahan, Mr T Callan (2), Mr I Callanan, Dr P L Callaway, Mr D Callaway, Mr I A Callender, Mrs K Calley, Mr R M Callf, Mr J C Callion, Mr S Callis, Mr S Callister, Mr J D Callow, Mr M Callow, Mr R Callow, Mr T Calloway, Mr A Calson, Miss J Calvert, Mr I Calvert, Mrs D Calvert, Mr R Calvin, Mrs C Calvin, Cambridgeshire Bird Club, Dr C H Cameron, Miss R Cameron, Mr A Cameron, Mr A J Cameron, Mr D Cameron (2), Mr E D Cameron, Mr N Cameron (2), Mr P Cameron, Ms J M Cameron, Mr P J Cammack, Andrew I Camp, Mr D A Camp, Mr J Camp, Mr J P Camp, Mrs S Camp, Dr Hazel Campbell, Dr J Campbell, Dr L H Campbell, Miss J E Campbell, Mr A Campbell, Mr B & Mrs S Campbell, Mr C Campbell, Mr D Campbell, Mr D E D Campbell, Mr D K Campbell, Mr G Campbell (3), Mr I Campbell, Mr J M Campbell, Mr M D Campbell, Mr N Campbell, Mr N G Campbell, Mr S Campbell, Mr T Campbell, Mr & Mrs J Campbell, Mrs A Campbell, Mrs D Campbell, Mrs F Campbell, Mrs L Campbell, Mrs Y Campbell, Ms A Campbell, Ms B Campbell, Ms L A Campbell, Revnd W M Campbell, Mr B Campbell-Taylor, Mr S Campion, Ms R Campling, Mr R Candeland, Mr G W Candelin, Miss R Candy, Mrs I Cann, Mrs K Cannatella, Mr A R Cannon, Gerry Cannon, Mr P Cannon, Mr G Canny, C Cant & C Walker, Mr J E Cant, Mr N Cant, Mr J Cantelo, Mrs K Canterbury, Miss S Caola, Cape Clear Bird Observatory, Mrs M J Caplin, Mr K A Capps, Mr P Capsey, Dr R Carden (2), Mr M A Cardus, Mrs P Cardus, Mr Cardy, Mr C M Careless, Miss H Carey, Mr D Carey, Mr J Carey (2), Mr K Carey, Mr N Carey, Steven Carey, Dr R Carleton, Mrs W Carlin, Miss H Carlton (2), Stuart Carlton, Mr W Carlyle, Carmarthenshire Bird Club, Mr A Carmichael, Mr I Carmichael, Mr S Carmichael,

Ms F Carmichael, Dr M Carmody, Mrs V Carnell, Dr J Carnie, Mr B R Carpenter, Mr C Carpenter, Mr R J Carpenter, Mr T Carpenter, Mrs E Carpenter, Mrs J Carpenter, Mr & Mrs Jeff Carpenter, Ms J Carpenter, Miss D G Carr, Mr Alastair Robert Carr, Mr D L Carr, Mr G M Carr, Mr I Carr, Mr M J Carr, Mr P Carr (2), Mr R J F Carr, Mrs J D Carr, Ms M Carr, Ms R Carr, Miss A Carrick, Mrs T Carrick, Mr M F Carrier, S & K Carriere, Mr A Carrington, Mr D G Carrington, Roger Carrington, Anne Carrington-Cotton, Dr M Carroll, Mr A Carroll (2), Mr F Carroll, Mr J Carroll, Mr M J & Mrs H Carroll, Mr P Carroll, Ms N Carroll, Mr B Carruthers, Miss J Carson, M Carson, Mrs M J Carson, Mrs S Carstairs, Dr I Carter, Nick Carter, Miss J K Carter, Miss L Carter, Miss S Carter, Miss W Carter, Mr A G Carter, Carter family, Mr B Carter (2), Mr C Carter, Colin Carter, Mr D Carter, D & M Carter, Mr G Carter (2), Mr I Carter, Mr J Carter (2), Jimmy Carter, Mr N Carter, Mr P Carter, Paul Carter, Mr R Carter, Mr S H Carter, Mr S Carter, Mr S J Carter, Mr T Carter (3), Mrs A Carter, Mrs C Carter, Mrs D Carter, Mrs H M Carter, Mrs M Carter, Mrs M D Carter, Mrs P Carter, Mrs V Carter, Mr P Cartlidge, Davy Cartwright, Mr V Cartwright, Mrs S J Cartwright, Miss H Carty, Mr P Carty, Mr D Cary, Dr R L Case, Mr N Case, Mr M Casey (2), Mr S Casey, Mrs A Cashford, Patrick Cashman, Mr J Cason, Mr C Cassells, Dr J Cassels & Arran Natural History Society, Miss C Cassidy, Mr M Cassidy, Mr M J P Cassidy, Ms J Cassidy, Casson family, Joy & Ian Castle, Mr J S Castle, Mr P Castle, Mr P E Castle, Castle Close Wildlife & Heritage Group, Castle Semple Countryside Rangers Pat, Denys, Hayley & Andy, Mr M Catchesides, Mr I Cater, Mr G Catley, Mr B R Catlin, Mr R C Catlin, Dr L W Catlow, Mr D C Catt, Ms F Cattanach, Mrs J Catterall, Mr M Catterwell, Mr C M Catto, Ms D Caudery, Mr F B Caudwell, Mr S Caudwell, Mr A Caulkett, Mrs Karen Cavalier, Mr P Cavalli, J & H Cavanagh, Mr M S Cavanagh, Mrs B Cavanagh, Mr D Cave, Mr J Caveney, Mr G R Caw, Mr D Cawdron, Mr S Cawkwell, P J Cawley, Dr D Cawthorne, Mrs J Cawthorne, Cayton & Flixton Carrs Wetland Project, CCW staff, Mrs D Cearns, Mr R Chace, Dr R Chadburn, Mr W Chadfield, Miss Y Chadwick, Mr A Chadwick, Mr D Chadwick, Mr J Chadwick, M G Chadwick, Martin Chadwick, Michael Chadwick, Mr M Chadwick (3), Mr P J W Chadwick, Mr S Chadwick, Mr T Chadwick, Mr J Chaffe, Mrs J Chaffe, M N Chaffe, Chailey Commons Society, Mr & Mrs J P Challis, Mr M L Chalmers, Mr S Chalmers, Mrs D Chalmers, Mrs S Chalmers, Ms G Chalmers, Mr J Chamberlain, Mr M Chamberlain (2), Mrs J Chamberlain, Mrs L Chamberlain, Mrs V Chamberlain, Mr A P Chamberlin, Miss C Chambers, Miss D E Chambers, Miss P Chambers, Miss S Chambers, Mr A Chambers, Mr A L Chambers, Mr D Chambers, Mr G Chambers (2), Keith Chambers, Mr N Chambers (2), Mr V A Chambers, Mrs M Chambers, Mr A Champion, Mr B Champion, Mr G D Champken, Miss N Chan, Miss C Chandler, Mr D Chandler, Mr N S Chandler, Mr P Chandler, Mike Chandler, Mr D Chaney, Mr R Changleng, Mrs C Chant, Mr J Chantler, Mr E Chaplin, Mrs M Chaplin, Dr Jason Chapman, Dr P Chapman, Miss J Chapman, Kate Chapman, Mr A Chapman (4), Mr B Chapman, Mr C Chapman, Chris Chapman, Mr H Chapman, Mr I Chapman, Mr J Chapman (2), Ken Chapman, Mr M Chapman (2), Mr N Chapman, Mr N J Chapman, Mr P Chapman (2), Mr R A Chapman (2), Mr R E Chapman, Stephen Chapman, Mrs M A Chapman, Mrs Michelle Chapman, S & J Chapman, Mrs W R Chapman, Mr N D Chappell, Mr R I Chappell, Celia Chapple, K & J Chapple, K G Charity, Dr K Charles, Dr R G Charles, Peter John Charles, Mrs A Charles, Mr E A Charleson, Mr P Charleston, Phil Charleton, J Howick & Netherside, Miss C Charlton, Mr C Charlton, Mr D Charlton, C Charlton & F Musgrave, Ms P Charlton, Mr H J Charlton-Jones, Drs Elisabeth C & Tom G Charman, Ms E Charman, Mr A Charnock, Mr P Charnock, Mr J Charteris, Dr R Chase, Mr S M Chastell, Dr C Chatham, Mr I Chattaway, Joanne Chattaway, Mrs S M Chatten, Mr V Chatterton, Mr P Chattle, Mr R Chatwin, Mr D Chavner, Mr J Chaytor, Mr G Checkley, Mr P R Cheek, Mrs T M Cheek, Mr A Cheeseman, Mr R A Cheetham, Mrs G Cheetham, Mr A S Cheke, Prof R A Cheke, Cheltenham Bird Club, Mr M Cherrill, Dr Peter C Cherry, Mr M S Cherry, Mrs J Cherry, Mr J Chesbrough, Cheshire Wirral Ornithological Society, Mrs S A Chesson, Mr D Chester, Mr J Chester (2), Mr S Chester, Mrs J Chester, Mr D Chester - Master, Mrs E Chestney, Mr G Chesworth, Miss M Cheung, Lt Cdr J M Cheverton R N, Mr D Chevins, Mr A P Chick, Mrs S J Chicken, Mrs J Chidlow, Mr S Chidwick, Mr B M Child, Mr J Childs, Mr K Childs, Mr P J Childs, Ms J Childs, Mr P Chillingworth, Mr S Chilton (2), Mr W Chilton, Ms J Chilton, Miss C Chilvers, Mr M Chinery, Mr K R & Mrs S E Chinn, Mr T J Chinn, Mr P Chinniah, Mr A Chipchase, Mr M Chippendale, Stephen Chippendale, Mr D Chisholm, Mr D Chivers, Mr J L Chivers & Mrs S R Chivers, Lady U Cholmeley, Mr M J Chorley, Mr D J Chown, Mr M Chown, Chown family, Ms J Chowne, Mr M Chris, Mr G Christer, Dr P Christian, Mr E Christian, Mr J Christian, Mr A Christie, Mr C J Christie, Mr C R Christie, Mr D Christie, Ms E R Christie, Dr S E Christmas, Dr R Christopher, Mr T Christopher, Mr B Church, Mr G Church, Mr P Church, Mr B Churches, Miss T Churchill, Miss W Churchill, Mr D J Churchill, Mr T Churchyard, Mr I Churm, Mr R Chuter, Mr & Mrs R N Cinderey, City of Edinburgh Council Countryside Ranger Service, Mr M Claassen, Mr G Clack, Mr K Claiden-Yardley, H & M Claire, Mr A Clamp, Mr S Clamp, Ms T G Clamp, Mr M Clancy, Mr P Clancy, Mr B Clapham, Rachel Clapham, Prof C S Clapham, Mr P Clapton, Mr R J Claridge, Dr H Clark, Dr N A Clark, Miss J Clark, Miss S Clark, Miss T Clark, Dr Bernard Clark, Mr B Clark, Mr C Clark, Mr D Clark (2), Mr D J Clark, Mr Dave Clark, Mr F C Clark, Mr H Clark, Mr J Clark (4), Mr & Mrs J E Clark, Mr J J Clark, J M Clark, Mr J R Clark, Mr K G Clark, Mr L Clark, Mr N Clark, Mr O Clark, Mr P Clark (2), Mr P L Clark, Mr R Clark (2), Mr R J Clark, Mr R T Clark, Mr R Clark, Mr S Clark (3), Mr W Clark (2), Mrs A Clark, Mrs C Clark, Mrs J Clark, Mrs L A Clark, Mrs S B Clark, Mrs S F Clark, Mrs T M Clark, Ms D Clark, Dr B Clarke, Miss M Clarke, Mr A J Clarke, Mr C S Clarke, Mr D Clarke (2), Derek WJ Clarke, Gordon Clarke, Graham Clarke, Mr G Clarke (2), Mr J Clarke (2), Mr J A Clarke, Mr J E Clarke, Mr J Clarke, Mr P Clarke (3), Mr R Clarke (2), Richard M Clarke, Mr R M Clarke, Mr S Clarke (5), Mr T Clarke, Clarke family, Mrs D Clarke, Mrs H Clarke, Mrs J Clarke, Mrs J M Clarke, Mrs Y R Clarke, Ms J R Clarke, Ms S Clarke, Ms T Clarke, Prof P Clarke, Mrs R Clarke-Simon, Mr E Clarkin, Dr K Clarkson, Dr Stephen Clarkson, Graham Clarkson, Mr J Clarkson, Miss C Clay, Mr A P Clay, Mr B R Clay, Mr G M Clay, Mr M Clay, Mr R M Clay, Mr B Clayden, K & M Claydon, K Claydon, Mr M Clayson, Mr N Clayson, Mr T Clayson, Mr D Clayton, Mr P J Clayton, Mr W Clayton, Ms C Clayton, Mr D Cleal, Miss V Cleaver, Ms V J Cleaver, Mr J Cleeter, Mrs G Cleeve, Mr T R Cleeves, Mr D Clegg, Ms A S Cleland, S Clement, Dr Andy Clements, Mr B Clements, John & Joy Clements, K & G Clements, Maurice Clements, Mr R J Clements, Mrs E C Clements, Ms A Clements, Mr A Clements-Partridge, Mr S Cleminson, Mr K J & Mrs A E Clench, Mr A Clerici, Mr G Clerkin, Mr B D Clews, Mrs H Clews, Mrs R Clews-Roberts, Mr A Cliff, Mr P Cliff, Mr A Clifford, Ms D Clifford, Mrs B Clifton, Mr W Clinch, Mrs J H Clinch, Mr S Clinton, Liz Clinton, Mrs J M Clissold, Mr J H Clitherow, Mrs G E K Clode, Mr P J Cloke, Miss S Clolse, Ms H Cloney (2), Mr N Clooney, Mr Jon Close, Ms E Close, Cameron Clotworthy, Mr R Cloude, Dr J Clough, Mr J Clough, Mr S Clough, Mrs C Clough, Cloughwood School, Mr D Clow, Mrs L Clow, C Clowes, Mr J M Cloyne, CLS & PJS, Mr E W Clubley, Mr D L Clugston, Mr A J Clunas, Mrs P A Clutterbuck, Clwyd Bird Recording Group, Clyde Muirshiel Ranger Service, Mr M Clydesdale, Miss S Coates, Mr L Coates, Mr S Coates (2), Mrs A Coates, Mrs J Coates, J Coatesworth, Mr D C Coath, Mr M Coath, Mr N Coats, Ms H F Coats, Ms Y Coats, Mr R Coatsworth, Mr T Coatsworth, Mr A Cobb, Mr J Cobb, Mrs F Cobb, Dr A Cobley, Mr M Coburn, Mr S Cochrane, Ms G Cochrane, Mr M Cock, Mr R P & Mrs C A Cockbain, Mr C Cockburn, Mr F Cockburn, Mrs S Cocker, Mrs S Cockerham, Mr J Cockram, Mr A Cockroft, Mr R Cockroft, Dr J A Cocks, Mr T D Codlin, Mr E Cody, Mr C Coe, Mr H C Coe, Mr J C Coe, Mrs S Coe, Coetir Mynydd, Mr P Coffey, Mr P M Cogan, David Cogger, Mrs A Coggon, Mrs S Coghlan, Mr A Cohen, Mr D A Cohen, Mr K Cohen, Mr S H Cohen, Ms Y Cohen, Mr J Coit, Mr D G Coker, Mrs A Colam-Ainsworth, Dr P Colbourn, Miss C H Colchester, Mr K Colclough, Dr H Cole, Mr A Cole, Mr C Cole (2), Mr J Cole, Mr M Cole, Mr M A Cole, Mr P Cole, Mr R C Cole, Celia & Graham Cole, Mrs D Cole, Mrs F A Cole, Mrs C Coleclough, Mr J Cole-Hamilton, Dr D Coleman, Dr P Coleman, Miss E Coleman, Mr B Coleman, Mr D Coleman, Mr J B Coleman, Mr J C Coleman, Mr J D Coleman, Mr J R Coleman, Mr P Coleman, Mr R Coleman, Mrs C Coleman, Mrs N Coleman, Mrs P Coleman, Ms F G Coleman, Ms T Coleman, Ms M Coleman-Cooke, Mr J Coleridge, Mr P J Coles, Mr T E J Coles, Mrs A Coles, Mrs R Coles, Mrs W Coles, Mr S Colgate, Mr M C Collard, Mr J H Colledge, Mr J N Collen, Mr A Collens, Mr T Collett, Mr T A Collett, Mr L Colley, Mr P Colley, Mr I Collier, Mr J Collier, Mr M P Collier, Mr R Collier, Ms J Collier, Mr P N Collin, Mr R Collingridge, Mr H T G Collings, Mr M Collings, Mr S Collings, Dr J Collins, Miss J Collins, Miss L Collins, Mr A Collins, Mr C Collins (2), Mr C B Collins, Mr C J Collins, Mr D R Collins, Mr E Collins, Mr F Collins, Mr J N Collins, Mr J Collins, Mr K Collins, Mr M Collins (2), Mr M C Collins, Mr P Collins, Mr R Collins, R E C Collins, Mr S Collins (2), Mr S N Collins, Mr S P Collins, Mr T S Collins, Mrs C W Collins, Mrs J Collins, Ms A Collins, Ms C Collins, Ms F Collins, Ms M L Collins, Mr J Collinson, Lee Collinson, Mrs C Collinson, Dr G Collis, Mr M Collis, Mr R Collishaw, Mr R Collister, Mr J R Collman, Catherine Collop, Mr S Colsell, Mr S Colton, Mr P Combridge, Mr R Comerford, Ms D M Comerford, Mr M Comley, Common family, Mr J Common (2), Ms A Compton, Mr M Concannon, Mr S Concannon, Mr J Condron, Mr D Coney, Mr S P Coney, Miss R Coneybeer, Mr J Conlin, Mr F Conn, Mr T Conn, Mr A Connell, Mr J Connell, Mr T Connell, Mr E P Connelly, Mrs J M Connelly, Mrs L Connelly, Mrs N F Connelly, Dr J Conner, Mr J Conner, Dr C N Connolly, Mr G Connolly, Mr J G Connolly, Mr S Connolly, Mrs M E Connolly, Mr J Connor, Mr N Connor, Mr C R Conroy, Mr D Constantine, Convent of Poor Clares, Mrs C Convery, Mrs M D Convery, Dr G J Conway, Mr D A Conway, Mr M Conway, Mr S T Conyers, Lisa Neale, Miss S Cook, Mr A Cook, Mr A J Cook, Alasdair Cook, Mr A S Cook, Mr C J G Cook, Mr C Cook, David Cook & Juliet Owen, Mr D Cook (2), Mr D L Cook, Mr D Cook, Team Cook, Mr G Cook, Henry Cook, Mr J Cook (3), Mr K Cook, Mr M Cook (2), Mr M J Cook, Mrs C A Cook & Mr A Beeton, Mr M J H Cook, Mr M Cook, Nick Cook, Mr N Cook, Mr P Cook (2), Mr R Cook (2), Mr S Cook, Mr S R Cook, Mrs A Cook (2), Mrs E A Cook, Mrs G M Cook, Mrs M P Cook, Mrs R Cook, Mrs S Cook, Miss M Cooke, Mr C Cooke (2), Mr D Cooke (2), Mr D C Cooke, Mr G Cooke, Mr I Cooke, Mr J D Cooke, Mr M Cooke (2), Mr M R Cooke, Mr N J Cook & Mrs W Cook, Mr P Cooke, Mr R Cooke, Mr & Mrs Cooke, Mrs J Cooke, Mr D J & Mrs K Cooke, Nicola Cooke, Mrs S Cooke (2), Ms C Cooke, Prof F Cooke, Mr A R & Mrs J Cookney, Mr M Cooksley, Mr D A Cookson, D & D Cooling, Miss R Coombes, Mr D Coombes, Mr J Coombes, Mr R Coombes, Mrs E Coombes, Ms R Coombes, Mr R Coombs, Ms G Coombs, Mr M J Coon, Mr P C Cooney, Mr T Cooney, Dr A J Cooper, Graham Cooper, Miss H Cooper & Mr D Nolan, Miss Cooper, Miss S Cooper, Miss V Cooper, Andrew Cooper, Tony Cooper, Barry J Cooper, Mr C Cooper, Mr D Cooper (2), Dennis Cooper, Mr E Cooper, Mr G L Cooper, Mr G W Cooper, J E J Cooper, Mr J Cooper (3), John Cooper, Mr J R Cooper, Mr M Cooper, Mr M C Cooper, Mr M S Cooper, Neil Cooper, Mr N Cooper, Philip Cooper, Mr P F Cooper, Mr R Cooper (2), Simon Cooper, Mr S Cooper (2), Mr S D Cooper, Mr T Cooper (2), Mr T J Cooper, Mrs B Cooper, Charmaine Cooper, E & A Cooper, Mrs L Cooper, Mrs M D Cooper, Mrs B Cooper-Poole, Mr S Cooter, Miss T E Cope, Mr C J Cope, Mr N J C Cope, Mr R D Cope, Simon Cope, Judy Copeland, Mr P G Copestake, Mr W Copestake, Mrs K Copestake, Dr A S Copland, Ms A Copley, Jeff Copner, Mr J D Copp, Ms A Copping, Chris Coppock, Mr S Copsey, Mr P Corbet, Mr T I Corbett, Mrs T & Mr A Corbett, Mr G Corbett-Marshall, Corbridge WEA, Mr T Corcoran, Mrs T Corcoran, Mr R Cordery, Mr D Cording, Mr G W Corfield, Mr P Corfield, Mr R A Corfield, Mrs S Corfield, Mr A Cormack, Mr J Cormack, Mr P Cornelius, Mr A Corner, Mr P Corner, Ms B Cornes, Mr G Cornfield, Mr G Cornick, Mr C Cornish, Mr D Cornish, Mr D M Cornish, Mr R J Cornish, Mr Len Cornwell, Mr L Cornwell, Mr R D Corran, Mrs L Corrie, Chris Corrigan, Mr M Corrigan, Mr C J Corse, Cortachy Primary School, N Cory-Wright, Cosford Conservation Group, Dr P J Cosgrove, Mr D Cosh, Mr B L Cosnette, Dr L Costa, Mr S Costa, Mr B Costello, Mr E Costello, Ms L Costello, Mr T Coster, Mr P Costigan, Mr A Cottam, Miss F Cottee, Mr D W Cottee, Mr G Cotter, Mr N R Cotter, Mr I Cotterell, Mr F Cottington, Mr R Cottis, N W Cottle, Mrs C Cottle, Dr D Cotton, Mrs C V Cotton, Ms S Cotton, Mr J Cottrell, Mr V Cottrell, Mrs P Cottrell, Mr R T Cottrill, Mr P Couch, Couchman family, Miss L Coughlan, Miss N Coulson, Mr A Coulson (2), Mr E Coulson, Mr M Coulstock, Mr S Coulter, Mr G Coultrip, Mr R Council, Ms S Counsell, Countryside Services, Mrs C Coupe (2), Mr B Couper, Bob Coursey & Ryedale Naturalists, Mr I Court, Mr P Courtney, Mrs M Courtney, Dr R Cousins, Dougal Cousins, Mr R Cousins, Mr G Cousquer, Mr H B Coventry, Mr M Coventry, Ms L Coventry, Mike Coverdale, Mr S Covey, Dr A V Cowan, Miss C Cowan, Mr J Cowan, Mr J M Cowan, Mr T C Cowan, Mrs D Cowan, Mr C Coward, Mrs A Cowdrey, Mr J Cowe, Mr C Cowell, Canon B Cowen, Mr R Cowen, Dr R Cowie, Mr J Cowie, E Cowley, Mr M Cowley, Mrs J Cowley, Mr J A Cowlin, Micheal & Mick Cowming, Mr S H Cowper, Mr J R Cowser, Mr A J Cowx, Dr AFJ Cox, Dr R A F Cox, Dr S Cox, Miss B Cox, Miss J Cox (2), Mr A Cox, Andrew Cox, Mr B Cox, Mr C Cox, Mr C J Cox, Chris Cox, Dave Cox, Mr G Cox, Mr I Cox, Mr I N Cox, Jeff Cox, Mr K Cox, Mr M Cox, Mr M & Mrs H Cox, Mr Richard Cox, Mr R Cox (3), Mr S Cox, Mrs F Cox, Mrs G L Cox, Mrs K M Cox, Mrs T Cox, Ms L Cox, Ms M Cox, Dr A Coxon, Dr D Coxon, Mr J Coxon, Mr Jim Coyle, Shaun P Coyle, Mr J Coyne, Mr S Coyne, Miss J Cozens, Helen Crabtree, Mr E Crabtree, Mr N Crabtree, Mrs S Crabtree, Ms E Crabtree, Mr B Cracknell, Mr J Cracknell, Mr C Crafford, Dr B Cragg, Mr D Cragg, Mr D A Cragg, Mr J Cragg, Mr P Cragg, Master R J Crago, Dr A Craig, Mr C Craig, Mr D Craig, Mr J J Craig & Ms J Morrison, Mr M Craig, Mrs E Craig, Mrs G Craig, Mr W A Craigie, Mr M S Cram, Mr K Cramer, Mr D J Cramp, Mr E Crandley, Miss E Crane, Mr K Crane, Mrs L Crankshaw, Mr G Cranston, Rebecca Cranston, Mr W Cranstoun, PA Cranswick, Ms J Crapnell, Brian & Liz Crathorne, Mr P Craven, Miss A Crawford, Miss E Crawford, Mr A D Crawford, Mr David A Crawford, Mr R Crawford (2), Mrs Ann Crawford, Mr D E Crawley, Mr &Mrs P Crawley, Mr D Crawshaw, Mr J R Crawshaw, Mrs P M Crawshaw, Mr R Crayk, Miss A Creamer, Mrs L Creamer, Major A J Crease, Dr Michael R Creasy, Mr J C Crebbin, Nicky & Tony Credland, Mr I Cree, Mr M Creegan, Mr B Cregan, Mr P Cregg, Mr M Creighton, Sam Creighton, Mr R H Creighton, Mr R Crellin, Mr A Cremin, Mr P Cremins, Dr W R L Cresswell, Mr M Cresswell, Mr N C Crighton, Mr T L Crilley, Miss S Cripps, Mr R Cripps, Bob Cripps, Mr K Crisp, Mr M Crisp, Mr P Crispin, Mr A Cristinacce, Mrs M Criswell, Mr M B Critchley, Mr M A Critchlow, Dr G C Crittenden, John Crix, Miss S Cro, Mr R Croasdaile, Mr N J Crocker, Mr P Crockett, Mrs R Crockett, Mr S Crockford, Mr K G Croft, Mr S Croft, Mrs C Croft, Ms L Croft, Ms Pat Crofton, Miss R E Croger, Mrs H Cromack, Mr J Cromie, Dr N J R Crompton, Mr R Crompton, David & Judith Crompton, Mr C Cronin, Mr D Cronin, Mr K Cronin, Mr S Cronin, Mrs A Cronin, Dr S J Crook, Mr D Crook, Mr H B Crook, Mr J Crook, Mr P F Crook, Mrs C Crook, Mrs H J Crook, Ms L S M Crook, Rt Revd J M Crook, Mr B Crookes, Dr Peter Crooks, Mr A Cropper, Mr P Cropper, Mr A Crory, Mr J Crosby, Mrs J Crosher, Dr J Cross, Michelle Elizabeth Cross, Mr A Cross, Mr A V Cross, Mr D Cross, Mr M Cross (2), Mr S Cross, Mrs A R Cross, Mrs C Cross, Mrs G E Cross, Mrs S Cross, Mrs S K Cross, Ms T Cross, Ms O Crossan, Mr R Crossen, Mike Crossley, Mr M Crossley, Mr R Crossley, Mr A Crossman, Mr N Crossman, Mr N A Croton, Miss V Crouch, Mr N C Crouch, Ms L Crouch, Mr A Croucher, Mr J D Crowder, Miss E Crowe, Mr A Crowe, Mr S Crowe, Ms O Crowe, Mr J Crowley, Dr P Crowley, Mr P Crowley, Ms M Crowley, Miss R Crowther, Mr N Crowther, Mr D Croxson, Mrs J Croxson, Mr J Croxton, Mr Robert J Croxton, Miss H Crozier, Susan Crozier, Mr J Crum, Mrs J Crummy, Michael G Crutch, Mr A N Crutchley, Mr C Cryer, Ms J Crystal, Mr K R Cservenka, Mr T Cubbin, Mr M G Cubitt, Mr A Cudmore, Mr T Cue, Mr M J Cuff, Mrs S Cuff, Mr M Cuffe, Mr T Cuffe, Mr D A Culham, Mr P Culkin, Dr J Patrick Cullen, Mr C Cullen, Mr D Cullen (2), Mr G Cullen, Mrs C Cullen, John Culleton, Mr R Culley, Mr S Culley, Mr J Culligan, Cullip family, Dr A J Culshaw, Mr D Culverhouse, Cumbria Bird Club, Mr I Cumming, Mr S Cumming, Ms M J Cumming, Mrs A C Cummings, Mrs P Cummings, Dr S Cummins, Mr G C Cundale, Mr A W Cundall, Mr G J Cundall, David Cunliffe, Mr J Cunliffe, Mr K C Cunnane, Miss L Cunningham, Mr A Cunningham, Mr A J Cunningham, Mr D Cunningham (2), Mr I Cunningham, Mr T Cunningham, Ms T Cunningham, Mrs C Cupit, Mr R Curd, Debbie & Maddy Ford, Dr P Cureton, Mrs J Cureton, Mr J Curley, Mr R Curley, Miss A J Curnow, Mr S Curnow, Mrs J Curnow, Mr E Curran, Mr G Curran, Mr K Curran, Mrs L Curran-Dumez, Mrs K Curran-White, Mr F A Currie,

Mr J Currie (2), Mr N Currie, R Currier, Mr A Curry, Mr M Curry, Mr M C Curry, Mr R Curry, Mr J Curson, Mr L Curson, Mr S Curson, Mr J Cursons, Mr M Curtin, Mr A Curtis, David Curtis, Mr J Curtis (2), Mr M Curtis, Mr Roger Curtis, Mr R Curtis, Mr S Curtis, Mr W F Curtis, Mrs D Curtis, C R Curtis, Mrs L Curwen, Mr J Cusworth, Mr C P Cuthbert, Mr C R Cuthbert, Mr C Cuthbert, Mr A Cuthbertson, Mr A Cutler, Mr P Cutler, Mrs C Cutler, Mr B Cuttell, Liz Cutting, Mr A Cutts (2), Mr M A J Cutts, Mr K E Cypher, Mr J Cyprus, Mr S Czornyj, Mr M J Czuczman

D Dr S R D Da Prato, Mr D Da Silva Grave, Mrs Y Daams, Andrew Dacre, Mr M Dadds, Mr R Dafydd, Dr J R Dagley, Dr G Dagnall, Mrs D Dahl, Ms L Dahl, Dr J N Daisley, Dr B Dale, Miss J Dale, Mr A Dale (2), Chris Dale, Mr M Dale, Mr P Dale, Mr R Dale, Mr V A Dalkin, Mr T Dallimore, Mr M B & Mrs L E Dalling, Mr D R Dallinger, Mrs H Dallinger-Mills, Miss L A Dallmeier, Mr A J A Dally, Andrew Dally, Mr R Dalrymple, Ms S Dalrymple, Dr B Dalton, Mr T Dalton, Mr F Daly, Mr G Daly, Mr M O Daly, Mr N P Daly, Mr R H Daly, Mrs S Daly, Ms H Daly, Mr R Dalziel, Dr A Damant, Mr D Damant, Miss C Damon, Miss N Dan, Mr D Dana, Mr B Danaher, Mr I Danby, Mr M Dancer, Dr M Dancey, Stoycho Danev, Ms S Dangerfield, Mr R Daniel, Miss L J Daniells, John Daniels, Mr J P Daniels, Daniels family, Mr P Daniels, Mr T P Daniels, Mrs B Daniels, Mrs J M Daniels, Ms E Daniels, Ms R Daniels, Mr I Danielson, Mr M Dannatt, Mrs L Dantinnes, Mr R Daplyn, Mr P Darby, Mrs V Darby, Mr J Darbyshire, Mr S Darbyshire, Mr J D'arcy, Dr P J Dare, Mrs C M Dare, Mr P J Dargue, Mr L Dark, Mr M Darke, Mr M Darlaston, Mr C Darling, Mr M Darlow, Dr J N Darroch, Mr J A Darroch, Mr R R & Mrs N Darsley, Darvell Group, Mr G C Darvill, Hugh Darwen, Miss P Dashwood, Mr L B Daunt, Ms A C Davage, Mr M Davenport, Miss C Davey, Mr G Davey, Mr P Davey, Mr S Davey, Mr F David, The Davids, Dr A V Davidson, Mr A Davidson, Drew Davidson, Mr E Davidson, Mr H Davidson, Mr I Davidson, Mr J Davidson (2), Mrs A Davidson, Mrs D Davidson (2), Mr T R Davie, Dr A D Davies, Dr C E Davies, Dr I Davies, Dr R Davies, Miss C Davies, Miss J Davies, Miss K Davies, Miss L Davies, Mr A Davies (2), Mr B Davies, Carlos Davies, Mr C Davies (2), Mr D Davies (2), Mr D C Davies, Mr D P Davies, Mr D Davies, Mr E Davies, Mr G Davies (3), Mr Gareth Davies, Mr G Davies, I Davies & family, Mr I Davies, Mr J Davies (2), Mr J A Davies, Mr J D Davies, Mr J G Davies (2), Mr J Davies, Mr K Davies, Keith & Judith Davies, Mr L Davies, Mr M Davies, Mr M H Davies, Mr M J Davies, Mr N Davies, Mr O Davies, P Davies family, Mr P Davies, Mr P N Davies, Mr R Davies (2), Mr Russell Davies, Mr R J Davies, Mr R T Davies, Mr S Davies (9), Mr S J Davies (2), Mr S Davies, Mr T Davies, A Davies family, Mrs D Davies, Heather Davies, T & H Davies, Mrs J Davies (3), Mrs L Davies, Mrs M Davies, Mrs P Davies, Mrs S Davies, Mrs S A E Davies, Ms G Davies, Ms J Davies, Ms L E Davies, Ms S I Davies, Ms S M Davies, Sir R Davies, Dr G Davis, Mr John Davis, Miss C Davis, Miss J Davis, Miss R Davis, Mr Andrew H Davis, Mr A M Davis, Mr B Davis, Mr G Davis, Mr J Davis (3), Mr J A Davis, Mr K J Davis, Mr M Davis, Mr & Mrs N Davis, Mr P Davis, Mr P E Davis MBE, Mr S W Davis, Tim Davis & Tim Jones, Mr T Davis, Mrs A Davis, Mrs P L Davis, Mrs S Davis, Ms A Davis, Ms C Davis, Ms L Davis, Miss L Davison, Mr A Davison, Mr B Davison, Mr C N Davison, Mr I Davison, Mr J Davison, Mr M Davison & Mrs A Middleton, Mr N Davison, Mr P Davison, Mrs V Davison, Mr J Davy, Mr A Daw, Mr P C Daw & Argyll Bird Club, Mr A Dawes, Mr A P Dawes, Mr G R Dawes, Mrs R Dawes, Ms J Dawes, Miss C Dawkins, Miss M Dawkins, Mr R Dawkins, Dawlish Warren NNR Staff, Dawlish Warren Recording Group, Mr D Dawney, Dr D G Dawson, Bob Dawson, Miss E Dawson, Miss J Dawson (2), Miss K Dawson, Mr K Dawson (2), Mr L Dawson, Mr M D Dawson, Mr M J Dawson, Dawson family, Mr P Dawson, Mr R Dawson, Mr T Dawson, Mrs A Dawson, Ms J Dawson, Ms K Dawson, Mr B Dawton, Mr A Day, Mr J C Day, Mr J P Day, Mr M Day, Mrs A Day (2), Mrs I Day, Mrs J Day, Mrs S Day, Mr R A Dazley, Mr G De Boer, Dr D De Freitas, Ms S De Jong, Mr S De La Haye, Mrs A De Muynck, Miss A F de Potier, Mr M A de Retuerto, Mrs Jean de Selincourt, Dr R De Souza, Mrs C De Wilde, Mr P A Deacon, Ms L Deacon, John & Rosemary Deag, Mr B Deakin, Mr P & Mrs B Deakin, Bill Deakins, M Dean & R Hill, Mr A R Dean, Mr C H Dean, Mr J Dean, Mr P Dean (2), Mr R J Dean, Mr S Dean, Mr T R Dean & Mrs J Dean, Mrs P Dean, Mr J Deans, Mr M Deans, Mr P G Deans, Mr D Dear, Dr C Dearden, Mr H T Dearden, Mr R A Dearden, Mrs P Debarr, Mr P K Dedicoat, John & Janet Dedman, Chris W Dee, Mrs M Dee, Miss A Deegan, Mr P Deegan, Mr T Deegan, Denise Deegan, Mrs A Deem, Mr J Deen, Defence Estates (MOD Bird Count), Mrs S Defferary, Mr J Degg, Mr L J Degnan, Mr M Degnan, Mr J Deighton, Mr R Deignan, Mr P Delahunty, Mr P Delaloye, Mr N Deller, Mr J B Dellow, Mr M Delong, S J Delve, Mr J R Delve, Mrs L E Demaine, Mr B Demby, Mr A Denard, Mrs B Dendy, Mrs S E Denham, Ms J Denham, Mr A Denholm, A & J Denison, Mr A E Denker, Mrs B Denman, Mr Denis Dennehy, Mr D F Denney, Tom Denney, Mr C Dennis, Mr G Dennis, Mr P Dennis (2), Mr R H Dennis, Mr S Dennis, Mrs C Dennis, Miss M Dennison, J & J Dennison, Mr R Dennison, Mr W Dennison, Dr M J H Denny, Mr G R J Dent, Mr M D Dent, Mr P Dent, Mr S Dent, Mr M L Denton, Mr R Denton, Ms T Denton, Mr P Denyer, Mr R A Denyer, Mr A W A Derks, Dr K J Derrett, Mr C Derri, Miss J Derricott, Mr M Derrington, Mrs L Derry, Mr M Derwin, Mr N Desmond, Dr N Detert, Steve Dettmar, Mrs F Devery, Mr R Deville, Mr AB Devitt, Mr R H Devitt, Mrs F Devon, DBWPS, Mrs G Devonshire, Mr Rafe Dewar, Mr R Dewar, Mr S Dewar, Mr T G Dewdney, Mr R A Dewey, Mr S Dewey, Mr O Dewhurst, Mr R Dewhurst, Mrs B Dewhurst, Mrs S L De-Wit, Mr J Dews, Mr D J Dewsbury, Mr J Dexter, Mrs L J Dezonie, Mr J R Diamond, Mr I Dibble, Mrs J Dibnah, Mr A Dick, Mr H Dick, Mr R Dick, Mr T D Dick, Mrs A Dick, Mrs K A Dick, Miss R Dickens, Ellie Dickens, Mr D S Dicker, Mr T Dickerson, Roger Dickey, Mr J Dickie, Mr B H B Dickinson, Mr D F Dickinson, Mr K Dickinson, Mr P K Dickinson, Mrs M Dickinson, Dr C Dickson, Miss W E Dickson, Mr J Dickson (2), Mrs B Dickson, Mrs D M Dickson, Mr F Dignan, Prof D Dilks, Mr D Dillon, Mr I A Dillon, Milko Dimitrov, Megan C Dimitrov & Maira Hall, Mr D Dine, Dr B Dines, Dr T Dines, Mr D Dines, Mr J Dingemans, Mr T & Mrs G Dingwall, Mrs S Dinham, Mr R Dinnes, Ms C E Diprose, Mr S Diserens, Mr G Ditchburn, Ms E Diver, Mr J Dix, Mr S Dix, Mr T J Dix, Mrs J Dix, Miss E Dixon, Miss J Dixon, Mr A Dixon (2), Mr G Dixon, Mr J M Dixon, Mr L Dixon, Mr M Dixon, Mr M J Dixon (2), Mr T E Dixon, Mrs A Dixon (2), Miss H D'mello, Mr R Dobbins, G E Dobbs, Mr R Dobbs, H Dobie, Roger Doble, Mr F Dobson, Ian Dobson, Mr I Dobson, Mr P Dobson, Mr R Dobson (3), Mr S Dobson, Angela Dobson, Mrs S Dobson (2), Ms G Dobson, Mr A Docherty, Mrs M Docherty, Mrs R Docherty, Mr B J H Dodd, Mr D E J Dodd, Nigel Dodd, Peter Dodd, Mr S G Dodd, Mr S P Dodd, Tony Dodd, Mr G W Dodds, Mr M Dodds, Mrs V A Dodds, Mr T Dodman, Mr B Dodson, Ms A D Dodson, Mr F R Dodsworth, Mr P Dodsworth, Mr J Doe, Mr C Doherty, Mr J Doherty, Mr P A Doherty, T A Doherty, Ms D Doherty Ryan, Mr G Doig, Mr J Dolan, Mrs H Dolan, Mr K Dolby, Miss E M Doley, Mr H Dolphin, Peter Dolton, Miss M Dominey, Richard Donaghey, Dr A Donaghy, Miss K Donaghy, Dr P Donald, Chris Donald, Mr C Donald, Mr D Donald, Mr B Donaldson, Mr D Donaldson, Mr H Donaldson, Mr J Donaldson, Mrs G Donaldson, Mr B Donato, Mr B Donders, Mr A Done, Mr J Done, Mr D Donnell, Anne Donnelly, Mr K Donnelly, Nicholas J Donnithorne FLS FRES, Mr S Donockley, Mr T Donoghue, Miss T Donohue, Miss B Donovan, Mr P Donovan, Dr P Doody, Mr O Doody, Mrs A Doore, Mr T M J Doran, Ms H Doran, Clare Dore, Mr M Dorling, Mr A Dorman, Mr K Dorman, Mr J Dormer, Mr H Dorrington, Mr J Doswell, Mr H E M Dott, Mr M D Douch, Mr P Douch, Dr T W Dougall, Ms A Dougall, Mr S Dougan, Alan & Carol Dougherty, Dr R W Doughty, Ross Doughty, Mark Doughty, Ms L Doughty, David Douglas, Miss H Douglas, Mr B Douglas, Mr E W Douglas, Mr F Douglas, Mr G Douglas, Mr M Douglas, Mr R P Douglas, Mrs E Douglas, S & I Douglas, Dounby School, Dr A F G Douse, Mr C Doust, Mr P Douthwaite, Mr R J Douthwaite, Dr S J & Mrs H J Dove, Mr M J Dove, Mr P Dove, Mr & Mrs T Dove, Mrs T Dove, Ms M J Dove, Miss M E Dover, Mr M Doveston, Miss C Dow, Ms L Dow, Mr J Dowding, Mr Dowding, Mr A Dowell, Mr F Dowell, Miss A Dower, Richard & Elaine Dowle, Mr C L & Mrs E I Dowling, Mr J Dowling, Ms B Dowling, Mr G J Down, Mr J Down, Malcolm Down, Mrs J P Down, Miss K Downes, Mr J A Downes, Mr P Downes, Mrs M Downes, Mr B Downey, Mr F Downey, Mr H Downham, Dr Iain Downie, Miss A Downing, Mr R E Downing, Mr S Downing, Mrs H Downing-Emms, Vince Downs, Andy Dowse, Mrs W Dowse, Mr J M Dowson, Malcolm Dowty, Rosemary Doyle, Mr C Doyle, Mr D Doyle, Mr F Doyle, Mr J Doyle, Mr P A T Doyle, Mr T Doyle (2), Mrs J Doyle, Mrs K E Doyle, Mr M J D'Oyly, Ms J Drage, Mr C Drakard, Mr I P Drake, Mr J Drake, Robert Drake, Mr S Drake, Mr T Drakeford, Mr J M Dransfield, Mr A Draper, Mr A J Draper, Bill Draper, Mr R Dray, Mrs P P Dray, Mr I Draycott, Miss M Draye, Mr P Drayton, Mr C H J Dredge, Miss E F Dresner, Mr Alan Drever, Miss K Drew, Mr E Drew, Mr M Drew, Mr P Drew, John Drewett, Mr A Drewitt, Ed Drewitt, Mr J Driessen, Mr A C Drinkel, Mrs S Drinkwater, Mr K Driske, Mr C Driver, Mr P Driver (2), Mr M Droy, Mrs J Druett, Mrs R Drumm, Mr J Drury, Mr R Drury, Ms C Drybrough, Mr P Dryburgh, Mr B Dryden, Mr D Dryden, Mr M Dryden, Mr R A Dryden, Miss A Drysdale & Forvie NNR, Ms P Drysdale, Mr C du Feu, Leo du Feu, Mr R Du Feu, Mr A S Duckels, Miss K Duckers, Dr D P Duckett, Mr M Duckham, Mr B Duckhouse, Mrs L Duckworth, Ms N A Duckworth, Christine Dudgeon, Mrs P R Dudgeon, Mr C Dudley, Mr S P Dudley, Mrs C Dudley, Mr C Duerdoth, Miss F Duff, Mr A Duff, Mr K Duff, Mr N Duff, Mr C Duffey, Mr S Duffield, Mr John F Duffy, Mr K Duffy, Mr R Duffy, Mrs J Duffy, Mrs G Duffy-Jentz, Mr D Dugan, Mr R Dugan, Mr C Dugdell, Mr D Duggan, Ms M Duggan, Ms O Duggan, Ms R Duggan, Mr A Dugmore, Mr K Dugmore, Mr S Duke, Mr F J Dummigan, Mr K A Dummigan, Mr P Dumpleton, Mr J E L Dunbar, I G Duncan, Dr L Duncan, Mr A Duncan, Mr J Duncan, Mr J R Duncan, Mr KP Duncan, Mr K Duncan, Mr R Duncan, Mr W Duncan, Mrs S Duncan, Ms A E Duncan, Ms K Duncan, Mrs P Duncanson, Mrs S Duncanson, Mrs J Dunch, Dunderrow National School, Dr E Dunford, Mr D Dunford, Dungeness Bird Observatory (2), Mr K Dunham, Mr C J Dunkerley, Mrs A Dunkley, Mr C R Dunlop, Mr J Dunlop, Dr J C Dunn, Robert Dunn, Mr J Dunn, Mrs K Dunn, Mrs M H Dunn, Ms N Dunn, Ms S Dunn, Prof G Dunn, Mr B Dunne (2), Mr D Dunne, Mrs N Dunne, Ms M Dunne, Mr S Dunnett, Ms N Dunphy, Mr D Dunstan, Mr R Dunstan, Mr S Dunstan, A B Dunstan, Mr J Dunster, Ms R Dunston, Mr P Dunworth, Mr B J Durham, Mr M Durham, S A Durham, Durham Bird Club, Mr H Durka, Mr A Durkan, Mr K Durman, Mrs J P Durman, Mr J Durnell, Mr P Durnell, Mr M Durney, Mr J Durrant, Mr R Durrant, Mrs J Durrant, Dursley Bird Watch Pres Society, Mr J Dustow, Mr C Dutton, Mr S Dutton, Mr T P Dutton, Dwane family, Mr M Dwyer, Mr P R Dwyer, Mrs H Dyas, Miss L Dyckhoff, Mr J Dyda, Mrs S Dyde, Mrs B Dye, Mr D Dyer, Mr J Dyer, Mr M P Dyer, Mr M Dyer, Mr R Dyer, Brian Dyke, Mrs P E Dykes, Mr J N Dymond, Ms L Dymond, Mr G P Dyne, Mr P Dyne, David & Ros Dyson, Mr B Dyson, Mr C D Dyson, Mr C J Dyson, Ms D Dyson

E Mr S Eachus, Mr R Eade, Ray Eades, Mrs P Eades, Mr R Eadie, Eae Ltd, Joan & Roger Eagles, Mr R Eagles, Mrs M Eagles, Mrs M M Eaglestone, Mr R Eames, D Eames, Mr R Eardley, Ms M J Earle, Ms J Earls, Mr A Earnshaw, Mr K Earnshaw, Lynn Earnshaw, Mr W J Earp, Les Easom, Mr M J East, Mr P R East, East Ayrshire Countryside Services, East Lancashire Ornithologist Club, East London Birders Forum, Mr M Easterbrook, Mr T G Easterbrook, Mrs J L Eastlake, Miss J Eastmead, Dr J Easton, Andrew C Easton, Mr Jim Easton, Mr J Easton, Mr M Easton, Mr M M Easton, Mrs M Easton, Mrs W Easton, Dr M Eaton, Mr A Eaton, Mr B Eaton, Mr D Eaton, Mr D R Eaton, Mr J Eaton, Mrs R Eaton, Mr T Eaton Walker, Mr R Eaves, Mr S Eaves, Mr A Eaves, Mr A Ebanks, Ian Ebbage, Mr R A & Mrs L N Ebbs, Mrs L Ebbs, Mr A Eberst, Mr R J Ebrey, Mr K G Ebsworthy, Ebworth Estate, Martin & Julie Eccles, Mr Howard W Eccles, Mr J Eccles, Mr T Eccles, Mr P Eccleshare, Mr A Eccleston, Mr A Eckersall, Ecological Survey & Assessment LTD, Gill Edden, Mr R Eddleston, Ms L Eddleston, Dr M J Eddowes, Mr J Ede, Mrs D Ede, Mr G Eden, Mr C Edgar, Mr M Edgar, Mr R D M Edgar, Mrs S Edgar, Mr J Edge, Mr M Edgeller, Mr J Edis, Mr W Edmond, Miss N Edmonds, Mr L Edmonds, Mr P Edmonds, Mrs M M L B Edmonds, Mr P Edmondson, Mr S Edmunds, Mrs B Edmunds, Mr A Edward, Dr J Edwards, Dr J P Edwards, A Edwards-Jones, Mr A H Edwards, Mr A J B Edwards, Leybourne-Edwards family, Mr B Edwards, Mr E J Edwards, Eirwen Yemn Edwards, Mr G Edwards (2), Mr J Edwards (2), Mr J H Edwards, Mr M Edwards, Mr N M Edwards, Mr R Edwards (2), Mr R L Edwards, Mr S Edwards (2), Mr S B Edwards, Mr S J Edwards, Mr S Edwards, Mr T Edwards, Mr T W Edwards, Mrs A Edwards, Mrs C Edwards (2), Mrs E A Edwards, Mrs J Edwards, Mrs M V Edwards, Mr B & Mrs P Edwards, Ms A Edwards, Ms E Edwards, Mr L Edwins, The Edyveans, Mr P Eele, D R Eele, Mr P Egan, Mr G Egarr, Mr H R Egelstaff, Miss C Eggleton, Dr S M Eglington, EHS, Ms S Einschutz, Mr B Eke, Mr G R Ekins, Mr R E Ekins, Mr A Elberse, Mr T P Elborn, Elcoate family, Mr M D Elcoate, Mr S Elcoate, Miss A Elder, Mr C Elder, Elder family, Mrs J Elderton, Mr M Eldridge, Mr P J Eldridge, Ms J Eldridge, Ms K Eldridge, Mr P J Elford, Mr S Elford, Mr T Elias, Mrs A Elias, Mr J Elkin, Mr R Elkington, Norman Elkins, Mrs M E Elkins, Mr A A Ellard, Ms N Ellard, Mr J Ellegiers, Mr J Elliot, Ms L Elliot, Dr S C Elliott, Miss S Elliott, Mr A Elliott, Mr B Elliott, Mr C Elliott, Mr D I F Elliott, Mr G D Elliott, Mr J C Elliott, Mr J W Elliott, Mr L Elliott, Mr M Elliott, Mr P Elliott, Mr S W Elliott, Mrs J Elliott, Mrs Jan Elliott, Miss N Elliott-Charles, Mr K Elliottt, Dr D Ellis, Miss C Ellis, Miss M Ellis, Miss V Ellis, Mr A J C Ellis, B R Ellis, Mr B Ellis (2), Mr D Ellis, Mr G Ellis, Ian D Ellis, Mr I G Ellis, Mr Ian S Ellis, Mr J Ellis (2), Mr K B Ellis, Mr N J Ellis, Mr P Ellis (3), Mr P J N Ellis, Mr P M Ellis, Mrs E M Ellis, Mrs J Ellis, Mrs P Ellis, Mrs R Ellis, Ms C Ellis, Ms L S Ellis & Ms G L White, Mr D Ellisdon, Mr R Ellison, Mrs B Ellison, Mrs G Ellison, Mrs V Ellison, Mr C Elmer, Mr D Elphick, Ian Elphick, Mrs A Elphinstone, Mrs M Else, N J C Elsey, Mr K Elsmore, Mrs J Elson, Mrs M Elson, Ms S Elstub, Mr D M Elsworth, Dr D Elton, Mr G Elton, Mrs J S Elvin, Miss J Elwell, Mr M Elwell, Ms J Elworthy, Mr A R Ely, Mr S Ely, Mr C Emary, Ms M Emberson, H N Embleton, Mr B C Embling, Mr L Embry, Ms C Embury, Mr J Emerson, Miss V Emery, Mr A Emery, Mr David Emery, Jon Emery, Mr R J Emery, Stephen David Emery, Mr & Mrs D W Emley, Mrs S Emlyn-Jones, Mr R P Emmens, Mr J E Emms, Mr S Empsall, Mr N Emsden, Mr P Emsley, Mrs D Endersby, Mr C England, Mr S England, Mrs P England, Dr M English, Mr A English, Mr J English, Mrs L English, Mr I Enlander, Mr R E Enoch, Mr M Enright, Seamus Enright, Mr John & Mrs Helen Ensor, Mrs J Enston, Environment & Heritage Sevice, Epa Lakes Team 1 -Dublin, EPA River Biologists, Mr P Eric, Mrs V Erith, Mr D C Errey, Mr R Errington, Mr E Escott, Mr P M J Espin, Mrs A Essex, Mr K Esslemont, ESTC Bird Club, Mr D Esther, Mr B Etheridge, Mr P Etheridge, Mr R C Etheridge, Dr G J Etherington, Mr P Etherington, Mrs J Etherington, Mr P Ettinger, Mr J Eustace, Mr D Eva, Mr D P P Eva, Dr A Evans, Dr A D Evans, Dr D Evans, Dr I Evans, Dr K A Evans, Dr K L Evans, Dr L Evans, Dr S Evans, Clare Evans, Miss J Evans, Mr A Evans (2), Mr B Evans, Mr C Evans (2), Chris Evans, Mr D Evans (3), Mr D B Evans, Mr D S C Evans, Mr D W Evans, Mr D Evans, Mr G Evans (5), Glynne & Sheila Evans, Mr G D Evans, Gordon Evans, Mr H W Evans, Mr I B Evans (2), Mr & Mrs J Evans, Mr J Evans (3), Mr Lyn Evans, Mr L G R Evans, Mr Mark Evans, M I Evans, Michael Evans, Mr M Evans (2), Mr M J Evans (2), Mr M Evans, Mr R & Mrs J Evans, Mr R Evans (5), Mr R E Evans, Mr R J Evans (2), Mr R Evans (2), Mr S Evans (3), Mr S P Evans, Mr T Evans, Mr T D Evans, Mr V Evans, Mr Evans & Mr Bevan, Mrs A Evans, Mrs C Evans, Mrs E Evans, Mrs F Evans, Mrs F M Evans, J & D Evans, Mrs J Evans, Mrs L Evans, Mrs L J K Evans, Mrs M Evans (3), Mrs P Evans (2), Rhian Evans, Ms G Evans, Ms J Evans, Ms K Evans, A H Eveleigh, Mr T Everden, Colin M Everett, Mr M J Everett, Alison Everett, Mr A Everhale, Mr A Everingham, Mr J Everitt, Mr L Everitt, Mr P Everitt, Mrs C Everson, Mr J Evry, Mr R Ewing, Mr A D Ewins, Miss E Exelby, Ms J Exell, Exeter University Ecosoc, Mr S Exton, Mr T Eyles, John Eyre, Mr M Eyre, Mrs S Eyre, Mr A Ezard, Mrs K Ezard

F Mr G Facer, Mr R Facey, Mr K Fackrell, Mr A Fagan, Mr B Fahey, Mr F Fahy, Mrs E Fahy, Mr T Faichen, Mr N F G Faiers, Mr Michael Faint, Fair Isle Bird Observatory, Richard Fairbank, Vic Fairbrother, Ms L Fairchild, J Fairclough, Mr J Fairclough, Mr K Fairclough, Mr M Fairclough, Miss E Fairfield, Mr S Fairfull, Mr L Fairhall, Mr R Fairhead, Mr D Fairhurst, Mr D Fairlamb, Malcolm Fairley, Dr L Fairman, Mr N P Fairney, J Fairweather, Mr H & Mrs L Fairweather, Ms C Faith, Mr R Falango, Falcon Works London, Dr H Falconer, Mrs J Falconer, A Falconer, Mr M H Fallon, Mr P Fallon, Ms M Fanning, Mrs A Farley, Mr A Farmer (2), Mr M Farmer, Mr P

Farmer, P Farmer, Mr R J Farmer, Mr D P Farncombe, Mr C F Farnell, John & Fiona Farnsworth, B Farquharson, Mr D Farragher, Mr D Farrant, Dr Andre R Farrar, Mr D Farrar, Mr M Farrar, Mrs J Farrar, Charles Farrell, Mrs M Farrell, Mrs S Farrell, Mrs J Farrell-Brooks, Mr J Farren, The Farrers, Mrs B Farrer, Mrs W A Farrer, Mr M Farrier, Mr T Farrington, Mr D Farrow, Mr G Farwell, Mr D Faulkner, Mr F J & Mrs S P Faulkner, Mr P D Faulkner, Mr P J & Mrs J Faulkner, Mr P Fava, Mr L F Favret, Miss S L J Fawcett, Mr N A Fawcett, Mr P F Fawcett, George & Wendy Fawcett, Sue Fawcett, Chris T Fawdington, Dr H Fay, Miss L Fay, Mrs K Feachem, Mrs E M I Fear, Mr H Fearn, Mrs C M Fearnley, Mr J D Fearnside, Mr J Feasey, Feathers Wild Bird Care, Mr G Featherstone, Mr J Featherstone, Mr R Featherstone, Mr M Feaviour, Mrs J Federici, Mr M Feely, Mrs G Feely, Mr D Feeney, Séamus Feeney, Mrs M J Feeney, Ms B Feeney, Mr D C Feige, Feld family, Mr L Feldwick, Dr J Fell, Ms J E Fell, Dr E C Fellowes, Dr M Fellowes, Dr B J Fellows, Mrs L Fellows, Mr R Feltham, Mr R C Felton, Mr B Fendley, Dr D Fenlon, Miss T Fenn, Mr A Fenn, Colin Fenn, Mr P Fenn, Mr R Fenn, Mrs J Fenna, L Fennell, Mr R Fennelly, Mr K Fenner, Mr M G Fenner, Ms Á Fenner, Mr J Fenteman-Bladen, Mr A Fenton, Mr J Fenton, Mr R Fenton, Mr Terry Fenton, Mrs A Fenton, Mrs U Fenton, Milford National School, Mr R A Ferbrache, Mrs P A Fereday, Miss H Ferguson, Mr A Ferguson, Mr C Ferguson, David Ferguson, Mr K Ferguson, Ferguson family, Mr R Ferguson, Mr R L N Ferguson, Jane Ferguson, Mrs S Ferguson, Dr D Fergusson, Mr P Fernandes, Ms M Fernandes, Mr D Fernández Bellon, Mr D Fernleigh, Mr D Fernley, Ms S Fernley, Dr P N Ferns, Mr P Ferns, Mr D Ferrier, C Ferriroli & J Jennings, Mr M Ferris, Mrs S Ferris, A J Ferris, Miss D Ferry, Ms C D Ferry, Mr C Few, Miss L Fewtrell, Mr M Ffrench-Constant, Andy Field, Mr A Field, Mr B Field, Mr M Field, Mr P Field, Mrs A Field, Mrs E Field, Mrs K G Field, Field Studies Council, Dame P Fielding, Mr J E Fields, Mr S Fields, Mr D Fieldsend, Mr M N Fielker, Ms J H Fife, Fife Bird Club, Fife Coast & Countryside Trust, Fife Ranger Service, Mr G A Figg, Ms Y Filbey, Filey Bird Observatory, Mr A Filmer, Mr J Filowiat, Mr I Filtness, Mr L Filtness, Mr G Finbow, Mr C Finch, Mr D Finch, Finch-Labinger, Mr S Finch, Mr T Finch, Mrs J Finch, K Finch, Mr & Mrs N Fincham, Mr S Findell, Mr K Findlater, Dr R Findlay, Mr D Findlay, Mr G Findlay, Mr I H Findlay, Mr R Findlay-Robinson, Mrs A Finegan, Dr I Fingland, Mr A Finlayson, Mr J Finn, Mr M Finn, Mr P D Finn, Mr A Finnegan, Ms T Finnen, Mr S Finnerty, Miss K Finney, Mrs J Finney, Mr R J Firmin, Miss V Firminger, Mr A Firth, Mr S Firth, Ms S Firth, Mrs Ann Marie Firth-Bernard, Miss C E Fischbacher, Mr J A Fish, Mr J L R T Fish, Miss J Fisher, Miss S Fisher, Mr C P Fisher, Mr D Fisher, Mr D J Fisher, Mr G Fisher, Mr IJ Fisher, Mr J Fisher (2), Mr J W Fisher, Mr K Fisher, Mr M Fisher (2), Mr S Fisher, Stuart Fisher, Mr T L Fisher, Mrs A L Fisher, Mrs D Fisher & QMU Birdlife, Mrs D Fisher, Janice Fisher, Mrs P Fisher, Mrs P R Fisher, Mrs R Fisher, Mrs S M Fisher, Ms G Fisher, Mr L Fishlock, Mrs A Fishwick, Mr B Fisk, Mr M Fitch, Mr S Fitch, Mr A Fitchett, Mr G J Fitchett, Mr S J Fitt, Miss M Fitter, Mr K Fitton, Mr L Fitt-Savage, Mr D Fitzgerald, Mrs A Fitzgerald, S Fitzgerald, Mrs S Fitzgibbon, Mr M Fitzpatrick, Mr R Fitzwilliam, Mr S Flack, Ms D Flack, Miss M Flaherty, Mr D P Flaherty MBE, Flamingo Land Theme Park & Zoo, Mr N J Flanagan, Mr T C Flanagan, Mr Alastair W Flannagan, Mr P Flatters, Mr B R Flavell, Miss S Flax, Mrs J Fleck, Fleet family, Dr J Fleming, Dr V Fleming, Mr A Fleming, Mr C Fleming, Mr D Fleming, Mr I Fleming, Mr J Fleming, Mr J A Fleming, Kim Fleming, Mr P Fleming, Mrs E Fleming, Mrs L Fleming, Mrs S Fleming, Ms C Fleming, Mrs S Flenk, Mr D Flenley, Mr R Flesher, Miss K L Fletcher, Mr A G Fletcher, Mr C Fletcher, Mr E W Fletcher, Mr I Fletcher, Mr J Fletcher (2), Mr J D Fletcher, Mr K Fletcher, Mr M R Fletcher, Mr N Fletcher, Mr P Fletcher, Mr P C Fletcher, Mr T Fletcher, Mrs M Fletcher, Mr R Flight, Mr C C Flindall, Mrs G Flinn, Miss V Flint, Nick Flint, Mr P Flint, Mrs S F Flint, Dr D Flockhart, Mr G L Flood, Mr J M Flood, Ms K Flood, Mr S Flory, Mr T E Flower & Ms R Rogers, The Flowers, Pat Flowerday, Dr J Flowers, Mr L Flowers, Mr M J Flowers, Mr W Flowers, Mr L Floyd, Mr R Floyd (2), Mr D Flumm, Mrs J Flux, Mr C Flynn (2), Mr C D Flynn, Mr D J Flynn, Mr H Flynn, Mr I Flynn, Mr J E Flynn, Mr O Flynn, Phil Flynn, Mr S G Flynn, Ms A Flynn, Mr I Foakes, Dr B Foex, Mrs J Fogden, Mr D N Fogg, Mr J Foley, Mr K Foley, Mr M Foley, Mr M F Foley, Mr P Foley, Mr S Foley (2), Mrs M Foley, Ms M Foley, Tristan D Folland, Mr P Follett, Gordon W Follows, Mr R Follows, Mr A Font, Mr S Foord, Miss C Foot, Mr D Foot, Ms E Foot, Miss B Foote, Mrs J Foote, Mrs K Footner, Ms J G Foott, Mr G Forbes, Mr I Forbes, Mr K Forbes, Mr M Forbes (2), Mr N Forbes, Mr S Forbes, Mr T Forbes, Mrs A Forbes, Mrs G Forbes, Rev T J Forbes Turner, Dr I Ford, Dr J E Ford, Mr A J Ford, Mr B Ford, Mr C Ford, Mr E Ford, Mr I Ford, Mr J Ford, Mr K Ford, Mr M G Ford, Mr R Ford, Mr R A Ford, Mr S B & Mrs H E W Ford, Mr S Ford, Trevor Ford, Mrs A Ford, Mrs J Ford, Mrs Jennifer & Martin Ford, Margaret Ford, Mrs T Ford, Mrs T D Ford, Jane Ford & David Harman, Ms N Ford, A A Ford, D R Ford, Miss N Forder, Mr D A Forder, Mr M Fordham, Mr & Mrs T Fordham, Mr W E Fordham, Mr S P Foreman, Mr S Forest, Forestry Commission Staff, Mr D B Forgham, Mr J Forgham, Forrest family, Mr G W Forrest, Mr M Forrest, Mr P Forrest, Mrs L Forrest, Claire Forrest, Mr J Forrester (2), Mr R J W Forrester, Mr R W Forrester, Mrs J A Forrester, Mr P Forsdick, Mr M Forshaw, Mrs J M Forshaw, C J Forss, Dr T J Forster, Mr R Forster, Mr S Forster, Mr T Forster, Prof M J C Forster, Alastair Forsyth, Mr C Forsyth, Mr E M Forsyth, Mr I Forsyth, Mr L Forsyth, Mr T Forsyth, Mrs D Forsyth, Mrs P Forsyth, Ms J M Forsyth, Mr L Fortune, Mr T J Forward, Mr T Forward, Mr A J Fossey, Mrs A Fossey, Mrs B Fossey, Mrs J Fossey, Miss C Foster, Christopher W Foster, Mr D Foster, Mr G Foster, Mr J Foster, Mr L M Foster, Mr M A Foster, Mr N & Mrs J Foster, Mr R Foster (2), Mr R J Foster, The Fosters, Mr S Foster, Mr S J Foster, Mr W Foster, Mrs A Foster (2), Mrs D Foster, Mrs S M Foster, Ms J Foster, Ms C M Fotheringham, Ms S Fotheringham, Dr M Foulkes, Miss K Foulkes, Mr N Foulkes, Mr P Fountain, Mr T Fountain, Mr D Fouracre, Mr B Fowkes, Dr J A Fowler, Miss C Fowler, Miss P M Fowler, Mr D Fowler, Mr G R Fowler, Mr N Fowler, Mr P Fowler, Mr R Fowler, Mrs B Fowler, Mrs P Fowler, Mrs S J Fowler, Mr N B Fowles, Mr R Fowles, Dr M K Fowlie, Mr R Fowling, Mr O Fox, Miss T Fox, Mr A Fox (2), Mr A S Fox, Mr C Fox, Mr D J Fox, Mr H Fox, Mr I Fox, Mr J Fox (3), Mr M & Mrs J Fox, Mr P Fox (2), Mr R Fox (3), Mr R J Fox, Mr T Fox, Mr T P Fox, Mrs A Fox, Ms P Fox, Mr E G Foxcroft, Dr A Foxton, Dr N Foxton, Mr C Foxton, Mr C Foy, Mr D Foy, Mr E L Foyle, Miss A Frame, Mrs G Frame, Mr J D France, Dr I S Francis, Miss A Francis, Mr G Francis (2), Mr J Francis, Mr P Francis (2), Francis family, Mr W G Francis, Mrs D Francis, Mr G Franck, Mr S Franck, Mrs S Frank, Mr M P Frankis, Mrs K Frankland (2), Mr M Franklin (2), Mr N Franklin (2), Mr P Franklin, Mr R D Franklin, Mrs H Y Franklin, Mrs E Franks, Mr R Frankum, Mrs V Franz, T J Fraser, Mr A Fraser (2), Mr B Fraser, Mr D S Fraser, Mr C Fraser, Mr G Fraser (2), Mr I Fraser, Mr P Fraser, Mr R Fraser, Mrs A Fraser, Mrs H Fraser, Mrs J Fraser, Mrs P Fraser, Mrs S Fraser, Mrs J Fraser-Martin, Michelle Frater, Mr C P Fray, Mr R M Fray, Miss Z Frayne, Mr L Frayne, Mrs S D Frearson, Mr S Frederick, Dr L M Free, Mr D Free, Mr S Free, Mrs D Free, Mr S Freear, Mr T H Freeborn, Dr K Freeman, Dr S N Freeman, Mr C Freeman, Mr D Freeman, Keith & Anne Freeman, Mr K R Freeman, Shirley Freeman, Mrs S Freeman, Mr P J Freestone, Mr R E Freestone, Miss A Fremeaux, Ms J Fremlin, Dr G French, Major A J French, Miss H French, Miss R French, Mr N French (2), Mr Paul R French, Mr T French, Mrs C D French, Dr J V Freshwater, Mr D J Fricker, Ms E Fricker, Mr R Friedlander, Mr D Friel, Mr A Friend, Friends Of West Ham Park, Miss C Friers, Mr J Friers, Mr J D Friese, Mr T Frieze, Dr N J Friswell, Ms C L Frith, Ms J Frith-Orr, Mr V E Froome, Mr A Frost, Mr A C Frost, The Frosts, Mr P Frost, Mr R Frost, Mr R A Frost, Mr T Frost, Mrs M Frost, Mrs P Frost, Mrs W Frost, Ms J Frost, Ms S Frost, Mr P Frostick, Mr M R Froud, Mr D Fry, Mr J Fry (3), Mr S Fry, Mr S J Fry, Mrs J Fry (2), Prof C H Fry, Miss M Fryer, Mrs A C Fryer, Mrs L Fryer, Fsc Volunteers, Mr M A Fuchter, Mr D Fuegi, Mrs B Fulbrook, Mr A J C Fulford, Dr R Fuller, Miss K Fuller, Mr C Fuller, Mr D J Fuller, Mr N F Fuller, Mr R J Fuller, Mr R Fuller, Fuller/Kimpton family, Mr W W Fuller, Rob Fuller, Pam Fullforth, Valerie Fullforth, Mrs J Fullwood, Mr B Fulton, Mr D Fulton, Mr I Fulton, Mrs J Fulton, Dr S Weir, Mrs E Funnell, Miss J Furber, Miss M Furber, Mrs S Furber, Alison Fure, Mr C Furlepa, Mr Christopher Furley, L Furness, Prof R W Furness, Miss J Furnival, Mr J S Furphy OBE, Mr V Fusco, Roy Fussell, Mrs L Fussell, Mr R Fyfe, Alasdair M M Fyffe, Fylde Bird Club

G Mr R G Gabb, Mrs V Gabbay, Mr J D Gaches, Mr T Gadsdon, Dr D Gaffney, Mr D Gaffney, Mr F Gaffney, David J Gains, Miss K Gaisford, Steve Gaites, Mr B Galbraith, Mr A Gale (2), Mr J Gale, Mr S Gale, Mr T J Gale, Mr N Galilee, Brian Gall, Mrs V Gall, Mr C Gallacher, J & S Gallacher, S & J Gallacher, Mr C Gallagher, Mr R Gallagher, Ms A Gallagher, Ms L Gallagher, Mr R D Galliers, Ms S Gallifent, Dr C H Gallimore, Mrs B C Gallivan, Dr A Galloway, Mr B Galloway, David & Mary Galloway, Mr N Galloway, Mr P E W Galloway, Mrs L Galloway, Barrie Galpin, Mrs J Galpin, Mrs J E Galpin, Sir Anthony Galsworthy, Mr J Galton, Mr N Galton, Mr M Galtry, Mr DJS Gamble, The Gambles, Mr J D Gambles, Game & Wildlife Conservation Trust, Mrs Carole Gammond, Mrs K Gander, Mrs S Ganderton-Jackson, Mr J Gandy, Mr S Gann, Howard Gannaway, Garbutt family, Dr E F J Garcia, Ms S Garcia, Mr H Garcia Rueda, Mr A H Gardiner, Mr C Gardiner, Mr G Gardiner, Mr I Gardiner, Mr J Gardiner, Mr S Gardiner, Mr J Gardner, Mr P Gardner, Mr R Gardner (2), Mr S Gardner, Mrs J Gardner, Mrs M Gardner, Miss C Garfitt, Miss C Garforth, Miss S Garforth, Mr B Garland, J & S Garlick, Mrs V Garlick, Mr A Garner, Mr D Garner, Mr G Garner, Mr M Garner, Mrs L Garner, P Garner, Mr S Garnett, Mr R D Garrett, Ms A Garrie, Mr J J Garrigan, Mr A Garrod, Ms G Garry, Mr N Garside, Dr B Garson, Mr A Garton, Mr D Gartside, Mr K Gartside, Mrs P Gartside, Mr L Garvey, Mr M Garwood, Mr J Gascoyne, Mr M R Gash, Mrs A Gashi, Mr D R Gaskell, Mr J Gaskell, Mr S Gaskell, Mr I Gasper, Ms A Gasston, Mr R Gatehouse, Ms F Gatens, Mr J Gates, Mr M Gates, Mr N J F Gates, Mr S J Gates, Mr C Gath, Mr P Gathercole, Mrs D Gathercole, Mr & Mrs E Gathercole, Mr D Gatland, Mr N Gatward, Ms G Gauden, Mr P Gaughran, Allan Gaunt, Mr R Gaunt, Mr R L Gaunt, Mr G N Gavaghan, Mr E Gavin, Mr A Gay, Mr C A Gay, Mr N Gay, Mrs K Gay, Mrs R Gay, Mrs C Gaydon, Dr R Gayer, Mr M A Gayer, Mr D Gayton, Mr R Gaze, Mr A Gear, Mrs S C Gear, Miss J E Gearing, John Gearing, Dr M Geary, Mr S Gebbett, Mrs O Geddes, Mr P Gee, Mrs M Gee, Dr G R Geen, J & J Geeson, Mr C Gent, Mr S Gent, Mrs P M Gentleman, Mr A George, Mr A J George, Mr M George, Mr W George, Mrs C George, Mrs C M George, Mrs Kay George, Mrs V George, Ms Z George, Mr M J Gerard, Lesley Gerber, Mr K Gerhardsen, Mr K F German, Ms I German, Miss L Gerrard, Mr I Gerrard, Mrs N Gerwitz, The Gethings, Mrs C Gethings, Dr N Ghazal Asswad, Mr S B Ghilks, Ms A Giacomelli, Mr A Gibb, Mr A S Gibb, Mrs B Gibb, Mrs C Gibbins, Mr P J Gibbon, Dr D W Gibbons, Mr L J Gibbons, Mr N Gibbons, Dr R G Gibbs, Mr A Gibbs, Mr B Gibbs, Mr G Gibbs, Mr P Gibbs, Mr W Gibbs, Mrs J Gibbs, Mrs K J Gibbs, Mrs P Gibby, Ms P Gibby, Mr A B Gibney, Mr W Gibney, Dr C C & Mrs M Gibson, Miss G Gibson, Mr A Gibson, Mr A H Gibson, Mr D Gibson, Mr I P Gibson, Mr J Gibson (2), Mr J F Gibson, Mike Gibson, Mr P Gibson, Mr R J Gibson, Floss Gibson, Mr S Gibson, Mr T E Gibson, Mrs L Gibson (2), Mrs M Gibson, Mrs S Gibson, Ms S Gibson (2), Mr T Gibson-Poole, Mr A Gicquel, Mrs S Giddens, Mrs C Giddy, D L Gifford, Mr L Gifford, Mr D Gilbert (2), Mr I Gilbert, Mr J Gilbert, Mr K Gilbert, Mr N Gilbert, Robert Gilbert, Mr Stephen A Gilbert, Mrs L Gilbert, Mr P Gilbertson, Mr S Gilbertson, Mr J Gilbody, Derek Gilby, Ms J Gilby, Mr M A Gilchrist, Mr J Gilder, Mr M Gilder, Gilding family, Mr I Gilea, Dr T Giles, Mr P Giles, Mr V R Giles, Mrs L Giles, Mr C Gilhooley, Mr A Gill, Mr A P Gill, Mr D J Gill, Mr R Gill (2), Mrs E M Gill, G W Gill & J E Gill, Miss B Gillam MBE, Mr K B Gillam, Mr D Gillan, Mr T Gilland, Mr S Gillard, Miss E Gillatt, Mr W Gillatt, Mr R Gillespie, Mrs L Gillespie, Mrs M Gillet, Dr A Gillham, Mr K Gillies, Mr N Gilligan, D & C Gillingham, Mr J K C Gillingham, Dr S Gillings, Mrs B Gillings, Gilliver family, Mr D Gilmore, Mr H Gilmour, Mr A Giloney, Mr D Gilroy, Mr R Gimson, Mr H B Ginn, Mr S Ginnaw, Mr M C Ginns, Mr S Ginty, Mr P Giovannini, Gipping Valley Birding, Mr J Girdley, Mr T M Girling, Mr J G Gissing, Mr M Gitsham, Dr G J Gittens, Mr K J Gittens, Dr T Gittings, Mr J W Gittins, Mr M Given, Dr P Glading, C Gladman, Mr M Gladstone, Mr W Gladwell, Mrs E Gladwell, Revnd T W Gladwin, Glamorgan Bird Club, Mr M Glanville, Glasgow NHS Hill-Walking Club, Mr D Glass, Mr D J Glaves, Mr T Glavin, Mr C Glaysher, Mr J A Glazebrook, Mr R Glazebrook, Glazier Elizabeth, Ron Gleadle & Dick Barton, Mr R Gleason, Mr B Gleeson, Mr F Gleeson, Mr J Glen, Mr P Glencross, Mr J M Glendenning, Glenisla Primary School, Mrs C Glenister, Mr N M Glenn, Mr Andrew Glenn-Craigie, Mr S Glinn, Mr T Gloster, Gloucestershire Ornithological Co-Ordinating Committee, Mr D Glover, Paul Glover, Mrs C Glover, Mrs G Glover, Mr J C Gloyn, Mr D Glue, Mr P Gluth, Mr B Glynn, Mr M Glynn, Mr J Goacher, Mr P Goacher, Mr C Goatcher, Mr D J Gobbett, Dr C Goble, Phillip Goble, Lynne Goble, Mr A Goddard, Mr E Goddard, Mr S H Goddard, Steve Goddard, Mrs A E Goddard, Mr A Godden, Mr N Godden, Mr R Godden, Mr D Godfrey, Mr M F Godfrey, Mr T Godfrey (2), Ms S Godfrey, Mr U Godo, Mr T D Godson, Mr N Godwin, Mr D Goff, Mr R Goff, Dr P Golborn, Mrs P Goldberg, Mrs J Goldberry, Mr K Golden, Golden Eagle Trust Data, Mrs G Goldfarb, Mr E Goldingay, Mr D J Golds, Dr P Goldschmidt, Mr D Goldsmith, Mrs S G Goldthorpe, Mrs C Golightly, Mr D Golson, Mr F R Gomes, Mr F R Gomm, Dr D Gompertz, Miss E Gonzalez, Mr J Gooch, Mr R & Mrs P M Gooch, Mrs SM Gooch, Mr C Good, Mr A Goodall, Mr G Goodall (2), Mr I Goodall, Mrs A Goodall, Mr C Goodey, Mrs J Goodey, Mr P F Goodfellow, Mr L Goodhand, Mr A A Goodhead, Mr A Gooding, Mr J Gooding, Mr M Goodlad, Mr R Goodlad, Mr J Goodley, Miss F Goodliffe, Goodliffe family, Mr C Goodman, Mr P Goodman (2), Dr N G Goodship, Mr H M Goodship, Mr J R Goodspeed, Mr R Goodwillie, Dr B J Goodwin, Mr D Goodwin, Mr M Goodwin, Mr R Goodwin, Mr S Goodwin (2), Mrs B Goodwin, Mrs P Goodwin, Dr Sally Gordon, Major J J Gordon, Mr A Gordon, James Gordon, Mr N E Gordon, Mr N Gordon, Mr P R Gordon, Mr R L Gordon, Mr T Gordon (2), Mr D Gordon, Mr A Gordon-Rogers, Mrs A Goreham, Mr P Goriup, Mr J J S Gorman, Brian B Gormley, Mr B Gormley, Miss G Gornall, Mr R Gorringe, Mr H J & Mrs D J Gorringe, Mr E K Gorton, Dr A G Gosler, Dr R Gosling, Tony Gosling, Mr R Gosnay, Mr D Gosney, Mr M R Goss, Mrs J Goss, Ms E Gottuli, Mr A Gough, Mr D Gough, Mr I Gough, B M & M R Gough, Mrs J Gough, Mr A F Gould, Mr J Gould, Mr M Gould, Mr P Gould, Mr R A Gould, Ms S Gould, Ms C Goulden, Mr D Goulder, Mr P Goulding, Mrs R D Govan, Mrs A Govier, John Gowenlock, Mrs E Gower, J S Gowers, Mrs C Gowing, Mrs K Gowthorpe, John T Goy, Mr K Grabe, Mr M Grabham, Mr A Grace, Mr E Grace, Mr K Grace, Mr A Gracie, Mr M Grady, Mr T Graeme, Miss C C Graham, Miss L Graham, Mr A Graham, Mr A N Graham, Mr D Graham, Mr J Graham (2), Mr K L Graham, Mr R Graham (2), Mr R J Graham, Mr S Graham, Mr S A Graham, Mr T Graham, Mrs J Graham, Ms S Graham, Mr J D Grainger, Mr P D Grainger, Grange Academy, Mrs M Granger, Miss J Grant, Mr D Grant, Mr D W Grant, Mr E Grant, Mr G Grant, Mr I Grant, Mr J Grant, Mr K Grant (3), Mr K R Grant, Mr K Grant, Mr M G Grant, Mr P V Grant, Mr R Grant, Mr T Grant, Mr T C R Grant, Mrs A Grant, Mrs M Grant, Sue J Grant, Ms S Grant, R Grant, Mr B Grantham, Mr M J Grantham, Mr V Grantham (2), Mr M Granville, Mr R Gratton, Miss L Gravely, David Gravenor, Mr C Graves, Mr O W Graves, Mr R Graves, Mrs J Graves, Mr T Gravett, Ms S Gravil, Dr B Gray, Dr C M Gray, Dr R Gray, The Grays, Mr A Gray (3), Mr C Gray, Mr D Gray, Mr D B Gray, Mr I J Gray, Mr J Gray, Mr M Gray, Mr M P Gray, Mr N Gray, Mr P Gray, Mr R Gray, Mr R W Gray, Mr S Gray, Mr W Gray (2), Mrs B Gray (2), Mrs B E Gray, Mrs C Gray, Mrs C C Gray, Jean Gray, Mrs K Gray, Dr G H Grayer, Mrs S Grayston, Mr W Greacen, Mr D N Greasley, Great Bustard Project, Great Garden Council, Greater Manchester Bird Recording Group, Miss F Greaves, Mr B Greaves, Mr M Greaves, Mr R P Greaves, Mrs M A Grech, Ms Y Grech, Dr A E Green, Dr D Green, Dr J Green, Miss C Green, Miss G Green, Mr A Green (2), Mr B Green, Mr C Green (2), Mr D E Green, Mr D Green, Mr G H Green MBE, Mr I Green, Mr J Green (2), Mr M Green (5), Mr M G Green, Mr N Green, Mr N C Green, Mr P Green (3), Mr R Green, Mr R A Green, Stephen Green, Mr S Green, Mr T T Green, Mrs H Green, Janet Green, K Green, Mrs K Green, Mrs M Green, Mrs W Green, Ms N Green, Ms R Green, Mr B H Greenacre, The Greenaways, Mr N Greenaway, Mrs P Greenaway, Mr M C Greene, Mr J Greener, Mr C Greenfield, The Greenfields, Greenfields Community Primary School, Mr P Greenhalgh, Mr S Greenhalgh, Mr M Greening, Mr M T Greening, Miss S Greenland, Martin Greenland, Mr B Greenough, Mr R Greenshields, Mr P Greenslade, Rev T Greenslade, Mrs C Greenway, M & B Greenwood, Dr J G Greenwood, Miss L Greenwood, Miss M Greenwood, Mr A Greenwood, Mr C I Greenwood, Mr R Greenwood (2), Prof J J D Greenwood CBE, Mr R Greer, Mr S J Greer, Mrs K Greevy, Miss R V Gregory, Mr C Gregory, Mr J Gregory, Mr L V Gregory, Mr M Gregory, Mr Paul A Gregory, Mr P W Gregory, Mr R D Gregory, Mr S Gregory, Mr W H Gregory, Gillian Margaret

Gregory, Mrs K E Gregory, Mrs R S Gregory, Mrs S Gregory, Mrs C Greig, Mr H E Grenfell, Mr P Grennard, Mr A Gretton, Mr R D Grew, Mr B E Grey, Mr H Grey, Mr K Grey, Mr M Grey, Mr F C Gribble MBE, Mr R J Gribble, Miss R Grice, Miss T Grice, Mr N Grice, Mr G Grieco, Mr D Grierson, Mr A Grieve, Mrs A Grieve, Dr M Griffin, Miss B M Griffin, Mr D G Griffin, Mr I Griffin, Mike Griffin, Mr M A Griffin, Mr R V J Griffin, Mrs C Griffin, Mrs E G Griffin, Mrs J Griffin, Mrs M Griffin, Mr D J Griffith, Mr J E Griffith, Mr M Griffith, Pete Wynn Griffith, Mr R Griffith, Mrs E Griffith, Dr M Griffiths, Dr P Griffiths (2), Dr R Griffiths, Miss D Griffiths, Helen J Griffiths, Miss K Griffiths, Miss L Griffiths, Mr A Griffiths, Mr B Griffiths, Mr D Griffiths (2), Mr D A Griffiths, Mr D N Griffiths, Mr E Griffiths, Mr G Griffiths, Mr H Griffiths, Mr J Griffiths, Mr J O Griffiths, Mr M Griffiths, Mr M J Griffiths, Mr M P Griffiths, Mr Robin D Griffiths, Mr S Griffiths (2), Mr T B Griffiths, Mrs L Griffiths, Mrs M Griffiths, Mrs N Griffiths, Mr R Grimes, Mr R Grimmond, Mr A Grimsey, Mr C Grimshaw, Mr R Grimshaw, Mr S Grimshaw, Miss S Grimwood, Ms S Grimwood, Ms C Grindley, Ms C Grindrod, Mr A L Grinter, Mr C Grisedale, Mr L Grisedale, Mrs C Grisely, Mr D I Griss, Dr D M Grist, Mr J Grist, Mr R Gritton, Mr T Grogan, Mr J D Groom, Mrs R D Gross, Mr P S Grosse, Mr M J Grounds, Mr N Grounds, Mr P Grout, Mr Tim Grout-Smith, Major P N Grove, Miss L Grove, Mr T Grove, Mr S Grover, Mrs E Grover, Mr N N Groves, Mr Richard Groves, Mr D J Gruar, Mr P J Gruar, Mr N Grubb, Mr A Grubb, Mr M Grubb, Mr J P B Grundy, Mr S Grundy, Mrs J Grundy, Miss S Grunsell, Mr S Guallar, Mr D Guard, Mr J R Guard, Mr P Guarnaccio, Mr A J Guest, Ms B M Guest, Prof J E Guest, Miss R Guillem, Ms E Guinan, Mr J Guiver, Prof K Gull, Mr B Gulliford, Mrs M Gulliver, Mrs M Gully, Mr D G Gumn, The Gunbys, Dr D A & Ms C Gunn, Dr M Gunn, Miss T Gunn, Mr D Gunn, Mr I D M Gunn, Mr R Gunn, Mr & Mrs Gunn, Ms C Gunn, Mrs S Gunnee, Ms M Gunning, Mr M Gunstone, Mr M Gurney, Mr R Gurowich, D & Y Gurr-Gearing, Miss A Guthrie, Miss J Guthrie, Mr A Guthrie, Mr G Guthrie, Ms K Guthrie, Tony Gutteridge, Mr J Guy, Mrs J Guy, Dr D Guyoncourt, Mr C Gwilliam, Mrs D M Gwilliams, Ms M Gwyn, Henderson-Gwynnes, Mrs E Gwynne, David Gynn, Gypps family

H Mr P Hacker, Mr D Hackett, Mr J Hackett, Mrs J Hackman, Miss T Haddon, Mr P Hadfield, Mr P Hadfield & Miss W Gawne, Mrs C A Hadfield, Mrs S M Hadfield, Paul Hadland & Anne Dale, Mr P R Hadley, Mrs C Hadley, Mr I Hadwin, Mr C Haffenden, Mr J Hagger, Mr J M Hague, Mrs A E Hague, Mr C Haigh, Mr P Haigh, Mr R Haigh, Mrs A Haigh, David &Avril Haines, Mr D Haines, Bill Haines, Mrs N Haines, Mrs S Haines, Dr I R Hainsworth, Mr M Hainsworth, Mr J Hakeman, Mr R Halahan, Mr F Halbert, Ms V Halcrow, Mr D S Halder, Dr P A Hale, Miss A Hale, Allan Hale, Mr A Hale, Mr D B Hale, John Hale, Mr R Hale, Mr S Hale, Mr D Hales, Mr P Hales (2), Mr R J Hales, Mr S Hales, Mr & Mrs Hale-Sutton, Mr L Haley, Mr P Halford, Miss G Haliday, Dr C N Hall, Dr E Hall, Dr M Hall, Maggie Hall, Dr R L Hall, Miss G Hall, Mr A Hall, Allan G Hall, Mr A S Hall, Mr C J Hall, Mr D Hall (2), Hall family, Mr G Hall (2), Mr G C Hall, Mr I Hall, Mr J Hall, Mr J J Hall, Mr K Hall, Mr K J Hall, Mr K S D Hall, Mr M Hall (2), Mr M P Hall, Mr P Hall (2), Mr P T Hall, Mr R Hall (3), Mr R I Hall, Mr S Hall (2), Mr T Hall (3), Mrs A Hall, Mrs A H Hall, Mrs C Hall, Mrs C P Hall, Di Hall, Mrs J Hall, Sally Hall, Mrs T Hall, Mrs W E Hall, Ms C Hall, Ms J Hall, Rosie Hall, Ms R Hall, Ms S Hall (2), Mr A R Hallam, Mr B Hallam (2), Mr R Hallam, Hallam family, Ms L Hallam, Mr N Hallas, Dr Richard J Hallett, Mark Hallett, The Halletts, Mrs K Halley, Dr T D Halliday, Miss C Halliday, Mr JB Halliday, Mr K Halliday, Mr R G Hallier, Revnd W L Halling, Mr Mark Halliwell, Mr R M Halliwell, Mr J M Halls, Mr P J Halpin, Mr T Halpin, Mrs C Halpin, M & P Halpin, Mr P Halsall, Mr T J Halsey, Mrs J Halsey, Ms S A Halsey, Mr R Halstead, Mrs J Halstead, Mr R G Hamar, Miss E Hamblin, Mrs M Hamblin, Mr R A Hambling, Miss B Hamill, Mr B Hamill, Dr P Hamilton, Mr B Hamilton (2), Mr D F Hamilton, Mr D R Hamilton, Mr F Hamilton, Mr G Hamilton, Mr J Hamilton, Mr K Hamilton, Mr M Hamilton, Mr P Hamilton & Ms M White, Mr P Hamilton, Mr T Hamilton, W Alex H Hamilton, Chris & John Hamilton, Ms R Hamilton (2), Ms V Hamilton, Mr N Hamilton Fletcher, Hamilton Grammar Ecoschool, Dr J C Hamlet, Mr M Hamlett, Mr I Hamlin, Mr P J Hamling, Mr P Hammersley, Mrs L Hammersley, Mr D Hammocks, Mr B Hammond, Mr J Hammond, Mr M Hammond, Mr P Hammond, Mrs L Hammond, Mrs L C Hammond, Ms M Hammond, Mr M Hammonds, Mr M Hampden-Smith, Hampshire Ornithological Society, Mr D Hampson, Mr I Hampson, Mr C Hampton, Mr M Hampton, Mr S Hampton, Mr M J Hamzij, Mr M Hanafin, Mrs D Hanan, Dr C Hancock, Miss S Hancock, Mr B Hancock, Mr B J & Mrs C Hancock, Mr J Hancock, Mr K Hancock, Mrs J Hancock, Mrs K J Hancock, Ms R Hancock, Mr P J Hancocks, Mr J Hancox, Dr J Hand, Mr K Hand, Mrs A Hand, Mrs L Hand, Handa Island, Mr H Handley, Mr J Handley, Mrs R Handyside, Mr S C & Mrs A Hankin, Mr D A & Mrs L Hankinson, Mr T Hankinson, Ms C Hanks, Miss K Hanley, Mr H Hanley, Mrs C Hanley, Deborah Deveney, Mr A Hanlon, Graham Hanlon, Mr H J Hanmer, Mr G Hann, Miss C Hanna, Mr G Hanna, Mr J C Hanna, Mr A C Hannaford, Mr D Hannaford-Hill, Mr K Hanney, Mrs M Hannis, Miss B S Hansell, Miss C Hansell, Mr J Hansford, Mr N Hansford, Mr N Hanshaw, Mr P Hanslip, Mr C P Hanson, Mrs J Hanson, Richard Harbird, Mr S Harborne, Mr A G Harbott, Mr R Harbott, Dr D Harbour, Mr M F Harcup, Mr D Hardgrave, Mr A J Hardie, Dr N J Harding, Mr A V Harding, Mr C D Harding, Mr I Harding, Mr M Harding, Mr M D M Harding, Mr W Harding, Mrs J Harding, Mrs M Harding, Mrs S Harding, Ms J Harding, Mrs A Hardman, Mrs S Hardman, Mr R Hards, Mr T Hardware, Mr P G Hardwick, Miss H Hardy, Mr G Hardy, Roger Hardy, Mrs C Hardy, Prof A R Hardy, Mr P Hare, Stewart Hares, Mr P Harford, Mrs K Hargreave, Dr K Hargreaves, Mr G Hargreaves, Mr R Hargreaves, Mr R A Hargreaves, Mrs E Hargreaves, Mrs M Hargreaves, Mrs J Harkness, B & G Harland, Ken Harland, Mr A Harler, Mr F J Harley, Mr R Harley, Mrs C Harlow, Dr A Harmer, Mr F Harmer, Miss F Harms, Ms A Harnden, Mr J Harnett, Mr J Harold, Mr R Harold, Dr D G C Harper, Chris Harper, Mr G Harper, John Harper, Mr K Harper, Mr M Harper, Marianne & Steve Harper, Mrs L Harper, Mary Harper, Mrs N Harper, Mrs R Harper, Mr R Harrald, Mr G Harries, Mrs H R F Harries, Mr P D Harrigan, Síle Harrington, Mr B Harrington, Mr B R Harrington, Mr J Harrington, Mr P Harrington, Mrs Y Harrington, Dr J C Harris, Dr M P Harris, Dr R J Harris, Miss A Harris, Miss C Harris, Miss E Harris (2), Miss M Harris (2), Miss S Harris, Mr A Harris, Mr A B J Harris, Mr A T Harris, Mr D A Harris, Mr D G Harris, Mr D P Harris, Harris family (2), Mr G Harris, G M Harris, Mr G O Harris, Mr I Harris, Mr J Harris (2), Mr J C Harris, Mr K Harris (3), Mr M Harris, Nigel G Harris, Mr O Harris, Mr P Harris (2), Mr P A Harris, Mr P W Harris, Mr R Harris (2), R G Harris, Mr R J Harris, Mr S M Harris, Mr T Harris, Mr W D Harris, Mrs E M Harris, Mrs J Harris (4), Mrs M Harris, Mrs S Harris, Mrs V Harris, Ms J Harris, Ms M Harris, Duncan Harris, Dr C Harrison, Miss C Harrison, Miss K Harrison, Miss S Harrison, Andrew Harrison, Michael Cooper, Mr A Harrison, Mr D Harrison (2), Mr D W Harrison, Mr G Harrison, Mr G R Harrison, Mr Ian Harrison, Mr J Harrison (2), Mr J M Harrison, Mr K Harrison, Mr L Harrison, Mr M Harrison (3), Mr M C Harrison, Mr R Harrison, Mr R L P & Mrs M E Harrison, Mr S Harrison (3), Mr T A Harrison, Mrs B A Harrison, Elizabeth Harrison, Mrs P Harrison, Pete & Valerie Harrison, J Harrison & J Margetts, Mr Ian E Harrison-Gammons, Mr G Harrop, Hugh Harrop, Mr J Harrop, Mr B Harrower, MR P A Harrup, Mr J Harry, Ms S M Harry, Alistair Hart, Miss E M Hart, Miss H Hart, Miss J Hart, Mr D Hart (3), Mr J M Hart, Mr M J Hart (2), Mr M P Hart, Mr P R Hart, Robin A Hart, Mr R Hart (2), Mrs E Hart, Mrs M Hart, Ms D Hart, Ms H Hart, Ms J Hart, Mrs S Harte, Ms Rachel Hartgrove, Mrs R Hartland, Dr A Hartley, Dr I R Hartley, Mr C Hartley, Mr W W Hartley, The Hartleys, Mrs P J Hartley, Mr N Hartnell, Miss J Hartshorn, Mr P Hartshorn, Mr M I Harvey, Michael Harvey & Tuesday Birders, Mr N Harvey (2), Mr P C Harvey, Mr P V Harvey, Mr R Harvey (3), Mr S C Harvey, Mr W G Harvey, Mrs E Harvey, Mrs J Harvey, Mrs M Harvey, Mr A Harwood, Mr D Harwood, Mr E Harwood, Mr M Harwood, Mr N Harwood, Mr S Harwood, Ms S J Harwood, Rev M Harwood, Mr & Mrs T A Hasdell, Mr J Haselhan, A Hashmi, Mr Philip Haskell, Mrs J Haskell, Mr D Haslam, Mr M Hassan, Dr V A Hassell, Mr C Hastings, Mr R Hastings, Mr J J Hatch, Mr C M Hatch, Mr N Hatch, Mr S R Hatch, Mrs S Hatch, Mrs M Hatcher, Mrs J Hatchett, Mr M Hatfield, Mrs E M Hatfield, Ms J Hatfield, Ms L Hatfield, Hatfield Moor Bird Study Group, Mrs L Hatfull, Mr C Hatherill, Dr P R Hatherley, Mr C Hatsell, Miss C S A Hattersley, Mr D Hatton, Mr P L Hatton, Mrs W A Hatton, Dr T Hatton-Ellis, Mr A Haugaard, Havergate Island Group, Miss K Haw, Mr J Haw, Mrs J P Hawes, Gavin Hawgood, Mr J Hawke, Mrs L Hawke, Mr D M Hawker, Mrs A Hawker, Mr & Mrs Hawker, Mr S Hawkes, Mrs S Hawkes, Mr L Hawkings, Mr A Hawkins, Clive Hawkins, Mr D Hawkins, Mr G Hawkins, Mr I Hawkins, Mr J Hawkins, Mr K Hawkins, Mr M Hawkins, Mr P Hawkins (2), E & H Hawkins, Mrs J Hawkins (2), Hawkwood Organic Nursery, Mrs C Hawley, Mrs J E Hawley, Mr J Haworth, Mr C N Haworth-Booth, Mr M Haworth-Booth, Mr B Hawthorne, Miss V Hawtin, Mr S Hawtin, Mr J N Hawtree, Haxby & Wigginton U3A Bird Group, Mr M Hay, Mr R Hay, Mrs S Hay, Mr R J & Mrs A Haycock, Mrs A N Haycock, Mr N Haycox, Mr S Haycox, Mr A Hayden, Mr J Hayden, Mr R M Haydon, Mr B Hayes, Mr D E Hayes, Mr I Hayes, Mr J Hayes, Mr L Hayes, Mr N Hayes, Mr P Hayes, Mr R Hayes (2), Valentine Hayes, Mr D Hayhow, Mr S J Hayhow, Linda Haylock, Mr T Hayman, Bill Haynes, Mr J Haynes, Mr P Haynes, Mr S Haynes, Dr S L Haysom, Mrs M E Hayter, Mr R L Haythornthwaite, T & W Hayward, Mr B M Hayward, Mr C Hayward, Mr I M Hayward, Mr R Hayward, T & K Hayward, Mr W Hayward, Mrs M Hayward, Prof M D Hayward, K S Hayward, Mr D Haywood, Mr J Hazell (2), Mr P J Hazelwood, Dr Grant A Hazlehurst, Miss T Hazlehurst, A Hazlehurst, Mr M Hazleton, Mr J Hazzard, Lady Head, Mr G Head, Mr J Head, Mrs N Heading, Mr J Headon, Dr K Heal, Mr J H R & Mrs M E Heal, Peter J Heald, Ms D Heales, Mr M Healey, Mr S Healey, Sue Healey, Miss K Healy, Mark G Healy, Mrs M Healy, Miss A Heaney, Ms J Heard, Clare Heardman, Mr E C Hearn, Mr P J Hearn, Mr R Hearn, Mr G W Hearnden, Mr D Hearne, Mr A W Heath, Mr G R Heath, Mr J Heath, Mr M Heath, Mr R Heath, Richard & Kay Heath, Mr S D Heath, Mrs S Heath, Peter & Pauline Heathcote, Mrs S A Heathcote, Ms C Heathcote, Mr S M J Heather, Ms V Heather, Miss Helen Heathman, Mr P Heatley, Miss J Heaton, A P Heaton, Mr G B Heaton, Mr J Heaton, Mr P Heaton, Mr R Heaton (2), Mrs B Heaton, Mrs J Heaton, Alan Heavisides, Dr J & Mrs M L Heber, Dr R Heckingbottom, Mrs J Hector, Miss H Hedderman, Mr S Hedditch, Ms S Hedger, Mr B Hedley, Miss N Heeley, Prof D W Heeley, Mr S Heery, Mr B Hegarty, Mr F Hegarty, Mrs P M Heighway, Miss I Heinaru, Dr C Heintzen, Miss S Helbling, Mrs L Hellam, Mrs Rebekah Hellard, Mr D W Helliar, Mr A Helm, Mr S Helm, Mr J Helps, Mrs D Helps, Mr J Hemingway, Mr R Hemming, Mrs J M Hemming, Mr G Hemmings, Ms S Hempel, Mr D Hemsley, Dr I G Henderson, Miss A Henderson, Andrew Henderson, Mr A J K Henderson, David Ross Henderson, Mr E Henderson, Mr F N G Henderson, Mr G Henderson, John Henderson, Mr J Henderson, Martin Henderson, Mr M Henderson, Mr R Henderson, Mr S Henderson, Ms R Henderson, Ms M Hendrick, Mr A Hendry, Mr C Hendry, Mr J Hendry, Mr T Hendry, Mr E Hendy, Mr P Henebury, Henfield Birdwatch, Mr D Hennessy, Mr B Henning, Mrs C Henriques, Dr M D Henry, Mr B L Henry, Mr M Henry, Mr R Henry, Henry family (2), Ms B Henry, Michael John Henshaw, Mr D Henshilwood, Miss K Hensman, Mr P Henson, Ms P Henton, Dr C J Henty, Miss E Henwood, Mr T Heppell, Mr D Heptinstall, The Heptinstalls, Mr D Hepworth, Mr G Hepworth, Mr K J Herber, Mr A Herbert, Mr I Herbert, Mr M Herbert, Mrs H Herbert, Ms J Herbert, Mr D Hercock, Mr James Herd, Hereford Ornithological Group, Herefordshire Ornithological Club, Herefordshire Ornithological Society, Mr P Herkenrath, Mr A Herman, Mrs B Herniman, Mr K Heron, Mr P Heron, Ms J Herridge, Mr J Herring, Ms A Herrod, Mr B Herschell, Mr G Hersey-Green, Herts Bird Club, Dr Russ Heselden, Mr R Hesketh (2), Hesketh Ecology, Mr G W Heslop, Mr I Heslop, Mr R E F Heslop, Mr L Hesp, Mr J Hesslewood, Mr Alan J Hetherington, Mr & Mrs Hetherington, Mr M Heubeck, The Hewards, Mr D G Hewett, Mrs C Hewett, Mr A Hewins, Mr D Hewitt, Mr F Hewitt, Mr G Hewitt, Mr M Hewitt, Mr N D Hewitt (2), Mr P J Hewitt, Mr R Hewitt, Mr R F Hewitt, Mr S J Hewitt, Mrs E R Hewitt, Mrs J Hewitt, Mrs J Hewlett, Ms P Hewlett, Dr C Hewson, Dr G Hewson, Mr C Heyes, Mr J Heyes, Heysham Bird Observatory, Mr K Heywood, Mr M J Heywood, Mr S Heywood, Mrs A Heywood, Mr A Hibberd, Mr G F Hibberd, Mr A O Hibbert & Miss C A J Lake, Mr J Hibbert, Mr R Hibbert, Mrs J Hible, Mr J Hickerton, Ms J Hickey, Miss A Hickman, Miss J Hickman, Mr A Hickman, Mr C Hickman, Mr P Hickman, Mrs L Hickman, Mr C Hickman-hale, Miss W Hicks, Hicks family, Mr B Hicks, Mr M Hicks, R K Hicks, Mr T J Hicks, Mr W Hickson, Mr G Higgin, Mr C Higgins, Mr P Higgins, Philip Higgins, Mr R Higgins, Mrs E Higgins, Mr J K Higginson, Mrs P M Higginson, Mr S Higgs, Mrs M Higham, Miss M Highfield, Dr A J Highton, Mr J Higman, Mr C Higson, Mr P Higson, Mr J Hiles, Mr J Hiley, R & S Hiley, Dr D Hill, Dr M J Hill, Dr T R G Hill, Miss M Hill, Miss R Hill, Miss Hill, Mr A Hill, Mr A R Hill, Mr B J Hill, Mr C Hill (2), Mr D Hill (4), Mr D A Hill, D M Hill, Mr G Hill (2), Mr I Hill, Mr J Hill (2), Mr K Hill, Mr M Hill, Mr P Hill (3), Mr R Hill (2), Mr R & Mrs L Hill, Mr R Hill (3), Mr R J Hill, Mr S Hill, Mr T Hill (2), Mr T J Hill, Trevor M Hill, Mrs C Hill, Mrs L Hill, Ms C Hill, Ms P Hill, Ms S Hill, Mrs M J Hille, Mr I Hillery, Mr K L Hillery, Mr R A Hilliard, Alec & Felicity Hillier, Mr J Hillier, Mrs C Hillier, Mrs H Hillier, Mr P Hillion, Paul Hillis, Ms J Hillman, Mr J L Hillsdon, Hilly Fields Bird Champion Group, Mrs D Hilson, Mr N Hilton, Mr P Hilton (2), Mrs D Hilton, Mr Z Hinchcliffe, Ms S Hinchcliffe, Mr R Hinchliffe, Dr K Hinckley, Mr A Hind, Mr C M Hind, Mr M Hind, Dudley Hind, Mr S Hind, Dr F R J Hinde, Mr C H Hindle, David J Hindle, D & H Hindle, Mr M J Hindle, Dr M Hindley, Mr K Hindmarch, Miss E Hine, Prof J P Hine, Mr M Hines & Dr J Sartori, Mr M Hines, Mr J Hingley, Mr S J Hingston, Mr J Hinkley, Mr W Hinks, Ms C Hinks, Hinsley family, Mr G Hipkin, Mr J Hipkin, Mr M Hipkin, Ms H Hipperson, Mr G Hird, Mr D Hirst, Mr M Hirst, Mrs K Hirst, Roger Hissett, Dr C Histon, Historic Scotland Ranger Service, Dr M J Hitch, Mr J D Hitchcock, Mr R Hitchen, Mr S Hitchen, Mr D I Hitchin, Mr A E Hitchmough, HMS Collingwood Wildlife Conservation Group, HMS Raleigh, Mr D Hoad, Mr G D Hoad, Mrs N Hoar, Dr D Hoare, Mr B Hoare, Mr J J Hoare, Mr R T Hoare, Mrs K Hoare, MS S Hoare, The Hobans, John Hobbs, Mr N A J Hobbs, Mrs D Hobbs, Mr C K Hobley, Mr A Hobson, Mr D Hobson, Mr J A Hobson, Mr M Hobson, Mr P Hobson, Mr R W Hobson, Ms M Hockey, Mr R L Hockin, Mr I Hockings, Mr A B Hodder, Mr P Hodge, Mr T N Hodge, Mrs E Hodge, Ms V Hodge, Lt Col (Retd) M G Hodges, Miss F Hodges, Mr G Hodges, Hodges Clan, Mr W A Hodges, Mrs J F Hodges, Ms Debbie J Hodges, Ms J E Hodges MBE, Mrs J Hodgetts, Mr M Hodgkin, Mr C Hodgkins, Mr M J Hodgkins, Ms S Hodgkinson, Dr C Hodgson, Miss J Hodgson, Mr I P Hodgson, Mr K Hodgson, Mike S Hodgson, Mrs S Hodgson, Mr T Hodkin, Mr D Hodkinson, Mrs H Hodkinson, Mr B Hodson, Mr J R Hodson, Mr M Hodson, Mr H Hoff, Mr C Hogan, Mr D Hogan (2), Mr M T Hogan, Mr S Hogan, Mr Hogan, Mr C Hogarth, Angus Hogg, Mr C Hogg, Mr R Hogg, Mr S Hogg, Mrs A C Hogg, Mrs P Hogg, Ms J Hogg, Miss M Hoggett, Mrs P Holbourn-Williams, Harry Holbrook, Mr M Holbrook-Bull, Mr A Hold, Mr A Holden, Mr M Holden, Peter Holden, Mr J Holder, Mr J D Holder, Mr D J Holding, Mr N Holding, Mr M Holdsworth, Mrs A M Holdsworth, Mrs C Holdsworth, Mr J Hole, Mrs M E Holehouse, Mr A Holgate, Mr J Holgate, Mr T Holgate, Mrs G Hollamby, Dr D Holland, Dr J P Holland, P K Holland, Mr A N S Holland, Mr C Holland, Mr D G Holland, Mr J P Holland, Mr R Holland (3), Mr T Holland (2), Ms M Holland, Mr N Hollands, Mr J Hollender, Mary Holley, Mr A Hollick (2), Mr S Holliday, Mark Holling, Mr M Hollingsworth, Mr R R Hollingsworth, Dr R Hollingworth, Mr SP Hollinrake, Mark Hollinsworth, Mr J Hollis, Mr R Hollis, Ms E Hollis, Mrs E Hollister, Mr D Hollow, Miss S V Holloway, Mr D Holloway, Mr R D Holloway, Mr S Holloway, Mr R Holm, Mr S Holm, Mr T R F Holman, Mrs M Holman, Holme Bird Observatory, Dr B M Holmes, Andy Holmes, Mr A Holmes, Mr G Holmes, Mr J Holmes, Mr N Holmes, Mr P J Holmes, Mr R Holmes (2), Mr R M Holmes, Mr S Holmes, Mrs B Holmes, Mrs B A Holmes, Mrs P Holmes, Miss S Holoran, Mr S Holroyd, Dr C A Holt, The Holts, Mr A Holt, Mr B Holt, Mr D Holt (3), Mr D J Holt, Mr D Holt, Mr J C Holt, Mr K Holt, Mr P Holt, Mr S Holtam, Mrs J A Holter, Mr A Holt-Kentwell, Mr B Holtom, Mr T Holwill, Mr P C W Holyday, Mr R Homan, Mr J Homer, Mr & Mrs E Homer, Mr D Hone, Mr C D Honer, Mr W D Honeywell, Mrs J Hood, Dr A Hoodless, Mr R Hoodless, Mr Bob Hook, Mr I Hook, Mr K A Hook, Mrs L Hook, Dr P Hooper, Miss D Hooper, Alan Hooper, Mr J Hooper, Mary & Michael Hooper, Mr M Hooper, Mr R Hooper, Mr C E Hope, Mr M Hope-Urwin, Mrs J Hopewell, Mrs B A M Hopkins, Mr B Hopkins, Mr D Hopkins, Mr G Hopkins, Ian Hopkins, Mr R Hopkins, Mr S Hopkins, W D M Hopkins, Ms R Hopkins, Mrs J Hopkinson, Mr J Hopper, Mr R Hopson, Mr M R Hopton, Mr G Hopwood,

Mr D Mccabe, Mr E Mccabe, Mr K McCabe, Mr A Mccafferty, Mr S Mccafferty, Mrs J Mccaffrey, John McCaig, Mr W Mccaig, Ms M Mccain, Mr N Mccall, Mrs D McCallion, Mr J R McCallum, Mrs S Mccamley, Dr Norman V Mccanch, Mr D McCandlish, Dr T S Mccann, Miss K Mccann, Mr R Mccann, Mr S Mccann (2), Mrs A McCann, Tom Mccanna, Mr M Mccarrick, Mr P McCarron, Dr Barry McCarthy, Miss J Mccarthy, Mr F McCarthy, John McCarthy, Mrs A Mccarthy (2), Mrs C Mccarthy, Mrs M McCarthy, Mrs T Mccarthy, Mrs W Mccarthy, Ms M Mccarthy, Mr P McCartney, Mr C Mcclean, Mr M J McCleary, Mr O Mccleave, Mr G Mcclelland, Mr K Mcclintock, Mr A McClure, Stephen McClure, Mrs H J McClure, Mr G McColl, Mr S McCombie, Mrs G Mcconnachie, Mr K Mcconnell, Ms J McConnell, Mr R McConville, Mrs S Mcconville, Ms A McConville, Mr J Mccorkell, Ms S Mccormack, Mr K S McCormick, Mr M Mccormick, Mr S McCormick, Ms K McCormick, Mr G McCoubrey, Mr S Mccourt, Dr C M McCowen, Mr R Mccraight, Mr A McCubbin, Ms F Mccudden, Mr S F H McCullagh, Mr I McCulloch, Miss K McCullough, Mr A McCullough, Mr D McCullough, Bob Mc Curley, Mr J Mccurry, Mr J A McCutcheon, Mr N McCutcheon, Mr L McDaid, Mr A Mcdermid, Miss P Mcdonald (2), Mr P McDonald, Mr S Mcdonald, Mrs C J McDonald, Ms R Mcdonald, Miss E Mcdonnell, Mr B McDonnell, Mr C McDonnell, Mr M McDonnell, Mr P Mcdonnell, B McDonnell, Miss S Mcdougall, Lindsay McDougall, Mrs D Mcdougall, Mr M McDowall, Dr S Mcdowell, Mr M McDowell, Mr J McEachen, Mr L Mcelderry, Conor McElhinney, Dr H Mcelroy, Mr D Mcenroe, Derek Mcewan & Catherine McEwan, Mr T Mcewan, Mr J Mcewen, Mrs A Mcewen, Mrs K Mcewen, Mrs G Mcfarland, Mr K Mcfarlane, Mr S & Mrs B Mcfarlane, Mrs S Mcfarlane, Miss D Mcgahey, Mr T Mcgarrity, S Mcgarry & R Hunt, Oonagh McGarry, Ms S Mcgarry, Mr C Mcgeachie, Mr K Mcgee, Mr J McGeehin, Mrs S McGeeney, Mr I McGeorge, Mrs D M McGhee, Mr B H McGhie, Mr M Mcgill, Mr J McGillivary, Katy McGilvray, Mr J McGinlay, Mr C Mcginley, Mr M McGinty, Mr A Mcglashan, Mrs P Mcgough, Mr P Mcgovern, Dr A J McGowan & Dr C F Braban, Mr J McGowan, Mr K Mcgowan, Mr L Mcgowan, Mrs E Mcgowan, Mr A C O McGrath, Mr B Mcgrath, Declan McGrath, Mr D Mcgrath, Mr D Mcgrath-Grogan, Mr C McGregor, Mr I Mcgregor, Mr I J McGregor, Mr R McGregor, Mr T Mcgregor, Ms T Mcgregor, Chris Mcguigan, Miss K Mcguinness, Mrs J Mcguinness, Miss R Mcguire, Mrs K Mcguire, Mcgurk family, Dr P Mcgurn, Mr P Mchaffie, Mr P Mchale, Mr E McHugh, Mr P Mchugh, Mr S McHugh, Mrs F McHugh, Mr D Mcilwraith, Mr J Mcinerney, Ms B Mcinerney, Mr A Mcinnes, Mr J Mcinnes, Mr I Mcintosh, Dr E Mcintyre, Mr D Mcintyre, Mr R McIntyre (2), Mr M McKavett, Clive R McKay, Miss S Mckay, Mr D Mckay, Mr G E McKay, Mr G S McKay, Mr D Mckean, Mr A Mckechnie, Mr A Mckee, Mr D McKee, Mr M Mckee, Neville McKee, Mrs J McKee, Mr B & Mrs J McKee, Dick McKeever, Dr Dervilla McKeith, Mr R McKellican, Mr M Mckenna (2), Mr N Mckenna, Mr P Mckenna, Mr R Mckenna, Mrs E McKenna, Dr S A McKenzie, Mr I J McKenzie, Mr M Mckenzie, Pauline McKenzie Lloyd, Miss K Mckeon, Mr B McKeown, Mr P Mckeown, Ms D McKeown, Ms P Mckeown, Mr I McKerchar, Mr F Mckevitt, Alastair McKie, Mr M Mckiernan, Mr J McKillop, Mrs G Mckillop, Mr J McKinlay, Mr T Mckinney, Mrs J Mclain, Ms S McLaney, Mr G McLaren, Mr I D McLaren, Mr P McLaren, Mr G Mclaughlin, Rónán Mclaughlin, Mr S McLaughlin, Ms C Mclaughlin, Dr K McLean, Mr D Mclean, Mr I Mclean (2), Mr M Mclean, Mrs V Mclean, Ms G Mclean, Ms T Mclean, Dr A G McLee, Mr J McLeish, Mr T McLellan, Dr K McLennan, Colin R McLeod, Mr D N Mcleod, Jon McLeod, Mr R McLeod, Mrs T Mcleod, Ms K Mcleod, Ms P McLeod, Mr B L McLernon, Mr A McLevy, Dr D McLoughlin, Enda J Mcloughlin, Mr J McLoughlin, Miss A McLure, Barry McMahon, Dr D Mcmahon, Mr I McMahon, Mr S Mcmahon, Mr W McMahon, Mrs E K McMahon, Mrs J Mcmahon, Mrs S McMahon, Ms J McMahon, Ms N Mcmahon, Mrs C Mcmann, Miss J Mcmanus, Mark McManus, Mr M Mcmanus, Mr J M McMeeking MBE, Miss B McMillan, Mr A Mcmillan, Robert McMillan, Mrs C Mcmullan, Mr L Mcmullin, Mrs G Mcmullin, Mr P Mcnab, Miss L Mcnaghten, Mr D F McNair, Peter McNamara, Mr D Mcnamee, Mrs C Mcnamee, Dr D Mcnay, Mr A F McNee, Mr D Mcnee, Mrs H Mcneice, Miss J Mcneil, Mr A K McNeil, Mrs B Mcneil, Dr E McNeill, Mr D Mcneill, Ms J McNeilly, Mr D Mcnicholas, Mr A Mcniff, Mr D McNulty, Anna McOustra, Mr B McOwan, Mr G G McPhail, Mr P Mcphail, Mr I McPherson, Mr P McQuail, Mr T McRoberts, Mr S McRoyall, Mr B Mcsherry (2), Dr C Mcsorley, John McSweeney, Mrs I Mcvicar, Mr D McWalter, Mr A McWilliam, Mr C McWilliam, Mr I McWilliam, Ms A Mcwilliam, Mr S Mcwilliams, Mr K Meacher, David Mead, Mr G Mead, Miss R Meade, Mr C Meade, Mr J Meade, Mr P & Mrs G Meader, Mr B S Meadows, Mr J Meadows, Michael C Meadows, The Mewdows, Mr A Meads, Ms M Meagher, Dr K Meakin, Mr T Meakin, Mrs H Meakin, Mr C Meaney, Miss G Mear, Richard & Barbara Mearns, Mrs B Mearns, Mr A Mears, Ms P Mears, Mrs V Measom, Mr C Measures, Mr G Measures, Mr D Medcalf, Mr & Mrs R D L Meddick, Mrs L Meddick, Mr S Meddings, Jamie Medford, Miss S E Medforth, Mr A Medforth, Ms L Medine, Mr J M Medland, Mr R D Medland, Mrs H Medley, Mrs D Medlicott, Allan Mee, Mr T Mee, Mr D G J Meech, Mr J Meechan, Meehan family, Mr S Meehan, Mr E R Meek, Mr N R Meek, Mr W R Meek, Miss H Megraw, Mr G Megson, Katherine Meikle, Ms J Meikle (2), Dame E Meilleur, Mr C Melhuish, Mr S Melhuish, Mr P Mella, Miss H Mellan, Mr C Mellenchip, Mr D Melling, John Melling, Mr C Mellon, Mr M J Mellon, Miss J Mellor, Miss R Mellor, David Mellor, Mr I Mellor, Mr M Mellor, N & D Melsom, David J Melville, Richard Menari, Miss C Mendez, Stephen Menzie, Mr D Menzies, Dr J Mercer, Dr J H Mercer, Miss F J Mercer, Mrs L R Mercer, Mrs V J Mercer, Mr P Merchant, Mr A Meredith, Mr R Meredith, Mr O J Merne, Ms E Merrick, Mr T D H Merrie, Mrs J Merrigan, Mr I Merrill, Mr M Merriman, Mr T J Merrington, Mrs T Merriott, Mr A Merritt, Mr M Merritt, Mr G Merry, Mr S Message, P & L Message, Mr C R Messenger, Mr Douglas Messenger, Mrs P A Messent, Mr P Meston, Keith Metcalf, Mr M Metcalf, Ms M Metcalf, Mr S Metcalfe, Prof N B Metcalfe, Mr D Methven, Mr M Metson, Mr J A Mettam, Mr A Meurer, Dr R Meyer, Ms K Meyer, Mr S Michael, Dr G P Michaels, Mr L Michalek, Dr A Michel, Prof C C Michel, Mr G Michelmore, Ms A Mick, Mr J Micks, Mr R Middlemiss, Mr B Middleton, Mr G Middleton, J Middleton & H Campbell, Mr J E Middleton, Dr P Middleton, Mr R Middleton, Mrs C Middleton, Ms G Middleton, Mrs M Middleton, Ms A Middleton, Dr A Midgley, Miss K Midgley, Mrs P Midwinter, Mr R Mielcarek, Mr C Miers, Miss M Mierzejewska, Mr J S Mighell BVMS, Mr P Mignaud, Mrs J Mihell, Mrs M Milan, Mr G Milanese, Sir A Milbank, Mrs J Milborrow, Mr N R Milbourne, Mr C Milburn, Miss E M Mildner, Mr D Mileham, Mr M Mileham, Dr O Miler, Dr C J Miles, Lord Stephen Miles, Miss I Miles, Miss K Miles, Mr J Miles, Mr K Miles, Mr M Miles, Mr S J Miles, Mr W T S Miles, D & V Milford, Mr A Millar, Mr J Millar, Mr J C G Millar, Mike Millar, Mrs M Millar, Mrs S Millar, Ms M Millar, Mr W G Millard, Clive Millard, Mr M Millard, Mr R Millard, Mr W Millard, Mr A Millbank, Miss D Miller, Mr A Miller (2), Mr Alan K Miller, Mr A R Miller, Whittingehame Filter House, Mr A Miller, Mr B Miller, Mr D Miller, Mr D G Miller, Mr D J Miller, Mr H J Miller, Mr J Miller (2), Mr M Miller (2), Mr M J Miller, Mr P E Miller, Mr R Miller (4), Mr T Miller, Mr W Miller, Mrs C Miller, Mrs C A Miller, Mrs M Miller, Mrs M J Miller, L Miller, Ms S Miller, Mr J Millett, Dr N Milligan, Dr S P Milligan, K Milligan, Mr R Milligan, Miss S Millinger, Mr B Millington, Mrs M Millns, Dr A Millon, Miss J J Mills, Mr B Mills, Mr C Mills, Mr C H Mills, Mr G Mills, Mr J R Mills, Mr L J Mills, Mike Mills, Mr S Mills (4), Mr T Mills, Mrs J Mills & Mr K Haydock, Mrs J Mills (2), Mrs M Mills, Mrs S Mills, Mrs C Millward, Dr I Milne, Miss N Milne, Mr D Milne, Mr J Milne, Mr P Milne, Mrs H Milne, Mr D L Milner, Mr J Milner, Mr P Milner, Lynda Milner, Miss A Milopoulos, Mr J Milroy, Dr T P Milsom, Mr K Milsom, Mr E Milton, Mr J Milton, Mr R G Milton, Revnd F A R Minay, Mr F Miniken, Mr D Minns, Mr S Minns, Mr S Minton, Miss E Miscioscia, Mr C Miskimmin, Mr I P Misselbrook, Mr M Missin, Mr A Mitcham, Mr T Mitcham, Mr D Mitchel, Dr I Mitchell, Dr R J Mitchell, Miss C Mitchell, Miss R Mitchell, Mr A Mitchell, Andy Mitchell, Mr B Mitchell, Mr C M Mitchell, Mr D Mitchell (2), Mr & Mrs D Mitchell, Mr D G Mitchell, Mr D Mitchell, Mr F W Mitchell, Mr G K Mitchell, Mr G Mitchell, Mr H Mitchell, Mr J Mitchell (3), Mr Jolyon R Mitchell, Mr K Mitchell, Mr L Mitchell, Mr M G Mitchell, Mr N Mitchell (3), Mr P Mitchell, Mr R Mitchell, Tony Mitchell - Ormiston, Mrs D Mitchell, Mrs H Mitchell, Mrs L Mitchell, Mrs M Mitchell, Mrs S Mitchell, Mrs T Mitchell, Ms H Mitchell, Mitchell family, Ms R A Mitchell, Mr C R Mitchell, Mr P Mitchener, Mrs M Moar, Mr G K Moates, Mr G Mock, Mr R Mock, MOD Aberporth Site Conservation Group, Mr R Moden, Ms M Moeller - Holtkamp, Mr G Moffat, Mr I D Moffat, Ms S Moffat, Mr F Mogford, Mr Neil Moglione, Mr E Molden, Mr J Mole, Mr P E Mollatt, Mr P Molloy, Sandra Molloy, Ms A Molnar, Mr D Moloney, Mr M Moloney, Mr J Molyneux, Mr A Monaghan, Mr C Monaghan, Mr F Monaghan, Mr J Monaghan, Mr M Monahan, Miss H Moncrieff, Mr A Moncrieff, Ms P J Moncur, Dr S Money, Mr L Money, Mr N Money, Ms S Money, Monifieth High Biodiversity Group, Paul Monk, Monk family, Mrs J R Monnington, Mr C K A Monteith, Mr C R Monteith, Mr G B Monteith, Mr P Montgomery, Rev S Montgomery, Sqn Ldr G P F Montgomery, Mr D J Montier, Mr D Moodie, Mr T S Moodie, Miss F Moody, Mr A Moody, Mr C Moody, Mrs J O Moody, Dr J Moon, Mr A V Moon, Mr D Moon, Mr John Moon, Mr S J Moon, Mrs J Moon, Mr A Mooney, Mrs M E Mooney, Ms C Mooney, Mr C Moonstone, Ms C Moor, Mr Neil Moorcroft, Dr P D Moore, Dr R C Moore, Miss E Moore, Miss K N Moore, Miss P Moore, Miss R Moore, Miss T Moore, Mr A Moore (3), Mr B Moore, Mr C Moore (2), R & D Moore, Mr F Moore (2), Glen Moore, Mr I Moore, Mr J Moore, Mr J R Moore, Jeremy Moore, Mr O Moore, Mr P Moore (3), Mr P G Moore, Mr P W & Mrs D J Moore, Mr S Moore (2), Tom Moore, Mrs C Moore, Mrs C C Moore, Mrs J Moore, Mrs R Moore, Mrs S Moore, Ms H Moore, Mr R Moores, Mrs G Moores, Mr S Moore-Vale, Mr I D Moorhouse, Ms P Moorhouse, Mr D Moors, Dr N Moran, Miss F Moran, Nick Moran, Mr P Moran, Mr S Moran, Mr W Moran, Sian Moran, Moray Firth Counters, Mrs S M Mordue, Mr A R Moreton, Mrs C Moreton, Mrs J Moreton, Dr J E Morgan, Dr S Morgan, Mr F C Morgan, Mr G Morgan (3), Mr G J Morgan, Mr H Morgan, Mr I Morgan, Mr John Morgan, Mr K Morgan, Mr N Morgan (2), Mr P Morgan (2), Mr R Morgan, Mr S Morgan, Mr S J Morgan, Mrs A Morgan, Mrs E Morgan, Morgan family, Mrs L Morgan (2), Mrs M Morgan, Mrs S Morgan (2), Mr B & Mrs S A Morgan, Mrs U Morgan, Vicky & Chris Morgan, Prof K Morgan, Mr P Morgan-Capner, D Morgan-Smith, Miss J Morgen, T & J Moriarty, Mr D D Morison, Prof R Morledge, Mr A H Morley, Mr I Morley, Mr S Morley, Mrs V Morley, Mr P Morling, Dr M Morphy, J & C Morrell, Dr E Morris, Dr G Morris, Miss S Morris, Mr A Morris, Mr A J Morris, Mr C Morris (2), Mr D Morris (2), Mr D H Morris, Mr D Morris, Mr G Morris, Glenn E Morris, Mr I Morris (2), Mr J Morris, Mr K Morris, Mr N Morris, Mr P Morris (2), Mr P D Morris, Mr P R Morris, Ray Morris, Mr R Morris (4), Mr R E Morris, Mr R G W Morris, Mr S Morris, Mr Sean Morris, SNH Rum, Mr T Morris, Wayne Morris, Mrs A Morris, Mrs E Morris, Mrs J M Morris, Mr J A & Mrs P L Morris, Mrs R Morris, Mrs S & Mr N Morris, Ms M Morris, Ms S Morris, Mr R Morrisey, Ms J Morrish, B & T Morris-Hale, Dr W M Morrison, Miss L Morrison, Mr C R Morrison, Mr D Morrison (2), Mr I Morrison, Mr J Morrison, Peter & Sue Morrison, Mrs A Morrison, K & J Morrison, Mr C Morrow, Rev D Morrow, Mr M J Morse, Mr Doug Mortimer, Mr I Mortimer, Mr P Mortimer, Mr & Mrs K Mortimore, Dr P P Morton, Miss H Morton, Miss J H Morton, Mr A Morton, Mr C Morton, Mr D Morton, Mr G Morton (2), Mr I Morton, Mr K Morton, Mr M Morton, Mr R M Morton, Mr R Morton, Mr S Morton, Mr T Morton, Mrs J V Morton, Mrs M Morton, Mrs S D Morton, Ms S Morton-Williams, Mr J C Moseley, Dr M E Moser, Mr G M B Moser, Mr C Moses, Mr M Moses, Dr Ben Moss, Dr D Moss & Mrs U C Price Moss, Mr A P Moss, Mr K Moss, Martin Moss, Mr M W Moss, Mr N Moss, Mr N A Moss, Robert Moss, Mr S Moss, Mr W C W Moss, Mrs A Moss, Mrs C Moss, Mrs J J Moss, Ms V Moss, Mr A Mossman, Mr J Mossman, Mrs H Mossman, Mr A P Mossop, Keir Mottram, Mr A Mould, Mr J M Mould, Mr R C Mould, Mr T Moulton, Mr P Mouncer, Mr R Mouncher, Mr R R & Mrs Y Mounsey, Mount Mascal Pony Club Centre, Mr D Mountain, Mrs H Mountford, Mourne Heritage Trust, Mr J Mousley, Mrs S R Mousley, Tony Moverley, Miss C Mowat, Joan Mowl, Mr A D Mowle, Miss B Moynihan, Mrs S Moys, Mr R Moyse, Mr G Moyser, Mr G F Moysey, Ms M Muckersie, Mr K Mudd, Ms S Muddell, Mr R Mudhar, Mr N Mugan, Mr P Mugridge, Miss C Muir, Mrs L Muir, Ms D Muir, Mr A Muirhead, Lys Muirhead, Mr S Muldal, Mrs S Mulford, Mr C Mulhern, Mr E Mulholland, Mr M Mulholland, Mr N Mulholland, Mrs C Mulholland, Mrs E Mulholland, Ms C M Mulholland, Ms J Mulholland, Dr C Mullen, Mr J Mullen, James Mullen, Mr M F Mullen, Mr M Mullen, Mr P Mullen, Mr J R Muller, Mr G Mullett, Mr I Mulley, Mr D & Mrs D Mullier, Mrs A Mulligan, Mrs J F Mullinger, Mr A Mullins, Mr S Mullins, D Mullins & R Nixon, Mr M Mullis, Mr P J Mulvihill, Dr R Mumford, Mrs W Mumford, Mr R V Mummery, Dr Rod Muncey, Mr D Munday, Mr S Munday, Ms L Munday, Mr L Mundy, Mr R Mundy, Ms E Mundy, Mr B Mung, Mr P Munnings, Dr P I Munro, Mr A Munro, E A Munro, Mr G Munro, Mr R Munro (2), Mrs G Munro, Ms K Munro, Mrs Z Munroe, Mr A R Murch, Dr D A Murdoch, Mr I Murdoch, Mr A Murdock, Mrs M Murfin, Mr R C Murfitt, Mr J Murgatroy, Mr C Murgatroyd, Miss J Murphy, Miss M E Murphy, Mr B Murphy, Mr D Murphy, Mr F Murphy, Mr G Murphy, Mr I Murphy, Mr J Murphy, Kevin Murphy, Mr M Murphy (2), Mr M D Murphy, Noel Murphy, Mr N Murphy, Mr N L Murphy, Mr P Murphy, Mr P W Murphy, Mr S Murphy, Mrs M Murphy, Margaret Murphy, Mrs S Murphy, Ms A Murphy, Ms B Murphy, Ms J Murphy, Ms S Murphy (2), Dr W D Murray, Miss J Murray, Miss K Murray, Mr A Murray, Mr A E Murray, Mr A J Murray, Mr F A E & Mrs L Murray, Mr G Murray, Mr J Murray (3), Mr J B Murray, Mr L Murray, Mr R D Murray, Mr S Murray (2), Mr T Murray (2), Mr & Mrs R Murray, Mrs A Murray, Denise Murray, Mrs G Murray (2), Mrs J Murray, Mrs S Murray, Ms E Murray, Miss J Murrell, Miss L Murtagh, Mr G Murtagh (2), Tom Murtagh, Mrs N Murtagh, Miss J Muscott, Mrs M Muse-Hodgson, Dr A Musgrove, Dave Musgrove, Roger L Musgrove, Mr S Musgrove, Mr A J Muston, J N Mutch, Mr C Mutimer, Mr D Mutters, RW & MI Myatt, Mr J Mycock, Mr G W Myers, Mr T M Myerscough, Mr W R Myerscough, Mrs J F Mylne, Mrs M J Mylne, Mr S J Mynard, Dr R J Mynott, Miss L M Mynott, Mr C A Mynott

N Ms C Naden, Mr JS Nadin, Mr T Nagle, Mrs C Nagle, Mr A Nairn, Ms A Naish, Mrs D Napier, Mr P Napthine, Mrs L Napton, Nar Valley Ornithological Society, Mr A J Nash, Mr C Nash, Mr M Nash, Mr N Nash (2), Mr P Nash, Mr R Nash (2), Mr R G Nash, Maj R H J Nash, Thomas V Nash, Mrs M Nash, Miss R J Nason, Mrs A Nason, Mr M Nattrass, Maurice Nauta & Brian Chapman, Mr J Naven, Ms S Nayler, Naylor family, Mr D Naylor, Mr P Naylor, Mr S Naylor, John Neal, Mr P Neal, Mr S Neal, Mrs B Neal, Mrs S Neal, C & F Neale, Mr D Neale, Mr J Neale, Mr M Neale (3), Mr R Neale, Mrs E A Neale, Aidan Patrick Joseph Neary, Mr W D Neate, Mr B Neath, Mrs M Neath, M & P Neatherway, Mr Russell R P Neave, Mr D Neavyn, Mr M Needham, Ms A Needham, S & J Needham, Mr B Needle, Mr H Needle, Janet Needs, Miss K Neeson, Mr J S Neighbour, Mr K Neil, Mrs J Neil, Bill Neill, Ms J Neill, Mr R Neilson, Mr T A Neilson, Rev R Neilson, Ms A Neligan, Dr P Nellist, Miss C Nelson, Miss J Nelson, Mr A Nelson, Dr Brian Nelson, Mr D Nelson, Mr R Nelson, Mr T Nelson, Mrs B Nelson, Mrs L Nelson, Mrs S Nelson, Ms R Nelson, Mr D Nesbitt, Nesbitt family, Mr D Nesham, Ness family, Mr M Nestor, Mr M J Netherwood, Mr H R Netley, Mr A Nevard, The Nevards, Mr C Neve, David Neville, Mr P Neville, Mr A C Neville-Smith, Ms C E Nevison, Mr D R Nevitt, Mr K New, Mrs L Newbold, Peter Newbound, Dr M Newby, Miss S Newby, Mr S G Newby, Mrs J Newcombe, Dr R G Newell, Mr M A Newell, Mr R Newland, Dr D A Newman, Dr S P Newman, Mr C J Newman, Errol C B Newman, Mr M A H Newman, Mr P Newman (2), Mr R Newman (2), Mrs I Newman, Ms D Newman, Mr P A Newmark, Dr John Newnham, Mr & Mrs G Newport, Mr P Newport, Mr D G Newsam, Miss S Newsome, Mr D Newsome, Mr M Newsome, Dr S E Newson, Mrs D Newson, Newstead Village School, Dr J Newton, Dr S Newton, Dr S F Newton, Miss A Newton, Miss C D Newton, Mr C Newton, Mr J Newton, Mr J A Newton, Mr K Newton, Paul Newton, Mr S Newton, The Newtons, Mr T R Newton, Mrs G Newton, Newtown Nature Reserve, Ms Á Ní Mhuircheartaigh, Mr D Nicholas, Mr G Nicholas, Ms L L Nicholas, Dr P E Nicholls, Miss A J Nicholls, Miss L Nicholls, Mr B Nicholls, Mr D Nicholls, Mrs A C Nicholls, Mrs E Nicholls, Ms L Nicholls, Ms T Nicholls, Mr A J Nichols, Mr C Nichols, Mr H & Mrs M M Nichols, Mr P F Nichols, Mr S Nichols, Steven C Nichols, Mr G G Nicholson, Mr I P Nicholson, Jim Nicholson, Mrs A Nicholson, Mrs E Nicholson, Nicholson family, Mrs R A Nicholson, Mr D Nickeas, Ms L J Nickell, Mr P Nickless, Mr S Nickols, Miss C Nicol, Alexander Nicol, Mr J Nicol, Mr M Nicoll, Mr W A Nicoll, Mr G Nicolson, Ms B Nicpheadaran, Miss A

Nielsen, Miss V Nightingale, Mr B Nightingale, Mr D Nightingale, Mr Josh Nightingale, Mr Mark Nightingale, Mr F Nind, Mr R B Ninnes, Mr C Nisbet, Mr D Nisbet, Mr J Nisbet, Mr T Niven, Ms K Niven, Mr G Nix, Dr M Nixon, Miss L Nixon, Mr A Nixon, E Flavelle & R Nixon, Mr C Nixon, Mr P Nixon, Mr R C Nixon, Mrs V Nixon, Mrs V A Nixon, Dr P C Noakes, Mrs J Noakes, Mr N T Noall, Mr B T Nobbs, Mr J L Nobbs, Mrs R Nobbs, Mr C Noble, Mr G Noble, K E Noble, Mr K Noble, Mr M Noble, Mr R Noble, Simon Noble, Mr S D Noble, Mr Stuart Noble, Ms B Noble, Ms J Noble, Mr R Nocher, Miss R Nock, Mrs K Nock, Ms A Noel, Mr A Nolan, Mr M Nolan, Mr P Nolan, Mr T Nolan, Mrs A Nolan, Mr A Norburn, Mrs A Norbury, Ms M Norden, Mr D Norfolk, Mr P Norfolk, Norfolk & Norwich Naturalists' Society, Dr R Norman, Mr C G Norman, Mr J Norman, Michael J Norman, Mr R K Norman, Mr S C Norman, Mr W Norman, Mrs J Norman, The Norman's Burbage, Prof D Norman, Ms W Normanton, Mr J Norrey, Mr S Norrie, Cdr A Y Norris R N, FRGS, Miss C Norris, Mr D Norris, Mr G Norris, Ian Norris, Mr M Norris, Mr R Norris (3), Mrs A Norris, Mr D North, Mr M North, Mr R North, Mr Stephen North, Ms L A North, North East Scotland Bird Club, North Northumberland Bird Club, Mr P H Northcote, Mrs S Northcott, Northern Ireland Birdwatchers Association, Northern Ireland Environment Agency, Northern Ireland Raptor Study Group, Mr D Northover, Mr R H Northridge, Northumberland & Tyneside Bird Club, Northumberland National Park, Mr J Norton, Mr T Norton, Mrs L Norton, Mr D Nott, Mr A Nottage, Rob & Linda Nottage, Mrs L Nottage, Nottinghamshire Birdwatchers, Nottrodt family, Mr J Nourish, Mrs S Nouvet, Miss R Noyland, NPWS, Prof J H A Nugent, Mr J A Nundy MBE, A D Nunn, Miss G Nunn-Turner, Mr D I & Mrs J Nurney, Mr N Nuthall, Miss K Nutt, Mr A Nuttall (2), David Nuttall, Mr M D Nuttall, Mrs A Nuttall, Mr R C Nutton, NW London RSPB Group, NWWT Conservation Staff, Mr J Nyhan

O Mr J O'Boyle, Mr D O'Brien, Mr P O'Callaghan, Mr J O'Carroll, Mr C Ó Coisneacháin, Mr P O'Connell, Mr F O'Connor, Mr G O'Connor, Ms O O'Connor, Mr D O' Connor, Mr S O'Dea, Mr P O'Donnell (2), Mr A O' Donnell, Ms H O'Dowd, Mr K O'Driscoll, Ms E O'Driscoll, Mr R O' Driscoll, Mr R Ó Gormáin, Mr D O'Gorman, Mr D O'Hare, Mr P O'Hora, Dr M O'Mahony, Mr P Ó Meachair, Miss M O'Neill, Gavin Ó Sé, Mr D O'Sullivan, Mr L O'Toole, Mr N Oakes, Mr S & Mrs E J Oakes, Mrs G Oakes, Mr M Oakland, Mrs J Oakland, Mr G Oakley, Mr J Oakley, Mr J N Oakley, Peter Oakley, Ms B A Oakley, Mr D Oakley-Martin, Mr J G Oates, Steve Oates, Mr S P Oates, Dr I Oberski, Dr M Ó Briain, C Ó Briain, Dr M Obrien, Mr C D J O'Brien, Mr F O'Brien, Mr L O'Brien, Mr W O'Brien, Mrs D M O'Brien, J O'Brien, Ms I O'Brien, Miss C O'Brien Moran, Ms C O'Brien Shanahan, Mr J O'Callaghan, Michael O'Clery, Miss M Oconnell, Miss B O'Connell, Mr F O'Connell, Mr G O'Connell, Mr M O'Connell, Mr J B O'Connell-Davidson, Brian O' Connor, Dr I O'Connor, Mr C O'Connor, Mr D O'Connor, Mr P O'Connor, Mr S J O'Connor, Mr W J O'Connor, Ms A O'Connor, Mr D Odell, Mr N Odin, Sue O' Donnell, Dr P O'Donnell, Dr S O'Donnell, Mr D O'Donnell, Dr P O'Donoghue, Mr B O'Donoghue, Mrs A O'Donoghue, Mrs J Odonovan, Ms S O'Donovan, Mrs R O'Dowd, Ms B Odriscoll, Miss D O'Driscoll, Mr M O'Driscoll, Mr O'Driscoll, Mr F O'Duffy, Mr P K O'Duffy, Mr B Odwyer, Mr A O'Dwyer, Mr J O'Dwyer, Mr D C Offer, Mr I Offer, Mr K Offord, Mr S P Offord, Mr C O'Flynn, Mr P J Ogden, Mr P Ogden, Mr J Ogilvie, Ms K Ogilvie, Ms L M Ogilvie, Dr J R Ogle, Mrs R O'Gorman, Ms C O'Grady, Miss N J O'Hanlon, Mr J O'Hanlon, Mr D J O'Hara, Mr K O'Hara, Mr S O'Hara, Mr T O'Hara, Ms R O'Hara, Miss V O'Hare, Mr T P O'Hare, Dr B Oji, Dr D A O'Kane, Mr A D O'Keefe, Mr C O'Keeffe, Mr D O'Keeffe, Mr M O'Keeffe, Mrs I Okell, Mr J D Okill, Mrs G Okill, Mrs T Okon, Mr M Oksien, Mr R A Old, Old Down Conservation Group Basingstoke, Mr A Oldacre, Mr C Oldershaw, Ms L Oldershaw, Mr F C Oldfield, Mr P Oldfield, Mr D F Oldham, Mr B M & Mrs V F O'Leary, Mr M J & Mrs R O'Leary, Ms B Oleksy, Mr T Oliphant, Dr G Oliver, Miss K Oliver, Miss M Oliver, Andy Oliver, Mr D Oliver, Mr G Oliver (2), Ian Oliver, Mr C Oliver, Mr M Oliver, Mr N Oliver, Mr P J Oliver, Mr R Oliver, Mr S J Oliver, Mr T R Oliver, Joan Oliver, The Olivers, Ms R Oliver, Mr J Olley, Mr S Ollivier, Miss K Olsen, Mr P Olson, Mr P J Olson, Mr G G O'Mahoney, Mr B O'Mahony, Mr T O'Mahony, Mr D S Omand, Mrs M O'Meara, Ms M O'Meara, Mr D Ó Murchú, Mr C Omus, Mr D O'Neill, Mr E O'Neill, Mr J O'Neill, Mr K O'Neill, Mr L O'Neill, M O'Neill, Mr S O'Neill, Mr R Onions, Mrs J Onions, OPH Dunkeld, Mr A M Oram, Mr M J Orchard, Mr P Orchard, Ms N Orchard, Mr M A Orchard-Webb, John O'Reilly, Mr P O'Reilly (2), Jeff O'Reilly, Ms B O'Reilly, Mr D Orford, Ms E O'Riordan, Orkney Bird Club, Miss A Orland, Dr P R Orman, Mr C Orman, Mr E Ormand, Mr P Orme, Mrs J K S Dyson-Orme, T Peter Ormerod, Mr J Ormerod, Mr R Ormerod, Mr A Ormond, Prof R Ormond, Mr S Ormondroyd, Mr C M Ormston, Mr N C A O'Rourke, Mr T O'Rourke, Ms C Orpin, Mr A Orr, Mr J Orr, Mr D C Orr-Ewing, Mr H J Orridge, Mr S A Orridge, Mr N Orson, Dr R A Orton, Geoff Orton, Robert Orton, Mr E Osbaldeston, Miss R L Osborn, Mr K Osborn, Mr Peter Osborn, Mr T Osborn, Miss B Osborne, Miss K Osborne, Mr C Osborne, Mr C A Osborne, Mr G Osborne (2), Jonathan Osborne, Mr J Osborne (3), N F Osborne, Mr R T & Mrs G Osborne, Mr W A Osborne, Mrs F Osborne, Ms H Osborne, Mrs L Osgerby, Patricia O'Shaughnessy, Mr S O'Shea, Miss F Osis, Mr J Osmond, L Osmund, Mr C Osthoff & Ms F Wilson, Mr G Ostler, Mr T L Ostler, Mr D O'Sulivan, Mr J O'Sullivan, Mr M O'Sullivan (2), Mr O O'Sullivan, Mr P O'Sullivan, Valerie O'Sullivan, Ms O O'Sullivan, Kirstine Oswald, Mr R A Oswald, Mr S O'Toole, Mr T O'Toole, Mr P G Ottaway, Mr B Ottery, Mr G H C Ottley, Mr C Otty, H Ougham & S Thomas, Mr P J Ouless, Mrs L Outhwaite, Mr I Outlaw, Mr M Outlaw, Mrs W Outram, Ms K Outteridge, Mr C Ovel, Mr E Ovens, Mr R Overall, Mrs M E Overall, Mr C P Overton, Mrs B Overton, Mr M K Overy, Mrs D Overy, Miss J Owen, Miss M J Owen, Mr C Owen, Mr C M Owen, Mr D Owen (4), Mr D W Owen, Mr G Owen, Mr G A Owen, Mr H J Owen, Mr M Owen, Mr M K Owen, Mr S Owen, Mrs A Owen, Mrs B Owen, Mrs R Owen, Ms E Owen, Ms J Owen, Dr N W Owens, Miss K Owens, Mr E Owens, Mr I Owens, Mr R Owens, Mr & Mrs Owens, Jan Owens, Ms B Owsianka, Prof A Oxbrow, Dr W M Oxbury, J V Oxenham, Mr D Oxford, Mr J Oxford, Oxford Ornithological Society, Miss S Oxlade, Dainis & Wendy Ozols, Miss J Ozyer

P Mr B Pacey, Mr P Pacey, Mr D J Pack, Mr P Packer, Mr R Packham, Mr M Paddock, Mr A F Padel, Mrs C E Padfield, Mr M Padmore, Miss H Page, Mr A Page, Mr D I Page, Mr J Page, Mr N Page, Mr N G Page, Mr P Page, Mr R Page, S & J Page, Mrs C Page, Mr H Paget-Wilkes, Dr C Paice, Mrs G Pain, Dr K Paine, Dr T Paine, Mr M Paine, Mrs S Paine, Mark G Painter, Ms D Painter, Mr T Paintin, Mrs J Pakenham, Mrs A Palethorpe, Mr J Palfery, Mr M Paling, Mr S Paling, Mr D Palmar, Dr S Palmer, Miss L Palmer, Mr A Palmer (2), Mr B Palmer, Mr D Palmer, Mr G Palmer (2), Mr J Palmer (2), Palmer family, Mark J Palmer, Mr M Palmer, Mr N W Palmer, Mr R Palmer, Mr Roger Palmer, Mr R B Palmer, Mr S Palmer, Mr S J Palmer, Mr S M Palmer, Mrs A Palmer, Mrs E Palmer, Mrs J P Palmer, Mrs L E Palmer, Ms L Palmer, Mr J Pam, Mr M Pamment, Mr K Panchen, Ms S Pandor, Mrs J Panman, Mr R J Pannell, Mr P Panton, Mr N Papas, Mr S Papps, Mr S Pardoe, Mr A J Parfitt, R A K Parfitt, Mr S E Parfitt, Mr W J Parfitt, Mrs C Parfitt, Mrs S Parfitt, Ms A Parfitt, Miss C Parfrey, Ms Z Paris, Nicky Parish, Mr J C Parish, Miss C Park, Mr I Park (2), Mrs A M Park, Dr R Parker, Miss M E Parker, Miss P E Parker, Mr A Parker (2), Mr A B Parker, Mr B Parker, Mr B J Parker, Mr D Parker (3), Mr G R Parker, Mr J Parker (2), Mr K Parker, Mr N Parker (2), Mr Owen Parker, Mr P Parker (2), Stephen J Parker, Mr T Parker (3), Mr W J Parker, Mrs J Parker, Mrs M Parker, Mrs S Parker, Mrs S E Parker, Ms C Parker, Ms T Parker, Mrs S A Parker-Munn, Parkes family, Mr C Parkes, Mr M Parkes, Mr T Parkes, Mrs P Parkes, Mrs C Parkhouse, Mr D Parkin, Mr J Parkin (2), T Parkin & S Hodgkinson, Prof D T Parkin, Miss V Parkins, David Parkinson, Mr K Parkinson, Mrs R Parkinson, Prof D Parkinson, Mr M Parks, Mrs P Parkyn, Dr S Parlane, Mr J Parlby, Mr D Parmenter, Mr P Parmenter, Mr & Mrs T W Parmenter, Mrs S Parmenter, Mr D Parnaby, Mr E Parnell, Dr L Parr (2), Chris Parr, Mr M D Parr, Mr M Parr, Dr J D Parrack, Mr G A Parrish, The Parrotts, Mr D Parrott, Mr M Parrott, Mrs J E Parrott, Dr C Parry, Dr M Parry, Miss E Parry, Mr A Parry, Mr C J Parry, Mr D Parry, Mr H Parry, Mr J Parry, Mr K G Parry, Mr M Parry, Mr R Parry, Mr & Mrs R L Parry, Mr W Parry, Ms J Parry, C B Parry, Mr H Parry-Jones, Mrs J Parry-Williams, Miss B Parslow, Mr J L F Parslow, Mr A J Parsons, The Sustainability Centre, Mr J Parsons (2), Mr N Parsons, Mr R Parsons, Mr R C Parsons, Mr R J Parsons, Mrs M Parsons, Mrs M C Parsons, Mrs S Parsons, Mr R Partner, Dr J K Partridge, Partridge family, Mr R G Partridge, Mr S J Partridge, Mrs S Partridge, Mr I P & Mrs F A Pascoe, Mrs L Pascoe, Mrs S Pashby, Mr M Passman, Mrs I H Passway, Mr S Paston, Mrs M Patch, Ms D Pate, Mr S Patel, Ms H Pateman, Dr A Paterson, C Paterson, Mr C Paterson, Mr D Paterson, Mr H Paterson, Mr I Paterson, Mr S Paterson, Rev N Paterson, Mr J F Patient, Mr S Patient, Mr S Patmore, Mr H Paton, Ms O Patricia, Mr E T Patrick, Mr N Patrick, Mr S Patrick, Kath Patrick/Nick Brown, Mr D Patterson, Mr John Patterson, Mr K Patterson, Mr P Patterson, Mr T W Patterson, Mrs B Patterson, Ms A Patterson, Mr R S Pattimore, Miss E J Pattinson, Ms A Pattison, Dr R G Paul, Mr B Paul, Mr C Paul, Mr R Paul, Mr S Paul, Mrs L Pauli, David E Paull, Mr A Paulls, Mr R Paulson, Mr A Pavey, J & C Pavey, Jon Pavey / Alison Stuck, Mr B Pavier, Mr D Pawlett, Ms J Pawlyn, Valerie Pawsey, Mr A W Pawson, Mrs H Paxton, Ms J Payne, Mr A Payne, Mr B Payne, Dr David Payne, Mr D E E Payne, Mr D H & Mrs L Payne, Mr E Payne, Mr J Payne, Mr M Payne, Mr R Payne, Mr R J Payne, Mr S Payne, Mr T Payne, Anita Payne, Mrs G Payne, Payne family, Ms C M Payne, Ms K Payne, Mr D B Paynter, Mr K Peace, Mr N D Peace, Mr R K Peace, Mr T Peach, Dr C Peacock, Mr M Peacock (2), Mr M A Peacock, Mrs P Peacock, Mr P Peakall, Mr C Peake, Mr M Peaker, Mr A Pearce, Mr A K Pearce, Mr A R Pearce, Mr D Pearce (2), Mr D F Pearce, Mr D Pearce, Mr G Pearce, Mr K Pearce, Mr S L Pearce, Mr & Mrs L Pearce, Pearce family, Mr J W Pearce-Higgins, Dr A J Pearman, Mr D Pearson (3), Mr M Pearson, Mr R & Mrs K W Pearson, Mr R Pearson (2), Mr R M Pearson, Mr R S Pearson, Mr R W Pearson, Mr R Pearson, Mrs J Pearson, Mrs N Pearson, P Sumner & J Coulter, Mr D Pearsons, Mr R H Peart, Ms E A Peart, Mr R Pease, Mr R Peasland, Lizzy Peat, Mr J D Peatfield, Mr O Peavoy, Miss H Peck, Mr M Peck, Ronald Peck, Mrs M Peck, Mr R Pedder, Mrs B Peddle, Mr R Pedley, Mr S Pedley, Mr A Pedliham, Mr C Peel, Mr D Peel (2), Mr M Peel, Mr S Peel, Mrs K J Peerless, Mr C L Peers, Mr M F Peers, Mr D A Peet, Mr J Pegden, Dave Pegg, Mr N Pegler, Mr A Peilow, Dr T Pell, L & J Pell, Dr R Pell, Dr R Pelling, Mrs M Pelling, Mrs L Pemble, Mr R Pembridge, Pembrokeshire Bird Group, Ms S Pendarves, Dr C J Pendlebury, Mr C Pendleton, Ms K Pendreigh, Mr M Penford, Mr D Penman, Mr J Penman, Mrs E Penman, Mr M K Penn, Mr M Pennell, Mr C Pennells, Mr D Penney, Mr P Penning, Mr M G Pennington, Mrs A Pennock, Mr C Penny, Mrs C M Penny, Mrs R Penny, Mr R F Penticost, Pentland Hills Ranger Service, Mr G Penton, Mr G R M Pepler, Mr G Peplow, Mr W F Peplow, Mrs D Peppe, Mr A Pepper, Mr A S Pepper, Mr & Mrs C E Pepper, Mr R T Pepper, Mr S Pepper, Ms S C Pepper, Dr C J Peppiatt, Dr Roger Peppiatt, Dr S Percival, Mr T J L Percival, Mrs D Percival, Mrs L Percival, T & H Peregrine, Mr P Peretti, Mr S Perfect, Mrs C Perikleous, Mr R Perkin, Miss C Perkins, Mr A J Perkins, Mr B Perkins, Mrs C Perkins, Mr D Perks, Mr I Perks, Miss H Perrett, Mr D Perriman, Mr C A Perrin, Ms S Perrin, Miss S Perring, Dr K W Perry, Mr A Perry, Mr A S Perry, Mr D Perry (2), Mr J Perry, Mr J W Perry, Mr R Perry, Mr S J Perry, Perry family, Miss H Perryman, Mrs J Persson, Mrs L Peruffo, Ms S Pesante-Mullis, Mr O Pescott, Mr S Petar, Mr L Petch, Mrs E A Petch, Mr Peter Hastings, Mr S Peter, Mr H Peter Crispin, Peterborough Bird Club, Miss H Peters, Mr A Peters, Mr I Peters (2), Mr J N Peters, Mr R Peters, Mr S Peters, Mrs A Peters, Ms V Peters, Miss L Petherbridge, Mr A Petherbridge, Mr S W Petrek, Mr D Petrie, Mr G Petrie, Mrs K Petrovics, Mrs M E Petrovics, Jacky Pett, Mrs S Pettett, Mrs J Pettit, Ms J Pettit, Miss G Pettitt, Mrs D Pettitt, Dr D Petts, Dr S J Petty, Mr T Petty, Pevensey Levels, Mr C B Peverett, Mr J Pewtress, Mr M Pewtress, Mr B Phalan, Mr R A Phelan, Mrs A Phelan, Miss A Phelps, Miss E Phelps, Mr M Phelps, Mr R & Mrs C A Phelps, Miss A Phillip, Dr M F Phillips, Miss A Phillips, Miss A L Phillips, Mr D Phillips, Mr H Phillips, Mr I Phillips (2), Mr J Phillips, Mr J H Phillips, Mr L Phillips, Mr M J Phillips, Mr N Phillips, John & Viv Phillips, Mr R Phillips, Mr P M Phillips, Roy & Mary Phillips, Mr T Phillips (2), Mr T J Phillips, Mr W F Phillips, Mrs E Phillips, Mrs G M Phillips, Mrs M Phillips (2), Mrs S Phillips, Ms H Phillips, Ms J Phillips, Mr C Phillipson, Mr I Phillipson, Mr V Phillpot, Mr B Philp, Mr R Philpott, Mrs Z Philpott, Ms L Phipps, Mr A Phizacklea, Mr A Pick, Mr M Pick, Ms V B Picken, Mr G Pickerell, Miss C Pickering, Mr R H Pickering, Mrs H M Pickering, Mr & Mrs Pickering, Ms L Pickersgill, Dr S Pickett, Miss H Pickett, Mr David Pickett, Mr G Pickett (2), Mr R S Pickett, Phil Pickin, Dr K Pickles, Mr M Pickles, Mrs J Pickles, Dr J Pickup, Zoe Lynette Pickup, Miss L Pickup, N Picozzi, Mr G Picton, Miss L Pictor, Mr G D Pictor, Mr D C Piec, Miss B Pierce (2), Miss R Pierce, Mr D Pierce, Mr G Pierce, Mr L Pierce, Mr N Pierce, Mr D R Piercy, Mr R Piercy, Mrs A J Pierssene, Dr J M Piggins, Mr M Piggott, Mr M J Pigott, Dr W J Pigram, Mr A R Pike, Mr N Pike, Mrs J Pike, Mr S Pilbeam, Mr I Pilling, Mr M Pilsworth, Mr H Pimborough, Miss S E Pimm, Mr T Pinchen, Mr T Pinches, Mr J M Pinder, Mr N Pinder, Mr R K Pinder, Mr S Pinder, Mrs J A Pinder, Mr C Piner, Mr S G Piner, Debbie Ping, Valerie Pink, Miss J Pinkney, Mr B Pinnick, Mr J Pinnick, Ms K Pinnick, Mr D Pinniger, Mr G Pinnock, Mr M Piotrowski, Mr S Piotrowski, Dr J Piper, Miss C J Piper, Mr D Piper, Mr B Pirie (2), Miss T K Pirnie, Mrs M Piska, Pitcher-Baker, Mr A P Pitcher, Mr B Pitcher (2), Mr R Pitcher, Mr A G Pitches, Mr Martin J Pitt, Mrs L M Pitt, Mrs S A Pittam, Mr M Pittaway, Mr D Pittendreigh, Mr M F Pittman, Mr R Pitts, Dr J D Plackett, Mr J Plackett, Mr C Planck, Dr A R Plant, Dr P J Plant, Miss S Plant, Mr D Plant, Mr I Plant, Mr M Plant, Mr B Planterose, Mr D Platt, Mr Gareth D A Platt, Bob Platt, Mr B L Pleasance, Mr N Pleasance, Mrs E Pleasance, Mr D Pleasants, Mr M Pleasants, Ms D Pledge, Mrs C Pledger, Mrs C Plenderleith, Mr R G Plowman, A Plowright, Mr R Pluck, Mr A J Plumb, Mr M Plumley, Mr R Plummer, Mr S Plummer, Ms N Plummer, Mr J R Plumpton, Ms A Plumridge, Mr R A Pocklington, Dr M Pocock, Mr A Pocock, Mr A Podmore, Mrs J Podmore, Miss M Podsiad, Mr R Pointer, Mr T Pointer, Dr M Polak, Dr R H C Poland, Mr A Polkey, Mr S Polkinghorne, Mr D J Pollard, Mr M J Pollard, Mr A J Polley, Mr M V Polley, Mr G C Pollinger, Mr M S Pollitt, Mr J Pollock, Mr P Pollock, Mrs A Pollock, Mrs D Pollock, Mr D Pomeroy, Mr B Pomfret, Nick Pomiankowski, Mr J M Pomroy, Mr M S Ponsford, Ms E Ponsford, Mr M Pont, Mr E D Ponting, Miss A Poole, Clive R Poole, Mr L Poole, Mr M Poole, Mr W G Poole, Mr E J Pooley, Mr F Pooley, Mrs G Pooley, Mr S Poots, Mr M Pope, Mr N Pope, Mr P Pope, Ms P A Pope, Mrs W Pope, Dr M Popescu, Dr R G Pople, Mr G D Pople, Mrs J Popley, Mr R Poppleton, Mr M Porteous, Mr B Porter (2), Mr C Porter (2), Mr D E Porter, Mr N R Porter, Mr P Porter, Mr R Porter, Mr S Porter, Mr T Porter, Mrs R Porter, Mrs S Porter, Ms H Porter, Mr S Portlock, Mr J Portman, Mr S G Posen, Mr K R Postones, Mr J Pott, Mr A Potter, Mr B Potter, Mr B C Potter, Mr C Potter, Mr D Potter, Mr J Potter, Mr L Potter, Mr M Potter (2), Mr N Potter, Mr R Potter, Mrs A Potter (2), Mrs Y Potter, Potteric Carr, Potters Bar RSPB Members Group, Mr C & Mrs M Potts, Mr P M Potts, Mr P Potts, Mrs Potts, Poulsom family, Lynne Poulson, Mr Bruce Poulter, Mr D J Poulter, Mr A Poulton, Poulton & Pilmer Families, Mr M Pountain, Mrs S Pow, Dr C D Powell, Ian Powell, Major J M N Powell, Miss S E Powell, Miss W Powell, Mr C R Powell, Mr D Powell (2), Mr D G Powell, Mr Graham D Powell, Mr G W Powell, Mr J Powell (2), Mr J R Powell, Mr M Powell, Mr P Powell, Mr R Powell, Tony Powell, Mrs J C Powell, Mrs M Powell, Ms A Powell, Ms B Powell, Mrs W Powell-Jones, Mr A Power (2), Mr B Power (2), Mr D J Power, Mr G Power, Mr J Power (2), Mr P Power, Mrs S Powis, Mr G Powley, Mr R M Powley, Mr J D Powne & Mrs D Ludlow, Prof I R Poxton, Richard Poyer, Mr J Poyner, Mr R Poyner, Mr D Poyser, Miss L Praciak, Mr A Prasad, Dr B Prater, Dr A J Prater, Mr G Pratt, Mr M P Pratt, Mrs E M Pratt, Mr S Preddy, Mr M Preece, Ms C Preece, Ms L Preece, Mr I R Prentice, Mr L Prentice, Mr R J F Prentice, Ms P Prenton, Mr B Prescott, Mr D Prescott, Mr M Prescott, Ms C Prested, Miss S Preston, Mr A Preston, Mr D Preston, Mr J Preston (3), Mr M A Preston, Mr S Preston, Mrs K Preston, Ann Preston, Dr C R Price, Dr D Price, Miss C Price, Miss L Price, Mr D J Price, Mr G Price, Mr K D Price, Mr M Price (3), Mr R Price (2), Mr R B Price, Mrs C Price, Mrs L Price, Mrs N Price, Mrs S Price, Mrs S H Price, Gillian Price, Ms S

Price, Mr J Price-Stephens, Mr D J Priddis, Mr D J Priddle, Mrs P Pridham, Mrs B Priest, Mr M M Priestley, Mr N Priestley, Mr S Priestnall, Mr & Mrs R Priestnall & family, Miss B Primrose, Mr M S Prina, Mr B Prince, Mr D Prince, Mr M G Prince, John & Sue Prince, Mr R Pring, Miss H Pringle, Mrs L Pringle, Sir Charles Pringle, Ms A Pringle Davies, Mr A S & Mrs J G Prior, Mr J K Prior, Mr R Prior, Mr W Prior, Elizabeth & Bob Prior, Mrs M Priscott, Dr I Pritchard, Miss C E Pritchard, Mr A Pritchard, Mr D Pritchard, Mr E P Pritchard, Mr I Pritchard, Mr K V Pritchard, Mr R Pritchard, Mrs J S Pritchard, Revnd C N Pritchard, Mrs L Pritchett, Mr F Prittie, Mr R Probets, Mr M Procter, Mrs H Procter, Mr I J Proctor, Mr K Proctor, Mr R Proctor, Mr S J Proctor, Mrs K Proctor, Mr B Proffitt, Ms A Prole, Mrs C Prosser, Ms J Prosser, Mr K Proud, Mrs A Proud, Paul Proudfoot, Maura Noonan, Joe Proudfoot, Proudlock family, Mr S Prouse, Miss J Prout, Mr G K Prout, Dr A D Prowse, Mr F Prowse, Miss K Prowting, Mr J Proyer, Mr B G Prudden, Mr G Pryce, Mr K T P Pryce, Mr N Pryce-Jones, Mr L Pryde, Mr W Prylinski, Ms T Prynne, Mr D Pryor, Mr R J Prytherch, Mrs H Pude, Mr E R Pugh, Mr R Pugh, Mr S Pugh, Mr M Pugh-Davies & Mrs J Davies, Mr A Pugsley, Pulford Parish, Mr G Pullan, Mr J Pullen, Mr J R Pullen, Mr M Pullen, Mr N Pullen, Dr N Pumphrey, Mrs S Puplampu, Mr A Purcell, Mr N Purdew, Mr M D Purton, Mr R Purveur, Dr G Purvis, Mr D Purvis, Mr M Pusey, Mr J Putley, Robert H A Putnam, Mrs M M Putnam, Mr B Puttock, Mr Michael Puxley, Dr G Pyatt, Mr A Pycroft, Dr R Pye, Mr D Pye, Robert Pyefinch, Mr K Pyman, The Pymars, Mr J H Pymm, Mrs S Pyne, Rupert Pyrah, Mr N Pyzik

Q Mr A Quadling, Tim Quantrill, Bill Quantrill, Mr H J Quarterman, Miss L Quaye, Mrs P M Quayle, Quedgeley Health Walks, Miss C Quigley, Quigley family, Mr G Quigley, Mrs A Quigley, Miss L Quinn, Mr A Quinn, Mr C B Quinn, Mr P Quinn, Mr T Quinn, Mrs H Quinn, The Quinns, Mrs V Quinn, Mr S Quinney, Mr A Quinton, Mr B Quinton, Mrs J Quirie, Mel & Chris Quy

R Mr B Rabbitts, Ms S Rabie, Mr C Raby, Dr D Race, Mr R Rackliffe, Dr B J Rackstraw, Mr T Rackstraw, Mr S H Radcliffe, Radden family, Dr P Radford, Doug Radford, Mr G L & Mrs P J Radford, Mr R Radford, Mrs S Radford, B Radford, Mr M Radka, Mrs J Radlett, Mr C Rae (2), Rae family, Mr J Rae, Mr M Rae, Mr R S Rae, Mr S Rae, Mrs J Rae, RAF Boulmer, RAF Coningsby (John Harry), RAF Lakenheath Conservation Group, Royal Air Force Ornithological Society, Dr Richard Rafe, Mr D Rafferty, Mr A Rafinski, Mr M Rafter, Mr S Rahilly, Dr A Raine, Mr A Raine, Miss E Rainey, Mrs M A Rainey, Mr K M Rainford, Rainham RSPB, Mr K Rainio, Mr B Rains, Mr M H Rains, Mr D Ralfs, Janet Ralley, Dr C Ralph, Mr S E Ralph, Mrs L Ralph, Mrs S Ralph, Ian L Ralphs, Mr A & Mrs D Ramage, Dr S Ramalingam, Mr A D K Ramsay, Mr G Ramsay, Mr I Ramsay, Miss K Ramsbottom, Mr C G Ramsden, Mrs J E Ramsden, Mr A Ramsell, Miss V Ramsey, Mrs A Ramsey, Mrs M Ramskill, Mr C D Rand, Miss J Randall, Mr G Randall, Mr I Randall, John Randall, Mr K Randall, Mrs M Randall, Miss N Ranger, Ranger Service, Mr J Rankin, Mr W P Rankin, Mr C Rankine, Ranscombe family, Mr B Ransom, Mr T Ransom, Mr P Raper, Mr G Rapley, Rare Breeding Birds Panel, Mrs C Ratcliff, Miss L Ratcliffe & Mr B Davies, Mr D B Ratcliffe, Mr M Ratcliffe, N M Ratcliffe, Mr P H Rathbone, Mrs J Raum, Dr S Raven, Colin Raven, Mr T Raven, Sue Raven & Greensand Trust, Mr J Rawles, Dr N D Rawlings, Christopher Rawlings, Mr D Rawlings, Brigadier P P Rawlins, Mr I Rawlins, Mrs M Rawlins, Mr S Rawlinson, Mrs E Rawlinson, Paul Rawnsley, Mr G Rawson, Mr S Rawson, Miss A Ray, Mr A Ray, Mr D Ray, Mr H Ray, Mr S Rayburn, Mr M Rayment, Mr N W Rayment, Colin Raymond, Miss J Rayner, Mr M Rayner, Mr P G Raynes, Mr E M & Mrs P J Raynor, Mr R Raynor, Mrs P Raynor, Mr D Read, Mr I Read, Mr M Read, Mr P Read, Mrs L Reader, Mr R Ready, Mr P Reaney, Mr D Rear, Mr M Reardon, Miss E Reast, Mr J B Reavey & family, Mr M Reavley, Mr Chris Reay, Mr G Reay, Mr N Reay, Mr P Reay, Mr P J Reay, C J Reay, Mr G W Rebecca, Redbridge Birdwatching, Mr C Reddick, Mr P Reddish, Mr A J Redfern, Mr M Redfern (2), Mrs C Redgate, Mr R J Redhead, Ms S Redican, Mr M Redmond, Mrs J Redmond, Mrs J Redpath, Mr E J Redshaw, Mrs J A Redway, Mr S Redwood, T Reed & J Richardson, Barry Reed, Mr C Reed, Mr D K Reed, Mr J W Reed, Peter Reed, Mr P Reed, Mrs A Reed, Sir C Reed, Mr R Reedman, Mr S Reeds, D Gareth Rees, Dr J Rees, Mr D Rees (2), Mr E I S Rees, Mr G H Rees, Mr R Rees, Ms J Rees, Vicky & Ben Rees, Miss M S Reeve, Mr J Reeve, Mr J C Reeve, Mr T Reeve, Mrs B Reeve, Mrs E Reeve, Mrs K Reeve, C Reeve, Mr D J Reeves, Mr J Reeves, Mr M Reeves, Mr S Reeves, Mrs S Reeves, Mrs T Reeves, J Oxford, Mr L Reich, Mrs D Reich, Ms I Reich, Dr A Reid, Dr J M Reid, Miss C Reid, Miss M Reid, Mr A J Reid, Mr A Reid, Mr A R Reid, R & A Reid, Mr D Reid, Mr D B Reid, Mr H Reid (2), Iain Reid, Mr J Reid, Mr P Reid, Mr R Reid, Mr S Reid, Mrs G S Reid, Mrs K Reid, Mrs M Reid, Ms M Reid, Dr B Reidy, Mr V Reidy, Mr M Reilly, Mr P Reilly, Ms M Reilly, Mr D Rein, Ms S Reiverson, Ms E Relf, Mr R H Relph, Mr P A G Rendel, Mr D G Rendell, Mrs S J Rendell-Read, Mr G Rendle, Mr D Renham, Mr D G Renn, Mr D A Renney, Mr J Renny, Mr J Renold, Ms J Renowden, Mr K Renshaw, Mr R Renshaw, Mr E Rentzenbrink, Dr A C Renwick, Mrs C Renwick (2), Mr S M Resting, Miss C Revett, Mr J Revill, Mr O Reville, Mr I Revitt, Ms E Reyes, Mr S Reynaert, Mr P J Reynier, Dr D J Reynolds, Miss J Reynolds, Chris Reynolds, Mr D Reynolds, Mr F Reynolds, Mr M Reynolds (2), Neil Reynolds, Mr P Reynolds (3), Mr T Reynolds, Mrs C Reynolds, Mrs H Reynolds, Mrs P Reynolds, Mr P G Rhind, Mr G D Rhodda, Mr A Rhodes (2), Rhodes family, Mr G Rhodes, Mr P Rhodes, Abigail Rhodes, Mr I L Rhymes, Glyn Rhys-Parry, Mr J M L Riall, Mr G Ribas, Mr J B Ribbands, Miss E M Rice, Mr D J Rice, Mr E Rice, Rich family, Mrs J Richard, Miss G Richards (2), Miss G C Richards, Mr A Richards, Mr C Richards (2), Mr C M Richards, Mr D Richards, Mr G Richards, Mr J Richards, Mr Michael W Richards, Mr N G J Richards, Mr S Richards (3), Mr T Richards, Mrs A Richards, Mrs H Richards, Mrs J Richards (2), Mrs S Richards, Ms G M Richards, Ms J Richards, Sue Richards, Mrs G Richards-Jones, Mr C Richardson, Mr C J Richardson, Mr D Richardson, Mr D E Richardson, Mr G Richardson, Mr J Richardson (2), Mr J E Richardson, Mr L Richardson, Mr M J Richardson, Mr M Richardson (2), Mr P Richardson (2), Mr P C Richardson, Mr P W Richardson, Mr R Richardson, Mr T Richardson, Mrs C Richardson, Mr A Riche, Mr A C Riches, Mr J Riches, Mrs G Riches, Mrs P Riches, Mr D Richey, Mr O J Richings, Miss S J Richmond, Richmond family, Mr M Richter, Mrs R Rickard, Mr B Rickett, Mrs J Ricketts Hein, Mr L Rickman, Mr S Ricks, Miss F Riddell, Mr D N Riddell, Mr S Riddell, Mrs J Riddell, Dr R Riddington, Gordon Riddle, Trevor Riddle, Mr M Riddler, Miss S Rider, Mr S Rider, I M Ridett, M R Ridett & the National Trust, Mr M Ridgard, Mr J E Ridge, Mr T Ridgers-Steer, Mr A A Riding, Mat Ridley, Mr R Ridley, Mr T Ridley, P & IC Ridout, Mrs B Ridyard, Mr B Rigby, Mr D Rigby, Mr M Rigby, Mr M E Rigby, Mr R Rigby, Mr S Rigby, Mrs B Rigby, Mr R J Rigdon, Ms A Rigg, Dr R Righelato, Dr C Riley, Dr H Riley, Mr A Riley, Mr D Riley, Mr G Riley, Mr L Riley, Mr N Riley, Mr T Riley, Mrs E Riley, Mrs J Riley (2), Ms J Riley & Mr M Todd, Mr D Rimes, Mr M R Rimmell, Mr A Rimmer, Mr B Rimmer, Ms M Rimmer, Ms C Ring, Ms M Ring, Mr A Ringrose, Ms M Rinsler, Miss S Rintoul, Mr S Rippon, Kate Risely, Mr A Rishko, Dr P Rishton, Mr P Risley, Mr M Rist, Mr A Ritchie, Mr C Ritchie, Mr K Ritchie, Mr M Ritchie, Mrs S Ritchie, Mrs V Ritchie, Dr S Rivers, John Rivoire, Mr J B Rix, RMAS Conservation Group, Mr A Robb, Mr H Robb, Mr G Robbins, Dr B J Roberts, Dr ET Roberts, G Roberts, Dr S C Roberts, Miss C Roberts, Miss J R Roberts, Miss M Roberts, Miss R A Roberts, Miss S Roberts, Mr A E & Mrs M A M Roberts, Mr A H N Roberts, Mr A N Roberts, Mr A Roberts, Mr D Roberts (2), Mr D C Roberts, Mr D R Roberts, Mr G Roberts, Mr G C M Roberts, Mr Glyn N Roberts, Mr G P Roberts, Mr G R Roberts, Mr H Roberts, Mr I Roberts (2), Mr I J & Mrs V Roberts, Mr J Roberts (3), Mr L Roberts, Mr L F Roberts, Mr M Roberts (3), Mr N Roberts (2), Mr P Roberts (3), Mr P I Roberts, Mr P J Roberts, Mr P L & Mrs J V Roberts, Mr R Roberts (3), Mr S J Roberts, Mr W Roberts, Mrs C Roberts, Mrs D Roberts, Mrs G Roberts, Mrs H Roberts, Jean Roberts, Mrs J Roberts (2), Mrs J P L Roberts, Mrs L Roberts, Mrs Roberts, Mrs M Roberts, Ms A Roberts, Ms J Roberts, Ms M Roberts, Rev J H Roberts, J Roberts, Dr B D Robertson, Miss G Robertson, Janice Robertson, Miss L Robertson, Mr A Robertson, Mr A Y Robertson, Derek Robertson, Mr Derek Robertson, Mr D I Robertson, Mr F Robertson, Mr G Robertson, Mr G A Robertson, Grant Robertson, Mr K D Robertson, Mr M Robertson, Mr N Robertson, Mr R Robertson (3), Mr S Robertson (2), Mrs J Robertson, Mrs M A Robertson, Ms A Robertson, Ms G Robertson, Ms G M M Robertson, Ms H Robertson, Robertson-Lloyd, Mr B Roberts-Wray, Mr T Robilliard, Dr J A Robinson, Rob Robinson, Miss C E Robinson, Miss D Robinson, Miss L Robinson, Mr A Robinson, Barry & Jean Robinson, Mr B Robinson, Mr C Robinson, Mr C M Robinson, Robinson family, Mr D Robinson (4), Mr D L Robinson, Mr D R Robinson, Mr G Robinson, Mr I Robinson (2), Mr J Robinson, Mr M Robinson (2), Mr M C Robinson, Mike F Robinson, R Robinson & A Ball, Mr N Robinson, Mr P Robinson, Mr P A Robinson, Peter J Robinson, Mr R Robinson, Mr S Robinson, Mr S J G Robinson, Mr T Robinson, Mr Robinson, Mrs A Robinson, Mrs E Robinson, Jacky Robinson, Mrs J M Robinson, Mrs J M S Robinson, Mr C & Mrs M A Robinson, Geoff Robinson, Rosina Robinson & Les Mears, Mrs R Robinson, Mrs S Robinson, Ms L Robinson, Ms N Robinson, Prof M Robinson, Mrs C Robinson-Begg, Mr D Roblin, Dr J Robson, Dr M Robson, Miss H Robson, Brian Robson, Mr B J Robson, Barry Robson, Mr C Robson, Mr G G Robson, Mr I Robson, Mr J W T Robson, Les Robson, Mr N Robson, Mr S Robson, Mrs J Robson, Ms B J M Robson, Susanna Robson, Mr S Roby, Mr A Roch, Rochdale Field Naturalists' Society, Miss O Roche, Mr G Roche, Mr R Roche, Mrs C Roche, Mrs S Roche, Mr & Mrs D Rochester, Mr P Rocke, Mr S E Roddick, Mr S J Roddis, Mr M Rodel-Duffy, The Rodemarks, Mr H W Roderick, Mr T Roderick, Dr G Rodger, Mr C Rodger, Mr W P Rodger, Ms S Rodger, Mrs E Rodgers, Mrs J Rodgers, Mr P Rodham, Dr R Rodway, Mrs Linda M Rodway, Mr S Rodwell, Mr D M Roe, Mr I Roe, Mr M & Mrs T Roe, Roe family, Ms J Roe, Ms T Roe, Mr G Roebuck, Mr J Roex, Mr V Roffe, Mrs L Rogan, Dr D Rogers, Dr M Rogers, Shelagh M Rogers, Mr A C N Rogers, Mr B S J Rogers (2), Mr C Rogers, David Rogers, Mr D A Rogers, Mr D I Rogers, Mr D P Rogers, Mr G Rogers, Mr P Rogers (2), Mr P F Rogers, Mr R B & Mr A Rogers, Mr T Rogers (2), Mrs A Rogers, Mrs C Rogers, Mrs M J Rogers, Ms J Rogers, Ms L Rogers, Ms S Rogers, Mr A C Rogerson, Mr D G Rogerson, Mr J C Rogerson, Ms H J Rogerson, Miss S Rogge, Rogha Na Héan, Mrs E Rohan, Mr A Rohl, Mrs C Roissetter, Miss J Rolf, Miss R Rolfe, Mr D Rolfe, Mr M D Rolfe, Mr R Roll, Mr C Rollie, Mr T Rollins, Mrs E J Rollinson, Mr J C Rolls, Dr H A Rolton, Alan Roman, Ms J L Roman, Mr K Rome, Mrs S Rook, Mr A Rooke, Mrs G Rooke, Mr M Rookes, Mrs C Rookes, Mr C G K Roome, Miss Eimear Rooney, Miss S Rooney, Mr A Rooney, Mr B Rooney, Mr K Rooney, Mr M Rooney, Mr S Rooney (2), Dr S Roos & Miss L Coiffait, Mr S M Root, Miss J Roper, The Ropers, Mr L Rorke, Mr S Rosborough, Mr A Roscoe, Dr H E Rose, P D Rose, Mr A C Rose, Mr I Rose, Mr J E Rose, Laurence Rose, Mr T Rose, Mrs S H Rose, Mr M P Roser, Miss N Rosin, Miss A Roskilly, Mr A S Rosney, Dr B Ross, Dr J Ross, Mr C & Ms K Ross, Mr C M Ross, Mr G Ross, Mr J Ross (2), Mr L Ross, Mr M B Ross, Mr N & Mrs R Ross, Mr R Ross, Mr S Ross, Mrs B Ross, Mrs B M Ross, Mrs D Ross, Mrs G Ross, Mrs J Ross, Mrs L Ross, Ms J Ross, Ms N Ross, Prof J Ross, Mr M Rossell, Rossendale Ornithologists' Club, Miss B L Rosser, Mr J Rosser, Mr S E Rosser, Mrs B Rosser, Mr B N Rossiter, Mr G Rostron, Mr J W Rotheroe, Mr D Rothery, Mr H Rothwell, Mr R Rothwell, Mr P Roughley, Julian Roughton, Mr J A & Ms C I Roulstone, Miss M Rouncefield, Mr Craig Round, Pete Rouse, Mr C J S Routh, Mr C Routledge, Martin Routledge, Mr S Routledge, Mr P Routliff, Dr M G Rowan, Mr P Rowan, Mrs L Rowberry, Mrs Hazel Wainwright, George Alan Rowden, Dr D J Rowe, Mr D Rowe, Jim Rowe, Mr J Rowe (2), Mr P Rowe (2), Mr R Rowe, Mr T Rowe, Mrs A Rowe, Anne & Charlie Rowe, Elspeth & Christopher Rowe, Mrs H M Rowe, Ms S M Rowe, Mr A Rowell, Mr I Rowing, Mr G Rowland, Miss B Rowlands, Alan Rowley, Mr J Rowley, Mr J W Rowley, Graham Rowling, Mr D Rowlinson, Mr P A S Rowse, Mr E S Rowsell, Mr M Rowson, Mr R Rowthorn, Mr E Roxburgh, Mr J C Roy, Mr K Roy, Mrs J M Roy, Mrs P Royall, Dr D Roycroft, Mr K B Roylance, Miss S Royle, Mr K Royle, Mr P Royle, Mrs R S Royle, Mr K P Royles, RSPB Conservation Science, RSPB Croydon Local Group, RSPB Cymru, RSPB Epsom & Ewell, RSPB Guildford Group, RSPB Minsmere, RSPB Ouse Valley Reserves, RSPB Ramsey Island, RSPB Wildlife Enquiries, Ms S Rubalcava, Mr B Ruckwood, Mr J Rudd, Mr M Rudd, Ms L Rudd, Dr M Ruddock, Mr D Rudge, Mr D B Rudge, Mr J Rudge, Mr S P Rudge, Mrs M Rudge, Ms A Rudge, Mr P J Rudkin, Mr A Rudland, Mrs C A Rudram, Mr R D Ruffell, Ms I Ruffles, Mr D W Rugg, Mrs J H Ruggles, Mrs J Rumball, Miss N Rumney, Lorna Rusbridge, Mr J Ruscoe, Mr M P Ruscoe, Mr E Rush, Mr S Rush, Mr S A Rush, Mr L Rushall, Mr J M Rushforth, Mr D & Mrs A Rushton, John Rushton, Mr Michael Rushton, Mr M Rushton, Mrs C Rushton, Mrs H Rushton, Ms C Rushton, Mr J F Ruskin, Mr Russ, Mr J Russ, Dr J H Russell, Miss A Russell, Miss D M Russell, Miss F J Russell, Miss T Russell, Mr A Russell (2), Mr D Russell, Mr Gerard Russell, Mr G Russell, Mr M Russell (2), Mr M D Russell, Mr M G Russell, Mr P Russell (2), Mr & Mrs V F Russell, Mrs J Russell, Mrs R Russell, Ms L Russell, Mr G Russon, Mr R Russon, Mrs L Rust, Allan Rustell, Miss N D Rutherford, Mr H Rutherford, Mr S Rutherford, Mr G Ruthven, Rutland Natural History Society, Mr G P Rutt, Mr P J Rutt, Dr N Rutter, Dr P Rutter, Mr C Rutter, Mr G Rutter, Mr J W Rutter, Mrs R Ruttle, Dr C Ryall, J Ryall, Miss K Ryan, Mr A Ryan, Mr C Ryan, Mr C & Mrs C E Ryan, Mr F Ryan, Mr Fintan Ryan, Mr J Ryan (2), Mr K P Ryan, Mr P Ryan, Mr S Ryan, Mr T Ryan, Mr T A Ryan, Mrs R Ryan, Ms B Ryan, Ms M Ryan, Mr P Rydall, Mr J R S Ryder, Mr P Ryder, Mr P D Ryder, Mr I Ryding, Mrs S Rye, Rye Bay RG, Rye Harbour Nature Reserve, Mr K Rylands, Mrs A Ryles, Mr W J Ryman, Mrs A Rymell (2), Mr A Rymer, Christopher Rymer

S Ms H Sackett, Miss J Saddington, Dr J Sadler, Mr B Sadler, Mrs M E Sadler, Mike Saffery, Ms M Sagar, Mr J M Sage, Annie & Stephen Sage, J & R Best, Mr S Sahonta, Mr R Sains, Sainsburys family, Ms A Sainsbury, Miss V Saint, Mr D Sale, Mr G W Sales, Mr D G Salmon, Mr M J Salmon, Mr W Salmon, Ms N Salmon, Mr A J Salter, Mr D Salter, Mr Salter, Mrs R Salter, S Saltman, Mr J Salwey, Miss B Samé, Mr P Sammonds, Mr D Sampson, Mr M G Sampson, Mrs V Sampson, Miss L Samson, Mr W A Samson, Mr J R & Mrs M A Samuel, D J Samuel, Mrs S Samuel, Mr M Samuels, Mrs M Samways, Miss K Samyn, Mr C Sandbach, Mr A Sandeman, Mr A Sanders (2), Mr M Sanders, Mr R Sanders, Ms D Sanders, Miss G Sanderson, Mr D J Sanderson, Mr J Sanderson, Mr K Sanderson, Mr M Sanderson, Mr N Sanderson, Mr R F Sanderson, Mrs K Sanderson, Mrs A Sandfield, Mr R Sandford, Ms F Sandford, Mr G Sandham, Mr R Sandham, S D Sandham, Dr R Sandifer, E Sandilands, Mrs D Sandland, Dr D Gerry Sanger, Mr A Sangster, Mrs L Sank, Mr J Sankey, Mr F I Sant, Mr N Santy, Mr M Sargeant, Mr S Sargeant, Mrs S Sargeant, Mr D Sargent, Mr I Sargent, Mr P Sargent, Ms L Sargent, Dr S P Satterthwaite, Saturday Volunteers Group, Caroline & Martin Saul, A L Saunders, D Saunders H T, Mr H Saunders, Mr I Saunders & Ms H Bailey, Mr J Saunders, Mr M Saunders (2), Sandy Saunders, Mr R Saunders (2), Mr R C Saunders, Mr S Saunders, Mr S E Saunders, Mr W Saunders, Mrs E L Saunders, Mr O Saunderson, Mr P Sauter, Mr D Savage, Mr M L Savage, Mr S Savage, Mrs M Savage, Mr R Savidge, Dr G Savill, Mr G Saville, N & J Saville, Mr W Saville, Dr J N Savory, Mr C J Savory, Mrs M Sawers, Mr J Sawtell, Dr J B Sawyer, Mr D Sawyer, Mrs H Saxby, Mr D A Saxon, Ms W Sayell, Mr K Sayer, Mrs M A Sayer, SBSG, Mr I Scaife, Mr G E Scales, Mr V Scannella, Mr R L Scantlebury, Mr E V Scarfe & Ms P M Sweet, Oliver Clifford Scarffe, Mr R Scarlett, Mr S Scarlett, Mrs J Scattergood, Dr A Schalk, Mr G Scheller, Mr A Schofield, Mr C Schofield, Mr L Schofield, Mr M Schofield (2), Mrs E Schofield, Mr G D Scholey, Mrs G A Scholey, Ms N Schorbon, Dr P Scobie, Mrs P Scoffield, Mr J Scoggins, SCOIL IOSAEF NAOFA, Scoil Mháthair Dé, Dr A Derick Scott, Dr D A Scott, Dr G W Scott, Miss E M Scott, Mr A Scott (2), Mr A D Scott, Mr B Scott (3), Mr C Scott, Mr D Scott (3), Mr D W Scott, Mr G Scott, Mr H I Scott, Mr I Scott, Mr J Scott (2), Mr K N Scott, Mr L Scott, Mr L A Scott, Mr M Scott (3), Mr P Scott (3), Mr R D S Scott, Mr R E Scott, Mr S Scott, Mr W J Scott, Mrs A Scott, L & A Scott, Mrs M S Scott, Mrs P Scott, Mrs R Scott, Ms D Scott, Ms G Scott, Ms L Scott, Naomi Jane Scott,

Ms R Scott, Mr M Scott-Ham, Scottish Ornithologists' Club, Scottish Raptor Monitoring Scheme, Mr D Scott-Langley, Dr T J Scragg, Mr J Scragg, Mrs S Scrase, Scrimgeour family, Mr J Scrimger, Mr N J Scriven, Mr V Scriven, Mrs A L Scrivener, Mr P G Scudamore, Mr C Scully (2), Mr S Scutt, SDBWS, SE Scotland Blitz Group, Sea Larks Birders, Mr R Seabrook, Mr T Seabrook, Ms K Seabrooke, Mr D Seager, Trudy Seagon, Mr M Seagrave, Mrs J Seale, Ms C Seale, Mr J Sealy, Prof B J Sealy, Mr P Seaman, Mr R Seargent, Miss E Searle, Mrs H Searle, Ms Margaret Searle, Mrs D Searles, Mr K Searstone, Mr C Seaton, Mr J D Seaton, Mr K J Seaton, Mr D Seawright, Mr J Secker, SECOS, Miss G Seddon, Ms A Seddon, Ms R Seddon, Mr A Sedgwick, Mr C Sedgwick, Dr A Seed, Miss C Seeds, Mr D T Sefton, Mr I Sefton, Sefton Coast & Countryside, Mr G Segelbacher, Miss L J Seiersen, Mrs B Seifert, Seil Natural History Group, Mr M Selby, Mr Peter Selby, Mr P R Selby, Mr A Self, Mr R Self, Paul Seligman, Mr M Sell, Mr M R W Sell, Mr P Sellars, Ms N Sellars, Dr R M Sellers, Mr D Selley, Mrs M Sellors, Mr A Selstrom, Mr C Selway, Mr Barrie Senior, Mr D Senior, Ms G Senior, Mr M V Sennitt, Ms E Sentier, SERC Trainee Team, Mr T Sergeant, H Sergeant, Mrs G M Service, Dr R P Setchfield, Mrs R Setter, Mr A Seward, Mr D Q Seward, Mr D Sexton (2), Mr S Sexton, Mr T Sexton (2), Mrs A Sexton, Mrs S P Sexton, Miss C Seymour (2), Miss M Seymour, Mr A Seymour, Mr P Seymour, Mr D Seymour-Fairey, Dr M M Shackell, Mr J Shackleton, Mr J D Shackleton, Alan Shadrack, Mr M Shaft, Mrs T E Shailer, Mr M Shakespeare, Mrs J Shakespeare, Mr E Shangi, Dr S Shanks, Mr J Shanks, Mr R J Shanks, Mr S Shann, Mr J Shannon, Mr B Sharkey, Neil Sharkey, Mrs J Sharland, Ms M Sharland, Mr G Sharman, Mr J Sharman (2), Dr S P Sharp, Mr G Sharp, Mr J Sharp, Mr M A Sharp, Mr P Sharp, Mr R Sharp (2), Mr S Sharp, Mrs C Sharp, Mrs J Sharp, Mrs J C Sharp, Mrs P Sharp, Mrs P S Sharp, Mrs T Sharp, Ms L Sharp, Miss A Sharpe, Chris Sharpe, Mr M Sharpe, Mr S Sharpe, Mrs A Sharpe, Mrs M M Sharpe, Ms S Sharpe, Mr Garry Sharples, Mr G Sharples, Mr M Sharples, Mr P Sharples, Mr D Sharps, Mr C Sharratt, Mr R Sharratt, Dr J T R Sharrock, Mr K Sharrock, Mr D Sharrod, Mr M J & Mrs B Shave, Ms S Shave, Dr P Shaw, Mr A Shaw, Mr C Shaw, Mr C I Shaw, D Shaw & A Boden, Mr D J Shaw, Mr D N Shaw, Mr G Shaw, Mr G Shaw, I & J Shaw, Mr J Shaw (2), Mr K D Shaw, Matthew G D Shaw, Marc Shaw OBE, Mr R Shaw, Mr R M Shaw, Mr S Shaw, Mr W Shaw (2), Mrs A Shaw, Shaw family, Mrs L Shaw, Mrs S Shaw, Anne Shaw, Ms D Shaw, Sandra Shaw, Mr E Shawyer, Jill Shayler, Mr K Shea, Mr B Sheaff, Mr P Sheahan, Horringer MAF, Mr D Shearer, Mr P Shearer, Alan Shearman, Mr M Shearwood, Ms M Sheddan, Dr D Sheehan, Mr D Sheehan, Mr J Sheehan, Mr R Sheehan, Mr M Sheehy, Sheehy-Humphreys, Sheehy family, Ms B Sheepy, Sheffield Bird Study Group, Mr B Sheils, Mr P Sheils, Mr J Sheldon, Mr R Sheldon, Prof B Sheldon, Annette Shelford, Ms D Shelley, Mr J Shelliker, Mr T Shelton, Dr J E Shemilt, Mrs S Shemwell, Mr G Shentall, Shenton Primary School, Mr P Shenton (2), Mrs A J Shenton, Mr B Shephard, Mr E Shephard, Dr Michael Shepherd, Mr A Shepherd, Mr A J Shepherd, Mr D T Shepherd, Mr H R Shepherd, Mr I E H Shepherd, Mr I Shepherd, Mr J A Shepherd, Mr M Shepherd (2), Mr P Shepherd (2), Mr S Shepherd (2), Mr C Shepherdson, Ms A Shepherdson, Mr M Shepherd-Wells, Mrs P R Shepley, Mr A Sheppard (2), Mr D Sheppard, Mr D J Sheppard, Geoff Sheppard, Mr R Sheppard, Mrs D Sheppard, Mrs L Sheppard, Neil M G Shepperson, Mr D Sheridan, Paddy & Sandra Sheridan, Mrs M Sheridan, Mrs J Sheridan-Ros, Mr G Sherman, Mr M Sherman, Mr D Sherrington, The Sherringtons, Mr P Sherrington, Mr R Sherry, Mrs J Sherry, Mr B Sherwin, Mr J S Sherwin, Mr A Sherwood (2), Mrs A Sherwood, Shetland Bird Club, Miss F Shevlin, Miss A Sheward, Mr M Shewring, Mr R Shield, Mr D Shields, Mr J Shields, Mr M Shields, Tom Shields, Mrs A Shields, Ms Lee Shields, Mr J R D Shillitoe, Shilton family, Ms A Shimmens, Mrs S P Shipman, Mr R Shipp, Mrs A Shipton, Mr B Shirky, Mr P Shirley, Mr G A Shoebridge, Mr P Shone, Mr R Shone, Mrs M Shone, Mrs T Shoosmith, Mr A Shooter, Norena Shopland, Mr A M Shore, Mrs B Shorney, Mr A Shorrock, B Shorrock, Mr B Short, Mr D Short, Graham R Short, Mr J Short, Mr M Short, Mr P C Short, Mr P P Short, Mrs J M Short, Mr M Shorten, Mr G Shortt, M Shotter, Mr F Shovlin, Mr J Shovlin, Mrs C Showell, Mr J Showers, Shropshire Ornithological Society, Mr M Shrubb, Robin Shrubsole, Mr Michael Shurmer, Mr D A Shute, Mr M Shuter, Dr J Shutes, Mr J D Shutt, Stephen Shutt, Miss S Shuttleworth, Mr A Sibley, Mrs J M Sibson, Mr D Sidebotham, Mr A Sides, Ms L Sidwell, Miss S Siggins, Mr P C Silburn, Ms L Silcock, Mrs D Silk, Mrs G M Silk, Mrs H Silk, Mr L Silkowski, Mr M Sillence, Mr D A Sillett, Mr I Sillett, Mr N Sillett, Mr D Sillito, Mr T Sillitoe, Mr D Sills, Mr N Sills, Mr P Sills, Ms K Silvan, Miss A Silver, Mrs C Silversides, Mr P Sim, Mr W F Simcox, Mr D Sime, Ms A G M Sime, Mrs S Simey, Ms J Simister, Dr J Simkin, Mrs A Simm, Miss B Simmonds, Mr S Simmonds, Mr D Simmonite, Paul & Joyce Simmons, Mr A Simmons, Mr C Simmons, Mr J Simmons, Mr M Simmons (2), Mr R Simmons, Mrs J Simmons, Ms M Simmons, Mr O Simms, Ms G Simms, Mr T B Simon, Dr M Simons, Andrew Simons, Mr J Simons, Mr D Simpkin, Mr M Simpkin, Dr R M Simpson, Miss A E Simpson, Miss C Simpson, Mr C Simpson, Mr D W Simpson, D & J Simpson, Mr G Simpson, Mr J Simpson (2), Mr M Simpson, Mr M T Simpson, Mr N Simpson (2), Mr R Simpson (2), Mr R D H Simpson, Mr S P Simpson, Mrs D Simpson, Mrs E Simpson, Mrs L Simpson, Mrs L J Simpson, Ms C Simpson, Ms K Simpson, The Simpson-Edwards, Mr A Sims, Mr B Sims, Mr C Sims, Mr Ian Sims, Kent Sims family, Karen Sims & John Towers, Ms M Sims, Mr C Sinclair, Mr P Sinclair, Mr J Sines, The Singhs, Mr M R Singleton, Mr R Singleton, Mr P Sinnadurai, Dr A D Sinnott, Mr P Sinte, Mr J Sirrett, Mrs S Sissons, Mr B A Sivyer, Mr Dave Sivyer, Mr M Skeffington, Mr D Skehan, Ms K Skelhorn, Mr I Skelton, Mr S Skelton, Mr T Skerl, Mr M A Skerratt, Mr M Skevington, Miss P Skidmore, Mr D Skidmore, David Skidmore, Dr I O Skillicorn, Mr D Skilling, Mr J Skilling, Mr R Skillings, Skilton family, Dr N J Skinner, Miss E Skinner, Mr A Skinner, Mr D C Skinner, Mr F Skinner, Mr K Skinner, Mr P Skinner, Mr G Skipper, Miss I Skipworth, Dr M B Skirrow, Skomer Island, Mr A Skotnicki, Miss A Skull, Mr J Skuriat, Mr J Skuse, Mr R Slack, Stuart Slack, Ms E Slack, Mr A M Slade, Mr P Slade, Mrs H Slade, Mr R C Slaney, Ms E Slark, Dr A Slater, Dr B Slater, Dr F M Slater, Mr B Slater, Mr D M K Slater, Mr K Slater, Mr P Slater, Mr Roy Slater, Mrs A Slater, Slater family, Mrs S Slater, Prof P Slater, Mr C Slator, Mr H A Slatter, Ms J M Slavin, Mr P Slavin, SLC Ranger Service, Ms R Sleat, Mr J Slee, Gordon & Lesley Sleight, Mr J Slobodian, Mr J Slocombe, The Slocombes, Chris Small, Mr J P Small, Mr A C Smallbone, Mr H M & Mrs D M Smalley, Mr D Smallshire, Dr Jeremy Smallwood, Mr R Smallwood, Ms D Smallwood, Dr J Smart, Mr D Smart, Mr M Smart, Mr N Smart, Mr O Smart, Mr P J Smart, Mr R Smart, Mrs S Smart, Ms C Smart, Mr A Smedley, Mr M Smedley, Mr P Smiddy, John Smillie, Mr T D Smit, Dr A Smith (2), Dr A H V Smith, Dr G Smith, Dr J J Smith, Dr K Smith, Ken Smith & Linda Smith, Dr Mike Smith, Dr S C H Smith, Dr T Smith, Miss A Smith (2), Miss B Smith, Miss C Smith, Miss E Smith (3), Miss G Smith, Miss G J Smith, J Smith & family, Miss L Smith (3), Mr A Smith (2), Mr & Mrs A Smith, Mr A Smith (4), Mr A D Smith, Mr A M Smith, Mr B Smith (4), Mr B W Smith, Mr C Smith (3), Mr C J Smith, Mr Christopher J Smith, Mr D J Smith, Mr D Smith (9), Mr D A Smith, Mr D J Smith (2), Mr D P Smith, Mr D R Smith, Mr D Smith, Mr E J Smith, Mike Smith, Mr F Smith, Graham Smith, Mr G Smith (7), Mr G M Smith, Mr G P Smith, Mr I A Smith, Mr J W Smith, Mr J Smith (6), Mr J A Smith, Mr J M & Mrs S Smith, Mr J V & Mrs J Smith, Mr K Smith (5), Leo Smith, Shropshire roving recorders, Mr Mark C Smith, Smith family, Mr M Smith (8), Mr M A Smith, Mr Marcus Smith, Mr M Smith, Mr M C Smith, Mr M F Smith, Mr M G Smith, Martin Smith, Michael J Smith, Mr M P J Smith, Michael S Smith, Mr M W Smith, Mr N Smith, Mr N A Smith, Mr N R Smith, Mr PG Smith, Peter C Smith, Mr P Smith (5), Mr Philip Smith, Mr P Smith (2), Mr & Mrs P Smith, Mr P Smith, Mr Peter A Smith, Mr P A Smith, Mr P G Smith, Mr P J Smith (2), Mr P R Smith (2), Richard Smith, Mr R Smith (10), Mr R E Smith, Mr R G Smith, Bob & Val Smith, Mr R Smith, Mr R N Smith, Mr R P Smith, Mr R T Smith, Mr S Smith (5), Mr T Smith (4), Mr V Smith, Mr V D Smith, Mr W Smith, W L & M J Smith, Mrs A Smith, Mrs A Judith Smith, Mrs D Smith (2), Mrs E Smith (2), G M Smith, Mrs I Smith, Mrs J Smith (2), Mrs L H Smith, J R Smith, Mrs M Smith (2), Mrs P Smith (2), Mrs Ros Smith, Mrs S Smith (3), Mrs V Smith, Y Smith & family, Ms C Smith, Ms J Smith (2), Ms J H Smith, Ms L Smith (2), Ms M Smith, Ms S Smith, R G Smith, Steve Smithee, Mr A C Smithen, Mr N J Smithers, Clifford Smout, Chris & Anne-Marie Smout, Dr G Smyth, Mr D Smyth, Mr J R Smyth, Mr M Smyth, Mr P A Smyth, Mr W Smyth, Mrs C Smyth, Mr A Snape, Mrs A Snape-Batten, Mr D Sneap, Mr M Snelgrove, Miss K Snell, Mrs E Snell, Ms B Snell, Mr M Snelling, SNH Lerwick, SNH Loch Lomond NNR, Mr A M Snook, Dr D W Snow, Mr P Snow, Mr T G Snow, Mr L Snowball, Mr A Snowden, Mr A J Snowden, Mr P Snowden, Mr D Snowdon, Mr J Snowdon, Mr J V Soames, Mr G Sobala, Mr J Sobey, SOC (Highland), Mr E Soden, Mrs J Solari, Someries Infant School, Mr F Somerville, Dr A Sommerville, Somoskovi family, Ms J Sonntag, Prof P Soothill, Mr P Soper, Mrs A Soper, Ms S Sorrell, Mrs L J Sourbutts, Mr P A L Souster, Mr R Soutar, Dr R Souter, Mr M M Souter, South Lanarkshire Council Countryside Ranger Service, South Wales Peregrine Monitoring Group, Mrs S Southall, Mr M Southam, Mrs S Southam, Brian Southern, Mr T Southgate, Mr B Southon, Miss J Southwell, Mr M Southwood, Janet Southwood, Neil Southworth, Mr S R Sowden, Mr D L Sowerbutts, Mr W H Sowerby, Lee Sparey, Mr Derek Sparkes, Mr P J & Mrs S E Sparkes, Mrs M Sparkes, Dr T Sparks, Mr G Sparshott, Mr J R Spaull, Mr P W Speak, Dr P Speake, Mr M Speck, Revnd K Speck, Mr I Speechley, Mrs V Speed, Mr Andrew Speer, Mr D & Mrs B Speight, Mrs B A Speight, Mr M Speir, Mr A Speirs, Mr W Speirs, Mr N Spellacy, Mr A Spellman, Miss J Spence, Miss L Spence, Mr B E Spence, Mr G Spence, Ian M Spence, Spence family, Mrs S Spence, Mrs V Spence, Dr S R Spencer, Mr D P Spencer, Mr H Spencer, Mr J & Mrs J Spencer, Mr J A Spencer, Mr K G Spencer, Mr M Spencer, Mr Paul A Spencer, Mr R Spencer, Mr R A Spencer, Mr T Spencer, Mrs P Spencer, Mrs S Spencer, Mr C P Sperring MBE & family, Ms L V Spicer, Mr K Spicer, Mr N Spicer, Mr R Spicer, Mrs J Spicer, Mr S G Spick, S & G Spickett, Mr N Spiers, Mr J Spilsbury, The Spinners, Michael Spittles, Pamela Spokes, The Spongoes, Dr D Spooner, Mr S J Spooner, Mrs H Spooner, Mr A Spottiswood, Mr J Spottiswood, Ms J Spouse, Mr J Spowart, Dr D Spratt, Prof C J Spray, Mrs J Sprigens, Dr M A Spriggs, Mr J Spriggs, Mr N Spring, Springall family, Mrs J K Springett, Springhallow School, Mr N Sprowell, Miss E Spruce, Ms M Spruit, Mr S D Sprules-Wright, Prof C J F Spry, Mrs G Spurway, Anthony Spybey, Mr J Squire, Mr P C Squire, Mrs J Squire, Ms B Squire, Mr R Squires, Mr S M Squires, Mr M Sreeves, Mrs E St John, Mrs S St Ledger, St Marys School Ascot, Paul St Pierre, St. Cuthbert's Primary School, St. Monica's HS Wildlife Explorers, Mrs N Stabler, Mr A Stables, Mr B Stacey, Derek Stacey, Mr D Stacey, Mr Graham A F Stacey, Mr G Stacey, Mr J Stacey, Mr R K Stacey, Mrs J Stacey, Tony Stachnicki, Dr A Stackhouse, Dr L Stader, Mr M Staff, Staff & Volunteers at SWT Montrose Basin, Mr A Stafford, Mr B Stafford, Mr D Stafford, Mrs F H Stafford, Mrs J V Stafford, Mr J Stagg, Mr P Stagg, Mrs D L Stagg, Mr D E Stainer, Mr B Stainsby, Mr Wayne Stainthorpe, The Straits, Mr R Stakim, Staley family, Mr B Stalker, Mr B Stallard, Miss H Stammers, Mr B Stammers, Mr C Stamp (2), Mr P Stamp, Mr P B Stanbrook, Mr P A Stancliffe, Mr M Standring, Mr F G Stanford, Mr A Stanger, Miss C Stanley, Mr D Stanley (2), Ms J Stanley, Mr R Stansfield, Mr S D Stansfield, Mrs E J Stansfield, Mr J Stanton, Mr D J Stanyard, Mr M P Stanyer, Ms A Staples, Miss K Stapleton, Mr C F Stapleton, Ms S Star, Mr J Starbuck, Mr C Stares, Mr A Stark, Miss P Starkey, D Starkins, Mr L Starling, Mr S Starling, Mrs A Starling, Dr M Starr, Ms J Startin, Mr T J Staton, Mr E G Staunton, Mrs L Staves, Ms M Stead, Mr B Steadman, Mr J Stean, Mr T Steane, Mr Keith Stedman, Mr R Stedman, Ms N Stedman, Dr J Steel, Mr D Steel, Ms C Steel, Ms J L Steel, Mr E J Steele, Mr L Steele (2), Mr L D Steele, Mr N Steele, Miss I M Steenbergen, Mr P Steepe, Dr M Steer, Mr L Steer, Mr R Steer, Mrs A Steer, Mrs L Stein, Mrs N Stelling, Mr M Stentiford MBE, Dr John Stephens, Miss N Stephens, Mr J J Stephens, Mr R Stephens (2), Mrs N Stephens, Ms L Stephens, Mr G & Mrs P Stephenson, Mr G C Stephenson, Mr K Stephenson, Mr M Stephenson (2), Sarah Stephenson, Mr W Sterling, Mr K Sterry, Mrs A Sterry, Ms A Sterry, Mr M Steven, Dr T A Stevens, Miss C Stevens, Mr A Stevens, Mr D Stevens, Mr E G Stevens, Ted Stevens, Mr G Stevens, Mr J Stevens, Mr M Stevens, Mr P Stevens (2), Mr P J Stevens (2), Mr P R Stevens, Mr R J Stevens, Mr T Stevens, Mrs A Stevens, Mrs J Stevens, Mrs K Stevens, Mrs P Stevens, Mrs S Stevens, Ms R A Stevens, Dr J Stevenson, Mr A Stevenson (2), Mr D Stevenson (2), Mr I Stevenson, Mr J Stevenson, Mr J N Stevenson, Mr K M Stevenson, Mr T A Stevenson, Mrs D J Stevenson, Mrs E Stevenson, Ms M Stevenson, Miss A Steventon, Mr D J Steventon, Mr L Steward, Mr M M Steward, Mr P Steward, Dr Jean M Stewart, Miss A Stewart, Mr A Stewart, B J Stewart, Mr B Stewart, Mr C B Stewart, Mr C V Stewart, Mr D Stewart, Mr G Stewart (3), Graeme Stewart & Bob Lyle, Ken Stewart, Mr L Stewart, Mr Mark Stewart, Mr P Stewart (2), Mr R Stewart, Stewart family, Mr W Stewart, Mrs A Stewart, Mrs M Stewart, Mr J Stewart-Smith, Ms A Stiegler, Dr L Still, Mrs S Stille, Mrs A Stimpson, Mr Michael Stinson, Mr C Stinton, Miss G Stirling, Mr J R Stirling, Mr R Stock, Mr A Stocker, Mr R G Stockhausen, Stocks family, Mr P Stockton, Mr J Stockwell, Mr M Stockwell, Dave Stoddard, Dr P G P Stoddart, Mr M Stokeld, Mr J Stokes, Mr R Stokes, Mr M Stolworthy, Dr H E Stone, Dr R Stone, Miss E Stone, Miss P Stone, Mr B H Stone, Mr D A Stone, Mr G Stone, Mr G W Stone, Mr T Stone, Mr W Stone, Mrs S Stone, Miss S Stonebridge, Mr B Stoneham, Mr J R Stonehouse, Mr A Stones, Mr B Stopgate, Mr P Stoppard, Mr C Storey, Gary Storey, Mrs A Stormont, Mr J Storrar, Mrs D Storrow, Mr C Stott, Mr R Stott, Mr K Stouse, Mr M R Stout, Mr A Stow, Dr T J Stowe, Dr A Strachan, Dr F A Strachan, Dr I Strachan, Miss L Strachan, Mr R Strachan, Ms N Strachan, Mr J Straffon, Mr C Straker, Arnold Strand, Mr A Strand, Mr A Strang, Mr A Strange, Mr P J Strangeman, Mr J Strangeway, Mr I Strangward, Mr D Strath, Mr R Strathdee, Mr A Straton, Ms L Stratton, Mr D Stratton-Thomsett, Mrs L Straughair, Mr P Straughan, Mr R Straughan (2), Mrs P M Straughan, Mr A Straw, Julie M Straw, Mrs C Streater, Miss C Street, Lucas-Street, Mr D Street, Mr D K Street, Mr L Street, Mrs R Street, Mrs J Streeter, Tamasine Stretton, Ms J Stretton, Mrs A Stribley, Mr J Strickland, Mr M J Strickland, Mr L Stride, Mrs M Stride, Mr J Strike, Mr D Stringer, Mr W R Stringer, Mrs J Stringer, Ms L Stringer, Ms W Stringer, Dr N Stronach, Mr J Stronach, Mr P Stronach, Mr A Strong, Mr B Strong, Mr W Strong, Mr P Stronge, Mr G Strother, David, Judy, Tom & Rachel Stroud, Mr J Stroud, Mr J Strowger, Mr C J Strudwick, Mr T Strudwick, G E Strugnell, Mr R Strugnell, The Strugnells, Mr E Stubbings & Ms B Bueche, Stubbs family, Mr M Stubbs, Mr P Stubbs, Mrs S Stubbs, Mr B Stuckey, Mr R F Stuckey, Mr G Studd, Dr I Stuijts, Miss B Stundon, Miss J Sturgeon, Mr C J Sturgeon, Ms A Sturgeon, Miss G M Sturgess, Mr I Sturgess, Mrs A Sturt, Dr R A Stuttard, Mr Peter Stuttard, Mr P Styles (2), Mr D Suddaby, Mr D Suff, Ms A Suffolk, Suffolk Records, Suffolk Wildlife Trust Surveys, Mr A Sullivan (2), Mr B Sullivan, Mr J Sullivan, Mr M Sullivan, Mr P Sullivan, Mrs M J Sullivan, Ms S Sullivan, Ms W Sully, Mr M Sulway, Mr M G Sumbler, Mrs B Summerfield, Ms S Summerfield, Dr M P Summers, Dr R W Summers, Mr A Summers, Mr C Summers, Mr D Summers, Mrs G Summers, Mrs H C Summers, Mrs R Summers, Miss A Summersgill, Mr A & Mrs J Summerson, Dr A T Sumner, Dr B Sumner, Mr J Sumner, Mr N A Sumner, Prof J Sumpter, Mr J Sunderland, Mr R Sunderland, Sunningwell Conservation Group, Mrs D Suppan, Mr M Sur, Surbiton & District Bird Watching Soc, Mr R Surgenor, Mr D J Sussex, Sussex Ornithological Society, David Sutcliffe, Mr E Sutcliffe, Mr M Sutcliffe, Mr M A Sutcliffe, Mr S J Sutcliffe, Mrs R A Sutcliffe, Mr G Suter, The Simmons, Mr H & Mrs F Sutherland, Mr J A Sutherland, Mr Martin Sutherland, Mrs E J O Sutherland, Mrs J Sutherland, Mrs M Sutherland, Mr I S Suttie, Mr S M Suttill, Dr Robin D Sutton, Mr B Sutton, Mr D H Sutton, Mr J Sutton, Mr K Sutton, Mr L J Sutton, Mr M Sutton (2), Paul Sutton, Mr R Sutton, Mr R K Sutton, Mrs B Sutton, Sutton family, Mrs J Sutton, Ms K Sutton-Spence, Mr S M Swaby, Miss H Swain, Miss J A Swain, Miss G Swainson, Mr J J Swale, Swaledale Birds, Mr B T Swallow, Mr John L Swallow, Swallow family, P Swallow, Mr P M Swan, Mrs J C Swan, Mr A J Swann, Mr R Swann, Mr R L Swann, Mr Thomas S Swann, Mrs G Swann, Mrs J Swann White, Mrs M Swanson, Mr E Swarts, Mrs R Swatman, Mr J R Sweby, Dr J J Sweeney, Mr A Sweeney, Mr J Sweeney, Mr K Sweeney, Mr M J Sweeney, Mr O Sweeney (2), Mr S Sweeney (2), Mrs E M Sweeney, Mrs L Sweeney, D Sweet & D Free, Tony Sweetland, Mr K Sweetmore, Mr C R Swift, Mr J H Swift, Rob Swift, Ms F Swift, Ann & Peter Swindale, Mr J M Swindells, Mrs S Swindells, Ms J Swindells, Mr R Swinford, Peter Swinney, Mr Arne D Swithenbank, Mr T Sydney, Dr Mike Sykes, Miss T K Sykes, S Sykes & S Cassidy, Mr & Mrs P A Sykes, Mr T Sykes, Mrs D Sykes, Mrs S U Sykes, Ms Heather J Sykes, Mr N Symes, Mr D Symmons, Mr G Symons, Mr S Szary, Mr A Szopa-Comley

T Mrs E Tabb, Mr C Tack, Mrs P Tack, Mrs E H Tackley, Mr P Tadeusz, Dr W Tagg, Mr B Taggart, Paul & Judy Tainton, Mr I Tait, Dr R Talbot, Mrs S Talbot, Mrs A Talintyre, Miss N Tallach, Mr R E Tallack, Mr R M Tallack, Mr P Tallantire, Mrs S K Tallents, Mr I Tallon, Mr J R B Tallowin, Ms C Tan, Mr J Tandy, Mr P Tankard, Mr D Tanner, Mr M Tanner, Mrs C Tanner, Mr A Taplin, Brother G Tarbuck, Mr & Mrs N Tardivel, Mr P Tarling, Mr T Tarpey, Mr P L Tarrant, L & J Tarrant, Mr K D Tarratt, Mr S Tarry, K Tarsey & S Cooper, Mr L Tarver, Miss S Tash, Tashkandy & Banks, Mark Tasker, Mrs F Tasker, Mr L J Taswell, Sue Tate, Mr D Tate, Mr J E Tate, Dr P Tatner, Mrs S Tattersall, Mr D Tattersley, Miss J Tavener, Dr J Taverner, Mrs J Tayler, Mr J Tayleur, Dr D L Taylor, Dr J Taylor, Dr M Taylor, Moss Taylor, Miss C Taylor, Miss C M Taylor, Miss F Taylor, Miss K A Taylor, Megan Taylor, Natalie Todman, Alan M Taylor, Mr A Taylor, Mr A I Taylor, Mr A J Taylor, Mr A M Taylor, Mr B Taylor (3), Mr Chris Taylor, Chris Taylor, Mr C Taylor (3), Don Taylor, Mr G Taylor (2), Mr G J Taylor, Mr H Taylor, Ian Taylor, Mr I Taylor (2), Mr I M Taylor, Mr I R Taylor, John Taylor, Mr J Taylor (6), Mr J E Taylor (2), Mr K Taylor (2), Mr M Taylor (4), Mr M H J Taylor, Mr M J Taylor (2), Mr M J & Mrs G C Taylor, Mr N J Taylor, Mr N W Taylor, Paul Taylor, Mr P Taylor (2), Mr P D Taylor, Mr P Taylor, Mr P R Taylor (2), Mr R Taylor (2), Mr Ray Taylor, Roger Taylor, Mr R J Taylor, Mr S Taylor, Mr SM Taylor, Mr Stewart Taylor MBE, Mr S Taylor (2), Mr S R Taylor, Di & Dennis Taylor, Mr T Taylor, Bill Taylor, Mrs B Taylor, Mrs C Taylor (2), Mrs C P Taylor, Mrs E A Taylor, Mrs H Taylor (2), Mrs J Taylor (2), Mrs J E Taylor, Jillian Taylor, Mrs K Taylor, Mrs L Taylor, Mrs R Taylor, Mrs S L Taylor, Graeme Taylor, Mrs V Taylor, Ms B Taylor, Ms J Taylor (2), Ms J M Taylor, Taylor-Bleanch, Ms K Taylor, Ms L Taylor, Ms N Taylor, Mrs G Taylor Shaw, The Taylor-Hodges, Ms P Taylor-Pearce, Mr D Teague, Mr A Teale, Teall family, Team 46, Mrs J E Teare, Mr W Tebbutt, George Tedbury, Mrs D W Tedd, Miss M Tedder, Mrs A Tee, Mr S Teears, Mrs C Tees, Mr M Teesdale, Mr Ray Teesdale, Ms R Teesdale, Mr G D Tegerdine, Mr W Telfer, Dr S Temperley, Miss C Temple, Mr Andy Temple, Mr J Temple, Mr K H Temple, Mr R Temple, Mr W Temple, Mrs C Temple, Eve Templeton, Ms M ten Cate, Mr J H Terry, Mr M Terry, Mr S Terry, Mrs C Terry, Mrs L Tester, Mr A I Tetlow, Dr E Teuten, Mr I F Tew, Mr T H, Mrs J Thacker, Mrs D Thackeray, Mr J Thackway, Mr C Thain, Miss S Thairs, Mrs J Thake, Dr A Tharme, Mrs L A Thatcher, Mr B & Mrs L Thaxter, The American Gardens Project, The Reservoirs Nature Society, Revnd J H Theaker, Mr B Theakston, Mr R Theakston, Mr D A Thelwell, Roger & Imogen Theobald, Miss M Theresa, Mr D Theron, Mr E Thetford, R M Thewlis, Mr D B Thexton, John A Thickitt, Mr M Thiebaut, Mr A Thiel, Mrs C Thistlethwaite, Mrs P Thistlethwaite, Mr C Thody, Willie M Thom, Mrs J Thom, Dr C B Thomas, Dr D K Thomas, Dr L P Thomas, Miss C Thomas, Miss E Thomas, Miss K Thomas (2), Miss R Thomas, Andrew D Thomas, Mr A Thomas (2), Mr B Thomas, Mr B J Thomas, Mr C Thomas (3), Mr D Thomas (3), Mr D G Thomas, Dave Thomas, Mr D R Thomas, Mr D Thomas, Mr E Thomas, Mr F R Thomas, Mr G Thomas (5), Mr G H Thomas (2), Mr J Thomas, Mike Thomas, Mr M Thomas (2), Mr M V Thomas, Mr N Thomas, Mr N R Thomas, Mr P Thomas, Mr P D Thomas, Mr R Thomas (4), Mr T Thomas, Mrs A Thomas, Mrs B Thomas, Mrs D Thomas, Mrs E Thomas, Mr G F & Mrs E A Thomas, Mrs J Thomas, Mrs J M & Mr A B Thomas, Mrs J M K Thomas, Mrs K Thomas, Mrs L Thomas, Mrs M Thomas (2), Ms A Thomas, Ms C Thomas, Ms J Thomas (2), Prof M Thomas, Prof M J Thomas, Mr D Thomason, Mr E Thomason, Mr P C Thomason, Sister S Thomasson, Dr E G Thompson, Miss B J Thompson, Miss C E Thompson, Miss F Thompson, Miss S Thompson, Mr A Thompson, Mr A J Thompson, Mr A P Thompson, Mr B Thompson, Brian Thompson, Mr C A Thompson, Mr D Thompson (3), Mr D G Thompson, Mr D J Thompson, Mr G Thompson, Mr G L Thompson, Mr I Thompson (2), Mr I R Thompson, John Thompson, Mr J Thompson (3), Mr Julian P Thompson, Mr M Thompson (2), Mr M J Thompson, Mr N Thompson, Mr P Thompson (4), Mr P G L Thompson, Mr P Thompson (2), Mr R Thompson, Mr S Thompson (2), Mr S L Thompson, Mr T Thompson, Mrs C Thompson, Mrs D Thompson, Mrs J M Thompson, Mrs S Thompson, Mrs S M Thompson, Mrs V Thompson, Ms M Thompson, T Thompson & R Bell, Thomsen family, Dr B Thomson, Dr C Thomson, Dr I J Thomson, Miss J L Thomson, Miss K Thomson, Derek Thomson, Mr E Thomson, Mr H Thomson, I D S Thomson, Ian Thomson, Mr J Thomson (2), Mr J H Thomson, Mr L Thomson, Mr N Thomson, Mr S Thomson (2), Mrs E Thomson, F & E Thomson, Mrs L Thomson, Mrs S M Thomson, Ms W Thomson, K Thomson, Mr M Thomson-Tur, Mr S Thorley, Mr T Thorley, Miss M Thorman, Miss M Thorne, Mr G Thorne, Mrs D Thorne, Mrs K Thorne, Mrs R Thorne, A K Thorne, Mr R Thorneycroft, Mr A Thornhill, Mr R N Thornhill, Mr M Thornley, Mr V Thornley, Mr J Thornton, Mr M Thornton, Mr P Thornton, Ms G A Thornton, Mr J Thorogood (2), Mrs M Thorogood, Dr Ray G Thorp, Miss R Thorp, Mrs R E Thorp, Miss K Thorpe, Mr A W Thorpe, Mr D Thorpe, Mr J Thorpe, Mr R Thorpe, R Thorpe & A Stratford, Mr C Thorpe-Dixon, Dr R Threadgould, Mr B Threlfall, Mr D Throup, Miss E Thrussell, Ms J Thrussell, Mr H C Thurgate, Dr J Thurlow, Mr G Thurlow, Mr G Thurnham, Mr J Thurston, Mr M H Thurston, G Thyne, Ms S Tibbits, Mr R S Ticehurst, Mr M M Tickler, Mr Matthew B Tickner, Mrs V Tidball, Mr I M Tidmarsh, Mr S J Tidswell, Mr D Tierney, Mr G R Tierney, Niall Tierney, Mr N Tigg, Miss E M Tigwell, The Tilburys, Mr A Tildesley (2), Mr D Tillett, Mr J Tilley, Mr P Tilley, S C Tilley, Miss R Tilling, Mr A Tilmouth, Mr A Tilt, Mr J Tilt, Mrs G Timberlake, Miss M J Timmis, Miss C Timms, Mr N Timms, Mr R M Timms, Dr I Timson, Mr A D Tindale, Mrs R Tindall, Mr R Tingley, Mrs W Tinley, Trevor Tinlin, Mrs C Tinn, Mr M J Tinnelly, Mr A G Tinning, Mrs A J Tinston, Mr K Tinworth, Mrs A Tippett, Mrs B Tipping, Mrs M V Tipping, Mrs G Tipton, Mr J J Tirrell, Mr A Tissier, Mr C Titcombe, Mr P Tithecott, Mr R C E Titman, Mr D Tittensor, Mr R Tivendale, Phil Tizzard, Mr L Toal, Mr B Toale, Mr M Tod, Alan L Todd, Mr F Todd, Mr R Todd, Ralph Todd, Mrs D Todd, Mrs L Todd, Mrs V Todd, Ms C Todd, Mrs S Tokley, Mr R Tollervey, Mr B D Tollitt, Mr N Tomalin, Miss I Tománková, Mrs B F Tombrock, Mr A B Tomczynski, Mr D Tomes, Mr W V Tomkins, Mrs J Tomkinson, Mr R J Tomlin, Fr J Tomlinson, Alan & Christine Tomlinson, Mr C Tomlinson, Mr C G Tomlinson, David Tomlinson, Mr M Tomlinson, Mr S Tomlinson, Mr W Tomlinson, Ms A Tomlinson, Mr M J Tompkinson, Dr A J Tompsett, Mr C Toms, Mike Toms, Mr M R Toms, Mr D Tonge, Mr G G Tonge, Nick Tonge, Mr S Tonge, A D W Tongue, Mr S G Tonkin, Mr H Tonks, Mr J Tookey, Mrs H Y Toole, Dr D K Toomer, Mrs J Toomer, Mr E Toomey, Russell Toon, Mr B Toone, Mr M Toone, Mr A Tooth, Mr N D Tooth, Mr J S Toovey, Dr C Topping, Mr E Topping, Mr P Topping, Ms V Topping, Torbreck Twitchers, Dr G Tordoff, Mr M Tordoff, Jack Torney, Mr M Torrington, Miss M Toth, Mrs J Totty, Mr C Tout, Mr M Tout, Mr T Tovey, Margaret Towers, Mr D F Towie, James T M Towill, Mr C & Mrs K L Towler, Prof H Towill, Mr C Townend, Mr R H Townhill, Mr L Towns, Mr P F Towns, Lady A S W Townsend, Miss Y Townsend, Mr A Townsend, Mr C Townsend, Mr M Townsend, Mr P Townsend, Mr T J Townsend, Mrs S Townsend, Ms K Townsend, Ms L Townsend, Mr B Townson, Mr P E Toynton, Mr R B Tozer, Mrs N Tracey, Mr B Tragett, Mr J Trainor, Mr K Travers, Mr W Travers OBE, Travis family, Mrs C Travis, Dr Y Traynor, Mr D Traynor (2), Ian Traynor, Miss R Treacy, Mr B Treacy, Tree family, Mr R L Treen, Mr P R Tregenza, Mr P Treloar, Mr P Trend, Mr S Trenerry, G & R Trerise, Ms C R Tresadern, Mr D Trestain, Treswell Wood, Dr P & Mr B Treves, M & A Trew, Mr P Trew, Mr M Trewby, Dr N H Trewin, Mr Nick Tribe, Mr C N Tricker, Mr M Tricker, Mr J Trimble, Miss K Trimingham, Mrs K Trinder, Mrs A Tring, Mr P Troake, Mr C Trollen, Charles Trollope, Mr D H Trollope, Mr H Trollope, Mr R Trollope, Mrs R M Troman, Mr N Trotman, Mr D Trott, Mr D Trotter, Mrs C Trotter, Mrs M Trotter, Trotternish Casuals, Nigel Troup, Neil Trout, Mr D I Trowman, Mike Trubridge, Bill Truckle, Mr R Trundle, Mrs S Trundle, Mr M Trussell, Mr F Try, Dr P C Tubb, Mr G Tubb, Mr M J R Tubb, Mrs J M Tubbs, Mr B Tuck, Mr R Tuck, Mrs M Tuck, Mrs R Tuck, Dr M R Tucker, Miss K J Tucker, Andrew & Diane Tucker, Mr A J Tucker, Mr D Tucker, Mr I Tucker, John & LIz Tucker, Mr K Tucker, Mr M Tucker, Mr N Tucker, Peter Tucker, Mr R Tucker, Simon Tucker, Mr V Tucker, Mr W Tucker, Mrs M Tuckey, Mr C Tudge, Christopher Tudge, Mrs J A Tudor-Hughes, Dr F Tuffen, Miss C Tuffnell, Mr J Tuijl, Mr D Tulley, David Tully, Mr J Tully, Mr R N Tune, Mrs S Tune, Dr W R Tunnicliffe, Mr K S Tunnicliffe, Mr T Tuohy, Mr M Turley, Miss E Turnbull, Mr I Turnbull, Mr M Turnbull, Mrs B Turnbull, Mrs E E Turnbull, Dr A K Turner, Dr Jane Turner, Dr J Turner, Dr J R Turner, Miss K Turner, Miss L Turner, Miss S Turner (2), Miss V Turner, Mr A Turner (2), Daniel M Turner, Mr D Turner, Mr G Turner, Mr H Turner, Mr I Turner, Turner family, Mr J Turner (2), Mr J A Turner, Mr J C E Turner, Mr J E Turner, Mr J G Turner, Mr J L Turner, Mr L Turner, Michael Turner, Mr M Turner (3), Mike & Sue Turner, Mr P C Turner, Mr R Turner (3), Raymond Turner, Mr R P Turner, Mr S Turner (3), Audrey & Bob Turner, Mrs K Turner, The Turners (2), Mrs R Turner, Mrs S Turner, Revnd A E Turner, Mr D & Mrs S M Turnidge, Mr R Turrell, Mike Turton, Mr N Turvey, Mrs C Turvey, Mr F Tusa, Sash Tusa, Mr R Tushingham, Mr P V Tutalo & Ms K Wilkins, Mr G Tuthill, Miss K Tutin, Mr D Tutt, Mr R A Tutthill, Miss N Tutton, Mr G Tutton, Mr C Tweddle, Mr J Tweedie, Mrs C Tweedle, Mr G Twells, Mr C Tweney, Mr R S Twigg, Mr W Twigg, Mrs L Twigg, Dr S Twigger, Mrs S E E Twigger, Mr M A Twiggs, Mr T Twiggs, Mr G Twinberrow, Mr P Twine, Mr M Twinn, C & I Twissell, Mr P Twist, Miss W Twitchett, Mr M Twydell, Dr John Twyford, Mr A Twyman, Mrs A Twyman, Mr Christopher J Tyas, Mr N Tyers, Dr S J Tyler, Mr C Tyler, Mr Jonathan P Tyler, Mr M W Tyler, Mr P Tyler, Rosemary Tyler, Dr B J Tymons, Mr C J Tynan, Mrs J M Tyndall, Mr A Tyner, Mr John Tyner, Mr C Type, Mr S Tyrer, Ian Tyrrell, Ms R Tyrrell, Mr A Tysoe, Mr A Tysom, Mr D Tyson, Mrs R Tyson

U Caroline Uff, Uhi Perth College, Mr P W Ullrich, Mr D K Underwood, Mr J Underwood, Mr P Underwood, Mr W M Underwood, Mrs S Underwood, Mr G Uney, Miss J Unsworth, Mr D J Unsworth, Mr R Unsworth, Mr Brian Unwin, Mr E N Unwin, Mr N Unwin, Mr J Upchurch, Jeff Danger Upex, Upper Clun Bird Group, Mr C Upstone, Michael Upstone, Andrew Upton, Mr Andrew J Upton, Mr G Upton, Mr R Upton, Upton family, Mrs A Upton, Prof G Upton, Mr E T Urbanski, Mr N Urch, Mr J Ure, Mr J Uren, Mr P Urpeth, Mr D Urquhart, Mr E D Urquhart, Mr H R Urquhart, Ms N Urry, Miss S Urwin, Mr B Urwin, Tony Usher, Mr J Usher Smith, Dr J Uttley, Mrs J E Uttley, Mr B Utton

V Mr J Vafidis, Mr J Vale, Mrs J Vale, Mr S Valentine, Mr M Vallance, Mr J Vallas, Mrs A Van De Velde, Mr H Van De Ven, Mrs M R van de Wetering, Mr N Van Der Veen, Mr S Van Hear, Dame M Van Nuenen, Ms J Vanderhook, Mrs C Vandersteen, Miss J Vangendt, Mr & Mrs M Vann, Miss A L Vanstone, Mrs I Vanstone, Mr P Varkala, Mr K A Varley, Mr M Varley, Mrs J Varley, Mrs S A Varrie, Mr R A Vasey, Mrs L Vasey, Dr H Vaughan, D & K Vaughan, Mr D R Vaughan, Mr M Vaughan, Mr P O Vaughan, Mr T Vaughan (2), T C Vaughan, Miss A Vaux, Mr D Veal, Mrs L Veal, Mr M Veater, Mr T Vee, Mr D Veitch, Mr J Veldman, Dr W A Venables, Mr D Venables, Mr H Venables, Mrs J Vendy, Mr F B Venn, Mr R Venn, Dr D Venters, Mrs C Venters, Mr D Venus, Mrs R Verdon, Mr G Vernall, Ms T Vernelli, Dr R Vernon, Mr R Vernon, Miss M Vertigan, A S Vials, Mr T Vick, Mr D Vickers, I Vickers, D & I Overseal, Mr M Vickers, D & M Vickers, Mrs J Vickers, Mr H J Vilkaitis, Mr M Vincent, Mr P R Vincent, Mr R J Vincent, Mr S Vincent, The Vincents, Mr T Vincett, Mr A Vine, Mr C Vine, Graham Vine, Mr I Vine, Mr N Vipond, Mr & Mrs Virden, Mr W Visick, Mrs K Visocchi, Mrs J Vitkovskis, Mr A Vittery, Tony Vittery, Miss K Vivian, Mrs G Vivian, Mr P Vodden, Mrs C Vodden, Mrs J Voisey, Mr M G Vokes, Mrs Y Von Cramon, Ms J Vonhof, Mr T Vorsterman, Donna Vose, Mrs G A Vowles, VRO - Newwildife, Mr L Vyrva, Miss C Vyse, Mr C Vyvyan, Mrs J Vyvyan-Penney

W F & L Waddingham, Mr D Wade, Mr T P Wade, Mrs I Wade, Mrs E Wade-Brown, Dr A H Wadsworth, Dr E Wadsworth, Mrs Wadsworth, Mr M Wagemakers, Mr D Wagstaff, Mr I Wagstaff, Mr W H Wagstaff, Mrs E M Wagstaff, Mr J Wain, Mr P Wain, Mrs J Waine, Eddie Wainscott, Mr B Wainwright, Gareth Wainwright, Mr J Wainwright, Mr R Wainwright, Mr S Wainwright, Mrs S Wainwright, Mr D W Waistell, Dr J Waite, Dr R K Waite, Mr G F Waite, Mr K Waite, Mr M Waite, Mr S A D Waite, Anne Waite, Mr S Wake, Dave Wakefield, Mr K Wakefield, Jeremy Wakeford, Mr R J Wakeford, Robert Wakelam, Mr P Wakelin, Mr M Wakeman, Mr T Wakeman, Mr P Wald, Mr P Walder, Mr J M Waldon, Mr G Waldron, Mr J Waldron, Mr M Waldron, Mr M J Waldron, Mr D Walduck, Mrs L K Walduck, Mrs C Wales, Mr C Walford, Mr G L Walford, Mr M Walford, Mr C Waling, Miss J Walkden, Walker family (2), Dr C H Walker, Dr S Walker, Miss B W Walker, Miss D C Walker, Miss R H Walker, Andrew P Walker, Mr A Walker (2), Mr Andrew Walker, Mr B Walker (2), Mr B J Walker, Mr C Walker (2), Mr D Walker (2), D K Walker, Mr G Walker (2), Mr G J Walker, Henry Walker, Mr I Walker, Mr I D Walker, Mr I R Walker, Mr J Walker (2), Mr J R Walker, Mr M Walker, Mr N Walker, Mr P Walker (2), Mr P N Walker, Mr R Walker (2), R & C Walker, Mr R P Walker, Mr S Walker (3), Mr S W Walker, Mr T Walker, Mr T H Walker, Angus Walker, Mrs A Walker, Mrs C Walker, Mrs D Walker, Mrs V Walker, Mrs V E Walker, Ms P Walker, Mr T Walkington, Mrs S A Walkinshaw, Mr J Walkley-Cox, Mr D Wall (2), Mr D H Wall, Mr G M Wall, Mr L Wall, Mr M J Wall (2), Nick Wall, Tom Wall, Mrs G Wall, Ms M Wall, Dr A Wallace, Jenny Wallace, Kit Wallace, Mr D Wallace, Mr D I M Wallace, Mr I Wallace, Mr J Wallace (2), Mr M F Wallace, Mr N Wallace, Mr S Wallace, Mrs C Wallace, Mrs D Wallace, Mrs L J Wallace, Mr A T Wallbank, M Wallen, Mr W R Waller, L & A Walley, Revnd Peter Walley, Mr J J Walling, Ms M Wallington, Miss G Wallis, Mr C Wallis, Mr R Wallis, Stephen Wallis, Mr T Wallis, Mrs H Wallis, Mrs V A Wallis, Walls family, Mr R Walls, Mrs G Walls, Dr J Walmsley, Dr M Walmsley, Mr A P Walmsley, Mrs S Walmsley, Walney Bird Observatory, Dr J F Walsh, Dr M Walsh, Alyn Walsh, Mr A Walsh, Mr B Walsh, Mr C Walsh (2), Mr D Walsh (2), Mr G D Walsh, Mr G D J Walsh, Mr I R Walsh, Mr J Walsh, Mr J R Walsh, McGrath & Walsh, Mr P M Walsh, Walsh family, Mrs L Walsh, Mrs S F Walsh, Ms M Walsh, Mr J Walshe (2), Dr G Walsh-Kemmis, Mr M F Walter, Dr S Walters, Miss E Walters, Mr D Walters, J & G Walers, Mr J M Walters, Mr P Walters (2), Mr S Walters, Mr T Walters, Mrs A Walters, Mrs L Walters, Mr L Walther, George Walthew, Mr C M Waltho, Miss R Walton, Mr A Walton, Clive Walton & Lothian Ringing Group, Mr J Walton, Mr J W Walton, Mr L Walton, Mr M Walton, Mr P W Walton, Mr S Walton, Mr T Walton, Mrs S Walton, The Waltons, Mrs C Walton-Waters, Mr D Wanklyn, Mr G Wann, Wanstead Birders, Mr R J Wapshott, Miss A Warburton, Mr D Warburton (2), Mr M Warburton, Mrs J Warburton, Mrs M Warburton, Sally Ward, Dr S D Ward, Dr S T Ward, Miss C Ward, Miss K Ward, Mr C Ward, Mr C J Ward, Mr D Ward, John Ward, Mr J & Mrs R Ward, Mr J N Ward, Mr J Ward, Mr K Ward, Ward family, Mr M R Ward, Mr P Ward, Roger Ward, Mr R Ward (2), Mr R M Ward, Steven Ward, Mr T Ward, Mrs J Ward, Mrs M Ward, Ms K Ward, Ms M Ward, B Ward, Mrs C Ward-Campbell, Mr A Wardell (2), Mrs J B Wardell, Mrs L Wardell, Ms K J Warden, Mr D Warden, Mr P C Warden, Mr R Wardle (2), Ms E M Wardle, Ms J Wardle, Mr B Wardman, Dr M A Ware, Mr D Ware, Ms D Wareham, Mr S Warford, Mr J Warham, Mr M Warham, Mr A Waring, Chris Waring, Mr C Waring, Mr M Waring, Mrs P Warke, Mrs P Warman, Mr J Warne, Mr M Warne, Mrs Heather Warne, Mr D Warner, Mr P Warner, Mr J D Warnes, Mr N G Warnock, Dr M Warren, Dr R F Warren, Mr A Warren, Mr J E Warren, Mr J P Warren, Mr Michael J Warren, Mr N Warren, Roger S Warren, Ms R Warren, Mr G J Warrilow, Mr S Warrilow, D & G Warriner, Dr S Warrington, Mr A Warrington, Mr P M Warrington, Mr S Warry, Jill C Warwick, Mr J Warwicker, Mrs M W R Washington, Mr K P Watcham, Mr S Watchorn, Watchorn & Dodd Families, Mr A F Waterer, Mr K W Waterfall, Mr J Waterfield, Miss A Waterhouse, Mr G E C Waterhouse, Mrs S Waterhouse, Mr A M Waterman, Mr G M Waterman, Mr A Waters, Mr D Waters (2), Mr T Waters, Mrs J Waters, Mr A Waterson, Mr P Waterton, Mr L Waterworth, Mrs L Waterworth, Mr D R Watkins, Mr G Watkins, Mr I Watkins, Mr P Watkins, Mrs S Watkins, Mr W R & Mrs A H Watkins, Mr P D Watkinson, Miss A Watmore, Mr B R Watmough, Mr G Watola, Dr A B & Mrs A Watson, Dr D A Watson, C & H Watson, Mr A Watson, Mr A R Watson, Mr C Watson, Mr C F Watson, Watson family, Mr D Watson (5), Mr D L Watson, Mr G Watson (2),

Mr G F Watson, Mr H Watson, Mr I S Watson, Mr J Watson (3), Mr J W Watson, Mr M H E Watson, Michael Ivan Godfrey Watson, Mr M Watson, Mr N & Mrs D Watson, Mr O M Watson, Mr P Watson (3), Peter E W Watson, Mr R Watson, Mr R F Watson, Mr R K Watson, Simon Watson, Mr S Watson, Mr T P M Watson, Mrs A Watson, Mrs B J Watson, Mrs H Watson, Mrs M R Watson, Marie Yvette Watson & family, Mrs R Watson, Mrs Y Watson, Ms S Watson, Ms S A Watson, C Watson, Mrs S Watson-Quirk, Dr J W H Watt, Mr D Watt, Mr G Watt, Mr I Watt, Mr J A L Watt, Mr J T Watt, Watt/Duff, Mr R E L Watt, Miss B Watts, Mr A Watts, Mr A J Watts, Mr C Watts, Mr Jason Watts, Mr P J Watts, Mr P N Watts MBE, Mr P Watts, Mr S Watts, Mr T Watts, Mrs P Watts, Dr Denise Wawman, Mr C & Mrs A Waygood, Mr T Wayman, Mr M Weake, Mr T Weale, Mrs G Wearmouth, Revnd A Wearmouth, Mr B Weatherburn, Mr D Weaver, Mrs P Weaver, Mr R Weaving, Dr L Webb, Miss L S Webb, Mr A Webb, Mr B Webb, Mr D Webb, Mr G Webb, Mr J R Webb, Mr M Webb (2), Mr P Webb, Richard Webb, Mr R Webb, Mr S Webb, Webb family, Mr T Webb, Mr T B Webb, Mr W Webb, Mr W J & Mrs G M Webb, Mrs A Webb, Mrs C Webb, Ms G Webb, Mr D Webber, Mr J Webber, Mr W John Webber, Mr S Weber, Ms A Webley, Mr J H Websper, Dr M Webster, Miss E Webster, Susan Webster, Mr A Webster, Mr C Webster, Mr D J Webster, Mr G J Webster, Mr M Webster, Michael Webster, Mr P Webster, Mr P J Webster, Mr P N Webster, Mrs E Webster, Mrs S Webster, Mr D Wedd, Dave &Jackie Wedd, Ms J Wedd, Ms R Weedon, Mr P Weeks, Mr S Weeratunge, Mr A M Weir, Mr G Weir, Martin Weir, Weir family, Mrs L Weir Bingham Mcandrew, Mr P S Weisner, Mr S Welch, Mr W Welch, Daniel Weldon, Mr K Welfare, Mr J T Welford, Mrs H J Wellard, Mr C Wellavize, Mr R Welling, Mr P Wellington, Wellington College, Miss A M Wells, Mr A Wells, Mr A P Wells, Mr B Wells, Mr C A G Wells, Mr Colin E Wells, Mr D Wells, Mr E Wells, Mr G A H Wells, Mr J Wells (2), Mr J H Wells, Mr John N Wells & Mrs Sue W Wells, Mr M Wells, Mr M G Wells, Mr P Wells (2), Peter Wells, Mr T Wells (2), Mr T J Wells, Mrs A A Wells, Mrs F Wells, Mrs J Wells, Mrs K Wells, V Wells, Ms Judith Wells & family, Ms S J Wells, Wells U3A, Welney WWT, Mr G Welsby, Dr A Welsh, Miss A Welsh, Mr P A Welsh, Mr Stephen Wende, Mrs A J Wende, Mr C Went, Mr M S Werndly, Chris Wernham, Mr A Wesley, Mr T Wesley, Mr Mark Wessels, Mrs S R Wesson, Miss A West, Sarah L West, Mr A West, Mr C West, Mr D West (2), Mr Derek G H West, Mark West, Mr R West, Mr S West, Mrs C West, Mrs H West, Mrs J West, Mrs N West, West Cliff Primary School, West Midlands Bird Club (Staffs), Mr P Westbrook, Miss H Westen, Mrs C Westerback, Dr A E Westerberg, Mr S Westerberg, Mrs D Westerhoff, Miss G Western, Mr D Western, Mrs J Westfold, Mr D Westlake, Mr L Westmoreland, Mr R G Westoby, Miss L Weston, Mr A Weston, Barry Weston, Ian Weston, Mr J Weston, Mr K Weston, Mr T Weston, Mrs C Weston, Ms S Weston, Mrs J H Westover, Dr R J Westrop, Mr A D Westwood, Mr B Westwood, Mr D Westwood, Ms J Wetlaufer, Mr R Weyl, Mr W Wez, WFE, Mr C Whale, Mr D Whaley, Mr T W Whalin, Mr E J Whalley, Mr P D Whalley, Mr F Wharram, Mr M Wharton, Mrs S Wharton, Mr G Whatley, Mrs J A Whatmough, Miss C M Wheadon, Mr J M & Mrs S Wheat, Anne Wheatcroft, Miss H Wheatley, Mr D Wheatley, Jeffery J Wheatley, Mr S Wheatley, Mr J Wheel, Mr W Wheel, Mr R Wheeldon, Fiona Wheeldon, Dr G Wheeler, Dr M J Wheeler, Mr D Wheeler (2), Mr D J Wheeler, Mr D K Wheeler, Mr G Wheeler, John Wheeler, Mr K Wheeler, Mr R Wheeler, Mr W Wheeler, Mrs J Wheeler, Mr G Whelan, Mr M Whelan, Mr N Whelan, Mr R Whelan, Mrs J Whelehan, Mr B Whenman, Andrew Whewell, Miss J Whilde, Mr P Whillance, Mr D S Whitaker, Mr G & Mrs P Whitaker, Mr I N & Mrs H Whitaker, Mr K Whitaker, Mr M J Whitaker, Mr M S F Whitaker, Mr S F Whitaker, Mrs L M Whitaker, Mr G J Whitby & Mrs C A Whitby, Whitby Naturalists' Club, Mr P J Whitcomb, Miss E Whitcombe, Mr T Whitcombe, Dr S White, Miss C White, Mr A White, B & H White, Mr B White (2), Mr C White, Mr C E White MBE, Mr D White (3), Mr D A White, Mr Denis M White, Mr D R White, Mr D T White, Mr G White (2), Mr G J White, Mr J White (2), Mr J R White, Mr J White, Mr K White (2), Mr K G White, Mr M White (2), Mr N White, White family, Mr P White, Mr R White, Mr Richard N White, Mr S White (3), Mr S J White, Mr W J White, Mrs A White, Mrs K White, The Whites, Sarah Priest, Mrs S White, Ms A White, Prof P R White, Mr T Whiteford, Ms S Whiteford, Mr B Whitehall, Dr S Whitehead (2), Dr T Whitehead, Lt Cdr P A Whitehead, Mr A Whitehead, Mr D Whitehead, Mr R Whitehead (2), Mr A G Whitehead, Mr T Whitehead, Ms D Whitehead, E Whitehead, Dr A Whitehouse, Dr D Whitehouse, Adam Whitehouse, Mr A Whitehouse, Mr M Whitehouse, Mrs G P Whitehouse, Mr A Whitelaw, Mr D B Whitelaw, Mr P Whiteman, Mrs V Whiteman, Mr M R Whiteside, Mrs L Whiteside, Miss L Whitfield, Mr & Mrs A Whitfield, Mr C J Whitfield, Mrs J Whitfield & Mr P Attwood, Mrs J Whitham, Mrs S Whiting, Mr C Whitlock, Mr K Whitmore, Mr M Whitmore, Ms R Whitmore, Andy Whitney, Mr R Whitney, Whitstable Bird Group, Dr P Whittaker, Mr I Whittaker, Mr M Whittaker, Mr P Whittaker, Mr S Whittaker (2), Mr M Whittam, Mr P Whittam, J Whittington & C Hassall, Mr C Whittle, Mr M Whittle, K Hollidge, Mr D Whitton, Mr J Whitton, Mr B A Whitworth, Mr P & Mrs M M Whitworth, Mrs A Whorwood, Mr M Whybrow, Bill Whybrow, Mr R G Whyman, Miss C Whysall, Mr A Whyte (2), Mr C Whyte, Mr G J Whyte, Mr K A R Whyte, Niall Whyte, Mr W Whyte, Mr RC Whytock, Mr Rory Whytock, Mr J A Wichall, Mr M Wicken, Mr M Wickenden, Dr S M Wickham, Mr B Wickham, Mr C Wicks, Mr J E Wicks, Mr S Wicks, Barrie Widden, Mr J Widdick, Mrs A Widdows, Mr M Widdowson, John Widgery, Mr R Widgery, Mrs C Wieloch, Miss T J Wiffen, Mr P Wiggans, Mrs H Wiggins, Mr A D Wight, Mr J Wightman, Mr J B Wightman, Mr S Wightman, The Wigleys, Mr P Wigley, Mrs D M Wignall, Mr S Wignill, Bill Wigram, Ms B Wijnberg, Mrs V Wikman, Mrs K Wilbraham, Mr M Wilby, Mr R Wilby, Mr B Wilcock, Mr M Wilcock, Mr S J Wilcock, Mrs J Wilcockson, Joan Wilcox, Barry Wilcox, Mr C Wilcox, Mr J Wilcox, Mr T Wilcox, Dorothy Wilcox, Miss J Wilcoxon, Mr J P P Wilczur, Dr J Wild, Mr D Wild, Mr P J Wild, Mrs M Wild, Mrs R Wild, Wild Frontier Ecology Ltd, Mrs C Wildash, Mr E Wilde, Mr M Wilde (2), Mrs B M Wilden, Ms J Wilder, Ms V Wilders, Miss S Wildgoose, Mr J Wilding, Mr M F Wildish, Mr L Wildman, Mr A Wileman, Mr & Mrs D A Wileman, Ms C Wilken, Ms B Wilkens, Mr R Wilkes, Prof Andrew O M Wilkie, Dr A Wilkins, Miss E Wilkins, Mr A K Wilkins, Mr I Wilkins, Mr M A Wilkins, Mr P Wilkins, Dr E Wilkinson, Dr K Wilkinson, Dr R Wilkinson, Miss M F Wilkinson, A P Wilkinson, Mr A Wilkinson, Mr B Wilkinson, Mr C Wilkinson (4), David Wilkinson, Mr D Wilkinson (2), Mr G Wilkinson, Mr J Wilkinson (2), Mr J A Wilkinson, Mike Wilkinson, Mr N Wilkinson, Miss E Ashworth & Mr I Sim, Mr R Wilkinson, Mr R B Wilkinson, Mr S Wilkinson, Mr Wilkinson, Mr T Wilkinson, Mrs A Wilkinson, Mrs Wilkinson, Mrs F Wilkinson, Ms R Wilkinson, Mr C Wilks, Mrs M A Wilks, Mr B Willams, Mr J Willans, Dr C Willard, Mr J B Willcock, Mrs A Willcocks, Brian Willder, Mr J Willdridge, Mr J Willett, Mr S Willett, Mr S R Willey, Ms J Willi, Dr A J Williams, Dr B R Williams, Dr G Williams, Dr J Williams, Dr M Williams, Dr N Williams & Ms N Gunson, Dr P Williams, Dr R Williams, Miss H Williams, Miss J Williams, Miss L Williams, Miss T M Williams, Andy Williams, Mr A Williams, Mr A H Williams, Mr A J Williams, A K Williams, Mr B Williams (3), Mr C Williams (2), Mr C T Williams, Mr D Williams (4), Mr D A Williams, Mr D P L Williams, Mr D Williams (2), David Williams, Mr E Williams (3), Mr E J Williams, Mr F Williams, Graham Williams, Audrey & Gwyn Williams, Mr G Williams (4), Mr H Williams, Mr I Williams (2), Mr I J Williams, Mr I T Williams, Mr J Williams (2), Mr J E Williams, Mr J H Williams, Mr J N Williams, Mr K Williams (2), Mike Williams, Mr M Williams (2), Michael Williams, Mr M J Williams (3), Mr M Williams, Mr N Williams (2), Mr O Williams & The Woodcock Network, Mr P & Mrs M Williams, Mr P Williams (4), Mr P C Williams, Mr R E Williams, Mr R Williams (5), Mr R A Williams, Mr R D Williams, Mr R D R Williams, Mr R G Williams, Mr R J Williams, Mr S Williams (2), Mr S P Williams, Mr T & Mrs L Williams, Mr T Williams (2), Mrs C Williams (2), Mrs D Williams, Mrs D J Williams, Mrs M Williams (2), Elspeth Williams, Mrs E A Williams, Mrs H Williams, Mrs J Williams (3), Mrs P A Williams, Mrs R Williams, Mrs S Williams, Ms A Williams, Ms D Williams, Ms H Williams, Ms L Williams, Ms R Williams, Prof I Williams, Dr B Williamson, Miss N Williamson, Clive & Sheila Williamson, John Williamson, Mr M Williamson, Mr N Williamson, Mr R Williamson, Mrs H R Williamson, Ms K Williamson, Mrs K Williamson-Smith, Mr N Willimott, R & C Willis, Dr S Willis, Miss S Willis, Mr L Willis, Mr P Willis (3), Mr R Willis, Mr R G Willis, R & S Willis, Mr S Willis, Mrs J Willis, Mr N D Willits, Mr P Willmer, Mr J W Willmott, Mr M Willmott, Mr M J Willmott, Mrs H M Willmott, Mr P Willoughby, Dr R Willows, Mr C Wills, Mr D L Wills, Mr J M Wills, Mr K Wills, Mr R Wills, Mrs V Willsher, Mrs C Willson, Mrs D V M Wilshere, Mr J Wilshire, Dr H Wilson, Dr J Wilson, Dr M Wilson (2), Dr R H Wilson, Miss K Wilson, Miss L Wilson, Miss N Wilson, Mr A Wilson (2), Mr A D D Wilson, Mr A M Wilson (2), Mr A S Wilson, Mr B Wilson (2), Mr C Wilson (2), Mr Chris J Wilson , Mr E J Wilson, Mr G Wilson (4), Mr G E Wilson, Mr H Wilson, Ian J Wilson, Mr I Wilson, Jeremy Wilson, Jim Wilson, John Wilson, Mr J Wilson (6), Mr J D Wilson (2), Mr J L Wilson, Mr J M Wilson, Mr John W Wilson, Mr K M Wilson, Mr L Wilson, Mr M Wilson (3), Mark & Caroline Wilson, Mr M D Wilson, Mr P Wilson, Mr P E Wilson, Mr P H Wilson, Mr P K Wilson, Peter M Wilson, Mr Richard I Wilson CEnv MIEEM, Mr R Wilson (2), Mr R D S Wilson, Mr R J Wilson, David Wilson, Mr S Wilson (3), Mr T Wilson, Mrs C Wilson OBE, Mrs E Wilson (2), Mrs G Wilson, Mrs H Wilson, Mrs J Wilson, Mrs P Wilson, Wells family, Mrs V Wilson (2), Ms C M Wilson, Ms G Wilson, Ms J Wilson, Ryan Wilson-Parr, Mrs S Wilson-Town, Mr I J Wilton, Mr R Wilton, Revnd C Wilton, Mr I Wiltshire, Mr R Wiltshire, Mr S Wiltshire, Mr M & Mrs D Wiltshire, Wiltshire Ornithological Society, Mr Keith Wimbush, Mr C Wimlett, Mr R S Wimpress, Mr J S Winder, Mrs C Winder, Mr D Windle, Windmill Farmers, Windsor family, Mr I Winfield, Mr S Wing, Mr J Wingate, Mr A Wingrove, Mr M R Wingrove, Miss J Winkless, Mrs V Winks, Dr A Winn, Alan Winn, Miss S E Winnard, Mr D Winnard, Miss M Winney, Mr D A Winnie, Miss M Winsch, Mrs A D Winship, Mrs L Winship-Wardell, Mr M Winslow, Dr R J Winspear, Mr J L Winsper, Mr J R Winsper, Mr G Winstanley, Mr M Winstanley, Ms G Winstanley, Mr R P Winston, Miss K Winstone, John Wint, Miss J Winter, Miss K Winter, Mr C Winter, Mr D Winter, Leslie Winter, Mr P D Winter, Mrs B Winter, John Winterbottom, Joy Winyard, Mr M Wisby, Mr P & Mrs D Wise, Mrs D Wise, Mrs S Wise, Wiseman family, Mr E J Wiseman, Mrs J Wiseman, Mr J & Mrs L K Wiseman, Mrs T Wishart, Mr I Witham, Ms S J Witham, C Withers, G & B West, Mr B R Withers, Mr G Withers, Mr I Witheyman, Mr D Withrington, Mrs G Withrington, Mrs H A Witte, Antony Witts, Howard Wix, Mr K Woan, Mr I Woiwod, Mrs S Wolfendale, K Wollen, Mr R P S Wolsey, Mr L Wolstenholme, Mr P Wolstenholme, Mrs A F Wolstenholme, Dr A Wood, Dr J B Wood, Dr N Wood, Dr S Wood, Mr A Wood (2), Mr A D Wood, Mr A J Wood, Mr G Wood (3), Mr H Wood, Mr I H Wood, James A Wood, Maen Rock Farm, Mr J Wood, Mr J C Wood, Mr J H Wood, Mr & Mrs JKR Wood, Mr M Wood, Mark 'Fifer' Wood, Mr N Wood, Mr N A Wood, Mr P Wood, Mr P A Wood, Mr P R Wood, Mr Russell Wood, Mr R W G Wood, Mr S Wood (3), Simon D Wood, Mr T Wood, Terry Wood, Betty Wood, Mrs B J Wood, Mrs C A Wood, Wood family, Ms R Wood, Mr H G Wood Homer, Geoff Woodard, Mr R G Woodard, Mr G D Woodburn, Mr N J Woodcock, Mr S Woodcock (2), Mrs J P Woodcock, Woodcutters For Wildlife, Mrs R Woodell, Mr D Woodfield, Chris Woodham, Jerry Woodham, Mr C W Woodhead, Mr N Woodhead, Ms A Woodhill, Miss Z J Woodhouse, Mr B D Woodhouse, Mr K Woodhouse, Mrs E Woodhouse, Mr G T Woodin, John & Jen Woodland, Mrs H A Woodland, Woodlands Primary School P5, Mr J Woodley, Mr T Woodley, Ms P Woodley, Ms H Woodman, Mr D H Woodmansey, Mr G Woodroffe, Mr W Woodrow, Dr J Woodruff, Mr P Woodruff, Miss C Woods, Mr A E L Woods, Mr K Woods, Mr P H Woods, Mr R P Woods, Mrs J Woods, Mrs M Woods, Ms F Woods, Ms K Woods, Mr D Woodthorpe, Dr S Woodward, Mr A & Mrs S D Woodward, Mr G Woodward, Mr I D Woodward, Mr M W C Woodward, Mr R Woodward, Mr S F Woodward, Mrs S Woodward, Ms C Woodward, Dr C D Wooff, Mr G Wookey, Mr D Wooldridge, Mr S A Woolfries, Mr A Woollaston, Mr P Woollen, Mr R Woollen, Mr J C Woollett, Mr S Woolley, Mrs G Woolley, Mr J Woolliams, Mr T Woolmer, Mr J Woolston, Ms M Wootton, Miss D Worboys, Mrs J Worboys, Mr G Wordley, Mr A Wordsworth, Mr T Worfolk, Mr F J Workman, Robert Workman, Mrs L Worledge, Miss S Worrall, Mrs J A Worrall, Mr I Worsley, Mr P Wortham, Miss L Worthington, Mr G Worthington, Mr M Worthington, Mrs J Worthington, Ms J M Worthington Wilde, Mr P Worthy, Mr M D Wortley, Mr J & Mrs M Wotherspoon, Mr J Wotton, Mr M Wozniak, Mr D I Wragg, Mrs A Wragg, Mrs K Wragg, Mr S Wragge-Morley, Mr J Wray, Mr J & Mrs A P Wren, Mr P Wren, Mr S B Wren, Mr D Wrench, Wrexham Birdwatchers, Dr C Wright, Dr C M V Wright, Dr M Wright, Dr S A Wright, Miss A Wright, Miss R Wright, Miss S Wright, Miss T R C Wright, Mr A Wright, Mr B F Wright, Mr C E Wright, Mr C H Wright, Mr C J Wright, Mr D Wright, Mr & Mrs D Wright, Mr D Wright (2), Mr E Wright, Mr Graeme Wright, Mr G Wright, Mr I D Wright, Mr I H Wright, Jack Wright, Mr John Wright, Mr J Wright (3), Mr L Wright, Mr M Wright (3), Mr M R Wright, Mr M T Wright, Mr N Wright (2), Mr P Wright, Mr P A Wright, Mr P M Wright, Mr R Wright (2), Mr R M Wright, Simon J Wright, Mr S Wright, Mr T Wright, Mrs A Wright, Mrs G R Wright, J & B Wright, Mrs J Wright, Mrs K S Wright, Mrs W E Wright, Ms A M Wright, Ms B Wright, Karen Wright, Ms S Wright, G W Wright, Mrs G Wrightson, Miss D Wrigley, Mr S Wrigley, V Wrigley, Mrs H Wroe, Mr C Wroot, Mr D Wulff, Mr M Wurr, WWT Caerlaverock Staff, Mr D Wyatt, Mr M J Wyatt, Mrs B Wyatt, Ms C Wyatt, Mr D Wyer, Mr G Wylie, Mr M Wylie, C Wylie, Mr J Wyllie, David Wymer, Ms Freda Wyn, Mr A R Wynde, Mr J Wyness, Dr R Wynn, Mr A Wynn, Mr J Wynn (2), Mr C Wynne

Y Dr D W Yalden, Mr S Yandall, Ms C P Yarborough, Mr J Yardley, Mr R Yardley, Ms L Yarrow, Mr E Yarwood, Dr B J Yates, Dr L J Yates, Miss K Yates, Mr C D Yates, Mr R J Yates, Mr S Yates, Mrs C Yates, Mr R Yaxley, Mr D Yeardley, Mr D Yelland, Ms A Yellowley, Mr R Yeoman Walker, Mrs J Yeomans, Mrs W Yolland, Mr C H York, Mrs B A York, Dr R Yorke, Mr G Youdale, Mr M Youdale, Mr S Youell, Miss C Young, Mr A J Young, Mr A Young, Mr C Young, Mr D Young, Mr D Young & Mrs Lloyd-Jenkins, Mr F Young, Mr G Young, Alan Young, Mr G T Young, Mr H G Young, Young family, Mr J Young (2), Mr P Young, Mr S Young (2), Mr S A Young, Stephen Young, Mr S F Young, Mr W Young, Mrs G Young, Mrs H Young, Mrs K Young, M R C Young, Mr & Mrs Young, Mr D Younger, Mr E Younger, Mr M J Youngman, Mr R E Youngman, Mr T Youngs, Mr I J Yoxall, Ysgol Bodafon

Z Elisha I Zadok, Dr S Zamze, Mr J Zantboer, Ms Y Zellmann, The Zephyrs, Mrs F Zobole, Dr B Zonfrillo

APPENDIX 2 – SUMMARY STATISTICS

THIS TABLE SUMMARISES information on range size (number of 10-km squares) and relative abundance (birds per hour per tetrad) in 2007–11, and changes in range size since previous atlases. It covers 510 species and major subspecies recorded in winter in at least one 10-km square or in the breeding season in at least one 10-km square with at least *possible* breeding evidence. Separate statistics are given for 'Britain' (includes Britain, the Isle of Man and Channel Islands; see Section 1.2) and Ireland. The table includes the scientific names of all taxa (BOU 2013b; Gill & Donsker 2013) and the two-letter species code for those that have one is given in brackets after the scientific name. The infinity symbol (∞) indicates the 'zero to some' changes for which a percentage cannot be calculated. For some taxa, totals include records of individuals of non-native or captive origin; it is worth noting that some individuals of these taxa may have been present during previous atlases but were not reported. See Section 4.6 for further details and the additional notes below on the derivation and interpretation of certain columns.

a. Using data from the full winter period (1 November–end of February).

b. Using 2007–11 data from mid November to end of February to maintain consistency with the seasonal recording period used by the *1981–84 Winter Atlas* (see Section 4.3.1).

c. Calculated by taking the percentage of surveyed tetrads that were occupied on the early and/or late visit (first-hour only) in each 10-km square, then averaged across all occupied 10-km squares. For breeding-season figures, 'occupied' means those with at least *possible* breeding evidence. Ireland figures were calculated using only checkerboard High-priority 10-km squares.

d. Having added zeroes where necessary for taxa not recorded on visits (e.g. for TTVs returning zero of a given species despite breeding evidence in the 10-km squares), the first-hour counts from early and late visits were averaged for each tetrad, then averaged across tetrads in each 10-km square, then averaged across all occupied 10-km squares. As in (c), breeding-season figures were averaged across 10-km squares with at least *possible* breeding evidence and Ireland figures were calculated using only checkerboard High-priority 10-km squares.

e. The number of 10-km squares where the species was recorded in the breeding season but with no true breeding evidence (i.e. squares coded as flying, migrant, summering, or lacking any code).

f. Using breeding-season records plus selected out-of-season breeding evidence (see Section 4.2.2).

g. Using breeding-season records only to maintain consistency with the seasonal recording periods of previous breeding atlases (see Section 4.3.2).

h. An index of breeding-season range change, ranging between -1 for wholesale loss to +1 for wholesale gain, for the period 1988–91 to 2008–11. The figures in brackets are the 95% confidence limits (see Box 4.6).

Species name *scientific name and two-letter code*	Island	Winter				Breeding season								
		Range size: *present* (a)	Range change (since 1981–84) (b)	Mean % of tetrads occupied per 10-km (c)	Mean birds per hour per tetrad (d)	Range size: *non-breeding* (e)	Range size: *possible* (f)	Range size: *probable* (f)	Range size: *confirmed* (f)	Range change (since 1968–72) (g)	Range change (since 1988–91) (g)	Fixed-effort change index (since 1988–91) (h)	Mean % of tetrads occupied per 10-km (c)	Mean birds per hour per tetrad (d)
Mute Swan *Cygnus olor* (MS)	Britain	1929	+21%	18%	0.9	139	84	146	1541	+9%	+12%	+0.14 (+0.11,+0.16)	18%	0.7
	Ireland	681	+<1%	19%	0.9	24	57	131	424	-4%	+9%	-0.07 (-0.11,-0.04)	19%	0.8
Black Swan *Cygnus atratus* (AS)	Britain	280	+780%	1%	<0.1	151	31	41	37	∞	+10600%	+1.00 (+1.00,+1.00)	3%	<0.1
	Ireland	13	∞	6%	0.1	2	1		1	∞	∞			
Trumpeter Swan *Cygnus buccinator* (TJ)	Britain	2	∞	2%	<0.1	1								
	Ireland													
Bewick's Swan *Cygnus columbianus* (BS)	Britain	385	-12%	1%	0.3	21								
	Ireland	15	-90%											
Whooper Swan *Cygnus cygnus* (WS)	Britain	1301	+35%	5%	0.5	467	9	7	22	+375%	0%	+1.00 (+1.00,+1.00)	8%	0.1
	Ireland	547	+16%	12%	1.3	130	3	2	3	∞	-38%			
Magpie Goose *Anseranas semipalmata* (MJ)	Britain	2	∞											
	Ireland													
White-faced Whistling Duck *Dendrocygna viduata* (YX)	Britain	3	∞			3								
	Ireland													
Fulvous Whistling Duck *Dendrocygna bicolor* (YV)	Britain	3	∞			3		1		∞	∞			
	Ireland													
Lesser Whistling Duck *Dendrocygna javanica* (YU)	Britain					1								
	Ireland													
Cape Barren Goose *Cereopsis novaehollandiae*	Britain	1	∞											
	Ireland													
Swan Goose (Chinese Goose) *Anser cygnoides* (HN)	Britain	37	+3300%	2%	<0.1	16	9	5	2	+1500%	+1500%		3%	<0.1
	Ireland	4	∞	4%	<0.1	1		2		∞	∞			

Species name *scientific name and two-letter code*	Island	Winter: Range size: *present* (a)	Winter: Range change (since 1981–84) (b)	Winter: Mean % of tetrads occupied per 10-km (c)	Winter: Mean birds per hour per tetrad (d)	Breeding season: Range size: *non-breeding* (e)	Breeding season: Range size: *possible* (f)	Breeding season: Range size: *probable* (f)	Breeding season: Range size: *confirmed* (f)	Breeding season: Range change (since 1968–72) (g)	Breeding season: Range change (since 1988–91) (g)	Breeding season: Fixed-effort change index (since 1988–91) (h)	Breeding season: Mean % of tetrads occupied per 10-km (c)	Breeding season: Mean birds per hour per tetrad (d)
Bean Goose *Anser fabalis* (BE)	Britain	197	+57%	<1%	<0.1	30								
	Ireland	4	-56%			3								
Bean Goose (Taiga) *Anser fabalis fabalis* (XF)	Britain	55	∞	<1%	0.1	5								
	Ireland	2	∞											
Bean Goose (Tundra) *Anser fabalis rossicus* (XR)	Britain	141	∞	<1%	<0.1	15								
	Ireland	1	∞			1								
Pink-footed Goose *Anser brachyrhynchus* (PG)	Britain	1049	+93%	7%	14.4	573		1	2	∞	-80%			
	Ireland	67	+369%	2%	<0.1	16					-100%			
White-fronted Goose *Anser albifrons* (WG)	Britain	563	+36%	3%	1.3	133					-100%			
	Ireland	111	-10%	4%	2.5	27								
White-fronted Goose (European) *Anser albifrons albifrons* (EW)	Britain	236	∞	1%	0.2	35								
	Ireland	2	∞											
White-fronted Goose (Greenland) *Anser albifrons flavirostris* (NW)	Britain	194	∞	7%	3.1	56								
	Ireland	104	∞	3%	1.3	23								
Lesser White-fronted Goose *Anser erythropus* (LC)	Britain	17	+129%	1%	<0.1	8								
	Ireland													
Greylag Goose *Anser anser* (GJ)	Britain	1888	+89%	15%	6.6	337	162	309	1198	+733%	+132%	+0.62 (+0.59,+0.64)	19%	1.8
	Ireland	182	+111%	7%	1.6	35	28	25	46	+1125%	+326%	+0.56 (+0.42,+0.71)	9%	0.6
Bar-headed Goose *Anser indicus* (HD)	Britain	140	+505%	<1%	<0.1	135	9	2	6	∞	+300%		1%	<0.1
	Ireland	1	∞	13%	0.1	1								
Snow Goose *Anser caerulescens* (SJ)	Britain	162	+28%	<1%	<0.1	82	1	1	4	∞	-45%		1%	0.1
	Ireland	9	+300%	2%	<0.1	1								
Ross's Goose *Anser rossii* (RJ)	Britain	58	∞	<1%	<0.1	28								
	Ireland													
Emperor Goose *Chen canagica* (EM)	Britain	13	∞			4			1	∞	-50%		6%	0.6
	Ireland	1	∞	10%	0.2									
Canada Goose *Branta canadensis* (CG)	Britain	1696	+67%	12%	2.6	227	109	234	1447	+162%	+49%	+0.31 (+0.28,+0.33)	18%	1.4
	Ireland	48	+80%	5%	0.3	14	11	13	18	+600%	+121%	+0.15 (0.00,+0.36)	9%	0.4
Cackling Goose *Branta hutchinsii* (LQ)	Britain	18	∞			10								
	Ireland	2	∞											
Hawaiian Goose *Branta sandvicensis* (NE)	Britain	2	∞			2								
	Ireland													
Barnacle Goose *Branta leucopsis* (BY)	Britain	796	+79%	3%	7.8	410	19	14	49	∞	+88%	+1.00 (+1.00,+1.00)	4%	0.4
	Ireland	76	+7%	7%	10.3	19			1	-50%	-50%		11%	7.2
Brent Goose *Branta bernicla* (BG)	Britain	586	+71%	7%	8.7	304					-100%			
	Ireland	197	+67%	22%	9.1	123					-100%			
Brent Goose (Dark-bellied) *Branta bernicla bernicla* (DB)	Britain	345	∞	8%	10.9	155								
	Ireland	25	∞	2%	<0.1	14								
Brent Goose (Light-bellied) *Branta bernicla hrota* (PB)	Britain	326	∞	2%	0.6	107								
	Ireland	180	∞	20%	8	101								
Brent Goose (Black Brant) *Branta bernicla nigricans* (BB)	Britain	60	+743%	1%	<0.1	16								
	Ireland	17	+1500%			5								
Red-breasted Goose *Branta ruficollis* (EB)	Britain	57	+563%	1%	<0.1	23			1	∞	∞			
	Ireland													
Egyptian Goose *Alopochen aegyptiaca* (EG)	Britain	393	+423%	6%	0.1	144	34	45	153	+1172%	+163%	+0.54 (+0.45,+0.64)	8%	0.2
	Ireland	1	∞			1								
Ruddy Shelduck *Tadorna ferruginea* (UD)	Britain	80	+630%	<1%	<0.1	104	5	7	4	∞	+220%		2%	<0.1
	Ireland	3	∞	6%	<0.1	1								
Shelduck *Tadorna tadorna* (SU)	Britain	1144	+19%	11%	1.9	254	93	375	680	+42%	+19%	+0.06 (+0.02,+0.09)	18%	1
	Ireland	206	+7%	17%	2.2	15	29	92	110	+7%	+25%	+0.07 (-0.04,+0.16)	21%	0.9
Cape Shelduck *Tadorna cana* (UE)	Britain	15	∞	1%	<0.1	5		1	1	∞	∞		3%	0.1
	Ireland													

Species name *scientific name and two-letter code*	Island	Winter: Range size: *present* (a)	Winter: Range change (since 1981–84) (b)	Winter: Mean % of tetrads occupied per 10-km (c)	Winter: Mean birds per hour per tetrad (d)	Breeding season: Range size: *non-breeding* (e)	Breeding season: Range size: *possible* (f)	Breeding season: Range size: *probable* (f)	Breeding season: Range size: *confirmed* (f)	Breeding season: Range change (since 1968–72) (g)	Breeding season: Range change (since 1988–91) (g)	Breeding season: Fixed-effort change index (since 1988–91) (h)	Breeding season: Mean % of tetrads occupied per 10-km (c)	Breeding season: Mean birds per hour per tetrad (d)
Australian Shelduck *Tadorna tadornoides* (UA)	Britain	5	∞	1%	<0.1	2								
	Ireland													
Paradise Shelduck *Tadorna variegata* (UB)	Britain	3	∞			1								
	Ireland													
Muscovy Duck *Cairina moschata* (MY)	Britain	220	∞	2%	<0.1	90	58	25	21	+1920%	+1920%	+1.00 (+1.00,+1.00)	2%	<0.1
	Ireland	3	∞	7%	0.1	1	4	1		∞	∞		6%	0.1
Wood Duck *Aix sponsa* (DC)	Britain	75	+218%	<1%	<0.1	44	9	10	10	+190%	-3%		2%	<0.1
	Ireland	2	∞			1		1		∞	∞			
Mandarin Duck *Aix galericulata* (MN)	Britain	592	+335%	2%	0.1	142	72	142	283	+1144%	+121%	+0.35 (+0.25,+0.44)	4%	0.1
	Ireland	9	∞	3%	<0.1	3	1	2	2	∞	∞			
Maned Duck *Chenonetta jubata* (MQ)	Britain	4	∞			2								
	Ireland													
Ringed Teal *Callonetta leucophrys* (IE)	Britain	17	∞	1%	<0.1	9		2		∞	∞			
	Ireland													
Crested Duck *Lophonetta specularioides* (QD)	Britain					1								
	Ireland													
Chiloe Wigeon *Anas sibilatrix* (HL)	Britain	30	∞	1%	<0.1	11		3		∞	∞			
	Ireland													
Wigeon *Anas penelope* (WN)	Britain	1859	+27%	12%	7.7	575	103	154	101	+27%	-1%	-0.07 (-0.15,0.00)	7%	0.2
	Ireland	497	+5%	15%	5.2	81	15	9		∞	-4%		9%	0.3
American Wigeon *Anas americana* (AW)	Britain	33	+233%	2%	<0.1	22								
	Ireland	10	+200%			2								
Gadwall *Anas strepera* (GA)	Britain	1139	+90%	6%	0.6	201	91	261	362	+351%	+100%	+0.44 (+0.38,+0.50)	8%	0.3
	Ireland	152	+143%	6%	0.2	24	16	24	15	+293%	+120%	+0.20 (0.00,+0.40)	12%	0.4
Falcated Duck *Anas falcata* (FT)	Britain	5	+400%			5	1			∞	∞			
	Ireland	1	∞											
Baikal Teal *Anas formosa* (IK)	Britain					3								
	Ireland	1	∞											
Teal *Anas crecca* (T.)	Britain	2163	+13%	13%	3.1	451	238	670	279	-14%	+3%	-0.08 (-0.12,-0.04)	7%	0.2
	Ireland	687	+<1%	18%	2.7	87	73	106	29	-46%	+11%	-0.25 (-0.31,-0.19)	8%	0.2
Green-winged Teal *Anas carolinensis* (TA)	Britain	108	+531%	<1%	<0.1	55								
	Ireland	36	+289%	1%	<0.1	8								
Mallard *Anas platyrhynchos* (MA)	Britain	2671	+5%	40%	4.9	31	76	222	2377	+1%	+2%	+0.02 (+0.01,+0.04)	45%	2.6
	Ireland	876	+2%	29%	2.1	12	106	203	594	-1%	+9%	+0.04 (+0.02,+0.07)	30%	1.3
Black Duck *Anas rubripes* (BD)	Britain	3	0%			4	2			∞	∞			
	Ireland	5	∞			1	1			∞	∞			
Pintail *Anas acuta* (PT)	Britain	800	+34%	3%	1	251	24	27	15	-23%	-22%	-0.41 (-0.72,-0.07)	3%	<0.1
	Ireland	102	-4%	4%	0.2	17	1			-92%	-89%			
White-cheeked Pintail *Anas bahamensis* (PN)	Britain	20	+433%	<1%	<0.1	12	1	1		∞	∞			
	Ireland													
Red-billed Teal *Anas erythrorhyncha* (YR)	Britain	1	∞											
	Ireland													
Speckled Teal *Anas flavirostris* (KQ)	Britain	5	∞						2	∞	∞		2%	<0.1
	Ireland													
Yellow-billed Pintail *Anas georgica* (YL)	Britain					2								
	Ireland													
Garganey *Anas querquedula* (GY)	Britain	47	+371%	<1%	<0.1	299	68	98	33	+46%	+44%	0.00 (-0.26,+0.29)	1%	<0.1
	Ireland	8	∞			26	5	6	2	+160%	+63%			
Blue-winged Teal *Anas discors* (TB)	Britain	2	0%			11	2			∞	∞			
	Ireland	5	∞			3								
Cinnamon Teal *Anas cyanoptera* (QA)	Britain					3								
	Ireland													

Species name *scientific name* *and two-letter code*	Island	Winter				Breeding season								
		Range size: *present* (a)	Range change (since 1981–84) (b)	Mean % of tetrads occupied per 10-km (c)	Mean birds per hour per tetrad (d)	Range size: *non-breeding* (e)	Range size: *possible* (f)	Range size: *probable* (f)	Range size: *confirmed* (f)	Range change (since 1968–72) (g)	Range change (since 1988–91) (g)	Fixed-effort change index (since 1988–91) (h)	Mean % of tetrads occupied per 10-km (c)	Mean birds per hour per tetrad (d)
Australian Shoveler	Britain	2	∞											
Anas rhynchotis (VA)	Ireland													
Shoveler	Britain	1071	+42%	6%	0.5	312	92	220	200	-2%	+13%	-0.06 (-0.15,+0.02)	5%	0.1
Anas clypeata (SV)	Ireland	204	+14%	8%	0.4	16	13	25	11	-4%	+9%	-0.58 (-0.62,-0.50)	3%	<0.1
Bernier's Teal	Britain	1	∞			1								
Anas bernieri (MK)	Ireland													
Chestnut Teal	Britain	6	∞			2								
Anas castanea (QB)	Ireland													
Marbled Duck	Britain	3	+100%	1%	<0.1	5								
Marmaronetta angustirostris (MB)	Ireland													
Red-crested Pochard	Britain	274	+245%	1%	<0.1	100	9	16	25	+1150%	+317%	+0.42 (0.00,+0.70)	2%	0.1
Netta rufina (RQ)	Ireland	3	-71%								-100%			
Rosy-billed Pochard	Britain	3	∞			1								
Netta peposaca (QR)	Ireland													
Pochard	Britain	1253	-13%	5%	0.7	291	93	95	131	-38%	-38%	-0.45 (-0.52,-0.37)	5%	0.3
Aythya ferina (PO)	Ireland	206	-49%	8%	2.6	12	14	1	9	-53%	-40%	-0.45 (-0.45,-0.33)	6%	0.3
Ring-necked Duck	Britain	61	+273%	1%	<0.1	31	1			∞	-75%			
Aythya collaris (NG)	Ireland	39	+3700%	1%	<0.1	14								
New Zealand Scaup	Britain	2	∞	6%	<0.1	2								
Aythya novaeseelandiae (NZ)	Ireland													
Ferruginous Duck	Britain	39	+67%	<1%	<0.1	15	1			∞	∞			
Aythya nyroca (FD)	Ireland	4	∞											
Tufted Duck	Britain	1820	+10%	11%	1.5	142	147	576	996	+33%	+16%	+0.02 (-0.01,+0.05)	13%	0.7
Aythya fuligula (TU)	Ireland	377	-14%	16%	2.8	34	64	122	81	-20%	+6%	-0.14 (-0.19,-0.10)	13%	0.7
Scaup	Britain	646	+58%	1%	0.1	240	2			-75%	-89%		6%	0.1
Aythya marila (SP)	Ireland	133	+54%	5%	0.5	25	3			∞	0%			
Lesser Scaup	Britain	30	∞	1%	<0.1	17								
Aythya affinis (AY)	Ireland	11	∞											
Eider	Britain	655	-1%	17%	3.2	176	56	75	334	+2%	-5%	-0.09 (-0.15,-0.04)	26%	3.2
Somateria mollissima (E.)	Ireland	85	-1%	9%	1.8	7	10	16	40	+71%	+44%	+0.12 (-0.20,+0.41)	19%	1.7
King Eider	Britain	23	+120%	5%	<0.1	15					-100%			
Somateria spectabilis (KE)	Ireland	2	+100%			2								
Long-tailed Duck	Britain	485	+6%	8%	0.6	180				-100%	-100%			
Clangula hyemalis (LN)	Ireland	90	+12%	3%	0.1	13								
Common Scoter	Britain	589	+43%	4%	0.7	489	6	14	4	-43%	-53%	-0.45 (-0.59,-0.32)	7%	0.1
Melanitta nigra (CX)	Ireland	114	+20%	5%	3.1	39	4	6	1	-27%	-31%		3%	0.1
Black Scoter	Britain					1								
Melanitta americana (DX)	Ireland													
Surf Scoter	Britain	35	+200%			25								
Melanitta perspicillata (FS)	Ireland	15	+100%			2								
Velvet Scoter	Britain	287	+58%	1%	0.1	117	1			∞	-50%			
Melanitta fusca (VS)	Ireland	38	+147%			3					-100%			
Barrow's Goldeneye	Britain													
Bucephala islandica (VG)	Ireland	2	∞			2								
Goldeneye	Britain	1678	+<1%	11%	0.4	643	10	13	22	+137%	-74%	-0.51 (-0.59,-0.42)	9%	0.2
Bucephala clangula (GN)	Ireland	280	-11%	11%	0.7	36	8			∞	-38%		10%	0.9
Hooded Merganser	Britain	7	∞	2%	<0.1	7								
Lophodytes cucullatus (HO)	Ireland													
Smew	Britain	397	+70%	1%	<0.1	48					-100%			
Mergellus albellus (SY)	Ireland	25	+100%	1%	<0.1	8								
Red-breasted Merganser	Britain	1009	+15%	13%	0.4	310	92	252	192	-19%	-20%	-0.21 (-0.25,-0.16)	13%	0.3
Mergus serrator (RM)	Ireland	217	-20%	15%	0.6	46	33	72	20	-50%	-26%	-0.33 (-0.42,-0.24)	11%	0.2

Species name *scientific name and two-letter code*	Island	Winter: Range size: *present* (a)	Winter: Range change (since 1981–84) (b)	Winter: Mean % of tetrads occupied per 10-km (c)	Winter: Mean birds per hour per tetrad (d)	Breeding season: Range size: *non-breeding* (e)	Breeding season: Range size: *possible* (f)	Breeding season: Range size: *probable* (f)	Breeding season: Range size: *confirmed* (f)	Breeding season: Range change (since 1968–72) (g)	Breeding season: Range change (since 1988–91) (g)	Breeding season: Fixed-effort change index (since 1988–91) (h)	Breeding season: Mean % of tetrads occupied per 10-km (c)	Breeding season: Mean birds per hour per tetrad (d)
Goosander *Mergus merganser* (GD)	Britain	1879	+81%	7%	0.2	392	189	258	423	+112%	+29%	+0.06 (+0.01,+0.10)	7%	0.1
	Ireland	80	+1800%	2%	<0.1	8		2	2	+300%	+100%			
Ruddy Duck *Oxyura jamaicensis* (RY)	Britain	317	+26%	2%	0.1	87	60	107	65	+1121%	-21%	-0.45 (-0.55,-0.36)	3%	<0.1
	Ireland	16	+133%	3%	<0.1	2	2	2	1	∞	-38%			
Lake Duck *Oxyura vittata* (OI)	Britain							1		∞	∞			
	Ireland													
Helmeted Guineafowl *Numida meleagris* (FW)	Britain	104	∞	2%	0.1	105	43	27	10	∞	+7900%	+1.00 (+1.00,+1.00)	3%	0.1
	Ireland	2	∞			3	4	4		∞	∞		4%	0.2
California Quail *Callipepla californica*	Britain					3								
	Ireland													
Northern Bobwhite *Colinus virginianus* (OQ)	Britain					2	3	1		0%	+300%			
	Ireland													
Red Grouse *Lagopus lagopus* (RG)	Britain	862	+12%	23%	1.5	26	166	232	450	-22%	-11%	-0.22 (-0.25,-0.19)	19%	0.5
	Ireland	160	-2%	5%	0.1	4	68	58	37	-66%	+1%	-0.40 (-0.47,-0.34)	5%	0.1
Ptarmigan *Lagopus muta* (PM)	Britain	145	+64%	8%	0.2	2	36	43	96	-11%	0%	-0.20 (-0.26,-0.12)	7%	0.1
	Ireland													
Black Grouse *Tetrao tetrix* (BK)	Britain	318	-5%	9%	0.2	14	59	297	77	-29%	-1%	-0.24 (-0.30,-0.20)	6%	0.1
	Ireland													
Capercaillie *Tetrao urogallus* (CP)	Britain	35	-55%	5%	0.1		16	14	21	-73%	-24%	-0.39 (-0.55,-0.26)	3%	<0.1
	Ireland													
Chukar *Alectoris chukar* (KR)	Britain					2	1			∞	-97%			
	Ireland													
Red-legged Partridge *Alectoris rufa* (RL)	Britain	1591	+75%	19%	1.1	97	224	723	697	+78%	+35%	+0.21 (+0.19,+0.23)	23%	0.6
	Ireland	20	+500%	6%	0.1	12	19	13	4	∞	+200%		5%	0.1
Grey Partridge *Perdix perdix* (P.)	Britain	1210	-23%	9%	0.3	117	184	541	506	-39%	-26%	-0.47 (-0.49,-0.44)	9%	0.1
	Ireland	11	-61%	2%	<0.1	2	4		8	-95%	-66%		3%	<0.1
Quail *Coturnix coturnix* (Q.)	Britain	3	+100%			124	550	286	31	+114%	+8%	-0.12 (-0.22,-0.01)	1%	<0.1
	Ireland					8	14	8	1	-30%	-32%		3%	<0.1
Silver Pheasant *Lophura nycthemera* (PV)	Britain	1	∞			1								
	Ireland													
Reeves's Pheasant *Syrmaticus reevesii* (RV)	Britain	48	∞	1%	<0.1	19	19	6		+1150%	+733%		1%	<0.1
	Ireland													
Pheasant *Phasianus colchicus* (PH)	Britain	2377	+11%	55%	2.6	40	211	571	1610	+6%	+5%	+0.08 (+0.07,+0.08)	64%	1.8
	Ireland	875	+16%	34%	0.5	4	272	340	320	+10%	+10%	+0.12 (+0.10,+0.14)	61%	0.9
Golden Pheasant *Chrysolophus pictus* (GF)	Britain	50	-2%	1%	<0.1	24	18	5	3	0%	-45%	-0.33 (-0.69,-0.06)	4%	<0.1
	Ireland									-100%				
Lady Amherst's Pheasant *Chrysolophus amherstiae* (LM)	Britain	7	0%	1%	<0.1	11	3	4		-46%	-22%			
	Ireland													
Indian Peafowl *Pavo cristatus* (PX)	Britain	108	∞	2%	<0.1	153	72	42	22	∞	∞	+1.00 (+1.00,+1.00)	4%	<0.1
	Ireland	3	∞			3	6	2		∞	∞		3%	<0.1
Red-throated Diver *Gavia stellata* (RH)	Britain	755	+33%	8%	0.2	348	47	122	176	+11%	-9%	-0.20 (-0.26,-0.13)	14%	0.2
	Ireland	208	+26%	9%	0.1	74		2	6	+14%	-20%		2%	<0.1
Black-throated Diver *Gavia arctica* (BV)	Britain	471	+41%	3%	<0.1	224	52	63	103	+3%	+10%	-0.14 (-0.22,-0.07)	6%	0.1
	Ireland	58	+247%	4%	0.1	21					-100%			
Pacific Diver *Gavia pacifica* (KD)	Britain	7	∞											
	Ireland	1	∞	25%	0.1									
Great Northern Diver *Gavia immer* (ND)	Britain	773	+48%	9%	0.1	437	1			-88%	-98%			
	Ireland	290	+22%	17%	0.4	144					-100%			
White-billed Diver *Gavia adamsii* (WV)	Britain	15	+150%	2%	<0.1	29								
	Ireland					3								
Black-browed Albatross *Thalassarche melanophris* (AA)	Britain	1	-100%								-100%			
	Ireland													

Species name *scientific name and two-letter code*	Island	Winter				Breeding season								
		Range size: *present* (a)	Range change (since 1981–84) (b)	Mean % of tetrads occupied per 10-km (c)	Mean birds per hour per tetrad (d)	Range size: *non-breeding* (e)	Range size: *possible* (f)	Range size: *probable* (f)	Range size: *confirmed* (f)	Range change (since 1968–72) (g)	Range change (since 1988–91) (g)	Fixed-effort change index (since 1988–91) (h)	Mean % of tetrads occupied per 10-km (c)	Mean birds per hour per tetrad (d)
Fulmar *Fulmarus glacialis* (F.)	Britain	545	+9%	23%	16.6	209	25	19	500	+2%	-6%	-0.09 (-0.16,-0.03)	29%	14
	Ireland	167	+5%	19%	7.9	29	5	11	155	+8%	+6%	-0.11 (-0.27,+0.04)	28%	7.6
Cory's Shearwater *Calonectris diomedea* (CQ)	Britain	2	0%			29								
	Ireland					7		1		∞	∞			
Great Shearwater *Puffinus gravis* (GQ)	Britain	1	∞			4								
	Ireland	2				6								
Sooty Shearwater *Puffinus griseus* (OT)	Britain	56	+2700%	<1%	<0.1	70								
	Ireland	6	∞			24								
Manx Shearwater *Puffinus puffinus* (MX)	Britain	47	+222%	<1%	<0.1	400	5	5	19	-28%	-8%	-1.00 (-1.00,-0.60)	1%	0.8
	Ireland	11	+67%			129	3		10	-38%	-23%		21%	1
Balearic Shearwater *Puffinus mauretanicus* (YQ)	Britain	48	∞			82								
	Ireland	2	∞			12								
Macaronesian Shearwater *Puffinus baroli*	Britain							1		∞	∞			
	Ireland													
Wilson's Petrel *Oceanites oceanicus*	Britain					2								
	Ireland					2								
Storm Petrel *Hydrobates pelagicus* (TM)	Britain	44	+150%	<1%	<0.1	152	5	1	36	-18%	-23%		4%	<0.1
	Ireland	6	+100%			50	7		11	-31%	-39%		17%	0.8
Leach's Petrel *Oceanodroma leucorhoa* (TL)	Britain	94	+473%	<1%	<0.1	14	1	1	7	-11%	-20%			
	Ireland	6	∞			4		1		∞	0%			
Gannet *Morus bassanus* (GX)	Britain	487	+52%	8%	2.3	768	1		26	+64%	+28%	-0.91 (-0.98,-0.84)	24%	100.1
	Ireland	163	-1%	14%	0.9	239			6	+100%	+20%	-1.00 (-1.00,-1.00)		
Cormorant *Phalacrocorax carbo* (CA)	Britain	2353	+53%	18%	0.7	1545	232	56	254	+39%	+69%	-0.21 (-0.25,-0.17)	21%	0.9
	Ireland	781	+18%	25%	0.8	430	71	11	87	0%	+4%	-0.45 (-0.50,-0.40)	29%	0.9
Shag *Phalacrocorax aristotelis* (SA)	Britain	905	+12%	25%	2.2	298	93	22	362	-9%	-4%	+0.04 (-0.02,+0.10)	31%	2.6
	Ireland	287	+2%	31%	1.7	80	42	10	118	-10%	+5%	+0.13 (-0.02,+0.28)	32%	2
Great White Pelican *Pelecanus onocrotalus* (YP)	Britain	1	∞			1								
	Ireland													
Bittern *Botaurus stellaris* (BI)	Britain	496	+161%	<1%	<0.1	46	27	25	30	+134%	+531%	+0.71 (+0.25,+1.00)	2%	<0.1
	Ireland	4	-20%			2								
American Bittern *Botaurus lentiginosus* (AM)	Britain	1	-100%											
	Ireland													
Little Bittern *Ixobrychus minutus* (LL)	Britain		-100%			5			1	0%	∞			
	Ireland					1								
Night-heron *Nycticorax nycticorax* (NT)	Britain	5	-17%			24								
	Ireland	1				5								
Green Heron *Butorides virescens* (HR)	Britain	2	-100%											
	Ireland													
Squacco Heron *Ardeola ralloides* (QH)	Britain	3	∞			6								
	Ireland													
Cattle Egret *Bubulcus ibis* (EC)	Britain	167	+7300%	<1%	<0.1	99	3		2	∞	∞		3%	<0.1
	Ireland	43	∞	1%	<0.1	11								
Little Egret *Egretta garzetta* (ET)	Britain	1167	+112900%	9%	0.1	656	153	58	122	∞	+16350%	+1.00 (+1.00,+1.00)	13%	0.2
	Ireland	383	∞	14%	0.2	98	69	16	42	∞	+12600%	+1.00 (+1.00,+1.00)	17%	0.3
Great White Egret *Ardea alba* (HW)	Britain	179	+7800%	<1%	<0.1	134	2			∞	∞			
	Ireland	9	∞			12								
Grey Heron *Ardea cinerea* (H.)	Britain	2627	+7%	26%	0.3	677	747	170	1021	+15%	-17%	-0.07 (-0.09,-0.06)	24%	0.3
	Ireland	905	+10%	30%	0.3	112	412	127	262	+4%	+1%	+0.02 (0.00,+0.05)	31%	0.3
Great Blue Heron *Ardea herodias*	Britain	1	∞											
	Ireland													
Purple Heron *Ardea purpurea* (UR)	Britain	2				56			1	0%	∞			
	Ireland					2								

Species name scientific name and two-letter code	Island	Winter				Breeding season								
		Range size: *present* (a)	Range change (since 1981–84) (b)	Mean % of tetrads occupied per 10-km (c)	Mean birds per hour per tetrad (d)	Range size: *non-breeding* (e)	Range size: *possible* (f)	Range size: *probable* (f)	Range size: *confirmed* (f)	Range change (since 1968–72) (g)	Range change (since 1988–91) (g)	Fixed-effort change index (since 1988–91) (h)	Mean % of tetrads occupied per 10-km (c)	Mean birds per hour per tetrad (d)
Black Stork *Ciconia nigra* (OS)	Britain					30								
	Ireland													
White Stork *Ciconia ciconia* (OR)	Britain	14	+200%			126								
	Ireland					2								
Sacred Ibis *Threskiornis aethiopicus* (IS)	Britain	1	-89%	7%	<0.1	10								
	Ireland													
Glossy Ibis *Plegadis falcinellus* (IB)	Britain	28	+733%	<1%	<0.1	20								
	Ireland	4	+300%			3								
Spoonbill *Platalea leucorodia* (NB)	Britain	105	+669%	1%	<0.1	233	4	1	2	∞	+133%			
	Ireland	10	+900%			11								
Greater Flamingo *Phoenicopterus ruber* (FL)	Britain	1	-92%			7								
	Ireland													
Chilean Flamingo *Phoenicopterus chilensis* (FM)	Britain					2								
	Ireland													
Lesser Flamingo *Phoeniconaias minor* (FK)	Britain					1								
	Ireland													
Pied-billed Grebe *Podilymbus podiceps* (PJ)	Britain	1	∞			1								
	Ireland	4	∞											
Little Grebe *Tachybaptus ruficollis* (LG)	Britain	1855	+27%	9%	0.2	106	234	267	1188	+23%	+31%	+0.08 (+0.04,+0.12)	8%	0.1
	Ireland	525	+24%	12%	0.2	16	109	110	251	-9%	+39%	+0.15 (+0.11,+0.20)	12%	0.2
Great Crested Grebe *Podiceps cristatus* (GG)	Britain	1170	+23%	8%	0.3	179	70	134	797	+31%	+12%	-0.04 (-0.09,0.00)	10%	0.3
	Ireland	323	+19%	13%	0.3	31	30	90	124	+9%	+8%	-0.07 (-0.12,-0.03)	18%	0.6
Red-necked Grebe *Podiceps grisegena* (RX)	Britain	239	+8%	<1%	<0.1	60	2			∞	-75%			
	Ireland	21	+350%	2%	<0.1	1					-100%			
Slavonian Grebe *Podiceps auritus* (SZ)	Britain	502	+65%	3%	<0.1	159	2	6	8	-16%	-33%		6%	0.2
	Ireland	51	+118%	1%	0.1	7				-100%				
Black-necked Grebe *Podiceps nigricollis* (BN)	Britain	205	+69%	1%	<0.1	111	12	18	23	+382%	+51%		1%	<0.1
	Ireland	10	+80%			1			1	0%	∞			
Honey-buzzard *Pernis apivorus* (HZ)	Britain					186	26	17	38	+555%	+167%		1%	<0.1
	Ireland													
Black Kite *Milvus migrans* (KB)	Britain	3	∞			90								
	Ireland					2								
Red Kite *Milvus milvus* (KT)	Britain	988	+1250%	12%	0.2	840	298	134	281	+1971%	+728%	+0.91 (+0.90,+0.92)	17%	0.2
	Ireland	33	+967%	6%	0.1	11	10	1	11	∞	∞		9%	0.1
White-tailed Eagle *Haliaeetus albicilla* (WE)	Britain	291	+1429%	3%	<0.1	176	23	11	60	∞	+318%	+0.81 (+0.69,+0.92)	8%	0.1
	Ireland	33	∞	2%	<0.1	27	3	2		∞	∞		6%	<0.1
Marsh Harrier *Circus aeruginosus* (MR)	Britain	347	+529%	7%	0.1	472	64	50	134	+876%	+114%	+0.61 (+0.53,+0.68)	13%	0.1
	Ireland	13	+300%	2%	<0.1	48	2		1	+200%	-57%			
Hen Harrier *Circus cyaneus* (HH)	Britain	1296	+24%	5%	<0.1	402	169	80	232	+29%	-3%	-0.18 (-0.24,-0.13)	7%	0.1
	Ireland	441	+100%	4%	<0.1	66	54	39	85	-7%	+45%	-0.25 (-0.30,-0.20)	6%	<0.1
Pallid Harrier *Circus macrourus*	Britain	1	∞			1								
	Ireland					1								
Montagu's Harrier *Circus pygargus* (MO)	Britain					121	16	6	14	-28%	+13%		1%	<0.1
	Ireland					8								
Goshawk *Accipiter gentilis* (GI)	Britain	517	+344%	2%	<0.1	135	114	144	256	+1291%	+106%	+0.41 (+0.32,+0.51)	2%	<0.1
	Ireland	5	∞			3	11	5	3	∞	+1800%			
Sparrowhawk *Accipiter nisus* (SH)	Britain	2481	+17%	19%	0.1	104	515	370	1469	+29%	+7%	+0.06 (+0.04,+0.09)	14%	0.1
	Ireland	858	+25%	18%	0.1	18	368	133	287	-2%	+19%	+0.12 (+0.09,+0.15)	17%	0.1
Harris's Hawk *Parabuteo unicinctus* (HA)	Britain	59	∞	1%	<0.1	36		4	2	∞	∞			
	Ireland													
Red-tailed Hawk *Buteo jamaicensis* (JZ)	Britain	6	∞	1%	<0.1	8		1		∞	∞		4%	<0.1
	Ireland													

Species name *scientific name and two-letter code*	Island	Winter				Breeding season								
		Range size: *present* (a)	Range change (since 1981–84) (b)	Mean % of tetrads occupied per 10-km (c)	Mean birds per hour per tetrad (d)	Range size: *non-breeding* (e)	Range size: *possible* (f)	Range size: *probable* (f)	Range size: *confirmed* (f)	Range change (since 1968–72) (g)	Range change (since 1988–91) (g)	Fixed-effort change index (since 1988–91) (h)	Mean % of tetrads occupied per 10-km (c)	Mean birds per hour per tetrad (d)
Buzzard *Buteo buteo* (BZ)	Britain	2661	+74%	53%	0.7	85	135	415	2033	+81%	+67%	+0.50 (+0.49,+0.50)	52%	0.6
	Ireland	551	+885%	27%	0.2	37	136	185	206	+1976%	+458%	+0.85 (+0.82,+0.87)	26%	0.3
Rough-legged Buzzard *Buteo lagopus* (RF)	Britain	125	+15%	<1%	<0.1	49					-100%			
	Ireland													
Golden Eagle *Aquila chrysaetos* (EA)	Britain	416	+11%	11%	0.1	81	85	88	232	+4%	-1%	-0.17 (-0.23,-0.12)	9%	0.1
	Ireland	10	+900%	3%	<0.1	13	3	1	4	∞	∞			
Osprey *Pandion haliaetus* (OP)	Britain	22	∞			832	90	23	173	+1032%	+68%	+0.52 (+0.46,+0.58)	6%	0.1
	Ireland	2				28					-100%			
Kestrel *Falco tinnunculus* (K.)	Britain	2578	+4%	33%	0.3	95	356	340	1768	-6%	-1%	-0.10 (-0.11,-0.08)	27%	0.2
	Ireland	904	+19%	24%	0.2	16	352	181	361	-6%	+11%	+0.05 (+0.02,+0.08)	20%	0.1
American Kestrel *Falco sparverius*	Britain	1												
	Ireland													
Red-footed Falcon *Falco vespertinus* (FV)	Britain					59					-100%			
	Ireland					4					-100%			
Merlin *Falco columbarius* (ML)	Britain	1572	+46%	4%	<0.1	461	191	94	378	+11%	-4%	-0.18 (-0.24,-0.12)	4%	<0.1
	Ireland	400	+80%	7%	<0.1	75	80	25	43	-40%	-6%	-0.09 (-0.18,0.00)	5%	<0.1
Hobby *Falco subbuteo* (HY)	Britain	12	∞	1%	<0.1	321	354	216	463	+295%	+64%	+0.29 (+0.23,+0.35)	4%	<0.1
	Ireland	1				42					-100%			
Lanner *Falco biarmicus* (FB)	Britain	9	∞			5								
	Ireland													
Saker Falcon *Falco cherrug* (JF)	Britain	5	∞			5								
	Ireland													
Gyr Falcon *Falco rusticolus* (YF)	Britain	8	+100%			6								
	Ireland	1	0%			1								
Peregrine *Falco peregrinus* (PE)	Britain	2158	+105%	6%	<0.1	610	280	214	974	+184%	+39%	-0.05 (-0.10,0.00)	5%	<0.1
	Ireland	445	+51%	9%	0.1	82	130	63	217	+276%	+41%	+0.03 (-0.05,+0.12)	8%	0.1
Water Rail *Rallus aquaticus* (WA)	Britain	1290	+53%	4%	<0.1	79	253	184	245	+6%	+59%	-0.04 (-0.15,+0.06)	2%	<0.1
	Ireland	287	+59%	9%	0.1	6	138	53	44	-17%	+31%	-0.07 (-0.15,0.00)	7%	0.1
Spotted Crake *Porzana porzana* (AK)	Britain	5	-73%			10	22	17	2	+5%	+58%		1%	<0.1
	Ireland						2			∞	+100%			
Little Crake *Porzana parva* (JC)	Britain		-100%			2								
	Ireland													
Corncrake *Crex crex* (CE)	Britain	1	∞			21	48	120	16	-72%	+14%	+0.04 (-0.13,+0.23)	6%	0.1
	Ireland					6	29	39	9	-91%	-69%	-0.59 (-0.81,-0.42)	4%	<0.1
Moorhen *Gallinula chloropus* (MH)	Britain	2031	+3%	31%	1	35	140	97	1842	-8%	+1%	-0.05 (-0.06,-0.03)	31%	0.6
	Ireland	711	-11%	21%	0.4	4	178	110	442	-19%	+2%	-0.06 (-0.09,-0.03)	21%	0.3
Purple Swamphen *Porphyrio porphyrio*	Britain					1								
	Ireland													
Coot *Fulica atra* (CO)	Britain	1679	+3%	15%	2.4	53	98	72	1465	-4%	+1%	-0.06 (-0.08,-0.03)	18%	1
	Ireland	383	-18%	12%	1.3	19	74	56	230	-36%	+1%	-0.18 (-0.22,-0.14)	14%	0.5
American Coot *Fulica americana* (AO)	Britain													
	Ireland	1	∞			1								
Crowned Crane *Balearica regulorum/pavonina* (KC)	Britain					1								
	Ireland													
Crane *Grus grus* (AN)	Britain	77	+492%	1%	<0.1	219	3	9	9	∞	+950%		3%	0.1
	Ireland	5	∞			2								
Great Bustard *Otis tarda* (US)	Britain	20	∞			10			1	∞	∞			
	Ireland													
Oystercatcher *Haematopus ostralegus* (OC)	Britain	1687	+40%	23%	9	243	117	417	1442	+28%	+15%	+0.11 (+0.09,+0.12)	37%	2.5
	Ireland	352	-5%	40%	7	102	65	76	106	-5%	-6%	+0.07 (-0.01,+0.16)	37%	2.9
Black-winged Stilt *Himantopus himantopus* (IT)	Britain					24		1	1	∞	∞			
	Ireland					2								

Species name *scientific name and two-letter code*	Island	Winter: Range size: *present* (a)	Winter: Range change (since 1981–84) (b)	Winter: Mean % of tetrads occupied per 10-km (c)	Winter: Mean birds per hour per tetrad (d)	Breeding season: Range size: *non-breeding* (e)	Breeding season: Range size: *possible* (f)	Breeding season: Range size: *probable* (f)	Breeding season: Range size: *confirmed* (f)	Breeding season: Range change (since 1968–72) (g)	Breeding season: Range change (since 1988–91) (g)	Breeding season: Fixed-effort change index (since 1988–91) (h)	Breeding season: Mean % of tetrads occupied per 10-km (c)	Breeding season: Mean birds per hour per tetrad (d)
Avocet	Britain	153	+184%	4%	1.6	158	10	16	115	+1663%	+404%	+0.79 (+0.68,+0.88)	7%	1
Recurvirostra avosetta (AV)	Ireland	1	-75%			1								
Stone-curlew	Britain	14	+233%	1%	<0.1	54	4	1	49	-42%	0%	+0.43 (+0.21,+0.63)	6%	0.1
Burhinus oedicnemus (TN)	Ireland					1								
Collared Pratincole	Britain					7								
Glareola pratincola (KM)	Ireland													
Oriental Pratincole	Britain					3								
Glareola maldivarum (GM)	Ireland													
Black-winged Pratincole	Britain					3								
Glareola nordmanni (KW)	Ireland													
Little Ringed Plover	Britain	5	∞			178	65	123	409	+107%	+42%	-0.10 (-0.19,-0.01)	3%	<0.1
Charadrius dubius (LP)	Ireland					4		1	2	∞	+200%			
Ringed Plover	Britain	827	+2%	12%	1	279	121	217	624	+5%	-7%	-0.23 (-0.27,-0.19)	15%	0.6
Charadrius hiaticula (RP)	Ireland	253	-2%	18%	1.9	28	27	62	152	-23%	-2%	-0.14 (-0.26,-0.04)	16%	0.6
Killdeer	Britain	2	-71%			3								
Charadrius vociferus (KL)	Ireland	1	0%											
Kentish Plover	Britain	1	-75%			22				-100%				
Charadrius alexandrinus (KP)	Ireland	1	∞			1								
Greater Sand Plover	Britain					1								
Charadrius leschenaultii (DP)	Ireland													
Dotterel	Britain	8	∞			162	15	32	35	+78%	-17%	-0.16 (-0.28,-0.03)	4%	0.1
Charadrius morinellus (DO)	Ireland	1	∞			4								
American Golden Plover	Britain	22	∞	<1%	<0.1	12								
Pluvialis dominica (ID)	Ireland	13	∞			3								
Pacific Golden Plover	Britain					5								
Pluvialis fulva (IF)	Ireland					2								
Golden Plover	Britain	1889	+19%	10%	7.9	736	106	227	349	-20%	-13%	-0.13 (-0.16,-0.10)	22%	0.6
Pluvialis apricaria (GP)	Ireland	520	+8%	12%	11.9	226	5	12	16	-50%	-42%	-0.27 (-0.38,-0.18)	9%	0.5
Grey Plover	Britain	514	+3%	9%	2.3	361					-100%			
Pluvialis squatarola (GV)	Ireland	154	+1%	13%	0.6	28								
Sociable Plover	Britain	1	∞											
Vanellus gregarius (IP)	Ireland													
White-tailed Plover	Britain					4								
Vanellus leucurus	Ireland													
Lapwing	Britain	2309	+1%	20%	10.5	114	105	325	1697	-17%	-9%	-0.21 (-0.23,-0.19)	28%	1.6
Vanellus vanellus (L.)	Ireland	730	-12%	24%	9.8	43	55	129	146	-53%	-33%	-0.45 (-0.49,-0.42)	13%	0.6
Blacksmith Lapwing	Britain		-100%			1								
Vanellus armatus (KS)	Ireland													
Knot	Britain	495	+27%	8%	26.3	383					-100%			
Calidris canutus (KN)	Ireland	123	+58%	9%	15.1	40								
Sanderling	Britain	407	+31%	10%	2.8	528					-100%			
Calidris alba (SS)	Ireland	137	+21%	15%	3.4	67								
Semipalmated Sandpiper	Britain					3								
Calidris pusilla (PZ)	Ireland	1				4								
Little Stint	Britain	100	+80%	1%	<0.1	184					-100%			
Calidris minuta (LX)	Ireland	8	+50%			5								
Temminck's Stint	Britain	1	-50%			131	1			-75%	-67%			
Calidris temminckii (TK)	Ireland					2								
Least Sandpiper	Britain					1								
Calidris minutilla (EP)	Ireland													
White-rumped Sandpiper	Britain	5	∞			10								
Calidris fuscicollis (WU)	Ireland	3	∞			1								

Species name *scientific name and two-letter code*	Island	Winter				Breeding season								
		Range size: *present* (a)	Range change (since 1981–84) (b)	Mean % of tetrads occupied per 10-km (c)	Mean birds per hour per tetrad (d)	Range size: *non-breeding* (e)	Range size: *possible* (f)	Range size: *probable* (f)	Range size: *confirmed* (f)	Range change (since 1968–72) (g)	Range change (since 1988–91) (g)	Fixed-effort change index (since 1988–91) (h)	Mean % of tetrads occupied per 10-km (c)	Mean birds per hour per tetrad (d)
Baird's Sandpiper *Calidris bairdii* (BP)	Britain	2	0%											
	Ireland													
Pectoral Sandpiper *Calidris melanotos* (PP)	Britain	6	∞	2%	<0.1	66								
	Ireland	2	∞			7								
Curlew Sandpiper *Calidris ferruginea* (CV)	Britain	46	+275%			161								
	Ireland	11	+75%			18								
Stilt Sandpiper *Calidris himantopus* (MI)	Britain					3								
	Ireland					2								
Purple Sandpiper *Calidris maritima* (PS)	Britain	499	+13%	9%	0.9	235			2	∞	-33%			
	Ireland	113	-10%	8%	0.2	36								
Dunlin *Calidris alpina* (DN)	Britain	883	+1%	9%	13.2	815	98	149	175	-11%	-26%	-0.20 (-0.25,-0.15)	17%	1.6
	Ireland	261	+3%	15%	9	141	5	8	7	-69%	-71%	-0.81 (-1.00,-0.75)	12%	14.5
Broad-billed Sandpiper *Calidris falcinellus* (OA)	Britain					12								
	Ireland					1								
Buff-breasted Sandpiper *Calidris subruficollis* (BQ)	Britain					13	1			∞	∞			
	Ireland					2								
Ruff *Calidris pugnax* (RU)	Britain	271	+14%	2%	0.1	274	8	8		+14%	-62%	-1.00 (-1.00,-0.75)	<1%	<0.1
	Ireland	32	+12%			22								
Jack Snipe *Lymnocryptes minimus* (JS)	Britain	1165	+71%	2%	<0.1	206					-100%			
	Ireland	203	+29%	5%	<0.1	9								
Snipe *Gallinago gallinago* (SN)	Britain	2496	+16%	16%	0.4	445	402	793	347	-31%	-15%	-0.21 (-0.24,-0.19)	17%	0.3
	Ireland	920	+6%	30%	0.7	44	230	308	53	-34%	-8%	-0.19 (-0.22,-0.16)	17%	0.2
Wilson's Snipe *Gallinago delicata*	Britain	1												
	Ireland													
Great Snipe *Gallinago media* (DS)	Britain					3								
	Ireland													
Long-billed Dowitcher *Limnodromus scolopaceus* (LD)	Britain	11	+175%			4								
	Ireland	7	+500%			2								
Woodcock *Scolopax rusticola* (WK)	Britain	2555	+42%	10%	0.1	157	220	515	117	-50%	-30%	-0.55 (-0.60,-0.51)	2%	<0.1
	Ireland	578	+20%	9%	0.1	12	55	74	11	-73%	-23%	-0.43 (-0.50,-0.35)	3%	<0.1
Black-tailed Godwit *Limosa limosa* (BW)	Britain	450	+177%	5%	2	605	9	20	11	-7%	-32%	-0.04 (-0.29,+0.24)	8%	2.1
	Ireland	159	+55%	16%	5.4	116			1	-83%	-89%		13%	0.1
Bar-tailed Godwit *Limosa lapponica* (BA)	Britain	522	+26%	7%	3.6	540					-100%			
	Ireland	183	+2%	12%	3.7	69					-100%			
Whimbrel *Numenius phaeopus* (WM)	Britain	104	+291%	2%	<0.1	1119	10	11	21	-29%	-49%	-0.41 (-0.56,-0.26)	21%	0.3
	Ireland	45	+111%	2%	<0.1	311								
Curlew *Numenius arquata* (CU)	Britain	1697	+15%	24%	4	611	305	608	711	-17%	-14%	-0.19 (-0.21,-0.17)	33%	1
	Ireland	609	-23%	31%	4.2	317	77	63	41	-78%	-73%	-0.73 (-0.75,-0.71)	15%	0.3
Upland Sandpiper *Bartramia longicauda* (UP)	Britain		-100%			1								
	Ireland													
Terek Sandpiper *Xenus cinereus* (TR)	Britain					5								
	Ireland					3								
Common Sandpiper *Actitis hypoleucos* (CS)	Britain	289	+35%	1%	<0.1	892	147	387	677	-14%	-15%	-0.21 (-0.24,-0.19)	22%	0.3
	Ireland	55	+23%	4%	<0.1	121	83	107	77	-41%	-15%	-0.40 (-0.46,-0.35)	10%	0.1
Spotted Sandpiper *Actitis macularius* (PQ)	Britain	8	∞			10								
	Ireland	2	0%			2								
Green Sandpiper *Tringa ochropus* (GE)	Britain	884	+56%	2%	<0.1	878	1		3	+100%	-43%		4%	<0.1
	Ireland	81	+63%	5%	<0.1	44				-100%				
Solitary Sandpiper *Tringa solitaria* (I.)	Britain					2								
	Ireland													
Spotted Redshank *Tringa erythropus* (DR)	Britain	192	+40%	1%	<0.1	236					-100%			
	Ireland	33	-16%	1%	<0.1	25								

Species name scientific name and two-letter code	Island	Winter				Breeding season								
		Range size: *present* (a)	Range change (since 1981–84) (b)	Mean % of tetrads occupied per 10-km (c)	Mean birds per hour per tetrad (d)	Range size: *non-breeding* (e)	Range size: *possible* (f)	Range size: *probable* (f)	Range size: *confirmed* (f)	Range change (since 1968–72) (g)	Range change (since 1988–91) (g)	Fixed-effort change index (since 1988–91) (h)	Mean % of tetrads occupied per 10-km (c)	Mean birds per hour per tetrad (d)
Greenshank *Tringa nebularia* (GK)	Britain	405	+48%	5%	0.1	747	38	111	108	+2%	+6%	+0.03 (-0.02,+0.09)	18%	0.2
	Ireland	264	+13%	18%	0.3	123				-100%	-100%			
Lesser Yellowlegs *Tringa flavipes* (LY)	Britain	8	+25%			20								
	Ireland	4	-33%			2								
Marsh Sandpiper *Tringa stagnatilis* (MD)	Britain					2								
	Ireland													
Wood Sandpiper *Tringa glareola* (OD)	Britain	2	-100%			404	5	7	10	+22%	+175%		4%	<0.1
	Ireland					9								
Redshank *Tringa totanus* (RK)	Britain	1299	+4%	22%	3.4	487	171	320	469	-43%	-35%	-0.42 (-0.45,-0.39)	18%	0.8
	Ireland	407	-1%	26%	3.1	144	33	41	39	-55%	-47%	-0.62 (-0.68,-0.56)	12%	0.7
Turnstone *Arenaria interpres* (TT)	Britain	795	+12%	20%	2.3	660					-100%			
	Ireland	275	+2%	20%	1.4	124					-100%			
Wilson's Phalarope *Phalaropus tricolor* (WF)	Britain	2	∞			1								
	Ireland	2	∞											
Red-necked Phalarope *Phalaropus lobatus* (NK)	Britain	5	∞			50	3	2	9	-22%	+56%		4%	<0.1
	Ireland					5	1	1		-50%	+100%			
Grey Phalarope *Phalaropus fulicarius* (PL)	Britain	176	+197%	<1%	<0.1	9								
	Ireland	27	+850%											
Pomarine Skua *Stercorarius pomarinus* (PK)	Britain	190	+917%	<1%	<0.1	150								
	Ireland	7	0%			19								
Arctic Skua *Stercorarius parasiticus* (AC)	Britain	107	+65%	1%	<0.1	345	12	33	74	+7%	-6%	-0.36 (-0.45,-0.27)	29%	0.8
	Ireland	11	-50%			55								
Long-tailed Skua *Stercorarius longicaudus* (OG)	Britain	10	+300%			50		1		∞	∞		14%	0.1
	Ireland					9								
Great Skua *Stercorarius skua* (NX)	Britain	175	+138%	<1%	<0.1	271	21	43	96	+97%	+42%	+0.01 (-0.08,+0.10)	43%	4.7
	Ireland	17	+150%			39	1	4	7	∞	∞		21%	0.3
Ivory Gull *Pagophila eburnea* (IV)	Britain	3	+50%											
	Ireland													
Sabine's Gull *Xema sabini* (AB)	Britain	32	+67%			12								
	Ireland	6	+300%			9								
Kittiwake *Rissa tridactyla* (KI)	Britain	564	-31%	5%	0.5	409	18	7	210	-10%	-14%	-0.33 (-0.45,-0.22)	15%	17.1
	Ireland	156	-26%	10%	1	76	12	8	62	-8%	+15%	-0.21 (-0.50,+0.12)	24%	39.4
Bonaparte's Gull *Chroicocephalus philadelphia* (ON)	Britain	7	∞			19								
	Ireland	5	+300%	3%	<0.1	4								
Black-headed Gull *Chroicocephalus ridibundus* (BH)	Britain	2238	-4%	44%	15	1287	264	135	599	-11%	+9%	-0.37 (-0.38,-0.35)	34%	6.1
	Ireland	715	-19%	35%	8.9	384	51	24	103	-55%	-12%	-0.65 (-0.67,-0.63)	36%	6
Little Gull *Hydrocoloeus minutus* (LU)	Britain	337	+52%	<1%	<0.1	418	1	2		∞	∞			
	Ireland	57	+13%	2%	0.1	26								
Ross's Gull *Rhodostethia rosea* (QG)	Britain	1	-100%			1								
	Ireland		-100%											
Laughing Gull *Larus atricilla* (LF)	Britain	3	-57%			7								
	Ireland		-100%			2								
Franklin's Gull *Larus pipixcan* (FG)	Britain	5	0%			11								
	Ireland													
Mediterranean Gull *Larus melanocephalus* (MU)	Britain	687	+356%	2%	0.1	452	24	22	44	+6500%	+843%	+0.59 (+0.39,+0.80)	5%	0.1
	Ireland	130	+638%	1%	<0.1	57	2	1	8	∞	∞		4%	0.1
Audouin's Gull *Larus audouinii*	Britain					1								
	Ireland													
Common Gull *Larus canus* (CM)	Britain	2325	+3%	32%	6.5	1304	144	119	496	-7%	+7%	-0.24 (-0.26,-0.22)	38%	4.2
	Ireland	510	-5%	32%	4.2	261	27	36	102	-3%	+59%	-0.41 (-0.45,-0.37)	29%	2.2
Ring-billed Gull *Larus delawarensis* (IN)	Britain	75	+145%	<1%	<0.1	20		1		∞	∞			
	Ireland	67	+230%	2%	<0.1	26								

Species name *scientific name and two-letter code*	Island	Winter				Breeding season								
		Range size: *present* (a)	Range change (since 1981–84) (b)	Mean % of tetrads occupied per 10-km (c)	Mean birds per hour per tetrad (d)	Range size: *non-breeding* (e)	Range size: *possible* (f)	Range size: *probable* (f)	Range size: *confirmed* (f)	Range change (since 1968–72) (g)	Range change (since 1988–91) (g)	Fixed-effort change index (since 1988–91) (h)	Mean % of tetrads occupied per 10-km (c)	Mean birds per hour per tetrad (d)
Lesser Black-backed Gull	Britain	1838	+22%	15%	1.3	1455	256	167	556	+65%	+63%	-0.11 (-0.14,-0.09)	30%	3.3
Larus fuscus (LB)	Ireland	389	+55%	13%	0.7	552	50	29	93	+3%	+53%	-0.44 (-0.48,-0.39)	36%	4.3
Herring Gull	Britain	2420	+3%	40%	9.8	1190	178	144	885	+38%	+40%	-0.02 (-0.03,0.00)	52%	11.5
Larus argentatus (HG)	Ireland	478	-38%	39%	5.4	299	47	27	135	-28%	0%	-0.40 (-0.44,-0.36)	57%	5
Yellow-legged Gull	Britain	421	∞	1%	<0.1	307	3	2	1	∞	∞		2%	0.1
Larus michahellis (YG)	Ireland	32	∞	1%	<0.1	13		1		∞	∞			
Caspian Gull	Britain	169	∞			46								
Larus cachinnans (YC)	Ireland													
American Herring Gull	Britain	1	∞											
Larus smithsonianus	Ireland	10	∞			2								
Iceland Gull	Britain	471	+61%	1%	<0.1	231								
Larus glaucoides (IG)	Ireland	121	+28%	2%	<0.1	61								
Thayer's Gull	Britain													
Larus thayeri	Ireland	3	∞											
Glaucous-winged Gull	Britain	2	∞											
Larus glaucescens (JG)	Ireland													
Glaucous Gull	Britain	452	+9%	1%	<0.1	150								
Larus hyperboreus (GZ)	Ireland	102	-23%	3%	<0.1	41								
Great Black-backed Gull	Britain	1982	-5%	24%	1.5	1102	138	116	421	+5%	+9%	-0.16 (-0.20,-0.13)	39%	1.9
Larus marinus (GB)	Ireland	445	-20%	33%	1.7	270	37	24	109	-30%	-1%	-0.29 (-0.36,-0.23)	49%	3.5
Bridled Tern	Britain					1								
Onychoprion anaethetus	Ireland													
Little Tern	Britain					252	8	12	82	-30%	-14%	-0.50 (-0.62,-0.37)	11%	0.6
Sternula albifrons (AF)	Ireland					24	1	4	17	-62%	-44%		13%	3.3
Gull-billed Tern	Britain					7								
Gelochelidon nilotica (TG)	Ireland													
Caspian Tern	Britain					8								
Hydroprogne caspia (CJ)	Ireland													
Whiskered Tern	Britain					39								
Chlidonias hybrida (WD)	Ireland					2								
Black Tern	Britain	4	∞			397				-100%				
Chlidonias niger (BJ)	Ireland	1				6								
White-winged Black Tern	Britain					18					-100%			
Chlidonias leucopterus (WJ)	Ireland					2								
Sandwich Tern	Britain	75	+120%	1%	<0.1	620	18	13	37	-14%	+11%	-0.51 (-0.61,-0.40)	17%	2.1
Sterna sandvicensis (TE)	Ireland	18	+1400%	1%	<0.1	201	6	2	19	-38%	-43%	-0.74 (-0.85,-0.63)	28%	7.6
Royal Tern	Britain													
Sterna maxima (QT)	Ireland					1								
Lesser Crested Tern	Britain					1					-100%			
Sterna bengalensis (TF)	Ireland													
Forster's Tern	Britain													
Sterna forsteri (FO)	Ireland	8	+500%			3								
Common Tern	Britain	15	0%	<1%	<0.1	573	119	92	430	+13%	+21%	-0.10 (-0.15,-0.06)	11%	0.4
Sterna hirundo (CN)	Ireland	2	∞			120	23	11	81	-47%	-6%	-0.27 (-0.36,-0.18)	15%	1.4
Roseate Tern	Britain					99	2		10	-57%	-47%		1%	<0.1
Sterna dougallii (RS)	Ireland					18	1		5	-77%	-55%		13%	0.2
Arctic Tern	Britain	54	+1400%	<1%	<0.1	470	32	29	222	-29%	-16%	-0.10 (-0.18,-0.03)	27%	3.9
Sterna paradisaea (AE)	Ireland	2				47	9	3	52	-38%	-24%	-0.48 (-0.67,-0.29)	10%	2
Guillemot	Britain	531	-21%	6%	11	402	34	7	183	-17%	-13%	-0.12 (-0.26,+0.01)	17%	30
Uria aalge (GU)	Ireland	172	+7%	10%	2.8	105	10	2	54	-23%	-3%	-0.53 (-0.81,-0.24)	23%	233.7
Razorbill	Britain	457	-10%	5%	0.8	344	33	7	199	-17%	-14%	+0.07 (-0.07,+0.20)	19%	4
Alca torda (RA)	Ireland	172	0%	10%	0.3	113	4	3	55	-28%	-5%	-0.60 (-0.82,-0.33)	16%	24.5

Species name *scientific name and two-letter code*	Island	Winter				Breeding season									
		Range size: *present* (a)	Range change (since 1981–84) (b)	Mean % of tetrads occupied per 10-km (c)	Mean birds per hour per tetrad (d)	Range size: *non-breeding* (e)	Range size: *possible* (f)	Range size: *probable* (f)	Range size: *confirmed* (f)	Range change (since 1968–72) (g)	Range change (since 1988–91) (g)	Fixed-effort change index (since 1988–91) (h)	Mean % of tetrads occupied per 10-km (c)	Mean birds per hour per tetrad (d)
Black Guillemot *Cepphus grylle* (TY)	Britain	372	+2%	15%	0.4	82	90	107	163	-9%	-29%	-0.01 (-0.10,+0.08)	18%	0.8
	Ireland	187	+8%	11%	0.2	39	48	50	79	+2%	+42%	-0.03 (-0.23,+0.19)	13%	0.5
Little Auk *Alle alle* (LK)	Britain	314	-<1%	1%	<0.1	3								
	Ireland	18	-41%											
Puffin *Fratercula arctica* (PU)	Britain	74	-52%	1%	<0.1	188	22	15	120	-27%	-13%	-0.13 (-0.34,+0.09)	14%	6.1
	Ireland	3	-73%			36	3	3	18	-33%	-17%		23%	11.7
Rock Dove/Feral Pigeon *Columba livia* (DV)	Britain	2332	+21%	28%	3.7	156	408	675	1150	+48%	+6%	+0.05 (+0.03,+0.06)	28%	2.1
	Ireland	573	+57%	27%	1.7	36	171	191	213	+42%	+66%	+0.40 (+0.36,+0.44)	26%	1.3
Stock Dove *Columba oenas* (SD)	Britain	1878	+10%	24%	0.8	49	157	524	1247	-2%	+5%	+0.06 (+0.05,+0.08)	33%	0.6
	Ireland	231	-14%	17%	0.4	18	100	115	74	-47%	-21%	-0.18 (-0.22,-0.15)	16%	0.2
Woodpigeon *Columba palumbus* (WP)	Britain	2491	+3%	75%	34.8	60	140	325	2127	+<1%	+2%	+0.02 (+0.01,+0.02)	82%	11.7
	Ireland	939	-<1%	78%	11.3	3	59	318	586	+1%	+2%	+0.02 (+0.01,+0.04)	91%	6.7
Collared Dove *Streptopelia decaocto* (CD)	Britain	2310	+21%	44%	1.7	58	143	604	1649	+15%	+8%	+0.11 (+0.10,+0.12)	47%	1.5
	Ireland	781	+70%	32%	0.8	17	143	358	352	+48%	+52%	+0.40 (+0.37,+0.42)	36%	0.7
Turtle Dove *Streptopelia turtur* (TD)	Britain	15	+225%	1%	<0.1	230	196	272	154	-51%	-34%	-0.59 (-0.61,-0.56)	10%	0.1
	Ireland	4	∞			26	2			-94%	-90%			
Rufous Turtle Dove *Streptopelia orientalis*	Britain	1	∞			2								
	Ireland													
Mourning Dove *Zenaida macroura*	Britain	1												
	Ireland	1												
Sulphur-crested Cockatoo *Cacatua galerita* (OK)	Britain	1	∞			3								
	Ireland													
Cockatiel *Nymphicus hollandicus* (QL)	Britain	10	∞			35	2			∞	∞			
	Ireland													
Red-rumped Parrot *Psephotus haematonotus*	Britain					1								
	Ireland													
Budgerigar *Melopsittacus undulatus* (UG)	Britain	6	∞			15				-100%	-100%			
	Ireland					1								
Alexandrine Parakeet *Psittacula eupatria* (AX)	Britain	1	∞			2			1	∞	∞			
	Ireland													
Ring-necked Parakeet *Psittacula krameri* (RI)	Britain	239	+214%	10%	0.5	121	24	21	46	+4400%	+43%	+0.79 (+0.68,+0.88)	24%	1
	Ireland	1	∞			1								
Peach-faced Lovebird *Agapornis roseicollis* (LV)	Britain	2	∞			2					-100%			
	Ireland													
Grey Parrot *Psittacus erithacus*	Britain	1	∞			1								
	Ireland													
Blue-crowned Parakeet *Aratinga acuticaudata*	Britain	1	∞			1								
	Ireland													
Monk Parakeet *Myiopsitta monachus*	Britain	4	∞	2%	<0.1	2	1		2	∞	+50%		1%	<0.1
	Ireland													
White-cheeked Turaco *Tauraco leucotis*	Britain	1	∞				1			∞	∞			
	Ireland													
Violet Turaco *Musophaga violacea*	Britain	1	∞											
	Ireland													
Great Spotted Cuckoo *Clamator glandarius* (UK)	Britain		-100%			2								
	Ireland	1	∞											
Cuckoo *Cuculus canorus* (CK)	Britain					136	955	1009	405	-7%	-2%	-0.22 (-0.23,-0.20)	20%	0.2
	Ireland					21	462	198	42	-26%	-1%	-0.12 (-0.15,-0.09)	19%	0.1
Barn Owl *Tyto alba* (BO)	Britain	2007	+71%	3%	<0.1	88	213	103	1586	+5%	+67%	+0.26 (+0.22,+0.31)	4%	<0.1
	Ireland	252	+79%	<1%	<0.1	18	119	25	124	-47%	+43%	-0.44 (-0.53,-0.36)	<1%	<0.1
Snowy Owl *Bubo scandiacus* (SO)	Britain	13	+500%			14				-100%	-100%			
	Ireland	5	∞			9								

Species name *scientific name and two-letter code*	Island	Winter: Range size: *present* (a)	Winter: Range change (since 1981–84) (b)	Winter: Mean % of tetrads occupied per 10-km (c)	Winter: Mean birds per hour per tetrad (d)	Breeding season: Range size: *non-breeding* (e)	Breeding season: Range size: *possible* (f)	Breeding season: Range size: *probable* (f)	Breeding season: Range size: *confirmed* (f)	Breeding season: Range change (since 1968–72) (g)	Breeding season: Range change (since 1988–91) (g)	Breeding season: Fixed-effort change index (since 1988–91) (h)	Breeding season: Mean % of tetrads occupied per 10-km (c)	Breeding season: Mean birds per hour per tetrad (d)
Eurasian Eagle-Owl *Bubo bubo* (EO)	Britain	16	∞			14	5	1	2	∞	+700%		1%	<0.1
	Ireland													
Little Owl *Athene noctua* (LO)	Britain	1181	+13%	4%	<0.1	48	181	222	835	-11%	+<1%	-0.22 (-0.26,-0.17)	5%	<0.1
	Ireland		-100%											
Tawny Owl *Strix aluco* (TO)	Britain	2172	+25%	2%	<0.1	50	270	288	1625	-6%	+6%	-0.33 (-0.36,-0.30)	4%	<0.1
	Ireland													
Long-eared Owl *Asio otus* (LE)	Britain	444	+50%	1%	<0.1	115	139	44	314	-19%	+8%	-0.45 (-0.58,-0.33)	1%	<0.1
	Ireland	140	+52%	2%	<0.1	11	72	37	294	+12%	+70%	-0.44 (-0.54,-0.35)	1%	<0.1
Short-eared Owl *Asio flammeus* (SE)	Britain	871	-19%	1%	<0.1	335	169	81	164	-48%	-39%	-0.63 (-0.69,-0.58)	4%	<0.1
	Ireland	54	0%	2%	<0.1	29	7	1	4	+1100%	+9%		2%	<0.1
Nightjar *Caprimulgus europaeus* (NJ)	Britain					68	54	161	108	-43%	+18%	-0.63 (-0.77,-0.49)	<1%	<0.1
	Ireland					2	2	2	1	-95%	-55%			
Common Nighthawk *Chordeiles minor*	Britain													
	Ireland	1	∞											
Swift *Apus apus* (SI)	Britain	4	-100%			268	139	207	1790	-3%	-4%	-0.07 (-0.08,-0.06)	38%	1.7
	Ireland					110	171	167	309	-26%	-13%	-0.13 (-0.16,-0.11)	26%	0.8
Pallid Swift *Apus pallidus*	Britain					1	1			∞	∞			
	Ireland													
Pacific Swift *Apus pacificus*	Britain					1								
	Ireland													
Alpine Swift *Apus melba* (AI)	Britain					43								
	Ireland					1								
Little Swift *Apus affinis*	Britain					1								
	Ireland													
Kingfisher *Alcedo atthis* (KF)	Britain	1615	+49%	5%	<0.1	169	439	270	626	+1%	+8%	-0.10 (-0.13,-0.05)	5%	<0.1
	Ireland	344	+22%	6%	<0.1	16	202	75	115	-25%	+26%	-0.23 (-0.28,-0.19)	5%	<0.1
Blue-cheeked Bee-eater *Merops persicus*	Britain					3								
	Ireland													
Bee-eater *Merops apiaster* (MZ)	Britain					84								
	Ireland					4								
Roller *Coracias garrulus*	Britain					3								
	Ireland													
Hoopoe *Upupa epops* (HP)	Britain	12	+80%	1%	<0.1	203				-100%	-100%			
	Ireland	3	∞			9								
Wryneck *Jynx torquilla* (WY)	Britain	2	0%			87	12	1		-73%	+117%		1%	<0.1
	Ireland	1				4								
Green Woodpecker *Picus viridis* (G.)	Britain	1556	+14%	28%	0.2	57	352	243	1044	+1%	+5%	+0.08 (+0.06,+0.10)	32%	0.3
	Ireland													
Great Spotted Woodpecker *Dendrocopos major* (GS)	Britain	2337	+33%	40%	0.4	22	139	175	2028	+14%	+19%	+0.30 (+0.29,+0.32)	41%	0.4
	Ireland	41	∞	3%	<0.1	6	18	8	17	+4000%	∞		2%	<0.1
Lesser Spotted Woodpecker *Dendrocopos minor* (LS)	Britain	615	-20%	2%	<0.1	93	284	168	108	-41%	-34%	-0.59 (-0.66,-0.53)	2%	<0.1
	Ireland													
Golden Oriole *Oriolus oriolus* (OL)	Britain					93	27		2	+180%	-38%		1%	<0.1
	Ireland					5								
Brown Shrike *Lanius cristatus*	Britain	2	∞											
	Ireland													
Red-backed Shrike *Lanius collurio* (ED)	Britain	1				127	6	4	3	-88%	-13%			
	Ireland					3								
Lesser Grey Shrike *Lanius minor*	Britain					3								
	Ireland													
Great Grey Shrike *Lanius excubitor* (SR)	Britain	294	+3%	1%	<0.1	87					-100%			
	Ireland		-100%			1								

Species name scientific name and two-letter code	Island	Winter				Breeding season								
		Range size: *present* (a)	Range change (since 1981–84) (b)	Mean % of tetrads occupied per 10-km (c)	Mean birds per hour per tetrad (d)	Range size: *non-breeding* (e)	Range size: *possible* (f)	Range size: *probable* (f)	Range size: *confirmed* (f)	Range change (since 1968–72) (g)	Range change (since 1988–91) (g)	Fixed-effort change index (since 1988–91) (h)	Mean % of tetrads occupied per 10-km (c)	Mean birds per hour per tetrad (d)
Southern Grey Shrike *Lanius meridionalis*	Britain	2	∞											
	Ireland													
Woodchat Shrike *Lanius senator* (OO)	Britain					56								
	Ireland					13								
Chough *Pyrrhocorax pyrrhocorax* (CF)	Britain	114	+26%	26%	0.9	17	8	20	82	+37%	+23%	+0.22 (+0.08,+0.37)	27%	0.8
	Ireland	185	+10%	26%	0.6	14	13	43	116	+4%	+2%	-0.02 (-0.16,+0.10)	35%	0.9
Magpie *Pica pica* (MG)	Britain	2131	+10%	71%	2.7	57	132	184	1774	+7%	+6%	+0.04 (+0.03,+0.05)	66%	1.9
	Ireland	986	-<1%	90%	3.5	3	64	123	795	+3%	+2%	+0.01 (0.00,+0.02)	87%	2.6
Jay *Garrulus glandarius* (J.)	Britain	2088	+16%	38%	0.5	66	356	592	1049	+14%	+16%	+0.12 (+0.10,+0.13)	28%	0.3
	Ireland	559	+86%	17%	0.2	15	195	166	161	+31%	+94%	+0.39 (+0.36,+0.42)	14%	0.1
Jackdaw *Corvus monedula* (JD)	Britain	2400	+2%	65%	12	66	75	107	2162	-3%	-<1%	+0.02 (+0.01,+0.03)	67%	6.5
	Ireland	968	+<1%	83%	15.9	8	38	78	841	-1%	+1%	+0.01 (0.00,+0.02)	83%	9.5
House Crow *Corvus splendens*	Britain													
	Ireland	1				2								
Rook *Corvus frugilegus* (RO)	Britain	2293	+<1%	55%	16.3	141	90	25	2048	-4%	-3%	-0.03 (-0.04,-0.01)	49%	8.8
	Ireland	956	-1%	81%	21.8	24	70	38	806	-2%	-1%	0.00 (-0.02,+0.01)	81%	14
Carrion Crow *Corvus corone* (C.)	Britain	2471	+3%	84%	6.9	76	158	106	2166	+6%	+3%	+0.02 (+0.01,+0.02)	84%	4.8
	Ireland	60	-32%	12%	0.1	28	18	4	4	+25%	+19%	0.00 (-0.33,+0.25)	9%	0.1
Hooded Crow *Corvus cornix* (HC)	Britain	773	-13%	48%	1.5	125	139	163	377	-1%	-2%	-0.04 (-0.06,-0.02)	48%	1
	Ireland	1005	+<1%	89%	3.2	7	78	205	714	+1%	+3%	+0.02 (+0.01,+0.03)	86%	2.4
Raven *Corvus corax* (RN)	Britain	2212	+97%	35%	0.6	210	199	417	1378	+58%	+74%	+0.46 (+0.44,+0.47)	27%	0.4
	Ireland	929	+48%	36%	0.4	35	211	252	412	+95%	+27%	+0.26 (+0.23,+0.29)	26%	0.3
Goldcrest *Regulus regulus* (GC)	Britain	2568	+11%	34%	0.6	68	374	467	1626	+5%	+6%	+0.05 (+0.03,+0.06)	29%	0.6
	Ireland	935	+13%	60%	1.4	1	203	215	503	+1%	+8%	+0.09 (+0.07,+0.11)	61%	1.5
Firecrest *Regulus ignicapilla* (FC)	Britain	603	+112%	2%	<0.1	125	99	59	60	+935%	+109%	+0.36 (+0.15,+0.53)	2%	<0.1
	Ireland	7	-60%			1								
Penduline Tit *Remiz pendulinus* (DT)	Britain	13	+500%			2	1			∞	0%			
	Ireland													
Blue Tit *Cyanistes caeruleus* (BT)	Britain	2571	+3%	80%	5.4	8	61	47	2460	+1%	+3%	+0.02 (+0.02,+0.03)	77%	3.6
	Ireland	968	+2%	86%	3.1	1	18	36	911	+2%	+4%	+0.05 (+0.03,+0.06)	85%	2.6
Great Tit *Parus major* (GT)	Britain	2563	+5%	76%	3.3	12	42	55	2465	+4%	+4%	+0.05 (+0.04,+0.06)	74%	2.4
	Ireland	960	+12%	77%	2	1	36	61	859	+6%	+8%	+0.11 (+0.10,+0.13)	81%	1.9
Crested Tit *Lophophanes cristatus* (CI)	Britain	73	+50%	13%	0.3	5	13	8	38	+28%	+16%	+0.30 (+0.21,+0.39)	14%	0.2
	Ireland													
Coal Tit *Periparus ater* (CT)	Britain	2526	+10%	45%	1.2	28	151	150	2164	+3%	+6%	+0.08 (+0.07,+0.09)	39%	0.9
	Ireland	934	+19%	74%	1.9	2	59	93	773	+8%	+8%	+0.11 (+0.09,+0.13)	70%	1.7
Willow Tit *Poecile montana* (WT)	Britain	691	-43%	6%	0.1	103	175	136	247	-55%	-50%	-0.64 (-0.67,-0.61)	5%	0.1
	Ireland													
Marsh Tit *Poecile palustris* (MT)	Britain	1173	-4%	15%	0.2	69	202	208	659	-22%	-7%	-0.27 (-0.30,-0.24)	10%	0.1
	Ireland													
Bearded Tit *Panurus biarmicus* (BR)	Britain	156	+16%	3%	0.1	15	13	12	58	+82%	+34%	+0.10 (-0.13,+0.33)	4%	0.1
	Ireland	1	0%						1	∞	-50%			
Calandra Lark *Melanocorypha calandra*	Britain					1								
	Ireland													
Black Lark *Melanocorypha yeltoniensis*	Britain					1								
	Ireland													
Short-toed Lark *Calandrella brachydactyla* (VL)	Britain	5	∞	1%	<0.1	22								
	Ireland					5								
Crested Lark *Galerida cristata*	Britain	1	∞			1								
	Ireland													
Woodlark *Lullula arborea* (WL)	Britain	243	+296%	2%	<0.1	31	33	33	89	-23%	+107%	+0.47 (+0.36,+0.58)	7%	0.1
	Ireland													

Species name *scientific name and two-letter code*	Island	Winter				Breeding season								
		Range size: *present* (a)	Range change (since 1981–84) (b)	Mean % of tetrads occupied per 10-km (c)	Mean birds per hour per tetrad (d)	Range size: *non-breeding* (e)	Range size: *possible* (f)	Range size: *probable* (f)	Range size: *confirmed* (f)	Range change (since 1968–72) (g)	Range change (since 1988–91) (g)	Fixed-effort change index (since 1988–91) (h)	Mean % of tetrads occupied per 10-km (c)	Mean birds per hour per tetrad (d)
Skylark *Alauda arvensis* (S.)	Britain	2318	+8%	30%	1.8	15	169	613	1971	-1%	+<1%	-0.03 (-0.03,-0.02)	58%	2.3
	Ireland	545	-14%	22%	1.4	3	221	334	302	-14%	-8%	-0.12 (-0.14,-0.10)	43%	1.2
Shore Lark *Eremophila alpestris* (SX)	Britain	62	-4%	<1%	<0.1	26								
	Ireland	1	∞											
Sand Martin *Riparia riparia* (SM)	Britain	28	+667%			571	245	157	1354	-14%	+12%	+0.11 (+0.09,+0.14)	17%	1.4
	Ireland	2				74	177	138	481	-6%	+34%	+0.19 (+0.17,+0.22)	27%	1.8
Swallow *Hirundo rustica* (SL)	Britain	301	+53%	1%	<0.1	75	52	61	2585	+3%	+2%	+0.02 (+0.01,+0.03)	74%	4.3
	Ireland	40	+46%	2%	<0.1	1	14	28	958	+1%	+2%	+0.02 (+0.01,+0.02)	94%	9.1
House Martin *Delichon urbicum* (HM)	Britain	97	-57%	<1%	<0.1	107	60	75	2377	+3%	+4%	+0.03 (+0.02,+0.04)	41%	1.8
	Ireland	5	+100%			12	54	45	837	+4%	+15%	+0.16 (+0.14,+0.18)	52%	2
Red-rumped Swallow *Cecropis daurica* (VR)	Britain	6	∞			80								
	Ireland					5								
Cetti's Warbler *Cettia cetti* (CW)	Britain	374	+433%	4%	0.1	28	137	148	136	+6783%	+375%	+0.75 (+0.69,+0.81)	7%	0.1
	Ireland	1	∞											
Long-tailed Tit *Aegithalos caudatus* (LT)	Britain	2408	+14%	44%	1.9	31	85	118	2051	+2%	+6%	+0.10 (+0.08,+0.11)	35%	0.8
	Ireland	838	+52%	32%	1.1	10	143	111	506	+5%	+39%	+0.25 (+0.23,+0.27)	23%	0.5
Greenish Warbler *Phylloscopus trochiloides* (NP)	Britain	1				4								
	Ireland													
Arctic Warbler *Phylloscopus borealis* (AP)	Britain	1												
	Ireland													
Pallas's Warbler *Phylloscopus proregulus* (PA)	Britain	58	+450%	1%	<0.1	5								
	Ireland													
Yellow-browed Warbler *Phylloscopus inornatus* (YB)	Britain	125	+1667%	<1%	<0.1	6								
	Ireland	5												
Hume's Warbler *Phylloscopus humei*	Britain	11	∞			1								
	Ireland													
Radde's Warbler *Phylloscopus schwarzi*	Britain	3	∞											
	Ireland													
Dusky Warbler *Phylloscopus fuscatus* (UY)	Britain	9	+100%			1								
	Ireland													
Western Bonelli's Warbler *Phylloscopus bonelli* (IW)	Britain	1					1			∞	∞			
	Ireland													
Wood Warbler *Phylloscopus sibilatrix* (WO)	Britain					268	394	207	200	-35%	-37%	-0.51 (-0.54,-0.48)	8%	0.1
	Ireland					8	13	5	1	+138%	-32%		1%	<0.1
Chiffchaff *Phylloscopus collybita* (CC)	Britain	1196	+85%	4%	<0.1	57	398	429	1583	+20%	+14%	+0.14 (+0.13,+0.15)	57%	1.5
	Ireland	169	+68%	7%	0.1	3	194	336	366	-1%	+7%	+0.07 (+0.06,+0.09)	59%	1.3
Iberian Chiffchaff *Phylloscopus ibericus*	Britain					1	4	1		∞	∞			
	Ireland					1								
Willow Warbler *Phylloscopus trochilus* (WW)	Britain	29	+40%	<1%	<0.1	55	184	376	2111	+3%	+2%	-0.02 (-0.03,-0.01)	57%	2.2
	Ireland	2	∞			8	39	122	803	+2%	+4%	+0.06 (+0.04,+0.07)	84%	4.2
Blackcap *Sylvia atricapilla* (BC)	Britain	1471	+57%	3%	<0.1	49	282	446	1692	+25%	+17%	+0.19 (+0.18,+0.20)	55%	1.2
	Ireland	393	+225%	6%	<0.1	5	173	336	385	+249%	+150%	+0.66 (+0.64,+0.67)	54%	0.9
Garden Warbler *Sylvia borin* (GW)	Britain	18	+33%			115	540	709	821	+12%	+10%	+0.01 (-0.01,+0.03)	20%	0.2
	Ireland	1	-100%			10	44	9	6	+11%	+5%	-0.10 (-0.21,0.00)	7%	0.1
Barred Warbler *Sylvia nisoria* (RR)	Britain	11	∞			1								
	Ireland	4	+100%	6%	<0.1									
Lesser Whitethroat *Sylvia curruca* (LW)	Britain	58	+140%	<1%	<0.1	96	403	402	651	+32%	+14%	0.00 (-0.03,+0.03)	15%	0.1
	Ireland	5	∞			10	5	2		∞	+600%			
Whitethroat *Sylvia communis* (WH)	Britain	5	-50%			48	208	238	1908	-<1%	+7%	+0.07 (+0.06,+0.08)	49%	1.2
	Ireland		-100%			3	268	207	308	-7%	+24%	+0.20 (+0.17,+0.22)	34%	0.4
Spectacled Warbler *Sylvia conspicillata*	Britain						1			∞	∞			
	Ireland													

Species name *scientific name and two-letter code*	Island	Winter				Breeding season								
		Range size: *present* (a)	Range change (since 1981–84) (b)	Mean % of tetrads occupied per 10-km (c)	Mean birds per hour per tetrad (d)	Range size: *non-breeding* (e)	Range size: *possible* (f)	Range size: *probable* (f)	Range size: *confirmed* (f)	Range change (since 1968–72) (g)	Range change (since 1988–91) (g)	Fixed-effort change index (since 1988–91) (h)	Mean % of tetrads occupied per 10-km (c)	Mean birds per hour per tetrad (d)
Dartford Warbler *Sylvia undata* (DW)	Britain	215	+449%	5%	0.1	17	21	29	99	+352%	+198%	+0.48 (+0.32,+0.60)	6%	0.1
	Ireland													
Marmora's Warbler *Sylvia sarda* (MM)	Britain							1		∞	∞			
	Ireland													
Subalpine Warbler *Sylvia cantillans*	Britain					34	2			∞	∞			
	Ireland					9					-100%			
Sardinian Warbler *Sylvia melanocephala*	Britain	1	∞			1								
	Ireland													
Grasshopper Warbler *Locustella naevia* (GH)	Britain	1				162	805	601	258	-12%	+40%	+0.10 (+0.06,+0.15)	6%	0.1
	Ireland					15	409	244	59	+3%	+74%	+0.42 (+0.39,+0.45)	19%	0.1
River Warbler *Locustella fluviatilis* (VW)	Britain					1	2			∞	∞			
	Ireland													
Savi's Warbler *Locustella luscinioides* (VI)	Britain					4	10	8	1	+58%	-30%			
	Ireland										-100%			
Eastern Olivaceous Warbler *Iduna pallida*	Britain					1								
	Ireland													
Icterine Warbler *Hippolais icterina* (IC)	Britain		-100%			47	25	1	1	∞	+2600%		<1%	<0.1
	Ireland					1					-100%			
Melodious Warbler *Hippolais polyglotta* (ME)	Britain					8	3			∞	∞			
	Ireland													
Sedge Warbler *Acrocephalus schoenobaenus* (SW)	Britain	1	-100%			82	482	449	1214	+3%	+13%	-0.03 (-0.06,-0.01)	18%	0.4
	Ireland					4	227	266	366	+<1%	+26%	+0.20 (+0.18,+0.23)	34%	0.5
Paddyfield Warbler *Acrocephalus agricola* (PY)	Britain					1								
	Ireland													
Blyth's Reed Warbler *Acrocephalus dumetorum*	Britain	1	∞			1								
	Ireland													
Marsh Warbler *Acrocephalus palustris* (MW)	Britain	1				39	27	13	8	+133%	+227%		<1%	<0.1
	Ireland													
Reed Warbler *Acrocephalus scirpaceus* (RW)	Britain	13	+25%			98	189	188	719	+40%	+37%	+0.11 (+0.08,+0.15)	14%	0.4
	Ireland	1				6	9	19	8	∞	+177%		8%	0.1
Great Reed Warbler *Acrocephalus arundinaceus* (QW)	Britain					8	4	1		∞	∞			
	Ireland					1								
Fan-tailed Warbler *Cisticola juncidis* (FZ)	Britain	1	∞				1		1	∞	∞			
	Ireland													
Waxwing *Bombycilla garrulus* (WX)	Britain	1655	+903%	1%	<0.1	270	1			0%	0%			
	Ireland	147	+1857%	<1%	0.1	7								
Nuthatch *Sitta europaea* (NH)	Britain	1609	+38%	30%	0.4	53	163	177	1248	+35%	+25%	+0.19 (+0.17,+0.21)	28%	0.4
	Ireland													
Treecreeper *Certhia familiaris* (TC)	Britain	2278	+9%	20%	0.2	46	382	299	1510	-5%	+3%	-0.04 (-0.06,-0.02)	17%	0.2
	Ireland	707	+16%	21%	0.2	16	297	128	248	-6%	+18%	+0.13 (+0.10,+0.16)	20%	0.2
Short-toed Treecreeper *Certhia brachydactyla* (TH)	Britain	11	-25%	6%	0.1	2	2		6	-60%	-33%	-0.60 (-1.00,0.00)	12%	0.3
	Ireland													
Wren *Troglodytes troglodytes* (WR)	Britain	2774	+6%	77%	1.9	4	71	171	2545	+1%	+1%	0.01 (0.00,+0.01)	83%	4.1
	Ireland	996	-<1%	92%	3.7		34	85	884	+1%	+1%	0.01 (0.00,+0.02)	96%	7.5
Starling *Sturnus vulgaris* (SG)	Britain	2592	+2%	63%	31.9	45	79	59	2451	-5%	-2%	-0.05 (-0.06,-0.04)	55%	6.3
	Ireland	987	+1%	79%	37.5	5	23	18	944	+<1%	+3%	0.04 (+0.03,+0.06)	79%	8.6
Rose-coloured Starling *Pastor roseus* (OE)	Britain	13	+900%			40								
	Ireland		-100%			2								
Dipper *Cinclus cinclus* (DI)	Britain	1353	+14%	13%	0.1	48	226	164	906	-11%	-2%	-0.18 (-0.21,-0.16)	11%	0.1
	Ireland	445	+19%	12%	0.1	6	154	82	263	-21%	+14%	-0.06 (-0.10,-0.02)	12%	0.1
White's Thrush *Zoothera dauma*	Britain					1								
	Ireland													

Species name *scientific name and two-letter code*	Island	Winter				Breeding season								
		Range size: *present* (a)	Range change (since 1981–84) (b)	Mean % of tetrads occupied per 10-km (c)	Mean birds per hour per tetrad (d)	Range size: *non-breeding* (e)	Range size: *possible* (f)	Range size: *probable* (f)	Range size: *confirmed* (f)	Range change (since 1968–72) (g)	Range change (since 1988–91) (g)	Fixed-effort change index (since 1988–91) (h)	Mean % of tetrads occupied per 10-km (c)	Mean birds per hour per tetrad (d)
Ring Ouzel	Britain	159	+300%	<1%	<0.1	502	132	96	196	-43%	-22%	-0.47 (-0.51,-0.42)	7%	0.1
Turdus torquatus (RZ)	Ireland	5	+300%			14	7	4	4	-57%	-48%			
Blackbird	Britain	2742	+2%	82%	6.9	14	66	57	2596	-1%	+1%	+0.01 (0.00,+0.02)	82%	5.6
Turdus merula (B.)	Ireland	993	-<1%	94%	8.7	2	13	42	932	+<1%	+1%	+0.01 (0.00,+0.02)	94%	6.6
Black-throated Thrush	Britain	1	0%			1								
Turdus atrogularis (XC)	Ireland													
Fieldfare	Britain	2637	+4%	45%	9.8	1227	12	7	4	-32%	-78%	-0.55 (-0.78,-0.31)	3%	0.3
Turdus pilaris (FF)	Ireland	951	+5%	57%	7.2	66				-100%				
Song Thrush	Britain	2595	+7%	50%	0.9	35	114	118	2434	-1%	+1%	+0.01 (0.00,+0.01)	65%	1.2
Turdus philomelos (ST)	Ireland	993	+1%	80%	2.4	2	117	145	711	+<1%	+3%	+0.03 (+0.02,+0.05)	80%	1.8
Redwing	Britain	2605	+5%	46%	5.7	720	47	10	18	-32%	-45%	-0.46 (-0.56,-0.37)	5%	<0.1
Turdus iliacus (RE)	Ireland	973	+4%	71%	16.5	60					-100%			
Mistle Thrush	Britain	2491	+8%	40%	0.5	35	150	211	2096	0%	+2%	-0.01 (-0.02,+0.01)	35%	0.4
Turdus viscivorus (M.)	Ireland	949	+4%	61%	1	8	125	158	611	-5%	+4%	+0.03 (+0.02,+0.05)	50%	0.7
American Robin	Britain	1	-50%											
Turdus migratorius (AR)	Ireland		-100%											
Spotted Flycatcher	Britain	5	-100%			202	226	252	1730	-10%	-8%	-0.35 (-0.37,-0.33)	12%	0.1
Muscicapa striata (SF)	Ireland					19	204	130	377	-21%	-3%	-0.13 (-0.16,-0.10)	19%	0.2
Robin	Britain	2741	+5%	81%	3.9	37	65	59	2551	+2%	+1%	+0.02 (+0.01,+0.02)	80%	3.3
Erithacus rubecula (R.)	Ireland	994	-1%	95%	6.9		18	30	937	+1%	+2%	+0.01 (0.00,+0.02)	95%	5.8
White-throated Robin	Britain					1								
Irania gutturalis	Ireland													
Thrush Nightingale	Britain					7								
Luscinia luscinia (FN)	Ireland													
Nightingale	Britain					68	163	127	77	-43%	-20%	-0.44 (-0.51,-0.37)	5%	0.1
Luscinia megarhynchos (N.)	Ireland													
Bluethroat	Britain	11	-100%			61	1	1		+100%	+100%			
Luscinia svecica (BU)	Ireland	2	∞			1								
Red-flanked Bluetail	Britain	8				1								
Tarsiger cyanurus	Ireland	1	∞											
Red-breasted Flycatcher	Britain	10	-83%	1%	<0.1	17								
Ficedula parva (FY)	Ireland	1				2								
Collared Flycatcher	Britain					5								
Ficedula albicollis	Ireland													
Pied Flycatcher	Britain	1	∞			271	110	67	357	-2%	-27%	-0.44 (-0.48,-0.40)	9%	0.1
Ficedula hypoleuca (PF)	Ireland					5					-100%			
Black Redstart	Britain	605	+52%	1%	<0.1	294	79	11	42	+94%	+28%	-0.63 (-1.00,-0.33)	1%	<0.1
Phoenicurus ochruros (BX)	Ireland	99	+2%	4%	<0.1	12	2	1		∞	∞		4%	<0.1
Redstart	Britain	15	-89%	<1%	<0.1	470	269	175	700	-31%	-14%	-0.23 (-0.25,-0.20)	17%	0.3
Phoenicurus phoenicurus (RT)	Ireland					10		1	1	-78%	-82%			
Whinchat	Britain	29	-58%			644	186	150	544	-47%	-37%	-0.49 (-0.52,-0.47)	11%	0.2
Saxicola rubetra (WC)	Ireland	3	-50%			13	14	12	19	-76%	-64%	-0.63 (-0.68,-0.59)	7%	0.1
Siberian Stonechat	Britain	3	∞											
Saxicola maurus	Ireland													
Stonechat	Britain	2414	+151%	16%	0.2	153	144	259	1475	+52%	+78%	+0.42 (+0.40,+0.44)	19%	0.3
Saxicola rubicola (SC)	Ireland	742	+46%	34%	0.5	11	104	139	510	-5%	+32%	+0.23 (+0.20,+0.26)	31%	0.5
Wheatear	Britain	125	-7%	<1%	<0.1	981	217	297	1101	-14%	-7%	-0.12 (-0.15,-0.10)	31%	0.7
Oenanthe oenanthe (W.)	Ireland	15	-57%			103	117	87	209	-24%	-5%	-0.10 (-0.14,-0.05)	26%	0.6
Pied Wheatear	Britain	2	+100%											
Oenanthe pleschanka (PI)	Ireland		-100%											
Black-eared Wheatear	Britain					1								
Oenanthe hispanica	Ireland					1								

Species name *scientific name and two-letter code*	Island	Winter: Range size: *present* (a)	Winter: Range change (since 1981–84) (b)	Winter: Mean % of tetrads occupied per 10-km (c)	Winter: Mean birds per hour per tetrad (d)	Breeding season: Range size: *non-breeding* (e)	Breeding season: Range size: *possible* (f)	Breeding season: Range size: *probable* (f)	Breeding season: Range size: *confirmed* (f)	Breeding season: Range change (since 1968–72) (g)	Breeding season: Range change (since 1988–91) (g)	Breeding season: Fixed-effort change index (since 1988–91) (h)	Breeding season: Mean % of tetrads occupied per 10-km (c)	Breeding season: Mean birds per hour per tetrad (d)
Desert Wheatear	Britain	11	∞	2%	<0.1									
Oenanthe deserti	Ireland													
Dunnock	Britain	2649	+7%	66%	1.6	26	158	186	2289	+1%	+4%	+0.06 (+0.05,+0.06)	67%	1.7
Prunella modularis (D.)	Ireland	984	+3%	81%	2.2	1	100	211	663	+<1%	+3%	+0.04 (+0.02,+0.06)	81%	2
House Sparrow	Britain	2539	+4%	54%	5.5	11	34	53	2512	-3%	+2%	+0.02 (+0.01,+0.03)	60%	6.2
Passer domesticus (HS)	Ireland	946	+7%	63%	5.1	4	36	53	877	-<1%	+7%	+0.08 (+0.06,+0.10)	68%	5.6
Tree Sparrow	Britain	1168	-20%	15%	0.8	136	98	101	799	-41%	-26%	-0.23 (-0.26,-0.21)	21%	0.7
Passer montanus (TS)	Ireland	242	+71%	15%	0.8	13	53	45	120	+78%	+67%	+0.43 (+0.36,+0.49)	21%	0.7
Zebra Finch	Britain	2	∞			3		1		∞	∞			
Taeniopygia guttata (FI)	Ireland													
Red Avadavat	Britain					1								
Amandava amandava	Ireland													
Pin-tailed Whydah	Britain	1								-100%				
Vidua macroura	Ireland													
Yellow Wagtail	Britain	11	-86%			346	121	126	538	-32%	-25%	-0.37 (-0.40,-0.34)	13%	0.2
Motacilla flava (YW)	Ireland					12				-100%	-100%			
Yellow Wagtail (Blue-headed)	Britain					132	8	10	1	+533%	∞		1%	<0.1
Motacilla flava flava (EH)	Ireland					2								
Citrine Wagtail	Britain					10								
Motacilla citreola	Ireland													
Grey Wagtail	Britain	2163	+38%	13%	0.1	167	220	270	1701	+19%	+11%	-0.07 (-0.09,-0.05)	14%	0.1
Motacilla cinerea (GL)	Ireland	761	+10%	26%	0.2	17	209	163	396	-14%	-6%	-0.13 (-0.16,-0.11)	24%	0.2
Pied Wagtail	Britain	2355	+11%	41%	0.8	27	59	93	2593	+3%	+2%	+0.03 (+0.02,+0.04)	49%	0.7
Motacilla alba (PW)	Ireland	986	+1%	76%	1.6	5	38	62	882	-<1%	+1%	+0.02 (+0.01,+0.04)	69%	1.1
White Wagtail	Britain	79	∞	<1%	<0.1	581	4	2	8	-33%	∞		1%	<0.1
Motacilla alba alba (WB)	Ireland	10	∞			48				-100%				
Richard's Pipit	Britain	58	+414%	1%	<0.1	8								
Anthus richardi (PR)	Ireland	4	∞											
Tawny Pipit	Britain	1				16								
Anthus campestris (TI)	Ireland					4								
Olive-backed Pipit	Britain	8	0%											
Anthus hodgsoni (OV)	Ireland													
Tree Pipit	Britain	3				290	294	395	580	-29%	-17%	-0.25 (-0.28,-0.23)	15%	0.2
Anthus trivialis (TP)	Ireland					4					-100%			
Pechora Pipit	Britain	1	∞											
Anthus gustavi	Ireland													
Meadow Pipit	Britain	2651	+10%	31%	1	154	241	281	2060	-2%	+1%	-0.04 (-0.05,-0.02)	49%	3.6
Anthus pratensis (MP)	Ireland	955	0%	67%	2.6	6	93	162	718	-3%	+3%	+0.02 (+0.01,+0.03)	64%	2.9
Red-throated Pipit	Britain					7								
Anthus cervinus (VP)	Ireland	1	∞											
Rock Pipit	Britain	969	+3%	23%	0.6	98	83	108	535	-3%	+9%	-0.11 (-0.17,-0.06)	24%	0.6
Anthus petrosus (RC)	Ireland	293	+2%	28%	0.7	7	32	42	211	-1%	+10%	-0.03 (-0.13,+0.07)	30%	0.9
Water Pipit	Britain	316	+211%	1%	<0.1	128								
Anthus spinoletta (WI)	Ireland	24	∞	2%	<0.1	7								
Buff-bellied Pipit	Britain	3	∞											
Anthus rubescens	Ireland	5	∞	6%	<0.1									
Chaffinch	Britain	2694	+4%	80%	9.1	23	56	112	2511	+3%	+2%	+0.02 (+0.01,+0.02)	87%	6.3
Fringilla coelebs (CH)	Ireland	980	-<1%	91%	9.5	2	37	104	820	-<1%	+1%	+0.01 (0.00,+0.02)	93%	6.1
Brambling	Britain	2053	+21%	5%	0.3	720	5	2		+40%	-46%		2%	<0.1
Fringilla montifringilla (BL)	Ireland	321	+53%	4%	0.1	7								
Greenfinch	Britain	2529	+16%	48%	1.8	33	139	274	2150	+7%	+10%	+0.11 (+0.10,+0.12)	60%	1.9
Chloris chloris (GR)	Ireland	918	+24%	47%	1.3	4	146	238	545	+1%	+14%	+0.15 (+0.13,+0.17)	59%	1.4

Species name *scientific name and two-letter code*	Island	Winter				Breeding season								
		Range size: *present* (a)	Range change (since 1981–84) (b)	Mean % of tetrads occupied per 10-km (c)	Mean birds per hour per tetrad (d)	Range size: *non-breeding* (e)	Range size: *possible* (f)	Range size: *probable* (f)	Range size: *confirmed* (f)	Range change (since 1968–72) (g)	Range change (since 1988–91) (g)	Fixed-effort change index (since 1988–91) (h)	Mean % of tetrads occupied per 10-km (c)	Mean birds per hour per tetrad (d)
Serin *Serinus serinus* (NS)	Britain	12	+100%			33	4	1		-17%	-44%			
	Ireland													
Domestic Canary *Serinus canaria* (NR)	Britain	2	∞			6				-100%				
	Ireland													
Goldfinch *Carduelis carduelis* (GO)	Britain	2511	+23%	46%	2.1	55	84	212	2156	+16%	+10%	+0.14 (+0.13,+0.15)	65%	2.3
	Ireland	962	+49%	54%	1.7	1	74	219	660	+4%	+27%	+0.29 (+0.28,+0.31)	68%	1.9
Siskin *Carduelis spinus* (SK)	Britain	2469	+44%	19%	1.3	461	321	414	1047	+180%	+51%	+0.41 (+0.39,+0.43)	25%	0.8
	Ireland	744	+189%	21%	0.9	22	184	193	164	+127%	+90%	+0.43 (+0.40,+0.46)	22%	0.4
Linnet *Carduelis cannabina* (LI)	Britain	1995	+12%	17%	2.6	42	109	411	1899	+1%	+6%	+0.01 (0.00,+0.03)	42%	1.7
	Ireland	679	+9%	31%	3.4	8	112	290	540	-2%	+20%	+0.22 (+0.20,+0.24)	53%	1.6
Twite *Carduelis flavirostris* (TW)	Britain	656	+19%	9%	1.7	95	121	186	229	-19%	-18%	-0.28 (-0.33,-0.24)	19%	0.5
	Ireland	68	-30%	4%	0.1	11	7	5	14	-80%	-57%	-0.60 (-0.80,-0.40)	7%	0.2
Lesser Redpoll *Carduelis cabaret* (LR)	Britain	1941	+20%	7%	0.3	382	382	506	659	-22%	-12%	-0.15 (-0.18,-0.13)	18%	0.4
	Ireland	788	+61%	30%	0.9	19	210	336	267	-3%	+51%	+0.42 (+0.39,+0.44)	38%	0.7
Common Redpoll *Carduelis flammea* (FR)	Britain	521	∞	1%	<0.1	117	11	5	5	∞	∞		2%	<0.1
	Ireland	21	∞			2								
Arctic Redpoll *Carduelis hornemanni* (AL)	Britain	19	∞			7								
	Ireland					1								
Two-barred Crossbill *Loxia leucoptera* (PD)	Britain	1	0%			6					-100%			
	Ireland													
Parrot Crossbill *Loxia pytyopsittacus* (PC)	Britain	5	-56%				2		7	∞	∞		1%	<0.1
	Ireland													
Common/Scottish Crossbill *Loxia sp* (XB)	Britain	1316	+177%	11%	0.6	582	373	220	497	+253%	+40%	+0.42 (+0.39,+0.45)	11%	0.3
	Ireland	239	+1150%	11%	0.3	26	128	76	51	+8133%	+58%	+0.32 (+0.28,+0.38)	9%	0.2
Trumpeter Finch *Bucanetes githagineus*	Britain					3	1			∞	∞			
	Ireland													
Common Rosefinch *Carpodacus erythrinus* (SQ)	Britain	8	∞			53	25	2		∞	+440%		<1%	<0.1
	Ireland					1								
Bullfinch *Pyrrhula pyrrhula* (BF)	Britain	2392	+10%	32%	0.5	42	180	631	1499	-1%	+6%	0.00 (-0.01,+0.02)	25%	0.3
	Ireland	891	+6%	57%	0.9	1	63	364	475	+3%	+9%	+0.06 (+0.04,+0.08)	51%	0.7
Hawfinch *Coccothraustes coccothraustes* (HF)	Britain	358	+28%	1%	<0.1	102	52	24	37	-76%	-65%	-0.78 (-0.88,-0.68)	1%	<0.1
	Ireland	2	∞	11%	0.1									
Snow Bunting *Plectrophenax nivalis* (SB)	Britain	954	+33%	6%	0.4	108	18	3	11	+129%	-24%		2%	<0.1
	Ireland	123	+48%	3%	0.1	5								
Lapland Bunting *Calcarius lapponicus* (LA)	Britain	268	+151%	2%	<0.1	66	1			∞	∞			
	Ireland	15	+1300%	1%	<0.1	2								
White-crowned Sparrow *Zonotrichia leucophrys*	Britain	1	∞			1								
	Ireland													
White-throated Sparrow *Zonotrichia albicollis*	Britain	2	∞			3	1			∞	∞			
	Ireland													
Dark-eyed Junco *Junco hyemalis* (JU)	Britain	3	∞			2								
	Ireland													
Pine Bunting *Emberiza leucocephalos* (EL)	Britain	1				1								
	Ireland													
Yellowhammer *Emberiza citrinella* (Y.)	Britain	1866	-9%	29%	1.4	57	201	383	1355	-21%	-13%	-0.19 (-0.20,-0.18)	45%	1.3
	Ireland	299	-38%	24%	0.7	7	87	119	157	-61%	-38%	-0.37 (-0.40,-0.34)	37%	0.8
Cirl Bunting *Emberiza cirlus* (CL)	Britain	23	-12%	18%	0.7	3	7	6	15	-84%	-10%	0.00 (-0.29,+0.27)	21%	0.4
	Ireland													
Ortolan Bunting *Emberiza hortulana* (OB)	Britain	1	∞			15								
	Ireland					1					-100%			
Rustic Bunting *Emberiza rustica*	Britain	4	∞			8								
	Ireland													

Species name *scientific name and two-letter code*	Island	Winter				Breeding season								
		Range size: *present* (a)	Range change (since 1981–84) (b)	Mean % of tetrads occupied per 10-km (c)	Mean birds per hour per tetrad (d)	Range size: *non-breeding* (e)	Range size: *possible* (f)	Range size: *probable* (f)	Range size: *confirmed* (f)	Range change (since 1968–72) (g)	Range change (since 1988–91) (g)	Fixed-effort change index (since 1988–91) (h)	Mean % of tetrads occupied per 10-km (c)	Mean birds per hour per tetrad (d)
Little Bunting *Emberiza pusilla* (LJ)	Britain	12	+267%			2								
	Ireland	1				1								
Reed Bunting *Emberiza schoeniclus* (RB)	Britain	2283	+25%	17%	0.4	62	277	541	1534	-5%	+7%	+0.04 (+0.02,+0.06)	24%	0.4
	Ireland	863	+18%	37%	0.7	3	182	282	458	-3%	+11%	+0.10 (+0.08,+0.12)	38%	0.5
Black-headed Bunting *Emberiza melanocephala*	Britain					4								
	Ireland													
Corn Bunting *Emberiza calandra* (CB)	Britain	557	-27%	7%	0.5	83	158	194	244	-56%	-36%	-0.50 (-0.53,-0.47)	12%	0.3
	Ireland		-100%							-100%	-100%			
Brown-headed Cowbird *Molothrus ater*	Britain					1								
	Ireland													

APPENDIX 3 ~ SURVEY INSTRUCTIONS AND FORMS

A INTRODUCTION TO *BIRD ATLAS 2007–11*

This information sheet was widely circulated at the start of the Atlas project and made available free for download from www.birdatlas.net.

1.

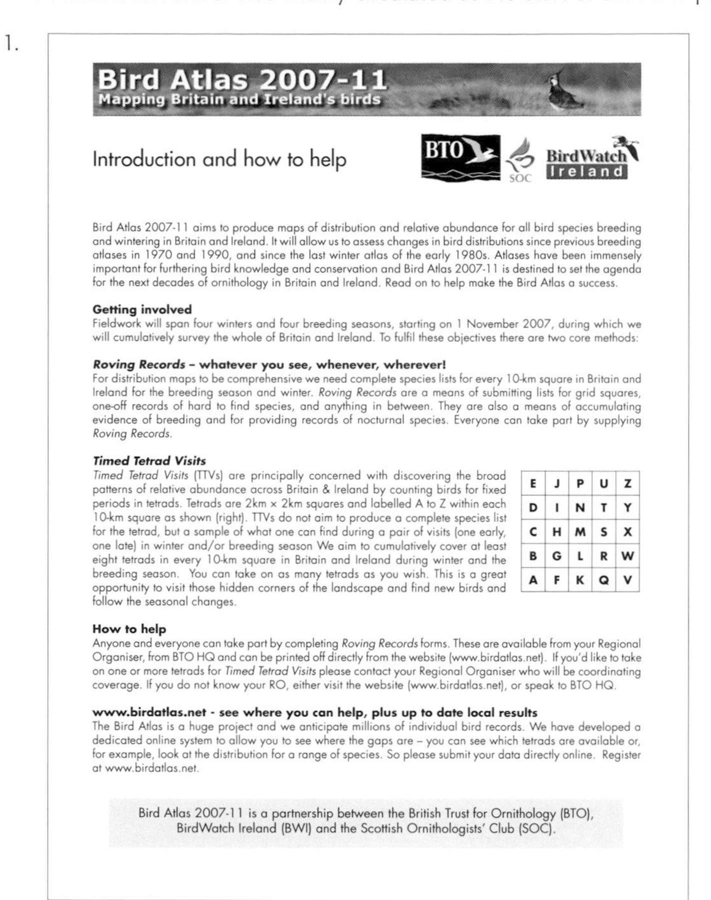

Bird Atlas 2007-11
Mapping Britain and Ireland's birds

Introduction and how to help

BTO SOC BirdWatch Ireland

Bird Atlas 2007-11 aims to produce maps of distribution and relative abundance for all bird species breeding and wintering in Britain and Ireland. It will allow us to assess changes in bird distributions since previous breeding atlases in 1970 and 1990, and since the last winter atlas of the early 1980s. Atlases have been immensely important for furthering bird knowledge and conservation and Bird Atlas 2007-11 is destined to set the agenda for the next decades of ornithology in Britain and Ireland. Read on to help make the Bird Atlas a success.

Getting involved
Fieldwork will span four winters and four breeding seasons, starting on 1 November 2007, during which we will cumulatively survey the whole of Britain and Ireland. To fulfil these objectives there are two core methods:

***Roving Records* – whatever you see, whenever, wherever!**
For distribution maps to be comprehensive we need complete species lists for every 10-km square in Britain and Ireland for the breeding season and winter. *Roving Records* are a means of submitting lists for grid squares, one-off records of hard to find species, and anything in between. They are also a means of accumulating evidence of breeding and for providing records of nocturnal species. Everyone can take part by supplying *Roving Records*.

Timed Tetrad Visits
Timed Tetrad Visits (TTVs) are principally concerned with discovering the broad patterns of relative abundance across Britain & Ireland by counting birds for fixed periods in tetrads. Tetrads are 2km × 2km squares and labelled A to Z within each 10-km square as shown (right). TTVs do not aim to produce a complete species list for the tetrad, but a sample of what one can find during a pair of visits (one early, one late) in winter and/or breeding season We aim to cumulatively cover at least eight tetrads in every 10-km square in Britain and Ireland during winter and the breeding season. You can take on as many tetrads as you wish. This is a great opportunity to visit those hidden corners of the landscape and find new birds and follow the seasonal changes.

E	J	P	U	Z
D	I	N	T	Y
C	H	M	S	X
B	G	L	R	W
A	F	K	Q	V

How to help
Anyone and everyone can take part by completing *Roving Records* forms. These are available from your Regional Organiser, from BTO HQ and can be printed off directly from the website (www.birdatlas.net). If you'd like to take on one or more tetrads for *Timed Tetrad Visits* please contact your Regional Organiser who will be coordinating coverage. If you do not know your RO, either visit the website (www.birdatlas.net), or speak to BTO HQ.

www.birdatlas.net - see where you can help, plus up to date local results
The Bird Atlas is a huge project and we anticipate millions of individual bird records. We have developed a dedicated online system to allow you to see where the gaps are – you can see which tetrads are available or, for example, look at the distribution for a range of species. So please submit your data directly online. Register at www.birdatlas.net.

Bird Atlas 2007-11 is a partnership between the British Trust for Ornithology (BTO), BirdWatch Ireland (BWI) and the Scottish Ornithologists' Club (SOC).

2.

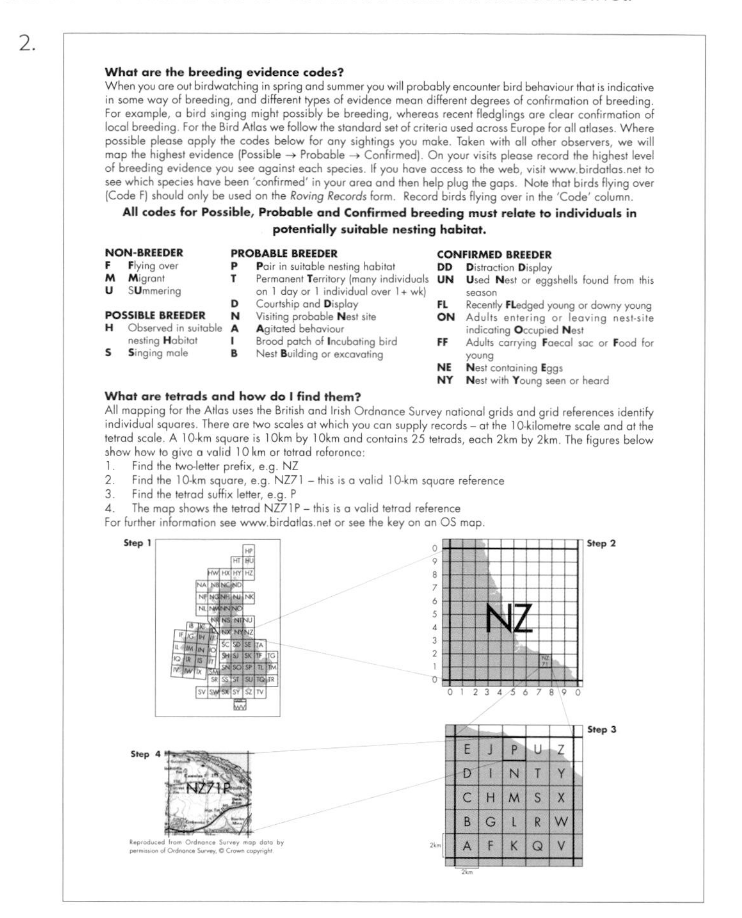

What are the breeding evidence codes?
When you are out birdwatching in spring and summer you will probably encounter bird behaviour that is indicative in some way of breeding, and different types of evidence mean different degrees of confirmation of breeding. For example, a bird singing might possibly be breeding, whereas recent fledglings are clear confirmation of local breeding. For the Bird Atlas we follow the standard set of criteria used across Europe for all atlases. Where possible please apply the codes below for any sightings you make. Taken with all other observers, we will map the highest evidence (Possible → Probable → Confirmed). On your visits please record the highest level of breeding evidence you see against each species. If you have access to the web, visit www.birdatlas.net to see which species have been 'confirmed' in your area and then help plug the gaps. Note that birds flying over (Code F) should only be used on the *Roving Records* form. Record birds flying over in the 'Code' column.

All codes for Possible, Probable and Confirmed breeding must relate to individuals in potentially suitable nesting habitat.

NON-BREEDER
F **F**lying over
M **M**igrant
U S**U**mmering

POSSIBLE BREEDER
H Observed in suitable nesting **H**abitat
S **S**inging male

PROBABLE BREEDER
P **P**air in suitable nesting habitat
T Permanent **T**erritory (many individuals on 1 day or 1 individual over 1+ wk)
D Courtship and **D**isplay
N Visiting probable **N**est site
A **A**gitated behaviour
I Brood patch of **I**ncubating bird
B Nest **B**uilding or excavating

CONFIRMED BREEDER
DD **D**istraction **D**isplay
UN **U**sed **N**est or eggshells found from this season
FL Recently **FL**edged young or downy young
ON Adults entering or leaving nest-site indicating **O**ccupied **N**est
FF Adults carrying **F**aecal sac or **F**ood for young
NE **N**est containing **E**ggs
NY **N**est with **Y**oung seen or heard

What are tetrads and how do I find them?
All mapping for the Atlas uses the British and Irish Ordnance Survey national grids and grid references identify individual squares. There are two scales at which you can supply records – at the 10-kilometre scale and at the tetrad scale. A 10-km square is 10km by 10km and contains 25 tetrads, each 2km by 2km. The figures below show how to give a valid 10 km or tetrad reference:

1. Find the two-letter prefix, e.g. NZ
2. Find the 10-km square, e.g. NZ71 – this is a valid 10-km square reference
3. Find the tetrad suffix letter, e.g. P
4. The map shows the tetrad NZ71P – this is a valid tetrad reference

For further information see www.birdatlas.net or see the key on an OS map.

B ROVING RECORDS FORM

This survey form was widely circulated at the start of the Atlas project and made available for free download from www.birdatlas.net.

1.

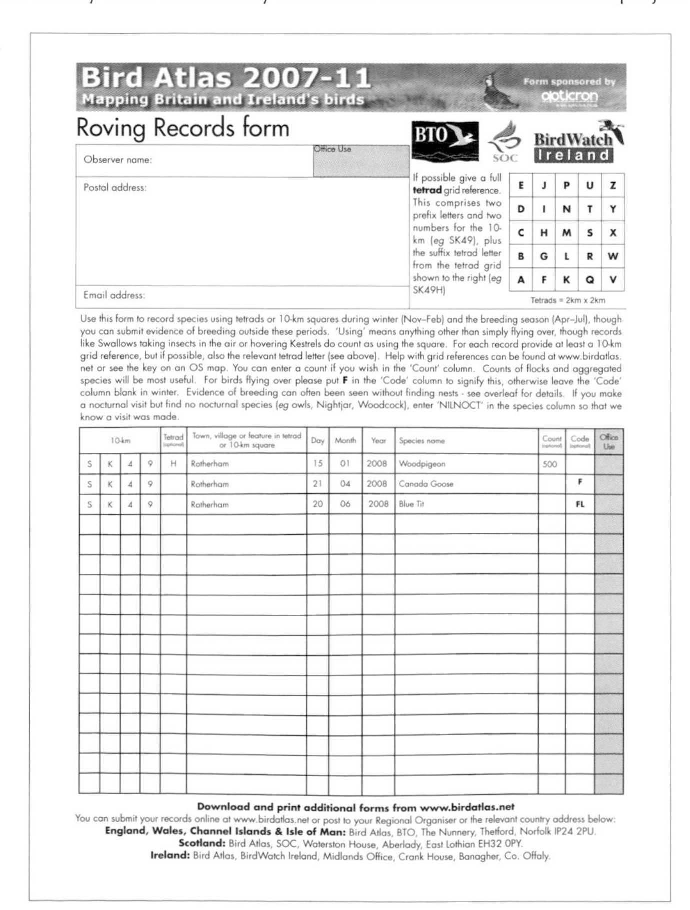

Bird Atlas 2007-11
Mapping Britain and Ireland's birds
Form sponsored by opticron

Roving Records form

BTO SOC BirdWatch Ireland

Observer name:
Office Use
Postal address:
Email address:

If possible give a full **tetrad** grid reference. This comprises two prefix letters and two numbers for the 10-km (*eg* SK49), plus the suffix tetrad letter from the tetrad grid shown to the right (*eg* SK49H)

E	J	P	U	Z
D	I	N	T	Y
C	H	M	S	X
B	G	L	R	W
A	F	K	Q	V

Tetrads = 2km x 2km

Use this form to record species using tetrads or 10-km squares during winter (Nov–Feb) and the breeding season (Apr–Jul), though you can submit evidence of breeding outside these periods. 'Using' means anything other than simply flying over, though records like Swallows taking insects in the air or hovering Kestrels do count as using the square. For each record provide at least a 10-km grid reference, but if possible, also the relevant tetrad letter (see above). Help with grid references can be found at www.birdatlas.net or see the key on an OS map. You can enter a count if you wish in the 'Count' column. Counts of flocks and aggregated species will be most useful. For birds flying over please put **F** in the 'Code' column to signify this, otherwise leave the 'Code' column blank in winter. Evidence of breeding can often been seen without finding nests - see overleaf for details. If you make a nocturnal visit but find no nocturnal species (*eg* owls, Nightjar, Woodcock), enter 'NILNOCT' in the species column so that we know a visit was made.

10-km				Tetrad (optional)	Town, village or feature in tetrad or 10-km square	Day	Month	Year	Species name	Count (optional)	Code (optional)	Office Use
S	K	4	9	H	Rotherham	15	01	2008	Woodpigeon	500		
S	K	4	9		Rotherham	21	04	2008	Canada Goose		**F**	
S	K	4	9		Rotherham	20	06	2008	Blue Tit		**FL**	

Download and print additional forms from www.birdatlas.net
You can submit your records online at www.birdatlas.net or post to your Regional Organiser or the relevant country address below:
England, Wales, Channel Islands & Isle of Man: Bird Atlas, BTO, The Nunnery, Thetford, Norfolk IP24 2PU.
Scotland: Bird Atlas, SOC, Waterston House, Aberlady, East Lothian EH32 0PY.
Ireland: Bird Atlas, BirdWatch Ireland, Midlands Office, Crank House, Banagher, Co. Offaly.

2.

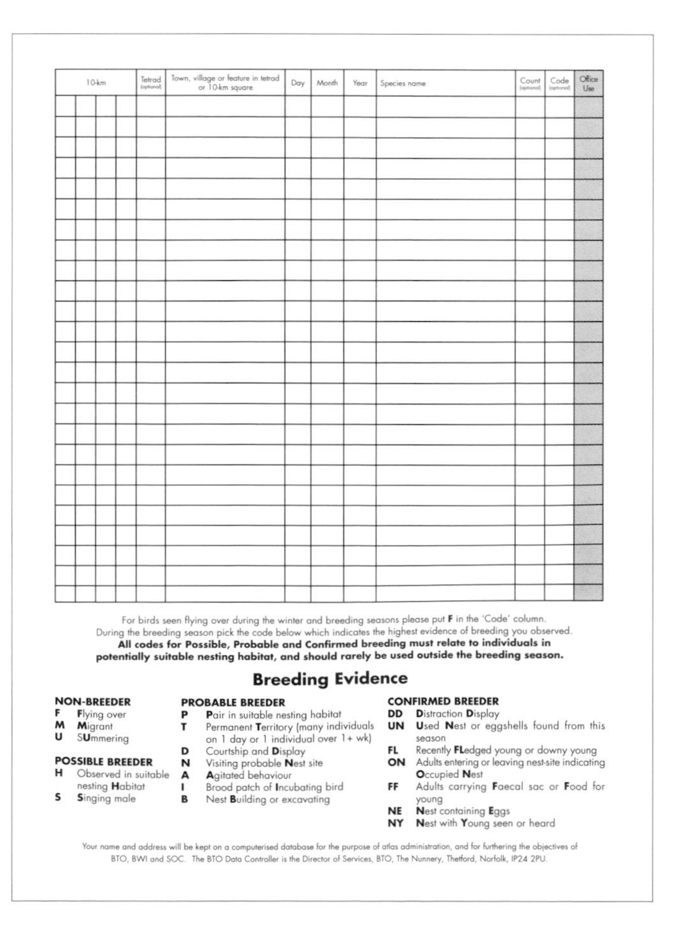

10-km				Tetrad (optional)	Town, village or feature in tetrad or 10-km square	Day	Month	Year	Species name	Count (optional)	Code (optional)	Office Use

For birds seen flying over during the winter and breeding seasons please put **F** in the 'Code' column.
During the breeding season pick the code below which indicates the highest evidence of breeding you observed.
All codes for Possible, Probable and Confirmed breeding must relate to individuals in potentially suitable nesting habitat, and should rarely be used outside the breeding season.

Breeding Evidence

NON-BREEDER
F **F**lying over
M **M**igrant
U S**U**mmering

POSSIBLE BREEDER
H Observed in suitable nesting **H**abitat
S **S**inging male

PROBABLE BREEDER
P **P**air in suitable nesting habitat
T Permanent **T**erritory (many individuals on 1 day or 1 individual over 1+ wk)
D Courtship and **D**isplay
N Visiting probable **N**est site
A **A**gitated behaviour
I Brood patch of **I**ncubating bird
B Nest **B**uilding or excavating

CONFIRMED BREEDER
DD **D**istraction **D**isplay
UN **U**sed **N**est or eggshells found from this season
FL Recently **FL**edged young or downy young
ON Adults entering or leaving nest-site indicating **O**ccupied **N**est
FF Adults carrying **F**aecal sac or **F**ood for young
NE **N**est containing **E**ggs
NY **N**est with **Y**oung seen or heard

Your name and address will be kept on a computerised database for the purpose of atlas administration, and for furthering the objectives of BTO, BWI and SOC. The BTO Data Controller is the Director of Services, BTO, The Nunnery, Thetford, Norfolk, IP24 2PU.

C TIMED TETRAD VISIT INSTRUCTIONS

So that observers could judge whether to request a TTV, these instructions were widely circulated and made available for free download from www.birdatlas.net.

1.

Timed Tetrad Visits - Instructions

Revised August 2008

SOC BirdWatch Ireland

Bird Atlas 2007-11 aims to produce maps of distribution and relative abundance for all bird species breeding and wintering in Britain and Ireland.

Timed Tetrad Visits are principally concerned with discovering the broad patterns of relative abundance by counting birds for fixed periods in a set of sample tetrads (2km × 2km squares). The aim, therefore, is to provide an indication of the number and diversity of birds seen in your tetrad(s) during either visits of one hour or two hours duration. The same methods apply for winter and the breeding season (with the exception of evidence of breeding).

Your Regional Organiser (RO) will be aiming to accumulate survey coverage of at least eight tetrads per 10-km square in winter and the breeding season – ideally the same eight tetrads. Please survey the tetrad(s) you have been allocated and if possible in both winter and the breeding season. **A tetrad requires only two visits in the winter and two in the breeding season and is then complete.** You can survey several tetrads if you wish but must coordinate this through your RO. Remember, you can also submit records of birds seen at other times, or in other areas, on a *Roving Records* form (available from your RO, BTO HQ, or download from www.birdatlas.net).

BIRD ATLAS ONLINE

The Bird Atlas is a huge project and we anticipate millions of individual bird records. To help handle this quantity of information, and to see how your results compare with others, we strongly encourage you to submit your data online. Registration is simple and opens up a world of immediate results. Visit www.birdatlas.net

WHICH TETRAD TO VISIT?

Liaise with your RO to decide which tetrad(s) you are to survey. Tetrads are identified from the full 10-km grid reference (*eg* SP24) plus a single letter suffix (*eg* J). The grid to the right shows the system of labelling tetrads within a 10-km square. For more information on tetrads and grid references see www.birdatlas.net, ask your RO, contact BTO HQ or BirdWatch Ireland.

E	J	P	U	Z
D	I	N	T	Y
C	H	M	S	X
B	G	L	R	W
A	F	K	Q	V

HOW LONG WILL IT TAKE? Choose from one hour, two hours or longer.

You have a choice. Everyone must spend the first hour of their *Timed Tetrad Visit* counting the number of individuals of each species they encounter (by sight and sound). You can then choose to spend a second hour counting. If you opt for the second hour your totals for the first hour and second hour should be recorded **separately** on the form.

Also, we have provided space on the form so that, if you wish, you can stay longer than two hours to list additional species or to check for evidence of breeding. Whichever you choose, ensure that you separate out the birds encountered during the first, second and subsequent hours in your notebook and when you submit the results online or on the recording form provided.

WHERE TO GO IN THE TETRAD?

Your visit should **cover the range of the habitats present in the tetrad.** We do not expect you to cover the whole tetrad in one hour or two hours, so choose a route that covers as many habitat types as possible. You can vary your route between visits. You may need to seek permission to visit some areas in order to cover key habitats in the tetrad. Look at an Ordnance Survey map or the Google Map on the www.birdatlas.net website to plan your route.

HOW MANY VISITS AND WHEN?

For the first time we are combining winter and breeding season fieldwork into one Atlas. If you have taken on a tetrad for winter you need to make two timed visits – one in early winter (1 November – 31 December) and

2.

one in late winter (1 January – 28 February). If you have taken on the tetrad for the breeding season you need to make two timed visits – one early (1 April – 31 May) and one late (1 June – 31 July). The optimal timing for TTVs in northern Britain are May (early visit) and June (late visit). If possible, offer to survey the same tetrad in one winter and one breeding season – it will give us interesting opportunities to make seasonal contrasts across Britain and Ireland. **Once a tetrad has received a pair of *Timed Tetrad Visits* in the breeding season and/or winter, it is complete.** It may still receive *Roving Records* effort, perhaps to pick up scarce and nocturnal species or to confirm breeding (see later).

WHAT TIME OF DAY?

In all seasons try to avoid periods of especially inclement weather when bird activity may be suppressed and the survey efficiency is reduced. During winter, avoid the first and last hour of daylight when birds can be moving from and to roosts and give rise to duplication across squares. During the breeding season make visits during mornings when bird activity is higher. If your square contains coastal habitats try to survey them from mid to high tide when birds on mudflats and on the sea will be more concentrated and closer, thus easier to identify and count.

WHAT TO DO IN THE FIELD

You have been provided with a *Timed Tetrad Visit* form for each tetrad you are surveying. This contains a full species list (tailored for winter or the breeding season) and is split for an early visit and a late visit. We recommend you record directly into a notebook in the field (see example below-right) and then transfer the totals to your form when you get home – or better still, enter them directly onto the Bird Atlas online system.

At the beginning of your visit note the time, or start a stopwatch. Walk around the tetrad, checking the main habitat types for birds. Keep a tally of the number of individuals you encounter (seen or heard) of each species. Continue until the first hour is completed. If you have opted to do a second hour, restart the clock and continue on your way around the tetrad, keeping a **separate** tally of the numbers encountered. Do not visit the same areas again, and try not to record the same individuals again. There is a third column (Extra) which you can use to list anything seen outside of the timed part of the visit – that might be to tick off a new species seen whilst walking back out of the square (*eg* the Great Tit: right), or if you decide to double back to check out an area in more detail and look for species you may have missed.

What to count: only count individuals that are using the square. Ignore individuals flying over the square except, for example, hunting raptors and hirundines that are effectively 'using' the square. During the breeding season exclude juveniles from your count. On the coast, count birds as far offshore as you are confident in identifying them, provided you think they are still in your tetrad. Anything offshore but not in the tetrad should be submitted as a *Roving Record* for the appropriate 10-km square. Count all introduced and feral species too.

Colonial nesting species (inland or coastal): (see TTV form for colonial species marked ★) if you encounter a colony (one or more nests) then please use the separate 'Colony Table' on the recording form. The minimum requirement is to tick that a colony is present on either of your visits. If it is possible to count the colony, please provide the maximum count of the number of Apparently Occupied Nests or the number of individuals in the colony. Shading on the Colony Table gives guidance on which method to use for specific species. Only rough estimates are required (see Tetrad Population Estimates overleaf). If you are unable to count the colony (too large, inaccessible etc) then by providing a tick to say a colony is present we can pursue other ways of achieving a colony count. If you encounter any of these species away from a colony, just count them as normal in the main part of the form.

Stopping the clock: if you encounter a flock of birds, a colony, or a particularly 'busy' area (*eg* a wood at dawn chorus) you can 'pause the clock' to do the counting. When counting large aggregations, or areas with high densities, we are looking for an approximate count so don't worry too much about getting the exact number. Remember to resume the clock before moving on.

3.

RECORDING EVIDENCE OF BREEDING ('BREED CODE')

During your tetrad visit note any behaviour that could be indicative of breeding using the coding system provided. You will see many of these activities during every-day birdwatching. Do not spend excessive time trying to confirm breeding – this will eat into your counting time too much – though you can pause the clock if you wish to try and confirm evidence of breeding. Evidence of breeding mainly applies to breeding season visits. However, some species do nest early (*eg* urban Blackbirds, Crossbills, Rooks) and you may encounter evidence of breeding on your late winter visit. Though the 'Breed Code' column is shaded out you can submit an evidence code. You only need to record the **highest** (Possible → Probable → Confirmed) level of breeding evidence you record, so if you see a pair of Blackbirds in suitable habitat (code **P**) and later a Blackbird carrying food for young (code **FF**) then you only need to record **FF** as this equates to Confirmed breeding rather than Probable breeding. So codes lower down the list should be used in preference to those higher up. You don't need to split up your counts by level of breeding evidence.

LAST BUT NOT LEAST, WHAT TO DO WHEN YOU GET BACK HOME

If you make notes directly into your notebook we advise you to transfer your records onto the *Timed Tetrad Visit* form as soon as possible, or better still, submit the records online at www.birdatlas.net. If you submit your records on paper forms, please ensure you provide your full contact details (email especially helpful). Please provide dates of visits and remember to indicate whether they were one hour or two hour visits.

TETRAD POPULATION ESTIMATES (TPE) (optional)

Having walked around part of the tetrad, you might be able to make some rough extrapolations of how many birds might have been present in the whole tetrad. For some species, these estimates, when used in conjunction with information from other sources, may help us calculate national population estimates. These estimates need only be approximate: estimates in the '1's need only be to the nearest 1, in the '10's to the nearest 10, in the '100's to the nearest 100, and so on. For some species this will be very easy – if you went to the only lake in the tetrad and saw 6 Tufted Ducks, your estimate will be 6. If you covered perhaps half the woodland in the square and recorded 33 Blue Tits, your estimate might be 70. In the breeding season, the TPE refers to the number of birds using the tetrad and not just those breeding there. Your TPE should based on the maximum count you make. Providing these Tetrad Estimates is entirely optional but would be extremely valuable.

Examples of completed forms

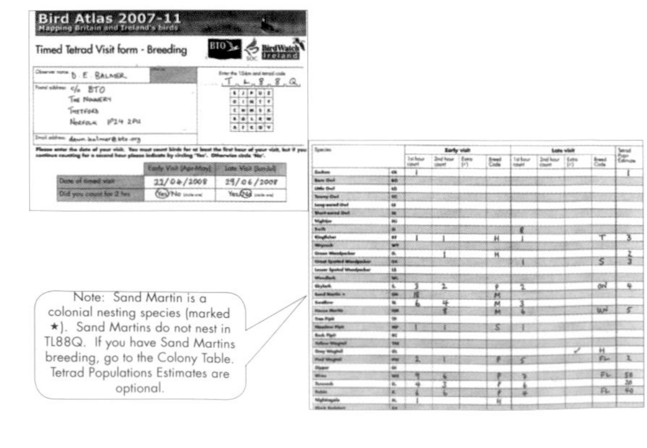

Bird Atlas 2007-11
Timed Tetrad Visit form - Breeding
Note: Sand Martin is a colonial nesting species (marked ★). Sand Martins do not nest in TL88Q. If you have Sand Martins breeding, go to the Colony Table. Tetrad Populations Estimates are optional.

4.

BREEDING EVIDENCE

For the Bird Atlas we follow the standard set of criteria used across Europe for all atlases. Where possible please apply the codes below for any sightings you make. Taken with all other observers, we will map the highest evidence (Possible → Probable → Confirmed). On your visits please record the highest level of breeding evidence you see against each species. If you have access to the web, visit www.birdatlas.net to see which species have been 'confirmed' in your area and then help plug the gaps.

All codes for Possible, Probable and Confirmed breeding must relate to individuals in potentially suitable nesting habitat.

NON-BREEDER
- **M** **M**igrant
- **U** S**U**mmering

POSSIBLE BREEDER
- **H** Observed in suitable nesting **H**abitat
- **S** **S**inging male

PROBABLE BREEDER
- **P** **P**air in suitable nesting habitat
- **T** Permanent **T**erritory (many individuals on 1 day or 1 individual over 1+ wk)
- **D** Courtship and **D**isplay
- **N** Visiting probable **N**est site
- **A** **A**gitated behaviour
- **I** Brood patch of **I**ncubating bird
- **B** Nest **B**uilding or excavating

CONFIRMED BREEDER
- **DD** **D**istraction **D**isplay
- **UN** **U**sed **N**est or eggshells found from this season
- **FL** Recently **FL**edged young or downy young
- **ON** Adults entering or leaving nest-site indicating **O**ccupied **N**est
- **FF** Adults carrying **F**aecal sac or **F**ood for young
- **NE** **N**est containing **E**ggs
- **NY** **N**est with **Y**oung seen or heard

What are tetrads and how do I find them?

All mapping for the Atlas uses the British and Irish Ordnance Survey national grids and grid references identify individual squares. There are two scales at which you can supply records – at the 10-kilometre scale and at the tetrad scale. A 10-km square is 10km by 10km and contains 25 tetrads, each 2km by 2km. The figures below show how to give a valid 10-km or tetrad reference:

1. Find the two-letter prefix, e.g. NZ
2. Find the 10-km square, e.g. NZ71 – this is a valid 10-km square reference
3. Find the tetrad suffix letter, e.g. P
4. The map shows the tetrad NZ71P – this is a valid tetrad reference

For further information see www.birdatlas.net or see the key on the OS map.

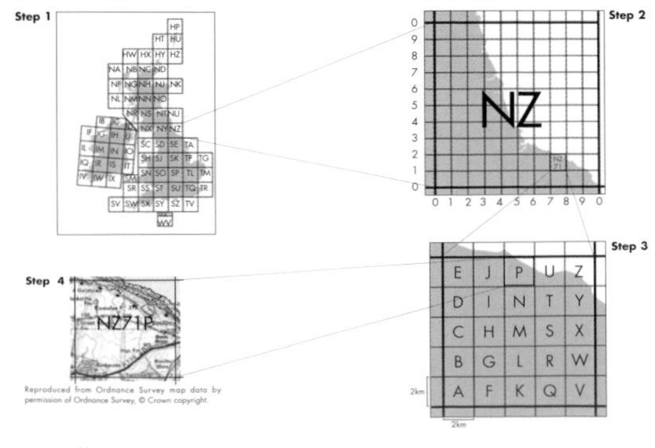

Reproduced from Ordnance Survey map data by permission of Ordnance Survey, © Crown copyright.

Contact addresses:

England, Wales, Channel Islands & Isle of Man: Bird Atlas, BTO, The Nunnery, Thetford, Norfolk IP24 2PU.
Scotland: Bird Atlas, SOC, Waterston House, Aberlady, East Lothian EH32 0PY.
Ireland: Bird Atlas, BirdWatch Ireland, Midlands Office, Crank House, Banagher, Co. Offaly.

D TIMED TETRAD VISIT FORM FOR WINTER

To limit opportunities for duplication of coverage, the winter TTV form was only made available, via post or download, to observers who had been allocated tetrads.

1.

Bird Atlas 2007-11
Mapping Britain and Ireland's birds

Timed Tetrad Visit form - Winter

BTO SOC BirdWatch Ireland

Observer name: | Office Use

Postal address:

Email address:

Enter the 10-km and tetrad code

Example S P 2 4 J

E	J	P	U	Z
D	I	N	T	Y
C	H	M	S	X
B	G	L	R	W
A	F	K	Q	V

Please enter the date of your visit. You must count birds for at least the first hour of your visit, but if you continue counting for a second hour please indicate by circling 'Yes'. Otherwise circle 'No'.

	Early Visit (Nov-Dec)	Late Visit (Jan-Feb)
Date of timed visit	dd/mm/yyyy	dd/mm/yyyy
Did you count for 2 hrs?	Yes/No (circle one)	Yes/No (circle one)

Overleaf: for each species encountered enter the number of individuals in the first hour (compulsory) and the second hour (optional) columns. If you stay longer still, enter a tick in the Extra column for additional species recorded. Optional: after your two visits provide an estimate of the number of individuals present in the whole tetrad. See instructions for full details.

Quick Fieldwork Instructions

- Make two visits to your tetrad in the winter season: early visit between 1 November – 31 December, late visit between 1 January – 28 February
- Avoid the first and last hour of the day when birds can be moving to and from roosts – try to avoid inclement weather. If your tetrad includes coastal habitat, try to visit from mid to high tide when birds will be closer and (hopefully!) easier to identify.
- Choose a route that will cover the range of habitats in the tetrad. Walk around keeping a tally of all the individuals you encounter (seen or heard) of each species for a minimum of one hour. If you want, you can continue for a second hour. Restart the clock and walk around other parts of the tetrad, keeping a **separate** tally of the numbers encountered. Keep a note of any species you see outside the timed visits and enter these on the form. You can vary the route taken on early and late visits.
- Only count individuals using the square – this includes raptors hunting.
- If you encounter a flock of birds or a very 'busy' area then you can pause the clock to do the counting. Resume the clock when you have finished counting.
- Although the Breed Code column is shaded out you may enter any breeding evidence you find for early nesting species such as Crossbill or Collared Dove.
- If possible, estimate the number of each species present in the whole tetrad, based on your counts in the part of the tetrad you covered. These estimates need only be approximate; estimate in the '1's need only be to the nearest 1, in the '10's to the nearest 10, in the '100's to the nearest 100, etc.
- Back at home, complete the form or enter your records online at www.birdatlas.net

2.

Species		Early visit				Late visit				Tetrad Popn Estimate
		1st hour count	2nd hour count	Extra (✓)	Breed Code	1st hour count	2nd hour count	Extra (✓)	Breed Code	
Mute Swan	MS									
Bewick's Swan	BS									
Whooper Swan	WS									
Bean Goose	BE									
Taiga Bean Goose	XF									
Tundra Bean Goose	XR									
Pink-footed Goose	PG									
White-fronted Goose	WG									
European W-fronted Goose	EW									
Greenland W-fronted Goose	NW									
Greylag Goose	GJ									
Canada Goose	CG									
Barnacle Goose	BY									
Brent Goose	BG									
Dark-bellied Brent Goose	DB									
Pale-bellied Brent Goose	PB									
Egyptian Goose	EG									
Shelduck	SU									
Mandarin	MN									
Wigeon	WN									
Gadwall	GA									
Teal	T.									
Mallard	MA									
Pintail	PT									
Shoveler	SV									
Red-crested Pochard	RQ									
Pochard	PO									
Tufted Duck	TU									
Scaup	SP									
Eider	E.									
Long-tailed Duck	LN									
Common Scoter	CX									
Velvet Scoter	VS									
Goldeneye	GN									
Smew	SY									
Red-breasted Merganser	RM									
Goosander	GD									
Ruddy Duck	RY									
Red Grouse	RG									
Ptarmigan	PM									
Black Grouse	BK									
Capercaillie	CP									
Red-legged Partridge	RL									
Grey Partridge	P.									
Pheasant	PH									
Golden Pheasant	GF									
Lady Amherst's Pheasant	LM									
Red-throated Diver	RH									
Black-throated Diver	BV									
Great Northern Diver	ND									
Little Grebe	LG									
Great Crested Grebe	GG									
Red-necked Grebe	RX									
Slavonian Grebe	SZ									
Black-necked Grebe	BN									
Fulmar	F.									
Gannet	GX									

3.

Species		Early visit				Late visit				Tetrad Popn Estimate
		1st hour count	2nd hour count	Extra (✓)	Breed Code	1st hour count	2nd hour count	Extra (✓)	Breed Code	
Cormorant	CA									
Shag	SA									
Bittern	BI									
Little Egret	ET									
Grey Heron	H.									
Red Kite	KT									
Marsh Harrier	MR									
Hen Harrier	HH									
Goshawk	GI									
Sparrowhawk	SH									
Buzzard	BZ									
Golden Eagle	EA									
Kestrel	K.									
Merlin	ML									
Peregrine	PE									
Water Rail	WA									
Moorhen	MH									
Coot	CO									
Oystercatcher	OC									
Avocet	AV									
Ringed Plover	RP									
Golden Plover	GP									
Grey Plover	GV									
Lapwing	L.									
Knot	KN									
Sanderling	SS									
Little Stint	LX									
Purple Sandpiper	PS									
Dunlin	DN									
Ruff	RU									
Jack Snipe	JS									
Snipe	SN									
Woodcock	WK									
Black-tailed Godwit	BW									
Bar-tailed Godwit	BA									
Whimbrel	WM									
Curlew	CU									
Spotted Redshank	DR									
Redshank	RK									
Greenshank	GK									
Green Sandpiper	GE									
Common Sandpiper	CS									
Turnstone	TT									
Grey Phalarope	PL									
Arctic Skua	AC									
Great Skua	NX									
Mediterranean Gull	MU									
Little Gull	LU									
Black-headed Gull	BH									
Ring-billed Gull	IN									
Common Gull	CM									
Lesser Black-backed Gull	LB									
Yellow-legged Gull	YG									
Herring Gull	HG									
Iceland Gull	IG									
Glaucous Gull	GZ									
Great Black-backed Gull	GB									

4.

Species		Early visit				Late visit				Tetrad Popn Estimate
		1st hour count	2nd hour count	Extra (✓)	Breed Code	1st hour count	2nd hour count	Extra (✓)	Breed Code	
Kittiwake	KI									
Sandwich Tern	TE									
Guillemot	GU									
Razorbill	RA									
Black Guillemot	TY									
Little Auk	LK									
Puffin	PU									
Feral Pigeon	FP									
Rock Dove	DV									
Stock Dove	SD									
Woodpigeon	WP									
Collared Dove	CD									
Ring-necked Parakeet	RI									
Barn Owl	BO									
Little Owl	LO									
Tawny Owl	TO									
Long-eared Owl	LE									
Short-eared Owl	SE									
Kingfisher	KF									
Green Woodpecker	G.									
Great Spotted Woodpecker	GS									
Lesser Spotted Woodpecker	LS									
Woodlark	WL									
Skylark	S.									
Shorelark	SX									
Swallow	SL									
House Martin	HM									
Meadow Pipit	MP									
Rock Pipit	RC									
Water Pipit	WI									
Grey Wagtail	GL									
Pied Wagtail	PW									
Waxwing	WX									
Dipper	DI									
Wren	WR									
Dunnock	D.									
Robin	R.									
Black Redstart	BX									
Stonechat	SC									
Blackbird	B.									
Fieldfare	FF									
Song Thrush	ST									
Redwing	RE									
Mistle Thrush	M.									
Cetti's Warbler	CW									
Dartford Warbler	DW									
Blackcap	BC									
Chiffchaff	CC									
Goldcrest	GC									
Firecrest	FC									
Bearded Tit	BR									
Long-tailed Tit	LT									
Marsh Tit	MT									
Willow Tit	WT									
Crested Tit	CI									
Coal Tit	CT									
Blue Tit	BT									

5.

Species		Early visit				Late visit				Tetrad Popn Estimate
		1st hour count	2nd hour count	Extra (✓)	Breed Code	1st hour count	2nd hour count	Extra (✓)	Breed Code	
Great Tit	GT									
Nuthatch	NH									
Treecreeper	TC									
Jay	J.									
Magpie	MG									
Chough	CF									
Jackdaw	JD									
Rook	RO									
Carrion Crow	C.									
Hooded Crow	HC									
Carrion/Hooded Crow	HB									
Raven	RN									
Starling	SG									
House Sparrow	HS									
Tree Sparrow	TS									
Chaffinch	CH									
Brambling	BL									
Greenfinch	GR									
Goldfinch	GO									
Siskin	SK									
Linnet	LI									
Twite	TW									
Common (Mealy) Redpoll	FR									
Lesser Redpoll	LR									
Redpoll (Common/Lesser)	FQ									
Common Crossbill	CR									
Scottish Crossbill	CY									
Parrot Crossbill	PC									
Unidentifield crossbill	XB									
Bullfinch	BF									
Hawfinch	HF									
Lapland Bunting	LA									
Snow Bunting	SB									
Yellowhammer	Y.									
Cirl Bunting	CL									
Reed Bunting	RB									
Corn Bunting	CB									

6.

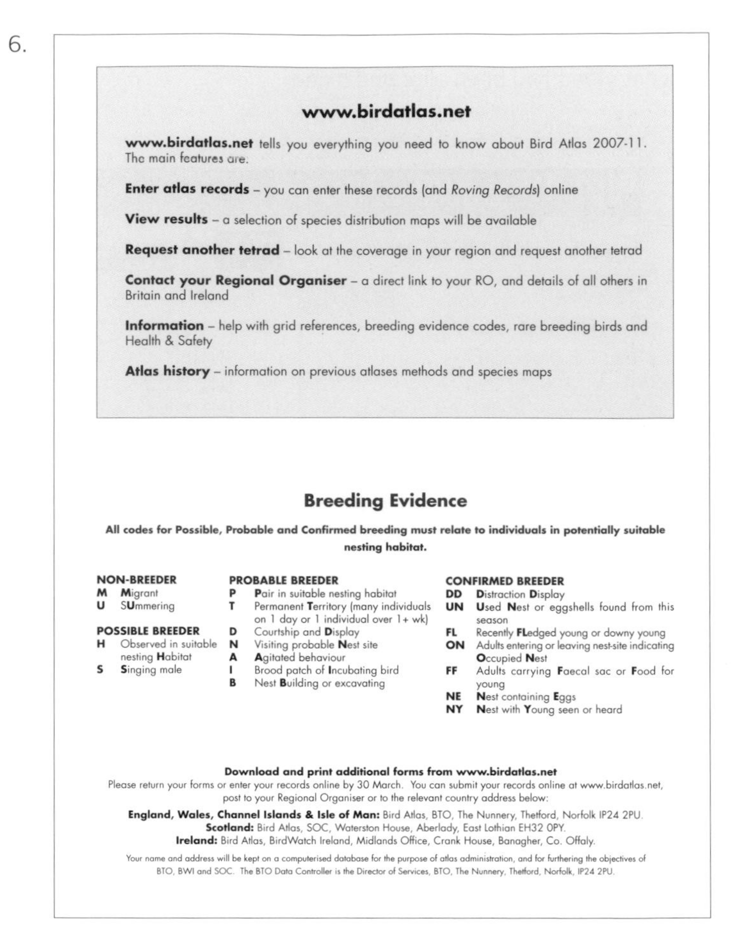

www.birdatlas.net

www.birdatlas.net tells you everything you need to know about Bird Atlas 2007-11. The main features are:

Enter atlas records – you can enter these records (and *Roving Records*) online

View results – a selection of species distribution maps will be available

Request another tetrad – look at the coverage in your region and request another tetrad

Contact your Regional Organiser – a direct link to your RO, and details of all others in Britain and Ireland

Information – help with grid references, breeding evidence codes, rare breeding birds and Health & Safety

Atlas history – information on previous atlases methods and species maps

Breeding Evidence

All codes for Possible, Probable and Confirmed breeding must relate to individuals in potentially suitable nesting habitat.

NON-BREEDER
- **M** **M**igrant
- **U** S**U**mmering

POSSIBLE BREEDER
- **H** Observed in suitable nesting **H**abitat
- **S** **S**inging male

PROBABLE BREEDER
- **P** **P**air in suitable nesting habitat
- **T** Permanent **T**erritory (many individuals on 1 day or 1 individual over 1+ wk)
- **D** Courtship and **D**isplay
- **N** Visiting probable **N**est site
- **A** **A**gitated behaviour
- **I** Brood patch of **I**ncubating bird
- **B** Nest **B**uilding or excavating

CONFIRMED BREEDER
- **DD** **D**istraction **D**isplay
- **UN** **U**sed **N**est or eggshells found from this season
- **FL** Recently **FL**edged young or downy young
- **ON** Adults entering or leaving nest-site indicating **O**ccupied **N**est
- **FF** Adults carrying **F**aecal sac or **F**ood for young
- **NE** **N**est containing **E**ggs
- **NY** **N**est with **Y**oung seen or heard

Download and print additional forms from www.birdatlas.net
Please return your forms or enter your records online by 30 March. You can submit your records online at www.birdatlas.net, post to your Regional Organiser or to the relevant country address below:

England, Wales, Channel Islands & Isle of Man: Bird Atlas, BTO, The Nunnery, Thetford, Norfolk IP24 2PU.
Scotland: Bird Atlas, SOC, Waterston House, Aberlady, East Lothian EH32 0PY.
Ireland: Bird Atlas, BirdWatch Ireland, Midlands Office, Crank House, Banagher, Co. Offaly.

Your name and address will be kept on a computerised database for the purpose of atlas administration, and for furthering the objectives of BTO, BWI and SOC. The BTO Data Controller is the Director of Services, BTO, The Nunnery, Thetford, Norfolk, IP24 2PU.

E TIMED TETRAD VISIT FORM FOR THE BREEDING SEASON

To limit opportunities for duplication of coverage, the breeding-season TTV form was only made available, via post or download, to observers who had been allocated tetrads.

1.

Bird Atlas 2007-11
Mapping Britain and Ireland's birds

Timed Tetrad Visit form - Breeding

BTO SOC BirdWatch Ireland

Observer name: Office Use

Postal address:

Email address:

Enter the 10-km and tetrad code

Example S P 2 4 J

E	J	P	U	Z
D	I	N	T	Y
C	H	M	S	X
B	G	L	R	W
A	F	K	Q	V

Please enter the date of your visit. You must count birds for at least the first hour of your visit, but if you continue counting for a second hour please indicate by circling 'Yes'. Otherwise circle 'No'.

	Early Visit (Apr-May)	Late Visit (Jun-Jul)
Date of timed visit	dd/mm/yyyy	dd/mm/yyyy
Did you count for 2 hrs?	Yes/No (circle one)	Yes/No (circle one)

Overleaf: for each species encountered enter the number of individuals in the first hour (compulsory) and the second hour (optional) columns. If you stay longer still, enter a tick in the Extra column for additional species recorded. Optional: after your two visits provide an estimate of the number of individuals present in the whole tetrad. See instructions for full details.

Quick Fieldwork Instructions

- Make two visits to your tetrad in the breeding season: early visit between 1 April – 31 May, late visit between 1 June – 31 July
- Make visits during the morning – try to avoid inclement weather. If your tetrad includes coastal habitat, try to visit from mid to high tide when birds will be closer and (hopefully!) easier to identify.
- Choose a route that will cover the range of habitats in the tetrad. Walk around keeping a tally of all the individuals you encounter (seen or heard) of each species for a minimum of one hour. If you want, you can continue for a second hour. Restart the clock and walk around other parts of the tetrad, keeping a **separate** tally of the numbers encountered. Keep a note of any species you see outside the timed visits and enter these on the form. You can vary the route taken on early and late visits.
- Look out for evidence of breeding but do not spend too long trying to confirm breeding. Codes are given on the back page.
- Only count individuals using the square – this includes raptors hunting and hirundines feeding. Only count full grown birds and where possible exclude juveniles.
- If you encounter a colony of colonial nesting species (species marked with ★) please tick the colony sheet to say a colony is present. If you can, estimate the number of individuals or apparently occupied nests (see Colony Table). Only one count per species is required - the maximum count recorded on the two visits.
- If you encounter a flock of birds, a colony or a very 'busy' area then you can pause the clock to do the counting. Resume the clock when you have finished counting.
- If possible, estimate the number of each species present in the whole tetrad, based on your counts in the part of the tetrad you covered. These estimates need only be approximate; estimate in the '1's need only be to the nearest 1, in the '10's to the nearest 10, in the '100's to the nearest 100, etc.
- Back at home, complete the form or enter your records online at www.birdatlas.net

2.

Species		Early visit				Late visit				Tetrad Popn Estimate
		1st hour count	2nd hour count	Extra (✓)	Breed Code	1st hour count	2nd hour count	Extra (✓)	Breed Code	
Mute Swan	MS									
Greylag Goose	GJ									
Canada Goose	CG									
Egyptian Goose	EG									
Shelduck	SU									
Mandarin	MN									
Wigeon	WN									
Gadwall	GA									
Teal	T.									
Mallard	MA									
Pintail	PT									
Garganey	GY									
Shoveler	SV									
Red-crested Pochard	RQ									
Pochard	PO									
Tufted Duck	TU									
Eider	E.									
Common Scoter	CX									
Goldeneye	GN									
Red-breasted Merganser	RM									
Goosander	GD									
Ruddy Duck	RY									
Red Grouse	RG									
Ptarmigan	PM									
Black Grouse	BK									
Capercaillie	CP									
Red-legged Partridge	RL									
Grey Partridge	P.									
Quail	Q.									
Pheasant	PH									
Golden Pheasant	GF									
Lady Amherst's Pheasant	LM									
Red-throated Diver	RH									
Black-throated Diver	BV									
Little Grebe	LG									
Great Crested Grebe	GG									
Slavonian Grebe	SZ									
Black-necked Grebe	BN									
Fulmar ★	F.									
Manx Shearwater ★	MX									
Storm Petrel ★	TM									
Leach's Petrel ★	TL									
Gannet ★	GX									
Cormorant ★	CA									
Shag ★	SA									
Bittern	BI									
Little Egret ★	ET									
Grey Heron ★	H.									
Honey Buzzard	HZ									
Red Kite	KT									
White-tailed Eagle	WE									
Marsh Harrier	MR									
Hen Harrier	HH									
Montagu's Harrier	MO									
Goshawk	GI									
Sparrowhawk	SH									
Buzzard	BZ									

3.

Species		Early visit				Late visit				Tetrad Popn Estimate
		1st hour count	2nd hour count	Extra (✓)	Breed Code	1st hour count	2nd hour count	Extra (✓)	Breed Code	
Golden Eagle	EA									
Osprey	OP									
Kestrel	K.									
Merlin	ML									
Hobby	HY									
Peregrine	PE									
Water Rail	WA									
Spotted Crake	AK									
Corncrake	CE									
Moorhen	MH									
Coot	CO									
Oystercatcher	OC									
Avocet	AV									
Stone Curlew	TN									
Little Ringed Plover	LP									
Ringed Plover	RP									
Dotterel	DO									
Golden Plover	GP									
Lapwing	L.									
Temminck's Stint	TK									
Dunlin	DN									
Ruff	RU									
Snipe	SN									
Woodcock	WK									
Black-tailed Godwit	BW									
Whimbrel	WM									
Curlew	CU									
Redshank	RK									
Greenshank	GK									
Wood Sandpiper	OD									
Common Sandpiper	CS									
Red-necked Phalarope	NK									
Arctic Skua *	AC									
Great Skua *	NX									
Mediterranean Gull *	MU									
Black-headed Gull *	BH									
Common Gull *	CM									
Lesser Black-backed Gull *	LB									
Herring Gull *	HG									
Great Black-backed Gull *	GB									
Kittiwake *	KI									
Sandwich Tern *	TE									
Roseate Tern *	RS									
Common Tern *	CN									
Arctic Tern *	AE									
Little Tern *	AF									
Guillemot *	GU									
Razorbill *	RA									
Black Guillemot	TY									
Puffin *	PU									
Feral Pigeon	FP									
Rock Dove	DV									
Stock Dove	SD									
Woodpigeon	WP									
Collared Dove	CD									
Turtle Dove	TD									
Ring-necked Parakeet	RI									

4.

Species		Early visit				Late visit				Tetrad Popn Estimate
		1st hour count	2nd hour count	Extra (✓)	Breed Code	1st hour count	2nd hour count	Extra (✓)	Breed Code	
Cuckoo	CK									
Barn Owl	BO									
Little Owl	LO									
Tawny Owl	TO									
Long-eared Owl	LE									
Short-eared Owl	SE									
Nightjar	NJ									
Swift	SI									
Kingfisher	KF									
Wryneck	WY									
Green Woodpecker	G.									
Great Spotted Woodpecker	GS									
Lesser Spoted Woodpecker	LS									
Woodlark	WL									
Skylark	S.									
Sand Martin *	SM									
Swallow	SL									
House Martin	HM									
Tree Pipit	TP									
Meadow Pipit	MP									
Rock Pipit	RC									
Yellow Wagtail	YW									
Grey Wagtail	GL									
Pied Wagtail	PW									
Dipper	DI									
Wren	WR									
Dunnock	D.									
Robin	R.									
Nightingale	N.									
Black Redstart	BX									
Redstart	RT									
Whinchat	WC									
Stonechat	SC									
Wheatear	W.									
Ring Ouzel	RZ									
Blackbird	B.									
Fieldfare	FF									
Song Thrush	ST									
Redwing	RE									
Mistle Thrush	M.									
Cetti's Warbler	CW									
Grasshopper Warbler	GH									
Savis Warbler	VI									
Sedge Warbler	SW									
Marsh Warbler	MW									
Reed Warbler	RW									
Dartford Warbler	DW									
Lesser Whitethroat	LW									
Whitethroat	WH									
Garden Warbler	GW									
Blackcap	BC									
Wood Warbler	WO									
Chiffchaff	CC									
Willow Warbler	WW									
Goldcrest	GC									
Firecrest	FC									
Spotted Flycatcher	SF									

5.

Species		Early visit				Late visit				Tetrad Popn Estimate
		1st hour count	2nd hour count	Extra (✓)	Breed Code	1st hour count	2nd hour count	Extra (✓)	Breed Code	
Pied Flycatcher	PF									
Bearded Tit	BR									
Long-tailed Tit	LT									
Marsh Tit	MT									
Willow Tit	WT									
Crested Tit	CI									
Coal Tit	CT									
Blue Tit	BT									
Great Tit	GT									
Nuthatch	NH									
Treecreeper	TC									
Golden Oriole	OL									
Jay	J.									
Magpie	MG									
Chough	CF									
Jackdaw	JD									
Rook *	RO									
Carrion Crow	C.									
Hooded Crow	HC									
Carrion/Hooded Crow	HB									
Raven	RN									
Starling	SG									
House Sparrow	HS									
Tree Sparrow	TS									
Chaffinch	CH									
Greenfinch	GR									
Goldfinch	GO									
Siskin	SK									
Linnet	LI									
Twite	TW									
Lesser Redpoll	LR									
Redpoll (Common/Lesser)	FQ									
Common Crossbill	CR									
Scottish Crossbill	CY									
Parrot Crossbill	PC									
Unidentified crossbill	XB									
Hawfinch	HF									
Bullfinch	BF									
Snow Bunting	SB									
Yellowhammer	Y.									
Cirl Bunting	CL									
Reed Bunting	RB									
Corn Bunting	CB									

Colony Table →

6.

Colony Table (enter maximum count over two visits)

Do not use this method for this species

Preferred count method for this species

This method is acceptable if the preferred method is not possible

Species		Colony present (Tick) MANDATORY	Estimated number of individuals at colony OPTIONAL	Estimated number of Apparently Occupied Nests OPTIONAL
Fulmar	F.			
Manx Shearwater	MX			
Storm Petrel	TM			
Leach's Petrel	TL			
Gannet	GX			
Cormorant	CA			
Shag	SA			
Little Egret	ET			
Grey Heron	H.			
Arctic Skua	AC			
Great Skua	NX			
Mediterranean Gull	MU			
Black-headed Gull	BH			
Common Gull	CM			
Lesser Black-backed Gull	LB			
Herring Gull	HG			
Great Black-backed Gull	GB			
Kittiwake	KI			
Sandwich Tern	TE			
Roseate Tern	RS			
Common Tern	CN			
Arctic Tern	AE			
Little Tern	AF			
Guillemot	GU			
Razorbill	RA			
Puffin	PU			
Sand Martin	SM			
Rook	RO			

Breeding Evidence

All codes for Possible, Probable and Confirmed breeding must relate to individuals in potentially suitable nesting habitat.

NON-BREEDER
- **M** **M**igrant
- **U** S**U**mmering

POSSIBLE BREEDER
- **H** Observed in suitable nesting **H**abitat
- **S** **S**inging male

PROBABLE BREEDER
- **P** **P**air in suitable nesting habitat
- **T** Permanent **T**erritory (many individuals on 1 day or 1 individual over 1+ wk)
- **D** Courtship and **D**isplay
- **N** Visiting probable **N**est site
- **A** **A**gitated behaviour
- **I** Brood patch of **I**ncubating bird
- **B** Nest **B**uilding or excavating

CONFIRMED BREEDER
- **DD** **D**istraction **D**isplay
- **UN** **U**sed **N**est or eggshells found from this season
- **FL** Recently **FL**edged young or downy young
- **ON** Adults entering or leaving nest-site indicating **O**ccupied **N**est
- **FF** Adults carrying **F**aecal sac or **F**ood for young
- **NE** **N**est containing **E**ggs
- **NY** **N**est with **Y**oung seen or heard

Download and print additional forms from www.birdatlas.net

Please return your forms by 31 August. You can submit your records online at www.birdatlas.net, post to your Regional Organiser or post to the relevant country address:

England, Wales, Channel Islands & Isle of Man: Bird Atlas, BTO, The Nunnery, Thetford, Norfolk IP24 2PU.
Scotland: Bird Atlas, SOC, Waterston House, Aberlady, East Lothian EH32 0PY.
Ireland: Bird Atlas, BirdWatch Ireland, Midlands Office, Crank House, Banagher, Co. Offaly.

Your name and address will be kept on a computerised database for the purpose of atlas administration, and for furthering the objectives of BTO, BWI and SOC. The BTO Data Controller is the Director of Services, BTO, The Nunnery, Thetford, Norfolk, IP24 2PU.

APPENDIX 4 – OUT-OF-SEASON BREEDING EVIDENCE USED IN CALCULATION OF SPECIES' RANGES

A SIGNIFICANT NUMBER of species have a breeding season that extends beyond the standard April–July period used for breeding atlas fieldwork and there is a strong case for the inclusion of out-of-season breeding evidence in the production of distribution maps (Section 2.2.2). This does not mean that all out-of-season breeding evidence should be used: for many passerine species, young birds can move considerable distances post-fledging, so using August records of fledged young could exaggerate the breeding range. Therefore, to use the data most appropriately to build the comprehensive species maps, a set of species-specific rules were created detailing which out-of-season breeding evidence codes would be used to define a species' range.

In total, 159 species were identified for which an aspect of breeding behaviour extended outside the April–July period. The table lists these species, with rows for March, August, September and October, and coloured shading in columns to indicate the breeding evidence codes (see Table 2.2) that were used in each month for the calculation of range. Distribution maps for breeding species not listed here were based solely on in-season breeding evidence (i.e. April–July inclusive).

It is unclear to what degree out-of-season breeding evidence was used in previous atlases. For this reason, no out-of-season breeding records were used in the calculation of distribution changes (Section 4.3.2).

SPECIES		H	S	P	T	D	N	A	I	B	DD	UN	FL	ON	FF	NE	NY
Mute Swan	MAR	■	■	■	■	■	■	■	■	■	■	■	■	■	■	■	■
	AUG			■	■	■	■	■	■	■	■	■	■	■	■	■	■
	SEPT																
	OCT																
Black Swan	MAR	■	■	■	■	■	■	■	■	■	■	■	■	■	■	■	■
	AUG			■	■	■	■	■	■	■	■	■	■	■	■	■	■
	SEPT																
	OCT																
Greylag Goose	MAR				■	■	■	■	■	■	■	■	■	■	■	■	■
	AUG			■	■	■	■	■	■	■	■	■	■	■	■	■	■
	SEPT																
	OCT																
Bar-headed Goose	MAR	■	■	■	■	■	■	■	■	■	■	■	■	■	■	■	■
	AUG			■	■	■	■	■	■	■	■	■	■	■	■	■	■
	SEPT																
	OCT																
Canada Goose	MAR	■	■	■	■	■	■	■	■	■	■	■	■	■	■	■	■
	AUG			■	■	■	■	■	■	■	■	■	■	■	■	■	■
	SEPT																
	OCT																
Barnacle Goose	MAR				■	■	■	■	■	■	■	■	■	■	■	■	■
	AUG			■	■	■	■	■	■	■	■	■	■	■	■	■	■
	SEPT																
	OCT																
Egyptian Goose	MAR	■	■	■	■	■	■	■	■	■	■	■	■	■	■	■	■
	AUG			■	■	■	■	■	■	■	■	■	■	■	■	■	■
	SEPT																
	OCT																
Shelduck	MAR	■	■	■	■	■	■	■	■	■	■	■	■	■	■	■	■
	AUG			■	■	■	■	■	■	■	■	■	■	■	■	■	■
	SEPT																
	OCT																
Muscovy Duck	MAR	■	■	■	■	■	■	■	■	■	■	■	■	■	■	■	■
	AUG			■	■	■	■	■	■	■	■	■	■	■	■	■	■
	SEPT																
	OCT																
Mandarin Duck	MAR	■	■	■	■	■	■	■	■	■	■	■	■	■	■	■	■
	AUG			■	■	■	■	■	■	■	■	■	■	■	■	■	■
	SEPT																
	OCT																
Mallard	MAR	■	■	■	■	■	■	■	■	■	■	■	■	■	■	■	■
	AUG			■	■	■	■	■	■	■	■	■	■	■	■	■	■
	SEPT																
	OCT																
Red Grouse	MAR	■	■	■	■	■	■	■	■	■	■	■	■	■	■	■	■
	AUG			■	■	■	■	■	■	■	■	■	■	■	■	■	■
	SEPT																
	OCT																

SPECIES		H	S	P	T	D	N	A	I	B	DD	UN	FL	ON	FF	NE	NY
Ptarmigan	MAR	■	■	■	■	■	■	■	■	■	■	■	■	■	■	■	■
	AUG			■	■	■	■	■	■	■	■	■	■	■	■	■	■
	SEPT																
	OCT																
Black Grouse	MAR	■	■	■	■	■	■	■	■	■	■	■	■	■	■	■	■
	AUG			■	■	■	■	■	■	■	■	■	■	■	■	■	■
	SEPT																
	OCT																
Capercaillie	MAR	■	■	■	■	■	■	■	■	■	■	■	■	■	■	■	■
	AUG	■	■	■	■	■	■	■	■	■	■	■	■	■	■	■	■
	SEPT																
	OCT																
Red-legged Partridge	MAR	■	■	■	■	■	■	■	■	■	■	■	■	■	■	■	■
	AUG			■	■	■	■	■	■	■	■	■	■	■	■	■	■
	SEPT																
	OCT																
Grey Partridge	MAR	■	■	■	■	■	■	■	■	■	■	■	■	■	■	■	■
	AUG			■	■	■	■	■	■	■	■	■	■	■	■	■	■
	SEPT																
	OCT																
Quail	MAR	■	■	■	■	■	■	■	■	■	■	■	■	■	■	■	■
	AUG			■	■	■	■	■	■	■	■	■	■	■	■	■	■
	SEPT																
	OCT																
Pheasant	MAR	■	■	■	■	■	■	■	■	■	■	■	■	■	■	■	■
	AUG			■	■	■	■	■	■	■	■	■	■	■	■	■	■
	SEPT																
	OCT																
Golden Pheasant	MAR	■	■	■	■	■	■	■	■	■	■	■	■	■	■	■	■
	AUG			■	■	■	■	■	■	■	■	■	■	■	■	■	■
	SEPT																
	OCT																
Indian Peafowl	MAR	■	■	■	■	■	■	■	■	■	■	■	■	■	■	■	■
	AUG			■	■	■	■	■	■	■	■	■	■	■	■	■	■
	SEPT																
	OCT																
Little Grebe	MAR	■	■	■	■	■	■	■	■	■	■	■	■	■	■	■	■
	AUG			■	■	■	■	■	■	■	■	■	■	■	■	■	■
	SEPT																
	OCT																
Great Crested Grebe	MAR			■	■	■	■	■	■	■	■	■	■	■	■	■	■
	AUG			■	■	■	■	■	■	■	■	■	■	■	■	■	■
	SEPT																
	OCT																
Fulmar	MAR																
	AUG			■	■	■	■	■	■	■	■	■		■		■	■
	SEPT			■	■	■	■	■	■	■	■	■		■		■	■
	OCT																
Manx Shearwater	MAR																
	AUG			■	■	■	■	■	■	■	■	■		■		■	■
	SEPT			■	■	■	■	■	■	■	■	■		■		■	■
	OCT											■		■		■	■
Storm Petrel	MAR																
	AUG											■		■		■	■
	SEPT											■		■		■	■
	OCT																
Leach's Petrel	MAR																
	AUG			■	■	■	■	■	■	■	■	■		■		■	■
	SEPT			■	■	■	■	■	■	■	■	■		■		■	■
	OCT																
Gannet	MAR																
	AUG											■		■		■	■
	SEPT											■		■		■	■
	OCT											■		■		■	■
Cormorant	MAR										■	■		■		■	■
	AUG			■	■	■	■	■	■	■	■	■		■		■	■
	SEPT																
	OCT																
Shag	MAR										■	■		■		■	■
	AUG			■	■	■	■	■	■	■	■	■		■		■	■
	SEPT																
	OCT																
Grey Heron	MAR	■	■	■	■	■	■	■	■	■	■	■		■		■	■
	AUG			■	■	■	■	■	■	■	■	■		■		■	■
	SEPT																
	OCT																
Spoonbill	MAR																
	AUG			■	■	■	■	■	■	■	■	■	■	■	■	■	■
	SEPT																
	OCT																
Honey-buzzard	MAR																
	AUG			■	■	■	■	■	■	■	■	■	■	■	■	■	■
	SEPT																
	OCT																

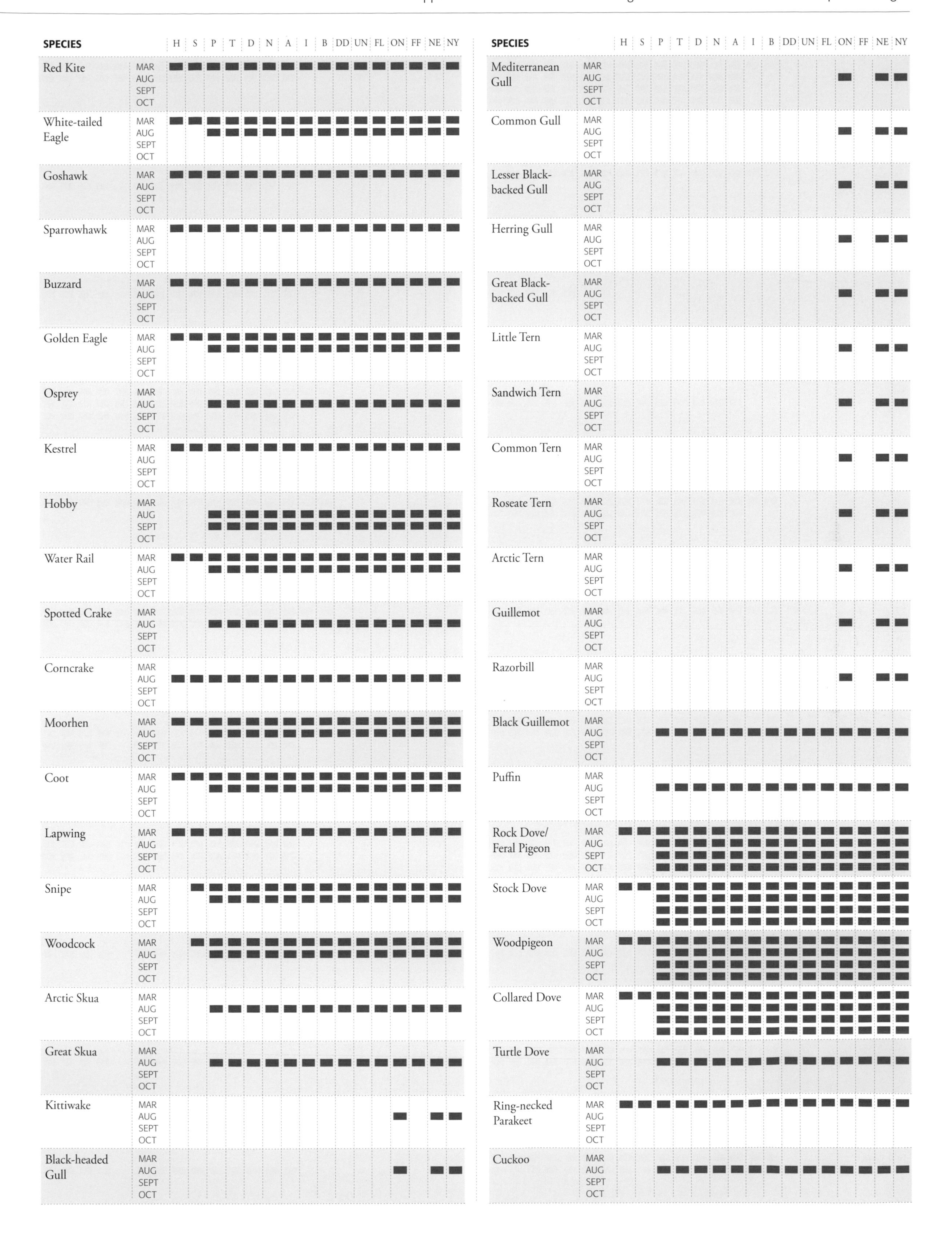

SPECIES		H	S	P	T	D	N	A	I	B	DD	UN	FL	ON	FF	NE	NY
Red Kite	MAR	■	■	■	■	■	■	■	■	■	■	■	■	■	■	■	■
	AUG																
	SEPT																
	OCT																
White-tailed Eagle	MAR	■	■	■	■	■	■	■	■	■	■	■	■	■	■	■	■
	AUG			■	■	■	■	■	■	■	■	■	■	■	■	■	■
	SEPT																
	OCT																
Goshawk	MAR	■	■	■	■	■	■	■	■	■	■	■	■	■	■	■	■
	AUG																
	SEPT																
	OCT																
Sparrowhawk	MAR	■	■	■	■	■	■	■	■	■	■	■	■	■	■	■	■
	AUG																
	SEPT																
	OCT																
Buzzard	MAR	■	■	■	■	■	■	■	■	■	■	■	■	■	■	■	■
	AUG																
	SEPT																
	OCT																
Golden Eagle	MAR	■	■	■	■	■	■	■	■	■	■	■	■	■	■	■	■
	AUG			■	■	■	■	■	■	■	■	■	■	■	■	■	■
	SEPT																
	OCT																
Osprey	MAR																
	AUG			■	■	■	■	■	■	■	■	■	■	■	■	■	■
	SEPT																
	OCT																
Kestrel	MAR	■	■	■	■	■	■	■	■	■	■	■	■	■	■	■	■
	AUG																
	SEPT																
	OCT																
Hobby	MAR																
	AUG			■	■	■	■	■	■	■	■	■	■	■	■	■	■
	SEPT			■	■	■	■	■	■	■	■	■	■	■	■	■	■
	OCT																
Water Rail	MAR	■	■	■	■	■	■	■	■	■	■	■	■	■	■	■	■
	AUG			■	■	■	■	■	■	■	■	■	■	■	■	■	■
	SEPT																
	OCT																
Spotted Crake	MAR																
	AUG			■	■	■	■	■	■	■	■	■	■	■	■	■	■
	SEPT																
	OCT																
Corncrake	MAR																
	AUG	■	■	■	■	■	■	■	■	■	■	■	■	■	■	■	■
	SEPT																
	OCT																
Moorhen	MAR	■	■	■	■	■	■	■	■	■	■	■	■	■	■	■	■
	AUG			■	■	■	■	■	■	■	■	■	■	■	■	■	■
	SEPT																
	OCT																
Coot	MAR	■	■	■	■	■	■	■	■	■	■	■	■	■	■	■	■
	AUG			■	■	■	■	■	■	■	■	■	■	■	■	■	■
	SEPT																
	OCT																
Lapwing	MAR	■	■	■	■	■	■	■	■	■	■	■	■	■	■	■	■
	AUG																
	SEPT																
	OCT																
Snipe	MAR		■	■	■	■	■	■	■	■	■	■	■	■	■	■	■
	AUG			■	■	■	■	■	■	■	■	■	■	■	■	■	■
	SEPT																
	OCT																
Woodcock	MAR		■	■	■	■	■	■	■	■	■	■	■	■	■	■	■
	AUG			■	■	■	■	■	■	■	■	■	■	■	■	■	■
	SEPT																
	OCT																
Arctic Skua	MAR																
	AUG			■	■	■	■	■	■	■	■	■	■	■	■	■	■
	SEPT																
	OCT																
Great Skua	MAR																
	AUG			■	■	■	■	■	■	■	■	■	■	■	■	■	■
	SEPT																
	OCT																
Kittiwake	MAR																
	AUG													■		■	■
	SEPT																
	OCT																
Black-headed Gull	MAR																
	AUG													■		■	■
	SEPT																
	OCT																
Mediterranean Gull	MAR																
	AUG													■		■	■
	SEPT																
	OCT																
Common Gull	MAR																
	AUG													■		■	■
	SEPT																
	OCT																
Lesser Black-backed Gull	MAR																
	AUG													■		■	■
	SEPT																
	OCT																
Herring Gull	MAR																
	AUG													■		■	■
	SEPT																
	OCT																
Great Black-backed Gull	MAR																
	AUG													■		■	■
	SEPT																
	OCT																
Little Tern	MAR																
	AUG													■		■	■
	SEPT																
	OCT																
Sandwich Tern	MAR																
	AUG													■		■	■
	SEPT																
	OCT																
Common Tern	MAR																
	AUG													■		■	■
	SEPT																
	OCT																
Roseate Tern	MAR																
	AUG													■		■	■
	SEPT																
	OCT																
Arctic Tern	MAR																
	AUG													■		■	■
	SEPT																
	OCT																
Guillemot	MAR																
	AUG													■		■	■
	SEPT																
	OCT																
Razorbill	MAR																
	AUG													■		■	■
	SEPT																
	OCT																
Black Guillemot	MAR																
	AUG			■	■	■	■	■	■	■	■	■	■	■	■	■	■
	SEPT																
	OCT																
Puffin	MAR																
	AUG			■	■	■	■	■	■	■	■	■	■	■	■	■	■
	SEPT																
	OCT																
Rock Dove/ Feral Pigeon	MAR	■	■	■	■	■	■	■	■	■	■	■	■	■	■	■	■
	AUG			■	■	■	■	■	■	■	■	■	■	■	■	■	■
	SEPT			■	■	■	■	■	■	■	■	■	■	■	■	■	■
	OCT			■	■	■	■	■	■	■	■	■	■	■	■	■	■
Stock Dove	MAR	■	■	■	■	■	■	■	■	■	■	■	■	■	■	■	■
	AUG			■	■	■	■	■	■	■	■	■	■	■	■	■	■
	SEPT			■	■	■	■	■	■	■	■	■	■	■	■	■	■
	OCT			■	■	■	■	■	■	■	■	■	■	■	■	■	■
Woodpigeon	MAR	■	■	■	■	■	■	■	■	■	■	■	■	■	■	■	■
	AUG			■	■	■	■	■	■	■	■	■	■	■	■	■	■
	SEPT			■	■	■	■	■	■	■	■	■	■	■	■	■	■
	OCT			■	■	■	■	■	■	■	■	■	■	■	■	■	■
Collared Dove	MAR	■	■	■	■	■	■	■	■	■	■	■	■	■	■	■	■
	AUG			■	■	■	■	■	■	■	■	■	■	■	■	■	■
	SEPT			■	■	■	■	■	■	■	■	■	■	■	■	■	■
	OCT			■	■	■	■	■	■	■	■	■	■	■	■	■	■
Turtle Dove	MAR																
	AUG			■	■	■	■	■	■	■	■	■	■	■	■	■	■
	SEPT																
	OCT																
Ring-necked Parakeet	MAR	■	■	■	■	■	■	■	■	■	■	■	■	■	■	■	■
	AUG																
	SEPT																
	OCT																
Cuckoo	MAR																
	AUG			■	■	■	■	■	■	■	■	■	■	■	■	■	■
	SEPT																
	OCT																

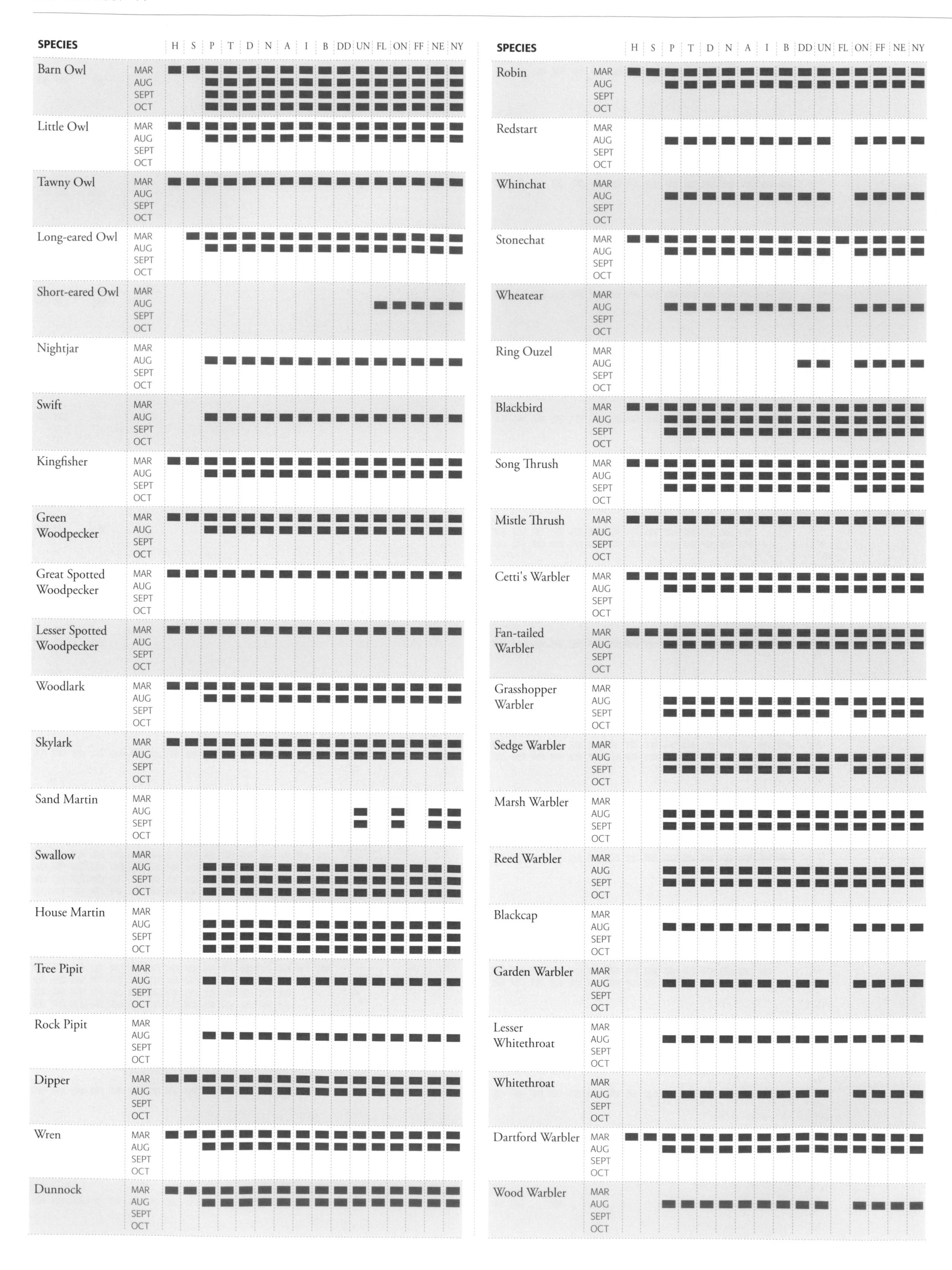
SPECIES
H S P T D N A I B DD UN FL ON FF NE NY
Barn Owl
Little Owl
Tawny Owl
Long-eared Owl
Short-eared Owl
Nightjar
Swift
Kingfisher
Green Woodpecker
Great Spotted Woodpecker
Lesser Spotted Woodpecker
Woodlark
Skylark
Sand Martin
Swallow
House Martin
Tree Pipit
Rock Pipit
Dipper
Wren
Dunnock
MAR
AUG
SEPT
OCT
SPECIES
H S P T D N A I B DD UN FL ON FF NE NY
Robin
Redstart
Whinchat
Stonechat
Wheatear
Ring Ouzel
Blackbird
Song Thrush
Mistle Thrush
Cetti's Warbler
Fan-tailed Warbler
Grasshopper Warbler
Sedge Warbler
Marsh Warbler
Reed Warbler
Blackcap
Garden Warbler
Lesser Whitethroat
Whitethroat
Dartford Warbler
Wood Warbler
MAR
AUG
SEPT
OCT

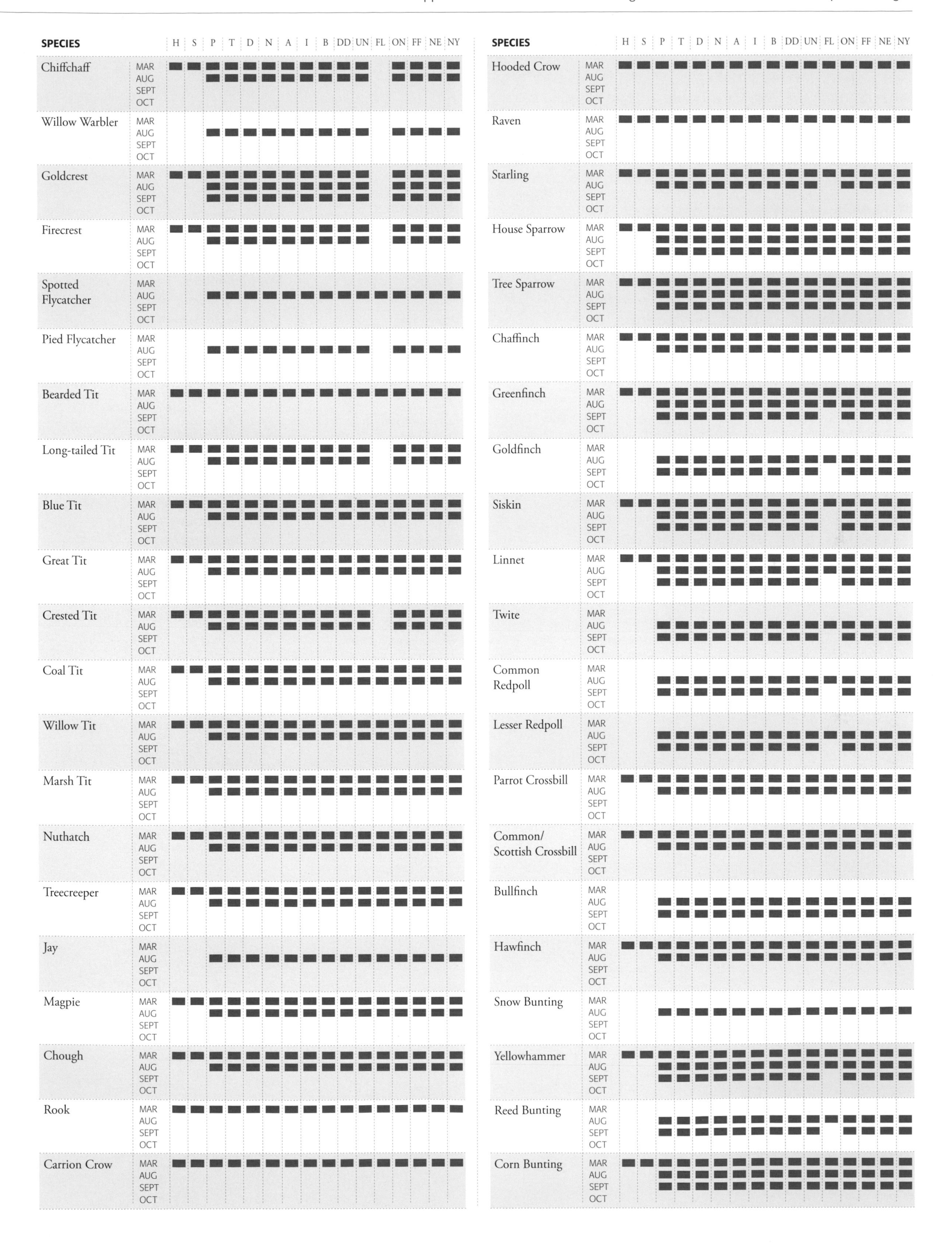

SPECIES	H	S	P	T	D	N	A	I	B	DD	UN	FL	ON	FF	NE	NY
Chiffchaff	MAR	MAR	MAR, AUG	MAR, AUG	MAR, AUG	MAR, AUG	MAR, AUG	MAR, AUG	MAR, AUG	MAR, AUG	MAR, AUG		MAR, AUG	MAR, AUG	MAR, AUG	MAR, AUG
Willow Warbler			AUG	AUG	AUG	AUG	AUG	AUG	AUG	AUG	AUG		AUG	AUG	AUG	AUG
Goldcrest	MAR	MAR	MAR, AUG, SEPT	MAR, AUG, SEPT	MAR, AUG, SEPT	MAR, AUG, SEPT	MAR, AUG, SEPT	MAR, AUG, SEPT	MAR, AUG, SEPT	MAR, AUG, SEPT	MAR, AUG, SEPT		MAR, AUG, SEPT	MAR, AUG, SEPT	MAR, AUG, SEPT	MAR, AUG, SEPT
Firecrest	MAR	MAR	MAR, AUG	MAR, AUG	MAR, AUG	MAR, AUG	MAR, AUG	MAR, AUG	MAR, AUG	MAR, AUG	MAR, AUG		MAR, AUG	MAR, AUG	MAR, AUG	MAR, AUG
Spotted Flycatcher			AUG	AUG	AUG	AUG	AUG	AUG	AUG	AUG	AUG	AUG	AUG	AUG	AUG	AUG
Pied Flycatcher			AUG	AUG	AUG	AUG	AUG	AUG	AUG	AUG	AUG		AUG	AUG	AUG	AUG
Bearded Tit	MAR	MAR	MAR	MAR	MAR	MAR	MAR	MAR	MAR	MAR	MAR	MAR	MAR	MAR	MAR	MAR
Long-tailed Tit	MAR	MAR	MAR, AUG	MAR, AUG	MAR, AUG	MAR, AUG	MAR, AUG	MAR, AUG	MAR, AUG	MAR, AUG	MAR, AUG		MAR, AUG	MAR, AUG	MAR, AUG	MAR, AUG
Blue Tit	MAR	MAR	MAR, AUG	MAR, AUG	MAR, AUG	MAR, AUG	MAR, AUG	MAR, AUG	MAR, AUG	MAR, AUG	MAR, AUG	MAR, AUG	MAR, AUG	MAR, AUG	MAR, AUG	MAR, AUG
Great Tit	MAR	MAR	MAR, AUG	MAR, AUG	MAR, AUG	MAR, AUG	MAR, AUG	MAR, AUG	MAR, AUG	MAR, AUG	MAR, AUG	MAR, AUG	MAR, AUG	MAR, AUG	MAR, AUG	MAR, AUG
Crested Tit	MAR	MAR	MAR, AUG	MAR, AUG	MAR, AUG	MAR, AUG	MAR, AUG	MAR, AUG	MAR, AUG	MAR, AUG	MAR, AUG		MAR, AUG	MAR, AUG	MAR, AUG	MAR, AUG
Coal Tit	MAR	MAR	MAR, AUG	MAR, AUG	MAR, AUG	MAR, AUG	MAR, AUG	MAR, AUG	MAR, AUG	MAR, AUG	MAR, AUG	MAR, AUG	MAR, AUG	MAR, AUG	MAR, AUG	MAR, AUG
Willow Tit	MAR	MAR	MAR, AUG	MAR, AUG	MAR, AUG	MAR, AUG	MAR, AUG	MAR, AUG	MAR, AUG	MAR, AUG	MAR, AUG	MAR, AUG	MAR, AUG	MAR, AUG	MAR, AUG	MAR, AUG
Marsh Tit	MAR	MAR	MAR, AUG	MAR, AUG	MAR, AUG	MAR, AUG	MAR, AUG	MAR, AUG	MAR, AUG	MAR, AUG	MAR, AUG	MAR, AUG	MAR, AUG	MAR, AUG	MAR, AUG	MAR, AUG
Nuthatch	MAR	MAR	MAR, AUG	MAR, AUG	MAR, AUG	MAR, AUG	MAR, AUG	MAR, AUG	MAR, AUG	MAR, AUG	MAR, AUG	MAR, AUG	MAR, AUG	MAR, AUG	MAR, AUG	MAR, AUG
Treecreeper	MAR	MAR	MAR, AUG	MAR, AUG	MAR, AUG	MAR, AUG	MAR, AUG	MAR, AUG	MAR, AUG	MAR, AUG	MAR, AUG	MAR, AUG	MAR, AUG	MAR, AUG	MAR, AUG	MAR, AUG
Jay			AUG	AUG	AUG	AUG	AUG	AUG	AUG	AUG	AUG	AUG	AUG	AUG	AUG	AUG
Magpie	MAR	MAR	MAR, AUG	MAR, AUG	MAR, AUG	MAR, AUG	MAR, AUG	MAR, AUG	MAR, AUG	MAR, AUG	MAR, AUG	MAR, AUG	MAR, AUG	MAR, AUG	MAR, AUG	MAR, AUG
Chough	MAR	MAR	MAR, AUG	MAR, AUG	MAR, AUG	MAR, AUG	MAR, AUG	MAR, AUG	MAR, AUG	MAR, AUG	MAR, AUG	MAR, AUG	MAR, AUG	MAR, AUG	MAR, AUG	MAR, AUG
Rook	MAR	MAR	MAR	MAR	MAR	MAR	MAR	MAR	MAR	MAR	MAR	MAR	MAR	MAR	MAR	MAR
Carrion Crow	MAR	MAR	MAR	MAR	MAR	MAR	MAR	MAR	MAR	MAR	MAR	MAR	MAR	MAR	MAR	MAR
Hooded Crow	MAR	MAR	MAR	MAR	MAR	MAR	MAR	MAR	MAR	MAR	MAR	MAR	MAR	MAR	MAR	MAR
Raven	MAR	MAR	MAR	MAR	MAR	MAR	MAR	MAR	MAR	MAR	MAR	MAR	MAR	MAR	MAR	MAR
Starling	MAR	MAR	MAR, AUG	MAR, AUG	MAR, AUG	MAR, AUG	MAR, AUG	MAR, AUG	MAR, AUG	MAR, AUG	MAR, AUG	MAR	MAR, AUG	MAR, AUG	MAR, AUG	MAR, AUG
House Sparrow	MAR	MAR	MAR, AUG, SEPT	MAR, AUG, SEPT	MAR, AUG, SEPT	MAR, AUG, SEPT	MAR, AUG, SEPT	MAR, AUG, SEPT	MAR, AUG, SEPT	MAR, AUG, SEPT	MAR, AUG, SEPT	MAR, AUG, SEPT	MAR, AUG, SEPT	MAR, AUG, SEPT	MAR, AUG, SEPT	MAR, AUG, SEPT
Tree Sparrow	MAR	MAR	MAR, AUG, SEPT	MAR, AUG, SEPT	MAR, AUG, SEPT	MAR, AUG, SEPT	MAR, AUG, SEPT	MAR, AUG, SEPT	MAR, AUG, SEPT	MAR, AUG, SEPT	MAR, AUG, SEPT	MAR, AUG, SEPT	MAR, AUG, SEPT	MAR, AUG, SEPT	MAR, AUG, SEPT	MAR, AUG, SEPT
Chaffinch	MAR	MAR	MAR, AUG	MAR, AUG	MAR, AUG	MAR, AUG	MAR, AUG	MAR, AUG	MAR, AUG	MAR, AUG	MAR, AUG	MAR, AUG	MAR, AUG	MAR, AUG	MAR, AUG	MAR, AUG
Greenfinch	MAR	MAR	MAR, AUG, SEPT	MAR, AUG, SEPT	MAR, AUG, SEPT	MAR, AUG, SEPT	MAR, AUG, SEPT	MAR, AUG, SEPT	MAR, AUG, SEPT	MAR, AUG, SEPT	MAR, AUG, SEPT	MAR, AUG	MAR, AUG, SEPT	MAR, AUG, SEPT	MAR, AUG, SEPT	MAR, AUG, SEPT
Goldfinch			AUG, SEPT	AUG, SEPT	AUG, SEPT	AUG, SEPT	AUG, SEPT	AUG, SEPT	AUG, SEPT	AUG, SEPT	AUG, SEPT	AUG	AUG, SEPT	AUG, SEPT	AUG, SEPT	AUG, SEPT
Siskin	MAR	MAR	MAR, AUG, SEPT	MAR, AUG, SEPT	MAR, AUG, SEPT	MAR, AUG, SEPT	MAR, AUG, SEPT	MAR, AUG, SEPT	MAR, AUG, SEPT	MAR, AUG, SEPT	MAR, AUG, SEPT	MAR	MAR, AUG, SEPT	MAR, AUG, SEPT	MAR, AUG, SEPT	MAR, AUG, SEPT
Linnet	MAR	MAR	MAR, AUG, SEPT	MAR, AUG, SEPT	MAR, AUG, SEPT	MAR, AUG, SEPT	MAR, AUG, SEPT	MAR, AUG, SEPT	MAR, AUG, SEPT	MAR, AUG, SEPT	MAR, AUG, SEPT	MAR, AUG	MAR, AUG, SEPT	MAR, AUG, SEPT	MAR, AUG, SEPT	MAR, AUG, SEPT
Twite			AUG, SEPT	AUG, SEPT	AUG, SEPT	AUG, SEPT	AUG, SEPT	AUG, SEPT	AUG, SEPT	AUG, SEPT	AUG, SEPT	AUG	AUG, SEPT	AUG, SEPT	AUG, SEPT	AUG, SEPT
Common Redpoll			AUG, SEPT	AUG, SEPT	AUG, SEPT	AUG, SEPT	AUG, SEPT	AUG, SEPT	AUG, SEPT	AUG, SEPT	AUG, SEPT	AUG	AUG, SEPT	AUG, SEPT	AUG, SEPT	AUG, SEPT
Lesser Redpoll			AUG, SEPT	AUG, SEPT	AUG, SEPT	AUG, SEPT	AUG, SEPT	AUG, SEPT	AUG, SEPT	AUG, SEPT	AUG, SEPT	AUG	AUG, SEPT	AUG, SEPT	AUG, SEPT	AUG, SEPT
Parrot Crossbill	MAR	MAR	MAR, AUG	MAR, AUG	MAR, AUG	MAR, AUG	MAR, AUG	MAR, AUG	MAR, AUG	MAR, AUG	MAR, AUG	MAR, AUG	MAR, AUG	MAR, AUG	MAR, AUG	MAR, AUG
Common/ Scottish Crossbill	MAR	MAR	MAR, AUG	MAR, AUG	MAR, AUG	MAR, AUG	MAR, AUG	MAR, AUG	MAR, AUG	MAR, AUG	MAR, AUG	MAR, AUG	MAR, AUG	MAR, AUG	MAR, AUG	MAR, AUG
Bullfinch			AUG, SEPT	AUG, SEPT	AUG, SEPT	AUG, SEPT	AUG, SEPT	AUG, SEPT	AUG, SEPT	AUG, SEPT	AUG, SEPT	AUG, SEPT	AUG, SEPT	AUG, SEPT	AUG, SEPT	AUG, SEPT
Hawfinch	MAR	MAR	MAR, AUG	MAR, AUG	MAR, AUG	MAR, AUG	MAR, AUG	MAR, AUG	MAR, AUG	MAR, AUG	MAR, AUG	MAR, AUG	MAR, AUG	MAR, AUG	MAR, AUG	MAR, AUG
Snow Bunting			AUG	AUG	AUG	AUG	AUG	AUG	AUG	AUG	AUG	AUG	AUG	AUG	AUG	AUG
Yellowhammer	MAR	MAR	MAR, AUG, SEPT	MAR, AUG, SEPT	MAR, AUG, SEPT	MAR, AUG, SEPT	MAR, AUG, SEPT	MAR, AUG, SEPT	MAR, AUG, SEPT	MAR, AUG, SEPT	MAR, AUG, SEPT	MAR, AUG	MAR, AUG, SEPT	MAR, AUG, SEPT	MAR, AUG, SEPT	MAR, AUG, SEPT
Reed Bunting			AUG, SEPT	AUG, SEPT	AUG, SEPT	AUG, SEPT	AUG, SEPT	AUG, SEPT	AUG, SEPT	AUG, SEPT	AUG, SEPT	AUG	AUG, SEPT	AUG, SEPT	AUG, SEPT	AUG, SEPT
Corn Bunting	MAR	MAR	MAR, AUG, SEPT	MAR, AUG, SEPT	MAR, AUG, SEPT	MAR, AUG, SEPT	MAR, AUG, SEPT	MAR, AUG, SEPT	MAR, AUG, SEPT	MAR, AUG, SEPT	MAR, AUG, SEPT	MAR, AUG, SEPT	MAR, AUG, SEPT	MAR, AUG, SEPT	MAR, AUG, SEPT	MAR, AUG, SEPT

APPENDIX 5 ~ SCIENTIFIC NAMES

SCIENTIFIC NAMES FOR all bird species recorded during *Bird Atlas 2007–11* are given in Appendix 2. Those for other taxa mentioned in the text are given below and follow the London Natural History Museum's *Dictionary of UK Species* (Natural History Museum 2013).

A PLANTS

birch	*Betula* spp.
Silver Birch	*Betula pendula*
Bramble	*Rubus fruticosus agg.*
Common Nettle	*Urtica dioica*
Dandelion	*Taraxacum officinale agg.*
Elephant Grass	*Miscanthus x giganteus*
elm	*Ulmus* spp.
gorse	*Ulex* spp.
heather	Ericaceae
iris	*Iris* spp.
Ivy	*Hedera helix*
larch	*Larix* spp.
Lodgepole Pine	*Pinus contorta*
oak	*Quercus* spp.
Sessile Oak	*Quercus petraea*
Oil-seed Rape	*Brassica napus* subsp. *oleifera*
poplar	*Populus* spp.
Potato	*Solanum tuberosum*
Rowan	*Sorbus aucuparia*
Scots Pine	*Pinus sylvestris*
Sitka Spruce	*Picea sitchensis*
Sweet Chestnut	*Castanea sativa*
wheat	*Triticum* spp.

B ANIMALS

American Mink	*Neovison vison*
ant	Formicidae
beetle	Coleoptera
Brown Rat	*Rattus norvegicus*
cranefly	Tipulidae
cricket	Orthoptera
deer	Cervidae
dragonfly	Odonate
European Rabbit	*Oryctolagus cuniculus*
Hedgehog (Western European)	*Erinaceus europaeus*
ladybird	Coccinellidae
Pike	*Esox lucius*
Pine Marten	*Martes martes*
Red Fox	*Vulpes vulpes*
Chinese Muntjac	*Muntiacus reevesi*
Roach	*Rutilus rutilus*
sandeel	Ammodytidae
Lesser Sandeel	*Ammodytes marinus*
Sheep (Feral)	*Ovis aries*
White-headed Duck	*Oxyura leucocephala*
Zebra Mussel	*Dreissena polymorpha*

REFERENCES

THE REFERENCE LIST below has been constructed using an 'author-date' approach as used in the *Migration Atlas*. The initials of the first author have been given priority over date of publication, so that the publications of each individual come together in the list, even if an author has used incomplete initials on their different publications.

For an individual author we have listed single-author references, then any with a second author, and thirdly any with three or more authors. The list then moves to the next individual in the alphabetised list. Single-author references are alphabetised, first by surname, second by the author's full initials and, third, by date of publication. Two-author references are alphabetised by first author, second author and then date. Where there are three or more authors, references in the text are in the form of 'Author *et al.* year'. So that these references can be quickly located, they are listed by first author and date, with the names of the second and subsequent authors, given lower priority in the alphabetisation process.

Adams, M.C. 1966. Firecrests breeding in Hampshire. *British Birds* 59: 240–246.

Aebischer, N.J., Green, R.E. & Evans, A.D. 2000. From science to recovery: four case studies of how research has been translated into conservation action in the UK. In *Ecology and Conservation of Lowland Farmland Birds* (eds Aebischer, N.J., Evans, A.D., Grice, P.V. & Vickery, J.A.), pp 43–54. BOU, Tring.

Aebischer, A., Nyffeler, P. & Arlettaz, R. 2010. Wide-range dispersal in juvenile Eagle Owls (*Bubo bubo*) across the European Alps calls for transnational conservation programmes. *Journal of Ornithology* 151: 1–9.

Ahola, M.P., Laaksonen, T., Eeva, T. & Lehikoinen, E. 2009. Great Tits lay increasingly smaller clutches than selected for: a study of climate- and density-related changes in reproductive traits. *Journal of Animal Ecology* 78: 1298–1306.

Alder, D. & Marsden, S. 2010. Characteristics of feeding-site selection by breeding Green Woodpeckers *Picus viridis* in a UK agricultural landscape. *Bird Study* 57: 100–107.

Allard, P. 2007. Population explosion of nesting gulls in Great Yarmouth. *Norfolk Bird & Mammal Report* 2006: 325–328.

Allen, D. & Tickner, M. 1996. Mediterranean Gull: a new breeding bird for Ireland. *Irish Birds* 5: 435–436.

Allen, D., Mellon, C., Enlander, I. & Watson, G. 2004. Lough Neagh ducks: recent changes in wintering populations. *Irish Birds* 7: 327–336.

Allen, D., Mellon, C., Mawhinney, K., Looney, D. & Milburne, J. 2005. The status of Red Grouse *Lagopus lagopus* in Northern Ireland 2004. *Irish Birds* 7: 449–460.

Amar, A., Redpath, S. & Thirgood, S. 2003. Evidence for food limitation in the declining Hen Harrier population on the Orkney Islands, Scotland. *Biological Conservation* 111: 377–384.

Amar, A., Picozzi, N., Meek, E.R., Redpath, S.M. & Lambin, X. 2005. Decline of the Hen Harrier *Circus cyaneus* population: do changes to demographic parameters and mating system fit a declining food hypothesis? *Bird Study* 52: 18–24.

Amar, A., Hewson, C.M., Thewlis, R.M., Smith, K.W., Fuller, R.J., Lindsell, J.A., Conway, G., Butler, S. & MacDonald, M.A. 2006. *What's Happening to Our Woodland Birds? Long-term changes in the populations of woodland birds.* BTO Research Report 169 & RSPB Research Report 19. BTO, Thetford.

Amar, A., Smith, K.W., Butler, S., Lindsell, J.A., Hewson, C.M., Fuller, R.J. & Charman, E.C. 2010a. Recent patterns of change in vegetation structure and tree composition of British broadleaved woodland: evidence from large-scale surveys. *Forestry* 83: 345–356.

Amar, A., Redpath, S., Sim, I. & Buchanan, G. 2010b. Spatial and temporal associations between recovering populations of Common Raven *Corvus corax* and British upland wader populations. *Journal of Applied Ecology* 47: 253–262.

Amar, A., Grant, M., Buchanan, G., Sim, I., Wilson, J., Pearce-Higgins, J.W. & Redpath, S. 2011. Exploring the relationships between wader declines and current land-use in the British uplands. *Bird Study* 58: 1–13.

Amar, A., Court, I.R., Davison, M., Downing, S., Grimshaw, T., Pickford, T. & Raw, D. 2012. Linking nest histories, remotely sensed land use data and wildlife crime records to explore the impact of grouse moor management on Peregrine Falcon populations. *Biological Conservation* 145: 86–94.

Anderson, B.J., Arroyo, B.E., Collingham, Y.C., Etheridge, B., Fernandez-de-Simon, J., Gillings, S., Gregory, R.D., Leckie, F.M., Sim, I.M.W., Thomas, C.D., Travis, J.M.J. & Redpath, S.M. 2009. Using distribution models to test alternative hypotheses about a species' environmental limits and recovery prospects. *Biological Conservation* 142: 488–499.

Anderson, O.R.J., Small, C.J., Croxall, J.P., Dunn, E.K., Sullivan, B.J., Yates, O. & Black, A. 2011. Global seabird bycatch in longline fisheries. *Endangered Species Research* 14: 91–106.

ap Rheinallt, T., McInerny, C.J., McGowan, R.Y. & Sweeney, J.J. 2012. Scottish Birds Records Committee report on rare birds in Scotland, 2010. *Scottish Birds* 32: 108–132.

Araújo, M.B. & New, M. 2007. Ensemble forecasting of species distributions. *Trends in Ecology and Evolution* 22: 42–47.

Araújo, M.B., Williams, P.H. & Fuller, R.J. 2002. Dynamics of extinction and the selection of nature reserves. *Proceedings of the Royal Society of London* Series B 269: 1971–1980.

Archer, M., Grantham, M., Howlett, P. & Stansfield, S. (eds). 2010. *Bird Observatories of Britain and Ireland.* T. & A.D. Poyser, London.

Armitage, M.J.S., Rehfisch, M.M. & Wernham, C.V. 1997. *The 1997 Breeding Sawbill Survey.* BTO Research Report 193. BTO, Thetford.

Askew, N.P., Searle, J.B. & Moore, N.P. 2007. Agri-environment schemes and foraging of Barn Owls *Tyto alba*. *Agriculture, Ecosystems & Environment* 118: 109–114.

Aspinall, S. & Aspinall, R. 2011. The Fair Isle Wren: population and territory occupancy 1950–2010. *British Birds* 104: 312–324.

Atkinson, P.W., Fuller, R.J. & Vickery, J.A. 2002. Large-scale patterns of summer and winter bird distribution in relation to farmland type in England and Wales. *Ecography* 25: 466–480.

Atkinson, P.W., Maclean, I.M.D. & Clark, N.A. 2010. Impacts of shellfisheries and nutrient inputs on waterbird communities in the Wash, England. *Journal of Applied Ecology* 47: 191–199.

Ausden, M. & Bolton, M. 2012. Breeding waders on wet grassland: factors influencing habitat suitability. In *Birds and Habitat: Relationships in Changing Landscapes* (ed Fuller, R.J.), pp 278–306. Cambridge University Press, Cambridge.

Ausden, M. & Fuller, R.J. 2009. Birds and habitat change in Britain. Part 2: past and future conservation responses. *British Birds* 102: 52–71.

Ausden, M. & Hirons, G.J.M. 2002. Grassland nature reserves for breeding wading birds in England and the implications for the ESA agri-environment scheme. *Biological Conservation* 106: 279–291.

Austin, G.E. & Rehfisch, M.M. 2005. Shifting nonbreeding distributions of migratory fauna in relation to climate change. *Global Change Biology* 11: 31–38.

Austin, G.E., Peachel, I. & Rehfisch, M.M. 2000. Regional trends in coastal wintering waders in Britain. *Bird Study* 47: 352–371.

Austin, G.E., Collier, M.P. & Rehfisch, M.M. 2007. *Non-estuarine Coastal Waterbird Survey: population estimates and broad comparisons with previous surveys.* BTO Research Report 501. BTO, Thetford.

Avery, M.I., Coulthard, N.D., Del Nevo, A.J., Leroux, A., Medeiros, F., Merne, O., Monteiro, L., Moralee, A., Ntiamoa-Baidu, Y., O'Briain, M. & Wallace, E. 1995. A recovery plan for Roseate Terns in the East Atlantic: an international programme. *Bird Conservation International* 5: 441–453.

Baatsen, R.G. 1990. Red-crested Pochard *Netta rufina* in the Cotswold Water Park. *Hobby* 16: 64–67.

Baillie, S.R. & Peach, W.J. 1992. Population limitation in Palearctic–African migrant passerines. *Ibis* 134 (Suppl. 1): 120–132.

Baines, D. 1988. The effects of improvement of upland, marginal grasslands on the distribution and density of breeding wading birds (Charadriiformes) in northern England. *Biological Conservation* 45: 221–236.

Baines, D., Moss, R. & Dugan, D. 2004. Capercaillie breeding success in relation to forest habitat and predator abundance. *Journal of Applied Ecology* 41: 59–71.

Baker, H., Stroud, D.A., Aebischer, N.J., Cranswick, P.A., Gregory, R.D., McSorley, C.A., Noble, D.G. & Rehfisch, M.M. 2006. Population estimates of birds in Great Britain and the United Kingdom. *British Birds* 99: 25–44.

Balk, L., Hägerroth, P.-Å., Åkerman, G., Hanson, M., Tjärnlund, U., Hansson, T., Hallgrimsson, G.T., Zebühr, Y., Broman, D., Mörner, T. & Sundberg, H. 2009. Wild birds of declining European species are dying from a thiamine deficiency syndrome. *Proceedings of the National Academy of the Sciences of the United States of America* 106: 12001–12006.

Balmer, D. & Vickery, J. 2010. Wildlife reports: birds. *British Wildlife* 21: 347–350.

Balmer, D. & Vickery, J. 2011. Wildlife reports: birds. *British Wildlife* 22: 196–199.

Balmer, D.E, Browne, S.J. & Rehfisch, M.M. 1996. A year in the life of Golden Pheasants *Chrysolophus pictus*. In *The Introduction and Naturalisation of Birds* (eds Holmes, J.S. & Simons, J.R.), pp 87–93. The Stationery Office, London.

Banks, A.N., Wright, L.J., Maclean, I.M.D., Hann, C. & Rehfisch, M.M. 2008. *Review of the Status of Introduced Non-native Waterbird Species in the Area of the African–Eurasian Waterbird Agreement: 2007 Update*. BTO Research Report 489. BTO, Thetford.

Banks, A.N., Burton, N.H.K., Calladine, J.R. & Austin, G.E. 2009. Indexing winter gull numbers in Great Britain using data from the 1953 to 2004 Winter Gull Roost Surveys. *Bird Study* 56: 103–119.

Banks, A.N., Crick, H.Q.P., Coombes, R., Benn, S., Ratcliffe, D.A. & Humphreys, E.M. 2010. The breeding status of Peregrine Falcons *Falco peregrinus* in the UK and Isle of Man in 2002. *Bird Study* 57: 421–436.

Barden, P. 2010. Breeding attempt by Ring-billed Gull in Scotland in 2009. *Scottish Birds* 30: 31–32.

Barrett, G., Silcocks, A., Barry, S., Cunningham, R. & Poulter, R. 2003. *The new atlas of Australian birds*. CSIRO Publishing, Collingwood, Victoria.

Bates, B., Etheridge, B., Elkins, N., Fox, J. & Summers, R.W. 2013. Pre-migratory change in mass and the migration track of a Common Sandpiper *Actitis hypoleucos* from Scotland. *Wader Study Group Bulletin* 119: 149–154.

Batten, L.A. 2001. European Honey-buzzard survey 2000 and 2001: preliminary results and request for further surveys. *British Birds* 94: 143–144.

Bauer, H.-G. & Woog, F. 2008. Nichteinheimische Vogelarten (Neozoen) in Deutschland, Teil 1: Auftreten, Bestande und Status. *Vogelwarte* 46: 157–194.

Bayly, N.J., Rumsey, S.J.R. & Clark, J.A. 2011. Crossing the Sahara desert: migratory strategies of the Grasshopper Warbler *Locustella naevia*. *Journal of Ornithology* 152: 933–946.

Beale, C.M., Burfield, I.J., Sim, I.M.W., Rebecca, G.W., Pearce-Higgins, J.W. & Grant, M. 2006. Climate change may account for the decline in British Ring Ouzels *Turdus torquatus*. *Journal of Animal Ecology* 75: 826–835.

Bealey, C.E., Green, R.E., Robson, R., Taylor, C.R. & Winspear, R. 1999. Factors affecting the numbers and breeding success of Stone Curlews *Burhinus oedicnemus* at Porton Down, Wiltshire. *Bird Study* 46: 145–156.

Bearhop, S., Fielder, W., Furness, R.W., Votier, S.C., Waldron, S., Newton, J., Bowen, G.J., Berthold, P. & Farnsworth, K. 2005. Assortative mating as a mechanism for rapid evolution of a migratory divide. *Science* 310: 502–504.

Beaugrand, G. 2004. Continuous Plankton Records: Plankton Atlas of the North Atlantic Ocean (1958–1999). Introduction and methodology. *Marine Ecology Progress Series Supplement* 2004: 3–10.

Bell, E., Boyle, D., Floyd, K., Garner-Richards, P., Swann, R., Luxmoore, R., Patterson, A. & Thomas, R. 2011. The ground-based eradication of Norway rats (*Rattus norvegicus*) from the Isle of Canna, Inner Hebrides, Scotland. In *Island Invasives: eradication and management* (eds Veitch, C.R., Clout, M.N. & Towns, D.R.), pp 269–274. IUCN, Gland.

Bellamy, P.E., Brown, N.J., Enoksson, B., Firbank, L.G., Fuller, R.J., Hinsley, S.A. & Schotman, A.G.M. 1998. The influences of habitat, landscape structure and climate on local distribution patterns of the Nuthatch (*Sitta europaea* L.). *Oecologia* 115: 127–136.

Bellamy, P.E. & Eaton, M.A. 2010. *2009 CSM bird monitoring of Caithness and Sutherland Peatlands SPA*. Unpublished RSPB report to SNH. RSPB, Sandy.

Benton, T.G., Bryant, D.M., Cole, L. & Crick, H.Q.P. 2002. Linking agricultural practice to insect and bird populations: a historical study over three decades. *Journal of Applied Ecology* 39: 673–687.

Berrow, S.D., Mackie, K.L., O'Sullivan, O., Shepherd, K.B., Mellon, C. & Coveney, J.A. 1993. The second International Chough Survey in Ireland, 1992. *Irish Birds* 5: 1–10.

Berthold, P. & Terrill, S.B. 1988. Migratory behaviour and population growth of Blackcaps wintering in Britain and Ireland: some hypotheses. *Ringing & Migration* 9: 153–159.

Bevington, A. 1991. Habitat selection in the Dunnock *Prunella modularis* in northern England. *Bird Study* 38: 87–91.

Bibby, C.J. 2003. Fifty years of *Bird Study*. *Bird Study* 50: 194–210.

Bircham, P.M.M. & Jordan, W.J. 1996. A consideration of some of the changes in distribution of "common birds" as revealed by *The New Atlas of Breeding Birds in Britain and Ireland*. *Ibis* 139: 177–179.

BirdLife International. 2004. *Birds in Europe: population estimates, trends and conservation status*. BirdLife Conservation Series No. 12. BirdLife International, Cambridge.

BirdLife International. 2013. *IUCN Red List for Birds*. URL: www.birdlife.org [31 July 2013].

Bivand, R.S., Pebesma, E.J. & Gómez-Rubio, V. 2008. *Applied Spatial Data Analysis with R*. Springer, New York.

Bland, R.L. & Dadds, M. 2012. *Avon Atlas 2007–11*. Published privately, Bristol.

Bodey, T.W., Smart, J., Smart, M.A. & Gregory, R.D. 2010. Reducing the impacts of predation on ground-nesting waders: a new landscape-scale solution? *Aspects of Applied Biology* 100: 167–174.

Böhning-Gaese, K. & Bauer, H.-G. 1996. Changes in species abundance, distribution, and diversity in a central European bird community. *Conservation Biology* 10: 175–187.

Boisseau, S. & Yalden, D.W. 1998. The former status of the Crane *Grus grus* in Britain. *Ibis* 140: 482–500.

Boland, H., McElwaine, J.G., Henderson, G., Hall, C., Walsh, A. & Crowe, O. 2010. Whooper *Cygnus cygnus* and Bewick's *C. columbianus bewickii* Swans in Ireland: results of the international census, January 2010. *Irish Birds* 9: 1–10.

Bolton, M., Tyler, G., Smith, K. & Bamford, R. 2007. The impact of predator control on Lapwing *Vanellus vanellus* breeding success on wet grassland nature reserves. *Journal of Applied Ecology* 44: 534–544.

Bonham, P.F. & Robertson, J.C.M. 1975. The spread of Cetti's Warbler in north-west Europe. *British Birds* 68: 393–408.

Bonter, D.N. & Cooper, C.B. 2012. Data validation in citizen science: a case study from Project FeederWatch. *Frontiers in Ecology and the Environment* 10: 305–307.

Both, C., Bouwhuis, S., Lessells, C.M. & Visser, M.E. 2006. Climate change and population declines in a long-distance migratory bird. *Nature* 441: 81–83.

BOU. 1980. Records Committee: Tenth Report (March 1980). *Ibis* 122: 564–568.

BOU. 1986. British Ornithologists' Union Records Committee: Twelfth Report (April 1986). *Ibis* 128: 601–603.

BOU. 2001. British Ornithologists' Union Records Committee: 27th Report (October 2000). *Ibis* 143: 171–175.

BOU. 2012. British Ornithologists' Union Records Committee: 40th Report (October 2011). *Ibis* 154: 212–215.

BOU. 2013a. The British List: a checklist of birds in Britain (8th edition). *Ibis* 155: 635–676.

BOU. 2013b. *The British List – The official list of birds recorded in Britain*. URL: www.bou.org.uk/british-list/ [31 July 2013].

Boyd, B. 1987. Black-winged Stilts breeding at Holme. *Norfolk Bird & Mammal Report* 1987: 89–96.

Bradbury, R.B. & Bradter, U. 2004. Habitat associations of Yellow Wagtails *Motacilla flava flavissima* on lowland wet grassland. *Ibis* 146: 241–246.

Bradbury, R.B., Pearce-Higgins, J.W., Wotton, S.R., Conway, G.J. & Grice, P.V. 2011. The influence of climate and topography in patterns of territory establishment in a range-expanding bird. *Ibis* 153: 336–344.

Branson, A. 2011. Weather for July and August 2011. *British Wildlife* 23: 42.

Breiman, L. 2001. Random forests. *Machine Learning* 45: 5–32.

Breiman, L. & Cutler, A. 2004. *Random Forests*. URL: www.stat.berkeley.edu/~breiman/RandomForests/ [31 July 2013].

Breiman, L., Friedman, J., Olshen, R. & Stone, C. 1984. *Classification and regression trees*. Wadsworth, Belmont.

Brenchley, A., Gibbs, G., Pritchard, R. & Spence, I. (eds). 2013. *The Breeding Birds of North Wales – Adar Nythu Gogledd Cymru*. Liverpool University Press, Liverpool. [In press].

Brickle, N.W. & Harper, D.G.C. 2002. Agricultural intensification and the timing of breeding of Corn Buntings *Miliaria calandra*. *Bird Study* 49: 219–228.

Brickle, N.W., Harper, D.G.C., Aebischer, N.J. & Cockayne, S.H. 2000. Effects of agricultural intensification on the breeding success of Corn Buntings *Miliaria calandra*. *Journal of Applied Ecology* 37: 742–755.

Bright, J.A., Anderson, G.Q.A., McArthur, T., Sage, R., Stockdale, J., Grice, P.V. & Bradbury, R.B. 2013. Bird use of establishment-stage Miscanthus biomass crops during the breeding season in England. *Bird Study* 60: 357–369.

Brindley, E., Norris, K., Cook, T., Babbs, S., Forster-Browne, C. & Yaxley, R. 1998. The abundance and conservation status of Redshank (*Tringa totanus*) nesting on saltmarshes in Great Britain. *Biological Conservation* 86: 289–297.

Brambilla, M., Rubolini, D. & Guidali, F. 2006. Eagle Owl *Bubo bubo* proximity can lower productivity of cliff-nesting Peregrines *Falco peregrinus*. *Ornis Fennica* 83: 20–26.

Brockie, K. 2011. Male Velvet Scoter paired with an Eider, Isle of May. *Scottish Birds* 31: 164.

Brooke, M. de L. & Davies, N.B. 1987. Recent changes in host usage by Cuckoos *Cuculus canorus* in Britain. *Journal of Animal Ecology* 56: 873–883.

Brooks, S.J., Jones, V.J., Telford, R.J., Appleby, P.G., Watson, E., McGowan, S. & Benn, S. 2012. Population trends in the Slavonian Grebe *Podiceps auritus* (L.) and Chironomidae (Diptera) at a Scottish loch. *Journal of Paleolimnology* 47: 631–644.

Broughton, R.K., Hill, R.A., Bellamy, P.E. & Hinsley, S.A. 2010. Dispersal, ranging and settling behaviour of Marsh Tits *Poecile palustris* in a fragmented landscape in lowland England. *Bird Study* 57: 458–472.

Broughton, R.K., Hill, R.A., Bellamy, P.E. & Hinsley, S.A. 2011. Nest-sites, breeding failure and causes of non-breeding in a population of British Marsh Tits *Poecile palustris*. *Bird Study* 58: 229–237.

Broughton, R.K., Hill, R.A., Freeman, S.N., Bellamy, P.E. & Hinsley, S.A. 2012. Describing habitat occupation by woodland birds using territory mapping and multiple vegetation models derived from remote sensing: an example using the Marsh Tit (*Poecile palustris*). *Condor* 114: 812–822.

Broughton, R.K., Hill, R.A. & Hinsley, S.A. 2013. Relationships between patterns of habitat cover and the historical distribution of the Marsh Tit, Willow Tit and Lesser Spotted Woodpecker in Britain. *Ecological Informatics* 14: 25–30.

Brown, A.F. & Atkinson, P.W. 1996. Habitat associations of coastal wintering passerines. *Bird Study* 43: 188–200.

Brown, A.F., Stillman, R.A. & Gibbons, D.W. 1995. Use of breeding bird atlas data to identify important bird areas: a northern England case study. *Bird Study* 42: 132–143.

Brown, A., Price, D., Slader, P., Booker, H., Lock, L. & Deveney, D. 2011. Seabirds on Lundy: their current status, recent history and prospects for the restoration of a once-important bird area. *British Birds* 104: 139–158.

Brown, A., Gilbert, G. & Wotton, S. 2012. Bitterns and Bittern conservation in the UK. *British Birds* 105: 58–87.

Browne, S.J. & Aebischer, N.J. 2001. *The role of agricultural intensification in the decline of the Turtle Dove* Streptopelia turtur. English Nature, Peterborough.

Browne, S.J. & Aebischer, N.J. 2004. Temporal changes in the breeding ecology of European Turtle Doves *Streptopelia turtur* in Britain, and implications for conservation. *Ibis* 146: 125–137.

Browne, S. & Aebischer, N. 2005. Studies of West Palearctic birds: Turtle Dove. *British Birds* 98: 58–72.

Bruce, K. 1997. Reed Warblers breeding in south west Scotland. *Scottish Birds* 19: 119–120.

Buckton, S.T. & Ormerod, S.J. 1997. Use of a new standardized habitat survey for assessing the habitat preferences and distribution of upland river birds. *Bird Study* 44: 327–337.

Bunn, D.S., Warburton, A.B. & Wilson, R.D.S. 1982. *The Barn Owl*. T. & A.D. Poyser, Calton.

Burfield, I.J. 2008. The conservation status and trends of raptors and owls in Europe. *Ambio* 37: 401–407.

Burnside, R.J., Carter, I., Dawes, A., Waters, D., Lock, L., Goriup, P. & Székely, T. 2012. The UK Great Bustard *Otis tarda* reintroduction trial: a 5-year progress report. *Oryx* 46: 112–121.

Burton, J.F. 1995. *Birds and Climate Change*. Christopher Helm Publishers Ltd, London.

Burton, N.H.K. 2007. Influences of restock age and habitat patchiness on Tree Pipits *Anthus trivialis* breeding in Breckland pine plantations. *Ibis* 149 (Suppl. 2): 193–204.

Burton, N.H.K. 2009. Reproductive success of Tree Pipits *Anthus trivialis* in relation to habitat selection in conifer plantations. *Ibis* 151: 361–372.

Burton, N.H.K., Watts, P.N., Crick, H.Q.P. & Edwards, P.J. 1999. The effects of preharvesting operations on Reed Buntings *Emberiza schoeniclus* nesting in oilseed rape *Brassica napus*. *Bird Study* 46: 369–372.

Burton, N.H.K., Banks, A.N., Calladine, J.R. & Austin, G.E. 2013. The importance of the United Kingdom for wintering gulls: population estimates and conservation requirements. *Bird Study* 60: 87–101.

Butler, C.J. 2003. *Population biology of the introduced Rose-ringed Parakeet* Psittacula krameri *in the UK*. PhD thesis, University of Oxford.

Buxton, J. & Durdin, C. 2011. *The Norfolk Cranes' Story*. Wren Publishing, Sheringham.

Byars, T. 2010. The status and distribution of the Lesser Whitethroat in Clyde and Ayrshire from 1983 to 2005. *Scottish Birds* 30: 316–325.

Cabot, D. 1996. Performance of the Roseate Tern population breeding in north-west Europe: Ireland, Britain and France, 1960–94. *Biology and Environment: Proceedings of the Royal Irish Academy* 96B: 55–68.

Cadbury, C.J. 2005. Bearded Tit and other reed-swamp passerine populations in the Hickling Broad complex 1980–2004. *Norfolk Bird & Mammal Report* 2004: 273–281.

Cadbury, C.J. & Olney, P.J.S. 1978. Avocet population dynamics in England. *British Birds* 71: 102–121.

Calladine, J. & Bray, J. 2012. The importance of altitude and aspect for breeding Whinchats *Saxicola rubetra* in the uplands: limitations of the uplands as a refuge for a declining, formerly widespread species? *Bird Study* 59: 43–51.

Calladine, J. & Morrison, N. 2010. The ranging behaviour and habitat selection by three Hawfinches *Coccothraustes coccothraustes* in late winter in Scotland. *Ornis Fennica* 87: 119–123.

Calladine, J., Baines, D. & Warren, P. 2002. Effects of reduced grazing on population density and breeding success of Black Grouse in northern England. *Journal of Applied Ecology* 39: 772–780.

Calladine, J.R., Park, K.J., Thompson, K. & Wernham, C.V. 2006. *Review of Urban Gulls and their Management in Scotland*. A report to the Scottish Executive (ENV/BTO/001/04). Scottish Executive, Edinburgh.

Calladine, J., Garner, G., Wernham, C. & Buxton, N. 2010. Variation in the diurnal activity of breeding Short-eared Owls *Asio flammeus*: implications for their survey and monitoring. *Bird Study* 57: 89–99.

Cambridgeshire Bird Club. 1997. *Cambridgeshire Bird Report 1996*. Cambridgeshire Bird Club, Cambridge.

Cambridgeshire Bird Club. 2001. *Cambridgeshire Bird Report 2000*. Cambridgeshire Bird Club, Cambridge.

Cambridgeshire Bird Club. 2011. *Cambridgeshire Bird Report 2009*. Cambridgeshire Bird Club, Cambridge.

Campbell, L.H. 1984. The impact of changes in sewage treatment on seaducks wintering in the Firth of Forth, Scotland. *Biological Conservation* 28: 173–180.

Camphuysen, C.J. 2011. Northern Gannets in the North Sea: foraging distribution and feeding techniques around the Bass Rock. *British Birds* 104: 60–76.

Carpenter, J., Smart, J., Amar, A., Gosler, A., Hinsley, S. & Charman, E. 2010. National-scale analyses of habitat associations of Marsh Tits *Poecile palustris* and Blue Tits *Cyanistes caeruleus*: two species with opposing population trends in Britain. *Bird Study* 57: 31–43.

Carrington, R. & Bradley, J. 2009. Common Gull *Larus canus* attempting to breed at Carsington Water. *Derbyshire Bird Report* 2008: 164.

Carter, I. & Newbery, P. 2004. Reintroduction as a tool for population recovery of farmland birds. *Ibis* 146 (Suppl. 2): 221–229.

Catt, D.C., Baines, D., Picozzi, N., Moss, R. & Summers, R.W. 1998. Abundance and distribution of Capercaillie *Tetrao urogallus* in Scotland 1992–1994. *Biological Conservation* 85: 257–267.

Central Statistics Office. 2013. *ICT Usage by Households*. URL: www.cso.ie/px/pxeirestat/Database/eirestat/ICT%20Usage%20by%20Households/ICT%20Usage%20by%20Households_statbank.asp?sp=ICT%20Usage%20by%20Households&Planguage=0 [31 July 2013].

CGIAR-CSI. 2008. *SRTM 90m Digital Elevation Data*. URL: http://srtm.csi.cgiar.org [31 July 2013].

Chabrzyk, G. & Coulson, J.C. 1976. Survival and recruitment in the Herring Gull *Larus argentatus*. *Journal of Animal Ecology* 45: 187–203.

Chamberlain, D.E. & Fuller, R.J. 2000. Local extinctions and changes in species richness of lowland farmland birds in England and Wales in relation to recent changes in agricultural land-use. *Agriculture, Ecosystems & Environment* 78: 1–17.

Chamberlain, D.E. & Fuller, R.J. 2001. Contrasting patterns of change in the distribution and abundance of farmland birds in relation to farming system in lowland Britain. *Global Ecology and Biogeography* 10: 399–409.

Chamberlain, D. & Siriwardena, G. 2000. The effects of agricultural intensification on Skylarks (*Alauda arvensis*): evidence from monitoring studies in Great Britain. *Environmental Reviews* 8: 95–113.

Chamberlain, D.E., Fuller, R.J., Bunce, R.G.H., Duckworth, J.C. & Shrubb, M. 2000. Changes in the abundance of farmland birds in relation to the timing of agricultural intensification in England and Wales. *Journal of Applied Ecology* 37: 771-788.

Chamberlain, D.E., Fuller, R.J., Garthwaite, D.G. & Impey, A.J. 2001. A comparison of farmland bird density and species richness in lowland England between two periods of contrasting agricultural practice. *Bird Study* 48: 245–251.

Chamberlain, D.E., Vickery, J.A., Glue, D.E., Robinson, R.A., Conway, G.J., Woodburn, R.J.W. & Cannon, A.R. 2005. Annual and seasonal trends in the use of garden feeders by birds in winter. *Ibis* 147: 563–575.

Chamberlain, D.E., Gosler, A.G. & Glue, D.E. 2007. Effects of the winter beechmast crop on bird occurrence in British gardens. *Bird Study* 54: 120–126.

Chamberlain, D., Arlettaz, R., Caprio, E., Maggini, R., Pedrini, P., Rolando, A. & Zbinden, N. 2012. The altitudinal frontier in avian climate impact research. *Ibis* 154: 205–209.

Chandler, R.J. 1981. Influxes into Britain and Ireland of Red-necked Grebes and other waterbirds during winter 1978/79. *British Birds* 74: 55–81.

Charles, D. 2009. Ring-billed Gull breeding with Common Gull on Copeland Islands Co. Down. The first confirmed breeding record for Ring-billed Gull in the Western Palearctic. *Northern Ireland Bird Report* 18: 122.

Charman, E.C., Smith, K.W., Gruar, D.J., Dodd, S. & Grice, P.V. 2010. Characteristics of woods used recently and historically by Lesser Spotted Woodpeckers *Dendrocopos minor* in England. *Ibis* 152: 543–555.

Charman, E.C., Smith, K.W., Dillon, I.A., Dodd, S., Gruar, D.J., Cristinacce, A., Grice, P.V. & Gregory, R.D. 2012. Drivers of low breeding success in the Lesser Spotted Woodpecker *Dendrocopos minor* in England: testing hypotheses for the decline. *Bird Study* 59: 255–265.

Chernetsov, N. & Huettmann, F. 2005. Linking global climate grid surfaces with local long-term migration monitoring data: spatial computations for the Pied Flycatcher to assess climate-related population dynamics on a continental scale. *Lecture Notes in Computer Science* 3482: 133–142.

Chisholm, K. 2007. History of the Wood Sandpiper as a breeding bird in Britain. *British Birds* 100: 112–121.

Clark, J.A., Balmer, D.E., Blackburn, J.R., Milne, L.J., Robinson, R.A., Wernham, C.V., Adams, S.Y. & Griffin, B.M. 2002a. Bird ringing in Britain and Ireland in 2000. *Ringing & Migration* 21: 25–61.

Clark, J.A., Balmer, D.E., Adams, S.Y., Grantham, M.J., Blackburn, J.R., Robinson, R.A., Wernham, C.V., Griffin, B.M. & Milne, L.J. 2002b. Bird ringing in Britain and Ireland in 2001. *Ringing & Migration* 21: 80–143.

Clark, J.M. & Eyre, J. 2012. Dartford Warblers on the Thames Basin and Wealden Heaths. *British Birds* 105: 308–317.

Clarke, R. 1995. *The Marsh Harrier*. Hamlyn, London.

Clarke, R. 1996. *Montagu's Harrier*. Arlequin Press, Chelmsford.

Clausen, P., Clausen, K.K., Fox, T., Pihl, S. & Vissing, M.S. 2012. East Atlantic Light-bellied Brent Goose population falls below 6,000 birds. *Goose News* (Autumn 2012). WWT/JNCC/Scottish Natural Heritage.

Clements, R. 2001. The Hobby in Britain: a new population estimate. *British Birds* 94: 402–408.

Clements, R. 2002. The Common Buzzard in Britain: a new population estimate. *British Birds* 95: 377–383.

Clements, R. 2005. Honey-buzzards in Britain. *British Birds* 98: 153–155.

Clements, R. 2008. The Common Kestrel population in Britain. *British Birds* 101: 228–234.

Clements, R. 2013. A UK population estimate for the Hawfinch. *British Birds* 106: 43–44.

Clode, D. & Macdonald, D.W. 2002. Invasive predators and the conservation of island birds: the case of American Mink *Mustela vison* and terns *Sterna* spp. in the Western Isles, Scotland. *Bird Study* 49: 118–123.

Coiffait, L., Clark, J.A., Robinson, R.A., Blackburn, J.R., Grantham, M.J., Leech, D.I., Marchant, J.H., Barber, L.J., De Palacio, D., Griffin, B.M., Moss, D. & Shephard, M.M. 2009. Bird ringing in Britain and Ireland in 2008. *Ringing & Migration* 24: 281–320.

Colhoun, K., Austin, G. & Newton, S. 2008. Wader populations on non-estuarine coasts in the Republic of Ireland: results of the 1997/98 Non-Estuarine Coastal Waterbird Survey (Ireland-NEWS). In *The European Non-Estuarine Coastal Waterbird Survey* (eds Burton, N.H.K., Rehfisch, M.M., Stroud, D.A. & Spray, C.J.), pp 39–48. International Wader Studies 18. International Wader Study Group, Thetford.

Colhoun, K., Mackie, K. & Gudmundsson, G.A. 2011. East Canadian Light-bellied Brent Goose autumn 2010 survey. *Goose News* (Autumn 2011). WWT/JNCC/ Scottish Natural Heritage.

Collins, K.P. 2008. Little Ringed Plover *Charadrius dubius* breeding in County Tipperary in 2008. *Irish Birds* 8: 435–436.

Collinson, M. 2006. Splitting headaches? Recent taxonomic changes affecting the British and Western Palearctic lists. *British Birds* 99: 306–323.

Collinson, J.M., Parkin, D.T., Knox, A.G., Sangster, G. & Svensson, L. 2008. Species boundaries in the Herring and Lesser Black-backed Gull complex. *British Birds* 101: 340–363.

Combridge, P., Christie, D.A. & Ferguson-Lees, J. 2003. Breeding European Honey-buzzards in Britain. *British Birds* 96: 258–260.

Combridge, P., Christie, D.A. & Ferguson-Lees, I.J. 2005. Breeding Honey-buzzards in Britain: fact, fiction and wishful thinking. *British Birds* 98: 488–489.

Concern for Swifts. 2013. *Concern for Swifts.* URL: www.concernforswifts.com [31 July 2013].

Conway, G.J. 2010. Fire in the Forest. *BTO News* 290: 23–24.

Conway, G.J. 2011. *Changes in migration strategy and wintering behaviour of Common Chiffchaff* Phylloscopus collybita. PhD thesis, University of East Anglia, Norwich.

Conway, G., Wotton, S., Henderson, I., Langston, R., Drewitt, A. & Currie, F. 2007. Status and distribution of European Nightjars *Caprimulgus europaeus* in the UK in 2004. *Bird Study* 54: 98–111.

Conway, G.J., Burton, N.H.K., Handschuh, M. & Austin, G.E. 2008. *UK population estimates from the 2007 Breeding Little Ringed Plover and Ringed Plover surveys*. BTO Research Report 510. BTO, Thetford.

Conway, G., Wotton, S., Henderson, I., Eaton, M., Drewitt, A. & Spencer, J. 2009. The status of breeding Woodlarks *Lullula arborea* in Britain in 2006. *Bird Study* 56: 310–325.

Cook, A. 1975. Changes in the Carrion/Hooded Crow hybrid zone and the possible importance of climate. *Bird Study* 22: 165–168.

Cook, M.J.H. 1982. Breeding status of the Crested Tit. *Scottish Birds* 12: 97–106.

Cooney, T. 1998. Ringed Plovers *Charadrius hiaticula* nesting on cut-away peat in County Offaly. *Irish Birds* 6: 283–284.

Copland, A. 2002. Delivering Corncrake *Crex crex* conservation in Ireland: past, present and future. *Irish Birds* 7: 33–42.

Copland, A. & Lusby, J. 2012. Lowland farmland. In *Bird Habitats in Ireland* (ed Nairn, R. & O'Halloran, J.), pp 124–137. The Collins Press, Wilton, Cork.

Copland, A., Crowe, O., Wilson, M.W. & O'Halloran, J. 2011. Habitat associations of Eurasian Skylarks *Alauda arvensis* breeding on Irish farmland and implications for agri-environment planning. *Bird Study* 59: 155–165.

Cornulier, T., Yoccoz, N.G., Bretagnolle, V., Brommer, J.E., Butet, A., Ecke, F., Elston, D.A., Framstad, E., Henttonen, H., Hörnfeldt, B., Huitu, O., Imholt, C., Ims, R.A., Jacob, J., Jędrzejewska, B., Millon, A., Petty, S.J., Pietiäinen, H., Tkadlec, E., Zub, K. & Lambin, X. 2013. Europe-wide dampening of population cycles in keystone herbivores. *Science* 340: 63–66.

Cosgrove, P. 2003. Mandarin Ducks in northern Scotland and the potential consequences for breeding Goldeneye. *Scottish Birds* 24: 1–10.

Coudrain, V., Arlettaz, R. & Schaub, M. 2010. Food or nesting place? Identifying factors limiting Wryneck populations. *Journal of Ornithology* 151: 867–880.

Coulson, J.C. 2010. A long-term study of the population dynamics of Common Eiders *Somateria mollissima*: why do several parameters fluctuate markedly? *Bird Study* 57: 1–18.

Coulson, J.C. 2012. *The Kittiwake*. T. & A.D Poyser, London.

Cowley, E. & Siriwardena, G. 2005. Long-term variation in survival rates of Sand Martins *Riparia riparia*: dependence on breeding and wintering ground weather, age and sex, and their population consequences. *Bird Study* 52: 237–251.

Craik, C. 1997. Long-term effects of North American Mink *Mustela vison* on seabirds in western Scotland. *Bird Study* 44: 303–309.

Craik, C. 1998. Recent Mink-related declines of gulls and terns in west Scotland and the beneficial effects of Mink control. *Argyll Bird Report* 14: 98–110.

Crick, H.Q.P. & Sparks, T.H. 1999. Climate change related to egg-laying trends. *Nature* 399: 423–424.

Cromie, J. 2002. Breeding status of Red-throated Diver *Gavia stellata* in Ireland. *Irish Birds* 7: 13–20.

Crowe, O. 2005. The Garden Bird Survey: monitoring birds of Irish gardens during winters between 1994/95 and 2003/04. *Irish Birds* 7: 475–482.

Crowe, O. 2011. *Recent patterns and trends in terrestrial bird populations in Ireland: their value in supporting bird conservation.* Unpublished PhD thesis, University College Cork.

Crowe, O., Austin, G.E., Colhoun, K., Cranswick, P., Kershaw, M. & Musgrove, A.J. 2008. Estimates and trends of waterbird numbers wintering in Ireland, 1994/95–2003/04. *Bird Study* 55: 66–77.

Crowe, O., Cummins, S., Gilligan, N., Smiddy, P. & Tierney, T.D. 2010a. An assessment of the current distribution and status of the Kingfisher *Alcedo atthis* in Ireland. *Irish Birds* 9: 41–54.

Crowe, O., Coombes, R.H., Lysaght, L., O'Brien, C., Choudhury, K.R., Walsh, A.J., Wilson, J.H. & O'Halloran, J. 2010b. Population trends of widespread breeding birds in the Republic of Ireland 1998–2008. *Bird Study* 57: 267–280.

Cumbria Raptor Study Group. 2002. *Birds of Prey in Cumbria: status, distribution and ecology 1992–2001: a 10 year review.* Cumbria Raptor Study Group.

Cummins, S.T. & O'Halloran, J. 2003. The breeding biology of the Stonechat *Saxicola torquata* in southwest Ireland. *Irish Birds* 7: 177–186.

Cummins, S., Bleasdale, A., Douglas, C., Newton, S., O'Halloran, J. & Wilson, H.J. 2010. The status of Red Grouse in Ireland and the effects of land use, habitat and habitat quality on their distribution. *Irish Wildlife Manuals, No. 50.* National Parks and Wildlife Service, Department of the Environment, Heritage and Local Government, Dublin.

Cunningham, P. 1987. Mink in the Outer Hebrides. *Scottish Bird News* 8: 4.

Cutler, D.R., Edwards Jr., T.C., Beard, K.H., Cutler, A., Hess, K.T., Gibson, J. & Lawler, J.J. 2007. Random forests for classification in ecology. *Ecology* 88: 2783–2792.

Dadam, D., Barimore, C.J., Shawyer, C.R. & Leech, D.I. 2011. *The BTO Barn Owl Monitoring Programme: final report 2000–2009.* BTO Research Report 577. BTO, Thetford.

Dadam, D., Clark, J.A., Robinson, R.A., Leech, D.I., Moss, D., Kew, A.J., Barber, L.J., Barimore, C.J., Blackburn, J.R., De Palacio, D.X., Griffin, B.M. & Schäfer, S. 2012. Bird ringing and nest recording in Britain and Ireland in 2011. *Bird Study* 27: 109–153.

Dagley, J.R. 1994. Golden Orioles in East Anglia and their conservation. *British Birds* 87: 205–219.

Davey, C.M. 2008. *The impact for game management for Pheasant (*Phasianus colchicus*) shooting on vertebrate biodiversity in British woodlands.* Unpublished PhD thesis, University of Bristol.

Davey, C.M., Chamberlain, D.E., Newson, S.E., Noble, D.G. & Johnston, A. 2011. Rise of the generalist: evidence for climate driven homogenization in avian communities. *Global Ecology and Biogeography* 21: 568–578.

Davis, A.H. & Vinicombe, K.E. 2011. The probable breeding of Ferruginous Ducks in Avon. *British Birds* 104: 77–83.

Davis, T.J. & Jones, T.A. 2007. *The Birds of Lundy.* Devon Birdwatching & Preservation Society and Lundy Field Society, Berrynarbor, Devon.

Dazley, R.A. & Trodd, P. 1994. *An Atlas of the Breeding Birds of Bedfordshire: 1988–92.* Bedfordshire Natural History Society, Bedford.

Dean, A. 2002. Wintering Glaucous, Iceland and Herring Gulls in the Midlands. *Birdwatch* 118: 12.

del Hoyo, J., Elliott, A. & Sargatal, J. (eds). 2002. *Handbook of the Birds of the World.* Volume 7: Jacamars to Woodpeckers. Lynx Edicions, Barcelona.

Delany, S. 1993. Introduced and escaped geese in Britain in summer 1991. *British Birds* 86: 591–599.

Delany, S., Scott, D., Dodman, T. & Stroud, D. (eds). 2009. *An Atlas of Wader Populations in Africa and Western Eurasia.* Wetlands International, Wageningen.

Dennis, M. 1992. Yellow-legged Herring Gulls in Essex. *British Birds* 85: 246.

Dennis, R. 2008. *A Life of Ospreys.* Whittles Publishing, Caithness.

Dennis, R. & Cook, M. 1996. Eagle Owl breeding in Moray. *Moray & Nairn Bird Report* 1995: 79–80.

Dennis, R.L.H., Sparks, T.H. & Hardy, P.B. 1999. Bias in butterfly distribution maps: the effects of sampling effort. *Journal of Insect Conservation* 3: 33–42.

Denny, R.E. & Summers, R.W. 1996. Nest-site selection, management and breeding success of Crested Tits *Parus cristatus* at Abernethy Forest, Strathspey. *Bird Study* 43: 371–379.

Devereux, C.L., McKeever, C.U., Benton, T.G. & Whittingham, M.J. 2004. The effect of sward height and drainage on Common Starlings *Sturnus vulgaris* and Northern Lapwings *Vanellus vanellus* foraging in grassland habitats. *Ibis* 146: 115–122.

Devon Birdwatching & Preservation Society. 2010. Devon Bird Report 2009. *Devon Birds* 63(3): 1–228.

Dickinson, J.L., Zuckerberg, B. & Bonter, D.N. 2010. Citizen science as an ecological research tool: challenges and benefits. *Annual Review of Ecology Evolution and Systematics* 41: 149–172.

Dierschke, J. 2001. *Die Überwinterungsökologie von Ohrenlerchen* Eremophila alpestris*, Schneeammern* Plectrophenax nivalis *und Berghänflingen* Carduelis flavirostris *im Wattenmeer.* PhD thesis, University of Oldenburg.

Dierschke, J. 2002. Food preferences of Shorelarks *Eremophila alpestris*, Snow Buntings *Plectrophenax nivalis* and Twites *Carduelis flavirostris wintering* in the Wadden Sea. *Bird Study* 49: 263–269.

Dillon, I.A., Smith, T.D., Williams, S.J., Haysom, S. & Eaton, M.A. 2009. Status of Red-throated Divers *Gavia stellata* in Britain in 2006. *Bird Study* 56: 147–157.

Dixon, A. & Haffield, J.P. 2013. Seed availability and timing of breeding of Common Crossbills *Loxia curvirostra* at Sitka Spruce *Picea sitchensis* dominated forestry plantations. *Ardea* 101: 33–38.

Dixon, A., Richards, C., Haffield, P., Roberts, G., Thomas, M. & Lowe, A. 2008. The National Peregrine Survey 2002: how accurate are the published results for Wales? *Welsh Birds* 5: 276–283.

Dobinson, H.M. & Richards, A.J. 1964. The effect of the severe winter of 1962/63 on birds in Britain. *British Birds* 57: 373–434.

Dobson, A.D.M., Clarke, M., Kjellén, N. & Clarke, R. 2012. The size and migratory origins of the population of Hen Harriers *Circus cyaneus* wintering in England. *Bird Study* 59: 218–227.

Dolman, P.M. 2012. Mechanisms and processes underlying landscape structure effects on bird populations. In *Birds and Habitat: Relationships in Changing Landscapes* (ed Fuller, R.J.), pp 93–124. Cambridge University Press, Cambridge.

Donaghy, A.M., Green, R.E. & O'Halloran, J. 2011. Timing of autumn migration of young Corncrake *Crex crex*. *Ibis* 153: 425–428.

Donald, P.F. 1997. The Corn Bunting *Miliaria calandra* in Britain: a review of current status, patterns of decline and possible causes. In *The Ecology and Conservation of Corn Buntings* Miliaria calandra (eds Donald, P.F. & Aebischer, N.J.), pp 11–26. UK Nature Conservation no. 13. JNCC, Peterborough.

Donald, P.F. 2004. *The Skylark.* T. & A.D. Poyser, London.

Donald, P.F. & Fuller, R.J. 1998. Ornithological atlas data: a review of uses and limitations. *Bird Study* 45: 129–145.

Donald, P.F. & Greenwood, J.J.D. 2001. Spatial patterns of range contraction in British breeding birds. *Ibis* 143: 593–601.

Dore, C.P., Ellis, P.M. & Stuart, E.M. 1996. Numbers of Whimbrel breeding in Shetland in 1989–1994 and previously. *Scottish Birds* 18: 193–196.

Dorset Bird Club. 2011. *Dorset Bird Report 2009.* Dorset Bird Club.

Dott, H.E.M. & Brown, A.W. 2000. A major decline in House Sparrows in central Edinburgh. *Scottish Birds* 21: 61–68.

Dougall, T.W., Holland, P.K. & Yalden, D.W. 2010. The population biology of Common Sandpipers in Britain. *British Birds* 103: 100–114.

Douglas, D.J.T., Newson, S.E., Leech, D.I., Noble, D.G. & Robinson, R.A. 2010. How important are climate-induced changes in host availability for population processes in an obligate brood parasite, the European Cuckoo? *Oikos* 119: 1834–1840.

Driver, J. & Dare, P.J. 2009. Population increase of Buzzards in Snowdonia, 1977–2007. *Welsh Birds* 6: 38–48.

Dudley, S.P. 2005. Changes to Category C of the British List. *Ibis* 147: 803–820.

Duncan, R. 2011. The 2010 Waxwing invasion. *Scottish Birds* 31: 71–76.

Eaton, M.A., Dillon, I.A., Stirling-Aird, P.K. & Whitfield, D.P. 2007. Status of the Golden Eagle *Aquila chrysaetos* in Britain in 2003. *Bird Study* 54: 212–220.

Eaton, M.A., Brown, A.F., Noble, D.G., Musgrove, A.J., Hearn, R.D., Aebischer, N.J., Gibbons, D.W., Evans, A. & Gregory, R.D. 2009. Birds of Conservation Concern 3: the population status of birds in the United Kingdom, Channel Isles and Isle of Man. *British Birds* 102: 296–341.

Eckerson, W.W. 1995. Three tier client/server architecture: achieving scalability, performance, and efficiency in client server applications. *Open Information Systems*

3: 44–50.

Eglington, S.M. & Pearce-Higgins, J.W. 2012. Disentangling the Relative Importance of Changes in Climate and Land-Use Intensity in Driving Recent Bird Population Trends. PLoS ONE 7(3): e30407. doi:10.1371/journal.pone.0030407.

Ekroos, J., Fox, A.D., Christensen, T.K., Petersen, I.K., Kilpi, M., Jonsson, J.E., Green, M., Laursen, K., Cervencl, A., de Boer, P., Nilsson, L., Meissner, W., Garthe, S. & Ost, M. 2012. Declines amongst breeding Eider *Somateria mollissima* numbers in the Baltic/Wadden Sea flyway. *Ornis Fennica* 89: 81–90.

Elith, J., Graham, C.H., Anderson, R.P., Dudík, M., Ferrier, S., Guisan, A., Hijmans, R.J., Huettmann, J.R., Leathwick, J.R., Lehmann, A., Li, J., Lohmann, L.G., Loiselle, B.A., Moritz, C., Nakamura, M., Makazawa, Y., Overton, J.M., Peterson, A.T., Phillips, S.J., Richardson, K., Scachetti-Pereira, R., Schapire, R.E., Soberón, J., Williams, S., Wisz, M.S. & Zimmermann, N.E. 2006. Novel methods improve prediction of species' distributions from occurrence data. *Ecography* 29: 129–151.

Ellis, P. 2004. *Living Shetland Biodiversity Action Plan: Red-necked Phalarope.* URL: www.livingshetland.org.uk/documents/RedNeckedPhalaropeSAP2004.pdf [31 July 2013].

Esselink, H. & Beekman, J.H. 1991. Between year variation and causes of mortality in the non-breeding population of the Mute Swan *Cygnus olor* in the Netherlands, with special reference to hunting. *Wildfowl* 42: 110–119.

Estrada, J., Pedrocchi, V., Brotons, L. & Herrando, S. (eds). 2004. *Atles des ocells nidificantes de Catalunya 1999–2002.* Institut Català d'Ornitologia (ICO)/Lynx Edicions, Barcelona.

Etheridge, B., Riley, H., Wernham, C., Holling, M., Stevenson, A. & Thompson, D. 2012. *Scottish Raptor Monitoring Scheme Report 2010.* Scottish Raptor Study Groups, Dunblane.

European Commission. 1994. *CORINE Land Cover Technical Guide.* EUR 12585 EN OPOCE Luxembourg.

Evans, A.D. 1992. The numbers and distribution of Cirl Buntings *Emberiza cirlus* breeding in Britain in 1989. *Bird Study* 39: 17–22.

Evans, A.D. & Smith, K.W. 1994. Habitat selection of Cirl Buntings *Emberiza cirlus* wintering in Britain. *Bird Study* 41: 81–87.

Evans, I.M., Summers, R.W., O'Toole, L., Orr-Ewing, D.C., Evans, R., Snell, N. & Smith, J. 1999. Evaluating the success of translocating Red Kites *Milvus milvus* to the UK. *Bird Study* 46: 129–144.

Evans, R.J., Wilson, J.D., Amar, A., Douse, A., MacLennan, A. & Whitfield, D.P. 2009. Growth and demography of a re-introduced population of White-tailed Eagles *Haliaeetus albicilla*. *Ibis* 151: 244–254.

Evans, R.J., O'Toole, L. & Whitfield, D.P. 2012. The history of eagles in Britain and Ireland: an ecological review of placename and documentary evidence from the last 1500 years. *Bird Study* 59: 335–349.

Ewald, J.A., Aebischer, N.J., Brickle, N.W., Moreby, S.J., Potts, G.R. & Wakeham-Dawson, A. 2002. Spatial variation in densities of farmland birds in relation to pesticide use and avian food resources. In *Avian Landscape Ecology: pure and applied issues in the large-scale ecology of birds* (eds Chamberlain, D. & Wilson, A.), pp 305–312. Proceedings of the 11th Annual IALE (UK) Conference, 10–13 September 2002. International Association for Landscape Ecology, Preston.

Ewing, S.R., Rebecca, G.W., Heavisides, A., Court, I.R., Lindley, P., Ruddock, M., Cohen, S. & Eaton, M.A. 2011. Breeding status of Merlins *Falco columbarius* in the UK in 2008. *Bird Study* 58: 379–389.

Ewing, S.R., Eaton, M.A., Poole, T.F., Davies, M. & Haysom, S. 2012. The size of the Scottish population of Capercaillie *Tetrao urogallus*: results of the fourth national survey. *Bird Study* 59: 126–138.

Ewins, P.J. & Tasker, M.L. 1985. The breeding distribution of Black Guillemots *Cepphus grylle* in Orkney and Shetland, 1982–84. *Bird Study* 32: 186–193.

Fahy, K. 2011. Irish Rare Bird Report 2010. *Irish Birds* 9: 283–314.

Ferguson, D. 2012. *The Birds of Buckinghamshire.* Buckinghamshire Bird Club, Beaconsfield.

Ferguson-Lees, I.J. & Sharrock, J.T.R. 1977. When will the Fan-tailed Warbler colonise Britain? *British Birds* 70: 152–158.

Ferguson-Lees, J., Castell, R. & Leech, D. 2011. *A field guide to monitoring nests.* BTO, Thetford.

Ferreras, P. & Macdonald, D.W. 1999. The impacts of American Mink *Mustela vison* on water birds in the upper Thames. *Journal of Applied Ecology* 36: 701–708.

Field, R.H. & Anderson, G.Q.A. 2004. Habitat use by breeding Tree Sparrows *Passer montanus*. *Ibis* 146 (Suppl. 2): 60–68.

Fielding, A.H. & Bell, J.F. 1997. A review of methods for the assessment of prediction errors in conservation presence/absence models. *Environmental Conservation* 24: 38–49.

Fielding, A., Haworth, P., Whitfield, P., McLeod, D. & Riley, H. 2011. *A Conservation Framework for Hen Harriers in the United Kingdom.* JNCC Report 441. JNCC, Peterborough.

Fife Bird Club. 2012. *Fife Bird Report 2010.* Fife Bird Club, Fife.

Fisher, J. 1952. *The Fulmar.* Collins, London.

Fisher, J. 1953. The Collared Turtle Dove in Europe. *British Birds* 46: 153–181.

Fletcher, K., Aebischer, N.J., Baines, D., Foster, R. & Hoodless, A.N. 2010. Changes in breeding success and abundance of ground-nesting moorland birds in relation to the experimental deployment of legal predator control. *Journal of Applied Ecology* 47: 263–272.

Foppen, R.P.B., ter Braak, C.J.F., Verboom, J. & Reijnen, R. 1999. Dutch Sedge Warblers *Acrocephalus schoenobaenus* and West-African rainfall: empirical data and simulation modelling show low population resilience in fragmented marshlands. *Ardea* 87: 113–127.

Forestry Commission. 2013. *Woodland Area, Planting and Restocking.* 2013 edition. URL: www.forestry.gov.uk/statistics [31 July 2013].

Forrester, R.W., Hopkins, I. & Menzies, D. 2012. *The birds of Bute: a bird atlas and local avifauna.* Bute Natural History Society, Rothesay, and Scottish Ornithologists' Club, Aberlady.

Fox, A.D. & Stroud, D.A. 2002. *Anser albifrons flavirostris* Dalgety & Scott 1948, Greenland White-fronted Goose. *Birds of the Western Palearctic Update* 4: 1-22.

Fox, A.D., Madsen, J., Boyd, H., Kuijken, E., Norriss, D.W., Tombre, I.M. & Stroud, D.A. 2005. Effects of agricultural change on abundance, fitness components and distribution of two arctic-nesting goose populations. *Global Change Biology* 11: 881–893.

Fox, R., Conrad, K.F., Parsons, M.S., Warren, M.S. & Woiwod, I.P. 2006. *The state of Britain's larger moths.* Butterfly Conservation and Rothamsted Research, Wareham, Dorset.

Francis, I. & Cook, M. (eds). 2011. *The Breeding Birds of North-East Scotland.* Scottish Ornithologists' Club, Aberdeen.

Franco, A.M.A., Anderson, B.J., Roy, D.B., Gillings, S., Fox, R., Moilanen, A. & Thomas, C.D. 2009. Surrogacy and persistence in reserve selection: landscape prioritization for multiple taxa in Britain. *Journal of Applied Ecology* 46: 82–91.

Franklin, J. 2009. *Mapping Species Distributions: Spatial Inference and Prediction.* Cambridge University Press, Cambridge.

Fraser, P.A. & Rogers, M.J. 2005. Report on scarce migrant birds in Britain in 2002. Part 2: American Wigeon to Ring-billed Gull. *British Birds* 98: 73–88.

Fraser, P.A. & Rogers, M.J. 2006a. Report on scarce migrant birds in Britain in 2003. Part 1: American Wigeon to Wryneck. *British Birds* 99: 74–91.

Fraser, P.A. & Rogers, M.J. 2006b. Report on scarce migrant birds in Britain in 2003. Part 2: Short-toed Lark to Little Bunting. *British Birds* 99: 129–147.

Fraser, P.A. & Ryan, J.F. 1995. Status of the Great Grey Shrike in Britain and Ireland. *British Birds* 88: 478–484.

Fray, R.M. 2013. Lesser Spotted Woodpecker, Scalloway, Shetland, October 2012 – the first record for Scotland. *Scottish Birds* 33: 78–81.

Fray, R., Davies, R., Gamble, D., Harrop, A. & Lister, S. 2009. *The Birds of Leicestershire and Rutland.* Christopher Helm, London.

Frederiksen, M., Wanless, S., Harris, M.P., Rothery, P. & Wilson, L.J. 2004. The role of industrial fisheries and oceanographic change in the decline of North Sea Black-legged Kittiwakes. *Journal of Applied Ecology* 41: 1129–1139.

Frederiksen, M., Edwards, M., Richardson, A.J., Halliday, N.C. & Wanless, S. 2006. From plankton to top predators: bottom-up control of a marine food web across four trophic levels. *Journal of Animal Ecology* 75: 1259–1268.

Frederiksen, M., Daunt, F., Harris, M.P. & Wanless, S. 2008. The demographic impact of extreme events: stochastic weather drives survival and population dynamics in a long-lived seabird. *Journal of Animal Ecology* 77: 1020–1029.

Frederiksen, M., Moe, B., Daunt, F., Phillips, R.A., Barrett, R.T., Bogdanova, M.I., Boulinier, T., Chardine, J.W., Chastel, O., Chivers, L.S., Christensen-Dalsgaard, S., Clément-Chastel, C., Colhoun, K., Freeman, R., Gaston, A.J., González-Solís, J., Goutte, A., Grémillet, D., Guilford, T., Jensen, G.H., Krasnov, Y., Lorentsen, S.-H., Mallory, M.L., Newell, M., Olsen, B., Shaw, D., Steen, H., Strøm, H., Systad, G.H., Thórarinsson, T.L. & Anker-Nilssen, T. 2012. Multicolony tracking reveals the winter distribution of a pelagic seabird on an ocean basin scale. *Diversity and Distributions* 18: 530–542.

Freeman, S.N. & Crick, H.Q.P. 2002. Population dynamics of House Sparrows *Passer domesticus* breeding in Britain: an integrated analysis. In *Investigation into the Causes of the Decline of Starlings and House Sparrows in Great Britain* (eds Crick, H.Q.P., Robinson, R.A., Appleton, G.F., Clark, N.A. & Rickard, A.D.), pp 193–212, BTO Research Report 290. BTO, Thetford.

Freeman, S.N. & Crick, H.Q.P. 2003. The decline of the Spotted Flycatcher *Muscicapa striata* in the UK: an integrated population model. *Ibis* 145: 400–412.

Freeman, S.N., Balmer, D.E. & Crick, H.Q.P. 2006. On the censusing of Tawny Owls *Strix aluco*. *Bird Census News* 19: 58–62.

Freeman, S.N., Robinson, R.A., Clark, J.A., Griffin, B.M. & Adams, S.Y. 2007. Changing demography and population decline in the Common Starling *Sturnus vulgaris*: a multisite approach to Integrated Population Monitoring. *Ibis* 149: 587–596.

Fuller, R.A., Warren, P.H., Armsworth, P.R., Barbosa, O. & Gaston, K.J. 2008. Garden bird feeding predicts the structure of urban avian assemblages. *Diversity and Distributions* 18: 131–137.

Fuller, R.J. 1982. *Bird Habitats in Britain*. T. & A.D. Poyser, Calton.

Fuller, R.J. 2001. Responses of woodland birds to increasing numbers of deer: a review of evidence and mechanisms. *Forestry* 74: 289–298.

Fuller, R.J. 2012a. *Birds and Habitat: relationships in changing landscapes*. Cambridge University Press, Cambridge.

Fuller, R.J. 2012b. Habitat quality and habitat occupancy by birds in variable environments. In *Birds and Habitat: Relationships in Changing Landscapes* (ed Fuller, R.J.), pp 37–62. Cambridge University Press, Cambridge.

Fuller, R.J. & Ausden, M. 2008. Birds and habitat change in Britain. Part 1: a review of losses and gains in the twentieth century. *British Birds* 101: 644–675.

Fuller, R.J., Gregory, R.D., Gibbons, D.W., Marchant, J.H., Wilson, J.D., Baillie, S.R. & Carter, N. 1995. Population declines and range contractions among lowland farmland birds in Britain. *Conservation Biology* 9: 1425–1441.

Fuller, R.J., Noble, D.G., Smith, K.W. & Vanhinsbergh, D. 2005. Recent declines in populations of woodland birds in Britain: a review of possible causes. *British Birds* 98: 116–143.

Fuller, R.J., Atkinson, P.W., Garnett, M.C., Conway, G.J., Bibby, C.J. & Johnstone, I.G. 2006. Breeding bird communities in the upland margins (ffridd) of Wales in the mid-1980s. *Bird Study* 53: 177–186.

Fuller, R.J., Gaston, K.J. & Quine, C.P. 2007. Living on the edge: British and Irish woodland birds in a European context. *Ibis* 149 (Suppl. 2): 53–63.

Fuller, R.J., Humphreys, E.M., Wilson, J.D., Hoccom, D.G. & Calladine, J. 2010. Changes in the breeding wader populations of the machair of the Western Isles, Scotland, between 2000 and 2007. *Bird Study* 57: 121–124.

Fulton, D. 2010. The breeding population of Northern Wheatears at Clee Hill, Shropshire, 1998–2009. *British Birds* 103: 223–228.

Furness, R.W. 2007. Responses of seabirds to depletion of food fish stocks. *Journal of Ornithology* 148 (Suppl. 2): S247–S252.

Galbraith, H., Murray, S., Rae, S., Whitfield, D.P. & Thompson, D.B.A. 1993. Numbers and distribution of Dotterel *Charadrius morinellus* breeding in Great Britain. *Bird Study* 40: 161–169.

Game & Wildlife Conservation Trust. 2013. *Red-legged Partridge*. URL: www.gwct.org.uk/research__surveys/wildlife_surveys_and_ngc/national_gamebag_census_ngc/birds__summary_trends/228.asp [31 July 2013].

Game Conservancy Trust. 2004. *Review of 2003*. Game Conservancy Trust, Fordingbridge.

Garden Bird Feeding Survey. 2013. *Results by species*. URL: www.bto.org/volunteer-surveys/gbfs/results/ [31 July 2013].

Garden BirdWatch. 2010. *Annual Results 2010*. URL: www.bto.org/volunteer-surveys/gbw/results/2010 [31 July 2013].

Garden BirdWatch. 2013. *BTO Garden BirdWatch Results*. URL: www.bto.org/volunteer-surveys/gbw/results [31 July 2013].

Gardiner, M.M., Allee, L.L., Brown, P.M.J., Losey, J.E., Roy, H.E. & Smyth, R.R. 2012. Lessons from lady beetles: accuracy of monitoring data from US and UK citizen-science programs. *Frontiers in Ecology and the Environment* 10: 471–476.

Garner, M. & Quinn, D. 1997. Identification of Yellow-legged Gulls in Britain. *British Birds* 90: 25–62.

Garner, M., Quinn, D. & Glover, B. 1997. Identification of Yellow-legged Gulls in Britain. Part 2. *British Birds* 90: 369–383.

Gaston, K.J. 2003. *The Structure and Dynamics of Geographic Ranges*. Oxford University Press, Oxford.

Gaston, K.J., Blackburn, T.M. & Lawton, J.H. 1997. Interspecific abundance-range size relationships: an appraisal of mechanisms. *Journal of Animal Ecology* 66: 579–601.

Gebhardt, H. 1996. Ecological and economic consequences of introductions of exotic wildlife (birds and animals) in Germany. *Wildlife Biology* 2: 205–211.

Gibbins, C., Small, B.J. & Sweeney, J. 2010. From the Rarities Committee's files. Identification of Caspian Gull. Part 1: typical birds. *British Birds* 103: 142–183.

Gibbins, C., Neubauer, G. & Small, B.J. 2011. From the Rarities Committee's files. Identification of Caspian Gull. Part 2: phenotypic variability and the field characteristics of hybrids. *British Birds* 104: 702–742.

Gibbons, D.W., Reid, J.B. & Chapman, R.A. 1993. *The New Atlas of Breeding Birds in Britain and Ireland: 1988–1991*. T. & A.D. Poyser, London.

Gibbons, D.W., Donald, P.F., Bauer, H.-G., Fornasari, L. & Dawson, I.K. 2007. Mapping avian distributions: the evolution of bird atlases. *Bird Study* 54: 324–334.

Gilbert, G. 2002. The status and habitat of Spotted Crakes *Porzana porzana* in Britain in 1999. *Bird Study* 49: 79–86.

Gilbert, G. 2012. Grasshopper Warbler *Locustella naevia* breeding habitat in Britain. *Bird Study* 59: 303–314.

Gilissen, N., Haanstra, L., Delany, S., Boere, G. & Hagemeijer, W. 2002. *Numbers and Distribution of Wintering Waterbirds in the Western Palearctic and Southwest Asia in 1997, 1998 and 1999. Results from the International Waterbird Census*. Wetlands International Global Series No. 11, Wageningen.

Gill, F. & Donsker, D. (eds). 2013. *IOC World Bird List (v 3.3)*. URL: www.worldbirdnames.org [31 July 2013].

Gill, J.A., Norris, K., Potts, P.M., Gunnarsson, T.G., Atkinson, P.W. & Sutherland, W.J. 2001. The buffer effect and large-scale population regulation in migratory birds. *Nature* 412: 436–438.

Gill, J.A., Langston, R.H.W., Alves, J.A., Atkinson, P.W., Bocher, P., Cidraes Vieira, N., Crockford, N.J., Gélinaud, G., Groen, N., Gunnarsson, T.G., Hayhow, B., Hooijmeijer, J., Kentie, R., Kleijn, D., Lourenço, P.M., Masero, J.A., Meunier, F., Potts, P.M., Roodbergen, M., Schekkerman, H., Schröder, J., Wymenga, E. & Piersma, T. 2007. Contrasting trends in two Black-tailed Godwit populations: a review of causes and recommendations. *Wader Study Group Bulletin* 114: 43–50.

Gillings, S. 2008. Designing a winter bird atlas field methodology: issues of time and space in sampling and interactions with habitat. *Journal of Ornithology* 149: 345–355.

Gillings, S. & Fuller, R.J. 2001. Habitat selection by Skylarks *Alauda arvensis* wintering in Britain in 1997/98. *Bird Study* 48: 293–307.

Gillings, S., Newson, S.E., Noble, D.G. & Vickery, J.A. 2005. Winter availability of cereal stubbles attracts declining farmland birds and positively influences breeding population trends. *Proceedings of the Royal Society Series B Biological Sciences* 272: 733–739.

Gillings, S., Austin, G.E., Fuller, R.J. & Sutherland, W.J. 2006. Distribution shifts in wintering Golden Plovers *Pluvialis apricaria* and Lapwing *Vanellus vanellus* in Britain. *Bird Study* 53: 274–284.

Gilroy, J.J. & Lees, A.C. 2003. Vagrancy theories: are autumn vagrants really reverse migrants? *British Birds* 96: 427–438.

Gilroy, J.J., Anderson, G.Q.A., Grice, P.V., Vickery, J.A., Bray, I., Watts, P.N. & Sutherland, W.J. 2008. Could soil degradation contribute to farmland bird declines? Links between soil penetrability and the abundance of Yellow Wagtails *Motacilla flava* in arable fields. *Biological Conservation* 141: 3116–3126.

Gilroy, J.J., Anderson, G.Q.A., Grice, P.V., Vickery, J.A. & Sutherland, W.J. 2010. Mid-season shifts in the habitat associations of Yellow Wagtails *Motacilla flava* breeding in arable farmland. *Ibis* 152: 90–104.

Golden Eagle Trust. 2013. *5th batch of Sea Eagles released in 2011*. URL: www.goldeneagle.ie [31 July 2013].

Goldeneye Study Group. 2010. *2010 Breeding season newsletter*. Goldeneye Study Group, Kingussie.

Goodenough, A.E., Elliot, S.L. & Hart, A.G. 2009. The challenges of conservation for declining migrants: are reserve-based initiatives during the breeding season appropriate for the Pied Flycatcher *Ficedula hypoleuca*? *Ibis* 151: 429–439.

Google Scholar. 2013. *Google Scholar search*. http://scholar.google.co.uk [31 July 2013].

Graham, I.M., Redpath, S.M. & Thirgood, S.J. 1995. The diet and breeding density of Common Buzzards *Buteo buteo* in relation to indices of prey abundance. *Bird Study* 42: 165–173.

Grant, M. & Pearce-Higgins, J.W. 2012. Spatial variation and habitat relationships in moorland bird assemblages: a British perspective. In *Birds and Habitat: Relationships in Changing Landscapes* (ed Fuller, R.J.), pp 207–236. Cambridge University Press, Cambridge.

Grant, M.C., Orsman, C., Easton, J., Lodge, C., Smith, M., Thompson, G., Rodwell, S. & Moore, N. 1999. Breeding success and causes of breeding failure of Curlew *Numenius arquata* in Northern Ireland. *Journal of Applied Ecology* 36: 59–74.

Grant, M.C., Cowie, N., Donald, C., Dugan, D., Johnstone, I., Lindley, P., Moncreiff, R., Pearce-Higgins, J.W., Thorpe, R. & Tomes, D. 2009. Black Grouse response to dedicated conservation management. *Folia Zoologica* 58: 195–206.

Graveland, J. 1999. Effects of reed cutting on density and breeding success of Reed Warbler *Acrocephalus scirpaceus* and Sedge Warbler *A. schoenobaenus*. *Journal of Avian Biology* 30: 469–482.

Gray, N., Thomas, G., Trewby, M. & Newton, S.F. 2003. The status and distribution of Choughs *Pyrrhocorax pyrrhocorax* in the Republic of Ireland 2002/03. *Irish Birds* 7: 147–156.

Great Bustard Project. 2013. *Bringing back the bustard*. URL: http://greatbustard.org/the-project/ [31 July 2013].

Great Crane Project. 2013. *The Great Crane Project*. URL: www.thegreatcraneproject.org.uk [31 July 2013].

Green, R.E. 1988. Effects of environmental factors on the timing and success of breeding in Common Snipe *Gallinago gallinago* (Aves: Scolopacidae). *Journal of Applied Ecology* 25: 79–93.

Green, R.E. 1996. Factors affecting the population density of the Corncrake *Crex crex* in Britain and Ireland. *Journal of Applied Ecology* 33: 237–248.

Green,R.E. & Griffiths, G.H. 1994.Use of preferred nesting habitat by Stone Curlews *Burhinus oedicnemus* in relation to vegetation structure. *Journal of Zoology* 233: 457–471.

Green, R.E. & Taylor, C.R. 1995. Changes in Stone Curlew *Burhinus oedicnemus* distribution and abundance and vegetation height on chalk grassland at Porton Down, Wiltshire. *Bird Study* 42: 177–181.

Green, R.E. & Williams, G. 1994. The ecology of the Corncrake *Crex crex* and action for its conservation in Britain and Ireland. In *Nature Conservation and Pastoralism in Europe: Proceedings of the Third European Forum* (eds Bignal, E. & Curtis, D.J.), pp 69–74. JNCC, Peterborough.

Green, R.E., Pienkowski, M.W. & Love, J.A. 1996. Long-term viability of the reintroduced population of the White-tailed Eagle *Haliaeetus albicilla* in Scotland. *Journal of Applied Ecology* 33: 357–368.

Green, R.E., Tyler, G.A., Stowe, T.J. & Newton, A.V. 1997. A simulation model of the effect of mowing of agricultural grassland on the breeding success of the Corncrake (*Crex crex*). *Journal of Zoology* 243: 81–115.

Green, R.E., Tyler, G.A. & Bowden, C.G.R. 2000. Habitat selection, ranging behaviour and diet of the Stone Curlew (*Burhinus oedicnemus*) in southern England. *Journal of Zoology* 50: 161–183.

Greenwood, J.J.D. 2007. Citizens, science and bird conservation. *Journal of Ornithology* 148: S77–S124.

Greenwood, J.G. & Tickner, M. 2002. How many Black Guillemots *Cepphus grylle* are there now in Northern Ireland? *Irish Naturalists' Journal* 27: 125–127.

Greenwood, J.J.D., Fuller, R.J. & Gibbons, D.W. 1997. Assessing changes in distribution from atlas data: reply to Bircham and Jordan. *Ibis* 139: 186–189.

Greenwood, J.J.D., Crick, H.Q.P. & Bainbridge, I.P. 2003. Numbers and international importance of raptors and owls in Britain and Ireland. In *Birds of Prey in a Changing Environment* (eds Thompson, D.B.A., Redpath, S.M., Fielding, A.H., Marquiss, M. & Galbraith, C.A.), pp 25–50. The Stationery Office, Edinburgh.

Gregory, J. 2008. The Black-winged Stilts nesting in Cheshire. *Birding World* 21: 250–251.

Gregory, R.D. & Marchant, J.H. 1996. Population trends of Jays, Magpies, Jackdaws and Carrion Crows in the United Kingdom. *Bird Study* 43: 28–37.

Gribble, F. 1983. Nightjars in Britain and Ireland in 1981. *Bird Study* 30: 157–176.

Gruar, D., Barritt, D. & Peach, W.J. 2006. Summer utilization of oilseed rape by Reed Buntings *Emberiza schoeniclus* and other farmland birds. *Bird Study* 53: 47–54.

Guilford, T.C., Meade, J., Freeman, R., Biro, D., Evans, T., Bonadonna, F., Boyle, D., Roberts, S. & Perrins, C.M. 2008. GPS tracking of the foraging movements of Manx Shearwaters *Puffinus puffinus* breeding on Skomer Island, Wales. *Ibis* 150: 462–473.

Guilford, T., Meade, J., Willis, J., Phillips, R.A., Boyle, D., Roberts, S., Collett, M., Freeman, R. & Perrins, C.M. 2009. Migration and stopover in a small pelagic seabird, the Manx Shearwater *Puffinus puffinus*: insights from machine learning. *Proceedings of the Royal Society B: Biological Sciences* 276: 1215–1223.

Guilford, T., Freeman, R., Boyle, D., Dean, B., Kirk, H., Phillips, R. & Perrins, C. 2011. A dispersive migration in the Atlantic Puffin and its implications for migratory navigation. *PLoS ONE* 6: e21336.

Gunnarsson, T.G., Gill, J.A., Potts, P.M., Atkinson, P.W., Croger, R.E, Gélinaud, G., Garðarsson, A. & Sutherland, W.J. 2005. Estimating population size in Black-tailed Godwits *Limosa limosa islandica* by colour-marking. *Bird Study* 52: 153–158.

Gurney, J.H. 1921. *Early Annals of Ornithology*. Witherby, London.

Gyimesi, A. & Lensink, R. 2010. *Risk analysis of the Egyptian Goose in The Netherlands*. Bureau Waardenburg bv, Culemborg.

Hafner, H. & Fasola, M. 1997. Long-term monitoring and conservation of herons in France and Italy. *Colonial Waterbirds* 20: 298–305.

Hall, C., Glanville, J.R., Boland, H., Einarsson, Ó., McElwaine, J.G., Holt, C.A., Spray, C.J. & Rees, E.C. 2012. Population size and breeding success of Icelandic Whooper Swans *Cygnus cygnus*: results of the 2010 international census. *Wildfowl* 62: 73–96.

Hallgrimsson, G.T., Summers, R.W., Etheridge, B. & Swann, R.L. 2012. The winter range of Nearctic Purple Sandpipers *Calidris maritima* on the East Atlantic Flyway. *Ardea* 100: 13–18.

Halupka, L., Dyrcz, A. & Borowiec, M. 2008. Climate change affects breeding of Reed Warblers *Acrocephalus scirpaceus*. *Journal of Avian Biology* 39: 95–100.

Hamer, K.C., Phillips, R.A., Hill, J.K., Wanless, S. & Wood, A.G. 2001. Contrasting foraging strategies of Gannets *Morus bassanus* at two North Atlantic colonies. *Marine Ecology Progress Series* 224: 283–290.

Hampshire Ornithological Society. 2011. *Hampshire Bird Report 2010*. Hampshire Ornithological Society.

Hancock, M. 2000. Artificial floating islands for nesting Black-throated Divers *Gavia arctica* in Scotland: construction, use and effect on breeding success. *Bird Study* 47: 165–175.

Hancock, M.H. & Wilson, J.D. 2003. Winter habitat associations of seed-eating passerines on Scottish farmland. *Bird Study* 50: 116–130.

Harding, B.D. 1979. *Bedfordshire Bird Atlas*. Bedfordshire Natural History Society, Bedford.

Harris, C.M., Calladine, J.R., Wernham, C.V. & Park, K.J. 2008. Impacts of piscivorous birds on salmonid populations and game fisheries in Scotland: a review. *Wildlife Biology* 14: 395–411.

Harris, M.P. & Wanless, S. 1991. The importance of the lesser sandeel *Ammodytes marinus* in the diet of the Shag *Phalacrocorax aristotelis*. *Ornis Scandinavica 22: 375–382.*

Harris, M.P. & Wanless, S. 1996. Differential responses of Guillemot *Uria aalge* and Shag *Phalacrocorax aristotelis* to a late winter wreck. *Bird Study* 43: 220–230.

Harris, M.P. & Wanless, S. 1997. Breeding success, diet and brood neglect in the Kittiwake (*Rissa tridactyla*) over an 11-year period. *ICES Journal of Marine Science* 54: 615–623.

Harris, M.P. & Wanless, S. 2011. *The Puffin*. T. & A.D. Poyser, London.

Harris, M.P., Wanless, S. & Rothery, P. 1998. Age-related effects of a non-breeding event and a winter wreck on Shags Phalacrocorax aristotelis. *Ibis* 140: 310–314.

Harris, M., Newell, M., Leitch, A., Bruce, W. & Hunt, J. 2009. Dramatic decline in numbers of Atlantic Puffins in the Firth of Forth. *Scottish Birds* 29: 132–134.

Harris, M.P., Daunt, F., Newell, M., Phillips, R.A. & Wanless, S. 2010. Wintering areas of adult Atlantic Puffins *Fratercula arctica* from a North Sea colony as revealed by geolocation technology. *Marine Biology* 157: 827–836.

Harrison, T.H. & Hollom, P.A.D. 1932. The Great Crested Grebe Enquiry, 1931. *British Birds* 26: 62–92, 102–131, 142–155 & 174–195.

Harrop, A. 1991. The status of Red-crested Pochard. *Birding World* 5: 171–175.

Harrop, A.H.J. 2002. The Ruddy Shelduck in Britain: a review. *British Birds* 95: 123–128.

Hartley, C. 2004. Little Gulls at sea off Yorkshire in autumn 2003. *British Birds* 97: 448–455.

Harvey, P.V. & Heubeck, M. 2012. Changes in the wintering population and distribution of Slavonian Grebes in Shetland. *British Birds* 105: 704–715.

Haylock, M.R., Hofstra, N., Klein Tank, A.M.G., Klok, E.J., Jones, P.D. & New, M. 2008. A European daily high-resolution gridded dataset of surface temperature and precipitation. *Journal of Geophysical Research: Atmospheres* 113, D20119.

Heaney, V., Lock, L., St Pierre, P. & Brown, A. 2008. Breeding seabirds on the Isles of Scilly. *British Birds* 101: 418–438.

Hebridean Mink Project. 2011. *Phase II Bulletin June 2011.* Hebridean Mink Project/Scottish Natural Heritage, Stornoway.

Helm, B., Fiedler, W. & Callion, J. 2006. Movements of European Stonechats *Saxicola torquata* according to ringing recoveries. *Ardea* 94: 33–44.

Henderson, B.D. 2012. The Great Grey Shrike in Dumfries & Galloway. *Scottish Birds* 32: 99–107.

Henderson, I.S. 2006. Recent measures to control Ruddy Ducks *Oxyura jamaicensis* in the United Kingdom. In *Waterbirds Around the World* (eds Boere, G.C., Galbraith, C.A. & Stroud, D.A.), pp 822–825. The Stationery Office, Edinburgh.

Henderson, I. 2009. Progress of the UK Ruddy Duck eradication programme. *British Birds* 102: 680–690.

Henderson, I. 2012. *UK Ruddy Duck Eradication Programme Project Bulletin.* Defra/Fera, York.

Henderson, I.G., Wilson, A.M., Steele, D. & Vickery, J.A. 2002. Population estimates, trends and habitat associations of breeding Lapwing *Vanellus vanellus*, Curlew *Numenius arquata* and Snipe *Gallinago gallinago* in Northern Ireland in 1999. *Bird Study* 49: 17–25.

Henderson, I.G., Fuller, R.J., Conway, G.J. & Gough, S.J. 2004. Evidence for declines in populations of grassland-associated birds in marginal areas of Britain. *Bird Study* 51: 12–19.

Henderson, I., Holt, C. & Vickery, J. 2007. National and regional patterns of habitat association with foraging Barn Swallows *Hirundo rustica* in the UK. *Bird Study* 54: 371–377.

Herbert, I.J. 1991. The status and habitat of the Garden Warbler at Crom Estate, Co. Fermanagh, and a review of its status in Ireland. *Irish Birds* 4: 369–376.

Heubeck, M. 2013. Censusing and monitoring breeding seabirds in Britain and Ireland. *British Birds* 106: 306–324.

Heubeck, M., Mellor, R.M. & Harvey, P.V. 1997. Changes in the breeding distribution and numbers of Kittiwakes *Rissa tridactyla* around Unst, Shetland, and the presumed role of predation by Great Skuas *Stercorarius skua*. *Seabird* 19: 12–21.

Heubeck, M., Aarvak, T., Isaksen, K., Johnsen, A., Petersen, I.K. & Anker-Nilssen, T. 2011. Mass mortality of adult Razorbills in the Skagerrak and North Sea area, autumn 2007. *Seabird* 24: 11–32.

Hewson, C.M. & Noble, D.G. 2009. Population trends of breeding birds in British woodlands over a 32-year period: relationships with food, habitat use and migratory behaviour. *Ibis* 151: 464–486.

Hewson, C.M., Amar, A., Lindsell, J.A., Thewlis, R.M., Butler, S., Smith, K. & Fuller, R.J. 2007. Recent changes in bird populations in British broadleaved woodland. *Ibis* 149 (Suppl. 2): 14–28

Hickling, R., Roy, D.B., Hill, J.K., Fox, R. & Thomas, C.D. 2006. The distributions of a wide range of taxonomic groups are expanding polewards. *Global Change Biology* 12: 450–455.

Highland SOC. 2011. *Highland Bird Report 2009.* Highland Branch of the Scottish Ornithologists' Club, Inverness.

Hill, M.O. 2012. Local frequency as a key to interpreting species occurrence data when recording effort is not known. *Methods in Ecology and Evolution* 3: 195–205.

Hillis, J.P. 2002. First Annual Report of the Irish Rare Breeding Birds Panel, 2002. *Irish Birds* 7: 375–384.

Hillis, J.P. 2007. Rare Breeding Birds in Ireland 2005 and 2006. *Irish Birds* 8: 249–262.

Hillis, J.P. 2010. Rare Irish breeding birds, 2009: the annual report of the Irish Rare Breeding Birds Panel (IRBBP). *Irish Birds* 9: 67–76.

Hillis, J.P. 2012. Rare Irish breeding birds, 2011: the annual report of the Irish Rare Breeding Birds Panel (IRBBP). *Irish Birds* 9: 411–420.

Hillis, J.P. & O'Sullivan, O. 2012. Rare grebes breeding in Ireland. *Irish Birds* 19: 511.

Hinsley, S.A., Bellamy, P.E., Newton, I. & Sparks, T.H. 1995. Habitat and landscape features influencing the presence of individual bird species in woodland fragments. *Journal of Avian Biology* 26: 94–104.

Hocart, R.P. 2010. *Guernsey's Countryside: an introduction to the history of the rural landscape.* La Société Guernesiaise.

Hole, D.G., Whittingham, M.J., Bradbury, R.B., Anderson, G.Q.A., Lee, P.L.M., Wilson, J.D & Krebs, J.R. 2002. Agriculture: widespread local House-Sparrow extinctions. *Nature* 418: 931–932.

Holland, P.K. & Yalden, D.W. 2002. Population dynamics of Common Sandpipers *Actitis hypoleucos* in the Peak District of Derbyshire – a different decade: a report of the failure of a population to recover from a catastrophic snow storm. *Bird Study* 49: 131–138.

Holland, R.A., Eigenbrod, F., Armsworth, P.R., Anderson, B.J., Thomas, C.D., Heinemeyer, A., Gillings, S., Roy, D.B. & Gaston, K.J. 2011. Spatial covariation between freshwater and terrestrial ecosystem services. *Ecological Applications* 21: 2034–2048.

Holling, M. & the Rare Breeding Birds Panel. 2007a. Rare breeding birds in the United Kingdom in 2003 and 2004. *British Birds* 100: 321–367.

Holling, M. & the Rare Breeding Birds Panel. 2007b. Non-native breeding birds in the United Kingdom in 2003, 2004 and 2005. *British Birds* 100: 638–649.

Holling, M. & the Rare Breeding Birds Panel. 2008. Rare breeding birds in the United Kingdom in 2005. *British Birds* 101: 276–316.

Holling, M. & the Rare Breeding Birds Panel. 2009. Rare breeding birds in the United Kingdom in 2006. *British Birds* 102: 158–202.

Holling, M. & the Rare Breeding Birds Panel. 2010a. Rare breeding birds in the United Kingdom in 2007. *British Birds* 103: 2–52.

Holling, M. & the Rare Breeding Birds Panel. 2010b. Rare breeding birds in the United Kingdom in 2008. *British Birds* 103: 482–538.

Holling, M. & the Rare Breeding Birds Panel. 2011a. Non-native breeding birds in the United Kingdom in 2006, 2007 and 2008. *British Birds* 104: 114–138.

Holling, M. & the Rare Breeding Birds Panel. 2011b. Rare breeding birds in the United Kingdom in 2009. *British Birds* 104: 476–537.

Holling, M. & the Rare Breeding Birds Panel. 2012. Rare breeding birds in the United Kingdom in 2010. *British Birds* 105: 352–416.

Holling, M. & the Rare Breeding Birds Panel. 2013. Rare breeding birds in the United Kingdom in 2011. *British Birds* [In press].

Hollom, P.A.D. 1957. The rare birds of prey, their present status in the British Isles: Goshawk. *British Birds* 50: 135–136.

Holloway, S. 1996. *The Historical Atlas of Breeding Birds in Britain and Ireland: 1875–1900.* T. & A.D. Poyser, London.

Holt, C. 2013. The changing status of the Great White Egret in Britain. *British Birds* 106: 246–257.

Holt, C.A., Fuller, R.J. & Dolman, P.M. 2010. Experimental evidence that deer browsing reduces habitat suitability for breeding Common Nightingales *Luscinia megarhynchos*. *Ibis* 152: 335–346.

Holt, C.A., Austin, G.E., Calbrade, N.A., Mellan, H.J., Mitchell, C., Stroud, D.A., Wotton, S.R. & Musgrove, A.J. 2011a. *Waterbirds in the UK 2009/10: The Wetland Bird Survey.* BTO/RSPB/JNCC, Thetford.

Holt, C.A., Fuller, R.J. & Dolman, P.M. 2011b. Breeding and post-breeding responses of woodland birds to modifications of habitat structure by deer. *Biological Conservation* 144: 2151–2162.

Holt, C.A., Hewson, C.M. & Fuller, R.J. 2012. The Nightingale in Britain: status, ecology and conservation needs. *British Birds* 105: 172–187.

Holt, C.A., Fuller, R.J. & Dolman, P.M. 2013. Deer reduce habitat quality for a woodland songbird: evidence from settlement patterns, demographic parameters and body condition. *Auk* 130: 13–20.

Hoodless, A.N. & Hirons, G.J.M. 2007. Habitat selection and foraging behaviour of breeding Eurasian Woodcock *Scolopax rusticola*: a comparison between contrasting landscapes. *Ibis* 149 (Suppl. 2): 234–249.

Hoodless, A.N. & Powell, A. 2010. Origins of wintering Woodcock: initial findings. *Game & Wildlife Conservation Trust Review of 2009*: 18–19. GWCT, Fordingbridge.

Hoodless, A.N., Lang, D., Aebischer, N.J., Fuller, R.J. & Ewald, J.A. 2009. Densities and population estimates of breeding Eurasian Woodcock *Scolopax rusticola* in Britain in 2003. *Bird Study* 56: 15–25.

Hope-Jones, P., Barrett, C.F., Mudge, G.P. & Harris, M.P. 1984. Physical conditions of auks beached in eastern Britain during the wreck of February 1983. *Bird Study* 31: 95–98.

Hõrak, P. & Lebreton, J.-D. 1998. Survival of adult Great Tits *Parus major* in relation to sex and habitat: a comparison of urban and rural populations. *Ibis* 140: 205–209.

Hornman, M., Hustings, F., Koffijberg, K., van Winden, E., SOVON Ganzen- en Zwanenwerkgroep & Soldaat, L. 2011. *Watervogels in Nederland in 2008/2009.* SOVON-monitoringrapport 2011/03, Waterdienst-rapport BM 10.24. SOVON Vogelonderzoek Nederland, Nijmegen.

Hornman, M., van Roomen, M., Hustings, F., Koffijberg, K., van Winden, E. & Soldaat, L. 2012a. Population trends in wintering and migrating waterbirds in the Netherlands in 1975–2010. *Limosa* 85: 97–116.

Hornman, M., Hustings, F., Koffijberg, K., Kleefstra, R., Klaassen, O., van Winden, E., SOVON Ganzen- en Zwanenwerkgroep & Soldaat, L. 2012b.

Watervogels in Nederland in 2009/2010. SOVON-monitoringrapport 2012/02, Waterdienst-rapport BM 12.06. SOVON Vogelonderzoek Nederland, Nijmegen.

Hudson, A.V. & Furness, R.W. 1989. The behaviour of seabirds foraging at fishing boats around Shetland. *Ibis* 131: 225–237.

Hudson, N. & the Rarities Committee. 2008. Report on rare birds in Great Britain in 2007. *British Birds* 101: 516–577.

Hudson, N. & the Rarities Committee. 2009. Report on rare birds in Great Britain in 2008. *British Birds* 102: 528–601.

Hudson, N. & the Rarities Committee. 2010. Report on rare birds in Great Britain in 2009. *British Birds* 103: 562–638.

Hudson, N. & the Rarities Committee. 2011. Report on rare birds in Great Britain in 2010. *British Birds* 104: 557–629.

Hudson, N. & the Rarities Committee. 2012. Report on rare birds in Great Britain in 2011. *British Birds* 105: 556–625.

Hudson, R. 1965. The spread of the Collared Dove in Britain and Ireland. *British Birds* 58: 105–139.

Hudson, R. & Marchant, J.H. 1984. *Population estimates for British breeding birds.* BTO Research Report 13. BTO, Tring.

Hughes, B., Bruce, J. & Ekins, G. 2000. *Movements and Distribution of Inland Breeding Cormorants in England.* Research Report 360. English Nature, Peterborough.

Hulme, M., Jenkins, G.J., Lu, X., Turnpenny, J.R., Mitchell, T.D., Jones, R.G., Lowe, J., Murphy, J.M., Hassell, D., Boorman, P., McDonald, R. & Hill, S. 2002. *Climate Change Scenarios for the United Kingdom: the UKCIP02 Scientific Report.* Tyndall Centre for Climate Change Research, School of Environmental Sciences, University of East Anglia, Norwich.

Hume, R.A. 1978. Variations in Herring Gulls at a Midland roost. *British Birds* 71: 338–345.

Hunt, J., Heffernan, M.L., McLoughlin, D., Benson, C. & Huxley, C. 2012. Common Scoter *Melanitta nigra* in Ireland: results from pre-breeding and brood surveys in 2012. *Irish Birds* 9: 341–348.

Hunter, J. & Crawford, W. 2001. *Java Servlet Programming.* 2nd edition. O'Reilly.

Huntley, B., Green, R.E., Collingham, Y.C. & Willis, S.G. 2007. *A Climatic Atlas of European Breeding Birds.* Lynx Edicions, Barcelona.

Hustings, F. 2002. Broedende Vuurgoudhanen *Regulus ignicapillus* in Nederland: hoe het verder ging. *Limosa* 75: 85–90.

Inglis, I.R., Isaacson, A.J., Smith, G.C., Haynes, P.J. & Thearle, R.J.P. 1997. The effect on the Woodpigeon (*Columba palumbus*) of the introduction of oilseed rape into Britain. *Agriculture, Ecosystems and Environment* 61: 113–121.

Irish Meteorological Service. 2013. Monthly summaries and bulletins. URL: www.met.ie/climate/monthly-weather-reports.asp [31 July 2013].

Iverson, R. & Prasad, A.M. 1998. Predicting abundance of 80 tree species following climate change in the eastern United States. *Ecological Monographs* 68: 465–485.

Jackson, D.B. 2005. Environmental correlates of lake occupancy and chick survival of Black-throated Divers *Gavia arctica* in Scotland. *Bird Study* 52: 225–236.

Jackson, D. 2009. *Interim Report on 2009 Breeding Whimbrel Surveys.* Unpublished report to Natural Research.

Jackson, D.B. & Green, R.E. 2000. The importance of the introduced Hedgehog (*Erinaceus europaeus*) as a predator of the eggs of waders (Charadrii) on machair in South Uist, Scotland. *Biological Conservation* 93: 333–348.

Jackson, D.B., Fuller, R.J. & Campbell, S.T. 2004. Long-term population changes among breeding shorebirds in the Outer Hebrides, Scotland, in relation to introduced Hedgehogs (*Erinaceus europaeus*). *Biological Conservation* 117: 151–166.

James, P. (ed). 1996. *Birds of Sussex.* Sussex Ornithological Society.

Jenkins, D. & Sparks, T.H. 2010. The changing bird phenology of Mid Deeside, Scotland. *Bird Study* 57: 407–414.

Jensen, F.P., Béchet, A. & Wymenga, E. (compilers). 2008. *International Single Species Action Plan for the Conservation of Black-tailed Godwit* Limosa l. limosa *&* L. l. islandica. AEWA Technical Series no. 37. Bonn, Germany.

JNCC. 2010. UK BAP Priority Species pages – *Phalaropus lobatus* (version 2). URL: http://jncc.defra.gov.uk/_speciespages/510.pdf [31 July 2013].

JNCC. 2012. *Seabird Population Trends and Causes of Change: 2012 Report.* URL: http://jncc.defra.gov.uk/page-3201 [31 July 2013].

Johnstone, I., Dyda, J. & Lindley, P. 2007a. The population status and hatching success of Curlews *Numenius arquata* in Wales in 2006. *Welsh Birds* 5: 78–87.

Johnstone, I., Thorpe, R., Moore, A. & Finney, S. 2007b. Breeding status of Choughs *Pyrrhocorax pyrrhocorax* in the UK and the Isle of Man in 2002. *Bird Study* 54: 23–34.

Johnstone, I., Mucklow, C., Cross, T., Lock, L. & Carter, I. 2011. The return of the Red-billed Chough to Cornwall: the first ten years and prospects for the future. *British Birds* 104: 416–431.

Jones, E. 1979. Breeding of the Short-eared Owl in south-west Ireland. *Irish Birds* 1: 377–380.

Jones, T., Smith, C., Williams, E. & Ramsay, A. 2008. Breeding performance and diet of Great Skuas *Stercorarius skua* and Parasitic Jaegers (Arctic Skuas) *S. parasiticus* on the west coast of Scotland. *Bird Study* 55: 257–266.

Jonsson, L. 1998. Yellow-legged Gulls and yellow-legged Herring Gulls in the Baltic. *Alula* 7: 2–11.

Kavanagh, B.P. 1987. The breeding density of Magpies *Pica pica* in Dublin city. *Irish Birds* 3: 387–394.

Kear, J. 2003. Cavity-nesting ducks: why woodpeckers matter. *British Birds* 96: 217–233.

Kehoe, C. & the British Birds Rarities Committee. 2006. Racial identification and assessment in Britain: a report from the RIACT subcommittee. *British Birds* 99: 619–645.

Keller, V. 2000. Winter distribution and population change of Red-crested Pochard *Netta rufina* in southwestern and central Europe. *Bird Study* 47: 176–185.

Keller, V. 2006. Population size and trend of the Red-crested Pochard *Netta rufina* in southwest/central Europe: an update. In *Waterbirds Around the World* (eds Boere, G.C., Galbraith, C.A. & Stroud, D.A.), pp 503–504. The Stationery Office, Edinburgh.

Keller, V. & Burkhardt, M. 2012. *Monitoring Hivernal des Oiseaux d'Eau 2010/11 en Suisse.* Station Ornithologique Suisse, Sempach.

Kelly, A., Leighton, K. & Newton, J. 2010. Using stable isotopes to investigate the provenance of an Eagle Owl found in Norfolk. *British Birds* 103: 213–222.

Kelsey, M.G., Green, G.H., Garnett, M.C. & Hayman, P.V. 1989. Marsh Warblers in Britain. *British Birds* 82: 239–256.

Kennedy, P.G., Ruttledge, R.F. & Scroope, C.F. 1954. *The Birds of Ireland.* Oliver & Boyd. London and Edinburgh.

Kent Ornithological Society. 1998. *Kent Bird Report 1996.* Kent Ornithological Society.

Kershaw, M. & Hughes, B. 2002. *The winter status and distribution of Ruddy Ducks in the UK 1966/67–1999/2000.* Report to the Department for Environment, Food and Rural Affairs. WWT Wetlands Advisory Service.

Kirk, G. 2010. Cold winter has dramatic effect. *BTO News* 286: 15.

Kirk, G.R. & Phillips, N.J. 2013. *The Birds of Gloucestershire.* Liverpool University Press. [In press].

Kjellén, N. & Roos, G. 2000. Population trends in Swedish raptors demonstrated by migration counts at Falsterbo, Sweden 1942–97. *Bird Study* 47: 195–211.

Kjellén, N. 2013. *Migration Counts at Falsterbo, SW Sweden.* URL: www.falsterbofagelstation.se [31 July 2013].

Knox, A.G. 1990. The sympatric breeding of Common and Scottish Crossbills *Loxia curvirostra* and *L. scotica* and the evolution of crossbills. *Ibis* 132: 454–466.

Knox, A.G., Helbig, A.J., Parkin, D.T. & Sangster, G. 2001. The taxonomic status of Lesser Redpoll. *British Birds* 94: 260–267.

Kramer, D. 1995. Inland spring passage of Arctic Terns in southern Britain. *British Birds* 88: 211–217.

Kubetzki, U., Garthe, S., Fifield, D., Mendel, B. & Furness, R.W. 2009. Individual migratory schedules and wintering areas of Northern Gannets. *Marine Ecology Progress Series* 394: 257–265.

Kujala, H., Vepsäläinen, V., Zuckerberg, B. & Brommer, J.E. 2013. Range margin shifts of birds revisited – the role of spatiotemporally varying survey effort. *Global Change Biology* 19: 420–430.

Kyrkos, A., Wilson, J.D. & Fuller, R.J. 1998. Farmland habitat change and abundance of Yellowhammers *Emberiza citrinella*: an analysis of Common Birds Census data. *Bird Study* 45: 232–246.

Lack, P.C. 1986. *The Atlas of Wintering Birds in Britain and Ireland.* T. & A.D. Poyser, Calton.

Lancashire & Cheshire Fauna Society. 2012. *Lancashire Bird Report 2011: the birds of Lancashire and North Merseyside.* Lancashire & Cheshire Fauna Society publication 115.

Langridge, A. 1996. *The status of the breeding population of Goldeneye* Bucephala clangula *in Scotland from 1974–1995.* Unpublished report. RSPB, Sandy.

Langston, R., Gregory, R. & Adams, R. 2002. The status of the Hawfinch in the UK

1975–1999. *British Birds* 95: 166–173.

Langston, R.H.W., Smith, T., Brown, A.F. & Gregory, R.D. 2006. Status of breeding Twite *Carduelis flavirostris* in the UK. *Bird Study* 53: 55–63.

Langston, R.H.W., Wotton, S.R., Conway, G.J., Wright, L.J., Mallord, J.W., Currie, F.A., Drewitt, A.L., Grice, P.V., Hoccom, D.G. & Symes, N. 2007. Nightjar *Caprimulgus europaeus* and Woodlark *Lullula arborea* – recovering species in Britain? *Ibis* 149 (Suppl. 2): 250–260.

Lauder, A. 2007. Trends in breeding duck populations at Loch Leven, Perth & Kinross. *Scottish Birds* 27: 55–63.

Lavers, C.P. & Haines-Young, R.H. 1997. Displacement of Dunlin *Calidris alpina schinzii* by forestry in the Flow Country and an estimate of value of moorland adjacent to plantations. *Biological Conservation* 79: 87–90.

Lawlor, M. & the Bailiwick of Guernsey Rarities Committee. 2010. *Report on rare birds in the Bailiwick of Guernsey in 2009.* La Société Guernesiaise. URL: www.guernseybirds.org.gg/documents/2009_Rarities_Report.pdf [31 July 2013].

Lawson, B., Robinson, R.A., Colvile, K.M., Peck, K.M., Chantrey, J., Pennycott, T.W., Simpson, V.R., Toms, M.P. & Cunningham, A.A. 2012. The emergence and spread of finch trichomonosis in the British Isles. *Philosophical Transactions of the Royal Society Series B* 367: 2852–2863.

Leathwick, J.R., Elith, J., Francis, M.P., Hastie, T. & Taylor, P. 2006. Variation in demersal fish species richness in the oceans surrounding New Zealand: an analysis using boosted regression trees. *Marine Ecology Progress Series* 321: 267–281.

Ledlie, R.C.B. & Pedler, E.G. 1938. Nesting of the Little Ringed Plover in Hertfordshire. *British Birds* 32: 90–102.

Lehikoinen, A. & Jaatinen, K. 2012. Delayed autumn migration in Northern European waterfowl. *Journal of Ornithology* 153: 563–570.

Lehikoinen, A., Jaatinen, K., Vahatalo, A., Clausen, P., Crowe, O., Deceuninck, B., Hearn, R.D., Holt, C.A., Hornman, M., Keller, V., Nilsson, L., Langendoen, T., Wahl, J. & Fox, A.D. 2013. Rapid climate-driven shifts in wintering distribution of three common waterbird species. *Global Change Biology* 19: 2071–2081.

Lensink, R. 1998. Temporal and spatial expansion of the Egyptian Goose *Alopochen aegyptiacus* in the Netherlands, 1967–94. *Journal of Biogeography* 25: 251–263.

Leonard, K. 2010. A survey of breeding Eiders *Somateria mollissima* in Northern Ireland in 2009. *Irish Birds* 9: 11–18.

Lever, C. 2005. *Naturalised Birds of the World.* T. & A.D. Poyser, London.

Lever, C. 2009. *The Naturalized Animals of Britain and Ireland.* New Holland Publishers, London.

Lewin-Koh, N.J. & Bivand, R. 2011. *maptools: tools for reading and handling spatial objects. R package version 0.8-10.* URL: http://CRAN.R-project.org/package=maptools [31 July 2013].

Lewis, A.J.G., Amar, A., Cordi-Piec, D. & Thewlis, R.M. 2007. Factors influencing Willow Tit *Poecile montanus* site occupancy: a comparison of abandoned and occupied woods. *Ibis* 149 (Suppl. 2): 205–213.

Lewis, A.J.G., Amar, A., Daniells, L., Charman, E.C., Grice, P. & Smith, K. 2009a. Factors influencing patch occupancy and within-patch habitat use in an apparently stable population of Willow Tits *Poecile montanus kleinschmidti* in Britain. *Bird Study* 56: 326–337.

Lewis, A.J.G., Amar, A., Charman, E.C. & Stewart, F.R.P. 2009b. The decline of the Willow Tit in Britain. *British Birds* 102: 386–393.

Lewis, M., Wilson, L.J., Söhle, I., Dean, B.J., Webb, A. & Reid, J.B. 2008. *Wintering sea ducks, divers and grebes in UK inshore areas: aerial surveys and shore-based counts 2006/07.* JNCC Report 414. JNCC, Peterborough.

Liaw, A. & Wiener, M. 2002. Classification and Regression by randomForest. *R News* 2: 18–22.

Liley, D. & Sutherland, W.J. 2007. Predicting the population consequences of human disturbance for Ringed Plovers *Charadrius hiaticula*: a game theory approach. *Ibis* 149: 82–94.

Lloyd, C., Tasker, M.L. & Partridge, K. 1991. *The Status of Seabirds in Britain and Ireland.* T. & A.D. Poyser, London.

Lloyd, J. & Friese, J. 2013. Little Ringed Plovers in southwest Wales. *British Birds* 106: 30–35.

Lock, L. 1994. *A review of the status and conservation of wintering Golden Plovers* Pluvialis apricaria *in Devon and Cornwall 1972–1991.* RSPB, Exeter.

Lock, L. & Cook, K. 1998. The Little Egret in Britain: a successful colonist. *British Birds* 91: 273–280.

Lourenço, P.M. & Piersma, T. 2008. Changes in the non-breeding distribution of Continental Black-tailed Godwits *Limosa limosa limosa* over 50 years: a synthesis of surveys. *Wader Study Group Bulletin* 115: 91–97.

Lovatt, J. & Madden, B. 2012. Impact of the 2009/10 severe winter on Skylark *Alauda arvensis*, Wren *Troglodytes troglodytes*, Stonechat *Saxicola torquatus* and Meadow Pipit *Anthus pratensis* populations at Bellacorick cutaway bog, Co. Mayo. *Irish Birds* 9: 505–507.

Lusby, J. 2012. *Exposure of Barn Owls* Tyto alba *in Ireland to second generation anticoagulant rodenticides (2006–2011).* Unpublished report to the National Parks and Wildlife Service and the Department of Agriculture, Food and the Marine.

Lusby, J., Fernández-Bellon, D., Norriss, D.W. & Lauder, A. 2011. Assessing the effectiveness of monitoring methods for Merlin *Falco columbarius* in Ireland: the Pilot Merlin Survey 2010. *Irish Birds* 9: 143–154.

Lynas, P., Newton, S.F. & Robinson, J.A. 2007. The status of birds in Ireland: an analysis of conservation concern 2008–2013. *Irish Birds* 8: 149–167.

Lysaght, L. 2002. *An atlas of the breeding birds of the Burren and the Aran Islands.* BirdWatch Ireland, Monkstown, Co. Dublin.

MacDonald, M.A., Cobbold, G., Mathews, F., Denny, M.J.H., Walker, L.K., Grice, P.V. & Anderson, G.Q.A. 2012. Effects of agri-environment management for cirl buntings on other biodiversity. *Biodiversity and Conservation* 21: 1477–1492.

Mackrill, T. 2013. *The Rutland Water Ospreys.* Bloomsbury, London.

Maclean, I.M.D., Burton, N.H.K. & Austin, G.E. 2006. *Declines in over-wintering diving ducks at Lough Neagh and Lough Beg: comparisons of within site, regional, national and European trends.* BTO Research Report 432. BTO, Thetford.

Maclean, I.M.D., Austin, G.E., Rehfisch, M.M., Blew, J., Crowe, O., Delany, S., Devos, K., Deceuninck, B., Gunther, K., Laursen, K., van Roomen, M. & Wahl, J. 2008. Climate change causes rapid changes in the distribution and site abundance of birds in winter. *Global Change Biology* 14: 2489–2500.

Madden, B., Hunt, J. & Norriss, D.W. 2009. The 2002 survey of the Peregrine *Falco peregrinus* breeding population in the Republic of Ireland. *Irish Birds* 8: 543–548.

Madders, M. 2000. Habitat selection and foraging success of Hen Harriers (*Circus cyaneus*) in west Scotland. *Bird Study* 47: 32–40.

Main, I.G. 1999. Overseas movements to and from Britain by Greenfinches *Carduelis chloris*. *Ringing & Migration* 19: 191–199.

Mallord, J.W., Dolman, P.M., Brown, A.F. & Sutherland, W.J. 2007. Linking recreational disturbance to population size in a ground-nesting passerine. *Journal of Applied Ecology* 44: 185–195.

Mallord, J.W., Orsman, C.J., Cristinacce, A., Butcher, N., Stowe, T.J. & Charman, E.C. 2012a. Mortality of Wood Warbler *Phylloscopus sibilatrix* nests in Welsh oakwoods: predation rates and the identification of nest predators using miniature nest cameras. *Bird Study* 59: 286–295.

Mallord, J.W., Charman, E.C. & Cristinacce, A. 2012b. Habitat associations of Wood Warblers *Phylloscopus sibilatrix* breeding in Welsh oakwoods. *Bird Study* 59: 403–415.

Marchant, J.H., Hudson, R., Carter, S.P. & Whittington, P. 1990. *Population Trends in British Breeding Birds.* BTO, Tring.

Marchant, J.H., Freeman, S.N., Crick, H.Q.P. & Beaven, L.P. 2004. The BTO Heronries Census of England and Wales 1928–2000: new indices and a comparison of analytical methods. *Ibis* 146: 323–334.

Marquiss, M. 2007. Seasonal pattern in hawk predation on Common Bullfinches *Pyrrhula pyrrhula*: evidence of an interaction with habitat affecting food availability. *Bird Study* 54: 1–11.

Marquiss, M. & Duncan, K. 1993. Variation in the abundance of Red-breasted Mergansers *Mergus serrator* on a Scottish river in relation to season, year, river hydrology, salmon density and spring culling. *Ibis* 135: 33–41.

Marquiss, M. & Newton, I. 1982. The Goshawk in Britain. *British Birds* 75: 243–260.

Marquiss, M. & Rae, R. 2002. Ecological differentiation in relation to bill size amongst sympatric, genetically undifferentiated crossbills *Loxia* spp. *Ibis* 144: 494–508.

Marquiss, M., Petty, S.J., Anderson, D.I.K. & Legge, G. 2003. Contrasting population trends of the Northern Goshawk (*Accipter gentilis*) in the Scottish/English Borders and north-east Scotland. In *Birds of Prey in a Changing Environment* (eds Thompson, D.B.A., Redpath, S.M., Fielding, A.H., Marquiss, M. & Galbraith, C.A.), pp 143–148. The Stationery Office, Edinburgh.

Marquiss, M., Hobson, K.A. & Newton, I. 2008. Stable isotope evidence for different regional source areas of Common Crossbill *Loxia curvirostra* irruptions into Britain. *Journal of Avian Biology* 39: 30–34.

Marquiss, M., Newton, I., Hobson, K.A. & Kolbeinsson, Y. 2012. Origins of irruptive migrations by Common Crossbills *Loxia curvirostra* into northwestern Europe revealed by stable isotope analysis. *Ibis* 154: 400–409.

Martin, A.R. 2002. *Ornithological fieldwork on Hermaness NNR, Shetland conducted by the Cambridge group, June/July 2002.* Report to Scottish Natural Heritage and Shetland Oil Terminal Environmental Advisory Group.

Martin, B. & Smith, J. 2007. A survey of breeding Black-necked Grebes in the UK: 1973–2004. *British Birds* 100: 368–378.

Mason, C.F. 2000. Thrushes now largely restricted to the built environment in eastern England. *Diversity and Distributions* 6: 189–194.

Mason, P. & Allsop, J. 2009. *The Golden Oriole.* T. & A.D. Poyser, London.

Massimino, D., Johnston, A., Pearce-Higgins, J. & Baillie, S. 2013. *Maps of population density and trends.* URL: www.bto.org/volunteer-surveys/bbs/latest-results/maps-population-density-and-trends [31 July 2013].

Mathers, R. 1993. Feral Mandarin Duck population breeding in County Down. *Irish Birds* 5: 76.

Matthysen, E. 2002. *The Nuthatches.* T. & A.D. Poyser, London.

Maxwell, J. 2002. Nest-site competition with Blue Tits and Great Tits as a possible cause of declines in Willow Tit numbers: observations in the Clyde area. *Glasgow Naturalist* 24: 47–50.

Maxwell, J. 2003. Scottish Willow Tits hanging on. *BTO News* 244: 24.

McDevitt, A.D., Kajtoch, Ł., Mazgajski, T.D., Carden, R.F., Coscia, I., Osthoff, C., Coombes, R.H. & Wilson, F. 2011. The origins of Great Spotted Woodpecker *Dendrocopos major* colonizing Ireland revealed by mitochondrial DNA. *Bird Study* 58: 361–364.

McGhee, H. 2000. Easter Ross rookeries in 1998–99. *Scottish Birds* 21: 36–42.

McGreal, E. 2007. The breeding status of the Raven *Corvus corax* in southwest Mayo. *Irish Birds* 8: 237–242.

McGreal, E. 2011. Census of inland breeding gulls in Counties Galway and Mayo. *Irish Birds* 9: 173–180.

McKenzie, A.J., Petty, S.J., Toms, M.P. & Furness, R.W. 2007. Importance of Sitka Spruce *Picea sitchensis* seed and garden bird-feeders for Siskins *Carduelis spinus* and Coal Tits *Periparus ater. Bird Study* 54: 236–247.

McLoughlin, D. & Cotton, D. 2008. The status of Twite in Ireland 2008. *Irish Birds* 8: 323–330.

McLoughlin, D., Benson, C., Williams, B. & Cotton, D. 2010. The movement patterns of two populations of Twites *Carduelis flavirostris* in Ireland. *Ringing & Migration* 25: 15–21.

McMahon, B.J. 2007. Irish agriculture and farmland birds, research to date and future priorities. *Irish Birds* 8: 195–206.

McMillan, R.L. 2005. *Skye Birds.* URL: www.skye-birds.com [31 July 2013].

McShea, W.J. & Rappole, J.H. 1997. Variable song rates in three species of passerines and implications for estimating bird populations. *Journal of Field Ornithology* 68: 367–375.

Mearns, R. & Mearns, B. 2012. The status and distribution of the Lesser Whitethroat in Dumfries & Galloway. *Scottish Birds* 32: 2–12.

Meek, W.R., Burman, P.J., Nowakowski, M., Sparks, T.H. & Burman, N.J. 2003. Barn Owl release in lowland southern England—a twenty-one year study. *Biological Conservation* 109: 271–282.

Meek, E.R., Bolton, M., Fox, D. & Remp, J. 2011. Breeding skuas in Orkney: a 2010 census indicates density-dependent population change driven by both food supply and predation. *Seabird* 24: 1–10.

Melling, T., Dudley, S. & Doherty, P. 2008. The Eagle Owl in Britain. *British Birds* 101: 478–490.

Melville, D.S. 1991. Yellow-legged Herring Gulls in Essex, 1973–74. *British Birds* 84: 342–343.

Mermod, M., Reichlin, T., Arlettaz, R. & Schaub, M. 2009. The importance of ant-rich habitats for the persistence of the Wryneck *Jynx torquilla* on farmland. *Ibis* 151: 731–742.

Messenger, D. 1993. Spring passage of Little Gulls across Northern England. *British Birds* 86: 397–406.

Messenger, A. & Roome, M. 2007. The breeding population of Hobby in Derbyshire. *British Birds* 100: 594–608.

Miles, W. 2011. The appearance and status of the St Kilda Wren. *British Birds* 104: 325–328.

Miles, W.T.S., Tallack, R.M., Thomason, B.H. & Okill, D.J. 2010. Leach's Storm-petrels *Oceanodroma leucorhoa* nesting on a new site in Shetland. *Seabird* 23: 145–150.

Millennium Ecosystem Assessment. 2005. *Ecosystems and Human Well-being: Synthesis.* Island Press, Washington D.C.

Milne, P. & McAdams, D.G. 2008. Irish Rare Bird Report 2006. *Irish Birds* 8: 395–416.

Milne, P. & McAdams, D.G. 2009. Irish Rare Bird Report 2007. *Irish Birds* 8: 583–610.

Milne, P. & O'Sullivan, O. 1997. Forty-fourth Irish Bird Report, 1996. *Irish Birds* 6: 61–90.

Milwright, R.D.P. 1998. Breeding biology of the Golden Oriole *Oriolus oriolus* in the Fenland basin of eastern England. *Bird Study* 45: 320–330.

Milwright, R.D.P. 2002. Redwing *Turdus iliacus* migration and wintering areas as shown by recoveries of birds ringed in the breeding season in Fennoscandia, Poland, the Baltic Republics, Russia, Siberia and Iceland. *Ringing & Migration* 21: 5–15.

Minot, E.O. & Perrins, C.M. 1986. Interspecific interference competition – nest sites for Blue and Great tits. *Journal of Animal Ecology* 55: 331–350.

Mitchell, C. 2010. *Status and Distribution of Icelandic-breeding Geese: results of the 2009 international census.* Wildfowl & Wetlands Trust Report. WWT, Slimbridge.

Mitchell, C. 2012. *Status and Distribution of Icelandic-breeding Geese: results of the 2011 international census.* Wildfowl & Wetlands Trust Report. WWT, Slimbridge.

Mitchell, C., Walsh, A., Hall, C. & Crowe, O. 2008. *Greenland Barnacle Geese* Branta leucopsis *in Britain and Ireland: results of the international census, spring 2008.* Internal report. Wildfowl & Wetlands Trust, Slimbridge.

Mitchell, C., Coulhoun, K., Fox, A., Griffin, L., Hall, C., Hearn, R., Holt, C. & Walsh, A. 2010. Trends in goose numbers wintering in Britain and Ireland, 1995 to 2008. *Ornis Svecica* 20: 128–143.

Mitchell, C., Hearn, R. & Stroud, D. 2012. The merging of populations of Greylag Geese breeding in Britain. *British Birds* 105: 498–505.

Mitchell-Jones, A.J., Amori, G., Bogdanowicz, W., Kryštufek, B., Reijnders, P.J.H., Spitzenberger, E., Stubbe, M., Thissen, J.B.M., Vohralik, V. & Zima, J. 1999. *The atlas of European mammals.* T. & A.D. Poyser, London.

Moisen, G.G., Freeman, E.A., Blackard, J.A., Frescino, T.S., Zimmermann, N.E. & Edwards Jr, T.C. 2006. Predicting tree species presence and basal area in Utah: a comparison of stochastic gradient boosting, generalized additive models, and tree-based methods. *Ecological Modelling* 199: 176–187.

Mollet, P., Zbinden, N. & Schmid, H. 2009. An increase in the population of woodpeckers and other bird species thanks to an increase in the quantities of deadwood? *Schweizerische Zeitschrift für Forstwesen* 160: 334–340.

Monaghan, P. 1992. Seabirds and sandeels: the conflict between exploitation and conservation in the northern North Sea. *Biodiversity & Conservation* 1: 98–111.

Moorcroft, D., Whittingham, M.J., Bradbury, R.B. & Wilson, J.D. 2002. The selection of stubble fields by wintering granivorous birds reflects vegetation cover and food abundance. *Journal of Applied Ecology* 39: 535–547.

Moorcroft, D., Wilson, J.D. & Bradbury, R.B. 2006. Diet of nestling Linnets *Carduelis cannabina* on lowland farmland before and after agricultural intensification. *Bird Study* 53: 156–162.

Morris, A., Burges, D., Fuller, R.J., Evans, A.D. & Smith, K.W. 1994. The status and distribution of Nightjars *Caprimulgus europaeus* in Britain in 1992. *Bird Study* 41: 181–191.

Morrison, C.A., Robinson, R.A., Clark, J.A. & Gill, J.A. 2010. Spatial and temporal variation in population trends in a long-distance migratory bird. *Diversity and Distributions* 16: 620–627.

Morrison, C.A., Robinson, R.A., Clark, J.A., Risely, K. & Gill, J.A. 2013. Recent population declines in Afro-Palaearctic migratory birds: the influence of breeding and non-breeding seasons. *Diversity and Distributions* 19: 1051–1058.

Moss, R., Picozzi, N., Summers, R.W. & Baines, D. 2000. Capercaillie *Tetrao urogallus* in Scotland – demography of a declining population. *Ibis* 142: 259–267.

Moss, R., Oswald, J. & Baines, D. 2001. Climatic change and breeding success: decline of the Capercaillie in Scotland. *Journal of Animal Ecology* 70: 47–61.

Murphy, C.W. 2007. Field Notes. *Down Democrat* (Downpatrick) 19 June 2007.

Murray, A.A. 2001. The Waxwing invasion of winter 2000/2001 – the biggest ever in Scotland? *Birding Scotland* 4: 91–93.

Murray, R.D. 1991. The first successful breeding of Nuthatch in Scotland. *Scottish Bird Report* 1989: 51–55.

Murray, R. 2013. Water Rails in Britain. *British Birds* 106: 229–231.

Murray, S., Shewry, M.C., Harden, J., Jamie, K. & Parsons, M. 2010. A survey of Leach's *Oceanodroma leucorhoa* and European Storm-petrel *Hydrobates pelagicus* populations on North Rona and Sula Sgeir, Western Isles, Scotland, in 2009. *Seabird*

23: 25–40.

Murray, T. & Cabot, D. 2002. Eider *Somateria mollissima*: a new breeding species in County Mayo. *Irish Birds* 7: 139–140.

Murton, R.K. & Porter, S.C. 1961. Greenland-type redpolls nesting in Inverness-shire. *British Birds* 54: 251–253.

Musgrove, A.J., Austin, G.E., Hearn, R.D., Holt, C.A., Stroud, D.A. & Wotton, S.R. 2011. Overwinter population estimates of British waterbirds. *British Birds* 104: 364–397.

Musgrove, A., Aebischer, N., Eaton, M., Hearn, R., Newson, S., Noble, D., Parsons, M., Risely, K. & Stroud, D. 2013. Population estimates of birds in Great Britain and the United Kingdom. *British Birds* 106: 64–100.

Nagle, T. 2007. The loss of Barn Owl *Tyto alba* breeding and roost sites in County Cork: a contributory factor in the species' decline. *Irish Birds* 8: 314–315.

Nairn, R. & O'Halloran, J. (eds). 2012. *Bird Habitats in Ireland*. The Collins Press, Wilton, Cork.

National Climate Information Centre. 2012. UK annual weather summary 2011. *Weather* 67: 43.

Natural History Museum. 2013. *Dictionary of UK Species*. URL: www.nhm.ac.uk/research-curation/scientific-resources/biodiversity/uk-biodiversity/uk-species/species/index.html [31 July 2013].

Nelson, S.H., Court, I., Vickery, J.A., Watts, P.N. & Bradbury, R.B. 2003. The status and ecology of the Yellow Wagtail in Britain. *British Wildlife* 14: 270–274.

Neubauer, G., Zagalska-Neubauer, M.M., Pons, J.-M., Crochet, P.-A., Chylarecki, P., Przystalski, A. & Gay, L. 2009. Assortative mating without complete reproductive isolation in a zone of recent secondary contact between Herring Gulls (*Larus argentatus*) and Caspian Gulls (*L. cachinnans*). *Auk* 126: 409–419.

Newnham, J. & Crabtree, H. 2012. *Sussex Bird Atlas 2007–11: the maps* [on CD]. Sussex Ornithological Society.

Newson, S.E., Marchant, J.H., Ekins, G.R. & Sellers, R.M. 2007. The status of inland-breeding Great Cormorants in England. *British Birds* 100: 289–299.

Newson, S.E., Evans, K.L., Noble, D.G., Greenwood, J.J.D. & Gaston, K.J. 2008a. Use of distance sampling to improve estimates of national population sizes for common and widespread breeding birds in the UK. *Journal of Applied Ecology* 45: 1330–1338.

Newson, S.E., Mitchell, P.I., Parsons, M., O'Brien, S.H., Austin, G.E., Benn, S., Black, J., Blackburn, J., Brodie, B., Humphreys, E., Leech, D.I., Prior, M. & Webster, M. 2008b. Population decline of Leach's Storm-petrels *Oceanodroma leucorhoa* within the largest colony in Britain and Ireland. *Seabird* 21: 77–84.

Newson, S.E., Johnston, A., Parrott, D. & Leech, D.I. 2011. Evaluating the population-level impact of an invasive species, Ring-necked Parakeet *Psittacula krameri*, on native avifauna. *Ibis* 153: 509–516.

Newson, S.E., Johnston, A., Renwick, A.R., Baillie, S.R. & Fuller, R.J. 2012. Modelling large-scale relationships between changes in woodland deer and bird populations. *Journal of Applied Ecology* 49: 278–286.

Newton, I. 1986. *The Sparrowhawk*. T. & A.D. Poyser, Calton.

Newton, I. 1997. Links between abundance and distribution of birds. *Ecography* 20: 137–145.

Newton, I. 1998. *Population Limitation in Birds*. Academic Press, London.

Newton, I. 2003. *The speciation and biogeography of birds*. Academic Press, London.

Newton, I. 2004. The recent declines of farmland bird populations in Britain: an appraisal of causal factors and conservation actions. *Ibis* 146: 579–600.

Newton, I. 2006. Movement patterns of Common Crossbills *Loxia curvirostra* in Europe. *Ibis* 148: 782–788.

Newton, I. 2013. Organochlorine pesticides and birds. *British Birds* 106: 189–205.

Newton, S. 2011. Ireland's Puffins – concerns over Puffin food supply. *Wings* 61: 21–23.

Nightingale, B. 2005. The status of Lady Amherst's Pheasant in Britain. *British Birds* 98: 20–25.

Nilsson, L. 2008. Changes in numbers and distribution of wintering waterfowl in Sweden during 40 years, 1967–2006. *Ornis Svecica* 18: 135–226.

Norman, D. (ed). 2008. *Birds in Cheshire and Wirral: a breeding and wintering atlas*. Liverpool University Press, Liverpool.

Norris, K., Brindley, E., Cook, T., Babbs, S., Forster Brown, C. & Yaxley, R. 1998. Is the density of Redshank *Tringa totanus* nesting on saltmarshes in Great Britain declining due to changes in grazing management? *Journal of Applied Ecology* 35: 621–634.

Norriss, D.W., Haran, B., Hennigan, J., McElheron, A., McLaughlin, D.J., Swan, V. & Walsh, A. 2010. Breeding biology of Merlins *Falco columbarius* in Ireland, 1986–1992. *Irish Birds* 9: 23–30.

Northern Ireland Birdwatchers' Association. 2008. *Northern Ireland Bird Report*. Volume 17. Northern Ireland Birdwatchers' Association, Belfast.

Northumberland and Tyneside Bird Club. 2010. *Birds in Northumbria 2010*. Northumberland and Tyneside Bird Club, Newcastle upon Tyne.

North West Norfolk Ringing Group. 2013. *Snow Bunting* Plectrophenax nivalis. URL: www.nwnrg.co.uk/research/snow_bunting/snowbunting.htm [31 July 2013].

O'Brien, M., Green, R.E. & Wilson, J. 2006. Partial recovery of the population of Corncrakes *Crex crex* in Britain 1993–2004. *Bird Study* 53: 213–224.

O'Brien, S.H., Wilson, L.J., Webb, A. & Cranswick, P.A. 2008. Revised estimate of numbers of wintering Red-throated Divers *Gavia stellata* in Great Britain. *Bird Study* 55: 152–160.

O'Connor, R.J. & Mead, C.J. 1984. The Stock Dove in Britain, 1930–1980. *British Birds* 77: 181–201.

O'Donoghue, B.G. 2012. Duhallow Hen Harriers *Circus cyaneus* – from stronghold to just holding on. *Irish Birds* 9: 349–356.

O'Donoghue, B., O'Donoghue, T.A. & King, F. 2011. The Hen Harrier in Ireland: conservation issues for the 21st century. *Biology and Environment: Proceedings of the Royal Irish Academy* 111: 1–11.

O'Donoghue, P.D., Cross, T.F. & O'Halloran, J. 1996. Carrion Crows in Ireland, 1969–1993. *Irish Birds* 5: 399–406.

O'Flynn, W.J. 1983. Population changes of the Hen Harrier in Ireland. *Irish Birds* 2: 337–343.

O'Halloran, R., Nairn, R., Smiddy, P. & Crowe, O. 2012. Bird habitats: a synthesis and future perspectives. In *Bird Habitats in Ireland* (eds Nairn, R. & O'Halloran, J.), pp 247–268. The Collins Press, Wilton, Cork.

O'Sullivan, O. 2010. Garden Bird Survey. e*Wings* 14: 22–26.

O'Sullivan, O. 2011. Garden Bird Survey. *eWings* 26: 22–24.

O'Sullivan, O. & Smiddy, P. 1992. Thirty-ninth Irish Bird Report, 1991. *Irish Birds* 4: 571–610.

O'Toole, L., Fielding, A.H. & Haworth, P.F. 2002. Re-introduction of the Golden Eagle *Aquila chrysaetos* into the Republic of Ireland. *Biological Conservation* 103: 303–312.

Ockendon, N., Hewson, C.M., Johnston, A. & Atkinson, P.W. 2012. Declines in British-breeding populations of Afro-Palearctic migrant birds are linked to bioclimatic wintering zone in Africa, possibly via constraints on arrival time advancement. *Bird Study* 59: 111–125.

Office for National Statistics. 2012. *Internet Access – Households and Individuals, 2012*. URL: www.ons.gov.uk/ons/publications/re-reference-tables.html?edition=tcm%3A77-270031 [31 July 2013].

Ogilvie, M. & the Rare Breeding Birds Panel. 1994. Rare breeding birds in the United Kingdom in 1991. *British Birds* 87: 366–393.

Ogilvie, M. & the Rare Breeding Birds Panel. 1995. Rare breeding birds in the United Kingdom in 1992. *British Birds* 88: 67–93.

Ogilvie, M. & the Rare Breeding Birds Panel. 1998. Rare breeding birds in the United Kingdom in 1995. *British Birds* 91: 417–447.

Ogilvie, M. & the Rare Breeding Birds Panel. 1999. Rare breeding birds in the United Kingdom in 1996. *British Birds* 92: 120–154.

Ogilvie, M. & the Rare Breeding Birds Panel. 2003a. Non-native birds breeding in the United Kingdom in 2001. *British Birds* 96: 620–625.

Ogilvie, M. & the Rare Breeding Birds Panel. 2003b. Rare breeding birds in the United Kingdom in 2001. *British Birds* 96: 476–519.

Ogilvie, M. & the Rare Breeding Birds Panel. 2004. Rare breeding birds in the United Kingdom in 2002. *British Birds* 97: 492–536.

Olsen, K.M. & Larsson, H. 2004. *Gulls of Europe, Asia and North America*. Christopher Helm, London.

Olsson, O., Bruun, M. & Smith, H.G. 2002. Starling foraging success in relation to agricultural land-use. *Ecography* 25: 363–371.

Orkney Bird Report Committee. 2009. *Orkney Bird Report 2008*. Orkney Bird Report Committee.

Ormerod, S.J. & Tyler, S.J. 1991. The influence of stream acidification and riparian land use on the feeding ecology of Grey Wagtails *Motacilla cinerea* in Wales. *Ibis* 133: 53–61.

Osborne, P.E. 2005. Key issues in assessing the feasibility of reintroducing the Great Bustard *Otis tarda* to Britain. *Oryx* 39: 22–29.

Özçift, A. 2011. Random forests ensemble classifier trained with data resampling strategy to improve cardiac arrhythmia diagnosis. *Computers in Biology and Medicine* 41: 265–271.

Palmer, P. 2013. Daylight hunting by Barn Owls – is England a special case? *British Birds* 106: 416.

Parkin, D.T. & Knox, A.G. 2010. *The Status of Birds in Britain and Ireland.* Christopher Helm, London.

Parkin, D.T., Collinson, M., Helbig, A.J., Knox, A.G. & Sangster, G. 2003. The taxonomic status of Carrion and Hooded Crows. *British Birds* 96: 274–290.

Parmesan, C. 2006. Ecological and evolutionary responses to recent climate change. *Annual Review of Ecology and Systematics* 37: 637–669.

Parrinder, E.D. 1989. Little Ringed Plovers *Charadrius dubius* in Britain in 1984. *Bird Study* 36: 147–153.

Parslow, J.L.F. 1973. *Breeding Birds of Britain and Ireland.* T. & A.D. Poyser, Berkhamsted.

Peach, W., Baillie, S. & Underhill, L. 1991. Survival of British Sedge Warblers *Acrocephalus schoenobaenus* in relation to west African rainfall. *Ibis* 133: 300–305.

Peach, W.J., Crick, H.Q.P. & Marchant, J.H. 1995a. The demography of the decline in the British Willow Warbler population. *Journal of Applied Statistics* 22: 905–922.

Peach, W., Du Feu, C. & McMeeking, J. 1995b. Site tenacity and survival rates of Wrens *Troglodytes troglodytes* and Treecreepers *Certhia familiaris* in a Nottinghamshire wood. *Ibis* 137: 497–507.

Peach, W.J., Siriwardena, G.M. & Gregory, R.D. 1999. Long-term changes in the abundance and demography of British Reed Buntings *Emberiza schoeniclus*. *Journal of Applied Ecology* 36: 798–811.

Peach, W.J., Lovett, L.J., Wotton, S.R. & Jeffs, C. 2001. Countryside stewardship delivers Cirl Buntings (*Emberiza cirlus*) in Devon, UK. *Biological Conservation* 101: 361–373.

Peach, W.J., Robinson, R.A. & Murray, K.A. 2004. Demographic and environmental causes of the decline of rural Song Thrushes *Turdus philomelos* in lowland Britain. *Ibis* 146: 50–59.

Peach, W.J., Vincent, K.E., Fowler, J.A. & Grice, P.V. 2008. Reproductive success of House Sparrows along an urban gradient. *Animal Conservation* 11: 493–503.

Pearce-Higgins, J.W., Grant, M.C., Robinson, M.C. & Haysom, S.L. 2007. The role of forest maturation in causing the decline of Black Grouse *Tetrao tetrix*. *Ibis* 149: 143–155.

Pearce-Higgins, J.W., Grant, M.C., Beale, C.M., Buchanan, G.M. & Sim, I.M.W. 2009a. International importance and drivers of change of upland bird populations. In *Drivers of Environmental Change in Uplands* (eds Bonn, A., Allot, T., Hubacek, K. & Stewart, J.), pp 209–227. Routledge, London.

Pearce-Higgins, J.W., Yalden, D.W., Dougall, T.W. & Beale, C.M. 2009b. Does climate change explain the decline of a trans-Saharan Afro-Palaearctic migrant? *Oecologia* 159: 649–659.

Pearce-Higgins, J.W., Dennis, P., Whittingham, M.J. & Yalden, D.W. 2010. Impacts of climate on prey abundance account for fluctuations in a population of a northern wader at the southern edge of its range. *Global Change Biology* 16: 12–23.

Penhallurick, R.D. 1978. Chiffchaffs wintering at a sewage-works in west Cornwall. *British Birds* 71: 183–186.

Pennington, M.G., Riddington, R. & Miles, W.T.S. 2012. The Lapland Bunting influx in Britain & Ireland in 2010/11. *British Birds* 105: 654–673.

Penteriani, V., Lourenço, R. & Delgado, M. 2012. Eagle Owls in Doñana: a conservation dilemma or not? *British Birds* 105: 88–95.

Perkins, A.J., Whittingham, M.J., Bradbury, R.B., Wilson, J.D., Morris, A.J. & Barnett, P.R. 2000. Habitat characteristics affecting use of lowland agricultural grassland by birds in winter. *Biological Conservation* 95: 279–294.

Perkins, A.J., Hancock, M.H., Butcher, N. & Summers, R.W. 2005. Use of time-lapse video cameras to determine cause of nest failure of Slavonian Grebes *Podiceps auritus*. *Bird Study* 52: 159–165.

Perkins, A.J., Maggs, H.E., Watson, A. & Wilson, J.D. 2011. Adaptive management and targeting of agri-environment schemes does benefit biodiversity: a case study of the Corn Bunting *Emberiza calandra*. *Journal of Applied Ecology* 48: 514–522.

Perring, F.H. & Walters, S.M. (eds). 1962. *Atlas of the British Flora.* Thomas Nelson & Sons, London.

Perry, K.W. & Agnew, P. 1993. Breeding Dipper populations in north-west Ireland, 1972–1992. *Irish Birds* 5: 45–48.

Perry, K.W. & Day, K.R. 1997. Numbers and distribution of Tree Sparrows *Passer montanus* around Lough Neagh, 1968–97. *Irish Birds* 6: 29–34.

Perry, K. & Kennedy, D. 2002. An observation of a female Goldeneye with a young duckling at Lough Neagh – the first Irish breeding record. *Northern Ireland Bird Report* 2000: 92–94.

Perry, K.W., Antoniazza, M. & Day, K.R. 1999. Abundance and habitat use by breeding Great Crested Grebes *Podiceps cristatus* at Lough Neagh (Northern Ireland) and at Lake Neuchâtel (Switzerland). *Irish Birds* 6: 269–276.

Petty, S.J., Anderson, D.I.K., Davison, M., Little, B., Sherratt, T.N., Thomas, C.J. & Lambin, X. 2003. The decline of Common Kestrels *Falco tinnunculus* in a forested area of northern England: the role of predation by Northern Goshawks *Accipiter gentilis*. *Ibis* 145: 472–483.

Phillips, R.A., Catry, P., Thompson, D.R., Hamer, K.C. & Furness, R.W. 1997. Inter-colony variation in diet and reproductive performance of Great Skuas *Catharacta skua*. *Marine Ecology Progress Series* 152: 285–293.

Piercy, D. 2007. Cumbria Bird Club – Long-eared Owl Winter Survey 2007. *Cumbria Bird Club News* 18: 34–36.

Pitches, A. 2006. Marsh Harriers soar to 200-year high. *British Birds* 99: 382.

Pitches, A. 2012. More on lead-shot poisoning. *British Birds* 105: 755–756.

Pollock, C., Reid, J.B. & White, R. 1996. The occurrences of Little Auks *Alle alle* off the east coast of Britain. *Sula* 10: 239–246.

Porter, R. 2012. Hearing tests for survey workers? *British Birds* 105: 152.

Potts, G.R. 2012. *Partridges*. Collins, London.

Prasad, A.M., Iverson, L.R. & Liaw, A. 2006. Newer classification and regression tree techniques: bagging and random forests for ecological prediction. *Ecosystems* 9: 181–199.

Prendergast, J.R., Wood, S.N., Lawton, J.H. & Eversham, B.C. 1993. Correcting for variation in recording effort in analyses of diversity hotspots. *Biodiversity Letters* 1: 39–53.

Preston, C.D. 2013. Following the BSBI's lead: the influence of the *Atlas of the British Flora*, 1962–2012. *New Journal of Botany* 3: 2–14.

Prince, P. & Clarke, R. 1993. The Hobby's breeding range in Britain. *British Wildlife* 4: 341–346.

Proffitt, F.M., Newton, I., Wilson, J.D. & Siriwardena, G.M. 2004. Bullfinch *Pyrrhula pyrrhula* breeding ecology in lowland farmland and woodland: comparisons across time and habitat. *Ibis* 146 (Suppl. 2): 78–86.

Project Parakeet. 2013. *Project Parakeet.* URL: www.projectparakeet.co.uk [31 July 2013].

Prytherch, R.J. 2013. The breeding biology of the Common Buzzard. *British Birds* 106: 264–279.

Public and Corporate Economic Consultants. 2006. *The Economic and Environmental Impact of Sporting Shooting in the UK.* PACEC, Cambridge.

Pyman, G.A. 1959. The status of the Red-crested Pochard in the British Isles. *British Birds* 52: 42–56.

Quinn, J.L., Still, L., Kirby, J.S., Carrier, M.C. & Lambdon, P. 1997. Scaup *Aythya marila* numbers and the Cockle *Cardium edule* fishery on the Solway Firth: are they related? *Wildfowl* 47: 187–193.

R Core Team. 2013. *R: a language and environment for statistical computing*. URL: www.r-project.org [31 July 2013].

Raine, A.F., Sowter, D.J., Brown, A.F. & Sutherland, W.J. 2006a. Migration patterns of two populations of Twite *Carduelis flavirostris* in Britain. *Ringing & Migration* 23: 45–52.

Raine, A.F., Sowter, D.J., Brown, A.F. & Sutherland, W.J. 2006b. Natal philopatry and local movement patterns of Twite *Carduelis flavirostris*. *Ringing & Migration* 23: 89–94.

Rajchard, J., Procházka, J. & Kindlmann, P. 2006. Long-term decline in Common Swift *Apus apus* annual breeding success may be related to weather conditions. *Ornis Fennica* 83: 66–72.

Rakhimberdiev, E.N., Verkuil, Y.I., Saveliev, A.A., Väisänen, R.A., Karagicheva, J.V., Soloviev, M.Y., Tomkovich, P.S. & Piersma, T. 2011. A global population redistribution in a migrant shorebird detected with continent-wide qualitative breeding survey data. *Diversity and Distributions* 17: 144–151.

Ramsden, D.J. 1998. Effect of barn conversions on local populations of Barn Owl *Tyto alba*. *Bird Study* 45: 68–76.

Ratcliffe, D.A. 2003. *The peregrine saga*. In: Thompson, D.B.A., Redpath, S.M.,Fielding, A.H., Marquiss, M.& Galbraith, C.A. (eds).*Birds of Prey in a Changing Environment*. pp 91-98. The Stationery Office, Edinburgh.

Ratcliffe, N., Pickerell, G. & Brindley, E. 2000. Population trends of Little and Sandwich Terns *Sterna albifrons* and *S. sandvicensis* in Britain and Ireland from 1969 to 1998. *Atlantic Seabirds* 2: 211–226.

Raven, M.J., Noble, D.G. & Baillie, S.R. 2007. *The Breeding Bird Survey 2006.* BTO Research Report 471. BTO, Thetford.

Raven, S.J. & Coulson, J.C. 1997. The distribution and abundance of *Larus* gulls nesting on buildings in Britain and Ireland. *Bird Study* 44: 13–34.

Redfern, C.P.F. & Clark, J.A. 2001. *Ringers' Manual.* BTO, Thetford.

Reed, T. 1981. The number of breeding landbird species on British islands. *Journal of Animal Ecology* 50: 613–624.

Rees, E.C. & Beekman, J.H. 2010. Northwest European Bewick's Swans: a population in decline. *British Birds* 103: 640–650.

Rees, G., Haycock, A., Haycock, B., Hodges, J., Sutcliffe, S., Jenks, P. & Dobbins, R. 2009. *Atlas of breeding birds in Pembrokeshire 2003–2007.* Pembrokeshire Bird Group.

Rehfisch, M.M., Wernham, C.V. & Marchant, J.H. (eds). 1999. *Population, Distribution, Movements and Survival of Fish-eating Birds in Great Britain.* Department of the Environment, Transport and the Regions, London.

Rehfisch, M.M., Holloway, S.J. & Austin, G.E. 2003. Population estimates of waders on the non-estuarine coasts of the UK and the Isle of Man during the winter of 1997–98. *Bird Study* 50: 22–32.

Rehfisch, M.M., Austin, G.E., Freeman, S.N., Armitage, M.J.S. & Burton, N.H.K. 2004. The possible impact of climate change on the future distribution and numbers of waders on Britain's non-estuarine coast. *Ibis* 146: 70–81.

Rehfisch, M.M., Allan, J.R. & Austin, G.E. 2010. The effect on the environment of Great Britain's naturalized Greater Canada *Branta canadensis* and Egyptian Geese *Alopochen aegyptiacus*. In *BOU Proceedings – The Impacts of Non-native Species* (eds Burton, N.H.K., Baker, H., Carter, I., Moore, N. & Clements, A.). URL: www.bou.org.uk/bouprocnet/impacts-of-non-native-species/ [31 July 2013].

Reid, J.M. & Riddington, R. 1998. Identification of Greenland and Iceland Redpolls. *Dutch Birding* 20: 261–269.

Reid, J.M., Bignal, E.M., Bignal, S., McCracken, D.I. & Monaghan, P. 2004. Identifying the demographic determinants of population growth rate: a case study of Red-billed Choughs *Pyrrhocorax pyrrhocorax*. *Journal of Animal Ecology* 73: 777–788.

Reneerkens, J., Benhoussa, A., Boland, H., Collier, M., Grond, K., Günther, K., Hallgrimsson, G.T., Hansen, J., Meissner, W., de Meulenaer, B., Ntiamoa-Baidu, Y., Piersma, T., Poot, M., van Roomen, M., Summers, R.W., Tomkovich, P.S. & Underhill, L.G. 2009. Sanderlings using African–Eurasian flyways: a review of current knowledge. *Wader Study Group Bulletin* 116: 2–20.

Renwick, A.R., Massimino, D., Newson, S.E., Chamberlain, D.E., Pearce-Higgins, J.W. & Johnston, A. 2012. Modelling changes in species' abundance in response to projected climate change. *Diversity and Distributions* 18: 121–132.

Riddiford, N. & Findley, P. 1981. *Seasonal Movements of Summer Migrants.* BTO, Tring.

Riddington, R. & Votier, S. 1997. Redpolls from Greenland and Iceland. *Birding World* 10: 147–149.

Ridgill, S.C. & Fox, A.D. 1990. *Cold weather movements of waterfowl in Western Europe.* IWRB Special Publication 13. International Waterfowl Research Bureau, Slimbridge.

Risely, K., Noble, D.G. & Baillie, S.R. 2009. *The Breeding Bird Survey 2008.* BTO Research Report 537. BTO, Thetford.

Risely, K., Renwick, A.R., Dadam, D., Eaton, M.A., Johnston, A., Baillie, S.R., Musgrove, A.J. & Noble, D.G. 2011. *The Breeding Bird Survey 2010.* BTO Research Report 597. BTO, Thetford.

Rivers, S.L. & Forsyth, A. 2012. The Lapland Bunting influx in autumn 2010. *Scottish Birds* 32: 258–269.

Roberts, J.L. & Jones, M.S. 2009. Spacing and breeding production of Buzzards *Buteo buteo* in north-east Wales, 1978–2008. *Welsh Birds* 6: 9–26.

Roberts, S.J., Lewis, J.M.S. & Williams, I.T. 1999. Breeding European Honey-buzzards in Britain. *British Birds* 92: 326–345.

Robertson, D. 2003. Eurasian Reed Warblers in Scotland: a review of probable breeding records. *Scottish Birds* 24: 36–39.

Robinson, H.W. 1917. Evidence for the breeding of the Green Sandpiper in Westmorland in 1917. *British Birds* 11: 103–104.

Robinson, M. & Becker, C.D. 1986. Snowy Owls on Fetlar. *British Birds* 79: 228–242.

Robinson, R.A. & Clark, J.A. 2012. *The Online Ringing Report: bird ringing in Britain & Ireland in 2011.* URL: www.bto.org/ringing-report [31 July 2013].

Robinson, R.A., Crick, H.Q.P. & Peach, W.J. 2003. Population trends of Swallows *Hirundo rustica* breeding in Britain. *Bird Study* 50: 1–7.

Robinson, R.A., Green, R.E., Baillie, S.R., Peach, W.J. & Thomson, D.L. 2004. Demographic mechanisms of the population decline of the Song Thrush *Turdus philomelos* in Britain. *Journal of Animal Ecology* 73: 670–682.

Robinson, R.A., Siriwardena, G.M. & Crick, H.Q.P. 2005a. Status and population trends of Starling *Sturnus vulgaris* in Great Britain. *Bird Study* 52: 252–260.

Robinson, R.A., Siriwardena, G.M. & Crick, H.Q.P. 2005b. Size and trends of the House Sparrow *Passer domesticus* population in Great Britain. *Ibis* 147: 552–562.

Robinson, R.A., Freeman, S.N., Balmer, D.E. & Grantham, M.J. 2007a. Cetti's Warbler *Cettia cetti*: analysis of an expanding population. *Bird Study* 54: 230–235.

Robinson, R.A., Baillie, S.R. & Crick, H.Q.P. 2007b. Weather-dependent survival: implications of climate change for passerine population processes. *Ibis* 149: 357–364.

Robinson, R.A., Balmer, D.E. & Marchant, J.H. 2008. Survival rates of hirundines in relation to British and African rainfall. *Ringing & Migration* 24: 1–6.

Robinson, R.A., Lawson, B., Toms, M.P., Peck, K.M., Kirkwood, J.K., Chantrey, J., Clatworthy, I.R., Evans, A.D., Hughes, L.A., Hutchinson, O.C., John, S.K., Pennycott, T.W., Perkins, M.W., Rowley, P.S., Simpson, V.R., Tyler, K.M. & Cunningham, A.A. 2010. Emerging infectious disease leads to rapid population declines of common British birds. *PLoS ONE* 5: e12215.

Rocchini, D., Hortal, J., Lengyel, S., Lobo, J.M., Jiménez-Valverde, A., Ricotta, C., Bacaro, G. & Chiarucci, A. 2011. Accounting for uncertainty when mapping species distributions: the need for maps of ignorance. *Progress in Physical Geography* 35: 211–226.

Rock, P. 2005. Urban gulls: problems and solutions. *British Birds* 98: 338–355.

Rooney, E. & Montgomery, W.I. 2013. Diet diversity of the Common Buzzard (*Buteo buteo*) in a vole-less environment. *Bird Study* 60: 147–155.

Rowell, H.E. & Spray, C.J. 2004. *The Mute Swan* Cygnus olor *(Britain and Ireland populations) in Britain and Northern Ireland 1960/61–2000/01.* Waterbird Review Series, WWT/JNCC, Slimbridge.

Roy, S. 2011. Strategies to improve landscape scale management of Mink populations in the west coast of Scotland: lessons learned from the Uists 2001–2006. In *Island Invasives: eradication and management* (eds Veitch, C.R., Clout, M.N. & Towns, D.R.), pp 114–117. IUCN, Gland.

RSPB. 2012. *SEEVIEWS: East Coast Sea Eagle Newsletter.* Issue 1. RSPB, Perth.

Ruddock, M., Dunlop, B.J., O'Toole, L., Mee, A. & Nagle, T. 2012. *Republic of Ireland National Hen Harrier Survey 2010.* Irish Wildlife Manual, no. 59. National Parks and Wildlife Service, Department of Arts, Heritage and the Gaeltacht, Dublin.

Russell, I.C., Dare, P.J., Eaton, D.R. & Armstrong, J.D. 1996. *Assessment of the Problem of Fish-eating Birds in Inland Fisheries in England and Wales.* Ministry of Agriculture, Fisheries and Food, London.

Saino, N., Rubolini, D., Lehikoinen, E., Sokolov, L.V., Bonisoli-Alquati, A., Ambrosini, R., Boncaraglio, G. & Møller, A.P. 2009. Climate change effects on migration phenology may mismatch brood parasitic cuckoos and their hosts. *Biology Letters* 5: 539–541.

Sanderson, F.J., Donald, P.F., Pain, D.J., Burfield, I.J. & van Bommel, F.P.J. 2006. Long-term population declines in Afro-Palearctic migrant birds. *Biological Conservation* 131: 93–105.

Sangster, G., Knox, A.G., Helbig, A.J. & Parkin, D.T. 2002. Taxonomic recommendations for European birds. *Ibis* 144: 153–159.

Sangster, G., Collinson, J.M., Helbig, A.J., Knox, A.G. & Parkin, D.T. 2005. Taxonomic recommendations for British birds: third report. *Ibis* 147: 821–826.

Sangster, G., Collinson, J.M., Knox, A.G., Parkin, D.T. & Svensson, L. 2007. Taxonomic recommendations for British birds: fourth report. *Ibis* 149: 853–857.

Sangster, G., Collinson, J.M., Crochet, P.-A., Knox, A.G., Parkin, D.T., Svensson, L. & Votier, S.C. 2011. Taxonomic recommendations for British birds: seventh report. *Ibis* 153: 883–892.

Sauer, J.R., Hines, J.E., Fallon, J.E., Pardieck, K.L., Ziolkowski Jr, D.J. & Link, W.A. 2011. *The North American Breeding Bird Survey, Results and Analysis 1966–2010.* Version 12 July 2011. United States Geological Survey, Patuxent Wildlife Research Center, Laurel, Maryland. URL: www.mbr-pwre.usgs.gov/bbs/bbs2010.html [31 July 2013].

Sauter, A., Korner-Nievergelt, F. & Jenni, L. 2010. Evidence of climate change effects on within-winter movements of European Mallards *Anas platyrhynchos*. *Ibis* 152: 600–609.

Scheiffarth, G., Wahls, S., Ketzenberg, C. & Exo, K.-M. 2002. Spring migration strategies of two populations of Bar-tailed Godwits, *Limosa lapponica*, in the

Wadden Sea: time minimizers or energy minimizers? *Oikos* 96: 346–354.

Schönn, S. 1986. Zu Status, Biologie, Ökologie und Schutz des Steinkauzes (*Athene noctua*) in der DDR. *Acta Ornithoecologica* 1: 103–133.

Schwemmer, P., Garthe, S. & Mundry, R. 2008. Area utilization of gulls in a coastal farmland landscape: habitat mosaic supports niche segregation of opportunistic species. *Landscape Ecology* 23: 355–367.

Scott, D., Scott, L. & McHaffie, P. 2009. Unexpected breeding of the Marsh Harrier *Circus aeruginosus* in County Down, during 2009. *Irish Birds* 8: 625–627.

Scott, R.E. 1978. Rough-legged Buzzards in Britain in 1973/74 and 1974/75. *British Birds* 71: 325–338.

Sellers, R.M. 2006. Breeding population estimate for Northern Wheatear in Britain. *British Birds* 99: 533–535.

Sellers, R.M., Ekins, G.R., Hughes, B. & Kirby, J.S. 1997. Population development of inland breeding Cormorants in Great Britain. *Ricerche di Biologia della Selvaggina* 26 (Suppl. 1): 11–21.

Sergio, F., Marchesi, L., Pedrini, P. & Penteriani, V. 2007. Coexistence of a generalist owl with its intraguild predator: distance-sensitive or habitat-mediated avoidance? *Animal Behaviour* 74: 1607–1616.

Sharpe, C. (ed). 2007. *Manx Bird Atlas: an atlas of breeding and wintering birds on the Isle of Man.* Liverpool University Press, Liverpool.

Sharrock, J.T.R. 1976. *The Atlas of Breeding Birds in Britain and Ireland.* T. & A.D. Poyser, Berkhamsted.

Shaw, G. 1995. Habitat selection by Short-eared Owls *Asio flammeus* in young coniferous forests. *Bird Study* 42: 158–164.

Shaw, L.M., Chamberlain, D.E. & Evans, M.R. 2008. The House Sparrow *Passer domesticus* in urban areas: reviewing a possible link between post-decline distribution and human socioeconomic status. *Journal of Ornithology* 149: 293–299.

Shawyer, C.R. 1987. *The Barn Owl in the British Isles: its past, present and future.* The Hawk Trust, London.

Sheldon, R., Bolton, M., Gillings, S. & Wilson, A. 2004. Conservation management of Lapwing *Vanellus vanellus* on lowland arable farmland in the UK. *Ibis* (Suppl. 2): 41–49.

Shetland Bird Club. 2007. *Shetland Bird Report 2006.* Shetland Bird Club, Shetland.

Shetland Bird Club. 2011. *Shetland Bird Report 2010.* Shetland Bird Club, Shetland.

Shrubb, M. 2004. The decline of the Kestrel in Wales. *Welsh Birds* 4: 65–66.

Shrubb, M. 2007. *The Lapwing.* T. & A.D. Poyser, London.

Shrubb, M. 2011. Some thoughts on the historical status of the Great Bustard in Britain. *British Birds* 104: 180–188.

Sim, I.M.W. 2003. Land use, Common Buzzards (*Buteo buteo*) and Rabbits (*Oryctolagus cuniculus*) in the Welsh Marches. In *Birds of Prey in a Changing Environment* (eds Thompson, D.B.A., Redpath, S.M., Fielding, A.H., Marquiss, M. & Galbraith, C.A.), pp 351–369. The Stationery Office, Edinburgh.

Sim, I.M.W., Cross, A.V., Lamacraft, D.L. & Pain, D.J. 2001. Correlates of Common Buzzard *Buteo buteo* density and breeding success in the West Midlands. *Bird Study* 48: 317–329.

Sim, I.M.W., Gregory, R.D., Hancock, M.H. & Brown, A.F. 2005. Recent changes in the abundance of British upland breeding birds. *Bird Study* 52: 261–275.

Sim, I.M.W., Eaton, M.A., Setchfield, R.P., Warren, P.K. & Lindley, P. 2008. Abundance of male Black Grouse *Tetrao tetrix* in Britain in 2005, and change since 1995–96. *Bird Study* 55: 304–313.

Sim, I., Rollie, C., Arthur, D., Benn, S., Booker, H., Fairbrother, V., Green, M., Hutchinson, K., Ludwig, S., Nicoll, M., Poxton, I., Rebecca, G., Smith, L., Stanbury, A. & Wilson, P. 2010. The decline of the Ring Ouzel in Britain. *British Birds* 103: 229–239.

Sim, I.M.W., Rebecca, G.W., Ludwig, S.C., Grant, M.C. & Reid, J.M. 2011. Characterizing demographic variation and contributions to population growth rate in a declining population. *Journal of Animal Ecology* 80: 159–170.

Siriwardena, G. 2004. Possible roles of habitat, competition and avian nest predation in the decline of the Willow Tit *Parus montanus* in Britain. *Bird Study* 51: 193–202.

Siriwardena, G.M. 2006. Avian nest predation, competition and the decline of British Marsh Tits *Parus palustris*. *Ibis* 148: 255–265.

Siriwardena, G.M., Baillie, S.R. & Wilson, J.D. 1998a. Variation in the survival rates of British farmland passerines with respect to their population trends. *Bird Study* 45: 276–292.

Siriwardena, G.M., Baillie, S.R., Buckland, S.T., Fewster, R.M., Marchant, J.H. & Wilson, J.D. 1998b. Trends in the abundance of farmland birds: a quantitative comparison of smoothed Common Birds Census indices. *Journal of Applied Ecology* 35: 24–43.

Siriwardena, G.M., Baillie, S.R. & Wilson, J.D. 1999. Temporal variation in the annual survival rates of six granivorous birds with contrasting population trends. *Ibis* 141: 621–636.

Siriwardena, G.M., Baillie, S.R., Crick, H.Q.P., Wilson, J.D. & Gates, S. 2000a. The demography of lowland farmland birds. In *Proceedings of the 1999 BOU Spring Conference: Ecology and Conservation of Lowland Farmland Birds* (eds Aebischer, N.J., Evans, A.D., Grice, P.V. & Vickery, J.A.), pp 117–133. BOU, Tring.

Siriwardena, G.M., Baillie, S.R., Crick, H.Q.P. & Wilson, J.D. 2000b. The importance of variation in the breeding performance of seed-eating birds in determining their population trends on farmland. *Journal of Applied Ecology* 37: 128–148.

Siriwardena, G.M., Freeman, S.N. & Crick, H.Q.P. 2001. The decline of the Bullfinch *Pyrrhula pyrrhula* in Britain: is the mechanism known? *Acta Ornithologica* 36: 143–152.

Siriwardena, G.M., Robinson, R.A. & Crick, H.Q.P. 2002. Status and population trends of the House Sparrow *Passer domesticus* in Great Britain. In *Investigation into the Causes of the Decline of Starlings and House Sparrows in Great Britain* (eds Crick, H.Q.P., Robinson, R.A., Appleton, G.F., Clark, N.A. & Rickard, A.D.). pp 33–52. BTO Research Report 290. BTO, Thetford.

Siriwardena, G.M., Stevens, D.K., Anderson, G.Q.A., Vickery, J.A., Calbrade, N.A. & Dodd, S. 2007. The effect of supplementary winter seed food on breeding populations of farmland birds: evidence from two large-scale experiments. *Journal of Applied Ecology* 44: 920–932.

Skórka, P., Wójcik, J.D. & Martyka, R. 2005. Colonization and population growth of Yellow-legged Gull *Larus cachinnans* in southeastern Poland: causes and influence on native species. *Ibis* 147: 471–482.

Skov, H., Heinänen, S., Žydelis, R., Bellebaum, J., Bzoma, S., Dagys, M., Durinck, J., Garthe, S., Grishanov, G., Hario, M., Kieckbusch, J.K., Kube, J., Kuresoo, A., Larsson, K., Luigujoe, L., Meissner, W., Nehls, H.W., Nilsson, L., Petersen, I.K., Roos, M.M., Pihl, S., Sonntag, N., Stock, A., Stipniece, A. & Wahl, J. 2011. *Waterbird Populations and Pressures in the Baltic Sea.* TemaNord 2011: 550. Nordic Council of Ministers, Copenhagen.

Slack, R. 2009. *Rare Birds Where and When: an analysis of status & distribution in Britain and Ireland.* Volume 1. Rare Bird Books, York.

Smart, J., Amar, A., O'Brien, M., Grice, P. & Smith, K. 2008. Changing land management of lowland wet grasslands of the UK: impacts on Snipe abundance and habitat quality. *Animal Conservation* 11: 339–351.

Smart, J., Amar, A., Sim, I.M., Etheridge, B., Cameron, D., Christie, G. & Wilson, J.D. 2010. Illegal killing slows population recovery of a re-introduced raptor of high conservation concern – the Red Kite *Milvus milvus*. *Biological Conservation* 143: 1278–1286.

Smiddy, P. 2002. Breeding of the Little Egret *Egretta garzetta* in Ireland, 1997–2001. *Irish Birds* 7: 57–60.

Smiddy, P. & Duffy, B. 1997. Little Egret *Egretta garzetta*: a new breeding bird for Ireland. *Irish Birds* 6: 55–56.

Smiddy, P. & O'Mahony, B. 1997. The status of the Reed Warbler *Acrocephalus scirpaceus* in Ireland. *Irish Birds* 6: 23–28.

Smith, B. 1988. Mink and waterways birds. *BTO News* 156: 10.

Smith, C., Culshaw, V., Turnbull, V. & Jones, T. 2010. *Handa Island Skua Project Annual Report 2010.* Unpublished report to SWT, SNH, JNCC, SOC and the Seabird Group.

Smith, D. 2004. Breeding and roosting Hawfinches (*Coccothraustes coccothraustes*) in Merioneth. *Welsh Birds* 4: 11–19.

Smith, K.W. 2005. Has the reduction in nest-site competition from Starlings *Sturnus vulgaris* been a factor in the recent increase of Great Spotted Woodpecker *Dendrocopos major* numbers in Britain? *Bird Study* 52: 307–313.

Smith, K.W. 2006. The implications of nest site competition from Starlings *Sturnus vulgaris* and the effect of spring temperatures on the timing and breeding performance of Great Spotted Woodpeckers *Dendrocopos major* in southern England. *Annales Zoologici Fennici* 43: 177–185.

Smith, K.W. 2007. The utilization of dead wood resources by woodpeckers in Britain. *Ibis* 149: 183–192.

Smith, K.W. & Charman, E.C. 2012. The ecology and conservation of the Lesser Spotted Woodpecker. *British Birds* 105: 294–307.

Smith, K.W. & Smith, L. 2013. The effect of supplementary feeding in early spring on the breeding performance of the Great Spotted Woodpecker *Dendrocopos major*.

Bird Study 60: 169–175.

Smith, K.W., Reed, J.M. & Trevis, B.E. 1992. Habitat use and site fidelity of Green Sandpipers *Tringa ochropus* wintering in Southern England. *Bird Study* 39: 155–164.

Smith, K.W., Reed, J.M. & Trevis, B.E. 1999. Nocturnal and diurnal activity patterns and roosting sites of Green Sandpipers *Tringa ochropus* wintering in southern England. *Ringing & Migration* 19: 315–322.

Smith, R.D. 1996. Racial composition of breeding and wintering Snow Buntings *Plectrophenax nivalis* in the north-east Scottish uplands. *Ringing & Migration* 17: 123–136.

Smith, R.D. & Summers, R.W. 2005. Population size, breeding biology and origins of Scottish Purple Sandpipers. *British Birds* 98: 579–588.

Snow, D.W. & Perrins, C.M. (eds). 1998. *The Birds of the Western Palearctic. Concise Edition* (2 volumes). Oxford University Press, Oxford.

SOVON Vogelonderzoek Nederland. 2002. *Atlas van de Nederlandse Broedvogels 1998–2000.* Nederlandse Fauna 5. Nationaal Natuurhistorisch Museum Naturalis, KNNV Uitgeverij & European Invertebrate Survey – Nederland, Leiden.

SOVON Vogelonderzoek Nederland. 2011. Risicoanalyse van geïntroduceerde ganzensoorten in Nederland. *SOVONinformatierapport 2010–06.* SOVON Vogelonderzoek Nederland, Nijmegen.

SOVON. 2013. *Netwerk Ecologische Monitoring.* URL: www.sovon.nl [31 July 2013].

Sparks, T. & Tryjanowski, P. 2007. Patterns of spring arrival dates differ in two hirundines. *Climate Research* 35: 159–164.

Sparks, T.H., Huber, K., Bland, R.L., Crick, H.Q.P., Croxton, P.J., Flood, J., Loxton, R.G., Mason, C.F., Newnham, J.A. & Tryjanowski, P. 2007. How consistent are trends in arrival (and departure) dates of migrant birds in the UK? *Journal of Ornithology* 148: 503–511.

Spencer, R. & the Rare Breeding Birds Panel. 1986. Rare breeding birds in the United Kingdom in 1983. *British Birds* 79: 53–81.

Squires, D., Grasse, J. & Falshaw, C. 2009. Northern Goshawks in mid/north Wales 1992–2006. *Welsh Birds* 6: 49–65.

Stanbury, A., Davies, M., Grice, P., Gregory, R. & Wotton, S. 2010. The status of the Cirl Bunting in the UK in 2009. *British Birds* 103: 702–711.

Stanbury, A. & the UK Crane Working Group. 2011. The changing status of the Common Crane in the UK. *British Birds* 104: 432–447.

Staton, J. 1945. The breeding of Black-winged Stilts in Nottinghamshire in 1945. *British Birds* 38: 322–328.

Stevens, D.K., Anderson, G.Q.A., Grice, P.V. & Norris, K. 2007. Breeding success of Spotted Flycatchers *Muscicapa striata* in southern England – is woodland a good habitat for this species? *Ibis* 149: 214–223.

Stevens, D.K., Anderson, G.Q.A., Grice, P.V., Norris, K. & Butcher, N. 2008. Predators of Spotted Flycatcher *Muscicapa striata* nests in southern England as determined by digital nest-cameras. *Bird Study* 55: 179–187.

Stevenson, A. 2005. Redpolls in the Outer Hebrides. *Birding World* 18: 124.

Stewart, J.R. 2007.The fossil and archaeological record of the Eagle Owl in Britain. *British Birds* 100: 481–486.

Stoddart, A.M. 2011. Iceland Redpoll at Kelling in 2009: new to Norfolk. *Norfolk Bird & Mammal Report* 2010: 41–42.

Stoddart, A. 2013. Redpolls in Britain – a review of their taxonomy, identification and status. *British Birds* [In press].

Stone, B.H., Sears, J., Cranswick, P.A., Gregory, R.D., Gibbons, D.W., Rehfisch, M.M., Aebischer, N.J. & Reid, J.B. 1997. Population estimates of birds in Britain and in the United Kingdom. *British Birds* 90: 1–22.

Stone, C.J., Webb, A., Barton, C., Ratcliffe, N., Reed, T.C., Tasker, M.L., Camphuysen, C.J. & Pienkowski, M.W. 1995. *An Atlas of Seabird Distribution in North-west European Waters.* JNCC, Peterborough.

Straughan, R. 2010. The return of the Red Kite *Milvus milvus* as a breeding bird to Northern Ireland. *Irish Birds* 9: 19–22.

Stroud, D.A., Fox, A.D., Urquhart, C. & Francis, I.S. (compilers). 2012a. *International Single Species Action Plan for the Conservation of the Greenland White-fronted Goose* Anser albifrons flavirostris, *2012–2022.* AEWA Technical Series 45. Bonn, Germany.

Stroud, D.A., Francis, I. & Stroud, R. 2012b. Spotted Crakes breeding in Britain and Ireland: a history and evaluation of current status. *British Birds* 105: 197–220.

Strubbe, D. & Matthysen, E. 2007. Invasive Ring-necked Parakeets *Psittacula krameri* in Belgium: habitat selection and impact on native birds. *Ecography* 30: 578–588.

Strubbe, D. & Matthysen, E. 2009a. Establishment success of invasive ring-necked and monk parakeets in Europe. *Journal of Biogeography* 36: 2264–2278.

Strubbe, D. & Matthysen, E. 2009b. Experimental evidence for nest-site competition between invasive Ring-necked Parakeets (*Psittacula krameri*) and native Nuthatches (*Sitta europaea*). *Biological Conservation* 142: 1588–1594.

Strubbe, D., Matthysen, E. & Graham, C.H. 2010. Assessing the potential impact of invasive Ring-necked Parakeets *Psittacula krameri* on native Nuthatches *Sitta europaea* in Belgium. *Journal of Applied Ecology* 47: 549–557.

Südbeck, P., Bauer, H.-G., Boschert, M., Boye, P. & Knief, W. 2007. Rote Liste der Brutvögel Deutschlands. Fourth edition, 30. November 2007. *Berichte zum Vogelschutz* 44: 23–81.

Suddaby, D., Shaw, K.D., Ellis, P.M. & Brockie, K. 1994. King Eiders in Britain and Ireland in 1958–90: occurrences and ageing. *British Birds* 87: 418–430.

Suddaby, D., Nelson, T. & Veldman, J. 2009. Resurvey and comparative changes of breeding wader populations of Irish machair and associated wet grasslands in 2009. *Irish Birds* 8: 533–542.

Sullivan, B.L., Wood, C.L., Iliff, M.J., Bonney, R.E., Fink, D. & Kelling, S. 2009. eBird: a citizen-based bird observation network in the biological sciences. *Biological Conservation* 142: 2282–2292.

Summers, R.W. 1998. Territory sizes of Crested Tits at Abernethy Forest, Strathspey. *Scottish Birds* 19: 177–179.

Summers, R.W. 1999. Numerical responses by crossbills *Loxia* spp. to annual fluctuations in cone crops. *Ornis Fennica* 76: 141–144.

Summers, R.W. 2000. The habitat requirements of the Crested Tit *Parus cristatus* in Scotland. *Scottish Forestry* 54: 197–201.

Summers, R.W. 2002. Parrot Crossbills breeding in Abernethy Forest, Highland. *British Birds* 95: 4–11.

Summers, R.W. & Buckland, S.T. 2011. A first survey of the global population size and distribution of the Scottish Crossbill *Loxia scotica. Bird Conservation International* 21: 186–198.

Summers, R.W. & Canham, M. 2001. The distribution of Crested Tits in Scotland during the 1990s. *Scottish Birds* 22: 20–27.

Summers, R.W., Mavor, R.A., Buckland, S.T. & MacLennan, A.M. 1999. Winter population size and habitat selection of Crested Tits *Parus cristatus* in Scotland. *Bird Study* 46: 230–242.

Summers, R.W. Jardine, D.C., Marquiss, M. & Rae, R. 2002. The distribution and habitats of crossbills *Loxia* spp. in Britain, with special reference to the Scottish Crossbill *Loxia scotica. Ibis* 144: 393–410.

Summers, R.W., Jardine, D.C. & Dawson, R.J.G. 2004. The distribution of the Scottish Crossbill, 1995–2003. *Scottish Birds* 24: 11–16.

Summers, R.W., Dawson, R.J.G. & Phillips, R.E. 2007. Assortative mating and patterns of inheritance indicate that the three crossbill taxa in Scotland are species. *Journal of Avian Biology* 38: 153–162.

Summers, R.W., Mavor, R.A. & Hancock, M.H. 2009. Correlates of breeding success of Horned Grebes in Scotland. *Waterbirds* 32: 265–275.

Summers-Smith, J.D. 2003. The decline of the House Sparrow: a review. *British Birds* 96: 439–446.

Summers-Smith, J.D. 2007. Is unleaded petrol a factor in urban House Sparrow decline? *British Birds* 100: 558–559.

Suorsa, P., Huhta, E., Jäntti, A., Nikula, A., Helle, H., Kuitunen, M., Koivunen, V. & Hakkarainen, H. 2005. Thresholds in selection of breeding habitat by the Eurasian Treecreeper (*Certhia familiaris*). *Biological Conservation* 121: 443–452.

Sutherland, W.J. & Allport, G. 1991. The distribution and ecology of naturalised Egyptian Geese *Alopochen aegyptiacus* in Britain. *Bird Study* 38: 128–134.

Swann, R.L. 2010. Icterine Warblers breeding in Sutherland in 2009. *Scottish Birds* 30: 126.

Taylor, A.J. & O'Halloran, J. 2002. The decline of the Corn Bunting, *Miliaria calandra*, in the Republic of Ireland. *Biology and Environment: Proceedings of the Royal Irish Academy* 102B: 165–175.

Taylor, B. & van Perlo, B. 1998. *Rails: a guide to the rails, crakes, gallinules and coots of the world.* Pica Press, Sussex.

Taylor, M. & Marchant, J.H. 2011. *The Norfolk Bird Atlas: summer and winter distributions 1999–2007.* BTO Books, Thetford.

Taylor, M., Seago, M., Allard, P. & Dorling, D. 1999. *The Birds of Norfolk.* Pica Press, Sussex.

Thaxter, C.B., Lascelles, B., Sugar, K., Cook, A.S.C.P., Roos, S., Bolton, M., Langston, R.H.W. & Burton, N.H.K. 2012. Seabird foraging ranges as a preliminary tool for identifying candidate Marine Protected Areas. *Biological Conservation* 156: 53–61.

Thirgood, S.J., Redpath, S.M., Haydon, D.T., Rothery, P., Newton, I. & Hudson, P.J. 2000. Habitat loss and raptor predation: disentangling long- and short-term causes of Red Grouse declines. *Proceedings of the Royal Society of London B* 267: 651–656.

Thomas, C.D. & Lennon, J.J. 1999. Birds extend their ranges northwards. *Nature* 399: 213.

Thomas, J. 2012. Two more new breeding species for Somerset. *The Bittern* 3: 1–3.

Thompson, D.B.A., MacDonald, A.J., Marsden, J.H. & Galbraith, C.A. 1995. Upland heather moorland in Great Britain: a review of international importance, vegetation change and some objectives for nature conservation. *Biological Conservation* 71: 163–178.

Thompson, W.L. 2002. Towards reliable bird surveys: accounting for individuals present but not detected. *Auk* 119: 18–25.

Thomson, D.L., Douglas-Home, H., Furness, R.W. & Monaghan, P. 1996. Breeding success and survival in the Common Swift *Apus apus*: a long-term study on the effects of weather. *Journal of Zoology* 239: 29–38.

Thomson, D.L., Baillie, S.R. & Peach, W.J. 1997. The demography and age-specific annual survival of Song Thrushes during periods of population stability and decline. *Journal of Animal Ecology* 66: 414–424.

Thorup, K. 1998. Vagrancy of Yellow-browed Warbler *Phylloscopus inornatus* and Pallas's Warbler *P. proregulus* in north-west Europe: misorientation on great circles? *Ringing & Migration* 19: 7–12.

Thorup, K., Sunde, P., Jacobsen, L.B. & Rahbek, C. 2010. Breeding season food limitation drives population decline of the Little Owl *Athene noctua* in Denmark. *Ibis* 152: 803–814.

Tierney, T.D., Dunne, J. & Callanan, T. 2000. The Common Scoter *Melanitta nigra nigra* breeding in Ireland, range expansion or site relocation? *Irish Birds* 6: 447–452.

Tierney, T.D., Hudson, J. & Casey, C. 2002. Survey of the breeding waders of the River Shannon Callows, 2002. *Irish Birds* 7: 21–32.

Tingley, M.W., Koo, M.S., Moritz, C., Rush, A.C. & Beissinger, S.R. 2012. The push and pull of climate change causes heterogeneous shifts in avian elevational ranges. *Global Change Biology* 18: 3279–3290.

Tompkins, D.M., Draycott, R.A.H. & Hudson, P.J. 2002. Field evidence for apparent competition mediated via the shared parasites of two gamebird species. *Ecology Letters* 3: 10–14.

Toms, M.P. 2009. Eagle Owls in Britain: origins and conservation implications. *British Wildlife* 20: 405–412.

Toms, M.P., Crick, H.Q.P. & Shawyer, C.R. 2001. The status of breeding Barn Owls *Tyto alba* in the United Kingdom 1995–97. *Bird Study* 48: 23–37.

Toon, S.J. 2007. Slavonian Grebe breeding with Great Crested Grebe. *British Birds* 100: 381.

Triplet, P., Overdijk, O., Smart, M., Nagy, S., Schneider-Jacoby, M., Karauz, E.S., Pigniczki, C., Baha El Din, S., Kralj, J., Sandor, A. & Navedo, J.G. (compilers). 2008. *International Single Species Action Plan for the Conservation of the Eurasian Spoonbill* Platalea leucorodia. AEWA Technical Series 35. Bonn, Germany.

Tryjanowski, P., Karg, M.K. & Karg, J. 2003. Diet composition and prey choice by the Red-backed Shrike *Lanius collurio* in western Poland. *Belgian Journal of Zoology* 133: 157–162.

Tryjanowski, P., Sparks, T.H. & Crick, H.Q.P. 2006. Red-backed Shrike (*Lanius collurio*) nest performance in a declining British population: a comparison with a stable population in Poland. *Ornis Fennica* 83: 181–186.

Tubbs, C.R. 1967. Numbers of Dartford Warblers in England during 1962–66. *British Birds* 60: 87–89.

Tucker, G.M. & Heath, M.F. (eds). 1994. *Birds in Europe: their conservation status.* BirdLife Conservation Series 3. BirdLife International, Cambridge.

Tully, J. 2006. A historical perspective of today's area Feral Pigeon population and the dovecote pigeons of c1780. *Avon Bird Report* 2005: 171–175.

Tyler, G.A., Green, R.E. & Casey, C. 1998. Survival and behaviour of Corncrake *Crex crex* clutches during the mowing of agricultural grassland. *Bird Study* 45: 35–50.

Tyler, S. & Ormerod, S. 1994. *The Dippers.* T. & A.D. Poyser, London.

Underhill-Day, J. 1998. Breeding Marsh Harriers in the United Kingdom, 1983–95. *British Birds* 91: 210–218.

Underwood, L.A. & Stowe, T.J. 1984. Massive wreck of seabirds in eastern Britain, 1983. *Bird Study* 31: 79–88.

van der Jeugd, H.P., Voslamber, B., van Turnhout, C., Sierdsema, H., Feige, N., Nienhuis, J. & Koffijberg, K. 2006. *Overzomerende ganzen in Nederland: grenzen aan de groei?* SOVON-onderzoeksrapport 2006/02. SOVON Vogelonderzoek Nederland, Beek-Ubbergen.

Van Nieuwenhuyse, D., Génot, J.-C. & Johnson, D.H. 2008. *The Little Owl: conservation, ecology and behaviour of* Athene noctua. Cambridge University Press, Cambridge.

van Turnhout, C.A.M., Hagemeijer, E.J.M. & Foppen, R.P.B. 2010. Long-term population developments in typical marshland birds in The Netherlands. *Ardea* 98: 283–299.

Vanhinsbergh, D.P. & Chamberlain, D.E. 2001. Habitat associations of breeding Meadow Pipits *Anthus pratensis* in the British uplands. *Bird Study* 48: 159–172.

Vanhinsbergh, D. & Evans, A. 2002. Habitat associations of the Red-backed Shrike (*Lanius collurio*) in Carinthia, Austria. *Journal für Ornithologie* 143: 405–415.

Venables, W.A., Baker, A.D., Clarke, R.M., Jones, C., Lewis, J.M.S., Tyler, S.J., Walker, I.R. & Williams, R.A. 2008. *The Birds of Gwent.* Christopher Helm, London.

Verkuil, Y.I., Karlionova, N., Rakhimberdiev, E.N., Jukema, J., Wijmenga, J.J., Hooijmeijer, J.C.E.W., Pinchuk, P., Wymenga, E., Baker, A.J. & Piersma, T. 2012. Losing a staging area: eastward redistribution of Afro-Eurasian ruffs is associated with deteriorating fuelling conditions along the western flyway. *Biological Conservation* 149: 51–59.

Vermeersch, G., Anselin, A., Devos, K., Herremans, M., Stevens, J., Gabriëls, J. & van der Krieken, B. 2004. *Atlas van de Vlaamse broedvogels 2000–2002.* Mededelingen van het Instituut voor Natuurbehoud 23, Brussels.

Veron, P.K. & Lawlor, M.P. 2010. The dispersal and migration of the Northern Gannet *Morus bassanus* from Channel Islands breeding colonies. *Seabird* 22: 37–47.

Vickery, J.A., Tallowin, J.R., Feber, R.E., Asteraki, E.J., Atkinson, P.W., Fuller, R.J. & Brown, V.K. 2001. The management of lowland neutral grasslands in Britain: effects of agricultural practices on birds and their food resources. *Journal of Applied Ecology* 38: 647–664.

Village, A. 1990. *The Kestrel.* T. & A.D. Poyser, London.

Vinicombe, K. 1982. Breeding and population fluctuations of the Little Grebe. *British Birds* 75: 204–218.

Vinicombe, K.E. 2000. Identification of Ferruginous Duck and its status in Britain and Ireland. *British Birds* 93: 4–21.

Vinicombe, K.E. 2002. Ruddy Shelducks in Britain. *British Birds* 95: 398–399.

Votier, S.C., Furness, R.W., Bearhop, S., Crane, J.E., Caldow, R.W.G., Catry, P., Ensor, K., Hamer, K.C., Hudson, A.V., Kalmbach, E., Klomp, N.I., Pfeiffer, S., Phillips, R.A., Prieto, I. & Thompson, D.R. 2004. Changes in fisheries discard rates and seabird communities. *Nature* 427: 727–730.

Walker, D. 1995. Status of Yellow-legged Gull at Dungeness, Kent. *British Birds* 88: 5–7.

Walker, L.A., Turk, A., Long, S.M., Wienburg, C.L., Best, J. & Shore, R.F. 2008. Second generation anticoagulant rodenticides in Tawny Owls (*Strix aluco*) from Great Britain. *Science of the Total Environment* 392: 93–98.

Wanless, S. & Harris, M. 2012. Scottish seabirds – past, present and future. *Scottish Birds* 32: 38–45.

Wanless, S., Murray, S. & Harris, M.P. 2005a. The status of Northern Gannet in Britain & Ireland in 2003/04. *British Birds* 98: 280–294.

Wanless, S., Harris, M.P., Redman, P. & Speakman, J.R. 2005b. Low energy values of fish as a probable cause of a major seabird breeding failure in the North Sea. *Marine Ecology Progress Series* 294: 1–8.

Wanless, S., Frederiksen, M., Harris, M.P. & Freeman, S.N. 2006. Survival of Gannets *Morus bassanus* in Britain and Ireland, 1959–2002. *Bird Study* 53: 79–85.

Ward, R.M., Cranswick, P.A., Kershaw, M., Austin, G.E., Brown, A.W., Brown, L.M., Coleman, J.C., Chisholm, H.K. & Spray, C.J. 2007. Numbers of Mute Swans *Cygnus olor* in Great Britain: results of the national census in 2002. *Wildfowl* 57: 3–20.

Warren, P. & Baines, D. 2008. Current status and recent trends in numbers and distribution of Black Grouse *Tetrao tetrix* in northern England. *Bird Study* 55: 94–99.

Waters, E. & Waters, D. 2005. The former status of Great Bustard in Britain. *British Birds* 98: 295–305.

Watson, A. & Francis, I. 2012. *Birds in North-east Scotland Then and Now.* Paragon Publishing, Rothersthorpe.

Watson, A. & Moss, R. 2004. Impacts of ski-development on Ptarmigan (*Lagopus mutus*) at Cairn Gorm, Scotland. *Biological Conservation* 116: 267–275.

Watson, A. & Smith, R. 1991. Scottish Snow Bunting numbers in summer 1970–87. *Scottish Birds* 16: 53–56.

Watson, A., Moss, R. & Rothery, P. 2000. Weather and synchrony in 10-year population cycles of Rock Ptarmigan and Red Grouse in Scotland. *Ecology* 81: 2126–2136.

Watson, M., Aebischer, N.J., Potts, G.R. & Ewald, J.A. 2007. The relative effects of raptor predation and shooting on overwinter mortality of Grey Partridges in the United Kingdom. *Journal of Applied Ecology* 44: 972–982.

Wearing, M. 2011. The decline and fall of the Golden Pheasant in east Hampshire. *Kingfisher* 6–11. Hampshire Ornithological Society.

Webb, A., Dean, B.J., O'Brien, S.H., Söhle, I., McSorley, C., Reid, J.B., Cranswick, P.A., Smith, L.E. & Hall, C. 2009. *The Numbers of Inshore Waterbirds using the Greater Thames during the Non-breeding Season; an Assessment of the Area's Potential for Qualification as a Marine SPA.* Report 374. JNCC, Peterborough.

Weir, D.N., Kitchener, A.C. & McGowan, R.Y. 2000. Hybridization and changes in the distribution of Iceland Gulls (*Larus glaucoides/kumlieni/thayeri*). *Journal of Zoology* 252: 517–530.

Weisshaupt, N., Arlettaz, R., Reichlin, T.S., Tagmann-Ioset, A. & Schaub, M. 2011. Habitat selection by foraging Wrynecks *Jynx torquilla* during the breeding season: identifying the optimal habitat profile. *Bird Study* 58: 111–119.

Welch, S. 2010. The influx of Quail to East Lothian in 2009. *Scottish Birds* 30: 147–152.

Welsh Kite Trust. 2011. Red Kite *Milvus milvus* distribution in Britain & Ireland 2011. *Boda Wennol* 26: 24–25.

Wenzel, M.A., Webster, L.M.I., Blanco, G., Burgess, M.D., Kerbiriou, C., Segelbacher, G., Piertney, S.B. & Reid, J.M. 2012. Pronounced genetic structure and low genetic diversity in European Red-billed Chough (*Pyrrhocorax pyrrhocorax*) populations. *Conservation Genetics* 13: 1213–1230.

Wesołowski, T. & Fuller, R.J. 2012. Spatial variation and temporal shifts in habitat use by birds at the European scale. In *Birds and Habitat: Relationships in Changing Landscapes* (ed Fuller, R.J.), pp 63–92. Cambridge University Press, Cambridge.

Wetlands International. 2013. *Waterbird Population Estimates.* URL: http://wpe.wetlands.org [31 July 2013].

Whilde, A. 1993. *Threatened Mammals, Birds, Amphibians and Fish in Ireland. Irish Red Data Book 2: Vertebrates.* HMSO, Belfast.

Whitehead, S.A., Wright, J. & Cotton, P.A. 1995. Winter field use by the European Starling *Sturnus vulgaris*: habitat preferences and the availability of prey. *Journal of Avian Biology* 26: 193–202.

Whitfield, D.P. 2002. Status of breeding Dotterel *Charadrius morinellus* in Britain in 1999. *Bird Study* 49: 237–249.

Whitfield, D.P., Fielding, A.H., McLeod, D.R.A. & Haworth, P.F. 2004. Modelling the effects of persecution on the population dynamics of Golden Eagles in Scotland. *Biological Conservation* 119: 319–333.

Whitfield, D.P., Fielding, A.H., McLeod, D.R.A., Haworth, P.F. & Watson, J. 2006. A conservation framework for the Golden Eagle in Scotland: refining condition targets and assessment of constraint influences. *Biological Conservation* 130: 465–480.

Whitfield, D.P., Fielding, A.H., McLeod, D.R.A., Morton, K., Stirling-Aird, P. & Eaton, M.A. 2007. Factors constraining the distribution of Golden Eagles *Aquila chrysaetos* in Scotland. *Bird Study* 54: 199–211.

Whitfield, D.P., Douse, A., Evans, R.J., Grant, J., Love, J., McLeod, D.R.A., Reid, R. & Wilson, J.D. 2009. Natal and breeding dispersal in a reintroduced population of White-tailed Eagles *Haliaeetus albicilla. Bird Study* 56: 177–186.

Whitfield, D.P., Marquiss, M., Reid, R., Grant, J., Tingay, R. & Evans, R.J. 2013. Breeding season diets of sympatric White-tailed Eagles and Golden Eagles in Scotland: no evidence for competitive effects. *Bird Study* 60: 67–76.

Wilkinson, N.I. & Wilson, J.D. 2010. Breeding ecology of Twite *Carduelis flavirostris* in a crofting landscape. *Bird Study* 57: 142–155.

Wilson, A.M., Henderson, A.C.B. & Fuller, R.J. 2002. Status of the Nightingale *Luscinia megarhynchos* in Britain at the end of the 20th century with particular reference to climate change. *Bird Study* 49: 193–204.

Wilson, A.M., Ausden, M. & Milsom, T.P. 2004. Changes in breeding wader populations on lowland wet grasslands in England and Wales: causes and potential solutions. *Ibis* 146 (Suppl. 2): 32–40.

Wilson, A.M., Vickery, J.A., Brown, A., Langston, R.H.W., Smallshire, D., Wotton, S. & Vanhinsbergh, D. 2005. Changes in the numbers of breeding waders on lowland wet grasslands in England and Wales between 1982 and 2002. *Bird Study* 52: 55–69.

Wilson, J. & Peach, W. 2006. Impact of an exceptional winter flood on the population dynamics of Bearded Tits (*Panurus biarmicus*). *Animal Conservation* 9: 463–473.

Wilson, J.D., Morris, A.J., Arroyo, B.E., Clark, S.C. & Bradbury, R.B. 1999. A review of the abundance and diversity of invertebrate and plant foods of granivorous birds in northern Europe in relation to agricultural change. *Agriculture, Ecosystems & Environment* 75: 13–30.

Wilson, J.D., Boyle, J., Jackson, D.B., Lowe, B. & Wilkinson, N.I. 2007. Effect of cereal harvesting method on a recent population decline of Corn Buntings *Emberiza calandra* on the Western Isles of Scotland. *Bird Study* 54: 362–370.

Wilson, J.D., Evans, A.D. & Grice, P.V. 2009. *Bird Conservation and Agriculture.* Cambridge University Press, Cambridge.

Wilson, M.W., Pithon, J., Gittings, T., Kelly, T.C., Giller, P.S. & O'Halloran, J. 2006. Effects of growth stage and tree species composition on breeding bird assemblages of plantation forests. *Bird Study* 53: 225–236.

Wilson, P. & Balmer, D. 2002. Yellow-legged Gulls in Norfolk: 1961–2001. *Norfolk Bird & Mammal Report* 2001: 273–280.

Winfield, D.K. & Winfield, I.J. 1994. Possible competitive interactions between overwintering Tufted Duck (*Aythya fuligula* (L.)) and fish populations of Lough Neagh, Northern Ireland: evidence from diet studies. *Hydrobiologia* 279: 377–386.

Winstanley, D., Spencer, R. & Williamson, K. 1974. Where have all the Whitethroats gone? *Bird Study* 21: 1–14.

Wiseman, E.J. 2004. Breeding Honey-buzzards in Britain. *British Birds* 97: 417.

Wiseman, E.J. 2012. Honey-buzzards in southern England. *British Birds* 105: 23–28.

Wood, S. 2007. *The Birds of Essex.* Christopher Helm, London.

Wotton, S.R. & Gillings, S. 2000. The status of breeding Woodlarks *Lullula arborea* in Britain in 1997. *Bird Study* 47: 212–224.

Wotton, S., Gibbons, D.W., Dilger, M. & Grice, P.V. 1998. Cetti's Warbler in the United Kingdom and the Channel Islands in 1996. *British Birds* 91: 77–89.

Wotton, S.R., Langston, R.H.W., Gibbons, D.W. & Pierce, A.J. 2000. The status of the Cirl Bunting *Emberiza cirlus* in the UK and the Channel Islands in 1998. *Bird Study* 47: 138–146.

Wotton, S.R., Carter, I., Cross, A.V., Etheridge, B., Snell, N., Duffy, K., Thorpe, R. & Gregory, R.D. 2002a. Breeding status of the Red Kite *Milvus milvus* in Britain in 2000. *Bird Study* 49: 278–286.

Wotton, S.R., Langston, R.H.W. & Gregory, R.D. 2002b. The breeding status of the Ring Ouzel *Turdus torquatus* in the UK in 1999. *Bird Study* 49: 26–34.

Wotton, S., Conway, G., Eaton, M., Henderson, I. & Grice, P. 2009. The status of the Dartford Warbler in the UK and the Channel Islands in 2006. *British Birds* 102: 230–246.

Wotton, S., Grantham, M., Moran, N. & Gilbert, G. 2011. Eurasian Bittern distribution and abundance in the UK during the 2009/10 winter. *British Birds* 104: 636–641.

Wright, L.J., Hoblyn, R.A., Green, R.E., Bowden, C.G.R., Mallord, J.W., Sutherland, W.J. & Dolman, P.M. 2009. Importance of climatic and environmental change in the demography of a multi-brooded passerine, the Woodlark *Lullula arborea. Journal of Animal Ecology* 78: 1191–1202.

Wright, L.J., Banks, A.N. & Rehfisch, M.M. 2010. The status of non-native waterbirds in Eurasia and Africa in 2007. In *BOU Proceedings – The Impacts of Non-native Species* (eds Burton, N.H.K., Baker, H., Carter, I., Moore, N. & Clements, A.). URL: www.bou.org.uk/bouprocnet/impacts-of-non-native-species/ [31 July 2013].

Wynn, R.B. 2009. Balearic Shearwaters in UK and Irish waters between 2004 and 2006. *British Birds* 102: 350–351.

Wynn, R.B., Ward, M. & Adams, M.C. 2012. The increasing Firecrest population in the New Forest, Hampshire. *British Birds* 105: 273–275.

Yésou, P. 2003. Recent changes in the summer distribution of Balearic Shearwater *Puffinus mauretanicus* off western France. *Scientia Marina* 67: 143–148.

Young, G., Lawlor, M., Paintin, A., Atkinson, M. & Watson, J. 2013. *A working list of the Birds of the Channel Islands.* URL: www.jerseybirds.co.uk [31 July 2013].

Zhang, Y., Cao, L., Barter, M., Fox, A.D., Zhao, M., Meng, F., Shi, H., Jiang, Y. & Zhu, W. 2011. Changing distribution and abundance of Swan Goose *Anser cygnoides* in the Yangtze River floodplain: the likely loss of a very important wintering site. *Bird Conservation International* 21: 36–48.

INDEX

ALL SPECIES MENTIONED in the introductory chapters and in species accounts are indexed here by their common and scientific names. Entries in bold indicate the start page of a species' account. Species that are listed only in Appendix 2 (pp 666–686) are not indexed.